Contents

Cast Irons 1

    Classification and Basic Metallurgy of Cast Iron 3
    Gray Iron 12
    Ductile Iron 33
    Compacted Graphite Iron 56
    Malleable Iron 71
    Alloy Cast Irons 85

Carbon and Low-Alloy Steels 105

    Steel Processing Technology 107
    Microstructures, Processing, and Properties of Steels 126
    Classification and Designation of Carbon and Low-Alloy
        Steels 140
    Physical Properties of Carbon and Low-Alloy Steels 195
    Carbon and Low-Alloy Steel Sheet and Strip 200
    Precoated Steel Sheet 212
    Carbon and Low-Alloy Steel Plate 226
    Hot-Rolled Steel Bars and Shapes 240
    Cold-Finished Steel Bars 248
    Steel Wire Rod 272
    Steel Wire 277
    Threaded Steel Fasteners 289
    Steel Springs 302
    Steel Tubular Products 327
    Closed-Die Forgings 337
    High-Strength Low-Alloy Steel Forgings 358
    Steel Castings 363
    Bearing Steels 380
    High-Strength Structural and High-Strength Low-Alloy
        Steels 389
    Dual-Phase Steels 424
    Ultrahigh-Strength Steels 430

Hardenability of Carbon and Low-Alloy Steels 449

    Hardenable Carbon and Low-Alloy Steels 451
    Hardenability of Carbon and Low-Alloy Steels 464
    Hardenability Curves 485

Fabrication Characteristics of Carbon and Low-Alloy Steels 571

    Sheet Formability of Steels 573
    Bulk Formability of Steels 581
    Machinability of Steels 591
    Weldability of Steels 603

Service Characteristics of Carbon and Low-Alloy Steels 615

    Elevated-Temperature Properties of Ferritic Steels 617
    Effect of Neutron Irradiation on Properties of Steels 653
    Low-Temperature Properties of Structural Steels 662
    Fatigue Resistance of Steels 673
    Embrittlement of Steels 689
    Notch Toughness of Steels 737

Specialty Steels and Heat-Resistant Alloys 755

    Wrought Tool Steels 757
    P/M Tool Steels 780
    Maraging Steels 793
    Ferrous Powder Metallurgy Materials 801
    Austenitic Manganese Steels 822
    Wrought Stainless Steels 841
    Cast Stainless Steels 908
    Elevated-Temperature Properties of Stainless Steels 930
    Wrought and P/M Superalloys 950
        Appendix: P/M Cobalt-Base Wear-Resistant Materials ... 977
    Polycrystalline Cast Superalloys 981
    Directionally Solidified and Single-Crystal Superalloys 995

Special Engineering Topics 1007

    Strategic Materials Availability and Supply 1009
        Appendix: Manganese Availability 1021
    Recycling of Iron, Steel, and Superalloys 1023

Metric Conversion Guide 1035

Abbreviations, Symbols, and Tradenames 1038

Index 1043

Metals Handbook®
———— TENTH EDITION ————

Volume 1
Properties and Selection: Irons, Steels, and High-Performance Alloys

Prepared under the direction of the
ASM INTERNATIONAL Handbook Committee

Joseph R. Davis, Manager of Handbook Development
Kathleen M. Mills, Manager of Book Production
Steven R. Lampman, Technical Editor
Theodore B. Zorc, Technical Editor
Heather F. Lampman, Editorial Supervisor
George M. Crankovic, Editorial Coordinator
Alice W. Ronke, Assistant Editor
Scott D. Henry, Assistant Editor
Janice L. Daquila, Assistant Editor
Janet Jakel, Word Processing Specialist
Karen Lynn O'Keefe, Word Processing Specialist

Robert L. Stedfeld, Director of Reference Publications

Editorial Assistance
Lois A. Abel
Robert T. Kiepura
Penelope Thomas
Nikki D. Wheaton

MATERIALS PARK, OH 44073

Copyright © 1990
by
ASM INTERNATIONAL®
All rights reserved

No part of this book may be reproduced, stored in a retrieval system, or transmitted, in any form or by any means, electronic, mechanical, photocopying, recording, or otherwise, without the written permission of the copyright owner.

First printing, March 1990

Metals Handbook is a collective effort involving thousands of technical specialists. It brings together in one book a wealth of information from world-wide sources to help scientists, engineers, and technicians solve current and long-range problems.

Great care is taken in the compilation and production of this Volume, but it should be made clear that no warranties, express or implied, are given in connection with the accuracy or completeness of this publication, and no responsibility can be taken for any claims that may arise.

Nothing contained in the Metals Handbook shall be construed as a grant of any right of manufacture, sale, use, or reproduction, in connection with any method, process, apparatus, product, composition, or system, whether or not covered by letters patent, copyright, or trademark, and nothing contained in the Metals Handbook shall be construed as a defense against any alleged infringement of letters patent, copyright, or trademark, or as a defense against liability for such infringement.

Comments, criticisms, and suggestions are invited, and should be forwarded to ASM INTERNATIONAL.

Library of Congress Cataloging-in-Publication Data

Metals handbook/prepared under the direction of the
ASM INTERNATIONAL Handbook Committee—10th ed.
p. cm.
Includes bibliographical references and index.
Contents: v. 1. Properties and selection—
irons, steels, and high-performance alloys.
ISBN 0-87170-377-7 (v. 1)
1. Metals—Handbooks, manuals, etc.
I. ASM International Handbook Committee.
TA459.M43   1990   90-115
620.1′6—dc20   CIP

SAN 204-7586

Printed in the United States of America

Foreword

For nearly 70 years the *Metals Handbook* has been one of the most widely read and respected sources of information on the subject of metals. Launched in 1923 as a single volume, it has remained a durable reference work, with each succeeding edition demonstrating a continuing upward trend in growth, in subject coverage, and in reader acceptance. As we enter the final decade of the 20th century, the ever-quickening pace of modern life has forced an increasing demand for timely and accurate technical information. Such a demand was the impetus for this, the 10th Edition of *Metals Handbook*.

Since the publication of Volume 1 of the 9th Edition in 1978, there have been significant technological advances in the field of metallurgy. The goal of the present volume is to document these advances as they pertain to the properties and selection of cast irons, steels, and superalloys. A companion volume on properties and selection of nonferrous alloys, special-purpose materials, and pure metals will be published this autumn. Projected volumes in the 10th Edition will present expanded coverage on processing and fabrication of metals; testing, inspection, and failure analysis; microstructural analysis and materials characterization; and corrosion and wear phenomena (the latter a subject area new to the Handbook series).

During the 12 years it took to complete the 17 volumes of the 9th Edition, the high standards for technical reliability and comprehensiveness for which *Metals Handbook* is internationally known were retained. Through the collective efforts of the ASM Handbook Committee, the editorial staff of the Handbook, and nearly 200 contributors from industry, research organizations, government establishments, and educational institutions, Volume 1 of the 10th Edition continues this legacy of excellence.

Klaus M. Zwilsky
President,
ASM INTERNATIONAL

Edward L. Langer
Managing Director,
ASM INTERNATIONAL

Policy on Units of Measure

By a resolution of its Board of Trustees, ASM INTERNATIONAL has adopted the practice of publishing data in both metric and customary U.S. units of measure. In preparing this Handbook, the editors have attempted to present data in metric units based primarily on Système International d'Unités (SI), with secondary mention of the corresponding values in customary U.S. units. The decision to use SI as the primary system of units was based on the aforementioned resolution of the Board of Trustees and the widespread use of metric units throughout the world.

For the most part, numerical engineering data in the text and in tables are presented in SI-based units with the customary U.S. equivalents in parentheses (text) or adjoining columns (tables). For example, pressure, stress, and strength are shown both in SI units, which are pascals (Pa) with a suitable prefix, and in customary U.S. units, which are pounds per square inch (psi). To save space, large values of psi have been converted to kips per square inch (ksi), where 1 ksi = 1000 psi. The metric tonne (kg $\times$ 10^3) has sometimes been shown in megagrams (Mg). Some strictly scientific data are presented in SI units only.

To clarify some illustrations, only one set of units is presented on artwork. References in the accompanying text to data in the illustrations are presented in both SI-based and customary U.S. units. On graphs and charts, grids corresponding to SI-based units appear along the left and bottom edges. Where appropriate, corresponding customary U.S. units appear along the top and right edges.

Data pertaining to a specification published by a specification-writing group may be given in only the units used in that specification or in dual units, depending on the nature of the data. For example, the typical yield strength of steel sheet made to a specification written in customary U.S. units would be presented in dual units, but the sheet thickness specified in that specification might be presented only in inches.

Data obtained according to standardized test methods for which the standard recommends a particular system of units are presented in the units of that system. Wherever feasible, equivalent units are also presented. Some statistical data may also be presented in only the original units used in the analysis.

Conversions and rounding have been done in accordance with ASTM Standard E 380, with attention given to the number of significant digits in the original data. For example, an annealing temperature of 1570 °F contains three significant digits. In this case, the equivalent temperature would be given as 855 °C; the exact conversion to 854.44 °C would not be appropriate. For an invariant physical phenomenon that occurs at a precise temperature (such as the melting of pure silver), it would be appropriate to report the temperature as 961.93 °C or 1763.5 °F. In some instances (especially in tables and data compilations), temperature values in °C and °F are alternatives rather than conversions.

The policy on units of measure in this Handbook contains several exceptions to strict conformance to ASTM E 380; in each instance, the exception has been made in an effort to improve the clarity of the Handbook. The most notable exception is the use of g/cm^3 rather than kg/m^3 as the unit of measure for density (mass per unit volume).

SI practice requires that only one virgule (diagonal) appear in units formed by combination of several basic units. Therefore, all of the units preceding the virgule are in the numerator and all units following the virgule are in the denominator of the expression; no parentheses are required to prevent ambiguity.

Preface

During the past decade, tremendous advances have taken place in the field of materials science. Rapid technological growth and development of composite materials, plastics, and ceramics combined with continued improvements in ferrous and nonferrous metals have made materials selection one of the most challenging endeavors for engineers. Yet the process of selection of materials has also evolved. No longer is a mere recitation of specifications, compositions, and properties adequate when dealing with this complex operation. Instead, information is needed that explains the correlation among the processing, structures, and properties of materials as well as their areas of use. It is the aim of this volume—the first in the new 10th Edition series of *Metals Handbook*—to present such data.

Like the technology it documents, the *Metals Handbook* is also evolving. To be truly effective and valid as a reference work, each Edition of the Handbook must have its own identity. To merely repeat information, or to simply make superficial cosmetic changes, would be self-defeating. As such, utmost care and thought were brought to the task of planning the 10th Edition by both the ASM Handbook Committee and the Editorial Staff.

To ensure that the 10th Edition continued the tradition of quality associated with the Handbook, it was agreed that it was necessary to:

- Determine which subjects (articles) not included in previous Handbooks needed to be added to the 10th Edition
- Determine which previously published articles needed only to be revised and/or expanded
- Determine which previously published articles needed to be completely rewritten
- Determine which areas needed to be de-emphasized
- Identify and eliminate obsolete data

The next step was to determine how the subject of properties selection should be addressed in the 10th Edition. Considering the information explosion that has taken place during the past 30 years, the single-volume approach used for Volume 1 of the 8th Edition (published in 1961) was not considered feasible. For the 9th Edition, three separate volumes on properties and selection were published from 1978 to 1980. This approach, however, was considered somewhat fragmented, particularly in regard to steels: carbon and low-alloy steels were covered in Volume 1, whereas tool steels, austenitic manganese steels, and stainless steels were described in Volume 3. After considering the various options, it was decided that the most logical and user-friendly approach would be to publish two comprehensive volumes on properties and selection. In the present volume, emphasis has been placed on cast irons, carbon and low-alloy steels, and high-performance alloys such as stainless steels and superalloys. A companion volume on properties and selection of nonferrous alloys and special-purpose materials will follow (see Table 1 for an abbreviated table of contents).

Principal Sections

Volume 1 has been organized into seven major sections:

- Cast Irons
- Carbon and Low-Alloy Steels

Table 1 Abbreviated table of contents for Volume 2, 10th Edition, Metals Handbook

Specific Metals and Alloys	Special-Purpose Materials
Wrought Aluminum and Aluminum Alloys	Soft Magnetic Materials
Cast Aluminum Alloys	Permanent Magnet Materials
Aluminum-Lithium Alloys	Metallic Glasses
Aluminum P/M Alloys	Superconducting Materials
Wrought Copper and Copper Alloys	Electrical Resistance Alloys
Cast Copper Alloys	Electric Contact Materials
Copper P/M Products	Thermocouple Materials
Nickel and Nickel Alloys	Low Expansion Alloys
Beryllium-Copper and Beryllium-Nickel Alloys	Shape-Memory Alloys
Cobalt and Cobalt Alloys	Materials For Sliding Bearings
Magnesium and Magnesium Alloys	Metal-Matrix Composite Materials
Tin and Tin Alloys	Ordered Intermetallics
Zinc and Zinc Alloys	Cemented Carbides
Lead and Lead Alloys	Cermets
Refractory Metals and Alloys	Superabrasives and Ultrahard Tool Materials
Wrought Titanium and Titanium Alloys	Structural Ceramics
Cast Titanium Alloys	
Titanium P/M Alloys	
Zirconium and Hafnium	**Pure Metals**
Uranium and Uranium Alloys	Preparation and Characterization of Pure Metals
Beryllium	Properties of Pure Metals
Precious Metals	
Rare Earth Metals	**Special Engineering Topics**
Germanium and Germanium Compounds	
Gallium and Gallium Compounds	Recycling of Nonferrous Alloys
Indium and Bismuth	Toxicity of Metals

Table 2 Summary of contents for Volume 1, 10th Edition, Metals Handbook

Section title	Number of articles	Pages	Figures(a)	Tables(b)	References
Cast Irons	6	104	155	81	108
Carbon and Low-Allow Steels	21	344	298	266	230
Hardenability of Carbon and Low-Alloy Steels	3	122	210	178	28
Fabrication Characteristics of Carbon and Low-Alloy Steels	4	44	56	10	85
Service Characteristics of Carbon and Low-Alloy Steels	6	140	219	22	567
Specialty Steels and Heat-Resistant Alloys	11	252	249	163	358
Special Engineering Topics	2	27	29	11	50
Totals	53	1033	1216	731	1426

(a) Total number of figure captions; some figures may include more than one illustration. (b) Does not include unnumbered in-text tables or tables that are part of figures

- Hardenability of Carbon and Low-Alloy Steels
- Fabrication Characteristics of Carbon and Low-Alloy Steels
- Service Characteristics of Carbon and Low-Alloy Steels
- Specialty Steels and Heat-Resistant Alloys
- Special Engineering Topics

Of the 53 articles contained in these sections, 14 are new, 10 were completely rewritten, and the remaining articles have been substantially revised. A review of the content of the major sections is given below; highlighted are differences between the present volume and its 9th Edition predecessor. Table 2 summarizes the content of the principal sections.

Cast irons are described in six articles. The introductory article on "Classification and Basic Metallurgy of Cast Irons" was completely rewritten for the 10th Edition. The article on "Compacted Graphite Iron" is new to the Handbook. Both of these contributions were authored by D.M. Stefanescu (The University of Alabama), who served as Chairman of Volume 15, *Casting*, of the 9th Edition. The remaining four articles contain new information on materials (for example, austempered ductile iron) and testing (for example, dynamic tear testing).

Carbon and Low-Alloy Steels. Key additions to this section include articles that explain the relationships among processing (both melt and rolling processes), microstructures, and properties of steels. Of particular note is the article by G. Krauss (Colorado School of Mines) on pages 126 to 139 and the various articles on high-strength low-alloy steels. Other highlights include an extensive tabular compilation that cross-references SAE-AISI steels to their international counterparts (see the article "Classification and Designation of Steels") and an article on "Bearing Steels" that compares both case-hardened and through-hardened bearing materials.

Hardenability of Carbon and Low-Alloy Steels. Following articles that introduce H-steels and describe hardenability concepts, including test procedures to determine the hardening response of steels, a comprehensive collection of hardenability curves is presented. Both English and metric hardenability curves are provided for some 86 steels.

Fabrication Characteristics. Sheet formability, forgeability, machinability, and weldability are described next. The article on bulk formability, which emphasizes recent studies on HSLA forging steels, is new to the Handbook series. The material on weldability was completely rewritten and occupies nearly four times the space allotted in the 9th Edition.

Service Characteristics. The influence of various in-service environments on the properties of steels is one of the most widely studied subjects in metallurgy. Among the topics described in this section are elevated-temperature creep properties, low-temperature fracture toughness, fatigue properties, and impact toughness. A new article also describes the deleterious effect of neutron irradiation on alloy and stainless steels. Of critical importance to this section, however, is the definitive treatise on "Embrittlement of Steels" written by G.F. Vander Voort (Carpenter Technology Corporation). Featuring more than 75 graphs and 372 references, this 48-page article explores the causes and effects of both thermal and environmental degradation on a wide variety of steels. Compared with the 9th Edition on the same subject, this represents a nearly tenfold increase in coverage.

Specialty Steels and Heat-Resistant Alloys. Eleven articles on wrought, cast, and powder metallurgy materials for specialty and/or high-performance applications make up this section. Alloy development and selection criteria as related to corrosion-resistant and heat-resistant steels and superalloys are well documented. More than 100 pages are devoted to stainless steels, while three new articles have been written on superalloys—including one on newly developed directionally solidified and single-crystal nickel-base alloys used for aerospace engine applications.

Special Engineering Topics. The final section examines two subjects that are becoming increasingly important to the engineering community: (1) the availability and supply of strategic materials, such as chromium and cobalt, used in stainless steel and superalloy production, and (2) the current efforts to recycle highly alloyed materials. Both of these subjects are new to the Handbook series. A second article on recycling of nonferrous alloys will be published in Volume 2 of the 10th Edition.

Acknowledgments

Successful completion of this Handbook required the cooperation and talents of literally hundreds of professional men and women. In terms of the book's technical content, we are indebted to the authors, reviewers, and miscellaneous contributors—some 200 strong—upon whose collective experience and knowledge rests the accuracy and authority of the volume. Thanks are also due to the ASM Handbook Committee and its capable Chairman, Dennis D. Huffman (The Timken Company). The ideas and suggestions provided by members of the committee proved invaluable during the two years of planning required for the 10th Edition. Lastly, we would like to acknowledge the efforts of those companies who have worked closely with ASM's editorial and production staff on this and many other Handbook volumes. Our thanks go to Byrd Data Imaging for their tireless efforts in maintaining a demanding typesetting schedule, to Rand McNally & Company for the care and quality brought to printing the Handbook, and to Precision Graphics, Don O. Tech, Accurate Art, and HaDel Studio for their attention to detail during preparation of Handbook artwork. Their combined efforts have resulted in a significant and lasting contribution to the metals industry.

The Editors

Officers and Trustees of ASM INTERNATIONAL

Klaus M. Zwilsky
President and Trustee
National Materials Advisory Board
National Academy of Sciences

Stephen M. Copley
Vice President and Trustee
Illinois Institute of Technology

Richard K. Pitler
Immediate Past President and Trustee
Allegheny Ludlum Corporation
(retired)

Edward L. Langer
Secretary and Managing Director
ASM INTERNATIONAL

Robert D. Halverstadt
Treasurer
AIMe Associates

Trustees

John V. Andrews
Teledyne Allvac

Edward R. Burrell
Inco Alloys International, Inc.

H. Joseph Klein
Haynes International, Inc.

Kenneth F. Packer
Packer Engineering, Inc.

Hans Portisch
VDM Technologies Corporation

William E. Quist
Boeing Commercial Airplanes

John G. Simon
General Motors Corporation

Charles Yaker
Howmet Corporation

Daniel S. Zamborsky
Consultant

Members of the ASM Handbook Committee (1990–1991)

Dennis D. Huffman
(Chairman 1986–; Member 1983–)
The Timken Company

Roger J. Austin (1984–)
ABARIS

Roy G. Baggerly (1987–)
Kenworth Truck Company

Robert J. Barnhurst (1988–)
Noranda Research Centre

Hans Borstell (1988–)
Grumman Aircraft Systems

Gordon Bourland (1988–)
LTV Aerospace and Defense Company

John F. Breedis (1989–)
Olin Corporation

Stephen J. Burden (1989–)
GTE Valenite

Craig V. Darragh (1989–)
The Timken Company

Gerald P. Fritzke (1988–)
Metallurgical Associates

J. Ernesto Indacochea (1987–)
University of Illinois at Chicago

John B. Lambert (1988–)
Fansteel Inc.

James C. Leslie (1988–)
Advanced Composites Products and Technology

Eli Levy (1987–)
The De Havilland Aircraft Company of Canada

William L. Mankins (1989–)
Inco Alloys International, Inc.

Arnold R. Marder (1987–)
Lehigh University

John E. Masters (1988–)
American Cyanamid Company

David V. Neff (1986–)
Metaullics Systems

David LeRoy Olson (1982–1988; 1989–)
Colorado School of Mines

Dean E. Orr (1988–)
Orr Metallurgical Consulting Service, Inc.

Elwin L. Rooy (1989–)
Aluminum Company of America

Kenneth P. Young (1988–)
AMAX Research & Development

Previous Chairmen of the ASM Handbook Committee

R.S. Archer
(1940–1942) (Member, 1937–1942)

L.B. Case
(1931–1933) (Member, 1927–1933)

T.D. Cooper
(1984–1986) (Member, 1981–1986)

E.O. Dixon
(1952–1954) (Member, 1947–1955)

R.L. Dowdell
(1938–1939) (Member, 1935–1939)

J.P. Gill
(1937) (Member, 1934–1937)

J.D. Graham
(1966–1968) (Member, 1961–1970)

J.F. Harper
(1923–1926) (Member, 1923–1926)

C.H. Herty, Jr.
(1934–1936) (Member, 1930–1936)

J.B. Johnson
(1948–1951) (Member, 1944–1951)

L.J. Korb
(1983) (Member, 1978–1983)

R.W.E. Leiter
(1962–1963) (Member, 1955–1958, 1960–1964)

G.V. Luerssen
(1943–1947) (Member, 1942–1947)

G.N. Maniar
(1979–1980) (Member, 1974–1980)

J.L. McCall
(1982) (Member, 1977–1982)

W.J. Merten
(1927–1930) (Member, 1923–1933)

N.E. Promisel
(1955–1961) (Member, 1954–1963)

G.J. Shubat
(1973–1975) (Member, 1966–1975)

W.A. Stadtler
(1969–1972) (Member, 1962–1972)

R. Ward
(1976–1978) (Member, 1972–1978)

M.G.H. Wells
(1981) (Member, 1976–1981)

D.J. Wright
(1964–1965) (Member, 1959–1967)

Authors and Reviewers

G. Aggen
 Allegheny Ludlum Steel Division
 Allegheny Ludlum Corporation
Frank W. Akstens
 Industrial Fasteners Institute
C. Michael Allen
 Adjelian Allen Rubeli Ltd.
H.S. Avery
 Consultant
P. Babu
 Caterpillar, Inc.
Alan M. Bayer
 Teledyne Vasco
Felix Bello
 The WEFA Group
S.P. Bhat
 Inland Steel Company
M. Blair
 Steel Founders' Society of America
Bruce Boardman
 Deere and Company Technical Center
Kurt W. Boehm
 Nucor Steel
Francis W. Boulger
 Battelle-Columbus Laboratories
 (retired)
Greg K. Bouse
 Howmet Corporation
John L. Bowles
 North American Wire Products
 Corporation
J.D. Boyd
 Metallurgical Engineering Department
 Queen's University
B.L. Bramfitt
 Bethlehem Steel Corporation
Richard W. Bratt
 Consultant
W.D. Brentnall
 Solar Turbines
C.R. Brinkman
 Oak Ridge National Laboratory
Edward J. Bueche
 USS/Kobe Steel Company
Harold Burrier, Jr.
 The Timken Company
Anthony Cammarata
 Mineral Commodities Division
 U.S. Bureau of Mines
A.P. Cantwell
 LTV Steel Company
M. Carlucci
 Lorlea Steels

Harry Charalambu
 Carr & Donald Associates
Joseph B. Conway
 Mar-Test Inc.
W. Couts
 Wyman-Gordon Company
Wil Danesi
 Garrett Processing Division
 Allied-Signal Aerospace Company
John W. Davis
 McDonnell Douglas
R.J. Dawson
 Deloro Stellite, Inc.
Terry A. DeBold
 Carpenter Technology Corporation
James Dimitrious
 Pfauter-Maag Cutting Tools
Douglas V. Doane
 Consulting Metallurgist
Mehmet Doner
 Allison Gas Turbine Division
Henry Dormitzer
 Wyman-Gordon Company
Allan B. Dove
 Consultant
 (deceased)
Don P.J. Duchesne
 Adjelian Allen Rubeli Ltd.
Gary L. Erickson
 Cannon-Muskegon Corporation
Walter Facer
 American Spring Wire Company
Brownell N. Ferry
 LTV Steel Company
F.B. Fletcher
 Lukens Steel Company
E.M. Foley
 Deloro Stellite, Inc.
R.D. Forrest
 Division Fonderie
 Pechinery Electrometallurgie
James Fox
 Charter Rolling Division
 Charter Manufacturing Company, Inc.
Edwin F. Frederick
 Bar, Rod and Wire Division
 Bethlehem Steel Corporation
James Gialamas
 USS/Kobe Steel Company
Jeffery C. Gibeling
 University of California at Davis
Wayne Gismondi
 Union Drawn Steel Co., Ltd.

R.J. Glodowski
 Armco, Inc.
Loren Godfrey
 Associated Spring
 Barnes Group, Inc.
Alan T. Gorton
 Atlantic Steel Company
W.G. Granzow
 Research & Technology
 Armco, Inc.
David Gray
 Teledyne CAE
Malcolm Gray
 Microalloying International, Inc.
Richard B. Gundlach
 Climax Research Services
I. Gupta
 Inland Steel Company
R.I.L. Guthrie
 McGill Metals Processing Center
 McGill University
P.C. Hagopian
 Stelco Fastener and Forging Company
J.M. Hambright
 Inland Bar and Structural Division
 Inland Steel Company
K. Harris
 Cannon-Muskegon Corporation
Hans J. Heine
 Foundry Management & Technology
W.E. Heitmann
 Inland Steel Company
T.A. Heuss
 LTV Steel Bar Division
 LTV Steel Company
Thomas Hill
 Speedsteel of New Jersey, Inc.
M. Hoetzl
 Surface Combustion, Inc.
Peter B. Hopper
 Milford Products Corporation
J.P. Hrusovsky
 The Timken Company
David Hudok
 Weirton Steel Corporation
S. Ibarra
 Amoco Corporation
J.E. Indacochea
 Department of Civil Engineering,
 Mechanics, and Metallurgy
 University of Illinois at Chicago
Asjad Jalil
 The Morgan Construction Company

William J. Jarae
  Georgetown Steel Corporation
Lyle R. Jenkins
  Ductile Iron Society
J.J. Jonas
  McGill Metals Processing Center
  McGill University
Robert S. Kaplan
  U.S. Bureau of Mines
Donald M. Keane
  LaSalle Steel Company
William S. Kirk
  U.S. Bureau of Mines
S.A. Kish
  LTV Steel Company
R.L. Klueh
  Metals and Ceramics Division
  Oak Ridge National Laboratory
G.J.W. Kor
  The Timken Company
Charles Kortovich
  PCC Airfoils
George Krauss
  Advanced Steel Processing and
    Products Research Center
  Colorado School of Mines
Eugene R. Kuch
  Gardner Denver Division
J.A. Laverick
  The Timken Company
M.J. Leap
  The Timken Company
P.W. Lee
  The Timken Company
B.F. Leighton
  Canadian Drawn Steel Company
R.W. Leonard
  USX Corporation
R.G. Lessard
  Stelpipe
  Stelco, Inc.
S. Liu
  Center for Welding and Joining
    Research
  Colorado School of Mines
Carl R. Loper, Jr.
  Materials Science & Engineering
    Department
  University of Wisconsin-Madison
Donald G. Lordo
  Townsend Engineered Products
R.A. Lula
  Consultant
W.C. Mack
  Babcock & Wilcox Division
  McDermott Company
T.P. Madvad
  USS/Kobe Steel Company
J.K. Mahoney, Jr.
  LTV Steel Company
C.W. Marshall
  Battelle Memorial Institute
G.T. Matthews
  The Timken Company
Gernant E. Maurer
  Special Metals Corporation

Joseph McAuliffe
  Lake Erie Screw Corporation
Thomas J. McCaffrey
  Carpenter Steel Division
  Carpenter Technology Corporation
J. McClain
  Danville Division
  Wyman-Gordon Company
T.K. McCluhan
  Elkem Metals Company
D.B. McCutcheon
  Steltech Technical Services Ltd.
Hal L. Miller
  Nelson Wire Company
K.L. Miller
  The Timken Company
Frank Minden
  Lone Star Steel
Michael Mitchell
  Rockwell International
R.W. Monroe
  Steel Founders' Society of America
Timothy E. Moss
  Inland Bar and Structural Division
  Inland Steel Company
Brian Murkey
  R.B. & W. Corporation
T.E. Murphy
  Inland Bar and Structural Division
  Inland Steel Company
Janet Nash
  American Iron and Steel Institute
Drew V. Nelson
  Mechanical Engineering Department
  Stanford University
G.B. Olson
  Northwestern University
George H. Osteen
  Chaparral Steel
J. Otter
  Saginaw Division
  General Motors Corporation
D.E. Overby
  Stelco Technical Services Ltd.
John F. Papp
  U.S. Bureau of Mines
Y.J. Park
  Amax Research Company
D.F. Paulonis
  United Technologies
Leander F. Pease III
  Powder-Tech Associates, Inc.
Thoni V. Philip
  TVP Inc.
Thomas A. Phillips
  Department of the Interior
  U.S. Bureau of Mines
K.E. Pinnow
  Crucible Research Center
  Crucible Materials Corporation
Arnold Plant
  Samuel G. Keywell Company
Christopher Plummer
  The WEFA Group
J.A. Pojeta
  LTV Steel Company

R. Randall
  Rariton River Steel
P. Repas
  U.S.S. Technical Center
  USX Corporation
M.K. Repp
  The Timken Company
Richard Rice
  Battelle Memorial Institute
William L. Roberts
  Consultant
G.J. Roe
  Bethlehem Steel Corporation
Kurt Rohrbach
  Carpenter Technology Corporation
A.R. Rosenfield
  Battelle Memorial Institute
James A. Rossow
  Wyman-Gordon Company
C.P. Royer
  Exxon Production Research Company
Mamdouh M. Salama
  Conoco Inc.
Norman L. Samways
  Association of Iron and Steel Engineers
Gregory D. Sander
  Ring Screw Works
J.A. Schmidt
  Joseph T. Ryerson and Sons, Inc.
Michael Schmidt
  Carpenter Technology Corporation
W. Schuld
  Seneca Wire & Manufacturing Company
R.E. Schwer
  Cannon-Muskegon Corporation
Kay M. Shupe
  Bliss & Laughlin Steel Company
V.K. Sikka
  Oak Ridge National Laboratory
Steve Slavonic
  Teledyne Columbia-Summerill
Dale L. Smith
  Argonne National Laboratory
Richard B. Smith
  Western Steel Division
  Stanadyne, Inc.
Dennis Smyth
  The Algoma Steel Corporation Ltd.
G.R. Speich
  Department of Metallurgical Engineering
  Illinois Institute of Technology
Thomas Spry
  Commonwealth Edition
W. Stasko
  Crucible Materials Corporation
  Crucible Research Center
Doru M. Stefanescu
  The University of Alabama
Joseph R. Stephens
  Lewis Research Center
  National Aeronautics and Space
    Administration
P.A. Stine
  General Electric Company
N.S. Stoloff
  Rensselaer Polytechnic Institute

John R. Stubbles
  LTV Steel Company
D.K. Subramanyam
  Ergenics, Inc.
A.E. Swansiger
  ABC Rail Corporation
R.W. Swindeman
  Oak Ridge National Laboratory
N. Tepovich
  Connecticut Steel
Millicent H. Thomas
  LTV Steel Company
Geoff Tither
  Niobium Products Company, Inc.
George F. Vander Voort
  Carpenter Technology Corporation

Elgin Van Meter
  Empire-Detroit Steel Division
  Cyclops Corporation
Krishna M. Vedula
  Materials Science & Engineering
    Department
  Case Western Reserve University
G.M. Waid
  The Timken Company
Charles F. Walton
  Consultant
Lee R. Walton
  Latrobe Steel Company
Yung-Shih Wang
  Exxon Production Research Company

S.D. Wasko
  Allegheny Ludlum Steel Division
  Allegheny Ludlum Corporation
J.R. Weeks
  Brookhaven National Laboratory
Charles V. White
  GMI Engineering and Management
    Institute
Alexander D. Wilson
  Lukens Steel Company
Peter H. Wright
  Chapparal Steel Company
B. Yalamanchili
  North Star Steel Texas Company
Z. Zimerman
  Bethlehem Steel Corporation

Contents

Cast Irons 1

    Classification and Basic Metallurgy of Cast Iron 3
    Gray Iron 12
    Ductile Iron 33
    Compacted Graphite Iron 56
    Malleable Iron 71
    Alloy Cast Irons 85

Carbon and Low-Alloy Steels 105

    Steel Processing Technology 107
    Microstructures, Processing, and Properties of Steels 126
    Classification and Designation of Carbon and Low-Alloy
        Steels 140
    Physical Properties of Carbon and Low-Alloy Steels 195
    Carbon and Low-Alloy Steel Sheet and Strip 200
    Precoated Steel Sheet 212
    Carbon and Low-Alloy Steel Plate 226
    Hot-Rolled Steel Bars and Shapes 240
    Cold-Finished Steel Bars 248
    Steel Wire Rod 272
    Steel Wire 277
    Threaded Steel Fasteners 289
    Steel Springs 302
    Steel Tubular Products 327
    Closed-Die Forgings 337
    High-Strength Low-Alloy Steel Forgings 358
    Steel Castings 363
    Bearing Steels 380
    High-Strength Structural and High-Strength Low-Alloy
        Steels 389
    Dual-Phase Steels 424
    Ultrahigh-Strength Steels 430

Hardenability of Carbon and Low-Alloy Steels 449

    Hardenable Carbon and Low-Alloy Steels 451
    Hardenability of Carbon and Low-Alloy Steels 464
    Hardenability Curves 485

Fabrication Characteristics of Carbon and Low-Alloy Steels 571

    Sheet Formability of Steels 573
    Bulk Formability of Steels 581
    Machinability of Steels 591
    Weldability of Steels 603

Service Characteristics of Carbon and Low-Alloy Steels 615

    Elevated-Temperature Properties of Ferritic Steels 617
    Effect of Neutron Irradiation on Properties of Steels 653
    Low-Temperature Properties of Structural Steels 662
    Fatigue Resistance of Steels 673
    Embrittlement of Steels 689
    Notch Toughness of Steels 737

Specialty Steels and Heat-Resistant Alloys 755

    Wrought Tool Steels 757
    P/M Tool Steels 780
    Maraging Steels 793
    Ferrous Powder Metallurgy Materials 801
    Austenitic Manganese Steels 822
    Wrought Stainless Steels 841
    Cast Stainless Steels 908
    Elevated-Temperature Properties of Stainless Steels 930
    Wrought and P/M Superalloys 950
        Appendix: P/M Cobalt-Base Wear-Resistant Materials ... 977
    Polycrystalline Cast Superalloys 981
    Directionally Solidified and Single-Crystal Superalloys ... 995

Special Engineering Topics 1007

    Strategic Materials Availability and Supply 1009
        Appendix: Manganese Availability 1021
    Recycling of Iron, Steel, and Superalloys 1023

Metric Conversion Guide 1035

Abbreviations, Symbols, and Tradenames 1038

Index 1043

Cast Irons

Classification and Basic Metallurgy of Cast Iron 3
Gray Iron 12
Ductile Iron 33
Compacted Graphite Iron 56
Malleable Iron 71
Alloy Cast Irons 85

Classification and Basic Metallurgy of Cast Iron

Doru M. Stefanescu, The University of Alabama

THE TERM CAST IRON, like the term steel, identifies a large family of ferrous alloys. Cast irons are multicomponent ferrous alloys, which solidify with a eutectic. They contain major (iron, carbon, silicon), minor (<0.1%), and often alloying (>0.1%) elements. Cast iron has higher carbon and silicon contents than steel. Because of the higher carbon content, the structure of cast iron, as opposed to that of steel, exhibits a rich carbon phase. Depending primarily on composition, cooling rate, and melt treatment, cast iron can solidify according to the thermodynamically metastable Fe-Fe$_3$C system or the stable Fe-Gr system. When the metastable path is followed, the rich carbon phase in the eutectic is the iron carbide; when the stable solidification path is followed, the rich carbon phase is graphite. Referring only to the binary Fe-Fe$_3$C or Fe-Gr system, cast iron can be defined as an iron-carbon alloy with more than 2% C. The reader is cautioned that silicon and other alloying elements may considerably change the maximum solubility of carbon in austenite (γ). Therefore, in exceptional cases, alloys with less than 2% C can solidify with a eutectic structure and therefore still belong to the family of cast iron.

The formation of stable or metastable eutectic is a function of many factors including the nucleation potential of the liquid, chemical composition, and cooling rate. The first two factors determine the graphitization potential of the iron. A high graphitization potential will result in irons with graphite as the rich carbon phase, while a low graphitization potential will result in irons with iron carbide. A schematic of the structure of the common types of commercial cast irons, as well as the processing required to obtain them, is shown in Fig. 1.

The two basic types of eutectics—the stable austenite-graphite or the metastable austenite-iron carbide (Fe$_3$C)—have wide differences in their mechanical properties, such as strength, hardness, toughness, and ductility. Therefore, the basic scope of the metallurgical processing of cast iron is to manipulate the type, amount, and morphology of the eutectic in order to achieve the desired mechanical properties.

Classification

Historically, the first classification of cast iron was based on its fracture. Two types of iron were initially recognized:

- *White iron*: Exhibits a white, crystalline fracture surface because fracture occurs along the iron carbide plates; it is the result of metastable solidification (Fe$_3$C eutectic)
- *Gray iron*: Exhibits a gray fracture surface because fracture occurs along the graphite plates (flakes); it is the result of stable solidification (Gr eutectic)

With the advent of metallography, and as the body of knowledge pertinent to cast iron increased, other classifications based on microstructural features became possible:

- *Graphite shape*: Lamellar (flake) graphite (FG), spheroidal (nodular) graphite (SG), compacted (vermicular) graphite (CG), and temper graphite (TG); temper graphite results from a solid-state reaction (malleabilization)
- *Matrix*: Ferritic, pearlitic, austenitic, martensitic, bainitic (austempered)

This classification is seldom used by the floor foundryman. The most widely used terminology is the commercial one. A first division can be made in two categories:

- *Common cast irons*: For general-purpose applications, they are unalloyed or low alloy
- *Special cast irons*: For special applications, generally high alloy

The correspondence between commercial and microstructural classification, as well as the final processing stage in obtaining common cast irons, is given in Table 1. A classification of cast irons by their commercial names and structure is also given in the article "Classification of Ferrous Casting Alloys" in Volume 15 of the 9th Edition of *Metals Handbook*.

Special cast irons differ from the common cast irons mainly in the higher content of alloying elements (>3%), which promote microstructures having special properties for elevated-temperature applications, corrosion resistance, and wear resistance. A classification of the main types of special cast irons is shown in Fig. 2.

Principles of the Metallurgy of Cast Iron

The goal of the metallurgist is to design a process that will produce a structure that will yield the expected mechanical proper-

Table 1 Classification of cast iron by commercial designation, microstructure, and fracture

Commercial designation	Carbon-rich phase	Matrix(a)	Fracture	Final structure after
Gray iron	Lamellar graphite	P	Gray	Solidification
Ductile iron	Spheroidal graphite	F, P, A	Silver-gray	Solidification or heat treatment
Compacted graphite iron	Compacted vermicular graphite	F, P	Gray	Solidification
White iron	Fe$_3$C	P, M	White	Solidification and heat treatment(b)
Mottled iron	Lamellar Gr + Fe$_3$C	P	Mottled	Solidification
Malleable iron	Temper graphite	F, P	Silver-gray	Heat treatment
Austempered ductile iron	Spheroidal graphite	At	Silver-gray	Heat treatment

(a) F, ferrite; P, pearlite; A, austenite; M, martensite; At, austempered (bainite). (b) White irons are not usually heat treated, except for stress relief and to continue austenite transformation.

4 / Cast Irons

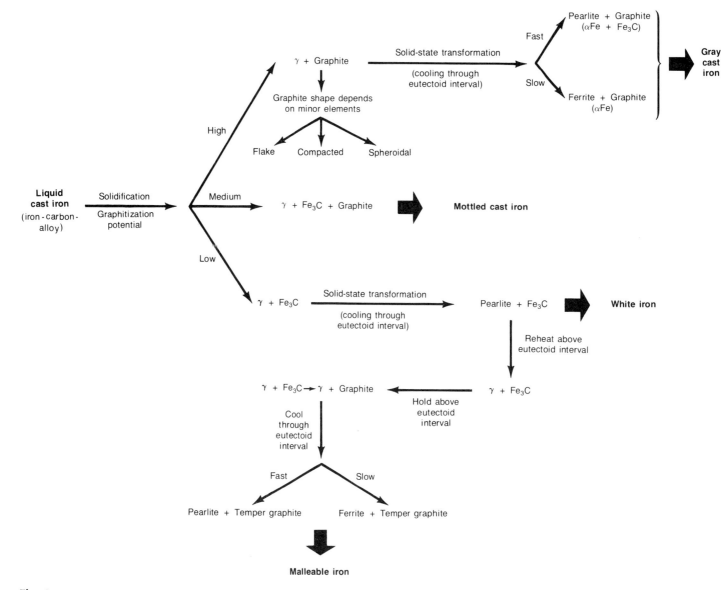

Fig. 1 Basic microstructures and processing for obtaining common commercial cast irons

ties. This requires knowledge of the structure-properties correlation for the particular alloy under consideration as well as of the factors affecting the structure. When discussing the metallurgy of cast iron, the main factors of influence on the structure that one needs to address are:

- Chemical composition
- Cooling rate
- Liquid treatment
- Heat treatment

In addition, the following aspects of combined carbon in cast irons should also be considered:

- In the original cooling or through subsequent heat treatment, a matrix can be internally decarburized or carburized by depositing graphite on existing sites or by dissolving carbon from them
- Depending on the silicon content and the cooling rate, the pearlite in iron can vary in carbon content. This is a ternary system, and the carbon content of pearlite can be as low as 0.50% with 2.5% Si
- The conventionally measured hardness of graphitic irons is influenced by the graphite, especially in gray iron. Martensite microhardness may be as high as 66 HRC, but measures as low as 54 HRC conventionally in gray iron (58 HRC in ductile)
- The critical temperature of iron is influenced (raised) by silicon content, not carbon content

The following sections in this article discuss some of the basic principles of cast iron metallurgy. More detailed descriptions of the metallurgy of cast irons are available in separate articles in this Volume describing certain types of cast irons. The Section "Ferrous Casting Alloys" in Volume 15 of the 9th Edition of *Metals Handbook* also contains more detailed descriptions on the metallurgy of cast irons.

Gray Iron (Flake Graphite Iron)

The composition of gray iron must be selected in such a way as to satisfy three basic structural requirements:

- The required graphite shape and distribution
- The carbide-free (chill-free) structure
- The required matrix

For common cast iron, the main elements of the chemical composition are carbon and silicon. Figure 3 shows the range of carbon and silicon for common cast irons as compared with steel. It is apparent that irons

Classification and Basic Metallurgy of Cast Iron / 5

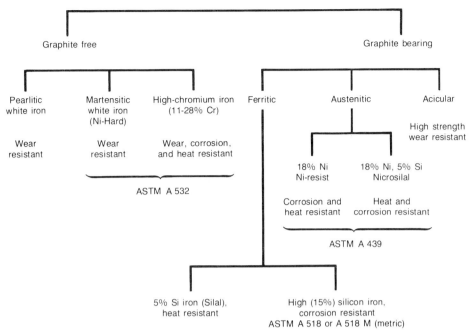

Fig. 2 Classification of special high-alloy cast irons. Source: Ref 1

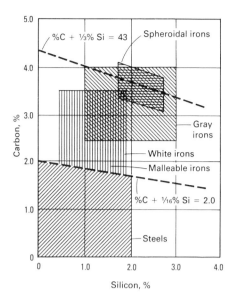

Fig. 3 Carbon and silicon composition ranges of common cast irons and steel. Source: Ref 2

have carbon in excess of the maximum solubility of carbon in austenite, which is shown by the lower dashed line. A high carbon content increases the amount of graphite or Fe_3C. High carbon and silicon contents increase the graphitization potential of the iron as well as its castability.

The combined influence of carbon and silicon on the structure is usually taken into account by the carbon equivalent (CE):

$$CE = \%\,C + 0.3(\%\,Si) + 0.33(\%\,P) - 0.027(\%\,Mn) + 0.4(\%\,S) \quad (Eq\ 1)$$

Additional information on carbon equivalent is available in the article "Thermodynamic Properties of Iron-Base Alloys" in Volume 15 of the 9th Edition of *Metals Handbook*. Although increasing the carbon and silicon contents improves the graphitization potential and therefore decreases the chilling tendency, the strength is adversely affected (Fig. 4). This is due to ferrite promotion and the coarsening of pearlite.

The manganese content varies as a function of the desired matrix. Typically, it can be as low as 0.1% for ferritic irons and as high as 1.2% for pearlitic irons, because manganese is a strong pearlite promoter.

From the minor elements, phosphorus and sulfur are the most common and are always present in the composition. They can be as high as 0.15% for low-quality iron and are considerably less for high-quality iron, such as ductile iron or compacted graphite iron. The effect of sulfur must be balanced by the effect of manganese. Without manganese in the iron, undesired iron sulfide (FeS) will form at grain boundaries. If the sulfur content is balanced by manganese, manganese sulfide (MnS) will form, which is harmless because it is distributed within the grains. The optimum ratio between manganese and sulfur for an FeS-free structure and maximum amount of ferrite is:

$$\%\,Mn = 1.7(\%\,S) + 0.15 \quad (Eq\ 2)$$

Other minor elements, such as aluminum, antimony, arsenic, bismuth, lead, magnesium, cerium, and calcium, can significantly alter both the graphite morphology and the microstructure of the matrix.

The range of composition for typical unalloyed common cast irons is given in Table 2. The typical composition range for low- and high-grade unalloyed gray iron (flake graphite iron) cast in sand molds is given in Table 3.

Both major and minor elements have a direct influence on the morphology of flake graphite. The typical graphite shapes for flake graphite are shown in Fig. 5. Type A graphite is found in inoculated irons cooled with moderate rates. In general, it is associated with the best mechanical properties, and cast irons with this type of graphite exhibit moderate undercooling during solidification (Fig. 6). Type B graphite is found in irons of near-eutectic composition, solidifying on a limited number of nuclei. Large eutectic cell size and low undercoolings are common in cast irons exhibiting this type of graphite. Type C graphite occurs in hypereutectic irons as a result of solidification with minimum undercooling. Type D graphite is found in hypoeutectic or eutectic irons solidified at rather high cooling rates, while type E graphite is characteristic for strongly hypoeutectic irons. Types D and E are both associated with high undercoolings during solidification. Not only graphite shape but also graphite size is important, because it is directly related to strength (Fig. 7).

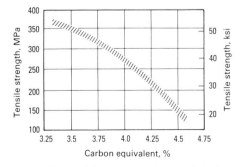

Fig. 4 General influence of carbon equivalent on the tensile strength of gray iron. Source: Ref 2

Table 2 Range of compositions for typical unalloyed common cast irons

Type of iron	Composition, %				
	C	Si	Mn	P	S
Gray (FG)	2.5–4.0	1.0–3.0	0.2–1.0	0.002–1.0	0.02–0.25
Compacted graphite (CG)	2.5–4.0	1.0–3.0	0.2–1.0	0.01–0.1	0.01–0.03
Ductile (SG)	3.0–4.0	1.8–2.8	0.1–1.0	0.01–0.1	0.01–0.03
White	1.8–3.6	0.5–1.9	0.25–0.8	0.06–0.2	0.06–0.2
Malleable (TG)	2.2–2.9	0.9–1.9	0.15–1.2	0.02–0.2	0.02–0.2

Source: Ref 2

6 / Cast Irons

Table 3 Compositions of unalloyed gray irons

ASTM 48 class	Carbon equivalent	C	Si	Mn	P	S
20B	4.5	3.1–3.4	2.5–2.8	0.5–0.7	0.9	0.15
55B	3.6	≤3.1	1.4–1.6	0.6–0.75	0.1	0.12

Alloying elements can be added in common cast iron to enhance some mechanical properties. They influence both the graphitization potential and the structure and properties of the matrix. The main elements are listed below in terms of their graphitization potential:

High positive graphitization potential (decreasing positive potential from top to bottom)

Carbon
Tin
Phosphorus
Silicon
Aluminum
Copper
Nickel

Neutral

Iron

High negative graphitization potential (increasing negative potential from top to bottom)

Manganese
Chromium
Molybdenum
Vanadium

This classification is based on the thermodynamic analysis of the influence of a third element on carbon solubility in the Fe-C-X system, where X is a third element (see the section "Influence of a Third Element on Carbon Solubility in the Fe-C-X System" in the article "Thermodynamic Properties of Iron-Base Alloys" in Volume 15 of the 9th Edition of *Metals Handbook*). Although listed as a graphitizer (which may be true thermodynamically), phosphorus also acts as a matrix hardener. Above its solubility level (probably about 0.08%), phosphorus forms a very hard ternary eutectic. The above classification should also include sulfur as a carbide former, although manganese and sulfur can combine and neutralize each other. The resultant manganese sulfide also acts as nuclei for flake graphite. In industrial processes, nucleation phenomena may sometimes override solubility considerations.

In general, alloying elements can be classified into three categories. Each is discussed below.

Silicon and aluminum increase the graphitization potential for both the eutectic and eutectoid transformations and increase the number of graphite particles. They form solid solutions in the matrix. Because they increase the ferrite/pearlite ratio, they lower strength and hardness.

Nickel, copper, and tin increase the graphitization potential during the eutectic transformation, but decrease it during the eutectoid transformation, thus raising the pearlite/ferrite ratio. This second effect is due to the retardation of carbon diffusion. These elements form solid solution in the matrix. Because they increase the amount of pearlite, they raise strength and hardness.

Chromium, molybdenum, tungsten, and vanadium decrease the graphitization potential at both stages. Thus, they increase the amount of carbides and pearlite. They concentrate in principal in the carbides, forming (FeX)$_n$C-type carbides, but also alloy the αFe solid solution. As long as carbide formation does not occur, these elements increase strength and hardness. Above a certain level, any of these elements will determine the solidification of a structure with both Gr and Fe$_3$C (mottled structure), which will have lower strength but higher hardness.

In alloyed gray iron, the typical ranges for the elements discussed above are as follows:

Element	Composition, %
Chromium	0.2–0.6
Molybdenum	0.2–1
Vanadium	0.1–0.2
Nickel	0.6–1
Copper	0.5–1.5
Tin	0.04–0.08

The influence of composition and cooling rate on tensile strength can be estimated using (Ref 3):

$$TS = 162.37 + 16.61/D - 21.78(\% C) \\ - 61.29(\% Si) - 10.59 (\% Mn - 1.7\% S) \\ + 13.80(\% Cr) + 2.05(\% Ni) + 30.66(\% Cu) \\ + 39.75(\% Mo) + 14.16 (\% Si)^2 \\ - 26.25(\% Cu)^2 - 23.83 (\% Mo)^2 \quad (Eq 3)$$

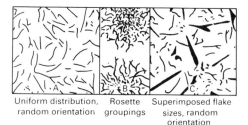
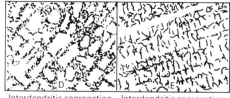

Fig. 5 Typical flake graphite shapes specified in ASTM A 247. A, uniform distribution, random orientation; B, rosette groupings; C, kish graphite (superimposed flake sizes, random orientation); D, interdendritic segregation with random orientation; E, interdendritic segregation with preferred orientation

where D is the bar diameter (in inches). Equation 3 is valid for bar diameters of 20 to 50 mm (⅛ to 2 in.) and compositions within the following ranges:

Element	Composition, %
Carbon	3.04–3.29
Chromium	0.1–0.55
Molybdenum	0.03–0.78
Silicon	1.6–2.46
Nickel	0.07–1.62
Sulfur	0.089–0.106
Manganese	0.39–0.98
Copper	0.07–0.85

The cooling rate, like the chemical composition, can significantly influence the as-cast structure and therefore the mechanical properties. The cooling rate of a casting is primarily a function of its section size. The dependence of structure and properties on section size is termed section sensitivity. Increasing the cooling rate will:

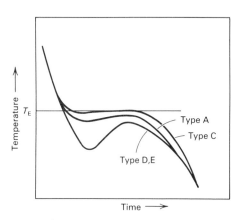

Fig. 6 Characteristic cooling curves associated with different flake graphite shapes. T_E, equilibrium eutectic temperature

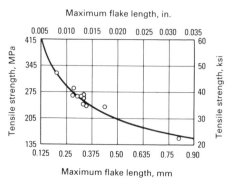

Fig. 7 Effect of maximum graphite flake length on the tensile strength of gray iron. Source: Ref 3

Classification and Basic Metallurgy of Cast Iron / 7

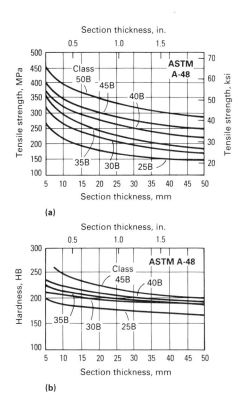

Fig. 8 Influence of section thickness of the casting on tensile strength (a) and hardness (b) for a series of gray irons classified by their strength as-cast in 30 mm (1.2 in.) diam bars. Source: Ref 2

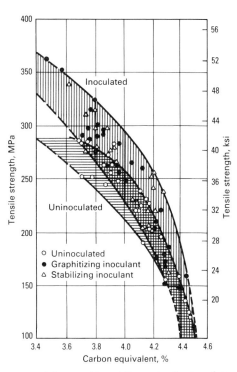

Fig. 9 Influence of inoculation on tensile strength as a function of carbon equivalent for 30 mm (1.2 in.) diam bars. Source: Ref 2

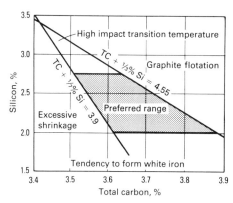

Fig. 10 Typical range for carbon and silicon contents in good-quality ductile iron. Source: Ref 2

- Refine both graphite size and matrix structure; this will result in increased strength and hardness
- Increase the chilling tendency; this may result in higher hardness, but will decrease the strength

Consequently, composition must be tailored in such a way as to provide the correct graphitization potential for a given cooling rate. For a given chemical composition and as the section thickness increases, the graphite becomes coarser, and the pearlite/ferrite ratio decreases, which results in lower strength and hardness (Fig. 8). Higher carbon equivalent has similar effects.

The liquid treatment of cast iron is of paramount importance in the processing of this alloy because it can dramatically change the nucleation and growth conditions during solidification. As a result, graphite morphology, and therefore properties, can be significantly affected. In gray iron practice, the liquid treatment used is termed inoculation and consists of minute additions of minor elements before pouring. Typically, ferrosilicon with additions of aluminum and calcium, or proprietary alloys are used as inoculants. The main effects of inoculation are:

- An increased graphitization potential because of decreased undercooling during solidification; as a result of this, the chilling tendency is diminished, and graphite shape changes from type D or E to type A
- A finer structure, that is, higher number of eutectic cells, with a subsequent increase in strength

As shown in Fig. 9, inoculation improves tensile strength. This influence is more pronounced for low-CE cast irons.

Heat treatment can considerably alter the matrix structure, although graphite shape and size remain basically unaffected. A rather low proportion of the total gray iron produced is heat treated. Common heat treatment may consist of stress relieving or of annealing to decrease hardness.

Ductile Iron (Spheroidal Graphite Iron)

Composition. The main effects of chemical composition are similar to those described for gray iron, with quantitative differences in the extent of these effects and qualitative differences in the influence on graphite morphology. The carbon equivalent has only a mild influence on the properties and structure of ductile iron, because it affects graphite shape considerably less than in the case of gray iron. Nevertheless, to prevent excessive shrinkage, high chilling tendency, graphite flotation, or a high impact transition temperature, optimum amounts of carbon and silicon must be selected. Figure 10 shows the basic guidelines for the selection of appropriate compositions.

As mentioned previously, minor elements can significantly alter the structure in terms of graphite morphology, chilling tendency, and matrix structure. Minor elements can promote the spheroidization of graphite or can have an adverse effect on graphite shape. The minor elements that adversely affect graphite shape are said to degenerate graphite shape. A variety of graphite shapes can occur, as illustrated in Fig. 11. Graphite shape is the single most important factor affecting the mechanical properties of cast iron, as shown in Fig. 12.

The generic influence of various elements on graphite shape is given in Table 4. The elements in the first group—the spheroidizing elements—can change graphite shape from flake through compacted to spheroidal. This is illustrated in Fig. 13 for magnesium. The most widely used element for the production of spheroidal graphite is magnesium. The amount of residual magnesium, Mg_{resid}, required to produce spheroidal graphite is generally 0.03 to 0.05%. The precise level depends on the cooling rate. A higher cooling rate requires less magnesium. The amount of magnesium to be added in the iron is a function of the initial sulfur level, S_{in}, and the recovery of magnesium, η, in the particular process used:

$$Mg_{added} = \frac{0.75\, S_{in} + Mg_{resid}}{\eta} \quad (Eq\ 4)$$

A residual magnesium level that is too low results in insufficient nodularity (that is, a low ratio between the spheroidal graphite and the total amount of graphite in the structure). This in turn results in a deterioration of the mechanical properties of the iron, as illustrated in Fig. 14. If the magnesium content is too high, carbides are promoted.

The presence of antispheroidizing (deleterious) minor elements may result in graphite shape deterioration, up to complete graphite degeneration. Therefore, upper limits are set on the amount of deleterious elements to be accepted in the composition of cast iron. Typical limits are given below (Ref 6):

8 / Cast Irons

Element	Composition, %
Aluminum	0.1
Arsenic	0.02
Bismuth	0.002
Cadmium	0.01
Lead	0.002
Antimony	0.002
Selenium	0.03
Tellurium	0.02
Titanium	0.1
Zirconium	0.1

These values can be influenced by the combination of various elements and by the presence of rare earths in the composition. Furthermore, some of these elements can be deliberately added during liquid processing in order to increase nodule count.

Alloying elements have in principle the same influence on structure and properties as for gray iron. Because a better graphite morphology allows more efficient use of the mechanical properties of the matrix, alloying is more common in ductile iron than in gray iron.

Cooling Rate. When changing the cooling rate, effects similar to those discussed for gray iron also occur in ductile iron, but the section sensitivity of ductile iron is lower. This is because spheroidal graphite is less affected by cooling rate than flake graphite.

The liquid treatment of ductile iron is more complex than that of gray iron. The two stages for the liquid treatment of ductile iron are:

- Modification, which consists of magnesium or magnesium alloy treatment of the melt, with the purpose of changing graphite shape from flake to spheroidal
- Inoculation (normally, postinoculation, that is, after the magnesium treatment) to increase the nodule count. Increasing the nodule count is an important goal, because a higher nodule count is associated with less chilling tendency (Fig. 15) and a higher as-cast ferrite/pearlite ratio

Heat treatment is extensively used in the processing of ductile iron because better advantage can be taken of the matrix structure than for gray iron. The heat treatments usually applied are as follows:

- Stress relieving
- Annealing to produce a ferritic matrix
- Normalizing to produce a pearlitic matrix

Table 4 Influence of minor elements on graphite shape

Element category	Element
Spheroidizer	Magnesium, calcium, rare earths (cerium, lanthanum, etc.), yttrium
Neutral	Iron, carbon, alloying elements
Antispheroidizer (degenerate shape)	Aluminum, arsenic, bismuth, tellurium, titanium, lead, sulfur, antimony

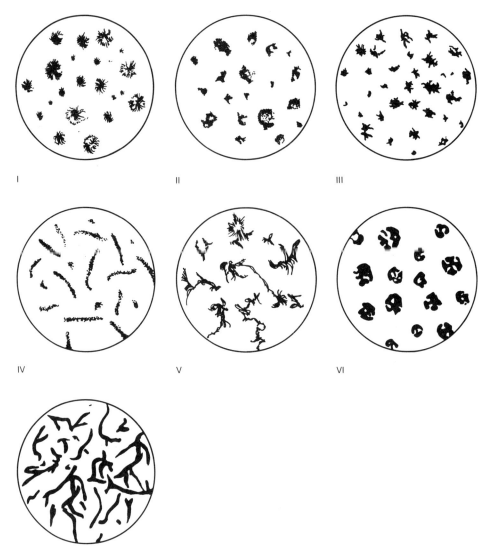

Fig. 11 Typical graphite shapes after ASTM A 247. I, spheroidal graphite; II, imperfect spheroidal graphite; III, temper graphite; IV, compacted graphite; V, crab graphite; VI, exploded graphite; VII, flake graphite

- Hardening to produce tempering structures
- Austempering to produce a ferritic bainite

The advantage of austempering is that it results in ductile irons with twice the tensile strength for the same toughness. A comparison between some mechanical properties of austempered ductile iron and standard ductile iron is shown in Fig. 16.

Compacted Graphite Irons

Compacted graphite irons have a graphite shape intermediate between spheroidal and flake. Typically, compacted graphite looks like type IV graphite (Fig. 11). Consequently, most of the properties of CG irons lie in between those of gray and ductile iron.

The chemical composition effects are similar to those described for ductile iron. Carbon equivalent influences strength less obviously than for the case of gray iron, but

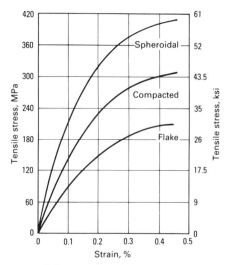

Fig. 12 Influence of graphite morphology on the stress-strain curve of several cast irons

Classification and Basic Metallurgy of Cast Iron / 9

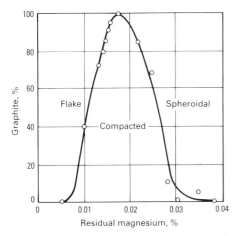

Fig. 13 Influence of residual magnesium on graphite shape

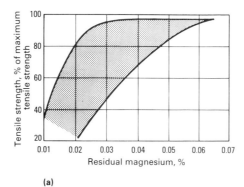

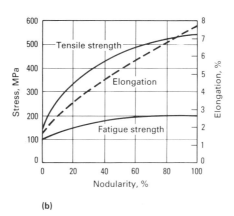

Fig. 14 Influence of residual magnesium (a) and nodularity (b) on some mechanical properties of ductile iron. Sources: Ref 4, 5

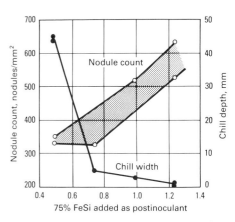

Fig. 15 Influence of the amount of 75% ferrosilicon added as a postinoculant on the nodule count and chill depth of 3 mm (0.12 in.) plates. Source: Ref 7

more than for ductile iron, as shown in Fig. 17. The graphite shape is controlled, as in the case of ductile iron, through the content of minor elements. When the goal is to produce compacted graphite, it is easier from the standpoint of controlling the structure to combine spheroidizing (magnesium, calcium, and/or rare earths) and antispheroidizing (titanium and/or aluminum) elements. Additional information is available in the article "Compacted Graphite Irons" in Volume 15 of the 9th Edition of *Metals Handbook*.

The cooling rate affects properties less for gray iron but more for ductile iron (Fig. 18). In other words, CG iron is less section sensitive than gray iron. However, high cooling rates are to be avoided because of the high propensity of CG iron for chilling and high nodule count in thin sections.

Liquid treatment can have two stages, as for ductile iron. Modification can be achieved with magnesium, Mg + Ti, Ce + Ca, and so on. Inoculation must be kept at a low level to avoid excessive nodularity.

Heat treatment is not common for CG irons.

Malleable Irons

Malleable cast irons differ from the types of irons previously discussed in that they have an initial as-cast white structure, that is, a structure consisting of iron carbides in a pearlitic matrix. This white structure is then heat treated (annealing at 800 to 970 °C, or 1470 to 1780 °F), which results in the decomposition of Fe_3C and the formation of temper graphite. The basic solid state reaction is:

$$Fe_3C \rightarrow \gamma + Gr \qquad (Eq\ 5)$$

The final structure consists of graphite and pearlite, pearlite and ferrite, or ferrite. The structure of the matrix is a function of the cooling rate after annealing. Most of the malleable iron is produced by this technique and is called blackheart malleable iron. Some malleable iron is produced in Europe by decarburization of the white as-cast iron, and it is called whiteheart malleable iron.

The composition of malleable irons must be selected in such a way as to produce a white as-cast structure and to allow for fast annealing times. Some typical compositions are given in Table 2. Although higher carbon and silicon reduce the heat treatment time, they must be limited to ensure a graphite-free structure upon solidification. Both tensile strength and elongation decrease with higher carbon equivalent. Nevertheless, it is not enough to control the carbon equivalent. The annealing time depends on the number of graphite nuclei available for graphitization, which in turn

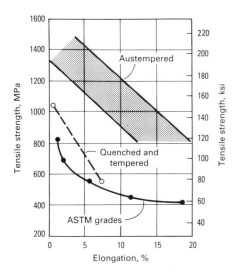

Fig. 16 Properties of some standard and austempered ductile irons. Source: Ref 8

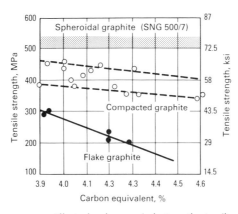

Fig. 17 Effect of carbon equivalent on the tensile strength of flake, compacted, and spheroidal graphite irons cast in 30 mm (1.2 in.) diam bars. Source: Ref 9

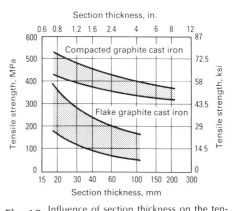

Fig. 18 Influence of section thickness on the tensile strength of CG irons. Source: Ref 10

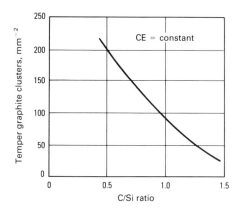

Fig. 19 Influence of C/Si ratio on the number of temper graphite clusters at constant carbon equivalent. Source: Ref 10

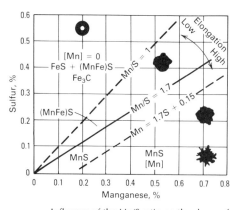

Fig. 20 Influence of the Mn/S ratio on the shape of temper graphite. Bracketed elements are dissolved in the matrix.

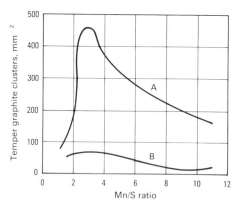

Fig. 21 Influence of the Mn/S ratio on the number of temper graphite clusters after annealing. A, low-temperature holding for 12 h at 350 °C (660 °F); B, no low-temperature holding

depends on, among other factors, the C/Si ratio. As shown in Fig. 19, a lower C/Si ratio (that is, a higher silicon content for a constant carbon equivalent) results in a higher temper graphite count. This in turn translates into shorter annealing times.

Manganese content and the Mn/S ratio must be closely controlled. In general, a lower manganese content is used when ferritic rather than pearlitic structures are desired. The correct Mn/S ratio can be calculated with Eq 2. Equation 2 is plotted in Fig. 20. Under the line described by Eq 2, all sulfur is stoichiometrically tied to manganese as MnS. The excess manganese is dissolved in the ferrite. In the range delimited by the lines given by Eq 2 and the line Mn/S = 1, a mixed sulfide, (Mn,Fe)S, is formed. For Mn/S ratios smaller than 1, pure FeS is also formed. It is assumed that the degree of compacting of temper graphite depends on the type of sulfides occurring in the iron (Ref 11). When FeS is predominant, very compacted, nodular temper graphite forms, but some undissolved Fe_3C may persist in the structure, resulting in lower elongations. When MnS is predominant, although the graphite is less compacted, elongation is higher because of the completely Fe_3C-free structure.

The Mn/S ratio also influences the number of temper graphite particles. From this standpoint, the optimum Mn/S ratio is about 2 to 4 (Fig. 21).

Alloying elements can be used in some grades of pearlitic malleable irons. The manganese content can be increased to 1.2%, or copper, nickel, and/or molybdenum can be added. Chromium must be avoided because it produces stable carbides, which are difficult to decompose during annealing.

Cooling Rate. Like all other irons, malleable irons are sensitive to cooling rate. Nevertheless, because the final structure is the result of a solid-state reaction, they are the least section sensitive irons. Typical correlations between tensile strength, elongation, and section thickness are shown in Fig. 22.

The liquid treatment of malleable iron increases the number of nuclei available for the solid-state graphitization reaction. This can be achieved in two different ways, as follows:

- By adding elements that increase undercooling during solidification. Typical elements in this category are magnesium, cerium, bismuth, and tellurium. Higher undercooling results in finer structure, which in turn means more γ-Fe_3C interface. Because graphite nucleates at the γ-Fe_3C interface, this means more nucleation sites for graphite. Higher undercooling during solidification also prevents the formation of unwanted eutectic graphite
- By adding nitrite-forming elements to the melt. Typical elements in this category are aluminum, boron, titanium, and zirconium

The heat treatment of malleable iron determines the final structure of this iron. It has two basic stages. In the first stage, the iron carbide is decomposed in austenite and graphite (Eq 5). In the second stage, the austenite is transformed into pearlite, ferrite, or a mixture of the two. Although there are some compositional differences between ferritic and pearlitic irons, the main difference is in the heat treatment cycle. When ferritic structures are to be produced, cooling rates in the range of 3 to 10 °C/h (5 to 18 °F/h) are required through the eutectoid transformation in the second stage. This is necessary to allow for a complete austenite-to-ferrite reaction. A typical annealing cycle for ferritic malleable iron is shown in Fig. 23. When pearlitic irons are to be produced, different schemes can be used, as shown in Fig. 24. The goal of the treatment is to

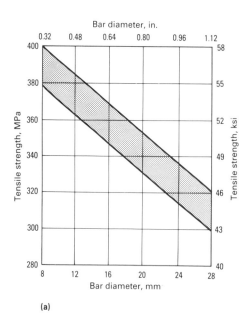

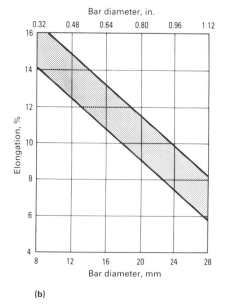

Fig. 22 Influence of bar diameter on the tensile strength (a) and elongation (b) of blackheart malleable iron. Source: Ref 13

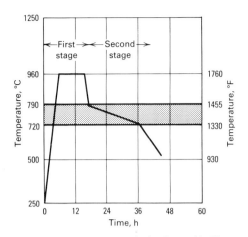

Fig. 23 Heat treatment cycle for ferritic blackheart malleable iron. Source: Ref 1

achieve a eutectoid transformation according to the austenite-to-pearlite reaction. In some limited cases, quenching-tempering treatments are used for malleable irons.

Special Cast Irons

Special cast irons, as previously discussed, are alloy irons that take advantage of the radical changes in structure produced by rather large amounts of alloying elements. Abrasion resistance can be improved by increasing hardness, which in turn can be achieved by either increasing the amount of carbides and their hardness or by producing a martensitic structure. The least expensive material is white iron with a pearlitic matrix. Additions of 3 to 5% Ni and 1.5 to 2.5% Cr result in irons with $(FeCr)_3C$ carbides and an as-cast martensitic matrix. Additions of 11 to 35% Cr produce

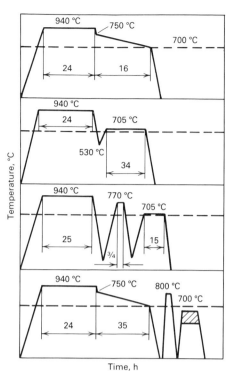

Fig. 24 Heat treatment cycles for pearlitic blackheart malleable irons

$(CrFe)_7C_3$ carbides, which are harder than the iron carbides. Additions of 4 to 16% Mn will result in a structure consisting of $(FeMn)_3C$, martensite, and work-hardenable austenite.

Heat resistance depends on the stability of the microstructure. Irons used for these applications may have a ferritic structure with graphite (5% Si), a ferritic structure with stable carbides (11 to 28% Cr), or a stable austenitic structure with either spheroidal or flake graphite (18% Ni, 5% Si). For corrosion resistance, irons with high chromium (up to 28%), nickel (up to 18%), and silicon (up to 15%) are used.

REFERENCES

1. R. Elliott, *Cast Iron Technology*, Butterworths, 1988
2. C.F. Walton and T.J. Opar, Ed., *Iron Castings Handbook*, Iron Castings Society, 1981
3. C.E. Bates, *AFS Trans.*, Vol 94, 1986, p 889
4. R. Barton, *B.C.I.R.A. J.*, No. 5, 1961, p 668
5. R.W. Lindsay and A. Shames, *AFS Trans.*, Vol 60, 1952, p 650
6. H. Morrogh, *AFS Trans.*, Vol 60, 1952, p 439
7. D.M. Stefanescu, *AFS Int. Cast Met. J.*, June 1981, p 23
8. J.F. Janowak and R.B. Gundlach, *AFS Trans.*, Vol 91, 1983, p 377
9. G.F. Sergeant and E.R. Evans, *Br. Foundryman*, May 1978, p 115
10. D.M. Stefanescu, *Metalurgia*, No. 7, 1967, p 368
11. K. Roesch, *Stahl Eisen*, No. 24, 1957, p 1747
12. R.P. Todorov, in *Proceedings of the 32nd International Foundry Congress* (Warsaw, Poland), International Committee of Foundry Technical Associations
13. K.M. Ankab, O.E. Shulte, and P.N. Bidulia, Isvestia Vishih Utchebnik Zavedenia-Tchornaia, *Metallurghia*, No. 5, 1966, p 168

Gray Iron

Revised by Charles V. White, GMI Engineering and Management Institute

CAST IRONS are alloys of iron, carbon, and silicon in which more carbon is present than can be retained in solid solution in austenite at the eutectic temperature. In gray cast iron, the carbon that exceeds the solubility in austenite precipitates as flake graphite. Gray irons usually contain 2.5 to 4% C, 1 to 3% Si, and additions of manganese, depending on the desired microstructure (as low as 0.1% Mn in ferritic gray irons and as high as 1.2% in pearlitics). Sulfur and phosphorus are also present in small amounts as residual impurities.

Certain important but low-tonnage specialty items in this family of cast metals (notably the austenitic and other highly alloyed gray irons) are not dealt with here; instead the emphasis is on the properties of gray irons used most often and in the largest tonnages. Information on the high-alloy gray irons is given in the article "Alloy Cast Irons" in this Volume. The basic metallurgy of gray cast irons is discussed in the article "Classification and Basic Metallurgy of Cast Iron" in this Volume.

Classes of Gray Iron

A simple and convenient classification of the gray irons is found in ASTM specification A 48, which classifies the various types in terms of tensile strength, expressed in ksi. The ASTM classification by no means connotes a scale of ascending superiority from class 20 (minimum tensile strength of 140 MPa, or 20 ksi) to class 60 (minimum tensile strength of 410 MPa, or 60 ksi). In many applications strength is not the major criterion for the choice of grade. For example, for parts such as clutch plates and brake drums, where resistance to heat checking is important, low-strength grades of iron are the superior performers. Similarly, in heat shock applications such as ingot or pig molds, a class 60 iron would fail quickly, whereas good performance is shown by class 25 iron. In machine tools and other parts subject to vibration, the better damping capacity of low-strength irons is often advantageous.

Generally, it can be assumed that the following properties of gray cast irons increase with increasing tensile strength from class 20 to class 60:

- All strengths, including strength at elevated temperature
- Ability to be machined to a fine finish
- Modulus of elasticity
- Wear resistance

On the other hand, the following properties decrease with increasing tensile strength, so that low-strength irons often perform better than high-strength irons when these properties are important:

- Machinability
- Resistance to thermal shock
- Damping capacity
- Ability to be cast in thin sections

Applications

Gray iron is used for many different types of parts in a very wide variety of machines and structures. Like parts made from other metals and alloys, parts intended to be produced as gray iron castings must be evaluated for the specific service conditions before being approved for production. Often a stress analysis of prototype castings helps establish the appropriate class of gray iron as well as any proof test requirements or other acceptance criteria for production parts.

Castability

Successful production of a gray iron casting depends on the fluidity of the molten metal and on the cooling rate (which is influenced by the minimum section thickness and on section thickness variations). Casting design is often described in terms of section sensitivity. This is an attempt to correlate properties in critical sections of the casting with the combined effects of composition and cooling rate. All these factors are interrelated and may be condensed into a single term, castability, which for gray iron may be defined as the minimum section thickness that can be produced in a mold cavity with given volume/area ratio and mechanical properties consistent with the type of iron being poured.

Fluidity. Scrap losses resulting from misruns, cold shuts, and round corners are often attributed to the lack of fluidity of the metal being poured.

Mold conditions, pouring rate, and other process variables being equal, the fluidity of commercial gray irons depends primarily on the amount of superheat above the freezing temperature (liquidus). As the total carbon (TC) content decreases, the liquidus temperature increases, and the fluidity at a given pouring temperature therefore decreases. Fluidity is commonly measured as the length of flow into a spiral-type fluidity test mold. The relation between fluidity and superheat is shown in Fig. 1 for four unalloyed gray irons of different carbon contents.

The significance of the relationships between fluidity, carbon content, and pouring temperature becomes apparent when it is realized that the gradation in strength in the ASTM classification of gray iron is due in large part to differences in carbon content (~3.60 to 3.80% for class 20; ~2.70 to 2.95% for class 60). The fluidity of these irons thus resolves into a measure of the practical limits of maximum pouring temperature as opposed to the liquidus of the iron being poured. These practical limits of maximum pouring temperature are largely determined by three factors:

- The ability of both mold and cores to withstand the impact of molten iron, an ability that decreases as the pouring temperature increases, thereby favoring low pouring temperatures
- The fact that metal tap temperatures seldom exceed 1550 °C (2825 °F). Because ladling and readling to the point of pouring generally accounts for temperature losses of 55 to 85 °C (100 to 150 °F), the final pouring temperatures seldom exceed 1450 to 1495 °C (2640 to 2720 °F), and in most instances maximum pouring temperatures in the range 1410 to 1450 °C (2570 to 2640 °F) are considered more realistic
- The necessity to control the overall thermal input to the mold in order to control the final desired microstructure

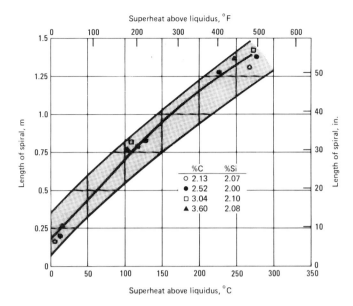

Fig. 1 Fluidity versus degree of superheat for four gray irons of different carbon contents

Table 1 Superheat above liquidus for 2% Si irons of various carbon contents poured at 1455 °C (2650 °F)

Carbon, %	Liquidus temperature °C	°F	Superheat above liquidus °C	°F
2.52	1295	2360	160	290
3.04	1245	2270	210	380
3.60	1175	2150	280	500

It can be seen from Table 1 that because of differences in liquidus temperature, the amount of superheat (and therefore fluidity) varies with carbon content when various compositions are cast from the same pouring temperature.

Microstructure

The usual microstructure of gray iron is a matrix of pearlite with graphite flakes dispersed throughout. Foundry practice can be varied so that nucleation and growth of graphite flakes occur in a pattern that enhances the desired properties. The amount, size, and distribution of graphite are important. Cooling that is too rapid may produce so-called chilled iron, in which the excess carbon is found in the form of massive carbides. Cooling at intermediate rates can produce mottled iron, in which carbon is present in the form of both primary cementite (iron carbide) and graphite. Very slow cooling of irons that contain large percentages of silicon and carbon is likely to produce considerable ferrite and pearlite throughout the matrix, together with coarse graphite flakes.

Flake graphite is one of seven types (shapes or forms) of graphite established in ASTM A 247. Flake graphite is subdivided into five types (patterns), which are designated by the letters A through E (see Fig. 2). Graphite size is established by comparison with an ASTM size chart, which shows the typical appearances of flakes of eight different sizes at 100× magnification.

Type A flake graphite (random orientation) is preferred for most applications. In the intermediate flake sizes, type A flake graphite is superior to other types in certain wear applications such as the cylinders of internal combustion engines. Type B flake graphite (rosette pattern) is typical of fairly rapid cooling, such as is common with moderately thin sections (about 10 mm, or ⅜ in.) and along the surfaces of thicker sections, and sometimes results from poor inoculation. The large flakes of type C flake graphite are typical of kish graphite that is formed in hypereutectic irons. These large flakes enhance resistance to thermal shock by increasing thermal conductivity and decreasing elastic modulus. On the other hand, large flakes are not conducive to good surface finishes on machined parts or to high strength or good impact resistance. The small, randomly oriented interdendritic flakes in type D flake graphite promote a fine machined finish by minimizing surface pitting, but it is difficult to obtain a pearlitic matrix with this type of graphite. Type D flake graphite may be formed near rapidly cooled surfaces or in thin sections. Frequently, such graphite is surrounded by a ferrite matrix, resulting in soft spots in the casting. Type E flake graphite is an interdendritic form, which has a preferred rather than a random orientation. Unlike type D graphite, type E graphite can be associated with a pearlitic matrix and thus can produce a casting whose wear properties are as good as those of a casting containing only type A graphite in a pearlitic matrix. There are, of course, many applications in which flake type has no significance as long as the mechanical property requirements are met.

Solidification of Gray Iron. In a hypereutectic gray iron, solidification begins with the precipitation of kish graphite in the melt. Kish grows as large, straight, undistorted flakes or as very thick, lumpy flakes that tend to rise to the surface of the melt because of their low relative density. When the temperature has been lowered sufficiently, the remaining liquid solidifies as a eutectic structure of austenite and graphite. Generally, eutectic graphite is finer than kish graphite.

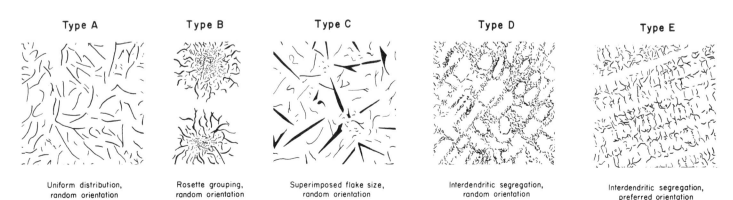

Fig. 2 Types of graphite flakes in gray iron (AFS-ASTM). In the recommended practice (ASTM A 247), these charts are shown at a magnification of 100×. They have been reduced to one-third size for reproduction here.

14 / Cast Irons

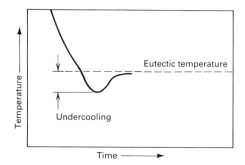

Fig. 3 Undercooling from rapid cooling of a eutectic composition

In hypoeutectic iron, solidification begins with the formation of proeutectic austenite dendrites. As the temperature falls, the dendrites grow, and the carbon content of the remaining liquid increases. When the increasing carbon content and decreasing temperature reach eutectic values, eutectic solidification begins. Eutectic growth from many different nuclei proceeds along crystallization fronts that are approximately spherical. Ultimately, the eutectic cells meet and consume the liquid remaining in the spaces between them. During eutectic solidification, the austenite in the eutectic becomes continuous with the dendritic proeutectic austenite, and the structure can be described as a dispersion of graphite flakes in austenite. After solidification, the eutectic cell structure and the proeutectic austenite dendrites cannot be distinguished metallographically except by special etching or in strongly hypoeutectic iron.

With eutectic compositions, obviously, solidification takes place as the molten alloy is cooled through the normal eutectic temperature range, but without the prior formation of a proeutectic constituent. During the solidification process, the controlling factor remains the rate at which the solidification is proceeding. The rapid solidification favored by thin section sizes or highly conductive molding media can result in undercooling. Undercooling can cause the solidification to start at a temperature lower than the expected eutectic temperature for a given composition (Fig. 3). This can result in a modification of the carbon form from A to E type or can completely suppress its formation and form primary carbides instead.

Room-Temperature Structure. Upon cooling from the eutectic temperature, the austenite will decompose, first by precipitating some of the dissolved carbon and then, at the eutectoid temperature, by undergoing complete transformation. The actual products of the eutectoid transformation depend on rate of cooling as well as on composition of the austenite, but under normal conditions the austenite will transform either to pearlite or to ferrite plus graphite.

Transformation to ferrite plus graphite is most likely to occur with slow cooling rates,

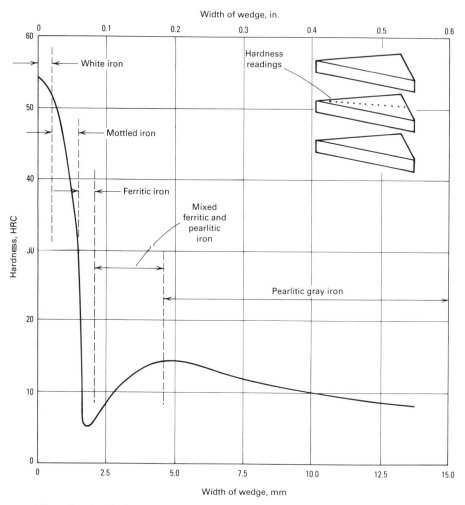

Fig. 4 Effect of section thickness on hardness and structure. Hardness readings were taken at increasing distance from the tip of a cast wedge section, as shown by inset. Composition of iron: 3.52% C, 2.55% Si, 1.01% Mn, 0.215% P, and 0.086% S. Source: Ref 2

which allow more time for carbon migration within the austenite; high silicon contents, which favor the formation of graphite rather than cementite; high values of carbon equivalent; and the presence of fine undercooled (type D) flake graphite. Graphite formed during decomposition is deposited on the existing graphite flakes.

When carbon equivalent values are relatively low or when cooling rates are relatively fast, the transformation to pearlite is favored. In some instances, the microstructure will contain all three constituents: ferrite, pearlite, and graphite. With certain compositions, especially alloy gray irons, it is possible to produce a martensitic matrix by oil quenching through the eutectoid transformation range or an austempered matrix by appropriate isothermal treatment (Ref 1). These treatments are often done deliberately in a secondary heat treatment where high strength or hardness is especially desired, such as in certain wear applications. The secondary heat treatment of gray iron castings is of great value in producing components that must be hard when machining requirements prohibit the use of components that are cast to final shape in white iron.

Section Sensitivity

In practice, the minimum thickness of section in which any given class of gray iron may be poured is more likely to depend on the cooling rate of the section than on the fluidity of the metal. For example, although a plate 300 mm (12 in.) square by 6 mm (0.24 in.) thick can be poured in class 50 as well as in class 25 iron, the former casting would not be gray iron because the cooling rate would be so rapid that massive carbides would be formed. Yet it is entirely feasible to use class 50 iron for a diesel engine cylinder head that has predominantly 6 mm (0.24 in.) wall sections in the water jackets above the firing deck. This is simply because the cooling rate of the cylinder head is reduced by the "mass effect" resulting from enclosed cores and the proximity (often less than 12 mm, or 0.47 in.) of one 6 mm (0.24 in.) wall to the other. Thus the shape of the casting has an important bearing on the choice of metal specification.

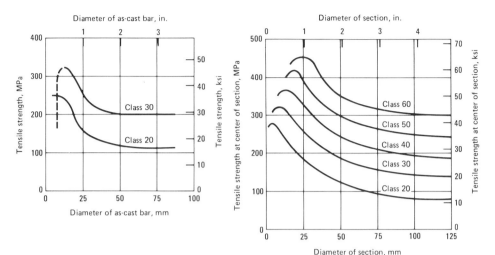

Fig. 5 Effect of section diameter on tensile strength at center of cast specimen for five classes of gray iron

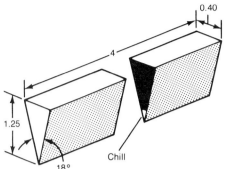

Fig. 6 Standard W2 wedge block used for measuring depth of chill (ASTM A 367). Dimensions given in inches

It should be recognized that the smallest section that can be cast gray, without massive carbides, depends not only on metal composition, but also on foundry practices. For example, by adjusting silicon content or by using graphitizing additions called inoculants in the ladle, the foundryman can decrease the minimum section size for freedom from carbides for a given basic composition of gray iron.

The mass effect associated with increasing section thickness or decreasing cooling rate is much more pronounced in gray iron than in cast steel. The mass effect in cast steel results in increased grain size in heavy sections. This also applies to gray iron, but the most important effects are on graphite size and distribution, and on amount of combined carbon.

For any given gray iron composition, the rate of cooling from the freezing temperature to below about 650 °C (1200 °F) determines the ratio of combined to graphitic carbon, which controls the hardness and strength of the iron. For this reason the effect of section size in gray iron is considerably greater than in the more homogeneous ferrous metals in which cooling rate does not affect the form and distribution of carbon throughout the metal structure.

Typical Effects of Section Size. When a wedge-shape bar with about a 10° taper is cast in a sand mold and sectioned near the center of the length, and Rockwell hardness determinations are made on the cut surface from the point of the wedge progressively into the thicker sections, the curves so determined show to what extent continually increasing section size affects hardness (Fig. 4).

Progressing along the curve from the left in Fig. 4, the following metallographic constituents occur. The tip of the wedge is white iron (a mixture of carbide and pearlite) with a hardness greater than 50 HRC. As the iron becomes mottled (a mixture of white iron and gray iron), the hardness decreases sharply. A minimum is reached because of the occurrence of fine type D flake graphite, which usually has associated ferrite in large amounts. With a slightly lower cooling rate, the structure becomes fine type A flake graphite in a pearlite matrix with the hardness rising to another maximum on the curve. This structure is usually the most desirable for wear resistance and strength. With increasing section thickness beyond this point, the graphite flakes become coarser, and the pearlite lamellae become more widely spaced, resulting in slightly lower hardness. With further increase in wedge thickness and decrease in cooling rate, pearlite decomposes progressively to a mixture of ferrite and graphite, resulting in softer and weaker iron.

The structures of most commercial gray iron castings are represented by the right-hand downward-sloping portion of the curve in Fig. 4, beyond 5 mm (0.2 in.) wedge thickness, and increasing section size is normally reflected by the gradual lowering of hardness and strength. However, thin sections may be represented by the left-hand downward-sloping portion.

Figure 5 shows the average tensile strength (up to ten tests per point) of two irons, for each of which six sizes of cylindrical round bars were cast and appropriate tensile specimens machined. With the class 20 iron, strength increases as the as-cast section decreases down to the 6 mm (0.24 in.) cast bar. However, for the class 30 iron, a section 6 mm (0.24 in.) in diameter is so small that the strength falls off sharply, because of the occurrence of type D flake graphite or mottled iron, or both. The other graph in Fig. 5 shows similar data for the same two classes of iron and for three higher classes.

Section sensitivity effects are used in the form of a wedge test in production control to judge the suitability of an iron for pouring a particular casting. In this test, a wedge-shape casting is poured and upon solidification is evaluated. The standard W2 wedge block specified in ASTM A 367 is shown in Fig. 6. The evaluation consists of measuring the length of the "chilled zone." The measurement, usually made in 0.8 mm (1/32 in.) increments, is related to empirically determined data obtained from a "good" casting. If the evaluation indicates an excessive sensitivity for a part, corrections are made to the molten metal prior to pouring.

Volume/Area Ratios. It is extremely difficult to predict with accuracy the cooling rate for castings other than fairly simple shapes. However, because minimum limita-

Table 2 Volume/area (V/A) ratios for round bars, square bars, and plates

Cast form and size	V/A ratio mm	in.
Bar, 13 mm (½ in.) diam × 533 mm (21 in.)	3.1	0.12
Bar, 13 mm (½ in.) square × 533 mm (21 in.)	3.1	0.12
Plate, 6.4 × 305 × 305 mm (¼ × 12 × 12 in.)	3.0	0.12
Bar, 30 mm (1.2 in.) diam × 533 mm (21 in.)(a)	7.4	0.29
Bar, 30 mm (1.2 in.) square × 533 mm (21 in.)	7.4	0.29
Plate, 16 × 305 × 305 mm (⅝ × 12 × 12 in.)	7.1	0.28
Bar, 50 mm (2 in.) diam × 560 mm (22 in.)	12.2	0.48
Bar, 50 mm (2 in.) square × 560 mm (22 in.)	12.2	0.48
Plate, 28.5 × 305 × 305 mm (1⅛ × 12 × 12 in.)	12.0	0.47
Bar, 100 mm (4 in.) diam × 460 mm (18 in.)	22.9	0.90
Bar, 100 mm (4 in.) square × 460 mm (18 in.)	22.9	0.90
Plate, 65 × 305 × 305 mm (2 9/16 × 12 × 12 in.)	22.8	0.90
Bar, 150 mm (6 in.) diam × 460 mm (18 in.)	32.7	1.29
Bar, 150 mm (6 in.) square × 460 mm (18 in.)	32.7	1.29
Plate, 114 × 305 × 305 mm (4½ × 12 × 12 in.)	32.7	1.29

(a) ASTM size B test bar. Source: Ref 3

Table 3 Bar and plate sizes of equivalent cooling rate

For 305 mm (12 in.) square plates, as recorded in Table 2

Bar diameter, in.	Plate thickness, in.	Ratio of bar diameter to plate thickness
½	¼	2.0
1.2	⅝	1.92
2	1⅛	1.78
4	2 9/16	1.56
6	4½	1.33

Table 4 Minimum prevailing casting sections recommended for gray irons

ASTM A48 class	Minimum thickness in.	mm	V/A ratio(a) in.	mm
20	⅛	3.2	0.06	1.5
25	¼	6.4	0.12	3.0
30	⅜	9.5	0.17	4.3
35	⅜	9.5	0.17	4.3
40	⅝	15.9	0.28	7.1
50	¾	19.0	0.33	8.4
60	1	25.4	0.42	10.7

(a) V/A ratios are for square plates.

tions are involved here, the problem can be resolved through comparisons of the casting design with ratios of volume to surface area or with minimum plate sections.

The volume/area (V/A) ratios for round, square, and plate sections provide a fairly accurate indication of the minimum casting sections possible in simple geometrical shapes (Table 2). The V/A ratios can be reported in either English or metric units and can be converted simply by treating them as length measurements.

Comparison of the ratios of volume to surface area for different shapes gives good agreement with the actual cooling rates of castings made in the same mold material. For long round bars and infinite flat plates, V/A is diameter/4 for bars and thickness/2 for plates; that is, a large plate casting would have the same cooling rate as a round bar with a diameter twice the plate thickness. Most castings, however, freeze somewhat faster than an infinite flat plate, and rather than establishing a 2-to-1 ratio of bar to plate, a smaller ratio will often give a better correlation with the cooling rate. The bar and plate sizes shown in Table 3 are nearly equivalent in cooling rate.

Similar comparisons have been made for production castings. In one study, the properties of a flat section from a 0.6 m (24 in.) cross pipe fitting having a nominal thickness of 29.5 mm (1.16 in.) were compared with the properties of a 50 mm (2 in.) diam cylindrical test bar cast from the same heat. The tensile strengths of the test bars were within about 16 MPa (2.3 ksi) of the tensile strengths of the cross pipe fittings for eight heats ranging in strength from about 205 to 310 MPa (30 to 45 ksi), an average variation of less than 8%. These results from production castings correlate well with the calculated equivalence given in Table 3. Other examples of this type of correlation are given in Ref 3.

Relationships developed for various specific castings are valid when an iron of controlled composition, and therefore of similar section sensitivity, is used consistently. For instance, with a copper-molybdenum iron of well-controlled composition, a tensile strength of 450 MPa (65 ksi) in the ASTM B test specimen has been found to ensure 345 MPa (50 ksi) tensile strength in a cast crankshaft 2.13 m (7 ft) long with sections thicker than 30.5 mm (1.2 in.). Such translation of properties of a small test bar to properties expected in a larger section cannot be done indiscriminately, because different irons may vary widely in section sensitivity.

Prevailing Sections

Although the ASTM size B test bar (30.5 mm, or 1.2 in., diam) is the bar most commonly used for all gray irons from class 20 to class 60, ASTM specification A 48 provides a series of bar sizes from which one that approximates the cooling rate in the critical section of the casting can be selected. In practice, it is customary to be somewhat more definite regarding the prevailing values of minimum casting section considered feasible for the various ASTM classes of cast iron. As summarized in Table 4, these minimum prevailing sections include the requirement of freedom from carbidic areas. In a platelike section, occasional thinner walls (such as ribs) are of no importance unless they are very thin or are appended to the outer edges of the casting.

Mechanical properties of class 30 and class 50 gray irons in various sections are shown in Fig. 7. For class 30 iron, the combined carbon content and hardness are still at a safe level in sections equivalent to a 10 mm (0.4 in.) plate, which has a V/A ratio of about 5 mm (0.20 in.). For class 50 iron, however, both combined carbon and Brinell hardness show marked increases when the thickness of the equivalent plate section is decreased to about 15 mm (0.6 in.), with V/A ratio around 7 mm (0.27 in.). These results are consistent with the minimum prevailing casting sections recommended in Table 4.

The hazards involved in pouring a given class of gray iron in a plate section thinner than recommended are discovered when the casting is machined. Typical losses as a result of specifying too high a strength for a prevailing section of 9.5 mm (⅜ in.) are given below (rejections were for "hard spots" that made it impossible to machine the castings by normal methods):

Class	Rejections, %
35	Negligible
45	25
55	80–100

In marginal applications, a higher class of iron may sometimes be used if the casting is cooled slowly (in effect, increasing the section thickness) by judicious placement of flow-offs and risers. An example is the successful production of a 25 mm (1 in.) diam single-throw crankshaft for an air compressor. This shaft was hard at the extreme ends when poured in class 50 iron. The difficulty was corrected by flowing metal through each end into flow-off risers that adequately balanced the cooling rate at the ends with the cooling rate at the center.

In sum, the selection of a suitable grade of gray iron for a specific casting necessarily requires an evaluation of the size and shape of the casting as related to its cooling rate, or volume/area ratio. For a majority of parts, this evaluation need be no more than a determination of whether or not the V/A ratio of the casting exceeds the minimum V/A ratio indicated for the grade considered.

Test Bar Properties

Mechanical property values obtained from test bars are sometimes the only available guides to the mechanical properties of the metal in production castings. When test bars and castings are poured from metal of the same chemical history, correlations can be drawn between the thermal history of the casting and that of the test bar. The strength of the test bar gives a relative strength of the casting, corrected for the cooling rate of the various section thicknesses. Through careful analysis of the critical sections of a casting, accurate predictions of mechanical behavior can be achieved.

Usual Tests. Tension and transverse tests on bars that are cast specifically for such tests are the most common methods used for evaluating the strength of gray iron.

Yield strength, elongation, and reduction of area are seldom determined for gray iron in standard tension tests. The transverse test measures strength in bending and has the additional advantage that a deflection value may be obtained readily. Minimum specification values are given in Table 5. Data can usually be obtained faster from the transverse test than from the tension test because machining of the specimen is unnecessary. The surface condition of the bar will affect the transverse test but not the tension test made on a machined specimen. Conversely, the presence of coarse graphite in the center of the bar, which can occur in an iron that is very section sensitive, will affect the tension test but not the transverse test.

Hardness tests, on either test bars or castings, are used as an approximate measure of strength and sometimes as an indication of relative machinability. Relationships between Brinell hardness and tensile strength generally follow the pattern repro-

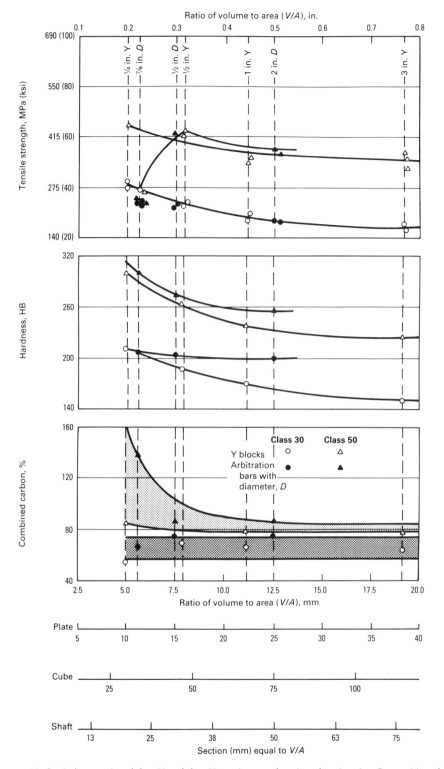

Fig. 7 Mechanical properties of class 30 and class 50 gray iron as a function of section size. Composition of the class 30 iron: 3.40% C, 2.38% Si, 0.71% Mn, 0.423% P, and 0.152% S; for the class 50 iron: 2.96% C, 1.63% Si, 1.05% Mn, 0.67% Mo, 0.114% P, and 0.072% S. Source: Ref 4

imum statistical variation between bars. By its nature gray iron behaves as a brittle material in tension, with no measurable elongation after fracture. This characteristic can be exaggerated by imposing a nonaxial load during tensile testing, resulting in statistical variations, which may not be a true measure of the quality of the iron. To overcome this tendency, many shops use self-aligning nonthreaded grips in the performance of tensile tests on gray iron test bars.

Typical Specifications. ASTM A 48 is typical of specifications based on test bars. In practice, one of three different standard sizes of separately cast test bars is used to evaluate the properties in the controlling section of the castings. After manufacturer and purchaser agree on a controlling section of the casting, the size of test bar that corresponds, approximately, to the cooling rate expected in that section is designated by a letter (see Table 6).

Most gray iron castings for general engineering use are specified as either class 25, 30, or 35. Specification A 48 is based entirely on mechanical properties, and the composition that provides the required properties can be selected by the individual producer. A manufacturer whose major production is medium-section castings of class 35 iron will find, for heavy-section castings for which the 50 mm (2 in.) test bar is required for qualifying, that the same composition will not meet the requirements for class 35. It will qualify only for some lower class, such as 25 or 30. As the thickness of the controlling section increases, the composition must be adjusted to maintain the same tensile strength.

SAE standard J431c for gray cast irons (see Tables 7 to 10) describes requirements that are more specific than those described in ASTM A 48. An iron intended for heavy sections, such as grade G3500, is specified to have higher strength and hardness in the standard test bar than does grade G2500, which is intended for light-section castings. Typical applications for the various grades are summarized in Table 11.

ASTM specifications other than A 48 include A 159 (automotive), A 126 (valves, flanges, and pipe fittings), A 74 (soil pipe and fittings), A 278 (pressure-containing parts for temperatures up to 340 °C, or 650 °F), A 319 (nonpressure-containing parts for elevated-temperature service) and A 436 (austenitic gray irons for heat, corrosion, and wear resistance). The austenitic gray irons are described in the article "Alloy Cast Irons" in this Volume. ASTM A 438 describes the standard method for performing transverse bending tests on separately cast, cylindrical test bars of gray cast iron.

Compressive Strength. When gray iron is used for structural applications such as machinery foundations or supports, the engineer, who is usually designing to support weight only, bases his calculations on the

duced in Fig. 8, which shows the variation of tensile strength with Brinell hardness for a series of gray irons produced by a single foundry. The data in Fig. 8 are from ASTM size A and B test bars poured in a series of inoculated gray irons. The successful use of Brinell hardness as a measure of strength depends on whether it can be proved suitable for the application, which may involve service tests or mechanical property tests on specimens cut from production castings.

Testing Precautions. In the assessment of mechanical properties for a series of heats, precautions should be taken to ensure min-

18 / Cast Irons

Table 5 Transverse breaking loads of gray irons tested per ASTM A 438

ASTM class(a)	Approximate tensile strength		Corrected transverse breaking load(a)					
			A bar(b)		B bar(c)		C bar(d)	
	MPa	ksi	kg	lb	kg	lb	kg	lb
20	138	20	408	900	816	1800	2720	6 000
25	172	25	465	1025	907	2000	3080	6 800
30	207	30	522	1150	998	2200	3450	7 600
35	241	35	578	1275	1089	2400	3760	8 300
40	276	40	635	1400	1179	2600	4130	9 100
45	310	45	699	1540	1270	2800	4400	9 700
50	345	50	760	1675	1361	3000	4670	10 300
60	414	60	873	1925	1542	3400	5670	12 500

(a) For separately cast test specimens produced in accordance with ASTM A 48, ASTM A 278, ASME SA278, FED QQ-I-652, or any other specification that designates ASME A438 as the test method. Included in specifications only by agreement between producer and purchaser. (b) 22.4 mm (0.88 in.) diam; 305 mm (12 in.) between supports. (c) 30.5 mm (1.20 in.) diam; 460 mm (18 in.) between supports. (d) 50.8 mm (2.00 in.) diam; 610 mm (24 in.) between supports

Table 6 Test bars designed to match controlling sections of castings (ASTM A 48)

Controlling section		Test bar	Diameter of as-cast test bar	
in.	mm		mm	in.
<0.25	<6	S	(a)	(a)
0.25–0.50	6–12	A	22.4	0.88
0.51–1.00	13–25	B	30.5	1.20
1.01–2.00	26–50	C	50.8	2.00
>2.00	>50	S	(a)	(a)

(a) All dimensions of test bar by agreement between manufacturer and purchaser

compressive strength of the material. Table 12, which summarizes typical values for mechanical properties of the various grades, shows the high compressive strength of gray irons. Figure 9 compares the stress-strain curves in tension and compression for a class 20 and a class 40 gray iron. The compressive strength of gray iron is typically three to four times that of the tensile strength. If loads other than dead weights are involved (unless these loads are constant), the problem is one of dynamic stresses, which requires the consideration of fatigue and damping characteristics.

Tensile strength is considered in selecting gray iron for parts intended for static loads in direct tension or bending. Such parts include pressure vessels, autoclaves, housings and other enclosures, valves, fittings, and levers. Depending on the uncertainty of loading, safety factors of 2 to 12 have been used in figuring allowable design stresses.

Transverse Strength and Deflection. When an ASTM arbitration bar is loaded as a simple beam and the load and deflection required to break it are determined, the resulting value is converted into a nominal index of strength by using the standard beam formula. The value that is determined is arbitrarily called the modulus of rupture. The values for modulus of rupture are useful for production control, but cannot be used in the design of castings without further analysis and interpretation. Rarely does a casting have a shape such that those areas subject to bending stress have a direct relationship to the round arbitration bar. A more rational approach is to use the tensile strength (or fatigue limit) and, after determining the section modulus of the actual shape, apply the proper bending formula. However, because the difficulty of obtaining meaningful values of tensile strength in tests of small specimens, the load computed in this manner is usually somewhat lower than the actual load required to rupture the part, unless unfavorable residual stresses are present in the finished part.

Elongation of gray iron at fracture is very small (of the order of 0.6%) and hence is seldom reported. The designer cannot use the numerical value of permanent elongation in any quantitative manner.

Torsional Shear Strength. As shown in Table 12, most gray irons have high torsional shear strength. Many grades have torsional strength greater than that of some grades of steel. This characteristic, along with low notch sensitivity, makes gray iron a suitable material for shafting of various types, particularly in the grades of higher tensile strength. Most shafts are subjected to dynamic torsional stresses, and the designer should carefully consider the exact nature of the loads to be encountered. For the higher-strength irons, stress concentration factors associated with changes of shape in the part are important for torque loads as well as for bending and tension loads.

Modulus of Elasticity. Typical stress-strain curves for gray iron are shown in Fig. 10. Gray iron does not obey Hooke's law, and the modulus in tension is usually determined arbitrarily as the slope of the line connecting the origin of the stress-strain curve with the point corresponding to one-fourth the tensile strength (secant modulus). Some engineers use the slope of the stress-strain curve near the origin (tangent modulus). The secant modulus is a conservative value suitable for most engineering work; design loads are seldom as high as one-fourth the tensile strength, and the deviation of the stress-strain curve from linearity is usually less than 0.01% at these loads. However, in the design of certain types of machinery, such as precision equipment, where design stresses are very low, the use of the tangent modulus may represent the actual situation more accurately.

The modulus of gray iron (see Table 13) varies considerably more than do the moduli for most other metals. Thus, in using observed strain to calculate stress, it is essential to measure the modulus of the particular gray iron specimen being considered. A significant range in modulus values is experienced because of both section size and chemical analysis variations (Fig. 11). In addition, the modulus experiences a linear reduction with increasing temperature. The rate of reduction can be reduced through alloy additions (Fig. 12). The numerical value of the modulus in torsion is always less than it is in tension, as is the case with steel.

Hardness of gray iron, as measured by Brinell or Rockwell testers, is an intermediate value between the hardness of the soft graphite in the iron and that of the harder metallic matrix. Variations in graphite size and distribution will cause wide variations in hardness (particularly Rockwell hardness) even though the hardness of the me-

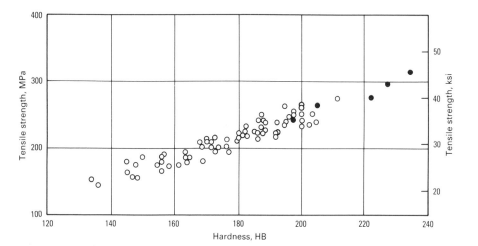

Fig. 8 Relationship between tensile strength and Brinell hardness for a series of inoculated gray irons from a single foundry. Open circles represent unalloyed gray iron, and closed circles represent alloy gray iron. Source: Ref 5

Table 7 Mechanical properties of SAE J431 automotive gray cast irons
Properties determined from as-cast test bar B (30.5 mm, or 1.2 in., diam)

SAE grade	Hardness, HB	Minimum transverse load		Minimum deflection		Minimum tensile strength	
		kg	lb	mm	in.	MPa	ksi
G1800	187 max	780	1720	3.6	0.14	124	18
G2500	170–229	910	2000	4.3	0.17	173	25
G3000	187–241	1000	2200	5.1	0.20	207	30
G3500	207–255	1110	2450	6.1	0.24	241	35
G4000	217–269	1180	2600	6.9	0.27	276	40

Table 8 Typical base compositions of SAE J431 automotive gray cast irons
See Table 7 for mechanical properties. If either carbon or silicon is on the high side of the range, the other should be on the low side.

UNS	SAE grade	Composition, %				
		TC(a)	Mn	Si	P	S
F10004	G1800(b)	3.40–3.70	0.50–0.80	2.80–2.30	0.15	0.15
F10005	G2500(b)	3.20–3.50	0.60–0.90	2.40–2.00	0.12	0.15
F10006	G3000(c)	3.10–3.40	0.60–0.90	2.30–1.90	0.10	0.15
F10007	G3500(c)	3.00–3.30	0.60–0.90	2.20–1.80	0.08	0.15
F10008	G4000(c)	3.00–3.30	0.70–1.00	2.10–1.80	0.07	0.15

(a) TC, total carbon. (b) Ferritic-pearlitic microstructure. (c) Pearlitic microstructure

Table 9 Mechanical properties of SAE J431 automotive gray cast irons for heavy-duty service
Properties determined from as-cast test bar B (30.5 mm, or 1.2 in., diam)

SAE grade	Hardness, HB	Minimum transverse load		Minimum deflection		Minimum tensile strength	
		kg	lb	mm	in.	MPa	ksi
G2500a	170–229	910	2000	4.3	0.17	173	25
G3500b	207–255	1090	2400	6.1	0.24	241	35
G3500c	207–255	1090	2400	6.1	0.24	241	35
G4000d	241–321(a)	1180	2600	6.9	0.27	276	40

(a) Determined on a specified bearing surface

Table 10 Typical base compositions of SAE J431 automotive gray cast irons for heavy-duty service
See Table 9 for mechanical properties.

UNS	SAE grade	Composition, %(a)				
		TC	Mn	Si	P	S
F10009	G2500a(b)	3.40 min	0.60–0.90	1.60–2.10	0.12	0.12
F10010	G3500b(c)	3.40 min	0.60–0.90	1.30–1.80	0.08	0.12
F10011	G3500c(c)	3.50 min	0.60–0.90	1.30–1.80	0.08	0.12
F10012	G4000d(d)	3.10–3.60	0.60–0.90	1.95–2.40	0.07	0.12

(a) If either carbon or silicon is on the high side of the range, the other should be on the low side. Alloying elements not listed in this table may be required. (b) Microstructure: size 2 to 4 type A graphite in a matrix of lamellar pearlite containing not more than 15% free ferrite. (c) Microstructure: size 3 to 5 type A graphite in a matrix of lamellar pearlite containing not more than 5% free ferrite or free carbide. (d) Alloy gray iron containing 0.85 to 1.25% Cr, 0.40 to 0.60% Mo, and 0.20 to 0.45% Ni or as agreed. Microstructure: primary carbides and size 4 to 7 type A or E graphite in a matrix of fine pearlite, as determined in a zone at least 3.2 mm (⅛ in.) deep at a specified location on a cam surface

Table 11 Automotive applications of gray cast iron

Grade	Typical uses	Grade	Typical uses
G1800	Miscellaneous soft iron castings (as-cast or annealed) in which strength is not a primary consideration	G3000	Automobile and diesel cylinder blocks, cylinder heads, flywheels, differential carrier castings, pistons, medium-duty brake drums and clutch plates
G2500	Small cylinder blocks, cylinder heads, air-cooled cylinders, pistons, clutch plates, oil pump bodies, transmission cases, gearboxes, clutch housings, and light-duty brake drums	G3500	Diesel engine blocks, truck and tractor cylinder blocks and heads, heavy flywheels, tractor transmission cases, heavy gearboxes
		G3500b	Brake drums and clutch plates for heavy duty service where both resistance to heat checking and higher strength are definite requirements
G2500a	Brake drums and clutch plates for moderate service requirements, where high-carbon iron is desired to minimize heat checking	G3500c	Brake drums for extra heavy duty service
		G4000	Diesel engine castings, liners, cylinders, and pistons
		G4000d	Camshafts

tallic matrix is constant. To illustrate this effect, the matrix microhardness of five types of hardened iron are compared with Rockwell C measurements on the same iron in Table 14.

If any hardness correlation is to be attempted, the type and amount of graphite in the irons being compared must be constant. Rockwell hardness tests are considered appropriate only for hardened castings (such as camshafts), and even on hardened castings, Brinell tests are preferred. Brinell tests must be used when attempting any strength correlations for unhardened castings.

Fatigue Limit in Reversed Bending

Because fatigue limits are expensive to determine, the designer usually has incomplete information on this property. Fatigue life curves at room temperature for a gray iron under completely reversed cycles of bending stress are shown in Fig. 13(a), in which each point represents the data from one specimen. The effects of temperature on fatigue limit and tensile strength are shown in Fig. 13(b) and 13(c), respectively.

Axial loading or torsional loading cycles are frequently encountered in designing parts of cast iron, and in many instances these are not completely reversed loads. Types of regularly repeated stress variation can usually be expressed as a function of a mean stress and a stress range. Wherever possible, the designer should use actual data from the limited information available. Without precisely applicable test data, an estimate of the reversed bending fatigue limit of machined parts may be made by using about 35% of the minimum specified tensile strength of the particular grade of gray iron being considered. This value is probably more conservative than an average of the few data available on the fatigue limit for gray iron.

An approximation of the effect of range of stress on fatigue limit may be obtained from diagrams such as Fig. 14. Tensile strength is plotted on the horizontal axis to represent fracture strength under static load (which corresponds to a 0 stress range). Reversed bending fatigue limit is plotted on the ordinate for 0 mean stress, and the two points are joined by a straight line. The resulting diagram yields a fatigue limit (maximum value of alternating stress) for any value of mean stress.

Few data available are applicable to design problems involving dynamic loading where the stress cycle is predominantly compressive rather than tensile. Some work done on aluminum and steel indicates that for compressive (negative) mean stress, the behavior of these materials could be represented by a horizontal line beginning at the fatigue limit in reversed bending, as indicated in Fig. 14. Gray iron is probably at least

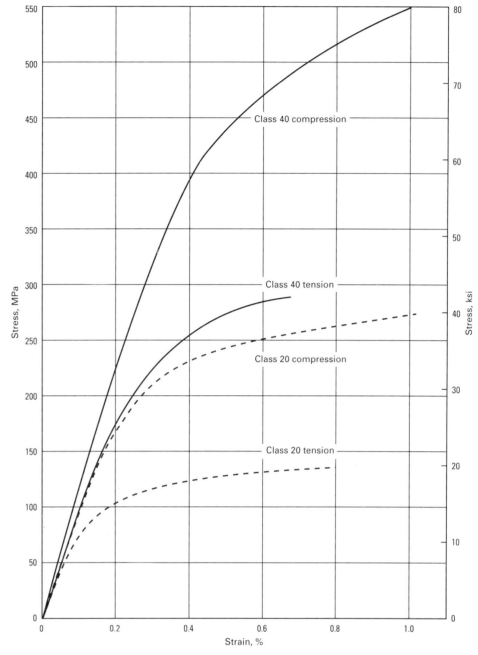

Fig. 9 A comparison of stress-strain curves in tension and compression for a class 20 and a class 40 gray iron. Source: Ref 6

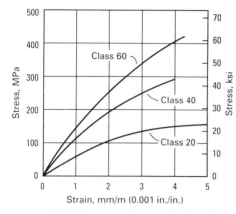

Fig. 10 Typical stress-strain curves for three classes of gray iron in tension

Table 13 Typical moduli of elasticity of as-cast standard gray iron test bars

ASTM A 48 class	Tensile modulus		Torsional modulus	
	GPa	10⁶ psi	GPa	10⁶ psi
20	66–97	9.6–14.0	27–39	3.9–5.6
25	79–102	11.5–14.8	32–41	4.6–6.0
30	90–113	13.0–16.4	36–45	5.2–6.6
35	100–119	14.5–17.2	40–48	5.8–6.9
40	110–138	16.0–20.0	44–54	6.4–7.8
50	130–157	18.8–22.8	50–55	7.2–8.0
60	141–162	20.4–23.5	54–59	7.8–8.5

enough data are available for a reliable S-N diagram for the gray iron proposed, the casting might be dimensioned to obtain a minimum safety factor of two based on fatigue strength. (Some uses may require more conservative or more liberal loading.) The approximate safety factor is best illustrated by point P in Fig. 14. The safety factor is determined by the distance from the origin to the fatigue limit line along a ray through the cyclic-stress point, divided by the distance from the origin to that point. In Fig. 14 this is OF/OP.

On this diagram, point P′ represents a stress cycle having a negative mean stress. In other words, the maximum compressive stress is greater than the tensile stress reached during the loading cycle. In this instance, the safety factor is the distance OF'/OP'. However, this analysis assumes that overloads will increase the mean stress and alternating stress in the same proportion. This may not always be true, particularly in systems with mechanical vibration in which the mean stress may remain constant. For this condition, the vertical line through P would be used; that is, DK/DP would be the factor of safety.

Most engineers use diagrams such as Fig. 14 mainly to determine whether a given condition of mean stress and cyclic stress results in a design safe for infinite life. The designer can also determine whether variations in the mean stress and the alternating stress that he anticipates will place his design in the unsafe zone. Usually the data required to analyze a particular set of con-

Table 12 Typical mechanical properties of as-cast standard gray iron test bars

ASTM A 48 class	Tensile strength		Torsional shear strength		Compressive strength		Reversed bending fatigue limit		Transverse load on test bar B		Hardness, HB
	MPa	ksi	MPa	ksi	MPa	ksi	MPa	ksi	kg	lb	
20	152	22	179	26	572	83	69	10	839	1850	156
25	179	26	220	32	669	97	79	11.5	987	2175	174
30	214	31	276	40	752	109	97	14	1145	2525	210
35	252	36.5	334	48.5	855	124	110	16	1293	2850	212
40	293	42.5	393	57	965	140	128	18.5	1440	3175	235
50	362	52.5	503	73	1130	164	148	21.5	1630	3600	262
60	431	62.5	610	88.5	1293	187.5	169	24.5	1678	3700	302

as strong as this for loading cycles resulting in negative mean stress, because it is much stronger in static compression than in static tension. It is therefore a natural assumption that the parallel behavior shown in Fig. 14 is conservative.

If, prior to design, the real stress cycle can be predicted with confidence and

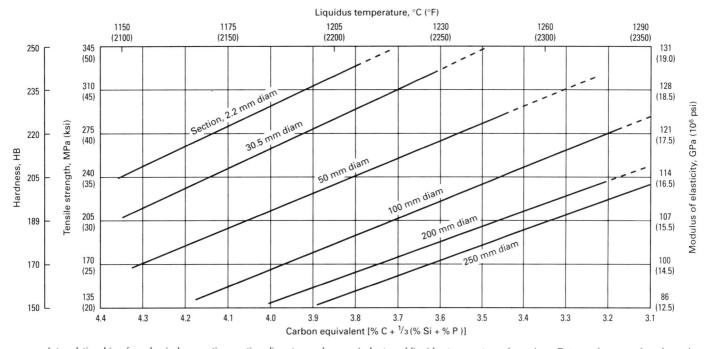

Fig. 11 Interrelationship of mechanical properties, section diameter, carbon equivalent, and liquidus temperature of gray iron. Data are from one foundry and are based on dry sand molding and ferrosilicon inoculation.

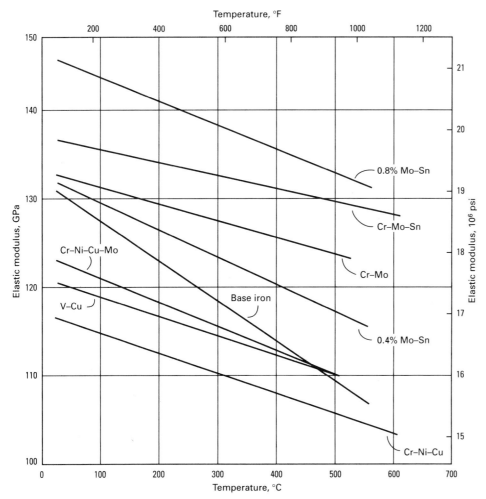

Fig. 12 Elastic modulus as a function of temperature in various alloyed gray cast irons. Source: Ref 7

Table 14 Influence of graphite type and distribution on the hardness of hardened gray irons

Type of graphite	Total carbon, %	Conventional hardness, HRC(a)	Matrix hardness, HRC(b)
A	3.06	45.2(c)	61.5
A	3.53	43.1	61.0
A	4.00	32.0	62.0
D	3.30	54.0	62.5
D	3.60	48.7	60.5

(a) Measured by conventional Rockwell C test. (b) Hardness of matrix, measured with superficial hardness tester and converted to Rockwell C. (c) Although this value was obtained in the specific test cited, it is not typical of gray iron of 3.06% C. Ordinarily the hardness of such iron is 48 to 50 HRC.

ditions are obtained experimentally. It is emphasized that the number of cycles of alternating stress implied in Fig. 14 is the number normally used to determine fatigue limits, that is, approximately 10 million. Fewer cycles, as encountered in infrequent overloads, will be safer than indicated by a particular point plotted on a diagram for infinite life. Too few data are available to draw a diagram for less than infinite life.

Fatigue Notch Sensitivity. In general, very little allowance need be made for a reduction in fatigue strength caused by notches or abrupt changes of section in gray iron members. The low-strength irons exhibit only a slight reduction in strength in the presence of fillets and holes. That is, the notch sensitivity index approaches 0; in other words, the effective stress concentration factor for these notches approaches 1. This characteristic can be explained by considering the graphite flakes in gray iron to be internal notches. Thus, gray iron can be thought of

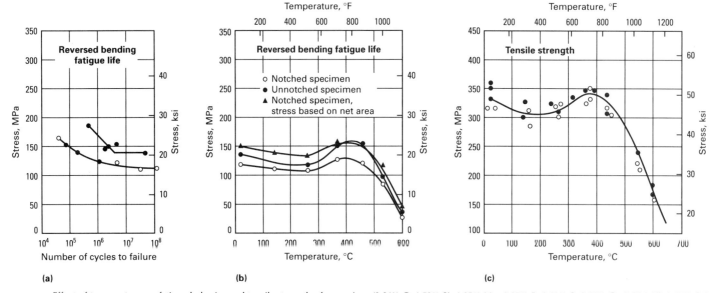

Fig. 13 Effect of temperature on fatigue behavior and tensile strength of a gray iron (2.84% C, 1.52% Si, 1.05% Mn, 0.07% P, 0.12% S, 0.31% Cr, 0.20% Ni, 0.37% Cu). (a) Reversed bending fatigue life at room temperature. (b) Reversed bending fatigue limit at elevated temperatures. (c) Tensile strength at elevated temperatures. Source: Ref 8

as a material that is already full of notches and that therefore has little or no sensitivity to the presence of additional notches resulting from design features. The strength-reducing effect of the internal notches is included in the fatigue limit values determined by conventional laboratory tests with smooth bars. High-strength irons usually exhibit greater notch sensitivity, but probably not the full theoretical value represented by the stress concentration factor. Normal stress concentration factors (see Ref 9) are probably suitable for high-strength gray irons.

Pressure Tightness

Gray iron castings are used widely in pressure applications such as cylinder blocks, manifolds, pipe and pipe fittings, compressors, and pumps. An important design factor for pressure tightness is uniformity of section. Parts of relatively uniform wall section cast in gray iron are pressure tight against gases as well as liquids. Most trouble with leaking castings is encountered when there are unavoidable, abrupt changes in section. Shrinkage, internal porosity, stress cracking, and other defects are most likely to occur at junctions between heavy and light sections.

Watertight castings are considerably less challenging to the foundryman than gastight castings. A slight sponginess or internal porosity at heavy sections will not usually leak water, and even if there is slight seepage, internal rusting will soon plug the passages permanently. For gastightness, however, castings must be quite sound.

Lack of pressure tightness in gray iron castings can usually be traced to internal porosity, which also is called internal shrinkage. In gray iron this seems to be a phenomenon distinctly different from the normal solidification shrinkage that often appears on the casting surface as a sink or draw, which can be cured by risering. Internal porosity or shrinkage (very difficult to prevent, even by the use of very heavy risers) is usually associated with poor feeding and lack of directional solidification. It can also be associated with heavy inoculation with calcium to promote a high graphite cell count. On the other hand, the use of strontium-bearing inoculants does not reduce cell size and can help control internal shrinkage in marginally gated castings.

Visible internal porosity may appear at centers of mass when the phosphorus content exceeds 0.25%. In critically gated castings, visible internal porosity may appear when the phosphorus content is as low as 0.09%. Chromium and molybdenum accentuate this effect of phosphorus, while nickel has a slight mitigating influence.

The effect of phosphorus may be caused in part by the fact that lowering phosphorus content also lowers the effective carbon equivalent. Lowering carbon equivalent by reducing carbon or silicon, or both, instead of phosphorus, might similarly reduce the leakage of pressure castings, but other foundry problems (such as increasing amounts of normal shrinkage) would be encountered. ASTM A 278 for pressure castings requires a carbon equivalent of 3.8% (max), a phosphorus content of 0.25% (max), and a sulfur content of 0.12% (max) for castings to be used above 230 °C (450 °F).

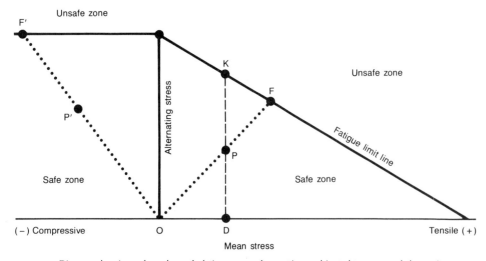

Fig. 14 Diagram showing safe and unsafe fatigue zones for cast iron subjected to ranges of alternating stress superimposed on a mean stress. Example point P shows conditions of tensile (positive) mean stress; P' shows compressive (negative) mean stress. The safety factor is represented by the ratio of OF to OP or OF' to OP'. For conditions of constant mean tensile stress, DK/DP is the safety factor.

Table 15 Machinability of gray iron

Microstructure	ASTM class	Tensile strength MPa	ksi	Hardness, HB	Cutting speed(a) m/min	ft/min
Acicular iron	50	407	59	263	46	150
Fine pearlite, alloy	40	310	45	225	95	310
Ferrite (annealed)	...	108	15.7	100	293	960
Coarse pearlite, no alloy	35	241	35	195	99	325

(a) Cutting speed at which removal of 3280 cm^3 (200 in.^{3}) produced 0.75 mm (0.030 in.) wear land on single-point carbide tools. Source: Ref 10

In addition to composition control, good overall foundry practice is required for consistently producing pressure-tight castings. Sand properties and gating must be controlled to avoid sand inclusions. Pouring temperature must be adequate for good fluidity, and heavy sections should be fed wherever possible.

Mold properties have been found to interact with composition to influence internal soundness. Generally, molds rammed tightly produce the soundest castings because of freedom from mold wall movement during solidification.

Impact Resistance

Where high impact resistance is needed, gray iron is not recommended. Gray iron has considerably lower impact strength than cast carbon steel, ductile iron, or malleable iron. However, many gray iron castings need some impact strength to resist breakage in shipment or use.

There is incomplete agreement on a standard method of impact testing for cast iron. Two methods that have been used successfully are given in ASTM A 327. Most impact testing of cast iron has been used as a research tool.

Most producers of cast iron pressure pipe use a routine pipe impact test as a control. Impact resistance in pipe helps avoid breakage in shipping and handling.

Machinability

The machinability of most gray cast iron is superior to that of most other cast irons of equivalent hardness, as well as to that of virtually all steel. The flake graphite introduces discontinuities in the metal matrix, which act as chip breakers. The graphite itself serves as a lubricant for the cutting tool. However, economical cutting depends on more than inherent machinability alone. Often, trouble in machining gray iron can be traced to one or more of several factors: the presence of chill at corners and in light sections, the presence of adhering sand on the surface of the casting, swells, usually the result of soft molds, shifted castings, shrinks, and phases included in the matrix as a result of melting practice.

Chill at corners and in light sections is more likely to be encountered with small castings, with higher-strength irons, and with designs that have light sections in the cope (or top) of the mold. Most foundries control iron with a chill test that gives an indication of the tendency of the iron to form white or mottled iron in light sections. The foundryman may treat his iron with a small amount (0.5 to 2.5 kg/Mg, or 1 to 5 lb/ton) of a graphitizing alloy such as calcium-bearing ferrosilicon or other proprietary inoculant, which effectively decreases chill. Inoculation to achieve control of the tendency to chill usually does not result in significant changes in the composition or the physical properties of the iron, although it does produce changes in the mechanical properties.

Light sections (5 mm, or 3/16 in.) usually cannot be cast in gray iron of higher than class 25. Class 30 iron can be cast in 6 mm (¼ in.) sections. These values are different for different designs, depending on how the casting is made and gated. The important thing to understand is that the cooling rate in the mold at the time of freezing determines whether the iron will be gray, white, or mottled. If the thin section is in the drag or near the gate, the flow of hot metal heats the mold, thereby decreasing the rate of cooling and enhancing the formation of the gray iron.

If chill is encountered, it is generally best to correct the trouble at its origin. It is usually uneconomical to anneal castings to remove chill because, in addition to heat treating for 2 h at 900 °C (1650 °F) for unalloyed irons, recleaning may be required for removing scale. Distortion beyond tolerance often occurs, and there are sacrifices in hardness and strength.

Adhering sand usually can be removed by effective cleaning, but sand present as the result of penetration of the iron into the mold wall is extremely difficult to blast clean. This is a foundry defect that is best corrected at the source. Slowing the speed of machining and increasing the rate of feed is the best approach to salvaging castings of this type. Carbide tools are better than high-speed tools for resisting the extreme abrasion.

Swells are most troublesome in operations such as broaching and in other setups tooled for high production. The additional metal often places an excessive load on the tool, which may chip or dull but not actually fail until some time after the troublesome parts have been machined. In highly automated machining centers, this casting defect can result in significant statistical variation in dimensions and high scrap rates.

Shifted castings are similar to swells in their action on cutting tools. Shifts or swells also may cause excessive tool loading if the locating points are affected. It is important to consider the positions of such locating points when designing the castings and also to avoid indiscriminate grinding of locating points in the foundry cleaning room.

Shrinks are not often present but can be troublesome when encountered in operations such as drilling. For example, the drill may tend to drift from its intended path to follow the shrink, which offers less resistance to the drill. Sometimes a drill may break because it encounters a region of higher hardness, which often is associated with an area of shrink. Cast iron is the easiest metal to cast without internal shrink. Eutectic freezing is accompanied by expansion due to the precipitation of low-density graphite, which aids in obtaining internal soundness.

Machinability rating is complex and is discussed in the article "Machinability of Steels" in this Volume. Criteria such as power per unit volume in unit time are of greatest importance in selecting a machine tool and the size of its motor. Machinability ratings based on tool life under standard test conditions are helpful but are not readily interpreted into the economics of machining.

One way to indicate the effect on machinability of changing from one grade of iron to another is shown in Table 15, which is based on an experiment in which metal was removed from four types of gray iron using single-point carbide tools. For each type of iron, cutting speed was varied until the removal of 3300 cm^3 (200 in.^{3}) of iron resulted in a 0.75 mm (0.030 in.) wear land on the tool. The values of cutting speed thus obtained serve as qualitative evaluations of machinability, but not as qualitative indices. Optimum cutting speeds, tool materials, cutting fluids, feeds, and finish requirements must be studied for any given machine tool setup. Tool life is an important factor, because the machine must be stopped to change tools and the tools must be resharpened. Progress has been made in decreasing both machine downtime for tool changes and resharpening cost by the use of solid carbide inserts or bits and by other means.

Annealing. The greatly improved machinability obtainable in gray iron by annealing has been advantageous to automotive and other industries for many years. Annealing is usually of the subcritical type, such as 1 h at 730 to 760 °C (1350 to 1400 °F). Some employ a cycle of heating to 785 to 815 °C (1450 to 1500 °F) and cooling at 22 °C/h (40 °F/h) to about 595 °C (1100 °F). These treatments graphitize the carbide in the pearlite and result in a ferritic matrix. Find-

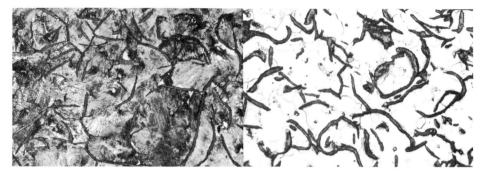

Fig. 15 Structure of class 35 iron, as-cast (left) and after annealing

Table 16 Effect of annealing on hardness and strength of class 35 gray iron

Condition	Tensile strength MPa	ksi	Hardness, HB
As-cast	268	38.9	217
Annealed	165	23.9	131

Composition of iron: 3.30% total C, 2.22% Si, 0.027% P, 0.18% S, 0.61% Mn, 0.03% Cr, 0.03% Ni, 0.14% Cu, Mo nil. Annealing treatment consisted of 1 h at 775 °C (1425 °F), followed by cooling in the furnace to 540 °C (1000 °F).

ing it uneconomical to graphitize primary carbide, most users try to avoid obtaining it. Annealing for improved machinability is most economical when the casting is small and the amount of machining is large.

The annealing treatments described result in sacrifices in hardness and strength. A typical class 35 iron will be downgraded to about class 20 in strength by this treatment. In applications in which wear resistance is important, such as cylinder blocks, gray iron is not annealed because of the unsatisfactory performance obtained with a ferritic matrix.

Table 16 and Fig. 15 show the changes in hardness, strength, and structure of class 35 gray iron obtained by annealing.

Wear

Gray iron is used widely for machine components that must resist wear. Different types of iron, however, exhibit great differences in wear characteristics. These differences do not correlate with the commonly measured properties of the iron.

In the discussion that follows, the general conclusions are related to metal-to-metal wear during sliding contact under conditions of normal lubrication. Although most of the supporting illustrations are for engine cylinders, the results have wider applicability.

The published data on gray iron wear are somewhat inconsistent; accelerated-wear tests often do not correlate with field service experience, nor does field experience in one application necessarily agree with that in another application. In many applications, properties of the surface are at least as important as properties of the metal; for instance, wear resistance may frequently be enhanced by lapping the wear areas. Relative hardness between mating parts also may be important to optimum wear resistance. Frequently, a hardness difference no greater than 10 points on the Brinell scale is considered optimum.

For components in sliding contact, such as engine cylinders, valve guides, and latheways, the recognized types of wear are cutting wear, abrasive wear, adhesive wear, and corrosive wear. In well-designed machinery, wear proceeds slowly and consists of combinations of very mild forms of these four types of wear. The predominant characteristic of this so-called normal wear is the development of a glazed surface during break-in. The primary objective in any wear application is to establish conditions that produce and maintain this glaze. A secondary objective is to minimize normal wear.

Cutting wear is caused by the mechanical removal of surface metal as a result of surface roughness and is similar to the action of a file. It usually occurs during the breaking in of new parts.

Abrasive wear is caused by the cutting action or loose, abrasive particles that get between the contacting faces and act like a lapping compound. Under some circumstances, abrasive particles embedded in one or both of the contacting faces can produce a similar action.

Adhesive wear is caused by metal-to-metal contact, resultant welding, and the breaking of those welds. When this happens on a large scale (a process known as galling), the metal is smeared, and severe surface damage results. Even in properly operating equipment, some wear occurs by adhesion on a microscopic scale. An intermediate form of adhesive wear called scuffing is less destructive than galling, but still results in an abnormally high rate of metal removal. In all probability, accelerated-wear tests between clean surfaces (with or without the presence of a lubricant) are predominantly tests of galling or scuffing.

Corrosive wear is a special type of wear that combines abrasion or adhesion with the chemical action of the environment. In engine cylinders, it is caused by condensed acidic products of combustion during low-temperature operation. Usually this kind of wear cannot be corrected by modifications in ordinary types of gray iron.

Resistance to Scuffing

Assuming reasonable design and operating conditions that minimize cutting and abrasive wear, scuffing or mild galling is the abnormal wear condition most likely to defeat the objective of obtaining and maintaining a glazed surface due to normal wear. Several alloy combinations with widely varied microstructures were tested by Shuck (Ref 11). A brake shoe type of specimen was held against a rotating gray iron drum for 1 h, after which both specimen and drum were checked for weight loss. The conditions of this test indicate that wear was caused primarily by adhesion (scuffing), although cutting wear may have had a contributing effect. The tests covered almost all conceivable microstructures and a wide variety of compositions having hardnesses below 300 HB. The conclusions, which basically agreed with other less comprehensive investigations, were:

- Microstructure determines wearing characteristics
- As the graphite structure becomes coarser and tends toward type A, scuffing decreases
- Interdendritic type D graphite and its associated ferrite give very poor results
- Secondary ferrite associated with random type A graphite is less damaging than that associated with type D
- Pearlitic, acicular, or tempered martensitic structures in the same hardness range are equal in wear resistance
- For a given type of graphite, as the matrix becomes more pearlitic and harder, wear resistance increases

These results were substantiated in scuff tests of cylinder sleeves in a diesel engine. The engine, equipped with the test sleeves, was operated at constant speed, and the horsepower was increased by increments above normal until scuffing occurred. The scuffing horsepower was expressed in a ratio with normal horsepower.

Effect of Graphite Structure. To eliminate differences in matrix, four types of gray iron were hardened and tempered at 205 °C (400 °F) to produce a uniform martensitic matrix. The data in Table 17 show that the greater the amount and flake size of graphite, the greater the resistance to scuffing. The changes in casting method in tests 3 and 4 caused no significant change in scuffing. The iron in test 5 is one used formerly in high-performance brake drums to give maximum resistance to scoring.

Table 17 Effect of graphite structure on resistance to scuffing

Test No.	Type of graphite(a)	Total carbon, %	Resistance to scuffing(b)
1	None (5150 steel)	...	<1(c)
2	100% type D, centrifugally cast	3.25 (average)	1.11
3	Type A, size 4 to 6, some type B, centrifugally cast	3.28	1.33
4	Same as 3 except cast in sand mold	3.35 (average)	1.30
5	Type A, size 3 to 4, some type C, sand cast	4.00	>1.45(d)

(a) Different chemical compositions were tested in two of the four types of iron. See Table 19 for compositions. Matrix of all specimens was tempered martensite. (b) Expressed as ratio: horsepower to produce scuffing divided by normal horsepower. (c) All the steel sleeves scuffed below normal horsepower. (d) Maximum available engine horsepower produced no scuffing.

Table 18 Effect of matrix microstructure on resistance to scuffing

Test No.	Microstructure of matrix	Hardness	Resistance to scuffing(a)
1	Pearlite, with ferrite occurring in graphitic areas	196–227 HB	<1(b)
2	Martensite tempered at 205 °C (400 °F)	53–56 HRC	1.06
3	Martensite tempered at 425 °C (800 °F)	44–47 HRC	1.22
4	Martensite tempered at 510 °C (950 °F)	39–41 HRC	1.39

(a) Expressed as a ratio: horsepower to produce scuffing divided by normal horsepower. (b) All sleeves scuffed below normal operating range.

Table 19 Compositions of irons reported in Table 17

Test No.	TC	Mn	Si	Cr	Ni	Mo	Cu	P	S
Type D graphite(a)									
1(b)	3.20	0.65	2.20	0.25	0.30	0.15	0.30	0.15	0.04
2	3.08	0.68	2.34	0.45	0.56	0.22	...	0.110	0.033
3	3.43	0.73	2.28	0.44	0.09	...	1.29	0.143	0.068
Type A fine graphite(c)									
1	3.38	0.61	1.99	0.45	0.59	...	1.63	...	...
2	3.28	0.70	2.46	0.24	0.27	...	0.94	0.23	0.068
3(b)	3.35	0.70	2.20	0.35	0.12	...	1.15	0.12	0.09
4	3.28	0.67	2.08	...	...	...	0.40	0.125	0.067
5	3.12	0.35	2.67	0.38	0.27	0.11	1.23	0.176	0.047
Type A coarse graphite(d)									
1	4.00	0.77	1.54	...	1.39	0.42	...	0.056	0.023

(a) Corresponds to test 2 in Table 17. (b) Typical composition. (c) Corresponds to tests 3 and 4 in Table 17. (d) Corresponds to test 5 in Table 17

Table 20 Effect of type of graphite on wear resistance

Test No.(a)	Type of graphite	Total carbon, %	Resistance to scuffing(a)	Sleeve wear, mm/1000 h (in./1000 h)	Ring wear (gap increase), mm/1000 h (in./1000 h)
2	100% type D	3.10–3.40	1.11	0.075 (0.003)	0.050 (0.002)
4	Type A, size 4 to 6, some type B	3.25–3.50	1.30	0.050 (0.002)	0.685 (0.027)
5	Type A, size 3 to 4, some type C	4.00	1.45	0.090 (0.0035)	2.15 (0.085)

(a) See Table 17.

Effect of Matrix Structure. Similar tests, in which the graphite was type D and the matrix was varied, gave the results shown in Table 18. The combination of type D graphite and ferrite in test 1 is especially poor even though the rest of the structure is pearlite. Pearlitic type A iron with small amounts of free ferrite gave ratios greater than 1.33 in the same tests, showing that the results obtained in test 1 are peculiar to the combination of type D graphite and ferrite. When type D graphite occurs in as-cast irons, it is usually associated with ferrite. Consequently, foundry practices that avoid the formation of type D graphite should be adopted when producing castings for wear applications.

Several investigators have noted similar, apparently inconsistent, results, which can be summarized as follows: increasing hardness from 160 to 260 HB increases resistance to scuffing; however, further increasing hardness above 260 HB decreases resistance to scuffing, even in heat-treated structures.

These tests, along with those of Ref 11, apparently apply only to scuffing and not to normal wear. As indicated in the section "Resistance to Normal Wear," wear resistance is different for different modes of wear.

Other investigators have observed that a fine network of steadite (iron phosphide-iron carbide eutectic) increases scuff resistance, while massive carbide reduces it.

Effect of Chemical Composition. The diesel cylinder scuff tests for evaluating the effect of graphite used five different compositions from five foundries for tests 3 and 4 and three compositions from two foundries for test 2. Despite the variations in source and in composition (Table 19), all fine type A irons performed nearly alike in the scuff tests. Similarly, the three type D materials performed essentially the same. No other composition was checked with coarse type A graphite. To cover the subject of scuffing more completely, other factors were evaluated during the cylinder sleeve tests. They help to clarify some of the additional controversial aspects of wear.

Surface finish effects were determined by running similar tests on cylinder sleeves with honed finishes of 0.75 and 2.3 μm (30 and 90 μin.) rms. In each, the sleeves were made from type D graphite iron, hardened, and tempered at 205 °C (400 °F).

The coarser finish gave about 20% greater scuff resistance; however, this effect was lost if the finish became finer under normal operating conditions. The loss of scuff resistance by the smoothing effect of normal wear may account for some of the failures that occur in certain applications after break-in has apparently been successful.

Other important factors affecting scuff resistance are the material and finish of the mating part and the type of lubricant used.

Resistance to Normal Wear

The scuff resistance described above should not be used as a basis for selecting a material that must resist normal wear. Resistance to normal wear, like resistance to scuffing, is affected by both graphite form and matrix microstructure (and possibly by the composition of the iron), but in a different manner.

Some of the same sleeve materials involved in the scuff tests were operated under conditions that produced normal wear with very little evidence of scuffing. Tests were run for approximately 1000 h using chromium-plated compression rings on the pistons. All sleeves were hardened and tempered at 205 °C (400 °F). Results are given in Table 20. Each of the three sleeve materials performed best when different types of wear were considered: the material in No. 2 test gave the lowest ring wear; in No. 4, the lowest sleeve wear; and in No. 5, the greatest resistance to scuffing. Thus, the choice of optimum sleeve material is a compromise.

Effect of Graphite Structure. It was concluded from the above tests that graphite produces a surface-roughening effect, which accounts for both the greater scuff

Table 21 Effect of matrix microstructure on resistance of gray iron to normal wear

Iron(a)	No. of tests	Type and size of graphite	Matrix	Rate of wear mm/1000 h	Rate of wear in./1000 h
A	28	A, size 3 to 4	Lamellar pearlite	0.14	0.0057
B	19	A, size 4	Lamellar pearlite plus phosphide	0.03	0.0012
C	4	A, size 5 to 6	Fine lamellar pearlite	0.022	0.0009
D	6	A, size 7 to 8	Very fine lamellar pearlite plus undetermined constituent	0.025	0.00097
E	30	A, size 5 to 6	Same as iron D	0.016	0.00062
F	3	A, size 3 to 4	Lamellar pearlite plus ferrite	0.050	0.002
G	2	Not shown	Martensite	0.012	0.00048

(a) Compositions are given in Table 22.

Table 22 Compositions of irons reported in Table 21

Iron	TC	Mn	Si	Cr	Mo	P	S	Others
A	3.00–3.30	0.90–1.10	1.15–1.35	...	...	0.20 max	0.12 max	0.80–1.10 Cu
B	3.40 max	0.90 max	1.50 max	...	...	0.35–0.50	0.13 max	...
C	2.85–3.30	1.00 max	1.25–1.75	0.30–0.40	0.25–0.35	0.20 max	0.12 max	1.00–1.50 Ni
D	2.80–3.10	0.80 max	4.6–5.0	1.90–2.20	...	0.20 max	0.12 max	...
E	3.10–3.40	0.80–1.00	3.30–3.70	1.20–1.50	...	0.40 max	0.12 max	...
F	3.30–3.40	0.90–1.10	1.75–2.00	...	0.40–0.50	0.20 max	0.12 max	...
G	3.00–3.20	0.90–1.10	1.15–1.35	...	...	0.20 max	0.12 max	...

resistance and the higher ring wear as the size and quantity of graphite are increased. In testing type D graphite sleeves with the rougher finish previously mentioned, improved scuff resistance was also obtained, but at the expense of greater ring wear, apparently from cutting. The larger quantity of graphite in test 5 leaves less load-carrying metal surface, and greater normal wear on the sleeve is the logical result.

The relatively high sleeve wear in test 2 was probably caused by slight scuffing, which gave high wear values for some sleeves tested, presumably because of the low safety margin of scuff resistance in this type of iron. Surprisingly, this mild scuffing did not seem to affect the chromium-plated rings, which apparently wore less severely in the presence of fine graphite. The plain gray iron oil-control rings, however, wore more in test 2 than in test 4. The principal effect of graphite on wear resistance is the elimination of scuffing, and when graphite is present in greater quantity and size than required for this purpose, it will reduce resistance to normal wear unnecessarily.

Effect of Matrix Microstructure. Gray iron is used for wear resistance in both the as-cast and hardened conditions. To show the effects of some of the matrix variations on wear resistance, comparable engine wear tests were made on seven types of gray iron cylinders, all of which apparently wore in a normal manner. These data are summarized in Tables 21 and 22.

By present standards, hardening is the process that can produce the most significant improvement in resistance to normal wear. Although iron E in Table 21 showed exceptionally good wear properties, it was less resistant than the hardened iron G and at the same time could not be cast readily into the required shape. Thus, when evaluated on the basis of combined properties, the hardened iron G was twice as wear resistant as iron D, the as-cast iron that had the best combination of wear resistance and castability. From other extensive testing, it has been concluded that hardened iron has as much as five times more wear resistance than pearlitic as-cast iron of the same general composition (Ref 12).

Another significant advantage of hardening is that it greatly reduces variations in matrix microstructure and thus provides a superior iron of more dependable performance. In most applications such as valve guides, latheways, and various sliding members in machines and engines, satisfactory wear resistance is obtained with properly specified and controlled as-cast gray iron. Hardened iron is used to obtain maximum wear resistance in severe wearing applications such as high-speed diesel cylinder sleeves, camshafts, gears, and similar heavily loaded wearing surfaces.

For more information on the effects of composition and microstructure on the wear resistance of cast iron, specifically that of white and chilled iron, see the article "Alloy Cast Irons" in this Volume.

Elevated-Temperature Properties

The most demanding applications of gray irons at elevated temperatures are those in which dimensional accuracy is important. The effect of elevated temperatures on dimensional stability is discussed below.

Less rigorous applications at high temperatures are those in which a load is to be carried but in which appreciable distortion can be tolerated. In this case, the important properties include tensile strength, fatigue properties, and creep-rupture strength at elevated temperatures. Figures 13(b) and (c) show the effects of temperature on the fatigue limit and tensile strength of a gray iron. Figure 16 shows the creep rupture properties of gray iron with molybdenum (Fig. 16a) and with molybdenum and chromium (Fig. 16b). As the figure shows, creep rupture strength is improved with the addition of chromium and molybdenum.

Dimensional Stability

Effect of Temperature. The dimensional stability of gray iron is degraded as temperature and exposure time increase. Dimensional stability at elevated temperatures is affected by factors such as growth, scaling, and creep rate.

Growth of the iron is the result of the breakdown of the pearlite to ferrite and graphite. To prevent this growth, alloying elements such as copper, molybdenum, chromium, tin, vanadium, and manganese must be added because they will stabilize the carbide structure. Examples of these effects are seen in Fig. 17. Alloying additions are generally recommended at temperatures above 400 °C (750 °F). Chromium and

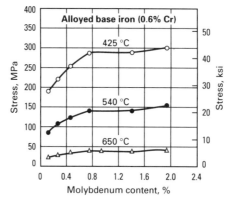

Fig. 16 Influence of chromium and molybdenum on stress to produce rupture in 100 h in gray iron at various temperatures. (a) Unalloyed base iron. (b) 0.6% Cr alloyed base iron. Source: Ref 13

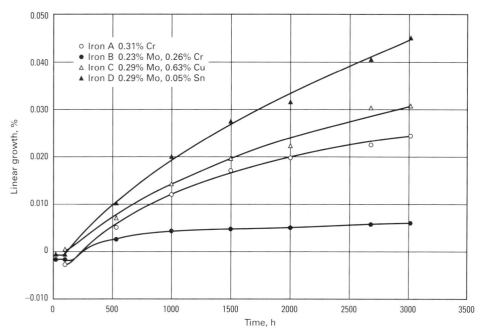

Fig. 17 Growth of four gray iron alloys produced from the same base iron (3.3% C, 2.2% Si) and tested at 455 °C (850 °F) in air. Source: Ref 7, 14

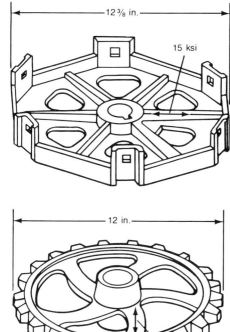

Fig. 18 Location and magnitude of residual stresses in two gray iron castings

chromium-nickel-molybdenum additions are most effective in retarding growth.

Scaling is another source of dimensional change. Table 23 gives the results of scaling tests. In general, differences in alloy content had only minor effects on scaling.

Creep. A gray iron part that operates at elevated temperature will deform by creep if the load is great enough. Table 24, which gives the creep characteristics of several gray irons at 370 °C (700 °F) and at stresses ranging from 72 to 165 MPa (10.5 to 24 ksi) suggests the usefulness of molybdenum additions for improving the creep properties of gray iron. Creep rupture properties are improved with additions of chromium and molybdenum (Fig. 16).

If the part is to operate below 480 °C (895 °F), two sources of dimensional inaccuracy must be considered: residual stresses and machining practice.

Residual stresses are present in all castings in the as-cast condition and are caused by:

- Differences in cooling rate between sections of the same casting because of different cross sections or locations in the mold
- Resistance of the mold to contraction of the casting during cooling
- High-energy cleaning, such as with shot, which can induce compressive stresses

Residual stresses as high as 220 MPa (32 ksi) have been reported in gray iron wheels. Stresses of 170 MPa (25 ksi) have been observed in localized areas of other gray iron castings. Residual stresses in engine cylinder blocks have been measured as high as 130 MPa (19 ksi). Figure 18 shows two examples of residual stress in light castings, as determined with resistance strain gages.

Only a small percentage of castings are stress relieved before machining, chiefly those requiring exceptional accuracy of dimensions or those with a combination of high or nonuniform stress associated with either low section stiffness or an abrupt change in section size. Castings of class 40, 50, and 60 iron are more likely to contain high residual stresses.

Figure 19 shows stress relief at seven temperatures between about 310 and 600 °C (600 and 1100 °F). The range from 480 to 600 °C (900 to 1100 °F) is recommended. For best results, castings should be held at temperature 1 h or more and then cooled in the furnace to below 450 °C (850 °F). High-strength and alloy gray irons require temperatures approaching 600 °C (1100 °F).

Machining Practice. In castings in the as-cast condition, residual tension and compression stresses are balanced, which makes the castings dimensionally stable at room temperature. However, when part of the surface is removed in machining, the balance of forces is altered, which can lead to distortion. If the casting is of relatively stiff section or has properly designed stiffness ribs, there may be no noticeable change in dimensions. Distortion will be most evident in castings of low stiffness from which a large volume of highly stressed metal has been removed.

Because the surface of a casting is often the principal site of residual stresses, a large

Table 23 Scaling of gray iron at elevated temperatures

Temperature exposure	Extent of scaling, mg/cm²
11.5 years	
At 350 °C (660 °F)	4
At 400 °C (750 °F)	4
64 weeks	
At 450 °C (840 °F)	3–4.5
At 500 °C (930 °F)	6–16

Table 24 Results of creep tests of gray iron at 370 °C (700 °F)

Iron composition, %(a)	Hardness, HB	Tensile strength MPa	ksi	Creep stress MPa	ksi	Deformation rate(b) 0–150 h	150–450 h	450–900 h	900–1200 h	1200–2000 h
3.4 C, 1.5 Si	197	217	31.5	72	10.5	12.1	12.1	5.9	3.4	3.0
2.95 C, 2.46 Si	237	310	45	72	10.5	5.3	5.3	5.3	3.4	0.0
3.2 C, 2.60 Si, 1.5 Ni	230	307	44.5	72	10.5	...	8.2(c)	5.3	0.0	0.0
2.75 C, 2.10 Si, 0.83 Mo	241	410	59.5	72	10.5	0.8	0.8	2.2	0.0	0.0
2.72 C, 2.5 Si, 0.83 Mo	241	362	52.5	119	17.3	10.0	0.0	0.0	0.0	...
2.72 C, 2.5 Si, 0.83 Mo	241	362	52.5	134	19.5	16.7	3.7	0.0	0.0	...
2.72 C, 2.5 Si, 0.83 Mo	241	362	52.5	165	24.0	30.0	6.4	1.6	0.0	...

(a) Also about 0.2% P (max), 0.1% S (max), and 0.7% Mn, except the 3.2C-2.60Si-1.5Ni iron, which had 0.15% S. (b) In mm/m · h, or 10⁻³ in./in. · h. (c) 0 to 500 h

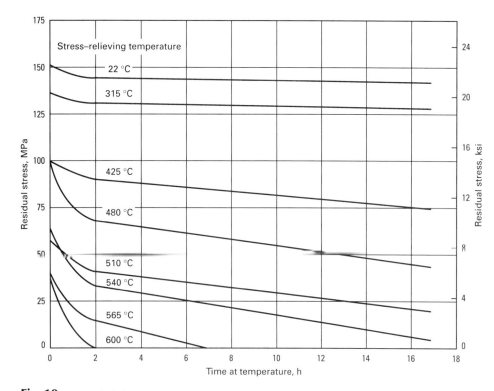

Fig. 19 Stress relief of gray iron at various temperatures. Source: Ref 17

proportion of the stress is relieved by rough machining, with consequent maximum distortion. If, before final machining, the casting is relocated carefully and properly supported in the machine tool fixtures, acceptable dimensional accuracy will usually be obtained in the finished piece.

In designing fixtures, it must be recognized that the usual gray iron (class 30) has a modulus of elasticity of about 97 GPa (14 × 10^6 psi), which means that deflection under tool bit loads will be approximately twice as great in a part made of this cast iron as in a steel part having the same section.

It is difficult to make general statements about the dimensional stability that can be achieved in a gray iron casting without stress relieving. However, it is well known that automotive engine blocks are taken directly from the foundry without stress relieving and are machined to tolerances of ±0.005 mm (±0.0002 in.) in parts such as crankshaft bearings, camshaft bearings, and cylinder bores. Therefore, if a casting is properly designed, is cast under controlled conditions, and is allowed to cool sufficiently in the mold before shakeout, and if proper machining practice is followed, extremely high dimensional stability can be obtained in many applications without stress relieving. On the other hand, it is frequently economical to stress relieve complex castings (other than engine blocks) that are produced in small quantities and that must be machined to precise dimensions.

Effect of Shakeout Practice

Shakeout practice may be influential in establishing the patterns of residual stresses in castings, because most residual stresses are basically caused by differences in cooling rate, and thus in contraction behavior, between light and heavy sections.

In addition to its effect on residual stress, shakeout practice may influence both microstructure and hardness of gray iron castings. If the iron is austenitic at the time the mold is dumped, higher hardness and residual stress may result. Many different microstructures may be obtained; the austenite in thin sections may transform in the mold, whereas in heavier sections, which cool more slowly, transformation may be delayed until air cooling after shakeout. In general, the effect of shakeout practice on hardness is negligible in unalloyed irons. Alloy irons are the most sensitive. However, the martensitic irons that contain enough alloying elements to reach full hardness with slow cooling in the mold will display little if any effect of shakeout practice.

Alloying to Modify As-Cast Properties

The term alloying as used here does not include inoculation because by definition the effect of inoculation on the mechanical properties of an iron is greater than can be explained by the change in chemical composition.

Strength and hardness, resistance to heat and oxidation, resistance to corrosion, electrical and magnetic characteristics, and section sensitivity can be changed by alloying, which can extend the application of gray iron into fields where costlier materials have traditionally been used.

There has also been considerable replacement of unalloyed gray iron by alloy iron to meet increased service demands and to provide increased safety factors.

The use of alloy iron often depends, in practice, on the relative production requirements in a given foundry. When the applications requiring alloy iron in a given plant are a small fraction of total production, manufacturing policy may dictate the use of unalloyed iron for all castings in order to achieve maximum uniformity of production practice. Continuous production of 450 to 1350 kg (1000 to 3000 lb) heats of alloy iron is usually needed for economical utilization.

Manganese, chromium, nickel, vanadium, and copper can also be used to strengthen cast irons. In many irons a combination of elements will provide the greatest increase in strength.

To develop resistance to the softening effect of heat and protect against oxidation, chromium is the most effective element. It stabilizes iron carbide and therefore prevents the breakdown of carbide at elevated temperatures; 1% Cr gives adequate protection against oxidation up to about 760 °C (1400 °F) in many applications. For temperatures above 760 °C (1400 °F), the chromium content should be greater than 15% for long-term protection against oxidation. This percentage of chromium suppresses the formation of graphite and makes the alloy solidify as white cast iron.

For corrosion resistance, chromium, copper, and silicon are effective. Additions of 0.2 to 1.0% Cr decrease the corrosion rate in seawater and weak acids. The corrosion resistance of iron to dilute acetic, sulfuric, and hydrochloric acids and to acid mine water can be increased by the addition of 0.25 to 1.0% Cu. For sulfur and acid corrosion, 15 to 30% Cr is effective. Silicon additions in the range of 14 to 15% give excellent corrosion resistance to sulfuric, nitric, and formic acids; however, both high-chromium and high-silicon irons are white, and the high-silicon irons are extremely brittle.

The electrical and magnetic properties of cast iron can be modified slightly by minor additions of alloying elements, but a major change in characteristics can be accomplished by the use of approximately 15% Ni or of nickel plus copper, which results in an austenitic iron that is virtually nonmagnetic. The austenitic gray irons also have good resistance to oxidation and growth at temperatures up to about 800 °C (1500 °F).

Molybdenum is an effective alloying addition for retaining strength in heavy sec-

Table 25 Hardness of quenched samples of gray iron

Condition	Plain cast iron		Cr-Ni-Mo cast iron	
	Combined carbon, %	Hardness, HB	Combined carbon, %	Hardness, HB
As-cast	0.69	217	0.70	255
After quenching from, °C (°F)				
650 (1200)	0.54	207	0.65	250
675 (1250)	0.38	187	0.63	241
705 (1300)	0.09	170	0.59	229
730 (1350)	0.09	143	0.47	217
760 (1400)	nil	137	0.45	197
790 (1450)	0.05	143	0.42	207
815 (1500)	0.47	269	0.60	444
845 (1550)	0.59	444	0.69	514
870 (1600)	0.67	477	0.76	601

Note: Specimens were 30.5 mm diam bars, 51 mm long (1.2 in. diam bars, 2 in. long) quenched in oil from temperatures shown. Source: Ref 18

Table 26 Hardenability data for gray irons quenched from 855 °C (1575 °F)
See Table 27 for compositions.

Distance from quenched end		Hardness, HRC					
mm	1/16 in. increments	Plain iron	Mo(A)	Mo(B)	Ni-Mo	Cr-Mo	Cr-Ni-Mo
3.2	2	54	56	53	54	56	55
6.4	4	53	56	52	54	55	55
9.5	6	50	56	52	53	56	54
12.7	8	43	54	51	53	55	54
15.9	10	37	52	50	52	55	53
19.0	12	31	51	49	52	54	53
22.2	14	26	51	46	52	54	52
25.4	16	26	49	45	52	54	53
28.6	18	25	46	45	52	53	52
31.8	20	23	46	44	51	50	51
34.9	22	22	45	43	47	50	50
38.1	24	22	43	44	47	49	50
41.3	26	21	43	44	47	47	49
44.4	28	20	40	41	45	47	48
47.6	30	19	39	40	45	44	50
50.8	32	17	39	40	45	41	47
54.0	34	18	36	41	44	38	46
57.2	36	18	40	40	45	36	45
60.3	38	19	38	37	45	34	46
63.5	40	22	38	36	42	35	46
66.7	42	20	35	35	42	32	45

Source: Ref 18

tions. It is normally added in amounts of 0.5 to 1.0%, but the low end of this range applies chiefly when molybdenum is added in combination with other elements. In the casting in thin sections, nickel is the most effective in combating the tendency to form chilled iron.

Base Irons. The selection of alloying elements to modify as-cast properties in gray iron depends to a large extent on the composition and method of manufacture of the base iron. For example, a foundry producing a base iron containing 2.3% Si and 3.4% total carbon for automotive castings might add 0.5 to 1.0% Cr if required to make heavier castings with the same hardness and strength as the normal castings. However, a foundry producing a base iron with 1.7% Si and 3.1% C for a heavy casting would add 0.5 to 0.8% Si to decrease hardness and chill when pouring this iron in light castings.

Depending on the strength desired in the final iron, the carbon equivalent of the base iron may vary from approximately 4.4% for weak irons to 3.0% for high-strength irons. The method of producing the base iron will affect mechanical properties and the alloy additions to be made, because factors such as type and percentage of raw materials in the metal charge, amount of superheat, and cooling rate of the iron after pouring all affect the properties. The base iron used for alloying will vary considerably from foundry to foundry, as will the alloying elements selected to give the desired mechanical properties. However, parts produced from different base irons and alloy additions can have the same properties and performance in service.

Heat Treatment

Gray iron, like steel, can be hardened by rapid cooling or quenching from a suitable elevated temperature. The quenched iron may be tempered by reheating in the range from 150 to 650 °C (300 to 1200 °F) to increase toughness and relieve stresses. The quenching medium may be water, oil, hot salt, or air, depending on composition and section size. Heating may be done in a furnace for hardening throughout the cross section, or it may be localized as by induction or flame so that only the volume heated above the transformation temperature is hardened. In the range of composition of the most commonly used unalloyed gray iron castings, that is, about 1.8 to 2.5% Si and 3.0 to 3.5% TC, the transformation range is about 760 to 845 °C (1400 to 1550 °F). The higher temperature must be exceeded in order to harden the iron during quenching. The proper temperature for hardening depends primarily on silicon content, not carbon content; silicon raises the critical temperature.

During the heating of unalloyed gray iron for hardening, graphitization of the matrix frequently begins as the temperature approaches 600 to 650 °C (1100 to 1200 °F) and may be entirely completed at a temperature of 730 to 760 °C (1350 to 1400 °F). This latter range is used for maximum softening. The changes in combined carbon content and hardness that occur upon heating and quenching of both alloyed and unalloyed gray iron are shown in Table 25.

Ordinarily, gray iron is furnace hardened from a temperature of 860 to 870 °C (1575 to 1600 °F). This results in a combined carbon content of about 0.7% and a hardness of about 45 to 52 HRC (415 to 514 HB) in the as-quenched condition. The actual hardness of the martensitic matrix is 62 to 67 HRC, but the presence of graphite causes a lower indicated hardness (Table 14). Temperatures much above this are not advisable because the as-quenched hardness will be reduced by retained austenite.

Oil is the usual quenching medium for through hardening. Quenching in water may be too drastic and may cause cracking and distortion unless the castings are massive and uniform in cross section. Hot oil and hot salt are sometimes used as quenching media to minimize distortion and quench cracking. Water is often used for quenching with flame or induction hardening if only the outer surface is to be hardened.

Hardenability of unalloyed gray iron is about equal to that of low-alloy steel. Hardenability can be measured using the standard end-quench hardenability test employed for steels. The hardenability of cast iron is increased by the addition of chromium, molybdenum, or nickel. Gray iron can be made air hardenable by the addition of the proper amounts of these elements. Some typical data on the hardenability of plain and alloy irons are shown in Table 26. Compositions of the irons are listed in Table 27.

Mechanical Properties. As-quenched gray iron is brittle. Tempering after quenching improves strength and toughness but decreases hardness. A temperature of about 370 °C (700 °F) is required before the toughness (impact strength) approaches the as-

30 / Cast Irons

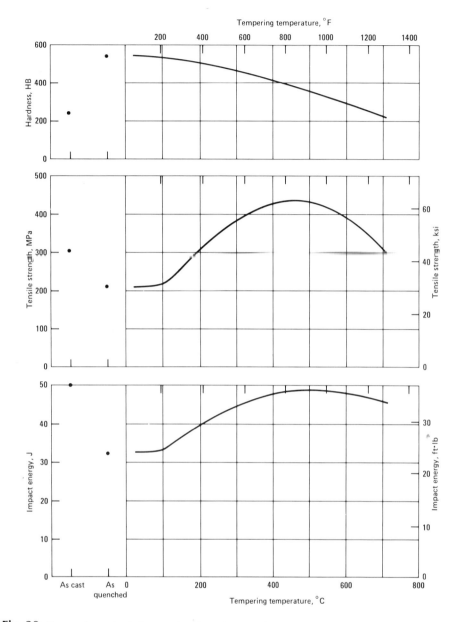

Fig. 20 Changes in mechanical properties of gray iron as a function of tempering temperature

Table 27 Compositions of irons for which hardenability data are given in Table 26

Iron	TC	CC(a)	GC(b)	Mn	Si	Cr	Ni	Mo	P	S
Plain	3.19	0.69	2.50	0.76	1.70	0.03	...	0.013	0.216	0.097
Mo(A)	3.22	0.65	2.57	0.75	1.73	0.03	...	0.47	0.212	0.089
Mo(B)	3.20	0.58	2.62	0.64	1.76	0.005	Trace	0.48	0.187	0.054
Ni-Mo	3.22	0.53	2.69	0.66	2.02	0.02	1.21	0.52	0.114	0.067
Cr-Mo	3.21	0.60	2.61	0.67	2.24	0.50	0.06	0.52	0.114	0.071
Cr-Ni-Mo	3.36	0.61	2.75	0.74	1.96	0.35	0.52	0.47	0.158	0.070

(a) CC, combined carbon. (b) GC, graphite carbon. Source: Ref 18

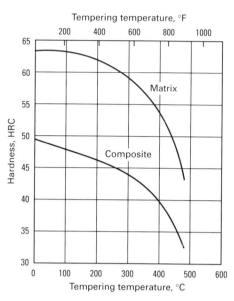

Fig. 21 Effect of tempering temperature on the hardness of quenched gray iron (3.30% C, 2.35% Si)

cast level. The tensile strength after tempering may be from 35 to 45% greater than the strength of the as-cast iron. Changes in properties brought about by quenching and tempering are shown in Fig. 20.

Heat treatment is not ordinarily used commercially to increase the overall strength of gray iron castings, because the strength of the as-cast metal can be increased at less cost by reducing the silicon and total carbon contents or by adding alloying elements. When gray iron is quenched and tempered, it usually is done to increase resistance to wear and abrasion by increasing hardness. A structure consisting of graphite embedded in a hard martensitic matrix is produced by heat treatment. Localized heat treatment such as flame or induction hardening can be used in some applications in which alloy iron or chilled iron has traditionally been used, often with a savings in cost. Heat treatment can be used when chilling is not feasible, as with complicated shapes or large castings, or when close tolerances that can be attained only by machining are required. Heat treatment extends the field of application of gray iron as an engineering material.

The hardness of quenched and tempered gray iron measured with either a Brinell or Rockwell C tester is a composite hardness that superimposes the effects of graphite type, distribution, and size on the hardness of the metal matrix. The true hardness of the matrix, measured with a microhardness test, is generally from 8 to 10 HRC points higher than the hardness indicated by the conventional Rockwell C test. This is shown in Fig. 21, which compares composite hardness with matrix hardness. The composite hardness was obtained by conventional Rockwell testing; the matrix hardness was measured with a Vickers indentor and a 200 g load, and the values were converted to equivalent Rockwell hardness. The hardness of the matrix and the change in matrix hardness with tempering temperature are about the same as in an alloy steel containing approximately 0.7% C.

After tempering at 370 °C (700 °F) for maximum toughness, the hardness of the metal matrix is still about 50 HRC. Where toughness is not required and a tempering temperature of 150 to 260 °C (300 to 500 °F) is acceptable, the matrix hardness is equivalent to 55 to 60 HRC. High matrix hardness and the presence of graphite result in a surface with good wear resistance for some

Table 28 Coefficients of thermal expansion for various gray irons compared with several steels and other irons

Metal			Thermal conductivity at indicated mean temperature, W/m · K		
			200 °F (95 °C)	400 °F (205 °C)	800 °F (425 °C)
Steels					
1020			51.9	48.5	41.5
4135			43.3	41.5	38.1
2335			38.1	38.1	36.3
Gray irons					
C, %	Si, %	Carbon equivalent, %			
3.50	2.25	4.25	57.1	51.9	
3.93	1.40	4.40	55.4	51.9	46.7
3.58	1.90	4.21	48.5		
3.16	1.54	3.67	46.7		41.5
2.92	1.75	3.50	36.3		
2.90	1.51	3.40 (Alloyed)	36.3	36.3	36.3
Ductile irons					
Ferritic, 2.35% Si			41.5	38.1	32.9
Pearlitic, alloyed			26.0	27.7	
High-alloy irons					
36% Ni steel (Invar)			10.4	...	...
36% Ni cast iron			39.8	...	...

Source: Ref 19

Table 29 Relative damping capacities of some common structural alloys

Material	Relative damping capacity
Gray iron, coarse flake	100–500
Gray iron, fine flake	20–100
Malleable iron	8–15
Ductile iron	5–20
Pure iron	5
Eutectoid steel	4
White iron	2–4
Aluminum	0.4

applications, for instance, farm implement gears, sprockets, diesel cylinder liners, and automotive camshafts.

Dimensional changes resulting from the hardening of gray iron are quite uniform and predictable if the prior structure and composition are uniform. Such dimensional changes can be allowed for in machining before heat treatment.

Complex shapes or nonuniform sections may be distorted as a result of the relief of residual casting stresses or a differential rate of cooling and hardening during quenching, or both. The former condition can be minimized by stress relieving at about 565 to 590 °C (1050 to 1100 °F) before machining. The latter can be minimized by either marquenching or austempering; both of these processes are used for cylinder liners where out-of-roundness must be held to a minimum. Through hardening is employed for gears, sprockets, hub bearings, and clutches.

Localized Hardening. In parts requiring only localized areas of hardness, conventional induction hardening or flame hardening may be used.

For flame hardening, it is generally desirable to alloy the iron with small amounts of either chromium or molybdenum to stabilize the iron carbide and thus prevent graphitization. Also, adequate localized hardening is enhanced when the structure contains little or no free ferrite prior to hardening. Compared to a pearlitic matrix, it may take from two to four times as long at temperature to condition a ferritic matrix for hardening. Even though the diffusion of carbon into austenite from the adjacent graphite is quite rapid, the attainment of a carbon concentration that can produce full hardness may require 1 to 10 min (or more) for a previously ferritic matrix, depending on ferrite composition, austenitizing temperature, and graphite distribution and spacing.

Water may be used as the quenching medium when the depth of hardening is shallow and the part is progressively heated and quenched. When only small areas are being hardened, the parts may be dropped into an oil quench. Localized hardening has been used for camshafts, gears, sprockets, and cylinder liners.

Physical Properties

Patternmakers' rules (shrink rules) allow 1% linear contraction upon solidification and cooling of gray iron. (The comparable allowances for other cast ferrous metals are 0% for annealed ductile iron, 0.7% for as-cast ductile iron, 1½ to 2% for white iron, 2% for cast carbon steel, and 2½% for cast alloy steel.) Often, in practice this allowance is inaccurate necessitating more exact evaluation of the effects of mass, composition, shape, mold, and core. Generally, contraction is less with an increase in mass and in gray irons increases with increasing tensile strength.

Three contraction phenomena are exhibited during the cooling of a cast iron from the molten state to room temperature: in the liquid, contraction of about 2.8% from 1425 °C (2600 °F) to the liquidus at about 1200 °C (2200 °F); contraction during solidification; and subsequent to solidification, contraction of about 3.0% from 1150 °C (2100 °F) to room temperature.

Shrinkage in volume during solidification ranges from negative shrinkage in "soft" irons to +1.94% in an iron containing about 0.90% combined carbon. The white irons undergo 4.0 to 5.5% contraction in volume within the same temperature range.

Coefficient of thermal expansion of gray irons is about 13 μm/m · °C (7.2 μin./in. · °F) in the range from 0 to 500 °C (32 to 930 °F) and about 10.5 μm/m · °C (5.8 μin./in. · °F) in the range from 0 to 100 °C (32 to 212 °F). The coefficient of thermal expansion is highly dependent on the matrix structure. The coefficients of thermal expansion of ferritic and martensitic irons are slightly higher than those of pearlitic irons (Table 28).

For the temperature range from 1070 °C (1960 °F) to room temperature, the coefficient of thermal expansion varies from 9.2 to 16.9 μm/m · °C (5.1 to 9.4 μin./in. · °F). At about room temperature, the commonly used figure of 10 μm/m · °C (5.5 μin./in. · °F) is accurate enough for moderate changes in temperature.

Density of gray irons at room temperature varies from about 6.95 Mg/m^3 for open-grained high-carbon irons to 7.35 Mg/m^3 for close-grained low-carbon irons. The density of white iron is about 7.70 Mg/m^3.

Thermal conductivity of gray iron is approximately 46 W/m · K.

Electrical and Magnetic Properties. The resistivity of gray iron, compared with that of other ferrous metals, is relatively high, apparently because of the amounts and distribution of the graphite. Increases in total carbon content and in silicon content increase resistivity.

The magnetic properties of gray iron may vary within wide limits, ranging from those of irons having low permeability and high coercive force (suitable for permanent magnets) to those of irons having high permeability, low coercive force, and low hysteresis loss (suitable for electrical machinery).

The highest magnetic induction and permeability are found in annealed white irons, such as malleable cast iron. Flake graphite, as in gray iron, does not affect hysteresis loss, but prevents the attainment of high magnetic induction by causing small demagnetizing forces.

Damping capacity is the capability of a material to quell vibrations and to dissipate the energy as heat, or simply the relative capability to stop vibrations or ringing. Because gray iron possesses high damping capacity, it is well suited for bases and supports, as well as for moving parts. It reduces or eliminates parasitic vibration. The noise level of a machine operating on a gray iron base is materially reduced. High damping capacity is especially desirable in structures and parts in which vibration can cause stresses in excess of those that result from direct loading. Table 29 compares the relative damping capacity of gray iron with the relative damping capacities of other common structural alloys.

REFERENCES

1. M.D. VanMaldegiam and K.B. Rundman, On The Structure and Properties of Austempered Gray Cast Iron, *Trans. AFS*, 1986, p 249
2. R. Schneidewind and R.G. McElwee, Composition and Properties of Gray Iron, Parts I and II, *Trans. AFS*, Vol 58, 1950, p 312-330
3. H.C. Winte, Gray Iron Castings Section Sensitivity, *Trans. AFS*, Vol 54, 1946, p 436-443
4. R.A. Flinn and R.W. Kraft, Improved Test Bars for Standard and Ductile Grades of Cast Iron, *Trans. AFS*, Vol 58, 1950, p 153-167
5. D.E. Krause, Gray Iron—A Unique Engineering Material, in *Gray, Ductile, and Malleable Iron Castings—Current Capabilities*, ASTM STP 455, American Society for Testing and Materials, 1969, p 3-28
6. C.F. Walton and T.J. Opar, *Iron Castings Handbook*, Iron Casting Society, 1981, p 235
7. R.B. Gundlach, The Effects of Alloying Elements on the Elevated-Temperature Properties of Gray Irons, *Trans. AFS*, 1983, p 389
8. W.L. Collins and J.O. Smith, Fatigue and Static Load Tests of a High-Strength Cast Iron at Elevated Temperatures, in *Proceedings of ASTM*, Vol 41, 1941, p 797-807
9. R.E. Peterson, *Stress Concentration Factors*, John Wiley, New York, 1974
10. *U.S. Air Force Machinability Report*, Vol 1, 1950, p 135
11. A.B. Shuck, A Laboratory Evaluation of Some Automotive Cast Irons, *Trans. AFS*, Vol 56, 1948, p 166-192
12. G.P. Phillips, Hardened Gray Iron—An Ideal Material for Diesel Engine Cylinder Sleeves and Liners, *Foundry*, Vol 80, Jan 1952, p 88-95, 222, 224, 226, 228
13. G.K. Turnbull and J.F. Wallace, Molybdenum Effect on Gray Iron Elevated-Temperature Properties, *Trans. AFS*, 1959, p 35
14. J.E. Bevan, Effect of Molybdenum on Dimensional Stability and Tensile Properties of Pearlitic Gray Irons at 600 to 850 °F (315 to 455 °C), Internal Report, Climax Molybdenum Company
15. D.G. White, Growth and Scaling Characteristics of Cast Irons With Undercooled and Normal Flake Graphite, *BCIRA J.*, 1963, p 223
16. K.B. Palmer, Design With Cast Irons at High Temperatures, 1, Growth and Scaling, Report 1248, British Cast Iron Research Association
17. J.H. Schaum, Stress Relief Heat Treatment of Gray Cast Iron, *Trans. AFS*, Vol 61, 1953, p 646-650
18. G.A. Timmons, V.A. Crosby, and A.J. Herzig, *Trans. AFS*, Vol 49, 1941, p 397
19. C.F. Walton, *Gray and Ductile Iron Castings Handbook*, Iron Founders' Society, 1971

SELECTED REFERENCES

- C.E. Bates, Alloy Element Effects on Gray Iron Properties, Part II, *Trans. AFS.*, 1986, p 889
- A.L. Boegehold, Present-Day Methods in Production and Utilization of Automotive Cast Iron, in *Proceedings of the ASTM Symposium on Developments in Automotive Materials*, American Society for Testing and Materials, 1930, p 5
- J.W. Grant, Comprehensive Mechanical Tests of Two Pearlitic Gray Irons, *BCIRA J.*, 1951, p 861
- R.B. Gundlach, Thermal Fatigue Resistance of Alloyed Gray Irons for Diesel Engine Components, *Trans. AFS.*, 1979, p 551
- K.B. Palmer, Creep Tests on a Flake Graphite Cast Iron at 400 °C, *BCIRA J.*, 1959, p 839

Ductile Iron

Revised by Lyle R. Jenkins, Ductile Iron Society;
and R.D. Forrest, Pechiney Electrometallurgie, Division Fonderie

DUCTILE CAST IRON, previously known as nodular iron or spheroidal-graphite (SG) cast iron (the international term is ductile iron), is cast iron in which the graphite is present as tiny spheres (nodules) (see Fig. 1). In ductile iron, eutectic graphite separates from the molten iron during solidification in a manner similar to that in which eutectic graphite separates in gray cast iron. However, because of additives introduced in the molten iron before casting, the graphite grows as spheres, rather than as flakes of any of the forms characteristic of gray iron. Cast iron containing spheroidal graphite is much stronger and has higher elongation than gray iron or malleable iron. It may be considered as a natural composite in which the spheroidal graphite imparts unique properties to ductile iron.

The relatively high strength and toughness of ductile iron give it an advantage over gray iron or malleable iron in many structural applications. Also, because ductile iron does not require heat treatment to produce graphite nodules (as does malleable iron to produce temper-carbon nodules), it can compete with malleable iron even though it requires a treatment and inoculation process. The mold yield is normally higher than with malleable iron. Ductile iron can be produced to x-ray standards because porosity stays in the thermal center. Malleable iron cannot tolerate porosity because voids migrate to the surface of hot spots such as fillets and appears as cracks.

Typically, the composition of unalloyed ductile iron differs from that of gray iron or malleable iron (Table 1). The raw materials used for ductile iron must be of higher purity. All cast irons can be melted in cupolas, electric arc furnaces, or induction furnaces. Ductile iron, as a liquid, has high fluidity, excellent castability, but high surface tension. The sands and molding equipment used for ductile iron must provide rigid molds of high density and good heat transfer.

The formation of graphite during solidification causes an attendant increase in volume, which can counteract the loss in volume due to the liquid-to-solid phase change in the metallic constituent. Ductile iron castings typically require only minimal use of risers (reservoirs in the mold that feed molten metal into the mold cavity to compensate for liquid contraction during cooling and solidification). Gray irons often do not require risers to ensure shrinkage-free castings. On the other hand, steels and malleable iron generally require heavy risering. Thus, the mold yield of ductile iron castings (the ratio of the weight of usable castings to the weight of metal poured) is

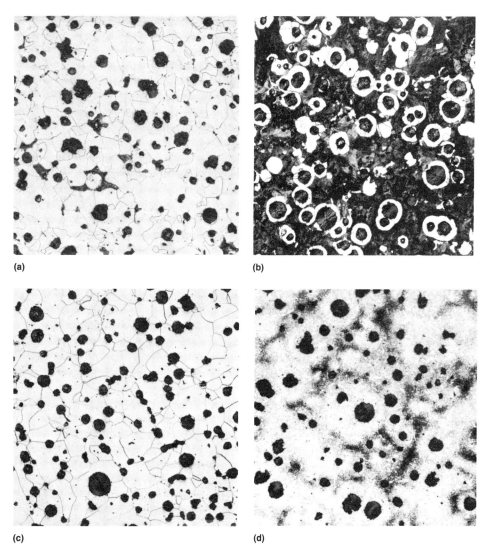

Fig. 1 Microstructures of ductile iron. (a) As-cast ferritic. (b) As-cast pearlitic; hardness, 255 HB. (c) Ferritic, annealed 3 h at 700 °C (1290 °F). (d) Pearlitic ductile iron oil quenched and tempered to 255 HB. All etched in 2% nital. 100×

Table 1 Composition of selected cast irons

Type	TC(a)	Mn	Si	Cr	Ni	Mo	Cu	P	S	Ce	Mg
Gray iron	3.25–3.50	0.50–0.90	1.80–2.30	0.05–0.45	0.05–0.20	0.05–0.10	0.15–0.40	0.12 max	0.15 max	...	...
Malleable iron	2.45–2.55	0.35–0.55	1.40–1.50	0.04–0.07	0.05–0.30	0.03–0.10	0.03–0.40	0.03 max	0.05–0.07	...	...
Ductile iron	3.60–3.80	0.15–1.00	1.80–2.80	0.03–0.07	0.05–0.20	0.01–0.10	0.15–1.00	0.03 max	0.002 max	0.005–0.20(b)	0.03–0.06

(a) TC, total carbon. (b) Optional

much higher than that of either steel castings or malleable iron castings, but not as high as that of gray iron. There are some cases of ductile iron castings being made without risers.

Often designers must compensate for the shrinkage of cast iron during both solidification and subsequent cooling to room temperature by making patterns with dimensions larger than those desired in the finished castings. Typically, ductile iron requires less compensation than any other cast ferrous metal. The allowances in patternmaker rules (shrink rules) are usually:

Type of cast metal	Shrinkage allowance, %
Ductile iron	0–0.7
Gray iron	1.0
Malleable iron	1.0
Austenitic alloy iron	1.3–1.5
White iron	2.0
Carbon steel	2.0
Alloy steel	2.5

Shrinkage allowance can vary somewhat from the percentages given above, and often different percentages must be used for different directions in one casting because of the influence of the solidification pattern on the amount of contraction that takes place in different directions. Shrinkage is volumetric, and the ratio of dimensions to volume influences each dimension. As ductile iron approaches a condition of shrinkage porosity, the graphite nodules tend to become aligned and can result in lower fatigue strength.

Most ductile iron castings are used as-cast, but in some foundries, some castings are heat treated before being shipped. Heat treatment varies according to the desired effect on properties. Any heat treatment, with the exception of austempering, reduces fatigue properties. Holding at subcritical (705 °C, or 1300 °F) temperature for no more than 4 h improves fracture resistance. Heating castings above 790 °C (1450 °F) followed by fast cooling (oil quench or air quench) significantly reduces fatigue strength and above-room-temperature fracture resistance. Ferritizing by heating to 900 °C (1650 °F) and slow cooling also reduces fatigue strength and above-room-temperature fracture resistance. Heating to above the critical temperature also reduces the combined carbon content of quenched and tempered microstructures and produces lower tensile strength and wear resistance than the same hardness produced as-cast. Some castings may be given hardening treatments (either localized surface or through hardened) that produce bainitic or martensitic matrices.

In recent years, considerable interest has been shown in the property improvements obtained in ductile iron by austempering heat treatments. Austempered ductile iron (ADI) has a matrix that is a combination of acicular (bainitic) ferrite and stabilized austenite. As the matrix structure is progressively varied from ferrite to ferrite plus pearlite to pearlite to bainite and finally to martensite, hardness, strength, and wear resistance increase, but impact resistance, ductility, and machinability decrease. Overall, however, this structure results in an exceptional combination of strength, ductility, and wear resistance. Table 2 shows the ASTM standards for austempered ductile iron. Some of the applications for austempered ductile iron include:

- Gears (including side and timing gears)
- Wear-resistant parts
- High-fatigue strength applications
- High-impact strength applications
- Automotive crankshafts
- Chain sprockets
- Refrigeration compressor crankshafts
- Universal joints
- Chain links
- Dolly wheels

Ductile iron can be alloyed with small amounts of nickel, molybdenum, or copper to improve its strength and hardenability. The addition of molybdenum is done with caution because of the tendency for intercellular segregation. Larger amounts of silicon, chromium, nickel, or copper can be added for improved resistance to corrosion, oxidation, or abrasion, or for high-temperature applications. The high-alloy ductile irons are covered in the article "Alloy Cast Irons" in this Volume. Representative microstructures of both unalloyed and alloyed ductile irons can be found in *Metallography and Microstructures*, Volume 9 of the 9th Edition of *Metals Handbook*.

Specifications

Most of the specifications for standard grades of ductile iron are based on properties; that is, strength and/or hardness is specified for each grade of ductile iron, and composition is either loosely specified or made subordinate to mechanical properties. Tables 3 and 4 list compositions, properties, and typical applications for most of the ductile irons that are defined by current standard specifications (except for the high-nickel, corrosion-resistant, and heat-resistant irons defined in ASTM A 439). As shown in Table 4, the ASTM system for designating the grade of ductile iron incorporates the numbers indicating tensile strength in ksi, yield strength in ksi, and elongation in percent. This system makes it easy to specify nonstandard grades that meet the general requirements of ASTM A 536. For example, grade 80-60-03 (80 ksi, or 552 MPa, minimum tensile strength, 60 ksi, or 414 MPa, yield strength, and 3% elongation) is widely used in applications for which relatively high ductility is not important. Grades 65-45-12 and 60-40-18 are used in areas requiring high ductility and impact resistance. Grades 60-42-10 and 70-50-05 are for special applications such as annealed

Table 2 ASTM standard A 897-90 and A 897M-90 mechanical property requirements of austempered ductile iron

Grade	Tensile (min) MPa	Tensile (min) ksi	Yield (min) MPa	Yield (min) ksi	Elongation, %	Impact(a) J	Impact(a) ft · lbf	Hardness, HB(c)
125-80-10	...	125	...	80	10	...	75	269–321
850-550-10	850	...	550	...	10	100	...	269–321
150-100-7	...	150	...	100	7	...	60	302–363
1050-700-7	1050	...	700	...	7	80	...	302–363
175-125-4	...	175	...	125	4	...	45	341–444
1200-850-4	1200	...	850	...	4	60	...	341–444
200-155-1	...	200	...	155	1	...	25	388–477
1400-1100-1	1400	...	1100	...	1	35	...	388–477
230-185	...	230	...	185	(b)	...	(b)	444–555
1600-1300	1600	...	1300	...	(b)	(b)	...	444–555

(a) Unnotched Charpy bars tested at 72 = 7 °F (22 = 4 °C). The values in the table are a minimum for the average of the highest three test values of four tested samples. (b) Elongation and impact requirements are not specified. Although grades 200-155-1, 1400-1100-1, 230-185, 1600-1300 are primarily used for gear and wear resistance applications, grades 200-155-1 and 1400-1100-1 have applications where some sacrifice in wear resistance is acceptable in order to provide a limited amount of ductility and toughness. (c) Hardness is not mandatory and is shown for information only.

Table 3 Compositions and general uses for standard grades of ductile iron

Specification No.	Grade or class	UNC	TC(a)	Si	Mn	P	S	Description	General uses
ASTM A 395; ASME SA395	60-40-18	F32800	3.00 min	2.50 max(b)	...	0.08 max	...	Ferritic; annealed	Pressure-containing parts for use at elevated temperatures
ASTM A 476; SAE AMS 5316C	80-60-03	F34100	3.00 min(c)	3.0 max	...	0.08 max	0.05 max	As-cast	Paper mill dryer rolls, at temperatures up to 230 °C (450 °F)
ASTM A 536	60-40-18(d)	F32800						Ferritic; may be annealed	Shock-resistant parts; low-temperature service
	65-45-12(d)	F33100						Mostly ferritic; as-cast or annealed	General service
	80-55-06(d)	F33800						Ferritic/pearlitic; as-cast	General service
	100-70-03(d)	F34800						Mostly pearlitic; may be normalized	Best combination of strength and wear resistance and best response to surface hardening
	120-90-02(d)	F36200						Martensitic; oil quenched and tempered	Highest strength and wear resistance
SAE J434	D4018(e)	F32800	3.20–4.10	1.80–3.00	0.10–1.00	0.015–0.10	0.005–0.035	Ferritic	Moderately stressed parts requiring good ductility and machinability
	D4512(e)	F33100						Ferritic/pearlitic	Moderately stressed parts requiring moderate machinability
	D5506(e)	F33800						Ferritic/pearlitic	Highly stressed parts requiring good toughness
	D7003(e)	F34800						Pearlitic	Highly stressed parts requiring very good wear resistance and good response to selective hardening
	DQ & T(e)	F30000						Martensitic	Highly stressed parts requiring uniformity of microstructure and close control of properties
SAE AMS 5315C	Class A	F33101	3.0 min	2.50 max(f)	...	0.08 max	...	Ferritic; annealed	General shipboard service

Note: For mechanical properties and typical applications, see Table 3. (a) TC, total carbon. (b) The silicon limit may be increased by 0.08%, up to 2.75 Si, for each 0.01% reduction in phosphorus content. (c) Carbon equivalent (CE), 3.8–4.5; CE = TC + 0.3 (Si + P). (d) Composition subordinate to mechanical properties; composition range for any element may be specified by agreement between supplier and purchaser. (e) General composition given under grade D4018 for reference only. Typically, foundries will produce to narrower ranges than those shown and will establish different median compositions for different grades. (f) For castings with sections 13 mm (½ in.) and smaller, may have 2.75 Si max with 0.08 P max, or 3.00 Si max with 0.05 P max; for castings with section 50 mm (2 in.) and greater, CE must not exceed 4.3

pipe or as-cast pipe fittings. Grades other than those listed in ASTM A 536 or mentioned above can be made to the general requirements of A 536, but with the mechanical properties specified by mutual agreement between purchaser and producer.

The Society of Automotive Engineers (SAE) uses a method of specifying iron for castings produced in larger quantities that is based on the microstructure and Brinell hardness of the metal in the castings themselves.

Both ASTM and SAE specifications are standards for tensile properties and hardness. The tensile properties are quasi-static and may not indicate the dynamic properties, such as impact or fatigue strength.

The International System of grade designation (ISO 1083) uses the tensile strength value, in MPa, and elongation percentage. Table 5 shows the property requirements of the ASTM, SAE, and ISO specifications. No standard for austempered ductile iron has yet been approved, although tentative specifications have been proposed, as shown in Table 6 and Fig. 2.

Ductile Iron Applications

Ductile iron castings are used for many structural applications, particularly those requiring strength and toughness combined with good machinability and low cost. The selection of casting, instead of mechanical

Table 4 Mechanical properties and typical applications for standard grades of ductile iron

Specification No.	Grade or class	Hardness, HB(a)	Tensile strength, min(b) MPa	ksi	Yield strength, min(b) MPa	ksi	Elongation in 50 mm (2 in.) (min), %(b)	Typical applications
ASTM A 395; ASME SA395	60-40-18	143–187	414	60	276	40	18	Valves and fittings for steam and chemical-plant equipment
ASTM A 476(c); SAE AMS 5316	80-60-03	201 min	552	80	414	60	3	Paper mill dryer rolls
ASTM A 536	60-40-18	...	414	60	276	40	18	Pressure-containing parts such as valve and pump bodies
	65-45-12	...	448	65	310	45	12	Machine components subject to shock and fatigue loads
	80-55-06	...	552	80	379	55	6	Crankshafts, gears, and rollers
	100-70-03	...	689	100	483	70	3	High-strength gears and machine components
	120-90-02	...	827	120	621	90	2	Pinions, gears, rollers, and slides
SAE J434	D4018	170 max	414	60	276	40	18	Steering knuckles
	D4512	156–217	448	65	310	45	12	Disk brake calipers
	D5506	187–255	552	80	379	55	6	Crankshafts
	D7003	241–302	689	100	483	70	3	Gears
	DQ&T	(c)	(d)	(d)	(d)	(d)	(d)	Rocker arms
SAE AMS 5315C	Class A	190 max	414	60	310	45	15	Electric equipment, engine blocks, pumps, housings, gears, valve bodies, clamps, and cylinders

Note: For compositions, descriptions, and uses, see Table 2. (a) Measured at a predetermined location on the casting. (b) Determined using a standard specimen taken from a separately cast test block, as set forth in the applicable specification. (c) Range specified by mutual agreement between producer and purchaser. (d) Value must be compatible with minimum hardness specified for production castings.

Table 5 Ductile iron property requirements of various national and international standards

Grade	Tensile strength MPa	ksi	0.2% offset yield strength MPa	ksi	Elongation (min, %)	Impact energy Mean(a) J	ft · lbf	Individual J	ft · lbf	Hardness, HB	Structure
ISO Standard 1083 (International)											
800-2	800	116	480	70	2	...	...	...	...	248–352	Pearlite or tempered
700-2	705	102	420	61	2	...	...	...	...	229–302	Pearlite
600-3	600	87	370	54	3	...	...	...	...	192–269	Pearlite + ferrite
500-7	500	73	320	46	7	...	...	...	...	170–241	Ferrite + pearlite
400-12	400	58	250	36	12	...	...	...	...	<201	Ferrite
370-17	370	54	230	33	17	13	9.5	11	8.1	<179	Ferrite
ASTM A 536 (United States)											
60-40-18	414	60	276	40	18	...	...	...	...	...	...
60-42-10	414	60	290	42	10	...	...	...	...	...	...
65-45-12	448	65	310	45	12	...	...	...	...	...	...
70-50-05	485	70	345	50	5	...	...	...	...	...	...
80-55-06	552	80	379	55	6	...	...	...	...	...	...
80-60-03	552	80	414	60	3	...	...	...	...	...	...
100-70-03	690	100	483	70	3	...	...	...	...	...	...
120-90-02	827	120	621	90	2	...	...	...	...	...	...
SAE J434 (United States)(b)											
D4018	414	60	276	40	18	...	...	...	...	170 max	Ferrite
D4512	448	65	310	45	12	...	...	...	...	156–217	Ferrite + pearlite
D5506	552	80	379	55	6	...	...	...	...	187–255	Ferrite + pearlite
D7003	690	100	483	70	3	...	...	...	...	241–302	Pearlite
DQ&T(c)	...	...	...	...	...	...	...	...	...	...	Martensite

(a) Mean value from three tests. (b) Specifications for these irons are primarily based on hardness and structure. Mechanical properties are given for information only. (c) Quenched and tempered grade; hardness subject to agreement between supplier and purchaser

Table 6 Some tentative specifications for austempered ductile iron

Grade	Tensile strength (min) MPa	ksi	0.2% offset yield strength (min) MPa	ksi	Elongation (min), %	Hardness, HB
Ductile Iron Society (United States)						
1	860	125	550	80	10	269–321
2	1035	165	690	100	7	302–363
3	1205	175	827	120	4	363–444
4	1380	200	965	140	2	388–477
Sulzer (Switzerland)						
GGG80 BAF	800	116	505	73	8	250–310
GGG100	1000	160	705	102	5	280–340
GGG120	1200	174	950	138	2	330–390
BCIRA (Great Britain)						
950/6	950	138	670	97	6	300–310
1050/3	1050	152	780	113	3	345–355
1200/1	1200	174	940	136	1	390–400

fabrication, as the production process often allows the designer to:

- Use to best advantage the combination of properties that is unique to ductile iron
- Combine several functions (or component shapes) in a single integrated configuration
- Realize the economic advantages inherent in casting, which is the simplest and most direct of the various production processes

Casting is like forming with a homogenous liquid that can flow smoothly into a wide variety of thicknesses, shapes, and contours. The material is consistent piece to

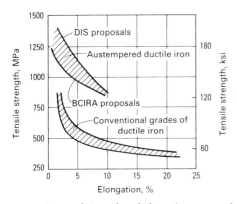

Fig. 2 Proposed strength and elongation ranges of austempered ductile iron compared to established criteria for other grades of ductile iron. DIS, Ductile Iron Society; BCIRA, British Cast Iron Research Association

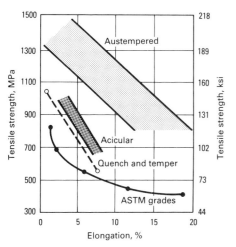

Fig. 3 Tensile strength and elongation relationship for various ductile iron grades. Source: Ref 1

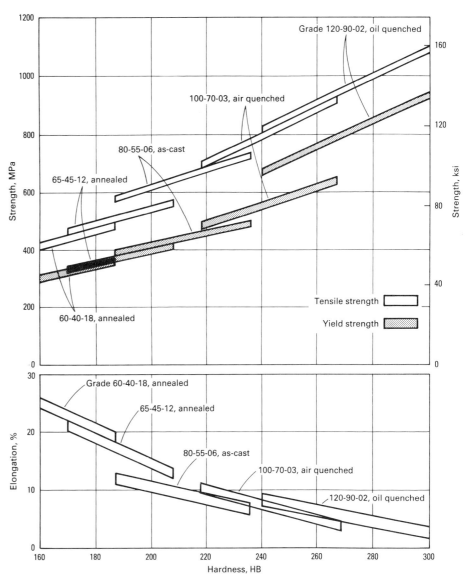

Fig. 4 Tensile properties of ductile iron versus hardness. Mechanical properties determined on specimens taken from a 25 mm (1 in.) keel block

piece, day after day because it is made from a pattern that has very low wear and dimensional change and that is produced to a specification. The microstructure is uniform, with neither directional flow lines nor variations from weld junctions or heat-affected zones. Any porosity is predictable and remains in the thermal center. Machining costs are low because there is less material to remove and to dispose of, castings are easier to machine, and cutting tools are subjected to less tool wear.

Properties unique to ductile iron include the ease of heat treating because the free carbon in the matrix can be redissolved to any desired level for hardness and strength control. Free carbon can be selectively hardened by flame, induction, laser, or electron beam. A 650 °C (1200 °F) anneal for 3 h can produce high toughness at low temperatures. Ductile iron can be austempered to high tensile strength, high fatigue strength, high toughness, and excellent wear resistance. Second, lower density makes ductile iron weigh 10% less than steel for the same section size. Third, the graphite content provides damping properties for quiet running gears. Also, the low coefficient of friction produces more efficient gear boxes. Furthermore, ductile iron has less tendency to gear seizures from the loss of lubricant.

Experience has proved that ductile iron works in applications in which experience and handbook data say it should not. This is because data in the literature do not describe the true properties of ductile iron, whereas it is very easy to make a prototype casting and try it in the field. This practice has resulted in very large cost savings and superior performance compared to the material it replaces. A ductile iron casting can be poured and shipped the same day. As-cast ductile iron castings are consistent in dimensions and weight because there is no distortion or growth due to heat treatment.

The automotive and agricultural industries are the major users of ductile iron castings. Almost 3×10^6 tonnes (2.9×10^6 t, or 3.2×10^6 tons) of ductile iron castings were produced in the United States in 1988, the majority of components being for automotive applications. Because of economic advantages and high reliability, ductile iron is used for such critical automotive parts as crankshafts, front wheel spindle supports, complex shapes of steering knuckles, disk brake calipers, engine connecting rods, idler arms, wheel hubs, truck axles, suspension system parts, power transmission yokes, high-temperature applications for turbo housings and manifolds, and high-security valves for many various applications. It can be rolled or spun into a desired shape or coined to an exact dimension.

The cast iron pipe industry is another major user of ductile iron.

Austempered ductile iron has resulted in many new applications for ductile iron. It is a high-strength, wear-resistant, heat-treated material. It has more than double the strength

38 / Cast Irons

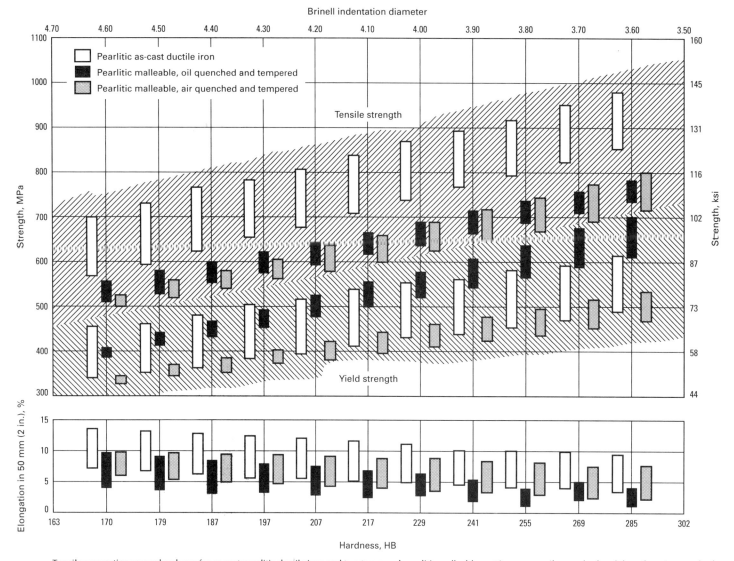

Fig. 5 Tensile properties versus hardness for as-cast pearlitic ductile iron and two tempered pearlitic malleable cast irons, one oil quenched and the other air quenched. Data are for 3000 kgf load. Source: Ref 2

of conventional ductile iron for a given level of ductility (Fig. 3). Austempered ductile iron gets its remarkable properties from a special austempering heat treatment. The resultant strength properties can be varied by controlling the heat treat cycle, which is described in this article in the section "Heat Treatment." In the austempering process, good-quality ductile iron can be transformed into a superior engineering material. It cannot transform poor-quality iron into a good-quality material.

Metallurgical Control in Ductile Iron Production

Greater metallurgical and process control is required in the production of ductile iron than in the production of other cast irons. Frequent chemical, mechanical, and metallurgical testing is needed to ensure that the required quality is maintained and that specifications are met.

The manufacture of high-quality ductile iron begins with the careful selection of charge materials that will give a relatively pure cast iron, free of the undesirable residual elements sometimes found in other cast irons. Carbon, manganese, silicon, phosphorus, and sulfur must be held at specified levels. Magnesium, cerium, and certain other elements must be controlled to attain the desired graphite shape and offset the deleterious effects of elements such as antimony, lead, titanium, tellurium, bismuth, and zirconium, which interfere with the nodulizing process and must be either eliminated or restricted to very low concentrations and neutralized by additions of cerium and/or rare earth elements. Alloying elements such as chromium, nickel, molybdenum, copper, vanadium, and boron act as carbide formers, as pearlite stabilizers, or as ferrite promoters. Alloys are controlled to the extent needed to obtain the required mechanical properties and/or microstructure in the critical section(s) of the casting.

A reduction of the sulfur content in the base iron to below 0.02% is necessary prior to the nodulizing process (except with certain pure magnesium treatment processes); this can be accomplished by basic melting alone or by desulfurization of the base metal (more commonly used today) before the magnesium-nodulizing alloy is added. If base iron sulfur is not reduced, excessive amounts of costly nodulizing alloys are required, and greater amounts of slag and dross are generated.

Graphite Shape and Distribution

There are three major types of nodulizing agents, all of which contain magnesium:

- Unalloyed magnesium metal
- Magnesium-containing ferrosilicons
- Nickel-base nodulizers

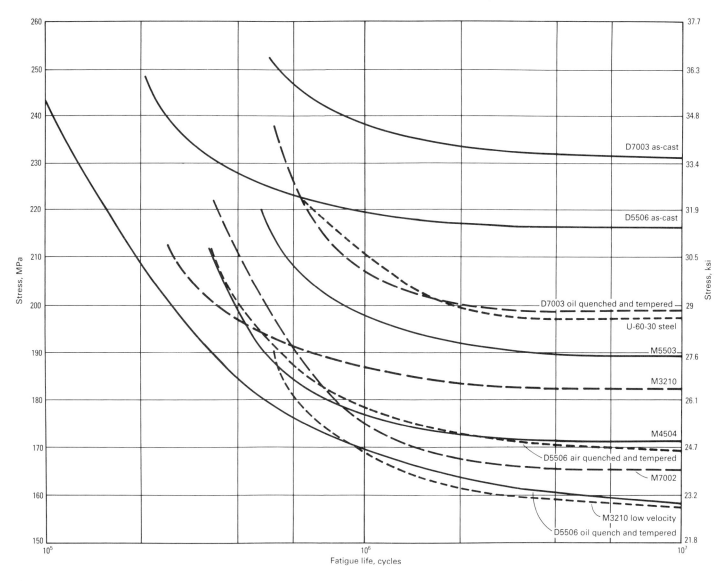

Fig. 6 Summary of fatigue life for ductile and malleable cast irons and one grade of cast steel. Source: Ref 2

Unalloyed magnesium metal has been added to molten iron as wire, ingots, or pellets; as briquets in combination with sponge iron; as pellets in combination with granular lime; or in the cellular pores of metallurgical coke.

Magnesium-Containing Alloys. The method of introducing magnesium alloys has varied from an open-ladle or covered-ladle method (in which the alloy is placed at the bottom of a ladle that has a height-to-diameter ratio of between 2:1 and 1.5:1 and iron is poured rapidly over the alloy) to a plunging method, or to a pressure-container method in which unalloyed magnesium is placed inside a container holding molten iron and the container is rotated so that the iron flows over the magnesium. The magnesium metal can also be plunged to the bottom of the iron in a pressure ladle. In all cases, magnesium is vaporized, and the vapors travel through the molten iron, lowering the sulfur content and promoting the formation of spheroidal graphite.

Nickel-Base Nodulizers. A nickel alloy with 14 to 16% Mg can be added to the ladle during filling or by plunging. The reaction is spectacular, but not violent, and a very consistent recovery is obtained. A disadvantage lies in the accompanying increase in nickel and the cost of the alloy. Other nickel alloys containing much lower magnesium contents (down to 4%) have also been used and involve a much quieter reaction.

Ferrosilicon-based inoculant is usually added to the nodulized iron when it is transferred to the pouring ladle. This produces a high nodule count and a matrix of preferred microstructure. With the proper control of shakeout time and temperature, certain ductile iron castings can meet grade specifications as-cast, without further heat treatment. Hardness uniformity is attained by the addition of pearlite-forming elements such as copper.

Testing and Inspection

Various tests are used to control the processing of ductile iron. The first is the analysis of raw materials and of the molten metal, both before and after the nodulizing treatment. Rapid thermal-arrest methods are used to confirm carbon, silicon, and carbon equivalence (CE) in the molten iron. Carbon equivalence is not used to allow broader carbon or silicon ranges; these elements still have definite individual ranges that must be held. Silicon content is determined by thermoelectric, spectrometric, and wet chemical analysis. Chill tests are used to control nucleation, indicate a carbide-forming tendency, and alert production control in the event that silicon has been omitted. The temperature of the molten iron in the furnace is measured by immersion thermocouple prior to the nodulizing process. It is also measured in the pouring ladle. Weight control is exercised for the amount of metal being treated, the

40 / Cast Irons

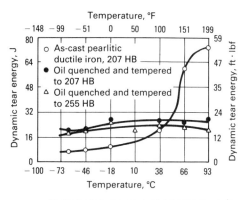

Fig. 7 Dynamic tear energy versus temperature for as-cast and heat-treated ductile irons. Specimen size: 190 mm (7½ in.) long, 130 mm (5⅛ in.) wide, and 41 mm (1⅝ in.) thick; 13 mm (½ in.) notch depth. Source: Ref 2

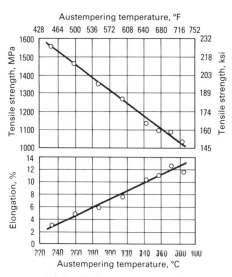

Fig. 8 Relationship between austempering temperature and the strength and ductility of a 1.5Ni-0.3Mo alloyed ductile iron. Austenitizing temperature was 900 °C (1650 °F). Source: Ref 3

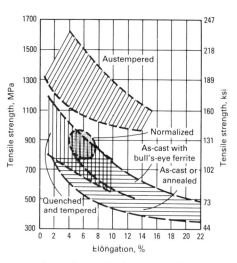

Fig. 9 Strength and ductility ranges of as-cast and heat-treated nodular irons. Source: Ref 4

amount of alloys being added, and the amount of metal being inoculated. Time control is used to ensure that all treated metal is poured within a given period, beginning with the nodulization treatment and including the inoculation. Minimum pouring temperatures are established, and any iron that does not meet the minimum standard is not allowed to be poured into molds.

A standard test coupon for microscopic examination must be poured from each batch of metal with the same iron that is poured into the last mold. This coupon is described in ASTM A 395. One ear of the test coupon is removed and polished to reveal graphite shape and distribution, as well as matrix structure. These characteristics are evaluated by comparison with standard ASTM/AFS photomicrographs, and acceptance or rejection of castings is based on this comparison.

Tensile-test specimens are machined from separately cast keel blocks, Y blocks, or modified keel blocks, as described in ASTM A 395. If the terms of purchase require tensile specimens to be taken from castings, the part drawing must identify the area of the casting and the size of the test specimen. A tensile specimen should never be cut from the centerline of a round section, but rather from the midradius or as near the surface as possible. These terms must be mutually acceptable to producer and purchaser.

The hardness testing of production castings is also used to evaluate conformance to specified properties. Some standard specifications, such as SAE J434b, relate strength and hardness, as shown in Fig. 4. A comparison of tensile properties with hardness of as-cast ductile iron and pearlitic malleable iron is shown in Fig. 5. This shows that the tensile strength of as-cast ductile iron is significantly higher than that of pearlitic malleable iron at all hardnesses. The elongation is higher than that of oil-quenched pearlitic malleable iron. Air-quenched pearlitic malleable iron has a higher elongation than oil-quenched iron because the air-quenched iron has a ferrite ring around the temper-carbon nodules. As oil-quenched pearlitic malleable iron is tempered to a lower hardness, the elongation increases. This is also true in the case of fatigue strength, as shown in Fig. 6, in which the endurance limit of 217 HB oil-quenched pearlitic malleable iron is higher than that of 255 HB. The yield strength of as-cast ductile iron is lower than oil-quenched and tempered pearlitic malleable iron when the hardness is above 217 HB. This is because the ferrite ring around the graphite nodules shows yielding at 0.2% offset. The yield strength of as-cast ductile iron can be increased to as high or higher than the yield strength of oil-quenched and tempered pearlitic malleable iron, and the hardness can be as high, but the tensile strength and elongation will be significantly lower. The fatigue strength will be lower, too, as shown in Fig. 6. When ductile iron is heated to 900 °C (1650 °F), oil quenched, and tempered, the peak of combined carbon between graphite nodules is reduced and spread out to the nodules, thereby eliminating the ring of ferrite. A loss in wear resistance is experienced, because of the lower combined carbon content at the same hardness as the as-cast iron. The producer and the purchaser must agree on a suitable location, usually indicated on the part drawing, for hardness testing. The preferred method for ductile iron hardness testing is the Brinell method. Surface preparation must be done carefully, and the casting must be cool before it is indented with the 10-mm tungsten carbide ball in a regularly calibrated machine. A proving ring is used to check the machine load of 30 kN (3000 kgf), and a stage micrometer is used to check the calibration of the Brinell scope. A hardness value may be listed as a Brinell indentation diameter (BID) or Brinell hardness number (HB). Other methods of measuring hardness with small indentors, such as the Rockwell method, have variations due to the small area of indentation that

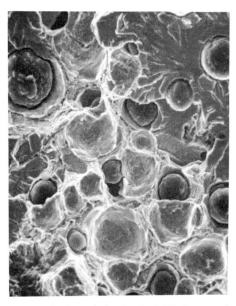

Fig. 10 Fractograph showing the brittle fracture of a ductile iron dynamic tear specimen. 355×

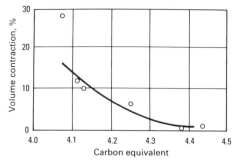

Fig. 11 Plot of carbon equivalent versus ductile iron feed metal requirement. Source: Ref 5

contains soft graphite nodules. Electromagnetic techniques may be used for the hardness testing of as-cast simple shapes, such as crankshafts (see the article "Electromagnetic Techniques for Residual Stress Measurements" in Volume 17 of the 9th Edition of *Metals Handbook*).

Verification of graphite nodularity in castings can be obtained using ultrasonic methods that measure the velocity of sound through the section being tested. Several companies have installed on-line systems to verify nodularity, particularly in automotive components such as brake calipers and steering knuckles. Nondestructive inspection of ductile iron castings, including ultrasonic verification of nodularity, is described in the articles "Castings" and "Ultrasonic Inspection" in Volume 17 of the 9th Edition of *Metals Handbook*. Sonic decay and resonant-frequency methods are also used.

Heat Treatment

The heat treatment of ductile iron castings produces a significant difference in mechanical properties from as-cast ductile iron. The beneficial result of heat treating ductile iron is the increase in impact resistance if the temperature is restricted to 705 °C (1300 °F) and the time is limited to 4 h. Most other forms of heat treatment, with the exception of austempering, cause a loss in fatigue strength, wear resistance, and room-temperature impact resistance, compared to as-cast ductile iron of the same hardness.

Stress Relieving. Occasionally ductile iron castings of large or nonuniform cross section are stress relieved at 540 to 595 °C (1000 to 1100 °F), which reduces warping and distortion during subsequent machining. Mechanical properties are reduced if the temperature selected causes a reduction in hardness.

Annealing. Full ferritizing annealing is used to remove carbides or stabilized pearlite in order to meet ASTM grade 60-40-18 requirements. This heat treatment usually involves heating to 900 °C (1650 °F), holding at temperature long enough to dissolve the carbides (possibly up to 3 h), then slow cooling at 85 °C/h (150 °F/h) to 705 °C (1300 °F) and still-air cooling to room temperature. This hypercritical anneal lowers the nil-ductility transition temperature (NDTT) to improve low-temperature fracture resistance, but results in a lower upper-shelf impact energy and a significant reduction in fatigue strength. In some cases, it may reduce the tensile and/or yield strength to levels below ASTM standards. The nil-ductility transition temperature is the temperature at which the material transforms from a partially ductile to a fully brittle material with no ductility.

Subcritical annealing produces either ASTM grade 60-40-18 or grade 65-45-12 for applications requiring high toughness and ductility. It produces greater low-temperature fracture resistance than as-cast ductile iron used for greater low-temperature fracture resistance than as-cast ductile iron used for steering knuckle applications.

Normalizing, quenching, and tempering produces ASTM grade 100-70-03, which is widely used for applications requiring good strength, wear resistance, and good response to localized hardening. Castings are heated to 900 °C (1650 °F), held at temperature for 3 h (allowing 1 h/25 mm, or 1 h/in., of section size to reach temperature), then air blasted or oil quenched, followed by tempering at 540 °C (1000 °F) or up to 675 °C (1250 °F), depending on the required final hardness. The yield strength may be higher than as-cast iron for the same hardness, but the ultimate tensile strength and elongation will be lower. The wear resistance at 228 HB (4.1 BID) will be lower than as-cast iron of the same hardness. At 255 HB (3.8 BID), the short-time fatigue strength (350 000 cycles) of D7003 oil quenched and tempered (OQT) may appear to be slightly higher, but the long life (10^6 cycles) will be significantly lower than that of as-cast

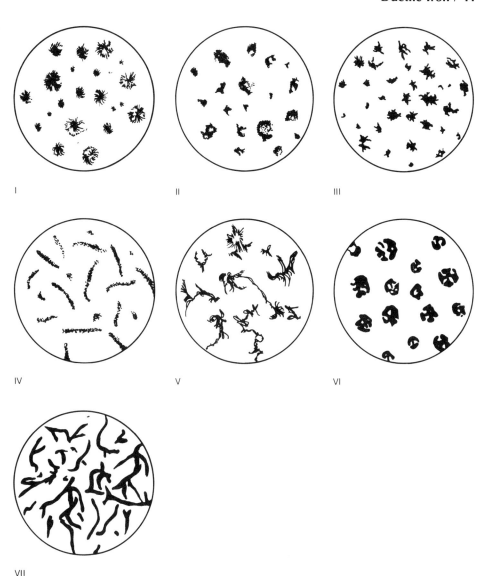

ASTM type(a)	Equivalent ISO form(b)	Description	ASTM type(a)	Equivalent ISO form(b)	Description
I	VI	Nodular (spheroidal) graphite	IV	III	Quasi-flake graphite
II	VI	Nodular (spheroidal) graphite, imperfectly formed	V	II	Crab-form graphite
III	IV	Aggregate, or temper carbon	VI	V	Irregular or open type nodules
			VII(c)	I	Flake graphite

(a) As defined in ASTM A 247. (b) As defined in ISO/R 945-1969 (E). (c) Divided into five subtypes: uniform flakes; rosette grouping; superimposed flake size; interdendritic, random orientation; and interdendritic, preferred orientation

Fig. 12 Seven graphite shapes used to classify cast irons

Table 7 Average mechanical properties of ductile irons heat treated to various strength levels

Determined for a single heat of ductile iron, heat treated to approximate standard grades. Properties were obtained using test bars machined from 25 mm (1 in.) keel blocks.

Nearest standard grade	Hardness, HB	Ultimate strength MPa	ksi	Yield strength MPa	ksi	Elongation in 50 mm (2 in.), %	Modulus GPa	10^6 psi	Poisson's ratio
Tension									
60-40-18	167	461	66.9	329(a)	47.7(a)	15.0	169	24.5	0.29
65-45-12	167	464	67.3	332(a)	48.2(a)	15.0	168	24.4	0.29
80-55-06	192	559	81.1	362(a)	52.5(a)	11.2	168	24.4	0.31
120-90-02	331	974	141.3	864(a)	125.3(a)	1.5	164	23.8	0.28
Compression									
60-40-18	167	...	...	359(a)	52.0(a)	...	164	23.8	0.26
65-45-12	167	...	...	362(a)	52.5(a)	...	163	23.6	0.31
80-55-06	192	...	...	386(a)	56.0(a)	...	165	23.9	0.31
120-90-02	331	...	...	920(a)	133.5(a)	...	164	23.8	0.27
Torsion									
60-40-18	167	472	68.5	195(b)	28.3(b)	...	63 / 65.5(c)	9.1 / 9.5(c)	...
65-45-12	167	475	68.9	207(b)	30.0(b)	...	64 / 65(c)	9.3 / 9.4(c)	...
80-55-06	192	504	73.1	193(b)	28.0(b)	...	62 / 64(c)	9.0 / 9.3(c)	...
120-90-02	331	875	126.9	492(b)	71.3(b)	...	63.4 / 64(c)	9.2 / 9.3(c)	...

(a) 0.2% offset. (b) 0.0375% offset. (c) Calculated from tensile modulus and Poisson's ratio in tension

iron. At 217 HB (4.1 BID), both the short-time and long-time fatigue strength are significantly lower. This is shown in Fig. 5. The impact resistance of oil-quenched and tempered ductile iron is lower than that of as-cast iron at 50 °C (125 °F) and above, but higher at temperatures under 45 °C (110 °F) (see Fig. 7).

Martensitic ductile iron ASTM grade 120-90-02 is produced by heating to 900 °C (1650 °F), holding to homogenize, quenching in agitated oil, and tempering at 510 to 565 °C (950 to 1050 °F).

Austempered ductile iron requires a two-stage heat treatment. The first stage, austenitizing, requires heating to and holding at about 900 °C (1650 °F). This is followed by the second stage, which requires quenching and isothermally holding at the required austempering temperature, usually in a salt bath.

Typically, austempered ductile iron is produced by heating the castings in a controlled atmosphere to an austenitizing temperature between 815 to 925 °C (1500 to 1700 °F). The castings are held at temperature for a long enough time to saturate the austenite with carbon in solution. The castings are then cooled at a rate sufficiently fast enough to avoid the formation of pearlite and other high-temperature transformation products to the appropriate transformation temperature (this may vary from 230 to 400 °C, or 450 to 750 °F, depending on the hardness and strength required). The castings are held at the selected transformation temperature for a long enough time to produce the desired properties. Austempered ductile iron transformed in the 370 °C (700 °F) range exhibits high ductility and impact resistance at a tensile strength of about 1035 MPa (150 ksi). When transformed at 260 °C (500 °F), it exhibits wear resistance comparable to case hardened steel and tensile strength in excess of 1380 MPa (200 ksi).

Thus, the selection of the austempering temperature and time of holding is critical. In general, austempering in the 240 to 270 °C (465 to 520 °F) range provides a component that has maximum strength but limited ductility, while austempering in the 360 to 380 °C (680 to 715 °F) range yields a component that exhibits maximum ductility and toughness, in combination with relatively high strength, albeit lower than that obtained at the lower austempering temperature (see Fig. 8). Figure 9 compares the strength and ductility of as-cast nodular irons and of nodular irons subjected to the heat treatments discussed in this section.

Surface Hardening

Ductile iron may be surface hardened to as high as 60 HRC which may have a microhardness of 62 HRC because the cone indentor averages the soft graphite nodules with the matrix hardness resulting in a lower reading than the actual metallic components of the matrix. The hardened zone produces a highly wear-resistant surface layer backed up by a core of tougher metal. Flame or induction methods can be used to heat the surface layer to about 900 °C (1650 °F) for a few seconds, after which the heated surface is quenched in a spray of water (which often contains a few percent of a water-soluble quenching aid). Surface hardening is most successful when the matrix is fully pearlitic and thus ASTM grades 100-7-03 and 120-90-02 respond best. However, as-cast hardnesses as low as 217 HB (4.1 BID) can be successfully surface hardened. Laser and electron-beam methods are also used for surface hardening.

Mechanical Properties

Most of the standard specifications for ductile iron require minimum strength and ductility, as determined by the use of separately cast, standard ASTM test bars described in ASTM A 395. The various specification limits have been established by the evaluation of the results from thousands of these test bars. The properties of test bars are useful approximations of the properties of finished castings. Test bar properties also make it possible to compare the metal from many different batches without having to account for the variations due to differences in the shapes being cast or in the production practices used in different foundries.

Test bars are machined from keel blocks, Y blocks, or modified keel blocks (see ASTM A 395 for details and dimensions). These test blocks are designed for ideal feeding from heavy molten metal heads over the mold and for controlled cooling at optimum rates. In practice, these characteristics may not be economically feasible or may be impossible to achieve because of the configuration of the casting. As a result, the properties of actual production castings may differ from those of test bars cast from the same heat of molten metal, a fact that is sometimes overlooked. Test bar properties are the most informative when their relationships to the properties of production castings have been previously established by the testing of bars machined from castings, by the selective overloading of cast-

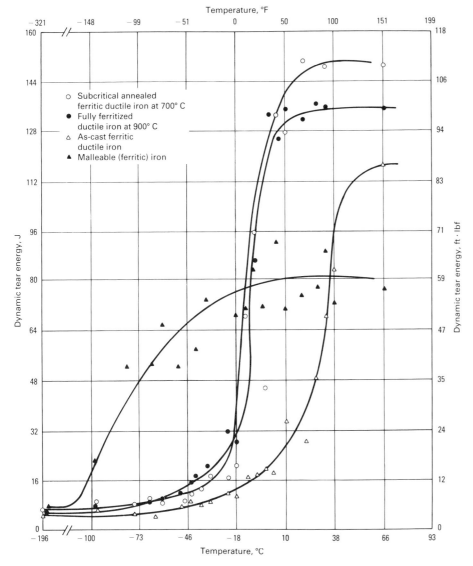

Fig. 13 Dynamic tear energy versus temperature for four ferritic cast irons. Specimen size: 190 mm (7½ in.) long, 130 mm (5⅛ in.) wide, and 41 mm (1⅝ in.) thick; 13 mm (½ in.) notch depth. Source: Ref 2

ings, by stress analysis, or by field testing. Test bars cast to near testing dimensions may be attached to the casting and removed for machining and testing. Dynamic properties such as impact resistance and fatigue strength may be tested by casting dynamic tear and reverse-bending paddle bar specimens. Figure 10 shows the brittle fracture appearance of a dynamic tear specimen.

Effect of Composition. The properties of ductile iron depend first on composition. Composition should be uniform within each casting and among all castings poured from the same melt. Many elements influence casting properties, but those of greatest importance are the elements that exert a powerful influence on matrix structure or on the shape and distribution of graphite nodules.

Carbon influences the fluidity of the molten iron and the shrinkage characteristics of the cast metal. Excess carbon not in solution, but in suspension, reduces fluidity. The volume of graphite is 3.5 times the volume of iron. As ductile iron solidifies, the carbon in solution precipitates out as graphite and causes an expansion of the iron, which can offset the shrinkage of the iron as it cools from liquid to solid. The amount of carbon needed to offset shrinkage and porosity is indicated in the following formula:

$$\% C + \tfrac{1}{7}\% Si \geq 3.9\% \qquad \text{(Eq 1)}$$

Carbon contents greater than this amount begin to decrease fatigue strength and impact strength before the effect is noticed on tensile strength. The size and the number of graphite nodules formed during solidification are influenced by the amount of carbon, the number of graphite nuclei, and the choice of inoculation practice. Normal graphite-containing ductile iron has 10% less weight than steel of the same section size. The graphite also provides lubricity for sliding friction, and the low coefficient of friction permits more efficiently running gears, which, furthermore, will not seize if a loss of lubricant is experienced in service. The graphite also produces ADI gears that are silent in operation. Graphite in the structure can provide good machinability of a ferritic material and then be available to redissolve into solution by heat treating to produce high strength and wear resistance.

The relationship between carbon content and silicon content in terms of the carbon equivalent, CE, is

$$CE = \% C + \frac{Si \%}{3} \qquad \text{(Eq 2)}$$

Figure 11 is a plot of the carbon equivalent versus the volume of feed metal required.

A chemical analysis for carbon content should be done prior to any formation of graphite. After graphite formation, the ability to obtain an accurate carbon content is limited because of the loss of carbon in sample preparation or, in some cases, because of segregation.

Silicon (Ref 6) is a powerful graphitizing agent. Within the normal composition limits, increasing amounts of silicon promote structures that have progressively greater amounts of ferrite; furthermore, silicon contributes to the solution strengthening and hardness of ferrite. Increasing the amount of ferrite reduces the yield strength and tensile strength, but increases the elongation and impact strength. The ferrite envelope surrounding the graphite nodule in pearlitic ductile iron reduces the indicated yield strength, but increases elongation, impact strength, and fatigue strength. Silicon reduces the impact strength of ferritic ductile iron both as-cast and subcritically annealed. To provide maximum resistance to fracture from room temperature down to −40 °C (−40 °F), silicon must be kept below 2.75% if the phosphorus content is below 0.02%. If the phosphorus content is 0.05%, the silicon content should be limited to 2.55%. High-temperature applications such as turbocharger housings require silicon contents from 3.75 to 4.25% and molybdenum contents up to 0.70%. This combination provides high-temperature oxidation resistance and dimensional stability. Even higher silicon contents are used for abrasion resistance.

Manganese (Ref 7). Among the alloying elements commonly used to improve the mechanical properties of ductile iron, manganese acts as a pearlite stabilizer and increases strength, but reduces ductility and machinability. It also promotes segregation to the cell boundaries and must be limited when making austempered ductile iron.

Nickel (Ref 7) is frequently used to increase strength by promoting the formation of fine pearlite and to increase hardenability, especially for surface-hardening appli-

44 / Cast Irons

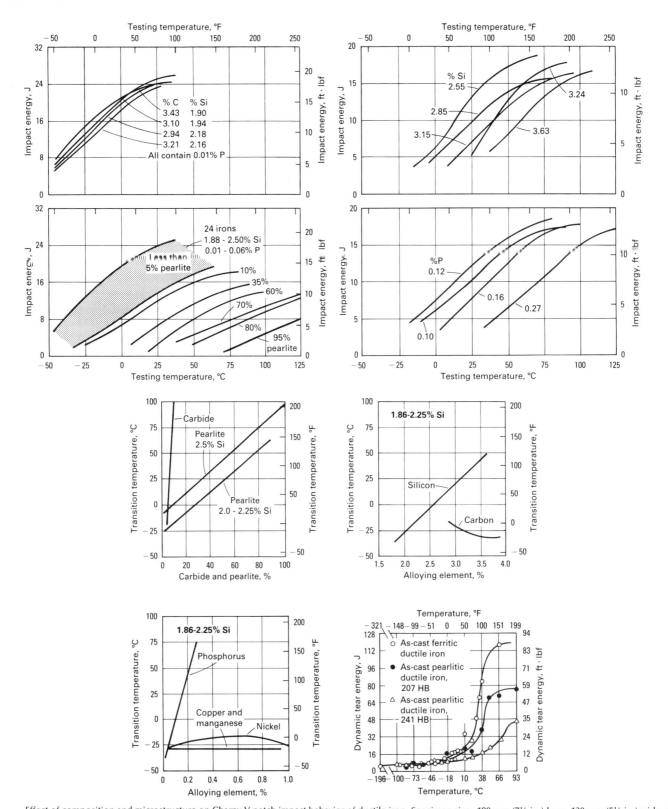

Fig. 14 Effect of composition and microstructure on Charpy V-notch impact behavior of ductile iron. Specimen size: 190 mm (7½ in.) long, 130 mm (5⅛ in.) wide, and 41 mm (1⅝ in.) thick; 13 mm (½ in.) notch depth. Source: Ref 8 for all except the dynamic tear energy graph

cations or for producing austempered ductile iron.

Copper (Ref 7) is used as a pearlite former for high strength with good toughness and machinability.

Molybdenum (Ref 7) is used to stabilize the structure at elevated temperatures. It is also used to add hardenability to heavy sections in producing austempered ductile iron. The amount must be controlled because of the tendency of molybdenum to segregate to the cell boundaries as stable carbides.

Effect of Graphite Shape. The presence of graphite in ductile cast iron in the shape of

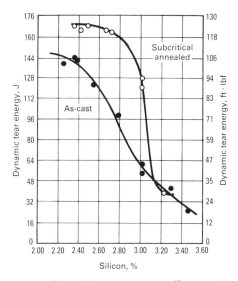

Fig. 15 Dynamic tear energy versus silicon content for two ferritic ductile irons. Test temperature was 25 °C (75 °F). Phosphorus content was 0.01%. Specimen size: 190 mm (7½ in.) long, 130 mm (5⅛ in.) wide, and 41 mm (1⅝ in.) thick; 13 mm (½ in.) notch depth. Source: Ref 2

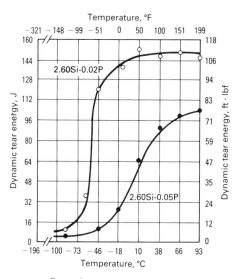

Fig. 16 Dynamic tear energy versus temperature to show effect of increasing phosphorus content in ferritic ductile irons. Specimen size: 190 mm (7½ in.) long, 130 mm (5⅛ in.) wide, and 41 mm (1⅝ in.) thick; 13 mm (½ in.) notch depth. Source: Ref 2

spheroidal nodules instead of sharp flakes such as those found in gray cast iron is caused by the addition of magnesium (or magnesium and cerium) to the molten iron, resulting in a fivefold to sevenfold increase in the strength of the cast metal. Shapes that are intermediate between a true nodular form (such as ASTM type 1, shown in Fig. 12) and a flake form (such as ASTM type VII, also shown in Fig. 12) yield mechanical properties that are inferior to those of ductile iron with a true nodular graphite. The size and uniformity of distribution of graphite nodules also influence properties, but to a lesser degree than graphite shape. Small, numerous nodules are usually accompanied by high tensile properties and tend to reduce the likelihood of the formation of chilled iron in thin sections or at edges. An optimum nodule density exists. Excessive nodules may weaken a casting to such a degree that it may not withstand the rigors of its intended application. Each material must be evaluated for its specific application.

A ductile iron composition can be converted to a compacted graphite iron composition. Minor alterations in composition produce a controlled microstructure consisting of more than 5% spheroidal and less than 20% spheroidal graphite, with the remainder of the graphite being a compacted, blunt, vermicular shape. This then becomes compacted graphite (CG) iron, which has lower strength than standard ductile iron and higher strength than gray iron, but which has better thermal transfer characteristics and less tendency for porosity than standard ductile iron.

Effect of Section Size. The variable chiefly affected by section size is the cooling rate. It, in turn, affects both the size of the graphite nodules and the microstructure of the matrix. The heavier the section, the more slowly it cools, and therefore the larger and fewer are the graphite nodules that form during solidification. When casting ductile iron in sections greater than about 64 mm (2½ in.), there is a possibility of producing centerline or inverse chill in the last section to solidify. This is usually controlled by inoculation techniques.

The structure of the matrix is essentially determined by the cooling rate through the eutectoid temperature range, although the specific effects of cooling rate are modified by the presence of alloying elements, as discussed previously in the section "Effect of Composition." Slow cooling rates prevalent in heavy sections promote the transformation to ferrite. If a pearlitic matrix is desired, pearlite formers (such as copper) are added to the molten iron. It is important that castings be allowed sufficient cooling time in the mold to allow the lamellar pearlite to be tempered in order to break up the plates partially and be rendered machinable at hardnesses up to 321 HB (3.40 BID). Insufficient cooling time may produce very fine pearlite, which reduces machinability. As-cast ductile iron anneals itself in the mold. Without a pearlite former, castings with variations in section thickness will have variations in hardness. Bainite and martensite are not found in as-cast structures because they are formed by heat treatment. Rapid cooling of thin sections may produce acicular structures, which are usually some form of combined carbon (carbides).

Bainitic ferrite and retained austenite are the main matrix constituents in austempered ductile iron. In order to minimize the harmful effects of segregation in medium- and thick-section ADI castings, the graphite nodule count must be maintained as high as possible, ideally at a level greater than 150 nodules/mm² (9.7×10^4 nodules/in.²).

Tensile Properties. The tensile properties of one heat of ductile iron heat treated to strength levels approximately equivalent to four standard ductile irons are given in Table 7. These values are not necessarily the average property values that can be expected for metal produced as-cast to the indicated grades. As-cast tensile and elongation values are higher than heat treated values; however, the yield strength values may be lower. Within each grade, strength and ductility vary somewhat with hardness, as shown in Fig. 4. In some instances, the ranges of expected strength and ductility overlap those for the next higher or lower grade.

As shown in Table 7, the modulus of elasticity in tension lies in the range of 162 to 170 GPa (23.5×10^6 to 24.5×10^6 psi) and does not vary greatly with grade. This value in tension should not be used in design for cantilever or three-point beam or torsion loading because the deflection is greater, and a value of 142 GPa (20.5×10^6 psi) should be used. The values of tensile modulus shown in Table 7 were determined using standard 12.83 mm (0.505 in.) diam tensile bars, with strain gages affixed to the reduced section (gage length).

Compressive Properties. The 0.20% offset yield strength of ductile iron in compression is generally reported to be 1.0 to 1.2 times the 0.2% offset yield strength in tension. The compressive properties shown in Table 7 were determined using specimens from the same single heat of ductile iron described in the section above.

Torsional Properties. Few data are available on the ultimate shear strength of ductile iron because it is very difficult to obtain accurate shear data on materials that exhibit some ductility. It is generally agreed that the ultimate shear strength of ductile cast iron is about 0.9 to 1.0 times the ultimate tensile strength. Table 7 gives data for shear strength and for 0.0375% offset yield strength in torsion for a single heat of ductile iron heat treated to strength levels approximately equivalent to those of four standard ductile irons.

Damping Capacity. The average damping capacity of ductile iron in the hardness range of 156 to 241 HB is about 6.6 times that of SAE 1018 and about 0.12 times that of ASTM class 30 gray cast iron. These data were obtained from resonant-frequency measurements:

Material	Log decay $\times 10^{-4}$
Ductile iron	8.316(a)
SAE 1018 steel	1.31(b)
	1.23(c)
Gray iron, class 30	68.67(a)

(a) Mean value. (b) Longitudinal direction. (c) Transverse direction

46 / Cast Irons

Austenitic ductile iron has a lower damping capacity than unalloyed ferritic ductile iron.

Impact Properties. Dynamic tear energy data for ferritic ductile iron are shown in Fig. 13. This figure compares the effect of heat treatment. Subcritical anneal was 3 h at 700 °C (1290 °F). Full ferritization was obtained by heating to 955 °C (1750 °F) and slow cooling in the same manner as for the ferritic malleable iron heat treatment. This data shows that ferritizing at temperatures above the critical temperature results in lower upper-shelf energy. It also reduces the fatigue strength of ductile iron, compared to a subcritical anneal. Figure 7 shows that heating to 900 °C (1650 °F) and then oil quenching and tempering reduces the upper-shelf energy, but raises the low-temperature energy to fracture.

Data from a comprehensive study of the impact properties of ductile iron are shown in Fig. 14. These data show that increasing pearlite decreases impact energy. Figure 15 shows that increasing silicon reduces impact energy at room temperature for ferritic ductile iron. Figure 16 shows that increasing phosphorus reduces impact energy and raises the nil-ductility transition temperature. The transition temperature is significantly affected by phosphorus and/or silicon content, but is affected little by other elements present within the normal variations in composition.

The effects of various heat treatments on Charpy V-notch impact energy are shown in Fig. 17 for a ductile iron alloyed with about 0.75% Ni. Curve F shows that austempering heat treatment not only improves elongation at high strength, but also produces the highest room-temperature impact energy of any of the conventional heat treatments, with the exception of surface-hardened as-cast ductile iron, curve A, with a hardness of 207 HB.

Fracture Toughness. Certain lower-strength grades of ductile iron do not fracture in a brittle manner when tested under nominal plane-strain conditions in a standard fracture toughness test. This behavior is contrary to the basic tenets of fracture mechanics and has been attributed to localized deformation in the ferrite envelope surrounding each graphite nodule. In the low-strength ductile irons, plane strain conditions are established only at temperatures low enough to embrittle the ferrite. Otherwise, an increase in the size of the fracture toughness test specimens does not provide the degree of mechanical constraint necessary to obtain a valid measurement of K_{Ic}.

Selected values of fracture toughness are given in Table 8. These values were determined using compact tension specimens 21 mm (0.83 in.) in width. All tests were made in accordance with ASTM E 399. The data given in Table 8 are for ductile iron with a nodule shape approximately corresponding to 50% ASTM type I. In this same study, cast iron with a graphite shape similar to that in ASTM type IV (vermicular graphite) exhibited significantly lower fracture toughness, although a lower ferrite content and the absence of a ferrite ring around each graphite particle were considered to be of far greater significance than the lack of nodularity.

Fatigue Strength. Figure 6 shows fatigue strength curves for pearlitic ductile iron of various grades and heat treatments compared with those of malleable iron and one grade of cast steel. The tests were made on a model VSP-150 variable-speed reverse-bending plate fatigue testing machine. The endurance limit for a given grade of ductile iron is influenced by surface condition. Figure 18 shows fatigue strength curves for ferritic and pearlitic ductile iron in both the notched and unnotched conditions. The tests were made on Wohler-type fatigue machines with polished specimens 10.6 mm (0.417 in.) in diameter. Figure 19 shows that the unnotched reversed-bending endurance limit for specimens with as-cast surfaces is considerably less than that for polished specimens of equivalent tensile strength. The absolute value of the as-cast endurance limit at any strength level depends on the surface or near-surface characteristics associated with a specific casting process. Shot blast cleaning contributes up to a 30% increase in the fatigue strength of the sand cast surface. This is controlled by the size of the shot, the size of the load, and the length of the cycle. The effect of shot blasting (shot peening) can be monitored by attaching an Almen block and strip to a sample casting and including it with the load. This should be done for any change in process.

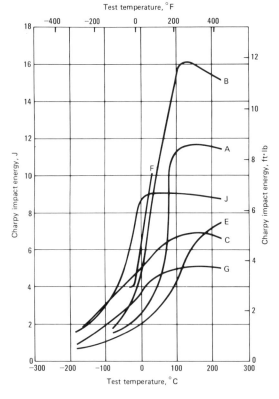

Sample(a)	Heat treatment
A	As-cast
B	Subcritical annealed
C	Quenched and tempered
J	Quenched and tempered
G	Quenched and tempered
E	Normalized and tempered
F(b)	Austempered (370 °C, or 700 °F)

Sample(a)	Tensile strength, MPA (ksi)	Yield strength, MPa (ksi)	Elongation in 50 mm (2 in.), %
A	535 (78)	340 (49)	13.0
B	430 (62)	315 (49)	23.6
C	635 (92)	525 (76)	8.8
J	750 (108)	580 (84)	9.4
G	1050 (152)	795 (115)	4.1
E	780 (113)	470 (68)	8.2
F(b)	930 (135)	655 (95)	11.5

(a) Analysis: 3.65% TC, 0.52% Mn, 0.065% P, 2.48% Si, 0.08% Cr, 0.78% Ni, 0.15% Cu. (b) Analysis not available; data from unpublished work conducted at International Harvester Company

Fig. 17 Effect of heat treatment on Charpy V-notch impact properties of ductile iron. Source: Ref 9

Table 8 Fracture toughness of ductile iron

Type of iron	Condition	Ultimate tensile strength MPa	ksi	Yield strength MPa	ksi	Elongation, %	K_{Ic}, MPa · m$^{0.5}$ (ksi · in.$^{1/2}$) at 20 °C (70 °F)	−40 °C (−40 °F)	−105 °C (−160 °F)
Ferritic									
3.0% Si	As-cast	521	75.6	427	62.0	11.0	...	35.1 (32.0)	30.2 (27.5)
3.5% Si	As-cast	547	79.4	471	68.3	9.0	...	27.0 (24.6)	...
Pearlitic									
2.5% Si	As-cast	703	102.0	374	54.2	7.5	...	37.1 (33.8)	...
	Normalized	918	133.2	552	80.0	3.6	45.3 (41.3)	...	...
	Austempered	...	...	620	90.0(a)	...	36.5 (33.3)	...	...

(a) Estimated. Source: Ref 10

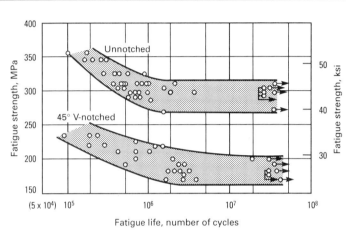

Grade	Tensile strength MPa	ksi	Unnotched Endurance limit MPa	ksi	Endurance ratio	Notched Endurance limit MPa	ksi	Endurance ratio	Stress concentration factor
60-40-18	480	70	205	30	0.43	125	18	0.26	1.67
80-55-06	680	99	275	40	0.40	165	24	0.24	1.67

Fig. 18 Plot of fatigue strength versus fatigue life for ductile iron in both the unnotched and 45° V-notched condition: (a) Ferritic (60-40-18 annealed). (b) Pearlitic (80-55-06 as-cast)

Ductile iron is cycle frequency sensitive. The cycle frequency used in testing should not exceed that which the part would experience in service.

The influence of tensile strength and matrix structure on the endurance ratio of ductile iron is shown in Fig. 20. Endurance ratio is defined as endurance limit divided by tensile strength. Because the endurance ratio of ductile iron declines as tensile strength increases, regardless of matrix structure, there may be little value in specifying a higher-strength ductile iron for a structure that is prone to fatigue failure; redesigning the structure to reduce stresses or strains may prove to be a better solution. For tempered martensitic ductile iron, the improvement in fatigue strength due to an increase in tensile strength is proportionately greater than for a ferritic or pearlitic grade; this is indicated by the shallower slope for martensitic ductile iron in Fig. 20.

Table 9 summarizes reversed-bending fatigue properties for three standard grades of ductile iron. The data were obtained using polished 10.6 mm (0.417 in.) diam test specimens.

Strain Rate Sensitivity. Ductile iron, like many steels, is strain rate sensitive. In a forming operation such as coining to a close dimension to remove a machining operation, rolling an idler arm ball socket, or thread rolling, the rate of material movement should be low enough to avoid cracking. In coining, a hydraulic press should be used when the operation is carried out at room temperature. When a stroke press is used, the part must be heated to a high enough temperature to avoid cracking during forming. When high rates of strain must be tolerated, the section size should be increased to reduce the degree of strain.

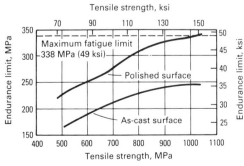

Fig. 19 Effect of surface condition on endurance limit of ductile iron. Tests made on 10.6 mm (0.417 in.) diam specimens. Fully reversed stress ($R = −1$)

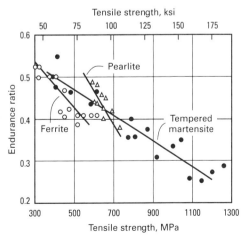

Fig. 20 Effect of tensile strength and matrix structure on endurance ratio for ductile iron. Source: Ref 11, 12

Table 9 Reversed-bending fatigue properties of three standard grades of ductile iron

	Tensile strength, S_t		Unnotched			45° V notched		
			Endurance limit, S_e		Endurance ratio(a)	Endurance limit, S_n		Endurance ratio(b)	Notch sensitivity factor(c)
Type	MPa	ksi	MPa	ksi		MPa	ksi		
64-45-12	490	71	210	30.5	0.43	145	21	0.30	1.4
80-55-06	620	90	275	40	0.44	165	24	0.27	1.7
120-90-02(d)	930	135	338	49	0.36	207	30	0.22	1.6

(a) S_e/S_t. (b) S_n/S_t. (c) S_e/S_n. (d) Oil quenched from 900 °C (1650 °F) and tempered at 595 °C (1100 °F)

Table 10 Oxide penetration of ductile iron and other materials at 705 °C (1300 °F)

	Oxide penetration					
	2000 h		3000 h		4000 h	
Material	mm/year	mil/year	mm/year	mil/year	mm/year	mil/year
Ductile iron						
80-55-06, 2.5% Si	1.35	53	...	...	1.05	41
60-40-18, 2.5% Si	1.4	56	...	...	0.7	28
Ferritic ductile						
4.0% Si	1.03	41	...	...	0.85	34
5.5% Si	0	0	...	...	0.08	3.3
Cast steel	5.1	201	5.05	199	...	...
Gray iron						
1.5% Si	3.3	131	2.8	110	...	...
1.65% Si	2.95	116	3.3	131	...	...
2.0% Si	4.6	181	3.3	131	...	...
2.5% Si	4.8	188	3.6	143	...	...
Pearlitic malleable	3.3	131	...	...	...	...

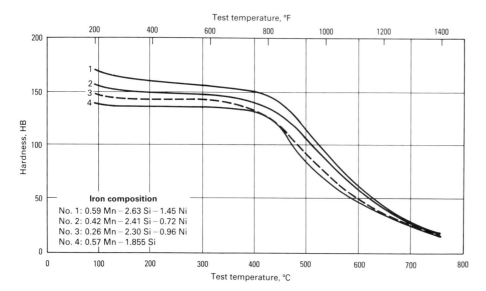

Fig. 21 Hot hardness of four annealed (ferritic) ductile irons. Source: Ref 13

Iron composition
No. 1: 0.59 Mn – 2.63 Si – 1.45 Ni
No. 2: 0.42 Mn – 2.41 Si – 0.72 Ni
No. 3: 0.26 Mn – 2.30 Si – 0.96 Ni
No. 4: 0.57 Mn – 1.855 Si

The **notch sensitivity** of ductile iron must be considered when designing crankshafts. The fillet in the crankpin must be a minimum of 1.65 mm (0.065 in.) thick, in contrast to 1.14 mm (0.045 in.) for the pearlitic malleable iron it replaces. It will have significantly greater fatigue strength than the part it replaces.

Mechanical Properties at Elevated Temperatures

The hardness and strength of all standard grades of ductile iron are relatively constant up to about 425 °C (800 °F). Hot hardness data for four annealed (ferritic) ductile irons are shown in Fig. 21. It should be noted that the hot hardness is lower with a lower silicon content.

The resistance to oxidation of ductile irons and other materials at 705 °C (1300 °F) is shown in Table 10. In ductile iron, resistance to oxidation increases with increasing silicon content. The data indicate that ductile iron of normal composition has much better oxidation resistance than does cast steel, gray iron, or pearlitic malleable iron.

The dimensional growth of ductile iron is much less than that of gray iron at elevated temperatures. Figure 22(a) compares the growth of ductile and gray irons at 900 °C (1650 °F). The compositions of the irons are shown in the table below the graph. Data on the growth of ductile and gray irons at 540 °C (1000 °F) for 6696 h are plotted in Fig. 22(b), which emphasizes the excellent performance of ASTM grade 60-40-18 ferritic ductile iron. Additional information on the properties of alloy ductile iron at elevated temperatures may be found in the article "Alloy Cast Irons" in this Volume.

The creep strength of ductile iron depends on composition and microstructure. As shown in Fig. 23(a), ferritic ductile iron has a creep resistance comparable to that of annealed low-carbon cast steel up to 650 °C (1200 °F). Table 11 shows the creep strength of ductile iron of various compositions. Representative minimum creep rates versus applied stress are shown in Fig. 23(b) and (c). The improvement in creep strength that can be achieved by adding molybdenum or copper is shown in Table 11 and Fig. 24. Figure 25 shows the stress necessary at a given temperature to produce a minimum creep rate of 0.0001%/h in ferritic and pearlitic ductile irons.

The stress-rupture properties of ferritic and pearlitic ductile irons are shown in Fig. 25 and 26. Addition of 2% Mo to a 4% Si ductile iron raises the rupture stress as shown in Fig. 27.

Master curves for stress rupture are given in Fig. 28. From these curves, it may be determined, for example, that unalloyed ductile iron stressed at 4 ksi (28 MPa) can be expected to endure for 10^4 h at 565 °C (1050 °F).

Hot-Tensile Properties. The short-time elevated-temperature tensile properties in Fig. 29(a) indicate the superior strength of pearlitic ductile iron. Figure 29(b) and (c) summarize the hot-tensile properties of annealed ductile iron compared to those of cast steel.

Hardenability

Ductile iron in various hardened conditions has become accepted for such applications as heavy-duty gears, spinning mandrels, pump liners, rolls, dies, clutch drums, pistons, brake drums, and agricultural-implement parts. Ductile iron has limited heat transfer characteristics because the graphite is in a spheroidal shape rather than the graphite flake form of cast iron. Spheroidal graphite (SG) iron provides an answer to the problem of heat transfer encountered with compacted graphite and yet SG iron has higher strength than gray cast iron. This is important in the case of brake drums or rotors, which may develop high braking surface temperatures and form hard spots of martensite. Alloy ductile iron can develop an acicular structure in the as-cast structure that may resemble bainite or martensite. However, the latter are structures developed by heat treatment. Most applications use quenching and tempering or isothermal

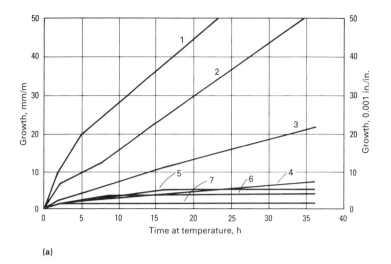
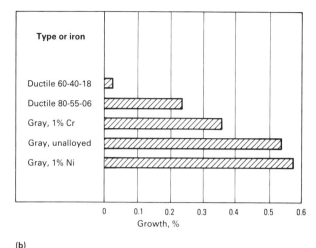

Type of iron	TC	Mn	Si	P	S	Cr	Ni	Mg
1 Class 25 gray iron	3.27	0.68	2.27	0.20	0.15	...	...	...
2 Class 35 alloy gray iron	3.19	0.93	2.10	0.16	0.092	0.37	1.12	...
3 Class 40 gray iron	3.15	0.90	1.28	0.092	0.10	...	...	...
4 Class 40 alloy gray iron	3.06	1.01	1.55	0.078	0.069	0.341	0.98	...
5 Ductile iron, as-cast(a)	3.43	0.47	2.19	0.13	0.009	...	1.95	0.077
6 Ductile iron, annealed(b)	3.50	0.54	2.58	0.12	0.010	...	1.50	0.055
7 Ductile iron, annealed(a)	3.43	0.47	2.19	0.13	0.009	...	1.95	0.07

(a) Cut from 115 mm (4.5 in.) wide keel block. (b) Cut from experimental piston after heat treating

Fig. 22 Growth of ductile and gray irons at elevated temperature. (a) Plot of growth versus time at 900 °C (1650 °F). (b) Growth of selected ductile and gray irons after 6696 h at 540 °C (1000 °F)

Table 11   Creep strength of several ductile irons at 425 °C (800 °F)

Chemical composition, %								Condition(a)	Pearlite, %	Stress for minimum creep rate of			
										0.0001%/h		0.00001%/h	
TC	Mn	P	Si	Ni	Mo	Cu	Mg			MPa	ksi	MPa	ksi
3.54	0.40	0.017	2.26	0.56	...	0.15	0.05	Ann, 950 °C (1740 °F)	1	48	7	...	...
3.54	0.40	0.017	2.26	0.56	...	0.15	0.05	Ann, 870 °C (1600 °F)	10	58.5	8.5	...	...
3.49	0.78	0.086	2.46	1.08	...	0.56	0.072	Ann, 870 °C (1600 °F)	20	141	20.5	...	...
3.49	0.78	0.086	2.46	1.08	...	0.56	0.072	Ann, 950 °C (1740 °F)	10	172	25	103	15
3.59	0.40	0.02	2.43	1.08	0.24	0.1	0.05	Ann, 870 °C (1600 °F)	2	124	18	86	12.5
3.61	0.47	0.02	2.43	1.19	0.81	0.1	0.05	Ann, 950 °C (1740 °F)	3	186	27	152	22
3.54	0.40	0.017	2.26	0.56	...	0.15	0.05	As-cast, stress relieved	30	79	11.5	...	...
3.49	0.37	0.085	2.50	1.22	...	0.1	0.064	As-cast, stress relieved	94	145	21	...	...
3.49	0.78	0.086	2.46	1.08	...	0.56	0.072	As-cast, stress relieved	100	172	25	114	16.5
3.61	0.47	0.02	2.43	1.19	0.81	0.1	0.05	As-cast, stress relieved	50	...	...	145	21

(a) Ann, annealed. Source: Ref 7

transformation treatments, which provide ductile iron with a hard surface overlaying a tough core. The hardenability of ductile irons covers a wide range, but a ductile iron generally has higher hardenability than a eutectoid steel with comparable alloy content (Ref 17).

Austenitizing temperatures between 845 and 925 °C (1550 and 1700 °F) affect hardenability according to the amount of combined carbon that is taken into solution. Higher temperature and longer time, up to that required for saturation of the austenite at the soaking temperature, increase the amount of combined carbon, which increases hardenability. Graphite has lower solubility in austenite than does pearlite (combined carbon); that is, it takes a higher temperature and a longer elapse of time to harden ferritic ductile than are required for pearlitic ductile iron. The time and temperature required for pearlitic ductile iron to reach maximum hardness depend on the combined carbon content, which is to some degree related to the hardness before heat treatment. An as-cast pearlitic ductile iron will normally have a higher combined carbon content than will normalized ductile iron of the same hardness. The silicon content of ductile iron re-

Table 12   Densities and thermal conductivities of the microconstituents in cast iron

Constituent	Density(a)		Thermal conductivity(b)	
	g/cm³	lb/in.³	W/m · K	Btu/ft² · h · °F
Ferrite	7.86	0.284	70–80	12.3–14.0
Austenite	7.84	0.283	...	...
Pearlite	7.78	0.281	50	8.8
Cementite	7.66	0.277	7	1.2
Martensite	7.63	0.276	...	...
Phosphide eutectic	7.32	0.264	...	...
Graphite	2.25	0.081	80–85(c)	14–15(c)
			285–425(d)	50–75(d)

(a) At 20 °C (68 °F). (b) 0 to 100 °C (32 to 212 °F). (c) Along C-axis. (d) Along basal plane

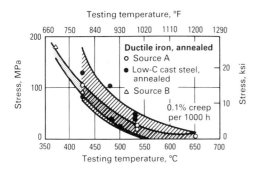

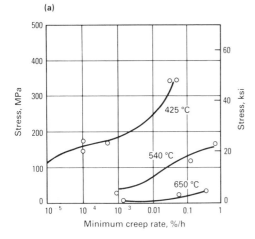

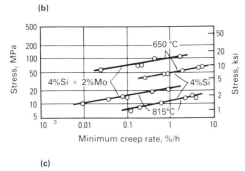

Fig. 23 Creep characteristics of ductile iron. (a) Stress versus testing temperature for annealed ferritic ductile irons. (b) Stress versus minimum creep rate for pearlitic ductile iron (Ref 14). (c) Stress versus minimum creep rate for 4% Si ductile iron and 4Si-2Mo ductile iron at test temperatures of 650 °C (1200 F) and 815 °C (1500 °F). Source: Ref 14

duces the amount of carbon taken into solution.

Insufficient time or temperature used in hardening ductile iron is made obvious by a microstructure that has the appearance of bainite in the original ferrite grain boundary and of transformation products (pearlite or martensite in the center of the grains).

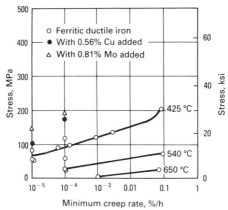

Fig. 24 Effect of molybdenum and copper additions on the creep characteristics of ferritic ductile iron. Source: Ref 14

Physical Properties

Cast irons are not homogeneous materials. Certain properties are affected more by the shape, size, and distribution of the graphite particles than by any other attribute of the structure, a behavior more like that of a composite material than of a homogeneous metal or alloy. Among the physical properties of ductile iron, density is affected only by the relative amounts of the microconstituents present, not by their form or distribution, whereas thermal and electrical conductivity are markedly affected by the form and distribution of the various phases, especially graphite, which has properties very different from those of the various metallic phases. Table 12 summarizes the densities and thermal conductivities of the various microconstituents of cast iron.

Density. For most ductile irons, density at room temperature is about 7.1 g/cm^3. Density is largely affected by carbon content and by the degree of graphitization and any amount of microporosity. A low-carbon pearlitic ductile iron might have a density as high as 7.4 g/cm^3, and a high-carbon ferritic ductile iron may have a density as low as 6.8 g/cm^3. Microporosity, when present, will produce a lower density, depending on the amount present. High-nickel compositions, such as the Ni-Resist ductile irons, have slightly greater density, about 7.4 to 7.7 g/cm^3.

Thermal Properties. Specific heat is relatively unaffected by composition, but of unalloyed ductile iron varies with temperature:

Temperature		Specific heat	
°C	°F	J/kg · K	Btu/lb · °F
20–200	70–390	461	0.110
20–300	70–570	494	0.118
20–400	70–750	507	0.121
20–500	70–930	515	0.123
20–600	70–1110	536	0.128
20–700	70–1290	603	0.144

The melting point varies with silicon content, and the melting range varies with carbon content. The closer the value of the carbon equivalent is to the eutectic composition, the narrower the melting range. Unalloyed and low-alloy ductile irons melt in the range of 1120 to 1160 °C (2050 to 2120 °F); austenitic high-nickel ductile irons melt at about 1230 °C (2250 °F).

The heat of fusion for all ferritic and pearlitic grades of ductile iron is about 210 to 230 kJ/kg (90 to 99 Btu/lb).

The coefficient of linear thermal expansion, like specific heat, is considered to be constant over a given temperature range even though the coefficient of thermal expansion actually varies with temperature. Because it varies with temperature, different values of expansion coefficient are needed for different temperature ranges. The value of the expansion coefficient is ordinarily determined by measuring the change in length of a standard-size specimen as it is heated and cooled over a specific temperature range. Coefficients of linear thermal expansion for ferritic and pearlitic ductile irons are given in Table 13. Values for highly alloyed ductile irons may be considerably different from those given. The coefficient of linear thermal expansion for austempered ductile iron is related to the proportion of ferrite and retained austenite in the austempered ductile iron.

The thermal conductivity of ferritic ductile iron is about 36 W/m · K (250 Btu · in./ft^2 · h · °F) over the temperature range of 20 to 500 °C (70 to 930 °F). There is a slight, negative temperature dependence for thermal conductivity, causing it to drop with increasing temperature. Graphite shape and alloy content have greater influences on thermal conductivity than does temperature; for instance, increasing the nickel plus silicon content of ferritic ductile iron reduces thermal conductivity, as shown in Fig. 30. Thermal conductivities of the various microconstituents found in ductile irons are given in Table 12.

Electrical and Thermal Relationship. The electrical and thermal conductivities of cast iron, like those of many metals, are related to each other in accordance with the Wiede-

Table 13 Coefficients of thermal expansion for ferritic and pearlitic ductile irons

Temperature range		Coefficient of thermal expansion			
		Ferritic ductile iron		Pearlitic ductile iron	
°C	°F	µm/m · K	µin./in. · °F	µm/m · K	µin./in. · °F
20–110	70–230	11.2	6.23	10.6	5.89
20–200	70–390	12.2	6.78	11.7	6.51
20–300	70–570	12.8	7.12	12.4	6.89
20–400	70–750	13.2	7.34	13.0	7.23
20–500	70–930	13.5	7.51	13.3	7.39
20–600	70–1110	13.7	7.62	13.6	7.56
20–700	70–1290	13.8	7.67	...	...
20–760	70–1400	14.8	8.23	14.8	8.23
20–870	70–1600	15.3	8.51	15.3	8.51

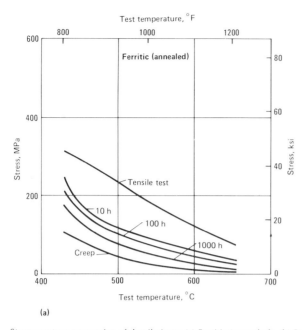
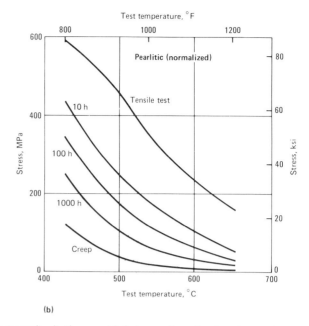

Fig. 25 Stress-rupture properties of ductile iron: (a) Ferritic (annealed). (b) Pearlitic (normalized). The curve labeled creep shows the stress-temperature combination that will result in a creep rate of 0.0001%/h. Source: Ref 15, 16

mann-Franz Law, which ascribes both properties to the mobility of free electrons. The shape of the graphite particles in cast iron greatly affects this relationship, as indicated in Fig. 31. The thermal conductivity of gray iron containing well-developed flake graphite is much higher than that of steel, and the electrical conductivity is much lower. As the shape of the graphite changes from flake to intermediate forms to fully spherical shapes, the difference between the thermal or electrical conductivity of the cast iron and that of steel becomes less. Accordingly, ductile irons have higher electrical conductivity and lower thermal conductivity than gray irons.

Electrical Resistivity. The reciprocal function of the electrical conductivity, or electrical resistivity, of ductile iron increases with temperature. For instance, irons with a specific resistance of 0.5 to 0.55 $\mu\Omega \cdot m$ at room temperature might have a resistance of 1.25 to 1.30 $\mu\Omega \cdot m$ at 650 °C (1200 °F). Increasing the amount of silicon in either a pearlite or ferritic matrix also increases electrical resistivity (see Fig. 32). Increasing the amount of graphite tends to increase electrical resistivity because graphite has a high resistance, but the graphitization of pearlite or cementite to produce a ferritic matrix often results in a net decrease in resistance because of the lower electrical resistivity of ferrite. Alloying elements that dissolve in the matrix normally increase resistance, but reversals of this tendency may be observed when the relative amounts of various microconstituents are changed by the presence of the alloying element.

Magnetic properties should be determined separately for each application in which they are important. Although the properties of cast irons do not approach those of permanent magnet alloys on the one hand, or of silicon steels on the other, cast irons often are used for parts that require known magnetic properties. Cast iron can be cast into intricate shapes and sections more easily than most permanent alloys. The magnetic properties of cast irons depend largely on structure; the influence of alloying elements is indirect, derived mainly from the influence exerted on structure. A ferritic structure exhibits low hysteresis loss and high permeability, whereas a pearlitic structure exhibits high hysteresis loss and low permeability. Free cementite produces an iron with low permeability, magnetic induction, and remanence, together with high coercive force and hysteresis loss. Spheroidal graphite slightly increases hysteresis loss in pearlitic irons, but considerably reduces hysteresis loss in ferritic irons. Low-phosphorus ferritic ductile iron that is essentially free of cementite and very low in combined carbon has the highest permeability and lowest hysteresis loss of all cast irons.

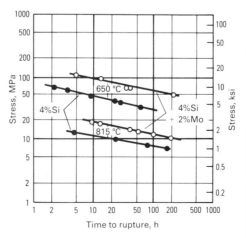

Fig. 26 Stress-rupture properties of 2.5Si-1.0Ni ductile irons. (a) Ferritic. (b) Pearlitic. Source: Ref 15, 16

Fig. 27 Effect of 2% Mo on the stress rupture of 4% Si ductile iron at 650 °C (1200 °F) and 815 °C (1500 °F). Source: Ref 7

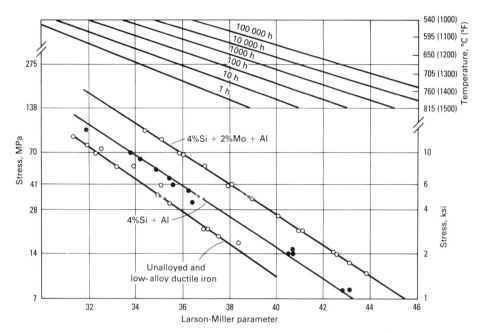

Fig. 28 Master curves for stress rupture of ductile iron. Larson-Miller parameter is $10^{-3}T(20 + \log t)$, where T is temperature in °R and t is time to rupture in hours. Source: Ref 14

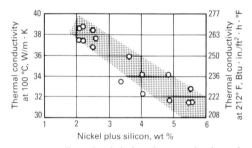

Fig. 30 Effect of nickel plus silicon on the thermal conductivity of ferritic ductile iron. Source: Ref 18

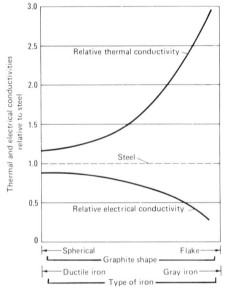

Fig. 31 Effect of graphite shape on the thermal and electrical conductivities of gray and ductile iron relative to those of steel. Source: Ref 14

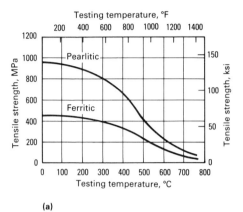

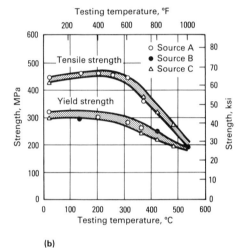

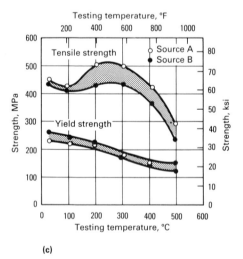

Fig. 29 Comparison of short-time tensile properties of cast products at elevated temperatures. (a) Pearlitic and ferritic ductile irons. (b) Ferritic ductile iron from three sources. (c) Low-carbon cast steel from two sources

In alloyed ductile irons, copper and nickel decrease permeability and increase hysteresis loss. When nickel is present in sufficient quantities to produce an austenitic matrix, ductile iron becomes paramagnetic. Manganese and chromium reduce magnetic induction, permeability, and remanent magnetism and increase hysteresis loss. Silicon has little effect on magnetic properties in pearlitic ductile iron, but slightly increases maximum permeability and reduces hysteresis loss in ferritic ductile iron.

Machinability

Ductile iron has been chosen in many instances on the basis of significantly lower machining costs, which resulted in lower overall cost of the part. Ductile iron has approximately the same machinability as gray iron of similar hardness. At low hardnesses (ductile iron with tensile strength up to about 550 MPa, or 80 ksi, the machinability of ductile iron is better than that of cast mild steel. At higher hardnesses, the difference in machinability between ductile iron and cast steel is less pronounced.

Recommended surface speeds for machining ductile iron castings are given in Table 14. Corresponding feeds for drilling and milling are given in Tables 15 and 16. More detailed data may be found in the article "Machining of Cast Irons" and other articles in *Machining*, Volume 16 of the 9th Edition of *Metals Handbook*. In general, the recommended cutting speeds given in Table 14 are based on a tool life of about 2 h of accumulated cutting time. Although the term recommended is used, the conditions should be considered nothing more than nominal starting conditions. In practice, recommended machining conditions may be altered as necessary to accommodate various part requirements, equipment characteristics, and manufacturing objectives.

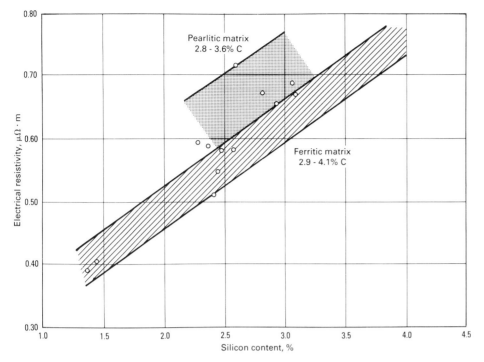

Fig. 32 The influence of the matrix structure and silicon content on the electrical resistivity of ductile iron at room temperature. Source: Ref 14

Experience has shown that when converting from malleable iron or pearlitic malleable iron castings to ductile iron, surface speeds should be increased by about 20%. This allows the chip to roll faster and break easier, improving tool life and thereby providing an increase in productivity. The higher elongation of ductile iron causes the chips to roll and produce more abrasion on the tool. This added abrasion is reduced if the chip breaks instead of rolling. Also, as-cast ductile iron does not have the surface decarburized layer normally found in malleable irons that decreases tool life.

Work hardening is produced when machining ductile iron, making it important to avoid light cuts. Coolants improve high-speed tool life, but are not as effective for carbide tool life until high-surface speeds are used.

A comparison can be made between the machinability of ASTM class 40 gray iron and that of ASTM grade 60-40-18 ductile iron. The gray iron has a tensile strength of about 275 MPa (40 ksi) and a hardness of 190 to 220 HBN. The ductile irons have a yield strength of about 275 MPa (40 ksi) and a hardness of about 187 HBN. The higher silicon content of ductile iron compared to malleable iron or gray cast iron causes the ferrite to be harder. The recommended cutting speed for rough turning using a single-point high speed steel tool at a feed of about 0.38 mm/rev (0.015 in./rev) is 20 m/min (70 sfm) for ASTM class 40 gray iron and 43 m/min (140 sfm) for ASTM grade 60-40-18 ductile iron. For a similar rough turning operation, cast mild steel with a tensile strength of about 415 to 485 MPa (60 to 70 ksi) requires a cutting speed of 33 m/min (110 sfm). Replacing the high speed steel tool with one made of sintered carbides allows increases in cutting speed for all three materials, but does not change the ratios of cutting speeds among the three materials. For ductile iron, the cutting speed, using carbide tools dry, increases to about 82 m/min (270 sfm).

Example 1: Comparison of the Machinability of Ferritic Ductile Iron With That of Selected Grades of Gray Cast Iron. For this test program, test castings representing a range of cooling conditions were produced (see Fig. 33). The five sections were then cut apart and centered, and the surfaces were turned off to depths of about 1.6 to 3.2 mm ($1/16$ to $1/8$ in.). The resulting diameters were 95, 70, 60, and 45 mm (3.75, 2.75, 2.37, and 1.75 in.). Test turning was done by making successive 0.508 mm (0.200 in.) cuts at a cutting speed of 76 m/min (250 sfm). Three cuts were made and timed for each test section. The rate of metal removal was calculated for each size, and the data are shown in Fig. 33(a).

Pearlitic Ductile Irons Compared With Pearlitic Gray Irons. The test described in Example 1 compared annealed ferritic ductile iron with pearlitic gray iron of slightly higher hardness. Other data comparing pearlitic gray irons with pearlitic ductile irons show no great machinability differences when the irons are of similar hardness. The comparison of tool life in the machining of ductile and gray irons is shown in Fig. 33(b).

Ductile Irons Compared With Malleable Cast Irons. In another investigation, a comparison was made between ductile iron castings and malleable iron castings. The machining operations included facing, boring, threading, and drilling. There was no significant difference in machining costs between these two materials when the work was done on a production basis.

Welding

Special materials and techniques are available for the repair welding of ductile

Table 14 Speeds for machining ductile iron

	High-speed steel tools, m/min (sfm)							Cemented carbide tools, sfm(a)		
Ductile iron	Turning(b)	Drilling(c)	Reaming(d)	Tapping	Thread chasing	Milling(e)	Shaping(f) and planing	Broaching	Turning(b)	Reaming(d)	Milling(e)
60-45-10 with full ferritic matrix	15–46 (50–150)	24–40 (80–130)	15–30 (50–100)	6–9 (20–30)	9–21 (30–70)	15–38 (50–125)	12–30 (40–100)	6–11 (20–35)	53–120 (175–400)	23–46 (75–150)	60–120 (200–400)
Semipearlitic matrix	12–27 (40–90)	15–30 (50–100)	12–21 (40–70)	4.5–6 (15–20)	6–15 (20–50)	11–20 (35–65)	9–23 (30–75)	4.5–7.6 (15–25)	30–90 (100–300)	15–27 (50–90)	53–107 (175–350)
80-60-03 with full pearlitic matrix	12–27 (40–90)	15–21 (50–70)	12–21 (40–70)	4.5–6 (15–20)	6–15 (20–50)	11–20 (35–65)	9–23 (30–75)	4.5–7.6 (15–25)	30–90 (100–300)	15–27 (50–90)	53–107 (175–350)

(a) In all instances, the longest tool life between regrinds will result when tools are operated at minimum to median speeds for the range given. (b) Feeds of 0.25 to 0.50 mm/rev (0.010 to 0.020 in./rev). Maximum speed is for cuts not more than 1.6 mm ($1/16$ in.) deep. (c) Feeds should be commensurate with drill diameter: light feeds for small-diameter drills and heavier feeds for larger drills. For recommended feeds, see Table 15. It is good practice to reduce speed as drill diameter increases; speeds at or near the maximum values given in this table are for drills 12 mm ($1/2$ in.) in diameter or less. (d) Use feeds three to four times those used for drills of similar size. An allowance of 0.30 to 0.38 mm (0.012 to 0.015 in.) is sufficient for reaming. (e) Speeds cited are principally for face milling; however, they may be used as a guide for plain milling. For recommended feeds, see Table 16. (f) Depth of cut and feed vary with sturdiness of the setup. The operating speed for roughing should approach the minimum value cited.

54 / Cast Irons

Table 15 Starting recommendations for drilling ductile irons with high-speed steel tools

Types classified by microstructure	Typical hardness, HB	Condition	Speed m/min (sfm)	Feed, mm/rev (in./rev) at nominal hole diameter, mm (in.)								High-speed steel tool material
				1.5 (1/16)	3 (1/8)	6 (1/4)	1 (1/2)	18 (3/4)	25 (1)	35 (1½)	50 (2)	
Ferritic 60-40-18, D4018, 65-45-12, D4512	140–190	Annealed	26–35 (85–115)	0.025 (0.001)	0.075 (0.003)	0.15 (0.006)	0.25 (0.010)	0.33 (0.013)	0.40 (0.016)	0.55 (0.021)	0.65 (0.026)	M10, M7, M1, S2, S3
Ferritic-pearlitic 80-55-06, D5506	190–225	As-cast	21 (70)	0.025 (0.001)	0.075 (0.003)	0.15 (0.006)	0.25 (0.010)	0.33 (0.013)	0.40 (0.016)	0.55 (0.021)	0.65 (0.025)	M10, M7, M1, S2, S3
	225–260		15 (50)	0.025 (0.001)	0.050 (0.002)	0.102 (0.004)	0.18 (0.007)	0.25 (0.010)	0.30 (0.012)	0.40 (0.015)	0.45 (0.018)	T15, M42, S9, S11(a)
Pearlitic martensitic 100-70-03, D7003	240–300	Normalized and tempered	14 (45)	0.025 (0.001)	0.050 (0.002)	0.102 (0.004)	0.18 (0.007)	0.20 (0.008)	0.25 (0.010)	0.33 (0.013)	0.40 (0.015)	T15, M42, S9, S11(a)
Martensitic 120-90-02, DQ&T	270–330	Quenched and tempered	9 (30)	...	0.025 (0.001)	0.050 (0.002)	0.102 (0.004)	0.13 (0.005)	0.15 (0.006)	0.18 (0.007)	0.20 (0.008)	T15, M42, S9, S11(a)
Austenitic (Ni-Resist) D2, D2-C, D-3A, D5, D-2M	120–200	Annealed	11 (35)	0.25 (0.001)	0.050 (0.002)	0.13 (0.005)	0.18 (0.007)	0.25 (0.010)	0.30 (0.012)	0.40 (0.016)	0.45 (0.020)	T15, M42, S9, S11(a)

(a) Any premium high-speed steel (T15, M33, M41-47, or S9, S10, S11, S12). Source: Ref 19

Table 16 Feeds for milling ductile iron

	Feed per tooth, when using			
	High-speed steel tools		Cemented carbide tools	
Type of cutter	mm	in.	mm	in.
Face mill	0.15–0.30	0.006–0.012	0.20–0.40	0.008–0.015
Plain mill	0.13–0.23	0.005–0.009	0.15–0.30	0.006–0.012
End mill	0.08–0.20	0.003–0.008	0.08–0.25	0.003–0.010
Circular saw	0.05–0.10	0.002–0.004	0.02–0.10	0.001–0.004

Note: For milling speeds, see Table 14.

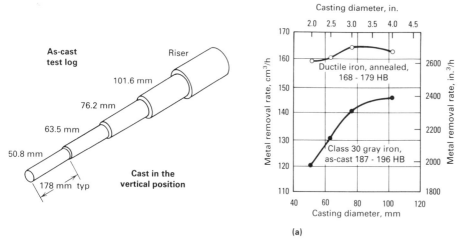

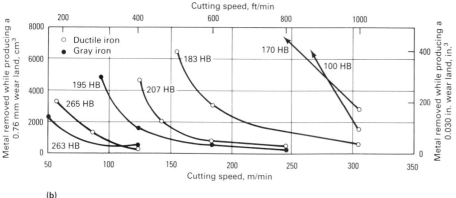

Fig. 33 Comparison of the machinability of ductile and gray irons. (a) Metal removal rates. (b) Tool life. Source: Ref 20

iron castings, or for joining ductile iron to itself or to other ferrous materials such as steel, gray iron, or malleable iron. Like the welding of other cast irons, the welding of ductile iron requires special precautions to obtain optimum properties in the weld metal and adjacent heat-affected zone. The main objective is to avoid the formation of cementite in the matrix material, which makes the welded region brittle; but in ductile iron an additional objective, that of retaining a nodular form of graphite, is of almost equal importance. The formation of martensite or fine pearlite can be removed by tempering.

A technique developed and patented by Oil City Iron Works uses a special ductile iron filler metal and a special welding flux that is introduced through a powder spray-type oxyacetylene welding torch. In this technique, which is used predominantly for the cosmetic repair of ductile iron castings, parent metal is puddled under a neutral region of slightly reducing flame, and filler metal is added as the special flux is sprayed into the puddle through the torch. A properly executed weld will be essentially free of eutectic carbides in the weld metal and will have a pearlitic matrix with bull's-eye ferrite surrounding the particles of spheroidal graphite.

Typically, the mechanical properties of the weld metal are very similar to those of the parent metal for all heat-treated conditions: as-welded, annealed, normalized, or quenched and tempered. The only possible exception is that the ductility of the weld metal may be slightly lower than that of equivalent parent metal at the lower hardnesses, such as those typical of the as-welded or annealed conditions.

Among other advantages, this process obtains a perfect color match with the cast metal; yields weld metal with composition, microstructure, and properties very close to those of the original casting; and minimizes the transition zone, the heat-affected zone, and any residual stresses.

As an alternative to the Oil City process, ductile iron can be welded with a high-nickel alloy, using the flux-cored arc welding (FCAW) process. In flux-cored arc welding, a hollow wire with the composition 50Ni-44Fe-4.25Mn-1.0C-0.6Si and containing a special flux is used as the electrode in standard FCAW equipment. This method is used to join ductile iron to itself or to steel or other types of cast iron more often than it is used to effect the cosmetic repair of castings. The mechanical properties of the high-nickel weld metal and of the adjacent heat-affected zone are usually equivalent to the properties of ASTM grade 65-45-12 ductile iron. A major disadvantage of welding with the high-nickel alloy is that it does not respond to heat treatment, and thus weldments made with this alloy cannot be heat treated to obtain uniformly high strength levels, as can weldments made using the Oil City process.

Low-temperature welding rods and wire that have high wetting properties on cast iron base metals, are available. This effects the joining of metals at such low temperatures that the base metal does not melt. The composition of the weld metal is such that it has dimensional changes with temperature similar to those of ductile iron, thereby reducing stresses. The color match is perfect, the hardness is low, the tensile strength is greater than 390 MPa (57 ksi), and the elongation is 25 to 30%. The weld metal is suitable for tungsten inert gas welding operations. The welding rods or wire are produced by Shichiho Metal Industrial Company Ltd.

Other methods for welding ductile iron include submerged arc welding and the use of austenitic consumable materials (see Ref 21-23 and the article "Arc Welding of Cast Irons," Volume 6, 9th Edition, *Metals Handbook*).

REFERENCES

1. R.B. Gundlach and J.F. Janowak, Approaching Austempered Ductile Iron Properties by Controlled Cooling in the Foundry, in *Proceedings of the First International Conference on Austempered Ductile Iron: Your Means to Improved Performance, Productivity, and Cost*, American Society for Metals, 1984
2. Lyle Jenkins, Ductile Iron—An Engineering Asset, in *Proceedings of the First International Conference on Austempered Ductile Iron: Your Means to Improved Performance, Productivity, and Cost*, American Society for Metals, 1984
3. R.B. Gundlach and J.F. Janowak, A Review of Austempered Ductile Iron Metallurgy, in *Proceedings of the First International Conference on Austempered Ductile Iron: Your Means to Improved Performance, Productivity, and Cost*, American Society for Metals, 1984
4. P.A. Blackmore and R.A. Harding, The Effects of Metallurgical Process Variables on the Properties of Austempered Ductile Irons, in *Proceedings of the First International Conference on Austempered Ductile Irons: Your Means to Improved Performance, Productivity, and Cost*, American Society for Metals, 1984
5. C.R. Loper, P. Banerjee, and R.W. Heine, Risering Requirements for Ductile Iron Castings in Greensand Moulds, *Gray Iron News*, May 1964, p 5-16
6. "The Fatigue Life of Cast Surface of Malleable and Nodular Iron," Bulletin 177, Metals Research and Development Foundation
7. D.L. Sponseller, W.G. Scholz, and D.F. Rundle, Development of Low-Alloy Ductile Irons for Service at 1200-1500 F, *AFS Trans.*, Vol 76, 1968, p 353-368
8. W.S. Pellini, G. Sandoz, and H.F. Bishop, Notch Ductility of Nodular Irons, *Trans. ASM*, Vol 46, 1954, p 418-445
9. C. Vishnevsky and J.F. Wallace, The Effect of Heat Treatment on the Impact Properties of Ductile Iron, *Gray Iron News*, July 1962, p 5-10
10. R.K. Nanstad, F.J. Worzala, and C.R. Loper, Jr., Static and Dynamic Toughness of Ductile Cast Iron, *AFS Trans.*, Vol 83, 1975
11. G.N.J. Gilbert, Tensile and Fatigue Tests on Normalized Pearlitic Nodular Irons, *J. Res.*, Vol 6 (No. 10), Feb 1957, p 498-504
12. R.C. Haverstraw and J.F. Wallace, Fatigue Properties of Ductile Iron, *Gray Ductile Iron News*, Aug 1966, p 5-19
13. H.D. Merchant and M.H. Moulton, Hot Hardness and Structure of Cast Irons, *Br. Foundryman*, Vol 57 (Part 2), Feb 1964, p 62-73
14. C.F. Walton, Ed., *Gray and Ductile Iron Castings Handbook*, Gray and Ductile Founders' Society, 1971
15. C.R. Wilks, N.A. Matthews, and R.W. Kraft, Jr., Elevated Temperature Properties of Ductile Cast Irons, *Trans. ASM*, Vol 47, 1954
16. F.B. Foley, Mechanical Properties at Elevated Temperatures of Ductile Cast Iron, *Trans. ASME*, Vol 78, 1956, p 1435-1438
17. C.C. Reynolds, W.T. Whittington, and H.F. Taylor, Hardenability of Ductile Iron, *AFS Trans.*, Vol 63, 1955, p 116-122
18. H.T. Angus, *Cast Iron: Physical and Engineering Properties*, 2nd ed., Butterworths, 1976
19. *Machining Data Handbook*, Vol 1, 3rd ed., Metcut Research Associates, 1980
20. "Machinability Report," U.S. Air Force, 1950
21. D.L. Olson, "Investigation of the MnO-SiO_2-Oxides and MnO-SiO_2-Fluorides Welding Flux Systems," DAAG29-77-G-0097, U.S. Army Research Office, June 1978
22. M.A. Davila, D.L. Olson, and T.A. Freese, Submerged Arc Welding of Ductile Iron, *Trans. AFS*, 1977
23. M.A. Davila and D.L. Olson, The Development of Austenitic Filler Materials for Welding Ductile Iron, Paper 23, Welding Institute Reprint, Welding Institute, 1978

Compacted Graphite Iron

Doru M. Stefanescu, The University of Alabama

COMPACTED GRAPHITE (CG) cast iron is also referred to as vermicular graphite, upgraded, or semiductile cast iron (Ref 1). It has been inadvertently manufactured in the past in the process of producing ductile iron, as a result of undertreatment with magnesium or cerium. Since 1965, after R.D. Schelleng obtained a patent for its production, CG iron has occupied its rightful place in the family of cast irons.

The graphite morphology of CG iron is rather complex. A typical scanning electron microscope photomicrograph of a compacted graphite particle etched out of the matrix is shown in Fig. 1(a). It is seen that compacted graphite appears in clusters that are interconnected within the eutectic cells. Classical optical metallography (Fig. 1b) exhibits graphite that is similar to type IV ASTM A 247 graphite (see the article "Classification and Basic Metallurgy of Cast Iron" in this Volume). Compacted graphite appears as thicker, shorter-flake graphite. In general, an acceptable CG iron is one in which at least 80% of the graphite is compacted graphite, there is a maximum of 20% spheroidal graphite (SG), and there is no flake graphite (FG).

This graphite morphology allows better use of the matrix, yielding higher strength and ductility than flake graphite cast iron. Similarities between the solidification patterns of flake and compacted graphite iron explain the good castability of the latter, compared to ductile iron (ductile iron, which is also termed nodular iron or spheroidal graphite iron, is called SG iron in this article). Also the interconnected graphite provides better thermal conductivity and damping capacity than spheroidal graphite.

Chemical Composition

The range of acceptable carbon and silicon contents for the production of CG iron is rather wide, as shown in Fig. 2. Nevertheless, the optimum carbon equivalent (CE) must be selected as a function of section thickness, in order to avoid carbon flotation when too high a CE is used, or excessive chilling tendency, when too low a CE is used. The manganese content can vary between 0.1 and 0.6%, depending on whether a ferritic or a pearlitic structure is desired. Phosphorus content should be kept below 0.06% in order to obtain maximum ductility from the matrix. The initial sulfur level should be below 0.025%, although techniques for producing CG iron from base irons with higher sulfur levels are now available (see the article "Compacted Graphite Irons" in Volume 15 of the 9th Edition of *Metals Handbook*). Residual sulfur after liquid treatment is typically in the range of 0.01 to 0.02%.

The change in graphite morphology from the flake graphite in the base iron to the compacted graphite in the final iron is achieved by liquid treatment with different minor elements. These elements may include one or more of the following: magnesium, rare earths (cerium, lanthanum, praseodymium, and so on), calcium, titanium, and aluminum. The amounts and combinations to be used are a function of the method of liquid treatment, base sulfur, section thickness, and so forth, and are discussed in Volume 15 of the 9th Edition of *Metals Handbook*. For example, Fig. 3 shows some typical correlations between treatment method (level of minor elements), initial sulfur level, and graphite shape when producing CG irons.

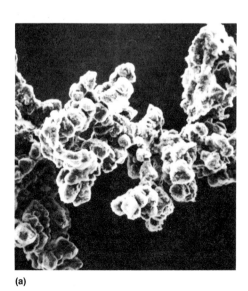

Fig. 1 Compacted graphite. (a) SEM photomicrograph showing deep-etched specimen. 200×. b) Optical photomicrograph

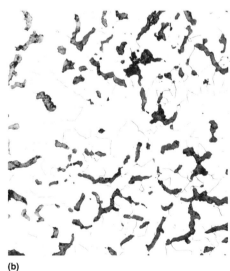

Fig. 2 Optimum range for carbon and silicon contents for CG iron. Source: Ref 2

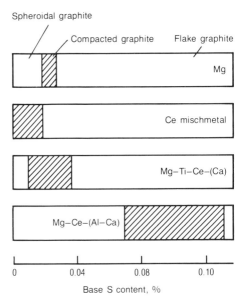

Spheroidizer	ΔS range(a) for compacted iron, %
Mg + Ce	−0.0155 to −0.032
Mg + Ti > 0.10%	−0.0155 to −0.042
Mg + Ti 0.05 − 0.1% + Al 0.2 − 0.3%	−0.0110 to −0.055
Mg + Al > 0.35%	−0.0060 to −0.35

(a) ΔS: final % S − 0.34 (% residual elements) − 1.33 (% Mg)

Fig. 3 Optimum range of initial sulfur level as a function of type (figure) and amount (table) of minor elements used for graphite compaction. The above table shows the sulfur range, ΔS, for compacted iron formation with different spheroidizers in an iron composition of 3.5% C, 2.1% Si, 0.75% Mn, and 0.03 to 0.08% P. Source: Ref 1, 3

Compacted graphite iron has a strong ferritization tendency. Copper, tin, molybdenum, and even aluminum can be used to increase the pearlite/ferrite ratio. Again, the optimum amounts of these elements for a particular matrix structure are largely a function of section size.

Castability

The fluidity of cast iron is a function of its pouring temperature, composition, and eutectic morphology. A higher temperature and higher CE result in better fluidity. Everything else being equal, the fluidity of CG iron is intermediate between that of FG (highest) and SG (lowest) iron (Ref 1). However, because CG iron has a higher strength than FG iron for the same CE, high-CE compositions of CG iron can be used for the pouring of thin castings.

Shrinkage Characteristics. With CG irons, obtaining sound castings free from external and internal shrinkage porosity is easier than with SG irons and slightly more difficult than with FG irons. This is because the tendency for mold wall movement also lies between that of SG and FG irons.

In relative numbers, solidification expansion has been found to be 4.4 for SG iron

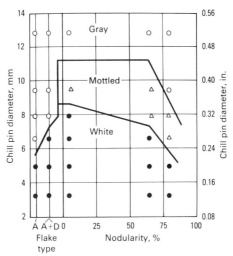

Fig. 4 Influence of graphite shape over the chilling tendency of cast iron. Type A graphite flake: uniform distribution and random orientation. Type D graphite flake: interdendritic segregation and random orientation (see the article "Classification and Basic Metallurgy of Cast Iron" in this Volume). Source: Ref 5

and 1 to 1.8 for CG iron if FG iron is 1 (Ref 4). Because of the rather low shrinkage of CG iron, it can sometimes be cast riserless. Expensive pattern changes are therefore not necessary when converting from gray iron to CG iron because the same gating and risering techniques can be applied.

Chilling Tendency. Although many believe that the chilling tendency of CG iron is also intermediate between that of FG (lowest) and SG (highest) irons, this is not true. Figure 4 shows the influence of nodularity on the structure of chill pins cast in air set molds (Ref 5). It can be seen that the highest chilling tendency is achieved for irons with nodularities between 6 and 64%. In other words, the chilling tendency of CG iron is higher than that of both SG and FG iron. This correlates with cooling curve data (see the article "Compacted Graphite Irons" in Volume 15 of the 9th Edition of *Metals Handbook*) and is explained by the combination of a low nucleation rate and low growth rate occurring during the solidification of CG iron.

Mechanical Properties at Room Temperature

The in-service behavior of many structural parts is a function not only of their mechanical strength, but also of their deformation properties. Thus it is not surprising to find that many castings fail not because of insufficient strength, but because of a low capacity for deformation. This is especially true under conditions of rapid loading and/or thermal stress. Particularly sensitive to such loading are casting zones that include some defects or abrupt changes in section thickness. The elongation values of about

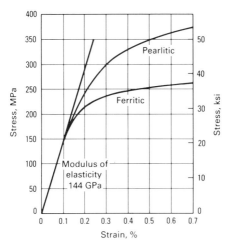

Fig. 5 Typical stress-strain curves for CG irons. The pearlitic iron had a tensile strength of 410 MPa (59.5 ksi) and an elongation of 1%. The ferritic iron had a tensile strength of 320 MPa (46.5 ksi) and an elongation of 3.5%. Source: Ref 1

1% obtainable with high-strength gray iron are insufficient for certain types of applications such as diesel cylinder heads (Ref 6). Compacted graphite irons have strength properties close to those of SG irons, at considerably higher elongations than those of FG iron, and with intermediate thermal conductivities. Consequently, they can successfully outperform other cast irons in a number of applications.

The main factors affecting the mechanical properties of CG irons both at room temperatures and at elevated temperatures are:

- Composition
- Structure (nodularity and matrix)
- Section size

In turn, the structure is heavily influenced by processing variables such as the type of raw materials, preprocessing of the melt (superheating temperature, holding time, desulfurization), and liquid treatment (graphite compaction and postinoculation).

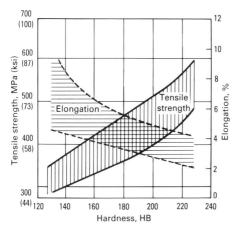

Fig. 6 Correlation between hardness, tensile strength, and elongation in CG irons. Source: Ref 8

Table 1 Comparison of properties of cerium-treated CG iron with FG iron of the same chemical composition, high-strength pearlitic FG iron, and ferritic SG iron in the as-cast condition

Property	High-strength pearlitic FG iron (100% pearlite, 100% FG)(a)	FG iron (100% pearlite, 100% FG)(b)	Ce-treated CG iron (>95% ferrite, >95% CG)(b)	SG iron (100% ferrite, 80% SG, 20% poor SG)(b)
Chemical composition, %	3.10 C, 2.10 Si, 0.60 Mn	3.61 C, 2.49 Si, 0.05 Mn	36.1 C, 2.54 Si, 0.05 Mn	3.56 C, 2.72 Si, 0.05 Mn
Tensile strength, MPa (ksi)	317 (46)	110 (16)	336 (48.7)	438 (63.5)
0.2% proof stress, MPa (ksi)	...	...	257 (37.3)	285 (41.3)
Elongation, %	...	...	6.7	25.3
Modulus of elasticity, GPa (10^6 psi)	108 (15.7)	96.9 (14.05)	158 (22.9)	176 (25.5)
Brinell hardness, HB	200	156	150	159
Charpy V-notched-bar impact toughness, J (ft · lbf)				
at 20 °C (68 °F)	...	...	9.32 (6.87)	24.5 (18.1)
at −20 °C (−4 °F)	...	...	6.57 (4.85)	9.81 (7.23)
at −40 °C (−40 °F)	...	...	7.07 (5.21)	6.18 (4.56)
Charpy impact bend toughness, J (ft · lbf)				
at 20 °C (68 °F)	4.9	2.0	32.07 (23.7)	176.5 (130.2)
at −20 °C (−4 °F)	...	...	26.48 (19.5)	148.1 (109.2)
at −40 °C (−40 °F)	...	...	26.67 (19.7)	121.6 (89.7)
Rotating-bar fatigue strength, MPa (ksi)	127.5 (18.5)	49.0 (7.1)	210.8 (30.6)	250.0 (36.3)
Thermal conductivity, W/(cm · K)	0.419	0.423	0.356	0.327

(a) Mechanical properties determined from a sample with a section size 30 mm (1.2 in.) in diameter. (b) Mechanical properties determined from a Y block 23 mm (0.9 in.) section. Source: Ref 7

Table 2 Tensile properties, hardness, and thermal conductivity of various CG irons at room temperature

Structural condition(a)	Degree of saturation, S_C(b)	Graphite type	Tensile strength MPa	Tensile strength ksi	0.2% proof stress MPa	0.2% proof stress ksi
Irons treated with additions of cerium						
As-cast ferrite (>95% F)	1.04	95% CG, 5% SG	336	48.7	257	37.3
Ferritic-pearlitic (>5% P)	1.04	95% CG, 5% SG	298	43.2	224	32.5
As-cast ferrite (90% F, 10% P)	1.00	85% CG, 15% SG	371	53.8	267	38.7
100% ferrite	1.00	85% CG, 15% SG	338	49.0	245	35.5
100% ferrite	1.04–1.09	CG	365 ± 63	53 ± 9	278 ± 42	40 ± 6
Ferritic-pearlitic (>90% F, <10% P)	...	>90% CG	300–400	43–58	250–300	36–43
Ferritic-pearlitic (85% F)	1.04	70% CG, 30% SG	320	46.4	242	35
Pearlitic (90% P, 10% F)	...	90% CG	400–550	58–80	320–430	46–62
Pearlitic (95% P, 5% F)	1.02	80% CG, 20% SG	410	59.5	338	49
Irons treated with combinations of Mg + Ti (+Ce)						
As-cast ferrite (0.004% Ce, <0.01% Mg, 0.28% Ti)	0.99	95% CG, 5% SG	319	46.3	264	38.3
100% ferrite (annealed) (0.018% Mg, 0.089% Ti, 0.032% As)	0.97	CG	292	42.3	225	32.6
As-cast ferrite (0.017% Mg, 0.062% Ti, 0.036% As)	1.01	CG	380	55	272	39.4
As-cast ferrite (0.024% Mg, 0.084% Ti, 0.030% As)	1.01	CG	388	56.3	276	40
As-cast pearlite (0.016% Mg, 0.094% Ti, 0.067% As)	0.99	CG	414	60	297	43.1
As-cast pearlite (0.026% Mg, 0.083% Ti, 0.074% As)	1.02	CG + SG	473	68.6	335	48.6
As-cast pearlite (70% P, 30% F)	...	CG	386	56.0	278	40.3

Structural condition(a)	Elongation, %	Static modulus of elasticity(c) GPa	Static modulus of elasticity(c) 10^6 psi	Dynamic (plastomat) modulus of elasticity GPa	Dynamic (plastomat) modulus of elasticity 10^6 psi	Thermal conductivity W/m · K	Hardness, HB
Irons treated with additions of cerium							
As-cast ferrite (>95% F)	6.7	...	...	158	22.9	35.6	150
Ferritic-pearlitic (>5% P)	5.3	...	...	144	20.9	38.5	128
As-cast ferrite (90% F, 10% P)	5.5	137	19.9	...	...	...	...
100% ferrite	8.0	...	...	...	...	...	140
100% ferrite	7.2 ± 4.5	...	...	...	...	...	138–156
Ferritic-pearlitic (>90% F, <10% P)	3–7	...	...	150	21.8	38.5	...
Ferritic-pearlitic (85% F)	3.5	140	20.3	...	...	...	164
Pearlitic (90% P, 10% F)	0.5–1.5	...	...	161	23.4	29.3	...
Pearlitic (95% P, 5% F)	1	144	20.9	...	...	...	220
Irons treated with combinations of Mg + Ti (+Ce)							
As-cast ferrite (0.04% Ce, <0.01% Mg, 0.28% Ti)	4	138	20.0	...	...	...	143
100% ferrite (annealed) (0.018% Mg, 0.089% Ti, 0.032% As)	6	...	...	...	...	...	129
As-cast ferrite (0.017% Mg, 0.062% Ti, 0.036% As)	2	145	21.0	147	21.3	...	179
As-cast ferrite (0.024% Mg, 0.084% Ti, 0.030% As)	2.5	...	...	151	21.9	...	184
As-cast pearlite (0.016% Mg, 0.094% Ti, 0.67% As)	2	...	...	150	21.8	...	205
As-cast pearlite (0.026% Mg, 0.083% Ti, 0.074% As)	2	...	...	161	23.4	...	217
As-cast pearlite (70% P, 30% F)	2	...	...	152	22.0	41.9	...

(a) F, ferrite; P, pearlite. (b) S_C = %C/4.25 − 0.3 (%Si + %P); see the article "Thermodynamic Properties of Iron-Base Alloys in Volume 15 of the 9th Edition of *Metals Handbook*. (c) Modulus of elasticity at zero stress (E_0). Source: Ref 6

Tensile Properties and Hardness. A comparison between some properties of FG, CG, and SG irons is given in Table 1. A listing of tensile properties of various CG irons produced by different methods is given in Table 2.

Compacted graphite irons exhibit linear elasticity for both pearlitic and ferritic matrices, but to a lower limit of proportionality than does SG iron (Fig. 5). The ratio of yield strength to tensile strength ranges from 0.72 to 0.82, which is higher than that for SG iron of the same composition. This makes possible a higher loading capacity. The limit of proportionality is 125 MPa (18 ksi) for both ferritic and pearlitic CG irons. It is slightly lower than that of SG iron. This can be explained by the higher notching effect from the sharper-edged morphology of the compacted graphite compared to spheroidal graphite. Consequently, for the same stress, plastic deformation occurs sooner around the graphite in CG iron than in SG iron, causing earlier divergence from proportionality (Ref 1).

As the hardness increases, tensile strength increases but elongation decreases, this being the effect of a higher pearlite/ferrite ratio (see Fig. 6). The ratio of tensile strength to Brinell hardness is somewhat higher for CG iron than for FG iron. In general, CG iron has lower hardness than an FG iron of equivalent strength because of the higher amount of ferrite in the structure. For the same elongation, CG iron has considerably less yield strength than SG iron, as shown in Fig. 7.

Effect of Composition. It has been demonstrated that the tensile properties of CG irons are much less sensitive to variations in carbon equivalent than are those of FG irons. Even at CE near the eutectic value of 4.3, both pearlitic and ferritic CG irons have higher strengths than does low-CE, high-duty, unalloyed FG cast iron (Fig. 8).

Although increasing the silicon content decreases the pearlite to ferrite ratio in the as-cast state, both the strength and hardness of as-cast and annealed CG irons improve. This is because of the hardening of ferrite by silicon. For the same reasons, elongation in the annealed condition decreases, but increases for the as-cast state (Ref 10). Although increasing the phosphorus content slightly improves strength, a maximum of 0.04% P is desirable to avoid lower ductility and impact strength.

The pearlite/ferrite ratio, and thus the strength and hardness of CG irons, can be increased by the use of a number of alloying elements such as copper, nickel, molybdenum, tin, manganese, arsenic, vanadium, and aluminum (Ref 6, 14). The effect of copper and molybdenum on the tensile properties of CG irons is shown in Fig. 9. After annealing to a fully ferritic structure, it is possible to increase the yield point of CG iron by 24% when using 1.5% Ni (Table 3). This is because of the strengthening of the solid solution by nickel (Ref 6). The reader is cautioned, however, that additions of copper, nickel, and molybdenum may increase nodularity (Ref 6).

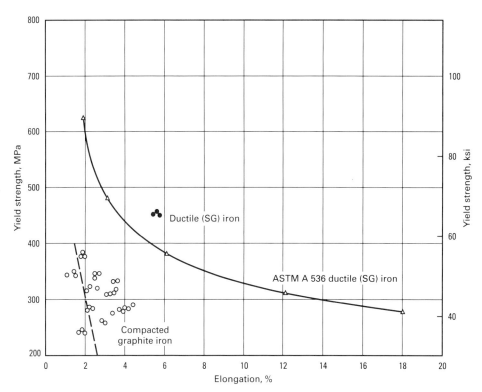

Fig. 7 Correlation between yield strength and elongation for CG and SG irons. Source: Ref 8

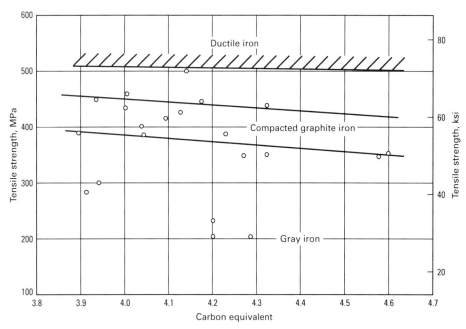

Fig. 8 Effect of carbon equivalent on the tensile strength of flake, compacted, and spheroidal graphite irons cast into 30 mm (1.2 in.) diam bars. Source: Ref 10

In order to compare the quality of different types of irons, several quality indexes can be used, such as the product of tensile strength and elongation (TS × El) or the ratio of tensile strength to Brinell hardness

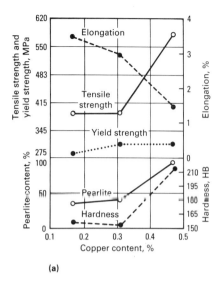
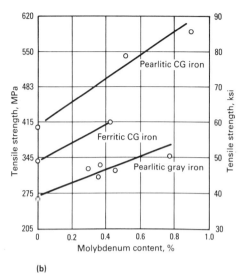

Fig. 9 Effect of (a) copper and (b) molybdenum on the tensile properties of CG iron. Source: Ref 13, 14

Fig. 10 Quality indexes for Fe-C-Si and Fe-C-Al irons. TS, tensile strength; El, elongation. Source: Ref 11

(TS/HB). Higher values of these indexes will characterize a better iron. Using the data given in Ref 10, some typical values were calculated for these indexes for unalloyed CG irons. Figure 10 compares the TS × El product and the TS/HB ratio for unalloyed and aluminum-alloyed CG irons. It can be seen that when 2% Si is replaced by 2% Al, a much better quality CG iron is produced (Ref 11).

Effect of Structure. One of the most important variables influencing the tensile properties of CG irons is nodularity. As nodularity increases, higher strength and elongation are to be expected, as shown in Table 4 and Fig. 11, although nodularity must be maintained at levels under 20% for the iron to qualify as CG iron. However, spheroidal graphite contents of up to 30% and even more must be expected in thin sections of castings with considerable variation in wall thickness.

As previously discussed, the pearlite/ferrite ratio can be increased by using alloying elements. Another way of increasing or decreasing this ratio is by using heat treatment. The influence of heat treatment on mechanical properties of CG irons with 20% nodularity is given in Table 3.

Effect of Section Size. Like all other irons, CG irons are rather sensitive to the influence of cooling rate, that is, to section size, because it affects both the pearlite/ferrite ratio and graphite morphology. As mentioned before, a higher cooling rate promotes more pearlite and increased nodularity. A typical example of the influence of section size on the microstructure of CG iron is provided in Fig. 12. Although CG iron is less section sensitive than FG iron (as shown in Fig. 13 for tensile strength), the influence of cooling rate may be quite significant (see the article "Compacted Graphite Irons" in Volume 15 of the 9th Edition of *Metals Handbook*). When the section size decreases until it is below 10 mm (0.4 in.), the tendency to increased nodularity and for higher chilling must be considered. This is particularly true for overtreated irons. While it is possible to eliminate the carbides that result from chilling by heat treatment, it is impossible to change the graphite shape, which remains spheroidal, with the associated consequences. Other factors influencing the cooling of castings, such as shakeout temperature, can also influence properties.

Compressive Properties. The stress-strain diagram for compression and tensile tests of CG iron is shown in Fig. 14. It can be seen that an elastic behavior occurs up to a compression stress of 200 MPa (30 ksi). Some compressive properties of the 179 HB as-cast ferritic CG iron in Table 2 are compared with those of SG iron in Table 5. It can be seen that the 0.1% proof stress in compression for CG iron is 76 MPa (11 ksi)

Table 3 Effect of heat treatment and alloying with nickel on the tensile properties of CG iron measured on a 25 mm section size

Heat treatment	Iron matrix(a)	Tensile strength MPa	ksi	Yield strength MPa	ksi	Elongation, %	Hardness, HB	Nickel, %
As-cast	60% F	325	47.1	263	38.1	2.8	153	0
Annealed(b)	100% F	294	42.6	231	33.5	5.5	121	0
Normalized(c)	90% P	423	61.3	307	44.5	2.5	207	0
As-cast	...	427	61.9	328	47.6	2.5	196	1.53
Annealed(b)	100% F	333	48.3	287	41.6	6.0	137	1.53
Normalized(c)	90% P	503	73	375	54.4	2.0	235	1.53

(a) F, ferrite; P, pearlite. (b) Annealed, 2 h at 900 °C (1650 °F), cooled in furnace to 690 °C (1275 °F), held 12 h, cooled in air. (c) Normalized, 2 h at 900 °C (1650 °F), cooled in air

Table 4 Properties of CG iron as a function of nodularity

Nodularity, %	Tensile strength MPa	ksi	Elongation, %	Thermal conductivity, W/(m·K)	Shrinkage, %
10–20	320–380	46–55	2–5	50–52	1.8–2.2
20–30	380–450	55–65	2–6	48–50	2.0–2.6
40–50	450–500	65–73	3–6	38–42	3.2–4.6

Source: Ref 15

Table 5 Comparison of tensile and compressive properties of CG and SG irons

Property	CG iron	SG iron	
Tensile strength, MPa (ksi)	380 (55)	370 (54)	420 (61)
0.1% proof stress, MPa (ksi)	246 (35.7)	224 (32.5)	261 (37.8)
0.2% proof stress, MPa (ksi)	242 (35.1)	236 (34.2)	273 (39.6)
Compressive stress			
0.1% proof stress, MPa (ksi)	322 (46.7)	247 (35.8)	284 (41.2)
0.2% proof stress, MPa (ksi)	350 (50)	250 (36.3)	287 (41.6)

Source: Ref 6

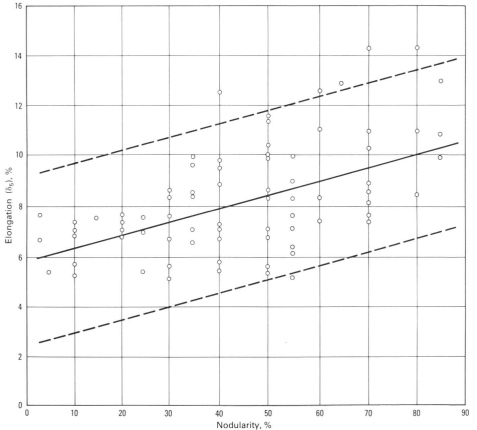

Fig. 11 Correlation between nodularity and elongation for ferritic CG iron. Source: Ref 16

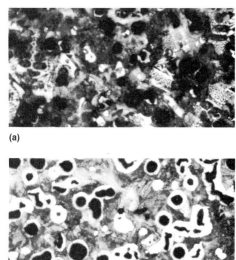

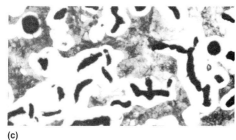

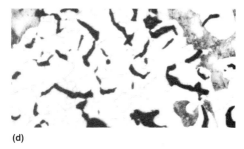

Fig. 12 Influence of section size on the microstructure of CG iron produced by inmold. Samples from series E3.3. (a) 3.2 mm (0.125 in.). (b) 6.4 mm (0.250 in.). (c) 12.7 mm (0.500 in.). (d) 25.4 mm (1.000 in.). (e) 50.8 mm (2.000 in.). 100×

higher than the 0.1% proof stress in tension, while for SG iron the difference is only 23 MPa (3.3 ksi). Compressive strengths up to 1400 MPa (203 ksi) have been reported for ferritic annealed CG irons (Ref 8).

Shear Properties. For a pearlitic CG iron, the shear strength on 20 mm (0.8 in.) diam specimens was measured at 365 MPa (53 ksi), with a shear-to-tensile strength ratio of 0.97 (Ref 17). Ratios of 0.90 for SG iron and of 1.1 to 1.2 for FG iron have been reported. Materials exhibiting some ductility have ratios lower than 1.0 (Ref 8).

Modulus of Elasticity. As is evident from Fig. 5 and 14, CG irons exhibit a clear zone of proportionality, both in tension and in compression. Typical values for both static and dynamic (resonance frequency method) measurements are given in Table 2. Dynamic tests give slightly higher numbers. In general, the moduli of elasticity for CG iron are similar to those of high-strength FG irons and can even be higher as nodularity increases.

The elasticity modulus measured by the tangent method depends on the level of stress, as shown in Fig. 15. A comparison of the stress dependency of the elasticity modulus for different types of cast irons is shown in Fig. 16. Poisson's ratios of 0.27 to 0.28 have been reported for CG irons (Ref 8).

Impact Properties. While SG iron exhibits substantially greater toughness at low pearlite contents, pearlitic CG irons have impact strengths equivalent to those of SG irons (Fig. 17). Charpy impact energy measurements at 21 °C (70 °F) and −41 °C (−42 °F) showed that CG irons produced from an SG-base iron absorbed greater energy than those made from gray iron-base iron (Ref 10). This is attributed to the solute hardening effects of tramp elements in the gray iron.

The results from dynamic tear tests were similar, although greater temperature dependence was observed. A comparison of the dyamic tear energies of CG cast irons is presented in Fig. 18. It is noted that significant differences in the values obtained occur in the ferritic condition, but that equivalent values are obtained when the matrix structure is primarily pearlitic.

Studies on crack initiation and growth under impact loading conditions showed that, in general, the initiation of matrix cracking was preceded by graphite fracture at the graphite-matrix interface, or through the graphite, or both. The most dominant form of graphite fracture appeared to be that occurring along the boundaries between graphite crystallites (Ref 18). Matrix cracks were usually initiated in the ferrite by transgranular cleavage (graphite was nearly always surrounded by ferrite), although in some instances intergranular ferrite fracture appeared to be the initiating mechanism. Matrix crack propagation generally occurred by a brittle cleavage mechanism, transgranular in ferrite, and interlamellar in pearlite. In general, the impact resistance of CG irons increases with carbon equivalent and decreases with phosphorus or increasing pearlite.

As may be seen in Table 6, cerium-treated CG irons seem to exhibit a higher impact energy than magnesium-titanium-treated irons. It is thought that this may be attributed to TiC and TiCN inclusions present in the matrix of magnesium titanium treated CG irons (Ref 6).

Fatigue Strength. Because the notching effect of graphite in CG iron is considerably lower than that in FG irons, it is expected that CG iron will have higher fatigue strengths than FG iron (Table 1). Table 7 lists the fatigue strengths of five CG irons from Table 2. The as-cast ferritic (>95% ferrite) CG with a hardness of 150 HB had the highest fatigue strength and the highest fatigue-endurance ratio (fatigue strength/tensile strength). Fatigue properties for three of these CG irons with comparable endurance ratios are shown in Fig. 19. It is evident that pearlitic structures, higher nodularity, and unnotched samples resulted in better fatigue strength. The fatigue-endurance ratio was 0.46 for a ferritic matrix, 0.45 for a pearlitic matrix, and 0.44 for a pearlitic higher-nodularity CG iron (Ref 17). With fatigue notch factors (ratio of unnotched to notched fatigue strength) of 1.71 to 1.79, CG iron is almost as notch sensitive as SG iron (>1.85). Gray iron is considerably less notch sensitive, with a notch factor of less than 1.5 (Ref 6).

Statistical analysis of a number of experimental data allowed the calculation of a relationship between fatigue strength (FS) and tensile strength (TS) of CG irons (Ref 6):

$$FS \text{ (in MPa)} = (0.63 - 0.00041 \cdot TS) \cdot TS \text{ (in MPa)} \quad \text{(Eq 1)}$$

Values calculated with this equation fit well between those of FG and SG irons. The intermediate position of CG irons from this standpoint is also shown in Fig. 20.

Table 6 Impact toughness of a cerium-treated CG cast iron and two magnesium-titanium-treated CG cast irons

Iron	Structural condition and graphite type	Test temperature °C	°F	Impact bend toughness(a) J	ft · lbf	Notched-bar(b) impact toughness J	ft · lbf
Cerium-treated (150 HB iron in Table 2)	>95% ferrite (as-cast); 95% CG, 5% SG	20	68	32.1	23.7	6.5	4.8
		−20	−4	26.5	19.5	4.6	3.4
		−40	−40	26.7	19.7	5.0	3.7
Magnesium (0.018%) and titanium (0.089%) treated	100% ferrite (annealed); CG	20	68	13.5–19	10–14	5.4	4.0
Magnesium (0.017%) and titanium (0.062%) treated	Ferritic (as-cast); CG	20	68	6.8–10.2	5–7.5	3.4	2.5

(a) Unnotched 10 × 10 mm (Charpy) testpiece. (b) V-notched 10 × 10 mm (Charpy) testpiece. Source: Ref 6

Table 7 Fatigue strengths and endurance ratios for five CG irons from rotating bending tests

Matrix structure	Graphite type	Tensile strength MPa	ksi	Fatigue strength MPa	ksi	Fatigue-endurance ratio	Hardness, HB
As-cast ferrite (>95% ferrite)	95% CG	336	48.7	211	30.6	0.63	150
As-cast ferrite	CG	388	56.3	178	25.8	0.46	184
As-cast pearlite	CG	414	60	185	26.8	0.45	205
As-cast pearlite	CG + SG	473	68.6	208	30.2	0.44	217
As-cast pearlite (70% pearlite)	CG	386	56.0	186	27	0.48	...

Source: Ref 6

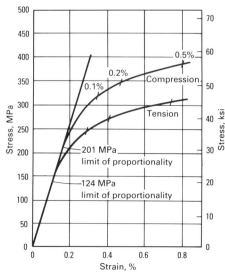

Fig. 13 Influence of section size on the tensile strength of CG iron. Source: Ref 15

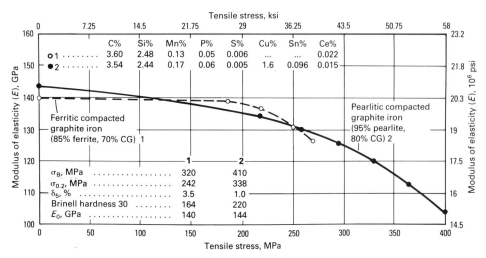

Fig. 15 Stress dependency by E-modulus for two heat-treated CG irons. Source: Ref 6

Fig. 14 Stress-strain curves in compression and tension for CG iron with 4.35 carbon equivalent. Source: Ref 9

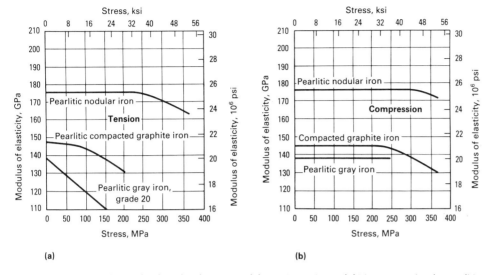

Fig. 16 Influence of stress level on the elasticity modulus (a) in tension and (b) in compression for pearlitic FG, CG, and SG irons. Source: Ref 6

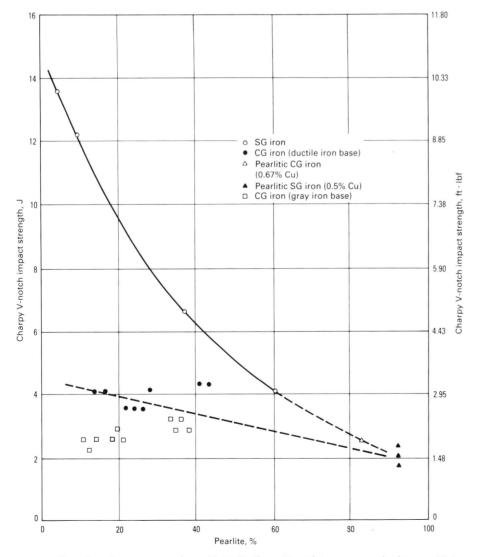

Fig. 17 Effect of pearlite content on the 21 °C (70 °F) Charpy V-notch impact strength of as-cast CG irons compared to that of SG iron. Source: Ref 10

A 36% decrease of the alternating bending fatigue strength was observed on unnotched bars with casting skin compared to machined bars. This compares with a 50% reduction for comparable strength FG iron and a 32% reduction for ferritic SG iron.

Elevated-Temperature Properties

Tensile Properties. The variation of tensile properties with temperature for CG iron produced with cerium-mischmetal treatment alloys is similar to that typical for SG iron (Fig. 21), but the values are somewhat lower (Ref 19). Similar results are reported for CG irons produced with Mg-Ti-ferrosilicon alloys shown in Fig. 22. As expected, a slight increase in nodularity led to higher tensile strength values at all temperatures.

Growth and Scaling. Tests conducted for 32 weeks in air have shown that at 500 °C (930 °F) the growth and scaling of CG iron was not significantly different from that exhibited by FG irons of similar composition. However, at 600 °C (1110 °F), the growth of CG irons was less than that of FG iron, and scaling resistance was superior (Fig. 23).

In other oxidation studies of cast irons conducted at 600 °C (1110 °F), it was concluded that weight gains due to oxidation are 10 to 15% higher for CG irons than for SG irons, but 30 to 60% lower for CG irons than for FG irons (Ref 21).

Thermal Fatigue. When castings are used in an environment where frequent changes in temperature occur, or where temperature differences are imposed on a part, thermal stresses occur in castings and may result in elastic and plastic strains and finally in crack formation. The casting can thus be destroyed as a result of thermal fatigue. Changes in microstructure, associated with stress-inducing volume changes, as well as surface and internal oxidation, may also be associated with temperature difference induced stresses.

The interpretation of thermal fatigue tests is complicated by the many different test methods employed by various investigators. The two widely accepted methods are constrained thermal fatigue and finned-disk thermal shock tests (Ref 22, 23).

In the constrained thermal fatigue test, a specimen (see Fig. 24a for dimensions) is mounted between two stationary plates that are held rigid by two columns, heated by high frequency (450 kHz) induction current, and cooled by conduction of heat to water-cooled grips (Fig. 24b). The thermal stress that develops in the test specimen is monitored by a load cell installed in one of the grips holding the specimen. During thermal cycling, compressive stresses develop upon heating, and tensile stresses develop upon cooling. As thermal cycling continues, the

Table 8 Thermal conductivity of structural constituents in iron-base alloys

Structural constituents	Thermal conductivity, W/(cm · K)		
	0–100 °C (32–212 °F)	500 °C (930 °F)	1000 °C (1830 °F)
Graphite			
Parallel to basal plane	2.93–4.19	0.84–1.26	0.42–0.63
Perpendicular to basal plane	~0.2		
Matrix			
Ferrite	0.71–0.80	0.42	0.29
Pearlite	0.50	0.44	
Cementite	0.071–0.084		

Source: Ref 1

Table 9 Thermal conductivities of FG, CG, and SG irons at various temperatures

Graphite shape	Carbon equivalent	Thermal conductivity, W/m · K (Btu/ft · h · °F)				
		100 °C (212 °F)	200 °C (390 °F)	300 °C (570 °F)	400 °C (750 °F)	500 °C (930 °F)
Flake	3.8	50.24 (29.02)	48.99 (28.30)	45.22 (26.12)	41.87 (24.19)	38.52 (22.25)
	4.8	53.39 (30.84)	50.66 (29.27)	47.31 (27.33)	43.12 (24.91)	38.94 (22.49)
Compacted	3.9	38.10 (22.01)	41.0 (23.69)	39.40 (22.76)	37.30 (21.55)	35.20 (20.34)
	4.1	43.54 (25.15)	43.12 (24.91)	40.19 (23.22)	37.68 (21.77)	35.17 (20.32)
Spheroidal	4.2	32.34 (18.68)	34.75 (20.08)	33.08 (19.11)	31.40 (18.14)	29.31 (16.93)

Source: Ref 9

specimen accumulates fatigue damage in a fashion similar to that in mechanical fatigue testing; ultimately, the specimen fails by fatigue. Initially the specimen develops compressive stress upon heating due to constrained thermal expansion (Fig. 25). Some yielding and stress relaxation occur during holding at 540 °C (1000 °F), and upon subsequent cooling the specimen develops residual tensile stress. During subsequent thermal cycling, the maximum compressive stress that has developed upon heating decreases continuously, and the maximum tensile stress upon cooling increases, as shown for six different irons in Fig. 26.

Experimental results (Fig. 27) point to higher thermal fatigue for CG iron than for FG iron and also indicate the beneficial effect of molybdenum. In fact, regression analysis of experimental results indicates that the main factors influencing thermal fatigue are tensile strength (TS) and molybdenum content:

$$\log N = 0.934 + 0.026 \cdot TS + 0.861 \cdot Mo \quad (Eq\ 2)$$

where N is the number of thermal cycles to failure, tensile strength is in kps per square inch (ksi), and molybdenum is in percent.

In the finned-disk thermal shock test, the specimen (see Fig. 28a for dimensions) is cycled between a moderate-temperature environment and a high-temperature environment, which causes thermal expansion and contraction. The thermal shock test apparatus is shown in Fig. 28(b). Because in this type of test thermal conductivity plays a significant role, FG iron showed much greater resistance to cracking than did CG iron. Major cracking occurred in less than 200 cycles in all CG iron specimens, while the unalloyed FG iron developed minor cracking after 500 cycles and major cracking after 775 cycles. The alloyed FG iron, because of its higher elevated-temperature strength, did not show any sign of cracking even after 2000 cycles (Ref 23). The CG iron containing more ferrite had a slightly better thermal fatigue resistance than the CG iron with less ferrite.

In general, for good resistance to thermal fatigue, cast irons must have high thermal conductivity; low modulus of elasticity; high strength at room and elevated temperatures; and, for use above 500 to 550 °C (930 to 1020 °F), resistance to oxidation and structural change. The relative ranking of irons varies with test conditions. When high cooling rates are encountered, experimental data and commercial experience show that thermal conductivity and a low modulus of elasticity are most important. Consequently, gray irons of high carbon content (3.6 to 4%) are superior (Ref 22, 23). When intermediate cooling rates exist, ferritic SG and CG irons have the highest resistance to cracking, but are subject to distortion. When low cooling rates exist, high-strength pearlitic SG irons or SG irons alloyed with silicon and molybdenum are best with regard to cracking and distortion (Fig. 29).

A rather detailed analysis of the behavior of various irons at elevated temperatures is given in Ref 6. Extensive experimental work on cylinder heads is reviewed. A critical analysis of most of the accepted criteria for assessing the quality of irons for castings used at elevated temperatures is also included.

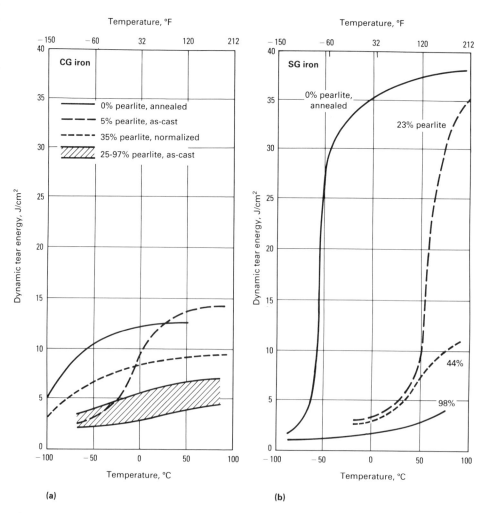

Fig. 18 Dynamic tear energy versus temperature for (a) CG and (b) SG irons. Source: Ref 10

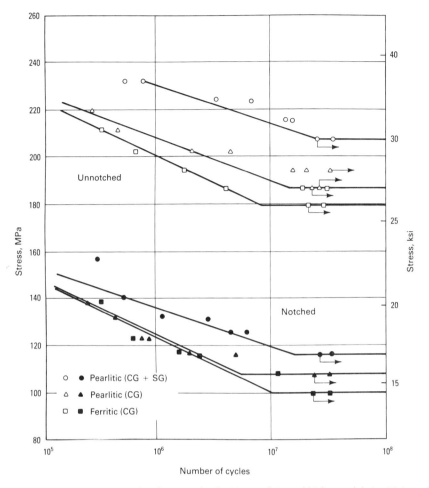

Fig. 19 Fatigue curves in rotating bending tests for ferritic, pearlitic, and higher nodularity CG irons from Table 7. Source: Ref 17

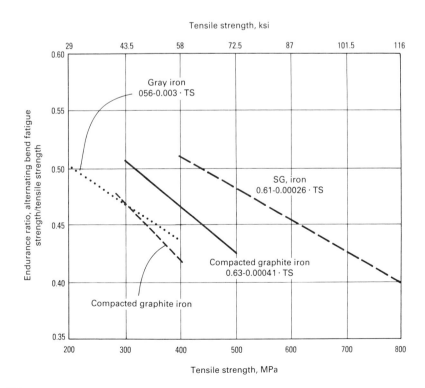

Fig. 20 Ratio of alternating bend fatigue strength/tensile strength of FG, CG, and SG irons. Source: Ref 6

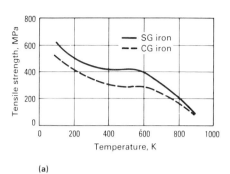

(a)

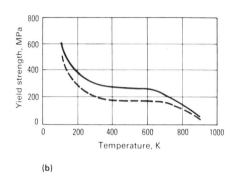

(b)

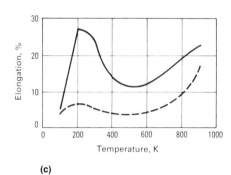

(c)

Fig. 21 Variation of tensile properties of Ce-treated CG and SG irons with temperature. Source: Ref 19

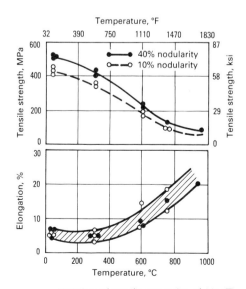

Fig. 22 Variation of tensile properties of Mg+Ti-treated CG irons. Source: Ref 20

66 / Cast Irons

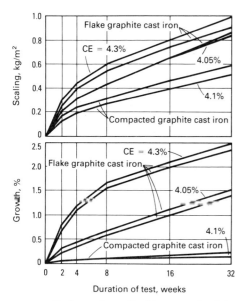

Fig. 23 Scaling and growth of heavy section flake and compacted graphite cast irons at 600 °C (1110 °F). Source: Ref 9

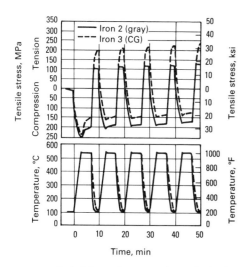

Fig. 25 Typical thermal stress cycles at the beginning of the test for FG and CG irons. Source: Ref 23

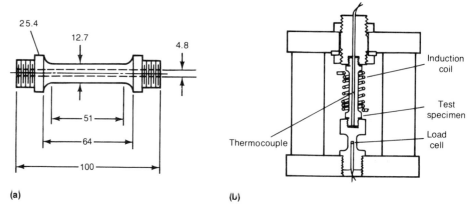

Fig. 24 (a) Dimensions of constrained fatigue test specimen. (b) Schematic of apparatus for constrained fatigue tests. Dimensions given in millimeters. Source: Ref 23

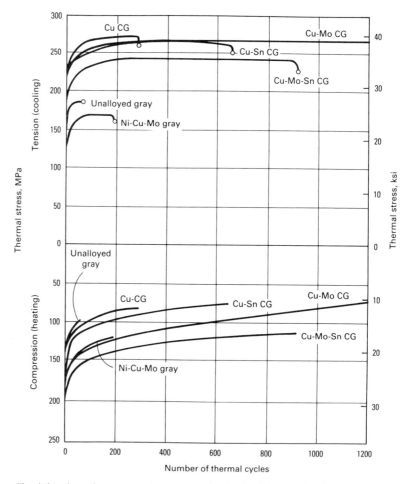

Fig. 26 The shift in thermal stress versus the number of cycles for six irons cycled between 100 and 540 °C (212 and 1000 °F). Source: Ref 23

Physical Properties

Thermal conductivity plays a significant role in structural components subjected to thermal stress. The higher the thermal conductivity, the lower the thermal gradients throughout the casting, and therefore the lower the thermal stresses. The microstructure of cast iron, and especially graphite morphology, greatly influence thermal conductivity, as implied by the data shown in Table 8. Graphite exhibits the highest thermal conductivity of all the metallographic constituents. The conductivity of graphite parallel to the basal plane is about four times higher than that perpendicular to its basal plane (Ref 24). Consequently, FG has higher thermal conductivity than SG (Fig. 30). It is therefore expected that FG iron will have higher thermal conductivity than SG iron, which in turn will be better than that of steel, as shown in Fig. 31 (Ref 26). Not unexpectedly, as the amount of graphite increases, thermal conductivity is also improved.

The thermal conductivity of ferrite is reduced by dissolved alloying elements. For steel, the conductivity of the matrix can be calculated by the equation (Ref 6):

$$\lambda = \lambda_0 - \ln\Sigma C \quad \text{(Eq 3)}$$

where λ is the thermal conductivity of alloyed steel, λ_0 is the thermal conductivity of unalloyed steel, and ΣC is the sum of alloying elements in %. This equation can also be used to estimate the influence of various

Compacted Graphite Iron / 67

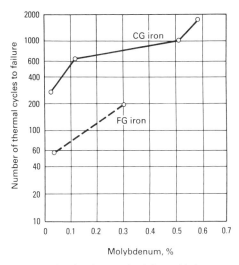

Fig. 27 Results of constrained thermal fatigue tests conducted between 100 and 540 °C (212 and 1000 °F). Source: Ref 23

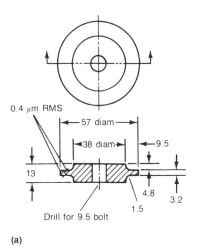

Fig. 28 (a) Dimensions (in millimeters) of finned-disk specimen. (b) Schematic of apparatus for finned-disk thermal shock test. Source: Ref 23

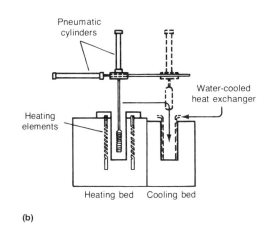

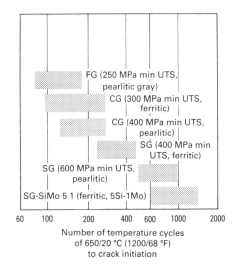

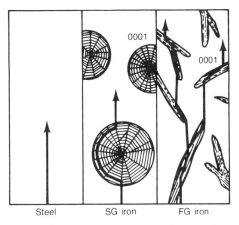

Fig. 30 Mechanism of heat conduction in various Fe-C alloys. Source: Ref 25

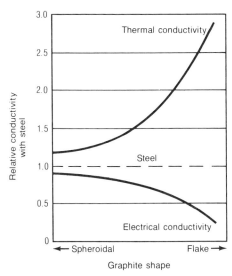

Fig. 31 Influence of graphite shape on the relative thermal and electrical conductivities of Fe-C alloys. Source: Ref 26

Composition					
C	Si	Mn	P	Mg	Other
2.96	2.90	0.78	0.66	...	0.12 Cr
3.52	2.61	0.25	0.051	0.015	...
3.52	2.25	0.40	0.054	0.015	1.47 Cu
3.67	2.55	0.13	0.060	0.030	...
3.60	2.34	0.50	0.053	0.030	0.54 Cu
3.48	4.84	0.31	0.067	0.030	1.02 Mo

Fig. 29 Results of thermal fatigue tests on various cast irons; specimens cycled between 650 and 20 °C (1200 and 70 °F). Source: Ref 22

alloying additions on the conductivity of cast irons.

Typical values for the thermal conductivity of CG iron at room temperature are given in Table 2, and various cast irons are compared in Table 9. From this last table it can be seen that the thermal conductivity of CG iron is very close to that of gray cast iron and considerably higher than that of SG iron (Ref 1, 9). This behavior is explained by the fact that much like flake graphite, compacted graphite is intercon- nected. As for FG irons, increasing the carbon equivalent results in higher thermal conductivity for CG iron. As the temperature is increased, the thermal conductivity reaches a maximum at about 200 °C (390 °F), an effect also shown by SG irons, but not by FG iron (Fig. 32). The thermal conductivity of a typical CG iron mold is compared in Fig. 33 with results for an ingot mold and bottom plate made from FG iron and a sample of ferritic SG iron (Ref 27). As previously implied, increased nodularity results in lower thermal conductivity (Ref 1, 28).

Thermal Expansion. For irons of similar chemical composition, there seems to be no difference in total expansion regardless of graphite shape (Ref 28). However, when different compositions are used in order for an iron to fall in a typical range for a given type of cast iron, the linear expansion of CG iron is between that of SG and FG irons (Fig. 34) (Ref 1).

Sonic and Ultrasonic Properties. Resonant frequency (sonic testing) and ultrasonic velocity measurements provide reliable methods for verifying the structure and properties of castings. As shown in Fig. 35, ultrasonic velocity is directly related to nodularity. Unfortunately it is rather difficult to distinguish between CG and low-nodularity SG irons. Better results seem to be obtained when ultrasonic velocity is related to tensile strength. Figure 36 shows the correlation between tensile strength and ultrasonic velocity or resonant frequency for test bars of 30 mm (1.2 in.) diameter.

When these tests are applied to castings, the ultrasonic velocity for CG structures is independent of the shape of the casting, but should be calibrated for the section thickness. Thus, for 30 mm (1.2 in.) diam bars, the range associated with CG is between 5.2 and 5.45 km/s, but for very large castings

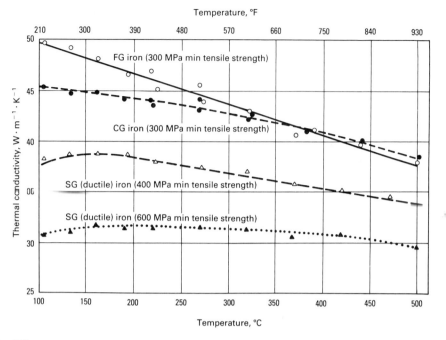

Fig. 32 Thermal conductivities of various cast irons. Source: Ref 1

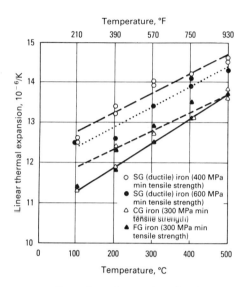

Fig. 34 Linear thermal expansion of various cast irons. Source: Ref 1

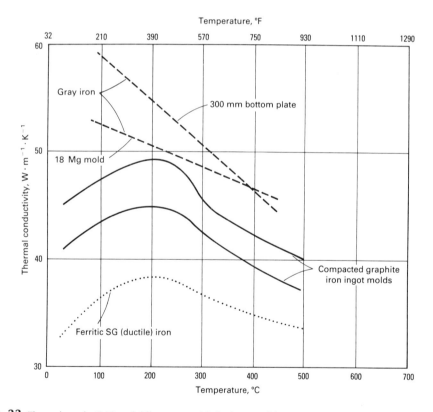

Fig. 33 Thermal conductivities of different materials for ingot molds. Source: Ref 27

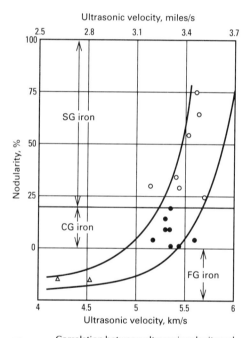

Fig. 35 Correlation between ultrasonic velocity and nodularity. Source: Ref 13

such as ingot molds, the ultrasonic velocity for good CG structures lies between 4.85 and 5.10 km/s. Sonic testing, on the other hand, must be calibrated for a particular design of casting, for which examples of satisfactory and unsatisfactory structures must be previously checked to provide a calibration range (Ref 9).

Other Properties

Corrosion Resistance. At room temperature, the corrosion rate of CG iron in 5% sulfuric acid is nearly half that of FG iron but higher than that of SG iron (Fig. 37). With increasing temperature, the difference becomes smaller. The pearlitic matrix has higher corrosion resistance than the ferritic one. As expected, corrosion accelerates when stress is applied (Ref 29). Detailed information on the corrosion resistance of cast irons is available in the article "Corrosion of Cast Irons" in Volume 13 of the 9th Edition of *Metals Handbook*.

Machinability. Standardized machinability tests comparing CG irons with other castings are difficult to find in the literature. The results of drill tests on castings shown in Fig. 38 seem to indicate that the machinability is similar to that of ductile iron and that the wear of the drill is greater than for FG iron.

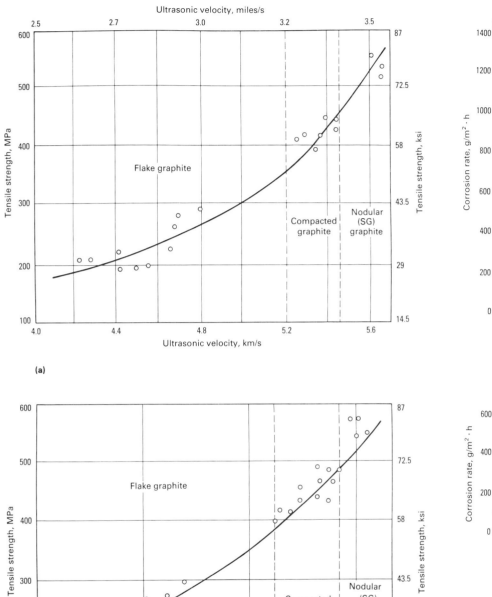

Fig. 36 Tensile strength related to (a) ultrasonic velocity and (b) resonant frequency for cast irons of varying graphite structures. Source: Ref 9

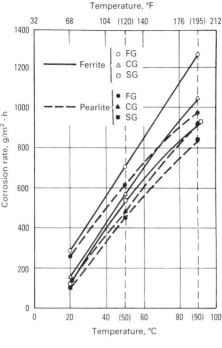

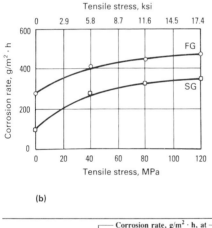

Iron type and matrix	Corrosion rate, g/m² · h, at		
	20 °C (68 °F)	50 °C (120 °F)	90 °C (195 °F)
FG iron			
Ferrite	280	587	1280
Pearlite	262	590	980
CG iron			
Ferrite	142	563	1037
Pearlite	139	478	920
SG iron			
Ferrite	123	536	923
Pearlite	99	480	842

Fig. 37 Influence of (a) temperature and (b) tensile stress on the corrosion behavior of various cast irons in 5% sulfuric acid. Source: Ref 29

Nevertheless, in general, from both experimental data and practical experience in machine shops, it can be concluded that for a given matrix the machinability of CG iron is between that of gray and ductile iron (Ref 1, 9, 21). The CG morphology makes the iron sufficiently brittle for machine swarf to break into small chips, yet strong enough to prevent the swarf from forming powdery chips. Neither large swarf nor fine, powdery swarf is ideal for high machinability (Ref 10).

Damping Capacity. The relative damping capacity of various irons, obtained by measuring the relative rates at which the amplitude of an imposed vibration decreases with time, Ref 17, is:

FG iron:CG iron:SG iron = 1.0:0.6:0.34 (Eq 4)

Apparently, changes in the carbon equivalent or matrix do not significantly influence the damping capacity, but heavier sections will produce higher damping capacities (Ref 9).

Applications

The applications of CG irons stem from their relative intermediate position between FG and SG irons. Compared to FG irons, CG irons have certain advantages:

- Higher tensile strength at the same carbon equivalent, which reduces the need for expensive alloying elements such as nickel, chromium, copper, and molybdenum
- Higher tensile strength to hardness ratio
- Much higher ductility and toughness, which result in a higher safety margin against fracture
- Lower oxidation and growth at high temperatures
- Less section sensitivity for heavy sections

Compared to SG irons, certain advantages can be claimed for CG irons:

- Lower coefficient of thermal expansion
- Higher thermal conductivity
- Better resistance to thermal shock
- Higher damping capacity
- Better castability, leading to higher casting yield, and the capability for pouring more intricate castings
- Improved machinability

CG iron can be substituted for FG iron in all cases in which the strength of FG iron has become insufficient, but in which a change to SG iron is undesirable because of the less favorable casting properties of the latter. Examples include bed plates for large diesel engines, crankcases, gearbox housings, turbocharger housings, connecting forks, bearing brackets, pulleys for truck servodrives, sprocket wheels, and eccentric gears.

Because the thermal conductivity of CG iron is higher than that of SG iron, CG iron is preferred for castings operating at elevated temperature and/or under thermal fatigue conditions. Applications include ingot molds, crankcases, cylinder heads, exhaust manifolds, and brake disks.

The largest industrial application by weight of CG iron produced is for ingot molds weighing up to 54 Mg (60 tons). According to a number of reports summarized in Ref 30, the life of ingot molds made of CG iron is 20 to 70% longer than the life of those made of FG iron.

In the case of cylinder heads, it was possible to increase engine output by 50% by changing from alloyed FG iron to ferritic CG iron (Ref 1). The specified minimum values for cylinder heads are 300 MPa (43 ksi) tensile strength, 240 MPa (35 ksi) yield strength, and 2% elongation.

Modern car and truck engines require that manifolds work at temperature ranges of 500 °C (930 °F). At this temperature, FG iron manifolds are prone to cracking, while SG iron manifolds tend to warp. CG iron manifolds warp and oxidize less and thus have a longer life. Other engineering applications are summarized in Ref 1 and 30.

REFERENCES

1. E. Nechtelberger, H. Puhr, J.B. von Nesselrode, and A. Nakayasu, Paper 1 presented at the 49th International Foundry Congress, International Committee of Foundry Technical Associations, Chicago, 1982
2. H.H. Cornell and C.R. Loper, Jr., *Trans. AFS*, Vol 93, 1985, p 435
3. R. Elliott, *Cast Iron Technology*, Butterworths, 1988
4. D.M. Stefanescu, I. Dinescu, S. Craciun, and M. Popescu, "Production of Vermicular Graphite Cast Irons by Operative Control and Correction of Graphite Shape," Paper 37 presented at the 46th International Foundry Congress, Madrid, 1979
5. D.M. Stefanescu, F. Martinez, and I.G. Chen, *Trans. AFS*, Vol 91, 1983, p 205
6. E. Nechtelberger, *The Properties of Cast Iron up to 500 °C*, Technicopy Ltd., 1980
7. J. Sissener, W. Thury, R. Hummer, and E. Nechtelberger, *AFS Cast Met. Res. J.*, 1972, p 178
8. C.F. Walton and T.J. Opar, Ed., *Iron Castings Handbook*, Iron Casting Society Inc., 1981
9. G.F. Sergeant and E.R. Evans, The British Foundryman, May 1978, p 115
10. K.P. Cooper and C.R. Loper, Jr., *Trans. AFS*, Vol 86, 1978, p 241
11. F. Martinez and D.M. Stefanescu, *Trans. AFS*, Vol 91, 1983, p 593
12. K.R. Ziegler and J.F. Wallace, *Trans. AFS*, Vol 92, 1984, p 735
13. J. Fowler, D.M. Stefanescu, and T. Prucha, *Trans. AFS*, Vol 92, 1984, p 361
14. R.B. Gundlach, *Trans. AFS*, Vol 86, 1978, p 551
15. *Spravotchnik po Tchugunomu Ljitiu (Cast Iron Handbook)*, 3rd ed., Mashinostrojenie, 1978
16. K.H. Riemer, *Giesserei*, Vol 63 (No. 10), 1976, p 285
17. K.B. Palmer, *BCIRA J.*, Report 1213, Jan 1976, p 31
18. A.F. Heiber, *Trans. AFS*, Vol 87, 1979, p 569
19. K. Hutterbraucker, O. Vohringer, and E. Macherauch, *Giessereiforschung*, No. 2, 1978, p 39
20. D.M. Stefanescu and G. Niculescu, unpublished research
21. I. Riposan, M. Chisamera, and L. Sofroni, *Trans. AFS*, Vol 93, 1985, p 35
22. K. Roehrig, *Trans. AFS*, Vol 86, 1978, p 75
23. Y.J. Park, R.B. Gundlach, R.G. Thomas, and J.F. Janowak, *Trans. AFS*, Vol 93, 1985, p 415
24. E. Mayer-Rassler, *Giesserei*, Vol 54 (No. 13), 1967, p 348
25. H. Kempers, *Giesserei*, Vol 53 (No. 1), 1966, p 15
26. K. Lohberg and J. Motz, *Giesserei*, Vol 44 (No. 11), 1957, p 305
27. P.A. Green and A.J. Thomas, *Trans. AFS*, Vol 87, 1979, p 569
28. R.W. Monroe and C.E. Bates, *Trans. AFS*, Vol 93, 1985, p 615
29. A.E. Krivosheev, B.V. Marintchenkov, and N.M. Fettisov, *Russ. Casting Prod.*, 1973, p 86
30. D.M. Stefanescu and C.R. Loper, Jr., *Giesserei-Prax.*, No. 5, 1981, p 74

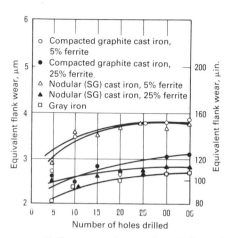

Fig. 38 Drill wear on various cast irons (drill speed, 780 rpm; rate of feed, 72 mm/min). Source: Ref 9

Malleable Iron

MALLEABLE IRON is a type of cast iron that has most of its carbon in the form of irregularly shaped graphite nodules instead of flakes, as in gray iron, or small graphite spherulites, as in ductile iron. Malleable iron is produced by first casting the iron as a white iron and then heat treating the white cast iron to convert the iron carbide into the irregularly shaped nodules of graphite. This form of graphite in malleable iron is called temper carbon because it is formed in the solid state during heat treatment.

Malleable iron, like ductile iron, possesses considerable ductility and toughness because of its combination of nodular graphite and a low-carbon metallic matrix. Consequently, malleable iron and ductile iron are suitable for some of the same applications requiring good ductility and toughness, with the choice between malleable and ductile iron based on economy and availability rather than properties. However, because solidification of white iron throughout a section is essential in the production of malleable iron, ductile iron has a clear advantage when the section is too thick to permit solidification as white iron. Malleable iron castings are produced in section thicknesses ranging from about 1.5 to 100 mm ($^{1}/_{16}$ to 4 in.) and in weights from less than 0.03 to 180 kg ($^{1}/_{16}$ to 400 lb) or more.

Ductile iron also has clear advantages over malleable iron when low solidification shrinkage is needed. In other applications, however, malleable iron has a distinct advantage over ductile iron. Malleable iron is preferred in the following applications:

- Thin-section casting
- Parts that are to be pierced, coined, or cold formed
- Parts requiring maximum machinability
- Parts that must retain good impact resistance at low temperatures
- Parts requiring wear resistance (martensitic malleable iron only)

Malleable iron (and ductile iron as well) also exhibits high resistance to corrosion, excellent machinability, good magnetic permeability, and low magnetic retention for magnetic clutches and brakes. The good fatigue strength and damping capacity of malleable iron are also useful for long service in highly stressed parts.

Metallurgical Factors

Although variations in heat treatment can produce malleable irons with different matrix microstructures (that is, ferritic, tempered pearlitic, tempered martensitic, or bainitic microstructures), the common feature of all malleable irons is the presence of uniformly dispersed and irregularly shaped graphite nodules in a given matrix microstructure. These graphite nodules, known as temper carbon, are formed by annealing white cast iron at temperatures that allow the decomposition of cementite (iron carbide) and the subsequent precipitation of temper carbon.

The desired formation of temper carbon in malleable irons has two basic requirements. First, graphite should not form during the solidification of the white cast iron, and second, graphite must also be readily formed during the annealing heat treatment. These two metallurgical requirements influence the useful compositions of malleable irons and the melting, solidification, and annealing procedures (see the article "Classification and Basic Metallurgy of Cast Iron" in this Volume for an introduction to the metallurgy of malleable iron). Metallurgical control is based on the following criteria:

- Produce solidified white iron throughout the section thickness
- Anneal on an established time-temperature cycle set to minimum values in the interest of economy
- Produce the desired graphite distribution (nodule count) upon annealing

Changes in melting practice or composition that would satisfy the first requirement listed above are generally opposed to satisfaction of the second and third, while attempts to improve annealability beyond a certain point may result in an unacceptable tendency for the as-cast iron to be mottled instead of white.

Composition. Because of the two metallurgical requirements described above, malleable irons involve a limited range of chemical composition and the restricted use of alloys. The chemical composition of malleable iron generally conforms to the ranges given in Table 1. Small amounts of chromium (0.01 to 0.03%), boron (0.0020%), copper (~1.0%), nickel (0.5 to 0.8%), and molybdenum (0.35 to 0.5%) are also sometimes present.

The common elements in malleable iron are generally controlled within about ±0.05 to ±0.15%. A limiting minimum carbon content is required in the interest of mechanical quality and annealability because decreasing carbon content reduces the fluidity of the molten iron, increases shrinkage during solidification, and reduces annealability. A limiting maximum carbon content is imposed by the requirement that the casting be white as-cast. The range in silicon content is limited to ensure proper annealing during a short-cycle high-production annealing process and to avoid the formation of primary graphite (known as mottle) during solidification of the white iron. Manganese and sulfur contents are balanced to ensure that all sulfur is combined with manganese and that only a safe, minimum quantity of excess manganese is present in the iron. An excess of either sulfur or manganese will retard annealing in the second stage and therefore increase annealing costs. The chromium content is kept low because of the carbide-stabilizing effect of this element and because it retards both the first-stage and second-stage annealing reactions.

A mixture of gray iron and white iron in variable proportions that produces a mottled (speckled) appearance is particularly damaging to the mechanical properties of the annealed casting, whether ferritic or pearlitic malleable iron. Primary control of mottle is achieved by maintaining a balance of carbon and silicon contents.

Because economy and castability are enhanced when the carbon and silicon contents of the base iron are in the higher

Table 1 Typical compositions for malleable iron

Element	Composition, % Ferritic	Pearlitic
Total carbon	2.2–2.9	2.0–2.9
Silicon	0.9–1.9	0.9–1.9
Manganese	0.2–0.6	0.2–1.3
Sulfur	0.02–0.2	0.05–0.2
Phosphorus	0.02–0.2	0.02–0.2

Source: Ref 1

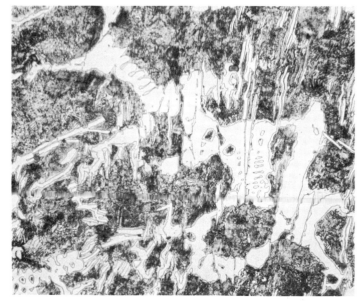

Fig. 1 Structure of as-cast malleable white iron showing a mixture of pearlite and eutectic carbides. 400×

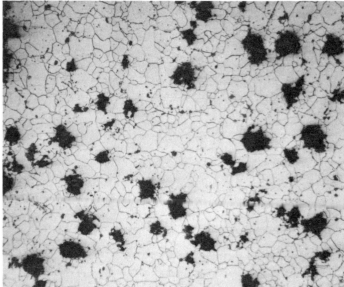

Fig. 2 Structure of annealed ferritic malleable iron showing temper carbon in ferrite. 100×

proportions of their respective ranges, some malleable iron foundries produce iron with carbon and silicon contents at levels that might produce mottle and then add a balanced, mild carbide stabilizer to prevent mottle during casting. Bismuth and boron in balanced amounts accomplish this control. A typical addition is 0.01% Bi (as metal) and 0.001% B (as ferroboron). Bismuth retards graphitization during solidification; small amounts of boron have little effect on graphitizing tendency during solidification, but accelerate carbide decomposition during annealing. The balanced addition of bismuth and boron permits the production of heavier sections for a given base iron or the utilization of a higher-carbon higher-silicon base iron for a given section thickness.

Tellurium can be added in amounts from 0.0005 to 0.001% to suppress mottle. Tellurium is a much stronger carbide stabilizer than bismuth during solidification, but also strongly retards annealing if the residual exceeds 0.003%. Less than 0.003% residual tellurium has little effect on annealing, but has a significant influence on mottle control. Tellurium is more effective if added together with copper or bismuth.

Residual boron should not exceed 0.0035% in order to avoid module alignment and carbide formation. Also, the addition of 0.005% Al to the pouring ladle significantly improves annealability without promoting mottle.

Melting Practices. (Ref 2). The iron for most present-day malleable iron is melted in coreless induction furnaces rather than the previous air furnace, cupola-air furnace, or cupola-electric furnace systems. The sulfur and nitrogen contents of the charge carbon used in melting must be high enough to provide 0.07 to 0.09% S and 80 to 120 ppm N in the iron. The sulfur reduces the surface tension and improves fluidity. The nitrogen increases the tensile strength without impairing elongation and toughness. Long holding periods in the molten state in the furnace and excessive superheating temperatures should be avoided, because they give rise to an unsatisfactory solidification structure, which in turn results in unsatisfactory heat-treated structures (Ref 2).

Melting can be accomplished by batch cold melting or by duplexing. Cold melting is done in coreless or channel-type induction furnaces, electric arc furnaces, or cupola furnaces. In duplexing, the iron is melted in a cupola or electric arc furnace, and the molten metal is transferred to a coreless or channel-type induction furnace for holding and pouring. Charge materials (foundry returns, steel scrap, ferroalloys, and, except in cupola melting, carbon) are carefully selected, and the melting operation is well controlled to produce metal having the desired composition and properties. Minor corrections in composition and pouring temperature are made in the second stage of duplex melting, but most of the process control is done in the primary melting furnace (Ref 2).

Molds are produced in green sand, silicate CO_2 bonded sand, or resin-bonded sand (shell molds) on equipment ranging from highly mechanized or automated machines to that required for floor or hand molding methods, depending on the size and number of castings to be produced. In general, the technology of molding and pouring malleable iron is similar to that used to produce gray iron.

Solidification. Molten iron produced under properly controlled melting conditions solidifies with all carbon in the combined form, producing the white iron structure fundamental to the manufacture of either ferritic or pearlitic malleable iron (Fig. 1). The base iron must contain balanced quantities of carbon and silicon to simultaneously provide castability, white iron in even the thickest sections of the castings, and annealability; therefore, precise metallurgical control is necessary for quality production. Thick metal sections cool slowly during solidification and tend to graphitize, producing mottled or gray iron. This is undesirable, because the graphite formed in mottled iron or rapidly cooled gray iron is generally of the type D configuration, a flake form in a dense, lacy structure, which is particularly damaging to the strength, ductility, and stiffness characteristics of both ferritic and pearlitic malleable iron.

After it solidifies and cools, the metal is in a white iron state, and gates, sprues, and feeders can be removed easily from the castings by impact. This operation, called spruing, is generally performed manually with a hammer because the diversity of castings produced in the foundry makes the mechanization or automation of spruing very difficult. After spruing, the castings proceed to heat treatment, while gates and risers are returned to the melting department for reprocessing.

First-Stage Anneal. Malleable iron castings are produced from the white iron by an annealing process that converts primary carbides into temper carbon. This initial anneal is then followed by additional heat treatments that produce the desired matrix microstructures. This section focuses on the initial (first-stage) anneal that produces the temper carbon in blackheart malleable iron. The additional heat treatments used to

Table 2 Properties of malleable iron castings
Microstructures and typical applications are given in Tables 3 and 4.

Specification No.	Class or grade	Tensile strength MPa	Tensile strength ksi	Yield strength MPa	Yield strength ksi	Hardness, HB	Elongation(a), %
Ferritic							
ASTM A 47 and A 338, ANSI G48.1, FED QQ-I-666c	32510	345	50	224	32	156 max	10
	35018	365	53	241	35	156 max	18
ASTM A 197	...	276	40	207	30	156 max	5
Pearlitic and martensitic							
ASTM A 220, ANSI G48.2, MIL-I-11444B	40010	414	60	276	40	149–197	10
	45008	448	65	310	45	156–197	8
	45006	448	65	310	45	156–207	6
	50005	483	70	345	50	179–229	5
	60004	552	80	414	60	197–241	4
	70003	586	85	483	70	217–269	3
	80002	655	95	552	80	241–285	2
	90001	724	105	621	90	269–321	1
Automotive							
ASTM A 602, SAE J158	M3210(b)	345	50	224	32	156 max	10
	M4504(c)	448	65	310	45	163–217	4
	M5003(c)	517	75	345	50	187–241	3
	M5503(d)	517	75	379	55	187–241	3
	M7002(d)	621	90	483	70	229–269	2
	M8501(d)	724	105	586	85	269–302	1

(a) Minimum in 50 mm (2 in.). (b) Annealed. (c) Air quenched and tempered. (d) Liquid quenched and tempered

Table 3 Grades of malleable iron specified according to hardness per ASTM A 602 and SAE J158
See Table 2 for mechanical properties.

Grade	Specified hardness, HB	Heat treatment	Microstructure	Typical applications
M 3210	156 max	Annealed	Ferritic	For low-stress parts requiring good machinability: steering-gear housings, carriers, and mounting brackets
M 4504	163–217	Air quenched and tempered	Ferrite and tempered pearlite(a)	Compressor crankshafts and hubs
M 5003	187–241	Air quenched and tempered	Ferrite and tempered pearlite(a)	For selective hardening: planet carriers, transmission gears, and differential cases
M 5503	187–241	Liquid quenched and tempered	Tempered martensite	For machinability and improved response to induction hardening
M 7002	229–269	Liquid quenched and tempered	Tempered martensite	For high-strength parts: connecting rods and universal-joint yokes
M 8501	269–302	Liquid quenched and tempered	Tempered martensite	For high strength plus good wear resistance: certain gears

(a) May be all tempered martensite for some applications

produce the desired matrix microstructure are discussed in the sections relating to ferritic, pearlitic, or martensitic microstructures.

During the first-stage annealing cycle, the carbon that exists in combined form, either as massive carbides or as a microconstituent in pearlite, is converted into nodules of graphite (temper carbon). The rate of annealing of a hard iron casting depends on chemical composition, nucleation tendency (discussed in the section "Control of Nodule Count" in this article), and annealing temperature. With the proper balance of boron content and graphitic materials in the charge, the optimum number and distribution of graphite nuclei are developed in the early portions of first-stage annealing, and growth of the temper carbon particles proceeds rapidly at any annealing temperature. An optimum iron will anneal completely through the first-stage reaction in approximately 3½ h at 940 °C (1720 °F). Irons with lower silicon contents or less-than-optimum nodule counts may require as much as 20 h for completion of first-stage annealing.

The temperature of first-stage annealing exercises considerable influence on the rate of annealing and the number of graphite particles produced. Increasing the annealing temperature accelerates the rate of decomposition of primary carbide and produces more graphite particles per unit volume. However, high first-stage annealing temperatures can result in excessive distortion of castings during annealing, which leads to straightening of the casting after heat treatment. Annealing temperatures are adjusted to provide maximum practical annealing rates and minimum distortion and are therefore controlled within the range of 900 to 970 °C (1650 to 1780 °F). Lower temperatures result in excessively long annealing times, while higher temperatures produce excessive distortion.

Annealing is done in high-production controlled-atmosphere continuous furnaces or batch-type furnaces, depending on production requirements. The furnace atmosphere for producing malleable iron in continuous furnaces is controlled so that the ratio of CO to CO_2 is between 1:1 and 20:1. In addition, any sources of water vapor or hydrogen are eliminated; the presence of hydrogen is thought to retard annealing, and it produces excessive decarburization of casting surfaces. Proper control of the gas atmosphere is important for avoiding an undesirable surface structure. A high ratio of CO to CO_2 retains a high level of combined carbon on the surface of the casting and produces a pearlitic rim, or picture frame, on a ferritic malleable iron part. A low ratio of CO to CO_2 permits excessive decarburization, which forms a ferritic skin on the casting with an underlying rim of pearlite. The latter condition is produced when a significant portion of the subsurface metal is decarburized to the degree that no temper carbon nodules can be developed during first-stage annealing. When this occurs, the dissolved carbon cannot precipitate from the austenite, except as the cementite plates in pearlite.

Control of Nodule Count. Proper annealing in short-term cycles and the attainment of high levels of casting quality require that controlled distribution of graphite particles be obtained during first-stage heat treatment. With low nodule count (few graphite particles per unit area or volume), mechanical properties are reduced from optimum, and second-stage annealing time is unnecessarily long because of long diffusion distances. Excessive nodule count is also undesirable, because graphite particles may become aligned in a configuration corresponding to the boundaries of the original primary cementite. In martensitic malleable iron, very high nodule counts are sometimes associated with low hardenability and nonuniform tempering. Generally, a nodule count of 80 to 150 discrete graphite particles per square millimeter of a photomicrograph magnified at 100× appears to be optimum. This produces random particle distribution, with short distances between the graphite particles.

Temper carbon is formed predominantly at the interface between primary carbide and saturated austenite at the first-stage annealing temperature, with growth around the nuclei taking place by a reaction involving diffusion and carbide decomposition. Although new nuclei undoubtedly form at the interfaces during holding at the first-stage annealing

Table 4 Grades of malleable iron specified according to minimum tensile properties
See Table 2 for hardness.

Specification No.	Class or grade(a)	ASTM metric equivalent class(b)	Microstructure	Typical applications
Ferritic				
ASTM A 47(c), ANSI G48.1, FED QQ-I-666c............	32510 35018	22010 24018	Temper carbon and ferrite	General engineering service at normal and elevated temperatures for good machinability and excellent shock resistance
ASTM A 338	(d)	...	Temper carbon and ferrite	Flanges, pipe fittings, and valve parts for railroad, marine, and other heavy-duty service to 345 °C (650 °F)
ASTM A 197, ANSI G49.1...	(e)	...	Free of primary graphite	Pipe fittings and valve parts for pressure service
Pearlitic and martensitic				
ASTM A 220(c), ANSI G48.2, MIL-I-11444B	40010 45008 45006 50005 60004 70003 80002 90001	280M10 310M8 310M6 340M5 410M4 480M3 560M2 620M1	Temper carbon in necessary matrix without primary cementite or graphite	General engineering service at normal and elevated temperatures. Dimensional tolerance range for castings is stipulated.

(a) The first three digits of the grade designation indicate the minimum yield strength (×100 psi), and the last two digits indicate minimum elongation (%). (b) ASTM specifications designated by footnote (c) provide a metric equivalent class where the first three digits indicate minimum yield strength in MPa. (c) Specifications with a suffix "M" utilize the metric equivalent class designation. (d) Zinc-coated malleable iron specified per ASTM A 47. (e) Cupola ferritic malleable iron

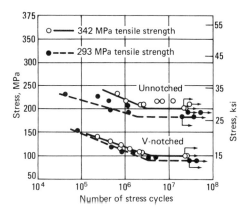

Fig. 3 Fatigue properties of two ferritic malleable irons (25 mm, or 1 in., diam bars) from bending fatigue tests on notched and unnotched specimens. The unnotched fatigue limit is about 200 MPa (29 ksi) for the iron with a 342 MPa (50 ksi) tensile strength and about 185 MPa (27 ksi) for the iron with a 293 MPa (42.5 ksi) tensile strength. Source: Ref 5

temperature, nucleation and graphitization are accelerated by the presence of nuclei that are created by appropriate melting practice. High silicon and carbon contents promote nucleation and graphitization, but these elements must be restricted to certain maximum levels because of the necessity that the iron solidify white.

Types and Properties of Malleable Iron

There are two basic types of malleable iron: blackheart and whiteheart. Blackheart malleable iron is the only type produced in North America and is the most widely used throughout the world. Whiteheart malleable iron is the older type and is essentially decarburized throughout in an extended heat treatment of white iron. This article considers only the blackheart type.

Malleable iron, like medium-carbon steel, can be heat treated to produce a wide variety of mechanical properties (Table 2). The different grades and mechanical properties are essentially the result of the matrix microstructure, which may be a matrix of ferrite, pearlite, tempered pearlite, bainite, tempered martensite, or a combination of these (all containing nodules of temper carbon). This matrix microstructure is the dominant factor influencing the mechanical properties. Other less significant factors include nodular count and the amount and compactness of the graphite (Ref 1). A higher nodular count may slightly decrease the tensile and yield strengths (Ref 3) as well as ductility (Ref 4). More graphite or a less compact form of graphite also tends to decrease strength (Ref 1).

The different microstructures of malleable irons are determined and controlled by variations in heat treatment and/or composition. Table 3, for example, lists various types of malleable irons used in automotive applications according to heat treatment and microstructure. The range of compositions for a ferritic or pearlitic microstructure is given in Table 1.

Because the mechanical properties of malleable iron are dominated by matrix microstructure, the mechanical properties may relate quite well to the relative hardness levels of different matrix microstruc-

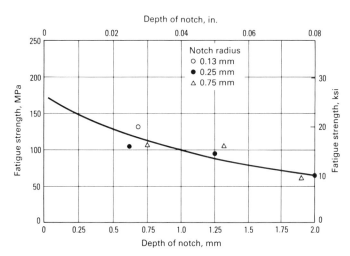

Fig. 4 Effects of notch radius and notch depth on the fatigue strength of ferritic malleable iron

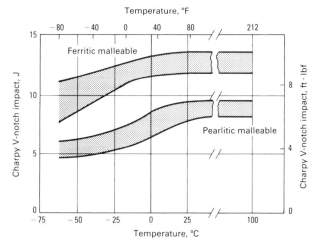

Fig. 5 Charpy V-notch transition curves for ferritic and pearlitic malleable irons. Source: Ref 1

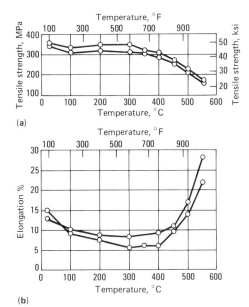

Fig. 6 Short-term high-temperature tensile properties of two ferritic malleable irons. (a) Tensile strength. (b) Elongation. Source: Ref 5

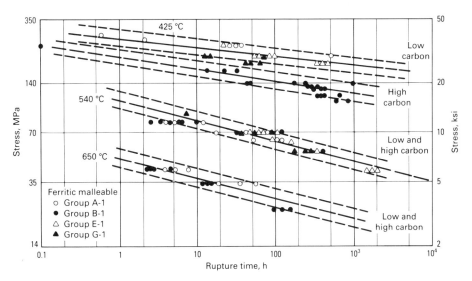

Group	Grade	Composition, %					
		C	Si	Mn	P	S	Cr
A-1	35018	2.21	1.14	0.35	0.161	0.081	...
B-1	32510	2.50	1.32	0.43	0.024	0.159	0.029
E-1	35018	2.16	1.17	0.38	0.137	0.095	0.017
G-1	35018	2.29	1.01	0.38	0.11	0.086	...

Fig. 7 Stress-rupture plot for various grades of ferritic malleable iron. The solid lines are curves determined by the method of least squares from the existing data and are least squares fit to the data. The dashed lines define the 90% symmetrical tolerance interval. The lower dashed curve defines time and load for 95% survivors, and the upper dashed curve is the boundary for 5% survivors. Normal distribution is assumed. Source: Ref 6

tures. This general effect of microstructure on malleable irons is similar to that of many other steels and irons. The softer ferritic matrix provides maximum ductility with lower strength, while increasing the amount of pearlite increases hardness and strength but decreases ductility. Martensite provides further increases in hardness and strength but with additional decreases in ductility.

The mechanical properties of pearlitic and martensitic malleable irons are closely related to hardness, as discussed in "Mechanical Properties" in the section "Pearlitic and Martensitic Malleable Irons" in this article. Therefore, grades of malleable irons are dependably specified by hardness and microstructure in ASTM A 602 and SAE J158 (Table 3). Malleable irons are also classified according to microstructure and minimum tensile properties (Table 4).

Table 2 summarizes some of the mechanical properties of the malleable irons listed in Tables 3 and 4. Additional information on the properties and heat treatment of ferritic, pearlitic, and martensitic malleable irons is provided in the following sections.

Ferritic Malleable Iron

The microstructure of ferritic malleable iron is shown in Fig. 2. A satisfactory structure consists of temper carbon in a matrix of ferrite. There should be no flake graphite and essentially no combined carbon in ferritic malleable iron. Because ferritic malleable iron consists of only ferrite and temper carbon, the properties of ferritic malleable castings depend on the quantity, size, shape, and distribution of temper carbon and on the composition of the ferrite.

Heat Treatment. Ferritic malleable iron requires a two-stage annealing cycle. The first stage converts primary carbides to temper carbon, and the second stage converts the carbon dissolved in austenite at the first-stage annealing temperature to temper carbon and ferrite.

After first-stage annealing, the castings are cooled as rapidly as practical to 740 to 760 °C (1360 to 1400 °F) in preparation for second-stage annealing. The fast cooling step requires 1 to 6 h, depending on the equipment used. Castings are then cooled slowly at a rate of about 3 to 10 °C (5 to 20 °F) per hour. During cooling, the carbon dissolved in the austenite is converted to graphite and deposited on the existing particles of temper carbon. This results in a fully ferritic matrix.

Composites. Fully annealed ferritic malleable iron castings contain 2.00 to 2.70% graphite carbon by weight, which is equivalent to about 6 to 8% by volume. Because the graphite carbon contributes nothing to the strength of the castings, those with the lesser amount of graphite are somewhat stronger and more ductile than those containing the greater amount (assuming equal size and distribution of graphite particles). Elements such as silicon and manganese in solid solution in the ferritic matrix contribute to the strength and reduce the elongation of the ferrite. Therefore, by varying base metal composition, slightly different strength levels can be obtained in a fully annealed ferritic product.

The mechanical properties that are most important for design purposes are tensile strength, yield strength, modulus of elasticity, fatigue strength, and impact strength. Hardness can be considered an approximate indicator that the ferritizing anneal was complete. The hardness of ferritic malleable iron almost always ranges from 110 to 156 HB and is influenced by the total carbon and silicon contents.

The tensile properties of ferritic malleable iron are usually measured on unmachined test bars. These properties are listed in Table 2.

The fatigue limit of unnotched ferritic malleable iron is about 50 or 60% of the tensile strength (see the two unnotched plots in Fig. 3). Figure 3 also plots the fatigue properties with notched specimens. Notch radius generally has little effect on fatigue strength, but fatigue strength decreases with increasing notch depth (Fig. 4).

The modulus of elasticity in tension is about 170 GPa (25×10^6 psi). The modulus in compression ranges from 150 to 170 GPa (22×10^6 to 25×10^6 psi); in torsion, from 65 to 75 GPa (9.5×10^6 to 11×10^6 psi).

Fracture Toughness. Because brittle fractures are most likely to occur at high strain rates, at low temperatures, and with a high restraint on metal deformation, notch tests such as the Charpy V-notch test are conducted over a range of test temperatures to

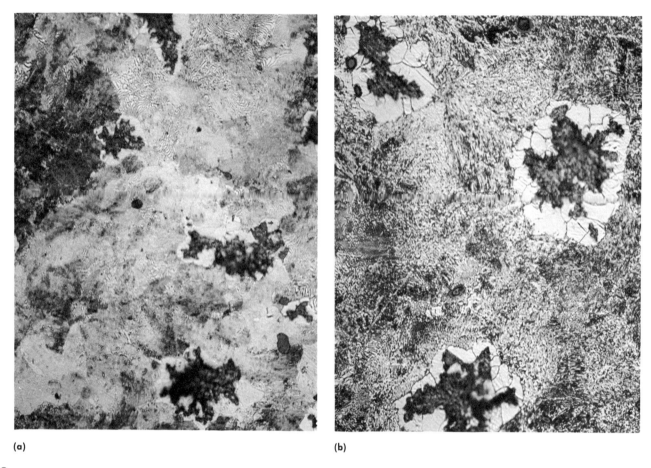

Fig. 8 Structure of air-cooled pearlitic malleable iron. (a) Slowly air cooled. 400×. (b) Cooled in an air blast. 400×

establish the toughness behavior and the temperature range of transition from ductile to a brittle fracture. Figure 5 illustrates the behavior of ferritic malleable iron and several types of pearlitic malleable iron in the Charpy V-notch test. This shows that ferritic malleable iron has a higher upper shelf energy and a lower transition temperature to a brittle fracture than pearlitic malleable iron. Additional information on the fracture toughness of malleable irons is available in the section "Pearlitic and Martensitic Malleable Iron" in this article.

Elevated-Temperature Properties. Short-term, high-temperature tensile properties typically show no significant change to 370 °C (700 °F). The short-term tensile properties of two ferritic malleable irons are shown in Fig. 6. Sustained-load stress-rupture data from 425 to 650 °C (800 to 1200 °F) are given in Fig. 7.

The corrosion resistance of ferritic malleable iron is increased by the addition of copper, usually about 1%, in certain applications, for example, conveyor buckets, bridge castings, pipe fittings, railroad switch stands, and freight-car hardware. One important use for copper-bearing ferritic malleable iron is chain links. Ferritic malleable iron can be galvanized to provide added protection. The effects of copper on the corrosion resistance of ferrous alloys are documented in Volume 13 of the 9th Edition of *Metals Handbook*.

Welding and Brazing. Welding of ferritic malleable iron almost always produces brittle white iron in the weld zone and the portion of the heat-affected zone immediately adjacent to the weld zone. During welding, temper carbon is dissolved, and upon cooling it is reprecipitated as carbide rather than graphite. In some cases, welding with a cast iron electrode may produce a brittle gray iron weld zone. The loss of ductility due to welding may not be serious in some applications. However, welding is usually not recommended unless the castings are subsequently annealed to convert the carbide to temper carbon and ferrite. Ferritic malleable iron can be fusion welded to steel without subsequent annealing if a completely decarburized zone as deep as the normal heat-affected zone is produced at the faying surface of the malleable iron part before welding. Silver brazing and tin-lead soldering can be satisfactorily used.

Pearlitic and Martensitic Malleable Iron

Pearlitic and martensitic-pearlitic malleable irons can be produced with a wide variety of mechanical properties, depending on heat treatment, alloying, and melting practices. The lower-strength pearlitic malleable irons are often produced by air cooling the casting after the first-stage anneal, while the higher-strength (pearlitic-martensitic) malleable irons are made by liquid quenching after the first-stage anneal. These two methods are discussed in the sections "Heat Treatment for Pearlitic Malleable Irons" and "Heat Treatment for Pearlitic-Martensitic Malleable Irons" in this article.

Given suitable heat treatment facilities, air cooling or liquid quenching after the first-stage anneal is generally the most economical heat treatment for producing pearlitic or martensitic-pearlitic malleable irons, respectively. Otherwise, ferritic iron produced from two-stage annealing is reheated to the austenite temperature and then quenched. This method is discussed in the section "Rehardened and Tempered Malleable Iron" in this article. Finally, the lower-strength pearlitic malleable irons can also be produced by alloying and a two-stage annealing process. The last method involves alloying during the melting process so that the carbides dissolved in the austenite do not decompose during cooling from the first-stage annealing temperature.

Heat Treatment for Pearlitic Malleable Irons. In the production of pearlitic mallea-

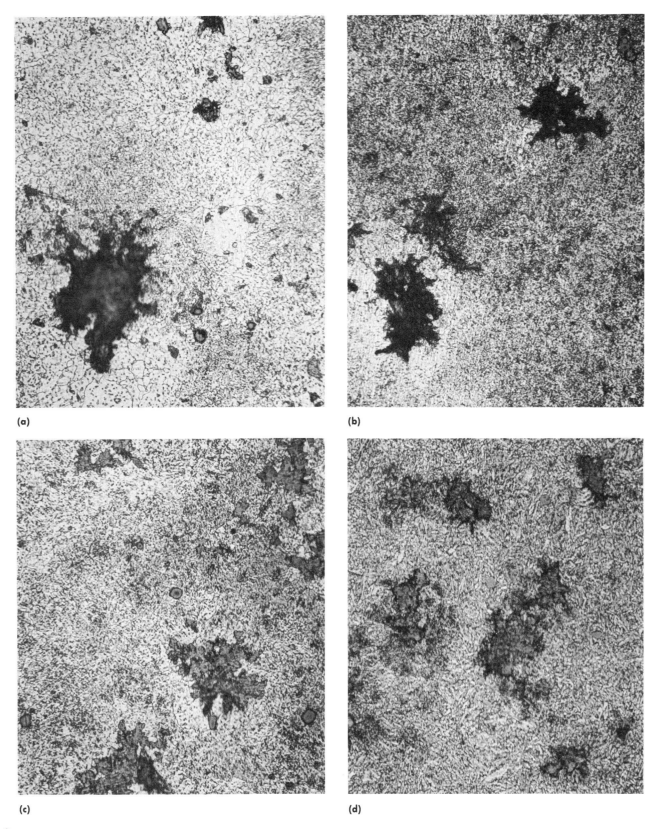

Fig. 9 Structure of oil-quenched and tempered martensitic malleable iron. (a) 163 HB. 500×. (b) 179 HB. 500×. (c) 207 HB. 500×. (d) 229 HB. 500×

ble iron, the first-stage anneal is identical to that used for ferritic malleable iron. After this, however, the process changes. Some foundries then slowly cool the castings to about 870 °C (1600 °F). During cooling, the combined carbon content of the austenite is reduced to about 0.75%, and the castings are then air cooled. Air cooling is accelerated by an air blast to avoid the formation of ferrite envelopes around the temper carbon particles (bull's-eye structure) and to produce a fine pearlitic matrix (Fig. 8). The castings are then tempered to specification, or they are reheated to reaustenitize at about 870 °C (1600 °F), oil quenched, and

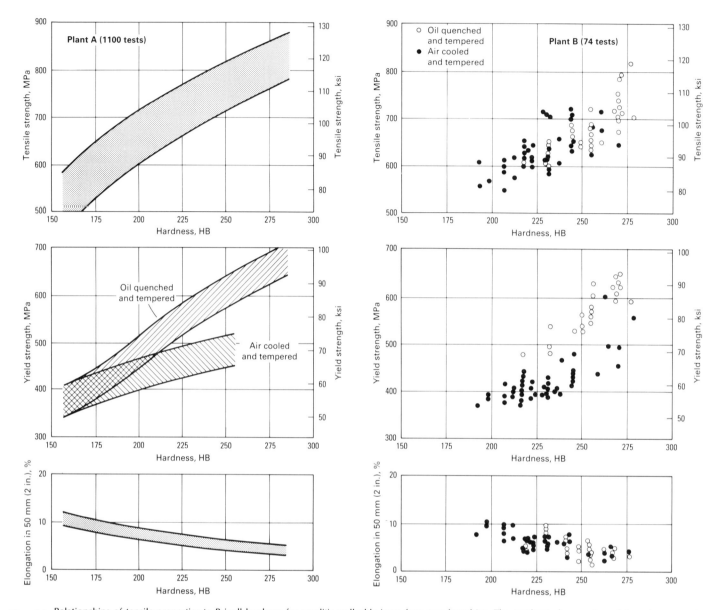

Fig. 10 Relationships of tensile properties to Brinell hardness for pearlitic malleable irons from two foundries. The mechanical properties of these irons vary in a substantially linear relationship with Brinell hardness, and in the low-hardness ranges (below about 207 HB), the properties of air-quenched and tempered material are essentially the same as those produced by oil quenching and tempering.

tempered to specification. Large foundries usually eliminate the reaustenitizing step and quench the castings in oil directly from the first-stage annealing furnace after stabilizing the temperature at 845 to 870 °C (1550 to 1600 °F).

The rate of cooling after first-stage annealing is important in the formation of a uniform pearlitic matrix in the air-cooled casting, because slow rates permit partial decomposition of carbon in the immediate vicinity of the temper carbon nodules, which results in the formation of films of ferrite around the temper carbon (bull's-eye structure). When the extent of these films becomes excessive, a carbon gradient is developed in the matrix. Air cooling is usually done at a rate not less than about 80 °C (150 °F) per minute.

Air-quenched malleable iron castings have hardnesses ranging from 269 to 321 HB, depending on casting size and cooling rate. Such castings can be tempered immediately after air cooling to obtain pearlitic malleable iron with a hardness of 241 HB or less.

Heat Treatment for Pearlitic-Martensitic Malleable Irons. High-strength malleable iron castings of uniformly high quality are usually produced by liquid quenching and tempering. The most economical procedure is direct quenching after first-stage annealing. In this procedure, the castings are cooled in the furnace to the quenching temperature of 845 to 870 °C (1550 to 1600 °F) and held for 15 to 30 min to homogenize the matrix. The castings are then quenched in agitated oil to develop a matrix microstructure of martensite having a hardness of 415 to 601 HB. Finally, the castings are tempered at an appropriate temperature between 590 and 725 °C (1100 and 1340 °F) to develop the specified mechanical properties. The final microstructure consists of tempered martensite plus temper carbon, as shown in Fig. 9. In heavy sections, higher-temperature transformation products such as fine pearlite are usually present.

Some foundries produce high-strength malleable iron by an alternative procedure in which the castings are forced-air cooled after first-stage annealing, retaining about 0.75% C as pearlite. The castings are then reheated at 840 to 870 °C (1545 to 1600 °F) for 15 to 30 min, followed by quenching and tempering as above for the direct-quench process.

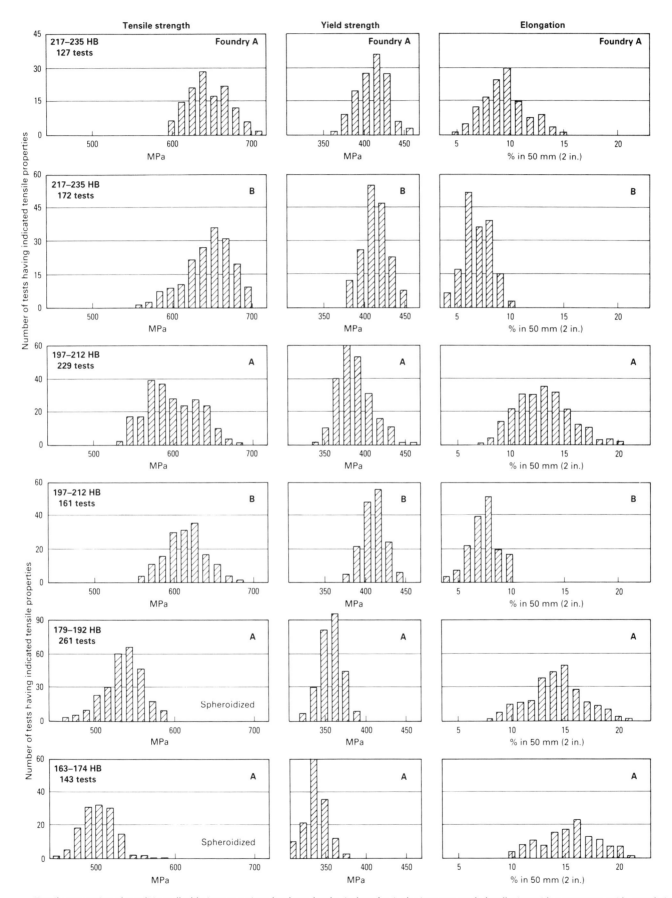

Fig. 11 Tensile properties of pearlitic malleable iron at various hardness levels. At foundry A, the iron was made by alloying with manganese, with completion of first-stage graphitization, air cooling under air blast from 938 °C (1720 °F), and subcritical tempering for spheroidizing.

80 / Cast Irons

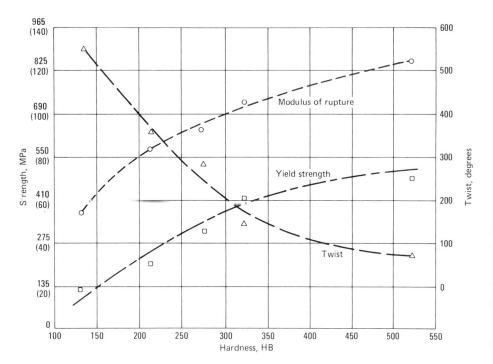

Fig. 12 Torsional properties of pearlitic malleable irons in relation to hardness. Source: Ref 1

Table 5 Fracture toughness of malleable irons

Malleable iron grade	Test temperature °C	°F	Yield strength MPa	ksi	K_{Ic} MPa $\sqrt{m}$	ksi $\sqrt{in.}$
Ferritic						
M3210	24	75	230	33	44	40
	−19	−3	240	35	42	38
	−59	−74	250	36	44	40
Pearlitic						
M4504 (normalized)	24	75	360	52	55	50
	−19	−2	380	55	48	44
	−57	−70	390	57	30	27
M5503 (quenched and tempered)	24	75	410	60	45	41
	−19	−3	440	64	52	47
	−58	−73	455	66	30	27
M7002 (quenched and tempered)	24	75	520	75	54	49
	−19	−3	550	80	38	35
	−58	−72	570	83	40	36

Source: Ref 7

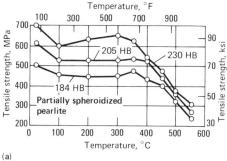

(a)

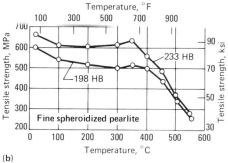

(b)

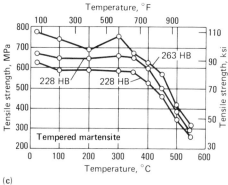

(c)

Fig. 13 Short-term elevated-temperature tensile strengths of (a) partially spheroidized pearlitic malleable irons produced by air cooling after the temper carbon anneal, (b) finely spheroidized pearlitic malleable irons produced by oil quenching after the temper carbon anneal, and (c) oil-quenched and tempered martensitic malleable irons. The two martensitic malleable irons with hardnesses of 228 HB were reheated (reaustenitized) after the temper carbon anneal (18 h soak at 950 °C, or 1740 °F) and then oil quenched. The 263 HB iron was oil quenched from 840 °C (1545 °F) after an anneal of 9.5 h at 950 °C (1740 °F). After oil quenching, all three martensitic irons were tempered. Source: Ref 5

Rehardened-and-tempered malleable iron can also be produced from fully annealed ferritic malleable iron with a slight variation in the heat treatment used for arrested-annealed (air-quenched) malleable. The matrix of fully annealed ferritic malleable iron is essentially carbon free, but can be recarburized by heating at 840 to 870 °C (1545 to 1600 °F) for 1 h. In general, the combined carbon content of the matrix produced by this procedure is slightly lower than that of arrested-annealed pearlitic malleable iron, and the final tempering temperatures required for the development of specific hardnesses are lower. Rehardened malleable iron made from ferritic malleable may not be capable of meeting certain specifications.

Tempering times of 2 h or more after either air cooling or liquid quenching are needed for uniformity. In general, the control of final hardness of the castings is precise, with process limitations approximately the same as those encountered in the heat treatment of medium- or high-carbon steels. This is particularly true when specifications require hardnesses of 241 to 321 HB where control limits of ±0.2 mm Brinell diameter can be maintained with ease. At lower hardnesses, a wider process control limit is required because of certain unique characteristics of the pearlitic malleable iron microstructure.

The mechanical properties of pearlitic and martensitic malleable iron vary in a substantially linear relationship with Brinell hardness (Fig. 10 and 11). In the low-hardness ranges, below about 207 HB, the properties of air-quenched and tempered pearlitic malleable are essentially the same as those of oil-quenched and tempered martensitic malleable. This is because attaining the low hardnesses requires considerable coarsening of the matrix carbides and partial second-stage graphitization. Either an air-quenched pearlitic structure or an oil-quenched martensitic structure can be coarsened and decarburized to meet this hardness requirement.

At higher hardnesses, oil-quenched and tempered malleable iron has higher yield strength and elongation than air-quenched and tempered malleable iron because of greater uniformity of matrix structure and finer distribution of carbide particles. Oil-

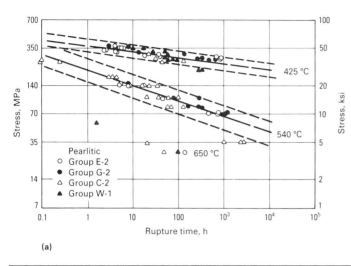
(a)

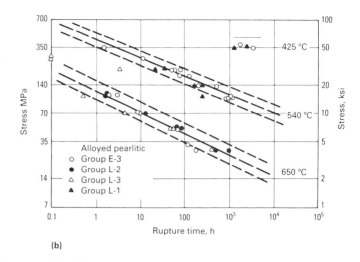
(b)

Material	Composition, %						
	C	Si	Mn	S	P	Cr	Others
Pearlitic (low carbon-high phosphorus)							
Group E-2	2.27	1.15	0.89	0.098	0.135	0.019	...
Group G-2	2.29	1.01	0.75	0.086	0.11	...	...
Pearlitic (high carbon-low phosphorus)							
Group C-2	2.65	1.35	0.41	0.15	...	0.018	0.0020 B
Group W-1	2.45	1.38	0.41	0.12	0.04	0.032	...
Alloyed pearlitic (low carbon-high phosphorus)							
Group E-3	2.21	1.13	0.88	0.110	0.122	0.021	0.47Mo,1.03Cu
Group L-1	2.16	1.18	0.72	0.120	0.128	...	0.34Mo,0.83 Ni
Group L-2	2.16	1.18	0.80	0.123	0.128	...	0.40Mo,0.62 Ni
Group L-3	2.32	1.14	0.82	0.117	0.128	...	0.38Mo,0.65 Ni

Fig. 14 Stress-rupture plot for pearlitic malleable iron (a) and alloyed pearlitic malleable iron (b). The solid lines are curves determined by the method of least squares from the existing data. The dashed lines define the 90% symmetrical tolerance interval. The lower dashed curve defines time and load for 95% survivors, and the upper dashed curve is the boundary for 5% survivors. Normal distribution is assumed. Source: Ref 6

quenched and tempered pearlitic malleable iron is produced commercially to hardnesses as high as 321 HB, while the maximum hardness for high-production air-quenched and tempered pearlitic malleable iron is about 255 HB. The lower maximum hardness is applied to the air-quenched material because:

- Hardness upon air quenching normally does not exceed 321 HB and may be as low as 269 HB; therefore, attempts to temper to a hardness range above 255 HB produce nonuniform hardness and make the process control limits excessive
- Very little structural alteration occurs during the tempering heat treatment to a higher hardness, and the resulting structure is more difficult to machine than an oil-quenched and tempered structure at the same hardness

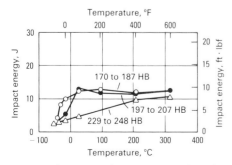

Fig. 15 Charpy V-notch impact energy of one heat of air-quenched and tempered pearlitic malleable iron

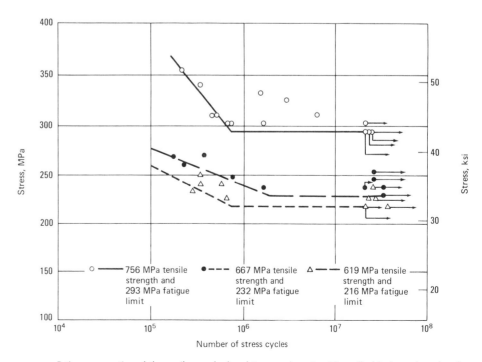

Fig. 16 Fatigue properties of three oil-quenched and tempered martensitic malleable irons from bending fatigue tests on unnotched 25 mm (1 in.) diam bars. Source: Ref 5

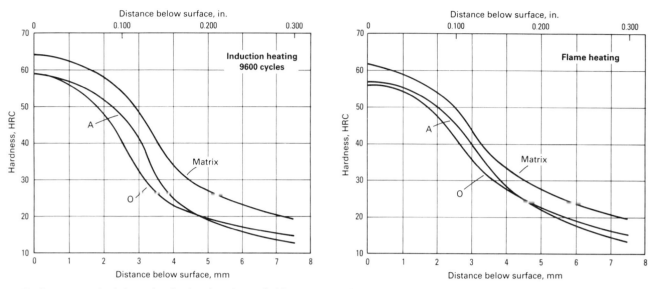

Fig. 17 Hardness versus depth for surface-hardened pearlitic malleable irons. Curves labeled "Matrix" show hardness of the matrix, converted from microhardness tests. O, oil quenched and tempered to 207 HB before surface hardening; A, air cooled and tempered to 207 HB before surface hardening

- There is only a slight improvement in other mechanical properties with increased hardness above 255 HB

Because of these considerations, applications for air-quenched and tempered pearlitic malleable iron are usually those requiring moderate strength levels, while the higher-strength applications need the oil-quenched and tempered material.

The tensile properties of pearlitic malleable irons are normally measured on machined test bars. These properties are listed in Table 2.

The compressive strength of malleable irons is seldom determined, because failure in compression seldom occurs. As a result of the decreased influence of the graphite nodules and the delayed onset of plastic deformation in compression, compressive yield strengths are characteristically slightly higher than tensile yield strengths for the same hardness (Ref 1, 7).

Shear and Torsional Strength. The shear strength of ferritic malleable irons is approximately 80% of the tensile strength, and for pearlitic iron it ranges from 70 to 90% of the tensile strength (Ref 7). The ultimate torsional strength of ferritic malleable irons is about 90% of the ultimate tensile strength. The yield strength in torsion is 75 to 80% of the value in tension (Ref 1). Torsional strengths for pearlitic grades are approximately equal to, or slightly less than, the tensile strength of the material. Yield strengths in torsion vary from 70 to 75% of the tensile yield strength (Ref 7). The characteristic torsional properties of ferritic and pearlitic malleable irons are related to hardness, as shown in Fig. 12. As expected, the amount of twist before failure decreases with increasing strength.

The modulus of elasticity in tension of pearlitic malleable iron is 176 to 193 GPa (25.5×10^6 to 28.0×10^6 psi). For automobile crankshafts, the modulus is important and must be determined with greater precision.

Fracture Toughness. The results of Charpy V-notch tests on pearlitic malleable iron are presented in Fig. 5. The fracture toughness of ferritic and pearlitic malleable irons has not been widely studied, but one researcher has estimated K_{Ic} values for these materials by using a J-integral approach (Ref 8). Table 5 summarizes the fracture toughness values obtained for the various grades of malleable iron at various temperatures. All of the materials exhibited stable crack extension prior to fracture for 25 mm (1 in.) wide compact-tension specimens.

As for ductile irons, fracture toughness testing indicates that malleable irons possess considerably more toughness than is indicated by Charpy impact toughness results. Although the fracture toughness values for pearlitic grades are similar to those obtained for ferritic grades, the higher yield strengths of the pearlitic grades indicate that their critical flaw sizes, which are proportional to $(K_{Ic}/\sigma_y)^2$, are less than those of the ferritic grades of malleable iron. Detailed information on the principles of fracture toughness and the nomenclature associated with fracture mechanics studies is available in the Section "Fracture Mechanics" and the article "Dynamic Fracture Testing" in Volume 8 of the 9th Edition of *Metals Handbook*.

Mechanical Properties at Elevated Temperatures. Figure 13 shows the short-term high-temperature tensile strength of five pearlitic malleable irons and three martensitic malleable irons. Generally, the room-temperature tensile strengths are related to hardness, while the tensile strengths at temperatures above about 450 °C (840 °F) exhibit asymptotic behavior.

Figure 13 also illustrates two exceptions of the general relationship between hardness and room-temperature tensile strength. The first exception is that the 230 HB pearlitic malleable iron in Fig. 13(a) has a slightly higher room-temperature tensile strength than the 233 HB pearlitic malleable iron in Fig. 13(b). This difference, however, diminishes at temperatures above 100 °C (210 °F).

The second exception is the difference in tensile strength for two malleable irons of the same hardness (Fig. 13c). This variation is perhaps attributable to the differences in heat treatment. Both of the martensitic malleable irons with hardnesses of 228 HB were annealed, cooled, reheated (reaustenitized), and then oil quenched. Before the reheat, however, the two irons underwent different cooling procedures. The 228 HB iron with

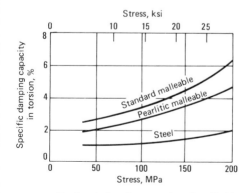

Fig. 18 Torsional damping capacity of malleable irons compared to steel. Source: Ref 1

Fig. 19 Examples of malleable iron automotive applications. (a) Driveline yokes. (b) Connecting rods. (c) Diesel pistons. (d) Steering gear housing. Courtesy of Central Foundry Division, General Motors Corporation

the higher strength was air cooled from 870 °C (1600 °F) after the temper carbon anneal (18 h soak at 950 °C, or 1740 °F), while the 228 HB martensitic iron with the lower strength was stabilized at 780 °C (1435 °F) for 6 h and then slow cooled to 700 °C (1290 °F) before reheating.

Sustained-load stress-rupture data for eight grades of pearlitic malleable iron are shown in Fig. 14. Results of high-temperature Charpy V-notch tests showing the effect of hardness on impact energy are given in Fig. 15.

The unnotched fatigue limits of tempered pearlitic malleable irons (air cooled or oil quenched) are about 40 to 50% of tensile strength. Tempered martensitic malleable irons (oil quenched) have an unnotched fatigue limit of about 35 to 40% of tensile strength (Fig. 16). The V-notched fatigue limits of the three irons in Fig. 16 ranged from 110 to 125 MPa (16 to 18 ksi) (Ref 5). Oil-quenched and tempered martensitic iron usually has a higher fatigue ratio than pearlitic iron made by the arrested anneal method.

Wear Resistance. Because of its structure and hardness, pearlitic and martensitic malleable irons have excellent wear resistance. In some moving parts where bushings are normally inserted at pivot points, heat-treated malleable iron has proved to be so wear resistant that the bushings have been eliminated. One example of this is the rocker arm for an overhead-valve automotive engine.

Welding and Brazing. Welding of pearlitic or martensitic malleable iron is difficult because the high temperatures used can cause the formation of a brittle layer of graphite-free white iron. Pearlitic and martensitic malleable iron can be successfully welded if the surface to be welded has been heavily decarburized.

Pearlitic or malleable iron can be brazed by various commercial processes. One application is the induction silver brazing of a pearlitic malleable casting and a steel shaft to form a planetary output shaft for an automotive transmission. In another automotive application, two steel shafts are induction copper brazed to a pearlitic malleable iron shifter shaft plate.

Selective Surface Hardening. Pearlitic malleable iron can be surface hardened by either induction heating and quenching or flame heating and quenching to develop high hardness at the heat-affected surface. Considerable research has been done to determine the surface-hardening characteristics of pearlitic malleable and its capability of developing high hardness over relatively narrow surface bands. In general, little difficulty is encountered in obtaining hardnesses in the range of 55 to 60 HRC, with the depth of penetration being controlled by the rate of heating and the surface temperature of the part being hardened (Fig. 17).

The maximum hardness obtainable in the matrix of a properly hardened pearlitic malleable part is 67 HRC. However, conventional hardness measurements made on castings show less than 67 HRC because of the presence of the graphite particles, which are averaged into the hardness. Generally, a casting with a matrix microhardness of 67 HRC will have about 62 HRC average hardness, as measured with the standard Rockwell tester. Similarly, a Rockwell or Brinell hardness test on softer structures will show less than matrix microhardness because of the presence of graphite.

Two examples of automobile production parts hardened by induction heating are rocker arms and clutch hubs. An example of a flame-hardened pearlitic malleable iron part is a pinion spacer used to support the cup of a roller bearing. To preclude service failures, the ends of the pinion spacer are flame hardened to a depth of about 2.3 mm (3/32 in.).

Malleable iron can be carburized, carbonitrided, or nitrided to produce a surface with improved wear resistance. In addition, heat treatments such as austempering have been used in specialized applications.

Damping Capacity

The good damping capacity and fatigue strength of malleable irons are useful for long service in highly stressed parts. Figure 18 compares the damping capacity of malleable irons to that of steels. The production of high internal stresses by quenching malleable iron can double the damping capacity, which is then gradually reduced as tempering relieves residual stresses (Ref 1).

Applications

Malleable iron castings are often selected because the material has excellent machinability in addition to significant ductility. In other applications, malleable iron is chosen because it combines castability with good toughness and machinability. Malleable iron is often chosen because of shock resistance alone. Tables 3 and 4 list some of the typical applications of malleable iron castings.

The requirement that any iron produced for conversion to malleable iron must solidify white places definite section thickness limitations on the malleable iron industry. Thick metal sections can be produced by melting a base iron of low carbon and silicon contents or by alloying the molten iron with a carbide stabilizer. However, when carbon and silicon are maintained at low levels, difficulty is invariably encountered in annealing, and the time required to convert primary and pearlitic carbides to temper carbon becomes excessively long. High-production foundries are usually reluctant to produce castings more than about 40 mm (1½ in.) thick. Some foundries, however, routinely produce castings as thick as 100 mm (4 in.).

After heat treatment, ferritic or pearlitic malleable castings are cleaned by shotblasting, gates are removed by shearing or grinding, and, where necessary, the castings are coined or punched. Close dimensional tolerances can be maintained in ferritic malleable iron and in the lower-hardness types of pearlitic malleable iron, both of which can be easily straightened in dies. The harder pearlitic malleable irons are more difficult to press because of higher yield strength and a greater tendency toward springback after die pressing. However, even the highest-strength pearlitic malleable can be straightened to achieve good dimensional tolerances.

Automotive and associated applications of ferritic and pearlitic malleable irons include many essential parts in vehicle power trains, frames, suspensions, and wheels. A partial list includes differential carriers, differential cases, bearing caps, steering-gear housings, spring hangers, universal-joint yokes, automatic-transmission parts, rocker arms, disc brake calipers, wheel hubs, and many other miscellaneous castings. Examples are shown in Fig. 19. Ferritic and pearlitic malleable irons are also used in the railroad industry and in agricultural equipment, chain links, ordnance material, electrical pole line hardware, hand tools, and other parts requiring section thicknesses and properties obtainable in these materials.

REFERENCES

1. C.F. Walton and T.J. Opar, Ed., *Iron Castings Handbook*, Iron Castings Society, 1981, p 297-321
2. L. Jenkins, Malleable Cast Iron, in *Encyclopedia of Materials Science and Engineering*, Vol 4, M.B. Bever, Ed., MIT Press, 1986, p 2725-2729
3. D.R. Askeland and R.F. Fleischman, Effect of Nodule Count on the Mechanical Properties of Ferritic Malleable Iron, *Trans. AFS*, Vol 86, 1978, p 373-378
4. J. Pelleg, Some Mechanical Properties of Cupola Malleable Iron, *Foundry*, Oct 1960, p 110-113
5. L.W.L. Smith et al., *Properties of Modern Malleable Irons*, BCIRA International Center for Cast Metals Technology, 1987
6. "Standard Specification for Malleable Iron Castings," A 47, *Annual Book of ASTM Standards*, American Society for Testing and Materials
7. G.N.J. Gilbert, *Engineering Data on Malleable Cast Irons*, British Cast Iron Research Association, 1968
8. W.L. Bradley, Fracture Toughness Studies of Gray, Malleable and Ductile Cast Iron, *Trans. AFS*, Vol 89, 1981, p 837-848

Alloy Cast Irons

Revised by Richard B. Gundlach, Climax Research Services; and Douglas V. Doane, Consulting Metallurgist

ALLOY CAST IRONS are considered to be those casting alloys based on the iron-carbon-silicon system that contain one or more alloying elements intentionally added to enhance one or more useful properties. The addition to the ladle of small amounts of substances (such as ferrosilicon, cerium, or magnesium) that are used to control the size, shape, and/or distribution of graphite particles is termed inoculation rather than alloying. The quantities of material used for inoculation neither change the basic composition of the solidified iron nor alter the properties of individual constituents. Alloying elements, including silicon when it exceeds about 3%, are usually added to increase the strength, hardness, hardenability, or corrosion resistance of the basic iron and are often added in quantities sufficient to affect the occurrence, properties, or distribution of constituents in the microstructure.

In gray and ductile irons, small amounts of alloying elements such as chromium, molybdenum, or nickel are used primarily to achieve high strength or to ensure the attainment of a specified minimum strength in heavy sections. Otherwise, alloying elements are used almost exclusively to enhance resistance to abrasive wear or chemical corrosion or to extend service life at elevated temperatures.

The strengthening effects of the various alloying elements in gray and ductile irons are dealt with in the articles "Gray Iron" and "Ductile Iron" in this Volume. This article discusses abrasion-resistant chilled and white irons, high-alloy corrosion-resistant irons, and medium-alloy and high-alloy heat-resistant gray and ductile irons. Table 1 lists approximate ranges of alloy content for various types of alloy cast irons covered in this article. Individual alloys within each type are made to compositions in which the actual ranges of one or more of the alloying elements span only a portion of the listed ranges; the listed ranges serve only to identify the types of alloys used in specific kinds of applications.

Classification of Alloy Cast Irons

Alloy cast irons can be classified as white cast irons, corrosion-resistant cast irons, and heat-resistant cast irons.

White cast irons, so named because of their characteristically white fracture surfaces, do not have any graphite in their microstructures. Instead, the carbon is present in the form of carbides, chiefly of the types Fe_3C and Cr_7C_3. Often, complex carbides such as $(Fe,Cr)_3C$ and $(Cr,Fe)_7C_3$, or those containing other carbide-forming elements, are also present.

Table 1 Ranges of alloy content for various types of alloy cast irons

Description	TC(b)	Mn	P	S	Si	Ni	Cr	Mo	Cu	Matrix structure, as-cast(c)
Abrasion-resistant white irons										
Low-carbon white iron(d)	2.2–2.8	0.2–0.6	0.15	0.15	1.0–1.6	1.5	1.0	0.5	(e)	CP
High-carbon, low-silicon white iron	2.8–3.6	0.3–2.0	0.30	0.15	0.3–1.0	2.5	3.0	1.0	(e)	CP
Martensitic nickel-chromium iron	2.5–3.7	1.3	0.30	0.15	0.8	2.7–5.0	1.1–4.0	1.0	...	M, A
Martensitic nickel, high-chromium iron	2.5–3.6	1.3	0.10	0.15	1.0–2.2	5–7	7–11	1.0	...	M, A
Martensitic chromium-molybdenum iron	2.0–3.6	0.5–1.5	0.10	0.06	1.0	1.5	11–23	0.5–3.5	1.2	M, A
High-chromium iron	2.3–3.0	0.5–1.5	0.10	0.06	1.0	1.5	23–28	1.5	1.2	M
Corrosion-resistant irons										
High-silicon iron(f)	0.4–1.1	1.5	0.15	0.15	14–17	...	5.0	1.0	0.5	F
High-chromium iron	1.2–4.0	0.3–1.5	0.15	0.15	0.5–3.0	5.0	12–35	4.0	3.0	M, A
Nickel-chromium gray iron(g)	3.0	0.5–1.5	0.08	0.12	1.0–2.8	13.5–36	1.5–6.0	1.0	7.5	A
Nickel-chromium ductile iron(h)	3.0	0.7–4.5	0.08	0.12	1.0–3.0	18–36	1.0–5.5	1.0	...	A
Heat-resistant gray irons										
Medium-silicon iron(j)	1.6–2.5	0.4–0.8	0.30	0.10	4.0–7.0	...	...	...	...	F
Nickel-chromium iron(g)	1.8–3.0	0.4–1.5	0.15	0.15	1.0–2.75	13.5–36	1.8–6.0	1.0	7.5	A
Nickel-chromium-silicon iron(k)	1.8–2.6	0.4–1.0	0.10	0.10	5.0–6.0	13–43	1.8–5.5	1.0	10.0	A
High-aluminum iron	1.3–2.0	0.4–1.0	0.15	0.15	1.3–6.0	...	20–25 Al	...	...	F
Heat-resistant ductile irons										
Medium-silicon ductile iron	2.8–3.8	0.2–0.6	0.08	0.12	2.5–6.0	1.5	...	2.0	...	F
Nickel-chromium ductile iron(h)	3.0	0.7–2.4	0.08	0.12	1.75–5.5	18–36	1.75–3.5	1.0	...	A
Heat-resistant white irons										
Ferritic grade	1–2.5	0.3–1.5	...	...	0.5–2.5	...	30–35	...	...	F
Austenitic grade	1–2.0	0.3–1.5	...	...	0.5–2.5	10–15	15–30	...	...	A

(a) Where a single value is given rather than a range, that value is a maximum limit. (b) Total carbon. (c) CP, coarse pearlite; M, martensite; A, austenite; F, ferrite. (d) Can be produced from a malleable-iron base composition. (e) Copper can replace all or part of the nickel. (f) Such as Duriron, Durichlor 51, Superchlor. (g) Such as Ni-Resist austenitic iron (ASTM A 436). (h) Such as Ni-Resist austenitic ductile iron (ASTM A 439). (j) Such as Silal. (k) Such as Nicrosilal

Fig. 1 Fracture surface of as-cast chilled iron. White, mottled, and gray portions are shown at full size, top to bottom.

White cast irons are usually very hard, which is the single property most responsible for their excellent resistance to abrasive wear. White iron can be produced either throughout the section (chiefly by adjusting the composition) or only partly inward from the surface (chiefly by casting against a chill). The latter iron is sometimes referred to as chilled iron to distinguish it from iron that is white throughout.

Chilled iron castings are produced by casting the molten metal against a metal or graphite chill, resulting in a surface virtually free from graphitic carbon. In the production of chilled iron, the composition is selected so that only the surfaces cast against the chill will be free from graphitic carbon (Fig. 1). The more slowly cooled portions of the casting will be gray or mottled iron. The depth and hardness of the chilled portion can be controlled by adjusting the composition of the metal, the extent of inoculation, and the pouring temperature.

White iron is a cast iron virtually free from graphitic carbon because of selected chemical composition. The composition is chosen so that, for the desired section size, graphite does not form as the casting solidifies. The hardness of white iron castings can be controlled by further adjustment of composition.

The main difference in microstructure between chilled iron and white iron is that chilled iron is fine grained and exhibits directionality perpendicular to the chilled face, while white iron is ordinarily coarse grained, randomly oriented, and white throughout, even in relatively heavy sections. (Fine-grain white iron can be produced by casting a white iron composition against a chill.) This difference reflects the effect of composition difference between the two types of abrasion-resistant iron.

Chilled iron is directional only because the casting, made of a composition that is ordinarily gray, has been cooled through the eutectic temperature so rapidly at one or more faces that the iron solidified white, growing inward from the chilled face. White iron, on the other hand, has a composition so low in carbon equivalent or so rich in alloy content that gray iron cannot be produced even at the relatively low rates of cooling that exist in the center of the heaviest section of the casting.

Corrosion-resistant irons derive their resistance to chemical attack chiefly from their high alloy content. Depending on which of three alloying elements—silicon, chromium, or nickel—dominates the composition, a corrosion-resistant iron can be ferritic, pearlitic, martensitic, or austenitic in its microstructure. Depending on composition, cooling rate, and inoculation practice, a corrosion-resistant iron can be white, gray, or nodular in both form and distribution of carbon.

Heat-resistant irons combine resistance to high-temperature oxidation and scaling with resistance to softening or microstructural degradation. Resistance to scaling depends chiefly on high alloy content, and resistance to softening depends on the initial microstructure plus the stability of the carbon-containing phase. Heat-resistant irons are usually ferritic or austenitic as-cast; carbon exists predominantly as graphite, either in flake or spherulitic form, which subdivides heat-resistant irons into either gray or ductile irons. There are also ferritic and austenitic white iron grades, although they are less frequently used and have no American Society for Testing and Materials (ASTM) designations.

Effects of Alloying Elements

In most cast irons, it is the interaction among alloying elements (including carbon and silicon) that has the greatest effect on properties. This influence is exerted largely by effects on the amount and shape of graphitic carbon present in the casting. For example, in low-alloy cast irons, depth of chill or the tendency of the iron to be white as-cast depends greatly on the carbon equivalent, the silicon in the composition, and the state of inoculation. The addition of other elements can only modify the basic tendency established by the carbon-silicon relationship.

On the other hand, abrasion-resistant white cast irons are specifically alloyed with chromium to produce fully carbidic irons. One of the benefits of chromium is that it causes carbide, rather than graphite, to be the stable carbon-rich eutectic phase upon solidification. At higher chromium contents (10% or more), M_7C_3 carbide becomes the stable carbon-rich phase of the eutectic reaction.

In general, only small amounts of alloying elements are needed to improve depth of chill, hardness, and strength. Typical effects on depth of chill are given in Fig. 2 for the alloying elements commonly used in low to moderately alloyed cast irons. High alloy contents are needed for the most significant improvements in abrasion resistance, corrosion resistance, or elevated-temperature properties.

Alloying elements such as nickel, chromium, and molybdenum are used, singly or in combination, to provide specific improvements in properties compared to unalloyed irons. Because the use of such elements means higher cost, the improvement in service performance must be sufficient to justify the increased cost.

Carbon. In chilled irons, the depth of chill decreases (Fig. 2c), and the hardness of the chilled zone increases, with increasing carbon content. Carbon also increases the hardness of white irons. Low-carbon white irons (~2.50% C) have a hardness of about 375 HB (Fig. 3), while white irons with fairly high total carbon (>3.50% C) have a hardness as high as 600 HB. In unalloyed white irons, high total carbon is essential for high hardness and maximum wear resistance. Carbon decreases transverse breaking strength (Fig. 4) and increases brittleness. It also increases the tendency for graphite to form during solidification, especially when the silicon content is also high. As a result, it is very important to keep the silicon content low in high-carbon white irons. The normal range of carbon content for unalloyed or low-alloy white irons is about 2.2 to 3.6%. For high-chromium white irons, the normal range is from about 2.2% to the carbon content of the eutectic composition, which is about 3.5% for a 15% Cr iron and about 2.7% for a 27% Cr iron.

The carbon content of gray and ductile alloy irons is generally somewhat higher than that of a white iron of similar alloy content. In addition, the silicon content is usually higher, so that graphite will be formed upon solidification.

Silicon is present in all cast irons. In alloy cast irons, as in other types, silicon is the chief factor that determines the carbon content of the eutectic. Increasing the silicon content lowers the carbon content of the eutectic and promotes the formation of graphite upon solidification. Therefore, the silicon content is the principal factor controlling the depth of chill in unalloyed or low-chromium chilled and white irons. This effect for relatively high carbon irons is summarized in Fig. 2(a).

In high-alloy white irons, silicon has a negative effect on hardenability; that is, it tends to promote pearlite formation in martensitic irons. However, when sufficient amounts of pearlitic-suppressing elements such as molybdenum, nickel, manganese, and chromium are present, increasing the

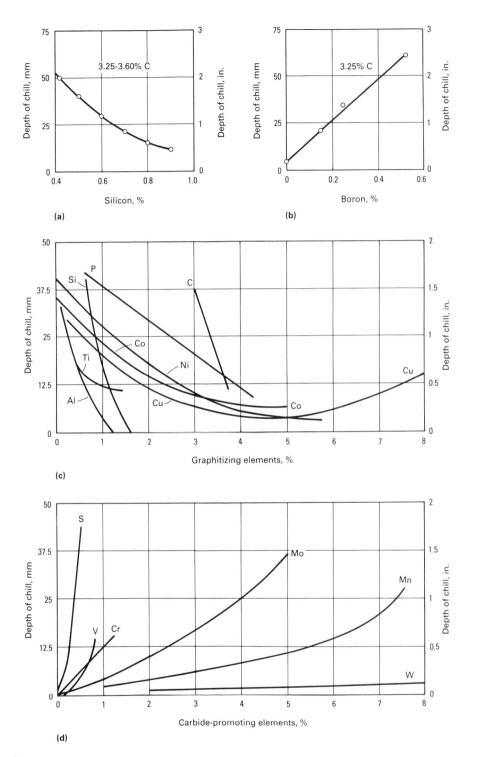

Fig. 2 Typical effects of alloying elements on depth of chill

silicon content raises the M_s temperature of the alloy, thus tending to increase both the amount of martensite and the final hardness.

The silicon content of chilled and white irons is usually between 0.3 and 2.2%. In martensitic nickel-chromium white irons, the desired silicon content is usually 0.4 to 0.9%. It is necessary to select carefully the charge constituents when melting a martensitic iron so that excessive silicon content is avoided. In particular, it is necessary to give special attention to the silicon content of the ferrochromium used in the furnace charge.

Silicon additions of 3.5 to 7% improve high-temperature properties by raising the eutectoid transformation temperature. The influence of silicon on the critical temperature is shown in Fig. 5. Elevated levels of silicon also reduce the rates of scaling and growth by forming a tight, adhering oxide scale. This occurs at silicon contents above 3.5% in ferritic irons and above 5% in 36% Ni austenitic irons. Additions of 14 to 17% (often accompanied by additions of about 5% Cr and 1% Mo) yield cast iron that is very resistant to corrosive acids, although resistance varies somewhat with acid concentration.

High-silicon irons (14 to 17%) are difficult to cast and are virtually unmachinable. High-silicon irons have particularly low resistance to mechanical and thermal shock at room temperature or moderately elevated temperature. However, above about 260 °C (500 °F), the shock resistance exceeds that of ordinary gray iron.

Manganese and sulfur should be considered together in their effects on gray or white iron. Alone, either manganese or sulfur increases the depth of chill, but when one is present, addition of the other decreases the depth of chill until the residual concentration has been neutralized by the formation of manganese sulfide. Generally, sulfur is the residual element, and excess manganese can be used to increase chill depth and hardness, as shown in Fig. 2(d). Furthermore, because it promotes the formation of finer and harder pearlite, manganese is often preferred for decreasing or preventing mottling in heavy-section castings.

Manganese, in excess of the amount needed to scavenge sulfur, mildly suppresses pearlitic formation. It is also a relatively strong austenite stabilizer and is normally kept below about 0.7% in martensitic white irons. In some pearlitic or ferritic alloy cast irons, up to about 1.5% Mn can be used to help ensure that specified strength levels are obtained. When manganese content exceeds about 1.5%, the strength and toughness of martensitic irons begin to drop. Abrasion resistance also drops, mainly because of austenite retention. Molten iron with a high manganese content tends to attack furnace and ladle refractories. Consequently, the use of manganese is limited in cast irons, even though it is one of the least expensive alloying elements.

The normal sulfur contents of alloy cast irons are neutralized by manganese, but the sulfur content is kept low in most alloy cast irons. In abrasion-resistant cast irons, the sulfur content should be as low as is commercially feasible, because several investigations have shown that sulfides in the microstructure degrade abrasion resistance. A sulfur content of 0.03% appears to be the maximum that can be tolerated when optimum abrasion resistance is desired.

Phosphorus is a mild graphitizer in unalloyed irons; it mildly reduces chill depth in chilled irons (Fig. 2c). In alloyed irons, the effects of phosphorus are somewhat obscure. There is some evidence that it reduc-

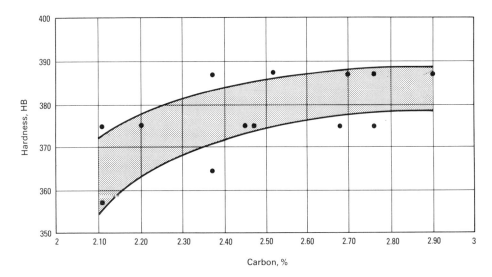

Fig. 3 Effect of carbon content on the hardness of low-carbon white iron

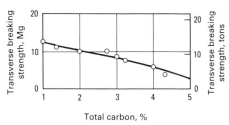

Fig. 4 Effect of total carbon on the transverse breaking strength of unalloyed white iron

es the toughness of martensitic white irons. The effect, if any, on abrasion resistance has not been conclusively proved. In heavy-section castings made from molybdenum-containing irons, high phosphorus contents are considered detrimental because they neutralize part of the deep-hardening effect of the molybdenum. It is considered desirable to keep the phosphorus content of alloy cast irons below about 0.3%, and some specifications call for less than 0.1%. In cast irons for high-temperature or chemical service, it is customary to keep the phosphorus content below 0.15%.

Chromium has three major uses in cast irons:

- To form carbides
- To impart corrosion resistance
- To stabilize the structure for high-temperature applications

Small amounts of chromium are routinely added to stabilize pearlite in gray iron, to control chill depth in chilled iron, or to ensure a graphite-free structure in white iron containing less than 1% Si. At such low percentages, usually no greater than 2 to 3%, chromium has little or no effect on hardenability, chiefly because most of the chromium is tied up in carbides. However, chromium does influence the fineness and hardness of pearlite and tends to increase the amount and hardness of the eutectic carbides. Consequently, chromium is often added to gray iron to ensure that strength requirements can be met, particularly in heavy sections. On occasion, it can be added to ductile iron for the same purpose. Also, relatively low percentages of chromium are used to improve the hardness and abrasion resistance of pearlitic white cast irons.

When the chromium content of cast iron is greater than about 10%, eutectic carbides of the M_7C_3 type are formed, rather than the M_3C type that predominates at lower chromium contents. More significantly, however, the higher chromium content causes a change in solidification pattern to a structure in which the M_7C_3 carbides are surrounded by a matrix of austenite or its transformation products. At lower chromium contents, the M_3C carbide forms the matrix. Because of the solidification characteristics, hypoeutectic irons containing M_7C_3 carbides are normally stronger and tougher than irons containing M_3C carbides.

The relatively good abrasion resistance, toughness, and corrosion resistance found in high-chromium white irons have led to the development of a series of commercial martensitic or austenitic white irons containing 12 to 28% Cr. Because much of the chromium in these irons is present in combined form as carbides, chromium is much less effective than molybdenum, nickel, manganese, or copper in suppressing the eutectoid transformation to pearlite and therefore has a lesser effect on hardenability than it has in steels. Martensitic white irons usually contain one or more of the elements molybdenum, nickel, manganese, and copper to give the required hardenability. These elements ensure that martensite will form upon cooling from above the upper transformation temperature either while the casting is cooling in the mold or during subsequent heat treatment.

It is difficult to maintain low silicon content in high-chromium irons because of the silicon introduced by high-carbon ferrochrome and other sources. Low silicon content is advantageous in that it provides for ready response to annealing and yields high hardness when the alloy is air quenched from high temperatures. High silicon content lessens response to this type of heat treatment. Although high-chromium white irons are sometimes used as-cast, their optimum properties are obtained in the heat-treated condition.

For developing resistance to the softening effect of heat and for protection against oxidation, chromium is the most effective element. It stabilizes iron carbide and therefore prevents the breakdown of carbide at elevated temperatures; 1% Cr gives ade-

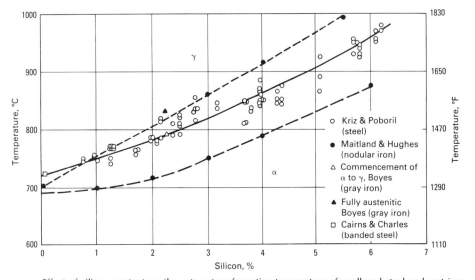

Fig. 5 Effect of silicon content on the α to γ transformation temperature of unalloyed steel and cast iron containing 0.09 to 4.06% C and 0.3 to 0.6% Mn

quate protection against oxidation up to about 760 °C (1400 °F) in many applications. For temperatures of 760 °C (1400 °F) and above, chromium contents up to 5.5% are found in austenitic ductile irons for added oxidation resistance. For long-term oxidation resistance at elevated temperatures, white cast irons having chromium contents of 15 to 35% are employed. This percentage of chromium suppresses the formation of graphite and makes the alloy solidify as white cast iron.

High levels of chromium stabilize the ferrite phase up to the melting point; typical high-chromium ferritic irons contain 30 to 35% Cr. Austenitic grades of high-chromium irons, which have significantly higher strength at elevated temperatures, contain 10 to 15% Ni, along with 15 to 30% Cr.

Nickel is almost entirely distributed in the austenitic phase or its transformation products. Like silicon, nickel promotes graphite formation, and in white and chilled irons, this effect is usually balanced by the addition of about one part chromium for every three parts nickel in the composition. If fully white castings are desired, the amount of chromium can be increased. Some low- and medium-alloy cast irons have a ratio as low as one part chromium to 1.3 parts nickel. In high-chromium irons, the nickel content may be as high as 15% to stabilize the austenite phase.

When added to low-chromium white iron in amounts up to about 2.5%, nickel produces a harder and finer pearlite in the structure, which improves its abrasion resistance. Nickel in somewhat larger amounts—up to about 4.5%—is needed to completely suppress pearlite formation, thus ensuring that a martensitic iron results when the castings cool in their molds. This latter practice forms the basis for production of the Ni-Hard cast irons (which are usually identified in standard specification as nickel-chromium martensitic irons). With small castings such as grinding balls, which can be shaken out of the molds while still hot, air cooling from the shakeout temperature will produce the desired martensitic structure even when the nickel content is as low as 2.7%. On the other hand, an excessively high nickel content (more than about 6.5%) will so stabilize the austenite that little martensite, if any, can be formed in castings of any size. Appreciable amounts of retained austenite in Ni-Hard cast irons can be transformed to martensite by refrigerating the castings at −55 to −75 °C (−70 to −100 °F) or by using special tempering treatments.

One of the Ni-Hard family of commercial alloy white irons (type IV Ni-Hard) contains 1.0 to 2.2% Si, 5 to 7% Ni, and 7 to 11% Cr. In the as-cast condition, it has a structure of M_7C_3 eutectic carbides in a martensitic matrix. If retained austenite is present, the martensite content and hardness of the alloy can be increased by refrigeration treatment or by reaustenitizing and air cooling. Ni-Hard IV is often specified for pumps and other equipment used for handling abrasive slurries because of its combination of relatively good strength, toughness, and abrasion resistance.

Nickel is used to suppress pearlite formation in large castings of high-chromium white iron (12 to 28% Cr). The typical amount of nickel is about 0.2 to 1.5%, and it is usually added in conjunction with molybdenum. Nickel contents higher than this range tend to excessively stabilize the austenite, leading to austenite retention. Control of composition is especially important for large castings that are intended to be martensitic, because their size dictates that they cool slowly regardless of whether they are to be used as-cast or after heat treatment.

Nickel additions of more than 12% are needed for optimum resistance to corrosion or heat. High-nickel gray or ductile irons usually contain 1 to 6% Cr and may contain as much as 10% Cu. These elements act in conjunction with the nickel to promote resistance to corrosion and scaling, especially at elevated temperatures. All types of cast iron with nickel contents above 18% are fully austenitic.

Copper in moderate amounts can be used to suppress pearlite formation in both low- and high-chromium martensitic white irons. The effect of copper is relatively mild compared to that of nickel, and because of the limited solubility of copper in austenite, copper additions probably should be limited to about 2.5% or less. This limitation means that copper cannot completely replace nickel in Ni-Hard-type irons. When added to chilled iron without chromium, copper narrows the zone of transition from white to gray iron, thus reducing the ratio of the mottled portion to the clear chilled portion.

Copper is most effective in suppressing pearlite when it is used in conjunction with about 0.5 to 2.0% Mo. The hardenability of this combination is surprisingly good, which indicates that there is a synergistic effect when copper and molybdenum are added together to cast iron. Combined additions appear to be particularly effective in the martensitic high-chromium irons. Here, copper content should be held to 1.2% or less; larger amounts tend to induce austenite retention.

Copper is used in amounts of about 3 to 10% in some high-nickel gray and ductile irons that are specified for corrosion or high-temperature service. Here, copper enhances corrosion resistance, particularly resistance to oxidation or scaling.

Molybdenum in chilled and white iron compositions is distributed between the eutectic carbides and the matrix. In graphitic irons, its main functions are to promote deep hardening and to improve high-temperature strength and corrosion resistance. In chilled iron compositions, molybdenum additions mildly increase depth of chill (they are about one-third as effective as chromium; see Fig. 2d). The primary purpose of small additions (0.25 to 0.75%) of molybdenum to chilled iron is to improve the resistance of the chilled face to spalling, pitting, chipping, and heat checking. Molybdenum hardens and toughens the pearlitic matrix.

Where a martensitic white iron is desired for superior abrasion resistance, additions of 0.5 to 3.0% Mo effectively suppress pearlite and other high-temperature transformation products (Fig. 6). Molybdenum is even more effective when used in combination with copper, chromium, nickel, or both chromium and nickel. Molybdenum has an advantage over nickel, copper, and manganese in that it increases depth of hardening without appreciably overstabilizing austenite, thus preventing the retention of undesirably large amounts of austenite in the final structure. Figure 6 illustrates the influence of different amounts of molybdenum on the hardenability of high-chromium white irons and shows that the hardenability (measured as the critical diameter for air hardening) increases as the ratio of chromium to carbon increases.

The pearlite-suppressing properties of molybdenum have been used to advantage in irons of high chromium content. White irons with 12 to 18% Cr are used for abrasion-resistant castings. The addition of 1 to 4% Mo is effective in suppressing pearlite formation, even when the castings are slowly cooled in heavy sections.

Molybdenum can replace some of the nickel in the nickel-chromium type of martensitic white irons. In heavy-section castings in which 4.5% Ni would be used, the addition of 1% Mo permits a reduction of nickel content to about 3%. In light-section castings of this type, where 3% Ni would normally be used, the addition of 1% Mo permits a reduction of nickel to 1.5%.

Molybdenum, in quantities of about 1 to 4%, is effective in enhancing corrosion resistance, especially in the presence of chlorides. In quantities of ½ to 2%, molybdenum improves high-temperature strength and creep resistance in gray and ductile irons with ferritic or austenitic matrices. Figure 7 illustrates the influence of molybdenum on the strength and creep resistance of high-silicon (4% Si) ferritic ductile iron at 705 °C (1300 °F).

Vanadium is a potent carbide stabilizer and increases depth of chill. The magnitude of the increase of depth of chill depends on the amount of vanadium and the composition of the iron as well as on section size and conditions of casting. The powerful chilling effect of vanadium in thin sections can be balanced by additions of nickel or copper, by a large increase in carbon or silicon, or both. In addition to its carbide-stabilizing

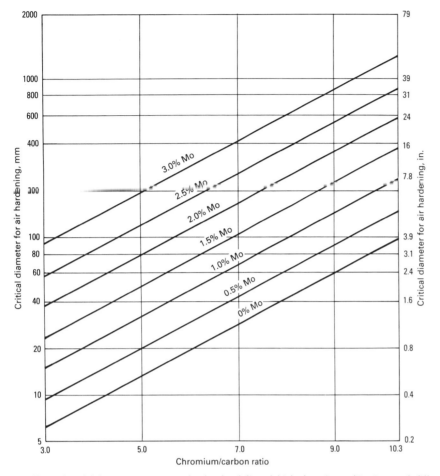

Fig. 6 Effect of molybdenum content on the hardenability of high-chromium white irons of different chromium-to-carbon (Cr/C) ratios

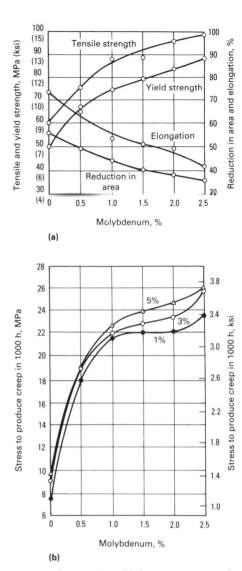

Fig. 7 Influence of molybdenum on (a) tensile properties and (b) creep resistance of 4% Si ductile iron at 705 °C (1300 °F)

influence, vanadium in amounts of 0.10 to 0.50% refines the structure of the chill and minimizes coarse columnar grain structure. Because of its strong carbide-forming tendency, vanadium is rarely used in gray or ductile irons for corrosion or elevated-temperature service.

Effects of Inoculants

Certain elements, when added in minute amounts in the pouring ladle, have relatively strong effects on the size, shape, and distribution of graphite in graphitic cast irons. Other elements are equally powerful in stabilizing carbides. These elements, called inoculants, appear to act more as catalysts than as participants in the reactions.

The main graphitizing inoculant is ferrosilicon, which is often added in detectable amounts (several kilograms per tonne) as a final adjustment of carbon equivalent in gray or ductile irons. In ductile irons, it is essential that the graphite be present in the final structure as nodules (spherulites) rather than as flakes. Magnesium, cerium, rare-earth elements, and certain proprietary substances are added to the molten iron just before pouring to induce the graphite to form in nodules of the desired size and distribution.

In white irons, tellurium, bismuth, and sometimes vanadium are the principal carbide-inducing inoculants. Tellurium is extremely potent; an addition of only about 5 g/t (5 ppm) is often sufficient. Tellurium has one major drawback. It has been found to cause tellurium halitosis in foundry workers exposed to even minute traces of its fumes; therefore, its use as an inoculant has been discouraged and sometimes prohibited.

Bismuth, in amounts of 50 to 100 g/t (50 to 100 ppm), effectively suppresses graphite formation in unalloyed or low-alloy white iron. In particular, bismuth is used in the low-carbon compositions destined for malleabilizing heat treatment. It has been reported that bismuth produces a fine-grain microstructure free from spiking, a condition that is sometimes preferred in abrasion-resistant white irons.

Vanadium, in amounts up to 0.5%, is sometimes considered useful as a carbide stabilizer and grain refiner in white or chilled irons. Nitrogen- and boron-containing ferroalloys have also been used as inoculants with reported beneficial effects. In general, however, the economic usefulness of inoculants in abrasion-resistant white irons has been inconsistent and remains unproved. Inoculants other than appropriate graphitizing or nodularizing agents are used rarely, if ever, in high-alloy corrosion-resistant or heat-resistant irons.

Abrasion-Resistant Cast Irons

It should be presumed that parts subjected to abrasion will wear out and will therefore need to be replaced from time to time. Also, for many applications, there will be one or more types of relatively low-cost material that will have adequate wear resistance and one or more types of higher-cost material that will have measurably superior wear resistance. For both situations, the ratio of wear rate to replacement cost should be evaluated; this ratio can be a very

Table 2 Chemical composition of standard martensitic white cast irons

Certain specific compositions of alloys II-B, II-C, II-D, and II-E are covered by U.S. Patent 3,410,682.

Class	Type	Designation	Composition, wt%(a)							
			TC(b)	Mn	P	S	Si	Cr	Ni	Mo	Cu
I	A	Ni-Cr-HC	3.0–3.6	1.3	0.30	0.15	0.8	1.4–4.0	3.3–5.0	1.0	...
I	B	Ni-Cr-LC	2.5–3.0	1.3	0.30	0.15	0.8	1.4–4.0	3.3–5.0	1.0	...
I	C	Ni-Cr-GB	2.9–3.7	1.3	0.30	0.15	0.8	1.1–1.5	2.7–4.0	1.0	...
I	D	Ni-Hi Cr	2.5–3.6	1.3	0.10	0.15	1.0–2.2	7–11	5–7	1.0	...
II	A	12% Cr	2.4–2.8	0.5–1.5	0.10	0.06	1.0	11–14	0.5	0.5–1.0	1.2
II	B	15% Cr-Mo-LC	2.4–2.8	0.5–1.5	0.10	0.06	1.0	14–18	0.5	1.0–3.0	1.2
II	C	15% Cr-Mo-HC	2.8–3.6	0.5–1.5	0.10	0.06	1.0	14–18	0.5	2.3–3.5	1.2
II	D	20% Cr-Mo-LC	2.0–2.6	0.5–1.5	0.10	0.06	1.0	18–23	1.5	1.5	1.2
II	E	20% Cr-Mo-HC	2.6–3.2	0.5–1.5	0.10	0.06	1.0	18–23	1.5	1.0–2.0	1.2
III	A	25% Cr	2.3–3.0	0.5–1.5	0.10	0.06	1.0	23–28	1.5	1.5	1.2

(a) Where a single value is given rather than a range, that value is a maximum limit. (b) Total carbon. Source: ASTM A 532-75a

effective means of evaluating the most economical use of materials. It is often more economical to use a less wear-resistant material and replace it more often. However, in some cases, such as when frequent occurrences of downtime cannot be tolerated, economy is less important than service life. Total cost-effectiveness must take into account the actual cost of materials, heat treatment, time for removal of worn parts and insertion of new parts, and other production time lost.

In general, chilled iron and unalloyed white iron are less expensive than alloy irons; they are also less wear resistant. However, the abrasion resistance of chilled or unalloyed white iron is entirely adequate for many applications. It is only when a clear performance advantage can be proved that alloy cast irons will show an economic advantage over unalloyed irons. For example, in a 1-year test in a mill for grinding cement clinker, grinding balls made of martensitic nickel-chromium white iron had to be replaced only about one-fifth as often as forged and hardened alloy steel balls. In another test of various parts in a brickmaking plant, martensitic nickel-chromium white iron was found to last three to four times as long as unalloyed white iron, in terms of both tonnage handled and lifetime in days. In both cases, martensitic nickel-chromium white iron showed a clear economic advantage as well as a clear performance advantage over the alternative materials.

Typical Compositions. The first two lines of Table 1 list the composition ranges for the typical commercial unalloyed and low-alloy grades of white and chilled irons used for abrasion-resistant castings. These are nominally classed as pearlitic white irons. Historically, most of the early white iron castings produced for abrasion resistance were cast from low-carbon, 1.0 to 1.6% Si unalloyed compositions, which were also used for malleable iron castings. As changes have occurred in demand and specific uses, the trend has been to produce a more abrasion-resistant 2.8 to 3.6% C, low-silicon grade, which is usually alloyed with chromium to suppress graphite and to increase the fineness and hardness of the pearlite. Other alloying elements such as nickel, molybdenum, copper, and manganese are used primarily to increase hardenability in order to obtain austenitic or martensitic structures.

Martensitic white irons have largely displaced pearlitic white irons for making many types of abrasion-resistant castings, with the possible exception of chilled iron rolls and grinding balls. Although martensitic white irons cost more than pearlitic irons, their much superior abrasion resistance, combined with the increasing costs of all castings, makes martensitic alloy white irons economically attractive. The better strength and toughness of martensitic irons favor their use, and the trend toward replacing cupola melting with electric furnace melting makes martensitic white irons relatively easy to produce.

Table 2 lists the composition ranges of commercial martensitic white cast irons. The iron alloys of class I are designed to be largely martensitic as-cast; the only heat treatment commonly applied is tempering.

The iron alloys of classes II and III are either pearlitic or austenitic as-cast, except in slow-cooling heavy sections, which may be partially martensitic. The iron alloys of classes II and III are usually heat treated as described below. There are several situations in which the abrasion resistance of the as-cast austenitic casting is very good; no heat treatment is applied in such cases.

Heat Treatment. Various high- and low-temperature heat treatments can be used to improve the properties of white and chilled iron castings. For the unalloyed or low-chromium pearlite white irons, heat treatment is performed primarily to relieve the internal stresses that develop in the castings as they cool in their molds. Generally, such heat treatments are used only on large castings such as mill rolls and chilled iron car wheels. Temperatures up to about 705 °C (1300 °F) can be used without severely reducing abrasion resistance. In some cases, the castings can be removed from their molds above the pearlitic-formation temperature and can then be isothermally transformed to pearlite (or to ferrite and carbide) in an annealing furnace. As the tempering or annealing temperature is increased, the time at temperature must be reduced to prevent graphitization. Results of reheating tests on unalloyed chilled iron with a composition of 3.25 to 3.60% C, 0.50 to 0.55% Si, 0.55 to 0.60% Mn, 0.33% P, and 0.13% S are given in Fig. 8.

Residual stresses in large castings result from volume changes during the transformation of austenite and during subsequent cooling of the casting to room temperature. Because these volume changes may not occur simultaneously in each part of the casting, they tend to set up residual stresses, which may be very high and may therefore cause the casting to crack in the foundry or in service. A quantitative indication of changes in expansion coefficient and specific heat with changes in temperature is shown in Fig. 9.

The nickel-chromium martensitic white irons, containing up to about 7% Ni and 11% Cr, are usually put into service after only a low-temperature heat treatment at 230 to 290 °C (450 to 550 °F) to temper the martensite and to increase toughness. If retained austenite is present and the iron therefore has less than optimum hardness, a subzero treatment down to liquid nitrogen temperature can be employed to transform much of the retained austenite to martensite. Subzero treatment substantially raises the hardness, often as much as 100 Brinell points. Following subzero treatment, the castings are almost always tempered at 230 to 260 °C (450 to 500 °F). The austenite-martensite microstructures produced in nickel-alloyed irons are often desirable for their intrinsic toughness.

It is possible to transform additional retained austenite by heat treating nickel-chromium white irons at about 730 °C (1350 °F). Such a treatment decreases matrix carbon and therefore raises the M_s temperature. However, high-temperature treatments are usually less desirable than subzero treatments because the former are more costly and more likely to induce cracking due to transformation stresses.

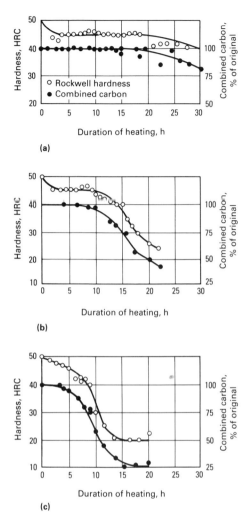

Fig. 8 Effect of annealing on hardness and combined carbon content in chilled iron. Effect of heating at (a) 815 °C (1500 °F), (b) 845 °C (1550 °F), (c) 870 °C (1600 °F) on hardness and combined carbon content of chilled portion of a chilled iron casting. See text for composition.

The high-chromium martensitic white irons (>12% Cr) must be subjected to a high-temperature heat treatment to develop full hardness. They can be annealed to soften them for machining, then hardened to develop the required abrasion resistance. Because of their high chromium content, there is no likelihood of graphitization while the castings are held at the reaustenitizing temperature.

The usual reaustenitizing temperature for high-chromium irons ranges from about 955 °C (1750 °F) for a 15Cr-Mo iron to about 1065 °C (1950 °F) for a 27% Cr iron. An appreciable holding time (3 to 4 h minimum) at temperature is usually mandatory to permit precipitation of dispersed secondary carbide particles in the austenite. This lowers the amount of carbon dissolved in the austenite to a level that permits transformation to martensite during cooling to room temperature. Air quenching is usually used, although small, simply shaped castings can be quenched in oil or molten salt without producing quench cracks. Following quenching, it is advisable to stress relieve (temper) the castings at about 205 to 260 °C (400 to 500 °F). Figure 10 is a continuous-cooling time-temperature-transformation diagram for a typical high-chromium iron designed for use in moderately heavy sections.

Microstructure. With rapid solidification, such as that which occurs in thin-wall castings or when the iron solidifies against a chill, the austenite dendrites and eutectic carbides are fine grained, which tends to increase fracture toughness. In low-chromium white irons, rapid solidification will also reduce any tendency toward formation of graphite. The presence of graphite severely degrades abrasion resistance. Chills in the mold can be used to promote directional solidification (Fig. 11) and therefore reduce shrinkage cavities in the casting. Certain inoculants, notably bismuth, may beneficially alter the solidification pattern by reducing spiking or by producing a finer as-cast grain size.

Immediately after solidification, the microstructure of unalloyed or low-chromium white irons consists of austenite dendrites, containing up to about 2% C, surrounded by M_3C carbides (Fig. 12). When the chromium content of the iron exceeds about 7 wt%, the structure contains M_7C_3 eutectic carbides surrounded by austenite (Fig. 13). This reversal of the continuous phase in the structure tends to increase the fracture toughness of white irons, but only those irons that have a hypoeutectic or eutectic carbon equivalent. All hypereutectic white irons are relatively brittle and are seldom used commercially.

After a white iron casting has solidified and begins to cool to room temperature, the carbide phase may decompose into graphite plus ferrite or austenite. This tendency to form graphite can be suppressed by rapid cooling or by the addition of carbide-stabilizing alloying elements—usually chromium, although inoculating with tellurium or bismuth is also very effective. Austenite in the solidified white iron structure normally undergoes several changes as it cools to ambient temperature. If it is cooled slowly enough, it tends to reject hypereutectoid carbon, either on existing eutectic carbide particles or as particles, platelets, or spines within the austenite grains. This precipitation occurs principally between about 1040 and 760 °C (1900 and 1400 °F). The rate of precipitation depends on both time and temperature.

As the austenite cools further, through the range of 705 to 540 °C (1300 to 1000 °F),

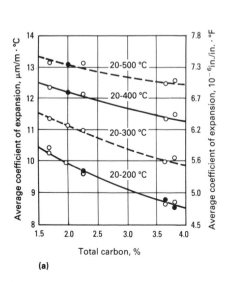
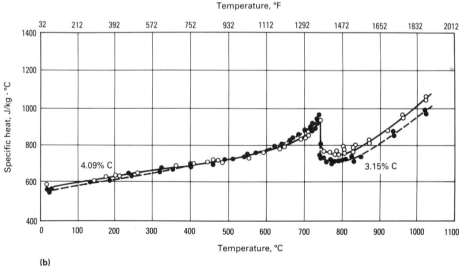

Fig. 9 Thermal expansion (a) and specific heat (b) of white iron

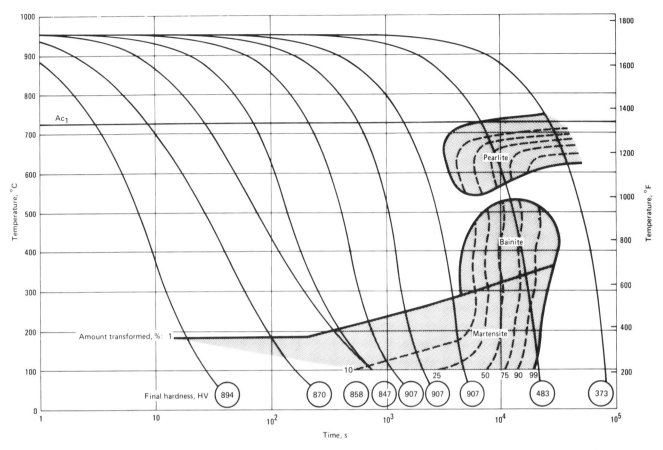

Fig. 10 Continuous-cooling transformation diagram for a white cast iron. Composition: 2.96TC-0.93Si-0.79Mn-17.5Cr-0.98Cu-1.55Mo; austenitized at 955 °C (1750 °F) for 2.5 h. Ac_1 is the temperature at which austenite begins to form upon heating.

it tends to transform to pearlite. This transformation, however, can be suppressed by rapid cooling and/or by the use of pearlite-suppressing elements in the iron.

Nickel, manganese, and copper are the principal pearlite-suppressing elements. Chromium does not contribute significantly to pearlitic suppression (hardenability) in many white irons, because most of the chromium is tied up in carbides. Molybdenum, a potent carbide former, is also tied up in carbides; however, in high-chromium irons, there is enough chromium and molybdenum remaining in the matrix to contribute significantly to hardenability.

Upon cooling below about 540 °C (1000 °F), the austenite may transform to bainite or martensite, thus producing martensitic

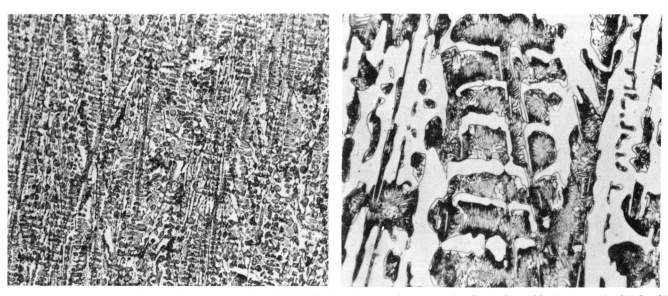

Fig. 11 Structure of unalloyed chill-cast white iron. Composition: 3.6TC-0.7Si-0.8Mn. Structure shows coarse lamellar pearlite and ferrite in a matrix of M_3C carbides. Left: 4% picral etch, 100×. Right: 4% picral etch, 1000×

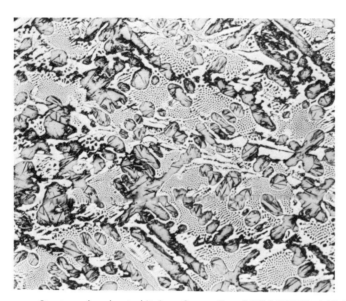 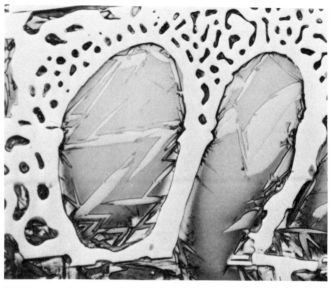

Fig. 12 Structure of sand-cast white iron. Composition: 3.4TC-0.4Si-0.6Mn-1.4Cr-3.0Ni. Structure shows austenite-martensite in a matrix of M_3C carbides. Left: Etched in 1% picric acid plus 5% HCl in methanol, 100×. Right: Etched in picral and a hot aqueous solution of picric acid, 1000×

white iron, which is currently the most widely used type of abrasion-resistant white iron. Martensitic white irons usually contain some retained austenite, which is not considered objectionable unless it exceeds about 15%. Retained austenite is metastable and may transform to martensite when plastically deformed at the wearing surface of the casting.

Silicon has a substantial influence on the microstructure of any grade of white iron. Normally, silicon content exceeds 0.3%, and it may range as high as 2.2% in some of the high-chromium grades. During the solidification of unalloyed or low-alloy irons, silicon tends to promote the formation of graphite, an effect that can be suppressed by rapid solidification or by the addition of carbide-stabilizing elements. After solidification, either while the casting is cooling to ambient temperature or during subsequent heat treatment, silicon tends to promote the formation of pearlite in the structure if it is the only alloy present. However, in the presence of chromium and molybdenum, both of which suppress ferrite, silicon has a minimal effect on ferrite and substantially suppresses bainite. In certain alloy white irons with high retained-austenite contents, increasing the silicon content raises the M_s temperature of the austenite, which in turn promotes the transformation of austenite to martensite. Silicon is also used to enhance the hardening response when the castings are cooled below ambient temperature. The microstructure of heat-treated high-chromium, high-molybdenum white iron is shown in Fig. 14.

Mechanical Properties. Hardness is the principal mechanical property of white iron that is routinely determined and reported. Other (nonstandard) tests to determine strength, impact resistance, and fracture toughness are sometimes employed by individual users, producers, or laboratories. Because of the difficulty of preparing test specimens, especially from heavy-section castings, these nonstandard tests are sel-

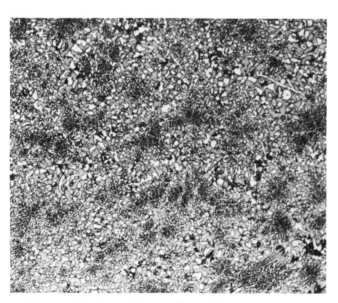 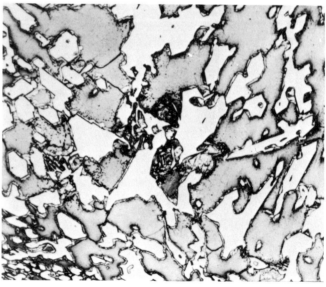

Fig. 13 Structure of sand-cast white iron. Composition: 3.5TC-0.4Si-0.8Mn-16.0Cr-3.0Mo. Structure shows M_7C_3 carbides in a matrix of austenite containing small amounts of pearlite and martensite. Left: Etched in 1% picric acid plus 5% HCl in methanol, 100×. Right: Etched in 1% picric acid plus 5% HCl in methanol, 1000×

Table 3 Mechanical properties of standard martensitic white cast irons
Compositions are given in Table 2.

Class	Type	Designation	Hardness, HB Minimum value Sand cast	Hardness, HB Minimum value Chill cast	Hardness, HB Maximum value Hardened	Hardness, HB Maximum value Annealed	Typical maximum section thickness mm	Typical maximum section thickness in.
I	A	Ni-Cr-HC	550	600	...	...	203	8
I	B	Ni-Cr-LC	550	600	...	...	203	8
I	C	Ni-Cr-GB	550	600	...	...	76(a)	3(a)
I	D	Ni-Hi-Cr	550	500	600	...	305	12
II	A	12% Cr	550	...	600	400	25(a)	1(a)
II	B	15% Cr-Mo-LC	450	...	600	400	102	4
II	C	15% Cr-Mo-HC	550	...	600	400	76	3
II	D	20% Cr-Mo-LC	450	...	600	400	203	8
II	E	20% Cr-Mo-HC	450	...	600	400	305	12
III	A	25% Cr	450	...	600	400	203	8

(a) Ball diameter. Source: ASTM A 532-75a

Table 4 Hardness conversions for white cast irons (from averaged data)

HB	HV	HRC	Scleroscope
High-chromium irons			
815	1000	68.5	...
800	975	68	...
790	950	67.5	...
775	925	67	...
760	900	66	...
745	875	65.0	...
730	850	64.5	...
720	825	63.5	...
700	800	62.5	...
680	775	61.5	...
660	750	61.0	...
640	725	59.5	...
625	700	58	...
610	675	57	...
585	650	56	...
560	625	54.5	...
540	600	53	...
520	575	51.5	...
490	550	50	...
475	525	48.5	...
440	500	47	...
420	475	45.5	...
395	450	43.5	...
370	425	41.7	...
...	400	40	...
Nickel-chromium irons			
750	830–860	...	90–93
700	740–770	...	84–87
650	690–720	...	79–82
600	630–660	...	75–78
550	570–610	...	70–73
500	510–540	...	67–70

dom used for routine quality control. Two exceptions are the tumbling-breakage test and the repeated-drop test, which have been routinely used by certain producers for testing grinding balls.

Minimum hardness values for pearlitic white irons are 321 HB for the low-carbon grade and 400 HB for the high-carbon grade. A chill cast high-carbon 2% Cr white iron may reach a hardness of about 550 HB. A typical hardness range for a sand cast high-carbon grade is about 430 to 500 HB.

Table 3 lists minimum specified hardness levels, together with maximum annealed hardness and typical maximum annealed hardness, and typical maximum section thickness, for the martensitic compositions listed in Table 2. The minimum hardness specified for the hardened (heat-treated) class II castings is well below the average expected hardness. These irons, when fully hardened so that they are free from high-temperature products of austenite transformation, will have hardness values ranging from about 800 to 950 HV (depending on retained austenite content), as for example the 847 to 907 HV range in Fig. 10. The 800 to 950 HV range is equivalent to 700 to 790 HB or 62.5 to 67.5 HRC. For optimum abrasion resistance of the class I irons in Table 3, the minimum Brinell hardness, as measured with a tungsten carbide ball or converted from HV or HRC values, should be 700 HB.

Hardness conversions for white irons are somewhat different from the published data for steel. Table 4 lists data for two classes of white irons: high-chromium irons and nickel-chromium irons. Because of inherent variations in structure for many cast irons, hardness conversion must be made cautiously. For example, Brinell hardness tests are more consistent and reliable for coarse structures such as those typical of heavy sections.

The tensile strength (in reality, the fracture strength) of pearlitic white irons normally ranges from about 205 MPa (30 ksi) for high-carbon grades to about 415 MPa (60 ksi) for low-carbon grades. The tensile strength of martensitic irons with M_3C carbides ranges from about 345 to 415 MPa (50 to 60 ksi), while high-chromium irons, with their M_7C_3-type carbides, usually have tensile strengths of 415 to 550 MPa (60 to 80 ksi). Limited data indicate that the yield strengths of white irons are about 90% of their tensile strengths. These data are extremely sensitive to variations in specimen alignment during testing. Because of the near-zero ductility of white irons, the use-

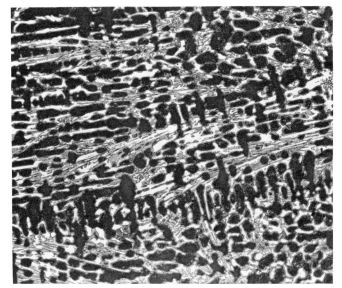
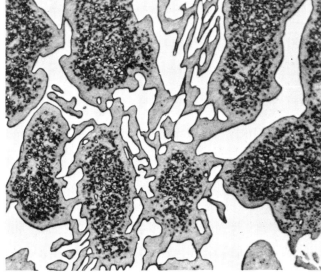

Fig. 14 Structure developed by the heat treatment of a high-chromium iron. Specimen from a chute liner 51 mm (2 in.) thick, containing 2.6% TC, 1.5% Si, 1.1% Mn, 14.3% Cr, and 3.0% Mo; reheated to 1040 to 1065 °C (1900 to 1950 °F) and air cooled. Structure is M_7C_3 eutectic carbides in a matrix of martensite and fine spheroidized M_3C carbides. Left: Hot aqueous picric acid etch, 100×. Right: Hot aqueous picric acid etch, 1000×

fulness of tensile test data for design or quality assurance is very limited.

Transverse strength, which is an indirect measurement of tensile strength and tensile ductility, can be determined with a moderate degree of accuracy on unmachined cast test bars. The product of transverse strength and deflection provides one measure of toughness. Table 5 lists typical values of transverse strength, deflection, and toughness for as-cast test bars. These values should be considered very general; the wide spread of these data emphasizes that properties depend to a marked degree on composition and the conditions under which the castings were produced.

Typical mechanical properties of alloyed white irons are given in Fig. 15 (Ref 1). Hardness, yield strength in compression, and impact energy absorbed by unnotched Charpy specimens are shown for as-cast austenitic and heat-treated martensitic irons. The chromium-molybdenum iron conforms to ASTM class II type E (see Table 2), the high-chromium iron conforms to ASTM class III type A, and the Ni-Hard IV iron conforms to ASTM class I type D.

The elastic modulus of a white iron is considerably influenced by its carbide structure. An iron with M_3C eutectic carbides has a tensile modulus of 165 to 195 GPa (24 to 28 × 10^6 psi), irrespective of whether it is pearlitic or martensitic, while an iron with M_7C_3 eutectic carbides has a modulus of 205 to 220 GPa (30 to 32 × 10^6 psi).

Physical Properties. The density of white irons ranges from 7.50 to 7.75 g/cm^3 (0.271 to 0.280 lb/in.^{3}). Increasing carbon content tends to decrease density; increasing the amount of retained austenite in the structure tends to increase density. Other physical properties are summarized in Table 6 for low-carbon white iron and martensitic nickel-chromium white iron.

Abrasion Resistance. The relative abrasion resistance of various types of white iron has been extensively studied in commercial service and in many types of laboratory abrasion tests. In general, martensitic white irons have substantially better abrasion resistance than pearlitic or austenitic white irons. There can be substantial differences in abrasion resistance among the various martensitic irons. The degree of superiority of one type over another can also vary considerably, depending on the application and also on whether abrasive wear is due to gouging, high-stress (grinding) abrasion, or low-stress scratching or erosion. In addition, performance in a wet environment may be quite different from that in a dry environment.

For nickel-chromium martensitic white irons, there are conflicting data as to the relative serviceability of sand cast and chill cast parts subjected to abrasive wear. This is not particularly surprising, because many of the data were obtained in tests using abrasive ores where the nature of the gangue was incompletely defined or largely ignored. As discussed in the article "Wear Failures" in Volume 11 of the 9th Edition of *Metals Handbook*, it is very important to completely characterize any abrasive substance, particularly

Table 5 Transverse strengths and relative toughness of various pearlitic and martensitic white irons in the as-cast condition
Data from as-cast 30.5 mm (1.2 in.) diam test bars broken over a 457 mm (18 in.) span

Type of iron	Basic composition	Transverse strength kg	Transverse strength lb	Deflection mm	Deflection in.	Toughness(a) kg · m	Toughness(a) lb · in.
Sand cast pearlitic	3.2–3.5 C, 1–2 Cr	635–815	1400–1800	2.0–2.3	0.080–0.092	1.29–1.87	112–162
Sand cast martensitic	2.8–3.6 C, 1.4–4 Cr, 3.3–5 Ni	1810–2490	4000–5500	2.0–3.0	0.08–0.12	3.68–7.60	320–660
	2.5–3.6 C, 7–11 Cr, 4.5–7 Ni	2270–2720	5000–6000	2.0–2.8	0.08–0.11	4.6–7.60	400–660
	2.8–3.4 C, 12–16 Cr, 2–4 Mo	1015–1370	2235–3015	3.2–3.6	0.125–0.14	3.21–4.93	279–422
	3.5–4.1 C, 12–16 Cr, 2.5–3 Mo	800–1000	1760–2200	2.0–2.8	0.08–0.110	1.60–2.80	140–240
Chill cast martensitic	2.8–3.6 C, 1.4–4 Cr, 3.3–5 Ni	2040–3180	4500–7000	2.0–3.0	0.08–0.12	4.15–9.68	360–840
	2.5–3.6 C, 7–11 Cr, 4.5–7 Ni	2500–3180	5500–7000	2.5–3.8	0.10–0.15	6.34–12.1	550–1050
	3.2–3.4 C, 12–16 Cr, 1.5–3 Mo	1980–2295	4360–5060	5.1–6.5	0.202–0.26	10–15.2	870–1320
	3.5–4.1 C, 12–16 Cr, 2.5–3 Mo	1270–1575	2800–3470	3.6–3.8	0.140–0.15	4.52–6.0	392–520

(a) Relative toughness evaluated as product of transverse strength times deflection

Table 6 Physical properties of selected alloy cast irons

Description(a)	Density g/cm^3	Density lb/in.^{3}	Coefficient of thermal expansion(b) μm/m · °C	Coefficient of thermal expansion(b) 10^{-6} in./in. · °F	Electrical resistivity, μΩ · m	Thermal conductivity W/m · K	Thermal conductivity Btu/ft · h · °F
Abrasion-resistant white irons							
Low-carbon white iron	7.6–7.8	0.275–0.282	12(c)	6.7(c)	0.53	22(d)	13(d)
Martensitic nickel-chromium iron	7.6–7.8	0.275–0.282	8–9(c)	4.4–5(c)	0.80	30(d)	17(d)
Corrosion-resistant irons							
High-silicon iron	7.0–7.05	0.252–0.254	12.4–13.1	6.9–7.3	0.50	...	...
High-chromium iron	7.3–7.5	0.264–0.271	9.4–9.9	5.2–5.5	...	...	...
High-nickel gray iron	7.4–7.6	0.267–0.275	8.1–19.3	4.5–10.7	1.0(d)	38–40	22–23
High-nickel ductile iron	7.4	0.267	12.6–18.7	7.0–10.4	1.0(d)	13.4	7.75
Heat-resistant gray irons							
Medium-silicon iron	6.8–7.1	0.246–0.256	10.8	6.0	...	37	21
High-nickel iron	7.3–7.5	0.264–0.271	8.1–19.3	4.5–10.7	1.4–1.7	37–40	21–23
Nickel-chromium-silicon iron	7.33–7.45	0.265–0.269	12.6–16.2	7.0–9.0	1.5–1.7	30	17
High-aluminum iron	5.5–6.4	0.20–0.23	15.3	8.5	2.4	...	...
Heat-resistant white iron							
High-chromium iron (ferritic)	7.3–7.5	0.264–0.271	9.3–9.9	5.2–5.5	...	20	12
Heat-resistant ductile irons							
Medium-silicon ductile iron	7.1	0.257	10.8–13.5	6.0–7.5	0.58–0.87	...	...
High-nickel ductile (20% Ni)	7.4	0.268	18.7	10.4	1.02	13	7.7
High-nickel ductile (23% Ni)	7.4	0.268	18.4	10.2	1.0(d)	...	...
High-nickel ductile (30% Ni)	7.5	0.270	12.6–14.4(e)	7.0–8.0(e)	...	...	...
High-nickel ductile (36% Ni)	7.7	0.278	7.2(e)	4(e)	...	...	...

(a) For compositions, see Table 1. (b) At 21 °C (70 °F). (c) 10–260 °C (50–500 °F). (d) Estimated. (e) 20–205 °C (70–400 °F)

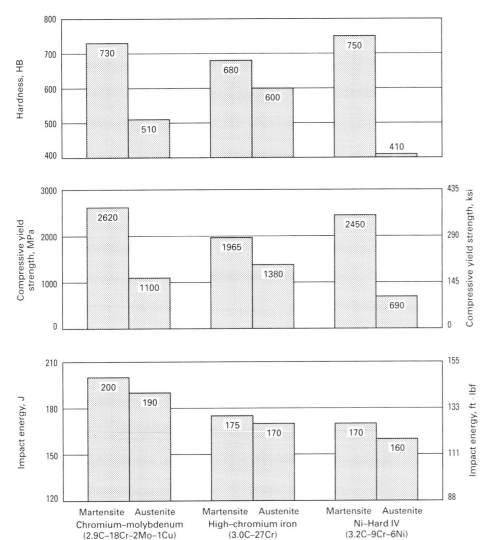

Fig. 15 Typical mechanical properties of white cast irons. Source: Ref 1

Table 7 Hardness of various abrasive minerals as related to hardness of microconstituents in white iron

	Mohs' scale	Constituent hardness HRC	Constituent hardness HV
Diamond	10	...	8000
Silicon carbide	...	...	2500
Tungsten carbide	...	...	2400
Corundum (Al_2O_3)	9	...	2000
M_7C_3-type carbides in high-chromium irons	...	...	1500–1800
Topaz	8	...	1300
M_3C-type carbides in low-chromium irons	...	...	1060–1240
Quartz (silica)	7	...	800–1000
Garnet	7	...	800–1000
High-carbon martensite	6½	62	770–800
Feldspar	6	...	500–600
Austenite in high-chromium irons	...	35–40	350–400
High-carbon pearlite	...	20–43	240–425

ores and other earthy mixtures, so that the results of tests and service evaluations can be properly analyzed.

The hardness of the abrasive material has a marked influence on relative abrasion rates. For example, when the abrasive is silicon carbide, which is hard enough to scratch M_3C and M_7C_3 carbides as well as martensite and pearlite, there may be little difference in relative wear rates among any of the white irons. However, with silica (the abrasive most commonly encountered in service), which is not hard enough to scratch M_7C_3 carbides but may scratch M_3C carbides and definitely will scratch martensite and pearlite, high-chromium white irons, with their M_7C_3 carbides, tend to provide superior performance. If the abrasive mineral is a silicate of intermediate hardness such as feldspar, which (theoretically) will not scratch fully hard martensite but will scratch pearlite, any of the martensitic white irons should perform much better than any of the pearlitic white irons. An indication of relative abrasion resistance can be obtained by comparing the hardness of the commonly encountered abrasive minerals with the hardness of the microconstituents in white irons, as indicated in Table 7.

The relatively low hardness of the retained austenite in high-chromium irons (Table 7) deserves special consideration. Because this austenite tends to work harden rapidly and may also transform to martensite, it is quite abrasion resistant when severely loaded. However, most abrasion tests and field experience indicate that irons containing considerable retained austenite are not as abrasion resistant as those put into service with fully martensitic microstructures.

The relative wear rates of several types of white iron, as determined in various laboratory tests, are listed in Table 8 (Ref 2). Field test data for grinding mill liners are also given in Table 8. For comparison, the abrasion resistance of a fully hardened, high-carbon martensitic steel is included, when available.

The laboratory tests include:

- A jaw crusher test, which provides a *gouging wear ratio* between the weight losses of the test material and a hardened and tempered structural steel plate (lower ratios indicate greater resistance to abrasion)
- Weight-loss data for a *pin test*, in which a test pin is continually abraded by fresh garnet abrasive cloth for a designated distance
- Weight-loss data for a *rubber-wheel test*, in which the specimen is abraded by a sand slurry forced between the specimen and a rotating rubber wheel under a designated load

Table 8 Abrasion resistance and wear rates of selected white irons

Material designation	ASTM designation Class	ASTM designation Type	Abrasion resistance Gouging wear ratio	Abrasion resistance Weight loss, g Pin test	Abrasion resistance Weight loss, g Rubber-wheel test	Relative wear rate in grinding mill liners(a)
Ni-Cr-HC	I	A	...	...	0.035	105–109
Ni-Cr-LC	I	B	0.151	...	0.055	...
15% Cr-Mo-LC	II	B	0.076	...	0.065	88–90
15% Cr-Mo-HC	II	C	0.044	...	0.036	...
20% Cr-Mo-HC	II	E	0.081	0.042	0.054	...
25% Cr	III	A	0.127	...	0.099	98–100
Martensitic steel (1C-5Cr-1Mo)	...	...	...	0.077	...	100

(a) Martensitic steel used as a reference and assigned a value of 100. Values < 100 indicate a more wear-resistant material; >100 indicates a less wear-resistant material. Source: Ref 3

98 / Cast Irons

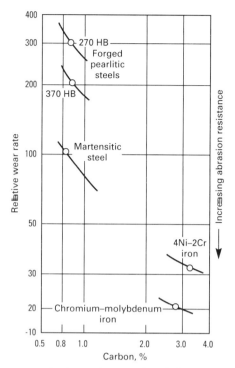

Fig. 16 Service performance in grinding cement clinker

The gouging wear test simulates rock-crushing service; the pin test provides high-stress abrasion, as found in various types of grinding mills; and the rubber-wheel test simulates low-stress abrasion, as found in classifiers and other equipment handling abrasive slurries.

The relative wear rates in grinding mill liners were obtained from field tests in which liner segments of various materials were weighed before and after service (Ref 3). The martensitic steel weight loss was assigned a value of 100, rates less than 100 indicate a more wear-resistant material, and rates greater than 100 indicate a material less resistant to wear.

Components of mills for grinding cement clinker have been made from several different irons and steels. The relative wear rates of mills in this service as determined during a field test are shown in Fig. 16.

Corrosion-Resistant Cast Irons

The corrosion resistance of gray cast iron is enhanced by the addition of appreciable amounts of nickel, chromium, and copper, singly or in combination, or silicon in excess of about 3%. Chemical composition ranges bracketing some of the more widely used corrosion-resistant cast irons are given in Table 1. Typical mechanical property ranges for the four basic types of corrosion-resistant cast iron are given in Table 9; physical property ranges for these irons are given in Table 6.

Up to 3% Si is normally present in all cast irons; in larger percentages, silicon is considered an alloying element. It promotes the formation of a strongly protective surface film under oxidizing conditions such as exposure to oxidizing acids. Relatively small amounts of molybdenum and/or chromium can be added in combination with high silicon. The addition of nickel to gray iron improves resistance to reducing acids and provides high resistance to caustic alkalies. Chromium assists in forming a protective oxide that resists oxidizing acids, although it is of little benefit under reducing conditions. Copper has a smaller beneficial effect on resistance to sulfuric acid.

High-silicon irons are the most universally corrosion-resistant alloys available at moderate cost. They are widely used for handling the corrosive media common in chemical plants, even when abrasive conditions are also encountered. When the silicon content is 14.2% or higher, these irons exhibit a very high resistance to boiling sulfuric acid (Fig. 17a). They are especially useful when the concentration of sulfuric acid is above 50%, at

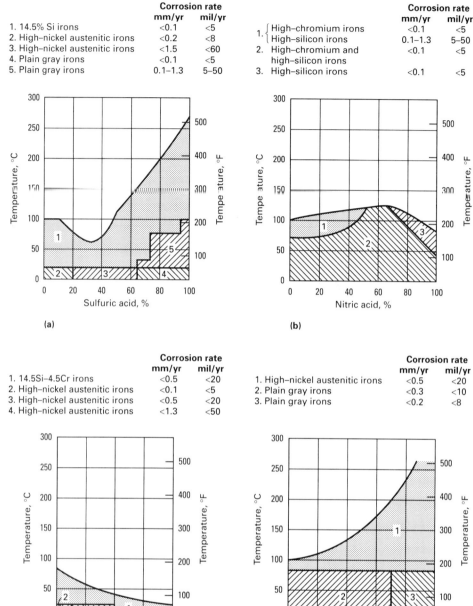

Fig. 17 Useful life of plain and alloyed cast irons in acid and alkaline media as a function of temperature and concentration of the corrodents. (a) Sulfuric acid. (b) Nitric acid. (c) Hydrochloric acid. (d) Sodium hydroxide

which point they are virtually immune to attack. As shown in Fig. 17(b), the high-silicon irons are also very resistant to nitric acid. Increasing the silicon content to 16.5% makes the alloy quite resistant to corrosion in boiling nitric and sulfuric acids at nearly all concentrations, but because this is accompanied by a reduction of mechanical strength, it is not ordinarily done in the United States.

The 14.5% Si iron is less resistant to the corrosive action of hydrochloric acid, but this resistance can be improved by additions of chromium and molybdenum and can be further enhanced by increasing the silicon content to 17%. The chromium-bearing silicon irons are very useful in contact with solutions containing copper salts, free wet chlorine, or other strongly oxidizing contaminants.

The high-silicon irons are very resistant to organic acid solutions at any concentration or temperature. However, their resistance to strong hot caustics is not satisfactory for most purposes. They are resistant to caustic solutions at lower temperatures and concentrations, and (although they are no better than unalloyed gray iron in this regard) they can be used where caustics and other corrosives are mixed or alternately handled. They have no useful resistance to hydrofluoric or sulfurous acids.

High-silicon irons have poor mechanical properties and particularly low thermal and mechanical shock resistance. These alloys are typically very hard and brittle, with a tensile strength of about 110 MPa (16 ksi) and a hardness of 480 to 520 HB. They are difficult to cast and are virtually unmachinable. Their considerable use stems from their outstanding resistance to acids such as those mentioned above. They are widely used for drain pipe in chemical plants, laboratories, hospitals, and schools. High-silicon iron towers, tubes, and fittings are standard equipment for concentrating sulfuric and nitric acids in the explosives and fertilizer industries.

High-silicon iron pumps, valves, mixing nozzles, tank outlets, and steam jets are widely used for handling severe corrodents such as chromic acid, sulfuric-acid slurries, bleach solutions, and acid-chloride slurries, which are frequently encountered in plants that manufacture paper, pigments, or dye stuffs or that use electroplating solutions. High-silicon irons are also widely used for anodes in impressed-current cathodic-protection systems, especially where aggressive environments such as seawater or chloride soils are encountered.

The mechanical strength and shock resistance of high-silicon irons can be improved by lowering the silicon content to 12% or slightly less; this practice is occasionally followed in the United States and Europe. However, reducing the silicon content to 12% causes a significant reduction in corrosion resistance and therefore is feasible only in applications where the loss in corrosion resistance has a minimal effect on service life or is offset by the benefit derived from the increase in strength.

High-chromium irons containing 20 to 35% Cr give good service with oxidizing acids, particularly nitric, but are not resistant to reducing acids. These irons are also reliable for use in weak acids under oxidizing conditions, in numerous salt solutions, in organic acid solutions, and in marine or industrial atmospheres.

The corrosion resistance of high-chromium cast iron to nitric acid is exceptional; it resists all concentrations of this acid up to 95% at room temperature. Its corrosion rate is less than 0.13 mm (0.005 in.) per year at all temperatures up to the boiling point for concentrations up to 70%. In handling nitric acid, the chromium irons are complementary to high-silicon irons. The former exhibit excellent corrosion resistance to all concentrations and temperatures, except for boiling concentrated acids, while the latter give better results in stronger acid. Data relating corrosion resistance to temperatures and acid concentrations are shown in Fig. 17(b) for both high-chromium and high-silicon irons.

The low-carbon, high-chromium irons are satisfactory for annealing pots; lead, zinc, or aluminum melting pots; conveyor links; and other parts exposed to corrosion at high temperature. Because the corrosion resistance is imparted by chromium present in solid solution in the ferritic matrix, this element must be present in sufficient quantity to combine with carbon as chromium carbide and still remain in the desired amount in the ferrite. Chromium contents of 30 to 33% are common in irons for use under conditions of severe acid corrosion.

High-chromium irons are resistant to all concentrations of sulfurous acid at temperatures up to 80 °C (175 °F), to sulfite liquors used in the papermaking industry, to hypochlorite bleaching liquors at room temperature, to cold aluminum sulfate in concentrations up to 5%, and to some salts that hydrolyze to give acid solutions. They resist all concentrations of phosphoric acid up to 60% at temperatures up to the boiling point

Table 9 Typical mechanical properties of corrosion-resistant cast irons

Type of iron(a)	Hardness, HB	Tensile strength MPa	ksi	Compressive strength MPa	ksi	Impact energy J	ft · lbf	Transverse breaking load(b) kg	lb	Transverse deflection(b) mm	in.
High-silicon iron	480–520	90–180	13–26	690	100	2.7–5.4(c) 0.1–3(d)	2–4(c) 0.1–2(d)	545–1000 910–1590	1200–2200 2000–3500	0.65 1.5–3.8	0.026 0.06–0.15
High-chromium iron	250–740	205–830	30–120	690	100	27–47(c)	20–35(c)	...	...	...	...
High-nickel gray iron	120–250	170–310	25–45	690–1100	100–160	80–200(c)	60–150(c)	820–1590	1800–3500	5–25	0.20–1.00
High-nickel ductile iron	130–240	380–480	55–70	1240–1380	180–200	14–40(d)	10–30(d)	...	...	...	...

(a) For composition ranges, see Table 1. (b) For as-cast 30.5 mm (1.2 in.) diam bar broken over a 457 mm (18 in.) span. (c) Unnotched 30.5 mm (1.2 in.) diam test bar broken over a 152 mm (6 in.) span in a Charpy testing machine. (d) Standard Charpy

Table 10 Typical mechanical properties of heat-resistant alloy cast irons

Type of iron(a)	Hardness, HB	Tensile strength MPa	ksi	Compressive strength MPa	ksi	Impact energy J	ft · lbf	Transverse breaking load(b) kg	lb	Transverse deflection(b) mm	in.
Medium-silicon gray iron	170–250	170–310	25–45	620–1040	90–150	20–31(c)	15–23(c)	455–1090	1000–2400	4.6–8.9	0.18–0.35
High-chromium gray iron	250–500	210–620	30–90	690	100	27–47(c)	20–35(c)	910–1590	2000–3500	1.5–3.8	0.06–0.15
High-nickel gray iron	130–250	170–310	25–45	690–1100	100–160	80–200(c)	60–150(c)	820–1360	1800–3000	5–25	0.2–1.0
Ni-Cr-Si gray iron	110–210	140–310	20–45	480–690	70–100	110–200(c)	80–150(c)	820–1130	1800–2500	7–35	0.3–1.4
High-aluminum gray iron	180–350	235–620	34–90	...	...	...	...	...	...	...	...
Medium-silicon ductile iron	140–300	415–690	60–100(c)	...	...	7–155(d)	5–115(d)	...	...	...	...
High-nickel ductile iron (20% Ni)	140–200	380–415	55–60(e)	1240–1380	180–200	16(f)	12(f)	...	...	...	...
High-nickel ductile iron (23% Ni)	130–170	400–450	58–65(g)	...	...	38(f)	28(f)	...	...	...	...

(a) For composition ranges, see Table 1. (b) Unnotched 30.5 mm (1.2 in.) diam test bar broken on 152 mm (6 in.) supports in a Charpy testing machine. (c) Yield strength, 310–520 MPa (45–75 ksi); elongation, 0.2%. (d) Standard Charpy test on 10 mm (0.4 in.) unnotched specimen. (e) Yield strength, 210–240 MPa (30–35 ksi); elongation, 8–20%. (f) Standard Charpy test on 10 mm (0.4 in.) notched specimen. (g) Yield strength, 195–240 MPa (28–35 ksi); elongation, 20–40%

100 / Cast Irons

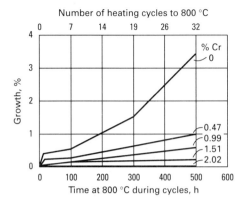

Fig. 18 Effect of chromium on the growth of cyclically heated gray iron. Growth is plotted as a function of the number of heating cycles and time at 800 °C (1470 °F).

and 85% concentrations up to 80 °C (175 °F). They also have good resistance to aerated seawater and most mine waters, including acidic types.

Chromium cast irons have better mechanical properties than high-silicon irons and respond readily to heat treatment when chromium and carbon contents are suitably balanced. Tensile strength as high as 480 MPa (70 ksi) is obtained with a hardness of 290 to 340 HB. These alloys are generally resistant to shock and can be machined; both of these properties are improved when carbon content is lowered to about 1.2%. As carbon is increased, machinability in the annealed condition decreases; consequently, irons with carbon contents of 3% or more should be used only when no machining is required. The maximum service temperature for high-chromium irons is generally 815 to 1095 °C (1500 to 2000 °F).

High-Nickel Irons. High-nickel austenitic cast irons are produced in several compositions, depending on desired properties and end use. Austenitic gray irons containing large percentages of nickel and copper are fairly resistant to mildly oxidizing acids, including dilute to concentrated sulfuric acid at room temperature. Their resistance to sulfuric acid is compared with that of other cast irons in Fig. 17(a). They are also fairly resistant to hydrochloric and some phosphoric acids at slightly elevated temperatures. Their useful range in hydrochloric acids is indicated in Fig. 17(c).

The corrosion behavior of high-nickel irons is similar to that of unalloyed gray iron in the presence of nitric acid. Although the nickel-containing iron exhibits better corrosion resistance than an 18-8 stainless steel, high-silicon irons are much better for both sulfuric and hydrochloric acids under the conditions shown in Fig. 17(a) and (c).

High-nickel irons exhibit fair resistance to some organic acids (such as acetic, oleic, and stearic acids) and to red oils. Irons with nickel contents of 18% or more are nearly immune to the effects of weak or strong alkalies and caustics, although subject to stress corrosion in strong hot caustics at stresses over 70 MPa (10 ksi). The resistance of these irons to various concentrations of sodium hydroxide at temperatures up to 260 °C (500 °F) is indicated in Fig. 17(d).

High-nickel irons are the toughest of all cast irons containing flake graphite. Although their tensile strength is relatively low, ranging from 140 to 275 MPa (20 to 40 ksi), they have satisfactory toughness and excellent machinability. High-nickel ductile irons, which are specially treated so that the graphite forms as spheroids rather than as flakes, have essentially the same corrosion resistance as high-nickel gray irons, but have much higher strength and ductility. A similar treatment applied to high-silicon iron provides no improvement in mechanical properties.

High-nickel irons provide satisfactory corrosion resistance at elevated temperatures up to about 705 to 815 °C (1300 to 1500 °F). Above this range, high-chromium irons are preferred.

Heat-Resistant Cast Irons

Heat-resistant cast irons are basically alloys of iron, carbon, and silicon having high-temperature properties markedly improved by the addition of certain alloying elements, singly or in combination, principally chromium, nickel, molybdenum, aluminum, and silicon in excess of 3%. Silicon and chromium increase resistance to heavy scaling by forming a light surface oxide that is impervious to oxidizing atmospheres. Both elements reduce the toughness and thermal shock resistance of the metal. Although nickel does not appreciably affect oxidation resistance, it increases strength and toughness at elevated temperatures by promoting an austenitic structure that is significantly stronger than ferritic structures above 540 °C (1000 °F). Molybdenum increases high-temperature strength in both ferritic and austenitic iron alloys. Aluminum additions are very potent in raising the equilibrium temperature (A_1) and in reducing both growth and scaling, but they adversely affect mechanical properties at room temperature.

The chemical composition ranges of some of the more widely used heat-resistant irons (both gray and ductile types) suitable for elevated-temperature service are given in Table 1. Typical physical properties are summarized in Table 6, and typical mechanical property ranges are given in Table 10. The performance of alloy gray irons at elevated temperatures is determined by a number of related properties, such as resistance to growth and oxidation, resistance to thermal shock, response to cyclic heating, creep resistance, rupture strength, and high-temperature fatigue strength.

Growth is the permanent increase in volume that occurs in some cast irons after prolonged exposure to elevated temperature or after repeated cyclic heating and cooling. It is produced by the expansion that accompanies graphitization, expansion, and contraction at the transformation temperature, combined with internal oxidation of the iron. Gases can penetrate the surface of hot cast iron at the graphite flakes and oxidize the graphite as well as the iron and silicon. The occurrence of fine cracks, or crazing, may accompany repeated heating and cooling through the transformation

Table 11 Growth of high-nickel irons in superheated steam at 480 °C (900 °F)

Type of iron	Growth, mm/m or 0.001 in./in., after		
	500 h	1000 h	2500 h
Gray iron (unalloyed)	2.3	5.2	14
High-nickel gray iron (20% Ni)	0.5	1.0	1.5
High-nickel gray iron (30% Ni)	0.3	0.45	0.48
High-nickel ductile iron (20% Ni)	0.3	0.5	0.5
High-nickel ductile iron (30% Ni)	0.3	nil	nil

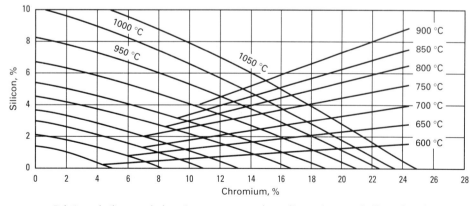

Fig. 19 Relation of silicon and chromium contents to the scaling resistance of silicon-chromium irons. Indicated are the temperatures at which various irons can be used with very little or insignificant scaling in sulfur-free oxidizing atmospheres.

temperature ranges because of thermal and transformational stresses.

Silicon contents of less than about 3.5% increase the rate of growth by promoting graphitization, but silicon contents of 4% or more retard growth. Both manganese and phosphorus decrease growth by acting as carbide stabilizers.

The carbide-stabilizing alloying elements, particularly chromium, effectively reduce growth in gray irons at 455 °C (850 °F) or above. Growth is not a problem below 400 °C (750 °F), except in the presence of superheated steam, where it can occur in coarse-grain irons at about 315 °C (600 °F). Even small amounts of chromium, molybdenum, and vanadium produce marked reductions in growth at the higher temperatures. The influence of chromium in reducing growth in gray iron at 800 °C (1470 °F) is shown in Fig. 18. These data indicate that in cyclic heating, including a total of 500 h at the upper temperature, a chromium content of about 2% serves to eliminate growth.

Chromium irons containing 24 to 34% Cr show no appreciable growth at 1095 °C (2000 °F). In general, cast irons containing 20 to 35% Cr can be used regularly at about 980 °C (1800 °F) and for short periods up to 1095 °C (2000 °F) with satisfactory resistance to growth and scaling.

High-nickel gray and ductile irons are also quite resistant to elevated-temperature growth. For example, the growth of gray and ductile versions of high-nickel cast iron is compared to that of unalloyed gray iron in Table 11; these data are for continuous exposure to superheated steam at 480 °C (900 °F). In addition to resistance to growth, the austenitic gray and ductile irons are resistant to warpage and cracking in cyclic elevated-temperature service. This resistance is attributed to the absence of phase transformation, to moderate elastic moduli, and to good mechanical properties at about 595 to 760 °C (1100 to 1400 °F).

Scaling. In addition to the internal oxidation that contributes to growth, a surface scale forms on unalloyed gray iron after exposure at sufficiently high temperature. The scale formed in air consists of a mixture of iron oxides. The important factor in scale formation is whether the scale, first, is essentially adherent and protective to the base metal or, second, tends to flake and permit continued oxidation of the metal.

Silicon, chromium, and aluminum increase the scaling resistance of cast iron by forming a light surface oxide that is impervious to oxidizing atmospheres. Unfortunately, these elements tend to reduce toughness and thermal shock resistance. The presence of nickel improves the scale resistance of most alloys containing chromium and, more important, increases their toughness and strength at elevated temperatures.

Carbon has a somewhat damaging effect above 705 °C (1300 °F) as a result of the mechanism of decarburization and the evolution of carbon monoxide and carbon dioxide. When these gases are evolved at the metal surface, the formation of protective oxide layers is hindered, and cracks and blisters may develop in the scale. Typical oxide penetration rates for unalloyed and alloyed cast irons and certain stainless steels in air, in a slightly reducing furnace atmosphere, and in sulfur-bearing flue gas are given in Tables 12 to 14.

Figure 19 indicates the temperatures at which various silicon-chromium irons can be used with only slight or insignificant scaling in sulfur-free oxidation atmospheres. Greater scaling rates can be toler-

Table 12 Oxidation of plain and alloy cast irons and one stainless steel

Iron	Composition, %(a)				Oxide penetration				Growth at 815 °C (1500 °F)(c)	
					At 760 °C (1400 °F)(b)		At 815 °C (1500 °F)(c)			
	TC	Si	Cr	Ni	mm/yr	mil/yr	mm/yr	mil/yr	mm/yr	mil/yr
Austenitic(d)	2.69	1.96	2.05	13.96	4.7	184	9.5	374	0.4	15
Austenitic(e)	2.97	2.38	4.87	(14.0)	2.4	96	5.9	232	0.2	8
Austenitic	2.40	1.57	2.98	30.28	2.1	83	6.3	249	0.2	8
Austenitic	(1.8)	(6.0)	(5.0)	(30.0)	0.05	2	1.3	53	0.2	8
Austenitic	(2.8)	(1.7)	(2.0)	(20.0)	4.2	166	7.9	312	0.2	8
Austenitic	(2.7)	(2.5)	(5.0)	(20.0)	1.9	74	3.6	143	0.4	15
Plain ferritic	(3.2)	(2.2)	...	...	>20(f)	>800(f)	>85(f)	>3300(f)	2.0	78
Low-alloy ferritic	(3.3)	(1.5)	(0.6)	(1.5)	>20(f)	>800(f)	>90(f)	>3500(f)	1.4	54
Low-alloy ferritic	(3.3)	(2.2)	(1.0)	(1.0)	5.8	228	25.9	1020	1.2	47
Low-alloy ferritic	(3.1)	(2.2)	(0.9)	(1.5)	7.2	284	29.0	1140	1.6	62
Type 309 stainless	...	...	(25.0)	(12.0)	nil	nil	nil	nil	nil	nil

(a) Parenthetical values are estimates. Phosphorus and sulfur contents in all iron samples were about 0.10%. (b) Exposure of 2000 h in electric furnace at 760 °C (1400 °F) with air atmosphere containing 17–19% O. (c) Exposed for 492 h in gas-fired heat-treating furnace at 815 °C (1500 °F). (d) 6.05% Cu. (e) 6.0% Cu. (f) Specimen completely burned. Source: Ref 4

Table 13 Oxidation of ferritic and austenitic cast irons and one stainless steel

Iron	Composition, %(a)				Growth		Oxide penetration	
	TC	Si	Cr	Ni	mm/yr	mil/yr	mm/yr	mil/yr
After 3723 h at 745–760 °C (1375–1400 °F) in electric furnace, air atmosphere								
Ferritic	3.05	2.67	0.90	1.55	2.0	78	(b)	(b)
Austenitic	2.97	1.63	1.89	20.02	0.8	31	6.9	270
Austenitic	2.52	2.67	5.16	20.03	nil	nil	0.2	6
Austenitic	2.32	1.86	2.86	30.93	nil	nil	2.0	78
Austenitic	1.86	5.84	5.00	29.63	nil	nil	<0.1	<3
309 stainless	...	...	(25.0)	(12.0)	nil	nil	<0.1	<3
After 1677 h at 815–925 °C (1500–1700 °F) in gas-fired furnace, slightly reducing atmosphere								
Ferritic	(3.2)	(2.2)	...	...	3.2	125	(b)	(b)
Austenitic(c)	(3.0)	(2.4)	(5.0)	(14.0)	0.4	15	8.4	330
Austenitic	(2.7)	(2.5)	(5.0)	(20.0)	0.4	15	5.6	220
Austenitic	(2.4)	(1.6)	(3.0)	(30.0)	0.4	15	6.9	270
Austenitic	(1.8)	(6.0)	(5.0)	(30.0)	0.4	15	0.1	5
309 stainless	...	...	(25.0)	(12.0)	nil	nil	0.1	5

(a) Parenthetical values are estimates. Phosphorus and sulfur contents in all iron samples were about 0.10%. (b) Sample was completely burned. (c) 6.0% Cu (est). Source: Ref 4

Table 14 Oxide penetration in ductile irons and one stainless steel

Iron	Estimated composition, %					Oxide penetration			
						Test 1		Test 2	
	TC	Si	Cr	Ni	Cu	mm/yr	mil/yr	mm/yr	mil/yr
After 15 months at 870 °C (1600 °F)(a)									
Austenitic	1.80	6.0	5.0	30.0	...	1.1	44	...	...
18-8 stainless	...	...	18.0	8.0	...	0.2	9	...	...
25-12 stainless	...	...	25.0	12.0	...	0.1	3	...	...
Austenitic	2.90	2.0	2.0	14.0	6.0	(b)	(b)	...	...
Austenitic	2.8	1.7	2.0	20.0	...	(b)	(b)	...	...
Austenitic	2.7	2.5	5.0	20.0	...	(b)	(b)	...	...
Air atmosphere, 400 h at 705 °C (1300 °F) (test 1) air atmosphere(c) (test 2)									
2.5 Si ductile	3.40	2.50	...	...	...	1.1	42	12.7	500
5.5 Si ductile	2.6	5.50	...	...	...	0.1	4	1.3	51
Austenitic ductile	2.3	2.5	1.7	20.0	...	1.1	42	4.4	175
Austenitic ductile	2.3	2.0	...	22.0	...	1.8	70	...	...
Austenitic ductile	2.1	5.5	5.0	30.0	...	1.1	4	nil	nil
Austenitic	3.0	1.6	1.9	20.0	...	2.5	98	7.6	300
309 stainless	...	...	25.0	12.0	...	nil	nil	nil	nil

(a) Exposed to flue gases from powdered coal containing 1.25–2.00% S. (b) Completely oxidized. (c) Exposed to a heat cycle, 600 h at 870–925 °C (1600–1700 °F), 600 h at 870–925 °C (1600–1700 °F) and 425–480 °C (800–900 °F), 600 h at 425–480 °C (800–900 °F). Source: Ref 5

ated in some applications, so that higher useful temperatures are possible. The presence of large amounts of nickel, as in Nicrosilal, or aluminum would increase the temperature limits shown for the various silicon-chromium irons. For example, a 7% Al iron has adequate scale resistance to about 900 °C (1650 °F), while 16 to 25% Al irons are virtually scale free at 1095 °C (2000 °F).

High-Temperature Strength. Measurements of short-time tensile strength, creep strength, and rupture strength provide a basis for evaluating the performance of metals at elevated temperatures. Creep rate increases with temperature and becomes an important design factor at elevated temperatures.

Creep is ordinarily reported in terms of strain for a specified period of time at a given tensile stress and temperature. Because cast irons can grow at elevated temperatures without the application of external stress, the measured increase in length is the sum of growth resulting from metallurgical causes and the mechanical elongation of creep.

Creep in gray iron is appreciably influenced by microstructure and composition. An unalloyed gray iron with a carbon equivalent of about 4% can usually be subjected to a tensile stress of 70 MPa (10 ksi) at 400 °C (750 °F) without exceeding a creep rate of 1% in 10 000 h. Low-alloy irons exhibit even less creep under similar conditions. Ductile irons may sustain stresses up to 185 MPa (27 ksi) at 425 °C (800 °F) without exceeding a creep rate of 1% in 10 000 h. Some austenitic ductile irons have about the same creep strength at 540 °C (1000 °F) as the unalloyed ductile irons display at 425 °C (800 °F).

The short-time high-temperature strength of a metal is taken as the stress that is sufficient to break a standard tension test specimen in a short period of time at elevated temperature. Often, the correlation between the high-temperature strength and load-carrying capacity of a metal over long periods of time is poor or nonexistent. Short-time tensile data for several alloy irons are reported in Fig. 20.

Because of the inadequacy of the short-time tension test, creep rupture test data are used more frequently in the evaluation of high-temperature properties. Although creep tests are performed at stresses that will not break the specimen, rupture tests are run to failure. Typical stress rupture data for the austenitic 20Ni, 20Ni-1Mo, 30Ni, and 30Ni-1Mo ductile irons are given in Fig. 21, along with data for ferritic 4Si and 4Si-1Mo. For comparison, stress rupture data for a cast 19Cr-9Ni stainless steel (ASTM A 297, grade HF) are included.

High-Silicon Irons. Although intermediate amounts of silicon increase the rate of growth in cast iron by increasing the rate of graphitization, additions of 4.5 to 8% Si greatly reduce both scaling and growth. Silicon also has the advantage of raising the transformation temperature to about 900 °C (1650 °F), thus increasing the operating temperature range that may be employed without encountering a phase change (Ref 6).

Ferritic high-silicon gray iron is rather brittle and has a low resistance to thermal shock at room temperature. However, it is superior to ordinary gray iron above about 260 °C (500 °F). An austenitic gray iron containing 5% Si, 18% Ni, and 2 to 5% Cr

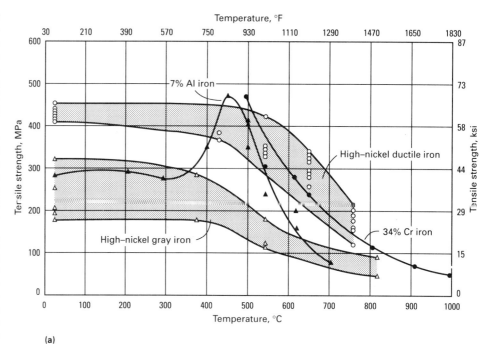

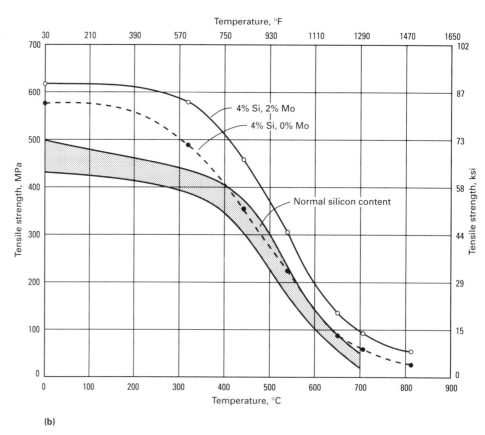

Fig. 20 Short-time elevated-temperature tensile strength of several (a) alloy irons and (b) ferritic nodular irons

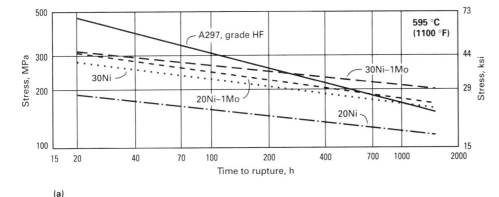

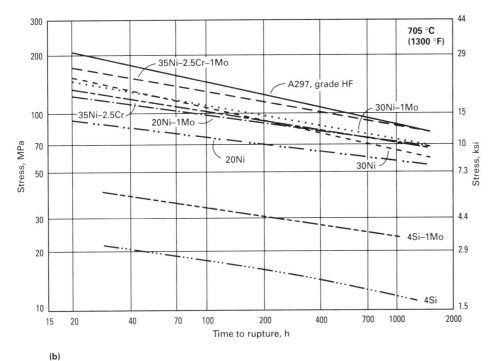

Fig. 21 Typical stress-rupture properties of high-nickel heat-resistant ductile irons. (a) At 595 °C (1100 °F). (b) At 705 °C (1300 °F). Source: Ref 6

exhibits considerably better toughness and thermal shock resistance. Both the plain silicon and the Ni-Cr-Si irons are British developments, the former known commercially as Silal and the latter as Nicrosilal. They exhibit excellent resistance to scaling in air up to 815 °C (1500 °F), and the Ni-Cr-Si iron can be successfully employed in sulfurous atmospheres. The maximum temperature for use of these irons is reported to be 900 °C (1650 °F) for the plain silicon iron and 955 °C (1750 °F) for the Ni-Cr-Si iron. Silicon-containing compositions are also available in ferritic ductile iron.

High-Chromium Irons. Chromium is widely used in heat-resistant irons because of its stabilizing influence on carbides, which deters growth, and its tendency to form a tight, protective oxide. Substantial improvement in oxidation resistance is obtained by the addition of 0.5 to 1% Cr for many applications up to 760 °C (1400 °F). Further improvement in resistance to scaling and growth at 760 °C (1400 °F) without excessive loss in toughness and machinability is reported for irons with up to 2% Cr.

The effect of chromium additions on cyclic heating characteristics and resistance to scaling is discussed in the section "Growth" in this article. Pertinent data are given in Fig. 18 and 19. Machinable castings with considerable heat resistance can be obtained with rather small additions of both chromium and nickel to cast iron. Chromium additions of 15 to 35% are employed for excellent oxidation and growth resistance at about 980 °C (1800 °F) and even up to 1095 °C (2000 °F) in oxidizing atmospheres or in the presence of certain chemicals.

The high-chromium irons exhibit a characteristically white structure and can be produced with fair machinability and good strength. Low silicon and carbon contents are desirable when toughness and thermal shock resistance are required. The thermal shock resistance of these irons is good, but their toughness is quite limited, even when carbon and silicon contents are low.

High-Nickel Irons. The austenitic cast irons containing 18 to 36% Ni, up to 7% Cu, and 1.75 to 4% Cr are used for both heat-resistant and corrosion-resistant applications. Known as Ni-Resist, this type of iron exhibits good resistance to high-temperature scaling and growth up to 815 °C (1500 °F) in most oxidizing atmospheres, good performance in steam service up to 530 °C (990 °F), and can handle sour gases and liquids up to 400 °C (750 °F). The maximum temperature of use is 540 °C (1000 °F) if appreciable sulfur is present in the atmosphere. Austenitic cast iron can be employed at temperatures as high as 950 °C (1740 °F). Austenitic irons have the advantage of considerably greater toughness and thermal shock resistance than the other heat-resistant alloy irons, although their strength is rather low.

High-nickel ductile irons are considerably stronger and tougher than the comparable gray irons. Tensile strengths of 400 to 470 MPa (58 to 68 ksi), yield strengths of 205 to 275 MPa (30 to 40 ksi), and elongations of 10 to 40% can be realized in high-nickel ductile irons.

High-Aluminum Irons. Alloy cast irons containing 6 to 7% Al, 18 to 25% Al, or 12 to 25% Cr plus 4 to 16% Al are reported to have considerably better resistance to scaling than several other alloy irons, including the high-silicon type. These irons have been little used commercially because of brittleness and poor castability.

Alloy Ductile Irons. Various corrosion-resistant and heat-resistant alloy irons can also be cast as ductile iron, with the graphite in the form of spheroids rather than in the normal flake-graphite shape characteristic of gray iron. When the graphite is present as spheroids, the metal exhibits improved elastic behavior, higher modulus of elasticity, a definite yield point, higher tensile strength, and improved ductility and toughness. Where service applications require these improved properties, alloy ductile irons can be efficiently utilized, although their metallurgical and production characteristics are more complex than those of comparable gray irons.

Alloy Cast Irons for Automotive Service

The most important automotive application of alloy cast irons is in brake drums and disks, for which the material is required to have a combination of high heat capacity, good thermal conductivity, and high emissivity so that it can dissipate a large amount of heat per unit volume. Also, to maintain strength and dimensional stability during cyclic heating and cooling, the material must have adequate high-temperature strength and resistance to thermal shock and must resist growth due to changes in structure.

Low-silicon chromium-molybdenum gray cast iron is generally preferred for automotive disk or drum brakes. Composition and graphite content are closely controlled to maintain adequate high-temperature strength, along with the graphite size and distribution that give adequate thermal conductivity and machinability. In general, decreasing silicon content from 2.5 to 1.5% increases thermal conductivity by about 10%. At the same time, the carbon equivalent must be kept relatively high to ensure solidification as gray iron rather than as white iron. Although most alloying elements decrease the thermal conductivity of gray iron, their effect is not as great as that of silicon. Gray irons alloyed with chromium and molybdenum are reported to have better thermal conductivity than unalloyed gray irons of comparable silicon and carbon contents. The alloy gray irons also have more stable structures and better stress rupture properties than the unalloyed irons.

REFERENCES

1. "Chrome-Moly White Cast Irons," Publication M-630, AMAX Inc., 1986
2. D.E. Diesburg and F. Borik, Optimizing Abrasion Resistance and Toughness of Steels and Irons for the Mining Industry, in *Materials for the Mining Industry*, Symposium Proceedings, Climax Molybdenum Company, 1974, p 26
3. T.E. Norman, A Review of Materials for Grinding Mill Liners, in *Materials for the Mining Industry*, Symposium Proceedings, Climax Molybdenum Company, 1974, p 208
4. R.J. Greene and F.G. Sefing, "Cast Irons in High-Temperature Service," National Association of Corrosion Engineers, March 1954
5. *Engineering Properties and Applications of the Ni-Resists and Ductile Ni-Resists*, No. 1231, Nickel Development Institute, International Nickel Company, 1975
6. W. Fairhurst and K. Roehrig, High Silicon Nodular Irons, *Foundry Trade J.*, Vol 146, 1979, p 657–681

SELECTED REFERENCES

- J. Dodd and J.L. Parks, Factors Affecting the Production and Performance of Thick Section High Chromium-Molybdenum Alloy Iron Castings, Publication M-383, AMAX Inc.
- *Engineering Properties and Applications of Ni-Hard*, The International Nickel Company, Inc.
- R.B. Gundlach, High-Alloy Graphitic Irons, in *Castings*, Vol 15, *Metals Handbook*, ASM INTERNATIONAL, 1988, p 698-701
- C.F. Walton and T.J. Opar, Ed., *Iron Castings Handbook*, Iron Castings Society, Inc., 1981

Carbon and Low-Alloy Steels

Steel Processing Technology 107
Microstructures, Processing, and Properties of Steels 126
Classification and Designation of Carbon and Low-Alloy Steels 140
Physical Properties of Carbon and Low-Alloy Steels 195
Carbon and Low-Alloy Steel Sheet and Strip 200
Precoated Steel Sheet 212
Carbon and Low-Alloy Steel Plate 226
Hot-Rolled Steel Bars and Shapes 240
Cold-Finished Steel Bars 248
Steel Wire Rod 272
Steel Wire 277
Threaded Steel Fasteners 289
Steel Springs 302
Steel Tubular Products 327
Closed-Die Forgings 337
High-Strength Low-Alloy Steel Forgings 358
Steel Castings 363
Bearing Steels 380
High-Strength Structural and High-Strength Low-Alloy Steels 389
Dual-Phase Steels 424
Ultrahigh-Strength Steels 430

Steel Processing Technology

R.I.L. Guthrie and J.J. Jonas, McGill Metals Processing Center, McGill University

OVER THE LAST THIRTY YEARS, remarkable advances have been made in the technology of steel processing operations. The first half of this article describes current ironmaking and steelmaking practices (melt, or liquid processing) and discusses the evolution of these processes and their effects on steel properties. The second half of this article describes solid processing of steel, with emphasis on rolling, thermomechanical processing, and annealing of flat steel products. Additional information on the processing of steel can be found in the references provided in this article as well as in the articles that follow in this Section.

Liquid Processing of Steel

The physical chemistry of steelmaking may appear deceptively simple for integrated steel mill operations where ore from the ground is converted into steel. The central reaction merely involves the reduction of iron oxide by carbon:

$$Fe_2O_3 \text{ (iron oxide)} + 2C \text{ (carbon)} \xrightarrow{\substack{1600\ °C \\ (2910\ °F)}} 2Fe \text{ (molten iron)} + CO/CO_2 \text{ gases} \quad \text{(Eq 1)}$$

The final reduction of oxide to liquid iron requires high temperatures of the order of 1600 °C (2910 °F), to overcome the chemical barrier to oxide reductions and the physical, or thermal, barrier of fusing iron. However, to yield a final steel product with the correct chemistry, quality, and property characteristics, the series of processes depicted in Fig. 1 is typically required.

Ironmaking

The first step in processing liquid iron into high-quality steel involves an ironmaking blast furnace, which has evolved over the centuries to become an efficient countercurrent exchanger of heat and of mass, or oxygen (Fig. 2). Iron oxide (in pellet or sinter form), coke, and limestone are successively charged through the top of the furnace. The charge slowly descends through the shaft (an 8-h journey) and is gradually heated by hot ascending gases (CO, CO_2, N_2, H_2, H_2O) with a transit time of about 3 s. Because the gas that is lower in the furnace is richer in carbon monoxide, it has a more reducing effect on iron oxides. Thus, the pellets are gradually reduced as a result of mass transfer of carbon monoxide (and hydrogen) from the gas phase into the pellet:

$$3Fe_2O_3 \text{ (hematite)} + CO \rightarrow 2Fe_3O_4 \text{ (magnetite)} + CO_4 \quad \text{(Eq 2)}$$

$$Fe_3O_4 \text{ (magnetite)} + CO \rightarrow 3FeO \text{ (wustite)} + CO_2 \quad \text{(Eq 3)}$$

Final deoxidation is accomplished down in the cohesive zone (Fig. 2), where high temperatures and highly reducing conditions result in the reduction of wustite (FeO) to iron. Impurities such as silica, sulfur, alumina, and magnesia, which are present in the original pellets and coke, associate with the lime/dolomite and are removed as a molten slag. To ensure that this slag is fluid, a composition of about 40% SiO_2, 50% CaO (+MgO) and 10% Al_2O_3 is desired, thereby placing it within the temperature valley, or well, of a ternary eutectic region. The final reduction of the charged pellets, ore, or sinter takes place either by:

$$FeO + CO \rightarrow Fe + CO_2 \quad \text{(Eq 4)}$$

or

$$CO_2 + C \text{ (coke)} \rightarrow 2CO \quad \text{(Eq 5)}$$

The reaction in Eq 4 is termed indirect reduction because the iron oxide is reduced through the intervention of a gaseous reductant. The reaction in Eq 5 is termed direct reduction because the direct contact of wustite with coke leads to droplets of iron that fall through the dripping zone into the hearth.

The CO_2 of Eq 4 reacts immediately with the carbon of the hot coke to form more CO as follows:

$$CO_2 + C \text{ (coke)} \rightarrow 2CO$$

This CO_2/CO reaction is often termed the solution loss reaction because it involves the dissolution of coke by CO_2.

Although the obvious purpose of a coke layer is to act as a reductant, the descending coke also plays another critical role. Part of the coke (known as the dead man) forms a supporting pillar for the overlaying burden (the ratio of iron and flux to coke and other fuels in the charge). In the region below the cohesive, or sticky, zone (Fig. 2), the remainder of the charge either is molten or is melting (that is, it is composed of slag and pig iron). The final role of the coke is to burn with hot air entering the coke raceways through the tuyeres, thereby generating the high-temperature heat needed for smelting.

Cokemaking. The production of coke required for the tasks described above is also a formidable, capital-intensive operation. The process involves the destructive distillation of metallurgical-grade coals in the coking chambers of the by-product coke ovens. The heat that is needed to distill the volatiles is transferred through the brickwork from adjacent vertical flues by combustion of enriched blast furnace off-gases. After an induction time of approximately 17 h, the incandescent coke is pushed out of the slot ovens into transfer railway cars. During its fall the column of coke breaks apart, forming large lumps that are then transferred to the quenching tower, where an intense and normally intermittent water spray quenches them for subsequent charging into the blast furnace. Retained moisture is kept to a minimum because of the endothermic character of the moisture and consequent thermal load in the blast furnace.

Blast Furnace Stove Use. To achieve overall thermal efficiency, and to generate the high temperatures required for the reduction to iron in the hearth region of the blast furnace, the incoming blast air is preheated to about 1000 °C (1830 °F) prior to its entry through the water-cooled copper tuyeres. This is accomplished by passing the cold-air blast through a stacked vertical column of preheated (hot) bricks in one of three blast furnace stoves. Because the cold air gradually extracts the stored heat, a separate heating phase is also necessary. This is effected by shutting off the cold-air blast to the stove, opening up the gas valve, and burning enriched blast furnace off-gas

108 / Carbon and Low-Alloy Steels

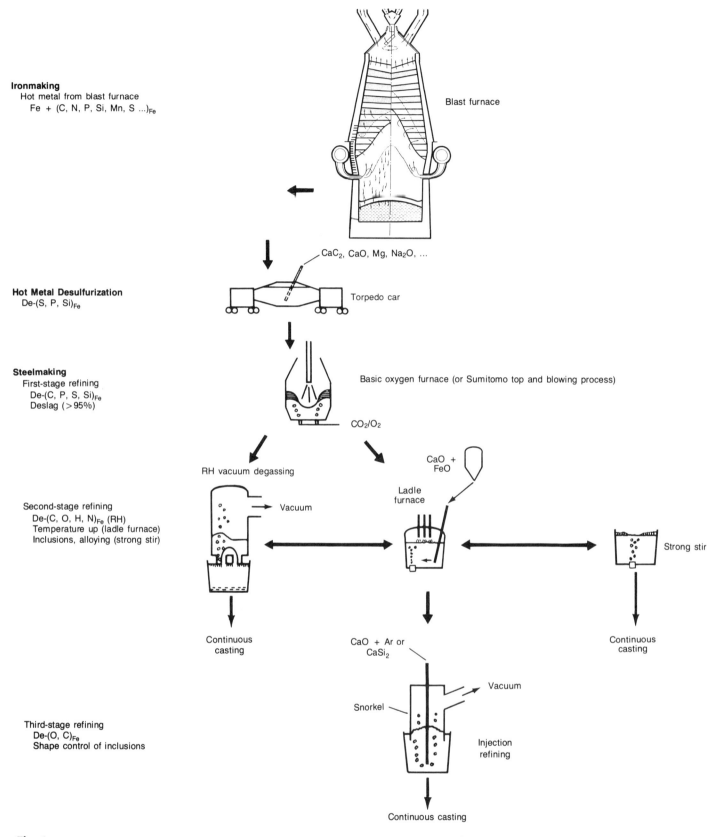

Fig. 1 Major steps in processing liquid iron into high-quality steels. RH, Ruhrstahl Hereaus process

(cleaned with water scrubbers, and electrostatic precipitators) to bring the cooled checkerwork of bricks back up to temperature. Because higher preheat temperatures translate directly into lower coke rates per net tonne of hot metal (NTHM), this heating and cooling cycle requires careful optimization.

Current Blast Furnace Technology. Over the years, significant improvements in burden preparation (such as the development of uniformly sized pellets) and burden lay-

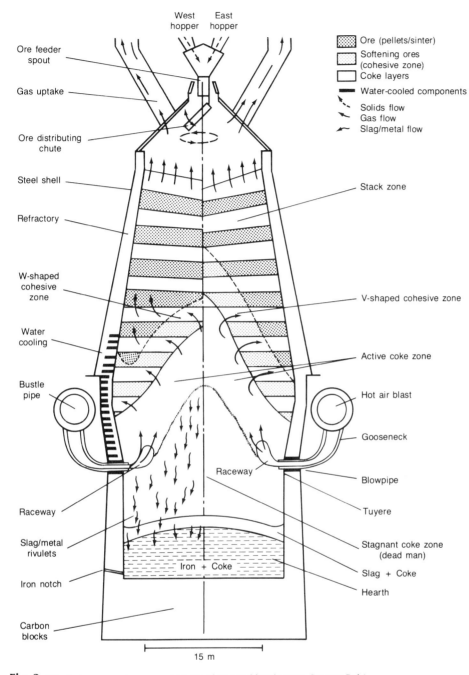

Fig. 2 Principal zones and component parts of an iron blast furnace. Source: Ref 1

to remove sulfur from the iron. The reactions taking place can be written as:

$$CaO \text{ (lime)} + (S)_{Fe} \rightarrow CaS + (O)_{Fe} \quad \text{(Eq 7)}$$

$$CaC_2 \text{ (calcium carbide)} + (S)_{Fe} \rightarrow CaS + 2(C)_{Fe} \quad \text{(Eq 8)}$$

$$(Mg)_{Fe} + (S)_{Fe} \rightarrow MgS \quad \text{(Eq 9)}$$

Enhanced desulfurization can be carried out in a blast furnace by using increased slag volumes to absorb the sulfur, but this method requires higher coke rates. Therefore, such practices were abandoned in the 1960s in favor of desulfurization external to the blast furnace.

It is important to remember that calcium and magnesium oxides are much more stable than their sulfide counterparts, calcium sulfide and magnesium sulfide. Consequently, these desulfurizing operations are only effective if dissolved oxygen levels within the iron are low. The presence of iron saturated with carbon ensures this condition; the fundamental interrelation between dissolved carbon and dissolved oxygen in high-carbon molten iron is (Ref 2):

$$C + O = CO \text{ (gas)} \quad \text{(Eq 10)}$$

$$K^{eq}_{1600 °C} = \frac{P_{CO} \text{ (atm)}}{\text{wt\% C wt\% O}} \approx 660 \quad \text{(Eq 11)}$$

where K^{eq} is the thermodynamic equilibrium constant for Eq 10.

The insertion of wt%$(C)_{Fe}$ = 4.4 wt% for hot metal would show wt%$(O)_{Fe} \sim$ 3 ppm for P_{CO} at atmospheric pressure if equilibrium applies. It is for this reason that desulfurization is so effective in hot metal. Calcia-rich slags have very high sulfur partition ratios with iron (~400). By contrast, sulfur partitioning in the steelmaking step is at best about 4 to 1 between a basic oxygen furnace (BOF) slag and oxygen-rich steel. Consequently, as much as possible of the sulfur-rich product that floats on the hot metal needs to be scraped or slagged off to prevent the sulfur from reverting to the metal during subsequent (low-carbon) steelmaking steps.

Current Hot-Metal Desulfurization Technology. The well-advanced process technology for desulfurization generally involves the submerged pneumatic injection of, for example, calcium carbide powder that is carried by nitrogen through a deeply submerged refractory-coated steel pipe of about a 25 mm (1-in.) inside diameter into hot metal contained within the torpedo car. This vessel (Fig. 1) is customarily used to transport hot metal from the ironmaking facilities to steelmaking operations downstream. Typical industrial practices reduce residual sulfur levels down to 0.01% to 0.02% $(S)_{Fe}$. The desulfurized hot metal usually is transported in the torpedo car from the blast furnace to the steelmaking shop, where it is emptied into the transfer ladle. As mentioned, any slag carryover

ering techniques have enhanced the kinetic efficiency of gas/solid and heat/mass transfer interactions. Higher air blast preheat temperatures and improved coke properties have also helped to reduce coke requirements from about 910 kg (2000 lb) per NTHM in the 1950s to current levels of 455 kg (1000 lb) per NTHM.

The iron that is tapped from the blast furnace is saturated with about 4.4% (or 22 at.%) C. It also contains other impurities that have been reduced from the oxides contained within the iron ore charge. Consequently, the hot metal also contains about 0.3 to 1.3 wt% $(Si)_{Fe}$, 0.5 to 2 wt% $(Mn)_{Fe}$, 0.1 to 1.0 wt% $(P)_{Fe}$ and 0.02 to 0.08 wt% $(S)_{Fe}$. The dissolved sulfur is largely derived from sulfur contained in the coking coal. Dissolved nitrogen levels of the order of 100 ppm would be typical from the air blast. To meet the stringent requirements for high-quality steels, these impurities [(C, S, N, P...)$_{Fe}$] must be brought to very low residual levels using the sequence of operations described below.

Hot Metal Desulfurization

Hot metal from the blast furnace is usually treated with lime, calcium carbide, magnesium, or mixtures of these substances

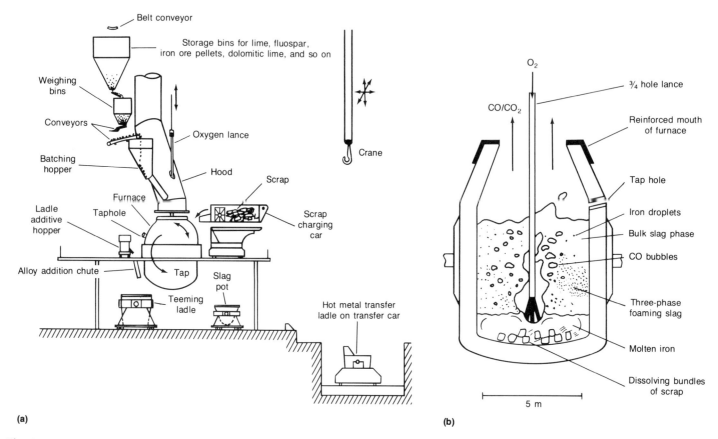

Fig. 3 Principal zones and component parts of a basic oxygen furnace for the production of steel in a melt shop. (a) Typical plant layout. (b) BOF vessel

into the hot metal transfer ladle needs to be removed prior to charging hot metal into the BOF in order to prevent sulfur reversion.

Japanese manufacturers can produce steels with residual hot metal levels of 1 to 2% P; they achieve dephosphorization ahead of the steelmaking step by using injections of sodium carbonate. Because strong compound-forming tendencies exist between phosphorus and sodium, as they do for sulfur and sodium, simultaneous desulfurization and dephosphorization is possible, provided the hot metal has first been desiliconized.

Steelmaking

First-Stage Refining. Because the blast furnace has produced hot metal saturated with carbon and containing other elements, the next operation requires that these impurities (particularly phosphorus) be removed to the required degree. Integrated steel plants normally rely on pneumatically blown oxygen vessels to accomplish these reactions. In a typical BOF, high-velocity (supersonic) jets of pure oxygen are blown onto the hot metal (Fig. 3). Dissolved carbon is oxidized and escapes as carbon monoxide (primarily) and carbon dioxide from the mouth of the vessel, while the other oxidized impurities (Si, Mn, P)$_{Fe}$ enter the slag by fluxing with additions of burnt lime (CaO).

To compensate for the vast amounts of heat liberated during these oxidation reactions, about 30% of the total charge to the furnace comprises steel scrap as coolant. The scrap coolant is required to prevent the temperature of the molten steel from exceeding 1650 °C (3000 °F) and thereby causing unnecessary refractory erosion. Once again, highly complex heat, mass, and fluid transport mechanisms are involved. For example, mass transfer of bath carbon to the scrap metal surfaces effectively dissolves light-section scrap, even though bath temperatures are well below the melting point of the scrap (1500 °C, or 2730 °F) during the major portion of a blow (Fig. 4). Once the bath temperature exceeds the scrap melting range (1500 to 1540 °C, or 2730 to 2800 °F), normal thermal processes that involve turbulent heat transfer will melt the scrap, which finally becomes assimilated into the molten bath. The removal of dissolved carbon as gas and the removal of dissolved silicon, manganese, and phosphorus to an upper slag phase takes place sequentially (Fig. 4), according to:

$$(Si)_{Fe} + O_2 \rightarrow (SiO_2)_{slag} \quad \text{(Eq 12)}$$

$$2(C)_{Fe} + O_2 \rightarrow 2CO \quad \text{(Eq 13)}$$

$$(Mn)_{Fe} + \tfrac{1}{2}O_2 \rightarrow (MnO)_{slag} \quad \text{(Eq 14)}$$

$$2(P)_{Fe} + \tfrac{5}{2}O_2 \rightarrow (P_2O_5)_{slag} \quad \text{(Eq 15)}$$

It should be emphasized that the exact transfer mechanisms are obscure and tend to remain so, due both to the opacity of the system and to the experimental difficulties and restrictions involved in direct measurements of important process variables at 1600 °C (2910 °F). However, the fact that the carbon drops linearly with time during the blow (following silicon elimination) indicates that the rate of oxygen supply controls the rate of decarburization; this is evident except at very low carbon levels, where the curve in Fig. 4 tails off with time.

Thus, towards the end of a BOF blow, the transport of dissolved carbon up to the fire point, where the oxygen jets impinge on the metal bath, has difficulty keeping up with the supply of oxygen. As a result, oxygen begins to dissolve in the steel bath at an increasing rate as the carbon-oxygen reaction heads away from the equilibrium curve for (C)$_{Fe}$ and (O)$_{Fe}$ in contact with a carbon monoxide environment at a partial pressure of 0.1 MPa (1 atm). Figure 5 illustrates the trajectory of the carbon-oxygen evolution as a function of process. The BOF-related curves start moving sharply higher as carbon levels drop below about 0.07 wt% C. The rapid increases in dissolved oxygen imply dirtier steels because greater amounts of deoxidizers (Al, Fe-Si) are needed to remove this oxygen, which is in the form of condensed oxide inclusions.

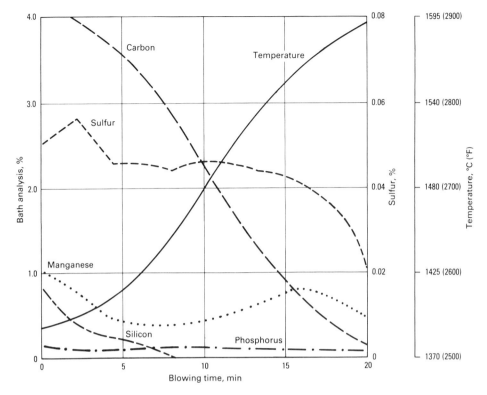

Fig. 4 Removal of elements from the bath in a BOF process. Source: Ref 3

Second-Stage Refining and Technology Advances. The recognition that the stirring being provided by the top-blown jet of a BOF furnace toward the end of the refining process was inadequate, together with the development of the Savarde-Lee shrouded tuyere (Ref 4), triggered a remarkable change in the technology of these oxygen-blown vessels. The tuyere development work made possible and practical the bottom blowing of low-pressure oxygen at high flow rates through a series (typically eight) of tuyeres set in the bottom of the furnace. Each tuyere consists of a central pipe for the oxygen jet and an annular space for injecting a hydrocarbon (such as methane) to form a solid mushroom of steel (Fig. 6). This mushroom protects the refractory base from the fluxing effects of FeO and has allowed the revived use of the Bessemer vessel of 1856 (Ref 5), except that pure oxygen rather than air is injected.

The first North American licensee named this process the quick-quiet basic oxygen process, or Q-BOP. The bottom-blown oxygen jets provide better mixing, lower turndown carbons (of the order of 0.01 wt% C), higher yields (less FeO in slag), and shorter processing times (for example, 14 versus 17 min/blow). One drawback, however, is higher levels of turndown hydrogen in the steel. This is caused by the endothermic cracking of the methane that is needed for the formation of the protective thermal accretions, or mushrooms (Ref 6). Higher levels of dissolved hydrogen can be deleterious for heavy-section products such as pipeline steels and ship plate products; postrefining stir with argon is sometimes favored for steels with these applications.

Another feature of these bottom-blown vessels is the need to inject a fine powdered lime simultaneously with oxygen. Top charging of lime particles or lumps in a similar manner to BOF operations leads to unacceptable foaming and slopping.

A wide variety of other processes have been spawned that take advantage of some features of both top- and bottom-blown vessels. In the Kawasaki basic oxygen process (K-BOP) operation, 30% of the oxygen is soft blown from a multihole lance set high above the steel bath, with the remainder injected through the base of the vessel using shrouded tuyere technology. This allows low turndown carbons (of the order of 0.02 to 0.04% C), together with higher scrap-melting capabilities (for example, 33% versus 30% of the charge). Other similar technologies, such as the German Kloeckner metallurgy scrap (KMS) process, are also in use.

The improved scrap-melting capability of such vessels is enabled by the burning of a higher proportion of effluent carbon monoxide to carbon dioxide within the upper reaches of the vessel itself. Part of the attendant heat can be usefully transferred back to the metal bath, allowing more scrap to be melted. Because scrap generally represents a less expensive source of iron units versus hot metal from the blast furnace, such operations can be profitable, even though they are more technically complex to operate.

Practically all BOF (or oxygen-blown method (OBM) or Linze-Donovitz (LD) method) steelmaking operations in North America now use bottom-blown gas injections to at least stir the steel bath. For example, nitrogen, argon, or carbon dioxide can be blown through submerged injector ports, plugs, or nozzles of various proprietary designs. The Sumitomo top and bottom blowing (STB) process, in which CO_2/N_2 mixtures are bottom blown at about 5% of the flow of the top-blown oxygen in a BOF-like vessel, is a good example of this concept. The STB process increases yields and lowers turndown carbons, thus approaching the performance of Q-BOP vessels.

Electric Furnace Steelmaking. Although integrated steel plants use oxygen-blown steelmaking vessels, many smaller steelmaking operations rely on return scrap steel (versus iron ore) as a primary source of material. For such operations, electric arc furnaces offer economic and technological advantages. These furnaces were originally considered appropriate for the production of tool and alloy steels, but they are also able to produce low-carbon steels of high quality. Currently, 30% of the steel production in North America derives from scrap recycling through remelting and refining operations in electric arc furnaces. One difficulty is that residuals, such as copper and tin in return scrap, are not diluted with a virgin hot-metal source in electric furnace steelmaking. However, with the introduction of prereduced ores (Ref 7) of low gange, or impurity levels (for example, >2% SiO_2) such problems can be mitigated.

Recent technological advances have stressed the role of the furnace as a melter rather than a refiner. Water-cooled panels are required to carry the ultrahigh-power kVA levels of modern furnaces.

Ferroalloy/Deoxidizer Additions. No matter which process is used, the raw steel poured from a furnace into a teeming ladle is too highly oxidized for immediate use because it contains about 0.04 to 0.1 wt% O. This level would cause blowholes in the steel if it were then solidified. Steel deoxidants such as aluminum, ferrosilicon, or carbon are therefore required to bring dissolved oxygen contents down to acceptable levels through precipitation of condensed oxides as inclusions. At the same time, additions of other ferroalloys (for example, Fe-Mn, Fe-Nb, Si-Mn, Fe-V) are made as needed to meet the chemical specifications required for the variety of steel grades that are commonly produced by any integrated steel company.

These bulk additions (13 to 100 mm, or ½ to 4 in., in diameter) either melt quickly (~40 to 120 s) or dissolve slowly (~60 to 360 s), depending on whether their melting ranges are below or above the steel bath temperature (typically 1570 to 1600 °C, or 2860 to 2910 °F) (Ref 8). Some are buoyant

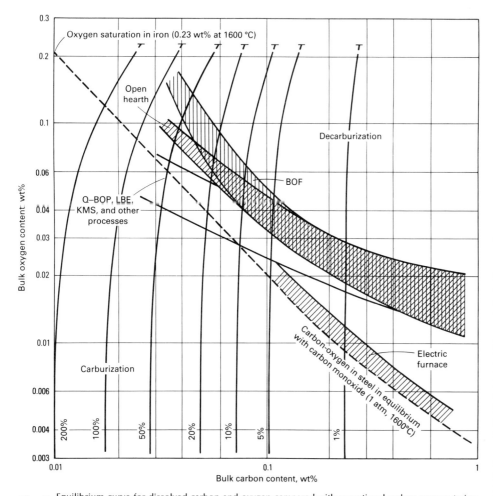

Fig. 5 Equilibrium curve for dissolved carbon and oxygen compared with operational carbon-oxygen trajectories for various steelmaking processes. The isopercentage error lines illustrate the relative importance of carbon to oxygen diffusion on carbon-oxygen kinetics. Q-BOP, quick-quiet basic oxygen process; LBE, lance bubbling equilibrium; KMS, Kloeckner Metallurgy Scrap; BOF, basic oxygen furnace. Source: Ref 1

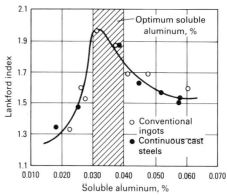

Fig. 7 Lankford index of aluminum-killed steels. Cold reduction 71.4%, anneal at 700 °C (1290 °F), 5h, furnace cool at 20 °C/h (36 °F/h)

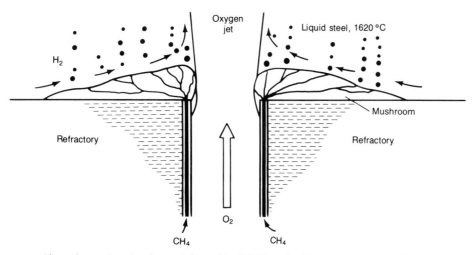

Fig. 6 Thermal accretions (mushrooms) formed in Q-BOP steelmaking operations using the Savarde-Lee shrouded tuyere

(for example, aluminum and ferrosilicon) and tend to float, while others, such as ferroniobium and ferrotungsten, sink rapidly (Ref 1). In either case, thorough metal mixing throughout the teeming ladle is needed (Ref 1). These large bulk additions are commonly added via alloy addition chutes during the last half of a 4 to 8 min furnace-tapping operation. Carryover of slag from the BOF into the ladle can make the recoveries of aluminum and ferrosilicon to the steel highly variable because slag deoxidation as well as metal deoxidation can occur. For these reasons, alloy addition sequencing is important, as are slag control techniques, to limit the net carryover of slag.

Ladle Steelmaking. The increasing need to produce quality products that meet much tighter chemical and physical specifications has led to major changes in steelmaking practices during the last two decades. These changes have centered on modifications to liquid steel within the ladle; therefore, this area of technology is known as ladle steelmaking.

To illustrate the critical nature of correct chemistry, aluminum-killed steels for deep-drawing operations require dissolved aluminum levels that range between 0.03 and 0.04% $(Al)_{Fe}$. The aluminum precipitates with dissolved nitrogen as aluminum nitride during subsequent batch-annealing operations. This precipitation controls grain growth and leads to steel with a fine grain structure and good deep-drawing qualities. Higher or lower levels of dissolved aluminum lead to poor performance indices (Fig. 7).

Even tighter specifications were required for high-strength low-alloy steels, which were introduced to compensate for weight reductions (that is, thinner gages) on automobile parts during the energy crises of the 1970s. Specifications called for dissolved niobium levels of 0.03%, a difficult target without close control of steel deoxidation procedures.

The production of interstitial-free steels for deep drawing (which are described in the article "High-Strength Structural and High-Strength Low-Alloy Steels" in this Volume) require carbon and nitrogen levels less than 50 ppm and controlled additions of titanium and/or niobium to scavenge carbon and nitrogen. To meet such stringent demands, secondary steelmaking processes, focusing on the teeming ladle, have been developed. Of these, the ladle furnace is used for melt reheating and temperature control. The Ruhrstahl Hereaus (RH) degasser, or tank degasser, is used to reduce

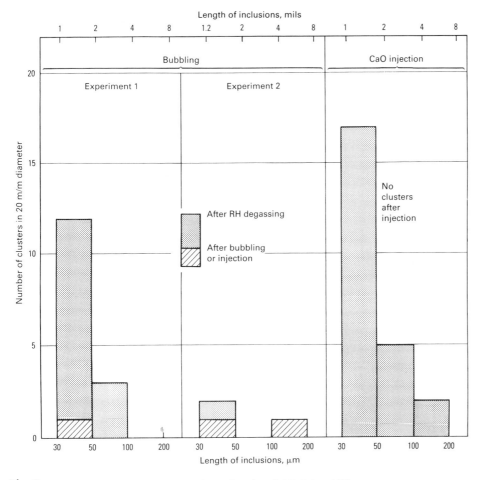

Fig. 8 Effect of lime powder injections on the quality of steel. RH, Ruhrstahl Hereaus process

dissolved $(C,O,H,N)_{Fe}$ levels. A third type of ladle station provides strong stir facilities by using argon and porous plugs set in the base of each teeming ladle, slag rake-off equipment, and wire feeding that allows precise additions of alloying elements, such as aluminum.

Third-stage refining, although still novel, has been conducted by, among others, Sumitomo Metals Industries; it is also known as the injection refining (IR) process (Fig. 1). First lime and then calcium silicide are fed pneumatically through a vertical lance into the teeming ladle. A refractory-lined hood placed over the surface of the steel prevents ingress of atmospheric oxygen. As the relatively large lime particles rise through the melt, they cleanse it by collecting the essentially stationary smaller-diameter (~1 to 10 μm, or 40 to 400 μin.) products of deoxidation. The results of ternary refining are shown in Fig. 8. The number of clusters is greatly reduced after RH degassing followed by strong bubbling. The clusters are totally eliminated with strong bubbling and lime additions.

A final injection of calcium silicide can be used to convert any remaining solid aluminum products of deoxidation into liquid calcium aluminate (preferably 12CaO · 7Al$_2$O$_3$) inclusions (Ref 9). Such inclusions pass easily through metering nozzles into the tundish and from there into the mold of a continuous casting machine.

By the end of these ladle-refining operations, the total residuals within the steel can be brought down to very low levels (~50 ppm total residuals for $(S,O,N,H,P)_{Fe}$) (Ref 10). The difficulty in the final liquid metal processing steps is to maintain this level of physical and chemical quality prior to final solidification in the continuous casting machines.

Tundish Metallurgy and Continuous Casting. The flow of steel from the tundish into the caster is shown in Fig. 9. Figure 10 illustrates the potential sources of contamination of purified steel emptied from the teeming ladle into the tundish. Using a sliding gate nozzle, metal is metered from the bottom of the teeming ladle into a tundish. This nozzle has to be shrouded with argon to avoid air infiltration, steel reoxidation, and the consequent generation of inclusions.

The tundish, in addition to acting as a metal distributor to two or more casters, serves as a further cleansing unit for inclusion removal. Therefore, current practices often use dam and weir combinations to modify the flow of steel within the tundish to enhance inclusion separation. This has led to a trend toward tundishes with larger volumes and thus longer residence times for a given throughput (for example, 60 tonne, or 66 ton, tundishes with a 7-min residence time for a 320 tonne, or 350 ton, ladle full of steel). A typical velocity field for a single-port water model tundish, using the computational fluid dynamic code METFLO (Ref 11), is illustrated in Fig. 11. The associated inclusion separation ratios (defined as the number of inclusions leaving per the number of inclusions entering a tundish) as a function of inclusion rise velocity are also given in Fig. 12. Flow modifiers have no influence on the very small inclusions collected by ternary refining (or by filters), but they can help clean the steel of midsize inclusions in the 50 to 200 μm (2 to 8 mils) range (Fig. 12). For larger inclusions with Stokes rising velocities greater than 5 mm/s (0.2 in./s), these flow controls are not needed for the set of operating conditions noted.

Tundishes are normally fitted with insulating covers to conserve heat. For highly deoxidized steels, they are protected with an argon gas cover to reduce reoxidation and inclusion formation. An artificial slag can also be added to absorb those inclusions that are floating out.

Contrary to popular belief, many inclusion clusters can reach large sizes within the tundish. Because they can be made up totally of alumina, large clusters are most likely the agglomerated products of deoxidation. Figure 13 presents data analyzing the large inclusions present in an aluminum-killed steel in a 60 tonne (66 ton) slab casting tundish not fitted with flow modifiers. A typical histogram of the inclusions, based on an on-line electric sensing technique using a Liquid Metal Cleanness Analyzer (LiMCA) (Ref 12), is compared with data from Japan for a wire quality steel (Ref 13). The slime extraction analysis technique (dissolution of large sample of steel by ferrous chloride, with elutriation to collect unreacted inclusions of alumina and/or silicates) was used for the Japanese data.

Microscopic techniques are inappropriate for the size range shown in Fig. 13, and slime extraction techniques require three days to complete. Nevertheless, such analysis is important because large inclusions can have a deleterious effect on the surface quality, paintability, and zinc-coating characteristics of steel sheet. Similarly, as such inclusions (of alumina or manganese silicates, and so on) are rolled out into long stringers, the transverse properties of steel sheet or plate, such as percent elongation and ultimate tensile strength, are severely compromised, as is metal formability. Consequently, the modification of these inclusions into calcium aluminate inclusions, which are refractory at rolling temperatures

114 / Carbon and Low-Alloy Steels

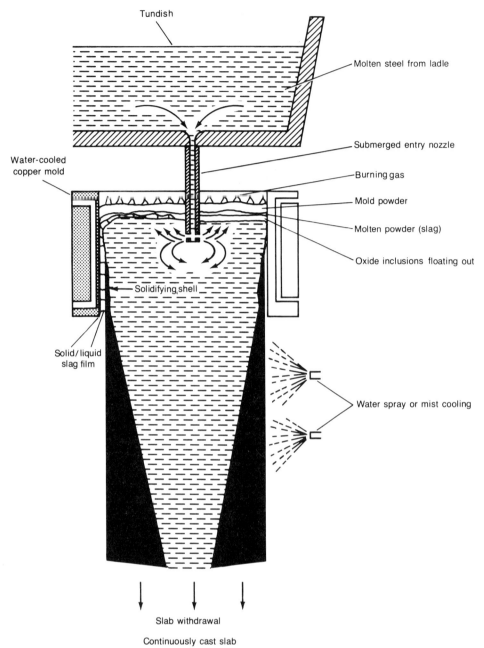

Fig. 9 Typical continuous casting operation

and retain their original spherical shape following rolling, is much preferred (Ref 9).

For other critical applications, the presence of inclusions with a diameter greater than about 50 μm (2 mils) needs to be prevented. Figure 14 shows a break in a steel wire fabricated for a steel-belted automobile tire (Ref 13). There is a move in the industry to filter steel for such applications to help eliminate inclusions under about 50 μm (2 mil) in size, which are not susceptible to flow modifiers.

Mold Metallurgy. The last opportunity for inclusions to be removed is in the mold. Metal enters the mold of a continuous caster through a submerged entry nozzle (Fig. 9); the ports of the nozzle are often angled upward in order to direct the exiting jets of metal up toward the steel surface. There, a layer of lubricating slag from fused mold powder further assimilates inclusions while it simultaneously protects the steel from reoxidation and provides lubrication between the forming shell of the steel and the surfaces of the oscillating mold.

It is preferred that the final structure of the solid steel be equiaxed rather than columnar so that cracking of the billet, slab, or bloom during unbending operations is less likely. Precise control of the metal superheat temperature is needed to prevent dendrite tips that are broken from the advancing columnar freezing front of steel from remelting. The dendrite tips are needed to act as nuclei for grain growth within the remaining melt. Electromagnetic stirring is also used to enhance uniformity of chemistry and structure, and to eliminate centerline segregation of solute-rich material. The cast steel is then cut with travelling oxytorches into slabs, billets, or blooms of appropriate length for further processing. The slabs are about 4 m (13 ft) long, 1 m (3.3 ft) wide, and 100 mm (4 in.) thick. These slabs are inspected and then charged to a slab reheating furnace for subsequent hot-rolling operations. Alternatively, in plants with advanced steelmaking practices where slab surface quality is guaranteed to be acceptable (that is, no scarfing is required), the slabs can be directly charged into the slab reheat furnace.

Future Technology for Liquid Steel Processing Operations

Because of the high capital cost of the blast furnace, melt shop, and hot-rolling mill complex, major research and development efforts are being made within the industry, with the objective of eliminating the number of process steps needed to produce a final product. Figure 15 shows past, present, and possible future process steps for the production of flat-rolled sheet. The object is to reduce the number of major processes down to two: direct steelmaking and direct, or near-net shape, casting. In direct steelmaking, the aim is to feed coal (rather than coke), together with iron ore pellets and lime flux, into an autogeneous reactor to produce iron that contains perhaps 2% C. In direct casting, the aim is to develop the technology needed to directly cast steel sheet perhaps 5 to 10 mm (0.2 to 0.4 in.) in thickness, at tonnage rates of 100 to 200 tonnes/h/m width (35 to 70 tons/h/ft width). Such performance characteristics would match those of the big slab casters of the present, but would have a dramatic impact on the capital and operating costs of the integrated steel plant of the future.

Processing of Solid Steel

As with liquid steel, several processing operations are required to convert steel into its wide variety of finished forms. Figure 15 shows the sequence of operations for flat rolling. After continuous casting and inspection, followed by slab reheating in the reheat furnace, the slab is prepared for the roughing and tandem hot strip mills. Rolled hot strip is then cooled on runout cooling tables and coiled. For thinner gages, hot-rolled strip is cold-rolled, which is followed by annealing and by various coating processes to protect against corrosion; coatings include zinc, tin, zincalume, paint, enamel, and so on. Slabs cut from the continuous casting machine are reheated to bring the steel to about 1200 °C (2190 °F).

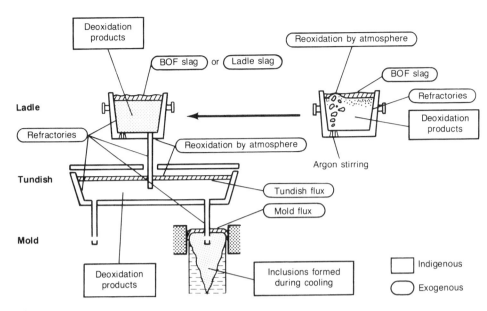

Fig. 10 Potential sources of contamination in the continuous casting process

Hot Rolling

Hot rolling is carried out with the steel in its γ, or austenite phase. Steel is evidently plastic and particularly malleable at the temperatures employed, which range from 1200 °C (2150 °F) to as low as 800 °C (1470 °F). This allows large reductions in thickness (for example, from 250 mm thick slab to 2 mm thick hot strip) with relatively small force. Following hot rolling, the steel transforms into its low-temperature α, or ferrite, phase (plus other constituents). The characteristics of this transformation, which has a significant effect on the mechanical properties of the product, depend on the cooling rates used on the runout cooling tables.

In early integrated steel mills, hot rolling traditionally began with the breakdown of cast ingots into rectangular slabs about 200 mm (8 in.) thick or square billets about 200 × 200 mm (8 × 8 in.) in cross section. With the gradual replacement of ingot casting by continuous casting (except for tool steels and other specialty or low-tonnage grades), the slabbing, or breakdown, stage of hot rolling has gradually disappeared from most mills. Layout of a modern hot strip mill is shown in Fig. 16. Hot rolling begins with roughing, which occurs at temperatures from 1200 °C (2190 °F) down to about 1100 °C (2010 °F). During roughing, slabs about 6 to 8 m (20 to 26 ft) long and 250 mm (10 in.) thick are converted into transfer bars about 30 to 50 mm (1.2 to 2 in.) thick and up to 40 or 50 m (130 or 165 ft) long. Round or square billets are transformed by analogous steps into bars about 100 m (330 ft) in length. Finish rolling, or finishing, is then carried out in a five-, six-, or seven-stand hot mill (Fig. 16), with finishing stand temperatures as low as 900 or 800 °C (1650 or 1470 °F). By this means, the transfer bars are converted into strips about 2 to 3 mm (0.08 to 0.12 in.) in thickness and 600 m (2000 ft) long, at a productivity level of about 100 tons/h per meter of steel strip width (Ref 14). The hot strip is coiled, while plate grades are retained in their rectangular form at thicknesses of about 10 to 30 mm (0.4 to 1.2 in.). In a similar manner, long products are either coiled if they are round and of small cross section (about 6 mm, or 0.24 in., in diameter), or cut to length if they are thicker or of irregular cross section, such as angle and channel shapes.

Controlled Rolling of Microalloyed Steels. The traditional concern of hot rolling was simply to reduce the cross-sectional size of the steel, as described above. However, with the use of microalloying elements such as niobium, vanadium, and/or titanium, hot-rolling at controlled temperatures is also used to condition the austenite so that a fine ferrite grain size is produced during cooling. This method of hot rolling, known as controlled rolling, relies on the precipitation of carbonitrides of the microalloying elements (niobium, vanadium, and/or titanium) to control austenite grain growth and recrystallization.

Over the last 20 years, there has been a gradual introduction of various types of controlled rolling (Ref 15), which presently include conventional controlled rolling, recrystallization controlled rolling, and dynamic recrystallization controlled rolling. With the various methods of controlled rolling, attractive properties can be imparted to materials in the as-hot-rolled condition, thereby eliminating the need for separate (and costly) heat treatment later. The use of controlled rolling leads to the production of steels with nearly twice the yield strength of the commodity grades produced by traditional rolling methods. This increase in yield strength is accompanied by an increase in fracture toughness (see the article "High-Strength Structural and High-Strength Low-Alloy Steels" in this Volume).

The increase in fracture toughness is a direct result of the considerable ferrite grain refinement caused by controlled rolling. Both the fracture properties and the yield strength depend directly on the inverse square root of the ferrite grain size, as given by the Hall-Petch relations (Ref 16):

$$\sigma_y = \sigma_0 + \kappa_y d^{-1/2} \quad \text{(Eq 16)}$$

and

$$\beta T = \alpha - \ln d^{-1/2} \quad \text{(Eq 17)}$$

where σ_y is the yield strength of the polycrystal, σ_0 is the yield strength of a single crystal of equivalent purity and condition, κ_y is the grain boundary strengthening coefficient, d is the mean grain size, T is the impact transition temperature, and α and β are constants. By reducing the ferrite grain size from 57 μm (2.25 mils), or ASTM grain size No. 5, for example, to 5 μm (0.2 mil), or ASTM grain size No. 12, yield strength increments of greater than 210 MPa (30 ksi) can be produced, and the impact transition temperature can be reduced by as much as 100 °C (180 °F).

Critical Temperatures. The three varieties of controlled rolling that have been developed to date rely on the recognition of the three critical temperatures of steel rolling. The first of the three temperatures is the no-recrystallization temperature, or T_{nr}. At temperatures above T_{nr}, the austenite recrystallizes between mill passes; as a result, the grain size is refined, and the work hardening accumulated within the roll pass is eliminated. This temperature can be detected by carrying out pilot rolling studies, by analyzing mill data on rolling loads, or by torsion testing (Fig. 17). Below the T_{nr}, the recrystallization of austenite no longer takes place between rolling passes. Work hardening, or strain, accumulates as a result, and the flow resistance or rolling load begins to increase more sharply with decreasing temperature (Fig. 17 and 18).

The second critical temperature, the upper critical temperature, or Ar_3, defines the start of the austenite-to-ferrite transformation on cooling (the r comes from the French *refroidissement*). The third temperature, the lower critical temperature, defines the end of the austenite-to-ferrite-plus-pearlite transformation and is known as Ar_1. It should be noted that these two critical temperatures do not correspond to the Ar_3 and Ar_1 values determined on annealed samples using classical dilatometry, for example, because the deformation introduced by rolling modifies the transformation behavior of the steel. Instead, Ar_3 and Ar_1 can be determined with the aid of a deformation dilatometer, which applies a compressive strain to the sample prior to

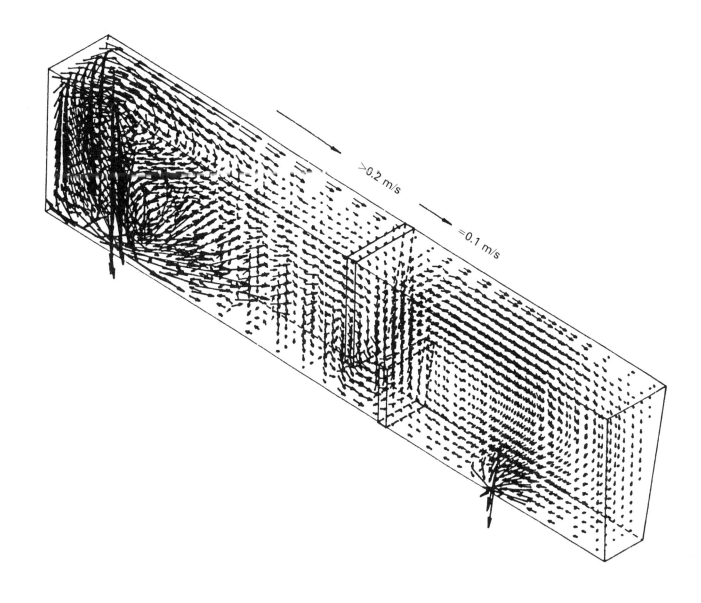

Fig. 11 Isometric view of flow fields predicted in a longitudinally bisected single-port water model slab casting tundish with a flow modification device consisting of a weir-dam arrangement placed at ⅔L. Flow rate of 0.007 m³/s (0.25 ft³/s length, 5.2 m (17 ft); depth, 111 m (3.6 ft); width (surface), 1.07 m (3.5 ft); flow modification device, 0.5L weir and dam arrangement

the initiation of cooling, or by the torsion simulation of rolling (Fig. 17 and 18).

Precipitation of Carbonitrides and Sulfides. Many of the particles that are precipitated during cooling after continuous casting are redissolved during reheating in the slab reheat furnace. These include AlN, MnS, Nb(C,N), Ti(C,N), $Ti_4C_2S_2$, TiS, and VN. During subsequent hot rolling, these reprecipitate fairly readily because the dislocations introduced during rolling act as nucleation sites for strain-induced precipitation. These particles are only about 2.5 nm (0.1 μin.) in diameter when they appear, but they can grow or coarsen up to diameters of 10 to 20 nm (0.4 to 0.8 μin.). The particles take several seconds to form; therefore, they are not produced during rolling operations of short duration. In general, these particles only play a role at temperatures below 1000 °C (1830 °F); that is, during finishing. The strain-induced precipitation of MnS is important in the processing of electrical steels, while that of Nb(C,N), $Ti_4C_2S_2$, and TiS is important in the rolling of interstitial-free steels, and that of Nb(C,N) (and to a lesser extent TiC and VN) is important in the controlled rolling of microalloyed or high-strength low-alloy steels (Ref 15, 18).

Precipitation is necessary because recrystallization can only be arrested during finish rolling, and only if a copious number of precipitates form during passage of the strip between successive mill stands. A high density of precipitates is promoted by the occurrence of cooling between passes (which increases the driving force for precipitation) and interpass intervals of about 10 s or more. As a result, a T_{nr} is only displayed by steels containing niobium, titanium, or vanadium and, furthermore, only when sufficient time is provided during finish rolling for the precipitates to nucleate and grow (Ref 18). This means that recrystallization is most readily arrested during rolling in slow reversing mills, such as plate and Steckel mills, while precipitation plays a much smaller role in tandem mills, such as hot strip, rod, and other mill installations where interpass times are short (~0.5 s or less).

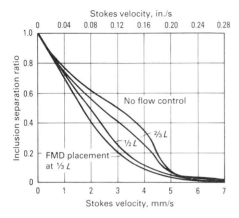

Fig. 12 Relationship between the inclusion separation ratio and Stokes rising velocities predicted for a full-scale water model of a slab casting tundish. FMD, flow-modification device consisting of a weir-dam arrangement (Fig. 11).

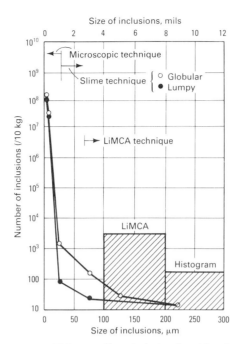

Fig. 13 Histogram of large inclusions in a slab casting tundish, based on the LiMCA technique, versus data for a wire quality steel, based on the slime extraction technique

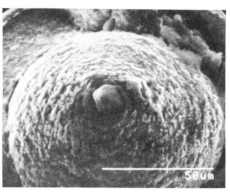

Fig. 14 Scanning electron micrographs of inclusions at a break in a steel cord wire

Conventional controlled rolling (CCR) was the first type of controlled rolling to come into regular commercial use. About 8 to 10% of the total steel tonnage rolled annually is now produced in this way. This process was originally developed for the production of plate grades for the manufacture of oil and gas pipelines, for which the required minimum yield strengths were 350 MPa (50 ksi), 420 MPa (60 ksi), and 490 MPa (70 ksi) (Ref 15). Because of the need for good weldability, low concentrations of carbon and carbon equivalents were specified. These were readily obtained by reducing carbon concentrations to 0.06 or 0.07%; small amounts of niobium (about 0.04%), in combination with vanadium (up to 0.1%) and molybdenum (up to 0.30%), were added for higher-strength grades.

During roughing operations, the coarse reheated austenite grains in a slab are first refined by repeated recrystallization, bringing the grain sizes down to about 20 μm (0.8 mil) or less. The transfer bar can then cool below the T_{nr} during transfer from roughing to the finishing facilities. When rolling is restarted or continued below the T_{nr}, recrystallization is no longer possible, and the austenite structure is progressively flattened in an operation known as pancaking. For pancaking to be successful, the accumulated reductions applied in this temperature range must add up to at least 80%. Finally, when the flattened austenite grains go through their transformation to ferrite, the ferrite produced has a very fine grain structure because of the large number of nucleation sites available on the expanded surfaces of the pancaked austenite grains. This leads to ferrite grain sizes in the range of 5 to 8 μm (0.2 to 0.3 mil). The fine-grain ferrite is responsible for the attractive combination of good toughness properties and high yield strengths (Ref 15, 18). It should be stressed that austenite pancaking is only possible in the absence of recrystallization, and its arrest is caused by the copious precipitation of Nb(C,N) during delays between mill passes.

Recrystallization Controlled Rolling (RCR). As described above, controlled rolling is generally based on the use of low finishing temperatures (that is, in the vicinity of 800 to 900 °C, or 1470 to 1650 °F), with the result that fine ferrite grain sizes appear after transformation. However, such finishing is inappropriate for certain products, such as heavy plates and thick-walled seamless tubes (Fig. 19), that cannot be finished at such low temperatures in the hot-rolling range due to excessive rolling loads. For such applications, it is possible to produce the fine microstructures required by carefully controlling the recrystallization of austenite and arranging for it to occur at successively lower temperatures during finish rolling (Ref 19). These temperatures are nevertheless above 900 °C (1650 °F) and thus are higher than those employed in CCR.

Two requirements must be met for the RCR process to be successful. One is that the recrystallization not be sluggish, so that the times required are not too long. This is achieved by employing vanadium rather than niobium as an alloying element. Vanadium acts as a grain refiner without bringing recrystallization to a complete stop, as niobium is inclined to do. The second requirement is that grain growth be prevented after each cycle of recrystallization; this grain growth can negate the refining effect of recrystallization at lower and lower temperatures. For this purpose, sufficient titanium is added to have about 0.01% available for the formation of fine particles of TiN during cooling after continuous casting (Ref 20). When this dispersion has an appropriate size and frequency distribution, it can completely prevent grain growth of the austenite after each cycle of recrystallization. The fine austenite grains, in turn, transform into relatively fine-grain ferrite, for example, 8 to 10 μm (0.3 to 0.4 mil) in diameter, leading to mechanical properties in the hot-rolled product that are acceptable for many purposes.

Dynamic Recrystallization Controlled Rolling (DRCR). When the interpass time is short, as in the case of rod, hot strip, and certain other rolling processes (Fig. 20), insufficient time is available for conventional recrystallization during the interpass delay. The amount of carbonitride precipitation that can take place is also severely limited. As a result, an alternative form of recrystallization is initiated. This is known as dynamic recrystallization, and it involves the nucleation and growth of new grains during (as opposed to after) deformation (Ref 21, 22). This also requires the accumulation of appreciable reductions, of the order of 100%, to enable the recrystallization process to spread completely through the microstructure inherited from the roughing process. Austenite grain sizes as small as 10 μm (0.4 mil) can be achieved with DRCR (Ref 23).

Low-temperature finishing by DRCR has the advantage of producing finer ferrite grain sizes after transformation than CCR; that is, 3 to 6 μm (0.12 to 0.24 mil), as

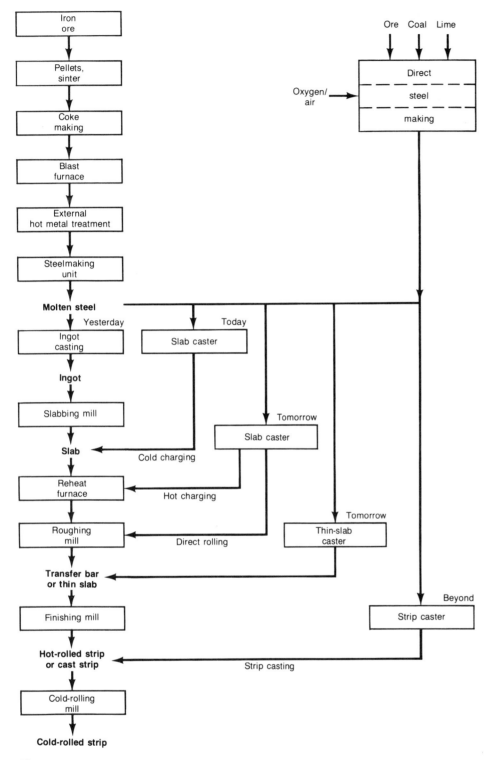

Fig. 15 Past, present, and future steel processing steps

tween the head and tail of a transfer bar, for example). Such a gradual decrease in temperature can lead to gage and flatness problems, as well as to a gradient in microstructure characteristics (and therefore in the mechanical properties) along the workpiece. One solution to this problem has involved the introduction of a coil box between the roughing and finishing stands of a hot strip mill (Fig. 21). When the transfer bar, which is 30 to 50 mm (1.2 to 2 in.) thick, arrives at the coil box, the leading edge is deflected and curled into a circular shape, and the entire bar is wound into a coil without a mandrel or spool being required. The coiled shape of the workpiece enables it to cool much more slowly as the bar is slowly fed (tail end first) into the hot strip mill. This technology has led to a significant improvement in the uniformity of the dimensions and properties of the final product, as well as to a decrease in the energy required for hot rolling.

Cooling Beds, Runout Table Cooling, and Coiling. Following hot rolling, the workpiece is generally cooled down to room temperature. For plates and bars this is carried out on cooling beds. For strip it is carried out on runout tables and during holding, which follows coiling. Figures 22(a) and (b) show a laminar flow cooling system used on a hot mill. For rods it is carried out in water boxes and along cooling lines employing large volumes of forced air (Fig. 23). Except for certain grades of stainless and electrical steels, such cooling always involves the transformation of austenite into ferrite, as well as into a number of other transformation products, such as pearlite, bainite, and martensite. The particular transformation product that forms, as well as its general characteristics, depends on the cooling rate that is achieved. The faster the cooling rate within a given product range, the stronger the microstructure that is produced. As a result, there is considerable interest at present in the use of accelerated cooling to promote the formation of appropriate microstructures. For most steel products, in fact, this process is the least expensive way to increase the strength.

When the transformation product consists largely of ferrite, rapid cooling decreases the ferrite grain size obtained from a fixed hot-rolling schedule. This is because of the hysteresis involved in phase transformations, as a result of which the actual (as opposed to equilibrium) transformation temperature displayed on cooling decreases as the cooling rate is increased. Lower transformation temperatures, in turn, lead to finer ferrite grain sizes, in part because the growth rates are lower at lower temperatures, but also because the ferrite nucleus density increases with increasing supercooling below the equilibrium transformation temperature.

opposed to 5 to 8 μm (0.2 to 0.3 mil) for the latter process (Ref 21). However, such low-temperature finishing increases the rolling load, and it can also make mill control more difficult because of the load drop associated with the initiation of dynamic recrystallization. It is important to note that under industrial rolling conditions, CCR, RCR, and DRCR can all occur to different degrees during a given operation. This can happen when the processing parameters have not been optimized so as to favor only strain-induced precipitation and austenite pancaking in the case of CCR, conventional recrystallization in the case of RCR, and dynamic recrystallization in the case of DRCR.

The Stelco Coil Box. One of the problems associated with batch processes such as the rolling of both long and flat products is the temperature rundown that develops (be-

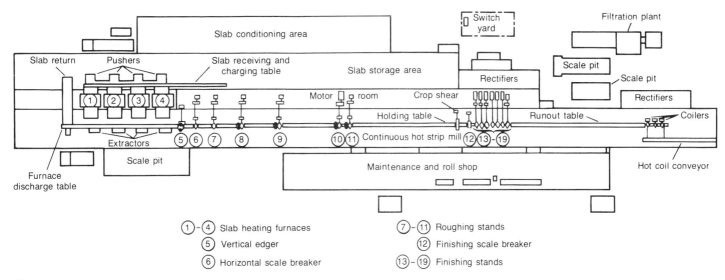

Fig. 16 Layout of a 2130 mm (84 in.) continuous hot strip mill. Source: Ref 14

When the transformation product consists largely of pearlite or contains appreciable volume fractions of pearlite, more rapid cooling leads to the formation of finer pearlite, which is associated with higher strength. However, if the cooling rate is too rapid for a given chemistry, some bainite or even martensite can form. Unless these transformations are carefully controlled, such structures lead to a lack of toughness and ductility and are therefore generally avoided. Nevertheless, B-modified bainitic steels are employed for the production of high-strength heavy plate (700 to 900 MPa, or 100 to 130 ksi, yield strength in the as-rolled condition); the carbon level in these steels is reduced to about 0.02% to improve toughness. Similarly, fully martensitic steels can be produced by quenching (a separate and therefore fairly expensive operation), in which case the brittleness is reduced by an appropriate tempering treatment.

Precipitation During Cooling and Coiling. The solubility of all the precipitate-forming elements in steel decreases as cooling progresses (Ref 18). Thus, carbonitrides and sulfides such as AlN, Fe_3C, MnS, Nb(C,N), Ti(C,N), $Ti_4C_2S_2$, TiS, and V(C,N) all tend to form, either on the runout table or cooling bed, or after coiling. Because precipitate nucleation and growth take time (for example, 1 to 10 s for nucleation, and 10 to 100 or 1000 s or more for growth, depending on the temperature), the amount and type of precipitation that takes place after hot rolling are sensitive functions of the cooling rate and conditions of holding. Thus, rapid cooling on the runout table suppresses precipitation, although some particles are inevitably formed during the austenite-to-ferrite transformation, because the solubilities in ferrite are appreciably lower than the respective levels that pertain to the austenite. The coiling temperature is also of considerable importance. Relatively high coiling temperatures, of the order of 750 °C (1380 °F), followed by the slow cooling rates associated with the geometry of coils, favor the precipitation of the carbonitrides (for example, AlN). Low

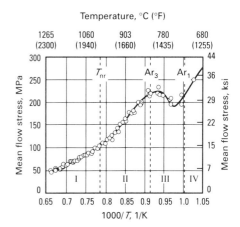

Fig. 18 Mean flow stress as a function of $1000/T$, where T is the absolute pass temperature. Region I corresponds to the temperature range where recrystallization occurs after each pass; region II falls between the no-recrystallization temperature and Ar_3, the upper critical temperature; region III is the intercritical temperature range; region IV lies below the Ar_1 of lower critical temperature. Source: Ref 17

Fig. 17 Stress-strain curves of torsional simulation for an average schedule of a Steckel hot mill. The pass number is shown above each flow curve. Source: Ref 17

120 / Carbon and Low-Alloy Steels

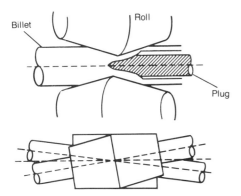

Fig. 19 Piercing stage in the manufacture of seamless tubes. Source: Ref 19

coiling temperatures, of the order of 550 °C (1020 °F), on the other hand, prevent AlN formation and keep these elements in solution for precipitation during annealing after cold rolling. The coiling temperature also determines the mean size and number of the particles that form; the former decreases and the latter increases as the temperature is lowered. These considerations are important because the final grain size after recrystallization and grain growth is directly proportional to the mean particle size for a given chemistry (that is, containing a given volume fraction of precipitate).

Warm Rolling

In recent years a renewed interest has developed in warm rolling, that is, the finish hot rolling of steel in the high-ferrite, as opposed to the low-austenite, temperature range. Warm rolling is possible because ferrite is actually softer than austenite at a given hot-rolling temperature. From the example shown in Fig. 24 it is evident that the ferrite in an interstitial-free (IF) steel has as little as half the flow resistance of the austenite prior to transformation. Such a large difference in flow stress can lead to serious gage and control problems when rolling is carried out in the vicinity of the γ-to-α transformation. These problems are avoided, however, if rolling is suspended during cooling through the intercritical range and resumed only when the steel has cooled below Ar_1.

The example given above is extreme because the low carbon level of an IF steel (about 30 ppm) means that the intercritical temperature range is reduced to as little as 30 °C (85 °F), and therefore the flow stress drop associated with passage through this range is very sharp. In conventional steels, the difference between the Ar_3 and Ar_1 temperatures is in the range of 100 °C (180 °F) or more. In such cases, the fully ferritic material is significantly colder than its fully austenitic counterpart, and thus the ferrite has a resistance to flow that is only moderately less than that of the austenite.

Fig. 20 Finishing stands of a 2130 mm (84 in.) hot strip mill. Courtesy of Mesta Machine Company

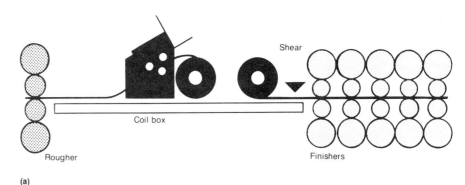

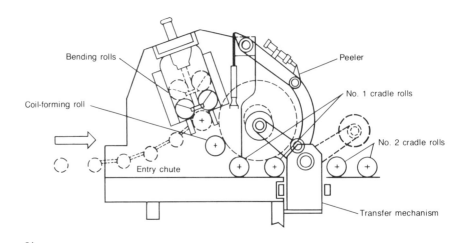

Fig. 21 Coil box of hot strip mill. (a) Position on delay table. (b) Schematic of coil box function. Source: Ref 14, 24

The warm rolling of IF steels is of commercial interest because lower reheating and rolling temperatures can be used, leading to lower scale losses and energy consumption rates in the slab reheat furnace. Furthermore, the textures developed during the warm rolling of ferrite do not differ appreciably from those produced during the cold rolling of the same phase, and this rolling step can therefore be employed for the production of steels with excellent formability characteristics.

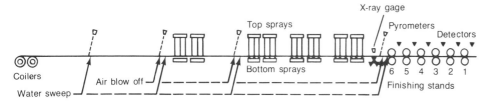

Fig. 22(a) Schematic showing the strip-cooling arrangement on a 2050 mm (82 in.) hot strip mill. See also Fig. 22(b). Source: Ref 14

Cold Rolling

About half of all rolled steel products are sold in the as-hot-rolled condition. This includes such obvious items as rails and structural sections (I beams, channels, angles, and so on), and plates and seamless tubes, as well as certain relatively thick grades of sheet and strip that are employed in the as-rolled condition (for example, for the forming of bumpers and car wheels, or for the manufacture of pipe). Nevertheless, numerous products require much more reduction in thickness or cross section and considerably better surface quality than can be produced by hot rolling; this is where cold rolling plays a role. In the cold rolling of flat products, numerous advances have been made in equipment and processes that have allowed for improved flatness and consistency of gage along the length of a coil. These advances include hydraulically inflatable rolls for control of the amount of crown, six-high mills with tapered backup rolls that can be inserted and withdrawn laterally as required, as well as a variety of roll-bending techniques (Fig. 25).

Cold rolling increases the hardness and yield strength, reduces the ductility and formability, and also introduces the specific texture components associated with thickness reduction at a constant width (known as plane-strain deformation) that can be

Fig. 22(b) Laminar flow cooling sprays on a hot strip mill. Courtesy of J.A. Pajeta, LTV Steel Company

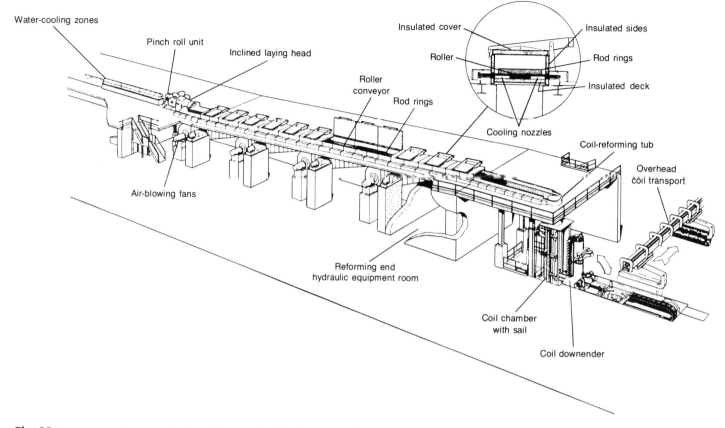

Fig. 23 Arrangement of a processing line with retarded cooling. Courtesy of Morgan Construction Company

later enhanced by annealing. Some products are sold and used in such a strengthened condition, particularly when little further deformation or shaping is involved. However, when appreciable further forming operations are to be carried out, such as in the manufacture of cans or the deep drawing of auto body parts, cold rolling must be followed by annealing. Annealing removes the work hardening introduced by rolling and thus restores the formability of the material. When employed after the appropriate processing of selected grades (for example, aluminum-killed drawing quality, also known as AKDQ, or IF steels), annealing also brings about an increase in the normal anisotropy (R value), which leads to significant further increases in the deep drawability.

Annealing

Batch Annealing. The annealing of coils weighing 9 to 27 tonnes (10 to 30 tons) is necessarily a very time-consuming operation. Typical heating and cooling times are of the order of 2 and 3 days, respectively, so that the rate of temperature change is about 12 °C/h (22 °F/h) during heating and −8 °C/h (−14 °F/h) during cooling. Consequently, a total process time of nearly 5 days is involved. Because of the considerable capital, maintenance, and energy costs associated with such heating operations (Fig. 23), the optimization of heating and cooling schedules to improve the productive capacity of these units is a subject of great interest at present.

The principal mechanism associated with batch annealing is recrystallization, which eliminates the strain hardening introduced in the previous rolling operations. Once initiated, at temperatures of 600 to 650 °C (1110 to 1200 °F), it takes only minutes to spread through the material and replace the grain structure flattened by rolling with equiaxed, strain-free grains. Although precipitate coarsening and particle solution and reprecipitation also occur, they are not of particular importance for most grades. For AKDQ steels, on the other hand, the characteristics of precipitation and of particle coarsening must be carefully controlled so that the grain orientations associated with high formability (that is, the $\{111\}\langle 110\rangle$ and to a lesser extent the $\{111\}\langle 112\rangle$ texture components) are favored during recrystallization and can in this way replace the other sets of grain orientations inherited from the rolling operation. This is brought about by pinning the undesirable $\{100\}$-oriented grains during heating through their optimum growth temperature range by means of AlN particles. These form during heating from the supersaturated Al and N held in solution as a result of the relatively low coiling temperatures (550 °C or 1020 °F) employed for the AKDQ grades. The AlN particles gradually redissolve and coarsen as the temperature is further increased during annealing, permitting growth of the $\{111\}$-oriented grains within the temperature range that favors this orientation.

Continuous Annealing. With the trend toward more and more continuous processing, batch annealing is gradually being replaced by continuous annealing (Fig. 26). The prime advantage of this process is the considerable increase in product uniformity along the length of a given coil. This is of increasing importance as tolerances and allowable property variabilities are reduced as a result of the increasing automatization of forming processes. Because of the much shorter process times involved (for example, 2 min instead of 5 days), the heating and cooling rates are much higher (15 °C/s, or 27 °F/s). Negligible hold times are required at the maximum temperatures of 700 to 800 °C (1290 to 1470 °F), which are significantly higher than in the batch annealing process.

For commodity grades, the temperatures and heating and cooling rates are not of critical importance. By contrast, for drawing quality grades such as the niobium- and titanium-base IF steels, the process parameters must again be carefully adjusted for compatibility with the particular chemistries employed. For these steels, the favored texture components are $\{554\}\langle 225\rangle$ and

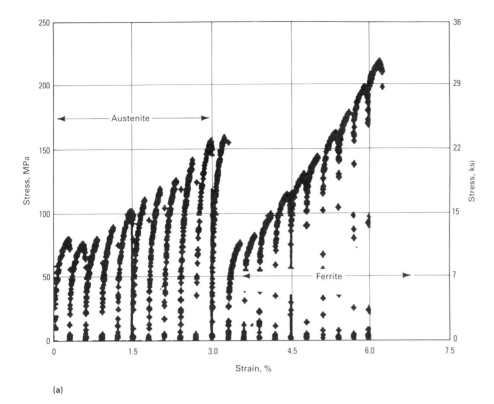

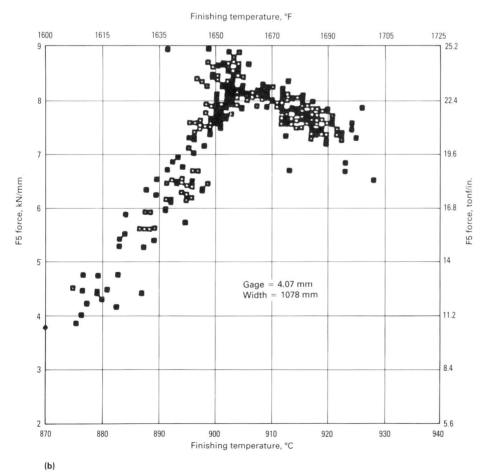

Fig. 24 Factors influencing the warm rolling of steel. (a) Stress-strain curves for interstitial-free steel rolled according to an idealized schedule A ($\epsilon \sim 2/s$). (b) Dependence of stand 5 separating force on rolling temperature. Source: Ref 25

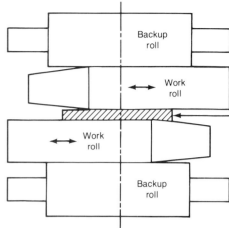

Fig. 25 Taper-adjusting method for crown control. Source Ref 24

$\{111\}\langle112\rangle$ and, to a lesser extent, $\{111\}\langle110\rangle$. These components differ slightly from those associated with the AKDQ steels and lead to still higher R values. As for the AKDQ grades, the appearance and disappearance of precipitates (in this case, Nb(C,N) and $Ti_4C_2S_2$), the mean sizes and volume fractions of these precipitates, and the amounts of niobium, titanium, carbon, and nitrogen remaining in solution must be carefully controlled if optimum product properties are to be produced. By tying up the carbon and nitrogen present in the form of carbonitrides and eliminating the yield drop in this way, low yield stresses are obtained, which lead to high initial work-hardening rates and n values. These, in association with the high R values that follow from the presence of the texture components described above, are responsible for the excellent drawability properties of these grades.

REFERENCES

1. R.I.L. Guthrie, *Engineering in Process Metallurgy*, Oxford Science Publications, Clarendon Press, 1989
2. E.T. Turkdogan, Physical Chemistry of Oxygen Steelmaking, Thermochemistry and Thermodynamics, in *B.O.F. Steelmaking*, Vol 2, *Theory*, Iron and Steel Society, 1975
3. A. Jackson, *Oxygen Steelmaking for Steelmakers*, 2nd ed., George Newnes Ltd., 1969
4. G. Savarde and R. Lee, French Patent 1,450,718, 1966
5. J.R. Stubbler, *The Original Steelmakers*, Iron and Steel Society, 1984
6. Y. Sahai and R.I.L. Guthrie, The Formation and Growth of Thermal Accretions in Bottom/Combination Blown Steelmaking Operations, *Iron Steelmaker*, April 1984, p 34-38
7. R.L. Reddy, Use of DRI in Steelmaking, in *Direct Reduced Iron—Technology and Economics of Produc-*

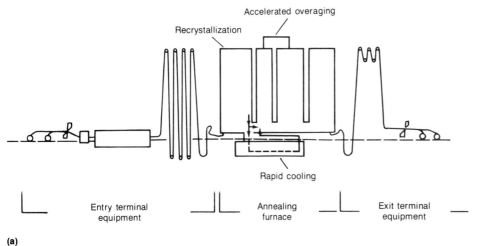

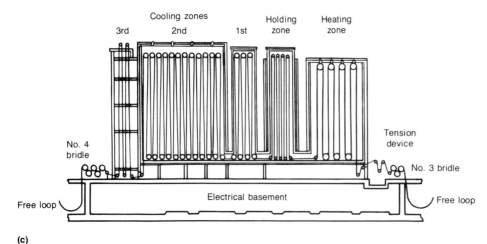

Fig. 26 Examples of continuous annealing configurations. (a) Continuous annealing line. (b) Entry looping section. (c) Heating, holding, and cooling zones. Source: Ref 24

tion and Use, Iron and Steel Society, 1980, p 104-118
8. R.I.L. Guthrie, *Addition Kinetics in Steelmaking*, in *Electric Furnace Proceedings*, Vol 35, Iron and Steel Society, 1977, p 30-41
9. R.I.L. Guthrie, The Use of Fluid Mechanics in Ladle Metallurgy, *Iron Steelmaker*, Vol 9 (No. 1), 1982 p 41-45
10. G.M. Faulring, J.W. Farrell, and D.C. Hilty, Steel Flow Through Nozzles: The Influence of Calcium, in *Continuous Casting*, Vol 1, L.J. Heaslip, A. McLean, and I.D. Sommerville, Ed., Iron and Steel Institute, 1983, p 57-66; see also p 23-42 for reoxidation inclusions
11. T. Emi and Y. Lida, Impact of Injection Metallurgy on the Quality of Steel Products, in *Scaninject III*, Part 1, Proceedings of a joint MEFOS/JERNKONTORET Conference (Lulea, Sweden), 1983, p 1-1 to 1-31
12. S. Joo and R.I.L. Guthrie, Mathematical Models and Sensors as an Aid to Steel Quality Assurance for Direct Rolling Operations, in *Proceedings of the Metals Society of the Canadian Institute of Mining and Metallurgy*, Vol 10, *Proceedings of an International Symposium on Direct Rolling and Hot Charging of Strand Cast Billets*, J.J. Jonas, R.W. Pugh, and S. Yue, Ed., Pergamon Press, 1989, p 193-209
13. H. Ichihashi, Sumitomo Metals Internal Report; see D.H. Nakajima, "On the Detection and Behaviour of Second Phase Particles in Steel Melts," Ph.D. thesis, McGill University, 1986
14. W.L. Roberts, *Hot Rolling of Steel*, Marcel Dekker, 1983
15. *Microalloying '75*, Proceedings of the International Symposium on High Strength Low Alloy Steels, Union Carbide, 1977
16. F.B. Pickering, *Physical Metallurgy and the Design of Steels*, Applied Science, 1978
17. S. Yue, F. Boratto, and J.J. Jonas, Designing an Industrial Controlled Rolling Schedule using Simple Statistical Process Analysis and Laboratory Modelling, in *Proceedings of the Conference on Hot and Cold-Rolled Sheet Steels*, R. Pradhan and G. Ludkovsky, Ed., The Metallurgical Society of the American Institute of Mining, Metallurgical, and Petroleum Engineers, 1988, p 349-359
18. W.J. Liu and J.J. Jonas, Ti(CN) Precipitation in Microalloyed Austenite During Stress Relaxation, *Metall. Trans. A*, Vol 19A, 1988, p 1415-1424; Calculation of the $Ti(C_yN_{1-y})$-$Ti_4C_2S_2$-MnS-Austenite Equilibrium in Ti-Bearing Steels, *Metall. Trans. A*, Vol 20A, 1989, p 1361-1371
19. S. Yue, R. Barbosa, J.J. Jonas, and P.J. Hunt, Manufacture of Seamless Tubing by Means of Recrystallized Controlled Rolling and Accelerated Cooling, in *30th Mechanical Working and Steel Processing Conference*, Iron and Steel Society of the American Institute of Mining, Metallurgical, and Petroleum Engineers, 1988, p 37-45
20. W. Roberts, in *HSLA Steels: Technology and Applications*, American Society for Metals, 1984, p 33, 67
21. F.H. Samuel, S. Yue, J.J. Jonas, and K.R. Barnes, Effect of Dynamic Recrystallization on Microstructural Evo-

lution During Strip Rolling, I.S.I.J. Int., in press
22. L.N. Pussegoda, S. Yue, and J.J. Jonas, Laboratory Simulation of Seamless Tube Piercing and Rolling Using Dynamic Recrystallization Schedules, *Metall. Trans.*, in press
23. J.J. Jonas and T. Sakai, A New Approach to Dynamic Recrystallization, in *Deformation, Processing and Structure*, G. Krauss, Ed., American Society for Metals, 1984 p 185-243
24. W.L. Roberts, *Flat Processing of Steel*, Marcel Dekker, 1988
25. F.H. Samuel, S. Yue, B.A. Zbinden, and J.J. Jonas, "Recrystallization Characteristics of a Ti-Containing Interstitial-Free Steel During Hot Rolling," Paper presented at the AIME Symposium on Metallurgy of Vacuum-Degassed Carbon Steel Products (Indianapolis, IN), American Institute of Mining, Metallurgical, and Petroleum Engineers, Oct 1989

Microstructures, Processing, and Properties of Steels

George Krauss, Advanced Steel Processing and Products Research Center, Colorado School of Mines

THE PERFORMANCE of steels depends on the properties associated with their microstructures, that is, on the arrangements, volume fractions, sizes, and morphologies of the various phases constituting a macroscopic section of steel with a given composition in a given processed condition. Because all the phases in steels are crystalline, steel microstructures are made up of various crystals, sometimes as many as three or four different types, which are physically blended by solidification, solid-state phase changes, hot deformation, cold deformation, and heat treatment. Each type of microstructure and product is developed to characteristic property ranges by specific processing routes that control and exploit microstructural changes. Thus, processing technologies not only depend on microstructure but are also used to tailor final microstructures. For example, sheet steel formability depends on the single-phase ferritic microstructures of low-carbon cold-rolled and annealed steel, while high strength and wear resistance are enhanced by carefully developed microstructures of very fine carbides in fine martensite in fine-grain austenite of high-carbon hardened steels.

This article describes microstructures and microstructure-property relationships in steels. An important objective is to relate microstructural evolution to the different processing schedules to which various types of steel are subjected. For example, low-carbon sheet and plate steels are subjected to much different processing schedules and develop significantly different microstructures and properties from those of medium-carbon forged and hardened steels. This article emphasizes the correlation of microstructure and properties as a function of carbon content and processing in relatively low-alloy steels. More highly alloyed steels, such as tool steels and stainless steels, are discussed in detail in the Section "Specialty Steels and Heat-Resistant Alloys" in this Volume.

Iron-Carbon Phase Diagram

The major component of steel is iron, which exists in two crystal forms below its melting point. One is the body-centered cubic (bcc) form, which is stable from below room temperature to 912 °C (1675 °F) and from 1394 °C (2540 °F) to the melting point of 1530 °C (2785 °F). In the former temperature range, bcc iron is known as α-ferrite, while in the higher temperature range, it is known as δ-ferrite. The other crystal form, which is stable between 912 and 1394 °C (1675 and 2540 °F), is the face-centered cubic (fcc) form, known as austenite or γ-iron.

Steels also contain carbon in amounts ranging from very small, of the order of 0.005 wt% in ultralow-carbon, vacuum-degassed sheet steels, to a maximum of 2.00 wt% in the highest-carbon tool steels. Carbon profoundly changes the phase relationships, microstructure, and properties in steels. Generally, carbon content is kept low in steels that require high ductility, high toughness, and good weldability, but is maintained at higher levels in steels that require high strength, high hardness, fatigue resistance, and wear resistance.

Figure 1 shows the iron-carbon phase diagram and the changes that carbon induces in the phase equilibria of pure iron. Carbon is an austenite stabilizer and expands the temperature range of stability of austenite. Its solubility is much higher in austenite (a maximum of 2.11 wt% in equilibrium with cementite at 1148 °C, or 3000 °F) than in ferrite (a maximum of 0.0218 wt% in equilibrium with cementite at 727 °C, or 1340 °F). The solubility of carbon in ferrite and austenite is a function of temperature; when the carbon atoms can no longer be accommodated in the octahedral interstitial sites between the iron atoms, a new phase that can accommodate more carbon atoms in its crystal structure will form (Ref 2). This phase is designated as cementite or iron carbide (Fe_3C) and has an orthorhombic crystal structure. Cementite formation and the temperature-dependent solubility of carbon in austenite and ferrite, as controlled by alloying and processing, account for the great variety of microstructures and properties produced in steels.

Alloys of iron and carbon that contain up to 2.00 wt% C are classified as steels, while those containing over 2.00 wt% C are classified as cast irons. Graphite is a more stable carbon-rich phase than cementite; its formation is promoted by a high carbon concentration and the presence of large amounts of such elements as silicon. Therefore, graphite is an important phase in cast irons but is rarely found in steels. When graphite does form, solubility limits and temperature ranges of phase stability are changed slightly, as indicated by the dashed lines in Fig. 1.

The austenite phase field shown in Fig. 1 is the basis for the hot workability and heat treatability of carbon steels. Single-phase austenite is readily hot worked; therefore, massive sections of steel can be hot reduced to smaller sections and structural shapes. Austenite, upon cooling, must transform to other microstructures. If slow cooled under conditions approximating equilibrium, austenite will change to mixtures of ferrite and cementite (Fig. 1); if cooled rapidly, it will change to martensite. Such transformations provide the basis of the heat treatments applied to steels. Thus, the fortuitous high-temperature stability of austenite in iron and iron-carbon alloys and the solid-state transformation of austenite upon cooling create many opportunities to optimize shape, section size, microstructure, and properties for many different applications.

Processing temperatures for the formation and transformation of austenite are set by the critical temperatures that mark the boundaries between the various phase fields of Fig. 1. The critical temperatures as a function of carbon content, which were initially identified by changes in slope or thermal arrests in heating and cooling curves, are given the designation "A." If

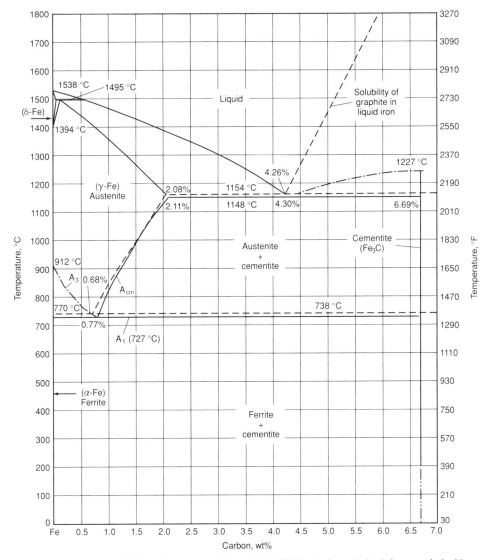

Fig. 1 Iron-carbon equilibrium diagram up to 6.67 wt% C. Solid lines indicate Fe-Fe$_3$C diagram; dashed lines indicate iron-graphite diagram. Source: Ref 1

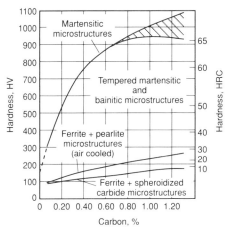

Fig. 2 Hardness as a function of carbon content for various microstructures in steels. Cross-hatched area shows effect of retained austenite. Source: Ref 1

Carbon Content and Properties

Figure 2 shows hardness as a function of carbon content for various types of microstructures. Hardness is readily measured and generally is directly proportional to strength and inversely proportional to ductility and toughness. Although the hardness-carbon relationships are shown as lines in Fig. 2, they are in fact better represented as bands because many factors may cause variations in hardness in a given microstructure. For example, the strength of low-carbon ferritic microstructures is very sensitive to grain size, while that of largely pearlitic microstructures is very sensitive to the interlamellar spacing of cementite and ferrite.

All types of microstructures increase in strength with increasing carbon content, but martensitic microstructures show the most dramatic increases. Because of the low solubility of carbon in ferrite (except for as-quenched martensite), the carbon is primarily concentrated in carbide phases. Therefore, much of the higher strength of medium- and high-carbon steels is due to higher volume fractions and finer dispersions of carbides in ferrite. Ferritic matrix grain sizes and morphology also significantly affect mechanical behavior at any given carbon level.

Figure 2 shows that all types of microstructures could be produced in a steel of a given carbon content. There are, however, practical limits to this observation. Low-carbon steels do not have sufficient hardenability to form martensite except in the thinnest sections and are therefore produced primarily with ferritic microstructures, which have excellent ductility for cold-working and forming operations. At the other extreme, medium- and high-carbon steels alloyed with chromium, nickel, and/or molybdenum may have such high hardenability for the formation of martens-

equilibrium conditions apply, the designations Ae$_1$, Ae$_3$ and Ae$_{cm}$, or simply A$_1$, A$_3$, and A$_{cm}$, are used to indicate the upper boundary of the ferrite-cementite phase field, the boundary between the ferrite-austenite and austenite phase fields, and the boundary between the austenite and austenite-cementite phase fields (Fig.1). If heating conditions apply (which raise the critical temperatures relative to equilibrium), then Ac$_1$, Ac$_3$, and Ac$_{cm}$ are used for the critical temperatures, with the c being derived from the French *chauffant*. If cooling conditions apply (which lower critical temperatures relative to equilibrium), then the designations Ar$_1$, Ar$_3$, and Ar$_{cm}$ are used, with the r being derived from the French *refroidissant*. There is hysteresis in the transformation temperatures (that is, Ac temperatures are higher than Ae temperatures and Ar temperatures are lower than Ae temperatures) because continuous heating and cooling leave insufficient time for complete diffusion-controlled transformation at the true equilibrium temperatures.

In addition to iron and carbon, steels contain many other elements that shift the boundaries of the iron-carbon phase diagram. Elements such as manganese and nickel are austenite stabilizers, which lower critical temperatures. Elements such as silicon, chromium, and molybdenum are ferrite stabilizers and carbide formers, which raise critical temperatures and shrink the austenite phase field (Ref 3). Other elements, such as titanium, niobium, and vanadium, may form temperature-dependent dispersions of nitrides, carbides, or carbonitrides in the austenite. These effects must be taken into account when setting processing temperature ranges for commercial alloys. For well-established grades, the optimum temperature ranges for hot work are listed in Ref 4.

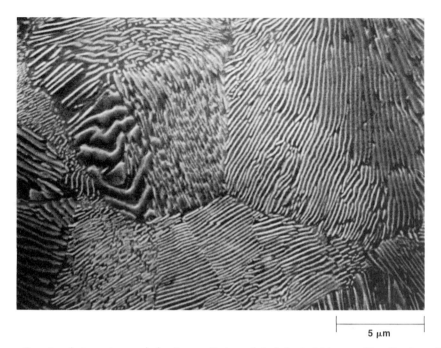

Fig. 3 Scanning electron micrograph showing pearlite in a rail steel of eutectoid composition. Courtesy of F. Zia-Ebrahimi

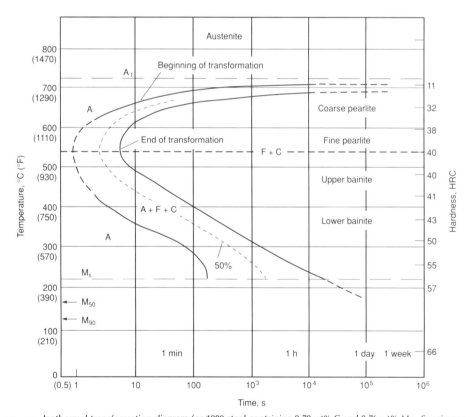

Fig. 4 Isothermal transformation diagram for 1080 steel containing 0.79 wt% C and 0.76 wt% Mn. Specimens were austenitized at 900 °C (1650 °F) and had an austenitic grain size of ASTM No. 6. The M_s, M_{50}, and M_{90} temperatures are estimated. Source: Ref 1

in this article describe in detail several important ferrous microstructures and the steels and processing methods used to produce them.

Pearlite and Bainite

Pearlite is the name given to the microstructure produced from austenite (A) during cooling of a steel by the following solid-state reaction:

$$A\ (0.77\ wt\%\ carbon) \underset{heating}{\overset{cooling}{\rightleftarrows}} F\ (0.02\ wt\%\ carbon)$$
$$+\ Fe_3C\ (6.67\ wt\%\ carbon) \qquad (Eq\ 1)$$

Such a reaction, in which one solid phase transforms to two other solid phases, is generically termed a eutectoid reaction. In the case of steels, the ferrite (F) and cementite (Fe_3C) form as roughly parallel lamellae, or platelets, to produce a composite lamellar two-phase structure. Figure 3 shows an example of pearlite formed in a rail steel that contains close to the eutectoid carbon content, 0.77 wt% C. In this scanning electron micrograph the cementite lamellae appear light and ferrite appears recessed because it has etched more deeply than the cementite.

The parallel lamellae act as diffraction gratings in the light microscope and diffract light of various wavelengths to produce the colors and the luster characteristic of pearls. Thus, early metallographers used the name pearlite for the unique lamellar structure they observed in steels. Within a given colony of pearlite, it has been shown that all the ferrite and cementite have largely the same crystallographic orientations. Therefore pearlite may be described as two interpenetrating single crystals (Ref 5).

Under equilibrium conditions, best approximated by very slow cooling, the pearlite reaction must occur at 727 °C (1340 °F), because austenite is not stable below that temperature (Fig. 1). However, pearlite formation is accomplished by diffusion, a time-dependent process. Carbon atoms diffuse away from regions that become ferrite to regions that become cementite. Thus, rapid cooling, which gives less time for diffusion, depresses pearlite formation to lower temperatures.

Figure 4 shows an isothermal transformation diagram for 1080 steel, which contains 0.79 wt% C and therefore transforms entirely to pearlite over a range of temperatures well below 727 °C (1340 °F). Figure 4 was produced by cooling the steel rapidly to a series of temperatures below A_1, holding at those temperatures, and then following, as a function of time, the transformation of austenite to pearlite. An incubation time is required for the initiation of transformation, and this time, as shown by the curve that marks the beginning of transformation, decreases with decreasing temperature. This

ite or bainite that other microstructures are formed only by special annealing treatments.

The preceding comments should help explain why various types of steels have evolved based on the most readily attainable microstructures and the property requirements that they satisfy for certain types of applications. Alloy design and processing approaches have also evolved and are still evolving to exploit the best features of each type of steel. The following sections

Fig. 5 Light micrograph showing patches of upper bainite (dark) formed in 4150 steel partially transformed at 460 °C (860 °F). Courtesy of F.A. Jacobs

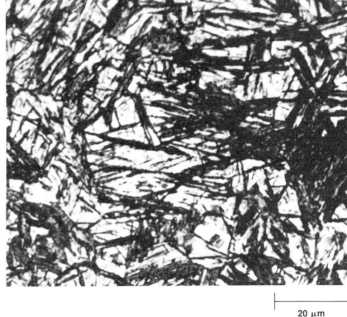

Fig. 6 Light micrograph showing lower bainite (dark plates or needles) formed in 4150 steel. Courtesy of F.A. Jacobs

acceleration is associated with increased undercooling, which provides greater thermodynamic driving force, increasing nucleation rates and decreasing interlamellar spacing. The latter, in turn, enhances growth rates because of reduced diffusion distances.

The kinetics of pearlite formation, based on the nucleation and growth of spherical pearlite colonies, are described by (Ref 6):

$$f(t) = 1 - \exp[-\pi NG^3 t^4/3] \quad \text{(Eq 2)}$$

where $f(t)$ is the volume fraction of pearlite formed at any time t at a given temperature, N is the nucleation rate of the colonies, and G is the growth rate of the colonies. Equation 2 describes well the slow initial and final transformation rates and the more rapid intermediate transformation rates observed for isothermal pearlite formation. With decreasing temperature below A_1, N and G increase, and the transformation of austenite to pearlite accelerates, as shown in Fig. 4.

During pearlite formation, in addition to carbon atom diffusion, iron atoms must also transfer across the interface between the austenite and pearlite. This short-range iron atom transfer is necessary to accomplish the crystal structure changes among the austenite, ferrite, and cementite. At a critical low temperature, this atom-by-atom short-range diffusion is no longer possible, and the iron atoms accomplish the crystal structure change by shearing or cooperative displacement (Ref 7). This change in transformation mechanism results in a new type of microstructure, referred to as bainite. The ferrite crystals assume elongated morphologies, and the cementite is no longer continuous and lamellar (Ref 8).

The bainite that forms at temperatures just below those at which pearlite forms (Fig. 4) is termed upper bainite. In medium- and high-carbon steels, it typically consists of groups of ferrite laths with coarse cementite particles between the laths. The bainite that forms at lower temperatures is termed lower bainite and consists of large needlelike plates that contain high densities of very fine carbide particles. Figures 5 and 6 show examples of upper and lower bainite, respectively. In these micrographs the bainite morphology is dominated by the ferrite phase, and the carbides are too fine to be resolved by the light microscope. In some low- and medium-carbon steels (generally, those alloyed with manganese, molybdenum, and silicon), bainitic microstructures with ferrite and austenite (or martensite formed from the austenite) will form instead of the classic ferrite-carbide bainitic structures (Ref 9).

Proeutectoid Ferrite and Cementite

Steels with a lower carbon content than the eutectoid composition (hypoeutectoid steels) and a higher carbon content than the eutectoid composition (hypereutectoid steels) form ferrite and cementite, respectively, prior to pearlite. The structures formed upon cooling between Ar_3 and Ar_1, and Ar_{cm} and Ar_1 are referred to as proeutectoid ferrite and proeutectoid cementite, respectively.

Figure 7 shows a low-carbon steel ferrite-pearlite microstructure that formed during air cooling from the austenite phase field (Fig. 1). The sequence of transformation to this microstructure is described below. When the temperature of the specimen reaches the Ar_3 temperature, proeutectoid ferrite begins to form. The ferrite crystals, or grains, as they are referred to by metallurgists, nucleate on austenite grain boundaries and grow, by rearrangement of iron atoms, from the fcc austenite structure into the bcc ferrite structure at the austenite-ferrite interface. Carbon atoms, because of their low solubility in the ferrite, are rejected into the untransformed austenite. When the steel reaches the Ar_1 temperature, most of the microstructure has transformed to proeutectoid ferrite, and the carbon content of the remaining austenite has been enriched to about 0.77 wt%, which is exactly the composition required for the pearlite reaction. Thus, the balance of the austenite transforms to pearlite, as described in the preceding section.

In Fig. 7, the ferrite appears white, and the boundaries between ferrite grains of different orientation appear as dark lines. The pearlite, in contrast to Fig. 3, appears uniformly black because the interlamellar spacing of the pearlite in this example is too fine to be resolved by the light microscope.

Figure 8 shows an example of proeutectoid cementite in a hypereutectoid steel. The cementite has formed as a thin network along the grain boundaries of the austenite, and the balance of the microstructure is martensite, which formed when the specimen was quenched from a temperature between A_{cm} and A_1. The cementite and its interfaces are preferred sites for fracture initiation and propagation, and as a result, proeutectoid cementite networks make hypereutectoid steels extremely brittle. Inter-

130 / Carbon and Low-Alloy Steels

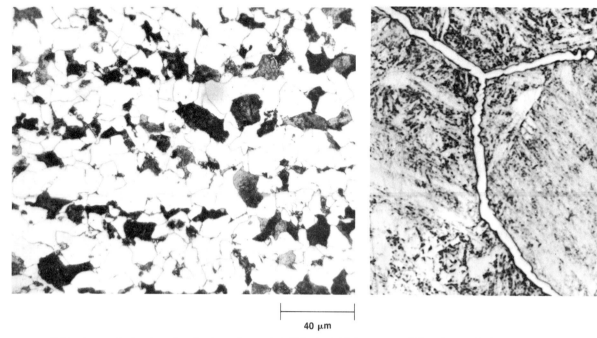

Fig. 7 Light micrograph showing microstructure of proeutectoid ferrite (white) and pearlite (dark) in a 0.17C-1.2Mn-0.19Si steel

Fig. 8 Light micrograph showing cementite network on prior-austenite grain boundaries in an Fe-1.12C-1.5Cr alloy. Courtesy of T. Ando

critical annealing treatments that break up and spheroidize the cementite are therefore used to increase toughness.

Processing: Ferrite-Pearlite Microstructures

Although some steel products are directly cast to shape, most are wrought or subjected to significant amounts of hot and/or cold work during their manufacture. Figure 9 depicts some of the primary processing steps that result in ferrite-pearlite microstructures. In some cases, processing produces finished products, such as plate and hot-rolled strip. In other cases, as discussed in the following section, further processing is applied; for example, steel bars are subsequently forged and heat treated, and hot-rolled strip is cold rolled and annealed.

The thermomechanical processing temperature ranges are shown in Fig. 9 relative to the critical temperatures identified in the Fe-C phase diagram (Fig. 1). All the hot rolling is done with steels in the austenitic condition, and because of the large section sizes and equipment design, the austenite transforms to microstructures of ferrite and pearlite. Depending on the casting technique, some steel products undergo several cycles of austenite to ferrite-pearlite transformation, as shown in Fig. 9. Each step in a cycle requires the nucleation and growth of new phases and offers the possibility of controlling grain size and the distribution of various microstructural components.

Traditionally, hot working has been used to reduce large ingots to products with reduced cross sections and special shapes, but there has been little attempt to control microstructure other than the use of slightly reduced finish hot-rolling temperatures. This approach has been substantially modified in recent years by the implementation of two major changes in process design: the casting of smaller and smaller sections, and the use of hot rolling to control microstructure and properties as well as to reduce section size.

The progression of casting technology from ingot to continuous (or strand) to thin slab to direct strip casting is shown in Fig. 9. Continuous casting eliminates the soaking and breakdown hot rolling of large ingots. Thin slab casting eliminates the roughing hot work applied to thick slabs that are produced by either ingot or continuous casting. Direct strip casting, which is still under development, would eliminate all hot work (Ref 10). A considerable savings of time and energy and improved surface quality result from the new casting techniques.

The control of microstructure and properties during hot-rolling involves a thermomechanical processing technique known as controlled rolling. Controlled rolling is used to enhance the toughness and strength of

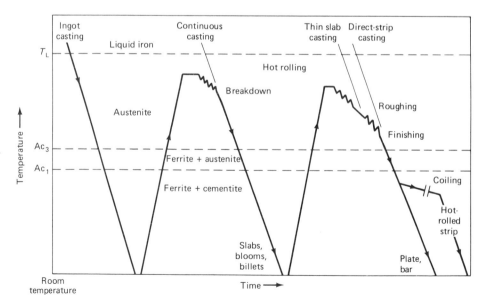

Fig. 9 Temperature-time schedules for primary processing of steels cast by various technologies

Microstructures, Processing, and Properties of Steel / 131

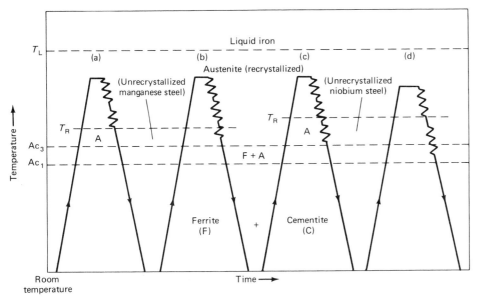

Fig. 10 Temperature-time schedules for thermomechanical processing of steels. (a) Normal processing. (b) Controlled rolling of carbon-manganese steel. (c) Controlled rolling of niobium-containing steel, finishing above Ac_3. (d) Controlled rolling of niobium-containing steel, finishing below Ac_3. Source: Ref 11

microalloyed low-carbon plate and strip steels by grain refinement (see Ref 11 to 13 and the article "High-Strength Structural and High-Strength Low-Alloy Steels" in this Volume). The fine grain sizes produce significantly increased strength, from the 210 MPa (30 ksi) yield strengths that are typical of conventionally hot-rolled low-carbon steels to yield strengths between 345 and 550 MPa (50 and 80 ksi).

The key to the use of controlled rolling is the formation of fine austenite grains that transform upon cooling to very fine ferrite grains. Deformation of austenite induces strains, which, at high temperatures, are rapidly eliminated by recrystallization, followed by grain growth of the austenite. However, at low deformation temperatures, grain growth is considerably retarded. If the temperature is low enough, even recrystallization is suppressed, especially in steels to which small amounts of alloying elements such as niobium have been added. The niobium, which is soluble at high temperatures, precipitates out as fine niobium carbonitrides at low austenitizing temperatures. These fine precipitate particles stabilize the deformation substructure of the deformed austenite and prevent recrystallization. Upon cooling, ferrite grains nucleate on the closely spaced grain boundaries of the unrecrystallized austenite and form very fine grain microstructures.

A number of thermomechanical processing schedules have been developed to produce low-carbon steels of high strength and toughness. Figure 10 shows schematically a number of treatments developed by Japanese steelmakers (Ref 12). Similar controlled rolling schedules are used worldwide to produce fine-grain low-carbon steels. A processing parameter introduced in Fig. 10 is the austenite recrystallization temperature (T_R), which is primarily dependent on the amount of deformation and alloying.

Controlled rolling is generally performed on microalloyed high-strength low-alloy (HSLA) steels, which have small amounts (generally less than 0.10%) of carbide- and nitride-forming elements such as niobium, titanium, and/or vanadium. The various microalloying elements have different temperature-dependent solubility products (Ref 11, 13), and hot-rolling parameters must be adjusted to fit specific alloy compositions (see the section "Controlled Rolling" in the article "High-Strength Structural and High-Strength Low-Alloy Steels" in this Volume).

Processing: Ferritic Microstructures

Large tonnages of hot-rolled steel strip are further processed to produce highly deformable sheet. As shown in Fig. 11, processing involves cold rolling to reduce thickness and improve surface quality, followed by annealing to produce microstructures consisting of ductile ferrite grains (Ref 14,15). Cold-rolled and annealed sheet steels have low carbon contents, usually less than 0.10%, and therefore contain little pearlite in the slow-cooled hot-rolled condition. However, if pearlite is present, it is deformed during cold rolling, and the cementite of the pearlite rapidly spheroidizes during annealing as the strained, cold-rolled ferrite recrystallizes to unstrained, equiaxed grains. These microstructural changes are shown in Fig. 12 for a 0.08% C steel containing 1.5% Mn and 0.21% Si (Ref 16).

A recently developed type of steel, made possible by the introduction of vacuum degassing into the steelmaking process, contains very low carbon, less than 0.008% (Ref 17). These steels are referred to as interstitial-free or ultralow-carbon steels and may contain small additions of niobium or titanium to tie up nitrogen and carbon residual from the steelmaking process. Interstitial-free steels have excellent deep-drawing properties (see the article "High-Strength Structural and High-Strength Low-Alloy Steels" in this Volume). The carbon content of the interstitial-free steels is below that of the solubility limit of carbon in bcc ferrite (Fig. 1 and 13); therefore, no pearlite forms in these steels. Also, the low

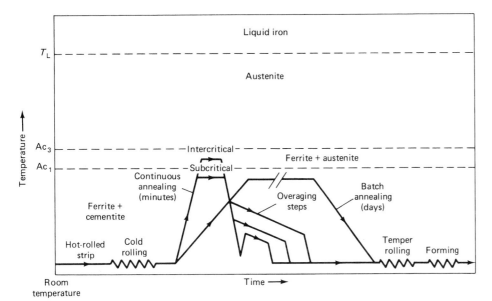

Fig. 11 Temperature-time processing schedule for cold-rolled and annealed low-carbon sheet steel. Continuous and batch annealing are schematically compared. Source: Ref 14

132 / Carbon and Low-Alloy Steels

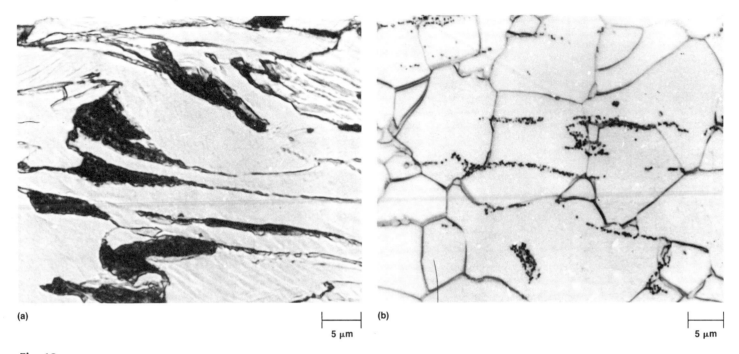

Fig. 12 Light micrographs of a 0.08C-1.5Mn-0.21Si steel. (a) After cold rolling 50%. (b) After cold rolling 50% and annealing at 700 °C (1290 °F) for 20 min. Source: Ref 16

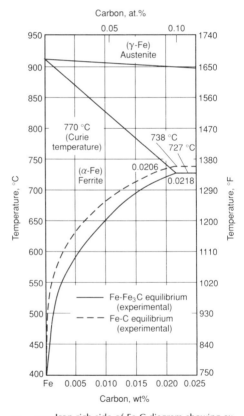

Fig. 13 Iron-rich side of Fe-C diagram showing extent of ferrite-phase field and decrease of carbon solubility in ferrite with decreasing temperature. Source: Ref 1

The formability of cold-rolled and annealed sheet steels, especially in stamping operations that require deep drawing, is significantly improved by the development of crystallographic textures that defer necking and fracture in thin sheets to higher strains. The preferred orientations, {111} planes parallel to the plane of the sheets and ⟨110⟩ directions in the rolling direction (that is, {111} ⟨110⟩ annealing textures), are promoted by aluminum deoxidation. The aluminum-killed steels, if finish hot rolled at high temperatures and coiled at low temperatures, retain aluminum and nitrogen in solid solution in hot-rolled strip and through cold rolling (Ref 15). During batch annealing, aluminum nitride particles precipitate and suppress the nucleation and growth of recrystallized grains in orientations other than the preferred orientations for good formability.

The traditional method of annealing cold-rolled sheet has been to heat stacks of coils in a batch process (Fig. 11). Batch annealing requires several days. Recently, continuous annealing lines, in which the sheet is uncoiled and rapidly passed through high-temperature zones in continuous annealing furnaces, have been installed and used to anneal cold-rolled sheet steels (Ref 14). Continuous annealing requires only minutes to recrystallize a section of sheet as it passes through the hot zone of a furnace.

Although cold-rolled and annealed sheet steels have low carbon contents, during annealing at temperatures close to the A_1 temperature, some carbon and nitrogen are always taken into solution (unless the steels are ultralow-carbon or interstitial-free steels). Figure 13 shows the carbon-rich side of the Fe-C diagram. Carbon has its maximum solubility at the A_1 temperature, and its solubility decreases with temperature to a negligible amount at room temperature. Nitrogen shows a similar relationship. Thus, if a steel is cooled from around A_1 at a rate that prevents gradual relief of supersaturation by cementite formation during cooling, the ferrite at room temperature may be highly supersaturated with respect to carbon and nitrogen. These interstitial elements then may segregate to dislocations in strained structures, a process referred to as strain aging, or they may precipitate out as fine carbide or nitride particles, a process referred to as quench aging (Ref 15). The aging processes may occur at room temperature or during heating at temperatures just above room temperature because of the high diffusivity of carbon and nitrogen in the bcc ferrite structure.

Strain aging and quench aging raise the yield strength of ferritic microstructures by pinning dislocations. When yielding does occur, new dislocations are generated and the stress drops to a lower level at which localized plastic deformation propagates across a specimen. The localized deformation is referred to as a Lüders band, and the process is described as discontinuous yielding. In deformed sheet steels, Lüders bands are called stretcher strains and, if present, result in unacceptable surface appearance in formed parts. In order to eliminate stretcher strains, cold-rolled and annealed sheet steels are temper rolled (Fig. 11). The temper rolling introduces just enough strain to exceed the Lüders strain. Beyond this point, sufficient dislocations are introduced, and all parts of a specimen or deformed sheet will strain uniformly or continuously.

interstitial content and the addition of stabilizing elements such as titanium or niobium eliminate strain aging and quench aging, as discussed below.

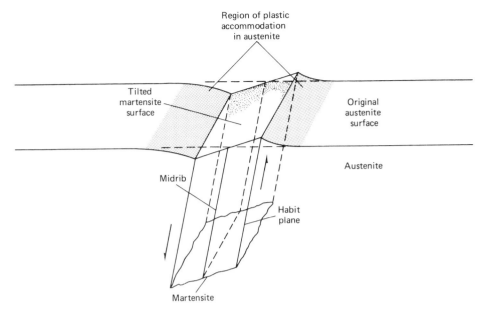

Fig. 14 Schematic of shear and surface tilt associated with the formaton of a martensite plate. Courtesy of M.D. Geib

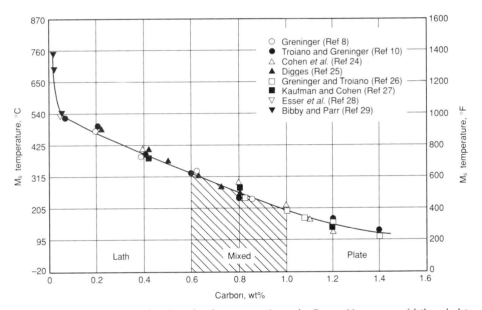

Fig. 15 M_s temperature as a function of carbon content in steels. Composition ranges of lath and plate martensite in iron-carbon alloys are also shown. Source: Ref 2; investigations referenced are identified by their number in that reference.

Martensite

Martensite is the phase that produces the highest hardness and strength in steels (Fig. 2). The martensitic transformation is diffusionless and occurs upon cooling at rates rapid enough to suppress the diffusion-controlled transformation of austenite to ferrite, pearlite, and bainite. Neither the iron atoms nor the carbon atoms diffuse. Therefore, the transformation occurs by shearing or the cooperative motion of large numbers of atoms. Figure 14 shows schematically the formation of a martensitic crystal. Macroscopically, the shears act parallel to a fixed crystallographic plane, termed the habit plane, and produce a uniformly tilted surface relief on a free surface. Not only is the crystal structure change from austenite (fcc) to martensite (bcc) (referred to as the lattice deformation) accomplished by the transformation, but also the product martensite is simultaneously deformed because of the constraints created by maintaining an unrotated and undistorted habit plane within the bulk austenite (Ref 2, 21). The deformation of the martensite is referred to as the lattice invariant deformation, and it produces a high density of dislocations or twins in martensite. This fine structure, together with the carbon atoms trapped within the octahedral interstitial sites of the body-centered tetragonal structure, produce the very high strength of as-quenched martensite (Ref 22).

Martensite begins to form at a critical temperature, defined as the martensite start (M_s) temperature. The transformation is accomplished by the nucleation and growth of many crystals. Because of the matrix constraints, the width of the martensitic units is limited, and the transformation proceeds primarily by the successive nucleation of new crystals. This process occurs only upon cooling to lower temperatures and is therefore independent of time. The latter type of transformation kinetics is termed athermal and is characterized by (Ref 23):

$$f = 1 - \exp - [0.01](M_s - T_q) \qquad \text{(Eq 3)}$$

where f is the fraction of martensite formed after quenching to any temperature, T_q, below M_s. Thus, the amount of martensite that forms at room temperature, for example, is a function only of M_s.

The M_s temperature is a function of the carbon and alloy content of steel, and a number of equations from which M_s can be calculated based on composition have been developed (Ref 2). Figure 15 shows that M_s decreases sharply with increasing carbon content in iron-carbon alloys. Almost all other alloying elements also lower M_s. A major effect of low M_s temperatures is incomplete martensite formation at room temperature. Therefore, in all martensitic structures, some austenite is retained, the

Continuously annealed sheet steels are very susceptible to aging effects, because the thin sheet cools rapidly from the annealing temperature in contrast to batch annealing. As a result, various types of overaging treatments are applied to continuously annealed steels, as shown in Fig. 11. These treatments are designed to remove carbon and nitrogen from solid solution by the precipitation of relatively coarse carbide and nitride particles.

Most cold-rolled steels are subcritically annealed; that is, they are annealed below the A_1 temperature. However, continuous annealing lines have made possible intercritical heating into the ferrite-austenite field, with cooling that is rapid enough to cause the austenite to transform to martensite. The martensite formation introduces a dislocation density exceeding that which can be pinned by the available carbon. As a result, early yielding is continuous and occurs with high rates of strain hardening. Intercritically annealed steels with ferrite-martensite microstructures are referred to as dual-phase steels and offer another approach to producing high strength levels that range from 345 to 550 MPa (50 to 80 ksi) in low-carbon steels (Ref 18-20). Dual-phase steels are discussed in the article "High-Strength Structural and High-Strength Low-Alloy Steels" in this Volume.

134 / Carbon and Low-Alloy Steels

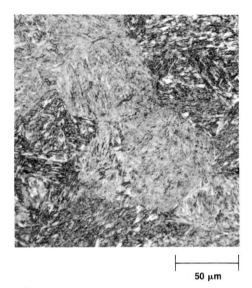

Fig. 16 Light micrograph of lath martensite in 4340 steel quenched from 940 °C (1725 °F) and tempered at 350 °C (660 °F). Packets of parallel laths are below resolution of light microscope. Source: Ref 24

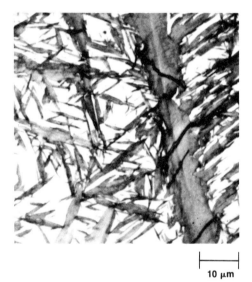

Fig. 17 Light micrograph of plate martensite and retained austenite in an Fe-1.39C alloy. Source: Ref 25

exact amount of which depends sensitively on composition.

Two morphologies of martensitic microstructures form in iron-carbon alloys and steels (Fig. 15). In low- and medium-carbon alloys, lath martensite forms. Lath martensite is characterized by parallel board or lath-shaped crystals. The laths have an internal structure consisting of tangled dislocations, and the microstructure contains small amounts of retained austenite between the laths. The groups of parallel laths are termed packets; many of the laths are too fine to be resolved in the light microscope. Figure 16 shows an example of lath martensite in 4340 steel.

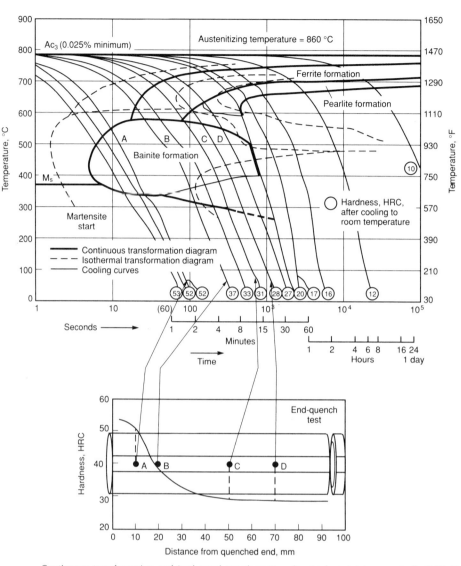

Fig. 18 Continuous transformation and isothermal transformation for steel containing nominally 0.4% C, 1.0% Cr, and 0.2% Mo. Several cooling rates are related to positions and hardness on a Jominy end-quench specimen. Source: Ref 1

In high-carbon steels, plate martensite forms. The martensitic crystals have the shape of plates, and adjacent units tend to be nonparallel. The fine structure associated with the plates often consists of fine transformation twins, and large amounts of retained austenite are present because of the low M_s temperatures. Figure 17 shows plate martensite and austenite in an Fe-1.36C alloy.

Martensite can form only if the diffusion-controlled transformations of austenite can be suppressed. On a practical level, this is accomplished by rapid quenching, for example, in water or brine baths. However, such drastic cooling introduces high surface tensile residual stresses and may cause quench cracking. Therefore, medium-carbon steels are alloyed with elements such as nickel, chromium, and molybdenum, which make it more difficult for the diffusion-controlled transformations to occur. As a result, martensite can be formed with less drastic cooling, such as oil quenching. The design of steels and cooling conditions to produce required amounts of martensite is the subject of the technology referred to as hardenability (Ref 26, 27).

The application of hardenability concepts characterize not only the conditions that produce martensite but also those under which other microstructures form. Thus, hardness gradients in bars of various diameters, cooled at various rates, can be estimated. Continuous cooling diagrams, in which cooling conditions that produce various microstructures are defined for a given steel, are often related to hardness gradients measured on Jominy end-quenched specimens, as shown in Fig. 18.

Tempering of Martensite

As-quenched martensite has very high strength, but has very low fracture resistance, or toughness. Therefore, almost all

Microstructures, Processing, and Properties of Steel / 135

Table 1 Tempering reactions in steel

Temperature range °C	°F	Reaction and symbol (if designated)	Comments
−40 to 100	−40 to 212	Clustering of two to four carbon atoms on octahedral sites of martensite (A1); segregation of carbon atoms to dislocations and boundaries	Clustering is associated with diffuse spikes around fundamental electron diffraction spots of martensite
20 to 100	70 to 212	Modulated clusters of carbon atoms on (102) martensite planes (A2)	Identified by satellite spots around electron diffraction spots of martensite
60 to 80	140 to 175	Long period ordered phase with ordered carbon atoms (A3)	Identified by superstructure spots in electron diffraction patterns
100 to 200	212 to 390	Precipitation of transition carbide as aligned 2 nm (0.08 μin.) diam particles (T1)	Recent work identifies carbides as η (orthorhombic, Fe_2C); earlier studies identified the carbides as ε (hexagonal, $Fe_{2.4}C$).
200 to 350	390 to 660	Transformation of retained austenite to ferrite and cementite (T2)	Associated with tempered-martensite embrittlement in low- and medium-carbon steels
250 to 700	480 to 1290	Formation of ferrite and cementite; eventual development of well-spheroidized carbides in a matrix of equiaxed ferrite grains (T3)	This stage now appears to be initiated by χ-carbide formation in high-carbon Fe-C alloys.
500 to 700	930 to 1290	Formation of alloy carbides in chromium-, molybdenum-, vanadium- and tungsten-containing steels. The mix and composition of the carbides may change significantly with time (T4).	The alloy carbides produce secondary hardening and pronounced retardation of softening during tempering or long-time service exposure around 500 °C (930 °F).
350 to 550	660 to 1020	Segregation and cosegregation of impurity and substitutional alloying elements	Responsible for temper embrittlement

Source: Ref 28

steels that are quenched to martensite are also tempered, or heated, to some temperature below A_1 in order to increase toughness. Depending on time and temperature, tempering treatments can produce a wide variety of microstructures and properties.

As-quenched martensitic microstructures are supersaturated with respect to carbon, have high residual stresses, contain a high density of dislocations, have a very high lath or plate boundary area per unit volume, and contain retained austenite. All these factors make martensitic microstructures very unstable and drive various phase transformations and microstructural changes during tempering. Table 1 lists the various reactions that develop during tempering. The most important changes are a result of aging and precipitation phenomena, which are caused by the supersaturation of carbon, and range from carbon atom clustering to transition carbide precipitation to cementite formation and spheroidization. Figure 19 shows a plate of high-carbon martensite tempered at 150 °C (300 °F). The dark basketweave structure is due to contrast associated with rows of very fine particles that are 2 nm (0.08 μin.) in size, of the orthorhombic transition carbide, η. The carbides must be imaged by other techniques (Ref 29, 30). Tempering at temperatures between 150 and 200 °C (300 and 390 °F) retains high hardness with increased toughness relative to as-quenched martensite.

Tempering at higher temperatures causes the formation of cementite, and, if strong

Fig. 19 Transmission electron micrograph showing effects of transition carbide precipitation (dark contrast) in fine structure of a plate of martensite in an Fe-1.22C alloy tempered at 150 °C (300 °F). Source: Ref 29

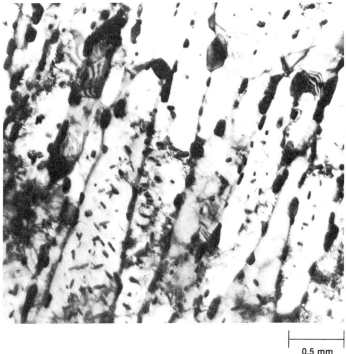

Fig. 20 Transmission electron micrograph showing the microstructure of 4130 steel water quenched from 900 °C (1650 °F) and tempered at 650 °C (1200 °F) Courtesy of F. Woldow

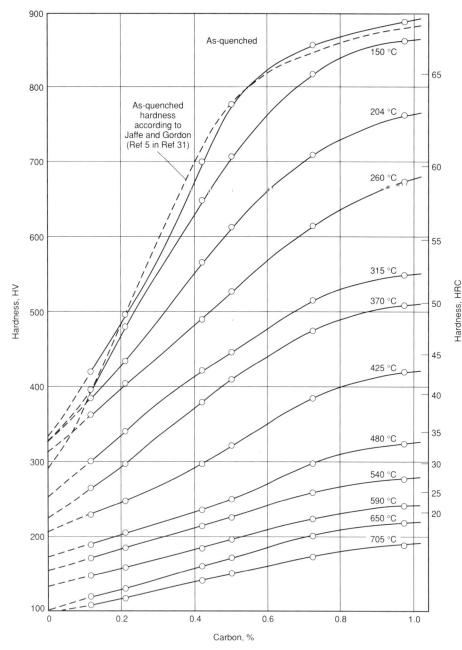

Fig. 21 Hardness as a function of carbon content in iron-carbon alloys quenched to martensite and tempered at various temperatures. Source: Ref 31

carbide-forming elements are present, alloy carbides. Concurrently, the laths or plates coarsen and the dislocation density is reduced by recovery mechanisms (Ref 2). In addition, retained austenite transforms to mixtures of cementite and ferrite between martensite laths and plates. The carbides formed by high-temperature tempering are much coarser than the transition carbides and are present at residual martensite interfaces and dispersed within the ferrite of the tempered martensite (Fig. 20).

Figure 21 shows the range of hardness levels which may be obtained by tempering at various temperatures as a function of the carbon content of the steel. The highest hardnesses for engineering applications are associated with the transition carbide microstructures produced by tempering at 150 °C (300 °F). These microstructures have excellent fatigue and wear resistance and are used for such applications as shafts, gears, and bearings. The lowest hardnesses are associated with microstructures of spheroidized carbides in a matrix of equiaxed ferrite. Steels with these microstructures are used when very high toughness or corrosion resistance (for example, resistance to H_2S in oil field applications) is required.

Toughness, or fracture resistance, generally increases with tempering temperature, but various types of embrittlement or reduced toughness can develop (Ref 2). Figure 22 shows impact toughness as a function of tempering temperature for selected sets of steels with high and low levels of phosphorus. Carbon content has a major influence on toughness. Medium-carbon tempered steels are quite tough, but high-carbon steels show very low impact toughness, which limits the application of hardened and tempered high-carbon steels to conditions of compressive loading without impact, such as in bearings. The effect of carbon on the toughness of low-temperature tempered specimens correlates with increasing densities of transition carbides and associated high strain hardening rates as carbon content increases (Ref 2).

Toughness reaches its peak in specimens tempered at 200 °C (390 °F); it drops to a minimum in specimens tempered around 300 °C (570 °F). This drop is referred to as tempered martensite embrittlement and is associated with the transformation of retained austenite to coarse carbide structures. Tempered martensite embrittlement is exacerbated by phosphorus segregation to prior-austenite grain boundaries and carbide interfaces, but this effect appears to be constant over the entire tempering range (Fig. 22). At higher tempering temperatures, between 350 and 550 °C (660 and 1020 °F), another embrittlement phenomenon may develop in steels containing phosphorus, antimony, or tin (Ref 34). This embrittlement is referred to as temper embrittlement, and requires long holding times or slow cooling through the embrittling temperature range. Alloy steels are most susceptible, and the cosegregation of the alloying elements with the impurities to prior austenite grain boundaries has been documented (Ref 35).

Processing: Quenched and Tempered Microstructures

Hardened steels with tempered martensitic microstructures are most frequently used in machine components that require high strength and excellent fatigue resistance under conditions of cyclic loading. Figure 23 shows a typical processing sequence for these components. Hot-rolled bars are received and forged, generally at high temperatures where deformation into complex shapes is readily accomplished. The forgings are air cooled, and ferrite-pearlite microstructures develop upon cooling to room temperature. A normalizing treatment to refine the coarse microstructures that originated because of high-temperature forging may be required, or a spheroidizing treatment to produce a microstructure of ferrite and spheroidized cementite may be applied if extensive machining prior to hardening is required. The forgings are then austenitized, quenched to martensite, and tempered to the properties

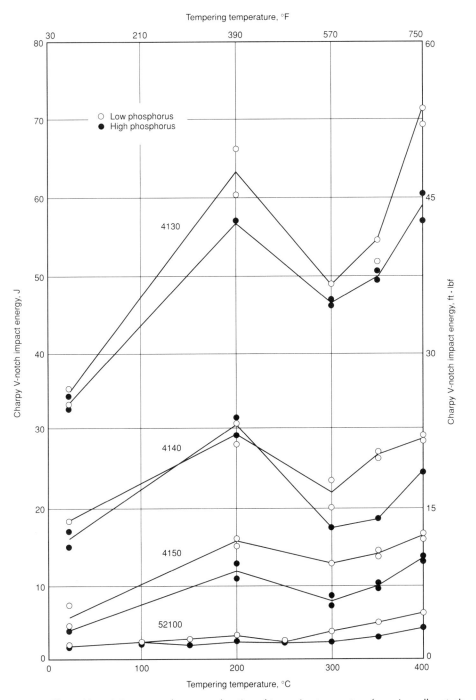

Fig. 22 Charpy V-notch impact toughness as a function of tempering temperature for various alloy steels. High phosphorus levels are about 0.02% and low phosphorus levels range between 0.002 and 0.009%. Source: Ref 32, 33

small amounts of vanadium and niobium and transform to precipitation-hardened microstructures of ferrite and pearlite. The hardness produced by rapid air cooling ranges from 25 to 30 HRC depending on the extent of precipitation and pearlite in the microstructure; ultimate strength values are over 690 MPa (100 ksi). Thus, the hardness and strength levels are not as high as can be produced by quenching and low-temperature tempering, but they are more than adequate for many automotive applications that require intermediate strengths (Ref 36).

The fatigue resistance of direct-cooled microalloyed steels is comparable to that of quenched and tempered steels of the same hardness, but the impact toughness is much lower. This reduced toughness is due to the well-known increase in the ductile-to-brittle temperature in steels with ferrite-pearlite microstructures as pearlite content increases (Fig. 25). In order to improve the toughness of direct-cooled forging steels, steels that transform to bainitic structures and forging steels with lower carbon concentrations and finer ferrite-pearlite microstructures are being developed (Ref 38).

Summary

This article has briefly described the major microstructures and the phase transformations by which these microstructures are developed in carbon and low-alloy steels. Each type of microstructure and product is developed to characteristic property ranges by specific processing routes that control and exploit microstructural changes. The incorporation of steel carbon content into microstructure has a profound effect on microstructure and properties, and steels fall naturally into low-strength/high ductility/high toughness or high-strength/high fatigue resistant/low toughness groups with increasing carbon content. The use of new casting techniques, microalloying, and thermomechanical processing are being used increasingly to reduce processing steps and to improve steel product microstructures and quality.

REFERENCES

1. G. Krauss, Physical Metallurgy and Heat Treatment of Steel, in *Metals Handbook Desk Edition*, H.E. Boyer and T.L. Gall, Ed., American Society for Metals, 1985, p 28-2 to 28-10
2. G. Krauss, *Steels: Heat Treatment and Processing Principles*, ASM INTERNATIONAL, 1989
3. J.S. Kirkaldy, B.A. Thompson, and E.A. Baganis, Prediction of Multicomponent Equilibrium and Transformation Diagrams for Low Alloy Steels, in *Hardenability Concepts with Applications to Steel*, D.V. Doane and J.S. Kirkaldy, Ed., The Metallurgical Society, 1978

described in the preceding section. Straightening and stress relieving operations may be applied if required.

Processing: Direct-Cooled Forging Microstructures

To reduce the number of processing steps associated with producing quenched and tempered microstructures, new alloying approaches have been developed to produce high-strength microstructures directly during cooling after forging. Figure 24 shows a schematic of such a processing approach and an alternate processing sequence that cold finishes hot-rolled bars. Eliminating heat treatment processing steps by direct cooling relative to quenching and tempering has obvious advantages.

One group of steels that has been developed for direct cooling is microalloyed medium-carbon steels (see Ref 36, 37 and the article "High-Strength Low-Alloy Steel Forgings" in this Volume). These steels contain

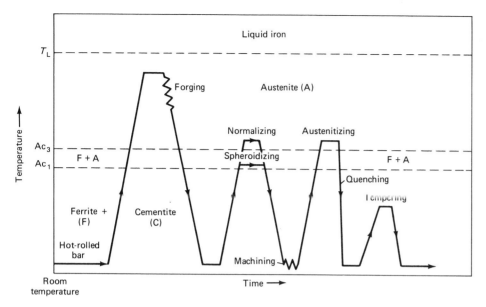

Fig. 23 Temperature-time processing schedules for producing quench and tempered forgings

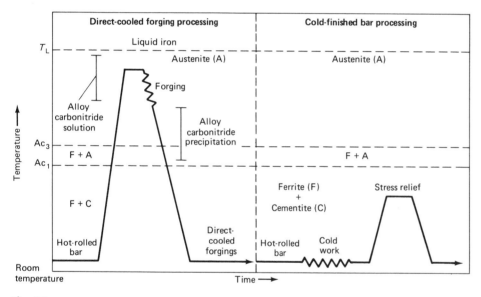

Fig. 24 Temperature-time schedule for producing direct-cooled forgings and cold-finished bars

4. *Heat Treaters Guide*, P.M. Unterweiser, H.E. Boyer, and J.J. Kubbs, Ed., American Society for Metals, 1982
5. M. Hillert, The Formation of Pearlite, in *Decomposition of Austenite by Diffusional Processes*, V.F. Zackay and H.I. Aaronson, Ed., Interscience, 1962, p 197-247
6. W.A. Johnson and R.F. Mehl, Reaction Kinetics in Processes of Nucleation and Growth, *Trans. AIME*, Vol 135, 1939, p 416-458
7. J.W. Christian and D.V. Edmonds, The Bainite Transformation, in *Phase Transformations and Ferrous Alloys*, A.R. Marder and J.I. Goldstein, Ed., The Metallurgical Society, 1984, p 293-325
8. R.F. Hehemann, Ferrous and Nonferrous Bainite Structures, in *Metals Handbook*, 8th ed., Vol 8, American Society for Metals, 1973, p 194-196
9. B.L. Bramfitt and J.G. Speer, A Perspective on the Morphology of Bainite, *Metall. Trans. A*, to be published in 1990
10. A.W. Cramb, New Steel Casting Processes for Thin Slabs and Strip, *Iron Steelmaker*, Vol 15 (No. 7), 1988, p 45-60
11. I. Tamura, H. Sekine, T. Tanaka, and C. Ouchi, *Thermomechanical Processing of High-Strength Low-Alloy Steels*, Butterworths, 1988
12. *Thermomechanical Processing of Microalloyed Austenite*, A.J. DeArdo, G.A. Ratz, and P.J. Wray, Ed., The Metallurgical Society, 1982
13. *Microalloyed HSLA Steels: Proceedings of Microalloying '88*, ASM INTERNATIONAL, 1988
14. P.R. Mould, An Overview of Continuous-Annealing Technology, in *Metallurgy of Continuous-Annealed Sheet Steel*, B.L. Bramfitt and D.L. Mangonon, Jr., Ed., The Metallurgical Society, 1982, p 3-33
15. W.C. Leslie, *The Physical Metallurgy of Steels*, McGraw-Hill, 1981
16. D.Z. Yang, E.L. Brown, D.K. Matlock, and G. Krauss, Ferrite Recrystallization and Austenite Formation in Cold-Rolled Intercritically Annealed Steel, *Metall. Trans. A*, Vol 11A, 1985, p 1385-1392
17. *Metallurgy of Vacuum-Degassed Steel Products*, R. Pradhan, Ed., The Metallurgical Society, to be published in 1990
18. *Structure and Properties of Dual-Phase Steels*, R.A. Kot and J.M. Morris, Ed., The Metallurgical Society, 1979
19. *Fundamentals of Dual-Phase Steels*, R.A. Kot and B.L. Bramfitt, Ed., The Metallurgical Society, 1981
20. D.K. Matlock, F. Zia-Ebrahimi, and G. Krauss, Structure, Properties and Strain Hardening of Dual-Phase Steels, in *Deformation, Processing and Structure*, G. Krauss, Ed., ASM INTERNATIONAL, 1984
21. B.A. Bilby and J.W. Christian, The Crystallography of Martensite Transformations, Vol 197, 1961, p 122-131
22. M. Cohen, The Strengthening of Steel, *Trans. TMS-AIME*, Vol 224, 1962, p 638-657
23. D.P. Koistinen and R.E. Marburger, A General Equation Prescribing the Extent of the Austenite-Martensite Transformation in Pure Iron-Carbon Alloys and Plain Carbon Steels, *Acta Metall.*, Vol 7, 1959, p 59-60
24. J.P. Materkowski and G. Krauss, Tempered Martensite Embrittlement in SAE 4340 Steel, *Metall. Trans. A*, Vol 10A, 1979, p 1643-1651
25. A.R. Marder, A.O. Benscoter, and G. Krauss, Microcracking Sensitivity in Fe-C Plate Martensite, *Metall. Trans.*, Vol 1, 1970, p 1545-1549
26. *Hardenability Concepts with Applications to Steel*, D.V. Doane and J.S. Kirkaldy, Ed., American Institute of Mining, Metallurgical, and Petroleum Engineers, 1978
27. C.A. Siebert, D.V. Doane, and D.H. Breen, *The Hardenability of Steels: Concepts, Metallurgical Influences, and Industrial Applications*, American Society for Metals, 1977
28. G. Krauss, Tempering and Structural Change in Ferrous Martensites, in *Phase Transformations in Ferrous Alloys*, A.R. Marder and J.I. Goldstein, Ed., The Metallurgical Society, 1984
29. D.L. Williamson, K. Nakazawa, and G.

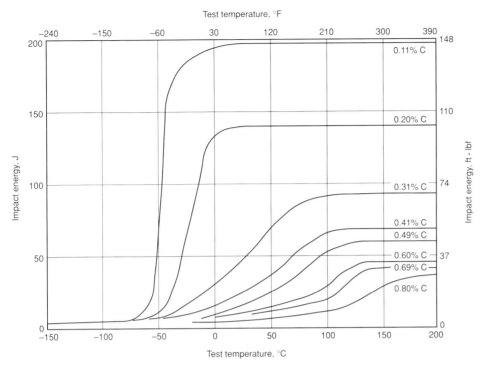

Fig. 25 Impact transition curves as a function of carbon content in normalized steels. Increase in ductile-to-brittle transition temperatures with increasing carbon content is due to increasing amounts of pearlite. Source: Ref 1

Krauss, A Study of the Early Stages of Tempering in an Fe-1.22 pct C Alloy, *Metall. Trans. A*, Vol 10A, 1979, p 1351-1363

30. Y. Hirotsu and S. Nagakura, Crystal Structure and Morphology of the Carbide Precipitated in Martensitic High Carbon Steel During the First Stage of Tempering, *Acta Metall.*, Vol 20, 1972, p 645-655

31. R.A. Grange, C.R. Hibral, and L.F. Porter, Hardness of Tempered Martensite in Carbon and Low Alloy Steels, *Metall. Trans. A*, Vol 8A, 1977, p 1775-1785
32. D.L. Yaney, "The Effects of Phosphorus and Tempering on the Fracture of AISI 52100 Steel," M.S. thesis, Colorado School of Mines, 1981
33. F. Zia-Ebrahimi and G. Krauss, Mechanisms of Tempered Martensite Embrittlement in Medium-Carbon Steels, *Acta Metall.*, Vol 32, 1984, p 1767-1777
34. C.J. McMahon, Jr., Temper Brittleness: An Interpretive Review, in *Temper Embrittlement in Steel*, STP 407, American Society for Testing and Materials, 1968, p 127-167
35. M. Guttman, P. Dumonlin, and M. Wayman, The Thermodynamics of Interactive Co-Segregation of Phosphorus and Alloying Elements in Iron and Temper-Brittle Steels, *Metall. Trans. A*, Vol 13A, 1982, p 1693-1711
36. *Fundamentals of Microalloying Forging Steels*, G. Krauss and S.K. Banerji, Ed., The Metallurgical Society, 1987
37. G. Krauss, Microalloyed Bar and Forging Steels, in *29th Mechanical Working and Steel Processing Conference Proceedings*, Vol XXV, Iron and Steel Society, 1988, p 67-77
38. K. Grassl, S.W. Thompson, and G. Krauss, "New Options for Steel Selection for Automotive Applications," SAE Technical Paper 890508, Society of Automotive Engineers, 1989

Classification and Designation of Carbon and Low-Alloy Steels

STEELS constitute the most widely used category of metallic material, primarily because they can be manufactured relatively inexpensively in large quantities to very precise specifications. They also provide a wide range of mechanical properties, from moderate yield strength levels (200 to 300 MPa, or 30 to 40 ksi) with excellent ductility to yield strengths exceeding 1400 MPa (200 ksi) with fracture toughness levels as high as 110 MPa$\sqrt{m}$ (100 ksi$\sqrt{in.}$).

This article will review the various systems used to classify carbon and low-alloy steels*, describe the effects of alloying elements on the properties and/or characteristics of steels, and provide extensive tabular data pertaining to designations of steels (both domestic and international). More detailed information on the steel types and product forms discussed in this article can be found in the articles that follow in this Section.

Classification of Steels

Steels can be classified by a variety of different systems depending on:

- *The composition*, such as carbon, low-alloy, or stainless steels
- *The manufacturing methods*, such as open hearth, basic oxygen process, or electric furnace methods
- *The finishing method*, such as hot rolling or cold rolling
- *The product form*, such as bar, plate, sheet, strip, tubing, or structural shape
- *The deoxidation practice*, such as killed, semikilled, capped, or rimmed steel
- *The microstructure*, such as ferritic, pearlitic, and martensitic (Fig. 1)
- *The required strength level*, as specified in ASTM standards
- *The heat treatment*, such as annealing, quenching and tempering, and thermomechanical processing
- *Quality descriptors*, such as forging quality and commercial quality

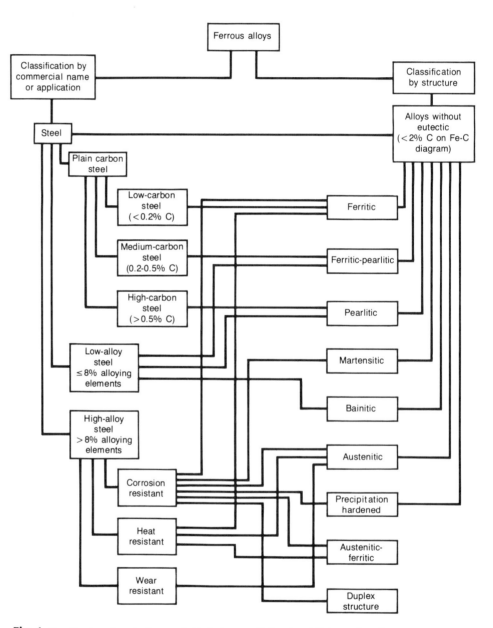

Fig. 1 Classification of steels. Source: D.M. Stefanescu, University of Alabama, Tuscaloosa

---

*The term low-alloy steel rather than the more general term alloy steel is being used to differentiate the steels covered in this article from high-alloy steels. High-alloy steels include steels with a high degree of fracture toughness (Fe-9Ni-4Co), which are described in the article "Ultrahigh-Strength Steels" in this Section of the Handbook. They also include maraging steels (Fe-18Ni-4Mo-8Co), austenitic manganese steels (Fe-1C-12Mn), tool steels, and stainless steels, which are described in separate articles in the Section "Specialty Steels and Heat-Resistant Alloys" in this Volume.

Classification and Designation of Carbon and Low-Alloy Steels / 141

Table 1 Carbon steel cast or heat chemical limits and ranges
Applicable only to semifinished products for forging, hot-rolled and cold-finished bars, wire rods, and seamless tubing

Element	Maximum of specified element, %	Range, %
Carbon(a)	≤0.12	...
	>0.12–0.25 incl	0.05
	>0.25–0.40 incl	0.06
	>0.40–0.55 incl	0.07
	>0.55–0.80 incl	0.10
	>0.80	0.13
Manganese	≤0.40	0.15
	>0.40–0.50 incl	0.20
	>0.50–1.65 incl	0.30
Phosphorus	>0.040–0.08 incl	0.03
	>0.08–0.13 incl	0.05
Sulfur	>0.050–0.09 incl	0.03
	>0.09–0.15 incl	0.05
	>0.15–0.23 incl	0.07
	>0.23–0.35 incl	0.09
Silicon (for bars)(b)(c)	≤0.15	0.08
	>0.15–0.20 incl	0.10
	>0.20–0.30 incl	0.15
	>0.30–0.60 incl	0.20
Copper	When copper is required, 0.20% minimum is commonly used	
Lead(d)	When lead is required, a range of 0.15–0.35 is generally used	

Note: Boron-treated fine-grain steels are produced to a range of 0.0005–0.003% B. Incl, inclusive. (a) The carbon ranges shown customarily apply when the specified maximum limit for manganese does not exceed 1.10%. When the maximum manganese limit exceeds 1.10%, it is customary to add 0.01 to the carbon range shown. (b) It is not common practice to produce a rephosphorized and resulfurized carbon steel to specified limits for silicon because of its adverse effect on machinability. (c) When silicon is required for rods the following ranges and limits are commonly used: 0.10 max; 0.07–0.15, 0.10–0.20, 0.15–0.35, 0.20–0.40, or 0.30–0.60. (d) Lead is reported only as a range of 0.15–0.35% because it is usually added to the mold or ladle stream as the steel is poured. Source: Ref 1

Table 2 Carbon steel cast or heat chemical limits and ranges
Applicable only to structural shapes, plates, strip, sheets, and welded tubing

Element	Maximum of specified element, %	Range, %
Carbon(a)(b)	≤0.15	0.05
	>0.15–0.30 incl	0.06
	>0.30–0.40 incl	0.07
	>0.40–0.60 incl	0.08
	>0.60–0.80 incl	0.11
	>0.80–1.35 incl	0.14
Manganese	≤0.50	0.20
	>0.050–1.15 incl	0.30
	>1.15–1.65 incl	0.35
Phosphorus	≤0.08	0.03
	>0.08–0.15 incl	0.05
Sulfur	≤0.08	0.03
	>0.08–0.15 incl	0.05
	>0.15–0.23 incl	0.07
	>0.23–0.33 incl	0.10
Silicon	≤0.15	0.08
	>0.15–0.30 incl	0.15
	>0.30–0.60 incl	0.30
Copper	When copper is required, 0.20% minimum is commonly specified	

Incl, inclusive. (a) The carbon ranges shown in the range column apply when the specified maximum limit for manganese does not exceed 1.00%. When the maximum manganese limit exceeds 1.00%, add 0.01 to the carbon ranges shown in the table. (b) Maximum of 0.12% C for structural shapes and plates. Source: Ref 1

Of the aforementioned classification systems, chemical composition is the most widely used internationally and will be emphasized in this article. Classification systems based on deoxidation practice and quality descriptors will also be reviewed. Information pertaining to the microstructural characteristics of steels can be found in the article "Microstructures, Processing, and Properties of Steels" in this Volume and in Volume 9, *Metallography and Microstructures*, of the 9th Edition of *Metals Handbook*.

Chemical Analysis

Chemical composition is often used as the basis for classifying steels or assigning standard designations to steels. Such designations are often incorporated into specifications for steel products. Users and specifiers of steel products should be familiar with methods of sampling and analysis.

Chemical analyses of steels are usually performed by wet chemical analysis methods or spectrochemical methods. Wet analysis is most often used to determine the composition of small numbers of specimens or of specimens composed of machine tool chips. Spectrochemical analysis is well-suited to the routine determination of the chemical composition of a large number of specimens, as may be necessary in a steel mill environment. Both classical wet chemical and spectrochemical methods for analyzing steel samples are described in detail in Volume 10, *Materials Characterization*, of the 9th Edition of *Metals Handbook*.

Heat and Product Analysis. During the steelmaking process, a small sample of molten metal is removed from the ladle or steelmaking furnace, allowed to solidify, and then analyzed for alloy content. In most steel mills, these heat analyses are performed using spectrochemical methods; as many as 14 different elements can be determined simultaneously. The heat analysis furnished to the customer, however, may include only those elements for which a range or a maximum or minimum limit exists in the appropriate designation or specification.

A heat analysis is generally considered to be an accurate representation of the composition of the entire heat of metal. Producers of steel have found that heat analyses for carbon and alloy steels can be consistently held within ranges that depend on the amount of the particular alloying element desired for the steel, the product form, and the method of making the steel. These ranges have been published as commercial practice, then incorporated into standard specifications. Standard ranges and limits of heat analyses of carbon and alloy steels are given in Tables 1 through 4.

Because segregation of some alloying elements is inherent in the solidification of an ingot, different portions will have local chemical compositions that differ slightly from the average composition. Many lengths of bar stock can be made from a single ingot; therefore, some variation in composition between individual bars must be expected. The compositions of individual bars might not conform to the applicable specification, even though the heat analysis does. The chemical composition of an individual bar (or other product) taken from a large heat of steel is called the product analysis or check analysis. Ranges and limits for product analyses are generally broader and less restrictive than the corresponding ranges and limits for heat analyses. Such limits used in standard commercial practice are given in Tables 5, 6, and 7.

Residual elements usually enter steel products from raw materials used to produce pig iron or from scrap steel used in steelmaking. Through careful steelmaking practices, the amounts of these residual elements are generally held to acceptable levels. Sulfur and phosphorus are usually considered deleterious to the mechanical properties of steels; therefore, restrictions are placed on the allowable amounts of these elements for most grades. The amounts of sulfur and phosphorus are invariably reported in the analyses of both carbon and alloy steels. Other residual alloying elements generally exert a lesser influence than sulfur and phosphorus on the properties of steel. For many grades of steel, limitations on the amounts of these residual elements are either optional or omitted entirely. Amounts of residual alloying elements are generally not reported in either heat or product analyses, except for special reasons.

Silicon Content of Steels. The composition requirements for many steels, particularly plain carbon steels, contain no specific restriction on silicon content. The lack of a silicon requirement is not an omission, but instead indicates recognition that the amount of silicon in a steel can often be traced directly to the deoxidation practice employed in making it (further information can be found in the section "Types of Steel Based on Deoxidation Practice" in this article).

Rimmed and capped steels are not deoxidized; the only silicon present is the residual amount left from scrap or raw materials, typically less than 0.05% Si. Specifications and orders for these steels customarily indicate that the steel must be made rimmed or capped, as required by the purchaser; restrictions on silicon content are not usually given.

The extent of rimming action during the solidification of semikilled steel ingots must be carefully controlled by matching the amount of deoxidizer with the oxygen content of the molten steel. The amount of silicon required for deoxidation may vary from heat to heat. Thus, the silicon content of the solid metal can also vary slightly from heat to heat. A maximum silicon content of

Table 3 Alloy steel heat composition ranges and limits for bars, blooms, billets, and slabs

Element	Maximum of specified element, %	Range, % Open hearth or basic oxygen steels	Range, % Electric furnace steels
Carbon	≤0.55	0.05	0.05
	>0.55–0.70 incl	0.08	0.07
	>0.70–0.80 incl	0.10	0.09
	>0.80–0.95 incl	0.12	0.11
	>0.95–1.35 incl	0.13	0.12
Manganese	≤0.60	0.20	0.15
	>0.60–0.90 incl	0.20	0.20
	>0.90–1.05 incl	0.25	0.25
	>1.05–1.90 incl	0.30	0.30
	>1.90–2.10 incl	0.40	0.35
Sulfur(a)	≤0.050	0.015	0.015
	>0.050–0.07 incl	0.02	0.02
	>0.07–0.10 incl	0.04	0.04
	>0.10–0.14 incl	0.05	0.05
Silicon	≤0.15	0.08	0.08
	>0.15–0.20 incl	0.10	0.10
	>0.20–0.40 incl	0.15	0.15
	>0.40–0.60 incl	0.20	0.20
	>0.60–1.00 incl	0.30	0.30
	>1.00–2.20 incl	0.40	0.35
Chromium	≤0.40	0.15	0.15
	>0.40–0.90 incl	0.20	0.20
	>0.90–1.05 incl	0.25	0.25
	>1.05–1.60 incl	0.30	0.30
	>1.60–1.75 incl	(b)	0.35
	>1.75–2.10 incl	(b)	0.40
	>2.10–3.99 incl	(b)	0.50
Nickel	≤0.50	0.20	0.20
	>0.50–1.50 incl	0.30	0.30
	>1.50–2.00 incl	0.35	0.35
	>2.00–3.00 incl	0.40	0.40
	>3.00–5.30 incl	0.50	0.50
	>5.30–10.00 incl	1.00	1.00
Molybdenum	≤0.10	0.05	0.05
	>0.10–0.20 incl	0.07	0.07
	>0.20–0.50 incl	0.10	0.10
	>0.50–0.80 incl	0.15	0.15
	>0.80–1.15 incl	0.20	0.20
Tungsten	≤0.50	0.20	0.20
	>0.50–1.00 incl	0.30	0.30
	>1.00–2.00 incl	0.50	0.50
	>2.00–4.00 incl	0.60	0.60
Copper	≤0.60	0.20	0.20
	>0.60–1.50 incl	0.30	0.30
	>1.50–2.00 incl	0.35	0.35
Vanadium	≤0.25	0.05	0.05
	>0.25–0.50 incl	0.10	0.10
Aluminum	≤0.10	0.05	0.05
	>0.10–0.20 incl	0.10	0.10
	>0.20–0.30 incl	0.15	0.15
	>0.30–0.80 incl	0.25	0.25
	>0.80–1.30 incl	0.35	0.35
	>1.30–1.80 incl	0.45	0.45

Element	Steelmaking process	Lowest maximum, %(c)
Phosphorus	Basic open hearth, basic oxygen, or basic electric furnace steels	0.035(d)
	Basic electric furnace E steels	0.025
	Acid open hearth or electric furnace steel	0.050
Sulfur	Basic open hearth, basic oxygen, or basic electric furnace steels	0.040(d)
	Basic electric furnace E steels	0.025
	Acid open hearth or electric furnace steel	0.050

Incl, inclusive. (a) A range of sulfur content normally indicates a resulfurized steel. (b) Not normally produced by open hearth process. (c) Not applicable to rephosphorized or resulfurized steels. (d) Lower maximum limits on phosphorus and sulfur are required by certain quality descriptors. Source: Ref 2

---

idized steels require no silicon; a requirement for minimum silicon content in such steel is unnecessary. A maximum permissible silicon content is appropriate for all killed plain carbon steels; a minimum silicon content implies a restriction that the steel must be silicon killed. Silicon is intentionally added to some alloy steels, for which it serves as both a deoxidizer and an alloying element to modify the properties of the steel. An acceptable range of silicon content would be appropriate for these steels.

Users and specifiers of steel mill products must realize that the silicon content of these items cannot be established independently of deoxidation practice. In ordering mill products, it is often desirable to cite a standard specification (such as an ASTM specification) where the various ramifications of restrictions on silicon content have already been considered in preparing the specification. In some instances, such as the forming of low-carbon steel sheet, the choice of deoxidation practice can significantly affect the performance of the steel; in such cases, it is appropriate to specify the desired practice.

Types of Steel Based on Deoxidation Practice (Ref 3)

Steels, when cast into ingots, can be classified into four types based on the deoxidation practice employed or, alternatively, by the amount of gas evolved during solidification. These types are killed, semikilled, rimmed, or capped steels (Fig. 2).

Killed steel is a type of steel from which there is only a slight evolution of gases during solidification of the metal after pouring. Killed steels are characterized by more uniform chemical composition and properties as compared to the other types. Alloy steels, forging steels, and steels for carburizing are generally killed.

Killed steel is produced by various steel-melting practices involving the use of certain deoxidizing elements which act with varying intensities. The most common of these are silicon and aluminum; however, vanadium, titanium, and zirconium are sometimes used. Deoxidation practices in the manufacture of killed steels are normally left to the discretion of the producer.

Semikilled steel is a type of steel wherein there is a greater degree of gas evolution than in killed steel but less than in capped or rimmed steel. The amount of deoxidizer used (customarily silicon or aluminum) will determine the amount of gas evolved. Semikilled steels generally have a carbon content within the range of 0.15 to 0.30%; they are used for a wide range of structural shape applications.

Semikilled steels are characterized by variable degrees of uniformity in composition, which are intermediate between those of killed and rimmed steels. Semikilled steel

0.10% is sometimes specified for semikilled steel, but this requirement is not very restrictive; for certain heats, a silicon addition sufficient to leave a residue of 0.10% may be enough of an addition to kill the steel.

Killed steels are fully deoxidized during their manufacture; deoxidation can be accomplished by additions of silicon, aluminum, or both, or by vacuum treatment of the molten steel. Because it is the least costly of these methods, silicon deoxidation is frequently used. For silicon-killed steels, a range of 0.15 to 0.30% Si is often specified, providing the manufacturer with adequate flexibility to compensate for variations in the steelmaking process and ensuring a steel acceptable for most applications. Aluminum-killed or vacuum-deox-

Classification and Designation of Carbon and Low-Alloy Steels / 143

Table 4 Alloy steel heat composition ranges and limits for plates

Element	Maximum of specified element, %	Range, % Open hearth or basic oxygen steels	Range, % Electric furnace steels
Carbon	≤0.25	0.06	0.05
	>0.25–0.40 incl	0.07	0.06
	>0.40–0.55 incl	0.08	0.07
	>0.55–0.70 incl	0.11	0.10
	>0.70	0.14	0.13
Manganese	≤0.45	0.20	0.15
	>0.45–0.80 incl	0.25	0.20
	>0.80–1.15 incl	0.30	0.25
	>1.15–1.70 incl	0.35	0.30
	>1.70–2.10 incl	0.40	0.35
Sulfur	≤0.060	0.02	0.02
	>0.060–0.100 incl	0.04	0.04
	>0.100–0.140 incl	0.05	0.05
Silicon	≤0.15	0.08	0.08
	>0.15–0.20 incl	0.10	0.10
	>0.20–0.40 incl	0.15	0.15
	>0.40–0.60 incl	0.20	0.20
	>0.60–1.00 incl	0.30	0.30
	>1.00–2.20 incl	0.40	0.35
Copper	≤0.60	0.20	0.20
	>0.60–1.50 incl	0.30	0.30
	>1.50–2.00 incl	0.35	0.35
Nickel	≤0.50	0.20	0.20
	>0.50–1.50 incl	0.30	0.30
	>1.50–2.00 incl	0.35	0.35
	>2.00–3.00 incl	0.40	0.40
	>3.00–5.30 incl	0.50	0.50
	>5.30–10.00 incl	1.00	1.00
Chromium	≤0.40	0.20	0.15
	>0.40–0.80 incl	0.25	0.20
	>0.80–1.05 incl	0.30	0.25
	>1.05–1.25 incl	0.35	0.30
	>1.25–1.75 incl	0.50	0.40
	>1.75–3.99 incl	0.60	0.50
Molybdenum	≤0.10	0.05	0.05
	>0.10–0.20 incl	0.07	0.07
	>0.20–0.50 incl	0.10	0.10
	>0.50–0.80 incl	0.15	0.15
	>0.80–1.15 incl	0.20	0.20
Vanadium	≤0.25	0.05	0.05
	>0.25–0.50 incl	0.10	0.10

Note: Boron steels can be expected to contain a minimum of 0.0005% B. Alloy steels can be produced with a lead range of 0.15–0.35%. A heat analysis for lead is not determinable because lead is added to the ladle stream while each ingot is poured. Incl, inclusive. Source: Ref 3

has a pronounced tendency for positive chemical segregation at the top-center of the ingot (Fig. 2).

Rimmed Steels. In the production of rimmed steels, no deoxidizing agents are added in the furnace. These steels are characterized by marked differences in chemical composition across the section and from the top to the bottom of the ingot (Fig. 2). They have an outer rim that is lower in carbon, phosphorus, and sulfur than the average composition of the whole ingot, and an inner portion, or core, that has higher levels than the average of those elements. The typical structure of the rimmed steel ingot results from a marked gas evolution during solidification of the outer rim.

During the solidification of the rim, the concentration of certain elements increases in the liquid portion of the ingot. During solidification of the core, some increase in segregation occurs in the upper and central portions of the ingot. The structural pattern of the ingot persists through the rolling process to the final product (rimmed ingots are best suited for steel sheets).

The technology of manufacturing rimmed steels limits the maximum content of carbon and manganese, and those maximums vary among producers. Rimmed steels do not retain any significant percentages of highly oxidizable elements such as aluminum, silicon, or titanium.

Capped steels have characteristics similar to those of rimmed steels but to a degree intermediate between those of rimmed and semikilled steels. A deoxidizer may be added to effect a controlled rimming action when the ingot is cast. The gas entrapped during solidification is in excess of that needed to counteract normal shrinkage, resulting in a tendency for the steel to rise in the mold. The capping operation limits the time of gas evolution and prevents the formation of an excessive number of gas voids within the ingot.

Mechanically capped steel is cast in bottle-top molds using a heavy metal cap.

Chemically capped steel is cast in open-top molds. The capping is accomplished by adding aluminum or ferrosilicon to the top of the ingot, causing the steel at the top surface to solidify rapidly. The top portion of the ingot is discarded.

The capped ingot practice is usually applied to steel with carbon contents greater than 0.15% that is used for sheet, strip, wire, and bars.

Quality Descriptors

The need for communication among producers and between producers and users has resulted in the development of a group of terms known as fundamental quality descriptors. These are names applied to various steel products to imply that the particular products possess certain characteristics that make them especially well suited for specific applications or fabrication processes. The fundamental quality descriptors in common use are listed in Table 8.

Some of the quality descriptors listed in Table 8 such as forging quality or cold extrusion quality are self-explanatory. The meaning of others is less obvious: for example, merchant quality hot-rolled carbon steel bars are made for noncritical applications requiring modest strength and mild bending or forming, but not requiring forging or heat treating. The descriptor for one particular steel commodity is not necessarily carried over to subsequent products made from that commodity—for example, standard quality cold-finished bars are made from special quality hot-rolled bars.

The various mechanical and physical attributes implied by a quality descriptor arise from the combined effects of several factors, including:

- The degree of internal soundess
- The relative uniformity of chemical composition
- The relative freedom from surface imperfections
- The size of the discard cropped from the ingot
- Extensive testing during manufacture
- The number, size, and distribution of nonmetallic inclusions
- Hardenability requirements

Control of these factors during manufacture is necessary to achieve mill products having the desired characteristics. The extent of the control over these and other related factors is another piece of information conveyed by the quality descriptor.

Some, but not all, of the fundamental descriptors may be modified by one or more additional requirements, as may be appropriate: special discard, macroetch test, restricted chemical composition, maximum incidental (residual) alloy, special hardenability or austenitic grain size. These restrictions could be applied to forging quality alloy steel bars, but not to merchant quality bars.

Understanding the various quality descriptors is complicated by the fact that most of the requirements that qualify a steel

Table 5 Product analysis tolerances for carbon and alloy steel plates, sheet, piling, and bars for structural applications

Element	Upper limit or maximum specified value, %	Tolerance, % Under minimum limit	Tolerance, % Over maximum limit
Carbon	≤0.15	0.02	0.03
	>0.15–0.40 incl	0.03	0.04
Manganese(a)	≤0.60	0.05	0.06
	>0.60–0.90 incl	0.06	0.08
	>0.90–1.20 incl	0.08	0.10
	>1.20–1.35 incl	0.09	0.11
	>1.35–1.65 incl	0.09	0.12
	>1.65–1.95 incl	0.11	0.14
	>1.95	0.12	0.16
Phosphorus	≤0.04	...	0.010
	>0.04–0.15 incl	...	(b)
Sulfur	≤0.05	...	0.010
Silicon	≤0.30	0.02	0.03
	>0.30–0.40 incl	0.05	0.05
	>0.40–2.20 incl	0.06	0.06
Nickel	≤1.00	0.03	0.03
	>1.00–2.00 incl	0.05	0.05
Chromium	≤0.90	0.04	0.04
	>0.90–2.10 incl	0.06	0.06
Molybdenum	≤0.20	0.01	0.01
	>0.20–0.40 incl	0.03	0.03
	>0.40–1.15 incl	0.04	0.04
Copper	0.20 minimum only	0.02	...
	≤1.00	0.03	0.03
	>1.00–2.00 incl	0.05	0.05
Titanium	≤0.10	0.01(c)	0.01(c)
Vanadium	≤0.10	0.01(c)	0.01(c)
	>0.10–0.25 incl	0.02	0.02
	Minimum only specified	0.01	...
Boron	Any	(b)	(b)
Niobium	≤0.10	0.01(c)	0.01(c)
Zirconium	≤0.15	0.03	0.03
Nitrogen	≤0.030	0.005	0.005

Incl, inclusive. (a) Manganese product analyses tolerances for bars and bar size shapes: ≤0.90, ±0.03; >0.90–2.20 incl, ±0.06. (b) Product analysis not applicable. (c) If the minimum of the range is 0.01%, the under tolerance is 0.005%. Source: Ref 4

for a particular descriptor are subjective. Only nonmetallic inclusion count, restrictions on chemical composition ranges and incidental alloying elements, austenitic grain size, and special hardenability are quantified. The subjective evaluation of the other characteristics depends on the skill and experience of those who make the evaluation. Although the use of these subjective quality descriptors might seem imprecise and unworkable, steel products made to meet the requirements of a particular quality descriptor can be relied upon to have those characteristics necessary for that product to be used in the indicated application or fabrication operation.

Effects of Alloying Elements (Ref 6)

Steels form one of the most complex group of alloys in common use. The synergistic effect of alloying elements and heat treatment produce a tremendous variety of microstructures and properties (characteristics). Given the limited scope of this article, it would be impossible to include a detailed survey of the effects of alloying elements on the iron-carbon equilibrium diagram. This complicated subject, which is briefly reviewed in the article "Microstructures, Processing, and Properties of Steels" in this Volume, lies in the domain of ferrous physical metallurgy and has also been reviewed extensively in the literature (Ref 7-11). In this section, the effects of various elements on steelmaking (deoxidation) practices and steel characteristics will be briefly outlined. It should be noted that the effects of a single alloying element are modified by the influence of other elements. These interrelations must be considered when evaluating a change in the composition of a steel. For the sake of simplicity, however, the various alloying elements listed below are discussed separately.

Carbon. The amount of carbon required in the finished steel limits the type of steel that can be made. As the carbon content of rimmed steels increases, surface quality becomes impaired. Killed steels in approximately the 0.15 to 0.30% C content level may have poorer surface quality and require special processing to attain surface quality comparable to steels with higher or lower carbon contents. Carbon has a moderate tendency to segregate, and carbon segregation is often more significant than the segregation of other elements. Carbon, which has a major effect on steel properties, is the principal hardening element in all steel. Tensile strength in the as-rolled condition increases as carbon content increases (up to about 0.85% C). Ductility and weldability decrease with increasing carbon.

Manganese has less of a tendency toward macrosegregation than any of the common elements. Steels above 0.60% Mn cannot be readily rimmed. Manganese is beneficial to surface quality in all carbon ranges (with the exception of extremely low carbon rimmed steels) and is particularly beneficial in resulfurized steels. It contributes to strength and hardness, but to a lesser degree than does carbon; the amount of increase is dependent upon the carbon content. Increasing the manganese content decreases ductility and weldability, but to a lesser extent than does carbon. Manganese has a strong effect on increasing the hardenability of a steel.

Phosphorus segregates, but to a lesser degree than carbon and sulfur. Increasing phosphorus increases strength and hardness and decreases ductility and notch impact toughness in the as-rolled condition. The decreases in ductility and toughness are greater in quenched and tempered higher-carbon steels. Higher phosphorus is often specified in low-carbon free-machining steels to improve machinability (see the article "Machinability of Steels" in this Volume).

Sulfur. Increased sulfur content lowers transverse ductility and notch impact tough-

Table 6 Product analysis tolerances for carbon and high-strength low-alloy steel bars, blooms, billets, and slabs

Element	Limit or maximum of specified range, %	Tolerance over the maximum limit or under the minimum limit, % ≤0.065 m² (100 in.²)	>0.065–0.129 m² (100–200 in.²) incl	>0.129–0.258 m² (200–400 in.²) incl	>0.258–0.516 m² (400–800 in.²) incl
Carbon	≤0.25	0.02	0.03	0.04	0.05
	>0.25–0.55 incl	0.03	0.04	0.05	0.06
	>0.55	0.04	0.05	0.06	0.07
Manganese	≤0.90	0.03	0.04	0.06	0.07
	>0.90–1.65 incl	0.06	0.06	0.07	0.08
Phosphorus(a)	Over maximum only, ≤0.40	0.008	0.008	0.010	0.015
Sulfur(a)	Over maximum only, ≤0.050	0.008	0.010	0.010	0.015
Silicon	≤0.35	0.02	0.02	0.03	0.04
	>0.35–0.60 incl	0.05	...	...	...
Copper	Under minimum only	0.02	0.03	...	...
Lead(b)	0.15–0.35 incl	0.03	0.03	...	...

Note: Rimmed or capped steels and boron are not subject to product analysis tolerances. Product analysis tolerances for alloy elements in high-strength low-alloy steels are given in Table 7. Incl, inclusive. (a) Because of the degree to which phosphorus and sulfur segregate, product analysis tolerances for those elements are not applicable for rephosphorized and resulfurized steels. (b) Product analysis tolerance for lead applies, both over and under the specified range. Source: Ref 2

Table 7 Product analysis tolerances for alloy steel bars, blooms, billets, and slabs

Element	Limit or maximum of specified range, %	Tolerance over the maximum limit or under the minimum limit for size ranges shown, %			
		≤0.065 m² (100 in.²)	>0.065–0.129 m² (100–200 in.²) incl	>0.129–0.258 m² (200–400 in.²) incl	>0.258–0.516 m² (400–800 in.²) incl
Carbon	≤0.30	0.01	0.02	0.03	0.04
	>0.30–0.75 incl	0.02	0.03	0.04	0.05
	>0.75	0.03	0.04	0.05	0.06
Manganese	≤0.90	0.03	0.04	0.05	0.06
	>0.90–2.10 incl	0.04	0.05	0.06	0.07
Phosphorus	Over max only	0.005	0.010	0.010	0.010
Sulfur	Over max only(a)	0.005	0.010	0.010	0.010
Silicon	≤0.40	0.02	0.02	0.03	0.04
	>0.40–2.20 incl	0.05	0.06	0.06	0.07
Nickel	≤1.00	0.03	0.03	0.03	0.03
	>1.00–2.00 incl	0.05	0.05	0.05	0.05
	>2.00–5.30 incl	0.07	0.07	0.07	0.07
	>5.30–10.00 incl	0.10	0.10	0.10	0.10
Chromium	≤0.90	0.03	0.04	0.04	0.05
	>0.90–2.10 incl	0.05	0.06	0.06	0.07
	>2.10–3.99 incl	0.10	0.10	0.12	0.14
Molybdenum	≤0.20	0.01	0.01	0.02	0.03
	>0.20–0.40 incl	0.02	0.03	0.03	0.04
	>0.40–1.15 incl	0.03	0.04	0.05	0.06
Vanadium	≤0.10	0.01	0.01	0.01	0.01
	>0.10–0.25 incl	0.02	0.02	0.02	0.02
	>0.25–0.50 incl	0.03	0.03	0.03	0.03
	Min value specified, check under min limit(b)	0.01	0.01	0.01	0.01
Tungsten	≤1.00	0.04	0.05	0.05	0.06
	>1.00–4.00 incl	0.08	0.09	0.10	0.12
Aluminum(c)	≤0.10	0.03	...	...	...
	>0.10–0.20 incl	0.04	...	...	...
	>0.20–0.30 incl	0.05	...	...	...
	>0.30–0.80 incl	0.07	...	...	...
	>0.80–1.80 incl	0.10	...	...	...
Lead(c)	0.15–0.35 incl	0.03(d)	...	...	...
Copper(c)	≤1.00	0.03	...	...	...
	>1.00–2.00 incl	0.05	...	...	...
Titanium(c)	≤0.10	0.01(b)	...	...	...
Niobium(c)	≤0.10	0.01(b)	...	...	...
Zirconium(c)	≤0.15	0.03	...	...	...
Nitrogen(c)	≤0.030	0.005	...	...	...

Note: Boron is not subject to product analysis tolerances. Incl, inclusive. (a) Resulfurized steels are not subject to product analysis limits for sulfur. (b) If the minimum of the range is 0.01%, the under tolerance is 0.005%. (c) Tolerances shown apply only to 0.065 m² (100 in.²) or less. (d) Tolerance is over and under. Source: Ref 2

ness but has only a slight effect on longitudinal mechanical properties. Weldability decreases with increasing sulfur content. This element is very detrimental to surface quality, particularly in the lower-carbon and lower-manganese steels. For these reasons, only a maximum limit is specified for most steels. The only exception is the group of free-machining steels, where sulfur is added to improve machinability; in this case a range is specified (see the article "Machinability of Steels" in this Volume). Sulfur has a greater segregation tendency than any of the other common elements. Sulfur occurs in steel principally in the form of sulfide inclusions. Obviously, a greater frequency of such inclusions can be expected in the resulfurized grades.

Silicon is one of the principal deoxidizers used in steelmaking; therefore, the amount of silicon present is related to the type of steel. Rimmed and capped steels contain no significant amounts of silicon. Semikilled steels may contain moderate amounts of silicon, although there is a definite maximum amount that can be tolerated in such steels. Killed carbon steels may contain any amount of silicon up to 0.60% maximum.

Silicon is somewhat less effective than manganese in increasing as-rolled strength and hardness. Silicon has only a slight tendency to segregate. In low-carbon steels, silicon is usually detrimental to surface quality, and this condition is more pronounced in low-carbon resulfurized grades.

Copper has a moderate tendency to segregate. Copper in appreciable amounts is detrimental to hot-working operations. Copper adversely affects forge welding, but it does not seriously affect arc or oxyacetylene welding. Copper is detrimental to surface quality and exaggerates the surface defects inherent in resulfurized steels. Copper is, however, beneficial to atmospheric corrosion resistance when present in amounts exceeding 0.20%. Steels containing these levels of copper are referred to as weathering steels and are described in the article "High-Strength Structural and High-Strength Low-Alloy Steels" in this Volume; they are also included in the descriptions of high-strength low-alloy steels given later in this article.

Lead is sometimes added to carbon and alloy steels through mechanical dispersion during teeming for the purpose of improving the machining characteristics of the steels. These additions are generally in the range of 0.15 to 0.35% (see the article "Machinability of Steels" in this Volume for details).

Boron is added to fully killed steel to improve hardenability. Boron-treated steels are produced to a range of 0.0005 to 0.003%. Whenever boron is substituted in part for other alloys, it should be done only with hardenability in mind because the lowered alloy content may be harmful for some applications. Boron is most effective in lower carbon steels. Boron steels are discussed in the Section "Hardenability of Carbon and Low-Alloy Steels" in this Volume.

Chromium is generally added to steel to increase resistance to corrosion and oxidation, to increase hardenability, to improve high-temperature strength, or to improve abrasion resistance in high-carbon compositions. Chromium is a strong carbide former. Complex chromium-iron carbides go into solution in austenite slowly; therefore, a sufficient heating time before quenching is necessary.

Chromium can be used as a hardening element, and is frequently used with a toughening element such as nickel to pro-

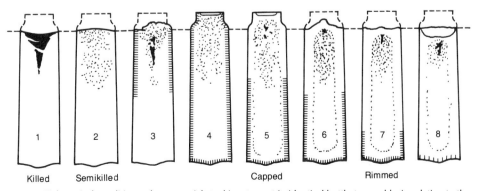

Fig. 2 Eight typical conditions of commercial steel ingots, cast in identical bottle-top molds, in relation to the degree of suppression of gas evolution. The dotted line indicates the height to which the steel originally was poured in each ingot mold. Depending on the carbon and, more importantly, the oxygen content of the steel, the ingot structures range from that of a fully killed ingot (No. 1) to that of a violently rimmed ingot (No. 8). Source: Ref 5

Table 8 Quality descriptions of carbon and alloy steels

Carbon steels			Alloy steels
Semifinished for forging 　Forging quality 　　Special hardenability 　　Special internal soundness 　　Nonmetallic inclusion 　　　requirement 　　Special surface **Carbon steel structural sections** 　Structural quality **Carbon steel plates** 　Regular quality 　Structural quality 　Cold-drawing quality 　Cold-pressing quality 　Cold-flanging quality 　Forging quality 　Pressure vessel quality **Hot-rolled carbon steel bars** 　Merchant quality 　Special quality 　　Special hardenability 　　Special internal soundness 　　Nonmetallic inclusion 　　　requirement 　　Special surface 　　Scrapless nut quality 　　Axle shaft quality 　　Cold extrusion quality 　　Cold-heading and cold-forging 　　　quality **Cold-finished carbon steel bars** 　Standard quality 　　Special hardenability 　　Special internal soundness 　　Nonmetallic inclusion 　　　requirement 　　Special surface 　　Cold-heading and cold-forging 　　　quality 　　Cold extrusion quality	**Hot-rolled sheets** 　Commercial quality 　Drawing quality 　Drawing quality special killed 　Structural quality **Cold-rolled sheets** 　Commercial quality 　Drawing quality 　Drawing quality special killed 　Structural quality **Porcelain enameling sheets** 　Commercial quality 　Drawing quality 　Drawing quality special killed **Long terne sheets** 　Commercial quality 　Drawing quality 　Drawing quality special killed 　Structural quality **Galvanized sheets** 　Commercial quality 　Drawing quality 　Drawing quality special killed 　Lock-forming quality **Electrolytic zinc coated sheets** 　Commercial quality 　Drawing quality 　Drawing quality special killed 　Structural quality **Hot-rolled strip** 　Commercial quality 　Drawing quality 　Drawing quality special killed 　Structural quality **Cold-rolled strip** 　Specific quality descriptions are not 　　provided in cold-rolled strip 　　because this product is largely 　　produced for specific end use	**Tin mill products** 　Specific quality descriptions are not 　　applicable to tin mill products **Carbon steel wire** 　Industrial quality wire 　Cold extrusion wires 　Heading, forging, and roll-threading 　　wires 　Mechanical spring wires 　Upholstery spring construction wires 　Welding wire **Carbon steel flat wire** 　Stitching wire 　Stapling wire **Carbon steel pipe** **Structural tubing** **Line pipe** **Oil country tubular goods** **Steel specialty tubular products** 　Pressure tubing 　Mechanical tubing 　Aircraft tubing **Hot-rolled carbon steel wire rods** 　Industrial quality 　Rods for manufacture of wire intended 　　for electric welded chain 　Rods for heading, forging, and 　　roll-threading wire 　Rods for lock washer wire 　Rods for scrapless nut wire 　Rods for upholstery spring wire 　Rods for welding wire	**Alloy steel plates** 　Drawing quality 　Pressure vessel quality 　Structural quality 　Aircraft physical quality **Hot-rolled alloy steel bars** 　Regular quality 　Aircraft quality or steel subject to 　　magnetic particle inspection 　Axle shaft quality 　Bearing quality 　Cold-heading quality 　Special cold-heading quality 　Rifle barrel quality, gun quality, shell 　　or A.P. shot quality **Alloy steel wire** 　Aircraft quality 　Bearing quality 　Special surface quality **Cold-finished alloy steel bars** 　Regular quality 　Aircraft quality or steel subject to 　　magnetic particle inspection 　Axle shaft quality 　Bearing shaft quality 　Cold-heading quality 　Special cold-heading quality 　Rifle barrel quality, gun quality, shell 　　or A.P. shot quality **Line pipe** **Oil country tubular goods** **Steel specialty tubular goods** 　Pressure tubing 　Mechanical tubing 　Stainless and head-resisting pipe, 　　pressure tubing, and mechanical 　　tubing 　Aircraft tubing 　Pipe

Source: Ref 6

duce superior mechanical properties. At higher temperatures, chromium contributes increased strength; it is ordinarily used for applications of this nature in conjunction with molybdenum.

Nickel, when used as an alloying element in constructional steels, is a ferrite strengthener. Because nickel does not form any carbide compounds in steel, it remains in solution in the ferrite, thus strengthening and toughening the ferrite phase. Nickel steels are easily heat treated because nickel lowers the critical cooling rate. In combination with chromium, nickel produces alloy steels with greater hardenability, higher impact strength, and greater fatigue resistance than can be achieved in carbon steels.

Molybdenum is added to constructional steels in the normal amounts of 0.10 to 1.00%. When molybdenum is in solid solution in austenite prior to quenching, the reaction rates for transformation become considerably slower as compared with carbon steel. Molybdenum can induce secondary hardening during the tempering of quenched steels and enhances the creep strength of low-alloy steels at elevated temperatures. Alloy steels that contain 0.15 to 0.30% Mo display a minimized susceptibility to temper embrittlement (see the article "Embrittlement of Steels" in this Volume for a discussion of temper embrittlement and other forms of thermal embrittlement).

Niobium. Small additions of niobium increase the yield strength and, to a lesser degree, the tensile strength of carbon steel. The addition of 0.02% Nb can increase the yield strength of medium-carbon steel by 70 to 100 MPa (10 to 15 ksi). This increased strength may be accompanied by considerably impaired notch toughness unless special measures are used to refine grain size during hot rolling. Grain refinement during hot rolling involves special thermomechanical processing techniques such as controlled rolling practices, low finishing temperatures for final reduction passes, and accelerated cooling after rolling is completed (further discussion of controlled rolling can be found in the article "High-Strength Structural and High-Strength Low-Alloy Steels" in this Volume).

Aluminum is widely used as a deoxidizer and for control of grain size. When added to steel in specified amounts, it controls austenite grain growth in reheated steels. Of all the alloying elements, aluminum is the most effective in controlling grain growth prior to quenching. Titanium, zirconium, and vanadium are also effective grain growth inhibitors; however, for structural grades that are heat treated (quenched and tempered), these three elements may have adverse effects on hardenability because their carbides are quite stable and difficult to dissolve in austenite prior to quenching.

Titanium and Zirconium. The effects of titanium are similar to those of vanadium and niobium, but it is only useful in fully killed (aluminum-deoxidized) steels because of its strong deoxidizing effects.

Classification and Designation of Carbon and Low-Alloy Steels / 147

Table 9 Raw steel production by type of furnace, grade, and cast

Year	Total all grades, net tons × 10³				Total production						Production by type of cast, net tons × 10³		
	Carbon	Alloy	Stainless	Total	By grade, %			By type of furnace, %			Ingots	Continuous castings	Steel castings
					Carbon	Alloy	Stainless	Open hearth	Basic oxygen process	Electric			
1988	86 823	10 902	2199	99 924	86.9	10.9	2.2	5.1	58.0	36.9	38 615	61 232	77
1987	77 976	9 147	2028	89 151	87.5	10.2	2.3	3.0	58.9	38.1	35 802	53 284	65
1986	71 413	8 505	1689	81 606	87.5	10.4	2.1	4.1	58.7	37.2	36 487	45 064	55
1985	76 699	9 877	1683	88 259	86.9	11.2	1.9	7.3	58.8	33.9	49 035	39 161	63
1984	79 918	10 838	1772	92 528	86.4	11.7	1.9	9.0	57.1	33.9	55 787	36 669	74

Source: Ref 12

Table 10 Net shipments of United States steel mill products, all grades

Steel products	1988		1987	
	Net tons × 10³	%	Net tons × 10³	%
Ingots and steel for castings	385	0.5	381	0.5
Blooms, slabs, and billets	1 542	1.8	1 212	1.6
Skelp	(a)	...	22	...
Wire rods	4 048	4.8	3 840	5.0
Structural shapes (≥75 mm, or 3 in.)	4 860	5.8	4 839	6.3
Steel piling	349	0.4	280	0.4
Plates cut in lengths	5 044	6.0	4 048	5.3
Plates in coils	2 284	2.7	(b)	...
Rails				
standard (>27 kg, or 60 lb)	460	0.5	351	0.5
all other	37	0.0	15	...
Railroad accessories	118	0.1	62	0.1
Wheels (rolled and forged)	(a)	...	58	0.1
Axles	(a)	...	29	...
Bars				
hot rolled	6 460	7.7	6 048	7.9
bar-size light shapes	1 373	1.6	1 190	1.6
reinforcing	5 091	6.1	4 918	6.4
cold finished	1 499	1.8	1 361	1.8
Tool steel	64	0.1	58	0.1
Pipe and tubing				
standard	1 238	1.5	969	1.3
oil country goods	1 130	1.3	919	1.2
line	808	1.0	620	0.8
mechanical	901	1.1	767	1.0
pressure	59	0.1	72	0.1
structural	178	0.2	180	0.2
pipe for piling	74	0.1	(c)	...
stainless	55	0.1	42	0.1
Wire				
drawn	1 073	1.3	800	1.0
nails and staples	(a)	...	218	0.3
barbed and twisted	(a)	...	49	0.1
woven wire fence	(a)	...	13	...
bale ties and baling wire	(a)	...	25	...
Black plate	283	0.3	205	0.3
Tin plate	2 806	3.3	2 765	3.6
Tin free steel	899	1.1	939	1.2
Tin coated sheets	81	0.1	79	0.1
Sheets				
hot rolled	12 589	15.0	13 048	17.0
cold rolled	13 871	16.5	13 859	18.1
Sheets and strip				
galvanized, hot dipped	8 115	9.7	7 660	10.0
galvanized, electrolytic	2 134	2.5	1 432	1.9
all other metallic coated	1 262	1.5	1 228	1.6
electrical	524	0.6	465	0.6
Strip				
hot rolled	1 203	1.4	657	0.9
cold rolled	941	1.1	929	1.2
Total steel mill products	**83 840**	**100.0**	**76 654**	**100.0**
Carbon	77 702	92.7	68 116	88.9
Stainless and heat resisting	1 586	1.9	1 418	1.8
Alloy (other than stainless)	4 552	5.4	7 120	9.3

(a) Effective 1 January 1988, these products are no longer classified as steel mill products by AISI. Consequently, comparable shipment tonnage is now included in applicable semifinished forms or drawn wire. (b) Prior to 1988 included in sheets hot rolled. (c) Prior to 1988 included in structural pipe and tubing. Source: Ref 12

Carbon Steels

The American Iron and Steel Institute defines carbon steel as follows (Ref 2, 3):

Steel is considered to be carbon steel when no minimum content is specified or required for chromium, cobalt, columbium [niobium], molybdenum, nickel, titanium, tungsten, vanadium or zirconium, or any other element to be added to obtain a desired alloying effect; when the specified minimum for copper does not exceed 0.40 per cent; or when the maximum content specified for any of the following elements does not exceed the percentages noted: manganese 1.65, silicon 0.60, copper 0.60.

Carbon steel can be classified, according to various deoxidation practices, as rimmed, capped, semikilled, or killed steel. Deoxidation practice and the steelmaking process will have an effect on the characteristics and properties of the steel (see the article "Steel Processing Technology" in this Volume). However, variations in carbon have the greatest effect on mechanical properties, with increasing carbon content leading to increased hardness and strength (see the article "Microstructures, Processing, and Properties of Steels" in this Volume). As such, carbon steels are generally categorized according to their carbon content. Generally speaking, carbon steels contain up to 2% total alloying elements and can be subdivided into low-carbon steels, medium-carbon steels, high-carbon steels, and ultrahigh-carbon steels; each of these designations is discussed below.

As a group, carbon steels are by far the most frequently used steel. Tables 9 and 10 indicate that more than 85% of the steel produced and shipped in the United States is carbon steel. Chemical compositions for carbon steels are provided in the tables referenced in the section "SAE-AISI Designations" in this article (see Tables 11 to 22).

Low-carbon steels contain up to 0.30% C. The largest category of this class of steel is flat-rolled products (sheet or strip) usually in the cold-rolled and annealed condition. The carbon content for these high-formability steels is very low, less than 0.10% C, with up to 0.4% Mn. Typical uses are in automobile body panels, tin plate, and wire products.

Zirconium can also be added to killed high-strength low-alloy steels to obtain improvements in inclusion characteristics, particularly sulfide inclusions where changes in inclusion shape improve ductility in transverse bending.

Table 11 SAE-AISI system of designations

Numerals and digits	Type of steel and nominal alloy content, %
Carbon steels	
10xx(a)	Plain carbon (Mn 1.00 max)
11xx	Resulfurized
12xx	Resulfurized and rephosphorized
15xx	Plain carbon (max Mn range: 1.00–1.65)
Manganese steels	
13xx	Mn 1.75
Nickel steels	
23xx	Ni 3.50
25xx	Ni 5.00
Nickel-chromium steels	
31xx	Ni 1.25; Cr 0.65 and 0.80
32xx	Ni 1.75; Cr 1.07
33xx	Ni 3.50; Cr 1.50 and 1.57
34xx	Ni 3.00; Cr 0.77
Molybdenum steels	
40xx	Mo 0.20 and 0.25
44xx	Mo 0.40 and 0.52
Chromium-molybdenum steels	
41xx	Cr 0.50, 0.80, and 0.95; Mo 0.12, 0.20, 0.25, and 0.30
Nickel-chromium-molybdenum steels	
43xx	Ni 1.82; Cr 0.50 and 0.80; Mo 0.25
43BVxx	Ni 1.82; Cr 0.50; Mo 0.12 and 0.25; V 0.03 min
47xx	Ni 1.05; Cr 0.45; Mo 0.20 and 0.35
81xx	Ni 0.30; Cr 0.40; Mo 0.12
86xx	Ni 0.55; Cr 0.50; Mo 0.20
87xx	Ni 0.55; Cr 0.50; Mo 0.25
88xx	Ni 0.55; Cr 0.50; Mo 0.35
93xx	Ni 3.25; Cr 1.20; Mo 0.12
94xx	Ni 0.45; Cr 0.40; Mo 0.12
97xx	Ni 0.55; Cr 0.20; Mo 0.20
98xx	Ni 1.00; Cr 0.80; Mo 0.25
Nickel-molybdenum steels	
46xx	Ni 0.85 and 1.82; Mo 0.20 and 0.25
48xx	Ni 3.50; Mo 0.25
Chromium steels	
50xx	Cr 0.27, 0.40, 0.50, and 0.65
51xx	Cr 0.80, 0.87, 0.92, 0.95, 1.00, and 1.05
Chromium (bearing) steels	
50xxx	Cr 0.50 } C 1.00 min
51xxx	Cr 1.02 } C 1.00 min
52xxx	Cr 1.45 } C 1.00 min
Chromium-vanadium steels	
61xx	Cr 0.60, 0.80, and 0.95; V 0.10 and 0.15 min
Tungsten-chromium steel	
72xx	W 1.75; Cr 0.75
Silicon-manganese steels	
92xx	Si 1.40 and 2.00; Mn 0.65, 0.82, and 0.85; Cr 0 and 0.65
High-strength low-alloy steels	
9xx	Various SAE grades
Boron steels	
xxBxx	B denotes boron steel
Leaded steels	
xxLxx	L denotes leaded steel

(a) The xx in the last two digits of these designations indicates that the carbon content (in hundredths of a percent) is to be inserted.

For rolled steel structural plates and sections, the carbon content may be increased to approximately 0.30%, with higher manganese up to 1.5%. These latter materials may be used for stampings, forgings, seamless tubes, and boiler plate.

Medium-carbon steels are similar to low-carbon steels except that the carbon ranges from 0.30 to 0.60% and the manganese from 0.60 to 1.65%. Increasing the carbon content to approximately 0.5% with an accompanying increase in manganese allows medium-carbon steels to be used in the quenched and tempered condition. The uses of medium carbon-manganese steels include shafts, couplings, crankshafts, axles, gears, and forgings. Steels in the 0.40 to 0.60% C range are also used for rails, railway wheels, and rail axles.

High-carbon steels contain from 0.60 to 1.00% C with manganese contents ranging from 0.30 to 0.90%. High-carbon steels are used for spring materials and high-strength wires.

Ultrahigh-carbon steels are experimental alloys containing approximately 1.25 to 2.0% C. These steels are thermomechanically processed to produce microstructures that consist of ultrafine, equiaxed grains of ferrite and a uniform distribution of fine, spherical, discontinuous proeutectoid carbide particles (Ref 13). Such microstructures in these steels have led to superplastic behavior (Ref 14). Properties of these experimental steels are described in Volume 14 of the 9th Edition of *Metals Handbook* (see the Appendix to the article "Superplastic Sheet Forming," entitled "Superplasticity in Iron-Base Alloys").

High-Strength Low-Alloy Steels

High-strength low-alloy (HSLA) steels, or microalloyed steels, are designed to provide better mechanical properties and/or greater resistance to atmospheric corrosion than conventional carbon steels. They are not considered to be alloy steels in the normal sense because they are designed to meet specific mechanical properties rather than a chemical composition (HSLA steels have yield strengths of less than 275 MPa, or 40 ksi). The chemical composition of a specific HSLA steel may vary for different product thicknesses to meet mechanical property requirements. The HSLA steels have low carbon contents (0.05 to ~0.25% C) in order to produce adequate formability and weldability, and they have manganese contents up to 2.0%. Small quantities of chromium, nickel, molybdenum, copper, nitrogen, vanadium, niobium, titanium, and zirconium are used in various combinations.

The HSLA steels are commonly furnished in the as-rolled condition. They may also be supplied in a controlled-rolled, normalized, or precipitation-hardened condition to meet specific property requirements. Primary applications for HSLA steels include oil and gas line pipe, ships, offshore structures, automobiles, off-highway equipment, and pressure vessels.

HSLA Classification. The types of HSLA steels commonly used include (Ref 15):

- *Weathering steels*, designed to exhibit superior atmospheric corrosion resistance
- *Control-rolled steels*, hot rolled according to a predetermined rolling schedule designed to develop a highly deformed austenite structure that will transform to a very fine equiaxed ferrite structure on cooling
- *Pearlite-reduced steels*, strengthened by very fine-grain ferrite and precipitation hardening but with low carbon content and therefore little or no pearlite in the microstructure
- *Microalloyed steels*, with very small additions (generally <0.10% each) of such elements as niobium, vanadium, and/or titanium for refinement of grain size and/or precipitation hardening
- *Acicular ferrite steel*, very low carbon steels with sufficient hardenability to transform on cooling to a very fine high-strength acicular ferrite (low-carbon bainite) structure rather than the usual polygonal ferrite structure
- *Dual-phase steels*, processed to a microstructure of ferrite containing small uniformly distributed regions of high-carbon martensite, resulting in a product with low yield strength and a high rate of work hardening, thus providing a high-strength steel of superior formability

The various types of HSLA steels may also have small additions of calcium, rare-earth elements, or zirconium for sulfide inclusion shape control. Compositions, properties, and applications of these steels can be found in the articles "High-Strength Structural and High-Strength Low-Alloy Steels," "Dual-Phase Steels," and "High-Strength Low-Alloy Steel Forgings" in this Volume.

Table 12 Carbon steel compositions
Applicable to semifinished products for forging, hot-rolled and cold-finished bars, wire rods, and seamless tubing

UNS number	SAE-AISI number	Cast or heat chemical ranges and limits, %(a)			
		C	Mn	P max	S max
G10050	1005	0.06 max	0.35 max	0.040	0.050
G10060	1006	0.08 max	0.25–0.40	0.040	0.050
G10080	1008	0.10 max	0.30–0.50	0.040	0.050
G10100	1010	0.08–0.13	0.30–0.60	0.040	0.050
G10120	1012	0.10–0.15	0.30–0.60	0.040	0.050
G10130	1013	0.11–0.16	0.50–0.80	0.040	0.050
G10150	1015	0.13–0.18	0.30–0.60	0.040	0.050
G10160	1016	0.13–0.18	0.60–0.90	0.040	0.050
G10170	1017	0.15–0.20	0.30–0.60	0.040	0.050
G10180	1018	0.15–0.20	0.60–0.90	0.040	0.050
G10190	1019	0.15–0.20	0.70–1.00	0.040	0.050
G10200	1020	0.18–0.23	0.30–0.60	0.040	0.050
G10210	1021	0.18–0.23	0.60–0.90	0.040	0.050
G10220	1022	0.18–0.23	0.70–1.00	0.040	0.050
G10230	1023	0.20–0.25	0.30–0.60	0.040	0.050
G10250	1025	0.22–0.28	0.30–0.60	0.040	0.050
G10260	1026	0.22–0.28	0.60–0.90	0.040	0.050
G10290	1029	0.25–0.31	0.60–0.90	0.040	0.050
G10300	1030	0.28–0.34	0.60–0.90	0.040	0.050
G10350	1035	0.32–0.38	0.60–0.90	0.040	0.050
G10370	1037	0.32–0.38	0.70–1.00	0.040	0.050
G10380	1038	0.35–0.42	0.60–0.90	0.040	0.050
G10390	1039	0.37–0.44	0.70–1.00	0.040	0.050
G10400	1040	0.37–0.44	0.60–0.90	0.040	0.050
G10420	1042	0.40–0.47	0.60–0.90	0.040	0.050
G10430	1043	0.40–0.47	0.70–1.00	0.040	0.050
G10440	1044	0.43–0.50	0.30–0.60	0.040	0.050
G10450	1045	0.43–0.50	0.60–0.90	0.040	0.050
G10460	1046	0.43–0.50	0.70–1.00	0.040	0.050
G10490	1049	0.46–0.53	0.60–0.90	0.040	0.050
G10500	1050	0.48–0.55	0.60–0.90	0.040	0.050
G10530	1053	0.48–0.55	0.70–1.00	0.040	0.050
G10550	1055	0.50–0.60	0.60–0.90	0.040	0.050
G10590	1059	0.55–0.65	0.50–0.80	0.040	0.050
G10600	1060	0.55–0.65	0.60–0.90	0.040	0.050
G10640	1064	0.60–0.70	0.50–0.80	0.040	0.050
G10650	1065	0.60–0.70	0.60–0.90	0.040	0.050
G10690	1069	0.65–0.75	0.40–0.70	0.040	0.050
G10700	1070	0.65–0.75	0.60–0.90	0.040	0.050
G10740	1074	0.70–0.80	0.50–0.80	0.040	0.050
G10750	1075	0.70–0.80	0.40–0.70	0.040	0.050
G10780	1078	0.72–0.85	0.30–0.60	0.040	0.050
G10800	1080	0.75–0.88	0.60–0.90	0.040	0.050
G10840	1084	0.80–0.93	0.60–0.90	0.040	0.050
G10850	1085	0.80–0.93	0.70–1.00	0.040	0.050
G10860	1086	0.80–0.93	0.30–0.50	0.040	0.050
G10900	1090	0.85–0.98	0.60–0.90	0.040	0.050
G10950	1095	0.90–1.03	0.30–0.50	0.040	0.050

(a) When silicon ranges or limits are required for bar and semifinished products, the values in Table 1 apply. For rods, the following ranges are commonly used: 0.10 max; 0.07–0.15%; 0.10–0.20%; 0.15–0.35%; 0.20–0.40%; and 0.30–0.60%. Steels listed in this table can be produced with additions of lead or boron. Leaded steels typically contain 0.15–0.35% Pb and are identified by inserting the letter L in the designation (10L45); boron steels can be expected to contain 0.0005–0.003% B and are identified by inserting the letter B in the designation (10B46). Source: Ref 1

Low-Alloy Steels

Low-alloy steels constitute a category of ferrous materials that exhibit mechanical properties superior to plain carbon steels as the result of additions of such alloying elements as nickel, chromium, and molybdenum. Total alloy content can range from 2.07% up to levels just below that of stainless steels, which contain a minimum of 10% Cr. For many low-alloy steels, the primary function of the alloying elements is to increase hardenability in order to optimize mechanical properties and toughness after heat treatment. In some cases, however, alloy additions are used to reduce environmental degradation under certain specified service conditions.

As with steels in general, low-alloy steels can be classified according to:

- *Chemical composition*, such as nickel steels, nickel-chromium steels, molybdenum steels, chromium-molybdenum steels, and so on, as described in the section "SAE-AISI Designations" in this article and as shown in Table 11
- *Heat treatment*, such as quenched and tempered, normalized and tempered, annealed, and so on
- *Weldability*, as described in the article "Weldability of Steels" in this Volume

Because of the wide variety of chemical compositions possible and the fact that some steels are used in more than one heat-treated condition, some overlap exists among the alloy steel classifications. In this article, four major groups of alloy steels are addressed: (1) low-carbon quenched and tempered (QT) steels, (2) medium-carbon ultrahigh-strength steels, (3) bearing steels, and (4) heat-resistant chromium-molybdenum steels.

Low-carbon quenched and tempered steels combine high yield strength (from 350 to 1035 MPa, or 50 to 150 ksi) and high tensile strength with good notch toughness, ductility, corrosion resistance, or weldability. The various steels have different combinations of these characteristics based on their intended applications. The chemical compositions of typical QT low-carbon steels are given in Table 23. Many of the steels are covered by ASTM specifications. However, a few steels, such as HY-80 and HY-100, are covered by military specifications. The steels listed are used primarily as plate. Some of these steels, as well as other similar steels, are produced as forgings or castings. More detailed information on low-carbon QT steels can be found in the articles "Hardenable Carbon and Low-Alloy Steels" and "High-Strength Structural and High-Strength Low-Alloy Steels" in this Volume.

Medium-carbon ultrahigh-strength steels are structural steels with yield strengths that can exceed 1380 MPa (200 ksi). Table 23 lists typical compositions. Many of these steels are covered by SAE-AISI designations or are proprietary compositions. Product forms include billet, bar, rod, forgings, sheet, tubing, and welding wire. A review of the heat treatments and resulting properties of these steels can be found in the article "Ultrahigh-Strength Steels" in this Volume.

Bearing steels used for ball and roller bearing applications are comprised of low-carbon (0.10 to 0.20% C) case-hardened steels and high carbon (~1.0% C) through-hardened steels (Table 23). Many of these steels are covered by SAE-AISI designations. Selection and properties of these materials are discussed in the article "Bearing Steels" in this Volume.

Chromium-molybdenum heat-resistant steels contain 0.5 to 9% Cr and 0.5 to 1.0% Mo. The carbon content is usually below 0.20%. The chromium provides improved oxidation and corrosion resistance, and the molybdenum increases strength at elevated temperatures. They are generally supplied in the normalized and tempered, quenched and tempered, or annealed condition. Chromium-molybdenum steels are widely used in the oil and gas industries and in fossil fuel and nuclear power plants. Various product forms and corresponding ASTM specifications for these steels are given in Table 24. Nominal chemical compositions are provided in Table 25. High-temperature property data for chromium-molybdenum steels are reviewed extensively in the article "Elevated-Temperature Properties of Ferritic Steels" in this Volume.

Table 13 Carbon steel compositions
Applicable only to structural shapes, plates, strip, sheets, and welded tubing

UNS number	SAE-AISI number	Cast or heat chemical ranges and limits, %(a)			
		C	Mn	P max	S max
G10060	1006	0.08 max	0.45 max	0.040	0.050
G10080	1008	0.10 max	0.50 max	0.040	0.050
G10090	1009	0.15 max	0.60 max	0.040	0.050
G10100	1010	0.08–0.13	0.30–0.60	0.040	0.050
G10120	1012	0.10–0.15	0.30–0.60	0.040	0.050
G10150	1015	0.12–0.18	0.30–0.60	0.040	0.050
G10160	1016	0.12–0.18	0.60–0.90	0.040	0.050
G10170	1017	0.14–0.20	0.30–0.60	0.040	0.050
G10180	1018	0.14–0.20	0.60–0.90	0.040	0.050
G10190	1019	0.14–0.20	0.70–1.00	0.040	0.050
G10200	1020	0.17–0.23	0.30–0.60	0.040	0.050
G10210	1021	0.17–0.23	0.60–0.90	0.040	0.050
G10220	1022	0.17–0.23	0.70–1.00	0.040	0.050
G10230	1023	0.19–0.25	0.30–0.60	0.040	0.050
G10250	1025	0.22–0.28	0.30–0.60	0.040	0.050
G10260	1026	0.22–0.28	0.60–0.90	0.040	0.050
G10300	1030	0.27–0.34	0.60–0.90	0.040	0.050
G10330	1033	0.29–0.36	0.70–1.00	0.040	0.050
G10350	1035	0.31–0.38	0.60–0.90	0.040	0.050
G10370	1037	0.31–0.38	0.70–1.00	0.040	0.050
G10380	1038	0.34–0.42	0.60–0.90	0.040	0.050
G10390	1039	0.36–0.44	0.70–1.00	0.040	0.050
G10400	1040	0.36–0.44	0.60–0.90	0.040	0.050
G10420	1042	0.39–0.47	0.60–0.90	0.040	0.050
G10430	1043	0.39–0.47	0.70–1.00	0.040	0.050
G10450	1045	0.42–0.50	0.60–0.90	0.040	0.050
G10460	1046	0.42–0.50	0.70–1.00	0.040	0.050
G10490	1049	0.45–0.53	0.60–0.90	0.040	0.050
G10500	1050	0.47–0.55	0.60–0.90	0.040	0.050
G10550	1055	0.52–0.60	0.60–0.90	0.040	0.050
G10600	1060	0.55–0.66	0.60–0.90	0.040	0.050
G10640	1064	0.59–0.70	0.50–0.80	0.040	0.050
G10650	1065	0.59–0.70	0.60–0.90	0.040	0.050
G10700	1070	0.65–0.76	0.60–0.90	0.040	0.050
G10740	1074	0.69–0.80	0.50–0.80	0.040	0.050
G10750	1075	0.69–0.80	0.40–0.70	0.040	0.050
G10780	1078	0.72–0.86	0.30–0.60	0.040	0.050
G10800	1080	0.74–0.88	0.60–0.90	0.040	0.050
G10840	1084	0.80–0.94	0.60–0.90	0.040	0.050
G10850	1085	0.80–0.94	0.70–1.00	0.040	0.050
G10860	1086	0.80–0.94	0.30–0.50	0.040	0.050
G10900	1090	0.84–0.98	0.60–0.90	0.040	0.050
G10950	1095	0.90–1.04	0.30–0.50	0.040	0.050

(a) When silicon ranges or limits are required, the following ranges and limits are commonly used: up to SAE 1025 inclusive, 0.10% max, 0.10–0.25%, or 0.15–0.35%. Over SAE 1025, 0.10–0.25% or 0.15–0.35%. Source: Ref 1

Table 14 Composition ranges and limits for merchant quality steels

SAE-AISI number	Cast or heat chemical ranges and limits, %(a)			
	C	Mn	P max	S max
M1008	0.10 max	0.25–0.60	0.04	0.05
M1010	0.07–0.14	0.25–0.60	0.04	0.05
M1012	0.09–0.16	0.25–0.60	0.04	0.05
M1015	0.12–0.19	0.25–0.60	0.04	0.05
M1017	0.14–0.21	0.25–0.60	0.04	0.05
M1020	0.17–0.24	0.25–0.60	0.04	0.05
M1023	0.19–0.27	0.25–0.60	0.04	0.05
M1025	0.20–0.30	0.25–0.60	0.04	0.05
M1031	0.26–0.36	0.25–0.60	0.04	0.05
M1044	0.40–0.50	0.25–0.60	0.04	0.05

(a) Merchant quality steel bars are not produced to any specified silicon content. Source: Ref 1

other attribute, such as strength level or surface smoothness.

In ASTM specifications, however, these terms are used somewhat interchangeably. In ASTM A 533, for example, type denotes chemical composition, while class indicates strength level. In ASTM A 515, grade identifies strength level; the maximum carbon content permitted by this specification depends on both plate thickness and strength level. In ASTM A 302, grade denotes requirements for both chemical composition and mechanical properties. ASTM A 514 and A 517 are specifications for high-strength quenched and tempered plate for structural and pressure vessel applications, respectively; each contains several compositions that can provide the required mechanical properties. However, A 514 type A has the identical composition limits as A 517 grade. Additional information can be found in the section "ASTM (ASME) Specifications" in this article.

Chemical composition is by far the most widely used basis for classification and/or designation of steels. The most commonly used system of designation in the United States is that of the Society of Automotive Engineers (SAE) and the American Iron and Steel Institute (AISI). The Unified Numbering System (UNS) is also being used with increasing frequency. Each of these designation systems is described below.

Designations for Steels

A designation is the specific identification of each grade, type, or class of steel by a number, letter, symbol, name, or suitable combination thereof unique to a particular steel. Grade, type, and class are terms used to classify steel products. Within the steel industry, they have very specific uses: grade is used to denote chemical composition; type is used to indicate deoxidation practice; and class is used to describe some

SAE-AISI Designations

As stated above, the most widely used system for designating carbon and alloy steels is the SAE-AISI system. As a point of technicality, there are two separate systems, but they are nearly identical and have been carefully coordinated by the two groups. It should be noted, however, that AISI has discontinued the practice of designating steels. Therefore, the reader should consult Volume 1, *Materials*, of the *SAE Handbook* for the most up-to-date information.

The SAE-AISI system is applied to semifinished forgings, hot-rolled and cold-finished bars, wire rod and seamless tubular

Table 15 Free-cutting (resulfurized) carbon steel compositions
Applicable to semifinished products for forging, hot-rolled and cold-finished bars, wire rods, and seamless tubing

UNS number	SAE-AISI number	Cast or heat chemical ranges and limits, %(a)			
		C	Mn	P max	S
G11080	1108	0.08–0.13	0.50–0.80	0.040	0.08–0.13
G11100	1110	0.08–0.13	0.30–0.60	0.040	0.08–0.13
G11170	1117	0.14–0.20	1.00–1.30	0.040	0.08–0.13
G11180	1118	0.14–0.20	1.30–1.60	0.040	0.08–0.13
G11370	1137	0.32–0.39	1.35–1.65	0.040	0.08–0.13
G11390	1139	0.35–0.43	1.35–1.65	0.040	0.13–0.20
G11400	1140	0.37–0.44	0.70–1.00	0.040	0.08–0.13
G11410	1141	0.37–0.45	1.35–1.65	0.040	0.08–0.13
G11440	1144	0.40–0.48	1.35–1.65	0.040	0.24–0.33
G11460	1146	0.42–0.49	0.70–1.00	0.040	0.08–0.13
G11510	1151	0.48–0.55	0.70–1.00	0.040	0.08–0.13

(a) When lead ranges or limits are required, or when silicon ranges or limits are required for bars or semifinished products, the values in Table 1 apply. For rods, the following ranges and limits for silicon are commonly used: up to SAE 1110 inclusive, 0.10% max; SAE 1117 and over, 0.10% max, 0.10–0.20%, or 0.15–0.35%. Source: Ref 1

Classification and Designation of Carbon and Low-Alloy Steels / 151

Table 16 Free-cutting (rephosphorized and resulfurized) carbon steel compositions
Applicable to semifinished products for forging, hot-rolled and cold-finished bars, wire rods, and seamless tubing

UNS number	SAE-AISI number	Cast or heat chemical ranges and limits, %(a)				
		C max	Mn	P	S	Pb
G12110	1211	0.13	0.60–0.90	0.07–0.12	0.10–0.15	...
G12120	1212	0.13	0.70–1.00	0.07–0.12	0.16–0.23	...
G12130	1213	0.13	0.70–1.00	0.07–0.12	0.24–0.33	...
G12150	1215	0.09	0.75–1.05	0.04–0.09	0.26–0.35	...
G12144	12L14	0.15	0.85–1.15	0.04–0.09	0.26–0.35	0.15–0.35

(a) When lead ranges or limits are required, the values in Table 1 apply. It is not common practice to produce the 12xx series of steels to specified limits for silicon because of its adverse effect on machinability. Source: Ref 1

Table 17 High-manganese carbon steel compositions
Applicable only to semifinished products for forging, hot-rolled and cold-finished bars, wire rods, and seamless tubing

UNS number	SAE-AISI number	Cast or heat chemical ranges and limits, %(a)			
		C	Mn	P max	S max
G15130	1513	0.10–0.16	1.10–1.40	0.040	0.050
G15220	1522	0.18–0.24	1.10–1.40	0.040	0.050
G15240	1524	0.19–0.25	1.35–1.65	0.040	0.050
G15260	1526	0.22–0.29	1.10–1.40	0.040	0.050
G15270	1527	0.22–0.29	1.20–1.50	0.040	0.050
G15360	1536	0.30–0.37	1.20–1.50	0.040	0.050
G15410	1541	0.36–0.44	1.35–1.65	0.040	0.050
G15480	1548	0.44–0.52	1.10–1.40	0.040	0.050
G15510	1551	0.45–0.56	0.85–1.15	0.040	0.050
G15520	1552	0.47–0.55	1.20–1.50	0.040	0.050
G15610	1561	0.55–0.65	0.75–1.05	0.040	0.050
G15660	1566	0.60–0.71	0.85–1.15	0.040	0.050

(a) When silicon, lead, and boron ranges or limits are required, the values in Tables 1 and 2 apply. Source: Ref 1

Table 18 High-manganese carbon steel compositions
Applicable only to structural shapes, plates, strip, sheets, and welded tubing

UNS number	SAE-AISI number	Cast or heat chemical ranges and limits, %(a)				Former SAE number
		C	Mn	P max	S max	
G15240	1524	0.18–0.25	1.30–1.65	0.040	0.050	1024
G15270	1527	0.22–0.29	1.20–1.55	0.040	0.050	1027
G15360	1536	0.30–0.38	1.20–1.55	0.040	0.050	1036
G15410	1541	0.36–0.45	1.30–1.65	0.040	0.050	1041
G15480	1548	0.43–0.52	1.05–1.40	0.040	0.050	1048
G15520	1552	0.46–0.55	1.20–1.55	0.040	0.050	1052

(a) When silicon ranges or limits are required, the values shown in Table 2 apply. Source: Ref 1

goods, structural shapes, plates, sheet, strip, and welded tubing. Table 11 summarizes the numerical designations used in both SAE and AISI.

Carbon steels contain less than 1.65% Mn, 0.60% Si, and 0.60% Cu; they comprise the 1xxx groups in the SAE-AISI system and are subdivided into four distinct series as a result of the difference in certain fundamental properties among them. Plain carbon steels in the 10xx group are listed in Tables 12 and 13; note that ranges and limits of chemical composition depend on the product form. Designations for merchant quality steels, given in Table 14, include the prefix M. A carbon steel designation with the letter B inserted between the second and third digits indicates the steel contains 0.0005 to 0.003% B. Likewise, the letter L inserted between the second and third digits indicates that the steel contains 0.15 to 0.35% Pb for enhanced machinability. Resulfurized carbon steels in the 11xx group are listed in Table 15, and resulfurized and rephosphorized carbon steels in the 12xx group are listed in Table 16. Both of these groups of steels are produced for applications requiring good machinability. Tables 17 and 18 list steels having nominal manganese contents of between 0.9 and 1.5% but no other alloying additions; these steels now have 15xx designations in place of the 10xx designations formerly used.

Certain steels have hardenability requirements in addition to the limits and ranges of chemical composition. They are distinguished from similar grades that have no hardenability requirement by the use of the suffix H. Limits and ranges of chemical composition for all carbon steel products reflect the restrictions on heat and product analyses given in Tables 1, 2, and 5. Hardenability characteristics of carbon steels and the carbon and carbon-boron H steels are discussed in the article "Hardenable Carbon and Low-Alloy Steels" in this Volume. Corresponding hardenability bands for these steels are given in the article "Hardenability Curves." Except where indicated, all of these designations for carbon steels are both AISI and SAE designations.

Alloy steels contain manganese, silicon, or copper in quantities greater than those listed for the carbon steels, or they have specified ranges or minimums for one or more of the other alloying elements. In the AISI-SAE system of designations, the major alloying elements in a steel are indicated by the first two digits of the designation (Table 11). The amount of carbon, in hundredths of a percent, is indicated by the last two (or three) digits. The chemical compositions of AISI-SAE standard grades of alloy steels are given in Table 19. For alloy steels that have specific hardenability requirements, the suffix H is used to distinguish these steels from corresponding grades that have no hardenability requirement (see the article "Hardenable Carbon and Low-Alloy Steels" in this Volume for chemical compositions of alloy H steels). As with carbon steels, the letter B inserted between the second and third digits indicates that the steel contains boron. The prefix E signifies that the steel was produced by the electric furnace process. Limits and ranges of chemical composition for all alloy steel products reflect the restrictions on heat and product analyses given in Tables 1 through 7. The designations in Table 19 are both AISI and SAE designations unless otherwise indicated.

Potential standard steels are listed in SAE J1081 and Table 20. These are experimental grades to which no regular AISI-SAE designations have been assigned. Some were developed to minimize the nickel content; others were devised to improve a particular attribute of a standard grade of alloy steel.

HSLA Steels. Several grades of HSLA steel are described in SAE Recommended Practice J410; their chemical compositions and minimum mechanical properties are listed in Table 21. These steels have been developed as a compromise between the convenient fabrication characteristics and low cost of plain carbon steels and the high strength of heat-treated alloy steels. These steels have excellent strength and ductility as-rolled.

Formerly Listed SAE Steels. A number of grades of carbon and alloy steels have been deleted from the list of SAE standard steels due to lack of use. For the convenience of those who might encounter an application for one of these grades, they are listed in Table 22.

UNS Designations

The Unified Numbering System (UNS) has been developed by ASTM and SAE and several other technical societies, trade associations, and United States government agencies. A UNS number, which is a designation of chemical composition and not a specification, is assigned to each chemical

152 / Carbon and Low-Alloy Steels

Table 19 Low-alloy steel compositions applicable to billets, blooms, slabs, and hot-rolled and cold-finished bars
Slightly wider ranges of compositions apply to plates. The article "Carbon and Low-Alloy Steel Plate" in this volume lists SAE-AISI plate compositions

UNS number	SAE number	Corresponding AISI number	Ladle chemical composition limits, %(a)								
			C	Mn	P	S	Si	Ni	Cr	Mo	V
G13300	1330	1330	0.28–0.33	1.60–1.90	0.035	0.040	0.15–0.35	...	...	...	...
G13350	1335	1335	0.33–0.38	1.60–1.90	0.035	0.040	0.15–0.35	...	...	...	...
G13400	1340	1340	0.38–0.43	1.60–1.90	0.035	0.040	0.15–0.35	...	...	...	...
G13450	1345	1345	0.43–0.48	1.60–1.90	0.035	0.040	0.15–0.35	...	...	...	...
G40230	4023	4023	0.20–0.25	0.70–0.90	0.035	0.040	0.15–0.35	...	...	...	...
G40240	4024	4024	0.20–0.25	0.70–0.90	0.035	0.035–0.050	0.15–0.35	...	...	0.20–0.30	...
G40270	4027	4027	0.25–0.30	0.70–0.90	0.035	0.040	0.15–0.35	...	...	0.20–0.30	...
G40280	4028	4028	0.25–0.30	0.70–0.90	0.035	0.035–0.050	0.15–0.35	...	...	0.20–0.30	...
G40320	4032	...	0.30–0.35	0.70–0.90	0.035	0.040	0.15–0.35	...	...	0.20–0.30	...
G40370	4037	4037	0.35–0.40	0.70–0.90	0.035	0.040	0.15–0.35	...	...	0.20–0.30	...
G40420	4042	...	0.40–0.45	0.70–0.90	0.035	0.040	0.15–0.35	...	...	0.20–0.30	...
G40470	4047	4047	0.45–0.50	0.70–0.90	0.035	0.040	0.15–0.35	...	...	0.20–0.30	...
G41180	4118	4118	0.18–0.23	0.70–0.90	0.035	0.040	0.15–0.35	...	0.40–0.60	0.08–0.15	...
G41300	4130	4130	0.28–0.33	0.40–0.60	0.035	0.040	0.15–0.35	...	0.80–1.10	0.15–0.25	...
G41350	4135	...	0.33–0.38	0.70–0.90	0.035	0.040	0.15–0.35	...	0.80–1.10	0.15–0.25	...
G41370	4137	4137	0.35–0.40	0.70–0.90	0.035	0.040	0.15–0.35	...	0.80–1.10	0.15–0.25	...
G41400	4140	4140	0.38–0.43	0.75–1.00	0.035	0.040	0.15–0.35	...	0.80–1.10	0.15–0.25	...
G41420	4142	4142	0.40–0.45	0.75–1.00	0.035	0.040	0.15–0.35	...	0.80–1.10	0.15–0.25	...
G41450	4145	4145	0.41–0.48	0.75–1.00	0.035	0.040	0.15–0.35	...	0.80–1.10	0.15–0.25	...
G41470	4147	4147	0.45–0.50	0.75–1.00	0.035	0.040	0.15–0.35	...	0.80–1.10	0.15–0.25	...
G41500	4150	4150	0.48–0.53	0.75–1.00	0.035	0.040	0.15–0.35	...	0.80–1.10	0.15–0.25	...
G41610	4161	4161	0.56–0.64	0.75–1.00	0.035	0.040	0.15–0.35	...	0.70–0.90	0.25–0.35	...
G43200	4320	4320	0.17–0.22	0.45–0.65	0.035	0.040	0.15–0.35	1.65–2.00	0.40–0.60	0.20–0.30	...
G43400	4340	4340	0.38–0.43	0.60–0.80	0.035	0.040	0.15–0.35	1.65–2.00	0.70–0.90	0.20–0.30	...
G43406	E4340(b)	E4340	0.38–0.43	0.65–0.85	0.025	0.025	0.15–0.35	1.65–2.00	0.70–0.90	0.20–0.30	...
G44220	4422	...	0.20–0.25	0.70–0.90	0.035	0.040	0.15–0.35	...	...	0.35–0.45	...
G44270	4427	...	0.24–0.29	0.70–0.90	0.035	0.040	0.15–0.35	...	...	0.35–0.45	...
G46150	4615	4615	0.13–0.18	0.45–0.65	0.035	0.040	0.15–0.25	1.65–2.00	...	0.20–0.30	...
G46170	4617	...	0.15–0.20	0.45–0.65	0.035	0.040	0.15–0.35	1.65–2.00	...	0.20–0.30	...
G46200	4620	4620	0.17–0.22	0.45–0.65	0.035	0.040	0.15–0.35	1.65–2.00	...	0.20–0.30	...
G46260	4626	4626	0.24–0.29	0.45–0.65	0.035	0.04 max	0.15–0.35	0.70–1.00	...	0.15–0.25	...
G47180	4718	4718	0.16–0.21	0.70–0.90	...	...	...	0.90–1.20	0.35–0.55	0.30–0.40	...
G47200	4720	4720	0.17–0.22	0.50–0.70	0.035	0.040	0.15–0.35	0.90–1.20	0.35–0.55	0.15–0.25	...
G48150	4815	4815	0.13–0.18	0.40–0.60	0.035	0.040	0.15–0.35	3.25–3.75	...	0.20–0.30	...
G48170	4817	4817	0.15–0.20	0.40–0.60	0.035	0.040	0.15–0.35	3.25–3.75	...	0.20–0.30	...
G48200	4820	4820	0.18–0.23	0.50–0.70	0.035	0.040	0.15–0.35	3.25–3.75	...	0.20–0.30	...
G50401	50B40(c)	...	0.38–0.43	0.75–1.00	0.035	0.040	0.15–0.35	...	0.40–0.60	...	...
G50441	50B44(c)	50B44	0.43–0.48	0.75–1.00	0.035	0.040	0.15–0.35	...	0.40–0.60	...	...
G50460	5046	...	0.43–0.48	0.75–1.00	0.035	0.040	0.15–0.35	...	0.20–0.35	...	...
G50461	50B46(c)	50B46	0.44–0.49	0.75–1.00	0.035	0.040	0.15–0.35	...	0.20–0.35	...	...
G50501	50B50(c)	50B50	0.48–0.53	0.75–1.00	0.035	0.040	0.15–0.35	...	0.40–0.60	...	...
G50600	5060	...	0.56–0.64	0.75–1.00	0.035	0.040	0.15–0.35	...	0.40–0.60	...	...
G50601	50B60(c)	50B60	0.56–0.64	0.75–1.00	0.035	0.040	0.15–0.35	...	0.40–0.60	...	...
G51150	5115	...	0.13–0.18	0.70–0.90	0.035	0.040	0.15–0.35	...	0.70–0.90	...	...
G51170	5117	5117	0.15–0.20	0.70–0.90	0.040	0.040	0.15–0.35	...	0.70–0.90	...	...
G51200	5120	5120	0.17–0.22	0.70–0.90	0.035	0.040	0.15–0.35	...	0.70–0.90	...	...
G51300	5130	5130	0.28–0.33	0.70–0.90	0.035	0.040	0.15–0.35	...	0.80–1.10	...	...
G51320	5132	5132	0.30–0.35	0.60–0.80	0.035	0.040	0.15–0.35	...	0.75–1.00	...	...
G51350	5135	5135	0.33–0.38	0.60–0.80	0.035	0.040	0.15–0.35	...	0.80–1.05	...	...
G51400	5140	5140	0.38–0.43	0.70–0.90	0.035	0.040	0.15–0.35	...	0.70–0.90	...	...
G51470	5147	5147	0.46–0.51	0.70–0.95	0.035	0.040	0.15–0.35	...	0.85–1.15	...	...
G51500	5150	5150	0.48–0.53	0.70–0.90	0.035	0.040	0.15–0.35	...	0.70–0.90	...	...
G51550	5155	5155	0.51–0.59	0.70–0.90	0.035	0.040	0.15–0.35	...	0.70–0.90	...	...
G51600	5160	5160	0.56–0.64	0.75–1.00	0.035	0.040	0.15–0.35	...	0.70–0.90	...	...
G51601	51B60(c)	51B60	0.56–0.64	0.75–1.00	0.035	0.040	0.15–0.35	...	0.70–0.90	...	...
G50986	50100(b)	...	0.98–1.10	0.25–0.45	0.025	0.025	0.15–0.35	...	0.40–0.60	...	...
G51986	51100(b)	E51100	0.98–1.10	0.25–0.45	0.025	0.025	0.15–0.35	...	0.90–1.15	...	...
G52986	52100(b)	E52100	0.98–1.10	0.25–0.45	0.025	0.025	0.15–0.35	...	1.30–1.60	...	...
G61180	6118	6118	0.16–0.21	0.50–0.70	0.035	0.040	0.15–0.35	...	0.50–0.70	...	0.10–0.15
G61500	6150	6150	0.48–0.53	0.70–0.90	0.035	0.040	0.15–0.35	...	0.80–1.10	...	0.15 min
G81150	8115	8115	0.13–0.18	0.70–0.90	0.035	0.040	0.15–0.35	0.20–0.40	0.30–0.50	0.08–0.15	...
G81451	81B45(c)	81B45	0.43–0.48	0.75–1.00	0.035	0.040	0.15–0.35	0.20–0.40	0.35–0.55	0.08–0.15	...

(continued)

(a) Small quantities of certain elements that are not specified or required may be found in alloy steels. These elements are to be considered as incidental and are acceptable to the following maximum amount: copper to 0.35%, nickel to 0.25%, chromium to 0.20%, and molybdenum to 0.06%. (b) Electric furnace steel. (c) Boron content is 0.0005–0.003%. Source: Ref 16

Table 19 (continued)

UNS number	SAE number	Corresponding AISI number	Ladle chemical composition limits, %(a)								
			C	Mn	P	S	Si	Ni	Cr	Mo	V
G86150	8615	8615	0.13–0.18	0.70–0.90	0.035	0.040	0.15–0.35	0.40–0.70	0.40–0.60	0.15–0.25	...
G86170	8617	8617	0.15–0.20	0.70–0.90	0.035	0.040	0.15–0.35	0.40–0.70	0.40–0.60	0.15–0.25	...
G86200	8620	8620	0.18–0.23	0.70–0.90	0.035	0.040	0.15–0.35	0.40–0.70	0.40–0.60	0.15–0.25	...
G86220	8622	8622	0.20–0.25	0.70–0.90	0.035	0.040	0.15–0.35	0.40–0.70	0.40–0.60	0.15–0.25	...
G86250	8625	8625	0.23–0.28	0.70–0.90	0.035	0.040	0.15–0.35	0.40–0.70	0.40–0.60	0.15–0.25	...
G86270	8627	8627	0.25–0.30	0.70–0.90	0.035	0.040	0.15–0.35	0.40–0.70	0.40–0.60	0.15–0.25	...
G86300	8630	8630	0.28–0.33	0.70–0.90	0.035	0.040	0.15–0.35	0.40–0.70	0.40–0.60	0.15–0.25	...
G86370	8637	8637	0.35–0.40	0.75–1.00	0.035	0.040	0.15–0.35	0.40–0.70	0.40–0.60	0.15–0.25	...
G86400	8640	8640	0.38–0.43	0.75–1.00	0.035	0.040	0.15–0.35	0.40–0.70	0.40–0.60	0.15–0.25	...
G86420	8642	8642	0.40–0.45	0.75–1.00	0.035	0.040	0.15–0.35	0.40–0.70	0.40–0.60	0.15–0.25	...
G86450	8645	8645	0.43–0.48	0.75–1.00	0.035	0.040	0.15–0.35	0.40–0.70	0.40–0.60	0.15–0.25	...
G86451	86B45(c)	...	0.43–0.48	0.75–1.00	0.035	0.040	0.15–0.35	0.40–0.70	0.40–0.60	0.15–0.25	...
G86500	8650	...	0.48–0.53	0.75–1.00	0.035	0.040	0.15–0.35	0.40–0.70	0.40–0.60	0.15–0.25	...
G86550	8655	8655	0.51–0.59	0.75–1.00	0.035	0.040	0.15–0.35	0.40–0.70	0.40–0.60	0.15–0.25	...
G86600	8660	...	0.56–0.64	0.75–1.00	0.035	0.040	0.15–0.35	0.40–0.70	0.40–0.60	0.15–0.25	...
G87200	8720	8720	0.18–0.23	0.70–0.90	0.035	0.040	0.15–0.35	0.40–0.70	0.40–0.60	0.20–0.30	...
G87400	8740	8740	0.38–0.43	0.75–1.00	0.035	0.040	0.15–0.35	0.40–0.70	0.40–0.60	0.20–0.30	...
G88220	8822	8822	0.20–0.25	0.75–1.00	0.035	0.040	0.15–0.35	0.40–0.70	0.40–0.60	0.30–0.40	...
G92540	9254	...	0.51–0.59	0.60–0.80	0.035	0.040	1.20–1.60	...	0.60–0.80	...	...
G92600	9260	9260	0.56–0.64	0.75–1.00	0.035	0.040	1.80–2.20	...	...	...	...
G93106	9310(b)	...	0.08–0.13	0.45–0.65	0.025	0.025	0.15–0.35	3.00–3.50	1.00–1.40	0.08–0.15	...
G94151	94B15(c)	...	0.13–0.18	0.75–1.00	0.035	0.040	0.15–0.35	0.30–0.60	0.30–0.50	0.08–0.15	...
G94171	94B17(c)	94B17	0.15–0.20	0.75–1.00	0.035	0.040	0.15–0.35	0.30–0.60	0.30–0.50	0.08–0.15	...
G94301	94B30(c)	94B30	0.28–0.33	0.75–1.00	0.035	0.040	0.15–0.35	0.30–0.60	0.30–0.50	0.08–0.15	...

(a) Small quantities of certain elements that are not specified or required may be found in alloy steels. These elements are to be considered as incidental and are acceptable to the following maximum amount: copper to 0.35%, nickel to 0.25%, chromium to 0.20%, and molybdenum to 0.06%. (b) Electric furnace steel. (c) Boron content is 0.0005–0.003%. Source: Ref 16

Table 20 SAE potential standard steel compositions

SAE PS number(a)	Ladle chemical composition limits, wt%								
	C	Mn	P max	S max	Si	Ni	Cr	Mo	B
PS 10	0.19–0.24	0.95–1.25	0.035	0.040	0.15–0.35	0.20–0.40	0.25–0.40	0.05–0.10	...
PS 15	0.18–0.23	0.90–1.20	0.035	0.040	0.15–0.35	...	0.40–0.60	0.13–0.20	...
PS 16	0.20–0.25	0.90–1.20	0.035	0.040	0.15–0.35	...	0.40–0.60	0.13–0.20	...
PS 17	0.23–0.28	0.90–1.20	0.035	0.040	0.15–0.35	...	0.40–0.60	0.13–0.20	...
PS 18	0.25–0.30	0.90–1.20	0.035	0.040	0.15–0.35	...	0.40–0.60	0.13–0.20	...
PS 19	0.18–0.23	0.90–1.20	0.035	0.040	0.15–0.35	...	0.40–0.60	0.08–0.15	0.0005–0.003
PS 20	0.13–0.18	0.90–1.20	0.035	0.040	0.15–0.35	...	0.40–0.60	0.13–0.20	...
PS 21	0.15–0.20	0.90–1.20	0.035	0.040	0.15–0.35	...	0.40–0.60	0.13–0.20	...
PS 24	0.18–0.23	0.75–1.00	0.035	0.040	0.15–0.35	...	0.45–0.65	0.20–0.30	...
PS 30	0.13–0.18	0.70–0.90	0.035	0.040	0.15–0.35	0.70–1.00	0.45–0.65	0.45–0.60	...
PS 31	0.15–0.20	0.70–0.90	0.035	0.040	0.15–0.35	0.70–1.00	0.45–0.65	0.45–0.60	...
PS 32	0.18–0.23	0.70–0.90	0.035	0.040	0.15–0.35	0.70–1.00	0.45–0.65	0.45–0.60	...
PS 33(b)	0.17–0.24	0.85–1.25	0.035	0.040	0.15–0.35	0.20 min	0.20 min	0.05 min	...
PS 34	0.28–0.33	0.90–1.20	0.035	0.040	0.15–0.35	...	0.40–0.60	0.13–0.20	...
PS 36	0.38–0.43	0.90–1.20	0.035	0.040	0.15–0.35	...	0.45–0.65	0.13–0.20	...
PS 38	0.43–0.48	0.90–1.20	0.035	0.040	0.15–0.35	...	0.45–0.65	0.13–0.20	...
PS 39	0.48–0.53	0.90–1.20	0.035	0.040	0.15–0.35	...	0.45–0.65	0.13–0.20	...
PS 40	0.51–0.59	0.90–1.20	0.035	0.040	0.15–0.35	...	0.45–0.65	0.13–0.20	...
PS 54	0.19–0.25	0.70–1.05	0.035	0.040	0.15–0.35	...	0.40–0.70	0.05 min	...
PS 55	0.15–0.20	0.70–1.00	0.035	0.040	0.15–0.35	1.65–2.00	0.45–0.65	0.65–0.80	...
PS 56	0.080–0.13	0.70–1.00	0.035	0.040	0.15–0.35	1.65–2.00	0.45–0.65	0.65–0.80	...
PS 57	0.08 max	1.25 max	0.040	0.15–0.35	1.00 max	...	17.00–19.00	1.75–2.25	...
PS 58	0.16–0.21	1.00–1.30	0.035	0.040	0.15–0.35	...	0.45–0.65	...	...
PS 59	0.18–0.23	1.00–1.30	0.035	0.040	0.15–0.35	...	0.70–0.90	...	...
PS 61	0.23–0.28	1.00–1.30	0.035	0.040	0.15–0.35	...	0.70–0.90	...	...
PS 63	0.31–0.38	0.75–1.10	0.035	0.040	0.15–0.35	...	0.45–0.65	...	0.0005–0.003
PS 64	0.16–0.21	1.00–1.30	0.035	0.040	0.15–0.35	...	0.70–0.90	...	...
PS 65	0.21–0.26	1.00–1.30	0.035	0.040	0.15–0.35	...	0.70–0.90	...	...
PS 66(c)	0.16–0.21	0.40–0.70	0.035	0.040	0.15–0.35	1.65–2.00	0.45–0.75	0.08–0.15	...
PS 67	0.42–0.49	0.80–1.20	0.035	0.040	0.15–0.35	...	0.85–1.20	0.25–0.35	...

(a) Some PS steels may be supplied to a hardenability requirement. (b) Supplied to a hardenability requirement of 15 HRC points within the range of 23–43 HRC at J4 (4/16 in. distance from quenched end), subject to agreement between producer and user. (c) PS 66 has a vanadium content of 0.10–0.15%. Source: Ref 17

composition of a metallic alloy. Available UNS designations are included in the tables in this article.

The UNS designation of an alloy consists of a letter and five numerals. The letters indicate the broad class of alloys; the numerals define specific alloys within that class. Existing designation systems, such as the AISI-SAE system for steels, have been incorporated into UNS designations. UNS is described in greater detail in SAE J1086 and ASTM E 527.

AMS Designations

Aerospace Materials Specifications (AMS), published by SAE, are complete specifications that are generally adequate for procurement purposes. Most of the AMS designations pertain to materials in-

154 / Carbon and Low-Alloy Steels

Table 21 Composition ranges and limits for SAE HSLA steels

SAE designation(b)	Heat composition limits, %(a)		
	C max	Mn max	P max
942X	0.21	1.35	0.04
945A	0.15	1.00	0.04
945C	0.23	1.40	0.04
945X	0.22	1.35	0.04
950A	0.15	1.30	0.04
950B	0.22	1.30	0.04
950C	0.25	1.60	0.04
950D	0.15	1.00	0.15
950X	0.23	1.35	0.04
955X	0.25	1.35	0.04
960X	0.26	1.45	0.04
965X	0.26	1.45	0.04
970X	0.26	1.65	0.04
980X	0.26	1.65	0.04

(a) Maximum contents of sulfur and silicon for all grades: 0.050% S, 0.90% Si. (b) Second and third digits of designation indicate minimum yield strength in ksi. Suffix X indicates that the steel contains niobium, vanadium, nitrogen, or other alloying elements. A second suffix K indicates that the steel is produced fully killed using fine-grain practice; otherwise, the steel is produced semi-killed. Source: Ref 18

tended for aerospace applications; the specifications may include mechanical property requirements significantly more severe than those for grades of steel having similar compositions but intended for other applications. Processing requirements, such as for consumable electrode remelting, are common in AMS steels. Chemical compositions for AMS grades of carbon and alloy steels are given in Tables 26 and 27, respectively.

Specifications for Steels

A specification is a written statement of the requirements, both technical and commercial, that a product must meet; it is a document that controls procurement. There are nearly as many formats for specifications as there are groups writing them, but any reasonably adequate specification will provide information about the items listed below:

- *Scope* may cover product classification, including size range when necessary, condition, and any comments on product processing deemed helpful to either the supplier or user. An informative title plus a statement of the required form may be used instead of a scope clause
- *Chemical composition* may be detailed, or it may be indicated by a well-recognized designation based on chemical composition. The SAE-AISI designations are frequently used
- *The quality statement* includes any appropriate quality descriptor and whatever additional requirements might be necessary. It may also include the type of steel and the steelmaking processes permitted
- *Quantitative requirements* identify allowable ranges of the composition and all physical and mechanical properties necessary to characterize the material. Test methods used to determine these properties should also be included, at least by reference to standard test methods. For reasons of economy, this section should be limited to those properties that are germane to the intended application
- *Additional requirements* can include special tolerances, surface preparation, and edge finish on flat-rolled products, as well as special identification, packaging, and loading instructions

Engineering societies, associations, and institutes whose members make, specify, or purchase steel products publish standard specifications, many of which have become well known and highly respected. Some of the important specification-writing groups are listed below. It is obvious from the names of some of these that the specifications prepared by a particular group may be limited to its own specialized field:

Organization	Acronym
Association of American Railroads	AAR
American Bureau of Shipbuilding	ABS
American Petroleum Institute	API
American Railway Engineering Association	AREA
American Society of Mechanical Engineers	ASME
American Society for Testing and Materials	ASTM
Society of Automotive Engineers	SAE
Aerospace Material Specification (of SAE)	AMS

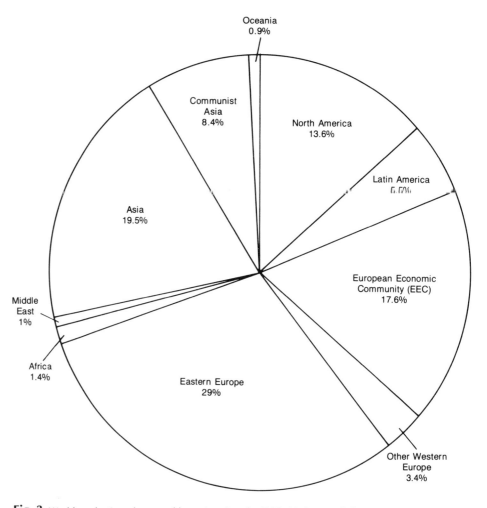

Fig. 3 World production of raw steel by region. See also Table 30. Source: Ref 12

ASTM (ASME) Specifications

The most widely used standard specifications for steel products in the United States are those published by ASTM. These are complete specifications, generally adequate for procurement purposes. Many ASTM specifications apply to specific products, such as A 574 for alloy steel socket head cap screws. These specifications are generally oriented toward performance of the fabricated end product, with considerable latitude in chemical composition of the steel used to make the end product.

ASTM specifications represent a consensus among producers, specifiers, fabricators, and users of steel mill products. In many cases, the dimensions, tolerances, limits, and restrictions in the ASTM specifications are similar to or the same as the corresponding items of the standard practices in the AISI Steel Products Manuals. Many of the ASTM specifications have been adopted by the American Society of Mechanical Engineers (ASME) with little or no modification; ASME uses the prefix S and the ASTM designation for these specifications. For example, ASME-SA213 and ASTM A 213 are identical.

Steel products can be identified by the number of the ASTM specification to which

Table 22 Composition ranges and limits for former standard SAE steels

SAE number	AISI number	UNS number	Composition, wt%									Date of obsolescence
			C	Mn	P max(b)	S max(b)	Si	Cr	Ni	Mo	V min	
1009	1009	...	0.15 max	0.60 max	0.040	0.050	...	...	...	...	...	1965
1011	...	G10110	0.08–0.13	0.60–0.90	0.040	0.050	...	...	...	...	...	1977
1033	1033	...	0.30–0.36	0.70–1.00	0.040	0.050	...	...	...	...	...	1965
1034	C1034	...	0.32–0.38	0.50–0.80	0.040	0.050	...	...	...	...	...	1968
1059(a)	...	...	0.55–0.65	0.50–0.80	0.040	0.050	...	...	...	...	...	1968
1062	C1062	...	0.54–0.65	0.85–1.15	0.040	0.050	...	...	...	...	...	1953
1086(a)	...	G10860	0.80–0.94	0.30–0.50	0.040	0.050	...	...	...	...	...	1977
1109	1109	G11090	0.08–0.13	0.60–0.90	0.040	0.08–0.13	...	...	...	...	...	1977
1111	B1111	...	0.13 max	0.60–0.90	0.07–0.12	0.10–0.15	...	...	...	...	...	1969
1112	B1112	...	0.13 max	0.70–1.00	0.07–0.12	0.16–0.23	...	...	...	...	...	1969
1113	B1113	...	0.13 max	0.70–1.00	0.07–0.12	0.24–0.33	...	...	...	...	...	1969
1114	C1114	...	0.10–0.16	1.00–1.30	0.040	0.08–0.13	...	...	...	...	...	1952
1115	1115	...	0.13–0.18	0.60–0.90	0.040	0.08–0.13	...	...	...	...	...	1965
1116	C1116	...	0.14–0.20	1.10–1.40	0.040	0.16–0.23	...	...	...	...	...	1952
1119	1119	G11190	0.14–0.20	1.00–1.30	0.040	0.24–0.33	...	...	...	...	...	1977
1120	1120	...	0.18–0.23	0.70–1.00	0.040	0.08–0.13	...	...	...	...	...	1965
1126	1126	...	0.23–0.29	0.70–1.00	0.040	0.08–0.13	...	...	...	...	...	1965
1132	1132	G11320	0.27–0.34	1.35–1.65	0.040	0.08–0.13	...	...	...	...	...	1977
1138	1138	...	0.34–0.40	0.70–1.00	0.040	0.08–0.13	...	...	...	...	...	1965
1145	1145	G11450	0.42–0.49	0.70–1.00	0.040	0.04–0.07	...	...	...	...	...	1977
1320	A1320	...	0.18–0.23	1.60–1.90	0.040	0.040	0.20–0.35	...	...	...	...	1956
1518	...	G15180	0.15–0.21	1.10–1.40	0.040	0.050	...	...	...	...	...	1977
1525	...	G15250	0.23–0.29	0.80–1.10	0.040	0.050	...	...	...	...	...	1977
1547	...	G15470	0.43–0.51	1.35–1.65	0.040	0.050	...	...	...	...	...	1977
1572	...	G15720	0.65–0.76	1.00–1.30	0.040	0.050	...	...	...	...	...	1977
2317	A2317	...	0.15–0.20	0.40–0.60	0.040	0.040	0.20–0.35	...	3.25–3.75	...	...	1956
2330	A2330	...	0.28–0.33	0.60–0.80	0.040	0.040	0.20–0.35	...	3.25–3.75	...	...	1953
2340	A2340	...	0.38–0.43	0.70–0.90	0.040	0.040	0.20–0.35	...	3.25–3.75	...	...	1953
2345	A2345	...	0.43–0.48	0.70–0.90	0.040	0.040	0.20–0.35	...	3.25–3.75	...	...	1952
2512	E2512	...	0.09–0.14	0.45–0.60	0.025	0.025	0.20–0.35	...	4.75–5.25	...	...	1953
2515	A2515	...	0.12–0.17	0.40–0.60	0.040	0.040	0.20–0.35	...	4.75–5.25	...	...	1956
2517	E2517	...	0.15–0.20	0.45–0.60	0.025	0.025	0.20–0.35	...	4.75–5.25	...	...	1959
3115	A3115	...	0.13–0.18	0.40–0.60	0.040	0.040	0.20–0.35	0.55–0.75	1.10–1.40	...	...	1953
3120	A3120	...	0.17–0.22	0.60–0.80	0.040	0.040	0.20–0.35	0.55–0.75	1.10–1.40	...	...	1956
3130	A3130	...	0.28–0.33	0.60–0.80	0.040	0.040	0.20–0.35	0.55–0.75	1.10–1.40	...	...	1956
3135	3135	...	0.33–0.38	0.60–0.80	0.040	0.040	0.20–0.35	0.55–0.75	1.10–1.40	...	...	1960
X3140	A3141	...	0.38–0.43	0.70–0.90	0.040	0.040	0.20–0.35	0.70–0.90	1.10–1.40	...	...	1947
3140	3140	...	0.38–0.43	0.70–0.90	0.040	0.040	0.20–0.35	0.55–0.75	1.10–1.40	...	...	1964
3145	A3145	...	0.43–0.48	0.70–0.90	0.040	0.040	0.20–0.35	0.70–0.90	1.10–1.40	...	...	1952
3150	A3150	...	0.48–0.53	0.70–0.90	0.040	0.040	0.20–0.35	0.70–0.90	1.10–1.40	...	...	1952
3215	...	...	0.10–0.20	0.30–0.60	0.040	0.050	0.15–0.30	0.90–1.25	1.50–2.00	...	...	1941
3220	...	...	0.15–0.25	0.30–0.60	0.040	0.050	0.15–0.30	0.90–1.25	1.50–2.00	...	...	1941
3230	...	...	0.25–0.35	0.30–0.60	0.040	0.050	0.15–0.30	0.90–1.25	1.50–2.00	...	...	1941
3240	A3240	...	0.35–0.45	0.30–0.60	0.040	0.040	0.15–0.30	0.90–1.25	1.50–2.00	...	...	1941
3245	...	...	0.40–0.50	0.30–0.60	0.040	0.040	0.15–0.30	0.90–1.25	1.50–2.00	...	...	1941
3250	...	...	0.45–0.55	0.30–0.60	0.040	0.040	0.15–0.30	0.90–1.25	1.50–2.00	...	...	1941
3310	E3310	...	0.08–0.13	0.45–0.60	0.025	0.025	0.20–0.35	1.40–1.75	3.25–3.75	...	...	1964
3312	...	...	0.08–0.13	0.45–0.60	0.025	0.025	0.20–0.35	1.40–1.75	3.25–3.75	...	...	1948
3316	E3316	...	0.14–0.19	0.45–0.60	0.025	0.025	0.20–0.35	1.40–1.75	3.25–3.75	...	...	1956
3325	...	...	20–30	0.30–0.60	0.040	0.050	0.15–0.30	1.25–1.75	3.25–3.75	...	...	1936
3335	...	...	30–40	0.30–0.60	0.040	0.050	0.15–0.30	1.25–1.75	3.25–3.75	...	...	1936
3340	...	...	35–45	0.30–0.60	0.040	0.050	0.15–0.30	1.25–1.75	3.25–3.75	...	...	1936
3415	...	...	0.10–0.20	0.30–0.60	0.040	0.050	0.15–0.30	0.60–0.95	2.75–3.25	...	...	1941
3435	...	...	0.30–0.40	0.30–0.60	0.040	0.050	0.15–0.30	0.60–0.95	2.75–3.25	...	...	1936
3450	...	...	0.45–0.55	0.30–0.60	0.040	0.050	0.15–0.30	0.60–0.95	2.75–3.25	...	...	1936
4012	4012	G40120	0.09–0.14	0.75–1.00	0.035	0.040	0.15–0.30	...	...	0.15–0.25	...	1977
4053	4053	...	0.50–0.56	0.75–1.00	0.040	0.040	0.20–0.35	...	...	0.20–0.30	...	1956
4063	4063	G40630	0.60–0.67	0.75–1.00	0.040	0.040	0.20–0.35	...	...	0.20–0.30	...	1964
4068	A4068	...	0.63–0.70	0.75–1.00	0.040	0.040	0.20–0.35	...	...	0.20–0.30	...	1957
4119	A4119	...	0.17–0.22	0.70–0.90	0.040	0.040	0.20–0.35	0.40–0.60	...	0.20–0.30	...	1956
4125	A4125	...	0.23–0.28	0.70–0.90	0.040	0.040	0.20–0.35	0.40–0.60	...	0.20–0.30	...	1950
4317	4317	...	0.15–0.20	0.45–0.65	0.040	0.040	0.20–0.35	0.40–0.60	1.65–2.00	0.20–0.30	...	1953
4337	4337	G43370	0.35–0.40	0.60–0.80	0.040	0.040	0.20–0.35	0.70–0.90	1.65–2.00	0.20–0.30	...	1964
4419	4520	...	0.18–0.23	0.45–0.65	0.035	0.040	0.15–0.30	...	...	0.45–0.60	...	1977
4419H	4419H	...	0.17–0.23	0.35–0.75	0.035	0.040	0.15–0.30	...	...	0.45–0.60	...	1977
4608	4608	...	0.06–0.11	0.25–0.45	0.040	0.040	0.025 max	...	1.40–1.75	0.15–0.25	...	1956
46B12(c)	46B12(c)	...	0.10–0.15	0.45–0.65	0.040	0.040	0.20–0.35	...	1.65–2.00	0.20–0.30	...	1957
X4620	X4620	...	0.18–0.23	0.50–0.70	0.040	0.040	0.20–0.35	...	1.65–2.00	0.20–0.30	...	1956
4621	4621	G46210	0.18–0.23	0.70–0.90	0.035	0.040	0.15–0.30	...	1.65–2.00	0.20–0.30	...	1977
4621H	4621H	...	0.17–0.23	0.60–1.00	0.035	0.040	0.15–0.30	...	1.55–2.00	0.20–0.30	...	1977
4640	A4640	...	0.38–0.43	0.60–0.80	0.040	0.040	0.20–0.35	...	1.65–2.00	0.20–0.30	...	1952
4812	4817	...	0.10–0.15	0.40–0.60	0.040	0.040	0.20–0.35	...	3.25–3.75	0.20–0.30	...	1956

(continued)

(a) These grades remain standard for wire rods. (b) Limits apply to semifinished products for forgings, bars, wire rods, and seamless tubing. (c) Boron content 0.0005–0.003%. (d) Contains 12.00–15.00% W. (e) Contains 15.00–18.00% W. (f) Contains 1.50–2.00% W. Source: Ref 19

Table 22 (continued)

SAE number	AISI number	UNS number	Composition, wt%									Date of obsolescence
			C	Mn	P max(b)	S max(b)	Si	Cr	Ni	Mo	V min	
5015	5015	G50150	0.12–0.17	0.30–0.50	0.035	0.040	0.15–0.30	0.30–0.50	...	...	...	1977
5045	5045	...	0.43–0.48	0.70–0.90	0.040	0.040	0.20–0.35	0.55–0.75	...	...	...	1953
5145	5145	G51450	0.43–0.48	0.70–0.90	0.035	0.040	0.15–0.30	0.70–0.90	...	...	...	1977
5145H	5145H	H51450	0.42–0.49	0.60–1.00	0.035	0.040	0.15–0.30	0.60–1.00	...	...	...	1977
5152	5152	...	0.48–0.55	0.70–0.90	0.040	0.040	0.20–0.35	0.90–1.20	...	...	...	1956
6115	...	...	0.10–0.20	0.30–0.60	0.040	0.050	0.15–0.30	0.80–1.10	...	...	0.15	1936
6117	6117	...	0.15–0.20	0.70–0.90	0.040	0.040	0.20–0.35	0.70–0.90	...	...	0.10	1956
6120	6120	...	0.17–0.22	0.70–0.90	0.040	0.040	0.20–0.35	0.70–0.90	...	...	0.10	1961
6125	...	...	0.20–0.30	0.60–0.90	0.040	0.050	0.15–0.30	0.80–1.10	...	...	0.15	1936
6130	...	...	0.25–0.35	0.60–0.90	0.040	0.050	0.15–0.30	0.80–1.10	...	...	0.15	1936
6135	...	...	0.30–0.40	0.60–0.90	0.040	0.050	0.15–0.30	0.80–1.10	...	...	0.15	1941
6140	...	...	0.35–0.45	0.60–0.90	0.040	0.050	0.15–0.30	0.80–1.10	...	...	0.15	1936
6145	6145	...	0.43–0.48	0.70–0.90	0.040	0.050	0.20–0.35	0.80–1.10	...	...	0.15	1956
6195	...	...	0.90–1.05	0.20–0.45	0.030	0.035	0.15–0.30	0.80–1.10	...	...	0.15	1936
71360(d)	...	...	0.50–0.70	0.30 max	0.035	0.040	0.15–0.30	3.00–4.00	...	...	...	1936
71660(e)	...	...	0.50–0.70	0.30 max	0.035	0.040	0.15–0.30	3.00–4.00	...	...	...	1936
7260(f)	...	...	0.50–0.70	0.30 max	0.035	0.040	0.15–0.30	0.50–1.00	...	...	...	1936
8632	8632	...	0.30–0.35	0.70–0.90	0.040	0.040	0.20–0.35	0.40–0.60	0.40–0.70	0.15–0.25	...	1951
8635	8635	...	0.33–0.38	0.75–1.00	0.040	0.040	0.20–0.35	0.40–0.60	0.40–0.70	0.15–0.25	...	1956
8641	8641	...	0.38–0.43	0.75–1.00	0.040	0.040–0.060	0.20–0.35	0.40–0.60	0.40–0.70	0.15–0.25	...	1956
8653	8653	...	0.50–0.56	0.75–1.00	0.040	0.040	0.20–0.35	0.50–0.80	0.40–0.70	0.15–0.25	...	1956
8647	8647	...	0.45–0.50	0.75–1.00	0.040	0.040	0.20–0.35	0.40–0.60	0.40–0.70	0.15–0.25	...	1948
8715	8715	...	0.13–0.18	0.70–0.90	0.040	0.040	0.20–0.35	0.40–0.60	0.40–0.70	0.20–0.30	...	1956
8717	8717	...	0.15–0.20	0.70–0.90	0.040	0.040	0.20–0.35	0.40–0.60	0.40–0.70	0.20–0.30	...	1956
8719	8719	...	0.18–0.23	0.60–0.80	0.040	0.040	0.20–0.35	0.40–0.60	0.40–0.70	0.20–0.30	...	1952
8735	8735	...	0.33–0.38	0.75–1.00	0.040	0.040	0.20–0.35	0.40–0.60	0.40–0.70	0.20–0.30	...	1952
8742	8742	G87420	0.40–0.45	0.75–1.00	0.040	0.040	0.20–0.35	0.40–0.60	0.40–0.70	0.20–0.30	...	1964
8745	8745	...	0.43–0.48	0.75–1.00	0.040	0.040	0.20–0.35	0.40–0.60	0.40–0.70	0.20–0.30	...	1953
8750	8750	...	0.48–0.53	0.75–1.00	0.040	0.040	0.20–0.35	0.40–0.60	0.40–0.70	0.20–0.30	...	1956
9250	9250	...	0.45–0.55	0.60–0.90	0.040	0.040	1.80–2.20	...	...	...	...	1941
9255	9255	G92550	0.51–0.59	0.70–0.95	0.035	0.040	1.80–2.20	...	...	...	...	1977
9261	9261	...	0.55–0.65	0.75–1.00	0.040	0.040	1.80–2.20	0.10–0.25	...	...	...	1956
9262	9262	G92620	0.55–0.65	0.75–1.00	0.040	0.040	1.80–2.20	0.25–0.40	...	...	...	1961
9315	E9315	...	0.13–0.18	0.45–0.65	0.025	0.025	0.20–0.35	1.00–1.40	3.00–3.50	0.08–0.15	...	1959
9317	E9317	...	0.15–0.20	0.45–0.65	0.025	0.025	0.20–0.35	1.00–1.40	3.00–3.50	0.08–0.15	...	1959
9437	9437	...	0.35–0.40	0.90–1.20	0.040	0.040	0.20–0.35	0.30–0.50	0.30–0.60	0.08–0.15	...	1950
9440	9440	...	0.38–0.43	0.90–1.20	0.040	0.040	0.20–0.35	0.30–0.50	0.30–0.60	0.08–0.15	...	1950
94B40(c)	94B40	G94401	0.38–0.43	0.75–1.00	0.040	0.040	0.20–0.35	0.30–0.50	0.30–0.60	0.08–0.15	...	1964
9442	9442	...	0.40–0.45	0.90–1.20	0.040	0.040	0.20–0.35	0.30–0.50	0.30–0.60	0.08–0.15	...	1950
9445	9445	...	0.43–0.48	0.90–1.20	0.040	0.040	0.20–0.35	0.30–0.50	0.30–0.60	0.08–0.15	...	1950
9447	9447	...	0.45–0.50	0.90–1.20	0.040	0.040	0.20–0.35	0.30–0.50	0.30–0.60	0.08–0.15	...	1950
9747	9747	...	0.45–0.50	0.50–0.80	0.040	0.040	0.20–0.35	0.10–0.25	0.40–0.70	0.15–0.25	...	1950
9763	9763	...	0.60–0.67	0.50–0.80	0.040	0.040	0.20–0.35	0.10–0.25	0.40–0.70	0.15–0.25	...	1950
9840	9840	G98400	0.38–0.43	0.70–0.90	0.040	0.040	0.20–0.35	0.70–0.90	0.85–1.15	0.20–0.30	...	1964
9845	9845	...	0.43–0.48	0.70–0.90	0.040	0.040	0.20–0.35	0.70–0.90	0.85–1.15	0.20–0.30	...	1950
9850	9850	G98500	0.48–0.53	0.70–0.90	0.040	0.040	0.20–0.35	0.70–0.90	0.85–1.15	0.20–0.30	...	1961
43BV12(c)	...	...	0.08–0.13	0.75–1.00	...	...	0.20–0.35	0.40–0.60	1.65–2.00	0.20–0.30	0.03	...
43BV14(c)	...	...	0.10–0.15	0.45–0.65	...	...	0.20–0.35	0.40–0.60	1.65–2.00	0.08–0.15	0.03	...

(a) These grades remain standard for wire rods. (b) Limits apply to semifinished products for forgings, bars, wire rods, and seamless tubing. (c) Boron content 0.0005–0.003%. (d) Contains 12.00–15.00% W. (e) Contains 15.00–18.00% W. (f) Contains 1.50–2.00% W. Source: Ref 19

they are made. The number consists of the letter A (for ferrous materials) and an arbitrary, serially assigned number. Citing the specification number, however, is not always adequate to completely describe a steel product. For example, A 434 is the specification for heat-treated (hardened and tempered) alloy steel bars. To completely describe steel bars indicated by this specification, the grade (SAE-AISI designation in this case) and class (required strength level) must also be indicated. The ASTM specification A 434 also incorporates, by reference, two standards for test methods (A 370 for mechanical testing and E 112 for grain size determination) and A 29, which specifies the general requirements for bar products.

SAE-AISI designations for the compositions of carbon and alloy steels are sometimes incorporated into the ASTM specifications for bars, wires, and billets for forging. Some ASTM specifications for sheet products include SAE-AISI designations for composition. The ASTM specifications for plates and structural shapes generally specify the limits and ranges of chemical composition directly, without the SAE-AISI designations. Table 28 includes a list of some of the ASTM specifications that incorporate AISI-SAE designations for compositions of the different grades of steel.

General Specifications. Several ASTM specifications, such as A 20 covering steel plate used for pressure vessels, contain the general requirements common to each member of a broad family of steel products. Table 29 lists general specifications for a number of product forms. As shown in Table 29, these general specifications are often supplemented by additional specifications describing a different mill form or intermediate fabricated product. Articles describing the compositions and properties of steels used for each of the product forms listed in Table 29 follow in this Section.

International Designations and Specifications

The steel industry has undergone major changes during the past 15 to 20 years. No

Table 23 Chemical compositions for typical low-alloy steels
Table 25 lists compositions of heat-resistant chromium-molybdenum low-alloy steels

Steel	C	Si	Mn	P	S	Ni	Cr	Mo	Other
Low-carbon quenched and tempered steels									
A 514/A 517 grade A	0.15–0.21	0.40–0.80	0.80–1.10	0.035	0.04	...	0.50–0.80	0.18–0.28	0.05–0.15 Zr(b) 0.0025 B
A 514/A 517 grade F	0.10–0.20	0.15–0.35	0.60–1.00	0.035	0.04	0.70–1.00	0.40–0.65	0.40–0.60	0.03–0.08 V 0.15–0.50 Cu 0.0005–0.005 B
A 514/A 517 grade R	0.15–0.20	0.20–0.35	0.85–1.15	0.035	0.04	0.90–1.10	0.35–0.65	0.15–0.25	0.03–0.08 V
A 533 type A	0.25	0.15–0.40	1.15–1.50	0.035	0.04	...	...	0.45–0.60	...
A 533 type C	0.25	0.15–0.40	1.15–1.50	0.035	0.04	0.70–1.00	...	0.45–0.60	...
HY-80	0.12–0.18	0.15–0.35	0.10–0.40	0.025	0.025	2.00–3.25	1.00–1.80	0.20–0.60	0.25 Cu 0.03 V 0.02 Ti
HY-100	0.12–0.20	0.15–0.35	0.10–0.40	0.025	0.025	2.25–3.50	1.00–1.80	0.20–0.60	0.25 Cu 0.03 V 0.02 Ti
Medium-carbon ultrahigh-strength steels									
4130	0.28–0.33	0.20–0.35	0.40–0.60	...	...	...	0.80–1.10	0.15–0.25	...
4340	0.38–0.43	0.20–0.35	0.60–0.80	...	...	1.65–2.00	0.70–0.90	0.20–0.30	...
300M	0.40–0.46	1.45–1.80	0.65–0.90	...	...	1.65–2.00	0.70–0.95	0.30–0.45	0.05 V min
D-6a	0.42–0.48	0.15–0.30	0.60–0.90	...	...	0.40–0.70	0.90–1.20	0.90–1.10	0.05–0.10 V
Carburizing bearing steels									
4118	0.18–0.23	0.15–0.30	0.70–0.90	0.035	0.040	...	0.40–0.60	0.08–0.18	...
5120	0.17–0.22	0.15–0.30	0.70–0.90	0.035	0.040	...	0.70–0.90	...	...
3310	0.08–0.13	0.20–0.35	0.45–0.60	0.025	0.025	3.25–3.75	1.40–1.75	...	...
Through-hardened bearing steels									
52100	0.98–1.10	0.15–0.30	0.25–0.45	0.025	0.025	...	1.30–1.60	...	...
A 485 grade 1	0.90–1.05	0.45–0.75	0.95–1.25	0.025	0.025	0.25	0.90–1.20	0.10	0.35 Cu
A 485 grade 3	0.95–1.10	0.15–0.35	0.65–0.90	0.025	0.025	0.25	1.10–1.50	0.20–0.30	0.35 Cu

(a) Single values represent the maximum allowable. (b) Zirconium may be replaced by cerium. When cerium is added, the cerium/sulfur ratio should be approximately 1.5/1, based on heat analysis.

Table 24 ASTM specifications for chromium-molybdenum steel product forms

Type	Forgings	Tubes	Pipe	Castings	Plate
½Cr-½Mo	A 182-F2	...	A 335-P2 A 369-FP2 A 426-CP2	...	A 387-Gr 2
1Cr-½Mo	A 182-F12 A 336-F12	...	A 335-P12 A 369-FP12 A 426-CP12	...	A 387-Gr 12
1¼Cr-½Mo	A 182-F11 A 336-F11/F11A A 541-C11C	A 199-T11 A 200-T11 A 213-T11	A 335-P11 A 369-FP11 A 426-CP11	A 217-WC6 A 356-Gr6 A 389-C23	A 387-Gr 11
2¼Cr-1Mo	A 182-F22/F22a A 336-F22/F22A A 541-C22C/22D	A 199-T22 A 200-T22 A 213-T22	A 335-P22 A 369-FP22 A 426-CP22	A 217-WC9 A 356-Gr10	A 387-Gr22 A 542
3Cr-1Mo	A 182-F21 A 336-F21/F21A	A 199-T21 A 200-T21 A 213-T21	A 335-P21 A 369-FP21 A 426-CP21	...	A 387-Gr 21
3Cr-1MoV	A 182-F21b	...	...	...	...
5Cr-½Mo	A 182-F5/F5a A 336-F5/F5A A 473-501/502	A 199-T5 A 200-T5 A 213-T5	A 335-P5 A 369-FP5 A 426-CP5	A 217-C5	A 387-Gr 5
5Cr-½MoSi	...	A 213-T5b	A 335-P5b A 426-CP5b	...	...
5Cr-½MoTi	...	A 213-T5c	A 335-P5c	...	...
7Cr-½Mo	A 182-F7 A 473-501A	A 199-T7 A 200-T7 A 213-T7	A 335-P7 A 369-FP7 A 426-CP7	...	A 387-Gr7
9Cr-1Mo	A 182-F9 A 336-F9 A 473-501B	A 199-T9 A 200-T9 A 213-T9	A 335-P9 A 369-FP9 A 426-CP9	A 217-C12	A 387-Gr9

Because steelmaking technology is available worldwide, familiarity with international designations and/or specifications for steels is critical. Table 31 cross-references SAE steels with those of a selected group of international designations and specifications, which are described below. Tables 32 to 43 provide chemical compositions for the non-SAE steels listed in Table 31. More detailed information on cross-referencing of steels can be found in Ref 20 and 21.

DIN standards are developed by Deutsches Institut für Normung in the Federal Republic of Germany. All West German steel specifications are preceded by the uppercase letters DIN followed by an alphanumeric or numeric code. The latter method, known as the Werkstoff number, uses numbers only with a decimal point after the first digit. Examples of both methods are given in Tables 31, 32, and 33, which cross-reference SAE and DIN designations and provide chemical compositions for DIN steels.

JIS standards are developed by the Japanese Industrial Standards Committee, which is part of the Ministry of International Trade and Industry in Tokyo. The JIS steel specifications begin with the uppercase letters JIS and are followed by an uppercase letter (G in the case of carbon and low-alloy steels) designating the division (product form) of the standard. This letter is followed by a series of numbers and

longer is the international steel marketplace dominated by the United States. In fact, the steel produced in all of North America accounts for less than 14% of total world production. The distribution of steel production among various nations is illustrated in Fig. 3; production figures are given in Table 30. With continuing advances in production by third world developing countries (Fig. 4), the international steel marketplace will continue to experience substantial changes during the 1990s.

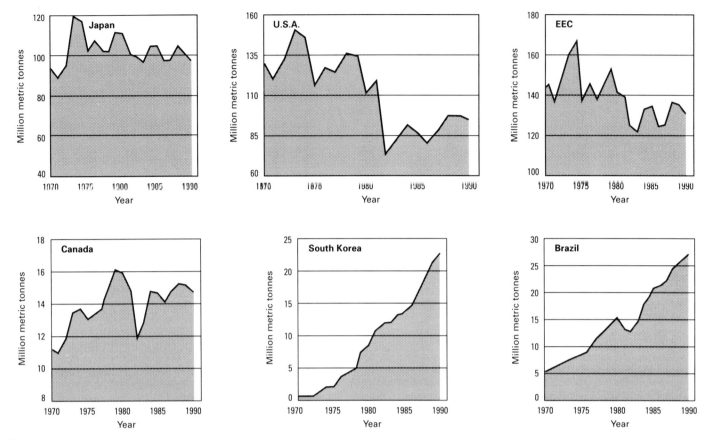

Fig. 4 Annual crude steel production for various industrialized and developing nations. Source: Wharton Econometric Forecasting Associates

Table 25 Nominal chemical compositions for heat-resistant chromium-molybdenum steels

Type	UNS designation	Composition, %(a)						
		C	Mn	S	P	Si	Cr	Mo
½Cr-½Mo	K12122	0.10–0.20	0.30–0.80	0.040	0.040	0.10–0.60	0.50–0.80	0.45–0.65
1Cr-½Mo	K11562	0.15	0.30–0.60	0.045	0.045	0.50	0.80–1.25	0.45–0.65
1¼Cr-½Mo	K11597	0.15	0.30–0.60	0.030	0.030	0.50–1.00	1.00–1.50	0.45–0.65
1¼Cr-½Mo	K11592	0.10–0.20	0.30–0.80	0.040	0.040	0.50–1.00	1.00–1.50	0.45–0.65
2¼Cr-1Mo	K21590	0.15	0.30–0.60	0.040	0.040	0.50	2.00–2.50	0.87–1.13
3Cr-1Mo	K31545	0.15	0.30–0.60	0.030	0.030	0.50	2.65–3.35	0.80–1.06
3Cr-1MoV(b)	K31830	0.18	0.30–0.60	0.020	0.020	0.10	2.75–3.25	0.90–1.10
5Cr-½Mo	K41545	0.15	0.30–0.60	0.030	0.030	0.50	4.00–6.00	0.45–0.65
7Cr-½Mo	K61595	0.15	0.30–0.60	0.030	0.030	0.50–1.00	6.00–8.00	0.45–0.65
9Cr-1Mo	K90941	0.15	0.30–0.60	0.030	0.030	0.50–1.00	8.00–10.00	0.90–1.10
9Cr-1MoV(c)	...	0.08–0.12	0.30–0.60	0.010	0.020	0.20–0.50	8.00–9.00	0.85–1.05

(a) Single values are maximums. (b) Also contains 0.02–0.030% V, 0.001–0.003% B, and 0.015–0.035% Ti. (c) Also contains 0.40% Ni, 0.18–0.25% V, 0.06–0.10% Nb, 0.03–0.07% N, and 0.04% Al

letters that indicate the specific steel; JIS designations are given in Tables 31, 34, and 35. For example, as indicated in Tables 31 and 34, JIS G3445 STKM11A is a low-carbon tube steel with the following chemical composition:

Element	Percentage/range
Carbon	0.12
Silicon	0.35
Manganese	0.60
Phosphorus	0.040
Sulfur	0.040

British standards (BS) are developed by the British Standards Institute in London, England. Tables 31, 36, and 37 list steels covered by British standards. Similar to the JIS standards, each British designation includes a product form and an alloy code. For example, as indicated in Tables 31 and 37, BS 979 708A37 is a low-alloy steel with the following chemical composition:

Element	Percentage/range
Carbon	0.35–0.40
Silicon	0.10–0.35
Manganese	0.70–1.00
Phosphorus	0.040
Sulfur	0.050
Nickel	...
Chromium	0.90–1.20
Molybdenum	0.15–0.25

AFNOR standards are developed by the Association Francaise de Normalisation in Paris, France. The correct format for reporting AFNOR standards, which are listed in Tables 31, 38, and 39, is as follows. An uppercase NF is placed to the left of the alphanumeric code. This code consists of an uppercase letter followed by a series of digits, which are subsequently followed by an alphanumeric sequence. For example, as indicated in Tables 31 and 38, NF A35-562 35MF6 is a resulfurized (free-cutting) steel with the following composition:

Element	Percentage/range
Carbon	0.33–0.39
Silicon	0.10–0.40
Manganese	1.10–1.70
Phosphorus	0.040
Sulfur	0.09–0.13

Table 26 Product descriptions and carbon contents for wrought carbon steels (AMS designations)

AMS designation	Product form	Carbon content	Nearest AISI-SAE grade	UNS number
5010H	Bars (screw machine stock)	...	1112	G12120
5020C	Bars, forgings, tubing	0.32–0.39(a)	11L37	G11374
5022K	Bars, forgings, tubing	0.14–0.20	1117	G11170
5024F	Bars, forgings, tubing	0.32–0.39(b)	1137	G11370
5032D	Wire (annealed)	0.18–0.23	1020	G10200
5036G	Sheet, strip (aluminum coated, low carbon)	...	...	...
5040H	Sheet, strip (deep-forming grade)	0.15 max	1010	G10100
5042H	Sheet, strip (forming grade)	0.15 max	1010	G10100
5044F	Sheet, strip (half-hard temper)	0.15 max	1010	G10100
5045E	Sheet, strip (hard temper)	0.25 max	1020	G10200
5046	Sheet, strip, plate (annealed)		1020	G10200
5047C	Sheet, strip (aluminum killed, deep forming grade)	0.08–0.13	1010	G10100
5050H	Tubing (seamless, annealed)	0.15 max	1010	G10100
5053F	Tubing (welded, annealed)	0.15 max	1010	G10100
5060F	Bars, forgings, tubing	0.13–0.18	1015	G10150
5061D	Bars, wire	Low	...	K00802
5062E	Bars, forgings, tubing, plate, sheet, strip	Low	...	K02508
5069D	Bars, forgings, tubing	0.15–0.20	1018	G10180
5070F	Bars, forgings	0.18–0.23	1022	G10220
5075E	Tubing (seamless, cold drawn, stress relieved)	0.22–0.28	1025	G10250
5077D	Tubing (welded)	0.22–0.28	1025	G10250
5080G	Bars, forgings, tubing	0.31–0.38	1035	G10350
5082D	Tubing (seamless, stress relieved)	0.31–0.38	1035	G10350
5085D	Plate, sheet, strip (annealed)	0.47–0.55	1050	G10500
5110E	Wire (carbon, spring temper, cold drawn)	0.75–0.88	1080	G10800
5112H	Wire (spring quality music wire, cold drawn)	0.70–1.00	1090	G10900
5115F	Wire (valve spring quality, hardened and tempered)	0.60–0.75	1070	G10700
5120J	Strip	0.68–0.80	1074	G10740
5121F	Sheet, strip	0.90–1.04	1095	G10950
5122F	Strip (hard temper)	0.90–1.04	1095	G10950
5132G	Bars	0.90–1.30	1095	G10950

(a) Contains 1.5% Mn and 0.025% Pb. (b) Contains 1.5% Mn

UNI standards are developed by the Ente Nazionale Italiano di Unificazione in Milan, Italy. Italian standards are preceded by the uppercase letters UNI followed by a four-digit product form code subsequently followed by an alphanumeric alloy identification (Tables 31, 40, and 41). For example, as indicated in Tables 31 and 40, UNI 5598 3CD5 is a low-carbon steel used for wire rod with the following composition:

Element	Percentage/range
Carbon	0.06
Silicon	...
Manganese	0.25–0.50
Phosphorus	0.035
Sulfur	0.035
Nitrogen	0.007

Swedish standards (SS_{14}) are prepared by the Swedish Standards Institution in Stockholm. Designations begin with the letters SS followed by the number 14 (all Swedish carbon and low-alloy steels are covered by SS_{14}). What subsequently follows is a four-digit numerical sequence similar to the German Werkstoff number. Swedish designations are listed in Tables 31, 42, and 43.

REFERENCES

1. "Chemical Compositions of SAE Carbon Steels," SAE J403, *1989 SAE Handbook*, Vol 1, *Materials*, Society of Automotive Engineers, p 1.08-1.10
2. "Alloy, Carbon and High Strength Low Alloy Steels: Semifinished for Forging; Hot Rolled Bars and Cold Finished Bars, Hot Rolled Deformed and Plain Concrete Reinforcing Bars," Steel Products Manual, American Iron and Steel Institute, March 1986
3. "Plates; Rolled Floor Plates: Carbon, High Strength Low Alloy, and Alloy Steel," Steel Products Manual, American Iron and Steel Institute, Aug 1985
4. "Standard Specification for General Requirements for Rolled Steel Plates, Shapes, Sheet Piling, and Bars for Structural Use," ASTM A 6/A 6M, American Society for Testing and Materials
5. *The Making, Shaping and Treating of Steel*, 10th ed., United States Steel Corporation, 1985
6. "Carbon and Alloy Steels," SAE J411, *1989 SAE Handbook* Vol 1, *Materials*, Society of Automotive Engineers, p 2.01-2.03
7. G. Krauss, *Steels—Heat Treatment and Processing Principles*, ASM INTERNATIONAL, 1989
8. W.C. Leslie, *The Physical Metallurgy of Steels*, McGraw-Hill, 1981
9. E.C. Bain and H.W. Paxton, *Alloying Elements in Steel*, American Society for Metals, 1966
10. A.K. Sinha, *Ferrous Physical Metallurgy*, Butterworths, 1989
11. R.W.K. Honeycombe, *Steels—Microstructure and Properties*, Edward Arnold Ltd., 1982
12. *Annual Statistical Report*, American Iron and Steel Institute, 1988 (copyright 1989)
13. O.D. Sherby, B. Walser, C.M. Young, and E.M. Cady, *Scr. Metall.*, Vol 9, 1975, p 569
14. T. Oyama, J. Wadsworth, M. Korchynsky, and O.D. Sherby, in *Proceedings of the Fifth International Conference on the Strength of Metals and Alloys*, International Series on the Strength and Fracture of Materials and Structures, Pergamon Press, 1980, p 381
15. L.F. Porter, High-Strength Low-Alloy Steels, in *Encyclopedia of Materials Science and Engineering*, MIT Press, 1986, p 2157-2162
16. "Chemical Compositions of SAE Alloy Steels," SAE J404, *1989 SAE Handbook*, Vol 1, *Materials*, Society of Automotive Engineers, p 1.10-1.12
17. "Potential Standard Steels," SAE J1081, *1989 SAE Handbook*, Vol 1, *Materials*, Society of Automotive Engineers, p 1.14-1.15
18. "High Strength Low Alloy Steel," SAE J310, *1989 SAE Handbook*, Vol 1, *Materials*, Society of Automotive Engineers, p 1.142-1.144
19. "Former SAE Standard and Former SAE EX-Steels," SAE J1249, *1989 SAE Handbook*, Vol 1, *Materials*, Society of Automotive Engineers, p 1.15-1.17
20. D.L. Potts and J.G. Gensure, *International Metallic Materials Cross-Reference*, Genium Publishing, 1989
21. C.W. Wegst, *Stahlschlüssel (Key to Steel)*, Verlag Stahlschlüssel Wegst GmbH, 1989

160 / Carbon and Low-Alloy Steels

Table 27 Product descriptions and nominal compositions for wrought alloy steels (AMS designations)

AMS designation	Product form(a)	Nominal composition, %					Nearest proprietary or AISI-SAE grade	UNS number
		C	Cr	Ni	Mo	Other		
6250H	Bars, forgings, tubing	0.07–0.13	1.5	3.5	...	...	3310	K44910
6255	Bars, forgings, tubing (P, DVM)	0.16–0.22	1.45	...	1.0	1.1 Si, 0.08 Al	CBS 600	K21940
6256A	Bars, forgings, tubing (P, DVM)	0.10–0.16	1.0	3.0	4.5	0.08 Al, 0.38 V	CBS 1000M	K71350
6260K	Bars, forgings, tubing (carburizing)	0.07–0.13	1.2	3.2	0.12	...	9310	G93106
6263G	Bars, forgings, tubing (carburizing grade, aircraft)	0.11–0.17	1.2	3.2	0.12	...	9315	...
6264G	Bars, forgings, tubing (carburizing)	0.14–0.20	1.2	3.2	0.12	...	9317	K44414
6265F	Bars, forgings, tubing (P, CVM)	0.07–0.13	1.2	3.25	0.12	...	9310	G93106
6266F	Bars, forgings, tubing	0.08–0.13	0.50	1.82	0.25	0.003 B, 0.06 V	43BV12	K21028
6267C	Bars, forgings, tubing	0.07–0.13	1.2	3.25	0.12	...	9310	G93106
6270L	Bars, forgings, tubing	0.11–0.17	0.50	0.55	0.20	...	8615	G86150
6272G	Bars, forgings, tubing	0.15–0.20	0.50	0.55	0.20	...	8617	G86170
6274K	Bars, forgings, tubing	0.18–0.23	0.50	0.55	0.20	...	8620	G86200
6275E	Bars, forgings, tubing	0.15–0.20	0.40	0.45	0.12	0.002 B	94B17	G94171
6276F	Bars, forgings, tubing (CVM)	0.18–0.23	0.50	0.55	0.2	...	8620	G86200
6277D	Bars, forgings, tubing (VAR, ESR)	0.18–0.23	0.50	0.55	0.20	...	8620	G86200
6278	Bars, forgings, tubing (for bearing applications; P, DVM)	0.11–0.15	4.1	3.4	4.2	1.2 V	...	...
6280G	Bars, forgings, rings	0.28–0.33	0.50	0.55	0.20	...	8630	G86300
6281F	Tubing (mechanical)	0.28–0.33	0.5	0.55	0.20	...	8630	G86300
6282F	Tubing (mechanical)	0.33–0.38	0.50	0.55	0.25	...	8735	G87350
6290F	Bars, forgings (carburizing)	0.11–0.17	...	1.8	0.25	...	4615	G46150
6292F	Bars, forgings (carburizing)	0.15–0.20	...	1.8	0.25	...	4617	G46170
6294F	Bars, forgings	0.17–0.22	...	1.8	0.25	...	4620	G42600
6299C	Bars, forgings, tubing	0.17–0.23	0.50	1.8	0.25	...	4320H	H43200
6300C	Bars, forgings	0.35–0.40	...	...	0.25	...	4037	G40370
6302E	Bars, forgings, tubing (low alloy, heat resistant)	0.28–0.33	1.25	...	0.50	0.65 Si, 0.25 V	17-22A(S)	K23015
6303D	Bars, forgings	0.25–0.30	1.25	...	0.50	0.65 Si, 0.85 V	17-22A(V)	K22770
6304G	Bars, forgings, tubing (low alloy, heat resistant)	0.40–0.50	0.95	...	0.55	0.30 V	17-22A	K14675
MAM 6304(b)	Bars, forgings, tubing (low alloy, heat resistant)	0.40–0.50	0.95	...	0.55	0.30 V	17-22A	K14675
6305A	Bars, forgings, tubing (VAR)	0.40–0.50	0.95	...	0.55	0.30 V	17-22A	K14675
6308A	Bars, forgings (VAR, ESR)	0.07–0.13	1.0	2.0	3.2	2.0 Cu, 0.10 V, 0.90 Si	Pyrowear alloy 53	K71040
6312E	Bars, forgings, tubing	0.38–0.43	...	1.8	0.25	...	4640	K22440
6317E	Bars, forgings (heat treated; 860 MPa tensile strength)	0.38–0.43	...	1.8	0.25	...	4640	K22440
6320H	Bars, forgings, rings	0.33–0.38	0.50	0.55	0.25	...	8735	G87350
6321D	Bars, forgings, tubing	0.38–0.43	0.42	0.30	0.12	0.003 B	81B40	K03810
6322K	Bars, forgings, rings	0.38–0.43	0.50	0.55	0.25	...	8740	G87400
6323G	Tubing (mechanical)	0.38–0.43	0.50	0.55	0.25	...	8740	G87400
6324E	Bars, forgings, tubing	0.38–0.43	0.65	0.70	0.25	...	8740 mod	K11640
6325F	Bars, forgings (heat treated; 725 MPa tensile strength)	0.38–0.43	0.50	0.55	0.25	...	8740	G87400
6327G	Bars, forgings (heat treated; 860 MPa tensile strength)	0.38–0.43	0.50	0.55	0.25	...	8740	G87400
6328H	Bars, forgings, tubing	0.48–0.53	0.50	0.55	0.25	...	8750	K13550
6330D	Bars, forgings, tubing	0.33–0.38	0.65	1.25	...	...	3135	K22033
6342G	Bars, forgings, tubing	0.38–0.43	0.80	1.0	0.25	...	9840	G98400
6348A	Bars (normalized)	0.28–0.33	0.95	...	0.20	...	4130	G41300
6349A	Bars (normalized)	0.38–0.43	0.95	...	0.20	...	4140	G41400
6350G	Plate, sheet, strip	0.28–0.33	0.95	...	0.20	...	4130	G41300
6351D	Plate, sheet, strip (spheroidized)	0.28–0.33	0.95	...	0.20	...	4130	G41300
6352E	Plate, sheet, strip	0.33–0.38	0.95	...	0.2	...	4135	G41350
6354C	Plate, sheet, strip	0.10–0.17	0.62	...	0.2	0.75 Si, 0.10 Zr	NAX 9115-AC	K11914
6355K	Plate, sheet, strip	0.28–0.33	0.50	0.55	0.20	...	8630	G86300
6356C	Plate, sheet, strip	0.30–0.35	0.95	...	0.20	...	4132	K13247
6357F	Plate, sheet, strip	0.33–0.38	0.50	0.5	0.25	...	8735	G87350
6358F	Plate, sheet, strip	0.38–0.43	0.50	0.55	0.25	...	8740	G87400
6359E	Plate, sheet, strip	0.38–0.43	0.80	1.8	0.25	...	4340	G43400
6360H	Tubing (seamless, normalized or stress relieved)	0.28–0.33	0.95	...	0.20	...	4130	G41300
6361B	Tubing (seamless, round; 860 MPa tensile strength)	0.28–0.33	0.95	...	0.2	...	4130	G41300
6362C	Tubing (seamless; 1035 MPa tensile strength)	0.28–0.33	0.95	...	0.2	...	4130	G41300
6365G	Tubing (seamless, normalized or stress relieved)	0.33–0.38	0.95	...	0.20	...	4135	G41350
6370J	Bars, forgings, rings	0.28–0.33	0.95	...	0.2	...	4130	G41300
6371G	Tubing (mechanical)	0.28–0.33	0.95	...	0.20	...	4130	G41300
6372G	Tubing (mechanical)	0.33–0.38	0.95	...	0.20	...	4135	G41350
6373C	Tubing (welded)	0.28–0.33	0.95	...	0.20	...	4130	G41300

(continued)

(a) P, premium quality; CVM, consumable vacuum melted; CVAR, consumable vacuum arc remelted; ESR, electroslag remelted; DVM, double vacuum melted; VAR, vacuum arc remelted; CM, consumable electrode remelted; VM, vacuum melted. (b) MAM, metric aerospace material specifications

Table 27 (continued)

AMS designation	Product form(a)	Nominal composition, %					Nearest proprietary or AISI-SAE grade	UNS number
		C	Cr	Ni	Mo	Other		
6378C	Bars (die drawn, free machining; 895 MPa yield strength)	0.39–0.48	0.95	...	0.20	0.015 Te	4142H	K11542
6379A	Bars (die drawn and tempered; 1140 MPa yield strength)	0.40–0.53	0.95	...	0.20	0.05 Te	4140 mod	G41400
6381D	Tubing	0.38–0.43	0.95	...	0.20	...	4140	G41400
6382J	Bars, forgings	0.38–0.43	0.95	...	0.20	...	4140	G41400
6385D	Plate, sheet, strip	0.27–0.33	1.25	...	0.50	0.65 Si, 0.25 V	17–22A(S)	K23015
6386B(1)	Plate, sheet (heat treated; 620 and 690 MPa yield strength)	0.15–0.21	0.50–0.80	...	0.18–0.28	0.40–0.80 Si, 0.05–0.15 Zr	...	K11856
6386B(2)	Plate, sheet (heat treated; 620 and 690 MPa yield strength)	0.12–0.21	0.40–0.65	...	0.15–0.25	0.20–0.35 Si, 0.01–0.03 Ti, 0.03–0.08 V, 0.0005–0.005 B	...	K11630
6386B(3)	Plate, sheet (heat treated; 620 and 690 MPa yield strength)	0.10–0.20	...	...	1.10–1.50	0.15–0.30 Si, 0.001–0.005 B	...	K11511
6386(4)	Plate, sheet (heat treated; 620 and 690 MPa yield strength)	0.13–0.20	0.85–1.20	...	0.15–0.25	0.20–0.40 Cu, 0.20–0.35 Si, 0.04–0.10 Ti, 0.0015–0.005 B	...	K11662
6386B(5)	Plate, sheet (heat treated; 620 and 690 MPa yield strength)	0.12–0.21	...	...	0.45–0.70	0.20–0.35 Si, 0.001–0.005 B	...	K11625
6390B	Tubing (mechanical, special surface quality)	0.38–0.43	0.95	...	0.20	...	4140	G41400
6395C	Plate, sheet, strip	0.38–0.43	0.95	...	0.20	...	4140	G41400
6396B	Sheet, strip, plate (annealed)	0.49–0.55	0.80	1.8	0.25	...	...	K22950
6406C	Plate, sheet, strip (annealed)	0.41–0.46	2.1	...	0.58	1.6 Si, 0.05 V	X200	K34378
6407D	Bars, forgings, tubing	0.27–0.33	1.2	2.05	0.45	...	HS-220	K33020
6408	Bars, forgings, tubing (annealed, ESR, CVM, VAR, P)	0.35–0.45	5.2	...	1.5	1.0 V	...	T20813
6409	Bars, forgings, tubing (normalized and tempered, quality cleanliness)	0.38–0.43	0.80	1.8	0.25	...	4340	G43400
6411C	Bars, forgings, tubing (CM)	0.28–0.33	0.85	1.8	0.40	...	4330 mod	K23080
6412H	Bars, forgings	0.35–0.40	0.80	1.8	0.25	...	4337	G43370
6413G	Tubing (mechanical)	0.35–0.40	0.80	1.8	0.25	...	4337	G43370
6414E	Bars, forgings, tubing (CVM)	0.38–0.43	0.80	1.8	0.25	...	4340	G43400
6415L	Bars, forgings, tubing	0.38–0.43	0.80	1.8	0.25	...	4340	G43400
MAM 6415(b)	Bars, forgings, tubing	0.38–0.43	0.80	1.8	0.25	...	4340	G43400
6416B	Superseded by AMS 6419							
6417C	Bars, forgings, tubing (CM)	0.38–0.43	0.82	1.8	0.40	1.6 Si, 0.08 V	300M	K44220
6418F	Bars, forgings, tubing, rings	0.23–0.28	0.30	1.8	0.40	1.3 Mn, 1.5 Si	Hy-tuf	K32550
6419C	Bars, forgings, tubing (CVM)	0.40–0.45	0.82	1.8	0.40	1.6 Si, 0.08 V	300M	K44220
6421B	Bars, forgings, tubing	0.35–0.40	0.80	0.85	0.20	0.003 B	98B37 mod	...
6422E	Bars, forgings, tubing	0.38–0.43	0.80	0.85	0.20	0.003 B	98BV40 mod	K11940
6423C	Bars, forgings, tubing	0.40–0.46	0.92	0.75	0.52	0.003 B	...	K24336
6424B	Bars, forgings, tubing	0.49–0.59	0.80	1.8	0.25	...	...	K22950
6426C	Bars, forgings, tubing (CVM)	0.80–0.90	1.0	...	0.58	0.75 Si	52CB	K18597
6427G	Bars, forgings, tubing	0.28–0.33	0.85	1.8	0.42	0.08 V	4330 mod	K23080
6428D	Bars, forgings, tubing	0.32–0.38	0.80	1.8	0.35	0.20 V	4335 mod	K23477
6429C	Bars, forgings, tubing, rings (CVM)	0.33–0.38	0.78	1.8	0.35	0.20 V	4335 mod	K33517
6430C	Bars, forgings, tubing, rings (special grade)	0.32–0.38	0.78	1.8	0.35	0.20 V	4335 mod	K33517
6431G	Bars, forgings, tubing (CVM)	0.45–0.50	1.05	0.55	1.0	0.11 V	D6AC	K24728
6432A	Bars, forgings, tubing	0.43–0.49	1.05	0.55	1.0	0.12 V	D6A	K24728
6433C	Plate, sheet, strip (special grade)	0.33–0.38	0.80	1.8	0.35	0.20 V	4335 mod	K33517
6434C	Plate, sheet, strip	0.33–0.38	0.78	1.8	0.35	0.20 V	4335 mod	K33517
6435C	Plate, sheet, strip (P, CM, annealed)	0.33–0.38	0.78	1.8	0.35	0.20 V	4335 mod	K33517
6436B	Plate, sheet, strip (low alloy, heat resistant, annealed)	0.25–0.30	1.25	...	0.50	0.65 Si, 0.85 V	17–22A(V)	K22770
6437D	Plate, sheet, strip	0.38–0.43	5.0	...	1.3	0.5 V	H-11	T20811
6438C	Plate, sheet, strip (P, CM)	0.45–0.50	1.05	0.55	1.0	0.11 V	D6AC	K24728
6439B	Sheet, strip, plate (annealed, CVM)	0.42–0.48	1.05	0.55	1.0	0.12 V	D6AC	K24729
6440J	Bars, forgings tubing (for bearing applications)	0.98–1.10	1.45	...	...	...	52100	G52986
6441G	Superseded by AMS 6440							
6442E	Bars, forgings (for bearing applications)	0.98–1.10	0.50	...	...	...	50100	G50986
6443E	Bars, forgings, tubing (CVM)	0.98–1.10	1.0	...	...	...	51100	G51986
6444G	Bars, wire, forgings, tubing (P, CVM)	0.98–1.10	1.45	...	...	...	52100	G52986
6445E	Bars, wire, forgings, tubing (CVM)	0.92–1.02	1.05	...	...	1.1 Mn	51100 mod	K22097
6446C	Bars, forgings (ESR)	0.98–1.10	1.00	...	...	...	51100	G51986
6447C	Bars, forgings, tubing (ESR)	0.98–1.10	1.45	...	...	...	52100	G52986

(continued)

(a) P, premium quality; CVM, consumable vacuum melted; CVAR, consumable vacuum arc remelted; ESR, electroslag remelted; DVM, double vacuum melted; VAR, vacuum arc remelted; CM, consumable electrode remelted; VM, vacuum melted. (b) MAM, metric aerospace material specifications

162 / Carbon and Low-Alloy Steels

Table 27 (continued)

AMS designation	Product form(a)	C	Cr	Ni	Mo	Other	Nearest proprietary or AISI-SAE grade	UNS number
6448F	Bars, forgings, tubing	0.48–0.53	0.95	...	...	0.22 V	6150	G61500
6449C	Bars, forgings, tubing (for bearing applications)	0.98–1.10	1.0	...	...	...	51100	G51986
6450E	Wire (spring)	0.48–0.53	0.95	...	...	0.22 V	6150	G61500
6451A	Wire, spring (oil tempered)	0.51–0.59	0.65	...	...	1.4 Si	9254	G92540
6454	Sheet, strip, plate (P, CM)	0.38–0.43	0.80	1.8	0.25	...	4340	G43400
6455F	Plate, sheet, strip	0.48–0.53	0.95	...	...	0.22 V	6150	G61500
6470H	Bars, forgings, tubing (nitriding)	0.38–0.43	1.6	...	0.35	1.1 Al	135 mod	K24065
6471C	Bars, forgings, tubing (nitriding, CVM)	0.38–0.43	1.6	...	0.35	1.2 Al	135 mod	K24065
6472B	Bars, forgings (nitriding, heat treated; 770 MPa tensile strength)	0.38–0.43	1.6	...	0.35	1.1 Al	135 mod	K24065
6475E	Bars, forgings, tubing (nitriding)	0.21–0.26	1.1	3.5	0.25	1.25 Al	...	K52355
6485	Bars, forgings	0.38–0.43	5.0	...	1.3	0.50 V	H-11	T20811
6487	Bars, forgings (P, CVM)	0.38–0.43	5.0	...	1.3	0.50 V	H-11	T20811
6488D	Bars, forgings (P)	0.38–0.43	5.0	...	1.3	0.50 V	H-11	T20811
6490D	Bars, forgings, tubing (for bearing applications; P, CVM)	0.77–0.85	4.0	...	4.2	1.0 V	M-50	T11350
6491A	Bars, forgings, tubing (for bearing applications; P, DVM)	0.80–0.85	4.1	...	4.2	1.0 V	M-50	T11350
6512B	Bars, forgings, tubing, rings (annealed)	...	...	18	4.9	7.8 Co, 0.40 Ti, 0.10 Al	Maraging 250	K92890
6514B	Bars, forgings, tubing, rings (annealed, CM)	...	...	18.5	4.9	9.0 Co, 0.65 Ti, 0.10 Al	Maraging 300	K93120
6518A	Sheet, strip, plate (solution treated, DVM)	...	...	19.0	3.0	0.10 Al, 1.4 Ti	...	...
6519A	Bars, forgings, tubing, springs (annealed, DVM)	...	...	19.0	3.0	0.10 Al, 1.4 Ti	...	...
6520B	Plate, sheet, strip (solution heat treated, CM)	...	...	18	4.9	7.8 Co, 0.40 Ti, 0.10 Al	Maraging 250	K92890
6521A	Plate, sheet, strip (solution heat treated, CM)	...	...	18.5	4.9	9.0 Co, 0.65 Ti, 0.10 Al	Maraging 300	K93120
6522	Plate (P, VM)	...	2.0	10.0	1.0	14.0 Co	AF 1410	K92571
6523C	Sheet, strip, plate (annealed, CVM)	0.17–0.23	0.75	9.0	1.0	0.09 V, 4.5 Co	HP 9-4-20	K91472
6524B	Sheet, strip, plate (annealed, CVM)	0.29–0.34	1.0	7.5	1.0	0.09 V, 4.5 Co	HP 9-4-30	K91283
6525A	Bars, forgings, tubing, rings (CVM)	0.17–0.23	0.75	9.0	1.0	0.09 V, 4.5 Co	HP 9-4-20	K91283
6526C	Bars, forgings, tubing, rings (annealed, CVM)	0.29–0.34	1.0	7.5	1.0	4.5 Co, 0.09 V	HP 9-4-30	K91313
6527	Bars, forgings (P, VM)	0.13–0.17	2.0	10.0	1.0	14 Co	AF 1410	K92571
6528	Bars (normalized, special aircraft quality cleanliness)	0.28–0.33	0.95	...	0.20	...	4130	G41300
6529	Bars (normalized, special aircraft quality cleanliness)	0.38–0.43	0.95	...	0.20	...	4140	G41400
6530H	Tubing (seamless)	0.28–0.33	0.55	0.50	0.20	...	8630	G86300
6535G	Tubing (seamless)	0.28–0.33	0.50	0.55	0.20	...	8630	G86300
6543A	Bars, forgings (solution treated, DVM)	0.10–0.14	2.0	10.0	1.0	8.0 Co	...	K91970
6544	Plate (solution treated, VM)	0.10–0.14	2.0	10.0	1.0	8.0 Co	...	K92571
6546C	Plate, sheet, strip (annealed, P, CM)	0.24–0.30	0.48	8.0	0.48	4.0 Co, 0.09 V	HP 9-4-25	K91122
6550H	Tubing (welded)	0.28–0.33	0.55	0.50	0.20	...	8630	G86300

(a) P, premium quality; CVM, consumable vacuum melted; CVAR, consumable vacuum arc remelted; ESR, electroslag remelted; DVM, double vacuum melted; VAR, vacuum arc remelted; CM, consumable electrode remelted; VM, vacuum melted. (b) MAM, metric aerospace material specifications

Table 28 ASTM specifications that incorporate AISI-SAE designations

A 29 ...... Carbon and alloy steel bars, hot rolled and cold finished
A 108 ..... Standard quality cold-finished carbon steel bars
A 295 ..... High carbon-chromium ball and roller bearing steel
A 304 ..... Alloy steel bars having hardenability requirements
A 322 ..... Hot-rolled alloy steel bars
A 331 ..... Cold-finished alloy steel bars
A 434 ..... Hot-rolled or cold-finished quenched and tempered alloy steel bars
A 505 ..... Hot-rolled and cold-rolled alloy steel sheet and strip
A 506 ..... Regular quality hot-rolled and cold-rolled alloy steel sheet and strip
A 507 ..... Drawing quality hot-rolled and cold-rolled alloy steel sheet and strip
A 510 ...... Carbon steel wire rods and coarse round wire
A 534 ...... Carburizing steels for antifriction bearings
A 535 ...... Special quality ball and roller bearing steel
A 544 ...... Scrapless nut quality carbon steel wire
A 545 ...... Cold-heading quality carbon steel wire for machine screws
A 546 ...... Cold-heading quality medium high carbon steel wire for hexagon-head bolts
A 547 ...... Cold-heading quality alloy steel wire for hexagon-head bolts
A 548 ...... Cold-heading quality carbon steel wire for tapping or sheet metal screws
A 549 ...... Cold-heading quality carbon steel wire for wood screws
A 575 ...... Merchant quality hot-rolled carbon steel bars
A 576 ...... Special quality hot-rolled carbon steel bars
A 646 ...... Premium quality alloy steel blooms and billets for aircraft and aerospace forgings
A 659 ...... Commercial quality hot-rolled carbon steel sheet and strip
A 682 ...... Cold-rolled spring quality carbon steel strip, generic
A 684 ...... Untempered cold-rolled high-carbon steel strip
A 689 ...... Carbon and alloy steel bars for springs
A 711 ...... Carbon and alloy steel blooms, billets, and slabs for forging
A 713 ...... High-carbon spring steel wire for heat-treated components
A 752 ...... Alloy steel wire rods and coarse round wire
A 827 ...... Carbon steel plates for forging and similar applications
A 829 ...... Structural quality alloy steel plates
A 830 ...... Structural quality carbon steel plates

Table 29   General ASTM specifications (in boldface) with corresponding supplementary specifications used to describe primary carbon and low-alloy steel product forms

ASTM designation	Title of specification
ASTM A 6/A 6M: General requirements for rolled steel plates, shapes, sheet piling, and bars for structural use(a)	
A 36/A 36M	Structural steel
A 131/A 131M	Structural steel for ships
A 242/A 242M	High-strength low-alloy structural steel
A 283/A 283M	Low and intermediate tensile strength carbon steel plates, shapes, and bars
A 284/A 284M	Low and intermediate tensile strength carbon-silicon steel plates for machine parts and general construction
A 328/A 328M	Steel sheet piling
A 441/A 441M	High-strength low-alloy structural manganese-vanadium steel
A 514/A 514M	High yield strength, quenched and tempered alloy steel plate suitable for welding
A 529/A 529M	Structural steel with 290 MPa (42 ksi) minimum yield point (12.7 mm, or ½ in., maximum thickness)
A 572/A 572M	High-strength low-alloy niobium-vanadium steels of structural quality
A 573/A 573M	Structural carbon steel plates of improved toughness
A 588/A 588M	High-strength low-alloy structural steel with 345 MPa (50 ksi) minimum yield point to 100 mm (4 in.) thick
A 633/A 633M	Normalized high-strength low-alloy structural steel
A 656/A 656M	Hot-rolled structural steel, high-strength low-alloy plate with improved formability
A 678/A 678M	Quenched and tempered carbon steel plates for structural applications
A 690/A 690M	High-strength low-alloy steel H-piles and sheet piling for use in marine environments
A 709	Structural steel for bridges
A 710/A 710M	Low-carbon age-hardening nickel-copper-chromium-molybdenum-niobium and nickel-copper-niobium alloy steels
A 769/A 769M	Electric resistance welded steel shapes
A 786/A 786M	Rolled steel floor plates
A 808/A 808M	High-strength low-alloy carbon-manganese-niobium-vanadium steel of structural quality with improved notch toughness
A 827	Plates, carbon steel, for forging and similar applications
A 829	Plates, alloy steel, structural quality
A 830	Plates, carbon steel, structural quality, furnished to chemical composition requirements
A 852/A 852M	Quenched and tempered low-alloy structural steel plate
A 857	Steel sheet piling, cold formed, light gage
A 871/A 871M	High-strength low-alloy structural steel plate with atmospheric corrosion resistance
ASTM A 20/A 20M: General requirements for steel plates for pressure vessels(a)	
A 202/A 202M	Pressure vessel plates, alloy steel, chromium-manganese-silicon
A 203/A 203M	Pressure vessel plates, alloy steel, nickel
A 204/A 204M	Pressure vessel plates, alloy steel, molybdenum
A 225/A 225M	Pressure vessel plates, alloy steel, manganese-vanadium
A 285/A 285M	Pressure vessel plates, carbon steel, low and intermediate tensile strength
A 299/A 299M	Pressure vessel plates, carbon steel, manganese-silicon
A 302/A 302M	Pressure vessel plates, alloy steel, manganese-molybdenum and manganese-molybdenum-nickel
A 353/A 353M	Pressure vessel plates, alloy steel, 9% Ni, double normalized and tempered
A 387/A 387M	Pressure vessel plates, alloy steel, chromium-molybdenum
A 442/A 442M	Pressure vessel plates, carbon steel, improved transition properties
A 455/A 455M	Pressure vessel plates, carbon steel, high-strength manganese
A 515/A 515M	Pressure vessel plates, carbon steel, for intermediate- and higher-temperature service
A 516/A 516M	Pressure vessel plates, carbon steel, for moderate- and lower-temperature service
ASTM A 20/A 20M: General requirements for steel plates for pressure vessels(a) (continued)	
A 517/A 517M	Pressure vessel plates, alloy steel, high strength, quenched and tempered
A 533/A 533M	Pressure vessel plates, alloy steel, quenched and tempered manganese-molybdenum and manganese-molybdenum-nickel
A 537/A 537M	Pressure vessel plates, heat-treated carbon-manganese-silicon steel
A 538/A 538M	Pressure vessel plates, alloy steel, precipitation hardening (maraging), 18% Ni
A 542/A 542M	Pressure vessel plates, alloy steel, quenched and tempered chromium-molybdenum
A 543/A 543M	Pressure vessel plates, alloy steel, quenched and tempered nickel-chromium-molybdenum
A 553/A 553M	Pressure vessel plates, alloy steel, quenched and tempered 8 and 9% Ni
A 562/A 562M	Pressure vessel plates, carbon steel, manganese-titanium, for glass or diffused metallic coatings
A 605/A 605M	Pressure vessel plates, alloy steel, quenched and tempered nickel-cobalt-molybdenum-chromium
A 612/A 612M	Pressure vessel plates, carbon steel, high strength, for moderate- and lower-temperature service
A 645/A 645M	Pressure vessel plates, 5% Ni, alloy steel, specially heat treated
A 662/A 662M	Pressure vessel plates, carbon-manganese, for moderate- and lower-temperature service
A 724/A 724M	Pressure vessel plates, carbon steel, quenched and tempered, for welded layered pressure vessels
A 734/A 734M	Pressure vessel plates, alloy steel and high-strength low-alloy steel, quenched and tempered
A 735/A 735M	Pressure vessel plates, low-carbon manganese-molybdenum-niobium alloy steel, for moderate- and lower-temperature service
A 736/A 736M	Pressure vessel plates, low-carbon age-hardening nickel-copper-chromium-molybdenum-niobium alloy steel
A 737/A 737M	Pressure vessel plates, high-strength low-alloy steel
A 738/A 738M	Pressure vessel plates, heat-treated carbon-manganese-silicon steel, for moderate- and lower-temperature service
A 782/A 782M	Pressure vessel plates, quenched and tempered manganese-chromium-molybdenum-silicon-zirconium alloy steel
A 832/A 832M	Pressure vessel plates, alloy steel, chromium-molybdenum-vanadium-titanium-boron
A 841/A 841M	Pressure vessel plates, produced by the thermomechanical control process
A 844/A 844M	Pressure vessel plates 9% Ni alloy, produced by the direct-quenching process
ASTM A29/A 29M(b): General requirements for hot-rolled and cold-finished carbon and alloy steel bars(c)	
Hot-rolled carbon steel bars	
A 321	Steel bars, carbon, quenched and tempered
A 499	Steel bars and shapes rerolled from rail steel, hot-rolled carbon
A 575(b)	Steel bars, merchant quality, hot-wrought carbon
A 576(b)	Steel bars, carbon, hot wrought, special quality
A 663	Steel bars, merchant quality, hot-wrought carbon, subject to mechanical property requirements
A 675	Steel bars and bar-size shapes, carbon, hot wrought, special quality, subject to mechanical property requirements
A 689(b)	Steel bars for springs, carbon and alloy
A 695	Steel bars, carbon, hot wrought, special quality, for fluid power applications
Cold-finished carbon steel bars	
A 108(b)	Steel bars, carbon, cold finished, standard quality
A 311	Steel bars subject to mechanical property requirements, stress relief annealed cold-drawn carbon
Hot-rolled alloy steel bars	
A 322(b)	Steel bars, hot-wrought alloy
A 304(b)	Steel bars, alloy, subject to end-quench hardenability requirements

(continued)

(a) From *Annual Book of ASTM Standards*, Vol 01.04. (b) Specification incorporates SAE-AISI designations as indicated in Table 28. (c) From *Annual Book of ASTM Standards*, Vol 01.05. (d) From *Annual Book of ASTM Standards*, Vol 01.01. (e) From *Annual Book of ASTM Standards*, Vol 01.03

Table 29 (continued)

ASTM designation	Title of specification
ASTM A29/A 29M(b): General requirements for hot-rolled and cold-finished carbon and alloy steel bars(c) (continued)	
Hot-rolled alloy steel bars (continued)	
A 434(b)	Steel bars, alloy, hot wrought or cold finished, quenched and tempered
A 739	Steel bars, alloy, hot wrought, for elevated-temperature or pressure-containing parts, or both
Cold-finished alloy steel bars	
A 331(b)	Steel bars, alloy, cold finished
A 434(b)	Steel bars, alloy, hot rolled or cold finished, quenched and tempered
A 696	Steel bars, carbon, hot rolled and cold finished, special quality, for pressure piping components and other pressure-containing parts
ASTM A 450/A 450M: General requirements for carbon, ferritic-alloy, and austenitic-alloy steel tubes(d)	
A 161	Seamless low-carbon and carbon-molybdenum steel still tubes for refinery service
A 178/A 178M	Electric resistance welded carbon steel boiler tubes
A 179/A 179M	Seamless cold-drawn low-carbon steel heat exchanger and condenser tubes
A 192/A 192M	Seamless carbon steel boiler tubes for high-pressure service
A 199/A 199M	Seamless cold-drawn intermediate alloy steel heat exchanger and condenser tubes
A 200	Seamless intermediate alloy steel still tubes for refinery service
A 209/A 209M	Seamless carbon-molybdenum alloy steel boiler and superheater tubes
A 210/A 210M	Seamless medium-carbon steel boiler and superheater tubes
A 213/A 213M	Seamless ferritic and austenitic alloy steel boiler, superheater, and heat exchanger tubes
A 214/A 214M	Electric resistance welded carbon steel heat exchanger and condenser tubes
A 226/A 226M	Electric resistance welded carbon steel boiler and superheater tubes for high-pressure service
A 249/A 249M	Welded austenitic steel boiler, superheater, heat exchanger, and condenser tubes
A 250/A 250M	Electric resistance welded carbon-molybdenum alloy steel boiler and superheater tubes
A 268/A 268M	Seamless and welded ferritic stainless steel tubing for general service
A 269	Seamless and welded austenitic stainless steel tubing for general service
A 270	Seamless and welded austenitic stainless steel sanitary tubing
A 271	Seamless austenitic chromium-nickel steel still tubes for refinery service
A 334/A 334M	Seamless and welded carbon and alloy steel tubes for low-temperature service
A 423/A 423M	Seamless and electric welded low-alloy steel tubes
A 539	Electric resistance welded coiled steel tubing for gas and fuel oil lines
A 556/A 556M	Seamless cold-drawn carbon steel feedwater heater tubes
A 557/A 557M	Electric resistance welded carbon steel feedwater heater tubes
A 688/A 688M	Welded austenitic stainless steel feedwater heater tubes
A 692	Seamless medium-strength carbon-molybdenum alloy steel boiler and superheater tubes
A 771	Austenitic stainless steel tubing for breeder reactor core components
A 789/A 789M	Seamless and welded ferritic and austenitic stainless steel tubing for general service
A 791/A 791M	Welded unannealed ferritic stainless steel tubing
A 803/A 803M	Welded ferritic stainless steel feedwater heater tubes
A 822	Seamless, cold-drawn carbon steel piping tubing for hydraulic system service
A 826	Austenitic and ferritic stainless steel duct tubes for breeder reactor core components
A 851	High-frequency induction-welded unannealed austenitic steel condenser tubes
ASTM A 530/A 530M: General requirements for specialized carbon and alloy steel pipe(d)	
A 106	Seamless carbon steel pipe for high-temperature service
A 312/A 312M	Seamless and welded austenitic stainless steel pipe
A 333/A 333M	Seamless and welded steel pipe for low-temperature service
A 335/A 335M	Seamless ferritic alloy steel pipe for high-temperature service
A 358/A 358M	Electric fusion welded austenitic chromium-nickel alloy steel pipe for high-temperature service
A 369/A 369M	Carbon and ferritic alloy steel forged and bored pipe for high-temperature service
A 376/A 376M	Seamless austenitic steel pipe for high-temperature central station service
A 381	Metal arc welded steel pipe for high-pressure transmission systems
A 405	Seamless ferritic alloy steel pipe specially heat treated for high-temperature service
A 409/A 409M	Welded large-diameter austenitic steel pipe for corrosive or high-temperature service
A 426	Centrifugally cast ferritic alloy steel pipe for high-temperature service
A 430/A 430M	Austenitic steel forged and bored pipe for high-temperature service
A 451	Centrifugally cast austenitic steel pipe for high-temperature service
A 452	Centrifugally cast austenitic steel cold-wrought pipe for high-temperature service
A 524	Seamless carbon steel pipe for atmospheric and lower temperatures
A 608	Centrifugally cast iron-chromium-nickel high-alloy tubing for pressure application at high temperatures
A 660	Centrifugally cast carbon steel pipe for high-temperature service
A 671	Electric fusion welded steel pipe for atmospheric and lower temperatures
A 672	Electric fusion welded steel pipe for high-pressure service at moderate temperatures
A 691	Carbon and alloy steel pipe electric fusion welded for high-pressure service at high temperatures
A 731/A 731M	Seamless and welded ferritic stainless steel pipe
A 790/A 790M	Seamless and welded ferritic and austenitic stainless steel pipe
A 813/A 813M	Single- or double-welded austenitic stainless steel pipe
A 814/A 814M	Cold-worked welded austenitic stainless steel pipe
A 872	Centrifugally cast ferritic and austenitic stainless steel pipe for corrosive environments
ASTM A682: General requirements for cold-rolled spring quality high-carbon steel strip(e)	
A 680	Steel, carbon strip, cold rolled hard, untempered spring quality
A 684	Steel, carbon strip, cold rolled soft, untempered spring quality
ASTM A 749: General requirements for hot-rolled carbon and high-strength low-alloy steel strip(e)	
ASTM A 505(b): General requirements for hot-rolled and cold-rolled alloy steel sheet and strip(e)	
A 506(b)	Alloy steel sheet and strip, hot rolled and cold rolled, regular quality and structural quality
A 507(b)	Alloy steel sheet and strip, hot rolled and cold rolled, drawing quality
A 873/A 873M	Chromium-molybdenum alloy steel sheet and strip for pressure vessels
ASTM A 505(b): General requirements for carbon steel wire rods and coarse round wire(e)	
ASTM A 752(b): General requirements for alloy steel wire rods and coarse bound wire(e)	

(a) From *Annual Book of ASTM Standards*, Vol 01.04. (b) Specification incorporates SAE-AISI designations as indicated in Table 28. (c) From *Annual Book of ASTM Standards*, Vol 01.05. (d) From *Annual Book of ASTM Standards*, Vol 01.01. (e) From *Annual Book of ASTM Standards*, Vol. 01.03

Table 30 World production of raw steel by country

Country	1988 Net tons × 10³	%	1987 Net tons × 10³	%	1986 Net tons × 10³	%
North America						
United States(a)	99 924	11.6	89 151	11.0	81 606	10.3
Canada	16 597	1.9	16 118	2.0	15 419	2.0
Total	116 521	13.6	105 269	13.0	97 025	12.3
Latin America						
Argentina	3 993	...	3 979	...	3 566	...
Brazil	27 133	...	24 502	...	23 405	...
Chile	987	...	793	...	778	...
Columbia	799	...	762	...	697	...
Cuba(b)	496	...	468	...	459	...
Mexico	8 591	...	8 346	...	7 901	...
Peru	551	...	502	...	537	...
Venezuela	3 979	...	4 103	...	3 750	...
Others	745	...	632	...	558	...
Total	47 274	5.5	44 087	5.4	41 651	5.3
Western Europe (EEC)						
Belgium	12 390	...	10 827	...	10 707	...
Luxembourg	4 033	...	3 640	...	4 084	...
Denmark	716	...	667	...	697	...
France	20 947	...	19 500	...	19 684	...
West Germany	45 209	...	39 956	...	40 933	...
Ireland	298	...	243	...	229	...
Italy	26 090	...	25 153	...	25 223	...
Netherlands	6 109	...	5 602	...	5 823	...
United Kingdom	20 958	...	19 208	...	16 231	...
Greece	1 604	...	1 000	...	1 113	...
Portugal(c)	884	...	795	...	783	...
Spain(c)	12 880	...	13 021	...	13 098	...
Total EEC	151 578	17.6	139 612	17.2	138 605	17.6
Other Western Europe						
Austria	5 034	...	4 741	...	4 731	...
Finland	3 080	...	2 942	...	2 851	...
Norway	1 000	...	937	...	933	...
Sweden	5 268	...	5 065	...	5 198	...
Switzerland	909	...	959	...	1 185	...
Turkey	8 832	...	7 769	...	6 534	...
Yugoslavia(b)	4 927	...	4 814	...	4 982	...
Total other Western Europe	29 050	3.4	27 227	3.4	26 414	3.3
Total Western Europe	180 628	21.0	166 839	20.6	165 019	20.9
Eastern Europe (communist bloc)						
Bulgaria	3 307	...	3 355	...	3 194	...
Czechoslovakia	16 975	...	16 993	...	16 658	...
East Germany	9 094	...	9 086	...	8 782	...
Hungary	3 858	...	3 993	...	4 095	...
Poland	18 739	...	18 898	...	18 898	...
Rumania	16 535	...	16 493	...	15 736	...
U.S.S.R.	180 777	...	178 501	...	177 123	...
Total Eastern Europe	249 285	29.0	247 319	30.5	244 486	31.0
Total Europe	429 913	50.0	414 158	51.1	409 505	51.9
Africa						
Zimbabwe	664	...	658	...	743	...
Republic of South Africa	9 647	...	9 620	...	9 805	...
Others	2 041	...	2 036	...	2 046	...
Total	12 352	1.4	12 314	1.5	12 594	1.6
Middle East						
Egypt	1 874	...	1 764	...	1 102	...
Saudi Arabia	1 505	...	1 505	...	1 213	...
Iran	1 378	...	1 378	...	1 213	...
Others	928	...	891	...	901	...
Total	5 685	0.7	5 538	0.7	4 429	0.6
Asia						
India	15 652	...	14 438	...	13 445	...
Japan	116 482	...	108 591	...	108 329	...
South Korea	21 069	...	18 499	...	16 044	...
Taiwan	9 163	...	6 379	...	6 112	...
Others	5 456	...	5 462	...	4 802	...
Total	167 822	19.5	153 369	18.9	148 732	18.9
Asia (communist bloc)						
China	65 036	...	61 751	...	57 209	...
North Korea	7 441	...	7 418	...	10 130	...
Total	72 477	8.4	69 169	8.5	67 339	8.5
Oceania						
Australia	6 944	...	6 724	...	7 357	...
New Zealand	551	...	451	...	316	...
Total	7 495	0.9	7 175	0.9	7 673	1.0
Total free world	532 354	61.9	489 309	60.3	471 682	59.8
Total communist bloc	327 185	38.1	321 770	39.7	317 266	40.2
Total world production	859 539	100.0	811 079	100.0	788 948	100.0

(a) United States data exclude steel produced by foundries which report their output to the Bureau of Census but not to the American Iron and Steel Institute as follows (in net tons × 10³): 1988, 1015; 1987, 830; 1986, 829. (b) Communist bloc. (c) Prior to 1986 included in other Western Europe. Source: Ref 12

166 / Carbon and Low-Alloy Steels

Table 31 Cross-reference to steels

This table cross-references standard SAE carbon and low-alloy steels to selected chemically similar steels from specifications established by standards organizations from the Federal Republic of Germany, Japan, the United Kingdom, France, Italy, and Sweden. Chemical compositions of the non-SAE steels listed in this table can be found in Tables 32 to 43.

United States (SAE)	Fed. R. of Germany (DIN)	Japan (JIS)	United Kingdom (BS)	France (AFNOR NF)	Italy (UNI)	Sweden (SS_{14})
Carbon steels						
1005	1.0288, D5-2 1.0303, QSt32-2 1.0312, D5-1 1.0314, D6-2 1.0393, ED3 1.0394, ED4 1.1012, RFe120	...	970 015A03	...	5598 3CD5	1160
1006	1.0311, D7-1 1.0313, D8-2 1.0317, RSD4 1.0321, St23 1.0334, StW23 1.0335, StW24 1.0354, St14Cu3 1.0391, EK2 1.0392, EK4 1.1009, Ck7	...	970 030A04 970 040A04 970 050A04	A35-564 XC6FF	5598 3CD6 5771 C8	1147 1225
1008	1.0010, D9 1.0318, St28 1.0320, St22 1.0322, USD8 1.0326, RSt28 1.0330, St2, St12 1.0333, St3, St13 1.0331, RoSt2 1.0332, StW22 1.0336, USt4, USt14 1.0337, RoSt4 1.0344, St12Cu3 1.0347, RRSt13 1.0357, USt28 1.0359, RRSt23 1.0375, Feinstblech T57, T61, T65, T70 1.0385, Weissblech T57, T61, T65, T70 1.0744, 6P10 1.0746, 6P20 1.1116, USD6	G3445 STKM11A (11A)	1449 3CR 1449 3CS 1449 3HR 1449 3HS 1717 ERW101 3606 261	A35-551 XC10 XC6 XC6FF	5598 3CD8	1142 1146
1010	1.0204, UQSt36 1.0301, C10 1.0328, USD10 1.0349, RSD9 1.1121, Ck10 1.1122, Cq10	G4051 S10C G4051 S9Ck	1449 40F30, 43F35, 46F40, 50F45, 60F55, 68F62, 75F70 (available in HR, HS, CS conditions) 1449 4HR, 4HS, 4CR, 4CS 970 040A10 (En2A, En2A/1, En2B) 970 045A10, 045M10 (En32A) 970 050A10 970 060A10 980 CEW1	A33-101 AF34 CC10 C10	5331 C10 6403 C10 7065 C10 7846 C10 5598 1CD10 5598 3CD12 5771 C12 7356 CB10FF, CB10FU	1232 1265 1311
1012	1.0439, RSD13	G4051 S12C	1449 12HS, 12CS 1501 141-360 970 040A12 (En2A, En2A/1, En2B) 970 050A12 970 060A12	A33-101 AF37 A-35 551 XC12 C12	...	1332 1431
1013	1.0036, USt37-2 1.0037, St37-2 1.0038, RSt37-2 1.0055, USt34-1 1.0057, RSt34-1 1.0116, St37-3 1.0218, RSt41-2 1.0219, St41-3 1.0307, StE210.7 1.0309, St35.4 1.0315, St37.8 1.0319, RRStE210.7 1.0356, TTSt35 1.0417 1.0457, StE240.7	...	3059 360 3061 360 3603 360	A35-551 XC12 CC12	5869 Fe360-1KG, Fe360-2KW 6403 Fe35-2 7070 Fe34CFN 7091 Fe34	1233 1234 1330

(continued)

Table 31 (continued)

United States (SAE)	Fed. R. of Germany (DIN)	Japan (JIS)	United Kingdom (BS)	France (AFNOR NF)	Italy (UNI)	Sweden (SS₁₄)
Carbon steels (continued)						
1015	1.0401, C15 1.1132, CQ15 1.1135, Ck16Al 1.1140, Cm15 1.1141, Ck15 1.1144 1.1148, Ck15Al	G4051 F15Ck G4051 S15C	970 040A15 970 050A15 970 060A15 970 080A15, 080M15 970 173H16	XC15	5331 C16 7065 C16 7356 CB15 7846 C15	1370
1016	1.0419, RSt44.2 1.0467, 15Mn3 1.0468, 15Mn3Al 1.1142, GS-Ck16	. . .	3059 440 3606 440 970 080A15, 080M15 970 170H15 970 173H16	. . .	. . .	1370 2101
1017	. . .	G4051 S17C	1449 17HS, 17CS 970 040A17 970 050A17 970 060A17	A35-551 XC18 A35-552 XC18 A35-566 XC18 A35-553 XC18S A35-554 XC18S	. . .	1312
1018	1.0453, C16.8	. . .	970 080A17	A33-101 AF42 C20	. . .	. . .
1019	. . .	. . .	. . .	. . .	. . .	. . .
1020	1.0402, C22 1.0414, D20-2 1.0427, C22.3 1.0460, C22.8 1.1149, Cm22 1.1151, Ck22	G4051 S20C G4051 S20CK	970 040A20 970 050A20 (En2C, En2D) 970 060A20	A35-551 XC18 A35-552 XC18 A35-553 XC18 A35-553 C20 A35-553 XC18S A35-554 XC18S CC20	5598 1CD20 5598 3CD20 6922 C21 7356 CB20FF	1450
1021	. . .	. . .	970 070M20 970 080A20	A35-551 21B3 A35-552 21B3 A35-553 21B3 A35-557 21B3 A35-566 21B3	5332 C20 7065 C20	. . .
1022	1.0432, C21 1.0469, 21Mn4 1.0482, 19Mn5 1.1133, 20Mn5, GS-20Mn5 1.1134, Ck19	. . .	3111 Type 9 970 120M19 970 170H20	A35-551 20MB5 A35-552 20 MB5 A35-553 20MB5 A35-556 20MB5 A35-557 20MB5 A35 566 20MB5 A35-566 20M5	5771 20Mn4	. . .
1023	1.1150, Ck22.8 1.1152, Cq22	G4051 S22C	1449 2HS, 22CS 970 040A22 (En2C, En2D) 970 050A22 970 060A22 970 080A22	. . .	5332 C20 7065 C20	. . .
1025	1.0406, C25 1.0415, D25-2, D26-2 1.1158, Ck25	G4051 S25C	. . .	A35-552 XC25 A35-566 XC25	5598 1CD25 5598 3CD25	. . .
1026	1.1155, GS-Ck25 1.1156, GS-Ck24	. . .	970 070M26 970 080A25 970 080A27	. . .	7845 C25 7847 C25	. . .
1029	1.0562, 28Mn4	G3445 STKM15A (15A), STKM15C (15C) G4051 S28C	970 060A27 970 080A27 (En5A)	A33-101 AF50 CC28 C30	. . .	. . .
1030	1.0528, C30 1.0530, D30-2 1.1178, Ck30 1.1179, Cm30 1.1811, G-31Mn4	G4051 S30C	1449 30HS, 30CS 970 060A30 970 080A30 (En5B) 970 080M30 (En5)	A35-552 XC32 A35-553 XC32	5332 C30 6403 C30 7065 C30 7845 C30 7874 C30 5598 3CD30 6783 Fe50-3 7065 C31	. . .
1035	1.0501, C35 1.0516, D35-2 1.1172, Cq35 1.1173, Ck34 1.1180, Cm35 1.1181, Ck35	G4051 S35C	1717 CDS105/106 970 060A35 970 080A32 (En5C) 970 080A35 (En8A) 980 CFS6	A33-101 AF55 A35-553 C35 A35-553 XC38 A35-554 XC38 XC35 XC38TS C35	5333 C33 5598 1CD35 5598 3CD35 7065 C35 7065 C36 7847 C36 7356 CB35	1550 1572
1037	1.0520, 31Mn4 1.0561, 36Mn4	G4051 S35C	3111 type 10 970 080M36 970 170H36	. . .	. . .	. . .
1038	No international equivalents					

(continued)

Table 31 (continued)

United States (SAE)	Fed. R. of Germany (DIN)	Japan (JIS)	United Kingdom (BS)	France (AFNOR NF)	Italy (UNI)	Sweden (SS₁₄)
Carbon steels (continued)						
1039	1.1190, Ck42Al	...	970 060A40 970 080A40 (En8C) 970 080M40 (En8) 970 170H41	40M5 A35-552 XC38H2 A35-553 38MB5 A35-556 38MB5 A35-557 38MB5 A35-557 XC38H2 XC42, XC42TS	...	...
1040	1.0511, C40 1.0541, D40-2 1.1186, Ck40 1.1189, Cm40	G4051, S40C	1287 1449 40HS, 40CS 3146 Class 1 Grade C 3146 Class 8 970 060A40 970 080A40 (En8C) 970 080M40 (En8)	A33-101 AF60 C40	5598 1CD40 5598 3CD40 6783 Fe60-3 6923 C40 7065 C40 7065 C41	...
1042	1.0517, D45-2	G4051 S43C	970 060A42 970 080A42 (En8D)	A35-552 XC42H1 A35-553 C40 CC45 XC42, XC42TS	...	...
1043	1.0558, GS-60.3	G4051 S43C	970 060A42 970 080A42 (En8D) 970 080M46	A35-552 XC42H2	7847 C43	...
1044	1.0517, D45-2	...	...	...	...	...
1045	1.0503, C45 1.1184, Ck46 1.1191, Ck45, GS-Ck45 1.1192, Cq45 1.1194, Cq45 1.1201, Cm45 1.1193, Cf45	G4051 S45C G5111 SCC5	970 060A47 970 080A47 970 080M46	A33-101 AF65 A35-552 XC48H1 A35-553 XC45 A35-554 XC48 XC48TS C45	3545 C45 5332 C45 7065 C45 7845 C45 7874 C45 5598 1CD45 5598 3CD45 7065 C46 7847 C46	1672
1046	1.0503, C45 1.0519, 45MnAl 1.1159, GS-46Mn4	...	3100 AW2 970 080M46	45M4TS A35-552 XC48H1 A35-552 XC48H2 XC48TS		...
1049	...	G3445 STKM17A (17A) G3445 STKM17C (17C)	970 060A47 970 080A47	A35-552 XC48H1 A35-554 XC48 XC48TS	6403 C48 7847 C48	...
1050	1.0540, C50 1.1202, D53-3 1.1206, Ck50 1.1210, Ck53 1.1213, Cf53 1.1219, Cf54 1.1241, Cm50	G4051 S50C G4051 S53C	1549 50HS 1549 50CS 970 060A52 970 080A52 (En43C) 970 080M50 (En43A)	A35-553 XC50	5332 C50 7065 C50 7065 C51 7845 C50 7874 C50 5598 1CD50 5598 3CD50 6783 Fe70-3 7847 C53	1674
1053	1.1210 Ck53 1.1213 Cf53 1.1219 Cf54	G4051 S53C	970 080A52 (En43C)	52M4TS A35-553 XC54	7847 C53	1674
1055	1.0518, D55-2 1.0535, C35 1.1202, D53-3 1.1203, Ck55 1.1209, Cm55 1.1210, Ck53 1.1213, Cf53 1.1219, Cf54 1.1220, D55-3 1.1820, C55W	G4051 S53C G4051 S55C	3100 AW3 970 060A57 970 070M55 970 080A52 (En43C) 970 080A57	A33-101 AF70 A35-552 XC55H1 A35-552 XC55H2 A35-553 XC54 XC55 C55	5598 3CD55 7065 C55 7845 C55 7874 C55 7065 C56 7847 C53	...
1059	1.0609, D58-2 1.0610, D60-2 1.0611, D63-2 1.1212, D58-3 1.1222, D63-3 1.1228, D60-3	...	970 060A62	A35-553 XC60	...	...
1060	1.0601, C60 1.0642, 60Mn3 1.1221, Ck60 1.1223, Cm60 1.1740, C60W	G4051 S58C	1449 60HS 1449 60CS 970 060A57 970 080A57	A35-553 XC60	3545 C60 7064 C60 7065 C60 7845 C60 7874 C60 5598 3CD60 7065 C61	1678

(continued)

Table 31 (continued)

United States (SAE)	Fed. R. of Germany (DIN)	Japan (JIS)	United Kingdom (BS)	France (AFNOR NF)	Italy (UNI)	Sweden (SS14)
Carbon steels (continued)						
1064	1.0611, D63-2 1.0612, D65-2 1.0613, D68-2 1.1222, D63-3 1.1236, D65-3	...	970 060A62 970 080A62 (En43D)	...	5598 3CD65	...
1065	1.0627, C68 1.0640, 64Mn3 1.1230, Federstahldraht FD 1.1233 1.1240, 65Mn4 1.1250, Federstahldraht VD 1.1260, 66Mn4	...	970 060A67 970 080A67 (En43E)	XC65	...	...
1069	1.0615, D70-2 1.0617, D73-2 1.0627, C68 1.1232, D68-3 1.1237 1.1249, Cf70 1.1251, D70-3 1.1520, C70W1 1.1620, C70W2	...		A35-553 XC68 XC70	...	...
1070	1.0603, C67 1.0643, 70Mn3 1.1231, Ck67	...	1449 70HS, 70CS 970 060A72 970 070A72 (En42) 970 080A72	XC70	3545 C70	1770
1074	1.0605, C75 1.0645, 76Mn3 1.0655, C74 1.1242, D73-3	...	970 070A72 (En42) 970 080A72	A35-553 XC75 XC70	3545 C75 7064 C75	1774
1075	1.0614, D75-2 1.0617, D73-2 1.0620, D78-2 1.1242, D73-3 1.1252, D78-3 1.1253, D75-3	...	...	A35-553 XC75 XC70	3545 C75 7064 C75 5598 3CD70 5598 3CD75	...
1078	1.0620, D78-2 1.0622, D80-2 1.0626, D83-2 1.1252, D78-3 1.1253, D75-3 1.1255, D80-3 1.1262, D83-3 1.1525, C80W1	G4801 SUP3	970 060A78	XC80	5598 3CD80	...
1080	1.1259 80Mn4 1.1265 D85-2	...	1449 80HS, 80CS 970 060A78 970 060A83 970 070A78 970 080A78 970 080A83	XC80	5598 3CD80 5598 3CD85	...
1084	1.1830, C85W	...	970 060A86 970 080A86	XC85	...	...
1085	1.0647, 85Mn3 1.1273, 90Mn4 1.1819, 90Mn4	...	970 080A83	...	...	...
1086	1.0616, C85, D85-2 1.0626, D83-2 1.0628, D88-2 1.1262, D83-3 1.1265, D85-3 1.1269, Ck85 1.1272, D88-3	...	970 050A86	A35-553 XC90	5598 3CD85 5598 3CD90	...
1090	1.1273, 90Mn4 1.1819, 90Mn4 1.1282, D95S3	...	1449 95HS 1449 95CS 970 060A96	...	3545 C90 7064 C90 5598 3CD95	...
1095	1.0618, D95-2 1.1274, Ck101 1.1275, Ck100 1.1282, D95S3 1.1291, MK97 1.1545, C105W1 1.1645, C105W2	G4801 SUP4	1449 95HS 1449 95CS 970 060A99	A35-553 XC100	3545 C100 7064 C100	1870

(continued)

Table 31 (continued)

United States (SAE)	Fed. R. of Germany (DIN)	Japan (JIS)	United Kingdom (BS)	France (AFNOR NF)	Italy (UNI)	Sweden (SS$_{14}$)
Carbon-manganese steels						
1513	1.0424, Schiffbaustahl CS:DS 1.0479, 13Mn6 1.0496, 12Mn6 1.0513, Schiffbaustahl A32 1.0514, Schiffbaustahl B32 1.0515, Schiffbaustahl E32 1.0549 1.0579 1.0583, Schiffbaustahl A36 1.0584, Schiffbaustahl D36 1.0589, Schiffbaustahl E36 1.0599 1.8941, QStE260N 1.8945, QStE340N 1.8950, QStE380N	. . .	1449 40/30 HR 1449 40/30 HS 1449 40/30 CS 1453 A2 2772 150M12 970 125A15 970 130M15 970 130M15 (En201)	12M5 A33-101 AF50-S A35-501 E35-4 A35-501 E36-2 A35-501 E36-3	. . .	. . .
1522	1.0471, 21MnS15 1.0529, StE350-Z2 1.1120, GS-20Mn5 1.1138, GS-21Mn5 1.1169, 20Mn6 1.8970, StE385.7 1.8972, StE415.7 1.8978 1.8979	G4106 SMn21	1503 221-460 1503 223-409 1503 224-490 3146 CLA2 980 CFS7	A35-551 20MB5 A35-552 20M5 A35-556 20M5 A35-552 20MB5 A35-553 20MB5 A35-556 20MB5 A35-557 20MB5 A35-566 20MB5	4010 FeG52 6930 20Mn6 7660 Fe510	2165 2168
1524	1.0499, 21Mn6Al 1.1133, 20Mn5, GS-20Mn5 1.1160, 22Mn6	G4106 SMn21 G5111 SCMn1	1456 Grade A 970 150M19 (En14A, En14B) 970 175H23 980 CDS9, CDS10	. . .	. . .	. . .
1526	. . .	. . .	970 120M28	A35-566 25MS5		2130
1527	1.0412, 27MnSi5 1.1161, 26Mn5 1.1165, 30Mn5 1.1165, GS-30Mn5 1.1170, 28Mn6	G5111 SCMn2	1453 A3 1456 Grade B1, Grade B2 3100 A5 3100 A6 970 150M28 (En14A, En14B)	. . .	4010 FeG60 7874 C28Mn	. . .
1536	1.0561, 36Mn4 1.1165, 30Mn5 1.1165, GS-30Mn5 1.1166, 34Mn5 1.1167, 36Mn5, GS-36Mn5 1.1813, G-35Mn5	G4052 SMn1H G4052 SMn433H G4106 SMn1 G4106 SMn433 G5111 SCMn2 G5111 SCMn3	1045 3100 A5, A6 970 120M36 (En15B) 970 150M36 (En15)	A35-552 32M5 A35-552 38MB5 A35-553 38MB5 A35-556 38MB5 A35-557 38MB5	4010 FeG60	. . .
1541	1.0563, E 1.0564, N-80 1.1127, 36Mn6 1.1168, GS-40Mn5	G4106 SMn2, SMn438 G4052 SMn2H, SMn438H G4106 SMn3, SMn443 G4052 SMn3H, SMn443H G5111 SCMn5	970 135M44 970 150M40	40M5 45M5 A35-552 40M6	. . .	2120 2128
1548	1.1128, 46Mn5 1.1159, GS-46Mn4	. . .	. . .	. . .	. . .	. . .
1551	1.0542, StSch80	. . .	. . .	24M4TS	. . .	. . .
1552	1.0624, StSch90B 1.1226, 52Mn5	. . .	. . .	55M5	. . .	. . .
1561	1.0908, 60SiMn5	. . .	. . .	. . .	. . .	. . .
1566	1.1233 1.1240, 65Mn4 1.1260, 66Mn7	. . .	. . .	. . .	. . .	. . .
Resulfurized carbon steels						
1108	1.0700, U7S10 1.0702, U10S10	G4804 SUM12	. . .	A35-562 10F1	. . .	. . .
1110	1.0703, R10S10	G4804 SUM11	. . .	. . .	. . .	. . .
1117	. . .	G4804 SUM31	970 210A15 970 210M17 (En32M) 970 214A15 970 214M15 (En202)	. . .	. . .	. . .
1118	. . .	. . .	970 214M15 (En201)	. . .	. . .	. . .
1137	. . .	G4804 SUM41	970 212M36 (En8M) 970 216M36 (En15AM) 970 225M36	35MF4 A35-562 35MF6	4838 CF35SMn10	. . .
1139	1.0726, 35S20	. . .	970 212A37 (En8BM) 970 212M36(En8M) 970 216M36 (En15AM) 970 225M36	35MF4 A35-562 35MF6	. . .	1957

(continued)

Table 31 (continued)

United States (SAE)	Fed. R. of Germany (DIN)	Japan (JIS)	United Kingdom (BS)	France (AFNOR NF)	Italy (UNI)	Sweden (SS$_{14}$)
Resulfurized carbon steels (continued)						
1140............	No international equivalents					
1141............	...	G4804 SUM42	970 212A42 (En8DM) 970 216A42	A35-562 45MF4	...	...
1144............	1.0727, 45S20	G4804 SUM43	970 212A42 (En8DM) 970 212M44 970 216M44 970 225M44 970 226M44	A35-562 45MF6	4838 CF44SMn28	1973
1146............	1.0727, 45S20	...	970 212M44	45MF4	...	...
1151............	1.0728, 60S20 1.0729, 70S20	...	...	...	...	1973
Resulfurized/rephosphorized carbon steels						
1211............	No international equivalents					
1212............	1.0711, 9S20 1.0721, 10S20 1.1011, RFe160K	G4804 SUM21	...	10F2 12MF4 S200	4838 10S20 4838 10S22 4838 CF9S22	...
1213............	1.0715, 9SMn28 1.0736, 9SMn36 1.0740, 9SMn40	G4804 SUM22	970 220M07 (En1A) 970 230M07 970 240M07 (En1A)	A35-561 S250 S250	4838 CF9SMn28 4838 CF9SMn32	1912
1215............	1.0736, 9SMn36	G4804 SUM23	970 240M07 (En1B)	A35-561 S300	4838 CF9SMn32 4838 CF9SMn36	...
12L14...........	No international equivalents					
Alloy steels						
1330............	No international equivalents					
1335............	1.5069, 36Mn7	...	...	...	...	...
1340............	1.5223, 42MnV7	...	...	...	...	...
1345............	1.0625, StSch90C 1.0912, 46Mn7 1.0913, 50Mn7 1.0915, 50MnV7 1.5085, 51Mn7 1.5225, 51MnV7	...	...	...	...	...
4023............	1.5416, 20Mo3	...	...	...	...	...
4024............	1.5416, 20Mo3	...	...	...	...	...
4027............	1.5419, 22Mo4	...	...	...	...	...
4028............	...	...	970 605M30	...	...	...
4032............	1.5411	G5111 SCMnM3	970 605A32 970 605H32 970 605M30 970 605M36 (En16)	...	...	...
4037............	1.2382, 43MnSiMo4 1.5412, GS-40MnMo4 3 1.5432, 42MnMo7	...	3111 Type 2/1 3111 Type 2/2 970 605A37 970 605H37	...	...	...
4042............	1.2382, 43MnSiMo4 1.5432, 42MnMo7	...	...	...	...	...
4047............	No international equivalents					
4118............	1.7211, 23CrMoB4 1.7264, 20CrMo5	G4052 SCM15H G4105 SCM21H G4052 SCM418H G4105 SCM418H	970 708H20 970 708M20	...	7846 18CrMo4	...
4130............	...	G4105 SCM1 G4105 SCM432 G4105 SCM2 G4105 SCM430 G4106 SCM2	1717 CDS110 970 708A30	A35-552 30CD4 A35-556 30CD4 A35-557 30CD4	30CrMo4 6929 35CrMo4F 7356 34CrMo4KB 7845 30CrMo4 7874 30CrMo4	2233
4135............	1.2330, 35CrMo4 1.7220, 34CrMo4 1.7220, GS-34CrMo4 1.7226, 34CrMoS4 1.7231, 33CrMo4	G4054 SCM3H G4054 SCM435H G4105 SCM1 G4105 SCM432 G4105 SCM3 G4105 SCM435	970 708A37 970 708H37	35CD4 A35-552 35CD4 A35-553 35CD4 A35-556 35CD4 A35-557 34CD4	5332 35CrMo4 6929 35CrMo4F 7356 34CrMo4KB 7845 35CrMo4 7874 35CrMo4	2234
4137............	1.7225, GS-42CrMo4	G4052 SCM4H G4052 SCM440H G4105 SCM4	3100 type 5 970 708A37 970 708H37 970 709A37 (continued)	40CD4 42CD4 A35-552 38CD4 A35-557 38CD4	5332 40CrMo4 5333 38CrMo4 7356 38CrMo4KB	...

Table 31 (continued)

United States (SAE)	Fed. R. of Germany (DIN)	Japan (JIS)	United Kingdom (BS)	France (AFNOR NF)	Italy (UNI)	Sweden (SS14)
Alloy steels (continued)						
4140	1.3563, 43CrMo4 1.7223, 41CrMo4 1.7225, 42CrMo4 1.7225, GS-42CrMo4 1.7227, 42CrMoS4	G4052 SCM4H G4052 SCM440H G4103 SNCM4 G4105 SCM4 G4105 SCM440	3100 Type 5 4670 711M40 970 708A40 970 708A42 (En19C) 970 708H42 970 708M40 970 709A40 970 709M40	40CD4 A35-552 42CD4, 42CDTS A35-556 42CD4, 42CDTS A35-557 42CD4, 42CDTS	3160 G40CrMo4 5332 40CrMo4 7845 42CrMo4 7847 41CrMo4 7874 42CrMo4	2244
4142	1.3563, 43CrMo4 1.7223, 41CrMo4	...	970 708A42 (En19C) 970 708H42 970 709A42	40CD4 A35-552 42CD4, 42CDTS A35-553 42CD4, 42CDTS A35-556 42CD4, 42CDTS A35-557 42CD4, 42CDTS	7845 42CrMo4 7874 42CrMo4	2244
4145	1.2332, 47CrMo4	G4052 SCM5H G4052 SCM445H G4105 SCM5, SCM445	970 708H45	A35-552 45SCD6 A35-553 45SCD6	...	...
4147	1.2332, 47CrMo4 1.3565, 48CrMo4 1.7228, 50CrMo4 1.7228, GS-50CrMo4 1.7230, 50CrMoPb4 1.7238, 49CrMo4	G4052 SCM5H G4052 SCM445H G4105 SCM5, SCM445	970 708A47	A35-552 45SCD6 A35-553 45SCD6 A35-571 50SCD6	...	...
4150	1.3565, 48CrMo4 1.7228, 50CrMo4 1.7228, GS-50CrMo4 1.7230, 50CrMoPb4 1.7238, 49CrMo4	...	...	A35-571 50SCD6	...	...
4161	1.7229, 61CrMo4 1.7266, GS-58CrMnMo4 4 3	G4801 SUP13	3100 BW4 3146 CLA12 Grade C	...	...	...
4320	...	G4103 SNCM23 G4103 SNCM420 G4103 SNCM420H	...	20NCD7 A35-565 18NCD4 A35-565 20NCD7	3097 20NiCrMo7 5331 18NiCrMo7 7846 18NiCrMo7	2523 2523-02
4340	1.6565, 40NiCrMo6	G4103 SNCM8 G4103 SNCM439 G4108 SNB23-1-5 G4108 SNB24-1-5	4670 818M40 970 2S.119	...	5332 40NiCrMo7 6926 40NiCrMo7 7845 40NiCrMo7 7874 40NiCrMo7 7356 40NiCrMo7KB	...
E4340	1.6562, 40NiCrMo7 3	...	970 2S.119	...	...	...
4422	1.5419, 22Mo	...	...	23D5	3608 G20Mo5	
4427	No international equivalent					
4615	...	...	...	15ND8	...	
4617		...	970 665A17 970 665H17 970 665M17 (En34)	...	...	
4620		...	970 665A19 970 665H20 970 665M20	2ND8	...	
4626		...	970 665A24 (En35B)	...	...	
4718	No international equivalent					
4720	...	...	...	18NCD4	...	
4815	No international equivalent					
4817	No international equivalent					
4820	No international equivalent					
50B40	1.7003, 38Cr2 1.7023, 28CrS2	G4052 SMnC3H G4052 SMnC443H G4106 SMnC3 G4106 SMnC443 G5111 SCMnCr4	...	A35-552 38C2 A35-556 38C2 A35-557 38C2 A35-552 42C2 A35-556 42C2 A35-557 42C2	7356 41Cr2KB	...
50B44	...	...	...	45C2	7847 45Cr2	...
5046	1.3561, 44Cr2	...	...	...	...	...
50B46	No international equivalent					
50B50	1.7138, 52MnCrB3	...	...	55C2	...	...
5060	1.2101, 62SiMnCr4	...	970 526M60 (En11)	61SC7 A35-552 60SC7	...	...

(continued)

Table 31 (continued)

United States (SAE)	Fed. R. of Germany (DIN)	Japan (JIS)	United Kingdom (BS)	France (AFNOR NF)	Italy (UNI)	Sweden (SS$_{14}$)
Alloy steels (continued)						
5115	1.7131, 16MnCr5, GS-16MnCr5 1.7139, 16MnCrS5 1.7142, 16MnCrPb5 1.7160, 16MnCrB5	G4052 SCr21H G4052 SCr415H G4104 SCr21 G4104 SCr415	970 527A17 970 527H17 970 527M17	16MC5 A35-551 16MC5	7846 16MnCr5	2127
5117	1.3521, 17MnCr5 1.7016, 17Cr3 1.7131, 16MnCr5, GS-16MnCr5 1.7139, 16MnCrS5 1.7142, 16MnCrPb5 1.7168, 18MnCrB5	. . .	. . .	18Cr4 A35-551 16MC5	. . .	. . .
5120	1.2162, 21MnCr5 1.3523, 19MnCr5 1.7027, 20Cr4 1.7028, 20Cr5 4 1.7121, 20CrMnS3 3 1.7146, 20MnCrPb5 1.7147, GS-20MnCr5 1.7149, 20MnCrS5	G4052 SCr22H G4052 SCr420H G4052 SMn21H G4052 SMn421H G4104 SCr22 G4104 SCr420	. . .	A35-551 20MC5 A35-552 20MC5	7846 20MnCr5	. . .
5130	1.8401, 30MnCrTi4	G4052 SCr2H G4052 SCr430H G4104 SCr2 G4104 SCr430	970 530A30 (En18A) 970 530H30	28C4	. . .	. . .
5132	1.7033, 34Cr4 1.7037, 34CrS4	G4104 SCr3 G4104 SCr435	970 530A32 (En18B) 970 530A36 (En18C) 970 530H32	A35-552 32C4 A35-553 32C4 A35-556 32C4 A35-557 32C4	7356 34Cr4KB 7874 34Cr4	. . .
5135	1.7034, 37Cr4 1.7038, 37CrS4 1.7043, 38Cr4	G4052 SCr3H G4052 SCr435H	3111 Type 3 970 530A36 (En18C) 970 530H36	38C4 A35-552 38C4 A35-553 38C4 A35-556 38C4 A35-557 38C4	5332 35CrMn5 6403 35CrMn5 5333 36CrMn4 7847 36CrMn4 7356 38Cr4KB 7845 36CrMn5 7874 36CrMn5 7847 38Cr4	. . .
5140	1.7035, 41Cr4 1.7039, 41CrS4 1.7045, 42Cr4	G4052 SCr4H G4052 SCr440H G4104 SCr4 G4104 SCr440	3111 Type 3 970 2S.117 970 530A40 (En18D) 970 530H40 970 530M40	A35-552 42C4 A35-557 42C4 A35-556 42C4	5332 40Cr4 7356 41Cr4KB 7845 41Cr4 7874 41Cr4	2245
5147	1.7145, GS-50CrMn4 4	. . .	3100 BW2, BW3 3146 CLA 12 Grade A 3146 CLA 12 Grade B	50C4	. . .	. . .
5150	1.7145, GS-50CrMn4 4 1.8404, 60MnCrTi4	. . .	3100 BW2 3100 BW3 3146 CLA 12 Grade A 3146 CLA 12 Grade B	. . .	. . .	2230
5155	1.7176, 55Cr3	G4801 SUP11 G4801 SUP9	. . .	A35-571 55C3	. . .	. . .
5160	1.2125, 65MnCr4	G4801 SUP9A	970 527A60 (En48) 970 527H60	. . .	. . .	. . .
51B60	No international equivalent					
E50100	1.2018, 95Cr1 1.3501, 100Cr2	. . .	. . .	A35-565 100C2	. . .	. . .
E51100	1.2057, 105Cr4 1.2109, 125CrSi5 1.2127, 105MnCr4 1.3503, 105Cr4	. . .	. . .	. . .	3160 G90Cr4	. . .
E52100	1.2059, 120Cr5 1.2060, 105Cr5 1.2067, 100Cr6 1.3505, 100Cr6 1.3503, 105Cr4 1.3514, 101Cr6 1.3520, 100CrMn6	. . .	970 534A99 (En31) 970 535A99 (En31)	. . .	100C6 3097 100Cr6	2258
6118	No international equivalent					
6150	1.8159, GS-50CrV4	G4801 SUP10	970 735A50 (En47) 970 S.204	A35-552 50CV4 A35-553 50CV4 A35-571 50CV4	3545 50CrV4 7065 50CrV4 7845 50CrV4 7874 50Cr₁V4	2230

(continued)

Table 31 (continued)

United States (SAE)	Fed. R. of Germany (DIN)	Japan (JIS)	United Kingdom (BS)	France (AFNOR NF)	Italy (UNI)	Sweden (SS$_{14}$)
Alloy steels (continued)						
8115............	No international equivalents					
81B45...........	No international equivalents					
8615............	...	...	...	15NCD2 15NCD4	3097 16NiCrMo2 5331 16NiCrMo2 7846 16NiCrMo2	...
8617............	...	...	970 805A17 970 805H17 970 805M17 (En 361)	18NCD4 18NCD6	...	...
8620............	1.6522, 20NiCrMo2 1.6523, 21NiCrMo2 1.6526, 21NiCrMoS2 1.6543, 21NiCrMo2 2	G4052 SNCM21H G4052 SNCM220H G4103 SNCM21 G4103 SNCM220	2772 806M20 970 805A20 970 805H20 970 805M20 (En362)	18NCD4 20NCD2 A35-551 19NCDB2 A35-552 19NCDB2 A35-551 20NCD2 A35-553 20NCD2 A35-565 20NCD2 A35-566 20NCD2	5331 20NiCrMo2 6403 20NiCrMo2 7846 20NiCrMo2	2506-03 2506-08
8622............	1.6541, 23MnNiCrMo5 2	...	2772 806M22 970 805A22 970 805H22 970 805M22	23NCDB4 A35-556 23MNCD5 A35-556 23NCDB2 A35-566 22NCD2	...	...
8625............	...	...	970 805H25 970 805M25	25NCD4 A35-556 25MNCD6 A35-566 25MNDC6	...	...
8627............	No international equivalents					
8630............	1.6545, 30NiCrMo2 2	...	...	30NCD2	7356 30NiCrMo2KB	...
8637............	...	...	970 945M38 (En100)	40NCD3	5332 38NiCrMo4 7356 38NiCrMo4KB 7845 39NiCrMo3 7874 39NiCrMo3	...
8640............	1.6546, 40NiCrMo2 2	...	3111 Type 7, 2S.147 970 945A40 (En 100C)	40NCD2 40NCD2TS 40NCD3TS 40NCD3	5333 40NiCrMo4 7356 40NiCrMo2KB 7845 40NiCrMo2 7874 40NiCrMo2 7847 40NiCrMo3	...
8642............	No international equivalents					
8645............	No international equivalents					
86B45...........	No international equivalents					
8650............	No international equivalents					
8655............	No international equivalents					
8660............	...	...	970 805A60 970 805H60	...	...	...
8720............	No international equivalents					
8740............	1.6546, 40NiCrMo2 2	...	3111 Type 7, 2S.147	40NCD2 40NCD2TS 40NCD3TS	7356 40NiCrMo2KB 7845 40NiCrMo2 7874 40NiCrMo2	...
8822............	No international equivalents					
9254............	No international equivalents					
9260............	...	G4801 SUP7	970 250A58 (En45A) 970 250A61 (En45A)	60S7 61S7	...	...
E9310...........	1.6657, 14NiCrMo13 4	...	970 832H13 970 832M13 (En36C) S.157	16NCD13	6932 15NiCrMo13 9335 10NiCrMo13	...
94B15...........	No international equivalents					
94B17...........	No international equivalents					
94B30...........	No international equivalents					

Table 32 Chemical compositions of German (DIN) carbon, carbon-manganese, resulfurized, and resulfurized/rephosphorized steels referenced in Table 31

Nearest SAE grade	DIN number	C	Si	Mn	P	S	Others
1008	1.0010, D9	0.10	0.30	0.50	0.070	0.060	...
1013	1.0036, USt37-2	0.17	...	0.20–0.50	0.050	0.050	0.007N
1013	1.0037, St37-2	0.17	0.30	0.20–0.50	0.050	0.050	0.009N
1013	1.0038, RSt37-2	0.17	0.03–0.30	0.20–0.50	0.050	0.050	0.009N
1013	1.0055, USt34-1	0.17	...	0.20–0.50	0.080	0.050	...
1013	1.0057, RSt34-1	0.17	0.03–0.30	0.20–0.50	0.080	0.050	...
1013	1.0116, St37-3	0.17	0.03–0.30	0.20–0.50	0.040	0.040	...
1010	1.0204, UQSt36	0.13	...	0.25–0.50	0.040	0.040	...
1013	1.0218, RSt41-2	0.11–0.17	0.03–0.25	0.40–0.60	0.040	0.040	...
1013	1.0219, St41-3	0.11–0.17	0.03–0.30	0.40–0.60	0.040	0.040	0.020–0.050Al
1005	1.0288, D5-2	0.06	...	0.40	0.030	0.030	...
1010	1.0301, C10	0.07–0.13	0.15–0.35	0.30–0.60	0.045	0.045	...
1005	1.0303, QSt32-2	0.06	0.1	0.20–0.40	0.040	0.040	0.02Al
1013	1.0307, StE210.7	0.17	0.45	0.35	0.040	0.035	...
1013	1.0309, St35.4	0.17	0.10–0.35	0.40	0.050	0.050	0.007N
1006	1.0311, D7-1	0.08	...	0.45	0.060	0.050	...
1005	1.0312, D5-1	0.06	0.10	0.40	0.040	0.050	...
1006	1.0313, D8-2	0.08	...	0.45	0.040	0.040	...
1005	1.0314, D6-2	0.06	...	0.40	0.040	0.040	...
1013	1.0315, St37.8	0.17	0.10–0.35	0.40–0.80	0.040	0.040	...
1006	1.0317, RSD4	0.03–0.07	0.07–0.17	0.50–0.70	0.025	0.025	0.025Cu
1008	1.0318, St28	0.13	...	...	0.050	0.050	...
1013	1.0319, RRStE210.7	0.17	0.45	0.35	0.040	0.035	0.020Al
1008	1.0320, St22	0.10	...	0.20–0.45	0.035	0.035	0.007N
1006	1.0321, St23	0.08	...	0.20–0.40	0.025	0.025	0.007N
1008	1.0322, USD8	0.06–0.10	...	0.45–0.65	0.030	0.030	0.12Cr, 0.12Ni, 0.17Cu, 0.007N
1008	1.0326, RSt28	0.10	0.05	0.45	0.030	0.030	0.025Al
1010	1.0328, USD10	0.08–0.12	...	0.50–0.70	0.030	0.030	0.007N
1008	1.0330, St2, 1.0330, St12	0.10	...	...	...	...	0.007N
1008	1.0331, RoSt2	0.10	...	0.30–0.60	0.045	0.045	0.007N
1008	1.0332, StW22	0.10	...	0.20–0.45	0.035	0.035	0.007N
1008	1.0333, St3, 1.0333, St13	0.10	...	...	...	...	0.007N
1006	1.0334, StW23	0.08	...	0.20–0.40	0.025	0.025	0.007N
1006	1.0335, StW24	0.08	0.03–0.10	0.40	0.025	0.025	...
1008	1.0336, USt4, 1.0336, USt14	0.09	...	0.25–0.60	0.030	0.030	0.007N
1008	1.0337, RoSt4	0.10	0.10	0.30–0.60	0.030	0.035	0.007N
1008	1.0344, St12Cu3	0.10	...	0.20–0.45	0.050	0.050	0.25–0.35Cu, 0.008N
1008	1.0347, RRSt13	0.10	0.03–0.10	0.40	0.025	0.025	0.020Al, 0.007N
1010	1.0349, RSD9	0.08–0.12	0.03–0.08	0.50–0.70	0.025	0.025	0.007N
1006	1.0354, St14Cu3	0.08	0.03–0.10	0.40	0.025	0.025	0.02Al, 0.25–0.35Cu, 0.007N
1013	1.0356, TTSt35	0.17	0.35	0.40	0.045	0.045	...
1008	1.0357, USt28	0.10	...	0.45	0.030	0.030	...
1008	1.0359, RRSt23	0.10	0.03–0.10	0.40	0.025	0.025	0.007N
1008	1.0375, Feinstblech T57	0.10	...	0.25–0.45	0.040	0.040	0.007N
	1.0377, Feinstblech T61	0.10	...	0.25–0.45	0.050	0.040	0.007N
	1.0378, Feinstblech T65	0.10	...	0.25–0.50	0.050	0.040	0.007N
	1.0379, Feinstblech T70	0.10	...	0.25–0.50	0.060	0.040	0.007N
	1.0385, Weissblech T57	0.10	...	0.25–0.45	0.040	0.040	0.007N
	1.0387, Weissblech T61	0.10	...	0.25–0.50	0.050	0.040	0.007N
	1.0388, Weissblech T65	0.10	...	0.25–0.50	0.050	0.040	0.007N
	1.0389, Weissblech T70	0.10	...	0.25–0.50	0.060	0.040	...
1006	1.0391, EK2	0.08	...	...	...	...	0.007N
1006	1.0392, EK4	0.08	...	...	...	...	0.020Al
1005	1.0393, ED3	0.04	...	...	...	...	0.007N
1005	1.0394, ED4	0.04	...	...	...	...	0.020Al
1015	1.0401, C15	0.12–0.18	0.15–0.35	0.30–0.60	0.045	0.045	...
1020	1.0402, C22	0.18–0.25	0.15–0.35	0.30–0.60	0.045	0.045	...
1025	1.0406, C25	0.22–0.29	0.40	0.40–0.70	0.045	0.045	...
1527	1.0412, 27MnSi5	0.24–0.30	0.30–0.55	1.10–1.60	0.040	0.040	0.020–0.050Al
1020	1.0414, D20-2	0.18–0.23	0.10–0.30	0.30–0.60	0.040	0.040	...
1025	1.0415, D25-2	0.23–0.28	0.10–0.30	0.30–0.60	0.04	0.04	...
1025	1.0415, D26-2	0.23–0.28	0.10–0.30	0.30–0.60	0.04	0.04	...
1013	1.0417	0.17	0.35	0.70	0.045	0.045	...
1016	1.0419, RSt44.2	0.18	0.45	0.80	0.050	0.050	0.008N
1513	1.0424, Schiffbaustahl CS:DS	0.16	0.10–0.35	1.00–1.35	0.040	0.040	0.015–0.06Al
1020	1.0427, C22.3	0.18–0.23	0.15–0.35	0.30–0.60	0.045	0.045	0.30Cr
1012	1.0439, RSD13	0.10–0.16	0.03–0.08	0.65–0.85	0.025	0.025	0.007N
1018	1.0453, C16.8	0.14–0.19	0.15–0.35	0.40–0.80	0.040	0.040	0.30Cr
1013	1.0457, StE240.7	0.17	0.45	0.40	0.040	0.035	...
1016	1.0467, 15Mn3	0.12–0.18	0.10–0.20	0.70–0.90	0.040	0.040	...
1016	1.0468, 15Mn3Al	0.12–0.18	0.10–0.20	0.70–0.90	0.040	0.040	0.020–0.050Al
1522	1.0471, 21MnSi5	0.18–0.24	0.10–0.35	1.10–1.60	0.040	0.040	0.020–0.040Al
1513	1.0479, 13Mn6	0.08–0.14	0.30–0.45	1.35–1.65	0.025	0.025	...
1513	1.0496, 12Mn6	0.08–0.14	0.08–0.25	1.35–1.65	0.025	0.025	0.030Al
1524	1.0499, 21Mn6Al	0.08–0.25	0.15–0.25	1.40–1.80	0.035	0.030	0.20Cu, 0.030Al

(continued)

Table 32 (continued)

Nearest SAE grade	DIN number	C	Si	Mn	P	S	Others
1035	1.0501, C35	0.32–0.39	0.40	0.50–0.80	0.045	0.045	...
1045	1.0503, C45	0.42–0.50	0.40	0.50–0.80	0.045	0.045	...
1040	1.0511, C40	0.37–0.45	0.15–0.35	0.60–0.80	0.045	0.045	...
1513	1.0513, Schiffbaustahl A32, 1.0514, Schiffbaustahl B32, 1.0515, Schiffbaustahl E32	0.18	1.00–1.50	0.90–1.60	0.040	0.040	0.015Al, 0.05Nb, 0.10V
1035	1.0516, D35-2	0.33–0.38	0.10–0.30	0.30–0.60	0.040	0.040	...
1042, 1044	1.0517, D45-2	0.43–0.48	0.10–0.30	0.30–0.70	0.040	0.040	...
1055	1.0518, D55-2	0.53–0.58	0.10–0.30	0.30–0.70	0.040	0.040	...
1037	1.0520, 31Mn4	0.28–0.36	0.20–0.50	0.80–1.10	0.045	0.045	0.020Al
1030	1.0528, C30	0.27–0.34	0.40	0.50–0.80	0.045	0.045	...
1522	1.0529, StE350-2Z	0.25	...	1.50	0.040	0.040	...
1030	1.0530, D30 2	0.28–0.33	0.10–0.30	0.30–0.60	0.040	0.040	...
1055	1.0535, C35	0.52–0.60	0.40	0.60–0.90	0.045	0.045	...
1050	1.0540, C50	0.47–0.55	0.40	0.60–0.90	0.045	0.045	...
1040	1.0541, D40-2	0.38–0.43	0.10–0.30	0.30–0.60	0.040	0.040	...
1551	1.0542, StSch80	0.45–0.63	0.50	0.80–1.80	0.050	0.050	0.007N
1513	1.0549	0.18	0.50	1.00–1.50	0.035	0.035	...
1043	1.0558, GS-60.3	0.40–0.48	0.30–0.50	0.70–0.90	0.030	0.030	...
1037, 1536	1.0561, 36Mn4	0.32–0.40	0.25–0.50	0.90–1.20	0.050	0.050	0.007N
1029	1.0562, 28Mn4	0.24–0.32	0.15–0.40	0.90–1.20	0.050	0.050	0.007N
1541	1.0563, E; 1.0564, N-80	0.45	0.10–0.30	1.20	0.040	0.060	0.007N
1513	1.0579	0.18	0.10–0.50	1.60	0.035	0.035	0.02Nb
1513	1.0583, Schiffbaustahl A36, 1.0584, Schiffbaustahl D36, 1.0589, Schiffbaustahl E36	0.18	0.10–0.50	0.90–1.60	0.040	0.040	0.15Al, 0.02–0.05Nb, 0.05–0.10V
1513	1.0599	0.15	0.50	1.70	0.030	0.030	...
1060	1.0601, C60	0.57–0.65	0.15–0.35	0.60–0.90	0.045	0.045	...
1070	1.0603, C67	0.65–0.72	0.40	0.60–0.90	0.045	0.045	...
1074	1.0605, C75	0.70–0.80	0.15–0.35	0.60–0.80	0.045	0.045	...
1059	1.0609, D58-2	0.55–0.60	0.10–0.30	0.30–0.70	0.040	0.040	...
1059	1.0610, D60-2	0.58–0.63	0.10–0.30	0.30–0.70	0.040	0.040	...
1059, 1064	1.0611, D63-2	0.60–0.65	0.10–0.30	0.30–0.70	0.040	0.040	...
1064	1.0612, D65-2	0.63–0.68	0.10–0.30	0.30–0.70	0.040	0.040	...
1064	1.0613, D68-2	0.65–0.70	0.10–0.30	0.30–0.70	0.040	0.040	...
1075	1.0614, D75-2	0.73–0.78	0.10–0.38	0.30–0.70	0.040	0.040	...
1086	1.0616, C85, 1.0616, D85-2	0.83–0.90	0.10–0.30	0.30–0.70	0.040	0.040	...
1069, 1075	1.0617, D-73-2	0.70–0.75	0.10–0.30	0.30–0.70	0.040	0.040	...
1095	1.0618, D95-2	0.90–0.99	0.10–0.30	0.30–0.70	0.040	0.040	...
1075, 1078	1.0620, D78-2	0.75–0.80	0.10–0.30	0.30–0.70	0.040	0.040	...
1078	1.0622, D80-2	0.78–0.83	0.10–0.30	0.30–0.70	0.040	0.040	...
1552	1.0624, StSch90	0.50–0.70	0.50	1.30–1.70	0.050	0.050	...
1078, 1086	1.0626, D83-2	0.80–0.85	0.10–0.30	0.30–0.70	0.040	0.040	...
1065, 1069	1.0627, C68	0.65–0.72	0.25–0.50	0.60–0.80	0.045	0.045	...
1086	1.0628, D88-2	0.85–0.90	0.10–0.30	0.30–0.70	0.040	0.040	...
1065	1.0640, 64Mn3	0.60–0.68	0.20–0.40	0.50–0.80	0.050	0.050	0.007N
1060	1.0642, 60Mn3	0.57–0.65	0.20–0.40	0.70–0.90	0.050	0.050	0.007N
1070	1.0643, 70Mn3	0.65–0.75	0.20–0.40	0.60–0.90	0.050	0.050	0.007N
1074	1.0645, 76Mn3	0.70–0.80	0.20–0.40	0.60–0.90	0.050	0.050	0.007N
1085	1.0647, 85Mn3	0.80–0.90	0.15–0.35	0.70–0.90	0.050	0.050	0.007N
1074	1.0655, C74	0.70–0.80	0.20–0.40	0.60–0.90	0.030	0.030	...
1108	1.0700, U7S10	0.10	...	0.40–0.70	0.080	0.08–0.12	...
1108	1.0702, U10S10	0.15	...	0.40–0.70	0.050	0.08–0.12	...
1110	1.0703, R10S10	0.15	0.40	0.30–0.80	0.050	0.08–0.12	...
1212	1.0711, 9S20	0.13	0.05	0.60–1.20	0.100	0.18–0.25	...
1213	1.0715, 9SMn28	0.14	0.05	0.90–1.30	0.100	0.24–0.32	...
1212	1.0721, 10S20	0.07–0.13	0.10–0.40	0.50–0.90	0.060	0.15–0.25	...
1139	1.0726, 35S20	0.32–0.39	0.10–0.40	0.50–0.90	0.060	0.15–0.25	...
1146	1.0727, 45S20	0.42–0.50	0.10–0.40	0.50–0.90	0.060	0.15–0.25	...
1151	1.0728, 60S20	0.57–0.65	0.10–0.40	0.50–0.90	0.060	0.15–0.25	...
1151	1.0729, 70S20	0.60–0.72	0.15–0.25	0.50–0.70	0.070	0.15–0.25	...
1213, 1215	1.0736, 9SMn36	0.15	0.05	1.00–1.50	0.100	0.32–0.40	...
1213	1.0740, 9SMn40	0.14	0.10–0.40	1.00–1.50	0.070	0.35–0.45	...
1008	1.0744, 6P10	0.09	...	0.20–0.45	0.08–0.15	0.050	...
1561	1.0908, 60SiMn5	0.55–0.65	1.00–1.30	0.90–1.10	0.050	0.050	0.007N
1006	1.1009, Ck7	0.08	0.15	0.50	0.035	0.035	...
1212	1.1011, RFe160K	0.10	0.10	0.50–0.90	0.080	0.18–0.27	0.04–0.10Al
1005	1.1012, RFe120	0.05	...	0.20–0.35	0.030	0.035	0.04–1.0Al
1008	1.1116, USD6	0.06–0.10	...	0.45–0.65	0.020	0.020	0.12Cr, 0.12Ni, 0.17Cu
1522	1.1120, GS-20Mn5	0.23	0.60	1.00–1.50	0.025	0.020	0.30Cr, 0.20Ni, 0.10Mo
1010	1.1121, Ck10, 1.1122 Cq10	0.07–0.13	0.15–0.35	0.30–0.60	0.035	0.035	...
1541	1.1127, 36Mn6	0.34–0.42	0.15–0.35	1.40–1.65	0.035	0.035	...
1548	1.1128, 46Mn5	0.42–0.48	0.25–0.45	1.15–1.35	0.035	0.035	...
1015	1.1132, CQ15	0.12–0.18	0.15–0.35	0.25–0.50	0.035	0.035	...
1022, 1524	1.1133, 20Mn5, 1.1133, GS-20Mn5	0.17–0.23	0.30–0.60	1.00–1.30	0.035	0.035	...
1022	1.1134, Ck19	0.15–0.23	0.15	0.40–0.60	0.030	0.025	0.15Cr, 0.15Cu, 0.03–0.08Al
1015	1.1135, Ck16Al	0.13–0.18	0.15	0.40–0.60	0.025	0.025	0.15Cu, 0.007N, 0.03–0.08Al

(continued)

Classification and Designation of Carbon and Low-Alloy Steels / 177

Table 32 (continued)

Nearest SAE grade	DIN number	Composition, wt%					
		C	Si	Mn	P	S	Others
1522	1.1138, GS-21Mn5	0.17–0.23	0.65	1.00–1.30	0.025	0.020	0.30Cr
1015	1.1140, Cm15	0.12–0.18	0.15–0.35	0.30–0.60	0.035	0.020–0.035	...
1015	1.1141, Ck15	0.12–0.18	0.15–0.35	0.30–0.60	0.035	0.035	...
1016	1.1142, GS-Ck16	0.12–0.19	0.30–0.50	0.50–0.80	0.030	0.030	0.30Cr, 0.007N
1015	1.1144	0.12–0.18	0.15–0.35	0.25–0.50	0.035	0.035	0.007N
1015	1.1148, Ck15Al	0.12–0.18	0.15–0.35	0.25–0.50	0.035	0.035	0.0003Al, 0.007N
1020	1.1149, Cm22	0.17–0.24	0.40	0.30–0.60	0.055	0.020–0.035	...
1023	1.1150, Ck22.8	0.18–0.25	0.15–0.35	0.30–0.60	0.035	0.035	0.30Cr
1020	1.1151, Ck22	0.17–0.24	0.40	0.30–0.60	0.035	0.030	...
1023	1.1152, Cq22	0.18–0.24	0.15–0.35	0.30–0.60	0.035	0.035	...
1026	1.1155, GS-Ck25	0.20–0.28	0.30–0.50	0.50–0.80	0.035	0.035	...
1026	1.1156, GS-Ck24	0.20–0.28	0.30–0.50	0.50–0.80	0.030	0.030	0.30Cr, 0.007N
1025	1.1158, Ck25	0.22–0.29	0.40	0.40–0.70	0.035	0.035	...
1046	1.1159, GS-46Mn4	0.42–0.50	0.25–0.50	0.90–1.20	0.045	0.045	...
1548	1.1159, GS-46Mn4	0.42–0.50	0.25–0.50	0.90–1.20	0.035	0.035	...
1524	1.1160, 22Mn6	0.18–0.25	0.15–0.30	1.30–1.65	0.035	0.035	...
1527	1.1161, 26Mn5	0.22–0.29	0.15–0.30	1.20–1.50	0.035	0.035	...
1527	1.1165, GS-30Mn5	0.27–0.34	0.30–0.50	1.20–1.50	0.035	0.035	...
1527, 1536	1.1165, 30Mn5	0.27–0.34	0.15–0.40	1.20–1.50	0.035	0.035	0.30Cr
1536	1.1166, 34Mn5	0.30–0.37	0.15–0.30	1.20–1.50	0.035	0.035	...
1536	1.1167, 36Mn5, 1.1167, GS-36Mn5	0.32–0.40	0.15–0.35	1.20–1.50	0.035	0.035	...
1541	1.1168, GS-40Mn5	0.36–0.44	0.30–0.50	1.20–1.50	0.035	0.035	...
1522	1.1169, 20Mn6	0.17–0.23	0.30–0.60	1.30–1.60	0.035	0.035	...
1527	1.1170, 28Mn6	0.25–0.32	0.15–0.40	1.30–1.65	0.035	0.035	0.30Cr
1035	1.1172, Cq35	0.32–0.39	0.15–0.35	0.50–0.80	0.035	0.035	...
1035	1.1173, Ck34	0.31–0.38	0.20	0.45–0.55	0.025	0.025	0.007N
1030	1.1178, Ck30	0.27–0.34	0.40	0.50–0.90	0.035	0.030	...
1030	1.1179, Cm30	0.27–0.34	0.40	0.50–0.90	0.035	0.020–0.035	...
1035	1.1180, Cm35	0.32–0.39	0.40	0.50–0.80	0.035	0.020–0.035	...
1035	1.1181, Ck35	0.32–0.39	0.15–0.30	0.50–0.80	0.035	0.035	...
1045	1.1184, Ck46	0.42–0.50	0.15–0.35	0.50–0.80	0.025	0.025	...
1040	1.1186, Ck40	0.37–0.43	0.15–0.35	0.50–0.80	0.035	0.035	...
1040	1.1189, Cm40	0.37–0.44	0.40	0.50–0.80	0.035	0.020–0.035	...
1039	1.1190, Ck42Al	0.39–0.44	0.25–0.40	0.75–0.90	0.035	0.035	0.007N
1045	1.1191, Ck45, 1.1191, GS-Ck45	0.42–0.50	0.15–0.35	0.50–0.80	0.035	0.035	...
1045	1.1192, Cq45, 1.1194, Cq45, 1.1201, Cm45	0.42–0.50	0.15–0.35	0.50–0.80	0.035	0.035	...
1045	1.1193, Cf45	0.43–0.49	0.15–0.45	0.50–0.80	0.025	0.035	...
1050, 1055	1.1202, D53-3	0.50–0.55	0.10–0.30	0.30–0.70	0.030	0.030	...
1055	1.1203, Ck55	0.52–0.62	0.40	0.60–0.90	0.035	0.030	...
1050	1.1206, Ck50	0.47–0.55	0.40	0.60–0.90	0.035	0.030	...
1055	1.1209 Cm55	0.52–0.60	0.15–0.35	0.60–0.90	0.035	0.020–0.035	...
1050, 1053, 1055	1.1210, Ck53	0.50–0.57	0.15–0.35	0.40–0.70	0.035	0.035	0.007N
1059	1.1212, D58-3	0.55–0.60	0.10–0.30	0.30–0.70	0.030	0.030	...
1050, 1053, 1055	1.1213, Cf53	0.50–0.57	0.15–0.35	0.40–0.70	0.025	0.035	...
1050, 1053, 1055	1.1219, Cf54	0.50–0.57	0.35	0.60–0.90	0.035	0.035	0.30Cu
1055	1.1220, D55-3	0.53–0.58	0.10–0.30	0.30–0.70	0.030	0.030	...
1060	1.1221, Ck60	0.57–0.65	0.15–0.35	0.60–0.90	0.035	0.035	...
1059, 1064	1.1222, D63-3	0.60–0.65	0.10–0.30	0.30–0.70	0.030	0.030	...
1060	1.1223 Cm60	0.57–0.65	0.40	0.60–0.90	0.030	0.020–0.035	...
1552	1.1226, 52Mn5	0.47–0.55	0.15–0.30	1.20–1.50	0.035	0.035	...
1059	1.1228, D60-3	0.58–0.63	0.10–0.30	0.30–0.70	0.030	0.030	...
1065	1.1230 Federstahldraht FD	0.60–0.70	0.25	0.50–0.90	0.030	0.030	0.12Cr
1070	1.1231, Ck67	0.65–0.72	0.15–0.35	0.60–0.90	0.030	0.030	...
1069	1.1232, D68-3	0.65–0.70	0.10–0.30	0.30–0.70	0.030	0.030	...
1065	1.1233	0.60–0.70	0.20–0.65	0.70–1.20	0.035	0.035	...
1064	1.1236, D65-3	0.63–0.68	0.10–0.30	0.30–0.70	0.030	0.030	...
1069	1.1237	0.65–0.75	0.15–0.30	0.70	0.030	0.030	...
1065, 1566	1.1240, 65Mn4	0.60–0.70	0.25–0.50	0.90–1.20	0.035	0.035	...
1050	1.1241, Cm50	0.47–0.55	0.40	0.60–0.90	0.035	0.020–0.035	...
1074, 1075	1.1242, D73-3	0.70–0.75	0.10–0.30	0.30–0.70	0.030	0.030	...
1069	1.1249, Cf70	0.68–0.75	0.15–0.35	0.20–0.35	0.025	0.035	0.007N
1065	1.1250 Federstahldraht VD	0.60–0.70	0.25	0.50–0.90	0.030	0.020	0.060Cu
1069	1.1251, D70-3	0.68–0.73	0.10–0.30	0.30–0.70	0.030	0.030	...
1075, 1078	1.1252, D78-3	0.75–0.80	0.10–0.30	0.30–0.70	0.030	0.030	...
1075, 1078	1.1253, D75-3	0.73–0.78	0.10–0.30	0.30–0.70	0.030	0.030	...
1078	1.1255, D80-3	0.78–0.83	0.10–0.30	0.30–0.70	0.030	0.030	...
1080	1.1259, 80Mn4	0.75–0.85	0.25–0.50	0.90–1.20	0.035	0.035	...
1065	1.1260, 66Mn4	0.60–0.71	0.15–0.30	0.85–1.15	0.035	0.035	...
1566	1.1260, 66Mn7	0.60–0.71	0.15–0.30	0.85–1.15	0.035	0.035	...
1078, 1086	1.1262, D83-3	0.80–0.85	0.10–0.30	0.30–0.70	0.030	0.030	...
1080	1.1265, D85-2	0.83–0.88	0.10–0.30	0.30–0.70	0.030	0.030	...
1086	1.1265, D85-3	0.83–0.88	0.10–0.30	0.30–0.70	0.030	0.030	...
1086	1.1269, Ck85	0.80–0.90	0.15–0.35	0.45–0.65	0.035	0.035	0.007N
1086	1.1272, D88-3	0.85–0.89	0.10–0.25	0.35–0.55	0.030	0.030	...
1090	1.1273, 90Mn4	0.85–0.95	0.25–0.50	0.90–1.10	0.035	0.035	...

(continued)

Table 32 (continued)

Nearest SAE grade	DIN number	C	Si	Mn	P	S	Others
1095	1.1274, Ck101	0.95–1.05	0.15–0.35	0.40–0.60	0.035	0.035	0.007N
1095	1.1275, Ck100	0.98–1.05	0.15–0.25	0.25	0.025	0.025	0.007N
1090, 1095	1.1282, D95S3	0.90–0.99	0.10–0.30	0.30–0.70	0.030	0.030	...
1095	1.1291, Mk97	0.95–0.99	0.10–0.25	0.25–0.45	0.025	0.025	0.007N
1055	1.1820, C55W	0.50–0.58	0.15	0.30–0.50	0.030	0.030	...
1069	1.1520, C70W1	0.65–0.74	0.10–0.25	0.10–0.25	0.020	0.020	...
1078	1.1525, C80W1	0.75–0.85	0.10–0.25	0.10–0.25	0.020	0.020	...
1095	1.1545, C105W1, 1.645, C105W2	1.00–1.10	0.10–0.25	0.10–0.25	0.020	0.020	...
1069	1.1620, C70W2	0.65–0.74	0.10–0.25	0.10–0.25	0.020	0.020	...
1060	1.1740, C60W	0.55–0.65	0.15–0.40	0.60–0.80	0.030	0.035	...
1030	1.1811, G-31Mn4	0.28–0.33	0.40–0.60	0.80–1.00	0.035	0.035	...
1536	1.1813, G-35Mn5	0.32–0.40	0.20–0.50	1.20–1.50	0.035	0.035	...
1055	1.1820, C55W	0.50–0.58	0.15	0.30–0.50	0.030	0.030	...
1084	1.1830, C85W	0.80–0.90	0.25–0.40	0.50–0.80	0.025	0.020	...
1513	1.8941, QStE260N, 1.8945, QStE340N	0.16	0.50	1.20	0.030	0.030	...
		0.16	0.50	1.50	0.030	0.030	...
1513	1.8950, QStE380N	0.18	0.50	1.60	0.030	0.030	...
1522	1.8970, StE385.7, 1.8972, StE415.7	0.23	0.55	1.00–1.50	0.040	0.035	0.12V, 0.020Al
1522	1.8978, 1.8979	0.26	0.35	1.40	0.040	0.050	...

Table 33 Chemical compositions of German (DIN) alloy steels referenced in Table 31

Nearest SAE grade	DIN number	C	Si	Mn	P	S	Ni	Cr	Mo	Others
1345	1.0625, StSch90C	0.45–0.65	0.40	1.70–2.10	0.030	0.030	...	...	...	...
1345	1.0912, 46Mn7	0.42–0.50	0.15–0.35	1.60–1.90	0.050	0.050	...	...	...	...
1345	1.0913, 50Mn7	0.45–0.55	0.40	1.60–2.00	0.040	0.040	...	...	...	0.007N
1345	1.0915, 50MnV7	0.48–0.55	0.15–0.35	1.60–2.00	0.050	0.050	...	...	...	0.007N, 0.07–0.12V
E50100	1.2018, 95Cr1	0.90–1.00	0.15–0.30	0.20–0.40	0.025	0.025	...	0.30–0.40	...	...
E51100	1.2057, 105Cr4	1.00–1.10	0.15–0.35	0.20–0.40	0.030	0.030	...	0.90–1.10	...	...
E52100	1.2059, 120Cr5	1.10–1.25	0.20–0.40	0.20–0.40	0.030	0.030	...	1.20–1.50	...	...
E52100	1.2060, 105Cr5	1.00–1.10	0.20–0.40	0.20	0.030	0.030	...	1.20–1.50	...	...
E52100	1.2067, 100Cr6	0.95–1.10	0.15–0.35	0.25–0.40	0.030	0.025	0.30	1.35–1.65	...	0.30Cu
5060	1.2101, 62SiMnCr4	0.58–0.66	0.90–1.20	0.90–1.20	0.030	0.030	...	0.40–0.70	...	...
E51100	1.2109, 125CrSi5	1.20–1.30	1.05–1.25	0.60–0.80	0.035	0.035	...	1.10–1.30	...	...
5160	1.2125, 65MnCr4	0.60–0.68	0.30–0.50	1.00–1.20	0.035	0.035	...	0.60–0.80	...	...
E51100	1.2127, 105MnCr4	1.00–1.10	0.15–0.30	1.00–1.20	0.035	0.035	...	0.70–1.00	...	...
5120	1.2162, 21MnCr5	0.18–0.24	0.15–0.35	1.10–1.40	0.030	0.030	...	1.00–1.30	...	...
4135	1.2330, 35CrMo4	0.32–0.37	0.20–0.40	0.60–0.80	...	...	...	0.90–1.10	0.20–0.25	...
4145, 4147	1.2332, 47CrMo4	0.43–0.50	0.15–0.35	0.60–0.80	0.025	0.025	...	0.90–1.20	0.25–0.40	...
4037, 4042	1.2382, 43MnSiMo4	0.36–0.46	0.80–1.00	0.85–1.10	0.035	0.035	...	...	0.10–0.25	...
E50100	1.3501, 100Cr2	0.95–1.05	0.15–0.35	0.25–0.40	0.030	0.025	0.30	0.40–0.60	...	0.30Cu
E51100, E 52100	1.3503, 105Cr4	1.00–1.10	0.15–0.35	0.25–0.40	0.030	0.025	...	0.90–1.15	...	...
E52100	1.3505, 100Cr6	0.95–1.05	0.15–0.35	0.25–0.40	0.030	0.025	0.30	1.35–1.65	...	0.30Cu
E52100	1.3514, 101Cr6	0.95–1.05	0.15–0.35	0.25–0.45	0.015	0.015	0.40	1.35–1.65	...	...
E52100	1.3520, 100CrMn6	0.90–1.05	0.50–0.70	1.00–1.20	0.030	0.025	0.30	1.40–1.65	...	...
5117	1.3521, 17MnCr5	0.15–0.20	0.15–0.40	1.00–1.30	0.035	0.035	...	0.80–1.10	...	0.30Cu
5120	1.3523, 19MnCr5	0.17–0.22	0.40	1.10–1.40	0.035	0.035	...	1.10–1.30	...	0.30Cu
5046	1.3561, 44Cr2	0.42–0.48	0.40	0.50–0.80	0.025	0.035	...	0.40–0.60	...	0.30Cu
4140	1.3563, 43CrMo4	0.40–0.46	0.40	0.60–0.90	0.025	0.035	...	0.90–1.20	0.15–0.30	0.30Cu
4142	1.3563, 43CrMo4	0.40–0.46	0.40	0.60–0.90	0.025	0.035	...	0.90–1.20	0.15–0.30	0.30Cu
4147, 4150	1.3565, 48CrMo4	0.46–0.52	0.40	0.50–0.80	0.025	0.030	...	0.90–1.20	0.15–0.30	0.30Cu
1335	1.5069, 36Mn7	0.32–0.40	0.30–0.45	1.60–1.90	0.030	0.030	...	...	...	...
1345	1.5085, 51Mn7	0.50	0.60	1.60–2.00	0.035	0.035	...	...	...	0.05Al
1340	1.5223, 42MnV7	0.38–0.45	0.15–0.35	1.60–1.90	0.035	0.035	...	...	...	0.07–0.12V
1345	1.5225, 51MnV7	0.48–0.55	0.15–0.35	1.60–1.92	0.035	0.035	...	...	...	0.07–0.12V
4032	1.5411	0.32–0.38	0.30–0.50	1.10–1.40	0.035	0.035	...	...	0.15–0.25	...
4037	1.5412, GS-40MnMo4 3	0.36–0.43	0.30–0.50	0.90–1.20	0.035	0.035	...	...	0.25–0.35	...
4023, 4024	1.5416, 20Mo3	0.16–0.24	0.15–0.35	0.50–0.80	0.040	0.040	...	...	0.25–0.35	...
4027	1.5419, 22Mo4	0.18–0.25	0.20–0.40	0.40–0.70	0.035	0.035	...	0.30	0.30–0.40	...
4037, 4042	1.5432, 42MnMo7	0.38–0.45	0.20–0.35	1.55–1.85	0.040	0.040	...	...	0.15–0.25	...
8620	1.6522, 20NiCrMo2	0.17–0.23	0.10–0.25	0.60–0.90	0.025	0.025	0.40–0.70	0.35–0.65	0.15–0.25	0.02–0.05Al
8620	1.6523, 21NiCrMo2	0.17–0.23	0.15–0.35	0.60–0.90	0.035	0.035	0.40–0.70	0.35–0.65	0.15–0.25	...
8620	1.6526, 21NiCrMoS2	0.17–0.23	0.40	0.65–0.95	0.035	0.020–0.035	0.40–0.70	0.40–0.70	0.15–0.25	...
8622	1.6541, 23NiCrMo5 2	0.20–0.26	0.15–0.35	1.10–1.40	0.025	0.025	0.40–0.70	0.40–0.60	0.20–0.30	0.02–0.05Al
8620	1.6543, 21NiCrMo2 2	0.18–0.23	0.20–0.35	0.70–0.90	0.035	0.035	0.40–0.70	0.40–0.60	0.20–0.30	...
8630	1.6545, 30NiCrMo2 2	0.27–0.34	0.15–0.40	0.70–1.00	0.035	0.035	0.40–0.70	0.40–0.60	0.15–0.30	...
8640, 8740	1.6546, 40NiCrMo2 2	0.37–0.44	0.15–0.40	0.70–1.00	0.035	0.035	0.40–0.70	0.40–0.60	0.15–0.30	...
E4340	1.6562, 40NiCrMo7 3	0.37–0.44	0.40	0.70–0.90	0.020	0.015	1.65–2.00	0.70–0.90	0.30–0.40	...
4340	1.6565, 40NiCrMo6	0.35–0.45	0.15–0.35	0.50–0.70	0.035	0.035	1.40–1.70	0.90–1.40	0.20–0.30	...
E9310	1.6657, 14NiCrMo13 4	0.12–0.17	0.15–0.40	0.30–0.60	0.025	0.020	3.00–3.50	0.80–1.10	0.20–0.30	...
50B40	1.7003, 38Cr2	0.35–0.42	0.40	0.50–0.80	0.035	0.030	...	0.40–0.60	...	...
50B40	1.7023, 28CrS2	0.35–0.42	0.40	0.50–0.80	0.035	0.020–0.030	...	0.40–0.60	...	...

(continued)

Table 33 (continued)

Nearest SAE grade	DIN number	Composition, wt%								
		C	Si	Mn	P	S	Ni	Cr	Mo	Others
5117	1.7016, 17Cr3	0.14–0.20	0.15–0.40	0.40–0.70	0.035	0.035	...	0.60–0.90	...	...
5120	1.7027, 20Cr4	0.17–0.23	0.40	0.60–0.90	0.035	0.035	...	0.90–1.20	...	...
5120	1.7028, 20Cr5 4	0.17–0.23	0.40	0.60–0.90	0.035	0.020–0.035	...	0.90–1.20	...	...
5132	1.7033, 34Cr4	0.30–0.37	0.15–0.40	0.60–0.90	0.035	0.035	...	0.90–1.20	...	...
5135	1.7034, 37Cr4	0.34–0.41	0.15–0.40	0.60–0.90	0.035	0.035	...	0.90–1.20	...	...
5140	1.7035, 41Cr4	0.38–0.45	0.15–0.40	0.50–0.80	0.035	0.035	...	0.90–1.20	...	...
5132	1.7037, 34CrS4	0.30–0.37	0.40	0.60–0.90	0.035	0.020–0.035	...	0.90–1.20	...	...
5135	1.7038, 37CrS4	0.34–0.41	0.40	0.60–0.90	0.035	0.020–0.035	...	0.90–1.20	...	...
5140	1.7039, 41CrS4	0.38–0.45	0.40	0.60–0.90	0.035	0.020–0.035	...	0.90–1.20	...	...
5135	1.7043, 38Cr4	0.34–0.40	0.15–0.40	0.60–0.90	0.025	0.035	...	0.90–1.20	...	...
5140	1.7045, 42Cr4	0.38–0.44	0.15–0.40	0.50–0.80	0.025	0.035	...	0.90–1.20	...	...
5120	1.7121, 20CrMnS3 3	0.17–0.23	0.20–0.35	0.60–1.00	0.040	0.020	...	0.60–1.00	...	...
5115, 5117	1.7131, 16MnCr5	0.14–0.19	0.15–0.40	1.00–1.30	0.035	0.035	...	0.80–1.10	...	...
5115, 5117	1.7131, GS-16MnCr5	0.14–0.19	0.15–0.40	1.00–1.30	0.035	0.035	...	0.80–1.10	...	...
50B50	1.7138, 52MnCrB3	0.48–0.55	0.15–0.35	0.75–1.00	0.035	0.035	...	0.40–0.60	...	0.0005(min)B
5115, 5117	1.7139, 16MnCrS5	0.14–0.19	0.15–0.40	1.00–1.30	0.035	0.020–0.035	...	0.80–1.10	...	...
5115, 5117	1.7142, 16MnCrPb5	0.14–0.19	0.15–0.35	1.00–1.30	0.035	0.035	...	0.80–1.10	...	0.20–0.35Pb
5147, 5150	1.7145, GS-50CrMn4 4	0.46–0.54	0.30–0.50	0.80–1.20	0.035	0.035	...	0.80–1.10	...	...
5120	1.7146, 20MnCrPb5	0.17–0.22	0.15–0.35	1.10–1.40	0.035	0.035	...	1.00–1.30	...	0.20–0.35Pb
5120	1.7147, GS-20MnCr5	0.17–0.22	0.15–0.40	1.10–1.40	0.035	0.035	...	1.00–1.30	...	...
5120	1.7149, 20MnCrS5	0.17–0.22	0.15–0.35	1.10–1.40	0.035	0.020–0.035	...	1.00–1.30	...	...
5115	1.7160, 16MnCrB5	0.14–0.18	0.15–0.35	1.00–1.30	0.035	0.015–0.035	...	0.90–1.20	...	0.0005(min)B
5117	1.7168, 18MnCrB5	0.16–0.20	0.15–0.35	1.00–1.30	0.035	0.015–0.035	...	0.90–1.20	...	0.0005(min)B
5155	1.7176, 55Cr3	0.52–0.59	0.15–0.40	0.70–1.00	0.035	0.035	...	0.60–0.90	...	...
4118	1.7211, 23CrMoB4	0.20–0.25	0.15–0.35	0.50–0.80	0.035	0.035	...	0.90–1.20	0.10–0.20	0.0005(min)B
4135	1.7220, 34CrMo4	0.30–0.37	0.40	0.60–0.90	0.035	0.030	...	0.90–1.20	0.15–0.30	...
4135	1.7220, GS-34CrMo4	0.30–0.37	0.15–0.40	0.50–0.80	0.035	0.035	...	0.90–1.20	0.15–0.30	...
4140	1.7223, 41CrMo4	0.38–0.44	0.15–0.40	0.50–0.80	0.025	0.035	...	0.90–1.20	0.15–0.30	...
4142	1.7223, 41CrMo4	0.38–0.44	0.15–0.40	0.50–0.80	0.025	0.035	...	0.90–1.20	0.15–0.30	...
4140	1.7225, 42CrMo4	0.38–0.45	0.40	0.50–0.80	0.035	0.030	...	0.90–1.20	0.15–0.30	...
4137, 4140	1.7225, GS-42CrMo4	0.38–0.45	0.30–0.50	0.50–0.80	0.035	0.035	...	0.80–1.20	0.20–0.30	...
4135	1.7226, 34CrMoS4	0.30–0.37	0.40	0.60–0.90	0.035	0.020–0.035	...	0.90–1.20	0.15–0.30	...
4140	1.7227, 42CrMoS4	0.38–0.45	0.40	0.60–0.90	0.035	0.020–0.035	...	0.90–1.20	0.15–0.30	...
4147	1.7228, 50CrMo4	0.46–0.54	0.15–0.40	0.50–0.80	0.035	0.035	...	0.90–1.20	0.15–0.30	...
4150	1.7228, 50CrMo4	0.46–0.54	0.15–0.40	0.50–0.80	0.035	0.035	...	0.90–1.20	0.15–0.30	...
4147	1.7228, GS-50CrMo4	0.46–0.54	0.25–0.50	0.50–0.80	0.035	0.035	...	0.90–1.20	0.15–0.25	...
4150	1.7228, GS-50CrMo4	0.46–0.54	0.25–0.50	0.50–0.80	0.035	0.035	...	0.90–1.20	0.15–0.25	...
4161	1.7229, 61CrMo4	0.57–0.65	0.15–0.35	0.40–0.60	0.035	0.035	...	0.90–1.20	0.15–0.25	...
4147, 4150	1.7230, 50CrMoPb4	0.46–0.54	0.15–0.40	0.50–0.80	0.035	0.035	0.60	0.90–1.20	0.15–0.30	0.15–0.30Pb
4135	1.7231, 33CrMo4	0.30–0.37	0.40	0.50–0.80	0.035	0.035	...	0.90–1.20	0.15–0.30	...
4147, 4150	1.7238, 49CrMo4	0.46–0.52	0.15–0.40	0.50–0.80	0.035	0.025	...	0.90–1.20	0.15–0.30	...
4118	1.7264, 20CrMo5	0.18–0.23	0.15–0.35	0.90–1.20	0.035	0.035	...	1.10–1.40	0.20–0.40	...
4161	1.7266, GS-58CrMnMo4 4 3	0.54–0.62	0.30–0.50	0.80–1.20	0.035	0.035	...	0.80–1.20	0.20–0.30	...
6150	1.8159, GS-50CrV4	0.47–0.55	0.15–0.40	0.70–1.10	0.035	0.035	...	0.90–1.20	...	0.10–0.20V
5130	1.8401, 30MnCrTi4	0.25–0.35	0.15–0.35	0.90–1.20	0.025	0.025	...	0.80–1.00	...	0.10Al, 0.15–0.30Ti
5150	1.8404, 60MnCrTi4	0.45–0.60	0.15–0.35	0.90–1.20	0.025	0.025	...	0.80–1.00	...	0.10Al, 0.15–0.30Ti

Table 34 Chemical compositions of Japanese Industrial Standard (JIS) carbon, carbon-manganese, resulfurized, and rephosphorized/resulfurized steels referenced in Table 31

Nearest SAE grade	JIS number	Composition, wt%					
		C	Si	Mn	P	S	Others
	G3445 (tubes)						
1008	STKM11A (11A)	0.12	0.35	0.60	0.040	0.040	...
1029	STKM15A (15A), STKM15C (15C)	0.25–0.35	0.35	0.30–1.00	0.040	0.040	...
1049	STKM17A (17A), STKM17C (17C)	0.45–0.55	0.40	0.40–1.00	0.040	0.040	...
	G4051 (carbon steels)						
1015	F15Ck	0.13–0.18	0.15–0.35	0.30–0.60	0.025	0.025	0.20Cr, 0.20Ni, 0.25Cu, Ni + Cr = 0.30
1010	S9Ck	0.07–0.12	0.15–0.35	0.30–0.60	0.025	0.025	...
1010	S10C	0.08–0.13	0.15–0.35	0.30–0.60	0.030	0.035	0.20Cr, 0.20Ni, 0.30Cu
1012	S12C	0.10–0.15	0.15–0.35	0.30–0.60	0.030	0.035	...
1015	S15C	0.13–0.18	0.15–0.35	0.30–0.60	0.030	0.035	...
1017	S17C	0.15–0.20	0.15–0.35	0.30–0.60	0.030	0.035	...
1020	S20C	0.18–0.23	0.15–0.35	0.30–0.60	0.030	0.035	...
1020	S20Ck	0.18–0.23	0.15–0.35	0.30–0.60	0.025	0.025	0.20Cr, 0.20Ni, 0.25Cu, Ni + Cr = 0.30
1023	S22C	0.20–0.25	0.15–0.35	0.30–0.60	0.030	0.035	...
1025	S25C	0.22–0.28	0.15–0.35	0.30–0.60	0.030	0.035	...
1029	S28C	0.25–0.31	0.15–0.35	0.60–0.90	0.030	0.035	...
1030	S30C	0.27–0.33	0.15–0.35	0.60–0.90	0.030	0.035	...
1035, 1037	S35C	0.32–0.38	0.15–0.35	0.60–0.90	0.030	0.035	...
1040	S40C	0.37–0.43	0.15–0.35	0.60–0.90	0.030	0.035	...
1042, 1043	S43C	0.40–0.46	0.15–0.35	0.60–0.90	0.030	0.035	0.20Cr, 0.20Ni, 0.30Cu, Ni + Cr = 0.35
1045	S45C	0.42–0.48	0.15–0.45	0.60–0.90	0.030	0.035	...
1050	S50C	0.47–0.53	0.15–0.35	0.60–0.90	0.030	0.035	...
1050, 1053, 1055	S53C	0.50–0.56	0.15–0.35	0.60–0.90	0.030	0.035	0.20Cr, 0.20Ni, 0.30Cu, Ni + Cr = 0.35
1055	S55C	0.52–0.58	0.15–0.35	0.60–0.90	0.030	0.035	...
1060	S58C	0.55–0.61	0.15–0.35	0.60–0.90	0.030	0.035	...
	G4052 (structural steels with specified hardenability bands)						
1536	SMn1H, SMn433H	0.30–0.36	0.15–0.35	1.20–1.50	0.030	0.030	...
1541	SMn2H, SMn438H	0.35–0.41	0.15–0.35	1.35–1.65	0.030	0.030	...
1541	SMn3H, SMn443H	0.40–0.46	0.15–0.35	1.35–1.65	0.030	0.030	...
	G4106 (Mn and Mn-Cr steels)						
1536	SMn1	0.30–0.36	0.15–0.35	1.20–1.50	0.030	0.030	...
1541	SMn2	0.35–0.41	0.15–0.35	1.35–1.65	0.030	0.035	...
1541	SMn3	0.40–0.46	0.15–0.35	1.35–1.65	0.030	0.030	...
1522, 1524	SMn21	0.17–0.23	0.15–0.35	1.20–1.50	0.030	0.030	...
1536	SMn433	0.30–0.36	0.15–0.35	1.20–1.50	0.030	0.030	...
1541	SMn443	0.40–0.46	0.15–0.35	1.35–1.65	0.030	0.030	...
1541	SMn438	0.35–0.41	0.15–0.35	1.35–1.65	0.030	0.030	...
	G4801 (spring steels)						
1078	SUP3	0.75–0.90	0.15–0.35	0.30–0.60	0.035	0.035	...
1095	SUP4	0.90–1.10	0.15–0.35	0.30–0.60	0.035	0.035	...
	4804 (free-cutting steels)						
1110	SUM11	0.08–0.13	...	0.30–0.60	0.040	0.08–0.13	...
1108	SUM12	0.08–1.13	...	0.60–0.90	0.040	0.08–0.13	...
1212	SUM21	0.13	...	0.70–1.00	0.07–0.12	0.16–0.23	...
1213	SUM22	0.13	...	0.70–1.00	0.07–0.12	0.24–0.33	...
1215	SUM23	0.09	...	0.75–1.05	0.040–0.09	0.26–0.35	...
1117	SUM31	0.14–0.20	...	1.00–1.30	0.040	0.08–0.13	...
1137	SUM41	0.32–0.39	...	1.35–1.65	0.040	0.08–0.13	...
1141	SUM42	0.37–0.45	...	1.35–1.65	0.040	0.08–0.13	...
1144	SUM43	0.40–0.48	...	1.35–1.65	0.040	0.24–0.33	...
	5111 (not identified)						
1045	SCC5	0.40–0.50	0.30–0.60	0.50–0.80	0.050	0.050	...
1524	SCMn1	0.20–0.30	0.30–0.60	1.00–1.60	0.040	0.040	...
1527, 1536	SCMn2	0.25–0.35	0.30–0.60	1.00–1.60	0.040	0.040	...
1536	SCMn3	0.30–0.40	0.30–0.60	1.00–1.60	0.040	0.040	...
1541	SCMn5	0.40–0.50	0.30–0.60	1.00–1.60	0.040	0.040	...

Table 35 Chemical compositions of Japanese Industrial Standard (JIS) alloy steels referenced in Table 31

Nearest SAE grade	JIS number	C	Si	Mn	P	S	Ni	Cr	Mo	Others
	G4052 (structural steels with specified hardenability bands)									
5135	SCr3H, SCr435H	0.32–0.38	0.15–0.35	0.55–0.90	0.030	0.030	...	0.85–1.25	...	...
5140	SCr4H, SCr440H	0.37–0.44	0.15–0.35	0.55–0.90	0.030	0.030	...	0.85–1.25	...	...
5115	SCr21H, SCr415H	0.12–0.18	0.15–0.35	0.55–0.90	0.030	0.030	...	0.85–1.20	...	...
5130	SCr2H, SCr430H	0.27–0.34	0.15–0.35	0.55–0.90	0.030	0.030	...	0.85–1.25	...	...
5120	SCr420H	0.17–0.23	0.15–0.35	0.55–0.90	0.030	0.030	...	0.85–1.25	...	...
4137, 4140	SCM4H, SCM440H	0.37–0.44	0.15–0.35	0.55–0.90	0.030	0.030	...	0.85–1.25	0.15–0.35	...
4145, 4147	SCM5H	0.42–0.49	0.15–0.35	0.55–0.90	0.030	0.030	...	0.85–1.25	0.15–0.35	...
4118	SCM415H	0.12–0.18	0.15–0.35	0.55–0.90	0.030	0.030	...	0.85–1.25	0.15–0.35	...
4118	SCM418H	0.15–0.21	0.15–0.35	0.55–0.90	0.030	0.030	...	0.85–1.25	0.15–0.35	...
4145, 4147	SCM445H	0.42–0.49	0.15–0.35	0.55–0.90	0.030	0.030	...	0.85–1.25	0.15–0.35	...
5120	SMn21H, SMn421H	0.16–0.23	0.15–0.35	1.15–1.55	0.030	0.030	...	0.85–1.25	0.15–0.35	...
50B40	SMnC3H, SMnC443H	0.39–0.46	0.15–0.35	1.30–1.70	0.030	0.030	...	0.35–0.70	...	...
8620	SNCM21H, SNCM220H	0.17–0.23	0.15–0.35	0.60–0.95	0.030	0.035	0.35–0.75	0.35–0.65	0.15–0.30	...
	G4054 (not identified)									
4135	SCM3H, SCM435H	0.32–0.39	0.15–0.35	0.55–0.90	0.030	0.030	...	0.85–1.25	0.15–0.35	...
	G4103 (Ni-Cr-Mo steels)									
4140	SNCM4	0.38–0.43	0.15–0.35	0.65–0.85	0.030	0.030	...	0.90–1.00	0.15–0.30	...
4340	SNCM8, SNCM439	0.36–0.43	0.15–0.35	0.60–0.90	0.030	0.030	1.60–2.00	0.60–1.00	0.15–0.30	...
8620	SNCM21, SNCM220	0.17–0.23	0.15–0.35	0.60–0.90	0.030	0.030	0.40–0.70	0.40–0.65	0.15–0.30	...
4320	SNCM23, SNCM420, SNCM420H	0.17–0.23	0.15–0.35	0.40–0.60	0.030	0.030	1.60–2.00	0.40–0.65	0.15–0.30	...
	G4104 (Cr steels)									
5130	SCr2, SCr430	0.28–0.33	0.15–0.35	0.60–0.85	0.030	0.030	...	0.90–1.20	...	...
5132	SCr3, SCr435	0.33–0.38	0.15–0.35	0.60–0.85	0.030	0.030	...	0.90–1.20	...	...
5140	SCr4, SCr440	0.38–0.43	0.15–0.35	0.60–0.85	0.030	0.030	...	0.90–1.20	...	...
5115	SCr21, SCr415	0.13–0.18	0.15–0.35	0.60–0.85	0.030	0.030	...	0.90–1.20	...	...
5120	SCr22, SCr420	0.18–0.23	0.15–0.35	0.60–0.85	0.030	0.030	...	0.90–1.20	...	...
	G4105 (Cr-Mo steels)									
4130, 4135	SCM1	0.27–0.37	0.15–0.35	0.30–0.60	0.030	0.030	...	1.00–1.50	0.15–0.30	...
4130	SCM2	0.28–0.33	0.15–0.35	0.60–0.85	0.030	0.030	...	0.90–1.20	0.15–0.30	...
4135	SCM3	0.33–0.38	0.15–0.35	0.60–0.85	0.030	0.030	...	0.90–1.20	0.15–0.30	...
4137, 4140	SCM4	0.38–0.43	0.15–0.35	0.60–0.85	0.030	0.030	...	0.90–1.20	0.15–0.30	...
4145, 4147	SCM5	0.43–0.48	0.15–0.35	0.60–0.85	0.030	0.030	...	0.90–1.20	0.15–0.30	...
4118	SCM21H	0.12–0.18	0.15–0.35	0.55–0.90	0.030	0.030	...	0.85–1.25	0.15–0.35	...
4118	SCM418H	0.15–0.21	0.15–0.35	0.55–0.90	0.030	0.030	...	0.85–1.25	0.15–0.35	...
4130	SCM430	0.28–0.33	0.15–0.35	0.60–0.85	0.030	0.030	...	0.90–1.20	0.15–0.30	...
4130, 4135	SCM432	0.27–0.37	0.15–0.35	0.30–0.60	0.030	0.030	...	1.00–1.50	0.15–0.30	...
4135	SCM435	0.33–0.38	0.15–0.35	0.60–0.85	0.030	0.030	...	0.90–1.20	0.15–0.30	...
4140	SCM440	0.38–0.43	0.15–0.35	0.60–0.85	0.030	0.030	...	0.90–1.20	0.15–0.30	...
4145, 4147	SCM445	0.43–0.48	0.15–0.35	0.60–0.85	0.030	0.030	...	0.90–1.20	0.15–0.30	...
	G4106 (Mn and Mn-Cr steels)									
4130	SCM2	0.28–0.33	0.15–0.35	0.60–0.80	0.030	0.030	...	0.90–1.20	0.15–0.30	...
50B40	SMnC3, SMnC443	0.40–0.46	0.15–0.35	1.35–1.65	0.030	0.030	...	0.35–0.70	...	...
	G4108 (bolting material)									
4340	SNB23-1-5	0.35–0.46	0.18–0.37	0.56–0.99	0.030	0.030	1.50–2.05	0.60–1.00	0.18–0.32	...
4340	SNB24-1-5	0.35–0.46	0.18–0.37	0.66–0.94	0.030	0.030	1.60–2.05	0.60–1.00	0.28–0.42	...
	G4801 (spring steels)									
9260	SUP7	0.55–0.65	1.80–2.20	0.70–1.00	0.035	0.035	...	...	...	...
5155	SUP9	0.50–0.60	0.15–0.35	0.65–0.95	0.035	0.035	...	0.65–0.95	...	...
5160	SUP9A	0.55–0.65	0.15–0.35	0.70–1.00	0.035	0.035	...	0.70–1.00	...	...
6150	SUP10	0.45–0.55	0.15–0.35	0.65–0.95	0.035	0.035	...	0.80–1.10	...	0.15–0.25V
5155	SUP11	0.50–0.60	0.15–0.35	0.65–0.95	0.035	0.035	...	0.65–0.95	...	0.0005(min)B
4161	SUP13	0.56–0.64	0.15–0.35	0.70–1.00	0.035	0.035	...	0.70–0.90	0.25–0.35	...
	G5111 (steel castings)									
50B40	SCMnCr4	0.35–0.45	0.30–0.60	1.20–1.60	0.040	0.040	...	0.40–0.80	...	...
4032	SCMnM3	0.30–0.40	0.30–0.60	1.20–1.60	0.040	0.040	...	0.20	0.15–0.35	...

182 / Carbon and Low-Alloy Steels

Table 36 Chemical compositions of British Standard (BS) carbon, carbon-manganese, resulfurized, and rephosphorized/resulfurized steels referenced in Table 31. Alternate designations are in parentheses.

Nearest SAE grade	BS number	C	Si	Mn	P	S	Others
\multicolumn{8}{c}{970 (carbon and carbon-manganese steel, free-cutting steel)}							
1005	015A03	0.06	0.10–0.40	0.40	0.050	0.050	...
1006	030A04	0.08	0.10–0.40	0.20–0.40	0.050	0.050	...
1006	040A04	0.08	0.10–0.40	0.30–0.50	0.050	0.050	...
1010	040A10 (En2A,2A/1,2B)	0.08–0.13	0.40	0.30–0.50	0.050	0.050	...
1012	040A12 (En2A,En2A/1,En2B)	0.10–0.15	0.40	0.30–0.50	0.050	0.050	...
1015	040A15	0.13–0.18	0.40	0.30–0.50	0.050	0.050	...
1017	040A17	0.15–0.20	0.40	0.30–0.50	0.050	0.050	...
1020	040A20	0.18–0.23	0.40	0.30–0.50	0.050	0.050	...
1023	040A22 (En2C,En2D)	0.20–0.25	0.40	0.30–0.50	0.050	0.050	...
1010	045A10	0.08–0.13	0.40	0.30–0.50	0.050	0.050	...
1010	045M10 (En32A)	0.07–0.13	0.10–0.40	0.30–0.60	0.050	0.050	...
1006	050A04	0.08	0.10–0.40	0.40–0.60	0.050	0.050	...
1010	050A10	0.08–0.13	0.10–0.40	0.40–0.60	0.050	0.050	...
1012	050A12	0.10–0.15	0.40	0.40–0.60	0.050	0.050	...
1015	050A15	0.13–0.18	0.40	0.50–0.70	0.050	0.050	...
1017	050A17	0.15–0.20	0.40	0.40–0.60	0.050	0.050	...
1020	050A20 (En2C,En2D)	0.18–0.23	0.40	0.40–0.60	0.050	0.050	...
1023	050A22	0.22–0.25	0.40	0.40–0.60	0.050	0.050	...
1086	050A86	0.83–0.90	0.10–0.40	0.40–0.60	0.050	0.050	...
1010	060A10	0.08–0.13	0.10–0.40	0.50–0.70	0.050	0.050	...
1012	060A12	0.10–0.15	0.40	0.50–0.70	0.050	0.050	...
1015	060A15	0.13–0.18	0.40	0.50–0.70	0.050	0.050	...
1017	060A17	0.15–0.20	0.40	0.50–0.70	0.050	0.050	...
1020	060A20	0.18–0.23	0.40	0.50–0.70	0.050	0.050	...
1023	060A22	0.20–0.25	0.04	0.50–0.70	0.050	0.050	...
1029	060A27	0.25–0.30	0.10–0.40	0.50–0.70	0.050	0.050	...
1030	060A30	0.28–0.33	0.40	0.50–0.70	0.050	0.050	...
1035	060A35	0.33–0.38	0.10–0.40	0.50–0.70	0.050	0.050	...
1039	060A40	0.38–0.43	0.40	0.50–0.70	0.050	0.050	...
1040	060A40	0.38–0.43	0.40	0.50–0.70	0.050	0.050	...
1042,1043	060A42	0.40–0.45	0.40	0.50–0.70	0.050	0.050	...
1045	060A47	0.45–0.50	0.10–0.40	0.50–0.70	0.050	0.050	...
1049	060A47	0.45–0.50	0.10–0.40	0.50–0.70	0.050	0.050	...
1050	060A52	0.50–0.55	0.40	0.50–0.70	0.050	0.050	...
1055	060A57	0.55–0.60	0.40	0.50–0.70	0.050	0.050	...
1060	060A57	0.55–0.60	0.40	0.50–0.70	0.050	0.050	...
1059	060A62	0.60–0.65	0.10–0.40	0.50–0.70	0.050	0.050	...
1064	060A62	0.60–0.65	0.10–0.40	0.50–0.70	0.050	0.050	...
1065	060A67	0.65–0.70	0.40	0.50–0.70	0.050	0.050	...
1070	060A72	0.70–0.75	0.40	0.50–0.70	0.050	0.050	...
1078	060A78	0.75–0.82	0.40	0.50–0.70	0.050	0.050	...
1080	060A78	0.75–0.82	0.40	0.50–0.70	0.050	0.050	...
1080	060A83	0.80–0.87	0.40	0.50–0.70	0.050	0.050	...
1084	060A86	0.83–0.90	0.40	0.50–0.70	0.050	0.050	...
1090	060A96	0.93–1.00	0.10–0.35	0.50–0.70	0.050	0.050	...
1095	060A99	0.95–1.05	0.40	0.50–0.70	0.050	0.050	...
1055	070M55	0.50–0.60	...	0.50–0.90	0.050	0.050	...
1070	070A72 (En42)	0.70–0.75	0.10–0.35	0.60–0.80	0.050	0.050	...
1074	070A72 (En42)	0.70–0.75	0.10–0.35	0.60–0.80	0.050	0.050	...
1080	070A78	0.75–0.82	0.10–0.40	0.60–0.80	0.050	0.050	...
1021	070M20	0.16–0.24	...	0.50–0.90	0.050	0.050	...
1026	070M26	0.22–0.30	...	0.50–0.90	0.050	0.050	...
1055	070M55	0.50–0.60	...	0.50–0.90	0.050	0.050	...
1015	080A15	0.13–0.18	0.40	0.70–0.90	0.050	0.050	...
1016	080A15	0.13–0.18	0.40	0.70–0.90	0.050	0.050	...
1018	080A17	0.15–0.20	0.40	0.70–0.90	0.050	0.050	...
1021	080A20	0.18–0.23	0.10–0.40	0.70–0.90	0.050	0.050	...
1023	080A22	0.20–0.25	0.10–0.40	0.70–0.90	0.050	0.050	...
1026	080A25	0.23–0.28	0.10–0.40	0.70–0.90	0.050	0.050	...
1026	080A27	0.25–0.30	0.10–0.40	0.70–0.90	0.050	0.050	...
1029	080A27 (En5A)	0.25–0.30	0.40	0.70–0.90	0.050	0.050	...
1030	080A30 (En5B)	0.28–0.33	0.40	0.70–0.90	0.050	0.050	...
1035	080A32 (En5C)	0.30–0.35	0.10–0.40	0.70–0.90	0.050	0.050	...
1035	080A35 (En8A)	0.33–0.38	0.10–0.40	0.70–0.90	0.050	0.050	...
1039	080A40 (En8C)	0.38–0.43	0.40	0.70–0.90	0.050	0.050	...
1040	080A40 (En8C)	0.38–0.43	0.40	0.70–0.90	0.050	0.050	...
1042	080A42 (En8D)	0.40–0.45	0.40	0.70–0.90	0.050	0.050	...
1043	080A42 (En8D)	0.40–0.45	0.40	0.70–0.90	0.050	0.050	...
1045	080A47	0.45–0.50	0.10–0.40	0.70–0.90	0.050	0.050	...
1049	080A47	0.45–0.50	0.10–0.40	0.70–0.90	0.050	0.050	...
1050	080A52 (En43C)	0.50–0.55	0.15–0.35	0.70–0.90	0.050	0.050	...
1053	080A52 (En43C)	0.50–0.55	0.15–0.35	0.70–0.90	0.050	0.050	...
1055	080A52 (En43C)	0.50–0.55	0.15–0.35	0.70–0.90	0.050	0.050	...
1055	080A57	0.55–0.60	0.40	0.70–0.90	0.050	0.050	...
1060	080A57	0.55–0.60	0.40	0.70–0.90	0.050	0.050	...

(continued)

Table 36 (continued)

Nearest SAE grade	BS number	C	Si	Mn	P	S	Others
	970 (carbon and carbon-manganese steel, free-cutting steel) (continued)						
1064	080A62 (En43D)	0.60–0.65	0.40	0.70–0.90	0.050	0.050	...
1065	080A67 (En43E)	0.65–0.70	0.40	0.70–0.90	0.050	0.050	...
1070, 1074	080A72	0.70–0.75	0.40	0.70–0.90	0.050	0.050	...
1080	080A78	0.75–0.82	0.40	0.70–0.90	0.050	0.050	...
1080, 1085	080A83	0.80–0.87	0.40	0.70–0.90	0.050	0.050	...
1084	080A86	0.83–0.90	0.40	0.70–0.90	0.050	0.050	...
1015	080M15	0.13–0.18	0.40	0.70–0.90	0.050	0.050	...
1016	080M15	0.13–0.18	0.40	0.70–0.90	0.050	0.050	...
1030	080M30 (En5)	0.26–0.34	...	0.60–1.00	0.050	0.050	...
1037	080M36	0.32–0.40	...	0.60–1.00	0.050	0.050	...
1039	080M40 (En8)	0.36–0.44	...	0.60–1.00	0.050	0.050	...
1040	080M40 (En8)	0.36–0.44	...	0.60–1.00	0.050	0.050	...
1043	080M46	0.42–0.50	...	0.60–1.00	0.050	0.050	...
1045	080M46	0.42–0.50	...	0.60–1.00	0.050	0.050	...
1046	080M46	0.42–0.50	...	0.60–1.00	0.050	0.050	...
1050	080M50 (En43A)	0.45–0.55	...	0.60–1.00	0.050	0.050	...
1053	080M52 (En43C)	0.50–0.55	0.15–0.35	0.70–0.90	0.050	0.050	...
1022	120M19	0.15–0.23	...	1.00–1.40	0.050	0.050	...
1526	120M28	0.24–0.32	...	1.00–1.40	0.050	0.050	...
1536	120M36 (En15B)	0.32–0.40	...	1.00–1.40	0.050	0.050	...
1513	125A15	0.13–0.18	0.10–0.40	1.10–1.40	0.050	0.050	...
1513	130M15	0.12–0.18	0.10–0.40	1.10–1.50	0.050	0.050	...
1513	130M15 (En201)	0.12–0.18	0.10–0.40	1.10–1.50	0.050	0.050	...
1513	130M15	0.12–0.18	0.10–0.40	1.10–1.50	0.050	0.050	...
1541	135M44	0.40–0.48	0.10–0.40	1.20–1.50	0.050	0.050	...
1524	150M19 (En14A)	0.15–0.23	...	1.30–1.70	0.050	0.050	...
1524	150M19 (En14B)	0.15–0.23	...	1.30–1.70	0.050	0.050	...
1527	150M28 (En14A)	0.24–0.32	...	1.30–1.70	0.050	0.050	...
1527	150M28 (En14B)	0.24–0.32	...	1.30–1.70	0.050	0.050	...
1536	150M36 (En15)	0.32–0.40	...	1.30–1.70	0.050	0.050	...
1541	150M40	0.36–0.44	0.10–0.40	1.30–1.70	0.050	0.050	...
1016	170H15	0.12–0.18	0.10–0.40	0.80–1.10	0.060	0.03–0.06	0.0005–0.005B
1022	170H20	0.17–0.23	0.10–0.40	0.80–1.20	0.050	0.050	0.0005–0.005B
1037	170H36	0.32–0.39	0.10–0.40	0.80–1.10	0.050	0.050	0.0005–0.005B
1039	170H41	0.37–0.44	0.10–0.40	0.80–1.10	0.050	0.050	0.0005–0.005B
1015	173H16	0.13–0.19	0.10–0.40	1.10–1.40	0.060	0.03–0.06	0.0005–0.005B
1016	173H16	0.13–0.19	0.10–0.40	1.10–1.40	0.060	0.03–0.06	0.0005–0.005B
1524	175H23	0.20–0.25	0.10–0.40	1.30–1.60	0.060	0.03–0.06	0.0005–0.005B
1117	210A15	0.13–0.18	0.10–0.40	0.90–1.20	0.050	0.10–0.18	...
1117	210M17 (En32M)	0.12–0.18	0.10–0.40	0.90–1.30	0.050	0.10–0.18	...
1139	212A37 (En8BM)	0.35–0.40	0.25	1.00–1.30	0.060	0.12–0.20	...
1141	212A42 (En8DM)	0.40–0.45	0.25	1.00–1.30	0.060	0.12–0.20	...
1144	212A42 (En8DM)	0.40–0.45	0.25	1.00–1.30	0.060	0.12–0.20	...
1137, 1139	212M36 (En8M)	0.32–0.40	0.25	1.00–1.40	0.060	0.12–0.20	...
1137, 1139	216M36 (En15AM)	0.32–0.40	0.25	1.30–1.70	0.060	0.12–0.20	...
1137, 1139	212M36 (En8M)	0.32–0.40	0.25	1.00–1.40	0.060	0.12–0.20	...
1144	212M44	0.40–0.48	0.25	1.00–1.40	0.060	0.12–0.20	...
1146	212M44	0.40–0.48	0.25	1.00–1.40	0.060	0.12–0.20	...
1117	214A15	0.13–0.18	0.10–0.40	1.10–1.50	0.050	0.10–0.18	...
1117	214M15 (En202)	0.12–0.18	0.10–0.40	1.20–1.60	0.050	0.10–0.18	...
1118	214M15 (En201)	0.12–0.18	0.10–0.40	1.20–1.60	0.050	0.10–0.18	...
1141	216A42	0.40–0.45	0.25	1.20–1.50	0.060	0.12–0.20	...
1144	216M44	0.40–0.48	0.25	1.20–1.50	0.060	0.12–0.20	...
1213	220M07 (En1A)	0.15	...	0.90–1.30	0.070	0.20–0.30	...
1137, 1139	225M36	0.32–0.40	0.25	1.30–1.70	0.060	0.12–0.20	...
1144	225M44	0.40–0.48	0.25	1.30–1.70	0.060	0.20–0.30	...
1144	226M44	0.40–0.48	0.25	1.30–1.70	0.060	0.22–0.30	...
1213	230M07	0.15	0.05	0.90–1.30	0.070	0.25–0.35	...
1213	240M07 (En1A)	0.15	...	1.10–1.50	0.070	0.30–0.40	...
1215	240M07 (En1B)	0.15	...	1.10–1.50	0.070	0.30–0.60	...
	980 (not identified)						
1524	CDS9, CDS10	0.26	0.35	1.20–1.70	0.050	0.050	...
1010	CEW1	0.013	...	0.60	0.050	0.050	...
1035	CFS6	0.30–0.40	0.35	0.50–0.80	0.050	0.050	...
1522	CFS7	0.20–0.30	0.35	1.20–1.50	0.050	0.050	...
	1045 (not identified)						
1536	1045	0.40	0.30	1.30–1.70	0.050	0.050	...
	1287 (not identified)						
1040	1287	0.40–0.48	0.30	0.50–0.90	0.045	0.045	...
	1449 (plate, sheet, and strip)						
1023	22HS	0.20–0.25	...	0.40–0.60	0.050	0.050	...
1008	3CR, 3CS, 3HR, 3HS	0.10	...	0.50	0.040	0.040	...
1010	4CR, 4CS, 4HR, 4HS	0.12	...	0.60	0.050	0.050	...

(continued)

Table 36 (continued)

Nearest SAE grade	BS number	C	Si	Mn	P	S	Others
	1449 (plate, sheet, and strip) (continued)						
1012	12CS, 12HS	0.10–0.15	...	0.40–0.60	0.050	0.050	...
1017	17CS, 17HS	0.15–0.20	...	0.40–0.60	0.050	0.050	...
1023	22CS	0.20–0.25	...	0.40–0.60	0.050	0.050	...
1030	30CS, 30HS	0.25–0.35	0.05–0.35	0.50–0.90	0.045	0.045	...
1040	40CS, 40HS	0.35–0.45	0.05–0.35	0.50–0.90	0.045	0.045	...
1010	40F30	0.12	...	1.20	0.030	0.035	...
1513	40/30CS, 40/30HR, 40/30HS	0.15	...	1.20	0.040	0.040	...
1010	43F35, 46F40, 50F45, 60F55, 68F62	0.12	...	1.20	0.030	0.035	...
1060	60CS, 60HS	0.55–0.65	0.05–0.30	0.50–0.90	0.045	0.045	...
1070	70CS, 70HS	0.65–0.75	0.05–0.30	0.50–0.90	0.045	0.045	...
1010	75F70CS, 75F70HR, 75F70HS	0.12	...	1.20	0.030	0.035	...
1080	80CS, 80HS	0.75–0.85	0.05–0.35	0.50–0.90	0.045	0.045	...
1090	95CS, 95HS	0.90–1.00	0.05–0.35	0.30–0.90	0.040	0.040	...
	1453 (not identified)						
1513	A2	0.10–0.20	0.10–0.35	1.00–1.60	0.040	0.040	...
1527	A3	0.25–0.30	0.30–0.50	1.30–1.60	0.050	0.050	0.25Cr, 0.25Ni
	1456 (not identified)						
1524	Grade A	0.18–0.25	0.50	1.20–1.60	0.050	0.050	...
1527	Grade B1, Grade B2	0.25–0.33	0.50	1.20–1.60	0.050	0.050	...
	1501 (plates for pressure vessels)						
1012	141–360	0.16	...	0.50	0.050	0.050	0.25Cr, 0.30Ni, 0.30Cu, 0.10Mo
	1503 (forgings for pressure vessels)						
1522	221–460	0.23	0.10–0.40	0.90–1.70	0.040	0.040	...
1522	223–409	0.25	0.10–0.40	0.90–1.70	0.040	0.040	0.01–0.06Nb
1522	224–490	0.25	0.10–0.40	0.90–1.70	0.040	0.040	0.15Al
	1549 (not identified)						
1050	50CS	0.45–0.55	0.05–0.35	0.50–0.90	0.045	0.045	...
1050	50HS	0.45–0.55	0.05–0.35	0.50–0.90	0.045	0.045	...
	1717 (tubes)						
1035	CDS105/106	0.30–0.40	0.35	0.30–0.90	0.050	0.050	...
1008	ERW101	0.10	...	0.60	0.060	0.060	...
	2772 (collier haulage and winding equipment)						
1513	150M12	0.10–0.15	0.10–0.35	1.30–1.70	0.050	0.050	...
	3059 (boiler and superheater tubes)						
1013	360	0.17	0.35	0.40–0.80	0.045	0.045	...
1016	440	0.12–0.18	0.10–0.35	0.90–1.20	0.040	0.035	...
	3601 (pipes and tubes)						
1013	360	0.17	0.35	0.40–0.60	0.045	0.045	...
	3100 (steel castings)						
1527	A5	0.25–0.33	0.60	1.20–1.60	0.050	0.050	...
1536	A5	0.25–0.33	0.60	1.20–1.60	0.050	0.050	...
1527	A6	0.25–0.33	0.60	1.20–1.60	0.050	0.050	...
1536	A6	0.25–0.33	0.60	1.20–1.60	0.050	0.050	...
1046	AW2	0.40–0.50	0.36	1.00	0.050	0.050	0.25Cr, 0.40Ni, 0.30Cu, 0.15Mo
1055	AW3	0.50–0.60	0.60	1.00	0.050	0.050	0.25Cr, 0.40Ni, 0.15Mo, 0.30Cu
	3111 (wire)						
1022	Type 9	0.17–0.23	0.15–0.35	0.80–1.10	0.040	0.040	0.02Al, 0.0008–0.005B
1037	Type 10	0.32–0.39	0.15–0.35	0.80–1.10	0.040	0.040	0.0008–0.005B, 0.02Al
	3146 (investment castings)						
1040	Class 1 Grade C	0.35–0.45	0.20–0.60	0.40–1.00	0.035	0.035	0.30Cr, 0.40Ni, 0.30Cu, 0.10Mo
1040	Class 8	0.37–0.45	0.20–0.60	0.50–0.80	0.035	0.035	0.30Cr, 0.40Ni, 0.30Cu, 0.10Mo
1522	CLA2	0.18–0.25	0.20–0.50	1.20–1.70	0.035	0.035	0.30Cr, 0.40Ni, 0.30Cu
	3601 (pipes and tubes)						
1013	360	0.17	0.35	0.40–0.80	0.045	0.045	...
	3603 (tubes for pressure vessels)						
1013	360	0.17	0.35	0.40–0.80	0.045	0.045	...
	3606 (tubes for heat exchangers)						
1008	261	0.06–0.10	0.10–0.35	0.60–0.80	0.020	0.020	0.20Cr, 0.40–0.60Mo, 0.06Al, 0.002–0.006B
1016	440	0.12–0.18	0.10–0.35	0.90–1.20	0.040	0.035	...

Table 37 Chemical compositions of British Standard (BS) alloy steels referenced in Table 31

Nearest SAE grade	BS number	C	Si	Mn	P	S	Ni	Cr	Mo	Others
	970 Part 1 (alloy steels)									
5115	527A17	0.14–0.19	0.10–0.35	0.70–0.90	0.035	0.040	...	0.70–0.90	...	...
5115	527H17, 527M17	0.14–0.20	0.10–0.35	0.70–1.00	0.035	0.040	...	0.60–0.90	...	...
5130	530A30 (En18A)	0.28–0.33	0.10–0.35	0.60–0.80	0.040	0.050	...	0.90–1.20	...	...
5132	530A32 (En18B)	0.30–0.35	0.10–0.35	0.60–0.80	0.040	0.050	...	0.90–1.20	...	...
5132, 5135	530A36 (En18C)	0.34–0.39	0.10–0.35	0.60–0.80	0.040	0.050	...	0.90–1.20	...	...
5140	530A40 (En18D)	0.38–0.43	0.10–0.35	0.60–0.80	0.040	0.050	...	0.90–1.20	...	...
5132	530H32	0.29–0.35	0.10–0.35	0.50–0.90	0.040	0.050	...	0.80–1.25	...	...
5135	530H36	0.33–0.40	0.10–0.35	0.50–0.90	0.040	0.050	...	0.80–1.25	...	...
5140	530H40	0.37–0.44	0.10–0.35	0.50–0.90	0.040	0.040	...	0.80–1.25	...	...
4032	605A32	0.30–0.35	0.10–0.35	1.30–1.70	0.040	0.050	...	...	0.22–0.32	...
4037	605A37	0.35–0.40	0.10–0.35	1.30–1.70	0.040	0.050	...	...	0.22–0.32	...
4032	605H32	0.29–0.35	0.10–0.35	1.25–1.75	0.035	0.040	...	...	0.22–0.32	...
4037	605H37	0.34–0.41	0.10–0.35	1.25–1.75	0.035	0.040	...	...	0.22–0.32	...
4032	605M36 (En16)	0.32–0.40	0.10–0.35	1.30–1.70	0.035	0.040	...	...	0.22–0.32	...
4617	665H17	0.14–0.20	0.10–0.35	0.35–0.75	0.035	0.040	1.50–2.00	...	0.20–0.30	...
4620	665H20, 665M20	0.17–0.23	0.10–0.35	0.35–0.75	0.035	0.040	1.50–2.00	...	0.20–0.30	...
4617	665M17 (En34)	0.14–0.20	0.10–0.35	0.35–0.75	0.035	0.040	1.50–2.00	...	0.20–0.30	...
4130	708A30	0.28–0.33	0.10–0.35	0.40–0.60	0.035	0.040	...	0.90–1.20	0.15–0.25	...
4135, 4137	708A37	0.35–0.40	0.10–0.35	0.70–1.00	0.040	0.050	...	0.90–1.20	0.15–0.25	...
4140	708A40	0.38–0.43	0.10–0.35	0.75–1.00	0.035	0.040	...	0.90–1.20	0.15–0.25	...
4140, 4142	708A42 (En19C)	0.40–0.45	0.10–0.35	0.70–1.00	0.040	0.050	...	0.90–1.20	0.15–0.25	...
4147	708A47	0.45–0.50	0.10–0.35	0.75–1.00	0.035	0.040	...	0.90–1.20	0.15–0.25	...
4118	708H20, 708M20	0.17–0.23	0.10–0.35	0.60–0.90	0.035	0.040	...	0.85–1.15	0.15–0.25	...
4135	708H37	0.34–0.41	0.10–0.35	0.65–1.05	0.040	0.050	...	0.80–1.25	0.15–0.25	...
4137	708H37	0.34–0.41	0.10–0.35	0.65–1.05	0.040	0.050	...	0.80–1.25	0.15–0.25	...
4140	708H42	0.39–0.46	0.10–0.35	0.65–1.05	0.040	0.050	...	0.80–1.25	0.15–0.25	...
4142	708H42	0.39–0.46	0.10–0.35	0.65–1.05	0.040	0.050	...	0.80–1.25	0.15–0.25	...
4145	708H45	0.42–0.49	0.10–0.35	0.65–1.05	0.035	0.040	...	0.80–1.25	0.15–0.25	...
4140	708M40	0.36–0.44	0.10–0.35	0.70–1.00	0.040	0.050	...	0.80–1.20	0.15–0.25	...
4137	709A37	0.35–0.40	0.10–0.35	0.75–1.00	0.035	0.040	...	0.90–1.20	0.25–0.35	...
4140	709A40	0.38–0.43	0.10–0.35	0.75–1.00	0.035	0.040	...	0.90–1.20	0.25–0.35	...
4142	709A42	0.40–0.45	0.10–0.35	0.75–1.00	0.035	0.040	...	0.90–1.20	0.25–0.35	...
4140	709M40	0.36–0.44	0.10–0.35	0.70–1.00	0.040	0.050	...	0.90–1.20	0.25–0.35	...
8617	805A17	0.15–0.20	0.10–0.35	0.70–0.90	0.035	0.040	0.40–0.70	0.40–0.60	0.15–0.25	...
8620	805A20	0.18–0.23	0.10–0.35	0.70–0.90	0.035	0.040	0.40–0.70	0.40–0.60	0.15–0.25	...
8622	805A22	0.20–0.25	0.10–0.35	0.70–0.90	0.035	0.040	0.40–0.70	0.40–0.60	0.15–0.25	...
8617	805H17, 805M17 (En361)	0.14–0.20	0.10–0.35	0.60–0.95	0.035	0.040	0.35–0.75	0.35–0.65	0.15–0.25	...
8620	805H20	0.17–0.23	0.10–0.35	0.60–0.95	0.035	0.040	0.35–0.75	0.35–0.65	0.15–0.25	...
8622	805H22, 805M22	0.19–0.25	0.10–0.35	0.60–0.90	0.035	0.040	0.35–0.75	0.35–0.65	0.15–0.25	...
E9310	832H13	0.10–0.16	0.10–0.35	0.35–0.60	(a)	(b)	3.00–3.75	0.70–1.00	0.10–0.25	...
E9310	832M13 (En36C)	0.10–0.16	0.10–0.35	0.35–0.60	(a)	(b)	3.00–3.75	0.70–1.00	0.10–0.25	...
8637	945M38 (En100)	0.34–0.42	0.10–0.35	1.20–1.60	0.040	0.050	0.60–0.90	0.40–0.60	0.15–0.25	...
	970 part 5 (spring steel)									
9260	250A58 (En45A)	0.55–0.62	1.70–2.10	0.70–1.00	0.050	0.050	...	...	...	...
9260	250A61 (En45A)	0.58–0.65	1.70–2.10	0.70–1.00	0.050	0.050	...	...	...	...
5160	527A60 (En48)	0.55–0.65	0.10–0.35	0.70–1.00	0.040	0.050	...	0.60–0.90	...	...
5160	527H60	0.55–0.65	0.10–0.35	0.65–1.05	0.040	0.050	...	0.55–0.90	...	...
6150	735A50 (En47)	0.46–0.54	0.10–0.35	0.60–0.90	0.040	0.050	...	0.80–1.10	...	...
8660	805A60	0.55–0.65	0.10–0.35	0.70–1.00	0.040	0.050	0.40–0.70	0.40–0.60	0.15–0.25	...
8660	805H60	0.55–0.65	0.10–0.35	0.65–1.05	0.040	0.050	0.35–0.65	0.35–0.65	0.15–0.25	...
	970 (unclassified alloy steels)									
5060	526M60 (En11)	0.55–0.65	0.10–0.35	0.50–0.80	0.040	0.050	...	0.50–0.80	...	...
5130	530H30	0.27–0.33	0.10–0.35	0.50–0.90	0.040	0.050	...	0.80–1.25	...	...
E52100	534A99 (En31)	0.95–1.10	0.10–0.35	0.25–0.40	0.040	0.050	...	1.20–1.60	...	...
E52100	535A99 (En31)	0.95–1.10	0.10–0.35	0.40–0.70	0.040	0.050	...	1.20–1.60	...	...
4028	605M30	0.26–0.34	0.10–0.35	1.30–1.70	0.040	0.050	...	...	0.22–0.32	...
4032	605M30	0.26–0.34	0.10–0.35	1.30–1.70	0.040	0.050	...	...	0.22–0.32	...
4617	665A17	0.15–0.20	0.10–0.35	0.45–0.65	0.040	0.050	1.60–2.00	0.25	0.20–0.30	...
4620	665A19	0.17–0.22	0.10–0.35	0.45–0.65	0.040	0.050	1.60–2.00	0.25	0.20–0.30	...
4626	665A24 (En35B)	0.22–0.27	0.10–0.35	0.45–0.65	0.040	0.050	1.60–2.00	0.25	0.20–0.30	...
8625	805H25, 805M25	0.22–0.28	0.10–0.35	0.60–0.95	0.040	0.050	0.35–0.75	0.35–0.65	0.15–0.25	...
8620	805M20 (En362)	0.17–0.23	0.10–0.35	0.60–0.95	0.035	0.040	0.35–0.75	0.35–0.65	0.15–0.25	...
8640	945A40 (En100C)	0.38–0.43	0.10–0.35	1.20–1.60	0.040	0.050	0.60–0.90	0.40–0.60	0.15–0.25	...
	1717 (not identified)									
4130	CDS110	0.26	0.35	0.40–0.80	0.050	0.050	...	0.80–1.20	0.15–0.30	...
	2772 (collier haulage and winding equipment)									
8622	806M22	0.19–0.25	0.10–0.35	0.60–0.95	0.040	0.050	0.35–0.75	0.35–0.65	0.15–0.25	...

(continued)

(a) Either ≦0.050P or ≦0.035P can be designated. (b) Either 0.025 to 0.050S or 0.015 to 0.040S can be designated.

Table 37 (continued)

Nearest SAE grade	BS (number)	C	Si	Mn	P	S	Ni	Cr	Mo	Others
	3100 (steel castings)									
5147, 5150	BW2	0.45–0.53	0.75	0.50–1.00	0.060	0.060	...	0.80–1.20	...	...
5147, 5150	BW3	0.45–0.53	0.75	0.50–1.00	0.060	0.060	...	0.80–1.20	...	...
4161	BW4	0.55–0.65	0.75	0.50–1.00	0.060	0.060	...	0.80–1.50	0.20–0.40	...
	3111 (wire)									
4037	Type 2/1	0.35–0.40	0.15–0.40	0.70–0.90	0.040	0.040	...	...	0.20–0.30	...
4037	Type 2/2	0.32–0.45	0.15–0.40	0.80–1.00	0.040	0.040	...	...	0.25–0.35	...
5135, 5140	Type 3	0.35–0.45	0.15–0.40	0.70–0.90	0.040	0.040	...	0.90–1.20	...	...
8640, 8740	Type 7	0.38–0.43	0.15–0.40	0.75–1.00	0.040	0.040	0.40–0.70	0.40–0.60	0.20–0.30	...
	3146 (investment castings)									
5147	CLA12 Grade A	0.45–0.55	0.30–0.80	0.50–1.00	0.035	0.035	0.40	0.80–1.20	0.10	0.30Cu
5147	CLA12 Grade B	0.45–0.55	0.30–0.80	0.50–1.00	0.035	0.035	0.40	0.80–1.20	0.10	0.30Cu
4161	CLA12 Grade C	0.55–0.65	0.30–0.80	0.50–1.00	0.035	0.035	0.40	0.80–1.50	0.20–0.40	0.30Cu
	4670 (alloy steel forgings)									
4140	711M40	0.36–0.44	0.10–0.35	0.60–1.10	0.040	0.040	0.40	0.90–1.50	0.25–0.40	...
4340	818M40	0.36–0.44	0.10–0.35	0.45–0.85	0.040	0.040	1.30–1.80	1.00–1.50	0.20–0.40	...
	(Aerospace materials: not identified)									
E9310	S.157	0.12–0.17	0.15–0.40	0.30–0.60	0.025	0.020	3.00–3.50	0.80–1.10	0.20–0.30	...
6150	S.204	0.46–0.54	0.10–0.35	0.60–0.90	0.025	0.020	...	0.80–1.10	...	0.15–0.25V
	970 (aerospace materials)									
5140	2S.117	0.35–0.45	0.10–0.35	0.60–0.90	0.040	0.040	...	0.90–1.20	...	...
4340, E4340	2S.119	0.36–0.44	0.15–0.35	0.45–0.70	0.025	0.020	1.30–1.70	1.10–1.40	0.20–0.35	...

Table 38 Chemical compositions of French (AFNOR NF) carbon, carbon-manganese, resulfurized, and rephosphorized/resulfurized steels referenced in Table 31

Nearest SAE grade	AFNOR NF number	C	Si	Mn	P	S	Others
	A33-101 (carbon steels)						
1010	AF34	0.12	0.30	0.30–0.60	0.040	0.040	...
1012	AF37	0.08–0.15	0.30	0.30–0.60	0.040	0.040	...
1018	AF42	0.14–0.21	0.10–0.40	0.50–0.80	0.040	0.040	...
1029	AF50	0.25–0.33	0.10–0.40	0.50–0.80	0.040	0.040	...
1513	AF50-S	0.20	0.55	1.50	0.040	0.040	...
1035	AF55	0.31–0.39	0.10–0.40	0.50–0.80	0.040	0.040	...
1040	AF60	0.37–0.45	0.10–0.40	0.50–0.80	0.040	0.040	...
1045	AF65	0.43–0.51	0.10–0.40	0.50–0.80	0.040	0.040	...
1055	AF70	0.50–0.58	0.10–0.40	0.50–0.80	0.040	0.040	...
	A35-501 (for general purpose)						
1513	E35-4	0.20	0.55	1.60	0.035	0.035	...
1513	E36-2	0.24	0.55	1.60	0.045	0.045	...
1513	E36-3	0.20	0.55	1.60	0.030	0.030	...
	A35-551 (cementation steels)						
1008	XC10	0.06–0.12	0.15–0.35	0.30–0.60	0.035	0.035	...
1012, 1013	XC12	0.10–0.16	0.15–0.35	0.30–0.60	0.035	0.035	...
1017, 1020	XC18	0.16–0.22	0.15–0.35	0.40–0.70	0.035	0.035	0.020 Al
1022, 1522	20MB5	0.16–0.22	0.10–0.40	1.10–1.40	0.035	0.035	0.0008–0.005B
1021	21B3	0.18–0.24	0.10–0.40	0.60–0.90	0.035	0.035	0.0008–0.005B
	A35-552 (semifinished products, bars, wire rods)						
1017, 1020	XC18	0.16–0.22	0.15–0.35	0.40–0.70	0.035	0.035	0.020 Al
1025	XC25	0.23–0.29	0.10–0.35	0.40–0.70	0.035	0.035	0.020 Al
1030	XC32	0.30–0.35	0.10–0.35	0.50–0.80	0.035	0.035	...
1039	XC38H2	0.35–0.40	0.15–0.35	1.20	0.035	0.035	...
1042	XC42H1	0.40–0.45	0.15–0.35	0.50–0.80	0.030	0.035	...
1043	XC42H2	0.40–0.45	0.15–0.35	1.20	0.035	0.035	...
1045	XC48H1	0.45–0.51	0.15–0.35	0.50–0.80	0.030	0.035	...
1046, 1049	XC48H1	0.45–0.51	0.15–0.35	0.50–0.80	0.030	0.035	...
1046	XC48H2	0.45–0.51	0.15–0.35	1.20	0.035	0.035	...
1055	XC55H1	0.52–0.60	0.15–0.35	0.50–0.80	0.030	0.035	...
1055	XC55H2	0.52–0.60	0.15–0.35	1.20	0.035	0.035	...
1522	20M5	0.16–0.22	0.15–0.35	1.10–1.40	0.035	0.035	...
1022	20MB5	0.16–0.22	0.10–0.40	1.10–1.40	0.035	0.035	0.0008–0.005B
1522	20MB5	0.16–0.22	0.10–0.40	1.10–1.40	0.030	0.035	0.0008–0.005B
1021	21B3	0.18–0.24	0.10–0.40	0.60–0.90	0.035	0.035	0.0008–0.005B
1536	32M5	0.32–0.38	0.15–0.35	1.10–1.40	0.035	0.035	...
1536	38MB5	0.34–0.40	0.10–0.40	1.10–1.40	0.035	0.035	0.0008–0.005B
1541	40M6	0.37–0.43	0.15–0.35	1.30–1.70	0.035	0.035	...

(continued)

Table 38 (continued)

Nearest SAE grade	AFNOR NF number	Composition, wt%					
		C	Si	Mn	P	S	Others
	A35-553 (strips)						
1020	C20	0.15–0.25	0.10–0.40	0.40–0.70	0.040	0.040	...
1035	C35	0.30–0.40	0.10–0.40	0.50–0.80	0.040	0.040	...
1042	C40	0.40–0.50	0.10–0.40	0.50–0.80	0.040	0.040	...
1017, 1020	XC18S	0.15–0.22	0.25	0.40–0.65	0.035	0.035	...
1030	XC32	0.30–0.35	0.10–0.35	0.50–0.80	0.035	0.035	...
1035	XC38	0.35–0.40	0.10–0.35	0.50–0.80	0.035	0.035	...
1045	XC45	0.42–0.48	0.15–0.35	0.50–0.80	0.035	0.035	...
1050	XC50	0.46–0.52	0.15–0.35	0.50–0.80	0.035	0.035	...
1053, 1055	XC54	0.50–0.57	0.15–0.35	0.40–0.70	0.035	0.035	...
1059, 1060	XC60	0.57–0.65	0.15–0.35	0.40–0.70	0.035	0.035	...
1069	XC68	0.65–0.73	0.15–0.35	0.40–0.70	0.035	0.035	...
1074, 1075	XC75	0.70–0.80	0.15–0.30	0.40–0.70	0.035	0.035	...
1086	XC90	0.85–0.95	0.15–0.30	0.30–0.50	0.030	0.025	...
1095	XC100	0.95–1.05	0.15–0.30	0.25–0.45	0.030	0.025	...
1022	20MB5	0.16–0.22	0.10–0.40	1.10–1.40	0.035	0.035	0.0008–0.005B
1522	20MB5	0.16–0.22	0.10–0.40	1.10–1.40	0.030	0.035	0.0008–0.005B
1021	21B3	0.18–0.24	0.10–0.40	0.60–0.90	0.035	0.035	0.0008–0.005B
1039, 1536	38MB5	0.34–0.40	0.10–0.40	1.10–1.40	0.035	0.035	0.0008–0.005B
	A35-554 (plate and universal plate)						
1017, 1020	XC18S	0.15–0.22	0.25	0.40–0.65	0.035	0.035	...
1035	XC38	0.35–0.40	0.10–0.35	0.50–0.80	0.035	0.035	...
1045, 1049	XC48	0.45–0.51	0.10–0.40	0.50–0.80	0.035	0.035	...
	A35-556 (locking bolts)						
1522	20M5	0.16–0.22	0.15–0.35	1.10–1.40	0.035	0.035	...
1022	20MB5	0.16–0.22	0.10–0.40	1.10–1.40	0.035	0.035	0.0008–0.005B
1522	20MB5	0.16–0.22	0.10–0.40	1.10–1.40	0.030	0.035	0.0008–0.005B
1039, 1536	38MB5	0.34–0.40	0.10–0.40	1.10–1.40	0.035	0.035	0.0008–0.005B
	A35-557 (high-performance bolts)						
1039	XC38H2	0.35–0.40	0.10–0.35	0.50–1.20	0.035	0.040	0.40Cr, 0.40Ni, 0.10Mo
1022	20MB5	0.16–0.22	0.10–0.40	1.10–1.40	0.035	0.035	0.0008–0.005B
1522	20MB5	0.16–0.22	0.10–0.40	1.10–1.40	0.030	0.035	0.0008–0.005B
1021	21B3	0.18–0.24	0.10–0.40	0.60–0.90	0.035	0.035	0.0008–0.005B
1039, 1536	38MB5	0.34–0.40	0.10–0.40	1.10–1.40	0.035	0.035	0.0008–0.005B
	A35-561 (free-cutting steels)						
1213	S250	0.14	0.08	0.90–1.50	0.110	0.250–0.320	...
1215	S300	0.15	0.09	1.00–1.60	0.110	0.300–0.400	...
	A35-562 (free-cutting steels for drawing)						
1108	10F1	0.07–0.13	0.10–0.40	0.60–0.90	0.040	0.09–0.13	...
1137, 1139	35MF6	0.33–0.39	0.10–0.40	1.10–1.70	0.040	0.09–0.13	...
1141	45MF4	0.42–0.49	0.10–0.40	0.80–1.10	0.040	0.09–0.13	...
1144	45MF6	0.41–0.48	0.10–0.40	1.30–1.70	0.040	0.24–0.33	...
	A35-564 (steels for cold forming)						
1006	XC6FF	0.04–0.08	0.10	0.25–0.40	0.030	0.030	...
	A35-566 (chain steels)						
1017, 1020	XC18	0.16–0.22	0.15–0.35	0.40–0.70	0.035	0.035	0.020Al
1025	XC25	0.23–0.29	0.10–0.35	0.40–0.70	0.035	0.035	0.020Al
1022	20M5	0.16–0.22	0.10–0.35	1.10–1.40	0.035	0.035	0.020Al
1022	20MB5	0.16–0.22	0.10–0.40	1.10–1.40	0.035	0.035	0.0008–0.005B
1522	20MB5	0.16–0.22	0.10–0.40	1.10–1.40	0.030	0.035	0.0008–0.005B
1021	21B3	0.18–0.24	0.10–0.40	0.60–0.90	0.035	0.035	0.0008–0.005B
1526	25MS5	0.24–0.30	0.30–0.55	1.10–1.60	0.035	0.035	...
	Nonstandardized steels						
1010	C10	0.12	0.30	0.30–0.60	0.040	0.040	...
1012	C12	0.08–0.15	0.30	0.30–0.60	0.040	0.040	...
1018	C20	0.14–0.21	0.10–0.40	0.50–0.80	0.040	0.040	...
1029	C30	0.25–0.33	0.10–0.40	0.50–0.80	0.040	0.040	...
1035	C35	0.31–0.39	0.10–0.40	0.50–0.80	0.040	0.040	...
1040	C40	0.37–0.45	0.10–0.40	0.50–0.80	0.040	0.040	...
1045	C45	0.43–0.51	0.10–0.40	0.50–0.80	0.040	0.040	...
1055	C55	0.50–0.58	0.10–0.40	0.50–0.80	0.040	0.040	...
1010	CC10	0.05–0.15	0.30	0.30–0.50	0.040	0.040	...
1013	CC12	0.06–0.18	0.30	0.40–0.70	0.050	0.050	...
1020	CC20	0.15–0.25	0.10–0.40	0.40–0.70	0.040	0.040	...
1029	CC28	0.25–0.30	0.10–0.40	0.40–0.70	0.050	0.050	...
1042	CC45	0.40–0.50	0.10–0.40	0.50–0.80	0.040	0.040	...
1212	S200	0.13	0.08	0.70–1.20	0.110	0.20–0.27	...
1213	S250	0.14	0.08	0.90–1.50	0.110	0.225–0.35	...
1008	XC6, XC6FF	0.04–0.09	0.10	0.25–0.45	0.030	0.030	...

(continued)

Table 38 (continued)

Nearest SAE grade	AFNOR NF number	C	Si	Mn	P	S	Others
	Nonstandardized steels (continued)						
1015	XC15	0.12–0.18	0.35	0.30–0.70	0.040	0.035	...
1035	XC35	0.32–0.38	0.10–0.40	0.50–0.80	0.040	0.035	...
1035	XC38TS	0.35–0.40	0.10–0.40	0.50–0.80	0.025	0.030	...
1039	XC42	0.14–0.45	0.10–0.40	0.50–0.80	0.025	0.035	...
1042	XC42	0.40–0.45	0.10–0.40	0.50–0.80	0.025	0.035	...
1039, 1042	XC42TS	0.40–0.45	0.10–0.40	0.50–0.80	0.025	0.035	...
1045, 1046, 1049	XC48TS	0.45–0.51	0.10–0.40	0.50–0.80	0.025	0.030	...
1055	XC55	0.55–0.57	0.15–0.30	0.40–0.70	0.035	0.035	...
1065	XC65	0.60–0.69	0.10–0.40	0.50–0.80	0.035	0.035	...
1069, 1070	XC70	0.68–0.77	0.10–0.40	0.50–0.80	0.035	0.035	...
1074, 1075	XC70	0.68–0.77	0.10–0.40	0.50–0.80	0.035	0.035	...
1078, 1080	XC80	0.75–0.85	0.10–0.40	0.50–0.80	0.035	0.035	0.12Cr
1084	XC85	0.80–0.98	0.20–0.40	0.40–0.70	0.040	0.040	...
1212	10F2	0.08–0.14	0.10–0.40	0.50–0.75	0.060	0.12–0.24	...
1513	12M5	0.10–0.15	0.40	0.90–1.40	0.040	0.035	...
1212	12MF4	0.09–0.15	0.10–0.40	0.90–1.20	0.060	0.12–0.24	...
1551	24M4TS	0.49–0.55	0.10–0.40	0.80–1.10	0.025	0.035	...
1137	35MF4	0.32–0.38	0.10–0.40	1.00–1.30	0.060	0.12–0.24	...
1139	35M4	0.42–0.49	0.10–0.40	0.80–1.10	0.040	0.09–0.13	...
1039, 1541	40M5	0.36–0.44	0.10–0.40	1.00–1.35	0.040	0.035	...
1146	45MF4	0.42–0.49	0.10–0.40	0.80–1.10	0.040	0.09–0.013	...
1046	45M4TS	0.43–0.49	0.10–0.40	0.80–1.10	0.035	0.035	...
1541	45M5	0.39–0.48	0.10–0.40	1.20–1.50	0.040	0.035	...
1053	52M4TS	0.49–0.55	0.10–0.40	0.80–1.10	0.025	0.035	...
1552	55M5	0.50–0.60	0.10–0.40	1.20–1.50	0.040	0.035	...

Table 39 Chemical compositions of French (AFNOR NF) alloy steels referenced in Table 31

Nearest SAE grade	AFNOR NF number	C	Si	Mn	P	S	Ni	Cr	Mo	Others
	A35-551 (cementation steels)									
5115, 5117	16MC5	0.14–0.19	0.10–0.40	1.00–1.30	0.035	0.035	...	0.80–1.10	...	...
8620	19NCDB2	0.17–0.23	0.10–0.40	0.65–0.95	0.035	0.035	0.40–0.70	0.40–0.65	0.15–0.25	0.0008–0.005B
5120	20MC5	0.17–0.22	0.10–0.40	1.10–1.40	0.035	0.035	...	1.00–1.30	...	...
8620	20NCD2	0.17–0.25	0.10–0.40	0.65–0.95	0.035	0.035	0.40–1.00	0.40–0.65	0.15–0.25	...
	A35-552 (semifinished products, bars, wire rods)									
8620	19NCDB2	0.17–0.23	0.10–0.40	0.65–0.95	0.035	0.035	0.40–0.70	0.40–0.65	0.15–0.25	0.0008–0.005B
5120	20MC5	0.17–0.22	0.10–0.40	1.10–1.40	0.035	0.035	...	1.00–1.30	...	...
4130	30CD4	0.27–0.33	0.10–0.40	0.60–0.90	0.035	0.035	...	0.90–1.20	0.15–0.25	...
5132	32C4	0.30–0.35	0.10–0.40	0.60–0.90	0.035	0.035	...	0.85–1.20	...	...
4135	35CD4	0.30–0.37	0.10–0.40	0.60–0.90	0.035	0.035	...	0.85–1.20	0.15–0.25	...
50B40	38C2	0.35–0.40	0.10–0.40	0.60–0.90	0.035	0.035	...	0.30–0.60	...	...
5135	38Cr4	0.35–0.40	0.10–0.40	0.60–0.90	0.035	0.035	...	0.85–1.20	...	...
4137	38CD4	0.35–0.41	0.10–0.40	0.60–0.90	0.035	0.035	...	0.90–1.20	0.15–0.25	...
50B40	42C2	0.40–0.45	0.10–0.40	0.60–0.90	0.035	0.035	...	0.30–0.60	...	...
5140	42C4	0.40–0.45	0.10–0.40	0.60–0.90	0.035	0.035	...	0.90–1.20	...	...
4140, 4142	42CD4	0.39–0.45	0.10–0.40	0.60–0.90	0.035	0.035	...	0.85–1.20	0.15–0.25	...
4140, 4142	42CDTS	0.39–0.45	0.10–0.40	0.60–0.90	0.035	0.035	...	0.85–1.20	0.15–0.25	...
4145, 4147	45SCD6	0.42–0.50	1.30–1.70	0.50–0.80	0.035	0.035	...	0.50–0.75	0.15–0.30	...
6150	50CV4	0.47–0.55	0.10–0.40	0.70–1.10	0.035	0.035	...	0.85–1.20	...	0.10–0.20V
5060	60SC7	0.55–0.65	1.30–1.80	0.60–0.90	0.035	0.035	...	0.45–0.70	...	...
	A35-553 (strips)									
8620	20NCD2	0.18–0.23	0.10–0.40	0.70–0.90	0.030	0.025	0.40–0.70	0.40–0.60	0.15–0.30	0.35Cu
5132	32C4	0.30–0.35	0.10–0.40	0.60–0.90	0.035	0.035	...	0.85–1.20	...	...
4135	35CD4	0.30–0.37	0.10–0.40	0.60–0.90	0.035	0.035	...	0.85–1.20	0.15–0.25	...
5135	38Cr4	0.35–0.40	0.10–0.40	0.60–0.90	0.035	0.035	...	0.85–1.20	...	...
4140	42CD4	0.39–0.46	0.20–0.50	0.50–0.80	0.030	0.025	...	0.95–1.30	0.15–0.30	...
4142	42CD4	0.39–0.45	0.10–0.40	0.60–0.90	0.035	0.035	...	0.85–1.20	0.15–0.25	...
4140, 4142	42CDTS	0.38–0.45	0.10–0.40	0.60–0.90	0.035	0.035	...	0.85–1.20	0.15–0.25	...
4145, 4147	45SCD6	0.42–0.50	1.30–1.70	0.50–0.80	0.035	0.035	...	0.50–0.75	0.15–0.30	...
6150	50CV4	0.47–0.55	0.10–0.40	0.70–1.10	0.035	0.035	...	0.85–1.20	...	0.10–0.20V
	A35-556 (locking bolts)									
8622	23NCDB2	0.20–0.25	0.10–0.35	0.65–0.95	0.030	0.025	0.40–0.70	0.40–0.65	0.15–0.25	0.0008(min)B, 0.020Al
8622	23MNCD5	0.20–0.26	0.10–0.35	1.10–1.40	0.030	0.025	0.40–0.70	0.40–0.60	0.20–0.30	0.020Al
8625	25MNCD6	0.23–0.28	0.10–0.35	1.40–1.70	0.020	0.020	0.40–0.70	0.40–0.70	0.20–0.30	0.020Al
4130	30CD4	0.27–0.33	0.10–0.40	0.60–0.90	0.035	0.035	...	0.90–1.20	0.15–0.25	...
5132	32C4	0.30–0.35	0.10–0.40	0.60–0.90	0.035	0.035	...	0.85–1.20	...	...
4135	35CD4	0.30–0.37	0.10–0.40	0.60–0.90	0.035	0.035	...	0.85–1.20	0.15–0.25	...

(continued)

Table 39 (continued)

Nearest SAE grade	AFNOR NF number	C	Si	Mn	P	S	Ni	Cr	Mo	Others
	A 35-556 (locking bolts) (continued)									
50B40	38C2	0.35–0.40	0.10–0.40	0.60–0.90	0.035	0.035	...	0.30–0.60	...	...
5135	38Cr4	0.35–0.40	0.10–0.40	0.60–0.90	0.035	0.035	...	0.85–1.20	...	...
50B40	42C2	0.40–0.45	0.10–0.40	0.60–0.90	0.035	0.035	...	0.30–0.60	...	...
5140	42C4	0.39–0.45	0.10–0.40	0.60–0.90	0.025	0.035	0.30	0.85–1.15	...	...
4140, 4142	42CD4	0.39–0.45	0.10–0.40	0.60–0.90	0.035	0.035	...	0.90–1.20	0.15–0.25	...
4140, 4142	42CDTS	0.39–0.45	0.10–0.40	0.60–0.90	0.035	0.035	...	0.90–1.20	0.15–0.25	...
	A35-557 (high-performance bolts)									
4130	30CD4	0.27–0.33	0.10–0.40	0.60–0.90	0.035	0.035	...	0.90–1.20	0.15–0.25	...
5132	32C4	0.30–0.35	0.10–0.40	0.60–0.90	0.035	0.035	...	0.85–1.20	...	...
4135	34CD4	0.31–0.37	0.10–0.40	0.60–0.90	0.035	0.035	...	0.90–1.20	0.15–0.25	...
50B40	38C2	0.35–0.40	0.10–0.40	0.60–0.90	0.035	0.035	...	0.30–0.60	...	...
5135	38Cr4	0.35–0.40	0.10–0.40	0.60–0.90	0.035	0.035	...	0.85–1.20	...	...
4137	38CD4	0.35–0.41	0.10–0.40	0.60–0.90	0.035	0.035	...	0.90–1.20	0.15–0.25	...
5140	42C4	0.40–0.45	0.10–0.40	0.60–0.90	0.035	0.035	...	0.90–1.20	...	...
4140, 4142	42CD4	0.39–0.45	0.10–0.40	0.60–0.90	0.035	0.035	...	0.85–1.20	...	...
4140, 4142	42CDTS	0.39–0.45	0.10–0.40	0.60–0.90	0.035	0.035	...	0.85–1.20	0.15–0.25	...
	A35-565 (ball- and roller-bearing steels)									
4320	18NCD4	0.16–0.22	0.20–0.35	0.50–0.80	0.030	0.025	0.90–1.20	0.35–0.55	0.15–0.30	0.35Cu
8620	20NCD2	0.17–0.25	0.10–0.20	0.65–0.95	0.035	0.035	0.40–1.00	0.40–0.65	0.15–0.25	...
4320	20NCD7	0.16–0.22	0.20–0.35	0.45–0.65	0.030	0.025	1.65–2.00	0.20–0.60	0.20–0.30	0.35Cu
E50100	100C2	0.95–1.10	0.15–0.35	0.20–0.40	0.030	0.025	...	0.40–0.60	...	...
	A35-566 (chain steels)									
8620	20NCD2	0.17–0.25	0.10–0.20	0.65–0.95	0.035	0.035	0.40–1.00	0.40–0.65	0.15–0.25	...
8622	22NCD2	0.20–0.25	0.10–0.35	0.65–0.95	0.030	0.025	0.40–0.70	0.40–0.65	0.15–0.25	...
8625	25MnDC6	0.23–0.26	0.10–0.35	1.40–1.70	0.020	0.020	0.40–0.70	0.40–0.60	0.20–0.30	0.020Al
	A35-571 (hot formed steel springs)									
6150	50CV4	0.47–0.55	0.10–0.40	0.70–1.10	0.035	0.035	...	0.85–1.20	...	0.10–0.20V
4147	50SCD6	0.46–0.54	1.40–1.80	0.70–1.10	0.025	0.020	...	0.80–1.10	0.20–0.35	...
5155	55C3	0.52–0.59	0.10–0.40	0.70–0.90	0.035	0.035	...	0.60–0.90	...	...
	Nonstandardized steels									
4620	2ND8	0.16–0.23	0.10–0.35	0.20–0.50	0.040	0.035	1.80–2.30	...	0.15–0.30	...
8615	15NCD2	0.13–0.18	0.10–0.40	0.70–0.90	0.040	0.035	0.40–0.70	0.40–0.60	0.15–0.25	...
8615	15NCD4	0.12–0.19	0.10–0.40	0.50–0.90	0.040	0.035	1.00–1.30	0.40–0.70	0.10–0.20	...
4615	15ND8	0.13–0.18	0.10–0.35	0.20–0.50	0.040	0.035	1.80–2.30	...	0.15–0.30	...
5115	16MC5	0.14–0.19	0.10–0.40	1.00–1.30	0.035	0.035	...	0.80–1.10	...	...
E9310	16NCD13	0.12–0.17	0.35	0.50	0.030	0.025	3.00–3.50	0.85–1.15	0.15–0.30	0.35Cu
5117	18Cr4	0.16–0.21	0.10–0.40	0.60–0.80	0.040	0.035	...	0.85–1.15	...	...
4720, 8620	18NCD4	0.16–0.22	0.20–0.35	0.50–0.80	0.030	0.025	0.90–1.20	0.35–0.55	0.15–0.30	0.35Cu
8617	18NCD4	0.16–0.20	0.20–0.35	0.50–0.80	0.030	0.025	0.90–1.20	0.35–0.55	0.15–0.30	0.35Cu
8617	18NCD6	0.14–0.20	0.10–0.40	0.60–0.90	0.035	0.035	1.20–1.60	0.85–1.15	0.15–0.30	...
8620	20NCD2	0.18–0.23	0.10–0.40	0.70–0.90	0.030	0.025	0.40–0.70	0.40–0.60	0.15–0.30	0.35Cu
4320	20NCD7	0.16–0.22	0.20–0.35	0.45–0.65	0.030	0.025	1.65–2.00	0.20–0.60	0.20–0.30	0.35Cu
4422	23D5	0.20–0.26	0.10–0.35	0.50–0.80	0.030	0.025	...	...	0.45–0.60	0.020Al
8622	23NCDB4	0.20–0.25	0.10–0.25	0.65–0.95	0.030	0.025	0.40–0.70	0.40–0.65	0.15–0.25	0.0008(min)B
8625	25NCD4	0.22–0.28	0.10–0.40	0.50–0.90	0.040	0.035	1.00–1.30	0.40–0.70	0.10–0.20	...
5130	28C4	0.25–0.30	0.40	0.60–0.90	0.040	0.035	...	0.85–1.15	...	...
8630	30NCD2	0.30–0.35	0.10–0.40	0.70–0.90	0.040	0.035	0.50–0.80	0.40–0.60	0.15–0.30	...
4135	35CD4	0.33–0.39	0.10–0.40	0.60–0.90	0.035	0.035	...	0.85–1.15	0.15–0.35	...
5135	38C4	0.35–0.40	0.10–0.40	0.60–0.90	0.035	0.035	...	0.85–1.15	...	...
4137, 4140, 4142	40CD4	0.39–0.46	0.20–0.50	0.50–0.80	0.030	0.025	...	0.95–1.30	0.15–0.30	...
8640, 8740	40NCD2	0.37–0.44	0.10–0.40	0.60–0.90	0.040	0.035	0.40–0.70	0.40–0.60	0.15–0.30	...
8640, 8740	40NCD2TS	0.38–0.44	0.10–0.40	0.70–1.00	0.025	0.030	0.40–0.70	0.40–0.60	0.15–0.30	...
8637, 8640	40NCD3	0.36–0.43	0.10–0.40	0.50–0.80	0.035	0.035	0.70–1.00	0.60–0.90	0.15–0.30	...
8640	40NCD3TS	0.38–0.44	0.10–0.40	0.70–1.00	0.025	0.030	0.40–0.70	0.40–0.60	0.15–0.30	...
8740	40NCD3TS	0.38–0.44	0.10–0.40	0.70–1.00	0.025	0.030	0.40–0.70	0.40–0.60	0.15–0.30	...
4137	42CD4	0.39–0.46	0.10–0.40	0.60–0.90	0.035	0.035	...	0.85–1.15	0.15–0.30	...
50B44	45C2	0.40–0.50	0.35	0.50–0.80	0.040	0.035	...	0.40–0.60	...	...
5147	50C4	0.46–0.54	0.10–0.40	0.60–0.90	0.040	0.035	...	0.85–1.15	...	...
50B50	55C2	0.50–0.60	0.35	0.60–0.90	0.040	0.035	...	0.40–0.60	...	...
9260	60S7	0.55–0.65	1.50–2.00	0.70–1.00	0.050	0.050	...	...	...	...
9260	61S7	0.57–0.64	1.60–2.00	0.60–0.90	0.035	0.035	...	0.45	...	...
5060	61SC7	0.57–0.64	1.60–2.00	0.60–0.90	0.035	0.035	...	0.20–0.45	...	...

Table 40 Chemical compositions of Italian (UNI) carbon, carbon-manganese, resulfurized, and rephosphorized/resulfurized steels referenced in Table 31

Nearest SAE grade	UNI number	C	Si	Mn	P	S	Others
	3545 (spring steels)						
1045	C45	0.42–0.50	0.15–0.40	0.40–0.90	0.035	0.035	...
1060	C60	0.57–0.65	0.15–0.40	0.50–0.90	0.035	0.035	...
1070	C70	0.65–0.72	0.15–0.40	0.50–0.90	0.035	0.035	...
1074, 1075	C75	0.70–0.80	0.15–0.30	0.40–0.70	0.035	0.035	...
1090	C90	0.85–0.95	0.15–0.30	0.40–0.70	0.035	0.035	...
1095	C100	0.98–1.05	0.15–0.30	0.35–0.60	0.035	0.035	...
	4010 (high strength castings)						
1522	FeG52	0.25	0.50	1.00–1.50	0.035	0.035	...
	4838 (free-cutting steels)						
1212	10S20	0.06–0.13	0.05	0.60–1.00	0.080	0.13–0.25	...
1212	10S22	0.06–0.12	0.10–0.40	0.50–0.90	0.070	0.18–0.26	...
1212	CF9S22	0.06–0.13	0.05	0.70–1.20	0.040–0.100	0.18–0.25	...
1213	CF9SMn28	0.06–0.13	0.05	0.90–1.30	0.040–0.100	0.24–0.32	...
1215	CF9SMn32	0.06–0.13	0.05	0.90–1.30	0.040–0.100	0.28–0.35	...
1215	CF9SMn36	0.06–0.13	0.05	1.00–1.50	0.040–0.100	0.32–0.40	...
1137	CF35SMn10	0.32–0.39	0.30	1.35–1.65	0.040	0.08–0.13	...
1144	CF44SMn28						
	5331 (not identified)						
1010	C10	0.07–0.12	0.35	0.35–0.60	0.035	0.035	...
1015	C16	0.12–0.18	0.35	0.30–0.70	0.035	0.035	...
	5332 (not identified)						
1021, 1023	C20	0.18–0.24	0.40	0.40–0.80	0.035	0.035	...
1030	C30	0.27–0.34	0.40	0.50–0.80	0.035	0.035	...
1045	C45	0.42–0.50	0.15–0.40	0.40–0.90	0.035	0.035	...
1050	C50	0.47–0.55	0.40	0.60–0.90	0.035	0.035	...
	5333 (not identified)						
1035	C33	0.30–0.36	0.40	0.60–0.90	0.035	0.035	0.25Cr, 0.25Cu
	5598 (wire rod for drawing)						
1010	1CD10	0.12	...	0.25–0.60	0.050	0.050	...
1020	1CD20	0.17–0.24	0.35	0.40–0.70	0.050	0.050	...
1025	1CD25	0.22–0.29	0.15–0.35	0.40–0.70	0.050	0.050	
1035	1CD35	0.32–0.39	0.15–0.35	0.40–0.70	0.050	0.050	
1040	1CD40	0.37–0.44	0.15–0.35	0.40–0.70	0.050	0.050	
1045	1CD45	0.42–0.49	0.15–0.35	0.40–0.70	0.050	0.050	
1050	1CD50	0.47–0.54	0.15–0.35	0.40–0.70	0.050	0.050	...
1005	3CD5	0.06	...	0.25–0.50	0.035	0.035	0.007N
1006	3CD6	0.08	...	0.25–0.50	0.035	0.035	0.007N
1008	3CD8	0.10	...	0.25–0.60	0.035	0.035	0.007N
1010	3CD12	0.08–0.13	...	0.30–0.60	0.035	0.035	0.007N
1020	3CD20	0.18–0.23	0.35	0.40–0.70	0.035	0.035	0.007N
1025	3CD25	0.23–0.28	0.15–0.35	0.40–0.70	0.035	0.035	0.008N
1030	3CD30	0.28–0.33	0.15–0.35	0.40–0.70	0.035	0.035	...
1035	3CD35	0.32–0.39	0.15–0.35	0.40–0.70	0.050	0.050	...
1040	3CD40	0.38–0.43	0.15–0.35	0.40–0.70	0.035	0.035	0.008N
1045	3CD45	0.43–0.48	0.15–0.35	0.40–0.70	0.035	0.035	0.008N
1050	3CD50	0.48–0.53	0.15–0.35	0.40–0.70	0.035	0.035	0.008N
1055	3CD55	0.53–0.58	0.15–0.35	0.40–0.70	0.035	0.035	0.008N
1060	3CD60	0.58–0.63	0.15–0.35	0.40–0.70	0.035	0.035	0.008N
1064	3CD65	0.63–0.68	0.15–0.35	0.40–0.70	0.035	0.035	0.008N
1075	3CD70	0.68–0.73	0.15–0.35	0.40–0.70	0.035	0.035	0.008N
1075	3CD75	0.73–0.78	0.15–0.35	0.40–0.70	0.035	0.035	0.008N
1078, 1080	3CD80	0.78–0.83	0.15–0.35	0.40–0.70	0.035	0.035	0.008N
1080	3CD85	0.83–0.88	0.15–0.35	0.40–0.70	0.035	0.035	0.008N
1086	3CD85	0.83–0.88	0.15–0.35	0.40–0.70	0.035	0.035	0.008N
1086	3CD90	0.88–0.93	0.15–0.35	0.40–0.70	0.035	0.035	0.008N
1090	3CD95	0.93–0.98	0.15–0.35	0.40–0.70	0.035	0.035	0.008N
	5771 (chain steels)						
1006	C8	0.08	...	0.35–0.45	0.035	0.035	...
1010	C12	0.12	0.30	0.35–0.60	0.035	0.035	...
1022	20Mn4	0.16–0.24	0.30	0.80–1.00	0.035	0.035	...
	5869 (plates for pressure vessels)						
1013	Fe360-1KG, Fe360-2KW	0.17	0.35	0.40	0.035	0.035	...
	6403 (tubes for use in motor vehicles)						
1030	C30	0.27–0.34	0.40	0.50–0.80	0.035	0.035	...
1049	C48	0.45–0.52	0.40	0.60–0.90	...	...	0.25Cr, 0.25Ni, 0.25Cu
1013	Fe35-2	0.17	0.10–0.35	0.40	0.035	0.035	...

(continued)

Table 40 (continued)

Nearest SAE grade	UNI number	Composition, wt%					
		C	Si	Mn	P	S	Others
	6783 (not identified)						
1030	Fe50-3	0.25–0.35	0.40	0.40–0.80	0.040	0.040	...
1040	Fe60-3	0.35–0.45	0.40	0.40–0.80	0.040	0.040	...
1050	Fe70-3	0.45–0.55	0.40	0.40–0.80	0.040	0.040	...
	6922 (aircraft material)						
1020	C21	0.18–0.24	0.10	0.30–0.60	0.035	0.035	...
	6923 (aircraft material)						
1040	C40	0.37–0.44	0.15–0.40	0.50–0.90	0.035	0.035	0.30Cr
	6930 (aircraft material)						
1522	20Mn6	0.17–0.25	0.10–0.35	1.30–1.70	0.030	0.025	0.40Ni, 0.10 Mo
	7064 (strip for springs)						
1060	C60	0.57–0.65	0.15–0.40	0.50–0.90	0.035	0.035	...
1074, 1075	C75	0.70–0.80	0.15–0.30	0.40–0.70	0.035	0.035	...
1090	C90	0.85–0.95	0.15–0.30	0.40–0.70	0.035	0.035	...
1095	C100	0.98–1.05	0.15–0.30	0.35–0.60	0.035	0.035	...
	7065 (strip)						
1010	C10	0.07–0.12	0.35	0.35–0.60	0.035	0.035	...
1015	C16	0.12–0.18	0.35	0.30–0.70	0.035	0.035	...
1021, 1023	C20	0.18–0.24	0.40	0.40–0.80	0.035	0.035	...
1030	C30, C31	0.27–0.34	0.40	0.50–0.80	0.035	0.035	...
1035	C35	0.33–0.38	0.35	0.40–0.65	0.020	0.020	
1035	C36	0.32–0.39	0.40	0.50–0.90	0.035	0.035	...
1040	C40	0.38–0.43	0.35	0.40–0.65	0.020	0.020	...
1040	C41	0.37–0.44	0.40	0.50–0.90	0.035	0.035	...
1045	C45	0.42–0.50	0.15–0.40	0.40–0.90	0.035	0.035	...
1045	C46	0.42–0.50	0.40	0.50–0.90	0.035	0.035	...
1050	C50	0.47–0.55	0.40	0.60–0.90	0.035	0.035	...
1050	C51	0.47–0.50	0.40	0.60–0.90	0.035	0.035	...
1055	C55	0.53–0.58	0.35	0.40–0.65	0.020	0.020	...
1055	C56	0.52–0.60	0.40	0.60–0.90	0.035	0.035	...
1060	C60	0.57–0.65	0.15–0.40	0.50–0.90	0.035	0.035	...
1060	C61	0.57–0.65	0.40	0.60–0.90	0.035	0.035	...
	7070 (for general purpose)						
1013	Fe34CFN	0.17	...	...	0.040	0.045	0.009N
	7091 (welded tubes)						
1013	Fe34	0.17	...	...	0.045	0.045	...
	7356 (steels for bolts, nuts, and rivets)						
1010	CB10FF, CB10FU	0.08–0.13	0.10	0.30–0.60	0.040	0.040	...
1015	CB15	0.12–0.18	0.10–0.40	0.30–0.60	0.035	0.035	...
1020	CB20FF	0.18–0.23	0.10	0.30–0.60	0.040	0.040	...
1035	CB35	0.34–0.39	0.15–0.40	0.50–0.80	0.035	0.035	...
	7660 (steels for pressure vessels)						
1522	Fe510	0.24	0.15–0.40	0.80–1.70	0.030	0.030	...
	7845 (heat-treatable steels)						
1030	C30	0.27–0.34	0.40	0.50–0.80	0.035	0.035	...
1045	C45	0.42–0.50	0.15–0.40	0.40–0.90	0.035	0.035	...
1050	C50	0.47–0.55	0.40	0.60–0.90	0.035	0.035	...
1055	C55	0.53–0.58	0.35	0.40–0.65	0.020	0.020	...
1060	C60	0.57–0.65	0.15–0.40	0.50–0.90	0.035	0.035	...
	7846 (cementation steels)						
1010	C10	0.07–0.12	0.35	0.35–0.60	0.035	0.035	...
1015	C15	0.12–0.18	0.15–0.35	0.30–0.60	0.035	0.035	...
	7847 (steels for superficial hardening)						
1035	C36	0.32–0.39	0.40	0.50–0.90	0.035	0.035	...
1043	C43	0.40–0.46	0.40	0.60–0.90	0.035	0.030	0.25Cr, 0.25Ni, 0.25Cu
1045	C46	0.42–0.50	0.40	0.50–0.90	0.035	0.035	...
1049	C48	0.45–0.52	0.15–0.40	0.50–0.80	0.030	0.030	...
1050, 1053, 1055	C53	0.50–0.57	0.15–0.40	0.50–0.80	0.030	0.030	...
	7874 (wrought steels)						
1030	C30	0.27–0.34	0.40	0.50–0.80	0.035	0.035	...
1045	C45	0.42–0.50	0.15–0.40	0.40–0.90	0.035	0.035	...
1050	C50	0.47–0.55	0.40	0.60–0.90	0.035	0.035	...
1055	C55	0.53–0.58	0.35	0.40–0.65	0.020	0.020	...
1060	C60	0.57–0.65	0.15–0.40	0.50–0.90	0.035	0.035	...

Table 41 Chemical compositions of Italian (UNI) alloy steels referenced in Table 31

Nearest SAE grade	UNI number	C	Si	Mn	P	S	Ni	Cr	Mo	Others
	3097 (ball and roller bearing steels)									
8615	16NiCrMo2	0.13–0.18	0.15–0.40	0.60–0.90	0.035	0.035	0.35–0.65	0.40–0.60	0.15–0.25	...
4320	20NiCrMo7	0.17–0.22	0.15–0.40	0.45–0.65	0.035	0.035	1.60–2.00	0.40–0.60	0.20–0.30	...
E52100	100Cr6	0.95–1.10	0.15–0.35	0.25–0.45	0.025	0.025	...	1.40–1.60	...	...
	3160 (wear-resistant sand castings)									
4140	G40CrMo4	0.37–0.42	0.20–0.35	0.75–1.00	0.035	0.035	...	0.80–1.20	0.15–0.25	...
E51100	G90Cr4	0.80–1.00	0.50	0.80	0.035	0.035	...	1.00–1.20	...	...
	3545 (spring steels)									
6150	50CrV4	0.47–0.55	0.20–0.40	0.70–0.90	0.035	0.035	...	0.80–1.20	...	0.10–0.20V
	3608 (steel casting)									
4422	G20Mo5	0.25	0.50	0.80	0.035	0.035	...	...	0.40–0.60	...
	5331 (not identified)									
8615	16NiCrMo2	0.13–0.18	0.15–0.40	0.60–0.90	0.035	0.035	0.35–0.65	0.40–0.60	0.15–0.25	...
4320	18NiCrMo7	0.15–0.21	0.35	0.40–0.70	0.035	0.035	1.50–1.80	0.40–0.70	0.20–0.30	...
8620	20NiCrMo2	0.18–0.23	0.35	0.70–0.90	0.030	0.025	0.40–0.70	0.40–0.60	0.15–0.25	...
	5332 (not identified)									
5135	35CrMn5	0.32–0.39	0.40	0.80–1.10	0.035	0.035	...	1.00–1.30	...	...
4135	35CrMo4	0.32–0.38	0.40	0.60–0.90	0.035	0.035	...	0.80–1.10	0.15–0.30	...
8637	38NiCrMo4	0.34–0.42	0.40	0.50–0.80	0.035	0.035	0.70–1.00	0.70–1.00	0.15–0.25	...
5140	40Cr4	0.37–0.44	0.40	0.50–0.80	0.035	0.035	...	0.90–1.20	...	...
4137, 4140	40CrMo4	0.37–0.44	0.40	0.70–1.00	0.035	0.035	...	0.90–1.20	0.15–0.25	...
4340	40NiCrMo7	0.37–0.43	0.40	0.50–0.80	0.035	0.035	1.60–1.90	0.60–0.90	0.20–0.30	...
	5333 (not identified)									
5135	36CrMn4	0.33–0.39	0.40	0.80–1.10	0.035	0.030	...	0.90–1.20	...	...
4137	38CrMo4	0.34–0.40	0.40	0.60–0.90	0.035	0.030	...	0.80–1.10	0.15–0.25	0.15–0.25Pb
8640	40NiCrMo4	0.37–0.43	0.40	0.50–0.80	0.035	0.030	0.70–1.10	0.60–0.90	0.15–0.25	...
	6932 (not identified)									
E9310	15NiCrMo13	0.12–0.17	0.15–0.40	0.30–0.60	0.030	0.025	3.00–3.50	0.80–1.10	0.20–0.30	...
	6403 (tubes for use in motor vehicles)									
8620	20NiCrMo2	0.18–0.23	0.35	0.70–0.90	0.030	0.025	0.40–0.70	0.40–0.60	0.15–0.25	...
5135	35CrMn5	0.32–0.39	0.40	0.80–1.10	0.035	0.035	...	1.00–1.30	...	...
	6926 (aircraft materials)									
4340	40NiCrMo7	0.37–0.43	0.40	0.50–0.80	0.035	0.035	1.60–1.90	0.60–0.90	0.20–0.30	...
	6929 (aircraft materials)									
4130, 4135	35CrMo4F	0.30–0.37	0.15–0.35	0.50–0.80	0.020	0.015	0.30	0.90–1.20	0.15–0.25	...
	7065 (strip)									
6150	50CrV4	0.47–0.55	0.20–0.40	0.70–0.90	0.035	0.035	...	0.80–1.20	...	0.10–0.20V
	7356 (steels for bolts, nuts, and rivets)									
8630	30NiCrMo2KB	0.27–0.34	0.15–0.40	0.70–1.00	0.035	0.035	0.40–0.70	0.40–0.60	0.15–0.30	...
4130, 4135	34CrMo4KB	0.30–0.37	0.15–0.40	0.50–0.80	0.035	0.035	...	0.90–1.20	0.15–0.30	...
5135	38Cr4KB	0.34–0.41	0.15–0.40	0.60–0.90	0.035	0.035	...	0.90–1.20	...	...
4137	38CrMo4KB	0.34–0.41	0.15–0.40	0.50–0.80	0.035	0.035	...	0.90–1.20	0.15–0.30	...
8637	38NiCrMo4KB	0.34–0.41	0.15–0.40	0.50–0.80	0.035	0.035	0.70–1.00	0.70–1.00	0.15–0.30	...
8640, 8740	40NiCrMo2KB	0.37–0.44	0.15–0.40	0.70–1.00	0.035	0.035	0.40–0.70	0.40–0.60	0.15–0.30	...
4340	40NiCrMo7KB	0.37–0.44	0.15–0.40	0.50–0.80	0.035	0.035	1.60–2.00	0.70–1.00	0.20–0.30	...
50B40	41Cr2KB	0.38–0.45	0.15–0.40	0.60–0.90	0.035	0.035	...	0.40–0.60	...	...
5140	41Cr4KB	0.38–0.45	0.15–0.40	0.50–0.80	0.035	0.035	...	0.90–1.20	...	...
5132	34Cr4KB	0.30–0.37	0.15–0.40	0.60–0.90	0.035	0.035	...	0.90–1.20	...	...
	7845 (heat-treatable steels)									
4130	30CrMo4	0.27–0.34	0.15–0.40	0.40–0.70	0.035	0.035	...	0.80–1.10	0.15–0.25	...
4135	35CrMo4	0.32–0.39	0.15–0.40	0.60–0.90	0.035	0.035	...	0.90–1.20	0.15–0.25	0.15–0.30Pb
5135	36CrMn5	0.33–0.40	0.15–0.40	0.80–1.10	0.035	0.035	...	1.00–1.30	...	...
8637	39NiCrMo3	0.35–0.43	0.15–0.40	0.50–0.80	0.035	0.035	0.70–1.00	0.60–1.00	0.15–0.25	...
8640, 8740	40NiCrMo2	0.37–0.44	0.15–0.40	0.70–1.00	0.035	0.035	0.40–0.70	0.40–0.60	0.15–0.30	...
4340	40NiCrMo7	0.37–0.43	0.40	0.50–0.80	0.035	0.035	1.60–1.90	0.60–0.90	0.20–0.30	...
5140	41Cr4	0.38–0.45	0.14–0.40	0.50–0.80	0.035	0.035	...	0.90–1.20	...	...
4140, 4142	42CrMo4	0.38–0.45	0.15–0.40	0.60–0.90	0.035	0.035	...	0.90–1.20	0.15–0.25	...
6150	50CrV4	0.47–0.55	0.20–0.40	0.70–0.90	0.035	0.035	...	0.80–1.20	...	0.10–0.20V
	7846 (cementation steels)									
8615	16NiCrMo2	0.13–0.18	0.15–0.40	0.60–0.90	0.035	0.035	0.35–0.65	0.40–0.60	0.15–0.25	...
5115	16MnCr5	0.13–0.19	0.15–0.40	1.00–1.30	0.035	0.035	...	0.80–1.20	...	...
4118	18CrMo4	0.15–0.21	0.15–0.40	0.60–0.90	0.035	0.035	...	0.85–1.15	0.15–0.25	...
4320	18NiCrMo7	0.15–0.21	0.35	0.40–0.70	0.035	0.035	1.50–1.80	0.40–0.70	0.20–0.30	...
5120	20MnCr5	0.17–0.22	0.15–0.40	1.10–1.40	0.035	0.035	...	1.00–1.30	...	...
8620	20NiCrMo2	0.18–0.23	0.35	0.70–0.90	0.030	0.025	0.40–0.70	0.40–0.60	0.15–0.25	...

(continued)

Table 41 (continued)

Nearest SAE grade	UNI number	Composition, wt%								
		C	Si	Mn	P	S	Ni	Cr	Mo	Others
	7847 (steels for superficial hardening)									
5135	36CrMn4	0.33–0.39	0.40	0.80–1.10	0.035	0.030	...	0.90–1.20	...	...
5135	38Cr4	0.34–0.40	0.15–0.40	0.60–0.90	0.030	0.030	...	0.90–1.20	...	...
8640	40NiCrMo3	0.37–0.43	0.15–0.40	0.50–0.80	0.030	0.030	0.70–1.00	0.60–1.00	0.15–0.25	...
4140	41CrMo4	0.38–0.44	0.15–0.40	0.50–0.80	0.030	0.030	...	0.90–1.20	0.15–0.25	...
50B44	45Cr2	0.42–0.48	0.15–0.40	0.50–0.80	0.030	0.030	...	0.40–0.60	...	...
	7874 (wrought steels)									
4130	30CrMo4	0.27–0.34	0.15–0.40	0.50–0.90	0.035	0.035	...	0.90–1.20	0.15–0.30	...
5132	34Cr4	0.30–0.37	0.15–0.40	0.60–0.90	0.035	0.035	...	0.90–1.20	...	...
4135	35CrMo4	0.32–0.39	0.15–0.40	0.50–0.90	0.035	0.035	...	0.90–1.20	0.15–0.30	0.15–0.30Pb
5135	36CrMn5	0.33–0.40	0.15–0.40	0.80–1.10	0.035	0.035	...	1.00–1.30	...	...
8637	39NiCrMo3	0.35–0.43	0.15–0.40	0.50–0.80	0.035	0.035	0.70–1.00	0.60–1.00	0.15–0.25	...
8640, 8740	40NiCrMo2	0.37–0.44	0.15–0.40	0.70–1.00	0.035	0.035	0.40–0.70	0.40–0.60	0.15–0.30	...
4340	40NiCrMo7	0.37–0.43	0.40	0.50–0.80	0.035	0.035	1.60–1.90	0.60–0.90	0.20–0.30	...
5140	41Cr4	0.38–0.45	0.14–0.40	0.50–0.80	0.035	0.035	...	0.90–1.20	...	...
4140, 4142	42CrMo4	0.38–0.45	0.15–0.40	0.50–0.90	0.035	0.035	...	0.90–1.20	0.15–0.30	...
6150	50CrV4	0.47–0.55	0.20–0.40	0.70–0.90	0.035	0.035	...	0.80–1.20	...	0.10–0.20V
	9335 (not identified)									
E9310	10NiCrMo13	0.08–0.13	0.12–0.35	0.40–0.70	0.030	0.025	3.00–3.50	1.00–1.40	0.08–0.15	0.35Cu
	Other									
4130	30CrMo4	0.27–0.33	0.35	0.40–0.70	0.035	0.035	...	0.80–1.10	0.15–0.25	...
E52100	100C6	0.95–1.10	0.35	0.30–0.50	0.030	0.030	...	1.40–1.65	...	P + S = 0.05

Table 42 Chemical compositions of Swedish Standard (SS) carbon, carbon-manganese, resulfurized, and rephosphorized/resulfurized steels referenced in Table 31

Nearest SAE grade	SS_{14} number	Composition, wt%					
		C	Si	Mn	P	S	Other
1008	1142, 1146	0.10	...	0.50	0.040	0.040	...
1006	1147	0.08	...	0.45	0.030	0.030	...
1005	1160	0.08	0.03	0.35	0.030	0.040	...
1006	1225	0.08	0.030	0.40–0.60	0.03	0.03	0.1Cr, 0.2Cu, 0.009N
1010	1232	0.13	...	0.30–0.70	0.050	0.050	0.25Cr, 0.30Cu, 0.009N
1013	1233, 1234	0.17	0.10–0.40	0.50	0.050	0.050	0.20Cr, 0.30Cu, 0.009N
1010	1265	0.07–0.13	0.30	0.25–0.45	0.030	0.040	...
1010	1311	0.14	0.20–0.45	0.40–0.70	0.060	0.050	...
1017	1312	0.20	0.05	0.40–0.70	0.050	0.050	0.009N
1013	1330	0.17	0.15–0.40	0.40–1.0	0.050	0.050	0.25Cr, 0.30Cu, 0.009N
1012	1332	0.10–0.14	0.15–0.40	0.50–0.80	0.040	0.040	0.25Cr, 0.40Cu, 0.009–0.015N
1015, 1016	1370	0.12–0.18	0.10–0.40	0.50–0.90	0.035	0.035	...
1012	1431	0.16	0.05–0.25	0.80–1.4	0.050	0.050	0.25Cr, 0.40Cu, 0.009–0.015N
1020	1450	0.25	0.10–0.40	0.40–0.80	0.050	0.050	...
1035	1550	0.28–0.40	...	0.40–0.90	0.050	0.050	...
1035	1572	0.32–0.39	0.15–0.40	0.50–0.80	0.035	0.035	...
1045	1672	0.43–0.50	0.15–0.40	0.50–0.80	0.035	0.035	...
1050, 1053	1674	0.48–0.55	0.15–0.40	0.60–0.90	0.035	0.035	...
1060	1678	0.57–0.65	0.15–0.40	0.60–0.90	0.035	0.035	...
1070	1770	0.55–0.80	0.15–0.80	0.50–0.90	0.035	0.035	...
1074	1774	0.60–0.95	0.15–0.40	0.30–0.80	0.035	0.035	...
1095	1870	0.94–1.06	0.10–0.35	0.30–0.60	0.030	0.030	...
1016	2101	0.20	0.10–0.50	0.80–1.60	0.050	0.050	0.25Cr, 0.30Cu, 0.009N
1213	1912	0.14	0.05	0.90–1.30	0.110	0.24–0.35	...
1139	1957	0.32–0.39	0.10–0.40	0.80–1.20	0.060	0.15–0.25	...
1144, 1151	1973	0.46–0.54	0.10–0.40	0.80–1.20	0.060	0.15–0.25	...
1541	2120	0.38–0.45	0.15–0.40	1.10–1.40	0.035	0.035	...
1541	2128	0.43	0.10–0.40	1.20–1.80	0.050	0.050	...
1526	2130	0.25–0.30	0.15–0.40	1.10–1.30	0.035	0.035	0.10–0.30Cr, 0.002B
1522	2165	0.24	0.6	1.6	0.06	0.050	...
1522	2168	0.28	0.6	1.6	0.06	0.050	...

Table 43 Chemical compositions of Swedish Standard (SS) alloy steels referenced in Table 31

Nearest SAE grade	S.S.$_{14}$ number	Composition, wt%								
		C	Si	Mn	P	S	Ni	Cr	Mo	Others
4130	2233	0.25–0.35	0.15–0.40	0.40–0.90	0.040	0.040	...	0.80–1.20	0.15–0.30	...
4135	2234	0.30–0.37	0.15–0.40	0.50–0.80	0.035	0.035	...	0.90–1.20	0.15–0.30	...
4140	2244	0.38–0.45	0.15–0.40	0.60–0.90	0.035	0.035	...	0.90–1.20	0.15–0.30	...
4142	2244	0.38–0.45	0.15–0.40	0.60–0.90	0.035	0.035	...	0.90–1.20	0.15–0.30	...
4320	2523, 2523-02	0.17–0.23	0.15–0.40	0.70–1.10	0.035	0.030–0.050	1.00–1.40	0.80–1.20	0.08–0.16	...
5115	2127	0.13–0.19	0.15–0.40	1.00–1.30	0.035	0.030–0.050	...	0.80–1.10	...	...
5140	2245	0.38–0.45	0.15–0.40	0.60–0.90	0.035	0.035	...	0.90–1.20	...	...
5150	2230	0.48–0.55	0.15–0.40	0.70–1.00	0.035	0.035	...	0.90–1.20	...	0.10–0.20V
6150	2230	0.48–0.55	0.15–0.40	0.70–1.00	0.035	0.035	...	0.90–1.20	...	0.10–0.20V
E52100	2258	0.95–1.10	0.15–0.35	0.25–0.45	0.030	0.025	...	1.35–1.65	...	...
8620	2506-03, 2506-08	0.17–0.23	0.15–0.40	0.60–0.95	0.035	0.030–0.50	0.35–0.75	0.35–0.65	0.15–0.25	...

Physical Properties of Carbon and Low-Alloy Steels*

Coefficients of linear thermal expansion for carbon and low-alloy steels

AISI-SAE grade	Treatment or condition	Average coefficient of expansion, μm/m · K, at °C (°F)(a)						
		20–100 (68–212)	20–200 (68–392)	20–300 (68–572)	20–400 (68–752)	20–500 (68–932)	20–600 (68–1112)	20–700 (68–1292)
1008	Annealed	12.6(b)	13.1(b)	13.5(b)	13.8(b)	14.2(b)	14.6(b)	15.0(b)
1008	Annealed	11.6	12.5	13.0	13.6	14.2	14.6	15.0
1010	Annealed	12.2(b)	13.0(b)	13.5(b)	13.9(b)	14.3(b)	14.7(b)	15.0(b)
1010	Unknown	11.9(c)	12.6	13.3	13.8	14.3	14.7	14.9
1010	Unknown	...	...	...	15.1(d)	...	...	...
1015	Rolled	11.9(b)	12.5(b)	13.0(b)	13.6(b)	14.2(b)	...	...
1015	Annealed	12.2(b)	...	...	13.4(b)	...	14.2(b)	...
1016	Annealed	12.0(b)	...	...	13.5(b)	...	14.4(b)	...
1017	Unknown	12.2(b)	...	...	13.5(b)	...	14.5(b)	...
1018	Annealed	12.0(b)	...	...	13.5(b)	...	14.6(b)	...
1019	Unknown	12.2(b)	...	...	13.5(b)	...	14.7(b)	...
1020	Annealed	11.7	12.1	12.8	13.4	13.9	14.4	14.8
1020	Unknown	12.2(b)	...	...	13.5(b)	...	14.2(b)	...
1021	Unknown	12.0(b)	...	...	13.5(b)	...	14.3(b)	...
1022	Annealed	12.2(b)	12.7(b)	13.1(b)	13.5(b)	13.9(b)	14.4(b)	14.9(b)
1023	Unknown	12.2(b)	...	...	13.5(b)	...	14.4(b)	...
1025	Annealed	12.0(b)	...	...	13.5(b)	...	14.4(b)	...
1026	Annealed	12.0(b)	...	...	13.5(b)	...	14.4(b)	...
1029	Annealed	12.0(b)	...	...	13.5(b)	...	14.4(b)	...
1030	Annealed	11.7(b)	...	...	13.5(b)	...	14.4(b)	...
1035	Annealed	11.1	11.9	12.7	13.4	14.0	14.4	14.8
1037	Annealed	11.1(b)	...	...	13.5(b)	...	14.6(b)	...
1039	Annealed	11.1(b)	...	...	13.5(b)	...	14.6(b)	...
1040	Annealed	11.3	12.0	12.5	13.3	13.9	14.4	14.8
1043	Annealed	11.3(b)	...	...	13.5(b)	...	14.6(b)	...
1044	Annealed	11.1(b)	12.0(b)	...	13.3(b)	...	...	...
1045	Annealed	11.6(b)	12.3(b)	13.1(b)	13.7(b)	14.2(b)	14.7(b)	15.1(b)
1045	Annealed	11.2(e)	11.9(e)	12.7(e)	13.5(e)	14.1(e)	14.5(e)	14.8(e)
1046	Unknown	11.1(b)	...	...	13.5(b)	...	...	...
1050	Annealed	11.1(b)	12.0(b)	...	13.5(b)	...	...	...
1052	Annealed	11.3(e)	11.8(e)	12.7(e)	13.7(e)	14.5(e)	14.7(e)	15.0(e)
1053	Unknown	11.1(b)	...	...	13.5(b)	...	...	...
1055	Annealed	11.0	11.8	12.6	13.4	14.0	14.5	14.8
1060	Annealed	11.1(e)	11.9(e)	12.9(e)	13.5(e)	14.1(e)	14.6(e)	14.9(e)
1064	Unknown	11.1(b)	...	...	13.5(b)	...	...	...
1065	Unknown	11.1(b)	...	...	13.5(b)	...	...	...
1070	Unknown	11.5(b)	...	...	13.3(b)	...	...	...
1078	Unknown	11.3(b)	...	...	13.3	...	...	...
1080	Annealed	11.0	11.6	12.4	13.2	13.8	14.2	14.7
1080	Unknown	11.7(b)	12.2(b)	...	...	...	...	...
1085	Annealed	11.1(b)	11.7(b)	12.5(b)	13.2(b)	13.6(b)	14.2(b)	14.7
1086	Unknown	11.1(b)	...	...	13.1(b)	...	...	...
1095	Unknown	...	...	...	...	...	...	14.6(d)
1095	Annealed	11.4	...	...	...	...	...	...
1095	Hardened	13.0(b)	...	...	...	...	...	...

(continued)

(a) To obtain coefficients in μin./in. · °F, multiply stated values by 0.556. (b) Stated value represents average coefficient between 0 °C (32 °F) and indicated temperature. (c) 10.3 μm/m · K from −100 to 20 °C (−148 to 68 °F); 9.8 μm/m · K from −150 to 20 °C (−238 to 68 °F). (d) Stated value represents average coefficient between 20 and 650 °C (68 and 1200 °F). (e) Stated value represents average coefficient between 25 °C (75 °F) and indicated temperature. (f) Stated value represents average coefficient between 20 and 95 °C (68 and 200 °F). (g) Stated value represents average coefficient between 20 and 370 °C (68 and 700 °F). (h) 8.6 μm/m · K from −195 to 20 °C (−320 to 68 °F); 10.0 μm/m · K from −130 to 20 °C (−200 to 68 °F). (i) Stated value represents average coefficient between 20 and 260 °C (68 and 500 °F). (j) Stated value represents average coefficient between 20 and 540 °C (68 and 1000 °F). (k) Stated value represents average coefficient between −18 and 95 °C (0 and 200 °F). (l) Stated value represents average coefficient between −18 and 650 °C (0 and 1200 °F). (m) 11.2 μm/m · K from −100 to 20 °C (−148 to 68°F); 10.4 μm/m · K from −150 to 20 °C (−238 to 68 °F). (n) Stated value represents average coefficient between 20 and 205 °C (68 and 400 °F). (o) Stated value represents average coefficient between 20 and 315 °C (68 and 600 °F). (p) Stated value represents average coefficient between 25 and 270 °C (77 and 518 °F). (q) Stated value represents average coefficient between 20 and 275 °C (68 and 525 °F). (r) Stated value represents average coefficient between −18 and 260 °C (0 and 500 °F). (s) Stated value represents average coefficient between −18 and 540 °C (0 and 1000 °F).

*These tabular data have been compiled from previous editions of *Metals Handbook* and other sources.

Coefficients of linear thermal expansion for carbon and low-alloy steels (continued)

AISI-SAE grade	Treatment or condition	Average coefficient of expansion, μm/m · K, at °C (°F)(a)						
		20–100 (68–212)	20–200 (68–392)	20–300 (68–572)	20–400 (68–752)	20–500 (68–932)	20–600 (68–1112)	20–700 (68–1292)
1117	Unknown	12.2(f)	...	...	13.1(g)	...	...	...
1118	Unknown	12.2(f)	...	...	13.3(g)	...	...	...
1132	Unknown	12.6(f)	...	...	...	...	...	...
1137	Unknown	12.8	...	...	...	...	...	...
1139	Unknown	12.6(f)	...	...	...	...	...	...
1140	Unknown	12.6(f)	...	...	...	...	...	...
1141	Unknown	...	12.6(b)	...	...	...	...	...
1144	Unknown	13.3	...	...	...	...	...	...
1145	Annealed	11.2(b)	12.1(b)	13.0(b)	13.6(b)	14.0(b)	14.6(b)	14.8(b)
1145	Annealed	11.6(b)	12.3(b)	13.1(b)	13.7(b)	14.2(b)	14.7(b)	15.1(b)
1146	Unknown	12.8	...	...	...	...	...	...
1151	Unknown	...	12.6(b)	...	...	...	...	...
1330	Unknown	12.0	12.8	13.3	...	...	...	...
1335	Unknown	12.2	12.8	13.3	...	...	...	...
1345	Unknown	12.0	12.6	13.3	...	...	...	...
1522	Annealed	12.0(b)	...	...	13.5(b)	...	14.4(b)	...
1524	Unknown	11.9	12.7	...	13.9	...	14.7	...
1524	Annealed	12.0(b)	...	...	13.5(b)	...	14.4(b)	...
1526	Annealed	12.0(b)	...	...	13.5(b)	...	14.4(b)	...
1541	Annealed	12.0(b)	...	...	13.5(b)	...	14.4(b)	...
1548	Unknown	11.9(b)	...	...	13.3(b)	...	14.6(b)	...
1551	Annealed	11.7(b)	...	...	13.5(b)	...	14.6(b)	...
1552	Unknown	11.1(b)	...	...	13.5(b)	...	...	...
1561	Annealed	11.1(b)	...	...	13.5(b)	...	14.6(b)	...
1566	Annealed	11.5(b)	...	...	13.5(b)	...	14.7(b)	...
2330	Annealed	10.9(e)	11.2(e)	12.1(e)	12.9(e)	13.4(e)	13.8(e)	...
2515	Unknown	10.9(f)(h)	...	12.6(i)	...	13.5(j)	...	...
3120	Unknown	11.3(k)	...	...	...	...	...	14.6(l)
3130	Unknown	11.3(k)	...	...	...	...	...	14.6(l)
3140	Unknown	11.3(k)	...	...	...	...	...	14.6(l)
3150	Unknown	11.3(k)	...	...	...	...	...	14.6(l)
4023	Unknown	11.7(k)	...	...	...	...	...	...
4027	Unknown	11.7(k)	...	...	...	...	...	...
4028	Unknown	11.9	12.4	12.9	...	...	...	...
4032	Unknown	11.9	12.4	12.9	...	...	...	...
4042	Unknown	11.9	12.4	12.9	...	...	...	...
4047	Unknown	11.9	12.4	12.9	...	...	...	...
4130	Unknown	12.2	...	...	13.7	...	14.6	...
4135	Unknown	11.7	12.2	12.8	...	...	...	...
4137	Unknown	11.7	12.2	12.8	...	...	...	...
4140	Oil hardened, tempered	12.3	12.7	...	13.7	...	14.5	...
4142	Unknown	11.7	12.2	12.8	...	...	...	...
4145	Oil hardened, tempered	11.7	12.2	12.8	...	...	...	...
4147	Unknown	11.7	12.2	12.8	...	...	...	...
4161	Unknown	11.5	12.2	12.9	...	...	...	...
4320	Unknown	11.3(k)	...	...	...	...	...	14.6(l)
4337	Unknown	11.3(k)	...	...	...	...	...	14.6(l)
4340	Oil hardened, tempered 600 °C (1110 °F)	12.3	12.7	...	13.7	...	14.5	...
4340	Oil hardened, tempered 630 °C (1170 °F)	(m)	12.4	...	13.6	...	14.3	...
4422	Unknown	11.7(k)	...	...	...	...	...	...
4427	Unknown	12.6	...	13.8	...	...	15.1	...
4615	Unknown	11.5	12.1	12.7	13.2	13.7	14.1	...
4617	Carburized and hardened	12.5	13.1	...	...	...	...	...
4626	Normalized and tempered	11.7(f)	...	12.6(i)	...	...	13.8(j)	...
4718	Unknown	11.3	12.2	13.1	...	...	...	...
4815	Unknown	11.5(f)	12.2(n)	13.1(o)	...	...	...	...
4820	Unknown	11.3(f)	12.2(n)	12.9(o)	...	...	...	...
5046	Unknown	11.9	12.4	12.9	...	...	...	...
50B60	Unknown	11.9	12.4	12.9	...	...	...	...
5117	Unknown	12.0	12.8	13.5	...	...	...	...
5120	Unknown	12.0	12.8	13.5	...	...	...	...
5130	Unknown	12.2	12.9	13.5	...	...	...	...
5132	Unknown	12.2	12.9	13.5	...	...	...	...
5135	Unknown	12.0	12.8	13.5	...	...	...	...
5140	Annealed	...	12.6	13.4	13.9	14.3	14.6	15.0
5145	Unknown	12.2	...	13.1(p)	...	...	...	...
5150	Unknown	12.8	...	13.7(p)	...	...	...	...

(continued)

(a) To obtain coefficients in μin./in. · °F, multiply stated values by 0.556. (b) Stated value represents average coefficient between 0 °C (32 °F) and indicated temperature. (c) 10.3 μm/m · K from −100 to 20 °C (−148 to 68 °F); 9.8 μm/m · K from −150 to 20 °C (−238 to 68 °F). (d) Stated value represents average coefficient between 20 and 650 °C (68 and 1200 °F). (e) Stated value represents average coefficient between 25 °C (75 °F) and indicated temperature. (f) Stated value represents average coefficient between 20 and 95 °C (68 and 200 °F). (g) Stated value represents average coefficient between 20 and 370 °C (68 and 700 °F). (h) 8.6 μm/m · K from −195 to 20 °C (−320 to 68 °F); 10.0 μm/m · K from −130 to 20 °C (−200 to 68 °F). (i) Stated value represents average coefficient between 20 and 260 °C (68 and 500 °F). (j) Stated value represents average coefficient between 20 and 540 °C (68 and 1000 °F). (k) Stated value represents average coefficient between −18 and 95 °C (0 and 200 °F). (l) Stated value represents average coefficient between −18 and 650 °C (0 and 1200 °F). (m) 11.2 μm/m · K from −100 to 20 °C (−148 to 68 °F); 10.4 μm/m · K from −150 to 20 °C (−238 to 68 °F). (n) Stated value represents average coefficient between 20 and 205 °C (68 and 400 °F). (o) Stated value represents average coefficient between 20 and 315 °C (68 and 600 °F). (p) Stated value represents average coefficient between 25 and 270 °C (77 and 518 °F). (q) Stated value represents average coefficient between 20 and 275 °C (68 and 525 °F). (r) Stated value represents average coefficient between 20 and 260 °C (0 and 500 °F). (s) Stated value represents average coefficient between −18 and 540 °C (0 and 1000 °F).

Coefficients of linear thermal expansion for carbon and low-alloy steels (continued)

AISI-SAE grade	Treatment or condition	Average coefficient of expansion, μm/m · K, at °C (°F)(a)						
		20–100 (68–212)	20–200 (68–392)	20–300 (68–572)	20–400 (68–752)	20–500 (68–932)	20–600 (68–1112)	20–700 (68–1292)
5155	Unknown	12.2	...	13.1(q)	...	...	...	...
52100	Annealed	11.9(b)	...	...	...	...	...	...
52100	Hardened	12.6(b)	...	...	...	...	...	...
6150	Annealed	12.2	12.7	13.3	13.7	14.1	14.4	...
6150	Hardened, tempered 205 °C (400 °F)	12.0	12.5	12.9	13.0	13.3	13.7	...
6150	Annealed	12.4(e)	12.6(e)	13.3(e)	13.8(e)	14.2(e)	14.5(e)	14.7(e)
8115	Unknown	11.9	12.6	13.3	...	...	...	...
81B45	Unknown	11.9	12.6	13.3	...	...	...	...
8617	Unknown	11.9(k)	...	12.8(r)	...	...	14.0(s)	...
8622	Unknown	11.1	12.2	12.9	...	...	...	...
8625	Unknown	11.1	12.2	12.9	...	...	...	...
8627	Unknown	11.3	12.2	12.9	...	...	...	...
8630	Unknown	11.3(k)	...	...	...	...	...	14.6(l)
8637	Unknown	11.3	12.2	12.8	...	...	...	...
8645	Oil hardened, tempered	11.7	12.2	12.8	...	...	...	...
8650	Oil hardened, tempered	11.7	12.2	12.8	...	...	...	...
8655	Oil hardened, tempered	11.7	12.2	12.8	...	...	...	...
8720	Unknown	11.3(k)	...	...	...	...	...	14.6(l)
8822	Unknown	11.3	12.2	12.9	...	...	...	...

(a) To obtain coefficients in μin./in. · °F, multiply stated values by 0.556. (b) Stated value represents average coefficient between 0 °C (32 °F) and indicated temperature. (c) 10.3 μm/m · K from −100 to 20 °C (−148 to 68 °F); 9.8 μm/m · K from −150 to 20 °C (−238 to 68 °F). (d) Stated value represents average coefficient between 20 and 650 °C (68 and 1200 °F). (e) Stated value represents average coefficient between 25 °C (75 °F) and indicated temperature. (f) Stated value represents average coefficient between 20 and 95 °C (68 and 200 °F). (g) Stated value represents average coefficient between 20 and 370 °C (68 and 700 °F). (h) 8.6 μm/m · K from −195 to 20 °C (−320 to 68 °F); 10.0 μm/m · K from −130 to 20 °C (−200 to 68 °F). (i) Stated value represents average coefficient between 20 and 260 °C (68 and 500 °F). (j) Stated value represents average coefficient between 20 and 540 °C (68 and 1000 °F). (k) Stated value represents average coefficient between −18 and 95 °C (0 and 200 °F). (l) Stated value represents average coefficient between −18 and 650 °C (0 and 1200 °F). (m) 11.2 μm/m · K from −100 to 20.2.C (−148 to 68 °F); 10.4 μm/m · K from −150 to 20 °C (−238 to 68 °F). (n) Stated value represents average coefficient between 20 and 205 °C (68 and 400 °F). (o) Stated value represents average coefficient between 20 and 315 °C (68 and 600 °F). (p) Stated value represents average coefficient between 25 and 270 °C (77 and 518 °F). (q) Stated value represents average coefficient between 20 and 275 °C (68 and 525 °F). (r) Stated value represents average coefficient between −18 and 260 °C (0 and 500 °F). (s) Stated value represents average coefficient between −18 and 540 °C (0 and 1000 °F).

Thermal conductivities of carbon and low-alloy steels

AISI-SAE grade	Treatment or condition	Conductivity, W/m · K, at °C (°F)(a)										
		0 (32)	100 (212)	200 (392)	300 (572)	400 (752)	500 (932)	600 (1112)	700 (1292)	800 (1472)	1000 (1832)	1200 (2192)
1008	Unknown	59.5	57.8	53.2	49.4	45.6	41.0	36.8	33.1	28.5	27.6	29.7
1008	Annealed	65.3(b)	60.3	54.9	...	45.2	...	36.4	...	28.5	27.6	...
1010	Unknown	...	46.7	...	...	...	...	...	...	...	...	...
1015	Annealed	51.9	51.0	48.9	...	...	...	...	...	...	...	...
1016	Annealed	51.9	50.2	47.6	...	...	...	...	...	...	...	...
1018	Annealed	51.9	50.8	48.9	...	...	...	...	...	...	...	...
1020	Unknown	51.9	51.0	48.9	...	...	...	...	...	...	...	...
1022	Annealed	51.9	50.8	48.8	...	...	...	...	...	...	...	...
1025	Annealed	51.9	51.1	49.0	46.1	42.7	39.4	35.6	31.8	26.0	27.2	29.7
1026	Annealed	51.9	50.1	48.4	...	...	...	...	...	...	...	...
1029	Annealed	51.9	50.1	48.4	...	...	...	...	...	...	...	...
1030	Annealed	...	51.0	...	...	...	...	...	...	...	...	...
1035	Annealed	...	50.8	...	...	...	...	...	...	...	...	...
1037	Annealed	...	51.0	...	...	...	...	...	...	...	...	...
1039	Annealed	...	50.7	...	...	...	...	...	...	...	...	...
1040	Annealed	...	50.7	...	...	...	...	...	...	...	...	...
1042	Annealed	51.9	50.7	48.2	45.6	41.9	38.1	33.9	30.1	24.7	26.8	29.7
1043	Annealed	...	50.8	...	...	...	...	...	...	...	...	...
1044	Annealed	...	50.8	...	...	...	...	...	...	...	...	...
1045	Annealed	...	50.8	...	...	...	...	...	...	...	...	...
1046	Unknown	51.2	49.7	...	...	...	...	...	...	...	...	...
1050	Annealed	51.2	49.7	46.8	...	...	...	...	...	...	...	...
1055	Unknown	51.2	49.7	...	...	...	...	...	...	...	...	...
1060	Unknown	50.5	...	46.8	...	...	...	...	...	...	...	...
1064	Unknown	51.2	49.7	...	...	...	...	...	...	...	...	...
1070	Unknown	49.9	48.4	...	...	...	...	...	...	...	...	...
1078	Annealed	47.8	48.2	45.2	41.4	38.1	35.2	32.7	30.1	24.3	26.8	30.1
1078	Unknown	49.6	48.1	...	...	...	...	...	...	...	...	...
1080	Unknown	50.5	...	46.8	...	...	...	...	...	...	...	...
1086	Unknown	49.9	48.4	...	...	...	...	...	...	...	...	...
1095	Unknown	...	46.7	...	...	...	...	...	...	...	...	...
1117	Unknown	51.9(b)	...	...	...	...	...	...	...	...	...	...
1118	Unknown	51.5(b)	...	...	...	...	...	...	...	...	...	...
1141	Unknown	...	50.5	47.6	...	...	...	...	...	...	...	...
1151	Unknown	...	50.5	47.6	...	...	...	...	...	...	...	...

(continued)

(a) To obtain conductivities in Btu/ft · h · °F, multiply values in table by 0.5777893; to obtain conductivities in Btu · in./ft² · h · °F, multiply values in table by 6.933472; to obtain conductivities in cal/cm · s · °C, multiply values in table by 0.0023884. (b) Thermal conductivity at 20 °C (68 °F). (c) Thermal conductivity at 260 °C (500 °F). (d) Thermal conductivity at 50 °C (120 °F)

Thermal conductivities of carbon and low-alloy steels (continued)

AISI-SAE grade	Treatment or condition	Conductivity, W/m · K, at °C (°F)(a)										
		0 (32)	100 (212)	200 (392)	300 (572)	400 (752)	500 (932)	600 (1112)	700 (1292)	800 (1472)	1000 (1832)	1200 (2192)
1522	Annealed	51.9	50.1	48.4	...	...	...	...	...	...	...	...
1524	Annealed	51.9	50.1	48.4	...	...	...	...	...	...	...	...
1526	Annealed	51.9	50.1	48.4	...	...	...	...	...	...	...	...
1541	Annealed	51.9	50.1	48.4	...	...	...	...	...	...	...	...
1548	Unknown	50.5	49.0	48.3	...	...	...	...	...	...	...	...
1551	Annealed	50.7	49.3	48.4	...	...	...	...	...	...	...	...
1561	Annealed	51.2	49.7	...	...	...	...	...	...	...	...	...
1566	Annealed	51.2	49.7	...	...	...	...	...	...	...	...	...
2515	Unknown	34.3(b)	...	...	38.2(c)	...	...	...	...	...	...	...
4037	Hardened and tempered	...	48.2	45.6	...	39.4	...	33.9	...	...	...	...
4130	Hardened and tempered	...	42.7	...	40.6	...	37.3	...	31.0	...	28.1	30.1
4140	Hardened and tempered	...	42.7	42.3	...	37.7	...	33.1	...	...	...	...
4145	Hardened and tempered	41.8(b)	...	...	...	...	...	...	...	...	...	...
4161	Hardened and tempered	42.7(b)	...	...	...	...	...	...	...	...	...	...
4427	Unknown	36.8(b)	...	...	...	...	...	...	...	...	...	...
4626	Unknown	...	44.1	...	...	...	...	...	...	...	...	...
5132	Unknown	48.6	46.5	44.4	42.3	38.5	35.6	31.8	28.9	26.0	28.1	30.1
5140	Hardened and tempered	...	44.8	43.5	...	37.7	...	31.4	...	...	...	...
8617	Unknown	...	43.3	...	...	...	...	...	...	...	...	...
8622	Unknown	...	37.5(d)	...	...	...	...	...	...	...	...	...
8627	Unknown	...	37.5(d)	...	...	...	...	...	...	...	...	...
8637	Unknown	...	37.5(d)	...	...	...	...	...	...	...	...	...
8822	Unknown	...	37.5(d)	...	...	...	...	...	...	...	...	...

(a) To obtain conductivities in Btu/ft · h · °F, multiply values in table by 0.5777893; to obtain conductivities in Btu · in./ft² · h · °F, multiply values in table by 6.933472; to obtain conductivities in cal/cm · s · °C, multiply values in table by 0.0023884. (b) Thermal conductivity at 20 °C (68 °F). (c) Thermal conductivity at 260 °C (500 °F). (d) Thermal conductivity at 50 °C (120 °F).

Specific heats of carbon and low-alloy steels

AISI-SAE grade	Treatment or condition	Mean apparent specific heat, J/Kg · K, at °C (°F)											
		50–100 (122–212)	150–200 (302–392)	200–250 (392–482)	250–300 (482–572)	300–350 (572–662)	350–400 (662–752)	450–500 (842–932)	550–600 (1022–1112)	650–700 (1202–1292)	700–750 (1292–1382)	750–800 (1382–1472)	850–900 (1562–1652)
1008	Annealed	481	519	536	553	574	595	662	754	867	1105	875	846
1010	Unknown	450	500	520	535	565	590	650	730	825	...	...	...
1015	Annealed	486	519	...	...	...	599	...	...	...	...	...	...
1016	Annealed	481	515	...	...	...	595	...	...	...	...	...	...
1017	Unknown	481(a)	...	...	...	...	...	...	...	...	...	...	...
1018	Annealed	486	519	...	...	...	599	...	...	...	...	...	...
1020	Unknown	486	519	...	...	...	599	...	...	...	...	...	...
1025	Annealed	486	519	532	557	574	599	662	749	846	1432	950	...
1030	Annealed	486	519	...	...	...	599	...	...	...	...	...	...
1035	Annealed	486	519	...	...	...	586	...	...	...	...	...	...
1040	Annealed	486	519	...	...	...	586	...	...	...	...	...	...
1042	Annealed	486	515	528	548	569	586	649	708	770	1583	624	548
1045	Annealed	486	519	...	...	...	586	...	...	...	...	...	...
1050	Annealed	486	519	...	...	...	590	...	...	...	...	...	...
1060	Unknown	502	544	...	...	...	...	...	...	...	...	...	...
1070	Unknown	490	532	...	...	...	...	...	...	...	...	...	...
1078	Annealed	490	532	548	565	586	607	670	712	770	2081	615	...
1086	Unknown	500	532	...	...	...	...	...	...	...	...	...	...
1095	Unknown	461(b)	...	...	...	...	...	...	...	...	...	...	...
1117	Unknown	481	...	...	...	...	...	...	...	...	...	...	...
1140	Unknown	461(b)	...	...	...	...	...	...	...	...	...	...	...
1151	Unknown	502(b)	...	...	...	...	...	...	...	...	...	...	...
1522	Annealed	486	519	...	...	...	599	...	...	...	...	...	...
1524	Annealed	477	511	528	544	565	590	649	741	837	1449	821	536
1561	Annealed	486	519	...	...	...	...	...	...	...	...	...	...
4032	Unknown	...	461(c)	...	...	...	...	...	...	...	...	...	...
4130	Hardened and tempered	477	515	...	544	...	595	657	737	825	...	833	...
4140	Hardened and tempered	...	473(d)	...	...	...	519(d)	...	561(d)	...	...	...	...
4142	Unknown	...	502(c)	...	...	...	...	...	...	...	...	...	...
4626	Normalized and tempered	335(e)	...	...	...	...	...	...	...	615(f)	...	...	...
4815	Unknown	481(b)	...	...	...	...	...	...	...	...	...	...	...
5132	Unknown	494	523	536	553	574	595	657	741	837	1499	934	574
5140	Hardened and tempered	452(d)	473(d)	...	...	...	519(d)	...	561(d)	...	...	...	...
8115	Unknown	461(b)	...	...	...	...	...	...	...	...	...	...	...
8617	Unknown	481(a)	...	...	...	...	...	...	...	...	...	...	...
8637	Unknown	...	502(c)	...	...	...	...	...	...	...	...	...	...

(a) Specific heat at 25–95 °C (75–200 °F). (b) Specific heat at 20–100 °C (68–212 °F). (c) Specific heat at 20–200 °C (68–392 °F). (d) Value presented is mean value of temperatures between 20 °C (68 °F) and the higher of the cited temperatures. (e) Specific heat at 10–25 °C (50–80 °F). (f) Average specific heat from 25–540 °C (80–1000 °F).

Electrical resistivities of carbon and low-alloy steels

AISI-SAE grade	Treatment or condition	Resistivity, μΩ · m, at °C (°F)											
		20 (68)	100 (212)	200 (392)	400 (752)	600 (1112)	700 (1292)	800 (1472)	900 (1652)	1000 (1832)	1100 (2012)	1200 (2182)	1300 (2372)
1008	Annealed	0.142	0.190	0.263	0.458	0.734	0.905	1.081	1.130	1.165	1.193	1.216	1.244
1010	Unknown	0.143	...	...	...	...	...	...	...	...	...	...	...
1015	Annealed	0.159(a)	0.219	0.292	...	...	...	...	...	...	...	...	...
1016	Annealed	0.160(a)	0.220	0.290	...	...	...	...	...	...	...	...	...
1018	Annealed	0.159(a)	0.219	0.293	...	...	...	...	...	...	...	...	...
1020	Unknown	0.159(a)	0.219	0.292	...	...	...	...	...	...	...	...	...
1022	Annealed	0.159(a)	0.219	0.293	...	...	...	...	...	...	...	...	...
1025	Annealed	0.159(a)	0.219	0.292	0.487	0.758	0.925	1.094	1.136	1.167	1.194	1.219	1.239
1029	Annealed	0.160(a)	0.220	0.290	...	...	...	...	...	...	...	...	...
1030	Annealed	0.166	...	...	...	...	...	...	...	...	...	...	...
1035	Annealed	0.163(a)	0.217	...	...	...	...	...	...	...	...	...	...
1040	Annealed	0.160(a)	0.221	...	...	...	...	...	...	...	...	...	...
1042	Annealed	0.171	0.221	0.296	0.493	0.766	0.932	1.111	1.149	1.179	1.207	1.230	...
1043	Annealed	0.163(a)	0.219	...	...	...	...	...	...	...	...	...	...
1045	Annealed	0.162(a)	0.223	...	...	...	...	...	...	...	...	...	...
1046	Unknown	0.163(a)	0.224	...	...	...	...	...	...	...	...	...	...
1050	Annealed	0.163(a)	0.224	0.300	...	...	...	...	...	...	...	...	...
1055	Unknown	0.163(a)	0.224	...	...	...	...	...	...	...	...	...	...
1060	Unknown	0.180	...	...	...	...	...	...	...	...	...	...	...
1065	Unknown	0.163(a)	0.224	...	...	...	...	...	...	...	...	...	...
1070	Unknown	0.168(a)	0.230	...	...	...	...	...	...	...	...	...	...
1078	Annealed	0.180	0.232	0.308	0.505	0.772	0.935	1.129	1.164	1.191	1.214	1.231	1.246
1080	Unknown	0.180	...	...	...	...	...	...	...	...	...	...	...
1095	Unknown	0.180	...	...	...	...	...	...	...	...	...	...	...
1137	Unknown	0.170	...	...	...	...	...	...	...	...	...	...	...
1141	Unknown	0.170	...	...	...	...	...	...	...	...	...	...	...
1151	Unknown	0.170	...	...	...	...	...	...	...	...	...	...	...
1524	Unknown	0.208	0.259	0.333	0.523	0.786	0.946	1.103	1.143	1.174	1.202	1.227	1.250
1524	Annealed	0.160(a)	0.220	0.290	...	...	...	...	...	...	...	...	...
1552	Unknown	0.163(a)	0.224	...	...	...	...	...	...	...	...	...	...
4130	Hardened and tempered	0.223	0.271	0.342	0.529	0.786	...	1.103	...	1.171	...	1.222	...
4140	Hardened and tempered	0.220	0.260	0.330	0.480	0.650	...	...	...	...	...	...	...
4626	Normalized and tempered	0.200(b)	...	...	...	...	...	...	...	...	...	...	...
4815	Unknown	0.260(a)	0.310	...	...	...	...	...	...	...	...	...	...
5132	Unknown	0.210	0.259	0.330	0.517	0.778	0.934	1.106	1.145	1.177	1.205	1.230	1.251
5140	Hardened and tempered	0.228	0.281	0.352	0.530	0.785	...	...	...	...	...	...	...
8615	Unknown	...	0.300(c)	...	...	...	...	...	...	...	...	...	...
8625	Unknown	...	0.300(c)	...	...	...	...	...	...	...	...	...	...
8720	Unknown	...	0.300(c)	...	0.480(d)	...	...	...	...	...	...	...	...

(a) Resistivity at 0 °C (32 °F). (b) Resistivity at −18 °C (0 °F). (c) Resistivity at 50 °C (120 °F). (d) Resistivity at 300 °C (570 °F)

Carbon and Low-Alloy Steel Sheet and Strip

Revised by the ASM Committee on Steel Sheet and Strip*

CARBON AND LOW-ALLOY STEEL SHEET AND STRIP have much in common with, but also some significant differences from, their flat-rolled plate counterparts. Categorizing these products on the basis of dimensions alone is impossible because of their interchangeability and the overlapping of sizes. Mechanical processing must also be taken into consideration when classifying these hot mill products. Generally, sheet and strip are produced as coils that are rolled in only one direction while using water to accelerate cooling. Plate products, on the other hand, are made from slabs that can be cross rolled (rolled parallel and perpendicular to the slab length) and that are air cooled. Plate is available in widths up to 5080 mm (200 in.). Strip refers to hot-rolled coils less than 305 mm (12 in.) wide or cold-rolled coils less than 608 mm (23^{15}/16 in.) wide. The maximum sheet width currently available from most manufacturers is 1830 mm (72 in.), although a few manufacturers have the capability of producing 1930 mm (76 in.) wide sheet in a 2130 mm (84 in.) mill facility.

Plain Carbon Steel

Plain carbon steel sheet and strip are used primarily in consumer goods. These applications require materials that are serviceable under a wide variety of conditions and that are especially adaptable to low-cost techniques of mass production into articles having good appearance. Therefore, these products must incorporate, in various degrees and combinations, ease of fabrication, adequate strength, excellent finishing characteristics to provide attractive appearance after fabrication, and compatibility with other materials and with various coatings and processes.

The steels used for these products are supplied over a wide range of chemical compositions (see Tables 1 and 2); however, the vast majority are unalloyed, low-carbon steels selected for stamping applications, such as automobile bodies and appliances. Thus, this section of the article will focus on low-carbon steel applications. For these major applications, typical compositions are 0.03 to 0.10% C, 0.15 to 0.50% Mn, 0.035% P (max), and 0.04% S (max).

Generally, rimmed (or capped) ingot cast steel has been used because of its lower price. More recently, these steels have been replaced by killed steels produced by the continuous casting process. This process is inherently suited to the production of killed steels. Where strain aging is to be avoided and/or when exceptional formability is required, steel killed with aluminum, regardless of the method of casting or manufacture, is preferred. Further details regarding steelmaking and deoxidation practice are given in the articles "Steel Processing Technology" and "Classification and Designation of Carbon and Low-Alloy Steels" in this Volume.

The width differentiation between sheet and strip made of plain carbon steel depends on the rolling process. It should be noted that both sheet and strip can be purchased as either cut lengths or coils. The standard dimensional tolerances for plain carbon steel strip are more restrictive than those for sheet. Standard size ranges of plain carbon steel sheet and strip are given in Table 3. Typical characteristics of the various qualities of these products are listed in Table 4(a) to 4(c).

Production of Carbon Steel Sheet and Strip

Carbon steel sheet and strip are available as hot-rolled and as cold-rolled products. Hot-rolled low-carbon steel sheet and strip are usually produced on continuous hot strip mills. The slab is heated and then passed through the mill, where the thickness is progressively reduced to the desired final dimension (see Fig. 3 and the corresponding text in the section "Direct Casting Methods" in this article). Some wide hot strip mills are capable of producing low-carbon steel sheet in thicknesses as low as 1.214 mm (0.0478 in.) (18 gage), but 1.897 mm (0.0747 in.) (14 gage) is considered a practical lower limit. Most narrow hot-strip mills are capable of producing low-carbon steel strip in thicknesses as low as 1.062 mm (0.0418 in.) (19 gage).

Cold-rolled low-carbon steel sheet and strip are produced from pickled hot-rolled coils by cold reduction to the desired thicknesses in either a continuous tandem mill or a reversing cold-reduction mill. The cold-rolling process allows thinner gages to be produced than can be obtained by hot rolling. Other advantages of cold-rolled steel are its better surface finish and dimensional control. The as-rolled steel is hard and has low ductility. Except when a fully work-hardened condition is desired, the steel is annealed to optimize its formability. This annealing can range from stress relieving through full recrystallization with ferrite grain growth and carbide agglomeration (see the article "Steel Processing Technology" in this Volume). After annealing, temper rolling (also called skin rolling or skin passing) is usually done to improve flatness and surface finish. Roller leveling or tension leveling can be used to improve flatness. Temper rolling, roller leveling, or tension leveling will also minimize the tendency of the material to develop stretcher strains during forming; this effect is permanent with killed steels and temporary with rimmed and capped steels (see the section "Surface Characteristics" in this article). Heating a killed steel, as in baking paint, may cause the steel to become susceptible to stretcher strains (see the article "Precoated Steel Sheet" in this Volume).

Most cold-rolled low-carbon steel sheet is available in two classes (Table 4a). Class 1 (temper rolled) is intended for applications where surface appearance is important and where specified surface and flatness re-

---

*David Hudok, Weirton Steel Corporation; J.K. Mahaney, Jr., S.A. Kish, and A.P. Cantwell, LTV Steel Company; Elgin Van Meter, Empire-Detroit Steel Division

Table 1 Compositions for hot-rolled and cold-rolled plain carbon steel sheet and strip

Grade designation	C	Mn	P(a)	S(a)	Cu(b)	Si
Hot-rolled sheet and strip: heavy thickness coils (ASTM A 635)						
1006	0.08(a)	0.45(a)	0.040	0.050	0.20(c)	...
1008	0.10(a)	0.50(a)	0.040	0.050	0.20(c)	...
1009	0.15(a)	0.60(a)	0.040	0.050	0.20(c)	...
1010	0.08–0.13	0.30–0.60	0.040	0.050	0.20(c)	...
1012	0.10–0.15	0.30–0.60	0.040	0.050	0.20(c)	...
1015	0.12–0.18	0.30–0.60	0.040	0.050	0.20(c)	...
1016	0.12–0.18	0.60–0.90	0.040	0.050	0.20(c)	...
1017	0.14–0.20	0.30–0.60	0.040	0.050	0.20(c)	...
1018	0.14–0.20	0.60–0.90	0.040	0.050	0.20(c)	...
1019	0.14–0.20	0.70–1.00	0.040	0.050	0.20(c)	...
1020	0.17–0.23	0.30–0.60	0.040	0.050	0.20(c)	...
1021	0.17–0.23	0.60–0.90	0.040	0.050	0.20(c)	...
1022	0.17–0.23	0.70–1.00	0.040	0.050	0.20(c)	...
1023	0.19–0.25	0.30–0.60	0.040	0.050	0.20(c)	...
1524	0.18–0.25	1.30–1.65	0.040	0.050	0.20(c)	...
Hot-rolled sheet and strip: commercial quality with 0.16–0.25% C (max) (ASTM A 659)						
1015	0.12–0.18	0.30–0.60	0.040	0.050	0.20(c)	...
1016	0.12–0.18	0.60–0.90	0.040	0.050	0.20(c)	...
1017	0.14–0.20	0.30–0.60	0.040	0.050	0.20(c)	...
1018	0.14–0.20	0.60–0.90	0.040	0.050	0.20(c)	...
1020	0.17–0.23	0.30–0.60	0.040	0.050	0.20(c)	...
1021	0.17–0.23	0.60–0.90	0.040	0.050	0.20(c)	...
1023	0.19–0.25	0.30–0.60	0.040	0.050		...
Cold-rolled, high-carbon, spring quality strip (ASTM A 682)						
1030	0.27–0.34	0.60–0.90	0.040(d)	0.050(d)	...	0.15–0.30(e)
1035	0.31–0.38	0.60–0.90	0.040(d)	0.050(d)	...	0.15–0.30(e)
1040	0.36–0.44	0.60–0.90	0.040(d)	0.050(d)	...	0.15–0.30(e)
1045	0.42–0.50	0.60–0.90	0.040(d)	0.050(d)	...	0.15–0.30(e)
1050	0.47–0.55	0.60–0.90	0.040(d)	0.050(d)	...	0.15–0.30(e)
1055	0.52–0.60	0.60–0.90	0.040(d)	0.050(d)	...	0.15–0.30(e)
1060	0.55–0.66	0.60–0.90	0.040(d)	0.050(d)	...	0.15–0.30(e)
1064	0.59–0.70	0.50–0.80	0.040(d)	0.050(d)	...	0.15–0.30(e)
1065	0.59–0.70	0.60–0.90	0.040(d)	0.050(d)	...	0.15–0.30(e)
1070	0.65–0.76	0.60–0.90	0.040(d)	0.050(d)	...	0.15–0.30(e)
1074	0.69–0.80	0.50–0.80	0.040(d)	0.050(d)	...	0.15–0.30(e)
1080	0.74–0.88	0.60–0.90	0.040(d)	0.050(d)	...	0.15–0.30(e)
1085	0.80–0.94	0.70–1.00	0.040(d)	0.050(d)	...	0.15–0.30(e)
1086	0.80–0.94	0.30–0.50	0.040(d)	0.050(d)	...	0.15–0.30(e)
1095	0.90–1.04	0.30–0.50	0.040(d)	0.050(d)	...	0.15–0.30(e)
Hot-rolled and cold-rolled sheet (ASTM A 568)						
1006	0.08(a)	0.45(a)	0.040	0.050	...	0.10(a)(f)
1008	0.10(a)	0.50(a)	0.040	0.050	...	0.10(a)(f)
1009	0.15(a)	0.60(a)	0.040	0.050	...	0.10(a)(f)
1010	0.08–0.13	0.30–0.60	0.040	0.050	...	0.10(a)(f)
1012	0.10–0.15	0.30–0.60	0.040	0.050	...	0.10(a)(f)
1015	0.12–0.18	0.30–0.60	0.040	0.050	...	0.10(a), 0.10–0.25, or 0.15–0.30(f)
1016	0.12–0.18	0.60–0.90	0.040	0.050	...	0.10(a), 0.10–0.25, or 0.15–0.30(f)
1017	0.14–0.20	0.30–0.60	0.040	0.050	...	0.10(a), 0.10–0.25, or 0.15–0.30(f)
1018	0.14–0.20	0.60–0.90	0.040	0.050	...	0.10(a), 0.10–0.25, or 0.15–0.30(f)
1019	0.14–0.20	0.70–1.00	0.040	0.050	...	0.10(a), 0.10–0.25, or 0.15–0.30(f)
1020	0.17–0.23	0.30–0.60	0.040	0.050	...	0.10(a), 0.10–0.25, or 0.15–0.30(f)
1021	0.17–0.23	0.60–0.90	0.040	0.050	...	0.10(a), 0.10–0.25, or 0.15–0.30(f)
1022	0.17–0.23	0.70–1.00	0.040	0.050	...	0.10(a), 0.10–0.25, or 0.15–0.30(f)
1023	0.19–0.25	0.30–0.60	0.040	0.050	...	0.10(a), 0.10–0.25, or 0.15–0.30(f)
1025	0.22–0.28	0.30–0.60	0.040	0.050	...	0.10(a), 0.10–0.25, or 0.15–0.30(f)
1026	0.22–0.28	0.60–0.90	0.040	0.050	...	0.10–0.25 or 0.15–0.30(f)
1030	0.27–0.34	0.60–0.90	0.040	0.050	...	0.10–0.25 or 0.15–0.30(f)
1033	0.29–0.36	0.70–1.00	0.040	0.050	...	0.10–0.25 or 0.15–0.30(f)
1035	0.31–0.38	0.60–0.90	0.040	0.050	...	0.10–0.25 or 0.15–0.30(f)
1037	0.31–0.38	0.70–1.00	0.040	0.050	...	0.10–0.25 or 0.15–0.30(f)
1038	0.34–0.42	0.60–0.90	0.040	0.050	...	0.10–0.25 or 0.15–0.30(f)
1039	0.36–0.44	0.70–1.00	0.040	0.050	...	0.10–0.25 or 0.15–0.30(f)
1040	0.36–0.44	0.60–0.90	0.040	0.050	...	0.10–0.25 or 0.15–0.30(f)
1042	0.39–0.47	0.60–0.90	0.040	0.050	...	0.10–0.25 or 0.15–0.30(f)
1043	0.39–0.47	0.70–1.00	0.040	0.050	...	0.10–0.25 or 0.15–0.30(f)
1045	0.42–0.50	0.60–0.90	0.040	0.050	...	0.10–0.25 or 0.15–0.30(f)
1046	0.42–0.50	0.70–1.00	0.040	0.050	...	0.10–0.25 or 0.15–0.30(f)
1049	0.45–0.53	0.60–0.90	0.040	0.050	...	0.10–0.25 or 0.15–0.30(f)
1050	0.47–0.55	0.60–0.90	0.040	0.050	...	0.10–0.25 or 0.15–0.30(f)
1055	0.52–0.60	0.60–0.90	0.040	0.050	...	0.10–0.25 or 0.15–0.30(f)
1060	0.55–0.66	0.60–0.90	0.040	0.050	...	0.10–0.25 or 0.15–0.30(f)
1064	0.59–0.70	0.50–0.80	0.040	0.050	...	0.10–0.25 or 0.15–0.30(f)
1065	0.59–0.70	0.60–0.90	0.040	0.050	...	0.10–0.25 or 0.15–0.30(f)
1070	0.65–0.76	0.60–0.90	0.040	0.050	...	0.10–0.25 or 0.15–0.30(f)
1074	0.69–0.80	0.50–0.80	0.040	0.050	...	0.10–0.25 or 0.15–0.30(f)
1078	0.72–0.86	0.30–0.60	0.040	0.050	...	0.10–0.25 or 0.15–0.30(f)
1080	0.74–0.88	0.60–0.90	0.040	0.050	...	0.10–0.25 or 0.15–0.30(f)
1084	0.80–0.94	0.60–0.90	0.040	0.050	...	0.10–0.25 or 0.15–0.30(f)
1085	0.80–0.94	0.70–1.00	0.040	0.050	...	0.10–0.25 or 0.15–0.30(f)
1086	0.80–0.94	0.30–0.50	0.040	0.050	...	0.10–0.25 or 0.15–0.30(f)
1090	0.84–0.98	0.60–0.90	0.040	0.050	...	0.10–0.25 or 0.15–0.30(f)
1095	0.90–1.04	0.30–0.50	0.040	0.050	...	0.10–0.25 or 0.15–0.30(f)
1524	0.18–0.25	1.30–1.65	0.040	0.050	...	0.10–0.25 or 0.15–0.30(f)
1527	0.22–0.29	1.20–1.55	0.040	0.050	...	0.10–0.25 or 0.15–0.30(f)
1536	0.30–0.38	1.20–1.55	0.040	0.050	...	0.10–0.25 or 0.15–0.30(f)
1541	0.36–0.45	1.30–1.65	0.040	0.050	...	0.10–0.25 or 0.15–0.30(f)
1548	0.43–0.52	1.05–1.40	0.040	0.050	...	0.10–0.25 or 0.15–0.30(f)
1552	0.46–0.55	1.20–1.55	0.040	0.050	...	0.10–0.25 or 0.15–0.30(f)

(a) Maximum. (b) Minimum. (c) When specified. (d) Ordinarily produced to 0.025% P (max) and 0.035% S (max) by cast or heat analysis. (e) When specified, silicon can be ordered at 0.10–0.20%. (f) When required. Source: Ref 1

quirements must be met. Class 2 is a product intended for applications where appearance is less important. Cold-rolled low-carbon steel strip is available in five hardness tempers ranging from full hard to dead soft (Table 5).

Quality Descriptors for Carbon Steels

The descriptors of quality used for hot-rolled plain carbon steel sheet and strip and cold-rolled plain carbon steel sheet include structural quality, commercial quality, drawing quality, and drawing quality, special killed (Table 4a). Some of the as-rolled material made to these qualities is subject to surface disturbances known as coil breaks, fluting, and stretcher strains; however, fluting and stretcher strains will not be produced during subsequent forming if the material is temper rolled and/or roller leveled immediately prior to forming. It should be noted that any beneficial effects of roller leveling deteriorate rapidly in nonkilled steel. In addition to the requirements listed below for the various qualities of plain carbon steel sheet and strip, special soundness can also be specified.

Commercial quality (CQ) plain carbon steel sheet and strip are suitable for moderate forming; material of this quality has sufficient ductility to be bent flat on itself in any direction in a standard room-temperature bend test. Commercial quality material is not subject to any other mechanical test requirements, and it is not expected to have exceptionally uniform chemical composition or mechanical properties. However, the hardness of cold-rolled CQ sheet is ordinarily less than 60 HRB at the time of shipment.

Drawing Quality. When greater ductility or more uniform properties than those afforded by commercial quality are required, drawing quality (DQ) is specified. Drawing quality material is suitable for the production of deep-drawn parts and other parts requiring severe deformation. When the deformation is particularly severe or resistance to stretcher strains is required, drawing quality, special killed (DQSK) is specified. When either type of drawing quality material is specified, the supplier usually guarantees that the material is capable of being formed into a specified part within an established breakage allowance. The identification of the part is included in the purchase order. Ordinarily, DQ or DQSK material is not subject to any other mechanical requirements, nor is it normally ordered to a specific chemical composition.

Special killed steel is usually an aluminum-killed steel, but other deoxidizers are sometimes used to obtain the desired characteristics. In addition to severe drawing applications, it is specified for applications requiring freedom from significant variations in mechanical properties or freedom from fluting and stretcher strains in temper-

202 / Carbon and Low-Alloy Steels

Table 2 Composition ranges and limits for plain carbon sheet and strip (ASTM specifications)

ASTM specification	Description(b)	C	Mn	P	S	Other
				Composition, %(a)		
A 611	CR SQ					
	Grades A, B, C, E	0.20	0.60	0.04	0.04	(c)
	Grade D { type 1	0.20	0.90	0.04	0.04	(c)
	{ type 2	0.15	0.60	0.20	0.04	(c)
A 366	CR CQ	0.15	0.60	0.035	0.04	(c)
A 109	CR strip					
	Tempers 1, 2, 3	0.25	0.60	0.035	0.04	(c)
	Tempers 4, 5	0.15	0.60	0.035	0.04	(c)
A 619	CR DQ	0.10	0.50	0.025	0.035	...
A 620	CR DQSK	0.10	0.50	0.025	0.035	(d)
A 570	HR SQ					
	Grades 30, 33, 36, and 40	0.25	0.90	0.04	0.05	(c)
	Grades 45, 50, and 55	0.25	1.35	0.04	0.05	(c)
A 569	HR CQ	0.15	0.60	0.035	0.040	(c)
A 621	HR DQ	0.10	0.50	0.025	0.035	...
A 622	HR DQSK	0.10	0.50	0.025	0.035	(d)
A 414	Pressure vessel					
	Grade A	0.15	0.90	0.035	0.040	(c)
	Grade B	0.22	0.90	0.035	0.040	(c)
	Grade C	0.25	0.90	0.035	0.040	(c)
	Grade D	0.25	1.20	0.035	0.040	(c)(e)
	Grade E	0.27	1.20	0.035	0.040	(c)(e)
	Grade F	0.31	1.20	0.035	0.040	(c)(e)
	Grade G	0.31	1.35	0.035	0.040	(c)(e)

(a) All values are maximum. (b) CR, cold rolled; HR, hot rolled; SQ, structural quality; DQ, drawing quality; DQSK, drawing quality, special killed; CQ, commercial quality. (c) Copper when specified as copper-bearing steel; 0.20% min. (d) Aluminum as deoxidizer usually exceeds 0.010% in the product. (e) Killed steel can be supplied upon request to the manufacturer for grades D through G. When silicon-killed steel is specified, a range of 0.15–0.30% Si shall be supplied. Source: Ref 1

strip are not ordinarily used in specifications unless special strength properties are required in the fabricated product. As a matter of general interest, however, the ranges of mechanical properties typical of sheet produced by three mills in these qualities are shown in Fig. 1. The bands would be wider if the product of the entire industry were represented.

It should be noted that the ranges are broader and the sheet harder for the hot-rolled than for the cold-rolled materials and that cold-rolled drawing quality, special killed sheet is produced to a narrower range of mechanical properties than cold-rolled drawing quality sheet, which is a rimmed steel grade. There is a great deal of overlapping in properties between commercial quality and drawing quality sheet.

Figure 2 shows the relationships among hardness, Olsen ductility, and sheet thickness in commercial quality and drawing quality hot-rolled low-carbon steel sheet, indicating the variations in properties that can occur in these materials. Stretchability, as measured by the Olsen value, is also shown to increase as sheet thickness increases.

In contrast to commercial and drawing quality materials, structural (physical) quality sheet and strip are produced in many grades having specific mechanical property minimums, seven of which are shown in Table 6. Cold-rolled low-carbon steel strip is not usually produced to specific strength requirements; typical mechanical property ranges for the various tempers of this product are listed in Table 5.

rolled material without subsequent roller leveling prior to forming. Special killed steels also have inherent characteristics that increase their formability.

Structural quality (SQ), formerly called physical quality (PQ), is applicable when specified strength and elongation values are required in addition to bend tests (Table 6). Minimum values of tensile strength ranging up to 690 MPa (100 ksi) in hot-rolled sheet and strip and up to 1035 MPa (150 ksi) in cold-rolled sheet are available. Cold-rolled strip, which does not have a quality descriptor, is available in five tempers that conform to specified Rockwell hardness ranges and bend test requirements (Table 5). It should be noted that steels with yield strengths exceeding 275 MPa (40 ksi) or tensile strengths greater than 345 MPa (50 ksi) are referred to as high-strength structural or high-strength low-alloy steels. These materials are described elsewhere in this Section of the Volume (see the articles "Classification and Designation of Carbon and Low-Alloy Steels" and "High-Strength Structural and High-Strength Low-Alloy Steels").

Mechanical Properties of Carbon Steels

The commonly measured tensile properties of plain carbon steel sheet and strip are not readily related to their performance in fabrication; the relationship between formability and values of the strain-hardening exponent, n, and the plastic strain ratio, r (determined in tensile testing), is discussed in the article "Sheet Formability of Steels" in this Volume. The mechanical properties of commercial quality, drawing quality, and drawing quality, special killed sheet and

Mill Heat Treatment of Cold-Rolled Products

Unless a hard temper is desired, cold-rolled carbon steel sheet and strip are always softened to improve formability. This is usually accomplished at the mill by a recrystallization heat treatment such as annealing or normalizing.

Annealing. Low-temperature recrystallization annealing, or process annealing, can be used to soften cold-rolled low-carbon

Table 3 Classification of hot-rolled and cold-rolled plain carbon steel sheet and strip product by size

Product	Thickness mm	Thickness in.	Width mm	Width in.	Other limitations	Specification symbol (ASTM No.) English units	Specification symbol (ASTM No.) Metric units
Hot-rolled sheet	1.2–6.0	0.045–0.230 incl	>300–1200 incl	>12–48 incl	Coils and cut lengths	A 569, A 621, or A 622	A 569M, A 621M, or A 622M
	1.2–4.5	0.045–0.180 incl	>1200	>48	Coils and cut lengths	A 569, A 621, or A 622	A 569M, A 621M, or A 622M
	6.0–12.5	0.230–0.500 incl	>300–1200 incl	>12–48 incl	Coils only	A 635	A 635M
	4.5–12.5	0.180–0.500 incl	>1200–1800 incl	>48–72 incl	Coils only	A 635	A 635M
Hot-rolled strip	1.2–5.0	0.045–0.203 incl	≤200	≤6	Coils and cut lengths	A 569, A 621, or A 622	A 569M, A 621M, or A 622M
	1.2–6.0	0.045–0.229 incl	>200–300 incl	>6–12 incl	Coils and cut lengths	A 569, A 621, or A 622	A 569M, A 621M, or A 622M
	6.0–12.5	0.230–0.500 incl	>200–300 incl	>8–12 incl	Coils only	A 635	A 635M
Cold-rolled sheet	0.35–2.0	0.014–0.082 incl	>50–300 incl	>2–12 incl	(a)	A 366, A 619, or A 620	A 366M, A 619M, or A 620M
	≥0.35	>0.014	>300	>12	(b)	A 366, A 619, or A 620	A 366M, A 619M, or A 620M
Cold-rolled strip	≤6.0	≤0.250	>12–600 incl	>0.50–23.9 incl	(c)	A 109	A 109M

Incl, inclusive. (a) Cold-rolled sheet, coils, and cut lengths, slit from wider coils with cut edge (only), thicknesses 0.356–2.08 mm (0.014–0.082 in.) and 0.25% C (max) by cost analysis. (b) When no special edge or finish (other than matte, commercial bright, or luster) is required and/or single-strand rolling of widths under 610 mm (24 in.) is not required. (c) Width 51–305 mm (2–12 in.) with thicknesses of 0.356–2.08 mm (0.014–0.082 in.) are classified as sheet when slit from wider coils, have a cut edge only, and contain 0.25% C (max) by cost analysis. Source: Ref 2

Carbon and Low-Alloy Steel Sheet and Strip / 203

Table 4(a) Summary of available types of hot-rolled and cold-rolled plain carbon steel sheet and strip

Quality or temper	Applicable basic specification number	AISI-SAE grade designation	Surface finish — Temper-rolled; for exposed parts(a) Description	Symbol	Annealed last; for unexposed parts(a) Description	Symbol	Edge(b) Description	Symbol
Hot-rolled sheet								
Commercial Quality .... A 569, A 635		1008–1012	As-rolled (black)	A	As-rolled (black)	A	Mill	M
			Pickled—dry	P	Pickled—dry	P	Mill	M
			Pickled and oiled	O	Pickled and oiled	O	Cut	C
Drawing quality .... A 621		1006–1008	As-rolled (black)	A	As-rolled (black)	A	Mill	M
			Pickled—dry	P	Pickled—dry	P	Mill	M
			Pickled and oiled	O	Pickled and oiled	O	Cut	C
Drawing quality, special killed ...... A 622		1006–1008	As-rolled (black)	A	As-rolled (black)	A	Mill	M
			Pickled—dry	P	Pickled—dry	P	Mill	M
			Pickled and oiled	O	Pickled and oiled	O	Cut	C
Hot-rolled strip								
Commercial quality .... A 569		1008–1012	As-rolled (black)	A	As-rolled (black)	A	Mill	M
			Pickled—dry	P	Pickled—dry	P	Mill	M
			Pickled and oiled	O	Pickled and oiled	O	Cut	M
Drawing quality .... A 621		1006–1008	As-rolled (black)	A	As-rolled (black)	A	Square	S
			Pickled—dry	P	Pickled—dry	P	Square	S
			Pickled and oiled	O	Pickled and oiled	O	Square	S
Drawing quality, special killed ...... A 622		1006–1008	As-rolled (black)	A	As-rolled (black)	A	Cut	C
			Pickled—dry	P	Pickled—dry	P	Cut	C
			Pickled and oiled	O	Pickled and oiled	O	Cut	C
Cold-rolled sheet								
Commercial quality .... A 366		1008–1012	Matte	E	Matte	U	Cut	(c)
			Commercial bright	B			Cut	(c)
			Luster	L			Cut	(c)
Drawing quality .... A 619		1006–1008	Matte	E	Matte	U	Cut	(c)
			Commercial bright	B			Cut	(c)
			Luster	L			Cut	(c)
Drawing quality, special killed ...... A 620		1006–1008	Matte	E	Matte	U	Cut	(c)
			Commercial bright	B			Cut	(c)
			Luster	L			Cut	(c)
Cold-rolled strip								
Temper description numbers 1, 2, 3, 4, 5 .... A 109		(d)	Matte	1	Matte	1	(b)	1, 2, 3, 4, 5, 6
			Regular bright	2	Regular bright	2	(b)	1, 2, 3, 4, 5, 6
			Best bright	3	Best bright	3	(b)	1, 2, 3, 4, 5, 6

(a) See Table 4(b). (b) See Table 4(c). (c) No symbol necessary; cut edge is standard. (d) Produced in five tempers with specific hardness and bend test limits; composition subordinate to mechanical properties. Source: Ref 2

steel. When done as a batch process, this type of annealing is known as box annealing. It is carried out by placing coils on a bottom plate and then enclosing them with a cover within which a protective gas atmosphere is maintained. A bell-type heating furnace is then placed over the atmosphere container. After heating to approximately 595 to 760 °C (1100 to 1400 °F), the charge is allowed to soak until the temperature is uniform throughout. The heating furnace is then removed, and the charge is allowed to cool in the protective atmosphere before being uncovered.

Cold-rolled steel can be batch annealed in coil form under a protective atmosphere. Some producers use a 100% hydrogen atmosphere in an effort to shorten annealing cycles.

Instead of box annealing, coils can also be treated by continuous annealing. With this process, which is usually intended to provide a fully recrystallized grain structure, coils are unwound and passed through an annealing furnace. The uncoiled steel strip passes through several different thermal zones of the furnace that serve to heat, soak, and cool the steel before it exits the furnace and is recoiled. This anneal cycle is very rapid and can be measured in seconds or minutes (as opposed to hours or days with a box anneal cycle). Generally, the rapid anneal cycle of a continuous anneal process results in material properties that are less ductile than those resulting from a box anneal cycle. However, continuous annealing results in more uniformity of properties throughout the length of a coil.

Open-coil annealing is used when uniform heating and/or gas contact across the entire

Table 4(b) Selection and specification of surface condition for plain carbon steel sheet

Specification symbol	Description of surface	Surface described applicable to
U(a)	Surface finish as normally used for unexposed automotive parts. Matte appearance. Normally annealed last	Cold-rolled sheet
E(b)..........	Surface finish as normally used for exposed automotive parts that require a good painted surface. Free from strain markings and fluting. Matte appearance. Temper rolled	Cold-rolled sheet
B.............	Same as above, except commercial bright appearance	Cold-rolled sheet
L.............	Same as above, except luster appearance	Cold-rolled sheet
1.............	No. 1 or dull finish (no luster). Especially suitable for lacquer or paint adhesion. Facilitates drawing by reducing the contact friction between the die and the metal	Cold-rolled strip
2.............	No. 2 or regular bright finish (moderately smooth). Suitable for many applications, but not generally applicable for parts to be plated, unless polished and buffed	Cold-rolled strip
3.............	No. 3 or best bright finish (relatively high luster). Particularly suitable for parts to be plated	Cold-rolled strip
A.............	As-rolled or black (oxide or scale not removed)	Hot-rolled sheet and strip
P.............	Pickled (scale removed), not oiled	Hot-rolled sheet and strip
O.............	Same as above, except oiled	Hot-rolled sheet and strip

(a) U, unexposed; also designated as class 2, cold-rolled sheet. (b) E, exposed; also designated as class 1, cold-rolled sheet. Source: Ref 2

Table 4(c)   Selection and specification of edge condition of plain carbon steel sheet and strip

Specification symbol	Description of edge	Edge described applicable to
None required	Cut edge	Cold-rolled sheet
1	No. 1 edge is a prepared edge of a specified contour (round, square, or beveled) supplied when a very accurate width is required or where the finish of the edge is required to be suitable for electroplating or both	Cold-rolled strip
2	No. 2 edge is a natural mill edge carried through the cold rolling from the hot-rolled strip without additional processing of the edge	Cold-rolled strip
3	No. 3 edge is an approximately square edge produced by slitting, on which the burr is not eliminated	Cold-rolled strip
4	No. 4 edge is a rounded edge produced by edge rolling the natural edge of hot-rolled strip or slit-edge strip. This edge is produced when the width tolerance and edge condition are not as exacting as for No. 1 edge	Cold-rolled strip
5	No. 5 edge is an approximately square edge produced by rolling or filing of a slit edge to remove burr only	Cold-rolled strip
6	No. 6 edge is a square edge produced by edge rolling the natural edge of hot-rolled strip or slit-edge strip, where the width tolerance and finish required are not as exacting as for No. 1 edge	Cold-rolled strip
M	Mill edge	Hot-rolled sheet and strip
C	Cut edge	Hot-rolled sheet and strip
S	Square edge (square and smooth, corners slightly rounded). Produced by rolling through vertical edging rolls during the hot-rolling operation	Hot-rolled strip

Source: Ref 2

width of the coil is required (for example, to obtain decarburization over the entire surface during production of material for porcelain enameling). In this process, the coils are loosely wound, permitting gas to flow freely between the coil convolutions. Annealing temperatures may be higher than those used in conventional box annealing.

Normalizing consists of heating the sheet or strip to a temperature above the Ac_3 point (~925 °C, or 1700 °F, for a steel that contains less than 0.15% C) in a continuous furnace containing an oxidizing atmosphere, then cooling to room temperature at a controlled rate (usually in still air). This treatment recrystallizes and refines the grain structure by phase transformation. Low-metalloid steel (enameling iron) for porcelain enameling is normalized rather than annealed because this steel will not readily recrystallize at box-annealing temperatures.

Surface Characteristics

The surface texture of low-carbon cold-rolled steel sheet and strip can be varied between rather wide limits. For chromium plating and similar finishes, a smooth, bright sheet or strip surface is necessary, but for porcelain enameling and many drawing operations, a rougher surface texture (matte finish) is preferred. In porcelain enameling, roughness tends to improve the adherence and uniformity of the coating; in certain drawing operations where heavy pressures are developed, the rougher type of surface is believed to retain more lubricant, thus aiding formation of the sheet by reducing friction and die galling.

Minor surface imperfections and slight strains are less noticeable on a dull surface than on a bright one. However, the surfaces of parts to be painted should not be so rough that the paint will not cover them adequately. A very smooth, bright surface can be obtained on sheet or strip by utilizing ground and polished rolling-mill rolls, and a dull (matte) surface can be obtained by either grit blasting or etching the rolls. For the purpose of evaluating surface roughness, an appropriate instrument is employed that measures the average height of surface asperities (peaks) in microinches and the number of peaks per inch that exceed a given height.

Cold-rolled sheet or strip can also be purchased with coined patterns that form a geometric design or that simulate such textures as leather grain. Such products are available in commercial quality, drawing quality, and drawing quality, special killed material. The texture is rolled into the steel surface after the sheet or strip has been annealed and thus has an effect on properties similar to that of a heavy temper-rolling pass. This effect, plus the notch effect of the pattern itself, somewhat reduces the formability of the sheet or strip.

Stretcher Strains. When loaded in tension, practically all hot-rolled or as-annealed cold-rolled plain carbon steels, whether rimmed, capped, or killed, exhibit a sharp upper yield point, a drop in load to the lower yield point, and subsequent plastic deformation at a nearly constant load (known as yield point elongation). The plastic deformation that occurs within this yield point elongation is accompanied by the formation of visible bands of deformation on the product surfaces. These bands are called stretcher strains or Lüders lines, and they can be aesthetically undesirable.

The tendency for stretcher strains to occur can be prevented through elimination of yield point elongation. In rimmed or capped steels, this is accomplished by subjecting the steel to small amounts of plastic deformation, usually by temper rolling, tension leveling, and/or roller leveling. Because overstraining the steel by these practices can increase strength and generally decrease ductility, it is usually desirable to strain the steel only by the amount required to eliminate yield point elongation. When properly processed, a killed steel, such as DQSK, provides a product with no yield point elongation.

Strain Aging. In rimmed or capped (but not killed) carbon steels, deformation (such as by temper rolling) followed by aging for several days or more at or slightly above room temperature will result in a return of the upper yield point and yield point elongation, increases in yield and tensile strengths, and a decrease in ductility. This treatment, called strain aging, may be desirable if the increase in strength can be used to advantage. However, strain aging often causes problems due to reduced formability and stretchability and the return of both yield point elongation and a propensity for stretcher strains. Further temper rolling may eliminate yield point elongation, but it will not restore stretchability. In applications where the appearance of stretcher strains is objectionable, killed steels, which

Table 5   Mechanical properties of cold-rolled low-carbon steel strip (ASTM A 109)

Temper	Hardness requirements, HRB	Bent test requirements(a)	Approximate tensile strength MPa	Approximate tensile strength ksi	Elongation in 50 mm (2 in.), %(b)
No. 1 (hard)	90 minimum(c), 84 minimum(d)	No bending in either direction	550–690	80–100	...
No. 2 (half-hard)	70–85(d)	90° bend across rolling direction around a 1t radius	380–520	55–75	4–16
No. 3 (quarter-hard)	60–75(e)	180° bend along rolling direction and 90° bend across rolling direction, both around a 1t radius	310–450	45–65	13–27
No. 4 (skin rolled)	65 maximum(e)	Bend flat on itself in any direction	290–370	42–54	24–40
No. 5 (dead soft)	55 maximum(e)	Bend flat on itself in any direction	260–340	38–50	33–45

(a) t = thickness of strip. (b) For strip 1.27 mm (0.050 in.) thick. (c) For strip of thickness 1.02–1.78 mm exclusive (0.040–0.070 in. exclusive). (d) For strip of thickness 1.78–6.35 mm exclusive (0.070–0.250 in. exclusive). (e) For strip of thickness 1.02–6.35 mm exclusive (0.040–0.250 in. exclusive)

Table 6 Tensile requirements for hot-rolled and cold-rolled plain carbon steel sheet and strip

Class or grade	Yield strength, minimum MPa	ksi	Tensile strength, minimum MPa	ksi	Elongation in 50 mm (2 in.), minimum, %
Structural quality hot-rolled sheet and strip in cut lengths or coils (ASTM A 570)(a)					
30	205	30	340	49	25.0(b)
33	230	33	360	52	23.0(b)
36	250	36	365	53	22.0(b)
40	275	40	380	55	21.0(b)
45	310	45	415	60	19.0(b)
50	345	50	450	65	17.0(b)
55	380	55	480	70	15.0(b)
Structural quality cold-rolled sheet in cut lengths or coils (ASTM A 611)(a)					
A	170	25	290	42	26
B	205	30	310	45	24
C	230	33	330	48	22
D, types 1 and 2	275	40	360	52	20
E	550(c)	80(c)	565	82	...
Hot-rolled sheet for pressure vessels (ASTM A 414)					
A	170(d)	25(d)	310	45	26(e)
B	205(d)	30(d)	345	50	24(e)
C	230(d)	33(d)	380	55	22(e)
D	240(d)	35(d)	415	60	20(e)
E	260(d)	38(d)	450	65	18(e)
F	290(d)	42(d)	485	70	16(e)
G	310(d)	45(d)	515	75	16(e)

(a) For coil products, testing by the producer is limited to the end of the coil. Results of such tests must comply with the specified values. However, design considerations must recognize that variation in strength levels may occur throughout the untested portions of the coil, but generally these levels will not be less than 90% of the minimum values specified. (b) At thickness, t, of 2.5–5.9 mm (0.097–0.230 in.). (c) On this full-hard product, the yield point approaches the tensile strength and because there is no halt in the gage or drop in the beam, the yield point shall be taken as the stress at 0.5% elongation, under load. (d) Yield strength determined by the 0.2% offset or 0.5% extension under load methods. (e) At thickness, t, of 3.7–5.9 mm (0.145–0.230 in.). Source: Ref 1

are resistant to aging, are preferable to rimmed and capped steels. For ingot casting, however, rimmed and capped steels are generally superior in inherent surface quality, are lower in cost, and are preferred over killed steel as long as the occurrence of stretcher strains is not a problem.

Strain aging is related to the presence of nitrogen in solid solution in the steel and is affected by time and termperature, with longer times and higher temperatures producing greater aging. The strain-aging rate is also dependent on the amount of deformation that has occurred and is increased when the deformation occurs at higher temperatures or lower strain rates. Another important variable that affects strain aging is the amount of nitrogen in solution. Killed carbon steels have very little susceptibility to strain aging because their nitrogen content is essentially chemically combined with aluminum. Rimmed and capped steels, however, tend to strain age because they contain greater amounts of nitrogen in solid solution (typically 6 to 30 ppm).

Control of Flatness

Plain carbon steel sheet is ordinarily sold to two standards of flatness:

- Commercial flatness, which is used where flatness is important but not critical
- The stretcher-level standard of flatness, which is required when little or no forming is to be done and the product is required to be flat and free from waves or oil can, or when flatness is necessary to ensure smooth automatic feeding of forming equipment

The permissible variations for the flatness of hot- and cold-rolled sheet have been established by the Technical Committee of the American Iron and Steel Institute and are given in the AISI Steel Products Manual. Commercial flatness can usually be produced by roller leveling or by temper rolling and roller leveling, but where very flat sheet is required, producers may have to resort to stretcher leveling, tension leveling, or other leveling processes.

In temper rolling, the steel is cold reduced, usually by ½ to 2%, which is also

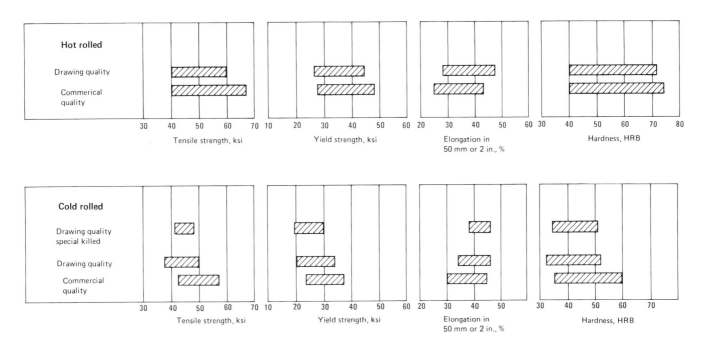

Fig. 1 Typical mechanical properties of low-carbon steel sheet shown by the range of properties in steel furnished by three mills. Hot-rolled sheet thickness from 1.519 to 3.416 mm (0.0598 to 0.1345 in., or 16 to 10 gage); cold-rolled sheet thickness from 0.759 to 1.519 mm (0.0299 to 0.0598 in., or 22 to 16 gage). All cold-rolled grades include a temper pass. All grades were rolled from rimmed steel except the one labeled special killed. See Table 5 for the mechanical properties of structural (physical) quality sheet.

206 / Carbon and Low-Alloy Steels

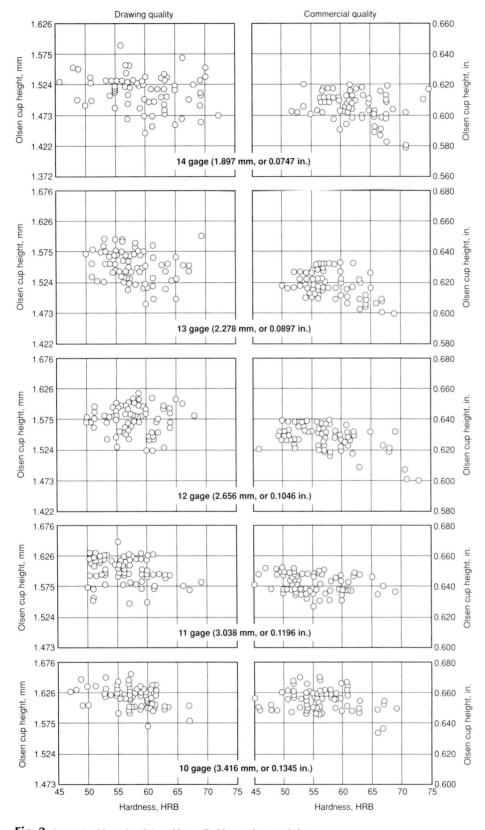

Fig. 2 Scatter in Olsen ductilities of hot-rolled low-carbon steel sheet

effective for removing yield point elongation and preventing stretcher strains.

In roller leveling, a staggered series of small-diameter rolls alternately flexes the steel back and forth. The rolls are adjusted so that the greatest deformation occurs at the entrance end of the rolls and less flexing occurs at the exit end. Stretcher strains can also be eliminated by roller leveling, as long as the deformation is great enough to remove yield point elongation. Dead-soft annealed sheet cannot be made suitable for production of exposed parts by roller leveling because the rolls kink the sheet severely, producing leveler breaks. The deformed areas or kinks will not deform further upon stretching and will appear as braised welts after forming.

Stretcher Leveling. Leveling by stretching cut lengths of the temper-rolled sheet lengthwise between jaws (stretcher leveling) is a more positive means of producing flatness. Elongation (stretching) during stretcher leveling may vary from about 1 to 3%, which exceeds the elastic limit of the steel and therefore results in some permanent elongation. The sheet must be of a killed or a capped steel having nearly uniform properties so that it will spring back uniformly across its full width and remain flat. It may be necessary to use killed steel having nearly uniform properties so that, after stretching, strain markings do not develop.

Tension Leveling. Another flattening process that is used for steel sheet is tension leveling, which combines the effects of stretcher and roller leveling. The sheet is pulled to a stress near its yield point while it is simultaneously flexed over small rolls; the combined tension and bending produce yielding at the flex points.

Modified Low-Carbon Steel Sheet and Strip

In addition to the low-carbon steel sheet and strip products already discussed in this article, there are numerous additional products available that are designed to satisfy specific customer requirements. These products are often made with low-carbon steels having chemical compositions slightly modified from those discussed earlier.

To be considered a plain low-carbon grade, a steel should contain no more than 0.25% C, 1.65% Mn, 0.60% S, and 0.60% Cu, but it may also contain small amounts of other elements, such as nitrogen, phosphorus, and boron, that are effective in imparting special characteristics when present singly or in combination. The modified low-carbon steel grades discussed below are designed to provide sheet and strip products having increased strength, formability, and/or corrosion resistance.

Carbon-Manganese Steels. Manganese is a solid-solution strengthening element in ferrite and is also effective in increasing hardenability. Manganese in amounts ranging from 1.0 to 1.5% is added to low-carbon steel (0.15 to 0.25% C) to provide enhanced strength (yield strength of about 275 MPa, or 40 ksi) with good ductility in hot-rolled and cold-rolled sheet and strip. Components fabricated from these higher-manganese steels can be heat treated by quenching

Table 7 Compositions for hot-rolled and cold-rolled low-alloy steel sheet and strip

AISI or SAE designation	Chemical composition ranges and limits, % (heat analysis)(a)								
	C	Mn	P	S	Si(b)	Ni	Cr	Mo	V
Regular quality and structural quality standard steels commonly produced (ASTM A 506)									
4118	0.18–0.23	0.70–0.90	0.035	0.040	0.15–0.30	...	0.40–0.60	0.08–0.15	...
4130	0.28–0.33	0.40–0.60	0.035	0.040	0.15–0.30	...	0.80–1.10	0.15–0.25	...
4140	0.38–0.43	0.75–1.00	0.035	0.040	0.15–0.30	...	0.80–1.10	0.15–0.25	...
4340	0.38–0.43	0.60–0.80	0.035	0.040	0.15–0.30	1.65–2.00	0.70–0.90	0.20–0.30	...
5140	0.38–0.43	0.70–0.90	0.035	0.040	0.15–0.30	...	0.70–0.90	...	...
5150	0.48–0.53	0.70–0.90	0.035	0.040	0.15–0.30	...	0.70–0.90	...	...
5160	0.55–0.65	0.75–1.00	0.035	0.040	0.15–0.30	...	0.70–0.90	...	...
8615	0.13–0.18	0.70–0.90	0.035	0.040	0.15–0.30	0.40–0.70	0.40–0.60	0.15–0.25	...
8620	0.18–0.23	0.70–0.90	0.035	0.040	0.15–0.30	0.40–0.70	0.40–0.60	0.15–0.25	...
Regular quality and structural quality standard steels not commonly produced (ASTM A 506)									
E3310	0.08–0.13	0.45–0.60	0.025	0.025	0.15–0.30	3.25–3.75	1.40–1.75	...	...
4012	0.09–0.14	0.75–1.00	0.040	0.040	0.15–0.30	...	...	0.15–0.25	...
4118	0.18–0.23	0.70–0.90	0.040	0.040	0.15–0.30	...	0.40–0.60	0.08–0.15	...
4135	0.33–0.38	0.70–0.90	0.040	0.040	0.15–0.30	...	0.80–1.10	0.15–0.25	...
4137	0.35–0.40	0.70–0.90	0.040	0.040	0.15–0.30	...	0.80–1.10	0.15–0.25	...
4142	0.40–0.45	0.75–1.00	0.040	0.040	0.15–0.30	...	0.80–1.10	0.15–0.25	...
4145	0.43–0.48	0.75–1.00	0.040	0.040	0.15–0.30	...	0.80–1.10	0.15–0.25	...
4147	0.45–0.50	0.75–1.00	0.040	0.040	0.15–0.30	...	0.80–1.10	0.15–0.25	...
4150	0.48–0.53	0.75–1.00	0.040	0.040	0.15–0.30	...	0.80–1.10	0.15–0.25	...
4320	0.17–0.22	0.45–0.65	0.040	0.040	0.15–0.30	1.65–2.00	0.40–0.60	0.20–0.30	...
E4340	0.38–0.43	0.65–0.85	0.025	0.025	0.15–0.30	1.65–2.00	0.70–0.90	0.20–0.30	...
4520	0.18–0.23	0.45–0.65	0.040	0.040	0.15–0.30	...	...	0.45–0.60	...
4615	0.13–0.18	0.45–0.65	0.040	0.040	0.15–0.30	1.65–2.00	...	0.20–0.30	...
4620	0.17–0.22	0.45–0.65	0.040	0.040	0.15–0.30	1.65–2.00	...	0.20–0.30	...
4718	0.16–0.21	0.70–0.90	0.040	0.040	0.15–0.30	0.90–1.20	0.35–0.55	0.30–0.40	...
4815	0.13–0.18	0.40–0.60	0.040	0.040	0.15–0.30	3.25–3.75	...	0.20–0.30	...
4820	0.18–0.23	0.50–0.70	0.040	0.040	0.15–0.30	3.25–3.75	...	0.20–0.30	...
5015	0.12–0.17	0.30–0.50	0.040	0.040	0.15–0.30	...	0.30–0.50	...	...
5046	0.43–0.50	0.75–1.00	0.040	0.040	0.15–0.30	...	0.20–0.35	...	...
5115	0.13–0.18	0.70–0.90	0.040	0.040	0.15–0.30	...	0.70–0.90	...	...
5130	0.28–0.33	0.70–0.90	0.040	0.040	0.15–0.30	...	0.80–1.10	...	...
5132	0.30–0.35	0.60–0.90	0.040	0.040	0.15–0.30	...	0.75–1.00	...	...
E51100	0.95–1.10	0.25–0.45	0.025	0.025	0.15–0.30	...	0.90–1.15	...	...
E52100	0.95–1.10	0.25–0.45	0.025	0.025	0.15–0.30	...	1.30–1.60	...	...
6150	0.48–0.53	0.70–0.90	0.040	0.040	0.15–0.30	...	0.80–1.10	...	0.15 min
8617	0.15–0.20	0.70–0.90	0.040	0.040	0.15–0.30	0.40–0.70	0.40–0.60	0.15–0.25	...
8630	0.28–0.33	0.70–0.90	0.040	0.040	0.15–0.30	0.40–0.70	0.40–0.60	0.15–0.25	...
8640	0.38–0.43	0.75–1.00	0.040	0.040	0.15–0.30	0.40–0.70	0.40–0.60	0.15–0.25	...
8642	0.40–0.45	0.75–1.00	0.040	0.040	0.15–0.30	0.40–0.70	0.40–0.60	0.15–0.25	...
8645	0.43–0.48	0.75–1.00	0.040	0.040	0.15–0.30	0.40–0.70	0.40–0.60	0.15–0.25	...
8650	0.48–0.53	0.75–1.00	0.040	0.040	0.15–0.30	0.40–0.70	0.40–0.60	0.15–0.25	...
8655	0.50–0.60	0.75–1.00	0.040	0.040	0.15–0.30	0.40–0.70	0.40–0.60	0.15–0.25	...
8660	0.55–0.65	0.75–1.00	0.040	0.040	0.15–0.30	0.40–0.70	0.40–0.60	0.15–0.25	...
8720	0.18–0.23	0.70–0.90	0.040	0.040	0.15–0.30	0.40–0.70	0.40–0.60	0.20–0.30	...
8735	0.33–0.38	0.75–1.00	0.040	0.040	0.15–0.30	0.40–0.70	0.40–0.60	0.20–0.30	...
8740	0.38–0.43	0.75–1.00	0.040	0.040	0.15–0.30	0.40–0.70	0.40–0.60	0.20–0.30	...
9260	0.55–0.65	0.70–1.00	0.040	0.040	1.80–2.20	...	...	...	...
9262	0.55–0.65	0.75–1.00	0.040	0.040	1.80–2.20	...	0.25–0.40	...	...
E9310	0.08–0.13	0.45–0.65	0.025	0.025	0.20–0.35	3.00–35.0	1.00–1.40	0.08–0.15	...
Drawing quality standard steels commonly produced (ASTM A 507)									
4118	0.18–0.23	0.70–0.90	0.035	0.040	0.15–0.30	...	0.40–0.60	0.08–0.15	...
4130	0.28–0.33	0.40–0.60	0.035	0.040	0.15–0.30	...	0.80–1.10	0.15–0.25	...
8615	0.13–0.18	0.70–0.90	0.035	0.040	0.15–0.30	0.40–0.70	0.40–0.60	0.15–0.25	...
8620	0.18–0.23	0.70–0.90	0.035	0.040	0.15–0.30	0.40–0.70	0.40–0.60	0.15–0.25	...
Drawing quality standard steels not commonly produced (ASTM M A 507)									
E3310	0.08–0.13	0.45–0.60	0.025	0.025	0.15–0.30	3.25–3.75	1.40–1.75	...	...
4012	0.09–0.14	0.75–1.00	0.040	0.040	0.15–0.30	...	...	0.15–0.25	...
4118	0.18–0.23	0.70–0.90	0.040	0.040	0.15–0.30	...	0.40–0.60	0.08–0.15	...
4320	0.17–0.22	0.45–0.65	0.040	0.040	0.15–0.30	1.65–2.00	0.40–0.60	0.20–0.30	...
4520	0.18–0.23	0.45–0.65	0.040	0.040	0.15–0.30	...	...	0.45–0.60	...
4615	0.13–0.18	0.45–0.65	0.040	0.040	0.15–0.30	1.65–2.00	...	0.20–0.30	...
4620	0.17–0.22	0.45–0.65	0.040	0.040	0.15–0.30	1.65–2.00	...	0.20–0.30	...
4718	0.16–0.21	0.70–0.90	0.040	0.040	0.15–0.30	0.90–1.20	0.35–0.55	0.30–0.40	...
4815	0.13–0.18	0.40–0.60	0.040	0.040	0.15–0.30	3.25–3.75	...	0.20–0.30	...
4820	0.18–0.23	0.50–0.70	0.040	0.040	0.15–0.30	3.25–3.75	...	0.20–0.30	...
5015	0.12–0.17	0.30–0.50	0.040	0.040	0.15–0.30	...	0.30–0.50	...	...
5115	0.13–0.18	0.70–0.90	0.040	0.040	0.15–0.30	...	0.70–0.90	...	...
5130	0.28–0.33	0.70–0.90	0.040	0.040	0.15–0.30	...	0.80–1.10	...	...
5132	0.30–0.35	0.60–0.90	0.040	0.040	0.15–0.30	...	0.75–1.00	...	...
8617	0.15–0.20	0.70–0.90	0.040	0.040	0.15–0.30	0.40–0.70	0.40–0.60	0.15–0.25	...
8630	0.28–0.33	0.70–0.90	0.040	0.040	0.15–0.30	0.40–0.70	0.40–0.60	0.15–0.25	...
8720	0.18–0.23	0.70–0.90	0.040	0.040	0.15–0.30	0.40–0.70	0.40–0.60	0.20–0.30	...
E9310	0.08–0.13	0.45–0.65	0.025	0.025	0.20–0.35	3.00–3.50	1.00–1.40	0.08–0.15	...

(a) The chemical ranges and limits shown are subject to product analysis tolerances. See ASTM A 505. (b) Other silicon ranges are available. Consult the producer. Source: Ref 1

Table 8 Standard sizes of hot-rolled and cold-rolled low-alloy steel sheet and strip: regular quality, structural quality, and drawing quality

Product	Applicable ASTM specification	Thickness range		Width range	
		mm	in.	mm	in.
Hot-rolled sheet	A 506, A 507	5.839–4.572 inclusive	0.2299–0.1800 inclusive	610–1220 inclusive	24–48 inclusive
		≤4.569	≤0.1799	>610	>24
Hot-rolled strip	A 506, A 507	≤5.156	≤0.2030	≤152	≤6
		≤5.839	≤0.2299	152–608 inclusive	>6–23$^{15}/_{16}$ inclusive
Cold-rolled sheet	A 506, A 507	≤5.839	≤0.2299	610–1220	24–48
		≤4.569	≤0.1799	>1220	>48
Cold-rolled strip	A 506, A 507	≤6.347	≤0.2499	≤608	≤23$^{15}/_{16}$

Source: Ref 1

and tempering to provide enhanced strength with good toughness (see the article "High-Strength Structural and High-Strength Low-Alloy Steels" in this Volume).

Carbon-Silicon Steels. Silicon, like manganese, is an effective ferrite-strengthening element and is sometimes added in amounts of about 0.5%, often in combination with 1.0 to 1.5% Mn, to provide increased strength in low-carbon hot-rolled and cold-rolled steel sheet and strip.

Nitrogenized and Rephosphorized Steels. Nitrogen is a strong interstitial strengthener, and phosphorus is an effective solid-solution strengthener in ferrite. Either about 0.010 to 0.015% N or 0.07 to 0.12% P is added to low-carbon steel to provide hot-rolled and cold-rolled sheet and strip with yield strengths in the range of 275 to 345 MPa (40 to 50 ksi) for low-cost structural components for buildings and automotive uses. Formed parts produced from nitrogenized steel can be further strengthened to yield strengths in the range of 415 to 485 MPa (60 to 70 ksi) as the result of strain aging that occurs at paint-curing temperatures.

Boron Steels. Boron is a strong carbide- and nitride-forming element and increases strength in quenched and tempered low-carbon steels through the formation of martensite and the precipitation strengthening of ferrite. Boron-containing killed carbon steels are available as low-cost replacements for the high-carbon and low-alloy steels used for sheet and strip. The low-carbon boron steels have better cold-forming characteristics and can be heat treated to equivalent hardness and greater toughness for a wide variety of applications, such as tools, machine components, and fasteners.

Copper Steels. Copper in amounts up to 0.5% is not only a mild solid-solution strengthener in ferrite, but it also provides enhanced atmospheric corrosion resistance together with improved paint retention in applications involving full exposure to the weather. Therefore, copper-bearing (0.20% Cu, minimum) steel is often specified by customers for use in sheet and strip for structures subject to atmospheric corrosion. Essentially all low-carbon steel sheet and strip products can be supplied in copper-bearing grades, if so specified. Copper-bearing steels, which are also referred to as weathering steels, are also described in the article "High-Strength Structural and High-Strength Low-Alloy Steels" in this Volume.

Low-Alloy Steel*

Low-alloy steel sheet and strip are used primarily for those special applications that require the mechanical properties normally obtained by heat treatment. A sizeable selection of the standard low-alloy steels are available as sheet and strip, either hot-rolled or cold rolled. The most commonly available alloys are listed in Table 7, along with their chemical compositions. In addition to standard low-alloy steels, high-strength low-alloy (HSLA) and dual-phase steels are available as sheet or strip for applications requiring tensile strengths in the range of 290 to 760 MPa (42 to 110 ksi), and ultrahigh-strength steels or maraging steels for applications requiring tensile strengths above 1380 MPa (200 ksi). These steels are discussed in the articles "High-Strength Structural and High-Strength Low-Alloy Steels," "Dual-Phase Steels," "Ultrahigh-Strength Steels" and "Maraging Steels" in this Volume.

Production of Sheet and Strip

As described earlier in this article, steel sheet and strip are flat-rolled products that can be rolled to finished thickness on either a hot mill or a cold mill. Hot-rolled steel sheet and strip are normally produced by passing heated slabs through a continuous mill consisting of a series of roll stands, where the thickness is progressively reduced to the desired final dimension.

Cold-rolled low-alloy steel sheet and strip are normally produced from pickled and annealed hot-rolled bands of intermediate thickness by cold reduction to desired thickness in a single-stand mill or tandem mill. Intermediate anneals may be required to facilitate cold reduction or to obtain the mechanical properties desired in the finished product. Cold rolling can produce thinner gages than can be obtained by hot rolling.

Low-alloy steel sheet and strip are produced in thicknesses similar to those typical of HSLA steel sheet and strip (Table 8). In general, tolerances similar to those given in the general requirements for hot-rolled and cold-rolled low-alloy and HSLA steel sheet and strip, ASTM A 505, apply to all low-alloy and HSLA steel sheet and strip. Available thicknesses and tolerances may vary among producers, due mainly to the interrelation between steel quality and rolling practice, as influenced by the equipment available for rolling the product.

Quality Descriptors

As it is used for steel mill products, the term quality relates to the general suitability of the mill product to make a given class of parts. For low-alloy steel sheet and strip, the various quality descriptors imply certain inherent characteristics, such as the degree of internal soundness and the relative freedom from harmful surface imperfections.

The quality descriptors used for alloy steel sheet and plate include regular quality, drawing quality, and aircraft quality, which are covered by ASTM specifications. The general requirements for these qualities include bearing quality and aircraft structural quality. Aircraft quality requirements are also defined in Aerospace Material Specifications (AMS).

Regular Quality. Low-alloy steel sheet and strip of regular quality are intended principally for general or miscellaneous applications where moderate drawing and/or bending is required. A smooth finish free of minor surface imperfections is not a primary requirement. Sheet and strip of this quality do not have the uniformity, the high degree of internal soundness, or the freedom from surface imperfections that are associated with other quality descriptors for low-alloy sheet and strip.

Regular quality low-alloy steel sheet and strip are covered by ASTM A 506. One or more of the following characteristics may be specified by the purchaser: chemical composition, grain size, or mechanical properties (determined by tensile and bend tests).

Drawing quality describes low-alloy steel sheet and strip for applications involving severe cold working such as deep-drawn or severely formed parts. Drawing quality low-alloy sheet and strip are rolled from steel produced by closely controlled steelmaking practices. The semifinished and finished mill products are subject to testing and inspection designed to ensure internal soundness, relative uniformity of chemical

---

*The term low-alloy steel rather than the more general term alloy steel is being used in this article as well as other articles in this Section of the Handbook. See the article "Classification and Designations of Carbon and Low-Alloy Steels" for definitions of various steel types.

Table 9 Tensile requirements of chromium-molybdenum alloy steel sheet and strip for pressure vessels (ASTM A 873)

Class	Yield strength, minimum MPa	ksi	Tensile strength, minimum MPa	ksi	Elongation in 50 mm (2 in.), minimum, at thickness t, %	
					t=3.8–5.9 mm (0.145–0.230 in.)	t=1.8–3.7 mm (0.070–0.144 in)
1	205	30	415	60	15	12
2	310	45	515	75	13	10
3	415	60	585	85	12	9
4	515	75	655	95	11	8
5	690	100	895	130	7	4

Source: Ref 1

composition, and freedom from injurious surface imperfections. Spheroidize annealing is generally specified so the mechanical properties and microstructure are suitable for deep drawing or severe forming. Drawing quality low-alloy steel sheet and strip are covered by ASTM A 507.

No standard test can fully evaluate resistance to breakage during deep drawing because successful drawing is affected by die clearances, die design, speed of drawing, lubricants, ironing, grade of steel, and any alteration of hardness, ductility, or surface condition that may develop during drawing. Thus, it cannot be assumed that merely specifying drawing quality steel will ensure a capability for drawing or forming a specific part under a given set of manufacturing conditions. Manufacturing trials may be necessary before purchase orders can be written for production material.

Bearing quality describes low-alloy steel sheet and strip intended for antifriction bearing parts. The steels are generally AISI-SAE alloy carburizing grades or AISI-SAE high-carbon chromium grades. These steels are produced using steelmaking and conditioning practices that are intended to optimize internal soundness and to provide a known size, shape, and distribution of nonmetallic inclusions. Standards of acceptance for microstructural quality are commonly reviewed and agreed upon between producer and purchaser for each order. Alternatively, internal soundness and microcleanliness can be determined by using immersion ultrasonic testing techniques to agreed-upon acceptance standards. More detailed information on low-alloy bearing steels can be found in the article "Bearing Steels" in this Volume.

Aircraft quality describes low-alloy steel sheet and strip for important or highly stressed parts of aircraft, missiles, and similar applications involving stringent performance requirements, especially in terms of internal cleanliness. The special mill practices required for producing aircraft quality sheet and strip include careful selection of the raw materials charged into the melting furnace, exceptionally close control of the steelmaking process, cropping and discarding more of the ingot than is normal during primary reduction, selection of specific heats or portions of heats for fulfillment of a given customer order, and using exceptionally close control over process variables during reheating and rolling. Aircraft quality low-alloy steel sheet and strip generally have an austenitic grain size predominantly ASTM No. 5 or finer, with grains as coarse as ASTM No. 3 permissible. Grain size tests are normally made on rerolling slabs or billets.

Aircraft quality low-alloy steel sheet and strip are covered by Aerospace Material Specifications (AMS 6454A, for example). Material of this quality is ordinarily certified that it has been produced as aircraft quality.

Aircraft structural quality low-alloy steel sheet and strip meet all the requirements of aircraft quality mill products described above. In addition, they meet specified requirements for mechanical properties, which may include tensile strength, yield strength, elongation, bend test results, or results of other similar tests. Many specimens from each heat must be tested to ensure compliance with the required mechanical properties.

Mill Heat Treatment

Hot-rolled regular quality low-alloy steel sheet and strip are normally available from the producer either as-rolled or heat treated. Standard mill heat-treated conditions are annealed, normalized, or normalized and tempered. Cold-rolled regular quality product is normally available only in the annealed condition.

Hot-rolled and cold-rolled drawing quality alloy steel sheet and strip are normally furnished by the producer in the spheroidize-annealed condition. They can be purchased in the as-rolled condition if they are to be spheroidize annealed by the user.

Aircraft quality products are normally furnished in a heat-treated condition. Hot-rolled products may be annealed, spheroidize annealed, normalized, or normalized and tempered by the producer. Cold-rolled products are normally furnished only in the annealed or spheroidize-annealed condition.

Annealing is done by heating the steel to a temperature near or below the lower critical temperature and holding at that temperature for a sufficient period, followed by slow cooling in the furnace. This process softens the sheet or strip for further processing, but not to the same degree as spheroidize annealing.

Spheroidize annealing involves prolonged heating at a temperature near or slightly below the lower critical temperature, followed by slow cooling. The objective of this process is to change the form of the carbides in the microstructure to a globular (spheroidal) shape, which produces the greatest degree of softening.

Normalizing consists of heating the sheet or strip to a temperature 55 to 70 °C (100 to 125 °F) above Ac_3 and then cooling to room temperature at a controlled rate (usually in still air). This treatment recrystallizes and refines the grains by phase transformation and can be used to obtain the desired mechanical properties.

Tempering consists of reheating steel to a predetermined temperature below the lower critical temperature, holding for a specified length of time, and then cooling under suitable conditions. When it is carried out as part of a mill heat treatment, tempering is done after normalizing to obtain the desired mechanical properties by modifying the as-normalized microstructure.

Quenching and tempering (or hardening) is normally reserved for the user to apply as one of the final steps in the fabricating process.

Mechanical Properties

In most instances, the mechanical properties of low-alloy steel furnished by the producer are of little consequence because they will be altered by heat treatment during fabrication. For low-alloy steel sheet and strip to be used in the mill condition, mechanical properties will vary, depending on both chemical composition and mill processing. Table 9 lists typical tensile properties for chromium-molybdenum low-alloy steel sheet and strip used for pressure vessels. Usually, low-alloy steel sheet and strip are custom produced to fulfill specific customer orders. Where necessary, any mechanical property requirements can be made part of the purchase order.

Because the chief benefits of low-alloy steel sheet and strip accrue to the user only after the finished part is heat treated, the mechanical properties of heat-treated low-alloy steels are the ones of greatest importance. These properties can be determined from hardenability curves (see the article "Hardenability Curves" in this Volume) and heat-treating guides such as those found in the articles "Hardenable Carbon and Low-Alloy Steels" and "Hardenability of Carbon and Low-Alloy Steels" in this Volume. In general, only those properties typical of through-hardened steel of the specific grade under consideration need to be considered. Except for the most shallow hardening grades used at thicknesses at or

210 / Carbon and Low-Alloy Steels

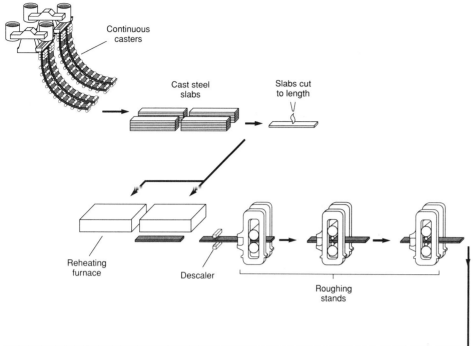

Fig. 3 Key components of a continuous casting operation. Source: SMS Engineering, Inc.

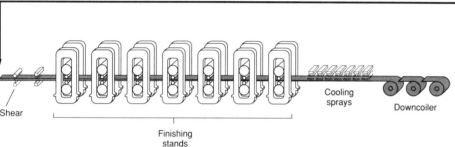

Fig. 4 Flowchart of operations for various strip casting processes. Source: Ref 3

Table 10 Alternative sheet, strip, and slab casting techniques

Country	Company	Caster type
Thin slab casting		
United States	Bethlehem-USX	Hazlett
	Nucor	Hazlett
		SMS-Concast
Great Britain	British Steel	Travelling block mold
Germany	SMS-Concast	Vertical static mold
	Krupp	Hazelett
	Mannesmann	Vertical static mold
Japan	Kawasaki Steel	Vertical twin belt
		Horizontal twin belt
	Sumitomo Metals	Hazlett
	Hitachi-Korf	Wheel and belt
	Nippon Steel	Twin belt
Switzerland	Alusuisse	Twin block mold
Austria	Hitachi-Korf	Wheel and belt
Thin strip casting		
United States	BSC-Armco-Inland-Weirton	Twin roll
	Armco	Single roll
	Allegheny-Ludlum	Single roll
	Argonne National Labs	Electromagnetic levitation
	United Technologies	Single roll
	LTV	Drum in drum
Japan	Nippon Steel	Twin roll
	Kawasaki Steel	Twin roll
	Nippon Kokan	Twin roll
	Nippon Metals	Twin roll
	Kobe	Twin roll
	Nippon Yakin	Twin roll
France	IRSID	Twin roll
Italy	CMS	Twin roll
	Danieli	Thin strip
Austria	Voest-Alpine	Single roll, Twin roll
Switzerland	Concast	Single roll
Germany	Mannesman-Battelle	Single substrate
Spray casting		
Great Britain	Osprey Metals Ltd.	Osprey process
	Sprayforming Developments	Spray forming
	Aurora Metals	Controlled spray deposition
	University of Swansea	Spray forming
Sweden	Sandviken	Osprey process
Germany	Mannesmann Demag	Spray forming sheet by Osprey process
Japan	Sumitomo	Osprey process
United States	M.I.T.	Dynamic liquid compaction
	Drexel University	Osprey process

Source: Ref 3

near the upper limit for sheet and strip, parts made of low-alloy steel sheet or strip will through harden when quenched. Many grades will through harden when quenched in a slow medium such as oil and may even through harden when air cooled. The possibility of oil quenching or air cooling should always be considered for hardening thin parts, especially when warping or distortion during hardening need to be minimized.

Parts made of low-alloy steel sheet and strip are sometimes carburized or carbonitrided to improve the mechanical properties or wear resistance of the surface layer. In some cases, parts that are difficult to form when made of a medium-carbon low-alloy steel can be formed from low-carbon low-

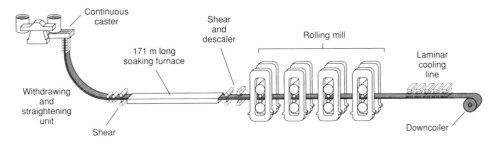

Fig. 5 Key components of a thin slab casting facility. Compare with Fig. 3. Source: SMS Engineering, Inc.

alloy steel and then carburized to a uniform but higher carbon content.

Direct Casting Methods

Because of the large investment needed to build conventional steelmaking casting and rolling facilities, the focus over the last ten years has been on reducing production costs and simplifying the overall steelmaking process. For the most part, cost savings have been achieved by the progression of casting technology from ingot to continuous casting, which eliminates soaking and breakdown hot rolling of large ingots. The following table compares the continuous cast share (in percent) for the United States, the European Economic Community (EEC), Japan, and the total world:

Country	1981, %	1989, %	1990, %
United States	20.3	63.7	66.2
EEC	42.5	73.7	74.6
Japan	70.7	94.6	95.1
Total world	24.3	44.4	46.8

Source: Wharton Econometric Forecasting Associates

Conventional continuous casting of steels requires the casting of a 150 to 250 mm (6 to 10 in.) thick by 800 to 2200 mm (31 to 86 in.) wide slab that is subsequently rolled down to a thickness of 1.5 to 25 mm (0.05 to 1.0 in.) utilizing a hot strip mill having both four-stand roughing and six- or seven-stand finishing mills (Fig. 3). This process requires a high degree of reduction and the equivalent input of energy.

Direct casting processes are alternatives to conventional slab casting processes. Direct casting processes for steel flat products could be defined as any casting process that produces a casting as close as possible to the final product dimensions of the next processing step. By this definition, direct casting could also be termed near-net shape casting because the final cast dimensions would approach the final product dimensions (Ref 3).

Presently, there are three direct casting alternatives. Listed in increasing order according to how close they come to producing near-net shape dimensions, these processes are (Ref 3):

- Thin slab casting
- Thin strip casting
- Spray casting

The flowcharts in Fig. 4 summarize the key operations involved in these three alternative direct casting processes and compare them with those of a continuous casting process in an integrated steel production facility.

Thin Slab Casting. Of the three direct casting processes listed above, only the thin slab casting process is being used commercially. In thin slab casting, a slab 40 to 60 mm (1.5 to 2.5 in.) is produced. Hot rolling is not completely eliminated in this process, but the amount of reduction necessary to produce strip is greatly reduced. However, the need for a heating furnace and a roughing mill is eliminated (Fig. 5). In addition, thin slab casting yields a finer grain structure and a better finish than that obtained with conventional continuous casting technology.

Table 10 lists some of the countries and specific firms engaged in research and development of thin slab casting worldwide. References 3 to 7 provide detailed information on the start-up of a thin slab casting minimill.

In thin strip casting, a strip that is generally less than 5 mm (0.2 in.) thick is cast. In this process, the most optimistic scenario is that the need for a hot strip mill will be eliminated altogether. As indicated in Table 10, there are three areas of concentration in thin strip casting:

- Single-roll process
- Twin-roll process
- Electromagnetic levitation

Strip casting is expected to be available for commercialization within the next five to ten years if significant advances in control and quality can be achieved. Currently, single-roll casting is closer to commercialization processes than twin-roll processes, especially in the area of stainless steel manufacture. Additional information on thin strip casting can be found in Ref 3.

In spray forming, a liquid metal is atomized and sprayed onto a substrate in an inert atmosphere to form a sheet (Ref 3). Because it eliminates conventional casting and hot rolling processes, spray forming is a true near-net shape casting technology. Compaction after forming is normally necessary to eliminate porosity and to achieve high density. This technology has been applied to the manufacture of rings, tubes, small billets, and pipes for both ferrous and nonferrous applications. Both centrifugal atomization processes such as controlled spray deposition and gas atomization processes are included in this category (Table 10).

The commercialization of spray casting for strip production is at least five to ten years in the future for bulk steelmaking. In addition, applying this technology to low-carbon aluminum-killed strip may be difficult because of surface quality and yield requirements.

REFERENCES

1. *Steel—Plate, Sheet, Strip, Wire*, Vol 01.03, *Annual Book of ASTM Standards*, American Society for Testing and Materials
2. *Materials*, Vol 1, *SAE Handbook*, Society of Automotive Engineers, 1989
3. A.W. Cramb, New Steel Casting Processes for Thin Slabs and Strip: A Historical Perspective, *Iron Steelmaker*, Vol 15 (No. 7), July 1988
4. W.D. Huskonen, Nucor Starts Up Thin Slab Mill, *33 Met. Prod.*, Aug 1989
5. G.J. McManus, Taking the Wraps off Nucor's Sheet Mill, *Iron Age*, June 1989
6. G. Flemming, F. Hollmann, M. Kolakowski, and H. Streubel, Continuous Casting of Strips, CSP: A Future Alternative for the Modernization of Slab Production, *Fachber. Hüttenprax. Metallweiterverarb.*, Vol 25 (No. 8), 1987
7. A. Collier, Hot Tech: Thin Slabs and Direct Steelmaking, *Iron Age*, July 1989

Precoated Steel Sheet

Revised by R.W. Leonard, USS Corporation, Division of USX Corporation

STEEL SHEET is often coated in coil form before fabrication either by the steel mills or by specialists known as coil coaters. This prefinished or precoated sheet is ready for fabrication and use without further surface coating. Precoated products yield lower production costs, improved product quality, shorter processing cycles, elimination of production hazards, conservation of energy, minimized ecological problems, and production expansion without a capital expenditure for new buildings and equipment.

Some precautions are necessary with precoated sheet. The product must be handled with more care to prevent scratches and damage to the prefinished surface. Metal finishing of damaged areas is more difficult than on uncoated sheet. Fabrication methods are more restrictive, bend radii must be more generous, and welding practices must be carefully chosen.

The basic types of precoating include metallic, pretreated, preprimed, and prepainted finishing. Metallic coating can be made up of zinc, aluminum, zinc-aluminum alloys, tin, and terne metal. Pretreatment coatings are usually phosphates, and preprimed finishes can be applied as a variety of organic-type coatings. These can be used as a primed-only coating, or a suitable paint topcoat can be applied. Prepainting consists of applying an organic paint system to steel sheet on a coil coating line either at a mill or at a coil coater. This article will address each of these coating processes. Emphasis will be placed on products that are galvanized by the hot dip process, although much of the discussion is equally applicable to electrogalvanizing and zinc spraying.

Zinc Coatings

Galvanizing is a process for rustproofing iron and steel by the application of a metallic zinc coating. It is applicable to products of nearly all shapes and sizes, ranging from nails, nuts, and bolts to large structural assemblies and steel sheet in coils and cut lengths. Other applications include roofing and siding sheets for buildings, silos, grain bins, heat exchangers, hot water tanks, pipe, culverts, conduits, air conditioner housings, outdoor furniture, and mail boxes. On all steel parts, galvanizing provides long-lasting, economical protection against a wide variety of corrosive elements in the air, water, or soil.

In the United States, more than 9×10^6 Mg (1×10^7 tons) of steel is produced annually by precoating. A large amount of this total is used by the automotive industry for both unexposed and exposed panels—from frames and floor pans to doors, fenders, and hoods (Fig. 1). Typically, 75% of the body, chassis, and power train components of one American automobile manufacturer's 1986 models consisted of galvanized precoated sheet (Fig. 2). Table 1 indicates that a typical 1986 American car utilized nearly 160 kg (350 lb) of zinc-coated steel components in its material composition. As indicated in Table 2, undervehicle test coupons evaluated after 2 years of exposure attest to the benefits of precoated steels in combating corrosion (additional information is available in the article "Corrosion in the Automobile Industry" in Volume 13 of the 9th Edition of *Metals Handbook*).

Metallic zinc is applied to iron and steel by three processes: hot dip galvanizing, electrogalvanizing, and zinc spraying. Most galvanized steel sheet is coated by the hot dip process, although there has been strong growth in electrogalvanizing capacity during the past few years.

Corrosion Resistance. The use of zinc is unique among methods for the corrosion protection of steel. The zinc coating serves a twofold purpose:

- It protects the steel from corrosive attack in most atmospheres, acting as a continuous barrier shield between the steel and the atmosphere
- It acts as a galvanic protector, slowly sacrificing itself in the presence of corrosive elements by continuing to protect the steel even when moderate-sized areas of bare metal have been exposed

This latter ability is possible because zinc is more electrochemically active than steel. This dual nature of zinc coatings is also available with some zinc/aluminum alloy coatings, but zinc coatings clearly offer the most galvanic protection. With most protective coatings that act only as a barrier, rapid attack commences when exposure of the base metal occurs.

The distance over which the galvanic protection of zinc is effective depends on the environment. When completely and continuously wetted, especially by a strong electrolyte (for example, seawater), relatively large areas of exposed steel will be protected as long as any zinc remains. In air, where the electrolyte is only superficially or discontinuously present (such as from dew or rain), smaller areas of bare steel are protected. The order of magnitude of this throwing power is nominally about 3.2 mm ($\frac{1}{8}$ in.), although this can vary significantly with the type of atmosphere. Nevertheless, galvanized parts exposed outdoors have remained rust free for many years, and the two basic reasons are the sacrificial protection provided by the zinc and the relatively stable zinc carbonate film that forms on the zinc surface to reduce the overall corrosion rate of the zinc coating.

The service life of zinc-coated steel is dependent on the conditions of exposure and on the coating thickness, as illustrated in Fig. 3. Although the coating process used to apply the zinc coating generally does not affect the service life, experience has shown that the corrosion resistance of galvanized coatings in the field cannot be accurately predicted from accelerated laboratory tests. Environmental factors such as atmospheric contaminants (sulfates, chlorides, and so on) and time of wetness have a large influence on the service life of galvanized steel. In polluted areas, such as severe industrial areas, the normally protective zinc carbonate film that forms on the surface of the zinc coating tends to be converted to soluble sulfates that are washed away by rain, thus exposing the zinc to further attack and accelerating the corrosion.

Coating Tests and Designations. The thickness (or weight), adhesion, and ductility of a zinc coating can have important effects on its service life and effectiveness against corrosion. Practical tests for these charac-

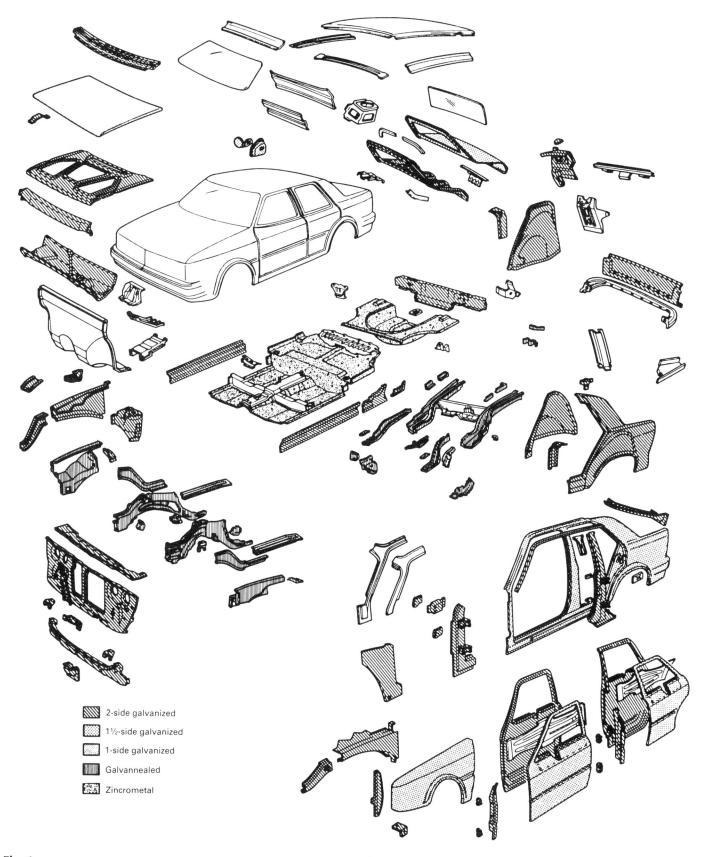

Fig. 1 Use of zinc-coated steels in a 1987 model by one U.S. automaker. Source: Ref 1

teristics are described in relevant specifications issued by the American Society for Testing and Materials (ASTM). Tests for coating thickness include microscopic measurement of the cross section, stripping tests in which the coating is removed from a given area (ASTM A 90), electrochemical stripping from a given area (ASTM B 504), and magnetic and electromagnetic methods

Table 1 Use of zinc-coated steel for a typical 1986 model U.S. car

Type	Amount of steel kg	lb	Amount of zinc kg	lb
One-side galvanized	33.5	74	0.55	1.21
Two-side galvanized	93	205	3.05	6.72
Zincrometal	29.5	65	0.19	0.41
Net total	156	344	3.8	8.34

Source: Ref 2

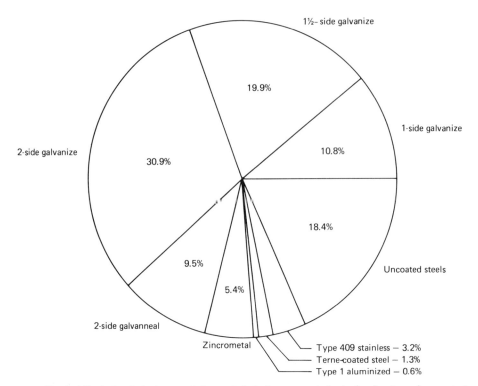

Fig. 2 Pie chart illustrating typical usage of zinc-coated steel components for body, chassis, and power train applications in a 1986 car manufactured by a U.S. automaker. Source: Ref 1

of measurement (ASTM E 376, A 123, A 754, B 499, and D 1186). Adhesion can be tested and rated by bend test methods described in ASTM A 525 and A 879. Other adhesion test methods include reverse impact and draw bend test.

Because the service life of a zinc-coated part in a given atmosphere is directly proportional to the thickness of zinc in the coating (Fig. 3), measurement of that amount is very important. The amount of coating is most often measured in terms of weight rather than thickness, usually by the method described in ASTM A 90. Specimens are cut from one or three spots in samples of the sheet, as described in ASTM A 525. These are weighed, the zinc is stripped (dissolved) in an acid solution, and the specimens are reweighed. The weight loss is reported in ounces per square foot of sheet or grams per square meter. When specimens from three spots are checked (triple-spot test), the value of weight loss is the average of the three specimens.

When the weight-loss method is used, the amount of coating measured is the total amount on both sides of the sheet. Ordinarily, the zinc coating is applied to both sides of the sheet. Therefore, a 2 oz/ft^2 coating has 305 g/m^2 (1 oz/ft^2) on each surface. This 28 g (1 oz) is equivalent to an average thickness of 43 μm (1.7 mils). When zinc-coated sheet is ordered, the minimum amount of coating can be specified as the weight determined by the triple-spot or single-spot test or by coating designations corresponding to these weights (Table 3).

Chromate Passivation. Several types of finishes can be applied to zinc-coated surfaces to provide additional corrosion resistance. The simplest type of finish applicable to fresh zinc surfaces is a chromate passivation treatment. This process is equally suitable for use on hot dip galvanized, electrogalvanized, zinc-sprayed, and zinc-plated articles. Usually, the treatment consists of simply cleaning and dipping the articles in a chromic acid or sodium dichromate solution at about 20 to 30 °C (68 to 86

Table 2 Corrosion of unpainted coated steel test coupons after 2 years of undervehicle exposure

Material	Coating weight per side g/m^2	oz/ft^2	Steel thickness mm	in.	Surface area showing base metal attack, % Vehicle 1	Vehicle 2	Average pit depth Vehicle 1 μm	mils	Vehicle 2 μm	mils
Hot dip										
Galvanized 1	120–150	0.39–0.49	0.71	0.028	0.6	14	0	0	15	0.6
Galvanized 2	100–120	0.33–0.39	0.90	0.035	0.3	27.3	11	0.43	56	2.2
Galvanized 3	55–90	0.18–0.30	0.45	0.018	0.5	5.0	0	0	15	0.6
Galvannealed 1	80–120	0.26–0.39	1.42	0.055	0	1.0	0	0	22	0.87
Galvannealed 2	75–85	0.25–0.28	0.89	0.035	0.3	32.8	11	0.43	86	3.4
One-side galvannealed	66	0.22	0.66	0.026	25	56.5	48	1.9	67	2.6
One-side electrodeposited										
Zn	90	0.30	0.88	0.035	61	86	64	2.5	120	4.7
Zn-15Ni-0.4Co	37	0.12	0.70	0.0275	46	67.5	75	3	81	3.2
Zn-16Ni	20	0.065	0.68	0.027	85	93.5	83	3.3	100	4
Zn-16Ni	40	0.13	0.68	0.027	38	79.3	73	2.9	128	5
Zn-16Al	25	0.08	0.68	0.027	59	84.3	64	2.5	97	3.8
Zn-22Al	40	0.13	0.68	0.027	54	76.5	64	2.5	90	3.5
Zinc-rich primer										
One-side Zincrometal	40	0.13	0.92	0.036	10.8	17.3	53	2.1	73	2.9
Uncoated										
Cold-rolled steel	...	...	0.51	0.020	100	100	>250	>10	>250	>10

(a) Vehicle 1, 660 days, 51 000 km (31 700 miles); vehicle 2, 660 days, 53 500 km (33 250 miles). Source: Ref 3

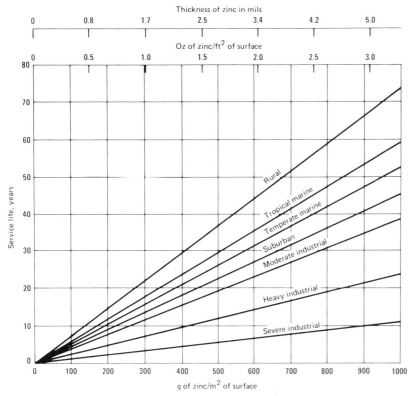

Fig. 3 Service life of zinc-coated steel sheet. Service life is measured in years to the appearance of first significant rusting.

Table 3 Designations and weights of zinc coating on hot dip galvanized sheet

Coating designation(b)	Minimum coating weights(a)			
	Triple-spot test		Single-spot test	
	g/m²	oz/ft²	g/m²	oz/ft²
G 235	717	2.35	610	2.00
G 210	640	2.10	549	1.80
G 185	564	1.85	488	1.60
G 165	503	1.65	427	1.40
G 140	427	1.40	366	1.20
G 115	351	1.15	305	1.00
G 90	275	0.90	244	0.80
G 60	183	0.60	152	0.50
G 01	No minimum		No minimum	
A 60	183	0.60	152	0.50
A 40	122	0.40	91	0.30
A 01	No minimum		No minimum	

(a) Total weight on both sides of sheet per unit area of sheet. (b) G, regular type of coating; A, zinc-iron alloy type of coating. Source: ASTM A 525

°F), followed by rinsing in cold fresh water and drying in warm air. The adherent chromate film may have a greenish or greenish-yellow iridescent appearance. Specification ASTM B 201 gives details of tests for measuring the adequacy and effectiveness of the chromate film. Chromate passivation helps prevent staining when galvanized sheet is stored under wet or humid conditions. Therefore, a thin, almost clear chromate or chromate/phosphate passivation film is often applied to the coating on hot dip coating lines.

Painting. The selection of galvanized steel as a material for barns, buildings, roofs, sidings, appliances, and many hardware items is based on the sacrificial protection and the barrier coating afforded the base metal by zinc coating. For additional protection and cosmetic appearance, paint coatings are often applied to the galvanized steel. The performance of these coatings is an important economic factor in the durability of this material (Table 4).

Galvanized steel, both new and weathered, can be painted with a minimum of preparation and with excellent adherence. On hot dip galvanized or zinc-plated steel, it is necessary to use special corrosion-inhibitive primers to prepare the surface before the paint is applied. This is partly because these types of zinc coating are too smooth to provide a mechanical key for the paint or lacquer and partly because the paint appears to react with the unprepared zinc surface in the presence of moisture to weaken the initially formed bond.

Many exposure tests have shown that zinc dust-zinc oxide paints (finely powdered zinc metallic and zinc oxide pigment in an oil or alkyd base) adhere best to galvanized steel surfaces under most conditions. Zinc dust-zinc oxide primers can be used over new or weathered galvanized steel and can be top coated with most oil or latex house paints or alkyd enamels. For the maintenance painting of galvanized steel, one or two coats of a zinc dust-zinc oxide paint are often used alone. The paint can be applied by brushing, rolling, or spraying.

Zinc sheets to be painted should not be treated at the mill with a chromate treatment, although they may be given a phosphate treatment to improve the adherence of the paint. Zinc-coated sheet steel is often prepainted in coil form by coil coating, as described in the section "Prepainted Sheet" in this article.

Packaging and Storage. Galvanized products in bundles, coils, or stacks of sheets must be protected from moisture, including condensation moisture, until openly exposed to the weather. They must be properly packaged and stored. Otherwise, wet-storage stain, a bulky white deposit that often forms on zinc surfaces stored under wet or humid conditions, may develop.

It is important to examine packages of galvanized products for damage and to take prompt action where cuts, tears, or other damage is evident. If the packaging is damaged or if moisture is present, the product should be dried at once and not repiled until thoroughly dry. Erection of materials should begin as soon as possible after the package arrives at the installation site.

If temporary storage of the galvanized product is absolutely necessary, it should be indoors. Where indoor storage is not possible, intact waterproof bundles can be stored at the site. The package should be slanted so that any condensation will drain

Table 4 Synergistic protective effect of galvanized steel/paint systems in atmospheric exposure

Type of atmosphere	Galvanized steel			Paint			Galvanized plus paint		
	Thickness		Service life(a), years	Thickness		Service life(a), years	Thickness		Service life(a), years
	μm	mils		μm	mils		μm	mils	
Heavy industrial	50	2	10	100	4	3	150	6	19
	75	3	14	150	6	5	225	9	29
	100	4	19	100	4	3	200	8	33
	100	4	19	150	6	5	250	10	36
Metropolitan (urban)	50	2	19	100	4	4	150	6	34
	75	3	29	150	6	6	225	9	52
	100	4	39	100	4	4	200	8	64
	100	4	39	150	6	6	250	10	67
Marine	50	2	20	100	4	4	150	6	36
	75	3	40	100	4	4	200	8	66
	100	4	40	150	6	6	250	10	69

(a) Service life is defined as time to about 5% red rust. Source: Ref 4

out, and it should be stored sufficiently high to allow air circulation beneath and to prevent rising water from entering. The stacks should be thoroughly covered with a waterproof canvas tarpaulin for protection from rain, snow, or condensation. The use of airtight plastic coverings should be avoided. To deter the formation of wet-storage stain, zinc-coated sheet can be purchased with a mill-applied chromate or chromate/phosphate film. Various proprietary mixtures are available.

Hot dip galvanizing is a process in which an adherent, protective coating of zinc and iron-zinc alloys is developed on the surfaces of iron and steel products by immersing them in a bath of molten zinc. Most zinc coated steel is processed by hot dip galvanizing.

One method of hot dip galvanizing is the batch process, which is used for fabricated steel items such as structurals or pipe. This method involves cleaning the steel articles, applying a flux to the surfaces, and immersing them in a molten bath of zinc for varying time periods to develop a thick alloyed zinc coating.

The most common form of hot dip galvanizing for steel sheet is done on a continuous galvanizing line. Coiled sheet is fed from pay-off reels through flatteners. It is then cleaned, bright annealed, and passed through the coating bath. After leaving the coating bath, the coating thickness is controlled by an air knife or steel rolls. The sheet is then cooled and recoiled or cut into lengths. The hot dip process normally coats both sides of the sheet. However, hot dip galvanized sheets can be coated on one side only for special uses, such as automotive exposed panels, by the use of special coating techniques. One-side coated sheet produced by the hot dip process is not commonly available. Continuous coating lines have to be specially modified to make one-side coated product.

A typical hot dip galvanized coating produced by the batch process consists of a series of layers (Fig. 4). Starting from the base steel at the bottom of the coating, each successive layer contains a higher proportion of zinc until the outer layer, which is relatively pure zinc, is reached. There is, therefore, no real line of demarcation between the iron and zinc, but a gradual transition through the series of iron-zinc alloys that provide a powerful bond between the base metal and the coating. These layers are identified in Table 5. The structure of the coating (the number and extent of the alloy layers) and its thickness depend on the composition and physical condition of the steel being treated as well as on a number of factors within the control of the galvanizer.

The ratio of the total thickness of the alloy layers to that of the outer zinc coating is affected by varying the time of immersion

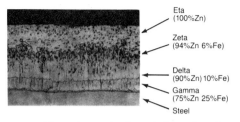

Fig. 4 Photomicrograph of a typical hot dip galvanized coating. The molten zinc is interlocked into the steel by the alloy reaction, which forms zinc-iron layers and creates a metallurgical bond. 250×

and the speed of withdrawal of the work from the molten zinc bath. The rate of cooling of the steel after withdrawal is another factor to be considered; rapid cooling gives small spangle size.

Sheet galvanizers operating continuous strip processes usually suppress the formation of alloy layers by adding 0.1 to 0.2% Al to the bath; this increases the ductility of the coating, thus rendering the sheet more amenable to fabrication (Fig. 5). Other elements can be added to galvanizing baths to improve the characteristics and appearance of the coating. Lead and antimony give rise to well-defined spangle effects.

During batch galvanizing, the zinc-iron alloy portion of the coating will represent 50 to 60% of the total coating thickness. However, certain combinations of elements may result in a coating that is either completely or almost completely alloyed. Visually, the zinc-iron alloy coating will have a gray, matte appearance because of the absence of the free-zinc layer. The free-zinc layer imparts the typical bright finish to a galvanized coating. Because of the greater percentage of the zinc-iron alloy present in the coating, the alloy-type coating may have a lower adherence than the regular galvanized coating.

The corrosion resistance of the zinc-iron and free zinc coating types is equal for all practical purposes. Steels containing carbon below 0.25%, phosphorus below 0.05%, and manganese below 1.35% (either individually or in combination) will normally develop regular galvanized coatings when conventional galvanizing techniques are used and when silicon is 0.05% or less or ranges between 0.15 and 0.3%. Fabricators and consumers should be aware that a gray

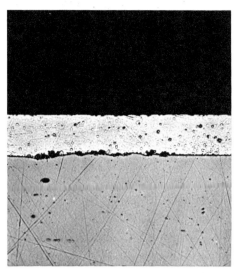

Fig. 5 Microstructure of continuously galvanized steel. In continuous hot dip galvanizing, the formation of various iron-zinc alloy layers is suppressed by the addition of 0.1 to 0.2% Al.

matte appearance may occur in batch galvanizing if silicon content exceeds 0.06%. This matte appearance does not reduce the long-term atmospheric corrosion protection of the galvanized coating.

Galvanized coatings on sheet products that are intended to be painted are frequently given treatments to make the spangle less obvious so that it does not show through the paint. A flat spangle without relief (suppressed spangle) can be obtained by small additions of antimony to the molten bath; smaller grain size (minimized spangle) can be produced by spraying the molten zinc with zinc dust, steam, air, or water just before it freezes. Finer grains are less visible through the paint and have narrower and smaller fractures on bending, often permitting the paint to bridge the gap and provide increased protection.

Galvanized steel sheet can be temper rolled to flatten surface irregularities such as dross and grain boundaries, thus providing an extra smooth surface more suitable for painting where critical surface requirements exist. At the galvanizing mill, galvanized steel sheet can be given a thermal treatment after coating, which converts all the free zinc to zinc-iron alloy, thus providing a spangle-free surface that is more suit-

Table 5 Properties of alloy layers of hot dip galvanized steels

Layer	Alloy	Iron, %	Melting point °C	Melting point °F	Crystal structure	Diamond pyramid microhardness	Alloy characteristics
Eta (η)	Zinc	0.03	419	787	Hexagonal	70–72	Soft, ductile
Zeta (ζ)	FeZn$_{13}$	5.7–6.3	530	986	Monoclinic	175–185	Hard, brittle
Delta (δ)	FeZn$_7$	7.0–11.0	530–670	986–1238	Hexagonal	240–300	Ductile
Gamma (Γ)	Fe$_3$Zn$_{10}$	20.0–27.0	670–780	1238–1436	Cubic	...	Thin, hard, brittle
Steel base metal	Iron	...	1510	2750	Cubic	150–175	...

Table 6 ASTM specifications for hot dip galvanized steel sheet
General requirements are given in A 525.

Specification	Application or quality
A 361	Sheet for roofing and siding
A 444	Sheet for culverts and underdrains
A 446	Structural (physical) quality sheet
A 526	Commercial quality sheet
A 527	Lock-forming quality sheet
A 528	Drawing quality sheet
A 642	Drawing quality, special killed sheet

Table 7 Bend test requirements for hot dip galvanized steel sheet
Table does not apply to structural (physical) quality sheet; see Table 8 instead.

Coating designation	Bend diameter for sheet thickness range(a)		
	0.33–0.97 mm (0.0131–0.0381 in.)	0.97–1.90 mm (0.0382–0.0747 in.)	1.90–4.46 mm (0.0748–0.1756 in.)
G 235	$2t$	$3t$	$3t$
G 210	$2t$	$2t$	$2t$
G 185	$2t$	$2t$	$2t$
G 165	$2t$	$2t$	$2t$
G 140	t	t	$2t$
G 115	0	0	t
G 90	0	0	t
G 60	0	0	0
G 01	0	0	0

(a) Value listed is the minimum diameter of rod (or mandrel), in multiples of the galvanized sheet thickness (t) around which the galvanized sheet can be bent 180° in any direction at room temperature without flaking of the coating on the outside of the bend. Source: ASTM A 525

able for painting. It can be painted without pretreatment (but not with all paints). As an added benefit, there is no spangle to show through the paint. However, the zinc-iron alloy coating is somewhat brittle and tends to powder if severely bent in fabrication.

Table 6 lists the seven ASTM specifications that cover hot dip galvanized steel sheet products. The general requirements for the products covered in these specifications are described in ASTM A 525. Included in this specification are the bend test requirements given in Table 7, but not included in these bend test requirements are those for structural (physical) quality sheet, which are given in Table 8.

The typical mechanical properties of commercial quality (CQ), drawing quality (DQ), and drawing quality, special killed (DQSK) hot dip galvanized steel sheet are listed in Table 9. Commercial quality sheet is satisfactory for applications requiring bending and moderate drawing. Drawing quality sheet has better ductility and uniformity than commercial quality and is excellent for ordinary drawing applications. Drawing quality, special killed sheet is superior to drawing quality and is excellent for applications requiring severe drawing. When higher strength is required, structural quality (SQ) sheet, also called physical quality (PQ) sheet, can be specified, although at some sacrifice in ductility (compare Tables 7 and 8). The minimum mechanical properties of structural quality sheet are presented in Table 10. Additional information is available in the article "Hot Dip Coatings" in Volume 13 of the 9th Edition of *Metals Handbook*.

Electrogalvanizing. Very thin formable zinc coatings ideally suited for deep drawing or painting can be obtained on steel products by electrogalvanizing. Zinc is electrodeposited on a variety of mill products: sheet, wire, and, in some cases, pipe. Electrogalvanizing the sheet and wire in coil form produces a thin, uniform coating of pure zinc with excellent adherence. The coating is smooth, readily prepared for painting by phosphatizing, and free of the characteristic spangles of hot dip zinc coatings. Electrogalvanizing can be used where a fine surface finish is needed. The appearance of the coating can be varied by additives and special treatments in the plating bath.

Electrodeposited zinc coatings are simpler in structure than hot dip galvanized coatings. They are composed of pure zinc, have a homogeneous structure, and are highly adherent. These coatings are not generally as thick as those produced by hot dip galvanizing. Electrogalvanized coating weights as high as 100 g/m² (0.3 oz/ft²) have been applied to one or both sides of steel sheet. The normal ranges of coating weights available are listed in ASTM Specifications A 591 and A 879. The coating thicknesses listed in A 591 are typically used when the application does not subject the steel sheet to very corrosive conditions or when the sheet is intended for painting. For more severe corrosion conditions, such as the need to protect cars from road salts and entrapped moisture, heavier coatings in the ranges listed in A 879 are used. These coating weights are applied to the steel sheets used for most body panels.

Electrodeposited zinc is considered to adhere to steel as well as any metallic coating. Because of the excellent adhesion of electrodeposited zinc, electrogalvanized coils of steel sheet and wire have good working properties, and the coating remains intact after severe deformation. Good adhesion depends on very close physical conformity of the coating with the base metal. Therefore, particular care must be taken during initial cleaning. Electrodeposition also affords a means of applying zinc coatings to finished parts that cannot be predipped. It is especially useful where a high processing temperature could damage the part. One advantage of electrodeposition is that it can be done cold and therefore does not change the mechanical properties of the steel.

Zincrometal is also used for outer body panels in automobiles. First introduced in 1972, Zincrometal is a coil coated product consisting of a mixed-oxide underlayer containing metallic zinc particles and a zinc-rich organic (epoxy) topcoat. It is weldable, formable, paintable, and compatible with commonly used adhesives. Zincrometal is primarily used in one-side applications to protect against inside-out corrosion. The corrosion resistance of Zincrometal is not as good as that of hot dip galvanized steels (Ref 1), and its use is declining substantially as more electrogalvanized steels and other types of coatings are employed.

Zinc alloy coated steels have also been developed. Coatings include zinc-iron (15 to 80% Fe) and zinc-nickel (10 to 14% Ni) alloys. These coatings are applied by electrodeposition. Zinc-iron coatings offer excellent corrosion resistance and weldability. Zinc-nickel coatings are more corrosion resistant than pure zinc coatings, but problems include brittleness from residual stresses and the fact that the coating is not completely sacrificial, as is a pure zinc coating. This can lead to accelerated corrosion of the steel substrate if the coating is damaged (Ref 5).

Multilayer coatings that take advantage of the properties of each layer have been developed in Europe. An example of this is Zincrox, a zinc-chromium-chromium oxide coating (Ref 5). The CrO_x top layer of this coating acts as a barrier to perforation and

Table 8 Bend test requirements for coating and base metal of structural (physical) quality hot dip galvanized steel sheet

Coating designations or base metal	Bend diameter for sheet grade(a)		
	A	B	C
G 235	$3t$	$3t$	$3t$
G 210	$2t$	$2t$	$2\frac{1}{2}t$
G 185	$2t$	$2t$	$2\frac{1}{2}t$
G 165	$2t$	$2t$	$2\frac{1}{2}t$
G 140	$2t$	$2t$	$2\frac{1}{2}t$
G 115	$1\frac{1}{2}t$	$2t$	$2\frac{1}{2}t$
G 90	$1\frac{1}{2}t$	$2t$	$2\frac{1}{2}t$
G 60	$1\frac{1}{2}t$	$2t$	$2\frac{1}{2}t$
G 01	$1\frac{1}{2}t$	$2t$	$2\frac{1}{2}t$
Base metal	$1\frac{1}{2}t$	$2t$	$2\frac{1}{2}t$

(a) Value listed is the minimum diameter of rod (or mandrel), in multiples of the galvanized sheet thickness (t), around which the sheet can be bent 180° in any direction at room temperature without flaking of the coating, or cracking of the base metal, on the outside of the bend. There are no bend test requirements for coatings and base metal of grades D, E, and F. Source: ASTM A 446

Table 9 Typical mechanical properties of hot dip galvanized, long terne, or aluminized steel sheet

Not temper rolled. See Table 10 for properties of structural (physical) quality galvanized sheet.

Quality	Tensile strength(a) MPa	ksi	Yield strength(a) MPa	ksi	Elongation in 50 mm or 2 in.(a), %	Hardness(a), HRB
CQ	310–385	45–56	235–290	34–42	30–38	47–68
	340	49	255	37	35	58
DQ	305–350	44–51	220–270	32–39	32–40	42–54
	325	47	250	36	37	50
DQ (postannealed)	310–340	45–49	215–260	31–38	37–42	40–52
	320	46	235	34	41	46
DQSK	310–375	45–54	205–270	30–39	34–42	46–58
	330	48	240	35	38	52
DQSK (postannealed)	310–345	45–50	180–230	26–33	38–45	42–55
	325	47	215	31	40	46

(a) Single values below ranges are average values.

Table 10 Minimum mechanical properties of structural quality hot dip galvanized steel sheet

Grade	Tensile strength MPa	ksi	Yield strength MPa	ksi	Elongation in 50 mm or 2 in., %
A	310	45	230	33	20
B	360	52	255	37	18
C	380	55	275	40	16
D	450	65	345	50	12
E(a)	570	82	550	80	...
F	480	70	345	50	12

(a) When the hardness value is 85 HRB or higher, no tensile test is required. Source: ASTM A 446

provides excellent paint adhesion and weldability (Ref 5).

Another relatively new development in zinc alloy coatings is Galfan, a Zn-5Al-mischmetal alloy coating applied by hot dipping. Applications in the United States are limited, but European automakers have used Galfan in such applications as brake servo housings, headlight reflectors and frames, and universal joint shrouds (Ref 6). Galfan is also being considered for oil pans, and heavily formed unexposed body panels. Detailed information is available in the article "Electroplated Coatings" in Volume 13 of the 9th Edition of *Metals Handbook*.

Zinc spraying consists of projecting atomized particles of molten zinc onto a prepared surface. Three types of spraying pistols are in commercial use: the molten metal pistol, the powder pistol, and the wire pistol. The sprayed coating is slightly rough and slightly porous; the specific gravity of a typical coating is approximately 6.35, compared to 7.1 for cast zinc. This slight porosity does not affect the protective value of the coating, because zinc is anodic to steel. The zinc corrosion products that form during service fill the pores of the coating, giving a solid appearance. The slight roughness of the surface makes it an ideal base for paint, when properly pretreated.

On-site zinc spraying can be performed on finished parts of almost any shape or size. When applied to finished articles, welds, ends, and rivets receive adequate coverage. Moreover, it is the only satisfactory method of depositing unusually thick zinc coatings (≥0.25 mm, or 0.01 in.).

Aluminum Coatings

Aluminized (aluminum-coated) steel sheet is used for applications where heat resistance, heat reflectivity, or barrier-layer resistance to corrosion is required. Aluminum coating of steel sheet is done on continuous lines similar to those used for hot dip galvanizing of steel sheet. Cold-rolled steel sheet is hot dipped into molten aluminum or an aluminum alloy containing 5 to 10% Si. The coating consists of two layers, the exterior layer of either pure aluminum or aluminum-silicon alloy and the steel base metal, with an aluminum-iron-silicon alloy layer in between. The thickness of this alloy can significantly affect the ductility, adhesion, uniformity, smoothness, and appearance of the surface and is controlled for optimum properties.

Aluminum-coated sheet steel combines the desirable properties of aluminum and steel. Steel has a greater load-bearing capacity, having a modulus of elasticity of about three times that of unalloyed aluminum. The thermal expansion of steel is approximately half as much as that of aluminum. The aluminum coating offers corrosion resistance, resistance to heat and oxidation, and thermal reflectivity. Typical applications are:

- Automotive mufflers and related components
- Catalytic converter heat shields
- Drying and baking ovens
- Industrial heating equipment
- Fireplaces
- Home incinerators and furnaces
- Fire and garage doors
- Kitchen and laundry appliances
- Metal buildings
- Agricultural equipment
- Silo roofs
- Playground equipment
- Outdoor furniture
- Signs, masts, and lighting fixtures
- Containers and wrappers

Coating Weight. Aluminum coatings on steel sheet are designated according to total coating weight on both surfaces in ounces per square foot of sheet, as indicated in Table 11. These coating categories are listed in ASTM Specification A 463. Type 1, Light Coating, is recommended for drawing applications and when welding is a significant portion of the fabrication. Type 1, Regular or Commercial, has approximately a 25 μm

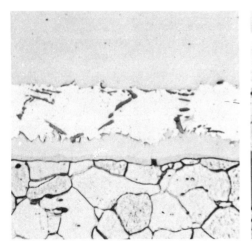
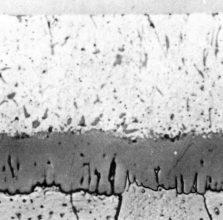

Fig. 6 Microstructure of aluminum coatings on steel. Left: Type 1 coating from top: a nickel filler, aluminum-silicon alloy, aluminum-silicon-iron alloy, and steel base metal. Right: Type 2 coating forms a layer of essentially pure aluminum (top) with scattered gray particles of aluminum-iron; the light gray center layer is aluminum-iron, and the bottom layer is the base steel. Both 1000×

Precoated Steel Sheet / 219

Table 11 Designations and weights of aluminum coating on aluminized steel sheet

Coating designation	Minimum coating weight(a)			
	Triple-spot test		Single-spot test	
	g/m²	oz/ft²	g/m²	oz/ft²
T1 25 (light)	80	0.25	60	0.20
T1 40 (regular)	120	0.40	90	0.30
T2 (regular)	230	0.75	200	0.65

(a) Total weight on both sides of sheet per unit area of sheet.
Source: ASTM A 463 and A 428

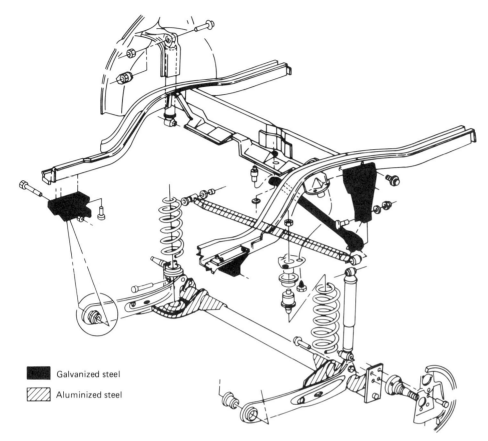

Fig. 7 Typical galvanized and aluminized steel rear suspension components used in American front-wheel drive automobiles. Source: Ref 1

(1 mil) thick coating on each surface (Fig. 6a). It is designated for applications requiring excellent heat resistance. Type 2 has a coating approximately 50 μm (2 mil) thick on each side (Fig. 6b). It is frequently used for atmospheric corrosion resistance. Coating weight on specimens from aluminum-coated sheet is determined by the test method in ASTM A 428. Figure 7 demonstrates how a typical rear suspension of a front-wheel drive vehicle utilizes type 1 aluminized steel components having a coating of Al-9Si-3Fe in conjunction with galvanized front pivot hangers, mounting brackets, and braces.

Base Metal and Formability. Aluminum coatings can be applied to CQ, DQ, or DQSK steel sheet, depending on the severity of the forming or drawing required. Only moderate forming and drawing are recommended for aluminized steel sheet, but there are numerous intricate components for heating, combustion, and other equipment being produced. Shallow crazing (hairline cracks) may occur in the coating if the bending and forming are too severe. To eliminate crazing, the radius of the bend should be increased. If the crazing is deep enough to expose the steel to the atmosphere during service, staining may occur. These stains generally have minimal effect on the serviceability of the product, because the corrosion stops at the crazed area after a relatively short exposure period. However, if water collects and does not drain off, corrosion products are dissolved and corrosion continues.

The mechanical properties of hot dip aluminized steel sheet are essentially the same as those of hot dip galvanized steel sheet (Table 9). When high strength is required, SQ aluminized steel sheet may be specified, although at some sacrifice in ductility.

Corrosion Resistance. The value of aluminum as a protective coating for steel sheet lies principally in its inherent corrosion resistance. In most environments, the long-term corrosion rate of aluminum is only about 15 to 25% that of zinc. Generally, the protective value of an aluminum coating on steel is a function of coating thickness. The coating tends to remain intact and therefore provides long-term protection.

Aluminum coatings do not provide sacrificial protection in most environments, particularly in atmospheric exposure. This is because a protective oxide film forms on the coating, which tends to passivate the aluminum and retard sacrificial protection. If the oxide film is destroyed, the aluminum will provide sacrificial protection to the base metal. In marine or salt-laden environments, the aluminum coating will protect sacrificially wherever chlorides destroy the surface oxide film.

Although staining or light rusting of the steel may occur at cut edges or crazing may occur where the aluminum does not protect, this action diminishes with further exposure time because of the self-sealing action of corrosion products. However, if insufficient slope or drainage permits water to pond or remain instead of running off freely, the corrosion products are dissolved and rusting will continue.

Heat Resistance. Aluminum-coated sheet steel has excellent resistance to high-temperature oxidation. At surface temperatures below about 510 °C (950 °F), the aluminum coating protects the steel base metal against oxidation without discoloration. Between 510 and 675 °C (950 and 1250 °F), the coating provides protection to the steel, but some darkening may result from the formation of aluminum-iron-silicon alloy. The alloy is extremely heat resistant, but upon long exposure at temperatures above 675 °C (1250 °F), the coating may become brittle and spall because of a different coefficient of expansion from that of the steel.

Because of their good resistance to scaling, combined with the structural strength of the steel base metal, type 1 coatings are used in automotive exhaust systems, heat exchangers, ovens, furnaces, flues, and tubing. The higher strength of the steel base metal, which melts at 1580 °C (2876 °F), enables steel sheet coated with either type 1 or type 2 coatings to perform for a longer time than aluminum alone in the event of fire.

Heat Reflection. The thermal reflectivity of aluminum-coated steel sheet is comparable to that of aluminum sheet. It is superior to galvanized steel sheet after a relatively short exposure time. All three sheet materials have thermal reflectivity of approximately 90% before exposure. However, after a few years, the value for galvanized steel decreases more than that for aluminized steel. Aluminum and aluminum-coated steel sheet retain 50 to 60% of their heat reflectivity. This is advantageous where heat must be confined, diverted, or excluded, as in heat transfer applications. When used for roofing and siding, alumi-

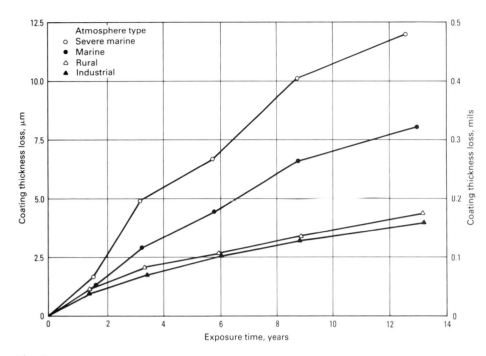

Fig. 8 Coating thickness loss of 55Al-Zn-coated steel in four atmospheres. Source: Ref 7

num-coated sheet keeps buildings cooler in summer and warmer in winter.

Weldability. Aluminum-coated steel sheet can be joined by electric resistance welding (spot welding or seam welding). It can also be metal arc welded, flash welded, or oxyacetylene welded. Thorough removal of grease, oil, paint, and dirt followed by wire brushing is recommended before joining. Special fluxes are required for metal arc and oxyacetylene welding. During spot welding, electrodes tend to pick up aluminum, and the tips must be dressed more frequently than during spot welding of uncoated steel. Also, higher current density is required.

Painting is generally unnecessary, but aluminum-coated sheet steel can be painted similarly to aluminum sheets. This includes removal of oil or grease and treatment with a phosphate, chromate, or proprietary wash-type chemical before painting.

Handling and Storage. The coating on aluminized steel sheet is soft, and care should be taken to avoid scratching and abrasion of the soft coating, which will mar the appearance and allow staining if the coating is removed. Wet-storage stains develop on aluminum-coated steel sheet that is continuously exposed to moisture. The sheet should be inspected for entrapped moisture when received and then stored indoors in a warm, dry place. Some added protection can be obtained by ordering the sheet oiled or chemically treated for resistance to wet-storage stain.

Aluminum-Zinc Alloy Coatings

In recent years, the desire and need to improve the corrosion resistance of galvanized coatings while retaining sacrificial galvanic corrosion behavior at sheared edges, and so on, have led to the commercial development of two types of hot dip aluminum-zinc alloy coatings. One type consists of about 55% Al and 45% Zn; the other type, zinc plus 5% Al. Both coating types contain small amounts of other alloying elements to improve wettability and/or coating adhesion during forming. Descriptions of these products are contained in ASTM A 792 (55Al-45Zn coating) and A 875 (Zn-5Al coating). These specifications include the general requirements, the coating categories available, and the product types that are available.

The 55% Al coating has been produced worldwide by a number of steel companies for more than 10 years. Its primary use is for preengineered metal buildings, especially roofing. In most environments, this coating has been found to have two to four times the corrosion resistance of galvanized coatings while retaining an adequate level of galvanic protection to minimize the tendency toward rust staining at edges and other breaks in the coating. Figure 8 illustrates the corrosion resistance of 55Al-Zn-coated steel exposed to four atmospheres. The coated sheet is available in similar grades (CQ, DQ, high strength, and so on) as hot dip galvanized and can be subjected to similar types of forming. It can also be

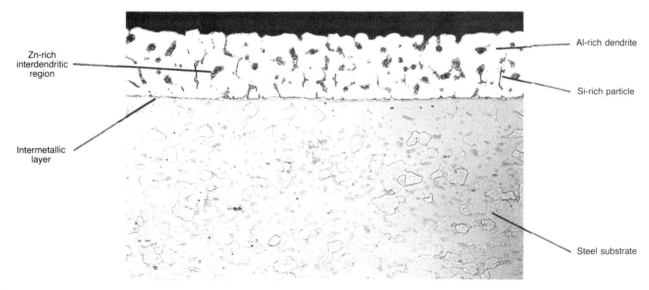

Fig. 9 Microstructure of an aluminum-zinc coated sheet. Etched with Amyl nital. 500×

painted either by coil-line painting methods or postpainting after fabrication.

The coating microstructure consists of an aluminum-iron intermetallic alloy bond between the steel and outer coating-metal layer (Fig. 9). This outer coating layer has a duplex microstructure, a matrix phase of an aluminum-rich composition, and a zinc-rich interdendritic phase. This zinc-rich phase corrodes preferentially to provide the galvanic corrosion protection. The coating contains about 2% Si, which is present in the microstructure as an elemental silicon phase. The silicon is added only to inhibit growth of the alloy layer during the hot dip coating operation.

Although this 55% Al coating is primarily used for metal-building applications, there are a variety of other applications, including appliances and automotive parts. It offers a level of heat-oxidation resistance intermediate between galvanized and aluminized coatings.

The Zn-5Al coating is also produced worldwide, but it is not as commonly available as the 55% Al coating. Its primary attribute is improved coating ductility compared to hot dip galvanized coatings.

This feature, along with a somewhat improved corrosion resistance, makes this coated-sheet product attractive for deep-drawn parts. Also, for prepainted sheets such as roll-formed metal-building panels, the improved coating ductility minimizes the tendency toward cracking of the paint along tension bends.

The Zn-5Al coated sheet is also available in similar grades (CQ, DQ, and so on) as hot dip galvanized. It is readily paintable, including coil-line prepainting.

Both types of aluminum-zinc coating have features that make them more attractive than galvanized for certain applications. Selection of either one should be based on consideration of the desired attributes and differences in fabricability, weldability, paintability, and so on, compared to the other coatings available.

Tin Coatings

Tin coatings are applied to steel sheet by electrolytic deposition or by immersion in a molten bath of tin (hot dip process). Hot dip tin coatings provide a nontoxic, protective, and decorative coating for food-handling, packaging, and dairy equipment, and they facilitate the soldering of components used in electronic and electrical equipment. In the United States, hot dip tin coating has been replaced by electrolytic tin coating.

Electrolytic tin coated steel sheet is used where solderability, appearance, or corrosion resistance under certain conditions is important, as in electronic equipment, food-handling and processing equipment, and laboratory clamps. It is generally produced with a matte finish formed by applying the coating to base metal sheet called black plate, which has a dull surface texture, and by leaving the coating unmelted. It can also be produced with a bright finish by applying the coating to base metal having a smooth surface texture and then melting the coating. Electrolytic tin coated sheet is usually produced in nominal thicknesses from 0.38 to 0.84 mm (0.015 to 0.033 in.) and in widths from 305 to 915 mm (12 to 36 in.).

Electrolytic tin coated steel sheet can be specified to one of the five coating-weight designations listed in Table 12. The coating weight is the total amount of tin on both surfaces, expressed in ounces per square foot of sheet area. Electrolytic coatings can be applied to CQ, DQ, or DQSK steel sheet, depending on the severity of the forming or drawing required. They can also be applied to SQ steel sheet when higher strength is required. Electrolytic tin coated steel sheet is covered in ASTM A 599. The mechanical properties of the steel sheet are unchanged by the electrolytic tin coating process.

Terne Coatings

Long terne steel sheet is carbon steel sheet continuously coated by the hot dip process with terne metal (lead with 3 to 15 wt% Sn). This coated sheet is duller in appearance than tin-coated sheet, hence the name (terne) from the French, which means dull or tarnished. The smooth, dull coating gives the sheet corrosion resistance, formability, excellent solderability, and paintability. The term long terne is used to describe terne-coated sheet, while short terne is used for terne-coated plate. Short terne, also called terneplate, is no longer produced in the United States.

Because of its unusual properties, long terne sheet has been adapted to a wide variety of applications. Its greatest use is in automotive gasoline tanks. Its excellent lubricity during forming, solderability and special corrosion resistance make the product well suited for this application. Other typical applications include:

- Automotive parts, such as air conditioners, air filters, cylinder head covers, distributor tubes, oil filters, oil pans, radiator parts, and valve rocker arm covers
- Electronic chassis and parts for radios, tape recorders, and television sets
- File drawer tracks
- Fire doors and frames
- Furnace and heating equipment parts
- Railroad switch lamps
- Small fuel tanks for lawn mowers, power saws, tractors, and outboard motors

Long terne sheet is often produced to ASTM A 308. The coatings are designated according to total coating weight on both surfaces in ounces per square foot of sheet area, as indicated in Table 13. For applications requiring good formability, the coating is applied over CQ, DQ, or DQSK low-carbon steel sheet. The terne coating acts as a lubricant and facilitates forming, and the strong bond of the terne metal allows it to be formed along with the base metal. When higher strength is required, the coating can be applied over SQ low-carbon steel sheet, although there is some sacrifice in ductility. In general, the mechanical properties of hot dip terne-coated steel are similar to those for cold-rolled steel. Terne coatings are applied by a flux-coating process that does not include in-line annealing. Therefore, the mechanical properties are obtained by pre-annealing using cycles comparable to those used for cold-rolled sheet.

Lead is well known for its excellent corrosion resistance, and terne metal is principally lead, with some tin added to form a tight, intermetallic bond with steel. The excellent corrosion resistance of terne sheet accounts for its wide acceptance as a material for gasoline tanks. However, because lead does not offer galvanic protection to the steel base metal, care must be exercised to avoid scratches and pores in the coating. Small openings may be sealed by corrosion products of iron, lead, and oxygen, but larger ones can corrode in an environment unfavorable to the steel base metal.

Long terne sheet can be readily soldered with noncorrosive fluxes using normal procedures because the sheet is already presoldered. This makes it a good choice for television and radio chassis and gasoline tanks, for which ease of solderability is important. It can also be readily welded by either resistance seam or spot welding

Table 12 Designations and weights of tin coating on electrolytic tin coated steel sheet

Coating designation	Minimum coating weight(a)			
	Triple-spot test		Single-spot test	
	g/m^2	oz/ft^2	g/m^2	oz/ft^2
25	3.7	0.012	2.8	0.009
50	7.3	0.024	5.6	0.018
75	11.0	0.036	8.2	0.027
100	14.6	0.048	11.0	0.036
125	18.3	0.060	13.8	0.045

(a) Total weight on both sides of sheet per unit area of sheet. Source: ASTM A 599

Table 13 Designations and weights of lead-tin coating on terne-coated steel sheet

Coating designation	Minimum coating weight(a)			
	Triple spot test		Single-spot test	
	g/m^2	oz/ft^2	g/m^2	oz/ft^2
LT01	No minimum		No minimum	
LT25	76	0.25	61	0.20
LT35	107	0.35	76	0.25
LT40	122	0.40	92	0.30
LT55	168	0.55	122	0.40
LT85	259	0.85	214	0.70
LT110	336	1.10	275	0.90

(a) Total weight on both sides of sheet per unit area of sheet. Source: ASTM A 308

methods. However, when the coating is subjected to high temperatures, significant concentrations of lead fumes can be released. Because of the toxicity of lead, the Occupational Safety and Health Administration and similar state agencies have promulgated standards that must be followed when welding, cutting, or brazing metals that contain lead or are coated with lead or lead alloys.

Long terne sheet has excellent paint adherence, which allows it to be painted using conventional systems, but this product is not usually painted. When painting is done, no prior special surface treatment or primer is necessary, except for the removal of ordinary dirt, oil, and grease. Oiled sheet, however, should be thoroughly cleaned to remove the oil.

Long terne sheet is normally furnished dry and requires no special handling. It should be stored indoors in a warm, dry place. Unprotected, outdoor storage of coils or bundles can result in white or gray staining of the terne coating, and if there are pores in the terne coating, rust staining can occur.

Phosphate Coatings

The phosphate coating of iron and steel consists of treatment with a dilute solution of phosphoric acid and other chemicals by which the surface of the metal, reacting chemically with the phosphoric acid, is converted into an integral layer of insoluble crystalline phosphate compound. This layer is less reactive than the metal surface and at the same time is more absorbent of lubricants or paints. Because the coating is an integral part of the surface, it adheres to the base metal tenaciously. The weight and crystalline structure of the coating, as well as the extent of penetration of the coating into the base metal, can be controlled by the method of cleaning before treatment, the method of applying the solution, the duration of treatment, and the changes in the chemical composition of the phosphating solution.

The two types of phosphate coatings in general use are zinc phosphate and iron phosphate. Within each type, chemical composition can be modified to suit various applications.

When zinc phosphate coatings are mill applied to galvanized sheets, the sheets are ready for immediate painting with the many paints readily available from industrial and retail suppliers. The zinc phosphate coated product is often referred to as phosphatized. Minor cleaning with mineral spirits, paint thinner, or naphtha may be necessary to remove fingerprints, oils, or dirt picked up during fabrication or handling. When mill-phosphatized sheets that are to be baked after painting are exposed to humid storage conditions for long periods of time, prebaking for several minutes at 150 °C (300 °F) prior to painting may be required to prevent blistering during baking.

The chief application for iron phosphate coatings is as a paint base for uncoated carbon steel sheet. Such a coating can be applied on coil coating lines.

The greatest tonnage of phosphate-coated steel is low-carbon flat-rolled material, which is used for applications such as sheet metal parts for automobiles and household appliances. Applications of the coatings range from simple protection to prepaint treatments for painted products, such as preengineered building panels and the side and top panels of washing machines, refrigerators, and ranges.

Phosphate coatings require a clean surface. The cleaning stage preceding phosphating removes foreign matter from the surface and makes uniformity of coating possible. This involves removal of oils, greases, and associated dirt by solvent degreasing or alkaline cleaning followed by thorough rinsing. Phosphate coatings are applied by spray, immersion, or roller coating.

A phosphate coating beneath a paint film helps prevent moisture and other corrosives that may penetrate the paint from reaching the metal. This prevents or delays the electrochemical reactions that lead to corrosion or rust. If the paint film sustains scratches or damage that exposes bare metal, the phosphate coating confines corrosion to the exposed metal surface, preventing the corrosion from spreading underneath the paint film. In painting applications, coarse or heavy phosphate coatings may be detrimental; they can absorb too much paint, thus reducing both gloss and adhesion, especially if deformation of the painted sheet steel is involved.

Preprimed Sheet

Primer paint coats are frequently applied to steel sheet at the mill or by a coil coater. Because their purpose is corrosion protection, they contain corrosion-inhibiting substances such as zinc powder, zinc chromate, or other compounds of zinc and/or chromium. Preprimed sheets are especially useful for parts that will have limited access after fabrication, rendering coating difficult. Parts made from preprimed sheet may receive a top coat after fabrication. The mill-applied phosphate coatings described in the previous section can also be considered prepriming treatments.

Formability (Ref 8). Preprimed steel offers advantages in forming metal fabrication through:

- Consistent surface morphology
- Reduced surface friction (reducing the flow over die surfaces) and reduced die wear, especially on the binder surfaces
- Reduced flaking and powdering (requiring less die maintenance), reduced need for metal finishing, and fewer surface defects
- Reduced galling

The painted surface acts as a cushion between substrate and stamping dies, which lessens the need for in-die lubrication and extends the life of the stamping die. The preprimed, prepainted surface can withstand severe forming and stretching. Thus, the need for lubricant is reduced or eliminated. This in turn provides a clean process environment and reduces the need for extensive cleaning along with phosphating and electrocoating.

Zinc Chromate Primers. Zinc chromate pigments are useful as corrosion inhibitors in paint. They are used as after-pickling coatings on steel and in primers. Federal specifications TT-P-57 and TT-P-645 cover zinc chromate paints. Zinc chromate pigments are unique; they are useful as corrosion inhibitors for both ferrous and nonferrous metals.

Zinc-Rich Primers. In recent years, manufacturers have developed various priming paints that will deposit films consisting mainly of metallic zinc and that have many properties in common with the zinc coatings applied by hot dip galvanizing, electroplating, metal spraying, or mechanical plating methods. Such films will provide some degree of sacrificial protection to the underlying steel if they contain 92 to 95% metallic zinc in the dry film and if the film is in electrical contact with the steel surface at a sufficient number of points. The type of zinc dust used in protective coatings is a heavy powder, light blue-gray in color, with spherically shaped particles having an average diameter of approximately 4 μm (160 μin.). Such powder normally contains 95 to 97% free metallic zinc with a total zinc content exceeding 99%. Many zinc-rich paints are air drying, although oven-curable primers containing a high content of zinc dust are available.

Depending on the nature of the binder, zinc-rich primers are classified as inorganic or organic. The inorganic solvent-base types are derived from organic alkyl silicates, which, upon curing, become totally inorganic. The organic zinc-rich coatings are formed by using zinc dust as a pigment in an organic binder. This binder may be any of the well-known coating vehicles, such as chlorinated rubber or epoxy. The zinc dust must be in sufficient concentration so that the zinc particles are in particle-to-particle contact throughout the film. In this way, zinc provides cathodic protection to the base metal. With the organic binder, there is no chemical reaction with the underlying surface, but the organic vehicle must wet the surface thoroughly to obtain mechanical adhesion.

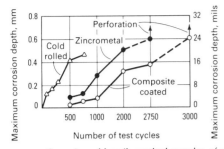

Fig. 10 Corrosion of heavily worked samples of a composite-coated steel, Zincrometal, and cold-rolled steel in a laboratory cyclic test. Test consisted of 28-min cycles of dipping in 5% saline solution at 40 °C (100 °F), humidifying at 50 °C (120 °F), and drying at 60 °C (140 °F). Source: Ref 9

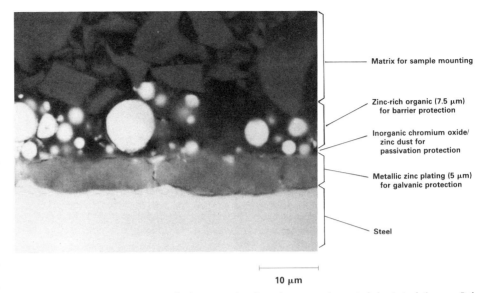

Fig. 11 Scanning electron micrograph of cross section through a composite-coated sheet steel. Source: Ref 12

The inorganic zinc coating forms its film and adheres to the steel surface by quite different means. The chemical activity during coating is quite similar for either water- or solvent-base inorganic binders. Zinc is the principal reactive element in the inorganic systems and is primarily responsible for the development of initial insolubility.

Zinc-rich primers offer a more versatile application of zinc to steel than galvanizing. Large, continuous, complex shapes and fabricated new or existing structures can be easily coated at manufacturing shops or in the field. The performance of zinc-rich primers has earned them a prominent place in the field of corrosion protection coatings. For example, zinc-rich primers are being preapplied to steel sheet as the first coat of a two-coat system for appliance applications such as refrigerator liners. However, the limitations of zinc-rich paints include cost and the required cleanliness of steel surfaces. They must be top coated in severe environments (pH under 6.0 and over 10.5).

The following comparisons should be helpful in selecting the binder system that is most suitable for an application. Inorganics have superior solvent and fuel resistance. They can withstand temperatures to 370 °C (700 °F) and are much easier to clean up after use. Inorganics do not blister upon exposure and are unaffected by weather, sunlight, or wide variations in temperature. They do not chalk, peel, or lose thickness over long periods of time. Also, they are easier to weld through and have excellent abrasion resistance and surface hardness. Organics use chlorinated rubber, epoxy, vinyl, phenoxy, or other coating vehicles, and the properties of the system are based on the characteristics of the vehicle used.

Organic Composite Coatings

Organic composite coated steels have been developed mainly by Japanese steelmakers in cooperation with automakers in that country, although development is underway in other countries as well. These coil coated products generally employ an electroplated zinc alloy base layer and a chemical conversion coating under a thin organic topcoat containing a high percentage of metal powder (Ref 9-11). The thinness of the organic topcoat allows for good formability without the risk of damaging the coating.

Figure 10 compares the corrosion resistance of one of these organic composite coated sheet steels to cold-rolled steel and to Zincrometal. Another of these products uses an organic-silicate composite topcoat only about 1 µm (40 µin.) thick and has corrosion resistance and weldability superior to that of Zincrometal (Ref 10). A bake-hardenable version of this material has also been developed (Ref 10). Researchers at a third Japanese steel company have developed a bake-hardenable organic composite coated sheet steel with a 0.8 to 1.5 µm (32 to 60 µin.) thick organic topcoat. The material possesses corrosion resistance, formability, and weldability equivalent to that of Zincrometal-KII, which uses a 7 µm (280 µin.) thick topcoat (Ref 11). Production of these composite-coated materials is increasing in anticipation of increased demand from Japanese automakers.

A similar material has been developed in the United States. This material has an electrodeposited zinc alloy base coat, a mixed intermediate layer of chromium oxide and zinc dust, and an organic topcoat for barrier protection (Ref 12). Figure 11 is a micrograph showing the cross section of the composite-coated steel. In salt spray tests comparing this material to electrodeposited zinc-nickel and Zincrometal, zinc-nickel failed after 216 h, Zincrometal at 480 h, and the composite coating at 960 h (Ref 12). This material was developed to have weldability, formability, and adhesive compatibility similar to that of Zincrometal. Developmental work is continuing.

Organic-Silicate Composite Coatings (Ref 13). Zinc-nickel electroplated steel sheet coated with an organic-silicate composite was developed by a Japanese steel company in an attempt to combine a highly corrosion resistant base zinc-nickel coating with a protective surface layer to prolong the coating life. With a view to forming a thin film with high corrosion resistance, the protective layer was designed as a two-layer protective film structure composed of a chromate film as a lower layer and the organic-silicate composite coating (the composite resin) as an upper layer. This protective film structure improves the corrosion resistance not only by the individual effects of each layer, such as the passivation of chromate film and the excellent corrosion resistance of the composite resin, but also by the suppression of excessive dissolution of Cr^{6+} from the lower chromate film layer by the sealing effect of the upper composite resin layer. This sealing effect sustains the passivation of chromate film more effectively in the corroding environment.

Prepainted Sheet

Prepainted steel sheet is a large and rapidly expanding market. The sheet is coated in coil form in a continuous coil-painting facility. Lower production costs, improved product quality, elimination of production hazards in the shop, customer satisfaction, conservation of energy, elimination of ecological problems, and the ability to expand production without capital expenditure for new buildings and equipment are some of the advantages of prepainted sheet over postpainting. Fabricated parts are readily joined by indirect projection welding, adhesives, tabs, and fasteners. Typical applications of prepainted steel sheet include tool sheds, preengineered buildings, swimming pools, automobiles, lighting fixtures, baseboard heaters, truck vans, mobile homes, home siding, metal awnings, air conditioners, freezers, refrigerators, ranges, washers, and dryers.

Selection of Paint System. A wide variety of paint systems are available on prepainted sheet. In selecting the proper system for a particular application, the user must consider fabrication requirements, the service life desired, and the service conditions that will be encountered. For example, in an aggressive environment a plastisol coating may be required. For a deep draw, a vinyl coating should be used instead of a polyester. For resistance to fading in sunlight, a silicone polyester may be suggested instead of a polyester or a vinyl paint.

In the preengineered building field, the paint system must be capable of being roll formed and still perform over the years under a wide variety of conditions without chalking, fading, cracking, or blistering. In the automotive field, the drawing properties of the coating must be considered in addition to corrosion protection from road salts. In the appliance industry, a high-gloss finish that will bend without cracking is important, along with resistance to such materials as detergents, solvents, mustard, ketchup, shoe polish, grape juice, and grease. Other product requirements frequently considered when selecting an appliance paint are color, hardness, adhesion, resistance to abrasion, corrosion, humidity, heat, and pressure marking.

For severe corrosion service and decorative effects, heavier coatings are supplied, often by laminating or bonding a solid film to the metal substrate. Typical applications include buildings, roofing and siding near pickling tanks, chemicals and other corrosive environments, and storm drains and culverts that are subjected to corrosive soils, mine acids, sewage, and abrasion. These culvert coatings can be a thermoplastic coal tar-base laminate 0.3 to 0.5 mm (0.012 to 0.020 in.) thick, or they can be a film of polyvinyl chloride.

Design Considerations. In using prepainted sheet, design should be considered. If necessary, bending radii, location of exposed edges, fastening methods, welding techniques, corner assembly, and other features should be modified to make them compatible with the base metal and paint system. For example, if a polyester paint is applied to bare steel sheet, a minimum bend radius of 3.2 mm (1/8 in.) is suggested to minimize cracking and crazing of the paint. If hot dip galvanized sheet is the substrate, a minimum bend radius of 6.4 mm (1/4 in.) should be used instead. Otherwise, the zinc coating may crack with sharper bending, and the paint may not be elastic enough to bridge the crack.

Paint is often cured at temperatures as high as 240 °C (465 °F). At the higher paint curing temperatures, the steel sheet may become fully aged and cause yield point elongation to return. The sheet is subject to the formation of stretcher strains during subsequent forming. Normally, return of yield point elongation is not objectionable in these applications. However, the formed part will sometimes be given a critical amount of strain, and strain lines may become visible. Frequently, this problem can be overcome by proper shop practices, particularly if the part has been roll formed. At times, however, it is necessary to use killed steels, which are considered essentially nonaging.

Shop Practices. Because a prepainted surface is composed of an organic material, the abuse that this surface can withstand is less than that of a metal sheet surface. Therefore, prior to using prepainted sheet for the first time, it is advisable to train shop personnel in proper handling practices and to examine shop equipment to eliminate sources of scratches. For example, dies, brake presses, and roll-forming equipment must have highly polished surfaces free of gouges, score marks, and so on. Clearances of the dies must be such that wiping of the paint film is avoided. Similarly, some care is needed when formed parts are put on carts or in containers for transfer from one location to another. It is not acceptable simply to pile one part on top of another. Good housekeeping is important to minimize the source of scratches. Frequent reexamination of shop equipment and parts containers is necessary to minimize scratches. Handling scratches can be refinished by retouching, which is costly and time consuming.

Packaging and Handling. Shop and field conditions should be considered when selecting packaging for prepainted sheet. Transit pickoff and job-site corrosion from entrapped moisture can be serious problems. For preengineered building sheets, for example, packaging after roll forming should include waterproof paper (no plastic wrapping), support sheets to prevent sagging, and pressure boards. Mixing sheets of different lengths in the bundle should be avoided. Once the bundle of formed prepainted sheets arrives at the job site, it should be inspected to determine if the packages are still intact and resistant to the weather.

Wherever possible, sheets should be erected on the day of delivery, or they should be protected from water condensation. Under-roof storage is desirable. However, if this is not possible, the waterproof bundles should be slanted so that any condensation will drain out. Damaged packages should be opened, inspected, and the sheets separated to allow complete drying. In addition to the prevention of moisture entrapment described above, chips from drilling operations should be brushed away to prevent rust spotting. Prepainted sheets should be installed with corrosion-resistant fasteners. The installation of sheets that are in contact with the soil should be avoided. Oil, grease, fingerprints, and other contaminants should be removed after installation.

REFERENCES

1. D.J. Bologna, Corrosion Resistant Materials and Body Paint Systems for Automotive Applications (SAE Paper 862015), in *Proceedings of the Automotive Corrosion and Prevention Conference*, P-188, Society of Automotive Engineers, 1986, p 69-80
2. "US Automotive Market for Zinc Coatings 1984-1986," Zinc Institute Inc.
3. R.J. Neville and K.M. DeSouza, Electrogalvanized or Hot Dip Galvanized—Results of Five Years of Undervehicle Corrosion Testing (SAE Paper 862010), in *Proceedings of the Automotive Corrosion and Prevention Conference*, P-188, Society of Automotive Engineers, 1986, p 31-40
4. J.F.H. van Eijnsbergen, Supplement (to Twenty Years of Duplex Systems), *Thermisch Verzinken*, Vol 8, 1979
5. M. Memmi et al., A Qualitative and Quantitative Evaluation of Zn + Cr-CrO_x Multilayer Coating Compared to Other Coated Steel Sheets (SAE Paper 862028), in *Proceedings of the Automotive Corrosion and Prevention Conference*, P-188, Society of Automotive Engineers, 1986, p 175-185
6. R.F. Lynch and F.E. Goodwin, "Galfan Coated Steel for Automotive Applications," SAE Paper 860658, Society of Automotive Engineers, 1986
7. H.E. Townsend and J.C. Zoccola, Atmospheric Corrosion Resistance of 55% Al-Zn Coated Sheet Steel: 13-Year Test Results, *Mater. Perform.*, Vol 18, 1979, p 13-20
8. B.K. Dubey, Prepainted Steel for Automotive Application, in *Corrosion-Resistant Automotive Sheet Steels*, L. Allegra, Ed., Proceedings of a Conference in conjunction with the 1988 World Materials Congress (Chicago), Sept 1988, ASM INTERNATIONAL, 1988
9. Y. Shindou et al., Properties of Organic Composite-Coated Steel Sheet for Automobile Body Panels (SAE Paper 862016), in *Proceedings of the Automotive Corrosion and Prevention Conference*, P-188, Society of Automotive Engineers, 1986, p 81-90
10. M. Yamashita, T. Kubota, and T. Adaniya, Organic-Silicate Composite Coated Steel Sheet for Automobile Body Panel (SAE Paper 862017), in *Proceedings of the Automotive Corro-*

sion and Prevention Conference, P-188, Society of Automotive Engineers, 1986, p 91-97
11. T. Mohri *et al.*, Newly Developed Organic Composite-Coated Steel Sheet With Bake Hardenability (SAE Paper 862030), in *Proceedings of the Automotive Corrosion and Prevention Conference*, P-188, Society of Automotive Engineers, 1986, p 199-208
12. T.E. Dorsett, Development of a Composite Coating for Pre-Coated Automotive Sheet Metal (SAE Paper 862027), in *Proceedings of the Automotive Corrosion and Prevention Conference*, P-188, Society of Automotive Engineers, 1986, p 163-173
13. T. Watanabe, T. Kubota, M. Yamashita, T. Urakawa, and M. Sagiyama, Corrosion-Resistant Precoated Steel Sheets for Automotive Body Panels, in *Corrosion-Resistant Automotive Sheet Steels*, L. Allegra, Ed., Proceedings of a Conference in conjunction with the 1988 World Materials Congress (Chicago), Sept 1988, ASM INTERNATIONAL, 1988

Carbon and Low-Alloy Steel Plate

Revised by F.B. Fletcher, Lukens Steel Company

STEEL PLATE is any flat-rolled steel product more than 200 mm (8 in.) wide and more than 6.0 mm (0.230 in.) thick or more than 1220 mm (48 in.) wide and 4.6 mm (0.180 in.) thick. The majority of mills for rolling steel plate have a working-roll width between 2030 and 5600 mm (80 and 220 in.). Therefore, the width of product normally available ranges from 1520 to 5080 mm (60 to 200 in.). Most steel plate consumed in North America ranges in width from 2030 to 3050 mm (80 to 120 in.) and ranges in thickness from 5 to 200 mm ($^{3}/_{16}$ to 8 in.). Some plate mills, however, have the capability to roll steel more than 640 mm (25 in.) thick.

Steel plate is usually used in the hot-finished condition, but the final rolling temperature can be controlled to improve both strength and toughness. Heat treatment is also used to improve the mechanical properties of some plate.

Steel plate is mainly used in the construction of buildings, bridges, ships, railroad cars, storage tanks, pressure vessels, pipe, large machines, and other heavy structures, for which good formability, weldability, and machinability are required. The impairment of these desirable characteristics with increasing carbon content usually limits the steel to the low-carbon and medium-carbon constructional grades, with the low-carbon grades predominating. Many alloy steels are also produced as plate. In the final structure, however, alloy steel plate is sometimes heat treated to achieve mechanical properties superior to those typical of the hot-finished product.

Steelmaking Practices

Steel plate is produced from continuously cast slabs or individually cast ingots or slabs. Preparing these steel slabs or ingots for subsequent forming into plates may involve requirements regarding deoxidation practices, austenite grain size, and/or secondary melting practices.

Deoxidation Practices. During the steelmaking process, segregation of carbon can occur when carbon reacts with the dissolved oxygen in the molten steel (a reaction that is favored thermodynamically at lower temperatures). Therefore, the practice of controlling dissolved oxygen in the molten metal before and during casting is an important factor in improving the internal soundness and chemical homogeneity of cast steel. Deoxidation is also important in lowering the impact transition temperatures. Deoxidation can be achieved by vacuum processing or by adding deoxidizing elements such as aluminum or silicon.

Steels are classified by their level of deoxidation: killed steel, semikilled steel, capped steel, and rimmed steel. The steel used for plates is usually either killed or semikilled. Semikilled steel is commonly used for casting ingots because it is more economical than killed steel. Continuously cast steels are normally fully killed to assure internal soundness.

Table 1 Available quality levels for carbon, HSLA, and low-alloy steel plate

Carbon steel plates	HSLA steel plates	Low-alloy steel plates
Regular quality	...	...
Structural quality	Structural quality	Structural quality
Drawing quality	Drawing quality	Drawing quality
Cold-drawing quality	Cold-drawing quality	Cold-drawing quality
Cold-pressing quality	...	...
Cold-flanging quality	Cold-flanging quality	Cold-flanging quality
Forging quality	Forging quality	...
Pressure vessel quality	Pressure vessel quality	Pressure vessel quality
...	...	Aircraft quality

Source: Ref 1

Table 2 Standard carbon steel plate compositions applicable for structural applications

When silicon is required, the following ranges and limits are commonly used for nonresulfurized carbon steel: 0.10% max, 0.07–0.15%, 0.10–0.20%, 0.15–0.30%, 0.35% max, 0.20–0.40, or 0.30–0.60%.

Steel designation		Chemical composition limits, %			
UNS	SAE or AISI No.	C	Mn	P(a)	S(a)
G10060	1006	0.08(a)	0.45(a)	0.040	0.050
G10080	1008	0.10(a)	0.50(a)	0.040	0.050
G10090	1009	0.15(a)	0.60(a)	0.040	0.050
G10100	1010	0.08–0.13	0.30–0.60	0.040	0.050
G10120	1012	0.10–0.15	0.30–0.60	0.040	0.050
G10150	1015	0.12–0.18	0.30–0.60	0.040	0.050
G10160	1016	0.12–0.18	0.60–0.90	0.040	0.050
G10170	1017	0.14–0.20	0.30–0.60	0.040	0.050
G10180	1018	0.14–0.20	0.60–0.90	0.040	0.050
G10190	1019	0.14–0.20	0.70–1.00	0.040	0.050
G10200	1020	0.17–0.23	0.30–0.60	0.040	0.050
G10210	1021	0.17–0.23	0.60–0.90	0.040	0.050
G10220	1022	0.17–0.23	0.70–1.00	0.040	0.050
G10230	1023	0.19–0.25	0.30–0.60	0.040	0.050
G10250	1025	0.22–0.28	0.30–0.60	0.040	0.050
G10260	1026	0.22–0.28	0.60–0.90	0.040	0.050
G10300	1030	0.27–0.34	0.60–0.90	0.040	0.050
G10330	1033	0.29–0.36	0.70–1.00	0.040	0.050
G10350	1035	0.31–0.38	0.60–0.90	0.040	0.050
G10370	1037	0.31–0.38	0.70–1.00	0.040	0.050
G10380	1038	0.34–0.42	0.60–0.90	0.040	0.050
G10390	1039	0.36–0.44	0.70–1.00	0.040	0.050
G10400	1040	0.36–0.44	0.60–0.90	0.040	0.050
G10420	1042	0.39–0.47	0.60–0.90	0.040	0.050
G10430	1043	0.39–0.47	0.70–1.00	0.040	0.050

(continued)

(a) Maximum

Table 2 (continued)

Steel designation		Chemical composition limits, %			
UNS	SAE or AISI No.	C	Mn	P(a)	S(a)
G10450	1045	0.42–0.50	0.60–0.90	0.040	0.050
G10460	1046	0.42–0.50	0.70–1.00	0.040	0.050
G10490	1049	0.45–0.53	0.60–0.90	0.040	0.050
G10500	1050	0.47–0.55	0.60–0.90	0.040	0.050
G10550	1055	0.52–0.60	0.60–0.90	0.040	0.050
G10600	1060	0.55–0.66	0.60–0.90	0.040	0.050
G10640	1064	0.59–0.70	0.50–0.80	0.040	0.050
G10650	1065	0.59–0.70	0.60–0.90	0.040	0.050
G10700	1070	0.65–0.76	0.60–0.90	0.040	0.050
G10740	1074	0.69–0.80	0.50–0.80	0.040	0.050
G10750	1075	0.69–0.80	0.40–0.70	0.040	0.050
G10780	1078	0.72–0.86	0.30–0.60	0.040	0.050
G10800	1080	0.74–0.88	0.60–0.90	0.040	0.050
G10840	1084	0.80–0.94	0.60–0.90	0.040	0.050
G10850	1085	0.80–0.94	0.70–1.00	0.040	0.050
G10860	1086	0.80–0.94	0.30–0.50	0.040	0.050
G10900	1090	0.84–0.98	0.60–0.90	0.040	0.050
G10950	1095	0.90–1.04	0.30–0.50	0.040	0.050
G15240	1524	0.18–0.25	1.30–1.65	0.040	0.050
G15270	1527	0.22–0.29	1.20–1.55	0.040	0.050
G15360	1536	0.30–0.38	1.20–1.55	0.040	0.050
G15410	1541	0.36–0.45	1.30–1.65	0.040	0.050
G15480	1548	0.43–0.52	1.05–1.40	0.040	0.050
G15520	1552	0.46–0.55	1.20–1.55	0.040	0.050

(a) Maximum

Table 3 Composition ranges and limits for AISI-SAE standard low-alloy steel plate applicable for structural applications

Boron or lead can be added to these compositions. Small quantities of certain elements not required may be found. These elements are to be considered incidental and are acceptable to the following maximum amounts: copper to 0.35%, nickel to 0.25%, chromium to 0.20%, and molybdenum to 0.06%.

AISI-SAE designation	UNS designation	Heat composition ranges and limits, %(a)					
		C	Mn	Si(b)	Cr	Ni	Mo
1330	G13300	0.27–0.34	1.50–1.90	0.15–0.30	...	...	...
1335	G13350	0.32–0.39	1.50–1.90	0.15–0.30	...	...	...
1340	G13400	0.36–0.44	1.50–1.90	0.15–0.30	...	...	...
1345	G13450	0.41–0.49	1.50–1.90	0.15–0.30	...	...	...
4118	G41180	0.17–0.23	0.60–0.90	0.15–0.30	0.40–0.65	...	0.08–0.15
4130	G41300	0.27–0.34	0.35–0.60	0.15–0.30	0.80–1.15	...	0.15–0.25
4135	G41350	0.32–0.39	0.65–0.95	0.15–0.30	0.80–1.15	...	0.15–0.25
4137	G41370	0.33–0.40	0.65–0.95	0.15–0.30	0.80–1.15	...	0.15–0.25
4140	G41400	0.36–0.44	0.70–1.00	0.15–0.30	0.80–1.15	...	0.15–0.25
4142	G41420	0.38–0.46	0.70–1.00	0.15–0.30	0.80–1.15	...	0.15–0.25
4145	G41450	0.41–0.49	0.70–1.00	0.15–0.30	0.80–1.15	...	0.15–0.25
4340	G43400	0.36–0.44	0.55–0.80	0.15–0.30	0.60–0.90	1.65–2.00	0.20–0.30
E4340(c)	G43406	0.37–0.44	0.60–0.85	0.15–0.30	0.65–0.90	1.65–2.00	0.20–0.30
4615	G46150	0.12–0.18	0.40–0.65	0.15–0.30	...	1.65–2.00	0.20–0.30
4617	G46170	0.15–0.21	0.40–0.65	0.15–0.30	...	1.65–2.00	0.20–0.30
4620	G46200	0.16–0.22	0.40–0.65	0.15–0.30	...	1.65–2.00	0.20–0.30
5160	G51600	0.54–0.65	0.70–1.00	0.15–0.30	0.60–0.90	...	...
6150(d)	G61500	0.46–0.54	0.60–0.90	0.15–0.30	0.80–1.15	...	...
8615	G86150	0.12–0.18	0.60–0.90	0.15–0.30	0.35–0.60	0.40–0.70	0.15–0.25
8617	G86170	0.15–0.21	0.60–0.90	0.15–0.30	0.35–0.60	0.40–0.70	0.15–0.25
8620	G86200	0.17–0.23	0.60–0.90	0.15–0.30	0.35–0.60	0.40–0.70	0.15–0.25
8622	G86220	0.19–0.25	0.60–0.90	0.15–0.30	0.35–0.60	0.40–0.70	0.15–0.25
8625	G86250	0.22–0.29	0.60–0.90	0.15–0.30	0.35–0.60	0.40–0.70	0.15–0.25
8627	G86270	0.24–0.31	0.60–0.90	0.15–0.30	0.35–0.60	0.40–0.70	0.15–0.25
8630	G86300	0.27–0.34	0.60–0.90	0.15–0.30	0.35–0.60	0.40–0.70	0.15–0.25
8637	G86370	0.33–0.40	0.70–1.00	0.15–0.30	0.35–0.60	0.40–0.70	0.15–0.25
8640	G86400	0.36–0.44	0.70–1.00	0.15–0.30	0.35–0.60	0.40–0.70	0.15–0.25
8655	G86550	0.49–0.60	0.70–1.00	0.15–0.30	0.35–0.60	0.40–0.70	0.15–0.25
8742	G87420	0.38–0.46	0.70–1.00	0.15–0.30	0.35–0.60	0.40–0.70	0.20–0.30

(a) Indicated ranges and limits apply to steels made by the open hearth or basic oxygen processes; maximum content for phosphorus is 0.035% and for sulfur 0.040%. For steels made by the electric furnace process, the ranges and limits are reduced as follows: C—0.01%; Mn—0.05%; Cr—0.05% (<1.25%), 0.10% (>1.25%); maximum content for either phosphorus or sulfur is 0.025%. (b) Other silicon ranges may be negotiated. Silicon is available in ranges of 0.10–0.20%, 0.20–0.30%, and 0.35% maximum (when carbon deoxidized) when so specified by the purchaser. (c) Prefix "E" indicates that the steel is made by the electric furnace process. (d) Contains 0.15% V minimum

Killed steel is fully deoxidized, and from the viewpoint of minimum chemical segregation and uniform mechanical properties, killed steel represents the best quality available. Therefore, killed steel is generally specified when homogeneous structure and internal soundness of the plate are required or when improved low-temperature impact properties are desired. Killed steel can be produced either fine or coarse grained without adversely affecting soundness, surface, or cleanliness. Generally, heavy-gage plate (thicker than 38 mm, or 1½ in.) is produced from killed steel to provide improved internal homogeneity.

Semikilled steel is deoxidized to a lesser extent than killed steel and therefore does not have the same degree of chemical uniformity or freedom from surface imperfections as killed steel. This type of steel is used primarily on lighter-gage plate, for which high reductions from ingot to plate thicknesses minimize the structural and chemical variations found in the as-cast ingot.

Austenitic Grain Size. Steel plate specifications for structural and pressure vessel applications may require a steelmaking process that produces a fine austenitic grain size. When a fine austenitic grain size is specified, grain-refining elements are added during steelmaking.

Aluminum is effective in retarding austenitic grain growth, resulting in improved toughness for heat-treated (normalized or quenched and tempered) steels. Steels used in high-temperature service normally contain only very small quantities of aluminum because aluminum may affect strain-aging characteristics and graphitization. However, the addition of aluminum may be necessary for some high-temperature steels (as well as most low-temperature steels) requiring good toughness. Other grain-refining elements, such as niobium, vanadium, and titanium, are used in high-strength low-alloy (HSLA) steels for grain refinement during rolling (see the ar-

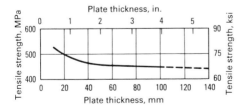

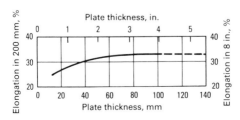

Fig. 1 Effect of thickness on tensile properties of 0.20% C steel plate

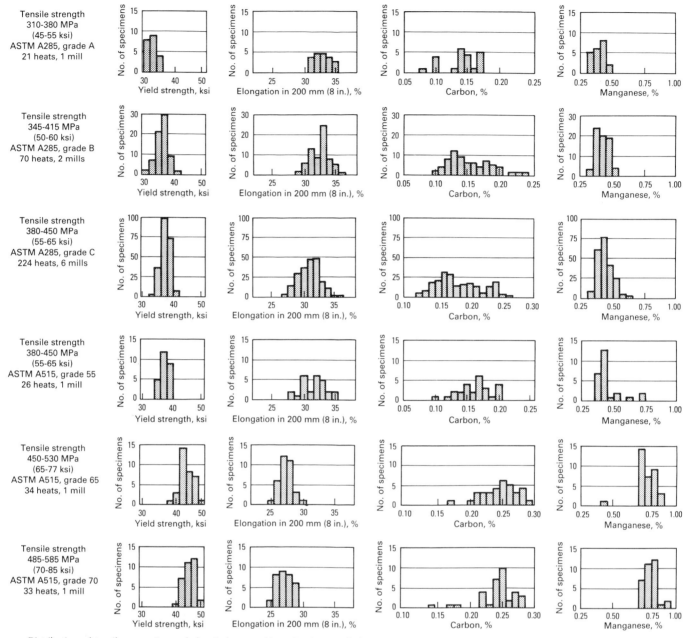

Fig. 2 Distribution of tensile properties and chemical composition of carbon steel plate. Data represent all the as-hot-rolled plate, 6 to 50 mm (¼ to 2 in.) thick, purchased to these specifications by one fabricator during a period of 8 years.

ticle "High-Strength Structural and High-Strength Low-Alloy Steels" in this Volume).

Melting Practices. The steel for plate products can be produced by the following primary steelmaking processes: open hearth, basic oxygen, or electric furnace. In addition, the steel can be further refined by secondary processes such as vacuum degassing or various ladle treatments for deoxidation or desulfurization.

Vacuum degassing is used to remove dissolved oxygen and hydrogen from steel, thus reducing the number and size of indigenous nonmetallic inclusions. It also reduces the likelihood of internal fissures or flakes caused when hydrogen content is higher than desired. A small cost premium is associated with the specification of vacuum degassing.

Desulfurization. By combining steel refining with the addition of ladle desulfurizing agents (for example, calcium or rare earth additions) immediately before casting or teeming, final plate steel sulfur content can be reduced to less than 0.005%. Lower sulfur content improves plate through-thickness properties and impact properties, but adds to the cost of the steel.

Platemaking Practices

As noted earlier, steel plates are produced from either continuous-cast slabs, pressure-cast slabs, or ingots. Steel ingots are typically between 380 and 1140 mm (15 and 45 in.) thick. These ingots first pass through a slabbing mill where they are reduced in thickness to make a slab. The slab is then inspected, and the surface is conditioned by grinding or scarfing to remove surface imperfections, and then reheated in furnaces prior to rolling to final plate thickness. Continuous-cast slabs and pressure-cast slabs are normally heated and rolled to final plate thickness in a single operation. The plate can then be roller leveled and cooled.

Microalloyed HSLA steels can be controlled rolled for grain refinement (see the article "High-Strength Structural and High-Strength Low-Alloy Steels" in this Vol-

Table 4 Composition of high-strength low-alloy steel plate

ASTM specification	Material grade or type	C	Mn	P	S	Si	Cr	Ni	Mo	Cu	V	Nb	Others
Structural quality													
A 131	AH32, DH32, EH32, AH36, DH36, EH36	0.18	0.90–1.60	0.04	0.04	0.10–0.50	0.25	0.40	0.08	0.35	0.10	0.05	...
A 242	1	0.15	1.00	0.15	0.05	...	...	...	...	0.20 min	...	...	(b)(e)
A 572	42	0.21	1.35	0.04	0.05	0.40(d)	...	...	...	...	(e)	(e)	(e)
	45	0.22	1.35	0.04	0.05	0.40(d)	...	...	...	...	(e)	(e)	(e)
	50	0.23	1.35	0.04	0.05	0.40(d)	...	...	...	...	(e)	(e)	(e)
	60	0.26	1.35	0.04	0.05	0.40(d)	...	...	...	...	(e)	(e)	(e)
	65	0.26(d)	1.65(d)	0.04	0.05	0.40	...	...	...	...	(e)	(e)	(e)
A 588	A	0.19	0.80–1.25	0.04	0.05	0.30–0.65	0.40–0.65	0.40	...	0.25–0.40	0.02–0.10	...	...
	B	0.20	0.75–1.35	0.04	0.05	0.15–0.50	0.50–0.70	0.50	...	0.20–0.40	0.01–0.10	...	...
	C	0.15	0.80–1.35	0.04	0.05	0.15–0.40	0.30–0.50	0.25–0.50	...	0.20–0.50	0.01–0.10	...	...
	D	0.10–0.20	0.75–1.25	0.04	0.05	0.50–0.90	0.50–0.70	...	...	0.30	...	0.04	Zr, 0.05-0.15
	K	0.17	0.50–1.20	0.04	0.05	0.25–0.50	0.40–0.70	0.40	0.10	0.30–0.50	...	0.005–0.05(f)	...
A 633	A	0.18	1.00–1.35	0.04	0.05	0.15–0.50	...	...	...	...	...	0.05	...
	B	0.18	1.00–1.35	0.04	0.05	0.15–0.50	...	...	...	...	0.10	...	...
	C	0.20	1.15–1.50	0.04	0.05	0.15–0.50	...	...	...	...	...	0.01–0.05	...
	D	0.20	0.70–1.60(d)	0.04	0.05	0.15–0.50	0.25	0.25	0.08	0.35	...	...	...
	E	0.22	1.15–1.50	0.04	0.05	0.15–0.50	...	...	...	...	0.04–0.11	(g)	N, 0.01–0.03(h)
A 656	3	0.18	1.65	0.025	0.035	0.60	...	...	...	...	0.08	0.005–0.15	N, 0.020
	7	0.18	1.65	0.025	0.035	0.60	...	...	...	...	0.005–0.015(i)	0.005–0.015(i)	N, 0.020
A 678	D	0.22	1.15–1.50	0.04	0.05	0.15–0.50	...	...	...	0.2 min(j)	0.04–0.11	(g)	N, 0.001–0.03
A 709	50	0.23	1.35(d)	0.04	0.05	0.15–0.40(d)	...	...	...	...	(e)	(e)	(e)
	50W					Identical to A 588 type A, B, or C (as specified)							
A 808	...	0.12	1.65	0.04	0.05	0.15–0.50	...	...	...	...	0.10	0.02–0.10	(Nb + V), 0.15
A 852	...	0.19	0.80–1.35	0.04	0.05	0.20–0.65	0.40–0.70	0.50	...	0.20–0.40	0.02–0.10	...	...
A 871	...	1.20	1.50	0.04	0.05	0.90	0.90	1.25	0.25	1.00	0.10	0.05	Zr, 0.15; Ti, 0.05
Pressure vessel quality													
A 734	B	0.17	1.60	0.035	0.015	0.40	...	0.35	0.25	0.25(j)	0.11	(k)	Al, 0.06; N, 0.030
A 737	B	0.20	1.15–1.50	0.035	0.030	0.15–0.50	...	...	...	...	...	0.05	...
	C	0.22	1.15–1.50	0.035	0.030	0.15–0.50	...	...	...	...	0.04–0.11	(k)	N, 0.03
A 841	...	0.20	0.70–1.60(d)	0.030	0.030	0.15–0.50	0.25	0.25	0.08	0.35	0.06	0.03	Al, 0.020 min

(a) Except as noted, when a single value is shown, it is a maximum limit. (b) Choice and amount of other alloying elements added to give the required mechanical properties and atmospheric corrosion resistance are made by the producer and reported in the heat analysis. (c) Elements commonly added include silicon, chromium, nickel, vanadium, titanium, and zirconium. (d) Limiting values vary with plate thickness. (e) For type 1, 0.005–0.05% Nb; for type 2, 0.01–0.15% V; for type 3, 0.05% Nb max + V = (0.02–0.15%); for type 4, N (with V) 0.015% max. (f) For plates under 13 mm (½ in.) thickness, the minimum niobium limit is waived. (g) Niobium may be present in the amount of 0.01–0.05%. (h) The minimum total aluminum content shall be 0.018% or the vanadium:nitrogen ratio shall be 4:1 minimum. (i) Niobium, or vanadium, or both, 0.005% min. When both are added, the total shall be 0.20% max. (j) Applicable only when specified. (k) 0.05% max Nb may be present.

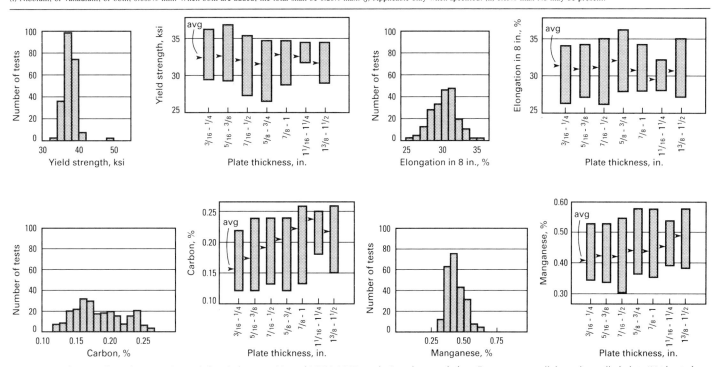

Fig. 3 Distribution of tensile properties and chemical composition of ASTM A 285, grade C, carbon steel plate. Data represent all the as-hot-rolled plate (224 heats from 6 mills) purchased to this specification by one fabricator during a period of 8 years.

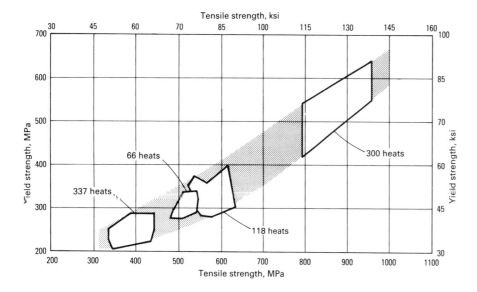

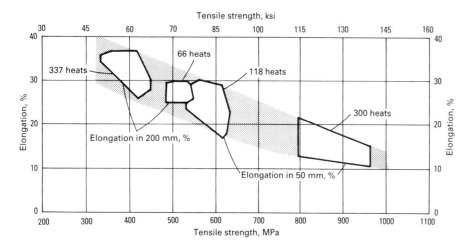

Fig. 4 Relation of tensile properties for hot-rolled carbon steel

Table 5 ASTM specifications for structural quality steel plate
General requirements for structural plate are covered in ASTM A 6.

ASTM specification(a)	Steel type and condition
Carbon steel	
A 36(b)	Carbon steel shapes, plates, and bars of structural quality
A 131(c)	Structural steel shapes, plates, bars, and rivets for use in ship construction (ordinary strength)
A 283(b)	Low and intermediate tensile strength carbon steel plates
A 284	Low and intermediate tensile strength carbon-silicon steel plates for machine parts and general construction
A 529(d)	Structural steel with 290 MPa (42 ksi) minimum yield point
A 573	Structural quality carbon-manganese-silicon steel plates with improved toughness
A 678	Quenched and tempered carbon and HSLA plates for structural applications
A 709	Carbon and HSLA steel structural shapes, plates, and bars, and quenched and tempered alloy steel for use in bridges
A 827(e)	Carbon steel plates for forging applications
A 830(e)	Structural quality carbon steel plates furnished to chemical requirements
(continued)	

(a) Also designated with the suffix "M" when the specification covers metric equivalents. (b) This specification is also published by the American Society of Mechanical Engineers, which uses the prefix "S" (for example, SA36). (c) See also Section 43 of the American Bureau of Shipping specifications and MIL-S-22698 (SH). (d) 13 mm (½ in.) maximum thickness. (e) See also Ref 1. (f) Tensile properties may also be specified when compatible. (g) Discontinued in 1989 and replaced by A 572. (h) Minimum yield point 345 MPa (50 ksi) to 100 mm (4 in.). Lower minimum yield points for thicker sections

ume). In this case, the reheating temperature is lower than usual, and the rolling practices are designed to impart heavy reductions at relatively low temperatures. This form of thermomechanically controlled processing (TMCP) is used for grain refinement, which results in plates with improved toughness and strength compared to conventional plate rolling. In some plate mills, controlled rolling is followed by accelerated cooling or direct quenching instead of air cooling. Attractive combinations of strength and toughness can be achieved by TMCP.

After cooling, plates are cut to size by shearing or thermal cutting. Following this operation, testing to confirm mechanical properties is customarily performed, and then the material is shipped to the fabricator. Certain plate products, however, require further processing such as heat treatment.

Plate Imperfections

Certain characteristic surface imperfections that can weaken the plate may appear on hot-finished steel; chemical segregation that can alter properties across the section may also be present. Some of these imperfections are discussed below.

Seams are the most common imperfections found in hot-finished steel. These longitudinal cracks on the surface are caused by blowholes and cracks in the original ingot that have been rolled closed, but not welded. For many plate applications, seams are of minor consequence. However, seams are harmful for applications involving heat treating or upsetting or in certain parts subjected to fatigue loading.

Decarburization, a surface condition common to all hot-finished steel, is produced during the heating and rolling operations when atmospheric oxygen reacts with the heated surface, removing carbon. This produces a soft, low-strength surface, which is often unsatisfactory for applications involving wear or fatigue. For this reason, critical parts or at least critical areas of parts are usually machined to remove this weakened surface.

Segregation. Alloying elements always segregate during the solidification of steel. Elements that are especially prone to segregation are carbon, phosphorus, sulfur, silicon, and manganese. The effect of segregation on mechanical properties and fabricability is insignificant for most plate steel applications. However, segregation may produce difficulties in subsequent operations such as forming, welding, punching, and machining.

Heat Treatment

Although most steel plate is used in the hot-finished condition, the following heat treatments are applied to plate that must meet special requirements.

Table 5 (continued)

ASTM specification(a)	Steel type and condition
Low-alloy steel	
A 514	Structural quality quenched and tempered alloy steel plates for use in welded bridges and other structures
A 709	See above under "Carbon steel"
A 710	Low-carbon age-hardening Ni-Cu-Cr-Mo-Nb, Ni-Cu-Nb, and Ni-Cu-Mn-Mo-Nb alloy steel plates, shapes, and bars for general applications
A 829(e)(f)	Structural quality alloy steel plates specified to chemical composition requirements
HSLA steel	
A 131(c)	Structural steel shapes, plates, bars, and rivets for use in ship construction (higher strength)
A 242	HSLA structural steel shapes, plates, and bars for welded, riveted, or bolted construction
A 441(g)	Mn-V HSLA steel plates, bars, and shapes
A 572	HSLA structural Nb-V steel shapes, plates, sheet piling, and bars for riveted, bolted, or welded construction of bridges, buildings, and other structures
A 588(h)	HSLA structural steel shapes, plates, and bars for welded, riveted, or bolted construction for use in bridges and buildings with atmospheric corrosion resistance approximately two times that of carbon steel with copper
A 633	Normalized HSLA structural steel for welded, riveted, or bolted construction suited for service at low ambient temperatures of −45 °C (−50 °F) or higher
A 656	Hot-rolled HSLA structural steel with improved formability for use in truck frames, brackets, crane booms, rail cars, and similar applications
A 678	See above under "Carbon steel"
A 709	See above under "Carbon steel"
A 808	Hot-rolled HSLA Mn-V-Nb structural steel plate with improved notch toughness
A 852	Quenched and tempered HSLA structural steel plate for welded, riveted, or bolted construction for use in bridges and buildings with atmospheric corrosion resistance approximately two times that of carbon steel with copper
A 871	HSLA structural steel plate in the as-rolled, normalized, or quenched and tempered condition with atmospheric corrosion resistance approximately two times that of carbon steel with copper

(a) Also designated with the suffix "M" when the specification covers metric equivalents. (b) This specification is also published by the American Society of Mechanical Engineers, which uses the prefix "S" (for example, SA36). (c) See also Section 43 of the American Bureau of Shipping specifications and MIL-S-22698 (SH). (d) 13 mm (½ in.) maximum thickness. (e) See also Ref 1. (f) Tensile properties may also be specified when compatible. (g) Discontinued in 1989 and replaced by A 572. (h) Minimum yield point 345 MPa (50 ksi) to 100 mm (4 in.). Lower minimum yield points for thicker sections

Normalizing consists of heating the steel above its critical temperature and cooling in air. This refines the grain size and provides improved uniformity of structure and properties of the hot-finished plate.

When toughness requirements are specified for certain thicknesses in some grades of normalized plate, accelerated cooling must be used in lieu of cooling in still air from the normalizing temperature. Such cooling is accomplished by fans to provide air circulation during cooling or by a water spray or dip. Accelerated cooling is used most often in plates with heavy thicknesses to obtain properties comparable to those developed by normalizing material in the lighter thicknesses.

Quenching consists of heating the steel to a suitable austenitizing temperature, holding at that temperature, and quenching in a suitable medium that depends on chemical composition and cross-sectional dimensions. As-quenched steels are hard, high in strength, and brittle. They are almost always tempered before being placed in service.

Tempering consists of reheating the steel to a predetermined temperature below the critical range, then cooling under suitable conditions. This treatment is usually carried out after normalizing or quenching to obtain desired mechanical properties. Those include a balance of strength and toughness to meet the designer's requirements.

Stress relieving consists of heating the steel to a subcritical temperature to release stresses induced by such operations as flattening or other cold working, shearing, or gas cutting. Stress relieving is not intended

Table 6 ASTM specifications of chemical composition for structural plate made of low-alloy steel or carbon steel

ASTM specification	Material grade or type	C	Mn	P	S	Si	Cr	Ni	Mo	Cu	Others
Low-alloy steel											
A 514	A	0.15–0.21	0.80–1.10	0.035	0.04	0.40–0.80	0.50–0.80	...	0.18–0.28	...	Zr, 0.05–0.15; B, 0.0025
	B	0.12–0.21	0.70–1.00	0.035	0.04	0.20–0.35	0.40–0.65	...	0.15–0.25	...	V, 0.03–0.08; Ti, 0.01–0.03; B, 0.0005–0.005
	C	0.10–0.20	1.10–1.50	0.035	0.04	0.15–0.30	...	...	0.15–0.30	...	B, 0.001–0.005
	E	0.12–0.20	0.40–0.70	0.035	0.04	0.20–0.40	1.40–2.00	...	0.40–0.60	...	Ti, 0.01–0.10 (b); B, 0.001–0.005
	F	0.10–0.20	0.60–1.00	0.035	0.04	0.15–0.35	0.40–0.65	0.70–1.00	0.40–0.60	0.15–0.50	V, 0.03–0.08; B, 0.0005–0.006
	H	0.12–0.21	0.95–1.30	0.035	0.04	0.20–0.35	0.40–0.65	0.30–0.70	0.20–0.30	...	V, 0.03–0.08; B, 0.0005–0.005
	J	0.12–0.21	0.45–0.70	0.035	0.04	0.20–0.35	...	...	0.50–0.65	...	B, 0.001–0.005
	M	0.12–0.21	0.45–0.70	0.035	0.04	0.20–0.35	...	1.20–1.50	0.45–0.60	...	B, 0.001–0.005
	P	0.12–0.21	0.45–0.70	0.035	0.04	0.20–0.35	0.85–1.20	1.20–1.50	0.45–0.60	...	B, 0.001–0.005
	Q	0.14–0.21	0.95–1.30	0.035	0.04	0.15–0.35	1.00–1.50	1.20–1.50	0.40–0.6	...	V, 0.03–0.08
	R	0.15–0.80	0.85–1.15	0.035	0.04	0.20–0.35	0.35–0.65	0.90–1.10	0.15–0.25	...	V, 0.03–0.08
	S	0.10–0.20	1.10–1.50	0.035	0.04	0.15–0.35	...	...	0.10–0.35	...	B, 0.001–0.005; Nb, 0.06 max(c)
	T	0.08–0.14	1.20–1.50	0.035	0.010	0.40–0.60	...	...	0.45–0.60	...	V, 0.03–0.08; B, 0.001–0.005

(continued)

Note: See Table 4 for the compositions of structural plate made of HSLA steel. (a) When a single value is shown, it is a maximum limit, except for copper, for which a single value denotes a minimum limit. (b) Vanadium can be substituted for part or all of the titanium on a one-for-one basis. (c) Titanium may be present in levels up to 0.06% to protect the boron additions. (d) Specification covers many AISI/SAE grades and chemistries. (e) Limiting values vary with plate thickness. (f) Minimum value applicable only if copper-bearing steel is specified. (g) Plates over 13 mm (½ in.) in thickness shall have a minimum manganese content not less than 2.5 times carbon content. (h) The upper limit of manganese may be exceeded provided C + 1/6 Mn does not exceed 0.40% based on heat analysis.

232 / Carbon and Low-Alloy Steels

Table 6 (continued)

ASTM specification	Material grade or type	C	Mn	P	S	Si	Cr	Ni	Mo	Cu	Others
Low-alloy steel (continued)											
A 709	100, 100W					(equivalent to A 514-A,B, C, E, F, H, J, M, P, Q)					
A 710	A	0.07	0.40–0.70	0.025	0.025	0.40	0.60–0.90	0.70–1.00	0.15–0.25	1.00–1.30	Nb, 0.02 min
	B	0.06	0.40–0.65	0.025	0.025	0.15–0.40	...	1.20–1.50	...	1.00–1.30	Nb, 0.02 min
	C	0.07	1.30–1.65	0.25	0.25	0.04	...	0.70–1.00	0.15–0.25	1.00–1.30	Nb, 0.02 min
A 829	(d)						(See Table 3.)				
Carbon steel											
A 36	...	0.29(e)	0.80–1.20(e)	0.04	0.05	0.15–0.40(e)	...	...	...	0.20(f)	...
A 131	A	0.26(e)	(g)	0.05	0.05	...	...	...	...	...	...
	B	0.21	0.80–1.10(h)	0.04	0.04	0.35	...	...	...	...	...
	D	0.21	0.70–1.35(e)(h)	0.04	0.04	0.10–0.35	...	...	...	...	...
	E	0.10	0.70–1.35(h)	0.04	0.04	0.10–0.35	...	...	...	...	...
	CS, DS	0.16	1.00–1.35(h)	0.04	0.04	0.10–0.35	...	...	...	...	...
A 283	A	0.14	0.90	0.04	0.05	0.04(e)	...	...	...	0.20(f)	...
	B	0.17	0.90	0.04	0.05	0.04(e)	...	...	...	0.20(f)	...
	C	0.24	0.90	0.04	0.05	0.04(e)	...	...	...	0.20(f)	...
	D	0.27	0.90	0.04	0.05	0.04(e)	...	...	...	0.20(f)	...
A 284	C	0.36(e)	0.90	0.04	0.05	0.15–0.40	...	...	...	...	...
	D	0.35(e)	0.90	0.04	0.05	0.15–0.40	...	...	...	...	...
A 529	...	0.27	1.20	0.04	0.05	...	...	...	...	0.20(f)	...
A 573	58	0.23	0.60–0.90(h)	0.04	0.05	0.10–0.35	...	...	...	...	...
	65	0.26(e)	0.85–1.20	0.04	0.05	0.15–0.40	...	...	...	...	...
	70	0.28(e)	0.85–1.20	0.04	0.05	0.15–0.40	...	...	...	...	...
A 678	A	0.16	0.90–1.50	0.04	0.05	0.15–0.50	0.25	0.25	0.08	0.20(f)–0.35	...
	B	0.20	0.70–1.60(e)	0.04	0.05	0.15–0.50	0.25	0.25	0.08	0.20(f)–0.35	...
	C	0.22	1.00–1.60	0.04	0.05	0.20–0.50	0.25	0.25	0.08	0.20(f)–0.35	...
A 709	36	0.27(e)	0.80–1.20(e)	0.04	0.05	0.15–0.40(e)	...	...	...	...	...
A 827	(d)						(See Table 11.)				
A 830	(d)						(See Table 2.)				

Note: See Table 4 for the compositions of structural plate made of HSLA steel. (a) When a single value is shown, it is a maximum limit, except for copper, for which a single value denotes a minimum limit. (b) Vanadium can be substituted for part or all of the titanium on a one-for-one basis. (c) Titanium may be present in levels up to 0.06% to protect the boron additions. (d) Specification covers many AISI/SAE grades and chemistries. (e) Limiting values vary with plate thickness. (f) Minimum value applicable only if copper-bearing steel is specified. (g) Plates over 13 mm (½ in.) in thickness shall have a minimum manganese content not less than 2.5 times carbon content. (h) The upper limit of manganese may be exceeded provided C + 1/6 Mn does not exceed 0.40% based on heat analysis.

to significantly modify the microstructure or to obtain desired mechanical properties.

Types of Steel Plate

Steel plate is classified according to composition, mechanical properties, and steel quality. The three general categories of steel plate considered in this article are carbon steel plate, low-alloy steel plate, and high-strength low-alloy (HSLA) steel plate. These three categories of steel plate are available in the steel plate quality levels given in Table 1. Further discussion on these various quality levels is provided in the section "Steel Plate Quality" in this article.

General Categories

Carbon steel plate is available in all quality levels except aircraft quality (Table 1) and is available in many grades. Generally, carbon steel contains carbon up to about 2% and only residual quantities of other elements except those added for deoxidation, with silicon usually limited to 0.60% and manganese to about 1.65%. The chemical composition requirements of standard carbon steel plate are listed in Table 2. These steels may be suitable for some structural applications when furnished according to ASTM A 830 and A 6. In addition to the carbon steels listed in Table 2, other carbon steel plates are also classified according to

Table 7 ASTM specifications of mechanical properties for structural plate made of carbon steel, low-alloy steel, and HSLA steel

ASTM specification	Material grade or type	Tensile strength(a) MPa	ksi	Yield strength(a) MPa	ksi	Minimum elongation(b) in 200 mm (8 in.), %	Minimum elongation(b) in 50 mm (2 in.), %
Carbon steel							
A 36	...	400–500	58–80	220–250(b)	32–36(b)	20	23
A 131	A, B, D, E, CS, DS	400–490	58–71	220(b)	32(b)	21(b)	24
A 283	A	310–415	45–60	165	24	27	30
	B	345–405	50–65	185	27	25	28
	C	380–485	55–70	205	30	22	25
	D	415–515(b)	60–75(b)	230	33	20	23
A 284	C	415	60	205	30	21	25
	D	415	60	230	33	21	24
A 529	...	415–585	60–85	290	42	19	...
A 573	58	400–490	58–71	220	32	21	...
	65	450–530	65–77	240	35	20	...
	70	485–620	70–90	290	42	18	...
A 678	A	485–620	70–90	345	50	...	22
	B	550–690	80–100	415	60	...	22
	C	585–793(b)	85–115(b)	450(b)	65(b)	...	19
A 709	36	400–550	58–80	250	36	20	23
A 827(c)	(d)	(See the section "Forging Quality Plates" in this article.)					
A 830(c)	(d)	(See text.)					
Low-alloy steel							
A 514	All	690–895(b)	100–130(b)	620(b)	90(b)	...	16
A 709	100, 100W	700–915	100–130	635(b)	90(b)	...	15(c)
A 710	A (class 1)	585(b)	85(b)	515(b)	75(b)	...	20
	A (class 2)	485(b)	70(b)	415(b)	60(b)	...	20
	A (class 3)	485(b)	70(b)	415(b)	60(b)	...	20
	B	605(b)	88(b)	515(b)	75(b)	...	18
	C (class 1)	690	100	620	90	...	20
	C (class 3)	620(b)	90(b)	550(b)	80(b)	...	20
A 829(c)	(d)	(See text.)					
(continued)							

(a) Where a single value is shown, it is a minimum. (b) Minimum and/or maximum values depend on plate width and/or thickness. (c) Specification does not specify mechanical properties. (d) Includes several AISI/SAE grades more specific requirements in various ASTM specifications (see the section "Steel Plate Quality" in this article).

Low-Alloy Steel Plate. Steel is considered to be low-alloy steel when either of the following conditions is met:

- The maximum of the range given for the content of alloying elements exceeds one or more of the following limits: 1.65% Mn, 0.60% Si, and 0.60% Cu
- Any definite range or definite minimum quantity of any of the following elements is specified or required within the limits of the recognized field of constructional alloy steels: aluminum, boron, chromium up to 3.99%, cobalt, niobium, molybdenum, nickel, titanium, tungsten, vanadium, zirconium, or any other alloying element added to obtain the desired alloying effect

Alloying elements are added to hot-finished plates for various reasons, including improved corrosion resistance and/or improved mechanical properties at low or elevated temperatures. Alloying elements are also used to improve the hardenability of quenched and tempered plate.

Low-alloy steels generally require additional care throughout their manufacture. They are more sensitive to thermal and mechanical operations, the control of which is complicated by the varying effects of different chemical compositions. To secure the most satisfactory results, consumers normally consult with steel producers regarding the working, machining, heat treating, or other operations to be employed in fabricating the steel; mechanical operations to be employed in fabricating the steel; mechanical properties to be obtained; and the conditions of service for which the finished articles are intended.

The chemical composition requirements of standard low-alloy steel plate are listed in Table 3. These low-alloy steels may be suitable for some structural applications when furnished according to ASTM A 6 and

Table 7 (continued)

ASTM specification	Material grade or type	Tensile strength(a) MPa	ksi	Yield strength(a) MPa	ksi	Minimum elongation(b) in 200 mm (8 in.), %	Minimum elongation(b) in 50 mm (2 in.), %
HSLA steels							
A 131	AH32, DH32, EH32	470–585	65–85	315	46	19	22
	AH36, DH36, EH36	490–620	71–90	...	51	19	22
A 242	...	435(b)	63(b)	290(b)	42(b)	18	21
A 572	42	415	60	290	42	20	24
	50	450	65	345	50	18	21
	60	520	75	415	60	16	18
	65	550	80	450	65	15	17
A 588	All	435(b)	63(b)	290(b)	42(b)	18	21
A 633	A	430–570	63–83	290	42	18	23
	C, D	450–590(b)	65–85(b)	315(b)	46(b)	18	23
	E	515–655(b)	75–95(b)	380(b)	55(b)	18	23
A 656	50	415	60	345	50	20	...
	60	485	70	415	60	17	...
	70	550	80	485	70	14	...
	80	620	90	550	80	12	...
A 678	D	620–760	90–110	515	75	...	18
A 709	50	450	65	345	50	18	21
	50W	485	70	345	50	18	21
A 808	...	415(b)	60(b)	290(b)	42(b)	18	22
A 852	...	620–760	90–110	485	70	...	19
A 871	60	520	75	415	60	16	18
	65	550	80	450	65	15	17

(a) Where a single value is shown, it is a minimum. (b) Minimum and/or maximum values depend on plate width and/or thickness. (c) Specification does not specify mechanical properties. (d) Includes several AISI/SAE grades

Table 8 ASTM specifications for pressure vessel quality steel plate
General requirements for pressure vessel plate are covered in ASTM A 20

Specification	Steel type and condition
Carbon steel	
A 285(a)	Carbon steel plates of low or intermediate tensile strength
A 299(a)	Carbon-manganese-silicon steel plates
A 442(a)	Carbon steel plates for applications requiring low transition temperature
A 455(a)	Carbon-manganese steel plates of high tensile strength
A 515(a)	Carbon-silicon steel plates for intermediate- and higher-temperature service
A 516(a)	Carbon steel plates for moderate- and lower-temperature service
A 537(a)	Heat-treated carbon-manganese-silicon steel plates
A 562	Titanium-bearing carbon steel plates for glass or diffused metallic coatings
A 612(a)	Carbon steel plates of high tensile strength for moderate- and lower-temperature service
A 662(a)	Carbon-manganese steel plates for moderate- and lower-temperature service
A 724	Quenched and tempered carbon steel plates for layered pressure vessels not subject to postweld heat treatment
Carbon steel (continued)	
A 738(a)	Heat-treated carbon-manganese-silicon steel plates for moderate- and lower-temperature service
Low-alloy steel	
A 202(b)	Cr-Mn-Si alloy steel plates
A 203(b)	Nickel alloy steel plates
A 204(b)(c)	Molybdenum alloy steel plates
A 225(b)	Mn-V alloy steel plates
A 302(b)(c)	Mn-Mo and Mn-Mo-Ni alloy steel plates
A 353(b)	Double normalized and tempered 9% Ni steel plates for cryogenic service
A 387(b)(c)	Cr-Mo alloy steel plates for elevated-temperature service
A 517(a)	Quenched and tempered alloy steel plates of high tensile strength
A 533(a)	Quenched and tempered Mn-Mo and Mn-Mo-Ni alloy steel plates
A 542(a)	Quenched and tempered Cr-Mo alloy steel plates
A 543(b)(a)	Quenched and tempered Ni-Cr-Mo alloy steel plates
A 553(a)	Quenched and tempered 8% and 9% Ni alloy steel plates
Low-alloy steel (continued)	
A 645(a)	Specially heat treated 5% Ni alloy steel plates for low- or cryogenic-temperature service
A 734	Quenched and tempered alloy and HSLA steel plates for low-temperature service
A 735	Low-carbon Mn-Mo-Nb alloy steel plates for moderate- and lower-temperature service
A 736	Age-hardening low-carbon Ni-Cu-Cr-Mo-Nb alloy steel plates
A 782	Quenched and tempered Mn-Cr-Mo-Si-Zr alloy pressure vessel steel plates
A 832	Cr-Mo-V-Ti-B alloy pressure vessel steel plates
A 844	9% Ni alloy pressure vessel steel plates produced by the direct-quenching process
HSLA steel	
A 734	See under "Alloy steel"
A 737(a)	HSLA steel plates for applications requiring high strength and toughness
A 841	Steel pressure vessel plate produced by thermomechanical control processes

(a) This specification is also published by the American Society of Mechanical Engineers, which adds an "S" in front of the "A" (for example, SA285). (b) Discontinued in 1989

Table 9 ASTM specifications of chemical composition for pressure vessel plate made of carbon and low-alloy steel

See Table 4 for the compositions of pressure vessel plate made of HSLA steel. The maximum limits per ASTM A 20 on unspecified elements are 0.40% Cu, 0.40% Ni, 0.30% Cr, 0.12% Mo, 0.03% V, and 0.02% Nb.

ASTM specification	Material grade or type	C	Mn	P	S	Si	Cr	Ni	Mo	Cu	Others
Carbon steel											
A 285	A	0.17	0.90	0.035	0.04	...	...	...	...	...	...
	B	0.22	0.90	0.035	0.04	...	...	...	...	...	...
	C	0.28	0.90	0.035	0.04	...	...	...	...	...	...
A 299	...	0.30(b)	0.90–1.50(b)	0.035	0.04	0.15–0.40	...	...	...	...	...
A 442	55	0.24(b)	0.80–1.10(b)	0.035	0.04	0.15–0.40	...	...	...	...	...
	60	0.27(b)	0.80–1.10(b)	0.035	0.04	0.15–0.40	...	...	...	...	...
A 455	...	0.33	0.85–1.20	0.035	0.04	0.10	...	...	...	...	...
A 515	55	0.28(b)	0.90	0.035	0.04	0.15–0.40	...	...	...	...	...
	60	0.31(b)	0.90	0.035	0.04	0.15–0.40	...	...	...	...	...
	65	0.33(b)	0.90	0.035	0.04	0.15–0.40	...	...	...	...	...
	70	0.35(b)	1.20	0.035	0.04	0.15–0.40	...	...	...	...	...
A 516	55	0.26(b)	0.60–1.20(b)	0.035	0.04	0.15–0.40	...	...	...	...	...
	60	0.27(b)	0.60–1.20(b)	0.035	0.04	0.15–0.40	...	...	...	...	...
	65	0.29(b)	0.85–1.20	0.035	0.04	0.15–0.40	...	...	...	...	...
	70	0.31(b)	0.85–1.20	0.035	0.04	0.15–0.40	...	...	...	...	...
A 537	Class 1, 2	0.24	0.70–1.60(b)	0.035	0.04	0.15–0.50	0.25	0.25	0.08	0.35	...
A 562	...	0.12	1.20	0.035	0.04	0.15–0.50	...	...	...	0.15 min	Ti min, 4 × C
A 612	...	0.29(b)	1.00–1.50(b)	0.035	0.04	0.15–0.50(b)	0.25	0.25	0.08	0.35	V, 0.08
A 662	A	0.14	0.90–1.35	0.035	0.04	0.15–0.40	...	...	...	...	...
	B	0.19	0.85–1.50	0.035	0.04	0.15–0.40	...	...	...	...	...
	C	0.20	1.00–1.60	0.035	0.04	0.15–0.50	...	...	...	...	...
A 724	A	0.18	1.00–1.60	0.035	0.04	0.55	0.25	0.25	0.08	0.35	V, 0.08
	B	0.20	1.00–1.60	0.035	0.04	0.50	0.25	0.25	0.08	0.35	V, 0.08
	C	0.22	1.10–1.60	0.035	0.04	0.20–0.60	0.25	0.25	0.08	0.35	B, 0.005; V, 0.008
A 738	A	0.24	1.60(b)	0.035	0.04	0.15–0.50	0.25	0.50	0.08	0.35	...
	B	0.20	0.90–1.50	0.030	0.025	0.15–0.55	0.25	0.25	0.08	0.35	V, 0.08
	C	0.20	1.60(b)	0.030	0.025	0.15–0.55	0.25	0.25	0.08	0.35	V, 0.08
Low-alloy steel											
A 202	A	0.17	1.05–1.40	0.035	0.040	0.60–0.90	0.35–0.60	...	...	...	...
	B	0.25	1.05–1.40	0.035	0.040	0.60–0.90	0.35–0.60	...	...	...	...
A 203	A	0.23(b)	0.80(b)	0.035	0.040	0.15–0.40	...	2.10–2.50	...	...	...
	B	0.25(b)	0.80(b)	0.035	0.040	0.15–0.40	...	2.10–2.50	...	...	...
	D	0.20(b)	0.80(b)	0.035	0.040	0.15–0.40	...	3.25–3.75	...	...	...
	E,F	0.23(b)	0.80(b)	0.035	0.040	0.15–0.40	...	3.25–3.75	...	...	...
A 204	A	0.25(b)	0.90	0.035	0.040	0.15–0.40	...	...	0.45–0.60	...	...
	B	0.27(b)	0.90	0.035	0.040	0.15–0.40	...	...	0.45–0.60	...	...
	C	0.28(b)	0.90	0.035	0.040	0.15–0.40	...	...	0.45–0.60	...	...
A 225	C	0.25	1.60	0.035	0.040	0.15–0.40	...	0.40–0.70	...	...	V, 0.13–0.18
	D	0.20	1.70	0.035	0.40	0.10–0.40	...	0.40–0.70	...	...	V, 0.10–0.18
A 302	A	0.25(b)	0.95–1.30	0.035	0.040	0.15–0.40	...	...	0.45–0.60	...	...
	B	0.25(b)	1.15–1.50	0.035	0.040	0.15–0.40	...	...	0.45–0.60	...	...
	C	0.25(b)	1.15–1.50	0.035	0.040	0.15–0.40	...	0.40–0.70	0.45–0.60	...	...
	D	0.25(b)	1.15–1.50	0.035	0.040	0.15–0.40	...	0.70–1.00	0.45–0.60	...	...
A 353	...	0.13	0.90	0.035	0.040	0.15–0.40	...	8.50–9.50	...	...	...
A 387	2	0.21	0.55–0.80	0.035	0.040	0.15–0.40	0.50–0.80	...	0.45–0.60	...	...
	5	0.15	0.30–0.60	0.040	0.030	0.50	4.00–6.00	...	0.45–0.65	...	...
	7	0.15	0.30–0.60	0.030	0.030	1.00	6.00–8.00	...	0.45–0.65	...	...
	9	0.15	0.30–0.60	0.030	0.030	1.00	8.00–10.00	...	0.90–1.10	...	...
	11	0.17	0.40–0.65	0.035	0.040	0.50–0.80	1.00–1.50	...	0.45–0.65	...	...
	12	0.17	0.40–0.65	0.035	0.040	0.15–0.40	0.80–1.15	...	0.45–0.60	...	...
	21	0.15(b)	0.30–0.60	0.035	0.035	0.50	2.75–3.25	...	0.90–1.10	...	...
	22	0.15(b)	0.30–0.60	0.035	0.035	0.50	2.00–2.50	...	0.90–1.10	...	...
	91	0.08–0.12	0.30–0.60	0.020	0.010	0.20–0.50	8.00–9.50	0.40	0.85–1.05	...	V, 0.18–0.25; Nb, 0.06–0.10; N, 0.03–0.07; Al, 0.04
A 517	A	0.15–0.21	0.80–1.10	0.035	0.040	0.40–0.80	0.50–0.80	...	0.18–0.28	...	B, 0.0025
	B	0.15–0.21	0.70–1.00	0.035	0.040	0.20–0.35	0.40–0.65	...	0.15–0.25	...	B, 0.0005–0.005
	C	0.10–0.20	1.10–1.50	0.035	0.040	0.15–0.30	...	...	0.20–0.30	...	B, 0.001–0.005
	E	0.12–0.20	0.40–0.70	0.035	0.040	0.20–0.35	1.40–2.00	...	0.40–0.60	...	B, 0.0015, 0.005
	F	0.10–0.20	0.60–1.00	0.035	0.040	0.15–0.35	0.40–0.65	0.70–1.00	0.40–0.60	...	B, 0.0005–0.006
	H	0.12–0.21	0.95–1.30	0.035	0.040	0.20–0.35	0.40–0.65	0.30–0.70	0.20–0.30	...	B, 0.0005
	J	0.12–0.21	0.45–0.70	0.035	0.040	0.20–0.35	...	...	0.50–0.65	...	B, 0.001–0.005
	M	0.12–0.21	0.45–0.70	0.035	0.040	0.20–0.35	...	1.20–1.50	0.45–0.60	...	B, 0.001–0.005
	P	0.12–0.21	0.45–0.70	0.035	0.040	0.20–0.35	0.85–1.20	1.20–1.50	0.45–0.60	...	B, 0.001–0.005
	Q	0.14–0.21	0.95–1.30	0.035	0.040	0.15–0.35	1.00–1.50	1.20–1.50	0.40–0.60	...	V, 0.03–0.08
	S	0.10–0.20	1.10–1.50	0.035	0.040	0.15–0.40	...	...	0.10–0.35	...	Ti, 0.06; Nb, 0.06
	T	0.08–0.14	1.20–1.50	0.035	0.010	0.40–0.60	...	...	0.45–0.60	...	B, 0.001–0.005; V, 0.03–0.08
A 533	A	0.25	1.15–1.50	0.035	0.040	0.15–0.40	...	...	0.45–0.60	...	...
	B	0.25	1.15–1.50	0.035	0.040	0.15–0.40	...	0.40–0.70	0.45–0.60	...	...
	C	0.25	1.15–1.50	0.035	0.040	0.15–0.40	...	0.70–1.00	0.45–0.60	...	...

(continued)

(a) When a single value is shown, it is a maximum limit, except where specified as a minimum limit. (b) Limiting values may vary with plate thickness. (c) When specified

Table 9 (continued)

ASTM specification	Material grade or type	Composition, %(a)									
		C	Mn	P	S	Si	Cr	Ni	Mo	Cu	Others
Low-alloy steel (continued)											
	D	0.25	1.15–1.50	0.035	0.040	0.15–0.40	...	0.20–0.40	0.45–0.60	...	...
A 542	A	0.15	0.30–0.60	0.025	0.025	0.50	2.00–2.50	0.40	0.90–1.10	0.40	V, 0.03
	B	0.11–0.15	0.30–0.60	0.015	0.15	0.50	2.00–2.50	0.25	0.90–1.10	0.25	V, 0.02
	C	0.10–0.15	0.30–0.60	0.025	0.025	0.13	2.75–3.25	0.25	0.90–1.10	0.25	V, 0.2–0.3; Ti, 0.015–0.035; B, 0.001–0.003
A 543	B	0.23	0.40	0.035	0.040	0.20–0.40	1.50–2.00	2.60–3.25(b)	0.45–0.60	...	V, 0.03
	C	0.23	0.40	0.020	0.020	0.20–0.40	1.20–1.80	2.25–3.25(b)	0.45–0.60	...	V, 0.03
A 553	I	0.13	0.90	0.035	0.040	0.15–0.40	...	8.50–9.50	...	...	...
	II	0.13	0.90	0.035	0.040	0.15–0.40	...	7.50–8.50	...	...	...
A 645	...	0.13	0.30–0.60	0.025	0.025	0.20–0.40	...	4.75–5.25	0.20–0.35	...	Al, 0.02–0.12; N, 0.020
A 734	A	0.12	0.45–0.75	0.035	0.015	0.40	0.90–1.20	0.90–1.20	0.25–0.40	...	Al, 0.06
A 735	...	0.06	1.20–2.20(b)	0.04	0.025	0.40	...	...	0.23–0.47	0.20–0.35(c)	Nb, 0.03–0.09
A 736	A	0.07	0.40–0.70	0.025	0.025	0.40	0.60–0.90	0.70–1.00	0.15–0.25	1.00–1.30	Nb, 0.02 min
	C	0.07	1.30–1.65	0.025	0.025	0.40	...	0.70–1.00	0.15–0.25	1.00–1.30	Nb, 0.02 min
A 782	...	0.20	0.7–1.20	0.035	0.040	0.40–0.80	0.50–1.00	...	0.20–0.60	...	Zr, 0.04–0.12
A 832	...	0.10–0.15	0.30–0.60	0.025	0.025	0.10	2.75–3.25	...	0.90–1.10	...	V, 0.20–0.30; Ti, 0.015–0.035; B, 0.001–0.003
A 844	...	0.13	0.90	0.020	0.020	0.15–0.40	...	8.50–9.50	...	...	...

(a) When a single value is shown, it is a maximum limit, except where specified as a minimum limit. (b) Limiting values may vary with plate thickness. (c) When specified

Table 10 ASTM specifications of mechanical properties for pressure vessel plate made of carbon steel, HSLA steel, or low-alloy steel

ASTM specification	Material grade or type	Tensile strength(a)		Yield strength(a)		Minimum elongation(b) in 200 mm (8 in.), %	Minimum elongation(b) in 50 mm (2 in.), %
		MPa	ksi	MPa	ksi		
Carbon steel							
A 285	A	310–450	45–65	165	24	27	30
	B	345–485	50–70	185	27	25	28
	C	380–515	55–75	205	30	23	27
A 299	...	515–655	75–95	275(b)	40(b)	16	19
A 442	55	380–515	55–75	205	30	21	26
	60	415–550	60–80	220	32	20	23
A 455	...	485–620(b)	70–90(b)	240(b)	35(b)	15	22
A 515	55	380–515	55–75	205	30	23	27
	60	415–550	60–80	220	32	21	25
	65	450–585	65–85	240	35	19	23
	70	485–620	70–90	260	38	17	21
A 516	55	380–515	55–75	205	30	23	27
	60	415–550	60–80	220	32	21	25
	65	450–585	65–85	240	35	19	23
	70	485–620	70–90	260	38	17	21
A 537	1	450–585(b)	65–85(b)	310(b)	45(b)	18	22
	2	485–620(b)	70–90(b)	315(b)	46(b)	...	20
	2	515–655(b)	75–95(b)	380(b)	55(b)	...	22
A 562	...	380–515	55–75	205	30	22	26
A 612	...	560–695(b)	81–101(b)	345	50	16	22
A 662	A	400–540	58–78	275	40	20	23
	B	450–585	65–85	275	40	20	23
	C	485–620	70–90	295	43	18	22
A 724	A, C	620–760	90–110	485	70	...	19
	B	655–795	95–115	515	75	...	17
A 738	A	515–655	75–95	310	45	...	20
	B	585–705	85–102	415	60	...	20
	C	485–620	70–90	315	46	...	20
HSLA steel							
A 734	B	530–670	77–97	450	65	...	20
A 737	B	485–620	70–90	345	50	18	23
	C	550–690	80–100	415	60	18	23
A 841	...	450–585(b)	65–85(b)	310(b)	45(b)	18	22
Low-alloy steel							
A 202	A	515–655	75–95	310	45	16	19
	B	585–760	85–110	325	47	15	18
A 203	A, D	450–585	65–85	255	37	19	23
	B, E	485–620	70–90	275	40	17	21
	F	515–655	75–95	345	50	...	20

(continued)

(a) Where a single value is shown, it is a minimum. (b) Minimum and/or maximum values depend on plate thickness. (c) As-rolled class 1 plate is limited to 25 mm (1 in.) thickness. (d) As-rolled and aged class 2 plate is limited to 25 mm (1 in.) thickness.

A 829. The effect of residual alloying elements on the mechanical properties of hot-finished steel plate is discussed in the section "Mechanical Properties" in this article. The effect of alloying elements on the hardenability and mechanical properties of quenched and tempered steels is discussed in the articles "Hardenable Carbon and Low-Alloy Steels" and "High-Strength Structural and High-Strength Low-Alloy Steels" in this Volume.

In addition to the low-alloy steels listed in Table 3, other low-alloy steel plates are also classified according to more specific requirements in various ASTM specifications. The chemical composition requirements and mechanical properties of low-alloy steel plate in ASTM standards are discussed in the section "Steel Plate Quality" in this article.

High-strength low-alloy steels offer higher mechanical properties and, in certain of these steels, greater resistance to atmospheric corrosion than conventional carbon structural steels. The HSLA steels are generally produced with emphasis on mechanical property requirements rather than the chemical composition limits. They are not considered alloy steels as described in the American Iron and Steel Institute (AISI) steel products manuals, even though utilization of any intentionally added alloy content would technically qualify as such.

There are two groups of compositions in this category:

- Vanadium and/or niobium steels, with a manganese content generally not exceeding 1.35% maximum and with the addition of 0.2% minimum copper when specified
- High-strength intermediate-manganese steels, with a manganese content in the range of 1.10 to 1.65% and with the

Table 10 (continued)

ASTM specification	Material grade or type	Tensile strength(a) MPa	Tensile strength(a) ksi	Yield strength(a) MPa	Yield strength(a) ksi	Minimum elongation(b) in 200 mm (8 in.), %	Minimum elongation(b) in 50 mm (2 in.), %
Low-alloy steel (continued)							
A 204	A	450–585	65–85	255	37	19	23
	B	485–620	70–90	275	40	17	21
	C	515–655	75–95	295	43	16	20
A 225	A	485–620	70–90	275	40	17	21
	B	515–655	75–95	295	43	16	20
	C	725–930	105–135	485	70	...	20
	D	515–690	75–100	380	55	...	19
A 302	A	515–655	75–95	310	45	15	19
	B	550–690	80–100	345	50	15	18
	C, D	550–690	80–100	345	50	17	20
A 353	...	690–825	100–120	515	75	...	20
A 387	2, 12 (class 1)	380–550	55–80	230	33	18	22
	11 (class 1)	415–585	60–85	240	35	19	22
	22, 21, 5, 7, and 9 (class 1)	415–585	60–85	205	30	...	18
	2 (class 2)	485–620	70–90	310	45	18	22
	11 (class 2)	515–690	75–100	310	45	18	22
	12 (class 2)	450–585	65–85	275	40	19	22
	22, 21, 5, 7, and 9 (class 2)	515–690	75–100	310	45	...	18
	91	585–760	85–110	415	60	...	18
A 517	A, B, C, F, H, J, M, S, T	795–930	115–135	690	100	...	16
	E, P, Q	725–930(b)	105–135(b)	620(b)	90(b)	...	14
A 533	1	550–690	80–100	345	50	...	18
	2	620–795	90–115	485	70	...	16
	3	690–860	100–125	570	83	...	16
A 542	1	725–860	105–125	585	85	...	14
	2	795–930	115–135	690	100	...	13
	3	655–795	95–115	515	75	...	20
	4	585–760	85–110	380	55	...	20
	4a	585–760	85–110	415	60	...	18
A 543	1	725–860	105–125	585	85	...	14
	2	795–930	115–135	690	100	...	14
	3	620–795	90–115	485	70	...	16
A 553	I, II	690–825	100–120	585	85	...	20
A 645	...	655–795	95–115	450	65	...	20
A 734	A	530–670	77–97	450	65	...	20
A 735	1(c)	550–690	80–100	450	65	12	18
	2(d)	585–720	85–105	485	70	12	18
	3	620–750	90–110	515	75	12	18
	4	655–790	95–115	550	80	12	18
A 736	A1	620–760	90–110	550	80	...	20
	A2	415–550(b)	60–80(b)	345(b)	50(b)	...	20
	A3	485–620(b)	70–90(b)	415(b)	60(b)	...	20
	C1	690–825	100–120	620	90	...	20
	C3	620–760(b)	90–110	550(b)	80(b)	...	20
A 782	1	670–820	97–119	550	80	...	18
	2	740–890	107–129	620	90	...	17
	3	795–940	115–136	690	100	...	16
A 832	...	585–760	85–110	415	60	...	18
A 844	...	690–825	100–120	585	85	...	20

(a) Where a single value is shown, it is a minimum. (b) Minimum and/or maximum values depend on plate thickness. (c) As-rolled class 1 plate is limited to 25 mm (1 in.) thickness. (d) As-rolled and aged class 2 plate is limited to 25 mm (1 in.) thickness.

Table 11 Compositions of forging quality steel plate specified in ASTM A 827

Grade UNS	Grade AISI	C	Mn	P(a)	S(a)	Si
G10090	1009	0.15(a)	0.60(a)	0.035	0.040	0.15–0.40
G10200	1020	0.17–0.23	0.30–0.60	0.035	0.040	0.15–0.40
G10350	1035	0.31–0.38	0.60–0.90	0.035	0.040	0.15–0.40
G10400	1040	0.36–0.44	0.60–0.90	0.035	0.040	0.15–0.40
G10450	1045	0.42–0.50	0.60–0.90	0.035	0.040	0.15–0.40
G10500	1050	0.47–0.55	0.60–0.90	0.035	0.040	0.15–0.40

(a) Maximum

addition of 0.2% minimum copper when specified

Other elements commonly added to HSLA steels to yield the desired properties include silicon, chromium, nickel, molybdenum, titanium, zirconium, boron, aluminum, and nitrogen. The chemical compositions of ASTM structural quality and pressure vessel quality plates made of HSLA steel are listed in Table 4. More information on HSLA steels is provided in the article "High-Strength Structural and High-Strength Low-Alloy Steels" in this Volume.

Steel Plate Quality

Steel quality, as the term applies to steel plate, is indicative of many conditions, such as the degree of internal soundness, relative uniformity of mechanical properties and chemical composition, and relative freedom from injurious surface imperfections. The various types of steel plate quality are indicated in Table 1.

The three main quality descriptors used to describe steel plate are regular quality, structural quality, and pressure vessel quality. Special qualities include cold-drawing quality, cold-pressing quality, cold-flanging quality, and forging quality carbon steel plate, along with drawing quality and aircraft quality alloy steel plate. Quality descriptors that have been used in the past include flange quality and firebox quality carbon and alloy steel plate and marine quality carbon steel plate. However, use of these descriptors has been discontinued in favor of pressure vessel quality.

Regular quality is the most common quality of carbon steel, which is applicable to plates with a maximum carbon content of 0.33%. Plates of this quality are not expected to have the same degree of chemical uniformity, internal soundness, or freedom from surface imperfections that is associated with structural quality or pressure vessel quality plate. Regular quality is usually ordered to standard composition ranges and is not customarily produced to mechanical property requirements. Regular quality is analogous to merchant quality for bars because there are normally no restrictions on deoxidation, grain size, check analysis, or other metallurgical factors. Also, this quality plate can be satisfactorily used for applications similar to those of merchant quality bars, such as those involving mild cold bending, mild hot forming, punching, and welding for noncritical parts of machinery.

Structural quality steel plate is intended for general structural applications such as bridges, buildings, transportation equipment, and machined parts. The various ASTM specifications for structural quality steel plate are given in Table 5. Most of the structural steel plate listed in Table 5 is furnished to both chemical composition limits (Table 6) and mechanical properties (Table 7). However, some structural steel plate (ASTM A 829 and A 830 in Table 5) is produced from the standard steels listed in Tables 2 and 3. These steels can be furnished only according to the chemical compositions specified by SAE/AISI steel designations. Factors affecting the mechanical properties of hot-finished carbon steel are discussed in the section "Mechanical Properties" in this article.

Pressure Vessel Plate. Steel plate intended for fabrication into pressure vessels must conform to specifications different from those of similar plate intended for structural applications. The major differences between the two groups of specifications are that pressure vessel plate must meet requirements for notch toughness and has more stringent limits for allowable surface and edge imperfections.

Table 8 lists the various ASTM specifications for pressure vessel steel plate. All of these steel plate specifications are furnished according to both chemical composition limits and mechanical properties.

The chemical composition limits of pressure vessel steel plate include a maximum phosphorus content of 0.035% and a maximum sulfur content of 0.040% by product analysis. The chemical compositions of various types of pressure vessel steel plate are given in Table 9.

Mechanical tests of pressure vessel steel plate involve a minimum of one tensile test for each as-rolled plate or a minimum of two tensile tests for quenched and tempered plates. The mechanical property requirements given in ASTM specifications for pressure vessel steel plate are listed in Table 10.

Aircraft quality plates are used for important or highly stressed parts of aircraft, missiles, and other applications involving stringent requirements. Plates of this quality require exacting steelmaking, conditioning, and process controls and are generally furnished from electric furnace steels in order to meet the internal cleanliness requirements outlined in Aerospace Materials Specification AMS-2301. The primary requirements of this quality are a high degree of internal soundness, good uniformity of chemical composition, good degree of cleanliness, and a fine austenitic grain size. Aircraft quality plates can be supplied in the hot-rolled or thermally treated condition.

Forging quality plates are intended for forging, quenching and tempering, or similar purposes or when uniformity of composition and freedom from injurious imperfections are important (see ASTM A 827). Plates of this quality are produced from killed steel and are ordinarily furnished with the phosphorus content limited to 0.035% maximum and the sulfur content limited to 0.040% maximum by heat analysis. Table 11 lists some AISI/SAE steels suitable for forging quality plate. Plates of this quality can be produced to chemical ranges and limits and mechanical properties. When mechanical properties are specified, two tension tests from each heat are taken from the same locations as tests for structural quality. Factors affecting mechanical properties are discussed in the section "Mechanical Properties" in this article.

Mechanical Properties

Of the various mechanical properties normally determined for steel plate, yield strength is an important design criterion in structural applications. Tensile strength is also an important design consideration in many design codes in the United States, but is useful primarily as an indication of fatigue properties. Yield strength is a design criterion in most design codes when the ratio of yield to tensile strength is less than 0.5. Ductility, as measured by tensile elongation and reduction in area, is seldom in itself a valuable design criterion, but is sometimes used as an indication of toughness and suitability for certain applications.

The mechanical properties of steel plate in the hot-finished condition are influenced by several variables, of which chemical composition is the most influential. Other factors include deoxidation practice, finishing temperature, plate thickness, and the presence of residual elements such as nickel, chromium, and molybdenum. For steels used in the hot-finished condition (such as plate), carbon content is the single most important factor in determining mechanical properties.

The static tensile properties of the various grades, types, and classes of steel plate covered by ASTM specifications are listed in Tables 7 and 10. It should be noted that some of these values vary with plate thickness and/or width. An example of the variation of tensile strength and elongation with thickness is shown in Fig. 1, which presents the minimum expected values for 0.20% C steel plate from 13 to 125 mm (½ to 5 in.) thick. Plate under 13 mm (½ in.) thick would show even slightly higher tensile strength and lower elongation because of the increased amount of hot working during rolling and the faster cooling rates after rolling.

The distribution of the tensile properties obtained for a larger number of heats of A 285, A 515, and A 516 steel plate is illustrated in Fig. 2, which also shows the distribution of the carbon and manganese content. The use of the carbon and manganese contents to control mechanical properties is clearly shown in Fig. 2; higher carbon and manganese contents accompany higher yield strengths.

Figure 3 repeats the data in Fig. 2 for the 224 heats of A 285, grade C plate. However, the data are augmented by the individual distributions for the various ranges of plate thickness included in the investigation. When steel is produced to a mechanical property requirement, it is common practice to vary the carbon and manganese to compensate for size influence. The use of higher-than-average carbon (and manganese) content to maintain yield strength as plate thickness increases is evident in Fig. 3.

The common mechanical properties of hot-finished steel, including plate, reliably related to each other, and this relation is relatively free from influence of composition for most of these properties. Figure 4 shows the relationship between yield strength, elongation, and tensile strength over a wide range of tensile strengths for various hot-rolled carbon steels.

Residual alloying elements generally have a minor strengthening effect on hot-finished

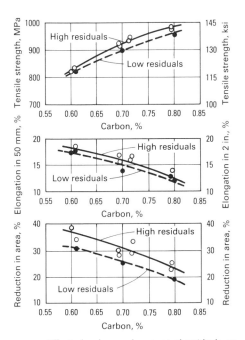

Fig. 5 Effect of carbon and amount of residuals on tensile properties of hot-finished carbon steel. Curves marked high residuals represent steel containing 0.06 to 0.12% Ni, 0.06 to 0.13% Cr, and 0.08 to 0.13% Mo. Curves marked low residuals represent steel containing 0.05% Ni max, 0.05% Cr max, and 0.04% Mo max. Total of 58 heats tested

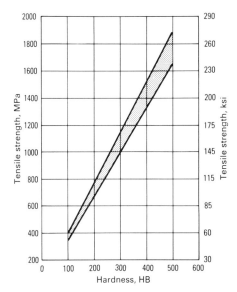

Fig. 6 Relation between hardness and tensile strength of steel. Range up to 300 HB is applicable to the hot-finished steel discussed in this article.

steels, such as plate. This effect cannot be considered in design because residuals vary greatly among the different steel producing plants. This influence is shown in Fig. 5, which demonstrates that the effect is minor.

Hardness is a relatively simple test to perform and is closely related to tensile strength, as shown in Fig. 6. A simple hardness test, used in conjunction with the data in Fig. 4, can be used to estimate yield strength, elongation, and tensile strength.

Fatigue Strength. The high-cycle (>1 million) fatigue properties of hot-finished steel, often called the fatigue limit, are more or less directly related to tensile strength and are greatly affected by the surface condition. The fatigue limit of machined specimens is about 40% of the tensile strength, depending on the surface finish. In contrast, unmachined hot-rolled steel, when loaded so that fatigue stresses are concentrated at the surface, will have a considerably lower fatigue limit because of decarburization, surface roughness, and other surface imperfections. For this reason, the location of maximum fatigue stresses should be carefully considered; for structural members designed in hot-finished steel, the surface should be machined off from critically stressed areas or an allowance made for the weakness of the hot-finished surface.

The presence of inclusions in hot-finished steel may also have an adverse effect on the fatigue limit. Large inclusions are considered harmful under the dynamic stresses of impact or fatigue, and the effect is greater in the harder steels.

Low-Temperature Impact Energy. When notch toughness is an important consideration, satisfactory service performance can be ensured by proper selection of the steel that will behave in a tough manner at its lowest operating temperature. The Charpy V-notch tests and crack-starter drop-weight tests provide a fairly reliable indication of the tendency toward brittle fracture in service. The transition temperatures of hot-finished steels are controlled principally by their chemical composition and ferrite grain size. For the steels considered in this article, carbon is of primary importance because of its effect in raising the transition temperature, lowering the maximum energy values, and widening the temperature range between completely tough and completely brittle behavior.

Manganese (up to about 1.5%) improves low-temperature properties. Also, as mentioned previously, the transition temperature is affected by the deoxidation practice used. The transition temperature decreases and the energy absorption before fracture at normal temperatures increases in the order of rimmed, capped, semikilled, and killed steels. In addition, killed steels contain larger amounts of silicon or aluminum than semikilled steels, and these elements improve low-temperature toughness and ductility. Because of variations in finishing temperatures and cooling rates, plate thickness influences the grain size and therefore the transition temperature. Extensive data on the impact properties of hot-finished steel are given in the article "Notch Toughness of Steels" in this Volume.

Elevated-Temperature Properties. The steel plate used in pressure vessel applications is often subjected to long-term elevated temperatures. Of the carbon and low-alloy steels used for pressure vessel plate, the behavior of 2.25Cr-1Mo steel (ASTM A 387, Class 22, in Table 9) at elevated temperatures has been studied more thoroughly than any other steel and has become the reference for comparing the elevated-temperature properties of low-alloy steels. Further information on the elevated-temperature properties of 2.25Cr-1Mo steel can be found in the article "Elevated-Temperature Properties of Ferritic Steels" in this Volume.

Directional Properties. An important characteristic of steel plate, known as directionality or fibering, must be considered. During the rolling operations, many inclusions, which are in a plastic condition at rolling temperatures, are elongated in the direction of rolling. At the same time, localized chemical segregates that have formed during solidification of the steel are also elongated. These effects reduce the ductility and impact properties transverse to the rolling direction, but have little or no effect on strength.

Fabrication Considerations

Formability. The cold formability of steel plate is directly related to the yield strength and ductility of the material. The lower the yield strength, the smaller the load required to produce permanent deformation; high ductility allows large deformation without fracture. Therefore, the lower-carbon grades are most easily formed.

Operations such as shearing and blanking are usually limited by the lack of the available facilities as the plate thickness increases. This also applies to bending operations. Of course, an adequate bend radius must be used to avoid fracture. Because of fibering effects, the direction of bend is also important; when the axis of a bend is parallel to the direction of rolling, small bend radii are usually difficult to form because of the danger of cracking.

Machinability. Machining operations are usually performed with little difficulty on most plate steels up to about 0.50% C. Higher-carbon steels can be annealed for softening. Steels with low carbon and manganese content, such as 1015, with large quantities of free ferrite in the microstructure may be too soft and gummy for good machining. Increasing the carbon content (to a steel such as 1025) improves the machinability.

Machining characteristics can be improved by factors that break up the chip as it is removed. This is usually accomplished by the introduction of large numbers of inclusions such as manganese sulfides or complex oxysulfides. These "free-machining" steels are somewhat more expensive, but are cost-effective when extensive machining is involved.

Weldability is a relative term that describes the ease with which sound welds possessing good mechanical properties can be produced in a material. The chief weldability factors are composition, heat input, and rate of cooling. These factors produce various effects, such as grain growth, phase changes, expansion, and contraction, which in turn determine weldability. Heat input and cooling rate are characteristics of the specific process and technique used and the thickness of the metal part being welded. Therefore, weldability ratings should state the conditions under which the rating was determined and the properties and soundness obtained.

For carbon steels, the carbon and manganese contents are the primary elements of the composition factor that determine the effect of the steel of given heating and cooling conditions. The great tonnage of steel used for welded applications consists of low-carbon steel, up to 0.30% C.

Generally, steels with a carbon content less than 0.15% are readily weldable by any method. Steel with a carbon range of 0.15 to 0.30% can usually be welded satisfactorily without preheating, postheating, or special electrodes. For rather thick sections (>25 mm, or 1 in.), however, special precautions such as 40 °C (100 °F) minimum preheat, 40 °C (100 °F) minimum temperature between weld passes, and a 540 to 675 °C (1000 to 1250 °F) stress relief may be necessary.

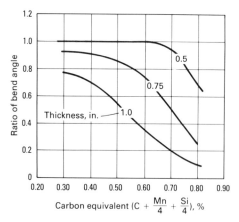

Fig. 7 Ratio (welded to unwelded) of bend angle for normalized steel plate. A high value of the ratio indicates high weldability. Source: Ref 2

Higher-carbon and higher-manganese grades can often be welded satisfactorily if preheating, special welding techniques, and postheating and peening are used. In the absence of such precautions to control the rate of cooling and to eliminate high stress gradients, cracks may occur in the weld and base metal. In addition, base metal properties such as strength, ductility, and toughness may be greatly reduced.

All comments about the effect of carbon and manganese on weldability must be qualified in terms of section size because of its relationship to heat input and cooling rate. In welding thicker sections, such as plate, the relatively cold base metal serves to greatly accelerate the cooling rate after welding with the result that plate thickness is a very important consideration. Figure 7 shows the effect of plate thickness and carbon equivalent on weldability as expressed in terms of a notch bend test.

REFERENCES

1. *Plates; Rolled Floor Plates: Carbon, High Strength Low Alloy, and Alloy Steel*, AISI Steel Products Manual, American Iron and Steel Institute, 1985
2. *Weldability of Steels*, Welding Research Council, 1953

Hot-Rolled Steel Bars and Shapes

Revised by Timothy E. Moss, J.M. Hambright, and T.E. Murphy, Inland Bar and Structural, Division of Inland Steel Company; and J.A. Schmidt, Joseph T. Ryerson and Sons, Inc.

HOT-ROLLED STEEL BARS and other hot-rolled steel shapes are produced from ingots, blooms, or billets converted from ingots or from strand cast blooms or billets and comprise a variety of sizes and cross sections. Bars and shapes are most often produced in straight lengths, but bars in some cross sections in smaller sizes are also produced in coils.

The term "bar" includes:

- Rounds, squares, hexagons, and similar cross sections 9.5 mm (3/8 in.) and greater across
- Flats greater than 5.16 mm (0.203 in.) in thickness and 152 mm (6 in.) and less in width, or 5.84 mm (0.230 in.) and greater in thickness and 203 mm (8 in.) and less in width
- Small angles, channels, tees, and other standard shapes less than 76 mm (3 in.) across
- Concrete-reinforcing bars

The term "shape" includes structural shapes and special shapes. Structural shapes are flanged, are 76 mm (3 in.) or greater in at least one cross-sectional dimension, and are used in structures such as bridges, buildings, ships, and railroad cars. Special shapes are those designed by users for specific applications.

Dimensions and Tolerances

The nominal dimensions of hot-rolled steel bars and shapes are designated in inches or millimeters with applicable tolerances, as shown in ASTM A 6 and A 29. Bars with certain quality descriptors have size limitations; these are covered in discussions of individual product qualities later in this article.

Bars or shapes can be cut to length in the mill by a number of methods, such as hot or cold shearing or sawing. The method used is determined by cross section, grade, and customer requirements. Some end distortion is to be expected from most methods. When greater accuracy in length or freedom from distortion is required, bars or shapes can be cut overlength, then recut on one or both ends by cold sawing or equivalent means.

If a bar or shape requires straightening, prior annealing is sometimes necessary, depending on the grade of steel and the cross-sectional shape of the part. The processing necessary to meet straightness tolerances is not intended to improve either the surface finish or accuracy of cross-sectional shape and may result in increased surface hardness. Length and straightness tolerances for bars and shapes are found in ASTM A 6 and A 29.

Surface Imperfections

Most carbon steel and alloy steel hot-rolled bars and shapes contain surface imperfections with varying degrees of severity. In virtually all cases, these defects are undesirable and may in some applications affect the integrity of the finished product.

Included in the manufacturing process for hot-rolled bars and shapes are various steps designed to minimize or eliminate surface defects. These steps include inspection of both the semifinished and the finished product and either subsequent removal of the defects or rejection of the material if defect removal is not possible. Inspection techniques range from visual inspection of the semifinished material to sophisticated electronic inspection of the finished product. Defects found in the semifinished product can be removed by hot scarfing, grinding, or chipping. Defects in the finished products are generally removed by grinding, turning, or peeling and, to a lesser degree, by chipping.

Currently, it is not technically feasible to produce defect-free hot-rolled bars. With the current demand for high-quality bar products, it is becoming increasingly common to subject hot-rolled bars to a cold-finishing operation, such as turning or grinding, coupled with a sensitive electronic inspection. With this process route, it becomes possible to significantly reduce both the frequency and the severity of surface defects.

Seams, Laps, and Slivers

Seams, laps, and slivers are probably the most common defects in hot-rolled bars and shapes.

Seams are longitudinal defects that can vary greatly in length and depth. It is quite common for steel users to refer to any longitudinal defect as a seam regardless of the true nature of the defect. However, there is a classical definition of a seam, as follows. Gas comes out of the solution as the liquid steel solidifies. This gas is trapped as bubbles or blowholes by the solidifying steel and appears as small holes under the surface of the steel. When the steel is reheated, some areas of the surface may scale off, exposing and oxidizing the interior of these blowholes. This oxidation prevents the blowholes from welding shut during rolling. This rolling then elongates the steel, resulting in a longitudinal surface discontinuity—a seam. As viewed in the cross section, seams are generally characterized as being perpendicular to the surface, completely surrounded by decarburization, and associated with disperse oxides.

Laps are mechanical defects that occur during the hot rolling of both semifinished and finished material. Laps are nothing more than a folding over of the material, resulting, for example, from gouging during the rolling process or misalignment of the pass lines or rolls. As viewed in the cross section, laps are characterized as being at an angle from the steel surface; they have decarburization on one side only of the defect and often contain entrapped scale.

Slivers usually appear as a scablike defect, adhering on one end to the surface of the hot-rolled steel. They are normally pressed into the surface during hot rolling. They can originate from short, rolled out defects such as torn corners that are not removed during conditioning. They can also result from conditioning gouges or mechanical gouges during rolling. Although there is no specific metallographic definition of slivers, metallographic examination can be used to determine the origin of these defects.

Decarburization

Another condition that could be considered a surface defect is decarburization. This condition is present to some degree on all hot-rolled steel. Decarburization occurs at very high temperatures when the surface carbon of the steel reacts with the oxygen in the furnace atmosphere. This loss of surface carbon results in a surface that is softer and unsuitable for any application involving wear or fatigue. Because of the existence of this condition, steel ordered for critical applications can be produced oversize and then ground to desired size.

Allowance for Surface Imperfections in Machining Applications

Experience has shown that when purchasers order hot-rolled or heat-treated bars that are to be machined, it is advisable for the purchaser to make adequate allowances for the removal of surface imperfections and to specify the sizes accordingly. These allowances depend on the way the surface metal is removed, the length and size of the bars, the straightness, the size tolerance, and the out-of-round tolerance. Bars are generally straightened before machining. For special quality carbon steel bars and regular quality alloy steel bars, either resulfurized or nonresulfurized (see the article "Cold-Finished Steel Bars" in this Volume), it is advisable that allowances for centerless-turned or centerless-ground bars be adequate to permit stock removal of not less than the amount shown below:

Bar diameter		Recommended minimum machining allowance per side, % of specific size	
mm	in.	Nonresulfurized	Resulfurized
≤51	≤2	2.6	3.4
>51	>2	1.6	2.4

Source: Ref 1

Note that these allowances are based on bars within straightness tolerance. Also, because straightness is a function of length, additional machining allowance may be required for turning long bars on centers. For steel bars subject to magnetic particle inspection, additional stock removal is recommended, as indicated in Table 1.

The allowances described above are usually more than sufficient to remove surface imperfections and result in considerable loss of material. Therefore, most experienced fabricators remove considerably less stock than recommended and take their chances on occasional difficulties. In conventional practice, depth of machining for hot-rolled bars is 1.6 mm (¹⁄₁₆ in.) for bars 38 to 76 mm (1½ to 3 in.) in diameter, and 3.2 mm (⅛ in.) for bars over 76 mm (3 in.) in diameter.

Table 1 Recommended minimum stock removal for steel bars subject to magnetic particle inspection

Hot-rolled size		Minimum stock removal from the surface(a)	
mm	in.	mm	in.
Up to 12.7	Up to ½	0.76	0.030
>12.7–19	>½–¾	1.14	0.045
>19–25	>¾–1	1.52	0.060
>25–38	>1–1½	1.90	0.075
>38–51	>1½–2	2.29	0.090
>51–64	>2–2½	3.18	0.125
>64–89	>2½–3½	3.96	0.156
>89–114	>3½–4½	4.75	0.187
>114–152	>4½–6	6.35	0.250
>152–191	>6–7½	7.92	0.312
>191–229	>7½–9	9.52	0.375
>229–254	>9–10	11.10	0.437

(a) The minimum reduction in diameter of rounds is twice the minimum stock removal from the surface.

Surface Treatment

It is uncommon for hot-rolled steel bars and shapes to be descaled by the producer or protected from the weather during transit. Most cleaning and coating operations are done either by the customer or by an intermediate processor.

Descaling of hot-rolled bars and shapes is generally done by either pickling or blasting, depending on the end use. There are several subsequent coatings that can be used. Oil is both the simplest and the least expensive to use and acts as a temporary rust preventive. Lime, in addition to serving as a rust preventive, can serve as a carrier for lubricants used during cold drawing or cold forging. A more sophisticated system includes descaling, followed by a zinc phosphate coating, coupled with a dry lubricant. This system provides some rust protection and serves as a lubricant for cold-forming operations.

Heat Treatment

Hot-rolled low-carbon and medium-carbon steel bars and shapes are often used in the as-rolled condition, but hot-rolled bars of higher-carbon steel and most hot-rolled alloy steel bars must be heat treated in order to attain the hardness and microstructure best suited for the final product or to make them suitable for processing. Such heat treatment consists of one or more of the following: some form of annealing, stress relieving, normalizing, quenching, and tempering.

Ordinary annealing is the term generally applied to heat treatment used to soften steel. The steel is heated to a suitable temperature, held there for some period of time, and then cooled; specific times, temperatures, and cooling rates vary. Maximum hardness compatible with common practice can be specified.

Annealing for specified microstructures can be performed to obtain improved machinability or cold-forming characteristics. The structures produced may consist of lamellar pearlite or spheroidized carbides. Special control of the time and temperature cycles is necessary. A compatible maximum hardness can be specified.

Stress relieving involves heating to a subcritical temperature and then cooling. For hot-rolled bars, the principal reason for stress relieving is to minimize distortion in subsequent machining. It is used to relieve the stresses resulting from cold-working operations, such as special machine straightening.

Normalizing involves heating to a temperature above the critical temperature range and then cooling in air. A compatible maximum hardness can be specified.

Hardening by quenching consists of heating steel to the correct austenitizing temperature, holding at that temperature for a sufficient time to produce homogeneous austenite, and quenching in a suitable medium (water, oil, synthetic oil or polymer, molten salts, or low-melting metals) depending on chemical composition and section thickness.

Tempering is an operation performed on normalized or quenched steel bars. In this technique, the bars are reheated to a predetermined temperature below the critical range and then cooled under suitable conditions.

When a hardness requirement is specified for normalized and tempered bars, the bars are ordinarily produced to a range of hardnesses equivalent to a 0.4 mm range of Brinell impression diameters. Quenched and tempered bars are ordinarily produced to a 0.3 mm range of Brinell impression diameters. Quenched and tempered bars can also be produced to minimum mechanical property requirements.

Product Requirements

Hot-rolled steel bars and shapes can be produced to chemical composition ranges or limits, mechanical property requirements, or both. The mechanical testing of hot-rolled steel bars and shapes can include tensile, Brinell or Rockwell hardness, bend, Charpy impact, fracture toughness, and short-time elevated-temperature tests, as well as tests for elastic limit, proportional limit, and offset yield strength, which require the use of an extensometer or plotting of a stress-strain curve. These tests are covered by ASTM A 370 and other ASTM standards.

Other tests sometimes required include the measurement of grain size and hardenability. Austenitic grain size is determined by the McQuaid-Ehn test, which is described in ASTM E 112. This test involves metallographic examination of a carburized specimen to observe prior austenitic grain boundaries. Hardenability can be measured by several methods, the most common be-

Table 2 Lowest maximum hardness that can be expected for hot-rolled steel bars, billets, and slabs with ordinary mill annealing

Steel grade	Maximum hardness, HB(a) Straightened	Nonstraightened	Steel grade	Maximum hardness, HB(a) Straightened	Nonstraightened	Steel grade	Maximum hardness, HB(a) Straightened	Nonstraightened
Carbon steels			**Alloy steels**			**Alloy steels**		
1141	201	192	4150	235	223	5155	229	217
1144	207	197	4161	241	229	5160	235	223
1151	207	201	4320	207	197	51B60	235	223
1541	207	197	4340	235	223	6118	163	156
1548	212	207	4419	170	163	6150	217	207
1552	212	207				81B45	201	192
15B41	207	197	4615	174	167			
15B48	212	207	4620	179	170	8615	163	156
			4621	179	170	8617	163	156
Alloy steels			4626	187	179	8620	170	163
1330	197	179	4718	179	170	8622	179	170
1335	197	187				8625	179	170
1340	201	192	4720	170	163			
1345	212	201	4815	223	192	8627	183	174
4012	149	143	4817	229	197	8630	187	179
			4820	229	197	8637	201	192
4023	156	149	5015	156	149	8640	207	197
4024	156	149				8642	212	201
4027	170	163	50B44	207	197			
4028	170	163	50B46	217	201	8645	217	207
			50B50	217	201	8655	235	223
4037	192	183	50B60	229	217	8720	170	163
4047	212	201	5120	170	163	8740	212	201
4118	170	163				8822	187	179
4130	183	174	5130	183	174			
4137	201	192	5132	187	179	9254	241	229
			5135	192	183	9255	241	229
4140	207	197	5140	197	187	9260	248	235
4142	212	201	5145	229	197	94B17	156	149
4145	217	207	5147	217	207	94B30	183	174
4147	223	212	5150	212	201			

(a) Specific microstructure requirements may necessitate modification of these hardness numbers.

ing the Jominy end-quench test, as described in ASTM A 255 (see the article "Hardenability of Carbon and Low-Alloy Steels" in this Volume).

Soundness and homogeneity can be evaluated by fracturing. The fracture test is commonly applied only to high-carbon bearing quality steel. Location of samples, number of tests, details of testing technique, and acceptance limits based on the test should be established in each instance.

Testing for nonmetallic inclusions consists of careful microscopic examination (at 100×) of prepared and polished specimens. The specimens should be taken on a longitudinal plane midway between the center and surface of the product. Location of specimens, number of tests, and interpretation of results should be established in each instance. Typical testing procedures are described in ASTM E 45. Nonmetallic inclusion content can also be measured on the macroscopic scale by magnetic particle tests such as those described in AMS 2300 and 2301. These tests involve the measurement of inclusion frequency and severity in a sampling scheme that represents the interior of the material.

Surface and subsurface nonuniformities are revealed by magnetic particle testing. This test was developed for, and is used on, fully machined or ground surfaces of finished parts. When the magnetic particle test is to be applied to bar stock, short-length samples should be heat treated and completely machined or ground.

Tensile and hardness tests are the most common mechanical tests performed on hot-rolled steel bars and shapes. Hardness is a relatively simple property to measure, and it is closely related to tensile strength, as shown in Fig. 1. When Fig. 2 is used together with Fig. 1, a simple hardness test can give an estimate of yield strength and elongation, as well as tensile strength.

It is not practicable to set definite limitations on tensile strength or hardness for carbon or alloy steel bars in the as-rolled condition. For mill-annealed steel bars, there is a maximum tensile strength or a maximum hardness (Table 2) that can be expected for each grade of steel. For steel bars in the normalized condition, maximum hardness, maximum tensile strength, minimum hardness, or minimum tensile strength can be specified. For normalized and tempered bars and for quenched and tempered bars, either maximum and minimum hardness or maximum and minimum tensile strength can be specified; for either property, the range that can be specified varies with tensile strength and is equivalent to a 0.4 mm range of Brinell indentation diameters at any specified location for normalized and tempered bars and to a 0.3 mm range for quenched and tempered bars.

It is essential that the purchaser specify the positions at which hardness readings are to be taken. When both hardness and tensile strength are specified at the same position, the limits should be consistent with each other. When hardness limits are specified as surface values, they may be inconsistent with tensile-test values, which of necessity are properties of the bulk metal; the inconsistency will vary according to the size of the bar and the hardenability of the steel. The purchaser should specify limits that take this inconsistency into account.

If the locations of hardness readings are not specified on the purchaser's order or specification, the hardness values are applicable to the bar surface after removal of decarburization. Hardness correction factors for bars of various diameters as described in ASTM E 18 should be employed if a flat area is not available on the bar tested.

Generally, yield strength, elongation, and reduction in area are specified as minimums for steel only in the quenched and tempered or the normalized and tempered condition, and they should be consistent with ultimate tensile strength or hardness. When quenched and tempered bars are cold worked by cold straightening, stress relieving may be required to restore elastic properties and to improve ductility.

Product Categories

Hot-rolled carbon steel bars are produced to two primary quality levels: merchant qual-

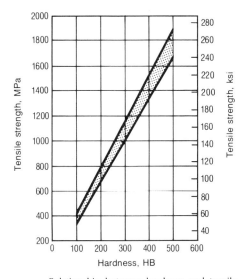

Fig. 1 Relationship between hardness and tensile strength of steel. Range up to 300 HB is applicable to the hot-finished steel discussed in this article. Source: Ref 2

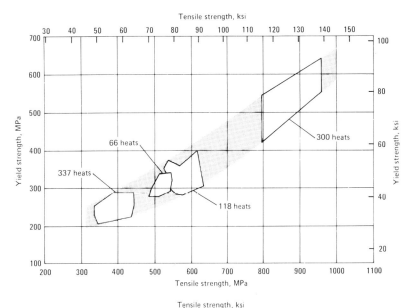

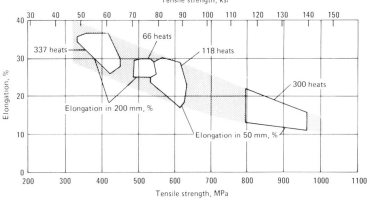

Fig. 2 Relation of tensile properties for hot-rolled carbon steel

ity and special quality. Merchant quality is the lower quality level and is not suitable for any operation in which internal soundness or freedom from surface imperfections is of primary importance. Special quality includes all bar categories with end-use-related and restrictive quality requirements.

The mechanical properties of hot-rolled carbon steel bars in the as-rolled condition are influenced by:

- Chemical composition
- Thickness or cross-sectional area
- Variables in mill design and mill practice

Carbon content is the dominant factor. The minimum expected mechanical properties of commonly used grades of hot-rolled carbon steel bars are shown in Fig. 3.

Quality descriptors for hot-rolled alloy steel bars are related to suitability for specific applications. Characteristics considered include inclusion content, uniformity of chemical composition, and freedom from surface imperfections.

Carbon steel and alloy steel structural shapes and special shapes do not have specific quality descriptors but are covered by several ASTM specifications (Table 3). In most cases, these same specifications also cover structural quality steel bars. The ASTM specifications covering other qualities of hot-rolled bars are listed in Table 4. The various categories of hot-rolled steel bar products and their characteristics are described in the following sections.

Merchant Quality Bars

Merchant quality is the least restrictive descriptor for hot-rolled carbon steel bars. Merchant quality bars are used in the production of noncritical parts of bridges, buildings, ships, agricultural implements, road-building equipment, railway equipment, and general machinery. These applications require only mild cold bending, mild hot forming, punching, and welding. Mild cold bending is bending in which a generous bend radius is used and in which the axis of the bend is at right angles to the direction of rolling.

Merchant quality bars should be free from visible pipe; however, they may contain pronounced chemical segregation, and for this reason, product analysis tolerances are not appropriate. Internal porosity, surface seams, and other surface irregularities may be present and are generally expected in bars of this quality. Consequently, merchant quality bars are not suitable for applications that involve forging, heat treating, or other operations in which internal soundness or freedom from surface imperfections is of primary importance.

Grades. Merchant quality bars can be produced to meet both chemical composition (heat analysis only) and mechanical properties. These steels can be supplied to chemical compositions within the ranges of 0.50% C (max), 0.60% Mn (max), 0.04% P (max), and 0.05% S (max), but are not produced to meet any specific silicon content, grain size, or any other requirement that would dictate the type of steel produced.

Merchant quality steel bars do not require the chemical ranges typical of standard steels. They are produced to wider carbon and manganese ranges and are designated by the prefix "M."

When ordering merchant quality bars to meet mechanical properties, the following strength ranges are to be used up to a maximum of 655 MPa (95 ksi):

- 70 MPa (10 ksi) for minimums up to but not including 415 MPa (60 ksi)
- 80 MPa (12 ksi) for minimums from 415 MPa (60 ksi) up to but not including 460 MPa (67 ksi)
- 100 MPa (15 ksi) for minimums from 460 to 550 MPa (67 to 80 ksi)

Specification ASTM A 663 defines the requirements for hot-wrought merchant quality carbon steel bars and bar-size shapes intended for noncritical constructional applications.

Sizes. Merchant quality steel rounds are not produced in diameters greater than 76 mm (3 in.).

244 / Carbon and Low-Alloy Steels

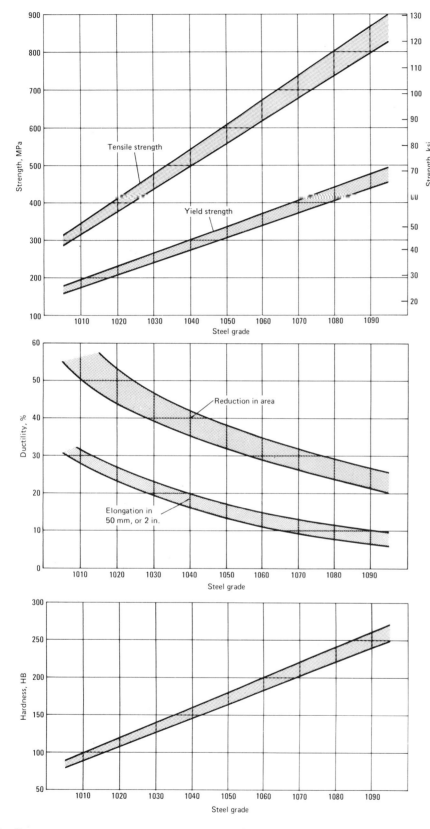

Fig. 3 Estimated minimum tensile properties of selected hot-rolled carbon steel bars

Table 3 Typical ASTM specifications for structural quality steel bars and steel structural shapes
Covered in ASTM A 6

Specification	Steel type and condition
Carbon steels	
A 36(a)(b)................	Carbon steel plates, bars, and shapes
A 131(c)..................	Carbon and HSLA steel plates, bars, shapes, and rivets for ships
A 529	Carbon steel plates, bars, shapes, and sheet piling with minimum yield strength of 290 MPa (42 ksi)
A 709	Carbon, alloy, and HSLA steel plates, bars, and shapes for bridges
Alloy steel	
A 710	Age-hardening low-carbon Ni-Cu-Cr-Mo-Nb and Ni-Cu-Nb alloy steel plates, bars, and shapes
High-strength low-alloy (HSLA) steels	
A 131(c)..................	See above under Carbon Steel
A 242	HSLA steel plates, bars, and shapes
A 572	Nb-V HSLA steel plates, bars, shapes, and sheet piling
A 588	HSLA steel plates, bars, and shapes with minimum yield point of 345 MPa (50 ksi)
A 633	Normalized HSLA steel plates, bars, and shapes
A 690	HSLA steel H-piles and sheet piling for use in marine environments

(a) This ASTM specification is also published by the American Society of Mechanical Engineers, which adds an S in front of the A. (b) See also Canadian Standards Association (CSA) specification G40.8. (c) See also Section 39 of the ABS specifications.

Special Quality Bars

Special quality bars are employed when end use, method of fabrication, or subsequent processing treatment requires characteristics not available in merchant quality bars. Typical applications, including many structural uses, require hot forging, heat treating, cold drawing, cold forming, and machining.

Special quality bars are required to be free from visible pipe and excessive chemical segregation. Also, they are rolled from billets that have been inspected and conditioned, as necessary, to minimize surface imperfections. Frequency and degree of surface imperfections are influenced by chemical composition, type of steel, and bar size. Resulfurized grades, certain low-carbon killed steels, and boron-treated steels are most susceptible to surface imperfections.

Some end uses or fabricating procedures can necessitate one or more extra requirements. These requirements include special hardenability, internal soundness, nonmetallic inclusion rating, and surface condition and are described in the AISI manual covering hot-rolled bars. The quality descriptor

Hot-Rolled Steel Bars and Shapes / 245

Table 4 Typical ASTM specifications for hot-rolled steel bars
See Table 3 for ASTM specifications for structural quality bars and structural shapes.

Specification	Steel type and condition
Carbon steels	
A 321(a)	Quenched and tempered carbon steel bars
A 575(a)	Merchant quality carbon steel bars
A 576(a)	Special quality carbon steel bars
A 663(a)	Merchant quality carbon steel bars subject to mechanical property requirements
A 675(a)	Special quality carbon steel bars subject to mechanical property requirements
Alloy steels	
A 295	Bearing quality high-carbon chromium steel billets, forgings, tube rounds, bars, rods, and tubes
A 304(a)	Alloy steel bars subject to end-quench hardenability requirements
A 322(a)	Alloy steel bars for regular constructional applications
A 434(a)	Quenched and tempered alloy steel bars, hot rolled or cold finished
A 485	Bearing quality high-carbon chromium steel billets, tube rounds, bars, and tubes modified for high hardenability
A 534	Carburizing alloy steel billets, tube rounds, bars, rods, wire, and tubes of bearing quality
A 535	Special quality alloy steel billets, bars, tube rounds, rods, and tubes for the manufacture of antifriction bearings

(a) Covered in ASTM A 29

for bars to which only one of these special requirements is applied is Restrictive Requirement Quality A. When a single special restriction other than the four mentioned above is applied, the quality descriptor is Restrictive Requirement Quality B. Multiple Restrictive Requirement Quality bars are those to which two or more restrictive requirements are applied.

Special quality steel bars can be produced using rimmed, capped, semikilled, or killed deoxidation practice. The appropriate type is dependent on chemical composition, quality, and customer specifications. Killed steels can be produced to coarse or fine austenitic grain size.

Special quality steel bars are produced to product chemical composition tolerances and can be purchased on the basis of heat composition. Special quality steel bars can also be produced to meet mechanical property requirements. The tensile strength ranges are identical to those presented in the section "Merchant Quality Bars" in this article. Additional information on mechanical property requirements and test frequencies is available in the appropriate ASTM specifications.

Sizes. Special quality steel bars are commonly produced in the following sizes:

- *Rounds*: 6.4 to 254 mm (¼ to 10 in.)
- *Squares*: 6.4 to 154 mm (¼ to 6¹/₁₆ in.)
- *Round-cornered squares*: 9.5 to 203 mm (⅜ to 8 in.)
- *Hexagons*: 9.5 to 103 mm (⅜ to 4¹/₁₆ in.)
- *Flats*: greater than 5.16 mm (0.203 in.) in thickness and 152 mm (6 in.) and less in width, or 5.84 mm (0.230 in.) and greater in thickness and 203 mm (8 in.) and less in width

Common size ranges have not been established for special quality bars of other shapes, including bar-size shapes, ovals, half-ovals, half-rounds, octagons, and special bar-size shapes.

Carbon Steel Bars for Specific Applications

Cold-working quality is the descriptor (replacing the older terminology of scrapless nut, cold forging, cold heading, and cold extrusion qualities) for hot-rolled bars used in the production of solid or hollow shapes by means of severe cold plastic deformation, such as (but not limited to) upsetting, heading, forging, and forward or backward extrusion involving movement of metal by expansion and/or compression. Such processing normally involves special inspection standards and requires sound steel of special surface quality and uniform chemical composition. If steel of the type or chemical composition specified does not have adequate cold-forming characteristics in the as-rolled condition, a suitable heat treatment, such as annealing or spheroidize annealing, may be necessary.

Axle Shaft Quality. Bars of axle shaft quality are produced for the manufacture of power-driven axle shafts for cars, trucks, and other vehicles. Because of their design or method of manufacture, these axles either are not machined all over or undergo less than the recommended amount of stock removal for proper cleanup of normal surface imperfections. Therefore, it is necessary to minimize the presence of injurious surface imperfections in bars of axle shaft quality through the use of special rolling practices, special billet and bar conditioning, and selective inspection.

Cold-Shearing Quality. There are limits to the sizes of hot-rolled steel bars that can normally be cold sheared without specially controlled production procedures. When the cold shearing of larger bars is desirable, it is recommended that cold-shearing quality bars be ordered. Bars of this quality have characteristics that prevent cracking even in these larger sizes. Cold-shearing quality bars are not produced to specific requirements such as

Table 5 AISI-SAE grades of hot-rolled alloy steel bars in ASTM specifications

ASTM specification	AISI-SAE grades
A 295	52100, 51100, 50100
A 304	All H grades except 4626H and 86B30H
A 322	All standard grades except 4032, 4042, 4135, 4422, 4427, 4617, 50B40, 5046, 5060, 5115, 5117, 50100, 8115, 86B45, 8650, 8660, 9310, and 94B15
A 434	By agreement
A 534	4023, 4118, 4320, 4620, 4720, 5120, 8620, E-3310, E-9310
A 535	3310, 4320, 4620, 4720, 4820, 52100, 52100 Mod. 1, 52100 Mod. 2, 52100 Mod. 3, 52100 Mod. 4, 8620, 9310

hardness, microstructure, shear life, or productivity. Maximum size (cross-sectional area) limitations for the cold shearing of hot-rolled steel bars without the specially controlled production procedures, and of cold-shearing quality bars, are given in the AISI manual that covers hot-rolled bars. If even larger bars are to be cold sheared, cold-shearing behavior can be further improved by suitable prior heat treatment.

Structural quality is the descriptor for hot-rolled bars used in the construction of bridges and buildings by riveting, bolting, or welding and for general structural purposes. The general requirements for bars of this quality are given in ASTM A 6; individual ASTM specifications are listed in Table 3.

Additional qualities of carbon steel bars are available for specific requirements. Such qualities are related to application and processing. They include:

- File quality
- Gun barrel quality
- Gun receiver quality
- Shell steel quality A
- Shell steel quality B
- Shell steel quality C
- Shell steel quality D
- Standard tube round quality

Alloy Steel Bars

Hot-rolled alloy steel bars are commonly produced in the same size as special quality steel bars. Common size ranges have not been established for other shapes of hot-rolled alloy steel bar, including bar-size shapes, ovals, half-ovals, half-rounds, octagons, and special bar-size shapes.

Hot-rolled alloy steel bars are covered by several ASTM specifications (Tables 3 and 4). Many of the alloys covered in these specifications are standard AISI-SAE grades (Table 5).

Hot-rolled alloy steel bars are also covered by several quality descriptors, which are discussed below. As with all quality descriptors, these descriptors differentiate bars on the basis of characteristic properties required to meet the particular conditions encountered during fabrication or use.

Regular quality is the basic or standard quality for hot-rolled alloy steel bars, such as those covered by ASTM A 322. Steel for this quality are killed, are usually produced to fine grain size, and are melted to chemical composition limits. Bars of this quality are inspected, conditioned, and tested to meet the normal requirements for regular construction applications for which alloy steel is used.

Axle Shaft Quality. Alloy steel bars of axle shaft quality are similar to carbon steel bars of the same quality (see the discussion of axle shaft quality bars in the section "Carbon Steel Bars for Specific Applications" in this article).

Ball and roller bearing quality and bearing quality apply to alloy steel bars intended for antifriction bearings. These bars are usually made from steels of the AISI-SAE standard alloy carburizing grades and the AISI-SAE high-carbon chromium series. These steels can be produced in accordance with ASTM A 534, A 535, A 295, or A 485 (Table 4). Bearing quality steel bars require restricted melting and special teeming, heating, rolling, cooling, and conditioning practices to meet rigid quality standards. Steelmaking practices may include vacuum treatment. The foregoing requirements include thorough examination for internal imperfections by one or more of the following methods: macroetch testing, microscopic examination for nonmetallic inclusions, ultrasonic inspection, and fracture testing.

It is not practical to furnish bearing quality steel bars in sizes exceeding 64 500 mm^2 (100 in.^{2}) in cross-sectional area to the same rigid requirements as those for bars in smaller sizes because of insufficient hot working in the larger bars. Usually, bars over 102 mm (4 in.) in thickness are forged to 102 mm (4 in.) square or smaller for testing.

Cold-Shearing Quality. Alloy steel bars of cold-shearing quality are similar to carbon steel bars of the same quality (see the discussion of cold-shearing quality bars in the section "Carbon Steel Bars for Specific Applications" in this article).

Cold-working quality, which replaces the older terminologies cold-heading quality and special cold-heading quality, is the descriptor for hot-rolled bars used in the production of solid or hollow shapes by means of severe cold plastic deformation, such as (but not limited to) upsetting, heading, forging, and forward or backward extrusion involving movement of metal by expansion and/or compression. Such processing normally involves special inspection standards and requires sound steel of special surface quality and uniform chemical composition. If steel of the type or chemical composition specified does not have adequate cold-forming characteristics in the as-rolled condition, a suitable heat treatment, such as annealing or spheroidize annealing, may be necessary.

Aircraft quality and magnaflux quality are the descriptors used for alloy steel bars for critical or highly stressed parts of aircraft and for other similar or corresponding purposes involving additional stringent requirements such as magnetic particle inspection, additional discard, macroetch tests, and hardenability control. To meet these requirements, exacting steelmaking, rolling, and testing practices must be employed. These practices are designed to minimize detrimental inclusions and porosity. Phosphorus and sulfur are usually limited to 0.025% maximum each.

Many parts for aircraft, missiles, and rockets require aircraft quality alloy steel bars. Magnetic particle testing as in AMS 2301 is sometimes specified for such applications. Some very critical aircraft, missile, and rocket applications require alloy steel bars of a quality attained only by vacuum melting or by an equivalent process. The requirements of AMS 2300 are sometimes specified for such applications.

Aircraft quality alloy steel bars are ordinarily made to Aerospace Material Specifications published by the Society of Automotive Engineers. Typical examples of parts for aircraft engines and airframes made from bars covered by AMS specifications are given in Table 6.

Structural quality is the descriptor for hot-rolled bars used in the construction of bridges and buildings by riveting, bolting, or welding and for general structural purposes. The general requirements for bars of this quality are given in ASTM A 6; the only individual ASTM specification referred to in A 6 that pertains to alloy steel bars is A 710.

Additional Qualities. The quality designations shown below apply to alloy steel bars intended for rifles, guns, shell, shot, and similar applications. They may involve requirements for amount of discard, macroetch testing, surface quality, or magnetic particle testing, as indicated in the product specification:

- AP shot quality
- AP shot magnaflux quality
- Gun quality
- Rifle barrel quality
- Shell quality
- Shell magnaflux quality

High-Strength Low-Alloy Steel Bars

In addition to the carbon steel and alloy steel bars of structural quality discussed in preceding sections of this article, ASTM A 6 also lists several specifications covering high-strength low-alloy (HSLA) steel bars of structural quality (Table 3). High-strength low-alloy steel bars are also covered in SAE J 1442.

Bars of these steels offer higher strength than that of carbon steel bars and are frequently selected for applications in which weight saving is important. They also offer increased durability, and many offer increased resistance to atmospheric corrosion. Additional information on HSLA steels is available in the articles "High-Strength Structural and High-Strength Low-Alloy Steels," "High-Strength Low-Alloy Steel Forgings" and "Bulk Formability of Steels" in this Volume.

Microalloyed steel bars constitute a class of special quality carbon steels to which small amounts of alloying elements such as vanadium, niobium, or titanium have been added. Microalloyed steels in the as-hot-rolled condition are capable of developing strengths higher than those of the base carbon grades through precipitation hardening. In some cases, strength properties comparable to those of the quenched and tempered base grade can be attained. These steels are finding increased application in shafting and automotive forgings.

Concrete-Reinforcing Bars

Concrete-reinforcing bars are available as either plain rounds or deformed rounds. Deformed reinforcing bars are used almost exclusively in the construction industry to furnish tensile strength to concrete structures. The surface of the deformed bar is provided with lugs, or protrusions, which inhibit longitudinal movement relative to the surrounding concrete. The lugs are hot

Table 6 Specifications and grades of alloy steel bars for aircraft parts

Part	AMS specification	AISI-SAE grade or approximate grade
Aileron, rudder, and elevator hinge pins	6415	E4340
Airframe parts (tubing, fittings, and braces)	6370	4130
	6280	8630
	6382	4140
	6322	8740
	6415	E4340
Bearings	6440	E52100
Bolts, studs, and nuts	6322	8740
Connecting rods	6415	E4340
Crankcases	6342	9840
	6382	4140
	6322	8740
Crankshafts	6415	E4340
Gears and shafts	6415	E4340
	6448	6150
	6274	8620
Landing gears	6322	8740
	6382	4140
	6415	E4340
Propellers, spiders, hubs, and barrels	6415	E4340
Springs	6450	6150

formed in the final roll pass by passing the bars between rolls into which patterns have been cut. Plain reinforcing bars are used more often for dowels, spirals, structural ties, and supports than as substitutes for deformed bars. Concrete-reinforcing bars are supplied either straight and cut to proper length, or bent or curved in accordance with plans and specifications.

Grades. Deformed and plain concrete-reinforcing bars rolled from billet steel are produced to the requirements of ASTM A 615 or A 706. For special applications that require deformed bars with a combination of strength, weldability, ductility, and improved bending properties, ASTM A 706 is specified, which is an HSLA steel. Deformed and plain concrete-reinforcing bars are also available rolled from railroad rails (ASTM A 616) and from axles for railroad cars (ASTM A 617), Specification ASTM A 722 covers deformed and plain uncoated high-strength steel bars for prestressing concrete structures.

Sizes. Numbers indicating sizes of reinforcing bars correspond to nominal bar diameter in eighths of an inch for sizes 3 through 8; this relationship is approximate for sizes 9, 10, 11, 14, and 18. The nominal values of bar diameter, cross-sectional area, and weight per unit length corresponding to these size numbers are given in Table 7. The nominal cross-sectional area and the nominal diameter of a deformed bar are the same as those of a plain bar of equal weight per foot.

Structural Shapes

Structural shapes, as stated previously, are flanged shapes 76 mm (3 in.) and greater in at least one cross-sectional dimension (smaller shapes are referred to as bar-size shapes) and are used in the construction of structures such as bridges, buildings, ships, and railroad cars. Included in this product category are regular structural shapes (see ASTM A 6), such as standard beams, wide-flange beams, columns, light beams, joists, stanchions and bearing piles, and certain tees, along with special structural shapes, which are those designed for specialized applications and that have dimensions and/or values of weight per foot that do not

Table 7 Dimensions of deformed and plain concrete-reinforcing bars of standard sizes

Bar size	Nominal diameter mm	in.	Cross-sectional area mm^2	in.^{2}	Nominal weight kg/m	lb/ft
3	9.52	0.375	71	0.11	0.560	0.376
4	12.70	0.500	129	0.20	0.994	0.668
5	15.88	0.625	200	0.31	1.552	1.043
6	19.05	0.750	284	0.44	2.235	1.502
7	22.22	0.875	387	0.60	3.042	2.044
8	25.40	1.000	510	0.79	3.973	2.670
9	28.65	1.128	645	1.00	5.059	3.400
10	32.26	1.270	819	1.27	6.403	4.303
11	35.81	1.410	1006	1.56	7.906	5.313
14	43.00	1.693	1452	2.25	11.384	7.65
18	57.33	2.257	2581	4.00	20.238	13.60

conform to regular shapes. Bar-size structural shapes (angles, channels, tees, and zees with greatest cross-sectional dimension less than 76 mm, or 3 in.) are considered to be in the merchant quality bar category rather than the structural shape category.

The common method of designating sizes of structural shapes is as follows:

- *Beams and channels*: By depth of cross section and weight per foot
- *Angles*: By length of legs and thickness in fractions of an inch or, more commonly, by length of legs and weight per foot. The longer leg of an unequal angle is commonly stated first
- *Tees*: By width of flange, overall depth of stem, and weight per foot, in that order
- *Zees*: By depth, width of flanges, and thickness in fractions of an inch or by depth, flange width, and weight per foot
- *Wide-flange shapes*: By depth, width across flange, and weight per foot, in that order

Most structural shapes are produced to meet specific standard specifications, such as those listed in Table 3. Structural shapes are generally furnished to chemical composition limits and mechanical property requirements.

Special requirements are sometimes specified for structural shapes to adapt them to conditions they will encounter during fabrication or service. These requirements may include specific deoxidation practices, additional mechanical tests, or nondestructive testing.

Special Shapes

Special shapes are hot-rolled steel shapes made with cross-sectional configurations uniquely suited to specific applications. Examples of custom-designed shapes are track shoes for tractors or tanks and sign-post standards.

The only type of standard shape in high production that falls in this classification is rail. Railroad rails of the standard American tee rail shape are produced from carbon steel to the dimensional, chemical, and other requirements of the American Railway Engineering Association (AREA). The sizes of railroad rails are designated in pounds per yard of length; rails are furnished in 40 to 64 kg (90 to 140 lb) sizes. The most common sizes are 52, 60, 62, and 64 kg (115, 132, 136, and 140 lb). The ordinary length of railroad rails is 12 m (39 ft). Carbon steel tee rails for railway track are covered by ASTM A 1; rail-joint bars and tie plates are covered in ASTM A 3, A 4, A 5, A 49, A 67, and A 241.

Light rails are available for light duty, such as in mines and amusement park rides, in sizes from 6.8 to 39 kg (15 to 85 lb). Light rails are covered by specifications of the American Society of Civil Engineers (ASCE).

Crane rails generally have heavier heads and webs than those of railroad rails in order to withstand the heavy loads imposed on them in service. Crane rails in sizes from 18 to 79 kg (40 to 175 lb) are furnished to ASCE, ASTM, and producers' specifications.

REFERENCES

1. *Alloy, Carbon and High Strength Low Alloy Steels: Semifinished for Forging; Hot Rolled Bars, Cold Finished Bars; Hot Rolled Deformed and Plain Concrete Reinforcing Bars*, AISI Steel Products Manual, American Iron and Steel Institute, 1986
2. *Materials*, Vol 1, *1989 SAE Handbook*, Society of Automotive Engineers, 1989

Cold-Finished Steel Bars

Revised by the ASM Committee on Cold-Finished Steel Bars*

COLD-FINISHED STEEL BARS are carbon and alloy steel bar products (round, square, hexagonal, flat, or special shapes) that are produced by cold finishing previous hot-wrought bars by means of cold drawing, cold forming, turning, grinding, or polishing (singly or in combination) to yield straight lengths or coils that are uniform throughout their length. Not covered in this article are flat-rolled products such as sheet, strip, or plate, which are normally cold finished by cold rolling, or cold-drawn tubular products.

Cold-finished bars fall into five classifications:

- Cold-drawn bars
- Turned and polished (after cold draw or hot roll) bars
- Cold-drawn, ground, and polished (after cold draw) bars
- Turned, ground, and polished bars
- Cold-drawn, turned, ground, and polished bars

Cold-drawn bars represent the largest tonnage production and are widely used in the mass production of machined and other parts. They have attractive combinations of mechanical and dimensional properties.

Turned and polished bars have the mechanical properties of hot-rolled products but have greatly improved surface finish and dimensional accuracy. These bars are available in sizes larger than those that can be cold drawn. Turned bars are defect and decarb free.

Cold-drawn, ground, and polished bars have the increased machinability, tensile strength, and yield strength of cold-drawn bars together with very close size tolerances. However, cold-drawn, ground, and polished bars are not guaranteed to be defect free.

Turned, ground, and polished bars have superior surface finish, dimensional accuracy, and straightness. These bars find application in precision shafting and in plating, where such factors are of primary importance.

Cold-drawn, turned, ground, and polished bars have improved mechanical properties, close size tolerances, and a surface free of imperfections.

Bar Sizes

Cold-finished steel bars are available in a wide variety of sizes and cross-sectional shapes. Normally, they are furnished in straight lengths, but in some sizes and cross sections they may be furnished in coils. Cold-finished steel bars are available with nominal dimensions designated in either inches or millimeters. Cold-finished product is available in standard size increments, which vary by size range. Special sizes can be negotiated depending on hot mill increments and cold-finish tooling. The sizes in which they are commonly available in bar and coil form are given in Table 1.

Product Types

In the manufacture of cold-finished bars, the steel is first hot rolled oversize to appropriate shape and is then subjected to mechanical operations (other than those intended primarily for scale removal) that affect its machinability, straightness, and end-cut properties. The two common methods of cold finishing bars are:

- Removal of surface material by turning or grinding, singly or in combination
- Drawing the material through a die of suitable configuration

Pickling or blasting to remove scale may precede turning or grinding and must always precede drawing. For bar products, cold rolling has been almost superseded by cold drawing. Nevertheless, cold-finished bars and special shapes are sometimes incorrectly described as cold rolled.

Commercial Grades. Any grade of carbon or alloy steel that can be hot rolled can also be cold finished. The choice of grade is

Table 1   Common commercially available sizes of cold-finished steel bars and coils

	Bars(a)								Coils(b), sizes	
	Minimum thickness or diameter		Maximum thickness or diameter		Size increments		Normal length			
Configuration	mm	in.	mm	in.	mm	in.	m	ft	mm	in.
Round	3.2	0.125	305	12	0.8–25 1.6–75 3.2–152	32nds to 1 in., 16ths to 3 in., 8ths to 6 in.	3.0–3.7 or 6.1–7.3	10–12 or 20–24	≤25	≤1
Square	3.2	0.125	152	6	1.6–38 3.2–70	16ths to 1½ in., 8ths to 2¾ in.	3.0–3.7	10–12	≤16	≤⅝
Hexagonal	3.2	0.125	102	4	1.6–50 6.4–102	16ths to 2 in., 4ths to 4 in.	3.0–3.7	10–12	≤16	≤⅝
Flat	3.2 thick × 6.4 wide	0.125 thick × 0.25 wide	76 × 371	3 thick × 14⅝ wide	1.6–17 3.2–44 6.4–76	16ths to ¹¹⁄₁₆ in., 8ths to 1¾ in., 4ths to 3 in.	3.0–3.7	10–12	≤14.3 × 15.9(c)	≤⁹⁄₁₆ × ⅝(c)

(a) Ref 1. (b) Ref 2. (c) Or other sections having cross-sectional areas ≤194 mm² (≤0.30 in.²)

*K.M. Shupe, Bliss & Laughlin Steel Company; Richard B. Smith, Stanadyne Western Steel; Steve Slavonic, Teledyne Columbia-Summerill; B.F. Leighton, Canadian Drawn Steel Company; W. Gismondi, Union Drawn Steel Company, Ltd.; John R. Stubbles, LTV Steel Company; Kurt W. Boehm, Nucor Steel; Donald M. Keane, LaSalle Steel Company

Table 2 Size tolerances for cold-finished carbon steel bars, cold drawn or turned and polished

This table includes tolerances for bars that have been annealed, spheroidize annealed, normalized, normalized and tempered, or quenched and tempered before cold finishing. This table does not include tolerances for bars that are annealed, spheroidize annealed, normalized, normalized and tempered, or quenched and tempered after cold finishing; the producer should be consulted for tolerances for such bars.

Size		Size tolerance — Maximum carbon (C) range, %									
		C ≤ 0.28		0.28 < C ≤ 0.55		C ≤ 0.55 including stress relief or annealed after cold finishing		C > 0.55		All grades quenched and tempered or normalized before cold finishing	
mm	in.	mm	in.	mm	in.	mm	in.	mm	in.	mm	in.
Rounds—cold drawn (to 102 mm, or 4 in., in size) or turned and polished											
To 38 inclusive	To 1½ inclusive	−0.05	−0.002	−0.08	−0.003	−0.10	−0.004	−0.13	−0.005	−0.13	−0.005
>38–64 inclusive	>1½–2½ inclusive	−0.08	−0.003	−0.10	−0.004	−0.13	−0.005	−0.15	−0.006	−0.15	−0.006
>64–102 inclusive	>2½–4 inclusive	−0.10	−0.004	−0.13	−0.005	−0.15	−0.006	−0.18	−0.007	−0.18	−0.007
>102–152 inclusive	>4–6 inclusive	−0.13	−0.005	−0.15	−0.006	−0.18	−0.007	−0.20	−0.008	−0.20	−0.008
>152–203 inclusive	>6–8 inclusive	−0.15	−0.006	−0.18	−0.007	−0.20	−0.008	−0.23	−0.009	−0.23	−0.009
>203–229 inclusive	>8–9 inclusive	−0.18	−0.007	−0.20	−0.008	−0.23	−0.009	−0.25	−0.010	−0.25	−0.010
Hexagons—cold drawn											
To 19 inclusive	To ¾ inclusive	−0.05	−0.002	−0.08	−0.003	−0.10	−0.004	−0.15	−0.006	−0.15	−0.006
>19–38 inclusive	>¾–1½ inclusive	−0.08	−0.003	−0.10	−0.004	−0.13	−0.005	−0.18	−0.007	−0.18	−0.007
>38–64 inclusive	>1½–2½ inclusive	−0.10	−0.004	−0.13	−0.005	−0.15	−0.006	−0.20	−0.008	−0.20	−0.008
>64–80 inclusive	>2½–3⅛ inclusive	−0.13	−0.005	−0.15	−0.006	−0.18	−0.007	−0.23	−0.009	−0.23	−0.009
>80–102 inclusive	>3⅛–4 inclusive	−0.13	−0.005	−0.15	−0.006	...	...	...	...	...	...
Squares—cold drawn(a)											
To 19 inclusive	To ¾ inclusive	−0.05	−0.002	−0.10	−0.004	−0.13	−0.005	−0.18	−0.007	−0.18	−0.007
>19–38 inclusive	>¾–1½ inclusive	−0.08	−0.003	−0.13	−0.005	−0.15	−0.006	−0.20	−0.008	−0.20	−0.008
>38–64 inclusive	>1½–2½ inclusive	−0.10	−0.004	−0.15	−0.006	−0.18	−0.007	−0.23	−0.009	−0.23	−0.009
>64–102 inclusive	>2½–4 inclusive	−0.15	−0.006	−0.20	−0.008	−0.23	−0.009	−0.28	−0.011	−0.28	−0.011
>102–127 inclusive	>4–5 inclusive	−0.25	−0.010	...	...	...	...	...	...	...	...
>127–152 inclusive	>5–6 inclusive	−0.36	−0.014	...	...	...	...	...	...	...	...
Flats—cold drawn(a)(b)											
To 19 inclusive	To ¾ inclusive	−0.08	−0.003	−0.10	−0.004	−0.15	−0.006	−0.20	−0.008	−0.20	−0.008
>19–38 inclusive	>¾–1½ inclusive	−0.10	−0.004	−0.13	−0.005	−0.20	−0.008	−0.25	−0.010	0.25	−0.010
>38–75 inclusive	>1½–3 inclusive	−0.13	−0.005	−0.15	−0.006	−0.25	−0.010	−0.30	−0.012	−0.30	−0.012
>75–102 inclusive	>3–4 inclusive	−0.15	−0.006	−0.20	−0.008	−0.28	−0.011	−0.40	−0.016	−0.40	−0.016
>102–152 inclusive	>4–6 inclusive	−0.20	−0.008	−0.25	−0.010	−0.30	−0.012	−0.50	−0.020	−0.50	−0.020
>152	>6	−0.33	−0.013	−0.38	−0.015	...	...	...	...	...	...

(a) Tolerances can be ordered all plus, or distributed plus and minus with the sum equivalent to the tolerances listed. (b) Width governs the tolerance for both width and thickness of flats, for example, when the maximum of carbon range is 0.28% or less for a flat 50 mm (2 in.) wide and 25 mm (1 in.) thick. The width tolerance is 0.13 mm (0.005 in.), and the thickness is the same, nearly 0.13 mm (0.005 in.).
Source: Ref 4

based on the attainable cold-finished and/or hardenability and tempering characteristics necessary to obtain the required mechanical properties.

Production methods vary widely among cold-finished cold-drawn suppliers. For example, one supplier currently anneals and cold draws grades 1070, 1090, and 5160, and in the future plans to do the same with grade 9254. Grade 1070 is a high-volume item, and cold drawing is required for precision sizing and subsequent nondestructive testing of the bar, using a rotating-probe eddy current device (see the articles "Eddy Current Inspection," "Remote-Field Eddy Current Inspection," and "Steel Bar, Wire, and Billets" in Volume 17 of the 9th Edition of *Metals Handbook*) for detecting surface seams. Cold drawing is also necessary because the smallest hot-rolled size typically available for some applications is not small enough for customer use. Thus, a supplier whose smallest hot-rolled bar size is 11.1 mm (0.437 in.) cold draws this diameter to as small as 9.98 mm (0.393 in.).

Carbon steels containing more than 0.55% C must be annealed prior to being cold drawn so that the hardness will be sufficiently low to facilitate the cold-drawing operation. For carbon steels containing up to 0.65% C, this will normally be a lamellar pearlitic anneal; for carbon steels containing more than 0.65% C, a spheroidize anneal is required. The type of structure required is normally reached by agreement between the steel producer and the customer.

Alloy steels containing more than 0.38% C are usually annealed before cold drawing.

Machined Bars. Bar products that are cold finished by stock removal can be:

- Turned and polished
- Turned, ground, and polished
- Cold drawn, ground, and polished
- Cold drawn, turned, and polished
- Cold drawn, turned, ground, and polished

Turning is done in special machines with cutting tools mounted in rotating heads, thus eliminating the problem of having to support long bars as in a lathe. Grinding is done in centerless machines. Polishing can be done in a roll straightener of the crossed-axis (Medart) type with polished rolls to provide a smooth finish. Polishing by grinding with an organic wheel or with a belt is of increasing interest (see the article "Grinding Equipment and Processes" in Volume 16 of the 9th Edition of *Metals Handbook*) because it is cost effective to grind and polish the bars on the same machine simply by using grinding wheels or belts of different grit size. Grinding produces a smoother finish than turning; polishing improves the surface produced by either technique. Turned, ground, and polished rounds represent the highest degree of overall accuracy, concentricity, straightness, and surface perfection attainable in commercial practice (Ref 3).

The surface finish desired is specified by using the process names given above because the industry has not developed standard numerical values for roughness, such as microinch or root mean square (rms) numbers. However, surface finish with respect to rms (root mean square deviation from the mean surface) as determined with a profilometer can be negotiated between the producer and a customer. This could be done for such critical-finish applications as turned and polished bars used to produce shafting as well as stock used to produce

Table 3 Size tolerances for cold-finished alloy steel bars, cold drawn or turned and polished
This table includes tolerances for bars that have been annealed, spheroidize annealed, normalized, normalized and tempered, or quenched and tempered before cold finishing. This table does not include tolerances for bars that are annealed, spheroidize annealed, normalized, normalized and tempered, or quenched and tempered after cold finishing; the producer should be consulted for tolerances for such bars.

Size		Size tolerance — Maximum carbon (C) range, %								
		C ≤ 0.28		0.28 < C ≤ 0.55		C ≤ 0.55 including stress relief or annealed after cold finishing		C > 0.55 with or without stress relieving or annealing after cold finishing		All carbons quenched and tempered (heat treated) or normalized and tempered before cold finishing
mm	in.	mm	in.	mm	in.	mm	in.	mm	in.	mm	in.
Rounds—cold drawn (to 102 mm, or 4 in., in size) or turned and polished											
In coils: To 25 inclusive	To 1 inclusive	0.05	0.002	0.08	0.003	0.10	0.004	0.13	0.005	0.13	0.005
Cut lengths: To 38 inclusive	To 1½ inclusive	0.08	0.003	0.10	0.004	0.13	0.005	0.15	0.006	0.15	0.006
>38–64 inclusive	>1½–2½ inclusive	0.10	0.004	0.13	0.005	0.15	0.006	0.18	0.007	0.18	0.007
>64–102 inclusive	>2½–4 inclusive	0.13	0.005	0.15	0.006	0.18	0.007	0.20	0.008	0.20	0.008
>102–152 inclusive	>4–6 inclusive	0.15	0.006	0.18	0.007	0.20	0.008	0.23	0.009	0.23	0.009
>152–203 inclusive	>6–8 inclusive	0.18	0.007	0.20	0.008	0.23	0.009	0.25	0.010	0.25	0.010
>203–229 inclusive	>8–9 inclusive	0.20	0.008	0.23	0.009	0.25	0.010	0.28	0.011	0.28	0.011
Hexagons—cold drawn											
To 19 inclusive	To ¾ inclusive	0.08	0.003	0.10	0.004	0.13	0.005	0.18	0.007	0.18	0.007
>19–38 inclusive	>¾–1½ inclusive	0.10	0.004	0.13	0.005	0.15	0.006	0.20	0.008	0.20	0.008
>38–64 inclusive	>1½–2½ inclusive	0.13	0.005	0.15	0.006	0.18	0.007	0.23	0.009	0.23	0.009
>64–79 inclusive	>2½–3⅛ inclusive	0.15	0.006	0.18	0.007	0.20	0.008	0.25	0.010	0.25	0.010
>79–102 inclusive	>3⅛–4 inclusive	0.15	0.006	...	...	...	...	...	...	...	...
Squares—cold drawn											
To 19 inclusive	To ¾ inclusive	0.08	0.003	0.13	0.005	0.15	0.006	0.20	0.008	0.20	0.008
>19–38 inclusive	>¾–1½ inclusive	0.10	0.004	0.15	0.006	0.18	0.007	0.23	0.009	0.23	0.009
>38–64 inclusive	>1½–2½ inclusive	0.13	0.005	0.18	0.007	0.20	0.008	0.25	0.010	0.25	0.010
>64–102 inclusive	>2½–4 inclusive	0.18	0.007	0.23	0.009	0.25	0.010	0.30	0.012	0.30	0.012
>102–127 inclusive	>4–5 inclusive	0.28	0.011	0.23	...	0.25	...	0.30	...	0.30	...
Flats—cold drawn(a)											
To 19 inclusive	To ¾ inclusive	0.10	0.004	0.13	0.005	0.18	0.007	0.23	0.009	0.23	0.009
>19–38 inclusive	>¾–1½ inclusive	0.13	0.005	0.15	0.006	0.23	0.009	0.28	0.011	0.28	0.011
>38–76 inclusive	>1½–3 inclusive	0.15	0.006	0.18	0.007	0.28	0.011	0.33	0.013	0.33	0.013
>76–102 inclusive	>3–4 inclusive	0.18	0.007	0.23	0.009	0.30	0.012	0.43	0.017	0.43	0.017
>102–152 inclusive	>4–6 inclusive	0.23	0.009	0.28	0.011	0.33	0.013	0.52	0.021	0.52	0.021
>152	>6	0.36	0.014	...	...	...	...	...	...	...	...

(a) Width governs the tolerance for both width and thickness of flats, for example, when the maximum of carbon range is 0.28% or less for a flat 50 mm (2 in.) wide and 25 mm (1 in.) thick. The width tolerance is 0.13 mm (0.005 in.), and the thickness is the same, nearly 0.13 mm (0.005 in.).
Source: Ref 4

machined parts for which a superior finish is required on surfaces not machined.

The published range of diameters both for turned and for turned and ground bars is 13 to 229 mm (½ to 9 in.) inclusive; for cold-drawn and ground bars, it is 3.2 to 102 mm (⅛ to 4 in.) inclusive. These are composites of size ranges throughout the industry; an individual producer may be unable to furnish a full range of sizes.

For example, one well-known producer supplies turned rounds from 13 to 229 mm (½ to 9 in.), another from 29 to 203 mm (1⅛ to 8 in.)—all finished sizes. Yet another producer supplies sizes up to and including 152 mm (6 in.) that are turned on special turning machines and ground on centerless grinders; larger sizes are lathe turned and ground on centers. Because turning and grinding do not alter the mechanical properties of the hot-rolled bar, this product can be machined asymmetrically with practically no danger of warpage (Ref 3).

Stock removal is usually dependent on American Iron and Steel Institute (AISI) seam allowances (Ref 2). Stock removal in turning, or turning and grinding, measured on the diameter, is normally 1.6 mm (¹⁄₁₆ in.) for sizes up to 38 mm (1½ in.), 3.2 mm (⅛ in.) for the 38 to 76 mm (1½ to 3 in.) range, 4.8 mm (³⁄₁₆ in.) for the 76 to 127 mm (3 to 5 in.) range, and 6.4 mm (¼ in.) for 127 mm (5 in.) diameter and larger.

Cold-drawn round bars are available in a range of diameters from 3.2 to 152 mm (⅛ to 6 in.). The maximum diameters available from individual producers, however, may vary from 76 to 152 mm (3 to 6 in.). The reduction in diameter in cold drawing, called draft, is commonly 0.79 mm (¹⁄₃₂ in.) for finished sizes up to 9.5 mm (⅜ in.) and 1.6 mm (¹⁄₁₆ in.) for sizes over 9.5 mm (⅜ in.). Some special processes use heavier drafts followed by stress relieving. One producer employs heavy drafting at elevated temperature. With this exception, drawing operations are begun with the material at room temperature to start, and the only elevated temperature involved is that developed in the bar as a result of drawing; this temperature rise is small and of little significance.

Originally, cold finishing, whether by turning or by cold rolling, was employed only for sizing to produce a bar with closer dimensional tolerances and a smoother surface. As cold-finished bar products were developed and improved, increased attention was paid to the substantial enhancement of mechanical properties that could be obtained by cold working. This additional advantage is now more fully appreciated, as evidenced by the fact that increased mechanical properties are an important consideration in about 40% of the applications. In approximately half of these applications, or 20% of the total, cold drawing is used only to increase strength; in the other 20%, close tolerances and better surface finish are desired in addition to increased strength.

As-rolled microalloyed high-strength low-alloy (HSLA) steels or microalloyed HSLA steels in various combinations of controlled drafting and furnace treatment provide an extension of property attainment. A high percentage of free-machining steels are cold drawn for the combination of size accuracy and improved machinability. Recent developments in microalloyed steels provide hot-rolled turned bars, under certain circumstances, having mechanical properties

Table 4 Size tolerances for cold-finished carbon and alloy steel round bars cold drawn, ground, and polished or turned, ground, and polished

Size				Tolerances from specified size	
Cold drawn, ground, and polished		Turned, ground, and polished			
mm	in.	mm	in.	mm	in.
To 38 incl	To 1½ incl	To 38 incl	To 1½ incl	−0.03	−0.001
>38–64 incl	>1½–2½ excl	>38–64 incl	>1½–2½ excl	−0.04	−0.0015
>64–76 incl	≥2½–3 incl	>64–76 incl	≥2½–3 incl	−0.05	−0.002
>76–102 incl	>3–4 incl	>76–102 incl	>3–4 incl	−0.08	−0.003
...	...	>102–152 incl	>4–6 incl	−0.10(a)	−0.004(a)
...	...	>152	>6	−0.13(a)	−0.005(a)

incl, inclusive; excl, exclusive. (a) For nonresulfurized steels (steels specified to maximum sulfur limits under 0.08% or for steels thermally treated, the tolerance is increased by 0.03 mm (0.001 in.). Source: Ref 4

Table 5 Straightness tolerances for cold-finished carbon and alloy steel bars

All grades quenched and tempered or normalized and tempered to ≤HB 302 before cold finishing; all grades stress relieved or annealed after cold finishing. Straightness tolerances are not applicable to bars having Brinell hardness exceeding 302. The tolerances are based on the following method of measuring straightness. Departure from straightness is measured by placing the bar on a level table so that the arc or departure from straightness is horizontal, and the depth of the arc is measured with a feeler gage and a straightedge.

It should be recognized that straightness is a perishable quality and may be altered by mishandling. The preservation of straightness in cold-finished bars requires the utmost care in subsequent handling. Specific straightness tolerances are sometimes required for carbon and alloy steels, in which case the purchaser should inform the manufacturer of the straightness tolerances and the methods to be used in checking the straightness.

Size		Length		Straightness tolerances (maximum deviation) from straightness in any 3 m (10 ft) portion of the bar							
				Carbon range, ≤0.28%				Carbon range >0.28% and all grades thermally treated			
				Rounds		Squares, hexagons, and octagons		Rounds		Squares, hexagons, and octagons	
mm	in.	m	ft	mm	in.	mm	in.	mm	in.	mm	in.
<16	<⅝	<4.6	<15	3.2	⅛	4.8	3/16	4.8	3/16	6.4	¼
<16	<⅝	≥4.6	≥15	3.2	⅛	7.9	5/16	7.9	5/16	9.5	⅜
≥16	≥⅝	<4.6	<15	3.2	⅛	3.2	⅛	3.2	⅛	4.8	3/16
≥16	≥⅝	≥4.6	≥15	3.2	⅛	4.8	3/16	4.8	3/16	6.4	¼

Source: Ref 4

Table 6 Minimum stock removal for cold-finished steel bars subject to magnetic particle inspection

For turned and polished alloy steel bars subject to magnetic particle inspection, the recommended minimum total stock removal from the surface (the amount removed by the producer plus the amount removed by the purchaser) is based on the hot-rolled alloy steel bar size used by the producer.

Cold-finished size		Minimum stock removal from the surface(a)	
mm	in.	mm	in.
To 11.1 incl	To 7/16 incl	0.76	0.030
>11.1–17.5 incl	>7/16–11/16 incl	1.14	0.045
>17.5–23.8 incl	>11/16–15/16 incl	1.52	0.060
>23.8–36.5 incl	>15/16–17/16 incl	1.91	0.075
>36.5–49.2 incl	>17/16–1 15/16 incl	2.29	0.090
>49.2–61.9 incl	>1 15/16–2 7/16 incl	3.18	0.125
>61.9–85.7 incl	>2 7/16–3⅜ incl	3.96	0.156
>85.7–111 incl	>3⅜–4⅜ incl	4.75	0.187

incl, inclusive. (a) For example, the minimum reduction in diameter of rounds is twice the minimum stock removal from the surface. Source: Ref 2

similar to cold-drawn nonmicroalloyed steels.

An appreciable fraction of all applications of cold finishing to carbon steel bars utilizes cold drawing to improve mechanical properties. For alloy steel, however, cold finishing is commonly used to improve surface finish and dimensional accuracy, and not for additional mechanical strength. When additional mechanical strength is desired, alloy steel bars may be heat treated (quenched and tempered) and then cold drawn and stress relieved. Elevated-temperature or warm-drawn steels are also available with increased mechanical strength and improved machinability.

Heavily drafted and strain-tempered carbon and alloy steels subjected to induction hardening of the surface provide many additional property combinations. The extra cost of using alloy steel in cold-finished bars can be justified only when heat treatment (quenching and tempering) is necessary for meeting the required strength level. Because work-hardening effects are removed during heating prior to quenching, the benefit of increased mechanical strength due to cold finishing is eliminated from the finished product.

Turning Versus Cold Drawing. Basic differences exist between bars finished by turning and those finished by cold drawing. First, it is obvious that turning and centerless grinding are applicable only to round bars, while drawing can be applied to a variety of shapes. Drawing, therefore, is more versatile than turning.

Second, there is a difference in the number and severity of the surface imperfections that may be present. Because stock is removed in turning and grinding, shallow surface imperfections and decarburization may be completely eliminated. When material is drawn, stock is only displaced, and surface imperfections are only reduced in depth (in the ratio of the change in bar diameter or section thickness). The length of these imperfections may be slightly increased because in the drawing operation an increase in length accompanies the reduction in cross section.

Cold-drawn bars can approach the freedom from surface imperfections obtained in turned or turned and ground bars if the hot-rolled bars from which they are produced are rolled from specially conditioned billets. Quality conditions such as cold-working quality are available from producers of hot-rolled bars. The depth limits of the surface imperfections are as agreed to between the producer and the customer. However, if maximum freedom from surface imperfections is the controlling factor, turned bars have an advantage.

Different size tolerances are applicable to cold-finished products, depending on shape, carbon content, and heat treatment. Listed in Tables 2 to 4 are the tolerances for cold-finished carbon and alloy steel bars published in ASTM A 29. These tables include cold-drawn bars; turned and polished rounds; cold-drawn, ground, and polished rounds; and turned, ground, and polished rounds. From the data in Tables 2 to 4, certain generalizations can be stated. The tolerances for cold-drawn and for turned and polished rounds, for example, are the same for sizes up to and including 102 mm (4 in.). There are differences, however, between the tolerances that apply to carbon steel and those that apply to alloy steels. Tolerances for several finishes also vary with certain levels of carbon content. Broader tolerances are applicable to bars that have been heat treated before cold finishing. In contrast, tolerances are closer when bars are ground, and these tolerances are independent of carbon content.

In addition to the size-tolerance requirements for all cold-finished steel bars, straightness tolerances are also of major importance for bars intended for use in automatic screw machines. Table 5 (also from ASTM A 29) details the straightness requirements for rounds, squares, hexagons, and octagons, which are the same for both carbon and alloy steel bars. As indicated in Table 6, special provisions are also made for bars subject to magnetic particle inspection.

Product Quality Descriptors

The term quality relates to the suitability of a mill product to become an acceptable part. When used to identify cold-finished steel bars, the various quality descriptors are indicative of many characteristics, such as degree of internal soundness, relative uniformity of chemical composition, and relative freedom from detrimental surface imperfections.

Because of the characteristic surface finish of cold-drawn bars, close visual inspection cannot identify detrimental surface imperfections. Therefore, for applications that do not allow surface imperfections on the finished surfaces of standard quality cold-drawn carbon steel bars and regular quality cold-drawn alloy steel bars, the user should recognize that some stock removal is necessary to eliminate such imperfections as seams. The recommended stock removal per side for all nonresulfurized grades is 0.025 mm (0.001 in.) per 1.6 mm (¹⁄₁₆ in.) of cross section, or 0.25 mm (0.010 in.), whichever is greater. For example, for a 25 mm (1 in.) bar, recommended stock removal is 0.41 mm (0.016 in.) per side. For the resulfurized grades, recommended stock removal is 0.038 mm (0.0015 in.) per 1.6 mm (¹⁄₁₆ in.), or 0.38 mm (0.015 in.), whichever is greater. Therefore, for a 25 mm (1 in.) bar, recommended stock removal is 0.61 mm (0.024 in.) per side.

Occasionally, some bars in a shipment may have imperfections that exceed the recommended stock removal limits. Therefore, for critical applications, inspection of finished parts is recommended, or more restrictive quality and/or additional inspection methods can be specified by agreement of both supplier and customer.

To minimize pitting, the recommended stock removal per side for cold-drawn bars that are to be decorative chromium plated is as follows:

Size, mm (in.)	Stock removal per side, mm (in.)
Through 7.9 (⁵⁄₁₆)	0.15 (0.006)
Over 7.9 (⁵⁄₁₆) through 11.1 (⁷⁄₁₆)	0.20 (0.008)
Over 11.1 (⁷⁄₁₆)	0.25 (0.010)

Carbon Steel Quality Descriptors

Standard quality is the descriptor applied to the basic quality level to which cold-finished carbon steel bars are produced. Standard quality cold-finished bars are produced from hot-rolled carbon steel of special quality (the standard quality for hot-rolled bars for cold finishing). Steel bars of standard quality must be free from visible pipe and excessive chemical segregation. They may contain surface imperfections. In general, the size of surface imperfections increases with bar size.

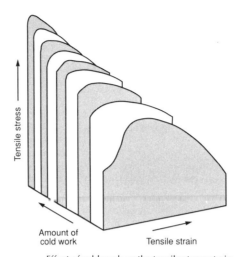

Fig. 1 Effect of cold work on the tensile stress-strain curve for low-carbon steel bars

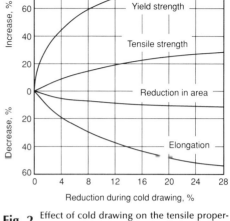

Fig. 2 Effect of cold drawing on the tensile properties of steel bars. Data are for bars up to 25 mm (1 in.) in cross section having a tensile strength of 690 MPa (100 ksi) or less before cold drawing.

Restrictive requirement quality A (RRA) incorporates all the features of standard quality carbon steel bars described above, plus any one of the following restrictive requirements.

Special surface bars are produced with special surface preparation to minimize the frequency and size of seams and other surface imperfections. These bars are used for applications in which machining allowances do not allow sufficient surface removal to clean up the detrimental imperfections that occur in standard quality bars.

Special internal soundness bars have greater freedom from chemical segregation and porosity than standard quality bars.

Special hardenability bars are produced to hardenability requirements other than those of standard H-steels.

Cold-finished carbon steel bars are also produced to inclusion ratings as determined by standard nonmetallic inclusion testing.

Restrictive requirement quality B (RRB) incorporates all the features of standard quality carbon steel bars, plus any one of the following.

Special discard is specified when minimized chemical segregation, special steel cleanliness, or internal soundness requirements dictate that the product be selected from certain positions in the ingot.

Minimized decarburization is specified whenever decarburization is important, as in heat treating for surface hardness requirements.

Single restrictions other than those noted above, such as special chemical limitations, special processing techniques, and other special characteristics not previously anticipated, are also covered by this quality level.

Multiple restrictive requirement quality (MRR) applies when two or more of the above-described restrictive requirements are involved.

Cold-forging quality A and cold-extrusion quality A apply to cold-finished carbon steel bars used in the production of solid or hollow shapes by means of cold plastic deformation involving the movement of metal by compression with no expansion of the surface and not requiring special inspection standards. For an individual application, if the type of steel or chemical composition specified does not provide adequate cold-forming characteristics in the as-drawn condition, a suitable heat treatment to provide proper hardness or microstructure may be necessary.

Cold-heading quality, cold-extrusion quality B, cold-upsetting quality, and cold-expansion quality apply to cold-finished carbon steel bars used in production of solid or hollow shapes by means of severe cold plastic deformation by cold heading, cold extrusion, cold upsetting, or cold expansion involving movement of metal by expansion and/or compression. Such bars are obtained from steel produced by closely controlled steelmaking practices and are subject to special inspection standards for internal soundness and surface quality and uniform chemical composition. For grades of steel with a maximum specified carbon content of 0.30% or more, an anneal or spheroidize anneal heat treatment may be required to obtain the proper hardness and microstructure for cold working.

Restrictive cold-working quality applies to cold-finished carbon steel bars used in the production of solid or hollow shapes by means of very severe cold plastic deformation involving cold working by expansion and/or compression. This degree of cold working normally involves restrictive inspection standards and requires steel that is exceptionally sound, of uniform chemical composition, and virtually free of detrimental surface imperfections. Such severe cold-forming operations

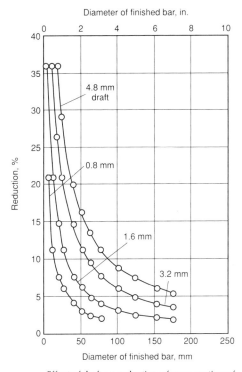

Fig. 3 Effect of draft on reduction of cross section of steel bars

- Axle shaft quality
- Shell steel quality A
- Shell steel quality C
- Rifle barrel quality
- Spark plug quality

Alloy Steel Quality Descriptors

Regular quality is the descriptor applied to the basic, or standard, quality level to which cold-finished alloy steel bars are produced. Steels for this quality are killed and are usually produced to a fine grain size. They are melted to chemical ranges and limits and are inspected and tested to meet normal requirements for regular constructional alloy steel applications. Regular quality cold-finished alloy steel bars may contain surface imperfections to the depths mentioned in the opening paragraphs of the section "Product Quality Descriptors" in this article. In general, the size of detrimental surface imperfections increases with bar size.

Cold-heading quality applies to cold-finished alloy steel bars intended for applications involving cold plastic deformation by such operations as upsetting, heading, or forging. Bars are supplied from steel produced by closely controlled steelmaking practices and are subject to mill testing and inspection designed to ensure internal soundness, uniformity of chemical composition, and freedom from detrimental surface imperfections. Proper control of hardness and microstructure by heat treatment and cold working is important for cold forming. Most cold-heading quality alloy steels are low- and medium-carbon grades. Typical low-carbon alloy steel parts, made by cold heading, include fasteners (cap screws, bolts, eyebolts), studs, anchor pins, and rollers for bearings. Examples of medium-carbon alloy steel cold-headed parts are bolts, studs, and hexagon-headed cap screws.

Special cold-heading quality applies to cold-finished alloy steel bars for applications involving severe cold plastic deformation when slight surface imperfections may cause splitting of a part. Bars of this quality are produced by closely controlled steelmaking practices to provide uniform chemical composition and internal soundness. Also, special processing (such as grinding) is applied at intermediate stages to remove detrimental surface imperfections. Proper control of hardness and microstructure by heat treatment and cold working is important for cold forming. Typical applications of alloy steel bars of this quality are front suspension studs, socket screws, and some valves.

Axle shaft quality applies to cold-finished alloy steel bars intended for the manufacture of automotive or truck-type, power-driven axle shafts, which by their design or method of manufacture are either not machined all over or undergo less than the recommended amount of stock removal for proper cleanup of normal surface imperfections. Axle shaft quality bars require special rolling practices, special billet and bar conditioning, and selective inspection techniques.

Ball and roller bearing quality and bearing quality apply to cold-finished alloy steel bars used for the manufacture of antifriction bearings. Such bars are usually produced from alloy steels of the AISI-SAE standard alloy carburizing grades and the AISI-SAE high-carbon chromium series. These steels can be produced in accordance with ASTM A 534, A 295, and A 485. Bearing quality steels are subjected to restricted melting and special teeming, heating, rolling, cooling, and conditioning practices to meet rigid

normally require suitable heat treatment to obtain proper hardness and microstructure for cold working.

Other Carbon Steel Qualities. The quality descriptors listed below are some of those that apply to cold-finished carbon steel bars intended for specific requirements and applications. They may have requirements for surface quality, amount of discard, macroetch tests, mechanical properties, or chemical uniformity as indicated in product specifications:

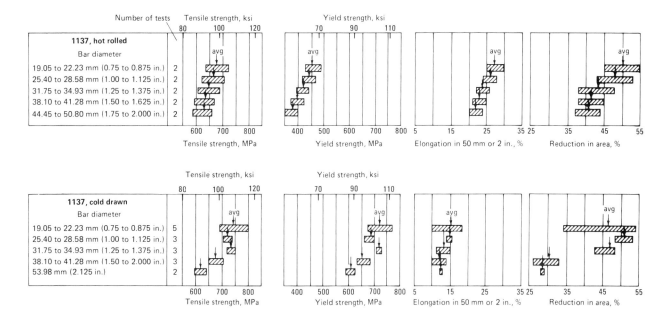

Fig. 4 Mechanical properties of hot-rolled and cold-drawn 1137 bars

254 / Carbon and Low-Alloy Steels

Table 7 Estimated minimum mechanical properties of cold-drawn carbon steel rounds, squares, and hexagons

Estimated minimum mechanical properties for sizes under 16 mm (⅝ in.) can be obtained from individual producers. The data in this table are not applicable to turned and polished or turned and ground bars, which have mechanical properties corresponding to those of hot-rolled steel bars of the same size and grade. The size of a square or hexagon is the distance between opposite sides.

Steel designation and size range		As-cold-drawn							Cold drawn followed by low-temperature stress relief							Cold drawn followed by high-temperature stress relief						
		Tensile Strength		Yield		Elongation in 50 mm (2 in.), %	Reduction in area, %	Hardness, HB	Tensile Strength		Yield		Elongation in 50 mm (2 in.), %	Reduction in area, %	Hardness, HB	Tensile Strength		Yield		Elongation in 50 mm (2 in.), %	Reduction in area, %	Hardness, HB
mm	in.	MPa	ksi	MPa	ksi				MPa	ksi	MPa	ksi				MPa	ksi	MPa	ksi			
1018, 1025																						
16–22 inclusive	⅝–⅞ inclusive	483	70	413	60	18	40	143	...	...	...	...	...	...	...	448	65	310	45	20	45	131
22–32 inclusive	>⅞–1¼ inclusive	448	65	379	55	16	40	131	...	...	...	...	...	...	...	414	60	310	45	20	45	121
32–51 inclusive	>1¼–2 inclusive	414	60	345	50	15	35	121	...	...	...	...	...	...	...	379	55	310	45	16	40	111
51–76 inclusive	>2–3 inclusive	379	55	310	45	15	35	111	...	...	...	...	...	...	...	345	50	276	40	15	40	101
1117, 1118																						
16–22 inclusive	⅝–⅞ inclusive	517	75	448	65	15	40	149	552	80	483	70	15	40	163	483	70	345	50	18	45	143
22–32 inclusive	>⅞–1¼ inclusive	483	70	414	60	15	40	143	517	75	448	65	15	40	149	448	65	345	50	16	45	131
32–51 inclusive	>1¼–2 inclusive	448	65	379	55	13	35	131	483	70	414	60	13	35	143	414	60	345	50	15	40	121
51–76 inclusive	>2–3 inclusive	414	60	345	50	12	30	121	448	65	379	55	12	35	131	379	55	310	45	15	40	111
1035																						
16–22 inclusive	⅝–⅞ inclusive	586	85	517	75	13	35	170	621	90	552	80	13	35	179	552	80	414	60	16	45	163
22–32 inclusive	>⅞–1¼ inclusive	552	80	483	70	12	35	163	586	85	517	75	12	35	170	517	75	414	60	15	45	149
32–51 inclusive	>1¼–2 inclusive	517	75	448	65	12	35	149	552	80	483	70	12	35	163	483	70	414	60	15	40	143
51–76 inclusive	>2–3 inclusive	483	70	414	60	10	30	143	517	75	448	65	10	30	149	448	65	379	55	12	35	131
1040, 1140																						
16–22 inclusive	⅝–⅞ inclusive	621	90	552	80	12	35	179	655	95	586	85	12	35	187	586	85	448	65	15	45	170
22–32 inclusive	>⅞–1¼ inclusive	586	85	517	75	12	35	170	621	90	552	80	12	35	179	552	80	448	65	15	45	163
32–51 inclusive	>1¼–2 inclusive	552	80	483	70	10	30	163	586	85	517	75	10	30	170	517	75	414	60	15	40	149
51–76 inclusive	>2–3 inclusive	517	75	448	65	10	30	149	552	80	483	70	10	30	163	483	70	379	55	12	35	143
1045, 1145																						
16–22 inclusive	⅝–⅞ inclusive	655	95	586	85	12	35	187	689	100	621	90	12	35	197	621	90	483	70	15	45	179
22–32 inclusive	>⅞–1¼ inclusive	621	90	552	80	11	30	179	655	95	586	85	11	30	187	586	85	483	70	15	45	170
32–51 inclusive	>1¼–2 inclusive	586	85	517	75	10	30	170	621	90	552	80	10	30	179	552	80	448	65	15	40	163
51–76 inclusive	>2–3 inclusive	552	80	483	70	10	30	163	586	85	517	75	10	25	170	517	75	414	60	12	35	149
1050, 1137, 1151																						
16–22 inclusive	⅝–⅞ inclusive	689	100	621	90	11	35	197	724	105	655	95	11	35	212	655	95	517	75	15	45	187
22–32 inclusive	>⅞–1¼ inclusive	655	95	586	85	11	30	187	689	100	621	90	11	30	197	621	90	517	75	15	40	179
32–51 inclusive	>1¼–2 inclusive	621	90	552	80	10	30	179	655	95	586	85	10	30	187	586	85	483	70	15	40	170
51–76 inclusive	>2–3 inclusive	586	85	517	75	10	30	170	621	90	552	80	10	25	179	552	80	448	65	12	35	163
1141																						
16–22 inclusive	⅝–⅞ inclusive	724	105	655	95	11	30	212	758	110	689	100	11	30	223	689	100	552	80	15	40	197
22–32 inclusive	>⅞–1¼ inclusive	689	100	621	90	10	30	197	724	105	655	95	10	30	212	655	95	552	80	15	40	187
32–51 inclusive	>1¼–2 inclusive	655	95	586	85	10	30	187	689	100	621	90	10	25	197	621	90	517	75	15	40	179
51–76 inclusive	>2–3 inclusive	621	90	552	80	10	20	179	655	95	581	85	10	20	187	586	85	483	70	12	30	170
1144																						
16–22 inclusive	⅝–⅞ inclusive	758	110	689	100	10	30	223	793	115	724	105	10	30	229	724	105	586	85	15	40	212
22–32 inclusive	>⅞–1¼ inclusive	724	105	655	95	10	30	212	758	110	689	100	10	30	223	689	100	586	85	15	40	197
32–51 inclusive	>1¼–2 inclusive	689	100	621	90	10	25	197	724	105	655	95	10	25	212	655	95	552	80	15	35	187
51–76 inclusive	>2–3 inclusive	655	95	586	85	10	20	187	689	100	621	90	10	20	197	621	90	517	75	12	30	179

Source: Ref 2

quality requirements. The steelmaking operations may include vacuum treatment. The foregoing requirements include thorough examination for internal imperfections by one or more of the following methods: macroetch testing, microscopic or ultrasonic examination for nonmetallic inclusions, and fracture testing.

Aircraft quality and magnaflux quality apply to cold-finished alloy steel bars for important or highly stressed parts of aircraft and for other similar or corresponding purposes involving additional stringent requirements, such as magnetic particle inspection, additional discard, macroetch tests, and hardenability control. To meet these requirements, exacting steelmaking, rolling, and testing practices must be employed. These practices are designed to minimize detrimental inclusions and porosity. Phosphorus and sulfur are usually limited to 0.025% maximum. There are many aircraft parts and many parts for missiles and other rockets that require aircraft quality steel. The magnetic particle testing requirements given in AMS 2301 are sometimes specified for such applications.

Other Alloy Steel Qualities. The quality descriptors listed below apply to cold-finished alloy steel bars intended for rifles, guns, shell, shot, and similar applications. They may have requirements for amount of discard, macroetch testing, surface requirements, or magnetic particle testing as indicated in the product specifications:

- Armor-piercing (AP) shot quality
- AP shot magnaflux quality
- Gun quality
- Rifle barrel quality
- Shell quality
- Shell magnaflux quality

Mechanical Properties

A major difference between machined and cold-drawn round bars is the improvement in tensile and yield strengths that

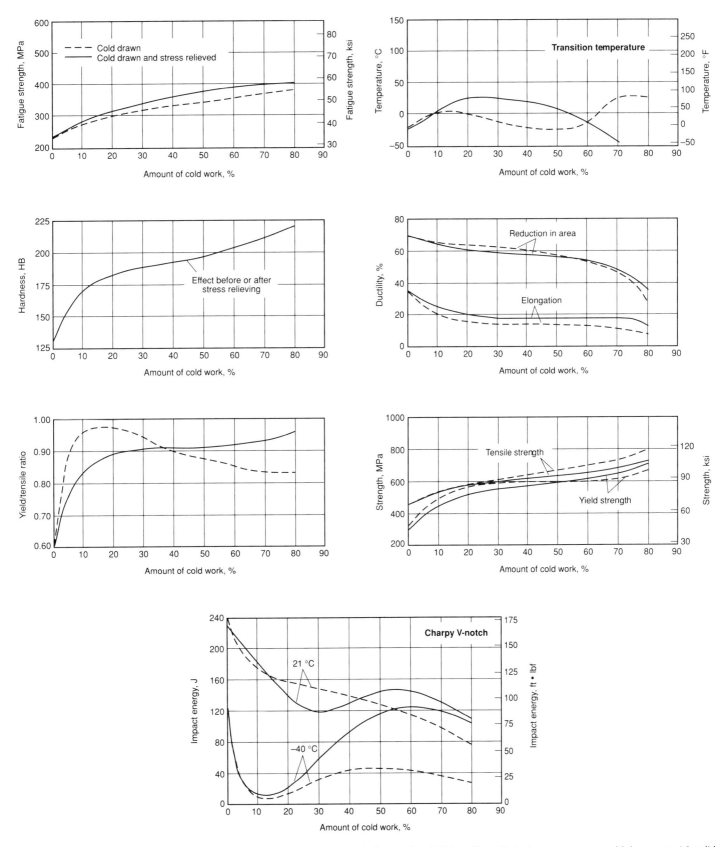

Fig. 5 Effects of cold drawing and of cold drawing and stress relieving on mechanical properties of 1016 steel bars. Dashed curves represent cold-drawn material; solid curves, material cold drawn and stress relieved. All bars were from a single heat. The bars were hot reduced to a diameter of 51 mm (2 in.) by conventional practice, then normalized and cold drawn to the reductions indicated. Note that the larger reductions are well beyond commercial ranges. Test specimens were taken from half-radius positions in the bars. The bars that were stress relieved after cold drawing were treated as follows: carbon steel bars, 2 h at 480 °C (900 °F); 8630 bars, 2 h at 540 °C (1000 °F). All bars were aged for 4 h at 100 °C (212 °F) after cold drawing, or after cold drawing and stress relieving, to simulate the condition of steel after several months of storage at room temperature. Source: Ref 5

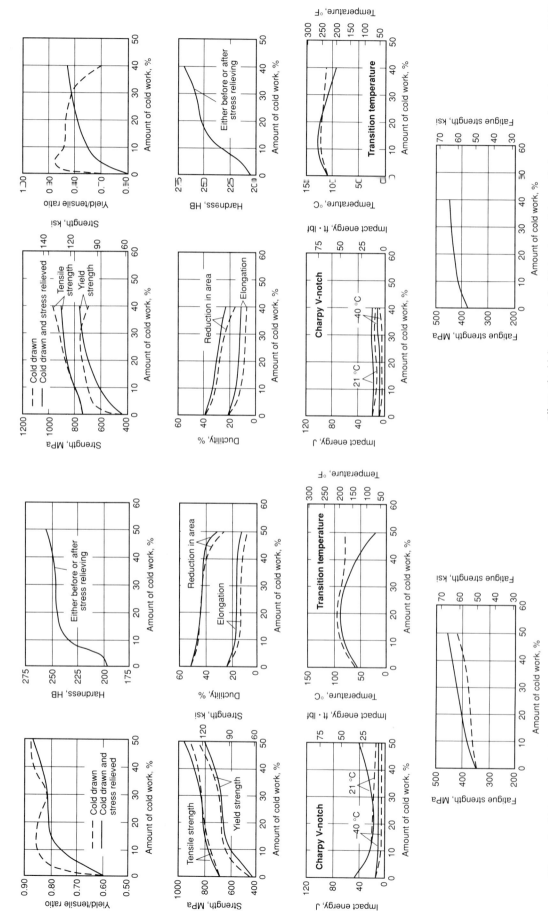

Fig. 6 Effects of cold drawing and of cold drawing and stress relieving on mechanical properties of 1040 steel bars. Dashed curves represent cold-drawn material; solid curves, material cold drawn and stress relieved. All bars were from a single heat. The bars were hot reduced to a diameter of 51 mm (2 in.) by conventional practice, then normalized and cold drawn to the reductions indicated. Note that the larger reductions are well beyond commercial ranges. Test specimens were taken from half-radius positions in the bars. The bars that were stress relieved after cold drawing were treated as follows: carbon steel bars, 2 h at 480 °C (900 °F); 8630 bars, 2 h at 540 °C (1000 °F). All bars were aged for 4 h at 100 °C (212 °F) after cold drawing, or after cold drawing and stress relieving, to simulate the condition of steel after several months of storage at room temperature. Source: Ref 5

Fig. 7 Effects of cold drawing and of cold drawing and stress relieving on mechanical properties of 1060 steel bars. Dashed curves represent cold-drawn material; solid curves, material cold drawn and stress relieved. All bars were from a single heat. The bars were hot reduced to a diameter of 51 mm (2 in.) by conventional practice, then normalized and cold drawn to the reductions indicated. Note that the larger reductions are well beyond commercial ranges. Test specimens were taken from half-radius positions in the bars. The bars that were stress relieved after cold drawing were treated as follows: carbon steel bars, 2 h at 480 °C (900 °F); 8630 bars, 2 h at 540 °C (1000 °F). All bars were aged for 4 h at 100 °C (212 °F) after cold drawing, or after cold drawing and stress relieving, to simulate the condition of steel after several months of storage at room temperature. Source: Ref 5

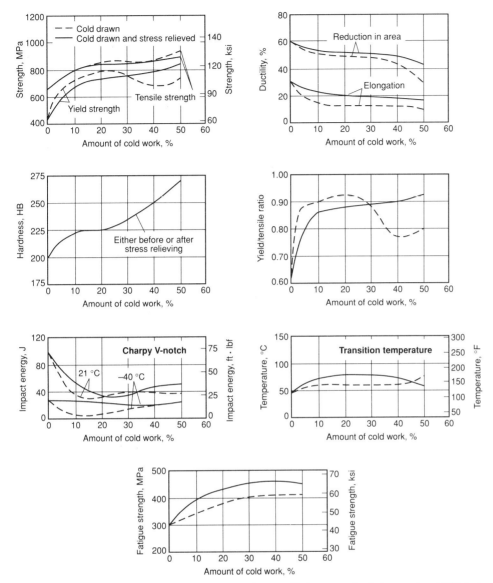

Fig. 8 Effects of cold drawing and of cold drawing and stress relieving on mechanical properties of 8630 steel bars. Dashed curves represent cold-drawn material; solid curves, material cold drawn and stress relieved. All bars were from a single heat. The bars were hot reduced to a diameter of 51 mm (2 in.) by conventional practice, then normalized and cold drawn to the reductions indicated. Note that the larger reductions are well beyond commercial ranges. Test specimens were taken from half-radius positions in the bars. The bars that were stress relieved after cold drawing were treated as follows: carbon steel bars, 2 h at 480 °C (900 °F); 8630 bars 2 h at 540 °C (1000 °F). All bars were aged for 4 h at 100 °C (212 °F) after cold drawing, or after cold drawing and stress relieving, to simulate the condition of steel after several months of storage at room temperature. Source: Ref 5

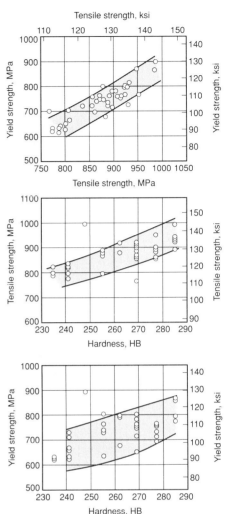

Fig. 9 Mechanical properties of 1144 steel bars cold drawn and stress relieved at 565 °C (1050 °F). Range of composition for 41 heats was 0.41 to 0.52% C, 1.33 to 1.68% Mn, and 0.220 to 0.336% S.

results from the cold work of drawing. Cold work also changes the shape of the stress-strain diagram, as shown in Fig. 1. Within the range of commercial drafts, cold work markedly affects certain mechanical properties (Fig. 2). The variations in percentage of reduction of cross section for bars drawn with normal commercial drafts of 0.8 and 1.6 mm (1/32 and 1/16 in.) and with heavy drafts of 3.2 and 4.8 mm (1/8 and 3/16 in.) are shown in Fig. 3. Normal reductions seldom exceed 20% and are usually less than 12%. According to Fig. 2, the more pronounced changes in significant tensile properties occur within this range of reductions (up to about 15%).

The minimum mechanical properties of several cold-drawn carbon steel bars in a range of sizes are presented in Table 7. In addition, the effects of both low- and high-temperature stress relief on the as-cold-drawn mechanical properties are noted. The mechanical property ranges and average values for one of the steels listed in Table 7 (1137 resulfurized steel) are presented in Fig. 4, which also shows the advantage in strength of cold-drawn over hot-rolled material.

Measurements of the changes in mechanical properties of three specific carbon steels (1016, 1040, and 1060) and one alloy steel (8630) as a result of cold drawing are shown in Fig. 5 to 8. Some of these data pertain to large reductions, well beyond commercial ranges. Data plotted as solid lines in Fig. 5 to 8 are for bars that were cold drawn and artificially aged, but not stress relieved. After drawing, these bars were aged for 4 h at 100 °C (212 °F) to simulate the natural aging resulting from several months of storage at room temperature. Data plotted as dashed lines in Fig. 5 to 8 are for cold-drawn and stress-relieved bars; carbon steels were stress relieved for 2 h at 480 °C (900 °F), and 8630 steel for 2 h at 540 °C (1000 °F), after cold drawing.

Tensile and Yield Strengths. The data in Fig. 2 and 5 to 8 indicate that as cold work increases up to about 15%, yield strength increases at a greater rate than tensile strength. The greatest improvement in strength results from the first 5% of reduction. Stress relieving modifies this pattern appreciably. The yield/tensile ratio is markedly affected by cold drawing. In this con-

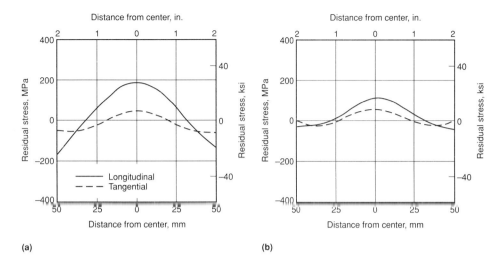

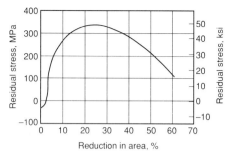

Fig. 11 Effect of increasing single-draft reduction on residual longitudinal stress at the surface of drawn steel wire

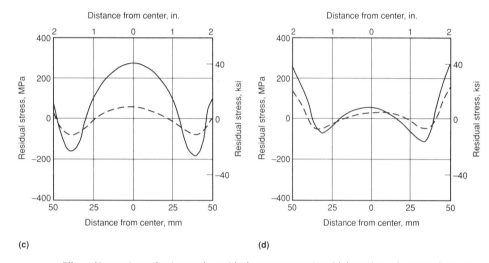

Fig. 10 Effect of increasing reduction on the residual stress patterns in cold-drawn bars of 1050 steel. (a) 4.1% reduction in area. (b) 6.2% reduction in area. (c) 8.3% reduction in area. (d) 12.3% reduction in area. Source: Ref 6

dition (cold drawn, not stress relieved; see Fig. 5 to 8), the data for yield/tensile ratio indicate a somewhat erratic behavior. However, the ratio follows a consistent upward trend with increased cold work and subsequent stress relief.

The hardness increases with increased cold work and, in most cases, is affected by stress relieving. There is considerable scatter in the relations between hardness and tensile strength and hardness and yield strength, as indicated by the data in Fig. 9 for 41 heats of cold-drawn and stress-relieved 1144 steel. However, there is a relationship between hardness and tensile strength or hardness and yield strength, because published tables allow approximation of the hardness or tensile strength (or yield strength) when one of the other values is known.

Impact Properties. Available data are limited on the effect of cold work on notched-bar impact properties. The results of one of the more important studies are included in Fig. 5 to 8, which show the effect of cold work over a wide range of drafts on three carbon steels with increasing carbon con-

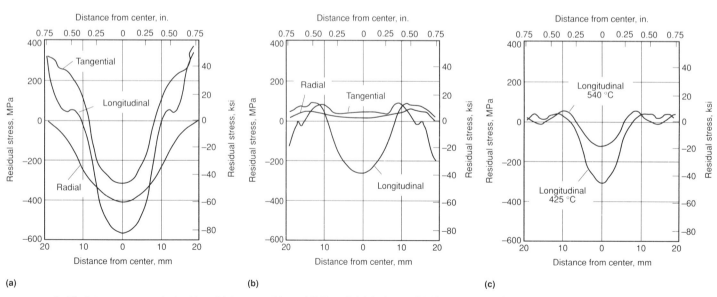

Fig. 12 Residual stress patterns obtained in cold-drawn steel bars of 1045 steel. (a) As-drawn. (b) After rotary straightening. (c) After stress relieving. Bars were cold drawn 20% from 43 to 38 mm (1 11/16 to 1 1/2 in.).

tents and the effect of cold work on 8630 alloy steel. Within the range of commercial drafts, energy absorbed (breaking strength) falls rapidly for the 1016 steel and less rapidly for 8630 steel. At any level of cold work, energy absorbed decreases with increased carbon content.

In the stress-relieved condition, the fracture transition temperature generally rises with increasing amounts of cold work up to 20 to 30% reduction. Beyond this commercial range of reductions, the transition temperature falls. For 1016 steel, extremely heavy drafts lower the transition temperature to below that of the original hot-rolled material. Increasing carbon content raises the transition temperature.

Residual Stresses

The stress pattern produced by cold drawing depends on the amount of reduction and the shape of the die, as well as the microstructure, hardness, and grade of steel. Figure 10 illustrates the effect of reduction in area on the magnitude and distribution of stresses in bars of 1050 steel reduced by the amounts shown. Cold drawing of the bars to 4.1% reduction resulted in surface compressive stresses, while increasing the amount of cold drawing to 12.3% reduction resulted in a change of the surface stresses from compressive to tensile. The variation in longitudinal stress over a much wider range of reduction values is shown in Fig. 11 for steel wire (the effect is qualitatively similar for bars). The greatest effect on the residual stress is caused by the first 10% reduction. The effect of a very light draft is to produce compressive stress at the surface, which rapidly changes to tensile stress with a relatively small increase in reduction.

Both straightening and stress relieving after cold drawing have significant effects on the residual stress pattern of the resulting product. Figure 12 shows the longitudinal, tangential, and radial stress patterns that result when a 43 mm (1 11/16 in.) diam carbon steel bar is drawn to 38 mm (1 1/2 in.), a reduction of 20%. These data indicate that the surface of the bar is in tension, the center is in compression, and both longitudinal and tangential stresses vary over wide ranges.

Straightening in a skewed-rolls (Medart) machine significantly reduces residual stress, particularly at the surface, as shown in Fig. 12(b). It is of interest to compare the longitudinal stress curve shown in Fig. 12(b) with that in Fig. 12(a). Figure 12(c) shows the effect of two stress-relieving temperatures (425 and 540 °C, or 800 and 1000 °F) on residual longitudinal stress. Stress relieving at these temperatures is only slightly more effective than straightening in reducing the residual stress level. This phenomenon may be accounted for by an analysis of the nature of the stresses that are developed in cold drawing.

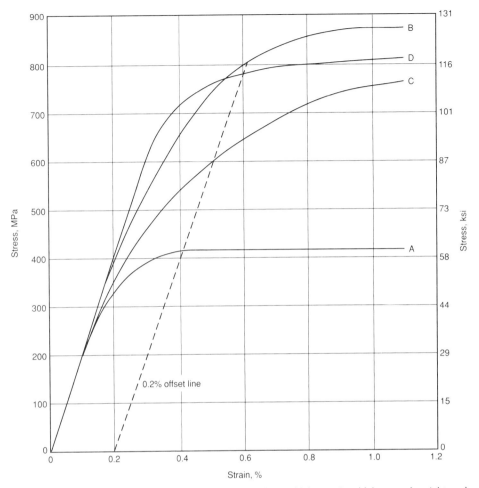

Fig. 13 Various production stages of 1144 steel. A, hot rolled; B, cold drawn; C, cold drawn and straightened; D, cold drawn, straightened, and strain relieved

The stress applied in cold drawing is sufficient to deform the material both elastically and plastically. Because the initiation of plastic strain depends on the development of maximum elastic strain, the ratio of these two strains after release of the deforming stress may be highly variable. If the deformation caused by cold drawing were uniform across the section, as in pure stretching, the elastic stress would be released by the release of the deforming stress. Because the plastic strain is not uniform, as shown by the dishing of the ends of drawn bars, neither is the accompanying elastic strain. When the deforming stress is removed from such a system, the remaining nonuniform elastic-strain energy cannot be released completely, because the resistance of low-strain regions prevents the complete recovery of regions of high strain. A pattern of residual stress results from this unequal adjustment.

Stress Relieving. The inevitable residual stresses in as-drawn bars can be relieved mechanically or thermally (Ref 7). Mechanical relief may take two forms. One involves the introduction of stresses of opposite sign, which can be accomplished by shot peening. A second approach is to plastically deform the material further, thus affording additional opportunity for the relief of nonuniform residual stresses. The data on rotary straightening in Fig. 12(b) demonstrate this effect.

The thermal stress relieving of cold-drawn bars—also known as strain drawing, strain annealing, strain relieving, preaging, and stabilizing—is probably the most widely used thermal treatment applied to cold-drawn bars. Its purpose is to modify the magnitude and distribution of residual stresses in the cold-finished bar and thus produce a product with the desired combination of mechanical properties for field service. Thermal stress relief temporarily reduces the strength level of the material (at the stress-relieving temperature) and enables the elastic-strain energy to find release in small but significant amounts of plastic deformation. After stress relieving, the maximum residual stress that can remain is equal to the yield strength of the material at the stress-relieving temperature.

Temperatures up to about 650 °C (1200 °F) are commonly used for the stress relieving of cold-drawn bars. The upper limit for

Table 8 Machinability ratings and recommended feeds and speeds for cold-drawn carbon steel bars

All cutting speeds and feeds based on cutting with high-speed steel tools. For cutting conditions for other machining operations, see the article "Machinability of Steels" in this Volume.

Steel designation AISI or SAE	Machinability rating(a), %	Form turning Width of cut, mm (in.)	Speed, m/min (sfm)	Feed, mm/rev (in./rev)	Single-point turning Depth of cut, mm (in.)	Speed, m/min (sfm)	Feed, mm/rev (in./rev)	Drilling Size of hole, mm (in.)	Speed, m/min (sfm)	Feed, mm/rev (in./rev)
12L14	170	12.7 (0.500)	85.3 (280)	0.089 (0.0035)	3.18 (0.125)	85.3 (280)	0.236 (0.0093)	6.35 (0.250)	50.3 (165)	0.160 (0.0063)
		25.4 (1.000)	79.2 (260)	0.076 (0.0030)	6.35 (0.250)	79.2 (260)	0.224 (0.0088)	12.7 (0.500)	50.3 (165)	0.175 (0.0069)
		38.1 (1.500)	79.2 (260)	0.074 (0.0029)	9.53 (0.375)	77.7 (255)	0.180 (0.0071)	19.05 (0.750)	54.9 (180)	0.206 (0.0081)
		50.8 (2.000)	76.2 (250)	0.053 (0.0021)	12.7 (0.500)	76.2 (250)	0.152 (0.0060)	25.4 (1.000)	54.9 (180)	0.234 (0.0092)
		63.5 (2.500)	71.6 (235)	0.043 (0.0017)	...	...	...	31.75 (1.250)	56.4 (185)	0.267 (0.0105)
1213, 1215	136	12.7 (0.500)	68.6 (225)	0.076 (0.0030)	3.18 (0.125)	68.6 (225)	0.216 (0.0085)	6.35 (0.250)	38.1 (125)	0.137 (0.0054)
		25.4 (1.000)	64.0 (210)	0.064 (0.0025)	6.35 (0.250)	64.0 (210)	0.203 (0.0080)	12.7 (0.500)	38.1 (125)	0.152 (0.0060)
		38.1 (1.500)	64.0 (210)	0.064 (0.0025)	9.53 (0.375)	62.5 (205)	0.165 (0.0065)	19.05 (0.750)	42.7 (140)	0.178 (0.0070)
		50.8 (2.000)	62.5 (205)	0.046 (0.0018)	12.7 (0.500)	61.0 (200)	0.140 (0.0055)	25.4 (1.000)	42.7 (140)	0.203 (0.0080)
		63.5 (2.500)	61.0 (200)	0.033 (0.0013)	...	...	...	31.75 (1.250)	44.2 (145)	0.229 (0.0090)
1119, 1212	100	12.7 (0.500)	50.3 (165)	0.064 (0.0025)	3.18 (0.125)	50.3 (165)	0.178 (0.0070)	6.35 (0.250)	32.0 (105)	0.114 (0.0045)
		25.4 (1.000)	48.8 (160)	0.051 (0.0020)	6.35 (0.250)	48.8 (160)	0.165 (0.0065)	12.7 (0.500)	32.0 (105)	0.127 (0.0050)
		38.1 (1.500)	48.8 (160)	0.046 (0.0018)	9.53 (0.375)	47.2 (155)	0.140 (0.0055)	19.05 (0.750)	35.0 (115)	0.152 (0.0060)
		50.8 (2.000)	47.2 (155)	0.038 (0.0015)	12.7 (0.500)	45.7 (150)	0.114 (0.0045)	25.4 (1.000)	35.0 (115)	0.178 (0.0070)
		63.5 (2.500)	45.7 (150)	0.030 (0.0012)	...	...	...	31.75 (1.250)	36.6 (120)	0.203 (0.0080)
1211	94	12.7 (0.500)	47.2 (155)	0.058 (0.0023)	3.18 (0.125)	47.2 (155)	0.168 (0.0066)	6.35 (0.250)	30.2 (99)	0.107 (0.0042)
		25.4 (1.000)	45.7 (150)	0.048 (0.0019)	6.35 (0.250)	45.7 (150)	0.155 (0.0061)	12.7 (0.500)	30.2 (99)	0.119 (0.0047)
		38.1 (1.500)	45.7 (150)	0.043 (0.0017)	9.53 (0.375)	44.5 (146)	0.132 (0.0052)	19.05 (0.750)	32.9 (108)	0.142 (0.0056)
		50.8 (2.000)	44.5 (146)	0.036 (0.0014)	12.7 (0.500)	43.0 (141)	0.107 (0.0042)	25.4 (1.000)	32.9 (108)	0.168 (0.0066)
		63.5 (2.500)	43.0 (141)	0.028 (0.0011)	...	...	...	31.75 (1.250)	34.4 (113)	0.193 (0.0076)
1117, 1118	91	12.7 (0.500)	45.7 (150)	0.056 (0.0022)	3.18 (0.125)	45.7 (150)	0.163 (0.0064)	6.35 (0.250)	29.0 (95)	0.104 (0.0041)
		25.4 (1.000)	44.2 (145)	0.046 (0.0018)	6.35 (0.250)	44.2 (145)	0.150 (0.0059)	12.7 (0.500)	29.0 (95)	0.114 (0.0045)
		38.1 (1.500)	44.2 (145)	0.041 (0.0016)	9.53 (0.375)	43.0 (141)	0.127 (0.0050)	19.05 (0.750)	32.0 (105)	0.140 (0.0055)
		50.8 (2.000)	43.0 (141)	0.036 (0.0014)	12.7 (0.500)	41.5 (136)	0.104 (0.0041)	25.4 (1.000)	32.0 (105)	0.163 (0.0064)
		63.5 (2.500)	41.5 (136)	0.028 (0.0011)	...	...	...	31.75 (1.250)	36.3 (119)	0.185 (0.0073)
1144, annealed	85	12.7 (0.500)	42.7 (140)	0.053 (0.0021)	3.18 (0.125)	42.7 (140)	0.150 (0.0059)	6.35 (0.250)	27.1 (89)	0.102 (0.0040)
		25.4 (1.000)	41.5 (136)	0.043 (0.0017)	6.35 (0.250)	41.5 (136)	0.140 (0.0055)	12.7 (0.500)	27.1 (89)	0.114 (0.0045)
		38.1 (1.500)	41.5 (136)	0.038 (0.0015)	9.53 (0.375)	40.2 (132)	0.119 (0.0047)	19.05 (0.750)	29.9 (98)	0.140 (0.0055)
		50.8 (2.000)	40.2 (132)	0.033 (0.0013)	12.7 (0.500)	38.7 (127)	0.102 (0.0040)	25.4 (1.000)	29.9 (98)	0.163 (0.0064)
		63.5 (2.500)	38.7 (127)	0.025 (0.0010)	...	...	...	31.75 (1.250)	31.1 (102)	0.179 (0.0070)
1141, annealed	81	12.7 (0.500)	41.1 (135)	0.051 (0.0020)	3.18 (0.125)	41.1 (135)	0.145 (0.0057)	6.35 (0.250)	26.2 (86)	0.102 (0.0040)
		25.4 (1.000)	39.6 (130)	0.043 (0.0017)	6.35 (0.250)	39.6 (130)	0.135 (0.0053)	12.7 (0.500)	26.2 (86)	0.114 (0.0045)
		38.1 (1.500)	39.6 (130)	0.038 (0.0015)	9.53 (0.375)	38.7 (127)	0.114 (0.0045)	19.05 (0.750)	28.6 (94)	0.137 (0.0054)
		50.8 (2.000)	38.7 (127)	0.030 (0.0012)	12.7 (0.500)	37.2 (122)	0.094 (0.0037)	25.4 (1.000)	28.6 (94)	0.160 (0.0063)
		63.5 (2.500)	37.2 (122)	0.025 (0.0010)	...	...	...	31.75 (1.200)	29.9 (98)	0.183 (0.0072)
1016, 1018, 1022	78	12.7 (0.500)	39.6 (130)	0.048 (0.0019)	3.18 (0.125)	39.6 (130)	0.140 (0.0055)	6.35 (0.250)	25.0 (82)	0.096 (0.0038)
		25.4 (1.000)	38.1 (125)	0.041 (0.0016)	6.35 (0.250)	38.1 (125)	0.130 (0.0051)	12.7 (0.500)	25.0 (82)	0.109 (0.0043)
		38.1 (1.500)	38.1 (125)	0.036 (0.0014)	9.53 (0.375)	36.9 (121)	0.110 (0.0043)	19.05 (0.750)	27.4 (90)	0.132 (0.0052)
		50.8 (2.000)	36.9 (121)	0.030 (0.0012)	12.7 (0.500)	35.7 (117)	0.090 (0.0035)	25.4 (1.000)	27.4 (90)	0.152 (0.0060)
		63.5 (2.500)	35.7 (117)	0.023 (0.0009)	...	...	...	31.75 (1.250)	28.7 (94)	0.173 (0.0068)

(continued)

(a) Based on a machinability rating of 100% for 1212 steel. Source: Ref 2

---

the stress-relief temperature for a particular cold-worked steel is the recrystallization or lower critical temperature of that steel, because if this temperature is exceeded, the strengthening effect of cold work is lost. The temperatures used in commercial practice frequently range from 370 to 480 °C (700 to 900 °F). When stress relieving is performed at relatively low temperatures (for example, 290 °C, or 550 °F), yield strength of most cold-drawn steels is increased. At higher temperatures, however, hardness, tensile strength, and yield strength are reduced, while elongation and reduction in area are increased. The choice of a specific time and temperature is dependent on chemical composition, cold-drawing practice, and the final properties required in the bar.

The various categories of stress relief can be divided into three groups:

- Group 1: Complete relief of all cold-working stresses
- Group 2: Relief of cold-working stresses to a limited degree to increase ductility and stability in the material
- Group 3: Relief of stresses in heavily drafted steels to develop high yield strength

Group 1 treatment is conducted above 540 °C (1000 °F). It removes all residual stresses that otherwise would cause objectionable distortion in machining.

Group 2 Treatment. With group 2 processing, lower temperatures in the range of 370 to 540 °C (700 to 1000 °F) are used, and the stresses are partially relieved to bring the mechanical properties within the limits of individual specifications. Applications falling into this class are those that may require ductility close to that of hot-rolled steel, along with good surface finish and close control of dimensions and stability during machining.

Group 3 stress relief is used for bars with heavy drafts. These drafts raise the tensile and yield strengths to high levels, but reduce elongation and reduction in area. Heating to 260 to 425 °C (500 to 800 °F) restores the ductility while retaining or increasing the strength and hardness imparted by the cold work.

The effect of stress relieving at two temperatures on the residual stress pattern of 38 mm (1.5 in.) diam 0.45% C steel bars that have been cold drawn 20% is shown in Fig. 12. The estimated minimum mechanical properties for cold-drawn carbon steel bars as-cold-drawn and as-cold-drawn followed by both a low- and a high-temperature stress-relieving treatment are given in Table 7. The cumulative effects of cold drawing, straightening, and stress relieving on the yield and tensile strength of 1144 steel are shown in Fig. 13.

Heat Treatment

Heat treatment by quenching and tempering, followed by scale removal and then

Table 8 (continued)

Steel designation AISI or SAE	Machinability rating(a), %	Form turning - Width of cut, mm (in.)	Form turning - Speed, m/min (sfm)	Form turning - Feed, mm/rev (in./rev)	Single-point turning - Depth of cut, mm (in.)	Single-point turning - Speed, m/min (sfm)	Single-point turning - Feed, mm/rev (in./rev)	Drilling - Size of hole, mm (in.)	Drilling - Speed, m/min (sfm)	Drilling - Feed, mm/rev (in./rev)
1144	76	12.7 (0.500)	38.1 (125)	0.048 (0.0019)	3.18 (0.125)	38.1 (125)	0.132 (0.0052)	6.35 (0.250)	24.1 (79)	0.094 (0.0037)
		25.4 (1.000)	36.9 (121)	0.038 (0.0015)	6.35 (0.250)	36.9 (121)	0.124 (0.0049)	12.7 (0.500)	24.1 (79)	0.107 (0.0042)
		38.1 (1.500)	36.9 (121)	0.036 (0.0014)	9.53 (0.375)	35.7 (117)	0.104 (0.0041)	19.05 (0.750)	26.5 (87)	0.127 (0.0050)
		50.8 (2.000)	35.7 (117)	0.028 (0.0011)	12.7 (0.500)	34.4 (113)	0.086 (0.0034)	25.4 (1.000)	26.5 (87)	0.147 (0.0058)
		63.5 (2.500)	34.4 (113)	0.023 (0.0009)	...	...	...	31.75 (1.250)	27.7 (91)	0.168 (0.0066)
1020, 1137, 1045, annealed	72	12.7 (0.500)	36.6 (120)	0.046 (0.0018)	3.18 (0.125)	36.6 (120)	0.127 (0.0050)	6.35 (0.250)	23.2 (76)	0.089 (0.0035)
		25.4 (1.000)	35.1 (115)	0.036 (0.0014)	6.35 (0.250)	35.1 (115)	0.119 (0.0047)	12.7 (0.500)	23.2 (76)	0.102 (0.0040)
		38.1 (1.500)	35.1 (115)	0.033 (0.0013)	9.53 (0.375)	34.1 (112)	0.102 (0.0040)	19.05 (0.750)	25.3 (83)	0.119 (0.0047)
		50.8 (2.000)	34.1 (112)	0.028 (0.0011)	12.7 (0.500)	32.9 (108)	0.081 (0.0032)	25.4 (1.000)	25.3 (83)	0.140 (0.0055)
		63.5 (2.500)	32.9 (108)	0.023 (0.0009)	...	...	...	31.75 (1.250)	26.2 (86)	0.163 (0.0064)
1035, 1141, 1050, annealed	70	12.7 (0.500)	35.1 (115)	0.043 (0.0017)	3.18 (0.125)	35.1 (115)	0.124 (0.0049)	6.35 (0.250)	22.3 (73)	0.086 (0.0034)
		25.4 (1.000)	34.1 (112)	0.036 (0.0014)	6.35 (0.250)	34.1 (112)	0.114 (0.0045)	12.7 (0.500)	22.3 (73)	0.097 (0.0038)
		38.1 (1.500)	34.1 (112)	0.033 (0.0013)	9.53 (0.375)	32.9 (108)	0.097 (0.0038)	19.05 (0.750)	24.4 (80)	0.114 (0.0045)
		50.8 (2.000)	32.9 (108)	0.028 (0.0011)	12.7 (0.500)	32.0 (105)	0.079 (0.0031)	25.4 (1.000)	24.4 (80)	0.135 (0.0053)
		63.5 (2.500)	32.0 (105)	0.020 (0.0008)	...	...	...	31.75 (1.250)	25.6 (84)	0.157 (0.0062)
1040	64	12.7 (0.500)	32.0 (105)	0.038 (0.0015)	3.18 (0.125)	32.0 (105)	0.112 (0.0044)	6.35 (0.250)	20.4 (67)	0.081 (0.0032)
		25.4 (1.000)	30.8 (101)	0.030 (0.0012)	6.35 (0.250)	30.8 (101)	0.104 (0.0041)	12.7 (0.500)	20.4 (67)	0.089 (0.0035)
		38.1 (1.500)	30.8 (101)	0.038 (0.0011)	9.53 (0.375)	29.9 (98)	0.086 (0.0034)	19.05 (0.750)	22.3 (73)	0.107 (0.0042)
		50.8 (2.000)	29.9 (98)	0.023 (0.0009)	12.7 (0.500)	29.0 (95)	0.071 (0.0028)	25.4 (1.000)	22.3 (73)	0.124 (0.0049)
		63.5 (2.500)	29.0 (95)	0.018 (0.0007)	...	...	...	31.75 (1.250)	23.2 (76)	0.142 (0.0056)
1045	57	12.7 (0.500)	29.0 (95)	0.036 (0.0014)	3.18 (0.125)	29.0 (95)	0.102 (0.0040)	6.35 (0.250)	18.3 (60)	0.071 (0.0028)
		25.4 (1.000)	27.7 (91)	0.030 (0.0012)	6.35 (0.250)	27.7 (91)	0.094 (0.0037)	12.7 (0.500)	18.3 (60)	0.079 (0.0031)
		38.1 (1.500)	27.7 (91)	0.025 (0.0010)	9.53 (0.375)	26.8 (88)	0.079 (0.0031)	19.05 (0.750)	19.8 (65)	0.094 (0.0037)
		50.8 (2.000)	26.8 (88)	0.023 (0.0009)	12.7 (0.500)	25.9 (85)	0.066 (0.0026)	25.4 (1.000)	19.8 (65)	0.112 (0.0044)
		63.5 (2.500)	25.9 (85)	0.018 (0.0007)	...	...	...	31.75 (1.250)	20.7 (68)	0.127 (0.0050)
1050	54	12.7 (0.500)	27.4 (90)	0.036 (0.0014)	3.18 (0.125)	27.4 (90)	0.097 (0.0038)	6.35 (0.250)	17.4 (57)	0.071 (0.0028)
		25.4 (1.000)	26.5 (87)	0.028 (0.0011)	6.35 (0.250)	26.5 (87)	0.089 (0.0035)	12.7 (0.500)	17.4 (57)	0.079 (0.0031)
		38.1 (1.500)	26.5 (87)	0.025 (0.0010)	9.53 (0.375)	25.6 (84)	0.076 (0.0030)	19.05 (0.750)	18.9 (62)	0.094 (0.0037)
		50.8 (2.000)	25.6 (84)	0.020 (0.0008)	12.7 (0.500)	24.7 (81)	0.061 (0.0024)	25.4 (1.000)	18.9 (62)	0.112 (0.0044)
		63.5 (2.500)	24.7 (81)	0.018 (0.0007)	...	...	...	31.75 (1.250)	19.8 (65)	0.127 (0.0050)

(a) Based on a machinability rating of 100% for 1212 steel. Source: Ref 2

cold drawing, can also be used as a method of producing stronger cold-finished bars in those grades amenable to quench hardening. Heat treatment provides the required increase in strength, and cold drawing provides the size and finish, with a minimal increase in the mechanical properties obtained by quench hardening. Alternatively, quenched and tempered bars can be cold finished by turning and polishing. When bars are cold finished by turning and polishing, there is no increase in the mechanical properties obtained by quench hardening.

For the cold drawing of quenched and tempered bars to be economically justifiable, the minimum strength level produced must be above that obtainable by conventional cold-drawn practices. The upper strength limit is not clearly defined, but for most applications it is the upper limit of machinability. The cold drawing of quenched and tempered bars is applicable to both carbon and alloy steels; however, for the process to be economically justifiable, alloy steel is used only in those sizes above which carbon steel will not respond satisfactorily to liquid quenching. Typically, quenched and tempered product offers superior ductility and heat-resistant properties. Other heat treatments—principally normalizing, full annealing, spheroidizing, and thermal stress relieving—can be applied to suitable grades of hot-rolled steel before or after cold drawing or turning and polishing as required by the end product.

Control of microstructure is frequently important, a good example being annealing for machinability. A controlled rate of continuous cooling through the pearlite transformation range (so-called cycle annealing) is employed. Isothermal cycles are also used. In the cycle-annealing process, the rate of cooling of the furnace charge is adjusted so that the time required to traverse the pearlite temperature interval is sufficient to allow completion of that transformation. By regulating the dwell in the transformation temperature range, the carbide distribution in the product can be varied from partly spheroidal to fully pearlitic, and the pearlite from coarse to fine. In this manner, the optimum machining structure can be obtained for the grade and the machining practice being used.

Spheroidize annealing thermal treatment is given to cold-finished bars that are to be used for severe cold-forming operations. The aim of this treatment is to develop a microstructure consisting of globular carbides in a ferrite matrix. The rate of spheroidizing depends to some degree on the original microstructures. Prior cold work also increases the rate of spheroidizing, particularly for subcritical spheroidizing treatments.

The spheroidized structure is desirable when minimum hardness and maximum ductility are important. Low-carbon steels are seldom annealed for machining because, in the annealed condition, they are very soft and gummy, which tends to produce long, stringy chips that cause handling problems at the machine tool and contribute to a rough surface finish on the machined part. When such steels are spheroidized, it is usually to permit severe cold deformation.

Carbon Restoration. During the hot-working operations involved in the production of bar products—the reduction of cast ingots, blooms, or billets and subsequent conversion in bar mills—decarburization of the bar surface takes place because of exposure to ambient oxygen at high temperatures throughout these operations. A specialized variant of full annealing, called carbon restoration or carbon correction, is utilized to compensate for the loss of carbon due to decarburization.

Carbon restoration for alloy steels is limited because vanadium carbide and molybdenum are not recovered. By heating the descaled hot-rolled bars to approximately 870 to 925 °C (1600 to 1700 °F) in a controlled atmosphere, it is possible to restore surface carbon to the required level. A modern controlled-atmosphere furnace is used for this purpose. Methane or other light hydrocarbons are burned with a controlled amount of air in an endothermic

Table 9 Estimated mechanical properties and machinability ratings of nonresulfurized carbon steel bars

Steel designation SAE and/or AISI No.	UNS No.	Type of processing(a)	Tensile strength MPa	ksi	Yield strength MPa	ksi	Elongation in 50 mm (2 in.), %	Reduction in area, %	Hardness, HB	Average machinability rating(b)
Manganese 1.00% maximum										
1006	G10060	Hot rolled	300	43	170	24	30	55	86	
		Cold drawn	330	48	280	41	20	45	95	50
1008	G10080	Hot rolled	303	44	170	24.5	30	55	86	
		Cold drawn	340	49	290	41.5	20	45	95	55
1010	G10100	Hot rolled	320	47	180	26	28	50	95	
		Cold drawn	370	53	300	44	20	40	105	55
1012	G10120	Hot rolled	330	48	180	26.5	28	50	95	
		Cold drawn	370	54	310	45	19	40	105	55
1015	G10150	Hot rolled	340	50	190	27.5	28	50	101	
		Cold drawn	390	56	320	47	18	40	111	60
1016	G10160	Hot rolled	380	55	210	30	25	50	111	
		Cold drawn	420	61	350	51	18	40	121	70
1017	G10170	Hot rolled	370	53	200	29	26	50	105	
		Cold drawn	410	59	340	49	18	40	116	65
1018	G10180	Hot rolled	400	58	220	32	25	50	116	
		Cold drawn	440	64	370	54	15	40	126	70
1019	G10190	Hot rolled	410	59	220	32.5	25	50	116	
		Cold drawn	460	66	380	55	15	40	131	70
1020	G10200	Hot rolled	380	55	210	30	25	50	111	
		Cold drawn	420	61	350	51	15	40	121	65
1021	G10210	Hot rolled	420	61	230	33	24	48	116	
		Cold drawn	470	68	390	57	15	40	131	70
1022	G10220	Hot rolled	430	62	230	34	23	47	151	
		Cold drawn	480	69	400	58	15	40	137	70
1023	G10230	Hot rolled	370	56	210	31	25	50	111	
		Cold drawn	430	62	360	52.5	15	40	121	65
1025	G10250	Hot rolled	400	58	220	32	25	50	116	
		Cold drawn	440	64	370	54	15	40	126	65
1026	G10260	Hot rolled	440	64	240	35	24	49	126	
		Cold drawn	490	71	410	60	15	40	143	75
1030	G10300	Hot rolled	470	68	260	37.5	20	42	137	
		Cold drawn	520	76	440	64	12	35	149	70
1035	G10350	Hot rolled	500	72	270	39.5	18	40	143	
		Cold drawn	550	80	460	67	12	35	163	65
1037	G10370	Hot rolled	510	74	280	40.5	18	40	143	
		Cold drawn	570	82	480	69	12	35	167	65
1038	G10380	Hot rolled	520	75	280	41	18	40	149	
		Cold drawn	570	83	480	70	12	35	163	65
1039	G10390	Hot rolled	540	79	300	43.5	16	40	156	
		Cold drawn	610	88	510	74	12	35	179	60
1040	G10400	Hot rolled	520	76	290	42	18	40	149	
		Cold drawn	590	85	490	71	12	35	170	60
1042	G10420	Hot rolled	550	80	300	44	16	40	163	
		Cold drawn	610	89	520	75	12	35	179	60
		NCD	590	85	500	73	12	45	179	70
1043	G10430	Hot rolled	570	82	310	45	16	40	163	
		Cold drawn	630	91	530	77	12	35	179	60
		NCD	600	87	520	75	12	45	179	70
1044	G10440	Hot rolled	550	80	300	44	16	40	163	
1045	G10450	Hot rolled	570	82	310	45	16	40	163	
		Cold drawn	630	91	530	77	12	35	179	55
		ACD	590	85	500	73	12	45	170	65
1046	G10460	Hot rolled	590	85	320	47	15	40	170	
		Cold drawn	650	94	540	79	12	35	187	55
		ACD	620	90	520	75	12	45	179	65
1049	G10490	Hot rolled	600	87	330	48	15	35	179	
		Cold drawn	670	97	560	81.5	10	30	197	45
		ACD	630	92	530	77	10	40	187	55
1050	G10500	Hot rolled	620	90	340	49.5	15	35	179	
		Cold drawn	690	100	580	84	10	30	197	45
		ACD	660	95	550	80	10	40	189	55
1055	G10550	Hot rolled	650	94	360	51.5	12	30	192	
		ACD	660	96	560	81	10	40	197	55
1060	G10600	Hot rolled	680	98	370	54	12	30	201	
		SACD	620	90	480	70	10	45	183	60
1064	G10640	Hot rolled	670	97	370	53.5	12	30	201	
		SACD	610	89	480	69	10	45	183	60
1065	G10650	Hot rolled	690	100	380	55	12	30	207	
		SACD	630	92	490	71	10	45	187	60
1070	G10700	Hot rolled	700	102	390	56	12	30	212	
		SACD	640	93	500	72	10	45	192	55
1074	G10740	Hot rolled	720	105	400	58	12	30	217	
		SACD	650	94.5	500	73	10	40	192	55

(continued)

(a) ACD, annealed cold drawn; NCD, normalized cold drawn; SACD, spheroidized annealed cold drawn. (b) Cold drawn 1212 = 100%. Source: Ref 8

Table 9 (continued)

Steel designation SAE and/or AISI No.	UNS No.	Type of processing(a)	Tensile strength MPa	ksi	Yield strength MPa	ksi	Elongation in 50 mm (2 in.), %	Reduction in area, %	Hardness, HB	Average machinability rating(b)
Manganese 1.00% maximum (continued)										
1078	G10780	Hot rolled	690	100	380	55	12	30	207	
		SACD	650	94	500	72.5	10	40	192	55
1080	G10800	Hot rolled	770	112	420	61.5	10	25	229	
		SACD	680	98	520	75	10	40	192	45
1084	G10840	Hot rolled	820	119	450	65.5	10	25	241	
		SACD	690	100	530	77	10	40	192	45
1085	G10850	Hot rolled	830	121	460	66.5	10	25	248	
		SACD	690	100.5	540	78	10	40	192	45
1086	G10860	Hot rolled	770	112	420	61.5	10	25	229	
		SACD	670	97	510	74	10	40	192	45
1090	G10900	Hot rolled	840	122	460	67	10	25	248	
		SACD	700	101	540	78	10	40	197	45
1095	G10950	Hot rolled	830	120	460	66	10	25	248	
		SACD	680	99	520	76	10	40	197	45
Manganese, maximum >1.00%										
1524	G15240	Hot rolled	510	74	280	41	20	42	149	
		Cold drawn	570	82	480	69	12	35	163	60
1527	G15270	Hot rolled	520	75	280	41	18	40	149	
		Cold drawn	570	83	480	70	12	35	163	65
1536	G15360	Hot rolled	570	83	310	45.5	16	40	163	
		Cold drawn	630	92	530	77.5	12	35	187	55
1541	G15410	Hot rolled	630	92	350	51	15	40	187	
		Cold drawn	710	102.5	600	87	10	30	207	45
		ACD	650	94	550	80	10	45	184	60
1548	G15480	Hot rolled	660	96	370	53	14	33	197	
		Cold drawn	730	106.5	620	89.5	10	28	217	45
		ACD	640	93.5	540	78.5	10	35	192	50
1552	G15520	Hot rolled	740	108	410	59.5	12	30	217	
		ACD	680	98	570	83	10	40	193	50

(a) ACD, annealed cold drawn; NCD, normalized cold drawn; SACD, spheroidized annealed cold drawn. (b) Cold drawn 1212 = 100%. Source: Ref 8

generator to produce a gas with a mixed ratio of CO to CO_2. By controlling the CO/CO_2 ratio of the endothermic gas, an atmosphere can be generated that will be in equilibrium with the carbon content of the steel to be treated. Low-ratio gas is in equilibrium with lower-carbon steels, and high-ratio gas is in equilibrium with higher content steels. The actual ratio used depends on the type of anneal and the grade of steel to be annealed. This ratio must be closely controlled, or the atmosphere will

Table 10 Estimated mechanical properties and machinability ratings of resulfurized carbon steel bars
All 1100 and 1200 series steels are rated on the basis of 0.10% maximun or coarse-grain melting practice.

Steel designation SAE and/or AISI No.	UNS No.	Type of processing	Tensile strength MPa	ksi	Yield strength MPa	ksi	Elongation in 50 mm (2 in.), %	Reduction in area, %	Hardness, HB	Average machinability rating(b)
1108	G11080	Hot rolled	340	50	190	27.5	30	50	101	
		Cold drawn	390	56	320	47	20	40	121	80
1117	G11170	Hot rolled	430	62	230	34	23	47	121	
		Cold drawn	480	69	400	58	15	40	137	90
1132	G11320	Hot rolled	570	83	310	45.5	16	40	167	
		Cold drawn	630	92	530	77	12	35	183	75
1137	G11370	Hot rolled	610	88	330	48	15	35	179	
		Cold drawn	680	98	570	82	10	30	197	70
1140	G11400	Hot rolled	540	79	300	43.5	16	40	156	
		Cold drawn	610	88	510	74	12	35	170	70
1141	G11410	Hot rolled	650	94	360	51.5	15	35	187	
		Cold drawn	720	105.1	610	88	10	30	212	70
1144	G11440	Hot rolled	670	97	370	53	15	35	197	
		Cold drawn	740	108	620	90	10	30	217	80
1146	G11460	Hot rolled	590	85	320	47	15	40	170	
		Cold drawn	650	94	550	80	12	35	187	70
1151	G11510	Hot rolled	630	92	340	50.5	15	35	187	
		Cold drawn	700	102	590	86	10	30	207	65
1211	G12110	Hot rolled	380	55	230	33	25	45	121	
		Cold drawn	520	75	400	58	10	35	163	95
1212	G12120	Hot rolled	390	56	230	33.5	25	45	121	
		Cold drawn	540	78	410	60	10	35	167	100
1213	G12130	Hot rolled	390	56	230	33.5	25	45	121	
		Cold drawn	540	78	410	60	10	35	167	135
12L14	G12144	Hot rolled	390	57	230	34	22	45	121	
		Cold drawn	540	78	410	60	10	35	163	170

(a) Cold drawn 1212 = 100%. Source: Ref 8

become decarburizing or carburizing to the steel. Modern instruments, such as oxygen probes, are available to maintain this close control.

After carbon restoration, bars are cold drawn. Material processed in this manner is useful when parts must have full hardness on the cold-drawn (unmachined) surface after heat treatment. Many induction-hardened parts make use of a carbon-restored material as a means of eliminating machining.

Machinability

Cold drawing significantly improves the machinability of the steels discussed in this article. The increase in hardness due to cold work causes the chips formed by a cutting tool to tear away from the workpiece more readily, and to be harder and more brittle, so that they break up easily and are less likely to build up on the tool edge. Deformation extends a shorter distance above the edge of the tool, giving a sharper cleavage at that point. These factors contribute to improvements in power consumption, tool wear, and surface finish. They result from the addition of the major contributors to improvements in machinability: phosphorus, sulfur, nitrogen (diatomic), lead, bismuth, tellurium, selenium, calcium, and so on, in various combinations.

In addition, the accuracy of size and section of cold-finished bars minimizes collet troubles and requires less surface removal to obtain concentricity. The freedom from scale on the cold-finished bar also improves tool life and may permit the surface of the bar to be used as the finished surface of the completed part.

Cold drawing generally improves the machinability of low-carbon steels because the high ductility of these materials in the hot-rolled condition can be lowered considerably without raising strength excessively. In contrast, a steel such as 1144, which is inherently low in ductility because of its higher carbon content, shows little improvement in machining after cold drawing. The increased hardness that results from cold drawing can be deleterious to the machinability of the higher-carbon steels; it may be helpful to stress relieve after cold drawing to reduce hardness.

Another approach to maximum machinability with the higher-carbon grades is to anneal before cold drawing. This puts the carbide in a form that is less abrasive to the cutting tool. Lamellar anneal and spheroidize annealing are used depending on carbon level, machinability requirements, and heat treat response requirements. The trade-off values must be decided for each individual application. Compared with hot-rolled steel, uniformity of hardness and structure are improved.

One of the conventional indexes of machinability is the ratio of tool life to that

Table 11 Hardness and machinability ratings of cold-drawn alloy steels

Steel designation AISI and/or SAE No.	UNS No.	Machinability rating(a)	Condition	Range of typical hardness, HB	Microstructure type(b)
1330	G13300	55	Annealed and cold drawn	179/235	A
1335	G13350	55	Annealed and cold drawn	179/235	A
1340	G13400	50	Annealed and cold drawn	183/241	A
1345	G13450	45	Annealed and cold drawn	183/241	A
4023	G40230	70	Cold drawn	156/207	C
4024	G40240	75	Cold drawn	156/207	C
4027	G40270	70	Annealed and cold drawn	167/212	A
4028	G40280	75	Annealed and cold drawn	167/212	A
4032	G40320	70	Annealed and cold drawn	174/217	A
4037	G40370	70	Annealed and cold drawn	174/217	A
4042	G40420	65	Annealed and cold drawn	179/229	A
4047	G40470	65	Annealed and cold drawn	179/229	A
4118	G41180	60	Cold drawn	170/207	C
4130	G41300	70	Annealed and cold drawn	187/229	A
4135	G41350	70	Annealed and cold drawn	187/229	A
4137	G41370	70	Annealed and cold drawn	187/229	A
4140	G41400	65	Annealed and cold drawn	187/229	A
4142	G41420	65	Annealed and cold drawn	187/229	A
4145	G41450	60	Annealed and cold drawn	187/229	A
4147	G41470	60	Annealed and cold drawn	187/235	A
4150	G41500	55	Annealed and cold drawn	187/241	A, B
4161	G41610	50	Spheroidized and cold drawn	187/241	B, A
4320	G43200	60	Annealed and cold drawn	187/229	D, B, A
4340	G43400	50	Annealed and cold drawn	187/241	B, A
E4340	G43406	50	Annealed and cold drawn	187/241	B, A
4422	G44220	65	Cold drawn	170/212	C
4427	G44270	65	Annealed and cold drawn	170/212	A
4615	G46150	65	Cold drawn	174/223	C
4617	G46170	65	Cold drawn	174/223	C
4620	G46200	65	Cold drawn	183/229	C
4626	G46260	70	Cold drawn	170/212	C
4718	G47180	60	Cold drawn	187/229	C
4720	G47200	65	Cold drawn	187/229	C
4815	G48150	50	Annealed and cold drawn	187/229	D, B
4817	G48170	50	Annealed and cold drawn	187/229	D, B
4820	G48200	50	Annealed and cold drawn	187/229	D, B
50B40	G50401	65	Annealed and cold drawn	174/223	A
50B44	G50441	65	Annealed and cold drawn	174/223	A
5046	G50460	60	Annealed and cold drawn	174/223	A
50B46	G50461	60	Annealed and cold drawn	174/223	A
50B50	G50501	55	Annealed and cold drawn	183/235	A
5060	G50600	55	Spheroidized annealed and cold drawn	170/212	B
50B60	G50601	55	Spheroidized annealed and cold drawn	170/212	B
5115	G51150	65	Cold drawn	163/201	C
5120	G51200	70	Cold drawn	163/201	C
5130	G51300	70	Annealed and cold drawn	174/212	A
5132	G51320	70	Annealed and cold drawn	174/212	A
5135	G51350	70	Annealed and cold drawn	179/217	A
5140	G51400	65	Annealed and cold drawn	179/217	A
5147	G51470	65	Annealed and cold drawn	179/229	A
5150	G51500	60	Annealed and cold drawn	183/235	A, B
5155	G51550	55	Annealed and cold drawn	183/235	A, B
5160	G51600	55	Spheroidized annealed and cold drawn	179/217	B
51B60	G51601	55	Spheroidized annealed and cold drawn	179/217	B
50100	G50986	40	Spheroidized annealed and cold drawn	183/241	B
51100	G51986	40	Spheroidized annealed and cold drawn	183/241	B
52100	G52986	40	Spheroidized annealed and cold drawn	183/241	B
6118	G61180	60	Cold drawn	179/217	C
6150	G61500	55	Annealed and cold drawn	183/241	B, A
8115	G81150	65	Cold drawn	163/202	C
81B45	G81451	65	Annealed and cold drawn	179/223	A
8615	G86150	70	Cold drawn	179/235	C
8617	G86170	70	Cold drawn	179/235	C
8620	G86200	65	Cold drawn	179/235	C
8622	G86220	65	Cold drawn	179/235	C
8625	G86250	60	Annealed and cold drawn	179/223	A
8627	G86270	60	Annealed and cold drawn	179/223	A
8630	G86300	70	Annealed and cold drawn	179/229	A
8637	G86370	65	Annealed and cold drawn	179/229	A
8640	G86400	65	Annealed and cold drawn	184/229	A
8642	G86420	65	Annealed and cold drawn	184/229	A
8645	G86450	65	Annealed and cold drawn	184/235	A

(continued)

(a) Based on cutting with high-speed tool steels and a machinability rating of 100% for 1212 steel. (b) Type A is predominantly lamellar pearlite and ferrite. Type B is predominantly spheroidized. Type C is a hot-rolled structure that depends on grade, size, and rolling conditions of the producing mill. The structure may be coarse or fine pearlite or bainite. The pearlite at low magnification may be blocky or acicular. For descriptive information, see Ref 9. Type D is a structure resulting from a subcritical anneal or temper anneal. It is usually a granular or spheroidized carbide condition confined to the hot-rolled grain pattern, which may be blocky or acicular. Source: Ref 8

Table 11 (continued)

Steel designation AISI and/or SAE No.	UNS No.	Machinability rating(a)	Condition	Range of typical hardness, HB	Microstructure type(b)
86B45	G86451	65	Annealed and cold drawn	184/235	A
8650	G86500	60	Annealed and cold drawn	187/248	A, B
8655	G86550	55	Annealed and cold drawn	187/248	A, B
8660	G86600	55	Spheroidized annealed and cold drawn	179/217	B
8720	G87200	65	Cold drawn	179/235	C
8740	G87400	65	Annealed and cold drawn	184/235	A
8822	G88220	55	Cold drawn	179/223	B
9254	G92540	45	Spheroidized annealed and cold drawn	187/241	B
9260	G92600	40	Spheroidized annealed and cold drawn	184/235	B
9310	G93106	50	Annealed and cold drawn	184/229	D
94B15	G94151	70	Cold drawn	163/202	C
94B17	G94171	70	Cold drawn	163/202	C
94B30	G94301	70	Annealed and cold drawn	170/223	A

(a) Based on cutting with high-speed tool steels and a machinability rating of 100% for 1212 steel. (b) Type A is predominantly lamellar pearlite and ferrite. Type B is predominantly spheroidized. Type C is a hot-rolled structure that depends on grade, size, and rolling conditions of the producing mill. The structure may be coarse or fine pearlite or bainite. The pearlite at low magnification may be blocky or acicular. For descriptive information, see Ref 9. Type D is a structure resulting from a subcritical anneal or temper anneal. It is usually a granular or spheroidized carbide condition confined to the hot-rolled grain pattern, which may be blocky or acicular. Source: Ref 8

Table 12 Typical mechanical properties of special-die-drawn carbon and alloy steel bars of medium-carbon content

Steel grades	Bar size range mm	Bar size range in.	Tensile strength MPa	Tensile strength ksi	Yield strength MPa	Yield strength ksi	Elongation in 50 mm (2 in.), %(a)	Reduction in area, %(a)	Hardness, HB(b)
Carbon steels									
1541, 1045	Up to 76 round	Up to 3 round	825	120	690	100	10.0	25.0	241–321
1052, 1141	6.4 through 89 round	¼ through 3½ round	825	120	690	100	10.0	25.0	241–321
1144, 1151	6.4 through 114 round and 6.4 through 51 hexagon	¼ through 4½ round and ¼ through 2 hexagon	825	120	690	100	10.0	25.0	248–321
1144	6.4 through 6.4 round and 6.4 through 38 hexagon	¼ through 2½ round and ¼ through 1½ hexagon	965	140	860	125	5.0	15.0	280
Alloy steels									
41xx, 51xx(c)	11.1 through 89 hexagon	7/16 through 3½ hexagon	860	125	725	105	14.0	45.0	269
41xx(c), 51xx(c)	11.1 through 89 hexagon	7/16 through 3½ hexagon	103.5	150	895	130	10.0	35.0	302
41xx(c)	11.1 through 89 hexagon	7/16 through 3½ hexagon	1170	170	1140	155	5.0	20.0	355

(a) Typical minimum. (b) Typical hardness range, subject to negotiation. Hardness is taken on a flat below decarburization or on the midradius. In case of a disagreement between hardness and tensile or yield strength, the latter properties govern. (c) May contain lead or tellurium or other additives for improved machinability. Source: Ref 8

Table 13 Typical tensile properties of cold-drawn and stress-relieved 1144 grade (UNS G11440) carbon steel bars subjected to normal and heavy drafts

Diameter of round, thickness of square, or distance between parallel faces of hexagon or flat mm	in.	Round or hexagon(a) mm	Round or hexagon(a) in.	Tensile strength minimum MPa	Tensile strength minimum ksi	Yield strength minimum MPa	Yield strength minimum ksi	Elongation in 50 mm (2 in.) minimum, %	Reduction in area minimum, %
Normal draft									
To 22 incl	To ⅞ incl	...	...	725	105	655	95	10	30
>22–32 incl	>⅞–1¼ incl	...	...	690	100	620	90	10	30
>32–51 incl	>1¼–2 incl	...	...	655	95	585	85	10	25
>51–76 incl	>2–3 incl	...	...	620	90	550	80	10	20
>76–114 incl	>3–4½ incl	...	...	585	85	520	75	10	20
Heavy draft									
...	...	To 22 incl	To ⅞ incl	795	115	690	100	8	25
...	...	>22–32 incl	⅞–1¼ incl	795	115	690	100	8	25
...	...	>32–51 incl	1¼–2 incl	795	115	690	100	8	25
...	...	>51–76 incl	2–3 incl	795	115	690	100	8	20
...	...	>76–114 incl	3–4½ incl	795	115	690	100	7	20

incl, inclusive. (a) Maximum size for hexagons is 38 mm (1½ in.).

encountered with 1212 cold-drawn bars. The average machinability ratings for cold-drawn carbon steel bars, nonresulfurized and resulfurized carbon steel bars, and alloy steel bars, based on a value of 100% for 1212 bars, are given in Tables 8 to 11. The relative machinabililty data listed in Tables 8 to 11 represent results obtained from experimental studies and actual shop production information on the general run of parts. Any extraordinary features of the part to be machined or physical conditions of the steel should be taken into consideration, and speeds and feeds altered accordingly. In addition, machinability is influenced by various metallurgical factors, such as degree of cold reduction, mechanical properties, grain size, and microstructure. Therefore, the data in Tables 8 to 11 are presented only as a starting point from which proper speeds and feeds for specific parts can be determined. Further discussion of the machinability of cold-drawn steel is included in the article "Machinability of Steels" in this Volume.

Strength Considerations

Cold drawing increases the tensile and yield strengths of hot-rolled carbon steel bars by about 10 and 70%, respectively. For example, hot-rolled low-carbon steel bars with yield-to-tensile-strength ratios of about 0.55 will have ratios of about 0.85 after cold drawing. Elongation, reduction in area, and impact strength will be decreased, but these changes are relatively insignificant in many structural and engineering applications.

Improvements in strength resulting from cold drawing are of interest to the design engineer seeking a better strength-to-weight ratio or a reduction in costs by elimination of alloy additions and heat treatment. They may also be useful in applications involving threads, notches, cutouts, size stability, and other design factors that might affect strength adversely.

It should be remembered, however, that the spread in mechanical properties within a single grade of cold-drawn steel may not lend itself to as close control as may be obtained by heat treating to the same strength level. Some variation in properties may result from differences in mill technique and amount of cold work. It should also be remembered that turned and polished bars and turned, ground, and polished bars have the same mechanical properties as hot-rolled bars.

Tensile property values that may be expected for cold-drawn carbon steel bars processed with normal drafts are shown in Fig. 14, which covers bars from 19 to 32 mm (¾ to 1¼ in.) in diameter. For smaller bars, tensile and yield strengths will be slightly higher; for larger bars, they will be slightly lower. The values of yield strength shown in Fig. 14 are conservative and are very close to the expected minimums.

266 / Carbon and Low-Alloy Steels

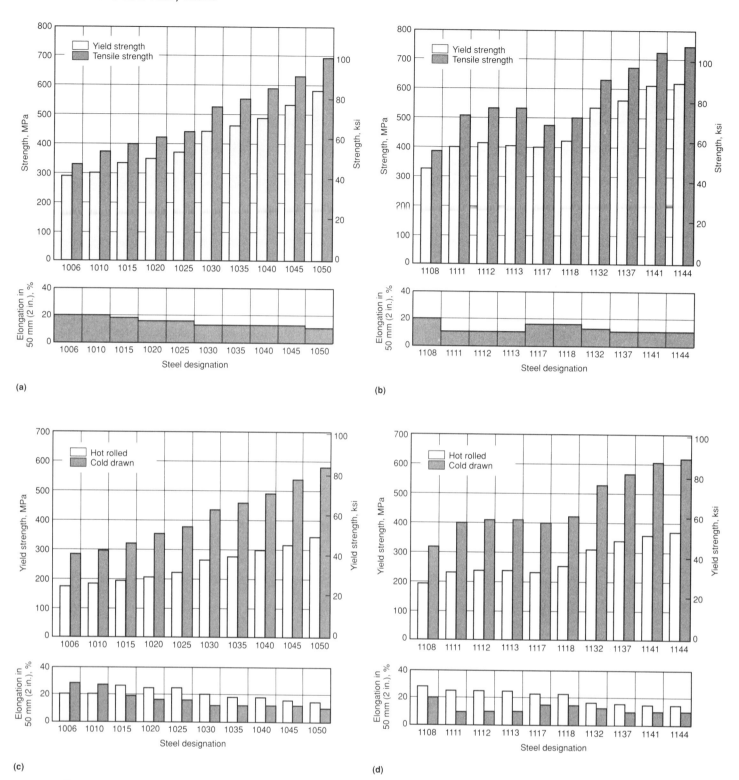

Fig. 14 Effect of cold drawing on the tensile properties of plain carbon and resulfurized carbon steels. Comparison of tensile strength and yield strength of 25 mm (1 in.) diam cold-drawn (a) plain carbon steels and (b) resulfurized carbon steels. Comparison of yield strengths of 25 mm (1 in.) diam hot-rolled and 25 mm (1 in.) diam cold-drawn (c) plain carbon and (d) resulfurized carbon steels. Source: Ref 10

Figures 15 to 19 present actual tensile test data for five grades of steel that are commonly cold drawn. In these charts, most of the minimum values for tensile and yield strengths are considerably higher than those shown in Fig. 14. Numerous factors may significantly affect these values, the principal ones being:

- The variation in composition from the low to the high limit of the specification. For example, in 1045 steel, carbon may vary from 0.40 to 0.53%, and manganese from 0.57 to 0.93%, on the basis of bar-to-bar checks (heat range plus allowable variations in product analysis)
- The effects of solidification rate, type of solidification pattern, and segregation

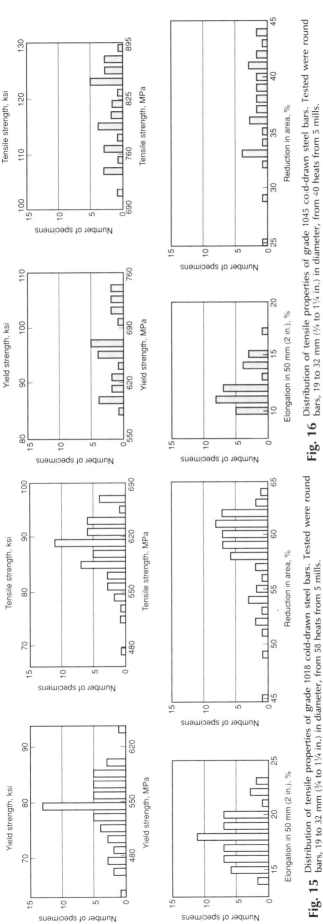

Fig. 15 Distribution of tensile properties of grade 1018 cold-drawn steel bars. Tested were round bars, 19 to 32 mm (¾ to 1¼ in.) in diameter, from 58 heats from 5 mills.

Fig. 16 Distribution of tensile properties of grade 1045 cold-drawn steel bars. Tested were round bars, 19 to 32 mm (¾ to 1¼ in.) in diameter, from 40 heats from 5 mills.

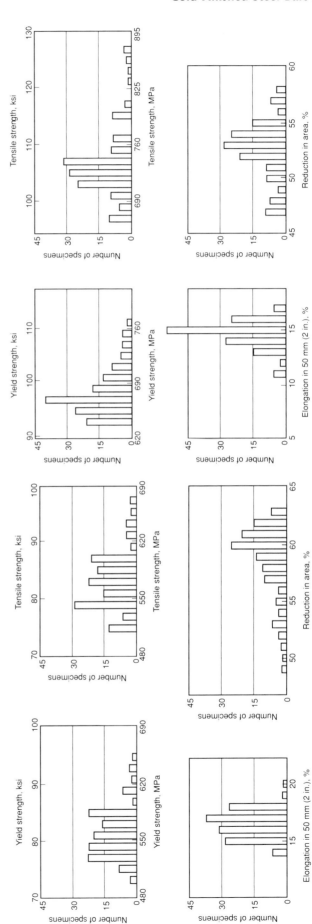

Fig. 17 Distribution of tensile properties of grade 1117 cold-drawn steel bars. Tested were round bars, 25 mm (1 in.) in diameter, from 25 heats from 2 mills.

Fig. 18 Distribution of tensile properties of grade 1137 cold-drawn steel bars. Tested were round bars, 25 mm (1 in.) in diameter, from 25 heats from 2 mills.

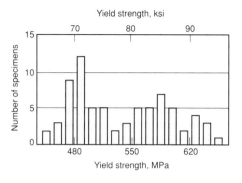
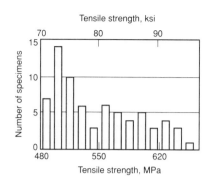
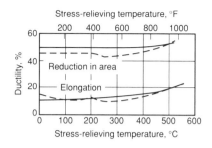
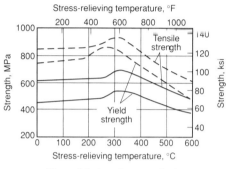

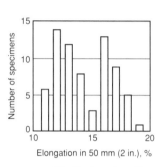
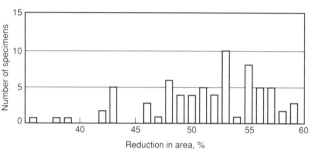

Fig. 19 Distribution of tensile properties of grade 12L14 cold-drawn steel bars. Tested were round bars, 19 to 38 mm (¾ to 1½ in.) in diameter, and hexagon bars, 14 to 25 mm (9/16 to 1 in.) in diameter, from 64 heats from 1 mill.

Fig. 20 Effect of draft and stress-relieving temperature on the tensile properties of cold-drawn carbon steel bars. Solid curves are for bars given a normal draft; dashed curves are for bars given a heavy draft.

- Temperature of the bar in the last pass in the hot mill and rate of cooling after rolling. These factors affect hardness, particularly when carbon content is 0.30% or more. With the same draft, bars of higher hardness will show higher strength and lower ductility after cold drawing than will bars of lower hardness
- Hot roll reduction
- Macro- and microstructure. Finish temperature has a dramatic effect on grain size

Therefore, the minimum property values for conventionally cold-drawn material must be conservatively low.

Special Die Drawing

Two special die-drawing processes have been developed to give improved properties over those offered by standard drawing practices. These processes are cold drawing using heavier-than-normal drafts, followed by stress relieving; and drawing at elevated temperatures. In the production of steel bars by these special processes, drafts of approximately 10 to 35% reduction in cross-

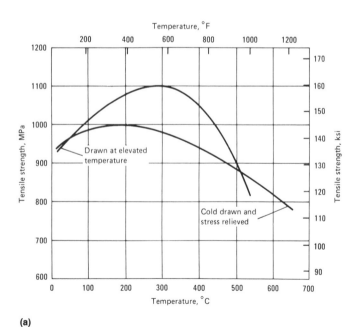
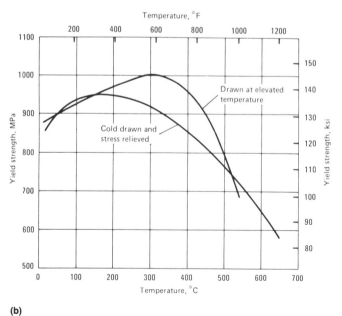

(a)   (b)

Fig. 21 Effects of stress-relieving or drawing temperature on the (a) tensile strength and (b) yield strength of cold-drawn and stress-relieved bars and on hot-drawn bars of 1144 steel. Bars, all from the same heat of steel and approximately 25 mm (1 in.) in diameter before drawing, were subjected to a draft of about 20%. Source: Ref 11

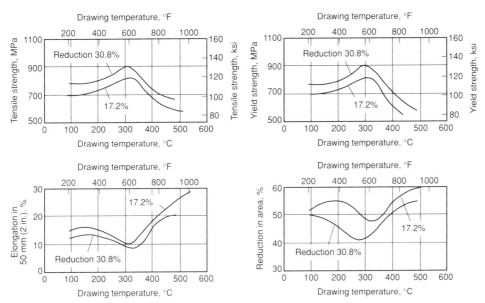

Fig. 22 Effects of drawing temperature and percentage reduction on mechanical properties of 19 mm (¾ in.) diam cold-drawn 1018 steel bars. Source: Ref 12

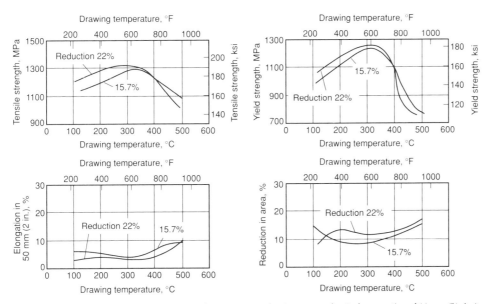

Fig. 23 Effects of drawing temperature and percentage reduction on mechanical properties of 14 mm (⁹⁄₁₆ in.) diam cold-drawn 1080 steel bars. Source: Ref 12

sectional area are employed at room or elevated temperature, depending on the practices and facilities of the individual producer. Stress-relieving temperatures vary over a similarly wide range, depending on producer facilities and end-product requirements. Typical tensile properties of plain carbon and alloy steel bars of medium carbon content subjected to special die-drawing processing are given in Table 12.

Heavy Drafts. Because of the engineering and economic advantages of cold-finished steels, a considerable effort has been made to improve the uniformity of mechanical properties after cold drawing. This has been accomplished by using heavier-than-normal drafts (10 to 35% reduction) followed by subcritical annealing. Stepwise or tandem drawing has been resorted to in order to avoid the formation of internal transverse fissures (cupping or bambooing) that may result from heavy drafts taken in one pass. The trend of property improvement resulting from this special practice can be seen in the comparison of normal and heavy drafts shown in Fig. 20.

Heavier drafts produce higher tensile and yield strengths. Elongation can be substantially improved by stress relieving at 510 °C (950 °F), and this treatment still provides tensile and yield strengths higher than those obtainable with normal drafts. The combination of properties resulting from heavier drafts and higher stress-relieving temperatures is most desirable from the design standpoint. Such processing is most effective when applied to medium-carbon steels of either normal or higher manganese content. In the medium-manganese range, 1045 and 1050 respond most favorably; 1137, 1141, 1144, 1527, 1536, and 1541 show good response for the higher-manganese grades. The sulfur content of the 11xx steels improves machinability without lowering mechanical properties. Typical tensile properties for 1144 bars subjected to normal and heavy drafts are given in Table 13.

Steel bars that have been cold drawn using heavier-than-normal drafts and then stress relieved are often used in place of quenched and tempered bars, and as already noted, several resulfurized grades (1137, 1141, and 1144) respond readily to this process with resulting high strength values. Because the microstructures of these steels are still pearlitic, they machine more easily than their quenched and tempered counterparts. Therefore, although these grades cost more than nonresulfurized grades, they can provide significant savings in manufacturing costs, chiefly through the elimination of heat treating. However, even though the strength of the special cold-drawn and stress-relieved bars may be equal to that of quenched and tempered steel, it is not advisable to translate other properties from one condition to the other.

The torsional strength and endurance limits of these special-process bars are similar to those of quenched and tempered bars at the same strength level. The same is true for the wear resistance of bars of the same surface hardness. The impact test values of the process bars, as measured by Izod or Charpy notched-bar test, are lower than those of quenched and tempered carbon steel bars and are significantly lower than those of quenched and tempered alloy steel bars. Failures of machine components usually result from fatigue, corrosion, wear, or shock loading. With the possible exception of the latter, there is no known correlation between instances of failure and the notched-bar impact test. When low temperatures or high pressures are involved and where there is doubt as to the suitability of these special-process bars, the design of the part should be reviewed.

Drawing at elevated temperatures between 95 and 540 °C (200 and 1000 °F), a special proprietary process, can, under optimum conditions, produce steel bars that have higher tensile and yield strengths than those of bars cold drawn with the same degree of reduction. The relative effects of cold drawing followed by stress relieving and of drawing at elevated temperature can be seen in Fig. 21. Both processes were carried out with 20% draft on 25 mm (1 in.) diam bars of 1144 steel. As shown in Fig. 21, elevated-temperature drawing affects tensile strength considerably, giving values

270 / Carbon and Low-Alloy Steels

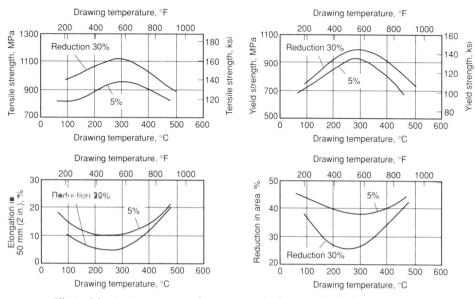

Fig. 24 Effects of drawing temperature and percentage reduction on mechanical properties of 16 mm (⅝ in.) diam cold-drawn 1144 steel bars. Source: Ref 12

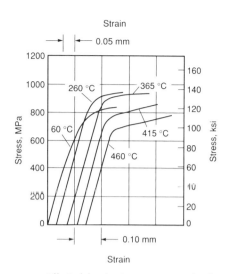

Fig. 26 Effect of drawing temperature on the shape of the tensile stress-strain curve for hot-drawn 1144 steel bars. Bars were reduced 7.2% to a diameter of 21.4 mm (²⁷⁄₃₂ in.). Source: Ref 12

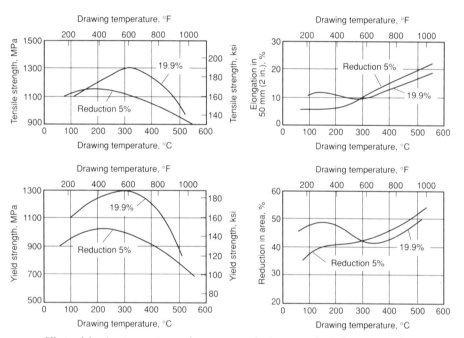

Fig. 25 Effects of drawing temperature and percentage reduction on mechanical properties of 16 mm (⅝ in.) diam cold-drawn 4140 steel bars. Source: Ref 12

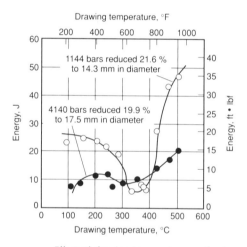

Fig. 27 Effect of drawing temperature on the impact properties of hot-drawn bars of two steels. Source: Ref 12

somewhat greater than those for cold-drawn stress-relieved bars. For yield strength, the same general effects are evident, but the difference between the two processes is not as pronounced. The elongation values are quite similar for both processes.

Figures 22 to 25 show the effects of two drafts and increasing drawing temperatures on bars of each of four steels. Between 260 and 315 °C (500 and 600 °F), tensile and yield strengths reach maximums and elongation and reduction in area reach minimums. Strength for any given drawing temperature increases with increased draft, with minor exceptions. Depending on chemical composition, a certain minimum draft is required to fully develop the effect on strength properties of drawing at elevated temperature. In general, the draft taken at room temperature would have to be doubled in order to match the strength developed in drawing at elevated temperature. The yield strengths of all four steels were increased from 12 to 35%, depending on percentage reduction in drawing and on drawing temperature. Carbon steels were less affected by increased draft than alloy steel 4140 when drawn at elevated temperatures.

The pronounced effect of drawing at elevated temperatures changes the shape of the stress-strain curve from round to sharp-kneed, as shown in Fig. 26. Hot drawing reverses the effect of cold work, that is, automatically stress relieves the steel. This effect on the stress-strain curve is significant in applications in which minimum plastic deformation is permissible, such as a stud that requires a proof load high in relation to its tensile strength.

Typical yield strength minimums for 1144 bars before and after drawing are as follows:

	MPa	ksi
As-hot-rolled	345	50
Cold drawn (normal draft) and stress relieved at low temperature	550	80
Cold drawn (heavy draft) and stress relieved at high temperature	690	100
Drawn at elevated temperature	860	125

The effect of drawing temperature on notched-bar impact values is shown in Fig. 27. Although impact values decrease with the first increase in drawing temperature, they later rise significantly. Thus, it may be possible to select a drawing temperature that will provide both good static strength and satisfactory notched-bar impact strength.

REFERENCES

1. J.G. Bralla, *Handbook of Product Design for Manufacturing*, McGraw-Hill, 1986
2. *Alloy, Carbon, and High Strength Low Alloy Steels, Semifinished for Forging; Hot Rolled Bars; Cold Finished Steel Bars; Hot Rolled Deformed and Plain Concrete Reinforcing Bars*, AISI Steel Products Manual, American Iron and Steel Institute, 1986
3. *Handbook of Machining Data for Cold Finished Steel Bars*, LTV Steel Flat Rolled and Bar Company, 1985
4. *Steel—Bars, Forgings, Bearing, Chain, Springs*, Vol 1.05, *Annual Book of ASTM Standards*, American Society for Testing and Materials, 1989
5. L.J. Ebert, Report WAL 310/90-85 to Watertown Arsenal, 1955
6. H. Buhler and H. Bucholz, Influence of Cold Drawn Reduction Upon Stresses in Round Bars, *Arch. Eisenhüttenwes.*, Vol 7, 1934, p 427-430
7. E. Dieter, *Mechanical Metallurgy*, McGraw-Hill, 1976
8. *Materials*, Vol 1, *1989 SAE Handbook*, Society of Automotive Engineers, 1989
9. "U.S. Air Force Machinability Report," Vol 2, Curtiss-Wright Corporation, 1951
10. "Estimated Properties and Machinability of Plain Carbon and Re-sulfurized Plain Carbon Steel Bars," SAE J414, SAE Handbook Supplement HS30, Society of Automotive Engineers, 1976, p 3.14
11. E.S. Nachtman and E.B. Moore, *J. Met.*, April 1955
12. E.S. Nachtman and E.B. Moore, *J. Met.*, April 1958

Steel Wire Rod

Revised by R.J. Glodowski, Armco, Inc.

WIRE ROD is a semifinished product rolled from billet on a rod mill and is used primarily for the manufacture of wire. The steel for wire rod is produced by all the modern processes, including the basic oxygen, basic open hearth, and electric furnace processes. Steel wire rod is usually cold drawn into wire suitable for further drawing; for cold rolling, cold heading, cold upsetting, cold extrusion, or cold forging; or for hot forging.

Although wire rod may be produced in several regular shapes, most is round in cross section. Round rod is usually produced in nominal diameters of 5.6 to 18.7 mm ($7/32$ to $47/64$ in.), advancing in increments of 0.4 mm ($1/64$ in.).* As the rod comes off the rolling mill, it is formed into coils. These coils are usually about 760 mm (30 in.) in inside diameter and weigh up to 2000 Kg (4400 lb). The dimensions and maximum weight of a single coil are determined by the capabilities of the rolling mill. Coil weights that exceed the capabilities of the rolling mill sometimes can be obtained by welding two or more coils together. The standard tolerances are ±0.4 mm (±$1/64$ in.) on the diameter and 0.64 mm (0.025 in.) maximum out-of-roundness.

Producers of wire rod may market their product as rolled, as cleaned and coated, or as heat treated, although users generally prefer to do such preparations themselves. These operations are explained in the following sections, along with the several recognized quality and commodity classifications applicable to steel wire rods.

Cleaning and Coating

Mill scale is cleaned from steel wire rods by pickling or caustic cleaning followed by water rinsing, or by mechanical means such as shot blasting with abrasive particles or reverse bending over sheaves. The chemical cleaning of steel wire rods is always followed by a supplementary coating operation. Lime, borax, or phosphate coating is applied to provide a carrier for the lubricant necessary for subsequent processing into wire. In lime coating, practices may be varied in order to apply differing amounts of lime on the rods depending on the customer requirements. Phosphate-coated rods may have a supplementary coating of lime, borax, or water-soluble soap. Mechanically descaled rods may be drawn without coating using only wire drawing soaps, or may be coated in a fashion similar to that used for chemically cleaned rods.

Heat Treatment

The heat treatments commonly applied to steel wire rod, either before or during processing into wire, include annealing, spheroidize annealing, patenting, and controlled cooling. Annealing commonly involves heating to a temperature near or below the lower critical temperature and holding at that temperature for a sufficient period of time, followed by slow cooling. This process softens the steel for further processing, but not to the same degree as does spheroidize annealing. Spheroidize annealing involves prolonged heating at a temperature near or slightly below the lower critical temperature (or thermal cycling at about the lower critical temperature), followed by slow cooling, with the object of changing the shape of carbides in the microstructure to globular (spheroidal), which produces maximum softness.

Patenting is a heat treatment usually confined to medium-high-carbon and high-carbon steels. In this process, individual strands of rod or wire are heated well above the upper critical temperature and then are cooled comparatively rapidly in air, molten salt, molten lead, or a fluidized bed. The object of patenting is to develop a microstructure of homogeneous, fine pearlite. This treatment generally is employed to prepare the material for subsequent wire drawing.

Controlled cooling is a heat treatment performed in modern rod mills in which the rate of cooling after hot rolling is carefully controlled. The process imparts uniformity of properties and some degree of control over scale, grain size, and microstructure.

Carbon Steel Rod

Carbon steels are those steels for which no minimum content is specified or required for chromium, nickel, molybdenum, tungsten, vanadium, cobalt, niobium, titanium, zirconium, aluminum, or any other element added to obtain a desired alloying effect; for which specified minimum copper content does not exceed 0.40%; for which specified maximum manganese content does not exceed 1.65%; and for which specified maximum silicon or copper content does not exceed 0.60%. In all carbon steels, small quantities of certain residual elements, such as chromium, nickel, molybdenum, and copper, are unavoidably retained from raw materials. These elements are considered incidental, although maximum limits are commonly specified for specific end-uses.

Carbon steel rods are produced in various grades, or compositions:

- Low-carbon steel wire rods (maximum carbon content ≤0.15%)
- Medium-low-carbon steel wire rods (maximum carbon content >0.15%, but ≤0.23%)
- Medium-high-carbon steel wire rods (maximum carbon content >0.23%, but ≤0.44%)
- High-carbon steel wire rods (maximum carbon content >0.44%)

Ordinarily, sulfur and phosphorus contents are kept within the usual limits for each grade of steel, while carbon, manganese, and silicon contents are varied according to the mechanical properties desired. Occasionally, sulfur and/or phosphorus may be added to the steel to improve the machinability.

Qualities and Commodities of Carbon Steel Rod

Rod for the manufacture of carbon steel wire is produced with manufacturing controls and inspection procedures intended to ensure the degree of soundness and free-

---

*Because steel wire rod manufactured in the United States is customarily produced to fractional-inch sizes, rather than decimal-inch or millimeter sizes, the millimeter conversion for wire rod sizes may not be a multiple of the 0.4 mm ($1/64$ in.) increment size.

dom from injurious surface imperfections necessary for specific applications. The various quality descriptors and commodities applicable to carbon steel wire rod are described below.

Industrial quality rod is manufactured from low-carbon or medium-low-carbon steel and is intended primarily for drawing into industrial quality wire. Rod of this quality is available in the as-rolled or heat-treated conditions. Practical limitations for drawing are: low-carbon rod 5.6 mm (7/32 in.) in diameter can be drawn without intermediate annealing to 2.0 mm (0.080 in.) by five conventional drafts; medium-low-carbon rod 5.6 mm (7/32 in.) in diameter can be drawn without intermediate annealing to 2.69 mm (0.106 in.) by four conventional drafts.

Chain Quality Rod. Rod for the manufacture of wire to be used for resistance welded chain is made from low-carbon and medium-low-carbon steel produced by practices that ensure their suitability for drawing into wire for this end-use. Good butt welding uniformity characteristics and internal soundness are essential for this application. Rod for the manufacture of wire to be used for fusion welded chain can be produced from specially selected low-carbon rimmed steel, but is more often made from continuous cast steel.

Fine wire quality rod is suitable for drawing into small-diameter wire either without intermediate annealing treatments or with only one such treatment. Rod 5.6 mm (7/32 in.) in diameter can be direct drawn into wire as fine as 0.9 mm (0.035 in.) without intermediate annealing. Wire finer than 0.9 mm (0.035 in.), for such products as insect-screen wire, weaving wire, and florist wire, is usually drawn in two steps: reducing to an intermediate size no smaller than 0.9 mm (0.035 in.), followed by annealing and redrawing to final size.

Fine wire quality rod is generally rolled from steel of grade 1005 or 1006 produced using techniques to provide good surface finish and internal cleanliness. In addition to these precautions, the producer may subject the rod to tests such as fracture or macroetch tests.

Cold finishing quality rod is intended for drawing into cold finished bars; the manufacture of such rod is controlled to ensure suitable surface conditions.

Heading, Cold Extrusion, or Cold Rolling Quality Rod. Rod used for the manufacture of heading, forging, cold extrusion or cold rolling quality wire is produced by closely controlled manufacturing practices. It is subject to mill testing and inspection to ensure internal soundness and freedom from injurious surface imperfections. Heat treatment as a part of wire mill processing is very important in the higher carbon grades of steel. For common upsetting, represented by the production of standard trimmed hexagon-head cap screws, 1016 to 1038 steel wire drawn from annealed rod is suitable. Wire for moderate upsetting, also produced from 1016 to 1038 steel, should be drawn from spheroidize annealed rod or should be in-process annealed. Wire for severe heading and forging, produced from rod of 1016 to 1541 steel, should be spheroidize annealed in process or at finished size. Rod of this quality is not intended for recessed-head or similar special-head applications.

In the production of rod for heading, forging, or cold extrusion in killed carbon steels with nominal carbon contents of 0.16% or more (AISI grades 1016 or higher), both austenitic grain size and decarburization should be controlled. Such steels can be produced with either fine or coarse austenitic grains, depending on the type of heat treatment and end-use. The maximum allowable amounts of decarburization as defined by the average value for the depth of the layer of free ferrite plus the layer of partial decarburization (the total affected depth) and the average depth of the layer of free ferrite alone are given below:

Nominal rod diameter, mm (in.)	Average depth of decarburization			
	Free ferrite		Total affected	
	mm	in.	mm	in.
≤9.6 (≤3/8)	0.10	0.004	0.30	0.012
10–12.8 (25/64–1/2)	0.13	0.005	0.36	0.014
13–18.7 (33/64–47/64)	0.15	0.006	0.41	0.016

If decarburization limits closer than these standard limits are required in a given manufactured product, it is sometimes necessary to incorporate means for carbon restoration in the annealing process. When there are discrepancies in decarburization test results, it is customary to make heat-treatment tests of the finished product to determine suitability for the particular application.

Wood screw quality rod includes low-carbon resulfurized and nonresulfurized wire rod for drawing into wire for the manufacture of slotted-head screws only, not for recessed-head or other special-head screws.

Scrapless Nut Quality Rod. Rod to be drawn into wire for scrapless nuts is produced by specially controlled manufacturing practices. It is subjected to mill tests and inspection designed to ensure internal soundness; freedom from injurious segregation and injurious surface imperfections; and satisfactory performance during cold heading, cold expanding, cold punching, and thread tapping.

Rod for scrapless nut wire commonly is made from low-carbon, resulfurized steels. Nonresulfurized steels are also used; these steels ordinarily are furnished only in grades containing more carbon than the resulfurized grades and with phosphorus content not exceeding 0.035% and sulfur content not exceeding 0.045% by heat analysis.

In making resulfurized steel for scrapless nut quality rod, either an ingot or continuous casting process can be used. In an ingot manufacturing process, sometimes the sulfur content is obtained through delayed mold additions to a conventional nonresulfurized rimming steel. The purpose of such a practice is to produce a steel consisting of a rim of low-sulfur steel suitable for expansion during nut forming around a high-sulfur interior section suitable for the piercing and threading operations involved in making scrapless nuts. When high sulfur content is secured through such mold additions, sulfur analyses are made on the solid billets rather than by heat analysis. It is customary to produce these steels to a specified sulfur range of either 0.08 to 0.13% or 0.04 to 0.09%. Because of the practice used in making the steel and the degree to which sulfur segregates, the sulfur content at various locations in a billet may vary from the indicated range.

When a continuous casting process is used to make resulfurized steel, the sulfur content is typically more uniform than the ingot process. However, continuous casting precludes the rimming practice described in the above paragraph.

Severe cold heading, severe cold extrusion, or severe scrapless nut quality rod is used for severe single-step or multiple-step cold forming where intermediate heat treatment and inspection are not possible. Rod of this quality is produced with carefully controlled manufacturing practices and rigid inspection practices to ensure the required degree of internal soundness and freedom from surface imperfections. A fully killed fine-grain steel is usually required for the most difficult operations. Normally, the wire made from this quality rod is spheroidize annealed, either in process or after drawing finished sizes. Decarburization limits and the steels to which they apply are the same as those described in the section "Heading, Cold Extrusion, or Cold Rolling Quality Rod" in this article.

Welding-quality rod is used to make wire for gas or electric-arc welding filler metal. Welding-quality rod can be made from selected ingots or billets of low-carbon rimmed, capped, or killed steel, but is preferably made from continuous cast steel. It is produced to several restricted ranges and limits of chemical composition; an example of the restricted ranges and limits for low-carbon, arc welding wire rod is shown below:

Element	Heat analysis, %
Carbon	0.10–0.15
Manganese	0.40–0.60
Phosphorus	0.025 max
Sulfur	0.035 max
Silicon	0.030 max

Table 1 Tensile strength of 5.6 mm (7/32 in.) diam hot-rolled low-carbon steel rod
Data obtained from rod produced with controlled cooling

Steel grade	Rimmed		Capped		Aluminum killed fine-grain steel		Silicon killed fine- or coarse-grain steel	
	MPa	ksi	MPa	ksi	MPa	ksi	MPa	ksi
1005	350	51	...	...	380	55	395	57
1006	360	52	365	53	395	57	405	59
1008	370	54	385	56	405	59	425	62
1010	385	56	400	58	420	61	440	64
1012	405	59	420	61	435	63	455	66
1015	425	62	440	64	450	65	475	69
1017	450	65	455	66	455	66	495	72
1018	...	...	475	69	490	71	525	76
1020	470	68	475	69	485	70	510	74
1022	...	...	...	...	520	75	565	82

Rod for welding-quality wire constitutes an exception to the general practice that rimmed or capped steel is not commonly subject to product analysis. Experience to date has shown the necessity for close control of composition, and therefore only billets from those portions of the ingot that conform to the applicable ranges and limits are used for welding-quality rod. For the majority of welding-quality rod that is made from continuous cast steel, these product checks may not be necessary.

Medium-high-carbon and high-carbon quality rod is wire rod intended for drawing into such products as strand wire, lockwasher wire, tire bead wire, upholstery spring wire, rope wire, screen wire (for heavy aggregate screens), aluminum cable steel reinforced core wire, and prestressed concrete wire.

These wire qualities are normally drawn directly from patented or control-cooled rod. When drawing to sizes finer than 2.0 mm (0.080 in.) (from 5.6 mm, or 7/32 in., rod), it is customary to employ in-process heat treatment before drawing to finish size. Medium-high-carbon and high-carbon quality rod is not intended for the manufacture of higher-quality wires such as music wire or valve spring wire.

Rod for Special Purposes. In addition to the carbon steel rod commodities described above, which have specific quality descriptors, several other commodities are produced, each having the characteristics necessary for a specific application, but for which no specific quality descriptor exists. Some of these commodities are made to standard specifications; the others are made to proprietary specifications that are mutually acceptable to both producer and user.

Rod for music wire, valve spring wire, and tire cord wire is rolled and conditioned to ensure the lowest possible incidence of imperfections. Surface imperfections are objectionable because they lower the fatigue resistance that is important in many of the end products made from these wires. Internal imperfections are objectionable because they make the rod unsuitable for cold drawing to high strength levels and the extremely fine sizes required.

Rod for concrete reinforcement is nondeformed rod produced from steel chemical compositions selected to provide the mechanical property requirements for grade 40 and grade 60, as described in ASTM A 615. This quality rod is produced in coils.

Rod for telephone and telegraph wire is produced by practices and to chemical compositions intended for the manufacture of wire having electrical and mechanical properties that will meet the requirements of the various grades of this type of wire.

Special Requirements for Carbon Steel Rod

Some of the quality descriptors discussed above imply special requirements for the manufacture and testing of wire rod. A few of the more common requirements are listed below. For some applications, it may be appropriate to add one or more special requirements to those implied by the quality descriptor.

Macroetch testing is deep-etch testing to evaluate internal soundness. A representative cross section is etched in a hot acid solution.

Fracture Testing. In fracture testing, a specimen is fractured to evaluate soundness and homogeneity.

Austenitic Grain Size Requirements. For applications involving carburizing or heat treatment, austenitic grain size for killed steels may be specified as either coarse (grain size 1 through 5) or fine (grain size 5 through 8 inclusive), in accordance with ASTM E 112.

Heat-Treating Requirements. When heat-treating requirements must be met in the purchaser's end product, all heat treatment procedures and mechanical property requirements should be clearly specified.

Nonmetallic inclusion testing comprises a microscopic examination of longitudinal sections of the rod to determine the nature and frequency of nonmetallic inclusions. Methods B or C of ASTM E 45 are commonly used.

Decarburization limits are specified for special applications when required. A specimen is polished so that the entire cross-sectional area is in a single plane, with no rounded edges. After etching with a suitable etchant, the specimen is examined microscopically (usually at 100 diameters), and the results are reported in hundredths of a millimeter or thousandths of an inch. The examination includes the entire periphery, and the results reported should include the amount of free ferrite and the total depth of decarburization. Further details of this microscopic method are contained in SAE Recommended Practice J419, Methods of Measuring Decarburization.

Mechanical Properties of Carbon Steel Rod

In the older mills, where rod was coiled hot, there was considerable variation within each coil because of the effect of varying cooling rates from the center to the periphery of the coil. Therefore, as-hot-rolled rod was seldom sold to specific mechanical properties because of the inherent variations of such properties. These properties for a given grade of steel varied from mill to mill and were influenced by both the type of mill and the source of steel being rolled.

In new rod mills, which are equipped with controlled cooling facilities, this intracoil variation is kept to a minimum. In such mills, finishing temperature, cooling of water, cooling air, and conveyor speed all are balanced to produce rod with the desired scale and microstructure. This structure, in turn, is reflected in the mechanical properties of the rod and permits the rod to be drawn directly for all but the most demanding applications. The primary source of intracoil variation on these new mills is the overlapping of the coiled rings on the conveyor. These overlapped areas cool at a slower rate than the majority of the ring.

Table 1 lists typical values of tensile strength for 5.6 mm (7/32 in.) low-carbon steel rod rolled on a modern rod mill equipped with controlled cooling facilities. The values shown are for rods rolled with full air cooling. Tensile strength values for larger-diameter rod are lower, decreasing by approximately 1.9 MPa (270 psi) for each 0.4 mm (1/64 in.) increment by which rod diameter exceeds 5.6 mm (7/32 in.). Similar analyses of rods rolled without full air cooling or rods rolled on an older mill, where the steel is coiled hot, would be expected to reveal lower tensile strength.

Table 2 shows typical expected tensile strength values for 5.6 mm (7/32 in.) medium-high-carbon and high-carbon steel rods rolled on a mill utilizing controlled cooling. The microstructure of such rod approximates that obtained by patenting. The strength generally falls between those obtained by air patenting and lead patenting. Most high-carbon steel wire is drawn from such rods without prior patenting. Tensile strengths for large-diameter rods have been

Table 2 Tensile strength of 5.6 mm (7/32 in.) diam hot-rolled medium-high-carbon and high-carbon steel rod
Data obtained from rod produced with controlled cooling

Carbon content of steel, %	Tensile strength for steel with manganese content of					
	0.60%		0.80%		1.00%	
	MPa	ksi	MPa	ksi	MPa	ksi
0.30	641	93	676	98	717	104
0.35	689	100	731	106	793	115
0.40	745	108	779	113	820	119
0.45	793	115	834	121	869	126
0.50	848	123	883	128	931	135
0.55	896	130	938	136	972	141
0.60	951	138	986	143	1020	148
0.65	1000	145	1041	151	1076	156
0.70	1055	153	1089	158	1124	163
0.75	1103	160	1138	165	1179	171
0.80	1151	167	1193	173	1227	178
0.85	1207	175	1241	180	1282	186

found to average 5.2 MPa (750 psi) lower for each 0.4 mm (1/64 in.) increment in diameter over 5.6 mm (7/32 in.). Additional strength in control-cooled high-carbon rods can be achieved by using microalloying techniques. These procedures should be used with caution because they may affect other properties such as ductility and durability.

Alloy Steel Rod

Alloy steels are those steels for which maximum specified manganese content exceeds 1.65% or maximum specified silicon or copper content exceeds 0.60%; or for which a definite range or definite minimum quantity of any other element is specified in order to obtain desired effects on properties. Detailed information on composition ranges and limits of alloy steels can be found in the article "Classification and Designation of Carbon and Low-Alloy Steels" in this Volume.

Qualities and Commodities for Alloy Steel Rod

The various qualities of alloy steel wire rod possess characteristics that are adapted to the particular conditions typically encountered during fabrication or service. Manufacture of these steels normally includes careful selection of raw materials for melting, exacting steelmaking practices, selective discard (when the steel is produced in ingots), extensive billet preparation, and extensive testing and inspection.

Occasionally, alloy steel of a special quality is specified for the manufacture of wire rod. Aircraft quality alloy steel may be specified for wire rods intended for processing into critical or highly stressed aircraft parts or for similar purposes. Bearing quality alloy steel may be specified for wire rods intended for processing into balls and rollers for antifriction bearings. Bearing quality alloy steel is usually specified when purchasing the standard carburizing grades, such as 4118, 4320, 4620, 4720, and 8620, or the through-hardening, high-carbon chromium grades such as E51100 and E52100. The various standard qualities and commodities available in alloy steel wire rod are described below.

Cold heading quality alloy steel rod is used for the manufacture of wire for applications involving cold plastic deformation by such operations as upsetting, heading, forging, or extrusion. Typical parts are fasteners (cap screws, bolts, eyebolts), studs, anchor pins, and balls and rollers for antifriction bearings.

Special cold heading quality alloy steel rod is used for wire for applications involving severe cold plastic deformation. Surface quality requirements are more critical than for cold heading quality. Steel with very uniform chemical composition and internal soundness, as well as special surface preparation of the semifinished steel, are required. Typical applications are ball joint suspension studs, socket head screws, recessed-head screws, and valves.

Welding quality alloy steel rod is used for the manufacture of wire used as filler metal in electric arc welding or for building up hard wearing surfaces of parts subjected to wear. The heat analysis limits given below for phosphorus and sulfur apply to this quality rod:

Steelmaking process	Maximum percent	
	P	S
Basic electric	0.025	0.025
Basic open hearth or basic oxygen	0.025	0.035

Special Requirements for Alloy Steel Rod

Alloy steel rod can be produced with special requirements in addition to those implied by the quality descriptors discussed above. These special requirements include those given below.

Special surface entails a product with minimal frequency and severity of seams and other surface imperfections.

Decarburization limits can be specified for special applications. An example of such limits are those shown in the table below for alloy steel rod for wire for heading, forging, roll threading, extrusion, lockwasher, and screwdriver applications. Listed below are the maximum allowable amounts of decarburization as defined by the average value for the depth of the layer of free ferrite plus the layer of partial decarburization (the total affected depth) and the average depth of the layer of free ferrite alone:

Nominal rod diameter, mm (in.)	Average depth of decarburization			
	Free ferrite		Total affected	
	mm	in.	mm	in.
≤6.4 (≤1/4)	0.08	0.003	0.20	0.008
6.8–9.6 (17/64–3/8)	0.08	0.003	0.25	0.010
10–12.8 (25/64–1/2)	0.10	0.004	0.30	0.012
13–18 (33/64–45/64)	0.13	0.005	0.36	0.014
18.2–25 (23/32–1)	0.15	0.006	0.41	0.016

When limits closer than those given above are required for the end product, it is sometimes appropriate to incorporate carbon restoration in the fabrication process. For some applications, the rod producer can include carbon restoration in the mill heat treatment. The method of measuring decarburization is the same as that described for carbon steel rods.

Heat-Treating Requirements. When the end product must be heat treated, the heat treatment and mechanical properties should be clearly defined.

Hardenability requirements are customarily specified by H-steel designations and hardenability bands. These steels and hardenability bands are discussed in the article "Hardenability of Carbon and Low-Alloy Steels" in this Volume.

Austenitic Grain Size Determination. Most alloy steels are produced using fine-grain practice. Fine-grain steels are useful in carburized parts, especially when direct quenching is involved, and are less sensitive than coarse-grain steels to variations in heat-treating practices. Coarse-grain steels are deeper hardening and are generally considered more machinable. Austenitic grain size is specified as either coarse (grain sizes 1 through 5) or fine (grain sizes 5 through 8), determined in accordance with ASTM E 112.

Nonmetallic-Inclusion Testing. When the nonmetallic-inclusion test is specified, it is commonly done on billets. Prepared and polished specimens are examined microscopically at 100 diameters. Sample locations, number of tests, and limits of acceptability should be established in each instance. Test procedures are described in ASTM E 45.

Magnetic-Particle Inspection. For alloy steel rod and wire products subject to magnetic-particle inspection, it is customary for the producer to test the product in a semifinished form, such as billets (using specimens properly machined from billets), to ensure that the heat conforms to the magnetic-particle inspection requirements, prior to further processing.

The method of inspection consists of suitably magnetizing the steel and applying a prepared magnetic powder, either dry or suspended in a suitable liquid, that adheres to the steel along lines of flux leakage. On properly magnetized steel, flux leakage develops along surface or subsurface discontinuities. The results of the inspection will vary with the degree of magnetization, the inspection procedure (including such conditions as relative location of surfaces tested), the method and sequence of magnetizing and applying the powder, and the interpretation. The testing procedure and standards of acceptance for magnetic-particle inspection are described in Aerospace Materials Specification 2301.

Macroetch Testing. Soundness and homogeneity of alloy steel rod are sometimes evaluated macroscopically by examining a properly prepared cross section of the product after it has been immersed in a hot acid solution. It is customary to use hydrochloric acid for this purpose.

ACKNOWLEDGMENT

The helpful suggestions provided by Zeev Zimerman, Bethlehem Steel Corporation, and Bhaskar Yalamanchili, North Star Steel Texas Company, are greatly appreciated.

Steel Wire

Revised by Allan B. Dove, Consultant

WIRE can be cold drawn from any of the types of carbon steel or alloy steel rod described in the article "Steel Wire Rod" in this Volume. For convenience, the various grades of carbon steel wire can be divided into the same four classes used for carbon steel rod. Based on carbon content, these classes are:

- Low-carbon steel wire (0.15% C max)
- Medium-low-carbon steel wire (>0.15 to 0.23% C)
- Medium-high-carbon steel wire (>0.23 to 0.44% C)
- High-carbon steel wire (>0.44% C)

The conventional four-digit or five-digit American Iron and Steel Institute—Society of Automotive Engineers (AISI-SAE) designation is used to specify the carbon or alloy steel used to make the wire. Carbon and alloy steel wire can be produced in qualities suitable for cold rolling, cold drawing, cold heading, cold upsetting, cold extrusion, cold forging, hot forging, cold coiling, heat treatment, or carburizing and for a wide variety of fabricated products.

Wire Configurations and Sizes

Shapes of Wires. Although wire is ordinarily thought of as being only round, it may have any one of an infinite number of sectional shapes, as required by end use. After ordinary round wire, the most common shapes are square, hexagonal, octagonal, oval, half-oval, half-round, triangular, keystone, and flat. In addition to these regular (symmetrical) shapes, wire is also made in various odd and irregular shapes for specific purposes.

Flat wire, as defined by AISI, is wire that has been cold rolled or drawn, has a prepared edge, is rectangular in shape, 25 mm (1 in.) or less in width, and less than 9.5 mm (3/8 in.) in thickness. Flat wire is generally produced from hot-rolled rods or specially prepared round wire by one or more cold-rolling operations intended primarily for the purpose of obtaining the size and section desired and for improving surface finish, dimensional accuracy, and mechanical properties.

Low-carbon steel flat wire can also be produced by slitting cold-rolled flat sheet or strip steel to the desired width. The width-to-thickness ratio and the specified type of edge generally determine the process that will be necessary to produce a specific flat wire item. The edges, finishes, and tempers obtainable in flat wire are similar to those furnished in cold-rolled strip. It should be noted that a product having an approximately rectangular section, rolled from carbon steel round wire of selected size, without edge, is also known as carbon steel flat wire.

Sizes of Wire. The size limits for the product commonly known as wire range from approximately 0.13 mm (0.005 in.) to (but not including) 25.4 mm (1 in.) for round sections and from a few tenths of a millimeter to approximately 16 mm (5/8 in.) for square sections. Larger rounds and squares (if passed through a die or rolled) and all sizes of hexagonal and octagonal sections are commonly known as cold-drawn bars.

The size (diameter) of round wire is expressed in decimal units or by gage numbers. In the United States, the conventional unit is the inch, and wire diameter is determined with micrometers capable of making measurements accurate to at least one thousandth of an inch. Sizes specified are expressed in ten thousandths of an inch, which should be followed by the metric dimension in brackets to two decimal places. There are several different systems of gage numbers that can be used for the measurement of wire, but in general these systems have fallen into disuse and have been replaced by sizes in thousandths of an inch or by metric dimensions. The size of music wire is usually expressed in music wire gage (MWG), which is the standard for this wire application. For iron and steel telephone and telegraph wire, the standard is the Birmingham wire gage (BWG) system. The system commonly used by manufacturers of steel wire (other than the exceptions noted) is the United States steel wire gage (USSWG) or, more commonly, the steel wire gage (SWG) system, and all unidentified gage numbers used in this article will refer to this system. The use of gage numbers for steel wire measurements is falling from favor, and the use of absolute units is gaining acceptance. Table 1 lists decimal equivalents in inches and millimeters for steel wire gage numbers from 7/0 (12.45 mm, or 0.490 in.) to 50 (0.112 mm, or 0.0044 in.).

Wire 20 gage and smaller in size is usually regarded as fine; wire of these sizes is normally drawn and coiled on 203 mm (8 in.) diam blocks. Larger blocks are used as finished wire diameter increases. For example, 2.34 or 0.092 in. (13 gage) wire is generally drawn on 559 mm (22 in.) blocks. Table 2 indicates the usual block sizes by gages for wires between 0.889 and 12.70 mm (0.035 and 0.500 in.).

Wire in gages 17 to 19 may be regarded as either fine or coarse, depending on manufacturing process and end use. The most commonly used wires are 8 to 20 gages in size and may be drawn from 5.6 mm (7/32 in.) diam rod. Tables 3 and 4 list some standard size tolerances for coarse and fine carbon steel round wire.

Wiremaking Practices

Wiredrawing. Steel wire is produced from coils of wire rod, after removal of scale, by one or more cold reduction processes intended primarily for the purpose of obtaining the desired size. Wiredrawing, which improves surface finish and dimensional accuracy, is the most common cold reduction process. A natural result of cold drawing is that the resultant wire develops mechanical and physical properties different from those of hot-rolled steel of like composition. By varying the amount of cold reduction and other wire mill practices, including thermal treatment, a wide variety of properties and finishes can be obtained.

The mechanical characteristics of steel wire (tensile strength, stiffness, ductility, hardness, and so on) result from wire mill treatment as well as from chemical composition; therefore, the mechanical properties of wire are less dependent on chemical composition than those of other steel mill forms. In most cases, the purchaser of wire is interested in suitable mechanical proper-

278 / Carbon and Low-Alloy Steels

Table 1 Steel wire gage sizes

Gage No.	Wire diameter mm	in.	Gage No.	Wire diameter mm	in.
7/0	12.45	0.490	16½	1.47	0.058
6/0	11.73	0.462	17	1.37	0.054
5/0	10.92	0.430	17½	1.30	0.051
4/0	10.01	0.394	18	1.22	0.048
3/0	9.19	0.362	18½	1.12	0.044
2/0	8.41	0.331	19	1.04	0.041
1/0	7.77	0.306	19½	0.97	0.038
1	7.19	0.283	20	0.89	0.035
1½	6.91	0.272	21	0.805	0.0317
2	6.65	0.262	22	0.726	0.0286
2½	6.43	0.253	23	0.655	0.0258
3	6.20	0.244	24	0.584	0.0230
3½	5.94	0.234	25	0.518	0.0204
4	5.72	0.223	26	0.460	0.0181
4½	5.49	0.216	27	0.439	0.0173
5	5.26	0.207	28	0.411	0.0162
5½	5.08	0.200	29	0.381	0.0150
6	4.88	0.192	30	0.356	0.0140
6½	4.67	0.184	31	0.335	0.0132
7	4.50	0.177	32	0.325	0.0128
7½	4.32	0.170	33	0.300	0.0118
8	4.11	0.162	34	0.264	0.0104
8½	3.94	0.155	35	0.241	0.0095
9	3.76	0.148	36	0.229	0.0090
9½	3.61	0.142	37	0.216	0.0085
10	3.43	0.135	38	0.203	0.0080
10½	3.25	0.128	39	0.191	0.0075
11	3.05	0.120	40	0.178	0.0070
11½	2.87	0.113	41	0.168	0.0066
12	2.69	0.106	42	0.157	0.0062
12½	2.51	0.099	43	0.152	0.0060
13	2.34	0.092	44	0.147	0.0058
13½	2.18	0.086	45	0.140	0.0055
14	2.03	0.080	46	0.132	0.0052
14½	1.93	0.076	47	0.127	0.0050
15	1.83	0.072	48	0.122	0.0048
15½	1.70	0.067	49	0.117	0.0046
16	1.57	0.062	50	0.112	0.0044

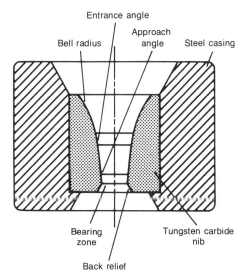

Fig. 1 Typical wiredrawing die

ties for a given application rather than in chemical composition.

Prior to drawing, the scale is removed from the material by acid pickling or mechanical descaling. If chemically cleaned, the coil is then rinsed with water, dipped in a vat containing lime in suspension or other material in solution, and in some cases baked to dry the coating and to liberate the mobile hydrogen that may have been absorbed by the steel during pickling.

In the cold drawing of wire, coiled rod or bar is drawn through the tapered hole of a die or through a series of dies; the number of dies used depends on the finished diameter required. In the past, the smallest rod section used was 5.6 mm (7/32 in.), but a few rod producers now provide 4 mm (0.157 in.) material. To begin drawing, one end of the rod is swaged to a point, inserted through the die, and attached to a power-driven reel (block). The block then pulls the material through the die and coils the drawn wire.

The common design of a wiredrawing die (Fig. 1) consists of a supporting ring of steel encasing a hard, wear-resistant nib. The nib consists of one or more carbides (such as tungsten, tantalum, or titanium) mixed with a bonding agent (such as cobalt), pressed into the desired shape, and sintered into its hardened rough form, after which it is mounted, sized, and polished. Diamond dies are sometimes used instead of carbide dies for special fine wire applications. The finest sizes, less than 0.51 mm (0.020 in.) in diameter, are generally drawn in diamond dies.

The amount of reduction during drawing is expressed as a percentage of the original cross-sectional area and is known as the overall draft or total reduction. When a rod has been reduced by cold reduction (or drawing), it is called a wire, even though many more reductions (drafts) may be necessary to reduce it to final size. This usually requires a thermal treatment between drafts. Wire produced by the single-draft drawing method refers to rod that is drawn through one die at a time, with the wire being removed from the block after each draft, until the desired diameter is achieved. Continuous drawing, which is more widely used, designates wire that is drawn through a series of dies with power-driven blocks between dies.

Table 2 Wiredrawing block sizes and corresponding coil diameters for coarse round wire

Wire size mm	in.	Coil weight <270 kg (600 lb) Common block size(a) mm	in.	Approximate inside coil diameter(b) mm	in.	Coil weight 270–680 kg (600–1500 lb) Common block size(a) mm	in.	Approximate inside coil diameter(b) mm	in.	Coil weight 680–1360 kg (1500–3000 lb) Common block size(a) mm	in.	Approximate inside coil diameter(b) mm	in.	Coil weight ≈ 1810 kg (4000 lb) Common block size(a) mm	in.	Approximate inside coil diameter(b) mm	in.
0.889 to <1.83	0.035 to <0.072	400	16	250–375	10–15	550	22	350–475	14–19	...	...	...	...	...	...	...	...
1.83 to <2.49	0.072 to <0.098	550	22	400–500	16–20	670	26	425–575	17–23	760	30	450–600	18–24	760	30	425–575	17–23
2.49 to <4.88	0.098 to <0.192	550	22	400–500	16–20	670	26	425–575	17–23	760	30	450–600	18–24	760	30	425–575	17–23
		670	26	450–600	18–24	760	30	475–625	19–25								
4.88 to <9.52	0.192 to <0.375	760	30	550–700	22–28	760	30	500–650	20–26	760	30	475–625	19–25	900	36	625–825	25–32
9.52 to <12.70	0.375 to <0.500	760	30	550–700	22–28	760	30	500–650	20–26	760	30	475–625	19–25	900	36	625–825	25–32
≥12.70	≥0.500	760	30	675–875	27–34	900	36	700–900	28–35	900	36	650–850	26–33	900	36	650–850	26–33

(a) Nearest actual metric block size commonly manufactured. (b) Metric sizes shown were rounded off to the nearest 25 mm. The figures on inside coil diameter are for guidance only. The inside diameter of the coil of wire after stripping from the block varies depending on the coil weight, grade of steel, type of riding stripper used, and size of wire. Generally, the inside diameter is slightly less than block diameter. The coils of the larger-size wire and higher-chemistry grades may have a tendency to expand.

Table 3(a) Size tolerances of galvanized coarse round wire available in coils

Wire diameter		Tolerances(a)					
		Regular or Type 1 coating		Type 2 coating		Type 3 coating	
mm	in.	mm	in.	mm	in.	mm	in.
0.889 to <1.93	0.035 to <0.076	±0.05	±0.002	±0.05	±0.002	±0.05	±0.002
1.93 to <3.76	0.076 to <0.148	±0.08	±0.003	±0.08	±0.003	±0.10	±0.004
3.76 to 6.35	0.148 to 0.250	±0.08	±0.003	±0.08	±0.003	±0.10	±0.004
>6.35	>0.250	±0.08	±0.003	±0.10	±0.004	±0.13	±0.005

(a) It is recognized that the surface of heavy zinc coating, particularly those produced by hot galvanizing, are not perfectly smooth and devoid of irregularities. If the tolerances shown above are rigidly applied to such irregularities that are inherent in the product, unjustified rejections of wire that would actually be satisfactory for use could occur. Therefore, it is intended that these tolerances be used in gaging the uniform areas of the galvanized wire. For certain galvanized wire commodities, different size tolerances are applicable, as shown under the specific commodity descriptions.

Table 3(b) Size tolerances of uncoated coarse round wire available in coils

Wire diameter		Diameter tolerance		Out-of-round tolerance	
mm	in.	mm	in.	mm	in.
0.889 to <1.93	0.035 to <0.076	±0.03	±0.001	±0.03	±0.001
1.93 to <12.70	0.076 to <0.500	±0.05	±0.002	±0.05	±0.002
12.70 to ≤25.37	0.500 to ≤0.999	±0.08	±0.003	±0.08	±0.003

The drawing speed increases with each draft. The wire is coiled on the last block as in single-draft drawing or it may be taken up on a spool. With either single-draft or continuous drawing, the term direct drawing is used when the wire is drawn from acid-cleaned or mechanically descaled rod directly to finished size without thermal treatment.

Lubricants. In the drawing operation, various materials are used as lubricants to produce different finishes on the surface of the wire and to minimize die wear. Drawing can be done either with dry lubricants or wet lubricants. In dry drawing, the base for lubrication is lime, borax, phosphate, or a combination of these, applied to the surface of the cleaned rod or wire. The lubricant, soap, grease, or oil is applied in the die box or die container at the time of drawing.

Wet drawing is used for manufacturing some classes of fine wire, for the light drafting of coarse wire, and for producing special finishes. The rod is generally dry drawn to a suitable size before wet drawing. The wire, with or without a preparatory heat treatment, is then cleaned and immersed in a solution of copper phosphate, iron or zinc phosphate, or a combination of these salts, depending on the finish desired. The wire thus prepared is drawn wet using a liquid lubricant.

Welds. In the continuous drawing of wire, industry practice is to weld rod coils together prior to the first draft. Ordinarily, the welds are difficult to identify unless indicated by painting, and the indicator paint can be removed from the finished wire if the customer so specifies. The welded joints will have mechanical properties that differ somewhat from those of the unaffected base metal. Therefore, samples taken for testing should not contain welds. If a test specimen is found to contain a weld, the specimen should be discarded as nonrepresentative of the wire coil and a new sample taken to establish compliance with the properties specified. If a variation in properties characteristic of welded joints between coils is objectionable, weld-free wire should be specified. It should be recognized that this restriction may limit finished wire coil weight based on the weight of the initial rod coil applied.

Finishes. The following finishes are specified to obtain smooth and clean surfaces, which may be required for functional or cosmetic reasons.

Common dry-drawn finish, sometimes referred to as bright finish, is ordinarily obtained by conventional dry-drawing practice. Many wiredrawing lubricants available in powder or viscous form consist primarily of calcium or sodium stearates.

Clean bright wire finish, which applies to low-carbon and medium-low-carbon dry-drawn wire, is a finish that is sufficiently clean and bright to ensure satisfactory performance in applications requiring spot welding and painting. Although not requiring additional drafting, clean bright wire finish does require special care in processing to provide a surface free of excess lime and lubricant. For some applications, further cleaning by the user or wire manufacturer may be necessary.

Extra smooth clean bright wire finish applies to low-carbon and medium-low-carbon wire given an especially smooth, clean, bright surface by dry drawing from rod selected with respect to surface characteristics and by varying lubrication and drafting practices. This finish is intended for use where a smooth, clean surface is of primary importance in some subsequent operation, such as electroplating. A special wiredrawing practice, which requires additional inspection, selection, and, in some cases, additional draft, is necessary to produce this finish.

Coppered finish and liquor finishes are obtained by drawing rod or wire that has been immersed in a copper sulfate solution or a copper-tin sulfate solution. The solution selected depends on whether copper-colored finish, straw- or brass-colored finish, or white liquor finish is required. Coppered finish and liquor finishes can be produced by dry or wet drawing. Extra coppered finish and extra liquor finishes are normally produced with an extra smooth clean finish.

Coppered finish and liquor finishes are extremely thin. Because of the nature and thinness of these finishes, they do not provide protection against corrosion. The coppered and liquor finishes are intended for appearance and for smooth clean surfaces only. Even handling with bare hands can cause discoloration; cotton gloves should be worn. Copper or brass finishes with heavier or more resistant coatings can be provided by electroplating. Prior to draft-

Table 4(a) Size tolerances for galvanized fine round wire available in coils

These tolerances are applicable to fine wire that is coated by conventional galvanizing procedures without the introduction of a prior annealing operation and are not appropriate for wire that is annealed by a separate operation following cold drawing and before the galvanizing operation.

Wire diameter		Tolerance(a)								
		Regular and type 1 coating				Type 3 coating				
		Diameter		Out-of-round		Diameter		Out-of-round		
mm	in.	mm	in.	mm	in.	mm	in.	mm	in.	
≤0.254	≤0.0100	+0.013, −0.008	+0.0005, −0.0003	0.010	0.0004	...	...	...	...	
>0.254 to 0.381	>0.0100 to 0.0150	+0.020, −0.010	+0.0008, −0.0004	0.015	0.0006	...	...	...	...	
>0.381 to 0.505	>0.0150 to 0.0199	+0.025, −0.013	+0.0010, −0.0005	0.020	0.0008	...	...	...	...	
>0.505 to 0.686	>0.0199 to 0.0270	+0.033, −0.013	+0.0013, −0.0005	0.023	0.0009	+0.038, −0.025	+0.0015, −0.0010	0.033	0.0013	
>0.686 to 0.881	>0.0270 to 0.0347	+0.038, −0.025	+0.0015, −0.0010	0.033	0.0013	+0.051, −0.025	+0.0020, −0.0010	0.038	0.0015	
>0.881 to 1.59	>0.0347 to 0.0625	+0.038, −0.038	+0.0015, −0.0015	0.038	0.0015	+0.051, −0.051	+0.0020, −0.0020	0.051	0.0020	

(a) Tolerances shown are generally followed by the producers.

Table 4(b) Coating weights for galvanized fine round wire available in coils

Wire diameter		Minimum coating weight for uncoated wire surface					
		Regular coating		Type 1 coating		Type 3 or class A coating	
mm	in.	g/m²	oz/ft²	g/m²	oz/ft²	g/m²	oz/ft²
≤0.254	≤0.0100	(a)	(a)	9.0	0.03	...	...
>0.254 to <0.381	>0.0100 to <0.0150	(a)	(a)	15	0.05	...	...
0.381 to <0.508	0.0150 to <0.020	(a)	(a)	21	0.07	...	...
0.508 to <0.889	0.020 to <0.035	(a)	(a)	31	0.10	92	0.30

(a) No specified minimum coating weight

Table 4(c) Size tolerances for annealed and galvanized fine wire available in coils and suitable for regular and type 1 coatings

Annealed and galvanized means that annealing is performed as a separate operation prior to galvanizing and is not a part of the galvanizing operation.

Wire diameter		Tolerance(a)					
		Diameter				Out-of-round	
mm	in.	mm		in.		mm	in.
≤0.254	≤0.0100	+0.013, −0.013		+0.0005, −0.0005		0.013	0.0005
>0.254 to 0.381	>0.0100 to 0.0150	+0.020, −0.015		+0.0008, −0.0006		0.018	0.0007
>0.381 to 0.505	>0.0150 to 0.0199	+0.025, −0.018		+0.0010, −0.0007		0.020	0.0008
>0.505 to 0.686	>0.0199 to 0.0270	+0.033, −0.018		+0.0013, −0.0007		0.025	0.0010
>0.686 to 0.881	>0.0270 to 0.0347	+0.038, −0.030		+0.0015, −0.0012		0.036	0.0014
>0.881 to 1.59	>0.0347 to 0.0625	+0.051, −0.051		+0.0020, −0.0020		0.051	0.0020

(a) Tolerances shown are generally followed by the producers

ing, it is necessary to remove the iron oxide scale from the rod. There are two common techniques used: chemical cleaning and mechanical descaling.

Cleaning and Coating. In chemical cleaning, scale and other surface contaminants are cleaned from steel wire by acid pickling followed by water rinsing. Cleaning is always followed by a supplementary coating operation (lime, borax, phosphate, or combinations of these).

Mechanical descaling by bending, wet abrasive blasting, or shotblasting can be used. This can be followed by coating with lime, borax, or soap solution, but the coating must be dry before reaching the drawing lubricant.

Lime, borax, and phosphate coating operations are performed to provide a carrier for the lubricant necessary for subsequent processing. In lime coating, practices can be varied to apply differing amounts of lime, depending on the end use. Phosphate-coated wire may have a supplementary coating of lime, borax, or a water-soluble soap. Wire is frequently cleaned and coated as a final operation, particularly where the wire has been thermally treated at finished size.

Thermal treatments for steel wire include stress relief, annealing, normalizing, patenting, and oil tempering. All thermal treatment of ferrous material involves time and temperature to provide the three phases: recovery, recrystallization, and grain growth.

Annealing is the general term applied to a variety of heat treatments for the purpose of softening the wire. Annealing commonly involves heating to a temperature above, near, or below the upper critical temperature, Ac_3 or Ac_{cm}. A number of processes are employed, all of which influence the surface finish obtained. If a particular finish is required on wire annealed at finish size, the producer should be consulted. Specific annealing processes are described below:

- Regular annealing (black annealing) is performed by heating coils of wire in a furnace, followed by slow cooling, without attempting to produce a specific microstructure or a specific surface finish
- Salt annealing is performed by immersing coils of wire in a molten salt bath, holding at the required temperature for a predetermined time, cooling in air, and removing the salt. Salt-annealed wire is not as soft as regular annealed wire and is used when maximum softness is not required or when wire slightly stiffer than regular annealed wire is desired. Salt annealing can be used as an intermediate procedure before final drawing
- Strand annealing is performed by passing wire in single strands through a bath of molten lead (lead annealing), a fluidized bed, or through open-fired furnaces (some with controlled atmospheres), followed by air cooling. Strand-annealed wire is not as soft as regular annealed wire and is used for products requiring greater strength or stiffness or as an in-process procedure in continuous operations such as wire galvanizing
- Bright annealing of coppered finished and liquor finished wire is performed by heating and cooling the wire in a controlled atmosphere to minimize oxidation
- Lime bright annealing is performed by heating and cooling bright dry-drawn wire in a controlled atmosphere to minimize oxidation
- Spheroidize annealing involves prolonged heating at a temperature near or slightly below the lower critical temperature, Ac_1, followed by very slow cooling, with the object of making the carbide particles in the microstructure globular (spheroidal) in order to obtain maximum softness

Annealing in-process is performed on dry low-, medium-, and high-carbon steel wire. There are two primary levels of annealing in this type: spheroidized anneal in-process and annealed in-process. The differentiation is based on a metallographic examination of grain structure after thermal treatment. If 80% or more of the grains are spheroid, then the product is designated spheroidized in-process. This requires a longer furnace time than annealed in-process.

In producing annealed in-process wire, an annealing heat treatment (followed by a separate cleaning and coating operation) is performed at an intermediate stage of wiredrawing to produce a softer wire for applications in which direct-drawn wire would be too hard or too stiff. Annealing in-process can also be used when controlled mechanical properties are required for a specific application.

It should be noted that it is normal to clean scaled rod or wire before thermal treatment in order to avoid difficult-to-remove secondary scale formation. Such practices employ cleaned lime-coated wire and rod.

Patenting is a thermal treatment that is usually confined to medium-high-carbon and high-carbon steels. In this process, individual strands of rod or wire are heated well above the upper critical temperature (generally accepted as 900 °C, or 1650 °F) rapidly in air, molten salt, fluidized beds, or molten lead. This treatment is generally used to produce a fine pearlitic grain structure to enhance the tensile properties in subsequent wiredrawing. Patenting can be used in a continuous line in which the material is cleaned and coated.

Oil tempering is a heat treatment for high-carbon steel wire in which strands of the wire at finished size are continuously heated to an appropriate temperature above the critical temperature range, oil quenched, and subsequently passed through a stress-relieving bath. Oil tempering is used in the production of such commodities as oil-tempered spring wire, which is used in certain types of mechanical springs that are not subjected to a final heat treatment after forming. Oil-tempered wire is intended primarily for the manufacture of products that are required to withstand stresses close to the limit of the elastic range of the wire. The mechanical proper-

Table 5 Common tensile strength ranges for coarse, round specification wire

Wire diameter		Tensile strength range(a)	
mm	in.	MPa	ksi
0.889 to <2.69	0.035 to <0.106	207	30
2.69 to 4.50	0.106 to 0.177	172	25
>4.50	>0.177	138	20

(a) Difference between specified minimum and maximum values

Table 6 Tensile strength ranges and limits of galvanized wire

Wire diameter		Tensile strength					
		Soft temper (maximum)(a)		Medium temper		Hard temper	
mm	in.	MPa	ksi	MPa	ksi	MPa	ksi
0.889 to <2.03	0.035 to <0.080	515	75	483–690	70–100	620–825	90–120
2.03 to <2.69	0.080 to <0.106	515	75	483–655	70–95	585–795	85–115
2.69 to 4.47	0.106 to 0.176	483	70	448–620	65–90	550–760	80–110
>4.47	>0.176	483	70	414–585	60–85	515–720	75–105

(a) Generally produced from maximum 0.13% C grades

Table 7(a) Zinc coating weights on galvanized wire

Wire diameter		Regular coating	Minimum zinc coating weight(a)					
			Type 1 coating		Type 2 coating		Type 3 coating	
mm	in.		g/m²	oz/ft²	g/m²	oz/ft²	g/m²	oz/ft²
0.889 to <1.22	0.035 to <0.048	No specified	31	0.10	92	0.30	120	0.40
1.22 to <1.57	0.048 to <0.062	minimum weight	46	0.15	92	0.30	120	0.40
1.57 to <1.93	0.062 to <0.076	of coating	46	0.15	110	0.35	150	0.50
1.93 to <2.03	0.076 to <0.080		61	0.20	120	0.40	180	0.60
2.03 to <2.34	0.080 to <0.092		76	0.25	140	0.45	200	0.65
2.34 to <2.51	0.092 to <0.099		92	0.30	150	0.50	210	0.70
2.51 to <3.76	0.099 to <0.148		92	0.30	150	0.50	240	0.80
3.76 to <4.88	0.148 to <0.192		120	0.40	180	0.60	240	0.80
4.88 to <5.26	0.192 to <0.207		150	0.50	210	0.70	270	0.90
≥5.26	≥0.207		200	0.65	230	0.75	270	0.90

(a) Uncoated wire surface

Table 7(b) Mandrel diameters, tests for adherence of zinc coating on galvanized wire

Wire diameter		Mandrel diameters(a)		
mm	in.	Regular or type 1 coating	Type 2 coating	Type 3 coating
0.889 to <1.93	0.035 to <0.076	1D	1D	2D
1.93 to <3.76	0.076 to <0.148	1D	2D	3D
3.76 to 6.35	0.148 to 0.250	2D	2D	4D
>6.35	>0.250	2D	2D	4D

(a) D, wire diameter

ties of oil-tempered wire provide resistance to permanent set under repeated and continuous stress. Oil-tempered wire is also used in applications that require abrasion resistance, such as sieves and screens.

An oxide coating of controlled thickness can be developed during the tempering of spring wire. This coating is normally left on the wire to help retain oil-base rust preventives on the surface of the wire. It also serves to prevent metal-to-metal contact during coiling or forming.

Specification Wire

There are some applications for low-carbon and medium-low-carbon steel wire that involve special requirements, such as specific tensile strength ranges or hardness limitations, the attainment of which involves special selection of steel and modification of conventional wire mill practices and/or thermal treatment (for example, annealed in-process wire). Such wire is commonly designated specification wire.

A standard specification that covers the general requirements for coarse, carbon steel round wire is ASTM A 510 (Table 5). Specification wire can be furnished with special finishes for subsequent processing, such as spot welding, electroplating, and tinning.

Metallic Coated Wire

Metallic coatings can be applied to wire by various methods, including hot dip processes, the electrolytic process, and metal cladding by rolling metallic strip over the wire.

Aluminized wire (aluminum-coated wire) is produced by passing fluxed strands of wire through a bath of molten aluminum or aluminum alloy. The material is then rapidly cooled. This may, depending on the analysis, increase or decrease the tensile values.

Brass-plated wire is produced by passing strands or wire through an electrolytic cell containing a solution of both copper and zinc salts. Typical applications for this product are steel cord used for reinforcing automobile tires and rubber hoses and when a pleasing appearance is important. Brass-plated wire is not intended for applications requiring corrosion resistance.

Bronze-coated wire, used for tire bead reinforcement around the rim area of the tire, is produced in continuous lines by passing wires through thermal treatment to adjust physical properties, acid cleaning, and tin-copper solutions.

Galvanized wire (zinc-coated wire) is produced by passing strands of wire through a bath of molten zinc (hot dip galvanized) or through an electrolytic cell containing a solution of a zinc salt (electrogalvanized). Galvanizing gives corrosion protection to wire. The wire is usually annealed in the same operation by being passed through molten lead, molten salt, or a fluidized bed, followed by cleaning or pickling, prior to galvanizing. The general requirements for galvanized carbon steel wire are given in ASTM A 641.

The term temper as applied to galvanized wire is a reference to stiffness or resistance to bending, not to thermal treatment. It is customarily expressed as soft, medium, or hard. Tensile strengths corresponding to these three tempers are given in Table 6.

Because the wire is generally strand annealed prior to zinc coating, the temper of the wire can be controlled by using different strand-annealing temperatures. Different properties can also be obtained by varying the chemical composition for a given annealing practice. The user of galvanized wire should ensure that the manufacturer has a thorough understanding of the properties, including both tensile strength and ductility, required by the end use of the wire.

Coating weights for galvanized wire vary depending on the application. The type of wire generally used for simple wire forms that do not require a minimum coating weight is variously called regular, common, single, tight wiped, ordinary, silver bright, or plain galvanized. Galvanized wire is also available with zinc coatings of specified weights: types 1, 2, and 3 and classes A, B, and C. Type 3 coating weight is the same as class A and is also called extra galvanized or double galvanized. Class B coating weight is twice class A, and class C is three times class A, as indicated in Table 7.

The coating weights given in Table 7 apply to ordinary galvanized wire only; heavier coatings are applied to wire for structural applications, such as bridge wire, strand wire, and concrete-reinforcing wire. Size tolerances for galvanized wire are given in Table 4. Tests for coating adherence are performed by wrapping the wire on a mandrel of diameter indicated in Table 7.

Tinned wire is produced by passing strands of wire continuously through a molten tin bath and then through tightly compressed wipes as the strands emerge from the bath. Tinned wire is commonly manufactured in three tempers: soft, medium hard, and hard. These three tempers are obtained by using the following techniques:

- Soft tinned wire is tinned after being annealed at or near finished size
- Medium-hard tinned wire is produced from thermal-treated wire
- Hard tinned wire is obtained by tinning wire that has been cold drawn to final size, usually without intermediate thermal treatment

Quality Descriptions and Commodities

Many types of steel wire have been developed for specific components of machines and equipment and for particular end uses. The unique properties of each of these types of wire are obtained by employing a specific combination of steel composition, steel quality, process thermal treatment, and cold-drawing practice.

These wires are normally grouped into broad usage categories. These categories, as well as some items in each category, are described in the following sections under their quality descriptions or commodity names.

Low-Carbon Steel Wire for General Usage

Low-carbon steel wire forms for general usage include industrial or standard-quality wire, annealed low-carbon manufacturers' wire, and merchant wire.

Industrial or standard-quality wire is produced from low- or medium-low-carbon steel, manufactured under close control in steelmaking procedures to provide internal soundness and freedom from detrimental surface imperfections for industrial applications. It is usually drawn directly from hot-rolled rod. Applications may involve specific chemical composition, but do not require specific tempers or tensile or softness limitations. The terms industrial or standard-quality wire should not be confused with or used in connection with commodity descriptions that apply to types of wire that have been developed for specific applications. Some of these are described in the AISI Steel Products Manual.

Annealed low-carbon manufacturers' wire, either black annealed or lime bright annealed, has the typical properties given in Table 8.

Merchant wire is annealed or soft galvanized low-carbon steel wire supplied in coils of exact weights. Black annealed wire can be supplied oiled and wiped for short-term corrosion protection.

Wire for Packaging and Container Applications

Included in wires for packaging and container applications are those designed for tying bags, baling hay and straw, binding and stapling boxes, and strapping and reinforcing shipping packages.

Baling wire for use on automatic hay-baling machines is 1.93 mm (0.076 in.) or 14½ gage, diam wire made from annealed low-carbon or annealed medium-low-carbon steel. The tensile strength of baling wire is 345 to 485 MPa (50 to 70 ksi), and minimum elongation is 12% in 250 mm (10 in.). The wire is commonly produced as rewound coils having an oil-base protective coating that will not harden or become gummy when applied to the wire at the time of rewinding.

Strapping wire, sometimes called tying wire, is used for strapping and reinforcing packages such as boxes, crates, and cartons. Commonly, it is annealed wire, coppered wire, or galvanized wire. It is made to specified tensile strength requirements and must have sufficient ductility and toughness to withstand severe crimping or twisting without breaking.

Wire for Structural Applications Other Than Prestressed Concrete

This classification includes galvanized bridge wire, zinc-coated or aluminum-coated steel strand wires, and concrete-reinforcing wire.

Galvanized bridge wire is used as a component in the fabrication of strands or ropes for bridge cables, building structures, and guys. The wire is drawn from high-carbon steel wire rods with closely controlled composition ranges to achieve the desired properties. Wire size is dependent on the size of the strand or rope to be produced. Coating weight for galvanized bridge wire is class A, B, or C. These coating weights are greater than those for the corresponding designations for ordinary galvanized wire. The mechanical properties of galvanized bridge wire are given in Table 9.

Zinc-coated strand wire is used in the manufacture of galvanized steel wire strand for guy, messenger, and ground wires and for similar applications. Zinc-coated steel strand wire is commonly produced to the specified tensile strength, elongation, and weight properties of zinc coating. Each grade of strand wire is intended for use in one of the five grades of strand covered in ASTM A 475 and A 363 or in equivalent specifications.

Aluminum-coated strand wire is similar to zinc-coated steel strand wire except for the coating, and it is used for similar purposes. It is manufactured in conformance with ASTM A 474.

Concrete-reinforcing wire (plain, not deformed) is used either as wire or for fabri-

Table 8 Tensile strength of annealed manufacturers' wire available in coils

AISI/SAE grade	UNS grade	Maximum tensile strength MPa	ksi
1006	G10060	379	55
1008	G10080	414	60
1010	G10100	448	65

Table 9(a) Coating weights for galvanized bridge wire used as bridge strand, structural strand, bridge rope, and structural rope

Coated wire diameter		Minimum coating weight of uncoated wire surface(a)					
		Class A		Class B		Class C	
mm	in.	g/m²	oz/ft²	g/m²	oz/ft²	g/m²	oz/ft²
1.04 to 1.55	0.041 to 0.061	120	0.40	240	0.80	370	1.20
>1.55 to 2.01	>0.061 to 0.079	150	0.50	310	1.00	460	1.50
>2.01 to 2.34	>0.079 to 0.092	180	0.60	370	1.20	550	1.80
>2.34 to 2.62	>0.092 to 0.103	210	0.70	430	1.40	640	2.10
>2.62 to 3.02	>0.103 to 0.119	240	0.80	490	1.60	730	2.40
>3.02 to 3.61	>0.119 to 0.142	260	0.85	520	1.70	780	2.55
>3.61 to 4.75	>0.142 to 0.187	270	0.90	550	1.80	820	2.70
>4.75	>0.187	310	1.00	610	2.00	920	3.00

(a) Uncoated wire surface

Table 9(b) Mechanical properties of galvanized bridge wire used for bridge strand, structural strand, bridge rope, and structural rope

Diameter		Minimum tensile strength(a)		Minimum yield strength at 0.7% extension under load(a)		Total elongation under load, in 250 mm (10 in.), percent minimum
mm	in.	MPa	ksi	MPa	ksi	
1.02 to <2.82	0.040 to <0.111	1520	220	1030	150	2.0
≥2.82	≥0.111	1520	220	1100	160	4.0
≥2.29	≥0.090	1450	210	1030	150	4.0
≥2.29	≥0.090	1380	200	965	140	4.0

(a) For actual cross section including zinc coating

Table 10 Composition of uncoated round high-carbon steel wire for tensioning by mechanical methods

Element	Chemical composition, %		
	Class I	Class II	Class III
Carbon(a)	0.45–0.75	0.50–0.85	0.55–0.88
Manganese(b)	0.60–1.10	0.60–1.10	0.60–1.10
Phosphorus	0.035 max	0.035 max	0.035 max
Sulfur	0.045 max	0.045 max	0.045 max
Silicon	0.15–0.35	0.15–0.35	0.15–0.35

(a) Not varying more than 13 points in any one lot. (b) Not varying more than 30 points in any one lot

Table 11 Strength of uncoated round high-carbon steel wire for tensioning by mechanical methods

Wire diameter		Minimum tensile strength					
		Class I		Class II		Class III	
mm	in.	MPa	ksi	MPa	ksi	MPa	ksi
4.11	0.162	1380	200	1590	231	1810	262
4.88	0.192	1320	192	1530	222	1740	252
6.35	0.250	1250	182	1450	211	1650	240
7.92	0.312	1200	174	1390	201	...	...

cation into welded wire fabric. Generally, this wire is produced to conform to ASTM A 82, which covers mechanical property requirements, tolerances, and test procedures. Mechanical property requirements include tensile strength, yield strength, reduction in area, and bend testing. Concrete-reinforcing wire is generally furnished with a clean bright finish suitable for spot welding, except that fine wire for welded fabric is galvanized (regular coating weight) after drawing to final size. Sizes are designated by W numbers related to sectional areas.

The surface deformation lines in deformed concrete-reinforcing wire provide a mechanical anchorage in concrete, conforming to ASTM A 496. Sizes are designated by D numbers related to the average cross-sectional areas of the wires between 6.45 to 200 mm^2 (0.01 and 0.31 in.^{2}).

Wire for Prestressed Concrete

Two types of uncoated round high-carbon steel wire are produced for prestressed concrete applications: cold drawn and cold drawn/stress relieved. The wire is used for linear or circular pretensioning or posttensioning concrete structural members. The stress-relieved product can be applied as a single wire or as a strand that has been stress relieved after stranding.

This product is described in American Society for Testing and Materials (ASTM) specifications as follows:

- A 648—Wire for prestressed concrete pipe
- A 416—Wire for stress-relieved strand
- A 421—Wire uncoated for linear prestressing

In North America, wires for direct linear prestressing are stress relieved.

Stress-relieved uncoated high-carbon wire is commonly used for the linear prestressing of concrete structures. It is produced in diameters of 4.88, 4.98, 6.35, and 7.01 mm (0.192, 0.196, 0.250, and 0.276 in.) to tensile strengths of 1620 to 1725 MPa (235 to 250 ksi). This wire can also be made in a low-relaxation grade. Full requirements are given in ASTM A 421. Full requirements for the stranded product are given in ASTM A 416.

After stress relieving, the inside diameters of coils are generally larger than those of drawn wire in order to prevent stress set or reintroduction of coil stress. Coil inside diameters may be as large as 200 times the wire diameter.

High-carbon wire for mechanical tensioning is commonly used for circular prestressing in the manufacture of concrete pressure pipe. Its chemical composition, tensile strength requirements, and specified wrap tests are indicated in Tables 10 to 12. Full requirements are given in ASTM A 648.

Wire for Electrical or Conductor Applications

Aluminum conductor steel reinforced wire, support wire, and telephone/telegraph wire are wires designed for electrical or conductor applications.

Aluminum conductor steel reinforced wire is a special commodity for the steel reinforcement of aluminum conductor cable. It is used either as a single center wire or as a multiple-wire strand and is supplied in diameters ranging from 1.27 to 4.83 mm (0.050 to 0.190 in.). This commodity is produced from either aluminum coated (ASTM B 341) or galvanized (ASTM B 498). Minimum tensile strength requirements range from 1140 to 1450 MPa (165 to 210 ksi), depending on the type and class of coating specified. Minimum yield strength and elongation, minimum coating weight, and mandrel wrap testing are additional requirements.

Support wire is a high-strength zinc-coated steel wire with a class A coating. This wire is used to support telegraph and telephone distribution lines. It is usually furnished in diameters of 2.77 to 3.76 mm (0.109 to 0.148 in.). Physical requirements are minimum tensile strength of 1310 MPa (190 ksi), minimum elongation of 2% in 250 mm (10 in.), and wrap test for ductility and adherence.

Table 12 Wrap test requirements for uncoated round high-carbon steel wire for tensioning by mechanical methods

Wire diameter		Mandrel diameters(a)		
mm	in.	Class I	Class II	Class III
4.11 to 6.35	0.162 to 0.250	2D	2D	2D
>6.35 to 7.92	>0.250 to 0.312	2D	2D	3D

(a) D, wire diameter. The wire shall wind without breakage on a cylindrical mandrel of the diameter specified. When the wire diameter exceeds 7.92 mm (0.312 in.), the resistance to winding on conventional test equipment is so great that the wrap test is generally not practiced.

Telephone and telegraph wire is galvanized wire for signal transmission purposes. It is manufactured in several grades, for example, grades EBB and BB (ASTM A 111) and grades 85, 135, and 195 (ASTM A 326). The designations EBB and BB are abbreviations of the long-standing trade terms Extra Best Best and Best Best, which are applied to galvanized telephone and telegraph wire. All these grade designations, however, are now associated with definite values of strength and electrical resistivity. For example, EBB designates the steel wire with the lowest resistivity considered appropriate for signal transmission lines, and BB designates the steel wire with resistivity in the middle of the range appropriate for this purpose.

Rope Wire

Rope wire is produced primarily for use in the construction of wire rope. The principal grades of rope wire are level 1, 2, 3, and 4 wires, each with its own requirements with respect to tensile and torsional properties. The tensile strength requirements for these grades are shown in Fig. 2. The torsion testing of rope wire is usually done on a 203 mm (8 in.) length of the wire held between the jaws of a torsion testing machine. Minimum torsion requirements are based on the following formulas:

Level 1(a)(b)	$T = 34.5 - 25.0d$(c)
Level 2(d)(b)	$T = 32.0 - 25.0d$
Level 3(e)(b)	$T = 29.5 - 25.0d$
Level 4(f), uncoated	
0.25–2.01 mm (0.010–0.079 in.)	$T = 27.5 - 25.0d$
2.03–4.04 mm (0.080–0.159 in.)	$T = 24.0 - 25.0d$
≥4.06 mm (0.160 in.)	By agreement between consumer and producer
Level 4(f), drawn galvanized	
0.25–2.01 mm (0.010–0.079 in.)	$T = 23.5 - 25.0d$
2.03–3.56 mm (0.080–0.140 in.)	$T = 20.0 - 25.0d$
≥3.58 mm (0.141 in.)	By agreement between consumer and producer

(a) Level 1 was previously designated as mild plow steel. (b) Equation applies to both uncoated and drawn galvanized wire. (c) T is the number of twists per 100 wire diameters of length and d is the wire diameter (in inches). (d) Level 2, plow steel. (e) Level 3, improved plow steel. (f) Level 4, extra improved plow steel

The latest standard-rated breaking strengths in Fig. 2 are those published in Federal Specification, Wire Rope and Strand RR W-410. In the torsion test used above, it is necessary that a specific weight

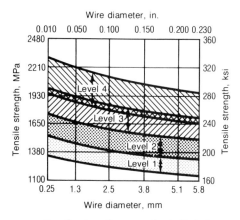

Fig. 2 Tensile strength ranges for indicated rope wire grades

be applied at the clamped end of the sample and that this weight and the revolutions per minute employed be mutually agreed upon by purchaser and supplier.

Galvanized rope wire can be galvanized either before or after drawing to finished size. When galvanizing is done at finished size, the strength requirements are 10% less than those for uncoated rope wire, and the torsion test requirements are reduced 30 to 50%, depending on size. Rope wire that is galvanized before being drawn to finished size meets the tensile and torsional properties of uncoated wire except as indicated in the above formulas for level 4 steel wire drawn after galvanizing. Wire galvanized at finished size can be produced on levels 1, 2, and 3.

Wire for Fasteners

Included in wire for fasteners is wire intended for such applications as bolts and cap screws, rivets, wood and self-tapping screws, and scrapless nuts. Depending on the application, such wire must be able to be forged, extruded, cold upset, roll or cut threaded, drilled, and hardened by suitable thermal treatment.

The type and grade of steel used in wire for fasteners depend on the requirements of the finished product and the nature of the required forming operations. Compositions range from that of 1006 grade steel, which is used for such items as common rivets, to that of a 0.55 to 0.65% C steel intended for lockwashers or screwdrivers. Low-carbon wire can be drawn from hot-rolled slow-cooled rod. Medium-carbon and high-carbon wire rod can be either annealed or spheroidized in-process.

The coating on the wire must provide sufficient lubrication in the header dies and must have the necessary lubricating qualities to prevent galling or undesirable die wear. Although lime-soap coatings are common, phosphate coatings are frequently used for more demanding applications. Producing phosphated wire may involve coating the cleaned rod or process wire with zinc phosphate and then coating with lime or borax to carry the lubricant during subsequent drawing.

It is important that the wire be internally homogeneous and free of seams and other surface imperfections. Decarburization must be held to a minimum for those products that are to be hardened by thermal treatment.

Mechanical Spring Wire for General Use

The several types of steel wire used for mechanical springs are produced in a variety of chemical compositions, but the primary consideration is that the wire have the specific properties necessary for the application. The required properties vary with the intended use of the spring and with the problems involved in its fabrication. The factors governing the selection of spring wire include:

- The load range through which the spring must operate
- The corresponding stress range for the wire
- Weight and space limitations
- Expected life of the spring
- Temperatures and other environmental conditions to be encountered in service
- Severity of deformation to be encountered during fabrication

As stress on the wire is increased, wire with higher strength is required. Because the surface of the wire is the most highly

Table 13 Tensile strength ranges for uncoated mechanical spring wire

Wire diameter		Tensile strength(a)									
		Hard-drawn wire						Oil-tempered wire			
		Class I		Class II		Clas III		Class I		Class II	
mm	in.	MPa	ksi	MPa	ksi	MPa	ksi	MPa	ksi	MPa	ksi
0.51	0.020	1950–2230	283–323	2230–2510	324–364	2410–2670	350–387	2020–2230	293–323	2230–2440	324–354
0.58	0.023	1920–2200	279–319	2210–2480	320–360	2360–2610	343–379	1990–2200	289–319	2210–2410	320–350
0.66	0.026	1900–2170	275–315	2180–2450	316–356	2320–2570	337–373	1970–2180	286–316	2190–2390	317–347
0.74	0.029	1870–2140	271–311	2150–2430	312–352	2280–2520	331–366	1950–2160	283–313	2160–2370	314–344
0.81	0.032	1830–2110	266–306	2120–2390	307–347	2250–2490	327–361	1930–2140	280–310	2140–2350	311–341
0.89	0.035	1800–2080	261–301	2080–2360	302–342	2220–2450	322–356	1890–2100	274–304	2100–2310	305–335
1.04	0.041	1760–2020	255–293	2030–2290	294–332	2160–2400	314–348	1830–2040	266–296	2050–2250	297–327
1.22	0.048	1710–1970	248–286	1980–2240	287–325	2110–2340	306–339	1790–1990	259–289	2000–2210	290–320
1.37	0.054	1680–1920	243–279	1930–2180	280–316	2070–2280	300–331	1740–1950	253–283	1960–2160	284–314
1.57	0.062	1630–1880	237–272	1880–2120	273–308	2020–2230	293–324	1700–1910	247–277	1920–2120	278–308
1.83	0.072	1600–1830	232–266	1840–2080	267–301	1980–2190	287–317	1660–1870	241–271	1880–2080	272–302
2.03	0.080	1570–1800	227–261	1810–2040	262–296	1940–2150	282–312	1620–1830	235–265	1830–2040	266–296
2.34	0.092	1520–1740	220–253	1750–1980	254–287	1900–2100	275–304	1590–1790	230–260	1800–2010	261–291
2.69	0.106	1490–1710	216–248	1720–1940	249–281	1850–2040	268–296	1550–1760	225–255	1770–1970	256–286
3.05	0.120	1450–1660	210–241	1670–1880	242–273	1810–2000	263–290	1520–1720	220–250	1730–1940	251–281
3.43	0.135	1420–1630	206–237	1640–1850	238–269	1780–1970	258–285	1480–1650	215–240	1660–1830	241–266
3.76	0.148	1400–1610	203–234	1620–1830	235–266	1740–1920	253–279	1450–1620	210–235	1630–1800	236–261
4.11	0.162	1380–1590	200–230	1590–1800	231–261	1720–1900	249–275	1410–1590	205–230	1590–1770	231–256
4.50	0.177	1340–1550	195–225	1560–1770	226–256	1690–1860	245–270	1380–1550	200–225	1560–1730	226–251
4.88	0.192	1320–1520	192–221	1530–1730	222–251	1660–1840	241–267	1340–1520	195–220	1520–1700	221–246
5.26	0.207	1310–1500	190–218	1510–1700	219–247	1640–1820	238–264	1310–1480	190–215	1490–1660	216–241
5.72	0.225	1280–1480	186–214	1480–1680	215–243	…	…	1300–1470	188–213	1480–1650	214–239
6.35	0.250	1250–1450	182–210	1450–1650	211–239	…	…	1280–1455	185–210	1450–1630	211–236
7.92	0.312	1200–1380	174–200	1390–1570	201–227	…	…	1260–1430	183–208	1440–1610	209–234
9.52	0.375	1150–1330	167–193	1340–1520	194–220	…	…	1240–1410	180–205	1420–1590	206–231
11.13	0.438	1140–1310	165–190	1320–1490	191–216	…	…	1210–1380	175–200	1390–1560	201–226
12.70	0.500	1080–1240	156–180	1250–1410	181–205	…	…	1170–1340	170–195	1350–1520	196–221
14.27	0.562	1050–1210	152–176	1220–1390	177–201	…	…	1140–1310	165–190	1320–1490	191–216
15.88	0.625	1010–1170	147–170	1180–1340	171–194	…	…	1140–1310	165–190	1320–1490	191–216

(a) For diameters other than those shown, it is customary to determine tensile strength ranges by interpolation. Not applicable to straightened and cut wire

Table 14 Tensile strength ranges for music spring steel wire

Wire diameter		Tensile strength		Wire diameter		Tensile strength	
mm	in.	MPa	ksi	mm	in.	MPa	ksi
0.10	0.004	3030–3340	439–485	1.40	0.055	2070–2280	300–331
0.13	0.005	2940–3250	426–471	1.50	0.059	2040–2250	296–327
0.15	0.006	2860–3160	415–459	1.60	0.063	2020–2230	293–324
0.18	0.007	2810–3100	407–449	1.70	0.067	2000–2210	290–321
0.20	0.008	2750–3040	399–441	1.83	0.072	1980–2190	287–317
0.23	0.009	2710–2990	393–434	1.93	0.076	1960–2160	284–314
0.25	0.010	2670–2950	387–428	2.03	0.080	1940–2150	282–312
0.28	0.011	2630–2910	382–422	2.16	0.085	1920–2120	279–308
0.30	0.012	2600–2880	377–417	2.29	0.090	1900–2100	276–305
0.33	0.013	2570–2840	373–412	2.41	0.095	1890–2090	274–303
0.36	0.014	2540–2810	369–408	2.54	0.100	1870–2070	271–300
0.38	0.015	2520–2790	365–404	2.59	0.102	1860–2060	270–299
0.41	0.016	2500–2760	362–400	2.72	0.107	1850–2040	268–296
0.46	0.018	2450–2710	356–393	2.79	0.110	1840–2030	267–295
0.51	0.020	2410–2670	350–387	2.84	0.112	1830–2030	266–294
0.56	0.022	2380–2630	345–382	3.07	0.121	1810–2000	263–290
0.61	0.024	2350–2600	341–377	3.18	0.125	1800–1990	261–288
0.66	0.026	2320–2570	337–373	3.30	0.130	1790–1970	259–286
0.71	0.028	2300–2540	333–368	3.43	0.135	1780–1960	258–285
0.76	0.030	2280–2520	330–365	3.56	0.140	1770–1950	256–283
0.81	0.032	2250–2490	327–361	3.68	0.145	1750–1940	254–281
0.86	0.034	2230–2470	324–358	3.81	0.150	1740–1920	253–279
0.91	0.036	2210–2450	321–355	3.96	0.156	1730–1910	251–277
0.97	0.038	2190–2430	318–352	4.11	0.162	1720–1900	249–275
1.02	0.040	2170–2410	315–349	4.50	0.177	1690–1860	245–270
1.07	0.042	2160–2390	313–346	4.88	0.192	1660–1840	241–267
1.14	0.045	2130–2360	309–342	5.26	0.207	1640–1820	238–264
1.22	0.048	2110–2340	306–339	5.72	0.225	1620–1790	235–260
1.30	0.051	2090–2310	303–335	6.35	0.250	1590–1760	230–255

Table 15 Typical sizes and tensile strength ranges for overhead door spring steel wire

Wire diameter		Tensile strength(a)			
		Hard drawn		Oil tempered	
mm	in.	MPa	ksi	MPa	ksi
3.76	0.148	1620–1900	235–276	1590–1760	230–255
4.11	0.162	1590–1870	231–271	1550–1720	225–250
4.50	0.177	1560–1830	226–266	1520–1690	220–245
4.88	0.192	1530–1800	222–261	1480–1650	215–240
5.26	0.207	1510–1770	219–257	1450–1620	210–235
5.72	0.225	1480–1740	215–253	1430–1610	208–233
6.35	0.250	1450–1720	211–249	1410–1590	205–230
7.92	0.312	1390–1620	201–237	1380–1550	200–225
9.52	0.375	1340–1590	194–230	1340–1520	195–220
11.13	0.438	1320–1560	191–226	1310–1480	190–215
12.70	0.500	1210–1450	176–210	1280–1450	185–210
14.27	0.562	1150–1390	167–201	1210–1380	175–200
15.88	0.625	1110–1340	161–194	1210–1380	175–200

(a) Not applicable to straightened and cut wire

stressed part of a spring and because spring motion gives rise to torsion effects, freedom from surface imperfections becomes increasingly important as maximum stress or required service life is increased. Surface condition is very important in music spring steel wire and is even more important in valve spring quality wire. The characteristics of springs are discussed in the article "Steel Springs" in this Volume.

There are three types of spring wire for general use:

- Hard-drawn spring wire (covered in ASTM A 227 and A 679)
- Oil-tempered spring wire (ASTM A 229)
- Spring steel wire for thermal-treated components (ASTM A 713)

Tensile strength ranges for the first two types are given in Table 13.

Music spring steel wire is used in springs that are subject to high stresses and require good fatigue properties. Final cold drawing is commonly performed by the wet white liquor method to develop a characteristic smooth bright luster. Phosphated music wire is also common. Manufacturers employ specialized coiling tests, twist tests, torsion tests, and bend tests to verify that the exacting requirements of this type of wire are met. Specification ASTM A 228 describes music spring steel wire in detail. Tensile strength ranges are given in Table 14.

Mechanical Spring Wire for Special Applications

Mechanical spring wire for such applications as automotive hood hinge springs, sewer augers, torque rods, and valve springs is specially designed for each application. One such material is oil-tempered valve spring quality carbon steel wire (ASTM A 230). Steel wire that is oil tempered for mechanical springs is defined in ASTM A 229 and A 679. Another is overhead door spring steel wire. Depending on design and conditions of service, making springs from the latter material may involve coiling with high initial tension and sharp bending to produce the required end formations, and the springs may be subjected to repetitive applications of high stress in service. Typical tensile strengths are shown in Table 15.

Upholstery Spring Construction Wire

Upholstery spring construction wire includes the several types of wire used for upholstery springs, plus the wires employed for borders and braces, hog rings, links, and frames. Upholstery spring wire is used in the manufacture of springs for mattresses, bed springs, and related applications. This type of spring wire is not intended for the manufacture of other types of springs. Upholstery spring wire is drawn from thermally treated or controlled cooled wire rod or wire. The producer determines manufacturing practice and selects specific carbon and manganese ranges within approximately 0.45 to 0.75% and 0.60 to 1.20%, respectively, to meet end-use requirements.

Upholstery spring wire is commonly produced with a dry-drawn finish and in sizes from 20 to 4 gage (0.89 to 5.72 mm, or 0.035 to 0.225 in.). Specification ASTM A 407 covers specifications for wire used in producing the various types of coil springs in common use. Table 16 lists the tensile strength ranges for one such type of wire.

Wire for Upholstery Springs. Zig-zag, square-form, and sinuous-type springs, which are extensively used in furniture, are made from hard-drawn carbon steel wire produced from controlled cooled rods, patented rods, or patented wire. These wires are commonly produced in 13 to 8 gage (2.34 to 4.11 mm, or 0.92 to 0.162 in.). Full requirements are covered in ASTM A 417. Tensile strength ranges for this type of wire are given in Table 17.

Upholstery spring wire is also produced for Marshall spring wire types (open wound, not knotted) and lacing helical springs (ASTM A 407). Border and brace wires, generally of lower tensile values, are produced as round wire reinforcements or for cold-rolling purposes.

Wire for Other Specific Applications

Steel wire for miscellaneous specific applications includes wire for such diverse products as brush handles, clothes hangers, chains, cotter pins, screens, staples, and steel wool. Two products for which substantial quantities of wire are used are chain link fences and construction wire for automobile tires.

Chain link fence wire is a low-carbon or medium-low-carbon steel wire suitable for the production of chain link fence fabric. It is produced in diameters of 3.05, 3.76, and 4.88 mm (0.120, 0.148, and 0.192 in.) and is supplied uncoated (for zinc coating after weaving), zinc coated prior to weaving, or aluminum coated prior to weaving.

Zinc-Coated Fence Wire. Requirements for zinc-coated chain link fence fabric, which is zinc coated either before or after weaving, are covered in ASTM A 392.

Aluminum-Coated Fence Wire. Requirements for aluminum-coated chain link fence fabric are covered in ASTM A 491. This wire is always aluminum coated before weaving.

Table 16 Tensile strength ranges for upholstery spring wire and lacing wire

Wire type	Diameter mm	in.	Gage	Tensile strength MPa	ksi
Upholstery spring wire					
Automatic coiling and knotting machines(a)	1.57	0.062	16	1620–1860	235–270
	1.83	0.072	15	1590–1830	230–265
	2.03	0.080	14	1550–1790	225–260
	2.34	0.092	13	1480–1720	215–250
	2.69	0.106	12	1410–1620	205–235
	3.05	0.120	11	1340–1550	195–225
	3.43	0.135	10	1310–1520	190–220
	3.76	0.148	9	1280–1480	185–215
	4.11	0.162	8	1240–1450	180–210
Marshall-type units	1.22	0.048	18	1760–2030	255–295
	1.37	0.054	17	1720–2000	250–290
	1.57	0.062	16	1720–2000	250–290
	1.83	0.072	15	1650–1930	240–280
	2.03	0.080	14	1590–1860	230–270
	2.34	0.092	13	1550–1830	225–265
	2.69	0.106	12	1520–1790	220–260
Lacing wire					
Regular	1.04	0.041	19	1620–1900	235–275
	1.22	0.048	18	1590–1860	230–270
	1.37	0.054	17	1550–1830	225–265
	1.57	0.062	16	1550–1830	225–265
Automatic	1.04	0.041	19	1720–2000	250–290
	1.22	0.048	18	1690–1970	245–285
	1.37	0.054	17	1650–1930	240–280

(a) Other sizes are available. Regular spring wire

Tire bead wire is used for reinforcing the beads of pneumatic tires. Uniformity in chemical and mechanical properties and a surface finish to which rubber will adhere are essential for satisfactory performance. Tire bead wire is most commonly produced 0.94 mm (0.037 in.) in diameter and brass plated. Mechanical property tests are conducted on wire samples that have been heated for 1 h at 150 °C (300 °F). Minimum breaking strength for wire 0.94 mm (0.037 in.) in diameter is 129 kg (285 lb). In torsion testing, the wire must withstand a minimum of 58 twists in a 203 mm (8 in.) gage length.

A considerable and growing quantity of wire between 0.18 and 0.38 mm (0.0069 and 0.015 in.) in diameter is used annually for the purpose of providing cord reinforcement for radial tires. This extremely fine, high-tensile wire is produced from controlled cooled rods by drawing, followed by drawing, patenting, and a second drawing and patenting. After the second patenting operation, the wire is brass plated wet drawn and then formed into strands to be embedded into the tire rubber.

Fine Wire

Fine wire is considered to include all wire less than 0.89 mm (0.035 in.) in diameter, as well as some coarser wire up to 1.57 mm (0.062 in.) in diameter when so designated. Fine wire is commonly produced with bright, liquor, coppered, or phosphate finishes; with galvanized, tin, or cadmium coatings; and in the hard-drawn, annealed or oil-tempered conditions. Aircraft cords, broom wire, brushes, fishhooks, florist wire, hose reinforcement, paper clips, insect screens, and safety pins are examples of items produced entirely or in part from various kinds of fine wire.

Aircraft cord wire is a hard-drawn, high-tensile-strength, high-carbon steel wire designed for the manufacture of flexible cords and multiple-wire strands for aircraft controls. This type of wire is usually either tin coated or zinc coated. Common sizes range from 0.18 to 0.81 mm (0.007 to 0.032 in.) in diameter.

Hose-reinforcing wire is a hard-drawn grade commonly furnished in three tensile ranges: 2070 to 2410 MPa (300 to 350 ksi), 2240 to 2590 MPa (325 to 375 ksi), and 2410 to 2760 MPa (350 to 400 ksi). Sizes up to 0.508 mm (0.020 in.) in diameter are provided in all ranges. Sizes over 0.508 mm (0.020 in.) are furnished in the lower two ranges. The wire should be sufficiently ductile to wrap around a mandrel four times the wire diameter without fracture.

Paper clip wire is low-carbon steel wire of relatively high strength. It can be tinned or drawn after galvanizing, or it can have a liquor or copper finish. Diameters of 1.02, 0.91, and 0.80 mm (0.040, 0.036, and 0.0315 in.) are most common. In these sizes, tensile strength is about 970 MPa (140 ksi).

Safety pin wire is supplied in two grades. The high-carbon grade has a tensile strength of approximately 2210 MPa (320 ksi), while the low-carbon grade has approximately 1280 MPa (185 ksi).

Alloy Wire

The chemical compositions, quality descriptions, requirements, and tests applicable to alloy wire are described in the article "Steel Wire Rod" in this Volume. Many alloy steel wires have been developed for specific applications, which include wires for bearings, chains, and springs, and for cold-heading and cold-forging applications. Several of these are described in the following sections.

Alloy steel wire for ball and roller bearings is produced by practices that ensure internal soundness, cleanliness, uniformity of chemical composition, and a good surface. High-carbon chromium steels, such as E51100 and E52100, are commonly used for ball bearings. Roller bearings are made from

Table 17 Tensile strength ranges for cold-drawn steel wire used for zig-zag, square-form, and sinuous types of upholstery spring units

Wire type	Wire diameter mm	in.	Wire gage	Tensile strength(a) Class I MPa	ksi	Class II MPa	ksi
Automobile seat and back spring units							
Type A, zig-zag	2.4	0.092	13	1520–1720	220–250	1590–1790	230–260
	2.6	0.106	12	1480–1690	215–245	1550–1760	225–255
	3.0	0.120	11	1450–1650	210–240	1480–1690	215–245
	3.5	0.135	10	1410–1620	205–235	1450–1650	210–240
	3.8	0.148	9	1380–1590	200–230	1450–1650	210–240
	4.2	0.162	8	1310–1520	190–220	1380–1590	200–230
Type B, square-form	2.4	0.092	13	1480–1690	215–245	1550–1760	225–255
	2.6	0.106	12	1450–1650	210–240	1520–1720	220–252
	3.0	0.120	11	1410–1620	205–235	1480–1690	215–245
	3.5	0.135	10	1380–1590	200–230	1450–1650	210–240
	3.8	0.148	9	1310–1520	190–220	1380–1590	200–230
	4.2	0.162	8	1240–1450	180–210	1310–1520	190–220
Furniture spring units							
Type C, sinuous	2.4	0.092	13	1620–1830	235–265	...	...
	2.6	0.106	12	1620–1830	235–265	...	...
	3.0	0.120	11	1590–1790	230–260	...	...
	3.5	0.135	10	1550–1760	225–255	...	...
	3.8	0.148	9	1520–1720	220–250	...	...
	4.2	0.162	8	1480–1690	215–245	...	...
	4.5	0.177	7	1450–1650	210–240	...	...
	5.0	0.192	6	1430–1630	207–237	...	...

(a) Tensile strength values for diameters not shown in this table shall conform to that shown for the next larger diameter. For example, for diameter 3.25 mm (0.128 in.), the value shall be the same as for 3.5 mm (0.135 in.).

Table 18(a) Tensile strength ranges for uncoated mechanical spring wire

Wire diameter		Tensile strength(a)									
		Hard-drawn wire						Oil-tempered wire			
		Class I		Class II		Class III		Class I		Class II	
mm	in.	MPa	ksi	MPa	ksi	MPa	ksi	MPa	ksi	MPa	ksi
0.51	0.020	1950–2230	283–323	2230–2510	324–364	2410–2670	350–387	2020–2230	293–323	2230–2440	324–354
0.58	0.023	1920–2200	279–319	2210–2480	320–360	2360–2610	343–379	1990–2200	289–319	2210–2410	320–350
0.66	0.026	1900–2170	275–315	2180–2450	316–356	2320–2570	337–373	1970–2180	286–316	2190–2390	317–347
0.74	0.029	1870–2140	271–311	2150–2430	312–352	2280–2520	331–366	1950–2160	283–313	2160–2370	314–344
0.81	0.032	1830–2110	266–306	2120–2390	307–347	2250–2490	327–361	1930–2140	280–310	2140–2350	311–341
0.89	0.035	1800–2080	261–301	2080–2360	302–342	2220–2450	322–356	1890–2100	274–304	2100–2310	305–335
1.04	0.041	1760–2020	255–293	2030–2290	294–332	2160–2400	314–348	1830–2040	266–296	2050–2250	297–327
1.22	0.048	1710–1970	248–286	1980–2240	287–325	2110–2340	306–339	1790–1990	259–289	2000–2210	290–320
1.37	0.054	1680–1920	243–279	1930–2180	280–316	2070–2280	300–331	1740–1950	253–283	1960–2160	284–314
1.57	0.062	1630–1880	237–272	1880–2120	273–308	2020–2230	293–324	1700–1910	247–277	1920–2120	278–308
1.83	0.072	1600–1830	232–266	1840–2080	267–301	1980–2190	287–317	1660–1870	241–271	1880–2080	272–302
2.03	0.080	1570–1800	227–261	1810–2040	262–296	1940–2150	282–312	1620–1830	235–265	1830–2040	266–296
2.34	0.092	1520–1740	220–253	1750–1980	254–287	1900–2100	275–304	1590–1790	230–260	1800–2010	261–291
2.69	0.106	1490–1710	216–248	1720–1940	249–281	1850–2040	268–296	1550–1760	225–255	1770–1970	256–286
3.05	0.120	1450–1660	210–241	1670–1880	242–273	1810–2000	263–290	1520–1720	220–250	1730–1940	251–281
3.43	0.135	1420–1630	206–237	1640–1850	238–269	1780–1970	258–285	1480–1650	215–240	1660–1830	241–266
3.76	0.148	1400–1610	203–234	1620–1830	235–266	1740–1920	253–279	1450–1620	210–235	1630–1800	236–261
4.11	0.162	1380–1590	200–230	1590–1800	231–261	1720–1900	249–275	1410–1590	205–230	1590–1770	231–256
4.50	0.177	1340–1550	195–225	1560–1760	226–256	1690–1860	245–270	1380–1550	200–225	1560–1730	226–251
4.88	0.192	1320–1520	192–221	1530–1730	222–251	1660–1840	241–267	1340–1520	195–220	1520–1700	221–246
5.26	0.207	1310–1500	190–218	1510–1700	219–247	1640–1820	238–264	1310–1480	190–215	1490–1660	216–241
5.72	0.225	1280–1480	186–214	1480–1680	215–243	...	...	1300–1470	188–213	1480–1650	214–239
6.35	0.250	1250–1450	182–210	1450–1650	211–239	...	...	1280–1450	185–210	1450–1630	211–236
7.92	0.312	1200–1380	174–200	1390–1570	201–227	...	...	1260–1430	183–208	1440–1610	209–234
9.52	0.375	1150–1330	167–193	1340–1520	194–220	...	...	1240–1410	180–205	1420–1590	206–231
11.13	0.438	1140–1310	165–190	1320–1490	191–216	...	...	1210–1380	175–200	1390–1560	201–226
12.70	0.500	1080–1240	156–180	1250–1410	181–205	...	...	1170–1340	170–195	1350–1520	196–221
14.27	0.562	1050–1210	152–176	1220–1390	177–201	...	...	1140–1310	165–190	1320–1490	191–216
15.88	0.625	1010–1170	147–170	1180–1340	171–194	...	...	1140–1310	165–190	1320–1490	191–216

(a) For diameters other than those shown, it is customary to determine tensile strength ranges by interpolation. Not applicable to straightened and cut wire

high-carbon wires or from low-carbon case-hardening steels such as 4620 and 8620.

Alloy steel wire for heading, forging, and roll threading is produced by specially controlled manufacturing practices and is subjected to mill tests and inspection to ensure internal soundness, uniformity of chemical composition, and freedom from surface imperfections. These precautions are necessary to provide satisfactory heading and forging performance and proper response to subsequent thermal treatment.

Alloy steel wire for upsetting or forging is commonly produced in one of the following conditions:

- Wire drawn from spheroidize-annealed hot-rolled coils
- Wire spheroidize annealed in process
- Wire spheroidize annealed at finished size

Generally, alloy steel wire for cold heading and cold forging is spheroidize annealed in-process. This permits a light finishing draft in which the wire is given a special coating finish to facilitate cold forming. There are cases, however (especially with high-carbon wire), in which the severity of cold work requires spheroidize annealing to be done at finished size. Specification ASTM A 547 covers alloy steel wire for trimmed hexagon-head bolts.

Alloy Steel Wire for Welded Chains. Chains made from alloy steel wire are generally used for slinging, hoisting, and load-binding purposes. The wire for these chains must have high strength and good abrasion resistance. In addition, the carbon content is generally held to a maximum of 0.25% to facilitate electric welding during fabrication of the chains. Typical grades of alloy steel that satisfy these requirements are 4615, 4620, 4320, 8620, and 8622. The wire is generally furnished in the spheroidize-annealed condition so that the amount of springback between the ends of the formed links does not interfere with welding.

Alloy steel spring wire is used for the manufacture of springs intended for operation at moderately elevated temperatures. Three grades in common use are 6150 chromium-vanadium steel, SAE 9260 silicon manganese, and 9254 chromium-silicon steel, which are covered in ASTM A 231, A 232, and A 401, respectively. ASTM A 877/A 877M also covers chromium-silicon grades; ASTM A 878/A 878M covers a grade of modified chromium-vanadium steel valve spring quality wire.

The wire is commonly produced in one of four conditions: oil tempered, spheroidize annealed, annealed in-process, and patented in-process. Oil-tempered alloy spring wire, commonly produced in diameters up to 12.7 mm (0.500 in.) to tensile strength or hardness requirements, is intended for very light forming and is generally used for coiling into common types of springs. Table 18 lists tensile strength ranges for several oil-tempered wires. Annealed alloy spring wire, produced in diameters up to 15.9 mm (0.625 in.), is used for severe cold forming or coiling (when annealed in process) or for very severe cold forming or coiling (when annealed at finished size) before hardening and tempering. Tensile strength is not commonly specified for annealed wire.

Mechanical Properties: Round Wire Versus Flat Wire

Most round wire is produced and tested to required tensile ranges. With flat wire, the calculation of tensile strength is inaccurate because of the difficulty of determining

Table 18(b) Mandrel diameters for wrap test of uncoated mechanical spring wire

Wire diameter		Mandrel diameters(a)			
		Hard drawn			Oil tempered,
mm	in.	Class I	Class II	Class III	Classes I and II
0.51 to 4.11	0.020 to 0.162	1D	2D	2D	1D
>4.11 to 7.92	>0.162 to 0.312	2D	4D	4D	2D

(a) D, wire diameter. The wire shall wind without breakage on a cylindrical mandrel of the diameter specified. For 1D mandrel diameter, the wire may be wound on itself. When the wire diameter exceeds 7.92 mm (0.312 in.), the resistance to winding on conventional test equipment is so great that the wrap test is generally not practiced.

the cross-sectional area as a result of the round edges and corners on the wire. Flat wire is therefore produced and tested to required hardness ranges.

Hardness and tensile properties are related, but a degree of caution must be exercised in converting one to the other. The standard conversion charts are not accurate for wire. Wire has very directional properties because its grain structure has been cold worked in only one direction. Generally, at a given hardness, wire has a higher tensile strength than that listed in the standard conversion charts.

ACKNOWLEDGMENT

The contributions of the following individuals were critical in the preparation of this article. T.A. Heuss, LTV Steel Bar Division; Bill Schuld, Seneca Wire and Manufacturing Company; and Walter Facer, American Spring Wire Company.

Threaded Steel Fasteners

Revised by the ASM Committee on Threaded Steel Fasteners*

THREADED FASTENERS for service between −50 and 200 °C (−65 and 400 °F) can be made from several different grades of steel, as long as the finished fastener meets the specified strength requirements. This article discusses the properties of threaded fasteners made from carbon and low-alloy steels containing a maximum of 0.55% C that are intended for use under the service conditions mentioned above. Guidelines are also discussed for the selection of steels for bolts (including cap screws), studs (including U-bolts), and nuts for service at the temperature range mentioned above and for service between 200 and 370 °C (400 and 700 °F). Steels used for threaded fasteners for service above 370 °C (700 °F) are also briefly discussed in this article.

Specification and Selection

Specifications are used to outline fastener requirements, to control the manufacturing process, and to establish functional or performance standards. Their goal is to ensure that fasteners will be interchangeable, dimensionally and functionally. The most common fastener specifications are product specifications that are set up to govern and define the quality and reliability of fasteners before they leave the manufacturer. These specifications determine what material to use; state the objectives for tensile strength, shear strength, and response to heat treatment; set requirements for environmental temperature or atmospheric exposure; and define methods for testing and evaluation. To ensure proper use of fasteners, additional specifications are necessary as a guide to proper design, application and installation. All American Society for Testing and Materials (ASTM) and Society of Automotive Engineers (SAE) specifications covering threaded fasteners require that the fastener be marked for grade identification (Table 1). Grade markings are a safety device that provide a positive check on selection, use, and inspection.

The selection and satisfactory use of a particular fastener are dictated by the design requirements and conditions under which the fastener will be used. Consideration must be given to the purpose of the fastener, the type and thickness of materials to be joined, the configuration and total thickness of the joint to be fastened, the operating environment of the installed fastener, and the type of loading to which the fastener will be subjected in service. A careful analysis of these requirements is necessary before a satisfactory fastener can be selected. The selection of the correct fastener or fastener system may simply involve satisfying a requirement for strength (static or fatigue) or for corrosion resistance. On the other hand, selection may be dictated by a complex system of specification and qualification controls. The extent and complexity of the system needed is usually dictated by the probable cost of a fastener failure. Failures of mechanical fasteners are discussed in Volume 11 of the 9th Edition of *Metals Handbook*.

Adequate testing is the most practical method of guarding against the failure of a new fastener system for a critical application. The designer must not extrapolate existing data to a different size of the same fastener, because larger-diameter fasteners have significantly lower fatigue endurance limits than smaller-diameter fasteners made from the same material and using the same manufacturing techniques and joint system.

Strength Grades and Property Classes

Although chemical composition can be an important factor when specifying and selecting threaded fasteners for many applications (particularly when the applications require elevated-temperature service, corrosion resistance, or good hardenability characteristics with adequate toughness properties), the primary criterion in selecting threaded fasteners involves the specification of strength levels. Consequently, the grade or class of bolts, studs, and nuts are widely used to designate the various strength or performance level of threaded fasteners in the specifications developed by the Society of Automotive Engineers, the International Organization for Standardization (ISO), the American Society for Testing and Materials, and/or the Industrial Fasteners Institute (IFI). This allows the purchaser of steel bolts, studs, and nuts to select the desired strength level by specifying a grade or class in the SAE, ISO, ASTM, or IFI specifications. The producer then selects a particular steel grade that meets the broad chemical composition ranges allowed in these specifications. This enables the producers to use the most economical material consistent with their equipment and production procedures to meet the specified properties. As a result, producers have adopted substantially the same manufacturing process for a given class of product, which has resulted in a certain degree of steel grade standardization. The strength and property designations of bolts and studs are typically based on minimum tensile (breaking) strength, while the grade designations of nuts are typically based on proof stress.

The following two sections describe the commonly used ISO and SAE grade designations for steel threaded fasteners made in metric sizes or in the U.S. system of inch dimensions. Of course, other specifications (such as some of the ASTM specifications listed in Table 1) may use their own grade designation systems.

Property class designations of metric fasteners are defined in ISO R898 and SAE J1199. The commonly used property classes of steel bolts and nuts made in metric sizes are given in Tables 2 and 3, respectively. In Tables 2 and 3, the property class numbers indicate the general level of tensile strength for bolts or proof stress for nuts in megapascals. In Table 3, for example, ISO class 9 nuts have a proof stress ranging from 900 to 920 MPa. In Table 2, the number following the decimal point in a class number for a

---

*Chairman, Frank W. Akstens, Industrial Fasteners Institute; James Gialamas, Edward J. Bueche, and T.P. Madvad, USS/Kobe Steel Company; Brian Murkey, RB&W Corporation; Joseph McAuliffe, Lake Erie Screw Corporation; Gregory D. Sander, Ring Screw Works; Edwin F. Frederick, Bethlehem Steel Corporation, Bar, Rod and Wire Division; Hal L. Miller, Nelson Wire Company; P.C. Hagopian, Stelco Fastener and Forging Company; and James Fox, Charter Rolling

bolt or stud indicates the ratio of the yield strength to the tensile strength; for example, an ISO class 5.8 stud has a tensile strength of 520 MPa and a yield strength of 420 MPa.

Some of the ISO property class designations given in Tables 2 and 3 are used in various specifications, such as:

- SAE J1199, "Mechanical and Material Requirements for Metric Externally Threaded Steel Fasteners"
- ASTM F 568, "Specification for Carbon and Alloy Steel Externally Threaded Metric Fasteners"
- ASTM A 325M, "Specification for High-Strength Bolts for Structural Steel Joints (Metric)"
- ASTM A 490M, "Specification for High-Strength Steel Bolts, Classes 10.9 and 10.9.3, for Structural Steel Joints (Metric)"
- ASTM A 563M, "Carbon and Alloy Steel Nuts (Metric)"

However, not all of the ISO property classes are used in these specifications for metric steel threaded fasteners. Specification SAE J1199, for example, no longer allows the high-hardness fasteners (ISO bolt classes 12.8 and 12.9), because these two classes are susceptible to delayed brittle fracture in corrosive environments. This change in SAE J1199 is in response to the stress-corrosion cracking of class 12.8 bolts in automobile rear suspensions after just 2 years of service in the Snow Belt of the United States (Ref 1).

Strength grades for the U.S. system of mechanical fasteners with inch dimensions are often defined per SAE specifications. The SAE strength grades for bolts and nuts specified in inch dimensions are given in Tables 2 and 3. As can be seen in Tables 2 and 3, the SAE strength grade numbers do not directly convert to a specific strength level, although they are generally organized by strength level, that is, the greater the number, the higher the strength level. A second number, following a decimal point, is sometimes added to represent a variation of the product with the general strength level. However, this number after the decimal point does not represent a strength ratio, as in the ISO system.

In addition to the SAE specification of strength grades for steel threaded fasteners in the U.S. system, other strength grades can also be specified in ASTM standards. Table 1 shows the bolt markings for the SAE strength levels, along with a partial list of some ASTM bolt grades. When the bolt markings are the same, the SAE grade is equivalent to the ASTM grade. In selecting approximate equivalents between the ISO classes and the various grades specified by SAE and/or ASTM, the following equivalents are suggested (for guidance purposes only) in SAE J1199:

- ISO class 4.6 is approximately equivalent to SAE J429, grade 1, and ASTM A 307, grade A
- ISO class 5.8 is approximately equivalent to SAE J429, grade 2
- ISO class 8.8 is approximately equivalent to SAE J429, grade 5, and ASTM A 449
- ISO class 9.8 has properties approximately 9% stronger than SAE J429-grade 5, and ASTM A 449
- ISO class 10.9 is approximately equivalent to SAE J429-grade 8, and ASTM A 354-grade BD

Steels for Threaded Fasteners

Many different low-carbon, medium-carbon, and alloy steel grades are used to make all the various strength grades and property classes of threaded steel fasteners suitable for service between −50 and 200 °C (−65 and 400 °F). In addition to the effects of steel composition on corrosion resistance and elevated-temperature properties, the hardenability of the steels used for threaded fasteners is important when selecting the chemical composition of the steel. As strength requirements and section size increase, hardenability becomes a major factor.

Grade 1022 steel is a popular low-carbon steel for threaded fasteners, although the low carbon content limits hardenability and therefore confines 1022 steel to the smaller diameter product sizes. For many product diameter sizes, grade 1038 steel is one of the most widely used steels for threaded fasteners up to the level of combined size and proof stress at which inadequate hardenability precludes further use. This medium-carbon steel has achieved its popularity because of excellent cold-heading properties, low cost, and availability.

Grade 1541 steel is extensively used for applications requiring hardenability greater than that of 1038 steel, but less than that of alloy steel. Figure 1 shows the depth of hardening when 19 mm (¾ in.) diam bolts made of 1541 steel are oil quenched. Figure 1 also shows the depth of hardening with 1038 steel but with a water quench. (Figure 2 shows a more direct comparison of 1541 and 1038 hardenability.)

Figure 2 shows cost-hardenability relationships for both oil- and water-quenched steels. An increase in hardenability does not necessarily mean an increase in cost per pound. Figure 2 is not intended to prescribe or imply the use of water quenching for alloy steels (which are normally oil quenched). These data are presented only to show the economic advantages of water quenching when it can be properly and successfully applied to the product being heat treated. Generally, the use of a water quench must be approached with caution, and a water quench practice is not necessarily recommended for some carbon steels (such as one of 1038 analysis) or requires careful analysis of part design, temperature control, and agitation to prevent quench cracking of low-carbon (SAE 1022) steel on an intermittent basis. Water quenching is precluded in most high-strength specifications due to uneven quenching and the resultant potential for quench cracking. For oil quenching, a large number of oil and

Table 1 Various ASTM and SAE grade markings for steel bolts and screws

Grade marking	Specification	Material
No mark	SAE grade 1	Low- or medium-carbon steel
	ASTM A 307	Low-carbon steel
	SAE grade 2	Low- or medium-carbon steel
(3 radial lines)	SAE grade 5, ASTM A 449	Medium-carbon steel, quenched and tempered
(1 radial line)	SAE grade 5.1	Low- or medium-carbon steel, quenched and tempered
(3 radial lines variant)	SAE grade 5.2	Low-carbon martensite steel, quenched and tempered
A 325	ASTM A 325, type 1	Medium-carbon steel, quenched and tempered
A 325	ASTM A 325, type 2	Low-carbon martensite steel, quenched and tempered
A 325	ASTM A 325, type 3	Atmospheric corrosion (weathering) steel, quenched and tempered
BB	ASTM A 354, grade BB	Low-alloy steel, quenched and tempered
BC	ASTM A 354, grade BC	Low-alloy steel, quenched and tempered
(6 radial lines)	SAE grade 7	Medium-carbon alloy steel, quenched and tempered. Roll threaded after heat treatment
(6 radial lines)	SAE grade 8	Medium-carbon alloy steel, quenched and tempered
(6 radial lines)	ASTM A 354, grade BD	Alloy steel, quenched and tempered
(radial lines variant)	SAE grade 8.2	Low-carbon martensite steel, quenched and tempered
A 490	ASTM A 490	Alloy steel, quenched and tempered

Table 2 ISO property classes and SAE strength grades for steel bolts and studs (including cap screws and U-bolts)
The minimum reduction in area for specimens machined from all the grades and classes of fasteners listed is 35%.

Strength grade or property class	Nominal diameter	Proof stress(a) MPa	ksi	Minimum tensile strength(b) MPa	ksi	Minimum yield strength(c)(d) MPa	ksi	Minimum elongation, %(c)	Maximum surface hardness, HR30N	Core hardness, HRC
ISO property classes(e) for metric sizes										
4.6	5–100 mm	225	33	400	58	240(f)	35(f)	22	...	67–95 HRB
4.8	1.6–16 mm	310	45	420	61	340	49	14	...	71–95 HRB
5.8(g)	5–24 mm	380	55	520	75	420	61	10	...	82–95 HRB
8.8	16–72 mm	600	87	830	120	660	96	12	53(h)	23–34
9.8	1.6–16 mm	650	94	900	131	720	104	10	56(h)	27–36
10.9	5–100 mm	830	120	1040	151	940	136	9	59(h)	33–39
12.8(i)	1.6–20 mm	940	136	1220	177	976	142	...	62(h)	38–43
12.9(i)	1.6–36 mm	970	141	1220	177	1100	160	8	63(h)	39–44
SAE strength grades(j) for sizes in U.S. system of inch dimensions										
1	¼–1½ in.	225	33	415	60	250(f)	36(f)	18	...	70–100 HRB
2	¼–¾ in.(k)	380	55	510	74	395	57	18	...	80–100 HRB
	>¾–1½ in.	225	33	415	60	250(f)	36(f)	18	...	70–100 HRB
4(l)	¼–1½ in.	450	65	795	115	690	100	10	...	22–32
5	¼–1 in.	585	85	830	120	635	92	14	54	25–34
	>1–1½ in.	510	74	725	105	560	81	14	50	19–30
5.1(m)	⅝–½ in.	585	85	830	120	...	...	...	50	25–40
5.2(n)	¼–1 in.	585	85	830	120	635	92	14	56	26–36
7(n)(o)	¼–1½ in.	725	105	915	133	795	115	12	54	28–34
8	¼–1½ in.	830	120	1035	150	895	130	12	58.6	33–39
8.1(l)	¼–1½ in.	830	120	1035	150	895	130	10	...	32–38
8.2(n)	¼–1 in.	830	120	1035	150	895	130	10	58.6	33–39

(a) Determined on full-size fasteners. (b) Determined on both full-size fasteners and specimens machined from fasteners. (c) Determined on specimens machined from fasteners. (d) Yield strength is stress to produce a permanent set of 0.2%. (e) Data from ASTM F 568 or SAE J1199. Values are for fasteners with coarse threads. (f) Yield point instead of yield strength for 0.2% offset. (g) Class 5.8 requirements apply to bolts and screws with lengths 150 mm (6 in.) and shorter and to studs of all lengths. (h) In SAE J1199, surface hardness shall not exceed base metal hardness by more than two points on the HRC scale or shall not exceed the maximums given. (i) As of Sept 1983, class 12.8 and 12.9 bolts were removed from SAE J1199 because of environmentally assisted cracking of 12.8 in automotive rear suspensions. Caution is advised when considering the use of class 12.8 or 12.9 bolts and screws. Capability of the bolt manufacturer, as well as the anticipated in-use environment, should be considered. High-strength products such as class 12.9 require rigid control of heat-treating operations and careful monitoring of as-quenched hardness, surface discontinuities, depth of partial decarburization, and freedom from carburization. Some environments may cause stress-corrosion cracking of nonplated as well as electroplated products. (j) Data from SAE J429. (k) For bolts and screws longer than 150 mm (6 in.), grade 1 requirements apply. (l) Studs only. (m) No. 6 screw and washer assemblies (sems) only. (n) Bolts and screws only. (o) Roll threaded after heat treatment

Table 3 ISO property classes and SAE strength grades for steel nuts
The following classes or grades do not normally include jam, slotted, castle, heavy, or thick nuts.

Strength grade or property class	Nominal diameter	Proof stress(a) MPa	ksi	Rockwell hardness, HRC Minimum	Maximum
ISO property classes(b) for metric nuts					
5(c)	1.6–2.5 mm	520	75	70 HRB	30
	3–4 mm	520	75	70 HRB	30
	5–6 mm	580	84	70 HRB	30
	8–10 mm	590	86	70 HRB	30
	12–16 mm	610	89	70 HRB	30
	20–36 mm	630	91	78 HRB	30
	42–100 mm	630	91	70 HRB	30
9(c)	3–4 mm	900	130	85 HRB	30
	5–6 mm	915	133	89 HRB	30
	8–10 mm	940	136	89 HRB	30
	12–16 mm	950	138	89 HRB	30
	20–36 mm	920	133	89 HRB	30
	42–100 mm	920	133	89 HRB	30
10(c)	1.6–10 mm	1040	151	26	36
	12–16 mm	1050	152	26	36
	20–36 mm	1060	154	26	36
12(c)	3–6 mm	1150	167	26	36
	8–10 mm	1160	168	26	36
	12–16 mm	1190	173	26	36
	20–36 mm	1200	174	26	36
	42–100 mm	1200	174	26	36
85 and 853(c)(d)	12–36 mm	1075	156	89 HRB	38
105 and 1053(c)(d)	12–36 mm	1245	181	26	38
SAE strength grades(e) for nuts in U.S. system of inch sizes					
2(f)	¼–1½ in.	620	90	...	32
5	¼–1 in.	830	120(g)	...	32
		750	109(h)	...	32
	>1–1½ in.	725	105(e)	...	32
		650	94(h)	...	32
8	¼–⅝ in.	1035	150	24	32
	>⅝–1 in.	1035	150	26	34
	>1–1½ in.	1035	150	26	36

(a) Determined on full-size nuts. (b) Data from ASTM A 563M. (c) For hex and hex-flange nuts only. (d) Classes 853 and 1053 are not recognized in ISO standards. Classes 853 and 1053 have atmospheric corrosion resistance comparable to that of the steels covered in ASTM A 588 and A 242 (see the section "Corrosion Protection" in this article). (e) Data from SAE J995. (f) Normally applicable only to square nuts, which are normally available only in grade 2. (g) For UNC, 9 UN thread series. (h) For UNF, 12 UN threaded series and finer

synthetic quenchants are available. Synthetic quenchants must be monitored carefully because they can rapidly change in terms of quenching speed.

Bolt Steels. Table 4 lists the compositions for the bolt steel grades given in Table 2. As previously noted, the producer of bolts is free to use any steel within the grade and class limitations of Table 4 to attain the properties of the specified grade or class in Table 2. However, specific applications sometimes require special characteristics, and the purchaser will consequently specify the steel composition. However, except where a particular steel is absolutely necessary, this practice is losing favor. A specific steel may not be well-suited to the fastener producer's processing facilities; specification of such a steel may result in unnecessarily high cost to the purchaser.

Most bolts are made by cold or hot heading. Resulfurized steels are used in the manufacture of nuts, but because of their tendency to split, these grades are not routinely used in the production of headed bolts. A more recent development relates to the use of calcium-treated steels instead of the C-1100 series steels for headed-bolt manufacture. Documented machinability data remain somewhat limited, but there are indications that the calcium-treated steels not only head well but also offer definite machinability benefits. Only a few bolts are machined from bars; these are usually of special design or the required quantities are extremely small. For such bolts, the extra

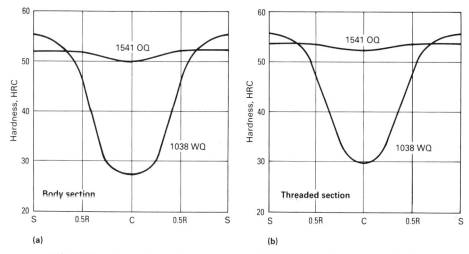

Fig. 1 Hardenability of 19 mm (¾ in.) diam bolts made of 1541 steel and oil quenched. (a) Body section. (b) Threaded section. The curves represent the average as-quenched hardnesses of fifteen 19 mm (¾ in.) diam bolts from one heat of each grade. C is center of the bolt, 0.5R is mid-radius, S is surface. The 1038 steel bolts were water quenched; the 1541 steel bolts, oil quenched.

cost for resulfurized grades of steel may be justified. For example, 1541 steel might be selected to make headed bolts of a specific size. If the same bolts were to be machined from bars, 1141 steel would be selected because of its superior machinability. Special bolts can usually be made more economically by machining from oversize upset blanks instead of from bars.

Stud Steels. The chemical compositions of studs (and U-bolts, which are basically studs formed into a U-shape) are given in Table 4; special modifications that apply to studs can be found in the footnotes. Because studs (and U-bolts) are not headed, it is not essential to restrict sulfur. It may be noted that grade 2 and class 5.8 permit 0.33% maximum sulfur, while grade 5 and classes 8.8 and 9.8 permit 0.13% maximum sulfur.

Stud (or U-bolt) threads, however, are not necessarily cut, but can be rolled for economy and good thread shape. A smaller-diameter rod must be used to roll a specific thread size than to cut the same thread size from rod. For example, a ½-13 thread could be cut from a rod 12.7 mm (0.500 in.) in diameter; a smaller diameter rod would be used to roll the same size threads. Grades 4 and 8.1 are made from a medium-carbon steel and obtain their mechanical properties not from quenching and tempering but from being drawn through a die with special processes. They are particularly suitable for studs because these materials cannot readily be formed into bolts.

Selection of Steel for Bolts and Studs. The following guidelines should be considered when selecting steel for bolts and studs (including cap screws and U-bolts):

- Depending on the capabilities of a facility, bolts up to 305 mm (12 in.) in length and 32 mm (1¼ in.) in diameter can be cold headed. For shops not having this or similar specialized equipment, bolts more than 150 mm (6 in.) in length or more than 19 mm (¾ in.) in diameter may have to be hot headed
- Strength requirements for steels for grade 1 bolts can be met with hot-rolled low-carbon steels
- Depending on the manufacturing method, the strength requirements for steels for grade 2 bolts ranging from 19 to 32 mm (¾ to 1¼ in.) or less in diameter can be met with cold-drawn low-carbon steels; sizes larger than this diameter range of 19 to 32 mm (¾ to 1¼ in.) require hot-rolled low-carbon steel only if the bolt is hot headed, but may be made of cold-finished material
- Grade 4 fasteners (studs only) require a cold-finished medium-carbon steel, specially processed to obtain higher-than-normal strength. Resulfurized steels are acceptable
- Grade 5 bolts and studs require quenched and tempered steel. The choice among carbon, 1541, and alloy steel will vary with the hardenability of the material, the size of the fastener, and the quench employed. Cost favors the use of carbon steel, including 1541; however, the possibility of quench cracking and excessive distortion determines the severity of the quench that can be used. The threading practice (before or after hardening) also determines the severity of quench that can be used if quench cracks in the threads are to be avoided. Generally, the use of a water quench must be approached with caution
- Fasteners made to grade 7 and 8 specifications normally require medium-carbon, fine-grain alloy steel. This steel is selected on a hardenability basis so a minimum of 90% martensite exists at the center after oil quenching. SAE J429 requires oil quenching of these two grades
- Fasteners of SAE grades 5.2 and 8.2 are made from low-carbon martensitic boron steels. These steels (and the low-carbon versions of ISO classes 8.8, 9.8, and 10.9 in Table 4) are readily formed because of the low carbon content, yet the boron gives them relatively high hardenability. Fasteners of these grades are hardened in oil or water, then tempered at minimum temperatures of 425 °C (800 °F) for the 8.2 grade and 340 °C (650 °F) for the 5.2 grade. Grades 5.2 and 8.2 are expected to offer the same mechanical properties as the corresponding nonboron grades 5 and 8 (Fig. 3), but grades 5.2 and 8.2 may have slightly better toughness and ductility than the medium carbon 5 and 8 grades at comparable hardness levels. *Grades 5.2 and 8.2 should be used with caution due to the potential for tempering (softening) at lower temperatures than grade 5 or 8 fasteners*
- *ISO bolt class 12.8 is also made from a low-carbon martensitic (boron) steel. However, this class of bolt is susceptible to stress-corrosion cracking and should be used with caution. This bolt class is no longer specified in SAE J1199 because of failures in automobiles after just two years of service (Ref 1).*
- *ISO bolt class 12.9 has also been removed from SAE J1199. Caution is advised when considering the use of class 12.9 bolts and screws because, like the 12.8 class, the 12.9 class is susceptible to stress-corrosion cracking. The capability of the bolt manufacturer, as well as the anticipated in-use environment, should be considered for both the 12.8 and 12.9 classes.* High-strength products such as class 12.9 require rigid control of the heat-treating operations and careful monitoring of as-quenched hardness, surface discontinuities, depth of partial decarburization, and freedom from carburization. Some environments may cause stress-corrosion cracking of nonplated as well as electroplated products
- For service temperatures of 200 to 370 °C (400 to 700 °F), specific bolt steels are recommended (Table 5) because relaxation is an influencing factor at these temperatures. Although other steels will fulfill requirements for the tabulated conditions, those listed are the commonly used grades. Only medium-carbon alloy steels are recommended; in all instances, they should be quenched and tempered

Nut Steels. The selection of steel for nuts is less critical than for bolts. The nut is usually not made from the same material as the bolt. Table 6 gives the chemical composition requirements for each property grade and class of steel nut shown in Table 3.

Lower-strength nuts (such as grades 2 and 5) are not heat treated. However, higher-grade nuts (such as grade 8) can be heat treated to attain specified hardness.

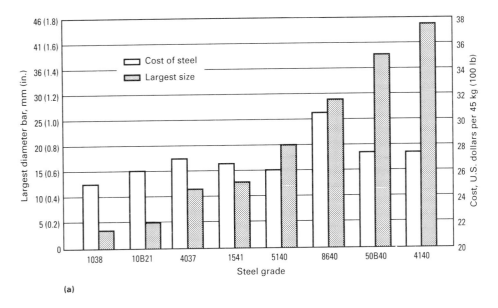

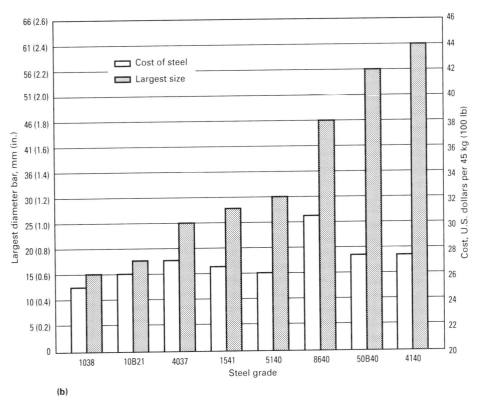

Steel	Cost, U.S. dollars/ton(a)	Largest size to quench to 42 HRC in center			
		Oil		Water	
		mm	in.	mm	in.
1038	500	3.8	0.15	15	0.6
10B21	520	5	0.2	18	0.7
4037	540	11	0.45	25	1
1541	530	13	0.5	28	1.1
5140	520	20	0.8	36	1.4
8640	610	29	1.15	46	1.8
50B40	550	38	1.5	56	2.2
4140	550	44	1.75	61	2.4

(a) As of 1989

Fig. 2 Cost and hardenability relations for oil-quenched (a) and water-quenched (b) steels for cold-formed fasteners

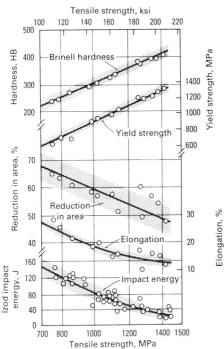

Fig. 3 Tensile and impact properties of fully quenched and tempered boron steels superimposed on normal expectancy bands for medium-carbon low-alloy steels without boron

Nuts are machined from bar stock, cold formed or hot formed, depending on configuration and production requirements. Size and configuration are usually more important than the material from which the nuts are made.

The bolt is normally intended to break before the nut threads strip. Regular hex nut dimensions are such that the shear area of the threads is greater than the tensile stress area of the bolt by more than 100%. Consequently, low-carbon steel nuts are customarily used even when the bolts are made of much higher strength material.

Low-carbon steel nuts are usually heat treated to provide mar resistance to the corners of the head or to the clamping face. Light case carburizing or carbonitriding is often employed to improve mar resistance.

When nuts are to be quenched and tempered, the steel must have the appropriate hardenability. Increasing the amounts of carbon and manganese or adding other alloying elements to provide increased hardenability will decrease the suitability of the material for cold forming. For this reason, low-carbon boron steels are widely used for quenched and tempered high-strength nuts. The low carbon content permits easy cold forming, while the boron enhances hardenability. Threading can be done before or after heat treatment, depending on the class of thread fit required and the hardness of the heat-treated nut.

Because the selection of steel for nuts is not critical, practice varies considerably. A common practice is to use steels such as

Table 4 Chemical compositions of steel bolts and studs (including cap screws and U-bolts)

Strength grade or property class	Nominal diameter	Material and treatment	C	P	S	Others
ISO property classes(b)						
4.6	5–100 mm	Low- or medium-carbon steel	0.55	0.048	0.058	...
4.8	1.6–16 mm	Low- or medium-carbon steel, partially or fully annealed as required	0.55	0.048	0.058	...
5.8	5–24 mm	Low- or medium-carbon steel, cold worked	0.13–0.55	0.048	0.058(c)	...
8.8	16–72 mm	Medium-carbon steel, quenched and tempered(d)(e)	0.25–0.55	0.048	0.058(f)	...
	16–36 mm	Low-carbon martensite steel, quenched and tempered(g)	0.15–0.40	0.048	0.058	(h)
8.8.3	16–36 mm	Atmospheric corrosion resistant steel, quenched and tempered	See ASTM F 568.			(i)
9.8	1.6–16 mm	Medium-carbon steel, quenched and tempered	0.25–0.55	0.048	0.058(f)	...
	1.6–16 mm	Low-carbon martensite steel, quenched and tempered(g)	0.15–0.40	0.048	0.058	(h)
10.9	5–20 mm	Medium-carbon steel, quenched and tempered(j)(k)	0.25–0.55	0.048	0.058	...
	5–100 mm	Medium-carbon alloy steel, quenched and tempered(j)	0.20–0.55	0.040	0.045	...
	5–36 mm	Low-carbon martensite steel, quenched and tempered(j)(g)	0.15–0.40	0.048	0.058	(h)
10.9.3	16–36 mm	Atmospheric corrosion resistant steel, quenched and tempered(j)	See ASTM F 568.			(i)
12.8(l)(m)	1.6–20 mm	Low-carbon martensite boron steel, quenched and tempered(j)(n)	0.16–0.27	0.038	0.048	(o)
12.9(m)	1.6–100 mm	Alloy steel, quenched and tempered(j)	0.31–0.65	0.045	0.045	(p)
SAE J429 strength grades						
1	¼–1½ in.	Low- or medium-carbon steel	0.55	0.048	0.058	...
2	¼–1½ in.	Low- or medium-carbon steel	0.55	0.048	0.058(c)	...
4	¼–1½ in.	Medium-carbon cold-drawn steel	0.55	0.048	0.13	...
5	¼–1½ in.	Medium-carbon steel, quenched and tempered	0.28–0.55	0.048	0.058(f)	...
5.1	⅝–½ in.	Low- or medium-carbon steel, quenched and tempered	0.15–0.30	0.048	0.058	...
5.2	¼–1 in.	Low-carbon martensitic steel, fully killed, fine grain, quenched and tempered	0.15–0.25	0.048	0.058	(h)
7	¼–1¼ in.	Medium-carbon alloy steel, quenched and tempered(q)(r)	0.28–0.55	0.040	0.045	...
8	¼–1¼ in.	Medium-carbon alloy steel, quenched and tempered(q)(r)	0.28–0.55	0.040	0.045	...
8.1	¼–1¼ in.	Drawn steel for elevated-temperature service: medium-carbon alloy steel or 1541 steel	0.28–0.55	0.048	0.058	...
8.2	¼–1 in.	Low-carbon martensitic steel, fully killed, fine grain, quenched and tempered(s)	0.15–0.25	0.048	0.058	(h)

(a) All values are for product analysis; where a single value is shown, it is a maximum. (b) Data from ASTM F 568. (c) For studs only, sulfur content may be 0.33% max. (d) For diameters through 24 mm, unless otherwise specified by the customer, the producer can use a low-carbon martensite steel with 0.15–0.40% C, 0.74% Mn (min), 0.048% P (max), 0.058% S (max), and 0.0005% B (min). (e) At producer's option, medium-carbon alloy steel can be used for diameters over 24 mm. (f) For studs only, sulfur content may be 0.13% max. (g) Requires special marking; see ASTM F 568. (h) 0.74% Mn (min) and 0.0005% B (min). (i) Available in six different types of compositions that include carbon, manganese, phosphorus, sulfur, silicon, copper, nickel, chromium, and vanadium or molybdenum in a few types. Selection of a type is at the option of the producer. (j) Steel for classes 10.9, 10.9.3, 12.8, and 12.9 products shall be fine grain and have a hardenability that will achieve a structure of approximately 90% martensite at the center of a transverse section one diameter from the threaded end of the product after oil quenching. (k) Carbon steel can be used at the option of the manufacturer for products of nominal thread diameters 12 mm and smaller. When approved by the purchaser, carbon steel can be used for products of diameters larger than 12 mm through 20 mm, inclusive. (l) No longer specified in SAE J1199. (m) Data obtained from the old (prior to Sept 1983) version of SAE J1199 and provided for information only. (n) Class 12.8 bolts required heat treatment in a continuous-type furnace having a protective atmosphere, and under no circumstances should heat treatment or carbon restoration be accomplished in the presence of nitrogen compounds, such as carbonitriding or cyaniding. (o) 0.74–1.46% Mn and 0.0005–0.003% B. (p) One or more of the alloying elements chromium, nickel, molybdenum, or vanadium shall be present in sufficient quantity to ensure that the specified strength properties are met after quenching and tempering. (q) Fine-grain steel with hardenability that will produce 47 HRC min at the center of a transverse section one diameter from the threaded end of the fastener after oil quenching (see SAE J407). (r) For diameters of ¼ through ¾ in., carbon steel can be used by agreement. At producer's option, 1541 steel, oil quenched and tempered, can be used for diameters through 7/16 in. (s) Steel with hardenability that will produce 38 HRC min at the center of a transverse section one diameter from the threaded end of the fastener after quenching

1108, 1109, 1110, 1113, or 1115, cold formed or machined from cold-drawn bars, for grade 2 nuts. Grade 5 nuts are commonly made from 1035 or 1038 steels, cold formed from annealed bars, cold drawn and stress relieved, or quenched and tempered. Grade 8 nuts are formed from low-carbon boron steels, then quenched and tempered.

Corrosion Protection

The most commonly used protective metal coatings for ferrous metal fasteners are zinc, cadmium, and aluminum. Tin, lead, copper, nickel, and chromium are also used, but only to a minor extent and for very special applications. In many cases, however, fasteners are protected by some means other than metallic coatings. They are sometimes sheltered from moisture or covered with a material that prevents moisture from making contact, thus drastically reducing or eliminating corrosion. For fasteners exposed to the elements, painting is universally used.

The low-alloy high-strength steel conforming to ASTM A 242 and A 588 forms its own protective oxide surface film. This type of steel, although it initially corrodes at the same rate as plain carbon steel, soon exhibits a decreasing corrosion rate, and after a few years, continuation of corrosion is practically nonexistent. The oxide coating formed is fine textured, tightly adherent, and a barrier to moisture and oxygen, effectively preventing further corrosion. Plain carbon steel, on the other hand, forms a coarse-textured flaky oxide that does not prevent moisture or oxygen from reaching the underlying noncorroded steel base.

The 853 and 1053 class nuts (Table 3) and bolt classes 8.8.3 and 10.9.3 (Table 4) have corrosion resistance characteristics similar to those of steels conforming to ASTM A 242 and A 588. These weathering steels are suitable for resisting atmospheric corrosion

Table 5 Recommended steels for bolts to be used at temperatures between 200 and 370 °C (400 and 700 °F)

All selections are based on a minimum tempering temperature of 455 °C (850 °F).

Bolt diameter		Steel recommended for a proof stress (at room temperature) of:		
mm	in.	520 MPa (75 ksi)	690 MPa (100 ksi)	860 MPa (125 ksi)
6.3–19	¼–¾	1038	4037	4037
19–32	¾–1¼	1038	4140	4140
32–50	1¼–2	4140	4140	4145

Table 6 Chemical compositions of steel nuts

Strength grade or property class	Composition, %(a)			
	C (max)	Mn (min)	P (max)	S (max)
SAE strength grades(b)				
2	0.47	...	0.12(c)	0.15(d)
5	0.55	0.30	0.05(e)(f)	0.15(d)(f)
8	0.55	0.30	0.04	0.05(g)
ISO property classes(h)				
5(i) and 9	0.55	...	0.04	0.15(c)(d)
8S	0.55	...	0.04	0.15
10	0.55	0.30	0.04	0.05(j)
853(k)	...	See ASTM A 563M.		
1053(k)	...	See ASTM A 563M.		
12	0.20–0.55	0.60	0.04	0.05(j)

(a) All values for heat analysis. (b) Data from SAE J995. (c) Resulfurized and rephosphorized material is not subject to rejection based on check analysis for sulfur. (d) If agreed, sulfur can be 0.23% max. (e) For acid bessemer steel, phosphorus can be 0.13% max. (f) If agreed, phosphorus can be 0.12% max and sulfur can be 0.35% max, provided manganese is 0.70% min. (g) If agreed, sulfur can be 0.33% max, provided manganese is 1.35% min. (h) Data from ASTM A 563M. (i) If agreed, free-cutting steel having maximums of 0.34% S, 0.12% P, and 0.35% Pb can be used. (j) If agreed, sulfur can be 0.15% max with a minimum of 1.35% Mn. (k) Corrosion-resistant grades are not included in ISO classifications. Class 853 is used with bolt grade 8.8.3 and has a selection of steel compositions at the option of the manufacturer. Class 1053 is used with bolt grade 10.9.3.

and have an atmospheric corrosion resistance approximately two times that of carbon structural steel with copper. However, these weathering steels are not recommended for exposure to highly concentrated industrial fumes or severe marine conditions, nor are they recommended for applications in which they will be buried or submerged. In these environments, the highly protective oxide does not form properly, and corrosion is similar to that for plain carbon steel.

Zinc Coating. Zinc is the coating material most widely used for protecting fasteners from corrosion. Electroplating and zinc phosphating are the two most frequently used method of application, followed by hot dipping and, to a minor extent, mechanical plating.

Hot dipping, as the name implies, involves immersing parts in a molten bath of zinc. Hot dip zinc coatings are sacrificial by electrochemical means, and these coatings for fasteners are covered in ASTM A 394. Zinc electroplating of fasteners is done primarily for appearance, where thread fit is critical, where corrosion is not expected to be severe, or where life expectancy is not great.

Specification ASTM B 633 for electrodeposited zinc coatings on steel specifies three coating thicknesses: GS, 25 μm (0.0010 in.); LS, 13 μm (0.0005 in.); and RS, 4 μm (0.00015 in.). These electrodeposited coatings are often given supplemental chromate coatings to develop a specific color and to enhance corrosion resistance. The corrosion life of a zinc coating is proportional to the amount of zinc present and chromate finish; therefore, the heaviest electrodeposited coating (GS) would have only about half the life of a hot dip galvanized coating.

Mechanical (nonelectrolytic) barrel plating is another method of coating fasteners with zinc. Coating weight can be changed by varying the amount of zinc used and the duration of barrel rotation. Such coatings are quite uniform and have a satisfactory appearance.

Cadmium coatings are also applied to fasteners by an electroplating process similar to that used for zinc. These coatings are covered in ASTM A 165. As is true for zinc, cadmium corrosion life is proportional to the coating thickness. The main advantage of cadmium over zinc is its much greater resistance to corrosion in marine environments and uniformity of torque-tension relationship. Cadmium-plated steel fasteners are also used in aircraft in contact with aluminum because the galvanic characteristics of cadmium are more favorable than those of zinc. Chromate coatings are also used over cadmium coatings for the reasons given for zinc-plated fasteners.

Aluminum coating on fasteners offers the best protection of all coatings against atmospheric corrosion. Aluminum coating also gives excellent corrosion protection in seawater immersion and in high-temperature applications.

Aluminum coatings are applied by hot dip methods at about 675 to 705 °C (1250 to 1300 °F). Aluminum alloy 1100 is usually used because of its general all-around corrosion resistance. As with any hot dip coating, a metallurgical bond is formed that consists of an intermetallic alloy layer overlaid with a coating of pure bath material.

Aluminum coatings do not corrode uniformly, as do zinc and cadmium coatings, but rather by pitting. In some cases, these pits may extend entirely through the coating to the base metal; in others, only through the overlay to the intermetallic layer. Pits, which may occur in a part soon after exposure, sometimes discolor the coated surface but cause little damage. The complex aluminum and iron oxide corrosion product seals the pits, and because the corrosion product is tightly adherent and impervious to attack, corrosion is usually limited. There is little tendency for corrosion to continue into the ferrous base, and there is none for undercutting and spalling of the coating.

Aluminum coatings will protect steel from scaling at temperatures up to about 540 °C (1000 °F); the aluminum coating remains substantially the same as when applied, and its life is exceptionally long. Above 650 °C (1200 °F), the aluminum coating diffuses into the steel to form a highly protective aluminum-iron alloy. This diffusing or alloying is time-temperature dependent; the higher the temperature, the faster the diffusion. However, scaling will not take place until all the aluminum is used up, which may take a thousand or more hours even at temperatures as high as 760 °C (1400 °F).

The prevention of galling at elevated temperatures is another characteristic of aluminum coatings. Stainless steel fasteners for use at 650 °C (1200 °F) have been aluminum coated just to prevent galling. Coated nuts can be removed with an ordinary wrench after many hours at these temperatures, which is impossible with uncoated nuts.

Fastener Performance at Elevated Temperatures

Selection of fastener material is perhaps the single most important consideration in elevated-temperature design. The basic design objective is to select a bolt material that will give the desired clamping force at all critical points in the joint.

Time- and Temperature-Related Factors. To achieve the basic design objective mentioned above, it is necessary to balance the three time- and temperature-related factors (modulus, thermal expansion, and relaxation) with a fourth factor—the amount of initial tightening or clamping force. These three time- and temperature-related factors affect the elevated-temperature performance of fasteners as follows.

Modulus of Elasticity. As temperature increases, the modulus of elasticity decreases; therefore, less load (or stress) is needed to impact a given amount of elongation (or strain) to a material than at lower temperatures. This means that a fastener stretched a certain amount at room temperature to develop preload will exert a lower clamping force at higher temperature.

Coefficient of Expansion. With most materials, the size of the part increases as the temperature increases. In a joint, both the structure and the fastener increase in size with an increase in temperature. If the coefficient of expansion of the fastener exceeds that of the joined material, a predictable amount of clamping force will be lost as temperature increases. Conversely, if the coefficient of expansion of the joined material is greater, the bolt may be stressed beyond its yield or even fracture strength, or cyclic thermal stressing may lead to thermal fatigue failure. Thus, matching of materials in joint design can ensure sufficient clamping force at both room and elevated temperatures without overstressing the fastener.

Relaxation. In a loaded bolt joint at elevated temperature, the bolt material will undergo permanent plastic deformation (creep) in the direction of the applied stress. This phenomenon, known as relaxation in loaded-joint applications, reduces the clamping force with time. Relaxation is the most important of the three time- and temperature-related factors and is discussed in more detail in the article "Elevated-Temperature Properties of Ferritic Steels" in this Volume.

Bolt Steels for Elevated Temperatures. Table 5 lists the recommended steels for bolts to be used at temperatures between 200 and 370 °C (400 and 700 °F). For higher temperatures up to 480 °C (900 °F), other alloy steels are used. For example, the medium-alloy chromium-molybdenum-vanadium steel conforming to ASTM A 193, grade B 16, is a commonly used bolt material in industrial turbine and engine applications to 480 °C (900 °F). An aircraft version of this steel, AMS 6304, is widely used in fasteners for jet engines. The 5% Cr tool steels, most notably H11, are also used for fasteners having a tensile strength of 1500 to 1800 MPa (220 to 260 ksi). They retain excellent strength through 480 °C (900 °F).

For temperatures above 480 °C (900 °F), heat-resistant alloys or superalloys are used for bolt materials. From 480 to 650 °C (900 to 1200 °F), corrosion-resistant alloy A-286 is used. Alloy 718, with a room-temperature tensile strength of 1240 MPa (180 ksi), has some applications in this temperature range. The nickel-base alloys René 41, Waspaloy, and alloy 718 can be used for most applications in the temperature range of 650 to 870 °C (1200 to 1600 °F).

Coatings for Elevated Temperatures. At moderate temperatures, where cadmium and zinc anticorrosion platings might normally be used, the phenomenon of stress alloying becomes an important consideration. Conventional cadmium plating, for example, is usable only to 230 °C (450 °F). At somewhat above that temperature, the cadmium is likely to melt and diffuse into the base material along the grain boundaries, causing cracking by liquid-metal embrittlement, which can lead to rapid failure. For corrosion protection of high-strength alloy steel fasteners used at temperatures between 230 and 480 °C (450 and 900 °F), special nickel-cadmium coatings such as that described in AMS 2416 are often used. At extremely high temperatures, coatings must be applied to prevent oxidation of the base material.

Effect of Thread Design on Relaxation. Fastener-manufacturing methods can also influence bolt performance at elevated temperature. The actual design and shape of the threaded fastener are also important, particularly the root of the thread. A radiused thread root is a major consideration in room-temperature design, being a requisite for good fatigue performance. However, at elevated temperature, a generously radiused thread root is also beneficial in relaxation performance. Starting at an initial preload of 483 MPa (70 ksi), a Waspaloy stud with square thread roots lost a full 50% of its clamping force after 20 h, with the curve continuing downward, indicating a further loss. A similar stud made with a large-radiused root lost only 36% of preload after 35 h.

Fastener Tests

The fastener manufacturer must perform periodic tests of the product to ensure that properties are maintained within specified limits. Guidelines for testing are provided in ASTM F 606 and in SAE J429 (bolts and studs in U.S. inch sizes), J995 (nuts in U.S. inch sizes), and J1216 (test methods for metric threaded fasteners). When requested in writing by the purchaser, the manufacturer will furnish a copy of a certified test report.

The most widely accepted method of determining the strength of full-size bolts and studs is a wedge tensile test for minimum tensile (breaking) strength. Testing of nuts involves a proof stress testing. Proof stress testing of bolts and studs is also performed before the test for ultimate wedge tensile strength.

The wedge tensile test of bolts accentuates the adverse conditions of bolts assembled under misalignment; therefore, it is also the most widely used quality control test for ultimate strength and head-to-body integrity and ductility. The test is performed by placing the wedge under the bolt head and, by means of suitable fixtures, stressing the bolt to fracture in a tensile test machine. To meet the requirements of the test, the tensile fracture must occur in the body, or threaded section, with no fracture at the

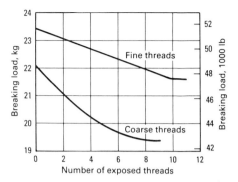

Fig. 4 Variation of breaking strength with number of exposed threads for ¾ in. diam SAE grade 5 bolts

junction of the body and head. In addition, the breaking strength should meet specified minimum strength requirements. Details of this test are given in ASTM F 606 and SAE J429.

The number of exposed threads between the bolt shank and the beginning of the nut or testing fixture influences the recorded tensile strength for both coarse-thread and fine-thread bolts. A typical variation of breaking strength with number of exposed threads is plotted in Fig. 4. The data are for ¾ in. diam SAE grade 5 bolts. Because of this variation, the number of exposed threads should be specified for wedge tensile testing of bolts and other threaded fasteners; generally, six exposed threads are specified.

The proof stress of a bolt or stud is a specified stress that the bolt or stud must withstand without detectable permanent set. For purposes of this test, a bolt or stud is deemed to have incurred no permanent set if the overall length after application and release of the proof stress is within ±0.013 mm (±0.0005 in.) of its original length. Length measurements are ordinarily made to the nearest 0.0025 mm (0.0001 in.). Because bolts and studs are manufactured in specific sizes, the proof stress values are commonly converted to equivalent proof load values, and it is the latter that are actually used in testing full-size fasteners. To compute proof load, the stressed area must first be determined. Because the smallest cross-sectional area is in the threads, the stressed area is computed as follows:

$$A_s(\text{in.}^2) = 0.7854\left(D - \frac{0.9743}{N}\right)^2 \quad (\text{Eq 1})$$

where A_s is the mean equivalent stress area in square inches, D is the nominal diameter in inches, and N is the number of threads per inch. The equivalent formula for metric threads is:

$$A_s(\text{mm}^2) = 0.7854(D - 0.9382P)^2 \quad (\text{Eq 2})$$

where A_s is the mean equivalent stress area in millimeters, D is the nominal diameter in

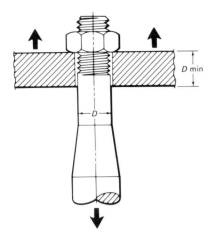

Fig. 5 Test setup for the proof testing of nuts

millimeters, and P is the thread pitch in millimeters.

For example, ½ in. diam bolts made to SAE grade 5 requirements have values of mean equivalent stress area for coarse threads (13 threads per inch) and fine threads (20 threads per inch) equal to 0.1419 in.^{2} and 0.1599 in.^{2}, respectively, according to Eq 1. These bolts must withstand a proof stress of 585 MPa (85 ksi) without a detectable difference between the initial length and the length after the proof load has been applied and released. For coarse threads, this requires a proof load of 53.6 kN (12 060 lb); for fine threads, the proof load is 60.5 kN (13 600 lb).

The proof stress of a nut is determined by assembling it on a hardened and threaded mandrel or on a test bolt conforming to the particular specification. The specified proof load for the nut is determined by converting the specified proof stress, using the mean equivalent stress area calculated for the mandrel or test bolt. This proof load is applied axially to the nut by a hardened plate, as shown in Fig. 5. The thickness of the plate is at least equal to the diameter of the mandrel or test bolt. The diameter of the hole in the plate is a specified small amount greater than that of the mandrel or test bolt diameter. To demonstrate acceptable proof stress, the nut must resist the specified proof load without failure by stripping or rupture, and it must be removable from the mandrel or test bolt by hand after initial loosening.

Details relating to the prescribed proof stress tests and other requirements for threaded steel fasteners of various sizes in both coarse and fine threads are available in SAE J429 and J995 for fasteners made to SAE strength grades or ISO 898 for fasteners made to ISO property classes. Additional information is also available in ASTM F 606.

Mechanical Properties

The major mechanical properties of fasteners include:

- Tensile (breaking) strength with a static load
- Hardness
- Fatigue strength with a dynamic loading
- Resistance to stress-corrosion cracking or other forms of environmentally induced cracking, such as hydrogen embrittlement and liquid-metal embrittlement

The following sections briefly review these mechanical properties of threaded steel fasteners.

Strengths With Static Loads. Tables 2 and 3 list the specified mechanical properties of the commonly used SAE strength grades and ISO property classes of steel bolts, studs, and nuts. As previously noted, the grades or classes of these specifications are based on tensile strength for bolts and studs or proof stress for nuts.

Grade 1038 steel is one of the most widely used steels for threaded fasteners. Typical distributions of tensile properties for bolts and cap screws made from 1038 steel, as evaluated by the wedge tensile test, are shown in Fig. 6. These data were obtained from one plant and represent tests from random lots of SAE grade 5 fasteners. The three histograms in Fig. 6 show three distributions typical of grade 5 fasteners. No significance should be attached to the apparent difference between average values, especially for the two hex-head bolts of different lengths. Specifications require only that bolt strength exceed a specified minimum value, not that bolts of different sizes have statistically equivalent average strengths.

Hardness Versus Tensile Strength. To ensure freedom from the effects of decarburization and nonrepresentative cooling rates during the quench, there is really only one preferred location for checking hardness. This location is the mid-radius of a transverse section taken one diameter from the threaded end of the bolt (Fig. 7). If the hardness tests are not taken at this location, then greater scatter will occur in the relation between hardness and tensile strength. For example, hardness tests taken at the bolt head will probably result in more scatter because the greater thermal mass of the bolt head produces differences in quench efficiency and may result in incomplete hardening.

The bolt shown in Fig. 7 was made from one heat of 1038 steel of the following composition: 0.38% C, 0.74% Mn, 0.08% Si, 0.025% P, 0.040% S, 0.08% Cr, 0.07% Ni, and 0.12% Cu. The bolt was quenched in water from 845 °C (1550 °F) and tempered at different temperatures in the range 365 to 600 °C (690 to 1110 °F) to produce a range of hardness of 20 to 40 HRC.

Figure 8 shows the results of a cooperative study in which eight laboratories tested a large number of bolts made from a single heat of 1038 steel. There were eight different lots of bolts, each heat treated to a different hardness level. Each of the eight laboratories tested bolts from all eight lots. Hardness readings were taken on a transverse section through the threaded portion of the bolt at the mid-radius of the bolt. The tranverse section was located one diameter from the threaded end of the bolt.

Fatigue Failures. Fatigue is one of the most common failure mechanisms of threaded fasteners, particularly if insufficient tightening of the fasteners results in flexing. To eliminate this cause of fatigue fracture at room temperature, the designer should specify as high an initial preload as

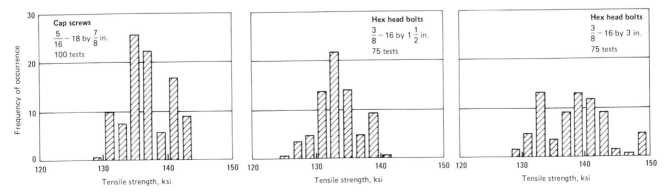

Fig. 6 Typical strength distributions for 10° wedge tests of 1038 steel cap screws and bolts

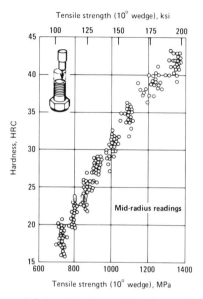

Fig. 7 Relation of hardness and tensile strength for SAE grade 5 bolts made of 1038 steel

practical. Higher clamping forces make more rigid joints and therefore reduce the rate of fatigue crack propagation from flexing. The optimum fastener-torque values for applying specific loads to the joint have been determined for many high-strength fasteners. However, these values should be used with caution, because the tension produced by a selected torque value depends directly on the friction between the contacting threads. The use of an effective lubricant on the threads may result in overloading of the fastener, while the use of a less effective lubricant may result in a loose joint. With proper selection of materials, proper design of bolt-and-nut bearing surfaces, and the use of locking devices, the assumption is that the initial clamping force will be sustained during the life of the fastened joint. This assumption cannot be made in elevated-temperature design. At elevated temperature, the induced bolt load will decrease with time as a result of creep.

Fatigue Strength. The factors affecting the fatigue strength of threaded fasteners include surface condition, mean stress, stress range, and the grain pattern at the head-to-shank fillet. If bolts made of two different steels have equivalent hardnesses throughout identical sections, their fatigue strengths will be similar (Fig. 9) as long as other factors such as mean stress, stress range, and surface condition are the same. If the results of fatigue tests on standard test specimens were interpreted literally, high-carbon steels would be selected for bolts. Actually, steels of high carbon content (>0.55% C) are unsuitable because they are notch sensitive.

The principal design feature of a bolt is the threaded section, which establishes a notch pattern inherent in the part because of its design. The form of the threads, plus any mechanical or metallurgical condition that also creates a surface notch, is much more important than steel composition in determining the fatigue strength of a particular lot of bolts. Some of these factors are discussed below.

Causes and Prevention of Fatigue Crack Initiation. The origin of a fatigue crack is usually at some point of stress concentration, such as an abrupt change of section, a deep scratch, a notch, a nick, a fold, a large inclusion, or a marked change in grain size. Fatigue failures in bolts often occur in the threaded section immediately adjacent to the edge of the nut (or mating part) on the washer side, at or near the first thread inside the nut (or mating part). This area of stress concentration occurs because the bolt elongates as the nut is tightened, thus producing increased loads on the threads nearest the bearing face of the nut, which add to normal service stresses. This condition is alleviated to some extent by using nuts of a softer material that will yield and distribute the load more uniformly over the engaged threads. Significant additional improvement in fatigue life is also obtained by rolling (cold working) the threads rather than cutting them and also by rolling threads after heat treatment rather than before.

Other locations of possible fatigue failure of a bolt under tensile loading are the thread runout and the head-to-shank fillet. Like the section of the bolt thread described in the previous paragraph, these two locations are also areas of stress concentration. Any measures that decrease stress concentration can lead to improved fatigue life. Typical examples of such measures are the use of UNJ increased root radius threads (see MIL-S-8879A) and the use of internal thread designs that distribute the load uniformly over a large number of bolt threads. Shape and size of the head-to-shank fillet are important, as is a generous radius from the thread runout to the shank. In general, the radius of this fillet should be as large as possible while at the same time permitting adequate head-bearing area. This requires a design trade-off between the head-to-shank radius and the head-bearing area to achieve optimum results. Cold working of the head fillet is another common method of preventing fatigue failure because it induces a residual compressive stress and increases the material strength.

Several other factors are also important in avoiding fatigue fracture at the head-to-shank fillet. The heads of most fasteners are formed by hot or cold forging, depending on the type of material and size of the bolt. In addition to being a relatively low-cost manufacturing method, forging provides smooth, unbroken grain flow lines through the head-to-shank fillet, which closely follows the external contour of the bolt (Fig. 10) and therefore minimizes stress raisers, which promote fatigue cracking. In the hot forging of fastener heads, temperatures must be carefully controlled to avoid overheating, which may cause grain growth. Several failures of 25 mm (1 in.) diam type H-11 airplane-wing bolts quenched and tempered to a tensile strength of 1800 to 1930 MPa (260 to 280 ksi) have been attributed to stress concentration that resulted from a large grain size in the shank. Other failures in these 25 mm (1 in.) diam bolts, as well as in other similarly quenched and tempered steel bolts, were the result of cracks in untempered martensite that formed as a result of overheating during finish grinding.

Influence of the Thread-Forming Method on Bolt Fatigue Strength. The method of forming the thread is an important factor influencing the fatigue strength of bolts. Specifically, there is a marked improvement when threads are rolled rather than either cut or ground, particularly when the threads are rolled after the bolt has been heat treated (Fig. 11).

Other factors being equal, a bolt with threads properly rolled after heat treatment—that is, free from mechanical imperfections—has a higher fatigue limit than one with cut threads. This is true for any strength category. The cold work of rolling increases the strength at the weakest section (the thread root) and imparts residual compressive stresses, similar to those imparted by shot peening. The larger and smoother root radius of the rolled thread also contributes to its superiority. Because of the fatigue life concern, all bolts and screws greater than grade 1 and less than 19 mm (¾ in.) in diameter and 150 mm (6 in.) in length are to be roll threaded. Studs and larger bolts and screws may have the threads rolled, cut, or ground.

Effect of Surface Treatment on Fatigue. Light cases, such as from carburizing or carbonitriding, are rarely recommended and should not be used for critical externally threaded fasteners, such as bolts, studs, or U-bolts. The cases are quite brittle and crack when the fasteners are tightened or bent in assembly or service. These cracks may then lead to fatigue cracking and possible fracture.

Chromium and nickel platings decrease the fatigue strength of threaded sections and should not be used except in a few applications, such as automobile bumper studs or similar fasteners that operate under conditions of low stress and require platings for appearance. Cadmium and zinc platings slightly reduce fatigue strength. Electroplated parts for high-strength applications should be treated after plating to eliminate or minimize hydrogen embrittlement (which is a strong contributor to fatigue cracking).

Installation. As noted at the beginning of this section on fatigue failures, bolt loading is a major factor in the fatigue failure of threaded fasteners. When placed into ser-

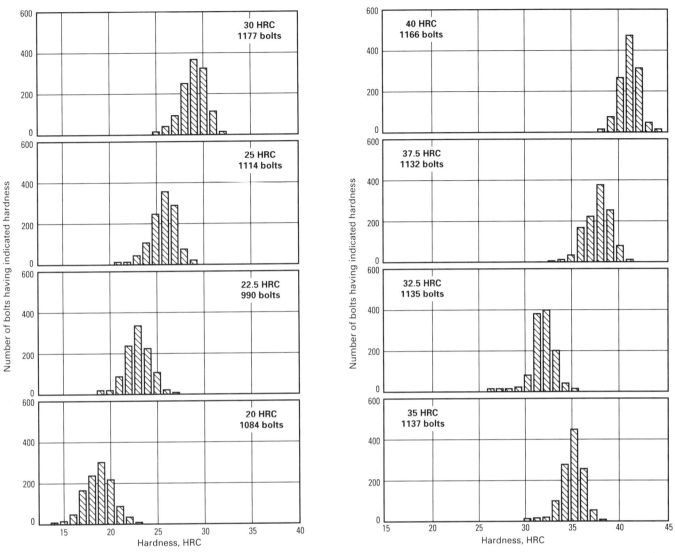

Fig. 8 Hardness distribution for eight lots of 1038 steel bolts. Bolts, 13 mm (½ in.) in diameter, were all made from one heat of steel. The bolts were heat treated in one plant to eight different levels of nominal hardness. Tests were made in the originating plant and in seven other laboratories.

vice, bolts are most likely to fail by fatigue if the assemblies involve soft gaskets or flanges, or if the bolts are not properly aligned and tightened. Fatigue resistance is also related to clamping force. In many assemblies, a certain minimum clamping force is required to ensure both proper alignment of the bolt in relation to other components of the assembly and proper preload on the bolt. The former ensures that the bolt will not be subjected to undue eccentric loading, and the latter that the correct mean stress is established for the application. In some cases, clamping stresses that exceed the yield strength may be desirable; experiments have shown that bolts clamped beyond the yield point have better fatigue resistance than bolts clamped below the yield point.

Decreasing the bolt stiffness can also reduce cyclic stresses. Methods commonly used are reduction of the cross-sectional area of the shank to form a waisted shank and rolling the threads further up the exposed shank to increase the "spring" length.

Stress-corrosion cracking (SCC) is an intergranular fracture mechanism that sometimes occurs in highly stressed fasteners after a period of time, and it is caused by a corrosive environment in conjunction with a sustained tensile stress above a threshold value. An adverse grain orientation increases the susceptibility of some materials to stress corrosion. Consequently, SCC can be prevented by excluding the corrodent, by keeping the static tensile stress of the fastener below the critical level for the material and grain orientation involved, or by changing to a less susceptible material or material condition. Because tensile loads (even residual tensile loads) are required to produce SCC, compressive residual stresses may prevent SCC.

As with the environmentally induced cracking from hydrogen embrittlement and liquid-metal embrittlement (see the article "Embrittlement of Steels" in this Volume), the understanding of SCC is largely phenomenological, without any satisfactory mechanistic model for predicting SCC or the other forms of environmentally induced cracking. This lack of mechanistic predictability of SCC is particularly unfortunate because measurable corrosion usually does not occur before or during crack initiation or propagation. Even when corrosion does occur, it is highly localized (that is, pitting, crevice attack) and may be difficult to detect. Moreover, SCC is a complex synergistic phenomenon resulting from the combined simultaneous interaction of mechanical and chemical conditions. Precorrosion followed by loading in an inert environment will not result in any observable crack propagation, while simultaneous environmental exposure and application of stress will cause time-dependent subcritical crack propagation.

The susceptibility of a metal to SCC depends on the alloy and the corrodent. The

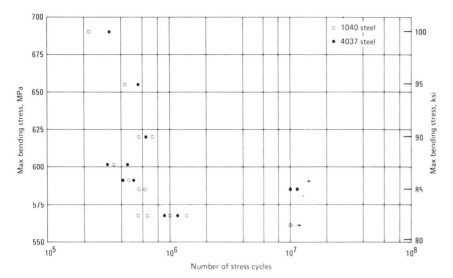

Fig. 9 Fatigue data for 1040 and 4037 steel bolts. The bolts (⅜ by 2 in., 16 threads to the inch) had a hardness of 35 HRC. Tensile properties of the 1040 steel at three-thread exposure were yield strength, 1060 MPa (154 ksi); tensile strength (axial), 1200 MPa (175 ksi); tensile strength (wedge), 1190 MPa (173 ksi). For the 4037 steel: yield strength, 1110 MPa (161 ksi); tensile strength (axial), 1250 MPa (182 ksi); tensile strength (wedge), 1250 MPa (182 ksi)

Fig. 10 Uniform, unbroken grain flow around the contours of the forged head of a threaded fastener. The uniform, unbroken grain flow minimizes stress raisers and unfavorable shear planes and therefore improves fatigue strength.

National Association of Corrosion Engineers, the Materials Technology Institute of the Chemical Process Industries, and others have published tables of corrodents known to cause SCC of various metal alloy systems (Ref 2-4). This literature should be used only as a guide for screening candidate materials for further in-depth investigation, testing, and evaluation. In general, plain carbon steels are susceptible to SCC by several corrodents of economic importance, including aqueous solutions of amines, carbonates, acidified cyanides, hydroxides, nitrates, and anhydrous ammonia. Carbon steels, low-alloy steels, and H-11 tool steels with ultimate tensile strengths above 1380 MPa (200 ksi) are susceptible to SCC in a seacoast environment. Of the various bolt steels, bolts made from ISO class 12.8 have experienced failures for SCC in automotive applications (Ref 1). Stainless steels are also susceptible to SCC in some environments.

Even though the micromechanistic causes of SCC are not entirely understood, some investigators consider SCC to be related to hydrogen damage and not strictly an active-path corrosion phenomenon. Although hydrogen can be a factor in the SCC of certain alloys (see Example 1), sufficient data are not available to generalize this concept. For example, SCC can be assisted by such factors as nuclear irradiation. More information on SCC is available in Volume 13 of the 9th Edition of *Metals Handbook*.

Example 1: Hydrogen-Assisted SCC Failure of Four AISI 4137 Steel Bolts. Figure 12 shows an example of hydrogen-assisted SCC failure of four AISI 4137 steel bolts having a hardness of 42 HRC. Although the normal service temperature (400 °C, or 750 °F) was too high for hydrogen embrittlement, the bolts were also subjected to extended shutdown periods at ambient temperatures. The corrosive environment contained trace hydrogen chloride and acetic acid vapors as well as calcium chloride if leaks occurred. The exact service life was unknown. The bolt surfaces showed extensive corrosion deposits. Cracks had initiated at both the thread roots and the fillet under the bolt head. Figure 12(b) shows a longitudinal section through the failed end of one bolt. Multiple, branched cracking was present, typical of hydrogen-assisted SCC in hardened steels. Chlorides were detected within the cracks and on the fracture surface. The failed bolts were replaced with 17-4 PH stainless steel bolts (Condition H1150M) having a hardness of 22 HRC (Ref 5).

Fabrication

Most bolts are made by cold heading. Other cold-forming methods, including cold extrusion, are used in the manufacture of threaded fasteners. Current technology is such that not only low-carbon steels but also medium-carbon and even low-alloy steels can be successfully impact extruded.

Parts having heads that are large in relation to the shank diameter can be hot headed or produced cold on a two-die, three-blow machine. Hot heading is also more practical for bolts with diameters larger than about 32 mm (1¼ in.) because of equipment limitations and increased probability of tool failures with cold heading.

Platings and Coatings. Most carbon steel fasteners are plated or coated. Common coatings are zinc, cadmium, and phosphate and oil. Other supplementary finishes are gaining popularity, especially in critical applications. The principal reason for fastener plating or coating is corrosion resistance (see the section "Corrosion Protection" in this article), although the appearance and the installation torque-tension relationship are also improved.

Plating is the deposition of metal onto the surface of the base metal. For commercial applications, plating is achieved by electroplating, hot dipping, or mechanical application. In general, the addition of plating increases the dimensions of the fastener by two times the plating thickness and by almost four times the plating thickness in the thread dimensions. The thread assembly may be affected by the increase in fastener size due to plating.

High-strength fasteners, usually with higher carbon content, are susceptible to hydrogen embrittlement when being acid cleaned, electrocleaned, or electroplated. Hydrogen penetration into a fastener can be minimized by baking at about 190 to 200 °C (375 to 400 °F) for 3 to 24 h. For applications in which hydrogen embrittlement is a concern or for a critical application of high-strength fasteners, the mechanical application of plating should be considered.

Clamping Forces

To operate effectively and economically, threaded fasteners should be designed to be torqued near the proof stress, as dictated by the cross-sectional area of the load-carrying parts of the fastener and the desired clamping force. The actual clamping force attained in any assembly will be influenced by such factors as:

- Roughness of the mating surfaces
- Coatings, contaminants, or lubricants on the mating surfaces
- Platings or lubricants on the threads

Typically, torque values are established to result in a clamping force equal to about 75% of the proof load. For some applications, bolts are torqued beyond their proof stress with no detrimental results, provided they are permanent fasteners holding static loads.

Because it is difficult to measure bolt tension (clamping force) in production in-

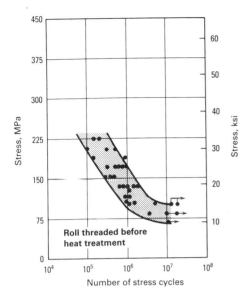

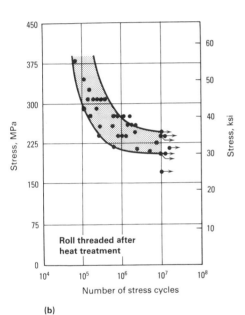

Fig. 11 Fatigue limits for roll-threaded steel bolts. (a) Four different lots of bolts that were roll threaded, then heat treated to average hardness of 22.7, 26.6, 27.6, and 32.6 HRC. (b) Five different lots that were heat treated to average hardnesses of 23.3, 27.4, 29.6, 31.7, and 33.0 HRC, then roll threaded. Bolts having higher hardnesses in each category had higher fatigue strengths.

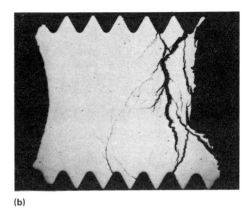

Fig. 12 4137 steel bolts (hardness: 42 HRC) that failed by hydrogen-assisted SCC caused by acidic chlorides from a leaking polymer solution. (a) Overall view of failed bolts. (b) Longitudinal section through one of the failed bolts in (a) showing multiple, branched hydrogen-assisted stress-corrosion cracks initiating from the thread roots. Source: Ref 5

stallations, torque values are used in most applications. Some critical joints in assembly-line processes are using torque-angle or torque-to-yield methods of tightening. In the torque-angle method, the bolt is torqued to a low seating torque level to mate all surfaces, then rotated a specific angle. The angle rotation has a linear relationship with extension because of the constant pitch and therefore with clamp load. In the torque-to-yield method, torque and angular rotation are monitored during installation by a microprocessor and bolt rotation continues until the relationship between the two is not linear. This point is defined as the yield point in torque tension.

The clamping forces generated at given torques are very dependent on the coefficient of friction at the threads and at the bearing face; therefore, they are highly dependent on fastener coatings. Common fastener coatings are zinc, cadmium, and phosphate and oil.

The maximum clamping force that can be effectively employed in any bolt is often limited by the compressive strength of the materials being bolted. If this value is exceeded, the bolt head or nut will be pulled into the parts being bolted, with a subsequent reduction in clamping force. The assembly then becomes loose, and the bolt is susceptible to fatigue failure. If high-tensile bolts are necessary to join low compressive strength materials, hardened washers should be used under the head of the bolt and under the nut to distribute bearing pressure more evenly and to avoid the condition described above.

The value of high clamping forces, apart from lessening the possibility of the nut loosening, is that the working stresses (against solid abutments) are always less than the clamping forces induced in a properly selected bolt. This ensures against cyclic stress and possible fatigue failure.

REFERENCES

1. T.J. Hughel, "Delayed Fracture of Class 12.8 Bolts in Automotive Rear Suspensions," SAE Technical Paper Series 820122, Society of Automotive Engineers, 1982
2. *Corrosion Data Survey—Metals Section*, 5th ed., National Association of Corrosion Engineers, 1974, p 268-269
3. D.R. McIntyre and C.P. Dillon, *Guidelines for Preventing Stress Corrosion Cracking in the Chemical Process Industries*, Publication 15, Materials Technology Institute of the Chemical Process Industries, 1985, p 8-14
4. *The Role of Stainless Steels in Petroleum Refining*, American Iron and Steel Institute, 1977, p 41
5. D. Warren, Hydrogen Effects on Steel, in *Process Industries Corrosion*, National Association of Corrosion Engineers, 1986, p 31-43

Steel Springs

Revised by Loren Godfrey, Associated Spring/Barnes Group, Inc.

STEEL SPRINGS are made in many types, shapes, and sizes, ranging from delicate hairsprings for instrument meters to massive buffer springs for railroad equipment. The major portion of this article discusses those relatively small steel springs that are cold wound from wire. Relatively large, hot-wound springs are quite different from cold-wound springs and are treated in a separate section. Flat and leaf springs are also treated separately to the extent that they differ from wire springs in material and fabrication.

Wire springs are of four types: compression springs (including die springs), extension springs, torsion springs, and wire forms. Compression springs are open wound with varying space between the coils and are provided with plain, plain and ground, squared, or squared and ground ends. The spring can be cylindrical, conical, barrel, or hour glass shaped. Extension springs are normally close wound, usually with specified initial tension, and, because they are used to resist pulling forces, are provided with hook or loop ends to fit the specific application. Ends may be integral parts of the spring or specially inserted forms. Torsion springs are usually designed to work over an arbor and to resist a force that causes the spring to wind. Wire forms are made in a wide variety of shapes and sizes.

Flat springs are usually made by stamping and forming of strip material into shapes such as spring washers. However, there are other types, including motor springs (clock type), constant-force springs, and volute springs, that are wound from strip or flat wire.

Chemical composition, mechanical properties, surface quality, availability, and cost are the principal factors to be considered in selecting steel for springs. Both carbon and alloy steels are used extensively.

Steels for cold-wound springs differ from other constructional steels in four ways.

- They are cold worked more extensively
- They are higher in carbon content
- They can be furnished in the pretempered condition
- They have higher surface quality

The first three items increase the strength of the steel, and the last improves fatigue properties. For flat, cold-formed springs made from steel strip or flat wire, narrower ranges of carbon and manganese are specified than for cold-wound springs made from round or square wire.

Where special properties are required, spring wire or strip made of stainless steel, a heat-resistant alloy, or a nonferrous alloy can be substituted for the carbon or alloy steel, provided that the design of the spring is changed to compensate for the differences in properties between the materials (see the section "Design" in this article).

Table 1 lists grade, specification, chemical composition, properties, method of manufacture, and chief applications of the materials commonly used for cold-formed springs. Hot-formed carbon and alloy steel springs are discussed in this article.

Mechanical Properties

Steels of the same chemical composition may perform differently because of different mechanical and metallurgical characteristics. These properties are developed by the steel producer through cold work and heat treatment or by the spring manufacturer through heat treatment. Selection of round wire for cold-wound springs is based on minimum tensile strength for each wire size and grade (Fig. 1) and on minimum reduction in area (40% for all sizes).

Rockwell hardness and tensile strength for any grade of spring steel strip depend on thickness. The same properties in different thicknesses can be obtained by specifying different carbon contents. The relationship between thickness of spring steel strip containing 0.50 to 0.95% C and Rockwell hardness is shown in Fig. 2. The optimum hardness of a spring steel increases gradually with decreasing thickness.

The hardness scale that can be used for thin metal depends on the hardness and the thickness of the metal (see Table 3 in the article "Rockwell Hardness Testing" in Volume 8 of the 9th Edition of *Metals Handbook*). For testing spring steel strip, which has a minimum hardness of 38 HRC, the Rockwell C scale is used for metal thicker than 0.89 mm (0.035 in.). For thickness ranges of 0.89 to 0.64 mm (0.035 to 0.025 in.), 0.64 to 0.5 mm (0.025 to 0.020 in.), and 0.5 to 0.33 mm (0.020 to 0.013 in.), the Rockwell 45N, 30N, and 15N scales are preferred. For thickness less than 0.33 mm (0.013 in.), the Vickers (diamond pyramid) scale is recommended. As the strip hardness increases, the thickness that can be safely tested decreases. It has been found that the readings obtained with the Vickers indentor are less subject to variation in industrial circumstances than those obtained with the Knoop indentor. The 500 g load Vickers test is used for spring steel strip in thicknesses as low as 0.08 mm (0.003 in.).

If readings are made using the proper hardness scale for a given thickness and hardness, they can be converted to HRC values using charts like those in the appendix to the article "Miscellaneous Hardness Tests" in Volume 8 of the 9th Edition of *Metals Handbook*. Similar charts appear in ASTM A 370 and in the cold-rolled flat wire section of the Steel Products Manual of the American Iron and Steel Institute (AISI). Chart No. 60 published by Wilson Instrument Division, American Chain & Cable Company, Inc., can also be used for this conversion. For specific steel springs, hardness can be held to within 4 points on the Rockwell C scale.

Note that in Table 1 and in the section "Design" in this article, design-stress values are given as percentages of minimum tensile strength. These values apply to springs that are coiled or formed and then stress relieved, which are used in applications involving relatively few load cycles. If each spring is coiled or formed so as to allow for some set, and then deflected beyond the design requirements, higher design stresses can be used. This is discussed in the section "Residual Stresses" in this article.

As a further aid in selecting steels for springs, Table 2 lists the suitable choices for cold-wound helical springs in various com-

Table 1 Common wire and strip materials used for cold-formed springs

Material type	Grade and specification	Nominal composition, %	Tensile properties		Torsion properties		Hardness, HRC(c)	Max allowable temperature, °C (°F)	Method of manufacture, chief applications, special properties
			Minimum tensile strength(a), MPa (ksi)	Modulus of elasticity (E), GPa (psi × 10^6)	Design stress, % of minimum tensile strength(b)	Modulus of rigidity (G), GPa (psi × 10^6)			
Cold-drawn wire									
High-carbon steel	Music wire, ASTM A 228	C 0.70–1.00 Mn 0.20–0.60	1590–2750 (230–399)	210 (30)	45	80 (11.5)	41–60	120 (250)	Drawn to high and uniform tensile strength; for high-quality springs and wire forms
	Hard drawn, ASTM A 227	C 0.45–0.85 Mn 0.30–1.30	Class I 1010–1950 (147–283) Class II 1180–2230 (171–324)	210 (30)	40	80 (11.5)	31–52	120 (250)	For average-stress applications; lower-cost springs and wire forms
	High-tensile hard drawn, ASTM A 679	C 0.65–1.00 Mn 0.20–1.30	1640–2410 (238–350)	210 (30)	45	80 (11.5)	41–60	120 (250)	For higher-quality springs and wire forms
	Oil tempered, ASTM A 229	C 0.55–0.85 Mn 0.30–1.20	Class I 1140–2020 (165–294) Class II 1320–2330 (191–324)	210 (30)	45	80 (11.5)	42–55	120 (250)	Heat treated before fabrication; for general-purpose springs
	Carbon VSQ(d), ASTM A 230	C 0.60–0.75 Mn 0.60–0.90	1480–1650 (215–240)	210 (30)	45	80 (11.5)	45–49	120 (250)	Heat treated before fabrication; good surface condition and uniform tensile strength
Alloy steel	Chromium vanadium, ASTM A 231, A 232(d)	C 0.48–0.53 Cr 0.80–1.10 V 0.15 min	1310–2070 (190–300)	210 (30)	45	80 (11.5)	41–55	220 (425)	Heat treated before fabrication; for shock loads and moderately elevated temperature; ASTM A 232 for valve springs
	Modified chromium vanadium VSQ(d), ASTM A 878	C 0.60–0.75 Cr 0.35–0.60 V 0.10–0.25	1410–2000 (205–290)	...	...	...	...	...	Heat treated before fabrication; for valve springs and moderately elevated temperatures
	Chromium silicon, ASTM A 877(d), A 401	C 0.51–0.59 Cr 0.60–0.80 Si 1.20–1.60	1620–2070 (235–300)	210 (30)	45	80 (11.5)	48–55	245 (475)	Heat treated before fabrication; for shock loads and moderately elevated temperature; ASTM A 877 for valve springs
Stainless steel	Type 302(18-8), ASTM A 313	Cr 17–19 Ni 8–10	860–2240 (125–325)	190 (28)	30–40	69 (10.0)	35–45	290 (550)	General-purpose corrosion and heat resistance; magnetic in spring temper
	Type 316, ASTM A 313	Cr 16–18 Ni 10–14 Mo 2–3	760–1690 (110–245)	190 (28)	40	69 (10.0)	35–45	290 (550)	Good heat resistance; greater corrosion resistance than 302; magnetic in spring temper
	Type 631 (17-7 PH), ASTM A 313	Cr 16–18 Ni 6.50–7.75 Al 0.75–1.50	Condition CH-900 1620–2310 (235–335)	200 (29.5)	45	76 (11.0)	38–57	340 (650)	Precipitation hardened after fabrication; high strength and general-purpose corrosion resistance; magnetic in spring temper
Nonferrous alloys	Copper alloy 510 (phosphor bronze A), ASTM B 159	Cu 94–96 Sn 4.2–5.8	720–1000 (105–145)	100 (15)	40	43 (6.25)	98–104(e)	90 (200)	Good corrosion resistance and electrical conductivity
	Copper alloy 170 (beryllium copper) ASTM B 197	Cu 98 Be 1.8–2.0	1030–1590 (150–230)	130 (18.5)	45	50 (7.0)	35–42	200 (400)	Can be mill hardened before fabrication; good corrosion resistance and electrical conductivity; high mechanical properties
	Monel 400, AMS 7233	Ni 66 Cu 31.5	1000–1240 (145–180)	180 (26)	40	65 (9.5)	23–32	230 (450)	Good corrosion resistance at moderately elevated temperature
	Monel K-500, QQ-N-286(f)	Ni 65 Cu 29.5 Al 2.8	1100–1380 (160–200)	180 (26)	40	65 (9.5)	23–35	290 (550)	Excellent corrosion resistance at moderately elevated temperature
High-temperature alloys	A-286 alloy	Fe 53 Ni 26 Cr 15	1100–1380 (160–200)	200 (29)	35	72 (10.4)	35–42	510 (950)	Precipitation hardened after fabrication; good corrosion resistance at elevated temperature
	Inconel 600, QQ-W-390(f)	Ni 76 Cr 15.8 Fe 7.2	1170–1590 (170–230)	215 (31)	40	76 (11.0)	35–45	370 (700)	Good corrosion resistance at elevated temperature
	Inconel 718	Ni 52.5 Cr 18.6 Fe 18.5	1450–1720 (210–250)	200 (29)	40	77 (11.2)	45–50	590 (1100)	Precipitation hardened after fabrication; good corrosion resistance at elevated temperature

(continued)

(a) Minimum tensile strength varies within the given range according to wire diameter (see the applicable specification). Maximum tensile strength is generally about 200 MPa (30 ksi) above the minimum tensile strength. (b) For helical compression or extension springs; design stress of torsion and flat springs taken as 75% of minimum tensile strength. (c) Correlation between hardness and tensile properties of wire is approximate only and should not be used for acceptance or rejection. (d) Valve-spring quality. (e) HRB values. (f) Federal specification. Source: Ref 1

Table 1 (continued)

Material type	Grade and specification	Nominal composition, %	Tensile properties — Minimum tensile strength(a), MPa (ksi)	Modulus of elasticity (E), GPa (psi × 10⁶)	Torsion properties — Design stress, % of minimum tensile strength(b)	Modulus of rigidity (G), GPa (psi × 10⁶)	Hardness, HRC(c)	Max allowable temperature, °C (°F)	Method of manufacture, chief applications, special properties
Cold-drawn wire (continued)									
	Inconel X-750, AMS 5698, 5699	Ni 73 Cr 15 Fe 6.75	No. 1 temper 1070 (155) Spring temper 1310–1590 (190–230)	215 (31)	40	83 (12.0)	No. 1 34–39 Spring 42–48	400–600 (750–1100)	Precipitation hardened after fabrication; good corrosion resistance at elevated temperature
Cold-rolled strip									
Carbon steel	Medium carbon (1050), ASTM A 682	C 0.47–0.55 Mn 0.60–0.90	Tempered 1100–1930 (160–280)	210 (30)	...	...	Annealed 85 max(e), tempered 38–50	120 (250)	General-purpose applications
	"Regular" carbon (1074), ASTM A 682	C 0.69–0.80 Mn 0.50–0.80	Tempered 1100–2210 (160–320)	210 (30)	...	...	Annealed 85 max(e), tempered 38–50	120 (250)	Most popular material for flat springs
	High carbon (1095), ASTM A 682	C 0.90–1.04 Mn 0.30–0.50	Tempered 1240–2340 (180–340)	210 (30)	...	...	Annealed 88 max(e), tempered 40–52	120 (250)	High-stress flat springs
Alloy steel	Chromium vanadium, AMS 6455	C 0.48–0.53 Cr 0.80–1.10 V 0.15 min	1380–1720 (200–250)	210 (30)	...	...	42–48	220 (425)	Heat treated after fabrication; for shock loads and moderately elevated temperature
	Chromium silicon, AISI 9254	C 0.51–0.59 Cr 0.60–0.80 Si 1.20–1.60	1720–2240 (250–325)	210 (30)	...	...	47–51	245 (475)	Heat treated after fabrication; for shock loads and moderately elevated temperature
Stainless steel	Type 301	Cr 16–18 Ni 6–8	1655–2650 (240–270)	190 (28)	...	...	48–52	150 (300)	Rolled to high yield strength; magnetic in spring temper
	Type 302 (18-8)	Cr 17–19 Ni 8–10	1280–1590 (185–230)	190 (28)	...	...	42–48	290 (550)	General-purpose corrosion and heat resistance; magnetic in spring temper
	Type 316	Cr 16–18 Ni 10–14 Mo 2–3	1170–1590 (170–230)	190 (28)	...	...	38–48	290 (550)	Good heat resistance; greater corrosion resistance than 302; magnetic in spring temper
	Type 631 (17-7 PH), ASTM A 693	Cr 16–18 Ni 6.50–7.75 Al 0.75–1.50	Condition CH-900 1655 (240)	200 (29)	...	...	46 min	340 (650)	Precipitation hardened after fabrication; high strength and general-purpose corrosion resistance; magnetic in spring temper
Nonferrous alloys	Copper alloy 510 (phosphor bronze A), ASTM B103	Cu 94–96 Sn 4.2–5.8	650–750 (95–110)	100 (15)	...	...	94–98(e)	90 (200)	Good corrosion resistance and electrical conductivity
	Copper alloy 170 (beryllium copper), ASTM B 194	Cu 98 Be 1.6–1.8	1240–1380 (180–200)	130 (18.5)	...	...	39 min	200 (400)	Can be mill hardened before fabrication; good corrosion resistance and electrical conductivity; high mechanical properties
	Monel 400, AMS 4544	Ni 66 Cu 31.5	690–970 (100–140)	180 (26)	...	...	98 min(e)	230 (450)	Good corrosion resistance at moderately elevated temperature
	Monel K-500, QQ-N-286(f)	Ni 65 Cu 29.5 Al 2.8	1170–1380 (170–200)	180 (26)	...	...	34 min	290 (550)	Excellent corrosion resistance at moderately elevated temperature
High-temperature alloys	A-286 alloy, AMS 5525	Fe 53 Ni 26 Cr 15	1100–1380 (160–200)	200 (29)	...	...	30–40	510 (950)	Precipitation hardened after fabrication; good corrosion resistance at elevated temperature
	Inconel 600, ASTM B 168, AMS 5540	Ni 76 Cr 15.8 Fe 7.2	1000–1170 (145–170)	215 (31)	...	...	30 min	370 (700)	Good corrosion resistance at elevated temperature
	Inconel 718, AMS 5596, AMS 5597	Ni 52.5 Cr 18.6 Fe 18.5	1240–1410 (180–204)	200 (29)	...	...	36	590 (1100)	Precipitation hardened after fabrication; good corrosion resistance at elevated temperature
	Inconel X-750, AMS 5542	Ni 73 Cr 15 Fe 6.75	1030 (150)	215 (31)	...	...	30 min	400–590 (750–1100)	Precipitation hardened after fabrication; good corrosion resistance at elevated temperature

(a) Minimum tensile strength varies within the given range according to wire diameter (see the applicable specification). Maximum tensile strength is generally about 200 MPa (30 ksi) above the minimum tensile strength. (b) For helical compression or extension springs; design stress of torsion and flat springs taken as 75% of minimum tensile strength. (c) Correlation between hardness and tensile properties of wire is approximate only and should not be used for acceptance or rejection. (d) Valve-spring quality. (e) HRB values. (f) Federal specification. Source: Ref 1

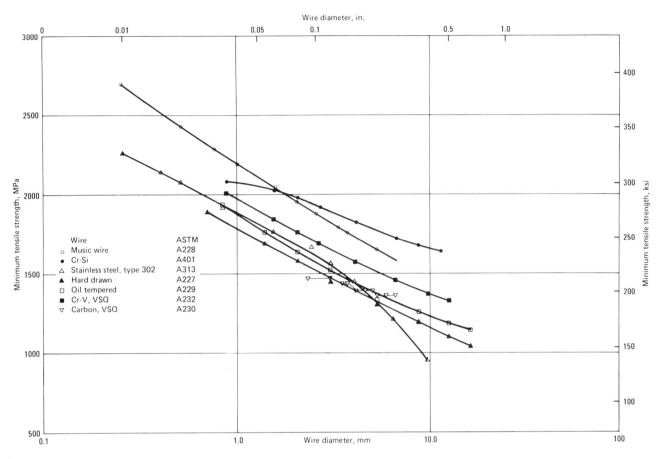

Fig. 1 Minimum tensile strength of steel spring wire. VSQ, valve-spring quality

binations of size, stress, and service. Each recommendation is the most economical steel that will perform satisfactorily under the designated conditions and that is commercially available in the specific size.

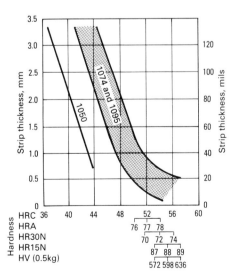

Fig. 2 Effect of strip thickness on the optimum hardness of spring steel strip for high-stress use. Hardness on HRC scale may be lowered 3 to 4 points for greater toughness. Instability of ductility is sometimes encountered above 57 HRC.

Fatigue strength is another important mechanical property of steel springs. However, this property is affected by many factors, and because of this complexity, fatigue is discussed in a separate section of this article.

Flat Springs. Figure 3 illustrates the different working stresses allowable in flat and leaf springs of 1095 steel that are to be loaded in each of three different ways: statically, variably, and dynamically. (These three types of loading are dealt with separately in the selection table for cold-wound springs, Table 2.) The stresses given in Fig. 3 are the maximum stresses expected in service. These data apply equally well to 1074 and 1050 steels if the stress values are lowered 10 and 20%, respectively. Except for motor or power springs and a few springs involving only moderate forming, most flat springs, because of complex forming requirements, are formed soft, then hardened and tempered.

For the optimum combination of properties, hypereutectoid spring steel in coil form should be held at hardening temperature for the minimum period of time. The presence of undissolved carbides indicates proper heat treatment. The extent of decarburization can be determined by microscopic examination of transverse sections or by microhardness surveys using Vickers or Knoop indentors with light loads (usually 100 g).

Power (clock) springs made from pretempered stock have a longer service life if controlled heat treatment can produce a fine, tempered martensitic structure with uniform distribution of excess carbide. If carbides are absent and the tempered martensitic structure is relatively coarse grained, the springs will have a smaller maximum free diameter after having been tightly wound in the barrel or retainer for a long time.

A recent development for lower-stressed flat springs is a hardened, 0.04 to 0.22% plain carbon strip steel, which is blanked, formed, and used with only a low-temperature stress-relief treatment. Thickness tolerances, however, are not as close as for spring steel. This material is available in tensile strengths of 900 to 1520 MPa (130 to 220 ksi).

Characteristics of Spring Steel Grade

General Spring Quality Wire. The three types of wire used in the greatest number of applications of cold-formed springs are:

- Hard-drawn spring wire
- Oil-tempered wire
- Music wire

Hard-Drawn Spring Wire (ASTM A 227). Among the grades of steel wire used for

Table 2 Recommended ASTM grades of carbon and alloy steel wire for cold-wound helical springs
See Table 1 for the type of wire and the composition of the ASTM grades given below.

Corrected maximum working stress(a)		Diameter of spring wire(b)					
MPa	ksi	0.13–0.51 mm (0.005–0.020 in.)	0.51–0.89 mm (0.020–0.035 in.)	0.89–3.18 mm (0.035–0.125 in.)	3.18–6.35 mm (0.125–0.250 in.)	6.35–12.70 mm (0.250–0.500 in.)	1270–15.88 mm (0.500–0.625 in.)
Compression springs, static load (set removed, springs stress relieved)(c)							
550	80	A 228(d)	A 227(d)	A 227(d)	A 227	A 227	A 227(e), A 229
690	100	A 228(d)	A 227(d)	A 227(d)	A 227	A 227(f), A 229	A 229
825	120	A 228(d)	A 227(d)	A 227	A 227(g), A 229	A 229(h)	...
965	140	A 228	A 227	A 227(i), A 229	A 229	A 401(h)	...
1100	160	A 228	A 227	A 228, A 229	A 229(j), A 228	A 401(k)(h)	...
1240	180	A 228	A 228	A 228	A 228(l)(h)	...	...
1380	200	A 228	A 228	A 228(m)(h)	...	...	...
1515	220	A 228	A 228(n)(h)	...	...	...	...
1655	240	A 228(o)(h)	...	...	...	...	...
Compression springs, variable load, designed for minimum life of 100 000 cycles (set removed, springs stress relieved)(p)							
550	80	A 228(d)	A 227(d)	A 227(d)	A 227(q), A 229	A 229(r), A 401	A 401
690	100	A 228(d)	A 227	A 227, A 229	A 229(s), A 401	A 401	A 401
825	120	A 228(d)	A 227	A 227(t), A 229	A 229(u), A 401	A 401(v)	...
965	140	A 228	A 229	A 229(w), A 228	A 228(x)	...	...
1100	160	A 228	A 228	A 228(w)(h), A 401	...	...	...
1240	180	A 228	A 228(y)(h)	...	...	...	...
1380	200	A 228(z)(h)	...	...	...	...	...
Compression springs, dynamic load, designed for minimum life of 10 million cycles (set removed, springs stress relieved)(p)							
415	60	A 228(d)	A 227(d)	A 227(d)	A 227(u), A 230	A 229(k)(h), A 230	...
550	80	A 228(d)	A 228(d)	A 229(t), A 228	A 230	A 230	...
690	100	A 228(d)	A 228	A 228(aa), A 230	A 230(bb)(h)	...	...
825	120	A 228(d)	A 228	A 230(cc)	...	...	...
Compression and extension springs, static load (set not removed, compression springs stress relieved)(c)							
550	80	A 228	A 227	A 227	A 227	A 227(dd), A 229	A 229
690	100	A 228	A 227	A 227(t), A 229	A 401	A 401(h)	...
825	120	A 228	A 227	A 227(ee), A 229	A 401	A 401(k)(h)	...
965	140	A 228	A 228	A 228(ff), A 401	A 401(h)	...	...
1100	160	A 228(h)	...	...	...	...	...
1240	180	A 228(o)(h)	...	...	...	...	...
Compression and extension springs, designed for minimum life of 100 000 cycles (set not removed, compression springs stress relieved)(p)							
415	60	A 228	A 227	A 227	A 227	A 227	A 229
550	80	A 228	A 227	A 227	A 227(x), A 229	A 229	A 401
690	100	A 228	A 229	A 229(cc), A 228	A 228(s), A 401	A 401(h)	...
825	120	A 228	A 228	A 228(w)	A 401(h)	...	...
965	140	A 228(h)	...	...	...	...	...
1100	160	A 228(o)(h)	...	...	...	...	...
Compression and extension springs, designed for minimum life of 10 million cycles (set not removed, compression springs stress relieved)(p)							
275	40	A 228	A 227	A 227	A 227	A 227(m), A 229	A 229
415	60	A 228	A 227	A 227(aa), A 230	A 230(bb)(h)	...	...
550	80	A 228	A 228	A 228(aa)(h), A 230	...	...	...
690	100	A 228(o)	...	...	...	...	...
Torsion springs (springs not stress relieved)(p)							
690	100	A 228	A 227	A 227	A 227(bb), A 229	A 229(v), A 401	...
825	120	A 228	A 227	A 227(w), A 229	A 229(bb), A 228	A 401(r)(h)	...
965	140	A 228	A 229	A 229(i), A 228	A 228(q)(h)	...	...
1100	160	A 228	A 228	A 228(w)(h)	...	...	...
1240	180	A 228	A 228(h)	...	...	...	...
1380	200	A 228(h)	...	...	...	...	...

(a) Stress corrected by the Wahl factor. See the section "Wahl Correction" in this article. (b) Where more than one steel is shown for an indicated range of wire diameter, the first is recommended up to the specific diameter listed in the footnote referred to; the last steel listed in any multiple choice is recommended for the remainder of the indicated wire diameter range. (c) Shot peening is not necessary for statically loaded springs. (d) Set removal not required in this range. (e) To 14.29 mm (0.563 in.). (f) To 10.32 mm (0.406 in.). (g) To 4.11 mm (0.162 in.). (h) Yielding likely to occur beyond this limit. (i) To 1.83 mm (0.072 in.). (j) To 3.81 mm (0.150 in.). (k) To 11.11 mm (0.437 in.). (l) To 5.33 mm (0.210 in.). (m) To 2.29 mm (0.090 in.). (n) To 0.81 mm (0.032 in.). (o) To 0.20 mm (0.008 in.). (p) Shot peening is recommended for wire diameter greater than 1.57 mm (0.062 in.) and smaller where obtainable. (q) To 3.94 mm (0.155 in.). (r) To 7.77 mm (0.306 in.). (s) To 4.76 mm (0.187 in.). (t) To 2.34 mm (0.092 in.). (u) To 3.76 mm (0.148 in.). (v) To 9.19 mm (0.362 in.). (w) To 1.57 mm (0.062 in.). (x) To 4.50 mm (0.177 in.). (y) To 0.74 mm (0.029 in.). (z) To 0.36 mm (0.014 in.). (aa) To 1.37 mm (0.054 in.). (bb) To 5.26 mm (0.207 in.). (cc) To 2.03 mm (0.080 in.). (dd) To 7.19 mm (0.283 in.). (ee) To 1.12 mm (0.044 in.). (ff) To 2.69 mm (0.105 in.).

cold-formed springs (Table 1), hard-drawn spring wire is the least costly. Its surface quality is comparatively low with regard to such imperfections as hairline seams. This wire is used in applications involving low stresses or static conditions.

Oil-tempered wire (ASTM A 229) is a general-purpose wire, although it is more susceptible to the embrittling effects of plating than hard-drawn spring wire. Its spring properties are obtained by heat treatment. Oil-tempered wire is slightly more expensive than hard-drawn wire; it is significantly superior in surface smoothness, but not necessarily in seam depth. Most cold-wound automotive springs are made of oil-tempered wire, although a small percentage are made of music wire and hard-drawn spring wire.

Music wire (ASTM A 228) is the carbon steel wire used for small springs. It is the least subject to hydrogen embrittlement by electroplating (see the section "Plating of Springs" in this article) and is comparable to valve-spring wire in surface quality.

Chromium-silicon and chromium-vanadium steel spring wire and strip are suitable for moderately elevated temperature service. The chromium-silicon steel spring wire, which has better relaxation resistance than the chromium-vanadium alloy,

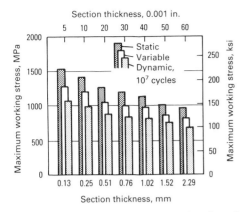

Fig. 3 Maximum working stress for bending flat and leaf springs made of 1095 steel

Table 3 Typical stress-relieving temperatures for steel spring wire
Applicable only for stress relieving after coiling and not valid for stress relieving after shot peening

Steel	Temperature(a) °C	°F
Music wire	230–260	450–500
Hard-drawn spring wire	230–290	450–550
Oil-tempered spring wire	230–400	450–750(b)
Valve spring wire	315–400	600–750
Cr-V spring wire	315–400	600–750
Cr-Si spring wire	425–455	800–850
Type 302 stainless	425–480	800–900
Type 631 stainless	480 ± 6(c)	900 + 10(c)

(a) Based on 30 min at temperature. (b) Temperature is not critical and can be varied over the range to accommodate problems of distortion, growth, and variation in wire size. (c) Based on 1 h at temperature

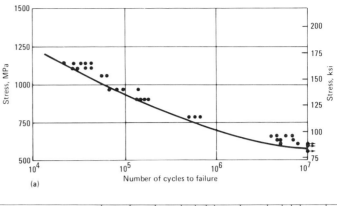
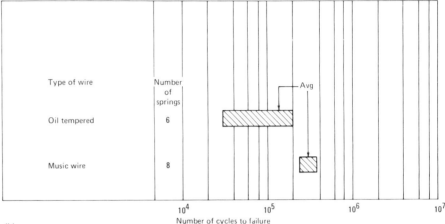

Fig. 4 Fatigue lives of compression coil springs made from various steels. (a) S-N diagram for springs made of minimum-quality music wire 0.56 mm (0.022 in.) in diameter. Spring diameter was 5.21 mm (0.205 in.); D/d was 8.32. Minimum stress was zero. Stresses corrected by Wahl factor (see the section "Wahl Correction" in this article). (b) Life of springs used in a hydraulic transmission. They were made of oil-tempered wire (ASTM A 229) and music wire (ASTM A 228). Wire diameter was 4.75 mm (0.187 in.), outside diameter of spring was 44.45 mm (1.750 in.), with 15 active coils in each spring. The springs were fatigue tested in a fixture at a stress of 605 MPa (88 ksi), corrected by the Wahl factor.

can be used at temperatures as high as 230 °C (450 °F). The cold-drawn spring wires of the chromium-vanadium and chromium-silicon alloys (ASTM A 231 and A 401, respectively) are heat treated before fabrication, while cold-rolled chromium-vanadium (AMS 6455) and chromium-silicon (AISI 9254) strip steels (and generally carbon strip steel as well) are heat treated after rolling and spring fabrication. The chromium-vanadium and chromium-silicon steel spring wires (ASTM A 231 and A 401) can be in either the annealed or oil-tempered condition before spring fabrication. Annealing can be performed before and after drawing, while oil tempering is performed after cold drawing.

High-tensile hard-drawn wire fills the gap where high strength is needed but where the quality of music wire is not required.

Valve-Spring Quality (VSQ) Wire. All valve-spring wires have the highest surface quality attainable in commercial production, and most manufacturers require that the wire conform to aircraft quality as defined in the AISI Steel Products Manual. Most VSQ wire producers remove the surface of the wire rod before drawing to final size. This practice improves the surface quality and eliminates decarburization.

Carbon steel spring wire is the least costly of the VSQ wires. The requirements for carbon VSQ wire in an oil-tempered condition are specified in ASTM A 230.

Chromium-vanadium steel wire of valve-spring quality (ASTM A 232) is superior to the same quality of carbon steel wire (ASTM A 230) for service at 120 °C (250 °F) and above. A modified chromium-vanadium steel of valve-spring quality is also specified in ASTM A 878. This modified chromium-vanadium wire has a smaller range of preferred diameters than ASTM A 232 and a lower minimum and maximum tensile strength than ASTM A 232 for a given wire diameter.

Chromium-silicon steel VSQ wire (ASTM A 877) can be used at temperatures as high as 230 °C (450 °F). The elevated-temperature behavior of this and other steel spring wires is discussed in the section "Effect of Temperature" in this article.

Annealed Spring Wire. Carbon steel wire of valve-spring quality, as well as chromium-vanadium and chromium-silicon steel wire of both spring and valve-spring quality, can be supplied in the annealed condition. This will permit severe forming of springs with a low spring index (ratio of mean coil diameter to wire diameter) and will also permit sharper bends in end hooks. (Although a sharp bend is never desired in any spring, it is sometimes unavoidable.)

Springs made from annealed wire can be quenched and tempered to spring hardness after they have been formed. However, without careful control of processing, such springs will have greater variations in dimensions and hardness. This method of making springs is usually used only for springs with special requirements, such as severe forming, or for small quantities, because springs made by this method may have less uniform properties than those of springs made from pretempered wire and are higher in cost. The amount of cost increase depends largely on design and required tolerances, but the cost of heat treating (which often involves fixturing expense)

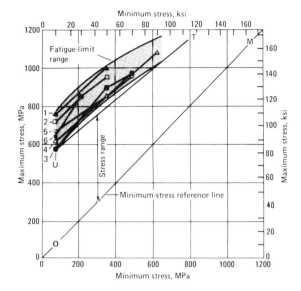

Fig. 5 Fatigue limits for compression coil springs made of music wire. Data are average fatigue limits from S-N curves for 185 unpeened springs of various wire diameters run to 10 million cycles of stress. All stresses were corrected for curvature using the Wahl correction factor. The springs were automatically coiled, with one turn squared on each end, then baked at 260 °C (500 °F) for 1 h, after which the ends were ground perpendicular to the spring axis. The test load was applied statically to each spring and a check made for set three times before fatigue testing. The springs were all tested in groups of six on the same fatigue testing machine at ten cycles per second. After testing, the unbroken springs were again checked for set and recorded. Number 4 springs, tested at 1070 MPa (155 ksi) maximum stress, had undergone about 2½% set after 10 million stress cycles, but the stresses were not recalculated to take this into account. None of the other springs showed appreciable set. The tensile strengths of the wires were according to ASTM A 228.

Spring No.	Wire diameter mm	in.	Spring outside diameter mm	in.	Spring index
1	0.81	0.032	9.52	0.375	10.7
2	0.81	0.032	6.35	0.250	6.8
3	1.22	0.048	15.88	0.625	12.0
4	2.59	0.102	22.22	0.875	7.6
5	3.07	0.121	22.22	0.875	6.2
6	4.50	0.177	22.22	0.875	4.9

Free length mm	in.	Total turns	Active turns	Total tested
22.10	0.87	6.0	4.2	16
26.97	1.062	7.0	5.2	28
44.45	1.75	7.0	5.2	38
60.20	2.37	7.0	5.2	43
57.15	2.25	7.5	5.7	35
57.15	2.25	7.5	5.7	25

and handling can increase total cost by more than 100%.

Stainless Steel Spring Wire. Cold-drawn type 302 stainless steel spring wire (specified in ASTM A 313) is high in heat resistance and has good corrosion resistance. The surface quality of type 302 stainless steel spring wire occasionally varies, seriously affecting fatigue resistance. Type 316 stainless is superior in corrosion resistance to type 302, particularly against pitting in salt water, but is more costly and is not considered a standard spring wire. Type 302 is readily available and has excellent spring properties in the full-hard or spring-temper condition. It is more expensive than any of the carbon steel wires for designs requiring a diameter larger than about 0.30 mm (0.012 in.) but is less expensive than music wire for sizes under about 0.30 mm (0.012 in.). In many applications, type 302 stainless can be substituted for music wire with only slight design changes to compensate for the decrease in modulus of rigidity.

For example, a design for a helical compression spring was based on the use of 0.25 mm (0.010 in.) diam music wire. The springs were cadmium plated to resist corrosion, but they tangled badly in the plating operation because of their proportions. A redesign substituted type 302 stainless steel wire of the same diameter for the music wire. Fewer coils were required because of the lower modulus of rigidity, and the springs did not require plating for corrosion resistance. The basic cost of this small-diameter stainless wire was, at the time, 20% less than the cost of the music wire. Elimination of plating and reduction of handling resulted in a total savings of 25%.

Wire Quality

Specification requirements for the spring materials listed in Table 1 include twist, coiling, fracture, or reduction-in-area tests, in addition to dimensional limits and minimum tensile strength. Such tests ensure that the wire has the expected ductility and has not been overdrawn (which could produce internal splits or voids).

In dynamic applications, in which fatigue strength is an important factor, the performance differences of spring materials depends on surface quality when materials are of similar composition and tensile strength. Because the initiation and growth of fatigue cracks is strongly affected by surface quality, seams and surface decarburization are important factors in dynamic applications of spring quality wire and especially valve-spring quality wire. Freedom from surface imperfections is of paramount importance in some applications of highly stressed springs for shock and fatigue loading, especially where replacement of a broken spring would be difficult and much more costly than the spring itself or where spring failure could cause extensive damage to other components.

Seams are evaluated visually, often after etching with hot 50% muriatic acid. The depth of metal removed can vary from 0.006 mm (¼ mil) to 1% of wire diameter. Examination of small-diameter etched wire requires a stereoscopic microscope, preferably of variable power so that the sizes of seams can be observed in relation to the diameter of the wire. The least expensive wires can have seams that are quite pronounced. Hard-drawn and oil-tempered wires occasionally have seams as deep as 3.5% of wire diameter, but usually not deeper than 0.25 mm (0.010 in.). On the other hand, wires of the highest quality (music wire, valve-spring wire) have only small surface imperfections, generally not deeper than ½% of wire diameter. Some grades can be obtained at moderate cost, with seam depth restricted to 1% of wire diameter.

Decarburization. There is no general numerical limit on decarburization, and phrases such as "held to a minimum consistent with commercial quality" are very elastic. It is usual for seams present during hot rolling to be partly decarburized to the full depth of the seam or slightly deeper.

For valve-spring quality, however, decarburization limits are more severe. Some manufacturers permit loss of surface carbon only if it does not drop below 0.40% for the first 0.025 mm (0.001 in.) and, within the succeeding 0.013 mm (0.0005 in.), becomes equal to the carbon content of the steel. As noted previously, most manufacturers of VSQ wire eliminate surface decarburization by removing the surface of the wire rod prior to drawing to final size.

General decarburization can be detrimental to the ability to maintain load. For hot-wound springs made directly from hot-rolled bars, it

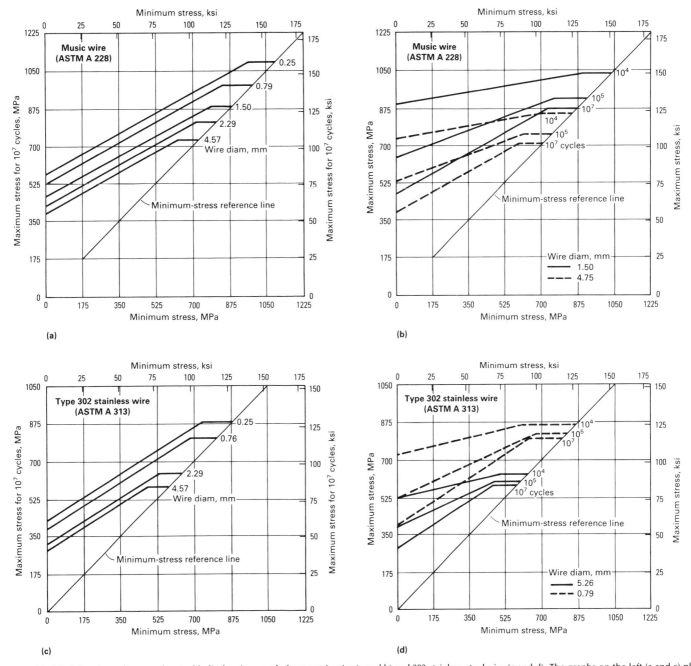

Fig. 6 Modified Goodman diagrams for steel helical springs made from music wire (a and b) and 302 stainless steel wire (c and d). The graphs on the left (a and c) plot maximum allowable stresses for 10 million cycles for a similar group of wire diameters. All stresses were corrected by the Wahl factor. See text for discussion.

is common practice to specify a torsional modulus of 72 GPa (10.5×10^6 psi) instead of the 80 GPa (11.5×10^6 psi) used for small spring wires. In part, this compensates for the low strength of the surface layer. Total loss of carbon from the surface during a heat-treating process is infrequent in modern wire mill products. Partial decarburization of spring wire is often blamed for spring failures, but quench cracks and coiling-tool marks are more frequently the actual causes. In wires of valve-spring or aircraft quality, a decarburized ferritic ring around the wire circumference is a basis for rejection. The net effects of seams and decarburization are described in the section "Fatigue" in this article.

Magnetic Particle and Eddy Current Testing. Inspection for seams and other imperfections in finished springs is generally carried out by magnetic particle inspection. In its various forms, this inspection method has proved to be the most practical nondestructive method for the inspection of springs that may affect human safety or for other reasons must not fail as a result of surface imperfections. For compression and extension springs, the inspection is always concentrated selectively on the inside of the coil, which is more highly stressed than the outside and is the most frequent location of start of failure.

Valve-spring wire is often 100% eddy current tested for seams and point defects by the wire mill. The defects are painted to ensure that they are not fabricated into finished springs.

Residual Stresses

Residual stresses can increase or decrease the strength of a spring material, depending on their direction. For example,

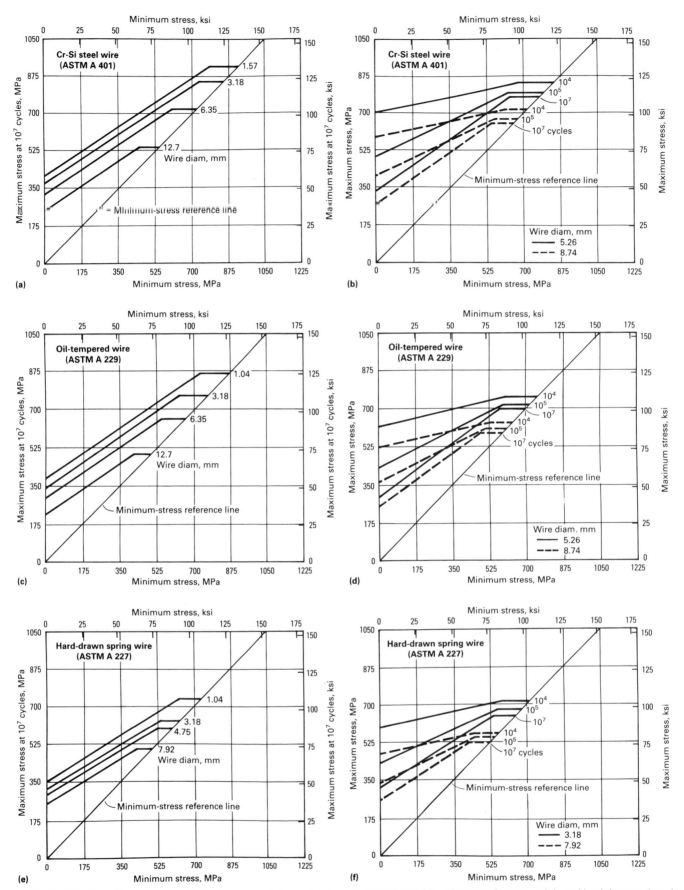

Fig. 7 Modified Goodman diagrams for steel helical springs made from chromium-silicon steel (a and b), oil-tempered wire (c and d), and hard-drawn spring wire (e and f). The graphs on the left (a, c, and e) plot maximum allowable stress for 10 million cycles for 3.18 mm (0.125 in.) diam wires and various other size wires. All stresses were corrected by the Wahl factor. See text for discussion.

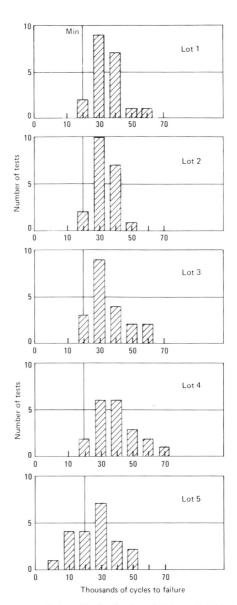

Fig. 8 Fatigue-life distribution of 0.50 mm (0.020 in.) diam music wire. Tested in a rotating-beam machine at a maximum stress of 1170 MPa (170 ksi) and a mean stress of zero

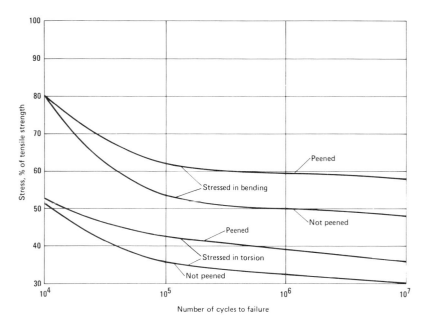

Fig. 9 Fatigue curves for peened and unpeened steel spring wires

residual stresses induced by bending strengthen wire for deflection in the same direction while weakening it for deflection in the opposite direction. In practice, residual stresses are either removed by stress relieving or induced to the proper direction by cold setting and shot peening.

Compression springs, torsion springs, flat springs, and retaining rings can be stress relieved and cold set. The treatment used depends on the design and application requirements of the individual spring.

Many compression springs are preset for use at higher stress. They are then known as springs with set removed. Compression springs can be made to close solid, without permanent set and without presetting, if the shear stress is less than a specified proportion of the tensile strength of the wire in the fully compressed spring (about 45% for music wire) and if the springs are properly stress relieved. The maximum shear stress in a fully compressed preset spring is about 33% higher, or approximately 60% of the tensile strength of the wire. Therefore, presetting to a maximum stress will permit the use of up to 45% less steel than is otherwise required, a savings greater than the cost of presetting for wire larger than about 3 mm (⅛ in.) in diameter. Also, the smaller, equally strong spring requires less space.

When the calculated uncorrected stress at solid height is greater than about 60% of the tensile strength (or, for cold-set springs, greater than the proportional limit stress), the spring can be neither cold set nor compressed to its solid height without taking a permanent set. Several types of springs are in this category, where the maximum permissible deflection must be calculated and positive stops provided to avoid permanent set in service.

Compression springs, cold wound from pretempered or hard-drawn high-carbon spring wire, should always be stress relieved to remove residual stresses produced in coiling. Extension springs are usually given a stress-relieving treatment to relieve stresses induced in forming hooks or other end configurations, but such treatment should allow retention of stresses induced for initial tension.

The treatment of wire retaining rings depends on whether the loading tends to increase or decrease the relaxed diameter of the spring. Most rings contain residual stresses in tension on the inside surface. For best performance, rings that are reduced in size in the application should not be stress relieved, while expanded rings should be. This consideration applies equally to torsion springs. It is common practice to give these springs a low-temperature heat treatment to provide dimensional stability.

Stress relieving affects the tensile strength and elastic limit, particularly for springs made from music wire and hard-drawn spring wire. The properties of both types of wire are increased by heating in the range of 230 to 260 °C (450 to 500 °F). Oil-tempered spring wire, except for the chromium-silicon grade, shows little change in either tensile strength or elastic limit after stress relieving below 315 °C (600 °F). Both properties then drop because of temper softening. Wire of chromium-silicon steel temper softens only above about 425 °C (800 °F).

The properties of spring steels are usually not improved by stress relieving for more than 30 min at temperature, except for age-hardenable alloys such as 631 (17-7 PH) stainless steel, which requires about 1 h to reach maximum strength. Typical stress-relief temperatures for steel spring wire are given in Table 3.

When springs are to be used at elevated temperatures, the stress-relieving temperatures should be near the upper limit of the range to minimize relaxation in service. Otherwise, lower temperatures are better.

Plating of Springs

Steel springs are often electroplated with zinc or cadmium to protect them from corrosion and abrasion. In general, zinc has been found to give the best protection in atmospheric environments, but cadmium is better in marine and similar environments involving strong electrolytes. Electroplating increases the hazards of stress raisers and residual tensile stresses because the hydrogen released at the surface during acid or cathodic electrocleaning or during plating can cause a time-dependent brittleness, which can act as though added tensile stress had been applied and can result in sudden fracture after minutes, hours, or hundreds of hours. Unrelieved tensile stresses can result in fracture during plating. Such stresses occur most severely at the inside of small-radius bends. Parts with such bends should always be stress relieved before plating. However, because even large index springs have been found to be cracked, general stress relief is always good practice.

Preparation for plating is also very important because hydrogen will evolve from any inorganic or organic material on the metal until the material is thoroughly covered. Such contaminants may be scarcely noticeable before plating except by their somewhat dark appearance. Thorough sandblasting or tumbling may be required to remove such layers.

Hydrogen Relief Treatment. If stress relieving has been attended to, and the springs are truly clean before plating, then the usual baking treatment of around 200 °C (400 °F) for 4 h should lessen the small amount of hydrogen absorbed and redistribute it to give blister-free springs, which will not fail.

Mechanical Plating. Another technique that solves the hydrogen problem is mechanical plating, which involves cold welding particles of zinc or other soft-metal powder to an immersion copper flash plate on the spring. While some hydrogen may be absorbed during acid dipping before plating, it does not result in a time-dependent embrittlement because the plated layer is inherently porous, even though it has a shiny appearance. The hydrogen easily diffuses through the pores within 24 h, leaving the steel ductile.

Fatigue

For those springs that are dynamically loaded, it is common practice to obtain basic mechanical data from *S-N* fatigue curves. A typical *S-N* diagram is shown in Fig. 4(a). For each cycle of fatigue testing, the minimum stress is zero and the maximum stress is represented by a point on the chart. An alternative method of presenting data on the fatigue life of springs is shown in Fig. 4(b).

Stress Range. In most spring applications, the load varies between initial and final positive values. For example, an automotive valve spring is compressed initially during assembly, and during operation it is further compressed cyclically each time the valve opens.

The shear-stress range (that is, the difference between the maximum and minimum

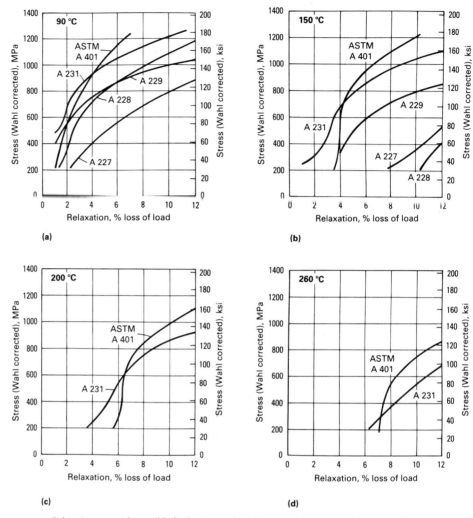

Fig. 10 Relaxation curves for steel helical springs of music wire (ASTM A 228), chromium-silicon spring wire (ASTM A 401), oil-tempered spring wire (ASTM A 229), chromium-vanadium spring wire (ASTM A 231), and hard-drawn spring wire (ASTM A 227) at (a) 90 °C (200 °F) and (b) 150 °C (300 °F). Relaxation curves for the low-alloy steel spring wires (ASTM A 231 and A 401) are also plotted at (c) 200 °C (400 °F) and (d) 260 °C (500 °F). All curves represent relaxation after exposure for 72 h at indicated temperatures.

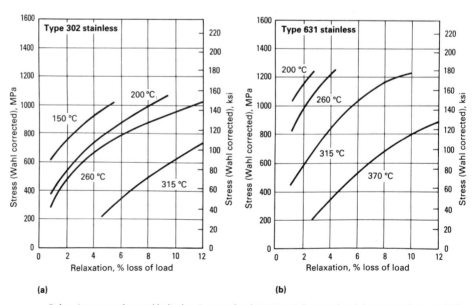

Fig. 11 Relaxation curves for steel helical springs made of (a) 302 stainless steel and (b) 631 stainless steel. The curves represent relaxation after exposure for 72 h at the indicated temperatures.

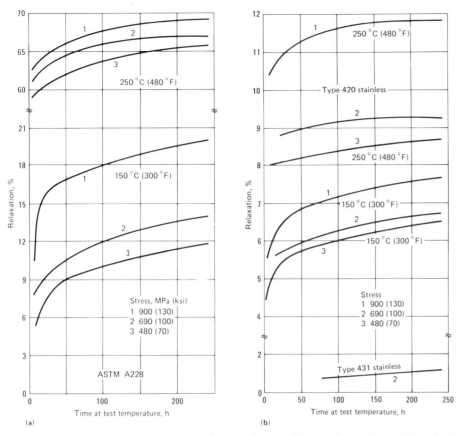

Fig. 12 Effect of time and temperature on the relaxation of ten-turn helical springs made from (a) music wire per ASTM A 228 and (b) 420 and 431 stainless steel wire. Wire diameter, 2.69 mm (0.106 in.); spring diameter, 25.4 mm (1.00 in.); free length, 76.2 mm (3.00 in.). Stresses were corrected by the Wahl factor.

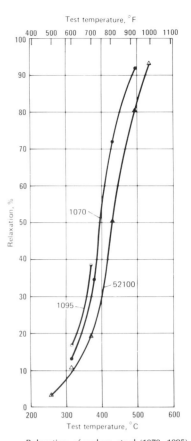

Fig. 13 Relaxation of carbon steel (1070, 1095) and SAE 52100 alloy steel circular flat springs (piston rings) at elevated temperatures. Spring hardness was 35 HRC. Springs were exposed to the indicated temperatures for 3 to 4 h.

of the stress cycle to which a helical steel spring may be subjected without fatigue failure) decreases gradually as the mean stress of the loading cycle increases. The allowable maximum stress increases up to the point where permanent set occurs. At this point, the maximum stress is limited by the occurrence of excessive set.

Figure 5 shows a fatigue diagram for music wire springs of various wire diameters and indexes. This is a modified Goodman diagram and shows the results of many fatigue-limit tests on a single chart. In Fig. 5, the 45° line OM represents the minimum-stress of the cycle, while the plotted points represent the fatigue limits for the respective minimum stresses used. The vertical distances between these points and the minimum-stress reference line represent the stress ranges for the music wire springs.

In fatigue testing, some scatter may be expected. The width of the band in Fig. 5 may be attributed partly to the normal changes in tensile strength with changes in wire diameter. There appears to be a trend toward higher fatigue limits for the smaller wire sizes. Line UT is usually drawn so as to intersect line OM at the average ultimate shear strength of the various sizes of wire.

Modified Goodman diagrams for helical springs made of several steels are shown in Fig. 6 (music wire and 302 stainless steel wire) and Fig. 7 (hard-drawn steel spring wire, oil-tempered wire, and chromium-silicon steel wire). In all instances, the plotted stress values were corrected by the Wahl factor (see the section "Wahl Correction" in this article). The data were obtained from various sources, including controlled laboratory fatigue tests, spot tests on production lots of springs, and correlation between rotating-beam fatigue tests on wire and unidirectional-stress fatigue tests on compression and extension helical springs.

The graphs on the left side of Fig. 6 and 7 plot the allowable stresses at 10 million cycles taken from *S-N* curves for various wire sizes diameters. The graphs on the right side of Fig. 6 and 7 show the allowable stresses at 10 000, 100 000, and 10 million cycles for two different wire diameters. In Fig. 6 and 7, the stress range is the vertical distance between the 45° line and the lines for the several wire sizes. The allowable maximum stress increases to a point of permanent set, indicated by the horizontal sections of the lines on the diagrams. For equal wire sizes, these diagrams show that music wire (Fig. 6a and b) is the most fatigue-resistant wire, with fatigue limits 50% greater than those of the least resistant wires (hard-drawn wires, Fig. 7e and f).

This difference is largely maintained under high-stress, short-life conditions. Figures 6 and 7 represent normal quality for each grade. Because of variations in production conditions, however, quality is not constant.

Figure 8 shows the range of fatigue life for five lots of music wire, all 0.50 mm (0.020 in.) in diameter. Wire was tested on a rotating-beam machine at a maximum stress of 1170 MPa (170 ksi) and a mean stress of zero. Results were correlated with fatigue tests on torsion springs as follows. A minimum fatigue life of 50 000 cycles was required of each spring; a minimum life of 20 000 cycles for the wire in the rotating-beam machine at 1170 MPa (170 ksi) gave satisfactory correlation with the 50 000 cycle service life of springs made from the wire. Lot 5 in Fig. 8 was rejected because it failed to meet the fatigue requirement. Subsequent fatigue tests on a pilot lot of springs made from lot 5 wire confirmed the inability of these springs to meet the fatigue requirement of 50 000 cycles.

Shot peening of springs improves fatigue strength by prestressing the surface in compression. It is usually applied to wire 1.6 mm (1/16 in.) or more in diameter. The type

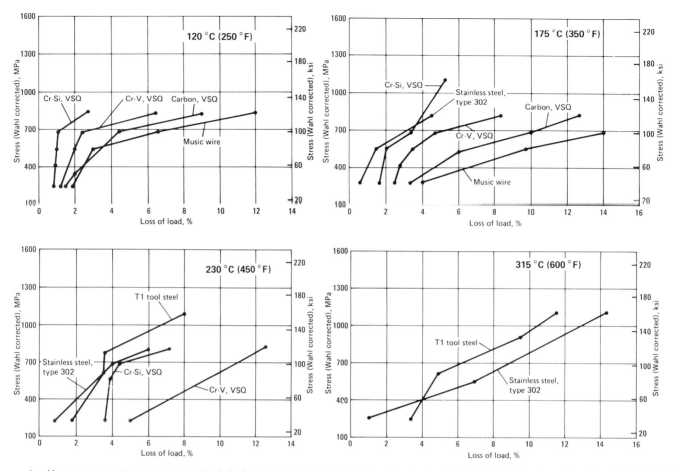

Fig. 14 Load-loss curves at various temperatures for helical compression steel springs made of music wire (ASTM A 228), VSQ carbon steel spring wire (ASTM A 230), VSQ chromium-vanadium steel spring wire (ASTM A 232), type 302 stainless steel spring wire (ASTM A 313), VSQ chromium-silicon steel spring wire (AISI 9254), and T1 high-speed tool steel. Results are based on tests of thousands of springs. All springs were made of pretempered wire and were stress relieved after coiling; none were shot peened, and all curves are for exposure of at least 72 h at the indicated temperatures.

of shot used is important; better results are obtained with carefully graded shot having no broken or angular particles. Shot size may be optimum at 10 to 20% of the wire diameter. However, for larger wire, it has been found that excessive roughening during peening with coarse shot lessens the benefits of peening, apparently by causing minute fissures. Also, peening thin material too deeply leaves little material in residual tension in the core; this negates the beneficial effect of peening, which requires internal tensile stress to balance the surface compression.

Shot peening is effective in largely overcoming the stress-raising effects of shallow pits and seams. Proper peening intensity is an important factor, but more important is the need for both the inside and outside surfaces of helical springs to be thoroughly covered. An Almen test strip necessarily receives the same exposure as the outside of the spring, but to reach the inside, the shot must pass between the coils and is thus greatly restricted. As a result, for springs with closely spaced coils, a coverage of 400% on the outside may be required to achieve 90% coverage on the inside. Cold-wound steel springs are normally stress relieved after peening to restore the yield point. A temperature of 230 °C (450 °F) is common because higher temperatures degrade or eliminate the improvement in fatigue strength.

The extent of improvement in fatigue strength to be gained by shot peening, according to one prominent manufacturer of cold-wound springs, is shown in Fig. 9. The bending stresses apply to flat springs, power springs, and torsion springs; the torsional stresses apply to compression and extension springs.

Effect of Temperature

The effect of elevated temperatures on the mechanical properties and performance of fabricated springs is shown in Fig. 10 to 14. The effect is reported as amount of load loss (relaxation), which is a function of chemical composition, maximum stress, and spring processing.

Figures 10 and 11 show some results from testing helical compression springs to determine the maximum relaxation at a given static working stress, temperature, and time. A specific spring height was determined for a given corrected stress. The spring was clamped at this height and placed in a convection oven for 72 h. It was removed and cooled, and the new free height was measured. Load loss was determined and amount of relaxation calculated. For example, a music wire spring designed at a corrected stress of 690 MPa (100 ksi) will relax a maximum of 3.8% when held at 90 °C (200 °F) for 72 h.

From the results shown in Fig. 10(a) and (b), springs made from music wire (ASTM A 228) are equal in performance to those made from oil-tempered wire (ASTM A 229) at 90 °C (200 °F) but are inferior at 150 °C (300 °F). The springs made from low-alloy steel wire such as chromium-silicon (ASTM A 401) or chromium-vanadium (ASTM A 231) provide improved relaxation resistance at elevated temperatures (Fig. 10), while the stainless steel springs (Fig. 11) exhibit further improvements in relaxation resistance. Percentage of load loss increases with shear stresses.

The effect of time and temperature on relaxation of springs is shown in Fig. 12. Rate of relaxation is greatest during the first

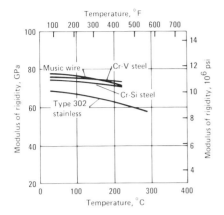

Fig. 15 Effect of temperature on modulus of rigidity of spring steels

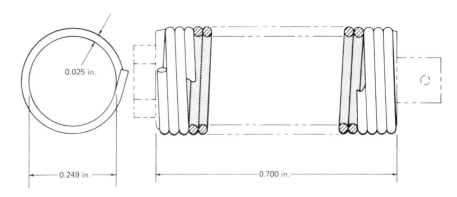

Fig. 16 Helical instrument spring that required a constant modulus of rigidity up to 90 °C (200 °F)

50 to 75 h. For longer periods of time, the rate is lower, decreasing as the logarithm of time when no structural change or softening occurs.

A significant effect of temperature on the relaxation of piston rings is indicated in Fig. 13. The rings were confined in test cylinders to maintain the outside ring diameter and were exposed to test temperature for 3 to 4 h. The load required to deflect the ring to working diameter was measured before and after each test to calculate the amount of relaxation.

Tests on thousands of springs under various loads at elevated temperatures are summarized in Fig. 14. Plain carbon spring steels of valve-spring quality are reliable at stresses up to 550 MPa (80 ksi) (corrected) and temperatures no higher than 175 °C (350 °F), in wire sizes no greater than 9.5 mm (3/8 in.). Slightly more severe applications may be successful if springs are preset at the operating temperature with loads greater than those of the application. Plain carbon spring steels of valve-spring quality should not be used above 200 °C (400 °F). To achieve an acceptable balance between relaxation and stress level, it is recommended that plain carbon steels be used at 120 °C (250 °F) or below (Table 1).

Except for high-speed steel, these tests revealed no advantage in springs heat treated after coiling compared with those at the same hardness made of pretempered wire and properly stress relieved. Deflection of a spring under load is inversely proportional to the modulus of rigidity, G, of the material. Variation with temperature is shown in Fig. 15.

A uniform deflection under load over a range of temperatures sometimes must be maintained. The instrument spring shown in Fig. 16 required a constant modulus up to 90 °C (200 °F) and, when made of music wire, drifted 5% in service. It was replaced with a satisfactory spring made of a nickel alloy with constant modulus (Fe-42Ni-5.4Cr-2.40Ti-0.60Al-0.45Mn-0.55Si-0.06C).

In another example, a spring was originally fabricated from oil-tempered wire (ASTM A 229) and performed satisfactorily when tested at room temperature. However, in service it was immersed in oil that attained a temperature slightly above 90 °C (200 °F), which was high enough to cause excessive relaxation over a period of 2 to 3 h. Chromium-silicon steel spring wire (ASTM A 401) was substituted for the oil-tempered wire, and at the identical operating temperature service was satisfactory. Design data for these springs are given in Table 4.

Hot-Wound Springs

Although some hot-wound springs are made of steels that are also used for cold-wound springs, hot-wound springs are usually much larger, which results in significant metallurgical differences.

Hardenability Requirements. Steels for hot-wound springs are selected mainly on the basis of hardenability. Carbon steels with about 0.70 to 1.00% C (1070 to 1095) are suitable and widely used for statically and dynamically loaded springs in the smaller sizes. Carbon steels are also used for larger springs in the lower stress range, where some hardenability can be sacrificed safely. In most cases, alloy steels are usually required for the larger sizes because of the need for hardenability. Most specifications for hot-wound alloy steel springs require 0.50 to 0.65% C and a minimum hardness of 50 HRC at the center after oil quenching from about 815 °C (1500 °F) and before tempering. (The austenitizing temperature will vary, depending on the specific steel.)

Springs subjected to lower stress ranges may not require this high hardenability and can therefore be made of the lower-priced carbon steels. Typical minimum hardness and hardenability values are given in Table 5 for hot-wound springs used under specific conditions of stress. Here it can be seen that lower as-quenched surface hardness and lower hardenability are permitted in the lower stress ranges for both statically and variably loaded springs. These requirements gradually increase as the stress increases. After tempering, a hardness of 50 HRC is often specified for a static stress of 1240 MPa (180 ksi) or for springs dynamically loaded at 825 MPa (120 ksi).

Table 4 Comparison of two spring materials for elevated-temperature service

	Condition at elastic limit			
	Load		Total deflection	
Grade of wire(a)	kg	lb	mm	in.
Oil-tempered (A229) ..	122.7	270.5	19.3	0.76
Cr-Si steel (A401)	139.7	308.0	21.8	0.86

(a) Design data (the same for both types of wire): mean diameter, 15.88 mm (0.625 in.); inside diameter, 12.12 mm (0.477 in.); wire diameter; 3.76 mm (0.148 in.); spring index, 4.22; total number of coils, 10; number of active coils, 8; spring rate, 6.36 kg/mm (356 lb/in.); working load, 5.17 kg (114 lb); deflection at working load, 7.92 mm (0.312 in.); stress at working load, 383 MPa (55.6 ksi); solid height, 37.6 mm (1.48 in.) max; free height, ~49.0 mm (1.93 in.); set removed, plain finish, variable type of load, environment of about 90 °C (200 °F)

Table 5 Typical minimum hardness and hardenability for steel used for hot-wound helical springs

As-oil-quenched, prior to tempering. Normal hardness, as-tempered, 44–49 HRC at surface

Corrected maximum solid stress		Hardness, HRC	
MPa	ksi	At surface	At center
Static load			
690	100	45	35
825	120	50	45
965	140	60	50
1100	160	60	50
1240	180	60	50
Variable load, designed for a minimum life of 50 000 cycles (set removed, 2.5% probability of failure, mean stress 515 MPa, or 75 ksi)			
690	100	45	35
825	120	50	45
965	140	60	50
1100	160	60	50
1240	180	60	50
Dynamic load, designed for a minimum life of 2 million cycles (set removed, shot peened, 2.5% probability of failure, mean stress 515 MPa, or 75 ksi)			
690	100	45	35
825	120	60	50

Table 6 Recommended steels for hot-wound helical springs

Where more than one steel is recommended for a specific set of conditions, they are arranged in the order of increasing hardenability. The first steel listed applies to the lower end of the designated wire diameter range and the last to the upper end of the range.

Corrected maximum solid stress		Diameter of spring wire (hot rolled)		
MPa	ksi	9.5–25.4 mm (3/8–1 in.)	25.4–50.8 mm (1–2 in.)	50.8–76.2 mm (2–3 in.)
Static load (set removed, static stress up to 80% of maximum solid stress)				
825	120	1070 1095	1095	1095
965	140	1070 1095	51B60H 4161	4161
1100	160	5150H 5160H 50B60H	51B60H 4161	4161
1240	180	5150H 5160H 50B60H	51B60H 4161	4161
Variable load, designed for minimum life of 50 000 cycles (set removed, 2.5% probability of failure, operating stress range not over 50% of solid stress)				
690	100	1095	1095	1095
825	120	1095	51B60H 4161	4161
965	140	5150H 5160H 50B60H	51B60H 4161	4161
1100	160	5150H 5160H 50B60H	51B60H 4161	4161
1240	180	5150H 5160H 50B60H	51B60H 4161	4161
Dynamic load, designed for minimum life of 2 million cycles (set removed, shot peened, 2.5% probability of failure, mean stress 515 MPa, or 75 ksi, operating stress range not over 50% of solid stress)				
690	100	1095	1095	1095
825	120	1095	51B60H 4161	4161

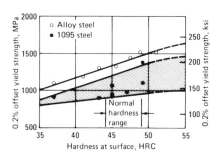

Fig. 17 Relationship between surface hardness and yield strength of steel bars after tempering. Alloy steels were martensitic throughout the section, as-quenched; 1095 bars were 12.7 to 50.8 mm (1/2 to 2 in.) in diameter.

Table 7 Steels with sufficient hardenability for torsion bar springs

Thickness of section		Minimum hardenability required	Steel
mm	in.		
29.21	1.150	J50 at 8	8650H, 5152H
33.91	1.335	J50 at 9	8655H, 50B60H
36.32	1.430	J50 at 8½	50B60H
39.88	1.570	J50 at 10	51B60H
46.23	1.820	J50 at 11	8660H
48.26	1.900	J50 at 14	4150H
57.15	2.250	J50 at 22	9850H

Recommended steels for hot-wound helical springs are given in Table 6, covering variations in stress range, type of loading, and wire size. Hardenability requirements increase as required strength and/or wire diameter increases.

The strengths obtained in bars with different hardenabilities are shown in Fig. 17. The band for 1095 steel demonstrates the variation in yield strength that results from variations in the thickness of the bar and the severity of quenching and from limited hardenability. Smaller bars, as well as those quenched more rapidly from the austenitizing temperature, follow the top of the band, while larger bars and those quenched less drastically fall in the lower half of the band. For example, a 13 mm (½ in.) round bar quenched in 5% caustic solution will have maximum properties, and a 50 mm (2 in.) round bar quenched in still oil will have yield strength near the minimum shown. The scatter for alloy steels in Fig. 17 is much narrower because the points only represent steels with sufficient hardenability to have a martensitic structure throughout the bar section, as-quenched.

The distribution of hardness test results at surface and center, as-quenched and as-tempered, is shown in Fig. 18 for a multiplicity of heats of 1095 steel and five alloy steels commonly used for hot-wound springs. The results of testing specimens 300 mm (12 in.) long correlate with those of testing production springs. For the alloy steels, a minimum of 50% martensite at the center of the quenched section was specified. Table 7 lists steels that meet hardenability requirements for torsion bar springs with section thicknesses from 29.21 to 57.15 mm (1.150 to 2.250 in.).

Surface Quality. Hot-wound springs fabricated from bars of large diameter will normally show much deeper surface decarburization, in the range from 0.13 to 0.38 mm (0.005 to 0.015 in.), unless special material preparation and processing techniques are used. Although decarburization is not desirable, in some design situations the use of the favorable residual stress pattern due to shallow hardening can reduce the negative effects of decarburization on fatigue performance (Table 8). The detrimental effects of decarburization are less noticeable on hot-wound than on cold-wound springs because other weaknesses, such as the surface imperfections and irregularities typical of a hot-rolled surface, create additional focal points for fatigue cracks.

Specifications for hot-wound springs and torsion bar springs usually include maximum seam depth. For example, the specifications used by a manufacturer of railway equipment allow seams with a maximum depth of 0.025 mm (0.001 in.) per 1.6 mm (1/16 in.) of wire diameter up to 0.41 mm (0.016 in.) for any bar size above 25 mm (1 in.). Another manufacturer allows seam depths of 0.41 mm (0.016 in.) for bars 25 mm (1 in.) in diameter, 0.81 mm (0.032 in.) for bars 25 to 44 mm (1.00 to 1.75 in.) in diameter, and 1.22 mm (0.048 in.) for bars 44 to 63.5 mm (1.75 to 2.50 in.) in diameter.

Design Stress. It should be noted that some organizations specify a much more conservative approach to stress than that presented in Table 6. The Spring Manufacturers Institute has adopted a chart from the Manufacturers Standardization Society of the Valve and Fittings Industry, for essentially static service, that calls for an admittedly uncorrected stress for alloy steel of 760 MPa (110 ksi) for a bar diameter of 13 mm (½ in.), reducing to 590 MPa (86 ksi) for a bar diameter of 92 mm (3⅝ in.). Their curve for 1095 steel is roughly 140 MPa (20 ksi) lower.

The desirability of conservative design in cyclical service is illustrated in Fig. 19, in which the minimum stress used was low. Such data on springs hot wound from bars with as-rolled surfaces are limited, and interpretation is therefore difficult. The value of peening, however, is made quite apparent. Surface imperfections can be removed by grinding, and this is normal practice, where the increased cost can be borne, in order to increase reliability at higher stresses.

With all of the limitations discussed above, there can be both cost and reliability advantages in using 1095 steel at low stresses. For example, railroad freight car

Steel Springs / 317

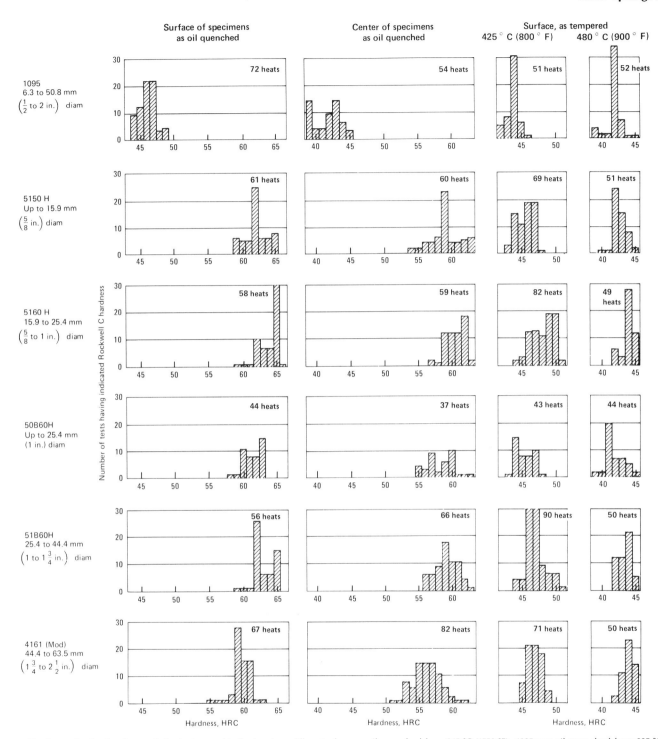

Fig. 18 Hardness distribution for steels for hot-wound helical springs. Alloy steels were oil quenched from 845 °C (1550 °F); 1095 was oil quenched from 885 °C (1625 °F). Data were obtained from hot-rolled, heat-treated laboratory test coupons, 305 mm (12 in.) long. Specimens were sectioned from the center of the coupons after heat treatment. These results on bars correlate with those on production springs. For the alloy steels, a minimum of 50% martensite at the center of the quenched section was specified.

springs, which are subject to severe corrosion pitting, are made of 1095 steel and designed for very low stress, and very little difficulty is encountered.

Costs

The relative costs of various spring steels in the form of round wire are given in Fig. 20. Base price may be outweighed by other costs, as indicated in Table 9, which lists possible extras for two alloy spring steels of about the same base price. Also, base price and other costs will vary somewhat with time.

The amount of material in a spring can be minimized by designing for the highest safe stress level. From the equation for volume of material:

$$V = E\left(\frac{4G}{S^2}\right) \qquad \text{(Eq 1)}$$

in which V is the volume of material in the spring (in cubic inches), E is the spring energy (in inch-pounds), G is the torsion modulus of elasticity (in pounds per square inch), and S is the stress (in pounds per square inch). A reduction of 5% in stress

Table 8 Properties of shallow-hardened and through-hardened hot-wound (wound from bar 32 mm, or 1¼ in., in diameter) helical compression springs

Heat treatment	Hardness, HRC Distance below surface				Fatigue tests(a) Millions of cycles to failure		
	1.6 mm (1/16 in.)	3.2 mm (⅛ in.)	6.35 mm (¼ in.)	Center	High	Low	Average
1045 steel, decarburized 0.38 mm (0.015 in.)							
As-quenched	60	50	45	30	...	...	...
Tempered	54.5	49	37.5	25	1.28	0.46	0.772
8655H steel, decarburized 0.03 mm (0.001 in.)							
As-quenched	60	60	60	60	...	...	...
Tempered	42	43	43	42	0.789	0.443	0.572

(a) 485 MPa (70 ksi) stress range (corrected), 415 MPa (60 ksi) mean stress (corrected); eight springs tested in each group

level increases required volume of material by 11%.

Cost of Springs. Tolerances are often the most important factors in spring cost and should be selected with cost in mind. Tolerances superimposed on tolerances are especially costly. For example, if a free-length tolerance were specified in addition to load tolerances at each of two deflections, the cost of manufacturing the spring could be much higher than if free-length tolerance were unspecified or indicated as a reference dimension. Important dimensions can sometimes be held to close tolerances at no additional cost by allowing wider tolerances elsewhere. Springs are resilient structures, and the tight tolerances held by the machining industry are not applicable. Spring diameter tolerances can often be held to within ±1 to 3%, while tolerances on load generally vary from ±5 to ±15%.

Design

All spring design is based on Hooke's Law: When a material is loaded within its elastic limit, the resulting strain in the material is directly proportional to the stress produced by the load. For springs, this means that the amount of deflection is proportional to the load or other force producing it.

Charts and formulas are available to aid in the design of springs. Charts that recommend categorically a design stress for a given spring steel can be reliable guides for springs used in static service only, unless otherwise qualified. Charts of any sort are of value only when it is known to which wire size(s) they apply, whether the data are for springs with or without presetting and with or without shot peening and whether the cited nominal stresses include a correction factor or a safety factor. The operating environment can have a pronounced effect on spring performance and is an important factor to consider in evaluating design data and in the material selection process.

The spring-design formulas given in Table 10 are valid for extension and compression springs of round and square wire below the stress at which yielding begins. For rectangular wire, the numerical coefficients vary with the ratio of width to thickness. Detailed information on these calculations is given in the SAE Handbook, Supplement J795 on Spring Design, and in the Spring Manufacturers Institute Handbook on Spring Design. Much of the labor of computation in design can be eliminated by the use of commercially available spring design software.

Wahl Correction. In a straight torsion bar spring of circular section, twisting produces a shear stress uniform at every point on the surface. However, in a helical spring coiled from round wire, the stress at the inside of the coil is higher.

The first formulas for stress in helical springs shown in Table 10 do not include additional stresses that exist due to curvature and direct shear. The magnitude of these stresses varies with the amount of curvature and resultant shift in the neutral axis. A set of formulas developed by A. M. Wahl, involving a correction factor known as the Wahl factor, takes into account these two effects (Fig. 21).

Table 11 shows a typical comparison between uncorrected and corrected stresses for two spring indexes and five grades of spring wire. The values in Table 11 indicate

Table 9 Pricing of automotive coiled spring steel

Item		Cost per 45 kg (100 lb)(a)	Item		Cost per 45 kg (100 lb)(a)	Item		Cost per 45 kg (100 lb)(a)
Alloy steel 6150			**Alloy steel 5160**			**Alloy steel 5160 (continued)**		
Base price	Alloy steel	$25.60	Base price	Alloy steel	$25.60	Quantity	Heat lots	...
Grade	6150	6.75	Grade	5160, standard fine grained	2.90	Marking	Paint one end of the bundle green; attach metal tag; show heat number, weight, part number, grade, length, and size	...
Quality	Electric furnace	2.30	Quality	Open hearth	...			
Restrictions	Restricted carbon and manganese	5.50	Restrictions	None, standard steel	...			
Size	13 mm (½ in.) diam	3.50	Size	16.48 mm (0.649 in.)	2.75			
Straightness	Special straightness	3.00		Close tolerance (½ standard)	1.50			
Treatment	Hot rolled, precision ground	8.95	Straightness	½ standard tolerance, machine straightened	2.25	Packaging	1600–1800 kg (3500–4000 lb) lifts wired with 4 or 5 wires, wrapped in waterproof paper with three steel bands outside paper for magnet unloading	...
Cleaning and coating	Pickled and oiled	4.10						
Preparation	None	...	Treatment, coating and cleaning	Hot rolled, pickled and oiled	3.25			
Testing	Restricted hardenability	3.85	Preparation	No surface conditioning allowed	...			
Cutting	1.5–2.4 m (5–8 ft) abrasive	1.00					1600–1800 kg (3500–4000 lb) lifts	0.50
Length	4.6 m (15 ft) dead length	0.50	Testing	Special decarburization, standard chemistry and hardenability	...		Paper wrap	0.25
Quantity	<900 kg (2000 lb)	10.00				Loading	Gondola cars blocked to maintain straightness; not less than 18 × 10³ kg (20 tons) per car	...
Marking	Continuous line	4.35						
		0.25						
Packaging	Paper wrap 1130 kg (2500 lb) or less box with runners, bar 4.6–6.7 m (15–22 ft)	5.60	Cutting	Machine cut	1.00			
			Length	Dead length 3.63 m (143 in.)	0.75			
Loading	Box car loading	0.25						
Total		**$85.50**				**Total**		**$40.75**

(a) 1989 prices

Steel Springs / 319

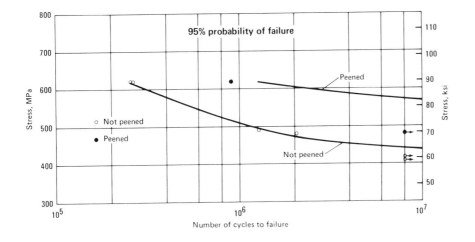

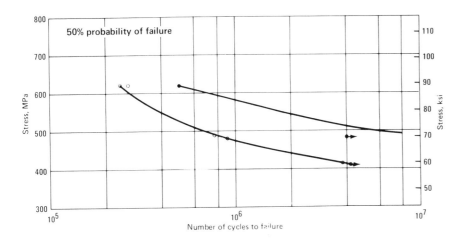

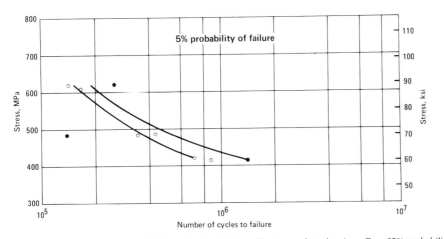

Fig. 19 Effect of peening on the probability of fatigue failure of hot-wound steel springs. Top: 95% probability of failure. Middle: 50% probability of failure. Bottom: 5% probability of failure. Springs were made from 16 to 27 mm (⅝ to 1 1/16 in.) diam 8650 and 8660 hot-rolled steel and heat treated to between 429 and 444 HB. Springs were shot peened to an average arc height of 0.2 mm (0.008 in.) on the type C Almen strip at 90% visual coverage.

that the stress in the spring is higher because of the shift in the neutral axis at the lower index. Therefore, in the first line, if the elastic limit is less than 1080 MPa (157 ksi), the spring will take a permanent set.

For springs with an index greater than 12, the Wahl correction is small; however, correction is needed for all springs subject to cyclic conditions.

Stress Range. Safe maximum stresses for a minimum of 10 million cycles of stress will generally be lower than the fatigue limits shown in Goodman diagrams (Fig. 5 to 7) because of variations in the wire (especially with regard to surface condition) and variations in the spring, manufacturing process, and application. Commercial tolerances on spring wire will also affect the selection of allowable stress for a production spring. For example, the 0.013 mm (0.0005 in.) tolerance on 0.81 mm (0.032 in.) diam music wire gives rise to a stress variation of 4½%. In addition, stresses vary proportionately with load tolerance. Tolerances on number of coils, squareness, and spring diameter also permit stresses different from those calculated. For these reasons, stresses near the upper fatigue limits may be unsafe for some applications.

Life. Unless required, it is wasteful to design a spring for infinite life. A fatigue curve or *S-N* diagram provides approximations of allowable stresses for designing to a desired life. For reliable operation, springs made of hard-drawn or oil-tempered wire, which may contain seams or scratches as deep as 3.5% of wire diameter or 0.25 mm (0.010 in.), should be limited to static applications or subjected to less than 10 000 loadings.

Ordnance springs typify long-stroke springs that have limited space and life requirements and that are subjected to shock loads. Stranded-wire springs have sometimes been used because of damping or design considerations. However, these springs are costly and chafe between coils. The stress level, stress range, and life (until set becomes excessive) of some springs used at high stresses, including some stranded-wire springs, are listed in Table 12. The life of these springs is not entirely dependent on maximum operating stress and stress range, but is also affected by the shock loading that is exerted. However, a fracture in one strand does not significantly affect the functioning of such a spring, which may develop several distributed breaks before becoming seriously impaired.

Compression Springs

The usual types of ends for compression springs are shown in Fig. 22, in order of increasing cost. An additional increase in the cost of grinding may be anticipated with close tolerances on squareness and length.

When springs are designed to work at maximum stress, allowance should be made for the effect on stress of the specified type of end. For example, in a compression spring with squared, or squared and ground, ends, the number of active coils will vary slightly throughout the deflection of the spring, and this results in spring rate and stress different from those originally calculated. This is especially important in springs having fewer than seven coils, because changes in the number of active coils result in high percentage changes in stress. The number of active coils changes because part of the active coil adjacent to the squared

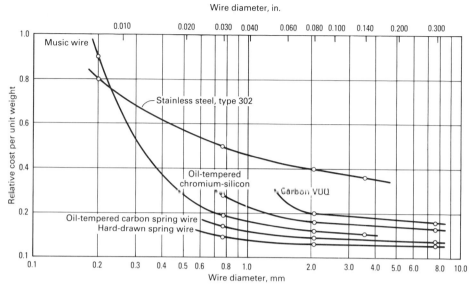

Fig. 20 Relative cost of spring steel wire

Table 10 Design formulas for helical extension and compression springs

Round wire	Square wire

Stress for statically loaded springs (psi)

$$S = \frac{2.55PD}{d^3} \qquad S = \frac{2.4PD}{t^3}$$

$$S = \frac{FGd}{\pi D^2 N} \qquad S = \frac{FGt}{2.32D^2 N}$$

Stress for round wire corrected by the Wahl factor (Fig. 21) (psi)

$$S_K = \frac{2.55PD}{d^3} \left[\frac{4C-1}{4C-4} + \frac{0.615}{C} \right]$$

Stress for square wire corrected by Wahl factor (Fig. 21) (psi)

$$S_K = \frac{2.4PD}{t^3} \left[1 + \frac{1.2}{C} + \frac{0.56}{C^2} + \frac{0.50}{C^3} \right]$$

Wire size for given load and stress (in.)

$$d^3 = \frac{2.55PD}{S} \qquad t^3 = \frac{2.4PD}{S}$$

Deflection of the spring under load (in.)

$$F = \frac{8PD^3 N}{Gd^4} \qquad F = \frac{5.58PD^3 N}{Gt^4}$$

Spring loading at a given deflection (lbf)

$$P = \frac{FGd^4}{8D^3 N} \qquad P = \frac{FGt^4}{5.58D^3 N}$$

$$P = \frac{Sd^3}{2.55D} \qquad P = \frac{St^3}{2.4D}$$

Note: S is the nominal shear stress (in pounds per square inch); P is the spring load or force (in pounds); D is the mean diameter of the spring (in inches) (outside diameter of the spring minus the wire diameter); d is the diameter of the wire (in inches); t is the thickness of the square wire before coiling (in inches); F is the deflection (in inches) under load P; G is the modulus of elasticity in shear (modulus of rigidity), equal to 11.5×10^6 psi for ordinary steels and 10×10^6 psi for stainless steel near room temperature; N is the number of active coils; C is the spring index, D/d; and K is the Wahl correction factor.

end closes down solid and becomes inactive as the spring approaches the fully compressed condition. In designs where this is objectionable, loads at two compressed lengths should be specified on the manufacturing print in order to ensure loads, rates, and stresses no higher than those calculated. Controlling loads in this way also controls the number of active coils.

For springs that will be operated at high stresses, tolerances that do not increase calculated stresses should be liberal so that tolerances on other dimensions that might increase stresses may be held closely. Therefore, if the number of coils must be held closely in order to keep spring stresses in control, tolerance on free length with ends ground should be large.

Active Coils. The number of active coils equals the total number of coils less those that are inactive at each end. For plain, unground ends the number of inactive coils depends on the length of contact between the end coils and the mating part. The number of active coils in a spring with plain, ground ends would approximately equal the number of turns of the wire untouched by grinding. One inactive coil on each end should be allowed in springs with squared ends or squared and ground ends. For springs of fewer than seven coils, this rule should be applied with care; in some springs, the degree of squareness should be considered.

Solid heights for the types of ends shown in Fig. 22(a) and (b) are computed by multiplying the wire diameter by the total number of coils plus one. Solid height for the type of end in Fig. 22(c) equals wire diameter times total number of coils plus one-half the wire diameter, if grinding at each end removes at least one-quarter of the wire thickness. For the end shown in Fig. 22(d), solid height is equal to the wire diameter times the total number of coils. A spring should be designed so that the coils do not touch or clash in service, because such contact will induce wear, fretting corrosion, and noise. The maximum solid height, specified primarily for inspection, should be determined after consideration of clearances and of tolerances on wire size to allow spring manufacturers to meet economically the requirements established for solid height.

Wire that is square or rectangular before coiling will upset at the inside of the coil and become trapezoidal in section during coiling. This limits deflection per coil because solid height is predicated on wire thickness at the inside of the coil. An approximate formula for the upset thickness, t_1, at the inside compared to the original thickness, t, of the rectangular or square wire is:

$$t_1 = t\left(1 + \frac{k}{c}\right) \qquad \text{(Eq 2)}$$

where $c = (D_o + D_i)/(D_o - D_i)$; D_o is the outside diameter of the spring; D_i is the inside diameter of the spring; and k is 0.3 for cold-wound springs and 0.4 for hot-wound springs and annealed materials. With wire that is keystone shaped before coiling, this difficulty is not encountered, but such wire is costly and is not readily available, particularly in small lots.

Effect of Modulus Change. A change in the value of G (modulus of rigidity) resulting from a change in material must be compensated for by a change in wire diameter, mean diameter, or number of active coils. If springs made from different steels must have precisely the same spring rate, the number of active coils, the wire diameter, or the mean diameter must differ. A typical example follows:

Parameter	Music wire	Type 302
Wire diameter, mm (in.)...	0.25 (0.010)	0.25 (0.010)
Coil diameter, mm (in.)...	5 (0.20)	5 (0.20)
Number of coils...	10	8.5
Spring rate, N/cm (lb/in.)...	0.313 (0.179)	0.313 (0.179)

Extension Springs

Extension springs are usually designed with some initial tension at the discretion of the spring maker. Occasionally, the pitch or space between coils is specified, or the spring must be close wound with no initial tension. When initial tension is specified, it is specified as a load with a tolerance of approximately ±15%. In making stress calculations, initial tension is included in the load, P, and if P is measured, initial tension is included in the measurement obtained.

The necessity for fastening extension springs to some other part may require secondary operations or costly fasteners. End hooks with sharp bends may have

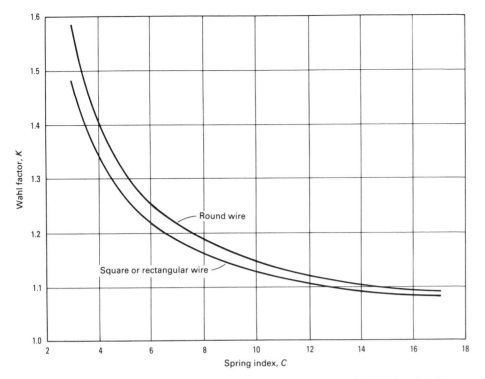

Fig. 21 Wahl correction factors for helical compression or extension springs. Spring index, C, is the mean diameter of the spring, D, divided by the diameter of the wire, d. When square wire, or rectangular wire coiled on the flat side, is used, the thickness of the wire, t, is substituted for d. When rectangular wire is coiled on edge, the width, b, is substituted for d. Source: Ref 2

Table 11 Effect of spring index on calculated spring working stress

Steel wire	Spring index	Wahl correction factor	Calculated spring working stress			
			Uncorrected		Corrected	
			MPa	ksi	MPa	ksi
Music	4.60	1.33	814	118	1082	157
	8.38	1.16	820	119	951	138
Hard drawn spring	4.60	1.33	634	92	841	122
	8.50	1.16	641	93	745	108
Oil-tempered spring	4.75	1.32	814	118	1075	156
	8.68	1.15	779	113	896	130
Chromium-vanadium spring	4.75	1.32	869	126	1145	166
	8.65	1.15	814	118	938	136
Chromium-silicon spring	4.75	1.32	1020	148	1344	195
	8.26	1.16	1076	156	1248	181

Table 12 Performance of highly stressed, shock-loaded music wire springs

Maximum stress		Stress range		Wire diameter		Spring index(a), D/d	Cycles to excessive set
MPa	ksi	MPa	ksi	mm	in.		
1103	160	862	125	2.29(b)	0.090(b)	3.9	2000
965	140	689	100	2.49	0.098	6.8	4000
1082	157	696	101	0.94(c)	0.037(c)	3.4	5000
1110	161	793	115	3.76	0.148	6.3	4000
1124	163	572	83	1.35	0.053	6.6	6000
1172	170	600	87	0.97	0.038	6.5	6000
945	137	558	81	1.57(b)	0.062(b)	4.8	7500
910	132	614	89	1.40	0.055	5.1	5000

(a) d, strand diameter. (b) Seven-strand wire. (c) Three-strand wire

stress concentrations that increase stresses by a factor of two or three. There are many common designs for end hooks.

End Hooks. Stresses in end hooks are combinations of torsional and bending stresses. Bending stresses diminish toward the body of the spring, where torsional stresses prevail. There is a region where both tensile and torsional stresses are present and are difficult to compute accurately. If the last coil of the spring adjacent to the end hook is reduced in diameter, the stress in the last coil and the end hook will be lower than that in the other coils.

The bending stresses in an end hook may be calculated from:

$$S_b = \frac{32PR}{\pi d^3}\left(\frac{r_c}{r_i}\right) \quad \text{(Eq 3)}$$

where PR is the bending moment (load times moment arm from centerline of load application to centerline of wire), d is the diameter of the wire, r_c is the radius of the sharp bend measured to centerline of wire, and r_i is the inside radius of the sharp bend. The ratio r_c/r_i is an approximate evaluation of the stress-concentrating effect of the configuration these radii describe. The ratio indicates the destructive effect of a small radius, r_i. The smaller this radius, the higher the stress at the bend, other values being the same.

Because of the possibilities for overextension and the stress concentrations in the ends, allowable stresses are more likely to be exceeded in extension springs than in compression springs. Therefore, allowable stresses that are 20% lower than those for compression springs are sometimes recommended for extension springs. However, when stresses in the ends are accurately known and the stroke of the spring is controlled in design of the spring or of the associated parts, the same values for allowable stresses can be used for extension springs as for compression springs that have not been preset, provided the initial tension is relatively low.

Leaf Springs*

Leaf springs, like all other springs, serve to absorb and store energy and then to release it. During this energy cycle, the stress in the spring must not exceed a certain maximum in order to avoid settling or premature failure. This consideration limits the amount of energy that can be stored in any spring.

In this regard, leaf springs are somewhat less efficient than helical springs in terms of energy storage per unit mass of material, but they are widely used in automotive applications because they may also function as structural members. Table 13 lists values of energy that may be stored in the active parts of leaf springs designed for maximum stress of 1100 MPa (160 ksi). If consideration of the inactive part of the spring required for axle anchorage, spring eyes, and so on, is included, the energy per pound of the total spring weight will be less than shown. For comparison, the stored energy in the active material of a helical spring made of round wire is 510 J/kg (2050 in. · lbf/lb) at 1100 MPa (160 ksi) and for a

*Abstracted by A.R. Shah, Senior Metallurgist, Automotive Operations, Rockwell International, from the "Manual on Design and Application of Leaf Springs," SAE Information Report J788a, 1982. Used with permission of the copyright holder, Society of Automotive Engineers, Inc.

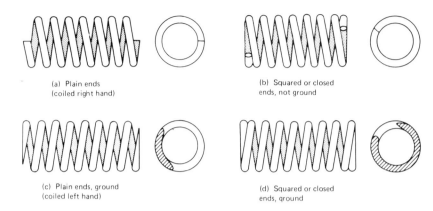

Fig. 22 Types of ends for compression springs. Cost increases from (a) to (d).

Table 13 Energy stored by steel leaf springs

Springs designed for maximum stress of 1100 MPa (160 ksi)

Leaf type(a)	Spring design	Energy per unit mass of spring	
		J/kg	in. · lbf/lb
F-1	Single leaf	43	173
	Multileaf, with all leaves full length	43	173
F-2	Multileaf, with properly stepped leaves	95	380
T-2	Single leaf	106	426
T-1	Single leaf	109	438
P-2	Single leaf	122	488
F-4	Single leaf	123	493

(a) See Table 14 for descriptions of leaf types

torsion bar of round cross section is 390 J/kg (1570 in. · lbf/lb) at 965 MPa (140 ksi). This comparison shows that a leaf spring is inherently heavier than other types of springs. Balancing this weight disadvantage is the fact that leaf springs can also be used as attaching linkage or structural members. In order to be economically competitive, leaf springs must therefore be so designed that this advantage is fully utilized.

A leaf spring can be constructed of a single leaf, of several leaves of equal (full) length, or of several leaves of decreasing (stepped) lengths. The various types of leaves used in single-leaf springs are described in Table 14. Of course, flat-profile, or type F, leaves can be used to construct multileaf springs. A leaf spring made entirely of full-length leaves of constant thickness (see type F-1) is much heavier and less efficient than a leaf spring made of properly stepped leaves (see type F-2) or a single-leaf spring (see types F-4, P-2, T-1, and T-2).

The maximum permissible leaf thickness for a given deflection is proportional to the square of the spring length. By choosing too short a length, the designer often makes it impractical for the spring maker to build a satisfactory spring, although the requirements for normal load, deflection, and stress can be fulfilled.

When a stepped spring is made of type F-2 leaves, its length should be chosen so that the spring will have no less than three leaves. Springs with many leaves are sometimes used for heavy loads, but they are economical only where use of short springs leads to definite savings in the supporting structure.

Leaf Springs for Vehicle Suspension. Leaf springs are most frequently used in vehicle suspension. The characteristics of a suspension system are affected chiefly by the spring rate and the static deflection.

The rate, or more accurately the load rate, of a spring is the change of load per inch of deflection. This is not the same at all positions of the spring and is different for the spring alone and for the spring as installed. The static deflection of a spring equals the static load divided by the rate at static load; it determines the stiffness of the suspension and the ride frequency of the vehicle. In most cases, the static deflection differs from the actual deflection of the spring between zero load and static load due to influences of spring camber and shackle effect.

A soft ride generally calls for a large static deflection of the suspension. There are, however, other considerations and limits, such as the following:

- A more flexible spring will have a larger total deflection and will be heavier
- In most applications, a more flexible spring will cause more severe striking through or will require a larger ride clearance (the spring travel on the vehicle from the design load position to the metal-to-metal contact position), disregarding rubber bumpers
- The change of standing height of the vehicle due to a variation of load is larger with a more flexible spring

The static deflection to be used also depends on the available ride clearance. Further, the permissible static deflection depends on the size of the vehicle because of considerations of stability in braking, accelerating, cornering, and so on.

Table 15 lists typical static deflections and ride clearances for various types of vehicles. These values are approximate and are meant to be used only as a general indication of current practice in suspension system design.

The weight of a spring for a given maximum stress is determined by the energy that is to be stored; this energy is represented by the area under a load-deflection diagram. The effects of changes in rate and clearance on the weight of the spring can easily be seen in this type of diagram.

Figure 23 shows theoretical load-deflection diagrams for two springs designed for the same load and ride clearance; the spring represented by the solid line in Fig. 23(a) is stiffer than the optimum spring for these design conditions, while the spring represented by the solid line in Fig. 23(b) is too flexible. Each spring, when fully deflected, stores the same amount of energy (1020 J, or 9000 in. · lbf), and the two springs will weigh almost the same if made of the same kind of material. The optimum spring (minimum energy and weight) for this design load and ride clearance will have stiffness intermediate between those of the two springs represented in Fig. 23. For this optimum spring, static deflection is equal to ride clearance, as indicated by the dashed lines in Fig. 23; the stored energy for this spring is only 900 J (8000 in. · lbf). Figure 23 also indicates that a change in ride clearance will affect the stored energy, and therefore the required weight, of the stiff spring much more than those of the flexible spring.

Steel Grades. The basic requirement of a leaf-spring steel is that it have sufficient hardenability relative to leaf thickness to ensure a fully martensitic structure throughout the entire cross section. Nonmartensitic transformation products detract from the fatigue properties. Automotive chassis leaf springs have been made from various fine-grain alloy steels such as grades 9260, 4068, 4161, 6150, 8660, 5160, and 51B60.

In the United States, almost all leaf springs are currently made of chromium steels such as 5160 or 51B60, or their H equivalents. With 5160, the chemical composition is specified as an independent variable (while the hardenability of the steel is a dependent variable that will vary with the composition); 5160H is essentially the same steel except that the hardenability is specified as an independent variable (while the composition is a dependent variable that can be adjusted to meet the hardenability-band requirement).

In general terms, higher alloy content is necessary to ensure adequate hardenability when thicker leaf sections are used. When considering a grade of steel, it is recommended that hardenability be either calculated from chemical composition or (for the

Table 14 Types of leaf springs

With flat profile: "F" types

F-1: rectangular cantilever. Constant in thickness (t_0) and in width (w_0). Under load (P), the stress is greatest at line of encasement and decreases at a constant rate to zero at line of load application. The elastic curve in bending (from an initially flat spring) has its smallest radius at line of encasement; that is, the rate of change of curvature is greatest at this line. This design is inefficient.

F-2: trapezoidal cantilever. Constant in thickness (t_0). Width decreases at a constant rate from w_0 at line of encasement to a specified dimension (w_e) at line of load application. Under load (P), the stress is greatest at line of encasement. This design is more efficient than type F-1.

F-3: triangular cantilever. Constant in thickness (t_0). Width decreases at a constant rate from w_0 at line of encasement to zero at point of load application. Under load (P), the stress is constant throughout the length; the elastic curve in bending is circular; that is, the rate of change of curvature is constant throughout the length. Although this design is highly efficient, it is impractical, because no material is provided for load application.

F-4: modified triangular cantilever. Same as type F-3 except for an end portion of length c with constant cross section ($t_0 \times w_c$) to facilitate load application. This design is slightly less efficient than type F-3.

With tapered profile: "T" types

T-1: tapered cantilever. Constant in width (w_0). Thickness decreases at a constant rate from t_0 at line of encasement to a specified dimension (t_e) at line of load application. Under load (P), the stress is greatest at line of encasement when the t_e/t_0 ratio is 0.50 or more. When the t_e/t_0 ratio is less (between 0.49 and 0.24), a higher degree of efficiency is obtained, with the line of peak stress some distance away from the line of encasement. The highest efficiency (approaching but not equaling that of the triangular F-3 and of the parabolic P-1 cantilevers) occurs when t_e/t_0 equals 0.357, with the peak stress (8.9% greater than at encasement) located at a distance from the line of encasement equal to 44.5% of the cantilever length (l).

T-2: modified tapered cantilever. Same as type T-1 except for an end portion of length c with constant cross section ($t_c \times w_0$) for material strength required for eye or load-bearing area. This design is slightly less efficient than type T-1.

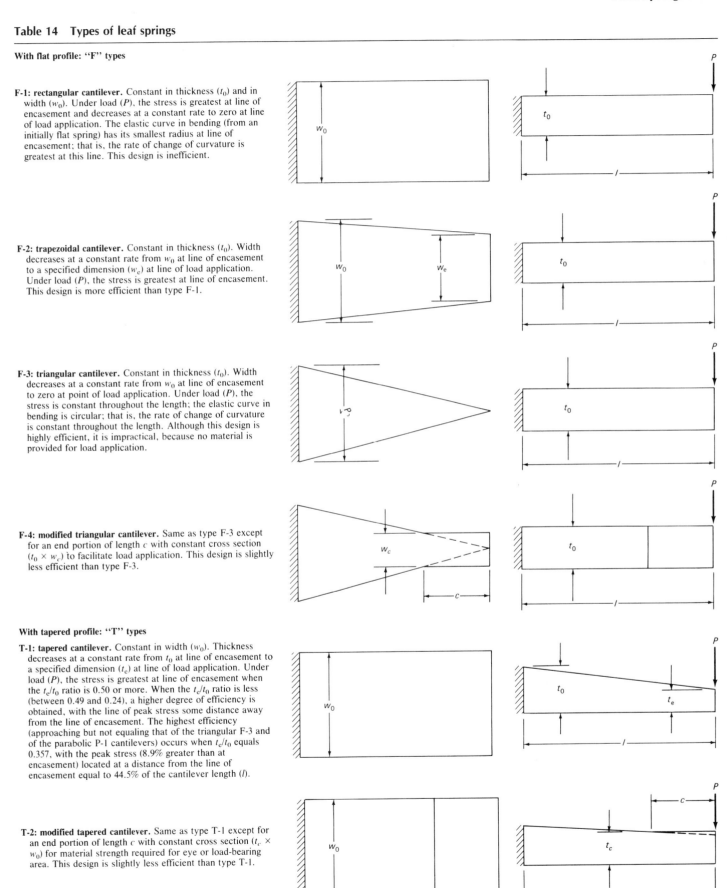

(continued)

324 / Carbon and Low-Alloy Steels

Table 14 (continued)

With tapered profile: "T" types (continued)

T-3: tapered-trapezoidal cantilever. Thickness decreases at a constant rate from t_0 at line of encasement to a specified dimension (t_e) at line of load application. Width increases at a substantially constant rate from w_0 at line of encasement to w_e at line of load application. This design approximates type T-1 for efficiency.

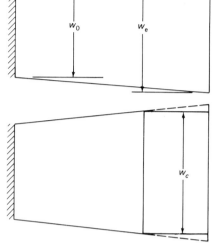
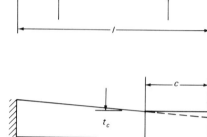

T-4: modified tapered-trapezoidal cantilever. Same as type T-3 except for an end portion of length c with constant cross section ($t_c \times w_c$) for material strength required for eye or load-bearing area. This design is slightly less efficient than type T-3.

With parabolic profile: "P" types

P-1: parabolic cantilever. Constant in width (w_0). Thickness decreases from t_0 at line of encasement in a parabolic profile that terminates in zero thickness at line of load application. Under load (P), the stress is constant throughout the length. The elastic curve in bending (from an initially flat spring) has its smallest radius at line of load application; that is, the rate of change of curvature is greatest at this line. Although this design is highly efficient, it is impractical because no material is provided for load application.

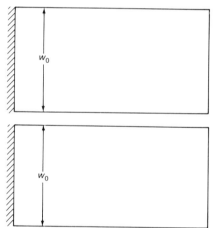
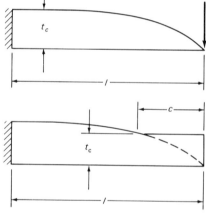

P-2: modified parabolic cantilever. Same as type P-1 except for an end portion of length c with constant cross section ($t_c \times w_0$) to facilitate load application. This design is slightly less efficient than type P-1.

P-3: parabolic-trapezoidal cantilever. Thickness decreases from t_0 at line of encasement in an approximately parabolic profile that terminates in zero thickness at line of load application. Width increases at a substantially constant rate from w_0 at line of encasement to w_e at line of load application. Under load (P), the stress is constant throughout the length. Although this design is highly efficient, it is impractical, because no material is provided for load application.

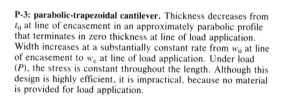
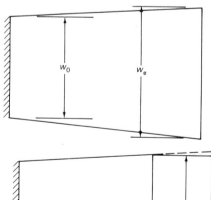
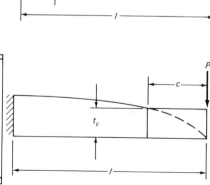

P-4: modified parabolic-trapezoidal cantilever. Same as type P-3 except for an end portion of length c with constant cross section ($t_c \times w_c$) to facilitate load application. This design is slightly less efficient than type P-3.

Table 15 Static deflections and ride clearances of various types of vehicles with steel leaf springs

Type of vehicle and load	Static deflection mm	in.	Ride clearance mm	in.
Passenger automobiles, at design load	100–300	4–12	75–125	3–5
Motor coaches, at maximum load	100–200	4–8	50–125	2–5
Trucks, at rated load				
For highway operation	75–200	3–8	75–125	3–5
For off-road operation	25–175	1–7	50–125	2–5

various H steels) determined from the hardenability-band charts in the article "Hardenability Curves" in this Volume. The following rule of thumb may be useful for correlating section size and steel grade:

Maximum section thickness		
mm	in.	Steel
8.2	0.323	5160
15.9	0.625	5160H
36.6	1.440	51B60H

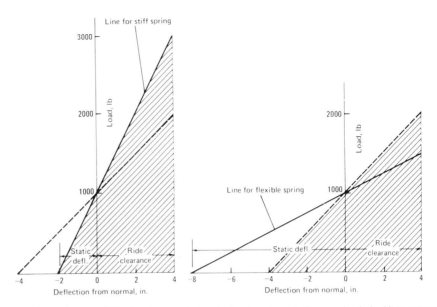

Fig. 23 Theoretical load-deflection diagrams for two leaf springs. In each diagram, the dashed line represents the minimum-energy spring having the same design load and ride clearance as the spring represented by the solid line.

Mechanical Properties. Steels of the same hardness in the tempered martensitic condition have approximately the same yield and tensile strengths. Ductility, as measured by elongation and reduction in area, is inversely proportional to hardness. Based on experience, the optimum mechanical properties for leaf-spring applications are obtained within the hardness range 388 to 461 HB. This range contains the six standard Brinell hardness numbers 388, 401, 415, 429, 444, and 461 (corresponding to the ball-indentation diameters 3.10, 3.05, 3.00, 2.95, 2.90, and 2.85 mm obtained with an applied load of 3000 kgf). A specification for leaf springs usually consists of a range covered by four of these hardness numbers, such as 415 to 461 HB (for thin section sizes). Typical mechanical properties of leaf-spring steel are as follows:

Tensile strength, MPa (ksi)	1310–1690 (190–245)
Yield strength (0.2% offset), MPa (ksi)	1170–1550 (170–225)
Elongation, %	7 minimum
Reduction in area, %	25 minimum
Hardness	42–49 HRC, or 338–461 HB for 3000 kg load (3.10–2.85 mm indentation diameter)

Mechanical Prestressing. Presetting, shot peening, and/or stress peening at ambient temperatures produces large increases in fatigue durability without increasing the size of the spring. These prestressing methods are more effective in increasing the fatigue properties of a spring than are changes in material.

When a load is applied to a leaf spring, the surface layers are subjected to the maximum bending stress. One surface of each leaf is in tension, and the other surface is in compression. The surfaces that are concave in the free position are generally tension surfaces under load, while the convex surfaces are generally in compression. Fatigue failures of the leaves start at or near the surface on the tension side. Because residual stresses are algebraically additive to load stresses, the introduction of residual compressive stresses in the tension surface by prestressing reduces operating stress level, thus increasing fatigue life.

Presetting (also known as cold setting, bulldozing, setting-down, scragging) produces residual compressive stresses in the tension surface and residual tensile stresses in the compression surface by forcing the leaves to yield or take a permanent set in the direction of subsequent service loading. Although this operation is beneficial to fatigue life, its primary effect is the reduction of settling (load loss) in service. Presetting is usually done after assembly.

Shot peening introduces compressive residual stresses by subjecting the tension side of each individual leaf to a high-velocity stream of shot. The SAE Manual on Shot Peening, HS-84, deals with the control

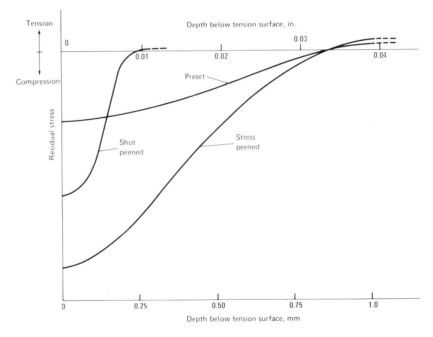

Fig. 24 Beneficial stress patterns induced by presetting and peening

of process variables, while techniques for control of peening effectiveness and quality are explained in Procedures for Using Standard Shot Peening Test Strip, SAE J443. Cut wire shot, sizes CW-23 to CW-41, and cast steel shot, sizes S-230 to S-390, are generally used for this purpose. The intensity of shot peening applied to light-weight and medium-weight springs is usually in the range of 0.25 to 0.50 mm (0.010 to 0.020 in.) as measured by Almen "A" strip. For heavy springs, single-leaf springs, and stress-peened springs, the intensity is usually 0.15 to 0.35 mm (0.006 to 0.014 in.), as measured by Almen "C" strip. Coverage in both cases should be at least 90%.

Stress peening (strain peening) is a means of introducing higher residual compressive stresses than are possible with shot peening with the leaf in the free (unloaded) position. Stress peening is done by shot peening the leaf while it is loaded (under stress) in the direction of subsequent service loading.

Curvature of a leaf spring will be changed by mechanical prestressing. The magnitude of the changes due to shot peening and presetting can be calculated using the formulas given in SAE Information Report J788a. The stress patterns induced by these processes are compared schematically in Fig. 24. The effects of presetting and shot peening are cumulative to some extent, but the results may be influenced by the sequence of operations.

Surface Finishes and Protective Coatings. Surface finish is defined as the surface condition of the spring leaves after the steel has been formed and heat treated, and prior to any subsequent coating treatment. Normally, automotive leaves are utilized as heat-treated or in the shot peened condition. An as-heat-treated finish will be a tight oxide produced by the quenching and tempering operations and will exhibit a blue or blue-black appearance. A shot-peened finish is the result of removing the blue or blue-black color by the peening operation and is characterized by a matte luster appearance.

A protective coating is a material added to the surfaces of individual leaves or to the exposed areas of leaf-spring assemblies. Its primary purpose is to prevent corrosion both in storage and in operational environments. All exposed surfaces to be coated must be free of loose scale and dirt. Peened surfaces should be coated as soon as possible to prevent the formation of corrosion. It is important that an enveloping coat be applied. An unprotected area or a break in the coating may contribute to localized corrosion and a reduction in fatigue life. Before a coating is specified, whether grease, oil, paint, or plastic, it should be evaluated for effects that its application might have on the fatigue life of the spring steel. The thickness and adhesion characteristics of the coating must be within the tolerances that have been established for the type of material being used in order to provide adequate corrosion protection and to ensure satisfactory performance.

REFERENCES

1. *Handbook of Spring Design*, Spring Manufacturers Institute, 1988
2. A.M. Wahl, *Mechanical Springs*, 2nd ed., McGraw-Hill, 1963

Steel Tubular Products

Revised by Dennis Smyth, The Algoma Steel Corporation Ltd.; R.G. Lessard, Stelpipe, a Unit of Stelco, Inc.; and Frank Minden, Lone Star Steel

STEEL TUBULAR PRODUCTS is the term used to cover all hollow steel products. Although these products are usually produced in cylindrical form, they are often subsequently altered by various processing methods to produce square, oval, rectangular, and other symmetrical shapes. Such products have applications that are almost innumerable, but they are most commonly used as conveyors of fluids and as structural members.

The products discussed in this article are those made from wrought carbon or alloy constructional steels and designated by the terms pipe, specialty tubing, and oil country tubular goods. Excluded are stainless steel items in the categories mentioned above and hollow steel products used as containers for storage and shipping.

In discussing the various tubular items in this article, emphasis is given to product classifications, available specifications, chemical compositions, sizes, and other dimensional characteristics. For more information on many of the manufacturing practices used in making steel tubular products, the reader may refer to Ref 1 and 2 and to Volume 14 of the 9th Edition of *Metals Handbook*.

Product Classification

The two simplest and broadest commercial classifications of steel tubular products are tube and pipe. (Although the application of the terms pipe and tube is not always consistent, the term pipe is commonly used to describe cylindrical tubular products made to standard combinations of outside diameter and wall thickness.) These two broad classifications are subdivided into several named use groups. For example, the term tube covers three such groups: pressure tubes, structural tubing, and mechanical tubing.

The term pipe covers five such groups: standard pipe, line pipe, oil country tubular goods, water well pipe, and pressure pipe. There is also pipe for special applications, such as conduit pipe and tubular piling, that do not fit any of these classifications. Each of these use groups, in turn, has a number of uses or named-use subdivisions. These are shown in Table 1 and described in a subsequent section of this article.

The named uses in Table 1 have been developed without regard to method of manufacture, size range, wall thickness, or degree of finish. For example, the names do not distinguish between those products commonly called pipe that have size and wall thicknesses the same as standard pipe, and those having different basic dimensions; all are termed pipe.

On a use basis, pressure pipe is distinguished from pressure tubes in that the latter are suitable for those applications in which heat is applied externally. The principal use groups and types of pressure tubes are shown in Table 2.

Structural and mechanical tubing do not follow this system of nomenclature. Instead, their names are derived from the method of fabrication and degree of finish, such as cold formed welded or seamless hot finished.

Steel tubular products can be made by forming a flat skelp, sheet, strip, or plate into a hollow cylinder and welding the resulting longitudinal seam together or by generating an opening in a solid cylinder by piercing and elongating the resultant hollow cylinder. The first group of products is called welded, and the second group, seamless. The processes for producing these products are discussed in the following sections.

Welding Processes

In producing welded steel tubular products, flat-rolled skelp, strip, sheet, or plate is formed into cylinders that are then joined at the longitudinal or spiral seam by one of the processes described below.

Electric resistance welding employs a series of operations, in the first of which the flat-rolled steel is cold shaped into tubular form. Welding is effected by the application of pressure and heat generated by induction or by an electric current through the seam. The welding pressure is generated by constricting rolls and the electromagnetic effects of the high welding current. Electric resistance welded tubular products having longitudinal seams are usually made in sizes ranging from 3.2 mm ($\frac{1}{8}$ in.) nominal to 0.6 m (24 in.) actual outside diameter, but larger sizes are also available. Electric resistance methods are also sometimes used for making tubular products by spiral welding. This

Table 1 Major types and uses of pipe

Type of pipe	Uses
Standard	Industrial or residential water steam, oil, or gas transmission; distribution or service lines; structural uses
Special	Conduit, piling, pipe for nipples, and so on
Line	Oil- or gas-transmission pipe, water-main pipe
Oil country tubular goods	Drill pipe, casing, tubing
Water well	Drive pipe, driven-well pipe, casing, reamed, and drifted pipe
Pressure	Pipe for elevated-temperature or pressure service

Table 2 Principal uses and types of pressure tubes

Use groups	Types
Water tube boilers	Generating tubes, superheater tubes, economizer tubes, circulator tubes, furnace wall tubes
Fire tube boilers	Boiler flues, superheater flues, superheater tubes, arch tubes, stay tubes, safe ends
Others	Feedwater heater tubes, oil still tubes

method is frequently used for large-diameter products.

Continuous Welding. In continuous (or furnace butt) welding, skelp with square or slightly beveled edges is furnace heated to the welding temperature. The heated stock is roll formed into cylindrical shape as it emerges from the furnace; additional heat is usually provided by an oxygen or air jet impinging on the seam edges, and the tube passes through constricting rolls where the seam edges are welded by the pressure of the rolls. Continuous welded products are available in nominal diameters from 3.2 to 100 mm (⅛ to 4 in.).

Fusion Welding. In fusion welding, the flat-rolled steel, with edges suitably prepared, is formed into tubular shape by either hot or cold shaping. The flat-rolled steel may be shaped longitudinally (straight seam) or bent into helical form (spiral welded). The edges are welded with or without simultaneously depositing filler metal in a molten or molten-and-vapor state. Mechanical pressure is not required to effect welding. Fusion may be accomplished by either electric arc or gas heating, or a combination of both. The upper limit of outside diameter is determined mainly by the forming method and the fusion welding process used.

Double submerged arc welding is a special method of shielded-arc welding in which the seam is submerged under a solid flux while being welded. The weld is made in two passes, one from inside the pipe and the other from outside the pipe, with approximately half the wall thickness being welded in each pass. The maximum diameter is limited only by the capability of the forming equipment, and double submerged-arc welded pipe is currently available in outside diameters of up to 3 m (120 in.).

Seamless Processes

Steel tubular products produced by the seamless processes are made in diameters up to ⅔ m (26 in.) by the rotary piercing method and up to 1.2 m (48 in.) by hot extrusion.

Rotary Piercing. In rotary piercing, rounds of the necessary diameter and length are first heated to rolling temperature. Each hot round is fed into a set of rolls having crossed axes and surface contours that pull it through the rolls, thus rupturing it longitudinally. The force of the rolls then causes the metal to flow around a piercing point, enlarging the axial hole, smoothing the inside surface, and forming a tube. After being pierced, the rough tube is usually hot rolled to final dimensions by means of plug mills or mandrel mills, which may be followed by stretch reducing mills.

Hot extrusion is a hot working process for making hollows, suitable for processing into finished tubing of regular and irregular form, by forcing hot, prepierced billets through a suitably shaped orifice formed by an external die and internal mandrel. The outside of the hollow acquires the size and contour imposed by the die; the inside conforms to the size and contour of the mandrel. Extruded hollows may be further worked into tubular products by cold-finishing methods.

Cupping and Drawing. In the cupping and drawing method of making seamless tubing, a circular sheet or plate is hot cupped in a press through several pairs of conical dies, each successive pair being deeper and more nearly cylindrical than the previous set. The rough tube then is drawn to finished size.

Cold Finishing

Pipe in suitable sizes and most products classified as tubing, both seamless and welded, may be cold finished. The process may be used to increase or decrease the diameter, to produce shapes other than round, to produce a smoother surface or closer dimensional tolerances, or to modify mechanical properties. The process most commonly used is cold drawing, in which the descaled hot-worked tube is plastically deformed by drawing it through a die and over a mandrel (mandrel drawing) to work both exterior and interior surfaces. Cold drawing through the die only (without a mandrel) is called sink drawing or sinking.

Tube Reducing and Swaging. In tube reducing by rotorolling or pilgering and in swaging, a reducing die works the tube hollow over a mandrel; swaging may, however, be done without a mandrel. The commercial importance of tube reducing is, first, that very heavy reductions (up to 85%) can be applied to mill length tubes, and second, that the process can be applied to the refractory alloys that are difficult to cold draw because of high power requirements.

Cold Finishing. Tubular products of circular cross section may be cold finished on the outside by turning, grinding, or polishing, or by any combination of these processes. They may be bored, skived, or honed on the inside diameter. Because these operations involve stock removal only, with negligible plastic deformation, there is no enhancement of mechanical properties.

Many of the standard specifications involving strength are based on the properties

Table 3 ASTM, API, and CSA specifications for carbon and alloy steel pipe

Specification	Product
ASTM specifications	
A 53(a)	Welded and seamless steel pipe, black and hot dipped, zinc coated
A 106(a)	Seamless carbon steel pipe for high-temperature service
A 134(a)	Arc-welded steel-plate pipe (sizes 400 mm, or 16 in., and over)
A 135(a)	Resistance-welded steel pipe
A 139	Arc-welded steel pipe (sizes 100 mm, or 4 in., and over)
A 211	Spiral-welded steel or iron pipe
A 252	Welded and seamless steel pipe piles
A 333(a)	Welded and seamless steel pipe for low-temperature service
A 335(a)	Seamless ferritic alloy steel pipe for high-temperature service
A 381	Double submerged-arc welded steel pipe for high-pressure transmission systems
A 405	Seamless ferritic alloy steel pipe, specially heat treated for high-temperature service
A 523	Resistance-welded or seamless steel pipe (plain end) for high-pressure electric cable conduit
A 524(a)	Seamless carbon steel pipe for atmospheric service and lower temperatures
A 587(a)	Resistance-welded low-carbon steel pipe for the chemical industry
A 589	Welded and seamless carbon steel water well pipe
A 671(a)	Arc-welded steel pipe for atmospheric service and lower temperatures
A 672(a)	Arc-welded steel pipe for high-pressure service at moderate temperatures
A 691	Arc-welded carbon or alloy steel pipe for high-pressure service at high temperatures
A 714	Welded and seamless HSLA steel pipe
A 795	Black and hot-dipped zinc-coated (galvanized) welded and seamless steel pipe for fire protection use
API specifications	
2B	Specification for fabricated structural steel and pipe
5CT	Specification for casing and tubing
5D	Specification for drill pipe
5L	Specification for line pipe
CSA standard	
CAN3-Z245.1-M86	Steel line pipe

(a) This ASTM specification is also published by ASME, which adds an "S" in front of the "A" (for example, SA 53).

Steel Tubular Products / 329

Table 4 Compositions of carbon steel pipe (ASTM, API, and CSA)

Specification	Pipe manufacturing process	Material grade	C	Mn	P	S	Si
ASTM specifications							
A 53 (type F)	Continuous or furnace butt welded	(b)	...	...	0.08	0.06	...
A 53 (type E)	Electric resistance welded	A(b)	0.25	0.95	0.05	0.06	...
		B(b)	0.30	1.20	0.05	0.06	...
A 53 (type S)	Seamless	A(b)	0.25	0.95	0.05	0.06	...
		B(b)	0.30	1.20	0.05	0.06	...
A 106	Seamless	A(b)(c)	0.25	0.27–0.93	0.025	0.025	0.10
		B(b)(c)	0.30	0.29–1.06	0.025	0.025	0.10
		C(b)(c)(d)	0.35	0.29–1.06	0.025	0.025	0.10
A 134	Arc welded	...	See material grade specifications.				
A 139	Arc welded	A(b)	...	1.00	0.040	0.050	...
		B(b)	0.30	1.00	0.040	0.050	...
		C(b)	0.30	1.20	0.040	0.050	...
		D(b)	0.30	1.30	0.040	0.050	...
		E(b)	0.30	1.40	0.040	0.050	...
A 211	Spiral welded	...	See material grade specifications.				
A 252	Welded or seamless	(b)	...	...	0.050	...	...
A 333	Welded or seamless	1(b)	0.30	0.40–1.06	0.05	0.06	...
		6(e)	0.30	0.29–1.06	0.048	0.058	0.10
A 381	Submerged arc welded	(f)	0.26	1.40	0.040	0.050	...
A 523	Electric resistance welded	A(b)	0.21	0.90	0.040	0.050	...
		B(b)	0.26	1.15	0.040	0.050	...
	Seamless	A(b)	0.22	0.90	0.040	0.050	...
		B(b)	0.27	1.15	0.040	0.050	...
A 524	Seamless	I, II(b)(c)	0.21	0.90–1.35	0.048	0.058	0.10–0.40
A 587	Electric resistance welded	(b)(c)(g)	0.15	0.27–0.63	0.048	0.050	...
A 589	Welded(h) or seamless	(b)	...	...	0.050	0.060	...
A 671	Arc welded	(f)	See material grade specifications.				
A 672	Arc welded	(f)	See material grade specifications.				
A 691	Arc welded	(f)	See material grade specifications.				
A 795	Welded or seamless	A	0.25	0.95	0.050	0.060	...
		B	0.30	1.20	0.050	0.060	...
API specifications							
2B	Arc welded	...	Specified by purchaser				
5CT	Welded(i) or seamless	H40, J55, K55, N80(b)	...	...	0.040	0.060	...
		C75 type 2(b)	0.43	1.50	0.040	0.060	0.45(j)
5CT	Seamless	P105, P110(b)	...	...	0.040	0.060	...
5L	Electric welded(k)	A25, class I(b)	0.21	0.30–0.60	0.045	0.06	...
		A25, class II(b)	0.21	0.30–0.60	0.045–0.080	0.06	...
		A(b)	0.21	0.90	0.04	0.05	...
		B(b)	0.26	1.15	0.04	0.05	...
	Furnace butt welded	A25, class I(b)	0.21	0.30–0.60	0.045	0.06	...
		A25, class II(b)	0.21	0.30–0.60	0.045–0.080	0.06	...
	Submerged arc welded	A(b)	0.21	0.90	0.04	0.05	...
		B(b)	0.26	1.15	0.04	0.05	...
	Seamless	A25, class I(b)	0.21	0.30–0.60	0.045	0.06	...
		A25, class II(b)	0.21	0.30–0.60	0.045–0.080	0.06	...
		A(b)	0.22	0.90	0.04	0.05	...
		B(b)	0.27	1.15	0.04	0.05	...
	Spiral welded(l)	A(b)	0.21	0.90	0.04	0.05	...
		B(b)	0.26	1.15	0.04	0.05	...
		X42(b)(m)	0.28	1.25	0.04	0.05	...
		X46, X52(b)(m)	0.30	1.35	0.04	0.05	...
	Welded(n)	X42(b)(m)	0.28	1.25	0.04	0.05	...
	Welded/cold expanded	X46, X52(b)(m)	0.28	1.25	0.04	0.05	...
	Welded, nonexpanded	X46, X52(b)(m)	0.30	1.35	0.04	0.05	...
	Seamless, cold expanded	X42(n), X46, X52(b)(m)	0.29	1.25	0.04	0.05	...
	Seamless, nonexpanded	X46, X52(b)(m)	0.31	1.35	0.04	0.05	...
5D	Seamless	(b)	...	...	0.04	0.06	...
CSA specification							
CAN3-Z245.1	Seamless or welded(o)	172 to 359(p)(q)	0.26	(r)	0.04	0.04	0.50 max
		386 to 550(p)(q)	0.26	(r)	0.03	0.035	0.50 max

(a) Where a single value is shown, it is a maximum limit, except for silicon, where a single value denotes a minimum limit. (b) Open hearth, electric furnace, or basic oxygen. (c) Killed. (d) Grade C is a special product supplied only by agreement between manufacturer and purchaser. (e) Open hearth or electric furnace. (f) Carbon steel grade(s) only. (g) Also Al, 0.02 min. (h) Furnace butt or electric resistance. (i) Electric resistance. (j) Maximum. (k) Electric resistance or electric induction. (l) Submerged arc. (m) Nb, Ti, V, or a combination, may be used by agreement between purchaser and manufacturer. (n) Cold expanded or nonexpanded. (o) Butt weld, electric weld, and submerged arc weld. (p) Pipe meeting the mechanical property requirements but deviating from the above limits may be supplied per agreement between purchaser and manufacturer. (q) Alloys may be added as stated in Table 5. (r) Maximum content may be subject to agreement between the manufacturer and purchaser.

of hot-rolled or cold-worked material. Some high-strength oil country goods are heat treated to achieve the combination of high strength, ductility, and sulfide stress corrosion cracking resistance required by the intended application.

Cold drawing may be employed to improve surface finish and dimensional accuracy and to increase the strength of tubular products. Some customer specifications prescribe strength levels that can best be attained by cold working.

Pipe Sizes and Specifications

Pipe is distinguished from tubing by the fact that it is produced in relatively few sizes and, therefore, in comparatively large quantities of each size. Size of pipe

Table 5 Compositions of alloy steel pipe (ASTM, API, and CSA)

Specification	Pipe manufacturing process	Material grade	C	Mn	P	S	Si	Cr	Ni	Mo	Cu	Others
ASTM specifications												
A 333	Welded or seamless	3(b)	0.19	0.31–0.64	0.05	0.05	0.18–0.37	...	3.18–3.82	...	...	...
		4(b)	0.12	0.50–1.05	0.04	0.04	0.08–0.37	0.44–1.01	0.47–0.98	...	0.40–0.75	0.04–0.30 Al
		7(b)	0.19	0.90	0.04	0.05	0.13–0.32	...	2.03–2.57	...	...	...
		8(b)	0.13	0.90	0.045	0.045	0.13–0.32	...	8.40–9.60	...	...	...
		9(b)	0.20	0.40–1.06	0.045	0.050	...	...	1.60–2.24	...	0.75–1.25	...
		10(b)	0.20	1.15–1.50	0.035	0.015	0.10–0.35	0.15 max	0.25 max	0.05	0.15 max	0.66 Al, 0.12V, and 0.05 Nb
A 335	Seamless	...	See Table 6.									
A 381	Submerged arc welded	(c)	Any adequate high-strength low-alloy steel									
A 405	Seamless	...	See Table 6.									
A 714	Welded(d) or seamless	I(b)	0.22	1.25	...	0.05	...	...	...	...	0.20 min	...
		II(b)	0.22	0.85–1.25	0.04	0.05	0.30	...	...	...	0.20 min	0.02 V min
		III(b)	0.23	1.35	0.04	0.05	0.30	...	...	...	0.20 min	(e)
		IV(b)	0.10	0.60	0.03–0.08	0.05	...	0.80–1.20	0.20–0.50	...	0.25–0.45	...
		V(b)	0.16	0.40–1.01	0.035	0.040	...	...	1.65 min	...	0.80 min	...
		VI(b)	0.15	0.50–1.00	0.035	0.045	...	0.30	0.40–1.10	0.10–0.20	0.30–1.00	...
		VII(b)	0.12	0.20–0.50	0.07–0.15	0.05	0.25–0.75	0.30–1.25	0.65	...	0.25–0.55	...
		VIII(b)	0.19	0.80–1.25	0.04	0.05	0.30–0.65	0.40–0.65	0.40	...	0.25–0.40	0.02–0.10 V
API specifications												
5CT	Welded(f)(b) or seamless(b)	C75 type 1	0.50	1.90	0.040	0.060	0.45	(g)	(g)	0.15–0.30	(g)	...
		C75 type 2(h)	0.43	1.50	0.040	0.060	0.45	...	...	...	...	...
		C75 type 3	0.38–0.48	0.75–1.00	0.040	0.040	...	0.80–1.10	...	0.15–0.25	...	...
		L80 type 1	0.43	1.90	0.040	0.060	0.45	...	0.25	...	0.35 max	...
		N80(h)	...	...	0.040	0.060	...	...	...	...	...	...
		C95(b)(h)	0.45(i)	1.90	0.040	0.060	0.45	...	...	...	...	...
	Seamless(b)	C75, L80 type 9 Cr	0.15	0.30–0.60	0.020	0.010	1.00	8.0–10.0	0.5	0.90–1.10	0.25	...
		C75, L80 type 13 Cr	0.15–0.22	0.25–1.00	0.020	0.010	1.00	12.0–14.0	0.5	...	0.25	...
		C90 type 1	0.35	1.00	0.020	0.010	...	1.20	0.99	0.75	...	...
		C90 type 2	0.50	1.90	0.030	0.010	...	(j)	0.99	(j)	...	...
		P-105, P-110(h)	...	...	0.04	0.06	...	...	...	...	...	...
	Welded(b)(f) or seamless(b)	Q125 type 1	0.35	1.00	0.020	0.010	...	1.20	0.99	0.75	...	...
		Q125 type 2	0.35	1.00	0.020	0.020	...	(j)	0.99	(j)	...	...
		Q125 type 3	0.50	1.90	0.030	0.010	...	(j)	0.99	(j)	...	...
		Q125 type 4	0.50	1.90	0.030	0.020	...	(j)	0.99	(j)	...	...
5L	Welded(b)(k)(l)	X56, X60	0.26	1.35	0.04	0.05	...	...	...	...	...	(e)
		X65	0.26	1.40	0.04	0.05	...	...	...	...	...	(e)
		X70	0.23	1.60	0.04	0.05	...	...	...	...	...	...
	Seamless(k)(b)	X56, X60	0.26	1.35	0.04	0.05	...	...	...	...	...	(e)
		X65, X70, X80	By agreement									
CSA specification												
CAN3-Z245.1	Seamless or welded(m)	172 to 359(n)	0.26	(o)	0.04	0.04	0.50	(o)	(o)	(o)	(o)	(p)
		386 to 550(n)	0.26	(o)	0.03	0.035	0.50	(o)	(o)	(o)	(o)	(p)

(a) Except for copper, where a single value is shown, it is a maximum limit. (b) Open hearth, electric furnace, or basic oxygen. (c) Alloy steel grade(s) only. (d) Furnace butt or electrical resistance. (e) 0.005 min Nb, 0.02 min V, or a combination thereof shall be used at the discretion of the manufacturer. (f) Electric resistance or electric flash. (g) Nickel, chromium, and copper combined shall not exceed 0.5%. (h) Alloys are typically added in various amounts, even though it is not specified in API Specification 5CT. (i) Carbon content may be increased to 0.55% max if the product is oil quenched. (j) No limits, but must be reported in analysis. (k) Cold expanded or nonexpanded. (l) Submerged arc, electric resistance, gas metal arc, spiral weld submerged-arc, and double seam. (m) Butt weld, electric weld, and submerged-arc weld. (n) Pipe meeting the mechanical property requirements but deviating from the above limits may be supplied subject to agreement of purchaser and manufacturer. (o) Maximum content may be subject to agreement by purchaser and manufacturer. (p) Maximums are 0.11% Nb, 0.11% V, 0.06% Ti, 0.001% B, and 0.020% (product analysis only) Ce.

is designed according to one of two methods.

It is now recommended that pipe less than 50 mm (2 in.) in diameter have a nominal or named size roughly equal to the inside diameter of standard-weight pipe. The outside diameter of each size is standard, regardless of weight, in order to allow the use of standard thread sizes. Therefore, the increase in wall thickness necessary to produce extra strong and double extra strong weights results in a decrease in inside diameter. The size designation, however, remains the same as for standard-weight pipe of the same outside diameter.

It is further recommended that for diameters of 50 mm (2 in.) and larger the pipe size be designated by outside diameter, wall thickness, and weight per unit length. In previous practice, the lower limits for designation by outside diameter were 90 mm (3½ in.) for line pipe with plain ends and 355 mm (14 in.) for all threaded pipe. The size of oil country goods is always designated by outside diameter.

For a reasonably complete list of the standardized sizes and weights of pipe for the major named uses, the AISI Steel Products Manual should be consulted. For oil country tubular goods, the specifications of the American Petroleum Institute (API) govern. Table 3 lists the current ASTM, API, and Canadian Standards Association (CSA) specifications covering pipe. Some of these involve several grades. The specifications listed cover carbon and alloy steels other than stainless, all methods of manufacture, and a wide range of service temperature.

Steel Tubular Products / 331

Table 6 Compositions of P and T grades of alloy steel pressure pipe and pressure tubes (ASTM)
See Table 10 for compositions of other alloy steel pressure tubes.

Grade(a)	Specification(b)	C	Mn	P	S	Si	Cr	Mo	V
P1, T1	A161, A209, A250, A335	0.10–0.20	0.30–0.80	0.045	0.045	0.10–0.50	...	0.44–0.65	...
T1a	A209, A250	0.15–0.25	0.30–0.80	0.045	0.045	0.10–0.50	...	0.44–0.65	...
T1b	A209, A250	0.14	0.30–0.80	0.045	0.045	0.10–0.50	...	0.44–0.65	...
P2, T2	A213, A335	0.10–0.20	0.30–0.61	0.045	0.045	0.10–0.30	0.50–0.81	0.44–0.65	...
T2	A250	0.10–0.20	0.30–0.61	0.030	0.020	0.10–0.30	0.50–0.81	0.446–0.65	...
T3b	A199, A200, A213	0.15	0.30–0.60	0.030	0.030	0.50	1.65–2.35	0.44–0.65	...
T4	A199, A200	0.15	0.30–0.60	0.030	0.030	0.50–1.00	2.15–2.85	0.44–0.65	...
P5, T5	A199, A200, A213, A335	0.15	0.30–0.60	0.030	0.030	0.50	4.00–6.00	0.45–0.65	...
P5b, T5b	A213, A335	0.15	0.30–0.60	0.030	0.030	1.00–2.00	4.00–6.00	0.45–0.65	...
P5c, T5c	A213, A335	0.12(d)	0.30–0.60	0.030	0.030	0.50	4.00–6.00	0.45–0.65	...
P7, T7	A199, A200, A213, A335	0.15	0.30–0.60	0.030	0.030	0.50–1.00	6.00–8.00	0.45–0.65	...
P9, T9	A199, A200, A213, A335	0.15	0.30–0.60	0.030	0.030	0.25–1.00	8.00–10.00	0.90–1.10	...
P11, T11	A199, A200, A213	0.15	0.30–0.60	0.030	0.030	0.50–1.00	1.00–1.50	0.44–0.65	...
T11	A250	0.15	0.30–0.60	0.030	0.020	0.50–1.00	1.00–1.50	0.44–0.65	...
P11	A335	0.15	0.30–0.60	0.025	0.025	0.50–1.00	1.00–1.50	0.44–0.65	...
P12, T12	A213, A335	0.15	0.30–0.61	0.045	0.045	0.50	0.80–1.25	0.44–0.65	...
P15	A335	0.15	0.30–0.60	0.030	0.030	1.15–1.65	...	0.44–0.65	...
T17	A213	0.15–0.25	0.30–0.61	0.045	0.045	0.15–0.35	0.80–1.25	...	0.15
P21, T21	A199, A200, A213, A335	0.15	0.30–0.60	0.030	0.030	0.50	2.65–3.35	0.80–1.06	...
P22, T22	A199, A200, A213	0.15	0.30–0.60	0.030	0.030	0.50	1.90–2.60	0.87–1.13	...
P22	A335	0.15	0.30–0.60	0.025	0.025	0.50	1.90–2.60	0.87–1.13	...
P24	A405	0.15	0.30–0.60	0.030	0.030	0.10–0.35	0.80–1.25	0.87–1.13	0.15–0.25
P91(e), T91	A199, A200, A213, A335	0.08–0.12	0.30–0.60	0.020	0.010	0.20–0.50	8.00–9.50	0.85–1.05	0.18–0.25
18Cr-2Mo(f)	A213	0.025	1.00	0.040	0.030	1.00	17.5–19.5	1.75–2.50	...

(a) Grades listed are found in one or more of the specifications listed adjacent to the grade grouping. (b) Specifications A 161, A209, A 250, and A 405 call for open hearth, electric furnace, or basic oxygen steel; specifications A 199, A 200, A 213 and A 335 call for electric furnace steel or steel made by other primary processes approved by the purchaser. (c) Where a single value is shown, it is a maximum limit, except for vanadium, where a single value denotes a minimum limit. (d) Grade P5c shall have a titanium content of not less than four times the carbon content and not more than 0.70%, or a niobium content of eight to ten times the carbon content; Grade T5c shall have a titanium content of not less than four times the carbon content and not more than 0.70%. (e) 0.030–0.070 N; 0.40 max Ni; 0.040 max Al; 0.06–0.10 Nb. (f) 0.035 max N; 1.00 max Ni + Cu; grade 18Cr-2Mo shall have Ti + Nb = 0.20 + 4(C + N) min, 0.80 max.

The situation with respect to the composition and strength required for the various end-uses may be summarized as follows: The simplest specification for pipe is ASTM A 53, which covers material that is produced for all ordinary purposes such as conveying fluids under low pressure that may be bent, coiled, or flanged. Chemical composition is specified, and strength requirements consist of a tension test in addition to the hydrostatic test or optional nondestructive test for seamless pipe.

The chemical composition limits for ASTM, API, and CSA carbon and alloy steel pipe are shown in Table 4 (carbon steel) and Table 5 (alloy steel). The composition limits for the alloy steel pressure pipe covered in several ASTM specifications are shown in Table 6. It should be noted that all of the specifications in Table 3 involve carbon (Table 4) or alloy (Table 5) steels with no more than a medium carbon content (0.50% max). The strength requirements (Table 7) are based on as-rolled material in the common qualities, with the classes being graded upward to annealed, normalized, normalized and tempered, or quenched and tempered properties for the higher qualities. Because ductility is considered desirable at higher strength levels, carbon content is restricted. To achieve such strength levels, alloying elements may be added or heat treatment may be employed. When welding is to be used as a method of joining, there is additional reason for carbon restriction.

Common Types of Pipe

The following brief descriptions concern the end-uses of some of the more common types of pipe.

Standard pipe is standard weight, extra strong, and double extra strong welded or seamless pipe of ordinary finish and dimensional tolerances, produced in sizes up to 660 mm (26 in.) in nominal diameter, inclusive. This pipe is used for fluid conveyance and some structural purposes.

Conduit pipe is welded or seamless pipe intended especially for fabrication into rigid conduit, a product used for the protection of electrical wiring systems. Conduit pipe is not subjected to hydrostatic testing unless so specified. It may be galvanized or bare, as specified. It is furnished in standard weight pipe sizes from 6 to 150 mm (¼ to 6 in.) in lengths of approximately 3 to 6 m (10 or 20 ft), with plain ends or threaded ends, as specified. Conduit pipe comes under the scope of Underwriters Laboratories Inc. Specification UL6 or ANSI standards.

Piling pipe is welded or seamless pipe for use as piles, with the cylinder section acting as a permanent load-carrying member or as a shell to form cast-in-place concrete piles. ASTM A 252 includes three grades, which have different minimum tensile strengths, a variety of diameters, ranging from 150 to 610 mm (6 to 24 in.), and a variety of wall thicknesses. Ends may be plain or beveled for welding.

Pipe for nipples is standard weight, extra strong, or double extra strong welded or seamless pipe, produced for the manufacture of pipe nipples. Pipe for nipples is commonly produced in random lengths with plain ends, in nominal sizes from 3 to 300 mm (⅛ to 12 in.). Close outside-diameter tolerances, sound welds, good threading properties, and surface cleanliness are essential in this product. Pipe for oil country tubular good couplings must be manufactured from seamless pipe. It is commonly coated with oil or zinc and is well protected in shipment.

Transmission or line pipe is welded or seamless pipe currently produced in sizes ranging from 3 mm (⅛ in.) nominal to 1.2 m (48 in.) actual outside diameter and is used principally for conveying gas or oil. Transmission pipe is produced with ends plain, threaded, beveled, grooved, flanged, or expanded, as required for various types of mechanical couplers or for welded joints. When threaded ends and couplings are required, recessed couplings are supplied. Transmission pipe is covered by API specification 5L and CSA specification Z245.1.

Water main pipe is welded or seamless steel pipe used for conveying water for municipal and industrial purposes. Pipe lines for such purposes are commonly designated as flow mains, transmission mains, force mains, water mains, or distribution mains. The mains are generally laid underground. Sizes range from 40 to 2450 mm (1½ to 96 in.) in nominal diameter in a variety of wall thicknesses. Pipe is produced with ends suitably prepared for mechanical couplers, with plain ends beveled

Table 7 Minimum tensile properties of carbon and alloy steel pipe (ASTM, API, and CSA)

Minimum elongation values may be different for different test directions, test specimen sizes, or pipe sizes; applicable specification gives additional details.

Specification	Pipe manufacturing process	Grade, class	Tensile strength MPa	ksi	Yield strength MPa	ksi
ASTM specifications						
A 53 (type F)	Furnace butt welded	...	310	45	170	25
A 53 (type E or S)	Welded(a) or seamless	A	330	48	205	30
		B	415	60	240	35
A 106	Seamless	A	330	48	205	30
		B	415	60	240	35
		C	485	70	275	40
A 134	Arc welded	...	See material grade specifications.			
A 135	Welded(a)	A	330	48	205	30
		B	415	60	240	35
A 139	Arc welded	A	330	48	205	30
		B	415	60	240	35
		C	415	60	290	42
		D	415	60	317	46
		E	455	66	360	52
A 211	Spiral arc welded	...	See material grade specifications.			
A 252	Welded or seamless	1	345	50	205	30
		2	415	60	240	35
		3	455	66	310	45
A 333	Welded or seamless	1	380	55	205	30
		3	450	65	240	35
		4	415	60	240	35
		6	415	60	240	35
		7	450	65	240	35
		8	690	100	515	75
		9	435	63	315	46
		10	550	80	450	65
A 335	Seamless	See Table 12.				
A 381	Submerged arc welded	Y35	415	60	240	35
		Y42	415	60	290	42
		Y46	435	63	316	46
		Y48(b)	460	67	330	48
		Y48(c)	430	62	330	48
		Y50(b)	475	69	345	50
		Y50(c)	440	64	345	50
		Y52(b)	495	72	360	52
		Y52(c)	455	66	360	52
		Y56(b)	515	75	385	56
		Y56(c)	490	71	385	56
		Y60(b)	540	78	415	60
		Y60(c)	515	75	415	60
		Y65(b)	550	80	450	65
		Y65(c)	535	77	450	65
A 405	Seamless	See Table 12.				
A 523	Welded(a) or seamless	A	330	48	205	30
		B	415	60	240	35
A 524	Seamless	I(b)	415	60	240	35
		II(c)	380	55	205	30
A 587	Welded(a)	...	330	48	207	30
A 589	Furnace butt welded	...	310	45	170	25
A 589	Welded(a) or seamless	A	330	48	207	30
		B	415	60	240	35
A 671	Arc welded	...	See material grade specifications.			
A 672	Arc welded	...	See material grade specifications.			
A 691	Arc welded	...	See material grade specifications.			
A 714	Welded or seamless	I, II	482	70	345	50
		III	450	65	345	50
		IV	400	58	250	36
		V	380	55	275	40
A 714 (type F)	Furnace butt welded					
A 714 (types E or S)	Welded(c) or seamless	V, VI	450	65	315	46
		VII	450	65	310	45
		VII	482	70	345	50
API specifications						
2B	Arc welded	...	Specified by purchaser			
5CT (casing or tubing)	Welded(a) or seamless	H-40	415	60	275	40
		J-55	517	75	380	55
		N-80	690	100	552	80
5CT (casing)	Welded(a) or seamless	K-55	655	95	380	55
5CT	Seamless	C-75	655	95	517	75
5CT	Welded(d) or seamless	L-80	655	95	552	80
		C90	690	100	620	90
		C-95	724	105	655	95
5CT (casing)	Welded(d) or seamless	Q125	930	135	860	125
5CT (casing)	Seamless	P-110	860	125	760	110
5CT (tubing)	Seamless	P-105	825	120	725	105
5D (drill pipe)	Seamless	X-95	725	105	655	95
		G-105	793	115	725	105
		S-135	1000	145	930	135
		E-75	690	100	517	75
5L	Welded(e) or seamless	A25	310	45	172	25
5L	Welded(f) or seamless	A	330	48	207	30
		B	415	60	240	35
		X42	413	60	289	42
		X46	435	63	317	46
		X52	455	66	358	52
		X56	489	71	385	56
		X60	517	75	413	60
		X65	530	77	450	65
		X70	565	82	482	70
		X80	620	90	552	80
CSA specification						
CAN3-Z245.1	Welded(g) or seamless	172	310	45	172	25
		207	331	48	207	30
		241	415	60	240	35
		290	415	60	290	42
		317	435	63	317	46
		359	455	66	360	52
		386	490	71	385	56
		414	517	75	415	60
		448	530	77	450	65
		483	565	82	485	70
		550	620	90	550	80

(a) Electric resistance. (b) For wall thicknesses 9.5 mm (3/8 in.) and under. (c) For wall thicknesses over 9.5 mm (3/8 in.). (d) Electric resistance or electric flash. (e) Furnace butt, electric resistance, electric flash, or electric induction. (f) Submerged arc, electric resistance, electric flash, electric induction, or gas metal arc. (g) Furnace butt, electrical resistance, or submerged-arc weld

---

for welding, or with bell and spigot joints for field connection. Pipe is produced in double random lengths of about 12 m (40 ft), single random lengths of about 6 m (20 ft), or in specified lengths. When required, it is produced with a specified coating or lining, or both. Steel water main pipe 150 mm (6 in.) in nominal diameter and larger is covered by Standard C200 of the American Water Works Association.

Oil country tubular goods is a collective term applied in the oil and gas industries to three kinds of pipe used in oil wells: drill pipe, casing, and tubing. These products conform to API specifications 5CT (casing and tubing) and 5D (drill pipe). The chemical composition requirements are contained in Tables 4 and 5, and the strength requirements are found in Table 7.

Drill pipe is used to transmit power by rotary motion from ground level to a rotary drilling tool below the surface and to convey flushing media to the cutting face of the tool. Drill pipe is produced in sizes ranging from 60 to 170 mm (2 3/8 to 6 5/8 in.) in outside diameter. Size designations refer to actual outside diameter and weight per foot. Drill pipe is usually upset, either internally or externally, or both, and is prepared to accommodate welded-on types of joints.

Casing is used as a structural retainer for the walls of oil or gas wells, to exclude undesirable fluids and to confine and conduct oil or gas from productive subsurface strata to ground level. Casing is produced in sizes from 115 to 500 mm (4 1/2 to 20 in.)

Table 8 Specifications for carbon and alloy steel pressure tubes (ASTM)

Specification	Product
A 161	Seamless low-carbon or carbon-molybdenum steel still tubes for refinery service
A 178(a)	Resistance-welded carbon and carbon-manganese steel boiler tubes
A 179(a)	Seamless cold-drawn low carbon steel heat-exchanger or condenser tubes
A 192(a)	Seamless carbon steel boiler tubes for high-pressure service
A 199(a)	Seamless cold-drawn intermediate alloy steel heat-exchanger or condenser tubes
A 200	Seamless intermediate alloy steel still tubes for refinery service
A 209(a)	Seamless carbon-molybdenum alloy steel boiler and superheater tubes
A 210(a)	Seamless medium-carbon steel boiler and superheater tubes
A 213(a)	Seamless ferritic and austenitic alloy steel boiler, superheater, and heat-exchanger tubes
A 214(a)	Resistance-welded carbon steel heat-exchanger and condenser tubes
A 226(a)	Resistance-welded carbon steel boiler and superheater tubes for high-pressure service
A 250(a)	Resistance-welded ferritic alloy steel boiler or superheater tubes
A 254	Copper-brazed steel tubes for general engineering use as fluid lines
A 334(a)	Welded and seamless carbon and alloy steel tubes for low-temperature service
A 423(a)	Resistance-welded or seamless low-alloy steel tubes for applications where corrosion resistance is important
A 539	Resistance-welded coiled steel tubes for gas or fuel oil lines
A 556(a)	Seamless cold-drawn carbon steel feed water heater tubes
A 557(a)	Resistance-welded carbon steel feed water heater tubes
A 692	Seamless medium-strength carbon-molybdenum alloy steel boilers and superheater tubes

(a) This specification is also published by ASME, which adds an "S" in front of the "A" (for example, SA 178)

Table 9 Compositions of carbon steel pressure tubes (ASTM)

Specification	Material grade	C	Mn	P	S	Si
A 161	Low carbon(b)(c)	0.10–0.20	0.30–0.80	0.048	0.058	0.25 max
A 178	A(b)	0.06–0.18	0.27–0.63	0.050	0.060	...
	C(b)	0.35	0.80	0.050	0.060	...
	D(b)	0.27	1.00–1.50	0.030	0.015	0.10 min
A 179	(b)	0.06–0.18	0.27–0.63	0.048	0.058	...
A 192	(b)(c)	0.06–0.18	0.27–0.63	0.048	0.058	0.25 max
A 210	A-1(b)(c)	0.27	0.93	0.048	0.058	0.10 min
	C(b)(c)	0.35	0.29–1.06	0.048	0.058	0.10 min
A 214	(b)	0.18	0.27–0.63	0.050	0.060	...
A 226	(b)	0.06–0.18	0.27–0.63	0.050	0.060	0.25 max
A 254	(b)	0.05–0.15	0.27–0.63	0.050	0.060	...
A 334	1(b)	0.30	0.40–1.06	0.05	0.06	...
	6(b)	0.30	0.29–1.06	0.048	0.058	0.10 min
A 539	(b)	0.15	0.63	0.050	0.060	...
A 556	A2(b)	0.18	0.27–0.63	0.048	0.058	...
	B2(b)	0.27	0.29–0.93	0.048	0.058	0.10 min
	C2(b)	0.30	0.29–1.06	0.048	0.058	0.10 min
A 557	A2(b)	0.18	0.27–0.63	0.050	0.060	...
	B2(b)	0.30	0.27–0.93	0.050	0.060	...
	C2(b)	0.35	0.27–1.06	0.050	0.060	...

(a) Maximum limit, except for silicon, where a single value is shown. (b) Open hearth, electric furnace, or basic oxygen. (c) Killed

in outside diameter. Size designations refer to actual outside diameter and weight per foot. Ends are commonly threaded and furnished with couplings, but may be prepared to accommodate other types of joints.

Tubing is used within the casing of oil wells to conduct oil and gas to ground level. It is produced in sizes from 26 to 114 mm (1.05 to 4.50 in.) in outside diameter, in several weights per foot. Ends are threaded for special integral-type joints or fitted with couplings and may or may not be upset externally.

Water well pipe is a collective term applied to four types of pipe that are used in water wells and that conform to ASTM A 589: type I, drive pipe; type II, reamed and drifted pipe; type III, driven well pipe; and type IV, casing pipe. The chemical composition and strength requirements are given in Tables 4 and 7, respectively.

Drive pipe is used to transmit power from ground level to a rotary drill tool below the surface and to convey flushing media to the cutting face of the tool. The lengths of pipe have specially threaded ends that permit the lengths to butt inside the coupling. Drive pipe is produced in nominal sizes of 150, 200, 300, 350, and 400 mm (6, 8, 12, 14, and 16 in.) in outside diameter.

Driven well pipe is threaded pipe in short lengths used for the manual driving of a drill tool or for use with short rigs. It may be furnished in random lengths ranging from 0.9 to 1.8 m (3 to 6 ft) or in random lengths ranging from 1.8 to 3.0 m (6 to 10 ft).

Casing is used both to confine and conduct water to ground level and as a structural retainer for the walls of water wells. It is produced as threaded pipe in random lengths from 4.9 to 6.7 m (16 to 22 ft) and in sizes from 90 to 220 mm (3½ to 8⅝ in.) in outside diameter. In western water well practice, welded strings are sometimes used.

Reseamed and drifted pipe is similar to casing, but is manufactured and inspected in a manner that assures the well driller that the pipe string will have a predetermined minimum diameter sufficient to permit unrestricted passage of pumps or other equipment through the string. The pipe is threaded and is produced in a wider range of sizes—25 to 300 mm (1 to 12 in.) in outside diameter—than standard water well casing.

Pressure pipe, as distinguished from pressure tubes, is a commercial term for pipe that is used to convey fluids at elevated temperature or pressure, or both, but that is not subjected to the external application of heat. This commodity is not differentiated from other types of pipe by ASTM, and the applicable specifications are listed with the other types in Table 3. Pressure pipe ranges in size from 3 mm (⅛ in.) nominal to 660 mm (26 in.) actual outside diameter in various wall thicknesses. Pressure pipe is furnished in random lengths, with threaded or plain ends, as required.

Pressure Tubes

Pressure tubes are given a separate classification by both ASTM (Table 8) and the producers. The chemical composition limits are shown in Table 9 for carbon steel and in Tables 6 and 10 for alloy steel; the strength requirements are shown in Tables 11 and 12. Pressure tubes are distinguished from pressure pipe in that they are suitable for the application of external heat while conveying pressurized fluids.

The principal named uses of pressure tubing are given in Table 2. These tubings are produced to actual outside diameter and minimum or average wall thickness (as specified by the purchaser) and may be hot finished or cold finished, as specified.

Double-wall brazed tubing is a specialty tubing confined to small sizes (see ASTM A 254). It is used in large quantities by the automotive industry for brake lines and fuel lines, and by the refrigeration industry for refrigerant lines. It is made by forming copper-coated strip into a tubular section with double walls, using either single-strip or double-strip construction. The tubing is then heated in a reducing atmosphere to join all mating surfaces completely. The resulting product is thus copper coated both inside and outside. When required by the intended service, a tin coating may be supplied. Available sizes range from 3 to 15 mm (⅛ to ⅝ in.) in outside diameter (OD) with wall thickness from 0.64 mm (0.025 in.) for 3 mm (⅛ in.) OD to 0.9 mm (0.035 in.) for 15

Table 10 Compositions of alloy steel pressure tubes (ASTM)
Table 6 gives compositions of the T grades of alloy steel pressure tubes.

Specification	Material grade	Composition, %(a)							
		C	Mn	P	S	Si	Ni	Cu	Others
A 334	3	0.19	0.31–0.64	0.05	0.05	0.18–0.37	3.18–3.82	...	...
	7	0.19	0.90	0.04	0.05	0.13–0.32	2.03–2.57	...	...
	8	0.13	0.90	0.045	0.045	0.13–0.32	8.40–9.60	...	...
	9	0.20	0.40–1.06	0.045	0.050	...	1.60–2.24	0.75–1.25	...
A 423	1	0.15	0.55	0.06–0.16	0.060	0.10	0.20–0.70	0.20–0.60	0.24–1.31 Cr
	2	0.15	0.50–1.00	0.04	0.05	...	0.40–1.10	0.30–1.00	0.10 Mo
A 692(b)	...	0.17–0.26	0.46–0.94	0.045	0.045	0.18–0.37	...	...	0.42–0.68 Mo

(a) Where a single value is shown, it is a maximum limit, except for silicon and molybdenum, where a single value denotes a minimum limit. (b) Hot or cold formed; seamless

mm (⅜ in.) OD (SAE Handbook). The product is generally made to fractional-inch sizes for use with standard compression fittings. It may be sink drawn for the improvement of surface finish and tolerances.

Structural Tubing

Structural tubing is used for the welded, riveted, or bolted construction of bridges and buildings and for general structural purposes. It is available as round, square, rectangular, or special-shape tubing, as well as tapered tubing. These products are covered by ASTM specifications (see Table 13). Structural tubing is produced with a maximum wall thickness of 13 mm (0.500 in.) and with maximum circumferences of 810 mm (32 in.) for seamless tubes and 1220 mm (48 in.) for welded tubes. The chemical composition limits and strength requirements are given in Tables 14 and 15, respectively.

Table 11 Minimum tensile properties of carbon and alloy steel pressure tubes (ASTM)
Table 12 gives minimum tensile properties of the T grades of alloy steel pressure tubes. Minimum elongation values may be different for different test directions, test specimen sizes, or tube sizes; see applicable specifications for details.

Specification	Grade	Tensile strength		Yield strength	
		MPa	ksi	MPa	ksi
A 178	A	No minimum requirements			
	C	415	60	255	37
	D	485	70	280	40
A 179	...	No minimum requirements			
A 192	(a)	325	47	180	26
A 210	A-1	415	60	255	37
	C	485	70	275	40
A 214	...	No minimum requirements			
A 226	(a)	325	47	180	26
A 254	...	290	42	170	25
A 334	1	380	55	207	30
	3	450	65	240	35
	6	415	60	240	35
	7	450	65	240	35
	8	690	100	520	75
	9	435	63	315	46
A 423	...	415	60	255	37
A 539	...	310	45	240	35
A 556, A 557	A2	322	47	180	26
	B2	410	60	255	37
A 692	C2	480	70	280	40
	...	441	64	290	42

(a) For design information only, not specification requirement

Mechanical Tubing

Mechanical tubing includes welded and seamless tubing used for a wide variety of mechanical purposes. It is usually produced to meet specific end-use requirements and therefore is produced in many shapes, to a variety of chemical compositions and mechanical properties, and with hot-rolled or cold-finished surfaces. Most mechanical tubing is ordered to ASTM specifications (see Table 13). Even when customer specifications are used, they usually reference portions of the ASTM Standard.

Mechanical tubing is not produced to specified standard sizes; instead, it is produced to specified dimensions, which may be anything the customer requires within the limits of production equipment or processes. Controlling tolerances are placed on outside diameter and wall thickness for hot-finished tubing and on outside diameter, inside diameter, and wall thickness for cold-finished tubing. Size is usually expressed in inches and decimals, but metric dimensions can be specified. Specifications for size may include any two of the controlling dimensions—outside diameter, inside diameter, and wall thickness—but never all three.

The chemical compositions commonly available in steel mechanical tubing cover a wide variety of standard AISI/SAE grades. In addition to the standard grades, numerous high-strength low-alloy grades and unique chemistries are produced to customer specifications. When the steel used, either carbon or alloy steel, requires normalizing or annealing after welding, such operations become a part of the specification. For example, a type of welded structural tubing made from carbon steel with nominal carbon content of 0.50% is usually furnished normalized.

Welded mechanical tubing is usually made by electric resistance welding, but some is made by the various fusion welding processes. In all instances, the exterior welding flash may be removed (if necessary) by cutting, grinding, or hammering.

Electric resistance welded (ERW) mechanical tubing is made from hot-rolled or cold-rolled carbon steel or from alloy steel strip. The welded tubing can be supplied as-welded, hot finished, or cold finished. Hot-finishing operations usually consist of either a stretch reducing mill or a hot reducing mill (hot sinking). Microstructural and hardness variations associated with the welding are modified by either seam annealing the weld zone or full body normalizing the entire tube. Sizes produced by electric resistance welding range in outside diameter from 6.4 to 400 mm (¼ to 16 in.) and in wall thickness from 1.65 to 17 mm (0.065 to 0.680 in.) for hot-rolled steel and 0.65 to 4.2 mm (0.025 to 0.165 in.) for cold-rolled steel. Hydraulically or electrically driven stretch reducing mills accomplish tube elongation, reduction in diameter, and control of wall thickness on very long mill lengths in essentially a continuous process. Walls as thick

Table 12 Minimum tensile properties of P and T grades of alloy steel pressure pipe and pressure tubes (ASTM)
Table 11 gives minimum tensile properties of other grades of alloy steel pressure tubes. Minimum elongation values may be different for different test directions, test specimen sizes, or tube sizes; see applicable specifications for details.

Specification	Grade(a)	Tensile strength		Yield strength	
		MPa	ksi	MPa	ksi
A 161, A 209, A 250, A 335	P1, T1, P2	380	55	207	30
A 209, A 250	T1a	415	60	220	32
A 209, A 250	T1b	365	53	193	28
A 199, A 200	(b)	415	60	170	25
A 213, A 355	(c)	415	60	207	30
A 405	P24	415	60	205	30
A 405	P24(d)	552	80	345	50

(a) Grades listed are found in one or more of the specifications listed adjacent to the grade grouping. (b) Grades T3b, T4, T5, T7, T9, T11, T21, and T22. (c) Grades T2, T3b, P5, T5, P5b, T5b, P5c, T5c, P7, T7, P9, T9, P11, T11, P12, T12, P15, T17, P21, T21, P22, T22. (d) Heat treated

Table 13 Specifications for carbon and alloy steel structural and mechanical tubing (ASTM)

Specification	Product
Structural tubing	
A 500	Cold formed welded or seamless carbon steel structural tubing in rounds and shapes
A 501	Hot-formed welded or seamless carbon steel structural tubing
A 595	Welded carbon and alloy steel tapered structural tubes
A 618	Hot-formed welded or seamless high-strength low-alloy structural tubing
A 847	Cold-formed welded and seamless high-strength, low-alloy structural tubing with improved atmospheric corrosion resistance
Mechanical tubing	
A 512	Cold-drawn butt-welded carbon steel mechanical tubing
A 513	Resistance-welded carbon and alloy steel mechanical tubing
A 519	Seamless carbon and alloy steel mechanical tubing

Table 15 Minimum tensile properties of carbon and alloy steel structural tubing (ASTM)

Minimum elongation values may be different for different test directions, test specimen sizes, or tube sizes.

Specification	Grade	Tensile strength MPa	ksi	Yield strength MPa	ksi
A 500	A(a)	310	45	228	33
	A(b)	310	45	268	39
	B(a)	400	58	290	42
	B(b)	400	58	317	46
	C(a)	427	62	317	46
	C(b)	427	62	345	50
A 501	...	400	58	250	36
A 595	A	450	65	380	55
	B, C	480	70	410	60
A 618	I, II	485	70	345	50
	III	450	65	345	50
A 847	...	485	70	345	50

(a) Round tubing. (b) Tube shapes other than round

as 18 mm (0.72 in.) are commercially available within limited outside diameter ranges.

Continuous-welded cold-finished mechanical tubing, as its name implies, is tubing that has been hot formed by furnace butt welding and cold finished. It is furnished sink drawn or mandrel drawn and is available in outside diameters up to 90 mm (3½ in.) and wall thicknesses from 0.9 to 13 mm (0.035 to 0.500 in.). The material is low-carbon steel, and the product is, in effect, a form of cold-drawn pipe. Although furnished in a narrower size range than electric resistance welded tubing, it has two advantages: within the available size range, heavier walls are available, and there is no problem with flash.

Seamless mechanical tubing, both hot and cold finished, is available in a wide variety of finishes and mechanical properties. It is made from carbon and alloy steels in sizes up to and including 325 mm (12¾ in.) in outside diameter.

Hot-finished seamless tubing is produced by rotary piercing or extrusion processes. Therefore, it has surfaces similar to the surface regularly produced on hot-rolled steel and, in general, cannot be held to dimensional tolerances as close as those of tubing produced by cold finishing. It is produced in sizes as small as 38 mm (1½ in.) in outside diameter.

Cold-finished mechanical tubing can be produced by means of surface removal or by cold working. Surface removal includes turning, polishing, grinding, or machining. Cold working involves cold reducing to effect changes in cross-sectional dimensions. Drawing over a mandrel (DOM) is the most common method of cold working mechanical tubing. Tubing is prepared for drawing by first pointing it. The end of the tube is mechanically reduced in outside diameter to allow the end to pass through the die for gripping. Tubing must be pickled and lubricated before drawing. Pickling is typically accomplished by hydrochloric or sulfuric acid immersion. A subsequent phosphate immersion coating on the steel surface acts as a binder for the soaplike lubricant. Complete coverage of both the inside and outside diameter is required to prevent galling and chatter during drawing. Several draw benches with 4×10^6 N (1×10^6 lbf) pulling force exist in the industry. Reductions in cross-sectional area of 10 to 30% are common. Drawing is usually followed by a thermal treatment, straightening, and nondestructive inspection.

Cold working and surface removal are used primarily for the purpose of obtaining smaller outside diameters (down to 3.2 mm, or ⅛ in.), better finishes, thinner walls, and closer dimensional accuracy than is possible in hot-finished tubing. In addition, cold-worked tubing offers improved mechanical properties and machinability. Cold working can also be used to produce tubes having cross-sectional shapes other than round. Cold-drawn tubing can be provided in the as-drawn, cold-worked condition, or thermally treated to the desired combination of mechanical properties or microstructure. Typical thermal treatments available are stress relief annealed, normalized, soft annealed, or quenched and tempered.

Square, rectangular, and special-shape sections are produced in welded or seamless tubing, starting with either round tubing of the required diameter and wall thickness or square, unwelded tubing.

Squaring is done in a Turk's head or by other cold-working methods. A Turk's head consists of a frame in which are mounted four rolls with their axes at 90° and adjusted so that the roll surfaces form an opening of the same shape as the section to be formed. The Turk's head is mounted on a draw bench, and the round tubes are passed through the rolls in the same manner as they are passed through dies.

When Turk's head shaping will not provide close enough tolerances on either the outside or inside of uniformly rounded corners, or close enough diagonal dimensions, the forming is customarily accom-

Table 14 Composition of carbon and alloy steel structural tubing (ASTM)

Specification	Tube manufacturing process	Material grade(b)	C	Mn	P	S	Si	Cu	Others
A 500	Cold formed; welded or seamless	A, B	0.26	...	0.04	0.05	...	0.20(c)	...
		C	0.23	1.35	0.04	0.05	...	0.20(c)	...
A 501	Hot formed; welded(d) or seamless	...	0.26	...	0.04	0.05	...	0.20(c)	...
A 595	Welded	A	0.15–0.25	0.30–0.90	0.04	0.05	0.04	...	...
		B	0.15–0.25	0.40–1.35	0.04	0.05	0.04	...	...
		C	0.12	0.20–0.50	0.07–0.15	0.05	0.25–0.75	0.25–0.55	(e)
A 618	Hot formed; welded(d) or seamless	Ia	0.15	1.00	0.15	0.05	...	0.20	...
		Ib	0.20	1.35	0.04	0.05	...	0.20(c)	...
		II	0.22	0.85–1.25	0.04	0.05	0.30	0.20	0.02V
		III	0.23	1.35	0.04	0.05	0.30	...	0.02V(f)
A 847	Cold formed; welded(g) or seamless	...	0.20	1.35	0.04	0.05	0.15	0.20(c)	(h)

(a) Where a single value is shown, it is a maximum limit, except for copper and "others," where a single value denotes a minimum limit. (b) Open hearth, electric furnace, or basic oxygen. (c) If Cr and Si contents are each 0.50% min, the Cu min does not apply. (d) Furnace butt or electric resistance. (e) 0.65 max Ni and 0.30–1.25 Cr; chemistry requirements may also conform to those in Table 1 of specification A 588/588M. (f) 0.05% min Nb may also be added. (g) Electric-resistance welding or electric-fusion. (h) Other elements commonly added include chromium, nickel, vanadium, titanium, and zirconium.

plished by means of die-and-mandrel shaping. The corners of sections processed by this means are approximately 90° arcs and have greater uniformity throughout than is provided by Turk's head shaping. Sections that can be processed in this manner are somewhat limited with respect to diameter, wall thickness, and outside and inside corner radii.

In addition to providing square and rectangular tubing, many producers of welded or seamless tubing supply a variety of special sections, such as oval, streamline, hexagonal, octagonal, round inside and hexagonal or octagonal outside, ribbed inside or out, triangular, round-ended rectangular, and D-shape. Available manufacturing equipment limits the size range and sections available from the various producers. These special sections may be made by passing round tube through Turk's head rolls or through a die with or without the use of a mandrel. Because the sections are special, dies and other tools are not kept available. Therefore, when inquiring about shapes other than square or rectangular, it is essential to give full details regarding dimensions and finish.

REFERENCES

1. T. Altan, S. Oh, and H. Gegel, *Metal Forming Fundamentals and Applications*, American Society for Metals, 1983
2. William T. Lankford, Jr. *et al.*, *The Making, Shaping and Treating of Steel*, 10th ed., Association of Iron and Steel Engineers, 1985

Closed-Die Forgings

Revised by James A. Rossow, Wyman-Gordon Company

FORGING is the process of working hot metal between dies, usually under successive blows and sometimes by continuous squeezing. Closed-die forgings, hot upset parts, and extrusions are shaped within a cavity formed by the closed dies.

Justification for selecting forging in preference to other, sometimes more economical, methods of producing useful shapes is based on several considerations. Mechanical properties in wrought materials are maximized in the direction of major metal flow during working. For complex shapes, only forging affords the opportunity to direct metal flow parallel to major applied service loads and to control, within limits, the refinement of the original ingot structures. Refinement of microstructure is a function of the temperature, the direction, and the magnitude of reduction from the cast ingot to the forged shape. Maximizing the structural integrity of the material permits refinement of design configuration, which in turn permits reduction of weight.

Adequate control of metal flow to optimize properties in complex forging configurations generally requires one or more upsetting operations prior to die forging and may require hollow forging or back extrusion to avoid flash formation at die parting lines. The additional operations and equipment required for hollow forging involve significant cost considerations, which must be justified by improved load capability of the forged part.

Types of Forgings

Forgings are classified in several ways, beginning with the general classifications open die and closed die. They are also classified in terms of the close-to-finish factor, or the amount of stock (cover) that must be removed from the forging by machining to satisfy the dimensional and detail requirements of the finished part (Fig. 1). Finally, forgings are further classified in terms of the forging equipment required for their manufacture, such as, for example, hammer upset forgings, ring-rolled forgings, and multiple-ram press forgings.

Of the various classifications, those based on the close-to-finish factor are most closely related to the inherent properties of the forging, such as strength and resistance to stress corrosion. In general, the type of forging that requires the least machining to satisfy finished-part requirements has the best properties. Thus, a finished part machined from a blocker-type forging usually exhibits mechanical properties and corrosion characteristics inferior to those of a part made from a close-tolerance, no-draft forging.

It should be anticipated that decreasing the amount of stock that must be removed from the forging by machining will almost invariably result in increased die costs. Also, equipment capacity requirements can be increased to produce a forging that is essentially net forged, or closer to finished dimensions. For example, when a window-frame forging was made as a conventional forging, requiring extensive subsequent machining, the frame could be readily produced by blocking and finishing in a 45 MN (5000 tonf) press. However, when the window-frame forging was produced as a close-tolerance, no-draft forging requiring no subsequent machining other than the drilling and re-arming of fastener holes, a 73 MN (8000 tonf) press was required.

Selection of Steel

Selection of a steel for a forged component is an integral part of the design process, and acceptable performance is dependent on this choice. A thorough understanding of the end use of the finished part will serve to define the required mechanical properties, surface finish requirements, tolerance to nonmetallic inclusions, and the attendant inspection methods and criteria.

Steels of forging quality are produced to a wide range of chemical compositions by electric furnace, open hearth, or pneumatic steelmaking processes. With each of the melting and rolling practices, a level of testing and evaluation of quality is exercised. The details of testing and quality evaluation may vary from producer to producer and should be a point of inquiry when forging stock is ordered. Should the designer require it, one or more special quality restrictions can be specified. These will bring into effect additional qualification testing by the producing mill.

Although electric furnaces have the ability to produce forging-quality material by a single slag melting practice, they are generally associated with the production of aircraft-quality alloy steels. High-strength alloy steels, because of their intolerance to inclusions, require this level of refinement to reduce the occurrence of nonmetallics (see the article "High-Strength Low-Alloy Steel Forgings" in this Volume). Occasionally, for higher-reliability applications, it is necessary to specify vacuum arc remelt or electroslag remelt steel. Steels produced to these quality levels are subject to highly restrictive evaluation procedures.

Microalloyed High-Strength Low-Alloy (HSLA) Steels. The use of microalloyed steels has evolved in recent years as an alternative to iron castings for applications such as automotive crankshafts. These steels typically have small additions of vanadium or niobium (0.05 to 0.10%) and can achieve acceptable properties in the non-heat-treated (as-forged) condition. Consequently, these alloys retain the advantages of the forging process while being economically competitive with castings because of the elimination of the heat-treat cycle.

Precipitation-Hardenable Stainless Steels. At the more demanding end of the spectrum for steel forgings is the series of precipitation-hardenable stainless steels such as 15-5PH and PH 13-8Mo. These materials provide an excellent combination of high strength, toughness, and corrosion resistance. These alloys are generally produced as vacuum induction melted plus consumable electrode remelted products. The hard-

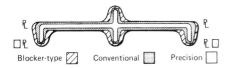

Fig. 1 Schematic composite of cross sections of blocker-type, conventional, and precision forgings

Table 1 Factors to be considered in the design of efficient forgings

Pattern of applied loads

Uniaxial loads. Tensile or compressive, or reversible with changes in operating conditions

Multiaxial or combined loads. Tensile, compressive, shear, bending, torsion, and bearing. These can be either parallel to a central axis or at an angle. Stress concentration should be minimized in design by specifying smooth, contoured fillets at changes of configuration. Where stress concentration cannot be avoided, notch toughness is usually important in material selection.

Cyclic loads. These can be either high- or low-cycle loads.

Sustained loads. If these loads are tensile, they may accelerate stress corrosion. Interference fits and residual stresses may give rise to sustained loading.

Thermal loads. These are caused by changes in temperature.

Load magnitudes and conditions of loading

Magnitudes

Rate of load application. Gradual or impact

Temperature. The major time accumulations should be estimated for minimum, normal, and maximum temperatures.

Environment. Cyclic periods of atmospheric condensation, chemical composition of environment, circumstances of corrosion, abrasion, erosion, or other wear

Life expectancy or reliability

Service life, including downtime, should be estimated. Repairability should be considered.

Special mechanical, physical, or chemical requirements, if any

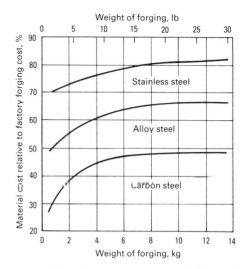

Fig. 2 Cost of steel as a percentage of total cost of forgings

ening mechanism includes solution heat treating and cooling at a rate sufficient to retain solute atoms or compounds in the supersaturated state. The forgings are usually supplied in this condition to facilitate subsequent machining operations. At the final stages of fabrication and assembly, the material is then aged or precipitation hardened to develop its desired strength, toughness, and corrosion resistance. Forgings produced from this family of alloys are typically utilized in the most demanding applications, such as aerospace and marine environments.

Forgeability is the relative ability of a steel to flow under compressive loading without fracturing. Except for resulfurized and rephosphorized grades, most carbon and low-alloy steels are usually considered to have good forgeability. Differences in forging behavior among the various grades of steel are small enough that selection of the steel is seldom affected by forging behavior. However, the choice of a resulfurized or rephosphorized steel for a forging is usually justified only if the forging must be extensively machined; because one of the principal reasons for considering manufacture by forging is the avoidance of subsequent machining operations, this situation is uncommon.

Design Requirements. The selection of a steel for a forged part usually requires some compromise between opposing factors—for example, strength versus toughness, stress-corrosion resistance versus weight, manufacturing cost versus useful load-carrying ability, production cost versus maintenance cost, and the cost of the steel raw material versus the total manufacturing cost of the forging. Material selection also involves consideration of melting practices, forming methods, machining operations, heat treating procedures, and deterioration of properties with time in service, as well as the conventional mechanical and chemical properties of the steel to be forged.

An efficient forging design obtains maximum performance from the minimum amount of material consistent with the loads to be applied, producibility, and desired life expectancy. To match a steel to its design component, the steel is first appraised for strength and toughness and then qualified for stability to temperature and environment. Optimum steels are then analyzed for producibility and finally for economy. Table 1 lists typical requirements needed to produce a cost-effective forging that meets customer specifications. Property requirements can be matched to candidate steels by consulting references on material properties available for standard steels.

Failure analyses are a useful data source for matching the properties of steels to requirements. Failure of a component can occur during operation within the design stress range. One cause of premature failure is lack of proper orientation of a critical design stress with the preferred grain flow of a forging (see the section "Mechanical Properties" in this article).

Unpredicted failure may also occur because of the deterioration of properties with time and service. For example, stress-corrosion cracking, which results from sustained tensile stress, may occur even in a typical ambient atmosphere. Under these conditions, failure is most likely to occur at locations in the forging that coincide with exposed end grain. Failure analyses may uncover other causes of premature failure, such as excessive grain growth, inclusions of nonmetallic impurities, grain flow folding from improper forging practice, lack of a wrought metallurgical structure, and the inadvertent production of stress raisers by machining to an overly sharp fillet or by poor fit in assembly.

The cost of steel as a percentage of the total manufacturing cost of forgings is shown in Fig. 2. These curves are based on an average of many actual forgings that are different in number of forging and heat-treating operations required, cost of steel, quantity, and setup cost. It should not be inferred from these data that an average 14 kg (30 lb) stainless steel forging will cost 34% more than an average carbon steel forging of the same weight.

Assuming a rating of 100% for the production attainable with the most forgeable steel, carbon steels gradually decrease in rating from 100% for 0.35% C to 75% for 0.95% C. The ratings for free-machining steels decrease almost proportionally with increasing carbon content, from 98% for C1117 to 93% for C1141, with one exception: B1112 and C1120 have a rating of 100% for attainable production. Ratings for alloy steels range from 94% for those with the least hardenability to 77% for those with the greatest.

Other processing characteristics likely to influence fabricability and finished-part costs should be considered when selecting steel for forgings. Most forgings require some machining, so the machining characteristics of the steel chosen may be a pertinent cost factor. Depending on mechanical property requirements, response to heat treatment may also be expected to affect costs.

Material Control

After completion of a forging design, there remains the responsibility of ensuring and verifying that the delivered forging will have all the properties and characteristics specified on the forging drawing. This responsibility is vested in material control, which controls all processes employed in the production of a forging, from the selection of raw material to final heat treatment and finish machining. It also establishes manufacturing standards to ensure reproducibility in processing and product uniformity. Material control depends on the proper application of drawings, specifications, manufacturing process controls, and quality assurance programs to satisfy all requirements for metallurgical integrity, mechanical properties, and dimensional accuracy. It can also provide for identification and certification of the forging so that it can be traced throughout processing and in service.

Responsibility for material control is subject to agreement between the purchaser and forging supplier. In many such agreements, the purchaser is responsible for design, material selection, and controls during manufacture; the forging supplier is responsible for maintaining adequate process control and inspection. The liability of the forging supplier is often limited to the replacement of forgings that have failed or have not met specifications.

Material Specification. As part of his responsibility for material selection, the purchaser (or designer) frequently prepares specifications or drawings that indicate the specific locations at which test specimens shall be removed from the forging and that will prescribe the nature and frequency of tests. Standard specifications are used for established materials with well-known mechanical properties. When advanced materials or processes, for which standard specifications are unavailable, are selected, it is the designer's responsibility to ensure that appropriate quality assurance procedures are established and incorporated into the purchase agreement.

Quality Assurance and Quality Control. For the forging supplier, material control encompasses quality assurance and quality control. Quality assurance consists of the tests and surveillance required to assure the user that the properties and attributes required by the design drawings and specifications are attained in all production parts. Quality control is the systematic monitoring of manufacturing variables and the control of these variables to maintain the final property variations within the customer's specification.

A critical forging is so designated on the forging drawing and receives very rigid material control. Forgings not designated as critical are subject to less stringent material control. The criticality of a forging is often limited to one location or to a direction of critical stress, such as an axis, because the remaining portions of the forging are employed at relatively low stress. For such forgings, material control is greater for the portion of the forging under the highest and most critical stress.

Identification. Material control for parts designated as critical includes a system for identification. Typically, each piece is given a serial number, which is recorded and provides a means of identifying the properties of the material, the forging and manufacturing processes, and the results of material-control surveillance. Marking of critical forgings for identification usually includes the part number, serial number, heat number, and forging supplier's symbol. The marking is located where it will remain visible on the part after assembly. If forged-on identification is removed by machining, critical parts are provided with metal tags to maintain identification during shop process-

Table 2 Testing plan for determining mechanical properties of forging material

	Number of tests for(a):							
	9Ni-4Co-0.30C steel (1520–1660 MPa, or 220–240 ksi)(b)				9Ni-4Co-0.45C steel (1790–1930 MPa, or 260–280 ksi)(c)			
Temperature and test	L	LT	ST	Total	L	LT	ST	Total
At −80 °C (−110 °F)								
Tension	2	3	2	7	2	4	2	8
Compression	1	3	1	5	1	4	...	5
Shear	1	3	1	5	1	3	1	5
Bearing e/D = 1.5(d)	1	2	1	4	1	1	1	3
Bearing e/D = 2.0(d)	1	2	1	4	1	4	1	6
At room temperature								
Tension	12	12	12	36	18	17	18	53
Compression	3	3	3	9	2	4	3	9
Shear	3	3	3	9	3	4	2	9
Bearing e/D = 1.5(d)	3	3	3	9	3	3	3	9
Bearing e/D = 2.0(d)	3	3	3	9	3	3	3	9
Modulus of elasticity	1	1	...	2	1	1	...	2
At 150 °C (300 °F)								
Tension	1	3	1	5	2	4	2	8
Compression	1	1	1	3	1	4	...	5
Shear	1	1	1	3	1	3	1	5
Bearing e/D = 1.5(d)	1	1	1	3	1	2	1	4
Bearing e/D = 2.0(d)	...	...	...	0	1	4	1	6
At 260 °C (500 °F)								
Tension	2	3	2	7	...	...	...	0
Compression	1	3	1	5	...	...	...	0
Shear	1	3	1	5	...	...	...	0
Bearing e/D = 1.5(d)	1	2	1	4	...	...	...	0
Bearing e/D = 2.0(d)	2	2	1	5	...	...	...	0
Total number of tests	42	57	40	139	42	65	39	146

(a) L, longitudinal; LT, long transverse; ST, short transverse. (b) Three heats. (c) Four heats. (d) D, hole diameter; e, edge distance measured from the hole center to the edge of the material in direction of applied stress

ing and until the parts are attached to an assembly that is stenciled or otherwise identified.

Routine Production. Material control for the routine production of critical forgings begins by grouping the forgings into lots of a specified maximum size. Results of testing are recorded for the entire lot. Typically, a lot consists of forgings of the same design produced by the same forging producer from the same heat of alloy, heat treated in the same batch, and offered for inspection by (or shipped to) the purchaser at the same time.

Combined Specifications. Material specifications define material properties or requirements. However, they often include one or more sections on processing (such as heat treatment, forging operations, and inspection). Processing specifications may include sections on materials. Therefore, material and process specifications are closely allied and are often combined.

Tests and Test Coupons. Tests contained in the material specifications are intended to provide correlation with, and interpretation of, the behavior of the material in actual use. The dynamic behavior of a full-size structural component can seldom be accurately predicted from simple room-temperature tests on small specimens. Analytical studies coupled with model or full-scale testing can augment simple tests in interpreting the complex behavior of materials.

The types of test specimens and tests specified for quality assurance depend on the conditions imposed on the final component in service. If, for example, a critical forging is to be subjected to large tensile loads, the designer would specify tests to measure fracture toughness and tensile yield strength. For components for elevated-temperature service, tests measuring strength, ductility, and creep at appropriate temperatures would be specified.

The first forging representative of the impression dies, following preliminary trials, is destructively tested for compliance with the specification. After the designer is satisfied that the test results conform to design parameters, the forging design is released for production.

The need to destroy entire forgings for quality assurance during a production run can sometimes be eliminated by designing test coupons, or prolongs (abbreviation for prolongation), into the forging. A prolong is a section of the forging that is removed from the finished forging; it provides test specimens that reflect the forging conditions experienced by the finished part. However, to accurately represent the mechanical properties of the finished part, a prolong must be designed so that it receives the same amount of deformation during forging as the region of the forging it is meant to represent.

The designer must sometimes compromise in selecting the location of test coupons, because of producibility considerations. He must then establish that the test coupons taken from alternative locations will provide data that are meaningful for the intended purpose. Destructive testing permits the designer to determine the degree of correlation between test values measured at

Table 3 Testing plan for determining forging material characteristics

	Number of tests for(b):							
	9Ni-4Co-0.30C steel (1520–1660 MPa, or 220–240 ksi)(c)				9Ni-4Co-0.45C steel (1790–1930 MPa, or 260–280 ksi)(d)			
Temperature and test(a)	L	LT	ST	Total	L	LT	ST	Total
At −80 °C (−110 °F)								
K_{Ic}	...	2	...	2	...	2	...	2
F_{tN} (K_t = 5)	...	2	...	2	...	2	...	2
At −55 °C (−65 °F)								
K_{Ic}	6	6	6	18	9	6	6	21
Exposed K_{Ic}	...	2	...	2	...	2	...	2
At room temperature								
K_{Ic}	...	6	...	6	3	6	2	11
F_{tN} (K_t = 5)	2	2	...	2	...	2	...	2
F_{tN} (K_t = 3)	...	2	...	2	2	...	...	2
Notched fatigue								
(K_t = 2.5, R = 0.06)	...	9	...	9	...	9	...	9
(K_t = 3.0, R = 0.06)	...	15	3	18	9	...	3	12
(K_t = 3.0, R = 0.04)	...	9	...	9	9	...	...	9
Stress corrosion								
K_{Ii}	...	13	...	13	2	15	...	17
Unnotched	...	9	...	9	...	6	...	6
Exposed(e)								
F_t	...	2	...	2	...	2	...	2
K_{Ic}	...	2	...	2	...	2	...	2
K_{Ii}	...	3	...	3	...	3	...	3
At 150 °C (300 °F)								
K_{Ic}	...	2	...	2	...	2	...	2
Total number of tests	6	86	9	101	34	59	11	104

(a) K_{Ic}, plane-strain fracture toughness; K_{Ii}, delayed fracture in salt water; F_{tN}, notched tensile specimen; R, minimum stress divided by maximum stress; K_t, theoretical stress-concentration factor; F_t, tensile specimen. (b) L, longitudinal; LT, long transverse; ST, short transverse. (c) Three heats. (d) Four heats. (e) Exposure for the 9Ni-4Co-0.30C steel: 5000 h, 275 MPa (40 ksi), 230 °C (450 °F); for the 9Ni-4Co-0.45C steel: 5000 h, 275 MPa (40 ksi), 150 °C (300 °F)

plan encompasses a total of 285 individual tests.

Table 3 outlines a plan for tests of special properties, including plane-strain fracture toughness, axial fatigue, notched tensile strength, and resistance to stress corrosion. The plan applies to the testing of seven heats of 9Ni-4Co steel at four different temperatures. The tests are designed to evaluate the effects of grain direction; some are conducted on specimens previously exposed under load (275 MPa, or 40 ksi) to temperatures of 150 to 230 °C (300 to 450 °F) for 5000 h. In total, 205 specimens are tested.

Wrought Structure and Ductility. Another aspect of material control ensures that the final forging has undergone sufficient plastic deformation to achieve the wrought structure necessary for development of the mechanical properties on which the design was based. Although some plastic deformation is achieved during the breakdown of a cast ingot into a forging billet, far more is imparted during the closed-die forging process. Material control for high-strength forgings may require determination of the mechanical properties of the forging billet, as well as those of the forging.

A measure of ductility or toughness is determined by measuring the reduction in area obtained in transverse tension test specimens. When corresponding tests are made of transverse and longitudinal specimens taken from forgings heat treated to the same strength level, it is possible to compare the mechanical properties of billet stock and forgings and to estimate the proportion of the final wrought metallurgical structure contributed by each.

Ductility and the Amount of Forging Reduction. A principal objective of material control is to ensure that optimum mechanical properties will be realized in the finished forging. The amount of reduction achieved in forging has a marked effect on ductility, as shown in Fig. 3, which compares ductility in the cast ingot, the wrought (rolled) bar or billet, and the forging. The curves in Fig. 3(a) indicate that when a wrought bar or billet is flat forged in a die, an increase in forging reduction does not affect longitudinal ductility, but does result in a gradual increase in transverse ductility. When a similar bar or billet is upset forged in a die, an increase in forging reduction results in a gradual decrease in axial ductility and a gradual increase in radial ductility.

The ductility of cast ingots varies with chemical composition, melting practice, and ingot size. The ductility of steel ingots of the same alloy composition also varies, depending on whether they were poured from air-melted or vacuum arc remelted steel. When starting with a large ingot of a particular alloy, it is at times practical to roll portions of the ingot to various billet or bar sizes with varying amounts of forging re-

critical locations and those measured at the locations to be monitored on each forging in production. Testing of coupons from each production forging should indicate not only whether the forging has exceeded the minimum values assigned to it but also the magnitudes of the values attained at critical locations.

Test Plans. Frequently, specifications are prepared from the results of tests on laboratory specimens, because the cost and time required for full-scale testing are usually prohibitive. Test plans for evaluation of the mechanical properties of two high-strength 9Ni-4Co steels used in aircraft service at temperatures ranging from −45 °C (−50 °F) to more than 205 °C (400 °F) are given in Tables 2 and 3, which illustrate the range and number of tests required for a very extensive type of evaluation.

As indicated in Table 2, test plans for mechanical properties include tension, compression, shear, and bearing strength tests; the effect of grain orientation is evaluated by testing specimens representative of the longitudinal, long-transverse, and short-transverse directions, as required. In addition to room-temperature tests, specimens are tested at −80, 150, and 260 °C (−110, 300, and 500 °F). The

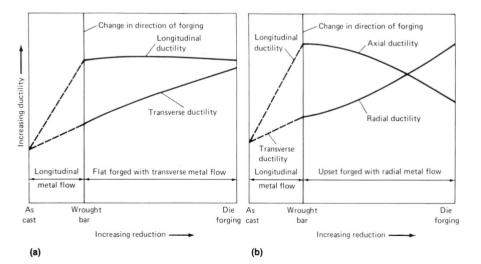

Fig. 3 Metal flow in forging. Effect of extent and direction of metal flow during forging on ductility. (a) Longitudinal and transverse ductility in flat-forged bars. (b) Axial and radial ductility in upset-forged bars

duction. The minimum amount of reduction is not standard, but is seldom less than 2:1 (ratio of ingot section area to billet section area). Reduction of steel ingot to billet is usually much greater than 2:1. In contrast, some heat-resisting alloy forgings are forged directly from a cast ingot. Often, it is not feasible to prepare billets for forgings that are so large that they require the entire weight of an ingot.

The amount of forging reduction represented by wrought metallurgical structures is best controlled by observation and testing of macroetch and tension test samples taken from completed forgings. These samples permit exploration of critical areas and, generally, of the entire forging. They are selected from the longitudinal, long-transverse, and short-transverse grain directions, as required. Etch tests permit visual observation of grain flow. Mechanical tests correlate strength and toughness with grain flow.

Grain Flow. Macroetching permits direct observation of grain direction and contour and also serves to detect folds, laps, and reentrant flow. By macroetching suitable specimens, grain flow can be examined in the longitudinal, long-transverse, and short-transverse directions. Macroetching also permits evaluation of complete sections, end-to-end and side-to-side, and a review of uniformity of macrograin size. Figure 4 illustrates grain flow in a representative forged part.

Grain Size and Microconstituents. Metallographic examination, using a microscope, is best suited for examining questionable areas revealed by macroetching, for measuring grain size, and for determining the nature and amount of microconstituents.

Fatigue Strength. Fatigue tests are used in material control under the following conditions and for the following purposes:

- Laboratory testing of small samples for the development or qualification of material
- Laboratory testing of complete components or subassemblies for design development
- Surveillance of components or assemblies in the field to ensure their continuing reliability in service

The laboratory fatigue testing of small samples for the qualification or development of material is done by standard methods. Test specimens are obtained either from mill products or from closed-die forgings, as required. Standard samples are small enough to permit selection from many locations within a forging and to correlate with various directions of grain flow. Testing is generally done at room temperature in air, although testing at higher or lower temperatures and in special atmospheres is feasible.

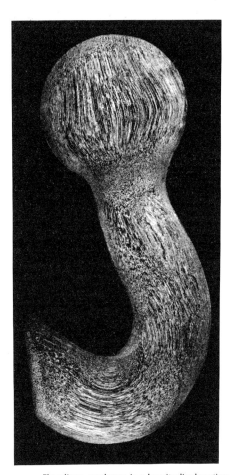

Fig. 4 Flow lines, as shown in a longitudinal section taken through a hook that was forged from 4140 steel

The application of small-scale laboratory fatigue testing to the analysis of components or assemblies introduces additional variables. One is the effect of surface condition. The curves in Fig. 5(a) demonstrate that the fatigue strength of steel specimens varies markedly, depending on whether the surface is polished, machined, hot rolled, or as-forged. The steel tested was an unidentified wrought low-alloy steel heat treated to 269 to 285 HB, equivalent to a tensile strength of 876 MPa (127 ksi) and a yield strength of 696 MPa (101 ksi). Sample preparation required that the specimens be machined and polished after heat treatment and that rolling or forging precede heat treatment. For a fatigue life of 10^6 cycles, the fatigue limit was 395 MPa (57 ksi) for the ground specimens, 315 MPa (46 ksi) for the machined specimens, 205 MPa (30 ksi) for the as-rolled specimens, and only 150 MPa (22 ksi) for the as-forged specimens.

The curves in Fig. 5(b) apply to steels with tensile strength ranging from 345 to 2070 MPa (50 to 300 ksi) and are approximations from several independent investigations. Sample preparation for as-forged or decarburized specimens at the 965 MPa (140 ksi) tensile strength level include 4140-type steels rough machined from bar stock, heated to approximately 900 °C (1650 °F) in a gas-fired muffle for 20 to 30 min, very lightly swaged from an original 7.47 mm (0.294 in.) diameter to a final diameter of 7.16 mm (0.282 in.), and air cooled. Heat treatment consisted of austenitizing in a salt bath at approximately 830 °C (1525 °F) for 45 min, oil quenching, tempering in air for 1 h at approximately 620 °C (1150 °F), and water quenching. Forging and heat treating produced a surface decarburized to a depth of about 0.064 mm (0.0025 in.). These specimens exhibited a fatigue strength, at 10^6 cycles, of about 310 MPa (45 ksi), compared with 470 MPa (68 ksi) for samples that were not forged but were machined or polished and free of decarburization. Decarburization lowers the strength levels obtained by heat treatment.

Laboratory control of surface condition is difficult to duplicate in the quantity production of forged components. Therefore, the fatigue strength of full-size components varies over a wider range than that of small specimens, because of variations in surface condition.

Fracture Toughness. The brittle fracture of forgings and other components as the result of crack propagation at stress levels considerably below the yield strength of the metal has led to widespread investigation of fracture characteristics and methods of evaluating fracture toughness. Results of these investigations are of major importance to material control, especially with respect to the development of tests for evaluating fracture toughness on which standards for material control can be based.

In the area of laboratory tests and analytical techniques, major emphasis has been placed on the development of dependable methods for evaluating the strength of metals that contain cracks or cracklike defects. Specifically, interest has centered on methods for determining plane-strain fracture toughness. Forged components are evaluated by testing small specimens removed from selected locations on the forging that are representative of the various grain directions. One test procedure comprises the bend testing of notched and fatigue-cracked specimens in a neutral environment (ASTM E 399). The objective of this test is to obtain a lower limiting value of fracture toughness that can be used to estimate the relationship between stress and defect size in a metal under service conditions in which high constraint would be expected.

In the test procedure referred to, a test specimen with a chevron notch is suitably precracked in fatigue. It is then tested in a bend test fixture provided with support rolls that rotate and move apart slightly to permit rolling contact and virtually eliminate the friction effect. The specimen is subjected to three-point bending, and the imposed load versus displacement change

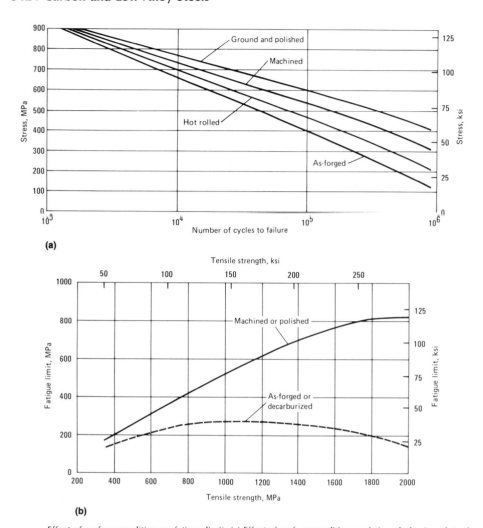

Fig. 5 Effect of surface condition on fatigue limit. (a) Effect of surface condition on fatigue behavior of steels that were hardened and tempered to 269 to 285 HB. (b) Effect of tensile strength level and surface condition of steel on fatigue limit; strengths are given for 10^6 cycle fatigue life.

across the notch is recorded on an autographic recorder. Fracture toughness is rated by a calculated parameter, the critical stress intensity.

End-Grain Exposure. Lowered resistance to stress-corrosion cracking in the long-transverse and short-transverse directions is related to end-grain exposure. A long, narrow test specimen sectioned so that the grain is parallel to the longitudinal axis of the specimen has no exposed end grain, except at the extreme ends, which are not subjected to loading. In contrast, a corresponding specimen cut in the transverse direction has end-grain exposure at all points along its length. End grain is especially pronounced in the short-transverse direction on die forgings designed with a flash line. Consequently, forged components designed to reduce or eliminate end grain have better resistance to stress-corrosion cracking.

Residual Stress. The sustained tensile stress at the surface of a forging that contributes to stress-corrosion cracking is the total of applied and residual stresses. When the residual stress constitutes a significant percentage of the total stress, it should be reduced or eliminated.

Common sources of residual tensile stresses include quenching, machining, and poor fit in assembly. Each can be suitably modified to reduce or eliminate tensile stresses, especially those present in an exposed surface. For example, drastic quenching places the surface of a heat-treatable alloy in a state of compression and the core in a state of tension. Furthermore, the compressed surface may be entirely removed during rough machining, exposing the tension-stressed core material. This hazard can be avoided by quenching after, rather than before, rough machining. In some applications, a surface in tension is placed in compression by shot peening.

Hydrogen-stress cracking occurs without corrosion; therefore, its initiation is not confined to exterior surfaces in contact with a corrosive medium. It can start at any suitable nucleus, such as an inclusion or void, as well as at a surface notch or other irregularity. Hydrogen-stress cracking at the interior is described as hydrogen embrittlement or hydrogen flaking.

Hydrogen-stress cracking has been observed, studied, and brought under control in most high-strength steels. The modern practice of vacuum melting can reduce residual hydrogen to negligible amounts. A hydrogen content of 3 to 6 ppm in an air-melted steel can be readily lowered to 0.6 to 1 ppm by vacuum arc remelting.

Provided that the initial hydrogen content of the steel is acceptably low, material control procedures must ensure that hydrogen pickup is avoided in all subsequent processing, including forging, heat treating, hot salt bath descaling, pickling, and plating. During forging, steels develop a surface scale and a decarburized surface layer, both of which are subsequently removed by grit-blasting and machining. Unless the steel is acid pickled, there is no possibility of hydrogen pickup.

Many of the critical parts made from steel forgings are protected by a coating of cadmium. Steel parts heat treated to strength levels higher than 1655 MPa (240 ksi) are especially sensitive to hydrogen pickup; if they are coated with cadmium, the coating is deposited in vacuum. Parts heat treated to strength levels lower than 1655 MPa (240 ksi) can be cadmium plated electrolytically, provided that a titanium-containing plating bath is used and the parts are subsequently baked at about 190 °C (375 °F) for 12 h.

Mechanical Properties

A major advantage of shaping metal parts by rolling, forging, or extrusion stems from the opportunities such processes offer the designer with respect to the control of grain flow. The strength of these and similar wrought products is almost always greatest in the longitudinal direction (or equivalent) of grain flow, and the maximum load-carrying ability in the finished part is attained by providing a grain-flow pattern parallel to the direction of the major applied service loads when, in addition, sound, dense, good-quality metal of sufficiently fine grain size has been produced throughout.

Grain Flow and Anisotropy. Metal that is rolled, forged, or extruded develops and retains a fiberlike grain structure aligned in the principal direction of working. This characteristic becomes visible on external and sectional surfaces of wrought products when the surfaces are suitably prepared and etched. The fibers are the result of elongation of the microstructural constituents of the metal in the direction of working. Therefore, the phrase "direction of grain flow" is commonly used to describe the dominant direction of these fibers within wrought metal products.

In wrought metal, the direction of grain flow is also evidenced by measurements of mechanical properties. Strength and ductil-

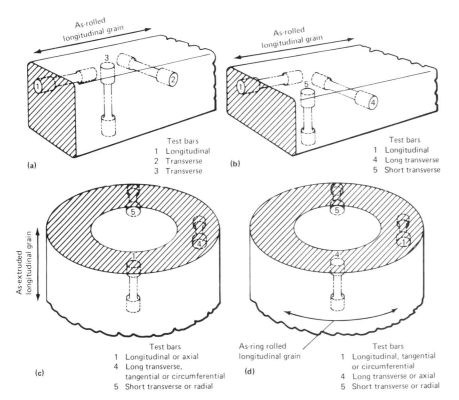

Fig. 6 Anisotropy and mechanical properties in forgings. Schematic views of sections from (a) square rolled stock, (b) rectangular rolled stock, (c) a cylindrical extruded section, and (d) a ring-rolled section, illustrating the effect of section configuration or forging process, or both, on the longitudinal direction in a forging

ity are almost always greater in the direction parallel to that of working. The characteristic of exhibiting different strength and ductility values with respect to the direction of working is referred to as mechanical anisotropy and is exploited in the design of wrought products. Although the best properties in wrought metals are most frequently the longitudinal (or equivalent), properties in other directions may yet be superior to those in products not wrought, that is, in cast ingots or in forging stock taken from a lightly worked ingot.

The square rolled section shown schematically in Fig. 6(a) is anisotropic with respect to the average mechanical properties of test bars positioned as shown. The average mechanical properties of longitudinal bar 1 are superior to those of transverse bars 2 and 3. Mechanical properties are equivalent for bars 2 and 3 because the section is square, which implies equal reduction in section in both transverse directions.

Mechanical anisotropy is also found in rectangular sections (Fig. 6b), in cylinders (Fig. 6c), and in rolled rings (Fig. 6d). Again, the best strength properties are, on the average, those of the longitudinal, as in test bar 1. Flat rolling of a section such as Fig. 6(a) to a rectangular section such as Fig. 6(b) enhances the average long-transverse properties of test bar 4 when compared with the short-transverse properties of test bar 5. Thus, such rectangular sections exhibit anisotropy among all three principal directions—longitudinal, long transverse, and short transverse. A design that employs a rectangular section such as Fig. 6(b) involves the properties in all these directions, not just the longitudinal. Thus, the longitudinal, long-transverse, and short-transverse service loads of rectangular sections are analyzed separately. The same concept can be applied to cylinders, whether extruded or rolled; longitudinal direction changes with the forging process used, as indicated in Fig. 6(c) and (d).

Anisotropy in High-Strength Steel. Although all wrought metals are mechanically anisotropic, the effects of anisotropy on mechanical properties vary among different metals and alloys. For example, a vacuum-melted steel of a given composition is generally less mechanically anisotropic than a conventionally killed, air-melted steel of the same composition. Response to etching to reveal the grain flow characteristic of anisotropy also varies. Metals with poor corrosion resistance are readily etched, while those with good corrosion resistance require more corrosive etchants and extended etching times to reveal grain flow.

In general, fatigue properties are markedly affected by the relation of flow-line direction to direction of stresses from applied loads: When flow lines are perpendicular to load stresses, a stress-raising effect is produced.

Closed-Die Forgings / 343

Table 4 gives test results from tensile specimens for two different kinds of alloy steel taken from a heat-treated forging. In another study of these directional effects on 4340 and 4330 modified steel, at different strength levels, large closed-die forgings approximately 305 mm (12 in.) in diameter and 1.2 m (4 ft) in length were bored to leave a wall thickness of 28.5 to 38 mm (1⅛ to 1½ in.) and then heat treated. These forgings had been formed from a 380 mm (15 in.) square billet by a 22 kN (5000 lbf) blacksmith hammer, a 110 kN (25 000 lbf) blocking hammer, and a 160 kN (35 000 lbf) finishing hammer.

Three sets of specimens were cut, as indicated in Table 5. One set was longitudinal, or parallel to the forging axis; another was transverse, or normal to the forging axis away from the parting line; and the third set was vertical across the parting line, with the critical section in the center of the parting-line region. The results of tests summarized in Table 5 show differences in properties resulting from orientation. Differences can be expected in all forgings regardless of the steel used or its strength after heat treatment.

A total of 7 tensile, 10 notch-tensile, 15 impact, and 5 bend specimens were tested for each steel in each orientation. Standard 12.83 mm (0.505 in.) round tensile specimens were used. Notch-tensile specimens of 7.62 mm (0.300 in.) nominal outside diameter had a circumferential 60° V-notch that removed 50% of the section area. The root radius of the notch was less than 0.025 mm (0.001 in.). Bend specimens 11.1 mm (0.438 in.) in diameter were 143 mm (5⅝ in.) long. The restricted-bend method was used; the specimen was supported at each end and centrally loaded.

Design Stress Calculations

Example 1: Redesign of a Forged Manual Gear-Shift Lever Mechanism to Overcome Unacceptable Fatigue Failure of Original Forging. In a hand-operated gear-shift lever mechanism, field performance was generally satisfactory, but some fatigue failures occurred in the 3.18 mm (⅛ in.) radius shown in Fig. 7. The lever is pivoted in a ball socket, and the maximum load transmitted at end A is 1.8 kN (400 lbf). The steel is 1049 hardened to 269 to 285 HB, with properties as given in Table 6.

As a simple lever, the stress at section A would be calculated as follows:

Bending moment = M
= (force at A)(lever arm)
= (400 lbf)(7 in.)
= 2800 lbf · in. or 315 N · m   (Eq 1)

$$\text{Stress} = \frac{\text{Bending moment}}{\text{Section modulus}}$$

= 262 MPa or 38 ksi   (Eq 2)

Table 4 Effect of orientation of test specimen on mechanical properties of steel forgings

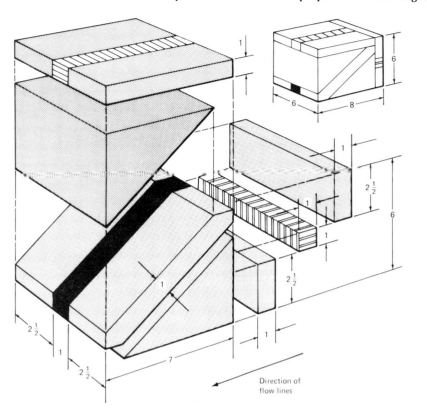

Note: Dimensions in inches.

Longitudinal | Diagonal | Transverse

Specimen orientation(a)	Yield strength MPa	ksi	Tensile strength MPa	ksi	Elongation, in 50 mm (2 in.), %	Reduction in area, %	Hardness, HB
Steel 5046							
Longitudinal	585	85	820	119	25.5	64.1	255
Diagonal	565	82	820	119	22.5	52.0	255
Transverse	600	87	825	120	11.5	20.5	262
Steel 4340							
Longitudinal	1005	146	1095	159	19.0	54.5	341
Diagonal	1000	145	1090	158	17.5	49.3	331
Transverse	1000	145	1095	159	13.5	29.9	341

(a) Specimen orientation described in illustration

Because an abrupt change in size occurs at section A, a stress concentration factor must be applied. This value is obtained from Fig. 8. Values applied to the curves are as follows: from the dimensions in Fig. 7, $r = 3.18$ mm (0.125 in.), $d = 22.2$ mm (0.875 in.), and $h = 4.75$ mm (0.187 in.). Therefore:

$$\frac{r}{d} = \frac{3.18}{22.2} = 0.143 \qquad (Eq\ 3)$$

and

$$\frac{h}{r} = \frac{4.75}{3.18} = 1.5 \qquad (Eq\ 4)$$

The stress concentration factor for these values, or r/d and h/r, is $K_f = 1.52$, from Fig. 8. Therefore, the calculated mean stress would be 1.52×262 MPa = 400 MPa (or 1.52×38 ksi = 58 ksi). This stress is well below the yield strength (696 MPa, or 101 ksi). However, the occurrence of failures under service conditions suggested fatigue as the mode of failure.

To investigate actual stresses, a strain gage was installed in the radius, and observations made during operation revealed stresses from 275 to 345 MPa (40 to 50 ksi). However, when gears were clashed, transient load impulses as long as 2½ s in duration resulted in stress as high as 620 MPa (90 ksi) at a frequency of 80 Hz. Therefore, about 200 stress reversals could occur several times during each hour the machine was in operation. This condition explains the possibility of the occurrence of 10^6 cycles in a few thousand hours of actual operation. It was necessary, therefore, to consider a means of reducing the stress because of fatigue.

A draftsman's layout of the related parts revealed that the diameter could be increased to 25 mm (1 in.) and the radius to 4.75 mm (0.187 in.). Also, instead of the as-forged surface, it was expedient to machine the radius as shown in the following stress analysis (Fig. 7).

In Eq 2, bending moment is unchanged, but the section modulus changes because of increased diameter. Then, the stress after redesigning is:

$$\text{Stress} = \frac{\text{Bending moment}}{\text{Section modulus}}$$
$$= 175\ \text{MPa or 25 ksi} \qquad (Eq\ 5)$$

In using Fig. 8:

$$\frac{r}{d} = \frac{4.75}{25} = 0.187 \qquad (Eq\ 6)$$

and

$$\frac{h}{r} = \frac{4.75}{4.75} = 1 \qquad (Eq\ 7)$$

Therefore, the stress concentration factor with the 4.75 mm (0.187 in.) radius would be reduced slightly, to 1.43, and the calculated mean stress after redesigning would be 1.43×175 MPa = 250 MPa (or 1.43×25 ksi = 36 ksi).

Life Expectancy. The foregoing computations have shown the calculated stresses for two different configurations at section A. It was determined experimentally that peak stresses of 620 MPa (90 ksi) were observed in the original condition. The peak stresses with the enlarged diameter and radius are reduced proportionally to the calculated stresses. Therefore, the new peak stress would be:

$$\frac{250}{400}(620) = 390\ \text{MPa or 57 ksi} \qquad (Eq\ 8)$$

The following discussion compares the effect of these maximum stresses on the life expectancy of the part.

In Fig. 5, the values for completely reversed cycles of stress for steel at a hardness range of 269 to 285 HB are plotted against life in cycles for as-forged, hot-rolled, machined, and ground surfaces. The problem at hand concerns as-forged versus machined surfaces. The forged surface would be decarburized and subject to the usual surface defects. At 10^6 cycles, as before, the allowable stress for a forged surface as read from Fig. 5 is 150 MPa (22 ksi), while for the machined surface it is 315 MPa (46 ksi)—both considerably less than the yield strength (690 MPa, or 100 ksi).

Figure 9 shows the relationship between maximum and mean stress for forged and machined surfaces of the example part. At

Table 5 Effect of location and orientation on mechanical properties of specimens cut from production forgings

All values shown are averages. A total of 7 tensile, 10 notch-tensile, 15 impact, and 5 bend specimens were tested for each steel. Standard 12.8 mm (0.505 in.) round tensile specimens were used. Notch-tensile specimens of 7.6 mm (0.300 in.) nominal outside diameter had a circumferential 60° V-notch that removed 50% of the section area. The root radius was less than 0.025 mm (0.001 in.). Bend specimens were 11.11 mm (0.4375 in.) in diameter by 143 mm (5⅝ in.) long. The data were obtained with the restricted-bend method, in which the specimen is supported at each end and the load is applied at the center of the specimen. Locations of specimens were longitudinal, transverse, and transverse to parting line (see cut).

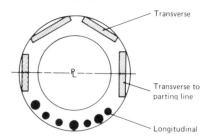

Property	4340 at 41 HRC			4330 (modified) at 43 HRC			4380 (modified) at 48 HRC		
	Longitudinal	Transverse	Transverse to parting line	Longitudinal	Transverse	Transverse to parting line	Longitudinal	Transverse	Transverse to parting line
Tension test with standard unnotched specimens									
Tensile strength									
MPa	1315	1315	1295	1360	1365	1360	1625	1650	1655
ksi	191	191	188	197	198	197	236	239	240
Yield strength, 0.2% offset									
MPa	1240	1250	1215	1260	1270	1270	1400	1400	1420
ksi	180	181	176	183	184	184	203	203	206
Elongation in 50 mm (2 in.), %	14.6	8.2	4.6	14.5	11.0	6.3	10.9	7.9	3.2
Reduction in area, %	48.9	17.1	9.8	47.7	28.0	13.3	44.3	27.2	7.5
Tension test with notched specimens									
Notch strength at									
20 °C, MPa	1935	1840	1710	1935	1915	1880	2040	1935	1750
68 °F, ksi	281	267	248	281	278	273	296	281	254
Notch strength at									
−54 °C, MPa	1990	1930	1730	1980	1970	1910	2015	1655	1315
−65 °F, ksi	289	280	251	287	286	277	292	240	191
Notch ductility at 20 °C (68 °F), %	3.46	2.74	1.76	3.00	2.58	1.72	2.00	1.90	1.36
Notch ductility at −54 °C (−65 °F), %	2.82	2.20	1.22	1.98	2.16	1.58	1.54	1.36	1.04
Charpy impact test									
Impact energy at									
20 °C, J	37	21	15	27	22	16	24	19	12
68 °F, ft · lbf	27.5	15.4	10.9	20.1	16.1	12.2	17.5	14.3	8.5
Impact energy at									
0 °C, J	27	20	13	26	22	17	23	14	10
32 °F, ft · lbf	20.0	14.5	9.6	19.0	16.6	12.8	17.2	13.8	7.2
Impact energy at									
−54 °C, J	23	15	12	22	22	12	19	15	8
−65 °F, ft · lbf	17.3	11.3	8.6	16.4	15.9	8.8	14.0	11.4	5.9
Restricted bend test									
Bend load									
kg	2530	2430	2140	2600	2500	2250	2990	3000	2770
lb	5570	5350	4710	5730	5520	4950	6600	6620	6110
Outside bend angle, degrees	180	38	15	180	67	27	147	51	18

Table 6 Fatigue strength of heat-treated wrought steel of various Brinell hardness ranges

Hardness, HB	Nominal tensile strength		Nominal yield strength		Fatigue strength at 10⁶ cycles — Type of surface							
					Ground		Machined		Hot rolled		Forged	
	MPa	ksi	MPa	ksi	MPa	ksi	MPa	ksi	MPa	ksi	MPa	ksi
160–187	530	77	330	48	225	33	205	30	165	24	125	18
187–207	615	89	415	60	270	39	235	34	170	25	130	19
207–217	680	99	475	69	305	44	255	37	185	27	135	20
217–229	710	103	510	74	315	46	260	38	185	27	135	20
229–241	750	109	550	80	340	49	275	40	195	28	135	20
241–255	785	114	595	86	350	51	290	42	200	29	145	21
255–269	835	121	650	94	380	55	305	44	205	30	145	21
269–285	875	127	695	101	395	57	315	46	205	30	150	22
285–302	930	135	760	110	420	61	340	49	215	31	150	22
302–321	980	142	825	120	440	64	350	51	220	32	160	23
321–352	1040	151	895	130	470	68	365	53	225	33	160	23
352–375	1145	166	1015	147	510	74	395	57	235	34	165	24
375–401	1215	176	1090	158	535	78	405	59	240	35	170	25
401–429	1295	188	1180	171	565	82	425	62	240	35	165	24
429–461	1395	202	1260	183	595	86	440	64	240	35	165	24
461–495	1495	217	1350	196	605	88	450	65	235	34	150	22
495–514	1605	233	1450	210	615	89	455	66	220	32	135	20
514–555	1660	241	1495	217	615	89	450	65	215	31	130	19

zero mean working stress (complete reversal), the values are the same as in Fig. 5 for 10⁶ cycles. At the point in the upper right-hand corner of Fig. 9 where mean and maximum stresses are equal, both stresses are equal to the tensile strength of the steel; therefore, the load is static.

In Fig. 9, the mean working stress of the original forging is represented by line 1, drawn from mean stress 395 MPa (57 ksi) vertically to its intersection with line 3 to predict a maximum stress of 475 MPa (69 ksi) for the forged surface, and drawn to its intersection with line 4 to predict a maximum stress of 565 MPa (82 ksi) for the machined surface. The stress of 565 MPa (82 ksi) is nearly equal to the maximum observed by strain gage analysis and therefore is indicative of failure. The vertical line 2 in Fig. 9 represents stresses in the rede-

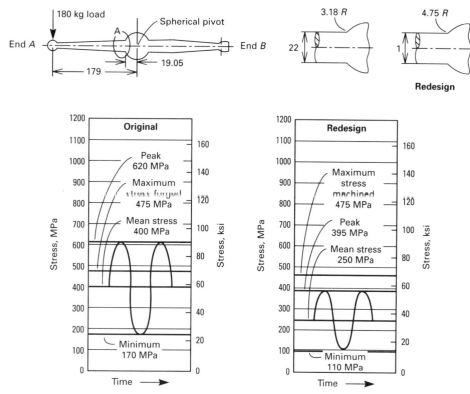

Fig. 7 Original and revised lever designs showing plot of stresses in the original and redesigned part. Dimensions given in millimeters

signed forging for which the mean stress previously calculated is 250 MPa (36 ksi). The values on the appropriate vertical line 2 of Fig. 9 show maximum working stresses of 360 MPa (52 ksi) for the forged surface and 475 MPa (69 ksi) for the machined.

The as-forged surface at 360 MPa (52 ksi) maximum working stress would not ensure satisfactory life, because the recalculated maximum stress is 390 MPa (57 ksi). However, the machined surface with a maximum working stress of 475 MPa (69 ksi) gives a safe margin above the 390 MPa (57 ksi) requirement for design stress. Interpreting these values, the forged surface should have a life expectancy of 10^6 cycles of stress. However, because the load cycle was somewhat uncertain, the machined radius was chosen to obtain a greater margin of safety. The foregoing discussion is diagrammed in Fig. 7 to show the values of calculated mean, peak, and minimum stresses in the problem on the original and the redesigned levers. Redesigning eliminated the failures.

Fundamentals of Hammer and Press Forgings

Many small forgings are made in a die that has successive cavities to preshape the stock progressively into its final shape in the last, or finish, cavity. Dies for large forgings are usually made to perform one operation at a time. The upper half of the die, having the deeper and more intricate cavity, is keyed or dovetailed into the hammer or press ram. The lower half is keyed to the sow block or bed of the hammer or press in precise alignment with the upper die. After being heated, the forging stock is placed in one cavity after another and is thus forged progressively to the final shape.

The parting line is the plane along the periphery of the forging where the striking faces of the upper and lower dies come together. Usually, the die has a gutter or recess just outside of the parting line to receive overflow metal or flash forced out between the two dies in the finish cavity (Fig. 10). More complex forgings may have other parting lines around holes and other contours within the forging that may or may not be in the same plane as the outer parting line.

For the greatest economy, the outer parting line should be in a single plane. When it must be along a contour, either step or locked dies (that is, dies with mating faces that lie in more than one plane) may be necessary to equalize thrust, as shown in Fig. 11. This may increase costs as much as 20%, because of the increased cost of dies and cost increases from processing difficulties in forging and trimming. Sharp steps or drops in the parting line should be limited to about 15° from the vertical in small parts and 25° in large forgings to prevent a tearing instead of a cutting action in trimming off the flash. Locked dies can sometimes be avoided by locating the parting line as shown in Fig. 12(b).

The specification of optional parting lines on forgings to be made in different shops allows these lines to vary from shop to shop. Unless the draft has been removed, this variation may cause difficulties in locating forgings when they are being chucked for subsequent machining. However, shearing the draft is not always an adequate remedy if trimming angles vary. Forgings made in different shops are likely to be more consistent in quality and to have less variation in shape when a definite parting line is specified.

Draft on the sides of a forging is an angle or taper necessary for releasing the forging from the die and is desirable for long die life and economical production. Draft requirements vary with the shape and size of the forging. The effect of part size on the amount of metal needed for draft is illustrated by Fig. 13.

Inside draft is draft on surfaces that tightens on the die as the forging shrinks during cooling; examples are cavities such as narrow grooves or pockets. Outside draft is draft on surfaces such as ribs or bosses that shrink away from the die during cooling. Both are illustrated in Fig. 14, which shows inside draft to be greater than outside draft—the usual relation. Recommended draft angles and tolerances are given in Table 7.

Increased draft, called blend draft or matched draft, may be needed on a side that is not very deep below the parting line in order to blend with a side of the forging of greater height above the parting line, as shown in Fig. 15. Increased draft is sometimes desirable or required in locked dies in order to strengthen the dies or trimmer so as to reduce breakage and cost. This can often be anticipated by sketching the die needed to shape a given forging. Cylindrical, spherical, square, rectangular, and some irregular sections can be forged without draft when the parting line is specified as shown in Fig. 16, but with some additional risk of breakage of dies. Other parts, such as the ends of cylinders, can be forged in locked dies at an angle so as to avoid draft on the ends.

Ribs and Bosses. Forgings that have ribs or bosses at or near the maximum heights recommended in Fig. 10 are usually forged at higher-than-normal temperatures (1230 to 1260 °C, or 2250 to 2300 °F) to ensure flow of the metal into the die cavities. Ribs are more readily formed in the upper die, where the temperature is higher; the lower die extracts heat from the forging, which is in continuous contact with it. The ribs formed by the upper die have better surface quality than those in the lower die, because scale left by the part is more easily removed.

The maximum height of a rib depends on its width at the base and on blocking operations that preshape the stock. Fillets of

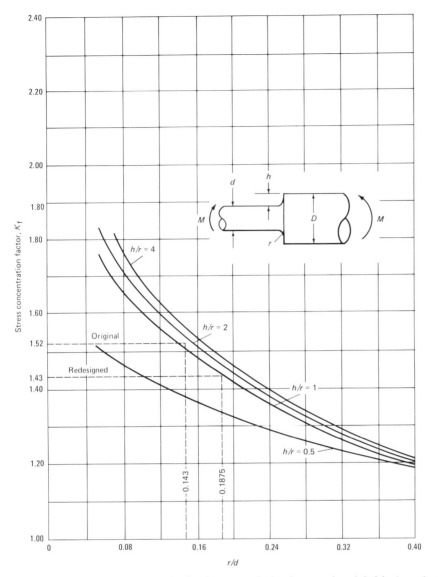

Fig. 8 Stress concentration factor obtained in bending a quenched and tempered steel shaft having a circular fillet

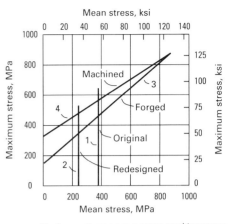

Fig. 9 Maximum stress versus mean working stress for steel with forged and machined surfaces. See Example 1 in text.

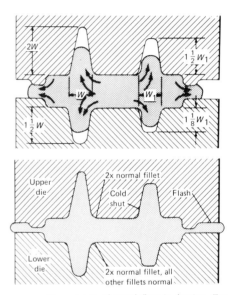

Fig. 10 Two stages of metal flow in forging. Top diagram shows limitations on height of ribs above and below the parting line.

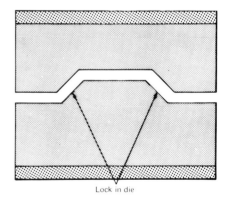

Fig. 11 Locked or stepped dies used to equalize thrust

Table 7 Draft and draft tolerances for steel forgings

Height or depth of draft		Commercial standard		Special standard	
mm	in.	Draft, degree	Tolerance(a), degree	Draft, degree	Tolerance(a), degree
Outside draft					
6.35–12.7	¼–½	...	...	3	+2
19–25	¾–1	5	+3	...	...
>12.7–25	>½–1	...	...	5	+2
>25–76	>1–3	7	+3	5	+3
>76	>3	7	+4	7	+3
Inside draft					
6.35–25.4	¼–1	7	+3	5	+3
>25.4	>1	10	+3	10	+3

(a) The minus tolerance is zero.

minimum size cannot be used at the base of a rib of maximum height if the rib is to be sound and completely filled. Twice the minimum fillet size should be used, a full radius is preferable at the crest of the rib, and draft should be increased if possible.

Fillets and Radii. In forging, some radii wear and grow greater; others become sharper under the combined effects of the pressure of the press or of repeated hammer blows and abrasion. Radii that are too small give the forge die a shearing action and develop high resistance to the flow of the metal, thus increasing die wear and reducing its life. Radii should be as large as the design will permit. Sharp radii in a forging die set up strains that cause the die to

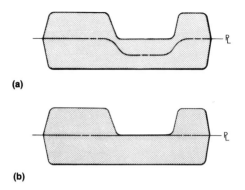

Fig. 12 Orientation of a forging in the die to avoid counterlocked dies and to eliminate draft. Workpiece forged with (a) and without (b) lock in die

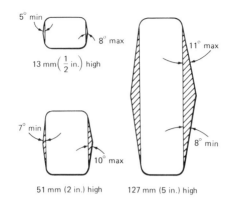

Fig. 13 Effect of part size on the amount of metal needed for draft in a forging

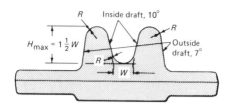

Fig. 14 Definition of inside and outside draft and limitations on the depth of the cavities between the ribs. Typically, inside-draft angles exceed outside-draft angles.

check, thus reducing die life and increasing cost. Very little material can be saved by producing a design that includes sharp internal (fillet) radii.

The effects of small and large radii are illustrated in Fig. 10, which shows a forging during two stages of the operation. In small steel forgings (<0.9 kg, or 2 lb), 3.2 mm (1/8 in.) radii in fillets are considered the absolute minimum. Common practice is to make the fillet radii twice the size of the corner radii. These radii increase in proportion to the size and weight of the forgings, as shown in Fig. 17. For steel forgings of average size (1.4 to 3.6 kg, or 3 to 8 lb), 6.4 mm (1/4 in.) fillet radii are normal. Recommended fillet and corner radii for various heights of rib or boss are given in Fig. 18.

Holes and Cavities. Holes should not obstruct the natural flow of the metal in the forging operation. If cavities lie perpendicular to the direction of metal flow, it may be necessary to add breakdown or blocking operations on the forging billet before it is placed in the dies for forging. Such operations add cost and so must be justified economically. In almost all cases in which a hole is to be punched, a forged cavity is provided to displace the metal in order to relieve the work load of the punch in the later operation. Holes and cavities should not be higher or deeper than the base of the widest cross section when normal fillets and radii are used. If a full radius or a hemispherical shape is allowed at the bottom of a cavity, the maximum depth of the cavity may be 1½ times the width (diameter), as shown in Fig. 14.

On shallow cavities, a draft angle of 7° and the required normal radii can be used. On cavities of maximum depth, the draft should be increased to 10 to 12°.

Minimum Web Thickness. The web in a forging is limited to the thickness at which it gets too cold before forging is completed. If the web gets cold enough in forging to look black, it prevents the part from being brought down to size. Figure 19 shows the limiting minimum web dimension as a function of web size. The minimums shown are generally accomplished on various metals with more or less difficulty—particularly in a problem forging that, in a forging hammer, requires more than a few blows for completion. Many web thicknesses that fall into the band between the two curves can be made only with difficulty and at extra cost. Some that fall below the curve are regularly produced, while the web thicknesses that fall above the upper curve are almost always made without extra cost in forgings that can be completed rapidly. When a web thinner than recommended is required, some advantage may be gained by tapering it 5 to 8° toward the thinnest section, at the center, but the average minimum thickness of the web must be retained to meet strength requirements.

Lightening Holes in Webs. Holes are not always desirable as a means of reducing weight, because of the effect on the strength of the part and stress concentration. Lightening holes are almost always produced by an added operation, and the expense involved is often unjustified. These holes should be used only in neutral or low-stress areas or to reduce cooling cracks and warpage caused by uneven cooling. Holes should be kept away from edges and provided with a strengthening bead to reduce stress concentration, as shown in Fig. 20. A hole near the edge of a forging usually

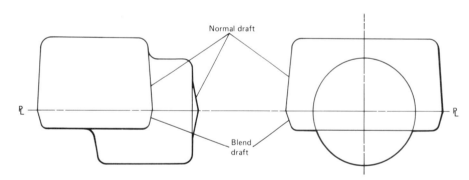

Fig. 15 Normal draft and blend draft in a forging

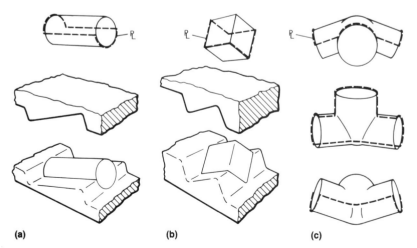

Fig. 16 Selection of parting lines to eliminate draft in (a) cylindrical, (b) square, and (c) tube-shaped forgings

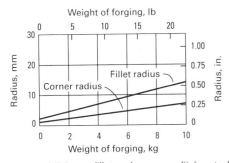

Fig. 17 Minimum fillet and corner radii for steel forgings

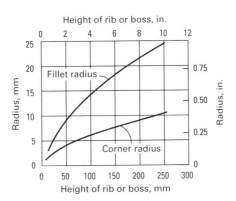

Fig. 18 Recommended fillet and corner radii in relation to height of rib or boss

leaves inadequate material in the highly stressed area around the hole, thus increasing the possibility of crack nucleation and severe distortion.

Scale Control. The reduction of scale formation and the removal of scale during the forging process are important considerations in meeting design requirements economically. Gas- and oil-fueled preheat furnaces should be adjusted to produce a reducing atmosphere and thus minimize the creation of scale. Induction heating can be a useful alternative to fossil fuel furnaces. In some cases, the weight loss due to scaling can be reduced from 2 to 3% for gas or oil furnaces to 0.2% for induction heating.

Tolerances

Forging tolerances, based on area and weight, that represent good commercial practice are listed in Tables 8 and 9. These tolerances apply to the dimensions shown in the illustration in Table 8. In using Tables 8 and 9 to determine the size of the forging, the related tolerances, such as mismatch, die wear, and length, should be added to the allowance for machining plus machined dimensions. On average, the tolerances listed in Tables 8 and 9 conform to the full process tolerance of actual production parts and yield more than 99% acceptance of any dimension specified from Tables 8 and 9. In particular, as shown in Table 10, instances may be found of precise accuracy or rarely as much as ±50% error in the tolerances recommended in Table 9. The values in Table 10 represent the product of a die for one run and not the full range of product between successive resinkings of the die.

Shift or mismatch tolerance allows for the misalignment of dies during forging, as shown in Fig. 21. All angular or flat surfaces of the die will erode or wear away and increase the volume of the forging, depending on the extent to which the forged metal flows over them. This increase is called spread, or die wear, and it must be included in the forging dimensions.

The characteristics of die wear are illustrated in Fig. 22. The part represented was made of 4140 steel, using ten blows in an 11 kN (2500 lbf) board hammer. Tolerances were commercial standard, and the part was later coined to a thickness tolerance of +0.25 mm, −0.000 mm (+0.010 in., −0.000 in.). The die block, 255 by 455 by 455 mm (10 by 18 by 18 in.), was hardened to 42 HRC. After 30 000 forgings had been produced, the die wore as indicated and the dies were resunk.

A range of tolerance is given for mismatch in Table 9. The higher values are to be added to tolerances for forgings that need locked dies or involved side thrust on the dies during forging. On forgings heavier than 23 kg (50 lb), it is sometimes necessary to grind out mismatch defects up to 3.2 mm (1/8 in.) maximum.

Length tolerance in Table 8 refers to variations in shrinkage that occur when forgings are finished at different temperatures. Length tolerance should be applied to overall lengths of forgings as well as to the locations of bosses, ribs, and holes.

Areas of a forging can be coined to hold closer tolerances, provided the metal is free to flow into an adjacent open area of the part. Under these circumstances, the toler-

Table 8 Recommended commercial tolerances on length and location

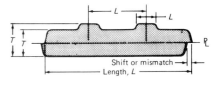

Maximum length of forging		Tolerance on length or location	
mm	in.	mm	in.
150	6	+1.19, −0.79	+0.047, −0.031
380	15	+1.57, −1.19	+0.062, −0.047
610	24	+3.18, −1.57	+0.125, −0.062
910	36	+3.18, −1.57	+0.125, −0.062
1220	48	+3.18, −3.18	+0.125, −0.125
1520	60	+4.75, −3.18	+0.187, −0.125
1830	72	+5.56, −3.18	+0.219, −0.125

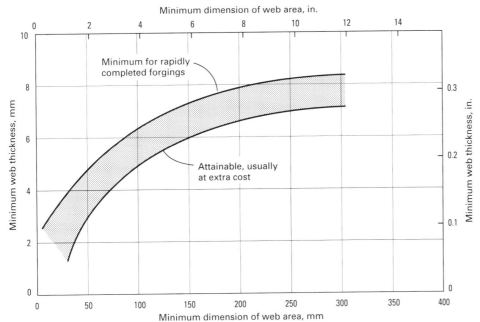

Fig. 19 Recommended minimum web thickness in relation to web dimensions

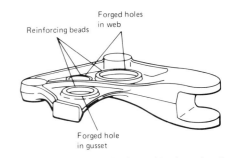

Fig. 20 Lightening holes located in the webs of a forging

350 / Carbon and Low-Alloy Steels

Table 9 Recommended commercial tolerances for steel forgings

Forging size				Tolerance					
Area		Weight		Thickness(a)		Mismatch(a)		Die wear	
10^3 mm²	in.²	kg	lb	mm	in.	mm	in.	mm	in.
3.2	5.0	0.45	1	+0.79, −0.41	+0.031, −0.016	+0.41 to +0.79	+0.016 to +0.031	+0.79	+0.031
4.5	7.0	3.2	7	+1.57, −0.79	+0.062, −0.031	+0.41 to +0.79	+0.016 to +0.031	+1.57	+0.062
6.5	10.0	0.7	1.5	+0.79, −0.79	+0.031, −0.031	+0.41 to +0.79	+0.016 to +0.031	+0.79	+0.031
7.7	12.0	5.5	12	+1.57, −0.79	+0.062, −0.031	+0.41 to +0.79	+0.016 to +0.031	+1.57	+0.062
12.9	20.0	0.9	2	+1.57, −0.79	+0.062, −0.031	+0.41 to +0.79	+0.016 to +0.031	+1.57	+0.062
12.9	20.0	14	30	+1.57, −0.79	+0.062, −0.031	+0.51 to +1.02	+0.020 to +0.040	+1.57	+0.062
24.5	38.0	2	4.5	+1.57, −0.79	+0.062, −0.031	+0.41 to +0.79	+0.016 to +0.031	+1.57	+0.062
24.5	38.0	36	80	+1.57, −0.79	+0.062, −0.031	+0.64 to +1.27	+0.025 to +0.050	+1.57	+0.062
32.3	50.0	3	8	+1.57, −0.79	+0.062, −0.031	+0.51 to +1.02	+0.020 to +0.040	+1.57	+0.062
32.3	50.0	27	60	+1.57, −0.79	+0.062, −0.031	+0.51 to +1.02	+0.020 to +0.040	+1.57	+0.062
32.3	50.0	45	100	+1.57, −0.79	+0.062, −0.031	+0.64 to +1.27	+0.025 to +0.050	+1.57	+0.062
61.3	95.0	5	11	+1.57, −0.79	+0.062, −0.031	+0.51 to +1.02	+0.020 to +0.040	+1.57	+0.062
85.2	132.0	8	17	+1.57, −0.79	+0.062, −0.031	+0.64 to +1.27	+0.025 to +0.050	+1.57	+0.062
107	166.0	33	73	+2.39, −0.79	+0.094, −0.031	+0.76 to +1.52	+0.030 to +0.060	+2.39	+0.094
113	175.0	68	150	+2.39, −0.79	+0.094, −0.031	+0.76 to +1.52	+0.030 to +0.060	+2.39	+0.094
130	201.0	18	40	+1.57, −0.79	+0.062, −0.031	+0.64 to +1.27	+0.025 to +0.050	+1.57	+0.062
155	240.0	23	51.5	+2.39, −0.79	+0.094, −0.031	+0.76 to +1.52	+0.030 to +0.060	+2.39	+0.094
161	250.0	114	250	+2.39, −0.79	+0.094, −0.031	+0.76 to +1.52	+0.030 to +0.060	+2.39	+0.094
171	265.0	27	60	+2.39, −0.79	+0.094, −0.031	+0.76 to +1.52	+0.030 to +0.060	+2.39	+0.094
177	275.0	30	65	+3.18, −0.79	+0.125, −0.031	+1.19 to +2.39	+0.047 to +0.094	+3.18	+0.125
194	300.0	34	75	+3.18, −1.57	+0.125, −0.062	+1.19 to +2.39	+0.047 to +0.094	+3.18	+0.125
194	300.0	159	350	+2.39, −0.79	+0.094, −0.031	+0.76 to +1.52	+0.030 to +0.060	+2.39	+0.094
242	375.0	205	450	+3.18, −0.79	+0.125, −0.031	+1.19 to +2.39	+0.047 to +0.094	+3.18	+0.125
268	415.0	139	306	+3.18, −1.57	+0.125, −0.062	+1.19 to +2.39	+0.047 to +0.094	+3.18	+0.125
339	525.0	340	750	+3.18, −1.57	+0.125, −0.062	+1.19 to +2.39	+0.047 to +0.094	+3.18	+0.125
580	900.0	455	1000	+3.18, −1.57	+0.125, −0.062	+1.19 to +2.39	+0.047 to +0.094	+3.18	+0.125

(a) The illustration in Table 8 shows locations of thickness and mismatch.

ances shown in Fig. 23 can be held in production without difficulty. Over a hot-sheared surface, the coining operation will bring the high points of the serrations within tolerance without removing all the depressions.

On average, Table 9 represents full process tolerance. Figure 24 indicates the relationship of the number of acceptable parts to the process tolerance. In application, the full process tolerance must be derived from the process capability, the full value of which is represented on the chart as full process tolerance within which 100% (theoretically 99.7%) acceptability will result.

For a given process and tolerance, if the designer chooses to narrow the tolerance to two-thirds of its full value, the acceptability will be reduced to 95%. Similarly, a reduction to one-third of the full tolerance would result in acceptability of 68%. Such reduction in tolerance incurs added expense because of the cost of rejected forgings and of 100% inspection to separate acceptable from rejected parts. Although it is possible to control the forging process more closely so as to increase the percentage of acceptable parts, the added expense will also increase the cost of the parts.

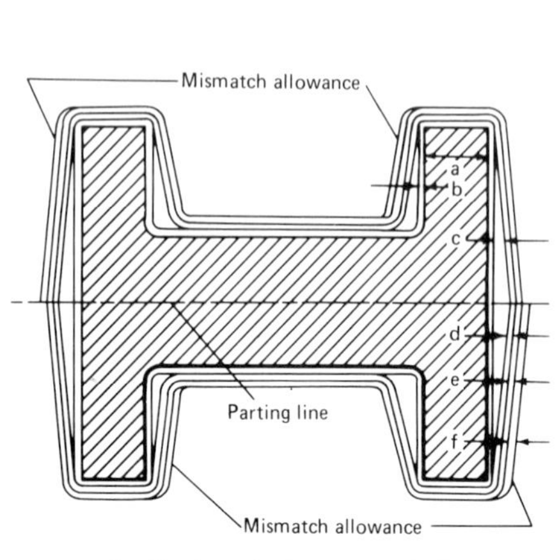

Table 10 Comparison of quality-control data with recommended tolerances for seven production forgings

Values represent the product of a die for one run and not the full range of product between successive resinkings of the die. All tolerances are plus; negative tolerances, zero. Plus signs in the last column indicate that the recommended tolerance is conservative compared with production experience. σ represents standard deviation in distribution of measured dimensions.

Part	Recommended tolerance (from Table 9)		Range of observed variation in length for specific quality-control limits				Difference between tolerance (from Table 9) and 6σ control limit	
			4σ		6σ			
	mm	in.	mm	in.	mm	in.	mm	in.
A	2.39	0.094	1.07	0.042	1.60	0.063	+0.79	+0.031
B	2.39	0.094	1.30	0.051	1.93	0.076	+0.46	+0.018
C	2.39	0.094	1.52	0.060	2.29	0.090	+0.10	+0.004
D	3.18	0.125	1.35	0.053	2.03	0.080	+1.15	+0.045
E	3.18	0.125	2.06	0.081	3.10	0.122	+0.08	+0.003
F	3.18	0.125	2.64	0.104	4.22	0.166	−1.04	−0.041
G	3.18	0.125	2.82	0.111	4.22	0.166	−1.04	−0.041

Fig. 21 Application of tolerances and allowances to forgings. The dimensions are not to scale. a, finish machined; b, machine allowance; c, draft allowance; d, die wear tolerance; e, shrink or length tolerance; f, mismatch allowance

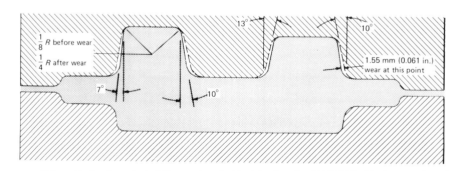

Fig. 22 Schematic showing extent of die wear in a die block hardened to 42 HRC. The block was evaluated for die wear after producing 30 000 forgings of 4140 steel at a rate of 10 blows/workpiece with an 11 kN (2500 lbf) hammer.

Closed-Die Forgings / 351

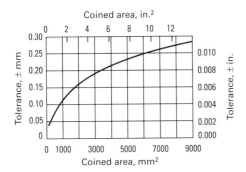

Fig. 23 Tolerances for coining unconfined areas of forgings

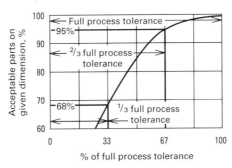

Fig. 24 Relationship of percentage of acceptable parts to process tolerance specified on a dimension

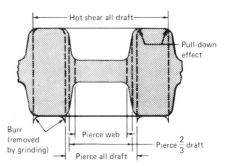

Fig. 25 Pull-down effect in hot shearing, a commonly specified operation for removing the draft from forgings. Pull-down effect, which results from the force of the shearing operation, marks the point where shear begins; burr, where shear ends.

Flash is trimmed in a press with a trimming die shaped to suit the plan view, outline, and side view contour of the parting line. The forging can be trimmed with a stated amount of burr or flash left around the periphery at the parting line. It can also be trimmed flush to the side face of the forging, or some of the draft can be trimmed off, provided the serrations or score marks left by the shearing operation are not an objectionable feature. In most commercial forgings, some draft is sheared away.

Trimming Tolerance. When the trim must cut through the flash only and leave the side of the forging untouched, it is necessary to use a trim dimension that includes burr tolerance, mismatch, draft tolerance, and die wear plus shrink tolerance. When it is satisfactory to trim draft partially, a closer trim tolerance can be held. Burr tolerance, listed in Table 11, applies to the amount of flash that should remain between the side of the forging at the parting line and the outside edge of the trim cut.

Draft tolerance depends on the height of the face having the draft and applies to the dimension across the forging at the parting line. Draft tolerance (plus) is listed in Table 12 for six different heights of the draft face on forgings.

Die wear tolerance allows for an economical life of tools by providing for acceptability of parts after the die has made a quantity of pieces. The tolerance to be added to trimming and draft tolerances to allow for die wear are given in Table 13. The fourth part of the total tolerance is 0.003 mm/mm (0.003 in./in.) of the greatest dimension across the forging at the trim line, to be added as shrink tolerance.

Example 2: Trim Tolerance as Summation of Four Individual Tolerances. The use of Tables 9 and 11 to 13 may be illustrated by the following example. Assume a 2.3 kg (5 lb) forging 127 mm (5 in.) high and 127 mm (5 in.) across at the minimum shearing dimension. The tolerance is the sum of 1.1 mm (0.045 in.) burr tolerance, 4.4 mm (0.175 in.) draft tolerance, 1.0 mm (0.040 in.) die wear tolerance, and 0.38 mm (0.015 in.) shrink tolerance, which equals 7.0 mm (0.275 in.) trim tolerance. Thus, 134.0 mm (5.275 in.) is the largest dimension allowed for trimming when the side of the forging must not be cut, and 127.0 mm (5.000 in.) is the smallest dimension to be allowed. This tolerance is most economical, but in many forgings the tolerance can be held as close as that shown in Table 3 by close control or extra operations.

Hot shearing removes the draft from forgings with a vertical cut that improves dimensional accuracy and leaves a serrated surface. This characteristic surface and accuracy represent an economical preparation for machining, broaching, coining, and accurate chucking in standard chucks. The surface is a substitute for rough machining or flame cutting.

The least expensive trimming operation on forgings is the cold trimming of small parts made from carbon steel of less than 0.50% C or from alloy steel of less than 0.30% C. However, to hot shear off about two-thirds of the draft is economical because the special locating tools required for shearing off all the draft are not needed. The holes and outside of a forging can be sheared in one operation with a combination trimmer and punch for parts where the sheared burr of the holes and the outside are not required to be on opposite sides of the forging.

The force of the shearing operation results in a rounded contour, called pull down, where the shear begins, and in sharp burr edges on the opposite side of the forging where the shear ends, as shown in Fig. 25. The trimming and piercing forces are sometimes great enough to crush or distort the forging. For example, the force of piercing a hole in a thin-wall hub may expand the outside dimension if there is no outer flange to support it. If the wall has a flange around the outside, the piercing forces may crush the lower ends of the hub.

Piercing. Expansion may occur in hot piercing a hole in a cylindrical piece when the unsupported outside wall is 1.5 times the diameter of the pierced hole. Parts with an outside diameter only 1.2 times the diameter of the pierced hole will be distorted. In some parts in which the outer surface is sheared in the same operation that pierces the hole, an outside diameter 1.4 times the diameter of the hole can sometimes be used with good results. Generally, however, 1.57 mm (0.062 in.) should be added to the end of the piece for such parts to allow for crushing of the ends if the ratio of outside diameter to inside diameter is less than 1.5. If the ratio is about 1.6, an allowance of 0.81 mm (0.032 in.) is satisfactory; for ratios of outside to inside diameter greater than 1.75, this allowance is seldom necessary. All of

Table 11 Tolerance on burr for steel forgings

Weight		Trim size(a)		Tolerance	
kg	lb	mm	in.	mm	in.
0.45	1	50	2	+0.79, −0.000	+0.031, −0.000
4.5	10	150	6	+1.57, −0.000	+0.062, −0.000
11	25	200	8	+3.18, −0.000	+0.125, −0.000
45	100	625	25	+6.35, −0.000	+0.250, −0.000

(a) The trim size refers to the greatest distance across the forging at the trim line.

Table 12 Draft increment of trim tolerance for steel forgings

Height of draft face		Tolerance	
mm	in.	mm	in.
6.35	¼	+0.38, −0.00	+0.015, −0.000
12.70	½	+0.51, −0.00	+0.020, −0.000
25	1	+0.9, −0.00	+0.035, −0.000
51	2	+1.5, −0.00	+0.060, −0.000
127	5	+4.5, −0.00	+0.175, −0.000
254	10	+8.9, −0.00	+0.350, −0.000

Table 13 Die wear increment of the trim tolerance for steel forgings

Weight		Trim size(a)		Tolerance	
kg	lb	mm	in.	mm	in.
0.45	1	50	2	+0.79, −0.00	+0.031, −0.000
4.5	10	150	6	+1.19, −0.00	+0.047, −0.000
11	25	200	8	+1.57, −0.00	+0.062, −0.000
45	100	625	25	+3.18, −0.00	+0.125, −0.000

(a) The trim size refers to the greatest distance across the forging at the trim line.

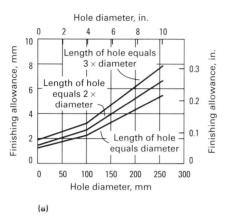
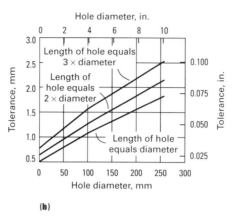

Fig. 26 Allowances (a) and tolerances (b) for hot-pierced holes that are to be broached

Table 15 Typical decarburization limits for steel forgings

Range of section size		Typical depth of decarburization	
mm	in.	mm	in.
<25	<1	0.8	0.031
25–100	1–4	1.2	0.047
100–200	4–8	1.6	0.062
>200	>8	3.2	0.125

given in Table 14. The allowances listed are based on weight and shape, which are significant principally as an index of the amount of probable warpage. In all cases, the depths of cut given in Table 14 will also remove permissible amounts of decarburization. The tolerance and allowance for hot piercing of holes that are to be broached are given in Fig. 26.

The precise effect of changes in section on the amount of distortion in heat treatment is unknown. For small parts, the usual method of overcoming distortion is by jig quenching or marquenching.

Decarburization. Because the loss of carbon greatly lowers the resistance to fatigue, the decarburized skin should be removed by machining in highly stressed areas. Table 15 lists the general limits for forgings of various sizes. The machining allowance is usually greater than the depths listed.

Design for Tooling Economy. Considerable reduction in cost may result from forging designs that incorporate provisions for location of the forging for machining and inspection. Forgings that will be clamped to a faceplate or a machine-tool table should be provided with three bosses under the proposed clamping points to locate the part and to avoid both distortion and the tendency to rock (Fig. 28). When the quantity of parts is too small to justify tooling and holes are to be drilled instead of hot sheared in forging, the holes can be spotted with a cone angle steeper than the drill and about one-half to two-thirds of its diameter. This procedure enables accuracy of locations comparable with length tolerances and squareness predictable from thickness tolerances if the part is not specially located for machining.

Long parts of irregular cross section that may have bow or camber should often be provided with steady-rest locations in the forging design for economical machining (Fig. 28). Machining economy and accuracy often result when a forging that presents a problem in machining is designed as a Siamese forging with multiple parts on the same forgings to be saw-cut apart after machining is completed.

Special opportunities for economy are present in the design of right- and left-hand forgings to be transformed later (Fig. 28). The method shown for avoiding right- and left-hand forging tooling should always be used for the manufacture of small quanti-

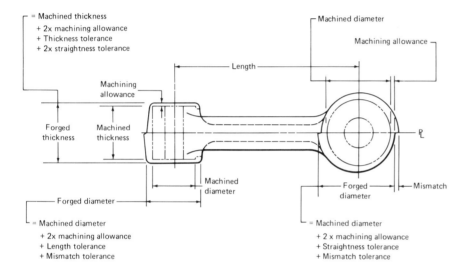

Fig. 27 Computation of surface stock allowances for forgings that are to be machined

Table 14 Machining allowance on each surface for typical steel forgings

Maximum weight of forging		Machining allowance per surface					
		Tail forgings		Flat forgings		Long forgings	
kg	lb	mm	in.	mm	in.	mm	in.
6.8	15	1.52–2.29	0.060–0.090	1.52–3.05	0.090–0.120	3.05	0.120
34	75	2.29–3.05	0.090–0.120	3.05	0.120	3.05	0.120
450	1000	3.05–4.83	0.120–0.190	3.05–6.35	0.120–0.250	4.83–6.35	0.190–0.250

the above ratios refer to dimensions after piercing.

Broaching Allowance. The amount of metal to be removed in broaching a hole in a forging can be controlled economically by hot piercing the hole to close tolerances or, less economically, by machining the hole before broaching. Figure 26 is a graph of the forging stock and tolerances to be allowed in a hole that is to be broached after hot piercing. In using Fig. 26, the allowance value at a given diameter is subtracted from the minimum broach diameter. This is the high limit for the size of the hot-pierced hole. The hole dimension is then specified as the high limit with a tolerance of plus zero, minus the tolerance read from Fig. 26.

Allowance for Machining

Surfaces that are to be machined must be forged oversize externally and undersize internally by an amount equal to the sum of applicable tolerances and machining allowance, as shown in Fig. 27. On machined surfaces parallel to the parting line, the stock allowance is affected by the tolerances for thickness and straightness. Machined surfaces perpendicular or nearly perpendicular to the parting plane are affected by the length tolerances, straightness, and mismatch tolerances. The minimum machining allowance, set somewhat arbitrarily at 1.52 mm (0.060 in.) on small forgings and as high as 6.35 mm (0.250 in.) on large forgings, is

Closed-Die Forgings / 353

Table 16 Length tolerances for unforged stems of upset forgings

Maximum length		Minimum tolerance	
mm	in.	mm	in.
150	6	+1.59, −0.00	+0.062, −0.000
250	10	+2.39, −0.00	+0.094, −0.000
500	20	+3.18, −0.00	+0.125, −0.000
>500	>20	+3.96, −0.00	+0.156, −0.000

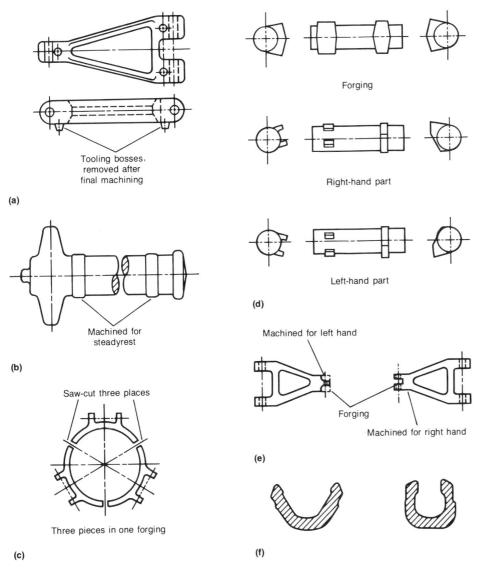

Fig. 28 Workpiece configurations that emphasize cost-effectiveness and versatility in terms of tooling design. (a), (b), and (c) Design for economy in tooling. (d) and (e) Combination of right-hand and left-hand parts in the same forging. (f) An unforgeable part that was bent to its final shape after preliminary forging

heading tool. Parting-line clearance is required in gripper dies for tangential clearance in order to avoid undercut and difficulty in the removal of the forging from the dies (Fig. 29a).

Tolerances for shear-cut ends have not been established. Figure 29(b) shows a shear-cut end on a 32 mm (1¼ in.) diam shank. Straight ends can be produced by torch cutting, hacksawing, or abrasive wheel cutoff at a higher cost than shearing.

Corner radii should follow the contours of the finished part, with a minimum radius of 1.6 mm (¹⁄₁₆ in.). Radii are not required at the outside diameter of the upset face, but can be specified as desired. Variations in thickness of the upset require variations in radii, as shown in Fig. 29(c), because the origin of the force is farther removed and the die cavity is more difficult to fill. When a long upset is only slightly larger than the original bar size, a taper is advisable instead of a radius.

Fillets can conform to the finished contour in most cases. The absolute minimum should be 3.2 mm (⅛ in.) on simple upsets (Fig. 29c).

Tolerances for all upset-forged diameters are generally +1.6 mm, −0 mm (+¹⁄₁₆ in., −0 in.) except for thin sections of flanges and upsets relatively large in ratio to the stock sizes used, where they are +2.4 mm, −0 mm (+³⁄₃₂ in., −0 in.). The increase in tolerances over the standard +1.6 mm, −0 mm (+¹⁄₁₆ in., −0 in.) is sometimes a necessity because of variations in size of hot-rolled mill bars, extreme die wear, or complexity of the part. Tolerances for unforged stem lengths are given in Table 16.

Draft angles may vary from 1 to 7°, depending on the characteristics of the forging design. Draft is needed to release the forging from the split dies; it also reduces the shearing of face surfaces in transfer from impression to impression. For an upset-forged part that requires several operations or passes, the dimensioning of lengths is determined on the basis of the design of each individual pass or operation.

Design of Specific Parts. A study of designs already being manufactured by the hot upset method of forging may serve as a guide for the development of similar applications. The following examples illustrate some typical exceptions to general design rules.

It can be seen from Fig. 30 that the size of stock required to produce the part deter-

ties, but may not be economical for large production. Figure 28 also shows a part that could not be forged without a secondary bending operation.

Design of Hot Upset Forgings

Hot heading, upset forging, or, more broadly, machine forging consists primarily of holding a bar of uniform cross section, usually round, between grooved dies and applying pressure on the end in the direction of the axis of the bar by using a heading tool so as to upset or enlarge the end into an impression of the die. The shapes generally produced include a variety of enlargements of the shank or multiple enlargements of the shank and reentrant angle configurations. Transmission cluster gears, pinion blanks, shell bodies, and many other shaped parts are adapted to production by the upset machine forging process. This process produces a looped grain flow of major importance for gear teeth. Simple, headed forgings can be completed in one step, while some that have large, configured heads or multiple upsets may require as many as six steps. Upset forgings weighing less than 0.45 to about 225 kg (1 to 500 lb) have been produced.

Machining Stock Allowances. The standard for machining stock allowance on any upset portion of the forging is 2.39 mm (0.094 in.), although allowances vary from 1.57 to 3.18 mm (0.062 to 0.125 in.), depending on the size of the upset, the material, and the shape of the part (Fig. 29a). Mismatch and shift of dies are each limited to 0.41 mm (0.016 in.) maximum. Mismatch is the location of the gripper dies with respect to each other, as shown in Fig. 29(a). Shift refers to the relation of the dies to the

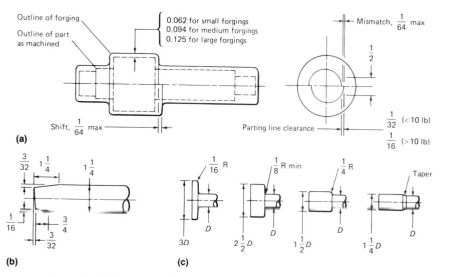

Fig. 29 Machining stock allowances for hot upset forgings. (a) Hot upset forging terminology and standards. (b) Probable shape of shear-cut ends. (c) Variation of corner radius with thickness of upset. These parts are the simplest forms of upset forgings. Dimensions given in inches

Table 17 Recommended size limitations for hot extrusions made of carbon and alloy steel

Inside diameter		Maximum depth of hole		Outside diameter range	
mm	in.	mm	in.	mm	in.
11	7/16	50	2	14–19	9/16–3/4
13	1/2	100	4	17–24	21/32–15/16
14	9/16	125	5	19–27	47/64–1 1/16
16	5/8	140	5 1/2	21–29	13/16–1 5/32
17.5	11/16	145	5 3/4	22–32	7/8–1 1/4
19	3/4	150	6	24–35	5/16–1 3/8
22	7/8	155	6 1/8	27–40	1 1/16–1 9/16
25	1	160	6 1/4	31–44	1 7/32–1 23/32
38	1 1/2	160	6 3/8	44–57	1 3/4–2 1/4
50	2	165	6 1/2	57–76	2 1/4–3
64	2 1/2	170	6 5/8	71–95	2 25/32–3 3/4
76	3	170	6 3/4	84–115	3 5/16–4 1/2
100	4	180	7	110–145	4 3/8–5 3/4
125	5	180	7 1/8	140–180	5 7/16–7
150	6	185	7 1/4	165–210	6 29/64–8 1/4
180	7	185	7 3/8	190–240	7 15/32–9 1/2
205	8	190	7 1/2	215–275	8 31/64–10 3/4
230	9	195	7 5/8	240–305	9 1/2–12
255	10	195	7 3/4	265–335	10 1/2–13 1/4

mines the allowances required for finish machining, thickness of upset, diameter tolerances, and corner radii. The amount of upset stock required depends on bar size and determines whether the stock can be sheared, flame cut, or torch cut, or separated by another method. Figure 29(c) illustrates a few of the simplest upset parts.

Figure 31(a) shows a variation from the straight axle-shaft type of design in which the beveled head of the upset is confined in the heading tool. This method usually requires that the design recognize a position in the forging where a flash, or excess metal, must be trapped between the dies and heading tool. This is indicated in Fig. 31(a) by the 3.2 mm (1/8 in.) minimum dimension. Another problem encountered in designs of the type shown in Fig. 31(a) is the filling of the barrel section at the point of transition from original stock size to slightly increased diameter. As noted, an additional amount of finish is required, along with a generous radius.

The same problem, shown in Fig. 31(b), can be overcome by a taper blending from the bar size to the shoulder diameters. This type of design is expedient where the two diameters are within 9.5 mm (3/8 in.) of each other.

Figure 31(c) is basically an axle-shaft type of forging with a long pilot. Because the pilot part of the forging must be carried in the heading tools, draft is required for withdrawal from the tool and usually should be no less than 1/2°. The length of the pilot determines the amount of draft, which may range to a maximum taper of 3°. Another design rule to be recognized is that the pilot diameter in the heading tool should be 1.6 mm (1/16 in.) larger than the bar diameter to allow the stock to bottom in the heading tool. Contingent on the number of passes required in producing the forging, plus the mill tolerance for the particular bar size, the pilot end diameter may require a maximum of 3.2 mm (1/8 in.) over the bar diameter.

Figure 31(d) illustrates a typical transmission cluster gear forging. The drafts specified are a requirement for this type of forging because the part must be carried in the die after being partially produced. Of necessity, the neck diameter is determined by the stock size required to produce the part, plus an allowance to make a fit with the heading tool similar to Fig. 31(c).

Figure 31(e) shows the radius required when the pilot end must pass into the header. A small radius on the heading tool can scrape off metal along part of the length of the bar end and forge the loose chips into the face of the forged flange.

Figure 31(f) illustrates the minimum tapers and radii required for depressions in upset forgings. A larger draft angle or radius, or both, decreases the possibility of cold shuts.

Figures 31(g) and (j) show variations of size and design of forgings that are pierced or punched. In such forgings, the allowance for machining of the holes varies from 1.0 to 2.0 mm (0.040 to 0.080 in.), according to size, for parts that are to be broached.

In Fig. 31(g), the draft allowances required are similar to those of Fig. 31(d) to facilitate removal of the part from the die during and after forging. In Fig. 31(g), the large diameter shows a tolerance 0.8 mm (1/32 in.) larger than standard because the large flange fabricated in the first pass must be carried in the die while the front flange is being upset. Carrying the flange

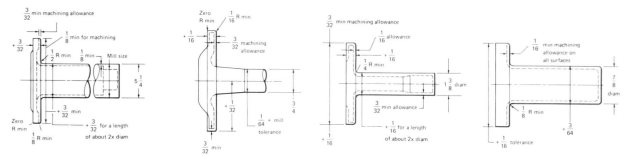

Fig. 30 Design practice for upset forgings with specifications determined by raw material stock diameter. Tolerances (shown with + or − sign), allowances, and design rules for upset forgings of various typical or common shapes. Dimensions given in inches

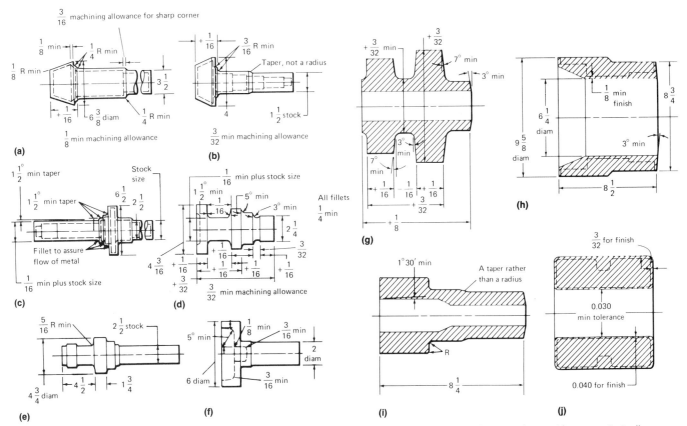

Fig. 31 Design practice for upset forgings in which specifications depend on position of flash in workpiece. Tolerances (shown with + or − sign), allowances, and design rules for upset forgings. See text for discussion. Dimensions given in inches

Table 18 Tolerances for straight extrusions of steel

Location of dimensions	Weight of forging kg	lb	Inside diameter, A mm	in.	Outside diameter, B mm	in.	Bottom thickness, C mm	in.	Total height, D mm	in.	Outside draft, degree	Inside draft, degree
	0.1	0.2	±0.13	±0.005	±0.13	±0.005	±0.25	±0.010	+0.79	+1/32	0–1	0–1
	0.2	0.4	±0.13	±0.005	±0.13	±0.005	±0.25	±0.010	+0.79	+1/32	0–1	0–1
	0.3	0.6	±0.13	±0.005	±0.13	±0.005	±0.25	±0.010	+0.79	+1/32	0–1	0–1
	0.4	0.8	±0.13	±0.005	±0.13	±0.005	±0.25	±0.010	+0.79	+1/32	0–1	0–1
	0.5	1.0	±0.15	±0.006	±0.15	±0.006	±0.30	±0.012	+1.19	+3/64	0–1	0–1
	0.9	2.0	±0.20	±0.008	±0.20	±0.008	±0.38	±0.015	+1.19	+3/64	0–2	0–2
	1.8	4.0	±0.25	±0.010	±0.25	±0.010	±0.51	±0.020	+1.19	+3/64	0–2	0–2
	2.7	6.0	±0.30	±0.012	±0.30	±0.012	±0.64	±0.025	+1.59	+1/16	0–2	0–2
	4.1	9.0	±0.38	±0.015	±0.38	±0.015	±0.79	±0.031	+1.59	+1/16	0–2	0–2
	5.5	12.0	±0.46	±0.018	±0.46	±0.018	±0.19	±0.036	+1.59	+1/16	0–3	0–3
	6.8	15.0	±0.51	±0.020	±0.51	±0.020	±1.02	±0.040	+3.18	+1/8	0–3	0–3
	9.1	20.0	±0.64	±0.025	±0.64	±0.025	±1.22	±0.048	+3.18	+1/8	0–3	0–3
	13.6	30.0	±0.76	±0.030	±0.76	±0.030	±1.52	±0.060	+3.18	+1/8	0–3	0–3

Table 19 Tolerances for controlled extrusions of steel

Location of dimensions	Weight of forging kg	lb	Inside diameter, A mm	in.	Outside diameter, B mm	in.	Bottom thickness, C mm	in.	Total height, D mm	in.	Outside draft, degree	Inside draft, degree
	0.1	0.2	±0.25	±0.010	±0.25	±0.010	±0.25	±0.010	±0.25	±0.010	0–1	0–3
	0.2	0.4	±0.25	±0.010	±0.25	±0.010	±0.25	±0.010	±0.25	±0.010	0–1	0–3
	0.3	0.6	±0.25	±0.010	±0.25	±0.010	±0.25	±0.010	±0.25	±0.010	0–1	0–3
	0.4	0.8	±0.25	±0.010	±0.25	±0.010	±0.25	±0.010	±0.25	±0.010	0–1	0–3
	0.5	1.0	±0.30	±0.012	±0.30	±0.012	±0.30	±0.012	±0.30	±0.012	0–2	1–4
	0.9	2.0	±0.41	±0.016	±0.41	±0.016	±0.38	±0.015	±0.38	±0.015	0–2	1–4
	1.8	4.0	±0.51	±0.020	±0.51	±0.020	±0.51	±0.020	±0.51	±0.020	0–2	2–5
	2.7	6.0	±0.61	±0.024	±0.61	±0.024	±0.64	±0.025	±0.64	±0.025	0–2	2–5
	4.1	9.0	±0.71	±0.028	±0.71	±0.028	±0.79	±0.031	±0.79	±0.031	0–2	3–6
	5.5	12.0	±0.79	±0.031	±0.79	±0.031	±0.91	±0.036	±0.91	±0.036	0–3	4–7
	6.9	15.0	±0.86	±0.034	±0.86	±0.034	±1.02	±0.040	±1.02	±0.040	0–3	5–8
	9.1	20.0	±0.97	±0.038	±0.97	±0.038	±1.22	±0.048	±1.22	±0.048	0–3	6–9
	13.6	30.0	±1.14	±0.045	±1.14	±0.045	±1.52	±0.060	±1.52	±0.060	0–3	7–10

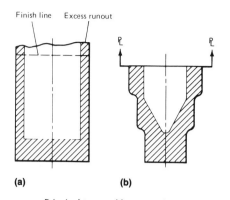

Fig. 32 Principal types of hot extrusion processes (a) Straight extrusion. (b) Controlled extrusion

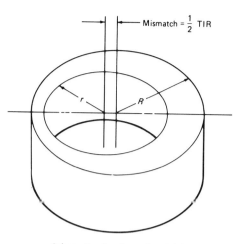

Fig. 33 Schematic showing mismatch in an extrusion. Definition of mismatch tolerance in terms of total indicator reading for values given in text

the punch to be withdrawn from the forging.

In addition to the various types of symmetrical forgings shown, some asymmetrical parts are readily forged. These include bolts of many sizes and designs, rod ends, trunnion forgings, steering sectors, universal joints, and a number of other parts with ends having various contours and dimensions.

Design of Hot Extrusion Forgings

Hot extrusion is used for making regular or irregular cup-shaped parts. The two principal types of hot extrusions are called straight extrusion, in which the metal is unconfined (Fig. 32a), and controlled extrusion, in which the metal is confined (Fig. 32b). More draft is allowed in a controlled extrusion because the flow of metal is restrained, and more stock is allowed because a flash is formed.

Size limitations for a range of small hot extrusions are given in Table 17. Tables 18 and 19 list the recommended dimensional tolerances applicable to straight extrusions and controlled extrusions, respectively.

Mismatch tolerances for 0.09 to 0.36 kg (0.2 to 0.8 lb) forgings are 0.20 mm (0.008 in.); 0.45 to 1.81 kg (1.0 to 4.0 lb), 0.25 to 0.30 mm (0.010 to 0.012 in.); 4.5 kg (10 lb), 0.41 mm (0.016 in.); 9.1 kg (20 lb), 0.56 mm (0.022 in.); and 14 kg (30 lb), 0.660 mm (0.026 in.). Mismatch is defined in Fig. 33. Mismatch tolerance must be added to other tolerances.

from pass to pass requires clearance of the flange in the die, and as the punch enters the forging, there is some upsetting action of the back flange, creating a slightly oval condition. The additional tolerance reflects not only die wear but also ovality. This also holds true for the neck diameter.

Figure 31(i) shows the amount of taper required when punching relatively deep holes. The length of the taper determines the amount of draft required to permit

Table 20 Properties of hot extrusion forgings at various locations and orientations (4340 steel)

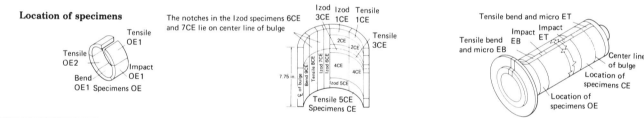

Code	Location	Orientation	Yield strength MPa	ksi	Tensile strength MPa	ksi	Gage length(a) mm	in.	Elongation, %	Reduction in area, %
Tensile properties										
EB	Open end	Longitudinal	1235	179	1310	190	50	2	14.5	46.7
OE1-1	Open end	Transverse	1200	174	1295	188	25	1	12.5	37.8
OE1-2	Open end	Transverse	1180	171	1310	190	25	1	10.0	38.3
OE1-3	Open end	Transverse	1205	175	1310	190	25	1	14.0	38.0
OE2-1	Open end	Transverse	1215	176	1325	192	25	1	12.0	38.5
OE2-2	Open end	Transverse	1215	176	1310	190	25	1	12.5	37.8
OE2-3	Open end	Transverse	1200	174	1295	188	25	1	12.0	40.3
ET	Closed end(b)	Longitudinal	1220	177	1310	190	50	2	14.0	47.0
8CE	Closed end(b)	Longitudinal	1240	180	1310	190	50	2	14.0	47.5
3CE	Closed end(b)	Transverse	1215	176	1305	189	25	1	12.0	36.7
1CE	Closed end	Transverse	1275	185	1310	190	25	1	11.0	19.0
5CE	Closed end	Transverse	1220	177	1305	189	25	1	12.5	34.7

Code	Location	Orientation	Izod impact energy J	ft · lbf	Bend angle, degree
Izod impact and bend properties					
EB	Open end	Longitudinal	30, 30	22, 22	70
OE	Open end	Transverse	22, 20, 23, 22	16.5, 15, 17, 16.5	35, 37, 37, 37
ET	Closed end	Longitudinal	30, 30	22, 22	...
6CE	Closed end(b)	Longitudinal	32	24	74
7CE-9CE	Closed end(b)	Longitudinal	33	24.5	...
3CE	Closed end(b)	Transverse	26	19.5	56
1CE	Closed end	Transverse	22	16.0	...
5CE	Closed end	Transverse	20	15.0	...

(a) Specimens with 50 mm (2 in.) gage length were standard 12.83 mm (0.505 in.) in diameter. Specimens with 25 mm (1 in.) gage length were subsize 6.35 mm (0.250 in.) in diameter. (b) At bulge

Machining allowance for forgings up to 6 m (20 ft) long, up to 305 mm (12 in.) OD, and 127 mm (5 in.) ID are 13 mm (½ in.) on length, 4.0 mm (5/32 in.) on the OD, and 3.2 mm (⅛ in.) on the ID.

Mechanical properties are dealt with in Table 20. A part of the extrusion was die forged and then extruded in the barrel section. In the area where the effect of the extrusion operation began at the end of the die-forged or hammered area, the flow lines were very irregular and curved or bulged to the outside. The extrusion was approximately 203 mm (8 in.) in diameter, 1.2 m (4 ft) in length, and had a 19 mm (¾ in.) wall thickness.

Transverse specimens show a marked decrease in impact strength and ductility as measured by bend angle. In straight extrusion, there is no flash line; therefore, transverse flash-line tests could not be conducted.

High-Strength Low-Alloy Steel Forgings

Peter H. Wright, Chapparal Steel Company

HIGH-STRENGTH LOW-ALLOY (HSLA) steels with excellent strength and toughness were originally developed as flat-rolled products for the Alaskan pipeline in the late 1960s.

Two HSLA families are acicular-ferrite steels (Ref 1) and pearlite-reduced steels (Ref 2). These steels contain microalloying additions of vanadium and niobium, and they have a yield strength of 480 MPa (70 ksi) and a Charpy V-notch fracture appearance transition temperature (FATT) of −60 °C (−75 °F). Typical thermomechanical processing of these materials involves reheating slabs to 1260 °C (2300 °F), with hot work continuing to 815 °C (1500 °F) and water spray cooling to 650 °C (1200 °F).

The objective of the researchers responsible for developing microalloy forging steels was to obtain the enhanced mechanical properties of hot-formed steel parts while simultaneously eliminating the need for heat treating of the steel. Elimination of the heat-treating operation reduces energy consumption and processing time as well as the material inventories resulting from intermediate processing steps.

Heat-treated forgings currently outperform alternative materials where strength, toughness, and reliability are primary considerations. However, methods must be found that achieve these benefits at lower cost. The goal of ferrous metallurgists, therefore, has been to achieve similar properties in steel forgings in the as-forged condition, that is, without subsequent heat treatment.

Unfortunately, the thermomechanical processing used for HSLA flat-rolled products cannot be readily transferred to hot forging. Typical hot forgings are produced from bars that are induction or gas heated to 1260 °C (2300 °F), then hot worked to 1120 °C (2050 °F). Grain growth and precipitate coarsening are rapid at these temperatures and are exacerbated by the mass effect of forgings compared with strip. Attempts to forge at temperatures low enough to optimize as-forged steel properties result in decreased equipment efficiency and excessive die wear.

Microalloying Elements

All microalloy steels contain small concentrations of one or more strong carbide- and nitride-forming elements. Vanadium, niobium, and titanium combine preferentially with carbon and/or nitrogen to form a fine dispersion of precipitated particles in the steel matrix. Table 1 summarizes the effects of the various elements.

Vanadium. Precipitation strengthening is one of the primary contributors to microalloy steels; it is most readily achieved with vanadium additions in the 0.03 to 0.10% range. Vanadium also improves toughness by stabilizing dissolved nitrogen, which, in electric furnace melted steel, may be as high as 0.012%. There is a linear relationship between vanadium content and yield and tensile strength up to 0.15% V; it can exceed this level if sufficient nitrogen is present. The impact transition temperature also increases when vanadium is added (Fig. 1).

Niobium can also have a strong precipitation-strengthening effect provided it is taken into solution at reheat and is kept in solution during forging. Its main contributions, however, are to form precipitates above the transformation temperature and to retard the recrystallization of austenite, thus promoting a fine-grain microstructure having improved strength and toughness. Concentrations vary from 0.020 to 0.10% Nb.

Molybdenum is included in Table 1 because of its use in second-generation microalloy steels. It has the effect of greatly simplifying the process controls necessary in a forge shop for the application of these materials.

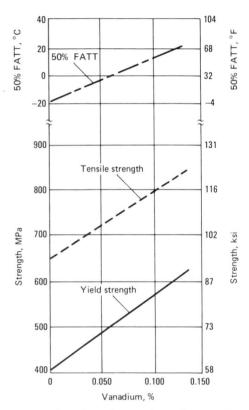

Fig. 1 Effect of vanadium on the tensile properties and FATT of a first-generation medium-carbon microalloy steel

Table 1 Effects of selected elements on the properties of microalloying steels

Element	Precipitation strengthening	Ferrite grain refinement	Nitrogen fixing	Structure modification
Vanadium	Strong	Weak	Strong	Moderate
Niobium	Moderate	Strong	Weak	None
Molybdenum	Weak	None	None	Strong
Titanium	None (<0.02% Ti) and strong (>0.05% Ti)	Strong	Strong	None

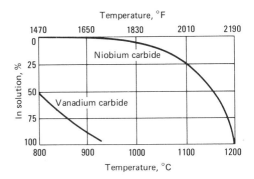

Fig. 2 Precipitation and dissolution characteristics of vanadium and niobium carbides in austenite

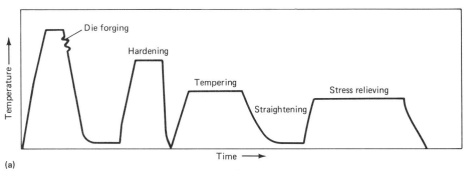

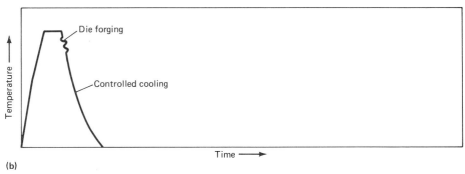

Fig. 3 Thermal cycles for conventional (a) and microalloy (b) steels. Source: Ref 4

Titanium can behave as both a grain refiner and a precipitate strengthener, depending on its content (Ref 3). At compositions greater than 0.050%, titanium carbides begin to exert a strengthening effect. However, at this time, titanium is used commercially to retard austenite grain growth and thus improve toughness. Typically, titanium concentrations range from 0.010 to 0.020%.

Metallurgical Effects

The precipitation and dissolution characteristics of vanadium and niobium compounds in austenite differ significantly. Upon cooling from the forging temperature, niobium carbide begins to precipitate at about 1205 °C (2200 °F), as shown in Fig. 2. Without subsequent hot work, the precipitates continue to form and coarsen as the temperature falls to 925 °C (1700 °F). Continued hot working into the 900 °C (1650 °F) temperature range, however, retards austenite recrystallization and precipitation, resulting in the development of a refined austenite grain size.

Vanadium carbonitride precipitation begins at about 950 °C (1740 °F) and becomes complete during transformation. Because vanadium carbonitride is a relatively low-temperature product, precipitate coarsening during accelerated cooling from the forge is minimal, and the maximum precipitation-strengthening effect is achieved.

The toughness of all hot-worked microalloy steels improves when ferrite grain refinement is maximized by controlling the finish hot-working temperature. For example, the 20 J (15 ft · lbf) transition temperature for a niobium-containing low-carbon steel is reduced from 30 to −80 °C (85 to −110 °F) by reducing the finishing temperature from 1050 to 900 °C (1920 to 1650 °F). The improvement is independent of the reheat temperature.

The solubility of niobium carbide in austenite is mainly a function of temperature and carbon content. The solid solubility limit of niobium in austenite can be readily exceeded at normal forging temperature. The same is true of titanium nitride precipitates. The opportunity therefore exists to avoid abnormal grain growth at forging temperatures by using a suitable concentration of either titanium or niobium.

The toughness of the acicular-ferrite steels originally designed for use on the Alaskan pipeline was a result of their complex microstructures, which were brought about by the addition of molybdenum to a 0.06% C (max), 1.8% Mn steel. Pearlite transformation is suppressed in these steels, and a ferrite microstructure is obtained over a wide range of cooling rates. This same microstructure can be achieved in small forgings. Figure 3 shows typical thermal cycles for conventional quench and temper and for microalloy process routes.

The metallurgical fundamentals described above were first applied to forgings in the early 1970s. A West German composition, 49MnVS3 (nominal composition: 0.47C-0.20Si-0.75Mn-0.060S-0.10V), was successfully used for automotive connecting rods. The steel was typical of the first generation of microalloy steels, with a medium carbon content (0.35 to 0.50% C) and additional strengthening through vanadium carbonitride precipitation. The parts were subjected to accelerated air cooling directly from the forging temperature. The AISI grade 1541 microalloy steel with either niobium or vanadium has been used in the United States for similar automotive parts for many years.

First-generation microalloy forging steels generally have ferrite-pearlite microstructures, tensile strengths above 760 MPa (110 ksi), and yield strengths in excess of 540 MPa (78 ksi). The room-temperature Charpy V-notch toughness of first-generation forgings is typically 7 to 14 J (5 to 10 ft · lbf), ambient. It became apparent that toughness would have to be significantly improved to realize the full potential of microalloy steel forgings.

Second-generation microalloy steels were introduced in about 1986 (Ref 5). These are typified by the West German grade 26MnSiVS7 (nominal composition: 0.26C-0.70Si-1.50Mn-0.040S-0.10V-0.02Ti). The carbon content of these steels was reduced to between 0.10 and 0.30%. They are produced with either a ferrite-pearlite microstructure or an acicular-ferrite structure. The latter results from the suppression of pearlite transformation products by an addition of about 0.10% Mo. Titanium additions have also been made to these steels to improve impact toughness even further. Two typical second-generation microalloy steel compositions used in the United States are listed in Table 2.

One of the primary concerns of any steel user is the consistency of finished-part properties. Heat treatment has successfully addressed this concern, and a method must be found to ensure the consistency of finished-part properties for microalloy steels. One disadvantage of ferrite-pearlite microalloy steels is that the finished strength and hardness are functions of the cooling rate. Cooling rate can vary because of either process changes or part geometry.

Figure 4 shows the effect of cooling rate on the yield strength of a forging made from

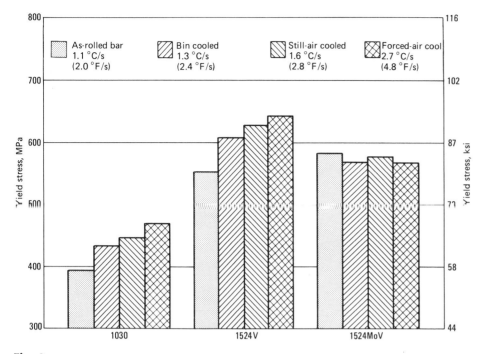

Fig. 4 Effect of cooling rate on the yield strength of three grades of forgings: 1030, 1524V, and 1524MoV

Table 2 Typical second-generation microalloy steel compositions used in the United States

Grade	C	Mn	P	S	Si	Mo	V	N
1524MoV	0.22	1.54	0.014	0.036	0.35	0.11	0.11	0.011
1524V	0.22	1.44	0.013	0.018	0.46	0.022	0.10	0.010

three grades of steel (Ref 6). The 1030 grade has been microalloyed with 0.03% Nb, the other two grades have compositions as shown in Table 2. Forging temperature was 1260 °C (2300 °F); the cooling conditions examined were traditional bin cooling after trimming the hot forging, still-air cooling of forgings isolated from each other, and forced-air cooling on a conveyor followed by final tub cooling from 595 °C (1100 °F).

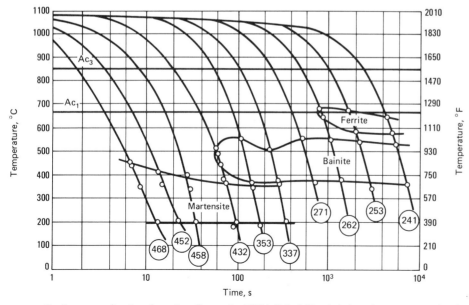

Fig. 6 Continuous cooling transformation diagram for 1524MoV steel. The circled numbers correspond to the HV hardness of microstructures produced by cooling at the rates shown. Source: Ref 7

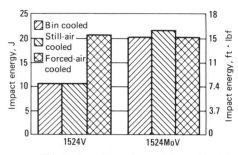

Fig. 5 Effect of cooling rate on Charpy V-notch impact toughness of two second-generation microalloy steels. Cooling rates ranged from 1.3 °C/s (2.4 °F/s) (bin cooling) to 2.7 °C/s (4.8 °F/s) (forced-air cooling). The molybdenum-containing steel is unaffected by cooling conditions.

The 1524V grade performs as predicted for ferrite-pearlite steels: Strength increases with cooling rate. The 1524MoV steel is neither as sensitive to cooling rate nor does it develop the same strength as the ferrite-pearlite steel.

Ambient-temperature Charpy V-notch impact tests on forgings in the same three cooling conditions show that the 1524MoV steel has a consistent 20 J (15 ft · lbf) of energy absorbed independent of the cooling rate. The 1524V grade must be conveyor cooled to achieve this toughness (Fig. 5). The molybdenum-treated steel removes the variable of accelerated cooling from the process, an advantage to the hot forger and end user because conveyors are unnecessary. It also makes possible the use of microalloy steels in forgings with cross sections up to at least 75 mm (3 in.) thick. It should be noted that a reduction in forging temperature would result in improved toughness in these steels. Figure 6 shows the continuous cooling transformation diagram for the 1524MoV steel.

Titanium-treated microalloy steels are currently in production in the United States, Germany, and Japan. The resistance to grain coarsening imparted by titanium nitride precipitation increases the toughness of the forgings. The ultimate strength of first- and second-generation microalloy steels is adequate for many engineering applications, but these steels do not achieve the toughness of conventional quenched and tempered alloys under normal hot-forging conditions.

Third-generation microalloy steels went into commercial production in the United States in 1989. These steels differ from their predecessors in that they are direct quenched from the forging temperature to produce microstructures of lath martensite with uniformly distributed temper carbides (Fig. 7). Without subsequent heat treatment, these materials achieve properties, including toughness, similar to those of standard quenched and tempered steels (Fig. 8 and 9). The metallurgical principles behind this development are as follows:

High-Strength Low-Alloy Steel Forgings / 361

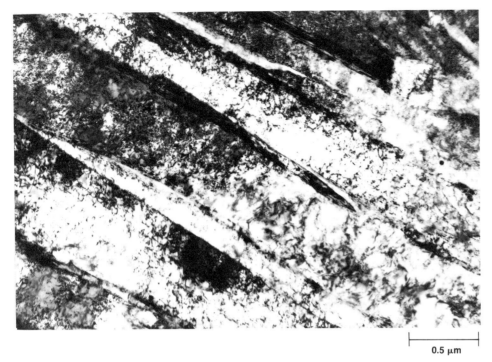

Fig. 7 Microstructure of a third-generation microalloy forging steel, which consists of lath martensite and uniformly distributed auto-tempered carbides

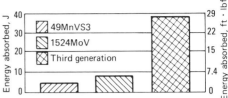

Fig. 8 Bar graph showing that direct-quenched third-generation steels absorb much more energy in Charpy V-notch impact testing at −30 °C (−20 °F) than earlier microalloy grades. The 20 J (15 ft · lbf) energy absorbed criterion used to qualify older alloys no longer applies.

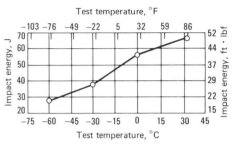

Fig. 9 Impact transition temperature profile of a third-generation microalloy steel in Charpy V-notch testing. New alloys maintain adequate toughness to −60 °C (−76 °F) and below.

- Niobium additions sufficient to exceed the solubility limit at the forging temperature. Undissolved Nb(CN) retards the recrystallization and grain growth of austenite during forging, trimming, and entry into the quenchant
- Composition control to ensure that the martensite finish temperature is above 205 °C (400 °F)
- A fast cold-water quench is performed on a moving conveyor through a spray chamber or by other appropriate equipment
- The relatively high martensite finish temperature, combined with the mass effect of a forging, results in an auto-tempered microstructure with excellent toughness and a hardness of 38 to 43 HRC. Section thicknesses of up to 50 mm (2 in.) have been successfully produced

The newest generation of microalloy forging steels (1989) has five to six times the toughness at −30 °C (−20 °F) and twice the yield strength of second-generation materials. No special forging practices are required except for the use of a water-cooling system. In a comparison of a third-generation microalloy steel to two standard quenched and tempered carbon and alloy steels (1040 and 4140), properties of the direct quenched microalloy grade were very similar to those of quenched and tempered 4140. The hardness of all three materials tested was 40 HRC (Fig. 10). Additional information on the effect of various elements on the metallurgy of steels is available in the article "Bulk Formability of Steels" in this Volume.

Future Outlook

Recent advances in titanium-treated and direct-quenched microalloy steels provide new opportunities for the hot forger to produce tough, high-strength parts without special forging practices. Product evaluations of these microalloy steels indicate that they are comparable to conventional quenched and tempered steels.

Warm forging continues to make steady progress as a cost-effective, precision manufacturing technique because it significantly reduces machining costs. Microalloy steels

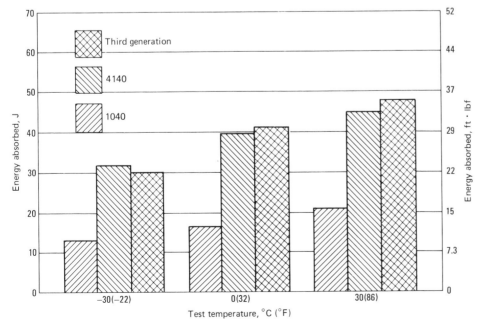

Fig. 10 Comparison of impact transition properties for third-generation microalloy steel to standard quenched and tempered carbon (1040) and alloy (4140) steel grades. All materials tested at 40 HRC hardness.

austenitized at 1040 °C (1900 °F), cooled to a warm forging temperature of 925 °C (1700 °F), forged, and cooled by air or water (depending on composition), will produce a range of physical properties. The resulting cost savings has the potential to improve the competitive edge that forging has over other manufacturing techniques.

REFERENCES

1. A.P. Coldren and J.L. Mihelich, Acicular Ferrite HSLA Steels for Line Pipe, in *Molybdenum Containing Steels for Gas and Oil Industry Applications*, No. M-321, Climax Molybdenum Company, 1976, p 14-28
2. G. Tither and W.E. Lauprecht, Pearlite Reduced HSLA Steels for Line Pipe, in *Molybdenum Containing Steels for Gas and Oil Industry Applications*, No. M-321, Climax Molybdenum Company, 1976, p 29-41
3. C. Webnscuan *et al.*, Recrystallizations of Austenite in Titanium Treated HSLA Steels, in *Proceedings of the International Conference on HSLA Steels* (Beijing, China), 1985, American Society for Metals, 1985, p 199-206
4. J. Stoeter and J. Kneller, Recent Developments in the Drop Forging of Crankshafts, *Met. Prog.*, March 1985, p 61
5. Huchteman and Schuler, 26MnSiVS7—ein nener mikrolegierter perlitischer Stahl mit hoher Zahigkeit, *Thyssen Edelstahl Tech. Ber.*, 1987
6. P.H. Wright *et al.*, What the Forger Should Know About Microalloy Steels, in *Fundamentals of Microalloying Forging Steels*, G. Krauss and S.K. Banerji, Ed., The Metallurgical Society, 1987, p 541-565
7. K.J. Grassl *et al.*, Thesis in Progress, Advanced Steel Processing and Products Research Center, Colorado School of Mines, 1988

Steel Castings

Revised by Malcolm Blair, Steel Founders' Society of America

STEEL CASTINGS are produced by pouring molten steel of the desired composition into a mold of the desired configuration and allowing the steel to solidify. The mold material may be silica, zircon, chromite sand, olivine sand, graphite, metal, or ceramic. The choice of mold material depends on the size, intricacy, dimensional accuracy of the casting, and cost. While the producible size, surface finish, and dimensional accuracy of castings vary widely with the type of mold, the properties of the cast steel are not affected significantly.

Steel castings can be made from any of the many types of carbon and alloy steel produced in wrought form. Those castings produced in any of the various types of molds and wrought steel of equivalent chemical composition respond similarly to heat treatment, have the same weldability, and have similar physical and mechanical properties. However, cast steels do not exhibit the effects of directionality on mechanical properties that are typical of wrought steels. This nondirectional characteristic of cast steel mechanical properties may be advantageous when service conditions involve multidirectional loading.

Another difference between steel castings and wrought steel is the deoxidation required during steelmaking. Cast steels are made only from fully killed (deoxidized) steel in a foundry, while wrought products can be made from rimmed, semikilled, or killed steel ingots in a mill. The method of producing the killed steel used for a casting may also differ from that used for a wrought product because of differences in the tapping temperatures required in casting and ingot production.

However, the salient features of producing killed steel in a casting foundry are the same as those features important to the production of fully killed steel ingots. For the deoxidation of carbon and low-alloy steels (that is, for control of their oxygen content), aluminum, titanium, and zirconium are used. Of these, aluminum is used more frequently because of its effectiveness and low cost. Unless otherwise specified, the normal sulfur limit for carbon and low-alloy steels is 0.06%, and the normal phosphorus limit is 0.05%.

Classifications and Specifications

The steel castings discussed in this article are divided into four general groups according to composition:

- Low-carbon steel castings
- Medium-carbon steel castings
- High-carbon steel castings
- Low-alloy steel castings

Other types of steel castings are discussed in separate articles on heat-resistant castings, stainless (corrosion-resistant) steel castings, and austenitic manganese steels in this Volume.

Carbon steels contain only carbon as the principal alloying element. Other elements are present in small quantities, including those added for deoxidation. Silicon and manganese in cast carbon steels typically range from 0.25 to about 0.80% Si and 0.50 to about 1.00% Mn.

Like wrought steels, carbon cast steels can be classified according to their carbon content into three broad groups:

- *Low-carbon steels*: 0.20% C or less
- *Medium-carbon steels*: 0.20 to 0.50% C
- *High-carbon steels*: 0.50% C or more

Carbon content is a principal factor affecting the mechanical properties of these steels. The other important factor is the method of heat treatment.

Figure 1 shows the basic trends of the mechanical properties of cast carbon steels as a function of carbon content for four different heat treatments. For a given heat treatment, a higher carbon content generally results in higher hardness and strength levels with lower ductility and toughness values. Further information on the properties of carbon cast steels is contained in subsequent sections of this article.

Low-alloy steels contain, in addition to carbon, alloying elements up to a total alloy content of 8%. Cast steels containing more than the following amounts of a single alloying element are considered low-alloy cast steel:

Element	Amount, %
Manganese	1.00
Silicon	0.80
Nickel	0.50
Copper	0.50
Chromium	0.25
Molybdenum	0.10
Vanadium	0.05
Tungsten	0.05

Aluminum, titanium, and zirconium are used for the deoxidation of low-alloy steels. Of these elements, aluminum is used most frequently because of its effectiveness and low cost.

Numerous types of cast low-alloy steel grades exist to meet the specific requirements of the end-use, such as structural strength and resistance to wear, heat, and corrosion. The designations of the American Iron and Steel Institute (AISI) and the Society of Automotive Engineers (SAE) have historically been used to identify the various types of steel by their principal alloy content. Cast steels, however, do not precisely follow the compositional ranges specified by AISI and SAE designations for wrought steels. In most cases, the cast steel grades contain 0.30 to 0.65% Si and 0.50 to 1.00% Mn, unless otherwise specified. The principal low-alloy cast steel designations, their AISI and SAE equivalents, and their alloy type are:

Cast steel designation	Nearest wrought equivalent	Alloying elements
1300	1300	Manganese
8000, 8400	8000, 8400	Manganese, molybdenum
80B00	80B00	Manganese, molybdenum, boron
2300	2300	Nickel
8600, 4300	8600, 4300	Nickel, chromium, molybdenum
9500	9500	Manganese, nickel, chromium, molybdenum
4100	4100	Chromium, molybdenum

The 8000, 8400, 2300, and 9500 alloy types are used extensively as cast steels. There are additional alloy types that are infrequently specified as cast steels, that is, 3100 (nickel-chromium), 3300 (nickel-chro-

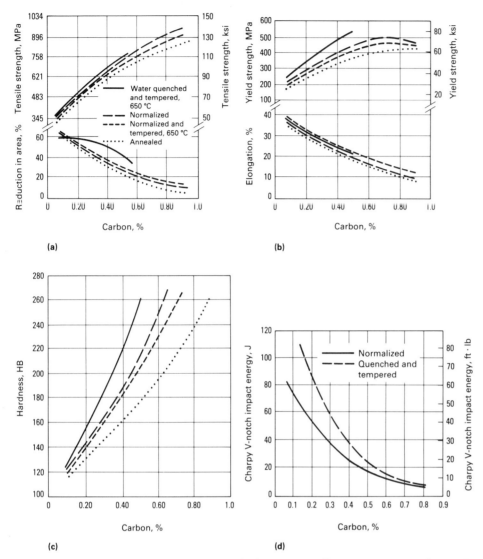

Fig. 1 Properties of cast carbon steels as a function of carbon content and heat treatment. (a) Tensile strength and reduction of area. (b) Yield strength and elongation. (c) Brinell hardness. (d) Charpy V-notch impact energy

mium), 4000 (molybdenum), 5100 (chromium), 6100 (chromium-vanadium), 4600 (nickel-molybdenum), and 9200 (silicon). Further information on the elements used in alloy steel castings is provided in the section "Low-Alloy Cast Steels" of this article.

Specifications. Steel castings are usually purchased to meet specified mechanical properties, with some restrictions on chemical composition. Tables 1 and 2 list the requirements given in various specifications of the American Society of Testing and Materials (ASTM) and in SAE J435c. Table 1 primarily lists carbon steel castings (with some comparable low-alloy types), while Table 2 lists several low-alloy cast steels and some cast steels with chromium contents up to 10.0%.

In the low-strength ranges, some specifications limit carbon and manganese content, usually to ensure satisfactory weldability. In SAE J435c, carbon and manganese are specified to ensure that the minimum desired hardness and strength are obtained after heat treatment. For special applications, other elements may be specified either as maximum or minimum, depending on the characteristics desired.

The ASTM specifications that include carbon and low-alloy grades of steel castings are A 216, A 217, A 352, A 356, A 389, A 487, and A 757. The ASTM specifications with grades of carbon steel castings are listed in Table 1. Table 2 lists the requirements for the low-alloy classes of steel castings given in some of the ASTM specifications mentioned above. In addition, ASTM specifications may address common requirements of all steel castings for a particular type of application. For example, ASTM A 703 specifies the general requirements of steel castings for pressure-containing parts.

If only mechanical properties are specified, the chemical composition of castings for general engineering applications is usually left to the discretion of the casting supplier. For specific applications, however, certain chemical composition limits have been established to ensure the development of specified mechanical properties after proper heat treatment, as well as to facilitate welding, uniform response to heat treatment, or other requirements. Hardness is specified for most grades of SAE J435c to ensure machinability, ease of inspection for high production rate items, or certain characteristics pertaining to wear.

SAE J435c includes three grades, HA, HB, and HC, with specified hardenability requirements. Figure 2 plots hardenability requirements, both minimum and maximum, for these steels. Hardenability is determined by the end-quench hardenability test described in the article "Hardenability of Carbon and Low-Alloy Steels" in this Volume. Other specifications require minimum hardness at one or two locations on the end-quench specimen. In general, hardenability is specified to ensure a predetermined degree of transformation from austenite to martensite during quenching, in the thickness required. This is important in critical parts requiring toughness and optimum resistance to fatigue.

Among the most commonly selected grades of steel castings are, first, a medium-carbon steel corresponding to ASTM A 27 65-35 or SAE 0030 and second, a higher-strength steel, often alloyed and fully heat treated, similar to ASTM A 148 105-85 or SAE 0105.

Particularly when the purchaser heat treats a part after other processing, a casting will be ordered to compositional limits closely equivalent to the AISI-SAE wrought steel compositions, with somewhat higher silicon permitted. As in other steel castings, it is best not to specify a range of silicon, but to permit the foundry to utilize the silicon and manganese combination needed to achieve required soundness in the shape being cast. The silicon content is frequently higher in cast steels than for the same nominal composition in wrought steel. Silicon above 0.80% is considered an alloy addition because it contributes significantly to resistance to tempering.

Railroad equipment manufacturers and other major users of steel castings may prefer their own or industry association specifications. Users of steel castings for extremely critical applications, such as aircraft, may use their own, industry association, or special-purpose military specifications. Foundries frequently make nonstandard grades for special applications or have their own specification system to meet the needs of the purchaser. Savings may be realized by using a grade that is standard with a foundry, especially for small quantities.

Steel Castings / 365

Table 1 Summary of specification requirements for various carbon steel castings.
Unless otherwise noted, all the grades listed in this table are restricted to a phosphorus content of 0.040% max and a sulfur content of 0.045% max.

Class or grade	Tensile strength(a) MPa	ksi	Yield strength(a) MPa	ksi	Minimum elongation in 50 mm (2 in.), %	Minimum reduction in area, %	Chemical composition(b), % C	Mn	Si	Other requirements	Condition or specific application
ASTM A 27: carbon steel castings for general applications											
N-1	...	...	...	...	...	...	0.25(c)	0.75(c)	0.80	0.06% S, 0.05% P	Chemical analysis only
N-2	...	...	...	...	...	...	0.35(c)	0.60(c)	0.80	0.06% S, 0.05% P	Heat treated but not mechanically tested
U60-30	415	60	205	30	22	30	0.25(c)	0.75(c)	0.80	0.06% S, 0.05% P	Mechanically tested but not heat treated
60–30	415	60	205	30	24	35	0.30(c)	0.60(c)	0.80	0.06% S, 0.05% P	Heat treated and mechanically tested
65–35	450	65	240	35	24	35	0.30(c)	0.70(c)	0.80	0.06% S, 0.05% P	Heat treated and mechanically tested
70–36	485	70	250	36	22	30	0.35(c)	0.70(c)	0.80	0.06% S, 0.05% P	Heat treated and mechanically tested
70–40	485	70	275	40	22	30	0.25(c)	1.20(c)	0.80	0.06% S, 0.05% P	Heat treated and mechanically tested
ASTM A 148: carbon steel castings for structural applications(d)											
80–40	550	80	275	40	18	30	(e)	(e)	(e)	0.06% S, 0.05% P	Composition and heat treatment necessary to achieve specified mechanical properties
80–50	550	80	345	50	22	35	(e)	(e)	(e)	0.06% S, 0.05% P	Composition and heat treatment necessary to achieve specified mechanical properties
90–60	620	90	415	60	20	40	(e)	(e)	(e)	0.06% S, 0.05% P	Composition and heat treatment necessary to achieve specified mechanical properties
105–85	725	105	585	85	17	35	(e)	(e)	(e)	0.06% S, 0.05% P	Composition and heat treatment necessary to achieve specified mechanical properties
SAE J435c: see Table 2 for alloy steel castings specified in SAE J435c											
0022	...	...	...	...	...	...	0.12–0.22	0.50–0.90	0.60	187 HB max	Low-carbon steel suitable for carburizing
0025	415	60	207	30	22	30	0.25(c)	0.75(c)	0.80	187 HB max	Carbon steel welding grade
0030	450	65	241	35	24	35	0.30(c)	0.70(c)	0.80	131–187 HB	Carbon steel welding grade
0050A	585	85	310	45	16	24	0.40–0.50	0.50–0.90	0.80	170–229 HB	Carbon steel medium-strength grade
0050B	690	100	485	70	10	15	0.40–0.50	0.50–0.90	0.80	207–255 HB	Carbon steel medium-strength grade
080	550	80	345	50	22	35	...	...	...	163–207 HB	Medium-strength low-alloy steel
090	620	90	415	60	20	40	...	...	...	187–241 HB	Medium-strength low-alloy steel
HA, HB, HC(f)	...	...	...	...	...	...	0.25–0.34	(f)	(f)	See Fig. 2.	Hardenability grades (Fig. 2)
ASTM A 216: carbon steel castings suitable for fusion welding and high-temperature service											
WCA	415–585	60–85	205	30	24	35	0.25	0.70(c)	0.60	(g)	Pressure-containing parts
WCB	485–655	70–95	250	36	22	35	0.30	1.00(c)	0.60	(g)	Pressure-containing parts
WCC	485–655	70–95	275	40	22	35	0.25	1.20(c)	0.50	(g)	Pressure-containing parts
Other ASTM cast steel specifications with carbon steel grades(h)											
A 352-LCA	415–585	60–85	205	30	24	35	0.25	0.70(c)	0.60	(g)(i)(j)	Low-temperature applications
A 352-LCB	450–620	65–90	240	35	24	35	0.30	1.00	0.60	(g)(j)(k)	Low-temperature applications
A 356-grade 1	485	70	250	36	20	35	0.35	0.70(c)	0.60	0.035% P max, 0.030% S max	Castings for valve chests, throttle valves, and other heavy-walled components for steam turbines
A 757-A1Q	450	65	240	35	24	35	0.30	1.00	0.60	(j)(k)(l)	Castings for pressure-containing applications at low temperatures

(a) Where a single value is shown, it is a minimum. (b) Where a single value is shown, it is a maximum. (c) For each reduction of 0.01% C below the maximum specified, an increase of 0.04% Mn above the maximum specified is permitted up to the maximums given in the applicable ASTM specifications. (d) Grades may also include low-alloy steels; see Table 2 for the stronger grades of ASTM A 148. (e) Unless specified by purchaser, the compositions of cast steels in ASTM A 148 are selected by the producer in order to achieve the specified mechanical properties. (f) Purchased on the basis of hardenability, with manganese and other elements added as required. (g) Specified residual elements include 0.30% Cu max, 0.50% Ni max, 0.50% Cr max, 0.20% Mo max, and 0.03% V max, with the total residual elements not exceeding 1.00%. (h) These ASTM specifications also include alloy steel castings for the general type of applications listed in the Table. (i) Testing temperature of −32 °C (−25 °F). (j) Charpy V-notch impact testing at the specified test temperature with an energy value of 18 J (13 ft · lbf) min for two specimens and an average of three. (k) Testing temperature of −46 °C (−50 °F). (l) Specified residual elements of 0.03% V, 0.50% Cu, 0.50% Ni, 0.40% Cr, and 0.25% Mo, with total amount not exceeding 1.00%. Sulfur and phosphorus content, each 0.025% max

Mechanical Properties

Carbon and low-alloy steel castings are produced to a great variety of properties because composition and heat treatment can be selected to achieve specific combinations of properties, including hardness, strength, ductility, fatigue, and toughness. Although selections can be made from a wide range of properties, it is important to recognize the general interrelationships among these properties. Specifically, higher hardness, lower toughness, and lower ductility values are generally associated with higher strength values.

This general relationship between strength, ductility, and toughness is illustrated in Fig. 3 and 4 for carbon and low-alloy cast steels, respectively. Because yield strength is a primary design criterion for structural applications, Fig. 3 and 4 plot tensile strength, ductility (as measured by elongation), and toughness (based on Charpy V-notch impact energy) versus yield strength. The remainder of this section compares these properties of carbon and low-alloy cast steels with those of wrought steels. This section also discusses fatigue properties and the effects of section size and heat treatment on the mechanical properties of castings.

Tensile and Yield Strengths. If ferritic steels are compared at a given level of hardness and hardenability, the tensile and yield strengths of cast, rolled, forged, and welded metal are virtually identical, regardless of alloy content. Consequently, where tensile and yield properties are controlling criteria, the designer can interchange rolled, forged, welded, and cast steel.

Ductility. The ductility of cast steels is nearly the same as that of forged, rolled, or welded steels of the same hardness. The longitudinal properties of rolled or forged steel are somewhat higher than the properties of cast steel or weld metal. However, the transverse properties are lower by an amount that depends on the amount of working. When service conditions involve multidirectional loading, the nondirectional characteristic of cast steels may be advantageous.

The toughness and impact resistance of steel depend on metal temperature, soundness, strength, and microstructure, which in turn are controlled by chemical composition and heat treatment. This is true for both cast and wrought steels. Gross inclusions, segregation, and high gas content, singly or in combination, will cause large variations

366 / Carbon and Low-Alloy Steels

Table 2 Summary of specification requirements for various alloy steel castings with chromium contents up to 10%

Material class(a)	Tensile strength(b) MPa	ksi	Yield strength(b) MPa	ksi	Minimum elongation in 50 mm (2 in.), %	Minimum reduction in area, %	Composition(c), % C	Mn	Si	Cr	Ni	Mo	Other
ASTM A 148: steel castings for structural applications(d)													
115–95	795	115	655	95	14	30	...	...	...	...	...	...	(e)
135–125	930	135	860	125	9	22	...	...	...	...	...	...	(e)
150–135	1035	150	930	135	7	18	...	...	...	...	...	...	(e)
160–145	1105	160	1000	145	6	12	...	...	...	...	...	...	(e)
165–150	1140	165	1035	150	5	20	...	...	...	...	...	...	(f)
165–150L	1140	165	1035	150	5	20	...	...	...	...	...	...	(f)
210–180	1450	210	1240	180	4	15	...	...	...	...	...	...	(f)
210–180L	1450	210	1240	180	4	15	...	...	...	...	...	...	(f)
260–210	1795	260	1450	210	3	6	...	...	...	...	...	...	(f)
260–210L	1795	260	1450	210	3	6	...	...	...	...	...	...	(f)
SAE J435c: see Table 1 for the carbon steel castings specified in SAE J435c													
0105	725	105	586	85	17	35	...	...	...	...	...	...	(h)
0120	827	120	655	95	14	30	...	...	...	...	...	...	(h)
0150	1035	150	862	125	9	22	...	...	...	...	...	...	(h)
0175	1207	175	1000	145	6	12	...	...	...	...	...	...	(h)
ASTM A 217: alloy steel castings for pressure-containing parts and high-temperature service													
WC1	450–620	65–90	240	35	24	35	0.25	0.50–0.80	0.60	0.35(i)	0.50(i)	0.45–0.65	(i)(j)
WC4	485–655	70–95	275	40	20	35	0.20	0.50–0.80	0.60	0.50–0.80	0.70–1.10	0.45–0.65	(j)(k)
WC5	485–655	70–95	275	40	20	35	0.20	0.40–0.70	0.60	0.50–0.90	0.60–1.00	0.90–1.20	(j)(k)
WC6	485–655	70–95	275	40	20	35	0.20	0.50–0.80	0.60	1.00–1.50	0.50(i)	0.45–0.65	(i)(j)
WC9	485–655	70–95	275	40	20	35	0.18	0.40–0.70	0.60	2.00–2.75	0.50(i)	0.9–1.20	(i)(j)
WC11	550–725	80–105	345	50	18	45	0.15–0.21	0.50–0.80	0.30–0.60	1.00–1.75	0.50(i)	0.45–0.65	(i)(l)
C5	620–795	90–115	415	60	18	35	0.20	0.40–0.70	0.75	4.00–6.50	0.50(i)	0.45–0.65	(i)(j)
C12	620–795	90–115	415	60	18	35	0.20	0.35–0.65	1.00	8.00–10.00	0.50(i)	0.90–1.20	(i)(j)
ASTM A 389: alloy steel castings (NT) suitable for fusion welding and pressure-containing parts at high temperatures													
C23	485	70	275	40	18	35	0.20	0.30–0.80	0.60	1.00–1.50	...	0.45–0.65	(h)(m)
C24	550	80	345	50	15	35	0.20	0.30–0.80	0.60	1.00–1.25	...	0.90–1.20	(h)(m)
ASTM A 487: alloy steel castings (NT or QT) for pressure-containing parts at high temperatures													
1A (NT)	585–760	85–110	380	55	22	40	0.30	1.00	0.80	0.35(n)	0.50(n)	0.25(n)(o)	0.5 Cu(h)(n)
2B (QT)	620–795	90–115	450	65	22	45	0.30	1.00	0.80	0.35(n)	0.50(n)	0.25(n)(o)	0.5 Cu(h)(n)
1C (NT or QT)	620	90	450	65	22	45	0.30	1.00	0.80	0.35(n)	0.50(n)	0.25(n)(o)	0.5 Cu(h)(n)
2A (NT)	585–760	85–110	365	53	22	35	0.30	1.10–1.40	0.80	0.35(i)	0.50(i)	0.10–0.30	(i)(p)
2B (QT)	620–795	90–115	450	65	22	40	0.30	1.10–1.40	0.80	0.35(i)	0.50(i)	0.10–0.30	(i)(p)
2C (NT or QT)	620	90	450	65	22	40	0.30	1.10–1.40	0.80	0.35(i)	0.50(i)	0.10–0.30	(i)(p)
4A (NT or QT)	620–795	90–115	415	60	20	40	0.30	1.00	0.80	0.40–0.80	0.40–0.80	0.15–0.30	(k)(p)
4B (QT)	725–895	105–130	585	85	17	35	0.30	1.00	0.80	0.40–0.80	0.40–0.80	0.15–0.30	(k)(p)
4C (NT or QT)	620	90	415	60	20	40	0.30	1.00	0.80	0.40–0.80	0.40–0.80	0.15–0.30	(k)(p)
4D (QT)	690	100	515	75	17	35	0.30	1.00	0.80	0.40–0.80	0.40–0.80	0.15–0.30	(k)(p)
4E (QT)	795	115	655	95	15	35	0.30	1.00	0.80	0.40–0.80	0.40–0.80	0.15–0.30	(k)(p)
6A (NT)	795	115	550	80	18	30	0.38	1.30–1.70	0.80	0.40–0.80	0.40–0.80	0.30–0.40	(k)(p)
6B (QT)	825	120	655	95	15	35	0.38	1.30–1.70	0.80	0.40–0.80	0.40–0.80	0.30–0.40	(k)(p)
7A (QT)(p)	795	115	690	100	15	30	0.20	0.60–1.00	0.80	0.40–0.80	0.70–1.00	0.40–0.60	(k)(p)(r)
8A (NT)	585–760	85–110	380	55	20	35	0.20	0.50–0.90	0.80	2.00–2.75	...	0.90–1.10	(k)(p)
8B (QT)	725	105	585	85	17	30	0.20	0.50–0.90	0.80	2.00–2.75	...	0.90–1.10	(k)(p)
8C (QT)	690	100	515	75	17	35	0.20	0.50–0.90	0.80	2.00–2.75	...	0.90–1.10	(k)(p)
9A (NT or QT)	620	90	415	60	18	35	0.33	0.60–1.00	0.80	0.75–1.10	0.50(i)	0.15–0.30	(i)(p)
9B (QT)	725	105	585	85	16	35	0.33	0.60–1.00	0.80	0.75–1.10	0.50(i)	0.15–0.30	(i)(p)
9C (NT or QT)	620	90	415	60	18	35	Composition same as 9A (NT or QT) but with a slightly higher tempering temperature						
9D (QT)	690	100	515	75	17	35	0.33	0.60–1.00	0.80	0.75–1.10	0.50(i)	0.15–0.30	(i)(p)
10A (NT)	690	100	485	70	18	35	0.30	0.60–1.00	0.80	0.55–0.90	1.40–2.00	0.20–0.40	(k)(p)
10B (QT)	860	125	690	100	15	35	0.30	0.60–1.00	0.80	0.55–0.90	1.40–2.00	0.20–0.40	(k)(p)
11A (NT)	485–655	70–95	275	40	20	35	0.20	0.50–0.80	0.60	0.50–0.80	0.70–1.10	0.45–0.65	(p)(s)
11B (QT)	725–895	105–130	585	85	17	35	0.20	0.50–0.80	0.60	0.50–0.80	0.70–1.10	0.45–0.65	(p)(s)
12A (NT)	485–655	70–95	275	40	20	35	0.20	0.40–0.70	0.60	0.50–0.90	0.60–1.00	0.90–1.20	(p)(s)
12B (QT)	725–895	105–130	585	85	17	35	0.20	0.40–0.70	0.60	0.50–0.90	0.60–1.00	0.90–1.20	(p)(s)
13A (NT)	620–795	90–115	415	60	18	35	0.30	0.80–1.10	0.60	0.40(t)	1.40–1.75	0.20–0.30	(p)(t)
13B (QT)	725–895	105–130	585	85	17	35	0.30	0.80–1.10	0.60	0.40(t)	1.40–1.75	0.20–0.30	(p)(t)
14A (QT)	825–1000	120–145	655	95	14	30	0.55	0.80–1.10	0.60	0.40(t)	1.40–1.75	0.20–0.30	(p)(t)
16A (NT)(u)	485–655	70–95	275	40	22	35	0.12(v)	2.10(v)	0.50	0.20(s)	1.00–1.40	0.10(s)	(s)(w)

(a) NT, normalized and tempered; QT, quenched and tempered. (b) When a single value is shown, it is a minimum. (c) When a single value is shown, it is a maximum. (d) Unless specified by the purchaser, the compositions of cast steels in ASTM A 148 are selected by the producer and therefore may include either carbon or alloy steels; see Table 1 for the lower-grade steels specified in ASTM A 148. (e) 0.06% S (max), 0.05% P (max). (f) 0.020% S (max), 0.020% P (max). (g) Similar to the cast steel in ASTM A 148. (h) 0.045% S (max), 0.040% P (max). (i) When residual maximums are specified for copper, nickel, chromium, tungsten, and vanadium, their total content shall not exceed 1.00%. (j) 0.50% Cu (max), 0.10% W (max), 0.045% S (max), 0.04% P (max). (k) When residual maximums are specified for copper, nickel, chromium, tungsten, and vanadium, their total residual content shall not exceed 0.60%. (l) 0.35% Cu (max), 0.03% V (max), 0.015% S (max), 0.020% P (max). (m) 0.15–0.25% V. (n) The specified residuals of copper, nickel, chromium, and molybdenum (plus tungsten), shall not exceed a total content of 1.00%. (o) Includes the residual content of tungsten. (p) 0.50% Cu (max), 0.10% W (max), 0.03% V (max), 0.045% S (max), 0.04% P (max). (q) Material class 7A is a proprietary steel and has a maximum thickness of 63.5 mm (2½ in.). (r) Specified elements include 0.15–0.50% Cu, 0.03–0.10% V, and 0.002–0.006% B. (s) When residual maximums are specified for copper, nickel, chromium, tungsten, molybdenum, and vanadium, their total shall not exceed 0.50%. (t) When residual maximums are specified for copper, nickel, chromium, tungsten, and vanadium, their total content shall not exceed 0.75%. (u) Low-carbon grade with double austenitization. (v) For each reduction of 0.01% C below the maximum, an increase of 0.04% Mn is permitted up to a maximum of 2.30%. (w) 0.20% Cu (max), 0.10% W (max), 0.02% V (max), 0.02% S (max), 0.02% P (max)

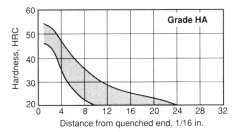 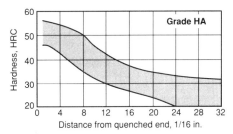 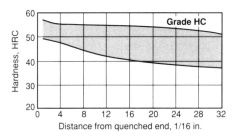

Fig. 2 End-quench hardenability limits for the hardenability grades of cast steel specified in SAE J435c. The nominal carbon content of these steels is 0.30% C (see Table 1). Manganese and other alloying elements are added as required to produce castings that meet these limits.

in impact resistance and may decrease it to a dangerously low level.

Several test methods are available for evaluating the toughness of steels and the resistance to sudden or brittle fracture. These include the Charpy V-notch impact test, the drop-weight test, the dynamic tear test, and specialized procedures to determine plane-strain fracture toughness. Each of these methods offers specific advantages that are unique to the test method. When notch toughness, drop-weight test results, or fracture toughness considerations are expected to be important in a particular casting application, it is often helpful to discuss these needs with the potential suppliers of the casting.

Charpy V-Notch Impact Tests. The notched-bar impact test is used to evaluate the toughness of materials from the impact energy required for fracture. Charpy V-notch impact testing at various temperatures is also particularly useful for determining the transition temperature from ductile to brittle fracture.

Because cast steels are nondirectional, their impact properties usually fall somewhere between the longitudinal and transverse properties of wrought steels with similar composition. The impact properties of wrought steels are usually listed for the longitudinal direction; the values shown are higher than those for cast steels of equivalent composition and thermal treatment. The transverse impact properties of wrought steels are usually 50 to 70% of those in the longitudinal direction above the transition temperature and, in some conditions of composition and degrees of working, even lower.

Impact properties are controlled by microstructure and, in general, are not significantly affected by microshrinkage or hydrogen. The effect of hardness, microstructure, and composition on the Charpy V-notch impact resistance at −40 °C (−40 °F) for cast modified 8630 steel and a cast molybdenum-boron steel are shown in Fig. 5. At the same hardness, both steels exhibit better impact resistance with a tempered martensitic microstructure than with a pearlitic one. The difference is of considerable magnitude. With either microstructure or steel composition, impact resistance decreases with higher hardness. However, increasing hardness has less effect with a pearlitic microstructure. For applications requiring good impact resistance, microstructure must be considered in selection as well as chemical composition and hardness. Further information on the effect of microstructure as controlled by composition and heat treatment is discussed in the article "Notch Toughness of Steels" in this Volume. Curves of impact energy versus temperature for casting steels are also presented in the section "Low-Temperature Toughness" in this article.

Drop-Weight Tests. Continuing efforts to provide a closer correlation between laboratory and service conditions have led to the development of the drop-weight test (DWT). Instead of a notched specimen, the DWT uses a rectangular plate with a crack-inducing bead of hardfacing metal on one face. The plate is supported, bead down, as a simple beam and is struck by a falling weight on the top face. The test piece and procedure are fully described in ASTM E 208. There is evidence that the results of this test provide a truer picture of the transition temperature range than is possible to obtain from the standard notched-bar impact test.

Nil ductility transition temperatures (NDTT) ranging from 38 °C (100 °F) to as low as −90 °C (−130 °F) have been recorded in tests on normalized and tempered cast carbon and low-alloy steels in the yield strength range of 207 to 655 MPa (30 to 95 ksi) (Fig. 6). Comparison of the data in Fig.

Fig. 3 Room-temperature properties of cast carbon steels after various heat treatments

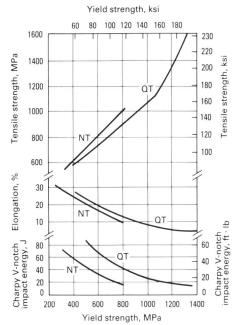

Fig. 4 Room-temperature properties of cast low-alloy steels. QT, quenched and tempered; NT, normalized and tempered

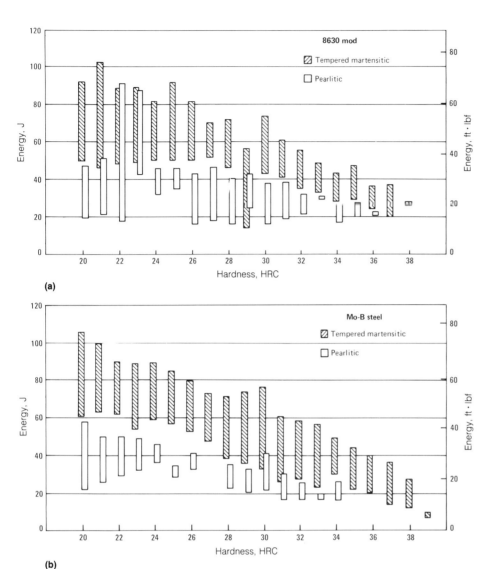

Fig. 5 Effect of hardness, microstructure, and composition on Charpy V-notch properties at −40 °C (−40 °F) for (a) modified 8630 alloy cast steel and (b) molybdenum-boron cast steel

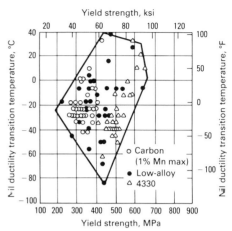

Fig. 6 Nil ductility transition temperatures and yield strengths of normalized and tempered commercial cast steels

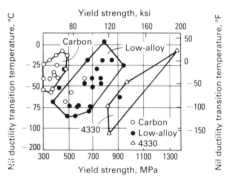

Fig. 7 Nil ductility transition temperatures and yield strengths of quenched and tempered commercial cast steels

6 with those of Fig. 7 shows the superior toughness values at equal strength levels that low-alloy steels offer compared to carbon steels. Depending on alloy selection, NDTT values of as high as 10 °C (50 °F) to as low as −107 °C (−160 °F) can be obtained in the yield strength range of 345 to 1345 MPa (50 to 195 ksi) (Fig. 7).

Testing for NDTT is described in ASTM E 208. An approximate relationship exists between the Charpy V-notch impact energy-temperature behavior and the NDTT value. The NDTT value frequently coincides with the ductile-brittle transition temperature determined in Charpy V-notch tests.

Fracture Mechanics Tests. Another toughness test that has found increasing use is the plane-strain fracture toughness test, which uses a precracked specimen. The plane-strain fracture toughness parameter, K_{Ic}, measured by this test allows a designer to determine how large a crack can exist in a critical section of a particular thick-walled casting without producing sudden brittle fracture when subjected to the maximum design load.

Fracture mechanics tests have the advantage over conventional toughness tests of being able to yield material property values that can be used in design equations. If the plane-strain fracture toughness, K_{Ic}, of a material is known at the temperature of interest, a designer can determine the critical combination of flaw size and stress required to cause failure in one load application. In addition, a designer can calculate the remaining life of a component having a discontinuity, or compute the largest acceptable flaw, using knowledge of the crack growth rate da/dN of a material and other fracture mechanics parameters.

Plane-strain fracture data on steel castings are available for a variety of alloys. Table 3 gives the room-temperature plane-strain fracture toughness for various low-alloy cast steels. Plane-strain fracture toughness of nickel-chromium-molybdenum cast steels is shown in Fig. 8, which indicates high K_{Ic} values of about 110 MPa√m (100 ksi√in.) at a 0.2% offset yield strength level of 1034 MPa (150 ksi). At a yield strength level of 1655 MPa (240 ksi), K_{Ic} values level off to about 66 MPa√m (60 ksi√in.). Data are also plotted in Fig. 8 for wrought plates made of comparable steel of somewhat higher carbon content. Plain-strain fracture toughness at low temperatures is discussed in the section "Low-Temperature Toughness."

Test methods for determining plane-strain fracture toughness are described in ASTM E 399. Because of the cost and complexity of plain-strain fracture toughness testing, approximations of K_{Ic} are also made on the basis of results obtained from various tensile, impact, and fatigue tests. For example, the possible relationship between K_{Ic} and the results of more simple and economic tests, such as notch toughness energy, have been investigated by Barson and Rolfe (Ref 1) and Groves and Wallace (Ref 2). These investigations correlated K_{Ic} values with Charpy V-notch ener-

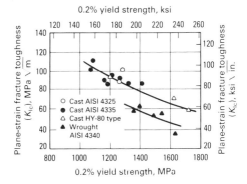

Fig. 8 Plane-strain fracture toughness K_{Ic} and strength relationships at room temperature for quenched and tempered nickel-chromium-molybdenum steels

gy measurements and reported the equations:

$$\left(\frac{K_{Ic}}{YS}\right)^2 = 5\left(\frac{CVN}{YS}\right) - 0.05 \quad (Eq\ 1)$$

$$\left(\frac{K_{Ic}}{YS}\right)^2 = 2.786\left(\frac{CVN}{YS}\right) + 0.09 \quad (Eq\ 2)$$

where the values of the fracture toughness yield strength, and impact energy are in conventional English units of ksi√in., ksi, and ft · lbf. Although the plots of these equations intersect at 0.24 ksi√in. and 0.06 ft · lbf, the Barson and Rolfe equation (Eq 1) has a slightly steeper slope. The differences in slope may have been caused by the differences in fracture toughness specimen size; Barson and Rolfe used 25 mm (1 in.) thick specimens of wrought steel, which were too thin to provide valid values for all but 4 of the 11 steels tested, while Groves and Wallace used 75 and 125 mm (3 and 5

Table 3 Plane-strain fracture toughness of cast low-alloy steels at room temperature

Alloy type	Heat treatment(a)	Yield strength, 0.2% offset MPa	ksi	Plane-strain fracture toughness, K_I MPa √m	ksi √in.
Fe-1.25 Cr-0.5 Mo	SRANTSR	275	40	88	80
Cast 1030	NT	303	44	127	116
Fe-0.5 Cr-0.5 Mo-0.25 V	NT	367	53	55	50
Fe-0.5 C-1.5 Mn	NT	406	59	107	97
Fe-0.5 C-1 Cr	NT	413	60	58	53
Fe-0.5 C	NT	425	61	65	59
Cast 9535	NT	614	89	67	61
Fe-0.35 C-0.6 Ni-0.7 Cr-0.4 Mo	NT	683	99	64	58
Cast 4335	SLQT	747	108	69	63
Cast 9536	NT	752	109	59	54
Fe-0.3 C-1 Ni-1 Cr-0.3 Mo	NT	787	114	66	60
Cast 4335	SLQT	814	118	96	87
Cast 4335	SLQT	903	131	105	96
Cast 4335	QT	1090	158	115	105
Cast 4335	QT	1166	169	92	84
Ni-Cr-Mo	QT	1207	175	98	89
Cast 4340	QT	1207	175	115	105
Cast 4325	QT	1263	183	90	82
Cast 4325	QT	1280	186	104	95
Cr-Mo	QT	1379	200	84	76
Cast 4340	QT	1450	210	67	61

(a) SR, stress relieved; A, annealed; N, normalized; T, tempered; Q, quenched; SLQ, slack quenched

in.) thick specimens of cast steel. It is important to remember that these relationships are empirical in nature because the small size of the Charpy specimen causes plane-strain conditions to develop in specimens of extremely brittle materials only.

Fatigue Properties. The most basic method of presenting engineering fatigue data is by means of the *S-N* curve. The *S-N* curve shows the life of the fatigue specimen in terms of the number of cycles to failure, N, and the maximum applied stress, S. Additional tests have been used, and the principal findings for cast steels are highlighted in the following sections.

Constant-Amplitude Tests of Smooth Bars. The endurance ratio (fatigue endurance limit divided by tensile strength) of cast carbon and low-alloy steels, determined in Moore rotating-beam bending fatigue tests (mean stress = 0), is generally taken to be about 0.40 to 0.50 for smooth bars. The data in Table 4 show that the endurance ratio is largely independent of strength, alloying elements, or heat treatment. It is also largely unaffected by low or high temperatures. In designing cast steel structures based on the fatigue ratio, it is advisable to use 40% of the tensile strength for a smooth bar when actual fatigue test values cannot be obtained. A factor of safety is added to this approximate figure.

Fatigue notch sensitivity, q, determined in rotating-beam bending fatigue tests, is related to the microstructure of the steel (composition and heat treatment) and to strength. Table 5 indicates that q generally increases with strength—from 0.23 for annealed carbon steel at a tensile strength of 576 MPa (83.5 ksi) to 0.68 for the higher-strength normalized and tempered low-alloy steels. The quenched and tempered steels with a martensitic structure are less notch sensitive than the normalized and tempered

Table 4 Fatigue endurance limit of cast carbon and low-alloy steels

Class(a) and heat treatment(b)	Tensile strength MPa	ksi	Fatigue endurance limit MPa	ksi	Endurance ratio
Carbon steels					
60 A	434	63	207	30	0.48
65 N	469	68	207	30	0.44
70 N	517	75	241	35	0.47
80 NT	565	82	255	37	0.45
85 NT	621	90	269	39	0.43
100 QT	724	105	310	45	0.47
Low-alloy steels(c)					
65 NT	469	68	221	32	0.47
70 NT	510	74	241	35	0.47
80 NT	593	86	269	39	0.45
90 NT	655	95	290	42	0.44
105 NT	758	110	365	53	0.48
120 QT	883	128	427	62	0.48
150 QT	1089	158	510	74	0.47
175 QT	1234	179	579	84	0.47
200 QT	1413	205	607	88	0.43

(a) Class of steel based on tensile strength. (b) A, annealed; N, normalized; NT, normalized and tempered; QT, quenched and tempered. (c) Below 8% total alloy content

Table 5 Fatigue notch sensitivity of several cast carbon and low-alloy steels

Steel	Tensile strength MPa	ksi	Endurance limit Unnotched MPa	ksi	Notched MPa	ksi	Fatigue endurance ratio Unnotched	Notched	Fatigue notch sensitivity factor, q
Normalized and tempered									
1040	648	94	260	37.7	193	28	0.40	0.30	0.29
1330	685	99.3	334	48.4	219	31.7	0.49	0.32	0.44
1330	669	97	288	41.7	215	31.2	0.43	0.32	0.28
4135	777	112.7	353	51.2	230	33.3	0.45	0.30	0.45
4335	872	126.5	434	63	241	34.9	0.50	0.28	0.68
8630	762	110.5	372	54	228	33.1	0.49	0.30	0.53
Quenched and tempered									
1330	843	122.2	403	58.5	257	37.3	0.48	0.31	0.48
4135	1009	146.4	423	61.3	280	40.6	0.42	0.28	0.43
4335	1160	168.2	535	77.6	332	48.2	0.46	0.29	0.51
8630	948	137.5	447	64.9	266	38.6	0.47	0.27	0.57
Annealed									
1040	576	83.5	229	33.2	179	26	0.40	0.31	0.23

Note: Notched tests run with theoretical stress concentration factor of 2.2. (b) $q = (K_f - 1)/(K_t - 1)$ where K_f is the endurance limit notched divided by endurance limit unnotched and K_t is the theoretical stress concentration factor

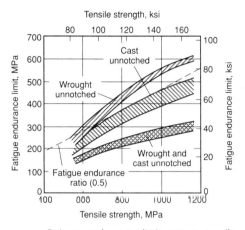

Fig. 9 Fatigue endurance limit versus tensile strength for notched and unnotched cast and wrought steels with various heat treatments. Data obtained in R.R. Moore rotating-beam fatigue tests; theoretical stress concentration factor = 2.2.

steels with a ferrite-pearlite microstructure. Similar results and trends have been reported on notch sensitivity for tests with sharper notches.

Cast steels suffer less degradation of fatigue properties due to notches than equivalent wrought steels. When laboratory test conditions are replaced with more realistic service conditions, the cast steels show much less notch sensitivity than do wrought steels. Table 6 indicates that wrought steels are 1.5 to 2.3 times as notch sensitive as cast steels. Under laboratory conditions with prepared specimens, the unnotched fatigue endurance limit of wrought steels is higher (Fig. 9). However, wrought and cast steels have similar fatigue endurance limits with notched specimens (Fig. 9).

Constant load amplitude fatigue crack growth properties for load ratios of $R = 0$ (Fig. 10a) indicate comparable properties for cast and wrought steel and slightly better properties for normalized and tempered cast carbon steel (SAE 1030) under load ratios of $R = -1$ (Fig. 10b). These tests were conducted in air at 10 to 30 Hz, depending on load ratio, initial stress intensity, and crack length. Figure 11 shows fatigue crack growth rates for various cast steels at room temperature and at low climatic temperatures.

Variable load amplitude fatigue tests indicate equal total life for cast and wrought carbon steels (cast 1030 and wrought 1020, respectively) (Fig. 12). The slower crack growth rate in the cast material compensated for the longer crack initiation life of the wrought carbon steel.

Section Size and Mass Effects. The size of a cast coupon or casting can have a marked effect on its mechanical properties. This effect reflects the influence of section size on the cooling rates achieved during heat treatment; a larger section has more mass, which slows the cooling rates within the section and thus affects the microstructure

Table 6 Fatigue notch sensitivity factors for cast and wrought carbon and low-alloy steels at a number of strength levels

Steel	Tensile strength MPa	ksi	Fatigue notch sensitivity factor, q
Annealed			
1040 cast	576	83.5	0.23
1040 wrought	561	81.4	0.43
Normalized and tempered			
1040 cast	649	94.2	0.29
1040 wrought	620	90.0	0.50
1330 cast	669	97.0	0.28
1340 wrought	702	101.8	0.65
4135 cast	777	112.7	0.45
4140 wrought	766	111.1	0.81
4335 cast	872	126.5	0.68
4340 wrought	859	124.6	0.97
8630 cast	762	110.5	0.53
8640 wrought	748	108.5	0.85
Quenched and tempered			
1330 cast	843	122.2	0.48
1340 wrought	836	121.2	0.73
4135 cast	1009	146.4	0.43
4140 wrought	1012	146.8	0.93
4335 cast	1160	168.2	0.51
4340 wrought	1161	168.4	0.92
8630 cast	948	137.5	0.57
8640 wrought	953	138.2	0.90

and mechanical properties achieved during cooling. The effect of increasing section size on the mechanical properties of a medium-carbon cast steel in the annealed and as-cast condition is shown in Fig. 13. Because of section size effects, the results of tests on specimens taken from very heavy sections and from large castings are helpful in predicting minimum properties in cast steel parts.

Mass effects are common to steels, whether rolled, forged, or cast, because the cooling rate during heat treating varies with section size and because the microstructure constituents, grain size, and nonmetallic inclusions increase in size from surface to center. However, the section size or mass effect is of particular importance in steel castings because mechanical properties are typically assessed from test bars machined from standardized coupons that are cast separately from or attached to the castings. The removal of test bars from the casting is impractical because the removal of material for testing would destroy the usefulness of the component.

Test specimens removed from a casting will not routinely exhibit the same properties as test specimens machined from the standard test coupon designs for which minimum properties are established in specifications. The mass effect discussed above, that is, the difference in cooling rate between the test coupons and the part being produced, is the fundamental reason for this situation. Several specifications provide for the mass effect by permitting the testing of coupons that are larger than normal and that have cooling rates more representative of those experienced by the part being produced. Among these specifications are ASTM E 208, A 356, and A 757.

Specimens. Mechanical properties are determined from test bars machined from coupons cast as a part of the casting, from test coupons separately cast from the same melt of steel used to produce the castings, or from test bars prepared from specified areas of the casting. Even though test specimens of the first two types are heat treated in the same operation as the castings they represent, their mechanical properties can differ significantly from those of similar specimens taken from the actual casting. Coupons machined from sections of a casting may show properties that also differ markedly from those exhibited by separately cast or cast-on test bars. The following factors contribute to the disparity: pouring and solidification conditions for separately cast or cast-on test bars can be made practically ideal, and response to heat treatment may be vastly different because of differences in cross section or thickness at the time of heat treatment unless the composition has been adequately adjusted to ensure hardening throughout the casting thickness.

For these reasons, separately cast test bars and cast-on test bars have been replaced in some instances by test bars prepared from specified areas for those castings that require a minimum weight or minimum design factor, as in aeronautical applications, or by heavy test bars representative of the casting sections, such as test bars with section T × 3T × 3T (where T is the nominal thickness of the casting). The preparation of test specimens from actual castings is costly, both from the standpoint of the casting destroyed and from the fact that much work and expense are involved in preparing a specimen. However, this type of specimen, used on a statistical basis, provides a high degree of reliability.

Heat treatment can provide the wide range of mechanical properties required for different applications. Lowest strength and hardness are obtained with low-carbon steel in the annealed condition, and highest strength is achieved with liquid-quenched alloy steel. Chemical composition and mechanical properties are usually specified as a range. Mechanical properties may be specified as minimums because there is a variation in composition and mechanical properties within any one grade and even within any one heat of steel. Variation in mechanical properties will also depend on specific response to heat treatment. Figures 3 and 4 show the effect of various heat treatments on the tensile properties of carbon and low-alloy cast steels, respectively. As Fig. 3 and 4 show, quenched and tempered steels exhibit higher ductility values for a given yield strength level than do normalized, normalized and tempered, or annealed steels. Moreover, when cast steels

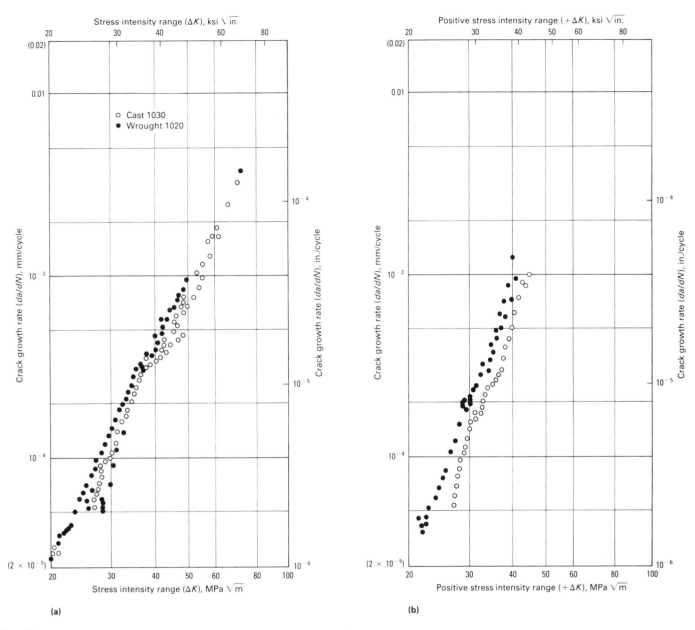

Fig. 10 Constant-amplitude fatigue behavior of normalized and tempered cast and wrought carbon steels. (a) Load ratio $R = 0$. (b) $R = -1$

are quenched and tempered, the range of strength and toughness is broadened (Fig. 4).

Low-Carbon Cast Steels

Low-carbon cast steels are those with a carbon content of less than 0.20%. Most of the tonnage produced in the low-carbon classification contains between 0.16 and 0.19% C, with 0.50 to 0.80% Mn, 0.05% P (max), 0.06% S (max), and 0.35 to 0.70% Si. In order to obtain high magnetic properties in electrical equipment, the manganese content is usually held between 0.10 and 0.20%. The properties of these dynamo steels may be slightly below those of typical low-carbon cast steels because of their manganese content.

Figure 1 includes the mechanical properties of carbon cast steels with low-carbon contents within the range of about 0.10 to 0.20%. There is very little difference between the properties of the low-carbon steels resulting from the use of normalizing heat treatments, and the properties of those that are fully annealed. In cast steels, as in rolled steels of this composition, increasing the carbon content increases strength and decreases ductility. Although the mechanical properties of low-carbon cast steels are nearly the same in the as-cast condition as they are after annealing, low-carbon steel castings are often annealed or normalized to relieve stresses and refine the structure.

Low-carbon steel castings are made in two important classes. One may be termed railroad castings, and the other miscellaneous jobbing castings. The railroad castings consist mainly of comparatively symmetrical and well-designed castings for which adverse stress conditions have been carefully studied and avoided.

Miscellaneous jobbing castings present a wide variation in design and frequently involve the joining of light and heavy sections. Varying sections make it more difficult to avoid high residual stress in the as-cast shape. Because residual stresses of large magnitude cannot be tolerated in many service applications, stress relieving becomes necessary. Therefore, the annealing of those castings is decidedly beneficial even though it may cause little improvement of mechanical properties. Castings for electrical or magnetic equipment are usually

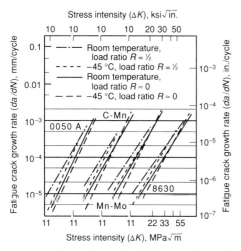

Fig. 11 Composite graph of fatigue crack growth rates for four cast steels at room temperature and −45 °C (−50 °F). Source: Ref 3

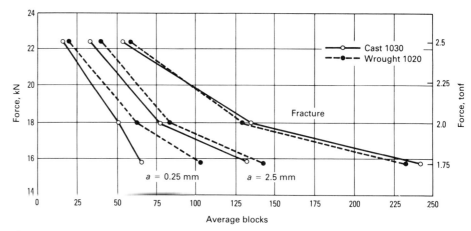

Fig. 12 Average blocks to specific crack lengths and fracture for comparable cast and wrought carbon steels in the normalized and tempered condition

fully annealed because this improves the electrical and magnetic properties.

An increase in mechanical properties can be obtained by quenching and tempering, provided the design of the casting is such that it can be liquid quenched without cracking. Impact resistance is improved by quenching and tempering, especially if a high tempering temperature is employed.

Uses. As has been mentioned, important castings for the railroads are produced from low-carbon cast steels. Some castings for the automotive industry are produced from this class of steel, as are annealing boxes, annealing bottoms, and hot-metal ladles. Low-carbon steel castings are also produced for case carburizing, by which process the castings are given a hard, wear-resistant exterior while retaining a tough, ductile core. The magnetic properties of this class of steel make it useful in the manufacture of electrical equipment. Free-machining cast steels containing 0.08 to 0.30% sulfur are also produced in low-carbon grades.

Medium-Carbon Cast Steels

The medium-carbon grades of cast steel contain 0.20 to 0.50% C and represent the bulk of steel casting production. In addition to carbon, they contain 0.50 to 1.50% Mn, 0.05% P (max), 0.06% S (max), and 0.35 to 0.80% Si. The mechanical properties at room temperature of cast steels containing from 0.20 to 0.50% C are included in Fig. 1. Steels in this carbon range are always heat treated, which relieves casting strains, refines the as-cast structure, and improves the ductility of the steel.

Unlike low-carbon castings, when medium-carbon steel castings are fully annealed, it is possible to increase the yield strength, the reduction of area, and the elongation over the entire range, compared to as-cast properties (Fig. 14). This increase is pronounced for steel with a carbon content between 0.25 and 0.50%. The hardness and tensile strength can be expected to fall off slightly following full annealing.

A very large proportion of steel castings of this grade are given a normalizing treatment, following by a tempering treatment. The improvement in mechanical properties of medium-carbon cast steel that may be expected after normalizing or normalizing and tempering is shown in Fig. 1.

If the design of a casting is suitable for liquid quenching, further improvements are possible in the mechanical properties. In fact, to develop mechanical properties to the fullest degree, steel castings should be heat treated by liquid quenching and tempering. Commercial procedure calls for tempering to obtain the desired strength level. Tempering temperatures of 650 to 705 °C (1200 to 1300 °F) are usually used to obtain higher ductility and impact properties.

High-Carbon Cast Steels

Cast steels containing more than 0.50% C are classified as high-carbon steels. This grade also contains 0.50 to 1.50% Mn, 0.05% P (max), 0.05% S (max), and 0.35 to 0.70% Si. The mechanical properties of high-carbon steels at room temperature are shown in Fig. 1. High-carbon cast steels are often fully annealed. Occasionally, a normalizing and tempering treatment is given, and for certain applications an oil quenching and tempering treatment may be used.

The microstructure of high-carbon steel is controlled by the heat treatment. Carbon also has a marked influence, for example, giving 100% pearlitic structure at eutectoid composition (~0.83% carbon). Higher proportions of carbon than eutectoid composition will increase the proeutectoid cementite, which is detrimental to the casting if it forms a network at the grain boundaries because of improper heat treatment (for example, slow cooling from above the A_{cm} temperature). Faster cooling will prevent the formation of this network and, hence, improve the properties.

Low-Alloy Cast Steels

Low-alloy cast steels contain a total alloy content of less than 8%. These steels have been developed and used extensively for meeting special requirements that cannot be met by ordinary plain carbon steels with low hardenability. The addition of alloys to plain carbon steel castings may be made for any of several reasons, such as to provide higher hardenability, increased wear resistance, higher impact resistance at increased strength, good machinability even at higher hardness, higher strength at elevated and low temperatures, and better resistance to corrosion and oxidation than the plain carbon steel castings. These materials are produced to meet tensile strength requirements of 485 to 1380 MPa (70 to 200 ksi), together with some of the above special requirements.

Alloy cast steels are used in machine tools; high-speed transportation units; steam turbines; valves and fittings; railway, automotive, excavating, and chemical processing equipment; pulp and paper machinery; refinery equipment; rayon machinery; and various types of marine equipment. They are also used in the aeronautics field.

Low-alloy cast steels may be divided into two classes according to use: those used for structural parts of increased strength, hardenability, and toughness, and those resistant to wear, abrasion, or corrosive attack under low- or high-temperature service conditions. There can be no sharp distinction between the two classes because many steels serve in both fields.

The present trend toward decreasing weight through the use of high-strength materials in lighter sections has had a marked effect on the development of low-alloy cast steels. Low-alloy grades, such as those in the 86xx, 41xx, and 43xx families, are capa-

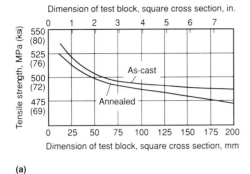
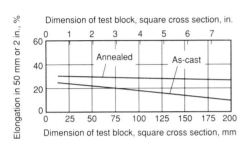
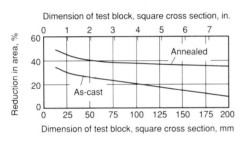
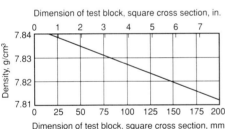
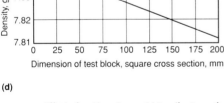

Fig. 13 Effect of section size on (a) tensile strength, (b) and (c) ductility, and (d) density of medium-carbon steel castings

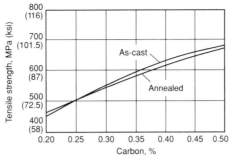
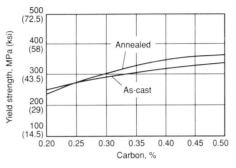
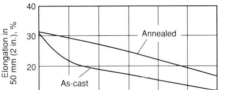
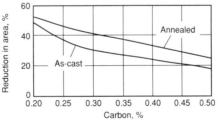

Fig. 14 Effect of annealing on the mechanical properties of medium-carbon steel castings

ble of producing mechanical properties with a yield strength 50% higher and a tensile strength 40% higher than carbon steels, with a ductility and impact resistance at least equal to unalloyed steels. Some 75 to 100 combinations of the available alloying materials have been regularly or occasionally used. It is doubtful that this many variations in composition are necessary or economical.

Alloying Elements. The compositions of low-alloy cast steels are primarily characterized by carbon contents under 0.45% and by small amounts of alloying elements, which are added to produce certain specific properties. Low-alloy steels are applied when strength requirements are higher than those obtainable with carbon steels. Low-alloy steels also have better toughness and hardenability than do carbon steels.

Carbon-Manganese Cast Steels. Manganese is the cheapest of the alloying elements and has an important effect in increasing the hardenability of steel. For this reason, many of the low-alloy cast steels now contain between 1 and 2% manganese. In the normalized steels in which grain refinement is also needed, vanadium, titanium, or aluminum is often added.

Carbon-manganese steels containing 1.00 to 1.75% Mn and 0.20 to 0.50% C have received considerable attention from engi-

neers in the past because of the excellent properties that can be developed with a single, relatively inexpensive alloying element and by a single normalizing or normalizing and tempering heat treatment. Carbon-manganese steels are also referred to as medium-manganese steels and are represented by the cast 1300 series of steels (1.60 to 1.90% Mn).

Manganese-molybdenum cast steels are very similar to the medium-manganese steels with the added characteristics of high yield strength at elevated temperatures, higher ratio of yield strength to tensile strength at room temperature, greater freedom from temper embrittlement, and greater hardenability. Therefore, these steels have replaced medium-manganese steel for certain applications.

There are two general grades of manganese-molybdenum cast steels:

- 8000 series (1.0 to 1.35% Mn, 0.10 to 0.30% Mo)
- 8400 series (1.35 to 1.75% Mn, 0.25 to 0.55% Mo)

For both of these alloy types, the selected carbon content is frequently between 0.20 and 0.35%, depending on the heat treatment employed and the strength characteristics desired.

Manganese-Nickel-Chromium-Molybdenum Cast Steels. The cast 9500 series low-alloy steels are primarily produced for their high hardenability. Sections exceeding 125 mm (5 in.) in thickness can be quenched and tempered to obtain a fully tempered martensitic structure. The composition range employed for the 9500 series is:

Element	Composition, %
Manganese	1.30–1.60
Nickel	0.40–0.70
Chromium	0.55–0.75
Molybdenum	0.30–0.40

Nickel or molybdenum with manganese refines the grain structure to a lesser extent than does vanadium, titanium, or aluminum, but each is important for increasing the ability of the steel to air harden. Chromium and vanadium impart considerable hardenability. Vanadium-containing steels are sometimes precipitation hardening and, therefore, may have higher tensile and yield strengths.

Nickel Cast Steels. Among the oldest alloy cast steels are those containing nickel. Nickel and nickel-vanadium steels are used for parts exposed to subzero conditions (such as return headers, valves, and pump castings in oil-refinery dewaxing processes) because of good notch toughness at lower temperatures. These steels are characterized by high tensile strength and elastic limit, good ductility, and excellent resistance to impact. The cast steels of the 2300 series contain 2.0 to 4.0% Ni, depending on the grade required.

Nickel-vanadium and nickel-manganese cast steels are used for structural purposes requiring wear resistance and high strength. The manganese-molybdenum cast steels are also used in these applications.

Nickel-Chromium-Molybdenum Cast Steels. The addition of molybdenum to nickel-chromium steel significantly improves hardenability and makes the steel relatively immune to temper embrittlement. Nickel-chromium-molybdenum cast steel is particularly well suited to the production of large castings because of its deep-hardening properties. In addition, the ability of these steels to retain strength at elevated temperatures extends their usefulness in many industrial applications.

Chromium-Molybdenum Cast Steels. Chromium contents of about 1.00% or more provide a nominal improvement in elevated-temperature properties. Cast steels containing chromium, molybdenum, vanadium, and tungsten have given good service in valves, fittings, turbines, and oil refinery parts, all of which are subjected to steam temperatures up to 650 °C (1200 °F).

The chromium cast steels (5100 series, 0.70 to 1.10% Cr) are not in common use in the steel casting industry. Although chromium leads the field as an alloying element for wear-resistant steels, it is seldom used alone. For example, the chromium-molybdenum steels are widely used.

Copper-Bearing Cast Steels. There are several types of copper-containing steels. Selection among these various types is primarily based on either their atmospheric-corrosion resistance (weathering steels) or the age-hardening characteristics that copper adds to steel.

High-strength cast steels cover the tensile strength range of 1200 to 2070 MPa (175 to 300 ksi). Cast steels with these strength levels and with considerable toughness and weldability were originally developed for ordnance applications. These cast steels can be produced from any of the above medium-alloy compositions by heat treating with liquid-quenching techniques and low tempering temperatures. Cast 4300 series steels or modifications thereof are usually employed.

Mechanical Properties. Figure 4 shows typical room-temperature mechanical properties of low-alloy steels plotted against yield strength. These properties are, of course, a function of alloy content, heat treatment, and section size.

Figure 15 shows the wide range of properties obtainable through changes in carbon and alloy content and heat treatment (note the properties for 0.30% C, 1.50% Mn, and 0.35% Mo steel). Figure 16 shows the variations in mechanical properties of a water-quenched cast 8630 steel as a function of tempering temperature. Section size effects were discussed in the section "Mechanical Properties" of this article.

Physical Properties

The physical properties of cast steel are generally similar to those of wrought steel.

Elastic constants of carbon and low-alloy cast steels as determined at room temperature are only slightly affected by changes in composition and structure. The modulus of elasticity, E, is about 200 GPa (30 × 10^6 psi), Poisson's ratio is 0.3, and the modulus of rigidity is 77.2 GPa (11.2 × 10^6 psi). Increasing temperature has a marked effect on the modulus of elasticity and the modulus of rigidity. A typical value of the modulus of elasticity at 200 °C (400 °F) is about 193 GPa (28 × 10^6 psi); at 360 °C (680 °F), 179 GPa (26 × 10^6 psi); at 445 °C (830 °F), 165 GPa (24 × 10^6 psi); and at 490 °C (910 °F), 152 GPa (22 × 10^6 psi). Above 480 °C (900 °F), the value of the modulus of elasticity decreases rapidly.

Density of cast steel is sensitive to changes in composition, structure, and temperature. The density of medium-carbon cast steel is about 7.8 Mg/m^3 (490 lb/ft^3). The density of cast steel is also affected somewhat by mass or size of section (Fig. 13d).

Electrical properties of carbon and low-alloy steel castings do not significantly affect usage. The only electrical property that may be regarded as having any importance is resistivity, which, for various annealed carbon steel castings with 0.07 to 0.20% C, is 0.13 to 0.14 μΩ · m. Resistivity increases with carbon content and is about 0.20 μΩ · m at 1.0% C.

Magnetic Properties. Steel castings form the housings for electrical machinery and magnetic equipment and carry only stray fluxes around the machines; hence, the magnetic properties of steel castings are less important than they were formerly when core material was manufactured from commercial cast iron and steel. Low-carbon cast dynamo steel has supplanted other cast metals for housings and frames for magnetic circuits.

The carbon content of the steel is very important in determining the magnetic properties. The maximum permeability and the saturation magnetization decrease, and the coercive force increases, as the carbon content increases. Manganese, phosphorus, sulfur, and silicon also increase the magnetic hysteresis loss in cast steels. This loss is equal to about 10 J/m^3 per cycle for B = 1 T for each 0.10% Mn, 0.01% S, and 0.01% P. Other factors being equal, the magnetic hysteresis loss is unaffected by more than 0.02% P. Magnetic properties change considerably, depending on the mechanical treatment and heat treatment of the steel.

Cast dynamo steels contain about 0.10% C, with other alloying elements held to a minimum; the castings are furnished in the annealed condition. Specifications require 0.05 to 0.15% C, 0.20% Mn, and 0.35 to 0.60 or 1.50 to 2.00% Si.

The magnetic properties of annealed cast dynamo steel that may normally be expected are:

Property	
Maximum permeability, mH/m	18.6
Hysteresis loss (induction for H = 11.9 kA/m), T	1.91
Saturation magnetization, T	2.14
Residual induction, T	1.10
Coercive force, A/m	29

As the carbon content is increased, maximum permeability and saturation magnetization decrease, and coercive force increases. Also, an increase in manganese and sulfur content increase the magnetic hysteresis loss.

Silicon and aluminum eliminate the allotropic transformation in iron and permit annealing at high temperature without recrystallization during cooling; thus, large grains can be obtained. These elements can be added in large quantities without affecting magnetic properties, but they do reduce the saturation value and increase the brittleness of the metal. Hysteresis loss varies directly with grain size number; therefore, the larger the grain size, the better the properties. Residual alloy content should be low because it lowers saturation value.

The factors that improve the machinability of dynamo steel decrease the magnetic properties. A disadvantage in the use of pure iron for dynamo steel is low resistivity. The iron must be rolled thin to keep eddy currents down; otherwise, the magnetic properties will be poor.

Volumetric Changes. In the foundry, all volume changes of a metal are pertinent, whether they occur in the liquid state, during solidification, or in the solid state. Of particular interest is the contraction that results when molten steel solidifies.

Volume changes that occur in the liquid state as the cast metal cools affect the planning for adequate metal to fill the mold. Contraction is of the order of 0.9% per 100 °C (180 °F) for a 0.30% C steel. The exact amount of contraction will vary with the chemical composition, but it is usually within the range of 0.8 to 1.0% per 100 °C (180 °F) for carbon and low-alloy steels. A larger contraction occurs upon solidification (2.2% for nearly pure iron to 4% for a 1.00% C steel). For cast carbon and low-alloy steels, a solidification contraction of 3.0% is generally assumed.

The greatest amount of contraction occurs as the solidified metal cools to room temperature. Solid-state contraction from the solidus to room temperature varies between 6.9 and 7.4% as a function of carbon content. Alloying elements have no signifi-

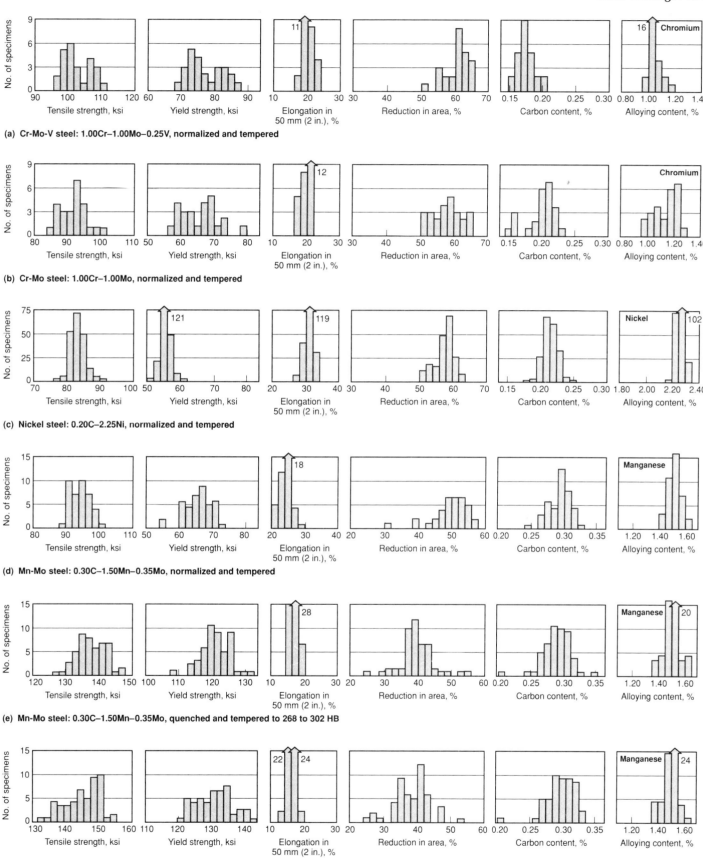

Fig. 15 Distribution of mechanical properties and carbon and alloy contents for alloy steel castings. (a) Cr-Mo-V steel, 1.00Cr-1.00Mo-0.25V, normalized and tempered; 25 heats. (b) Cr-Mo steel, 1.00Cr-1.00Mo, normalized and tempered; 25 heats. (c) Nickel steel, 0.20C-2.25Ni, normalized and tempered; 200 heats. (d) Mn-Mo steel, 0.30C-1.50Mn-0.35Mo, normalized and tempered; 40 heats. (e) Mn-Mo steel, 0.30C-1.50Mn-0.35Mo, quenched and tempered; 268 to 302 HB; 50 heats. (f) Mn-Mo steel, 0.30C-1.50Mn-0.35Mo, quenched and tempered; 300 to 321 HB; 50 heats

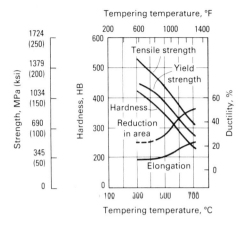

Fig. 16 Mechanical properties of water-quenched cast 8630 steel

cant effect on the amount of this contraction. The rigid form of the mold hinders contraction and results in the formation of stresses within the cooling casting that may be great enough to cause fracture or hot tears in the casting. The hot metal has low strength just after solidification. The rigidity of the mold makes the proper relation of casting configuration to accommodate this contraction one of the most important factors in producing a successful casting.

In commercial production, a combination of all three contraction components may operate simultaneously. Molten metal in contact with the mold wall solidifies quickly and proceeds to solidify toward the center of the casting. The solid envelope undergoes contraction in the solid state, while a portion of the still-molten metal is solidifying. The remaining molten metal contracts as its temperature decreases toward the freezing point. Because of contraction factors, many casting designs require considerable development to produce a sound casting.

Engineering Properties

Wear Resistance. Cast steels have wear resistance comparable to that of wrought steel of similar composition and condition. Chromium leads the field as an alloying element for wear-resistant steels but is seldom used alone. Nickel-vanadium, manganese-molybdenum, and nickel-manganese cast steels are used for numerous structural purposes requiring wear resistance and high strength.

Corrosion resistance of cast steel is similar to that of wrought steel of equivalent composition. Data published on the corrosion resistance of wrought carbon and low-alloy steels under various conditions may be applied to cast steels.

Low-alloy steels are generally not considered corrosion resistant, and casting compositions are not normally selected on the basis of corrosion resistance. In some environments, however, significant differences are observed in corrosion behavior such that the corrosion rate of one steel may be half that of another grade. In general, steels alloyed with small amounts of copper tend to have somewhat lower corrosion rates than copper-free alloys. As little as 0.05% Cu has been shown to exert a significant effect. In some environments, nominal levels of nickel, chromium, phosphorus, and silicon may also bring about modest improvements, but when these four elements are present, the addition of copper holds little if any advantage. Detailed information on the corrosion resistance of steels is available in the articles "Corrosion of Carbon Steels," "Corrosion of Alloy Steels," and "Corrosion of Cast Steels" in Volume 13 of the 9th Edition of *Metals Handbook*.

Soil Corrosion. Cast steel pipe has been tested for various periods up to 14 years in different types of soil. The results of these tests were compared directly with results from tests on wrought steel pipe of similar composition, and no significant difference in the corrosion of the two materials could be detected. However, the actual corrosion rate and rate of pitting of the cast pipe varied widely, depending on the soil and aeration conditions. Data on soil corrosion of pipe are summarized in the article "Soil Corrosion" in Volume 1 of the 9th Edition of *Metals Handbook*.

Elevated-Temperature Properties. Steels operating at temperatures above ambient are subject to failure by a number of mechanisms other than mechanical stress or impact. These include oxidation, hydrogen damage, sulfide scaling, and carbide instability, which manifests itself as graphitization.

The environmental factors involved in elevated-temperature service (370 to 650 °C, or 700 to 1200 °F) require that steels used in this temperature range be carefully characterized. As a consequence, four ASTM specifications have been developed for cast carbon and low-alloy steels for elevated-temperature service. One of these specifications, ASTM A 216, describes carbon steels; the other three, A 217, A 356, and A 389, cover low-alloy steels.

The two alloying elements common to nearly all the steel compositions used at

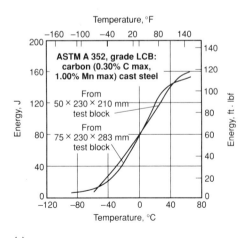
(a)

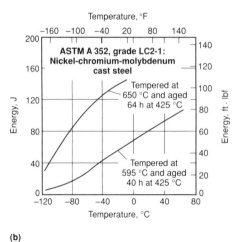
(b)

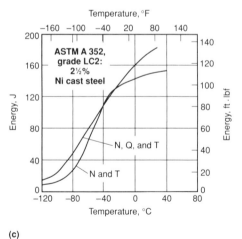
(c)

Fig. 17 Effect of temperature on the Charpy V-notch impact energy of a carbon steel and two low-alloy cast steels specified in ASTM A 352 for low-temperature service. (a) Charpy V-notch energies for a carbon cast steel, 0.30% C (max) with 1.00% Mn (max), quenched, tempered, and stress relieved, taken from a 50 × 230 × 210 mm (2 × 9 × 8¼ in.) test block and from a 75 × 230 × 283 mm (3 × 9 × 11⅛ in.) test block. (b) Charpy V-notch energies for nickel-chromium-molybdenum cast steel specimens (taken from 50 × 230 × 210 mm, or 2 × 9 × 8¼ in., test block) from steel with two different tempering and aging treatments after being air cooled from 955 °C (1750 °F), reheated to 900 °C (1650 °F), and then water quenched. (c) Charpy V-notch energies for 2½% Ni cast steel specimens (taken from 75 × 230 × 283 mm, or 3 × 9 × 11⅛ in., test block) after being air cooled (normalized, N) from 900 °C (1650 °F) and either tempered (T) at 620 °C (1150 °F) or reheated to 900 °C (1650 °F), water quenched (Q) and then tempered at 620 °C (1150 °F). All specimens were taken at locations greater than one-fourth the thickness in from the surface of test blocks having an ASTM grain size of 6 to 8. The curves represent average values for several tests at each test temperature.

Steel Castings / 377

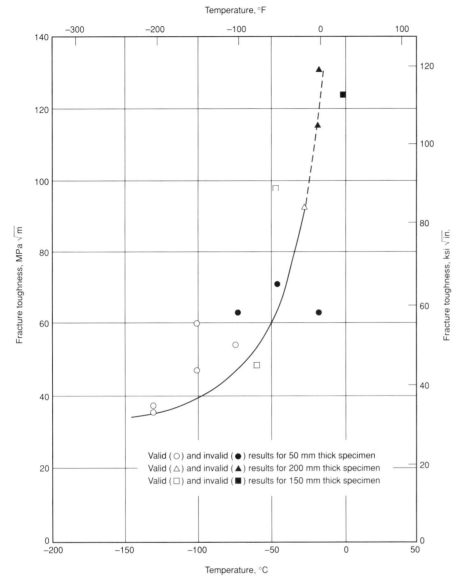

Fig. 18 Temperature dependence of plane-strain fracture toughness of a carbon steel casting (grade WCC of ASTM A 216). Test blocks were 508 × 508 × 1219 mm (20 × 20 × 48 in.), annealed 8 h at 900 °C (1650 °F), furnace cooled to 315 °C (600 °F), tempered 8 h at 605 °C (1125 °F), furnace cooled to 315 °C (600 °F), reheated to 955 °C (1750 °F) and held 8 h, furnace cooled to 900 °C (1650 °F), equalized, accelerated cooled to 95 °C (200 °F), final tempered 8 h at 650 °C (1200 °F), and air cooled. Compact tension specimens of three thicknesses as indicated were used. The open data points are the only symbols that indicate valid test results. Source: Ref 4

treatment on the impact resistance of grade LC2-1.

Fracture Toughness. Figure 18 shows the effect of temperature on the fracture toughness of a carbon (0.25% C, max) cast steel specified in ASTM A 216. Table 7 shows the effect of lower temperatures on the fracture toughness of five cast steels. As described below, the fracture toughness values in Table 7 are reported in terms of either J_{Ic} or J_c when circumstances prevented the direct determination of K_{Ic}.

For cost-effective testing, the J-integral approach was used to estimate the fracture toughness of some of the steels in Table 7. The J-integral approach can be used to estimate the fracture toughness of steels having considerable elastic-plastic behavior (Ref 5) without using large specimens. Four of the steels exhibited appreciable ductility and low yield strength in smooth-specimen tensile tests, which could require that specimen thickness and crack length range from about 50 to 500 mm (2 to 20 in.) for valid K_{Ic} determinations. The larger dimensions are unreasonable for cost-effective testing. Therefore, the J-integral was used to estimate fracture toughness characteristics without the use of large specimens. For linear elastic plane strain, J_{Ic} is related to K_{Ic} for linear elastic plane strain:

$$J_{Ic} = G_{Ic} = \frac{K_{Ic}^2}{E}(1 - \nu^2) \qquad \text{(Eq 3)}$$

where G is the strain energy release rate per unit crack extension, E is Young's modulus, and ν is Poissons' ratio.

Valid values of J_{Ic}, and hence conservative estimates of K_{Ic}, were obtained for four of the cast steels at room temperature and two cast steels at −45 °C (−50 °F) (Table 7). Only J_c, and hence estimates of K_c, were obtained for the other material and temperature conditions. The tests that did not produce valid J_{Ic} values were in the Charpy V-notch transition temperature energy region. Landes *et al.*, using A 471 wrought steel, showed that J_c fracture toughness values obtained in this region from small J-integral specimens had substantial scatter and that the lower limit of this J_c scatter band was similar to that of equivalent toughness measured on larger specimens (Ref 6). Landes *et al.* suggested that the lower boundary value of the J_c scatter may be reasonable to use for conservative-design criteria in the Charpy V-notch transition temperature region.

As shown in Table 7, Mn-Mo cast steel exhibited the highest fracture toughness (J_{Ic} or K_{Ic}) at both room temperature and −45 °C (−50 °F), while the 0050A cast steel showed the lowest fracture toughness (J_c or K_c) at both temperatures. The three martensitic cast steels (C-Mn, Mn-Mo, and 8630) had better fracture toughness at room temperature than the two ferritic-pearlitic cast

elevated temperatures are molybdenum and chromium. Molybdenum contributes greatly to creep resistance. Depending on microstructure, it has been shown that 0.5% Mo reduces the creep rate of steels by a factor of at least 10^3 at 600 °C (1110 °F).

Chromium also reduces the creep rate, although modestly, at levels to approximately 2.25%. At higher chromium levels, creep resistance is somewhat reduced. Vanadium improves creep strength and is indicated in some specifications. Other elements that improve creep resistance include tungsten, titanium, and niobium. The effect of tungsten is similar to that of molybdenum, but on a weight percent basis more tungsten is required in order to be equally beneficial. Titanium and niobium have been shown to improve the creep properties of carbon-free alloys, but because they remove carbon from solid solution, their effect tends to be variable. None of the latter three elements appears in U.S. specifications for cast steels for elevated-temperature service.

Low-Temperature Toughness. In addition to the soundness, strength, and microstructure of a metal, toughness, too, is strongly affected by temperature. Steel castings suitable for low-temperature service are specified in ASTM A 352 and A 757. Figure 17 shows the effect of temperature on the impact resistance of three grades of cast steels conforming to ASTM A 352. Figure 17(b) also shows the effect of heat

Table 7 Fracture toughness values for five cast steels at room temperature and −45 °C (−50 °F)

Steel	Room temperature								
	J_{Ic}		J_c		K_{Ic}		K_c		
	kJ/m²	in. · lbf/in.²	kJ/m²	in. · lbf/in.²	MPa √m	ksi √in.	MPa √m	ksi √in.	
0030	73	415	...	...	130	118	...	...	
0050A	...	...	37(a)	209	...	...	92(a)	84	
	...	...	25(b)	145	...	...	77(b)	70	
C-Mn	84	479	...	...	138	126	...	...	
Mn-Mo	139	794	...	...	179	163	...	...	
8630	80	456	...	...	135	123	...	...	

Steel	−45 °C (−50 °F)								Percent decrease
	J_{Ic}		J_c		K_{Ic}		K_c		
	kJ/m²	in. · lbf/in.²	kJ/m²	in. · lbf/in.²	MPa √m	ksi √in.	MPa √m	ksi √in.	
0030	...	...	49(a)	282	...	...	108(a)	98	32
	...	...	92(b)	...	...	...	92(b)	85	40
0050A	...	...	17(a)	95	...	...	61(a)	56	55
	...	...	14(b)	78	...	...	56(b)	51	46
C-Mn	75	428	...	...	132	120	...	...	11
Mn-Mo	118	674	...	...	166	151	...	...	15
8630	...	...	38(a)	218	...	...	94(a)	86	52
	...	...	30(b)	174	...	...	85(b)	77	62

(a) Average value. (b) Lowest value. Source: Ref 3

steels (0030 and 0050A). The 8630 cast steel had the largest decrease in fracture toughness at −45 °C (−50 °F) compared to room temperature. The C-Mn and Mn-Mo steels had ductile stable crack growth and the highest J values at both room temperature and −45 °C (−50 °F). Based on J-integral tests, they were the best steels at both temperatures.

Machinability. Extensive lathe and drilling tests on steel castings have not revealed significant differences in the machinability of steels made by different melting processes, nor of wrought and cast steel, provided strength, hardness, and microstructure are equivalent. The skin or surface on a sand mold casting often wears down cutting tools rapidly, possibly because of adherence of abrasive mold materials to the casting. Therefore, the initial cut should be deep enough to penetrate below the skin, or the cutting speed may be reduced to 50% of that recommended for the base metal.

Microstructure has considerable effect on the machinability of cast steels. It is sometimes possible to improve the machining characteristics of a steel casting by 100% through normalizing, normalizing and tempering, or annealing.

Weldability. Steel castings have welding characteristics comparable to those of wrought steel of the same composition, and welding these castings involves the same considerations.

The severe quenching effect produced when using a small welding rod to weld a large section results in the formation of martensite in the base metal area immediately adjacent to the weld (in the heat-affected zone). This can happen even in low-carbon steel, causing loss of ductility in the heat-affected zone. Usually cast steels with a maximum of 0.20% C and 0.50% Mn present fewer problems from this effect. However, it is essential that all of the carbon steels (with more than 0.20% C) and the air-hardening alloyed steels be preheated before welding at the standard recommended temperatures, maintaining a proper interpass temperature, and then postweld heat treated to produce sufficient ductility.

To prevent cracking in carbon and low-alloy cast steels, the hardness of the weld bead should not exceed 350 HV, except where conditions are such that only compressive forces result from the welding. This value may not be low enough in configurations in which extreme restraint is involved.

Virtually all castings receive a stress-relief heat treatment after welding, even composite fabrications in which steel castings are welded to wrought steel shapes.

The maximum compositional limits that have been placed by the industry on readily weldable grades of castings are 0.35% C, 0.70% Mn, 0.30% Cr, and 0.25% Mo (max) plus W, with a total of 1.00% undesirable elements, predicated on the widespread use of stress-relief heat treatment in the steel casting industry. For each 0.01% decrease in the specified maximum carbon content, most specifications permit an increase of 0.04% Mn above the maximum specified, up to a maximum of 1.00% (ASTM A 27, grade 70-40, and A 216, grade WCC, allow up to a maximum of 1.40% Mn). Specifications for weldable grades of cast steel are ASTM A 27, A 216, A 217, A 352, A 487, and A 757. Specifications covering control of weld quality are ASTM E 164 and E 390.

Many welds that fail do not fail in the weld but in the zone immediately adjacent to the weld. While the weld is being made, this zone is heated momentarily to a melting temperature. The temperature decreases as the distance from the weld increases. Such heating induces structural changes, specifically, the development of hard, brittle areas adjacent to the weld deposits, which reduce the toughness of the area and frequently cause cracking during and after cooling. Likewise, certain alloying elements other than carbon, such as nickel, molybdenum, and chromium, bring about air hardening of the base metal. For these reasons, the quantity of alloying elements to be used must be limited unless special precautions are taken, such as the preheating of the base metal to 150 to 315 °C (300 to 600 °F). Increased hardness in the heat-affected zone of the base metal can be removed by postweld heat treating the welded casting or by heating it for definite periods at 650 to 675 °C (1200 to 1250 °F). This also relieves stresses from welding.

For the arc welding of steel castings, a high-grade heavily coated electrode (AWS E7018 type), granular flux, or CO_2 atmosphere is generally desirable. These coatings contain little or no combustible material. Mineral coatings are often used to keep hydrogen absorption at a minimum and thereby limit underbead cracking. Selection of the number of passes and of welding conditions is similar to welding practice for wrought steels.

Welds in castings may be radiographed by gamma- or x-ray methods to ascertain the degree of homogeneity of the welded section. The most common imperfections are incomplete fusion, slag inclusions, and gas bubbles. Magnetic particle inspection is also useful in the detection of surface and near-surface cracks.

The mechanical properties of welds joining cast steel to cast steel and of welds joining cast steel to wrought steel are of the same order as similar welds joining wrought steel to wrought steel. Most tensile specimens machined across the weld will break outside the weld, in the heat-affected zone. This does not mean that the weld is stronger than the casting base metal. Closely controlled welding techniques and stress relieving are necessary to prevent brittleness in the heat-affected zone.

Nondestructive Inspection

Highly stressed steel castings for aircraft and for high-pressure or high-temperature service must pass rigid nondestructive inspection. ASTM specifications E 186, E 280, and E 446 cover radiographic standards for steel castings; E 94 covers ASTM recommended practice for radiographic testing; and E 142 covers the quality control of radiographic testing. Radiographic acceptance standards must be agreed upon by the user and producer before production begins. Critical areas to be radiographed may be identified on the casting drawing.

Magnetic particle inspection is used on highly stressed steel castings to detect surface discontinuities or imperfections at or just below the surface. ASTM E 709 should be consulted for establishing cause for rejection before the castings are produced. The imperfections are evaluated by using ASTM E 125, "Reference Photographs for Magnetic Particle Indications on Ferrous Castings." The extent of inspection and the acceptance standards must be agreed upon by the user and producer at the time of the order.

Liquid penetrant inspection can be used on steel castings, but it is primarily used to inspect nonmagnetic materials such as nonferrous metals and austenitic steels for possible surface discontinuities. ASTM E 165 is the specification used, and E 433 gives the standard reference photographs for liquid penetrant inspection.

Ultrasonic testing is sometimes used on steel castings to detect imperfections below the surface in heavy sections that are 0.3 to 8.5 m (1 to 28 ft) thick. Test surfaces of castings should be free of material that would interfere with the ultrasonic test. These castings may be as cast, blasted, ground, or machined. The standard specification for ultrasonic testing is ASTM A 609, "Longitudinal-Beam Ultrasonic Inspection of Carbon and Low-Alloy Steel Castings." This technique is intended to complement ASTM recommended practice E 94 for the use of radiographic testing in the detection of discontinuities.

Hydrostatic testing, or pressure testing, is used on valves and castings intended to contain steam or fluids, such as those made to ASTM specifications A 216, A 217, A 352, A 356, A 389, and A 487. If a casting must pass a pressure test, essential factors must be noted on the blueprint, and the details of the test should be understood by the buyer and producer.

REFERENCES

1. J.M. Barson and S.T. Rolfe, Correlation Between Charpy V-Notch Test Results in the Transition-Temperature Range, in *Impact Testing of Metals*, Special Technical Publication 466, American Society for Testing and Materials, 1970, p 280-302
2. M.T. Groves and J.F. Wallace, Cast Steel Plane Fracture Toughness, Charpy V-Notch and Dynamic Tear Test Correlations, *Steel Cast. Res.*, No. 70, March 1975, p 1-9
3. "Fatigue and Fracture Toughness of Five Carbon or Low-Alloy Steels at Room or Low Climatic Temperatures," Research Report 9A, Steel Founders' Society of America, Oct 1982
4. H.D. Greenberg and W.G. Clark, Jr., A Fracture Mechanics Approach to the Development of Realistic Acceptance Standards for Heavy Walled Steel Castings, *Met. Eng. Q.*, Vol 9 (No. 3), Aug 1969, p 30-39
5. J.R. Rice, A Path Independent Integral and the Approximate Analysis of Strain Concentration by Notches and Cracks, *J. Appl. Mech.*, Vol 35, June 1968, p 379
6. J.D. Landes and D.H. Shaffer, Statistical Characteristics of Fracture in the Transition Region, in *Fracture Mechanics*, STP 700, American Society for Testing and Materials, 1980, p 368

Bearing Steels

Harold Burrier, Jr., The Timken Company

ROLLING-ELEMENT BEARINGS, whether ball bearings or roller bearings with spherical, straight, or tapered rollers, are fabricated from a wide variety of steels. In a broad sense, bearing steels can be divided into classes intended for normal service, high-temperature service, or service under corrosive conditions.

Bearings for normal service conditions, a category that includes over 95% of all rolling-element bearings, are applicable when:

- Maximum temperatures are of the order of 120 to 150 °C (250 to 300 °F), although brief excursions to 175° C (350 °F) may be tolerated
- Minimum ambient temperatures are about −50 °C (−60 °F)
- The contact surfaces are lubricated with such materials as oil, grease, or mist
- The maximum Hertzian contact stresses are of the order of 2.1 to 3.1 GPa (300 to 450 ksi)

Bearings used under normal service conditions also experience the effects of vibration, shock, misalignment, debris, and handling. Therefore, the fabrication material must provide toughness, a degree of temper resistance, and microstructural stability under temperature extremes. The material must also exhibit the obvious requirement of surface hardness for wear and fatigue resistance.

High-Carbon or Low-Carbon Steels

Traditionally, bearings have been manufactured from both high-carbon (1.00%) and low-carbon (0.20%) steels. The high-carbon steels are used in either a through-hardened or a surface induction-hardened condition in special integral bearing configurations, such as the automotive wheel spindle shown in Fig. 1. Low-carbon bearing steels are carburized to provide the necessary surface hardness while maintaining other desirable properties in the core.

The parallel development of both high- and low-carbon steels for bearing applications is rooted in history. The early European manufacturers chose to use familiar chromium-type tool steels. The American bearing manufacturers, on the other hand, added a carburized case to their soft, plain steel bearings to meet the higher hardness requirements of more highly loaded rolling-element bearings.

Both high-carbon and low-carbon materials have survived because each offers a unique combination of properties that best suits the intended service conditions. For example, high-carbon steels:

- Can carry somewhat higher contact stresses, such as those encountered in point contact loading in ball bearings
- Can be quenched and tempered, which is a simpler heat treatment than carburizing
- May offer greater dimensional stability under temperature extremes because of their characteristically lower content of retained austenite

Carburizing steels, on the other hand, offer:

Fig. 1 Cross-section of a surface-hardened high-carbon steel automotive spindle

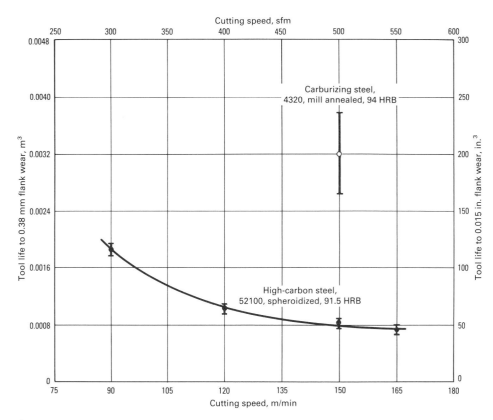

Fig. 2 Comparison of tool life at various cutting speeds in the machining of 52100 and 4320 steel tubing

- Greater surface ductility (because of their retained austenite content) to better resist the stress-raising effects of asperities, misalignment, and debris particles
- A higher level of core toughness to resist through-section fracture under severe service conditions
- A compressive residual surface stress condition to resist bending loads imposed on the ribs of roller bearings and reduce the rate of fatigue crack propagation through the cross section
- Easier machining of the base material in manufacturing

Figure 2 illustrates this last point in a comparison of tool life in machining tubing of a high-carbon steel (AISI 52100) and a carburizing steel (4320 mod) at similar hardness levels.

Table 1 compares selected mechanical properties of the core material in a quenched and tempered carburizing steel with those of the unhardened portion of an induction-hardened high-carbon steel component. The values for both strength and toughness are greater in the quenched and tempered low-carbon steel. The surface properties are largely a function of the hardness.

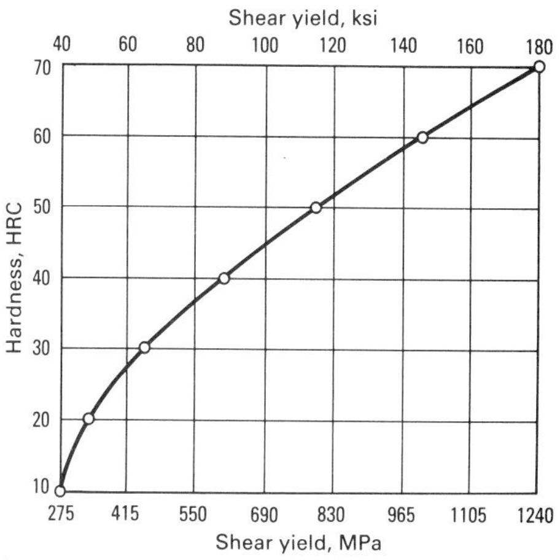

In rolling-contact bearings, it is essential to maintain an adequate strength throughout the region of maximum subsurface shear stresses. Figure 3 shows an estimated relationship between hardness and shear yield strength that is applicable to either steel type. The success of a given steel in a bearing application is not as much a function of the steel type as how it is treated. Fatigue resistance generally increases with hardness; the maximum depends on the steel type. Figure 4 compares the bending fatigue lives of through-carbon and carburized steels as a function of surface hardness. In bending fatigue, the combination of compressive residual surface stresses with a higher composite section toughness gives the advantage to the carburized steel.

Fig. 3 Estimated relationship between hardness and shear yield strength

Microstructure Characteristics

High-Carbon Bearing Steels. Figure 5 shows a typical microstructure of a hardened and tempered high-carbon bearing steel such as AISI 52100. The matrix is high-carbon martensite, containing primary carbides and 5 to 10% retained austenite. The hardness throughout the section is typically 60 to 64 HRC. Table 2 lists the compositions of selected high-carbon bearing steels currently in use. The first three grades are listed in order of increasing hardenability; they are applied to bearing sections of increasing thickness to ensure freedom from nonmartensitic transformations in hardening. Grade TBS-9 is a lower-chromium bearing steel, which, because of its residual alloy content, has a hardenability similar to that of AISI 52100. The remaining steels are representative of overseas steels applied to bearing components.

Carburizing Bearing Steels. Figure 6 shows typical case and core microstruc-

Table 1 Core properties of carburized versus induction-hardened components

Material	Hardness, HB	Yield strength MPa	ksi	Ultimate tensile strength MPa	ksi	Impact energy J	ft · lbf	Machining, %(a)
8620(b)	30–45 HRC	825–965	120–140	1105–1240	160–180	55–110	40–80	65
5160(c)	197	275	40	725	105	10	7	55
1095(c)	192	380	55	655	95	3	2	45
52100(c)	197	345	50	635	92	...	...	40

(a) Where 1212 represents 100%. (b) Quenched and tempered. (c) Annealed

Table 2 Nominal compositions of high-carbon bearing steels

Grade	C	Mn	Si	Cr	Ni	Mo
AISI 52100	1.04	0.35	0.25	1.45	...	...
ASTM A 485-1	0.97	1.10	0.60	1.05	...	...
ASTM A 485-3	1.02	0.78	0.22	1.30	...	0.25
TBS-9	0.95	0.65	0.22	0.50	0.25 max	0.12
SUJ 1(a)	1.02	<0.50	0.25	1.05	<0.25	<0.08
105Cr6(b)	0.97	0.32	0.25	1.52	...	...
SHKH15-SHD(c)	1.00	0.40	0.28	1.48	<0.30	...

(a) Japanese grade. (b) German grade. (c) Russian grade

382 / Carbon and Low-Alloy Steels

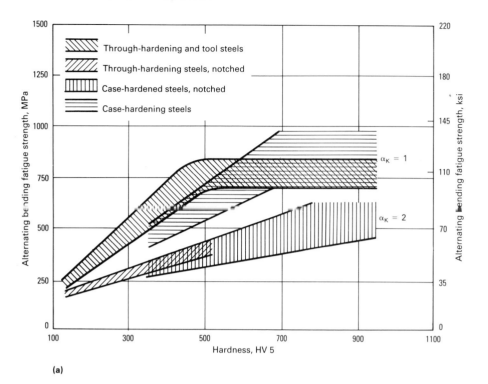

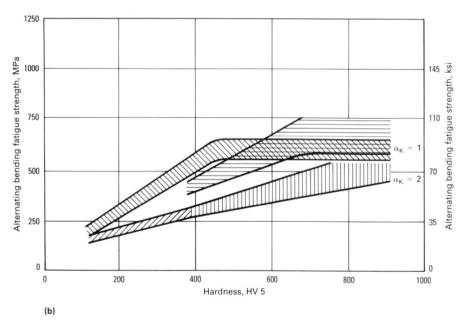

Fig. 4 Rotating-beam fatigue strength of case-hardening, through-hardening, and tool steels as a function of surface hardness. (a) Testpiece diameter of 6 mm (0.25 in.), triangular torque. (b) Testpiece diameter of 12 mm (0.5 in.), constant torque. Source: Ref 1

Table 3 Carburizing bearing steels

Grade	Composition, %					
	C	Mn	Si	Cr	Ni	Mo
4118	0.20	0.80	0.22	0.50	...	0.11
5120	0.20	0.80	0.22	0.80	...	...
8620	0.20	0.80	0.22	0.50	0.55	0.20
4620	0.20	0.55	0.22	...	1.82	0.25
4320	0.20	0.55	0.22	0.50	1.82	0.25
3310	0.10	0.52	0.22	1.57	3.50	...
SCM420	0.20	0.72	0.25	1.05	...	0.22
20MnCr5	0.20	1.25	0.27	1.15	...	...

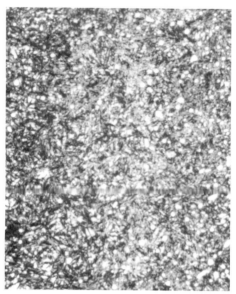

Fig. 5 Typical microstructure of a high-carbon through-hardened bearing component

tures of carburized bearing components. The case microstructure consists of high-carbon martensite with retained austenite in the range of 15 to 40%. Case hardness is typically 58 to 64 HRC. In the core of carburized bearings, the microstructure consists of low-carbon martensite; it also often contains variable amounts of bainite and ferrite. The core hardness may vary from 25 to 48 HRC.

Table 3 lists the compositions of typical carburizing bearing steels. The AISI grades are listed in approximate order of increasing hardenability or section size applicability. SCM420 and 20MnCr5 are Japanese and German grades, respectively, found in carburized bearing components. In addition to standard AISI grades, bearing steels can also be designed so that their hardenability matches the requirements of specific section thicknesses. Alloy conservation and a more consistent heat-treating response are benefits of using specially designed bearing steels.

The selection of a carburizing steel for a specific bearing section is based on the heat-treating practice of the producer, either direct quenching from carburizing or reheating for quenching, and on the characteristics of the quenching equipment. The importance of a proper case microstructure to the ability of a bearing to resist pitting fatigue is illustrated in Fig. 7. In particular, the presence of pearlite, resulting from a mismatch of quenching conditions and case hardenability, is shown to have a detrimental effect.

Bearing Steel Quality

Apart from a satisfactory microstructure, which is obtained through the proper combination of steel grade and heat treatment,

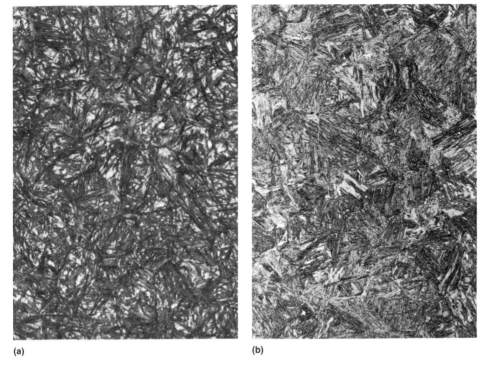

Fig. 6 Typical microstructures of carburized bearing components. (a) Case. (b) Core

the single most important factor in achieving high levels of rolling-contact fatigue life in bearings is the cleanliness, or freedom from harmful nonmetallic inclusions, of the steel. Bearing steels can be produced by one of these techniques:

- Clean-steel air-melt practices
- Electroslag remelting
- Air melting followed by vacuum arc remelting
- Vacuum induction melting followed by vacuum arc remelting

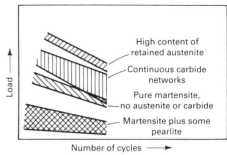

Fig. 7 Effect of surface microstructure on the shape of S-N curve for surface fatigue (pitting). Source: Ref 1

Cleanliness, cost, and reliability can increase depending on which practice is chosen.

Bearing steel cleanliness is most commonly rated by using microscopic techniques, such as those defined in ASTM A 295 for high-carbon steels and A 534 for carburizing steels. The worst fields found in metallographically prepared sections of the steel can be compared with rating charts (J-K charts) according to the type of inclusion: sulfides, stringer-type oxides, silicates, and globular-type oxides.

Table 4 lists the current levels of each of the inclusion types allowed for air-melted bearing quality steels. Several manufacturers produce bearings with significantly low-

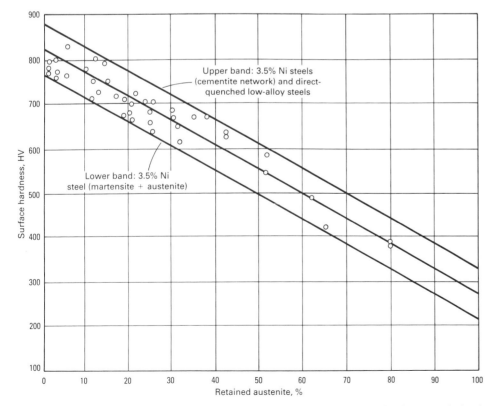

Fig. 8 Influence of retained austenite on the surface hardness of carburized alloy steels (reheat quenched and tempered at 150 to 185 °C, or 300 to 365 °F). Source: Ref 1

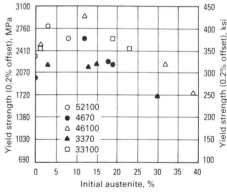

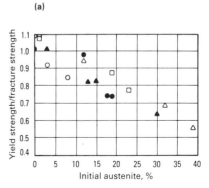

Fig. 9 Influence of retained austenite on surface ductility. (a) Yield strength data. (b) Yield-to-fracture-strength ratio

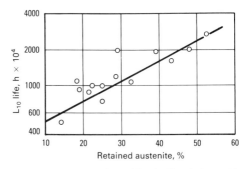

Fig. 10 Variation of 90% life of rolling fatigue with the amount of retained austenite. Source: Ref 2

er levels of nonmetallic inclusion content than allowed by rating charts.

Bearing steel cleanliness can also be rated by oxygen analysis, the magnetic particle method (AMS 2301, AMS 2300), and ultrasonic methods. Of these, the ultrasonic method appears to show a superior correlation with bearing fatigue life when the oxygen content of the steel is less than 20 ppm. This is due to the larger volume of material sampled by the technique.

Effect of Heat Treatment

The importance of matching the hardenability and quenching of a bearing steel has been pointed out above. However, within this restriction, other heat treatment variables have been found to affect the performance of bearings, particularly under the less-than-ideal conditions of debris contamination.

Retained austenite in the microstructure is known to reduce the surface hardness of

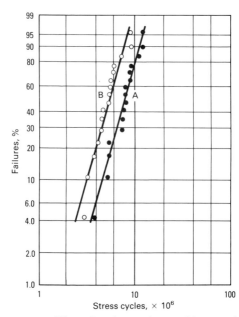

Fig. 13 Effect of surface carbon enrichment of 52100 steel on rolling-contact performance. A, carburized test bars; B, uncarburized test bars. Source: Ref 5

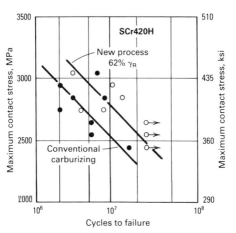

Fig. 11 Effect of carbonitriding to increase retained austenite on rolling-contact fatigue. Source: Ref 3

high-carbon steels, as shown in Fig. 8. The ductility of the surface, as expressed by the ratio of yield strength to fracture strength in Fig. 9, is improved by increasing amounts of retained austenite. This improved ductility often results in improved rolling-contact fatigue performance. Figure 10 illustrates this improvement by showing the results of a fatigue performance study of 14 carburizing steel compositions. The retained austenite level can be raised by adding nitrogen to the case (Fig. 11) or by increasing carburizing cycles (Fig. 12). Improved performance has been demonstrated in high-carbon steels when carburizing increased the carbon in solution and the austenite content. Figure 13 shows the improved performance of carburized AISI 52100 steel bars in comparison with uncarburized bars of the same material. Other means of increasing surface ductility, such as optimizing the austenitiz-

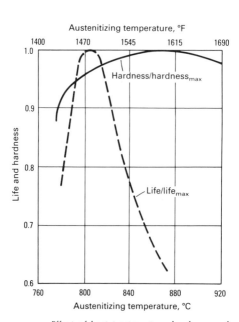

Fig. 14 Effect of heat treatment on hardness and life of 52100 steel. Source: Ref 6

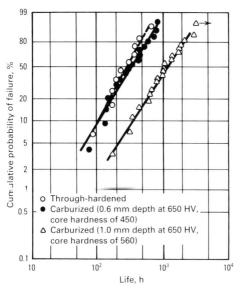

Fig. 12 Performance of through-hardened and carburized bearings in a debris environment with a load of 17.6 kN (1800 kgf) and a rotation speed of 2000 rev/min. Source: Ref 4

ing temperature (Fig. 14), controlling the quenching rate (Fig. 15), and austempering to produce a completely bainitic microstructure (Fig. 16), have been shown to improve the fatigue performance of high-carbon bearing steels for certain applications.

Special-Purpose Bearing Steels

When bearing service temperatures exceed about 150 °C (300 °F), common low-

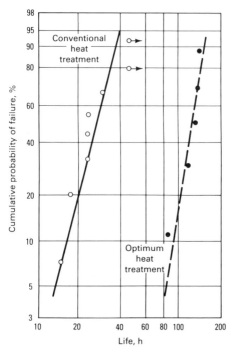

Fig. 15 Effect of controlling quench severity on the performance of 52100 steel bearings. Source: Ref 7

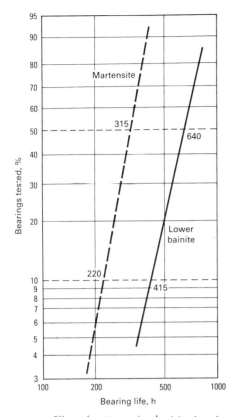

Fig. 16 Effect of austempering heat treatment on the performance of 52100 steel bearings. Source: Ref 8

alloy steels cannot maintain the necessary surface hardness to provide satisfactory fatigue life. The low corrosion resistance of these steels makes them susceptible to attack by environmental moisture, as well as aggressive gaseous or liquid contaminants. Therefore, specialized steels are often applied when these service conditions exist.

Table 5 lists the compositions of certain bearing steels suited for high-temperature service. These steels are typically alloyed with carbide-stabilizing elements such as chromium, molybdenum, vanadium, and silicon to improve their hot hardness and temper resistance. The listed maximum operating temperatures are those at which the hardness at temperature falls below a minimum of 58 HRC. Figure 17 compares the hot hardness behavior of some high-carbon tool and bearing steels to AISI 52100 steel. Table 6 indicates the effect of extended exposure to elevated temperatures on the recovered (room-temperature) hardness of various steels, both carburized and through-hardened.

An important application of the high-temperature bearing steels is aircraft and stationary turbine engines. Bearings made from M50 steel have been used in engine applications for many years. Jet engine speeds are being continually increased in order to improve performance and efficiency; therefore, the bearing materials used in these engines must have increased section toughness to withstand the stresses that result from higher centrifugal forces (Fig. 18). For this reason, the carburizing high-temperature bearing steels, such as M50-NiL and CBS-1000M, are receiving much attention. The core toughness of these steels is more than twice that of the through-hardening steels. Figure 19 compares the case and core fracture toughness of some of the common through-hardening and carburizing bearing steels. Figure 20

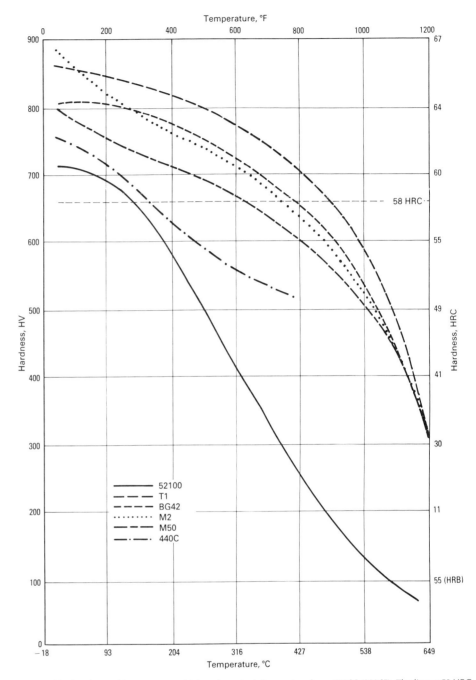

Fig. 17 Hot hardness of homogeneous high-carbon steels for service above 150 °C (300 °F). The line at 58 HRC indicates the maximum service temperature at which a basic dynamic load capacity of about 2100 MPa (300 ksi) can be supported in bearings and gears. Source: Ref 9

Table 4 Nonmetallic inclusion ratings

Specification	Type A		Type B		Type C		Type D	
	Thin	Heavy	Thin	Heavy	Thin	Heavy	Thin	Heavy
ASTM A 534-76 (carburizing steel)	3.0	2.0	3.0	2.5	2.5	1.5	2.0	1.5
ASTM A 295-84 (high-carbon steel)	2.5	1.5	2.0	1.0	0.5	0.5	1.0	1.0

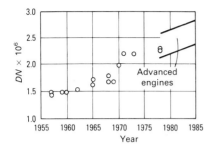

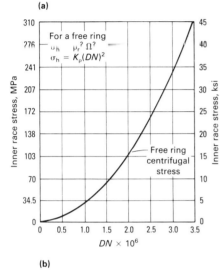

Fig. 18 Increasing section toughness of bearing materials used for jet engine applications. (a) Trend in aircraft engine main bearing in units of DN, the bearing bore diameter in millimeters multiplied by the rotation of the shaft in revolutions per minute. (b) Estimated inner race tangential stress versus bearing DN. Source: Ref 10

illustrates another benefit of the carburizing steels by comparing the compressive residual stress gradient present in carburized races with the tensile residual stresses found in through-hardened races. The presence of compressive residual stresses may help to retard the propagation of radial fatigue cracks through the race cross-section.

In general, high-temperature carburizing steels require more care in the carburizing process than conventional low-alloy carburizing steels. Because of the high content of chromium and silicon in the high-temperature steels, some precarburizing treatment, such as preoxidation, is always necessary to promote satisfactory carburizing.

Bearings that require the highest corrosion resistance necessitate the use of stainless grades with greater than 12% Cr. At this time, no satisfactory carburizing technique has been developed for these grades. Thus, all corrosion-resistant bearing steels are of the through-hardening type (Table 7). Steels such as the 440C modification, CRB-7, and BG42 also offer good high-temperature hardness. Figure 21 compares the hot hardness and hardness retention properties of selected corrosion-resistant steels to those of 52100 and M50 steels.

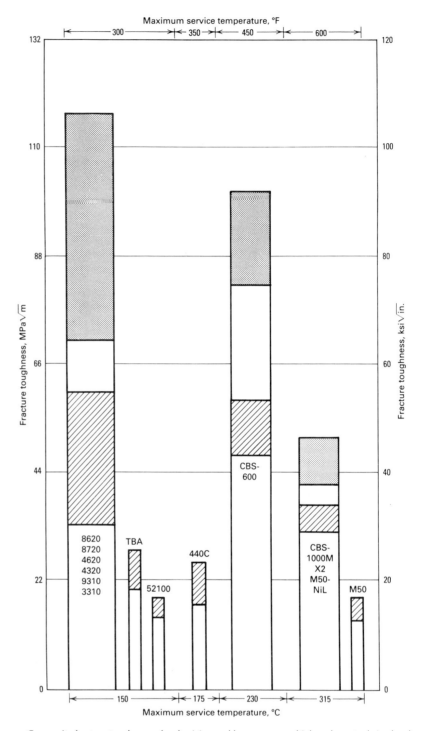

Fig. 19 Composite fracture toughness of carburizing and homogeneous high-carbon steels in slow bending. Case depth is 0.76 to 0.89 mm (0.030 to 0.035 in.) to 0.50% C level. Shaded areas indicate range of K values for cracks originating in core. Cross-hatched areas indicate range of K values for cracks originating in case. Charpy-sized specimens were carburized, hardened, tempered, and precracked to several depths in case and core regions before testing. As cracks progress inward, the fracture resistance of carburized composites improves significantly. Source: Ref 9

REFERENCES

1. *The Influence of Microstructure on the Properties of Case-Hardened Components*, American Society for Metals, 1980
2. H. Muro, Y. Sadaoka, S. Ito, and N. Tsushima, The Effect of Retained Austenite on the Rolling Fatigue of Carburized Steels, in *Proceedings of the Twelfth Japanese Congress on Materials Research* (Kyoto, Japan), Society of Materials Science, 1969
3. K. Nakamura, K. Mihara, Y. Kiba-

Table 5 Nominal compositions of high-temperature bearing steels

Steel	Composition, %								Maximum operating temperature(a)	
	C	Mn	Si	Cr	Ni	Mo	V	Other	°C	°F
M50	0.85	...	...	4.10	...	4.25	1.00	...	315	600
M50-NiL	0.13	0.25	0.20	4.20	3.40	4.25	1.20	...	315	600
Pyrowear 53	0.10	0.35	1.00	1.00	2.00	3.25	0.10	2.00 Cu	205	400
CBS-600	0.19	0.60	1.10	1.45	...	1.00	...	0.06 Al	230	450
Vasco X2-M	0.15	0.29	0.88	5.00	...	1.50	0.5	1.50 W	230	450
CBS-1000M	0.13	0.55	0.50	1.05	3.00	4.50	0.40	0.06 Al	315	600
BG42	1.15	0.50	0.30	14.5	...	4.00	1.20	...	370	700

(a) Maximum service temperature, based on a minimum hot hardness of 58 HRC

Table 6 Room-temperature hardness of CBS-600 and CBS-1000 after exposure up to 540 °C (1000 °F)

Steel type	Exposure time, 10^3 h	Hardness as heat treated, HRC	Minimum HRC after exposure for indicated time at °C (°F)						
			205 (400)	260 (500)	315 (600)	370 (700)	425 (800)	480 (900)	540 (1000)
Hardness of case layers (0.70–1.0% C)									
CBS-600	3	62	60	60	60	57	...	...	...
CBS-1000M	1	60	60	60	60	60	60	60	51
CBS-1000M(a)	1	60	59	59	59	58	58	...	...
9310	1	60	58	55	53	...	...	...	...
8620	1	60	58	56	53	47	...	...	...
52100	1	61	58	56	53	47	...	...	...
M50	1	62	62	62	62	62	62	60	52
Hardness of core regions(b)									
CBS-600	3	41	41	41	41	41	...	...	...
CBS-1000M	1	46	46	46	46	46	46	42	28
CBS-1000M(a)	1	44	44	44	44	44	44	40	27

(a) Oil quenched from 955 °C (1750 °F) rather than from the standard temperature of 1095 °C (2000 °F). (b) Core carbon 0.20% for CBS-600; 0.15% for CBS-1000M

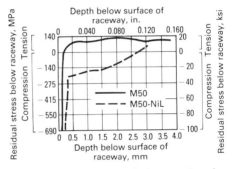

Fig. 20 Comparison of residual stresses in carburized versus through-hardened steel races. The higher residual compression of carburizing M50-NiL provides greater resistance to fracture, fatigue damage, and stress corrosion. Source: Ref 11

yashi, and T. Naito, Improvement on the Fatigue Strength of Case-Hardened Gears by a New Heat Treatment Process, in *Analysis and Design of Off-Highway Powertrains*, SP-522, Society of Automotive Engineers, 1982

4. N. Tsushima, H. Nakashima, and K. Maeda, "Improvement of Rolling Contact Fatigue Life of Carburized Tapered Roller Bearings," Paper 860725, presented at the *Earthmoving Industry Conference* (Peoria, IL), Society of Automotive Engineers, 1986
5. C.A. Stickels and A.M. Janotik, Controlling Residual Stresses in 52100 Bearing Steel by Heat Treatment, *Met. Prog.*, Sept 1981
6. H. Schlicht, Materials Properties Adapted to the Actual Stressing in a Rolling Bearing, *Ball Roller Bearing Eng.*, Vol 1, 1981
7. I. Sugiura, O. Kato, N. Tsushima, and H. Muro, "Improvement of Rolling Bearing Fatigue Life under Debris-Contaminated Lubrication by Decreasing the Crack-Sensitivity of the Material," Preprint 81-AM-1E-2, American Society of Lubrication Engineers, 1981
8. G.E. Hollox, R.A. Hobbs, and J.M. Hampshire, Lower Bainite Bearings for Adverse Environments, *Wear*, Vol 68,

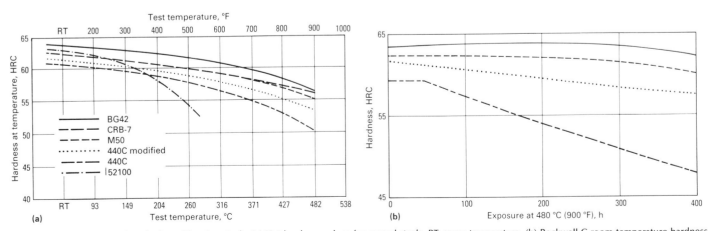

Fig. 21 Hardness properties of selected bearing steels. (a) Hot hardness values for several steels. RT, room temperature. (b) Rockwell C room-temperature hardness after exposure at 480 °C (900 °F). Source: Ref 12, 13

Table 7 Corrosion-resistant bearing steels

Grade	C	Mn	Si	Cr	Mo	W	V	Nb
BG42	1.15	0.50	0.30	14.50	4.00	...	1.2	...
440C	1.00	0.40	0.30	17.00	0.50	...	...	...
440C modified	1.05	0.40	0.30	14.00	4.00	...	...	...
CRB-7	1.10	0.35	0.30	14.00	2.0	...	1.0	0.25

9. C.F. Jatczak, Specialty Carburizing Steels for High Temperature Service, *Met. Prog.*, April 1978
10. E.N. Bamberger, B.L. Averbach, and P.K. Pearson, "Improved Fracture Toughness Bearings," AFWAL-TR-84-2103, Air Force Wright Aeronautical Laboratories, AFSC, Jan 1985
11. T.V. Philip, New Bearing Steel Beats Speed and Heat, *Power Transmission Des.*, June 1986
12. "LESCALLOY BG42," Data Sheet, Latrobe Steel Company
13. T.V. Philip, A New Bearing Steel; A New Hot Work Die Steel, *Met. Prog.*, Feb 1980

SELECTED REFERENCES

- W.F. Burd, A Carburizing Gear Steel for Elevated Temperatures, *Met. Prog.*, May 1985
- H.I. Burrier, Jr., Alloy Substitution for Flexibility and Performance, in *Proceedings of the Workshop on Conservation and Substitution Technology for Critical Metals in Bearings and Related Components for Industrial Equipment and Opportunities for Improved Bearing Performance*, United States Bureau of Mines/Vanderbilt University, 1984
- C.F. Jatczak, Hardenability in High Carbon Steels, *Metall. Trans.*, Vol 4, 1973
- J.D. Stover and R.V. Kolarik, Air-Melted Steel With Ultra-Low Inclusion Stringer Content Further Improves Bearing Fatigue Life, in *Proceedings of the 4th International Conference on Automotive Engineering*, SAE 871 20B, Society of Automotive Engineers

High-Strength Structural and High-Strength Low-Alloy Steels

HIGH-STRENGTH carbon and low-alloy steels have yield strengths greater than 275 MPa (40 ksi) and can be more or less divided into four classifications:

- As-rolled carbon-manganese steels
- As-rolled high-strength low-alloy (HSLA) steels (which are also known as microalloyed steels)
- Heat-treated (normalized or quenched and tempered) carbon steels
- Heat-treated low-alloy steels

These four types of steels have higher yield strengths than mild carbon steel in the as-hot-rolled condition (Table 1). The heat-treated low-alloy steels and the as-rolled HSLA steels also provide lower ductile-to-brittle transition temperatures than do carbon steels (Fig. 1).

These four types of high-strength steels have some basic differences in mechanical properties and available product forms. In terms of mechanical properties, the heat-treated (quenched and tempered) low-alloy steels offer the best combination of strength (Table 1) and toughness (Fig. 1). However, these steels are available primarily as bar and plate products and only occasionally as sheet and structural shapes. In particular, structural shapes (I-beams, channels, wide-flanged beams, or special sections) can be difficult to produce in the quenched and tempered condition because shape warpage can occur during quenching. Heat treating steels is also a more involved process than the production of as-rolled steels, which is one reason the as-rolled HSLA steels are an attractive alternative. The as-rolled HSLA steels are also commonly available in all the standard wrought product forms (sheet, strip, bar, plate, and structural shapes).

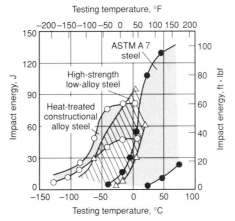

Fig. 1 General comparison of Charpy V-notch toughness for a mild-carbon steel (ASTM A 7, now ASTM A 283, grade D), an HSLA steel, and a heat-treated constructional alloy steel

This article considers four types of high-strength structural steel (which is defined here as those steels with yield strengths greater than 275 MPa, or 40 ksi): high-strength carbon steel, carbon-manganese steel, quenched and tempered low-alloy steel, and HSLA steel. Particular emphasis is placed on HSLA steels, which are an attractive alternative in structural applications because of their competitive price-per-yield strength ratios (generally, HSLA steels are priced from the base price of carbon steels but have higher yield strengths than as-rolled carbon steels). High-strength steels are used to reduce section sizes for a given design load, which allows weight savings. Reductions in section size may also be beneficial in obtaining the desired strength level during the production of structural steel. Whether steels are furnished in the as-hot-rolled or heat-treated condition, the strength levels tend to decrease as section size increases. In as-hot-rolled or normalized steel, this results from the coarser microstructure (larger grains and coarser pearlite) that develops from the slower cooling rates on the rolling mill for the thicker sections. In quenched and tempered steels, the lower strengths result because the transformation temperature increases as section thickness increases and the amount of martensite (the strongest microstructural constituent) progressively decreases. Thus, as the section size increases, it becomes more difficult to obtain the strength levels characteristic of a particular alloy.

Structural Carbon Steels

Structural carbon steels include mild steels, hot-rolled carbon-manganese steels, and heat-treated carbon steels. Mild steels and carbon-manganese steels are available in all the standard wrought forms: sheet, strip, plate, structural shapes, bar, bar-size shapes, and special sections. The heat-

Table 1 General comparison of mild (low-carbon) steel with various high-strength steels

Steel	C (max)	Chemical composition, %(a) Mn	Si	Other	Minimum yield strength MPa	ksi	Minimum tensile strength MPa	ksi	Minimum ductility (elongation in 50 mm, or 2 in.), %
Low-carbon steel	0.29	0.60–1.35	0.15–0.40	(b)	170–250	25–36	310–415	45–60	23–30
As-hot rolled carbon-manganese steel	0.40	1.00–1.65	0.15–0.40	...	250–400	36–58	415–690	60–100	15–20
HSLA steel	0.08	1.30 max	0.15–0.40	0.02 Nb or 0.05 V	275–450	40–65	415–550	60–80	18–24
Heat-treated carbon steel									
Normalized(b)	0.36	0.90 max	0.15–0.40	...	200	29	415	60	24
Quenched and tempered	0.20	1.50 max	0.15–0.30	0.0005 B min	550–690	80–100	660–760	95–110	18
Quenched and tempered low-alloy steel	0.21	0.45–0.70	0.20–0.35	0.45–0.65 Mo, 0.001–0.005 B	620–690	90–100	720–800	105–115	17–18

(a) Typical compositions include 0.04% P (max) and 0.05% S (max). (b) If copper is specified, the minimum is 0.20%.

treated grades are available as plate, bar, and, occasionally, sheet and structural shapes.

Mild (low-carbon) steels are normally considered to have carbon contents up to 0.25% C with about 0.4 to 0.7% Mn, 0.1 to 0.5% Si, and some residuals of sulfur, phosphorus, and other elements. These steels are not deliberately strengthened by alloying elements other than carbon; they contain some manganese for sulfur stabilization and silicon for deoxidation. Mild steels are mostly used in the as-rolled, forged, or annealed condition and are seldom quenched and tempered.

The largest category of mild steels is the low-carbon (<0.08% C, with ≤0.4% Mn) mild steels used for forming and packaging. Mild steels with higher carbon and manganese contents have also been used for structural products such as plate, sheet, bar, and structural sections. Typical examples include:

Steel	Minimum yield strength MPa	ksi
Hot-rolled SAE 1010 steel sheet	207	30
ASTM A 283, grade D	228	33
ASTM A 36	250	36

Before the advent of HSLA steels, these mild steels were commonly used for the structural parts of automobiles, bridges, and buildings. In automotive applications, for example, hot-rolled SAE 1010 sheet has long been used as a structural steel. However, as lighter weight automobiles became more desirable during the energy crisis, there was a trend to reduce weight by using higher-strength steels with suitable ductility for forming operations.

The trend for structural steels used in the construction of bridges and buildings has also been away from mild steels and toward HSLA steels. For many years, ASTM A 7 (now ASTM A 283, grade D) was widely used as structural steel. In about 1960, improved steelmaking methods resulted in the introduction of ASTM A 36, with improved weldability and slightly higher yield strength. Now, however, HSLA steels often provide a superior substitute for ASTM A 36, because HSLA steels provide higher yield strengths without adverse effects on weldability. Weathering HSLA steels also provide better atmospheric corrosion resistance than carbon steel.

Hot-Rolled Carbon-Manganese Structural Steels. For rolled structural plate and sections, one of the earliest approaches in achieving higher strengths involved the use of higher manganese contents. Manganese is a mild solid-solution strengthener in ferrite and is the principal strengthening element when it is present in amounts over 1% in rolled low-carbon (<0.20% C) steels. Manganese can also improve toughness properties (Fig. 2b).

Before World War II, strength in hot-rolled structural steels was achieved by the addition of carbon up to 0.4% and manganese up to 1.5%, giving yield strengths of the order of 350 to 400 MPa (50 to 58 ksi). The strengthening of these steels relies primarily on the increase in carbon content, which results in greater amounts of pearlite in the microstructure and thus higher tensile strengths. However, the high carbon contents of these steels greatly reduces notch toughness and weldability. Moreover, the increase of pearlite contents in hot-rolled carbon and alloy steels has little effect on yield strength, which, rather than tensile strength, has increasingly become the main strength criterion in structural design.

Nevertheless, carbon-manganese steels with suitable carbon contents are used in a variety of applications. Table 2 lists some high-strength carbon-manganese structural steels in the as-hot-rolled condition. If structural plate or shapes with improved toughness are required, small amounts of aluminum are added for grain refinement. Carbon-manganese steels are also used for stampings, forgings, seamless tubes, and boiler plates. Some of these steels are described according to product form in previous articles of this Volume.

High-strength structural carbon steels have yield strengths greater than 275 MPa (40 ksi) and are available in various product forms:

- Cold-rolled structural sheet
- Hot-rolled carbon-manganese steels in the form of sheet, plate, bar, and structural shapes
- Heat-treated (normalized or quenched and tempered) carbon steels in the form of plate, bar, and occasionally, sheet and structural shapes

This section focuses on the heat-treated carbon structural steels, which typically attain yield strengths of 290 to 690 MPa (42 to 100 ksi). Cold-rolled carbon steel sheet with yield strengths greater than 275 MPa (40 ksi) are discussed in the article "Carbon and Low-Alloy Steel Sheet and Strip" in this Volume. High-strength carbon-manganese steels in the as-hot-rolled condition are discussed in the previous section of this article.

The heat treatment of carbon steels consists of either normalizing or quenching and tempering. These heat treatments can be used to improve the mechanical properties of structural plate, bar, and occasionally, structural shapes. Structural shapes (such as I-beams, channels, wide-flange beams, and special sections) are primarily used in the as-hot-rolled condition because warpage is difficult to prevent during heat treatment. Nevertheless, some normalized or quenched and tempered structural sections can be produced in a limited number of section sizes by some manufacturers.

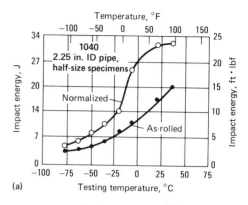

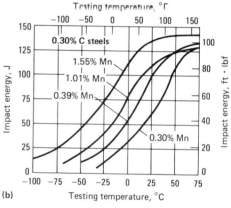

Fig. 2 Effect of (a) normalizing and (b) manganese content on the Charpy V-notch impact energy of normalized carbon steels. (a) Impact energy and transition temperature of 1040 steel pipe, deoxidized with aluminum and silicon. (b) Charpy V-notch impact energy for normalized 0.30% C steels containing various amounts of manganese

Normalizing involves air cooling from austenitizing temperatures and produces essentially the same ferrite-pearlite microstructure as that of hot-rolled carbon steel, except that the heat treatment produces a finer grain size. This grain refinement makes the steel stronger, tougher, and more uniform throughout. Typical product forms and tensile properties of normalized carbon structural steels are given in Table 2. Charpy V-notch impact energies at various temperatures are given in Fig. 2(b) for a normalized carbon steel with varying manganese contents.

Quenching and tempering, that is, heating to about 900 °C (1650 °F), water quenching, and tempering at temperatures of 480 to 600 °C (900 to 1100 °F) or higher, can provide a tempered martensitic or bainitic microstructure that results in better combinations of strength and toughness. An increase in the carbon content to about 0.5%, usually accompanied by an increase in manganese, allows the steels to be used in the quenched and tempered condition. For quenched and tempered carbon-manganese steels with carbon contents up to about 0.25% (Table 2), low hardenability restricts the section sizes to about 150 mm (6 in.).

Table 2 Typical compositions, tensile properties, and product sizes of high-strength structural carbon steels

Specification and grade or class	Product form	Product thickness(a) mm	in.	Carbon	Heat analysis composition, %(b) Manganese	Silicon	Copper	Yield strength MPa	ksi	Tensile strength MPa	ksi	Elongation in 200 mm (8 in.), %
As-hot-rolled carbon-manganese steels												
ASTM A 529	Bar, plate, and shapes	13	½	0.27	1.20	...	0.20(c)	290	42	415–585	60–85	19
ASTM A 612	Plate	13	½	0.25	1.00–1.35	0.15–0.40	0.35	345	50	570–725	83–165	16
		20	¾	0.25	1.00–1.35	0.15–0.40	0.35	345	50	560–695	81–101	16
		20–25	¾–1	0.25	1.00–1.50	0.15–0.50	0.35	345	50	560–695	81–101	16
ASTM A 570, grades 45, 50, 55	Sheet	6	0.229	0.25	1.35	...	0.20(c)	310–380	45–55	415–480	60–70	14–10
ASTM A 662, grade B	Plate	40	1½	0.19	0.85–1.50	0.15–0.40	...	275	40	450–585	65–85	20
ASTM A 662, grade C	Plate	40	1½	0.20	1.00–1.60	0.15–0.50	...	295	43	485–620	70–90	18
SAE J410, grade 945C	Sheet and strip	...	...	0.23	1.40	...	...	310	45	415	60	...
	Plate, bar, and shapes	13	½	0.23	1.40	...	...	310	45	450	65	18
	Plate, bar, and shapes	13–40	½–1½	0.23	1.40	...	...	290	42	427	62	19
	Plate, bar, and shapes	40–75	1½–3	0.23	1.40	...	...	275	40	427	62	19
SAE J410, grade 950C	Sheet and strip	...	...	0.25	1.60	...	...	345	50	483	70	...
	Plate, bar, and shapes	13	½	0.25	1.60	...	...	345	50	483	70	18
	Plate, bar, and shapes	13–40	½–1½	0.25	1.60	...	...	310	45	462	67	19
	Plate, bar, and shapes	40–75	1½–3	0.25	1.60	...	...	290	42	434	63	19
Normalized structural carbon-manganese steels												
ASTM A 537, class 1	Plate	40	1½	0.24	0.70–1.35	0.15–0.50	0.35	345	50	485–620	70–90	18
	Plate	40–65	1½–2½	0.24	1.00–1.60	0.15–0.50	0.35	345	50	485–620	70–90	18
	Plate	65–100	2½–4	0.24	1.00–1.60	0.15–0.50	0.35	310	45	450–585	65–85	18
ASTM A 612	Plate	Same as ASTM A 612 in the as-rolled condition, but can be normalized for improved impact toughness										
ASTM A 633, grade A	Plate	100	4	0.18	1.00–1.35	0.15–0.50	...	290	42	430–570	63–83	18
ASTM A 662, grade A	Plate	40–50	1½–2	0.14	0.90–1.35	0.15–0.40	...	275	40	400–540	58–78	20
ASTM A 662, grade B	Plate	40–50	1½–2	0.19	0.85–1.50	0.15–0.40	...	275	40	450–585	65–85	20
ASTM A 662, grade C	Plate	40–50	1½–2	0.20	1.00–1.60	0.15–0.50	...	295	43	485–620	70–90	18
ASTM A 738, grade A	Plate	65(d)	2.5(d)	0.24	1.50(d)	0.15–0.50	0.35	310	45	515–655	75–95	20(e)
ASTM A 737, grade B	Plate	100	4	0.20	1.15–1.50	0.15–0.50	...	345	50	485–620	70–90	18
Quenched and tempered structural carbon-manganese steels												
SAE J 368a, grade Q980	Plate	20	¾	0.20	1.35	...	...	550	80	655–795	95–115	18(e)
ASTM A 537, class 2	Plate	40	1½	0.24	0.70–1.35	0.15–0.50	0.35	415	60	550–690	80–100	22(e)
		40–65	1½–2½	0.24	1.00–1.60	0.15–0.50	0.35	415	60	550–690	80–100	22(e)
		65–100	2½–4	0.24	1.00–1.60	0.15–0.50	0.35	380	55	515–655	75–95	22(e)
		100–150	4–6	0.24	1.00–1.60	0.15–0.50	0.35	315	46	485–620	70–90	22(e)
ASTM A 678, grade A	Plate	40	1½	0.16	0.90–1.50	0.15–0.50	0.20(c)	345	50	485–620	70–90	22(e)
ASTM A 678, grade B	Plate	40	1½	0.20	0.70–1.35	0.15–0.50	0.20(c)	415	60	550–690	80–100	22(e)
		40–65	1½–2½	0.20	1.00–1.60	0.15–0.50	0.20(c)	415	60	550–690	80–100	22(e)
ASTM A 678, grade C	Plate	20	¾	0.22	1.00–1.60	0.20–0.50	0.20(c)	515	75	655–790	95–115	19(e)
		20–40	¾–1½	0.22	1.00–1.60	0.20–0.50	0.20(c)	485	70	620–760	90–110	19(e)
		40–50	1½–2	0.22	1.00–1.60	0.20–0.50	0.20(c)	450	65	585–720	85–105	19(e)
ASTM A 738, grade B	Plate	65	2.5	0.20	0.90–1.50	0.15–0.55	0.35	415	60	585–705	85–102	20(e)
ASTM A 738, grade C	Plate	65	2.5	0.20	1.50	0.15–0.50	0.35	415	60	550–690	80–100	22(e)
		65–100	2.5–4	0.20	1.62	0.15–0.50	0.35	380	55	515–655	75–95	22(e)
		100–150	4–6	0.20	1.62	0.15–0.50	0.35	315	46	485–620	70–90	20(e)

(a) Product thicknesses are a maximum unless a range is given. (b) Compositions are a maximum unless a range is given or if otherwise specified in footnotes. Residual amounts of sulfur and phosphorus are limited in all grades and have specified maximums of 0.035 to 0.04% P (max) and 0.04 to 0.05% S (max), depending on the specification. (c) Minimum amount of copper if specified. (d) Over 65 mm (2.5 in.), ASTM 738 grade A requires quenching and tempering and 1.62% Mn (max) to achieve the specified strength levels. (e) Elongation in 50 mm (2 in.).

The yield strength of quenched and tempered carbon-manganese steel plate varies from 315 to 550 MPa (46 to 80 ksi), depending on section thickness (Table 2). Minimum Charpy V-notch impact toughness may be as high as 27 to 34 J (20 to 25 ft · lbf) at temperatures as low as −68 °C (−90 °F) for quenched and tempered carbon steel having yield strengths of 345 MPa (50 ksi). However, for quenched and tempered carbon steel with 690 MPa (100 ksi) yield strengths (Table 1), the impact values are lower, normally about 20 J (15 ft · lbf) at −60 °C (−75 °F). All grades can be grain refined with aluminum to improve toughness.

In addition to high-strength plate applications, quenched and tempered carbon-manganese steels are used for shafts and couplings. Steels with 0.40 to 0.60% C are used for railway wheels, tires, and axles, while those with higher carbon contents can be used as high-strength wire laminated spring materials, often with silicon-manganese or chromium-vanadium additions. The higher-carbon steels are also used for rails (0.7% C) and, over a range of carbon contents (typically, 0.20-0.50% C), for reinforcing bar.

Quenched and Tempered Low-Alloy Steel

Alloy steels are defined as those steels that: (1) contain manganese, silicon, or copper in quantities greater than the maximum limits (1.65% Mn, 0.60% Si, and 0.60% Cu) of carbon steel; or (2) that have specified ranges or minimums for one or more other alloying additions. The low-alloy steels are those steels containing alloy elements, including carbon, up to a total alloy content of about 8.0%.

Except for plain carbon steels that are microalloyed with just vanadium, niobium, and/or titanium (see the section "Microalloyed Quenched and Tempered Grades" in this article), most low-alloy steels are suitable as engineering quenched and tempered steels and are generally heat treated for engineering use. Low-alloy steels with suitable alloy compositions have greater hardenability than structural carbon steel and, thus, can provide high strength and good toughness in thicker sections by heat treatment. Their alloy contents may also provide improved heat and corrosion resistance. However, as the alloy contents increase, alloy steels become more expensive and more difficult to weld. Quenched and tempered structural steels are primarily available in the form of plate or bar products.

Alloying Elements and Their Effect on Hardenability and Tempering. Quenched

Table 3 Typical compositions of quenched and tempered low-alloy steel plate
Additional grades can be found in the article "Carbon and Low-Alloy Steel Plate" in this Volume.

Specification or common designation	Grade, type, or class	C	Mn	P	S	Si	Cr	Ni	Mo	Cu	Others
SAE J368a	Q980B	0.20	1.50	0.04	0.05	...	...	...	...	...	0.0005 B, min
	Q990B	0.20	1.50	0.04	0.05	...	...	...	...	...	0.0005 B, min
	Q9100B	0.20	1.50	0.04	0.05	...	...	...	...	...	0.0005 B, min
ASTM A 514 or A 517	A	0.15–0.21	0.80–1.10	0.035	0.04	0.40–0.80	0.50–0.80	...	0.18–0.28	...	0.05–0.15 Zr, 0.0025 B
	B	0.12–0.21	0.70–1.00	0.035	0.04	0.20–0.35	0.40–0.65	...	0.15–0.25	...	0.03–0.08 V, 0.01–0.03 Ti, 0.0005–0.005 B
	C	0.10–0.20	1.10–1.50	0.035	0.04	0.15–0.30	...	...	0.20–0.30	...	0.001–0.005 B
	E	0.12–0.20	0.40–0.70	0.035	0.04	0.20–0.35	1.40–2.00	...	0.40–0.60	0.20–0.40	0.04–0.10 Ti(b), 0.0015–0.005 B
	F	0.10–0.20	0.60–1.00	0.035	0.04	0.15–0.35	0.40–0.65	0.70–1.00	0.40–0.60	0.15–0.50	0.03–0.08 V
ASTM A 533	A	0.25	1.15–1.50	0.035	0.040	0.15–0.30	...	...	0.45–0.60	...	...
	B	0.25	1.15–1.50	0.035	0.040	0.15–0.30	...	0.40–0.70	0.45–0.60	...	...
	C	0.25	1.15–1.50	0.035	0.040	0.15–0.30	...	0.70–1.00	0.45–0.60	...	...
	D	0.25	1.15–1.50	0.035	0.040	0.15–0.30	...	0.20–0.40	0.45–0.60	...	...
ASTM A 543	B	0.23	0.40	0.020	0.020	0.20–0.40	1.50–2.00	2.60–4.00(c)	0.45–0.60	...	0.03 V
	C	0.23	0.40	0.020	0.020	0.20–0.40	1.20–1.50	2.25–3.50	0.45–0.60	...	0.03 V
ASTM A 678	D	0.22	1.15–1.50	0.04	0.05	0.15–0.50	...	...	...	0.20(d)	0.04–0.11 V(e)
ASTM A 709	70W	0.19	0.80–1.35	0.04	0.05	0.20–0.65	0.40–0.70	0.50	...	0.25–0.40	0.02–0.10 V
ASTM A 709, grades 100 and 100W, Type	A	0.15–0.21	0.80–1.10	0.035	0.04	0.40–0.80	0.50–0.80	...	0.18–0.28	...	0.05–0.15 Zr, 0.0025 B (min)
	B	0.12–0.21	0.70–1.00	0.035	0.04	0.20–0.35	0.40–0.65	...	0.15–0.25	...	0.03–0.08 V, 0.01–0.03 Ti
	C	0.10–0.20	1.10–1.50	0.035	0.04	0.20–0.35	...	...	0.20–0.30	...	0.0005–0.005 B, 0.001–0.005 B
	E	0.12–0.20	0.40–0.70	0.035	0.04	0.20–0.40	1.40–2.00	...	0.40–0.60	...	0.04–0.10 Ti(b), 0.001–0.005 B
	F	0.10–0.20	0.60–1.00	0.035	0.04	0.15–0.35	0.40–0.65	0.70–1.00	0.40–0.60	0.15–0.50	0.03–0.08 V, 0.0005–0.006 B
	H	0.12–0.21	0.95–1.30	0.035	0.04	0.20–0.35	0.40–0.65	0.30–0.70	0.20–0.30	...	0.03–0.08 V, 0.0005–0.005 B
	J	0.12–0.21	0.45–0.70	0.035	0.04	0.04–0.20	...	...	0.50–0.65	...	0.001–0.005 B
	M	0.12–0.21	0.45–0.70	0.035	0.04	0.04–0.20	...	1.20–1.50	0.45–0.65	...	0.001–0.005 B
	P	0.12–0.21	0.45–0.70	0.035	0.04	0.04–0.20	0.85–1.20	1.20–1.50	0.45–0.65	...	0.001–0.005 B
HY-80	...	0.12–0.18	0.10–0.40	0.025	0.025	0.15–0.35	1.00–1.80	2.00–3.25	0.20–0.60	0.25	0.03 V, 0.02 Ti
HY-100	...	0.12–0.20	0.10–0.40	0.025	0.025	0.15–0.35	1.00–1.80	2.25–3.50	0.20–0.60	0.25	0.03 V, 0.02 Ti
HY-130	...	0.12	0.60–0.90	0.010	0.015	0.15–0.35	0.40–0.70	4.75–5.25	0.30–0.65	...	0.05–0.10 V

(a) When a single value is shown, it is a maximum limit. (b) Vanadium may be substituted for part or all of the titanium content on a one-for-one basis. (c) Limiting values vary with plate thickness. (d) Minimum when specified. (e) Niobium may be present in ASTM A 678, grade D, in the amount of 0.01 to 0.05%.

and tempered steels have carbon contents in the range of 0.10 to 0.45%, with alloy contents, either singly or in combination, of up to 1.5% Mn, 5% Ni, 3% Cr, 1% Mo, 0.5% V, 0.10% Nb; in some cases they contain small additions of titanium, zirconium and/or boron. Generally the higher the alloy content, the greater the hardenability, and the higher the carbon content, the greater the available strength. Some typical compositions of quenched and tempered low-alloy steel plate are shown in Table 3. The response to heat treatment is the most important function of the alloying elements in these steels.

Effect on Hardenability. The hardenability of a steel is the property that determines the depth and distribution of martensite induced by quenching. It is usually the single most important criterion for selecting a low-alloy steel. To ensure adequate hardenability, the alloys must be in solution in austenite so that they retard the diffusion-controlled transformation of austenite to ferrite-pearlite. This allows the slower cooling of a piece or the quenching of a larger piece in a given medium without subsequent transformation of austenite to undesirable ferrite-pearlite transformation products.

Hardenability is measured in terms of an ideal diameter (D_I), which facilitates the comparison of the hardening response of different steels to the same quenching medium. The ideal diameter, D_I, is affected by austenite grain size, carbon content, and alloy content; an increase in any of these factors reduces or eliminates diffusion-controlled transformations, thereby encouraging the formation of martensite. Carbon is the most potent alloy for increasing hardenability, but it can be undesirable in structural steels because of its adverse effects on weldability and toughness. The other two factors, grain size and alloy elements, are considered below and in the article "Hardenability of Carbon and Low-Alloy Steels" in this Volume.

Austenite grain size influences not only hardenability but also strength and toughness. Increases in austenite grain size reduce the strength of a given transformation product (see Fig. 3b for martensitic transformation products, for example), but these increases do allow hardening to a greater depth than fine-grained steels, all other factors being equal. For steels in which pearlite (or ferrite) limits the hardenability of the steel, a useful diagram relating grain size (ASTM grain size number) to ideal critical diameter (D_I) in steels was developed by Grossmann in 1952 and is shown in Fig. 3(a). For such steels, the influence of grain size can be considered independent of the steel composition. For alloy steels in which bainite, rather than pearlite, is the dominant structure limiting full hardening, the effect of austenitic grain size is not the same. Low-carbon (0.06%) bainites may not be greatly affected by prior-austenite grain size. Although a larger grain size improves hardenability, increases in grain size increase the possibility of quench cracking. The small effect of prior-austenite grain size on the strength of martensite is shown in Fig. 3(b) for two alloy steels. Larger grain size can also degrade toughness, although avoiding proeutectoid ferrite is the overriding concern in the maintenance of notch toughness.

The effect of alloying elements on hardenability was shown by Grossmann in 1942

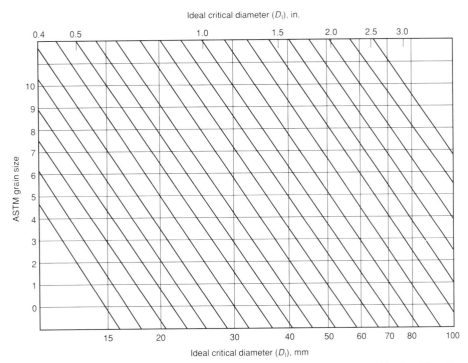

Fig. 3(a) Diagram showing direct relationship between ASTM grain size number and hardenability. For a grain size increment of one ASTM grain size number, multiply by 1.083. For a grain size increase of two ASTM size numbers, multiply by 1.172. For an increased grain size of three ASTM size numbers, multiply by 1.270. Source: Ref 1

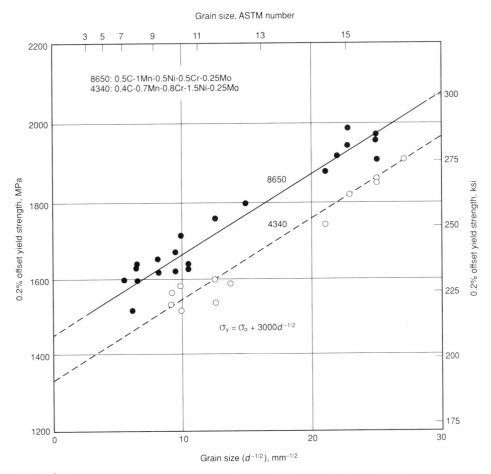

Fig. 3(b) Effect of prior-austenite grain size on the strength of martensite. Source: Ref 2

to be a multiplicative effect rather than an additive effect. During subsequent research, multiplying factors were developed with the realization that at times interaction effects occurred (that is, the multiplying factor for a given percentage of an element was not the same when added in conjunction with another element as it was when the element was used alone). For example, the multiplying factor for molybdenum varies with nickel content (Fig. 4). Also, there is a different set of multiplying factors for low-carbon alloys than for medium-carbon alloys. Because of this, Fig. 4 provides multiplying factors for steels having carbon contents similar to those shown in Table 3.

Some other interaction effects of alloying elements on hardenability are shown in Table 4. In general, alloying elements can be separated according to whether they are austenite stabilizers, such as manganese, nickel, and copper, or ferrite stabilizers (for example, γ-loop formers), such as molybdenum, silicon, titanium, vanadium, zirconium, tungsten, and niobium (Ref 4). Ferrite stabilizers require a much lower alloying addition than the austenite stabilizers for an equivalent increase in hardenability. However, with many of these ferrite stabilizers the competing process of carbide precipitation in the austenite depletes the austenite of both carbon and alloy addition, thus lowering hardenability. The precipitates also produce grain refinement, which further decreases hardenability.

Effects of Tempering. Because the hard martensite produced after quenching is also extremely brittle, virtually all hardened steels undergo a subcritical heat treatment referred to as tempering. Tempering improves the toughness of the as-quenched martensite, but also softens the steel, thus causing a decrease in strength and an increase in ductility. This softening is largely due to the rapid coarsening of cementite (Fe_3C) with increasing tempering temperature and a reduction in dislocation density.

Alloying elements can help retard the degree of softening during tempering, and certain elements are more effective than others. The alloys that act as solid-solution strengtheners (nickel, silicon, aluminum, and manganese) remain dissolved in the martensite and do not significantly retard the softening effect, although silicon (Table 4) does retard softening by inhibiting the coarsening of iron carbide (Fe_3C). The most effective elements in retarding the rate of softening during tempering are the strong carbide-forming elements such as molybdenum, chromium, vanadium, niobium, and titanium (Table 4). The metal carbides produced from these elements are harder than martensite (Fig. 5) and have a fine dispersion because the diffusion of the carbide-forming elements is more sluggish than the diffusion of carbon. The lower diffusion rate of the carbide-forming elements inhibits the

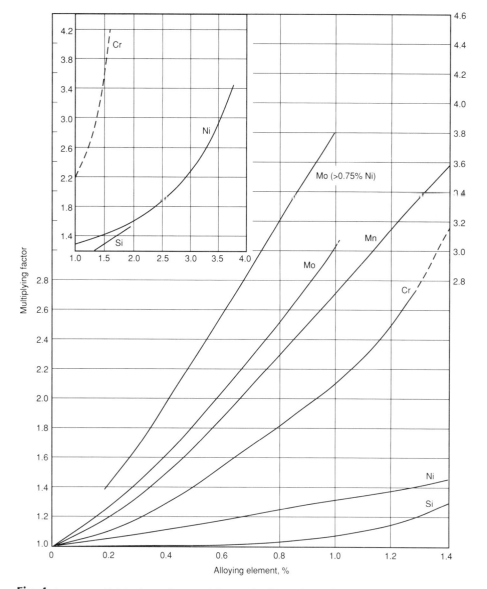

Fig. 4 Average multiplying factors for several elements in alloy steels containing 0.15 to 0.25% C. Source: Ref 1

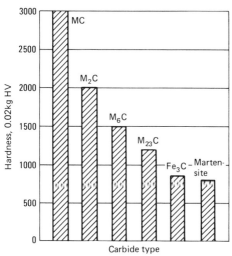

Carbide type	Alloying element	Composition, %
MC	C	13.0
	Fe	4.0
	W	23.0
	Mo	14.0
	V	43.0
	Cr	4.5
M_2C	C	6.0
	Fe	7.0
	W	41.0
	Mo	28.0
	V	11.0
	Cr	8.0
M_3C	C	6.7
	Fe	76.0
	W	5.0
	Mo	4.0
	V	2.0
	Cr	8.0
$M_{23}C_6$	C	4.0
	Fe	45.0
	W	25.0
	Mo	18.0
	V	4.0
	Cr	5.0
M_6C	C	3.0
	Fe	35.0
	W	35.0
	Mo	19.0
	V	3.3
	Cr	3.3

Fig. 5 Hardness of martensite and various carbides in an M2 tool steel. Representative analyses of carbide compositions are shown in the accompanying table. Source: Ref 5

coarsening of Fe_3C and thus retards the rate of softening at elevated temperatures.

The formation of carbides, which is a diffusion-controlled process dependent upon the migration of the carbide-forming elements, can reduce the rate of softening at all tempering temperatures. The degree of softening also depends on the quantity of the carbide-forming element. Figure 6, for example, shows reductions in softening at all tempering temperatures with various amounts of molybdenum. Similar effects occur for other carbide-forming elements, with the retardation in softening depending upon the type of carbide formed. In Fig. 5, for example, the maximum hardness obtainable in martensite is compared with the range of hardness for metal carbides (MC, M_2C, M_6C, M_3C, and $M_{23}C$), with representative analyses of carbide compositions shown in the table accompanying Fig. 5. In this case, the hardest carbide in steel, MC, is predominately a carbide of vanadium. The M_2C is a carbide of tungsten and molybdenum, and some vanadium.

If the carbide-forming elements are present in sufficient quantity, the metal carbides not only reduce softening but also produce a hardness increase at the higher tempering temperature (Fig. 6). This hardness increase is frequently referred to as secondary hardening. Given a sufficient level of carbide-forming elements, secondary hardening depends on a high enough temperature to allow a sufficient diffusion rate of the carbide-forming elements. Moreover, because the diffusion of the carbide formers is a more sluggish process than carbon diffusion, the metal carbides formed have a fine dispersion and are very resistant to coarsening. The latter characteristic of the fine metal carbides provides good creep resistance and is used to advantage in steels that must not soften during elevated-temperature exposure. See the articles "Wrought Tool Steels" and "Elevated-Temperature Properties of Ferritic Steels" in this Volume for additional information on secondary hardening steels with additions of molybdenum, vanadium, chromium, tungsten, and/or other carbide-forming elements.

Microalloyed Quenched and Tempered Grades. Although fittings with 0.6% Mn and induction bends use quenching and tempering as a standard practice, mild steels (plain, low-carbon steels with less than 0.7% Mn) with microalloying additions of vanadium, niobium, or titanium are seldom used as quenched and tempered steels. However, elements such as boron and vanadium are

Table 4 Effects of alloy elements on the heat treatment of quenched and tempered alloy steels

Effect of alloy on hardenability during quenching	Effect of alloy on tempering
Manganese contributes markedly to hardenability, especially in amounts greater than 0.8%. The effect of manganese up to 1.0% is stronger in low- and high-carbon steels than in medium-carbon steels.	Manganese increases the hardness of tempered martensite by retarding the coalescence of carbides, which prevent grain growth in the ferrite matrix. These effects cause a substantial increase in the hardness of tempered martensite as the percentage of manganese in the steel increases.
Nickel is similar to manganese at low alloy additions, but is less potent at the high alloy levels. Nickel is also affected by carbon content, the medium-carbon steels having the greatest effect. There is an alloy interaction between manganese and nickel that must be taken into account at lower austenitizing temperatures.	Nickel has a relatively small effect on the hardness of tempered martensite, which is essentially the same at all tempering temperatures. Because nickel is not a carbide former, its influence is considered to be due to a weak solid-solution strengthening.
Copper is usually added to alloy steels for its contribution to atmospheric-corrosion resistance and at higher levels for precipitation hardening. The effect of copper on hardenability is similar to that of nickel, and in hardenability calculations it has been suggested that the sum of copper plus nickel be used with the appropriate multiplying factor of nickel.	Copper is precipitated out when steel is heated to about 425–650 °C (800–1200 °F) and thus can provide a degree of precipitation hardening.
Silicon is more effective than manganese at low alloy levels and has a strengthening effect on low-alloy steels. However, at levels greater than 1% this element is much less effective than manganese. The effect of silicon also varies considerably with carbon content and other alloys present. Silicon is relatively ineffective in low-carbon steel but is very effective in high-carbon steels.	Silicon increases the hardness of tempered martensite at all tempering temperatures. Silicon also has a substantial retarding effect on softening at 316 °C (600 °F), and has been attributed to the inhibiting effect of silicon on the conversion of ε-carbide to cementite(a).
Molybdenum is most effective in improving hardenability. Molybdenum has a much greater effect in high-carbon steels than in medium-carbon steels. The presence of chromium decreases the multiplying factor, whereas the presence of nickel enhances the hardenability effect of molybdenum(b).	Molybdenum retards the softening of martensite at all tempering temperatures. Above 540 °C (1000 °F), molybdenum partitions to the carbide phase and thus keeps the carbide particles small and numerous. In addition, molybdenum reduces susceptibility to tempering embrittlement.
Chromium behaves much like molybdenum and has its greatest effect in medium-carbon steels. In low-carbon steel and carburized steel, the effect is less than in medium-carbon steels, but is still significant. As a result of the stability of chromium carbide at lower austenitizing temperatures, chromium becomes less effective.	Chromium, like molybdenum, is a strong carbide-forming element that can be expected to retard the softening of martensite at all temperatures. Also, by substituting chromium for some of the iron in cementite, the coalescence of carbides is retarded.
Vanadium is usually not added for hardenability in quenched and tempered structural steels (such as ASTM A 678, grade D) but is added to provide secondary hardening during tempering. Vanadium is a strong carbide former, and the steel must be austenitized at a sufficiently high temperature and for a sufficient length of time to ensure that the vanadium is in solution and thus able to contribute to hardenability. Moreover, solution is possible only if small amounts of vanadium are added (c).	Vanadium is a stronger carbide former than molybdenum and chromium and can therefore be expected to have a much more potent effect at equivalent alloy levels. The strong effect of vanadium is probably due to the formation of an alloy carbide that replaces cementite-type carbides at high tempering temperatures and persists as a fine dispersion up to the A_1 temperature.
Tungsten has been found to be more effective in high-carbon steels than in steels of low carbon content (less than 0.5%). Alloy interaction is important in tungsten-containing steels, with manganese-molybdenum-chromium having a greater effect on the multiplying factors than silicon or nickel additions.	Tungsten is also a carbide former and behaves like molybdenum in simple steels. Tungsten has been proposed as a substitute for molybdenum in reduced-activation ferritic steels for nuclear applications(d).
Titanium, niobium, and zirconium are all strong carbide formers and are usually not added to enhance hardenability for the same reasons given for vanadium. In addition, titanium and zirconium are strong nitride formers, a characteristic that affects their solubility in austenite and hence their contribution to hardenability.	Titanium, niobium, and zirconium should behave like vanadium because they are strong carbide formers.
Boron can considerably improve hardenability, the effect varying notably with the carbon content of the steel. The full effect of boron on hardenability is obtained only in fully deoxidized (aluminum-killed) steels.	Boron has no effect on the tempering characteristics of martensite, but a detrimental effect on toughness can result from the transformation to nonmartensitic products.

(a) Ref 3. (b) Fig. 3. (c) See the section "Microalloyed Quenched and Tempered Grades" in this article. (d) See the article "Elevated-Temperature Properties of Ferritic Steels" in this Volume. Source: Ref 2, 4

considered as substitutes for other elements that enhance hardenability. For example, the high cost of molybdenum in the late 1970s prompted considerable research in an effort to partially or completely replace molybdenum with microadditions of vanadium or of vanadium plus titanium (Ref 7-10). The titanium was added in order to form titanium nitride, thereby retaining an increased amount of vanadium in solution. This provided for a more efficient use of vanadium as a hardenability agent.

In terms of hardenability, the basic difficulty with vanadium (and other strong carbide formers such as titanium, niobium, and zirconium) is that the hardenability of steel can be increased only if small amounts are added and if the steel is austenitized at a high enough temperature and for a long enough time to ensure that the vanadium (or other strong carbide former) is in solution. Opinions vary as to the practical maximum amount of vanadium that can be added while still avoiding the nucleating effect of undissolved vanadium carbides, which would reduce hardenability. Complete solubility of vanadium during austenitization may not be the only factor in raising hardenability (Ref 11). The interaction of vanadium with other elements and the stabilization of nitrogen (with titanium) also influence hardenability. For example, Sandberg et al. (Ref 9, 10) investigated completely V-substituted variants of 4140-base series (0.4C-1Cr) with titanium additions, as well as partially V-substituted variants with and without titanium additions. The study concluded that:

- Complete substitution of molybdenum by vanadium does not increase the hardenability over standard 4140 (0.20% Mo) even when all the vanadium is dissolved during austenitization
- Steels containing 0.1 to 0.2% V and 0.04% Ti are characterized by significantly increased hardenability (10 to 25% in D_1) over standard 4140
- Microalloy combinations of V + Mo + Ti (~0.06-0.06-0.04%) provide very high hardenability, with D_1 being up to 60% greater than the D_1 in standard 4140 with 0.20% Mo. This effect is completely absent in a partially substituted steel without titanium (or aluminum as discussed below)

With regard to the third observation, however, Manganon (Ref 7, 8) has reported that for 4330-base steel (0.3C-0.5Cr-1.9Ni), 0.15% V (without titanium) can be substituted for 0.3% Mo without detriment to hardenability (although at present, com-

Table 5 Minimum tensile properties and maximum plate thickness for the quenched and tempered low-alloy steels listed in Table 3

Specification or common designation	Grade, type, or class	Plate thickness(a) mm	in.	Minimum yield strength MPa	ksi	Tensile strength MPa	ksi	Minimum elongation in 50 mm (2 in.), %
SAE J368a	Q980B	32	1¼	550	80	655–795	95–115	18
	Q990B	40	1½	620	90	690–895	100–130	18
	Q9100B	25	1	690	100	760 (min)	110 (min)	18
ASTM A 514 or A 517	A, B, C	32	1¼	690	100	760–895	110–130	18
	E, F	65	2½	690	100	760–895	110–130	18
	E	65–150	2½–6	620	90	690–895	100–130	16
ASTM A 533	Class 1 (type A, B, or C)	300	12	345	50	550–690	80–100	18
	Class 2 (type A, B, or C)	300	12	485	70	620–795	90–115	16
	Class 3 (type A, B, or C)	65	2½	570	83	690–860	100–125	16
ASTM A 543	B, C	(b)	(b)	485–585	70–85	620–930	90–135	14–16
ASTM A 678	D	75	3	515	75	620–760	90–110	18
ASTM A 709	70W	100	4	425	70	620–760	90–110	19
	100, 100W	65	2½	690	100	760–895	110–130	18
	100, 100W	65–100	2½–4	620	90	690–895	100–130	16
HY-80		20	¾	550–690	80–100	...	...	19
		20–200	¾–8	550–685	80–100	...	...	20
HY-100		20	¾	690–825	100–120	...	...	17
		20–150	¾–6	690–790	100–115	...	...	18
HY-130		9.5–14	⅜–9/16	895–1030	130–150	...	...	14
		14–100	9/16–4	895–1000	130–145	...	...	15

(a) Maximum plate thickness for the specified mechanical properties when a single value is shown. (b) Maximum plate thickness is not defined in ASTM A 543; plates in ASTM A 543 are intended for applications requiring plate thicknesses of 50 mm (2 in.) or more.

plete substitution is more expensive because the price of molybdenum is much lower than that of vanadium). Nevertheless, there also exists a synergism between molybdenum and vanadium such that the hardenability of a steel containing 0.15% V and 0.10% Mo is considerably superior to that of standard 4330 with 0.3% Mo. Additions of titanium to this partially substituted steel produced a further marginal increase in hardenability. Niobium also produces a niobium-molybdenum synergy in quenched and tempered steels; niobium can be present in amounts up to about 0.04% Nb without a decrease in the hardenability of carbon steels.

The pronounced effect on hardenability of molybdenum-vanadium combinations without titanium as observed by Manganon (Ref 7, 8) in 4330 steels, can probably be reconciled with the third result of Sandberg et al., in that the latter studied steels containing 0.06% Al, which would be expected to remove nitrogen to about the same extent as 0.04% Ti. Because small amounts of dissolved nitrogen promote the formation of VN (which removes vanadium from solution and thus detracts from the effectiveness of vanadium as a hardenability raiser), dissolved nitrogen must be limited via the presence of titanium or excess aluminum. By tying up the nitrogen as TiN or AlN, the vanadium can be in solution and thus increase hardenability.

Mechanical Properties. Quenched and tempered alloy steels can offer a combination of high strength and good toughness. Table 5 lists some typical tensile properties of the low-alloy plate steels given in Table 3. Figure 1 compares the low-temperature impact toughness of a heat-treated alloy steel with the impact toughness of a mild steel (ASTM A 7, which is now ASTM A 283, grade D) and an HSLA steel.

In addition, quenched and tempered alloy steel plate is available with ultrahigh strengths and enhanced toughness. Ultrahigh-strength steels with yield strengths above 1380 MPa (200 ksi) are described in the article "Ultrahigh-Strength Steels" in this Volume. Also within the category of ultrahigh-strength quenched and tempered steels is ball and roller bearing steel (see the article "Bearing Steels" in this Volume).

Enhanced toughness and high strength are achieved in the nickel-chromium-molybdenum alloys, which include steels such as ASTM A 543, HY-80, HY-100, and HY-130 (Table 3). These steels use nickel to improve toughness. The Charpy V-notch impact energies of the HY-80 and HY-100 grades are shown in Table 6. Figure 7 shows the Charpy V-notch impact energies of HY-130 at various temperatures.

High-Nickel Steels for Low-Temperature Service. For applications involving exposure to temperatures from 0 to −195 °C (32 to −320 °F), the ferritic steels with high nickel contents are typically used. Such applications include storage tanks for liquefied hydrocarbon gases and structures and machinery designed for use in cold regions. Properties of steels at low temperatures are discussed in the article "Low-Temperature Properties of Structural Steels" in this Volume.

The steels considered for the above applications also include the HY-130 steel in Table 3 and the various steels shown in Table 7. These steels utilize the effect of

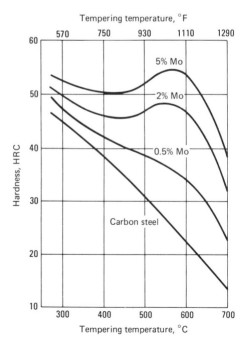

Fig. 6 Retardation of softening and secondary hardening during the tempering of a 0.35% C steel with various additions of molybdenum. Source: Ref 6

Table 6 Charpy V-notch impact strengths of two nickel-chromium-molybdenum steels

Alloy	Plate thickness mm	in.	Transverse impact strength at −18 °C (0 °F) J	ft·lbf	Transverse impact strength at −85 °C (−120 °F) J	ft·lbf	Longitudinal impact strength at −85 °C (−120 °F) J	ft·lbf
HY-80	<50	<2	81	60	47	35	68	50
	>50	>2	81	60	34	25	40	30
HY-100	<50	<2	75	55	40	30	68	50
	>50	>2	75	55	34	25	40	30

Table 7 Compositions of ferritic nickel steel plate for use at subzero temperatures

ASTM specification	Compositions of plates <50 mm (2 in.) thick, %(a)							
	C	Mn	P	S	Si	Ni	Mo	Others
A 203 A	0.17	0.70	0.035	0.040	0.15–0.30	2.10–2.50	...	...
A 203 B	0.21	0.70	0.035	0.040	0.15–0.30	2.10–2.50	...	...
A 203 C	0.17	0.70	0.035	0.040	0.15–0.30	3.25–3.75	...	...
A 203 D	0.20	0.70	0.035	0.040	0.15–0.30	3.25–3.75	...	...
A 645	0.13	0.30–0.60	0.025	0.025	0.20–0.35	4.75–5.25	0.20–0.35	0.02–0.12 Al, 0.020 N
A 353	0.13	0.90	0.035	0.040	0.15–0.30	8.5–9.5	...	...
A 553 I	0.13	0.90	0.035	0.040	0.15–0.30	8.5–9.5	...	...
A 553 II	0.13	0.90	0.035	0.040	0.15–0.30	7.5–8.5	...	...

(a) Single values are maximum limits.

Table 8 Typical tensile properties of ferritic nickel steels at low temperatures

Temperature		Tensile strength		Yield strength		Elongation, %	Reduction in area, %
°C	°F	MPa	ksi	MPa	ksi		
A 645 plate, longitudinal orientation(a)							
24	75	715	104	530	76.8	32	72
−168	−270	930	135	570	82.9	28	68
−196	−320	1130	164	765	111	30	62
A 353 plate, longitudinal orientation(b)							
24	75	780	113	680	98.6	28	70
−151	−240	1030	149	850	123	17	61
−196	−320	1190	172	950	138	25	58
−253	−423	1430	208	1320	192	18	43
−269	−452	1590	231	1430	208	21	59
A 553-I plate, longitudinal orientation(c)							
24	75	770	112	695	101	27	69
−151	−240	995	144	885	128	18	42
−196	−320	1150	167	960	139	27	38

(a) Quenched and tempered, and reversion annealed. (b) Double normalized and tempered: held at 900 °C (1650 °F) 1 h for each 25 mm (in.) of thickness, air cooled; 790 °C (1450 °F) 1 h for each 25 mm (1 in.) of thickness, air cooled; held at 570 °C (1050 °C) for 1 h for each 25 mm (1 in.) of thickness, air cooled or water quenched. (c) Quenched and tempered, 800 °C (1475 °F), water quenched; 570 °C (1050 °F) for 30 min for each 25 mm (1 in.) of thickness, air cooled or water quenched. Source: Ref 12–18

nickel content in reducing the impact transition temperature, thereby improving toughness at low temperatures. Carbon and alloy steel castings for subzero-temperature service are covered by ASTM standard specification A 757.

The 5% Ni alloys for low-temperature service include HY-130 in Table 3 and ASTM A 645 in Table 7. Typical Charpy V-notch impact energies of HY-130 at low temperatures are shown in Fig. 7. For steel purchased according to ASTM A 645, minimum Charpy V-notch impact requirements for 25 mm (1 in.) plate are designated at −170 °C (−275 °F) for hardened, tempered, and reversion-annealed plate. Minimum impact energies at this temperature range from 5 J (4 ft · lbf) for a 10 × 2.50 mm (0.4 × 0.1 in.) transverse specimen to 22 J (16 ft · lbf) for a 10 × 10 mm (0.4 × 0.4 in.) transverse specimen.

Double normalized and tempered 9% nickel steel is covered by ASTM A 353, and quenched and tempered 8% and 9% nickel steels are covered by ASTM A 553 (types I and II). For quenched and tempered material, the minimum lateral expansion in Charpy V-notch impact tests is 0.38 mm (0.015 in.). Charpy tests on 9% Ni steel (type I) are conducted at −195 °C (−320 °F); tests on 8% Ni steel (type II) are conducted at −170 °C (−275 °F). The transverse Charpy V-notch impact energies must not be less than 27 J (20 ft · lbf) at the specified temperature. Each impact test value must constitute the average value of three specimens, with not more than one value being below the specified minimum value of 27 J (20 ft · lbf) and no value being below 20 J (15 ft · lbf) for full-size specimens. Longitudinal Charpy impact properties must not be less than 34 J (25 ft · lbf) at the specified temperatures.

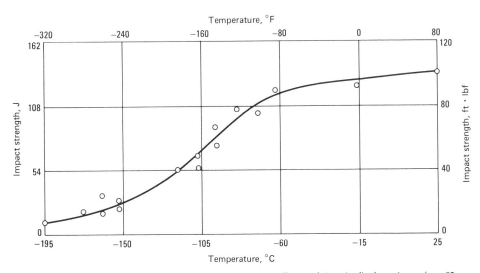

Fig. 7 Typical Charpy V-notch impact strengths of a 5% Ni low-alloy steel. Longitudinal specimens from 25 mm (1 in.) HY-130 steel plate were used.

Table 9 Fracture toughness of 5% and 9% Ni steel plate for compact tension specimens in transverse or longitudinal orientation

Alloy and condition	Yield strength(a)		Fracture toughness (K_{Ic}), J(b), at					
			−162 °C (−260 °F)		−196 °C (−320 °F)		−269 °C (−452 °F)	
	MPa	ksi	MPa√m	ksi√in.	MPa√m	ksi√in.	MPa√m	ksi√in.
5% Ni steel (A 645) quenched, tempered, and reversion annealed	534	77.5	196	178	87.1	79.3	58.4	53.2
9% Ni steel (A 553, type I) quenched and tempered	689	99.9	...	...	184	167	...	...

(a) At room temperature. (b) Fracture toughness determined from the J-integral method. Source: Ref 15–20

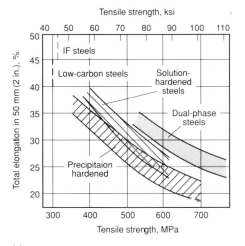
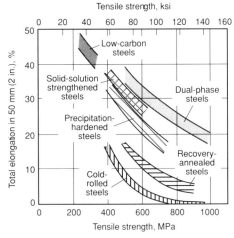
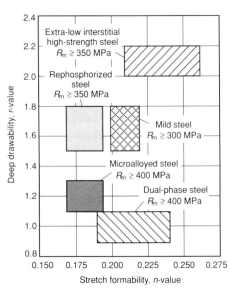

Fig. 8 Tensile and forming properties of dual-phase steels and interstitial-free (IF) steels. (a) Strength-elongation relationships for various hot-rolled sheet steels. (b) Strength-elongation relationships for various cold-rolled sheet steels. (c) Deep-drawing properties of steel sheet grades. Source: Ref 21

Typical tensile properties of 5% and 9% Ni steels at room temperature and at subzero temperatures are presented in Table 8. Yield and tensile strengths increase as testing temperature is decreased. These steels remain ductile at the lowest testing temperatures.

Ferritic nickel steels are too tough at room temperature for valid fracture toughness (K_{Ic}) data to be obtained on specimens of reasonable size, but limited fracture toughness data have been obtained on these steels at subzero temperatures by the J-integral method. Results of these tests are presented in Table 9. The 5% Ni steel retains relatively high fracture toughness at −162 °C (−260 °F), and the 9% Ni steel retains relatively high fracture toughness at −196 °C (−320 °F). These temperatures approximate the minimum temperatures at which these steels may be used.

HSLA Steels

High-strength low-alloy (HSLA) steels are a group of low-carbon steels that utilize small amounts of alloying elements to attain yield strengths greater than 275 MPa (40 ksi) in the as-rolled or normalized condition. These steels have better mechanical properties and sometimes better corrosion resistance than as-rolled carbon steels. Moreover, because the higher strength of HSLA steels can be obtained at lower carbon contents, the weldability of many HSLA steels is comparable to or better than that of mild steel (Ref 21).

High-strength low-alloy steels are primarily hot-rolled into the usual wrought product forms (sheet, strip, bar, plate, and structural sections) and are commonly furnished in the as-hot-rolled condition. However, the production of hot-rolled HSLA products may also involve special hot-mill processing that further improves the mechanical properties of some HSLA steels and product forms. These processing methods include:

- *The controlled rolling* of precipitation-strengthened HSLA steels to obtain fine austenite grains and/or highly deformed (pancaked) austenite grains, which during cooling transform into fine ferrite grains that greatly enhance toughness while improving yield strength
- *The accelerated cooling* of, preferably, controlled-rolled HSLA steels to produce fine ferrite grains during the transformation of austenite. These cooling rates cannot be rapid enough to form acicular ferrite, nor can they be slow enough so that high coiling temperatures result and thereby causing the overaging of precipitates
- *The quenching or accelerated air or water cooling* of low-carbon steels (≤0.08% C) that possess adequate hardenability to transform into low-carbon bainite (acicular ferrite). This microstructure offers an excellent combination of high yield strengths (275 to 690 MPa, or 60 to 100 ksi), excellent weldability and formability, and high toughness (controlled rolling is necessary for low ductile-brittle transition temperatures)
- *The normalizing* of vanadium-containing HSLA steels to refine grain size, thereby improving toughness and yield strength
- *The intercritical annealing* of HSLA steels (and also carbon-manganese steels with low carbon contents) to obtain a dual-phase microstructure (martensite islands dispersed in a ferrite matrix). This microstructure exhibits a lower yield strength but, because of rapid work-hardening capability, provides a better combination of ductility and tensile strength than conventional HSLA steels (Fig. 8) and improved formability

The usefulness and cost effectiveness of these processing methods are highly dependent on product form and alloy content, which are considered in more detail in the following sections.

In addition to hot-rolled products, HSLA steels are also furnished as cold-rolled sheet and forgings. Cold-rolled HSLA sheet and HSLA forgings are discussed in the section "Applications of HSLA Steels" in this article. HSLA forgings are also covered in the article "High-Strength Low-Alloy Steel Forgings" in this Volume. The main advantage of HSLA forgings (like as-hot-rolled HSLA products) is that yield strengths in the range of 275 to 485 MPa (40 to 70 ksi) or perhaps higher can be achieved without heat treatment. Base compositions of these microalloyed ferrite-pearlite forgings are typically 0.3-0.50% C and 1.4-1.6% Mn. Low-carbon bainitic HSLA steel forgings have also been developed.

HSLA Steel Categories and Specifications

High-strength low-alloy steels include many standard and proprietary grades designed to provide specific desirable combinations of properties such as strength, toughness, formability, weldability, and atmospheric-corrosion resistance. These steels are not considered alloy steels, even though their desired properties are achieved by the use of small alloy additions. Instead, HSLA steels are classified as a separate steel category, which is similar to as-rolled mild-carbon steel with enhanced mechanical properties obtained by the judicious (small) addition of alloys and, perhaps, special processing techniques such as controlled rolling. This separate product recognition of HSLA steels is reflected by the fact that HSLA steels are generally priced from the base price for carbon steels, not from the base price for alloy steels. Moreover, HSLA steels are often sold on the

Table 10 Summary of characteristics and intended uses of HSLA steels described in ASTM specifications

ASTM specification(a)	Title	Alloying elements(b)	Available mill forms	Special characteristics	Intended uses
A 242	High-strength low-alloy structural steel	Cr, Cu, N, Ni, Si, Ti, V, Zr	Plate, bar, and shapes ≤ 100 mm (4 in.) in thickness	Atmospheric-corrosion resistance four times that of carbon steel	Structural members in welded, bolted, or riveted constructions
A 572	High-strength low-alloy niobium-vanadium steels of structural quality	Nb, V, N	Plate, bar, shapes, and sheet piling ≤ 150 mm (6 in.) in thickness	Yield strengths of 290 to 450 MPa (42 to 65 ksi), in six grades	Welded, bolted, or riveted structures, but mainly bolted or riveted bridges and buildings
A 588	High-strength low-alloy structural steel with 345 MPa (50 ksi) minimum yield point ≤ 100 mm (4 in.) in thickness	Nb, V, Cr, Ni, Mo, Cu, Si, Ti, Zr	Plate, bar, and shapes ≤ 200 mm (8 in.) in thickness	Atmospheric-corrosion resistance four times that of carbon steel; nine grades of similar strength	Welded, bolted, or riveted structures, but primarily welded bridges and buildings in which weight savings or added durability is important
A 606	Steel sheet and strip, hot rolled and cold rolled, high strength low alloy with improved corrosion resistance	Not specified	Hot-rolled and cold-rolled sheet and strip	Atmospheric-corrosion resistance twice that of carbon steel (type 2) or four times that of carbon steel (type 4)	Structural and miscellaneous purposes for which weight savings or added durability is important
A 607	Steel sheet and strip, hot rolled and cold rolled, high strength low alloy niobium and/or vanadium	Nb, V, N, Cu	Hot-rolled and cold-rolled sheet and strip	Atmospheric-corrosion resistance twice that of carbon steel, but only when copper content is specified; yield strengths of 310 to 485 MPa (45 to 70 ksi) in six grades	Structural and miscellaneous purposes for which greater strength or weight savings is important
A 618	Hot-formed welded and seamless high-strength low-alloy structural tubing	Nb, V, Si, Cu	Square, rectangular, round, and special-shape structural welded or seamless tubing	Three grades of similar yield strength; may be purchased with atmospheric-corrosion resistance twice that of carbon steel	General structural purposes, included welded, bolted, or riveted bridges and buildings
A 633	Normalized high-strength low-alloy structural steel	Nb, V, Cr, Ni, Mo, Cu, N, Si	Plate, bar, and shapes ≤ 150 mm (6 in.) in thickness	Enhanced notch toughness; yield strengths of 290 to 415 MPa (42 to 60 ksi) in five grades	Welded, bolted, or riveted structures for service at temperatures ≥ −45 °C (−50 °F)
A 656	High-strength, low-alloy, hot-rolled structural vanadium-aluminum-nitrogen and titanium-aluminum steels	V, Al, N, Ti, Si	Plate, normally ≤ 16 mm (⅝ in.) in thickness	Yield strength of 552 MPa (80 ksi)	Truck frames, brackets, crane booms, rail cars, and other applications for which weight savings is important
A 690	High-strength low-alloy steel H-piles and sheet piling	Ni, Cu, Si	Structural-quality H-piles and sheet piling	Corrosion resistance two to three times greater than that of carbon steel in the splash zone of marine structures	Dock walls, sea walls, bulkheads, excavations, and similar structures exposed to seawater
A 709, grade 50 and 50W	Structural steel	V, Nb, N, Cr, Ni, Mo	All structural-shape groups and plate ≤ 100 mm (4 in.) in thickness	Minimum yield strength of 345 MPa (50 ksi). Grade 50W is a weathering steel.	Bridges
A 714	High-strength low-alloy welded and seamless steel pipe	V, Ni, Cr, Mo, Cu, Nb	Pipe with nominal pipe size diameters of 13 to 660 mm (½ to 26 in.)	Minimum yield strengths ≤ 345 MPa (50 ksi) and corrosion resistance two to four times that of carbon steel	Piping
A 715	Steel sheet and strip, hot rolled, high strength low alloy with improved formability	Nb, V, Cr, Mo, N, Si, Ti, Zr, B	Hot rolled sheet and strip	Improved formability(c) compared to A 606 and A 607; yield strengths of 345 to 550 MPa (50 to 80 ksi) in four grades	Structural and miscellaneous applications for which high strength, weight savings, improved formability, and good weldability are important
A 808	High-strength low-alloy steel with improved notch toughness	V, Nb	Hot-rolled steel plate ≤ 65 mm (2½ in.) in thickness	Charpy V-notch impact energies of 40–60 J (30–45 ft · lbf) at −45 °C (−50 °F)	Railway tank cars
A 812	High-strength low-alloy steels	V, Nb	Steel sheet in coil form	Yields strengths of 450–550 MPa (65–85 ksi)	Welded layered pressure vessels
A 841	Plate produced by thermomechanical controlled processes	V, Nb, Cr, Mo, Ni	Plates ≤ 100 mm (4 in.) in thickness	Yield strengths of 310–345 MPa (45–50 ksi)	Welded pressure vessels
A 847	Cold-formed welded and seamless high-strength low-alloy structural tubing with improved atmospheric-corrosion resistance	Cu, Cr, Ni, Si, V, Ti, Zr, Nb	Welded tubing with maximum periphery of 1625 mm (64 in.) and wall thickness of 16 mm (0.625 in.) or seamless tubing with maximum periphery of 810 mm (32 in.) and wall thickness of 13 mm (0.50 in.)	Minimum yield strengths ≤ 345 MPa (50 ksi) with atmospheric-corrosion resistance twice that of carbon	Round, square, or specially shaped structural tubing for welded, riveted, or bolted construction of bridges and buildings
A 860	High-strength butt-welding fittings of wrought high-strength low-alloy steel	Cu, Cr, Ni, Mo, V, Nb, Ti	Normalized or quenched and tempered wrought fittings	Minimum yield strengths ≤ 485 MPa (70 ksi)	High-pressure gas and oil transmission lines
A 871	High-strength low-alloy steel with atmospheric corrosion resistance	V, Nb, Ti, Cu, Mo, Cr	As-rolled plate ≤ 35 mm (1⅜ in.) in thickness	Atmospheric-corrosion resistance four times that of carbon structural steel	Tubular structures and poles

(a) For grades and mechanical properties, see Table 16. (b) In addition to carbon, manganese, phosphorus, and sulfur. A given grade may contain one or more of the listed elements, but not necessarily all of them; for specified compositional limits, see Table 13. (c) Obtained by producing killed steel, made to fine grain practice, and with microalloying elements such as niobium, vanadium, titanium, and zirconium in the composition

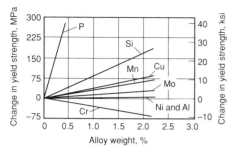

Fig. 9 Solid-solution strengthening of ferrite. Source: Ref 24

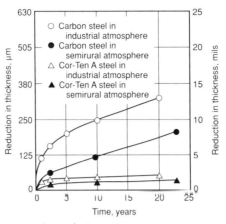

Fig. 10 Atmospheric-corrosion resistance of a proprietary high-phosphorus HSLA weathering steel (Cor-Ten A). Composition of weathering steel: 0.12% C (max), 0.20 to 0.50% Mn, 0.07 to 0.15% P, 0.05% S (max), 0.25 to 0.75% Si, 0.25 to 0.75% Cu, 0.30 to 1.25% Cr, and 0.65% Ni (max)

basis of minimum mechanical properties, with the specific alloy content left to the discretion of the steel producer.

Although HSLA steels are available in numerous standard and proprietary grades (see, for example, the listing of over 600 HSLA steels in Ref 23), HSLA steels can be divided into six categories:

- *Weathering steels*, which contain small amounts of alloying elements such as copper and phosphorus for improved atmospheric corrosion resistance and solid-solution strengthening
- *Microalloyed ferrite-pearlite steels*, which contain very small (generally, less than 0.10%) additions of strong carbide or carbonitride-forming elements such as niobium, vanadium, and/or titanium for precipitation strengthening, grain refinement, and possibly transformation temperature control
- *As-rolled pearlitic steels*, which may include carbon-manganese steels but which may also have small additions of other alloying elements to enhance strength, toughness, formability, and weldability
- *Acicular ferrite (low-carbon bainite) steels*, which are low-carbon (<0.08% C) steels with an excellent combination of high yield strengths, weldability, formability, and good toughness
- *Dual-phase steels*, which have a microstructure of martensite dispersed in a ferritic matrix and provide a good combination of ductility and high tensile strength (Fig. 8)
- *Inclusion shape controlled steels*, which provide improved ductility and through-thickness toughness by the small additions of calcium, zirconium, or titanium, or perhaps rare-earth elements so that the shape of the sulfide inclusions are changed from elongated stringers to small, dispersed, almost spherical globules
- *Hydrogen-induced cracking resistant steels* with low carbon, low sulfur, inclusion shape control, and limited manganese segregation, plus copper contents greater than 0.26%

These seven categories are not necessarily distinct groupings, in that an HSLA steel may have characteristics from more than one grouping. For example, all the above types of steels can be inclusion shape controlled. Microalloyed ferrite-pearlite steel may also have additional alloys for corrosion resistance and solid-solution strengthening. A separate category might also be considered for the HSLA 80 (Navy) nickel-copper-niobium steel (0.04% C, 1.5% Mn, 0.03% Nb, 1.0% Ni, 1.0% Cu, and 0.7% Cr). Table 10 describes some typical HSLA steels, their available mill forms, and their intended applications.

Weathering Steels. The first HSLA steels developed were the weathering steels. These steels contain copper and other elements that enhance corrosion resistance, solid-solution strengthening, and some grain refinement of the ferritic microstructure. The solid-solution strengthening effect of several alloying elements is shown in Fig. 9. Of these, copper, phosphorus, and silicon provide corrosion resistance in addition to solid-solution strengthening.

Several of the ASTM specifications listed in Table 10 cover weathering steels with enhanced atmospheric-corrosion resistance. The original architectural grade of weathering steel is covered by ASTM A 242, and the heavier structural grade is covered by ASTM A 588. These two steels reduce corrosion by forming their own protective oxide surface film. Although these steels initially corrode at the same rate as plain carbon steel, they soon exhibit a decreasing corrosion rate, and after a few years, continuation of corrosion is practically nonexistent. The protective oxide coating is fine textured, tightly adherent, and a barrier to moisture and oxygen, effectively preventing further corrosion. Plain carbon steel, on the other hand, forms a coarse-textured flaky oxide that does not prevent moisture or oxygen from reaching the underlying noncorroded steel base. Steels conforming to ASTM A 242 and A 588 are not recommended for exposure to highly concentrated industrial fumes or severe marine conditions, nor are they recommended for applications in which they will be buried or submerged. In these environments, the highly protective oxide does not form properly, and corrosion is similar to that for plain carbon steel.

The numerous grades of HSLA weathering steels can be classified into two groups:

- Weathering steels with normal low-phosphorus contents and multiple-alloy additions for solid-solution strengthening and enhanced corrosion resistance
- Proprietary weathering steels with high phosphorus contents (0.05 to 0.15%) for strengthening and corrosion resistance, together with multiple-alloy additions similar to those of the low-phosphorus weathering steels

The atmospheric-corrosion resistance of the low-phosphorus weathering steels (Ref 23) is two to six times that of carbon structural steel, while the proprietary weathering steels may have higher corrosion resistance. For example, an early (circa 1933) proprietary weathering steel, Cor-Ten A produced by United States Steel Corporation, has an atmospheric-corrosion resistance of five to eight times greater than that of carbon steel (Fig. 10), depending on the environment. This high-phosphorus weathering steel has a minimum yield strength of 345 MPa (50 ksi) in section thicknesses up to 13 mm (½ in.).

Microalloying with vanadium and/or niobium can improve the yield strength of weathering steels; the addition of niobium also improves toughness. Normalizing or controlled rolling-cooling can also refine the grain size (and thus improve toughness and yield strength). However, if normalizing or accelerated cooling is used to refine grain size, the effect of carbon content and alloy additions on hardenability and the potential for undesirable transformations to upper bainite and Widmanstätten ferrite must be considered.

Microalloyed ferrite-pearlite steels use additions of alloying elements such as niobium and vanadium to increase strength (and thereby increase load-carrying ability) of hot-rolled steel without increasing carbon and/or manganese contents. Extensive studies during the 1960s on the effects of niobium and vanadium on the properties of structural-grade materials resulted in the discovery that very small amounts of niobium and vanadium (<0.10% each) strengthen the standard carbon-manganese steels without interfering with subsequent processing. Carbon content thus could be reduced to improve both weldability and toughness because the strengthening effects of niobium and vanadium compensated for

Table 11 Compositions, mill forms, and characteristics of HSLA steels described in SAE J410c
See Table 17 for mechanical properties.

Grade(a)	C (max)	Mn (max)	P (max)	Other elements(c)	Available mill forms	Special characteristics
942X	0.21	1.35	0.04	Nb, V	Plate, bar, and shapes ≤ 100 mm (4 in.) in thickness	Similar to 945X and 945C except for better weldability and formability
945A	0.15	1.00	0.04	...	Sheet, strip, plate, bar, and shapes ≤ 75 mm (3 in.) in thickness	Excellent weldability, formability, and notch toughness
945C	0.23	1.40	0.04	...	Sheet, strip, plate, bar, and shapes ≤ 75 mm (3 in.) in thickness	Similar to 950C except that lower carbon and manganese content improve weldability, formability, and notch toughness
945X	0.22	1.35	0.04	Nb, V	Sheet, strip, plate, bar, and shapes ≤ 75 mm (3 in.) in thickness	Similar to 945C except for better weldability and formability
950A	0.15	1.30	0.04	...	Sheet, strip, plate, bar, and shapes ≤ 75 mm (3 in.) in thickness	Good weldability, notch toughness, and formability
950B	0.22	1.30	0.04	...	Sheet, strip, plate, bar, and shapes ≤ 75 mm (3 in.) in thickness	Fairly good notch toughness and formability
950C	0.25	1.60	0.04	...	Sheet, strip, plate, bar, and shapes ≤ 75 mm (3 in.) in thickness	Fair formability and toughness
950D	0.15	1.00	0.15	...	Sheet, strip, plate, bar, and shapes ≤ 75 mm (3 in.) in thickness	Good weldability and formability; phosphorus should be considered in conjunction with other elements
950X	0.23	1.35	0.04	Nb, V	Sheet, strip, plate, bar, and shapes ≤ 40 mm (1.5 in.) in thickness	Similar to 950C except for better weldability
955X	0.25	1.35	0.04	Nb, V, N	Sheet, strip, plate, bar, and shapes ≤ 40 mm (1.5 in.) in thickness	Similar to 945X and 950X except that progressively higher strengths are obtained by increasing the carbon and manganese contents, or by adding nitrogen ≤ 0.015%; formability and weldability generally decrease with increased strength; toughness varies with composition and mill practice
960X	0.26	1.45	0.04	Nb, V, N	Sheet, strip, plate, bar, and shapes ≤ 40 mm (1.5 in.) in thickness	
965X	0.26	1.45	0.04	Nb, V, N	Sheet, strip, plate, bar, and shapes ≤ 20 mm (0.75 in.) in thickness	
970X	0.26	1.65	0.04	Nb, V, N	Sheet, strip, plate, bar, and shapes ≤ 20 mm (0.75 in.) in thickness	
980X	0.26	1.65	0.04	Nb, V, N	Sheet, strip, plate, bar, and shapes ≤ 10 mm (0.38 in.) in thickness	

(a) Fully killed steel made to fine grain practice may be specified by adding a second suffix, K, for instance, 945XK. Steels made to K practice are normally specified only for applications requiring better toughness at low temperatures than steels made to normal semikilled practice. (b) 0.05% P (max) and 0.90 % Si (max), all grades. (c) Elements normally added singly or in combination to produce specified mechanical properties and other characteristics. Other alloying elements such as copper, chromium, and nickel may be added to enhance atmospheric-corrosion resistance.

the reduction in strength due to the reduction in carbon content.

The mechanical properties of microalloyed HSLA steels result, however, from more than just the mere presence of microalloying elements. Austenite conditioning, which depends on the complex effects of alloy design and rolling techniques, is also an important factor in the grain refinement of hot-rolled HSLA steels. Grain refinement by austenite conditioning with controlled rolling methods has resulted in improved toughness and high yield strengths in the range of 345 to 620 MPa (50 to 90 ksi) (Ref 25). This development of controlled-rolling processes coupled with alloy design has produced increasing yield strength levels accompanied by a gradual lowering of the carbon content. Many of the proprietary microalloyed HSLA steels have carbon contents as low as 0.06% or even lower, yet are still able to develop yield strengths of 485 MPa (70 ksi). The high yield strength is achieved by the combined effects of fine grain size developed during controlled hot rolling and precipitation strengthening that is due to the presence of vanadium, niobium, and titanium.

The various types of microalloyed ferrite-pearlite steels include:

- Vanadium-microalloyed steels
- Niobium-microalloyed steels
- Niobium-molybdenum steels
- Vanadium-niobium microalloyed steels
- Vanadium-nitrogen microalloyed steels
- Titanium-microalloyed steels
- Niobium-titanium microalloyed steels
- Vanadium-titanium microalloyed steels

These steels may also include other elements for improved corrosion resistance and solid-solution strengthening, or enhanced hardenability (if transformation products other than ferrite-pearlite are desired).

Some specifications of microalloyed HSLA steels do not specify the range of microalloying additions needed to achieve the desired strength level. These steels are often specified in terms of mechanical properties, with the amounts of microalloying elements left to the discretion of the steel producer. For example, SAE specification J410 covers various HSLA steels with varying carbon and manganese limits for sheet, strip, plate, bar, and shapes (Table 11). This specification limits only carbon and manganese contents (besides impurities such as sulfur), with the amounts of microalloying elements depending on the desired mechanical properties and corrosion resistance. Similar specifications based on mechanical properties without specific microalloying amounts also are given in SAE J1392 (hot- and cold-rolled HSLA sheet), SAE J1442 (hot-rolled HSLA plate), and some of the ASTM specifications in Table 10 (ASTM A 242, A 606, and A 715). The other ASTM specifications in Table 10 define specific ranges for microalloying additions.

Vanadium Microalloyed Steels. The development of vanadium-containing steels occurred shortly after the development of weathering steels, and flat-rolled products with up to 0.10% V are widely used in the hot-rolled condition. Vanadium-containing steels are also used in the controlled-rolled, normalized, or quenched and tempered condition.

Vanadium contributes to strengthening by forming fine precipitate particles (5 to 100 nm in diameter) of V(CN) in ferrite during cooling after hot rolling. These vanadium precipitates, which are not as stable as niobium precipitates, are in solution at all normal rolling temperatures and thus are very dependent on the cooling rate for their formation. Niobium precipitates, however, are stable at higher temperatures, which is beneficial for achieving fine-grain ferrite (see the section "Niobium Microalloyed Steels" in this article).

The strengthening from vanadium averages between 5 and 15 MPa (0.7 and 2 ksi) per 0.01 wt% V (Ref 26), depending on carbon content and rate of cooling from hot rolling (and thus section thickness). The cooling rate, which is determined by the hot-rolling temperature and the section thickness, af-

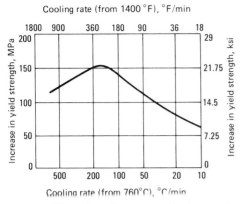

Fig. 11 Effect of cooling rate on the increase in yield strength due to precipitation strengthening in a 0.15% V steel. Source: Ref 26

fects the level of precipitation strengthening in a 0.15% V steel, as shown in Fig. 11. An optimum level of precipitation strengthening occurs at a cooling rate of about 170 °C/min (306 °F/min) (Fig. 11). At cooling rates lower than 170 °C/min (306 °F/min), the V(CN) precipitates coarsen and are less effective for strengthening. At higher cooling rates, more V(CN) remains in solution, and thus a smaller fraction of V(CN) particles precipitate and strengthening is reduced. For a given section thickness and cooling medium, cooling rates can be increased or decreased by increasing or decreasing, respectively, the temperature before cooling. Increasing the temperature results in larger austenite grain sizes, while decreasing the temperature makes rolling more difficult.

Manganese content and ferrite grain size also affect the strengthening of vanadium microalloyed steels. The effect of manganese on a hot-rolled vanadium steel is shown in Table 12. The 0.9% increase in manganese content increased the strength of the matrix by 34 MPa (5 ksi) because of solid-solution strengthening. The precipitation strengthening by vanadium was also enhanced because manganese lowered the austenite-to-ferrite transformation temperature, thereby resulting in a finer precipitate dispersion. This effect of manganese on precipitation strengthening is greater than its effect in niobium steels. However, the absolute strength in a niobium steel with 1.2% Mn is only about 50 MPa (7 ksi) less than that of vanadium steel but at a much lower alloy level (that is, 0.06% Nb versus 0.14% V).

The third factor affecting the strength of vanadium steels is the ferrite grain size produced after cooling from the austenitizing temperature. Finer ferrite grain sizes (which result in not only higher yield strengths but also improved toughness and ductility) can be produced by either lower austenite-to-ferrite transformation temperatures or by the formation of finer austenite grain sizes prior to transformation. Lowering the transformation temperature, which affects the level of precipitation strengthening as mentioned above, can be achieved by alloy additions and/or increased cooling rates. For a given cooling rate, further refinement of ferrite grain size is achieved by the refinement of the austenite grain size during rolling.

The austenite grain size of hot-rolled steels is determined by the recrystallization and grain growth of austenite during rolling. Vanadium hot-rolled steels usually undergo conventional rolling but are also produced by recrystallization controlled rolling (see the section "Controlled Rolling" in this article). With conventional rolling, vanadium steels provide moderate precipitation strengthening and relatively little strengthening from grain refinement. The maximum yield strength of conventionally hot-rolled vanadium steels with 0.25% C and 0.08% V is about 450 MPa (65 ksi). The practical limit of yield strengths for hot-rolled vanadium-microalloyed steel is about 415 MPa (60 ksi), even when controlled rolling techniques are used. Vanadium steels subjected to recrystallization controlled rolling require a titanium addition so that a fine precipitate of TiN is formed that restricts austenite grain growth after recrystallization. Yield strengths from conventional controlled rolling are limited to a practical limit of about 415 MPa (60 ksi) because of the lack of retardation of recrystallization. When both strength and impact toughness are important factors, controlled-rolled low-carbon niobium steel (such as X-60 hydrogen-induced cracking resistant plate) is preferable.

Niobium Microalloyed Steels. Like vanadium, niobium increases yield strength by precipitation hardening; the magnitude of the increase depends on the size and amount of precipitated niobium carbides (Fig. 12). However, niobium is also a more effective grain refiner than vanadium. Thus, the combined effect of precipitation strengthening and ferrite grain refinement makes niobium a more effective strengthening agent than vanadium. The usual niobium addition is 0.02 to 0.04%, which is about one-third the optimum vanadium addition.

Strengthening by niobium is 35 to 40 MPa (5 to 6 ksi) per 0.01% addition. This strengthening was accompanied by a considerable impairment of notch toughness until special rolling procedures were developed and carbon contents were lowered to avoid formation of upper bainite. In general, high finishing temperatures and light deformation passes should be avoided with niobium steels because that may result in mixed grain sizes or Widmanstätten ferrite, which impair toughness.

Niobium steels are produced by controlled rolling, recrystallization controlled rolling, accelerating cooling, and direct quenching. The recrystallization controlled rolling of niobium steel can be effective without titanium, while recrystallization rolling of vanadium steels requires titanium for grain refinement. Also, much less niobium is needed, and niobium-titanium steels can be recrystallization controlled rolled at higher temperatures. At present, offshore platform steels up to 75 mm (3 in.) thick with yield strengths of 345 to 415 MPa (50 to 60 ksi) are routinely produced.

Vanadium-Niobium Microalloyed Steels. Steels microalloyed with both niobium and vanadium provide a higher yield strength in the conventionally hot-rolled condition than that achievable with either element alone. As conventionally hot rolled, the niobium-vanadium steels derive almost all of their increased strength from precipitation strengthening and therefore have high ductile-brittle transition temperatures. If the steel is controlled rolled, the addition of both niobium and vanadium together is especially advantageous for increasing the yield strength and lowering ductile-brittle

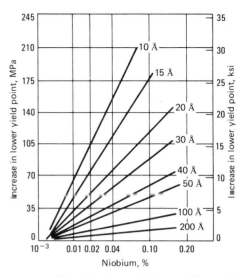

Fig. 12 Effect of niobium carbide on yield strength for various sizes of niobium carbide particles

Table 12 Effect of manganese content on the precipitation strengthening of a vanadium-microalloyed steel with a base composition of 0.08% C and 0.30% Si

Vanadium content, %	Yield strength		Change in yield strength	
	MPa	ksi	MPa	ksi
0.3% Mn				
0.00	297	43	0	0
0.08	352	51	55	8
0.14	380	55	83	12
1.2% Mn				
0.00	331	48	0	0
0.08	462	67	131	19
0.14	552	80	221	32

Source: Ref 23, 27

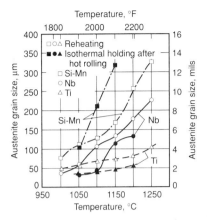

Fig. 13(a) Austenite grain coarsening during reheating and after hot rolling for a holding time of 30 min. Titanium contents were between 0.008 and 0.022% Ti. Source: Ref 25

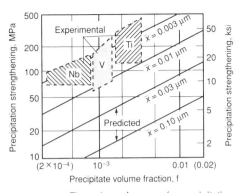

Fig. 13(b) The dependence of precipitation strengthening on average precipitate size ($\bar{x}$) and fraction according to theory and experimental observations for given microalloying additions. Source: Ref 22

transition temperatures by grain refinement.

Usually niobium-vanadium steels are made with relatively low carbon contents (<0.10% C). This reduces the amount of pearlite and improves toughness, ductility, and weldability. These steels are frequently referred to as pearlite-reduced steels.

Niobium-molybdenum microalloyed steels may have either a ferrite-pearlite microstructure or an acicular ferrite microstructure. Steels with the latter microstructure are discussed in the section "Acicular Ferrite (Low-Carbon Bainite) Steels" in this article.

In ferrite-pearlite niobium steels, the addition of molybdenum increases the yield strength and tensile strength by about 20 MPa (3 ksi) and 30 MPa (4.5 ksi), respectively, per 0.1% Mo, over an investigated range of 0% to 0.27% Mo (Ref 28). The principal effect of molybdenum on the microstructure is to alter the morphology of the pearlite and to introduce upper bainite as a partial replacement for pearlite. However, because the individual strength values of pearlite and bainite are somewhat similar, it has been proposed that the strength increase is due to solid-solution strengthening and enhanced Nb(CN) precipitation strengthening caused by a molybdenum-niobium synergism. The interaction between molybdenum and niobium (or vanadium) has been proposed as an explanation for the increase in precipitation strengthening by the addition of molybdenum. This effect has been attributed to the reduced precipitation in austenite from an increase in solubility arising from a decrease in carbon activity brought about by molybdenum (Ref 28). With less precipitation in austenite, more precipitates could form in the ferrite, resulting in enhanced strength. Also, molybdenum has been identified in the precipitates themselves; its presence may increase their strengthening effectiveness by increasing coherency strains and/or by increasing the volume fraction of precipitation (Ref 28). These metallurgical factors, when considered in conjunction with the effectiveness of controlled rolling to temperatures just below the Ar_3 temperature, have led to the development of a more economical X-70 molybdenum-niobium linepipe steel (Ref 28).

Vanadium-Nitrogen Microalloyed Steels. Vanadium combines more strongly with nitrogen than does niobium and forms VN precipitates in vanadium-nitrogen steel. Nitrogen additions to high-strength steels containing vanadium have become commercially important because the additions enhance precipitation hardening. Precipitation hardening may be accompanied by a drop in notch toughness, but this can often be overcome by lowering the carbon content. The precipitation of vanadium nitride also acts as a grain refiner.

Some producers use nitrogen additions to assist in the precipitation strengthening of controlled-cooled sheet and plate with thicknesses above 9.5 mm (0.375 in.). In one case (Ref 23), hot-rolled plates with vanadium and 0.018 to 0.022% N have been produced by controlled cooling in thicknesses up to 16 mm (0.625 in.) with yield strengths of 550 MPa (80 ksi). However, delayed cracking is a major problem in these steels. The use of nitrogen is not recommended for steels that will be welded because of its detrimental effect on notch toughness in the heat-affected zone.

Titanium-Microalloyed Steels. Titanium in low-carbon steels forms into a number of compounds that provide grain refinement, precipitation strengthening, and sulfide shape control. However, because titanium is also a strong deoxidizer, titanium can be used only in fully killed (aluminum deoxidized) steels so that titanium is available for forming into compounds other than titanium oxide. Commercially, steels precipitation strengthened with titanium are produced in thicknesses up to 9.5 mm (0.375 in.) in the minimum yield strength range from 345 to 550 MPa (50 to 80 ksi), with controlled rolling required to maximize strengthening and improve toughness.

Like niobium and/or vanadium steels, titanium microalloyed steels are strengthened by mechanisms that involve a combination of grain refinement and precipitation strengthening; the combination depends on the amount of alloy additions and processing methods. In reheated or continuously cast steels, small amounts of titanium (≤0.025% Ti) are effective grain refiners because austenite grain growth is retarded (Fig. 13a) by titanium nitride. Small amounts of titanium are also effective in recrystallization controlled rolling because titanium nitride retards the grain growth of recrystallized austenite. In conventional controlled rolling, however, titanium is a moderate grain refiner, causing less refinement than niobium but more than vanadium.

In terms of precipitation strengthening (Fig. 13b), a sufficient amount of titanium is required to form titanium carbide. Small percentages of titanium (<0.025% Ti) form mainly into TiN, which has an effect on austenite grain growth but little effect on precipitation strengthening because the precipitates formed in the liquid are too coarse. Increasing the titanium content leads to the formation first of titanium-containing manganese sulfide inclusions (Mn, Ti)S, and then of globular carbosulfides, $Ti_4C_2S_2$ (which provide sulfide shape control). The formation of $Ti_4C_2S_2$ is accompanied by and followed by titanium carbide (TiC) formation, which can be used for the precipitation strengthening of low-carbon steels. To determine the amount of titanium that is available for precipitation strengthening, the total titanium content must be adjusted for the formation of the coarse, insoluble titanium nitride and carbosulfides that do not participate in precipitation strengthening.

Experimentally observed strength increases from TiC precipitation have ranged up to 440 MPa (64 ksi) for very fine particles (less than 30 Å) and a relatively large fraction of precipitate (Fig. 13b). If sufficient amounts of titanium are used, titanium carbide can provide more precipitation strengthening than either niobium or vanadium. However, because higher levels of precipitation strengthening are generally associated with reduced toughness, grain refinement would be necessary to improve toughness. Titanium is a moderate grain refiner (compared to niobium and vanadium in hot-rolled steels), and the high levels of precipitation strengthening of titanium microalloyed steels result in a severe penalty in toughness. Using only titanium as a strengthener in high-strength hot-rolled strip has also resulted in unacceptable variability in mechanical properties (see the paper "Restricted Yield Strength Variation in High Strength Low Alloy Steels" by E.G. Hamburg in Ref 11).

Titanium-Niobium Microalloyed Steels. Although precipitation-strengthened titanium steels have limitations in terms of toughness and variability of mechanical properties, research has shown that an addition of titanium to low-carbon niobium steels results in an overall improvement in properties (see, for example, the article "Recent Developments in Automotive Hot-Rolled Strip Steels" by G. Tither in Ref 25). Titanium increases the efficiency of niobium because it combines with the nitrogen-forming titanium nitrides, thus preventing the formation of niobium nitrides. This allows for increased solubility of niobium in the austenite resulting in subsequent increased precipitation of Nb(C,N) particles in the ferrite. The addition of 0.04% titanium to steel strip containing various amounts of niobium consistently produced a yield strength increase of about 105 MPa (15 ksi) for a coiling temperature of 675 °C (1250 °F). Hot-rolled niobium-titanium steel strip is effective in achieving yield strengths of 550 MPa (80 ksi) in ferrite-pearlite steels. An addition of either vanadium or molybdenum can raise yield strengths to 690 MPa (100 ksi).

As-rolled pearlitic structural steels are a specific group of steels in which enhanced mechanical properties (and, in some cases, resistance to atmospheric corrosion) are obtained by the addition of moderate amounts of one or more alloying elements other than carbon. Some of these steels are carbon-manganese steels and differ from ordinary carbon steels only in having a greater manganese content. Other pearlitic structural steels contain small amounts of alloying elements, which are added to enhance weldability, formability, toughness, and strength.

The as-rolled pearlitic structural steels are characterized by as-rolled yield strengths in the range of 290 to 345 MPa (42 to 50 ksi). They are not intended for quenching and tempering and should not be subjected to such treatment. For certain applications, they may be annealed, normalized, or stress relieved, processes which may alter mechanical properties.

When these steels are used in welded structures, care must be exercised in grade selection and in the specification of the welding process details. Certain grades may be welded without preheating or postheating. The basic disadvantages of these steels is that the pearlitic microstructure raises the ductile-brittle transition temperature but does not improve yield strengths. In addition, the high carbon content (relative to other HSLA steels) reduces weldability.

Acicular Ferrite (Low-Carbon Bainite) Steels. Another approach to the development of HSLA steels is to obtain a very fine, high-strength acicular ferrite microstructure, instead of the usual polygonal ferrite microstructure, during the cooling transformation of ultralow carbon (<0.08%

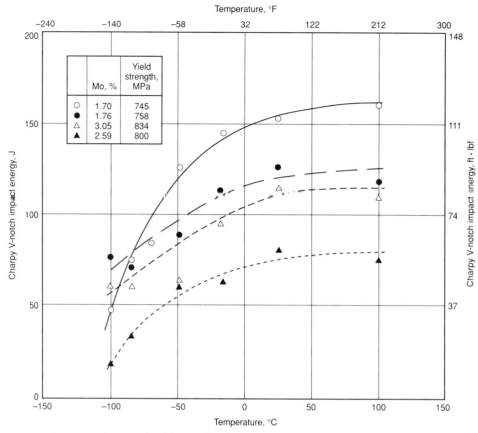

Fig. 14 Impact toughness and yield strengths of 25 mm (1 in.) thick, controlled-rolled, ultralow-carbon bainitic steel plate. In addition to the molybdenum contents shown, nominal contents of other alloying elements included 0.024 to 0.027% C, 0.92 to 1.00% Mn, 3.54 to 3.63% Ni, and 0.050 to 0.055% Nb. Source: Ref 29

C) steels with sufficient hardenability (by additions of manganese, molybdenum, and/or boron). Niobium can also be used for precipitation strengthening and grain refinement. The principal difference between the structure of acicular ferrite (which is also referred to as low-carbon bainite) and that of polygonal ferrite is that the former is characterized by a high dislocation density and fine, highly elongated grains that are not exhibited in polygonal ferrite.

Acicular ferrite steels can be obtained by quenching or, preferably, by air cooling with suitable alloys for hardenability. The principal advantage of this type of HSLA steel is the unusual combination of high yield strengths (415 to 690 MPa, or 60 to 100 ksi), high toughness (Fig. 14), and good weldability. A major application of these steels is linepipe in arctic conditions.

Typical properties of conventionally rolled, air-cooled 13 mm (0.5 in.) plate are (Ref 24):

- Yield strength: 470 to 530 MPa (68 to 77 ksi)
- Charpy V-notch 50% shear area fracture appearance transition temperature: −23 to 10 °C (−10 to 50 °F)
- Charpy V-notch ductile-shelf energy: greater than 136 J (100 ft · lbf)

Properties of a controlled-rolled acicular ferrite steel are summarized in Fig. 14.

The major application of acicular ferrite steel involves oil pipelines in arctic conditions. This application requires a combination of high strength, superior toughness, excellent resistance to hydrogen-induced cracking, and first-rate field weldability. In answer to these needs, Nippon Kokan K.K. developed a tough acicular ferrite steel for linepipe through the optimization of carbon and niobium content, the addition of boron, and/or the application of on-line accelerated cooling.

In this pipe, optimum carbon content ranges from 0.01 to 0.05%. Below 0.01% carbon, grain boundaries in the heat-affected zone (HAZ) are embrittled, resulting in intergranular, hydrogen-induced cracking and loss of toughness in the HAZ. The addition of boron and/or the application of on-line accelerated cooling ensures both high strength and superior toughness, along with desirable welding properties.

Three grades for arctic service are available: X-65, X-70, and X-80. An X-70 composition includes 0.03% C, 0.25% Si, 1.91% Mn, 0.008% P, 0.001% S, 0.048% N, plus titanium, boron, and calcium. Work at Bethlehem Steel Corporation to improve the properties of X-70 linepipe through the optimization of composition and processing parameters found that:

- An X-70 strength requirement, along with a 68 J (50 ft · lbf) minimum Charpy energy at 0 °C (32 °F) can be met with a controlled-rolled, low-carbon, low-sulfur vanadium-niobium composition. Full-curve Charpy data for the 10.9 mm (0.429 in.) wall linepipe indicate that with the chemistry and processing used, a 68 J (50 ft · lbf) Charpy requirement could also be met at −18 °C (0 °F) at this wall thickness
- There is a positive shift in both yield and tensile strength during pipe forming that is related to the amount of expansion and the corresponding stress-strain characteristics, such as work hardening and Lüders extension. As the percentage of expansion increases, the shift in yield strength rises continuously, but the shift in tensile strength levels off at about 27 to 35 MPa (4 to 5 ksi) after about 1.4% expansion
- Various Linde wire and flux combinations produce excellent toughness in the pipe body and HAZ, but only the Linde wire 83 and Linde flux 101 combinations provide excellent Charpy absorbed energies in the weld metal, such that the pipe body, heat-affected zone, and weld metal satisfy a 68 J (50 ft · lbf) requirement at 0 °C (32 °F)

Dual-phase steels have a microstructure with 80 to 90% polygonal ferrite and 10 to 20% martensite islands dispersed throughout the ferrite matrix. These steels have a low yield strength and continuous yielding behavior; therefore they form just like low-strength steel, but they can also provide high strength in the finished component because of their rapid work-hardening rate. Figure 8 compares the ductility and tensile strength of dual-phase steels with other HSLA steels. Typical as-shipped yield strength is 310 to 345 MPa (45 to 50 ksi).

Dual-phase steels can be produced from low-carbon steels in three ways (Ref 30):

- Intercritical austenitization of carbon-manganese steels followed by rapid cooling
- Hot rolling with ferrite formers such as silicon and transformation-delaying elements such as chromium, manganese, and/or molybdenum
- Continuous annealing of cold-rolled carbon-manganese steel followed by quenching and tempering

Additional information on dual-phase steels is contained in the article "Dual-Phase Steels" in this Volume.

Interstitial-Free (IF) Steels. Steels with very low interstitial contents exhibit excellent formability with low yield strength, high elongation, and good deep drawability (Fig. 8c). With the addition of carbonitride-forming elements, the deep drawability and the non-aging properties are further improved. The effect of niobium, unlike that of other microalloyed elements, is to improve the planar anisotropy, reducing earing. This is due to the finer grain size already in the hot strip material prior to cold rolling. Titanium is also added to improve the effectiveness of niobium (see the section "Titanium-Niobium Microalloyed Steels" in this article). Information on the processing and steelmaking of IF steels is given in the articles "Steel Processing Technology" and "Microstructures, Processing, and Properties of Steels" in this Volume.

Inclusion Shape Controlled Grades. An important development in microalloyed HSLA steels is the use of inclusion shape control. Sulfide inclusions, which are plastic at rolling temperatures and thus elongate and flatten during rolling, adversely affect ductility in the short transverse (through thickness) direction. The main objective of inclusion shape control is to produce sulfide inclusions with negligible plasticity at even the highest rolling temperatures.

The preferred method for sulfide shape control involves calcium-silicon ladle additions. However, sulfide shape control is also performed with small additions of rare-earth elements, zirconium, or titanium that change the shape of the sulfide inclusions from elongated stringers to small, dispersed, almost spherical globules. This change in the shape of sulfide inclusions substantially increases transverse impact energy and improves formability. Inclusion shape control was introduced with the advent of hot-rolled sheet and light plate having a yield strength of 550 MPa (80 ksi) in the as-rolled condition. This technology has also been extended to include grades with lower yield strengths ranging from 310 to 550 MPa (45 to 80 ksi). The improved formability of these grades is recognized in ASTM A 715. Inclusion shape control with rare-earth elements is seldom used because rare-earth elements produce relatively dirty steel.

Sulfide inclusion shape control performs several important roles in HSLA steels. It improves transverse impact energy, and it can minimize lamellar tearing in welded structures by improving through-thickness properties that are critical in constrained weldments.

Control of HSLA Steel Properties

Most HSLA steels are furnished in the as-hot-rolled condition with ferritic-pearlitic microstructure. The exceptions are the controlled-rolled steels with an acicular ferrite microstructure and the dual-phase steels with martensite dispersed in a matrix of polygonal ferrite. These two types of HSLA steels use the formation of eutectoid structures for strengthening, while the ferritic-pearlitic HSLA steels generally require strengthening of the ferrite. Pearlite is generally an undesirable strengthening agent in structural steels because it reduces impact toughness and requires higher carbon contents. Moreover, yield strength is largely unaffected by a higher pearlite content.

Strengthening Mechanisms in Ferrite. The ferrite in HSLA steels is typically strengthened by grain refinement, precipitation hardening, and, to a lesser extent, solid-solution strengthening. Grain refinement is the most desirable strengthening mechanism because it improves not only strength but also toughness.

Grain refinement is influenced by the complex effects of alloy design and processing methods. For example, the various methods of grain refinement used in the three different stages of hot rolling (that is, reheating, hot rolling, and cooling) include:

- The addition of titanium (Fig. 13a) or aluminum to retard austenite grain growth when the steel is reheated for hot deformation or subsequent heat treatment
- The controlled rolling of microalloyed steels to condition the austenite so that it transforms into fine-grain ferrite
- The use of alloy additions and/or faster cooling rates to lower the austenite-to-ferrite transformation temperature

The use of higher cooling rates for grain refinement may require consideration of its effect on precipitation strengthening and the possibility of undesirable transformation products.

Precipitation strengthening occurs from the formation of finely dispersed carbonitrides developed during heating and cooling. Because precipitation strengthening is generally associated with a reduction in toughness, grain refinement is often used in conjunction with precipitation strengthening to improve toughness.

Precipitation strengthening is influenced by the type of carbonitride, its grain size, and, of course, the number of carbonitrides precipitated. The formation of MC is the most effective metal carbide in the precipitation strengthening of microalloyed niobium, vanadium, and/or titanium steels. The number of fine MC particles formed during heating and cooling depends on the solubility of the carbides in austenite and on cooling rates (Fig. 11). Figure 13(b) shows the dependence of precipitation strengthening on precipitate size and volume fraction.

Steelmaking. Precise steelmaking operations are also essential in controlling the properties and chemistry of HSLA steels. Optimum property levels depend on such factors as the control of significant alloying elements and the reduction of impurities and nonmetallic inclusions.

Developments in secondary steelmaking such as desulfurization, vacuum degassing, and argon shrouding have enabled better control of steel chemistry and the effective use of microalloyed elements. In particular,

Table 13 Compositional limits for HSLA steel grades described in ASTM specifications

ASTM specification(a)	Type or grade	UNS designation	C	Mn	P	S	Si	Cr	Ni	Cu	V	Other
A 242	Type 1	K11510	0.15	1.00	0.45	0.05	...	...	...	0.20 min	...	...
A 572	Grade 42	...	0.21	1.35(c)	0.04	0.05	0.30(c)	...	...	0.20 min(d)	...	(e)
	Grade 50	...	0.23	1.35(c)	0.04	0.05	0.30(c)	...	...	0.20 min(d)	...	(e)
	Grade 60	...	0.26	1.35(c)	0.04	0.05	0.30	...	...	0.20 min(d)	...	(e)
	Grade 65	...	0.23(c)	1.65(c)	0.04	0.05	0.30	...	...	0.20 min(d)	...	(e)
A 588	Grade A	K11430	0.10–0.19	0.90–1.25	0.04	0.05	0.15–0.30	0.40–0.65	...	0.25–0.40	0.02–0.10	...
	Grade B	K12043	0.20	0.75–1.25	0.04	0.05	0.15–0.30	0.40–0.70	0.25–0.50	0.20–0.40	0.01–0.10	...
	Grade C	K11538	0.15	0.80–1.35	0.04	0.05	0.15–0.30	0.30–0.50	0.25–0.50	0.20–0.50	0.01–0.10	...
	Grade D	K11552	0.10–0.20	0.75–1.25	0.04	0.05	0.50–0.90	0.50–0.90	...	0.30	...	0.04 Nb, 0.05–0.15 Zr
	Grade K	...	0.17	0.5–1.20	0.04	0.05	0.25–0.50	0.40–0.70	0.40	0.30–0.50	...	0.10 Mo, 0.005–0.05 Nb
A 606	...	...	0.22	1.25	...	0.05	...	...	...	...	...	...
A 607	Grade 45	...	0.22	1.35	0.04	0.05	...	...	...	0.20 min(d)	...	(e)
	Grade 50	...	0.23	1.35	0.04	0.05	...	...	...	0.20 min(d)	...	(e)
	Grade 55	...	0.25	1.35	0.04	0.05	...	...	...	0.20 min(d)	...	(e)
	Grade 60	...	0.26	1.50	0.04	0.05	...	...	...	0.20 min(d)	...	(e)
	Grade 65	...	0.26	1.50	0.04	0.05	...	...	...	0.20 min(d)	...	(e)
	Grade 70	...	0.26	1.65	0.04	0.05	...	...	...	0.20 min(d)	...	(e)
A 618	Grade Ia	...	0.15	1.00	0.15	0.05	...	...	...	0.20 min	...	...
	Grade Ib	...	0.20	1.35	0.04	0.05	...	...	...	0.20 min(f)	...	...
	Grade II	K12609	0.22	0.85–1.25	0.04	0.05	0.30	...	...	...	0.02 min	...
	Grade III	K12700	0.23	1.35	0.04	0.05	0.30	...	...	...	0.02 min	0.005 Nb min(g)
A 633	Grade A	K01802	0.18	1.00–1.35	0.04	0.05	0.15–0.30	...	...	...	...	0.05 Nb
	Grade C	K12000	0.20	1.15–1.50	0.04	0.05	0.15–0.50	...	...	...	...	0.01–0.05 Nb
	Grade D	K02003	0.20	0.70–1.60(c)	0.04	0.05	0.15–0.50	0.25	0.25	0.35	...	0.08 Mo
	Grade E	K12202	0.22	1.15–1.50	0.04	0.05	0.15–0.50	...	...	...	0.04–0.11	0.01–0.05 Nb(d), 0.01–0.03 N
A 656	Type 3	...	0.18	1.65	0.025	0.035	0.60	...	...	...	0.08	0.020 N, 0.005–0.15 Nb
	Type 7	...	0.18	1.65	0.025	0.035	0.60	...	...	...	0.005–0.15	0.020 N, 0.005–0.10 Nb
A 690	...	K12249	0.22	0.60–0.90	0.08–0.15	0.05	0.10	...	0.40–0.75	0.50 min	...	...
A 709	Grade 50, type 1	...	0.23	1.35	0.04	0.05	0.40	...	...	...	...	0.005–0.05 Nb
	Grade 50, type 2	...	0.23	1.35	0.04	0.05	0.40	...	...	...	0.01–0.15	...
	Grade 50, type 3	...	0.23	1.35	0.04	0.05	0.40	...	...	...	(h)	0.05 Nb max
	Grade 50, type 4	...	0.23	1.35	0.04	0.05	0.40	...	...	...	(i)	0.015 N max
A 715	...	...	0.15	1.65	0.025	0.035	...	...	...	...	...	V, Ti, Nb added as necessary
A 808	...	...	0.12	1.65	0.04	0.05 max or 0.010 max	0.15–0.50	...	...	...	0.10	0.02–0.10 Nb, V+Nb = 0.15 max
A 812	65	...	0.23	1.40	0.035	0.04	0.15–0.50(j)	...	...	...	V+Nb = 0.02–0.15	0.05 Nb max
	80	...	0.23	1.50	0.035	0.04	0.15–0.50	0.35	...	...	V+Nb = 0.02–0.15	0.05 Nb max
A 841	...	...	0.20	(k)	0.030	0.030	0.15–0.50	0.25	0.25	0.35	0.06	0.08 Mo, 0.03 Nb, 0.02 Al total
A 871	...	...	0.20	1.50	0.04	0.05	0.90	0.90	1.25	1.00	0.10	0.25 Mo, 0.15 Zr, 0.05 Nb, 0.05 Ti

(a) For characteristics and intended uses, see Table 10; for mechanical properties, see Table 16. (b) If a single value is shown, it is a maximum unless otherwise stated. (c) Values may vary, or minimum value may exist, depending on product size and mill form. (d) Optional or when specified. (e) May be purchased as type 1 (0.005–0.05 Nb), type 2 (0.01–0.15 V), type 3 (0.05 Nb, max, plus 0.02–0.15 V) or type 4 (0.015 N, max, plus V ≥ 4 N). (f) If chromium and silicon are each 0.50% min, the copper minimum does not apply. (g) May be substituted for all or part of V. (h) Niobium plus vanadium, 0.02 to 0.15%. (i) Nitrogen with vanadium content of 0.015% (max) with a minimum vanadium-to-nitrogen ratio of 4:1. (j) When silicon-killed steel is specified. (k) For plate under 40 mm (1.5 in.), manganese contents are 0.70 to 1.35% or up to 1.60% if carbon equivalents do not exceed 0.47%. For plate over 40 mm (1 to 5 in.), ASTM A 841 specifies manganese contents of 1.00 to 1.60%.

the use of vacuum degassing equipment allows the production of interstitial-free (IF) steels. The IF steels exhibit excellent formability, high elongation, and good deep drawability (Fig. 8).

Compositions and Alloying Elements. Chemical compositions for the HSLA steels specified by ASTM standards (Table 10) are listed in Table 13. The principal function of alloying elements in these ferrite-pearlite HSLA steels, other than corrosion resistance, is strengthening of the ferrite by grain refinement, precipitation strengthening, and solid-solution strengthening. Solid-solution strengthening is closely related to alloy contents (Fig. 9), while grain refinement and precipitation strengthening depend on the complex effects of alloy design and thermomechanical treatment.

Alloying elements are also selected to influence transformation temperatures so that the transformation of austenite to ferrite and pearlite occurs at a lower temperature during air cooling. This lowering of the transformation temperature produces a finer-grain transformation product, which is a major source of strengthening. At the low carbon levels typical of HSLA steels, elements such as silicon, copper, nickel, and phosphorus are particularly effective for producing fine pearlite. Elements such as manganese and chromium, which are present in both the cementite and ferrite, also strengthen the ferrite by solid-solution strengthening in proportion to the amount dissolved in the ferrite.

Carbon markedly increases the amount of pearlite in the microstructure and is one of the more potent, as well as economical, strengthening elements. However, higher carbon contents reduce weldability and the impact toughness of steel. Increases in pearlite content are also ineffective in improving yield strength, which is often the main strength criterion in structural applications.

In the presence of alloying elements, the practical maximum carbon content at which HSLA steels can be used in the as-rolled condition is approximately 0.20%. Higher levels of carbon tend to form martensite or bainite in the microstructure of as-rolled steels, although some of the higher-strength low-alloy steels have carbon contents that approach 0.30%. Many of the proprietary microalloyed HSLA steels have carbon contents of 0.06% or even lower, yet are still able to develop yield strengths of 345 to 620 MPa (50 to 90 ksi). Carbon levels as low as 0.03% are utilized in some alloy designs. The required strength is developed by the combined effect of:

- Fine grain size developed during controlled hot rolling and enhanced by microalloyed elements (especially niobium)
- Precipitation strengthening caused by the presence of vanadium, niobium, and titanium in the composition

Nitrogen in amounts up to about 0.02% has been used to obtain strengths typical of HSLA steels and at reasonable cost. For carbon and carbon-manganese steels, such a practice is limited to light-gage products because the increase in strength is accompanied by a drop in notch toughness. In some applications, nitrogen contents are limited to 0.005%. Nitrogen additions to high-strength steels containing vanadium have become commercially important because such additions enhance precipitation hardening. The precipitation of vanadium nitride in vanadium-nitrogen steels also improves grain refinement because it has a lower solubility in austenite than vanadium carbide (Fig. 15). The total effect of nitrogen on yield strength is discussed in the section "Tensile Properties" in this article.

Manganese is the principal strengthening element in plain carbon high-strength structural steels when it is present in amounts over 1%. It functions mainly as a mild solid-solution strengthener in ferrite, but it also provides a marked decrease in the austenite-to-ferrite transformation temperature. In addition, manganese can enhance the precipitation strengthening of vanadium steels (Table 12) and, to a lesser extent, niobium steels.

Silicon. One of the most important applications of silicon is its use as a deoxidizer in molten steel. It is usually present in fully deoxidized structural steels in amounts up to 0.35%, which ensures the production of sound, dense ingots. Silicon has a strengthening effect in low-alloy structural steels. In larger amounts, it increases resistance to scaling at elevated temperatures. Silicon has a significant effect on yield strength enhancement by solid-solution strengthening (Fig. 9) and is widely used in HSLA steels for riveted or bolted structures. It can be used up to 0.30% in weldable steels; higher amounts produce a deterioration in notch toughness and weldability (Ref 23).

Copper. Approximately 0.20% copper is used to provide resistance to atmospheric

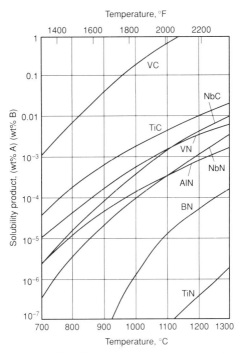

Fig. 15 Solubility product versus temperature for eight carbonitride compounds in austenite

corrosion. Its effect on resistance to corrosion is enhanced when phosphorus is present in amounts greater than about 0.05%.

Copper in levels in excess of 0.50% also increases the strength of both low- and medium-carbon steels by virtue of ferrite strengthening, which is accompanied by only slight decreases in ductility. In amounts over about 0.60%, copper can precipitate as ϵ-copper, which precipitation strengthens the ferrite. Copper can be retained in solid solution even at the slow rate of cooling obtained when large sections are normalized, but it is precipitated out when the steel is reheated to about 510 to 605 °C (950 to 1125 °F). At about 1% copper, the yield strength is increased by about 70 to 140 MPa (10 to 20 ksi), regardless of the effects of other alloying elements. Copper in amounts up to 0.75% is considered to have only minor adverse effects on notch toughness or weldability.

Copper precipitation hardening gives the steel the ability to be formed extensively (while still low in strength) and then precipitation hardened as a complex shape or welded assembly. This avoids the distortion or difficulty encountered by quenching compared with components made from conventionally heat-treated alloy steel.

Steels containing more than 0.50% Cu can exhibit hot shortness, with the result that cracks and a rough surface may develop during hot working. Careful control of oxidation during heating and prevention of overheating can minimize these conditions.

The addition of nickel in amounts equal to at least one-half the copper content is very beneficial to the surface quality of copper-bearing steels. Copper is also added in amounts between 0.25-0.35% to improve resistance to hydrogen-induced cracking in aqueous environments and H_2S environments.

Phosphorus is an effective solid-solution strengthener in ferrite (Fig. 9). It also enhances corrosion resistance but causes a decrease in ductility. Phosphorus at low levels (<0.05% P) can also cause embrittlement through segregation to the prior-austenite grain boundaries.

The atmospheric-corrosion resistance of steel is increased appreciably by the addition of phosphorus, and when small amounts of copper are present in the steel, the effect of the phosphorus is greatly enhanced. When both phosphorus and copper are present, there is a greater beneficial effect on corrosion resistance than the sum of the effects of the individual elements.

Chromium is often added with copper to obtain improved atmospheric-corrosion resistance. Upon exposure to the atmosphere, a steel composition with about 0.12% P, 0.85% Cr, and 0.40 Cu% develops a particularly adherent, dense oxide coating that is characteristic of weathering steels.

Nickel can be added in amounts up to about 1% in several HSLA steels and in amounts up to 5% for high-strength heat-treated alloy grades. It moderately increases strength by solution hardening of the ferrite. In HSLA steels, it enhances atmospheric-corrosion resistance and, when present in combination with copper and/or phosphorus, increases the seawater corrosion resistance of steels. Nickel is often added to copper-bearing steels to minimize hot shortness.

Molybdenum in hot-rolled HSLA steels is used primarily to improve hardenability when transformation products other than ferrite-pearlite are desired. For example, molybdenum is an essential ingredient for producing as-rolled acicular ferrite steels. In addition, it effectively enhances elevated-temperature properties.

Molybdenum (0.15 to 0.30%) in microalloyed steels also increases the solubility of niobium in austenite, thereby enhancing the precipitation of NbC(N) in the ferrite. This increases the precipitation-strengthening effect of Nb(C,N). Molybdenum has also been shown to join the Nb(C,N) precipitates, which further raises the yield strength.

Niobium. Small additions (0.03 to 0.05% Nb) of niobium increase yield strength by a combination of precipitation strengthening (Fig. 12) and grain refinement. Niobium is a more effective grain-refining element than vanadium because niobium carbide is more stable in austenite than vanadium carbide at typical rolling temperatures. The lower solubility of niobium carbide in austenite (Fig. 15) provides more stable precipitate particles, which pin the austenite grain boundaries and thus retard austenite grain growth.

Aluminum is widely used as a deoxidizer and was the first element used to control austenite grain growth during reheating. During controlled rolling, niobium and titanium are more effective grain refiners than aluminum. Titanium (Fig. 13a), niobium, zirconium, and vanadium are also effective grain growth inhibitors during reheating. However, for steels that are heat treated (quenched and tempered), these four elements may have adverse effects on hardenability because their carbides are quite stable and difficult to dissolve in austenite prior to quenching.

Vanadium strengthens HSLA steels by both precipitation hardening the ferrite and refining the ferrite grain size. The precipitation of vanadium carbonitride in ferrite can develop a significant increase in strength that depends not only on the rolling process used, but also on the base composition. Carbon contents above 0.13 to 0.15% and manganese content of 1% or more enhances the precipitation hardening, particularly when the nitrogen content is at least 0.01%. Grain size refinement depends on thermal processing (hot rolling) variables, as well as vanadium content.

Titanium is unique among common alloying elements in that it provides both precipitation strengthening and sulfide shape control. Small amounts of titanium (<0.025%) are also useful in limiting austenite grain growth (Fig. 13a). However, it is useful only in fully killed (aluminum deoxidized) steels because of its strong deoxidizing effects. The versatility of titanium is limited because variations in oxygen, nitrogen, and sulfur affect the contribution of titanium as a carbide strengthener.

Zirconium can also be added to killed high-strength low-alloy steels to improve inclusion characteristics, particularly in the case of sulfide inclusions, for which changes in inclusion shape improve ductility in transverse bending.

Boron has no effect on the strength of normal hot-rolled steel but can considerably improve hardenability when transformation products such as acicular ferrite are desired in low-carbon hot-rolled plate. Its full effect on hardenability is obtained only in fully deoxidized (aluminum-killed) steels.

Rare-earth elements, principally cerium, lanthanum, and praseodymium, can be used to provide shape control of sulfide inclusions. Sulfide inclusions, which are plastic at rolling temperatures and thus elongate and flatten during rolling, adversely affect ductility in the short transverse (through-thickness) direction. The chief role of rare-earth additives is to produce rare-earth sulfide and oxysulfide inclusions, which have negligible plasticity at even the highest rolling temperatures. Excessive amounts of cerium (>0.02%) and other rare-earth elements lead to oxide or oxysulfide stringers that may affect directionality. Treatment with rare-earth elements is seldom used because they produce relatively dirty steels. Treatment with calcium is preferred for sulfide inclusion shape control.

Controlled Rolling. The hot-rolling process has gradually become a much more closely controlled operation, and controlled rolling is now being increasingly applied to microalloyed steels with compositions carefully chosen to provide optimum mechanical properties at room temperature. Controlled rolling is a procedure whereby the various stages of rolling are temperature controlled, with the amount of reduction in each pass predetermined and the finishing temperature precisely defined. This processing is widely used to obtain reliable mechanical properties in steels for pipelines, bridges, offshore platforms, and many other engineering applications. The use of controlled rolling has resulted in improved combinations of strength and toughness and further reductions in the carbon content of microalloyed HSLA steels. This reduction in carbon content improves not only toughness but also weldability.

The basic objective of controlled rolling is to refine and/or deform austenite grains during the rolling process so that fine ferrite grains are produced during cooling. Controlled rolling can be performed on carbon steels but is most beneficial in steels with vanadium and/or niobium additions. During hot rolling, the undissolved carbonitrides of vanadium and niobium pin austenite grain boundaries and thus retard austenite grain growth. In carbon steels, however, the temperatures involved in hot rolling lead to marked austenite grain growth, which basically limits any benefit of grain refinement by controlled rolling. Controlled rolling is performed on strip, plate, and bar mills but not on continuous hot strip mills. On a hot strip mill, the water cooling on the runout table ensures a fine grain size.

The three methods of controlled rolling are:

- Conventional controlled rolling
- Recrystallization controlled rolling
- Dynamic recrystallization controlled rolling

These three methods use different techniques for grain refinement, but they are all preceded by a roughing operation to refine grain size by repeated recrystallization. In the roughing stage, stable carbonitride precipitates are desirable because they pin the grain boundaries of the recrystallized austenite and thus prevent their growth. Niobium is more effective than vanadium in preventing austenite grain growth during rolling because niobium forms precipitates that are less soluble than vanadium carbide in austenite (Fig. 15). Roughing can achieve austenite grain sizes on the order of 20 μm (0.8 mil). The austenite grains are then either deformed or further refined by controlled rolling during finishing operations.

Conventional controlled rolling is based on the deformation, or flattening (pancaking), of austenite grains so that a large number of nucleation sites exist on the deformed austenite grain boundaries and on the deformation bands with the austenite grains. These nucleation sites allow the formation of very fine-grain ferrite during transformation cooling. This process requires a total reduction of up to 80% in a temperature range where the austenite deforms but does not recrystallize.

Niobium is the most effective alloying element for grain refinement by conventional controlled rolling. During the rolling reductions at temperatures below 1040 °C (1900 °F), the niobium in solution suppresses recrystallization by solute drag or by strain-induced Nb(C,N) precipitation on the deformed austenite and slip planes. The strain-induced precipitates are too large to affect precipitation strengthening but are beneficial for two reasons: They allow additional suppression of recrystallization by preventing migration of austenite grain subboundaries, and they provide a large number of nuclei in the deformed austenite for the formation of fine ferrite particles during cooling. The strain-induced precipitates in the austenite detract from the precipitation-hardening potential of the ferrite by removing the available niobium from austenite solid solution. Nevertheless, a useful measure of precipitation strengthening is possible in controlled-rolled niobium steels.

The controlled rolling of niobium steels can lead to ferrite grain sizes in the range of 5 to 10 μm (ASTM grain size numbers 10 to 12). General combinations of yield strength and ductile-brittle transition temperatures for niobium steel with various compositions and section sizes are shown in Fig. 16. Because the precipitation of Nb(CN) in the austenite during hot rolling retards recrystallization and raises the temperature at which recrystallization of austenite ceases (the recrystallization stop temperature), a broader temperature range is possible for hot working the steel to produce highly deformed austenite. The optimum amount of niobium to suppress recrystallization between passes can be as little as 0.02%. Titanium, zirconium, and vanadium are not as effective as niobium in raising the recrystallization stop temperature. Titanium and zirconium nitride formed during solidification and upon cooling of the slab do not readily dissolve upon reheating to hot-rolling temperatures. Although these nitrides may prevent grain coarsening upon reheating, they are not effective in preventing recrystallization, because insufficient titanium or zirconium remains in solution at the

High-Strength Structural and High-Strength Low-Alloy Steels / 409

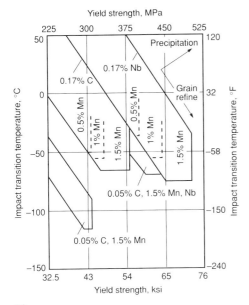

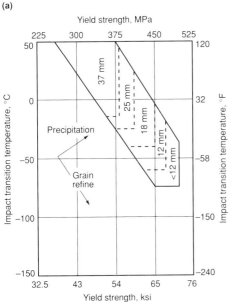

Fig. 16 General combinations of yield stress and impact transition temperatures available in controlled-rolled steels of various (a) compositions and (b) section sizes. Source: Ref 24

rolling temperature to precipitate on deformed austenite during boundaries during hot-rolling and thus suppress austenite recrystallization. Vanadium, on the other hand, is so soluble that precipitation does not readily occur in the austenite at normal hot-rolling temperatures. The concentrations of niobium, titanium, vanadium, carbon, and nitrogen; the degree of strain; the time between passes; the strain rate; and the temperature of deformation all influence recrystallization during hot working.

Recrystallization Controlled Rolling. Although conventional controlled rolling can lead to very fine ferrite grain sizes, the low finishing temperature (750 to 900 °C, or 1400 to 1650 °F) of this method leads to increased rolling loads for heavy plate and thick-walled seamless tube. For thicker sections, recrystallization controlled rolling is used to refine austenite grain size. This process can result in ferrite grain sizes on the order of 8 to 10 μm (0.3 to 0.4 mil).

Recrystallization controlled rolling involves the recrystallization of austenite at successively lower temperatures below roughing temperatures but still above 900 °C (1650 °F). Recrystallization should not be sluggish for this method to succeed, and thus vanadium can be beneficial because vanadium carbide is readily dissolved at rolling temperature and therefore unavailable for suppressing recrystallization. However, vanadium steels require stable carbonitrides, such as titanium nitride, to retard grain growth after recrystallization. Niobium steels, on the other hand, can undergo recrystallization controlled rolling at higher temperatures with Nb(C,N) precipitates eventually forming. This precipitation of Nb(C,N) will restrict austenite grain growth and may preclude the need for a titanium addition.

Dynamic recrystallization controlled rolling is used when there is insufficient time for recrystallization between rolling passes. This process initiates recrystallization during deformation and requires appreciable reductions (for example, 100%) to achieve an austenite grain size of 10 μm (0.4 mil). With low-temperature finishing, dynamic recrystallization can result in ferrite grain as fine as 3 to 6 μm (0.12 to 0.24 mil).

Interpass Cooling in Controlled Rolling. One of the major drawbacks with conventional controlled rolling is excessive processing time, especially on single-stand mills lacking facilities for removing plates off line during the hold period. Therefore, some investigations have examined the possibility of reducing processing time by the application of accelerated cooling both during the delay period and between individual roughing and finishing passes. In general, it would appear that the process time can be shortened by 30 to 40%, depending on exactly when forced cooling is applied in the rolling schedule. The as-rolled microstructure is not affected very much by intermittent cooling between passes but prolonged exposure to water during the hold period causes the temperature of the surface layers of the plate to fall below Ar₃. On resumption of rolling, this surface shell is reaustenitized by heat conduction from the interior of the plate; the surface austenite grain size is thereby considerably finer than that in the rest of the material, and this difference persists, of course, in the microstructure after rolling. Such nonuniformity of microstructure and properties in the through-thickness direction is hardly desirable, but it is unlikely to be directly detrimental to any particular bulk property of the plate (with the possible exception of bendability). Plates processed with shorter rolling times by the application of accelerated cooling between passes and/or under holding appear to exhibit equivalent strength and toughness properties to those controlled rolled in the conventional manner (Ref 11).

Normalizing. As with carbon steels, normalizing can be used to refine the ferrite-pearlite grain size in HSLA steels. However, because microalloyed HSLA steels may involve some strengthening by precipitation hardening, normalizing may cause a reduction in precipitation strengthening when stable carbonitride precipitates coarsen at typical austenitizing temperatures for normalizing. For example, normalizing a conventionally hot-rolled niobium steel reduces the yield strength considerably so that much of the precipitation strengthening increment that is due to niobium in the as-rolled condition is lost (Ref 23, 31). At the usual austenitizing temperatures for normalizing, most of the niobium has not been taken into solution and is present as relatively coarse precipitates. Thus, niobium inhibits grain coarsening, but provides little or no precipitation strengthening when the steel is normalized. However, the grain refining accounts for the improved low-temperature impact properties and most of the small increase in strength, compared to a niobium-free steel (Ref 23). Normalized niobium steels thus provide a fairly good combination of yield strength and ductile-brittle impact transition temperatures, even in semikilled steels (which contain free nitrogen that is detrimental to impact toughness). Niobium steels, along with titanium and vanadium steels, in the form of normalized, heavy-gage plate find important applications in offshore and general construction, shipbuilding, and pressure vessels in Europe. Normalized steels of this type are always fully aluminum killed (Ref 23). A typical example of a normalized niobium steel in the United States is the grade C steel in ASTM A 633 (Table 13). Minimum yield strengths of this steel range from 345 MPa (50 ksi) for plate thicknesses under 65 mm (2.5 in.) to 315 MPa (46 ksi) for plate thicknesses of 65 to 100 mm (2.5 to 4 in.). Charpy V-notch impact energies for the normalized steels in ASTM A 633 range from 41 J (30 ft · lbf) for transverse specimens at 0 °C (32 °F) to 20 J (15 ft · lbf) for transverse specimens at −50 °C (−60 °F).

Although normalized niobium steels have a finer ferrite grain size than do normalized carbon-manganese steels when strength is of paramount importance, controlled rolling of niobium steels is preferred over normalizing because controlled rolling also provides a measure of precipitation strengthening. Vanadium, on the other hand, causes marked precipitation strengthening with limited grain re-

Table 14 Mechanical properties of a representative sample of normalized HSLA steels produced in Europe

	Minimum yield strength, MPa (ksi), for a plate thickness of		Minimum ultimate tensile strength, MPa (ksi), for a plate thickness of		Minimum Charpy V-notch impact energy, J (ft · lbf) at			
					0 °C (32 °F)		−50 °C (−60 °F)	
Steel	≤16 mm (≤0.63 in.)	50<t≤80 mm (2<t≤3 in.)	≤16 mm (≤0.63 in.)	50<t≤80 mm (2<t≤3 in.)	Longitudinal	Transverse	Longitudinal	Transverse
Usiten 420-II	420 (61)	...	550 (80)	...	56 (41)	44 (32)	28 (21)	21 (15)
Usiten 460-I	460 (67)	...	590 (85)	...	48 (35)	36 (27)	...	...
Usiten 460-II	...	420 (61)	...	570 (82)	48 (35)	36 (27)	28 (21)	16 (12)
FG 43 T	420 (61)	...	530 (77)	...	...	...	27 (20)	27 (20)(a)
FG 47 CT	460 (67)	...	560 (81)	...	...	...	27 (20)	27 (20)(a)
FG 51 T	500 (72)	...	610 (88)	...	...	31 (23)	27 (20)(b)	...
Dillinal 55/43E	420 (61)	380 (55)	530 (77)	530 (77)	90 (66)	70 (52)	30 (22)	27 (20)
Dillinal 58/47E	460 (67)	420 (61)	560 (81)	560 (81)	90 (66)	70 (52)	30 (22)	27 (20)
Hyplus 29	450 (65)	400 (58)(d)	570 (82)	570 (82)	54 (40)	...	27 (20)(c)	...
BS 4360:55E	450 (65)	415 (60)(d)	550 (80)	550 (80)	...	61 (45)(a)	...	27 (20)
WSTE 500	480 (70)	450 (65)	610 (88)	610 (88)	≥44 (≥32)(a)	...	...	...

(a) 20 °C (-4 °F). (b) −40 °C (−40 °F). (c) −30 °C (−22 °F). (d) For plate thicknesses of 40 to 63 mm (1.5 to 2.5 in.)

finement upon normalizing because practically all of the vanadium is in solution at usual normalizing temperatures. The strengthening with vanadium in normalized steels is far greater than can be obtained in normalized niobium steels (Ref 23). In the normalized condition, vanadium semikilled structural steels also have impact properties superior to those of the semikilled niobium steels because vanadium combines with part of the nitrogen, thereby reducing the free-nitrogen content.

Table 14 lists some typical impact toughness values and yield strengths of aluminum-killed European vanadium steels in the normalized condition (see Table 15 for chemical compositions). In the United States, a typical example of a normalized vanadium steel is the grade E steel in ASTM A 633 (Table 13). Minimum yield strengths for this steel range from 415 MPa (60 ksi) for plate thicknesses under 100 mm (4 in.) to 380 MPa (55 ksi) for plate thicknesses of 100 to 150 mm (4 to 6 in.). Charpy V-notch impact energies are the same as for ASTM A 633, grade C, mentioned in the previous paragraph.

Precipitation Strengthening With Copper. Steels with about 0.6% or more copper can be precipitation strengthened by aging in a temperature range from about 425 to 650 °C (800 to 1200 °F). When reheated to 510 to 607 °C (950 to 1125 °F) after rolling, precipitation-hardened copper-bearing steels with more than 0.75% Cu have typical mechanical properties:

Tensile strength, MPa (ksi)	540 (78)
Yield strength, MPa (ksi)	485 (70)
Elongation in 50 mm (2 in.), %	20
Reduction of area, %	60

Mechanical Properties of Hot-Rolled HSLA Steels

The three major types of HSLA steel microstructures are dual-phase microstructures, acicular ferrite microstructures, and ferrite-pearlite microstructures. Typical tensile properties of dual-phase HSLA steels are summarized in Fig. 8 of this article and are discussed in more detail in the article "Dual-Phase Steels" in this Volume. Some limited yield strength and impact toughness values of various controlled-rolled acicular ferrite HSLA steels are given in Fig. 14.

Hot-rolled steels with ferrite-pearlite microstructures are the most common HSLA steels and are the primary focus of this section. Commercially available HSLA steels with ferrite-pearlite microstructures have yield strengths ranging up to 700 MPa (100 ksi), which is almost a fourfold increase over the 200 MPa (30 ksi) yield strength of conventional hot-rolled plain carbon steel. The increased strengths are developed by variations in composition and processing, but compositions having similar strengths are produced in different ways by different producers. There are three principal microalloying additions: niobium, vanadium, and titanium. Other additions can be made, depending on processing capabilities (particularly the cooling facilities) and on property requirements for the finished steel. In order to achieve good transverse properties, calcium-silicon ladle additions (or, less often, additions of rare-earth elements) are used in niobium- or vanadium-containing steels to control the shape of sulfide inclusions. For HSLA compositions containing titanium, ladle treatments for sulfide shape control are not required; titanium itself has the desired effect on the shape of sulfide inclusions. In recent years, calcium treatment has replaced rare-earth elements in sulfide inclusion shape control.

Tensile properties of some hot-rolled ferrite-pearlite HSLA steels are summarized in Tables 16 and 17. These properties are influenced by alloying elements and production methods.

In vanadium-microalloyed HSLA steels, yield strengths are influenced by manganese content (Table 12), nitrogen content (Fig. 17), and cooling rates (Fig. 11). The cooling rate, which determines the level of precipitation strengthening in vanadium steels, depends on the temperature, the product thickness, and the cooling medium. For thin products, temperature is the primary factor

Table 15 Chemical composition of a representative sample of normalized HSLA steels produced in Europe

		Composition, %										
Steel	Producer	C max	Mn max	Si max	P max	S max	Nb max	V	Ni	Cr max	Mo max	Cu
Usiten 420-II	Usinor	0.22	1.6	0.55	0.035	0.030	...	...	0.5–0.7	0.2	0.1	0.3 max
Usiten 460-I	Usinor	0.20	1.7	0.50	0.035	0.030	0.045	0.07–0.13	≤0.2	0.2	0.1	0.3 max
Usiten 460-II	Usinor	0.18	1.7	0.40	0.035	0.030	0.045	0.09–0.15	0.5–0.7	0.2	0.1	0.3 max
FG 43 T	Thyssen	0.18	1.7	0.5	0.030	0.030	...	0.10–0.18	0.7 max	...	...	...
FG 47 CT	Thyssen	0.15	1.5	0.5	0.030	0.030	...	0.08–0.18	0.5–0.7	...	...	0.5–0.7
FG 51 T	Thyssen	0.21	1.7	0.5	0.030	0.030	...	0.10–0.20	0.4–0.7	...	...	...
Dillinal 55/43E	Dillinger	0.18	1.7	0.5	0.025	0.015	...	0.10–0.18	0.7 max	...	...	...
Dillinal 58/47E	Dillinger	0.20	1.7	0.5	0.025	0.015	...	0.10–0.20	0.7 max	...	...	...
Hyplus 29	BSC	0.22	1.6	0.5	0.05	0.03	...	0.20 max	...	...	...	...
BS 4360:55E	BSC	0.22	1.6	0.6	0.04	0.04	0.10	0.20 max	...	...	...	...
WSTE 500	Creusot-Loire	0.18	1.6	...	0.015	0.010	...	0.10 max	0.3–0.7	...	...	...

Table 16 Tensile properties of HSLA steel grades specified in ASTM standards

ASTM specification(a)	Type, grade, or condition	Product thickness(b) mm	Product thickness(b) in.	Minimum tensile strength(c) MPa	Minimum tensile strength(c) ksi	Minimum yield strength(c) MPa	Minimum yield strength(c) ksi	Minimum elongation, %(c) in 200 mm (8 in.)	Minimum elongation, %(c) in 50 mm (2 in.)	Bend radius(c) Longitudinal	Bend radius(c) Transverse
A 242	Type 1	20	3/4	480	70	345	50	18	...		
		20–40	3/4–1 1/2	460	67	315	46	18	21		
		40–100	1 1/2–4	435	63	290	42	18	21		
A 572	Grade 42	150	6	415	60	290	42	20	24	(d)	...
	Grade 50	100	4	450	65	345	50	18	21	(d)	...
	Grade 60	32	1 1/4	520	75	415	60	16	18	(d)	...
	Grade 65	32	1 1/4	550	80	450	65	15	17	(d)	...
A 588	Grades A–K	100	4	485	70	345	50	18	21	(d)	...
		100–125	4–5	460	67	315	46	...	21	(d)	...
		125–200	5–8	435	63	290	42	...	21	(d)	...
A 606	Hot rolled	Sheet		480	70	345	50	...	22	t	2t–3t
	Hot rolled and annealed or normalized	Sheet		450	65	310	45	...	22	t	2t–3t
	Cold rolled	Sheet		450	65	310	45	...	22	t	2t–3t
A 607	Grade 45	Sheet		410	60	310	45	...	22–25	t	1.5t
	Grade 50	Sheet		450	65	345	50	...	20–22	t	1.5t
	Grade 55	Sheet		480	70	380	55	...	18–20	1.5t	2t
	Grade 60	Sheet		520	75	415	60	...	16–18	2t	3t
	Grade 65	Sheet		550	80	450	65	...	15–16	2.5t	3.5t
	Grade 70	Sheet		590	85	485	70	...	14	3t	4t
A 618	Ia, Ib, II	19	3/4	485	70	345	50	19	22	t–2t	...
	Ia, Ib, II, III	19–38	3/4–1 1/2	460	67	315	46	18	22	t–2t	...
A 633	A	100	4	430–570	63–83	290	42	18	23	(d)	...
	C, D	65	2.5	485–620	70–90	345	50	18	23	(d)	...
	C, D	65–100	2.5–4	450–590	65–85	315	46	18	23	(d)	...
	E	100	4	550–690	80–100	415	60	18	23	(d)	...
	E	100–150	4–6	515–655	75–95	380	55	18	23	(d)	...
A 656	50	50	2	415	60	345	50	20	...	(d)	...
	60	40	1 1/2	485	70	415	60	17	...	(d)	...
	70	25	1	550	80	485	70	14	...	(d)	...
	80	20	3/4	620	90	550	80	12	...	(d)	...
A 690	...	100	4	485	70	345	50	18	...	2t	...
A 709	50	100	4	450	65	345	50	18	21	...	...
	50W	100	4	485	70	345	50	18	21	...	...
A 715	Grade 50	Sheet		415	60	345	50	...	22–24	0	t
	Grade 60	Sheet		485	70	415	60	...	20–22	0	t
	Grade 70	Sheet		550	80	485	70	...	18–20	t	1.5t
	Grade 80	Sheet		620	90	550	80	...	16–18	t	1.5t
A 808	...	40	1 1/2	450	65	345	50	18	22	...	...
		40–50	1 1/2–2	450	65	315	46	18	22	...	...
		50–65	2–2 1/2	415	60	290	42	18	22	...	...
A 812	65	Sheet		585	85	450	65	...	13–15	...	...
	80	Sheet		690	100	550	80	...	11–13	...	...
A 841	...	65	2.5	485–620	70–90	345	50	18	22	...	...
		65–100	2.5–4	450–585	65–85	310	45	18	22	...	...
A 871	60, as-hot-rolled	5–35	3/16–1 3/8	520	75	415	60	16	18	...	...
	65, as-hot-rolled	5–20	3/16–3/4	550	80	450	65	15	17	...	...

(a) For characteristics and intended uses, see Table 10; for specified compositional limits, see Table 13. (b) Maximum product thickness except when a range is given. No thicknesses are specified for sheet products. (c) May vary with product size and mill form. (d) Optional supplementary requirement given in ASTM A 6

Table 17 Mechanical properties of HSLA steel grades described in SAE J410

Grade(a)	Minimum tensile strength(b) MPa	Minimum tensile strength(b) ksi	Minimum yield strength(b)(c) MPa	Minimum yield strength(b)(c) ksi	Minimum elongation, %(b) in 200 mm (8 in.)	Minimum elongation, %(b) in 50 mm (2 in.)	Bend diameter(b)(d)
942X	415	60	290	42	20	24	t–3t
945A	415–450	60–65	275–310	40–45	18–19	22–24	t–3t
945C	415–450	60–65	275–310	40–45	18–19	22–24	t–3t
945X	415	60	310	45	19	22–25	t–2.5t
950A	430–483	63–70	290–345	42–50	18–19	22–24	t–3t
950B	430–483	63–70	290–345	42–50	18–19	22–24	t–3t
950C	430–483	63–70	290–345	42–50	18–19	22–24	t–3t
950D	430–483	63–70	290–345	42–50	18–19	22–24	t–3t
950X	450	65	345	50	18	22	t–3t
955X	483	70	380	55	17	20	t–3.5t
960X	520	75	415	60	16	18	1.5t–3t
965X	550	80	450	65	15	16	2t–3t
970X	590	85	485	70	14	14	3t
980X	655	95	550	80	10	12	3t

(a) For compositions, available mill forms, and special characteristics of these steels, see Table 11. (b) May vary with product size and mill form; for specific limits, refer to SAE J410. (c) 0.2% offset. (d) 180° bend test at room temperature. Used for mill acceptance purposes only; not to be used as a basis for specifying fabricating procedures

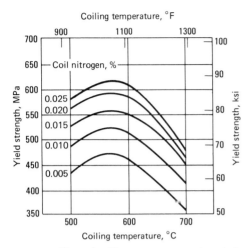

Fig. 17 Effect of coiling temperature and total nitrogen content on the yield strength of a vanadium microalloyed steel (0.13% C, 1.4% Mn, 0.5% Si, 0.12% V). Yield strengths are for a 6.3 mm (0.25 in.) thick sheet, finished at 900 °C (1650 °F) and then cooled at 17 °C/s (30 °F/s). Source: Ref 5

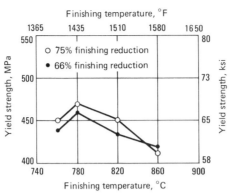

Fig. 18 Effect of finishing temperatures and reductions during finishing on the yield strength of a niobium-manganese steel (0.1% C, 1.25% Mn, 0.027% Nb, 0.2% Si, 0.03% Al). The reheat temperature was 1080 °C (1975 °F), with a cooling rate of 10 °C/s (18 °F/s) and a cooling temperature of 600 °C (1110 °F). Source: Ref 32

influencing cooling rates. For example, the strength of vanadium steel sheet is shown in Fig. 17 as a function of coiling temperature.

Yield strengths in niobium steels are also affected by rolling procedures and cooling rates. Figure 18 shows the effect of finishing temperature on the yield strength of a niobium-manganese steel after finish-rolling reductions of 75 and 66%. The range of yield strengths for controlled-rolled niobium steels of various compositions and section thicknesses is shown in Fig. 16. A typical controlled-rolled and accelerated-cooled HSLA steel is specified in ASTM A 841.

The notch toughness of microalloyed structural steels, as evaluated by Charpy-impact or drop-weight tests, is generally superior to that of structural carbon steels. The transition temperatures of the former also are lower. Figure 1 compares Charpy V-notch impact values for structural carbon steel (ASTM A 7), an HSLA steel, and a heat-treated constructional alloy steel. In the presence of a stress raiser, brittle failure is less likely to occur at subnormal temperatures in steels with lower transition temperatures. The transition temperatures of proprietary grades in the as-rolled or normalized conditions are controlled principally by chemical composition (particularly carbon) and ferrite grain size.

Notch toughness is reduced when the ferrite-pearlite HSLA steels are strengthened by precipitation hardening. The notch toughness of vanadium steels can be improved by normalizing or recrystallization controlled rolling. The notch toughness of niobium steels is normally improved by conventional controlled rolling, although the effects of recrystallization controlled rolling are under study. Figure 16 shows the range in notch toughness for controlled-rolled niobium steels, as evaluated by ductile-brittle transition temperatures.

Brittle Fracture. Most structural steels exhibit a transition in fracture behavior from ductile to brittle when the temperature is lowered to some critical temperature known as the nil-ductility transition temperature (NDTT). This temperature is defined as the temperature at which steel loses its ability to flow plastically in the presence of a sharp, cracklike discontinuity. At and below the NDTT, a brittle cleavage fracture will initiate from this discontinuity when stresses approaching the yield strength are reached in the volume of material surrounding the discontinuity. Once initiated, brittle fracture can propagate easily through regions of the structure that are subjected only to low levels of applied stress. In some steels, the transition from ductile to brittle fracture can occur at relatively high temperatures if a mechanical or metallurgical notch is present. If no sharp notch or crack is present, temperatures as low as −75 °C (−100 °F) are necessary to produce brittle fracture.

For welded structures, it is assumed that sharp notches are present; this makes brittle fracture at normal operating temperatures an important concern. Most structural members remain within elastic loadings, except for corners, cutouts, and similar locations where slight yielding may occur. For this type of service, brittle fracture is possible when certain conditions exist:

- The temperature is below the characteristic NDTT value for the steel being used
- A cracklike notch is present
- Load values are sufficiently large to raise the nominal stress level in the area of the notch to values close to or exceeding the yield strength

At and below the NDTT, the effects of residual stress may be considered. Weldments can contain residual stresses as high as 80% of the original yield strength. In these cases, an applied stress of only 20% of the yield strength would be sufficient to initiate brittle fracture. All three factors, high stress, low temperature, and cracklike flaws, must be present for brittle fracture to initiate. However, as flaw size increases, the stress required for crack initiation decreases. With larger flaw sizes, the levels can be well below the yield strength, and residual stresses will play an increasingly important role.

At temperatures slightly above or below the NDTT, localized plastic deformation is required to initiate brittle fracture, and residual stresses are less harmful. At temperatures well below the NDTT, design is critical; the nominal stress (the sum of applied stress and residual stress) must not exceed the yield strength at the locations of cracklike flaws. In general, the lower the temperature below NDTT, the less severe the notch must be to initiate brittle fracture. For very thin sections, stresses act in a manner that makes brittle fracture considerations less critical.

Each type of high-strength low-alloy steel has a range of NDTT values that depends on chemical composition, as-rolled ferritic grain size, and other variables; a range of 35 °C (60 °F) is not uncommon. On the basis of limited test data, expected NDTT ranges are estimated in Table 18. Grain coarsening may occur in different products because of rolling practice, which will increase the NDTT to above the maximum values shown in Table 18.

Directionality of Properties. In microalloyed HSLA steels, the changes in mechanical properties resulting from the use of niobium and vanadium, together with controlled rolling, result in improved yield strength, weldability, and toughness. The ferrite grain size is reduced, with an attendant increase in yield strength. Because of this increase, any reduction in toughness due to precipitation strengthening can generally be tolerated. The remaining proper-

Table 18 Typical nil-ductility transition temperature ranges for several HSLA steels

Type of steel	Nil-ductility transition temperature °C	°F
345 MPa (50 ksi) minimum yield strength; fully killed	−70 to −30	−90 to −25
345 and 550 MPa (50 and 80 ksi) minimum yield strength; fully killed; microalloyed, with low-carbon content and inclusion shape control	−70 to −30	−90 to −25
310 MPa (45 ksi) minimum yield strength; fully killed	−50 to −18	−60 to 0
310 MPa (45 ksi) minimum yield strength; silicon killed	−40 to −7	−40 to 20
290–450 MPa (42–65 ksi) minimum yield strength; semikilled	−25 to 10	−10 to 50

High-Strength Structural and High-Strength Low-Alloy Steels / 413

Table 19 Listing of SAE J410 grades in approximate order of decreasing toughness, formability, and weldability

Toughness	Formability	Weldability
945A	945A	945A
950A	950A	950A
950B	945C, 945X	950D
950D	950B, 950X, 942X	945X
945X, 950X	950D	950B, 950X
945C, 950C, 942X	950C	945C
955X	955X	955X, 950C, 942X
960X	960X	960X
965X	965X	965X
970X	970X	970X
980X	980X	980X

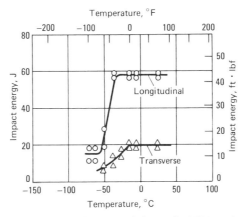

Fig. 19 Typical transition behavior for HSLA steel without inclusion shape control. Data determined on half-size Charpy V-notch test specimens

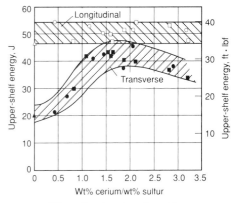

Fig. 20 Effect of cerium-to-sulfur ratio on upper-shelf impact energy for HSLA steel. Circles, steel treated with mischmetal; squares, steel treated with rare-earth silicides

ties, however, are typical only of samples tested in the direction of rolling. In the transverse direction, toughness is reduced considerably, and formability is inadequate because of the characteristic shape of nonmetallic inclusions which, during rolling, become elongated in the rolling direction. The effect of elongated inclusions on notch toughness is shown in Fig. 19. For the data presented, the impact energy for 100% shear fracture (upper-shelf energy) exceeds 55 J (40 ft · lbf) in the longitudinal direction, but is only 20 J (15 ft · lbf) in the transverse direction. For aluminum-killed steels, the low-shelf energy in the transverse direction is caused primarily by elongated sulfide inclusions. Reducing the sulfur content to 0.010% is not sufficient to eliminate directionality.

To prevent the sulfides from becoming excessively elongated during hot rolling, it is necessary to alter their composition. This can be done by adding elements such as zirconium, titanium, calcium, or rare earths, which form sulfides having high melting points. Sulfides with high melting points are less plastic at hot rolling temperatures and cannot be deformed readily. As indicated in Fig. 20, rare-earth additions can effectively enhance transverse toughness so that when the cerium-to-sulfur ratio is between 1.5 and 2.0, the transverse upper-shelf energy approaches that of the longitudinal direction. Of the elements mentioned above, calcium and rare earths are the most frequently used for improving transverse or through-thickness properties.

Fatigue Characteristics of HSLA Steels. Many structural applications for HSLA steels involve cyclic loading. The fatigue behavior of these steels therefore becomes important. In the listing below, some of the fatigue characteristics of HSLA steels are compared to similar characteristics of hot-rolled low-carbon steels:

- HSLA steels have fatigue properties equivalent or superior to those of hot-rolled low-carbon steel. Microalloyed HSLA steels of the three major compositional approaches (vanadium, niobium, and titanium) have similar properties
- Although the notch sensitivity is somewhat greater, HSLA steels have greater notch fatigue resistance than hot-rolled low-carbon steels
- Large plastic prestrains tend to impair the fatigue life of both HSLA and hot-rolled low-carbon steels, particularly the response of HSLA steels to cyclic loading at high-strain amplitudes. Tensile prestrains are more detrimental to fatigue resistance than are compressive prestrains for both classes of material
- Most of the gains in monotonic strength achieved by work hardening are not retained under cyclic loading for either HSLA or hot-rolled low-carbon steel

Forming of HSLA Steels

High-strength low-alloy steels are generally formed at room temperature using conventional equipment. Cold forming should not be done at temperatures below 10 °C (50 °F). As a class, high-strength steels are inherently less formable than low-carbon steels because of their greater strength and lower ductility. This reduces their ability to distribute strain. The greater strength makes it necessary to use greater forming pressure and to allow for more springback compared to low-carbon steels. However, high-strength steels have good formability, and straight bends can be made to relatively tight bend radii, especially with the grades having lower strengths and greater ductility. Further, high-strength steels can be stamped to relatively severe shapes such as automotive bumper facings, wheel spiders, and engine-mounting brackets. Table 19 ranks the formability of the various HSLA steels specified in SAE J410 (Table 11).

With the advent of inclusion shape control, cold formability has been substantially improved. Any grade produced with inclusion shape control can be more severely formed than a grade of the same strength level produced without inclusion shape control. Inclusion shape control enables the steel to be formed to nearly the same extent in both the longitudinal and transverse directions and is responsible for the moderately good formability of the higher-strength HSLA steels such as the grades having 550 MPa (80 ksi) yield strengths.

As with any metal, the bendability of HSLA steels is inversely proportional to strength and thickness, as shown by the suggested minimum bend radii in Table 20. These minimum bend radii are generally conservative, especially for inclusion shape controlled grades, as indicated by the bend test radii given in Table 21. The minimum radii in Table 20 represent safe production practice for parts made from HSLA steels

Table 20 Suggested minimum inside bend radii for SAE J410 steels of various strengths and thicknesses

Grade	Minimum bend radius for thickness of		
	<4.6 mm (0.180 in.)	4.6–6.4 mm (0.180–0.250 in.)	6.4–12.7 mm (0.250–0.500 in.)
942X	...	t	$2t$
945A, 945C	t	$2t$	$2.5t$
945X	t	t	$2t$
950A, 950B, 950C, 950D	t	$2t$	$3t$
950X	$1.5t$	$2.5t$	$2.5t$
955X	$2t$	$3t$	$3t$
960X	$2.5t$	$3.5t$	$3.5t$
965X	$3t$	$4t$	$4t$
970X	$3.5t$	$4.5t$	$4.5t$
980X(a)	$3.5t$	$4.5t$	$4.5t$

(a) Available only in thicknesses to 9.5 mm (0.375 in.)

Table 21 Specified bend test radii for inclusion shape controlled ASTM A 715 steel sheet

Grade	Bend test radius for Transverse bends(a)	Longitudinal bends(a)
50	0.5t	0
60	0.5t	0
70	0.75t	0.5t
80	0.75t	0.5t

(a) For sheet thicknesses up to 5.84 mm (0.2299 in.)

without inclusion shape control; more liberal bend radii than those shown in Table 21 may be allowed for production of parts made from the inclusion shape controlled grades.

High-strength low-alloy steels can be hot formed. However, hot forming usually alters mechanical properties, and a particular problem that arises in many applications is that some of the more recent thermomechanical processing techniques (such as controlled rolling) used for plates in particular are not suitable where hot forming is used during fabrication. The kind of property deterioration obtained is shown in Fig. 21 for a thermomechanically treated steel. Similar effects can be predicted for accelerated-cooled steels. This problem can be circumvented by the use of a rolling finishing temperature that coincides with the hot-forming temperature (900 to 930 °C, or 1650 to 1700 °F). Subsequent hot forming therefore simply repeats this operation, and deterioration in properties is then small or even absent provided that grain growth does not occur.

Producers should be consulted for recommendations of specific hot-forming temperatures and for comments on their effects on mechanical properties. According to one producer, satisfactory results can be obtained with certain grades by forming at temperatures between 815 and 900 °C (1500 and 1600 °F) without appreciable hardening after cooling. However, hot forming at these temperatures may result in material with mechanical properties equivalent to those of annealed or normalized material. Most producers advise against hot forming at temperatures below 650 °C (1200 °F). In the case of HSLA linepipe steels, hot bends are manufactured by induction heating followed by forming and (optional) quenching and tempering. The heating time is short (2 to 4 min), and limited grain growth occurs. Thus bending temperatures in the range of 850 to 1040 °C (1560 to 1900 °F) have been used successfully.

Welding of HSLA Steels

High-strength low-alloy steels are readily welded by any of the welding processes used for plain carbon structural steels, including shielded metal arc, submerged arc, flux-cored arc, gas metal arc, and electrical resistance methods. Welding can usually be performed without the need for preheat or postheat. Because the carbon content of HSLA steels is low, these steels exhibit hardening characteristics (in the HAZ) similar to those of plain carbon steels of similar low-carbon content. They do not exhibit the severe hardening in the HAZ that characterizes some plain carbon steels (those with sufficient carbon contents to attain yield and tensile strengths comparable to those of HSLA steels). Weldability generally decreases with increasing carbon content. For shielded metal arc welding, both mild steel covered electrodes and low-hydrogen electrodes can be used. Use of the latter is often recommended in order to preclude the need for preheating except when welding thick, highly restrained sections.

Submerged arc welding is often preferred because of the high production rates that can be attained. However, notch toughness of the weld metal in submerged arc welds can vary because of such factors as base metal composition, filler metal composition, type of welding flux, welding speed, current, voltage, and joint preparation. The choice of a particular steel is often greatly influenced by its ability to be used in welded construction, with less attention being given to the influence of steel composition on weld metal composition. In a submerged arc weld, the base metal may sometimes account for as much as 50 to 75% of the total weld metal. Base metal composition, therefore, can have an important effect on weld microstructure.

With constructional HSLA plate steels, the use of low-hydrogen electrodes and other precautions to minimize hydrogen pickup is advised. Various electrodes are available to match base metal strength and toughness. Preheating is generally required for all thicknesses over 25 mm (1 in.) and for highly restrained joints. Specific preheating temperatures depend largely on the grade, thickness, and welding process used and usually range from 40 to 200 °C (100 to 400 °F). Depending on design strength requirements for welded joints in heat-treated alloy

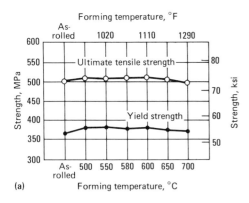

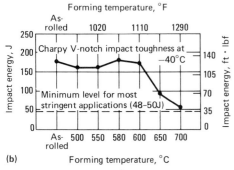

Fig. 21 Mechanical properties of a thermomechanically treated HSLA steel after reheating at different temperatures. (a) Yield and tensile strength. (b) Charpy V-notch impact toughness at −40 °C (−40 °F)

Table 22 Chronology of trends and developments in HSLA steels

Period	Focus	Advances in technology
1939–1960	Discovery	Microalloying elements were used in a minor way in plate and in structural shapes to ensure the attainment of minimum yield strength
1960–1965	Research and experimentation on strengthening mechanisms	Development of a physical metallurgical basis. Hot-rolled semikilled or fully killed steels. Emphasis on replacing heat-treated steel
1965–1972	Toughening; experimentation on grain refinement and desulfurization	Hot-rolled sheet, plate, pipeline steels, and structural shapes. Development of controlled rolling
1970–1976	Weldability; formability	Emphasis on lowering carbon and carbon equivalent, reduction of sulfur content, and control of inclusion shape. Improved bendability and stretchability
1972 to present	Secondary properties and steelmaking	Emphasis on brittle and ductile fracture control in pipelines and offshore structures, fatigue, resistance to H_2S corrosion, yielding behavior, formability of sheet steel, and development of continuous casting technology
1979 to present	Reformulation and reconstitution	Changes in availability of alloying elements such as molybdenum reawakened alloy developers. Previously inaccessible avenues opened by improvements in steelmaking. Focus on environmental degradation
1980 to present	Diffusion of technology	Cross-fertilization using development from other product areas
Future	Maturity; trend to higher strengths	Emphasis on reducing alloying costs and processing steps. Reconstitution of the steels to reflect direct hot-rolling and plate mill water-cooling technology

grades, low-hydrogen electrodes E70 (485 MPa, or 70 ksi, minimum tensile strength) through E120 (830 MPa, or 120 ksi, minimum tensile strength) are used. In applications involving low stress levels (as when high-strength steel is used chiefly for its wear-resisting properties), E70 class electrodes are preferred. For joints requiring maximum strength, E120 class electrodes can be used. Electrodes containing vanadium must be avoided for any weldments that are to be subsequently stress relieved because reheating at temperatures below the critical temperature causes precipitation of vanadium carbides and raises the transition temperature of the weld metal.

The ductility and toughness of high-strength heat-treated steels depend on a tempered martensitic type of microstructure. Therefore, welding heat and preheat must be controlled so that the weld zone is cooled rapidly enough to retain a martensitic microstructure rather than the upper transformation products such as pearlite and ferrite that would result from unrestricted high-heat inputs and and slow cooling. For this reason, the high heat usually generated by submerged arc and gas-shielded metal arc welding makes these processes unsuitable when maximum joint efficiency is required in high-strength heat-treated materials less than 13 mm (½ in.) thick. Many electrode-flux combinations have been developed to produce acceptable welds with a yield strength over 620 MPa (90 ksi) and an NDTT of about −100 °C (−150 °F) in both the as-welded and stress-relieved conditions. The best electrodes are based on either boron-molybdenum-titanium alloying or, more conventionally, nickel alloying.

Applications of HSLA Steels

High-strength low-alloy steels gained their first use as structural shapes and plates in the early 1960s because of their ability to be welded with ease. By the early 1970s, they were also used in pipelines in both elevated-temperature and severe arctic conditions. Later in the 1970s, concurrent with the energy crisis, another dominant application involved the use of HSLA steels to reduce the weight of parts and assemblies in automobiles and trucks. In the 1980s, bars, forgings, and castings have emerged as applications of particular interest. For example, weldable and nonweldable reinforcing bars are available, and high-strength forgings in the as-forged condition are being used to replace quenched and tempered steels. Shapes such as elbows and fittings for pipelines are also being cast out of microalloyed steel. A chronology of some of the important trends and developments in HSLA technology (as of 1984) is presented in Table 22.

High-strength low-alloy steels are used in a wide variety of applications, and properties can be tailored to specific applications by the combination of composition and structures obtained in processing at the mill. For example, low-carbon and closely controlled carbon equivalent values provide good toughness and weldability. Good yield strength and fracture toughness are the products of fine grain size. The effects of microalloying elements (mainly niobium, vanadium, and titanium, but also aluminum and, to a lesser degree, nitrogen) are fully exploited by such mill finishing processes as controlled rolling and accelerated cooling. The scope of HSLA applications and different product forms used at the international level are implied by the production statistics for 1986 in Table 23. A more complete listing of structural applications includes bridges, buildings, electricity pylons, pressure vessels, penstocks, steel piling, railway tracks, trucks, trailers, earthmoving equipment, mining equipment, tanks, and reinforcing bars. The numbers in Table 23 show percentages of use in various applications by country. In some applications use by country varies markedly; in some instances, the explanation is found in local conditions or requirements. Japan, for example, is the major user of HSLA for shipbuilding, while Europe is the leader in plate and section applications in offshore construction for the harsh material requirements of oil and gas production platforms on the European continental shelf.

Selection Guidelines. As a rule, it is necessary to look beyond comparative material costs to justify the use of HSLA steel. These steels almost always cost more than carbon steel and sometimes the alloy steel they compete with in a given application. High-strength low-alloy steels typically have an edge over other materials when their unique properties allow:

- Weight reductions
- New or more efficient designs with improved performance
- Attractive reductions in fabrication/manufacturing costs
- Reduced transportation costs from weight reductions

These benefits may come in combinations, depending on the application. In automobiles, for example, the initial advantage in substituting an HSLA grade for a mild steel part is in weight reduction. The gage of the metal can be reduced because HSLA steel is stronger. This weight reduction not only provides improved gas mileage but can also have multiplying effects in the design when lightening one component leads to a series of weight reductions in associated parts and assemblies. In some situations, however, it is possible to justify the specification of HSLA steel on the basis of comparative material costs alone. The technique, called cost-effective weight reduction, is shown in Fig. 22. The replacement of a mild steel part with HSLA steel is said to be cost effective when full advantage can be taken of its higher strength. The amount of weight savings needed to make HSLA steel cost competitive in terms of material costs is indicated in Fig. 23. Data based on substituting a grade with a yield strength of 340 to 550 MPa (50 to 80 ksi) for a mild steel indicate that a savings in weight of about 30% is needed to reduce material costs. Of course, the savings in a given application depends on relative material costs.

Another type of benefit involves the elimination of fabrication/manufacturing operations, such as those associated with the heat treating of quenched and tempered medium-carbon steels. For example, HSLA steels allow the production of high-strength forgings without any subsequent heat treatment (Fig. 24). By using a microalloyed steel in the as-forged condition, it is possible to

Table 23 Proportion of product forms made with HSLA steel (1986)

Product form	Europe	North America	Japan
Linepipe	95	95	95
Shipbuilding	40	20	75
Offshore steels			
Plate	90	30	70
Sections	70	20	10
Pressure vessels	30	25	85
Structural			
Sections	30	20	10
Sections, automotive	70	70	30
Sections, ships	15–30	20	10
Sheet piling	25	15	100
Rebar	100	>50	100
Plate	25	20	10–30
Sheet and coil (including galvanized)			
Automotive	20	10	20
Building (not rebar)	95	80	70

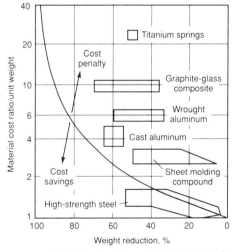

Fig. 22 Material cost ratio as a function of weight reduction potential for various materials compared to mild steel as the base. Solid line is break-even line for materials cost.

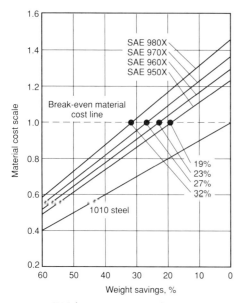

Fig. 23 Weight savings required to break even when substituting 340 to 550 MPa (50 to 80 ksi) yield strength HSLA steels for mild steel

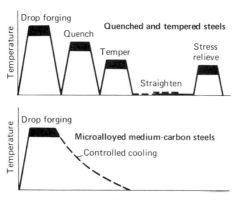

Fig. 24 Required manufacturing steps for producing crankshafts from quenched and tempered and microalloyed medium-carbon steels

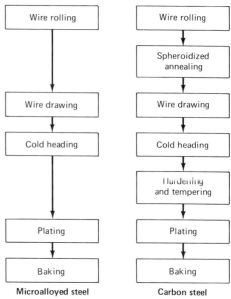

Fig. 25 Manufacturing sequence for class 8.8 automotive bolts from microalloyed and carbon steels

eliminate quenching, tempering, straightening, and stress-relieving operations. In applications such as the forging of crankshafts or connecting rods, final savings in part costs of the order of 10% have been gained. Another example of reducing manufacturing steps with HSLA steel involves the production of cold-headed automotive fasteners (Fig. 25). Microalloyed steel wire rod replaced heat-treated medium-carbon wire. Benefits were realized before drawing and cold heading (elimination of spheroidize annealing) and after these operations (elimination of quenching and tempering).

Other factors to consider in the selection of HSLA steel include:

- *Formability*: Limitations of microalloying technology in a given application may be overcome through part redesign and minor modifications in processing procedures
- *Notch toughness*: Microalloying in combination with controlled rolling generally provides better toughness properties than those of hot-rolled steel
- *Fatigue properties*: Adequate in most cases, with the possible exception being in the area of a welded joint
- *Weldability*: No longer considered a barrier to usage
- *Machinability*: At least similar and sometimes better than that of quenched and tempered steels. Control of alumina formation with calcium additions and substitutions of aluminum grain refinement by niobium have also been introduced. Free-machining grades (with added sulfur) are also available
- *Corrosion resistance*: Weathering HSLA steels can provide atmospheric-corrosion resistance four to eight times that of carbon steel, depending on the environment
- *Hydrogen-induced cracking and stress-corrosion cracking resistance*: Linepipe and pressure vessel steels have been developed with superior resistance to H_2S degradation

Oil and gas pipelines are a classic application of HSLA steel, and one of the first applications involved the use of acicular ferrite steel for pipelines in the Arctic Circle; see the section "Acicular Ferrite (Low-Carbon Bainite) Steels" in this article. The development of high-strength linepipe grades has permitted the use of large-diameter pipe operating at high pressure in excess of 11 MPa (1600 psi). Grades with minimum yield strengths up to 483 MPa (70 ksi) in thicknesses up to 25 mm (1 in.) are readily available. While tensile strength is a key requirement in linepipe, other properties are no less critical to the fabrication and operation of oil and gas pipelines. They include weldability, fracture toughness, and resistance to sour gas attack. In addition to higher strengths in greater thicknesses, HSLA steels can provide excellent toughness, good field weldability, resistance to ductile crack propagation, and, in some cases, resistance to sour gas and oil.

The girth weld is the main joining problem because it is a low heat input operation (HAZ cracking is not a problem at high welding temperatures) that is carried out at high speed. The problem is alleviated by taking maximum advantage of the lowest possible carbon equivalent steels made possible by advanced thermomechanical rolling or accelerated cooling.

Fracture toughness is another major requirement. The approach to avoiding brittle fracture involves a combination of lower carbon content and fine grain size by microalloying along with thermomechanical rolling or accelerated cooling.

The common technique being used to prevent the propagation of ductile cracks through a gas pipeline is to lower sulfur content to less than 0.005%. Low sulfur content, in this instance about 0.002%, also reduces reactions with sour gas and oil transported in pipelines, and calcium treatment is used to ensure that residual sulfide inclusions are of the modified round type rather than the elongated type. Also important are the elimination of all other nonmetallic inclusions, particularly oxides, and the reduction or elimination of segregation by a combination of modified casting practices at the mill and a reduction in the content of the main segregating elements, carbon, manganese, and phosphorus.

Automotive Applications. High-strength low-alloy steels used by the United States automobile industry fall into four major groups:

- *Structural grades*: These materials are produced by a combination of phosphorus and microalloying elements (mainly niobium) to develop yield strengths between 255 and 345 MPa (37 and 50 ksi). They have a cost advantage over other HSLA grades with equivalent strength and are available in the hot- or cold-rolled condition. Drawing characteristics are rated superior to those of other HSLA steels
- *Formable grades*: Alloying elements commonly used to obtain high strength are niobium, vanadium, titanium, or zirconium. Transverse ductility is improved with zirconium, calcium, or rare-earth additions by changing the morphology of sulfide inclusions. Strength levels for automotive applications vary from 275 to 550 MPa (40 to 80 ksi). Fully killed grades are used most frequently, but semikilled grades are also used

Table 24 Comparison of typical 1972 and 1986 chemical composition of offshore structural steel

Element	Typical, 1972(a)	Typical, 1986(b) U.S. mill	Typical, 1986(b) Foreign mill
Carbon	0.17	0.15	0.12
Manganese	1.30	1.34	1.44
Phosphorus	0.025	0.015	0.009
Sulfur	0.02	0.006	0.001
Silicon	0.40	0.30	0.38
Niobium	0.05	0.05	0.20
Aluminum, total	0.03	0.040	0.035
Nickel	...	0.17	0.18
Chromium	...	0.08	0.009
Molybdenum	...	0.056	0.001
Vanadium	...	0.002	0.001
Copper	...	0.032	0.16
Arsenic	...	...	0.003
Tin	...	...	0.001
Antimony	...	...	0.000
Carbon equivalent, max(c)	0.41	0.40	0.38

(a) Minimum yield strength of 290 MPa (42 ksi). (b) Minimum yield strength of 345 MPa (50 ksi). (c) Carbon equivalent (CE) = C + (Mn/6) + [(Cr + Mo + V)/5] + [(Ni + Cu)/15]

- *Dual-phase grades*: These steels are used because of their superior formability over equivalent-strength microalloyed steels. Their duplex structure (ferrite and martensite) gives them a high strain hardening capability. Yield strengths range from 275 to 550 MPa (40 to 80 ksi)
- *Ultrahigh-strength grades*: In this instance, yield strengths of 550 MPa (80 ksi) are obtained by recovery-annealing cold-rolled HSLA grades. The most common strength levels used are 690, 825, and 965 MPa (100, 120, and 140 ksi). Low ductility has limited their applications

Experience in the application of cold- and hot-rolled HSLA sheet in automotive applications indicates the importance of additional requirements with respect to stiffness, crash behavior, fatigue life, corrosion resistance, acoustic properties, and, of course, formability and weldability. For one manufacturer (Ref 33), the substitution of HSLA for unalloyed steel requires a gage reduction of at least one-third to avoid an increase in manufacturing costs. This manufacturer (Opel in Germany), found certain limitations of HSLA sheet in automotive applications (Ref 33):

- Limited formability must be taken into consideration
- Fatigue strength of electric resistance welded joints is not much better than that of mild steel
- Increased notch sensitivity, which means higher cut edge sensitivity, must be taken into consideration
- With the use of thinner gages, reduced rigidity, susceptibility to corrosion, and noise behavior must be taken into consideration

The function of the automotive part is also important in determining whether or not the application of HSLA steels is feasible. For applications in which yield strength is the major consideration, as in door beams and bumper beams, the efficient use of HSLA can be made, and weight savings of 40 to 60% are possible when substituting, for example, higher-strength grades, such as 340 to 550 MPa (50 to 80 ksi) grades for mild steel. However, for other parts, such as outer body panels, which are also subjected to bending and/or buckling, stiffness is more important than yield strength. Part geometry and modulus of elasticity, the latter of which is the same for mild steel and HSLA steel, are the key factors. In such instances, an economic case cannot be made for using HSLA steel. However, through component redesign, such as adding beads or ribs to upgrade stiffness properties, the performance of the part may be improved and allow the efficient use of a higher-strength HSLA grade.

In addition to improvements in gas mileage through weight reduction, benefits may be found in increasing payloads without a change in fuel consumption. Candidate applications here include trucks, rail cars, off-highway vehicles, and ships. Ship applications of HSLA, however, are also limited by buckling and stiffness considerations.

Offshore Applications. The essential characteristics of steels for these applications include:

- Yield strength in the region of 350 to 415 MPa (50 to 60 ksi)
- Good weldability
- High resistance to lamellar tearing
- Lean composition to minimize preheating requirements
- High toughness in the weld heat-affected zone
- Good fracture toughness at the designated operating temperatures

Some of these goals have been realized through a reduction in impurities such as

Table 25 Cold-forming strip steels

Steel type	Processing	Yield strength MPa (ksi)	Tensile strength MPa (ksi)	Carbon	Alloying element
Precipitation strengthened and grain refined	Hot rolled	280–600 (40–87)	400–700 (58–100)	0.05–0.15	≤ 1.5% Mn max plus Ti, Nb, or V
	Cold rolled	240–440 (35–64)	340–550 (49–80)	0.05–0.15	≤ 1.5% Mn max plus Ti, Nb, or V
Solid-solution strengthened	Hot or cold rolled C Mn	230–338 (33–49)	370–520 (54–75)	0.10–0.20	≤ 1.5% Mn max
	Cold rolled Al killed	220–300 (32–43)	340–435 (49–63)	<0.05	≤ 0.01% P
	Cold rolled, interstitial free	220–300 (32–43)	300–400 (43–58)	<0.005	≤ 0.1% P plus Ti, Nb, or B
Bake hardening	Cold rolled Al killed	200–230 max (29–33 max) 240–300(a) (35–43)(a)	310–380 (45–55) ...	0.01–0.02	≤ 0.1% max, possibly Si
	Cold rolled Nb	210 max (30 max) 240(a) (35)(a)	360–380 (52–55) ...	<0.005	≤ 0.1% P plus Nb
Dual phase	Hot and cold rolled	200–900 (29–130)	350–1300 (50–190)	0.02–0.15	≤ 1.5% Mn
Transformation strengthened	Hot and cold rolled	≤1400 (200)	≤1500 (215)	≤0.20	Mn ≤ 1.5% Mn
Cold-work strengthened	Cold rolled	≤1000 (145)	≤1100 (160)	≤0.20	≤ 1.5% Mn, with Ti, Nb, and V possible

(a) After baking

sulfur, nitrogen, and phosphorus in the steelmaking process for conventional steels. A major challenge, however, was to reduce carbon equivalents to improve weldability while still maintaining strength. This trend toward lower carbon equivalents and adequate strengths is shown in Table 24. Controlled rolling and accelerated cooling of niobium steels has also allowed reductions in carbon contents, which can be further reduced when accelerated cooling is employed.

Most offshore structures have been built using normalized carbon-manganese-niobium steel. Advances in computer control and rolling capability have led to the development of thermomechanically controlled processes that produce steels with higher strength, high fracture toughness, improved weldability, and lower cost. Thermomechanically controlled processes combine controlled rolling and accelerated cooling (with controlled water sprays) or direct quenching to room temperature. Very fine-grain steel (ASTM grain size numbers 10 to 12) is produced. These steels are characterized by low-carbon content (usually less than 0.10% C), which makes them less susceptible to increases in hardness caused by rapid cooling rates between 425 to 260 °C (800 to 500 °F) during welding. Potentially, these steels can be welded with little or no preheat.

Two approaches have been taken to eliminate lamellar tearing. One is to reduce sulfur to levels below 0.008%, while the other involves modification of the sulfide shape. The latter relies on the addition of calcium or rare-earth metals to form spheroidal calcium or rare-earth sulfides. This approach usually results in both the elimination of lamellar tearing and an improvement in transverse impact properties. Both sulfur reduction and sulfide shape control are often used to eliminate lamellar tearing. Calcium treatment is preferred for sulfide inclusion shape control.

In using continuously cast steels, centerline segregation is also a concern. The approach favored by users is to minimize segregation through steelmaking practice and to minimize the use of elements (carbon, manganese, sulfur, and phosphorus) known to segregate easily during solidification. Soft, in-line reduction practices have been introduced. Users also specify minimum slab soaking time and temperature to control centerline segregation.

Cold-Forming Strip. The attraction of HSLA in this instance, particularly in automotive applications, is to make lighter parts by reducing gage requirements. Formability is a problem that is dealt with in several ways; welding is not as much of a problem with strip as it is with plate. Generally, welding problems are dealt with through the process itself or with surface treatments, rather than with modification of the fundamental characteristics of the steels themselves.

A tabulation of the cold-forming strip steels presently being used (along with their yield and tensile properties, ranges of carbon content, and alloying additions) are listed in Table 25. This tabulation includes the important category of interstitial-free steels, which have excellent deep-drawing characteristics (Fig. 8c).

The larger tonnage of cold-rolled steels have yield strengths less than 345 MPa (50 ksi). The microalloyed grades derive their higher yield strength from a combination of grain refinement and precipitation strengthening. They are normally produced in the hot-rolled or cold-rolled and annealed condition. Because of their limited formability, they are typically used for structures and reinforcement parts for automobiles, buildings, and various forms of domestic equipment. Forming properties are summarized in Table 26.

Solid-solution strengthened grades have better formability than precipitation-strengthened grades. The most common alloying elements in this instance are manganese, silicon, and phosphorus. When phosphorus is above 0.10%, it tends to present problems in electric resistance welding. Solution-strengthened steels can be used in automotive body panels. These steels can be further strengthened by the addition of a small percentage of niobium (Table 25).

Bake-hardened grades are of interest because they are formed at a low strength level and subsequently strengthened during the paint-baking operation, at about 170 °C (340 °F). Advantage is taken of the quench aging mechanism. Carbon is quenched into solution at the cooling rates obtained after batch or continuous annealing and is later precipitated during the paint-baking operation.

Transformation-strengthened HSLA steels include dual-phase alloys in which the second phase may be martensite, bainite, or pearlite, depending on the strength requirements. These steels have good formability and weldability. They are used extensively in rolled form.

The highest strength strip with yield strengths up to 1600 MPa (230 ksi) are alloyed to ensure transformation to martensite. Applications include door impact beams and rear bumper supports. The formability of these grades is limited.

Corrosion protection is of concern because the corrosion resistance of nonweathering HSLA steel may not be significantly different from that of other steels. In this case, a number of coatings that inhibit corrosion and enhance the appearance of automobiles and buildings are available. They include zinc galvanized and aluminized coatings, which are applied by hot dip or continuous processes. A variety of colored organic coatings are available.

Structural Steels. Niobium, vanadium, and titanium microalloy additions are used, and chromium and molybdenum are added for elevated-temperature service. Processes used include controlled rolling, normalizing, quenching and tempering. For rolled shapes, direct accelerated cooling from the rolling temperature is a recent innovation. Steel frames for buildings are expected to be competitive with other building construction materials, including prestressed concrete. Benefits include an increase in the load bearing capacity of columns, weight savings, and gains in space. Composite construction (HSLA rolled sections and reinforcing bar bonded together with concrete) also appears to have potential. These composites offer great strength in compression, high building strength, and good resistance to fire.

Shipbuilding. Ship applications are limited by buckling and stiffness considerations. Growth in usage of hull plate is apparently held back by the fact that because HSLA and mild steel have the same elastic modulus, reducing thickness can lower elastic rigidity. This has caused concern regarding potential problems with hull rigidity.

One of the main challenges in this application area has been to reduce the need for welding preheat in the most common hull plate thicknesses. Some exceptions to this are reported. The Japanese are using readily welded hull structures employing HSLA grades that are processed by accelerated cooling and have yield strengths up to 400 MPa (58 ksi).

Crane and vehicle applications include trucks tankers, mobile cranes, dumper trucks, and heavy-duty material-handling equipment. Important properties include strength and toughness, resistance to brittle

Table 26 Forming properties of various steel types

Primary strengthening mechanism	Stretchability	Drawability	Strength range
Cold worked	Very poor	Poor	Large
Recovery annealed	Poor	Poor	Large
Transformation strengthened	Moderate	Moderate	Large
HSLA precipitation grain refined	Moderate	Moderate	Fairly large
Solid-solution aluminum killed	Good	Good	Limited
Bake hardenable	Good	Good	Limited
Dual phase	Good	Moderate	Large

fracture, fatigue strength, weldability, and formability. In crane construction, high minimum yield strengths are needed to obtain high load-bearing capacities to handle dead loads. For example, a quenched and tempered grade with a minimum yield strength of 960 MPa (140 ksi) offers both technical and economic advantages, especially with cranes subjected to lifting heavy loads. Acceptable brittle fracture properties are obtained, for example, with acicular ferrite steel.

Resistance to cyclic loading is extremely important; in particular the fatigue strength of welded joints which is typically not as strong as that of the base metal because of microstructural changes due to the heat from welding. Improvements can be realized through postweld treatments. In using controlled rolled grades, heat input must be limited because of the likelihood of softening in the HAZ during welding.

Microalloyed HSLA steels used in commercial vehicles are typically cold formed. Good forming properties in the longitudinal and transverse directions are required when thermomechanically rolled steels are used. The desired result is obtained by optimizing sulfide shape control by the addition of calcium and cerium.

Railway Tank Car Applications. Three improved grades of high-strength steel for railway tank cars are defined by the ASTM specifications below. All are specified to a minimum yield strength of 345 MPa (50 ksi) and a tensile strength of 485 MPa (70 ksi).

- ASTM A 633D is a lower-carbon version of the presently used ASTM A 612 (TC-128B) steel (carbon-manganese steel with nickel, chromium, molybdenum, and vanadium additions). It is produced as a normalized grade
- ASTM A 737B (LC) is representative of the first generation of niobium microalloyed steels, which is produced as a normalized steel because equivalent properties cannot be developed by controlled rolling
- ASTM A 808 is representative of second-generation (controlled-rolled) niobium-vanadium microalloyed steels. For low-temperature service, sulfur should be limited to 0.010% (max)

These grades were compared by preliminary weldability analysis and fracture analysis (Ref 34). General conclusions indicate that:

- ASTM A 633D and A 737B (LC) are expected to satisfy near-term requirements, but they require postweld heat treatment to meet objectives
- ASTM A 808 (CR) with 0.010% S (max) is expected to satisfy long-term requirements. Postweld heat treatment is not needed because a ductile HAZ is developed as-welded, a distinct advantage in making field repairs

Bars, forgings, and castings have emerged as the latest area of opportunity for microalloyed HSLA steels, and there has been rapid assimilation of the technology developed in other product areas. In particular, the search for higher strengths, elimination or reduction of heat treatments, simplified cold finishing or machining operations, and improved toughness have been combined with a simultaneous concern for other end-use properties such as toughness, machinability, and weldability. The result has been an integrated approach, including the proper selection of microstructure and transformation conditions, the application of thermomechanical processing and control of deoxidation and solidification practices (Ref 35).

The range of available steel types and compositions is indicated in Table 27. Carbon contents range from 0.05% in weldable castings and acicular ferrite bar used in cold heading to 0.60% for spring steels and high-strength wire. Tensile strengths range from 600 to 2000 MPa (87 to 290 ksi) and higher.

Microalloying is used to satisfy a wide variety of metallurgical and manufacturing objectives, such as the substitution of as-rolled or as-forged steels for parts that required heat treatment for the improvement of toughness, fatigue behavior, machinability, or response to cold rolling. Examples of how microalloying is used in combination with processing to develop or improve specific properties follow.

Vanadium is generally used when it is necessary to develop hardening in conjunction with low austenitizing temperatures, but it is often necessary to add titanium, which prevents grain growth during high-temperature forging applications. For normal high-temperature rolling conditions, higher levels of precipitation strengthening in vanadium (and/or niobium) steels may have a detrimental effect on toughness. When controlled rolling is used in combination with niobium (0.03 to 0.08%) or extralow carbon content, toughness improved dramatically, and the properties of niobium steels are superior to those obtained in vanadium steels.

Niobium is added to improve the toughness of as-forged components. Such steels are widely used in Europe for automobile components. Niobium-containing steels are accepted for safety related parts such as steering knuckles and suspension components.

Niobium or vanadium are also added to hot-rolled carbon-manganese steel bars to effect grain refinement and improve toughness. Steel grades include 1020 through 1090. Niobium and/or vanadium are also added to the 4100 series for the same reason.

Medium- and Low-Carbon Microalloyed Bars. Vanadium and niobium act as precipitation strengtheners in medium-carbon (0.40%) steels as used widely in concrete-reinforcing bars. If enhanced toughness is needed for low-temperature service conditions or if weldability is a problem, carbon is reduced to the 0.10 to 0.20% range, maximizing the contribution of microalloying, rather than pearlite, in the strengthening process. Further improvements are available through controlled rolling.

However, low-carbon bainitic microalloyed steels appear to have better potential for replacing high strength-high toughness quenched and tempered steels than do medium-carbon microalloyed steels (Ref 36). Medium-carbon microalloyed steels are not ideal candidates for using the benefits of microalloying technology because their carbon contents and precipitation-strengthening effects contribute to poor notch toughness. Low-carbon steels microalloyed with ferrite-bainite microstructures exhibit a better combination of strength and toughness properties (Ref 36).

Solute niobium is used to eliminate the quenching and tempering of forged crankshafts (Table 27). Forging is followed by high-temperature (1225 °C, or 2240 °F) induction heating and controlled air cooling, which results in a desirable microstructure consisting of 80% pearlite and 20% ferrite.

Microalloying of carbon-manganese steel bars can increase the strength of steel bars (0.35 to 0.45% C) used to reinforce concrete. Recent applications call for improvements in bendability, weldability, and toughness for low ambient service temperatures or for low-temperature structures, such as refrigerated warehouses, liquid natural gas storage facilities, foundations for nuclear power stations, and construction in seismic-prone areas. For these conditions, carbon content is lowered, and the strengthening properties of niobium and vanadium are used singly or in combination.

Microalloying elements as a substitute for aluminum in grain refinement can facilitate the production of desired inclusion compositions for applications in which machining or surface treatments such as carburizing or nitriding are involved. Such steels are designed to decrease the abrasive tool wear caused by hard alumina inclusions. Deoxidation practice produces soft, glassy inclusions that are relatively plastic at hot working temperatures. The elimination of aluminum requires the use of another strong deoxidant such as calcium and the addition of niobium or other suitable grain refiner.

Niobium or vanadium containing castings have applications that are given in Table 27. Large castings, for example, are used in offshore construction. Built-in reinforcements improve performance under cyclic loading conditions. Many low-carbon steels have inherently high ductility and can tolerate the presence of some potentially injurious second-phase particles such as

Table 27 Summary of microalloyed bar products, castings, and forgings

Steel product	Typical composition, %						Applications
	C	Mn	Si	Nb	V	Other	
Hot-rolled bar							
High carbon	0.58	0.80	0.30	0.04	...	...	Drawn wire having residual-deformation capability
Medium carbon							
Rebar	0.38	1.20	0.20	0.03 (or V)	...	...	Standard reinforcing bar
Structural	0.35	1.00	0.25	0.02–0.10	...	...	Machined automotive parts, or forged vanadium steel
Engineering	0.40	0.80	0.20	...	0.09 (or Nb)	...	Replaces quenched and tempered steel
Machined bar	0.30	1.20	0.30	0.03	...	Al free	Machined bar in automotive and engineering equipment
Low carbon	0.10–0.20	1.65	0.15	0.04	...	...	Weldable reinforcing bar, Arctic concrete structures, cryogenic storage tanks
Very low carbon	0.10	1.70	0	0.08	...	Boron	Cold-headed parts
Heat-treated bar							
Concast billets	0.25	1.30	0.20	0.03	...	...	Heat-treated parts for tractors, and so forth
Engineering	0.60	0.85	0.80	...	0.11	...	High-strength drawn wire
Rebar	0.18	0.60	0.05	...	0.05	...	Concrete structures
High carbon	0.58	0.85	2.10	0.15	0.12	...	
Forgings							
As-forged bar	0.30–0.40	0.80	0.20	...	0.09 (or Nb and 0.50 Cr)	...	Automotive components, universal couplings, stabilizer bars, wheel hubs, knuckle arms, connecting rods
Crankshafts	0.53	0.80	0.40	0.06 (or 0.10V)	...	...	Lightweight German automobiles
Small forgings	0.12	1.45	0.35	0.07	0.07 (or 0.015)	...	Automobile and off-highway equipment
Castings							
Nodes	0.15	1.40	0.30	0.03	0.07	0.48 CE max	Offshore platforms
Fittings	0.18	1.50	0.20	0.03	0.05	...	Pipeline elbows and fittings, and so on
Small castings	0.05	1.60	0.10	0.06	...	0.40 Mo	Small engineering castings, brake parts, compressors, and so on

carbide eutectics. Niobium and niobium-vanadium microalloyed low-carbon steels with manganese and molybdenum have been developed and applied in Europe. Castings are used in either the as-cast or as-quenched and tempered condition. In either case, weldability is reported to be excellent. A development for the future are low-carbon (0.02%) manganese-niobium-boron-titanium steels, originally developed as high heat input filler (weld) metals.

Cold-Rolled HSLA Steel. Although most HSLA steel is hot rolled, large tonnages of cold-rolled HSLA sheet and strip are produced. However, because many applications of cold-rolled steel already involve thin product forms, substituting a thinner, stronger steel for weight savings may not be feasible because of stiffness and deflection considerations.

The general types of cold-rolled HSLA steels utilize precipitation strengthening, solid-solution strengthening, or strengthening from a dual-phase (martensite in ferrite) microstructure. These types of cold-rolled HSLA steel are compared in Fig. 8(b) in terms of ductility and tensile strength. The cold-rolled solution-hardened steels have superior elongation and stretchability compared with cold-rolled precipitation-strengthened steels of the same tensile strength, but are limited to a tensile strength of about 600 MPa (87 ksi). The cold-rolled precipitation-strengthened steels allow higher tensile strengths compared with solution-strengthened steels, but these have less ductility at a given tensile strength (Fig. 8b). Because titanium provides sulfide shape control and precipitation strengthening, titanium steels have plastic-strain ratios up to 1.4 at tensile strengths of 550 to 650 MPa (80 to 95 ksi) (Ref 23).

Dual-phase steels offer improved ductility at higher tensile strengths (Fig. 8b). Sheets of dual-phase steel have very low yield strengths without sharp yield points (that is, continuous yielding), but the yield strengths are increased substantially during forming or by a tempering or aging treatment at 300 to 400 °C (570 to 750 °F) for about 5 min after air cooling. The formability (in terms of plastic-strain ratios and strain-hardening exponents) of a dual-phase steel is compared with other cold-rolled automotive sheet steels in Table 28 and Fig. 8(c). Although many of the high-strength cold-rolled steels have rather low plastic-strain ratios and, hence, are not deep drawn as readily, their drawability can be improved by a suitable precoating lubricant (Ref 23, 37). Interstitial-free steels offer the best deep-drawing properties (Fig. 8c).

Selection and Application Example. The following describes the use of structural and HSLA steels in an architectural application.

Example: Design and Construction of Retractable SkyDome Roof. Toronto, Canada's SkyDome, completed in June 1989, is a multipurpose structure that seats 54 000 for baseball, 55 000 for football, and up to 70 000 for special events (Ref 38). The SkyDome incorporates a patented rigid frame structural steel retractable roof. Within a span of 20 min, the movable roof transforms the SkyDome from an open-air stadium into an enclosed air-conditioned or heated domed stadium.

Roof Dimensions. The roof consists of four panels, three of which are movable, that stack and interlock on top of each other to provide a closed, sealed, and weather-tight structure (Fig. 26). Utilization of four parabolic arch panels in this arrangement exposes 91% of the seats in the open-air section and stacks all four panels within the building footprint (Ref 39). The roof covers 32 275 m^2 (347 420 ft^2, or 8 acres); it rises to a maximum height of 86 m (282 ft), which is high enough to accommodate a 31-story skyscraper within its confines. The roof has a steel dead weight of 69 000 kN (7760 tonf)

Table 28 Mechanical properties of cold-rolled, high-strength steels

Steel grade	Yield strength		Tensile strength		Elongation, %	Formability factors	
	MPa	ksi	MPa	ksi		Strain-hardening exponent (n)	Plastic-strain ratio (r)
Low carbon, killed	172	25	275	40	40	0.23	1.4
Structural grade, high strength	275	40	380	55	35	0.22	1.4
Formable grade, high strength	345	50	415	60	30	0.20	1.1
	550	80	620	90	18	0.14	0.8
Dual phase, high strength	345	50	550	80	25	0.24	0.8
Ultrahigh strength	825	120	860	125	8	...	...
	965	140	1000	145	4	...	...

(Ref 39, 40). The movable panels have the following dimensions:

Panel	Depth m	ft	Span m	ft
1	50	165	175	575
2	50	165	206	675
3	50	165	190	625

Source: Ref 38

Roof Opening Procedure (Ref 41). The roof of the SkyDome is closed most of the time. Panel 4 is fixed in place at the north end and has the configuration of a segment of a spherical shell or quarterdome. Panel 1 has a similar shape but is slightly larger. When the roof is in the closed position, panel 1 is located at the south end, directly opposite panel 4. As the roof opens, panel 1 travels along a 310 m (1020 ft) circular 4-rail track, carried by 22 steel-wheeled bogies (10 of which are fitted with two 7.5-kW, or 10-hp, dc electric-drive units), and parks itself directly above panel 4. Some of these roof-carrying bogies weigh up to 36 tonnes (40 tons) per unit.

Panels 2 and 3 sit adjacent to each other when the roof is closed. A total of 16 bogies are used to move panels 2 and 3. Eight of the bogies on panel 2 and six of the bogies on panel 3 are equipped with drive units. Panel 2 is slightly higher and larger than panel 3. Both move linearly northward when the roof is opening. Panel 2 initially moves along the parallel tracks and passes directly over panel 3, which then also starts its travel to the north end of the stadium.

All three movable panels are simultaneously in motion for a short portion of the 20-min operation when panel 1 begins its journey around the circular track. Panel 1 eventually comes to rest approximately 180° from its original position. When the SkyDome roof is open, panel 1 nests above the similarly shaped panel 4 but below panel 3, while panel 2 sits atop the other three panels (Fig. 26).

Structural Components. Figure 27 shows the SkyDome in its final stages of construction. Construction of the geometrically complex SkyDome roof involved the fabrication and erection of 7575 tonnes (8330 tons) of structural steel components, 975 tonnes (1070 tons) of bogies, and 930 tonnes (1025 tons) of rails, all to a high degree of precision (Ref 41).

The 7575 tonnes (8330 tons) of zinc-coated structural steels used in the roof structure are employed in a linear system of parallel arches with transverse trusses of hollow structural steel (HSS). These HSS trusses conform to Canadian Standards Association (CSA) specification CAN 3-G40.21-M81, which closely approximates ASTM A 572. CSA grade 300 W is equivalent to ASTM grade 42, while CSA grade 300 W is equivalent to ASTM grade 50. CSA type W steels are suitable for general welded construction where notch toughness at low temperatures is not a design requirement (Ref 42).

Grade 300 W, with a minimum yield strength of 300 MPa (44 ksi), is a standard general-purpose weldable hot-rolled structural steel with 130 to 170 HB hardness. In addition to being used for the joint connections and plates in the structural framework of the SkyDome, grade 300 W structural plate was used extensively in the construction of the bogies that move the roof panels.

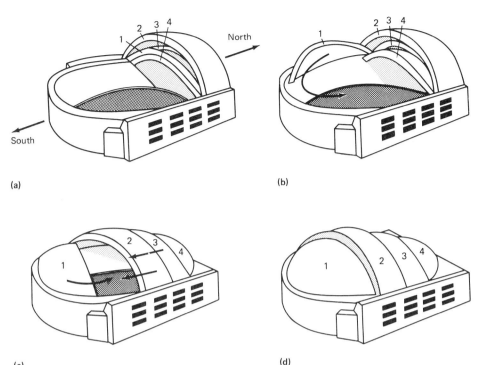

Fig. 26 Location and progressive movement of the three movable roof panels of the SkyDome as it is converted from an open-air facility into an enclosed environmentally controlled dome. (a) All four panels stacked at the north end of the SkyDome in the open position. Panel 4 remains stationary. (b) Panel 1 is first to move as it begins its circular motion. (c) All three movable panels are in motion as panels 2 and 3 begin their straight-line motion while panel 1 continues in its circular path. (d) Within 20 min, panel movement is complete and the structure is in its closed position. Source: Adapted from *Sports Illustrated*, 12 June 1989

Fig. 27 SkyDome construction nearing completion. Courtesy of Jane Welowszky, SkyDome Stadium Corporation of Ontario

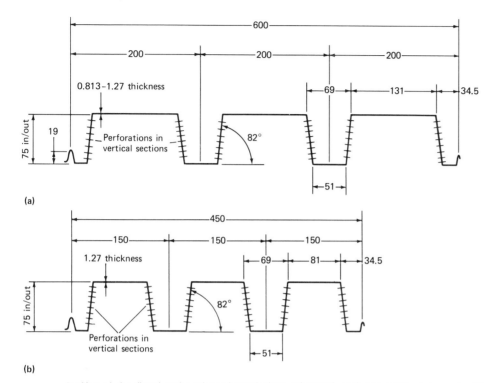

Fig. 28 Profiles of the fluted and perforated interlocking galvanized steel corrugated roof panels. All dimensions are given in millimeters. (a) D-200 roof panel, 600 mm (24 in.) wide coverage. (b) D-150 roof panel, 450 mm (18 in.) wide coverage

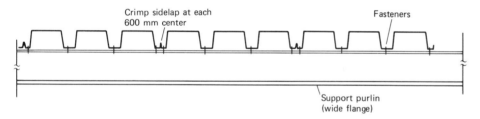

Fig. 29 Interlocking ends of corrugated roof panels

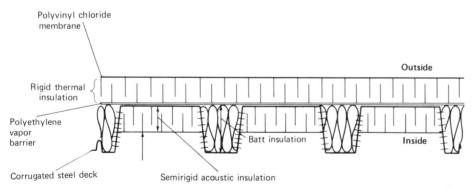

Fig. 30 Cross section of acoustic roof deck components used for the SkyDome. Setup shown is for the D-200 (600 mm, or 24 in., wide) corrugated steel panel; an identical setup is used for the D-150 (450 mm, or 18 in., wide) panel.

Grade 350 W, with a minimum yield strength of 350 MPa (50 ksi), is a cold-rolled and stress-relieved HSLA weldable structural steel with 130 to 180 HB hardness. Grade 350 W structural steel was seam welded to form a hollow square tube which serves as a roof tubular support. These HSS square tubes range in size from 150 × 150 mm (6 × 6 in.), with a thickness of 6.4 mm (¼ in.), to 300 × 300 mm (12 × 12 in.), with a thickness of 12 mm (½ in.). The use of standard structural steel hollow tubes has resulted in a roof that has great strength and stiffness characteristics and yet is relatively lightweight and has a depth relative to the large span, which is close to 206 m (675 ft) wide.

Trusses for the four panels were shop-fabricated in large sections and then bolted on-site using 140 000 ASTM A 325 galvanized 25 mm (1 in.) diam bolts and corresponding ASTM A 194 nuts. Both fastener types were made from quenched and tempered medium-carbon high-strength structural steels.

Acoustic Roof Deck. The actual roof covering consists of an acoustic steel roof deck weighing 500 tonnes (550 tons). The galvanized corrugated square-profile steel panels conform to ASTM A 446 grades B (255 MPa, or 37 ksi, minimum yield strength; 310 MPa, or 45 ksi, minimum tensile strength) and C (275 MPa, or 40 ksi, minimum yield strength; 380 MPa or 55 ksi minimum tensile strength); (Ref 42). The thickness of the steel strip and sheet used to fabricate the panels varies from 0.813 to 1.27 mm (0.032 to 0.050 in.) depending on location and load requirements.

The corrugated panels are fluted, perforated (Fig. 28), and provided with interlocking J-hook ends (Fig. 29) to provide a baffle that contains crowd noise within the confines of the structure when it is closed. Fabricated in lengths up to 6 m (20 ft), the corrugated panels have square profiles of 450 mm (18 in.) or 600 mm (24 in.). Both the inner and the outer exposed sections of the corrugated panels contain acoustic pads to absorb sound (Fig. 30). In addition, the roof sandwich also includes thermal insulation and a 1 mm (0.04 in.) thick single-ply polyvinyl chloride membrane.

ACKNOWLEDGMENT

The information presented on the SkyDome multipurpose stadium in this article was made possible by the data provided and reviewed by the following individuals: Harry Charalambu, Carr & Donald Associates, Toronto, Ontario (data on bogies); Don P.J. Duchesne and C. Michael Allen, Adjelian Allen Rubeli, Ltd., Ottawa, Ontario (data on roof structural framework); and Mike Carlucci, Lorlea Steels, Brampton, Ontario (data on steel deck corrugated panels).

REFERENCES

1. C.A. Siebert, D.V. Doane, and D.H. Breen, *The Hardenability of Steels—Concepts, Metallurgical Influences, and Industrial Applications*, American Society for Metals, 1977, p 66, 101
2. R.W.K. Honeycombe, *Steels—Microstructure and Properties*, Edward Arnold, London, 1982
3. R.A. Grange, C.R. Hribal, and L.F. Porter, Hardness of Tempered Martens-

ite in Carbon and Low-Alloy Steels, *Metall. Trans. A*, Vol 8A, 1977, p 1775-1785
4. A.R. Marder, Heat-Treated Alloy Steels, in Vol 3 of *Encyclopedia of Materials Science and Engineering*, M.B. Bever, Ed., Pergamon Press and MIT Press, 1986, p 2111-2116
5. W.C. Leslie, *The Physical Metallurgy of Steels*, McGraw-Hill, 1981, p 200, 370
6. E.C. Bain and H.W. Paxton, *Alloying Elements in Steel*, 2nd ed., American Society for Metals, 1962
7. P.L. Manganon, *J. Heat Treat.*, Vol 1, 1980, p 47-60
8. P.L. Manganon, *Metall. Trans. A*, Vol 13A, 1982, p 319-320
9. O. Sandberg, P. Westerhult, and W. Roberts, Report 1687, Swedish Institute for Metals Research, 1982
10. S. Gong, A. Sandberg, and R. Lagneborg, in *Mechanical Working and Steel Processing*, XIX, American Institute of Mining, Metallurgical, and Petroleum Engineers, 1981, p 563-582
11. W. Roberts, Recent Innovations in Alloy Design and Processing of Microalloyed Steels, in *HSLA Steels—Technology and Applications*, American Society for Metals, 1984, p 48, 55
12. K.R. Hanby, et al., "Handbook on Materials for Superconducting Machinery," MCIC-HB-04, Metals and Ceramics Information Center, Battelle Columbus Laboratories, Jan 1977
13. K.A. Warren and R.P. Reed, *Tensile and Impact Properties of Selected Materials From 20° to 300 °K*, Monograph 63, National Bureau of Standards, June 1963
14. J.P. Bruner and D.A. Sarno, An Evaluation of Three Steels for Cryogenic Service, in *Advances in Cryogenic Engineering*, Vol 24, K.D. Timmerhaus et al., Ed., Plenum Press, 1978, p 529-539
15. A.G. Haynes et al., *Strength and Fracture Toughness of Nickel Containing Steels*, STP 579, American Society for Testing and Materials, 1975, p 288-323
16. A.W. Pense and R.D. Stout, "Fracture Toughness and Related Characteristics of the Cryogenic Nickel Steels," Bulletin 205, Welding Research Council, May 1975
17. D.A. Sarno, J.P. Bruner, and G.E. Kampschaefer, Fracture Toughness of 5% Nickel Steel Weldments, *Weld. J.*, Vol 39 (No. 11), Nov 1974, p 486s-494s
18. H.I. McHenry and R.P. Reed, Fracture Behavior of the Heat-Affected Zone in 5% Ni Steel Weldments, *Weld. J.*, Vol 56 (No. 4), April 1977, p 104s-112s
19. R.L. Tobler et al., Low Temperature Fracture Behavior of Iron Nickel Alloy Steels, in *Properties of Materials for Liquified Natural Gas Tankage*, STP 579, American Society for Testing and Materials, Sept 1975, p 261-287
20. D.A. Sarno, D.E. McCabe, and T.G. Heberling, Fatigue and Fracture Toughness Properties of 9 Percent Nickel Steel at LNG Temperatures, in *J. Eng. Ind. (Trans. ASME)*, Series B, Vol 95 (No. 4), Nov 1973, p 1069-1075
21. *Welding Handbook*, Vol 4, *Metals and Their Weldability*, 7th ed., American Welding Society, 1982, p 24-32
22. P.E. Repas, Metallurgical Fundamentals for HSLA Steels, in *Microalloyed HSLA Steels*, ASM INTERNATIONAL, 1988, p 5
23. E.E. Fletcher, *High-Strength Low-Alloy Steels: Status Selection and Physical Metallurgy*, Battelle Press, 1979
24. F.B. Pickering, *Physical Metallurgy and Design of Steels*, Applied Science Publishers, 1978, p 65, 80
25. A.J. DeArdo, An Overview of Microalloyed Steels, in *Effects of Microalloying on the Hot-Working Behavior of Ferrous Alloys*, 8th Process Technology Development Proceedings, American Iron and Steel Institute, 1988
26. L.F. Porter, High-Strength Low-Alloy Steels, in *Encyclopedia of Materials Science and Engineering*, Vol 3, M.B. Bever, Ed., Pergamon Press and MIT Press, 1986, p 2157-2162
27. L. Meyer, F. Heisterkamp, and W. Mueschenborn, Columbium, Titanium, and Vanadium in Normalized, Thermo-Mechanically Treated and Cold-Rolled Steels, in *Microalloying '75*, Conference Proceedings (Washington, D.C., Oct 1975), Union Carbide Corporation, 1977, p 153-167
28. G. Tither, Developments in HSLA Steels Particularly for Pipe Line and Offshore Structural Applications, in *Metallurgical and Welding Advances in High Strength Low-Alloy Steels*, The Danish Corrosion Centre and The Danish Welding Institute, 1984
29. C.I. Garcia and A.J. DeArdo, Structure and Properties of ULCB Plate Steels for Heavy Section Application, in *Microalloyed HSLA Steels*, ASM INTERNATIONAL, 1988, p 291-299
30. A.K. Sinha, *Ferrous Physical Metallurgy*, Butterworths, 1989, p 628
31. W.B. Morrison, The Influence of Small Niobium Additions on Properties of Carbon-Manganese Steels, *J. Iron Steel Inst.*, Vol 201, 1963, p 317-325
32. P. Choquet et al., Optimization of the Hot Rolling of High Grade Pipeline Steels at the Hot Strip Mill, in *Microalloyed HSLA Steels*, ASM INTERNATIONAL, 1988, p 571-579
33. K.E. Richter, Cold- and Hot-Rolled Microalloyed Steel Sheet in Opel Cars—Experience and Applications, in *HSLA Steels—Technology and Applications*, American Society for Metals, 1984, p 485-491
34. D.H. Stone and W.S. Pellini, Application of HSLA Steels for Construction of Railroad Tank Cars, in *Microalloyed HSLA Steels*, ASM INTERNATIONAL, 1988, p 411-420
35. J.M. Gray et al., Property Improvements in Bars and Forgings Through Microalloying and Inclusion Engineering, *HSLA Steels—Technology and Applications*, American Society for Metals, 1984, p 967-979
36. C.I. Garcia, E.J. Palmiere, and A.J. DeArdo, An Alternative Approach to the Alloy Design and Thermomechanical Processing of Low-Carbon Microalloyed Bar Products, in *1987 Mechanical Working and Steel Processing Proceedings*, American Institute of Mining, Metallurgical, and Petroleum Engineers, 1987
37. S. Hayami and T. Furukawa, A Family of High-Strength, Cold-Rolled Steels, in *Microalloying '75*, Conference Proceedings (Washington, DC, Oct 1975), Union Carbide, 1977, p 311-320
38. The SkyDome Retractable Roof System, *Eng. Dig. (Canada)*, June 1989
39. "Trend 71: Depicting Steel in Modern Architecture," 30.5M/BP/8901/01, Stelco Steel, Jan 1989
40. The Design of the Retractable Roof System, *Eng. Dig. (Canada)*, June 1989
41. SkyDome Roof Fabrication and Construction, *Eng. Dig. (Canada)*, June 1989
42. "Structural Steels: Selection and Uses," 10M/SC/8344/6, 5th ed., Stelco Inc., Dec 1983

Dual-Phase Steels

G.R. Speich, Department of Metallurgical Engineering, Illinois Institute of Technology

DUAL-PHASE STEELS are a new class of high-strength low alloy (HSLA) steels. This class is characterized by a tensile strength value of approximately 550 MPa (80 ksi) and by a microstructure consisting of about 20% hard martensite particles dispersed in a soft ductile ferrite matrix. The term dual phase refers to the predominance in the microstructure of two phases, ferrite and martensite (Fig. 1). However, small amounts of other phases, such as bainite, pearlite, or retained austenite, may also be present.

In addition to high tensile strength, other unique properties of these steels include continuous yielding behavior, a low 0.2% offset yield strength, and a higher total elongation than other HSLA steels of similar strength.

Although limited research on dual-phase steels began in the early 1970s (Ref 2), an intense interest in these steels commenced with work done in 1975 and 1976. A 1975 study showed that continuous annealing in the intercritical temperature range produced steels with a ferrite-martensite microstructure (similar to that shown in Fig. 1) and a ductility superior to that of normal precipitation-hardened or solid solution hardened HSLA sheet steels (Ref 3). Some of these results are shown in Fig. 2 and in the article "High-Strength Structural and High-Strength Low-Alloy Steels" in this Volume. A 1976 study showed that intercritical heat treatment of vanadium-nitrogen HSLA steels resulted in a large increase in ductility at a strength level of 550 MPa (80 ksi) (Fig. 3). This study also showed that the increased ductility was accompanied by an increased formability of automotive parts (Ref 4). Prior to this time, steels with a strength of 550 MPa (80 ksi) that were produced without intercritical heat treatment were considered to have poor formability.

Large-scale efforts were begun at this time in many steel research laboratories to develop new steel compositions as well as alternative processing procedures for producing large tonnages of dual-phase steels. It was predicted that in the 1980s automobile manufacturers would need the large tonnages to produce more fuel-efficient cars. Gasoline mileage requirements were mandated by the government because of the gasoline shortage; it was expected that using thinner sheets of high-strength dual-phase steels for automobile bodies would save weight and enable the cars to meet these requirements. Much of this work has been reported in several review papers (Ref 1, 5, 6). Pertinent aspects of dual-phase steels, such as heat treatment, microstructures, mechanical properties, and new advances are discussed in this article.

Typical chemical compositions of present day dual-phase steels are given in Table 1. In general, these steels have a carbon content of less than 0.1%, which ensures that they can be spot welded. The carbon content also produces about 20% of the martensite in the microstructure after intercritical annealing and rapid cooling. Manganese in amounts of 1 to 1.5% is added to ensure sufficient hardenability so that martensite is formed upon rapid cooling. Chromium and molybdenum have also been added in amounts that are usually under 0.6%. Silicon is added to provide solid solution hardening. Small amounts of microalloying additions, such as vanadium, niobium, and titanium, may be added to provide precipitation hardening and/or grain size control. Nitrogen may be added to intensify the precipitation-hardening effects of vanadium.

Heat Treatment of Dual-Phase Steels

Annealing Techniques and Steel Compositions. Dual-phase steels can be produced by intercritical heat treatment with either con-

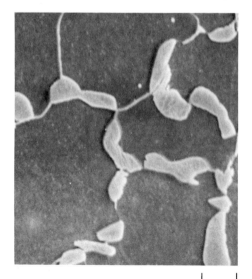

Fig. 1 Ferrite-martensite microstructure of a dual-phase steel (0.06% C, 1.5% Mn; water quenched from 760 °C, or 1400 °F). Source: Ref 1

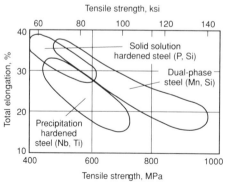

Fig. 2 Relation between tensile strength and total elongation for various HSLA sheet steels. Source: Ref 3

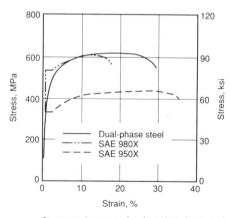

Fig. 3 Stress-strain curves for the HSLA sheet steels SAE 50X and SAE 80X (with yield strengths of 340 and 550 MPa, or 50 and 80 ksi, respectively) and a dual-phase steel (with a yield strength of 550 MPa, or 80 ksi). Source: Ref 4

Table 1 Typical dual-phase steel compositions

Production method	Composition, wt%						
	C	Mn	Si	Cr	Mo	V	N
Continuous annealing, hot-rolled gage	0.11	1.43	0.61	0.12	0.08	0.06	0.01
Continuous annealing, cold-rolled gage	0.11	1.20	0.40	...	...	...	...
Box annealing	0.12	2.10	1.40	...	...	...	...
As rolled	0.06	0.90	1.35	0.50	0.35	...	...

tinuous annealing or box annealing techniques. In continuous annealing, the steel sheet is heated for a short time (30 s) in the intercritical temperature range to form ferrite-austenite mixtures. This is followed by accelerated cooling (6 °C/s, or 10 °F/s) to transform the ferrite-austenite mixtures into ferrite-martensite mixtures (Fig. 4). The actual cooling rate depends on sheet thickness and the quenching conditions on a given annealing line.

Box annealing is similar to continuous annealing, but the heating times are much longer (3 h), and the cooling rates are much slower (10 °C/h, or 20 °F/h). Because of its slow cooling rates, box annealing requires steels with much higher alloy contents than does continuous annealing (Table 1). To achieve desired hardenability with box annealing, 2.5% Mn steels that can at times contain appreciable amounts of silicon and chromium have been proposed.

Formation of Austenite During Intercritical Annealing. The formation of austenite during intercritical annealing can be separated into several steps (Ref 7). The first is almost instantaneous nucleation of austenite at pearlite or grain boundary cementite particles, followed by rapid growth of austenite until the carbide or pearlite is dissolved. The next step is the slower growth of austenite into ferrite at a rate that is controlled by carbon diffusion in austenite at high temperatures (850 °C, or 1560 °F) and by manganese diffusion in ferrite (or along grain boundaries) at low temperatures (750 °C, or 1380 °F). Finally, there is a very slow equilibration of ferrite and austenite at a rate that is controlled by manganese diffusion in austenite, under conditions of very long time annealing at low temperatures.

An example of austenite formation in a 0.06C-1.5Mn steel after short-time annealing at 740 °C (1365 °F) is shown in Fig. 4. The kinetics of austenite formation at different temperatures in a 0.12C-1.5Mn steel are summarized in Fig. 5, with the time and controlling kinetic mechanism for each of the various steps indicated at each temperature.

Transformation of Austenite After Intercritical Annealing. Although the transformation of austenite after intercritical annealing is similar to the transformation of austenite after regular annealing, several features make the former process unique. First, because the carbon content of the austenite, C^γ, is fixed by the intercritical annealing temperature (Fig. 6), the hardenability of the austenite varies with this temperature. Second, because the martensite forms by a diffusionless transformation, the martensite phase inherits the carbon content of the austenite, which is much higher than that of the original steel C_O. Finally, because it is important to maintain the necessary carbon balance, using the lever rule will indicate that the amount of the austenite will vary with the intercritical temperature and the original carbon content of the steel (Fig. 7).

Upon water quenching at high cooling rates from the intercritical annealing temperature, practically all of the austenite transforms to martensite that has the classical high-carbon martensite substructure (Fig. 8). At slower cooling rates, such as in hot-oil quenching, some retained austenite remains in the form of very small particles (Fig. 9). This austenite is stable upon cooling to room temperature or subzero temperatures but transforms upon plastic straining (Fig. 10).

Changes in the Ferrite Phase During Intercritical Annealing. Because the ferrite is in contact with the austenite at the intercritical temperature, upon cooling from this temperature it may grow epitaxially into the austenite as part of the transformation mechanism (Ref 12). In cold-rolled steels,

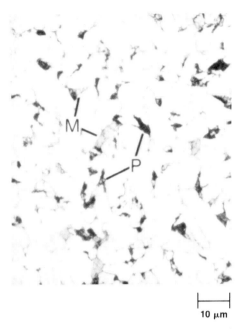

Fig. 4 Formation of austenite in 0.06C-1.5Mn steel from preexisting pearlite after short-time annealing in the intercritical temperature range (30 s at 740 °C, or 1365 °F). M, Martensite (austenite at intercritical temperature); P, pearlite (dissolution only partly complete). Source: Ref 7

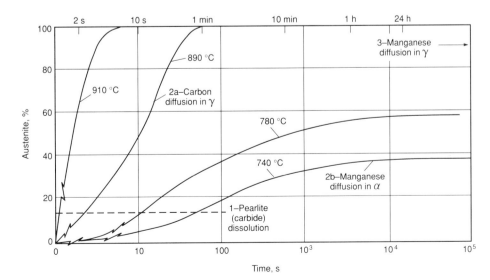

Fig. 5 Kinetics of austenite formation in 0.12C-1.5Mn steel. Source: Ref 7

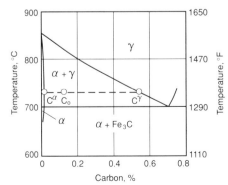

Fig. 6 Phase diagram for 1.5% Mn steel (paraequilibrium conditions). Source: Ref 7

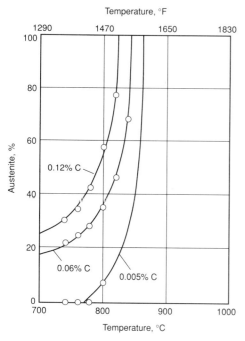

Fig. 7 Percentage of austenite formed at various intercritical temperatures for 1.5% Mn steels containing 0.005 to 0.12% C (paraequilibrium conditions). Source: Ref 8

Fig. 8 Transmission electron micrograph of martensite substructure and high dislocation density in ferrite in a 0.04C-1.5Mn steel intercritically annealed at 726 °C (1340 °F) and rapidly cooled. Source: Ref 9

recrystallization of the ferrite occurs rapidly and is generally complete before the steel reaches the intercritical temperature, even during the rapid heating encountered on most continuous annealing lines (Ref 12). Changes in the carbon content of the ferrite may occur during intercritical annealing. These changes arise from several different causes. First, the solubility of carbon in the ferrite may be lower at the intercritical temperature than it is in the as-received material. Second, the carbon content of the ferrite may decrease because of the increased alloy content. For example, the addition of 1.5% Mn under equilibrium conditions would lower the solubility of carbon at 740 °C (1365 °F) from 0.02 to 0.005%, as shown by the isothermal section of the iron-manganese-carbon phase diagram in Fig. 11. Additions of silicon are also reported to lower the carbon content of the ferrite (Ref 14). Variations in the cooling rate from the intercritical temperature can also affect the carbon content of the ferrite phase because of the precipitation of cementite from the ferrite.

Mechanical Properties of Dual-Phase Steels

Work Hardening and Yield Behavior. In general, ferrite-martensite steels do not show a yield point. The combination of high residual stresses and a high mobile dislocation density in the ferrite causes plastic flow to occur easily at low plastic strains (Ref 12). As a result, yielding occurs at many sites throughout the ferrite, and discontinuous yielding is suppressed (Fig. 3).

The work-hardening process in dual-phase steels can be separated into three stages (Ref 12). In the first stage (0.1 to 0.5% strain), rapid work hardening takes place because of the elimination of residual stresses and the rapid buildup of back stresses caused by the plastic incompatibility of the two phases. In the second stage (0.5 to 4% strain), transformation of retained austenite occurs. Finally, in the third stage (4 to 18% strain), dislocation cell structures are formed, and further deformation in the ferrite is governed both by dynamic recovery and cross slip and by eventual yielding of the martensite.

The work-hardening behavior of dual-phase steels is thus very complex, especially in the initial stages. However, the high initial work-hardening rate is believed to contribute to the good formability of these steels, compared to other HSLA steels of a similar strength level (Ref 4). Also, the lack of a yield point in these steels eliminates Lüders band formation and ensures that a good surface finish is obtained after forming.

Yield and Tensile Strength. Based on simple composite-strengthening theory, the strength of dual-phase steels should be expected to increase when either the volume fraction or strength (hardness) of the martensite phase is increased (Fig. 12). Actual data for 1.5% Mn steels support this idea (Fig. 13). In addition, the strength of the mixture increases as the strength of the ferrite or matrix phase increases. As al-

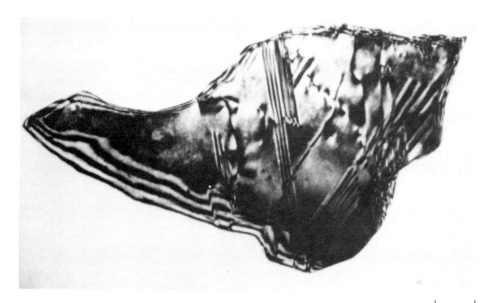

Fig. 9 Transmission electron micrograph of a retained austenite particle in a 0.072C-1.3Mn-0.08Cb-0.08V steel intercritically annealed 5 min at 900 °C (1650 °F) and hot-oil quenched. Source: Ref 10

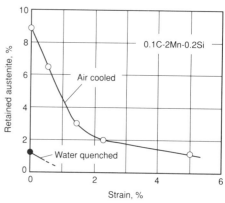

Fig. 10 Percentage of retained austenite after air cooling or water quenching followed by plastic straining. Source: Ref 11

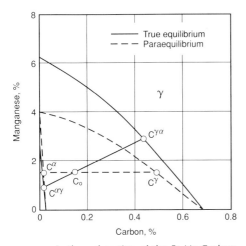

Fig. 11 Isothermal section of the Fe-Mn-C phase diagram at 740 °C (1365 °F) showing both the paraequilibrium and true equilibrium phase boundaries. Source: Ref 13

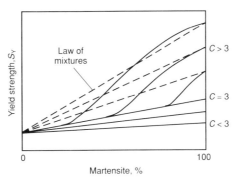

Fig. 12 Effect of the ratio of the yield strength of martensite to the yield strength of ferrite (defined as C) on the yield strength of ferrite-martensite mixtures. Source: Ref 2

ready stated, the strength of the martensite which depends primarily on the carbon content of the phase, is determined by the intercritical annealing conditions and the original carbon content of the steel. The strength of the ferrite phase depends on the grain size and solid solution hardening contributions of the alloying elements. A complete discussion of the factors controlling the strength of dual-phase steels requires the use of more sophisticated models and continuum mechanics.

Ductility. The enhanced ductility of dual-phase steels has been attributed to many causes, including the lower carbon content of the ferrite, the plasticity of the martensite phase, the amount of epitaxial ferrite, and the amount of retained austenite. A single cause for the enhanced ductility of dual-phase steels has not yet been isolated. Of these factors, however, the amount of retained austenite and the manner in which it transforms upon plastic straining has gained more acceptance as an explanation than the others. In cases where the amount of retained austenite is increased by controlling the annealing cycle, ductility has been found to reach a maximum value (Fig. 14).

Tempering and Strain Aging. Tempering of water-quenched dual-phase sheet steels can also be done during subsequent hot dip galvanizing operations or during paint-baking cycles, although in the latter case the part has been plastically strained, and the tempering actually represents a strain-aging process. In-line termpering may be used in commercial continuously annealed sheet to prevent changes in properties upon subsequent tempering.

Such tempering involves a combination of effects expected for tempering of each of the individual phases. Thus, in the high-carbon martensite phase, recovery of the defect structure, precipitation of carbides, and transformation of retained austenite are expected. Similarly, in the ferrite phase,

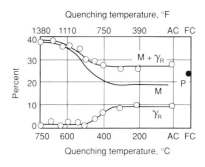

(a)

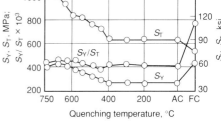

(b)

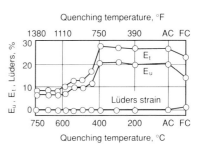

(c)

Fig. 14 Variation in selected properties in a 0.1C-0.2Si-2Mn steel cooled to various temperatures from 750 °C (1380 °F) and quenched. AC, air cool; FC, furnace cool. (a) Variation in retained austenite content, γ_R; P, pearlite; M, martensite. (b) Variation in yield strength, S_Y, and tensile strength, S_T. (c) Variation in total elongation, E_t, uniform elongation, E_u, and Lüders strain. Source: Ref 11

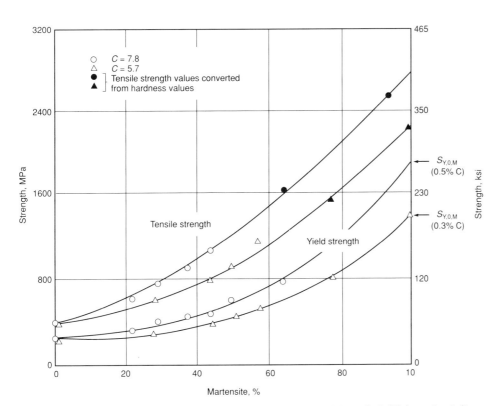

Fig. 13 Yield and tensile strength of ferrite-martensite mixtures in 1.5% Mn steels. Solid data points indicate tensile strength values converted from hardness values. $S_{Y,0,M}$, yield strength intercept of martensite; C, ratio of the yield strength of martensite to the yield strength of ferrite. Source: Ref 15

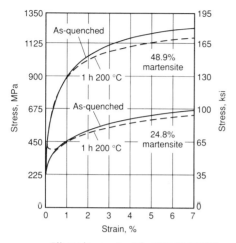

Fig. 15 Effect of tempering 1 h at 200 °C (390 °F) on the stress-strain behavior of 0.06 and 0.20% C steels containing 1.5% Mn with various percentages of martensite. Source: Ref 15

both carbon segregation to dislocations and precipitation of carbides are expected. In addition, there are synergistic effects that arise from the presence of both phases, such as the generation of residual stresses and a high dislocation density in the ferrite. Therefore, carbon segregation to the dislocations and the elimination of the residual stresses by the volume contraction of the ferrite are important parts of the tempering process. At temperatures near 200 °C (390 °F), the segregation of carbon to the dislocations and the elimination of residual stresses results in an increase in yield strength and a return of discontinuous yielding, but only if the martensite content is below 30% (Fig. 15).

In general, dual-phase steels are nonaging at room temperature (Ref 16), and they exhibit sluggish aging behavior at temperatures up to 270 °C (520 °F). However, deformation caused either by cold rolling or by tensile straining accelerates the aging process. One study has suggested that the more rapid strain aging by tensile straining is a result of making the dislocation density more uniform than that which is present in the temper-rolled condition (Ref 16). As a result, the diffusion distance for carbon is reduced, and carbon diffusion is faster. Figure 16 shows the effect of aging at 170 °C (340 °F) after temper rolling and tensile straining on the return of the yield point in the absence of strain.

New Advances in Dual-Phase Steels

Prior dual-phase steel developments have been focused primarily on applications involving cold- or hot-rolled sheet containing about 0.1% C. Carbon content was kept low because of the need for good spot weldability. More recent developments have been directed at producing dual-phase steels with

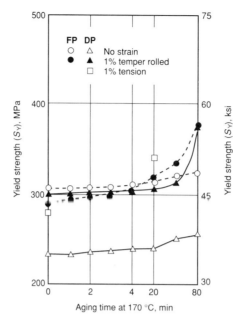

(a)

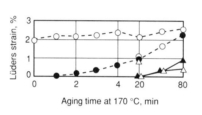

(b)

Fig. 16 Effect of aging at 170 °C (340 °F) for various times on (a) yield strength and (b) Lüders strain in a 0.08C-1.2Mn-0.5Cr steel in both ferrite-pearlite, FP, and dual-phase, DP, conditions. Source: Ref 16

higher carbon contents (0.2 to 0.4% C) for use in forging or bar applications (Ref 17, 18). Although such steels cannot be easily welded because of their high levels of carbon, they are suitable for forging or bar applications where spot welding is not required and where mechanical fastening is possible. A typical chemical composition of such steels is 0.40% C, 1.49% Si, 0.84% Mn, 0.005% P, 0.006% S, 0.035% Al, and 0.0024% N (total). In addition to a higher carbon content these steels also have a higher silicon content.

These newer high-carbon dual-phase steels have a unique combination of total elongation and tensile strength (Fig. 17). By increasing the carbon content, the tensile strength of the steel is increased up to 1040 MPa (150 ksi), yet the total elongation remains constant at 30%. This unusually good combination of tensile strength and total elongation is superior to that which can be obtained with conventional steels; it is associated with the high amounts of retained austenite, which vary from 10 to 20% as the carbon content is increased (Fig. 17). Producing such large amounts of austenite in these steels involves intercritical annealing followed by isothermal transformation in the bainite temperature range (Fig. 18). The high silicon content of these steel results in the displacement of almost all the carbon into the austenite and the suppression of cementite formation. As a result, the carbon content of the austenite at the isothermal transformation temperature is increased. This results in the stabilization of the austenite at room temperature, and ductility is enhanced because of the transformation of the metastable austenite upon plastic straining.

REFERENCES

1. G.R. Speich, Physical Metallurgy of Dual-Phase Steels, in *Fundamentals of Dual-Phase Steels*, The Metallurgical Society, 1981, p 1-45
2. I. Tamura, Y. Tomata, A. Okao, Y. Yamaoha, H. Ozawa, and S. Kaotoni, On the Strength and Ductility of Two-Phase Iron Alloys, *Trans Iron Steel Inst. Jpn.*, Vol 13, 1973, p 283-292
3. S. Hayami and T. Furakawa, A Family of High-Strength Cold-Rolled Steels, in *Microalloying 75*, Vanitech, 1975, p 78-87
4. M.S. Rashid, "A Unique High-Strength Sheet Steel with Superior Formability," Preprint 760206, Society of Automotive Engineers, 1976
5. *Formable HSLA and Dual-Phase Steels*, A.T. Davenport, Ed., American Institute of Mining, Metallurgical, and Petroleum Engineers, 1977
6. *Structure and Properties of Dual-Phase Steels*, R.A. Kot and J.W. Morris, Ed., American Institute of Mining, Metallurgical, and Petroleum Engineers, 1979
7. G.R. Speich, V.A. Demarest, and R.L. Miller, The Formation of Austenite During Intercritical Annealing of Dual-Phase Steels, *Metall. Trans. A*, Vol 12A, 1981, p 1419-1428
8. G.R. Speich and R.L. Miller, United States Steel Research Laboratory, private communication, 1975
9. P.R. Mould and C.C. Skena, Structure and Properties of Cold-Rolled Ferrite-Martensite (Dual-Phase) Steel Sheets, in *Formable HSLA and Dual-Phase Steels*, American Institute of Mining, Metallurgical, and Petroleum Engineers, 1979, p 183-205
10. J.M. Rigsbee and P.J. VanderArend, Laboratory Studies of Microstructures and Structure-Property Relationships in Dual-Phase Steels, in *Formable HSLA and Dual-Phase Steels*, American Institute of Mining, Metallurgical, and Petroleum Engineers, 1979, p 56-86
11. T. Furakawa, H. Morikawa, H. Takechi, and K. Koyama, Process Factors for Highly Ductile Dual-Phase Steel

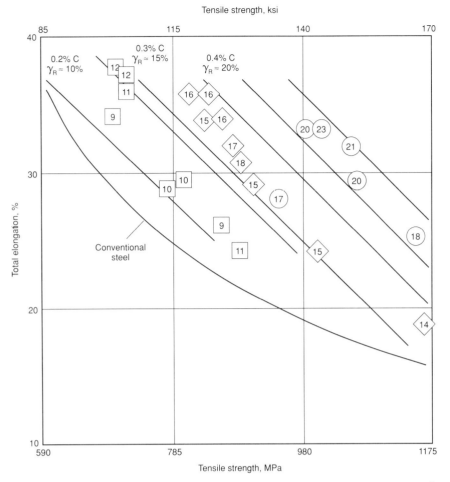

Fig. 17 Variation in total elongation and tensile strength of 0.2 to 0.4% C steels containing 1.2% Si and 1.2% Mn. Numbers in symbols indicate the percentage of retained austenite, γ_R. Source: Ref 17

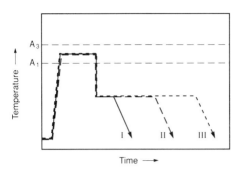

Fig. 18 Schematic of the heat treatment cycle used to produce ferrite-martensite-austenite-bainite steels. I, time required for a microstructure of martensite plus retained austenite; II, time required for a microstructure of retained austenite plus bainite; III, time required for a bainite microstructure. A_1, lower critical temperature; A_3, upper critical temperature. Source: Ref 18

Sheet, in *Structure and Properties of Dual-Phase Steels*, American Institute of Mining, Metallurgical, and Petroleum Engineers, 1979

12. D.K. Matlock, G. Krauss, L. Ramos, and G.S. Huppi, A Correlation of Processing Variables With Deformation Behavior of Dual-Phase Steels, in *Structure and Properties of Dual-Phase Steels*, American Institute of Mining, Metallurgical, and Petroleum Engineers, 1979, p 62-90

13. M. Hillert and M. Waldenstrom, Isothermal Sections of the Fe-Mn-C System in the Temperature Range 873-1373 K, *Calphad*, Vol 1, 1977, p 97-132

14. G. Thomas and J.Y. Koo, Developments in Strong, Duplex, Ferritic-Martensitic Steels, in *Structure and Properties of Dual-Phase Steels*, American Institute of Mining, Metallurgical, and Petroleum Engineers, 1979, p 183-201

15. G.R. Speich and R.L. Miller, Mechanical Properties of Ferrite-Martensite Steels, in *Structure and Properties of Dual-Phase Steels*, American Institute of Mining, Metallurgical, and Petroleum Engineers, 1979, p 145-182

16. T. Tanaka, M. Nishida, K. Nashiguchi, and T. Kato, Formation and Properties of Ferrite Plus Martensite Dual Phase Structures, in *Structure and Properties of Dual-Phase Steels*, American Institute of Mining, Metallurgical, and Petroleum Engineers, 1979, p 221-241

17. S. Tasuhara, M. Osamu, H. Takechi, Y. Ishii, and M. Usada, "Influence of Retained Austenite on the Bending Formability of Steel Sheet Containing Retained Austenite," Paper presented at the 114th Iron and Steel Institute of Japan Meeting, Tokyo, 1987

18. M. Matsumara, Y. Sakuma, and H. Takechi, Enhancement of Elongation by Retained Austenite in Intercritical Annealed 0.4C-1.5Si-0.8Mn Steel, *Trans. Iron Steel Inst. Jpn.*, Vol 27, 1987, p 570-579

SELECTED REFERENCES

- R.A. Kot and B.L. Bramfitt, Ed., *Fundamentals of Dual-Phase Steels*, American Institute of Mining, Metallurgical, and Petroleum Engineers, 1981

Ultrahigh-Strength Steels

Revised by Thoni V. Philip, TVP Inc.; and Thomas J. McCaffrey, Carpenter Steel Division of Carpenter Technology Corporation

STRUCTURAL STEELS with very high strength levels are often referred to as ultrahigh-strength steels. The designation ultrahigh-strength is arbitrary because no universally accepted strength level for the term has been established. Also, as structural steels with greater and greater strength have been developed, the strength range for which the term is applied has gradually increased. This article describes those commercial structural steels capable of a minimum yield strength of 1380 MPa (200 ksi).

In addition to the steels discussed in this article, many other proprietary and standard steels are used for essentially the same types of applications, but at strength levels slightly below the arbitrary, lower limit of 1380 MPa (200 ksi) established above for the ultrahigh-strength class of constructional steels. Medium-alloy steels such as 4330V and 4335V (vanadium-modified versions of the corresponding AISI standard steels) are among the more widely used steels for yield strengths of 1240 to 1380 MPa (180 to 200 ksi). Certain proprietary steels such as Hy Tuf (a silicon-modified steel similar to 300M) exhibit excellent toughness at strengths up to or slightly above 1380 MPa. The toughness of Hy Tuf is about the same as a maraging steel in this strength range. For properties and other information on steels and strength ranges not discussed here, the reader is referred to sources such as *Aerospace Structural Metals Handbook* (Ref 1) and to producer data sheets.

The ultrahigh-strength class of constructional steels is quite broad and includes several distinctly different families of steels. This article covers only medium-carbon low-alloy steels, medium-alloy air-hardening steels, and high fracture toughness steels. It does not cover 18Ni maraging steels, which are described in the article "Maraging Steels" in this Volume. Ultrahigh-strength steels of the stainless type (martensitic, martensitic precipitation hardenable, semiaustenitic precipitation hardenable, and cold-rolled austenitic steels) are covered in the Section "Specialty Steels and Heat-Resistant Alloys" in this Volume.

The effects of thermomechanical treatments such as ausforming and hot-cold working on the properties of many ultrahigh-strength steels have been investigated extensively. With adequate deformation while the steel is in a metastable condition, strength levels not attainable by standard quench and temper treatments have been obtained, quite often with higher ductility and fracture toughness than those normally expected at these very high strength levels. However, such thermomechanical treatments are not widely used commercially, presumably because of practical difficulties in adapting experimental techniques to actual production. Therefore, thermomechanical treatments and the properties that result from them are also excluded from this article.

Medium-Carbon Low-Alloy Steels

The medium-carbon low-alloy family of ultrahigh-strength steels includes AISI/SAE 4130, the higher-strength 4140, and the deeper hardening, higher-strength 4340. Several modifications of the basic 4340 steel have been developed. In one modification (300M), silicon content is increased to prevent embrittlement when the steel is tempered at the low temperatures required for very high strength. In AMS 6434, vanadium is added as a grain refiner to increase toughness, and the carbon is slightly reduced to promote weldability. Ladish D-6ac contains the grain refiner vanadium; slightly higher carbon, chromium, and molybdenum than 4340; and slightly lower nickel. Other less widely used steels that may be included in this family are 6150 and 8640. Chemical compositions are given in Table 1.

No new or distinctly different commercial steels have been added to this class of steels in recent years. Rather, developmental efforts have been primarily aimed at increasing ductility and toughness by improving melting and processing techniques and by using stricter process control and inspection. Steels with fewer and smaller nonmetallic inclusions and mill products with fewer internal and surface imperfections are produced by the use of selected raw materials as the melting charge and the employment of advanced melting techniques such as vacuum carbon deoxidation, vacuum degassing, electroslag remelting (ESR), vacuum arc remelting (VAR), and double vacuum melting (vacuum induction melting followed by vacuum arc remelting). These techniques yield less variation in properties from heat to heat and lot to lot; greater ductility and toughness, especially in the transverse direction; and greater in-service reliability.

Medium-carbon low-alloy ultrahigh-strength steels are readily hot forged, usually at temperatures ranging from 1065 to 1230 °C (1950 to 2250 °F). Specific forging temperatures are described in the subsequent sections for the various steels. To avoid stress cracks resulting from air hardening, the forged part should be slowly cooled in a furnace or embedded in lime, ashes, or other insulating material. Prior to machining, the usual practice is to normalize at 870 to 925 °C (1600 to 1700 °F) and temper at 650 to 675 °C (1200 to 1250 °F), or to anneal by furnace cooling from 815 to 845 °C (1500 to 1550 °F) to about 540 °C (1000 °F) if the steel is a deep air-hardening grade. These treatments impart moderately hard structures consisting of medium-to-fine pearlite. In this condition, the steel has a machinability rating of about one-half that of B1112. A very soft structure consisting of spheroidized carbides in a matrix of ferrite can be obtained by full annealing. Such a structure is not as well suited for machining as a normalized structure: The steel tends to tear, chips break away with difficulty, and metal tends to build up on the machining tool. However, the soft and ductile spheroidized structure is preferred for severe cold-forming operations such as spinning and deep drawing.

Medium-carbon low-alloy steels are cut, sheared, punched, and cold formed in the annealed condition. Cutting is commonly done with a saw or abrasive disk; if these steels are flame cut, most of them are preheated to about 315 °C (600 °F). After flame cutting, because the cut edge is hard, blanks

Ultrahigh-Strength Steels / 431

Table 1 Compositions of the ultrahigh-strength steels described in text

Designation or trade name	Composition, wt%(a)							
	C	Mn	Si	Cr	Ni	Mo	V	Co
Medium-carbon low-alloy steels								
4130	0.28–0.33	0.40–0.60	0.20–0.35	0.80–1.10	...	0.15–0.25	...	...
4140	0.38–0.43	0.75–1.00	0.20–0.35	0.80–1.10	...	0.15–0.25	...	...
4340	0.38–0.43	0.60–0.80	0.20–0.35	0.70–0.90	1.65–2.00	0.20–0.30	...	...
AMS 6434	0.31–0.38	0.60–0.80	0.20–0.35	0.65–0.90	1.65–2.00	0.30–0.40	0.17–0.23	...
300M	0.40–0.46	0.65–0.90	1.45–1.80	0.70–0.95	1.65–2.00	0.30–0.45	0.05 min	...
D-6a	0.42–0.48	0.60–0.90	0.15–0.30	0.90–1.20	0.40–0.70	0.90–1.10	0.05–0.10	...
6150	0.48–0.53	0.70–0.90	0.20–0.35	0.80–1.10	...	...	0.15–0.25	...
8640	0.38–0.43	0.75–1.00	0.20–0.35	0.40–0.60	0.40–0.70	0.15–0.25	...	...
Medium-alloy air-hardening steels								
H11 mod	0.37–0.43	0.20–0.40	0.80–1.00	4.75–5.25	...	1.20–1.40	0.40–0.60	...
H13	0.32–0.45	0.20–0.50	0.80–1.20	4.75–5.50	...	1.10–1.75	0.80–1.20	...
High fracture toughness steels								
AF1410(b)	0.13–0.17	0.10 max	0.10 max	1.80–2.20	9.50–10.50	0.90–1.10	...	13.50–14.50
HP 9-4-30(c)	0.29–0.34	0.10–0.35	0.20 max	0.90–1.10	7.0–8.0	0.90–1.10	0.06–0.12	4.25–4.75

(a) P and S contents may vary with steelmaking practice. Usually, these steels contain no more than 0.035 P and 0.040 S. (b) AF1410 is specified to have 0.008P and 0.005S composition. Ranges utilized by some producers are narrower. (c) HP 9-4-30 is specified to have 0.10 max P and 0.10 max S. Ranges utilized by some producers are narrower.

are annealed before being formed or machined.

Preferably, medium-carbon low-alloy steels are welded in the annealed or normalized condition and then heat treated to the desired strength. Such processes as inert-gas tungsten-arc, shielded metal-arc, inert-gas metal-arc, and pressure processes, as well as flash welding, can be used. Filler wire having the same composition as the base metal is preferred, but if such is not available, the filler wire should at least be of a composition that will produce a deposit that responds to heat treatment in approximately the same manner as the base metal. To avoid brittleness and cracking, preheating and interpass heating are used, and complex structures are stress relieved or hardened and tempered immediately after welding.

4130 Steel

AISI/SAE 4130 is a water-hardening alloy steel of low-to-intermediate hardenability. It retains good tensile, fatigue, and impact properties up to about 370 °C (700 °F); however, it has poor impact properties at cryogenic temperatures. This steel is not subject to temper embrittlement and can be nitrided. It usually is forged at 1100 to 1200 °C (2000 to 2200 °F); finishing temperature should never fall below 980 °C (1800 °F).

Available as billet, bar, rod, forgings, sheet, plate, tubing, and castings, 4130 steel is used to make automotive connecting rods, engine mounting lugs, shafts, fittings, bushings, gears, bolts, axles, gas cylinders, airframe components, hydraulic lines, and nitrided machinery parts.

Heat Treatments. The standard heat treatments that apply to 4130 steel are:

- *Normalize*: Heat to 870 to 925 °C (1600 to 1700 °F) and hold for a time period that depends on section thickness; air cool. Tempering at 480 °C (900 °F) or above is often done after normalizing to increase yield strength
- *Anneal*: Heat to 830 to 860 °C (1525 to 1575 °F) and hold for a time period that depends on section thickness or furnace load; furnace cool
- *Harden*: Heat to 845 to 870 °C (1550 to 1600 °F) and hold, then water quench; or heat to 860 to 885 °C (1575 to 1625 °F); hold and then oil quench. Holding time depends on section thickness
- *Temper*: Hold at least ½ h at 200 to 700 °C (400 to 1300 °F); air cool or water quench. Tempering temperature and time at temperature depend mainly on desired hardness or strength level
- *Spheroidize*: Heat to 760 to 775 °C (1400 to 1425 °F) and hold 6 to 12 h; cool slowly

Properties. Table 2 summarizes the typical properties obtained by tempering water-quenched and oil-quenched 4130 steel bars

Table 2 Typical mechanical properties of heat-treated 4130 steel

Tempering temperature		Tensile strength		Yield strength		Elongation in 50 mm (2 in.), %	Reduction in area, %	Hardness, HB	Izod impact energy	
°C	°F	MPa	ksi	MPa	ksi				J	ft·lbf
Water quenched and tempered(a)										
205	400	1765	256	1520	220	10.0	33.0	475	18	13
260	500	1670	242	1430	208	11.5	37.0	455	14	10
315	600	1570	228	1340	195	13.0	41.0	425	14	10
370	700	1475	214	1250	182	15.0	45.0	400	20	15
425	800	1380	200	1170	170	16.5	49.0	375	34	25
540	1000	1170	170	1000	145	20.0	56.0	325	81	60
650	1200	965	140	830	120	22.0	63.0	270	135	100
Oil quenched and tempered(b)										
205	400	1550	225	1340	195	11.0	38.0	450	...	...
260	500	1500	218	1275	185	11.5	40.0	440	...	...
315	600	1420	206	1210	175	12.5	43.0	418	...	...
370	700	1320	192	1120	162	14.5	48.0	385	...	...
425	800	1230	178	1030	150	16.5	54.0	360	...	...
540	1000	1030	150	840	122	20.0	60.0	305	...	...
650	1200	830	120	670	97	24.0	67.0	250	...	...

(a) 25 mm (1 in.) diam round bars quenched from 845 to 870 °C (1550 to 1600 °F). (b) 25 mm (1 in.) diam round bars quenched from 860 °C (1575 °F)

Table 3 Effects of mass on typical properties of heat-treated 4130 steel
Round bars oil quenched from 845 °C (1550 °F) and tempered at 540 °C (1000 °F)

Bar size(a)		Tensile strength		Yield strength		Elongation in 50 mm (2 in.), %	Reduction in area, %	Surface hardness, HB
mm	in.	MPa	ksi	MPa	ksi			
25	1	1040	151	880	128	18.0	55.0	307
50	2	740	107	570	83	20.0	58.0	223
75	3	710	103	540	78	22.0	60.0	217

(a) 12.83 mm (0.505 in.) diam tensile specimens were cut from center of 25 mm diam bar and from midradius of 50 and 75 mm diam bars.

4140 Steel

AISI/SAE 4140 is similar in composition to 4130 except for a higher carbon content. It is used in applications requiring a combination of moderate hardenability and good strength and toughness, but in which service conditions are only moderately severe. Because of its higher carbon content, 4140 steel has greater hardenability and strength than does 4130, but with some sacrifice in formability and weldability. Tensile strengths of up to 1650 MPa (240 ksi) are readily achieved in 4140 through conventional quench and temper heat treatments. This steel can be used at temperatures as high as 480 °C (900 °F), above which its strength decreases rapidly with increasing temperature. The material can be readily nitrided. Like other martensitic and ferritic steels, 4140 undergoes a transition from ductile to brittle behavior at low temperatures, the transition temperature varying with heat treatment and stress concentration. When 4140 is heat treated to high strength levels, it is subject to hydrogen embrittlement, such as that resulting from acid pickling or from cadmium or chromium electroplating. Ductility can be restored by baking for 2 to 4 h at 190 °C (375 °F).

The forging of 4140 steel can be done readily, usually at 1100 to 1200 °C (2000 to 2200 °F); the finishing temperature should not be below 980 °C (1800 °F). Parts should be cooled slowly after hot forming. This steel has good weldability using any of the standard welding methods. For welding, preheating at 150 to 260 °C (300 to 500 °F) and postheating at 600 to 675 °C (1100 to 1250 °F), followed by slow cooling, are recommended. Cold drawn 4140 has a machinability rating of 62% (B1112, 100%).

The steel 4140 is available as bar, rod, forgings, sheet, plate, strip, and castings. It is used for many high-strength machine parts (some of them nitrided) such as connecting rods, crankshafts, steering knuckles, axles, oil well drilling bits, piston rods, pump parts, high-pressure tubing, large industrial gears, flanges, collets, machine tool parts, wrenches, tong jaws, sprockets, and studs.

Heat Treatments. The standard heat treatments that apply to 4140 steel are:

- *Normalize*: Heat to 845 to 900 °C (1550 to 1650 °F) and hold for a time period that depends on section thickness; air cool
- *Anneal*: Heat to 845 to 870 °C (1550 to 1600 °F) and hold for a time period that depends on section thickness or furnace load; furnace cool
- *Harden*: Heat to 830 to 870 °C (1525 to 1600 °F) and hold; then oil quench. (For water quenching, which is rarely used, hardening temperatures are 815 to 845 °C, or 1500 to 1550 °F.) Holding time depends on section thickness
- *Temper*: Hold at least ½ h at 175 to 230 °C (350 to 450 °F) or 370 to 675 °C (700 to 1250 °F); air cool or water quench. Tempering temperature and time at temperature depend mainly on desired hardness. To avoid blue brittleness, 4140 is usually not tempered between 230 and 370 °C (450 and 700 °F); 4140 is not subject to temper embrittlement
- *Spheroidize*: Heat to 760 to 775 °C (1400 to 1425 °F) and hold 6 to 12 h; cool slowly

Properties. Table 4 summarizes the mechanical properties obtained by tempering oil-quenched 4140 steel at various temperatures. Because 4140 is not a deep-hardening steel, section size should be considered when specifying heat treatment, especially for high strength levels. The effects of mass on hardness and tensile properties are given in Table 5. As expected, 4140 steel has low impact strength at cryogenic temperatures.

4340 Steel

AISI/SAE 4340 steel is considered the standard by which other ultrahigh-strength steels are compared. It combines deep hardenability with high ductility, toughness, and strength. It has high fatigue and creep resistance. It is often used where severe service conditions exist and where high strength in heavy sections is required. In thin sections, this steel is air hardening; in practice, it is usually oil quenched. It is especially immune to temper embrittlement. It does not soften readily at elevated temperatures; that is, it exhibits good retention of strength. Hydrogen embrittlement is a problem for 4340 heat treated to tensile strengths greater than about 1400 MPa (200 ksi). Parts exposed to hydrogen, such as during pickling and plating, should be baked subsequently. This steel exhibits extremely poor resis-

Table 4 Typical mechanical properties of heat-treated 4140 steel
Round bars of 13 mm (½ in.) diameter, oil quenched from 845 °C (1550 °F)

Tempering temperature °C	°F	Tensile strength MPa	ksi	Yield strength MPa	ksi	Elongation in 50 mm (2 in.), %	Reduction in area, %	Hardness, HB	Izod impact energy J	ft · lbf
205	400	1965	285	1740	252	11.0	42	578	15	11
260	500	1860	270	1650	240	11.0	44	534	11	8
315	600	1720	250	1570	228	11.5	46	495	9	7
370	700	1590	231	1460	212	12.5	48	461	15	11
425	800	1450	210	1340	195	15.0	50	429	28	21
480	900	1300	188	1210	175	16.0	52	388	46	34
540	1000	1150	167	1050	152	17.5	55	341	65	48
595	1100	1020	148	910	132	19.0	58	311	93	69
650	1200	900	130	790	114	21.0	61	277	112	83
705	1300	810	117	690	100	23.0	65	235	136	100

Table 5 Effects of mass on typical properties of heat-treated 4140 steel

Diameter of bar(a) mm	in.	Tensile strength MPa	ksi	Yield strength MPa	ksi	Elongation in 50 mm (2 in.), %	Reduction in area, %	Surface hardness, HB
25	1	1140	165	985	143	15	50	335
50	2	920	133	750	109	18	55	302
75	3	860	125	655	95	19	55	293

(a) Round bars oil quenched from 845 °C (1550 °F) and tempered at 540 °C (1000 °F), 12.83 mm (0.505 in.) diam tensile specimens were cut from center of 25 mm (1 in.) diam bars and from midradius of 50 and 75 mm (2 and 3 in.) diam bars.

Table 6 Effects of mass on the mechanical properties of 4340 steel

Section diameter mm	in.	Tensile strength MPa	ksi	Yield strength MPa	ksi	Elongation in 50 mm (2 in.), %	Reduction in area, %	Hardness, HB
Oil quenched and tempered(a)								
13	½	1460	212	1380	200	13	51	...
38	1½	1450	210	1365	198	11	45	...
75	3	1420	206	1325	192	10	38	...
Water quenched and tempered(b)								
75	3	1055	153	930	135	18	52	340
100	4	1035	150	895	130	17	50	330
150	6	1000	145	850	123	16	44	322

(a) Austenitized at 845 °C (1550 °F); tempered at 425 °C (800 °F). (b) 75 mm (3 in.) diam bar austenitized at 800 °C (1475 °F); 100 and 150 mm (4 and 6 in.) diam bars austenitized at 815 °C (1500 °F). All sizes tempered at 650 °C (1200 °F). Test specimens taken at midradius. Source: Ref 2

Table 7 Typical mechanical properties of 4340 steel
Oil quenched from 845 °C (1550 °F) and tempered at various temperatures

Tempering temperature		Tensile strength		Yield strength		Elongation in 50 mm (2 in.), %	Reduction in area, %	Hardness		Izod impact energy	
°C	°F	MPa	ksi	MPa	ksi			HB	HRC	J	ft · lbf
205	400	1980	287	1860	270	11	39	520	53	20	15
315	600	1760	255	1620	235	12	44	490	49.5	14	10
425	800	1500	217	1365	198	14	48	440	46	16	12
540	1000	1240	180	1160	168	17	53	360	39	47	35
650	1200	1020	148	860	125	20	60	290	31	100	74
705	1300	860	125	740	108	23	63	250	24	102	75

Table 8 Notch toughness, fracture toughness, and K_{Iscc} for 4340 steel tempered to different hardnesses

Hardness, HB	Equivalent tensile strength(a)		Charpy V-notch impact energy		Plane-strain fracture toughness (K_{Ic})		K_{Iscc} in seawater	
	MPa	ksi	J	ft · lbf	MPa√m	ksi√in.	MPa√m	ksi√in.
550	2040	296	19	14	53	48	8	7
430	1520	220	30	22	75	68	30	27
380	1290	187	42	31	110	100	33	30

(a) Estimated from hardness

tance to stress-corrosion cracking when tempered to tensile strengths of 1500 to 1950 MPa (220 to 280 ksi). It can be readily nitrided, which often improves fatigue life.

The 4340 steel is usually forged at 1065 to 1230 °C (1950 to 2250 °F); after forging, parts may be air cooled in a dry place or, preferably, furnace cooled. The machinability rating of 4340 is 55% for cold-drawn material, and 45% for annealed material (cold-rolled B1112, 100%). A partly spheroidized structure obtained by normalizing and then tempering at 650 °C (1200 °F) is best for optimum machinability. The 4340 steel has good welding characteristics. It can be readily gas or arc welded, but welding rods of the same composition should be used. Because 4340 is air hardening, welded parts should be either annealed or normalized and tempered shortly after welding.

The steel 4340 is widely and readily available as billet, bar, rod, forgings, sheet, tubing, and welding wire. It is also produced as light plate and castings. Typical applications include bolts, screws, and other fasteners; gears, pinions, shafts, and similar machinery components; crankshafts and piston rods for engines; and landing gear and other critical structural members for aircraft.

Heat Treatments. The standard heat treatments that apply to 4340 steel are:

- *Normalize*: Heat to 845 to 900 °C (1550 to 1650 °F) and hold for a time period that depends on section thickness; air cool
- *Anneal*: Heat to 830 to 860 °C (1525 to 1575 °F) and hold for a time period that depends on section thickness or furnace load; furnace cool
- *Harden*: Heat to 800 to 845 °C (1475 to 1550 °F) and hold 15 min for each 25 mm (1 in.) of thickness (15 min, minimum); oil quench to below 65 °C (150 °F), or quench in fused salt at 200 to 210 °C (390 to 410 °F), hold 10 min, and then air cool to below 65 °C (150 °F)
- *Temper*: Hold at least ½ h at 200 to 650 °C (400 to 1200 °F); air cool. Temperature and time at temperature depend mainly on desired final hardness
- *Spheroidize*: The preferred schedule is to preheat to 690 °C (1275 °F) and hold 2 h, increase the temperature to 745 °C (1375 °F) and hold 2 h, cool to 650 °C (1200 °F) and hold 6 h, furnace cool to about 600 °C (1100 °F), and finally air cool to room temperature. An alternative schedule is to heat to 730 to 745 °C (1350 to 1375 °F), hold several hours, and then furnace cool to room temperature
- *Stress relieve*: After straightening, forming, or machining, parts may be stress relieved at 650 to 675 °C (1200 to 1250 °F)
- *Bake*: To avoid hydrogen embrittlement, plated parts must be baked at least 8 h at 185 to 195 °C (365 to 385 °F) as soon after plating as possible

Properties. Through-hardening of 4340 steel can be done by oil quenching, for round sections up to 75 mm (3 in.) in diameter, and by water quenching, for larger sections (up to the limit of hardenability). The influence of section size on tensile properties of oil-quenched and water-quenched 4340 is indicated in Table 6.

Variation in hardness of 4340 with the tempering temperature is shown in Fig. 1. Typical mechanical properties of oil-quenched 4340 are given in Table 7. Additional data on mechanical properties (impact strength, fracture toughness, and threshold stress intensity for stress-corrosion cracking) are given in Table 8.

Variations in tensile properties with test temperature are plotted in Fig. 2 and 3; low-temperature impact properties are plotted in Fig. 4.

Consumable electrode vacuum melting (commonly known as vacuum arc remelting, or VAR) and electroslag remelting

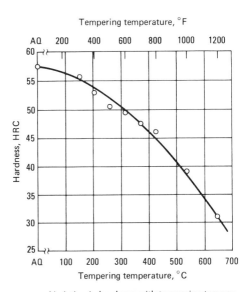

Fig. 1 Variation in hardness with tempering temperature for 4340 steel. All specimens oil quenched from 845 °C (1550 °F) and tempered 2 h at temperature. AQ, as quenched

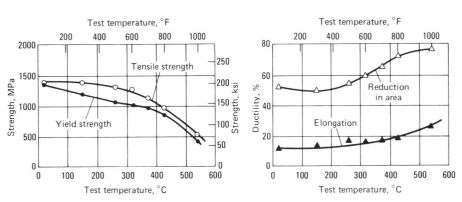

Fig. 2 Variation in tensile properties with temperature for 4340 steel. Properties determined using specimens heat treated to a room-temperature tensile strength of 1380 MPa (200 ksi). Source: Ref 1, 3

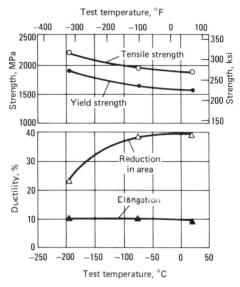

Fig. 3 Low-temperature tensile properties of 4340 steel. Properties determined for specimens oil quenched from 860 °C (1575 °F) and double tempered at 230 °C (450 °F). Source: Ref 1, 4

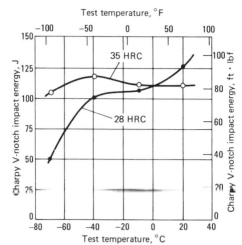

Fig. 4 Low-temperature toughness of 4340 steel. Determined for material heat treated to room-temperature hardnesses of 28 and 35 HRC

25 ppm O_2, and 100 ppm N_2; these can be reduced to about 0.9 ppm H_2, 4 ppm O_2, and 53 ppm N_2 by vacuum arc remelting. The remelted steels are more homogeneous than are the air-melted products. Table 9 compares average transverse tensile properties obtained on several heats of air melted and vacuum arc remelted 4340 steel tempered to different strength levels. Longitudinal and transverse mechanical properties, including R.R. Moore fatigue data and plane-strain fracture toughness data, of air-melted and vacuum arc remelted 4340 are compared in Table 10. Mechanical properties are significantly improved by VAR processing, especially in the transverse direction, although for plane-strain fracture toughness there appear to be no significant differences between longitudinal and transverse values for the same heat. Table 11 compares longitudinal properties of VAR- and ESR-processed 4340 bars. No significant differences in properties between VAR and ESR steels are evident; the two processes appear to give roughly equivalent improvements over air melting.

300M Steel

Alloy 300M is basically a silicon-modified (1.6% Si) 4340 steel, but it has slightly higher carbon and molybdenum contents and also contains vanadium. This steel exhibits deep hardenability and has ductility and toughness at tensile strengths of 1860 to 2070 MPa (270 to 300 ksi). Many of the properties of this steel are similar to those of 4340 steel, except that the increased silicon content provides deeper hardenability, increased solid-solution strengthening, and better resistance to softening at elevated temperatures. Compared to 4340 of similar strength, 300M can be tempered at a higher temperature, which provides greater relief of quenching stresses. The so-called 260 °C (500 °F) embrittlement is displaced to higher temperatures. Because of the high silicon and molybdenum contents, 300M is particularly prone to decarburization. During thermal processing, care should be exercised to avoid decarburization, or the decarburized layer should be removed after processing. When heat treated to strength levels higher than 1380 MPa (200 ksi), 300M is susceptible to hydrogen embrittlement. If the steel is properly baked after plating, the resulting improvement in properties is better than that for 4340 or D-6ac steel of equal strength.

The steel 300M is forged at 1065 to 1095 °C (1950 to 2000 °F). Forging should not be continued below 925 °C (1700 °F). After forging, it is preferred that parts be slowly cooled in a furnace, but they may be allowed to cool in air in a dry place. Although 300M can be readily gas or arc welded, welding is generally not recommended; welding rod of the same composition should be used. Because 300M is an air-hardening steel, parts should be either annealed or normalized and tempered after welding. The machinability rating of annealed 300M is about 45% (B1112, 100%). A partially spheroidized structure, obtained by normalizing and then tempering at 650 to 675 °C (1200 to 1250 °F), gives optimum machinability.

Table 9 Transverse tensile properties of air-melted and vacuum arc remelted 4340 steels
Properties listed are averages of several heats from the same producer; billet size and amount of hot reduction were not available.

Tempering temperature		Tensile strength		Yield strength		Elongation in 50 mm (2 in.), %	Reduction in area, %
°C	°F	MPa	ksi	MPa	ksi		
Air melted							
230	450	1945	282	1585	230	6.0	14.0
480	900	1380	200	1190	173	8.0	16.0
540	1000	1240	180	1125	163	10.0	22.0
Vacuum arc remelted							
230	450	1930	280	1635	237	6.5	17.0
480	900	1380	200	1210	175	9.0	20.0
540	1000	1240	180	1100	160	10.5	24.0

Table 10 Average mechanical properties of air-melted and vacuum arc remelted heats of 4340 steel
Hot reduced about 98% to round billets 100 to 115 mm (4 to 4½ in.) in diameter. Specimens were normalized at 900 °C (1650 °F), oil quenched from 845 °C (1550 °F), refrigerated, and tempered 2 h at 205 °C (400 °F).

Specimen direction	Tensile strength		Yield strength		Reduction in area, %	Plane-strain fracture toughness (K_{Ic})		Fatigue limit(a)	
	MPa	ksi	MPa	ksi		MPa√m	ksi√in.	MPa	ksi
Air melted									
Longitudinal	2005	291	1660	241	47.5	44.5	40.5	795	115
Transverse	2000	290	1655	240	8.9	45.8(b) 48.8(c)	41.7(b) 44.4(c)	540	78
Vacuum arc remelted									
Longitudinal	2035	295	1660	241	49.2	60.4	55.0	965	140
Transverse	2015	292	1650	239	40.2	61.5(b) 59.7(c)	56.0(b) 54.3(c)	715	104

(a) At 10^7 cycles. (b) WR orientation. (c) WW orientation. Source: Ref 5

Table 11 Longitudinal mechanical properties of bar stock made from remelted 4340 steel

Bars were normalized at 900 °C (1650 °F), oil quenched from 845 °C (1550 °F), and tempered 2 h at 541 °C (1005 °F). All specimens taken from midradius

Melting method	Tensile strength MPa	ksi	Yield strength MPa	ksi	Elongation in 4D, %	Reduction in area, %	Charpy V-notch impact energy at −12 °C (10 °F) J	ft · lbf	Hardness, HRC
Vacuum arc remelted(a)	1210	175	1120	163	16.4	61.2	65	48	37
Electroslag remelted(b)	1185	171	1090	158	16.1	59.0	64	47	37

(a) 91.9 mm (3.62 in.) round. (b) 117.4 mm (4.625 in.) round

Table 12 Typical mechanical properties of 300M steel

Round bars, 25 mm (1 in.) in diameter, oil quenched from 860 °C (1575 °F) and tempered at various temperatures

Tempering temperature °C	°F	Tensile strength MPa	ksi	Yield strength MPa	ksi	Elongation in 50 mm (2 in.), %	Reduction in area, %	Charpy V-notch impact energy J	ft · lbf	Hardness, HRC
90	200	2340	340	1240	180	6.0	10.0	17.6	13.0	56.0
205	400	2140	310	1650	240	7.0	27.0	21.7	16.0	54.5
260	500	2050	297	1670	242	8.0	32.0	24.4	18.0	54.0
315	600	1990	289	1690	245	9.5	34.0	29.8	22.0	53.0
370	700	1930	280	1620	235	9.0	32.0	23.7	17.5	51.0
425	800	1790	260	1480	215	8.5	23.0	13.6	10.0	45.5

Table 13 Effects of mass on tensile and impact properties of 300M steel

Round bars, normalized at 900 °C (1650 °F), oil quenched from 860 °C (1575 °F), and tempered at 315 °C (600 °F)

Bar diameter mm	in.	Tensile strength MPa	ksi	Yield strength MPa	ksi	Elongation in 50 mm (2 in.), %	Reduction in area, %	Charpy V-notch impact energy when tested at 21 °C (70 °F) J	ft · lbf	−46 °C (−50 °F) J	ft · lbf	−73 °C (−100 °F) J	ft · lbf
25	1	1990	289	1690	245	9.5	34.1	30	22	26	19	24	18
75	3	1940	281	1630	236	9.5	35.0	26	19	19	14	12	9
145	5¾	2120	308	1800	261	7.3	22.3	12	9	9	7	7	5

Table 14 Transverse tensile properties of air-melted and vacuum arc remelted 300M steel

Tempering temperature °C	°F	Tensile strength MPa	ksi	Yield strength MPa	ksi	Elongation in 50 mm (2 in.), %	Reduction in area, %
Air melted							
315	600	1960	284	1620	235	5.0	11
425	800	1760	255	1540	223	7.0	14
540	1000	1585	230	1480	215	9.0	22
Vacuum arc remelted							
260	500	2020	293	1620	235	7.0	25
425	800	1760	255	1550	225	10.0	34
540	1000	1585	230	1480	215	11.0	35

Typical applications of 300M, which is available as bar, sheet, plate, wire, tubing, forgings, and castings, are aircraft landing gear, airframe parts, fasteners, and pressure vessels.

Heat Treatments. The standard heat treatments that apply to 300M steel are:

- *Normalize*: Heat to 915 to 940 °C (1675 to 1725 °F) and hold for a time period that depends on section thickness; air cool. If normalizing to enhance machinability, charge into a tempering furnace at 650 to 675 °C (1200 to 1250 °F) before the steel reaches room temperature
- *Harden*: Austenitize at 860 to 885 °C (1575 to 1625 °F). Oil quench to below 70 °C (160 °F) or quench in salt at 200 to 210 °C (390 to 410 °F), hold 10 min, and then air cool to 70 °C (160 °F) or below
- *Temper*: Hold at 2 to 4 h at 260 to 315 °C (500 to 600 °F). Double tempering is recommended. This tempering procedure produces the best combination of high yield strength and high impact properties. Tempering outside this temperature range results in severe deterioration of properties
- *Spheroidize*: Heat to about 775 °C (1430 °F) and hold for a time period that depends on section thickness or furnace load. Cool to 650 °C (1200 °F) at a rate no faster than 5.5 °C/h (10 °F/h); then cool to 480 °C (900 °F) no faster than 10 °C/h (20 °F/h); finally, air cool to room temperature. The same schedule is recommended for annealing

Properties. Variations in hardness and mechanical properties of 300M with tempering temperature are presented in Table 12. Because 300M has deep hardenability, heat-treated bars 75 mm (3 in.) in diameter have essentially the same tensile properties as do bars 25 mm (1 in.) in diameter. Reductions in tensile ductility and impact strength, however, are observed in heat-treated bars 145 mm (5¾ in.) in diameter. Variations in properties of 300M (including impact strength at room and low temperatures) with section size are presented in Table 13.

In R.R. Moore fatigue tests of polished specimens, endurance limits of about 800 MPa (116 ksi) and 585 MPa (85 ksi) were found for longitudinal and transverse specimens, respectively, of air-melted 300M heat treated to a tensile strength of about 2025 MPa (294 ksi) by oil quenching from 870 °C (1600 °F) and tempering at 290 °C (550 °F). Vacuum arc remelting and electroslag remelting improve transverse ductility and impact strength by producing a cleaner microstructure. Transverse tensile properties of air-melted and vacuum arc remelted 300M are compared in Table 14. The data are average values from several heats.

In another study, transverse tensile properties were determined for specimens of 300M taken from 40 billets, each 125 mm (5 in.) square, representing seven vacuum arc remelted heats. Specimens were normalized for 1 h at 925 °C (1700 °F), reheated 1 h at 870 °C (1600 °F), oil quenched, and tempered 4 h at 315 °C (600 °F). Average properties were as follows: tensile strength, 1978 MPa (286.9 ksi); yield strength, 1671 MPa (242.4 ksi); elongation, 9.3%; and reduction in area, 36.6%. Property ranges were: tensile strength, 1896 to 2039 MPa (275.0 to 295.7 ksi); yield strength, 1581 to 1752 MPa (229.3 to 254.1 ksi); elongation, 9 to 10%; and reduction in area, 31.7 to 45.0%.

Samples from two heats of VAR 300M steel heat treated to an average tensile strength of 2025 MPa (294 ksi), average yield strength of 1710 MPa (248 ksi), average elongation of 11.5%, average reduction in area of 45.6%, and average Charpy V-notch impact value of 27 J (20 ft · lbf) showed an average K_{Ic} value of 60.2 MPa$\sqrt{m}$ (54.8 ksi$\sqrt{in.}$) and an average K_{Iscc} value of 15.9 MPa$\sqrt{m}$ (14.5 ksi$\sqrt{in.}$) in a 3.5% NaCl solution (Ref 6). Table 15 presents longitudinal and transverse tensile and plane-strain fracture toughness data for both air-melted and vacuum arc remelted 300M steel. R.R. Moore fatigue test results on transverse specimens indicate fatigue

Table 15 Average mechanical properties of air-melted and vacuum arc remelted heats of 300M steel

Hot reduced 96 to 98% to round-cornered square billets, about 100 × 100 mm (4 × 4 in.). Specimens were normalized at 925 °C (1700 °F), oil quenched from 870 °C (1600 °F), refrigerated, and double tempered, 2 + 2 h at 315 °C (600 °F).

Specimen direction	Tensile strength MPa	ksi	Yield strength MPa	ksi	Reduction in area, %	Plane-strain fracture toughness (K_{Ic}) MPa√m	ksi√in.
Air melted							
Longitudinal	2095	304	1805	262	44.8	49.3	44.9
Transverse	2035	295	1750	254	23.6	58.7(a)	53.4(a)
						61.4(b)	55.9(b)
Vacuum arc remelted							
Longitudinal	2080	302	1785	259	47.8	57.4	52.2
Transverse	2015	292	1760	255	33.6	64.0(a)	58.3(a)
						61.4(b)	55.9(b)

(a) WR orientation. (b) WW orientation

limits of about 580 MPa (84 ksi) for air-melted 300M and about 675 MPa (98 ksi) for VAR 300M; the latter represents an improvement of about 17% over air-melted 300M. In the same study, there was no significant difference in transverse fatigue limit between VAR 300M and ESR 300M that had been hot reduced only 75% from the ingot. In all these studies, remelted steels (whether VAR or ESR) had better ductility and toughness, particularly in the transverse direction, although there did not appear to be any significant difference between longitudinal and transverse plane-strain fracture toughness values.

D-6a and D-6ac Steel

Ladish D-6a is a low-alloy ultrahigh-strength steel developed for aircraft and missile structural applications. It is designed primarily for use at room-temperature tensile strengths of 1800 to 2000 MPa (260 to 290 ksi). The steel D-6a maintains a very high ratio of yield strength to tensile strength up to a tensile strength of 1930 MPa (280 ksi), combined with good ductility. It has good notch toughness, which results in high resistance to impact loading. It is deeper hardening than 4340 and does not exhibit temper embrittlement. It retains high strength at elevated temperature. Susceptibility of D-6a to stress-corrosion cracking and corrosion fatigue in moist and aqueous environments is comparable to that of 300M steel at the same strength level. The alloy is called D-6a when produced by air melting in an electric furnace and D-6ac when produced by air melting followed by vacuum arc remelting. The mechanical properties of D-6a and D-6ac differ somewhat as a result of the differences in melting practices. Other characteristics of the two steels, including processing behavior, are identical.

D-6a and D-6ac are available as bar, rod, billet, and forgings and can be made as flat-rolled products (sheet and plate) as well. These forms are used in landing gear and critical structural components for aircraft, motor cases for solid-fuel rockets, shafts, gears, springs, dies, dummy blocks, and backer blocks.

Processing. To forge D-6a, it should be heated to a maximum temperature of 1230 °C (2250 °F); forging should be finished above 980 °C (1800 °F). Finished forgings should be cooled slowly, either in a furnace or embedded in a suitable insulating medium such as ashes or lime. For maximum machinability, the parts should be charged into a 650 °C (1200 °F) furnace immediately after forging and held for 12 h; temperature should be increased to 900 °C (1650 °F) and held for a time period that depends on section size; parts should be cooled to 650 °C (1200 °F), held 10 h, and finally air cooled to room temperature.

The material D-6a, even in heavy sections, can be welded provided that the techniques and controls normally employed for welding medium-carbon, high hardenability alloy steels are used. Welding rod of the same composition should be used. For critical applications, gas tungsten arc welding is preferred; filler metal wire should be of vacuum-melted stock containing less carbon than is in the base metal and minimum amounts of phosphorus, sulfur, and dissolved gases. Welds made in this manner will have higher toughness than that of the base metal, but slightly lower strength. Preheat and interpass temperatures of 230 to 290 °C (450 to 550 °F) are recommended. Highly restrained weldments should be postheated 1½ h at 300 to 330 °C (575 to 625 °F) and cooled in still air. When the weldments reach 150 °C (300 °F), they should be charged immediately into a furnace for stress relieving at 650 to 700 °C (1200 to 1300 °F).

Annealed D-6a has a machinability rating of 50 to 55% (B1112, 100%). When the steel is to be severely cold formed, it is usually normalized and then spheroidized before working.

Heat Treatments. The standard heat treatments that apply to D-6a and D-6ac steels are:

- *Normalize*: Heat to 870 to 955 °C (1600 to 1750 °F) and hold for a time period that depends on section thickness; air cool
- *Anneal*: Heat to 815 to 845 °C (1500 to 1550 °F) and hold for a time period that depends on section thickness or furnace load; furnace cool to 540 °C (1000 °F) at a rate no greater than 28 °C/h (50 °F/h); and then air cool to room temperature
- *Harden*: Austenitize at 845 to 900 °C (1550 to 1650 °F) for ½ to 2 h. Sections no thicker than 25 mm (1 in.) may be air cooled. Sections thicker than 25 mm (1 in.) may be oil quenched to 65 °C (150 °F) or salt quenched to 205 °C (400 °F) and then air cooled. For optimum dimensional stability, "aus-bay" quench in a furnace at 525 °C (975 °F), equalize the temperature, and then quench in an oil bath held at 60 °C (140 °F) or quench in 205 °C (400 °F) salt and air cool. The cooling rate during quenching significantly affects fracture toughness. For high fracture toughness, especially in heavy sections, austenitize at 925 °C (1700 °F), aus-bay quench to 525 °C (975 °F), equalize, and oil quench to 60 °C (140 °F)
- *Temper*: Immediately after hardening, hold 2 to 4 h in the range of 200 to 700 °C (400 to 1300 °F), depending on desired strength or hardness
- *Spheroidize*: Heat to 730 °C (1350 °F) and hold 5 h; increase temperature to 760 °C (1400 °F) and hold 1 h; furnace cool to 690 °C (1275 °F) and hold 10 h; furnace cool to 650 °C (1200 °F) and hold 8 h; air cool to room temperature
- *Stress relieve*: Heat to an appropriate temperature in the range of 540 to 675 °C (1000 to 1250 °F) and hold for 1 to 2 h; air cool

Properties. The effect of tempering temperature on typical room-temperature hardness of D-6a steel bar is shown in Fig. 5; other mechanical properties of D-6a bar are given in Table 16. Tensile properties of heat-treated D-6ac billet material are given in Table 17. D-6a maintains impact resistance to very low temperatures (Fig. 6). Typical elevated-temperature tensile properties of D-6a are presented in Table 18, and in Table 19 room-temperature tensile properties after 10 h and 100 h exposures at 540 °C (1000 °F) are compared with properties of unexposed material. Data on stress-rupture life at 480 and 540 °C (900 and 1000 °F) are presented in Fig. 7. The effects of tempering temperature on smooth-bar and notched-bar tensile strengths are plotted in Fig. 8. As mentioned in the section on heat treatment, the rate of cooling during quenching has a significant effect on fracture toughness. When the steel is heat treated for high fracture toughness (as outlined above), a K_{Ic} value of 99 to 104 MPa√m (90 to 95 ksi√in.) can be obtained at a tensile strength of about 1650 MPa (240 ksi) (Ref 7).

Table 16 Typical mechanical properties of D-6a steel bar
Normalized at 900 °C (1650 °F), oil quenched from 845 °C (1550 °F), and tempered at various temperatures

Tempering temperature		Tensile strength		Yield strength		Elongation in 50 mm (2 in.), %	Reduction in area, %	Charpy V-notch impact energy	
°C	°F	MPa	ksi	MPa	ksi			J	ft · lbf
150	300	2060	299	1450	211	8.5	19.0	14	10
205	400	2000	290	1620	235	8.9	25.7	15	11
315	600	1840	267	1700	247	8.1	30.0	16	12
425	800	1630	236	1570	228	9.6	36.8	16	12
540	1000	1450	210	1410	204	13.0	45.5	26	19
650	1200	1030	150	970	141	18.4	60.8	41	30

Table 17 Tensile properties of double-tempered D-6ac billet material
Austenitized 1 h at 900 °C (1650 °F), quenched in fused salt at 205 °C (400 °F) for 5 min, then air cooled to room temperature. Tempered 1 h at 205 °C; second temper, 4 h at indicated temperature

Second tempering temperature		Tensile strength		Yield strength		Elongation in 50 mm (2 in.), %	Reduction in area, %
°C	°F	MPa	ksi	MPa	ksi		
480	900	1686.5	244.6	1540.3	223.4	11.1	40.0
510	950	1652.7	239.7	1519.7	220.4	13.2	44.1
540	1000	1613.4	234.0	1483.8	215.2	13.7	47.2

Table 18 Tensile properties of D-6a steel at elevated temperatures
Normalized at 900 °C (1650 °F), oil quenched from 845 °C (1550 °F), and tempered 28 °C (50 °F) above indicated test temperature

Test temperature		Tensile strength		Yield strength		Elongation in 50 mm (2 in.), %	Reduction in area, %
°C	°F	MPa	ksi	MPa	ksi		
260	500	1839.6	266.8	1294.9	187.8	15.2	55.0
315	600	1629.3	236.3	1256.3	182.2	15.8	57.4
370	700	1430.0	207.4	1164.6	168.9	14.7	55.0
425	800	1277.6	185.3	1096.3	159.0	15.2	55.0
480	900	1158.4	168.0	976.3	141.6	16.2	57.5
540	1000	957.7	138.9	831.5	120.6	19.8	64.5
595	1100	497.1	72.1	395.8	57.4	41.7	86.5

Table 19 Room-temperature properties of D-6a steel after various times at 540 °C (1000 °F)
Steel normalized at 900 °C (1650 °F), oil quenched from 845 °C (1550 °F), and tempered at 565 °C (1050 °F)

Time at 540 °C (1000 °F), h	Tensile strength		Yield strength		Elongation in 50 mm (2 in.), %	Reduction in area, %
	MPa	ksi	MPa	ksi		
0	1410	204	1340	195	14.8	50.5
10	1410	204	1330	193	14.5	52.0
100	1330	193	1260	183	14.8	51.0

Table 20 Typical fatigue limits for D-6a and D-6ac steel

Steel	Tempering temperature		Hardness, HRC	Fatigue limit(a)	
	°C	°F		MPa	ksi
D-6a(b)	540	1000	45.0	760	110
			46.0	750	108
			45.0	740	107
D-6ac(b)	575	1075	42.0	780	113
			42.5	795	115
			41.5	760	110
			43.5	750	108
D-6a(c)	315	600	50.0	740	107

(a) At 10^6 stress cycles in rotating-beam tests using polished specimens. (b) 1520 MPa (220 ksi) tensile strength. (c) 1790 MPa (260 ksi) tensile strength

Table 21 Room-temperature and elevated-temperature fatigue limits for D-6ac steel
Smooth specimens heat treated to a tensile strength of 1860 MPa (270 ksi); test speed, 186 kHz

Test temperature		Tension-tension(a)		Tension-compression(b)	
°C	°F	MPa	ksi	MPa	ksi
24	75	1035	150	690	100
232	450	930	135	550	80
288	550	930	135	575	75

(a) Mean stress equal to alternating stress ($R = 0$). (b) Mean stress equal to zero ($R = -1$)

The alloy is susceptible to stress-corrosion cracking; the K_{Iscc} value, in both water and 3.5% NaCl solution, appears to be less than 16 MPa $\sqrt{m}$ (15 ksi $\sqrt{in.}$). The steel has high fatigue strength; Table 20 presents data for both D-6a and D-6ac derived from rotating-beam fatigue tests, and Table 21 gives data for rapid-cycle (about 186 kHz) tension-tension and tension-compression fatigue tests at room and elevated temperatures.

6150 Steel

AISI/SAE 6150 is a tough, shock-resisting, shallow-hardening chromium-vanadium steel with high fatigue and impact resistance in the heat-treated condition. It can be nitrided for maximum surface hardness and abrasion resistance; nitriding characteristics are similar to those of 4140 and 4340 steels. The steel 6150 may be forged from temperatures up to 1200 °C (2200 °F), but the usual temperature range is 1175 to 950 °C (2150 to 1750 °F). Parts made of 6150 steel can readily be welded using any of the standard welding methods. After welding, parts should be normalized and then hardened and tempered to the desired hardness.

For best machinability, 6150 should be in the annealed condition. Machinability rating is about 55% (B1112, 100%). As with other low-alloy steels of about the same carbon content, the optimum microstructure for machining is coarse lamellar pearlite and/or coarse spheroidite. Chips are continuous and springy, which can make the steel difficult to machine.

Available as bar, rod, plate, sheet, strip, wire, and tubing, 6150 steel may be forged

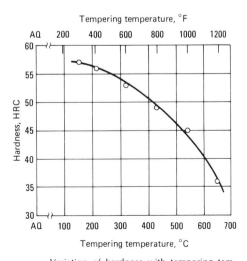

Fig. 5 Variation of hardness with tempering temperature for D-6a steel. All specimens oil quenched from 845 °C (1550 °F) and tempered 2 h at temperature. AQ, as quenched

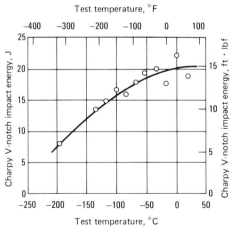

Fig. 6 Low-temperature toughness of a D-6a steel. All specimens heat treated to a room-temperature tensile strength of 1790 to 1860 MPa (260 to 270 ksi). Each data point is the average of three tests.

438 / Carbon and Low-Alloy Steels

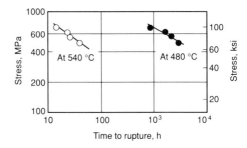

Fig. 7 Stress-rupture life for a D-6a steel. Determined for material heat treated to a room-temperature tensile strength of 1380 to 1520 MPa (200 to 220 ksi)

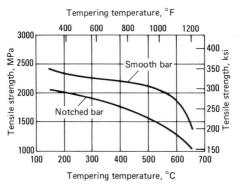

Fig. 8 Variation in smooth-bar and notched bar tensile strengths with tempering temperature for D-6a steel. Specimens oil quenched from 845 °C (1550 °F) and tempered 2 h at temperature. Notched bars had a stress-concentration factor, K_t, of 4.2.

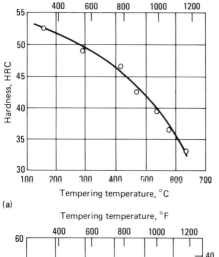

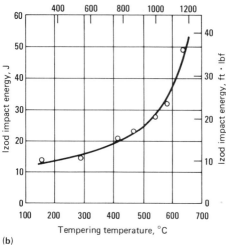

Fig. 9 Variation in (a) hardness and (b) impact energy with tempering temperature for 6150 steel. Specimens oil quenched from 885 °C (1625 °F) and tempered 2 h at temperature

or cast into shapes. Typical applications include gears, pinions, springs (both coiled and flat), shafts, axles, heavy duty pins, bolts, and machinery parts.

Heat Treatments. The heat treatments that apply to 6150 steel are:

- *Normalize*: Heat to 870 to 955 °C (1600 to 1750 °F) and hold for a time period that depends on section thickness; air cool
- *Anneal*: Heat to 845 to 900 °C (1550 to 1650 °F) and hold for a time period that depends on section thickness or furnace load; furnace cool
- *Harden*: Austenitize at 845 to 900 °C (1550 to 1650 °F); oil quench
- *Temper*: Hold at least ½ h at 200 to 650 °C (400 to 1200 °F). Tempering temperature and time at temperature primarily depend on desired final hardness
- *Austemper*: Austenitize in a salt bath at 845 to 900 °C (1550 to 1650 °F); quench in a salt bath at 230 to 315 °C (450 to 600 °F), hold 20 to 30 min, and quench or air cool to room temperature
- *Martemper*: Austenitize in a salt bath at 845 to 870 °C (1550 to 1600 °F); quench in a salt bath at 230 to 260 °C (450 to 500 °F), equalize, and then air cool or quench to room temperature. Temper to desired hardness
- *Spheroidize*: Heat to 800 to 830 °C (1475 to 1525 °F), hold until heated through, furnace cool to 650 °C (1200 °F), and hold several hours; then cool slowly to room temperature

Properties. Typical properties of small-diameter round sections of 6150 tempered at various temperatures are given in Table 22. Variations in hardness and Izod impact energy with tempering temperature are plotted in Fig. 9. The effects of section size on tensile properties and hardness are presented in Table 23.

8640 Steel

AISI/SAE 8640 was especially designed to provide the maximum hardenability and best combination of properties possible with minimum alloying additions. The material 8640 is an oil-hardening steel, but may be water hardened if precautions are taken to prevent cracking. It exhibits properties similar to those of 4340, except that its strength in large sections is not as high.

This steel is available as billets, bars, rods, forgings and castings. It is used to make gears, pinions, shafts, axles, studs, fasteners, machinery parts, and forged hand tools.

Processing. The 8640 steel may be forged at temperatures up to 1200 °C (2200 °F), but is usually forged in the range from 1175 to 950 °C (2150 to 1750 °F). Forged parts are cooled slowly from the forging temperature, then annealed prior to machining.

This steel can be welded by any of the standard welding methods. Because 8640 has some air-hardening tendencies, preheating to 150 to 260 °C (300 to 500 °F) before welding and postheating after welding are recommended. Stress relieving at 600 to 650 °C (1100 to 1200 °F) is quite satisfactory for most welded parts.

Cold-drawn 8640 has a machinability rating of 64% (B1112, 100%). Annealing prior to cold drawing can improve machinability by about 10%.

Table 22 Room-temperature tensile properties of heat-treated 6150 steel

Tempering temperature		Tensile strength		Yield strength		Elongation in 50 mm (2 in.), %	Reduction in area, %	Hardness, HB	Izod impact energy	
°C	°F	MPa	ksi	MPa	ksi				J	ft · lbf
Round bars, 14 mm (0.55 in.) in diameter(a)										
205	400	2050	298	1810	263	1	5	610	...	...
260	500	2070	300	1810	263	4	12	570	...	...
315	600	1950	283	1720	250	7	27	540	...	...
370	700	1770	257	1620	235	10	37	505	9	7
425	800	1585	230	1490	216	11	42	470	14	10
480	900	1410	204	1340	195	12	44	420	16	12
540	1000	1250	182	1210	175	13	46	380	20	15
595	1100	1150	167	1080	157	16	47	350	28	21
Round bars, 25 mm (1 in.) in diameter(b)										
425	800	1570	228	1450	210	10	37	461	...	...
480	900	1360	197	1210	175	11	41	401	...	...
540	1000	1180	171	1030	150	12	45	341	...	...
595	1100	1030	150	875	127	15	50	302	...	...
650	1200	920	133	760	110	19	55	262	...	...
705	1300	810	118	660	96	23	61	235	...	...

(a) Normalized at 870 °C (1600 °F), oil quenched from 860 °C (1575 °F), and tempered at various temperatures. (b) Oil quenched from 860 °C and tempered at various temperatures

Table 23 Effects of mass on properties of heat-treated 6150 steel

Specimen size(a)		Tensile strength		Yield strength		Elongation in 50 mm (2 in.), %	Reduction in area, %	Hardness, HB
mm	in.	MPa	ksi	MPa	ksi			
25	1	1185	172	1040	151	14	45	341
50	2	1170	170	1030	149	13	48	341
75	3	1090	158	950	138	13	47	331

(a) Round bars, oil quenched from 830 °C (1525 °F) and tempered at 540 °C (1000 °F); 12.83 mm (0.505 in.) diam tensile specimens taken from center of 25 mm (1 in.) bars and from midradius of 50 and 75 mm (2 and 3 in.) diam bars

Heat Treatments. The heat treatments that apply to 8640 steel are:

- *Normalize*: Heat to 870 to 925 °C (1600 to 1700 °F) and hold for a time period that depends on section thickness; air cool
- *Anneal*: Heat to 845 to 870 °C (1550 to 1600 °F) and hold for a time period that depends on section thickness or furnace load; furnace cool
- *Harden*: Austenitize at 815 to 845 °C (1500 to 1550 °F); quench in oil or water
- *Temper*: Hold at least ½ h at 200 to 650 °C (400 to 1200 °F)
- *Spheroidize*: Heat to 705 to 720 °C (1300 to 1325 °F) and hold several hours; furnace cool

Properties. Variations in properties of heat-treated round sections of 8640 with tempering temperature are given in Table 24. Variations in properties with section size (mass effect) are given in Table 25. Impact behavior is illustrated in Fig. 10.

Medium-Alloy Air-Hardening Steels

The ultrahigh-strength steels H11 modified (H11 mod) and H13, which are popularly known as 5% Cr hot-work die steels, are discussed in this section. Besides being extensively used in dies, these steels are widely used for structural applications, but not as widely as they once were, primarily because of the development of several other steels at essentially the same cost but with substantially greater fracture toughness at equivalent strength. Nonetheless, H11 mod and H13 possess some attractive features. Both can be hardened through in large sections by air cooling. The chemical compositions of these steels are given in Table 1.

H11 Modified

This steel is a modification of the martensitic hot-work die steel AISI H11, the significant difference being a slightly higher carbon content. The H11 mod steel can be heat treated to strengths exceeding 2070 MPa (300 ksi). It is air hardened, which results in minimal residual stress after hardening. Because H11 mod is a secondary hardening steel, it develops optimum properties when tempered at temperatures above 510 °C (950 °F). The high tempering temperatures used for this steel provide substantial stress relief and stabilization of properties so that the material can be used to advantage at elevated temperatures. This also enables heat-treated parts to be warm worked at temperatures as much as 55 °C (100 °F) below the prior tempering temperature or to be preheated for welding. At high strength levels (those exceeding 1800 MPa, or 260 ksi), H11 mod has good ductility, impact strength, notch toughness, and fatigue life, as well as high creep and rupture strength, at temperatures up to about 650 °C (1200 °F). It is used for parts requiring maximum levels of strength, ductility, toughness, fatigue resistance, and thermal stability at temperatures between −75 and 540 °C (−100 and 1000 °F). At elevated temperatures, parts should be protected from corrosion (oxidation) by an appropriate surface treatment. The material H11 mod has good formability in the annealed condition and is readily welded. It is subject to hydrogen embrittlement. Its fracture toughness is rather low; if it is used in critical applications at yield strengths above 1380 MPa (200 ksi), care should be taken to eliminate small discontinuities.

The H11 mod steel is available as bar, billet, rod, wire, plate, sheet, strip, forgings, and extrusions. It is used for parts requiring high strength combined with either strength retention at elevated temperatures or moderate corrosion resistance. Typical applications include aircraft landing gear components, airframe components, internal parts for steam and gas turbines, fasteners, springs, and hot-work dies.

Parts to be used at elevated temperatures are commonly protected by nickel-cadmium plating. If such plating is done, baking to avoid hydrogen-induced delayed cracking is recommended. Alternatively, part surfaces may be protected from oxidation by hot dipping in aluminum or by applying a heat-resistant paint.

Processing. The material H11 mod is readily forged from 1120 to 1150 °C (2050 to 2100 °F). Preferably, stock should be preheated at 790 to 815 °C (1450 to 1500 °F) and then heated uniformly to the forging tem-

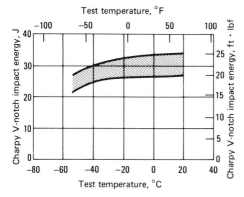

Fig. 10 Low-temperature toughness of 8640 steel. Determined for material oil quenched and tempered to room-temperature hardness of 401 to 415 HB

Table 24 Typical room-temperature mechanical properties of 8640 steel

Tempering temperature		Tensile strength		Yield strength		Elongation in 50 mm (2 in.), %	Reduction in area, %	Impact energy		Hardness	
°C	°F	MPa	ksi	MPa	ksi			J	ft · lbf	HB	HRC
Round bars, 13.5 mm (0.53 in.) in diameter(a)											
205	400	1810	263	1670	242	8.0	25.8	11.5(b)	8.5(b)	555	55
315	600	1585	230	1430	208	9.0	37.3	15.6(b)	11.5(b)	461	48
425	800	1380	200	1230	179	10.5	46.3	27.8(b)	20.5(b)	415	44
540	1000	1170	170	1050	152	14.0	53.3	56.3(b)	41.5(b)	341	37
650	1200	870	126	760	110	20.5	61.0	96.9(b)	71.5(b)	269	28
Round bars, 25 mm (1 in.) in diameter(a)											
425	800	1382	200.5	1230	179	10	46	27(c)	20(c)	415	44
480	900	1250	181	1120	162	13	51	41(c)	30(c)	388	42
540	1000	1070	155	940	137	17	56	54(c)	40(c)	331	36
595	1100	1020	148	910	132	16	57	73(c)	54(c)	302	32
650	1200	865	125.5	760	110.5	20	61	83(c)	61(c)	269	28

(a) Oil quenched from 830 °C (1525 °F) and tempered at indicated temperature. (b) Izod. (c) Charpy V-notch

Table 25 Effects of mass on properties of heat-treated 8640 steel

Bar size(a)		Tensile strength		Yield strength		Elongation in 50 mm (2 in.), %	Reduction in area, %	Surface hardness, HB
mm	in.	MPa	ksi	MPa	ksi			
25	1	1070	155	940	137	17	56	331
50	2	910	132	770	112	18	57	293
75	3	860	125	710	103	19	58	277

(a) Oil quenched from 830 °C (1525 °F) and tempered at 540 °C (1000 °F)

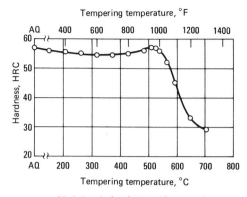

Fig. 11 Variation in hardness with tempering temperature for H11 mod steel. All specimens air cooled from 1010 °C (1850 °F) and double tempered, 2 + 2 h at temperature. AQ, as quenched

Table 26 Typical longitudinal mechanical properties of H11 mod steel
Air cooled from 1010 °C (1850 °F); double tempered, 2 + 2 h at indicated temperature

Tempering temperature		Tensile strength		Yield strength		Elongation in 50 mm (2 in.), %	Reduction in area, %	Charpy V-notch impact energy		Hardness, HRC
°C	°F	MPa	ksi	MPa	ksi			J	ft · lbf	
510	950	2120	308	1710	248	5.9	29.5	13.6	10.0	56.5
540	1000	2005	291	1675	243	9.6	30.6	21.0	15.5	56.0
565	1050	1855	269	1565	227	11.0	34.5	26.4	19.5	52.0
595	1100	1540	223	1320	192	13.1	39.3	31.2	23.0	45.0
650	1200	1060	154	855	124	14.1	41.2	40.0	29.5	33.0
705	1300	940	136	700	101	16.4	42.2	90.6	66.8	29.0

Table 27 Typical short-time elevated-temperature properties of H11 mod steel
Longitudinal specimens taken from bar stock air cooled from 1010 °C (1850 °F) and double tempered, 2 + 2 h at indicated tempering temperature

Test temperature		Tensile strength		Yield strength		Elongation in 50 mm (2 in.), %	Reduction in area, %	Charpy V-notch impact energy	
°C	°F	MPa	ksi	MPa	ksi			J	ft · lbf
Tempered at 540 °C (1000 °F)									
260	500	1860	270	1520	220	9.9	33.2	...	...
315	600	1840	267	1490	216	10.3	34.5	...	...
425	800	1670	242	1440	209	12.0	42.6	...	...
480	900	1580	229	1365	198	12.3	46.1	...	...
540	1000	1480	215	1255	182	13.7	48.2	...	...
650	1200	610	88	583	84.5	24.8	95.2	...	...
Tempered at 565 °C (1050 °F)									
Room	Room	1810	262	1480	215	9.8	35.4	...	...
150	300	1700	246	1365	198	10.1	36.1	29.4	21.7
260	500	1610	233	1340	195	10.2	35.8	41.2	30.4
315	600	1600	231.5	1330	193	10.3	36.0	42.7	31.5
425	800	1500	217	1270	184	11.4	38.8	40.0	29.5
480	900	1420	206	1140	166	12.2	39.3	39.7	29.3
540	1000	1240	180	970	141	12.2	41.3	41.4	30.5
595	1100	980	142	720	105	12.8	46.8	45.0	33.2
650	1200	590	85	440	64	19.0	66.8	80.0	59.0
Tempered at 595 °C (1100 °F)									
260	500	1340	195	1130	164	10.0	45.0	44	33
315	600	1310	190	1100	160	10.0	48.1	...	...
425	800	1230	178	1010	146	12.4	52.2	41	30
480	900	1130	164	900	131	13.5	56.0	...	...
540	1000	980	142	790	115	15.5	62.0	...	...

perature. Forging should not be continued below 925 °C (1700 °F). Stock may be reheated as often as necessary. Because H11 mod is air hardening, it must be cooled slowly after forging to prevent stress cracks. After forging, the part should be charged into a furnace at about 790 °C (1450 °F); soaked until the temperature is uniform; and then slowly cooled, either while retained in the furnace or buried in an insulating medium such as lime, mica, or a siliceous filler material such as silocel. When the forging has cooled, it should be annealed.

The steel H11 mod, even in heavy sections, is readily welded. Fusion welding generally is accomplished with an inert-gas process or with coated electrodes. Filler metal should be of the same general composition. Preheating at about 540 °C (1000 °F) is recommended, and during welding the temperature should be maintained above 315 °C (600 °F). Thin sheet can be welded without preheating, but should be postheated at about 760 °C (1400 °F). Weldments, especially heavy-section weldments, should be cooled slowly in a furnace or an insulating medium. All weldments should be fully annealed after welding. Weldments of H11 mod have shown weld metal strength and ductility equal to or greater than those of the base metal. The machinability rating for H11 mod is about 60% of the rating for 1% C steel, or about 45% of that for B1112.

Heat Treatments. The standard heat treatments that apply to H11 mod steel are:

- *Normalize*: Generally not necessary. For effective homogenization, heat to about 1065 °C (1950 °F), soak 1 h for each 25 mm (1 in.) of thickness; air cool. Anneal immediately after the part reaches room temperature. Note: There is a possibility that H11 mod may crack during this treatment
- *Anneal*: Heat to 845 to 885 °C (1550 to 1625 °F) and hold to equalize temperature; cool very slowly in the furnace to about 480 °C (900 °F) and then more rapidly to room temperature. This treatment should produce a fully spheroidized microstructure free of grain boundary carbide networks
- *Harden*: Preheat to 760 to 815 °C (1400 to 1500 °F) and then raise the temperature to 995 to 1025 °C (1825 to 1875 °F) and hold 20 min plus 5 min for each 25 mm (1 in.) of thickness; air cool. For a few applications, oil quenching from the low end of the hardening temperature range may be done. Air cooling, which produces less distortion than oil quenching, is more commonly employed
- *Temper*: Heat at the secondary hardening temperature of about 510 °C (950 °F) for maximum hardness and strength, or above the secondary hardening peak to temper back to a lower hardness or strength. A minimum of 1 h at temperature should be allowed, but preferably parts should be double tempered: Hold 2 h at temperature, cool to room temperature, and then hold 2 h more at temperature. Triple tempering is more desirable, especially for critical parts. For high-temperature applications, parts should be tempered at a temperature above the maximum service temperature to guard against unwanted changes in properties during service
- *Stress relieve*: Heat to 650 to 675 °C (1200 to 1250 °F); cool slowly to room temperature. This treatment is often used to achieve greater dimensional accuracy in heat-treated parts by stress relieving rough-machined parts, then finish machining, and finally heat treating to the desired hardness
- *Nitride*: Finish-machined and heat-treated parts should be gas or liquid nitrided at temperatures of about 525 °C (980 °F). The depth of the nitrided case depends on time at temperature. For example, gas nitriding in 20 to 30% dissociated ammonia for 8 to 48 h normally produces a case depth of about 0.2 to 0.35 mm (0.008 to 0.014 in.)

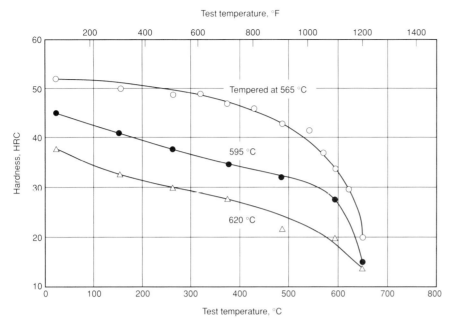

Fig. 12 Typical hot hardness of H11 mod steel. Specimens air cooled from 1010 °C (1850 °F) and double tempered, 2 + 2 h at indicated tempering temperature. Rockwell hardness converted from microhardness values

Table 28 Longitudinal room-temperature tensile properties of H11 mod steel after exposure to elevated temperature for 10 or 100 h
Air cooled from 1010 °C (1850 °F); double tempered

Exposure temperature		Exposure time, h	Tensile strength		Yield strength		Elongation in 50 mm (2 in.), %	Reduction in area, %
°C	°F		MPa	ksi	MPa	ksi		
510(a)	950(a)	100	1790	260	1760	255	11.5	42.8
540(b)	1000(b)	10	1650	239	1410	204	12.4	49.9
		100	1450	210	1300	188	13.7	52.9
540(c)	1000(c)	10	1385	201	1190	173	14.1	52.4
		100	1300	189	1100	160	15.2	58.2

(a) Tempered 2 + 2 h at 540 °C (1000 °F). (b) Tempered 2 + 2 h at 565 °C (1050 °F). (c) Tempered 2 + 2 h at 595 °C (1100 °F)

Table 29 Typical stress-rupture properties of H11 mod steel
Air cooled from 1010 °C (1850 °F) and tempered 1 h at temperature

Test temperature		Stress		Rupture life, h	Elongation(a), %	Reduction in area(a), %
°C	°F	MPa	ksi			
Tempered at 565 °C (1050 °F)						
480	900	1380	200	10.4	9.9	33.7
		1170	170	257.3	5.7	33.7
540	1000	1070	155	11.1	9.9	20.9
		760	110	58.0	12.4	24.9
		520	75	254.0	13.3	30.3
		480	70	318.0	17.3	35.4
Tempered at 650 °C (1200 °F)						
540	1000	655	95	6.9	24.5	32.3
		480	70	31.4	25.4	46.7
		275	40	435.4	31.3	68.7
595	1100	410	60	3.3	25.7	60.4
		205	30	73.1	49.3	64.6
		140	20	444.3	55.2	71.3
650	1200	205	30	2.3	28.1	75.0
		100	15	70.5	69.9	84.8

(a) At rupture

- *Bake*: After plating in an acid bath, or after other processing that might introduce hydrogen into the metal, parts should be baked 24 h or longer at 190 °C (375 °F) or above

Properties. Variations in hardness of H11 mod with tempering temperature are plotted in Fig. 11. Variations in typical room-temperature longitudinal mechanical properties with tempering temperature are given in Table 26. This steel is quite notch tough; the ratio of notched-bar tensile strength (K_t = 3.3) to smooth-bar tensile strength at room temperature is about 1.4 at a smooth-bar tensile strength of 1380 MPa (200 ksi) and about 1.15 at 1930 MPa (280 ksi).

Because of secondary hardening characteristics, H11 mod has good temper resistance, which results in high hardness and strength at elevated temperatures. Hot hardness of H11 mod tempered to three room-temperature hardness levels is plotted in Fig. 12. Regardless of the initial room-temperature hardness, the hot hardness drops precipitously to levels corresponding to the annealed state at temperatures above about 620 °C (1150 °F). Short-time elevated-temperature tensile properties of bar stock tempered at 540, 565, and 595 °C (1000, 1050, and 1100 °F) are given in Table 27. The high temper resistance of H11 mod is indicated in Table 28, which gives room-temperature tensile properties after prolonged exposure to elevated temperatures. This material has fairly good stress-rupture properties at temperatures below 650 °C (1200 °F), as shown in Table 29.

The steel H11 mod has comparatively low fracture toughness, but has good resistance to stress-corrosion cracking compared to other ultrahigh-strength steels heat treated to the same strength. In a four-point bend test, plate 13 mm (½ in.) thick heat treated to 1295 MPa (188 ksi) yield strength by tempering at 595 °C (1100 °F) had a room-temperature K_{Ic} value of 59 MPa$\sqrt{m}$ (54 ksi$\sqrt{in.}$). The K_{Iscc} value in 3.5% NaCl solution was 33 MPa$\sqrt{m}$ (30 ksi$\sqrt{in.}$). In another bend test (Ref 8), 12.7 mm plate with a yield strength of about 1450 MPa (210 ksi) resulting from tempering at 580 °C (1080 °F) had K_{Ic} values of 37 MPa$\sqrt{m}$ (34 ksi$\sqrt{in.}$) at room temperature and 60 MPa$\sqrt{m}$ (55 ksi$\sqrt{in.}$) at 93 °C (200 °F).

Significant improvements in H11 mod properties such as transverse ductility and toughness, particularly in large sections, can be achieved by vacuum arc remelting and electroslag remelting, both of which result in greater homogeneity and cleanliness of the steel. A comparison of midradius transverse tensile properties of air-melted and vacuum arc remelted billets is presented in Table 30. Each value is the average of four tests, two from the top of the ingot, and two from the bottom.

H13 Steel

AISI H13 is a 5% Cr ultrahigh-strength steel similar to H11 mod in composition, heat treatment, and many properties. The main difference in composition is the higher vanadium content of H13 (see Table 1); this leads to a greater dispersion of hard vanadium carbides, which results in higher wear resistance. Also, H13 has a slightly wider

Table 30 Effect of billet size and melting method on transverse strength and ductility of H11 mod steel
Air cooled from 1010 °C (1850 °F); triple tempered, 2 + 2 + 2 h at 540 °C (1000 °F)

Billet size	Melting method	Tensile strength MPa	ksi	Reduction in area, %
150 × 150 mm (6 × 6 in.)	Air	1965	285	16.1
	VAR	1985	288	25.7
300 × 300 mm (12 × 12 in.)	Air	1972	286	7.2
	VAR	2013	292	19.7

Table 31 Typical longitudinal room-temperature mechanical properties of H13 steel
Round bars, oil quenched from 1010 °C (1850 °F) and double tempered, 2 + 2 h at indicated temperature

Tempering temperature °C	°F	Tensile strength MPa	ksi	Yield strength MPa	ksi	Elongation in 4D gage length, %	Reduction in area, %	Charpy V-notch impact energy J	ft · lbf	Hardness, HRC
527	980	1960	284	1570	228	13.0	46.2	16	12	52
555	1030	1835	266	1530	222	13.1	50.1	24	18	50
575	1065	1730	251	1470	213	13.5	52.4	27	20	48
593	1100	1580	229	1365	198	14.4	53.7	28.5	21	46
605	1120	1495	217	1290	187	15.4	54.0	30	22	44

Table 32 Longitudinal short-time tensile properties of H13 steel bar
Oil quenched from 1010 °C (1850 °F) and double tempered to indicated hardness

Room-temperature hardness, HRC	Test temperature °C	°F	Tensile strength MPa	ksi	Yield strength MPa	ksi	Elongation in 4D gage length, %	Reduction in area, %
52(a)	425	800	1620	235	1240	180	13.7	50.6
	540	1000	1305	189	1000	145	13.9	54.0
	595	1100	1020	148	825	120	17.5	65.4
	650	1200	450	65	340	49	28.9	88.9
48(b)	425	800	1400	203	1150	167	15.0	59.9
	540	1000	1160	168	960	139	17.1	62.4
	595	1100	940	136	750	109	18.0	68.5
	650	1200	455	66	350	51	33.6	89.0
44(c)	425	800	1200	174	1005	146	17.0	64.1
	540	1000	995	144	820	119	20.6	70.0
	595	1100	827	120	690	100	22.6	74.0
	650	1200	450	65	350	51	28.4	87.6

(a) Tempered 2 + 2 h at 527 °C (980 °F). (b) Tempered 2 + 2 h at 575 °C (1065 °F). (c) Tempered 2 + 2 h at 605 °C (1120 °F).

range of carbon content than does H11 mod. Depending on the producer, the carbon content of H13 may be near the high or low side of the accepted range, with a corresponding variation in strength and ductility for a given heat treatment.

Like H11 mod, H13 is a secondary hardening steel. It has good temper resistance and maintains high hardness and strength at elevated temperatures. It is deep hardening, which allows large sections to be hardened by air cooling. H13 steel can be heat treated to strengths exceeding 2070 MPa (300 ksi); like H11 mod, it has good ductility and impact strength. With standard heat treatment, the fracture toughness of H13 appears to be even lower than that of H11 mod.

The material H13 has good resistance to thermal fatigue. Hot-work tooling made of H13 can be safely water cooled between hot-working operations. Its resistance to thermal fatigue, erosion, and wear has made it a preferred die material for aluminum and magnesium die casting, as well as for many other hot-work applications. However, H13 is subject to hydrogen embrittlement. It can be nitrided for additional wear resistance.

Although H13 has not been used as widely as H11 mod as an ultrahigh-strength constructional steel, the similarities in properties make H13 equally attractive for such applications. This is particularly true in noncritical service in which slightly higher wear resistance is an advantage.

The H13 steel is available as bar, rod, billet, and forgings. Typical hot-work applications include die casting dies, inserts, cores, ejector pins, plungers, sleeves, slides, forging dies, extrusion dies, dummy blocks, and mandrels. Other tooling and structural applications include punches, shafts, beams, torsion bars, shrouds, and ratchets.

Processing. For forging, H13 steel is heated slowly and uniformly to a temperature of 1090 to 1150 °C (2000 to 2100 °F), preferably after preheating at 760 to 815 °C (1400 to 1500 °F). The steel should be thoroughly heated before forging. Forging should not be done below 900 °C (1650 °F), but the parts may be reheated as often as necessary. Because H13 is air hardening, parts should be cooled slowly after forging. Simple forgings may be cooled in an insulating medium such as dry ashes, lime, or expanded mica. The best practice for large forgings is to place in a heated furnace at about 790 °C (1450 °F), soak until the temperature is uniform, shut off the furnace, and let cool slowly. Parts should then be given a full spheroidizing anneal.

When annealed H13 is welded, it should be preheated to 540 °C (1000 °F), or to as high a temperature as is practical, preferably in a furnace to ensure uniform, stress-free preheating. Uncoated rod, preferably of the same general composition, should be used, with shielded-arc equipment. The temperature of the part should be kept above 315 °C (600 °F), and the part reheated if necessary, until welding has been completed. After welding, the part should be cooled slowly in an insulating medium and given a full anneal. A heat-treated part such as a die can be welded using the same procedure, preferably preheating the part in a furnace to a temperature about 55 °C (100 °F) below the original tempering temperature. After welding, the part should be placed in a furnace at the preheating temperature and cooled slowly to room temperature. It is recommended that the part then be reheated to just below the original tempering temperature and air cooled. This helps to relieve welding stresses and blend the hardness of the weld area into that of the base metal. Regardless of the situation, adequate preheating and slow cooling are essential to minimize the risk of cracking during welding.

Fully annealed H13 has a machinability rating that is about 70% of the rating for 1% C tool steel, or about 45% of that for B1112.

Heat Treatments. The heat treatments that apply to H13 steel are:

- *Normalize*: Not recommended for H13. Some improvement in homogeneity can be obtained by preheating to about 790 °C (1450 °F), heating slowly and uniformly to 1040 to 1065 °C (1900 to 1950 °F), holding 1 h for each 25 mm (1 in.) of thickness, and then air cooling. Just before the part reaches room temperature, it should be recharged into a furnace and given a full anneal. Note: There is a risk of cracking during this treatment, especially if done in a furnace in which the atmosphere is not controlled to prevent surface decarburization
- *Anneal*: Heat uniformly to 860 to 900 °C (1575 to 1650 °F) in a furnace with controlled atmosphere, or with the part packed in a neutral compound, so that decarburization is prevented. Cool very slowly in the furnace to about 480 °C (900 °F); then cool more rapidly to room temperature. This treatment results in a fully spheroidized microstructure

Table 33 Longitudinal impact properties of H13 bar tempered at different temperatures

Tempering temperature(a)		Hardness(b), HRC	Charpy V-notch impact energy at test temperature of							
			−73 °C (−100 °F)		21 °C (70 °F)		260 °C (500 °F)		540 °C (1000 °F)		595 °C (1100 °F)	
°C	°F		J	ft · lbf	J	ft · lbf	J	ft · lbf	J	ft · lbf	J	ft · lbf
524	975	54	7	5	14	10	27	20	31	23	...	...
565	1050	52	7	5	14	10	30	22	34	25	34(c)	25(c)
607	1125	47	8	6	24	18	41	30	45	33	43	32
615	1140	43	9.5	7	24	18	52	38	60	44	57	42

(a) Air cooled from 1010 °C (1850 °F) and double tempered, 2 + 2 h at indicated temperature. (b) At room temperature. (c) At 565 °C (1050 °F)

Table 34 Longitudinal fracture toughness of H13 steel

Air cooled from 1050 °C (1920 °F) and tempered 2 h at temperature

Tempering temperature		Plane-strain fracture toughness(a)(K_{Ic})	
°C	°F	MPa√m	ksi√in.
400	750	47.7	43.4
475	885	33.0	30.0
500	930	27.4	24.9
530	985	24.3	22.1
550	1020	23.1	21.0
600	1110	33.2	30.2
625	1155	52.4	47.7
650	1200	77.7	70.7

(a) Values should not be used for design purposes because they represent material that was not heat treated in the usual manner, but was cooled from an austenitizing temperature higher than normal. Source: Ref 9

- *Harden*: Heat slowly and uniformly to 995 to 1025 °C (1825 to 1875 °F) and soak 20 min plus 5 min for each 25 mm (1 in.) of thickness; preheating at 790 to 815 °C (1450 to 1500 °F) is usually recommended for thick parts. Air cool in still air. Air cooling is usually done from the high side of the hardening temperature range. For a few applications, H13 may be oil quenched from the low side of the hardening temperature, but at the risk of distortion or cracking
- *Temper*: Heat at the secondary hardening peak of about 510 °C (950 °F) for maximum hardness and strength, or at higher temperatures to temper back to a lower level of hardness or strength. Double tempering, that is, 2 h at temperature, air cooling, then 2 h more at temperature, is recommended. Occasionally, triple tempering may be desirable
- *Stress relieve*: Heat to 650 to 675 °C (1200 to 1250 °F) and soak 1 h or more; cool slowly to room temperature. This treatment is often used to achieve greater dimensional accuracy in heat-treated parts by stress relieving rough-machined parts, then finish machining, and finally heat treating to the desired hardness
- *Nitride*: Finish-machined and heat-treated parts may be nitrided to produce a highly wear-resistant surface. Because it is carried out at the normal tempering temperature, nitriding can serve as the second temper in a double-tempering treatment. The depth of the nitrided case depends on the time at temperature. For example, gas nitriding at 510 °C (950 °F) for 10 to 12 h produces a case depth of 0.10 to 0.13 mm (0.004 to 0.005 in.) and for 40 to 50 h produces a case depth of about 0.3 to 0.4 mm (0.012 to 0.016 in.). Parts that have been deep nitrided are usually lapped or gently surface ground to remove the thin, brittle white layer. Selective nitriding is sometimes done to produce a nitrided case only where it is needed. Copper plating is preferred for stopping off areas that are not to be nitrided; stop-offs containing lead should be avoided, because lead embrittles H13 steel

Properties. The properties presented in this section are for H13 with a carbon content in the middle of the composition range (for composition, see Table 1). Somewhat different properties should be expected when the carbon content is near either the high end or the low end of the range.

The variation in hardness of H13 with tempering temperature is plotted in Fig. 13. Room-temperature longitudinal mechanical properties of bars tempered to different hardness levels are given in Table 31. Because it is a secondary hardening steel, H13 maintains high hardness and strength at elevated temperatures. Values of hot hardness for specimens tempered to three different hardness levels are presented in Fig. 14. Typical short-time tensile properties of H13 bar tempered to different hardness levels are listed in Table 32 for various service (test) temperatures. Regardless of initial room-temperature hardness, H13 attains essentially the same low properties at 650 °C (1200 °F). Like H11 mod, H13 has good

Table 35 Comparison of transverse mechanical properties of air-melted and electroslag-remelted H13 steel

Specimen property(a)	Air melted(b)	Electroslag remelted(c)
Tensile strength, MPa (ksi)	1615 (234)	1650 (239)
Yield strength, MPa (ksi)	1395 (202)	1425 (207)
Elongation in 4D gage length, %	2.9	5.6
Reduction in area, %	5.7	12.8
Charpy V-notch impact energy, J (ft · lbf)	5 (4)	7 (5)

(a) Specimens taken from midradius of round bar; oil quenched from 1010 °C (1850 °F); and double tempered, 2 + 2 h at 588 °C (1090 °F). Final hardness, 48 HRC. (b) Round section, 355 mm (14 in.) in diameter. (c) Round section, 457 mm (18 in.) in diameter

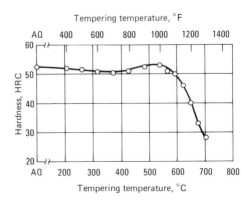

Fig. 13 Variation in hardness with tempering temperature for H13 steel. All specimens air cooled from 1025 °C (1875 °F) and tempered 2 h at temperature. AQ, as quenched

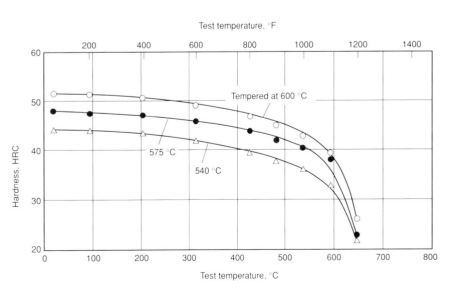

Fig. 14 Typical hot hardness values of H13 steel. Specimens oil quenched from 1010 °C (1850 °F) and double tempered, 2 + 2 h at indicated tempering temperature

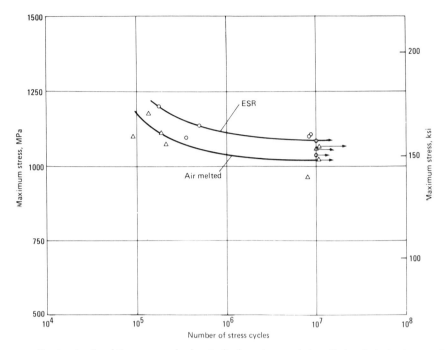

Fig. 15 Tension-tension fatigue curves for longitudinal specimens of air-melted and electroslag-remelted heats of H13 steel. Axial fatigue tests performed in an Ivy machine at a frequency of 60 Hz; the alternating stress was 67% of the mean stress for all tests ($R = 0.2$). Arrows signify tests terminated without any sign of fatigue cracking.

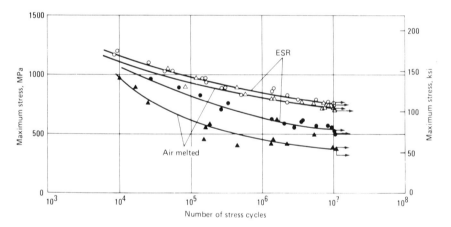

Fig. 16 Tension-compression fatigue curves for air-melted and electroslag-remelted heats of H13 steel. Axial fatigue tests performed in an Ivy machine at a frequency of 60 Hz; the stresses were fully reversed for all tests ($R = -1$). Open symbols indicate longitudinal fatigue data; filled symbols indicate transverse data. Arrows signify tests terminated without any sign of fatigue cracking. Source: Ref 10

Table 36 Room-temperature mechanical properties of HP 9-4-30 steel

Property	Typical value for hardness of 49–53 HRC(a)	44–48 HRC(b)	Minimum value(c)
Tensile strength, MPa (ksi)	1650–1790 (240–260)	1520–1650 (220–240)	1520 (220)
Yield strength, MPa (ksi)	1380–1450 (200–210)	1310–1380 (190–200)	1310 (190)
Elongation in 4D gage length, %	8–12	12–16	10
Reduction in area, %	25–35	35–50	35
Charpy V-notch impact energy, J (ft · lbf)	20–27 (15–20)	24–34 (18–25)	24 (18)
Fracture toughness (K_{Ic}), MPa$\sqrt{m}$ (ksi$\sqrt{in.}$)	66–99 (60–90)	99–115 (90–105)	...

(a) Oil quenched from 845 °C (1550 °F), refrigerated to −73 °C (−100 °F) and double tempered at 205 °C (400 °F). (b) Same heat treatment as (a) except double tempered at 550 °C (1025 °F). (c) For sections forged to 75 mm (3 in.) or less in thickness (or to less than 0.016 m², or 25 in.², in total cross-sectional area), quenched to martensite and double tempered at 540 °C (1000 °F)

impact properties at temperatures up to 540 °C (1000 °F). Charpy V-notch impact energies at subzero, room, and elevated temperatures are compared in Table 33.

Little work has been done to determine the fracture toughness of H13. It appears that with the standard heat treatment H13 has slightly lower fracture toughness than H11 mod. Available data on fracture toughness of H13 are given in Table 34.

Vacuum arc remelted (VAR) and electroslag remelted (ESR) H13 have better cleanliness and homogeneity than air melted H13. The VAR and ESR processes are used to produce steels with superior ductility, impact strength, and fatigue resistance, especially in the transverse direction, and particularly in large section sizes. Table 35 compares room-temperature transverse mechanical properties of air-melted and ESR H13 in large section sizes.

In axial (tension-tension) fatigue tests, the life of ESR H13 was superior to that of air-melted H13 (Fig. 15). When both were fatigue tested under fully reversed stresses, there was no significant difference in the life of longitudinal specimens. However, for transverse specimens, the life of ESR H13 was significantly better (Fig. 16).

High Fracture Toughness Steels

High-strength, high fracture toughness steels as described in this article are commercial structural steels capable of a yield strength of 1380 MPa (200 ksi) and a K_{Ic} of 100 MPa$\sqrt{m}$ (91 ksi$\sqrt{in.}$). (These steels also exhibit stress corrosion cracking resistance.) A number of developmental steels that are not fully commercial alloys are excluded from this discussion. The HP-9-4-30 and AF1410 steels, however, are discussed below.

Both these alloys are of the Ni-Co-Fe type and have a number of similar characteristics. Both are weldable, and the melt practice requires a minimum of VAR. Control of residual elements to low levels is required for optimum toughness. The machining practices used for 4340 steel are generally satisfactory; however, Ni-Co-Fe alloys are considered more difficult to machine than are alloy steels.

HP-9-4-30 Steel

During the 1960s, Republic Steel Corporation introduced a family of four weldable steels, all of which had high fracture toughness when heat treated to medium/high strength levels. Only HP-9-4-30 has been produced in significant quantities and is comparable to the high-strength, high fracture toughness steels of this article. The chemical composition is given in Table 1. The HP-9-4-30 steel is usually electric arc melted and then vacuum arc remelted. Forging temperatures should not exceed

Table 37 Mechanical properties of HP 9-4-30 steel at three test temperatures

All specimens normalized at 900 °C (1650 °F); oil quenched from 845 °C (1550 °F); and double tempered, 2 + 2 h at 550 °C (1025 °F)

Test temperature		Tensile strength		Yield strength		Elongation in 4D gage length, %	Plane-strain fracture toughness(a)(K_{Ic})	
°C	°F	MPa	ksi	MPa	ksi		MPa$\sqrt{m}$	ksi$\sqrt{in.}$
24	75	1530	222	1275	185	16	109.3	99.5
260	500	1380	200	1215	176	16	106.3	96.7
345	650	1325	192	1110	161	17	103.8	94.5

(a) WR orientation; average of two tests at each temperature. Source: Ref 11

Table 38 Room-temperature mechanical properties of HP 9-4-30 steel after 1000 h at various elevated temperatures

All specimens austenitized, quenched to martensite, and tempered at 540 °C (1000 °F)

Temperature of exposure		Tensile strength		Yield strength		Elongation in 25 mm (1 in.), %	Reduction in area, %	Charpy V-notch impact energy	
°C	°F	MPa	ksi	MPa	ksi			J	ft · lbf
Not exposed		1650	239	1350	196	14	52	39	29
205	400	1585	230	1405	204	16	60	41	30
345	650	1585	230	1440	209	15	56	38	28
425	800	1650	239	1400	203	14	50	34	25
480	900	1565	227	1395	202	15	51	26	19

Table 39 Effect of various heat treatments on mechanical properties of a cobalt-nickel steel (VIM/VAR plate of AF1410 steel)

Heat treatment(a)(b)	Ultimate strength		Yield strength		Elongation, %	Reduction in area, %	Charpy V-notch	
	MPa	ksi	MPa	ksi			J	ft · lbf
Plate of 15 mm (5/8 in.) thickness								
X + water quench per (c) + Z	1580	229	1515	220	16	60	91	67
X + refrigeration treatment per (d) + Z	1650	239	1550	225	17	69	83	61
X + vermiculite cool and refrigeration per (e) + Z	1620	235	1490	216	17	70	84	62
X + reaustenitization and refrigeration per (f) + Z	1660	241	1525	221	17	73	113	83
Average for several heats								
Heat treatment per (g)	1675	243	1590	231	...	...	92	68
Plate of 75 mm (3 in.) thickness								
Y + water quench per (c) + Z	1585	230	1540	223	16	66	65	48
Y + refrigeration treatment per (d) + Z	1680	244	1540	223	17	70	81	60
Y + vermiculite cool and refrigeration per (c) + Z	1480	215	1380	200	18	68	58	43
Y + reaustenitization, air cool, and refrigeration per (f) + Z	1670	242	1540	223	17	69	95	70

(a) Time at 900 °C (1650 °F) or 815 °C (1500 °F) is as follows: 1 h for the 15 mm (5/8 in.) plate or 3 h for the 75 mm (3 in.) plate. (b) Initial and final heat treatments: X = 900 °C (1650 °F) for 1 h with air cooling and 675 °C (1250 °F) for 8 h with air cooling; Y = 900 °C (1650 °F) for 3 h with air cooling and 675 °C (1250 °F) for 8 h with air cooling; Z = 510 °C (950 °F) for 5 h with air cooling. (c) 815 °C (1500 °F) for the time per (a) and water quenching. (d) 815 °C (1500 °F) for the time per (a) with air cooling and a refrigeration treatment of −73 °C (−100 °F). (e) 815 °C (1500 °F) for the time per (a) with vermiculite cool and a refrigeration treatment of −73 °C (−100 °F). (f) 900 °C (1650 °F) for time per (a) with air cooling, 815 °C (1500 °F) for time per (a) with air cooling, and refrigeration at −73 °C (−100 °F). (g) 900 °C (1650 °F) for time per (a) with water quench, 815 °C (1500 °F) for time per (a) with water quench, 815 °C (1500 °F) for time per (a) with water quench, and 510 °C (950 °F) for 5 h with air cooling. Source: Ref 1

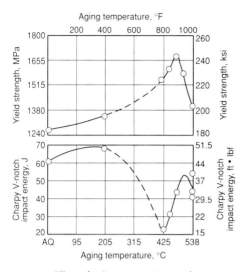

Fig. 17 Effect of aging temperature on impact energy (bottom) and yield strength (top) of AF1410 steel (VIM/VAR plate 15 mm, or 5/8 in., thick). Heat treatments: Heat at 900 °C (1650 °F) for ½ h and water quench; heat at 815 °C (1500 °F) for ½ h and water quench; age for 5 h at indicated temperatures and air cool. AQ, as quenched. Source: Ref 1

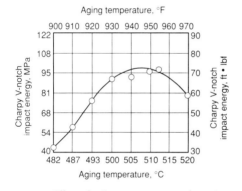

Fig. 18 Effect of aging temperature on impact energy of AF1410 steel (VIM/VAR plate 15 mm, or 5/8 in., thick). Heat treatments: See Fig. 17.

1120 °C (2050 °F). The HP-9-4-30 steel is capable of developing a tensile strength of 1520 to 1650 MPa (220 to 240 ksi) with a plain-strain fracture toughness of 100 MPa$\sqrt{m}$ (91 ksi$\sqrt{in.}$). This steel has deep hardenability and can be fully hardened to martensite in sections up to 150 mm (6 in.) thick. The HP-9-4-30 steel in the heat-treated condition can be formed by bending, rolling, or shear spinning. Heat-treated parts can readily be welded. Tungsten arc welding under inert-gas shielding is the preferred welding process. Neither postheating nor postweld heat treating is required. After welding, parts may be stress relieved at about 540 °C (1000 °F) for 24 h. This is a stress-relieving treatment and has no adverse effect on the strength or toughness of either the weld metal or the base metal. The HP-9-4-30 steel is available as billet, bar, rod, plate, sheet, and strip. It has been used for aircraft structural components, pressure vessels, rotor shafts for metal forming equipment, drop hammer rods, and high-strength shock-absorbing automotive parts.

Heat Treatments. The heat treatments that apply to HP 9-4-30 steel are:

- *Normalize*: Heat to 870 to 925 °C (1600 to 1700 °F) and hold 1 h for each 25 mm (1 in.) of thickness (1 h minimum); air cool
- *Anneal*: Heat to 620 °C (1150 °F) and hold 24 h; air cool
- *Harden*: Austenitize at 830 to 860 °C (1525 to 1575 °F) and hold 1 h for each 25 mm (1 in.) of thickness (1 h minimum); water or oil quench. Complete the martensitic transformation by refrigerating at least 1 h at −87 to −60 °C (−125 to −75 °F); allow to warm to room temperature
- *Temper*: Hold at 200 to 600 °C (400 to 1100 °F), depending on desired final strength; double tempering is preferred. The most widely used tempering treatment is double tempering, that is, 2 h at temperature, air cooling, and then 2 h more at temperature, at a temperature ranging from 540 to 580 °C (1000 to 1075 °F)
- *Stress relieve*: Usually required only after welding restrained sections. Heat to 540

°C (1000 °F) and hold 24 h; air cool to room temperature

Properties. Table 36 presents room-temperature mechanical properties of HP 9-4-30 double tempered at three different temperatures. The data for material double tempered at 540 °C (1000 °F) represent minimum mechanical properties for this condition: properties listed for the other conditions may be considered typical. Smooth-bar fatigue strengths at 10^7 cycles of 830 MPa (120 ksi) have been reported for material double tempered at 540 °C (1000 °F); the corresponding notched-bar fatigue strength was 380 MPa (55 ksi).

The HP 9-4-30 steels have good thermal stability, which makes them suitable for long-term service at temperatures up to at least 370 °C (700 °F). Table 37 gives short-time tensile properties and fracture toughness values for 25 mm (1 in.) plate tested at room temperature and at 260 and 345 °C (500 and 650 °F). In the same study, it was reported that the fatigue crack propagation rate was not affected by temperatures up to 345 °C (650 °F). Room-temperature mechanical properties of material exposed to elevated temperature for 1000 h are given in Table 38. Stresses to cause rupture in 100 h have been reported as 860 MPa (125 ksi) at 480 °C (900 °F) and 380 MPa (55 ksi) at 540 °C (1000 °F) for material double tempered at 540 °C (1000 °F). Corresponding values for 1000 h life are 585 MPa (85 ksi) at 480 °C (900 °F) and 195 MPa (28 ksi) at 540 °C (1000 °F).

AF1410 Steel

The steel AF1410 was an outgrowth of the U.S. Air Force sponsorship of the advanced submarine hull steels, the result of which was the development of the low-carbon Fe-Ni-Co type alloys. These alloys had significant stress corrosion cracking resistance. By raising the cobalt and carbon content, the ultimate tensile strength was increased to a typical 1615 MPa (235 ksi). This increase in strength was obtained while maintaining a K_{Ic} value of 154 MPa$\sqrt{m}$ (140 ksi$\sqrt{in.}$). This combination of strength and toughness exceeds that of other commercially available steels, and the alloy has been considered as a replacement for titanium in certain aircraft parts. The AF1410 material (see Table 1 for composition) is air hardenable in sections up to 75 mm (3 in.) thick. The preferred melting practice is presently vacuum induction melting followed by vacuum arc remelting (VIM/VAR). However, initially VIM and VIM/ESR practices were used. Melting practice requires that impurity elements be kept at very low levels to ensure high fracture toughness. Although forgeable to 1120 °C (2050 °F), at least a 40% reduction must be obtained below 900 °C (1650 °F) to attain maximum properties. Weldability is good using a continuous wave (CW)-gas tungsten arc welding (GTAW) process, provided that high-purity wire is used and oxygen contamination is avoided. Information on stress-corrosion cracking is incomplete; however, it has been determined that at 52 HRC AF1410 has a K_{Iscc} of 66 MPa$\sqrt{m}$ (60 ksi$\sqrt{in.}$). Obtainable as bar, billet, rod, plate, sheet, and strip, AF1410 has been used for aircraft structural components.

Table 40 Mechanical properties of a cobalt-nickel steel (AF1410) in various quenching media

Test specimens were 50 mm (2 in.) plate from VIM/VAR melt with the heat treatment: 675 °C (1250 °F) for 8 h with air cooling, 900 °C (1650 °F) for 1 h, quenching, 830 °C (1525 °F) for 1 h, quenching, refrigeration at −73 °C (−100 °F) for 1 h, 510 °C (950 °F) for 5 h, and air cooling. Source: Ref 1

Quench medium	Ultimate strength MPa	ksi	Yield strength MPa	ksi	Elongation, %	Reduction in area, %	Charpy V-notch J	ft · lbf	Plane-strain fracture toughness (K_{Ic}) MPa$\sqrt{m}$	ksi$\sqrt{in.}$
Air	1680	244	1475	214	16	69	69	51	174	158
Oil	1750	254	1545	224	16	69	65	48	154	140
Water	1710	248	1570	228	16	70	65	48	160	146

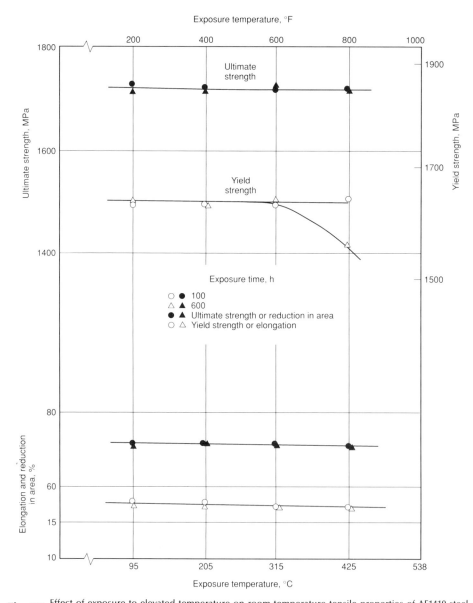

Fig. 19 Effect of exposure to elevated temperature on room-temperature tensile properties of AF1410 steel (VIM/VAR plate 30 mm, or 1¼ in., thick). Heat treatment: Heat at 900 °C (1650 °F) for ½ h and water quench, heat at 815 °C (1500 °F) for ½ h and water quench, and then heat at 510 °C (950 °F) for 5 h and air cool. Source: Ref 1

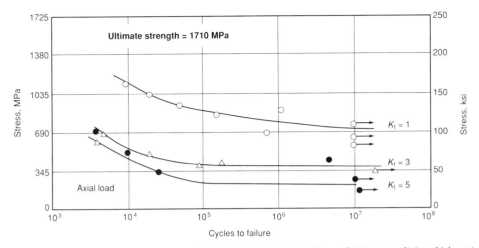

Fig. 20 S-N curves at room temperature for AF1410 steel (VIM/VAR plate with 20 mm, or ¾ in., thickness) tested in transverse direction at several stress concentrations for $R = -1$. Heat treatment: Heat at 900 °C (1650 °F) for 1 h and water quench, heat at 815 °C (1500 °F) for 1 h and water quench, and then heat at 510 °C (950 °F) for 5 h and air cool. Source: Ref 1

austenite, and these will weaken the matrix of the steel. The best combination of strength and ductility results from a 510 °C (950 °F) age.

Heat Treatment. The heat treatments that apply to AF1410 steel are:

- *Normalized and Overaged*: This is the condition the material is normally supplied for best machinability. Heat between 880 °C (1620 °F) and 900 °C (1650 °F), hold 1 h for each 25 mm (1 in.) of thickness; air cool and over age at 675 °C (1250 °F) for 5 h minimum
- *Anneal*: Usually, normalizing and overaging (above) are used to soften and stress relieve the product. A stress relief of 675 °C (1250 °F) may be applied to relieve mechanical stress
- *Harden*: Double austenitize, first at 870 to 900 °C (1600 to 1650 °F), hold 1 h for each 25 mm (1 in.) of thickness; oil, water, or air cool depending on section size; reaustenitize at 800 to 815 °C (1475 to 1500 °F); oil, water, or air cool. An alternative is to single austenitize at 800 to 815 °C (1475 to 1500 °F), hold 1 h for each 25 mm (1 in.) of thickness; oil, water, or air cool depending on section size
- *Quench*: Air cooling from the austenitizing temperature will produce tensile strength, toughness, and fatigue strength essentially equal to oil or water quenching in section sizes up to 75 mm (3 in.). Refrigeration treatment of −73 °C (−100 °F) is optional. The aim is to reduce the amount of retained austenite. There is no real evidence that such a treatment has any substantial effect on the material or the mechanical properties
- *Aging*: Age at 480 to 510 °C (900 to 950 °F) 5 to 8 h. Air cooling is normally employed

Properties. Tensile strength properties and impact energy for VIM/VAR melted plate quenched in air, water, or vermiculite and cooled following austenitizing are shown in Table 39. The heat treatments showed some effect on the tensile and impact properties for both 15 mm (⅝ in.) and 75 mm (3 in.) VIM/VAR melted plate.

Room-temperature tensile properties after exposure to long-term elevated temperatures show some degradation after 500 h (Fig. 19). Fracture toughness, tensile properties, and impact energy of VIM/VAR melted 50 mm (2 in.) plate quenched and aged at 510 °C (950 °F) (a premachining heat treatment had been applied) are shown after quenching in different media (Table 40). Axial load fatigue data are provided in Fig. 20. Maximum stress for 10^7 cycles is approximately 690 MPa (100 ksi) for $K_t = 1$. Comparison of fatigue crack growth rate in low-humidity air (Fig. 21) and in 3½% saltwater solution (Fig. 22) indicates the relative insensitivity of AF1410 to saltwater exposure.

The microstructure of AF1410 consists of Fe-Ni lath martensite, precipitation on which produces the strengthening mechanism. Quenching from the austenitizing temperature produces a highly dislocated lath martensite that has a high toughness, as measured by the Charpy V-notch impact test (Fig. 17). Aging produces a complex series of changes in carbide structure. At approximately 425 °C (800 °F), Fe_3C is precipitated. At 455 °C (850 °F), Fe-Cr-Mo M_2C carbide is obtained, which at 480 °C (900 °F) will begin to produce a pure Mo-Cr M_2C carbide. By raising the temperature to 510 °C (950 °F), the M_2C will begin to coarsen; at 540 °C (1000 °F) M_2C will begin to be replaced by M_6C, which has little strengthening effect. The steel is normally austenitized and aged. The secondary hardening, which is due to the aging, produces a maximum tensile strength when aged at 480 °C (900 °F) using a 5 h aging time and a minimum impact energy when aged at 425 °C (800 °F), as shown in Fig. 17. When aged in the temperature range between 425 °C (800 °F) and 540 °C (1000 °F), the impact energy exhibits a maximum at about 508 °C (947 °F), as shown in Fig. 18. At aging temperatures above 540 °C (1000 °F), both the tensile strength and the impact energy decrease rather rapidly.

The steel is subject to austenite reversion during aging. At normal aging temperatures, the retained austenite is generally less than 1% by volume. However, 540 °C (1000 °F) or higher will produce large amounts of

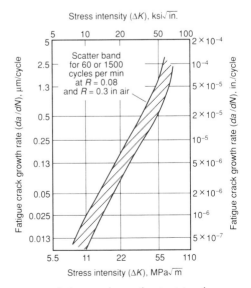

Fig. 21 Fatigue crack growth rate at two frequencies and two R ratios for AF1410 steel (VIM/VAR plate 25 mm, or 1 in., thick) at room temperature in low-humidity air. Heat treatment same as in Fig. 20. Source: Ref 1

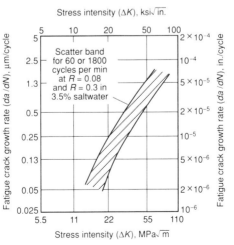

Fig. 22 Fatigue crack growth rate at two frequencies and two R ratios for AF1410 steel (VIM-VAR plate 25 mm, or 1 in., thick) at room temperature in 3½% saltwater. Heat treatment same as in Fig. 20. Source: Ref 1

REFERENCES

1. W.F. Brown, Jr., *Aerospace Structural Metals Handbook*, Code 1224, Metals and Ceramics Information Center, 1989, p 1-30
2. A.M. Hall, Sr., *Introduction to Today's Ultrahigh-Strength Structural Steels*, STP 498, American Society for Testing and Materials, 1971
3. G. Sachs, "Survey of Low-Alloy Aircraft Steels Heat Treated to High Strength Levels," WADC TR 53-254, Part 4, Wright Air Development Center, Dec 1953
4. R.L. McGee, J.E. Campbell, L.R. Carlson, and G.K. Manning, "The Mechanical Properties of Certain Aircraft Structural Metals at Very Low Temperature," WADC TR 58-386, Wright Air Development Center, Nov 1958
5. J.J. Hauser and M.G.H. Wells, The Effect of Inclusions on Fatigue Properties in High-Strength Steels, Mechanical Working and Steel Processing, in Vol 14, Iron and Steel Society of the American Institute of Mining, Metallurgical, and Petroleum Engineers, 1976, p 314-348
6. G.M. Waid, "Stress Corrosion and Hydrogen Embrittlement Studies of Ultrahigh-Strength Steels (HP 310, 4340, 300M)," Research Report, Republic Steel Corporation, Sept 1976
7. L. Peterman and R.L. Jones, Effects of Quenching Variables on Fracture Toughness of D-6ac Steel Aerospace Structures, *Met. Eng. Q.*, Vol 15 (No. 2), May 1975, p 59
8. "Plane Strain Fracture Toughness (K_{Ic}) Data Handbook for Metals," Army Materials and Mechanics Research Center, Dec 1973
9. K. Firth and R.D. Gargood, Fractography and Fracture Toughness of 5% Cr-Mo-V Ultrahigh-Strength Steels, "Fracture Toughness of High-Strength Materials: Theory and Practice," Publication 120, The Iron and Steel Institute, 1970, p 81
10. T.V. Philip, ESR: A Means of Improving Transverse Mechanical Properties in Tool and Die Steels, *Met. Technol.*, Dec 1975, p 554
11. H.I. McHenry, "Fatigue Crack Propagation in Steel Alloys at Elevated Temperature," Report EER-EW-1029, Convair Aerospace Division, General Dynamics Corporation, Sept 1970

Hardenability of Carbon and Low-Alloy Steels

Hardenable Carbon and Low-Alloy Steels 451
Hardenability of Carbon and Low-Alloy Steels 464
Hardenability Curves 485

Hardenable Carbon and Low-Alloy Steels

Revised by Eugene R. Kuch, Gardner Denver Division of Cooper Industries

CARBON STEELS are produced in greater tonnage and have wider usage than any other metal because of their versatility and low cost. For about ten years before World War II, there was a trend toward replacing carbon steels with alloy steels. The scarcity of alloying elements at that time caused a reappraisal of carbon steels, and in many instances, users reverted to carbon grades. With rapidly rising costs and the prospect of permanent shortages of various alloying elements, this trend is continuing.

Carbon steels proved to be satisfactory upon reappraisal because:

- Hardenability, although less than that of alloy steels, was adequate for many parts, and for some parts, shallow hardening was actually advantageous because it minimized quench cracking
- Refinements in heat-treating methods, such as induction hardening and flame hardening, made it possible to obtain higher properties from carbon steels than was previously possible
- Many new compositions were added to the carbon steel group, permitting a more discriminating selection

More than 50 grades are nominally available in the 10xx-series carbon steels, and more than 30 grades are available in the resulfurized 11xx and 12xx series. Also available are about 20 grades in the 15xx series (originally 10xx series steels with higher than normal manganese contents, usually in the range of 1.00 to 1.65%). The versatility of carbon steels has been further extended by the addition of lead for improved machinability and of boron for greater hardenability and for improvement in cold heading.

In addition to the standard carbon and alloy steels, there is a specific class of high-strength low-alloy (HSLA) steels. These materials are normally furnished to mechanical property requirements rather than to composition ranges. These steels are generally not intended for quenching and tempering by the user. Their properties and characteristics are covered in the articles "High-Strength Structural and High-Strength Low-Alloy Steels" and "High-Strength Low-Alloy Steel Forgings" in this Volume.

Generally, alloy steels are not used without appropriate heat treatment. By far the largest tonnage of alloy steels is of the type with nominal carbon contents between about 0.25 and 0.55%. If an alloy steel is to be used for carburized parts, the nominal carbon content usually does not exceed 0.20%. Various combinations and amounts of manganese, silicon, nickel, chromium, molybdenum, vanadium, and boron are commonly present in these steels to enhance the properties of the quenched and tempered alloy.

The microstructure (tempered martensite or bainite) produced by quenching and tempering alloy steels is characterized by good toughness, that is, the capacity to deform without rupture, at a given strength level. The basic phenomenon of obtaining a favorable microstructure by heat treatment is also possible in plain carbon steels, although primarily in thin sections. Thus, the most important effect of alloying elements in steel is to induce the formation of martensite or bainite, and accompanying superior properties, in large sections. The level of hardness or strength of these structures is a function of the carbon content of the transformation products rather than of the alloying elements present.

The general effect of alloying elements dissolved in austenite is to decrease the rate of austenite transformation at subcritical temperatures. Because the desirable products of transformation in these steels (martensite and lower bainite) are formed at low temperatures, this decreased transformation rate is essential. Consequently, pieces can be cooled more slowly, or larger pieces can be quenched in a given medium, without transformation of austenite to the undesirable high-temperature products (pearlite or upper bainite). This function, decreasing the rate of transformation and thereby facilitating ultimate transformation to martensite or lower bainite, is known as hardenability, the most important effect of alloying elements in hardenable steels. By increasing hardenability, alloying elements greatly extend the potential for enhanced properties to the large sections required for many applications.

Many of the curves in this article and in the article "Hardenability Curves" in this Volume serve to illustrate the principle that, regardless of composition, tempered steels of the same hardness have approximately the same tensile strength.

The maximum (as-quenched) hardness of heat-treated steel depends primarily on the carbon content. Alloying elements have little effect on the maximum hardness that can be developed in steel, but they profoundly affect the depth to which this maximum hardness can be developed in a part of specific size and shape. Thus, for a specific application, one of the first decisions to be made is what carbon level is required to obtain the desired hardness. The next step is to determine what alloy content will give the proper hardening response in the section size involved. This is not to imply that tempered martensitic steels are alike in every respect, regardless of composition, because the alloy content is responsible for differences in the preservation of strength at elevated temperatures; in abrasion resistance; in resistance to corrosion; and even, to a certain extent, in toughness. However, the similarities are sufficiently marked to permit reasonably accurate predictions of mechanical properties from hardness rather than from composition, thereby justifying the emphasis on hardenability as the most important function of the alloying elements.

Hardenability

Required hardenability is the most important factor influencing a choice between carbon- and alloy steel. It is now possible to purchase carbon steels produced to a spe-

Table 1 Composition of carbon and carbon-boron H steels

UNS No.	SAE or AISI No.	Ladle chemical composition, wt%				
		C	Mn	Si	P, maximum(b)	S, maximum(b)
H10380	1038H	0.34/0.43	0.50/1.00	0.15/0.35	0.040	0.050
H10450	1045H	0.42/0.51	0.50/1.00	0.15/0.35	0.040	0.050
H15220	1522H	0.17/0.25	1.00/1.50	0.15/0.35	0.040	0.050
H15240	1524H	0.18/0.26	1.25/1.75	0.15/0.35	0.040	0.050
H15260	1526H	0.21/0.30	1.00/1.50	0.15/0.35	0.040	0.050
H15410	1541H	0.35/0.45	1.25/1.75	0.15/0.35	0.040	0.050
H15211	15B21H(a)	0.17/0.24	0.70/1.20	0.15/0.35	0.040	0.050
H15281	15B28H(a)	0.25/0.34	1.00/1.50	0.15/0.35	0.040	0.050
H15301	15B30H(a)	0.27/0.35	0.70/1.20	0.15/0.35	0.040	0.050
H15351	15B35H(a)	0.31/0.39	0.70/1.20	0.15/0.35	0.040	0.050
H15371	15B37H(a)	0.30/0.39	1.00/1.50	0.15/0.35	0.040	0.050
H15411	15B41H(a)	0.35/0.45	1.25/1.75	0.15/0.35	0.040	0.050
H15481	15B48H(a)	0.43/0.53	1.00/1.50	0.15/0.35	0.040	0.050
H15621	15B62H(a)	0.54/0.67	1.00/1.50	0.40/0.60	0.040	0.050

(a) These steels contain 0.0005 to 0.003% B. (b) If electric furnace practice is specified or required, the limit for both phosphorus and sulfur is 0.025%, and the prefix E is added to the SAE or AISI number. Source: Ref 1

cific hardenability band. As of 1989, there were six carbon H steels (1038H, 1045H, 1522H, 1524H, 1526H, and 1541H), nine carbon-boron H steels (15B21H, 15B28H, 15B30H, 15B35H, 15B37H, 15B41H, 15B46H, 15B48H, and 15B62H) and 70 standard alloy H steels for which hardenability data had been compiled. This establishes a more definite distinction between carbon and alloy steels than can be done by actual hardenability values because the latter are available as H steels in most of the tonnage grades. Tables 1 and 2 list the compositions of all the H steels.

Carbon steels with low manganese (some grades as low as 0.30%) and a virtual absence of residual nickel, chromium, and molybdenum are the lowest in hardenability of all standard steels. This holds true for nearly all carbon levels, because in the absence of higher manganese or other alloying elements, carbon functions almost entirely to control maximum hardness and has only a minor effect on hardenability.

Most of the hardenable 1xxx-series carbon steels have 0.60 to 0.90% manganese, although several grades have higher and others have lower manganese contents. Those steels with a maximum manganese content between 1.00% and 1.65% have been renumbered and now constitute the 15xx series. The effect of manganese on the variation of strength with carbon content in as-hot-rolled carbon steels is shown in Fig. 1. Manganese has a marked effect on hardenability. Even a difference of 0.25% makes a significant difference on the end-quench hardenability of 0.50% carbon steel (Fig. 2).

Considering the range of manganese that is available in carbon steels, it follows that a wide range of hardenability can exist. For example, 1541 steel frequently shows end-quench hardenability values higher than the minimum of the hardenability band for 1340 steel (Fig. 3). Thus, there is a gradual transition in hardenability from carbon grades to alloy grades.

Steels are not necessarily better because they are higher in hardenability. They are better only when higher hardenability is required. There are many applications for which minimum, rather than maximum, hardenability is needed, which accounts for the many low-manganese grades melted. For example, it is often desirable to produce thin layers of maximum hardness on shaft bearings or cam contours. This is usually accomplished by induction or flame hardening; however, if the hardened zone is too deep, an unfavorable pattern of residual stresses will be established, with resultant cracking in quenching or premature failure in service. In one instance, cams were made from standard 1050 steel (0.60 to 0.90% Mn) and induction hardened to 60 HRC to a depth of about 1.6 mm (¹⁄₁₆ in.). If the hardened zone became as deep as 3.2 mm (¹⁄₈ in.), a significant number of parts cracked. Cracking was eliminated by using a modified grade of 1050 steel (0.30 to 0.60% Mn), which resulted in a shallower hardened zone after induction hardening.

It is more economical to use carbon steels whenever possible. The higher-manganese grades cost more than the lower-manganese grades, but are less expensive than the least expensive alloy grades.

Measuring Hardenability. Many tests have been proposed for measuring and comparing the hardenability of various grades of steel. The end-quench test (often referred to as the Jominy test) is used for testing the hardenability of carbon and alloy steels. This test involves heating a test bar 25.4 mm (1.0 in.) in diameter to the proper austenitizing temperature, placing the bar in a special hardenability fixture, and quenching only the end surface with water. Hardness is then measured along a plane just below the surface, at intervals of about 1.6 mm

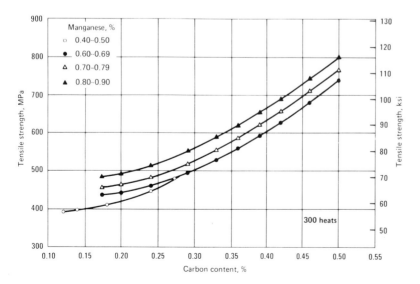

Fig. 1 Effect of manganese and carbon on tensile strength of as-hot-rolled carbon steels

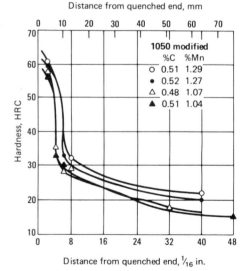

Fig. 2 Effect of carbon and manganese on end-quench hardenability of 1050 steel. The steels with 1.29 and 1.27% manganese contained 0.06% residual chromium. Steels with 1.07 and 1.04% manganese contained 0.06 and 0.08% residual chromium, respectively. No other residual elements were reported.

Table 2 Composition of standard alloy H steel

The ranges and limits on this table apply only to material not exceeding 1.3×10^5 mm² (200 in.²) in cross-sectional area, 460 mm (18 in.) in width, or 4.5 Mg (10 000 lb) in weight per piece. Ranges and limits are subject to the permissible variations for product analysis shown in Table 4 of SAE J409.

UNS No.	SAE or AISI No.	C	Mn	Si	Ni	Cr	Mo	V
H13300	1330H	0.27/0.33	1.45/2.05	0.15/0.35	...	...	...	...
H13350	1335H	0.32/0.38	1.45/2.05	0.15/0.35	...	...	...	...
H13400	1340H	0.37/0.44	1.45/2.05	0.15/0.35	...	...	...	...
H13450	1345H	0.42/0.49	1.45/2.05	0.15/0.35	...	...	...	...
H40270	4027H	0.24/0.30	0.60/1.00	0.15/0.35	...	...	0.20/0.30	...
H40280(c)	4028H(c)	0.24/0.30	0.60/1.00	0.15/0.35	...	...	0.20/0.30	...
H40320	4032H	0.29/0.35	0.60/1.00	0.15/0.35	...	...	0.20/0.30	...
H40370	4037H	0.34/0.41	0.60/1.00	0.15/0.35	...	...	0.20/0.30	...
H40420	4042H	0.39/0.46	0.60/1.00	0.15/0.35	...	...	0.20/0.30	...
H40470	4047H	0.44/0.51	0.60/1.00	0.15/0.35	...	...	0.20/0.30	...
H41180	4118H	0.17/0.23	0.60/1.00	0.15/0.35	...	0.30/0.70	0.08/0.15	...
H41300	4130H	0.27/0.33	0.30/0.70	0.15/0.35	...	0.75/1.20	0.15/0.25	...
H41350	4135H	0.32/0.38	0.60/1.00	0.15/0.35	...	0.75/1.20	0.15/0.25	...
H41370	4137H	0.34/0.41	0.60/1.00	0.15/0.35	...	0.75/1.20	0.15/0.25	...
H41400	4140H	0.37/0.44	0.65/1.10	0.15/0.35	...	0.75/1.20	0.15/0.25	...
H41420	4142H	0.39/0.46	0.65/1.10	0.15/0.35	...	0.75/1.20	0.15/0.25	...
H41450	4145H	0.42/0.49	0.65/1.10	0.15/0.35	...	0.75/1.20	0.15/0.25	...
H41470	4147H	0.44/0.51	0.65/1.10	0.15/0.35	...	0.75/1.20	0.15/0.25	...
H41500	4150H	0.47/0.54	0.65/1.10	0.15/0.35	...	0.75/1.20	0.15/0.25	...
H41610	4161H	0.55/0.65	0.65/1.10	0.15/0.35	...	0.65/0.95	0.25/0.35	...
H43200	4320H	0.17/0.23	0.40/0.70	0.15/0.35	1.55/2.00	0.35/0.65	0.20/0.30	...
H43400	4340H	0.37/0.44	0.55/0.90	0.15/0.35	1.55/2.00	0.65/0.95	0.20/0.30	...
H43406(d)	E4340H(d)	0.37/0.44	0.60/0.95	0.15/0.35	1.55/2.00	0.65/0.95	0.20/0.30	...
H46200	4620H	0.17/0.23	0.35/0.75	0.15/0.35	1.55/2.00	...	0.20/0.30	...
H47180	4718H	0.15/0.21	0.60/0.95	0.15/0.35	0.85/1.25	0.30/0.60	0.30/0.40	...
H47200	4720H	0.17/0.23	0.45/0.75	0.15/0.35	0.85/1.25	0.30/0.60	0.15/0.25	...
H48150	4815H	0.12/0.18	0.30/0.70	0.15/0.35	3.20/3.80	...	0.20/0.30	...
H48170	4817H	0.14/0.20	0.30/0.70	0.15/0.35	3.20/3.80	...	0.20/0.30	...
H48200	4820H	0.17/0.23	0.40/0.80	0.15/0.35	3.20/3.80	...	0.20/0.30	...
H50401(e)	50B40H(e)	0.37/0.44	0.65/1.10	0.15/0.35	...	0.30/0.70	...	...
H50441(e)	50B44H(e)	0.42/0.49	0.65/1.10	0.15/0.35	...	0.30/0.70	...	...
H50460	5046H	0.43/0.50	0.65/1.10	0.15/0.35	...	0.13/0.43	...	...
H50461(e)	50B46H(e)	0.43/0.50	0.65/1.10	0.15/0.35	...	0.13/0.43	...	...
H50501(e)	50B50H(e)	0.47/0.54	0.65/1.10	0.15/0.35	...	0.30/0.70	...	...
H50601(e)	50B60H(e)	0.55/0.65	0.65/1.10	0.15/0.35	...	0.30/0.70	...	...
H51200	5120H	0.17/0.23	0.60/1.00	0.15/0.35	...	0.60/1.00	...	...
H51300	5130H	0.27/0.33	0.60/1.10	0.15/0.35	...	0.75/1.20	...	...
H51320	5132H	0.29/0.35	0.50/0.90	0.15/0.35	...	0.65/1.10	...	...
H51350	5135H	0.32/0.38	0.50/0.90	0.15/0.35	...	0.70/1.15	...	...
H51400	5140H	0.37/0.44	0.60/1.00	0.15/0.35	...	0.60/1.00	...	...
H51470	5147H	0.45/0.52	0.60/1.05	0.15/0.35	...	0.80/1.25	...	...
H51500	5150H	0.47/0.54	0.60/1.00	0.15/0.35	...	0.60/1.00	...	...
H51550	5155H	0.50/0.60	0.60/1.00	0.15/0.35	...	0.60/1.00	...	...
H51600	5160H	0.55/0.65	0.65/1.10	0.15/0.35	...	0.60/1.00	...	...
H51601(e)	51B60H(e)	0.55/0.65	0.65/1.10	0.15/0.35	...	0.60/1.00	...	...
H61180	6118H	0.15/0.21	0.40/0.80	0.15/0.35	...	0.40/0.80	...	0.10/0.15
H61500	6150H	0.47/0.54	0.60/1.00	0.15/0.35	...	0.75/1.20	...	0.15
H81451(e)	81B45S(e)	0.42/0.49	0.70/1.05	0.15/0.35	0.15/0.45	0.30/0.60	0.08/0.15	...
H86170	8617H	0.14/0.20	0.60/0.95	0.15/0.35	0.35/0.75	0.35/0.65	0.15/0.25	...
H86200	8620H	0.17/0.23	0.60/0.95	0.15/0.35	0.35/0.75	0.35/0.65	0.15/0.25	...
H86220	8622H	0.19/0.25	0.60/0.95	0.15/0.35	0.35/0.75	0.35/0.65	0.15/0.25	...
H86250	8625H	0.22/0.28	0.60/0.95	0.15/0.35	0.35/0.75	0.35/0.65	0.15/0.25	...
H86270	8627H	0.24/0.30	0.60/0.95	0.15/0.35	0.35/0.75	0.35/0.65	0.15/0.25	...
H86300	8630H	0.27/0.33	0.60/0.95	0.15/0.35	0.35/0.75	0.35/0.65	0.15/0.25	...
H86301(e)	86B30H(e)	0.27/0.33	0.60/0.95	0.15/0.35	0.35/0.75	0.35/0.65	0.15/0.25	...
H86370	8637H	0.34/0.41	0.70/1.05	0.15/0.35	0.35/0.75	0.35/0.65	0.15/0.25	...
H86400	8640H	0.37/0.44	0.70/1.05	0.15/0.35	0.35/0.75	0.35/0.65	0.15/0.25	...
H86420	8642H	0.39/0.46	0.70/1.05	0.15/0.35	0.35/0.75	0.35/0.65	0.15/0.25	...
H86450	8645H	0.42/0.49	0.70/1.05	0.15/0.35	0.35/0.75	0.35/0.65	0.15/0.25	...
H86451(e)	86B45H(e)	0.42/0.49	0.70/1.05	0.15/0.35	0.35/0.75	0.35/0.65	0.15/0.25	...
H86500	8650H	0.47/0.54	0.70/1.05	0.15/0.35	0.35/0.75	0.35/0.65	0.15/0.25	...
H86550	8655H	0.50/0.60	0.70/1.05	0.15/0.35	0.35/0.75	0.35/0.65	0.15/0.25	...
H86600	8660H	0.55/0.65	0.70/1.05	0.15/0.35	0.35/0.75	0.35/0.65	0.15/0.25	...
H87200	8720H	0.17/0.23	0.60/0.95	0.15/0.35	0.35/0.75	0.35/0.65	0.20/0.30	...
H87400	8740H	0.37/0.44	0.70/1.05	0.15/0.35	0.35/0.75	0.35/0.65	0.20/0.30	...
H88220	8822H	0.19/0.25	0.70/1.05	0.15/0.35	0.35/0.75	0.35/0.65	0.30/0.40	...
H92600	9260H	0.55/0.65	0.65/1.10	1.70/2.20	...	...	...	...
H93100(d)	9310H(d)	0.07/0.13	0.40/0.70	0.15/0.35	2.95/3.55	1.00/1.45	0.08/0.15	...
H94151(e)	94B15H(e)	0.12/0.18	0.70/1.05	0.15/0.35	0.25/0.65	0.25/0.55	0.08/0.15	...
H94171(e)	94B17H(e)	0.14/0.20	0.70/1.05	0.15/0.35	0.25/0.65	0.25/0.55	0.08/0.15	...
H94301(e)	94B30H(e)	0.27/0.33	0.70/1.05	0.15/0.35	0.25/0.65	0.25/0.55	0.08/0.15	...

(a) Small quantities of certain elements may be found in alloy steel that are not specified or required. These elements are to be considered incidental and acceptable to the following maximum amounts: copper to 0.35%, nickel to 0.25%, chromium to 0.20%, and molybdenum to 0.06%. (b) For open hearth and basic oxygen steels, maximum sulfur content is to be 0.040%, and maximum phosphorus content is to be 0.035%. Maximum phosphorus and sulfur in basic electric furnace steels are to be 0.025% each. (c) Sulfur content range is 0.035/0.050%. (d) Electric furnace steel. (e) These steels contain 0.0005 to 0.003% B. Source: Ref 1

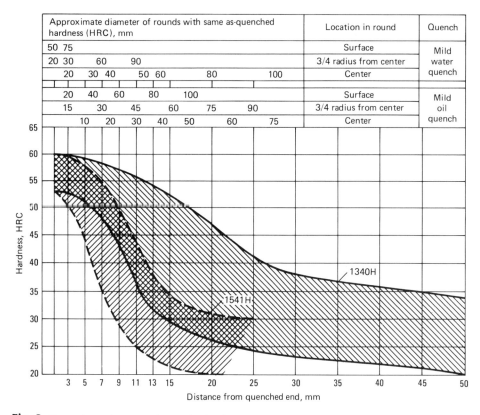

Fig. 3 Comparison of 1340H and 1541H hardenability ranges. Source: Ref 1

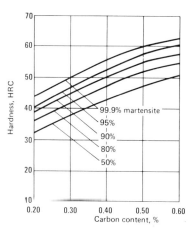

Fig. 4 Average relationship between carbon content and hardness for steels containing different amounts of untempered martensite in their microstructures

($1/16$ in.) from the water-quenched end, to determine how far from this end the hardness extends. Typical end-quench hardenability data are given in the article "Hardenability Curves" in this Volume.

Some industries have found the SAC (surface area center) test, also known as the Rockwell inch test, to be more discriminating than the end-quench test for determining the hardenability of shallow-hardening steels because of the sharp gradient on the end-quench curve.

Carbon Steels Containing up to 0.25% C

Carbon steels with nominal carbon contents below about 0.25% are the first of three arbitrary classifications of carbon steels. Three principal types of heat treatments are used for this first group of steels:

- Case-hardening treatments, which are used mainly to produce a hard, wear-resistant surface layer
- Conditioning treatments, such as process annealing, that prepare the steel for certain fabricating operations
- Quenching and tempering to improve machinability

The improvements in mechanical properties that can be gained by the quenching and tempering of low-carbon steels, however, are minimal and are usually not worth the cost.

Many industrial applications require a part that is hard and wear resistant, but in almost every instance these properties are of prime importance at the surface of the part. Usually, it is a significant advantage in processing or service (or both) to have a part that is hard and wear resistant at the surface, yet softer, more ductile, and tougher below the surface. Case-hardening processes such as carburizing, carbonitriding, nitriding, and cyaniding are intended to produce just such a combination of properties by changing the composition of the surface layer through the absorption and inward diffusion of carbon, nitrogen, or both. Case hardening has very definite economic advantages for parts that must be formed and/or machined to intricate shapes and must be wear resistant in service, with surface properties approximating those normally attributed to high-carbon steels.

The cold-heading properties of most steels are improved by process annealing, spheroidizing, or stress relieving. In general, process annealing is done at the steel mill. Additional heat treatment is not used unless required, for at least two reasons: First, the process could cost more than any savings realized in cold heading, and, second, cold-headed products often depend for their final strength on work hardening prior to and during the heading process, and if reannealed before cold heading, they may lose much of their potential strength.

Heat treatments are frequently employed to improve machinability. The generally poor machinability of the low-carbon steels, except those containing sulfur or other special alloying elements, results principally from the fact that the proportion of free ferrite to carbide is high. This situation cannot be changed fundamentally, but the machinability can be improved by putting the carbide in its most voluminous form, pearlite, and dispersing this pearlite evenly throughout the ferrite mass. Normalizing is commonly used with success, but best results are obtained by quenching the steel in oil from a temperature of about 815 to 870 °C (1500 to 1600 °F). In steels other than 1025 and 1524, no martensite is formed, and the parts do not require tempering.

Carbon Steels Containing 0.25 to 0.55% C

Because of their higher carbon content, carbon steels containing 0.25 to 0.55% C are usually used in the hardened and tempered condition. By varying the quenching medium (cooling rate) and tempering temperature, a wide range of mechanical properties can be produced. These steels are the most versatile of the three groups of carbon steels and are most commonly used for crankshafts, couplings, tie-rods, and many other machinery parts that have required hardness values in the range of 229 to 444 HB.

In this group of steels, which includes water-hardening and oil-hardening types, hardenability is very sensitive to changes in chemical composition, particularly to changes in the content of carbon, manganese, silicon, and residual elements. It is also sensitive to grain size. Figure 4 illustrates the general relationship between carbon content and hardness obtained in quenching for microstructures containing various amounts of martensite.

The rate of the heating of parts for quenching has a marked effect on hardenability under certain conditions. If the structure is nonuniform as a result of severe banding or lack of proper normalizing or annealing, extremely rapid heating (such as may be obtained in liquid baths) does not allow sufficient time for the diffusion of carbon and other elements in the austenite. Thus, nonuniform or low hardness will be produced unless the time at temperature is extended. In heating steels that contain free carbide (for example, spheroidized material), sufficient time must be allowed for the carbides to dissolve; otherwise, the austenite at the time of quenching will have a lower carbon content than is represented by the chemical composition of the steel, and disappointing results may be obtained. However, this condition may be produced deliberately (see the section "Carbon Steels Containing 0.55 to 1.00% C" below).

Usually these medium-carbon steels should be either normalized or annealed before hardening to obtain the best mechanical properties after hardening and tempering. Parts made from bar stock are frequently given no treatment prior to hardening, but it is common practice to normalize or anneal forgings.

Most bar stock, both hot finished and cold finished, is machined as received, except for the higher-carbon grades and small sizes that require annealing to reduce the as-received hardness. Forgings are usually normalized because this treatment avoids the extreme softening and consequent reduction of machinability that results from annealing.

In some instances, a cycle treatment is used. The parts are heated as for normalizing and are then cooled rapidly in the furnace to a temperature somewhat above the nose of the time-temperature transformation (TTT) curve, that is, within the transformation range that produces pearlite. The parts are then held at temperature or cooled slowly until the desired amount of transformation has taken place; thereafter, they are cooled in any convenient manner. Specially arranged furnaces are usually required. Details of the treatments vary widely and are frequently determined by the furnace equipment available.

Cold-headed products are commonly made from these steels, especially from those containing less than 0.40% C. Process treating before cold working is usually necessary because the higher carbon content decreases workability. For certain uses, these steels are normalized or annealed above the upper transformation temperature, but more frequently, a spheroidizing treatment is used. The degree of spheroidization required depends on the application. After shaping operations are finished, the parts may be heat treated by quenching and tempering.

These medium-carbon steels are widely used for machinery parts for moderate duty.

When such parts are to be machined after heat treatment, the maximum hardness is usually held to 321 HB and frequently much lower.

Water is the quenching medium most commonly used because it is the most economical and the easiest to install. Caustic soda solution (5 to 10% NaOH by weight) is used in many instances with improved results. Compared to water, it is a faster and therefore a more thorough and uniform quench, producing better mechanical properties in all but light sections. Because of its rapid action, more scale is removed from the parts.

When used hot (55 to 70 °C, or 130 to 160 °F), caustic soda solution frequently makes possible drastic quenching of parts that could not be water quenched without cracking. The disadvantages of caustic soda are that it can be used only in a closed system with provisions made for cooling, operators must be protected against contact with it, the solution must be checked frequently, and the proper concentration must be maintained. Woolen clothing disintegrates rapidly after contact with it; cotton clothing is not affected.

Salt solutions (brine) are often successfully used. Up to about 40 °C (100 °F), they produce almost as good results as caustic solutions, but are much less effective when hot. Like caustic solutions, they require a closed system. Salt solutions are not as dangerous to operators as are hot caustic ones, but their corrosive action on iron and steel equipment is very serious.

When the section is thin or the properties required after heat treatment are not high, oil quenching is often used. This usually eliminates cracking and is very effective in reducing distortion.

Of the four quenching media, oil tends to provide the slowest cooling of steel parts, and brine the most rapid (Ref 2). Furthermore, the effectiveness of each of these liquids increases if it is agitated. Sometimes it is desirable to use a quenching medium with a severity between that of water and oil. For this, the so-called polymer quenchants are used. These include polyvinyl alcohol (PVA) (the most common), polyalkylene glycol ethers (PAGs), polyvinylpyrrolidone (PVP), and polyacrylates (Ref 2).

A wide range of austenitizing temperatures is necessary for compatibility with the possible variations in composition, grain size, section size, and quenching medium or cooling rate. Lower temperatures should be used for the higher-manganese steels, light sections, coarse-grained material, and water quenching; higher temperatures are required for lower-manganese steels, heavy sections, fine-grain materials, and oil quenching. Excessive temperatures should be avoided; in many steels, the higher austenitizing temperatures promote the retention of austenite.

Many common hand tools, such as pliers, open-end wrenches, screwdrivers, and a few edged tools (for example, tin snips and brush knives) are made from these steels. The cutting tools are necessarily quenched locally on the cutting edges, in water, brine, or caustic solution, and are subsequently given suitable tempering treatments. In some instances, the edge is time quenched; the remainder of the tool is then oil quenched for partial strengthening. When made of these grades of steel, pliers, wrenches, and screwdrivers are usually quenched in water, either locally or completely, and then are properly tempered.

Carbon Steels Containing 0.55 to 1.00% C

Carbon steels with these higher carbon contents are more restricted in application than are the 0.25 to 0.55% C steels because of decreased machinability, poor formability, and poor weldability, which make them more difficult and more costly to fabricate. They are also less ductile after hardening and tempering. Higher-carbon steels such as 1070 to 1095 are especially suitable for springs that require resistance to fatigue and permanent set. They are also used in the nearly fully hardened condition (55 HRC and higher) for applications in which abrasion resistance is a primary requirement, as for agricultural tillage tools such as plowshares and knives for cutting hay or grain.

Forged parts should be annealed because the refinement of the forging structure is important in producing a high-quality hardened product and because the as-forged parts are too hard for cold trimming or for economical machining. Ordinary annealing practice, followed by furnace cooling to about 595 °C (1100 °F), is satisfactory for most parts.

Most of the parts made from steels in this group are hardened by conventional quenching. However, special techniques are necessary at times. Both oil and water quenching are used; water is used for heavy sections of the lower-carbon steels and for cutting edges, and oil is used for general purposes. Austempering and martempering are often applied successfully; the principal advantages from such treatments are considerably reduced distortion, elimination of breakage (in many instances), and greater toughness at high hardness.

For heavy machinery parts, such as shafts, collars, and the like, 1055 and 1561 steels may be used, either normalized and tempered for low strength or quenched and tempered for moderate strength. Other steels in the list may be used, but the combination of carbon and manganese in the two mentioned makes them particularly well adapted to such applications.

Even with favorable hardenability factors and with the use of a drastic quench, these steels are much shallower hardening than are alloy steels because carbon alone, or in combination with manganese in the amounts involved here, does not promote deep hardening. Therefore, the sections for which such steels are suited will definitely be limited. Even so, the danger of quench cracking is real and must be carefully guarded against when such parts are heat treated, especially parts of nonuniform section.

For hardening wrenches (except the Stillson type), screwdrivers, pliers, and similar tools, oil quenching is generally used, followed by tempering to the required hardness range. Even when no reduction of as-quenched hardness is desired, stress relieving at 150 to 190 °C (300 to 375 °F) is desirable to help prevent sudden service failures. In Stillson-type wrenches, the jaw teeth are really cutting edges and are nearly always quenched in water or brine to produce a hardness of 50 to 60 HRC. Either the jaws may be locally heated and quenched or the parts may be heated all over and the jaws locally time quenched in water or brine. The entire part is then quenched in oil for partial hardening of the remainder. In this way, considerable structural strength is obtained.

Hammers must possess high hardness on the striking face and somewhat lower hardness on the claws. They are usually locally hardened and tempered on each end, depending on type. The striking face is always quenched in water or brine. Satisfactory service depends on obtaining the proper depth of fully hardened (martensitic) surface on this face and then stress relieving at about 175 °C (350 °F). Final hardness is usually 50 to 58 HRC on the striking face and 40 to 47 HRC on claws.

Hand cutting tools, particularly axes and hatchets, must possess high hardness and high relative toughness in their cutting edges, as well as the ability to hold a keen edge. Because nothing is as effective as carbon in imparting the latter property, the carbon content is always higher than if hardness and toughness alone were required. Many such tools are given an ordinary furnace anneal after forging, but high-quality tools are prepared for hardening by spheroidization, which may be performed as a separate operation after regular annealing. Most frequently, however, the spheroidizing and hardening treatments are combined by heating to 870 °C (1600 °F), quenching in oil, and then tempering at 675 to 745 °C (1250 to 1375 °F).

For hardening, the cutting edges of such tools are usually heated in liquid baths to the lowest temperature at which the piece can be hardened and are then quenched in brine. The quick heating of the liquid bath and the low temperature fail to put all the spheroidal carbide into solution. As a result, the cutting edge of the tool consists of martensite with less carbon than indicated by the chemical composition of the steel; it also contains particles of cementite. In this condition, the tool is at its maximum toughness, relative to its hardness, and the cementite particles promote long life of the cutting edge. Final hardness may range from 55 to 60 HRC.

Agricultural Machinery Components. Manufacturers of agricultural implements make wide use of the steels in this group. Braces, control rods, shafts, and similar parts are often made of high-carbon steels to obtain increased strength without the additional cost of heat treating. The principal heat treated parts are plowshares, moldboards, coulters, cultivator shovels, disks for harrows and plows, mower and binder knives, ledger plates, and band knives. Those parts used for cutting or turning soil must be moderately tough, with the ability to resist abrasion. They are made of steels containing various combinations of carbon and manganese that will permit full hardening.

Grass-cutting and grain-cutting tools are usually made of 1090 or 1095 steel because the higher carbon level provides the desired edge to give a long service life. These parts are made from strip stock by blanking. They are not annealed below the transformation range, except for some process annealing that may be done at the rolling mill. As a result, spheroidization is obtained only occasionally. Local hardening is done either by induction heating or in continuous furnaces provided with fixtures that permit the pieces to be heated on the cutting edges only. Upon discharge, the parts are quenched all over in oil and are tempered at a low temperature. Final hardness on the cutting edges may range from 55 to 60 HRC.

Alloy Steels

Alloy steels are ordinarily quench hardened and tempered to the level of strength desired for the application. The tempered martensite or bainite thus obtained imparts toughness, the capability of deforming without rupture.

The several elements commonly dissolved in austenite prior to quenching increase hardenability in approximately the following ascending order: nickel, silicon, manganese, chromium, molybdenum, vanadium, and boron. The addition of several alloying elements in small amounts is more effective in increasing hardenability than the addition of much larger amounts of one or two alloying elements.

To increase hardenability effectively, it is essential that the alloying elements be in solution in the austenite. Steels containing the carbide-forming elements, that is, chromium, molybdenum, and vanadium, require special consideration in this respect. These elements are present predominantly in the carbide form in annealed steels; such carbides dissolve only at higher temperatures, and dissolution proceeds more slowly than for iron carbide. Hence, for effectiveness, heating schedules should be set up to dissolve an adequate proportion of these elements. It is uneconomical to heat at such low temperatures that the carbide-forming elements are largely undissolved.

Because the basic function of alloying in these steels is to increase hardenability, the selection of a steel and the choice of suitable austenitizing conditions should primarily be based on obtaining adequate hardenability. More than adequate hardenability is rarely a disadvantage, except in terms of cost.

Alloying Elements in Quenching

Because the sections treated are often relatively large and because the alloying elements have the general effect of lowering the temperature range at which martensite is formed, the thermal and transformational stresses set up during quenching tend to be greater in these alloy steel parts than those encountered in quenching the necessarily smaller sections of plain carbon steels. In general, the greater stresses result in distortion and risk of cracking.

Alloying elements, however, have two functions that tend to offset these disadvantages. First and probably most important is the capacity to permit the use of a lower carbon content for a given application. The decrease in hardenability accompanying the decrease in carbon content may be readily offset by the hardenability effect of the added alloying elements, and the lower carbon steel will exhibit a much lower susceptibility to quench cracking. This lower susceptibility results from the greater plasticity of the low-carbon martensite and from the generally higher temperature range at which martensite is formed in the lower-carbon materials. Quench cracking is seldom encountered in steels containing 0.25% C or less, and the susceptibility to cracking increases progressively with increasing carbon content.

The second function of the alloying elements in quenching is to permit slower rates of cooling for a given section because of increased hardenability, thereby generally decreasing the thermal gradient and, in turn, the cooling stress. It should be noted, however, that this is not altogether advantageous, because the direction, as well as the magnitude, of the stress existing after the quench is important in relation to cracking.

To prevent cracking, surface stresses after quenching should be either compressive or at a relatively low tensile level. In general, the use of a less drastic quench suited to the hardenability of the steel will result in lower distortion and greater freedom from cracking.

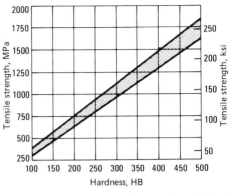

Fig. 5 Relation between tensile strength and Brinell hardness for steels in the as-rolled, normalized, or quenched and tempered condition. The tensile strength in ksi is approximately one-half the Brinell hardness number and in MPa is approximately 3½ times the Brinell hardness number.

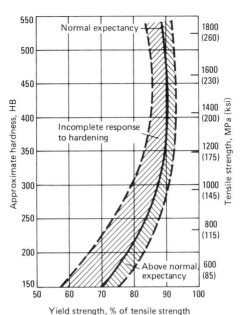

Fig. 6 Relation between tensile strength and yield strength for quenched and tempered steels. Source: Ref 1

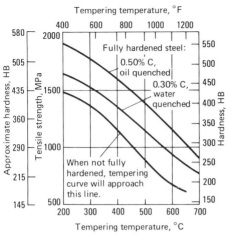

Fig. 7 Effect of tempering temperature on tensile strength and hardness of hardened carbon and alloy steels with carbon contents of 0.50 and 0.30%

Furthermore, the increased hardenability of these alloy steels may permit heat treatment by austempering or martempering, and therefore the level of adverse residual stress before tempering may be held to a minimum. In austempering, the workpiece is cooled rapidly to a temperature in the lower bainite region and is held at that temperature so that the section transforms completely to bainite. Because transformation occurs at a relatively high temperature and proceeds rather slowly, the stress level after transformation is quite low, and distortion is minimal.

In martempering, the workpiece is cooled rapidly to a temperature just above M_s, held there until the piece attains a uniform temperature throughout, and then is cooled slowly (usually by air cooling) through the martensite range. This procedure causes martensite to form more or less simultaneously throughout the entire section, thereby holding transformational stresses at a very low level, which minimizes distortion and the danger of cracking.

Mechanical Properties

When quenched to martensite and tempered to the same hardness, carbon and alloy steels have similar tensile properties in that portion of the cross section that reacts to the quench. If carbon steel has the hardenability required by the critical section of the part and the quench used, the resulting tensile strength, yield strength, and elongation in the fully hardened zone will be in the same range as in a similar zone in an alloy steel quenched and tempered to the same hardness. The similarity in properties of the hardened zones is maintained, regardless of the depth of hardening, but the strength of the pieces will be governed by the thickness of the respective hardened zone (depth of hardening).

Figure 5 shows the relationship between hardness and tensile strength for hardened and tempered, as-rolled, annealed, or normalized carbon and alloy constructional steels. Because of the effect of cold working, this relationship may not apply to cold-drawn steels. Figure 6 shows the relation between tensile strength and yield strength. The effect of tempering temperature on tensile strength and hardness is shown in Fig. 7.

An important exception to this similarity of properties is the relationship between tensile strength and the reduction in area. Figure 8 shows the direct relationship between ductility and hardness and illustrates that the reduction in area decreases as hardness increases and, for a given hardness, the reduction is higher for alloy steels than for plain carbon steels. Figures 5 to 8 further reinforce the contention that, despite some differences in certain properties, the major difference between carbon and alloy steels is hardenability.

One other and sometimes important difference between carbon and alloy steels is that, for the same hardness levels, fully quenched alloy steels require higher tempering temperatures than do carbon steels. This higher tempering temperature is presumed to reduce the stress level in the finished parts without impairing mechanical properties.

Property relations given in Fig. 5 to 8 illustrate general correlations among mechanical properties of steels. Normal variations in composition and grain size from heat to heat and even within one heat produce a considerable scatter of results in sections of the same size.

Changes in section size have a greater influence on the mechanical properties of carbon steels (particularly when quenched and tempered) than on the properties of alloy steels because of the lower hardenability of carbon steels. The section size of a heat-treated section affects not only specific properties, but also the relation of one property to another. As the section size increases, incomplete response to hardening lowers the ratio of yield strength to tensile strength. Figure 9 shows mechanical property relations for 1030 steel in both the as-rolled and quenched and tempered conditions, as a function of section size. The tensile strength decreases as the section size increases for a given composition and heat treatment, and there is some lowering of the ratio of yield to tensile strength.

As the section size increases and the hardness and strength levels increase, the hardenability of carbon steels is no longer adequate, and alloy steels must be used. The H-band alloy steels that should be considered for different design yield strength levels for highly stressed parts are given in Table 3 for round sections up to 102 mm (4 in.) in diameter. Either oil quenching or water quenching these alloys, as noted, should give 80% martensite at the indicated location within the section. For moderately stressed parts, a 50% martensite structure at the center is frequently adequate. The H-band alloy steels that can produce 50% martensite at the indicated location by oil or water quenching are given in Table 4 for round sections up to 102 mm (4 in.) in diameter. This information is intended to be a guide in the selection of an appropriate alloy steel; variations in equipment and techniques will greatly influence the final properties obtained. The grades are listed in the approximate order of increasing alloy cost to further aid in the initial screening of candidate alloys.

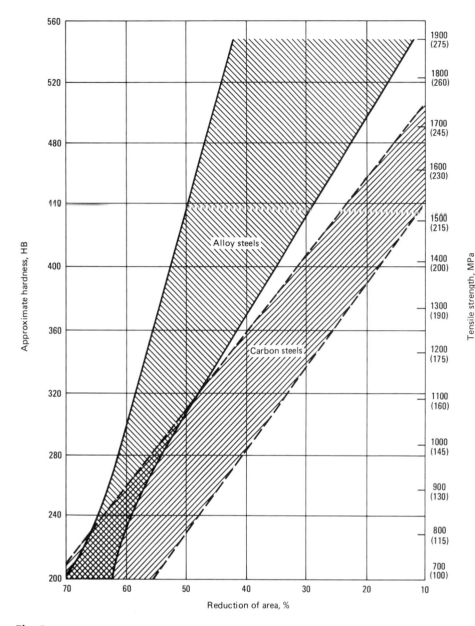

Fig. 8 Relation between tensile strength and the reduction in area for quenched and tempered steels

Figure 11 shows the response to tempering of four carbon and alloy steels containing 0.45% C. All steels were tempered for 1 h at the temperatures indicated. Somewhat shorter or longer intervals at temperature would affect hardness values to various degrees, depending on the tempering temperature.

The general effect of alloying is to retard the tempering rate, and therefore alloy steels require a higher tempering temperature to obtain a given hardness than does carbon steel of the same carbon content. However, the individual elements show significant differences in the magnitude of their retarding effect. Nickel, silicon, aluminum, and, to a large extent, manganese, all of which have little or no tendency to occur in the carbide phase and merely remain dissolved in ferrite, have only a minor effect on the hardness of the tempered steel, as would be expected from the general pattern of solid-solution hardening.

However, the carbide-forming elements, chromium, molybdenum, and vanadium, retard softening, particularly at higher tempering temperatures. These elements do not merely raise the tempering temperature; when they are present in higher percentages, the rate of tempering is no longer a continuous function of tempering temperature. That is, the tempering curves for these steels will show a range of tempering temperature in which the tempering is retarded or, with relatively high alloy content, in which the hardness may actually increase with an increase in tempering temperature. This characteristic behavior is known as secondary hardening and results from a delayed precipitation of fine alloy carbides. Secondary hardening is most often encountered in the higher-alloy tool steels.

As mentioned previously, the primary purpose of tempering is to impart plasticity or toughness to the steel, and the loss in strength is only incidental to this very important increase in toughness. The increase in toughness after tempering reflects two effects of tempering:

- The relief of residual stress induced during quenching
- The precipitation, coalescence, and spheroidization of iron and alloy carbides, resulting in a microstructure of greater plasticity

In addition to their effects on microstructure, the alloying elements have a secondary function. The higher tempering temperatures for a given hardness, which has been determined to be characteristic of alloy steels (particularly those containing carbide-forming elements), will presumably permit greater relaxation of residual stress and thereby improve properties. Furthermore, as discussed in the section "Alloying Elements in Quenching" in this article, the hardenability of these steels may permit the

Increasing carbon content consistently increases tensile and yield strength and decreases elongation and reduction in area, regardless of whether the steel is as-rolled or quenched and tempered (provided the ranges of tempering temperatures are the same). However, there is one major disadvantage to increasing the carbon content: Carbon steels show an increasing tendency to crack on quenching as the carbon content increases above about the 0.35% level. Consequently, parts to be made from steel having a carbon content greater than 0.35% should be tested for quench cracking before production is begun.

Variations in chemical composition within a specific grade contribute to the scatter of mechanical properties. This is illustrated by the test data in Fig. 10, where the properties for two heats of quenched and tempered 1050 steel are compared for a tempering range of 315 to 650 °C (600 to 1200 °F).

Tempering

Hardened steels are softened by reheating, although this effect may not be sought in tempering. The real need is to increase the capability of the steel to flow moderately without fracture, and this is inevitably accompanied by a loss of strength. The tensile strength is very closely related to hardness in this class of steels, as heat treated; thus, the effects of tempering can be followed by measuring the Brinell or Rockwell hardness.

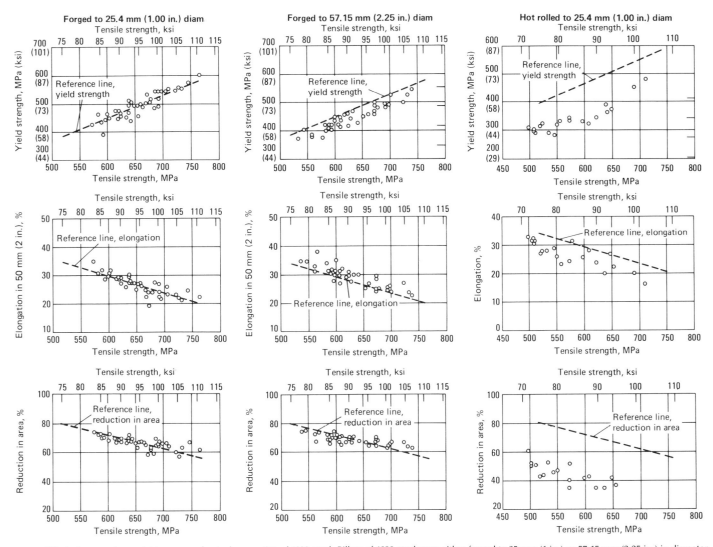

Fig. 9 Effect of processing variables on mechanical properties of 1030 steel. Billets of 1030 steel were either forged to 25 mm (1 in.) or 57.15 mm (2.25 in.) in diameter, then quenched and tempered, or they were hot rolled to 25 mm (1 in.) in diameter and not heat treated. Heat-treated specimens were water quenched from 870 °C (1600 °F) and tempered at 535 to 650 °C (1000 to 1200 °F). Specimens were taken from the center of 25 mm (1 in.) bars and at half-radius from the 57.15 mm (2.25 in.) bars. Reference lines represent mean values for the forged and heat-treated 25 mm (1 in.) diam bars.

use of less drastic quenching practices, so that the stress level before tempering will be lower, permitting these steels to be used at a higher level of hardness; this is because higher temperatures are not required for relief of quenching stresses. It should be noted, however, that this latter characteristic is only a secondary function of alloying elements in tempering; the effect primarily reflects the hardenability function of the alloying elements.

Another secondary function of alloying elements in tempering is to permit the use of steels with lower carbon content for a given level of hardness, because adequate tempering may be ensured by the retardation of softening caused by alloying. This results in greater freedom from cracking and generally improved plasticity at any given hardness. Here again, the function of alloying elements in tempering is a secondary function; their primary function is to increase hardenability sufficiently to offset the effect of a decreased carbon content.

The increase in plasticity upon tempering is discontinuous in those alloy steels that contain the carbide-forming elements; the behavior of notched specimens shows a characteristic irregularity at approximately 260 to 315 °C (500 to 600 °F). The quenched martensitic steel gains toughness, as reflected in a notched-bar impact test, by tempering at temperatures as high as 205 °C (400 °F). However, after tempering at higher temperatures, in the temper-brittle range, these types of steel lose toughness until they may be less tough than the same steels not tempered. Still higher tempering temperatures restore greater toughness (see Fig. 12).

The mechanism of this behavior is not fully understood, but it seems to be associated with the first precipitation of carbide particles and is presumably a grain boundary phenomenon; fractures of steels tempered in this region tend to follow intergranular paths. Thus, there is a range of tempering temperatures at about 205 to 370 °C (400 to 700 °F) never used for these steels; the tempering temperature is either below 205 °C (400 °F) or above 370 °C (700 °F). Although this phenomenon is common to all of these alloy steels, the alloying elements have a secondary function in this connection; a combination of carbon and alloy contents of suitable hardenability may be chosen that would permit tempering to the desired strength at temperatures outside this undesirable range.

Temper brittleness is another example of a discontinuous increase in plasticity subsequent to the tempering of steels containing the carbide-forming elements. This phe-

Table 3 Alloy steel selection guide for highly stressed parts

Unless otherwise indicated in the footnotes, any steel in this table may be considered for a lower strength level or a smaller section, or both.

Required yield strength MPa	ksi	As-tempered hardness HRC	HB	≤13 mm (½ in.) At center	13–25 mm (½–1 in.) At center	25–38 mm (1–1½ in.) At midradius	38–50 mm (1½–2 in.) At midradius	50–63 mm (2–2½ in.) At ¾ radius	63–75 mm (2½–3 in.) At ¾ radius	75–89 mm (3–3½ in.) At ¾ radius	89–102 mm (3½–4 in.) At ¾ radius
Oil quenched and tempered											
620–860(a)	90–125(b)	23–30	241–285	1330H 5132H 4130H 8630H		94B30H					
860–1030(c)	125–150	30–36(d)	285–341	1335H 5135H	4135H 8640H 94B30H 8740H	4137H		4142H	9840H		4337H 4340H
1030–1170(e)	150–170	36–41(f)	331–375	1340H 5140H 4135H 8637H 94B30H 3140H	50B40H 4137H 8642H 8645H 8742H	4140H 94B40H		4145H 9840H	86B45H 4337H	4340H	
1170–1275(g)	170–185	41–46(h)	375–429	50B46H 5145H 50B40H 4140H 8640H 8642H 8645H 8740H 8742H	5155H 50B44H 5147H 94B40H 6150H	81B45H 4142H 4145H 8650H 8655H 4337H	86B45H 9840H	4147H 4340H	4150H		9805H E4340H
>1275(i)	>185	46 min(j)	429 min	5150H 5155H 50B44H 5147H 9260H 81B45H 8650H 86B45H 6150H	5160H 50B50H 9262H 4147H 8655H	50B60H 51B60H 8660H	4150H			9850H	
Water quenched and tempered(k)											
620–860(a)	90–125	23–30(b)	241–285		5130H 5132H 4130H 8630H	5135H			4135H	94B30H	
860–1030(c)	125–150	30–36(d)	285–341		1330H 5135H	1335H	4135H(l) 8640H(l) 8740H(l) 3140H(l)	1340H(m) 8637H(m)	50B40H 8642H 94B30H	4137H 4140H	94B40H
1030–1170(e)	150–170	36–41(f)	331–375	1330H 1335H 5130H 5132H 5135H 4130H 8630H	4042H 4047H	1340H 50B46H 5140H 4135H 8637H 94B30H 3140H	50B40H(l) 4137H(l) 8642H(l) 8745H(l)	8640H(m) 8740H(m)	50B44H 5147H 4140H 8645H 8742H	94B40H	81B45H 4142H 4337H
1170–1275(g)	170–185	41–46(h)	375–429	5140H 4037H 4042H 4137H 8637H	1340H 50B46H 3140H	5145H 50B40H 8640H 8642H 8740H	50B44H(l) 5147H(l) 81B45H(l) 94B40H(l)	4140H(m) 8645H(m) 8742H(m)	4142H	81B45H 4337H	4145H 4147H 86B45H 9840H 4340H E4340H
>1275(i)	>185	46 min(j)	429 min	5046H 50B46H 5145H 4047H 4142H 8642H	5147H 4145H 8645H 86B45H	50B44H		81B45H(m)	4147H		

(a) Tensile strength, 790 to 940 MPa (115 to 138 ksi). (b) As-quenched hardness, 42 HRC, or 388 HB. (c) Tensile strength, 940 to 1100 MPa (136 to 160 ksi). (d) As-quenched hardness, 44 HRC, or 415 HB. (e) Tensile strength, 1100 to 1300 MPa (160 to 188 ksi). (f) As-quenched hardness, 48 HRC, or 461 HB. (g) Tensile strength, 1300 to 1530 MPa (188 to 222 ksi). (h) As-quenched hardness, 51 HRC, or 495 HB. (i) Tensile strength, over 1530 MPa (222 ksi). (j) As-quenched hardness, 55 HRC, or 555 HB. (k) Through steels with 0.47% C nominal. (l) May be substituted for steels listed under the 50 to 63 mm (2 to 2½ in.) column at same strength level or less. (m) Not recommended for applications requiring 80% martensite at midradius in sections 38 to 50 mm (1½ to 2 in.) in diameter because of insufficient hardenability. Source: Ref 3

Table 4 Alloy steel selection guide for moderately stressed parts

Unless otherwise indicated in the footnotes, any steel in this table may be considered for a lower strength level or a smaller section, or both.

Required yield strength MPa	ksi	As-tempered hardness HRC	As-tempered hardness HB	At center ≤13 mm (½ in.)	At center 13–25 mm (½–1 in.)	At midradius 25–38 mm (1–1½ in.)	At midradius 38–50 mm (1½–2 in.)	At ¾ radius 50–63 mm (2–2½ in.)	At ¾ radius 63–75 mm (2½–3 in.)	At ¾ radius 75–89 mm (3–3½ in.)	89–102 mm (3½–4 in.)
Oil quenched and tempered											
620–860(a)	90–125	23–30(b)	241–285	1330H 5132H 4130H 8630H	8737H	50B40H 8642H 94B30H 8740H 3140H	4140H 94B40H		4142H		
860–1030(c)	125–150	30–36(d)	285–341	1335H 4042H 4047H 5135H	4135H 8640H 94B30H 8740H 3140H	50B44H 5147H 4137H 8645H 8742H		4142H	4145H	4147H 86B45H 9840H	4337H 4340H
1030–1170(e)	150–170	36–41(f)	331–375	1340H 5140H 4135H 8637H 94B30H 3140H	5150H 50B40H 4137H 8642H 8645H 8742H	5160H 50B50H 4140H 94B40H 6150H	51B60H 8655H	4145H 9840H 4337H	4147H 86B45H	4150H 4340H	
1170–1275(g)	170–185	41–46(h)	375–429	5145H 50B40H 50B46H 4063H 4140H 8640H 8642H 8745H 8740H 8742H	5155H 50B44H 5147H 94B40H 6150H	81B45H 4142H 4145H 8650H 8655H 4337H	86B45H 9840H	4147H 8660H 4340H	4150H		9850H E4340H
>1275(i)	>185	46 min(j)	429 min	5150H 5155H 50B44H 5147H 9260H 81B45H 8650H 86B45H 6150H	5160H 50B50H 9262H 4147H 8655H	50B60H 51B60H 8660H	4150H			9850H	
Water quenched and tempered(k)											
620–860(a)	90–125	23–30(b)	241–285		4037H 5130H 5132H 4130H 8630H	5135H	8637H(l)	5140H(m)	4135H	50B40H 8642H 94B30H 3140H	4137H
860–1030(c)	125–150	30–36(d)	285–341		1330H 5135H	1335H	4135H(l)	1340H(m) 8637H(m)	50B40H 8640H 8642H 94B30H 8740H 3140H	50B44H 5147H 4137H 8645H 8742H	4140H 94B40H
1030–1170(e)	150–170	36–41(f)	331–375	1330H 1335H 5130H 5132H 5135H 4130H 8620H	4042H 4047H	1340H 50B46H 5140H 4135H 8637H 94B30H 3140H	50B40H(l) 4137H(l) 8642H(l)	5145H(m) 8640H(m) 8742H(m)	50B44H 8640H 4140H 8645H 8742H	94B40H	81B45H 4142H 4337H
1170–1275(g)	170–185	41–46(h)	375–429	5140H 4037H 4042H 4137H 8637H	1340H 50B46H 3140H	5145H 50B40H 8640H 8642H 8740H	50B44H(l) 5147H(l) 94B40H(l)	4140H(m) 8645H(m) 8742H(m)	4142H	81B45H 4337H	4145H 4147H 86B45H 9840H 4340H E4340H
>1275(i)	>185	46 min(j)	429 min	5046H 50B46H 5145H 4047H 4142H 8742H	5147H 4145H 8645H 86B45H	50B44H		81B45H(m)	4147H		

(a) Tensile strength, 790 to 940 MPa (115 to 136 ksi). (b) As-quenched hardness, 42 HRC, or 388 HB. (c) Tensile strength, 940 to 1100 MPa (136 to 160 ksi). (d) As-quenched hardness, 44 HRC, or 415 HB. (e) Tensile strength, 1100 to 1300 MPa (160 to 188 ksi). (f) As-quenched hardness, 48 HRC, or 461 HB. (g) Tensile strength, 1300 to 1530 MPa (188 to 222 ksi). (h) As-quenched hardness, 51 HRC, or 495 HB. (i) Tensile strength, over 1530 MPa (222 ksi). (j) As-quenched hardness, 55 HRC, or 555 HB. (k) Through steels with 0.47% C nominal. (l) May be substituted for steels listed under the 50 to 63 mm (2 to 2½ in.) column at same strength level or less. (m) Not recommended for applications requiring 50% martensite at midradius in sections 38 to 50 mm (1½ to 2 in.) in diameter because of insufficient hardenability. Source: Ref 3

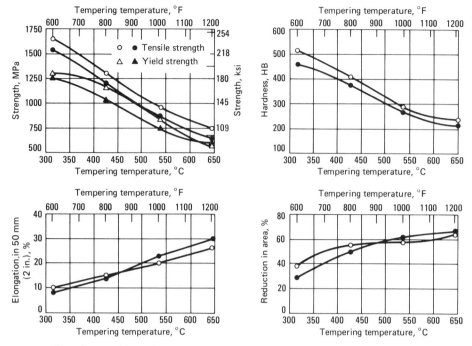

Fig. 10 Effect of composition and tempering temperature on mechanical properties of 1050 steel. Properties are summarized for two heats of 1050 steel that was forged to 38 mm (1.50 in.) in diameter, then water quenched and tempered at various temperatures. Open symbols are for heats containing 0.52 C and 0.93 Mn; closed symbols, for those containing 0.48 C and 0.57 Mn.

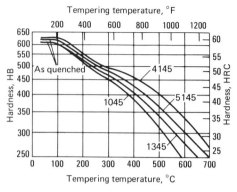

Fig. 11 Tempering characteristics of four 0.43% carbon and alloy steels tempered for 1 h

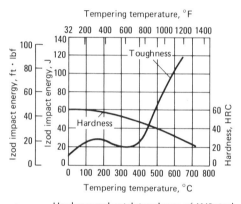

Fig. 12 Hardness and notch toughness of 4140 steel tempered for 1 h at various temperatures

nomenon is manifested as a loss of toughness, observed after slow cooling from tempering temperatures of 575 °C (1070 °F) or higher or after tempering in the temperature range between approximately 375 and 575 °C (700 and 1070 °F). Thus, a steel that is susceptible to temper embrittlement may lose much of its plasticity, as indicated by a notched-bar impact test, during slow cooling from a high tempering temperature, although the same steel will be very tough if it is quenched from the same tempering temperature. This expedient of quenching from the tempering temperature is often overlooked as a practical means for avoiding severe temper embrittlement in susceptible steels tempered at 575 °C (1070 °F) or higher. In steels susceptible to temper brittleness, embrittlement will also be observed after tempering at 375 to 575 °C (700 to 1070 °F), particularly if the tempering times are protracted. Under such circumstances, quenching from the tempering temperature will never restore the toughness.

High manganese, phosphorus, and chromium concentrations appear to accentuate the embrittling reaction; molybdenum has a definite retarding effect. Here again, the carbon and alloying elements may be chosen so that the susceptibility to temper embrittlement is minimized or the desired strength level is obtained by tempering either below 375 °C (700 °F) or above 575 °C (1070 °F) and then quenching. Temper brittleness is discussed in greater detail in the article "Embrittlement of Steels" in this Volume.

Distortion in Heat Treatment

Distortion during heat treatment may occur with almost any hardenable carbon or alloy steel, although distortion is usually more severe for carbon grades than for alloy grades of equivalent carbon content. Carbon steels distort more than alloy steels mainly because carbon steels require a water or brine quench to develop full hardness (at least in sections thicker than about 9.5 mm, or 3/8 in.). This often eliminates carbon steels from consideration for critical parts.

This distortion may be observed as a change in dimensions (size distortion) or a change in configuration or contour (shape distortion or warpage), or both. A more complete discussion of these types of distortion and the factors that influence them may be found in Ref 4 to 6.

Several factors contribute to the total distortion that occurs during heat treatment. These include residual stresses that may be present as a result of machining or other cold-working operations, the method of placing in the furnace, the rate of heating, nonuniform heating, and the normal volumetric changes that occur with phase transformations. However, the most important, single factor is uneven cooling during quenching, caused mainly by the configuration and by changes in cross-sectional area. Symmetrical parts with little or no variation in section may have almost no distortion, whereas complex parts with wide variations in section may distort so much that they cannot be used (or at least so much that they require excessive finishing operations to make them suitable for use).

Other factors being equal, the distortion in carbon steels will increase as the carbon content increases because of the gradual lowering of the martensite start (M_s) temperature with increasing carbon. There is also a significant variation in the magnitude of distortion and direction of dimensional change among different heats of the same grade of steel, even though other variables are minimal. This happens because of several factors, including minor variations in composition and grain size, but mainly because of the history of the steel with regard to hot working, cold working, and heat treatment.

Because of the different variables that contribute to total distortion, the prediction of distortion in actual parts is seldom reliable if it is based on the behavior of small test pieces. The most practical approach is to make studies on pilot lots of actual pieces that have been heat treated under production conditions. This procedure eliminates the shape variable so that the direction and magnitude of distortion can be plotted as ranges that incorporate most of the other variables. After a quantity of such data has

Induction and Flame Hardening

The relatively low hardenability of carbon steels is often a primary reason for choosing them in preference to alloy steels for parts that are to be locally heat treated by flame or induction hardening. One of the oldest rules for selecting steels for heat treating is to choose grades that are no higher in carbon or alloy content than is essential to develop required properties. This rule remains valid in the selection of steels to be heat treated by induction or flame processes.

When the peripheries of steel parts are heated rapidly and quenched, the tendency to crack depends mainly on a combination of four factors:

- Final surface hardness
- Temperature to which the surface has been heated
- Uniformity of heating
- Depth of hardened zone

The optimum heat pattern for either induction or flame heating depends on the type of steel and on the mass and shape of the part. The ideal heat pattern for any specific part will provide a hardened shell to a depth that will strengthen the part by establishing a favorable stress pattern. However, if the hardened zone is too deep for the specific section thickness, high tensile stresses are established in the surface layers, and these may either cause cracking or adversely affect service life.

Excessive depth of the hardened zone can be caused by improper processing (overheating, for instance) or by the choice of a steel with excessive hardenability. However, excessive carbon can aggravate other contributing factors and become the basic cause for cracking. The M_s temperature decreases as the carbon content increases. It is lowered further by higher austenitizing temperatures. In general, as the M_s temperature is lowered, the probability of surface cracking increases.

Fabrication of Parts and Assemblies

Fabrication processes are usually performed on hardenable carbon and alloy steels in the unhardened condition, that is, prior to heat treating. This is done primarily to avoid the high cost and difficulty of fabrication that are characteristic of high-strength materials. However, even in the unhardened condition, there are differences among the various grades in respect to formability, weldability, machinability, and forgeability properties. In many instances, difficulties arising during the fabrication of a given hardenable steel are directly related to the maximum hardness that can be developed and to hardenability.

REFERENCES

1. *1989 SAE Handbook*, Vol 1, *Materials*, Society of Automotive Engineers, 1989
2. S.L. Semiatin and D.E. Stutz, *Induction Heat Treatment of Steel*, American Society for Metals, 1986, p 24
3. *Republic Alloy Steels*, Republic Steel Corporation, 1961
4. B.S. Lement, *Distortion in Tool Steels*, American Society for Metals, 1959
5. *Properties and Selection of Tool Materials*, American Society for Metals, 1975
6. J.A. Ferrante, Controlling Part Dimensions During Fabrication and Heat Treatment, *Met. Prog.*, Vol 87 (No. 1), Jan 1965, p 87-90; reprinted in *Source Book on Heat Treating*, Vol I, American Society for Metals, 1975

Hardenability of Carbon and Low-Alloy Steels

Revised by Harold Burrier, Jr., The Timken Company

HARDENABILITY OF STEEL is the property that determines the depth and distribution of hardness induced by quenching. Steels that exhibit deep hardness penetration are considered to have high hardenability, while those that exhibit shallow hardness penetration are of low hardenability. Because the primary objective in quenching is to obtain satisfactory hardening to some desired depth, it follows that hardenability is usually the single most important factor in the selection of steel for heat-treated parts.

Hardenability should not be confused with hardness as such or with maximum hardness. The maximum attainable hardness of any steel depends solely on carbon content. Also, the maximum hardness values that can be obtained with small test specimens under the fastest cooling rates of water quenching are nearly always higher than those developed under production heat-treating conditions, because hardenability limitations in quenching larger sizes may result in less than 100% martensite formation. The effects of carbon and martensite content on hardness are shown in Fig. 1. Basically, the units of hardenability are those of cooling rate, for example, degrees per second. These cooling rates, as related to the continuous-cooling-transformation behavior of the steel, determine the hardness and microstructure outcome of a quench. In practice, these cooling rates are often expressed as a distance, with other factors such as the thermal conductivity of steel and the rate of surface heat removal being held constant. Therefore, the terms Jominy distance and ideal critical diameter can be used.

The hardenability of steel is governed almost entirely by the chemical composition (carbon and alloy content) at the austenitizing temperature and the austenite grain size at the moment of quenching. In some cases, the chemical composition of the austenite may not be the same as that determined by chemical analysis, because some carbide may be undissolved at the austenitizing temperature. Such carbides would be reflected in the chemical analysis, but because the carbides are undissolved in the austenite, neither their carbon nor alloy content can contribute to hardenability. In addition, by nucleating transformation products, undissolved carbides can actively decrease hardenability. This is especially important in high-carbon (0.50 to 1.10%) and alloy carburizing steels, which may contain excess carbides at the austenitizing temperature. Consequently, such factors as austenitizing temperature, time at temperature, and prior microstructure are sometimes very important variables when determining the basic hardenability of a specific steel composition. Certain ingot casting and hot reduction practices may also develop localized or periodic inhomogeneities within a given heat, further complicating hardenability measurements. The effects of all these variables are discussed in this article.

Hardenability Testing

The hardenability of a steel is best assessed by studying the hardening response of the steel to cooling in a standardized configuration in which a variety of cooling rates can be easily and consistently reproduced from one test to another.

The Jominy end-quench test fulfills the cooling rate requirements of hardenability testing of a broad range of alloy steels. The test specimen, a 25.4 mm (1.000 in.) diam bar 102 mm (4 in.) in length, is water quenched on one end face. The bar from which the specimen is made must be normalized before the test specimen is machined. The test involves heating the test specimen to the proper austenitizing temperature and then transferring it to a quenching fixture so designed that the specimen is held vertically 12.7 mm (0.5 in.) above an opening through which a column of water can be directed against the bottom face of the specimen (Fig. 2a). While the bottom end is being quenched by the column of water, the opposite end is cooling slowly in air, and intermediate positions along the specimen are cooling at intermediate rates. After the specimen has been quenched, parallel flats 180° apart are ground 0.38 mm (0.015 in.) deep on the cylindrical surface. Rockwell C hardness is measured at intervals of 1/16 in. (1.6 mm) for alloy steels and 1/32 in. (0.8 mm) for carbon steels, starting from the water-quenched end. A typical plot of these hardness values and their positions on the test bar, as shown in Fig. 2(b), indicates the relation between hardness and cooling rate, which in effect is the hardenability of the steel. Figure 2(b) also shows the cooling rate for the designated test positions. Details of the standard test method are available in ASTM A 255 and SAE J406.

The Carburized Hardenability Test. It is often necessary to determine the hardenability of the high-carbon case regions of carburized steels. Such information is important in controlling carburizing and quenching practice and in determining the ability of a specific steel to meet the microstructural and case depth requirements of the carburized component manufactured from the steel. As a general rule, adequate core hardenability does not ensure adequate case hardenability, especially when it is required to reheat for hardening after carburizing rather than to quench directly from the carburizing furnace. Two factors are responsible for this fact. The first is that equal alloying additions do not have the same effect on the hardenability of all carbon levels of alloyed steels. The second factor (as noted earlier) is that the high-carbon case regions do not always achieve full solution of alloy and carbides, as is normally achieved in the austenite of the low-carbon core region, prior to quenching. Accordingly, direct measurements of case hardenability are very important whenever a carburizing steel must be selected for a specific application.

Measurements of case hardenability are performed as follows. A standard end-quench bar is pack carburized for 9 h at 925 °C (1700 °F) and end quenched in the usual

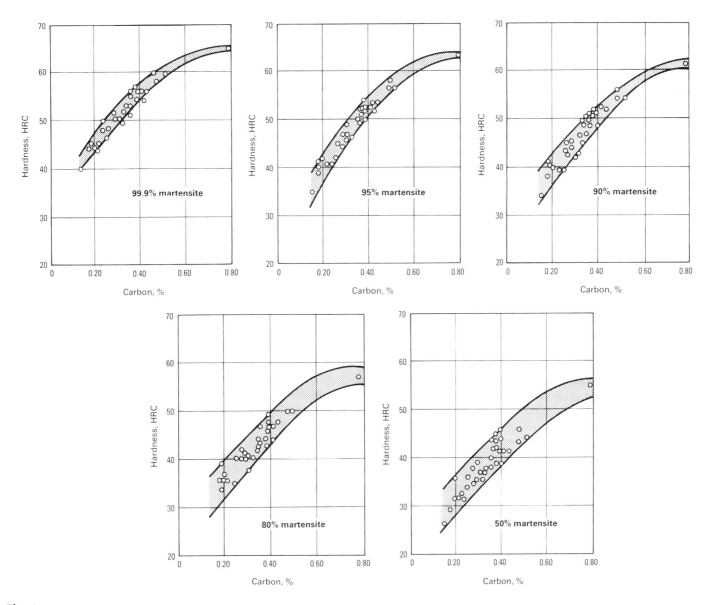

Fig. 1 Effect of carbon on the hardness of martensite structures

manner. A comparison bar is simultaneously carburized in the same pack to determine carbon penetration. Successive layers are removed from it and analyzed chemically to determine the carbon content at various depths. When a carbon-penetration curve is established, depths to various carbon levels can be determined in the Jominy bar, assuming that the distribution of carbon in the end-quench specimen is the same as in the carbon gradient bar. Longitudinal flats are then carefully ground to various depths on the end-quench bar (usually to carbon concentrations of 1.1, 1.0, 0.9, or 0.8%, and in some cases to as low as 0.6%), and hardenability is determined at these carbon levels by hardness traverses. In grinding, care must be exercised to avoid overheating and tempering, and in conducting hardness surveys, similar concern must be shown to ensure that the hardness level corresponds to a single carbon level by remaining in the exact center of the flat. Rockwell A hardness readings are preferable to Rockwell C readings because they minimize the depth of indentor penetration into softer subsurface layers. Rockwell A values are converted into Rockwell C values for plotting, as illustrated in Fig. 3, which shows the curves of carburized hardenability of an EX19 steel. In the higher-carbon layers of carburized specimens, the hardness will be influenced by the presence of retained austenite. Therefore, it is often useful to evaluate the microstructure/depth relationship by metallographically polishing and etching the ground flats. The Jominy distance to some chosen level of nonmartensitic transformation product can then be used as a measure of hardenability.

The case hardenability of steels that are carburized and then reheated for hardening at temperatures below 925 °C (1700 °F), such as 8620, 4817, and 9310, can also be determined by using a modification of this technique. The carburized end-quench specimens and companion gradient bars are oil quenched together from carburizing, but are then reheated in an atmosphere furnace to the desired austenitizing temperature for a total of 55 to 60 min, which should ensure at least 30 to 35 min at temperature. The hardenability specimen is then end quenched, and the carbon gradient bar is oil quenched and tempered to facilitate machining for carbon gradient determination, as described above. It is recommended that case hardenability tests be performed on no fewer than two test specimens. A more detailed description of the case hardenability measurement technique appears in SAE J406.

Air Hardenability Test. Occasionally, the hardening performance either of a steel cooled at a rate slower than that applied to the end-quench bar or of steels of very high hardenability must be determined. An air

466 / Hardenability of Carbon and Low-Alloy Steels

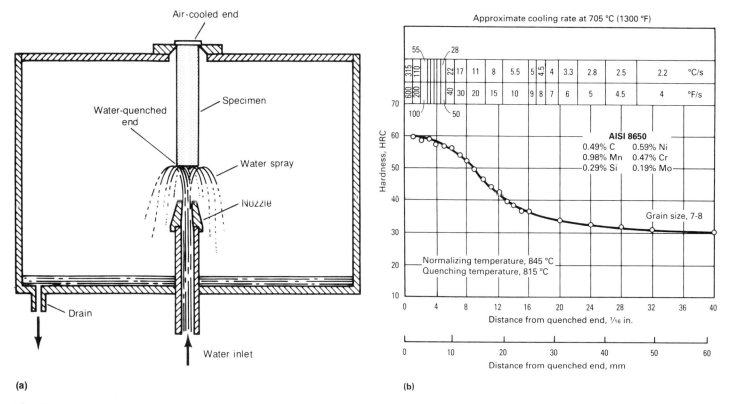

Fig. 2 Jominy end-quench apparatus (a) and method for presenting end-quench hardenability data (b)

hardenability test method described in Ref 1 can be employed for this purpose. In this test, a machined and partially threaded round test specimen, 25.4 mm (1.000 in.) in diameter and 254 mm (10 in.) long, is inserted to a depth of 152 mm (6 in.) in a hole drilled in a bar 152 mm (6 in.) in diameter and 381 mm (15 in.) long, thus leaving 102 mm (4 in.) of the test bar length exposed (Fig. 4). A second test specimen can be inserted at the opposite end of the bar holder to serve as a duplicate. With both test bars securely in place, the assembly is heated to the proper austenitizing temperature, after which it is transferred to a convenient location for cooling in still air. This cooling procedure results in very slow and ever decreasing cooling rates along the length of the test bars. Hardness is then measured at discrete intervals along each test bar and plotted against distance from the exposed end on charts specifically designed for this purpose.

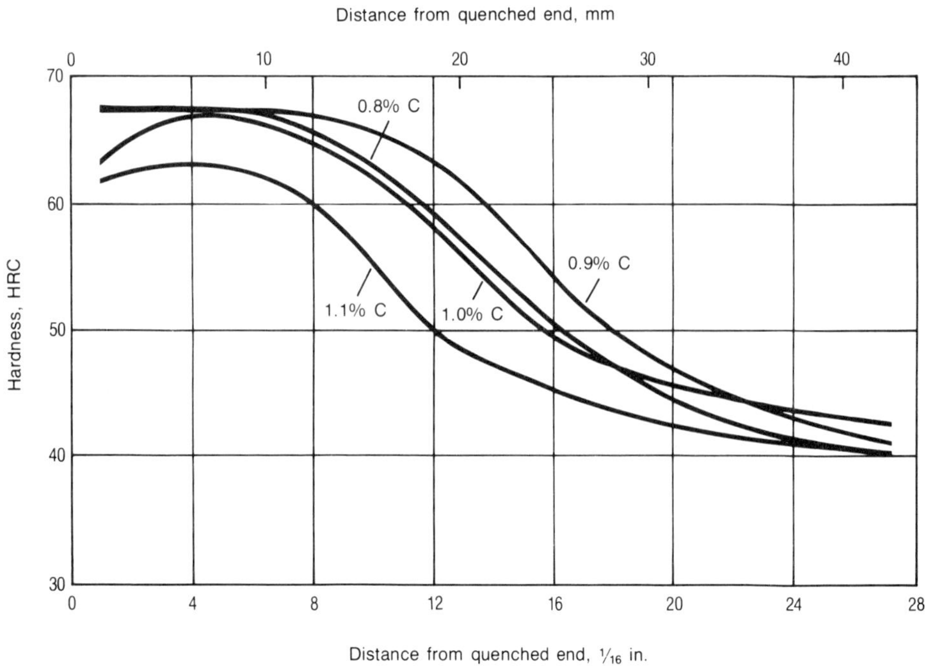

Continuous-Cooling-Transformation Diagrams. The use of continuous-cooling-transformation diagrams determined dilatometrically, for example, can also be helpful in evaluating the cooling behavior of high-hardenability steels.

Low-Hardenability Steels

In plain carbon and very low-alloy steels, the cooling rate at even the 1.6 mm (1/16 in.) position on a standard Jominy bar may not be fast enough to produce full hardening. Therefore, this test lacks discrimination between these steels. Tests that are more suited to very low hardenability steels include the hot-brine test and the surface-area-center (SAC) test.

In the hot-brine test proposed by Grange, coupons (Fig. 5) are quenched in brine maintained at a series of different temperatures. As shown in Fig. 6, the resulting hardnesses provide a very sensitive test of hardenability.

In the SAC test, a 25.4 mm (1.000 in.) round bar is normalized by cooling in air and then reaustenitized for water quench-

Fig. 3 Carburized hardenability, EX19 steel. Composition: 0.18 to 0.23% C, 0.90 to 1.20% Mn, 0.40 to 0.60% Cr, 0.08 to 0.15% Mo, 0.0005% B (min)

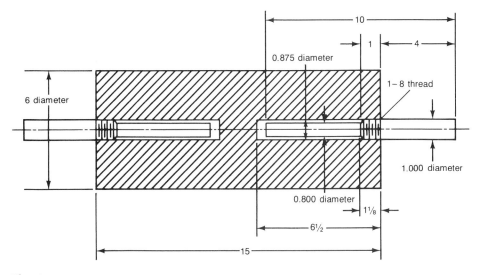

Fig. 4 Dimensions (given in inches) of components in air hardenability test setup

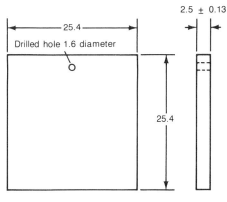

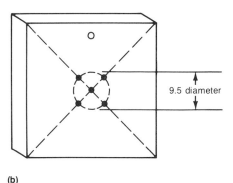

Fig. 5 Hot-brine hardenability test specimen. (a) Specimen dimensions. (b) Method of locating hardness impressions after heat treatment. Dimensions given in millimeters. Source: Ref 2

ing. Hardnesses are measured on a specimen cut from the center of the 100 mm (4 in.) length. Hardness is determined on the surface, the center, and at 1.6 mm (1/16 in.) intervals from surface to center. An area hardness is then computed as the sum of the average hardness in each interval × 1/16 (Fig. 7). The resulting set of three-digit numbers, for example, SAC No. 63-52-42, indicates a surface hardness of 63 HRC, a Rockwell-inch area of 52, and a center hardness of 42 HRC. Testing details are given in SAE J406.

Calculation of Hardenability

The hardenability of a steel is primarily a function of the composition (carbon, alloying elements, and residuals) and the grain size of the austenite at the instant of quenching. If this relationship can be determined quantitatively, it should be possible to predict the hardenability of a steel through a relatively simple calculation.

Such a technique was published by Grossmann in 1942, based on his observation that hardenability could be expressed as the product of a series of composition-related multiplying factors (Ref 3). The result of the calculation is an estimate of D_I, the ideal critical diameter of the steel. The multiplying-factor principle is still used today in several hardenability calculation techniques (see the article "High-Strength Structural and High-Strength Low-Alloy Steels" in this Volume for examples of multiplying factors for quench and tempered low-alloy steels). Other researchers have developed methods based on regression equations and on calculation from thermodynamic and kinetic first principles. To date, none of the hardenability prediction methods has proved to be universally applicable to all steel types; that is, different predictors are more suited to steels of given alloying systems, carbon contents, and hardenability levels. In addition, it is often necessary to fine-tune the predictions based on the characteristics (residuals, melt practice, and so on) of a particular steel producer. Some excellent discussions of current thinking on this subject are available in Ref 4 and 5. Properly used, hardenability calculations can provide a valuable tool for designing cost-effective alternative steels, for deciding the disposition of heats in the mill prior to rolling, and possibly for replacing the costly and time-consuming measurement of hardenability.

Effect of Carbon Content

Carbon has a dual effect in hardenable alloy steels: It controls maximum attainable hardness and contributes substantially to hardenability. The latter effect is enhanced by the quantity and type of alloying elements present. It might be concluded, therefore, that increasing the carbon content is the least expensive approach to improving hardenability. This is true to a degree, but several factors weigh against the use of large amounts of carbon:

- High carbon content generally decreases toughness at room and subzero temperatures
- It produces harder and more abrasive microstructures in the annealed conditions, which makes cold shearing, sawing, machining, and other forms of cold processing more difficult
- It makes the steel more susceptible to hot shortness in hot working
- It makes the steel more prone to cracking and distortion in heat treatment. Because of these disadvantages, more than 0.60% C is seldom used in steels for machine parts, except for springs and bearings, and steels with 0.50 to 0.60% C are used less frequently than those containing less than 0.50% C

Figure 8 shows the differences between minimum hardenability curves for six series of steels. In each series, alloy content is essentially constant, and the effect of carbon content on hardenability can be observed over a range from 0.15 to 0.60%. The hardness effect is shown by the vertical

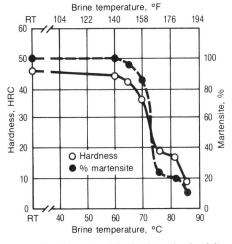

Fig. 6 Typical results of the hot-brine hardenability test. Steel composition: 0.18% C, 0.81% Mn, 0.17% Si, and 1.08% Ni. Austenitized at 845 °C (1550 °F). Grain size: 5 to 7. RT, room temperature. Source: Ref 2

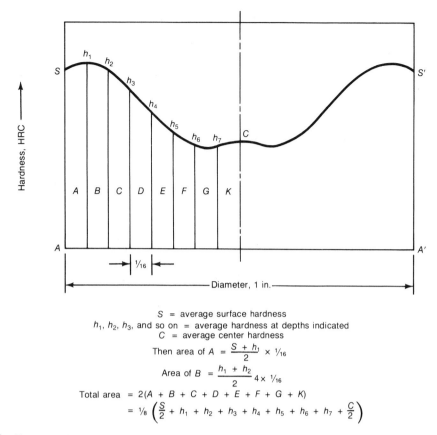

Fig. 7 Surface-area-center estimation of area

S = average surface hardness
h_1, h_2, h_3, and so on = average hardness at depths indicated
C = average center hardness

Then area of $A = \dfrac{S + h_1}{2} \times 1/16$

Area of $B = \dfrac{h_1 + h_2}{2} \times 1/16$

Total area $= 2(A + B + C + D + E + F + G + K)$
$= 1/8 \left(\dfrac{S}{2} + h_1 + h_2 + h_3 + h_4 + h_5 + h_6 + h_7 + \dfrac{C}{2} \right)$

distance between the curves at any position on the end-quench specimen, that is, for any cooling rate. This effect varies significantly, depending on the type and amounts of alloying elements. For example, referring to Fig. 8 (d) to (f), an increase in carbon content from 0.35 to 0.50% in each of the three series of steels causes hardness increases (in Rockwell C points) at four different end-quench positions, as shown below:

Series	Distance from quenched surface, in.			
	1/16	4/16	8/16	12/16
41xxH	8	10	17	20
51xxH	8	13	9	8
86xxH	8	12	18	12

The hardenability effect of carbon content is read on the horizontal axis in Fig. 8. If the inflection points of the curves are used to approximate the position of 50% martensite transformation, the effect of carbon content on hardenability in 8650 versus 8630 steel can be expressed as +4/16; that is, the inflection point is moved from the 5/16 position to the 9/16 position. Similarly, with nominal carbon contents of 0.35 and 0.50%, the hardenability effect of carbon is seen to be less (2/16) in 51xx series steels and more (6/16) in 41xx steels.

Considering the combined hardening and hardenability effects in terms of quenching speed, the cooling rate (or quenching speed) required to produce 45 HRC is affected more by 0.15% C with certain combinations of alloying elements than it is by other combinations. For example, in a steel containing 0.75 Cr and 0.15 Mo (a 41xxH series steel, for example), increasing the carbon content by 0.15% lowers the required or critical cooling rate to obtain 45 HRC from 25 to 4.6 °C (45 to 8.3 °F) per second, while in a steel containing 0.75% Cr and no molybdenum (51xxH series), the same increase in carbon content lowers the cooling rate from 47 to 21 °C (85 to 37 °F) per second.

The practical significance of the effect of carbon and alloy contents on cooling rate is considerable. In a 51 mm (2 in.) diam bar of 4150 steel, a hardness of 45 HRC can be obtained at half-radius using an oil quench without agitation. In a 4135 steel bar of the same diameter, to obtain the same hardness at half-radius would require a strongly agitated water quench. Comparing 32 mm (1¼ in.) diam bars of 5135 and 5150 steel, an agitated water quench will produce a hardness of 45 HRC at half-radius in the 5135 bar; the identical condition can be obtained in the 5150 bar using an oil quench with moderate agitation. Thus, an increase or decrease in carbon content or an alloying addition, such as 0.15% Mo, affects the results obtained both in terms of the quenching severity required and the section size in which the desired results can be obtained.

Figure 9 shows how steels are rated on the basis of ideal critical diameter by expressing the effect of carbon and alloy content on the section size that will harden to 50% martensite at the center, assuming an ideal quench. An ideal quench is defined as one that removes heat from the surface of the steel as fast as it is delivered to the surface. In general, the relation between hardness and carbon content that is important in practice is obscured in this rating method because the steel is rated to a constant microstructure. Hardness decreases continuously with lower carbon contents.

Alloying Elements

The most important function of the alloying elements in heat-treatable steel is to increase hardenability. Increased hardenability makes possible the hardening of larger sections and the use of an oil rather than a water quench to minimize distortion and to avoid quench cracking.

When the standard alloy steels are considered, it is found that, for practical purposes, all compositions develop the same tensile properties when quenched to martensite and tempered to the same hardness below 50 HRC. However, it should not be inferred that all tempered martensites of the same hardness are alike in all respects. For example, plain carbon martensites have lower reduction-in-area values than alloy martensites. A further difference, sometimes important, is that fully quenched alloy steels require, for the same hardness levels, higher tempering temperatures than carbon steels. This difference in tempering temperature may serve to reduce the residual stress level in finished parts. The stress reduction could be an advantage or a disadvantage, depending on whether a controlled compressive stress is desired in the part. Although tensile properties may not differ significantly from one alloy level to another, considerable differences may exist in fracture toughness and low-temperature impact properties. In general, steels with a higher nickel content, such as 4320, 3310, and 4340, offer much greater toughness at a given hardness level. In some applications, the toughness factor rather than hardenability may dictate steel selection, but hardenability is still important, because steels that can be fully quenched to 100% martensite are much tougher than those that cannot.

Usually, the least expensive means of increasing hardenability at a given carbon content is by increasing the manganese content. Chromium and molybdenum, already referred to as increasing hardenability, are also among the most economical elements per unit of increased hardenability. Nickel is the most expensive per unit, but is warranted when toughness is a primary consideration.

Hardenability of Carbon and Low-Alloy Steels / 469

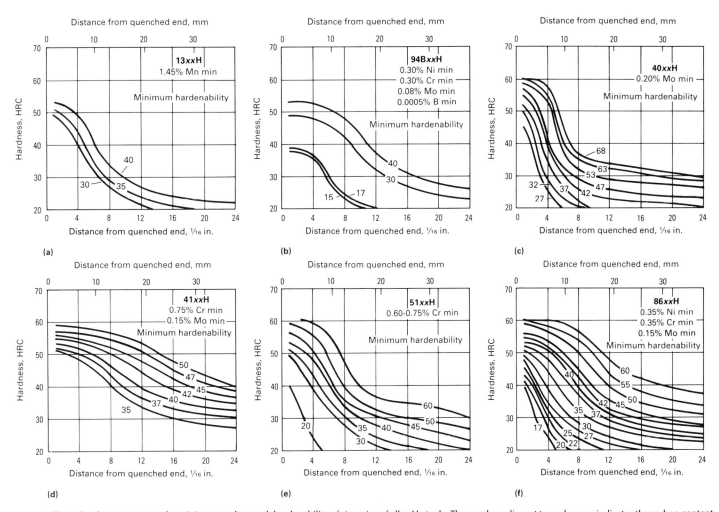

Fig. 8 Effect of carbon content on the minimum end-quench hardenability of six series of alloy H-steels. The number adjacent to each curve indicates the carbon content of the steel, to be inserted in place of *xx* in alloy designation.

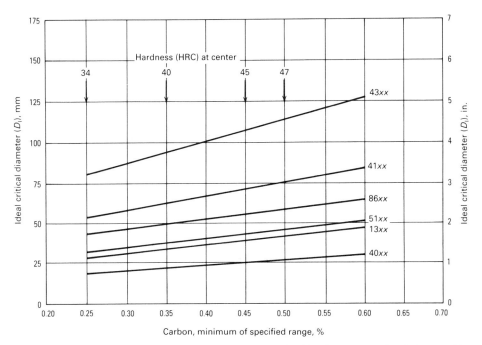

Fig. 9 Effect of carbon content on ideal critical diameter, calculated for the minimum chemical composition of each grade

Important synergistic effects, not yet fully defined, can also occur when combinations of alloying elements are used in place of single elements. Some examples of known synergistic combinations are nickel plus manganese, molybdenum plus nickel, and silicon plus manganese.

Boron. Another potent and economical alloying element is boron, which markedly increases hardenability when added to a fully deoxidized steel. The effects of boron on hardenability are unique in several respects:

- A very small amount of boron (about 0.001%) has a powerful effect on hardenability
- The effect of boron on hardenability is much less in high-carbon than in low-carbon steels
- Nitrogen and deoxidizers influence the effectiveness of boron
- High-temperature treatment reduces the hardenability effect of boron

Recommended austenitizing temperatures for boron H-steels are given with the H-bands.

470 / Hardenability of Carbon and Low-Alloy Steels

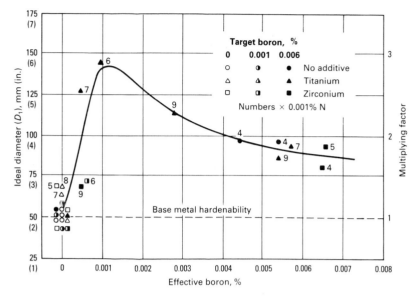

Fig. 10 Influence of effective boron content (β_{eff}) on the hardenability of an 8620 type steel. β_{eff} = B-[(N-0.002)-Ti/5-Zr/15] $\geq$ 0. Source: Ref 5

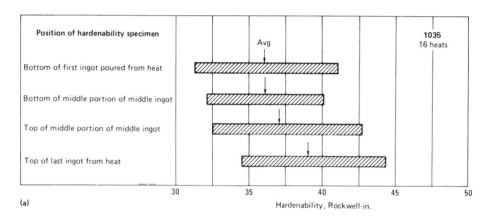

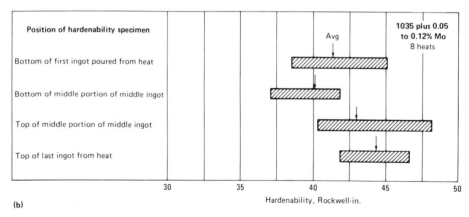

Fig. 11 Effect of test location of (a) 1035 steel and (b) 1035 steel with 0.05 to 0.12% Mo on SAC (Rockwell-in.) hardenability

Figure 10 illustrates the very small amount of boron required for an optimum hardenability effect when appropriate protection of the boron is afforded by additions of titanium or zirconium. In carburizing steels, the effect of boron on case hardenability may be completely lost if nitrogen is abundant in the carburizing atmosphere. The cost of boron is usually much less than that of other alloying elements having approximately the same hardenability effect.

Effect of Grain Size

The hardenability of a carbon steel may increase as much as 50% with an increase in austenite grain size from ASTM 8 (6 to 10) to ASTM 3 (1 to 4). The effect becomes more pronounced if the carbon content is increased at the same time. When the danger of quench cracking is remote (no abrupt changes in section thickness) and engineering considerations permit, it may sometimes appear to be more practical to use a coarser-grain steel rather than a fine-grain or more expensive alloy steel to obtain hardenability. However, this is not recommended, because the use of coarser-grain steels usually involves a serious sacrifice in notch toughness and may lead to other difficulties.

Variations Within Heats

Segregation of carbon, manganese, and other elements always occurs during ingot pouring and solidification. As a result, the hardenability of the steel in these segregated portions will differ from that in the remainder of the ingot. In general, specimens taken from the top of the ingot have higher hardenability than steel from the middle, and specimens from the bottom of the ingot will have lower hardenability than steel from the middle. This gradual increase in hardenability from the bottom of the first ingot to the top of the last ingot is illustrated in Fig. 11(a) for 16 heats of 1035 carbon steel. The hardenability spread for 8 heats of 1035 steel containing 0.05 to 0.12% Mo, plotted in Fig. 11(b), shows a similar trend. Comparison between Fig. 11(a) and (b) shows the effect of molybdenum on hardenability.

The same effect is observed in alloy steels. End-quench hardenability test results for one heat of 4028 steel (Fig. 12) show higher hardenability for a cast bar taken from the top of the last ingot of the heat than for a specimen from the melting floor and labeled cast end-quench specimen. The latter was taken from about the middle of the heat. After the heat of steel was rolled, the hardenability was slightly lower, as shown by the curve representing results on eight end-quench specimens. Data for 465 heats of ten other steels are summarized in Fig. 13.

Effect of Hot Working. Processing variables, such as the amount of hot working and the location of the test specimen in the semifinished section, have an effect on hardenability. A 330 mm (13 in.) square bloom of 1330 steel was forged progressively to bar sizes of 305, 255, 205, and 150 mm (12, 10, 8, and 6 in.) in diameter. Each bar size was evaluated by tests on end-quench specimens cut from five locations (center, quarter-radius, half-radius, three-quarter-radius, and just below the surface). Data in Fig. 14 show that the variation in harden-

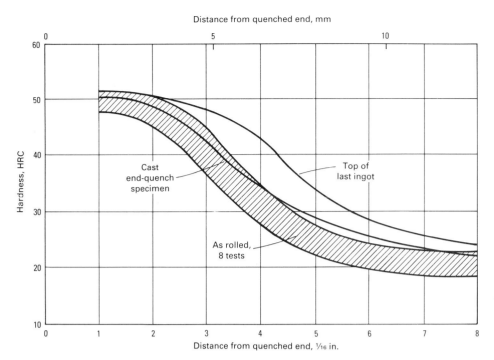

Fig. 12 Variation of hardenability within a heat of 4028 steel

ability narrows as the bar size is decreased by hot work.

Determining Hardenability Requirements

The basic information needed to specify a steel with adequate hardenability includes:

- The as-quenched hardness required prior to tempering to final hardness that will produce the best stress-resisting microstructure
- The depth below the surface to which this hardness must extend
- The quenching medium that should be used in hardening

As-Quenched Hardness. The Iron and Steel Technical Committee of the Society of Automotive Engineers (SAE) War Engineering Board approved and issued the relation shown in Fig. 15(a) as a recommendation for as-quenched hardness as a function of the hardness desired after tempering. Figure 15(a) does not specify the degree of hardening (percentage martensite) preferred in obtaining the as-quenched hardnesses indicated. It is possible, as shown in Fig. 15(b), to select steels that will produce these hardnesses with less than 90% martensite.

To ensure optimum properties, common practice is to select the steel with the lowest carbon content that will produce the indicated as-quenched hardness using the quenching medium available (or one that can be made available). Following this procedure, the structures possessing the indicated hardnesses would be fully hardened; that is, they would contain more than 90% martensite, which is a common and practical definition of full hardening and the one employed by the SAE committee. For components subjected to bending in service, it is considered adequate to have 90% martensite at the three-quarter-radius location. To ensure this, hardness levels are specified at half-radius.

Depth of Hardening. The depth and percentage of martensite to which parts are hardened may affect their serviceability, but it always affects the hardenability required and therefore the cost. In parts less highly stressed in bending, hardening to 80% martensite at three-quarter-radius of the part as finished may be sufficient; in other parts, even less depth may be required. The latter include principally those parts designed for low deflection under load, in which even the exterior regions are only moderately stressed. In contrast, some parts loaded principally in tension and others operating at high hardness levels, such as springs of all types, are usually hardened more nearly through the section. In automobile leaf springs, the leaves are designed with a low section modulus in the direction of loading. The allowable deflection is large, and most of the cross section is highly stressed.

In general, hardening need be no deeper than is required to provide the strength to sustain the load at a given depth below the surface. Therefore, parts designed to resist only surface wear, pure bending, or rolling contact often do not justify the cost of providing the hardenability required for hardening through the entire cross section.

When service requirements mandate that hardening must produce more than 80% martensite, the section size that can be hardened to a prescribed depth decreases rapidly as the percentage of martensite required increases. For example, let us assume that 95% martensite (51 HRC minimum hardness) is required in 8640H steel. Then the largest section size that can be hardened to the center in oil would be 16 mm (⅝ in.); a 25 mm (1 in.) section could be hardened to only three-quarter-radius. Again, on the basis of 95% martensite, the deepest hardening of standard steels, 4340H, will harden to the center of a 51 mm (2 in.) section; on the basis of 80% martensite (45 HRC), a 92 mm (3⅝ in.) round will harden to the center in oil.

The above examples emphasize the need for engineering judgment in requiring very deep hardening or unusually high percentages of martensite. When these requirements are not wholly justified, the results are overspecification of steel at higher cost and greater likelihood of distortion and quench cracking.

Quenching Media. The cooling potential of quenching media is a critical factor in heat-treating processes because of its contribution to attaining the minimum hardenability requirement of the part or section being heat treated. The cooling potential, a measure of quenching severity, can be varied over a rather wide range by:

- Selection of a particular quenching medium
- Control of agitation
- Additives that improve the cooling capability of the quenchant

Any or all of these variables can be employed to increase quenching severity and provide the following advantages:

- Permit the use of less expensive (lower-alloy) steels of lower hardenability
- Optimize the properties of the steel selected
- Permit the use of less expensive quenching media
- Improve productivity and achieve cost reductions as a result of shorter cycle times and higher production rates

In practice, however, two other considerations modify the selection of quenching medium and quenching severity: the amount of distortion that can be tolerated and the susceptibility to quench cracking.

In general, the more severe the quenchant and the less symmetrical the part being quenched, the greater the size and shape changes that result from quenching and the greater the risk of quench cracking. Consequently, although water quenching is less costly than oil quenching and water-quenched steels are less expensive than those requiring oil quenching, it is important that the parts to be hardened be care-

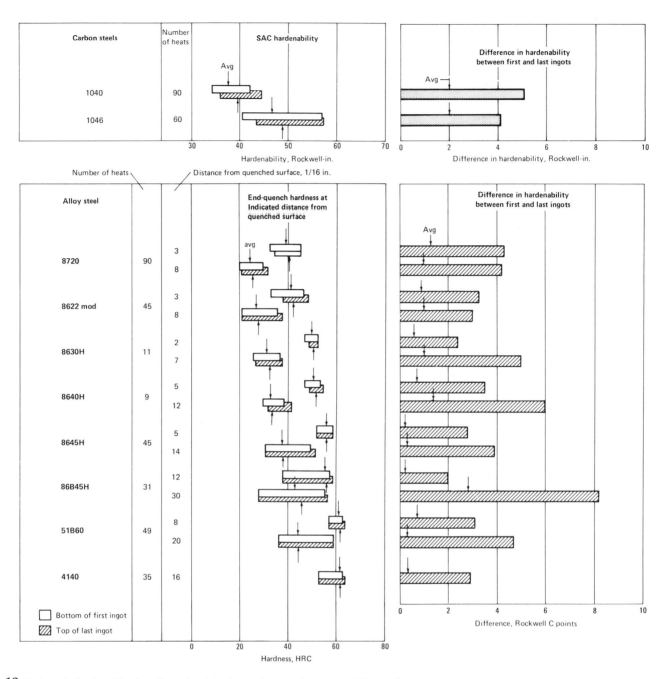

Fig. 13 Variation in hardenability from first to last ingot in heat for several carbon and alloy steels

Table 1 Quenching severities, H, for various media and quenching conditions

Quenchant agitation	Typical flow rates		Typical H values			
	m/min	sfm	Air	Mineral oil	Water	Brine
None	0	0	0.02	0.20–0.30	0.9–1.0	2.0
Mild	15	50	...	0.20–0.35	1.0–1.1	2.1
Moderate	30	100	...	0.35–0.40	1.2–1.3	...
Good	61	200	0.05	0.40–0.60	1.4–2.0	...
Strong	230	750	...	0.60–0.80	1.6–2.0	4.0

fully reviewed to determine whether the amount of distortion and the possibility of cracking as a result of water quenching will permit taking advantage of the lower cost of water quenching. Oil, salt, and synthetic water-polymer quenchants are alternatives, but their use often requires steels of higher alloy content to satisfy hardenability requirements.

A rule regarding selection of a steel and quenching medium for a given part is that the steel should have a minimum hardenability not exceeding that required by the quenching severity of the medium selected. The steel should also contain the lowest carbon content compatible with the required hardness and strength properties. This rule is based on the fact that the quench cracking susceptibility of steels increases with a decrease in M_s temperature and/or an increase in carbon content.

Table 1 lists typical quenching severity, or H, values for the common quenching media and conditions. These data are for media containing no additives. Figure 16 shows the effects of additives and of other quenching media. According to these data, considerable improvement in the cooling

Hardenability of Carbon and Low-Alloy Steels / 473

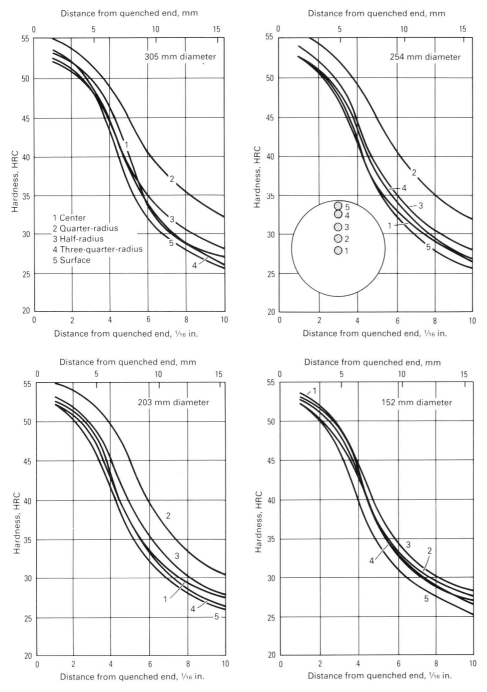

Fig. 14 Effect of hot working and location of test bars on end-quench hardenability of 1330 steel. A 330 mm (13 in.) bloom was progressively forged to bars of the diameters shown.

capability of quenchants can be obtained by such additions as water to hot salt, proprietary additives to oil, and polyalkylene glycol (polymer) to water. The polymer-water mixtures polyacrylamide gel (PAG), polyvinyl pyrrolidone (PVP), and polyvinyl alcohol (PVA) are gaining favor because they can be made to span the quenching severity range from oil to water by simple variation of the glycol (polymer) concentration in water. Also, because they are free of fire hazards and obnoxious environmental pollution agents, they have no adverse effect on working conditions. The quenching severity of these media should be tested at frequent intervals because dragout and thermal breakdown may affect their quenching efficiency.

Hardenability Versus Size and Shape. When end-quench data such as those shown in Fig. 2 are available, either of two methods can be used to estimate the hardenability a steel part of given size and configuration must have to achieve the desired hardness, strength, and microstructure at critical locations when quenched in various production media. These methods are:

- *Method 1*: The correlation of end-quench hardness data (J_{eh}) with equivalent hardness locations in variously quenched shapes
- *Method 2*: The correlation of end-quench cooling rate data (J_{ec}) with equivalent cooling rate locations in variously quenched production shapes

Method 1 (Fig. 17) is the more accurate and preferred method, because in practice it has been found that, when cooling at the same rates, large sections produce somewhat lower hardnesses than smaller sections, including end-quench and air hardenability bars. This difference has been attributed to two factors (Ref 6):

- Higher contraction stresses in large parts accentuate the transformation of austenite
- Quenching severity, H, decreases with an increase in section size

Also, in using the cooling rate method (method 2), it is difficult to determine cooling rates with a high degree of accuracy. Nevertheless, correlations that equate cooling conditions along the end-quench bar (J_{ec}) with those in production shapes quenched in various liquid media are also extremely useful when attempting to establish the required hardenability and/or quenching conditions for a production part.

General Hardenability Selection Charts

Figures 18 and 19 show correlations of J_{ec} equivalent cooling rates in end-quench hardenability specimens and round bars of up to 102 mm (4 in.) in diameter when quenched in oil, water, brine, and hot salt at various controlled agitation rates. They correlate bar diameter with equivalent positions on the end-quench hardenability specimen for ten modes of quenching, for both scaled and scale-free bars, and with data grouped according to bar location instead of by quenching mode.

Table 2 has been devised to work with the charts in Fig. 18 and 19 and includes most of the steels for which H-bands have been established, showing the location on the end-quenched specimen of the low limit of the H-band for six different hardness levels that might be specified for as-quenched hardness: 55, 50, 45, 40, 35, and 30 HRC. The last two levels apply primarily to the core hardness of carburized parts.

The use of Fig. 18 and Table 2 is described in the following example. This method substantially reduces the amount of chart hopping that has in the past been needed to examine all the available steels for the purpose of selecting one.

Table 2 Classification of H steels according to minimum hardness at various distances from quenched end

Distance from quenched end, 1/16th in.	H steels with a minimum hardenability curve that intersects the specified hardness at the indicated distance from the quenched end of the hardenability specimen	Typical values obtained by the use of Fig. 18(a) bar diameter (in.) for equivalent cooling rate at:					
		Three-quarter-radius		Half-radius		Center	
		Oil at 200 sfm; $H = 0.5$	Water at 200 sfm; $H = 1.5$	Oil at 200 sfm; $H = 0.5$	Water at 200 sfm; $H = 1.5$	Oil at 200 sfm; $H = 0.5$	Water at 200 sfm; $H = 1.5$
30 HRC							
2½	8617, 4118, 4620, 5120, 1038, 1522, 4419	0.4	1.5	...	1.1	...	0.8
3	4812, 4027, 1042, 1045, 1146, 1050, 1524, 1526, 4028, 6118	0.6	1.8	...	1.2	0.3	0.95
3½	4720, 6120, 8620, 4032	0.7	2.05	0.5	1.4	0.45	1.1
4	4815, 8720, 4621, 8622, 1050(b)	0.9	2.35	0.7	1.5	0.6	1.3
4½	46B12, 4817, 4320, 8625, 5046	1.05	2.6	0.8	1.6	0.7	1.45
5	4037, 1541, 4718, 8822	1.2	2.9	0.9	1.8	0.85	1.6
5½	94B15, 8627, 4042, 1541, 15B35	1.4	3.2	1.1	1.9	1.0	1.7
6	94B17	...	...	...	...	...	...
6½	4820, 1330, 4130, 8630, 1141	1.7	3.8	1.4	2.2	1.25	2.0
7	9130, 5130, 5132, 4047	1.85	...	1.5	2.4	1.35	2.1
7½	1335, 50B46, 15B37	2.0	...	1.7	2.5	1.5	2.2
8	5135	2.1	...	1.8	2.7	1.6	2.35
9½	1340	2.5	...	2.2	3.3	1.9	2.7
10	8635, 5140, 4053, 50B40	2.6	...	2.3	3.4	2.0	2.8
11	4640	2.8	...	2.4	3.7	2.15	3.0
12	8637, 1345, 50B44, 5145, 94B30	3.05	...	2.6	3.9	2.3	3.2
14	50B50	...	...	...	...	...	...
16	4135, 5147, 8645, 8740	3.85	...	3.3	...	2.8	3.85
20	4063	...	...	3.6	...	...	...
22	4068, 50B60, 5155, 86B30, 9260	...	...	3.7	...	...	...
24	4137, 5160, 6150, 81B45, 51B60, 8650	...	...	3.85	...	...	...
32	4140	...	...	...	...	...	...
35 HRC							
1½	8617	...	0.9	...	0.8	...	0.45
2	4812, 4118, 4620, 5120, 1038, 1522, 4419, 6118	...	1.2	...	0.9	...	0.65
2½	4028, 4720, 8620, 4027, 1042, 1045, 1146, 1050, 1524, 1526	0.4	1.5	...	1.1	...	0.8
3	9310, 46B12, 4320, 6120, 8720, 4621, 8622, 8625, 4032, 4815	0.6	1.8	...	1.2	0.3	0.95
3½	4815, 4817, 94B17, 5046, 1050(b), 4781, 8822	0.7	2.05	0.5	1.4	0.45	1.1
4	8627, 4037	0.9	2.35	0.7	1.5	0.6	1.3
4½	94B15, 4042, 1541	1.05	2.6	0.8	1.6	0.7	1.45
5	4820, 1330, 4130, 5130, 8630, 5132, 1141, 50B46, 4047, 15B35, 94B17	1.2	2.9	0.9	1.8	0.85	1.6
5½	1335	1.4	3.2	1.1	1.9	1.0	1.7
6	5135	1.55	3.5	1.2	2.1	1.1	1.85
6½	15B37	...	...	...	...	...	...
7	8635, 1340, 5140, 4053	1.85	...	1.5	2.4	1.35	2.1
8	4063, 1345, 5145	2.1	...	1.8	2.7	1.6	2.35
8½	8637	2.2	...	1.9	2.9	1.7	2.45
9	4640, 4068, 50B40	2.35	...	2.0	3.1	1.8	2.6
9½	8640, 50B44, 5150	2.5	...	2.2	3.3	1.9	2.7
10	8740, 9260	...	...	...	...	...	...
10½	4135, 50B50	2.7	...	2.35	3.5	2.1	2.9
13	4137	3.25	...	2.8	...	2.45	3.4
16	4140, 6150, 81B45, 86B30	3.85	...	3.3	...	2.8	3.85
40 HRC							
1	5120, 6120	...	0.65	...	0.6	...	0.3
1½	4118, 4620, 4320, 4720, 8620, 8720, 1038, 1522, 1526, 4621	...	0.9	...	0.8	...	0.45
2	8622, 8625, 4027, 1045, 1524, 4028, 4718	...	1.2	...	0.9	...	0.65
2¼	1146	0.3	1.3	...	1.0	...	0.7

(continued)

(a) If based on equivalent hardness, actual bar diameter will be less. (b) High residual alloy

Table 2 (continued)

Distance from quenched end, 1/16th in.	H steels with a minimum hardenability curve that intersects the specified hardness at the indicated distance from the quenched end of the hardenability specimen	Typical values obtained by the use of Fig. 18(a) bar diameter (in.) for equivalent cooling rate at:					
		Three-quarter-radius		Half-radius		Center	
		Oil at 200 sfm; H = 0.5	Water at 200 sfm; H = 1.5	Oil at 200 sfm; H = 0.5	Water at 200 sfm; H = 1.5	Oil at 200 sfm; H = 0.5	Water at 200 sfm; H = 1.5
40 HRC (continued)							
2½	4820, 8627, 4032, 1042, 1050	0.4	1.5	...	1.1	...	0.8
3	4037, 8822	0.6	1.8	...	1.2	0.3	0.95
3½	4130, 5130, 8630, 5046, 1050(b), 1541	0.7	2.05	0.5	1.4	0.45	1.1
4	1330, 5132, 4042	0.9	2.35	0.7	1.5	0.6	1.3
4½	5135, 1141, 4047	1.05	2.6	0.8	1.6	0.7	1.45
5	1335, 50B46, 15B35	1.2	2.9	0.9	1.8	0.85	1.6
5½	8635, 5140, 4053, 15B37	1.4	3.2	1.1	1.9	1.0	1.7
6	1340, 9260, 4063	1.55	3.5	1.2	2.1	1.1	1.85
6½	8637, 5145, 1345	1.7	3.8	1.4	2.2	1.25	2.0
7	4640, 4068	1.85	...	1.5	2.4	1.35	2.1
7½	8640, 5150	2.0	...	1.7	2.5	1.5	2.2
8	4135, 8740, 50B40	2.1	...	1.8	2.7	1.6	2.35
8½	6145, 9261, 50B44, 5155	2.2	...	1.9	2.9	1.7	2.45
9	4137, 8642, 5147, 50B50, 94B30	2.35	...	2.0	3.1	1.8	2.6
9½	8742, 8645, 5160, 9262	2.5	...	2.2	3.3	1.9	2.7
10½	6150, 50B60	2.7	...	2.35	3.5	2.1	2.9
11	4140	2.8	...	2.4	3.7	2.15	3.0
11½	81B45, 8650, 5152	2.9	...	2.5	3.8	2.25	3.1
12	86B30	...	...	...	...	...	...
13	51B60	3.25	...	2.8	...	2.45	3.4
14	8655	3.45	...	2.95	...	2.6	3.55
15	4142	3.65	...	3.1	...	2.7	3.7
15½	8750	3.75	...	3.2	...	2.75	3.8
18	4145, 8653, 8660	...	...	3.45	...	...	...
19	9840, 86B45	...	...	3.5	...	...	...
20	4147	...	...	3.6	...	...	...
24	4337, 4150	...	...	3.85	...	...	...
32	4340	...	...	...	...	...	...
36+	E4340, 9850	...	...	...	...	...	...
45 HRC							
1	4027, 4028, 8625	...	...	...	...	...	...
1½	8627, 1038	...	0.9	...	0.8	...	0.45
2	4032, 1042, 1146, 1045	...	1.2	...	0.9	...	0.65
2½	4130, 5130, 8630, 4037, 1050, 5132	0.4	1.5	...	1.1	...	0.8
3	1330, 5046, 1541	0.6	1.8	...	1.2	0.3	0.95
3¼	1050(b)	0.65	1.9	...	1.3	0.4	1.05
3½	1335, 5135, 4042, 4047	0.7	2.05	0.5	1.4	0.45	1.1
4	8635, 1141	0.9	2.35	0.7	1.5	0.6	1.3
5	8637, 1340, 5140, 50B46, 4053, 9260, 15B37	1.2	2.9	0.9	1.8	0.85	1.6
5½	5145, 4063	1.4	3.2	1.1	1.9	1.0	1.7
6	4135, 4640, 4068, 1345	1.55	3.5	1.2	2.1	1.1	1.85
6½	8640, 8740, 5150, 94B30	1.7	3.8	1.4	2.2	1.25	2.0
7	4137, 8642, 6145, 9261, 50B40	1.85	...	1.5	2.4	1.35	2.1
7½	8742, 50B44, 5155	2.0	...	1.7	2.5	1.5	2.2
8	8645, 5147	2.1	...	1.8	2.7	1.6	2.35
8½	4140, 6150, 5160, 9262, 50B50	2.2	...	1.9	2.9	1.7	2.45
9	50B60	2.35	...	2.0	3.1	1.8	2.6
9½	81B45, 8650, 86B30	2.5	...	2.2	3.3	1.9	2.7
10	5152	2.6	...	2.3	3.4	2.0	2.8
11	51B60, 8655	2.8	...	2.4	3.7	2.15	3.0
11½	4142	2.9	...	2.5	3.8	2.25	3.1
12	8750	3.05	...	2.6	3.9	2.3	3.2
13	8653, 8660	3.25	...	2.8	...	2.45	3.4
14	9840, 4145	3.45	...	2.95	...	2.6	3.55
16	86B45, 4147	3.85	...	3.3	...	2.8	3.85
17	4337	...	...	3.35	...	...	...
18	4150	...	...	3.45	...	...	...
22	4340	...	...	3.7	...	...	...
26	4161	...	...	...	...	...	...
30	E4340	...	...	...	...	...	...
36	9850	...	...	...	...	...	...
50 HRC							
1	4032, 5132, 1038	...	0.65	...	0.6	...	0.3

(continued)

(a) If based on equivalent hardness, actual bar diameter will be less. (b) High residual alloy

Table 2 (continued)

Distance from quenched end, 1/16th in.	H steels with a minimum hardenability curve that intersects the specified hardness at the indicated distance from the quenched end of the hardenability specimen	Typical values obtained by the use of Fig. 18(a) bar diameter (in.) for equivalent cooling rate at:					
		Three-quarter-radius		Half-radius		Center	
		Oil at 200 sfm; H = 0.5	Water at 200 sfm; H = 1.5	Oil at 200 sfm; H = 0.5	Water at 200 sfm; H = 1.5	Oil at 200 sfm; H = 0.5	Water at 200 sfm; H = 1.5
50 HRC (continued)							
1½	1335, 5135, 8635, 4037, 1042, 1146, 1045	...	0.9	...	0.8	...	0.45
2	4135, 1541, 15B35, 15B37	...	1.2	...	0.9	...	0.65
2¼	1050(b)	0.3	1.3	...	1.0	...	0.7
2½	4042	0.4	1.5	...	1.1	...	0.8
3	8637, 5140, 5046, 4047	0.6	1.8	...	1.2	0.3	0.95
3½	4137, 1141, 1340	0.7	2.05	0.5	1.4	0.45	1.1
4	4640, 5145, 50B46	0.9	2.35	0.7	1.5	0.6	1.3
4½	8640, 8740, 1053, 9260	1.05	2.6	0.8	1.6	0.7	1.45
5	8642, 4063, 1345, 50B40	1.2	2.9	0.9	1.8	0.85	1.6
5½	8742, 6145, 5150, 4068	1.4	3.2	1.1	1.9	1.0	1.7
6	4140, 8645	1.55	3.5	1.2	2.1	1.1	1.85
6½	9261, 50B44, 5155	1.7	3.8	1.4	2.2	1.25	2.0
7	5147, 6150	1.85	...	1.5	2.4	1.35	2.1
7½	5160, 9262, 50B50	2.0	...	1.7	2.5	1.5	2.2
8	4142, 81B45, 8650	2.1	...	1.8	2.7	1.6	2.35
8½	5152, 50B60	2.2	...	1.9	2.9	1.7	2.45
9½	4337, 8750, 8655	2.5	...	2.2	3.3	1.9	2.7
10	4145, 51B60	2.6	...	2.3	3.4	2.0	2.8
10½	9840	2.7	...	2.35	3.5	2.1	2.9
11	8653, 8660	2.8	...	2.4	3.7	2.15	3.0
11½	8645	2.9	...	2.5	3.8	2.25	3.1
12	86B45	...	...	...	...	...	...
13	4340, 4147	3.25	...	2.8	...	2.45	3.4
14	4150	3.45	...	2.95	...	2.6	3.55
20	E4340	...	...	3.6	...	...	...
22	9850, 4161	...	...	3.7	...	...	...
55 HRC							
1	1141, 1042, 4042, 4142, 1045, 1146, 1050(b), 8642	...	0.65	...	0.6	...	0.3
1½	50B46	...	0.9	...	0.8	...	0.45
2	8742, 5046, 4047, 5145	...	1.2	...	0.9	...	0.65
2½	6145	0.4	1.5	...	1.1	...	0.8
3	4145, 8645, 1345	0.6	1.8	...	1.2	0.3	0.95
3½	86B45, 5147, 4053, 9260	0.7	2.05	0.5	1.4	0.45	1.1
4½	5150, 4063	1.05	2.6	0.8	1.6	0.7	1.45
5	81B45, 6150, 9261, 5155	1.2	2.9	0.9	1.8	0.85	1.6
5½	8650, 5152, 4068	1.4	3.2	1.1	1.9	1.0	1.7
6	50B50	1.55	3.5	1.2	2.1	1.1	1.85
6½	5160, 9262	1.7	3.8	1.4	2.2	1.25	2.0
7	4147, 8750, 8655	1.85	...	1.5	2.4	1.35	2.1
7½	50B60	2.0	...	1.7	2.5	1.5	2.2
9	8653, 51B60, 8660	2.35	...	2.0	3.1	1.8	2.6
9½	4150	2.5	...	2.2	3.3	1.9	2.7
17	9850	...	...	3.35	...	...	...

(a) If based on equivalent hardness, actual bar diameter will be less. (b) High residual alloy

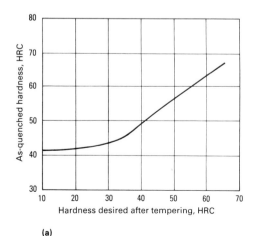
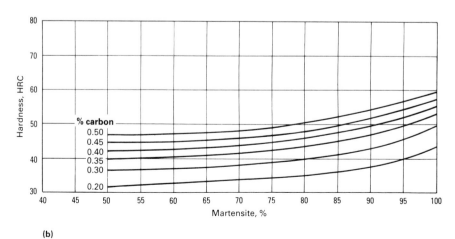

Fig. 15 Curves for steel selection based on hardness. (a) Minimum as-quenched hardness to produce various final hardnesses after tempering. (b) Dependence of as-quenched hardness on percentages of martensite and carbon

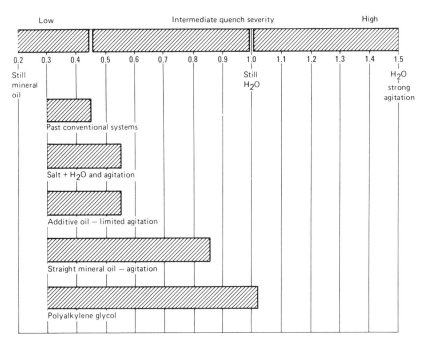

Fig. 16 Approximate quench severities for quenching media containing additives to improve cooling capacity

will be used for heating to the austenitizing temperature. Therefore, the chart for half-radius in Fig. 18(c) is applicable.

The following steps will then lead to the selection of a steel. First, trace horizontally at the level of 1½ in. diameter to the curve for oil quench at 60 m/min (200 sfm) (curve 5). From the point of intersection with this curve, trace vertically to the x-axis to determine the location on the end-quenched bar that has the same cooling rate as the point at half-radius in the 1½ in. round. This location is $6\cdot5/16$ from the quenched end of the bar. Then turn to the section of Table 2 that gives the location of 45 HRC on the end-quenched bar for the various H-steels. Here it is found that four steels will produce 45 HRC at $6\cdot5/16$ from the end of the bar: 8640, 8740, 5150, and 94B30. If some additional hardenability is not undesirable, steels that will produce 45 HRC at $7/16$ can be included—4137, 8642, 6145, and 50B40. Steel 9261 is also in the same category, but it would not be applicable, because it is a spring steel used only when the as-quenched hardness must be as high as 50 to 55 HRC. Therefore, eight steels are available that will meet the hardenability requirements of the stipulated specification. From knowledge of other characteristics of these steels, including machinability, forgeability, crackability, distortion, availability, and cost, the selector can decide which of these eight will be the most desirable for the part in question.

Example 1: Selection of a Steel with 38 mm (1½ in.) Diam Section Equivalent Having 45 HRC at Half-Radius. This example traces the steps needed to select a steel that will harden to 45 HRC at half-radius in a part having a significant section equivalent to a 38 mm (1½ in.) diam bar. First, it is assumed that, to prevent distortion, the quench will be in oil at 60 m/min (200 sfm) (H = 0.5) and that a nonscaling atmosphere

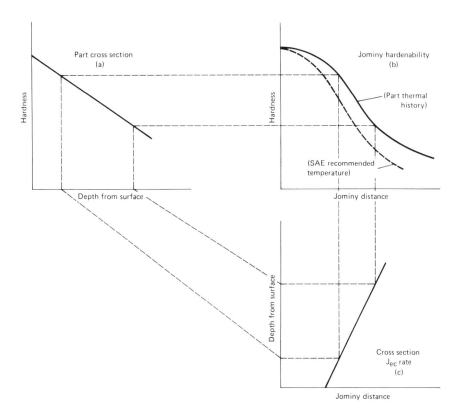

Fig. 17 Determination of Jominy equivalent hardness (J_{eh}) rates

Jominy equivalent hardness (J_{eh}) rates are determined by comparing the hardnesses of cross sections of parts receiving the established production heat treatment to hardnesses obtained on end-quenched bars of the same steel. A typical procedure is as follows:
1. Select hardening and quenching conditions that the production hardening equipment can easily fulfill.
2. Select a low-hardenability steel, such as 8620, 4023, or 1040, and manufacture a quantity of finished components: gears, bearings, shafts.
3. Quench a number of these components (in the uncarburized condition) in the production facility.
4. Measure the hardnesses obtained at all critical locations from the surface to the core.
5. Compare the measured hardness values at these locations with equivalent hardness values produced at some end-quench (J_{eh}) location on a Jominy bar made from the same heat and end quenched from the same thermal conditions.
6. The J_{eh} values obtained in this fashion define the equal hardness cooling conditions for each location in the production-quenched component.
7. Finally, select from available end-quench data a steel that will produce the hardnesses required at each critical J_{eh} location in the finished production part. If end-quench data are not available, calculate a suitable composition by one of the standard methods.

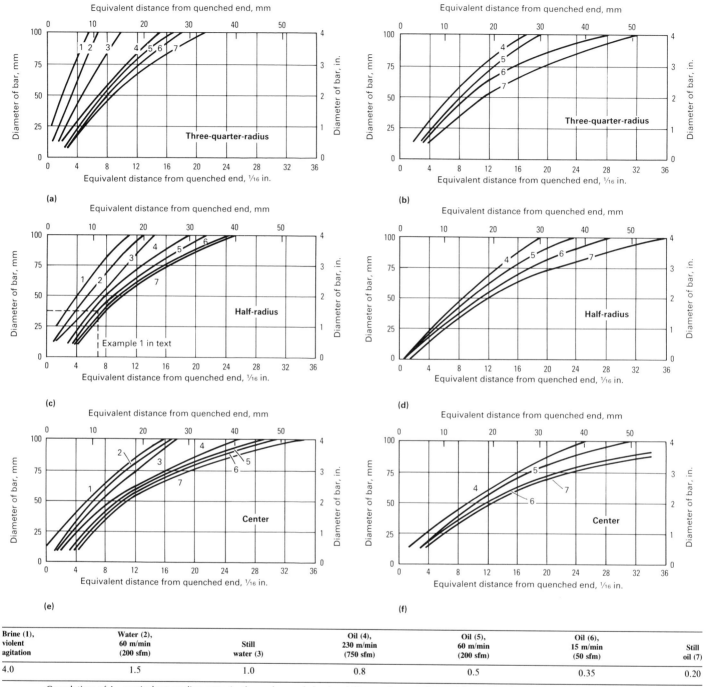

Fig. 18 Correlation of J_{ec} equivalent cooling rates in the end-quench hardenability specimen and round bars quenched in oil, water, and brine. (a), (c), and (e) Nonscaling austenitizing atmosphere. (b), (d), and (f) Austenitized in air

Scaled Rounds. When values for scaled round bars are desired, Fig. 18(b), (d), and (f) can be used. However, prediction of results for sizes less than 25 mm (1 in.) in diameter should not be the basis for important decisions involving costly purchases without further checking, because values for these sizes were obtained by extrapolation.

Figure 20 shows another correlation for rounds based on the equivalent hardness criterion. In Fig. 20, cooling conditions from the surface to the center of rounds of various sizes quenched in media ranging in quenching severity from 0.20 to infinity are correlated with J_{eh}; they are given in $\frac{1}{16}$ in. units producing the same hardness on the end-quench bar. Figure 20 is especially useful for estimating through-section strength, because the entire hardness profile of the prospective steel (and, to a degree, microstructure as well) can be predicted for rounds with different diameters from one set of end-quench data. Instructions for the procedure are given in the caption.

Rectangular or Hexagonal Bars and Plate. Except in critical or borderline applications, size relationships for rounds can be applied without correction to square or hexagonal sections. Figures 18 through 20 can also be used for rectangular bars in which the ratio of width to thickness (W/T) is less than 4, but the value 1.4 times the thickness should

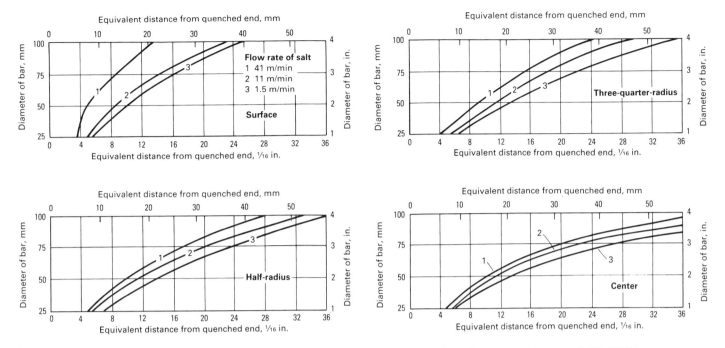

Fig. 19 Correlation of J_{ec} equivalent cooling rates in the end-quenched hardenability specimen and round bars quenched in salt at 205 °C (400 °F)

be used as the equivalent round. Large plates cool considerably more slowly than bars. The cooling rate relationships shown in Fig. 21 and 22 apply to these shapes.

Tubular Parts. The application of end-quench hardenability data to the selection of steel for hollow cylindrical sections is based largely on production experience with similar parts. There has been some progress in equating tubular sections to round bars and in developing dimensionless temperature-time charts for long hollow cylinders. Hollomon and Zener (Ref 7) determined by calculation the diameter of solid steel cylinders that, when quenched in a given medium, could be expected to have the same hardness at the center as the minimum hardness in the wall of hollow cylinders when quenched in the same medium. The rule of thumb of doubling the tube wall thickness to obtain the diameter of an equivalent solid bar is a useful first approximation.

Estimating Hardenability. When actual end-quench hardenability data are unavailable, the hardening performance of a steel of given chemical composition can be estimated from calculated hardenability data. The various methods proposed for calculating hardenability are given in the section "Calculation of Hardenability" in this article. Details can be found in Ref 3 and 8 to 13.

Use of the Charts

The true measure of applicability of any steel to a part requiring heat treatment is the relation of its hardenability to the critical cross section of the part at the time it is heat treated. The term critical cross section refers to that section of the part where service stresses are highest and therefore where the highest mechanical properties are required. For example, if the part is a rough forging 64 mm (2½ in.) in diameter at the critical cross section, which is later machined to 50 mm (2 in.) in diameter, and the finished part must be hardened to three-quarter-radius (that is, 6.4 mm, or ¼ in., deep), then the hardenability of the steel must be such that the rough forging will harden 13 mm (½ in.) deep.

Figure 23 shows the correlation between cooling rates along the end-quench hardenability specimen and at four locations in round bars up to 102 mm (4 in.) in diameter for both oil and water quenching at 60 m/min (200 sfm). The curves in Fig. 23 provide data that can be used directly in steel selection. Following is an example of their practical application to a specific problem of steel selection.

Example 2: Use of Hardenability Charts to Verify that 4140H Steel Will Fulfill Hardness Specifications for a 44.45 mm (1.75 in.) Diam Shaft. A shaft 44.45 mm (1.75 in.) in diameter and 1.1 m (3½ ft) long is required in a machine. The engineering analysis indicates that the torsion requirements will approach a maximum of 170 MPa (25 ksi) and that the bending stresses will reach a maximum of 550 MPa (80 ksi). Because several other parts in production in the same plant are being made from 4140H steel, it is desired to know whether 4140H has enough hardenability for this shaft.

Because the shear stress in torsion is about one-half that in bending, the latter will be the primary consideration. In bending, stresses approach zero in the neutral axis; therefore, the steel need not be hardened completely to the center. This is helpful because the distribution of stress in quenching will decrease the danger of quench cracking and, after tempering, should leave the exterior portion of the shaft in compression.

In order to withstand a fatigue load of 550 MPa (80 ksi) in bending, a minimum hardness of 35 HRC is required. For this example, it will be assumed that 35 HRC should be obtained by tempering a structure that, as-quenched, contains at least 80% martensite. From experience with similar parts, it is known that the 80% martensite structure should be present down to the three-quarter-radius position in the shaft.

Because 4140H has a minimum carbon content of 0.37%, the first operation on the charts (Fig. 24) is to find the as-quenched hardness that corresponds to 0.37% C in an 80% martensite structure. As shown in the top chart of Fig. 24—the same data as in Fig. 1(d)—this as-quenched hardness is 45 HRC.

The original question (whether 4140H is appropriate for this part) can now be rephrased to read: Will 4140H provide the required minimum as-quenched hardness of 45 HRC at three-quarter-radius in the 44.45 mm (1.75 in.) diam shaft? To determine the answer to this question, enter the middle chart of Fig. 24 (this is the same as Fig. 23b) at the diameter level of 44.45 mm (1.75 in.) and move horizontally to an intersection

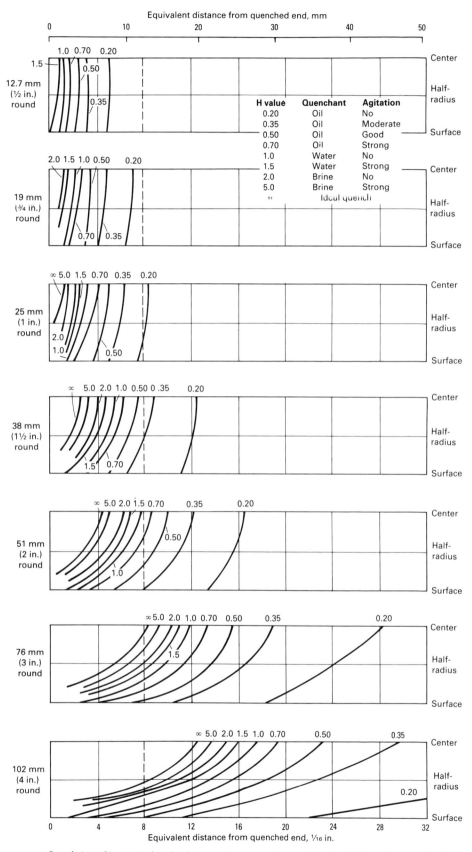

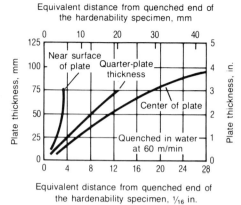

Fig. 20 Correlation of J_{eh} equivalent hardness positions in end-quench hardenability specimen and various locations in round bars quenched in oil, water, and brine. The dashed line shows the various positions in ½ to 4 in. diam rounds that are equivalent to the 8/16 in. distance on the end-quench bar. To determine cross-sectional hardnesses from results of end-quench tests, pick out the end-quench hardness at an appropriate point on the bottom line and extend an imaginary line upward to the curved line that corresponds to the quenching severity needed to obtain that hardness for the given diameter of round.

Fig. 21 Correlation of equivalent cooling rates in the end-quench specimen and quenched plates

with the ¾-radius curve. This intersection occurs at the 6.5/16 position on the specimen. Then, move down vertically into the bottom chart to an intersection with the curve for minimum hardenability of 4140H. The intersection occurs at 49 HRC. Because no more than 45 HRC is required, 4140H has more than enough hardenability for this part.

H-Steels

Hardenability bands are Jominy curves, based on much historical data, that describe the expected hardenability of many grades of carbon and alloy steels. The H-steels are guaranteed by the supplier to meet these limits for specific ranges of chemical composition. In general, the allowable composition ranges of H-steels are slightly wider than those of steels melted to composition specifications in order to accommodate differences in the residual levels and practices of different mills. These steels are designated by the letter "H" following the composition code or preceding the UNS designation.

The charts in the following article, "Hardenability Curves," in this Volume show composition limits and hardenability bands for SAE-AISI steels that can be purchased on a hardenability basis. The minimums of these bands are summarized in Table 2.

Current steelmaking technologies—ladle refining, for example—have permitted excellent control of composition. As a result, hardenability specifications much closer than those indicated by the H-bands can often be worked out between purchaser and supplier. The availability of fractional band hardenability is a function of both the base steel composition and the commonality of the grade.

When an H-steel is specified, the steel producer shows on the shipping papers or by some other acceptable means the hardenability characteristics of the heat involved. The heat hardenability is shown

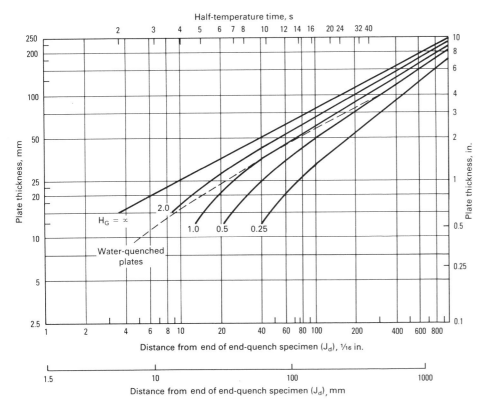

Fig. 22 Correlation between J_{ec} and center cooling rates in plates quenched at various severities

either by hardness values at specified reference points or at the following distances from the quenched end of the test specimen: 1/16, 1/8, 1/4, 1/2, 3/4, 1, 1 1/4, 1 1/2, 1 3/4, and 2 in. No reading below 20 HRC is reported. The heat hardenability is determined from either a cast or a forged end-quench test bar.

Use of Hardenability Limits

H-band limits are shown graphically in the article "Hardenability Curves" in this Volume for convenience in estimating the hardness values obtainable at various locations on the end-quench test bar and for quick comparison of the various H-grades. However, the graphs are not used for specification purposes. Tables appearing with the graphs in the article "Hardenability Curves" in this Volume show the minimum and maximum Rockwell C hardness values at the corresponding distances from the quenched end of the standard end-quenched test specimen for all H-steels.

When desirable, the maximum and minimum limits at the 1/16-in. position can be specified in addition to the other two points. When it is necessary to specify more than two points on the hardenability band (exclusive of the maximum and minimum limits at 1/16 in.), a tolerance of two points Rockwell C over a 3/16 in. portion of either curve (except at 1/16 in.) is permitted.

Maximum Hardenability Limits

As pointed out in the preceding paragraphs, maximum hardenability can sometimes be specified as well as minimum. Although minimum hardenability is significant in relation to the maximum section to be hardened, the maximum hardenability is related chiefly to minimum sections and their tendencies to distort or crack, especially when made from higher-carbon steels.

For example, assume that there is a part for which 4137 will not quite provide the necessary minimum hardenability. By changing to 4142, the minimum hardenability will be increased by 4/16 at 45 HRC, a worthwhile increase. The danger is that, by increasing carbon to provide greater hardenability, the maximum as-quenched hardness is also increased. Many parts have a thin section sensitive to both maximum hardenability and high as-quenched hardness, and because of this combined effect of higher carbon, such sections often break during quenching.

The higher-carbon steels transform to martensite at a lower temperature; that is, the M_s temperature is lower. At this temperature, the steel is less plastic and therefore less able to withstand the strains set up by the volume increase (about 1 1/2%) when austenite transforms to martensite. Also, the higher-carbon martensites are harder and more brittle and cannot withstand the severe strains set up in quenching as well; therefore, pieces with an unfavorable configuration, such as a shaft with a flange, may develop quench cracks.

Steels for Case Hardening

Many alloy steels for case hardening are now specified on the basis of core hardenability. Although the same considerations generally apply as for the selection of uncarburized grades, there are some peculiarities in carburizing applications.

First, in a case-hardened steel, the hardenability of both case and core must be considered. Because of the difference in carbon content, case and core have quite different hardenabilities, and this difference is much greater for some steels than for others. Moreover, the two regions have different functions to perform in service. Until the introduction of lean alloy steels such as the 86xx series, with and without boron, there was little need to be concerned about case hardenability, because the alloy content combined with the high carbon content always provided adequate hardenability. This is still generally true when the steels are direct quenched from carburizing, so that the carbon and alloying elements are in solution in the case austenite. In parts that are reheated for hardening and in heavy-sectioned parts, however, both case and core hardenability requirements should be carefully evaluated.

The hardenability of the steels as purchased will be the core hardenability. Because these low-carbon steels, as a class, are shallow hardening and because of the wide variation in the section sizes of case-hardened parts, the hardenability of the steel must be related to some critical section of the part, for example, the pitch line or the root of a gear tooth. This is best accomplished by making a part of a steel of known hardenability, heat treating it, and then, by means of equivalence of hardness, relating the hardenability in the critical section or sections to the proper positions on the end-quench hardenability specimens, both base carbon and carburized.

Finally, notice that the relationship between the thermal gradient and the carbon (hardenability) gradient during quenching of a carburized part can make a measurable difference in the case depth as measured by hardness. That is, an increase in base hardenability can produce a higher proportion of martensite for a given carbon level, yielding a deeper measured case depth. Therefore, a shallower carbon profile and shorter carburizing time could be used to attain the desired result in a properly chosen steel.

Core Hardness. A common mistake is to specify too narrow a range of core hardness. When the final quench is from a temperature high enough to allow the development of full core hardness, the hardness variation at any location will be that of the

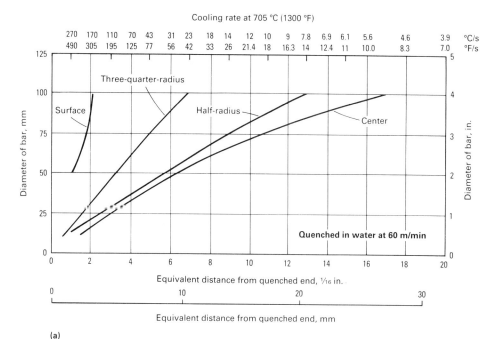

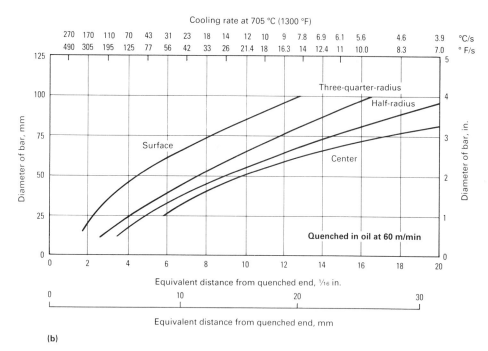

Fig. 23 Equivalent cooling rates for round bars quenched in water (a) and oil (b). Correlation of equivalent cooling rates in the end-quenched hardenability specimen and quenched round bars free from scale. Data for surface hardness are for mild agitation; other data are for 60 m/min (200 sfm).

hardenability band of the steel at the corresponding position on the end-quenched hardenability specimen. One way to alter this state of affairs is to use higher-alloy steels. In the commonly used alloy steels having a maximum of 2% total alloy content, the range for the core hardness of sections such as gear teeth is 12 to 15 points HRC. Higher-alloy steels exhibit a narrower range; for example, in 4815 the range is 10 points, while in 3310 it is 8 points. Such steels are justified only for severe service or special applications.

In standard steels purchased to chemical composition requirements rather than to hardenability, the range can be 20 points HRC or more; for example, 8620 may vary from 20 to 45 HRC at the ⁴⁄₁₆ in. position. This 25-point range emphasizes the advantage of purchasing to hardenability specifications to avoid the intolerable variation possible within the ranges for standard chemistry steels. Without resorting to the use of high-alloy steels, another way to control core hardness within narrow limits is to use a final quench from a lower temperature, so that full hardness will be developed in the case without the disadvantage of excessive core hardness.

Case-Hardened Steel Applications. In addition to the complexities already mentioned, there are highly variable conditions in heat treating and sometimes differences of opinion, even among qualified engineers. The subject can be simplified to some extent by dividing it into applications involving, first, gears and similar parts and, second, all others.

Gears are almost always oil quenched because distortion must be held to the lowest possible level. This means that alloy steels are usually selected—which particular alloy is much debated. The lower-alloy steels such as 4023, 5120, 4118, 8620, and 4620, with a carbon range between 0.15 and 0.25%, are widely used and generally satisfactory. The first choice usually would be made from the last two steels mentioned, either of which should be safe for all ordinary applications. The final choice, based on service experience or dynamometer testing, should be the least expensive steel that will do the job. To this list should be added 1524, which, although not classified commercially as an alloy steel, has sufficient manganese to make it oil hardening up to an end-quench correlation point of ³⁄₁₆.

For heavy-duty applications, higher-alloy grades such as 4320, 4817, and 9310 are justifiable if based on actual performance tests. The life testing of gears in the same mountings used in service, to prove both the design and the steel selection, is particularly important.

The carbonitriding process extends the use of carbon steels such as 1016, 1018, 1019, and 1022 into the field of light-duty gearing by permitting the use of oil quenching in teeth of eight diametral pitch and finer. Steels selected for such applications should be specified silicon-killed fine-grain in order to ensure uniform case hardness and dimensional control. The core of such gears will, of course, have the properties of low-carbon steel, oil quenched. In the thin sections of fine-pitch teeth, this may be up to 25 HRC. The carbonitriding process is usually limited, for economic reasons, to maximum case depths of approximately 0.6 mm (0.025 in.).

Nongear Applications. In other applications, when distortion is not a major factor, the carbon steels described above, water quenched, can be used up to a 50 mm (2 in.) diameter. In larger sizes, low-alloy steels, water quenched, such as 5120, 4023, and 6120, can be used, but possible distortion and quench cracking must be guarded against.

Steel Castings

Cast steels have about the same hardenability values as their wrought counterparts of the same composition. Because of their coarser grain, steels in the as-cast condition frequently show higher hardenability than the same steels in the wrought condition, but after they are normalized, hardenability will be more nearly equal. Variation in the hardenability of cast steels is caused by the same factors as in wrought steels.

Results from tests on standard end-quench specimens taken from various locations within a heavy casting reveal no significant effect of location on the hardenability of a modified 4032 steel (Fig. 25). Additional data on the hardenability of cast steels are given in SAE J434 and J435 and in Ref 14.

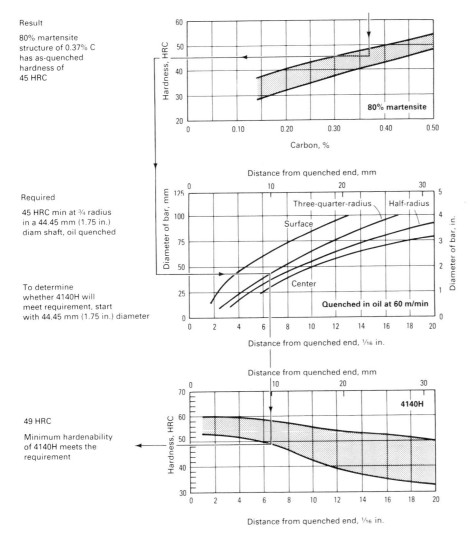

Fig. 24 Illustration of the use of hardenability data in steel selection

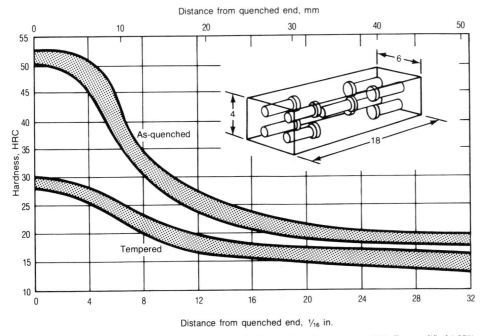

Fig. 25 Range of results for hardenability specimens cut from various locations in a 4032 alloy modified 1.35% Mn steel casting. Dimensions in illustration given in inches

REFERENCES

1. C.F. Jatczak, Effect of Microstructure and Cooling Rate on Secondary Hardening of Cr-Mo-V Steels, *Trans. ASM*, Vol 58, 1965, p 195
2. R.A. Grange, Estimating the Hardenability of Carbon Steels, *Metall. Trans.*, Vol 4, Oct 1973, p 2231
3. M.A. Grossmann, Hardenability Calculated from Chemical Composition, *Trans. AIME*, Vol 150, 1942, p 227
4. D.V. Doane and J.S. Kirkaldy, Ed., *Hardenability Concepts With Applications to Steel*, The Metallurgical Society, 1978
5. C.S. Siebert, D.V. Doane, and D.H. Breen, *The Hardenability of Steels*, American Society for Metals, 1977
6. D.J. Carney, Another Look at Quenchants, Cooling Rates and Hardenability, *Trans. ASM*, Vol 46, 1954, p 882
7. J.H. Hollomon and C. Zener, Quenching and Hardenability of Hollow Cylinders, *Trans. ASM*, Vol 33, 1944, p 1
8. J.M. Hodge and M.A. Orehoski, Relationship Between Hardenability and Percentage of Martensite in Some Low-Alloy Steels, *Trans. AIME*, Vol 167, 1946, p 627
9. I.R. Kramer, S. Siegel, and J.G. Brooks, Factors for the Calculation of Hardenability, *Trans. AIME*, Vol 167, 1946, p 670
10. A.F. de Retana and D.V. Doane, Hardenability of Carburizing Steels, *Met. Prog.*, Vol 100 (No. 3), Sept 1971, p 65
11. C.F. Jatczak, *Trans. AIME*, Vol 4, 1973
12. E. Just, New Formulas for Calculating Hardenability Curves, *Met. Prog.*, Vol 96 (No. 5), Nov 1969, p 87
13. J.S. Kirkaldy, *Metall. Trans.*, Vol 4, Oct 1973
14. *Materials*, Vol 1, *1989 SAE Handbook*, Society of Automotive Engineers, 1989

SELECTED REFERENCES

- A.L. Boegehold, Hardenability Control for Alloy Steel Parts, *Met. Prog.*, Vol 53 (No. 5), May 1948, p 697
- D.V. Doane and J.S. Kirkaldy, Ed., *Hardenability Concepts With Applications to Steel*, The Metallurgical Society, 1978
- C.F. Jatczak, Determining Hardenability From Composition, *Met. Prog.*, Vol 100 (No. 3), Sept 1971, p 60
- J.L. Lamont, *Iron Age*, 14 Oct 1943
- R.A. Rege, P.E. Hamill, and J.M. Hodge, The Effects of Geometry on the Cooling of Plates and Bars, *Trans. ASM*, Vol 62, 1969, p 333
- Report OSRD 3743, National Defense Research Committee
- J.M. Tartaglia and G.T. Eldis, Core Hardenability Calculations for Carburizing Steels, *Metall. Trans. A*, Vol 15A (No. 6), June 1984, p 1173-1183
- E.W. Weinman, R.F. Thomson, and A.L. Boegehold, Correlation of End-Quenched Test Bars and Rounds, *Trans. ASM*, Vol 44, 1952, p 803

Hardenability Curves

HARDENABILITY CURVES for more than 80 types of carbon and alloy H-band steels comprise this article (Ref 1, 2). The tabular data used to compile these curves are also included with each graph. Values from these tables are used for specification purposes, and SAE recommends choosing two points to designate the hardenability. The two points may be designated in any one of the ways listed below and illustrated in Fig. 1:

- The minimum and maximum hardness values at any desired distance. This method is illustrated as points A–A and is specified as J43 to J54 = 3/16 in. Obviously, the distance selected would be that distance on the end-quench specimen that corresponds to the section used by the consumer
- The minimum and maximum distances at which any desired hardness value occurs. This method is illustrated as points B–B and would be specified as J39 = 4/16 to 9/16 in.
- Two maximum hardness values at two desired distances, illustrated as points C–C and specified as J50 = 5/16 in. (max), J34 = 12/16 in. (max)
- Two minimum hardness values at two desired distances, illustrated as points D–D and specified as J35 = 5/16 in. (min), J21 = 16/16 in. (min)
- Any minimum hardness plus any maximum hardness, E–E, specified as J37 max = 10/16, J32 min = 6/16

It should be noted that each H-band hardenability limit curve is presented graphically and in tabular form, in both metric and English units.

REFERENCES

1. *Materials*, Vol 1, *1989 SAE Handbook*, Society of Automotive Engineers, 1989
2. "Alloy, Carbon and High Strength Low Alloy Steels: Semifinished for Forging; Hot Rolled Bars and Cold Finished Bars, Hot Rolled Deformed and Plain Concrete Reinforcing Bars," Steel Products Manual, American Iron and Steel Institute, March 1986

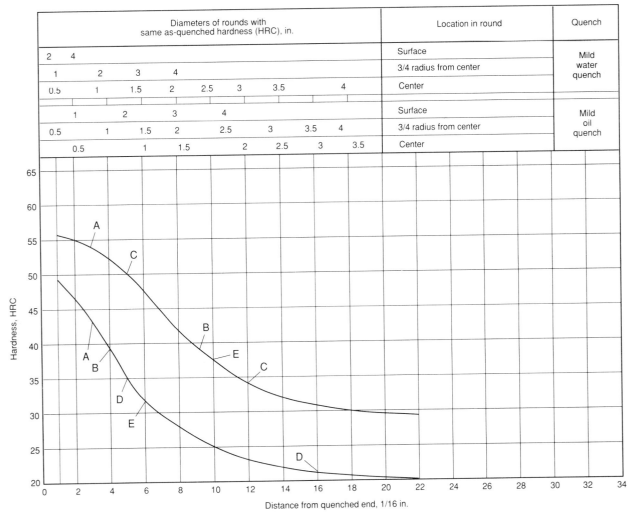

Fig. 1 Typical hardenability curve shown with English units. See text for discussion of designated points.

SAE/AISI 1038H UNS H10380

Heat-treating temperatures recommended by SAE
Normalize (for forged or rolled specimens only): 870 °C (1600 °F)
Austenitize: 845 °C (1550 °F)

Hardness limits for specification purposes

J distance, mm	Hardness, HRC Maximum	Hardness, HRC Minimum
1.5	58	51
3	56	37
5	49	25
7	33	22
9	29	20
11	27	...
13	26	...
15	25	...
20	24	...
25	22	...
35	...	...

Hardness limits for specification purposes

J distance, 1/16 in.	Hardness, HRC Maximum	Hardness, HRC Minimum
1	58	51
1.5	56	42
2	55	34
2.5	53	29
3	49	26
3.5	43	24
4	37	23
4.5	33	22
5	30	22
5.5	29	21
6	28	21
6.5	27	20
7	27	...
7.5	26	...
8	26	...
9	25	...
10	25	...
12	24	...
14	23	...
16	21	...

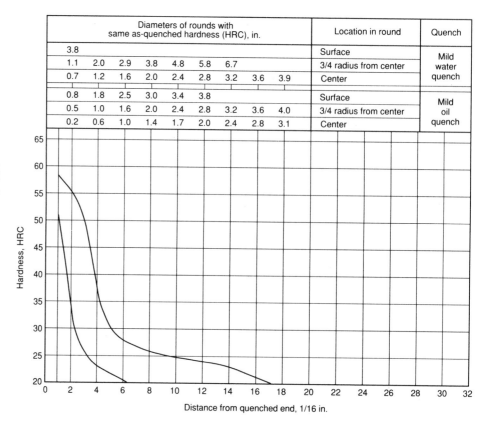

SAE/AISI 1045H   UNS H10450

Heat-treating temperatures recommended by SAE
Normalize (for forged or rolled specimens only): 870 °C (1600 °F)
Austenitize: 845 °C (1550 °F)

Hardness limits for specification purposes

J distance, mm	Hardness, HRC Maximum	Hardness, HRC Minimum
1.5	62	55
3	60	45
5	53	31
7	36	27
9	32	25
11	31	24
13	30	23
15	29	22
20	28	20
25	27	...
30	...	...

Hardness limits for specification purposes

J distance, 1/16 in.	Hardness, HRC Maximum	Hardness, HRC Minimum
1	62	55
1.5	61	52
2	59	42
2.5	56	34
3	52	31
3.5	46	29
4	38	28
4.5	34	27
5	33	26
5.5	32	26
6	32	25
6.5	31	25
7	31	25
7.5	30	24
8	30	24
9	29	23
10	29	22
12	28	21
14	27	20
16	26	...
18	25	...
20	23	...
22	22	...
24	21	...

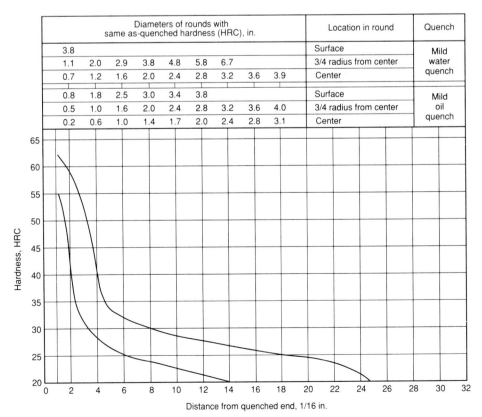

SAE/AISI 1522H UNS H15220

Heat-treating temperatures recommended by SAE
Normalize (for forged or rolled specimens only): 925 °C (1700 °F)
Austenitize: 925 °C (1700 °F)

Hardness limits for specification purposes

J distance, mm	Hardness, HRC Maximum	Hardness, HRC Minimum
1.5	50	41
3	48	35
5	45	23
7	39	20
9	32	...
11	27	...
13	...	...

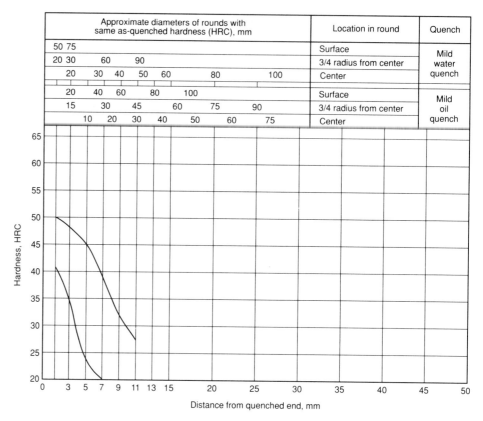

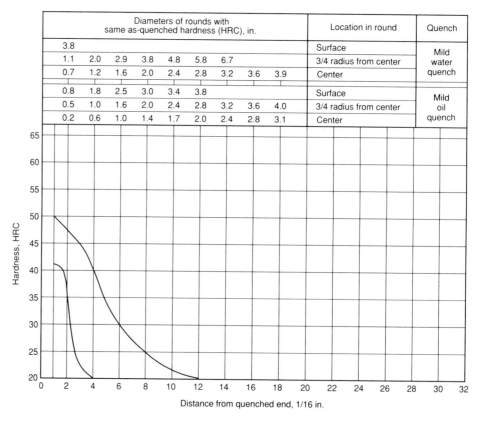

Hardness limits for specification purposes

J distance, 1/16 in.	Hardness, HRC Maximum	Hardness, HRC Minimum
1	50	41
1.5	48	41
2	47	32
2.5	46	27
3	45	22
3.5	42	21
4	39	20
4.5	37	...
5	34	...
5.5	32	...
6	30	...
6.5	28	...
7	27	...
7.5	...	...
8	25	...
9	23	...
10	22	...
12	20	...
14	...	...

SAE/AISI 1524H   UNS H15240

Heat-treating temperatures recommended by SAE
Normalize (for forged or rolled specimens only): 900 °C (1650 °F)
Austenitize: 870 °C (1600 °F)

Hardness limits for specification purposes

J distance, mm	Hardness, HRC Maximum	Hardness, HRC Minimum
1.5	51	42
3	49	39
5	44	26
7	38	21
9	34	...
11	30	...
13	27	...
15	25	...
20	23	...
25	...	...

Hardness limits for specification purposes

J distance, 1/16 in.	Hardness, HRC Maximum	Hardness, HRC Minimum
1	51	42
1.5	49	42
2	48	38
2.5	47	34
3	45	29
3.5	43	25
4	39	22
4.5	38	20
5	35	...
5.5	34	...
6	32	...
6.5	30	...
7	29	...
7.5	28	...
8	27	...
9	26	...
10	25	...
12	23	...
14	22	...
16	20	...

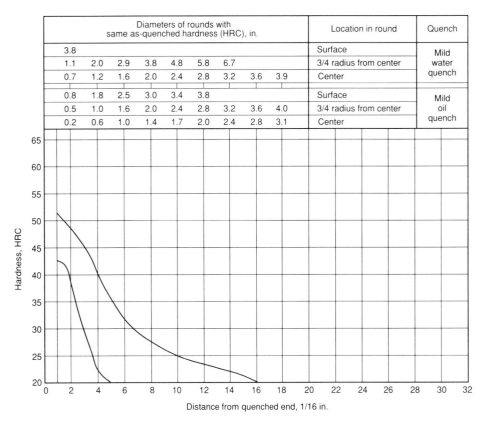

SAE/AISI 1526H UNS H15260

Heat-treating temperatures recommended by SAE
Normalize (for forged or rolled specimens only): 900 °C (1650 °F)
Austenitize: 870 °C (1600 °F)

Hardness limits for specification purposes

J distance, mm	Hardness, HRC Maximum	Hardness, HRC Minimum
1.5	53	44
3	50	39
5	44	24
7	37	20
9	32	...
11	28	...
13	25	...
15	24	...
20	...	...

Hardness limits for specification purposes

J distance, 1/16 in.	Hardness, HRC Maximum	Hardness, HRC Minimum
1	53	44
1.5	50	42
2	49	38
2.5	47	33
3	46	26
3.5	42	25
4	39	21
4.5	37	20
5	33	...
5.5	31	...
6	30	...
6.5	28	...
7	27	...
7.5	26	...
8	26	...
9	24	...
10	24	...
12	23	...
14	22	...
16	21	...
18	20	...

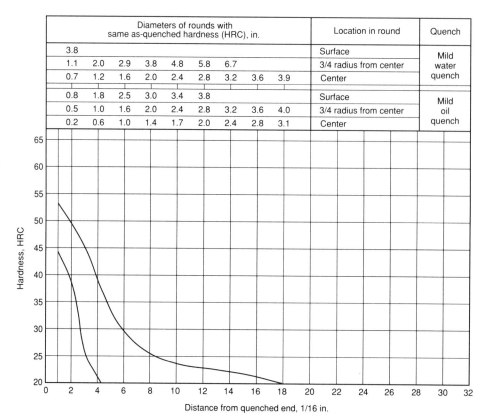

Hardenability Curves / 491

SAE/AISI 1541H UNS H15410

Heat-treating temperatures recommended by SAE
Normalize (for forged or rolled specimens only): 870 °C (1600 °F)
Austenitize: 845 °C (1550 °F)

Hardness limits for specification purposes

J distance, mm	Hardness, HRC Maximum	Hardness, HRC Minimum
1.5	60	53
3	59	50
5	57	43
7	53	36
9	49	29
11	44	25
13	38	23
15	35	22
20	32	20
25	30	...
30	...	...

Hardness limits for specification purposes

J distance, 1/16 in.	Hardness, HRC Maximum	Hardness, HRC Minimum
1	60	53
1.5	59	52
2	59	50
2.5	58	47
3	57	44
3.5	56	41
4	55	38
4.5	53	35
5	52	32
5.5	50	29
6	48	27
6.5	46	26
7	44	25
7.5	41	24
8	39	23
9	35	23
10	33	22
12	32	21
14	31	20
16	30	...
18	30	...
20	29	...
22	28	...
24	26	...

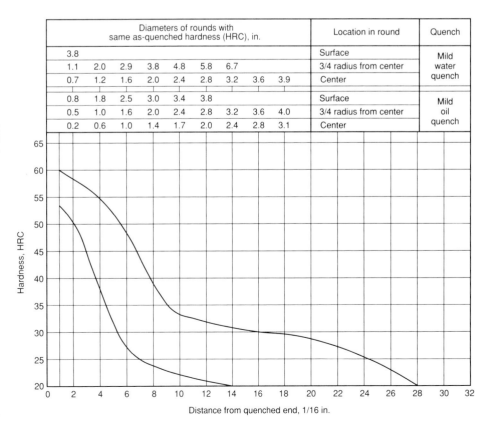

SAE/AISI 15B21H UNS H15211

Heat-treating temperatures recommended by SAE
Normalize (for forged or rolled specimens only): 925 °C (1700 °F)
Austenitize: 925 °C (1700 °F)

Hardness limits for specification purposes

J distance, mm	Hardness, HRC Maximum	Hardness, HRC Minimum
1.5	48	41
3	48	40
5	46	36
7	43	27
9	38	...
11	30	...
13	...	...

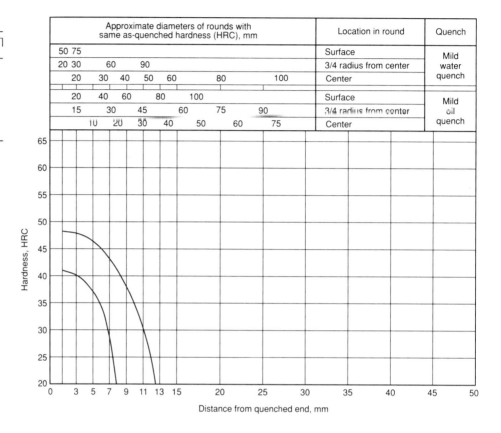

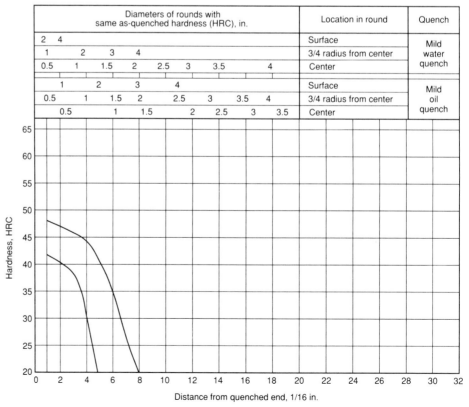

Hardness limits for specification purposes

J distance, 1/16 in.	Hardness, HRC Maximum	Hardness, HRC Minimum
1	48	41
1.5	48	41
2	47	40
2.5	47	39
3	46	38
3.5	45	36
4	44	30
4.5	42	23
5	40	20
5.5	38	...
6	35	...
6.5	32	...
7	27	...
7.5	22	...
8	20	...
9	...	...

SAE/AISI 15B28H   UNS H15281

Heat-treating temperatures recommended by SAE
Normalize (for forged or rolled specimens only): 900 °C (1650 °F)
Austenitize: 870 °C (1600 °F)

Hardness limits for specification purposes

J distance, mm	Hardness, HRC Maximum	Hardness, HRC Minimum
1.5	53	47
3	53	47
5	53	46
7	52	43
9	51	35
11	50	24
13	48	21
15	45	20
20	35	...
25	29	...
30	26	...
35	25	...
40	24	...
45	23	...
50	20	...

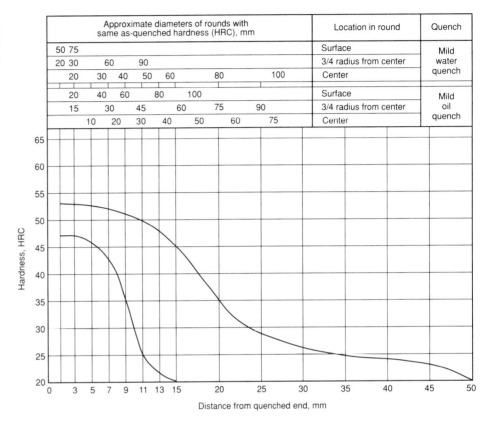

Hardness limits for specification purposes

J distance, 1/16 in.	Hardness, HRC Maximum	Hardness, HRC Minimum
1	53	47
2	53	47
3	52	46
4	51	45
5	51	42
6	50	32
7	49	25
8	48	21
9	46	20
10	43	...
11	40	...
12	37	...
13	34	...
14	31	...
15	30	...
16	29	...
18	27	...
20	25	...
22	25	...
24	24	...
26	23	...
28	22	...
30	21	...
32	20	...

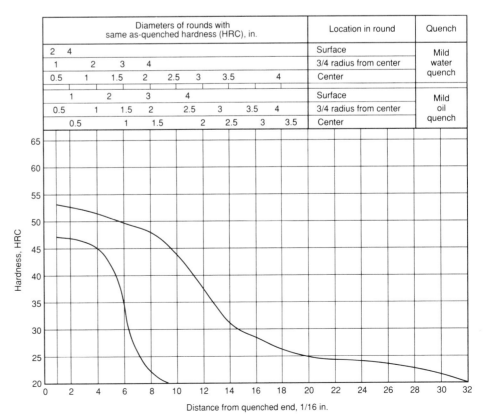

SAE/AISI 15B30H UNS H15301

Heat-treating temperatures recommended by SAE
Normalize (for forged or rolled specimens only): 900 °C (1650 °F)
Austenitize: 870 °C (1600 °F)

Hardness limits for specification purposes

J distance, mm	Hardness, HRC Maximum	Hardness, HRC Minimum
1.5	55	48
3	54	47
5	53	45
7	52	38
9	49	25
11	45	20
13	38	...
15	31	...
20	26	...
25	23	...
30	20	...
35	...	...

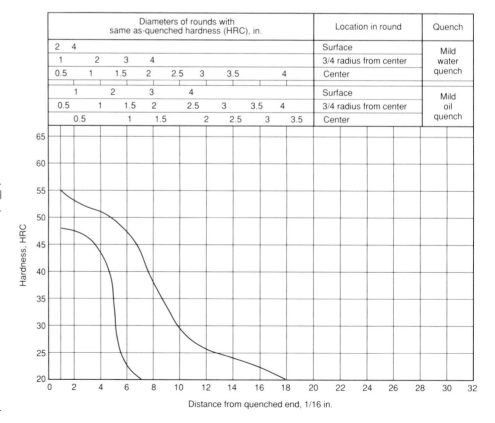

Hardness limits for specification purposes

J distance, 1/16 in.	Hardness, HRC Maximum	Hardness, HRC Minimum
1	55	48
2	53	47
3	52	46
4	51	44
5	50	32
6	48	22
7	43	20
8	38	...
9	33	...
10	29	...
11	27	...
12	26	...
13	25	...
14	24	...
15	23	...
16	22	...
18	20	...
20	...	...

SAE/AISI 15B35H UNS H15351

Heat-treating temperatures recommended by SAE
Normalize (for forged or rolled specimens only): 870 °C (1600 °F)
Austenitize: 845 °C (1550 °F)

Hardness limits for specification purposes

J distance, mm	Hardness, HRC Maximum	Hardness, HRC Minimum
1.5	58	51
3	57	50
5	56	49
7	54	45
9	52	32
11	47	24
13	39	21
15	32	20
20	27	...
25	25	...
30	24	...
35	23	...
40	22	...
45	20	...
50	...	...

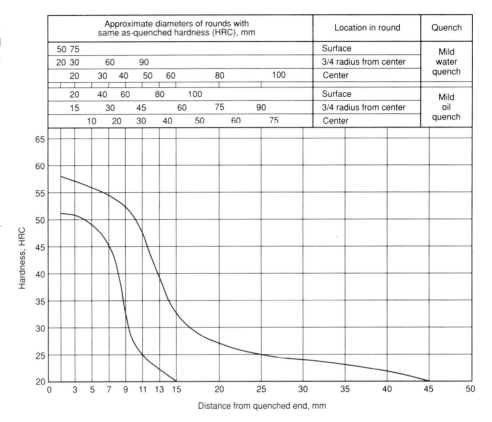

Hardness limits for specification purposes

J distance, 1/16 in.	Hardness, HRC Maximum	Hardness, HRC Minimum
1	58	51
2	56	50
3	55	48
4	54	48
5	53	39
6	51	28
7	47	24
8	41	22
9	...	...
10	30	20
11	...	...
12	27	...
13	...	...
14	26	...
15	...	...
16	25	...
18	...	...
20	24	...
22	...	...
24	22	...
26	...	...
28	20	...
30	...	...

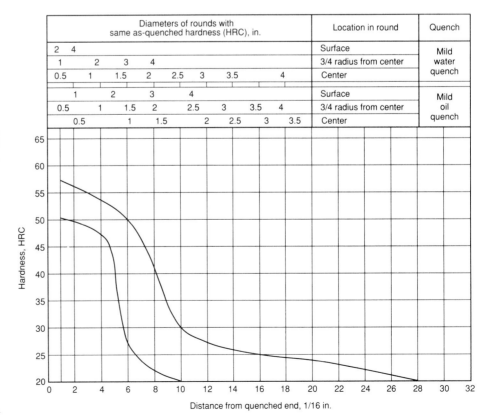

SAE/AISI 15B37H UNS H15371

Heat-treating temperatures recommended by SAE
Normalize (for forged or rolled specimens only): 870 °C (1600 °F)
Austenitize: 845 °C (1550 °F)

Hardness limits for specification purposes

J distance, mm	Hardness, HRC Maximum	Hardness, HRC Minimum
1.5	58	50
3	57	50
5	56	49
7	54	46
9	53	39
11	51	31
13	50	26
15	47	23
20	38	20
25	30	...
30	28	...
35	26	...
40	25	...
45	23	...
50	...	...

Hardness limits for specification purposes

J distance, 1/16 in.	Hardness, HRC Maximum	Hardness, HRC Minimum
1	58	50
2	56	50
3	55	49
4	54	48
5	53	43
6	52	37
7	51	33
8	50	26
9	...	...
10	45	22
11	...	...
12	40	21
13	...	...
14	33	20
15	...	...
16	29	...
18	...	...
20	27	...
22	...	...
24	25	...
26	...	...
28	23	...
30	...	...
32	21	...

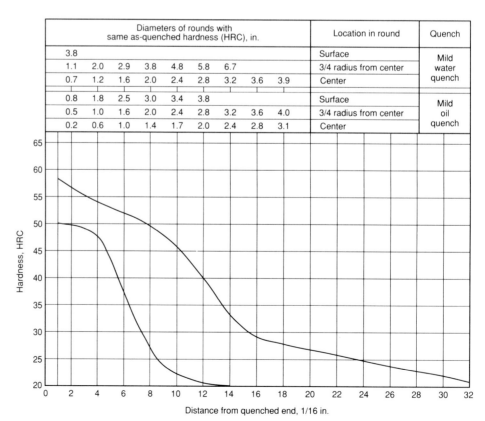

SAE/AISI 15B41H UNS H15411

Heat-treating temperatures recommended by SAE
Normalize (for forged or rolled specimens only): 870 °C (1600 °F)
Austenitize: 845 °C (1550 °F)

Hardness limits for specification purposes

J distance, mm	Hardness, HRC Maximum	Hardness, HRC Minimum
1.5	60	53
3	60	52
5	59	52
7	58	51
9	58	50
11	57	49
13	56	47
15	55	41
20	53	26
25	50	24
30	45	23
35	39	21
40	35	20
45	32	...
50	31	...

Hardness limits for specification purposes

J distance, 1/16 in.	Hardness, HRC Maximum	Hardness, HRC Minimum
1	60	53
2	59	52
3	59	52
4	58	51
5	58	51
6	57	50
7	57	49
8	56	48
9	55	44
10	55	37
11	54	32
12	53	28
13	52	26
14	51	25
15	50	25
16	49	24
18	46	23
20	42	22
22	39	21
24	36	21
26	34	20
28	33	...
30	31	...
32	31	...

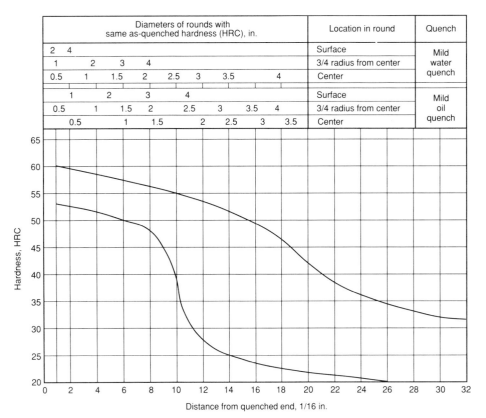

SAE/AISI 15B48H   UNS H15481

Heat-treating temperatures recommended by SAE
Normalize (for forged or rolled specimens only): 870 °C (1600 °F)
Austenitize: 845 °C (1550 °F)

Hardness limits for specification purposes

J distance, mm	Hardness, HRC Maximum	Hardness, HRC Minimum
1.5	63	56
3	63	55
5	62	55
7	61	54
9	60	53
11	59	45
13	57	33
15	56	30
20	49	27
25	39	25
30	33	24
35	31	23
40	30	22
45	29	...
50	28	...

Hardness limits for specification purposes

J distance, 1/16 in.	Hardness, HRC Maximum	Hardness, HRC Minimum
1	63	56
2	62	56
3	62	55
4	61	54
5	60	53
6	59	52
7	58	42
8	57	34
9	56	31
10	55	30
11	53	29
12	51	28
13	48	27
14	45	27
15	41	26
16	38	26
18	34	25
20	32	24
22	31	23
24	30	22
26	29	21
28	29	20
30	28	...
32	28	...

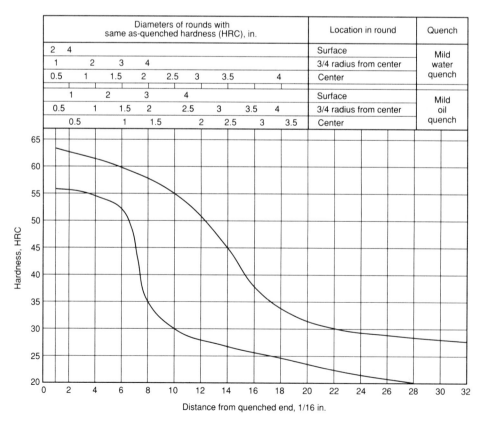

SAE/AISI 15B62H   UNS H15621

Heat-treating temperatures recommended by SAE
Normalize (for forged or rolled specimens only): 870 °C (1600 °F)
Austenitize: 845 °C (1550 °F)

Hardness limits for specification purposes

J distance, mm	Hardness, HRC Maximum	Hardness, HRC Minimum
1.5	...	60
3	...	60
5	65	60
7	65	59
9	65	58
11	65	56
13	64	50
15	64	42
20	63	34
25	60	32
30	56	31
35	48	30
40	42	29
45	37	27
50	34	26

Hardness limits for specification purposes

J distance, 1/16 in.	Hardness, HRC Maximum	Hardness, HRC Minimum
1	...	60
2	...	60
3	...	60
4	...	60
5	65	59
6	65	58
7	64	57
8	64	52
9	64	43
10	63	39
11	63	37
12	63	35
13	62	35
14	62	34
15	61	33
16	60	33
18	58	32
20	54	31
22	48	30
24	43	30
26	40	29
28	37	28
30	35	27
32	34	26

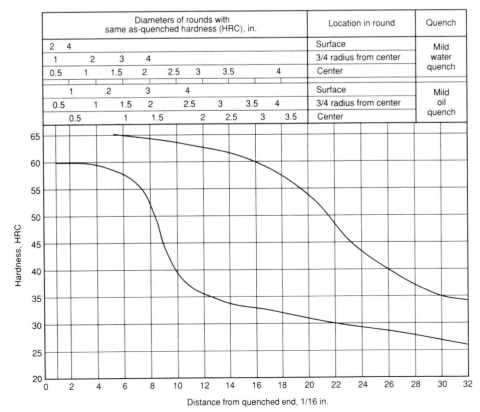

500 / Hardenability of Carbon and Low-Alloy Steels

SAE/AISI 1330H   UNS H13300

Heat-treating temperatures recommended by SAE
Normalize (for forged or rolled specimens only): 900 °C (1650 °F)
Austenitize: 870 °C (1600 °F)

Hardness limits for specification purposes

J distance, mm	Hardness, HRC Maximum	Hardness, HRC Minimum
1.5	56	49
3	56	47
5	55	44
7	53	38
9	51	32
11	48	28
13	45	25
15	43	24
20	39	20
25	35	...
30	33	...
35	32	...
40	31	...
45	31	...
50	30	...

Hardness limits for specification purposes

J distance, 1/16 in.	Hardness, HRC Maximum	Hardness, HRC Minimum
1	56	49
2	56	47
3	55	44
4	53	40
5	52	35
6	50	31
7	48	28
8	45	26
9	43	25
10	42	23
11	40	22
12	39	21
13	38	20
14	37	...
15	36	...
16	35	...
18	34	...
20	33	...
22	32	...
24	31	...
26	31	...
28	31	...
30	30	...
32	30	...

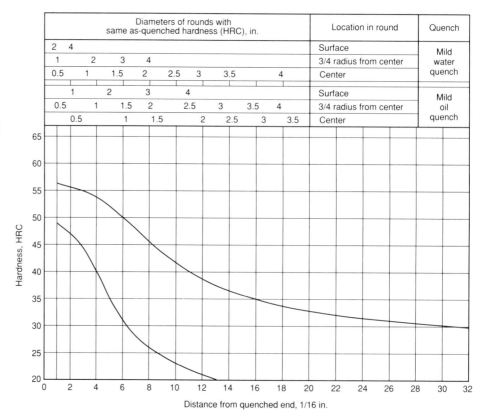

Hardenability Curves / 501

SAE/AISI 1335H UNS H13350

Heat-treating temperatures recommended by SAE
Normalize (for forged or rolled specimens only): 870 °C (1600 °F)
Austenitize: 845 °C (1550 °F)

Hardness limits for specification purposes

J distance, mm	Hardness, HRC Maximum	Hardness, HRC Minimum
1.5	58	51
3	58	49
5	57	46
7	55	42
9	53	36
11	50	31
13	47	28
15	45	27
20	41	23
25	37	21
30	35	...
35	33	...
40	32	...
45	31	...
50	30	...

Hardness limits for specification purposes

J distance, 1/16 in.	Hardness, HRC Maximum	Hardness, HRC Minimum
1	58	51
2	57	49
3	56	47
4	55	44
5	54	38
6	52	34
7	50	31
8	48	29
9	46	27
10	44	26
11	42	25
12	41	24
13	40	23
14	39	22
15	38	22
16	37	21
18	35	20
20	34	...
22	33	...
24	32	...
26	31	...
28	31	...
30	30	...
32	30	...

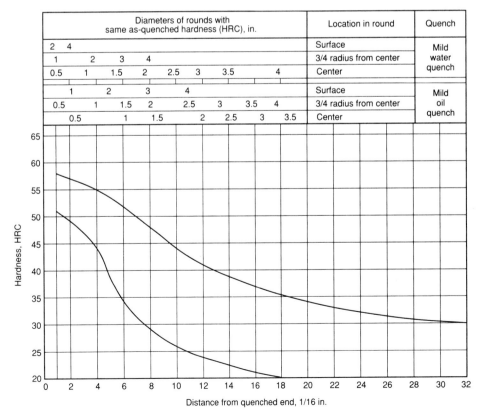

SAE/AISI 1340H   UNS H13400

Heat-treating temperatures recommended by SAE
Normalize (for forged or rolled specimens only): 870 °C (1600 °F)
Austenitize: 845 °C (1550 °F)

Hardness limits for specification purposes

J distance, mm	Hardness, HRC Maximum	Hardness, HRC Minimum
1.5	60	53
3	60	52
5	59	50
7	58	48
9	57	42
11	56	36
13	54	32
15	52	30
20	47	26
25	41	24
30	39	23
35	37	22
40	36	21
45	35	20
50	34	20

Hardness limits for specification purposes

J distance, 1/16 in.	Hardness, HRC Maximum	Hardness, HRC Minimum
1	60	53
2	60	52
3	59	51
4	58	49
5	57	46
6	56	40
7	55	35
8	54	33
9	52	31
10	51	29
11	50	28
12	48	27
13	46	26
14	44	25
15	42	25
16	41	24
18	39	23
20	38	23
22	37	22
24	36	22
26	36	21
28	35	21
30	34	20
32	34	20

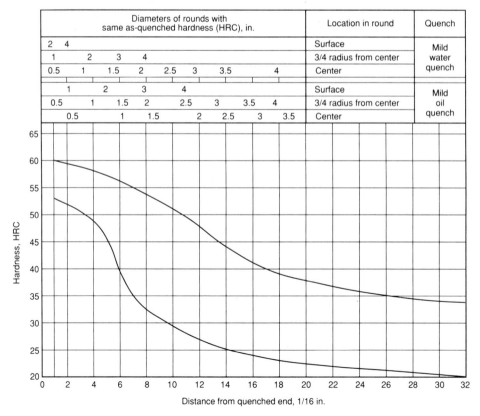

SAE/AISI 1345H UNS H13450

Heat-treating temperatures recommended by SAE
Normalize (for forged or rolled specimens only): 870 °C (1600 °F)
Austenitize: 845 °C (1550 °F)

Hardness limits for specification purposes

J distance, mm	Hardness, HRC Maximum	Hardness, HRC Minimum
1.5	63	56
3	63	56
5	63	54
7	62	52
9	61	46
11	60	38
13	59	35
15	58	31
20	55	29
25	51	27
30	48	26
35	47	25
40	46	24
45	45	24
50	45	24

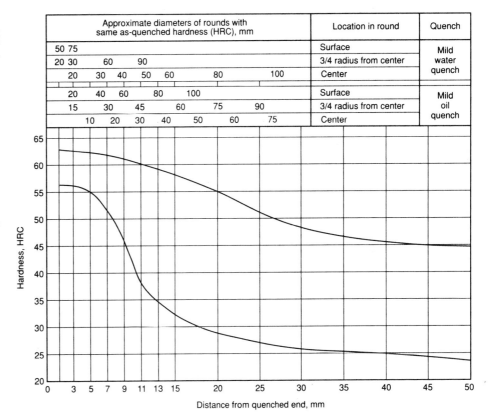

Hardness limits for specification purposes

J distance, 1/16 in.	Hardness, HRC Maximum	Hardness, HRC Minimum
1	63	56
2	63	56
3	62	55
4	61	54
5	61	51
6	60	44
7	60	38
8	59	35
9	58	33
10	57	32
11	56	31
12	55	30
13	54	29
14	53	29
15	52	28
16	51	28
18	49	27
20	48	27
22	47	26
24	46	26
26	45	25
28	45	25
30	45	24
32	45	24

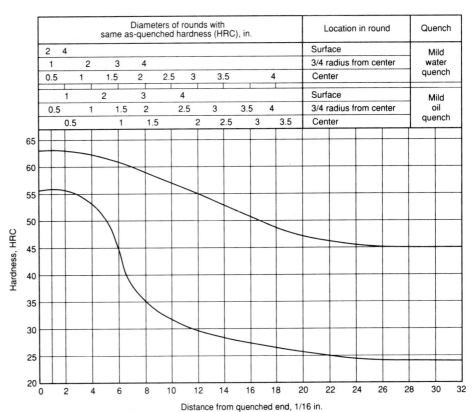

SAE/AISI 4027H   UNS H40270
SAE/AISI 4028H   UNS H40280

Heat-treating temperatures recommended by SAE
Normalize (for forged or rolled specimens only): 900 °C (1650 °F)
Austenitize: 870 °C (1600 °F)

Hardness limits for specification purposes

J distance, mm	Hardness, HRC Maximum	Hardness, HRC Minimum
1.5	52	45
3	51	41
5	45	32
7	40	23
9	32	20
11	29	...
13	26	...
15	25	...
20	23	...
25	22	...
30	21	...
35	...	...

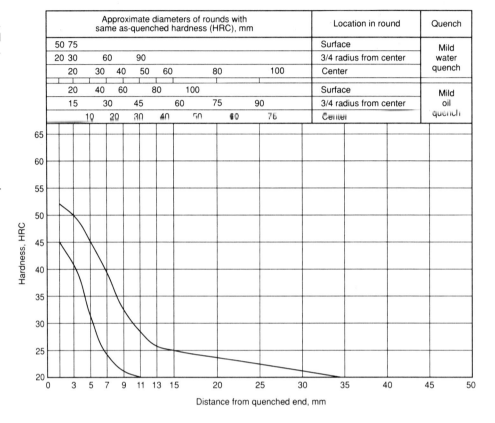

Hardness limits for specification purposes

J distance, 1/16 in.	Hardness, HRC Maximum	Hardness, HRC Minimum
1	52	45
2	50	40
3	46	31
4	40	25
5	34	22
6	30	20
7	28	...
8	26	...
9	25	...
10	25	...
11	24	...
12	23	...
13	23	...
14	22	...
15	22	...
16	21	...
18	21	...
20	20	...
22	...	...

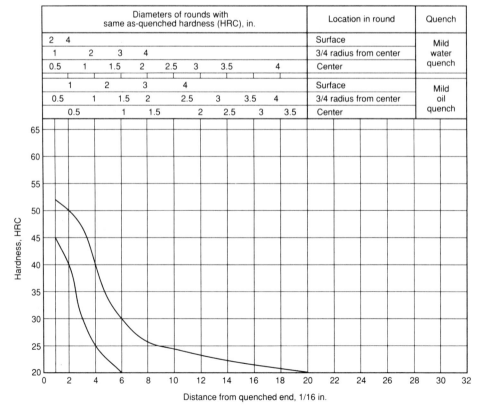

SAE/AISI 4032H   UNS H40320

Heat-treating temperatures recommended by SAE
Normalize (for forged or rolled specimens only): 900 °C (1650 °F)
Austenitize: 870 °C (1600 °F)

Hardness limits for specification purposes

J distance, mm	Hardness, HRC Maximum	Hardness, HRC Minimum
1.5	57	50
3	55	46
5	51	34
7	44	27
9	36	24
11	32	22
13	29	20
15	27	...
20	24	...
25	23	...
30	23	...
35	22	...
40	21	...
45	20	...
50	...	...

Hardness limits for specification purposes

J distance, 1/16 in.	Hardness, HRC Maximum	Hardness, HRC Minimum
1	57	50
2	54	45
3	51	36
4	46	29
5	39	25
6	34	23
7	31	22
8	29	21
9	28	20
10	26	...
11	26	...
12	25	...
13	24	...
14	24	...
15	23	...
16	23	...
18	23	...
20	22	...
22	22	...
24	21	...
26	21	...
28	20	...
30	...	...

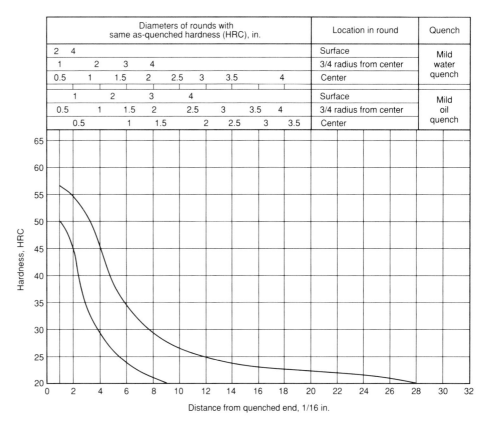

SAE/AISI 4037H UNS H40370

Heat-treating temperatures recommended by SAE
Normalize (for forged or rolled specimens only): 870 °C (1600 °F)
Austenitize: 845 °C (1550 °F)

Hardness limits for specification purposes

J distance, mm	Hardness, HRC Maximum	Hardness, HRC Minimum
1.5	59	52
3	57	50
5	54	42
7	49	32
9	41	27
11	35	24
13	32	21
15	30	20
20	27	...
25	26	...
30	25	...
35	25	...
40	25	...
45	24	...
50	23	...

Hardness limits for specification purposes

J distance, 1/16 in.	Hardness, HRC Maximum	Hardness, HRC Minimum
1	59	52
2	57	49
3	54	42
4	51	35
5	45	30
6	38	26
7	34	23
8	32	22
9	30	21
10	29	20
11	28	...
12	27	...
13	26	...
14	26	...
15	26	...
16	25	...
18	25	...
20	25	...
22	25	...
24	24	...
26	24	...
28	24	...
30	23	...
32	23	...

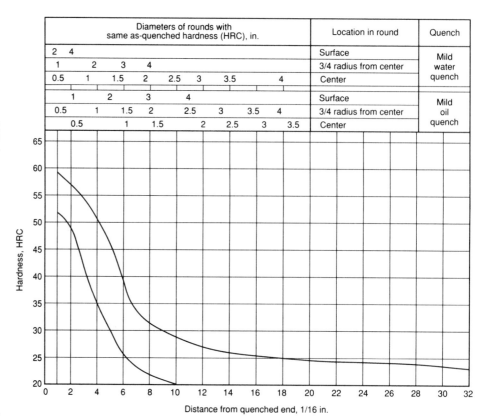

SAE/AISI 4042H   UNS H40420

Heat-treating temperatures recommended by SAE
Normalize (for forged or rolled specimens only): 870 °C (1600 °F)
Austenitize: 845 °C (1550 °F)

Hardness limits for specification purposes

J distance, mm	Hardness, HRC Maximum	Minimum
1.5	62	55
3	61	53
5	58	47
7	54	36
9	48	30
11	40	27
13	36	25
15	33	24
20	31	23
25	29	22
30	28	21
35	28	20
40	27	...
45	27	...
50	26	...

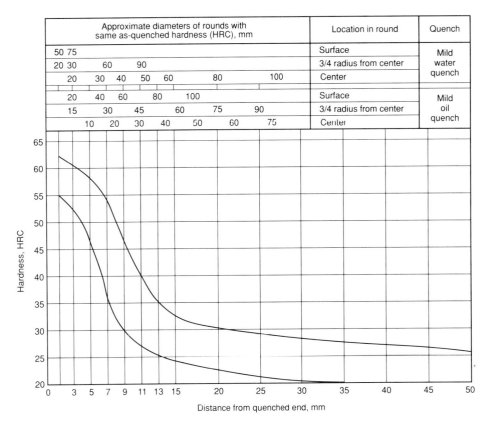

Hardness limits for specification purposes

J distance, 1/16 in.	Hardness, HRC Maximum	Minimum
1	62	55
2	60	52
3	58	48
4	55	40
5	50	33
6	45	29
7	39	27
8	36	26
9	34	25
10	33	24
11	32	24
12	31	23
13	30	23
14	30	23
15	29	22
16	29	22
18	28	22
20	28	21
22	28	20
24	27	20
26	27	...
28	27	...
30	26	...
32	26	...

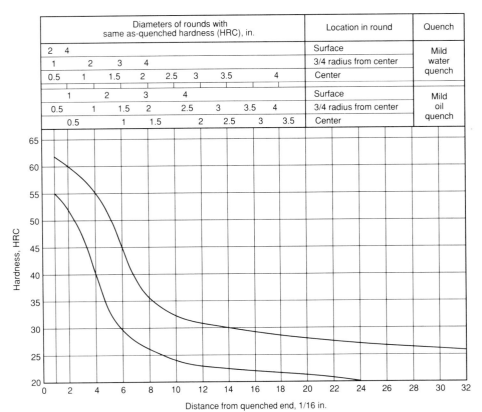

SAE/AISI 4047H UNS H40470

Heat-treating temperatures recommended by SAE
Normalize (for forged or rolled specimens only): 870 °C (1600 °F)
Austenitize: 845 °C (1550 °F)

Hardness limits for specification purposes

J distance, mm	Hardness, HRC Maximum	Hardness, HRC Minimum
1.5	64	57
3	63	55
5	60	49
7	57	39
9	53	33
11	48	30
13	43	28
15	39	27
20	34	25
25	33	24
30	31	24
35	30	23
40	30	23
45	29	22
50	29	21

Hardness limits for specification purposes

J distance, 1/16 in.	Hardness, HRC Maximum	Hardness, HRC Minimum
1	64	57
2	62	55
3	60	50
4	58	42
5	55	35
6	52	32
7	47	30
8	43	28
9	40	28
10	38	27
11	37	26
12	35	26
13	34	25
14	33	25
15	33	25
16	32	25
18	31	24
20	30	24
22	30	23
24	30	23
26	30	22
28	29	22
30	29	21
32	29	21

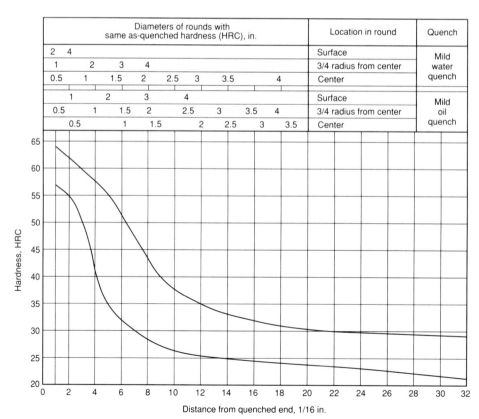

SAE/AISI 4118H   UNS H41180

Heat-treating temperatures recommended by SAE
Normalize (for forged or rolled specimens only): 925 °C (1700 °F)
Austenitize: 925 °C (1700 °F)

Hardness limits for specification purposes

J distance, mm	Hardness, HRC Maximum	Hardness, HRC Minimum
1.5	48	41
3	46	37
5	40	27
7	34	22
9	29	...
11	27	...
13	25	...
15	24	...
20	21	...
25	...	...

Hardness limits for specification purposes

J distance, 1/16 in.	Hardness, HRC Maximum	Hardness, HRC Minimum
1	48	41
2	46	36
3	41	27
4	35	23
5	31	20
6	28	...
7	27	...
8	25	...
9	24	...
10	23	...
11	22	...
12	21	...
13	21	...
14	20	...
15	...	...

SAE/AISI 4130H   UNS H41300

Heat-treating temperatures recommended by SAE
Normalize (for forged or rolled specimens only): 900 °C (1650 °F)
Austenitize: 870 °C (1600 °F)

Hardness limits for specification purposes

J distance, mm	Hardness, HRC Maximum	Hardness, HRC Minimum
1.5	56	49
3	55	46
5	53	40
7	51	36
9	48	32
11	44	28
13	41	26
15	39	25
20	34	24
25	33	23
30	33	22
35	32	20
40	31	...
45	31	...
50	30	...

Hardness limits for specification purposes

J distance, 1/16 in.	Hardness, HRC Maximum	Hardness, HRC Minimum
1	56	49
2	55	46
3	53	42
4	51	38
5	49	34
6	47	31
7	44	29
8	42	27
9	40	26
10	38	26
11	36	25
12	35	25
13	34	24
14	34	24
15	33	23
16	33	23
18	32	22
20	32	21
22	32	20
24	31	...
26	31	...
28	30	...
30	30	...
32	29	...

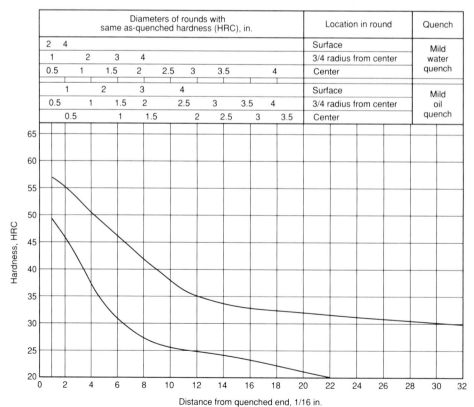

SAE/AISI 4135H   UNS H41350

Heat-treating temperatures recommended by SAE
Normalize (for forged or rolled specimens only): 870 °C (1600 °F)
Austenitize: 845 °C (1550 °F)

Hardness limits for specification purposes

J distance, mm	Hardness, HRC Maximum	Hardness, HRC Minimum
1.5	58	51
3	58	50
5	57	49
7	56	48
9	56	46
11	55	42
13	53	39
15	52	37
20	49	32
25	45	30
30	43	28
35	41	27
40	40	27
45	39	26
50	37	26

Hardness limits for specification purposes

J distance, 1/16 in.	Hardness, HRC Maximum	Hardness, HRC Minimum
1	58	51
2	58	50
3	57	49
4	56	48
5	56	47
6	55	45
7	54	42
8	53	40
9	52	38
10	51	36
11	50	34
12	49	33
13	48	32
14	47	31
15	46	30
16	45	30
18	44	29
20	42	28
22	41	27
24	40	27
26	39	27
28	38	26
30	38	26
32	37	26

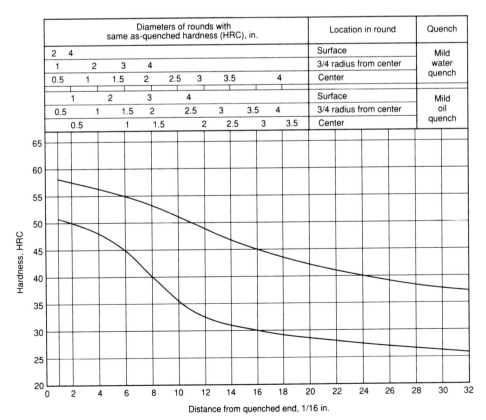

SAE/AISI 4137H UNS H41370

Heat-treating temperatures recommended by SAE
Normalize (for forged or rolled specimens only): 870 °C (1600 °F)
Austenitize: 845 °C (1550 °F)

Hardness limits for specification purposes

J distance, mm	Hardness, HRC Maximum	Hardness, HRC Minimum
1.5	59	52
3	59	51
5	58	50
7	58	49
9	57	48
11	56	45
13	55	42
15	55	39
20	52	35
25	48	33
30	46	31
35	44	30
40	43	29
45	42	29
50	41	29

Hardness limits for specification purposes

J distance, 1/16 in.	Hardness, HRC Maximum	Hardness, HRC Minimum
1	59	52
2	59	51
3	58	50
4	58	49
5	57	49
6	57	48
7	56	45
8	55	43
9	55	40
10	54	39
11	53	37
12	52	36
13	51	35
14	50	34
15	49	33
16	48	33
18	46	32
20	45	31
22	44	30
24	43	30
26	42	30
28	42	29
30	41	29
32	41	29

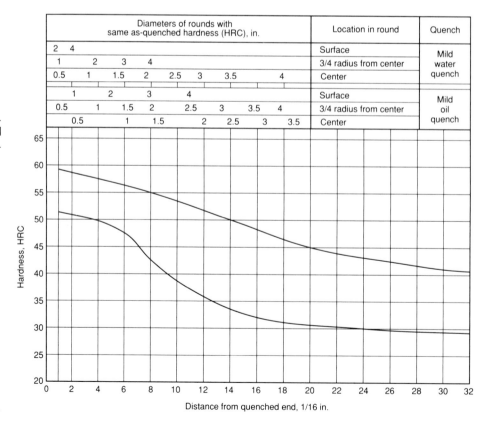

SAE/AISI 4140H   UNS H41400

Heat-treating temperatures recommended by SAE
Normalize (for forged or rolled specimens only): 870 °C (1600 °F)
Austenitize: 845 °C (1550 °F)

Hardness limits for specification purposes

J distance, mm	Hardness, HRC Maximum	Hardness, HRC Minimum
1.5	60	53
3	60	52
5	60	52
7	59	51
9	59	50
11	58	48
13	57	46
15	57	43
20	55	38
25	53	35
30	51	33
35	49	32
40	48	32
45	46	31
50	45	30

Hardness limits for specification purposes

J distance, 1/16 in.	Hardness, HRC Maximum	Hardness, HRC Minimum
1	60	53
2	60	53
3	60	52
4	59	51
5	59	51
6	58	50
7	58	48
8	57	47
9	57	44
10	56	42
11	56	40
12	55	39
13	55	38
14	54	37
15	54	36
16	53	35
18	52	34
20	51	33
22	49	33
24	48	32
26	47	32
28	46	31
30	45	31
32	44	30

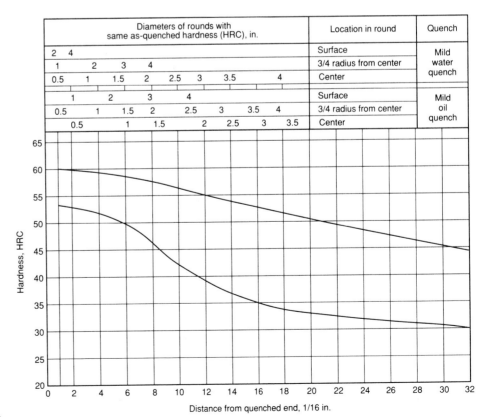

SAE/AISI 4142H   UNS H41420

Heat-treating temperatures recommended by SAE
Normalize (for forged or rolled specimens only): 870 °C (1600 °F)
Austenitize: 845 °C (1550 °F)

Hardness limits for specification purposes

J distance, mm	Hardness, HRC Maximum	Hardness, HRC Minimum
1.5	62	55
3	62	54
5	62	54
7	62	53
9	61	52
11	61	51
13	60	49
15	60	48
20	58	43
25	56	39
30	55	36
35	53	35
40	52	34
45	51	33
50	50	33

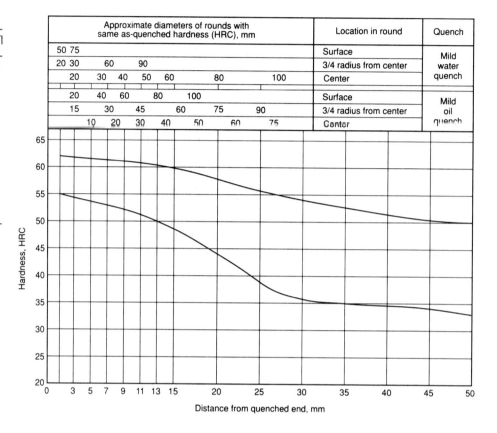

Hardness limits for specification purposes

J distance, 1/16 in.	Hardness, HRC Maximum	Hardness, HRC Minimum
1	62	55
2	62	55
3	62	54
4	61	53
5	61	53
6	61	52
7	60	51
8	60	50
9	60	49
10	59	47
11	59	46
12	58	44
13	58	42
14	57	41
15	57	40
16	56	39
18	55	37
20	54	36
22	53	35
24	53	34
26	52	34
28	51	34
30	51	33
32	50	33

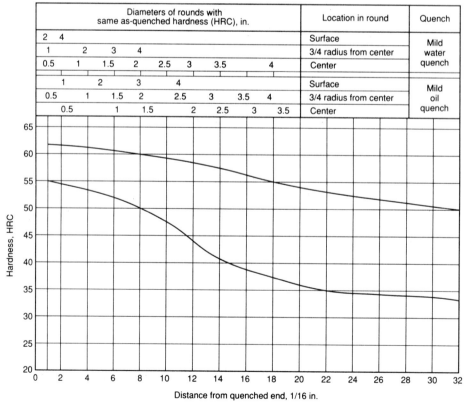

SAE/AISI 4145H   UNS H41450

Heat-treating temperatures recommended by SAE
Normalize (for forged or rolled specimens only): 870 °C (1600 °F)
Austenitize: 845 °C (1550 °F)

Hardness limits for specification purposes

J distance, mm	Hardness, HRC Maximum	Hardness, HRC Minimum
1.5	63	56
3	63	55
5	63	55
7	62	54
9	62	53
11	61	52
13	61	51
15	60	50
20	59	47
25	58	42
30	57	39
35	56	37
40	55	35
45	55	34
50	55	34

Hardness limits for specification purposes

J distance, 1/16 in.	Hardness, HRC Maximum	Hardness, HRC Minimum
1	63	56
2	63	55
3	62	55
4	62	54
5	62	53
6	61	53
7	61	52
8	61	52
9	60	51
10	60	50
11	60	49
12	59	48
13	59	46
14	59	45
15	58	43
16	58	42
18	57	40
20	57	38
22	56	37
24	55	36
26	55	35
28	55	35
30	55	34
32	54	34

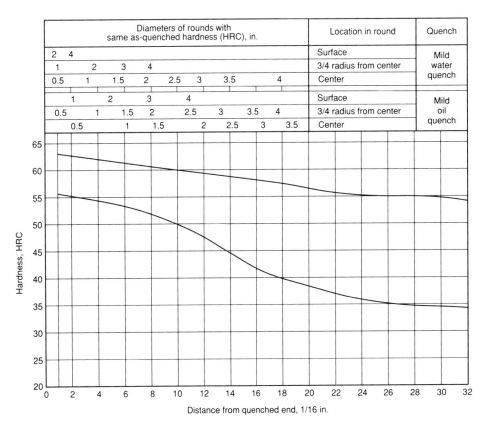

SAE/AISI 4147H UNS H41470

Heat-treating temperatures recommended by SAE
Normalize (for forged or rolled specimens only): 870 °C (1600 °F)
Austenitize: 845 °C (1550 °F)

Hardness limits for specification purposes

J distance, mm	Hardness, HRC Maximum	Minimum
1.5	64	57
3	64	57
5	64	56
7	64	55
9	63	55
11	63	55
13	63	54
15	63	53
20	62	50
25	60	45
30	59	42
35	58	39
40	57	37
45	57	36
50	56	36

Hardness limits for specification purposes

J distance, 1/16 in.	Hardness, HRC Maximum	Minimum
1	64	57
2	64	57
3	64	56
4	64	56
5	63	55
6	63	55
7	63	55
8	63	54
9	63	54
10	62	53
11	62	52
12	62	51
13	61	49
14	61	48
15	60	46
16	60	45
18	59	42
20	59	40
22	58	39
24	57	38
26	57	37
28	57	37
30	56	37
32	56	36

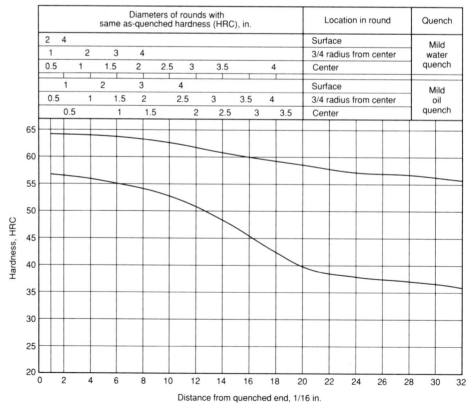

SAE/AISI 4150H   UNS H41500

Heat-treating temperatures recommended by SAE
Normalize (for forged or rolled specimens only): 870 °C (1600 °F)
Austenitize: 845 °C (1550 °F)

Hardness limits for specification purposes

J distance, mm	Hardness, HRC Maximum	Hardness, HRC Minimum
1.5	65	59
3	65	59
5	65	58
7	65	58
9	65	57
11	65	57
13	65	56
15	64	55
20	63	51
25	62	47
30	61	44
35	60	41
40	59	39
45	58	38
50	58	38

Hardness limits for specification purposes

J distance, 1/16 in.	Hardness, HRC Maximum	Hardness, HRC Minimum
1	65	59
2	65	59
3	65	59
4	65	58
5	65	58
6	65	57
7	65	57
8	64	56
9	64	56
10	64	55
11	64	54
12	63	53
13	63	51
14	62	50
15	62	48
16	62	47
18	61	45
20	60	43
22	59	41
24	59	40
26	58	39
28	58	38
30	58	38
32	58	38

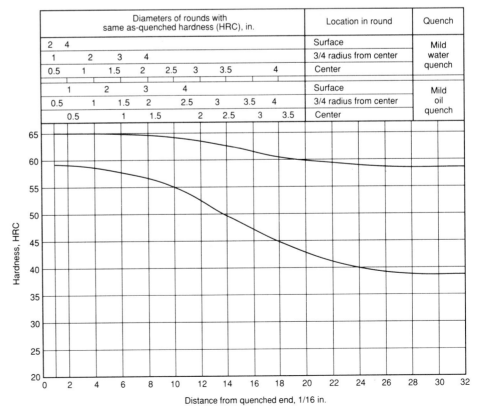

SAE/AISI 4161H UNS H41610

Heat-treating temperatures recommended by SAE
Normalize (for forged or rolled specimens only): 870 °C (1600 °F)
Austenitize: 845 °C (1550 °F)

Hardness limits for specification purposes

J distance, mm	Hardness, HRC Maximum	Hardness, HRC Minimum
1.5	65	60
3	65	60
5	65	60
7	65	60
9	65	60
11	65	60
13	65	60
15	65	60
20	65	58
25	64	56
30	63	53
35	63	50
40	63	46
45	63	43
50	63	41

Hardness limits for specification purposes

J distance, 1/16 in.	Hardness, HRC Maximum	Hardness, HRC Minimum
1	65	60
2	65	60
3	65	60
4	65	60
5	65	60
6	65	60
7	65	60
8	65	60
9	65	59
10	65	59
11	65	59
12	64	59
13	64	58
14	64	58
15	64	57
16	64	56
18	64	55
20	63	53
22	63	50
24	63	48
26	63	45
28	63	43
30	63	42
32	63	41

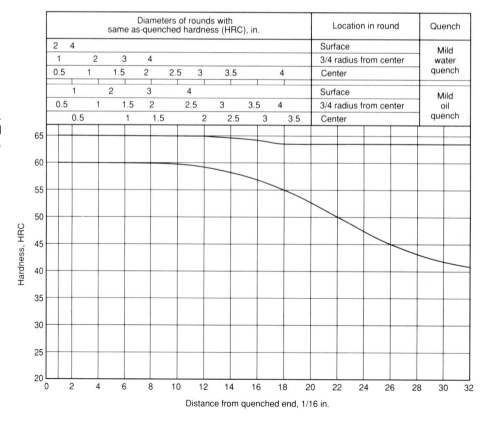

SAE/AISI 4320H   UNS H43200

Heat-treating temperatures recommended by SAE
Normalize (for forged or rolled specimens only): 925 °C (1700 °F)
Austenitize: 925 °C (1700 °F)

Hardness limits for specification purposes

J distance, mm	Hardness, HRC Maximum	Hardness, HRC Minimum
1.5	48	41
3	47	39
5	45	35
7	42	30
9	39	27
11	36	25
13	34	23
15	32	22
20	28	...
25	26	...
30	25	...
35	25	...
40	24	...
45	24	...
50	24	...

Hardness limits for specification purposes

J distance, 1/16 in.	Hardness, HRC Maximum	Hardness, HRC Minimum
1	48	41
2	47	38
3	45	35
4	43	32
5	41	29
6	38	27
7	36	25
8	34	23
9	33	22
10	31	21
11	30	20
12	29	20
13	28	...
14	27	...
15	27	...
16	26	...
18	25	...
20	25	...
22	24	...
24	24	...
26	24	...
28	24	...
30	24	...
32	24	...

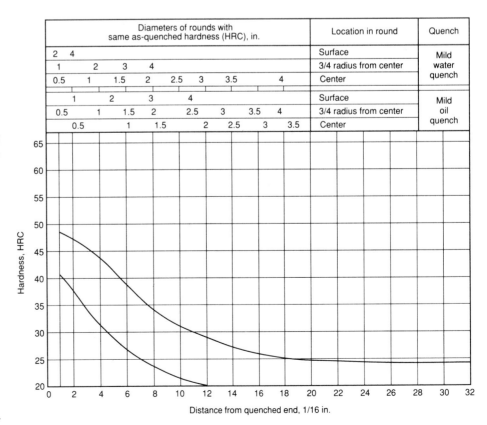

SAE/AISI 4340H UNS H43400

Heat-treating temperatures recommended by SAE
Normalize (for forged or rolled specimens only): 870 °C (1600 °F)
Austenitize: 845 °C (1550 °F)

Hardness limits for specification purposes

J distance, mm	Hardness, HRC Maximum	Hardness, HRC Minimum
1.5	60	53
3	60	53
5	60	53
7	60	53
9	60	53
11	60	53
13	60	52
15	60	52
20	59	50
25	58	48
30	58	46
35	57	44
40	57	43
45	56	42
50	56	40

Hardness limits for specification purposes

J distance, 1/16 in.	Hardness, HRC Maximum	Hardness, HRC Minimum
1	60	53
2	60	53
3	60	53
4	60	53
5	60	53
6	60	53
7	60	53
8	60	52
9	60	52
10	60	52
11	59	51
12	59	51
13	59	50
14	58	49
15	58	49
16	58	48
18	58	47
20	57	46
22	57	45
24	57	44
26	57	43
28	56	42
30	56	41
32	56	40

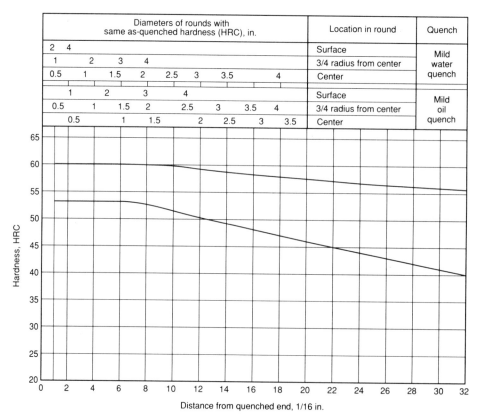

SAE/AISI E4340H UNS H43406

Heat-treating temperatures recommended by SAE
Normalize (for forged or rolled specimens only): 870 °C (1600 °F)
Austenitize: 845 °C (1550 °F)

Hardness limits for specification purposes

J distance, mm	Hardness, HRC Maximum	Hardness, HRC Minimum
1.5	60	53
3	60	53
5	60	53
7	60	53
9	60	53
11	60	53
13	60	53
15	60	53
20	60	52
25	59	51
30	58	50
35	58	49
40	57	47
45	57	46
50	57	44

Hardness limits for specification purposes

J distance, 1/16 in.	Hardness, HRC Maximum	Hardness, HRC Minimum
1	60	53
2	60	53
3	60	53
4	60	53
5	60	53
6	60	53
7	60	53
8	60	53
9	60	53
10	60	53
11	60	53
12	60	52
13	60	52
14	59	52
15	59	52
16	59	51
18	58	51
20	58	50
22	58	49
24	57	48
26	57	47
28	57	46
30	57	45
32	57	44

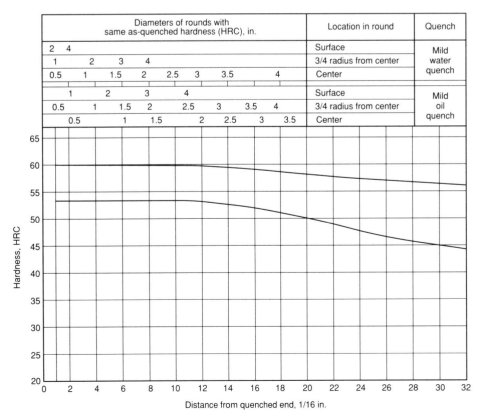

SAE/AISI 4620H   UNS H46200

Heat-treating temperatures recommended by SAE
Normalize (for forged or rolled specimens only): 925 °C (1700 °F)
Austenitize: 925 °C (1700 °F)

Hardness limits for specification purposes

J distance, mm	Hardness, HRC Maximum	Hardness, HRC Minimum
1.5	48	41
3	46	37
5	42	28
7	37	23
9	33	...
11	30	...
13	27	...
15	26	...
20	23	...
25	22	...
30	21	...
35	...	...

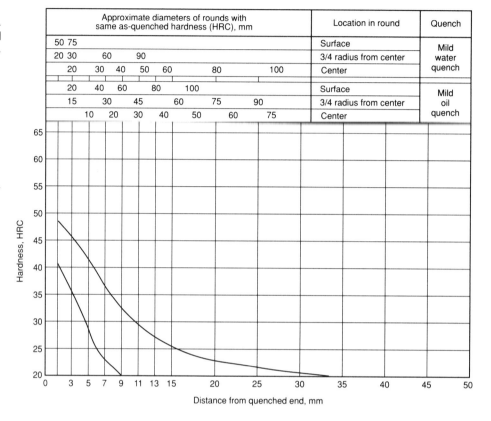

Hardness limits for specification purposes

J distance, 1/16 in.	Hardness, HRC Maximum	Hardness, HRC Minimum
1	48	41
2	45	35
3	42	27
4	39	24
5	34	21
6	31	...
7	29	...
8	27	...
9	26	...
10	25	...
11	24	...
12	23	...
13	22	...
14	22	...
15	22	...
16	21	...
18	21	...
20	20	...
22	...	...

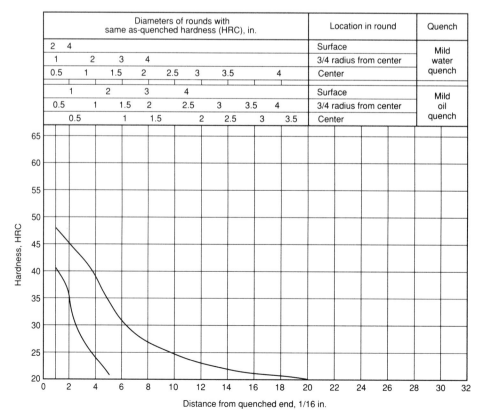

SAE/AISI 4626H   UNS H46260

Heat-treating temperatures recommended by SAE
Normalize (for forged or rolled specimens only): 925 °C (1700 °F)
Austenitize: 925 °C (1700 °F)

Hardness limits for specification purposes

J distance, mm	Hardness, HRC Maximum	Hardness, HRC Minimum
1.5	51	45
3	48	36
5	38	29
7	31	23
9	27	20
11	25	...
13	24	...
15	23	...
20	21	...
25	...	...

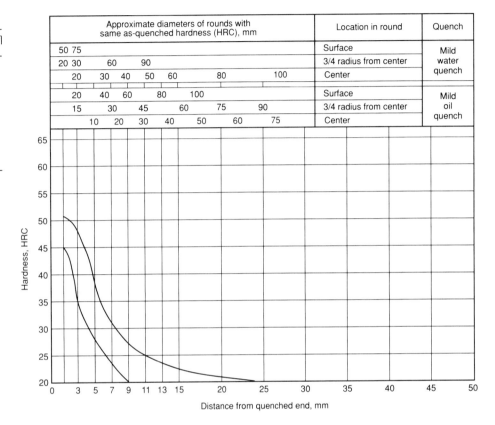

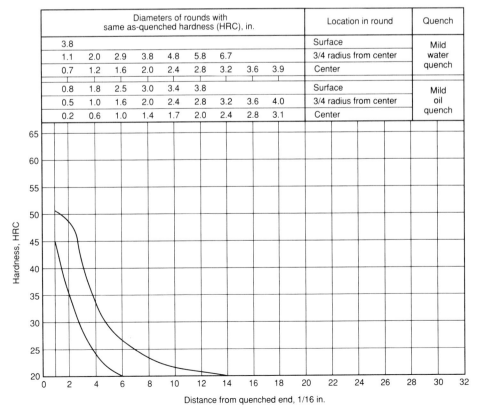

Hardness limits for specification purposes

J distance, 1/16 in.	Hardness, HRC Maximum	Hardness, HRC Minimum
1	51	45
2	48	36
3	41	29
4	33	24
5	29	21
6	27	...
7	25	...
8	24	...
9	23	...
10	22	...
11	22	...
12	21	...
13	21	...
14	20	...
15	...	...

SAE/AISI 4718H UNS H47180

Heat-treating temperatures recommended by SAE
Normalize (for forged or rolled specimens only): 925 °C (1700 °F)
Austenitize: 925 °C (1700 °F)

Hardness limits for specification purposes

J distance, mm	Hardness, HRC Maximum	Hardness, HRC Minimum
1.5	47	40
3	47	40
5	46	38
7	43	31
9	39	28
11	36	25
13	34	23
15	32	22
20	29	21
25	27	20
30	26	...
35	26	...
40	25	...
45	25	...
50	24	...

Hardness limits for specification purposes

J distance, 1/16 in.	Hardness, HRC Maximum	Hardness, HRC Minimum
1	47	40
2	47	40
3	45	38
4	43	33
5	40	29
6	37	27
7	35	25
8	33	24
9	32	23
10	31	22
11	30	22
12	29	21
13	29	21
14	28	21
15	27	20
16	27	20
18	27	...
20	26	...
22	26	...
24	25	...
26	25	...
28	24	...
30	24	...
32	24	...

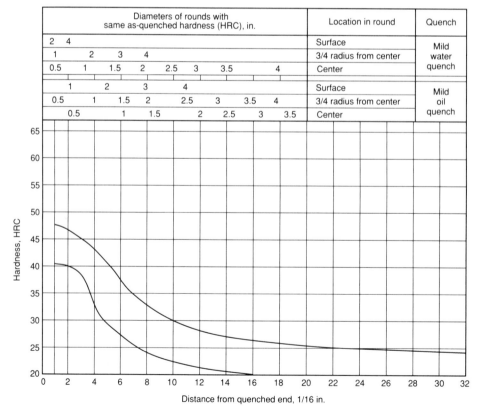

SAE/AISI 4720H   UNS H47200

Heat-treating temperatures recommended by SAE
Normalize (for forged or rolled specimens only): 925 °C (1700 °F)
Austenitize: 925 °C (1700 °F)

Hardness limits for specification purposes

J distance, mm	Hardness, HRC Maximum	Hardness, HRC Minimum
1.5	48	41
3	47	39
5	43	32
7	38	25
9	33	22
11	30	20
13	28	...
15	27	...
20	24	...
25	23	...
30	22	...
35	21	...
40	20	...
45	...	...

Hardness limits for specification purposes

J distance, 1/16 in.	Hardness, HRC Maximum	Hardness, HRC Minimum
1	48	41
2	47	39
3	43	31
4	39	27
5	35	23
6	32	21
7	29	...
8	28	...
9	27	...
10	26	...
11	25	...
12	24	...
13	24	...
14	23	...
15	23	...
16	22	...
18	21	...
20	21	...
22	21	...
24	20	...
26	...	...

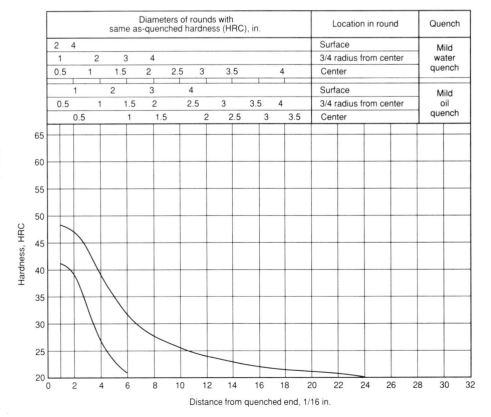

SAE/AISI 4815H   UNS H48150

Heat-treating temperatures recommended by SAE
Normalize (for forged or rolled specimens only): 925 °C (1700 °F)
Austenitize: 845 °C (1550 °F)

Hardness limits for specification purposes

J distance, mm	Hardness, HRC Maximum	Hardness, HRC Minimum
1.5	45	38
3	45	36
5	44	33
7	42	28
9	40	25
11	37	22
13	33	20
15	32	...
20	29	...
25	27	...
30	26	...
35	25	...
40	24	...
45	24	...
50	23	...

Hardness limits for specification purposes

J distance, 1/16 in.	Hardness, HRC Maximum	Hardness, HRC Minimum
1	45	38
2	44	37
3	44	34
4	42	30
5	41	27
6	39	24
7	37	22
8	35	21
9	33	20
10	31	...
11	30	...
12	29	...
13	28	...
14	28	...
15	27	...
16	27	...
18	26	...
20	25	...
22	24	...
24	24	...
26	24	...
28	23	...
30	23	...
32	23	...

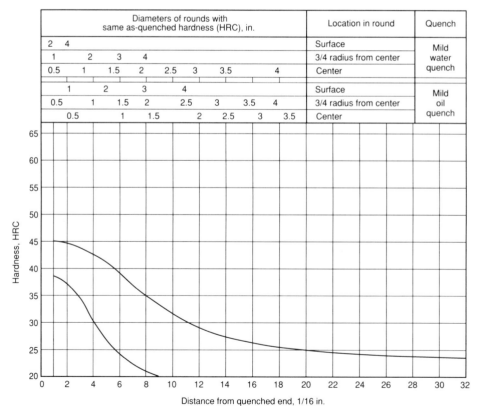

SAE/AISI 4817H UNS H48170

Heat-treating temperatures recommended by SAE
Normalize (for forged or rolled specimens only): 925 °C (1700 °F)
Austenitize: 845 °C (1550 °F)

Hardness limits for specification purposes

J distance, mm	Hardness, HRC Maximum	Hardness, HRC Minimum
1.5	46	39
3	46	38
5	45	35
7	44	31
9	42	28
11	39	25
13	37	23
15	34	21
20	31	...
25	28	...
30	27	...
35	26	...
40	25	...
45	25	...
50	25	...

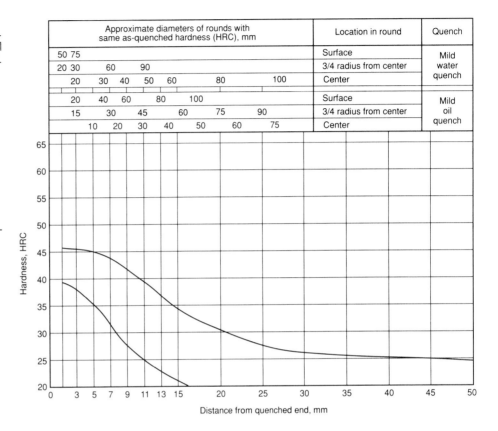

Hardness limits for specification purposes

J distance, 1/16 in.	Hardness, HRC Maximum	Hardness, HRC Minimum
1	46	39
2	46	38
3	45	35
4	44	32
5	42	29
6	41	27
7	39	25
8	37	23
9	35	22
10	33	21
11	32	20
12	31	20
13	30	...
14	29	...
15	28	...
16	28	...
18	27	...
20	26	...
22	25	...
24	25	...
26	25	...
28	25	...
30	24	...
32	24	...

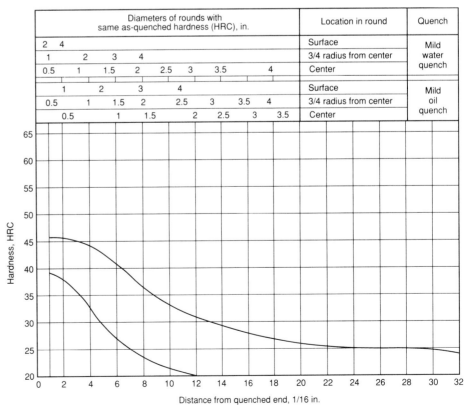

SAE/AISI 4820H UNS H48200

Heat-treating temperatures recommended by SAE
Normalize (for forged or rolled specimens only): 925 °C (1700 °F)
Austenitize: 845 °C (1550 °F)

Hardness limits for specification purposes

J distance, mm	Hardness, HRC Maximum	Hardness, HRC Minimum
1.5	48	41
3	48	40
5	48	39
7	46	36
9	45	32
11	43	29
13	40	27
15	39	25
20	35	22
25	32	21
30	29	20
35	28	...
40	27	...
45	26	...
50	26	...

Hardness limits for specification purposes

J distance, 1/16 in.	Hardness, HRC Maximum	Hardness, HRC Minimum
1	48	41
2	48	40
3	47	39
4	46	38
5	45	34
6	43	31
7	42	29
8	40	27
9	39	26
10	37	25
11	36	24
12	35	23
13	34	22
14	33	22
15	32	21
16	31	21
18	29	20
20	28	20
22	28	...
24	27	...
26	27	...
28	26	...
30	26	...
32	25	...

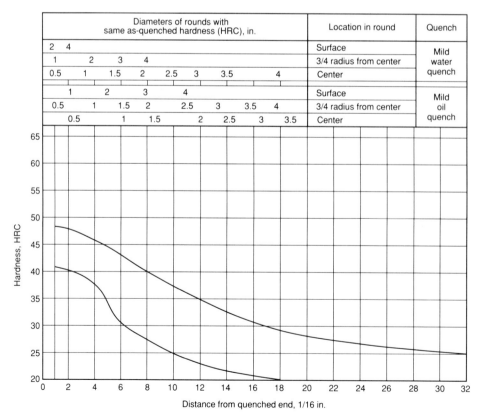

SAE/AISI 50B40H   UNS H50401

Heat-treating temperatures recommended by SAE
Normalize (for forged or rolled specimens only): 870 °C (1600 °F)
Austenitize: 845 °C (1550 °F)

Hardness limits for specification purposes

J distance, mm	Hardness, HRC Maximum	Hardness, HRC Minimum
1.5	60	53
3	60	53
5	60	52
7	59	51
9	59	49
11	58	44
13	57	38
15	56	33
20	50	27
25	43	24
30	37	22
35	35	...
40	34	...
45	32	...
50	30	...

Hardness limits for specification purposes

J distance, 1/16 in.	Hardness, HRC Maximum	Hardness, HRC Minimum
1	60	53
2	60	53
3	59	52
4	59	51
5	58	50
6	58	48
7	57	44
8	57	39
9	56	34
10	55	31
11	53	29
12	51	28
13	49	27
14	47	26
15	44	25
16	41	25
18	38	23
20	36	21
22	35	...
24	34	...
26	33	...
28	32	...
30	30	...
32	29	...

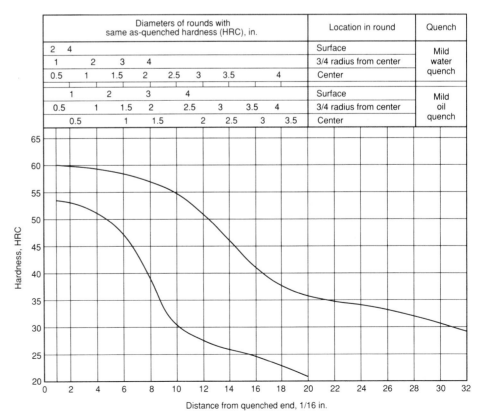

SAE/AISI 50B44H UNS H50441

Heat-treating temperatures recommended by SAE
Normalize (for forged or rolled specimens only): 870 °C (1600 °F)
Austenitize: 845 °C (1550 °F)

Hardness limits for specification purposes

J distance, mm	Hardness, HRC Maximum	Hardness, HRC Minimum
1.5	63	56
3	63	56
5	63	55
7	62	54
9	61	52
11	61	49
13	60	42
15	59	36
20	55	30
25	49	27
30	42	25
35	38	23
40	37	21
45	35	...
50	34	...

Hardness limits for specification purposes

J distance, 1/16 in.	Hardness, HRC Maximum	Hardness, HRC Minimum
1	63	56
2	63	56
3	62	55
4	62	55
5	61	54
6	61	52
7	60	48
8	60	43
9	59	38
10	58	34
11	57	31
12	56	30
13	54	29
14	52	29
15	50	28
16	48	27
18	44	26
20	40	24
22	38	23
24	37	21
26	36	20
28	35	...
30	34	...
32	33	...

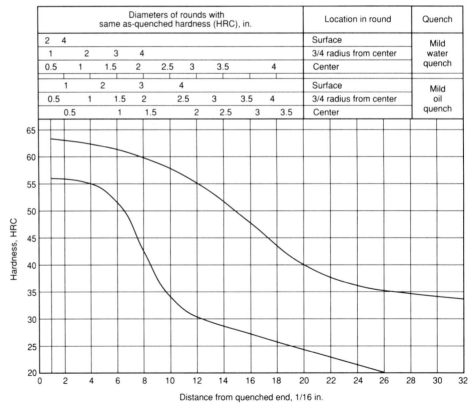

SAE/AISI 5046H   UNS H50460

Heat-treating temperatures recommended by SAE
Normalize (for forged or rolled specimens only): 870 °C (1600 °F)
Austenitize: 845 °C (1550 °F)

Hardness limits for specification purposes

J distance, mm	Hardness, HRC Maximum	Hardness, HRC Minimum
1.5	63	56
3	62	54
5	59	40
7	54	30
9	48	27
11	39	26
13	35	25
15	34	25
20	32	22
25	30	20
30	29	...
35	27	...
40	26	...
45	24	...
50	23	...

Hardness limits for specification purposes

J distance, 1/16 in.	Hardness, HRC Maximum	Hardness, HRC Minimum
1	63	56
2	62	55
3	60	45
4	56	32
5	52	28
6	46	27
7	39	26
8	35	25
9	34	24
10	33	24
11	33	23
12	32	23
13	32	22
14	31	22
15	31	21
16	30	21
18	29	20
20	28	...
22	27	...
24	26	...
26	25	...
28	24	...
30	23	...
32	23	...

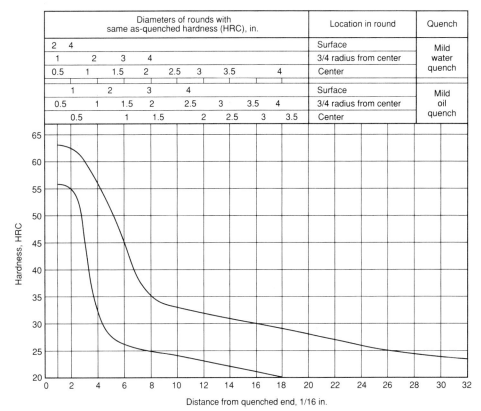

SAE/AISI 50B46H UNS H50461

Heat-treating temperatures recommended by SAE
Normalize (for forged or rolled specimens only): 870 °C (1600 °F)
Austenitize: 845 °C (1550 °F)

Hardness limits for specification purposes

J distance, mm	Hardness, HRC Maximum	Hardness, HRC Minimum
1.5	63	56
3	62	55
5	61	53
7	60	47
9	59	35
11	58	31
13	56	29
15	53	28
20	42	26
25	37	24
30	35	22
35	34	21
40	32	...
45	31	...
50	29	...

Hardness limits for specification purposes

J distance, 1/16 in.	Hardness, HRC Maximum	Hardness, HRC Minimum
1	63	56
2	62	54
3	61	52
4	60	50
5	59	41
6	58	32
7	57	31
8	56	30
9	54	29
10	51	28
11	47	27
12	43	26
13	40	26
14	38	25
15	37	25
16	36	24
18	35	23
20	34	22
22	33	21
24	32	20
26	31	...
28	30	...
30	29	...
32	28	...

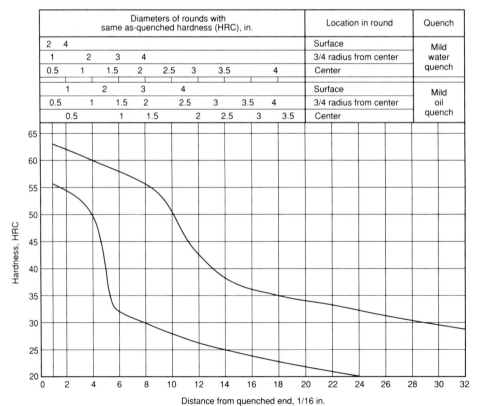

SAE/AISI 50B50H   UNS H50501

Heat-treating temperatures recommended by SAE
Normalize (for forged or rolled specimens only): 870 °C (1600 °F)
Austenitize: 845 °C (1550 °F)

Hardness limits for specification purposes

J distance, mm	Hardness, HRC Maximum	Hardness, HRC Minimum
1.5	65	59
3	65	59
5	65	59
7	64	57
9	63	55
11	63	52
13	62	46
15	62	39
20	59	32
25	54	29
30	49	27
35	44	26
40	40	24
45	38	22
50	37	20

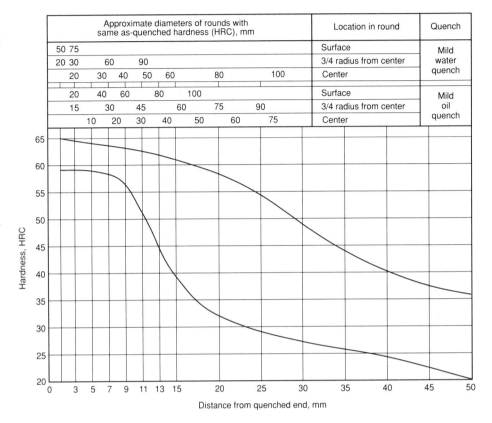

Hardness limits for specification purposes

J distance, 1/16 in.	Hardness, HRC Maximum	Hardness, HRC Minimum
1	65	59
2	65	59
3	64	58
4	64	57
5	63	56
6	63	55
7	62	52
8	62	47
9	61	42
10	60	37
11	60	35
12	59	33
13	58	32
14	57	31
15	56	30
16	54	29
18	50	28
20	47	27
22	44	26
24	41	25
26	39	24
28	38	22
30	37	21
32	36	20

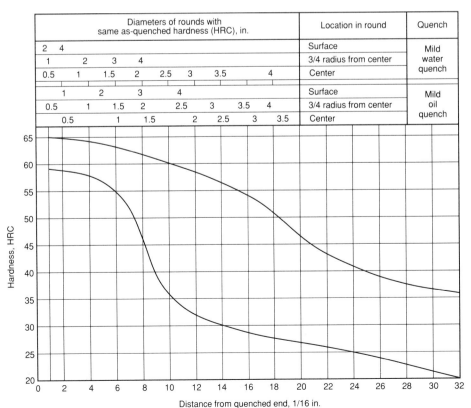

SAE/AISI 50B60H   UNS H50601

Heat-treating temperatures recommended by SAE
Normalize (for forged or rolled specimens only): 870 °C (1600 °F)
Austenitize: 845 °C (1550 °F)

Hardness limits for specification purposes

J distance, mm	Hardness, HRC Maximum	Hardness, HRC Minimum
1.5	...	60
3	...	60
5	...	60
7	...	60
9	...	59
11	...	57
13	65	51
15	65	44
20	65	36
25	62	34
30	59	32
35	56	30
40	52	28
45	48	27
50	45	25

Hardness limits for specification purposes

J distance, 1/16 in.	Hardness, HRC Maximum	Hardness, HRC Minimum
1	...	60
2	...	60
3	...	60
4	...	60
5	...	60
6	...	59
7	...	57
8	65	53
9	65	47
10	64	42
11	64	39
12	64	37
13	63	36
14	63	35
15	63	34
16	62	34
18	60	33
20	58	31
22	55	30
24	53	29
26	51	28
28	49	27
30	47	26
32	44	25

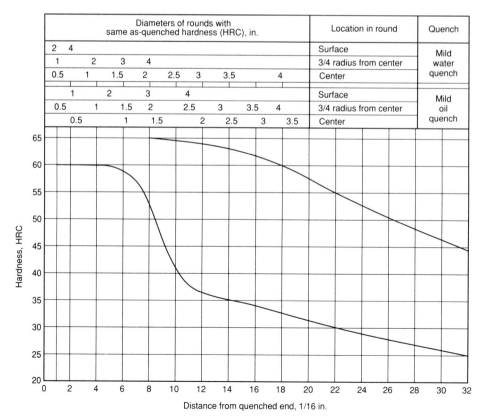

SAE/AISI 5120H UNS H51200

Heat-treating temperatures recommended by SAE
Normalize (for forged or rolled specimens only): 925 °C (1700 °F)
Austenitize: 925 °C (1700 °F)

Hardness limits for specification purposes

J distance, mm	Hardness, HRC Maximum	Hardness, HRC Minimum
1.5	48	40
3	46	34
5	41	27
7	34	22
9	31	20
11	29	...
13	27	...
15	25	...
20	22	...
25	...	...

Hardness limits for specification purposes

J distance, 1/16 in.	Hardness, HRC Maximum	Hardness, HRC Minimum
1	48	40
2	46	34
3	41	28
4	36	23
5	33	20
6	30	...
7	28	...
8	27	...
9	25	...
10	24	...
11	23	...
12	22	...
13	21	...
14	21	...
15	20	...
16	...	...

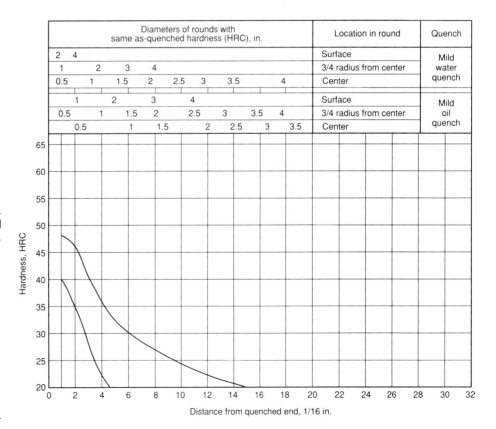

SAE/AISI 5130H UNS H51300

Heat-treating temperatures recommended by SAE
Normalize (for forged or rolled specimens only): 900 °C (1650 °F)
Austenitize: 870 °C (1600 °F)

Hardness limits for specification purposes

J distance, mm	Hardness, HRC Maximum	Hardness, HRC Minimum
1.5	56	49
3	55	46
5	53	42
7	51	37
9	48	33
11	45	30
13	42	27
15	39	25
20	35	21
25	33	...
30	31	...
35	30	...
40	28	...
45	26	...
50	24	...

Hardness limits for specification purposes

J distance, 1/16 in.	Hardness, HRC Maximum	Hardness, HRC Minimum
1	56	49
2	55	46
3	53	42
4	51	39
5	49	35
6	47	32
7	45	30
8	42	28
9	40	26
10	38	25
11	37	23
12	36	22
13	35	21
14	34	20
15	34	...
16	33	...
18	32	...
20	31	...
22	30	...
24	29	...
26	27	...
28	26	...
30	25	...
32	24	...

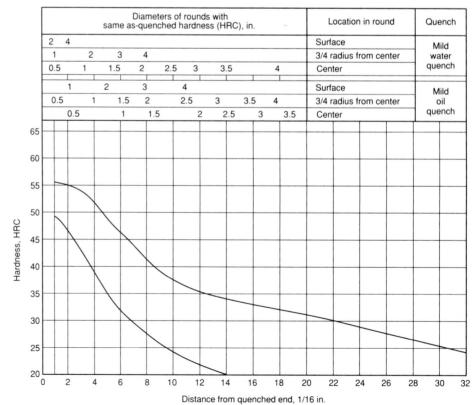

SAE/AISI 5132H   UNS H51320

Heat-treating temperatures recommended by SAE
Normalize (for forged or rolled specimens only): 870 °C (1650 °F)
Austenitize: 845 °C (1600 °F)

Hardness limits for specification purposes

J distance, mm	Hardness, HRC Maximum	Hardness, HRC Minimum
1.5	57	50
3	56	47
5	54	43
7	52	38
9	49	33
11	45	29
13	42	26
15	39	25
20	35	21
25	33	...
30	32	...
35	31	...
40	29	...
45	27	...
50	25	...

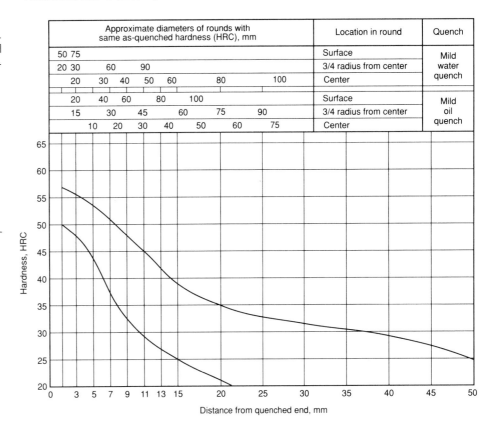

Hardness limits for specification purposes

J distance, 1/16 in.	Hardness, HRC Maximum	Hardness, HRC Minimum
1	57	50
2	56	47
3	54	43
4	52	40
5	50	35
6	48	32
7	46	29
8	42	27
9	40	25
10	38	24
11	37	23
12	36	22
13	35	21
14	34	20
15	34	...
16	33	...
18	32	...
20	31	...
22	30	...
24	29	...
26	28	...
28	27	...
30	26	...
32	25	...

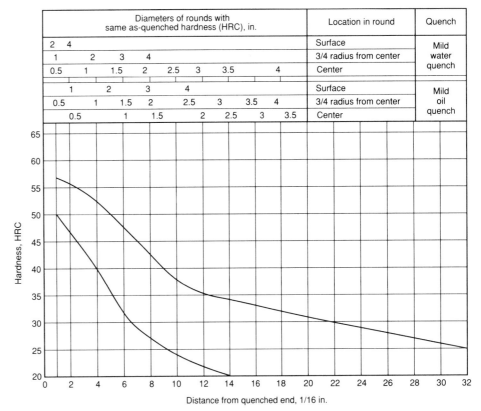

SAE/AISI 5135H UNS H51350

Heat-treating temperatures recommended by SAE
Normalize (for forged or rolled specimens only): 870 °C (1650 °F)
Austenitize: 845 °C (1600 °F)

Hardness limits for specification purposes

J distance, mm	Hardness, HRC Maximum	Hardness, HRC Minimum
1.5	58	51
3	58	49
5	56	46
7	54	41
9	53	36
11	50	32
13	47	30
15	44	27
20	40	23
25	37	21
30	35	...
35	34	...
40	33	...
45	32	...
50	31	...

Hardness limits for specification purposes

J distance, 1/16 in.	Hardness, HRC Maximum	Hardness, HRC Minimum
1	58	51
2	57	49
3	56	47
4	55	43
5	54	38
6	52	35
7	50	32
8	47	30
9	45	28
10	43	27
11	41	25
12	40	24
13	39	23
14	38	22
15	37	21
16	37	21
18	36	20
20	35	...
22	34	...
24	33	...
26	32	...
28	32	...
30	31	...
32	30	...

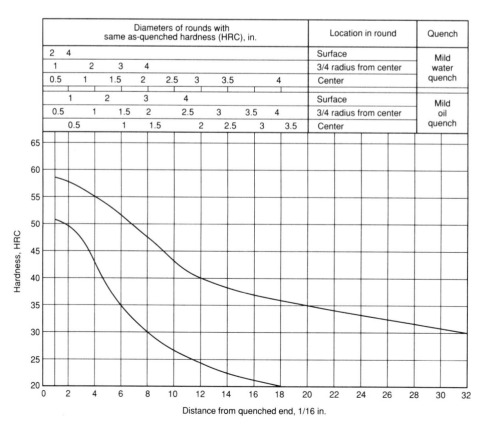

SAE/AISI 5140H   UNS H51400

Heat-treating temperatures recommended by SAE
Normalize (for forged or rolled specimens only): 870 °C (1600 °F)
Austenitize: 845 °C (1550 °F)

Hardness limits for specification purposes

J distance, mm	Hardness, HRC Maximum	Hardness, HRC Minimum
1.5	60	53
3	59	52
5	58	50
7	57	45
9	55	40
11	53	35
13	50	32
15	47	30
20	42	28
25	39	25
30	36	23
35	35	21
40	34	...
45	33	...
50	32	...

Hardness limits for specification purposes

J distance, 1/16 in.	Hardness, HRC Maximum	Hardness, HRC Minimum
1	60	53
2	59	52
3	58	50
4	57	48
5	56	43
6	54	38
7	52	35
8	50	33
9	48	31
10	46	30
11	45	29
12	43	28
13	42	27
14	40	27
15	39	26
16	38	25
18	37	24
20	36	23
22	35	21
24	34	20
26	34	...
28	33	...
30	33	...
32	32	...

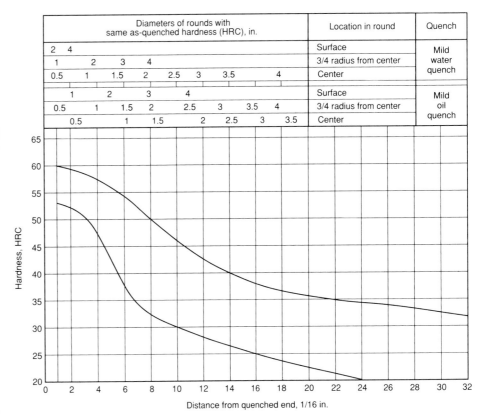

SAE/AISI 5147H UNS H51470

Heat-treating temperatures recommended by SAE
Normalize (for forged or rolled specimens only): 870 °C (1600 °F)
Austenitize: 845 °C (1550 °F)

Hardness limits for specification purposes

J distance, mm	Hardness, HRC Maximum	Hardness, HRC Minimum
1.5	64	57
3	64	56
5	64	55
7	63	53
9	62	52
11	61	49
13	60	44
15	60	39
20	58	33
25	57	31
30	55	29
35	53	27
40	52	25
45	50	23
50	49	21

Hardness limits for specification purposes

J distance, 1/16 in.	Hardness, HRC Maximum	Hardness, HRC Minimum
1	64	57
2	64	56
3	63	55
4	62	54
5	62	53
6	61	52
7	61	49
8	60	45
9	60	40
10	59	37
11	59	35
12	58	34
13	58	33
14	57	32
15	57	32
16	56	31
18	55	30
20	54	29
22	53	27
24	52	26
26	51	25
28	50	24
30	49	22
32	48	21

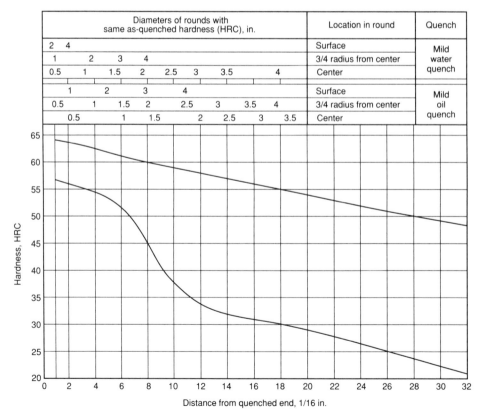

SAE/AISI 5150H   UNS H51500

Heat-treating temperatures recommended by SAE
Normalize (for forged or rolled specimens only): 870 °C (1600 °F)
Austenitize: 845 °C (1550 °F)

Hardness limits for specification purposes

J distance, mm	Hardness, HRC Maximum	Hardness, HRC Minimum
1.5	65	59
3	65	58
5	64	57
7	63	54
9	62	50
11	60	43
13	58	37
15	57	35
20	52	31
25	47	29
30	44	28
35	42	27
40	40	26
45	39	24
50	38	22

Hardness limits for specification purposes

J distance, 1/16 in.	Hardness, HRC Maximum	Hardness, HRC Minimum
1	65	59
2	65	58
3	64	57
4	63	56
5	62	53
6	61	49
7	60	42
8	59	38
9	58	36
10	56	34
11	55	33
12	53	32
13	51	31
14	50	31
15	48	30
16	47	30
18	45	29
20	43	28
22	42	27
24	41	26
26	40	25
28	39	24
30	39	23
32	38	22

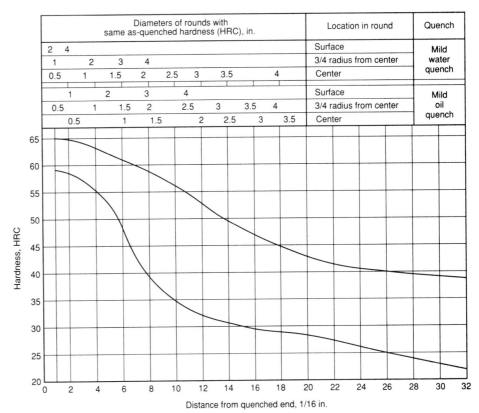

SAE/AISI 5155H UNS H51550

Heat-treating temperatures recommended by SAE
Normalize (for forged or rolled specimens only): 870 °C (1600 °F)
Austenitize: 845 °C (1550 °F)

Hardness limits for specification purposes

J distance, mm	Hardness, HRC Maximum	Hardness, HRC Minimum
1.5	...	60
3	65	60
5	65	59
7	64	56
9	64	53
11	63	48
13	61	40
15	60	37
20	56	34
25	50	32
30	46	30
35	44	29
40	43	28
45	42	27
50	41	25

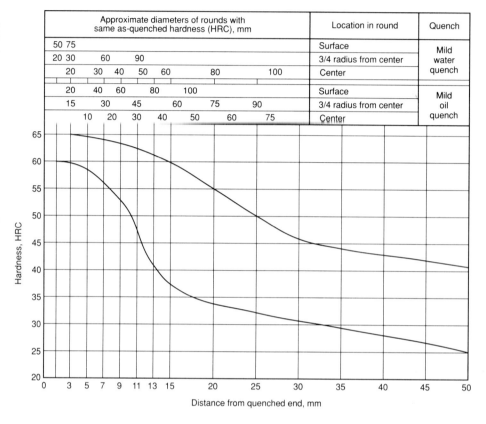

Hardness limits for specification purposes

J distance, 1/16 in.	Hardness, HRC Maximum	Hardness, HRC Minimum
1	...	60
2	65	59
3	64	58
4	64	57
5	63	55
6	63	52
7	62	47
8	62	41
9	61	37
10	60	36
11	59	35
12	57	34
13	55	34
14	52	33
15	51	33
16	49	32
18	47	31
20	45	31
22	44	30
24	43	29
26	42	28
28	41	27
30	41	26
32	40	25

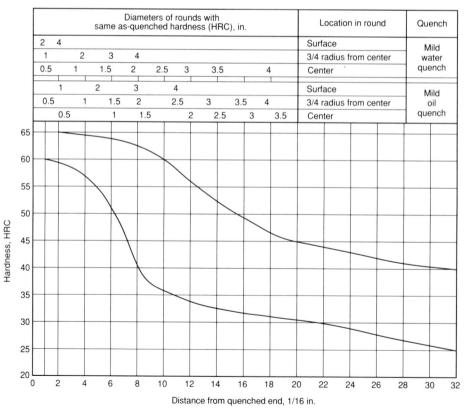

SAE/AISI 5160H   UNS H51600

Heat-treating temperatures recommended by SAE
Normalize (for forged or rolled specimens only): 870 °C (1600 °F)
Austenitize: 845 °C (1550 °F)

Hardness limits for specification purposes

J distance, mm	Hardness, HRC Maximum	Hardness, HRC Minimum
1.5	...	60
3	...	60
5	...	60
7	...	59
9	65	57
11	64	52
13	64	46
15	62	40
20	58	36
25	53	34
30	49	32
35	46	30
40	44	28
45	42	27
50	41	27

Hardness limits for specification purposes

J distance, 1/16 in.	Hardness, HRC Maximum	Hardness, HRC Minimum
1	...	60
2	...	60
3	...	60
4	65	59
5	65	58
6	64	56
7	64	52
8	63	47
9	62	42
10	61	39
11	60	37
12	59	36
13	58	35
14	56	35
15	54	34
16	52	34
18	48	33
20	47	32
22	46	31
24	45	30
26	44	29
28	43	28
30	43	28
32	42	27

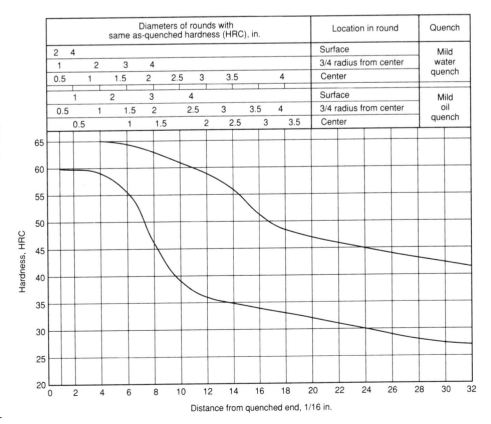

SAE/AISI 51B60H   UNS H51601

Heat-treating temperatures recommended by SAE
Normalize (for forged or rolled specimens only): 870 °C (1600 °F)
Austenitize: 845 °C (1550 °F)

Hardness limits for specification purposes

J distance, mm	Hardness, HRC Maximum	Hardness, HRC Minimum
1.5	...	60
3	...	60
5	...	60
7	...	60
9	...	59
11	...	58
13	...	55
15	...	51
20	65	40
25	63	37
30	61	35
35	57	32
40	54	30
45	51	28
50	47	25

Hardness limits for specification purposes

J distance, 1/16 in.	Hardness, HRC Maximum	Hardness, HRC Minimum
1	...	60
2	...	60
3	...	60
4	...	60
5	...	60
6	...	59
7	...	58
8	...	57
9	...	54
10	...	50
11	...	44
12	65	41
13	65	40
14	64	39
15	64	38
16	63	37
18	61	36
20	59	34
22	57	33
24	55	31
26	53	30
28	51	28
30	49	27
32	47	25

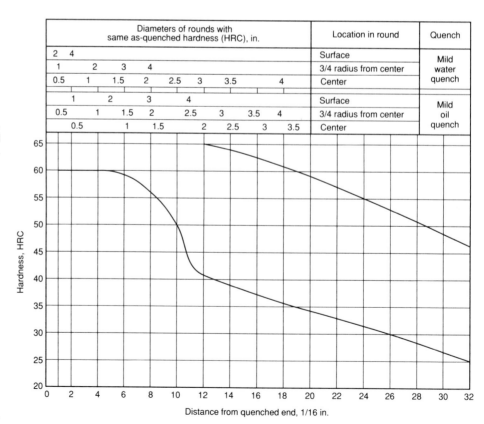

SAE/AISI 6118H UNS H61180

Heat-treating temperatures recommended by SAE
Normalize (for forged or rolled specimens only): 925 °C (1700 °F)
Austenitize: 925 °C (1700 °F)

Hardness limits for specification purposes

J distance, mm	Hardness, HRC Maximum	Hardness, HRC Minimum
1.5	46	39
3	44	36
5	39	28
7	32	23
9	30	20
11	28	...
13	27	...
15	25	...
20	24	...
25	23	...
30	22	...
35	21	...
40	20	...
45	...	...

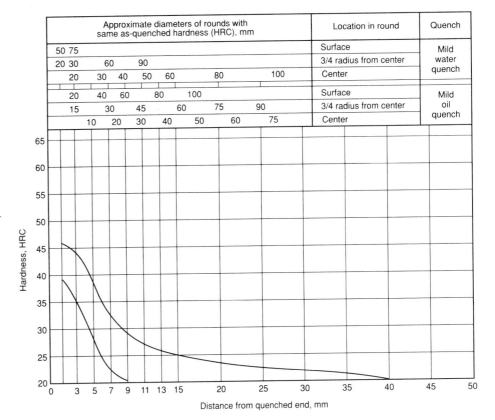

Hardness limits for specification purposes

J distance, 1/16 in.	Hardness, HRC Maximum	Hardness, HRC Minimum
1	46	39
2	44	36
3	38	28
4	33	24
5	30	22
6	28	20
7	27	...
8	26	...
9	26	...
10	25	...
11	25	...
12	24	...
13	24	...
14	23	...
15	23	...
16	22	...
18	22	...
20	21	...
22	21	...
24	20	...
26	...	...

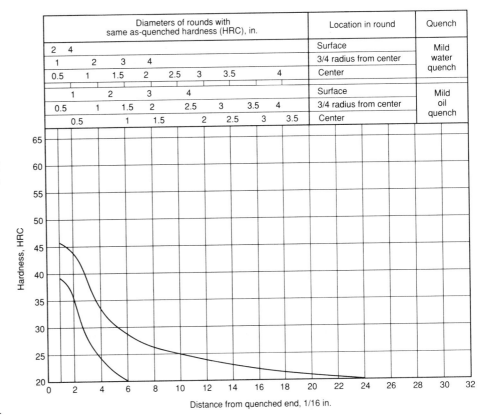

SAE/AISI 6150H UNS H61500

Heat-treating temperatures recommended by SAE
Normalize (for forged or rolled specimens only): 900 °C (1650 °F)
Austenitize: 870 °C (1600 °F)

Hardness limits for specification purposes

J distance, mm	Hardness, HRC Maximum	Hardness, HRC Minimum
1.5	65	59
3	65	58
5	65	57
7	64	55
9	63	53
11	63	50
13	61	46
15	60	42
20	58	37
25	53	35
30	50	33
35	47	31
40	45	29
45	44	27
50	43	25

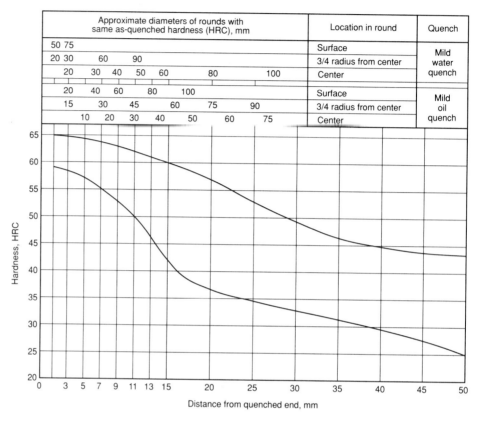

Hardness limits for specification purposes

J distance, 1/16 in.	Hardness, HRC Maximum	Hardness, HRC Minimum
1	65	59
2	65	58
3	64	57
4	64	56
5	63	55
6	63	53
7	62	50
8	61	47
9	61	43
10	60	41
11	59	39
12	58	38
13	57	37
14	55	36
15	54	35
16	52	35
18	50	34
20	48	32
22	47	31
24	46	30
26	45	29
28	44	27
30	43	26
32	42	25

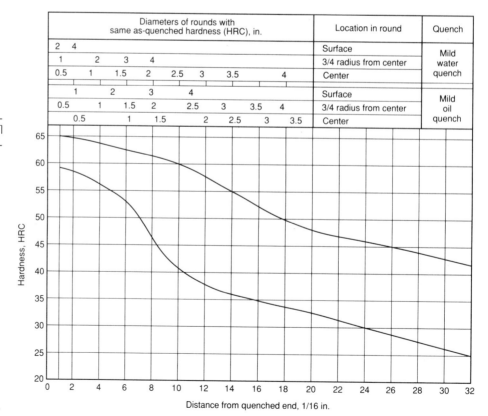

SAE/AISI 81B45H   UNS H81451

Heat-treating temperatures recommended by SAE
Normalize (for forged or rolled specimens only): 870 °C (1600 °F)
Austenitize: 845 °C (1550 °F)

Hardness limits for specification purposes

J distance, mm	Hardness, HRC Maximum	Hardness, HRC Minimum
1.5	63	56
3	63	56
5	63	56
7	63	56
9	63	55
11	63	53
13	62	49
15	61	47
20	59	38
25	57	35
30	55	33
35	52	31
40	50	29
45	47	28
50	44	27

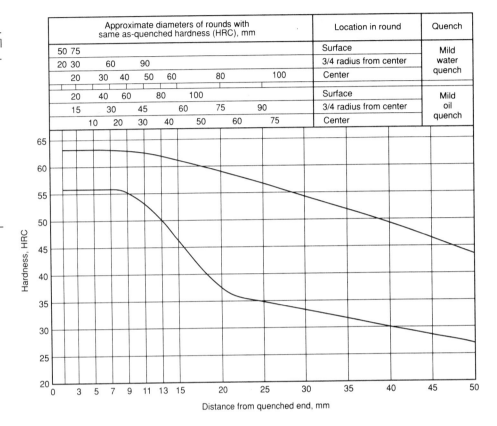

Hardness limits for specification purposes

J distance, 1/16 in.	Hardness, HRC Maximum	Hardness, HRC Minimum
1	63	56
2	63	56
3	63	56
4	63	56
5	63	55
6	63	54
7	62	53
8	62	51
9	61	48
10	60	44
11	60	41
12	59	39
13	58	38
14	57	37
15	57	36
16	56	35
18	55	34
20	53	32
22	52	31
24	50	30
26	49	29
28	47	28
30	45	28
32	43	27

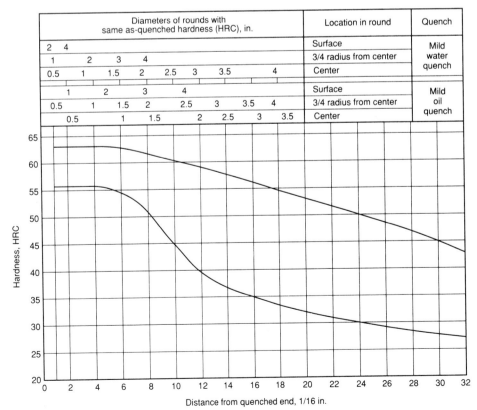

SAE/AISI 8617H UNS H86170

Heat-treating temperatures recommended by SAE
Normalize (for forged or rolled specimens only): 925 °C (1700 °F)
Austenitize: 925 °C (1700 °F)

Hardness limits for specification purposes

J distance, mm	Hardness, HRC Maximum	Hardness, HRC Minimum
1.5	46	39
3	44	33
5	42	27
7	37	23
9	32	20
11	29	...
13	27	...
15	25	...
20	23	...
25	22	...
30	20	...
35	...	...

Hardness limits for specification purposes

J distance, 1/16 in.	Hardness, HRC Maximum	Hardness, HRC Minimum
1	46	39
2	44	33
3	41	27
4	38	24
5	34	20
6	31	...
7	28	...
8	27	...
9	26	...
10	25	...
11	24	...
12	23	...
13	23	...
14	22	...
15	22	...
16	21	...
18	21	...
20	20	...
22	...	...

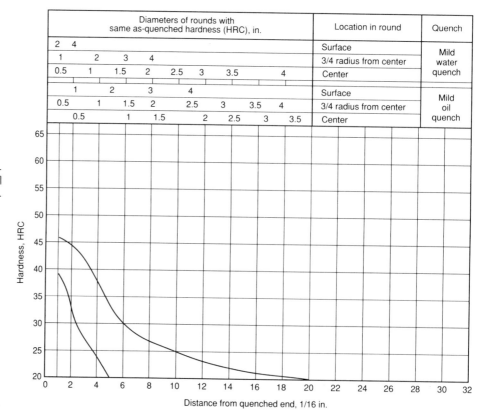

SAE/AISI 8620H   UNS H86200

Heat-treating temperatures recommended by SAE
Normalize (for forged or rolled specimens only): 925 °C (1700 °F)
Austenitize: 925 °C (1700 °F)

Hardness limits for specification purposes

J distance, mm	Hardness, HRC Maximum	Hardness, HRC Minimum
1.5	48	41
3	47	37
5	44	31
7	40	25
9	35	22
11	33	20
13	30	...
15	29	...
20	26	...
25	24	...
30	23	...
35	23	...
40	23	...
45	22	...
50	22	...

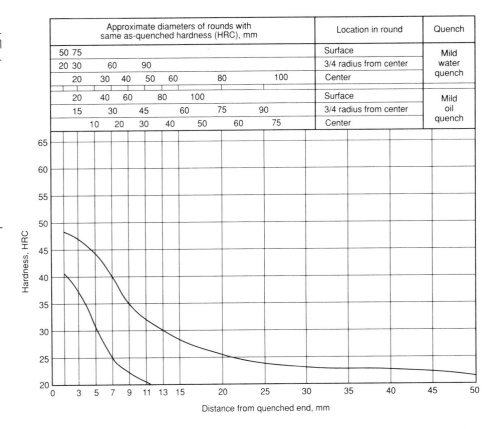

Hardness limits for specification purposes

J distance, 1/16 in.	Hardness, HRC Maximum	Hardness, HRC Minimum
1	48	41
2	47	37
3	44	32
4	41	27
5	37	23
6	34	21
7	32	...
8	30	...
9	29	...
10	28	...
11	27	...
12	26	...
13	25	...
14	25	...
15	24	...
16	24	...
18	23	...
20	23	...
22	23	...
24	23	...
26	23	...
28	22	...
30	22	...
32	22	...

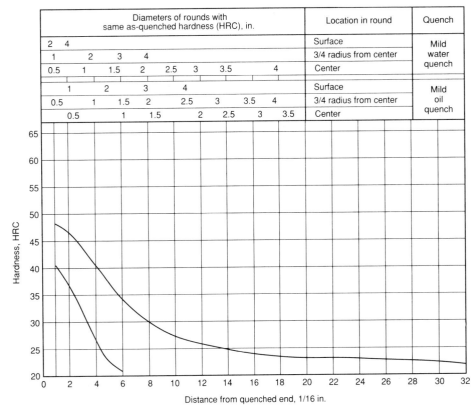

SAE/AISI 8622H UNS H86220

Heat-treating temperatures recommended by SAE
Normalize (for forged or rolled specimens only): 925 °C (1700 °F)
Austenitize: 925 °C (1700 °F)

Hardness limits for specification purposes

J distance, mm	Hardness, HRC Maximum	Hardness, HRC Minimum
1.5	50	43
3	50	39
5	47	34
7	43	28
9	39	25
11	35	22
13	32	20
15	31	...
20	28	...
25	26	...
30	25	...
35	24	...
40	24	...
45	24	...
50	24	...

Hardness limits for specification purposes

J distance, 1/16 in.	Hardness, HRC Maximum	Hardness, HRC Minimum
1	50	43
2	49	39
3	47	34
4	44	30
5	40	26
6	37	24
7	34	22
8	32	20
9	31	...
10	30	...
11	29	...
12	28	...
13	27	...
14	26	...
15	26	...
16	25	...
18	25	...
20	24	...
22	24	...
24	24	...
26	24	...
28	24	...
30	24	...
32	24	...

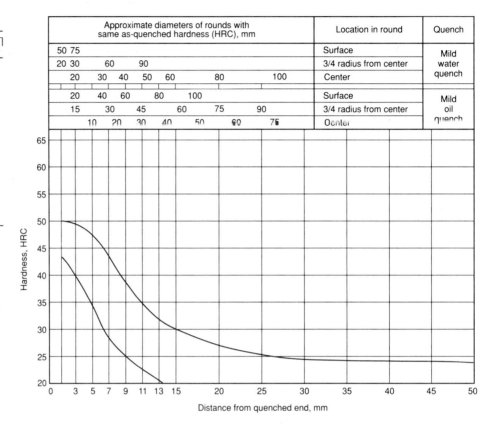

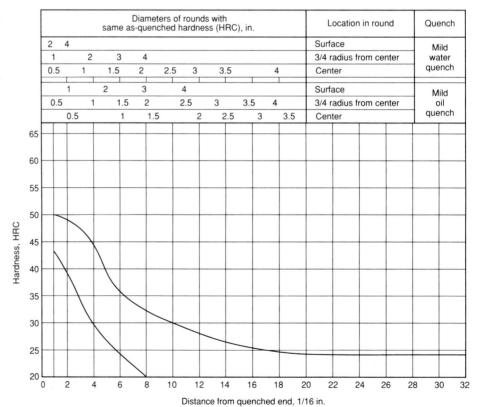

SAE/AISI 8625H UNS H86250

Heat-treating temperatures recommended by SAE
Normalize (for forged or rolled specimens only): 900 °C (1650 °F)
Austenitize: 870 °C (1600 °F)

Hardness limits for specification purposes

J distance, mm	Hardness, HRC Maximum	Hardness, HRC Minimum
1.5	52	45
3	51	40
5	48	35
7	45	31
9	41	28
11	38	25
13	35	23
15	33	21
20	29	...
25	28	...
30	27	...
35	26	...
40	26	...
45	26	...
50	25	...

Hardness limits for specification purposes

J distance, 1/16 in.	Hardness, HRC Maximum	Hardness, HRC Minimum
1	52	45
2	51	41
3	48	36
4	46	32
5	43	29
6	40	27
7	37	25
8	35	23
9	33	22
10	32	21
11	31	20
12	30	...
13	29	...
14	28	...
15	28	...
16	27	...
18	27	...
20	26	...
22	26	...
24	26	...
26	26	...
28	25	...
30	25	...
32	25	...

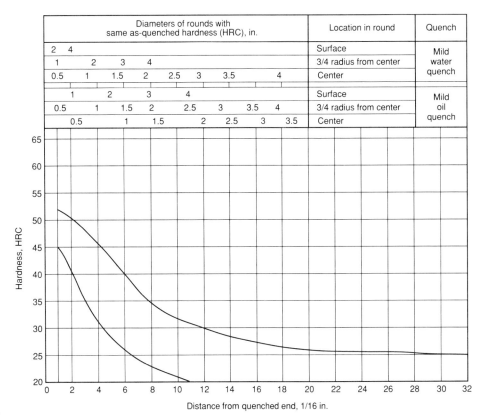

SAE/AISI 8627H   UNS H86270

Heat-treating temperatures recommended by SAE
Normalize (for forged or rolled specimens only): 900 °C (1650 °F)
Austenitize: 870 °C (1600 °F)

Hardness limits for specification purposes

J distance, mm	Hardness, HRC Maximum	Hardness, HRC Minimum
1.5	54	47
3	53	43
5	50	38
7	47	34
9	44	31
11	41	27
13	38	25
15	35	24
20	32	21
25	30	20
30	28	...
35	27	...
40	27	...
45	27	...
50	27	...

Hardness limits for specification purposes

J distance, 1/16 in.	Hardness, HRC Maximum	Hardness, HRC Minimum
1	54	47
2	52	43
3	50	38
4	48	35
5	45	32
6	43	29
7	40	27
8	38	26
9	36	24
10	34	24
11	33	23
12	32	22
13	31	21
14	30	21
15	30	20
16	29	20
18	28	...
20	28	...
22	28	...
24	27	...
26	27	...
28	27	...
30	27	...
32	27	...

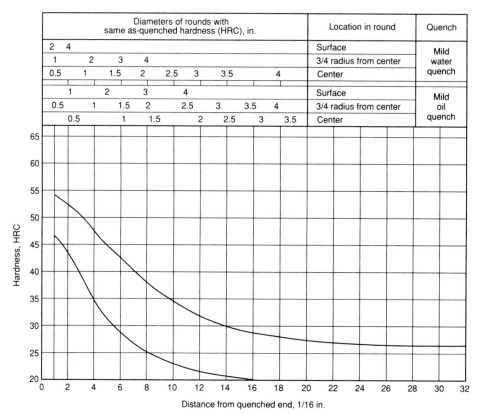

SAE/AISI 8630H   UNS H86300

Heat-treating temperatures recommended by SAE
Normalize (for forged or rolled specimens only): 900 °C (1650 °F)
Austenitize: 870 °C (1600 °F)

Hardness limits for specification purposes

J distance, mm	Hardness, HRC Maximum	Hardness, HRC Minimum
1.5	56	49
3	55	46
5	54	42
7	51	38
9	48	33
11	44	29
13	41	27
15	38	26
20	34	23
25	31	21
30	30	20
35	29	...
40	29	...
45	29	...
50	29	...

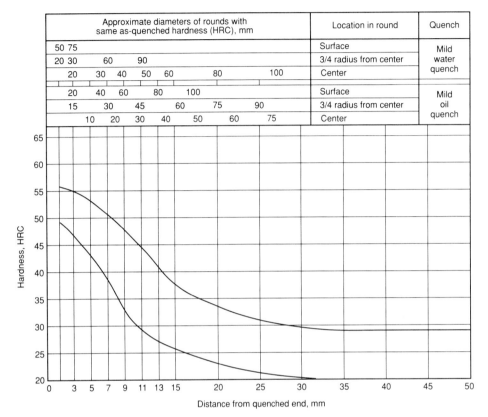

Hardness limits for specification purposes

J distance, 1/16 in.	Hardness, HRC Maximum	Hardness, HRC Minimum
1	56	49
2	55	46
3	54	43
4	52	39
5	50	35
6	47	32
7	44	29
8	41	28
9	39	27
10	37	26
11	35	25
12	34	24
13	33	23
14	33	22
15	32	22
16	31	21
18	30	21
20	30	20
22	29	20
24	29	...
26	29	...
28	29	...
30	29	...
32	29	...

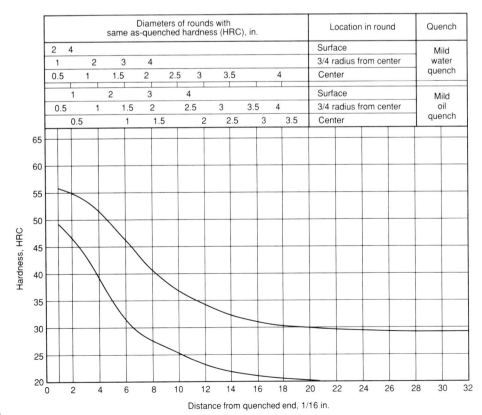

SAE/AISI 86B30H UNS H86301

Heat-treating temperatures recommended by SAE
Normalize (for forged or rolled specimens only): 900 °C (1650 °F)
Austenitize: 870 °C (1600 °F)

Hardness limits for specification purposes

J distance, mm	Hardness, HRC Maximum	Hardness, HRC Minimum
1.5	56	49
3	56	49
5	55	48
7	55	48
9	54	48
11	54	47
13	53	46
15	53	44
20	52	39
25	50	35
30	48	33
35	46	30
40	43	28
45	41	27
50	40	25

Hardness limits for specification purposes

J distance, 1/16 in.	Hardness, HRC Maximum	Hardness, HRC Minimum
1	56	49
2	55	49
3	55	48
4	55	48
5	54	48
6	54	48
7	53	48
8	53	47
9	52	46
10	52	44
11	52	42
12	51	40
13	51	39
14	50	38
15	50	36
16	49	35
18	48	34
20	47	32
22	45	31
24	44	29
26	43	28
28	41	27
30	40	26
32	39	25

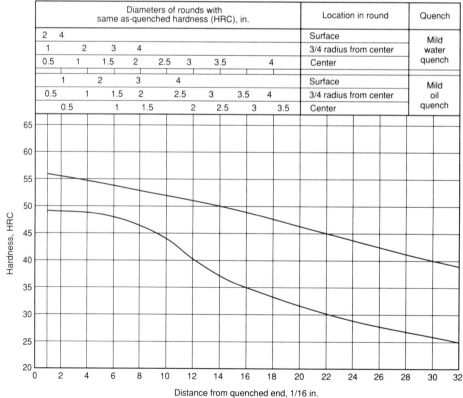

SAE/AISI 8637H   UNS H86370

Heat-treating temperatures recommended by SAE
Normalize (for forged or rolled specimens only): 870 °C (1600 °F)
Austenitize: 845 °C (1550 °F)

Hardness limits for specification purposes

J distance, mm	Hardness, HRC Maximum	Hardness, HRC Minimum
1.5	59	52
3	59	51
5	58	49
7	57	47
9	55	43
11	54	39
13	52	36
15	50	33
20	45	29
25	41	27
30	38	25
35	36	24
40	35	24
45	35	23
50	35	23

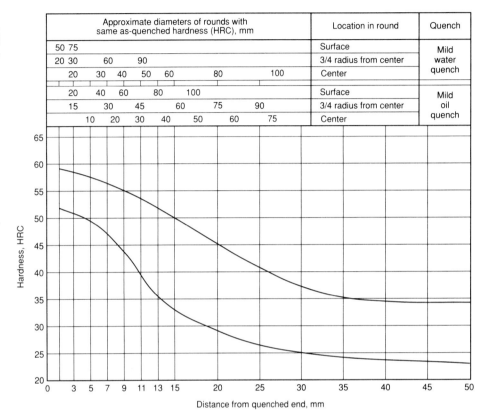

Hardness limits for specification purposes

J distance, 1/16 in.	Hardness, HRC Maximum	Hardness, HRC Minimum
1	59	52
2	58	51
3	58	50
4	57	48
5	56	45
6	55	42
7	54	39
8	53	36
9	51	34
10	49	32
11	47	31
12	46	30
13	44	29
14	43	28
15	41	27
16	40	26
18	39	25
20	37	25
22	36	24
24	36	24
26	35	24
28	35	24
30	35	23
32	35	23

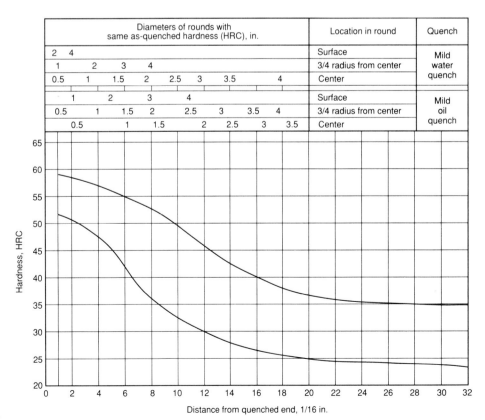

SAE/AISI 8640H UNS H86400

Heat-treating temperatures recommended by SAE
Normalize (for forged or rolled specimens only): 870 °C (1600 °F)
Austenitize: 845 °C (1550 °F)

Hardness limits for specification purposes

J distance, mm	Hardness, HRC Maximum	Hardness, HRC Minimum
1.5	60	53
3	60	53
5	60	52
7	60	50
9	58	47
11	57	42
13	55	30
15	54	36
20	48	31
25	43	27
30	40	26
35	39	25
40	38	24
45	37	24
50	37	24

Hardness limits for specification purposes

J distance, 1/16 in.	Hardness, HRC Maximum	Hardness, HRC Minimum
1	60	53
2	60	53
3	60	52
4	59	51
5	59	49
6	58	46
7	57	42
8	55	39
9	54	36
10	52	34
11	50	32
12	49	31
13	47	30
14	45	29
15	44	28
16	42	28
18	41	26
20	39	26
22	38	25
24	38	25
26	37	24
28	37	24
30	37	24
32	37	24

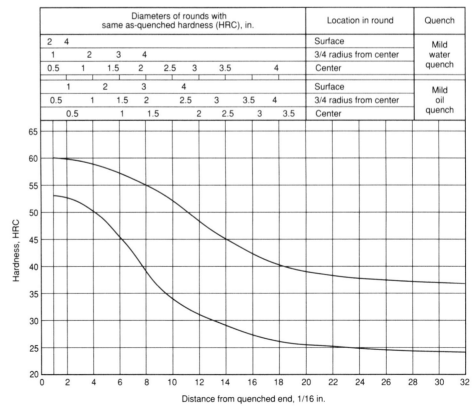

SAE/AISI 8642H   UNS H86420

Heat-treating temperatures recommended by SAE
Normalize (for forged or rolled specimens only): 870 °C (1600 °F)
Austenitize: 845 °C (1550 °F)

Hardness limits for specification purposes

J distance, mm	Hardness, HRC Maximum	Hardness, HRC Minimum
1.5	62	55
3	62	54
5	62	53
7	61	51
9	60	49
11	59	46
13	58	42
15	56	38
20	52	32
25	47	29
30	44	28
35	41	27
40	40	27
45	39	26
50	39	26

Hardness limits for specification purposes

J distance, 1/16 in.	Hardness, HRC Maximum	Hardness, HRC Minimum
1	62	55
2	62	54
3	62	53
4	61	52
5	61	50
6	60	48
7	59	45
8	58	42
9	57	39
10	55	37
11	54	34
12	52	33
13	50	32
14	49	31
15	48	30
16	46	29
18	44	28
20	42	28
22	41	27
24	40	27
26	40	26
28	39	26
30	39	26
32	39	26

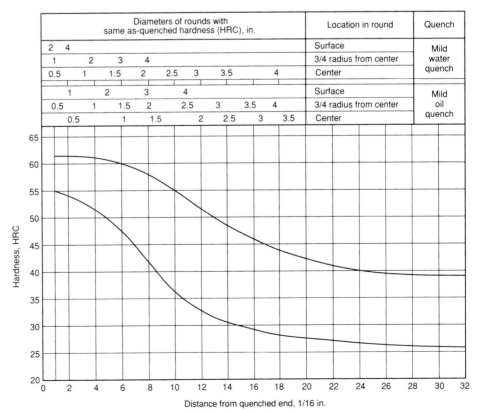

SAE/AISI 8645H UNS H86450

Heat-treating temperatures recommended by SAE
Normalize (for forged or rolled specimens only): 870 °C (1600 °F)
Austenitize: 845 °C (1550 °F)

Hardness limits for specification purposes

J distance, mm	Hardness, HRC Maximum	Hardness, HRC Minimum
1.5	63	56
3	63	56
5	63	55
7	63	53
9	62	51
11	61	48
13	59	45
15	58	41
20	54	34
25	49	31
30	46	29
35	43	28
40	42	27
45	42	27
50	41	27

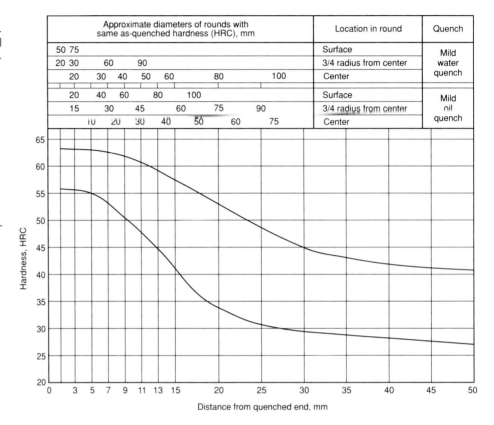

Hardness limits for specification purposes

J distance, 1/16 in.	Hardness, HRC Maximum	Hardness, HRC Minimum
1	63	56
2	63	56
3	63	55
4	63	54
5	62	52
6	61	50
7	61	48
8	60	45
9	59	41
10	58	39
11	56	37
12	55	35
13	54	34
14	52	33
15	51	32
16	49	31
18	47	30
20	45	29
22	43	28
24	42	28
26	42	27
28	41	27
30	41	27
32	41	27

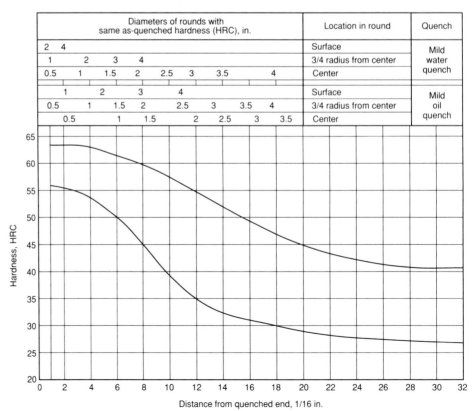

SAE/AISI 86B45H   UNS H86451

Heat-treating temperatures recommended by SAE
Normalize (for forged or rolled specimens only): 870 °C (1600 °F)
Austenitize: 845 °C (1550 °F)

Hardness limits for specification purposes

J distance, mm	Hardness, HRC Maximum	Hardness, HRC Minimum
1.5	63	56
3	63	56
5	63	55
7	62	54
9	62	53
11	61	52
13	61	51
15	60	51
20	59	49
25	58	45
30	58	40
35	57	36
40	57	33
45	56	32
50	56	31

Hardness limits for specification purposes

J distance, 1/16 in.	Hardness, HRC Maximum	Hardness, HRC Minimum
1	63	56
2	63	56
3	62	55
4	62	54
5	62	54
6	61	53
7	61	52
8	60	52
9	60	51
10	60	51
11	59	50
12	59	50
13	59	49
14	59	48
15	58	48
16	58	45
18	58	42
20	58	39
22	57	37
24	57	35
26	57	34
28	57	32
30	56	32
32	56	31

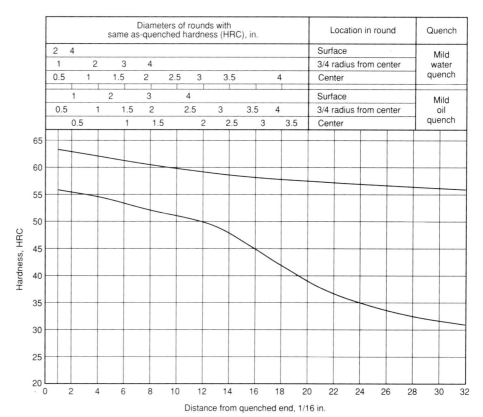

SAE/AISI 8650H UNS H86500

Heat-treating temperatures recommended by SAE
Normalize (for forged or rolled specimens only): 870 °C (1600 °F)
Austenitize: 845 °C (1550 °F)

Hardness limits for specification purposes

J distance, mm	Hardness, HRC Maximum	Hardness, HRC Minimum
1.5	65	59
3	65	59
5	65	58
7	65	56
9	64	55
11	63	53
13	62	50
15	61	46
20	59	38
25	57	34
30	54	32
35	52	31
40	49	30
45	47	29
50	46	29

Hardness limits for specification purposes

J distance, 1/16 in.	Hardness, HRC Maximum	Hardness, HRC Minimum
1	65	59
2	65	58
3	65	57
4	64	57
5	64	56
6	63	54
7	63	53
8	62	50
9	61	47
10	60	44
11	60	41
12	59	39
13	58	37
14	58	36
15	57	35
16	56	34
18	55	33
20	53	32
22	52	31
24	50	31
26	49	30
28	47	30
30	46	29
32	45	29

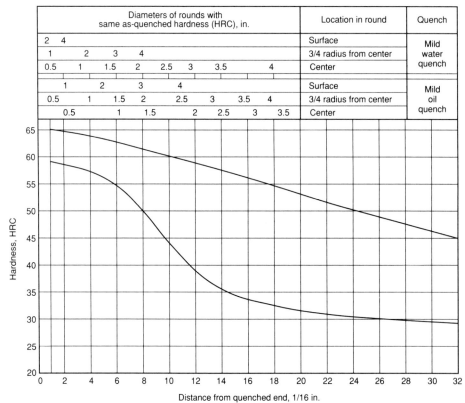

SAE/AISI 8655H   UNS H86550

Heat-treating temperatures recommended by SAE
Normalize (for forged or rolled specimens only): 870 °C (1600 °F)
Austenitize: 845 °C (1550 °F)

Hardness limits for specification purposes

J distance, mm	Hardness, HRC Maximum	Hardness, HRC Minimum
1.5	...	60
3	...	60
5	...	59
7	...	57
9	...	56
11	...	55
13	...	53
15	65	51
20	65	42
25	64	39
30	62	36
35	60	34
40	58	34
45	56	33
50	54	32

Hardness limits for specification purposes

J distance, 1/16 in.	Hardness, HRC Maximum	Hardness, HRC Minimum
1	...	60
2	...	59
3	...	59
4	...	58
5	...	57
6	...	56
7	...	55
8	...	54
9	...	52
10	65	49
11	65	46
12	64	43
13	64	41
14	63	40
15	63	39
16	62	38
18	61	37
20	60	35
22	59	34
24	58	34
26	57	33
28	56	33
30	55	32
32	53	32

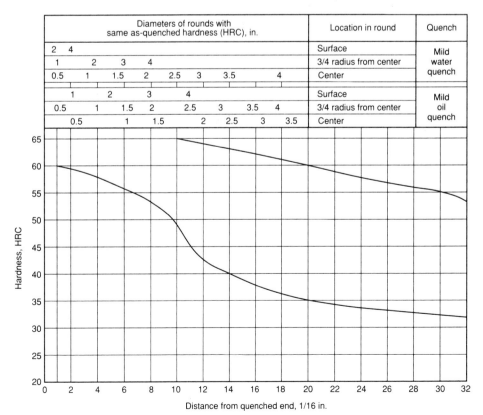

SAE/AISI 8660H UNS H86600

Heat-treating temperatures recommended by SAE
Normalize (for forged or rolled specimens only): 870 °C (1600 °F)
Austenitize: 845 °C (1550 °F)

Hardness limits for specification purposes

J distance, mm	Hardness, HRC Maximum	Hardness, HRC Minimum
1.5	...	60
3	...	60
5	...	60
7	...	60
9	...	59
11	...	58
13	...	56
15	...	53
20	...	46
25	...	42
30	65	39
35	64	38
40	62	37
45	61	36
50	60	35

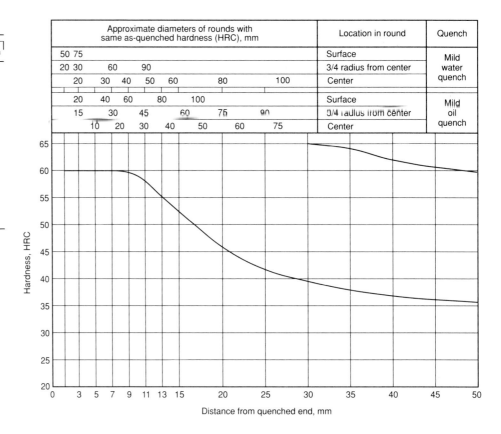

Hardness limits for specification purposes

J distance, 1/16 in.	Hardness, HRC Maximum	Hardness, HRC Minimum
1	...	60
2	...	60
3	...	60
4	...	60
5	...	60
6	...	59
7	...	58
8	...	57
9	...	55
10	...	53
11	...	50
12	...	47
13	...	45
14	...	44
15	...	43
16	65	42
18	64	40
20	64	39
22	63	38
24	62	37
26	62	36
28	61	36
30	60	35
32	60	35

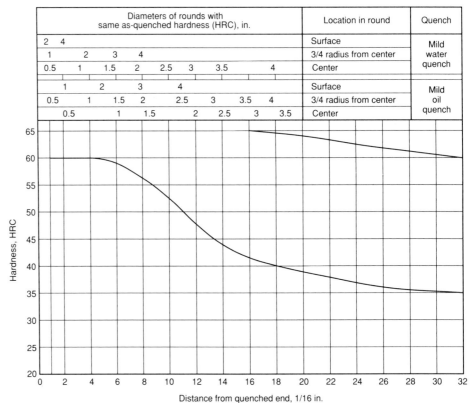

SAE/AISI 8720H UNS H87200

Heat-treating temperatures recommended by SAE
Normalize (for forged or rolled specimens only): 925 °C (1700 °F)
Austenitize: 925 °C (1700 °F)

Hardness limits for specification purposes

J distance, mm	Hardness, HRC Maximum	Hardness, HRC Minimum
1.5	48	41
3	47	39
5	45	35
7	41	29
9	37	25
11	33	22
13	31	21
15	29	...
20	27	...
25	25	...
30	24	...
35	23	...
40	23	...
45	23	...
50	22	...

Hardness limits for specification purposes

J distance, 1/16 in.	Hardness, HRC Maximum	Hardness, HRC Minimum
1	48	41
2	47	38
3	45	35
4	42	30
5	38	26
6	35	24
7	33	22
8	31	21
9	30	20
10	29	...
11	28	...
12	27	...
13	26	...
14	26	...
15	25	...
16	25	...
18	24	...
20	24	...
22	23	...
24	23	...
26	23	...
28	23	...
30	22	...
32	22	...

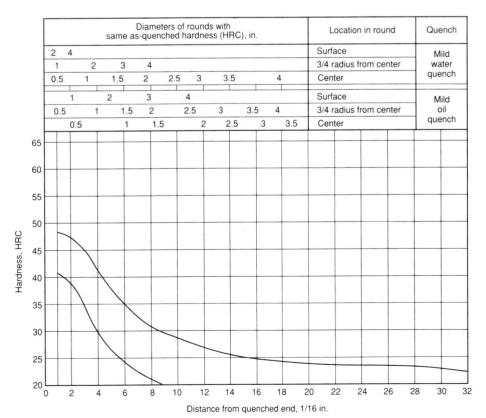

SAE/AISI 8740H UNS H87400

Heat-treating temperatures recommended by SAE
Normalize (for forged or rolled specimens only): 870 °C (1600 °F)
Austenitize: 845 °C (1550 °F)

Hardness limits for specification purposes

J distance, mm	Hardness, HRC Maximum	Hardness, HRC Minimum
1.5	60	53
3	60	52
5	60	51
7	60	49
9	59	46
11	58	43
13	56	39
15	54	36
20	50	31
25	45	29
30	43	28
35	41	27
40	40	27
45	39	26
50	38	26

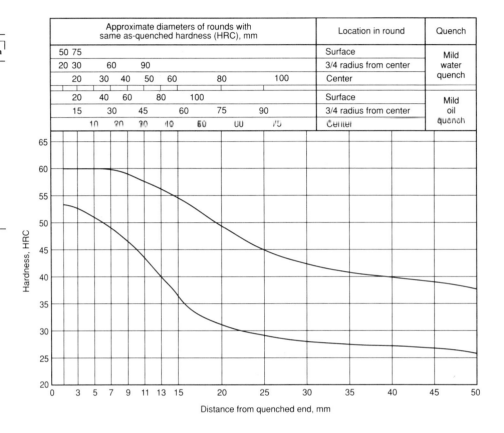

Hardness limits for specification purposes

J distance, 1/16 in.	Hardness, HRC Maximum	Hardness, HRC Minimum
1	60	53
2	60	53
3	60	52
4	60	51
5	59	49
6	58	46
7	57	43
8	56	40
9	55	37
10	53	35
11	52	34
12	50	32
13	49	31
14	48	31
15	46	30
16	45	29
18	43	28
20	42	28
22	41	27
24	40	27
26	39	27
28	39	27
30	38	26
32	38	26

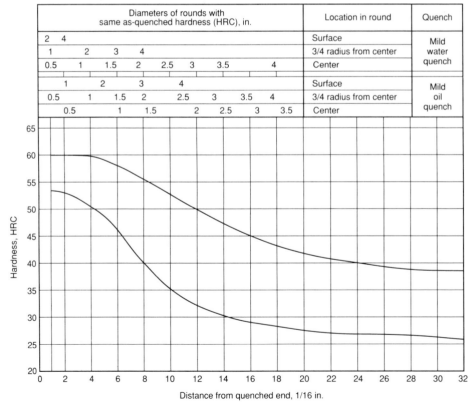

SAE/AISI 8822H   UNS H88220

Heat-treating temperatures recommended by SAE
Normalize (for forged or rolled specimens only): 925 °C (1700 °F)
Austenitize: 925 °C (1700 °F)

Hardness limits for specification purposes

J distance, mm	Hardness, HRC Maximum	Hardness, HRC Minimum
1.5	50	43
3	49	42
5	47	38
7	45	31
9	41	28
11	38	26
13	35	24
15	33	23
20	31	21
25	29	20
30	29	...
35	28	...
40	27	...
45	27	...
50	27	...

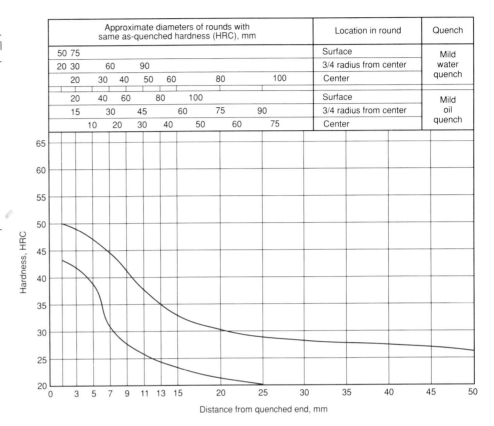

Hardness limits for specification purposes

J distance, 1/16 in.	Hardness, HRC Maximum	Hardness, HRC Minimum
1	50	43
2	49	42
3	48	39
4	46	33
5	43	29
6	40	27
7	37	25
8	35	24
9	34	24
10	33	23
11	32	23
12	31	22
13	31	22
14	30	22
15	30	21
16	29	21
18	29	20
20	28	...
22	27	...
24	27	...
26	27	...
28	27	...
30	27	...
32	27	...

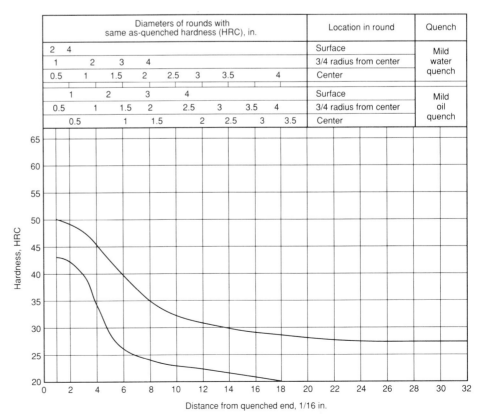

SAE/AISI 9260H UNS H92600

Heat-treating temperatures recommended by SAE
Normalize (for forged or rolled specimens only): 900 °C (1650 °F)
Austenitize: 870 °C (1600 °F)

Hardness limits for specification purposes

J distance, mm	Hardness, HRC Maximum	Hardness, HRC Minimum
1.5	...	60
3	...	60
5	65	58
7	63	50
9	62	42
11	60	38
13	58	36
15	54	35
20	47	33
25	40	32
30	38	31
35	37	30
40	36	29
45	35	28
50	35	28

Hardness limits for specification purposes

J distance, 1/16 in.	Hardness, HRC Maximum	Hardness, HRC Minimum
1	...	60
2	...	60
3	65	57
4	64	53
5	63	46
6	62	41
7	60	38
8	58	36
9	55	36
10	52	35
11	49	34
12	47	34
13	45	33
14	43	33
15	42	32
16	40	32
18	38	31
20	37	31
22	36	30
24	36	30
26	35	29
28	35	29
30	35	28
32	34	28

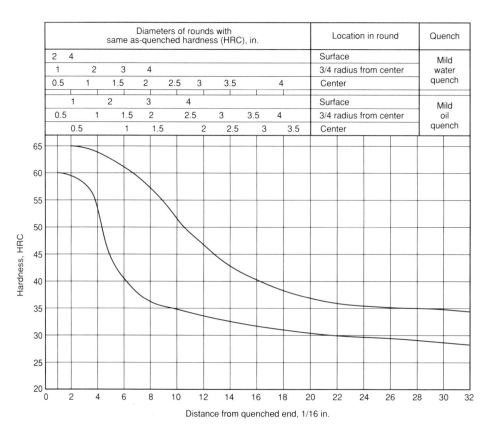

SAE/AISI 9310H UNS H93100

Heat-treating temperatures recommended by SAE
Normalize (for forged or rolled specimens only): 925 °C (1700 °F)
Austenitize: 845 °C (1550 °F)

Hardness limits for specification purposes

J distance, mm	Hardness, HRC Maximum	Hardness, HRC Minimum
1.5	43	36
3	43	35
5	43	34
7	43	33
9	43	31
11	42	30
13	41	28
15	40	27
20	38	26
25	36	25
30	35	25
35	35	25
40	34	25
45	34	24
50	33	24

Hardness limits for specification purposes

J distance, 1/16 in.	Hardness, HRC Maximum	Hardness, HRC Minimum
1	43	36
2	43	35
3	43	35
4	42	34
5	42	32
6	42	31
7	42	30
8	41	29
9	40	28
10	40	27
11	39	27
12	38	26
13	37	26
14	36	26
15	36	26
16	35	26
18	35	26
20	35	25
22	34	25
24	34	25
26	34	25
28	34	25
30	33	24
32	33	24

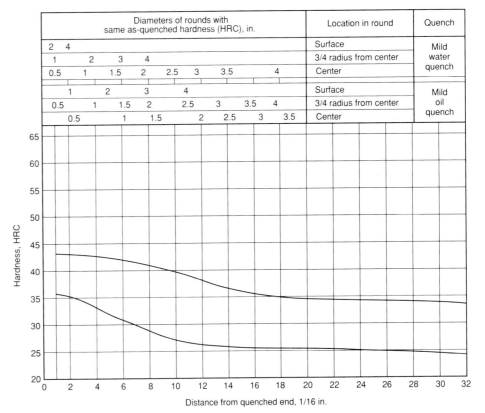

SAE/AISI 94B15H   UNS H94151

Heat-treating temperatures recommended by SAE
Normalize (for forged or rolled specimens only): 925 °C (1700 °F)
Austenitize: 925 °C (1700 °F)

Hardness limits for specification purposes

J distance, mm	Hardness, HRC Maximum	Hardness, HRC Minimum
1.5	45	38
3	45	38
5	45	37
7	44	34
9	42	30
11	40	26
13	38	22
15	36	20
20	31	...
25	28	...
30	26	...
35	24	...
40	23	...
45	22	...
50	22	...

Hardness limits for specification purposes

J distance, 1/16 in.	Hardness, HRC Maximum	Hardness, HRC Minimum
1	45	38
2	45	38
3	44	37
4	44	36
5	43	32
6	42	28
7	40	25
8	38	23
9	36	21
10	34	20
11	33	...
12	31	...
13	30	...
14	29	...
15	28	...
16	27	...
18	26	...
20	25	...
22	24	...
24	23	...
26	23	...
28	22	...
30	22	...
32	22	...

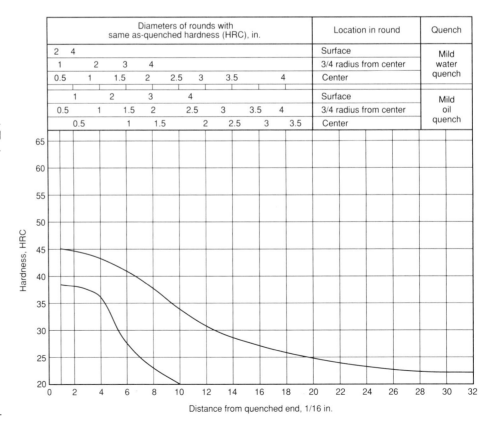

SAE/AISI 94B17H UNS H94171

Heat-treating temperatures recommended by SAE
Normalize (for forged or rolled specimens only): 925 °C (1700 °F)
Austenitize: 925 °C (1700 °F)

Hardness limits for specification purposes

J distance, mm	Hardness, HRC Maximum	Hardness, HRC Minimum
1.5	46	39
3	46	39
5	46	38
7	45	36
9	44	31
11	43	26
13	41	24
15	39	22
20	34	...
25	30	...
30	28	...
35	26	...
40	25	...
45	24	...
50	23	...

Hardness limits for specification purposes

J distance, 1/16 in.	Hardness, HRC Maximum	Hardness, HRC Minimum
1	46	39
2	46	39
3	45	38
4	45	37
5	44	34
6	43	29
7	42	26
8	41	24
9	40	23
10	38	21
11	36	20
12	34	...
13	33	...
14	32	...
15	31	...
16	30	...
18	28	...
20	27	...
22	26	...
24	25	...
26	24	...
28	24	...
30	23	...
32	23	...

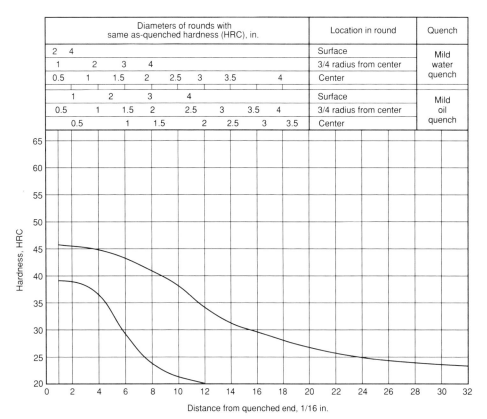

SAE/AISI 94B30H UNS H94301

Heat-treating temperatures recommended by SAE
Normalize (for forged or rolled specimens only): 900 °C (1650 °F)
Austenitize: 870 °C (1600 °F)

Hardness limits for specification purposes

J distance, mm	Hardness, HRC Maximum	Hardness, HRC Minimum
1.5	56	49
3	56	49
5	56	48
7	55	47
9	55	46
11	54	44
13	53	41
15	53	38
20	51	31
25	47	26
30	43	24
35	40	23
40	37	22
45	36	21
50	34	20

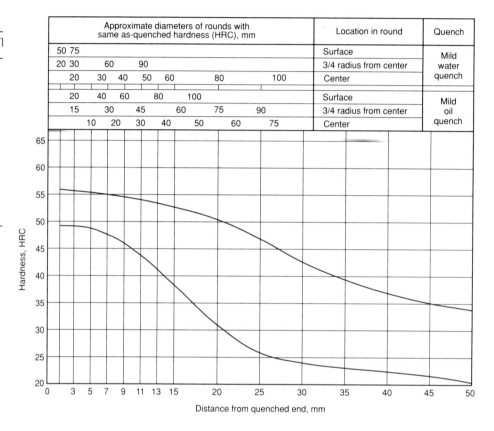

Hardness limits for specification purposes

J distance, 1/16 in.	Hardness, HRC Maximum	Hardness, HRC Minimum
1	56	49
2	56	49
3	55	48
4	55	48
5	54	47
6	54	46
7	53	44
8	53	42
9	52	39
10	52	37
11	51	34
12	51	32
13	50	30
14	49	29
15	48	28
16	46	27
18	44	25
20	42	24
22	40	23
24	38	23
26	37	22
28	35	21
30	34	21
32	34	20

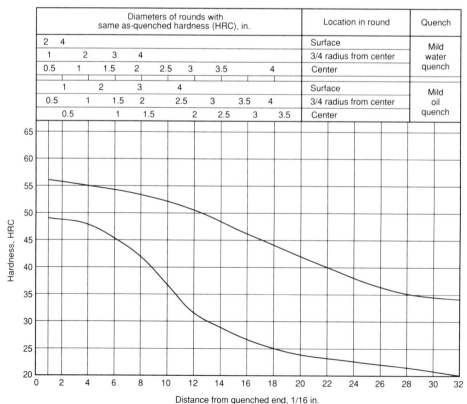

Fabrication Characteristics of Carbon and Low-Alloy Steels

Sheet Formability of Steels 573
Bulk Formability of Steels 581
Machinability of Steels 591
Weldability of Steels 603

Sheet Formability of Steels

Revised by W.G. Granzow, Armco, Inc., Research & Technology

STEEL SHEET is widely used for industrial and consumer products, partly because it is relatively strong, easily joined, and readily available at moderate cost. However, it is not these factors, but the formability of steel sheets that is primarily responsible for its wide use. The term formability is commonly used to describe the ability of a steel to maintain its structural integrity while being plastically deformed into various shapes. However, not all shapes require the same forming characteristics, and a steel that has excellent formability in one application may exhibit poor formability in another application. In practice, therefore, formability must be optimized by selecting a grade of steel that has the forming characteristics needed to make the required shape.

These forming characteristics are normally estimated from an analysis of the mechanical properties of steel, which are determined by uniaxial tensile tests. Although this type of test does not simulate any commercial forming operations, the test results have been universally used for many years to evaluate formability, and some understanding of them is essential to the understanding of sheet steel formability. Sheet metal forming methods are described in detail in Volume 14 of the 9th Edition of *Metals Handbook*.

Examples of formed parts that require different forming characteristics in the steel are shown in Fig. 1. Part A was formed by drawing; that is, all the metal that was required to form the part from a flat blank came from the flanges. This shape requires that the steel have a high plastic-strain ratio, or r value, which determines the resistance of steel sheet to thinning during forming operations. Part C was formed by stretching; the flange on the blank was clamped during forming, and all of the metal that was required to form the part came from reducing the thickness of the metal. This type of part requires good ductility in the steel. However, the r value should be low, because high r values can cause failures of stretched parts. Part B has failed in plane strain, which is a type of stretching. Parts that develop this strain condition, such as automotive panels, require good ductility.

This article discusses the mechanical properties and formability of steel sheet, the use of circle grid analysis to identify the properties of complicated shapes, and various simulative forming tests. It covers the effects of steel composition, steelmaking practices, and metallic coatings, as well as the correlation between microstructure and formability. A guide to the selection of steel sheet is also included.

Mechanical Properties and Formability

The mechanical properties of steel sheet that influence its forming characteristics, either directly or indirectly, can be measured by uniaxial tension testing, such as that described in ASTM E 8. The tensile test results of particular interest include the yield strength, ultimate tensile strength, total elongation, uniform elongation, yield point elongation, plastic-strain ratio, planar anisotropy, and the strain-hardening exponent. Uniaxial tensile tests may be made with specimens obtained from longitudinal, diagonal, transverse, or other orientations relative to the rolling direction. Typical mechanical properties for common grades of hot-rolled and cold-rolled steel sheets are given in Tables 1 and 2.

Yield strength of steel sheet is indicative of both formability and strength after forming. Several types of yielding behavior are observed in steel sheet (see Fig. 2). When yield point elongation occurs, the lowest value observed during discontinuous yielding is reported as the yield strength. In the absence of an abrupt change in the load-extension curve, the stress at 0.2% offset or 0.5% extension under load is reported as the yield strength.

In forming plain carbon steel sheet, a yield strength of 240 MPa (35 ksi) or more increases the likelihood of excessive springback and breakage during forming. However, the use of material having a yield strength of less than 140 MPa (20 ksi) may result in parts with insufficient strength levels.

High-strength, formable sheet steels have been developed for applications for which increased strength or reduced weight, in addition to moderate formability, are required. The yield strengths of these steels generally range from 345 to 690 MPa (50 to 100 ksi).

Total Elongation. After fracture, the tensile specimen is pieced together, and the length between gage marks is measured. In this manner, elongation is calculated and reported as a percentage of the original gage length, which is usually 50 mm (2 in.). (A gage length of 200 mm, or 8 in., may be used for heavier-gage metals.) Specimens of sheet metals used for tensile tests usually have short, parallel-sided, reduced sections, but slightly tapered reduced sections are sometimes used to control the location of necking and fracture. Values of elongation resulting from tests of different speci-

Fig. 1 Parts that required different forming characteristics in the steel sheet

574 / Fabrication Characteristics of Carbon and Low-Steel Alloys

Table 1 Typical mechanical properties of hot-rolled steel sheet

Type or quality	Special feature	Yield strength MPa	ksi	Tensile strength MPa	ksi	Elongation in 50 mm (2 in.), %	Hardness, HRB	Strain-hardening exponent, n	Plastic-strain ratio, r_m
Commercial	Standard properties	262	38	359	52	30	55	0.15	0.9
Drawing (rimmed)	Improved properties	241	35	345	50	35	50	0.18	1.0
Drawing (special killed)	Nonaging	241	35	345	50	40	50	0.20	1.0
Medium strength	Inclusion shape control	345	50	414	60	25	70	0.15	0.9
High strength	Inclusion shape control	552	80	620	90	15	90	...	...

Table 2 Typical mechanical properties of cold-rolled steel sheet

Type or quality	Special feature	Yield strength MPa	ksi	Tensile strength MPa	ksi	Elongation in 50 mm (2 in.), %	Hardness, HRB	Strain-hardening exponent, n	Plastic-strain ratio, r_m
Commercial	Standard properties	234	34	317	46	35	45	0.18	1.0
Drawing (rimmed)	Stretchable	207	30	310	45	42	40	0.22	1.2
Drawing (special killed)	Deep drawing	172	25	296	43	42	40	0.22	1.6
Interstitial free	Extra deep drawing	152	22	317	46	42	45	0.24	2.0
Medium strength	Formable	414	60	483	70	25	85	0.20	1.2
High strength	Moderately formable	689	100	724	105	10	25(a)	...	...

(a) HRC

mens of the same material may vary because of differences in gage length, sheet thickness, edge preparation and finish, test methods, or other factors.

Typical values of the amount of elongation in 50 mm (2 in.) are listed in Tables 1 and 2 for common formable grades of steel sheet. Generally, an elongation of 35 to 45% in 50 mm (2 in.) is normal for conventional low-carbon steels, with higher values indicating better formability.

Uniform Elongation. The total elongation of a sheet tensile specimen comprises two parts, uniform elongation and postuniform elongation. For a material that follows the power relationship for hardening ($\sigma = K\epsilon^n$), the uniform elongation (measured in true strain) is equal to the strain-hardening exponent, n. The postuniform elongation depends on both the strain-hardening behavior and the strain rate sensitivity response of the metal to the applied stress. When a neck forms, the strains and strain rate within the neck are greater than in the outside regions, and increased strain hardening may offset the weakening due to the reduced cross-sectional thickness, causing a shift of deformation to regions outside the neck.

The engineering elongation to maximum load, e_u, is related to the strain-hardening exponent, n, by the equation:

$$n = \ln(1 + e_u) \quad \text{(Eq 1)}$$

Typical values of e_u for low-carbon steels range from 20 to 30%. The e_u and associated n values indicate the work-hardening rate of sheet metals and, thus, the capability of the metal to deform in stretch, plane-strain, and bending deformation modes. Other factors, such as strain-rate sensitivity, can enhance or detract from the capability of a metal to be formed into a part. For example, the n values of low-carbon steel and 1100-O grade aluminum are about the same; however, both the total elongation and the forming limit of aluminum are considerably lower than those of low-carbon steel because aluminum has a negative value of m, the strain-rate sensitivity in response to the applied flow stress:

$$m = \frac{\ln(\sigma_1/\sigma_2)}{\ln(\dot{\epsilon}_1/\dot{\epsilon}_2)} \quad \text{(Eq 2)}$$

where σ is the flow stress and $\dot{\epsilon}$ is the strain rate.

For sheet metals that fail by local necking, uniform elongation may not give a true estimate of formability. Estimates based on total elongation are often considered more reliable.

Yield point elongation is the portion of total elongation that occurs during discontinuous yielding at the yield stress. It is accompanied by the formation of surface defects known as Lüders lines, or stretcher strains, which are considered imperfections in many applications of steel sheet because of their unsightly appearance. Yield point elongation during tensile testing indicates that Lüders lines are likely to occur during forming.

Yield point elongation requires the presence of interstitial residual alloying elements, particularly carbon or nitrogen; consequently, low-interstitial steels do not exhibit this effect. Yield point elongation can be suppressed by temper rolling the steel sheet at the mill. However, unless the nitrogen has been combined with another element (usually aluminum), the steel will age harden after a period that varies from a few hours to a year or more (depending on storage temperature and other factors). Aged steels can be used in most forming

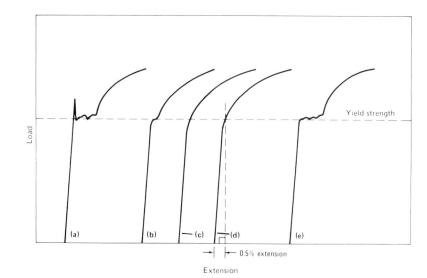

Fig. 2 Load-extension curves for steel sheet having the same yield strength, but different characteristic behavior. (a) Annealed soft-rimmed or aluminum-killed steel; yield strength is the lowest stress measured during yield point elongation. (b) Lightly temper-rolled rimmed steel; stress at the jog in the curve is reported as yield strength. (c) and (d) Temper-rolled low-carbon steel. May be rimmed, aluminum-killed, or interstitial-free steel with no detectable yield point. The yield strength is calculated from the load at 0.2% offset (c) or from the load at 0.5% extension (d). (e) Rimmed steel with a yield point elongation due to aging at room temperature for several months. The yield strength is the lowest stress measured during yield point elongation.

operations, provided they are roller leveled or flex leveled immediately before fabrication, although these methods are less effective than temper rolling.

Plastic-strain ratio, r, describes the resistance of steel sheet to thinning during forming operations. This is the ratio of the true strain in the width direction, ϵ_w, to the true strain in the thickness direction, ϵ_t, of plastically strained sheet metal:

$$r = \frac{\epsilon_w}{\epsilon_t} \qquad \text{(Eq 3)}$$

The plastic-strain ratio is related to the crystallographic orientation of low-carbon steels. A standard method for determining r by using the tension test is given in ASTM E 517. The value will vary with test direction (relative to the coil rolling direction) in anisotropic metals. An average value, r_m, (sometimes designated $\bar{r}$), represents the normal plastic anisotropy of the steel sheet:

$$r_m = \frac{r_0 + 2r_{45} + r_{90}}{4} \qquad \text{(Eq 4)}$$

Hot-rolled and normalized cold-rolled steels are generally isotropic (r_m of 1.0). Rimmed steels usually have an r_m of 1.2, but this value may be higher in special cases, as with some low-manganese low-sulfur products. Aluminum-killed steels will be more anisotropic, with an r_m of 1.6. Higher values (up to 2.5) may be attained by controlling composition and processing. The upper limit for commercial steels is about 3.0, although values near 3.0 are seldom achieved. Interstitial-free steels tend to have the highest r_m at approximately 2.0. The r_m value predicts the ability of metals to deform in draw.

Planar anisotropy may be reported as:

$$\Delta r = \frac{r_0 + r_{90} - 2r_{45}}{2} \qquad \text{(Eq 5)}$$

Planar anisotropy is a measure of the amount of high points, or ears, that will develop on the edges of deep-drawn cylindrical cups or similar parts. High points in the rolling and transverse directions are noted when Δr is positive (for low-carbon, drawing-quality, aluminum-killed steel sheet). For some high-strength low-alloy steels, Δr is negative, and earing occurs at 45° to the rolling direction. For most applications, values of Δr near 0 are preferred, because such values imply a minimal tendency to form ears when metals are drawn into cylindrical cups.

The strain-hardening exponent, n, is the slope of the true stress-true strain curve, when plotted on logarithmic coordinates. A significant portion of the curve is nearly a straight line for many low-carbon steels. The data are assumed to fit the equation:

$$\sigma = K\epsilon^n \qquad \text{(Eq 6)}$$

The n value will normally be approximately 0.22 for low-carbon steels used to form

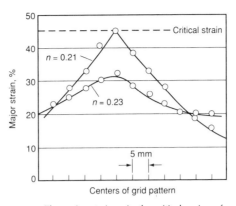

Fig. 3 The major strain ϵ_1 in the critical region of a formed part is more uniformly distributed for the steel having the higher value of n. One of these two parts (which are identical except for the n value of the steel selected) was strained to the point of excessive thinning; the other, made from steel with the higher n value, showed no inclination to fracture. Source: Ref 1

complex-shape parts. Higher values (up to 0.26) indicate improved capabilities to deform in stretch. Freshly rolled rimmed steels generally have n values comparable to those of aluminum-killed steels. After aging, values of n for rimmed steels are less than those for aluminum-killed steels. Some low-carbon steels that are not fully processed for formability, especially hot-rolled grades, will have n values as low as 0.10, but most of the formable grades will have n values above 0.14.

The effects of different n values on strain distribution in critical regions of a specific formed part are shown in Fig. 3. Parts formed from steel sheet with a low n value (0.21) may undergo excessive thinning and fracture in critical regions. Identical parts formed from sheet with a higher n value (0.23) frequently will be strong enough in the critical areas to transfer strain to adjacent areas, thereby avoiding failure during forming.

Circle Grid Analysis

Uniaxial tension tests determine the mechanical properties of steel sheet under closely controlled, frictionless conditions, which are different from the conditions that normally occur during sheet metal forming operations. However, experience has shown that test results often correlate with the ability of a steel to be successfully formed into parts. The relationship is not always a simple one. For example, the parts shown in Fig. 1 require the steel to exhibit different r and n values for successful processing. With practice, the mechanical properties that are required to form simple shapes like those shown can be estimated, but the properties required to form more complicated shapes can best be determined using circle grid analysis.

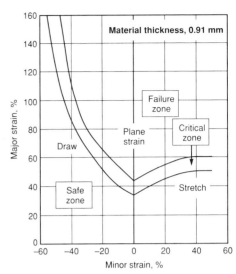

Fig. 4 A forming limit diagram (FLD). Strains in the critical zone and above it will result in excessive breakage. Strain conditions on the left side require high r values, while strain conditions on the center and right side require good ductility (a high percentage elongation in tensile tests).

To use circle grid analysis, a pattern consisting of uniformly spaced, uniformly sized circles (usually 2.5 or 5.0 mm, or 1.0 or 0.2 in., in diameter) is etched onto a flat sheet of steel that is to be formed, and the steel is then processed into the desired shape. The circles change into ellipses with the deformation of the gridded steel blank. The ellipses are then measured to determine the maximum dimension (major strain) and the minimum dimension (minor strain). These strains are then plotted on a forming limit diagram (Fig. 4). The location and magnitude of the plotted points on the diagram indicate the severity of the forming operation. Their position on the right (stretch) side or left (draw) side of the diagram indicates the strain condition in the steel, where 0% minor strain indicates a plane-strain condition. Experience has shown that breakage associated with critical strains on the right, or stretch, side may be resolved by increasing the total elongation in the steel. Breakage associated with strains on the left, or draw, side may be eliminated by increasing the r-value of the steel. The magnitude of all of the strains can be reduced by changing the sheet metal forming conditions.

The diagram in Fig. 4 is for 0.914 mm (0.036 in.) thick drawing-quality steel sheets. The critical area curves will be lower for thinner steels and higher for thicker steels. The curves can also be adjusted to accommodate commercial quality steels and other grades with lower n values than drawing-quality steels. The adjustments are made by moving the plane-strain intercept, or the point at which the lower curve crosses the 0% minor strain line, according to Fig. 5.

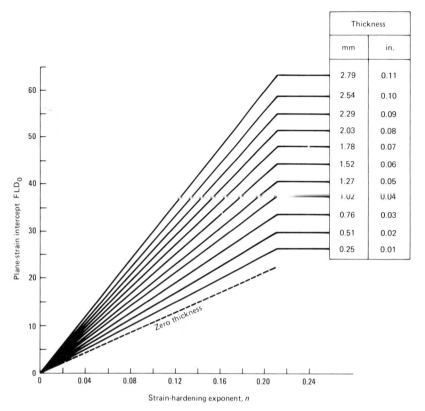

Fig. 5 Relationship between the plane-strain intercept on a forming limit diagram (FLD$_0$) and the strain-hardening exponent as a function of thickness. FLD$_0$ depends only on thickness for values of n greater than 0.21. Source: Ref 2

The plotted points on a forming limit diagram show the magnitude of the strains that develop in steel sheets that are processed into a particular shape. Although the magnitude of the strains will be slightly different in steels with different r values, the plotted points generally show the strain conditions that will occur in any steel that has been used successfully to make a particular part. Therefore, in theory, the quality of the steel that is used to perform the analysis should not be important. In practice, it is advantageous to determine the mechanical properties of the steel when the circle grid analysis is made because excessive strains exhibited by formable steel are a clear indication that changes in processing will be required.

Simulative Forming Tests

Numerous tests have been developed to observe and compare the performance of steel sheets under actual forming conditions. The best known are the Olsen, Erichsen, and Swift cup tests, and the limiting dome height (LDH) test. The Olsen and Erichsen cup tests measure steel-stretching performance, the Swift cup test measures steel-drawing performance, and the LDH test measures the performance of a steel under plane-strain conditions.

The Olsen cup test measures the maximum penetration of a 22 mm (0.875 in.) diam hemispherical punch into a clamped, flat blank of steel. The punch depth at failure is the Olsen cup value. The Erichsen test is similar, but the punch diameter is 20 mm (0.790 in.). Experience has shown that a conscientious operator using a specific testing machine can use the Olsen or Erichsen cup test results to evaluate the ductility of the steel. However, the correlation between the test results and the steel performance in many sheet metal forming operations has not been good.

The Swift cup test determines the maximum diameter circular blank that can be successfully drawn into a flat-bottom cup. The die is usually 100 mm (4 in.) in diameter, although other dimensions are also used. The results are expressed as the limiting draw ratio (LDR):

$$\text{LDR} = \frac{\text{Maximum diameter blank}}{\text{Cup diameter}}$$

The LDR is a measure of the drawability of the steel, and the test results correlate well with the r value.

The LDH test measures the resistance of a steel to failure under plane-strain conditions. The test sample is a 180 mm (7 in.) long rectangle with an experimentally determined width (~135 mm, or 5.25 in.) that is clamped and stretched to failure over a 100 mm (4 in.) diam hemispherical punch. The sample width is the width that is found to produce the minimum height in the test; this minimum height is the LDH value for that material.

The LDH value has been shown to correlate well with the performance of steel sheets in stretch-type automotive body panels, which normally fail under plane-strain conditions. However, it does not correlate with steel performance on deep-drawn panels.

All simulative forming tests are affected by friction between the sheet metal and the tooling. This creates problems with sample and tooling preparation because the factors that affect friction in sheet metal forming are poorly understood. In practice, the samples and tooling are usually cleaned with a light mineral oil and wiped dry before testing. This treatment provides sufficient lubricity to prevent a "stick-slip" effect, while minimizing the effect of the lubrication.

Effects of Steel Composition on Formability

Low-carbon sheet steels are generally preferred for forming. These steels typically contain less than 0.10% carbon and less than 1% total intentional and residual alloying elements. The amount of manganese, the principal alloying addition, normally ranges from 0.15 to 0.35%. Controlled amounts of silicon, niobium, titanium, or aluminum may be added either as deoxidizers or to develop certain properties. Residual elements, such as sulfur, chromium, nickel, molybdenum, copper, nitrogen, and phosphorus, are usually limited as much as possible. In steelmaking shops, these amounts are based on the quality of sheet being produced. Alloy sheet steels (including high-strength low-alloy grades) however, contain specified amounts of one or more of these elements.

Carbon content is particularly significant in steels that are intended for complex forming applications. An increase in the carbon content of steel increases the strength of the steel and reduces its formability. These effects are caused by the formation of carbide particles in the ferrite matrix and by the resulting small grain size. The amount of carbon in steel sheet is generally limited to 0.10% or less to maximize the formability of the sheet.

Manganese enhances the hot-working characteristics of the steel and facilitates the development of the desired grain size. Some manganese is also necessary to neutralize the detrimental effects of sulfur, particularly for hot workability. Typical manganese contents for low-carbon steel sheet range from 0.15 to 0.35%; manganese contents up to 2.0% may be specified in high-strength low-alloy steels. When the sulfur content of the steel is very low, the manganese content also can be low, which allows the steel to be processed to develop high r values.

Phosphorus and sulfur are considered undesirable in steel sheet intended for forming, drawing, or bending because their presence increases the likelihood of cracking or splitting. Allowable levels of phosphorus and sulfur depend on the desired quality level. For example, commercial-quality cold-rolled sheet must contain less than 0.035% P and 0.040% S. For some applications, phosphorus may be added to the steel to increase the strength. Sulfur usually appears as manganese sulfide stringers in the microstructure. These stringers can promote splitting, particularly whenever an unrestrained edge is deformed.

Silicon content in low-carbon steel varies according to the deoxidation practice employed during production. In rimmed steels (so called because of the rimming action caused by outgassing during solidification from the molten state), the silicon content is generally less than 0.10%. When silicon rather than aluminum is used to kill the rimming action, the silicon content may be as high as 0.40%. Silicon may cause silicate inclusions, which increase the likelihood of cracking during bending. Silicon also increases the strength of the steel and thus decreases its formability.

Chromium, nickel, molybdenum, vanadium, and other alloying elements are present in low-carbon steel only as residual elements. With proper scrap selection and control of steelmaking operations, these elements are generally held to minimum amounts. Each of these elements increases the strength and decreases the formability of steel sheet. High-strength low-alloy steels may contain specified amounts of one or more of these elements.

Copper is generally considered an innocuous residual element in steel sheet. The strengthening effect of copper is almost negligible in typical residual amounts of less than 0.10%. However, copper is added to steel in amounts exceeding 0.20% to improve resistance to atmospheric corrosion.

Niobium strengthens high-strength low-alloy steel through the formation of niobium carbides and nitrides. It can also be used either alone or in combination with titanium to develop high r values in interstitial-free steels. These alloying elements remove the interstitial elements carbon and nitrogen from solid solution. Consequently, the steel shows no yield point elongation.

Titanium is a strong carbide and nitride former. It helps develop high r values and eliminates yield point elongation and the aging of cold-rolled annealed steel sheet. Titanium streaks may be a problem in some grades, especially in the form of surface defects in exposed applications.

Aluminum is added to steel to kill the rimming action and thus produce a very clean steel known as an aluminum-killed, or special-killed, steel. Aluminum combines with both the oxygen and nitrogen to stop the outgassing of the molten steel when it is added to the ladle or mold. Aluminum also aids the development of preferred grain orientations to attain high r values in cold-rolled and annealed steel sheet. Elongated grains of an approximate ASTM 7 size are found in most well-processed aluminum-killed steels. Because the aluminum combines with the nitrogen, the steel is not subject to strain aging.

Nitrogen can significantly strengthen low-carbon steel. It also causes strain aging of the steel. The effects of nitrogen can be controlled by deoxidizing the melt with aluminum.

Cerium and other rare earth elements may be added to steel to change the shape of manganese sulfide inclusions from being needlelike or ribbonlike to being globular. Globular inclusions reduce the likelihood of cracking if the sheet is formed without restraining the edges.

Oxygen content of molten steel determines its solidification characteristics in the ingot. Excessive amounts of oxygen impede nitride formation and thus negate the effects of alloying elements added to minimize strain aging. Deoxidizers such as silicon, aluminum, and titanium will control the oxygen content. When oxygen combines with these deoxidants, complex nonmetallics are formed. Although most nonmetallics dissolve in the slag, some may become trapped in the steel, causing the surface defects of seams and slivers.

Effects of Steelmaking Practices on Formability

The formability of steel sheet is determined to a great extent by the steelmaking practices employed in manufacturing. The user of steel sheet normally specifies certain characteristics for the sheet, thus ensuring that the material can be formed in a predictable manner. Adherence to these specifications also implies that the producer of the sheet has observed whatever steelmaking practices are necessary to enable the product to perform as indicated. The user can specify either hot-rolled or cold-rolled sheet, and he must select an appropriate quality designation. Sometimes deoxidation practice is specified. The user has some latitude in choosing the surface finish of the sheet. Of course, the user specifies the dimensions and tolerances of the sheet and the type of edge to be supplied.

Hot-rolled steel is rolled to its final thickness in an elevated-temperature process. The finishing temperature is determined by the composition of the steel and the desired properties. In the as-hot-rolled condition, the steel has a dark-gray oxide coating on its surface, which offers limited corrosion protection as long as it is undisturbed. However, the oxide flakes off during forming and may be undesirable around the press. Because the oxide coating also interferes with steel surface lubricants, it should be removed before the final finishing of most formed parts. Hot-rolled steel may be ordered pickled (using either hot sulfuric or hydrochloric acid to remove the oxide) and oiled to inhibit in-transit rusting. Hot-rolled steel in the as-pickled condition will show stretcher strains or Lüders lines on the surface after forming. Whenever surface appearance is important, the steel should be ordered with a temper-rolled surface (skin pass of less than 2% cold reduction) to reduce this tendency. If aging is a problem because of storage requirements, special-killed hot-rolled steel should be ordered.

There is no preferred grain orientation providing high r values in hot-rolled steel, but improved grain size and resistance to longitudinal splitting may be attained by closely controlling chemical composition, which differs between commercial- and drawing-quality hot-rolled steel. Higher strength levels (when necessary for the part being formed) are obtained by alloy additions and processing controls to develop improved structure. Because higher strength is associated with forming problems such as lower ductility, increased springback, and longitudinal bend failures, only high-strength low-alloy steels designed for improved formability should be used in structural parts made by press operations. Mechanical properties of several types of hot-rolled sheet are given in Table 1.

Cold-rolled steel sheet for forming is produced by the cold reduction of hot-rolled pickled coils, followed by annealing and possibly additional processing, such as temper rolling. Class 1 (E, exposed) should be ordered when a controlled surface finish is required. Class 2 (U, unexposed) is intended for applications in which surface appearance is not of primary importance. Both classes are available as commercial-quality, drawing-quality, or drawing-quality special-killed cold-rolled steel. Mechanical properties of cold-rolled steel sheet are given in Table 2.

Most cold-rolled steels exhibit yield point elongation in the as-annealed condition. This appears as Lüders lines, or stretcher strains, on the surface of formed parts (for example, flat areas near the corners of pan-shape draws) that have been subjected to moderate forming operations. The yield point elongation may be removed by temper rolling the annealed coils. Because temper rolling strengthens the steel and reduces its ductility, it is usually limited to 0.5 to 1.5% elongation of the strip. Temper rolling under tension is more effective than flex rolling or roller leveling for eliminating yield point elongation because the steel is more uniformly strained through the thickness. These latter methods are sometimes used in the plants of fabrica-

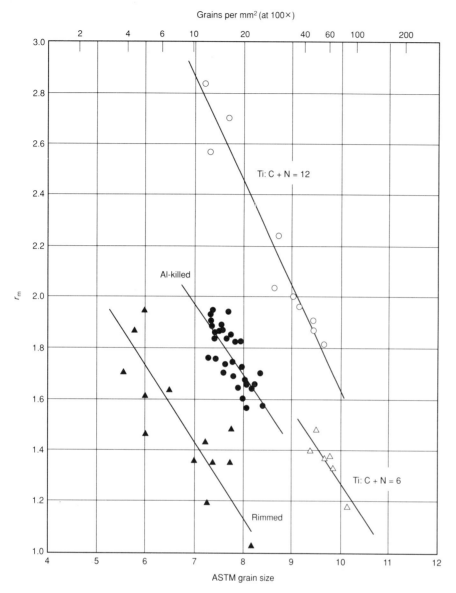

Fig. 6 Variation in r_m with grain size for four low-carbon sheet steels. Steels were cold reduced 70% and annealed. Source: Ref 4

tors because the equipment is less expensive and because it permits the use of aged coils of rimmed steels that may show strain on the surface of formed parts.

In addition to many as-processed surface finishes, cold-rolled sheet may be ordered with a metallic coating that provides corrosion protection or a decorative finish that reduces the manufacturing costs of parts such as appliances or building panels.

Rimmed steels are available as both hot-rolled and cold-rolled products. The rimming action caused by outgassing during solidification produces a relatively pure iron layer on the surface of the ingot. Thus, rimmed steel generally has a better surface finish than killed steel. After the annealing treatments used to regain ductility in the product following cold reduction to final thickness, rimmed steels must be temper rolled to prevent the formation of Lüders lines during forming.

The two available quality levels of rimmed steel are achieved by controlling chemical composition and annealing practice. Commercial quality is standard, whereas drawing quality is produced under stricter tolerance levels for impurities and is given a longer anneal to ensure uniformity throughout the coil, as well as good formability. Rimmed steels are more suited to stretch-type deformation than to deep drawing, for which aluminum-killed steels are generally recommended. Rimmed steels will age after a period of time following temper rolling. Consequently, there is a time limit on any performance guarantee on drawing-quality rimmed steel.

Aluminum-killed steels are deoxidized with aluminum and, possibly, with silicon. As already mentioned, use of aluminum results in a very clean steel, known as aluminum-killed or drawing-quality special-killed steel. Exceptional resistance to thinning through the sheet thickness (as measured by the plastic strain ratio, r) can be developed through the controlled processing of these steels. Because the pure iron skin characteristic of rimmed steel does not exist in aluminum-killed steel, surface imperfections may occasionally be encountered on aluminum-killed sheet. Both class 1 and class 2 drawing-quality aluminum-killed steels are produced. It should be noted that some aluminum-killed steels that cannot meet the formability requirements for drawing-quality sheet are sold as commercial-quality steel.

Interstitial-free steel is vacuum degassed to reduce the amounts of the interstitial elements carbon, nitrogen, and oxygen. It is usually processed to achieve high values of r_m (~2.00). This type of steel is not subject to strain aging at any stage of processing or manufacture; it exhibits no yield point elongation. Interstitial-free steel can withstand deeper draws with less breakage than other grades of steel sheet, and coated products made from it generally retain excellent formability.

Surface finish may be specified for cold-rolled steel sheet. The need for uniformity among parts that must have matching surface finishes (such as automobile fenders and hoods), even when made from different materials, often dictates the sheet finish. A surface roughness of 0.8 to 1.5 μm (30 to 60 μin.) for average peak height and two to six peaks per millimeter (50 to 150 peaks per inch) is considered standard for cold-rolled steel sheet. The surface finish is determined by the finish applied to the cold-mill rolls and the temper-mill rolls. Roll finishes are obtained by shot blasting or electroetching a ground roll surface so that the roll is roughened sufficiently to transfer the pattern to the sheet. As these rolls are used, their finish tends to become smoother; there may be a consequent change in appearance among coils, and press performance may vary slightly. A rougher sheet surface tends to hold lubricant better and resists galling and cold welding to die surfaces during forming. For parts requiring little forming, a smoother and often preferred finish can be attained when roughness is minimized.

Correlation Between Microstructure and Formability

The formability of steel sheet is related to various microstructural features of the sheet. For example, grain size and shape, grain orientation relative to the rolling direction, and the various microconstituents present in the steel are reflected in its forming behavior.

Grain size of steel sheet influences formability in two opposing ways. Petch (Ref 3)

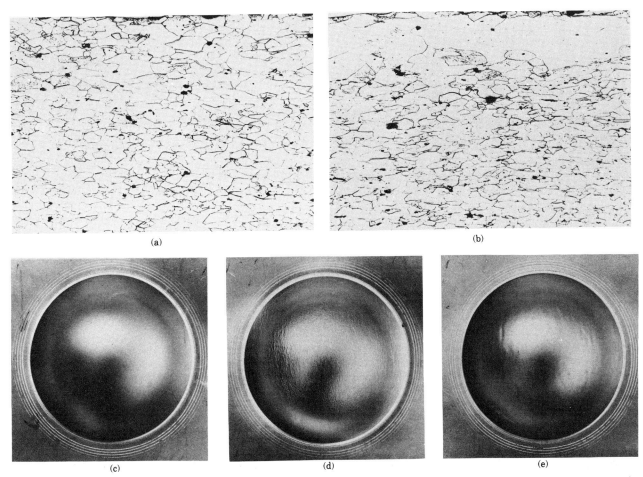

Fig. 7 Forming behavior of decarburized rimmed steel sheet (1.12 mm, or 0.044 in., thick) containing normal grain size distribution and abnormally large surface grains, which resulted from a change from normal manufacturing practice. (a) Cross section of test cup made from normally manufactured steel sheet. Grain size ASTM 6 throughout. (b) Cross section of test cup made from steel sheet containing abnormally large surface grains. Grain size ASTM 3 at one surface and ASTM 7 elsewhere. (a) and (b) both 100×, 3% nital etch. (c) Outside surface of test cup made from normally manufactured steel sheet. (d) Outside surface of test cup made from steel sheet having abnormally large grains on outside surface of cup. Note pronounced orange peel effect. (e) Outside surface of test cup made from steel sheet having abnormally large grains on inside surface of cup. (c), (d), and (e) all ⅔×

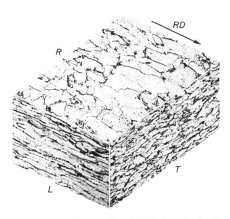

Fig. 8 Low-carbon steel, cold rolled 65%, showing the grain structure in the rolling plane (R), the longitudinal plane (L), and the transverse plane (T). RD, rolling direction

has shown that the yield strength of low-carbon steel varies inversely with the square root of the grain diameter. Fine-grained steels are quite strong, but they have low strain-hardening exponents and limited formability. Blickwede (Ref 4) has shown (Fig. 6) that r_m decreases as grain size decreases. Coarse-grained steels have better formability, but the roughened surface (called orange peel) that results from stretching steel with grain sizes that are below ASTM 5 is unacceptable for many applications. Grain sizes of ASTM 7 or 8 are usually a good compromise between formability and surface appearance. High-strength low-alloy steels, however, are usually produced with extremely small grain sizes (as small as ASTM 12) to increase both strength and toughness.

Figure 7 shows the effect of abnormally large surface grains on surface appearance after forming. It should be noted that the effect of the large surface grains is visible on the opposite surface of the sheet.

Grain shape of the ferrite can also affect sheet formability. Rimmed and hot-rolled aluminum-killed steels generally have equiaxed grains. Cold-rolled aluminum-killed steels, when properly processed, generally exhibit pancake-shape ferrite grains (Fig. 8). This grain shape is associated with the preferred grain orientation that is responsible for the excellent formability of aluminum-killed steels.

The microconstituents that are found in low-carbon steel at room temperatures include iron carbides and various nonmetallic inclusions. The most common inclusions are sulfides, silicates, and oxides. Aluminum-killed steels will also contain submicroscopic particles of aluminum nitrides.

These microconstituents can affect the formability of steel sheet by altering its strength. Alloying elements that dissolve in ferrite strengthen the steel appreciably, thereby reducing its formability. Nonmetallic inclusions may form a distinctive pattern of stringers that reflects the processing history from ingot to sheet. These elongated particles affect the formability of sheet primarily because they encourage cracking at the edge of a part during forming.

Effect of Metallic Coatings on Formability

Low-carbon steels are often coated with zinc, aluminum, or terne to improve their resistance to corrosion. The coating may be applied by submerging the steel in a container filled with the molten metal (hot-dipped) or by electroplating. The formability of these coated steels is the same as the formability of the base metal to which they are applied, modified by the frictional effects and handling problems imposed by the coating.

The base metal for electrogalvanized and terne-coated steels is normally processed in the same manner as uncoated steel base metal, and the mechanical properties are the same. However, zinc and aluminum hot-dipped coated steels are processed in continuous annealing lines, and this treatment affects the mechanical properties and the formability of the base metal. Conventional commercial-quality (CQ), drawing-quality (DQ), and drawing-quality special-killed (DQSK) hot-dipped coated steels will exhibit higher yield and tensile strengths, higher hardness, lower elongation, and lower r values than uncoated low-carbon steels, and they will generally have poorer formability. Decarburized hot-dipped coated steels (sometimes referred to as IF or DDQSK grades) are not similarly affected by the continuous annealing process, and these grades usually have somewhat better mechanical properties and formability than uncoated low-carbon steels.

The coatings can be slippery or abrasive. A terne coating is often applied on difficult parts because of its excellent lubricity. Because many hot-dipped zinc coatings are slippery, they may require slightly more blank holder pressure during press forming than do uncoated steels. Aluminum coatings and electrogalvanized coatings tend to develop higher friction during forming and thus generally require better lubrication than uncoated steels.

All steel coatings are softer than the steel base metal and can be scraped or gouged off the base metal surface with sharp burrs on blank edges or rough areas in the processing equipment. Coating that is scraped off tends to build up on the tooling (flaking), thereby producing a poor surface on the formed part. The solution is to maintain sharp cutting edges on blanking tools to minimize burr height, as well as a polished surface in die contact areas. Flaking can also be controlled with improved lubrication practice.

Gray cast iron dies, in particular, cause flaking problems. Ideally, cast tooling that is to be used to press form a coated steel should be made of cast steel or nodular cast iron. However, flaking problems in gray cast iron tooling can be eliminated by chrome plating or ion nitriding the tool contact surfaces.

Selection of Steel Sheet

Steel sheet selection should be based on an understanding of available grades of sheet and forming requirements. Other factors that should be considered when selecting a material for forming into a particular part include:

- Purpose of the part and its service requirements
- Thickness of the sheet metal and allowable tolerances
- Size and shape of blanks for the forming operation
- Equipment available for forming
- Quantities required
- Available handling equipment for sheets or coils
- Local availability of sheet products
- Surface characteristics of the steel sheet
- Special finishes or coatings for appearance or for corrosion resistance
- Aging propensity and its relation to time before use
- Strength of the steel sheets as-delivered
- Strength requirements in the formed part

Because these factors may be interdependent and large quantities are generally used in part manufacture, it is often desirable that steel selection be made after consultation with either the technical representatives of suppliers or the steel producer. Some parts require specialized low-carbon steel that has been processed to enhance a given mechanical property. These are other less critical formed parts that can use a wide selection of both hot-rolled and cold-rolled steel sheet. The user and producer should understand not only how steels are produced but what a steel mill can do to obtain specific properties in order to prevent the purchase of a steel possessing unwanted properties. The user should be aware of special steels that, although more costly, may reduce production costs and forming problems, resulting in per-part savings.

Low-carbon steels, coated and uncoated, are generally supplied as commercial-quality, drawing-quality, and drawing-quality special-killed grades. Some steel mills also offer specialized grades, such as interstitial-free deep-drawing steels and enameling steels. Some of the forming characteristics of the more commonly used formable grades are:

- *Commercial quality*: Available in hot-rolled, cold-rolled, and coated grades. The least expensive grade of sheet steel. Subject to aging (mechanical properties may deteriorate with time). Not intended for difficult-to-form shapes
- *Drawing quality*: Available in hot-rolled, cold-rolled, and coated grades. Exhibits better ductility than CQ grade steels, but has low r values. Subject to aging (mechanical properties may deteriorate with time). Has excellent base metal surface quality
- *Drawing-quality special-killed*: Available in hot-rolled, cold-rolled, and coated grades, with good forming capabilities. Not subject to aging (mechanical properties do not change with time)
- *Interstitial-free steels*: Available in cold-rolled and coated grades, with excellent forming capabilities. Not subject to aging (mechanical properties do not change with time)
- *Enameling steels*: Available in cold-rolled grades. Various types of processing are used to make a product that is satisfactory for porcelain enameling. All grades have good forming capabilities
- *Higher-strength steel sheets*: Available in hot-rolled, cold-rolled, and coated grades. Various types of processing are used to obtain the desired strength levels. In general, the formability of these grades decreases as yield strength increases. Springback may be a problem at lower sheet thicknesses

Steel sheet selection can be assisted by circle grid analysis, which provides a reliable description of the strain condition in press-formed shapes and which indicates whether the steel is capable of making the required shape or whether a more formable grade is required. Also, if a circle grid analysis shows severe strains with good-quality sheet steel that has normal mechanical properties, this is a strong indication that some modifications will have to be made in the forming process if production is to be maintained. By accurately identifying the problem areas in sheet metal forming operations, circle grid analysis can produce significant savings for both the manufacturers and the users of steel sheet.

REFERENCES

1. S.P. Keeler, Understanding Sheet Metal Formability, *Machinery*, Vol 74 (No. 6-11), Feb-July 1968
2. S.P. Keeler and W.G. Brazier, Relationship Between Laboratory Material Characterization and Press-Shop Formability, in *Microalloying 75*, Union Carbide Corporation, 1977, p 517-528
3. N.J. Petch, *The Ductile-Cleavage Transition in Alpha-Iron, Fracture*, B.L. Averback et al., Ed., Technology Press, 1959
4. D.J. Blickwede, Micrometallurgy by the Millions, *Trans. ASM*, Vol 61, 1968, p 653-679

SELECTED REFERENCES

- W.F. Hosford and R.M. Caddell, *Metal Forming—Mechanics and Metallurgy*, Prentice-Hall, 1983
- G. Sachs and H.E. Voegeli, Principles and Methods of Sheet Metal Fabricating, Reinhold Publishing, 1966
- "Sheet Steel Formability," Committee of Sheet Steel Producers, American Iron and Steel Institute, 1984

Bulk Formability of Steels

BULK FORMABILITY, also known as workability, refers to the relative ease with which a metal can be shaped through deformation processes such as forging, extrusion, or rolling. Bulk formability is related to sheet formability in only the broadest sense, in that both characteristics provide quantitative estimates of the strength and ductility of a metal. The latter property—ductility, or the resistance of the material to failure—is usually of primary concern in describing both bulk and sheet formability.

Formability Characteristics

Bulk Versus Sheet Formability. To clarify the distinction between bulk formability and sheet formability, it may be useful to compare and contrast the types of deformation that occur during typical bulk and sheet forming processes.

In both processes, the surfaces of the deforming metal are in contact with forming tools, and friction may have a major influence on material flow. In bulk forming, the surface-to-volume ratio of the formed part increases considerably under the action of largely compressive stresses. Plastic deformation is much more prevalent than elastic deformation; therefore, elastic recovery after deformation is negligible. Important material characteristics include flow stress, failure behavior, and the metallurgical transformations that characterize the alloy system in question.

In sheet forming operations, the metal is plastically deformed by tensile loads, often without significant changes in sheet thickness or surface characteristics. The magnitudes of plastic and elastic deformation may be similar, resulting in a significant amount of elastic recovery or springback. A key characteristic of a material is the plastic-strain ratio r, the resistance of the sheet to thinning during deformation.

The major emphasis in determining both the bulk formability and sheet formability of materials is on measuring and predicting the limits of deformation before fracture. A useful tool for graphically depicting the bulk formability of a material is the workability diagram, which indicates the locus of normal free-surface strains that result in fracture. The analogous concept for sheet formability is the forming limit diagram.

Tests for Bulk Formability. Both of these graphic depictions of formability rely on data gathered from laboratory formability tests. A wide variety of tests are used to determine bulk formability, ranging from general tests (tension and torsion tests, for example) to specialized tests that have a very narrow scope and range of application. Test procedures commonly used for determining the bulk formability of steels are covered in the section "Formability Tests" in this article.

Bulk Formability of Carbon and Alloy Steels. Despite the large number of available compositions, all the materials in this category exhibit essentially similar bulk formability characteristics. Exceptions to this are steels containing free-machining additives such as sulfides; these materials are not as receptive to bulk forming as nonfree-machining grades (Fig. 1).

Generally, the bulk formability of carbon and alloy steels improves as the deformation rate increases. The improvement has been primarily attributed to the increased heat of deformation generated at high deformation rates.

Because steels are the most commonly forged materials, a particularly important aspect of their bulk formability is forgeability, the ability to flow readily and fill forging die recesses without fracturing. An important measure of forgeability is flow stress, the amount of force required to deform the material at a specific temperature and strain rate. The section "Evaluating Forgeability" in this article includes information on measuring flow stress.

This article will present procedures for various formability tests used for carbon and alloy steels. The metallurgy and thermomechanical processing of high-strength low-alloy (microalloyed) steels will also be discussed.

Formability Tests

Tests for bulk formability can be divided into two main categories: primary tests and specialized tests. Primary tests such as tension and torsion tests have somewhat limited utility; specialized tests more closely simulate the deformation experienced in actual bulk forming processes and may give better indications of formability.

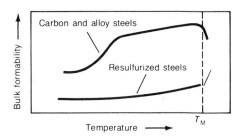

Fig. 1 Comparison of the bulk formability of carbon and low-alloy steels with the formability of resulfurized grades. T_M is the absolute melting temperature of the alloys. Source: Ref 1

Primary Tests

The primary tests for workability are those for which the stress state is well known and controlled. Generally, these are small-scale laboratory simulation tests.

The tension test is widely used to determine the mechanical properties of a material (Ref 1). Uniform elongation, total elongation, and reduction in area at fracture are frequently used as indices of ductility. However, the extent of deformation possible in a tension test is limited by the formation of a necked region in the tension specimen. This introduces a triaxial tensile stress state and leads to fracture. For carbon and alloy steels, tension tests are primarily used under special high strain rate, hot tension test conditions to establish the range of hot-working temperatures. The principal advantage of hot tension testing for carbon and alloy steels is that minimum and maximum hot-working temperatures are clearly established.

Most commercial hot tensile testing is done with a Gleeble unit, which is a high strain rate, high-temperature testing machine (Ref 2). A solid buttonhead specimen that has a reduced diameter of 6.35 mm (0.250 in.) and an overall length of 88.9 mm (3.5 in.) is held horizontally by water-cooled copper jaws (grips), through which electric power is introduced to resistance heat the test specimen (Fig. 2). Specimen temperature is monitored by a thermocouple welded to the specimen surface at the middle of its length. The thermocouple, with a function generator, controls the heat fed into the specimen according to a programmed cycle. Therefore, a specimen can be tested under

Fig. 2 Gleeble test unit used for hot tension and compression testing. (a) Specimen in grips showing attached thermocouple wires and linear variable differential transformer (LVDT) for measuring strain. (b) Close-up of a test specimen. Courtesy of Duffers Scientific, Inc.

time and temperature conditions that simulate hot-working sequences.

The specimen is loaded by a pneumatic-hydraulic system. The load can be applied at any time in the thermal cycle. Temperature, load, and crosshead displacement are measured as a function of time.

The percent reduction in area is the primary result obtained from the hot tension test. This measure of ductility is used to assess the ability of the material to withstand crack propagation. Reduction in area adequately detects small ductility variations in materials caused by composition or processing when the material is of low-to-moderate ductility. It does not reveal small ductility variations in materials of very high ductility.

In the torsion test, deformation is caused by pure shear, and large strains can be achieved without the limitations imposed by necking (Ref 3). Because the strain rate is proportional to rotational speed, high strain rates are easily obtained. Moreover, friction has no effect on the test, as it does in compression testing. The stress state in torsion may represent the typical stress in metalworking processes, but deformation in the torsion test is not an accurate simulation of metalworking processes, because of excessive material reorientation at large strains.

Fracture data from torsion tests are usually reported in terms of the number of twists to failure or the surface fracture strain to failure. Figure 3 shows the relative hot workability of two AISI carbon and alloy steels as indicated by the torsion test. The test identifies the optimal hot-working temperature for each of the two steels. The section "Evaluating Forgeability" in this article contains information on the use of torsion testing to evaluate the forgeability of carbon and alloy steels.

The compression test, in which a cylindrical specimen is upset into a flat pancake, is usually considered to be a standard bulk formability test. The average stress state during testing is similar to that in many bulk deformation processes, without introducing the problems of necking (in tension) or material reorientation (in torsion). Therefore, a large amount of deformation can be achieved before fracture occurs. The stress state can be varied over wide limits by controlling the barreling of the specimen through variations in geometry and by reducing friction between the specimen ends and the anvil with lubricants.

Compression testing has developed into a highly sophisticated test for formability in cold upset forging, and it is a common quality control test in the hot forging of carbon and alloy steels. Compression forging is a useful method of assessing the frictional conditions in hot working. The principal disadvantage of the compression test is that tests at a constant, true strain rate require special equipment.

Ductility Testing. The basic hot ductility test consists of compressing a series of cylindrical or square specimens to various thicknesses or to the same thickness with varying specimen length-to-diameter (length-to-width) ratios. The limit for compression without failure by radial or peripheral cracking is considered to be a measure of bulk formability. This type of test has been widely used in the forging industry. Longitudinal notches are sometimes machined into the specimens before compression. Because the notches apparently cause more severe stress concentrations, they enable the test to provide a more reliable index of the workability to be expected in a complex forging operation.

The bend test is useful for assessing the formability of thick steel sheet and plate. Generally, this test is most applicable to cold-working operations. The principal stress and strains developed during bending are defined in Fig. 4. The critical parameter is the width-to-thickness ratio, w/t. If $w/t > 8$, bending occurs under plane-strain conditions ($\epsilon_2 = 0$) and $\sigma_2/\sigma_1 = 0.5$. If $w/t > 8$, the bend ductility is independent of the exact w/t ratio. If $w/t < 8$, then the stress state and bend ductility depends strongly on the width-to-thickness ratio.

Specialized Tests

In the plane-strain compression test, the specimen is a thin plate or sheet that is compressed across the width of the strip by narrow platens that are wider than the strip. The elastic constraints of the undeformed shoulders of material on each side of the platens prevent extension of the strip in the width dimension; hence the term plane strain.

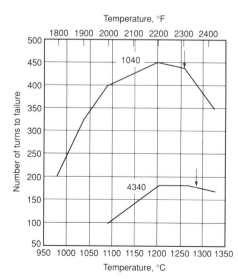

Fig. 3 Ductility of two AISI carbon and alloy steels determined in hot torsion tests. Arrows denote suitable hot-working temperatures.

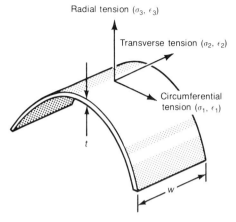

Fig. 4 Bend region defining the direction of principal stresses and strains in bend testing

Deformation occurs in the direction of platen motion and in the direction normal to the length of the platen. To ensure that lateral spread is negligible, the width of the strip should be at least six to ten times the breadth of the platens. To ensure that deformation beneath the platens is essentially homogeneous, the ratio of platen breadth to strip thickness, b/t, should be between 2 and 4 at all times. It may be necessary to change the platens during testing to maintain this condition. True strains of 2 can be achieved by carrying out the test in increments in order to provide good lubrication and to maintain the proper b/t ratio. Because of its geometry, this test is more applicable to rolling operations than to forging.

The partial-width indentation test is similar to the plane-strain compression test, but it does not subject the test specimen to true plane-strain conditions (Ref 4). In this test, a simple slab-shaped specimen is deformed over part of its width by two opposing rectangular anvils having widths smaller than that of the specimen. Upon penetrating the workpiece, the anvils longitudinally displace metal from the center, creating overhangs (ribs) that are subjected to secondary, nearly uniaxial tensile straining. The material ductility under these conditions is indicated by the reduction in the rib height at fracture. The test geometry has been standardized.

One advantage of this test is that it uses a specimen of simple shape; another is that as-cast materials can be readily tested. One edge of the specimen can contain original surface defects. The test can be conducted hot or cold.

The secondary-tension test is a modification of the partial-width indentation test. In this test, a hole or a slot is machined in the slab-type specimen adjacent to where the anvils indent the specimen. With this design, the ribs are sufficiently stretched to ensure fracture in even the most ductile materials. The fracture strain is based on reduction in area where the rib is cut out so that the fracture area can be photographed or traced on an optical comparator.

Ring Compression Test. When a flat ring-shaped specimen is upset in the axial direction, the resulting change in shape depends only on the amount of compression in the thickness direction and the frictional conditions at the die/ring interfaces. If the interfacial friction were zero, the ring would deform in the same manner as a solid disk, with each element flowing outward radially at a rate proportional to its distance from the center.

In the case of small, but finite, interfacial friction, the outside diameter is smaller than in the zero-friction case. If the friction exceeds a critical value, frictional resistance to outward flow becomes so high that some of the ring material flows inward to the center. Measurements of the inside diameters of compressed rings provide a particularly sensitive means of studying interfacial friction because the inside diameter increases if the friction is low and decreases if the friction is higher. The ring test, then, is a compression test with a built-in frictional measurement. Therefore, it is possible to measure the ring dimensions and compute both the friction value and the basic flow stress of the ring material at the strain under the given deformation conditions.

The ring compression test can be used to measure the flow stress under high-strain practical forming conditions. The only instrumentation required is that for measuring the force needed to produce the reduction in height.

Evaluating Forgeability

The hot forging of carbon and alloy steels into intricate shapes is rarely limited by forgeability aspects, with the exception of the free-machining grades mentioned earlier. Section thickness, shape complexity, and forging size are limited primarily by the cooling that occurs when the heated workpiece comes into contact with the cold dies. For this reason, equipment that has relatively short die contact times, such as hammers, is often preferred for forging intricate shapes in steel.

Because forging is a complex process, a single test cannot be relied on to determine forgeability. However, several testing techniques have been developed for predicting forgeability, depending on alloy type, microstructure, die geometry, and process variables. This section will summarize some of the common tests for determining formability in open-die and closed-die forgings.

Forgeability

Hot Twist Testing. One common means of measuring the forgeability of steels is the hot twist test. As the name implies, this test involves twisting heated bar specimens to fracture at a number of different temperatures selected to cover the possible hot-working temperature range of the test material. The number of twists to fracture and the torque required to maintain twisting at a constant rate are reported. The temperature at which the number of twists is the greatest, if such a maximum exists, is assumed to be the optimal hot-working temperature of the test material. Figure 5 shows the forgeabilities of several carbon steels as determined by hot twist testing. More information on the hot twist test is available in Ref 5 to 7.

Wedge-Forging Test. In this test, a wedge-shaped piece of metal is machined from a cast ingot or wrought billet and forged between flat, parallel dies (Fig. 6). The dimensions of the wedge must be selected so that a representative structure of the ingot is tested. Coarse-grain materials require larger specimens than fine-grain materials. The wedge-forging test is a gradient test in which the degree of deformation varies from a large amount at the thick end (h_2) to a small amount or no deformation at the thin end (h_1). The specimen should be used on the actual forging equipment in which production will occur to allow for the effects of deformation velocity and die chill on workability.

Tests can be made at a series of preheat temperatures, beginning at about nine-tenths of the solidus temperature or the incipient melting temperature. After testing at each temperature, the deformation that causes cracking can be established. In addition, the extent of recrystallization as a function of strain and temperature can be determined by performing metallographic examinations in the direction of the strain gradient.

The sidepressing test consists of compressing a cylindrical bar between flat, parallel dies where the axis of the cylinder is parallel to the surfaces of the dies. Because the cylinder is compressed on its side, this testing procedure is termed sidepressing. This test is sensitive to surface-related cracking and to the general unsoundness of the bar because high tensile stresses are created at the center of the cylinder.

For a cylindrical bar deformed against flat dies, the tensile stress is greatest at the start of deformation and decreases as the bar assumes more of a rectangular cross section. The degree of tensile stress can be reduced at the outset of the tests by changing from flat dies to curved dies that support the bar around part of its circumference.

The typical sidepressing test is conducted with unconstrained ends. In this case, failure occurs by ductile fracture on the expanding end faces. If the bar is constrained to deform in plane strain by preventing the ends from expanding, deformation will be in pure shear, and cracking will be less likely. Plain-strain conditions can be achieved if

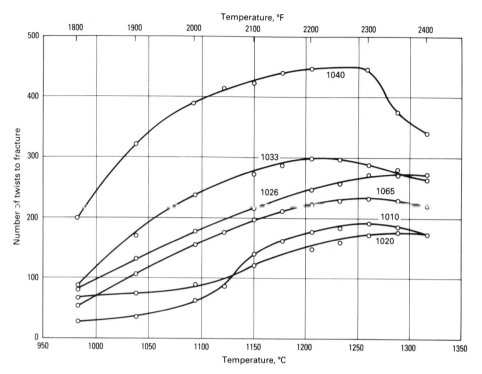

Fig. 5 Forgeabilities of various carbon steels as determined using hot twist testing. Source: Ref 5

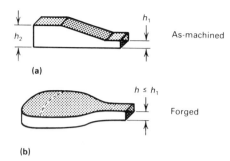

Fig. 6 Specimen for the wedge-forging test. (a) As-machined specimen. (b) Specimen after forging

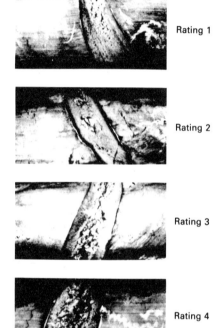

Fig. 7 Suggested rating system for notched-bar upset test specimens that exhibit progressively poorer forgeability. A rating of 0 indicates freedom from ruptures in the notched area.

the ends are blocked from longitudinal expansion by machining a channel or cavity into the lower die block.

The notched-bar upset test is similar to the conventional upset test, except that axial notches are machined into the test specimens. The notched-bar test is used with materials of marginal forgeability for which the standard upset test may indicate an erroneously high degree of workability. The introduction of notches produces high local stresses that induce fracture. The high levels of tensile stress in the test are believed to be more typical of those occurring in actual forging operations.

Test specimens are prepared by longitudinally quartering a forging billet, thus exposing center material along one corner of each test specimen. Notches with 1.0 or 0.25 mm (0.04 or 0.01 in.) radii are machined into the faces. A weld button is frequently placed on one corner to identify the center and surface material of alloys that are difficult to forge because of segregation.

Specimens are heated to predetermined temperatures and upset about 75%. The specimen is oriented with the grooves (notches) in the vertical direction. Because of the stress concentration effect, ruptures are most likely to occur in the notched areas. These ruptures can be classified according to the rating system shown in Fig. 7. A rating of 0 indicates that no ruptures are observed, and higher numbers indicate an increasing frequency and depth of rupture.

Truncated-Cone Indentation Test. This test involves the indentation of a cylindrical specimen by a conical tool. As a result of the indentation, cracking is made to occur beneath the surface of the testpiece at the tool/material interface. The reduction (measured at the specimen axis) at which cracking occurs can be used to compare the workability of different materials. Alternatively, the reduction (stroke) at which a fixed crack width is produced or the width of the crack at a given reduction can be used as a measure of workability.

The truncated cone was developed as a test that minimizes the effects of surface flaws and the variability they produce in workability (Ref 8). This test has been primarily used in cold forging.

Flow Localization

Complex forgings frequently develop regions of highly localized deformation. Shear bands may span the entire cross section of a forging and, in extreme cases, produce shear cracking. Flow localization can arise from constrained deformation due to die chill or high friction. However, flow localization can also occur in the absence of these effects if the metal undergoes flow softening or negative strain hardening.

Nonisothermal Upset Test. The simplest workability test for detecting the influence of heat transfer (die chilling) on flow localization is the nonisothermal upset test, in which the dies are much colder than the workpiece. Zones of flow localization must be made visible by sectioning and metallographic preparation.

The sidepressing test conducted in a nonisothermal manner can also be used to detect flow localization. Several test specimens are sidepressed between flat dies at several workpiece temperatures, die temperatures, and working speeds. The formation of shear bands is determined by metallography. Flow localization by shear band formation is more likely in the sidepressing test than in the upset test. This is due to the absence of a well-defined axisymmetric chill zone. In the sidepressing of round bars, the contact area starts out as 0 and builds up slowly with deformation. In addition, because the deformation is basically plane strain, surfaces of zero extension are present, along which block shearing can initiate and propagate. These are natural

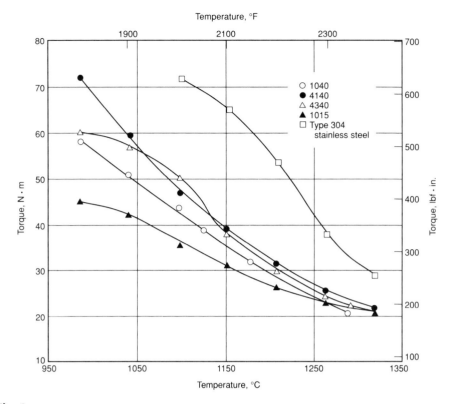

Fig. 8 Deformation resistance versus temperature for various carbon and alloy steels. Source: Ref 9

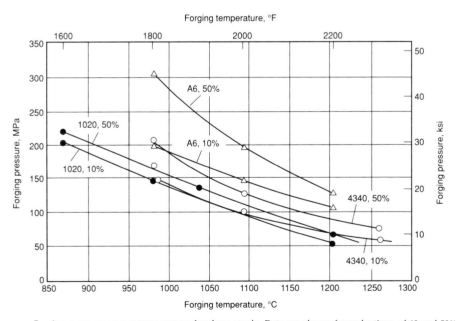

Fig. 9 Forging pressure versus temperature for three steels. Data are shown for reductions of 10 and 50%; strain rate was constant at 0.7 s^{-1}. Source: Ref 10

surfaces along which shear strain can concentrate into shear bands.

Flow Stress and Forging Pressure

Flow stresses and forging pressures can be obtained from torque curves generated in hot twist tests or from hot compression or tension testing. Figure 8 shows torque versus temperature curves for several carbon and alloy steels obtained from hot twist testing. These data show that the relative forging pressure requirements for this group of alloys do not vary widely at normal hot-forging temperatures. A curve for AISI type 304 stainless steel is included to illustrate the effect of higher alloy content on flow strength.

Figure 9 shows actual forging pressure measurements for 1020 and 4340 steels and AISI A6 tool steel for reductions of 10 and 50%. Forging pressures for 1020 and 4340 vary only slightly at identical temperatures and strain rates. Considerably greater pressures are required for the more highly alloyed A6 material, and this alloy also exhibits a more significant increase in forging pressure with increasing reduction.

Flow Stress in Compression. Ideally, the determination of flow stress in compression should be carried out under isothermal conditions (no die chilling) at a constant strain rate and with a minimum of friction in order to minimize barreling. These conditions can be met with conventional servohydraulic testing machines.

In flow stress determination, a specific load is applied to a cylindrical specimen of known height, the load is removed, and the new height is determined in order to calculate a true strain. Upon relubrication, the specimen is subjected to an increased load, unloaded, and measured. The cycle is then repeated.

Microalloyed Steels*

Microalloying—the use of small amounts of elements such as vanadium and niobium to strengthen steels—has been in practice since the 1960s to control the microstructure and properties of low-carbon steels. Most of the early developments were related to plate and sheet products in which microalloy precipitation, controlled rolling, and modern steelmaking technology were combined to increase strength significantly relative to that of other low-carbon steels.

The application of microalloying technology to forging steels has lagged behind that of flat-rolled products because of the different property requirements and thermomechanical processing of forging steels. Forging steels are commonly used in applications in which high strength, fatigue resistance, and wear resistance are required. These requirements are most often filled by medium-carbon steels. Thus, the development of microalloyed forging steels has centered on grades containing 0.30 to 0.50% C.

Regardless of product form (plate, bar, or forging), microalloyed steels are a classic example of a successful metallurgical innovation in which alloying additions and thermomechanical processing have been brought together effectively to attain improved combinations of engineering properties through microstructural control. This practice is relatively inexpensive because only small concentrations of the alloying elements (typically niobium, vanadium, or titanium) are needed to form carbides or carbonitrides. Where possible, the associated thermomechanical processing is intro-

---
*Portions of this section are adapted from S.S. Hansen, Microalloyed Plate and Bar Products: Production Technology, in *Fundamentals of Microalloying Forging Steels*, The Metallurgical Society, 1987, p 155-172

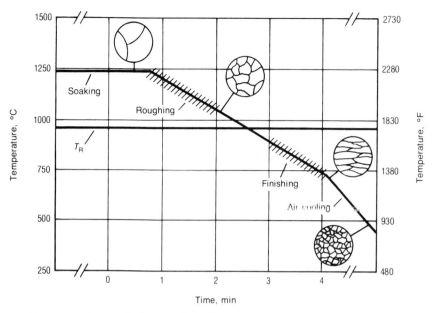

Fig. 10 Temperature-time profile for controlled rolling of 19 mm (¾ in.) thick microalloyed steel plate. T_R, recrystallization temperature

duced merely as a modification of the final hot-working operation. Although precipitation hardening with fine carbonitrides and substructural changes due to warm rolling of austenite-ferrite mixtures can contribute to the strengthening of microalloyed steels, the microstructural feature that ultimately provides a favorable balance of strength and toughness is a small ferritic grain size.

The following sections compare the processing of microalloyed plate and bar products. More information on the metallurgy and properties of microalloyed steel forgings is available in the article "High-Strength Low-Alloy Steel Forgings" in this Volume.

Processing of Microalloyed Plate Steels

Figure 10 shows a temperature-time profile for the rolling of 19 mm (¾ in.) microalloyed steel plate. Initially, slabs are reheated to temperatures in the range of 1100 to 1250 °C (2010 to 2280 °F). The rolling operation itself generally involves two distinct stages: high-temperature rolling or roughing and a lower-temperature series of deformation steps designated as finishing. If the roughing and finishing operations are continuous, the process is termed hot rolling, but if there is a delay between the two stages, as shown in Fig. 10, the process is referred to as controlled rolling. After rolling, the plate is usually air cooled, although a recently developed technology involves water spray cooling of the plate after finish rolling.

Overall, the plate-rolling operation lends itself to considerable control of the thermomechanical treatment. The slab reheat temperature can be reduced if desired. In fact, some rolling strategies involve only reheating to 960 °C (1760 °F) prior to rolling. Delays can be built into the rolling operation (although with some penalty in productivity), and a considerable range of finishing temperatures can be achieved. This operation can accommodate the most severe controlled-rolling schedules, including the deformation of austenite-ferrite mixtures.

In metallurgical terms, the controlled-rolling operation in microalloyed steels serves two purposes. The first is to refine the relatively coarse, as-reheated austenitic microstructure by a series of high-temperature rolling and recrystallization steps. The second purpose of the rolling operation is to impose a moderate-to-heavy reduction in a temperature regime where austenite recrystallization is inhibited between rolling passes (below the recrystallization temperature indicated in Fig. 10) such that the plastically deformed austenite grains remain pancaked. Subsequent transformation after rolling into ferritic microstructures results in the desired fine grain size and associated mechanical properties. The various parts of the rolling operation are discussed separately and in greater detail in the following sections of this article.

Thermomechanical Treatment (Rolling). Typically, the initial rolling passes are conducted at relatively high temperatures, just below the slab-reheating temperature. At these temperatures, each deformation step is usually followed by rapid recrystallization and grain growth. Recently, a thermomechanical processing procedure called recrystallization controlled rolling has been proposed. It combines repeated deformation and recrystallization steps with the addition of austenite grain-growth inhibitors such as titanium nitride to refine the starting austenitic grain size and to restrict grain growth after recrystallization. Such processing would obviate the need for low-temperature controlled rolling (Ref 11-13). However, even with optimum compositions and the adoption of rather difficult reduction schedules, there seems to be a limit to the degree of austenitic refinement that can be achieved by repeated recrystallization; the finest recrystallized austenitic grain sizes produced by this process are about 15 μm (600 μin.). Depending on the subsequent cooling rate, transformation can then result in a ferritic grain size of 6 to 8 μm (240 to 320 μin.) (Ref 13). This is useful degree of structural refinement and is appropriate in those cases where controlled rolling is not possible (for example, due to mill load constraints).

However, still finer grain sizes can be attained through the use of additional microalloying elements along with a controlled-rolling sequence in which austenite recrystallization is substantially retarded during the later rolling passes. This process develops a pancaked grain morphology and a much higher surface area per unit volume than are possible in recrystallized austenite (Ref 14). During this process, the austenite recrystallization and carbonitride precipitation reactions are coupled in the sense that each is greatly influenced by the other.

Transformation to Ferrite. The transformation of the austenite grains into ferritic microstructures determines the final grain size and associated mechanical properties of the microalloyed plate. The effects of austenitic morphology and the transformation temperature range (as governed by alloy content, rolling deformation, and cooling rate) are of the greatest importance. Even after the minimum austenitic grain thickness has been produced, the temperature range of the austenite-to-ferrite transformation must be controlled to determine the reaction kinetics. Increasing the ferritic nucleation rate and decreasing the ferritic growth rate can produce a finer ferritic grain size. These effects are generally achieved by alloying or controlled cooling.

Precipitation and Substructural Strengthening. Although grain refinement offers the best combination of strength and toughness, there is a practical limit to the yield strength level that can be achieved with this strengthening mechanism alone: about 450 MPa (65 ksi) for a grain size of 3 μm (120 μin.). For higher strength levels, additional strengthening mechanisms must be used; however, these mechanisms can have deleterious effects on toughness (Ref 15). For example, as vanadium is added to a controlled-rolled niobium-containing microalloyed plate steel, the yield strength is increased about 7 MPa (1 ksi) for every 0.01% increase in vanadium. This strengthening is due to the precipitation of vanadium-rich carbonitrides in the ferrite during air cooling of the plate after rolling. For a typical vanadium content of 0.10%, this precipita-

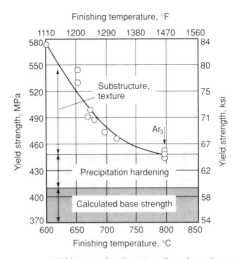

Fig. 11 Yield strength of a microalloyed steel as a function of finishing temperature. Grain size: 5 μm (200 μin.). Source: Ref 16

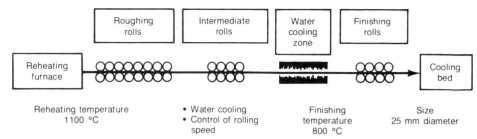

Fig. 12 Controlled-rolling process for microalloyed steel bar. Source: Ref 17

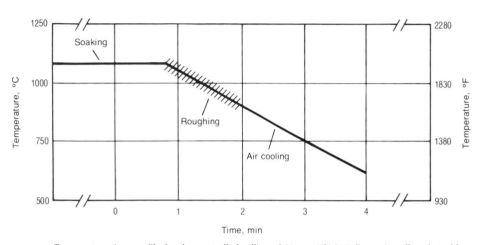

Fig. 13 Temperature-time profile for the controlled rolling of 44 mm (1¾ in.) diam microalloyed steel bar. Compare with Fig. 10.

tion strengthening can raise the yield strength to about 525 MPa (76 ksi). This results in the deformation of austenite-ferrite mixtures and the development of a warm-worked ferritic substructure. As shown in Fig. 11, strength increases progressively as rolling is continued to lower temperatures. At the same time, toughness is also reduced; therefore, this strengthening mechanism is used in microalloyed steels mainly as a way to reach the last increment of required strength.

High-Strength Low-Alloy (HSLA) Plate Products. The basic metallurgical principles discussed above are used to produce a range of as-rolled microalloyed plate steels at thicknesses to 102 mm (4 in.) and yield strengths as high as 550 MPa (80 ksi). In these steels, niobium is added to control the recrystallization temperature, manganese is often used to control austenite formation, and vanadium is introduced for precipitation strengthening. Finish rolling is carried out at temperatures low enough to produce substructural strengthening.

Processing of Microalloyed Bars

In contrast to the plate-rolling process, the thermomechanical treatment possible on a modern bar mill (Fig. 12) is somewhat limited in scope. For example, the temperature-time profile for the rolling of 44 mm (1¾ in.) diam bar shown in Fig. 13 can be compared to the plate-rolling profile shown in Fig. 10. There are clear differences between these rolling processes:

- Lower reheat temperatures, typically in the range of 1100 to 1200 °C (2010 to 2190 °F), are used in bar rolling. This lower temperature, in combination with the generally higher carbon levels in bar products, limits the amount of niobium that can be dissolved upon reheating. For example, in a 0.20% C steel, only about 0.01% Nb is soluble at 1100 °C (2010 °F). In contrast, vanadium is still readily soluble at bar reheat temperatures. Consequently, in HSLA bar grades, vanadium is the microalloying element commonly used to obtain the highest possible strength levels
- Even though the lower reheat temperatures typical of bar products place some limitations on the use of different microalloying elements, these lower temperatures do provide for a finer as-reheated austenitic grain size than is typical of slabs reheated for conversion to plate. With a small titanium addition and continuous casting, as-reheated austenitic grain sizes of 50 to 60 μm (0.0020 to 0.0024 in.) can be achieved in billets destined for bar
- Finishing temperatures in bar rolling are relatively high, even with the use of interstand cooling

Consequently, recrystallization controlled rolling becomes quite important in bar rolling, and the rolling strategy must be designed to produce the finest possible recrystallized austenitic grain size. Subsequent control of the austenite-to-ferrite transformation range is still important to maximize ferritic grain refinement. Nevertheless, as discussed earlier for plate, the ferritic grain size that can be produced on transformation from a recrystallized austenite is limited compared to the grain size that can be produced on transformation from austenitic grains that have been flattened by rolling below the recrystallization temperature. Thus, while moderate grain refinement can be achieved in an as-rolled microalloyed bar, this grain size will be somewhat coarser than the grain size of controlled-rolled HSLA plates.

Alternative Strengthening Mechanisms in Microalloyed Bar Steels. Because the degree of ferritic grain refinement possible in as-rolled microalloyed bar steels is somewhat limited, and because substructural strengthening is not possible, alternative strengthening mechanisms must be employed to reach yield strength levels comparable to those of plate grades. For example, in the alloy design of microalloyed bar steels, precipitation and pearlite strengthening must be relied on to a greater extent than in the design of plates. In view of the limited solubility of niobium or titanium at the reheat temperatures used in bar processing, vanadium is usually used to obtain the required level of precipitation strengthening in HSLA bar grades. Precipitation of V(C,N) during or after transformation can provide significant strengthening increments. In this regard, nitrogen level is also of importance. Judicious selection of both the vanadium and nitrogen levels is required to produce the desired level of precipitation strengthening. Similarly, an increase in the carbon content and thus the pearlite volume fraction of a bar steel can also be used to increase strength (Ref 18-20). In addition to moderate grain refinement, precipitation hardening with VN and an increase in the pearlite volume fraction can be used to produce yield strength levels up to 625 MPa (91 ksi) in microalloyed bars. Of course, these two strengthening mechanisms have very deleterious effects on toughness.

High-Strength Low-Alloy Bar Products. The alloy and process design principles discussed above are employed to produce a reasonable selection of as-rolled microalloyed bar steels. Yield strength levels up to 625 MPa (91 ksi) have been publicized, although the available yield strength level is influenced by bar thickness. In these steels, titanium is sometimes added to control austenitic grain growth (on reheating and after recrystallization during rolling), carbon and manganese can be balanced to control the transformation temperature range, vanadium and nitrogen are used for precipitation strengthening, and carbon is increased (as required) to raise the pearlite fraction of the ferrite and pearlite microstructural aggregate. Compared to the higher-carbon quenched and tempered grades that are currently used in competitive applications, these microalloyed bar steels offer comparable strength at lower cost because less alloy is required and heat-treating costs are eliminated. However, the toughness of the microalloyed bar grades developed to date is still somewhat lower than that of the higher-carbon, heat-treated steels. While the toughness levels currently available in commercial microalloyed bar steels may be adequate for many applications, considerable effort is being made at the present time to improve the toughness of microalloyed bar grades.

Comparison of Microalloyed Plate and Bar Products

The differing processing approaches for plate and bar are reflected in the microstructure and properties that are ultimately developed in these two product forms. Consider, as an example, the attributes of microalloyed plate and bar grades at the 550 MPa (80 ksi) yield strength level. The bar product (Fig. 14) has a coarser ferritic grain size and a significantly higher pearlite volume fraction (due to the higher carbon content) than the plate product. The comparison of compositions and yield-strengthening increments shown in Fig. 15 reflects these differences in microstructure. The most significant yield strength increment in the plate product is due to ferritic grain refinement, while the bar product must rely more on precipitation strengthening. In both products, however, small strengthening contributions are required beyond the grain size and precipitation hardening level to reach a yield strength of 550 MPa (80 ksi). In plate, this increment is obtained by rolling to develop a ferritic substructure, while in bar a small contribution from pearlite is necessary. Of course, these different strengthening mechanisms have an impact on the toughness that can be achieved in these two product forms (Fig. 16). Because it has a finer ferritic grain size and relies less on precipitation strengthening, the plate product exhibits significantly better toughness than the bar product.

Processing of Microalloyed Forging Steels

The driving force behind the development of microalloyed forging steels has been the need to reduce manufacturing costs. This is accomplished in these materials by means of a simplfied thermomechanical treatment (that is, a controlled cooling following hot forging) that achieves the desired properties without the separate quenching and tempering treatments required by conventional carbon and alloy steels.

Control of Properties. In order to realize the full strengthening potential of microalloying additions, it is necessary to use a soaking temperature prior to forging that is high enough to dissolve all vanadium-bearing precipitates. A soaking temperature above 1100 °C (2010 °F) is preferred. Rapid induction heating methods for bar and billet to conventional commercial forging temperatures of 1250 °C (2280 °F) are acceptable and allow sufficient time for the dissolution of the microalloying constituents.

Tensile strength decreases slightly as the finish forging temperature is reduced, but there is no significant effect on yield strength. Ductility and toughness show a significant increase with a reduction in finishing temperature; this is due to grain refinement of the austenite and increased ferrite content. Forgers are beginning to utilize this approach to enhancing the toughness of as-forged microalloyed steel; however, low finish forging temperatures are often avoided to minimize die wear. The specified properties of microalloyed forging steels can be achieved over a wide range of finishing temperatures.

One of the most important processing factors affecting the properties of as-forged microalloyed steels is the postforging cooling procedure. Increasing the cooling rate generally increases the yield and tensile strength because it enhances grain refinement and precipitation hardening. At high cooling rates, an optimum can be reached; above this rate the strength reduces due to the suppression of precipitation and the introduction of low-temperature transformation products.

The optimum cooling rate and maximum hardness are significantly influenced by the alloy and residual element content of the steel. Nevertheless, through control of the steel composition it is possible to ensure that the specified mechanical properties are achieved over a wide range of section sizes and cooling conditions.

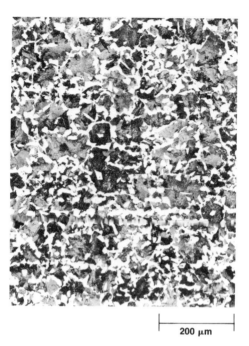

Fig. 14 Grain structure of a microalloyed steel bar product of composition Fe-0.38C-1.18Mn-0.16V-0.018N. Source: Ref 21

Properties of Forged Parts. Because of the improved strength and hardness of microalloyed bar steels, it is possible to use them for the production of many forged parts and eliminate the need for subsequent heat treatment. This is particularly true when air cooling is applied to the as-forged parts. Numerous forgings weighing as little as 1 kg (3 lb) to well over 11 kg (25 lb) have been produced by this approach. The parts produced from microalloyed steel include connecting rods and caps, stub yokes, weld yokes, wheel hubs, stabilizer bars, blower shafts, sucker rods, anchor bolts, and U-bolts.

An example of the improved properties that can be obtained is shown in Fig. 17, in which the cross-sectional hardness of an air-cooled microalloyed 1541 forged part is compared with a similar quenched and tempered 1043 part. The hardness is much more uniform in the microalloyed part, and as a result the fatigue life at this location in the part is an order of magnitude greater than that in the quenched and tempered part.

Effects of Hot Mill Finishing Temperature (Ref 23). In tests of a grade of 1541 steel microalloyed with 0.10% V, higher hardness and tensile strengths were obtained with higher hot mill finishing temperatures (Fig. 18). Reduction in area also increased slightly at higher finishing temperatures. Yield strength and percent elongation values did not vary with finishing temperature over the range investigated. The steel finished at the lowest temperature (970 °C, or 1780 °F) had the highest impact strength in subsequent Charpy V-notch testing (Fig. 19).

Effects of Forging Temperature. To achieve an optimum balance of strength and

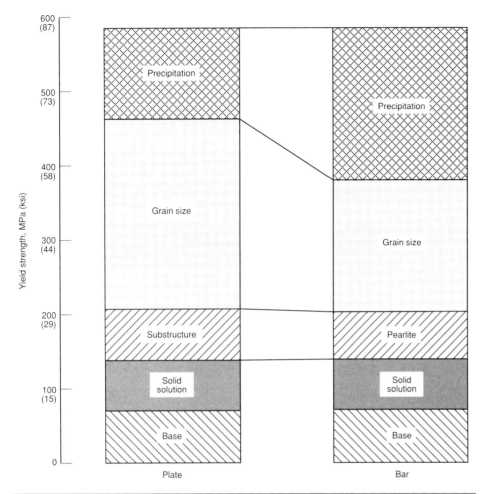

Product	Composition, %					
	C	Mn	Si	Nb	V	N
Plate	0.14	1.45	0.25	0.035	0.08	0.012
Bar	0.26	1.38	0.20	0.005	0.16	0.018

Fig. 15 Composition and yield strength increments of microalloyed plate and bar steels with yield strengths of 550 MPa (80 ksi)

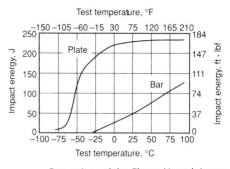

Fig. 16 Comparison of the Charpy V-notch impact toughness of microalloyed plate and bar steels with yield strengths of 550 MPa (80 ksi)

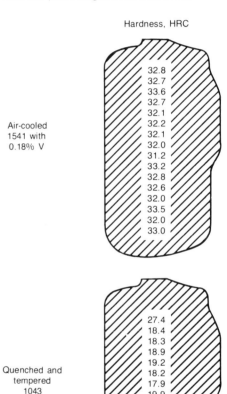

Fatigue test results		
Stress		
MPa	ksi	Cycles to failure
Air-cooled 1541 with 0.18% V		
520	75	219 700
550	80	113 200
Quenched and tempered 1043		
520	75	29 100
550	80	18 400

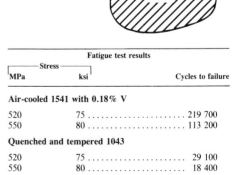

Fig. 17 Comparison of the cross-sectional Rockwell C hardness and fatigue strength of microalloyed steel forgings (air-cooled 1541 grade containing 0.18% V) and medium-carbon quenched and tempered steel forgings (1043 grade). Source: Ref 22

toughness properties, forged parts produced from microalloyed steel must be air-cooled through the transformation temperature. Slow cooling rates resulting from batch cooling must be avoided.

Because the forging process is the final thermal processing step in the production of parts from microalloyed chemistries, it is important that the forging operation be controlled in the same manner that the steelmaker controls the bar-rolling operation. The key forging variables that require process control are the reheating temperature and the postforging cooling rate (Ref 23).

REFERENCES

1. A.M. Sabroff, F.W. Boulger, and H.J. Henning, *Forging Materials and Practices*, Reinhold, 1968
2. E.F. Nippes, W.F. Savage, B.J. Bastian, and R.M. Curran, An Investigation of the Hot Ductility of High-Temperature Alloys, *Weld. J.*, Vol 34, April 1955, p 183-196s
3. M.J. Luton, Hot Torsion Testing, in *Workability Testing Techniques*, G.E. Dieter, Ed., American Society for Metals, 1984, p 95-133
4. S.M. Woodall and J.A. Schey, Development of New Workability Test Techniques, *J. Mech. Work. Technol.*, Vol 2, 1979, p 367-384
5. *Evaluating the Forgeability of Steel*, 4th ed., The Timken Company, 1974
6. H.K. Ihrig, The Effect of Various Elements on the Hot Workability of Steel, *Trans. AIME*, Vol 167, 1946, p 749-777
7. C.L. Clark and J.J. Russ, A Laboratory Evaluation of the Hot Working Characteristics of Metals, *Trans. AIME*, Vol 167, 1946, p 736-748
8. T. Okamoto, T. Fukuda, and H. Hagita, Material Fracture in Cold Forging—Systematic Classification of Working Methods and Types of Cracking in Cold Forging, *Sumitomo Search*,

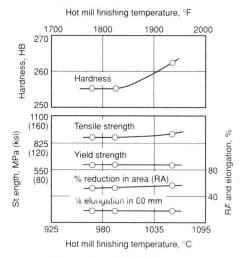

Fig. 18 Effect of hot mill finishing temperature on hardness and tensile properties of 1541 steel containing 0.10% V. Source: Ref 23

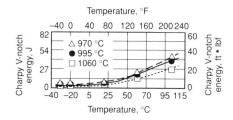

Fig. 19 Effect of hot mill finishing temperature on impact properties of 1541 steel containing 0.10% V. Source: Ref 23

No. 9, May 1973, p 46; *Source Book on Cold Forming*, American Society for Metals, 1975, p 216-226

9. C.T. Anderson, R.W. Kimball, and F.R. Cattoir, Effect of Various Elements on the Hot Working Characteristics and Physical Properties of Fe-C Alloys, *J. Met.*, Vol 5 (No. 4), April 1953, p 525-529

10. H.J. Henning, A.M. Sabroff, and F.W. Boulger, "A Study of Forging Variables," Technical Documentary Report ML-TDR-64-95, Battelle Memorial Institute, March 1964

11. W. Roberts, A. Sandberg, T. Siwecki, and T. Werlefors, Prediction of Microstructure Development During Recrystallization Hot Rolling of Ti-V Steels, in *HSLA Steels: Technology and Applications*, American Society for Metals, 1984, p 67

12. H. Sekine, T. Maruyama, H. Kageyama, and Y. Kawashima, in *Thermomechanical Processing of Microalloyed Austenite*, The Metallurgical Society, 1982, p 141

13. T. Siwecki, A. Sandberg, W. Roberts, and R. Lagneborg, in *Thermomechanical Processing of Microalloyed Austenite*, The Metallurgical Society, 1982, p 167

14. L.J. Cuddy, *Metall. Trans. A*, Vol 15A, 1984, p 87-98

15. F.B. Pickering, in *Microalloying '75*, Union Carbide Corporation, 1975, p 9

16. J.H. Little, J.A. Chapman, W.B. Morrison, and B. Mintz, *The Microstructure and Design of Alloys*, Vol 1, The Metals Society, 1974, p 80

17. T. Sampei, T. Abe, H. Osuzu, and I. Kosazu, Microalloyed Bar for Machine Structural Use, in *HSLA Steels: Technology and Applications*, American Society for Metals, 1984, p 1063-1070

18. H.J. Kouwenhoven, *Trans. ASM*, Vol 62, 1969, p 437-446

19. T. Gladman, I.D. McIvor, and F.B. Pickering, *J. Iron Steel Inst.*, Vol 210, 1972, p 916-930

20. F.B. Pickering, in *Hardenability Concepts with Applications to Steels*, The Metallurgical Society, 1978, p 179

21. B.L. Bramfitt, S.S. Hansen, D.P. Wirick, and W.B. Collins, Development of a Microalloyed Joint Bar, in *Microalloyed HSLA Steels*, ASM INTERNATIONAL, 1988, p 451-457

22. J.F. Held and B.A. Lauer, Development of Microalloyed Medium Carbon Hot Rolled Bar Products, in *HSLA Steels: Technology and Applications*, American Society for Metals, 1984, p 1071-1080

23. J.F. Held, Some Factors Influencing the Mechanical Properties of Microalloyed Steel, in *Fundamentals of Microalloying Forging Steels*, The Metallurgical Society, 1987, p 175-188

Machinability of Steels

Francis W. Boulger, Battelle-Columbus Laboratories (retired)

THE MACHINABILITY OF CARBON AND ALLOY STEELS is affected by many factors, such as the composition, microstructure, and strength level of the steel; the feeds, speeds, and depth of cut; and the choice of cutting fluid and cutting tool material. These machining characteristics, in turn, affect the cost of producing steel parts, particularly when the cost of machining represents a major part of the cost of the finished part. This article describes the influence of the various attributes of carbon and alloy steels on machining characteristics.

It should be recognized that the relative cost of cutting two steels in a particular operation, such as boring, is not necessarily the same as the relative ease of cutting the same two steels in another operation, such as broaching. Machining processes differ in operational metal removal characteristics; some place a greater premium on high machinability of the workpiece than others. Several common machining processes are listed in approximate decreasing order of machinability requirement, as follows:

- Internal broaching
- External broaching
- Tapping
- Generation of gear teeth
- Deep drilling
- Boring
- Screw machining with form tools
- High-speed light-feed screw machining
- Milling
- Shallow drilling
- Planing and shaping
- Turning with single-point tools
- Sawing
- Grinding

The designer's choice of part shape and dimensions largely determines the selection of the machining process. The mechanical properties needed for satisfactory service performance usually dictate the selection of the workpiece material and condition of heat treatment. Consequently, decisions about materials by the manufacturing engineer are generally reduced to choosing between similar grades of steel (for example, between 4140 and 8640) rather than between very different grades (for example, 4140 and 12L14).

Measures of Machinability

The term machinability is used to indicate the ease or difficulty with which a material can be machined to the size, shape, and desired surface finish. The terms machinability index and machinability rating are used as qualitative and relative measures of the machinability of a steel under specified conditions. There are no clear-cut or unambiguous meanings for these terms and no standard or universally accepted method of measuring machinability. Historically, machinability judgments have been based on one or more of the following criteria:

- *Tool life*: Measured by the amount of material that can be removed by a standard cutting tool under standard cutting conditions before tool performance becomes unacceptable or tool wear reaches a specified amount
- *Cutting speed*: Measured by the maximum speed at which a standard tool under standard conditions can continue to provide satisfactory performance for a specified period
- *Power consumption*: Measured by the power required to remove a unit volume of material under specified machining conditions
- Comparisons with a standard steel based on experience in machine shops
- Quality of surface finish
- Feeds resulting from a constant thrust force

Some of the test criteria are best suited to laboratory studies intended to elicit information about the effects of small changes in microstructure, composition, or processing history on machinability. Other types of tests are useful for studying the effects of geometry changes or cutting tool composition.

Years ago, machinability ratings were also relied upon as aids for choosing machining conditions to be used on materials unfamiliar to production personnel. This is rarely necessary now because detailed and reliable guides to suitable practices, such as the *Machining Data Handbook* (Ref 1), are readily available and widely used.

Tool life and cutting speed can be related by the equation:

$$V_c T^n = C_t \qquad \text{(Eq 1)}$$

where V_c is the cutting speed, T is the tool life, and n and C_t are empirical constants that reflect the cutting conditions under which the tests were made and the machinability of the material. In 1907, Taylor presented Eq 1 to describe single-point turning; the constant C_t is often called the Taylor constant. Because typical values of n for high-speed steel (HSS) tools range from 0.1 to 0.2, small variations in cutting speed are equivalent to enormous changes in tool life. Therefore, it is more practical to measure machinability as the cutting speed necessary to cause tool failure within a specified period, usually 60 min, than as tool life at a specified cutting speed. To determine the machinability of a particular steel, tool life for each of several cutting speeds (with standardized cutting conditions and tool shape) must be measured. Values of n and C_t can be determined from these data, and the cutting speed that corresponds exactly to the specified tool life can be calculated.

Tool life tests are used in laboratories to evaluate the effects of changes in tool materials, cutting variables, processing history, or workpiece compositions or microstructure on the ease of removing material. They are also useful for predicting tool life and choosing cutting speeds for industrial operations.

There are several criteria that can be used to define the failure of cutting tools. One criterion is the complete destruction of the cutting surface of the tool. A second criterion is the wear of the tool to the extent that the quality of the machined surface becomes unacceptable. Perhaps the most widely used criterion for tool failure is wear of the surface of the tool to some predetermined amount of flank wear. Sometimes, especially in screw machine tests, a specific increase in a part dimension is used to define tool life.

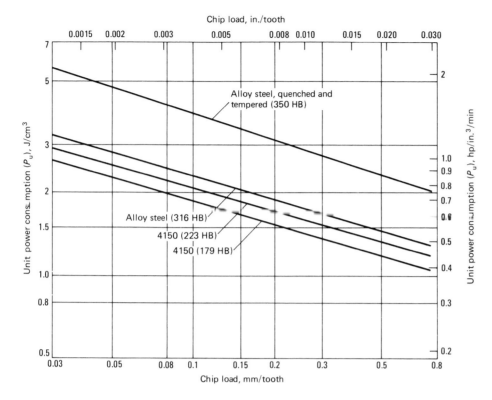

Fig. 1 Unit power consumption for surface broaching (HSS tools)

Regardless of the criterion adopted for tool failure, any machinability rating that depends on tool life measurements will be affected by the cutting tool. The choice of tool material, the configuration of the tool, the sharpness of the cutting edge, and the efficiency with which the tool is cooled can affect the machinability rating of the steel under test. For example, typical values of n for HSS tools range from 0.1 to 0.2; for carbide tools, typical values of n range from 0.2 to 0.4. Therefore, machinability testing should be carefully standardized (as described in Ref 2 and 3) so that the test reflects the machinability of the material rather than variations in the test procedures.

Power Consumption. The forces acting on a tool during cutting, as measured on a dynamometer, can be used to estimate the power consumed in metal cutting. The power consumption (expressed in watts) in cutting operations is approximately equal to the product of the cutting speed, V_c (expressed in units of meters per second), and the component of the cutting force parallel to the cutting direction, F_c (expressed in newtons). In English units, this relation for power consumption is:

$$P = \frac{V_c \times F_c}{33\,000} \quad \text{(Eq 2)}$$

where P is power consumption (at the spindle) in horsepower units, F_c is the cutting force in pounds, and V_c is the cutting speed in feet per minute.

To calculate the unit power consumption, which reflects the power requirements for cutting a given quantity of a particular material, it is necessary to divide the power consumption, P, by the metal removal rate, Q (which is typically expressed in units of either cubic centimeters or inches per minute). Therefore, the unit power consumption, P_u, is:

$$P_u = \frac{P}{Q} \quad \text{(Eq 3)}$$

In cutting operations, $Q \simeq d \cdot f \cdot V_c$, where d is the depth of cut and f is the feed rate. In metric units, the unit power consumption, P_u, is given by:

$$P_u = \frac{F_c}{fd} \quad \text{(Eq 4)}$$

where P_u is in joules per cubic centimeter, F_c is in newtons, and f and d are in millimeters. If F_c is given in pounds of force and f and d in inches, then P_u in horsepower per cubic inch per minute is given by:

$$P_u = \frac{2.525 \times 10^{-6} F_c}{fd} \quad \text{(Eq 4a)}$$

Typical data for unit power consumption of steels machined in different processes are given in Table 1 and Fig. 1. Note that the unit power consumption increases with increasing hardness, which reflects the resistance of the material to the deformation required in machining operations.

The ranges of values for unit power requirements for a particular hardness level cover the spread for sharp and dull tools. The energy is used in deforming metal in the chips and the surface layers of the workpiece and in overcoming friction. Figure 1 shows that decreasing the amount of metal removed by each tooth on a broach increases the unit power consumption, because the friction between the tool and chip or workpiece increases. The same effects occur when feeds are reduced in turning or milling operations. The choice of cutting tool shape and material and the application of coolants have comparatively little effect on the unit power consumption, except by altering the amount of power expended in friction. Unit power consumption cannot be easily correlated with tool life; the factors that affect the unit power consumption are primarily the inherent machinability of the material and the power consumed by friction, while the additional factors that affect tool life include the shape and material of the tool, the temperature of the interface between the tool and chip, and the extent of the abrasive action of the chip on the surface of the tool.

Quality of surface finish is another means of assessing the machinability of materials. Because the surface finish of a machined part may affect its performance in service, it is sometimes useful to rate the machining characteristics of candidate materials in terms of the surface finish that can be expected from machining under specified conditions. Such ratings are generally qualitative, although materials that have high machinability ratings, as determined by other rating methods, usually produce smooth

Table 1 Average unit power requirements for turning, drilling, and milling plain carbon and alloy steels

	Unit power, kW/cm³/min (hp/in.³/min)(a)		
Hardness, HRC	Turning with a feed of 0.12–0.50 mm/rev (0.005–0.020 in./rev)	Drilling with a feed of 0.05–0.20 mm/rev (0.002–0.008 in./rev)	Milling with a feed of 0.12–0.30 mm/tooth (0.005–0.012 in./tooth)
85–200 HB	0.050–0.064 (1.1–1.4)	0.046–0.059 (1.0–1.3)	0.050–0.064 (1.1–1.4)
35–40	0.064–0.077 (1.4–1.7)	0.064–0.077 (1.4–1.7)	0.068–0.086 (1.5–1.9)
40–50	0.068–0.086 (1.5–1.9)	0.077–0.096 (1.7–2.1)	0.082–0.100 (1.8–2.2)
50–55	0.091–0.114 (2.0–2.5)	0.096–0.118 (2.1–2.6)	0.096–0.118 (2.1–2.6)
55–58	0.155–0.191 (3.4–4.2)	0.118–0.146 (2.6–3.2)	0.118–0.146 (2.6–3.2)

(a) Power requirements at spindle-drive motor, corrected for 80% spindle-drive efficiency. Source: Ref 4

surfaces, partly because they are machined at high speeds.

Machinability Testing for Screw Machines. Although the principles described have been used for judging the machining characteristics of steels by short-time tests made on small amounts of material and for quantifying various characteristics, they are not useful for setting up screw machines. Therefore, engineers concerned with parts produced on automatic bar machines or with making steels for such applications have developed special procedures for determining machinability (Ref 5). Test parts are produced by simultaneous cutting with form tools and by finish-forming and cutoff operations. The conditions to be used and the records to be kept are described in ASTM E 618. Ratings for different lots are assigned on the basis of maximum production rates for parts meeting specified dimensional and surface roughness tolerances.

Scatter in Machinability Ratings

Considerable scatter in the machinability data for the steel chosen as the reference for machinability ratings, B1112, is illustrated in Fig. 2. Within the composition range permitted for that grade, it was found that unintentional variations in carbon, sulfur, and, principally, silicon contents cause the machinability index of B1112 to vary as much as 20% below or 60% above the nominal value of 100 assigned to it. The data on multiple heats of B1113 indicate a similarly large spread of values for that steel.

The effects of small variations in composition and grain size on machinability are sometimes greater than those from variations in hardness. Therefore, variations in performance should be expected when machining different lots of ostensibly similar material. The scatter in machinability data indicates that the precision of machinability ratings is limited, to some extent at least, by the concept of using the average behavior of a particular grade of steel as the standard for comparison.

In general, differences of 5% in machinability ratings are not likely to be significant or reproducible. It has been shown that such a scatter can result from variations in composition that meet the chemical ranges permitted by specifications for the grade (Ref 7).

Machinability Ratings of Steels

It is generally agreed that machinability, as defined by tool life, depends on or correlates with the following characteristics of the workpiece:

- Structure
- Chemical composition

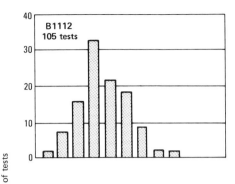
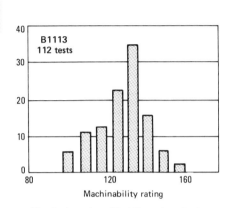

Fig. 2 Distribution of machinability ratings for B1112 and B1113 steels. Source: Ref 6

- Dispersion of second-phase particles
- Mechanical properties, such as strength and hardness
- Physical properties, such as thermal conductivity

Variations in these characteristics control the performance of workpieces machined under comparable conditions. Materials with superior machinability give better tool lives at equal cutting speeds or permit higher cutting speeds while maintaining equal tool lives. Adopting either alternative—that is, better tool lives or higher cutting speeds—improves productivity and lowers machining costs.

Most published machinability ratings have been based on the performance of steels in one type of operation (usually turning) and with one type of cutting tool. This is particularly true of data from laboratory tests. Consequently, it is of interest to determine whether data obtained with carbide or coated-carbide tools rank materials in the same order of machinability as ratings obtained with HSS tools. Similarly, it is important to determine whether steels exhibiting superiority in one type of operation, such as turning, will also perform better in drilling, boring, milling, or other operations.

In the following two sections, machinability rankings are compared for different cutting tool materials and different machining operations. In both sections, machinability is assessed in terms of cutting speeds. Assuming that the other machining parameters are comparable, the machinability of a material is reflected by its permissible cutting speed for a usefully long tool life. This provides an approach for determining the information mentioned above. Information on suitable cutting speeds, based on experience in many industrial shops, has been collected, evaluated, and published (Ref 1). Usually, the recommended cutting speed gave tool lives of about 2 h for HSS tools and about 1 h for carbide tools.

Order of Machinability Rankings With Different Cutting Tool Materials. The cutting speeds used for the cross plots in Fig. 3 are those recommended for turning metals, at cut depths of 1.0 mm (0.04 in.) and at appropriate feeds, with three types of cutting tools. The workpiece materials represented in Fig. 3 include 3 types of stainless steel, 14 grades of constructional steel, and 5 varieties of cast iron. Their hardnesses ranged from 100 to 325 HB. The plots show close and consistent relationships among the turning speeds recommended for the three types of cutting tools.

Table 2 lists the statistical attributes conventionally employed for judging the reliability of the correlations shown in Fig. 3. The goodness of fit between the points and the trend lines is indicated by the correlation coefficient or r value. For the two plots in Fig. 3, the r values are high enough to permit the conclusion that the relationships shown in the chart are statistically significant at the 99.9% level. The coefficients of determination (r^2 values) indicate the proportion of the total variation in the dependent variable explained by its relationship with the independent variable. High proportions of the variance among permissible cutting speeds (89 and 76%) were accounted for by the relationships between indexable carbides and the other two types of cutting tools. The coefficient of determination for the relationship between suggested cutting speeds for coated carbide and steel tools was smaller and less indicative of a strong correlation. The slope and intercept values in Table 2 describe the mean-square trend lines for the relationships when used in an equation with the form:

$$y = (\text{slope})x + (\text{intercept})$$

The correlation data confirm the opinion that workpiece characteristics that improve the ease of machining a metal will be effective in turning tests made with different types of cutting tools. The trend lines also indicate that changing to a better tool material does not raise the permissible cutting speed an equal or fixed amount for different metals.

Order of Machinability Rankings With Different Machining Operations. The cutting speeds prescribed for the same 22 types of ferrous metals were also used to obtain their machinability ratings in different types of

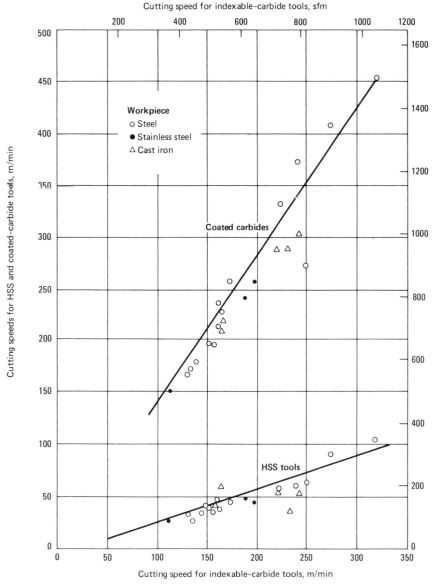

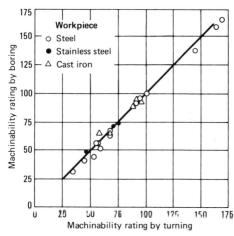

Fig. 4 Correlations among machinability ratings for different materials based on recommended speeds for turning and for boring with HSS tools. See text for details.

Turning	Boring	Turning	Boring
95	94	62	56
100	100	57	56
148	141	53	44
90	88	88	85
162	159	57	65
167	165	95	94
68	65	67	65
58	53	90	88
45	41	71	71
35	32	45	44
68	68	74	74

HSS		Indexable carbide		Coated carbide		HSS		Indexable carbide		Coated carbide	
m/min	sfm	m/min	sfm	m/min	sfm	m/min	sfm	m/min	sfm	m/min	sfm
60	200	240	790	365	1200	40	130	161	530	213	700
64	210	250	820	373	1225	37	120	155	510	229	750
90	300	275	900	411	1350	34	110	130	425	168	550
58	190	221	725	335	1100	56	185	221	725	290	950
106	350	320	1050	457	1500	37	120	236	775	290	950
41	135	150	500	198	650	60	200	165	540	213	700
35	115	142	465	180	600	43	140	165	540	213	700
27	90	134	440	175	575	58	190	245	800	305	1000
44	145	174	570	260	850	46	150	198	650	260	850
40	130	160	525	239	785	29	95	114	375	150	500
41	135	150	500	198	650	47	155	189	620	245	800

Fig. 3 Correlation among suggested cutting speeds for turning different ferrous metals with indexable-carbide, coated-carbide, and HSS tools. Cutting speeds are from Ref 1.

operations. The ratings were based on the speeds (in surface feet per minute) recommended for machining with HSS tools (Ref 1). The cutting speeds suggested for 1212 steel, at a hardness level of 150 to 200 HB, were used as the basis for comparison. The machinability rating of a material was taken to be 100 times the ratio of its recommended cutting speed to that for 1212 steel for otherwise comparable cutting conditions.

Figure 4 shows a crossplot of the machinability ratings calculated as above for boring and for turning with a cut depth of 1.0 mm (0.04 in.). The correlation chart shows that the agreement was very close between the two types of metal removal operations. Although not shown in the chart, the least-squares trend lines for machinability ratings based on rough reaming and those based on turning at a heavier cut depth (0.38 mm, or 0.15 in.) closely matched the one shown for boring; that is, they fell within the scatter band of points shown for boring.

Table 3 gives the correlation statistics comparing the machinability ratings calculated from speeds recommended for different types of operations. Ratings for the first three operations listed in Table 3 were closely related to those based on light turning cuts. They all had very high coefficients of correlation and determination, and the slopes and intercepts for the mean-square trend lines were similar. The correlation coefficients were similar for end milling and for face milling with turning. Nevertheless, they show that the data fit the trend lines quite well. The coefficients of determination indicate that over 73% of the variances were accounted for. On the other hand, the correlation between machinability ratings for drilling and turning was poor when the 22 types of materials were considered in the statistical analysis. Discrepancies of this type are often caused by differences in the ease with which chips can be removed from the scene of the action. In this case, excluding the data for the five cast irons from consideration resulted in a more meaningful

Table 2 Correlation data of recommended cutting speeds for turning operations with different types of cutting tool materials

The cutting speeds used for these correlation analyses are given in Fig. 3 and are those recommended in Ref 1 for turning 5 types of cast irons, 3 types of stainless steels, and 14 constructional steels.

Paired tool materials		Correlation statistic			
Dependent variable	Independent variable	Correlation coefficient	Coefficient of determination	Slope of trend line (Fig. 3)	Intercept on y-axis (Fig. 3)
HSS	Indexable carbide	0.8722	0.761	0.35	−34.9
Coated carbide	Indexable carbide	0.9427	0.890	1.41	−3.97
HSS	Coated carbide	0.8387	0.703	3.74	241

Table 3 Correlation data for the machinability ratings of different machining operations on ferrous materials

Correlation data are based on the machinability ratings(a) of 22 different types of ferrous materials.

Machining operation compared with turning at 1.0 mm (0.04 in.) depth of cut	Correlation statistic of the machinability ratings(a) for the machining operation in the left column and turning with a 1.0 mm (0.04 in.) depth of cut		Trend line	
	Correlation coefficient	Coefficient of determination	Slope	Intercept
Turning with a 3.8 mm (0.15 in.) depth of cut	0.9954	0.991	1.02	−2.26
Boring	0.9964	0.993	0.99	−1.13
Reaming	0.9681	0.973	1.08	−10.4
Face milling	0.8827	0.779	0.78	14.9
End milling	0.8583	0.737	0.67	22.5
Drilling(a)	0.6786	0.460	0.61	42.7
Drilling(b)	0.8623	0.740	0.56	37.5

(a) The ratings were calculated from speeds recommended in Ref 1 for machining 22 different types of material with HSS tools. Those speeds were compared with the recommendations for 1212 steel to be processed under similar conditions. The recommended cutting speed for 1212 steel, at 175 HB, was assigned a machinability rating of 100. (b) Ratings from 17 pairs of observations; ratings for cast iron not considered

correlation, as indicated in the last line in Table 3.

Venkatesh and Narayanan (Ref 8) used statistical methods to evaluate relationships among machinability ratings determined by drilling, turning, and milling processes. They found better correlations among ratings from different processes than among those determined with different tool materials.

Microstructure

In general terms, it is possible to differentiate between extremes of metallurgical structure that are easy or difficult to machine. However, when specific cases are encountered, such as differentiating between the machining characteristics of 5130 and 8630 in a particular operation, the presence of a certain microstructure is not, of itself, definitive for selecting one or the other of the two similar steels. This is particularly true when both steels can be heat treated to develop the same type of microstructure.

There are rough correlations among hardness, microstructure, and machinability. Experience teaches that, when machining high- or medium-carbon alloy steels such as 4140, the maximum tool life is obtained with workpieces in the annealed condition. Tool wear is accelerated by increases in hardness level. Based on many observations, machinability theory and practice indicate that the optimum conditions or microstructures for machining steels of different carbon contents are usually as follows:

Carbon, %	Optimum microstructure
0.06–0.20	As-rolled (most economical)
0.20–0.30	Less than 75 mm (3 in.) in diameter, normalized; 75 mm (3 in.) in diameter and over, as-rolled
0.30–0.40	Annealed to give coarse pearlite, minimum ferrite
0.40–0.60	Coarse lamellar pearlite to coarse spheroidite
0.60–1.00	100% spheroidite, coarse to fine

The above examples have only qualitative utility. An attempt to define the relative machinability of steels by a quantitative measure of microstructure has one additional disadvantage: Even assuming that an optimum structural combination could be predetermined and accurately measured, it would still be necessary to regularly achieve this desired combination on a production basis. This means that all parts would have to be heat treated accurately enough to produce an identical amount and type of microstructure in each. Using commercial steels and production heat treatment practices, such precision is usually impracticable.

Among normalized and annealed steels, those with lower hardness and smaller amounts of pearlite can be machined at higher speeds for equal tool lives. By assuming a direct relationship between carbon and pearlite contents, Kronenberg reported that the life of carbide cutting tools decreased as the carbon content of workpiece steels increased (Fig. 5). Araki and associates (Ref 9) made a similar analysis on data they obtained on 4135 steel that was heat treated to a variety of structures having different hardnesses. As illustrated in Fig. 6, the harder specimens caused tool failure in a shorter time than the softer specimens. These results are comparable to those reported by Armarego and Brown, shown in Fig. 7. An equation of the form:

$$V_c H^x = C \qquad (Eq\ 5)$$

where H is the Brinell hardness number and x and C are empirical constants, has been used by several investigators. Mayer and Lee (Ref 11) combined Eq 1 and 5 to obtain a relationship between relative tool life and hardness for different steels and tool materials, shown in Fig. 8.

The microconstituents frequently encountered in steels can be identified as contributing to or detracting from the machinability of the steel. Ferrite can be readily cut and causes little tool wear, but it also contributes to the formation of a built-up edge on the tool and a relatively poor surface finish on the workpiece. Spheroidized structures can behave similarly, but large quantities of massive carbide particles can cause significant wear on the tool. Pearlite is harder than ferrite and generally causes greater tool wear; the finer the pearlite plate spacing, the shorter the tool life. A built-up edge is less common when machining pearlite than when machining ferrite. Hard constituents, such as massive carbides or oxides, can be very abrasive to the cutting tool; such particles generally accelerate tool wear. Soft constituents, such as lead or manganese sulfide, generally improve the machinability of the steel.

As discussed above, there are correlations between machinability and several interrelated factors, such as composition, microstructure, and hardness; however, it is not clear that there is an exact causal relationship between any one of these factors and machinability. Furthermore, the effects on machining behavior of these factors acting together are generally less than those from differences in tool materials, tool configurations, or the choice of machining process.

Carbon Steels

Carbon steels nearly always have better machinability than alloy steels of comparable carbon content and hardness. Steels hardened and tempered to hardness levels greater than 300 HB are an exception to this observation; under such conditions, alloy steels have superior machinability, which is usually attributed to, first, the higher tempering temperature required to temper an alloy steel to a specified hardness level and, second, nonuniformity of microstructure due to limited hardenability in carbon steels.

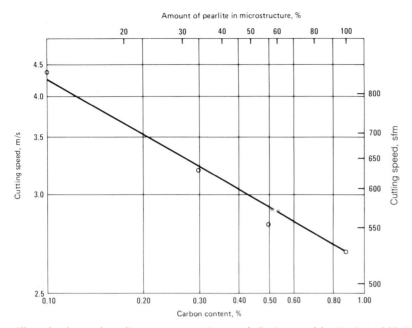

Fig. 5 Effect of carbon and pearlite content on cutting speed. Cutting speed for 60-min tool life in steels containing different amounts of carbon and pearlite; 0.65 mm² (0.001 in.²) cross-sectional cutting area; carbide tool

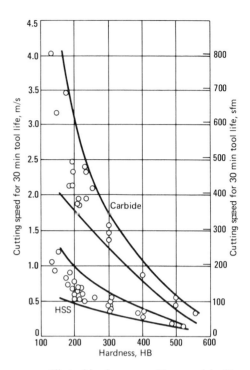

Fig. 7 Effect of hardness on cutting speed for 30-min tool life, using HSS and carbide tools. Source: Ref 10

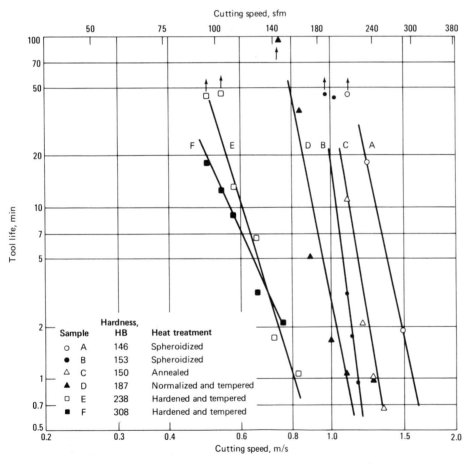

Fig. 6 Effect of hardness on tool life curves. Workpiece: 4135 steel. Tool material: cobalt-tungsten (10% Co, 10% W) high-speed steel per Japanese designation SKH57. Machining conditions: depth of cut = 2.0 mm (0.08 in.); feed rate = 0.2 mm/rev (0.008 in./rev). Source: Ref 9

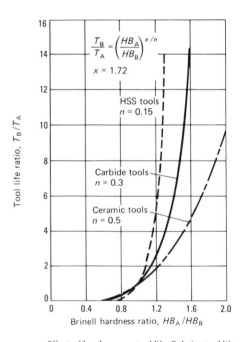

Fig. 8 Effect of hardness on tool life. Relative tool life as a function of relative hardness for three tool materials; the value of x, 1.72, used in constructing these curves is a conservatively estimated maximum. Source: Ref 11

Relative machinability ratings for some plain carbon steels are given in Table 4. Carbon content has a dominant effect on the machinability of carbon steels, chiefly because it governs strength, hardness, and ductility. Increasing the carbon content of steel increases its strength and the unit power consumption for cutting. Data in Fig. 9 show the effects of increasing carbon content and manganese content on unit power consumption.

Low-carbon steels containing less than 0.15% C are low in strength in the annealed condition; they machine poorly because they are soft and gummy and adhere to the cutting tools. The machinability of these

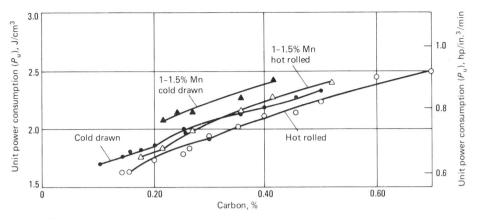

Fig. 9 Effect of carbon content on unit power consumption. Unit power consumption for hot-rolled and cold-drawn steels of two different manganese levels containing various amounts of carbon

Table 4 Machinability ratings of plain carbon steels
Machinability ratings are from the percentage of cutting speed for 1212 steel at a given tool life.

SAE/AISI grade(a)	Machinability rating	Hardness, HB	SAE/AISI grade(a)	Machinability rating	Hardness, HB
1212	100	175	1050	45	197
1005	45	95	1050	55(b)	189
1006	50	95	1065	60(b)	183
1010	55	105	1070	55(b)	187
1015	60	111	1075	48	192
1017	65	116	1085	45(c)	192
1019	70	131	1095	45(c)	197
1030	70	149	1524	60	163
1038	65	163	1536	55	187
1040	60	170	1541	45	207
1045	55	179	1547	40	207
1045	60(b)	170	1547	45(b)	187

(a) Values are for steels cold drawn from the hot-rolled condition, unless otherwise indicated. (b) Annealed, then cold drawn. (c) Spheroidized, then cold drawn

Table 5 Machinability ratings of resulfurized and rephosphorized carbon steels, percent of cutting speed for B1112/1212

Grade(a)	Machinability rating, %	Hardness, HB
1117	90	137
1118	85	143
1137	70	197
1140	70	170
1141	70	212
1144	80	217
1146	70	187
1151	65	207
1212	100	...
1213	136	...
1215	136	...
12L14	160	163
12L14(b)	190	137
12L14(c)	235	137
12L14(d)	295	137

(a) All values are for cold-drawn steels. (b) Proprietary free-machining variant of 12L14. (c) Proprietary free-machining variant of 12L14 containing bismuth. (d) Proprietary free-machining variants of 12L14 containing bismuth, selenium, or tellurium

grades can best be improved by work hardening to raise the strength level and lower the ductility.

Steels in the 0.15 to 0.30% C range are usually machined satisfactorily in the as-rolled, as-forged, annealed, or normalized condition with a predominantly pearlitic structure. The medium-carbon grades, containing up to about 0.55% C, machine best if an annealing treatment that produces a mixture of lamellar pearlite and spheroidite is utilized. If the structure is not partially spheroidized, the strength and hardness may be too high for optimum machinability. For steels with carbon content higher than about 0.55%, a completely spheroidized structure is preferred. Hardened and tempered structures are generally not desired for machining.

Selection of a carbon steel grade within the standard 10xx series is seldom based entirely on machinability, although it may be a factor in selection when other functional requirements can be satisfied by more than one grade. Both tool life and production rate are adversely affected by increases in carbon content. To minimize tool wear and maximize production rate, carbon content should be held to the lowest level consistent with mechanical property requirements.

Carbon content also affects surface finish in machining, although its effect can be greatly modified by the nature of the cutting operation or by the cutting conditions. Low values of surface roughness resulting from machining can be most easily achieved with carbon steels containing approximately 0.25 to 0.35% C.

The practical significance of carbon content to economy in machining is relatively slight in most selection problems involving hardenable carbon steels because the difference in carbon content between steels with equivalent mechanical properties is unlikely to exceed 0.10%. Therefore, assuming a need for comparable mechanical properties, a choice between 1045 and 1050 would be realistic, while a choice between 1030 and 1090 would not.

Resulfurized Carbon Steels

There is a significant improvement in machinability when a resulfurized carbon steel is substituted for a plain carbon steel of approximately the same carbon content. In carbon steels, the sulfur content is ordinarily restricted to a maximum of 0.05%. In the manufacture of resulfurized steels, sulfur is deliberately added to achieve the desired sulfur level. The most common range of sulfur content in resulfurized steels is 0.08 to 0.13%, but some grades permit sulfur content as high as 0.35%. Machinability ratings for standard resulfurized steels are given in Table 5.

Sulfur is added to steel for the sole purpose of decreasing machining costs, either by increasing productivity through greater machining speeds and improved tool life or by eliminating secondary operations through an improvement in finish. Sulfide inclusions, depending on their size, shape, and orientation, improve machining by causing the formation of a broken chip instead of a stringy or continuous chip and by providing a built-in lubricant that prevents the chips from sticking to the tool and undermining the cutting edge. By minimizing this adherence, less power is required, finish is improved, and the speed of machining can often be doubled, compared with machining a similar, nonresulfurized grade. A tightly curled chip that breaks readily is also particularly helpful in milling, deep drilling, tapping, slotting, and reaming because the chip is forced to move within a confined area in these operations.

The reduced friction, lower specific power requirements, and improved chip characteristics when machining resulfurized steels all contribute to increased production rates. The advantage of free-machining steels over carbon and alloy grades in terms of unit power consumption is shown in Fig. 10. The difference is important because almost all the energy of cutting is converted into heat in the cutting zone.

The manganese content of resulfurized steels must be high enough to ensure that all the sulfur is present in the form of manganese sulfide (MnS) particles. When a high sulfur content is accompanied by an increase in manganese content, a better surface finish is obtainable, which usually re-

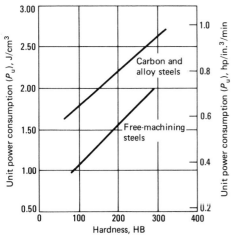

Fig. 10 Unit power consumption for free-machining and standard grades of carbon and alloy steels as a function of hardness

sults in an improvement in dimensional accuracy.

Control and Effect of Sulfide Morphology. The control of MnS particle shape, size, and distribution is a critical aspect of steelmaking (Ref 12). The particles may remain somewhat globular or may become elongated during rolling to form stringers parallel to the direction of rolling. Figure 11 shows that the size and shape of sulfide particles have pronounced effects on the machinability of steels having similar compositions. The two bars came from different ingots in the same heat. In this case, the difference in machinability was caused by variations in oxygen content and was reflected by the differences in silicon content. The presence of aluminum or other strong deoxidizers changes the shape of sulfide inclusions and may impair machinability. In theory at least, differences in rolling practice that affect the characteristics of sulfide inclusions also influence machinability.

For years, the beneficial effects of large, globular sulfides on machinability have been realized by controlling the silicon contents of semikilled steels (Ref 12). Recently, the effects of inclusion morphology on other characteristics have received considerable attention. Striking benefits resulting from changing inclusions have been reported, but there has been less agreement about the mechanisms through which they act.

Royer (Ref 13) reported that a 25% improvement in machinability resulted from increasing the manganese content of type 303 stainless steel. He attributed the improvement to softer, less abrasive sulfide inclusions. Presumably, the higher manganese level may also have affected the total volume and size of the manganese sulfides.

Yaguchi (Ref 14) measured thrust and cutting forces in tests on leaded, free-cutting steels. The life to catastrophic failure of the HSS tools improved considerably with inclusion size. The minima in cutting forces shifted toward higher speeds as the sulfides increased in size. Yaguchi concluded that the size of the inclusions influenced the temperature distribution at the tip of the tool and indirectly affected the formation of the built-up edge on the tool. It is widely known that the presence or absence of a built-up edge influences tool life.

Abeyama and associates (Ref 15) controlled the morphology of sulfides in resulfurized free-cutting steels through additions of tellurium. Treatments producing Te/S ratios of 0.2 were most effective in improving machinability, and these treatments also reduced the anisotropy of mechanical properties. The researchers attributed both effects to the influence of tellurium on the melting or softening temperatures of the inclusions. Raising the melting point of the sulfides by alloying would minimize the elongation of the mixed sulfides during hot rolling.

Katayama (Ref 16) found that sulfide shape also affected the machinability of continuously cast billets. This subject is of interest because steels made by that practice, rather than from ingots, have smaller as-cast inclusions that are not elongated as much during hot rolling because the reductions are smaller. However, this difference between ingots and continuously cast steel may not always manifest itself in terms of machinability. Welburn and Naylor (Ref 17) reported that the machinability of 1144 steel made from continuously cast billets is equivalent to that of bars made from ingots. Their conclusions were based on both laboratory tests and experience by customers.

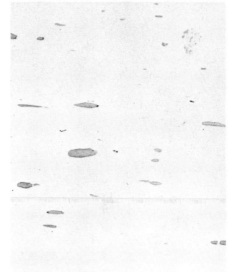

Chemical composition, %	Steel A	Steel B
Carbon	0.07	0.07
Manganese	0.94	0.94
Sulfur	0.200	0.200
Phosphorus	0.094	0.093
Silicon	0.009	0.044
Machinability rating, %	176	125

Fig. 11 Influence of size and shape of sulfide inclusions on machinability. Two steels, identical in composition except for silicon content, exhibited different machinability ratings that were traced to differences in the size and shape of MnS inclusions. Source: Ref 12

Economic factors influence the use of resulfurized steels because resulfurized grades cost more than plain carbon steels. For example, 25 mm (1 in.) round hot-rolled bars of 1117 steel cost approximately 8% more than plain carbon steel bars of a composition that is similar except for the sulfur content. Nevertheless, the economic benefits of machining resulfurized steels are large enough to justify annual purchases of 2 million tons of free-cutting steel in the United States.

The use of resulfurized grades depends on whether the higher steel cost can be offset by lower machining cost. Considering only the cost of removing chips, the use of free-cutting steels can seldom be justified if less than about 10% of the bar is removed in machining the parts. However, as the amount of metal machined off approaches or exceeds 20%, the resulfurized grades should be considered. When the required finish can be obtained in a primary operation with a free-cutting grade but a secondary operation would be necessary for a grade that is not free cutting, the 11xx series can be justified for a smaller amount of metal removal. In such parts, the savings in handling and the elimination of an operation usually more than offset the extra cost of the steel without consideration of machining speed or tool life.

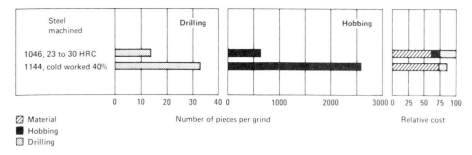

Fig. 12 Comparison between machining 1046 and 1144

Figure 12 illustrates how the substitution of heavily drawn and stress-relieved 1144 for heat-treated 1046 can significantly increase the number of parts machined between regrinds of the tools and therefore the cost of manufacturing the part.

Carbon Steels With Other Additives

Phosphorus, as well as sulfur, is often added to improve the machining characteristics of low-carbon steels. The phosphorus range for 12L14 and 1215 is 0.04 to 0.09%. The phosphorus limits are 0.07 to 0.12% for other steels in the 12xx series. The limits are set because phosphorus, like carbon, increases the hardness and strength of the steel. Consequently, excessive phosphorus contents impair machining characteristics and some other properties of steel.

Phosphorus is soluble in iron and increases the strength of ferrite, an effect that promotes chip breaking in cutting operations. The phosphorus helps to avoid the formation of long, stringy chips in some operations and may result in a better surface finish.

Nitrogen. The effects of nitrogen on the machining characteristics of 12L14 steel were studied by Watson and Davies (Ref 18). They found that nitrogen adversely affected the life of HSS tools used for turning and form cutting. This effect was attributed to strain age hardening, caused by nitrogen, before testing. No beneficial effects of nitrogen were detected.

Selenium and tellurium additions improve machinability but are not available in standard grades of steel. These additions are expensive (selenium treatment increases the cost of steel by about 15%). When they are used, they are often used in combination with sulfur or lead. Typical percentages of either element would be 0.04 or 0.05%. Both elements seem to exert beneficial effects by promoting the retention of globular-shaped sulfide-type inclusions. For the same reason, they are considered to have a less deleterious effect than sulfur on mechanical properties. The data in Fig. 13 show that the effect of tellurium on machinability can be appreciable. The data were obtained on steels with a nominal tensile strength of 1035 MPa (150 ksi) by turning with a form tool and measuring the diameters of successive parts. The presence of 0.042% Te quadrupled the number of parts made between tool changes and improved the surface finish. Tata and Sampsell (Ref 19) reported that selenium is even more effective than tellurium in improving the machinability of steels, particularly alloy steels.

Calcium additions improve the machining characteristics of steels fully deoxidized with aluminum. The cost of the special treatment is relatively modest. Steels made by aluminum deoxidation practices ordinarily contain small inclusions of aluminum silicate in quantities essentially independent of the amount of aluminum added to the steel. The inclusions are often assumed to be alumina, and the poorer machinability of aluminum-killed steels, compared to steels deoxidized with silicon, is often attributed to the supposedly abrasive effects of the inclusions. The validity of this explanation is debatable. Calcium additions result in larger inclusions consisting of calcium-aluminum silicates. Joseph and Tipnis (Ref 20) considered these inclusions to be softer and less abrasive than those in steels not treated with calcium and concluded that such attributes benefited machinability. Tests on a series of 1045 steels led Subramanian and Kay (Ref 21) to the same conclusions.

Abeyama and colleagues (Ref 15, 22) demonstrated that calcium treatments benefit the machining characteristics of several types of steel. Their studies on heavy-duty steels led them to believe that machinability is affected by the chemical composition of the inclusions and by temperatures at the tool point. Fombarlet (Ref 23) attributed the better machining properties of 1048, 4142, and 8620 steels to appropriate treatments with calcium.

Leaded Carbon and Resulfurized Steels

The addition of lead to carbon steels is another means of increasing the machinability of the steel and improving the surface finish of machined parts. Lead is added to the molten steel during teeming of ingots or, sometimes, to the ladle. Because lead is insoluble, or nearly so, in molten steel, a fine dispersion of lead particles develops as the steel solidifies. The lead is usually found near or surrounding the sulfide inclusions. On special order, nearly all carbon steels in the 10xx and 11xx series can be produced with 0.15 to 0.35% Pb. The grades are identified by inserting the letter "L" between the second and third numerals of the grade designation, for example, 10L45. It is generally believed that lead has a minimal effect on the yield or ultimate strength,

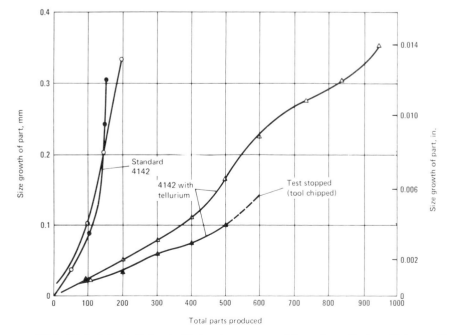

Fig. 13 Effect of tellurium on tool wear. Tool wear, as measured by part growth, in multiple-operation machined parts of quenched and tempered 4142 and a similar grade with tellurium. Cutting speed was 0.5 m/s (99 sfm). Source: Ref 11

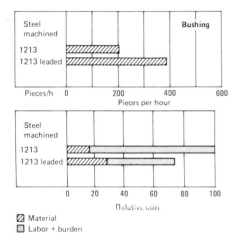

Fig. 14 Effect of lead content on multiple-operation machining. Graphs show the production rate and relative cost of machining bushings from 1113 and 12L13 steel with cutting speeds chosen for an 8-h tool life.

Table 6 Machinability ratings for alloy steels compared to 1212 steel
The machinability rating of 1212 steel is assigned at 100.

Grade	Machinability rating(a)	Hardness, HB	Grade	Machinability rating(a)	Hardness, HB
1330	55(b)	179–235	5140	65(b)	179–217
1340	50(b)	183–241	5160	55(d)	179–217
1345	45(c)	183–241	51B60	55(d)	179–217
4024	75(c)	156–207	50100	40(d)	183–240
4028	75(c)	167–212	51100	40(d)	183–240
4042	65(b)	179–229	52100	40(d)	183–240
4130	70(b)	187–229	8115	65(c)	163–202
4140	65(b)	187–229	81B45	65(b)	179–223
41L40	85(b)	185–230	8630	70(b)	179–229
4150	55(b)	187–240	8620	65(c)	179–235
4340	50(b)	187–240	86L20	85(c)	...
4620	65(b)	183–229	8660	55(d)	179–217
50B40	65(b)	174–223	8645	65(b)	184–217
50B60	55(d)	170–212	86B45	65(b)	184–217
5130	70(b)	174–212	8740	65(b)	184–217

(a) Ratings are for cold-finished bars. (b) Microstructure composed of ferrite and lamellar pearlite. (c) Microstructure composed mainly of acicular pearlite and bainite. (d) Microstructure composed primarily of spheroidite

ductility, or fatigue properties of steels at room temperature and moderate strength levels. Lead can also be added to alloy steels to improve machinability without sacrificing room-temperature mechanical properties. Environmental considerations may restrict the manufacture or use of leaded steels.

Leaded steels cost about 5% more than similar nonleaded compositions. Because requirements of machinability and finish are the only reasons for using leaded grades, the extra cost for these steels must be justified by either or both of these factors.

As with resulfurized grades of carbon steel, consideration must be given to the amount of stock being removed. For example, if only 10% of the bar weight is to be removed in machining, the extra cost for leaded grades may not be justifiable. However, if 20% or more of the bar is converted to chips, the leaded grades should be considered.

The problem of surface finish requires consideration of specific parts, and no general statements are valid. In a case study of one particular part, the finish obtained in drilling and reaming standard 1050 steel was marginal, and parts were frequently rejected and subsequently reworked. The use of leaded 1050 corrected this condition, and the added cost of the steel was justified by finish alone.

Most of the resulfurized grades can be produced with an addition of 0.15 to 0.35% Pb. The lead addition augments the effect of sulfur, permitting a further increase of machining speed and better finish. For screw machine parts where more than 50% of the bars become chips, there may be justification for the higher cost, especially where there are high finish requirements.

The increased cost of leaded steel is unjustified when machine tools are already being operated at maximum speed on plain resulfurized steels. On the other hand, machine tools that have been designed for higher speeds can take advantage of the leaded resulfurized grades.

In another case study, surface finish was the deciding factor in selecting a leaded steel for parts about 13 mm (½ in.) in diameter and 75 mm (3 in.) long made from 1141 steel on a multiple-operation machine. Finish requirements on the ends of the parts could not be met in the cutoff operation, and a secondary facing operation was necessary. A change to leaded 1141 steel provided just enough better finish from the cutoff operation so that the finish requirement could be met, and the elimination of the facing operation justified the cost of the leaded grade.

The example shown in Fig. 14 illustrates the importance of machinability of steel in determining the cost of small parts. A change from 1213 to 12L13 nearly doubled the production rate of these bushings and reduced the total cost per part, even though the 12L13 steel was more costly.

Reh and coworkers (Ref 24) described the characteristics of lead-free, high-sulfur steels containing bismuth. They found bismuth to be more efficient than lead in reducing cutting forces and in improving surface finish. They attributed the pronounced improvements caused by bismuth to the spheroidization of inclusions and to the liquid-metal embrittlement of grain boundaries, by liquid bismuth, during cutting.

Carburizing Steels

It is difficult to evaluate the relative machining economy of plain carbon carburizing steels, except in terms of specific parts, largely because these steels possess conflicting properties that may either promote or detract from economy in machining. On the one hand, their low carbon content may be beneficial to tool life and production rate. However, their relatively soft, gummy structure results in a tearing action in cutting, which is harmful to surface finish and dimensional accuracy. It is this conflict in properties that sometimes makes it advantageous to machine or partially machine these steels in the carburized condition. Although higher carbon content adversely affects tool life, the higher carbon areas are more controllable in terms of surface finish and dimensional accuracy. In operations such as gear cutting, the finish and accuracy may be more decisive factors than tool wear or production rate.

Through-Hardening Alloy Steels

The steels chosen for parts that must be hardened throughout must contain enough carbon to achieve the desired hardness after quenching and sufficient alloy content to obtain the desired percentage martensite in the thickest section of the part. The combination of carbon and alloy contents can make these steels difficult to machine. The extent of the difficulties encountered in machining these steels depends primarily on the microstructure and hardness of the steel and secondarily on its alloy content. Figure 15 illustrates the magnitude of the variations in tool life that may be expected from differences in hardness and microstructure in a single steel, 4340. A similar effect for 4135 steel is shown in Fig. 6. Machinability ratings for several alloy steels are given in Table 6.

A comparison of the machinability ratings with the compositions of these steels indicates that all of the alloying elements that increase the hardenability of the steel decrease machinability; ferrite-strengthening elements such as nickel and silicon decrease the machinability more than equivalent amounts of carbide-forming elements such as chromium and molybdenum.

It is not uncommon for heat-treating considerations to overshadow both machining

Machinability of Steels / 601

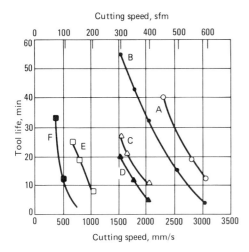

Sample	Hardness, HB	Heat treatment
○ A	206	Spheroidized
● B	221	Annealed
△ C	321	Normalized
▲ D	400	Hardened and tempered
□ E	500	Hardened and tempered
■ F	515	Hardened and tempered

Fig. 15 Effect of hardness on tool life curves. Workpiece: 4340 steel. Tool material: C6 carbide. Source: Ref 25

and material costs in the selection of steel. On occasion, heat-treating responses may dictate the selection of a less machinable or a more expensive steel so that the lowest total costs can be realized.

The sulfur content of through-hardening alloy steels can significantly affect machining behavior. Variations in residual sulfur level can account for unexplained differences in the machining behavior of different lots of the same material. Many grades of hardenable alloy steels can be obtained in the resulfurized condition. The differences in tool life and cutting speed between standard and high-sulfur 4150 steels are substantial. Tests by Field and Zlatin (Ref 26) showed that raising the sulfur content from 0.04 to 0.09% increased the cutting speed for 60-min tool life by 25%. Alloy steels containing lead are available and useful. As indicated in Table 6, the machinability rating of the leaded grade 41L40 is 85, while the rating for 4140 is only 65. The performance of these two grades in several machining operations is indicated in Table 7. The data are from a case study described in Ref 26.

Another important factor that can affect the choice of steel for a through-hardening application is the effect of alloying elements added for machinability on the mechanical properties of the steel. These steels are often used at high-strength levels, where the deleterious effects of inclusions, particularly on transverse properties, might not be permissible. The effect of sulfur, in the amounts usually specified for enhanced machinability, is generally considered to be

Table 7 Effect of lead on cutting speed and tool life in machining alloy steels

Operation	Standard 4140	Leaded 4140
Turning		
Hardness, HB	300	300
Cutting speed, rev/min	321	495
Feed, mm/rev (in./rev)	0.30 (0.012)	0.30 (0.012)
Tool life, parts per tool grind	4	18–20
Turning		
Cutting speed, rev/min	460	740
Feed, mm/rev (in./rev)	0.15 (0.006)	0.23 (0.009)
Drilling(a)		
Cutting speed, m/min (sfm)	8.76 (28.75)	10.55 (34.6)
Feed, mm/rev (in./rev)	0.10 (0.004)	0.15 (0.006)

(a) In drilling standard 4140 steel, the 19 mm (¾ in.) diam hole jammed with chips and the drill had to be removed frequently for cleaning. When using leaded 4140 steel, the entire depth was drilled without removing the tool.

more damaging than that of lead. For some applications, neither machinability additive can be tolerated.

Cold-Drawn Steel

Cold drawing generally improves the machinability of steels containing less than about 0.2% C. The improvement is most noticeable in plain carbon steels, as shown in Fig. 16. The machinability of higher-carbon steels, or alloy steels, is less affected by cold work. This improvement in machinability may be attributed to reduced cutting forces and/or the characteristics of chip removal. Kopalinsky and Oxley (Ref 28) found that cold drawing lowered the cutting forces and improved the tool life and surface finish of low-carbon steels. Screw machine tests by Yaguchi (Ref 29) showed that the workpiece surface finish improved continuously with increases in reduction in area up to 29%. These effects were not characteristic of steels with high nitrogen contents (Ref 18). The improved machinability of cold-drawn steels can also be attributed to the decrease in ductility that results from cold working; thus, the chips are generally not long and stringy.

Cold-finished bars have closer dimensional tolerances, better surfaces, and, usually, higher strength than hot-finished bars. The first two factors may be significant in the selection of steels to be machined in multiple-operation machines or other high-production equipment. These considerations are discussed in the article "Cold-Finished Steel Bars" in this Volume.

The machining characteristics of cold-drawn steels are only rarely a decisive criterion for selection. The extra strength obtained with cold-drawn steel may be more important from a cost standpoint, because it is often high enough to eliminate the need for heat treatment.

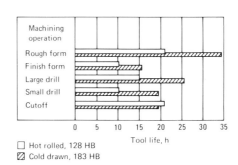

Fig. 16 Effect of cold drawing on tool life. Workpiece: 1016 steel, 25 mm (1 in.) in diameter. Machining conditions: multiple-operation machined with a cutting speed of 0.73 m/s (144 sfm). Source: Ref 27

REFERENCES

1. *Machining Data Handbook*, 3rd ed., Metcut Research Associates Inc., 1980
2. "Life Tests for Single-Point Tools of Sintered Carbide," B94.36-1956 (R 1971), American National Standards Institute
3. "Tool Life Testing With Single-Point Turning Tools," ANSI/ASME B94.55M-1985, American National Standards Institute
4. J.F. Kahles, Elements of the Machining Process, in *Metals Handbook; Desk Edition*, American Society for Metals, 1985, p 27.10
5. "Machining Performance of Ferrous Metals Using an Automatic Screw/Bar Machine," E 618-81-03.01, *Annual Book of ASTM Standards*, American Society for Testing and Materials
6. F.W. Boulger, *Influence of Metallurgical Properties on Metal-Cutting Operations*, Society of Manufacturing Engineers, 1958
7. F.W. Boulger and H.J. Grover, Machinability Can Be Related to Composition, *Tool Eng.*, Vol 40, March 1958
8. V.C. Venkatesh and V. Narayanan, Machinability Correlations Among Turning, Milling and Drilling Processes, *Ann. CIRP*, Vol 35 (No. 1), 1986, p 59-62
9. T. Araki et al., Some Results of Cooperative Research on the Effect of Heat Treated Structure on the Machinability of a Low Alloy Steel in *Influence of Metallurgy on Machinability*, V.A. Tipnis, Ed., American Society for Metals, 1975, p 381-395
10. E.J.A. Armarego and R.H. Brown, *The Machining of Metals*, Prentice-Hall, 1969
11. J.E. Mayer, Jr., and D.G. Lee, Influence of Machinability on Productivity and Machining Cost, in *Influence of Metallurgy on Machinability*, V.A. Tipnis, Ed., American Society for Metals, 1975, p 31-54
12. F.W. Boulger et al., Superior Machinability of MX Steel Explained, *Iron*

Age, Vol 167, 17 May 1951, p 90-95
13. W.E. Royer, Making Stainless More Machinable—303 Super X, *Autom. Mach.*, Vol 47 (No. 5), May 1986, p 47-49
14. H. Yaguchi, Effect of MnS Inclusion Size on Machinability of Low-Carbon, Leaded, Resulfurized Free-Machining Steel, *J. Appl. Metalwork.*, Vol 3 (No. 3), July 1986, p 214-225
15. S. Abeyama et al., Development of Free Machining Steel With Controlled-Shape Sulfides, *Bull. Jpn. Inst. Met.*, Vol 24 (No. 6), 1985, p 518-520
16. S. Katayama et al., Improvements in Machinability of Continuously-Cast, Low-Carbon, Free-Cutting Steels, *Trans. ISI*, Vol 25 (No. 9), Sept 1985, p B229
17. R.M. Welburn and D.J. Naylor, Production and Machinability of Billet-Cast Medium Carbon High Sulfur (Over 0.08%) Free-Machining Steels, in *Proceedings of the Conference on Continuous Casting*, Institute of Metals, 1985
18. J.D. Watson and R.H. Davies, The Effects of Nitrogen on the Machinability of Low-Carbon Free-Machining Steels, *J. Appl. Metalwork.*, Vol 3 (No. 2), 1984, p 110-119
19. H.J. Tata and R.E. Sampsell, Effects of Additions on Machinability and Properties of Alloy-Steel Bars, Paper 730114, *Trans. SAE*, Vol 82, 1973
20. R.A. Joseph and V.A. Tipnis, The Influence of Non-Metallic Inclusions on the Machinability of Free-Machining Steels, in *Influence of Metallurgy on Machinability*, V.A. Tipnis, Ed., American Society for Metals, 1975, p 55-72
21. S.V. Subramanian and D.A.R. Kay, Inclusions and Matrix Effects on the Machinability of Medium Carbon Steels, in Conference Proceedings, Ottawa, Ontario, Canada, Canadian Government Publishing Centre, 1985
22. T. Kato, S. Abeyama, A. Kimura, and S. Nakamura, The Effect of Ca Oxide Inclusions on the Machinability of Heavy Duty Steels, in *The Machinability of Engineering Materials*, R.W. Thompson, Ed., Conference Proceedings, 13-15 Sept (Rosemont, IL), American Society for Metals, 1983, p 323-337
23. J. Fombarlet, Improvement in the Machinability of Engineering Steels Through Modification of Oxide Inclusions, in *The Machinability of Engineering Materials*, 13–15 Sept (Rosemont, IL), R.W. Thompson, Ed., Conference Proceedings, American Society for Metals, 1983, p 366-382
24. B. Reh, U. Finger et al., Development of Bismuth-Alloyed High Performance Easy Machining Steel, Neue Hütte, Vol 31 (No. 9), Sept 1986, p 327-330
25. N. Zlatin and J. Christopher, Machining Characteristics of Difficult to Machine Materials, in *Influence of Metallurgy on Machinability*, V.A. Tipnis, Ed., American Society for Metals, 1975, p 296-307
26. M. Field and N. Zlatin, Evaluation of Machinability of Rolled Steels, Forgings and Cast Irons, *Machining—Theory and Practice*, American Society for Metals, 1950, p 341-376
27. J.D. Armour, Metallurgy and Machinability of Steels, *Machining—Theory and Practice*, American Society for Metals, 1950, p 123-168
28. E.M. Kopalinsky and P.L.B. Oxley, Predicting Effects of Cold Working on Machining Characteristics of Low-Carbon Steels, *J. Eng. Ind. (Trans. ASME)*, Vol 109 (No. 3), 1987, p 257-264
29. H. Yaguchi and N. Onodera, Effect of Cold Working on the Machinability of AISI 12L14 Steel, in *Strategies for Automation of Machining: Materials and Processes*, Proceedings of an International Conference (Orlando, FL), ASM INTERNATIONAL, 1987, p 15-26

Weldability of Steels

S. Liu, Center for Welding and Joining Research, Colorado School of Mines
J.E. Indacochea, Department of Civil Engineering, Mechanics, and Metallurgy, University of Illinois at Chicago

THE MAIN OBJECTIVE of this article is to survey the factors controlling the weldability of carbon and low-alloy steels in arc welding. A good understanding of the chemical and physical phenomena that occur in the weldment is necessary for welding modern steels. Therefore, the influence of operational parameters, thermal cycles, and metallurgical factors on weld metal transformations and the susceptibility to hot and cold cracking are discussed. Common tests to determine steel weldability are also described.

The carbon and low-alloy steels group comprises a large number of steels that differ in chemical composition, strength, heat treatment, corrosion resistance, and weldability. These steels can be further divided into subgroups:

- Carbon steels
- High-strength low-alloy (HSLA) steels
- Quenched and tempered (QT) steels
- Heat-treatable low-alloy (HTLA) steels
- Precoated steels

This article addresses only the basic principles that affect the weldability of carbon and low-alloy steels. More detailed information concerning the other aspects of welding, such as joint design, defects, and failure in weldments and the influence of these factors on different groups of steels, can be found in Volumes 1, 6, and 11 of the 9th Edition of *Metals Handbook* and in the "Selected References" at the end of this article.

Characteristic Features of Welds

Single-Pass Weldments. To understand weldability, it is necessary to recognize the various weld regions. In the case of a single-pass bead, the weldment is generally divided into two main regions: the fusion zone, or weld metal, and the heat-affected zone (HAZ), as shown in Fig. 1. Within the fusion zone, the peak temperature exceeds the melting point of the base metal, and the chemical composition of the weld metal will depend on the choice of welding consumables, the base metal dilution ratio, and the operating conditions.

Under conditions of rapid cooling and solidification in the weld metal, alloying and impurity elements segregate extensively to the center of the interdendritic or intercellular regions and to the center parts of the weld, resulting in significant local chemical inhomogeneities. Accordingly, the transformation behavior of the weld metal may be quite different from that of the base metal, even when the bulk chemical composition is not significantly changed by the welding process. The typical anisotropic nature of the solidified weld metal structure is also shown in Fig. 1.

The chemical composition remains largely unchanged in the HAZ because the peak temperature remains below the melting point of the parent plate. Nevertheless, considerable microstructural change takes place within the HAZ during welding as a result of the extremely harsh thermal cycles. The material immediately adjacent to the fusion zone is heated high into the austenitic temperature range. The microalloy precipitates that developed in the previous stages of processing will generally dissolve, and unpinning of austenite grain boundaries occurs with substantial growth of the grains, forming the coarse-grain HAZ. The average size of the austenite grains, which is a function of the peak temperature attained, decreases with increasing distance from the fusion zone. The cooling rate also varies from point to point in the HAZ; it increases with increasing peak temperature at constant heat input and decreases with increasing heat input at constant peak temperature. Because of varying thermal conditions as a function of distance from the fusion line, the HAZ is actually composed of coarse-grain zones (CGHAZ), fine-grain zones (FGHAZ), intercritical zones (ICHAZ), and subcritical zones (SCHAZ). The various HAZ regions of a single-pass low-carbon steel butt weld are shown in Fig. 2.

In multipass weldments, the situation is much more complex because of the presence of reheated zones within the fusion zone, as shown in Fig. 3. The partial refinement of the microstructure by subsequent weld passes increases the inhomogeneity of the various regions with respect to microstructure and mechanical properties. Reaustenitization and subcritical heating can have a profound effect on the subsequent structures and properties of the HAZ. Toughness property deterioration is related to small regions of limited ductility and low cleavage resistance within the CGHAZ that are known as the localized brittle zones (LBZ). Localized brittle zones consist of unaltered CGHAZ, intercritically reheated coarse-grain (IRCG) heat-affected zone, and subcritically reheated coarse-grain (SRCG) heat-affected zone. At an adjacent fusion line, the LBZs may be aligned, as shown in Fig. 3. The aligned LBZs offer short and easy paths for crack propagation. Fracture occurs along the fusion line.

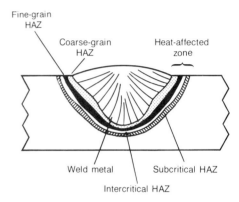

Fig. 1 Various regions of a bead-on-plate weld

Metallurgical Factors That Affect Weldability

Hardenability and Weldability. Hardenability in steels is generally used to indicate austenite stability with alloy additions. However, it has also been used as an indicator of weldability and as a guide for selecting a material and welding process to avoid excessive hardness and cracking in the HAZ. Steels with high hardness often contain a high volume fraction of martensite, which is extremely susceptible to crack-

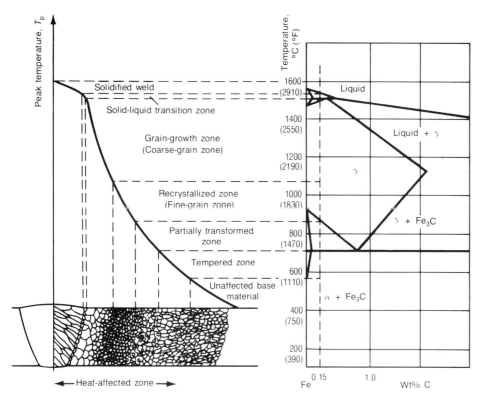

Fig. 2 Various regions of the HAZ of a single-pass low-carbon steel weld metal with 0.15 wt% C

ing during processing. Hardenability is also used to indicate the susceptibility of a steel to hydrogen-induced cracking.

Traditionally, empirical equations have been developed experimentally to express weldability. Carbon equivalent (CE) is one such expression; it was developed to estimate the cracking susceptibility of a steel during welding and to determine whether the steel needs pre- and postweld heat treatment to avoid cracking. Carbon equivalent equations do include the hardenability effect of the alloying elements by expressing the chemical composition of the steel as a sum of weighted alloy contents. To date, several CE expressions with different coefficients for the alloying elements have been reported. The International Institute of Welding (IIW) carbon equivalent equation is:

$$CE = C + \frac{Mn}{6} + \frac{Ni}{15} + \frac{Cu}{15} + \frac{Cr}{5}$$
$$+ \frac{Mo}{5} + \frac{V}{5} \qquad (Eq\ 1)$$

where the concentration of the alloying elements is given in weight percent. It can be seen in Eq 1 that carbon is the element that most affects weldability. Together with other chemical elements, carbon may affect the solidification temperature range, hot tear susceptibility, hardenability, and cold-cracking behavior of a steel weldment. Figure 4 summarizes the CE and weldability description of some steel families. Because of the simplification and generalization involved in Fig. 4, it should be used cautiously for actual welding situations.

The application of CE expressions is also empirical. For example, the IIW carbon equivalent equation has been used successfully with traditional medium-carbon low-alloy steels. Steels with lower CE values generally exhibit good weldability. When the CE of a steel is less than 0.45 wt%, weld cracking is unlikely, and no heat treatment is required. When the CE is between 0.45 and 0.60 wt%, weld cracking is likely, and preheat in the range of approximately 95 to 400 °C, (200 to 750 °F) is generally recommended. When the CE of a steel is greater than 0.60 wt%, there is a high probability that the weld will crack and that both preheat and postweld heat treatments will be required to obtain a sound weld.

However, Eq 1 does not accurately correlate with the microstructures and properties of newly developed low-carbon microalloyed steels over extended alloy ranges. Thus, new expressions based on solution thermodynamics and kinetic considerations were developed to obtain better predictions of the alloy behavior and weldability of low-carbon low-alloy steels. Complex interactive terms, rather than simple additive forms, are included in these equations. An example of one such expression is:

$$CE = k_1C[1 + k_2C + k_3Mn + \ldots$$
$$+ k_{11} \ln C + k_{22}C \ln C + k_{33}Mn \ln Mn$$
$$+ \ldots + k_{111}CMn + \ldots] \qquad (Eq\ 2)$$

where $k_1, k_2, \ldots$, and so on, are the weighted coefficients multiplied to the concentration of the alloying elements. Nonlinear terms such as $\ln X_i$, $X_i \ln X_i$, and X_iX_j represent the interaction effect among the alloying elements X_i and X_j. Equations with these nonlinear terms are more useful in predicting arc welding behavior.

Several expressions are also available for other steel groups with a wider range of alloying elements and with different prior heat treatments, hydrogen contents, and weld hardnesses. Recently, expressions that include fabrication conditions such as heat input, cooling rate, joint design and restraint conditions have also been proposed. An example of this type of equation is:

$$P_H = P_{cm} + 0.075 \log_{10} H + \frac{R_f}{40\ 000} \qquad (Eq\ 3)$$

where P_H is the cracking susceptibility parameter, H is the concentration of hydrogen (in parts per million), R_f is the restraint stress (in megapascals), and:

$$P_{cm} = C + \frac{Mn}{20} + \frac{Si}{30} + \frac{Cu}{20} + \frac{Ni}{60} + \frac{Cr}{20}$$
$$+ \frac{Mo}{15} + \frac{V}{10} + 5B \qquad (Eq\ 4)$$

The thickness of the part being welded can also be related to CE as a compensated carbon equivalent (CCE) as follows:

$$CCE = CE + 0.00254e \qquad (Eq\ 5)$$

where e is the thickness of the part (in millimeters). Equations 3 to 5 are valid only for specific ranges of chemical composition and welding conditions. Nevertheless, despite the different forms and terms included in the predictive equations, the main objective remains that of estimating the weldability and cracking susceptibility of the material.

Weld Metal Microstructure. Inherent in the welding process is the formation of a pool of molten metal directly below a moving heat source. The shape of this molten pool is determined by the flow of both heat and metal, with melting occurring ahead of the heat source and solidification occurring behind it. Heat input determines the volume of molten metal and therefore the dilution and weld metal composition, as well as the thermal conditions under which solidification takes place. Also important to solidification is the crystalline growth rate, which is geometrically related to weld travel speed and weld pool shape. Thus, weld pool shape, weld metal composition, cooling rate, and growth rate are all factors that are interrelated with heat input, which in turn will affect the solidification microstructure and the tolerance of the weldment to hot cracking.

Incipient melting at base metal grain boundaries immediately adjacent to the fu-

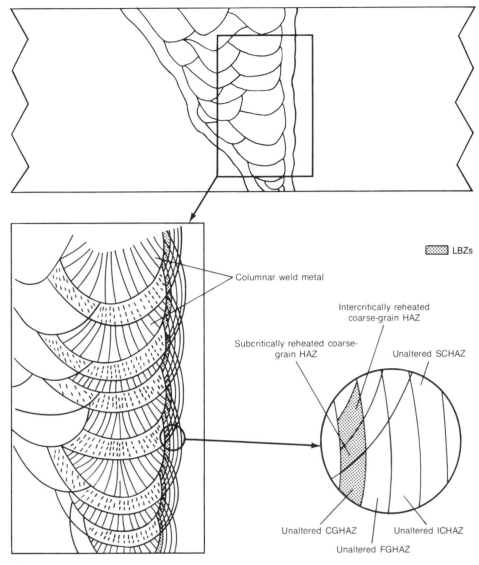

Fig. 3 Overlapping of HAZ to form localized brittle zones aligned along the fusion line

sion zone allows these grains to serve as seed crystals for epitaxial grain growth during weld metal solidification. The continuous growth of the epitaxial grains results in large columnar grains whose boundaries provide easy paths for crack propagation. An elongated weld pool will yield straight and broad columnar grains, which promote the formation of centerline cracking because of impurity segregation, mechanical entrapment of inclusions, and the shrinkage stresses that develop during solidification. Epitaxial columnar growth is particularly deleterious in multipass welds where grains can extend continuously from one weld bead to another.

Hot tears originate near the liquid/solid interface when strains from solidification shrinkage and thermal contraction cause rupture of the liquid films of low melting point located at grain boundaries. The susceptibility of an alloy to hot tearing is related to its inability to accommodate strain through dendrite interlocking as well as the tendency of tears to backfill with the remaining liquid. The time interval during which liquid films can exist in relation to the rate of strain generation may also play a role in hot tear susceptibility. Ferrous alloys can be hot tear sensitive depending on the amount of phosphorus and sulfur impurities they contain. Carbon and nickel are also known to influence hot cracking in steel welding.

When the solidified steel weld metal cools down, solid-state transformation reactions may occur. As in solidification, the two main factors that determine the final microstructure are the chemical composition and thermal cycle of the weld metal. In most structural steels, weld metal will solidify as δ-ferrite. At the peritectic temperature, austenite will form from the reaction between liquid weld metal and δ-ferrite, and subsequent cooling will lead to the formation of α-ferrite. During the austenite-to-ferrite transformation, proeutectoid ferrite forms first along the austenite grain boundaries; this is known as grain-boundary ferrite. Subsequent to grain-boundary ferrite formation, ferrite sideplates develop in the form of long needlelike ferrite laths that protrude from the allotriomorphs. A coarse austenite grain size and a low carbon content, in combination with a relatively high degree of supercooling, are found to promote ferrite sideplate formation. These laths can be properly characterized by their length-to-width aspect ratios; values above 10:1 are common.

As the temperature continues to drop, intragranular acicular ferrite will nucleate and grow in the form of short laths separated by high angle boundaries. The inclination between orientations of adjacent acicular ferrite laths is usually larger than 20°. The random orientation of these laths provides good resistance to crack propagation. Acicular ferrite laths have aspect ratios ranging from 3:1 to 10:1.

During proeutectoid ferrite formation, carbon is rejected continuously from the ferrite phase, enriching the remaining austenite, which later transforms into a variety of constituents, such as martensite (both lath and twinned), bainite, pearlite, and retained austenite. Because of the acicular nature of the bainite laths, they can also be described by their aspect ratio, with values similar to those of Widmanstätten sideplates. More frequently, however, bainite laths occur in the form of packets associated with grain boundaries. Figure 5 illustrates the microstructure of a low-carbon steel weld metal.

Heat-Affected Zone Microstructure. In terms of microstructure, long bainite laths with alternate layers of connected martensite islands are generally found in the CGHAZ of high-strength low-alloy steel weldments. Martensite islands (martensite-retained austenite constituents) are formed because of the enrichment of carbon in austenite in the intercritical zone. Coarse austenite grain size in the near-fusion region of the HAZ can suppress high-temperature transformation products in favor of martensite and bainite upon cooling. Upper bainite has a relatively high transformation temperature and is stable relative to the thermal cycles subsequent to those of the first pass. Fluctuation of the chemical composition of the microalloying elements could also contribute to carbon equivalent change and to the amount of hard martensite present in the CGHAZ.

In the FGHAZ, even though the peak temperature attained is above thermal cycle Ac_3, it is still well below the grain-coarsening temperature. The smaller prior-austenite grain size and subsequent ferrite transformation produce a refined microstrucure having grains smaller than those of the parent material. The microstructure is sim-

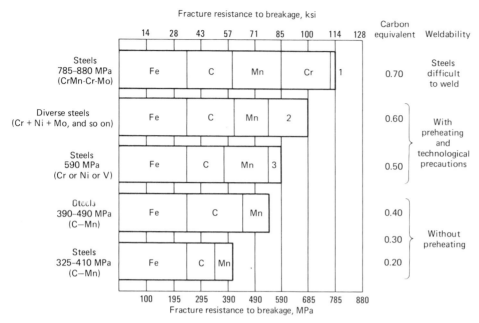

Fig. 4 Weldability of several families of steels as a function of carbon equivalent. 1, Mo; 2, Cr + Ni + Mo + Si, and so on; 3, Cr or V or Ni + Si, and so on

ilar to that of a normalized steel, with considerable toughness.

Only partial transformation takes place in the ICHAZ, resulting in a mixture of austenite and ferrite at the peak temperature of the thermal cycle. Upon cooling, the austenite in a matrix of soft ferrite decomposes, and the final microstructure depends on the bulk and local composition of the alloying elements. The cooling rate is also an important factor in determining the amount of martensite and bainite in the ferrite matrix. In the SCHAZ, no observable microstructural changes are observed. Some spheroidization of carbides may occur.

Upon reheating by subsequent weld passes, precipitates or preprecipitate clusters may form, reducing the toughness. Irregularly shaped particles may also coalesce and strain the surrounding matrix, further lowering the toughness. During HAZ thermal cycles between Ac_1 and Ac_3, the austenite becomes enriched with carbon, which, upon cooling, transforms to martensite islands. In the as-welded condition, this transformation affects the IRCG region more than the other reheated zone. Figure 6 illustrates the different phases that can be found in a low-carbon steel HAZ.

Chemical Composition Effect. The presence of a certain phase in the final microstructure of a weldment can be explained by means of a continuous cooling transformation (CCT) diagram, which is formed by two sets of curves: the percent transformation curves and the cooling curves. The percent transformation curves define the regions of stability of the different phases. The cooling curves represent the actual thermal conditions that the weld experienced. The intersection of these two sets of curves determines the final microstructure of the different weld zones. Figure 7 illustrates the use of a CCT diagram to determine the microstructure of a low-carbon low-alloy steel weld metal.

Hardenability elements, such as carbon, manganese, chromium, and molybdenum, suppress the start of austenite decomposition to lower temperatures. This is equivalent to pushing the transformation curves to the right side of the CCT diagram, resulting in a refined microstructure. Inclusion formers, such as oxygen and sulfur, accelerate the austenite-to-ferrite transformation by providing more nucleation sites for the reaction to initiate at higher temperatures.

Faster cooling has the same effect as an increase in hardenability elements, while a slower cooling rate acts in the same direction as a decrease in hardenability agents or an increase in nucleation site providers. Because the cooling rate varies from point to point in the HAZ, the microstructure also changes accordingly, with martensite and bainite in regions close to the fusion line.

Preweld and Postweld Heat Treatments. In the welding of carbon and low-alloy steels, the final microstructure of the weldment is primarily determined by the cooling rate from the peak temperature. Because the alloy level in carbon and low-alloy steel is low, the major physical properties of the steel are not affected. Thus, temperature gradient and heat input become the important parameters in weld metal microstructural evolution. A slower cooling rate decreases shrinkage stress, prevents excessive hardening, and allows time for hydrogen diffusion. Cooling rate (CR) is of particular importance and is a function of the difference in temperature, ΔT, as well as the thermal conductivity of the material, k. The cooling rate can be expressed for thin-plate and thick-plate welding, respectively, as:

$$CR \propto k\,\Delta T^3$$
$$CR \propto k\,\Delta T^2 \qquad \text{(Eq 6)}$$

During preheating, the initial temperature of the plate increases, decreasing the cooling rate and the amount of the hard phases, such as martensite and bainite, in the weld microstructure. For the welding of hardenable steels, it is important to determine the critical cooling rate (CCR) that the base metal can tolerate without cracking:

$$CCR(°F/s) = \frac{6.598}{(CE - 0.3074)} - 16.26 \qquad \text{(Eq 7)}$$

The higher the carbon equivalent of an alloy, the lower the critical or allowable cooling rate. The use of a low-hydrogen welding electrode also becomes more important. Preheating should be applied to adjust the cooling rate accordingly.

Weld Cracking

Most evidence indicates that a weld-cracking failure mechanism is microstructure related. In the case of cold cracking, recent crack tip opening displacement (CTOD) results show that the reduction in toughness of HSLA weldments is related to the CGHAZ and that cracks generally propagate along or near the fusion line. The CGHAZ of microalloyed steel welds generally exhibits the highest hardness of the entire HAZ. The high-carbon untempered martensite in this region is the major cause of embrittlement. The amount of precipitates (carbides, nitrides, and carbonitrides) is found to be the highest in the regions next to the SCHAZ and the lowest at or next to the fusion line. As a result, there is a slight increase in microalloying element in solution in the CGHAZ, which increases the hardenability of this region.

Hydrogen-Induced Cracking. The effect of hydrogen on weld cracking should also be mentioned. Moisture pickup from the atmosphere that is incorporated into the molten puddle, either directly or via the welding consumables, is the main source of hydrogen. The presence of hydrogen increases the HAZ cracking susceptibility of high-strength steel weldments. Also known as underbead, cold, or delayed cracking, it is perhaps the most serious and least understood of all weld-cracking problems. It generally occurs at a temperature below approximately 95 °C (200 °F) either immediately upon cooling or after a period of several hours. The crack can be both transgranular and intercrystalline in character, but mainly follows prior-austenite grain

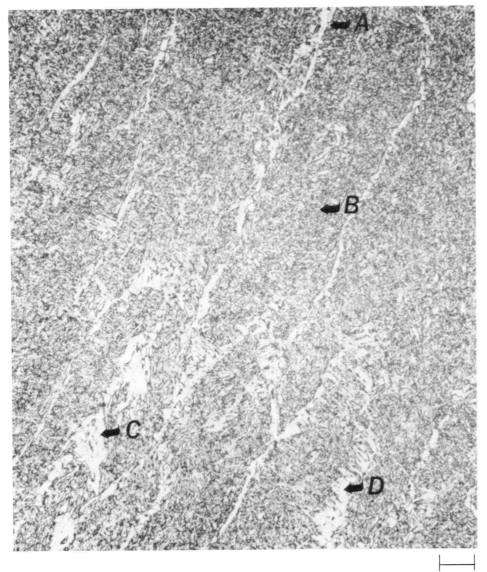

Fig. 5 Weld metal microstructure of HSLA steel. A, grain-boundary ferrite; B, acicular ferrite; C, bainite; D, sideplate ferrite

boundaries. The initiation of cold cracking is particularly associated with notches, such as the toe of the weld, or with inhomogeneities in microstructure that exhibit sudden changes in hardness, such as slag inclusions, martensite/ferrite interfaces, or even grain boundaries. Like most other crack growth phenomena, hydrogen-induced cracking is accentuated in the presence of high-restraint weld geometries and matrix hardening. Such cracking is associated with the combined presence of three factors:

- The presence of hydrogen in the steel (even very small amounts, measured in parts per million)
- A microstructure that is partly or wholly martensitic
- High residual stresses (generally as a result of thick material)

If one or more of these conditions is absent or at a low level, hydrogen-induced cracking will not occur. However, high cooling rates such as those found in manual processes further enhance the probability of weld HAZ cold cracking.

The tolerance of steels for hydrogen decreases with increasing carbon or alloy content. Hydrogen-induced cracking can be controlled by choosing a welding process or an electrode that produces little or no hydrogen. Postweld heat treatments can be used to decrease or eliminate the residual hydrogen or to produce a microstructure that is insensitive to hydrogen cracking. Finally, welding procedures that result in low restraining stresses will also reduce the risk of weld cracking.

Stress-relief cracking due to reheating is of concern when welding quenched and tempered grades and heat-resistant steels containing significant levels of carbide formers, such as chromium, molybdenum, and vanadium. When weldments of these steels are heated above approximately 510 °C (950 °F), intergranular cracking along the prior-austenite grain boundaries may take place in the CGHAZ. Also known as reheat cracking and stress-rupture cracking, stress-relief cracking is thought to be closely related to the phenomenon of creep rupture. Furthermore, during reheating, the reprecipitation of carbides is likely to occur, further increasing the hardness. The precipitation of carbides during stress relaxation alters the delicate balance between resistance to grain-boundary sliding and resistance to deformation within the coarse grains of the heat-affected zone.

Some procedures that can be used singly or in combination to decrease stress-relief cracking in steels include the selection of a more appropriate weld joint design, weld location, and sequence of assembly to minimize restraint and stress concentrations. Selecting a filler metal that will provide a weld metal that has significantly lower strength than that of the HAZ at the heat-treating temperature is another way to minimize stress-relief cracking. Peening each layer of weld metal to generate a surface compressive stress state that counteracts shrinkage stresses is also very effective.

Lamellar cracking, better known as lamellar tearing, is characterized by a steplike crack parallel to the rolling plane. Figure 8 shows a typical feature of lamellar tearing, the horizontal and vertical cracking of the base plate. The problem occurs particularly when making tee and corner joints in thick plates such that the fusion boundary of the weld runs parallel to the plate surface. High tensile stresses can develop perpendicular to the midplane of the steel plate, as well as parallel to it. This tearing, usually associated with inclusions in the steel, progresses from one inclusion to another.

There is some evidence that sensitivity to lamellar tearing is increased by the presence of hydrogen in the steel. Inclusions that contain low-melting compounds, such as those of sulfur and phosphorus, also increase the sensitivity of steel to lamellar tearing by wetting the prior-austenite grain boundaries; this makes them too weak and fragile to withstand the thermal stresses during cooling. Some approaches that can minimize lamellar tearing are:

- Changing the location and design of the welded joint to minimize through-thickness strains
- Using a lower-strength weld metal
- Reducing available hydrogen
- Buttering the surface of the plate with weld metal prior to making the weld
- Using preheat and interpass temperatures of at least 95 °C (200 °F)

Fig. 6 Heat-affected zone exhibiting a wide variety of microstructures in the intercritical and subcritical regions. A, spheroidized carbides; B, bainite and martensite

- Using base plates with inclusion shape control

Hot cracking, or solidification cracking, occurs at elevated temperatures and is usually located in the weld metal. Hot cracking also can be found in the HAZ, where it is known as liquation cracking. Solidification cracking in weld deposits during cooling occurs predominately at the weld centerline or between columnar grains. The fracture path of a hot crack is intergranular. The causes of solidification cracking are well understood. The partition and rejection of alloying elements at columnar grain boundaries and ahead of the advancing solid/liquid interface produce significant segregation. The elements of segregation form low-melting phases or eutectic structures to produce highly wetting films at grain boundaries. They weaken the structure to the extent that cracks form at the boundaries under the influence of the tensile residual stresses during cooling. Liquation cracking is also associated with grain-boundary segregation and is aggravated by the melting of these boundaries near the fusion line. These impurity-weakened boundaries tend to rupture as the weld cools because of the high residual stresses.

Inclusions. Large amounts of sulfur and phosphorus are added to some steels to provide free-machining characteristics. These steels have relatively poor weldability because of hot tearing in the weld metal caused by low-melting compounds of phosphorus and sulfur at the grain boundaries. Iron oxide and iron sulfide inclusions, if present, are also harmful because of their solubility change with temperature and their propensity to precipitate at grain boundaries, contributing to low ductility, cracking, and porosity. Laminations, which are flat separations or weaknesses that sometimes occur beneath and parallel to the surface of rolled products, have a slight tendency to open up if they extend to the weld joint.

Weldability of Steels

Low-carbon steels are mainly used in structural applications. Steels with less than 0.15 wt% C may harden to 30 to 40 HRC. Plain carbon steels containing less than 0.30 wt% C and 0.05 wt% S can be welded readily by most methods with little need for special measures to prevent weld cracking. The welding of sections that are more than 25 mm (1 in.) thick, particularly if the carbon content of the base metal exceeds 0.22 wt%, may require that the steel be preheated to approximately 40 °C (100 °F) and stress relieved at approximately 525 to 675 °C (1000 to 1250 °F).

For low-carbon steels, a low-alloy filler metal is generally recommended for meeting mechanical property requirements. The general procedure is to match the filler with the base metal in terms of strength or, for dissimilar welds, to match the lower-strength material. Often, however, higher-strength weld metal may actually require a softer HAZ to undergo a relatively large amount of strain when the joint is subjected to deformation near room temperature. Nevertheless, a low-strength filler metal should not be used indiscriminately as a remedy for cracking difficulties.

Medium-Carbon Steels. If steel containing about 0.5 wt% C is welded by a procedure commonly used for low-carbon steel, the heat-affected zone is likely to be hard, low in toughness, and susceptible to cold cracking. As indicated previously, preheating the base metal can greatly reduce the rate at which the weld area cools, thus reducing the likelihood of martensite formation. Postheating can further retard the cooling of the weld or can temper any martensite that might have formed.

The appropriate preheat temperature depends on the carbon equivalent of the steel, the joint thicknesses, and the welding procedure. With a carbon equivalent in the 0.45 to 0.60 wt% range, a preheat temperature in the range of approximately 95 to 100 °C (200 to 400 °F) is generally recommended. The minimum interpass temperature should be the same as the preheat temperature. A low-hydrogen welding procedure is mandatory with these steels. Modifications in welding procedure, such as the use of a larger V-groove or of multiple passes, also decrease the cooling rate and the probability of weld cracking.

Dilution can be minimized by depositing small weld beads or by using a welding procedure that provides shallow penetration. This is done to minimize carbon pick-

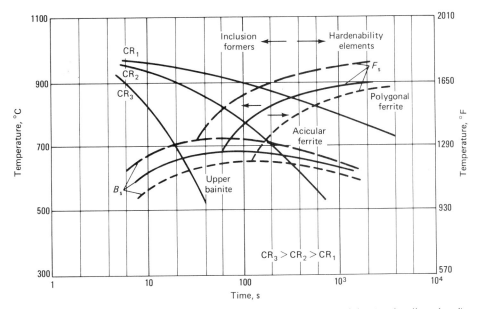

Fig. 7 Continuous cooling transformation diagram for an HSLA steel weld metal showing the effect of cooling rate and chemical composition on microstructure. CR, cooling rate

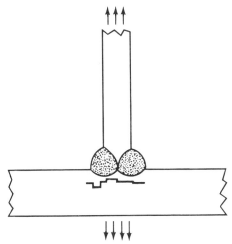

Fig. 8 Lamellar tear caused by thermal contraction strain

up from the base metal and the amount of hard transformation products in the fusion zone. Low heat input to limit dilution is also recommended for the first few layers in a multipass weld.

High-carbon steels generally contain over 0.60 wt% C and exhibit a very high elastic limit. They are often used in applications where high wear resistance is required. These steels have high hardenability and sensitivity to cracking in both the weld metal and the HAZ. A low-hydrogen welding procedure must be used for arc welding. Preheat and postheat will not actually retard the formation of brittle high-carbon martensite in the weld. However, preheating can minimize shrinkage stresses, and postheating can temper the martensite that forms. Successful welding of high-carbon steel requires the development of a specific welding procedure for each application. The composition, thickness, and configuration of the component parts must be considered in process and consumable selections.

High-strength low-alloy steels are designed to meet specific mechanical properties rather than a chemical composition. The alloy additions to HSLA steels strengthen the ferrite, promote hardenability, and help to control grain size. Weldability decreases as yield strength increases. For all practical purposes, welding these steels is the same as welding plain carbon steels that have similar carbon equivalents. Preheating may sometimes be required, but postheating is almost never required.

Quenched and tempered steels are furnished in the heat-treated condition with yield strengths ranging from approximately 350 to 1000 MPa (50 to 150 ksi), depending on the composition. The base metal is kept at less than 0.22% C for good weldability. Preheating must be used with caution when welding QT steels because it reduces the cooling rate of the weld HAZ. If the cooling rate is too slow, the reaustenitized zone adjacent to the weld metal can transform either to ferrite with regions of high-carbon martensite, or to coarse bainite, of lower strength and toughness. A moderate preheat, however, can ensure against cracking, especially when the joint to be welded is thick and highly restrained. A postweld stress-relief heat treatment is generally not required to prevent brittle fracture in weld joints in most QT steels.

Heat-treatable low-alloy steels. Examples of HTLA steels include AISI 4140, AISI 4340, AISI 5140, AISI 8640, and 300M. The high hardness of these steels requires that welding be conducted on materials in an annealed or overtempered condition, followed by heat treatment to counter martensite formation and cold cracking. However, high preheating is often used with a low-hydrogen process on these steels in a quenched and tempered condition, as in motor shaft applications. Preheating, or interpass heating, for both the weld metal and the HAZ are recommended. Hydrogen control is also essential to prevent weld cracking. Extremely clean vacuum-melted steels are preferred for welding.

Low sulfur and phosphorus, as described previously, are required to reduce hot cracking. Segregation, which occurs because of the extended temperature range at which solidification takes place, reduces high-temperature strength and ductility. Fillers of lower carbon and alloy content are highly recommended. Preheat and interpass temperatures of 315 °C (600 °F) or higher are very harsh environments for welders because of the physical discomfort and because an oxide layer forms at the weld joint. However, the cooling rate must be controlled to allow the formation of a bainitic microstructure instead of the hard martensite. The bainitic microstructure can be heat treated afterward to restore the original mechanical properties of the structure. Specifications and procedures should be followed rigorously for difficult-to-weld materials.

Precoated Steels. Thin plates and steel sheets are often precoated to protect them from oxidation and corrosion. The coatings commonly used are aluminum (aluminized), zinc (galvanized), and zinc-rich primers. As expected, the coating originally at the weld region is destroyed during fusion welding, and the effectiveness of the coating adjacent to the weld is significantly decreased by the welding heat. In the case of aluminized steels, the formation of aluminum oxide may adversely affect the wetting and weld pool shape. The welding electrode and filler metals should be selected carefully. A basic coating shielded metal arc (SMA) welding electrode is recommended.

For galvanized steels, weld cracking is generally attributed to intergranular penetration by zinc. Zinc dissolves considerably in iron to form an intermetallic compound at temperatures close to the melting temperature of zinc. Thus, molten zinc penetrates along the grain boundaries, leaving behind a brittle compound that fractures during cooling with the onset of a tensile stress state. Cracking occurs primarily at the throat region of a fillet weld, where shrinkage strain is more significant. The use of hot-dipped coatings results in more severe cracking, while thin electrogalvanized coatings are the least susceptible to cracking.

Low-silicon electrodes and rutile-base SMA welding rods are both good for galvanized steel welding. Specific welding and setup procedures should be followed, such as removing the zinc coating by an oxy-fuel

610 / Fabrication Characteristics of Carbon and Low-Alloy Steels

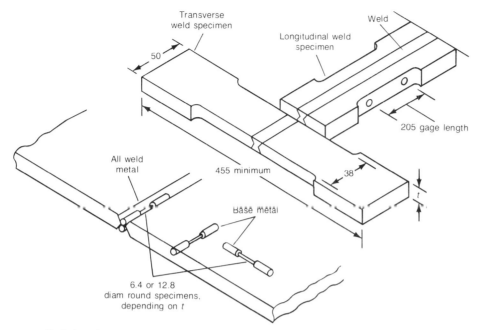

Fig. 9 Typical tension test specimens for evaluating welded joints. Both plate-type specimens have identical dimensions. All dimensions given in millimeters

process or by grinding, ensuring a large root opening, and using a slower welding speed to allow zinc vaporization and to prevent zinc entrapment in the weld metal. Adequate ventilation and fuel extraction should be mandatory in welding galvanized steels because of the health hazard of zinc fumes.

Weldability Tests

Weldability tests are conducted to provide information on the service and performance of welds. However, the data obtained in these tests can also be applied to the design of useful structures. Frequently these data are obtained from the same type of test specimens used in determining the base metal properties. Predicting the performance of structures from a laboratory-type test is very complex because of the nature of the joint, which is far from homogeneous, metallurgically or chemically. Along with the base metal, the weld joint consists of the weld metal and the HAZ. Thus, a variety of properties are to be expected throughout the welded joint. Careful interpretation and application of the test results are required.

There are currently many tests that evaluate not only the strength requirements of steel structures, but also the fracture characteristics and the effect of environmental conditions on early failure of the weldments. Selected major tests are described below.

Weld Tension Test. To obtain an accurate assessment of the strength and ductility of welds, several tension test specimens can be used; all weld metal specimens, transverse weld specimens, and longitudinal weld specimens are shown in Fig. 9. In the all weld metal test, base metal dilution should be minimized if the test is to be representative of the weld metal. However, the resulting properties may not be easy to translate into those properties achievable from welds made in an actual weld joint.

Interpreting test results for the transverse butt weld test is complicated by the different strengths and ductilities generally found in the various regions of the joint. The primary information gained from this test is the ultimate tensile strength. Yield strength and elongation requirements are generally not specified.

Tests of HAZ properties that are unaffected by the presence of either base metal or weld metal are not easy to conduct because it is practically impossible to obtain specimens made up entirely of the HAZ. In addition, as indicated earlier, the HAZ is composed of various regions, each with its own distinct properties. Simulated HAZ specimens that are generated and tested using a Gleeble thermomechanical testing system can be used to provide a more accurate assessment of the tensile properties of this region.

Bend Test. Different types of bend tests are used to evaluate the ductility and soundness of welded joints. Bend test results are expressed in various terms, such as percent elongation in outer fibers, minimum bend radius prior to failure, go/no-go for specific test conditions, and angle of bend prior to failure. Various specimen designs, both notched and unnotched, and testing techniques have been used. Today, unnotched specimens can be used in quality control tests, while notched specimens may be used to predict in-service behavior; however, most notched bend tests are used for research purposes and are not in common industrial use. Transverse bend tests are useful because they quite often reveal the presence of defects that are not detected in tension tests. However, the transverse specimen suffers from the same weakness as the transverse weld tension test specimen in that nonuniform properties along the length of the specimen can cause nonuniform bending, although this is often compensated for by the use of a wraparound bend fixture.

Hardness testing can be used to complement information gained through tension or bend tests by providing information about the metallurgical changes caused by welding. Routine methods for the hardness testing of metals are well established. In carbon and low-alloy steels, the hardness near the fusion line in the HAZ may be much higher than in the base metal because of the formation of martensite. In the HAZ areas where the temperature is low, the hardness may be lower than in the base metal because of tempering effects.

The drop-weight test design is based on service failures resulting from brittle fracture initiation at a small flaw located in a region of high stress. The drop-weight test can be considered a limited-deflection bend test that uses a crack starter to introduce a running crack in the specimen. The specimen is a bar on which a brittle crack starter weld is deposited. This overlay cracks when the bar is deflected by the drop weight. A series of tests is performed at different temperatures to determine the testing temperature below which the crack will propagate to the edges of the specimen. This critical temperature is also called the nil-ductility temperature (NDT), defined as the highest temperature at which the propagating crack reaches the edge of the specimen. Therefore, the drop-weight test is also known as the NDT test.

The Charpy V-notch (CVN) test is the most popular technique for evaluating the impact properties of welds. The energy absorbed by a sample at fracture determines the toughness of the specimen. In this test, specimens at different temperatures are broken using a pendulum hammer. A typical plot of CVN results for a carbon and low-alloy steel is illustrated in Fig. 10. The plot shows that there is a transition from low- to high-energy fracture over a narrow temperature range; this is associated with a change from transcrystalline to ductile fracture. Therefore, material quality can be defined in terms of this transition temperature.

In the CVN testing of welds, the notch is typically located at the weld centerline. For CVN testing of the heat-affected zone, the notch is more typically introduced at the CGHAZ. However, because precise location of a notch is never simple in the HAZ, simulated weld samples are used instead.

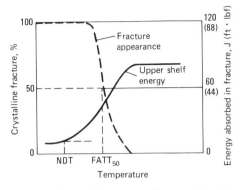

Fig. 10 Schematic impact transition curve for steel. FATT$_{50}$, fracture appearance transition temperature

The crack tip opening displacement test measures toughness, primarily for elastic-plastic conditions. In CTOD tests, the clip gage opening at the onset of fracture is measured and used to calculate the crack opening displacement at the crack tip. The critical value of CTOD at fracture, Δc, is a critical strain parameter that is analogous to the critical stress-intensity parameter. K_{Ic}. The CTOD test provides a useful method of determining the critical flaw size. Nevertheless, the test is very sensitive to changes in sample thickness, hardness, and strength, and it is difficult to obtain valid results in practical specimen thicknesses.

The application of fracture mechanics to the prevention of catastrophic failure in weldments is, however, complicated by the nature of the weldment. In addition to their metallurgical heterogeneity, weldments often contain high residual stresses. Consequently, it is inadequate to fracture test the base metal and assume that the critical crack length thus determined is valid when the base metal is made into a weldment. The fracture toughness criterion must be determined for the base metal, the HAZ, and the weld metal. By first determining the zone with the lowest toughness value, it is then possible to evaluate a more realistic critical flaw size. However, the plane-strain fracture toughness tests are preproduction or pilot plant type tests that provide a rational means for designs and engineers to estimate the effects of new designs, materials, or fabrication practices on the fracture-safe performance of structures. Other popular tests include compact tension (CT) and wedge opening load (WOL) tests, which are commonly used in the evaluation of structural weldments. Further discussion of CTOD and other fracture toughness testing of welds is available in the "Selected References" at the end of this article.

Stress-Corrosion Cracking Test. The presence of corrosive environments in a steel weldment may accelerate the initiation of a crack. Usually, the higher the strength of the steel, the more susceptible it becomes to stress-corrosion cracking. The steels considered in this article are not usually exposed to severely corrosive evironments, but rather to the atmosphere, moisture, hydrocarbons, fertilizers, and soils. Nevertheless, welding can lower corrosive resistance by the introduction of:

- Compositional differences that promote galvanic attack between weld metal, HAZ, and base metal when the joint in immersed in a conducting liquid
- Residual stresses that can cause stress-corrosion cracking
- Surface flaws that can act as sites for stress-corrosion cracking

Stress-corrosion cracking is generally delayed cracking, with longer time to failure at lower stresses. Most stress-corrosion tests are fairly long in terms of time because of the slow crack initiation that occurs in unnotched test bars. However, it has been found that the long initiation period can be eliminated by testing precracked specimens. Additional information on the stress-corrosion cracking test is available in the "Selected References" at the end of this article.

Fabrication Weldability Tests

There are various types of tests for determining the susceptibility of the weld joint to different types of cracking during fabrication. They are:

- Restraint tests
- Externally loaded tests
- Underbead cracking tests
- Lamellar tearing tests

Table 1 summarizes the applications, controllable test variables, and typical test data of several fabrication weldability tests to illustrate the differences among them. Of

Table 1 Comparison of weldability tests for fabrication

Test	Fields of use	Controllable variables	Type of data	Specific equipment	Cost
Lehigh restraint test	Weld metal hot and cold cracks, root cracks, HAZ hydrogen cracks, stress-relief cracks	Joint geometry, process, filler metal, restraint level, heat input, preheat, postweld heat treatment	Critical restraint, or % hindered control	None	Costly machining
Slot test	HAZ hydrogen cracks	Filler metal, interpass time, preheat	Time to crack, critical preheat	None	Low cost
Rigid restraint test	Weld metal hot and cold cracks, root cracks, HAZ hydrogen cracks	Joint geometry, process restraint level, filler metal, heat input, preheat	Critical restraint	Restraint jig	Costly machining and setup
Tekken test	Weld metal root cracks, HAZ hydrogen cracks	Joint geometry, process filler metal, heat input, preheat	Critical preheat	None	Low cost
Circular groove test	Weld metal hot and cold cracks, HAZ hydrogen cracks	Process, filler metal, preheat	Go/no-go	None	Costly preparation
Implant test	HAZ hydrogen cracks, stress-relief cracks	Process, filler metal, preheat, postweld heat treatment	Critical fracture stress, critical preheat	Loading jig	Intermediate cost
Tension restraint cracking test	HAZ hydrogen cracks	Process, filler metal, heat input, preheat	Critical fracture stress, critical preheat	Loading jig	Costly machining and setup
Varestraint test	Weld metal and HAZ hot cracks	Process, filler metal, heat input	Crack length, % strain	Loading jig	Costly preparation and analysis
Longitudinal bead-on-plate test	HAZ hydrogen cracks	Electrical type, heat input	% cracking	None	Low cost
Controlled thermal severity test	HAZ hydrogen cracks in fillet welds	Electrical type, cooling rate, preheat	Go/no-go (at two cooling rates)	None	Costly preparation
Cruciform test	HAZ hydrogen cracks, weld metal root cracks	Process, heat input, preheat, filler metal	Go/no-go	None	Costly preparation
Lehigh cantilever test	Lamellar tearing	Process, filler metal, heat input, preheat	Critical restraint stress and strain	Loading jig	Costly specimen preparation
Cranfield test	Lamellar tearing	Filler metal	Number of passes to crack	None	Low cost
Nick bend test	Weld metal soundness	Filler metal	Go/no-go	None	Low cost

Source: Ref 1

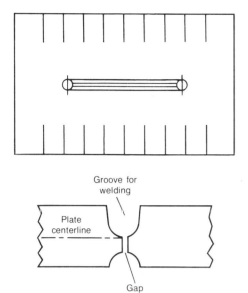

Fig. 11 Basic outline of a Lehigh test specimen

the many tests identified in Table 1, the Lehigh restraint test, the Varestraint test, and the controlled thermal severity test are described below.

The Lehigh restraint test (Fig. 11) is particularly useful for quantitatively rating the crack susceptibility of a weld metal as affected by electrode variables. This test provides a means of imposing a controllable severity of restraint on the root bead that is deposited in a butt weld groove with dimensions suitable to the application. Slots are cut in the sides and ends of a plate prior to welding. By changing the length of the slots, the degree of plate restraint on the weld is varied without significantly changing the cooling rate of the weld. Therefore, a critical restraint for cracking can be determined for given welding conditions. This sample is also useful for hydrogen cracking.

The Varestraint test (Fig. 12) determines the susceptibility of the welded joint to hot cracking. The test utilizes external loading to impose controlled plastic deformation in a plate while a weld bead is being deposited on the long axis of the plate. The specimen is mounted as a cantilever beam, and a pneumatically driven yoke is positioned to force the specimen downward when the welding arc reaches a predetermined position. By the choice of the radius to which the plate is bent, the severity of deformation causing cracking can be determined. Strain from 0 to 4% can be chosen according to the susceptibility of the joint to hot cracking. When the bending moment is applied transverse to the weld axis, the test is termed transvarestraint. A spot Varestraint test can also be conducted by keeping the arc stationary; bending is applied at the moment the arc is extinguished.

The controlled thermal severity test (Fig. 13) is designed to measure the cracking sensitivity of steels under cooling rates controlled by the thickness of the plates and the number of paths available for dissipating the welding heat. It is conducted with a plate bolted and anchor welded to a second plate in a position to provide two fillet (lap) welds. The fillet located at the plate edges has two paths of heat flow. The lap weld located near the middle of the bottom plate has three paths of heat flow, thus inducing faster cooling. The fillet welds are made first and allowed to cool, followed by the lap welds. After a holding time of 72 h at room temperature, the degree of cracking is determined by measuring the crack length on metallographic specimens.

A number of other tests have been developed that contain welds in a circular configuration. The circular patch test has probably the most severe testing conditions; the two varieties are the Navy circular patch restraint test and the segmented circular patch restraint test. Cracking is detected by visual, radiographic, and liquid penetrant inspection. The cracking susceptibility of a material is measured as the total crack length and expressed as a percentage of the weld length. These tests can be used to

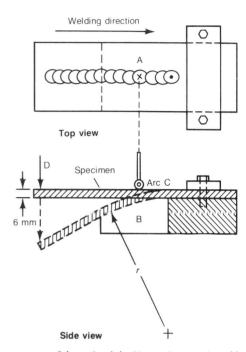

Fig. 12 Schematic of the Varestraint test. A, weld location; B, die; C, arc; D, load; r, radius of deformation

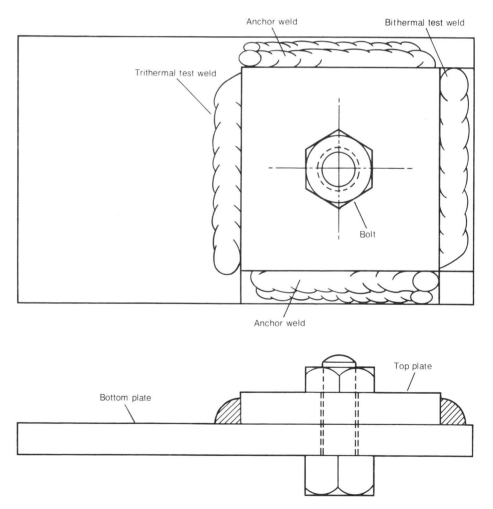

Fig. 13 Schematic of the controlled thermal severity test

determine both hot and cold cracking in the weld metal and the HAZ. Depending on the results, a go/no-go criteria is established for weld qualification. Detailed descriptions of these tests can be found in the "Selected References" at the end of this article.

REFERENCE

1. R. Stout, *Weldability of Steels*, 4th ed., Welding Research Council, 1987

SELECTED REFERENCES

- R.R. Barr and F.M. Burdekin, Design Against Brittle Failure, in *Rosenhain Centenary Conference Proceedings*, R.G. Baker and A. Kelly, Ed., The Royal Society, p 85, 1975
- O. Blodgett, "Why Preheat? An Approach to Estimating Correct Preheat Temperature," Brochure G-231, Lincoln Arc Welding Foundation, June 1970
- B.F. Brown, "Stress Corrosion Cracking and Corrosion Fatigue of High Strength Steels," Report 210, Defense Metals Information Center, 1964, p 91-102
- "Classification of Microstructures in Low Carbon Low Alloy Steel Weld Metal and Terminology," DOC IX-1282-83, International Institute of Welding, 1983
- J. Cornu, *Advanced Welding Systems: Fundamentals of Fusion Welding Technology*, IFS/Springer Verlag, 1988
- C.L.M. Cottrell, Hardness Equivalent May Lead to a More Critical Measure of Weldability, *Met. Constr.*, Vol 16 (No. 12), 1984, p 740-743
- C.E. Cross, Ø. Grong, S. Liu, and J.F. Capes, Metallography and Welding Process Control, in *Applied Metallography*, G. Vander Voort, Ed., Van Nostrand Reinhold, 1985
- G.J. Davies and J.G. Garland, Solidification Structures and Properties of Fusion Welds, *Int. Met. Rev.*, No. 20, 1975, p 83-106
- K. Easterling, *Introduction to the Physical Metallurgy of Welding*, Butterworths, 1983
- D.P. Fairchild, Brittle Zones in Structural Welds, in *Welding Metallurgy of Structural Steels*, J. Koo, Ed., The Metallurgical Society, 1987, p 303-318
- H. Granjon, "Notes on the Carbon Equivalent," DOC IX-555-67, International Institute of Welding, 1967
- J.D. Harrison, M.G. Davies, G.L. Archer, and M.S. Kamath, "The COD Approach and Its Application to Welded Structures," Report 55/1978, The Welding Institute, 1978
- Y. Ito and K. Bessyo, "Weldability Formula of High Strength Steels Related to Heat Affected Zone Cracking," DOC IX-576-68, International Institute of Welding, 1968
- J. Koo and A. Ozekan, Local Brittle Zone Microstructure, in *Welding Metallurgy of Structural Steels*, J. Koo, Ed., The Metallurgical Society, 1987, p 119-135
- S. Kou, *Welding Metallurgy*, Wiley Interscience, 1987
- S. Liu, D.L. Olson, and D.K. Matlock, A Thermodynamic and Kinetic Approach in the Development of Expressions for Alloy Behavior Prediction, *J. Heat Treat.*, Vol 4 (No. 4), 1986, p 309-316
- F. Matsuda, T. Hashimoto, and T. Senda, Fundamental Investigations on Solidification Structure, *Trans. Natl. Res. Inst. Met. (Jpn.)*, Vol 11, 1969, p 43-58
- H.G. Pisarski and J. Kudoh, Exploratory Studies on the Fracture Toughness of Multi-Pass Welds With Locally Embrittled Regions, in *Welding Metallurgy of Structural Steels*, J. Koo, Ed., The Metallurgical Society, 1987, p 263-276
- *Properties and Selection: Irons and Steels*, Vol 1, 9th ed., *Metals Handbook*, American Society for Metals, 1978
- S.T. Rolfe, "Development of a K_{Ic} Stress Corrosion Specimen," Technical Report, United States Steel Applied Research Laboratory, 1965
- A.B. Rothwell, CAN/MET Report 79-6, *Can. Weld. Fabr.*, Vol 20, 1980
- C.P. Royer, A User's Perspective on HAZ Toughness, in *Welding Metallurgy of Structural Steels*, J. Koo, Ed., The Metallurgical Society, 1987, p 255-262
- H. Suzuki, "Carbon Equivalent and Maximum Hardness," DOC IX-1279-83, International Institute of Welding, 1983
- *Welding, Brazing, and Soldering*, Vol 6, 9th ed., *Metals Handbook*, American Society for Metals, 1983
- *Welding Handbook*, Vol I and II, 7th ed., American Welding Society, 1983

Service Characteristics of Carbon and Low-Alloy Steels

Elevated-Temperature Properties of Ferritic Steels 617
Effect of Neutron Irradiation on Properties of Steels 653
Low-Temperature Properties of Structural Steels 662
Fatigue Resistance of Steels 673
Embrittlement of Steels 689
Notch Toughness of Steels 737

Elevated-Temperature Properties of Ferritic Steels

CARBON STEELS and low-alloy steels with ferrite-pearlite or ferrite-bainite microstructures are used extensively at elevated temperatures in fossil-fired power-generating plants, aircraft power plants, chemical-processing plants, and petroleum-processing plants. Carbon steels are often used up to about 370 °C (700 °F) under continuous loading, but also have allowable stresses defined up to 540 °C (1000 °F) in Section VIII of the ASME Boiler and Pressure Vessel Code. Carbon-molybdenum steels with 0.5% Mo are used up to 540 °C (1000 °F), while low-alloy with 0.5-1.0% Mo in combination with 0.5-9.0% Cr and sometimes other carbide formers (such as vanadium, tungsten, niobium, and titanium) are often used up to about 650 °C (1200 °F). For temperatures above 650 °C (1200 °F), austenitic alloys are generally used. However, these general maximum-use temperature limits do not necessarily apply in specific applications with different design criteria. Tables 1 and 2, for example, list maximum-use temperatures in two specific application areas with different design criteria.

This article covers some elevated-temperature properties of carbon steels and low-alloy steels with ferrite-pearlite and ferrite-bainite microstructures for use in boiler tubes, pressure vessels, and steam turbines. In these applications, the selection of steels to be used at elevated temperatures generally involves compromise between the higher efficiencies obtained at higher operating temperatures and the cost of equipment, including materials, fabrication, replacement, and downtime costs. The highly alloyed steels, which depend on an austenitic matrix for their high-temperature properties, generally have higher resistance to mechanical and chemical degradation at elevated temperatures than the low-alloy ferritic steels. However, a higher alloy content generally means higher cost. Therefore, carbon and low-alloy ferritic steels are extensively used in several forms (piping, pressure vessel plates, bolts, structural parts) in a variety of applications that involve exposure to elevated temperatures. In addition, interest in ferritic steels has increased recently because their relatively lower thermal expansion coefficient and higher thermal conductivity make them more attractive than austenitic steels in applications where thermal cycling is present.

To illustrate the tonnage requirements for carbon and low-alloy steels in industrial construction, 1360 Mg (1500 tons) of pressure tubing were required for the construction of a single 500 MW coal-fired generating plant. The quantities of the various carbon and low-alloy steels used in the pressure tubing were as follows:

Steel type	Tons	% of total tonnage
Carbon	540	36
C-½Mo	150	10
1¼Cr-1Mo	495	33
2¼Cr-1Mo	150	10
9Cr-1Mo	165	11

This list of carbon and low-alloy steels is for pressure tube applications and does not include the chromium-molybdenum-vanadium steels that are used for turbine rotors, high-temperature bolts, and pressure tubing.

Carbon and Low-Alloy Steels for Elevated-Temperature Service

The numerous types of steels used in elevated-temperature applications include the following:

- Carbon steels
- Low-alloy steels
- High steels
- Stainless steels
- Hot-work tool steels
- Iron-base superalloys

Within the context of this article, the low-alloy steels considered are the creep-resistant steels with 0.5 to 1.0% Mo combined with 0.5 to 9.0% Cr and perhaps other carbide formers (such as vanadium, tungsten, niobium, and titanium). High-strength low-alloy (HSLA) steels are not considered here because they typically have molybdenum contents below 0.5%, which limits their resistance against creep and temper embrittlement. However, HSLA steels, which are discussed in the article "High-Strength Structural and High-Strength Low-Alloy Steels" in this Volume, may be effective substitutes for carbon steels in elevated-temperature applications. Another category of ferritic steels for elevated-tem-

Table 1 Temperature limits of superheater tube materials covered in ASME Boiler Codes

	Maximum-use temperature			
	Oxidation/graphitization criteria, metal surface(a)		Strength criteria, metal midsection	
Material	°C	°F	°C	°F
SA-106 carbon steel	400–500	750–930	425	795
Ferritic alloy steels				
0.5Cr-0.5Mo	550	1020	510	950
1.2Cr-0.5Mo	565	1050	560	1040
2.25Cr-1Mo	580	1075	595	1105
9Cr-1Mo	650	1200	650	1200
Austenitic stainless steel				
Type 304H	760	1400	815	1500

(a) In the fired section, tube surface temperatures are typically 20–30 °C (35–55 °F) higher than the tube midwall temperature. In a typical U.S. utility boiler, the maximum metal surface temperature is approximately 625 °C (1155 °F).

Table 2 Suggested maximum temperatures in petrochemical operations for continuous service based on creep or rupture data

	Maximum temperature based on creep rate		Maximum temperature based on rupture	
Material	°C	°F	°C	°F
Carbon steel	450	850	540	1000
C-0.5 Mo steel	510	950	595	1100
2¼ Cr-1Mo steel	540	1000	650	1200
Type 304 stainless steel	595	1100	815	1500
Alloy C-276 nickel-base alloy	650	1200	1040	1900

618 / Service Characteristics of Carbon and Low-Alloy Steels

Table 3(a) Compositions of steels for elevated-temperature service

ASME specification	UNS designation	Nominal composition	Product form	C	Mn	Si	P	S	Cr	Ni	Mo	Others
SA-106A	K02501	C	Seamless carbon steel pipe	0.25(a)	0.27–0.93	0.10(b)	0.048(a)	0.058(a)	...	...	...	...
SA-106B	K01700	C-Si	Seamless carbon steel pipe	0.30(a)	0.29–1.06	0.10(b)	0.048(a)	0.058(a)	...	...	...	...
SA-285A	K03006	C	Carbon steel PV plate	0.17(a)	0.90(a)	...	0.035(a)	0.045(a)	...	...	...	...
SA-299	K02803	C-Mn-Si	C-Mn-Si steel PV plate	0.28(a)	0.90–1.40	0.15–0.30	0.035(a)	0.040(a)	...	...	...	0.25 Cu(a)
SA-204A	K11820	C-½Mo	Mo alloy steel PV plate	0.18(a)	0.90(a)	0.15–0.30	0.035(a)	0.040(a)	...	...	0.45–0.60	...
SA-302A	K12021	Mn-Mo	Mn-Mo-Mn and Mo-Ni alloy PV plate	0.20(a)	0.95–1.30	0.15–0.30	0.035(a)	0.040(a)	...	...	0.45–0.60	...
SA-533B2	K12539	Mn-Mo-Ni	Mn-Mo-Mn and Mo-Ni alloy steel PV plate	0.25(a)	1.15–1.50	0.15–0.30	0.035(a)	0.040(a)	...	0.40–0.70	0.45–0.60	0.10 Cu(a)
SA-517F	K11576	...	High-strength alloy steel PV plate	0.10–0.20	0.60–1.00	0.15–0.35	0.035(a)	0.040(a)	0.40–0.65	0.70–1.00	0.40–0.60	0.002–0.006 B, 0.15–0.050 Cu, 0.03–0.08 V
SA-335 P12	K11562	1Cr-½Mo	Seamless ferritic alloy steel pipe for high-temperature service	0.15(a)	0.30–0.61	0.50(a)	0.045(a)	0.045(a)	0.50–1.25	...	0.44–0.65	...
SA-217WC6	J12072	1¼Cr-½Mo	Alloy steel castings	0.20(a)	0.50–0.80	0.60(a)	0.04(a)	0.045(a)	1.00–1.50	...	0.45–0.65	...
SA-387Gr22	K21590	2¼Cr-1Mo	Cr-Mo alloy steel PV plate	0.15(a)	0.30–0.60	0.50(a)	0.035(a)	0.035(a)	2.0–2.5	...	0.90–1.10	...
SA-387Gr5	S50100	5Cr-½Mo	Cr-Mo alloy steel PV plate	0.15(a)	0.30–0.60	0.50(a)	0.040(a)	0.030(a)	4.0–6.0	...	0.45–0.65	...
SA-217C12	J82090	9Cr-1Mo	Alloy steel castings	0.02(a)	0.35–0.65	1.00(a)	0.04(a)	0.045(a)	8.0–1.0	...	0.90–1.20	...

(a) Maximum. (b) Minimum

Table 3(b) Room-temperature mechanical properties of steels for elevated-temperature service listed in Table 3(a)

ASME specification	Tensile strength MPa	Tensile strength ksi	Yield strength, minimum MPa	Yield strength, minimum ksi	Minimum elongation in 50 mm (2 in.), %	Minimum reduction in area, %
SA-106A	330	48(a)	207	30	35(b), 25(c)	...
SA-106B	415	60(a)	241	35	30(b), 16.5(c)	...
SA-285A	310–380	45–55	165	24	27(d), 30	...
SA-299	515–620	75–90	290	42	16(d)	...
SA-204A	445–530	65–77	255	37	19(d), 23	...
SA-302A	515–655	75–95	310	45	15(d), 19	...
SA-533B2	620–790	90–115	475	70	16	...
SA-517F	795–930	115–135	690	100	16	35–45
SA-335P12	415	60(a)	207	30	30(b), 20(c)	...
SA-217WC6	485–620	70–90	275	40	20	35
SA-387Gr22-1	415–585	60–85	207	30	18(d), 45	40
SA-387Gr5-2	515–690	75–100	310	45	18(d), 22	45
SA-217C12	620–795	90–115	415	60	18	35

(a) Minimum. (b) Longitudinal. (c) Transverse. (d) Elongation in 200 mm (8 in.)

perature service are manganese-molybdenum-nickel ferritic steels (ASTM A 302 and A 533), which are commonly used for pressure vessels in light-water reactors. High-alloy steels, stainless steels, hot-work tool steels, and the iron-base superalloys are discussed in the Section "Specialty Steels and Heat-Resistant Alloys" in this Volume.

Alloy Designations and Specifications

Carbon and low-alloy steels used for elevated-temperature service are usually identified by American Iron and Steel Institute (AISI) designations; aerospace material specification (AMS), American Society of Mechanical Engineers (ASME), or American Society for Testing and Materials (ASTM) specification number; nominal composition; or trade name. These steels have also been assigned numbers in the Unified Numbering System. In addition, there are Military and Federal specifications covering many of these steels.

Steel products manufactured for use under the ASME Boiler and Pressure Vessel Code must comply with provisions of the appropriate ASME specification. Each specification includes information on ranges and limits of composition, dimensions and tolerances, minimum mechanical properties, and other functional requirements. The designations applied to these products include the letters "SA," the number of the specification, and possibly other letters or numbers to distinguish among the various types, grades, and classes within a single specification. Most ASTM specifications are identical to the ASME specification of the same number except that the ASTM designations begin with the letter "A." Some examples of ASME specifications for elevated-temperature steels, as well as their compositions and typical room-temperature mechanical properties, are given in Tables 3(a) and (b).

Aerospace material specifications, as the name suggests, are specifications for products intended for the aerospace industry. The nominal compositions, typical applications, and typical mechanical properties of steels often identified by AMS numbers are given in Table 4.

The AISI designation for steels intended for elevated-temperature service is a three-digit number beginning with a 6, such as 601. The AISI designations are also included in Table 4.

Carbon Steels

Carbon steels are the predominant materials in pressure vessel fabrication because of their low cost, versatile mechanical properties, and availability in fabricated forms. They are the most common materials used in noncorrosive environments in the temperature range of −29 to 425 °C (−20 to 800 °F) in oil refineries and chemical plants. Although the ASME code gives allowable stresses for temperatures greater than 425 °C (800 °F), it also notes that prolonged exposure at these temperatures may result in the carbide phase of the carbon steel being converted to graphite. This phenomenon, known as graphitization, is a cumulative process dependent on the time the material is at or above 425 °C (800 °F). The result is a weakening of the steel after high-temperature exposure (Fig. 1). Carbon steels are also increasingly affected by creep at temperatures above 370 °C (700 °F). Figure 2 shows the effect of temperature on the stress-to-rupture life of a carbon steel.

Table 4 Compositions and mechanical properties of AISI steels for elevated-temperature service

AISI designation	AMS designations	Commercial designation	UNS designations	Typical applications	Nominal composition, %					
					C	Mn	Si	Cr	Mo	V
601	6304	...	K14675	Bolting and structural parts	0.46	0.60	0.26	1.00	0.50	0.30
602	6302, 6385, 6458	17-22 AS	K23015	Bolting and structural parts	0.30	0.55	0.65	1.25	0.50	0.25
603	6303, 6436	17-22 AV	K22770	Turbine rotors and aircraft parts	0.27	0.75	0.65	1.25	0.50	0.85
610	6437, 6485	H11 mod	T20811 K74015	Ultrahigh-strength components	0.40	0.30	0.90	5.00	1.30	0.50

AISI designation	Room-temperature tensile properties						Temperature at which 70 MPa (10 ksi) will cause rupture in				Temperature to produce min creep rate at 70 MPa (10 ksi)			
	Yield strength		Tensile strength		Elongation in 50 mm (2 in.), %	Reduction in area, %	1000 h		10 000 h		1 µm/m · h		0.1 µm/m · h	
	MPa	ksi	MPa	ksi			°C	°F	°C	°F	°C	°F	°C	°F
601	710	103	855	124	29	61	620	1150	595	1100	...	...	...	...
602	745–930	108–135	880–1060	128–154	16–21	53–63	625	1160	590	1090	...	...	555	1030
603	1000	145	1100	160	17	52	650	1200	613	1135	...	...	565	1050
610	1480	215	1805	262	10	36	630	1170	595	1100	560	1040	540	1000

Creep-Resistant Low-Alloy Steels

Creep-resistant low-alloy steels usually contain 0.5 to 1.0% Mo for enhanced creep strength, along with chromium contents between 0.5 and 9% for improved corrosion resistance, rupture ductility, and resistance against graphitization. Small additions of carbide formers such as vanadium, niobium, and titanium may also be added for precipitation strengthening and/or grain refinement. The effects of alloy elements on transformation hardening and weldability are, of course, additional factors.

The three general types of creep-resistant low-alloy steels are chromium-molybdenum steels, chromium-molybdenum-vanadium steels, and modified chromium-molybdenum steels. Chromium-molybdenum steels are used primarily for tube, pipe, and pressure vessels, where the allowable stresses may permit creep deformation up to about 5% over the life of the component. Typical creep strengths of various chromium-molybdenum steels are shown in Fig. 3. Figure 3 also shows the creep strength of a chromium-molybdenum steel with vanadium additions. Chromium-molybdenum-vanadium steels provide higher creep strengths and are used for high-temperature bolts, compressor wheels, or steam turbine rotors, where allowable stresses may require deformations less than 1% over the life of the component.

Chromium-molybdenum steels are widely used in oil refineries, chemical industries, and electrical power generating stations for piping, heat exchangers, superheater tubes, and pressure vessels. The main advantage of these steels is the improved creep strength from molybdenum and chromium additions and the enhanced corrosion resistance from chromium. The creep strength of chromium-molybdenum steels is derived mainly from two sources: solid-solution strengthening of the matrix ferrite by carbon, molybdenum, and chromium; and precipitation hardening by carbides. Creep strength generally, but not always, increases with higher amounts of molybdenum and chromium. The effects of chromium and molybdenum on creep strength are complex (see "Effects of Composition" in this article). In Fig. 3, for example, 2.25Cr-1Mo steel has a higher creep strength than 5Cr-0.5Mo steel.

Chromium-molybdenum steels are available in several product forms (see Table 24 in the article "Classification and Designation of Carbon and Low-Alloy Steels" in this Volume). In actual applications, boiler tubes are used mostly in the annealed condition, whereas piping is used mostly in the normalized and tempered condition. Bend sections used in piping, however, are closer to an annealed condition than to a normalized condition. As a result of the cooling rates employed in these treatments, the microstructures of chromium-molybdenum steels may vary from ferrite-pearlite aggregates to ferrite-bainite aggregates. Bainite microstructures have better creep resistance under high-stress, short-time conditions but degrade more rapidly at high temperatures than pearlitic structures. As a result, ferrite-pearlite material has better intermediate-term, low-stress creep resistance. Because both microstructures will eventually spheroidize, it is expected that over long service lives the two microstructures will converge to similar creep strengths.

The 0.5Mo steel with 0.15% C is used for piping and superheater tubes operating at metal temperatures to 455 °C (850 °F). Above this temperature, spheroidization and graphitization may increase the possibility of failure in service. Use of carbon-molybdenum steel has been largely discontinued for the higher temperatures because

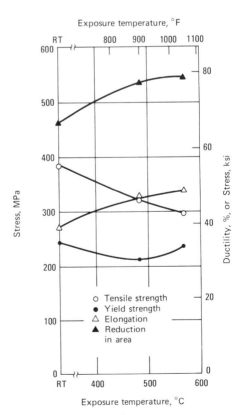

Fig. 1 Effect of elevated-temperature exposure on the room-temperature tensile properties of normalized 0.17% C steel after exposure (without stress) to indicated temperature for 83 000 h

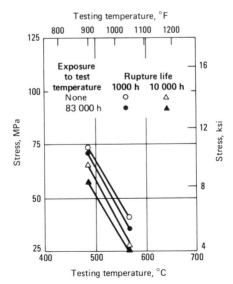

Fig. 2 Effect of exposure to elevated temperature on stress-to-rupture of carbon steel. Stress-to-rupture in 1000 and 10 000 h at the indicated temperature for specimens of normalized 0.17% C steel exposed to the test temperature (without stress) for 83 000 h and for similar specimens not exposed to elevated temperature prior to testing

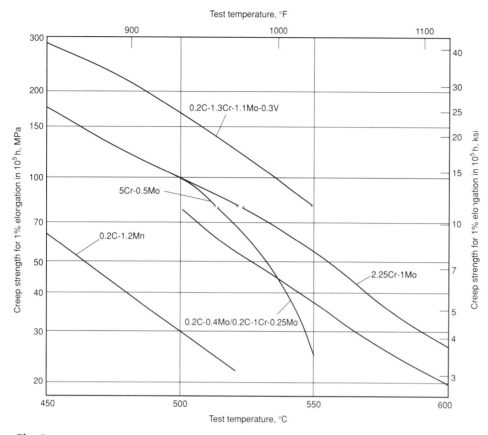

Fig. 3 General comparison of creep strengths of various creep-resistant low-alloy steels

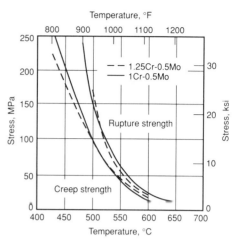

Fig. 4 Creep strength (0.01%/1000 h) and rupture strength (100 000 h) of 1Cr-0.5Mo and 1.25Cr-0.5Mo steel. Source: Ref 1

of graphitization. Chromium steels are highly resistant to graphitization and are therefore preferred for service above 455 °C (850 °F).

The *1.0Cr-0.5Mo steel* is used for piping, cracking-still tubes, and boiler tubes for service temperatures to 510 or 540 °C (950 or 1000 °F). The similar 1.25Cr-0.5Mo steel is used up to 590 °C (1100 °F) and has comparable stress-rupture and creep properties as that of the 1.0Cr-0.5Mo alloy (Fig. 4).

The *2.25Cr-1.0Mo steel* has better oxidation resistance and creep strength than the steels mentioned above. The 2.25Cr-1Mo steel is a highly favored alloy for service up to 650 °C (1200 °F) without the presence of hydrogen or 480 °C (900 °F) in a hydrogen environment. This steel, which has substantial documentation of its elevated-temperature properties (Ref 2 through 6), is discussed in more detail in the section "Elevated-Temperature Behavior of 2¼Cr-1Mo Steel" in this article.

The *5, 7, and 9% Cr steels* are generally lower in stress rupture and creep strength than the lower-chromium steels because the strength at elevated temperatures typically drops off with an increase in chromium. However, this may not always be the case, depending on the service temperature (Fig. 5) and the exposure (Fig. 6 and 7). Heat treatment is also an important factor. The main advantage of these steels is the improved oxidation resistance from the increased chromium content.

The chromium-molybdenum-vanadium steels are manufactured with higher carbon ranges (such as 0.28 to 0.33% and 0.40 to 0.50%) and are used in the normalized and tempered or quenched and tempered condition. Because of the relatively high yield strengths (Fig. 8) and creep strengths (Fig. 3), these steels are suitable for bolts, compressor wheels in gas turbines and steam turbine rotors, and other parts operating at temperatures up to 540 °C (1000 °F). The most common low-alloy composition contains 1% Cr, 1% Mo, and 0.25% V.

Bolt Applications. The basic compositions of low-alloy high-temperature bolt steels has evolved from chromium-molybdenum steels to chromium-molybdenum-vanadium steels. The chromium-molybdenum steels used until the late 1940s had creep strengths adequate for service at temperatures up to about 480 °C (895 °F). With the increasing need for a higher-strength steel, a 1Cr-1Mo-¼V steel strengthened by stable V_4C_3 precipitates was developed. This alloy was found to be adequate for steam temperatures up to 540 °C (1000 °F). When steam temperatures reached about 565 °C (1050 °F) in the mid-1950s, a 1Cr-1Mo-¾V steel, in which vanadium and carbon had been stoichiometrically optimized to get the largest volume fraction of V_4C_3 and hence the highest creep strength, was developed. Unfortunately, this development had overlooked the importance of rupture ductility, and many creep-rupture failures of bolts due to notch sensitivity occurred. The loss in rupture ductility was subsequently countered by grain refinement and by compositional modifications involving titanium and boron. Melting practice is also another factor in improving rupture ductility.

High-Temperature Rotor Applications. Since it was introduced in the 1950s, 1Cr-1Mo-0.025V steel has remained the industry standard in turbine rotor applications, although a few higher-alloy rotor steels (12% Cr) have been developed (see the article "Elevated-Temperature Properties of Stainless Steels" in this Volume). It is well recognized that 1Cr-1Mo-0.25V rotor steels are limited by their creep strength for service up to about 540 °C (1000 °F).

The desired properties in chromium-molybdenum-vanadium steel rotors is made possible by careful control of heat treatment and composition. In the United States, the usual practice has been to air cool the rotors from the austenitizing temperature in order to achieve a highly creep-resistant, but somewhat less tough, upper bainitic microstructure. In Europe, however, manufacturers have resorted to oil quenching of rotors from the austenitizing temperature to achieve a better compromise between creep strength and toughness. Oil quenching of 1Cr-1Mo-0.25V rotors may shift the transformation product increasingly toward lower bainite, but it is unlikely that the cooling rates needed for formation of martensite (that is, 10 000 °C/h or 20 000 °F/h) are ever encountered.

Comparative evaluation of creep properties of chromium-molybdenum-vanadium steels with martensite, bainite, and ferrite-pearlite as the principal microstructure have been conducted by numerous investigators, and the results have been reviewed elsewhere (Ref 8). There is consensus that upper bainitic structures provide the best creep resistance coupled with adequate ductility. Toughness properties are discussed in Ref 9.

Turbine Casing Applications. Chromium-molybdenum-vanadium steels are also used for turbine casings. The table below compares the maximum application temperatures of various low-alloy steels used for turbine casings (Ref 10):

Casing material	Maximum application temperature	
	°C	°F
C-½Mo (0.25C max, 0.20-0.50Si, 0.5-1.0Mn, 0.50-0.70Mo)	480	895
Cr-½Mo (0.15C max, 0.60Si max, 0.5-0.8Mn, 1.0-1.5Cr, 0.45-0.65Mo)	525	975
2¼Cr-1Mo (0.15C max, 0.45Si max, 0.4-0.8Mn)	540	1000
Cr-Mo-V (0.15C max, 0.15-0.30Si max, 0.4-0.6Mn, 0.7-1.2Cr, 0.7-1.2Mo, 0.25-0.35V)	565	1050
½Cr-Mo-V (0.1-0.15C, 0.45Si max, 0.4-0.7Mn, 0.4-0.6Cr, 0.4-0.6Mo, 0.22-0.28V)	565	1050

Modified Chromium-Molybdenum Steels. To achieve higher process efficiencies in future coal conversion plants, chemical-processing plants, and petrochemical-refining plants, several modified versions of chromium-molybdenum pressure vessel steels have been investigated for operation at higher temperatures and pressures than those currently encountered. The higher temperatures affect the elevated-temperature strength, the dimensional deformation, and the metallurgical stability of an alloy, while higher operating pressures require either higher-strength alloys or thicker sections.

Of the unmodified ferritic steels, SA-387 grade 22, class 2 (normalized and tempered 2¼Cr-1Mo unmodified steel) meets the requirements for the fabrication of large pressure vessels per Section VIII, Division 2 of the ASME Boiler and Pressure Vessel Code. Unfortunately, the thick-section hardenability is insufficient to prevent the formation of cementite, even with accelerated cooling procedures and lower tempering conditions (Ref 11). This persistence of cementite in thick-section SA-387 grade 22, class 2 is a concern regarding hydrogen attack (Ref 12). Other unmodified chromium-molybdenum steels such as 3Cr-1Mo and 5Cr-0.5Mo (SA-387 grades 21 and 5) resist hydrogen attack, but the design allowables are below those of 2¼Cr-1Mo steel at some temperatures of interest. Higher-chromium alloys, such as 7Cr-0.5Mo and 9Cr-1Mo, also have strengths below normalized and tempered 2¼Cr-1Mo and so have not been considered in the United States for heavy-wall vessels.

Therefore, several modified chromium-molybdenum alloys have been investigated for thick-section vessels in a hydrogen environment. These modified chromium-molybdenum alloys contain various microalloying elements such as vanadium, niobium, titanium, and boron. Three categories (Ref 11) of modified chromium-molybdenum steels investigated for thick-section applications in a hydrogen environment are:

- *3Cr-1Mo modified with vanadium, titanium, and boron* (Ref 13): This steel is approved for service up to 455 °C (850 °F), is fully hardenable, resists hydrogen attack, and has strengths capable of meeting the design allowables of normalized and tempered 2¼Cr-1Mo steel
- *9Cr-1Mo steel modified with vanadium and niobium* (Ref 14, 15): This steel has strengths exceeding those of 2.25Cr-1Mo and is approved for use to more than 600 °C (1110 °F) for steam and hydrogen service
- *2¼Cr-1Mo steel modified with vanadium, titanium, and boron* (Ref 4, 16): Vanadium-modified 2¼Cr-1Mo steel is fully hardenable, resists hydrogen attack, and exceeds the strength of normalized and tempered 2¼Cr-1Mo steel

Other modified alloys, such as 3Cr-1.5Mo-0.1V-0.1C, have also been investigated (Ref 13, 17). The modified alloys have improved hardenability over unmodified 2¼Cr-1Mo steel. However, these modified chromium-molybdenum steels with bainitic microstructures undergo a strain softening (Ref 18-21), which may be a limitation in applications with cyclic stresses.

The modified 9Cr-1Mo steel is an attractive alloy because it has strengths (Fig. 5) capable of meeting or exceeding the allowable stresses of stainless steel (Fig. 9). Microstructural work has indicated that the improved strength of the modified alloy derives from two factors. First, fine $M_{23}C_6$ precipitate particles nucleate on Nb(C,N), which first appears during the heat treatment. Second, the vanadium enters $M_{23}C_6$ and retards its growth at the service temperature. The finer distribution of $M_{23}C_6$ adds to the strength, and its retarded grain-size growth holds the strength for long periods of time at the service temperature. The grain-coarsening behavior of the modified 9Cr-1Mo steel as a function of normalizing temperature and time-temperature exposure is shown in Fig. 10.

The H11 die steels have very high yield strengths (Fig. 8) and are primarily used in aircraft and missiles when high strength-to-weight ratios are desired. The H11 die steels are basically medium-carbon, 5% Cr steels with molybdenum and vanadium added. This composition air hardens from the austenitizing temperature and is tempered to a tensile strength of 1500 to 2200 MPa (215 to 320 ksi) (45 to 58 HRC). These properties apply to thin sheet as well as heavy forgings because the hardenability is fairly constant up to 38 mm (2⁴⁄₁₆ in.) on a standard end-quench specimen.

Exposure of this steel at a temperature as close as 30 °C (50 °F) below the tempering temperature for 100 h or longer will have little effect on hardness and tensile strength. The high tempering temperature eliminates most residual stress. The retention of 70 to 80% of the room-temperature strength up to 540 °C (1000 °F) gives H11 steel a high strength-to-weight ratio at elevated temperatures. Additional information on H11 die steels is contained in the article "Ultrahigh-Strength Steels" in this Volume.

Mechanical Properties at Elevated Temperatures

The allowable design stresses for steels at elevated temperatures may be controlled by different mechanical properties, depending on the application and temperature exposure. For applications with temperatures below the creep-temperature range, tensile strength or the yield strength at the expected service temperature generally controls allowable stresses. For temperatures in the creep range, allowable stresses are determined from either creep-rupture properties or the degree of deformation from creep. In recent years, the worldwide interest in life extension of high-temperature components has also promoted considerably more interest in elevated-temperature fatigue. This effort has led to tests and methods for evaluating the effects of creep-fatigue interaction on the life of elevated-temperature components.

Ductility and toughness may also be important considerations, although ductility and toughness considerations commonly do not enter directly into the setting of allowable stresses. In elevated-temperature applications, ductility and toughness may not remain fixed in magnitude or character and often change with temperature and with time at temperature. The changes, which may be beneficial but are often deleterious, are of interest both at service temperature and, because of shutdowns, at ambient temperatures. Ductility is also an important factor that influences notch sensitivity and creep-fatigue interaction.

The types of tests used to evaluate the mechanical properties of steels at elevated temperatures include:

- Short-term elevated-temperature tests
- Long-term elevated-temperature tests
- Fatigue tests (including thermal fatigue and thermal shock tests)
- Time-dependent fatigue tests
- Ductility and toughness tests
- Short-term and long-term tests following long-term exposure to elevated temperatures

Several methods are used to interpret, interpolate, and extrapolate the data from

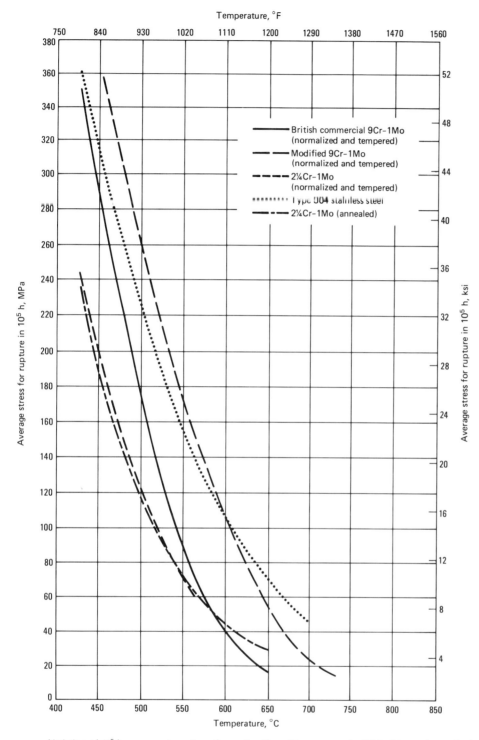

Fig. 5 Variation of 10^5-h creep-rupture strength as a function of temperature for 2¼Cr-1Mo steel, standard 9Cr-1Mo, modified 9Cr-1Mo, and 304 stainless steel. Source: Ref 7

some of these tests, as described in the section "Methods for Correlating, Interpolating, and Extrapolating Elevated-Temperature Mechanical Property Data" in this article.

Short-Term Elevated-Temperature Tests

Short-term elevated-temperature tests include the elevated-temperature tensile tests (described in ASTM E 21), a test for elastic modulus (ASTM E 231), compression tests, pin bearing load tests, and the hot hardness test. The mechanical properties determined by means of the tensile test include ultimate tensile strength, yield strength, percent elongation, and percent reduction in area. Because elevated-temperature tensile properties are sensitive to strain rate, these tests are conducted at carefully controlled strain rates. Tensile strength data obtained on specimens of annealed 2¼Cr-1Mo steel at various temperatures and at strain rates ranging from 2.7×10^{-6} s^{-1} to 144 s^{-1} are shown in Fig. 11.

In designing components that are to be produced from low-alloy steels and to be exposed to temperatures up to 370 °C (700 °F), the yield and ultimate strengths at the maximum service temperature can be used much as they would be used in the design of components for service at room temperature. Figure 8 compares the short-time elevated-temperature yield and tensile strengths of selected alloys. Certain codes require that appropriate factors be applied in calculating allowable stresses.

Elevated-temperature values of elastic modulus can be determined during tensile testing or dynamic testing by measuring the natural frequency of a test bar at the designated test temperature. Figure 12 shows values of elastic modulus at temperatures between room temperature and 650 °C (1200 °F) for several low-alloy steels, determined during static tensile loading and dynamic loading.

Compression tests and pin bearing load tests (ASTM E 209 and E 238) can be used to evaluate materials for applications in which the components will be subjected to these types of loading at elevated temperatures. Hot hardness tests can be used to evaluate materials for elevated-temperature service and can be applied to the qualification of materials in the same way in which room-temperature hardness tests are applied.

Components for many elevated-temperature applications are joined by welding. Elevated-temperature properties of both the weld metal and the heat-affected zones can be determined by the same methods used to evaluate the properties of the base metal.

Long-Term Elevated-Temperature Tests

Long-term elevated-temperature tests are used to evaluate the effects of creep, which is defined as the time-dependent strain that occurs under constant load at elevated temperatures. Creep is observed in steels at temperatures above about 370 °C (700 °F). In general, creep occurs at a temperature slightly above the recrystallization temperature of a metal or alloy; at such a temperature, atoms become sufficiently mobile to allow time-dependent rearrangement of the structure. In time, creep may lead to excessive deformation and even fracture at stresses considerably below those determined in room-temperature and elevated-temperature short-term tension tests.

Typical creep behavior consists of three distinct stages, as shown in Fig. 13. Following initial elastic-plastic strain resulting from the immediate effects of the applied load, there is a region of increasing plastic

Elevated-Temperature Properties of Ferritic Steels / 623

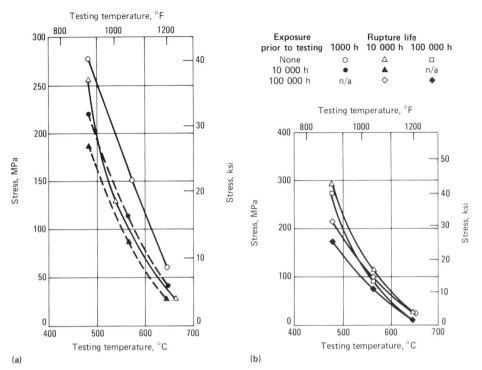

Fig. 6 Effect of elevated-temperature exposure on stress-rupture behavior of (a) normalized and tempered 2¼Cr-1Mo steel and (b) annealed 9Cr-1Mo steel. Exposure to stress-rupture testing was at the indicated test temperatures (without stress) and was 10 000 h long for the 2¼Cr-1Mo steel and 100 000 h long for the 9Cr-1Mo steel. n/a, data not available at indicated exposure and rupture life

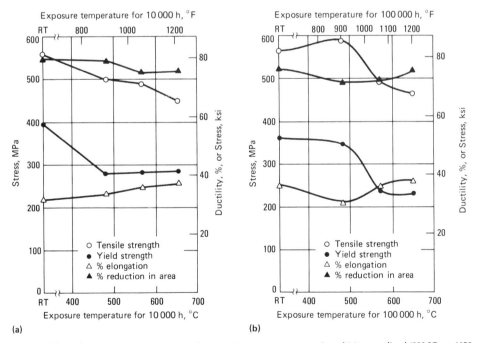

Fig. 7 Effect of temperature exposure on the room-temperature properties of (a) normalized (900 °C, or 1650 °F) and tempered (705 °C, or 1300 °F) 2¼Cr-1Mo steel after exposure (without stress) to indicated temperature for 10 000 h and (b) annealed 9Cr-1Mo steel after exposure (without stress) to indicated temperatures for 100 000 h

increased strain rate with rapid extension to fracture (third-stage, or tertiary creep). Tertiary creep has no distinct beginning but does refer to the region with an increasing rate of extension that is followed by fracture. Under certain conditions, some metals may not exhibit all three stages of plastic extension. For example, at high stresses or temperatures, the absence of primary creep is not uncommon, with secondary creep or, in extreme cases, tertiary creep following immediately upon loading.

Of all the parameters pertaining to the creep curve, the most important for engineering applications are the creep rate and the time to rupture. These parameters are determined from long-term elevated-temperature tests that include creep, creep-rupture, and stress-rupture tests (ASTM E 139) and notched-bar rupture tests (ASTM E 292). In addition, relaxation tests (ASTM E 328) are used to evaluate the effect of creep behavior on the performance of high-temperature bolt steels. These tests are described in Volume 8 (p 311 to 328) of the 9th Edition of *Metals Handbook*.

Creep Strength. When the rate or degree of deformation is the limiting factor, the design stress is based on the minimum (secondary) creep rate and design life after allowing for initial transient creep. The stress that produces a specified minimum creep rate of an alloy or a specified amount of creep deformation in a given time (for example, 1% total creep in 100 000 h) is referred to as the limiting creep strength or limiting stress. Typical creep strengths of various low-alloy steels are shown in Fig. 3. Table 2 also lists some suggested maximum service temperatures of various low-alloy steels based on creep rate. Figure 14 shows the 0.01%/1000 h creep strength of carbon steel as a function of room temperature tensile strength.

Stress Rupture. When fracture is a limiting factor, stress-rupture values are used in design. Stress-rupture values of various low-alloy chromium-molybdenum steels are shown in Fig. 4, 5, and 6. Figures 15 and 16 show typical creep-rupture values of carbon and 1Cr-1Mo-0.25V steel, respectively.

It should be recognized that long-term creep and stress-rupture values (for example, 100 000 h) are often extrapolated from shorter-term tests. Whether these property values are extrapolated or determined directly often has little bearing on the operating life of high-temperature parts. The actual material behavior is often difficult to predict accurately because of the complexity of the service stresses relative to the idealized, uniaxial loading conditions in the standardized tests and because of the attenuating factors such as cyclic loading, temperature fluctuations, or metal loss from corrosion.

For those alloys in which failure occurs before a well-defined start of tertiary creep,

strain at a decreasing strain rate (first-stage, or primary, creep). Following the primary creep region, there is a region where the creep strain increases at a minimum, and almost constant, rate of plastic strain (second-stage, or secondary creep). This nominally constant creep rate is generally known as the minimum creep rate and is widely employed in research and engineering studies. Finally, there is a region of drastically

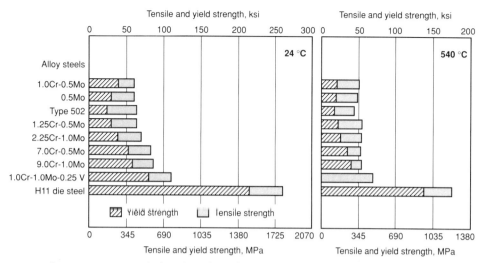

Fig. 8 Room-temperature and short-time elevated-temperature tensile strengths and yield strengths of selected steels containing less than 10% alloy. The 1.0Cr-0.5Mo steel, 0.5Mo steel, type 502, and 2.25Cr-1.0Mo steel were annealed at 843 °C (1550 °F). The 1.25Cr-0.5Mo steel was annealed at 815 °C (1500 °F). The 7.0Cr-0.5Mo and 9.0Cr-1.0Mo steels were annealed at 900 °C (1650 °F). The 1.0Cr-1.0Mo-0.25V steel was normalized at 955 °C (1750 °F) and tempered at 650 °C (1200 °F). H11, hardened 1010 °C (1850 °F), tempered 565 °C (1050 °F)

it is useful to use notched specimens or specimens with both smooth and notched test sections (with the cross-sectional area of the notch equal to that of the smooth test section). If the material is notch sensitive, the specimen will fail in the notch before failure occurs in the smooth section. It has been well recognized for many years that notch sensitivity is related to creep ductility. It has been suggested that a minimum smooth-bar creep ductility of about 10% in terms of reduction in area may be desirable for avoidance of notch sensitivity (Ref 25 and 26). Limited published data on notched stress-rupture properties of low-alloy ferritic steels for elevated temperatures indicate that these steels generally are not notch sensitive. Representative stress-rupture data for notched and unnotched specimens of AISI 603 steel are presented in Fig. 17.

Relaxation Tests. Creep tests on metals are usually carried out by keeping either the applied load or the stress constant and noting the specimen strain as a function of time. In another type of test, known as the stress relaxation test, a sample is first deformed to a given strain and then the stress is measured as a function of time such that the total strain remains constant. Stress relaxation tests are more difficult to carry out than ordinary creep tests and are more difficult to interpret. However, stress relaxation is an important elevated-temperature property in the design of bolts or other devices intended to hold components in contact under pressure. If the service temperature is high enough, the extended-time stress on the bolt causes a minute amount of creep, which results in a reduction in the restraining force.

Because of their low relaxed stresses, carbon steels are usually used only at temperatures below 370 °C (700 °F). Various low-alloy steels have been widely used up to metal temperatures of about 540 °C (1000 °F). Modified 12% Cr steels can be used for slightly higher temperatures. The common austenitic stainless steels are seldom used because of their low yield strength in the annealed condition, but are used in the cold-worked condition. The superstrength alloys are usually employed only at the highest temperatures. The comparative 1000-h relaxation strengths of these classes of alloys are shown in Fig. 18(a). More recent data are provided in Ref 29.

Carbon steel is not recognized as a high-temperature bolting material under ASTM standards or by the ASME Boiler Code. One of the most widely used low-alloy steels for moderately high-temperature bolting applications is quenched and tempered 4140, in accordance with ASTM A 193, grade B7. Its relaxation behavior is approximately indicated by the solid lines for 0.65-1.10Cr-0.10-0.30Mo steels in Fig. 18(b) and (c). The relaxation strength of 4140 is greater after normalizing and tempering than in the quenched and tempered condition. However, this steel is nearly always used in the quenched and tempered condition in order to obtain more consistent mechanical properties. Chromium-molybdenum steels similar to 4140 except that they contain approximately 0.50% Mo (A 193, grade B7A) have also been widely used. They have slightly higher relaxation strength than 4140 but are less readily available.

The strongest low-alloy steels are those with approximately 1% Cr, 0.5% Mo, and 0.25% V, in the normalized and tempered condition (A 193, grade B14) or the quenched and tempered condition (A 193, grade B16). Some of these grades are produced with rather high silicon contents (~0.75%), which seems to increase resistance to tempering. These grades have been satisfactory in service up to 540 °C (1000 °F) metal temperature in the absence of excessive follow-up or retightening. However, they are somewhat notch sensitive in creep rupture and in impact at room temperature, especially in the normalized and tempered condition.

Fatigue

At room temperature and in nonaggressive environments (and except at very high frequencies), the frequency at which loads are applied has little effect on the fatigue strength of most metals. The effects of frequency, however, become much greater as the temperature increases or as the presence of corrosion becomes more significant. At high temperatures, creep becomes more of a factor, and the fatigue strength seems to depend on the total time stress is applied rather than solely on the number of cycles. The behavior occurs because the continuous deformation (creep) under load at high temperatures affects the propagation of fatigue cracks. This effect is referred to as creep-fatigue interaction. The quantification of creep-fatigue interaction effects and the application of this information to life prediction procedures constitute the primary objective in time-dependent fatigue tests. Time-dependent fatigue tests are also used to assess the effect of load frequency on corrosion fatigue.

Effect of Load Frequency on Corrosion Fatigue. In aggressive environments, fatigue strength is strongly dependent on frequency. Corrosion fatigue strength (endurance limit at a prescribed number of cycles) will generally decrease as the cyclic frequency is decreased. This effect is most important at frequencies of less than 10 Hz.

The frequency dependence of corrosion fatigue is thought to result from the fact that the interaction of a material and its environment is essentially a rate-controlled process. Low frequencies, especially at low strain amplitudes or when there is substantial elapsed time between changes in stress levels, allow time for interaction between material and environment; high frequencies do not, particularly when high strain amplitude is also involved. At very high frequencies or in the plastic-strain range, localized heating may seriously affect the properties of the part. Such effects normally are not considered to be related to a corrosion fatigue phenomenon.

When environments have a deleterious effect on fatigue behavior, a critical range of frequencies of loading may exist in which the mechanical/environmental interaction is significant. Above this range the effect usually disappears, while below this range the effect may diminish.

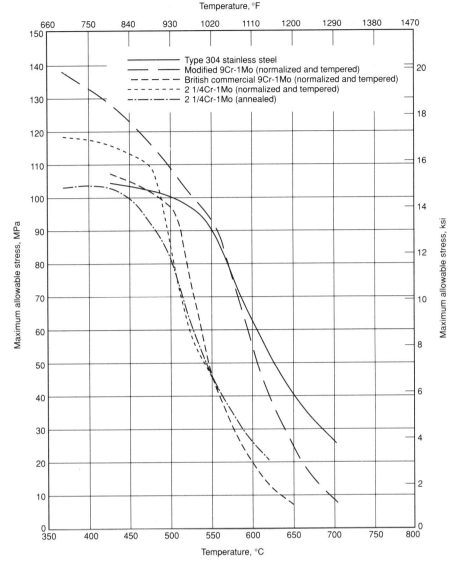

Fig. 9 Estimated design allowable stresses (Section VIII of ASME Boiler and Pressure Vessel Code) as a function of temperature for modified 9Cr-1Mo steel, standard 9Cr-1Mo, 2¼Cr-1Mo steel, and 304 stainless steel. Source: Ref 7

Creep-fatigue interaction is an elevated-temperature phenomenon that can seriously reduce fatigue life and creep-rupture strength. Figure 19 illustrates the effect of time-dependent fatigue when the elevated temperature is within the creep range of a material. Figure 19 shows a continuous strain cycling waveform (Fig. 19a) and a hold cycling waveform (Fig. 19b) for fatigue strength testing. Figure 19(c) shows the fatigue life from a continuous strain cycle and from cycling with two different hold times. This decrease in fatigue life with increasing hold time or decreasing frequency, which occurs at temperatures within the creep range, is referred to as time-dependent fatigue or creep-fatigue interaction. It has been attributed to a number of factors, including the formation of intergranular voids or classical creep damage (which permits intergranular crack propagation under cyclic loading conditions), environmental interaction (corrosion fatigue), mean stress effects, and microstructural instabilities of defects produced as a result of stress and/or thermal aging, irradiation damage, and fabrication processing.

Most of these changes can occur at elevated temperatures and are time and possibly waveform dependent. There is also ample evidence to show that rupture ductility has a major influence on creep-fatigue interaction. Because this effect is believed to be caused by the influence of rupture ductility on the creep-fracture component, endurance in continuous-cycle and in high-frequency or short-hold-time fatigue tests (where fracture is fatigue-dominated) will be relatively unaffected. As the frequency is decreased or as the hold time is increased, the effect of rupture ductility becomes more pronounced. Endurance data for several ferritic steels, in relation to the range of rupture ductility exhibited by them, are illustrated in Fig. 20. The lower the ductility, the lower the creep-fatigue endurance. In addition, long hold periods, small strain ranges, and low ductility favor creep-dominated failures, whereas short hold periods, intermediate strain ranges, and high creep ductility favor creep-fatigue-interaction failures. Similar results have been presented for austenitic stainless steels (see the article "Elevated-Temperature Properties of Stainless Steels" in this Volume).

To determine the effect of cyclic loading superimposed on a constant load at elevated temperatures, several types of fatigue testing can be employed: continuous alternating stress, continuous alternating strain, tension-tension loading with the stress ratios greater than 0, and special waveforms that provide specific holding times at maximum load. Results of these tests show which factors are most contributory to deformation and fracture of the specimens for the testing conditions employed. Further information on time-dependent fatigue is available in the article "Creep-Fatigue Interaction" in Volume 8 of the 9th Edition of *Metals Handbook*.

Fatigue-Crack Growth. Although the *S-N* curves have been used in the past as the basic tool for design against fatigue, their limitations have become increasingly obvious. One of the more serious limitations is the fact that they do not distinguish between crack initiation and crack propagation. Particularly in the low-stress regions, a large fraction of a component's life may be spent in crack propagation, thus allowing for crack tolerance over a large portion of the life. Engineering structures often contain flaws or cracklike defects that may altogether eliminate the crack-initiation step. A methodology that quantitatively describes crack growth as a function of the loading variables is, therefore, of great value in design and in assessing the remaining lives of components.

Because fatigue-crack growth rates are obtained at various ΔK ranges and temperature ranges, it is difficult to compare the various types of materials directly. At a constant ΔK (arbitrarily chosen as 30 MPa$\sqrt{m}$, or 27 ksi$\sqrt{in.}$), a clear trend of crack-growth-rate increase with increasing temperature can be seen as shown in Fig. 21. In this figure it can be seen that at temperatures up to about 50% of the melting point (550 to 600 °C, or 1020 to 1110 °F), the growth rates are relatively insensitive to temperature, but the sensitivity increases rapidly at higher temperatures. The crack-growth rates for all the materials at temperatures up to 600 °C (1110 °F) relative to the room-temperature rates can be estimated by a maximum correlation factor of 5 (2 for ferritic steels).

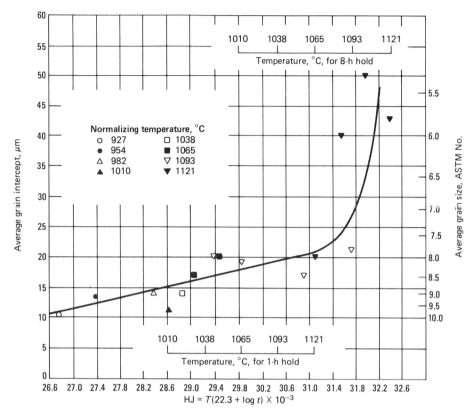

Fig. 10 Grain-coarsening behavior of a modified 9Cr-1Mo steel (9Cr-1Mo steel with 0.06 to 0.10% Nb and 0.18 to 0.25% V). Source: Ref 7

Thermal fatigue refers to the gradual deterioration and eventual cracking of a material from cyclic thermal transients. In the past, thermal fatigue traditionally has been treated as synonymous with isothermal low-cycle fatigue at the maximum temperature of the thermal cycle. Life-prediction techniques also have evolved from the low-cycle-fatigue (LCF) literature. More recently, advances in finite-element analysis and in servohydraulic test systems have made it possible to analyze complex thermal cycles and to conduct thermomechanical fatigue (TMF) tests under controlled conditions. The assumed equivalence of isothermal LCF tests and TMF tests has been brought into question as a result of a number of studies. It has been shown that for the same total strain range, the TMF test can be more damaging under certain conditions than the pure LCF test. Information on the thermal fatigue of materials is provided in Ref 33.

High-cycle thermal fatigue frequently results from intermittent wetting of a hot surface by a coolant having a considerably lower temperature. In this case, thermal fatigue cracks may initiate at the surface after a sufficient number of cycles. In other cases, the thermal cycling or ratcheting may result in plastic deformation. Thermal ratcheting is the progressive cyclic inelastic deformation that occurs as a result of cyclic strains caused by thermal or secondary mechanical stresses; sustained primary loading often contributes to thermal ratcheting. Salt pots used to contain heat-treating salt are subject to thermal ratcheting whenever the salt goes through a freeze-melt cycle.

Ductility and Toughness

Although steels typically have adequate ambient temperature toughness and excellent elevated-temperature ductility, several embrittling mechanisms can occur during elevated-temperature exposure. Consequently, ductility and toughness tests are useful in assessing embrittling mechanisms. Information on the toughness of steels is provided in the article "Notch Toughness of Steels" in this Volume.

Figure 22 shows that toughness may actually decrease if steels are tempered in the range of 260 to 370 °C (500 to 700 °F). This decrease in toughness is referred to as tempered martensite embrittlement, 350 °C embrittlement, or 500 °F embrittlement and is discussed in more detail in the article "Embrittlement of Steels" in this Volume. As a result of this embrittlement, the tempering range between 260 and 370 °C (500 and 700 °F) is generally avoided in commercial practice. Another type of embrittlement—temper embrittlement—may occur in certain alloy steels as a result of holding on slow cooling through certain temperature ranges (see the text below discussing Fig. 23).

Another method of assessing toughness is to estimate the ductile-to-brittle transition temperature by performing notched-specimen impact tests at various temperatures. Steels are susceptible to a lowering of absorbed impact energy with decreasing temperature of use or testing. This change in energy value is accompanied by a transition from a fibrous to a crystalline-appearing fracture. The temperature at which some specified level of energy absorption or fracture appearance occurs is defined as a transition temperature. Transition temperature is an important concept because it defines a change in the mode of fracture from one that is caused predominantly by a shear mechanism to one that propagates primarily by cleavage (or along the grain boundaries in the case of temper embrittlement).

Shifts in the ductile-to-brittle transition temperature are measured to detect the presence of temper embrittlement, as shown in Fig. 23. In this case, 3140 steel (containing nominally 1.15% Ni and 0.65% Cr) was embrittled by both isothermal tempering and slow furnace cooling through the critical temperature range of about 375 to 575 °C (706 to 1070 °F). Additional information on temper embrittlement is available in the article "Embrittlement of Steels" in this Volume.

A third method of assessing the effects of embrittlement mechanisms is by ductility (reduction of area) measurements. Creep embrittlement effects, for example, are usually reported in terms of a ductility minimum in stress-rupture tests, while temper embrittlement is usually recorded as an upward shift in Charpy V-notch transition temperature.

Creep embrittlement occurs in roughly the same temperature range as temper embrittlement, but is not reversible with heat treatment. Creep embrittlement also seems to depend on tempering reactions inside grains and on the presence of a carbide denuded zone at prior austenite grain boundaries, while segregation effects producing temper embrittlement occur at distances only a few atomic diameters from the grain boundary. Some investigators maintain that impurities known to produce temper embrittlement also contribute to the development of creep embrittlement. Some general characteristics of creep embrittlement are:

- Creep embrittlement has been shown to occur in the temperature range 425 to 595 °C (800 to 1100 °F) for alloy steels having ferrite plus carbide microstructures
- Creep embrittlement appears after longer times and becomes more severe the lower the position in the embrittling temperature range
- Creep embrittlement is manifested by a loss and then partial recovery of stress-rupture ductility with decreasing stress

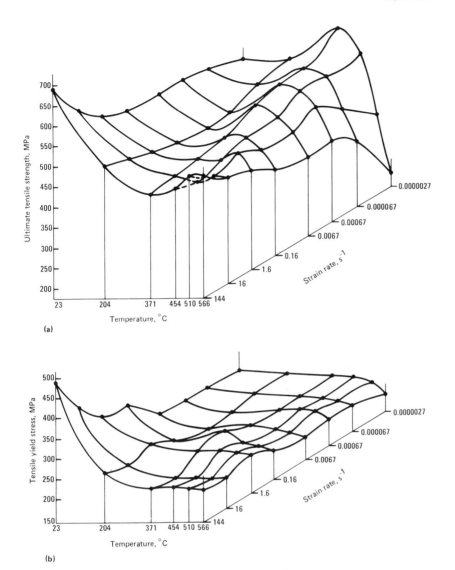

Fig. 11 Effect of test temperature and strain rate on the strength of annealed 2¼Cr-1Mo steel. Tensile strength (a) and yield strength (b) of 2¼Cr-1Mo steel tested at various temperatures and strain rates. Source: Ref 22

- The development of embrittlement is invariably associated with a transition from transgranular to intergranular fracture. Voids and microcracks are found throughout a creep embrittled microstructure. These voids form along prior austenite grain boundaries transverse to the tensile direction
- The mechanism for creep embrittlement appears to be closely associated with tempering reactions inside grains and at the grain boundaries during the creep process. The formation of fine, needlelike precipitates in grain interiors, accompanied by the development of a denuded zone and elongated alloy carbides at grain boundaries seems to contribute significantly to the embrittlement process
- Loss in toughness produced by creep embrittlement is largely unaffected by subsequent heat treatments; and void formation caused by creep is irreversible

Long-Term Exposure

Long-term exposure to elevated temperature may affect either short-term or long-term properties. For example, the initial microstructure of creep-resistant chromium-molybdenum steels consists of bainite and ferrite containing Fe_3C carbides, ϵ carbides, and fine M_2C carbides. Although a number of different carbides may be present, the principal carbide phase responsible for strengthening is a fine dispersion of M_2C carbides, where M is essentially molybdenum. With increasing aging in service, or tempering in the laboratory, a series of transformations of the carbide phases takes place that eventually transform M_2C into M_6C and $M_{23}C$ (where the M in the latter two metal carbides is mostly chromium). Such an evolution of the carbide structure results in coarsening of the carbides, changes in the matrix composition, and an overall decrease in creep strength. The effect of exposure on the stress-rupture strength of two chromium-molybdenum steels is shown in Fig. 6.

Other metallurgical changes (such as spheroidization and graphitization) and corrosion effects may also occur during long-term exposure at elevated temperature. Therefore, tests after long-term exposure may be useful in determining the effect of these metallurgical changes on short-term or long-term properties.

Data Presentation and Analysis

Presentation of Tensile and Yield Strength. One method for comparing steels of different strengths is to report elevated-temperature strength as a percentage of room-temperature strength; this method is illustrated in Fig. 24. The strength levels of the steels represented in Fig. 24 varied from 480 to 1100 MPa (70 to 160 ksi).

Presentation of Creep Data. Four different presentations of the same creep data for 2¼Cr-1Mo steel are given in Fig. 25. In Fig. 25(a) to (c), only the creep strain is plotted. In the isochronous stress-strain diagram (Fig. 25d), total strain is used. The overall format of Fig. 25(d) is particularly useful in design problems in which total strain is a major consideration.

Methods for Correlating, Interpolating, and Extrapolating Elevated-Temperature Mechanical Property Data. The behavior of steels at elevated temperatures can be affected by many variables, including time, temperature, stress, and environment. A variety of methods have been devised for correlating, interpolating, and extrapolating elevated-temperature mechanical property data. Further information on the analysis of elevated-temperature data is contained in Ref 32, Volume 8 of the 9th Edition of *Metals Handbook*, and MPC-7 of the American Society of Mechanical Engineers.

Larson-Miller Parameter. Several parameters have been used for comparison of and interpolation between stress-rupture data. The most widely used is the Larson-Miller parameter, P, defined by the equation:

$$P = T(C + \log t) \times 10^{-3} \quad \text{(Eq 1a)}$$

where T is the test temperature in degrees Rankine, t is the rupture time in hours, and C is a constant whose value is approximately 20 for low-alloy steels. If T is given in Kelvins, the equation is:

$$P = 1.8\,T(C + \log t) \times 10^{-3} \quad \text{(Eq 1b)}$$

The Larson-Miller parameter is used with an experimentally determined graph such as that shown in Fig. 26 to correlate stress, temperature, and rupture time. Each graph should include the ranges of time and temperature for which the data apply; extrapolation beyond these ranges is generally not appropriate.

A similar parameter was used by Smith (Ref 36) to describe the creep behavior of

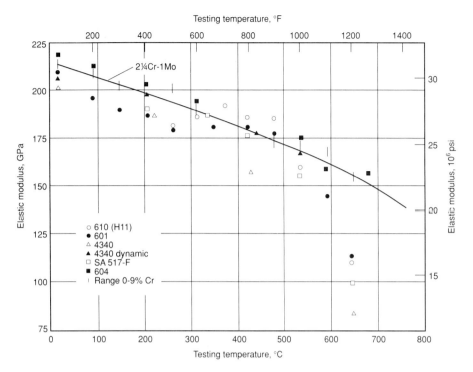
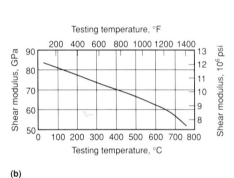
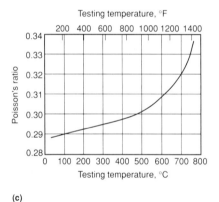

Fig. 12 Effect of test temperature on elastic modulus, shear modulus, and Poisson's ratio. (a) Effect of test temperature on elastic modulus for several steels commonly used at elevated temperatures. Dynamic measurements of elastic modulus were made by determining the natural frequencies of test specimens; static measurements were made during tensile testing. (b) Effect of test temperature on shear modulus of 2¼Cr-1Mo steel. (c) Effect of test temperature on Poisson's ratio of 2¼Cr-1Mo steel. Source: Ref 23

tion. These changes in slope are often indicative of microstructural changes. Marked differences in slope between curves representing temperatures separated by less than 100 °C (180 °F) should be regarded as evidence that the slope of the lower-temperature curve will change over the time period of extrapolation, indicating the need for longer tests or careful approximations of the probable influence of the change in slope. Such changes in slope are almost always in the direction of lower stress-rupture strength than would be predicted by straight-line extrapolation.

Because of microstructural instabilities, deviations from the ideal creep curve must also be considered. Primary creep may be virtually absent or may be excessive and extend over long periods of time. Secondary creep may persist only for very short time periods or may exhibit nonclassical behavior. The creep behavior of annealed 2.25Cr-1Mo steel, for example, exhibits creep curves that differ from a classical three-stage creep curve in that two steady-state stages occur (Ref 37). During the first steady-state stage, the creep rate is controlled by the motion of dislocations that contain atmospheres of carbon and molybdenum atom clusters, a process termed interaction solid-solution hardening (see the section "Strengthening Mechanisms" in this article). Eventually, the precipitation of Mo_2C removes molybdenum and carbon from solution, and the creep rate increases to a new steady state where the creep rate is controlled by atmosphere-free dislocations moving through a precipitate field. These nonclassical curves occur at intermediate stresses. As the stress decreases, the first steady-state stage disappears because the dislocation velocity decreases and the molybdenum-carbon atmosphere will be able to diffuse with the dislocations. At high stresses, a classical curve occurs when the creep rate is controlled by a combination of processes that operate in the two steady-state stages of the nonclassical curves (Ref 37). Such factors indicate the need to experimentally check values of deformation predicted by extrapolation of secondary creep data.

The extrapolation of stress-rupture ductility with parametric techniques has been considered a potential method for predicting long-term ductility from short-term tests (Ref 38). Because the stress-rupture ductility of many alloys used at elevated temperatures varies with temperature and stress, the objective is to develop a combined (stress, temperature) parameter that can be correlated to rupture ductility over a wide range of stresses and temperatures. Reference 38 compares the correlation between some parametric models and rupture ductility data for a 1¼Cr-½Mo steel in the temperature range of 510 to 620 °C (950 to 1150 °F).

9Cr-1Mo steel. The creep rate parameter, P', is given by:

$$P' = T(20 - \log r) \times 10^{-3} \quad \text{(Eq 2a)}$$

where T is the test temperature in degrees Rankine, and r is the minimum creep rate in percent per hour. If T is given in Kelvins, the equation is:

$$P' = 1.8 T(20 - \log r) \times 10^{-3} \quad \text{(Eq 2b)}$$

The creep parameter is used with an experimentally determined graph such as the one shown in Fig. 27 for 2.25Cr-1Mo steel.

Extrapolation of Creep and Rupture Data. It should be recognized that long-term creep and stress-rupture values (for example, 100 000 h) are often extrapolated from shorter-term tests. Whether these property values are extrapolated or determined directly often has little bearing on the operating life of high-temperature parts. The actual material behavior is often difficult to predict accurately because of the complexity of the service stresses relative to the idealized, uniaxial loading conditions in the standardized tests and because of the attenuating factors such as cyclic loading, temperature fluctuations, or metal loss from corrosion.

Marked changes in the slope of stress-rupture curves (see, for example, the lower plot in Fig. 15(b) near 480 °C, or 900 °F) must also be considered in data extrapola-

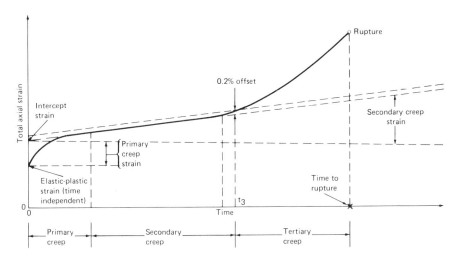

Fig. 13 Schematic representation of classical creep behavior

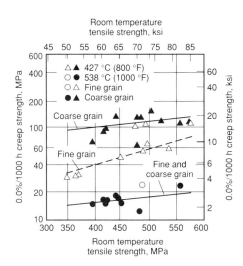

Fig. 14 Relationship between creep strength (0.01%/1000 h) and ultimate tensile strength of a carbon steel. Creep strength estimates made using isothermal lot constants. Source: Ref 24

Methods for Predicting Time-Dependent Fatigue (Creep-Fatigue Interaction) Behavior. Many methods have been employed to extrapolate available data to estimate the time-dependent fatigue life of materials. Development of a mathematical formulation for life prediction is one of the most challenging aspects of creep-fatigue interaction. It is complicated by the fact that any proposed formulation must account for strain rate, relaxation at constant strain, creep at constant load, the difference between tension and compression creep and/or relaxation, or combinations of all of these.

Linear damage summation is perhaps the most widely known and simplest of the many life prediction methods and has been used extensively in the evaluation of creep-fatigue interaction. It is based on a simple relation that fatigue damage can be expressed as a cycle fraction of damage and that creep damage can be expressed as a time fraction of damage. It is also assumed that these quantities can be added linearly to represent damage accumulation. Failure occurs when this summation reaches a certain value.

Other methods include the ductility exhaustion approach, the frequency modified approach, and strain range partitioning. These methods are reviewed in Ref 30, 32 and in Volume 8 of the 9th Edition of *Metals Handbook*.

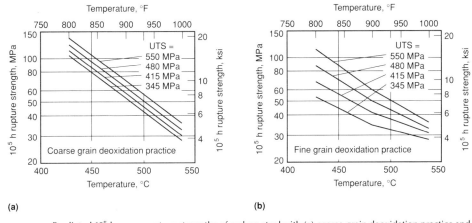

Fig. 15 Predicted 10^5-h creep-rupture strengths of carbon steel with (a) coarse-grain deoxidation practice and (b) fine-grain deoxidation practice. Source: Ref 24

Corrosion

Although the mechanical properties establish the allowable design-stress levels, corrosion effects at elevated temperatures often set the maximum allowable service temperature of an alloy. The following sections describe the three common forms of corrosion—oxidation, sulfidation, and hydrogen attack—that occur at elevated temperature. Corrosion considerations with liquid-metal environments are also summarized. More detailed information on corrosion and its prevention is available in Volume 13 of the 9th Edition of *Metals Handbook*.

Oxidation from steam or air is a serious problem that can occur at elevated temperatures. When metal is exposed to an oxidizing gas at elevated temperature, corrosion can occur by direct reaction with the gas. This type of corrosion is referred to as

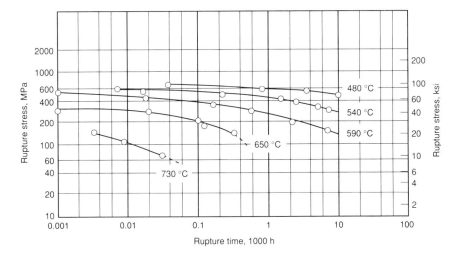

Fig. 16 Time-temperature-rupture data of a 1Cr-1Mo-0.25V steel

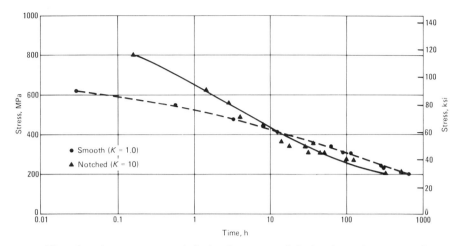

Fig. 17 Effect of notch on stress-rupture behavior. Stress-rupture behavior of smooth ($K = 1.0$) and notched specimens of AISI 603 steel tested at 595 °C (1100 °F). All specimens were normalized at 980 °C (1800 °F) and tempered 6 h at 675 °C (1250 °F). Source: Ref 27

tarnishing, high-temperature oxidation, or scaling. The rate of attack increases substantially with temperature. The surface film typically thickens as a result of reaction at the scale/gas or metal/scale interface due to cation or anion transport through the scale, which behaves as a solid electrolyte.

Alloys intended for high-temperature applications are designed to have the capability of forming protective oxide scales. Chromium provides oxidation resistance in alloy steels, and Fig. 28 compares the loss by scaling for alloys with varying levels of chromium. Silicon can also improve oxidation resistance (Fig. 28), although it also reduces creep strength and may promote temper embrittlement when other impurities are present.

The scaling data given in Fig. 28 were obtained in the presence of air. If other variables affecting oxidation are changed, such as gas composition, the heating method, temperature, pressure, or velocity, different rates of scaling can be expected. Elements such as sulfur, vanadium, and sodium can change the nature of metal oxidation, sometimes increasing it to a catastrophic level of several inches per year.

At elevated temperatures, steam decomposes at metal surfaces to hydrogen and oxygen and may cause steam oxidation of steel, which is somewhat more severe than air oxidation at the same temperature. Fluctuating steam temperatures tend to increase the rate of oxidation by causing scale to spall, which exposes fresh metal to further attack. Table 1 gives the maximum-use temperatures for several boiler alloys for which code standards exist. The strength criteria are based on the wall midsection temperatures, which are typically 25 °C (45 °F) lower than the outer surface temperature.

In a water environment, corrosion is significantly influenced by the concentrations of dissolved species, pH, temperature, suspended particles, and bacteria. Temperature plays a dual role with respect to oxygen corrosion. Increasing the temperature will reduce oxygen solubility. In open systems, in which oxygen can be released from the system, corrosion will increase up to a maximum at 80 °C (175 °F), where the oxygen solubility is 3 mg/L. Beyond this temperature, the reduced oxygen content limits the oxygen reduction reaction, preventing occurrence of the iron dissolution process. Thus, the corrosion rate of carbon steel decreases, and at boiling water conditions, the temperature effect is similar to room temperature with a high oxygen content. For closed systems, in which oxygen cannot escape, corrosion continues to increase linearly with temperature. The other factors affected by temperature are the diffusion of oxygen to the metal surface, the viscosity of water, and solution conductivity. Increasing the temperature will increase the rate of oxygen diffusion to the metal surface, thus increasing corrosion rate because more oxygen is available for the cathodic reduction process. The viscosity will decrease with increasing temperature, which will aid oxygen diffusion.

Sulfidation. Corrosion by various sulfur compounds at temperatures between 260 and 540 °C (500 and 1000 °F) is a common problem in many petroleum-refining processes and, occasionally, in petrochemical processes. When the sulfur activity (partial pressure and concentration) of the gaseous environment is sufficiently high, sulfide phases, instead of oxide phases, can be formed. In the majority of environments encountered in practice by oxidation-resistant alloys, Al_2O_3 or Cr_2O_3 should form in preference to any sulfides, and destructive sulfidation attack occurs mainly at sites where the protective oxide has broken down. The role of sulfur, once it has entered the alloy, appears to be to tie up the chromium and aluminum as sulfides, effectively redistributing the protective scale-forming elements near the alloy surface and thus interfering with the process of formation or re-formation of the protective scale. If sufficient sulfur enters the alloy so that all immediately available chromium or aluminum is converted to sulfides, then the less stable sulfides of the base metal may form because of morphological and kinetic reasons. It is these base metal sulfides that are often responsible for the observed accelerated attack, because they grow much faster than the oxides or sulfides of chromium or aluminum; in addition, they have relatively low melting points, so that molten slag phases are often possible.

Sulfur can also transport across continuous protective scales of Al_2O_3 and Cr_2O_3 under certain conditions, with the result that discrete sulfide precipitates can be observed immediately beneath the scales on alloys that are behaving in a protective manner. For the reasons indicated above, as long as the amount of sulfur present as sulfides is small, there is little danger of accelerated attack. However, once sulfides have formed in the alloy, there is a tendency for the sulfide phases to be preferentially oxidized by the encroaching reaction front and for the sulfur to be displaced inward, forming new sulfides deeper in the alloy, often in grain boundaries or at the sites of other chromium- or aluminum-rich phases, such as carbides. In this way, fingerlike protrusions of oxide/sulfide can be formed from the alloy surface inward, which may act to localize stress or otherwise reduce the load-bearing section.

The relative corrosivity of sulfur compounds generally increases with temperature. Depending on the process particulars, corrosion is in the form of uniform thinning, localized attack, or erosion-corrosion. Corrosion control depends almost entirely on the formation of protective metal sulfide scales that exhibit parabolic growth behavior. In general, nickel and nickel-rich alloys are rapidly attacked by sulfur compounds at elevated temperatures, while chromium-containing steels provide excellent corrosion resistance (as does aluminum). The combination of hydrogen sulfide and hydrogen can be particularly corrosive, and as a rule, austenitic stainless steels are required for effective corrosion control.

The effect of temperature and alloy compositions on sulfidic corrosions depends on the sulfur compounds present. Figure 29 shows the rates of sulfur corrosion of various steels as a function of temperature. These so-called McConomy curves can be used to predict the relative corrosivity of crude oils and their various fractions (Ref 40). Although this method relates corrosivity to total sulfur content, and therefore does not take into account the variable effects of different sulfur compounds, it can provide reliable corrosion trends if certain corrections are applied. Plant experience

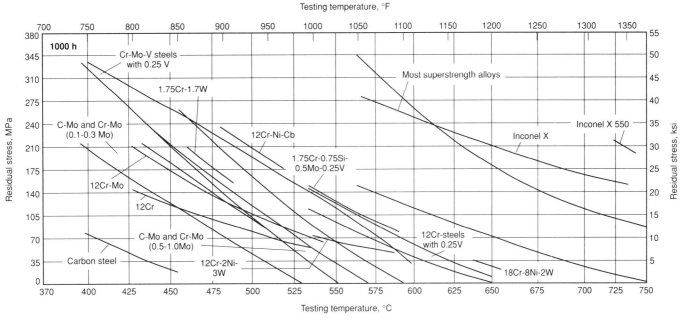

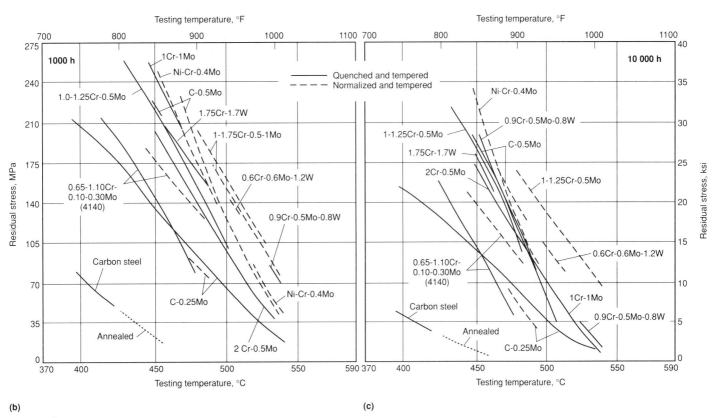

Fig. 18 Comparison of relaxation strengths (residual stress) of various steels. (a) Comparison of low-alloy steels with superstrength alloys. (b) Low-alloy steels at 1000 h. (c) Low-alloy steels at 10 000 h. Source: Ref 28

has shown that the McConomy curves, as originally published, tend to predict excessively high corrosion rates. The curves apply only to liquid hydrocarbon streams containing 0.6 wt% S (unless a correction factor given in Fig. 30 for sulfur content is applied) and do not take into account the effects of vaporization and flow regime. The curves can be particularly useful, however, for predicting the effect of operational changes on known corrosion rates.

Over the years, it has been found that corrosion rates predicted by the original McConomy curves should be decreased by a factor of roughly 2.5, resulting in the modified curves shown in Fig. 29. The curves demonstrate the beneficial effects of alloying steel with chromium in order to reduce corrosion rates. Corrosion rates are roughly halved when the next higher grade of low-alloy steel (for example, 2.25Cr-

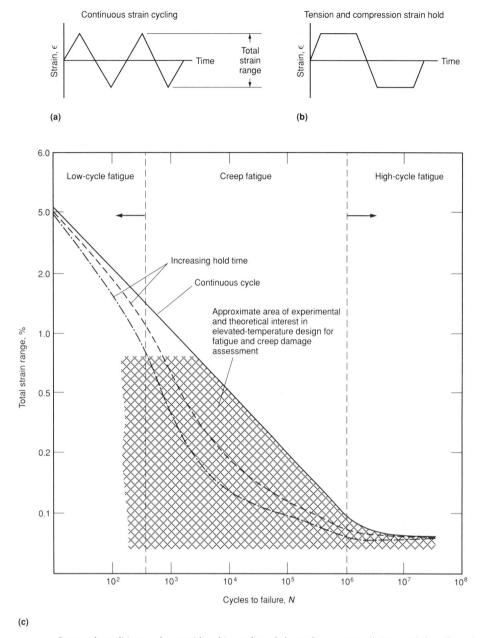

Fig. 19 Range of conditions to be considered in studies of elevated-temperature fatigue and the effect of continuous cycling (a) and strain hold (b) on elevated-temperature fatigue (c). Source: Ref 30

1Mo, 5Cr-0.5Mo, 7Cr-0.5Mo, or 9Cr-1Mo steel) is selected. Essentially, no corrosion occurs with stainless steels containing 12% or more chromium. Although few data are available, plant experience has shown that corrosion rates start to decrease as temperatures exceed 455 °C (850 °F). Two explanations frequently offered for this phenomenon are the possible decomposition of reactive sulfur compounds and the formation of a protective coke layer.

Sulfidic Corrosion With the Presence of Hydrogen. The presence of hydrogen increases the severity of high-temperature sulfidic corrosion. Hydrogen converts organic sulfur compounds to hydrogen sulfide; corrosion becomes a function of hydrogen sulfide concentration (or partial pressure).

A number of researchers have proposed various corrosion rate correlations for high-temperature sulfidic corrosion in the presence of hydrogen (Ref 41-47), but the most practical correlations seem to be the so-called Couper-Gorman curves. The Couper-Gorman curves are based on a survey conducted by National Association of Corrosion Engineers (NACE) Committee T-8 on Refining Industry Corrosion (Ref 48).

The Couper-Gorman curves differ from those previously published in that they reflect the influence of temperature on corrosion rates throughout a whole range of hydrogen sulfide concentrations. Total pressure was found not to be a significant variable between 1 and 18 MPa (150 and 2650 psig). It was also found that essentially no corrosion occurs at low hydrogen sulfide concentrations and temperatures above 315 °C (600 °F) because the formation of iron sulfide becomes thermodynamically impossible. Curves are available for carbon steel, 5Cr-0.5Mo steel, 9Cr-1Mo steel, 12% Cr stainless steel, and 18Cr-8Ni austenitic stainless steel. For the low-alloy steels, two sets of curves apply, depending on whether the hydrocarbon stream is naphtha or gas oil. The curves again demonstrate the beneficial effects of alloying steel with chromium to reduce the corrosion rate.

Modified Couper-Gorman curves are shown in Fig. 31. To facilitate the use of these curves, original segments of the curves were extended (dashed lines). In contrast to sulfidic corrosion in the absence of hydrogen, there is often no real improvement in corrosion resistance unless chromium content exceeds 5%. Therefore, the curves for 5Cr-0.5Mo steel also apply to carbon steel and low-alloy steels containing less than 5% Cr. Stainless steels containing at least 18% Cr are often required for essentially complete immunity to corrosion. Because the Couper-Gorman curves are primarily based on corrosion rate data for an all-vapor system, partial condensation can be expected to increase corrosion rates because of droplet impingement.

When selecting steels for resistance to high-temperature sulfidic corrosion in the presence of hydrogen, the possibility of high-temperature hydrogen attack should be considered. Conceivably, this problem arises when carbon steel and low-alloy steels containing less than 1% Cr are chosen for temperatures exceeding 260 °C (500 °F) and hydrogen partial pressures above 700 kPa (100 psia) and when corrosion rates are expected to be relatively low.

Hydrogen Damage. Because iron-base alloys are principal materials of construction, these alloys have been the focus of most of the studies relating to hydrogen effects. In addition, ferritic (body-centered cubic) steels have a particular sensitivity to hydrogen. For these reasons, hydrogen effects on steel are important. Such hydrogen effects have been thoroughly described in a review of hydrogen damage (Ref 49).

Hydrogen damage includes hydrogen embrittlement, hydrogen attack, and hydrogen blistering. In ferrous alloys, embrittlement by hydrogen is generally restricted to those alloys having a hardness of 22 HRC or greater. The other forms of hydrogen damage, such as hydrogen attack or hydrogen blistering, are associated with low-alloy or carbon steels.

Thermodynamic calculations by Odette (Ref 50) have also shown that the carbide M_3C is significantly less stable in hydrogen

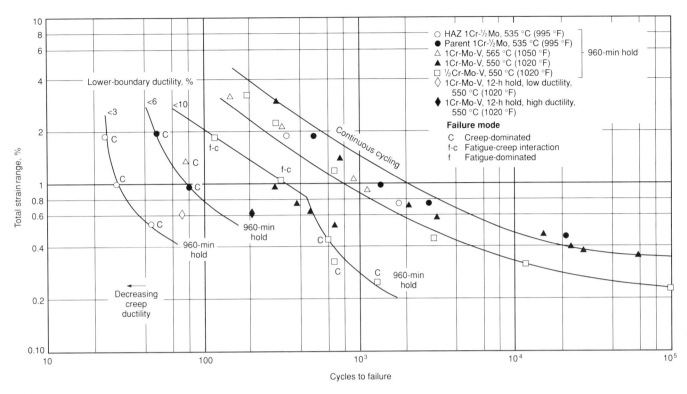

Fig. 20 Effect of ductility on endurance of ferritic steels. Source: Ref 31

environments than alloy carbides such as M_2C, M_7C_3, $M_{23}C_6$, and M_6C. As shown in Fig. 32, this M_3C carbide can persist in normalized 2.25Cr-1Mo steel for up to 50 h at 700 °C (1290 °F) and hence may be present during service. When alloy content is increased to 3Cr-1.5Mo, Ritchie et al. (Ref 51) have shown that the tempering kinetics are significantly accelerated and that M_3C can be eliminated from the microstructure within 1 h at 700 °C (1290 °F). Hydrogen exposures of 3Cr-1.0Mo-0.5Ni steel at 600 °C (1110 °F) and 17 MPa (2.5 ksi)

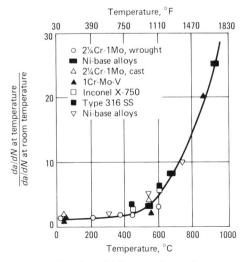

Fig. 21 Variation of fatigue-crack growth rates as a function of temperature at $\Delta K = 30$ MPa$\sqrt{m}$ (27 ksi$\sqrt{in}$.). Source: Ref 32

for 1000 h gave no indication of the voids characteristic of hydrogen attack on these steels (Ref 51).

Hydrogen attack is a damage mechanism that is associated with carbon and low-alloy steels exposed to hydrogen-containing environments at temperatures above 220 °C (430 °F) (Ref 49). With hydrogen at elevated temperatures and pressures, there is increasing availability of atomic hydrogen that can easily penetrate metal structures and react internally with reducible species. Exposure to the environment is known to result in a direct chemical reaction with the carbon in the steel. The reaction occurs between absorbed hydrogen and the iron carbide phase, resulting in the formation of methane:

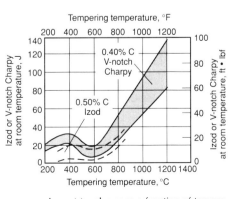

Fig. 22 Impact toughness as a function of tempering temperature of hardened, low-alloy, medium-carbon steels. Source: Ref 34

$$2H_2 + Fe_3C \rightarrow CH_4 + 3Fe$$

Unlike nascent hydrogen, the resulting methane gas does not dissolve in the iron lattice. Internal gas pressures develop, leading to the formation of voids, blisters, or cracks. The generated defects lower the strength and ductility of the steel. Because the carbide phase is a reactant in the mechanism, its absence in the vicinity of generated defects serves as direct evidence of the mechanism itself.

Hydrogen attacks occur in carbon steels and can lead to fissuring of the steel. Alloy steels with stable carbides, such as chromium and molybdenum carbides, are less susceptible to this form of attack. For example, 2.25Cr-1Mo suffers some decarburization in high-temperature high-pressure hydrogen, but is less likely to fissure than carbon steel. Hydrogen attack does not occur in austenitic stainless steels (Ref 49).

The susceptibility of steels to attack by hydrogen can be judged from the Nelson curves, which indicate the regions of temperature and pressure in which a variety of steels will suffer attack. Nelson curves for various alloy steels are shown in Fig. 33. In Fig. 33(b), the Nelson curve indicates only the operating limits for the surface decarburization of 2¼Cr-1Mo steel. There are, however, indications that the limiting condition for using quenched and tempered 2¼Cr-1Mo steel will not be decarburization but rather the loss of integrity through methane bubble growth (Ref 54). This may only be true if fatigue is not a significant

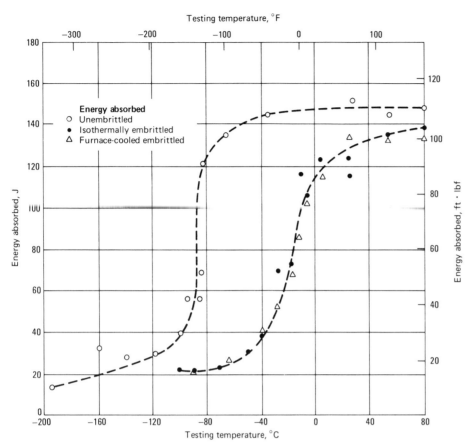

Fig. 23 Shift in impact transition curve to higher temperature as a result of temper embrittlement produced in SAE 3140 steel by isothermal holding and furnace cooling through the critical range. Source: Ref 34

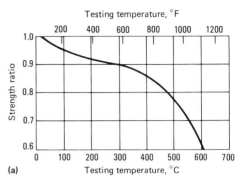

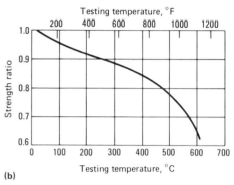

Fig. 24 Ratios of elevated-temperature strength to room-temperature strength for hardened and tempered 2¼Cr-1Mo steel tempered to room-temperature tensile strengths ranging from 480 to 1100 MPa (70 to 160 ksi). (a) Tensile strength. (b) Yield strength. Source: Ref 2

design consideration, because decarburization can affect fatigue strength. The top curve in Fig. 33(b) shows an estimate of the operating limits for the formation of methane bubbles (that is, hydrogen attack) in 2¼Cr-1Mo steel (Ref 53).

Hydrogen blistering is a mechanism that involves the hydrogen damage of unhardened steel near ambient temperature. It is known that the entry of atomic hydrogen into steel can result in its collection, as the molecular species, at internal defects or interfaces. If the entry kinetics are substantial (promoted by an acidic environment, high corrosion rates, and cathodic poisons), the resulting internal pressure will cause internal separation (fissuring or blistering) of the steel. Such damage typically occurs at large, elongated inclusions and results in delaminations known as hydrogen blisters. Field experience indicates that fully killed steels are more susceptible than semikilled steels (Ref 55), but the nature and size of the original inclusions appear to be the key factors with regard to susceptibility. Rimmed steels or free-machining grades with high levels of sulfur or selenium would most likely show a high susceptibility to blistering. Stepwise cracking at the ends of blisters indicates an effect of elongated inclusions in the delamination process (Ref 49, 55). Similar stepwise cracking occurs in the hydrogen-induced failure of low-alloy pipeline steels (Ref 56). Both stepwise cracking and blistering appear to be limited to environments in which acidic corrosion occurs and in which cathodic poisons, such as sulfide, are present to promote hydrogen entry.

Hydrogen embrittlement, unlike hydrogen attack or blistering, can occur without immediate and resolvable damage within the metal structure. In this respect, hydrogen embrittlement is a somewhat reversible process. For example, hydrogenation plant equipment, operating at about 540 °C (1000 °F) with absorbed hydrogen in the steel, is cooled from operating temperature at 30 to 40 °C (50 to 75 °F) per hour with no breakage. This cooling rate is apparently slow enough to allow most of the absorbed hydrogen to diffuse from the metal without causing excessive embrittlement. Another way of removing detrimental atomic hydrogen derives from its mobility at higher temperatures. A bake-out cycle, involving temperatures of 175 to 205 °C (350 to 400 °F), allows the diffusion and escape of hydrogen from the metal or alloy. If the hydrogen charging conditions were not severe enough to cause internal damage, the bake-out cycle (described in Federal Specification QQC-320) will restore full ductility.

Susceptibility to hydrogen embrittlement is strongly influenced by the strength level of the metal or alloy. In steels, untempered martensite is the most susceptible phase. Lamellar carbide structures are less desirable than those with spheroidized structures. In general, iron-base alloys with a ferritic or martensitic structure are restricted to a maximum hardness of 22 HRC. Most other alloys are restricted to a maximum hardness of 35 HRC. There are exceptions in both cases. The procedures for materials testing in a wet hydrogen sulfide environment, which is the most aggressive in promoting hydrogen entry, are discussed in NACE TM-01-77. Further information on hydrogen embrittlement is provided in the article "Embrittlement of Steels" in this Volume.

Resistance to Liquid-Metal Corrosion. The following sections describe the resistance of steels to various liquid-metal environments. Liquid metals can attack steels in various ways. One form of attack that can occur is intergranular penetration and/or dissolution by liquid metal. The unfortunate aspect of this mode of attack is that it can result in a loss of strength without any large weight loss or change in appearance. In this respect, it resembles the more familiar aqueous intergranular corrosion. Additional in-

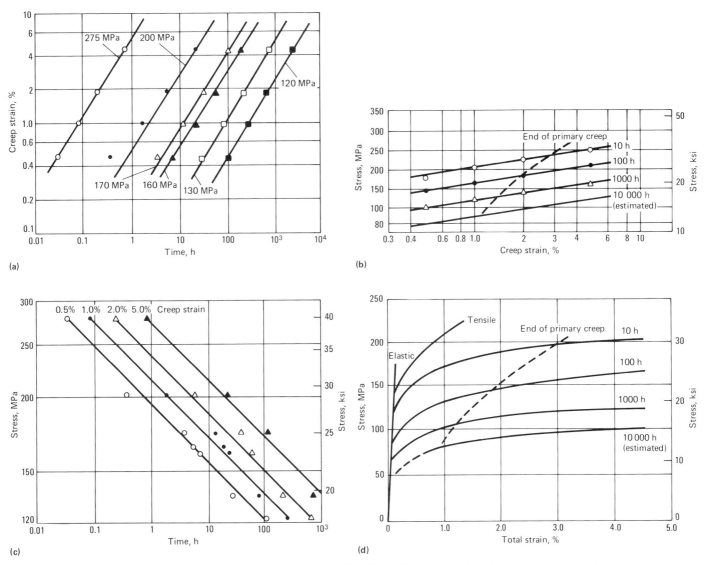

Fig. 25 Analysis of creep data. Creep behavior of 2¼Cr-1Mo steel tested at 540 °C (1000 °F). (a) Creep strain-time plot; constant-stress lines have been drawn parallel. (b) Stress-creep strain plot. (c) Stress-time plot; constant-strain lines have been drawn parallel. (d) Isochronous stress-strain curves. Source: Ref 35

formation on liquid-metal corrosion can be found in Volume 13 of the 9th Edition of *Metals Handbook*.

Another phenomenon is liquid-metal embrittlement, which requires the presence of both stress and a liquid metal. The cracks that occur during the embrittlement process may be intergranular or transgranular. The process seems to be similar in many ways to stress-corrosion cracking. Steels have been reported to undergo embrittlement by lithium, indium, cadmium, zinc, tellurium, and various lead-tin solders. Additional information on liquid-metal embrittlement can be found in the article "Embrittlement of Steels" in this Volume.

Sodium and Sodium-Potassium Alloys. Plain carbon and low-alloy steels are generally suitable for long-term use in these media at temperatures to 450 °C (840 °F). Beyond these temperatures, stainless steels are required.

The principal disadvantage of ferritic steels in sodium systems is the decarburization potential and its possible effect on the mechanical properties of the ferritic steel and the other system materials. Reference 5 considers aspects of low-carbon 2¼Cr-1Mo steel as the construction material for the sodium-heated steam generator of a liquid-metal fast breeder reactor.

Lithium is somewhat more aggressive to plain carbon steels than sodium or sodium-potassium. As a result, low-alloy steels should not be considered for long-term use above 300 °C (570 °F). At higher temperatures, the ferritic stainless steels show better results.

Cadmium. Low-alloy steels exhibit good serviceability to 700 °C (1290 °F).

Zinc. Most engineering metals and alloys show poor resistance to molten zinc, and carbon steels are no exception.

Antimony. Low-carbon steels have poor resistance to attack by antimony.

Mercury. Although plain carbon steels are virtually unattacked by mercury under nonflowing or isothermal conditions, the presence of either a temperature gradient or liquid flow can lead to drastic attack. The corrosion mechanism seems to be one of dissolution, with the rate of attack increasing rapidly with temperature above 500 °C (930 °F). Alloy additions of chromium, titanium, silicon, and molybdenum, alone or in combination, show resistance to 600 °C (1110 °F). Where applicable, the attack of ferrous alloys by mercury can be reduced to negligible amounts by the addition of 10 ppm Ti to the mercury; this raises the useful range of operating temperatures to 650 °C (1200 °F). Additions of metal with a higher affinity for oxygen than titanium, such as sodium or magnesium, may be required to prevent oxidation of the titanium and loss of the inhibitive action.

Aluminum. Plain carbon steels are not satisfactory for the long-term containment of molten aluminum.

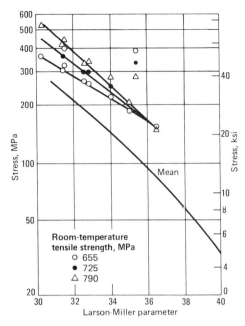

Fig. 26 Larson-Miller plot of stress-rupture behavior of 2¼Cr-1Mo steel. Variation in Larson-Miller parameter with stress to rupture for normalized and tempered and hardened and tempered specimens of 2¼Cr-1Mo steel tested between 425 and 650 °C (800 and 1200 °F) for rupture life to 10 000 h; the data are grouped according to the room-temperature tensile strength of the steel. Larson-Miller plot for annealed steel included for comparison. Source: Ref 2

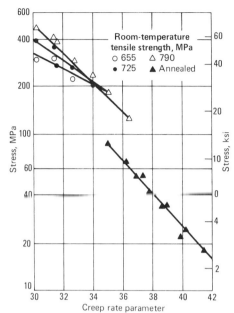

Fig. 27 Modified Larson-Miller plot of creep behavior of 2¼Cr-1Mo steel. Variation in creep rate parameter with creep stress for normalized and tempered and hardened and tempered specimens of 2¼Cr-1Mo steel tested between 425 and 650 °C (800 and 1200 °F) for test duration to 10 000 h; the data are grouped according to the room-temperature tensile strength of the steel. Creep rate data for annealed steel included for comparison. Source: Ref 2, 36

Gallium is one of the most aggressive of all liquid metals and cannot be contained by carbon or low-alloy steels at elevated temperatures.

Indium. Carbon and low-alloy steels have poor resistance to molten indium.

Lead, Bismuth, Tin, and Their Alloys. Low-alloy steels have good resistance to lead up to 600 °C (1110 °F), to bismuth up to 700 °C (1290 °F), and to tin only up to 150 °C (300 °F). The various alloys of lead, bismuth, and tin are more aggressive.

Factors Affecting Mechanical Properties

The factors affecting the mechanical properties of steels include the nature of the strengthening mechanisms, the microstructure, the heat treatment, and the alloy composition. This section describes these factors, with particular emphasis on chromium-molybdenum steels (especially 2¼Cr-1Mo) used for elevated-temperature service.

In addition, various service factors such as thermal exposure and environmental conditions can induce metallurgical changes, which may affect the mechanical properties of steels used at elevated temperatures. These metallurgical changes include spheroidization, graphitization, decarburization, and carburization. Depending on the temperature and exposure environment, ferritic steels used at elevated temperatures may also be susceptible to embrittlement phenomena such as temper embrittlement, temper martensite embrittlement, creep embrittlement, hydrogen embrittlement, and liquid-metal embrittlement. Embrittlement phenomena are discussed in the article "Embrittlement of Steels" in this Volume.

Strengthening Mechanisms. The creep strength of a steel is affected by the typical strengthening mechanisms—namely, grain refinement, solid-solution hardening, and precipitation hardening. Of these various strengthening mechanisms, the refinement of grain size is perhaps the most unique because it is the only strengthening mechanism that also increases toughness. Figure 14 shows the effect of grain size on the creep strength of a carbon steel.

The creep strength of chromium-molybdenum steels is mainly derived from a complex combination of solid-solution (primarily interaction solid-solution strengthening) and precipitation effects, as illustrated in

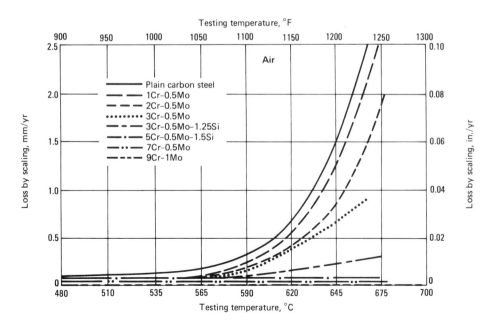

Fig. 28 Effect of temperature on metal loss from scaling for several carbon and alloy steels in air

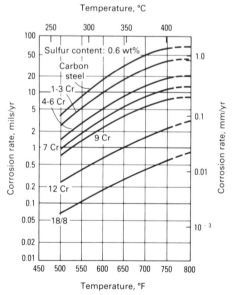

Fig. 29 Modified McConomy curves showing the effect of temperature on high-temperature sulfidic corrosion of various steels and stainless steels. Source: Ref 39

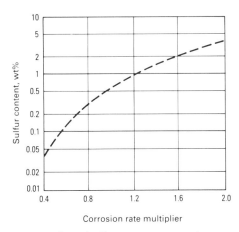

Fig. 30 Effect of sulfur content on corrosion rates predicted by modified McConomy curves in the temperature range of 290 to 400 °C (550 to 750 °F). Source: Ref 39

ing it easier for dislocations to move. In precipitation hardening, heating of the alloy to an excessively high temperature can cause solutionizing of the precipitates. At intermediate temperatures, the precipitates can coarsen and become less-effective impediments to dislocation motion. High stresses and high-strain cyclic loading also can lead to accelerated softening.

The solid-solution strengthening effect illustrated in Fig. 34 occurs primarily from a process termed interaction solid-solution hardening (or strengthening), which is a mechanism that involves the interaction of substitutional and interstitial solutes (Ref 57-59). This process occurs in ferritic alloys that contain in solid solution interstitial and substitutional elements that have an affinity for each other. As a result of this strong attraction, atom pairs or clusters could form dislocation atmospheres that hinder dislocation motion and therefore strengthen the steel. Other solid-solution effects from either a pure substitutional solute or a pure interstitial solute do not alone provide significant creep strengthening in carbon manganese steels and molybdenum steels. The addition of interstitial solutes to iron has no significant creep-strengthening effect above 450 °C (840 °F), while the substitutional solutes manganese, chromium, and molybdenum give rise to only modest increases in strength in the absence of interstitial solutes (Ref 59). However, when certain combinations of substitutional and interstitial solutes are present together (for example, manganese-nitrogen, molybdenum-carbon, and molybdenum-nitrogen), there is a substantial increase in creep strength (Ref 59). These combinations give rise to the strengthening process of interaction solid-solution hardening.

Precipitation-strengthening effects are probably negligible in the carbon-manganese steels typically used for elevated-temperature applications (Ref 57), although strengthening by fine NbC particles has been observed (Ref 60). Precipitation strengthening is more significant in molybdenum steels, for which the strengthening precipitates are mainly Mo_2C and Mo_2N. Further increases in precipitation strengthening can be achieved with additions of niobium or vanadium to chromium-molybdenum steels. The stability of the carbides increases in the following order of alloying elements: chromium, molybdenum, vanadium, and niobium. Fine and closely dispersed precipitates of NbC and VC are thus desirable, followed by the other carbides. This precipitation strengthening effect in creep-resistant chromium-molybdenum steels is related to secondary hardening, as discussed below.

Secondary Hardening. If the mechanical properties of tempered steels need to be maintained at elevated service temperatures, the problem is to reduce the amount of softening during tempering so that higher strength (hardness) can be achieved at higher temperatures. One way to reduce softening is with strong carbide formers such as chromium, molybdenum, and vanadium. These carbide formers induce an effect known as secondary hardening. Without these elements, iron-carbon alloys and low-

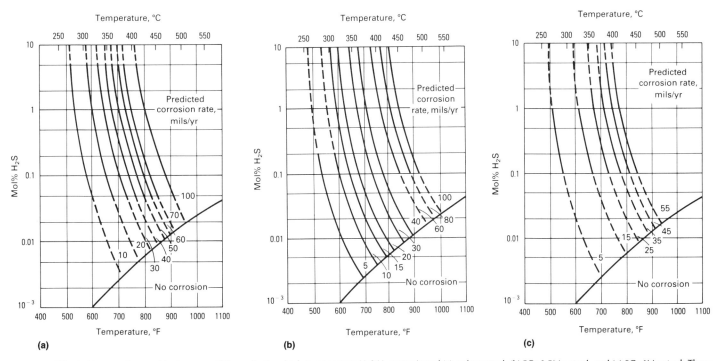

Fig. 31 Effect of temperature and hydrogen sulfide content on high-temperature H_2S/H_2 corrosion of (a) carbon steel, (b) 5Cr-0.5Mo steel, and (c) 9Cr-1Mo steel. These corrosion rates are based on the use of gas oil desulfurizers; corrosion rates with naphtha desulfurizers may be slightly less severe. Source: Ref 39

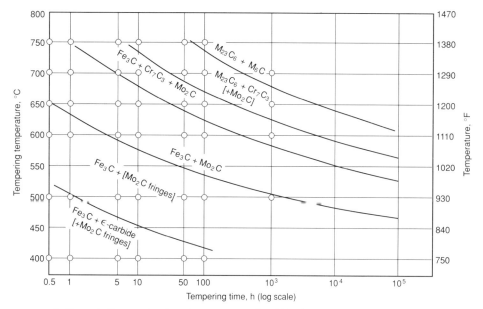

Fig. 32 Isothermal diagram showing the sequence of carbide formation on tempering of normalized 2¼Cr-1Mo steel. Source: Ref 12

carbon steels soften rapidly with increasing tempering temperature, as shown in Fig. 35. This softening is largely due to the rapid coarsening of cementite with increasing tempering temperature, a process dependent on the diffusion of carbon and iron. If present in a steel in sufficient quantity, however, the carbide-forming elements not only retard softening but also form fine alloy carbides that produce a hardness increase at higher tempering temperatures. This hardness increase is frequently referred to as secondary hardening. This hardening can also occur during elevated-temperature service and is related to creep strength, as shown in Fig. 36.

Secondary hardening allows higher tempering temperatures, and this increases the range of service temperatures. Figure 37 shows secondary hardening in a series of steels containing molybdenum. The secondary hardening peaks develop only at high tempering temperatures because alloy carbide formation depends on the diffusion of the carbide-forming elements, a more sluggish process than that of carbon and iron diffusion. As a result, not only is a finer dispersion of particles produced but also the alloy carbides, once formed, are quite resistant to coarsening. The latter characteristic of the fine alloy carbides is used to advantage in tool steels that must not soften even though high temperatures are generated by their use in hot-working dies or high-speed machining. Also, ferritic low-carbon steels containing chromium and molybdenum are used in pressure vessels and reactors operated at temperatures around 540 °C (1000 °F) because the alloy carbides are slow to coarsen at those temperatures.

The beneficial property changes from secondary hardening can be improved by increasing the intensity of secondary hardening, by decreasing the rate of overaging of the secondary-hardening carbide, or by increasing the temperature of secondary hardening. The intensity of secondary hardening can be increased by increasing the mismatch between the carbide precipitate and the matrix (Ref 62). Although this tends to cause more rapid overaging, the net effect can be beneficial, so that a higher strength after tempering is achieved. Increased mismatch is produced by:

- Increasing the lattice parameter of the carbide precipitate
- Decreasing the lattice parameter of the matrix

The carbide Mo_2C can dissolve both chromium and vanadium. Chromium, being a smaller atom than molybdenum, reduces the lattice parameter of Mo_2C, but vanadium increases it. Chromium therefore tends to decrease the intensity of secondary hardening. Moreover, chromium causes the Mo_2C carbide to be less stable (that is, to give maximum secondary hardening at lower tempering temperatures) and accelerates overaging (Fig. 38). On the other hand, vanadium increases the lattice parameter of Mo_2C and stabilizes the carbide. The result is a greater intensity of secondary hardening.

Effects of Microstructure. It is widely accepted that the strength and impact toughness of carbon and chromium-molybdenum steels with fully bainitic microstructures are better than those with a ferritic-bainitic microstructure. Bainitic microstructures also have better creep resistance under high-stress, short-time conditions, but degrade more rapidly at high temperatures than pearlitic structures. As a result, ferrite-pearlite material has better intermediate-term, low-stress creep resistance. Because both microstructures will eventually spheroidize, it is expected that over long service lives the two microstructures will converge to similar creep strengths. This convergence can be estimated to occur in about 50 000 h at 540 °C (1000 °F) for 2.25Cr-1Mo steel, based on the limited data presented in Fig. 39. Investigations of chromium-molybdenum steels for one application concluded that tempered bainite is the optimum microstructure for creep resistance (Ref 64). However, the carbide precipitates are also an important microstructural factor in achieving optimum creep behavior, and for some microstructures an untempered condition may be desirable (see the following section "Effects of Heat Treatment" in this article). Moreover, even though bainitic microstructures improve strength, toughness, and creep resistance, chromium-molybdenum steels with bainitic and tempered martensitic microstructures also undergo strain softening during mechanical cycling. This effect of strain softening of bainitic chromium-molybdenum steels has undergone several investigations (Ref 18-21).

Microstructure may also influence the carbide precipitation and strengthening mechanism of chromium-molybdenum steels. In 2¼Cr-1Mo steel, for example, precipitation reactions are known to occur much more rapidly in bainite than in proeutectoid ferrite (Ref 12). In addition, the interaction solid-solution strengthening of 2¼Cr-1Mo steel is influenced by microstructure. In tensile studies, it was concluded that interaction solid-solution hardening in bainitic (normalized and tempered) 2.25Cr-1Mo steel is due to chromium-carbon interactions, while it is due to molybdenum-carbon interactions in the proeutectoid ferrite of annealed steel (Ref 65).

Effects of Heat Treatment. Figure 40 shows the general effect of three heat treatments on the creep-rupture strength of 2.25Cr-1Mo steel. Like long-term exposure (Fig. 39), the creep-rupture strengths converge in Fig. 40.

The use of tempering is also an important factor that influences the level of precipitation strengthening and solid-solution strengthening in chromium-molybdenum steel (Fig. 34). In a normalized molybdenum steel (Fig. 34a), the initial contribution from solid-solution strengthening is greater than that of the normalized and tempered steel. In the normalized and tempered molybdenum steels (Fig. 34b), the initial contribution from precipitation strengthening will be larger than that from the normalized steel. In addition, the precipitation-strengthening effect in the normalized and tempered steel will reach a maximum and begin to decline at an earlier stage due to the earlier incidence of overaging in tempered material. This is a potential consideration in applica-

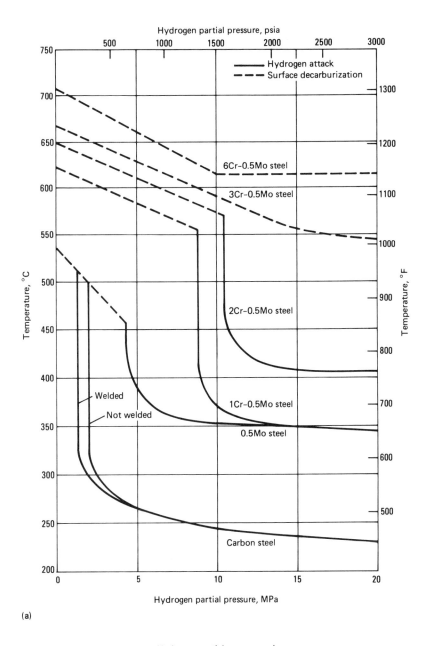

(a)

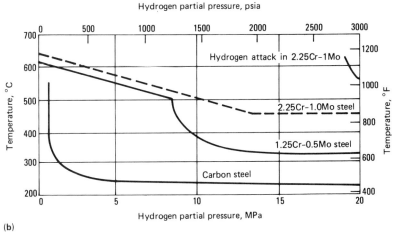

(b)

Fig. 33 (a) Nelson curves defining the operating limits of various alloys in a hydrogen environment. Curves adapted from the chart of Nelson. Source: Ref 52. (b) Nelson curves for three steels given in Ref 36 and an estimate of the operating limit (solid line) for the formation of methane bubbles in 2.25Cr-1Mo steel. Source: Ref 53

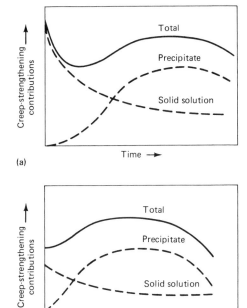

Fig. 34 Schematic of changes in creep strengthening contributions at 550 °C (1020 °F) in (a) normalized molybdenum steel and (b) normalized and tempered molybdenum steel. Source: Ref 57

tions requiring creep resistance over long times and at high temperatures.

As noted in the previous section "Effects of Microstructure," an investigation for one application concluded that tempered bainite resulted in the optimum creep resistance. In ferrite-pearlite or ferrite-bainite structures, however, it has been suggested that the best creep resistance at relatively high stresses is obtained in the untempered condition, because the dislocations introduced on loading are then able to nucleate a finer dispersion of particles in ferrite grains than is obtained by tempering in the absence of strain (Ref 57). Application of this concept does not apply to bainitic structures. In bainitic steels, where the dislocation density is already higher than that introduced upon straining of a ferrite/pearlite steel, the use of untempered structures is unlikely to prove beneficial to short-term creep strength (Ref 57). Ultimately, it is the balance of hardness (or strength) and toughness required in service that determines the conditions of tempering for a given application. Figure 41 shows the variation of properties from the tempering of a modified 9Cr-1Mo alloy (9Cr-1Mo with 0.06 to 0.10 wt% Nb and 0.18 to 0.25 wt% V).

Effects of Composition. The mechanical properties of carbon and low-alloy steels are determined primarily by composition and heat treatment. The effects of alloying elements in annealed, normalized and tempered, and quenched and tempered steels are discussed below.

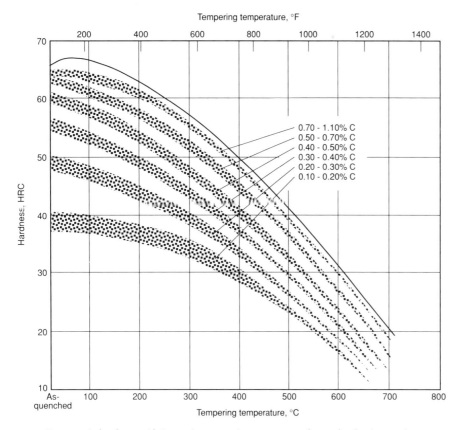

Fig. 35 Decrease in hardness with increasing tempering temperature for steels of various carbon contents. Source: Ref 61

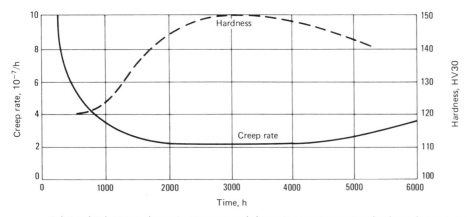

Fig. 36 Relationship between change in creep rate and change in room-temperature hardness during creep of normalized 1% Mo steel tested at 123 MPa (17.8 ksi) at 550 °C (1020 °F). Under these test conditions, secondary creep coincided with maximum precipitation hardening. Source: Ref 57

Carbon increases both the strength and hardenability of steel at room temperature but decreases the weldability and impact toughness. In plain carbon and carbon-molybdenum steels intended for elevated-temperature service, carbon content is usually limited to about 0.20%; in some classes of tubing for boilers, however, carbon may be as high as 0.35%. For chromium-containing steels, carbon content is usually limited to 0.15%. Carbon increases short-term tensile strength but does not add appreciably to creep resistance at temperatures above 540 °C (1000 °F) because carbides eventually become spheroidized at such temperatures.

Manganese, in addition to its normal function of preventing hot shortness by forming dispersed manganese sulfide inclusions, also appears to enhance the effectiveness of nitrogen in increasing the strength of plain carbon steels at elevated temperatures. Manganese significantly improves hardenability, but contributes to temper embrittlement.

Phosphorus and sulfur are considered undesirable because they reduce the elevated-temperature ductility of steel. This reduction in ductility is demonstrated by reductions in stress-rupture life and thermal fatigue life. Phosphorus contributes to temper embrittlement.

Silicon increases the elevated-temperature strength of steel. It also increases the resistance to scaling of the low-chromium steels in air at elevated temperatures. Silicon is one factor in temper embrittlement.

Chromium in small amounts (~0.5%) is a carbide former and stabilizer. In larger amounts (up to 9% or more), it increases the resistance of steels to corrosion. Chromium also influences hardenability.

The effect of chromium in ferritic creep-resistant steels is complex. By itself, chromium gives some enhancement of creep strength, although increasing the chromium content in lower-carbon grades does not increase resistance to deformation at elevated temperatures (Ref 59). When added to molybdenum steel, chromium generally leads to some reduction in creep strength (Ref 67) such as that shown in Fig. 42. For the 1.0Mo steel in Fig. 42, the optimum creep strength occurs with about 2.25% Cr. Chromium is most effective in strengthening molybdenum steels (0.5 to 1.0% Mo) when it is used in amounts of 1 to 2½%.

Figure 43 summarizes the effects of chromium content on the tensile and yield strengths of chromium-molybdenum steels containing 0.5 to 1.0% Mo and various amounts of chromium. The effect of temperature is reported as the test temperature at which strength is reduced to 60% of its room-temperature value. Chromium is most effective in strengthening these chromium-molybdenum steels when it is used in amounts of 1 to 2½%.

Molybdenum is an essential alloying element in ferritic steels where good creep resistance above 450 °C (840 °F) is required. Even in small amounts (0.1 to 0.5%), molybdenum increases the resistance of these steels to deformation at elevated temperatures. Much greater creep strength can be obtained by increasing the molybdenum level to about 1% but at the expense of greatly reduced rupture ductility (Ref 69). Additions of chromium can improve rupture ductility.

Molybdenum is a carbide stabilizer and prevents graphitization. For certain ranges of stress and temperature, the dissolving of iron carbide and the concurrent precipitation of molybdenum carbide cause strain hardening in these steels. Molybdenum in amounts of 0.5% or less also minimizes temper embrittlement.

Niobium and vanadium are added to improve elevated-strength properties. Vanadium is also added to some of the higher-carbon steels to provide additional resistance to tempering and to retard the growth of carbides at service temperatures. Niobium is sometimes added to these steels

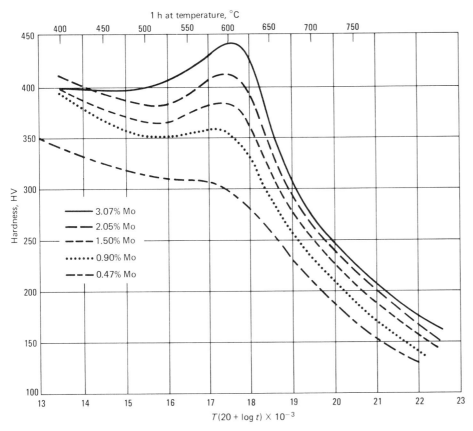

Fig. 37 Retardation of softening and secondary hardening during tempering of steels with various molybdenum contents. Source: Ref 61

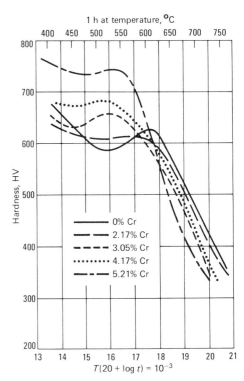

Fig. 38 Effect of chromium on the tempering characteristics of 0.45C-1.75Mo-0.75V steels. Source: Ref 62

to increase their strength through the formation of carbides. Niobium and vanadium improve resistance to hydrogen attack, but may promote hot (reheat) cracking.

Boron is added to increase hardenability. Boron can cause hot shortness and can impair toughness.

Tungsten behaves like molybdenum in simple steels and has been proposed for replacing molybdenum in nuclear applications (Ref 70-72).

Thermal Exposure and Aging. Thermal exposure over time is one of the main service conditions affecting mechanical properties because the metallurgical structure of steel changes with time at temperature. For example, a ferritic matrix may be either fine or coarse-grained initially, and the carbides may vary from lamellar to completely spheroidized. With increasing time at service temperatures, the metallurgical structure slowly approaches a more stable state. For example, there may be some increase in ferrite grain size, the carbides may spheroidize, and the structure of carbon and carbon-molybdenum steels may approach the graphitized condition, with large irregular nodules of graphite in a ferrite matrix and few, if any, remaining carbides.

The thermal exposure of molybdenum and molybdenum-chromium ferritic steels also contributes to complex aging phenomena, which are governed by the complicated carbide precipitation processes that occur in the steel. Figure 32, for example, shows the sequence of carbide formations in 2.25Cr-1Mo steel. The M_2C carbide (where M is primarily molybdenum) is the principal carbide for strengthening in this steel. The Mo_2C first precipitates during heat treatment and/or elevated-temperature exposure. The Mo_2C forms a high density of fine needles or platelets and thus contributes to strengthening by dispersion hardening. During thermal exposure, however, the unstable Mo_2C carbide eventually transforms into large globular particles of $M_{23}C$ and η carbide. These particles are thought to have little strengthening effect, although there

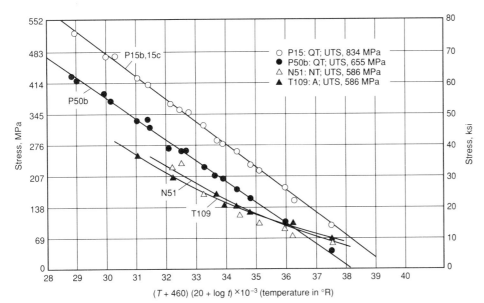

Fig. 39 Variation in stress-rupture strength of 2¼Cr-1Mo steel under different heat treatments. QT, quenched and tempered; NT, normalized and tempered; A, annealed; UTS, ultimate tensile strength. Source: Ref 63

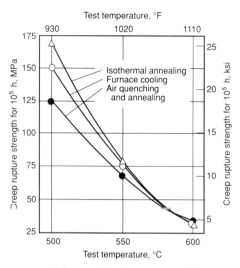

Fig. 40 Influence of heat treatment on 10^5-h creep-rupture strength of 2¼Cr-1Mo steel. Source: Ref 66

are some indications that $M_{23}C$ and η carbide present after long aging times in 2¼Cr-1Mo steel can enhance rupture strength (Ref 37). Precipitation kinetics also depend on microstructure. The strengthening carbide Mo_2C precipitates more rapidly in bainite than proeutectoid ferrite. Similarly, the Mo_2C is replaced more quickly by more stable carbides in bainite than in proeutectoid ferrite. In either case, these precipitation reactions influence the strength in a similar way, regardless of whether the microstructure is bainite or proeutectoid ferrite.

Spheroidization of the carbides in a steel occurs over time because spheroidized microstructures are the most stable microstructure found in steels. This spheroidization of carbides reduces strength and increases ductility.

The effect of spheroidization on the rupture strength of a typical carbon-molybdenum steel containing 0.17% C and 0.42% Mo, at 480 and 540 °C (900 and 1000 °F), is shown in Fig. 44 for several initial metallurgical structures (normalized or annealed,

fine or coarse grained). In these tests, the structure of the steel affected the rupture strength; for example, the stress for failure of a spheroidized structure in a given time was sometimes only half that of a normalized structure.

At 480 °C (900 °F), a coarse-grain normalized structure was the strongest for both short-time and long-time tests. The spheroidized structures were weaker than the normalized or annealed structures for short-time tests at both 480 and 540 °C (900 and 1000 °F). As the test time increased, the rupture values for all the structures tended to approach a common value.

The rate of spheroidization depends on the initial microstructure. The slowest spheroidizing is associated with pearlitic microstructures, especially those with coarse interlamellar spacings. Spheroidizing is more rapid if the carbides are initially in the form of discrete particles, as in bainite, and even more rapid if the initial structure is martensite.

Graphitization is a microstructural change that sometimes occurs in carbon or low-alloy steels subjected to moderate temperatures for long periods of time. The microstructure of carbon and carbon-molybdenum steels used for high-temperature applications such as vessels or pipes is normally composed of pearlite, which is a mixture of ferrite with some iron carbide (cementite). However, the stable form of carbon is graphite rather than cementite. Therefore, the pearlite can decompose into ferrite and randomly dispersed graphite, while the cementite will tend to disappear in these materials if they are in service long enough at metal temperatures higher than 455 °C (850 °F). This graphitization from the decomposition of pearlite into ferrite and carbon (graphite) can embrittle steel parts, especially when the graphite particles form along a continuous zone through a load-carrying member. Graphite particles that are randomly distributed throughout the microstructure cause only moderate loss of strength. Graphitization can be resisted by steels containing more than 0.7% Cr; such steels always contain at least 0.5% Mo as well, largely to impart elevated-temperature strength and resistance to temper embrittlement.

Graphitization and the formation of spheroidal carbides are competing mechanisms of pearlite decomposition. The rate of decomposition is temperature dependent for both mechanisms, and the mechanisms have different activation energies. As shown in Fig. 45, graphitization is the usual mode of pearlite decomposition at temperatures below about 550 °C (about 1025 °F), and the formation of spheroidal carbides can be expected to predominate at higher temperatures. Because graphitization involves prolonged exposure to moderate temperatures, it seldom occurs in boiling-surface tubing. Economizer tubing, steam piping, and other components that are exposed to temperatures from about 425 to 550 °C (800 to 1025 °F) for several thousand hours are more likely than boiler-surface tubing to be embrittled by graphitization.

The heat-affected zones adjacent to welds are among the most likely locations for graphitization to occur. Figure 46(a) shows a carbon-molybdenum steel tube that ruptured in a brittle manner along fillet welds after 13 years of service. Investigation of this failure revealed that the rupture was caused by the presence of chainlike arrays of embrittling graphite nodules (Fig. 46b and c) along the edges of heat-affected zones associated with each of the four welds on the tube. Arrays of graphite nodules were also found in the same locations on welds in several adjacent tubes, necessitating replacement of the entire tube bank.

Decarburization is a loss of carbon from the surface of a ferrous alloy as a result of heating in a medium (for example, hydrogen) that reacts with carbon. Unless special precautions are taken, the risk of losing carbon from the surface of steel is always present in any heating to high temperatures in an oxidizing atmosphere. A marked reduction in fatigue strength is noted in steels

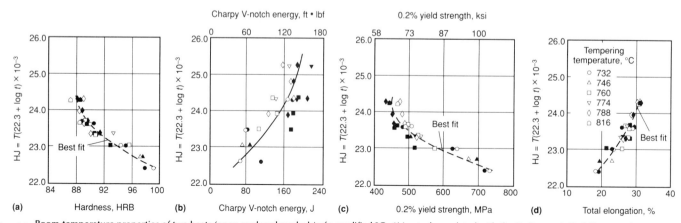

Fig. 41 Room-temperature properties of two heats (open or closed symbols) of a modified 9Cr-1Mo steel correlated with the Holloman-Jaffe (HJ) tempering parameter. (a) Hardness. (b) Charpy energy. (c) 0.2% yield strength. (d) Total elongation at room temperature. Source: Ref 7

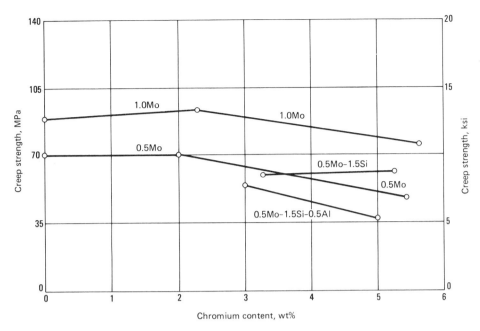

Fig. 42 Effect of chromium on the creep strength (stress to produce a minimum creep rate of 0.0001% per hour) of several steels containing small amounts of molybdenum, silicon, and aluminum at 540 °C (1000 °F). Source: Ref 68

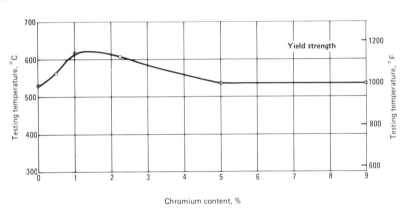

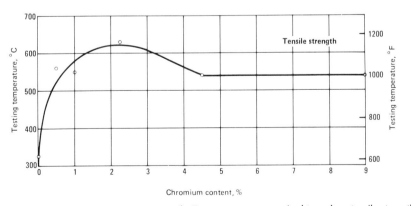

Fig. 43 Effect of chromium content on strength. Test temperature required to reduce tensile strength and yield strength to 60% of their room-temperature values for chromium-molybdenum steels containing 0.5 to 1.0% Mo and the indicated amount of chromium

with decarburized surfaces. The effect of decarburization is much greater on high tensile strength steels than on steels with low tensile strength.

Carburization. As in the case of sulfide penetration, carburization of high-temperature alloys is thermodynamically unlikely except at very low oxygen partial pressures, because the protective oxides of chromium and aluminum are generally more likely to form than the carbides. However, carburization can occur kinetically in many carbon-containing environments. Carbon transport across continuous nonporous scales of Al_2O_3 or Cr_2O_3 is very slow, and alloy pretreatments likely to promote such scales (for example, initially smooth surfaces or preoxidation) have generally been found to be effective in decreasing carburization attack.

The suitability of carburized metal for further service can be determined by evaluating its properties and condition. The mechanical properties of the carburized layer vary markedly from those of the unaffected metal. Room-temperature ductility and toughness are decreased, and hardness is increased greatly. This deterioration is important if the carburized layer is stressed in tension because cracking is quite likely to occur. Weldability is adversely affected. Welds in carburized materials frequently show cracks because of thermal tensile stresses, even with preheating and postheating. Ductility at temperatures above 400 °C (750 °F) is usually adequate. The corrosion resistance of the low chromium-molybdenum steels commonly used for elevated-temperature applications is reduced because of the reduction in effective chromium content. For the same reason, carburized stainless steels may have a relatively low resistance to general corrosion and to intergranular corrosion, particularly while the equipment is shut down. Minor amounts of carburization do not affect creep and rupture strengths significantly.

Factors Affecting Fatigue Strength. As described in the section "Creep-Fatigue Interaction" in this article, the hold times (dwell periods) and the waveform of cyclic strains influence the fatigue strength of metals at elevated temperatures. These factors affect the assessment of low-cycle fatigue and creep fatigue (Fig. 19). In addition, environmental effects and strain aging also influence fatigue strength.

Environmental Effects. It has long been recognized that oxidation at elevated temperatures can have a marked effect (usually an acceleration) on fatigue crack initiation and growth (Ref 74-80). In many alloys, intergranular oxidation initiates intergranular cracks at temperatures that depend on waveform or frequency but are near one-half the melting point (Ref 77). Penetration of oxygen along slip bands with subsequent localized embrittlement and cracking is another possibility. An example of the effects of environment on the fatigue strength of 2¼Cr-1Mo steel is shown in Fig. 47. These tests were conducted in bending at a frequency of 0.05 Hz.

Dynamic Strain Aging. In addition to the effects of environment and temperature, ferritic low-carbon and alloy steels, when

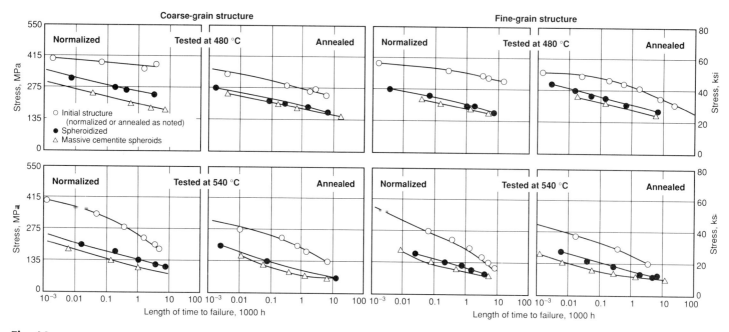

Fig. 44 Effect of spheroidization on the rupture strength of carbon-molybdenum steel (0.17C-0.88Mn-0.20Si-0.42Mo). Source: Ref 73

subjected to inelastic deformation in certain ranges of temperature, strain, and strain rate, undergo dynamic strain aging. Dynamic strain aging, which involves the interaction of interstitials and/or carbide or nitride formers such as chromium, molybdenum, and manganese with strain-induced dislocations, has been shown to markedly influence the cyclic strain rate dependent hardening characteristics, thus affecting both the initiation and growth of fatigue cracks in ferritic materials (Ref 81). Figure 48 shows the deleterious effect of a decreasing strain rate on the fatigue strength of an annealed 2¼Cr-1Mo steel at various temperatures. Typically, strength is increased and ductility is decreased over the temperature ranges where aging occurs; therefore, both low- and high-cycle fatigue properties can be influenced accordingly. Thus, understanding fatigue, creep fatigue, environment, and strain aging interactions in the intermediate- to high-cycle life region is important.

Air Hardening. The amount of air hardening of the various low-alloy steels depends on composition and austenitizing temperatures (Table 5). Steels that harden to more than 300 HB upon air cooling from 1065 °C (1950 °F) require close control of welding operations. Gas-shielded fusion

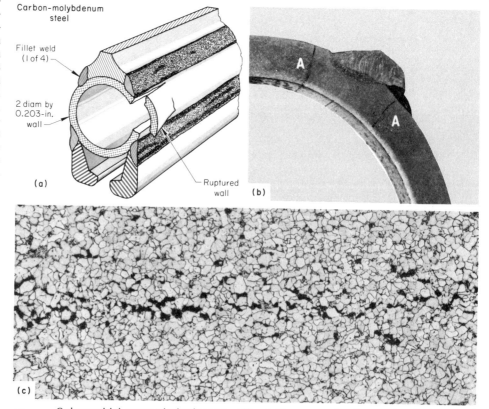

Fig. 45 Temperature-time plot of pearlite decomposition by the competing mechanisms of spheroidization and graphitization in carbon and low-alloy steels. The curve for spheroidization is for conversion of one-half of the carbon in 0.15% C steel to spheroidal carbides. The curve for graphitization is for conversion of one-half of the carbon in aluminum-deoxidized, 0.5% Mo cast steel to nodular graphite.

Fig. 46 Carbon-molybdenum steel tube that ruptured in a brittle manner after 13 years of service because of graphitization at weld heat-affected zones. (a) View of tube showing dimensions, locations of welds, and rupture. (b) Macrograph showing graphitization along edges of a weld heat-affected zone (at A); this was typical of all four welds. 2×. (c) Micrograph of a specimen etched in 2% nital showing chainlike array of embrittling graphite nodules (black) at the edge of a heat-affected zone. 100×

Elevated-Temperature Behavior of 2¼Cr-1Mo Steel

The elevated-temperature behavior of 2¼Cr-1Mo steel has been studied more thoroughly than that of any other steel. The available data on annealed and normalized and tempered 2¼Cr-1Mo steel are summarized in Ref 82 and 83. The rupture strength and creep ductility of 2¼Cr-1Mo steel in various heat-treated conditions are reviewed in Ref 82. The following conclusions were reached:

- The stress-rupture strength generally increases linearly with room-temperature tensile strength up to about 565 °C (1050 °F) for times up to 10 000 h
- At a given strength level, tempered bainite results in higher creep strength than tempered martensite or ferrite-pearlite aggregates for temperatures up to 565 °C (1050 °F) and times up to 100 000 h. For higher temperatures and times, the ferrite-pearlite structure is the strongest
- Rupture ductility generally decreases with rupture time, reaches a minimum, and then increases again. Test temperature, room-temperature tensile strength, austenitizing temperature, and impurity content increase the rate of decrease of ductility with time and cause the ductility minimum to occur at shorter times

In terms of application, this steel has an excellent service record in both fossil fuel and nuclear fuel plants for generating electricity. The severe operating conditions in these plants have justified extensive studies of the behavior of 2¼Cr-1Mo steel under complex loading conditions and in unusual environments. This steel has become a reference against which the performance of other steels can be measured.

Specifications, Steelmaking Practices, and Heat Treatments. Some of the specifications for 2¼Cr-1Mo steel in the ASME Boiler and Pressure Vessel Code are listed in Table 6, which also includes product forms and room-temperature mechanical property requirements. For some of these specifications, composition ranges and limits differ slightly from those given in Tables 3(a) and (b).

In the United States, 2¼Cr-1Mo steel is normally manufactured in an electric furnace. In Japan, basic oxygen processes are used. For certain critical applications, vacuum arc remelting or electroslag remelting is appropriate.

The austenitizing temperature for 2¼Cr-1Mo steel is about 900 °C (1650 °F). Heat treatments commonly employed with 2¼Cr-1Mo steel include:

- *Normalize and temper*: Austenitize at 910 to 940 °C (1650 to 1725 °F), cool in air, temper at 580 to 720 °C (1075 to 1325 °F)
- *Oil quench and temper*: Austenitize at 940 to 980 °C (1725 to 1800 °F), quench in

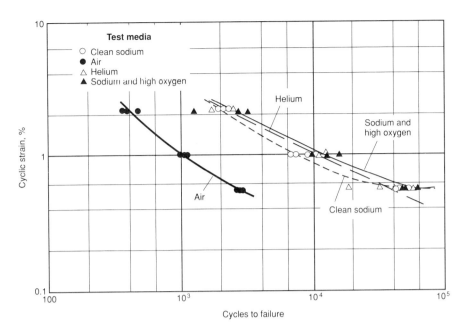

Fig. 47 Fatigue test results of 2¼Cr-1Mo steel in sodium, air, and helium at 593 °C (1100 °F). Source: Ref 81

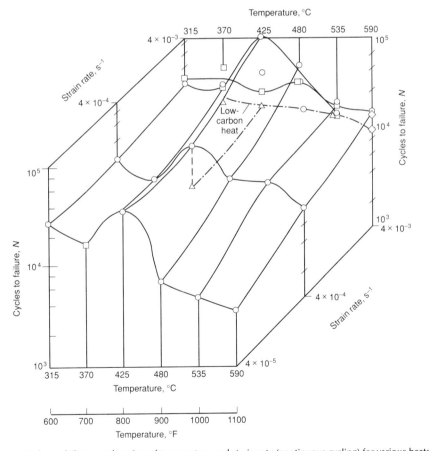

Fig. 48 Cycles to failure as a function of temperature and strain rate (continuous cycling) for various heats of isothermally annealed 2¼Cr-1Mo steel. Source: Ref 81

methods with nonconsumable tungsten electrodes and filler rods of parent metal are suitable for welding the 5% Cr steels having about 0.40% C. Except for the chromium-molybdenum-vanadium type, these steels are normally annealed at the mill, and no further heat treatment is necessary except stress relief after welding for the air-hardening steels. Heavy sections can be preheated.

Table 5 Air-hardening characteristics of low-alloy high-temperature steels

Specimens were cylinders 25 mm (1 in.) in diameter and 50 mm (2 in.) long. After air cooling from the designated temperatures they were sliced longitudinally, and the hardness determined. The values are influenced by chemical variations within the permissible limits of the specifications.

Steel	Annealed	Hardness, HB, after air cooling from						
		760 °C (1400 °F)	815 °C (1500 °F)	870 °C (1600 °F)	900 °C (1650 °F)	955 °C (1750 °F)	1010 °C (1850 °F)	1065 °C (1950 °F)
0.50Mo	137	149	149	149	163	170	187	187
1Cr-0.50Mo	137	149	149	149	170	181	187	187
1.25Cr-0.50Mo	156	149	156	179	187	229	223	212
2.25Cr-1Mo	140	149	149	235	311	311	321	285
Type 502	137	137	137	321	341	341	341	341
7Cr-0.5Mo	156	156	156	321	363	388	388	363
9Cr-0.5Mo	163	170	170	269	321	388	388	375

Source: Ref 9

oil, temper at 570 to 705 °C (1065 to 1300 °F)

Short-Term Elevated-Temperature Mechanical Properties of 2¼Cr-1Mo Steel. The effects of test temperature on the tensile and yield strengths of 2¼Cr-1Mo steel are illustrated in Fig. 11, 24, and 49. Data for annealed specimens and for hardened and tempered specimens are also included. The large variations in both tensile strength and yield strength with temperature and strain rate (Fig. 11) are caused by strain rate, temperature, and microstructure.

The effects of elevated temperatures, elongation, and reduction in area for annealed specimens tested at standard strain rates are illustrated in Fig. 50. Specimens tested at about 400 °C (750 °F) showed both an increase in strength and a reduction in ductility, both of which were caused by strain aging. However, the reduction in ductility was relatively small.

The effects of temperature on modulus of elasticity, shear modulus, and Poisson's ratio are shown in Fig. 12. The modulus of elasticity diminishes from 215 GPa (31 × 10^6 psi) at room temperature to 140 GPa (20.3 × 10^6 psi) at 760 °C (1400 °F); similarly, the shear modulus diminishes from 83 GPa (12.05 × 10^6 psi) at room temperature to 52.4 GPa (7.6 × 10^6 psi) at 760 °C (1400 °F). Poisson's ratio increases from 0.288 at room temperature to 0.336 at 760 °C (1400 °F).

Long-Term Elevated-Temperature Mechanical Properties of 2¼Cr-1Mo Steel. The creep and stress-rupture behavior of annealed specimens and hardened and tempered specimens of 2¼Cr-1Mo steel are illustrated in Fig. 25, 26, 27, and 49. With regard to rupture life and creep rate, the hardened and tempered specimens were able to withstand higher stresses than the annealed specimens.

The ductility exhibited by stress-rupture specimens can be roughly correlated with stress level or rupture life. In general, specimens tested at high stress levels have short rupture lives, and such specimens exhibit greater reduction in area than similar specimens tested at lower stress levels. These data show considerable scatter but no evidence of brittle behavior by this steel. The relaxation behavior of 2¼Cr-1Mo steel is illustrated in Fig. 51.

Long-term exposure to elevated temperature can reduce the room-temperature and elevated-temperature properties of 2¼Cr-1Mo steel. Some of these effects are illustrated in Fig. 6(a), 7(a), 52, and 53. Figure 7(a) shows the changes in room-temperature tensile properties caused by exposure (without stress) to elevated temperatures.

Figure 52 shows the effect of variations in aging time (without stress) at 455 °C (850 °F) on the ultimate tensile and yield strengths of two heats of 2¼Cr-1Mo steel tested at the same temperature. The difference in strength between these two heats was observed even before the tests; the differences were probably caused by variations in composition and microstructure. The same factors account for strength changes during aging because they affect both the size and distribution of carbides in the steel.

As shown in Fig. 53, prolonged aging without stress at 565 °C (1050 °F) can reduce time to rupture for annealed 2¼Cr-1Mo steel. Similarly, the data in Fig. 6(a) show that prolonged exposure to high temperatures without stress substantially reduces stress to rupture in a fixed time. The amount of reduction in stress to rupture is greatest for exposure at 480 °C (900 °F).

Elevated-Temperature Fatigue Behavior of 2¼Cr-1Mo Steel. The results of strain-controlled fatigue tests at 425, 540, and 595 °C (800, 1000, and 1100 °F) on specimens of annealed 2¼Cr-1Mo steel are shown in Fig. 54. Within this range, the test temperature had relatively little effect on the number of cycles to failure. Other strain-controlled fatigue tests (Fig. 48) have shown that reducing the carbon content to 0.03% decreases the fatigue strength. Furthermore, because of variations in strain-aging effect, specimens from one heat with a higher carbon content ran longer at 425 °C (800 °F) than at 315 °C (600 °F).

The crack growth rate data shown in Fig. 55 and 56 were obtained from precracked specimens subjected to cyclic loading at a constant maximum load. Crack extension was measured at intervals during testing. The stress intensity factor range increased as crack length was increased. Figure 55 illustrates the increase in crack growth rate with increasing test temperature. The data in Fig. 56 indicate that in elevated-temperature tests at a given stress intensity factor range, crack growth rate increases as cyclic frequency is decreased. These fracture mechanics data can be applied to the design of structural components that may contain un-

Table 6 Room-temperature mechanical properties of 2¼Cr-1Mo steel in various product forms

ASME specification	Grade	Product form	Mechanical properties				Minimum elongation in 50 mm (2 in.), %	Minimum reduction in area, %
			Yield strength		Ultimate tensile strength			
			MPa	ksi	MPa	ksi		
SA-182	F22	Pipe flanges, fillings, and valves	275	40	485	70	...	...
SA-199	T22	Seamless cold-drawn tubes	170	25	415	60	30	...
SA-213	T22	Seamless ferritic alloy steel tubes	170	25	415	60	30	...
SA-217	WC9	Alloy steel castings	275	40	480	70	20	33
SA-333	P22	Welded and seamless pipe	205	30	415	60	30	20
SA-336	F22	Alloy steel forgings	310	45	515–755	75–110	18	25
	F22a	Alloy steel forgings	205	30	415–585	60–85	20	35
SA-369	FP22	Ferritic alloy steel forged and bored pipe	205	30	415	60	20–30	...
SA-387	GR22, class 1	Chromium-molybdenum PV plate	205	30	415–585	60–85	18(a), 45	40
	GR22, class 2	Chromium-molybdenum PV plate	310	45	515–690	75–100	18(a), 45	40
SA-426	CP22	Centrifugally cast ferritic alloy steel pipe	275	40	480	70	20	35
SA-542	Class 1	Chromium-molybdenum alloy steel plate	585	85	725–860	105–125	14	...
	Class 2	Chromium-molybdenum alloy steel plate	690	100	790–930	115–135	15	...

(a) Elongation in 200 mm (8 in.)

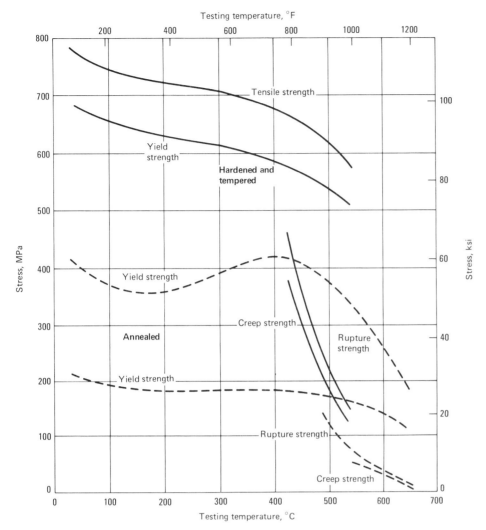

Fig. 49 Effect of test temperature on strength of 2¼Cr-1Mo steel. Effect of test temperature on tensile strength, yield strength, creep strength (for creep rate of 0.1 μm/m · h), and stress to rupture (for life of 100 000 h) of annealed specimens (dashed lines) and hardened and tempered specimens (solid lines) of 2¼Cr-1Mo steel. Source: Ref 2

In one investigation, the elevated-temperature tensile and creep-rupture properties of weldments in 2¼Cr-1Mo steel were measured (Ref 87, 88). Specimens were cut from the weld metal and the base metal; other specimens had transverse welds. All specimens were tempered at 705 °C (1300 °F) before testing. In all these tests, the weld metal was stronger than either the base metal or the specimens containing transverse welds. Specimens with transverse welds invariably fractured in the base metal. The high strength of the weld metal relative to that of the base metal was attributed to differences in microstructure. The base metal, which had been normalized and tempered, contained more ferrite and less bainite than the weld metal. In these tests, the base metal was the weakest part of the welded structure.

Thermal Expansion and Conductivity

Because of their higher thermal conductivity and lower thermal coefficient of expansion, ferritic steels may be more desirable than austenitic steels when thermal cycling occurs in service. Figures 57 and 58 indicate the thermal conductivity and expansion coefficient for carbon and low-alloy steels as a function of temperature.

ACKNOWLEDGMENT

ASM INTERNATIONAL would like to thank R.L. Klueh of Oak Ridge National Laboratory for providing literature to update this article. Thanks are also extended to Joseph Conway of Mar-Test, Inc., and R.W. Swindeman and C.R. Brinkman of Oak Ridge National Laboratory for their review.

detected discontinuities or that may develop cracks in service.

The introduction of a holding period at the peak strain of each fatigue cycle reduces fatigue life as described in the section "Creep-Fatigue Interaction." From studies conducted on annealed 2.25Cr-1Mo steel in air (Ref 6, 81, 86), it is possible to conclude the following (Ref 30):

- Compressive hold periods are more damaging than tensile hold periods, and hold periods imposed on the tension-going side of the hysteresis loop are more damaging in terms of reduced cycle life than hold periods on the compression-going side
- Linear damage summation of fatigue and creep damage does not sum to a unique value

The fact that the damage sums are less than 1 indicates apparent creep-fatigue interaction, but because the values shown do not trend toward a unique value, the linear damage summation method is highly questionable for data extrapolation. The primary reason that the damage sums are less than 1 is that significant environmental interaction or corrosion fatigue occurs in air. This oxidation can substantially reduce the time for crack initiation in smooth bar tests, depending on waveform, and is not adequately accounted for by the simple linear damage summation of fatigue and creep damage fractions. Environmental interaction is discussed in the section "Factors Affecting Fatigue Strength" in this article.

Properties of Welds in 2¼Cr-1Mo Steel. Welding is often required in the fabrication of pressure vessels, boilers, heat exchangers, and similar structures for use at elevated temperatures in power plants, refineries, chemical-processing plants, and similar applications. Therefore, in evaluating materials for these structures, it is important to consider the mechanical properties of welded joints.

REFERENCES

1. G.V. Smith, *Evaluation of the Elevated Temperature Tensile and Creep-Rupture Properties of ½Cr-½Mo, 1Cr-½Mo, and 1¼Cr-½Mo-Si Steels*, DS 50, American Society for Testing and Materials, 1973
2. G.V. Smith, *Supplemental Report on the Elevated-Temperature Properties of Chromium-Molybdenum Steels (An Evaluation of 2¼Cr-1Mo Steel)*, DS 6 S2, American Society for Testing and Materials, March 1971
3. G.S. Sangdahl and H.R. Voorhees, Quenched-and-Tempered 2¼Cr-1Mo Steel at Elevated Temperatures—Tests and Evaluation, in 2¼ Chrome-1 Molybdenum Steel in Pressure Vessels and Piping, American Society of Mechanical Engineers, 1972
4. G.S. Sangdahl and M. Semchyshen, Ed., *Application of 2¼Cr-1Mo for*

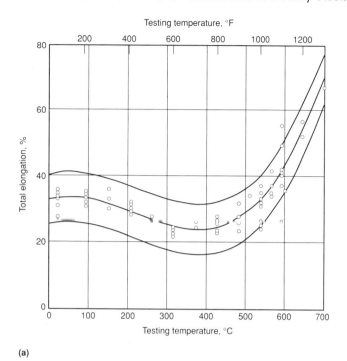
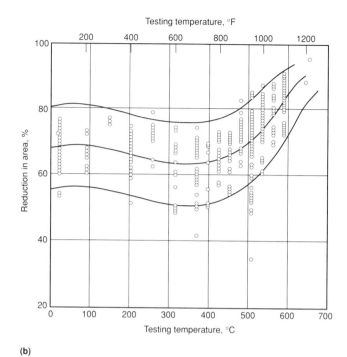

Fig. 50 Effect of test temperature on ductility. (a) Elongation in 50 mm (2 in.) and (b) reduction in area for annealed specimens of 2¼Cr-1Mo steel tested at the indicated temperatures. Source: Ref 84

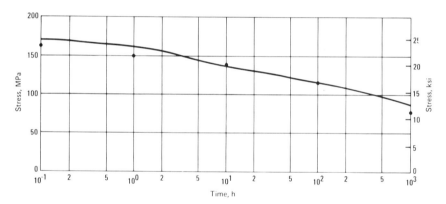

Fig. 51 Relaxation behavior of 2¼Cr-1Mo steel. Specimens were stressed to level indicated on ordinate of graph and exposed to elevated temperature for indicated duration; remaining stress indicated on graph. Source: Ref 84

Thick-Wall Pressure Vessels, STP 755, American Society for Testing and Materials, 1982

5. Low Carbon and Stabilized 2¼% Chromium 1% Molybdenum Steels, American Society for Metals, 1973
6. C.R. Brinkman et al., Time-Dependent Strain-Controlled Fatigue Behavior of Annealed 2¼Cr-1Mo Steel for Use in Nuclear Steam Generator Design, J. Nucl. Mater., Vol 62, 1976, p 181-204
7. V.K. Sikka, "Development of a Modified 9Cr-1Mo Steel for Elevated Temperature Service," in Proceedings of Topical Conference on Ferritic Alloys for Use in Nuclear Energy Technologies, The Metallurgical Society of AIME, 1984, p 317-327
8. R. Viswanathan, Strength and Ductility of CrMoV Steels in Creep at Elevated Temperatures, ASTM J. Test. and Eval., Vol 3 (No. 2), 1975, p 93-106
9. R. Viswanathan and R.I. Jaffee, Toughness of Cr-Mo-V Steels for Steam Turbine Rotors, ASME J. Eng. Mater. Tech., Vol 105, Oct 1983, p 286-294
10. R. Crombie, High Integrity Ferrous Castings for Steam Turbines—Aspects of Steel Development and Manufacture, Mater. Sci. Tech., Vol 1, Nov 1985, p 986-993
11. J.A. Todd et al., New Low Chromium Ferritic Pressure Vessel Steels, in Mi-Con 86: Optimization of Processing, Properties, and Service Performance Through Microstructural Control, STP 979, American Society for Testing and Materials, 1986, p 83-115
12. R.G. Baker and J. Nutting, J. Iron Steel Inst., Vol 192, 1959, p 257-268
13. T. Ishiguro et al., Research on Chrome Moly Steels, R.A. Swift, Ed., MPC-21, American Society of Mechanical Engineers, 1984, p 43-51
14. V.K. Sikka, M.G. Cowgill, and B.W. Roberts, Creep Properties of Modified 9Cr-1Mo Steel, in Conference on Ferritic Alloys for Use in Nuclear Energy Technologies, American Institute of Mining, Metallurgical and Petroleum Engineers, 1984, p 413-423
15. V.K. Sikka, G.T. Ward, and K.C. Thomas, in Ferritic Steels for High Temperature Applications, American Society for Metals, 1982, p 65-84
16. R.L. Klueh and R.W. Swindeman, The Microstructure and Mechanical Properties of a Modified 2.25Cr-1Mo Steel, Metall. Trans. A, Vol 17A, 1986, p 1027-1034
17. R.L. Klueh and A.M. Nasreldin, Metall. Trans. A, Vol 18A, 1987, p 1279-1290
18. W.B. Jones, Effects of Mechanical Cycling on the Substructure of Modified 9Cr-1Mo Ferritic Steel, in Ferritic Steels for High-Temperature Applications, A.K. Khare, Ed., American Society for Metals, 1983, p 221-235
19. J.L. Handrock and D.L. Marriot, Cyclic Softening Effects on Creep Resistance of Bainitic Low Alloy Steel Plain and Notched Bars, in Properties of High Strength Steels for High-Pressure Containments, E.G. Nisbett, Ed., MPC-27, American Society of Mechanical Engineers, 1986
20. R.W. Swindeman, Cyclic Stress-Strain-Time Response of a 9Cr-1Mo-V-Nb Pressure Vessel Steel at High Temper-

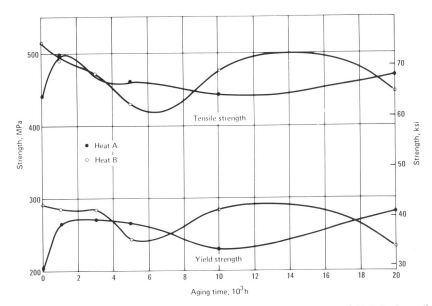

Fig. 52 Effect of exposure to elevated temperature on the strength of 2¼Cr-1Mo steel. Variation in tensile and yield strengths of two different heats of 2¼Cr-1Mo steel after exposure (without stress) to test temperature of 455 °C (850 °F)

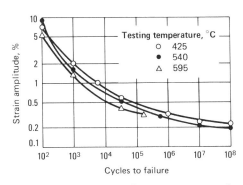

Fig. 54 Effect of elevated temperature on strain-controlled fatigue behavior of annealed 2¼Cr-1Mo steel. Strain rate was greater than 4 mm/m · s. Source: Ref 84

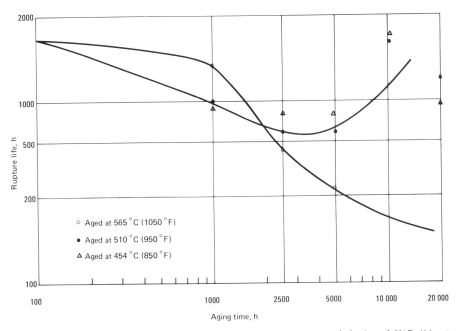

Fig. 53 Effect of exposure to elevated temperature on the stress-rupture behavior of 2¼Cr-1Mo steel. Variation in rupture life for specimens of annealed 2¼Cr-1Mo steel exposed to various elevated temperatures for the durations indicated. After aging, all specimens were stressed to 140 MPa (20 ksi) and tested at 565 °C (1050 °F).

ature, in *Low Cycle Fatigue*, STP 942, American Society for Testing and Materials, 1987, p 107-122
21. S. Kim and J.R. Weertman, Investigation of Microstructural Changes in a Ferritic Steel Caused by High Temperature Fatigue, *Metall. Trans. A*, Vol 19A, 1988, p 999-1007
22. R.L. Klueh and R.E. Oakes, Jr., High Strain Rate Tensile Properties of Annealed 2¼Cr-1Mo Steel, *J. Eng. Mater. Technol.*, Vol 98, Oct 1976, p 361-367
23. *Digest of Steels for High Temperature Service*, 6th ed., The Timken Roller Bearing Company, 1957
24. M. Prager, *Factors Influencing the Time-Dependent Properties of Carbon Steels for Elevated Temperature Pressure Vessels*, MPC 19, American Society of Mechanical Engineers, 1983, p 12, 13
25. R.M. Goldhoff, Stress Concentration and Size Effects in a CrMoV Steel at Elevated Temperatures, *Joint International Conference on Creep*, Institute of Mechanical Engineers, London, 1963
26. R. Viswanathan and C.G. Beck, Effect of Aluminum on the Stress Rupture Properties of CrMoV Steels, *Met. Trans. A*, Vol 6A, Nov 1975, p 1997-2003
27. "Aerospace Structural Metals Handbook," AFML-TR-68-115, Army Materials and Mechanics Research Center, 1977
28. J.W. Freeman and H. Voorhees, in *Relaxation Properties of Steels and Superstrength Alloys at Elevated Temperatures*, STP 187, American Society for Testing and Materials
29. H.R. Voorhees and M.J. Manjoine, *Compilation of Stress-Relaxation Data for Engineering Alloys*, DS 60, American Society for Testing and Materials, 1982
30. C.R. Brinkman, High-Temperature Time-Dependent Fatigue Behavior of Several Engineering Structural Alloys, *Int. Met. Rev.*, Vol 30 (No. 5), 1985, p 235-258
31. D.A. Miller, R.H. Priest, and E.G. Ellison, A Review of Material Response and Life Prediction Techniques Under Fatigue-Creep Loading Conditions, *High Temp. Mater. Proc.*, Vol 6 (No. 3 and 4), 1984, p 115-194
32. R. Viswanathan, *Damage Mechanisms and Life Assessment of High-Temperature Components*, ASM INTERNATIONAL, 1989
33. *Thermal Fatigue of Materials and Components*, STP 612, American Society for Testing and Materials, 1976
34. M.A. Grossman and E.C. Bain, *Principles of Heat Treatment*, 5th ed., American Society for Metals, 1964
35. J.B. Conway, Parametric Considerations Applied to Isochronous Stress-Strain Plots, in *The Generation of Isochronous Stress-Strain Curves*, A.O. Schaefer, Ed., American Society of Mechanical Engineers, 1972
36. G.V. Smith, *Evaluation of the Elevated Temperature Tensile and Creep-Rupture Properties of 3-9% Chromium-Molybdenum Steels*, DS 58, American So-

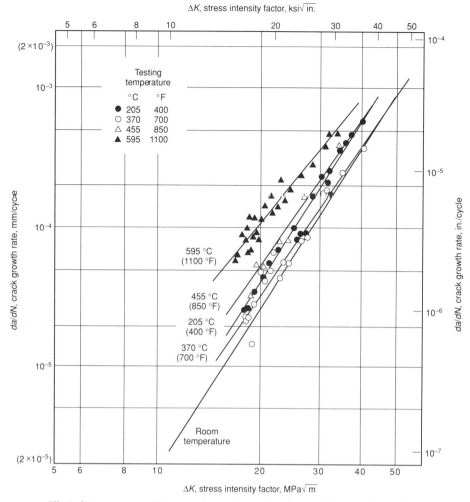

Fig. 55 Effect of temperature on fatigue crack growth rate. Variations in fatigue crack growth rate with test temperature for specimens of 2¼Cr-1Mo steel tested in air. Stress ratio was 0.05; cyclic frequency was 400 per minute. Source: Ref 85

ciety for Testing and Materials, 1975
37. R.L. Klueh, Interaction Solid Solution Hardening in 2.25Cr-1Mo Steel, *Mater. Sci. Eng.*, Vol 35, 1978, p 239-253
38. R. Viswanathan and R.D. Fardo, Parametric Techniques for Extrapolating Rupture Ductility, in *Ductility and Toughness Considerations in Elevated Temperature Service*, G.V. Smith, Ed., MPC-8, American Society of Mechanical Engineers, 1978
39. J. Gutzeit, High Temperature Sulfidic Corrosion of Steels, in *Process Industries Corrosion—The Theory and Practice*, National Association of Corrosion Engineers, 1986
40. H.F. McConomy, High-Temperature Sulfidic Corrosion in Hydrogen-Free Environment, *Proc. API*, Vol 43 (III), 1963, p 78-96
41. E.B. Backensto, R.D. Drew, and C.C. Stapleford, High Temperature Hydrogen Sulfide Corrosion, *Corrosion*, Vol 12 (No. 1), 1956, p 6t-16t
42. G. Sorell and W.B. Hoyt, Collection and Correlation of High Temperature Hydrogen Sulfide Corrosion Data, *Corrosion*, Vol 12 (No. 5), 1956, p 213t-234t
43. C. Phillips, Jr., High Temperature Sulfide Corrosion in Catalytic Reforming of Light Naphthas, *Corrosion*, Vol 13 (No. 1), 1957, p 37t-42t
44. G. Sorell, Compilation and Correlation of High Temperature Catalytic Reformer Corrosion Data, *Corrosion*, Vol 14 (No. 1), 1958, p 15t-26t
45. W.H. Sharp and E.W. Haycock, Sulfide Scaling Under Hydrorefining Conditions, *Proc. API*, Vol 39 (III), 1959, p 74-91
46. J.D. McCoy and F.B. Hamel, New Corrosion Data for Hydrosulfurizing Units, *Hydrocarbon Process.*, Vol 49 (No. 6), 1970, p 116-120
47. J.D. McCoy and F.B. Hamel, Effect of Hydrosulfurizing Process Variables on Corrosion Rates, *Mater. Prot. Perform.*, Vol 10 (No. 4), 1971, p 17-22
48. A.S. Couper and J.W. Gorman, Computer Correlations to Estimate High Temperature H2S Corrosion in Refinery Streams, *Mater. Prot. Perform.*, Vol 10 (No. 1), 1971, p 31-37
49. D. Warren, Hydrogen Effects on Steel, in *Process Industries Corrosion*, National Association of Corrosion Engineers, 1986, p 21-30
50. G.R. Odette, *Conference Proceedings on Materials for Coal Conversion and Utilization*, National Bureau of Standards, 1982
51. R.O. Ritchie et al., *J. Mater. Energy Sys.*, Vol 6 (No. 3), p 151-162
52. G.A. Nelson, Metals for High Pressure Hydrogenation Plants, *Trans. ASME*, Vol 73, 1951, p 205-213
53. I. Masaoka et al., Hydrogen Attack Limit of 2¼Cr-1Mo Steel, in *Current Solutions to Hydrogen Problems in Steel*, American Society for Metals, 1982, p 247
54. P.G. Shewmon et al., On the Nelson Curve for 2¼Cr-1Mo Steel, in *Research on Chrome-Moly Steels*, MPC-21, American Society of Mechanical Engineers, 1984, p 1-8
55. R.L. Schuyler III, Hydrogen Blistering of Steel in Anhydrous Hydrofluoric Acid, *Mater. Perform.*, Vol 18 (No. 8), 1979, p 9-16
56. G. Herbsleb et al., Occurrence and Prevention of Hydrogen Induced Stepwise Cracking and Stress Corrosion Cracking of Low Alloy Pipeline Steels, *Corrosion*, Vol 37 (No. 5), 1981, p 247-255
57. J.D. Baird et al., Strengthening Mechanisms in Ferritic Creep Resistant Steels, in *Creep Strength in Steel and High Temperature Alloys*, The Metals Society, 1974, p 207-216
58. J.D. Baird and A. Jamieson, *J. Iron Steel Inst.*, Vol 210, 1972, p 841
59. J.D. Baird and A. Jamieson, *J. Iron Steel Inst.*, Vol 210, 1972, p 847
60. B.B. Argent et al., *J. Iron Steel Inst.*, Vol 208, 1970, p 830-843
61. G. Krauss, *Principles of Heat Treatment of Steel*, American Society for Metals, 1980
62. F.B. Pickering, *Physical Metallurgy and the Design of Steels*, Applied Science, 1978
63. R. Viswanathan, Strength and Ductility of 2¼Cr-1Mo Steels in Creep, *Met. Tech.*, June 1974, p 284-293
64. J. Orr, F.R. Beckitt, and G.D. Fawkes, The Physical Metallurgy of Chromium-Molybdenum Steels for Fast Reactor Boilers, in *Ferritic Steels for Fast Reactor Steam Generators*, S.F. Pugh and E.A. Little, Ed., British Nuclear Energy Society, 1978, p 91
65. R.L. Klueh, *J. Nucl. Mater.*, Vol 68, 1977, p 294
66. J. Ewald, et al., Over 30 Years Joint Long-Term Research on Creep Resistant Materials in Germany, in *Advances in Material Technology for Fossil Power Plants*, R. Viswanathan and R.I.

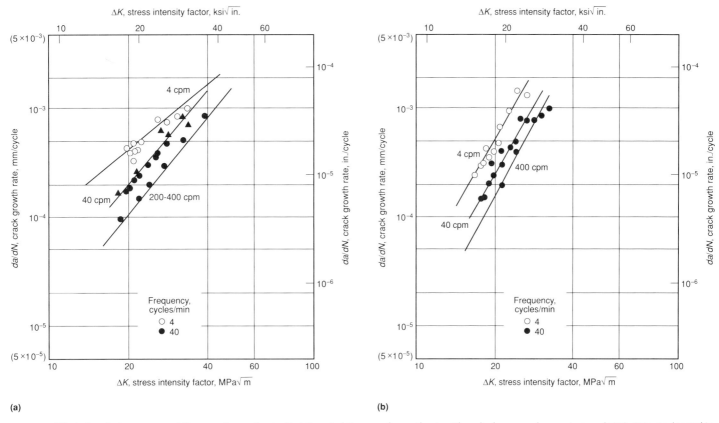

Fig. 56 Effect of cyclic frequency on fatigue crack growth rate. Variations in fatigue crack growth rate with cyclic frequency for specimens of 2¼Cr-1Mo steel tested in air. Stress ratio was 0.05. (a) Tested at 510 °C (950 °F). (b) Tested at 595 °C (1100 °F). Source: Ref 85

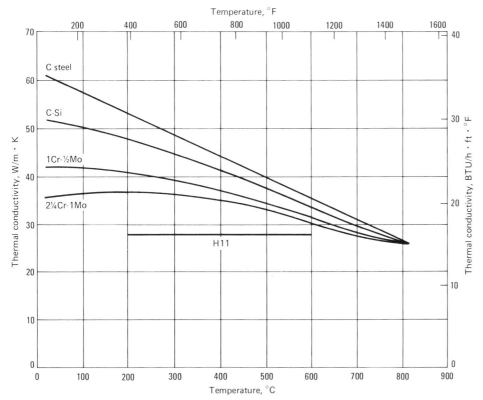

Fig. 57 Thermal conductivity of carbon and low-alloy steels at various temperatures

Jaffee, Ed., ASM INTERNATIONAL, 1987, p 33-39
67. A. Krisch, *Jernkontorets Ann.*, Vol 155, 1971, p 323-331
68. G.V. Smith, *Properties of Metals at Elevated Temperatures*, McGraw-Hill, 1950, p 231
69. J.D. Baird, *Jernkontorets Ann.*, Vol 151, 1971, p 311-321
70. R.L. Klueh and P.J. Maziasz, Reduced-Activation Ferritic Steels: A Comparison With Cr-Mo Steels, *J. Nucl. Mater.*, Vol 155-157, 1988, p 602-607
71. R.L. Klueh and E.E. Bloom, The Development of Ferritic Steels for Fast Induced-Radioactive Decay for Fusion Reactor Applications, in *Nuclear Engineering and Design/Fusion 2*, North-Holland, 1985, p 383-389
72. R.L. Klueh and P.J. Maziasz, Low-Chromium Reduced-Activation Ferritic Steels, in *Reduced-Activation Materials for Fusion Reactors*, STP 1046, American Society for Testing and Materials, to be published
73. S.H. Weaver, The Effect of Carbide Spheroidization Upon the Rupture Strength and Ductility of Carbon Molybdenum Steel, *Proc. ASTM*, Vol 46, 1946, p 856-866
74. L.F. Coffin, *Metall. Trans.*, Vol 3, 1972, p 1777-1788
75. L.A. James, *J. Eng. Mater. Technol.*, Vol 98, July 1976, p 235-243

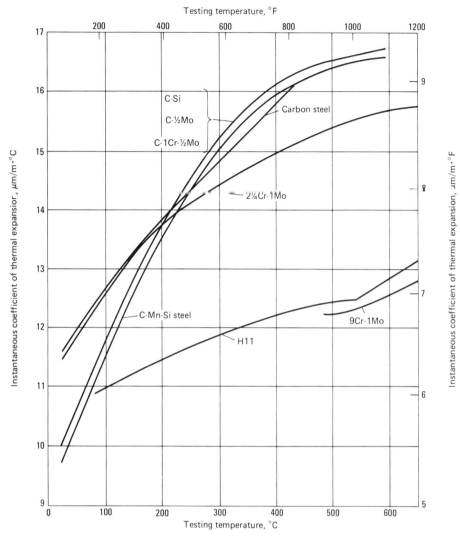

Fig. 58 Coefficients of thermal expansion for carbon and low-alloy steels at various temperatures. These are not mean values of the coefficient over a range of temperatures.

76. M. Gell and G.R. Leverant, in *Fatigue at Elevated Temperatures*, STP 520, American Society for Testing and Materials, 1973, p 37-66
77. J.C. Runkle and R.M. Pelloux, in *Fatigue Mechanisms*, STP 675, J.T. Fong, Ed., American Society for Testing and Materials, 1979, p 501-527
78. D.J. Duquette, Environmental Effects I: General Fatigue Resistance and Crack Nucleation in Metals and Alloys, in *Fatigue and Microstructure*, American Society for Metals, 1979, p 335-363
79. H.L. Marcus, Environmental Effects II: Fatigue-Crack Growth in Metals and Alloys, in *Fatigue and Microstructure*, American Society for Metals, 1979, p 365-383
80. P. Marshall, in *Fatigue at High Temperature*, R.P. Skelton, Ed., Applied Science, 1983, p 259-303
81. C.R. Brinkman et al., Time-Dependent Strain-Controlled Fatigue Behavior of Annealed 2¼Cr-1Mo Steel for Use in Nuclear Steam Generator Design, *J. Nucl. Mater.*, Vol 62, 1976, p 181-204
82. R. Viswanathan, Strength and Ductility of 2¼Cr-1Mo Steels in Creep at Elevated Temperatures, *Met. Technol.*, June 1974, p 284-294
83. G.V. Smith, Elevated Temperature Strength and Ductility of Q&T 2¼Cr-1Mo Steel, in *Current Evaluation of 2¼Cr-1Mo Steel in Pressure Vessels and Piping*, American Society of Mechanical Engineers, 1972
84. M.K. Booker, T.L. Hebble, D.O. Hobson, and C.R. Brinkman, Mechanical Property Correlations for 2¼Cr-1Mo Steel in Support of Nuclear Reactor Systems Design, *Int. J. Pressure Vessels Piping*, Vol 5, 1977
85. C.R. Brinkman, W.R. Corwin, M.K. Booker, T.L. Hebble, and R.L. Klueh, "Time Dependent Mechanical Properties of 2¼Cr-1Mo Steel for Use in Steam Generator Design," ORNL-5125, Oak Ridge National Laboratory, 1976
86. J.J. Burke and V. Weiss, in *Fatigue Environment and Temperature Effects*, Plenum Press, 1983, p 241-261
87. R.L. Klueh and D.A. Canonico, Microstructure and Tensile Properties of 2¼Cr-1Mo Steel Weldments With Varying Carbon Contents, *Weld. J.* (Research Supplement), Sept 1976
88. R.L. Klueh and D.A. Canonico, Creep-Rupture Properties of 2¼Cr-1Mo Steel Weldments With Varying Carbon Content, *Weld. J.* (Research Supplement), Dec 1976

Effect of Neutron Irradiation on Properties of Steels*

R.L. Klueh, Metals and Ceramics Division, Oak Ridge National Laboratory

DAMAGE TO STEELS from neutron irradiation affects the properties of steels and is an important factor in the design of safe and economical components for fission and fusion reactors. Damage occurs when high-energy neutrons displace metal atoms from their normal lattice positions to form interstitials and vacancies. It is the disposition of these defects that influences properties during and after irradiation. In addition to the formation of vacancies and interstitials, transmutation reactions can also occur when neutrons are absorbed by the atoms of an irradiated alloy. These transmutation reactions produce new metal atoms and gas atoms of hydrogen and/or helium within the alloy matrix. Of these various transmutation by-products, transmutation helium is considered the most significant in exacerbating property changes.

The effects of damage caused by neutron irradiation include swelling (volume increase), irradiation hardening, and irradiation embrittlement (the influence of irradiation hardening on fracture toughness). These effects are primarily associated with high-energy (>0.1 MeV) neutrons. Consequently, irradiation damage from neutrons is of considerable importance in fast reactors, which produce a significant flux of high-energy neutrons during operation.

However, irradiation damage from neutrons is also a factor in commercial light-water reactors, even though neutrons in a light-water reactor are moderated to reduce their energy (most neutrons in the spectrum of these reactors are thermal neutrons with energies much less than 1 eV). Such reactors produce a small flux of high-energy neutrons, and until recently, these neutrons were the only ones considered to cause the irradiation effects observed in power reactors. However, as discussed in the section "Irradiation Embrittlement" in this article, recent observations have indicated that thermal neutrons can also cause irradiation effects. Therefore, material damage from neutron irradiation is important not only in fast reactors, such as the experimental Fast Flux Test Facility (FFTF) in Hanford, Washington, and the Super Phenix fast-breeder electric power reactor in France, but also in the many commercial light-water power reactors. In addition, the future of economically viable fusion reactors may also depend on the development of irradiation-resistant alloys.

This article discusses the effects of high-energy neutrons on steels, with particular emphasis on the steels listed in Table 1. For the pressure vessels of light-water reactors the manganese-molybdenum-nickel ferritic steels (ASTM A 302-B and A 533-B) are commonly used. These steels are quenched and tempered, which produces a tempered martensite and/or tempered bainite microstructure. Austenitic steels such as type 316 stainless steel are proposed for fusion reactors and are used in fast reactors for fuel cladding, ducts, and other structural components. These steels are used in either the solution-annealed or the 20% cold-worked condition. Special irradiation-resistant austenitic steels have been developed for these applications. An example of such a new steel is the prime candidate alloy (PCA) for fusion (Table 1). For both the fast-breeder reactor and fusion reactors, chromium-molybdenum ferritic steels are being considered. Of special interest are the 9Cr-1MoVNb and 12Cr-1MoVW steels; 2¼Cr-1Mo steel is also considered for fusion reactors. These steels are used in a normalized and tempered condition, which gives a tempered martensite microstructure in the 9Cr-1MoVNb and 12Cr-1MoVW steels and a tempered bainite microstructure in the 2¼Cr-1Mo steel.

Irradiation Damage Processes

The current understanding of neutron irradiation effects has been obtained from studies on materials irradiated in fission test reactors, such as the FFTF, the Experimental Breeder Reactor (EBR-II), and the High Flux Isotope Reactor (HFIR) (Ref 1). Irradiation is described in terms of the flux of neutrons striking the steel being irradiated, which is measured as the number of neutrons per square meter per second (n/m² · s), and the fluence, which is the time-integrated flux in neutrons per square meter (n/m²). A typical flux for a fast reactor is ~5 × 10^{19} n/m² · s.

Displacement Damage. When a steel is irradiated in a high-energy neutron field, neutrons collide with atoms in the material and displace them from their lattice positions (Ref 2, 3). The first atom struck and displaced by a neutron is termed a primary knock-on atom. When this primary knock-on atom recoils from the impact, it collides with other atoms, which in turn recoil and collide with still other atoms. Therefore, an incoming neutron can produce a displacement cascade, by which a large number of atoms are displaced from lattice sites. The displaced atoms of a displacement cascade move into interstitial positions (termed interstitials) and leave behind a like number of vacant sites (vacancies).

This displacement of atoms by irradiation is described in terms of displacements per atom (dpa), which is a measure of the average number of times an atom is displaced from its lattice position. The dpa can be calculated from the neutron fluence received by the steel (Ref 2, 3). During 1 year of fast reactor operation, each atom in stainless steel is typically displaced more than 30 times (30 dpa/yr). For a light-water reactor, the displacement rate in steel is about 0.03 dpa/yr. For a fusion reactor, displacement rates of up to 60 dpa/yr might be expected.

The disposition of the defects—interstitials and vacancies—determines the effect of the atomic displacements on the properties of an irradiated material. Although the average number of atomic displacements is described by the dpa unit,

*Research sponsored by the Office of Fusion Energy, U.S. Department of Energy, under contract DE-AC05-84OR21400 with the Martin Marietta Energy Systems, Inc.

Table 1 Typical compositions for steels of interest for nuclear reactor applications

Steel	Chemical composition, wt%										
	Cr	Ni	Mo	Mn	Si	C	V	Nb	W	Ti	N
Austenitic stainless steels											
316 stainless steel	18.0	13.0	2.5	2.0	0.8	0.05	...	...	...	0.05	0.05
PCA steel	14.0	16.0	2.5	1.7	0.4	0.05	...	...	...	0.25	...
Ferritic steels											
A302	0.1	0.2	0.5	1.3	...	0.2	...	...	...	...	...
A533	...	0.5	0.5	1.3	...	0.2	...	...	...	...	...
2¼Cr-1Mo	2.25	0.25	1.0	0.5	0.3	0.12	...	...	...	...	...
9Cr-1MoVNb	9.0	0.2	1.0	0.5	0.3	0.10	0.2	0.08	0.5	...	0.05
12Cr-1MoVW	12.0	0.5	1.0	0.5	0.4	0.20	0.3	...	...	...	...

only a fraction of these displacements produce damage and property effects. In general, most (typically 95 to 99%) of the displaced atoms from a displacement cascade recombine with a vacancy. This is because the interstitials and vacancies produced in a displacement cascade are near each other and have a strong likelihood of recombining.

Therefore, displacement damage from neutron irradiation occurs from only a small portion of the atomic displacements. The interstitials and vacancies from this portion of the displacements do not recombine but instead migrate to sinks, where they are absorbed or accumulate. Sinks include surfaces, grain boundaries, dislocations, and existing cavities.

This migration of defects can also result in the formation of defect clusters; those consisting of interstitials can evolve into dislocation loops, while vacancy clusters can develop into microvoids or cavities (Fig. 1). Solute clusters can also form under certain conditions.

The type of cluster defect that forms depends on the irradiation temperature. Below about 0.35 T_m (where T_m is the melting point of the irradiated material in degrees kelvin), interstitials are mobile relative to vacancies, and the interstitials combine to form dislocation loops. This gives rise to an increase in strength and a decrease in ductility.

Vacancies become increasingly mobile above 0.35 T_m, and a dislocation and cavity structure results (Fig. 1). This microstructure occurs because certain sinks have a bias and do not accept vacancies and interstitials equally (Ref 2, 3). If all sinks accepted both defects equally, the vacancies and interstitials would annihilate at a sink, and no swelling would result. However, within a grain, interstitials are accepted preferentially by dislocations. This leaves an excess of vacancies to be absorbed by cavities, giving rise to the observed swelling.

Finally, at high irradiation temperatures (greater than about 0.6 T_m), defect clusters are unstable. That is, the high equilibrium vacancy concentration and rapid diffusion lead to vacancy-interstitial recombination, which thus reduces the number of defects and the effects of displacement damage on properties. However, at temperatures ≥0.5 T_m, any transmutation helium produced during irradiation can lead to problems.

Cavities. Two types of cavities form during irradiation—bubbles and voids. Bubbles contain gas at a pressure greater than or equal to the surface tension pressure. Voids have internal gas pressure below the equilibrium pressure. The origin of gas in irradiated material is described in the section "Transmutation Helium" in this article.

Radiation-Induced Segregation. Because certain alloying elements can be preferentially associated with or rejected by vacancies or interstitials, such elements can be transported to or from sinks when the defects migrate. This radiation-induced segregation can cause detrimental effects and must be considered when developing alloys for irradiation resistance (Ref 2, 3).

Transmutation Helium. In addition to displacement damage, a neutron can be absorbed by an atom of the irradiated alloy, resulting in a transmutation reaction that produces a new metal atom and hydrogen and/or helium gas atoms within the alloy being irradiated. Indications are that small amounts of new metal atoms have little effect on properties. Hydrogen will have little effect on properties because, at the operating temperatures of most reactors (250 to 550 °C, or 480 to 1020 °F), it should readily diffuse from the alloy. However, any helium produced can affect the properties; it is relatively insoluble in metals and will therefore be incorporated into the bubbles or voids that can form within the matrix and on grain boundaries and precipitate interfaces.

The displacement damage formed in fusion and fission reactors is similar, and fusion damage can be simulated by fission reactor irradiations. However, the much higher energy of the neutrons (up to 14 MeV) produced in a fusion reactor will lead to much more transmutation helium than occurs in most fission irradiations. Such a simultaneous development of displacement damage and helium can affect both the swelling behavior and the mechanical prop-

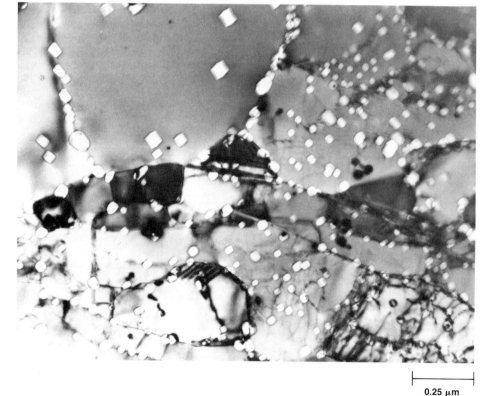

Fig. 1 Cavities (indicated as the white rectangles and circles) formed in type 316 stainless steel irradiated to 60 dpa at 600 °C (1110 °F) in the HFIR. Courtesy of P.J. Maziasz, Oak Ridge National Laboratory

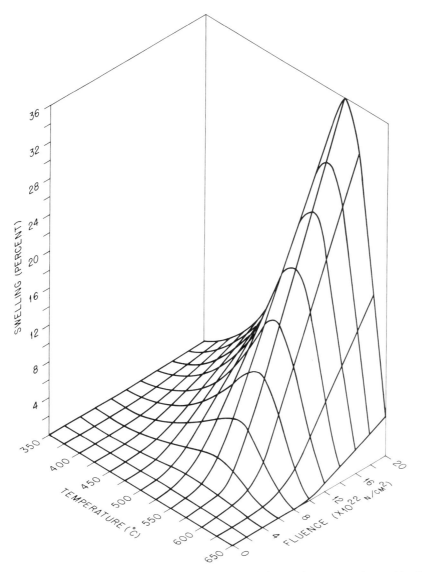

Fig. 2 Effect of temperature and neutron fluence on the swelling behavior of type 316 stainless steel irradiated in a fast reactor (EBR-II). Source: Ref 3

erties. Much recent research has been directed at determining the effect of helium on the properties of irradiated candidate structural alloys and on developing alloys that will withstand these effects (Ref 3).

Void Swelling

As mentioned earlier in the section "Displacement Damage" in this article, the bias of dislocations for interstitials causes an excess of vacancies to agglomerate at cavities (Fig. 1), which thus causes a volume increase or swelling of the irradiated alloy (bias-driven void swelling). Void swelling (measured as $\Delta V/V$, where ΔV is the change in volume of the irradiated material and V is the original volume) of several tens of percent is observed in some stainless steels. Large amounts of swelling cannot be tolerated in a reactor component, and considerable effort has been directed toward the development of swelling-resistant alloys for use in fast-breeder and fusion reactors. Void swelling is unimportant for light-water power reactors because of the low flux of high-energy neutrons in the neutron spectrum of such reactors.

Swelling in Austenitic Stainless Steels. Irradiations to fluences that produce displacement damage greater than 100 dpa have been conducted on types 304 and 316 stainless steel, and the swelling can be described as a function of temperature and fluence (Ref 2, 3), as illustrated in Fig. 2. For constant fluence, a peak swelling temperature is observed. At a constant temperature, there is an incubation time, after which swelling develops slowly with a power-law dependence on fluence. This transient regime is eventually replaced by a rapid-swelling regime characterized by a linear dependence (steady-state swelling) on fluence. Swelling in the steady-state regime for austenitic steels occurs at approximately 1%/dpa; the steady-state rate is essentially independent of composition and fabrication variables and is a weak function of temperature, irradiation rate, and stress (Ref 4, 5).

The high rate of swelling in the steady-state regime (~1%/dpa) suggested that the structural lifetime could be increased only by extending the transient regime (Ref 4, 5). Voids nucleate during the transient period, while their growth and coalescence occur during steady state. Void nucleation is aided by small amounts of dissolved gases (oxygen and nitrogen) or gases formed by transmutation reactions (helium). Gas can combine with irradiation-induced vacancies to form bubbles. These bubbles collect vacancies until a critical radius for a void is reached, after which growth is bias driven. If the conversion from bubbles to voids can be inhibited, the transient stage can be extended, and swelling resistance is improved.

Although void swelling occurs because of the slight bias of dislocations as sinks for interstitials, a very high dislocation density can provide sufficient sinks for both vacancies and interstitials, where they can recombine and annihilate. This information has led to the use of 20%-cold-worked type 316 stainless steel for fast-breeder reactor fuel cladding.

Small amounts of titanium were found to extend the transient regime (Ref 6). Titanium was thought to getter oxygen and other dissolved gases and thus reduce the nucleation rate, which was delayed until small amounts of transmutation helium could aid nucleation. The addition of titanium and niobium to form carbides and phosphorus to form phosphides has a similar effect on transient swelling (Ref 4). In this case, the fine dispersions of stable precipitate particles trap helium at their interfaces, forming a high density of small bubbles at these locations. Because the irradiation-produced vacancies are collected by a larger number of cavities, the formation of the critical radius for void growth is delayed, and the transient regime is extended. The high density of cavities can also become the dominant sink for both vacancies and interstitials and therefore sites for recombination. Such alloying techniques have been applied in the breeder reactor and fusion reactor alloy development programs in an attempt to develop irradiation-resistant stainless steels, such as the PCA steel for fusion (Table 1).

In a fusion reactor, the large amounts of helium generated simultaneously with displacement damage will give rise to irradiation effects different from those observed in a breeder reactor. The information available for large helium concentrations at relevant damage rates is from comparisons of the behavior of type 316 stainless steel irradiated in fast reactors, such as the EBR-II, and

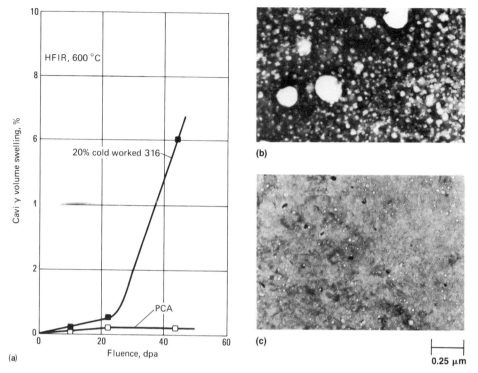

Fig. 3 Comparison of the swelling behavior and microstructure of cold-worked type 316 stainless steel and cold-worked PCA steel irradiated in the HFIR at 600 °C (1110 °F). (a) Cavity volume swelling versus neutron fluence. (b) Microstructure of 316 stainless steel after about 43 dpa. (c) Microstructure of the PCA steel after about 43 dpa. Source: Ref 4

in a mixed-spectrum reactor, such as the HFIR. In a mixed-spectrum reactor, the neutron spectrum contains both thermal and fast neutrons. When a nickel-containing steel is irradiated in the HFIR, the thermal neutrons react with ^{58}Ni to form helium, while the fast neutrons produce displacement damage just as they do in a fast reactor. Natural nickel contains approximately 68% ^{58}Ni.

Recent evidence indicates that increasing the He:dpa ratios from approximately 0.5 (EBR-II) to approximately 60 (HFIR) (the He concentration is in atomic parts per million, appm) may shorten or extend the transition regime, depending on the heat chemistry and the thermomechanical treatment (Ref 4). An extended transition regime is associated with a high density of bubbles, the inhibition of radiation-induced segregation, and the delayed conversion of bubbles to voids. For 20%-cold-worked type 316 stainless steel irradiated in the HFIR, helium is trapped on the high density of dislocations, leading to the nucleation of a high density of bubbles. Because of the large number of small bubbles present, they become the dominant sinks for both vacancies and interstitials, and swelling is inhibited. In solution-annealed steel, the opposite occurs. Helium is not effectively trapped, and bubble nucleation occurs on a coarser scale than in cold-worked material. This accelerates the transition of cavities from bubbles to voids and leads to greater swelling than that observed in the absence of high helium concentrations.

The precipitation of fine titanium-rich carbides (designated as MC, where the M indicates the carbide contains more than one type of metal atom) in the PCA steel for fusion (Table 1) enhances bubble nucleation per increment of generated helium (Ref 4). As long as fine dispersions of MCs are preserved during irradiation, the association of fine MCs with helium bubbles hinders bubble coarsening by coalescence. The resulting high density of bubble/precipitate sinks also suppresses radiation-induced segregation and thus further enhances MC stability. This is illustrated in Fig. 3, which compares the swelling behavior and microstructure of cold-worked type 316 stainless steel and PCA steel. The low-swelling PCA steel contains a high density of small bubbles, compared to the large voids in the stainless steel. Helium-enhanced MC stability and suppressed radiation-induced segregation are essential in extending the transient regime of swelling for fusion compared to fast-breeder reactor irradiation. The performance of PCA steel illustrated in Fig. 3 is for an He:dpa ratio of approximately 60, while the value for a fusion reactor is expected to be about 10 to 12. It still must be determined whether the metal carbide will remain stable under fusion reactor conditions.

Swelling in Ferritic Steels. When chromium-molybdenum ferritic steels such as 2¼Cr-1Mo, 9Cr-1MoVNb, and 12Cr-1MoVW (Table 1) were irradiated in fast reactors to neutron fluences of 17.6×10^{26} n/m^2 and a displacement-damage level of up to 80 dpa (helium concentrations <10 appm) at 300 to 600 °C (570 to 1110 °F), these steels swelled much less than austenitic stainless steels exposed to similar conditions (Ref 7-9). However, a significant amount of irradiation-induced precipitation occurred during these irradiations, and the subsequent influence of this precipitation on strength may prove to be the most significant effect for these steels (Ref 7-9).

As discussed above, the effects of simultaneous displacement damage and transmutation helium can be studied by irradiating nickel-containing alloys in HFIR. Standard 9Cr-1MoVNb steel contains approximately 0.2% Ni; 12Cr-1MoVW steel, approximately 0.5% Ni. These steels were irradiated in HFIR at 300 to 600 °C (570 to 1110 °F) to approximately 36 dpa to produce about 30 and 100 appm of helium for the 9Cr and 12Cr steels, respectively (Ref 9). Considerably more swelling was observed when helium was present than when irradiated in fast reactors to much higher damage levels with little helium present. For both steels, swelling was a maximum after irradiation at 400 °C (750 °F). Irradiation at 300, 500, and 600 °C (570, 930, and 1110 °F) resulted in some cavity formation, but negligible swelling. After irradiation at 300 and 400 °C (570 and 750 °F), the cavities were homogeneously distributed in the matrix. Irradiation at 500 and 600 °C (930 and 1110 °F) resulted in small bubbles at lath boundaries, prior-austenite grain boundaries, and dislocations. These were concluded to be helium-filled bubbles because of their small, uniform size and because they formed preferentially on structural defects. When compared to results in fast reactors, where no helium was produced, the HFIR results indicated that helium assists the nucleation of cavities (Ref 9).

To further study the effect of helium on swelling, 9Cr-1MoVNb and 12Cr-1MoVW steels doped with 2% Ni were irradiated in the HFIR. Examination of 9Cr-1MoVNb steel irradiated to 38 dpa to produce over 400 appm of helium indicated that the helium caused larger amounts of void and bubble formation over the range of 300 to 600 °C (570 to 1110 °F) than occurred during fast reactor irradiation (>10 appm He) (Ref 9). Furthermore, significantly greater amounts of precipitation occurred at 400 and 500 °C (750 and 930 °F) under conditions of increased void formation. It appeared that these precipitate changes were the result of irradiation-induced solute segregation, in which the migration of vacancies and helium to cavity surfaces is accompanied by the preferential migration of solute atoms away from these sites. This can result in irradiation-induced precipitate phases that are un-

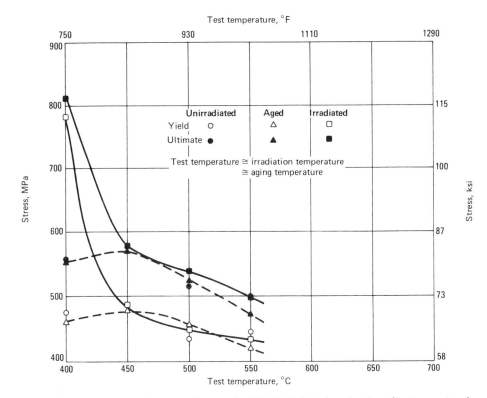

Fig. 4 0.2% yield stress and ultimate tensile strength of 9Cr-1MoVNb steel as a function of test temperature for irradiated specimens (12 dpa), as-heat-treated controls, and thermally aged controls. The test temperature equals the irradiation and aging temperatures; specimens were aged 5000 h, which corresponded to the time in-reactor. Source: Ref 10

detail. An attempt has been made to model the behavior observed in the EBR-II in terms of the defects produced during irradiation (Ref 16). Only minor differences were observed between stainless steels irradiated in the EBR-II and the HFIR up to approximately 600 °C (1110 °F), indicating that helium had little effect on tensile behavior under these conditions (Ref 17).

Elevated-Temperature Tensile Behavior—Helium Embrittlement. For elevated temperatures, displacement damage is no longer stable, and flow properties are basically unaffected by irradiation (Fig. 4). However, in certain irradiated alloys containing helium, the strength decreases upon irradiation at elevated temperatures, but the ductility also decreases (Ref 18). Total elongation measured in a tensile test drops to only a few tenths of a percent. Although a temperature of approximately $0.5\ T_m$ is often associated with helium embrittlement, the temperature will depend on the helium concentration and the tensile strain rate. As the helium concentration increases and/or the strain rate decreases, the temperature at which helium embrittlement occurs will decrease.

Elevated-temperature helium embrittlement is accompanied by intergranular fracture and is thought to be caused by helium on grain boundaries. For austenitic stainless steels, the effect can occur with the presence of only a few atomic parts per million of helium—even the small amounts formed during fast-reactor irradiation (Ref 18). Embrittlement is more severe for cold-worked than for solution-annealed material, although the effect generally appears at temperatures at which recrystallization or recovery of the cold-worked material occurs.

The large difference between the embrittlement of cold-worked and solution-annealed austenitic steels may indicate that some grain-boundary migration (by recrystallization or grain growth) during irradiation or testing may be necessary to obtain the extremely low elongations found in austenitic stainless steels. The difference between the cold-worked and solution-annealed type 316 stainless steels might be explained by grain boundaries collecting helium during recrystallization of the cold-worked structure.

Such embrittlement could impose an upper temperature limit on the use of austenitic steels. Again, alloying with titanium can improve resistance to embrittlement. The improvement is due to bubble refinement when helium is trapped on MC particles, as shown in Fig. 5 for PCA steel. The PCA-A1 was in a solution-annealed condition with no MC particles (Fig. 5c). The PCA-B3 was aged at 800 °C (1470 °F) prior to 25% cold working. When the helium was trapped in the fine bubbles formed on the metal carbides (Fig. 5b), a relatively small change in ductility occurred upon irradiation at 600 °C

Mechanical Properties

Although ferritic steels are more resistant to swelling than austenitic steels, irradiation may have a more critical effect on the mechanical properties of ferritic steels. In particular, the effect of irradiation on fracture behavior is of crucial importance in light-water reactors and may limit the use of ferritic steels in fusion reactors. As noted above, low-temperature irradiation can result in the formation of dislocation loops, solute clusters, vacancy clusters, precipitates, and microvoids. This microstructural alteration causes most of the changes in mechanical properties. Transmutation helium can also affect mechanical properties.

Low-Temperature Tensile Behavior. An example of the effect of fast reactor irradiation on the strength of a ferritic steel is shown in Fig. 4 for 9Cr-1MoVNb steel irradiated in the EBR-II at 390 to 550 °C (735 to 1020 °F) and tested at the irradiation temperature. An increase in the yield stress is observed for irradiation up to about 425 °C (800 °F). No hardening occurred at 450 °C (840 °F) and above. With increasing fluence below approximately 425 °C (800 °F), strength increases to a saturation level, after which it remains unchanged. Depending on the steel and the fluence, irradiation-enhanced softening is possible at temperatures above about 425 °C (800 °F) because of recovery and precipitate coarsening, which can be hastened by the irradiation. The change in the ultimate tensile strength is similar to that of the yield stress. Changes in ductility reflect the strength changes; an increase in strength results in a ductility decrease and vice versa. In general, adequate ductility is maintained for these irradiation conditions. Similar observations have been made for 12Cr-1MoVW (Ref 11) and 2¼Cr-1Mo steels (Ref 12).

To determine the effect of helium, nickel-doped 9Cr-1MoVNb and 12Cr-1MoVW steel tensile specimens were irradiated in the HFIR (Ref 13). The results were compared with results for undoped steels irradiated similarly and undoped and doped specimens irradiated in the EBR-II, in which little helium was generated. At 300 and 400 °C (570 and 750 °F), results indicated that the transmutation helium caused an increase in strength in addition to that caused by the displacement damage. No helium effect was apparent on specimens irradiated at 500 °C (930 °F).

A qualitatively similar behavior is observed when austenitic stainless steels are irradiated in the EBR-II (little helium). For type 316 stainless steel, both solution-annealed (Ref 14) and 20%-cold-worked (Ref 15) steels have been investigated in some

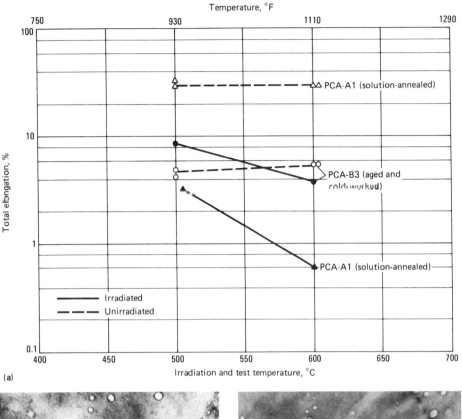

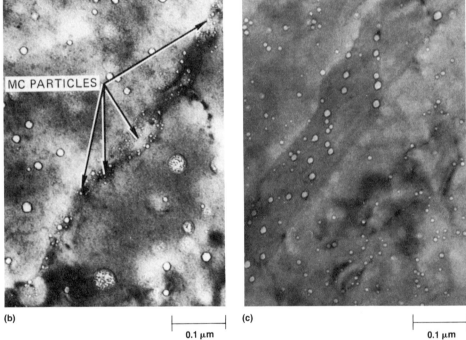

Fig. 5 Tensile ductility (a) of solution-annealed PCA steel and aged and cold-worked PCA steel. Irradiation caused a large decrease in the ductility of the solution-annealed PCA steel but not the cold-worked steel. This difference was correlated with fine bubbles on the MC precipitates that were present in the aged and cold-worked steel (b) but not in the solution-annealed steel (c). Source: Ref 17

(1110 °F) to approximately 22 dpa and 1750 appm of helium (Ref 17). Without helium trapping (Fig. 5c), the helium collected at grain boundaries, and a much larger decrease in ductility occurred. It should be noted, however, that irradiation to higher fluences will be required to determine if this resistance to helium embrittlement continues.

All indications are that the martensitic steels, such as 9Cr-1MoVNb and 12Cr-1MoVW, are much more immune to helium embrittlement than the austenitic steels (Ref 19). The reasons for such resistance are not completely understood. Immunity is not inherent in the body-centered cubic (bcc) crystal structure (compared to the face-centered cubic structure), because helium embrittlement occurs in vanadium and niobium alloys. It appears likely that the resistance to helium embrittlement is related to the martensitic microstructure.

In the normalized condition, martensite has a fine lath structure containing a high density of dislocations. After tempering, a ferrite matrix containing a high density of carbide particles and a lower dislocation density remains. However, the distinctive lath structure is still evident; long laths, separated mostly by low-angle boundaries, are grouped in packets. The packets and some laths are separated by high-angle boundaries. Prior-austenite grain boundaries are also present. This type of fine microstructure should allow the partitioning of helium atoms to the various boundaries, including the precipitate boundaries. Such a wide distribution of helium should effectively keep the helium concentration on a given high-angle grain boundary relatively low and should reduce the probability of intergranular failure (an intergranular failure is expected to propagate along high-angle boundaries).

If the above conclusions are correct, then a much larger effect of helium would be expected on pure iron or on a steel with a polygonal ferrite structure (as opposed to a martensitic or bainitic microstructure). The many collection sites available in martensite are not present in pure iron, and the grain boundaries in such precipitate-free microstructures are more mobile at elevated temperatures. Indeed, this may explain the reduced ductility values observed on steels with a ferrite microstructure. For example, the total elongation of types 405 and 430 ferritic stainless steels containing 40 appm of helium decreased from 52 to 33% and 89 to 48%, respectively, when tested at 700 °C (1290 °F) (Ref 20).

Irradiation Embrittlement. A major concern for bcc ferritic steels involves the effect of irradiation on fracture toughness (Ref 21, 22). Irradiation can cause large increases in the ductile-to-brittle transition temperature (DBTT) and decreases in the upper-shelf energy (USE), as measured by Charpy V-notch specimens. Even if the DBTT of the unirradiated steel is below room temperature, it can be well above room temperature after irradiation.

Irradiation embrittlement is related to the radiation-produced dislocation loops that form below $0.35\ T_m$; irradiation-induced precipitates can also have an effect. Loops are barriers to dislocation motion and give rise to the strengthening discussed above. The relationship of this increase in flow stress to irradiation embrittlement is shown

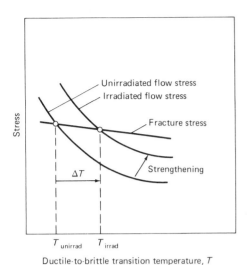

Fig. 6 Schematic of suggested mechanism by which a strength increase due to irradiation causes an upward shift in the DBTT. Source: Ref 21

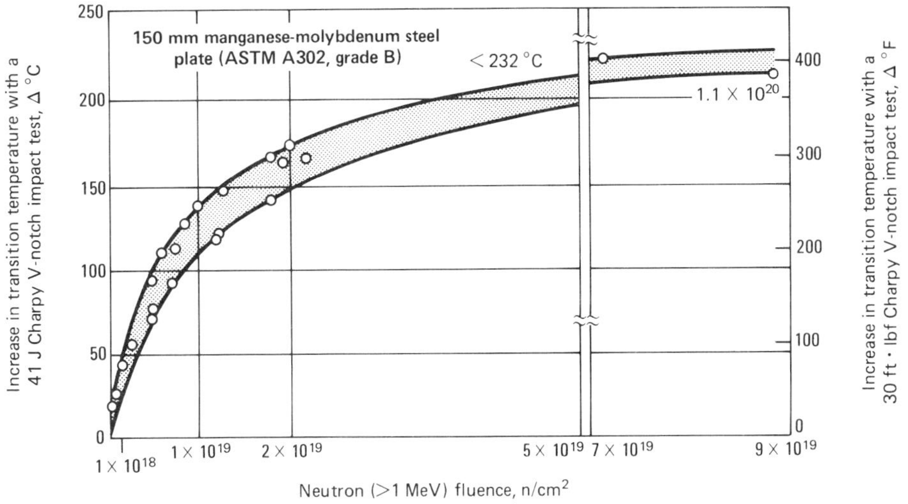

Fig. 7 Effect of neutron fluence on the 41 J (30 ft · lbf) transition temperature in Charpy impact tests at temperatures below 232 °C (450 °F). Test specimen: 150 mm (6 in.) thick manganese-molybdenum steel (ASTM A 302, grade B). Source: Ref 21

schematically in Fig. 6. Figure 6 shows how irradiation has shifted the flow stress upward. Under the assumption that the intersection of the fracture stress curve and the flow stress curve is the DBTT for the unflawed condition, the increase in flow stress is seen to cause a shift in the DBTT.

Although swelling and the other aspects of radiation damage do not play a role in light-water power reactors, irradiation embrittlement has been a major concern. Low-alloy pressure vessel steels specified by ASTM A 302-B and A 533-B are commonly used for this application.

Shifts in DBTT of over 200 °C (360 °F) have been observed in A 302-B irradiated at less than 232 °C (450 °F) to fluences of approximately 1×10^{24} n/m^2, which is less than 0.1 dpa (Fig. 7). Because displacement damage can be eliminated by annealing, the magnitude of the DBTT shift for a given fluence generally decreases with temperature. For a given temperature, embrittlement is rapid with increasing fluence at low fluence. A marked decrease in rate of embrittlement (as measured by the upward shift in DBTT) is then observed, and the rate appears to go to zero with increasing fluence (that is, saturation) (Fig. 7).

Irradiation embrittlement of the pressure vessel steels exposed to fluences typical of light-water reactors is affected by heat-to-heat variations, microstructure, and residual element content (Ref 21). Of the residual elements, copper and phosphorus have the greatest effect. One proposed mechanism for the effect of copper is that it enhances the formation of dislocation loops that lead to hardening (Ref 21). The effect of phosphorus has been attributed to a mechanism similar to the role it plays in temper embrittlement. This effect manifests itself under these conditions because of radiation-enhanced diffusion (Ref 21). Postirradiation heat treatment can restore the DBTT of low-alloy pressure vessel steels (Ref 21).

Neutrons are moderated in light-water reactors to produce thermal neutrons, but a considerable flux of fast neutrons is still present. It is these neutrons with energies above about 1 MeV that are generally considered to produce the damage that causes the irradiation embrittlement (Ref 21). However, accelerated embrittlement was recently observed in surveillance specimens for the pressure vessel of the HFIR. These ASTM A 212-B steel specimens were irradiated at about 50 °C (120 °F) in a high thermal-to-fast-flux ratio position, where the measured property change was about an order of magnitude larger than that expected on the basis of the fast neutron fluence (Ref 23).

Although thermal neutrons do not possess sufficient energy to dislodge an atom from the matrix, thermal neutrons can cause damage indirectly through transmutation reactions. In particular, a reaction between a thermal neutron and ^{56}Fe to form ^{57}Fe causes an atom recoil when a γ-ray is released. Displacement damage by this recoil can cause the embrittlement observed in the HFIR (Ref 24). Recoil from a reaction between a thermal neutron and boron to form an α-particle can also cause displacement damage. Damage from these transmutation reactions becomes important whenever the displacement-damage energy deposited by these reactions comes within an order of magnitude of that deposited by fast neutrons (Ref 24). The displacement damage produced by thermal neutrons is believed to be more efficient than that produced by fast neutrons in causing microstructural changes leading to embrittlement. This is because the displacement cascades from the recoil reactions are smaller than those for fast neutrons; consequently, less in-cascade recombination (loss) of vacancies and interstitials takes place for thermal neutrons. Recoil displacements are expected to be especially important at low temperatures, as is the case in the support structure of light-water reactors (Ref 24).

Irradiation embrittlement must also be considered in the development of ferritic steels for fast reactors and fusion reactors (Ref 22). The ferritic steels considered for use for fast-breeder reactors and fusion reactors are quite different from the low-alloy pressure vessel steels used for light-water reactors (Table 1). Furthermore, the types of fluences to which these steels will be exposed are considerably greater. While a light-water reactor steel will be irradiated to levels of the order of 1×10^{24} n/m^2, which produces a displacement-damage level of less than 0.1 dpa, fluences of two to three orders of magnitude higher and damage levels exceeding 200 dpa are expected for fast-breeder and fusion reactors.

Charpy impact tests for the pressure vessel steels are generally conducted using standard Charpy V-notch specimens. Because of space limitations in most test reactors used for the fast-breeder and fusion reactor programs, miniature Charpy V-notch (half- and third-size) specimens are used (Ref 25). Results from comparisons of the different specimen sizes have shown that the smaller specimens exhibit behavior relatively similar to that of large specimens (that is, increase in DBTT and decrease in USE). However, the magnitude of the USE is greatly reduced, and the DBTT is lower (Ref 25).

Because of the appearance of a saturation in the shift in DBTT of the pressure vessel steels after an irradiation to approximately 2×10^{23} n/m^2 (Fig. 7), miniature specimens of 12Cr-1MoVW steel were irradiated at 300 °C (570 °F) to typical light-water reactor fluences in the University of Buffalo reactor

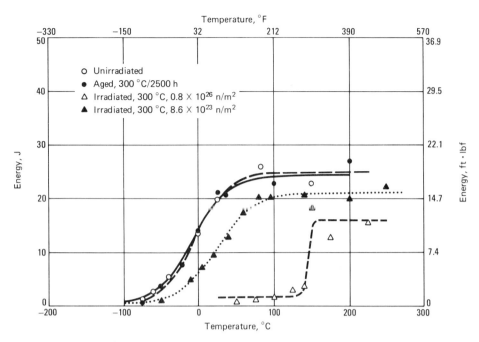

Fig. 8 Shift in DBTT for 12Cr-1MoVW steel irradiated at 300 °C (570 °F) to fluence of 8.6 × 10²³ and 8 × 10²⁵ n/m². Source: Ref 26

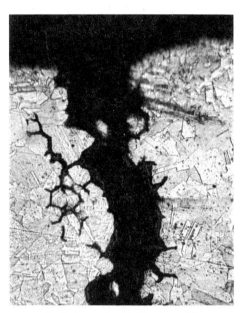

Fig. 9 Irradiation-assisted SCC of a stainless steel (Type 304) control blade absorber tube. 500×

(UBR) to 8.6 × 10²³ n/m² (<0.1 dpa) and compared with similar specimens irradiated in the HFIR to approximately 8 × 10²⁵ n/m² (4 to 9 dpa) (Fig. 8). After irradiation in the UBR, the DBTT increased by 36 °C (65 °F), and after irradiation in the HFIR, it increased by 164 °C (295 °F), indicating no saturation. Comparable decreases in the USE were also observed (Fig. 8).

Although the saturation observed for steels irradiated in light-water reactors did not apply to the HFIR, 9Cr-1MoVNb and 12Cr-1MoVW steels showed a saturation when irradiated in the EBR-II fast reactor (Ref 27). Above approximately 13 dpa, no further increase in the DBTT shift occurred when irradiated up to 26 dpa. However, when these steels were irradiated in the mixed neutron spectrum of the HFIR, the saturation observed in the EBR-II did not apply. A much larger shift in DBTT was observed (Ref 24). One possible explanation is that the helium generated in the HFIR played a role.

Another unusual effect occurred when 12Cr-1MoVW was irradiated at 300 and 400 °C (570 and 750 °F) in the HFIR: The shift in DBTT at 400 °C (750 °F) was 217 °C (390 °F), but was only 164 °C (295 °F) after irradiation at 300 °C (570 °F) (Ref 26). The shift at 400 °C (750 °F) was also higher than that after irradiation at 55 °C (130 °F) (Ref 28). These observations were contrary to observations of pressure vessel steels, where the amount of shift generally decreased with increasing irradiation temperature. Further, the shift in DBTT is expected to be proportional to hardening as measured in a tensile test where the hardening decreases with increasing temperature. When 9Cr-1MoVNb and 12Cr-1MoVW steels were irradiated above 400 °C (750 °F), the magnitude of the shift again decreased, and above approximately 450 °C (840 °F) the shift became quite small (Ref 27, 29). All indications are, therefore, that the shift in DBTT for 12Cr-1MoVW steel goes through a maximum with irradiation temperature (Ref 26).

Such a maximum in DBTT shift may arise from the dislocation and precipitation structure that results from irradiation near 400 °C (750 °F) (Ref 9, 30). This structure could have a maximum influence near 400 °C (750 °F), because at lower temperatures, precipitate formation is inhibited by unfavorable kinetics, and at higher temperatures, precipitate coarsening causes softening. An effect of helium cannot be ruled out, because the maximum shift in DBTT with temperature relies on data obtained from irradiations in the HFIR. It was also found that swelling in 9Cr-1MoVNb and 12Cr-1MoVW steels was a maximum at 400 °C (750 °F) (Ref 9). Because of the design of fast reactors in which irradiation temperatures below approximately 390 °C (735 °F) are not possible, no tests can be made to determine a maximum for irradiations in such reactors.

Other Properties. Irradiation effects on properties such as thermal creep (Ref 31) and fatigue (Ref 32) have been investigated. Two interesting phenomena that may be important for fast fission reactors and fusion reactors are irradiation creep (Ref 33) and the effect of irradiation on intergranular stress-corrosion cracking (ISCC) (Ref 34).

Irradiation Creep. If a specimen stressed to a load less than that required to cause thermal creep is irradiated in a fast reactor, deformation (creep) occurs under certain experimental conditions (temperature, stress, fluence). One explanation for irradiation creep involves the preferential absorption of radiation-produced interstitials by dislocations properly oriented relative to the stress such that they can climb under the influence of the irradiation-produced interstitials (Ref 33).

Irradiation-assisted SCC is an intergranular SCC phenomenon that has occurred in the stainless steel neutron source holders, control blade absorber tubes (Fig. 9), and nuclear instrument tube holders of commercial boiling-water reactors. Irradiation-assisted SCC has also occurred in commercial pressurized-water reactors (Ref 32).

Like the ISCC of stainless steels without any neutron irradiation, the irradiation-assisted SCC of stainless steels requires the presence of oxygenated high-temperature water. However, it is extremely important to note that irradiation-assisted SCC does not require the chromium depletion sensitization or high tensile stresses that are implicit in the ISCC of nonirradiated stainless steel components. There is some evidence that nonsensitized stainless steels become partially sensitized (that is, chromium depletion at the grain boundaries) during irradiation by the effect of radiation-induced segregation, thus reducing the resistance of the stainless steel to ISCC in aqueous environments (Ref 34). The mechanism of irradiation-assisted SCC appears to involve the simultaneous interaction of highly irradiated nonsensitized material with diffusion of impurities (sulfur, silicon, and phosphorus) to the grain boundaries, low stress (fabrication, irradiation creep), and high-temperature water with short-lived oxidizing species. In boiling-water reactors, no irradiation-assisted SCC has been observed at a neutron (>1 MeV) fluence below about 5×10^{20} n/cm^2 (see the article "Corrosion in the Nuclear Power Industry" in Volume 13 of the 9th Edition of *Metals Handbook*). High tensile stresses are not necessary for irradiation-assisted SCC, and cracks may occur at lower stresses for higher fluences. However, high tensile stress can exacerbate the problem.

REFERENCES

1. R.L. Klueh and E.E. Bloom, Radiation Facilities for Fusion Reactor First Wall and Blanket Structural Materials Development, *Nucl. Eng. Des.*, Vol 73, 1982, p 101-125
2. J.O. Stiegler and L.K. Mansur, Radiation Effects in Structural Materials, *Ann. Rev. Mater. Sci.*, Vol 9, 1979, p 405-454
3. L.K. Mansur and E.E. Bloom, Radiation Effects in Reactor Structural Alloys, *J. Met.*, Vol 34, 1982, p 23-31
4. P.J. Maziasz, Swelling and Swelling Resistance Possibilities of Austenitic Stainless Steel in Fusion Reactors, *J. Nucl. Mater.*, Vol 122 and 123, 1984, p 472-486
5. F.A. Garner, Recent Insights on the Swelling and Creep of Irradiated Austenitic Alloys, *J. Nucl. Mater.*, Vol 122 and 123, 1984, p 459-471
6. R.A. Weiner and A. Boltax, Comparison of High Fluence Swelling Behavior of Austenitic Stainless Steels, in *Effects of Irradiation on Materials: Tenth Conference*, STP 725, American Society for Testing and Materials, 1981, p 484-499
7. D.S. Gelles, Microstructural Examination of Several Commercial Ferritic Alloys Irradiated to High Fluence, *J. Nucl. Mater.*, Vol 103 and 104, 1981, p 975-980
8. E.A. Little and D.A. Stowe, Void-Swelling in Irons and Ferritic Steels: II. An Experimental Survey of Materials Irradiated in a Fast Reactor, *J. Nucl. Mater.*, Vol 87, 1979, p 25-39
9. P.J. Maziasz, R.L. Klueh, and J.M. Vitek, Helium Effects on Void Formation in 9Cr-1MoVNb and 12Cr-1MoVW Irradiated in HFIR, *J. Nucl. Mater.*, Vol 141-143, 1986, p 929-937
10. R.L. Klueh and J.M. Vitek, Elevated-Temperature Tensile Properties of Irradiated 9Cr-1MoVNb Steel, *J. Nucl. Mater.*, Vol 132, 1985, p 27-31
11. R.L. Klueh and J.M. Vitek, Tensile Behavior of Irradiated 12Cr-1MoVW Steel, *J. Nucl. Mater.*, Vol 137, 1985, p 44-50
12. R.L. Klueh and J.M. Vitek, Tensile Properties of 2¼Cr-1Mo Steel Irradiated to 23 dpa at 390 to 550 °C, *J. Nucl. Mater.*, Vol 140, 1986, p 140-148
13. R.L. Klueh and J.M. Vitek, Postirradiation Tensile Behavior of Nickel-Doped Ferritic Steels, *J. Nucl. Mater.*, Vol 150, 1987, p 272-280
14. R.L. Fish and J.J. Holmes, Tensile Properties of Annealed Type 316 Stainless Steel After EBR-II Irradiation, *J. Nucl. Mater.*, Vol 46, 1973, p 113-120
15. R.L. Fish, N.S. Cannon, and G.L. Wire, Tensile Property Correlations for Highly Irradiated 20 Percent Cold-Worked Type 316 Stainless Steel, in *Effects of Radiation on Structural Materials*, STP 683, American Society for Testing and Materials, 1979, p 450-465
16. G.D. Johnson, F.A. Garner, H.R. Brager, and R.L. Fish, Microstructural Interpretation of the Fluence and Temperature Dependence of the Mechanical Properties of Irradiated AISI 316, in *Effects of Radiation on Materials: Tenth Conference*, STP 725, American Society for Testing and Materials, 1981, p 393-412
17. P.J. Maziasz, A Perspective on Present and Future Alloy Development Efforts on Austenitic Stainless Steels for Fusion Application, *J. Nucl. Mater.*, Vol 133-134, 1985, p 134-140
18. E.E. Bloom, Irradiation Strengthening and Embrittlement, in *Radiation Damage in Metals*, American Society for Metals, 1976, p 295-329
19. R.L. Klueh and J.M. Vitek, The Resistance of 9Cr-1MoVNb and 12Cr-1MoVW Steels to Helium Embrittlement, *J. Nucl. Mater.*, Vol 117, 1983, p 295-302
20. D. Kramer, A.G. Pard, and C.G. Rhodes, A Survey of Helium Embrittlement of Various Alloy Types, in *Irradiation Embrittlement and Creep in Fuel Cladding and Core Components*, British Nuclear Energy Society, 1972, p 109-115
21. J.R. Hawthorne, Irradiation Embrittlement, in *Treatise on Materials Science and Technology*, Vol 25, Academic Press, 1983, p 461-524
22. G.E. Lucas and D.S. Gelles, The Influence of Irradiation on Fracture and Impact Properties of Fusion Reactor Materials, *J. Nucl. Mater.*, Vol 155-157, 1988, p 164-177
23. R.K. Nanstad, K. Farrell, D.N. Braski, and W.R. Corwin, Accelerated Neutron Embrittlement of Ferritic Steels at Low Fluence: Flux and Spectrum Effects, *J. Nucl. Mater.*, Vol 158, 1988, p 1-6
24. L.K. Mansur and K. Farrell, On Mechanisms by Which a Soft Neutron Spectrum May Induce Accelerated Embrittlement, *J. Nucl. Mater.*, to be published
25. W.R. Corwin and A.M. Hougland, Effect of Specimen Size and Material Condition on the Charpy Impact Properties of 9Cr-1Mo-V-Nb Steel, in *The Use of Small-Scale Specimens for Testing Irradiated Material*, STP 888, American Society for Testing and Materials, 1986, p 325
26. J.M. Vitek, W.R. Corwin, R.L. Klueh, and J.R. Hawthorne, On the Saturation of the DBTT Shift of Irradiated 12Cr-1MoVW With Increasing Fluence, *J. Nucl. Mater.*, Vol 141-143, 1986, p 948-953
27. W.L. Hu and D.S. Gelles, The Ductile-to-Brittle Transition Behavior of Martensitic Steels Neutron Irradiated to 26 dpa, in *Influence of Radiation on Material Properties: 13th International Symposium (Part II)*, STP 956, American Society for Testing and Materials, 1987, p 83-97
28. R.L. Klueh, J.M. Vitek, W.R. Corwin, and D.J. Alexander, Impact Behavior of 9-Cr and 12-Cr Ferritic Steels After Low-Temperature Irradiation, *J. Nucl. Mater.*, Vol 155-157, 1988, p 973-977
29. W.R. Corwin, J.M. Vitek, and R.L. Klueh, Effect of Nickel Content of 9Cr-1MoVNb and 12Cr-1MoVW Steels on the Aging and Irradiation Response of Impact Properties, *J. Nucl. Mater.*, Vol 149, 1987, p 312-320
30. D.S. Gelles and L.K. Thomas, Effects of Neutron Irradiation on Microstructure in Experimental and Commercial Ferritic Steels, in *Ferritic Alloys for Use in Nuclear Energy Technologies*, The Metallurgical Society, 1984, p 559-568
31. C. Wassilew, Influence of Helium Embrittlement on Post-Irradiation Creep Rupture Behavior of Austenitic and Martensitic Stainless Steels, in *Nuclear Technology and Applications of Stainless Steels at Elevated Temperatures*, The Metals Society, 1982, p 172-181
32. B. Van der Schaaf, The Effect of Neutron Irradiation on the Fatigue and Fatigue-Creep Behaviour of Structural Materials, *J. Nucl. Mater.*, Vol 155-157, 1988, p 156-163
33. W.A. Coghlan, Recent Irradiation Creep Results, *Int. Met. Rev.*, Vol 31, 1986, p 241-290
34. A.J. Jacobs and G.P. Wozadlo, Irradiation-Assisted Stress Corrosion Cracking as a Factor in Nuclear Power Plant Aging, in *Proceedings of the International Conference on Nuclear Power Plant Aging, Availability Factor, and Reliability Analysis*, American Society for Metals, 1985, p 173-180

Low-Temperature Properties of Structural Steels

Mamdouh M. Salama, Conoco Inc.

CRITICAL STRUCTURAL COMPONENTS must be fabricated from steels that exhibit adequate low-temperature fracture toughness because of the serious consequences of failure due to brittle fracture. Codes used for the design of offshore structures specify low-temperature toughness requirements, and steel specifications that satisfy these requirements have been developed. The need for steels with higher fracture toughness and better weldability, as well as lower cost, has prompted major advancements in structural steel technology. These advancements are highlighted by the development of controlled-rolled and accelerated-cooled steels. This article reviews fracture resistance assessment procedures for welded joints and includes discussions on fatigue crack growth and fracture toughness. Fracture toughness requirements specified by different design codes are presented, and American Petroleum Institute (API), British Standards Institution (BSI), and American Society for Testing and Materials (ASTM) specifications for offshore structural steels are summarized, and applications of these specifications are discussed. This article also focuses on advances made in steel technology and the impact of these advances on the fracture toughness of steel.

Design and Failure Criteria

Three major factors contribute to service failure of steel structures:

- Brittle failure due to the presence of fabrication defects
- Fatigue crack development
- Crack development as a result of accidental damage

It is not practical or economical to fabricate defect-free structures. Although the use of appropriate inspection and quality control procedures can limit the size of defects, it cannot eliminate defects entirely. Proper fatigue design practices and in-service inspection can control the growth of fatigue cracks; however, complete elimination of small fatigue cracks is unrealistic, particularly for complex welded structures. Ductile failure due to growth of fatigue cracks to a large, plastic collapse critical size is a rare event, but it is still more common than brittle fracture, especially in structures subjected to the turbulent North Sea environment. Ductile failure in the absence of cracklike defects is experienced only in cases of accidental overloads that grossly exceed normal design stresses.

In addition to catastrophic failures of ships, tankers, offshore structures, pipelines, bridges, and vessels (Ref 1-3), numerous minor brittle failures of structures under construction or in service have resulted in delays and expensive repairs. To minimize the probability of these failures, the design of modern structures is based on the combined use of the methods of both classical design and structural integrity design. Structural integrity design is employed to prevent structural failure due to brittle fracture or premature fatigue cracking. Integrity design provides a tool for assessing fracture resistance by integrating stress analysis with evaluations of fabrication quality and the mechanical properties of the steel. The mechanical properties that are evaluated include fatigue crack growth, fracture toughness, and basic tensile properties (for example, yield strength and tensile strength).

Currently, all design guidelines, codes, or standards for critical applications emphasize fracture control procedures that provide for the evaluation of properties such as fracture toughness, weldability, and strength. Stringent steel qualification criteria have contributed to the development of low-cost structural steels possessing superior mechanical properties. These structural steels combine desired properties such as higher strength, improved weldability, and higher fracture toughness in one product. These properties are vital in steels used for offshore structures because the inaccessibility of these structures makes in-service inspection and repair very difficult and extremely expensive.

Assessment of Fracture Resistance

The offshore industry has used several advanced fracture mechanics methodologies and tests to establish allowable final defect, a_f (Ref 4). These include crack tip opening displacement (CTOD), tests, and, to a lesser extent, crack growth resistance, J_R, and failure assessment diagram methods. The CTOD approach allows calculation of size of the allowable final defect, a_f, using the following expression for the ratio between the critical defect and the CTOD value (Ref 5):

$$\frac{a_f}{\text{CTOD}} = \frac{1}{2\pi\,(Y/E)\,[\text{SCF}\,(S/Y) + \alpha - 0.25]} \quad (\text{Eq 1})$$

where a_f is the half-length of a through-thickness rectilinear crack (for surface and buried cracks, and for crack geometries other than through-thickness rectilinear cracks, the equivalent through-thickness cracks, can be estimated by utilizing relationships available in fracture mechanics handbooks or from the literature (for example, Ref 6), Y/E is the ratio between the yield strength and the modulus of elasticity of the material, S/Y is the ratio between the nominal applied stress and the yield strength, SCF is the stress concentration factor, and α is the stress relief parameter, which equals 1.0 for no stress relief (that is, residual stress equals Y) and equals 0.0 for full stress relief (that is, no residual stress).

Ensuring against brittle fracture by specifying a blanket CTOD value is difficult without performing detailed fatigue life calculations. However, toughness specifications in terms of CTOD values are valuable when used in conjunction with fatigue crack growth rate data in the framework of fracture mechanics analysis; they can provide

Table 1 Yield strength and impact energy guidelines for low-temperature structural steels

	Minimum yield strength		Minimum Charpy V-notch impact energy							
			Average				Individual			
			Transverse		Longitudinal		Transverse		Longitudinal	
Specification	MPa	ksi	J	ft · lbf	J	ft · lbf	J	ft · lbf	J	ft · lbf
API RP 2A	<280	<40	20	15	...	...	13(a)	9.8(a)	...	...
	280–360	40–52	34	25	...	...	27(a)	19.8(a)	...	...
	>360	>52	48	35	...	...	41(a)	29.8(a)	...	...
United Kingdom DEn	Y(b)	Y(b)	$0.10Y$	$0.10Y$	...	...	$0.07Y$	$0.07Y$	...	...
DNV	$Y \leq 275$	$Y \leq 40$	18	13	27	20	14(c)	9.8(c)	20(c)	15(c)
	$Y > 275$	$Y > 40$	$0.07\,Y$	$0.07\,Y$	$0.10\,Y$	$0.10\,Y$	$0.053\,Y$(c)	$0.053\,Y$(c)	$0.075\,Y$(c)	$0.075\,Y$(c)

(a) Minimum individual impact energy = minimum average −7 J, or −5.2 ft · lbf. (b) Y, minimum yield strength of the thinnest plate. (c) Minimum individual CVN value = 0.75 × (minimum required average value)

useful information on tolerable defects, remaining product life, and allowable loading conditions. Because of the complexity of CTOD testing (Ref 7), most design codes still rely on Charpy V-notch (CVN) energy and transition temperature concepts as the main fracture toughness acceptance criteria. The CVN impact test is performed following international standards such as ASTM A 370 or BSI 131.

Fracture Toughness Requirements

Almost all design guidelines for critical structures specify a minimum fracture toughness requirement. This section summarizes toughness requirements given by three existing design guidelines used in the offshore industry. These guidelines have been developed by API, the United Kingdom Department of Energy (DEn), and Det Norske Veritas (DNV), a Norwegian ship classifier. The fracture toughness criteria used in these guidelines are based mainly on CVN impact energy and transition temperature criteria.

According to API RP 2A (Ref 8), underwater joints should meet notch toughness requirements as established by either the Naval Research Laboratory drop-weight test (ASTM E 208) or the CVN impact energy test. For the drop-weight test, the joints should be rated for no-break performance. The CVN test is performed on transverse test specimens. The minimum Charpy energy is specified as a function of the minimum yield strength of the steel (Table 1). The test temperature is specified as a function of the lowest anticipated service temperature (LAST) and the pipe diameter-to-thickness ratio, D/t:

	Test temperature(b)		Condition
D/t(a)	°C	°F	of testpiece
<20	LAST − 10	LAST − 18	As-fabricated
20–30	LAST − 30	LAST − 54	Flat plate
>30	LAST − 20	LAST − 36	Flat plate

(a) D/t, diameter-to-thickness ratio. (b) LAST, lowest anticipated service temperature

For a D/t less than 20, the Charpy specimen is machined from as-fabricated pipe, and the test temperature is 10 °C (18 °F) below LAST. On the other hand, for a higher D/t, testing is conducted on samples machined from flat plates. A lower temperature is specified for this test to account for possible toughness deterioration due to strains induced by forming (D/t of 20 is equivalent to 5% strain). No recommendations about the effect of plate thickness on toughness requirements are provided by the API specification.

The DEn specification (Ref 9) provides fracture toughness criteria for structures located in the North Sea. The toughness criteria depend on the minimum yield strength and thickness of the plate, and on the postweld heat treatment, stress concentration, and location of test specimens. The recommended minimum average CVN values of transverse specimens are listed in Table 1, and the recommended test temperatures are summarized in Table 2.

The DNV standard (Ref 10) provides fracture tougness requirements using both CVN and CTOD approaches. The recommended average minimum energy level depends on the yield strength as given in Table 1. Table 3 lists the recommended impact-testing temperature in terms of the design temperature, T_D, where T_D is defined as 5 °C (9 °F) below the most probable lowest monthly mean temperature. The DNV specifications include minimum CTOD value requirements for weld procedure qualifications for plates 50 mm (2 in.) and greater in thickness. The requirements at the minimum design temperature are 0.35 mm (0.014 in.) for as-welded or local postweld heat-treated conditions, and 0.25 mm (0.010 in.) after furnace postweld heat treatment.

There are some differences in toughness requirements among the design codes. Also, requirements continue to change with new issues of each code as more data become available. Table 4 highlights the importance of low-temperature toughness requirements for offshore structural steels by comparing the toughness requirements for two cases using the different codes. For thicker plates and for structures used in more severe environments such as the Arctic, higher fracture toughness values are required.

Fatigue Crack Growth in Structural Steel

Steel structures generally include complex welded joints that have large local stress concentrations and that are subject to fatigue loadings induced by environmental forces. This, in addition to fabrication defects that are often present in welded structures, will result in the early initiation of fatigue cracks. Eventually the crack grows to a size at which failure may occur. Therefore, the fatigue life of welded components can be estimated by integrating an appropriate crack growth equation such as Paris law between the allowable initial defect, a_i, and the final defect at which failure occurs, a_f. The size of the final defect depends on the fracture toughness of the material and the applied stress.

The Paris equation (Eq 2) is bounded by the threshold value, ΔK_o, and the critical value, K_{max}, which is a measure of the fracture toughness as shown in Fig. 1. The crack growth equation provides a relationship between the crack growth rate, da/dN, and the stress intensity factor range, ΔK in the following form:

$$\frac{da}{dN} = C\,(\Delta K)^m \qquad (\text{Eq 2})$$

The stress intensity factor range, ΔK, is defined by:

$$\Delta K = \Delta S\,F\,\sqrt{\pi a} \qquad (\text{Eq 3})$$

where ΔS is the cyclic stress range, F is a correction factor dependent on component and crack geometries, and a is the half-length of a through-thickness rectilinear crack.

The crack growth parameters C and m are experimentally determined constants that depend on the material, loading condition, and environment. Reference 6 provides a C value of 3×10^{-13} and an m value of 3, in units of N (load) and mm (length), for ferritic steels with yield strengths up to 600 MPa (87 ksi). These values are based on the upper limit of air fatigue data shown in Fig. 2 for a variety of weld metals and heat-affected zone (HAZ) microstructures (Ref 11). Since the development of these data, extensive fatigue

Table 2 Recommended CVN test temperatures according to United Kingdom Department of Energy specifications

| Thickness, t | | Test location | As-welded | | | | Postweld heat treatment | | | |
| mm | in. | | Highly stressed | | Others | | Highly stressed | | Others | |
			°C	°F	°C	°F	°C	°F	°C	°F
$t \leq 20$	$t \leq 25/32$	Subsurface	−20	−5	−20	−5	−20	−5	−10	15
$20 < t \leq 100$	$25/32 < t \leq 4$	Subsurface	−40	−40	−30	−20	−30	−20	−20	−5
$40 < t \leq 100$	$19/16 < t \leq 4$	Mid-thickness	−30	−20	−20	−5	−20	−5	−20	−5

crack growth data have been developed for offshore structural steels such as BS 4360 grade 50D. Figure 3 presents crack growth data for this steel in both an air environment and in free corrosion and cathodic protection (CP) conditions in seawater. The CP levels were between −800 and −1100 mV (with respect to a silver/silver chloride reference electrode). Figure 3 was developed using data from research done exclusively on rectangular through-thickness notched parent steel specimens (no weld metal or HAZ data are included) (Ref 12). The test frequency varied between 0.1 and 1.0 Hz, the temperature between 5 and 20 °C (40 and 70 °F) and the stress ratio, R, between 0.0 and 0.5. Based on these data, a C value of 2.3×10^{-12} and an m value of 3 (in units of N and mm) have been suggested (Ref 13). These values predict crack growth rates that are about one order of magnitude higher than crack growth rates calculated using the values recommended in BS PD6493. Fatigue lives are therefore reduced by one order of magnitude due to the interactive effect of crack size.

The Paris crack growth equation is generally valid within the ΔK range of 300 to 1800 MPa$\sqrt{mm}$ (9 to 52 ksi$\sqrt{in.}$). Values of K below about 300 MPa$\sqrt{mm}$ (9 ksi$\sqrt{in.}$) fall in the threshold range where crack propagation does not occur, and values above about 1800 MPa$\sqrt{mm}$ (52 ksi$\sqrt{in.}$) fall in the range where the static mode of fracture occurs as the fracture toughness limit of the material is approached.

The following relationship between ΔK_o and the applied stress ratio, R, is provided in BS PD6493:

$$\Delta K_o = 190 - 144 R \text{ MPa} \sqrt{mm} \quad \text{(Eq 4)}$$

In Eq 4, ΔK_o depends on R. The Paris equation (Eq 2) does not depend on R. The relationship presented in Eq 4 provides the lower bound to all published threshold data for grade 50D steel in air and seawater (Ref 9). It has been suggested that other data for similar steels and for austenitic steels lie below the PD6493 line (Ref 10). Including these data, the following relationship, based on a 97.7% probability of survival for the data in Fig. 4, has been recommended (Ref 10):

$$\Delta K_o = 170 - 214 R \text{ MPa} \sqrt{mm} \text{ for } 0 \leq R < 0.5$$
$$= 63 \quad \text{MPa} \sqrt{mm} \text{ for } R \geq 0.5 \quad \text{(Eq 5)}$$

Steel Specifications

Structural steel specifications are generally based on the appropriate national or industry standards such as ASTM, API, BSI, and so on. In most cases, standards provide mainly basic requirements such as limits on chemical composition and tensile properties. During the mid-1960s, several in-service and structural fabrication problems in Gulf of Mexico offshore structures were encountered. These problems indicated that common pipes such as API 5L B and structural steels such as ASTM A 7 and ASTM A 36 do not always meet the design or service needs of the offshore industry (Ref 3). Failure analysis studies on several salvaged structures have shown that low notch toughness, laminations in the steel, lamellar tearing, and poor weldability were major contributors to the failures. These results made offshore operators and certifying authorities cognizant of the need for more restrictive standards to ensure that the steels used for offshore applications are of high quality and satisfy strict fracture toughness and weldability requirements. Therefore, standards such as API 2H, 2Y, and 2W (Ref 14-16) were developed. The types of structural steels that are addressed in these standards include killed fine-grain normalized, controlled-rolled, and quenched and tempered steels, as well as controlled-rolled accelerated-cooled (referred to as thermomechanical control process, or TMCP), steel. In addition to the above API grades, special grades from general standards such as ASTM and BSI are also specified for offshore structures. Table 5 summarizes the chemical composition and mechanical properties of some offshore structural steels. There are several differ-

Table 3 Recommended CVN test temperature according to DNV specifications

| Thickness, t | | Special steel | | Primary steel | | Secondary steel | |
mm	in.	°C	°F	°C	°F	°C	°F
$t \leq 12.5$	$t \leq \frac{1}{2}$	T_D	T_D	T_D	T_D	...	...
$12.5 < t \leq 25.5$	$\frac{1}{2} < t \leq 1$	$T_D - 20$	$T_D - 36$	T_D	T_D	...	...
$25.5 < t \leq 50$	$1 < t \leq 2$	$T_D - 40$	$T_D - 72$	$T_D - 20$	$T_D - 36$	T_D	T_D
$t > 50$	$t > 2$	$T_D - 40$	$T_D - 72$	$T_D - 40$	$T_D - 72$	$T_D - 20$	$T_D - 36$

(a) T_D, design temperature

Table 4 Low-temperature fracture toughness requirements for 50 mm (2 in.) thick plate used in offshore structures

| Structure location | Minimum water temperature | | Specification | Minimum yield strength | | Minimum individual CVN impact energy | | Temperature | |
	°C	°F		MPa	ksi	J	ft · lbf	°C	°F
Gulf of Mexico	5	40	API	345	50	27	20	−25	−15
				415	60	41	30	−25	−15
	5	40	DEn	345	50	25	18	−40	−40
				415	60	30	22	−40	−40
	5	40	DNV	345	50	25	18	−35	−30
				415	60	30	22	−40	−40
North Sea	0	32	API	345	50	27	20	−30	−20
				415	60	41	30	−30	−20
	0	32	DEn	345	50	25	18	−40	−40
				415	60	30	22	−40	−40
	0	32	DNV	345	50	25	18	−40	−40
				415	60	30	22	−45	−50

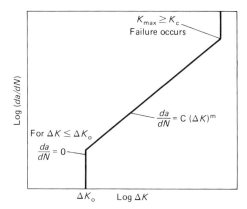

Fig. 1 Idealized fatigue crack growth model

ences among these specifications in the details they provide regarding limitations on steelmaking, chemical composition, mechanical properties, and quality.

Tables 6 and 7 compare BSI and API specifications for tensile strength and toughness properties of similar grades of steel. In addition to the differences in toughness values, there are differences in how each specification handles the effect of thickness on the yield strength. Furthermore, API 2W and 2Y provide not only minimum yield and tensile strength limits, but also an upper limit on the yield strength.

A limit on the maximum yield strength is very important to ensure a reasonable match between the strength of the weld metal and the base plate. In general, it is desirable to ensure that the weld metal strength is higher than the steel strength.

In addition to the CVN toughness requirements in Table 7, API provides two supplements using different toughness criteria. The first supplement is for toughness based on crack tip opening displacement tests of weld HAZ. Tests are performed in accordance with Section 3 of API RP 2Z (Ref 18); the heat input is 1.5 to 5 kJ/mm (38 to 125 kJ/in.), and the preheat is 100 to 250 °C (210 to 480 °F). For thicknesses up to and including 75 mm (3 in.), the required CTOD value is 0.25 mm (0.010 in.) at −10 °C (15 °F). For thicknesses greater than 75 mm (3 in.), the required CTOD value is 0.38 mm (0.015 in.) at −10 °C (15 F). The second supplement is for toughness of plates using the drop-weight test. The test is done in accordance with ASTM E 208 using P-3 specimens. The acceptable criterion is no-break performance at −35 °C (−30 °F).

Although standards for offshore structural steels are generally more restrictive than those used by other industries, they provide only minimum requirements for tensile properties, fracture toughness properties, control of chemical composition, and dimensional tolerances. Therefore, offshore operators often include additional requirements in the steel purchase specifications. These specifications generally include additional limitations on chemical composition along with requirements for higher toughness, a weldability evaluation, reduced tolerances, and an increased frequency of testing. Table 8 compares the chemical compositions of typical offshore structural steels and the composition allowed by the API 2H (Ref 14). In the typical steels, limits are imposed on more elements, and the maximum limits of carbon, sulfur, phosphorus, and carbon equivalent are reduced. These restrictions are intended to ensure improved toughness and weldability.

Advances in Steel Technology

Many significant advances in steelmaking processes have been made by steel companies to meet the demand for high-quality lower-cost structural steels with higher strength, improved weldability, and increased fracture toughness. These advances include the close monitoring of the supply of desulfurized iron in the blast furnace, the widespread use of continuous casting of thick slab for rolling to plate, the introduction of vacuum arc degassing, vacuum degassing, and argon stirring and injection

Table 5 Chemical composition and mechanical properties of selected offshore structural steels

Specification	Grade	Condition	C	Si	Mn	P	S	Nb	Al, total	N	Ni	Cr	Mo	Cu	Ti	Yield strength(b), MPa (ksi)	Tensile strength(b), MPa (ksi)
API 2H	42	Normalized	0.18	0.15–0.40	0.90–1.35	0.03	0.010	0.04	0.02–0.06	0.012	...	...	...	...	...	289 (42)	427–565 (62–82)
	50	Normalized	0.18	0.15–0.40	1.15–1.60	0.03	0.010	0.01–0.04	0.02–0.06	0.012	...	...	...	...	...	345 (50)	483–620 (70–90)
API 2W	42	TMCP	0.16	0.15–0.50	0.90–1.35	0.03	0.010	0.03	0.02–0.06	0.012	0.75	0.25	0.08	0.35	0.003–0.02	290–462 (42–67)	427 (62)
	50	TMCP	0.16	0.15–0.50	1.15–1.60	0.03	0.010	0.03	0.02–0.06	0.012	0.75	0.25	0.08	0.35	0.003–0.02	345–483 (50–75)	448 (65)
	60	TMCP	0.16	0.15–0.50	1.15–1.60	0.03	0.010	0.03	0.02–0.06	0.012	1.0	0.25	0.15	0.35	0.003–0.02	414–621 (60–90)	517 (75)
API 2Y	42	Q&T	0.16	0.15–0.50	0.90–1.35	0.03	0.010	0.08	0.02–0.06	0.012	0.75	0.25	0.08	0.35	0.003–0.02	290–462 (42–67)	427 (62)
	50	Q&T	0.16	0.15–0.50	1.15–1.60	0.03	0.010	0.08	0.02–0.06	0.012	0.75	0.25	0.08	0.35	0.003–0.02	345–517 (50–75)	448 (65)
	60	Q&T	0.16	0.15–0.50	1.15–1.60	0.03	0.010	0.08	0.02–0.06	0.012	1.0	0.25	0.15	0.35	0.003–0.02	414–621 (60–90)	517 (75)
BS 4360(c)	43D	Normalized	0.16	0.50	1.50	0.040	0.040	0.003–0.10(d)	...	...	...	...	...	...	...	270 (39)	430–510 (62–74)
	50D	Normalized	0.18	0.10–0.50	1.50	0.040	0.040	0.003–0.10(d)	...	...	...	...	...	...	...	345 (50)	490–620 (71–90)
	55E	Normalized	0.22	0.10–0.60	1.60	0.040	0.040	0.003–0.10(e)	...	...	...	...	...	...	...	430 (62)	550–700 (80–101)
ASTM A 633	C	Normalized	0.20	0.15–0.50	1.15–1.50	0.040	0.050	0.01–0.05	...	...	...	...	...	...	...	345 (50)	485–620 (70–90)
	D	Normalized	0.20	0.15–0.50	1.00–1.60	0.040	0.050	...	...	...	0.25	0.25	0.08	0.35	...	345 (50)	485–620 (70–90)
ASTM A 131	EH32	Normalized	0.18	0.10–0.50	0.90–1.60	0.04	0.04	0.05	...	...	0.40	0.25	0.08	0.35	...	315 (46)	470–585 (68–85)
	EH36	Normalized	0.18	0.10–0.50	0.90–1.60	0.04	0.04	0.05	...	...	0.40	0.25	0.08	0.35	...	360 (51)	490–620 (71–90)
ASTM A 710	A (class 3)	Quenched and precipitation heat treated	0.07	0.40	0.40–0.70	0.025	0.025	0.02 min	...	...	0.7	0.7–1.0	0.60–0.90	0.15–0.25	1.00–1.30	...	515 (75)	585 (85)

Q&T, quenched and tempered. (a) Heat analysis. (b) Values are selected for a thickness of 25 mm (1 in.); values may be reduced as thickness increases. (c) Ref 17. (d) 0.003–0.10% V. (e) 0.003–0.20% V

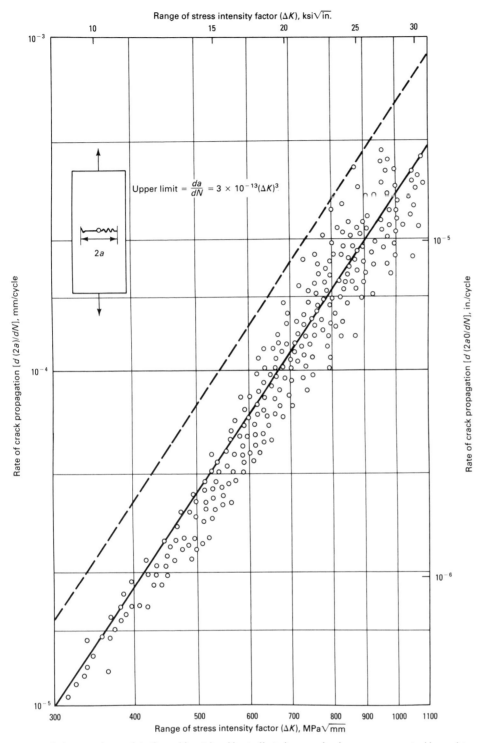

Fig. 2 Fatigue crack growth in the weld metal and heat-affected zones of carbon-manganese steel base plates in an air environment

techniques, along with almost exclusive use of basic oxygen process steelmaking (Ref 19). These advances in the steelmaking process have resulted in major improvements in structural steels, including significant control of alloying elements (for example, carbon, manganese, niobium, vanadium, and aluminum), major reductions in impurities (for example, sulfur, phosphorus, and nitrogen), and improved uniformity of composition and properties. Recent advances in computer control and rolling capacity have allowed the development of a new class of high-strength low-alloy steels, namely TMCP steels. The TMCP involves both controlled rolling and controlled (accelerated) cooling to produce steels with a very fine grain size (ASTM 10 to 12). The main aim of TMCP is to increase strength and fracture toughness and improve weldability by reducing the carbon equivalent and controlling the chemical composition (additional information is available in the article "Weldability of Steels" in this Volume). The API 2W specification (Ref 16) covers TMCP steel plates with minimum yield strengths between 290 and 415 MPa (42 and 60 ksi).

Strength in TMCP steels is maximized by reducing the ferrite grain size and increasing the volume fraction of the second phase. Accelerated cooling is used to achieve these effects. The influence of cooling rate on strength and toughness is shown in Fig. 5. A variation in cooling rate can be expected between surface and mid-thickness regions of thick plates. The addition of small amounts of niobium is very effective in strengthening the steel without affecting toughness (Fig. 6). However, the addition of more than 0.04% Nb is not desirable because it can cause a reduction in toughness, particularly in the subcritically reheated grain-coarsened heat-affected zone.

Fracture Toughness Characteristics of Structural Steels

Fracture toughness of steel has improved greatly as a result of advances in steel technology. Figure 7 compares CVN transition curves of old and new steels using transverse subsurface specimens. Steel A is a 60 mm (2⅜ in.) carbon-manganese steel that was used in 1975. It has a carbon level of 0.21%. Steel B is a modern 70 mm (2¾ in.) thick normalized carbon-manganese steel with a carbon level of 0.114% and some microalloying (0.29% Ni, 0.025% Nb, and 0.022% Cu). Steel C is a modern 50 mm (2 in.) thick controlled-rolled and accelerated-cooled TMCP steel with a carbon level of 0.11% and some microalloying (0.23% Ni, 0.03% Nb, and 0.24% Cu). The yield strengths of steels A, B, and C are 355, 369, and 506 MPa (51, 54, and 73 ksi), respectively. Figure 7 shows the improved fracture toughness of modern steels as indicated by a decrease in the transition temperature and an increase in the upper-shelf energy.

Figure 7 also presents nil-ductility transition temperature (NDTT) results for the three steels as determined by the drop-weight test (ASTM E 208). The results show that the NDTT for each steel corresponds to a different location on the CVN curve. The old steel, A, has an NDTT of −30 °C (−20 °F), which corresponds to a CVN transition temperature at 45 J (33 ft · lbf) and 85% crystallinity. The NDTT of B, the modern normalized steel, is −40 °C (−40 °F), which corresponds to a CVN transition temperature at 235 J (173 ft · lbf) and 8% crystallinity. The NDTT of C, the TMCP steel, is −60 °C (−75 °F), which corresponds to a

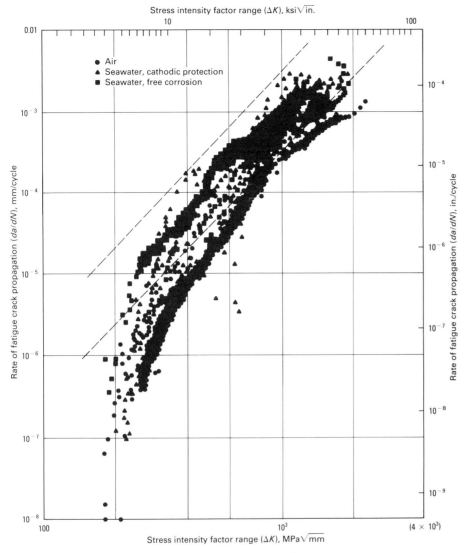

Fig. 3 Fatigue crack growth of BS 4360 grade 50D steel in air and in free corrosion and cathodic protection conditions in seawater

CVN transition temperature at 200 J (150 ft · lbf) and 15% crystallinity.

Test data obtained for API 2W grade 50, an 89 mm (3½ in.) thick TMCP steel, indicate a similar relationship between the NDTT and the CVN transition curve (Ref 23). The API 2W grade 50 steel has an NDTT of −60 °C (−75 °F), which corresponds to CVN impact energy at NDTT of about 200 J (150 ft · lbf); the CVN energy at a CTOD value of 0.25 mm (0.010 in) is about 150 J (110 ft · lbf) (Fig. 8).

These test results show that modern steels have higher fracture toughness and lower NDTT than older steels. However, the results also raise questions about the suitability of the current practice of assessing the transition temperature of modern steels based on a CVN energy level of 25 to 45 J (18 to 33 ft · lbf) as shown in Table 4. For older steels, CVN energy of 45 J (33 ft · lbf) corresponds to a temperature that is about the same as the NDTT, while for modern steels it corresponds to a temperature that is about 40 to 60 °C (105 to 140 °F) lower than the NDTT. This difference in steel behavior needs to be addressed in both design codes and steel specifications.

Fracture Toughness of Welded Structures

Assessing the fracture toughness of offshore structural steel involves evaluating not only base plate toughness but also HAZ and weld metal toughness. Although both HAZ toughness and weld metal toughness requirements are usually included in fabrication specifications, HAZ toughness requirements are sometimes specified in the steel purchase agreement. Because the small amount of material at the tip of the

Table 6 Tensile properties of offshore low-temperature structural steels

Specification	Grade	Thickness, t (mm)	Thickness, t (in.)	Minimum yield strength (MPa)	Minimum yield strength (ksi)	Minimum tensile strength (MPa)	Minimum tensile strength (ksi)
BS 4360	43D, 43E	$t \leq 16$	$t \leq 0.63$	280	41	430–510	62–74
		$16 < t \leq 40$	$0.63 < t \leq 1.6$	270	39	430–510	62–74
		$40 < t \leq 63$	$1.6 < t \leq 2.5$	255	37	430–510	62–74
		$63 < t \leq 100$	$2.5 < t \leq 4.0$	240	35	430–510	62–74
	50D, 50E, 50F	$t \leq 16$	$t \leq 0.63$	355	51	490–620	71–90
		$16 < t \leq 40$	$0.63 < t \leq 1.6$	345	50	490–620	71–90
		$40 < t \leq 63$	$1.6 < t \leq 2.5$	340	49	490–620	71–90
		$63 < t \leq 100$	$2.5 < t \leq 4.0$	(b)	(b)	490–620	71–90
	55E, 55F	$t \leq 16$	$t \leq 0.63$	450	65	550–700	80–101
		$16 < t \leq 40$	$0.63 < t \leq 1.6$	430	62	550–700	80–101
		$40 < t \leq 63$	$1.6 < t \leq 2.5$	415	60	550–700	80–101
	55E	$63 < t \leq 100$	$2.5 < t \leq 4.0$	400	58	550–700	80–101
	55F	$63 < t \leq 100$	$2.5 < t \leq 4.0$	(b)	(b)	550–700	80–101
API 2H	42	$t \leq 63$	$t \leq 2.5$	289	42	427–565	62–82
		$t > 63$	$t > 2.5$	289	42	427–565	62–82
	50	$t \leq 63$	$t \leq 2.5$	345	50	483–620	70–90
		$t > 63$	$t > 2.5$	324	47	483–620	70–90
API 2W, 2Y	42	$t \leq 25$	$t \leq 1.0$	290–462	42–67	427	62
		$t > 25$	$t > 1.0$	290–427	42–62	427	62
	50	$t \leq 25$	$t \leq 1.0$	345–517	50–75	448	65
		$t > 25$	$t > 1.0$	345–483	50–70	448	65
	60	$t \leq 25$	$t \leq 1.0$	414–621	60–90	517	75
		$t > 25$	$t > 1.0$	414–586	60–85	517	75

(a) Tensile strength is the same for all thicknesses. (b) By agreement

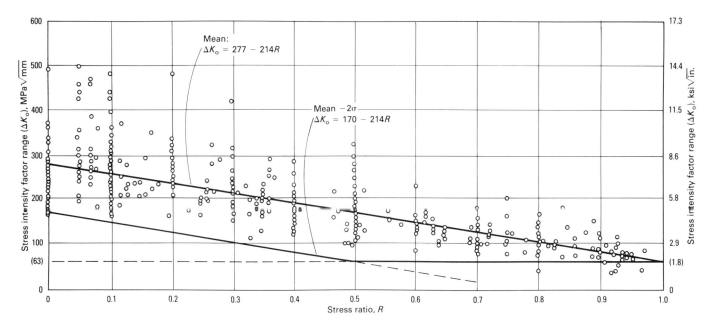

Fig. 4 Fatigue crack growth threshold data for ferritic steels with yield strengths up to 600 MPa (87 ksi)

sharp fatigue crack can be examined with CTOD testing, a detailed evaluation of the toughness of the different HAZ regions is possible (Ref 24). This positional accuracy allows for the identification of isolated regions in the HAZ with toughness substantially lower than that of the bulk material.

These local brittle zones (LBZs) have occurred during testing for crack extension at stress levels far lower than those at which cracks extended in the bulk material.

The presence of LBZs is not a new problem, nor is it limited to modern steels. In most steels, LBZs are associated with the

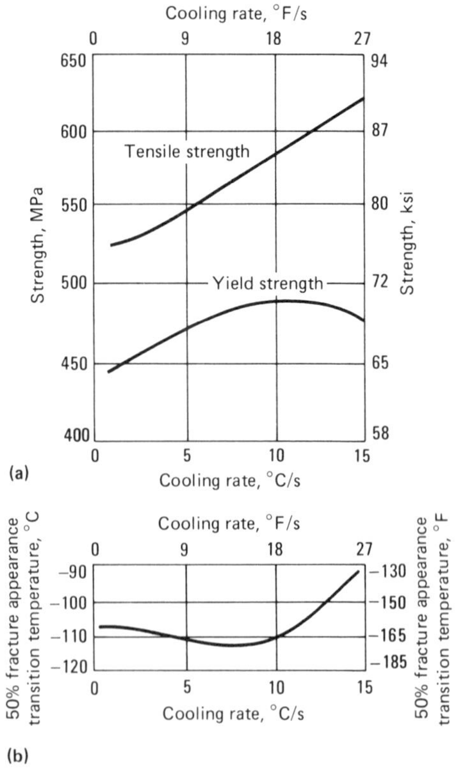

Table 7 Toughness requirements for offshore low-temperature structural steels

		Minimum average energy, CVN		Test temperature	
Specification	Grade	J	ft · lbf	°C	°F
BS 4360	43D, normalized	41	30	−10	15
		27	20	−20	−5
	43E, normalized	61	45	−20	−5
		27	20	−50	−60
	50D, normalized	41	30	−20	−5
		27	20	−30	−20
	50E, normalized	47	35	−30	−20
		27	20	−50	−60
	50F, quenched and tempered	47	35	−30	−20
		27	20	−50	−60
	55E, normalized	61	45	−20	−5
		27	20	−50	−60
	55F, quenched and tempered	41	30	−40	−40
		27	20	−60	−75
API 2H, 2W, 2Y	42	34(a)(b)	25(a)(b)	−40	−40
	42, S-2(c)	34(a)(b)	25(a)(b)	−60	−75
	50	41(a)(b)	30(a)(b)	−40	−40
	50, S-2(c)	41(a)(b)	30(a)(b)	−60	−75
API 2W, 2Y	60	48(a)(b)	35(a)(b)	−40	−40
	60, S-2(c)	48(a)(b)	35(a)(b)	−60	−75

(a) API CVN tests use transverse specimens. (b) API provides supplement S-7 for CVN tests using specimens uniformly strained 5% and aged at 250 °C (480 °F) for 1 h. (c) S-2 supplementary requirement

Fig. 5 Effect of cooling rate on selected properties of TMCP steels. (a) Strength. (b) Toughness. Source: Ref 20

Table 8 API composition specification (product analysis) for offshore low-temperature structural steels compared with typical United States and foreign specifications

	Composition, %															Carbon equivalent, maximum(a)
Specification	C	Mn	P	S	Si	Nb	Al, total	Ni	Cr	Mo	V	Cu	As	Sn	Sb	
API 2H (1988), grade 50	0.22, max	1.07–1.60	0.04 max	0.015 max	0.13–0.45	0.005–0.05	0.015–0.06	(b)	(b)	(b)	(b)	(b)	(b)	(b)	(b)	0.43
Typical United States mill	0.15	1.34	0.015	0.006	0.30	0.05	0.040	0.17	0.08	0.056	0.002	0.032	(c)	(c)	(c)	0.40
Typical foreign mill	0.12	1.44	0.009	0.001	0.38	0.020	0.035	0.18	0.009	0.001	0.001	0.16	0.003	0.001	0.000	0.38

(a) International Institute of Welding carbon equivalent (CE): CE = %C + (%Cr + %Mo + %V)/5 + (%Ni + %Cu)/15. (b) Not specified. (c) Not reported

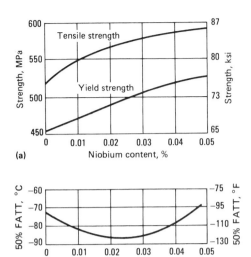

Fig. 6 Effect of niobium content on selected properties of TMCP steels. (a) Strength. (b) Toughness. Source: Ref 21

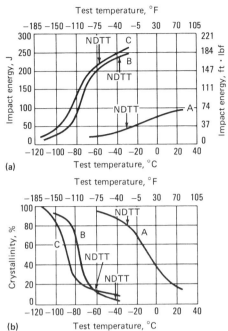

Fig. 7 Correlation of (a) Charpy V-notch impact energy and (b) crystallinity with nil-ductility transition temperature (NDTT) for three steels: A, 60 mm (2⅜ in.) thick old carbon-manganese steel (0.21% C) with a yield strength of 355 MPa (51 ksi); B, 70 mm (2¾ in.) thick modern carbon-manganese steel (0.114C-0.29Ni-0.025Nb-0.022 Cu) with a yield strength of 369 MPa (54 ksi); and C, 50 mm (2 in.) thick TMCP steel (0.11C-0.23Ni-0.03Nb-0.24Cu) with a yield strength of 506 MPa (73 ksi). Source: Ref 22

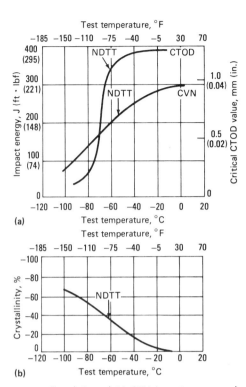

Fig. 8 Correlation of (a) CVN impact energy and crack tip opening displacement and (b) crystallinity with nil-ducility transition temperature for API 2W grade 50, a TMCP steel with a thickness of 89 mm (3½ in.)

grain-coarsened regions of the heat-affected zone (GCHAZ). Figure 9 identifies the different HAZ regions in a single-bevel multipass weld. Figure 9 also presents a plan view of a polished section illustrating a method for calculating the length and the percent of the grain-coarsened (GC) regions. Evaluation of wide plate tests suggests that fractures are likely to initiate from the GC areas where the grain size is greater than 80 μm (0.0024 in.), or ASTM No. 4 (Ref 22, 25).

There are several reasons for the current interest in LBZs. The need to reduce costs results in optimized structures that have less redundancy and a large number of highly stressed joints. To reduce welding costs, narrow groove preparation is used that may result in an HAZ that is normal to the loading direction. Also, unlike normalized steel in which the HAZ yield strength is higher than that of the base plate, TMCP steel sometimes has an HAZ yield strength that is lower than that of both the weld metal and the base plate (Fig. 10). Softening of the HAZ can also be expected if thermal cutting is used during fabrication. This behavior can limit the application of thermal cutting to TMCP steels.

Figure 11 shows the effect of weld metal strength on the CVN toughness of the HAZ for a carbon-manganese steel using the same narrow groove welding procedures. The yield strength of the steel is 458 MPa (66 ksi). Similar results were obtained using CTOD tests; the CGHAZ toughness decreases (that is, transition temperature increases) with increasing weld metal strength. Therefore, a highly overmatched weld metal (that is, a high ratio of weld metal yield strength, Y_w to base metal yield strength Y_b) is not desirable. The increase in weld metal strength appears to cause an unfavorable deformation state (constraint) that reduces toughness and enhances brittle crack initiation and propagation of GCHAZ. These results also suggest that care must be taken when correlating HAZ toughness values based on simulated microstructure with values obtained from actual welded joints. The combination of lower structural redundancy, higher stresses, and the location of lower-strength heat-affected zones normal to the loading directions can result in situations where fatigue cracks can propagate through more GC regions, thus increasing the possibility of brittle fracture. Toughness values supplied by ASTM and API are established by testing under ideal laboratory conditions and may not reflect actual HAZ toughness values experienced in the field. Customers should expect toughness values to be lower than those specified in standards.

Grain-coarsened regions will always exist in structural steels, but their size depends on both the steel and the welding process used. Therefore, it is necessary to identify an acceptable size for these regions. One study indicates that as long as the percentage of GC regions sampled by the crack front of CTOD specimens is less than 7%, no low toughness values can be measured (Fig. 12). Similar evaluations have shown that a major deterioration in CTOD values results only when the GC regions sampled by the crack fron exceed 10% (Ref 28).

Steel companies are responding to industry concerns about LBZs by working on the development of LBZ-free steels through appropriate microalloying (Ref 29). The development of an LBZ-free low aluminum-boron steel has been reported (Ref 30). The low aluminum-boron, medium nitrogen chemistry (0.009% Al, 0.0024% B, 0.0052% N, with no copper or niobium additions) was designed to nucleate the maximum amounts of boron nitride precipitates during the cooling of the initial thermal cycle in the GCHAZ. This is intended to promote the austenite-to-ferrite transformation and to prevent the bainite transformation. Figure 13 compares the LBZ-free steel with a conventional TMCP steel. The two steels are of the 415-MPa (60-ksi) yield strength class. The cumulative distribution of the critical heat-affected zone CTOD values of the two steels were compared using 27 mm (1¹/₁₆ in.) CTOD specimens. These specimens were machined from joints welded using submerged arc welding with a heat input of 5.0 kJ/mm (125 kJ/in.).

ACKNOWLEDGMENT

The author wishes to express his thanks to the management of Conoco Inc. for permission to publish this article.

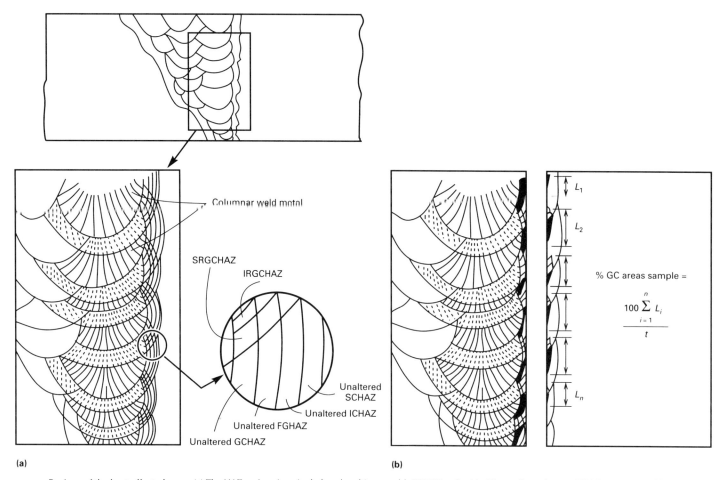

Fig. 9 Regions of the heat-affected zone. (a) The HAZ regions in a single-bevel multipass weld. SCHAZ, subcritical heat-affected zone; ICHAZ, intercritical heat-affected zone; FGHAZ, fine-grain heat-affected zone; SRGCHAZ, subcritically reheated grain-coarsened heat-affected zone; IRGCHAZ, intercritically reheated grain-coarsened heat-affected zone. (b) Plan view of a polished weld section showing a method for calculating the length and the percent of the GCHAZ. GC, grain-coarsened

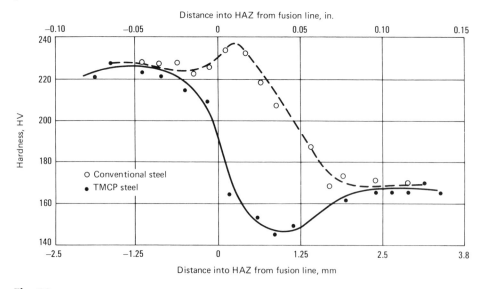

Fig. 10 Heat-affected zone hardness of conventional (normalized) steel and TMCP steel

REFERENCES

1. W.S. Pellini, *Guidelines for Fracture-Safe and Fatigue-Reliable Design of Steel Structures*, The Welding Institute, 1983
2. S.T. Rolfe and J.M. Barsom, *Fracture and Fatigue Control in Structures—Applications of Fracture Mechanics*, Prentice-Hall, 1977
3. M.L. Peterson, Steel Selection for Offshore Structures, *J. Petrol. Technol.*, Vol 27, 1975, p 274-282
4. H.C. Rhee and M.M. Salama, Application of Fracture Mechanics Method to Offshore Structural Crack Instability Analysis, *J. Ocean Eng. Technol.*, Vol 1, (No. 1), 1987, p 94-103
5. M.G. Dawes and M.S. Kamath, The CTOD Design Curve Approach to Crack Tolerance, in *Proceedings of the Conference on Tolerance of Flaws in Pressurized Components*, Institution of Mechanical and General Technician Engineers, 1978
6. "Guidance on Some Methods for the Derivation of Acceptance Levels for Defects in Fusion Welded Joints," PD6493, British Standards Institution, 1980
7. "Method for Crack Opening Displacement (COD) Testing," BS 5762, British Standards Institution, 1979
8. "API Recommended Practice for Planning, Designing and Constructing Fixed Offshore Platforms," RP 2A, American Petroleum Institute, 1989
9. "Offshore Installations: Guidance on Design and Construction," United Kingdom Department of Energy, 1985

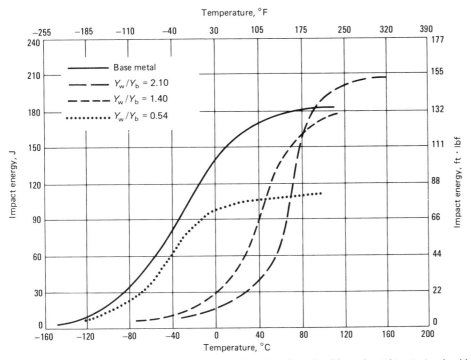

Fig. 11 Plot of impact energy versus temperature to show the effect of weld metal matching (ratio of weld metal yield strength, Y_w, to base plate yield strength, Y_b) on CVN transition curves for the heat-affected zone. Source: Ref 26

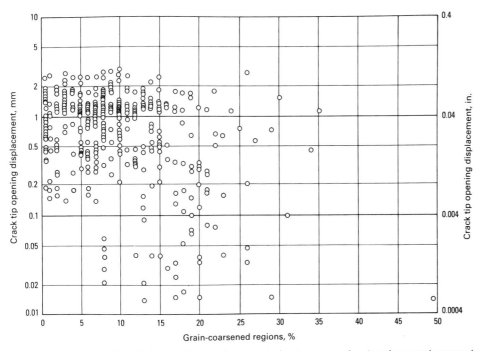

Fig. 12 Crack tip opening displacement versus the percent of grain-coarsened regions for several structural steels. Source: Ref 27

10. "Rules for the Design, Construction and Inspection of Offshore Structures," Det Norske Veritas, 1977
11. S.J. Maddox, Fatigue Crack Propagation Data Obtained from Parent Plate, Weld Metal and HAZ in Structural Steels, *Weld. Res. Int.*, Vol 4 (No. 1.), 1974
12. G.S. Booth and S.J. Dobbs, Corrosion Fatigue Crack Growth in BS 4360 Grade 50D Steel—An Analysis, *Weld. Inst. Res. Bull.* Vol 27 (No. 9), 1986, p 293-297
13. S.J. Maddox, Revision of the Fatigue Clauses in BS PD 6493, in *Proceedings of the International Conference on Weld Failures*, The Welding Institute, 1988, p P47-1 to P47-15
14. "Specification for Carbon Manganese Steel Plates for Offshore Platform Tubular Joints," API 2H, American Petroleum Institute, 1989
15. "Specifications for Steel Plates, Quenched-and-Tempered, for Offshore Structures," API 2Y, American Petroleum Institute, 1989
16. "Specification for Steel Plates for Offshore Structures, Produced by Thermomechanical Control Process (TMCP)," API 2W, American Petroleum Institute, 1989
17. "Specification for Weldable Structural Steels," BS 4360, British Standards Institution, 1979
18. "Recommended Practice for Preproduction Qualification for Steel Plates for Offshore Structures," RP 2Z, American Petroleum Institute, 1987
19. E.F. Walker, Steel Quality, Weldability and Toughness, in *Steel in Marine Structures*, C. Noordhoek and J. de Back, Ed., Elsevier Science Publishers, 1987, p 49-69
20. Y. Nakano, K. Amano, J. Kudo, E. Kobayashi, T. Ogawa, S. Kaihara, and A. Sato, "Preheat and PWHT-Free 150-mm Thick API 2W Grade 60 Steel Plate for Offshore Structures," in *Proceedings of the 7th International Conference on Offshore Mechanics and Arctic Engineering*, M.M. Salama *et al.*, Ed., Vol 3, American Society of Mechanical Engineers, 1988, p 89-101
21. T. Shiwaku, T. Shimohata, S. Takashima, H. Kaji, and K. Masubuchi, YS 420 and 460 MPa Class High Strength Steel Plates for Arctic Offshore Structures Manufactured by TMCP, in *Proceedings of the 7th International Conference on Offshore Mechanics and Arctic Engineering*, M.M. Salama *et al.*, Ed., Vol 3, American Society of Mechanical Engineers, 1988, p 95-101
22. A.C. de Koning, J.D. Harston, K.D. Nayler, and R.K. Ohm, Feeling Free Despite LBZ, in *Proceedings of the 7th International Conference on Offshore Mechanics and Arctic Engineering*, M.M. Salama *et al.*, Ed., Vol 3, American Society of Mechanical Engineers, 1988, p 161-179
23. M. Kurihara, H. Kagawa, and I. Watanbe, Coarse Grain HAZ Toughness Evaluation on Heavy Gauge TMCP Steel Plate By Wide Plate Test, in *Proceedings of the 8th International Conference on Offshore Mechanics and Arctic Engineering*, M.M. Salama *et al.*, Ed., Vol 3, American Society of Mechanical Engineers, 1989, p 649-656
24. S.E. Webster and E.F. Walker, The Significance of Local Brittle Zones to the Integrity of Large Welded Structures, in *Proceedings of the 7th International Conference on Offshore Mechanics and Arctic Engineering*, M.M.

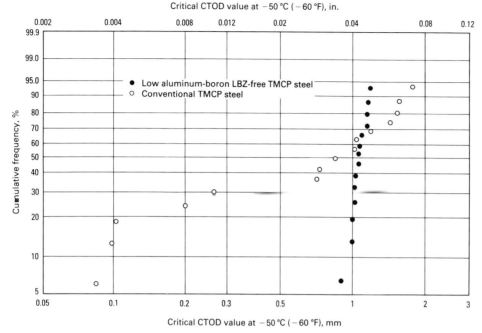

Fig. 13 Heat-affected zone toughness of low aluminum-boron LBZ-free TMCP steel and conventional TMCP steel. Heat input using submerged arc welding is 5.0 kJ/mm (125 kJ/in.). Source: Ref 30

Salama *et al.*, Ed., Vol 3, American Society of Mechanical Engineers, 1988, p 395-403

25. R.M. Denys, Fracture Control and Brittle Zones, A General Appraisal, in *Proceedings of the International Conference on Weld Failures*, The Welding Institute, 1988, p P44-1 to P44-17

26. M. Kocak, L. Chen, and G. Gnirss, Effects of Notch Position and Weld Metal Matching on CTOD of HAZ, in *Proceedings of the International Conference on Weld Failures*, The Welding Institute, 1988, p P7-1 to P7-10

27. D.P. Fairchild, Fracture-Toughness Testing of Weld Heat-Affected Zones in Structural Steel, in *Fatigue and Fracture Testing of Weldments*, STP 1058, H.I. McHenry and J.M. Potter, Ed., American Society for Testing and Materials, 1990

28. K. Hirabayashi, H. Harasawa, H. Kobayashi, T. Sakaurai, M. Hano, and T. Yasuoka, Welding Procedures of Offshore Structure to Achieve Toughness of the Welded Joint, in *Proceedings of the 6th International Conference on Offshore Mechanics and Arctic Engineering*, M.M. Salama *et al.*, Ed., Vol 3, American Society of Mechanical Engineers, 1987, p 151-157

29. K. Ohnishi, S. Suzuki, A. Inami, R. Someya, S. Sugisawa, and J. Furusawa, Advanced TMCP Steel Plates for Offshore Structures, in *Microalloyed HSLA Steels: Proceedings of Microalloying 88*, ASM INTERNATIONAL, 1988, p 215-224

30. S. Suzuki, K. Arimochi, J. Furusawa, K. Bessyo, and R. Someya, Development of LBZ-Free Low Al-B-Treated Steel Plates, in *Proceedings of the 8th International Conference on Offshore Mechanics and Arctic Engineering*, M.M. Salama *et al.*, Ed., American Society of Mechanical Engineers, 1989, p 657-663

Fatigue Resistance of Steels

Bruce Boardman, Deere and Company, Technical Center

FATIGUE is the progressive, localized, and permanent structural change that occurs in a material subjected to repeated or fluctuating strains at nominal stresses that have maximum values less than (and often much less than) the tensile strength of the material. Fatigue may culminate into cracks and cause fracture after a sufficient number of fluctuations. The process of fatigue consists of three stages:

- Initial fatigue damage leading to crack initiation
- Crack propagation to some critical size (a size at which the remaining uncracked cross section of the part becomes too weak to carry the imposed loads)
- Final, sudden fracture of the remaining cross section

Fatigue damage is caused by the simultaneous action of cyclic stress, tensile stress, and plastic strain. If any one of these three is not present, a fatigue crack will not initiate and propagate. The plastic strain resulting from cyclic stress initiates the crack; the tensile stress promotes crack growth (propagation). Careful measurement of strain shows that microscopic plastic strains can be present at low levels of stress where the strain might otherwise appear to be totally elastic. Although compressive stresses will not cause fatigue, compressive loads may result in local tensile stresses.

In the early literature, fatigue fractures were often attributed to crystallization because of their crystalline appearance. Because metals are crystalline solids, the use of the term crystallization in connection with fatigue is incorrect and should be avoided.

Fatigue Resistance

Variations in mechanical properties, composition, microstructure, and macrostructure, along with their subsequent effects on fatigue life, have been studied extensively to aid in the appropriate selection of steel to meet specific end-use requirements. Studies have shown that the fatigue strength of steels is usually proportional to hardness and tensile strength; this generalization is not true, however, for high tensile strength values where toughness and critical flaw size may govern ultimate load carrying ability. Processing, fabrication, heat treatment, surface treatments, finishing, and service environments significantly influence the ultimate behavior of a metal subjected to cyclic stressing.

Predicting the fatigue life of a metal part is complicated because materials are sensitive to small changes in loading conditions and stress concentrations and to other factors. The resistance of a metal structural member to fatigue is also affected by manufacturing procedures such as cold forming, welding, brazing, and plating and by surface conditions such as surface roughness and residual stresses. Fatigue tests performed on small specimens are not sufficient for precisely establishing the fatigue life of a part. These tests are useful for rating the relative resistance of a material and the baseline properties of the material to cyclic stressing. The baseline properties must be combined with the load history of the part in a design analysis before a component life prediction can be made.

In addition to material properties and loads, the design analysis must take into consideration the type of applied loading (uniaxial, bending, or torsional), loading pattern (either periodic loading at a constant or variable amplitude or random loading), magnitude of peak stresses, overall size of the part, fabrication method, surface roughness, presence of fretting or corroded surface, operating temperature and environment, and occurrence of service-induced imperfections.

Traditionally, fatigue life has been expressed as the total number of stress cycles required for a fatigue crack to initiate and grow large enough to produce catastrophic failure, that is, separation into two pieces. In this article, fatigue data are expressed in terms of total life. For the small samples that are used in the laboratory to determine fatigue properties, this is generally the case; but, for real components, crack initiation may be as little as a few percent or the majority of the total component life.

Fatigue data can also be expressed in terms of crack growth rate. In the past, it was commonly assumed that total fatigue life consisted mainly of crack initiation (stage I of fatigue crack development) and that the time required for a minute fatigue crack to grow and produce failure was a minor portion of the total life. However, as better methods of crack detection became available, it was discovered that cracks often develop early in the fatigue life of the material (after as little as 10% of total lifetime) and grow continuously until catastrophic failure occurs. This discovery has led to the use of crack growth rate, critical crack size, and fracture mechanics for the prediction of total life in some applications. Hertzberg's text (Ref 1) is a useful primer for the use of fracture mechanics methods.

Prevention of Fatigue Failure

A thorough understanding of the factors that can cause a component to fail is essential before designing a part. Reference 2 provides numerous examples of these factors that cause fracture (including fatigue) and includes high-quality optical and electron micrographs to help explain factors.

The incidence of fatigue failure can be considerably reduced by careful attention to design details and manufacturing processes. As long as the metal is sound and free from major flaws, a change in material composition is not as effective for achieving satisfactory fatigue life as is care taken in design, fabrication, and maintenance during service. The most effective and economical method of improving fatigue performance is improvement in design to:

- Eliminate or reduce stress raisers by streamlining the part
- Avoid sharp surface tears resulting from punching, stamping, shearing, and so on
- Prevent the development of surface discontinuities or decarburizing during processing or heat treatment
- Reduce or eliminate tensile residual stresses caused by manufacturing, heat treating, and welding
- Improve the details of fabrication and fastening procedures

674 / Service Characteristics of Carbon and Low-Alloy Steels

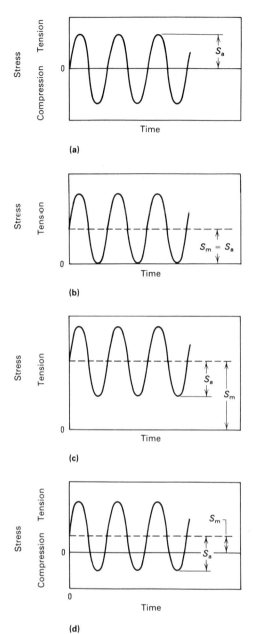

Fig. 1 Types of fatigue test stress. (a) Alternating stress in which $S_m = 0$ and $R = -1$. (b) Pulsating tensile stress in which $S_m = S_a$, the minimum stress is zero, and $R = 0$. (c) Fluctuating tensile stress in which both the minimum and maximum stresses are tensile stresses and $R = \frac{1}{3}$. (d) Fluctuating tensile-to-compressive stress in which the minimum stress is a compressive stress, the maximum stress is a tensile stress, and $R = -\frac{1}{3}$

Control of or protection against corrosion, erosion, chemical attack, or service-induced nicks and other gouges is an important part of proper maintenance of fatigue life during active service life. Reference 3 contains numerous papers pertaining to these subjects.

Symbols and Definitions

In most laboratory fatigue testing, the specimen is loaded so that stress is cycled

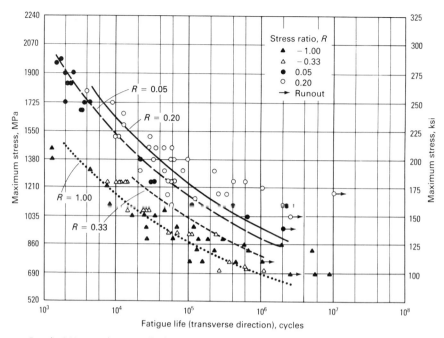

Fig. 2 Best-fit S-N curves for unnotched 300M alloy forging with an ultimate tensile strength of 1930 MPa (280 ksi). Stresses are based on net section. Testing was performed in the transverse direction with a theoretical stress concentration factor, K_t, of 1.0. Source: Ref 4

either between a maximum and a minimum tensile stress or between a maximum tensile stress and a specified level of compressive stress. The latter of the two, considered a negative tensile stress, is given an algebraic minus sign and called the minimum stress.

Applied Stresses. The mean stress, S_m, is the algebraic average of the maximum stress and the minimum stress in one cycle:

$$S_m = \frac{(S_{max} + S_{min})}{2} \quad \text{(Eq 1)}$$

The range of stress, S_r, is the algebraic difference between the maximum stress and the minimum stress in one cycle:

$$S_r = S_{max} - S_{min} \quad \text{(Eq 2)}$$

The stress amplitude, S_a, is one-half the range of stress:

$$S_a = \frac{S_r}{2} = \frac{(S_{max} - S_{min})}{2} \quad \text{(Eq 3)}$$

During a fatigue test, the stress cycle is usually maintained constant so that the applied stress conditions can be written $S_m \pm S_a$, where S_m is the static or mean stress and S_a is the alternating stress equal to one-half the stress range. The positive sign is used to denote a tensile stress, and the negative sign denotes a compressive stress. Some of the possible combinations of S_m and S_a are shown in Fig. 1. When $S_m = 0$ (Fig. 1a), the maximum tensile stress is equal to the maximum compressive stress; this is called an alternating stress, or a completely reversed stress. When $S_m = S_a$ (Fig. 1b), the minimum stress of the cycle is zero; this is called a pulsating, or repeated, tensile stress. Any other combination is known as an alternating stress, which may be an alternating tensile stress (Fig. 1c), an alternating compressive stress, or a stress that alternates between a tensile and a compressive value (Fig. 1d).

Nominal axial stresses can be calculated on the net section of a part (S = force per unit area) without consideration of variations in stress conditions caused by holes, grooves, fillets, and so on. Nominal stresses are frequently used in these calculations, although a closer estimate of actual stresses through the use of a stress concentration factor might be preferred.

Stress ratio is the algebraic ratio of two specified stress values in a stress cycle. Two commonly used stress ratios are A, the ratio of the alternating stress amplitude to the mean stress ($A = S_a/S_m$) and R, the ratio of the minimum stress to the maximum stress ($R = S_{min}/S_{max}$). The five conditions that R can take range from $+1$ to -1:

- Stresses are fully reversed: $R = -1$
- Stresses are partially reversed: R is between -1 and zero
- Stress is cycled between a maximum stress and no load: The stress ratio R becomes zero
- Stress is cycled between two tensile stresses: The stress ratio R becomes a positive number less than 1
- An R stress ratio of 1 indicates no variation in stress, and the test becomes a sustained-load creep test rather than a fatigue test

S-N Curves. The results of fatigue tests are usually plotted as the maximum stress or stress amplitude versus the number of cycles, N, to fracture, using a logarithmic scale for the number of cycles. Stress may be plotted on either a linear or a logarithmic scale. The resulting curve of data points is called an *S-N* curve. A family of *S-N* curves for a material tested at various stress ratios is shown in Fig. 2. It should be noted that the fully reversed condition, $R = -1$, is the most severe, with the least fatigue life. For carbon and low-alloy steels, *S-N* curves (plotted as linear stress versus log life) typically have a fairly straight slanting portion with a negative slope at low cycles, which changes with a sharp transition into a straight, horizontal line at higher cycles.

An *S-N* curve usually represents the median, or B_{50}, life, which represents the number of cycles when half the specimens fail at a given stress level. The scatter of fatigue lives covers a very wide range and can occur for many reasons other than material variability.

A constant-lifetime diagram (Fig. 3) is a summary graph prepared from a group of *S-N* curves of a material; each *S-N* curve is obtained at a different stress ratio. The diagram shows the relationship between the alternating stress amplitude and the mean stress and the relationship between maximum stress and minimum stress of the stress cycle for various constant lifetimes. Although this technique has received considerable use, it is now out of date. Earlier editions of the *Military Standardization Handbook* (Ref 5) used constant lifetime diagrams extensively, but more recent editions (Ref 4) no longer include them.

Fatigue limit (or endurance limit) is the value of the stress below which a material can presumably endure an infinite number of stress cycles, that is, the stress at which the *S-N* diagram becomes and appears to remain horizontal. The existence of a fatigue limit is typical for carbon and low-alloy steels. For many variable-amplitude loading conditions this is true; but for conditions involving periodic overstrains, as is typical for many actual components, large changes in the long-life fatigue resistance can occur (see the discussion in the section "Comparison of Fatigue Testing Techniques" in this article).

Fatigue strength, which should not be confused with fatigue limit, is the stress to which the material can be subjected for a specified number of cycles. The term fatigue strength is used for materials, such as most nonferrous metals, that do not exhibit well-defined fatigue limits. It is also used to describe the fatigue behavior of carbon and low-alloy steels at stresses greater than the fatigue limit.

Stress Concentration Factor. Concentrated stress in a metal is evidenced by surface discontinuities such as notches, holes, and scratches and by changes in microstructure

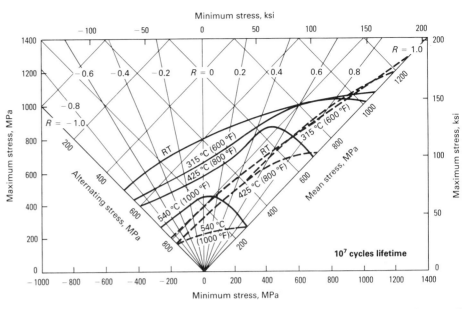

Fig. 3 Constant-lifetime fatigue diagram for AISI-SAE 4340 alloy steel bars, hardened and tempered to a tensile strength of 1035 MPa (150 ksi) and tested at various temperatures. Solid lines represent data obtained from unnotched specimens; dashed lines represent data from specimens having notches with $K_t = 3.3$. All lines represent lifetimes of ten million cycles. Source: Ref 5

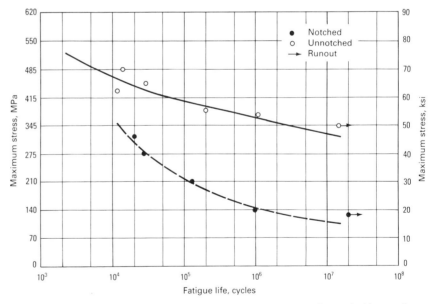

Fig. 4 Room temperature *S-N* curves for notched and unnotched AISI 4340 alloy steel with a tensile strength of 860 MPa (125 ksi). Stress ratio, R, equals -1.0. Source: Ref 4

such as inclusions and thermal heat affected zones. The theoretical stress concentration factor, K_t, is the ratio of the greatest elastically calculated stress in the region of the notch (or other stress concentrator) to the corresponding nominal stress. For the determination of K_t, the greatest stress in the region of the notch is calculated from the theory of elasticity or by finite-element analysis. Equivalent values may be derived experimentally. An experimental stress concentration factor is a ratio of stress in a notched specimen to the stress in a smooth (unnotched) specimen.

Fatigue notch factor, K_f, is the ratio of the fatigue strength of a smooth (unnotched) specimen to the fatigue strength of a notched specimen at the same number of cycles. The fatigue notch factor will vary with the life on the *S-N* curve and with the mean stress. At high stress levels and short cycles, the factor is usually less than at lower stress levels and longer cycles because of a reduction of the notch effect by plastic deformation.

Fatigue notch sensitivity, q, is determined by comparing the fatigue notch factor, K_f, and the theoretical stress concentration fac-

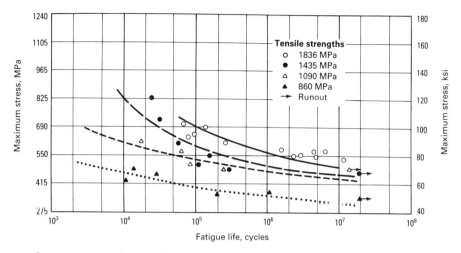

Fig. 5 Room temperature S-N curves for AISI 4340 alloy steel with various ultimate tensile strengths and with $R = -1.0$. Source: Ref 4

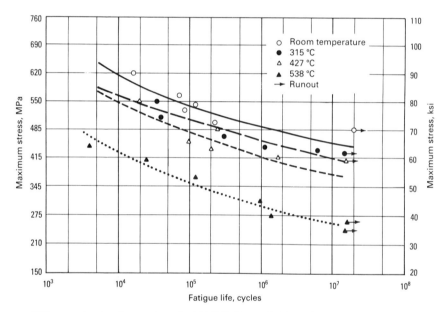

Fig. 6 S-N curves at various temperatures for AISI 4340 alloy steel with an ultimate tensile strength of 1090 MPa (158 ksi). Stress ratio, R, equals -1.0. Source: Ref 4

tor, K_t, for a specimen of a given size containing a stress concentrator of a given shape and size. A common definition of fatigue notch sensitivity is:

$$q = \frac{K_f - 1}{K_t - 1} \quad \text{(Eq 4)}$$

in which q may vary between 0 (where $K_f = 1$, no effect) and 1 (where $K_f = K_t$, full effect). This value may be stated as a percentage. As the fatigue notch factor varies with the position on the S-N curve, so does notch sensitivity. Most metals tend to become more notch sensitive at low stresses and long cycles. If they do not, it may be that the fatigue strengths for the smooth (unnotched) specimens are lower than they could be because of surface imperfections. Most metals are not fully notch sensitive under high stresses and a low number of cycles. Under these conditions, the actual peak stress at the base of the notch is partly in the plastic strain condition. This results in the actual peak stress being lower than the theoretical peak elastic stress used in the calculation of the theoretical stress concentration factor.

Stress-Based Approach To Fatigue

The design of a machine element that will be subjected to cyclic loading can be approached by adjusting the configuration of the part so that the calculated stresses fall safely below the required line on an S-N plot. In a stress-based analysis, the material is assumed to deform in a nominally elastic manner, and local plastic strains are neglected. To the extent that these approximations are valid, the stress-based approach is useful. These assumptions imply that all the stresses will essentially be elastic.

The S-N plot shown in Fig. 4 presents data for AISI-SAE 4340 steel, heat treated to a tensile strength of 1035 MPa (150 ksi) in the notched and unnotched condition. Figure 5 shows the combinations of cyclic stresses that can be tolerated by the same steel when the specimens are heat treated to different tensile strengths ranging from 860 to 1790 MPa (125 to 260 ksi).

The effect of elevated temperature on the fatigue behavior of 4340 steel heat treated to 1035 MPa (150 ksi) is shown in Fig. 6. An increase in temperature reduces the fatigue strength of the steel and is most deleterious for those applications in which the stress ratio, R, lies between 0.4 and 1.0 (Fig. 3). A decrease in temperature may increase the fatigue limit of steel; however, parts with preexisting cracks may also show decreased total life as temperature is lowered, because of accompanying reductions in critical crack size and fracture toughness.

Figure 7 shows the effect of notches on the fatigue behavior of the ultrahigh-strength 300M steel. A K_t value of 2 is obtained in a specimen having a notch radius of about 1 mm (0.040 in.). For small parts, such a radius is often considered large enough to negate the stress concentration associated with any change in section. The significant effect of notches, even those with low stress concentration factors, on the fatigue behavior of this steel is apparent.

Data such as those presented in Fig. 3 to 7 may not be directly applicable to the design of structures because these graphs do not take into account the effect of the specific stress concentration associated with reentrant corners, notches, holes, joints, rough surfaces, and other similar conditions present in fabricated parts. The localized high stresses induced in fabricated parts by stress raisers are of much greater importance for cyclic loading than for static loading. Stress raisers reduce the fatigue life significantly below those predicted by the direct comparison of the smooth specimen fatigue strength with the nominal calculated stresses for the parts in question. Fabricated parts in simulated service have been found to fail at less than 50 000 repetitions of load, even though the nominal stress was far below that which could be repeated many millions of times on a smooth, machined specimen.

Correction Factors for Test Data. The available fatigue data normally are for a specific type of loading, specimen size, and surface roughness. For instance, the R.R. Moore rotating-beam fatigue test machine uses a 7.5 mm (0.3 in.) diam specimen that is free of any stress concentrations (because of specimen shape and a surface that has been polished to a mirror finish), and that is subjected to completely reversed bending stresses. For the fatigue

Fatigue Resistance of Steels / 677

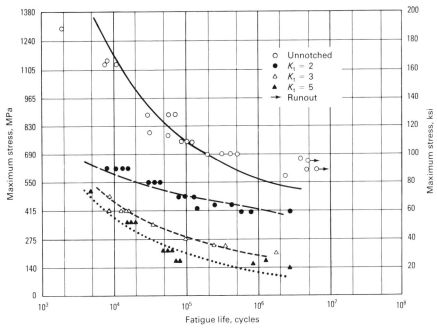

Fig. 7 Room-temperature S-N curves for a 300M steel with an ultimate tensile strength of 2000 MPa (290 ksi) having various notch severities. Stress ratio, R, equals 1.0. Source: Ref 4

Table 1 Correction factors for surface roughness (K_s), type of loading (K_l), and part diameter (K_d), for fatigue life of steel parts

Factor	Value for loading in Bending	Torsion	Tension
K_l	1.0	0.58	0.9(a)
K_d			
where $d \leq 10$ mm (0.4 in.)	1.0	1.0	1.0
where 10 mm (0.4 in.) $< d \leq 50$ mm (2 in.)	0.9	0.9	1.0
K_s	See Fig. 8.		

(a) A lower value (0.06 to 0.85) may be used to take into account known or suspected undetermined bending because of load eccentricity. Source: Ref 6

limits used in design calculations, Juvinall (Ref 6) suggests the correction of fatigue life data by multiplying the fatigue limit from testing, N_i, by three factors that take into account the variation in the type of loading, part diameter, and surface roughness:

$$\text{Design fatigue limit} = K_l \cdot K_d \cdot K_s \cdot N_i \quad (\text{Eq 5})$$

where K_l is the correction factor for the type of loading, K_d for the part diameter, and K_s for the surface roughness. Values of these factors are given in Table 1 and Fig. 8.

Strain-Based Approach To Fatigue

A strain-based approach to fatigue, developed for the analysis of low-cycle fatigue data, has proved to be useful for analyzing long-life fatigue data as well. The approach can take into account both elastic and plastic responses to applied loadings. The data are presented on a log-log plot similar in shape to an S-N curve; the value plotted on the abscissa is the number of strain reversals (twice the number of cycles) to failure, and the ordinate is the strain amplitude (half the strain range).

During cyclic loading, the stress-strain relationship can usually be described by a loop, such as that shown in Fig. 9. For purely elastic loading, the loop becomes a straight line whose slope is the elastic modulus, E, of the material. The occurrence of a hysteresis loop is most common. The definitions of the plastic strain range, $\Delta\epsilon_p$, the elastic strain range, $\Delta\epsilon_e$, the total strain range, $\Delta\epsilon_t$, and the stress range, $\Delta\sigma$, are indicated in Fig. 9. A series of fatigue tests, each having a different total strain range, will generate a series of hysteresis loops. For each set of conditions, a characteristic number of strain reversals is necessary to cause failure.

As shown in Fig. 10, a plot on logarithmic coordinates of the plastic portion of the strain amplitude (half the plastic strain range) versus the fatigue life often yields a straight line, described by the equation:

$$\frac{\Delta\epsilon_p}{2} = \epsilon'_f (2N_f)^c \quad (\text{Eq 6})$$

where ϵ'_f is the fatigue ductility coefficient, c is the fatigue ductility exponent, and N_f is the number of cycles to failure.

Because the conditions under which elastic strains have the greatest impact on fatigue behavior are the long-life conditions where stress-based analysis of fatigue is appropriate, the effects of elastic strain on fatigue are charted by plotting stress amplitude (half the stress range) versus fatigue life on logarithmic coordinates. As shown in Fig. 11, the result is a straight line having the equation:

$$\frac{\Delta\sigma}{2} = \sigma'_f (2N_f)^b \quad (\text{Eq 7})$$

where σ'_f is the fatigue strength coefficient and b is the fatigue strength exponent.

The elastic strain range is obtained by dividing Eq 7 by E:

$$\frac{\Delta\epsilon_e}{2} = \frac{\sigma'_f}{E} (2N_f)^b \quad (\text{Eq 8})$$

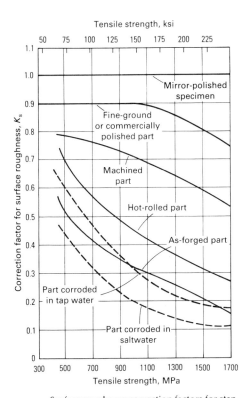

Fig. 8 Surface roughness correction factors for standard rotating-beam fatigue life testing of steel parts. See Table 1 for correction factors from part diameter and type of loading. Source: Ref 6

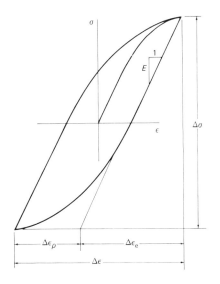

Fig. 9 Stress-strain hysteresis loop. Source: Ref 7

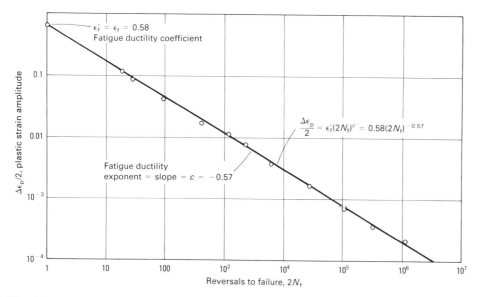

Fig. 10 Ductility versus fatigue life for annealed AISI-SAE 4340 steel. Source: Ref 8

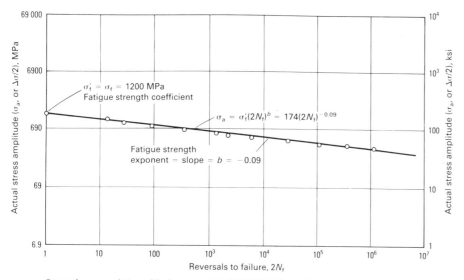

Fig. 11 Strength versus fatigue life for annealed AISI-SAE 4340 steel. The equation for the actual stress amplitude, σ_a, is shown in ksi units. Source: Ref 8

The total strain range is the sum of the elastic and plastic components, obtained by adding Eq 6 and 8 (see Fig. 12):

$$\frac{\Delta\epsilon}{2} = \epsilon_f'(2N_f)^c + \frac{\sigma_f'}{E}(2N_f)^b \quad (Eq\ 9)$$

For low-cycle fatigue conditions (frequently fewer than about 1000 cycles to failure), the first term of Eq 9 is much larger than the second; thus, analysis and design under such conditions must use the strain-based approach. For long-life fatigue conditions (frequently more than about 10 000 cycles to failure), the second term dominates, and the fatigue behavior is adequately described by Eq 7. Thus, it becomes possible to use Eq 7 in stress-based analysis and design.

Figure 13 shows the fatigue life behavior of two high-strength plate steels for which extensive fatigue data exist. ASTM A 440 has a yield strength of about 345 MPa (50 ksi); the other steel is a proprietary grade hardened and tempered to a yield strength of about 750 MPa (110 ksi). Under long-life fatigue conditions, the higher-strength steel can accommodate higher strain amplitudes for any specified number of cycles; such strains are elastic. Thus, stress and strain are proportional, and it is apparent that the higher-strength steel has a higher fatigue limit. With low-cycle fatigue conditions, however, the more ductile lower-strength steel can accommodate higher strain amplitudes. For low-cycle fatigue conditions (in which the yield strength of the material is exceeded on every cycle), the lower-strength steel can accommodate more strain reversals before failure for a specified strain amplitude. For strain amplitudes of 0.003 to 0.01, the two steels have the same fatigue life, 10^4 to 10^5 cycles. For this particular strain amplitude, most steels have the same fatigue life, regardless of their strength levels. Heat treating a steel to different hardness levels does not appreciably change the fatigue life for this strain amplitude (Fig. 14).

Fuchs and Stephens's text (Ref 9), *Proceedings of the SAE Fatigue Conference* (Ref 10), and the recently published update to the SAE *Fatigue Design Handbook* (Ref 11) provide much additional detail on the use of state-of-the-art fatigue analysis methods. In fact, the chapter outline for the latter work, shown in Fig. 15, provides an excellent checklist of factors to include in a fatigue analysis.

Metallurgical Variables of Fatigue Behavior

The metallurgical variables having the most pronounced effects on the fatigue behavior of carbon and low-alloy steels are strength level, ductility, cleanliness of the steel, residual stresses, surface conditions, and aggressive environments. At least partly because of the characteristic scatter of fatigue testing data, it is difficult to distinguish the direct effects of other variables such as composition on fatigue from their effects on the strength level of steel. Reference 3 addresses some excellent research in the area of microstructure and its effect on fatigue.

Strength Level. For most steels with hardnesses below 400 HB (not including precipitation hardening steels), the fatigue limit is about half the ultimate tensile strength. Thus, any heat treatment or alloying addition that increases the strength (or hardness) of a steel can be expected to increase its fatigue limit as shown in Fig. 5 for a low-alloy steel (AISI 4340) and in Fig. 16 for various other low-alloy steels as a function of hardness. However, as shown in Fig. 14 for medium-carbon steel, a higher hardness (or strength) may not be associated with improved fatigue behavior in a low-cycle regime ($<10^3$ cycles) because ductility may be a more important factor.

Ductility is generally important to fatigue life only under low-cycle fatigue conditions. Exceptions to this include spectrum loading where there is an occasional overload with millions of smaller cycles, or extremely brittle materials where crack propagation dominates. The fatigue-ductility coefficient, ϵ_f', can be estimated from the reduction in area occurring in a tension test.

Cleanliness of a steel refers to its relative freedom from nonmetallic inclusions. These inclusions generally have a deleterious effect on the fatigue behavior of steels, particularly for long-life applications. The type, number, size, and distribution of nonmetallic inclusions may have a greater effect on the fatigue life of carbon and alloy steel than will differences in composition, microstructure, or stress gradients. Nonmetallic inclu-

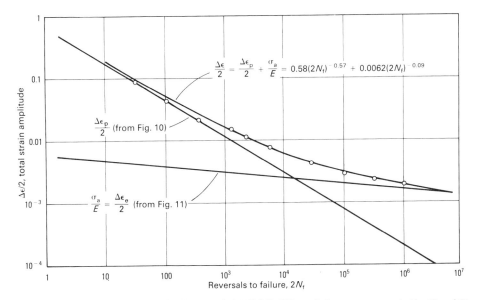

Fig. 12 Total strain versus fatigue life for annealed AISI-SAE 4340 steel. Data are same as in Fig. 10 and 11. Source: Ref 8

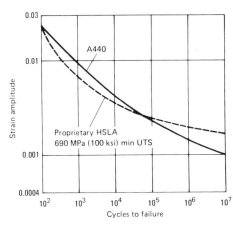

Fig. 13 Total strain versus fatigue life for two high-strength low-alloy (HSLA) steels. Steels are ASTM A 440 having a yield strength of about 345 MPa (50 ksi) and a proprietary quenched and tempered HSLA steel having a yield strength of about 750 MPa (110 ksi). Source: Ref 7

sions, however, are rarely the prime cause of the fatigue failure of production parts; if the design fatigue properties were determined using specimens containing inclusions representative of those in the parts, any effects of these inclusions would already be incorporated in the test results. Great care must be used when rating the cleanliness of a steel based on metallographic examination to ensure that the limited sample size (volume rated) is representative of the critical area in the final component.

Points on the lower curve in Fig. 17 represent the cycles to failure for a few specimens from one bar selected from a lot consisting of several bars of 4340H steel. Large spherical inclusions, about 0.13 mm (0.005 in.) in diameter, were observed in the fracture surfaces of these specimens. The inclusions were identified as silicate particles. No spherical inclusions larger than 0.02 mm (0.00075 in.) were detected in the other specimens.

Large nonmetallic inclusions can often be detected by nondestructive inspection; steels can be selected on the basis of such inspection. Vacuum melting, which reduces the number and size of nonmetallic inclusions, increases the fatigue limit of 4340 steel, as can be seen in Table 2. Improvement in fatigue limit is especially evident in the transverse direction.

Surface conditions of a metal part, particularly surface imperfections and roughness, can reduce the fatigue limit of the part. This effect is most apparent for high-strength steels. The interrelationship between surface roughness, method of producing the surface finish, strength level, and fatigue limit is shown in Fig. 8, in which the ordinate represents the fraction of fatigue limit relative to a polished test specimen that could be anticipated for the combination of strength level and surface finish.

Fretting is a wear phenomenon that occurs between two mating surfaces. It is adhesive in nature, and vibration is its essential causative factor. Usually, fretting is accompanied by oxidation. Fretting usually occurs between two tight-fitting surfaces that are subjected to a cyclic, relative motion of extremely small amplitude. Fretted regions are highly sensitive to fatigue cracking. Under fretting conditions, fatigue cracks are initiated at very low stresses, well below the fatigue limit of nonfretted specimens.

Decarburization is the depletion of carbon from the surface of a steel part. As indicated in Fig. 18, it significantly reduces the fatigue limits of steel. Decarburization of from 0.08 to 0.75 mm (0.003 to 0.030 in.)

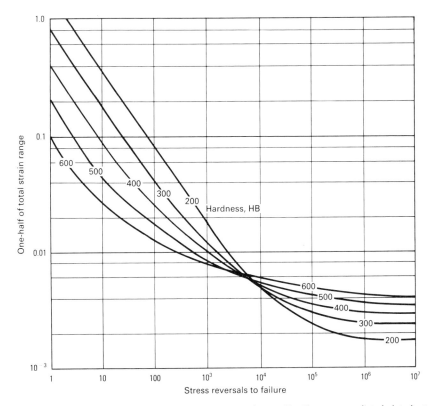

Fig. 14 Effect of hardness level on plot of total strain versus fatigue life. These are predicted plots for typical medium-carbon steel at the indicated hardness levels. The prediction methodology is described under the heading "Notches" in this article.

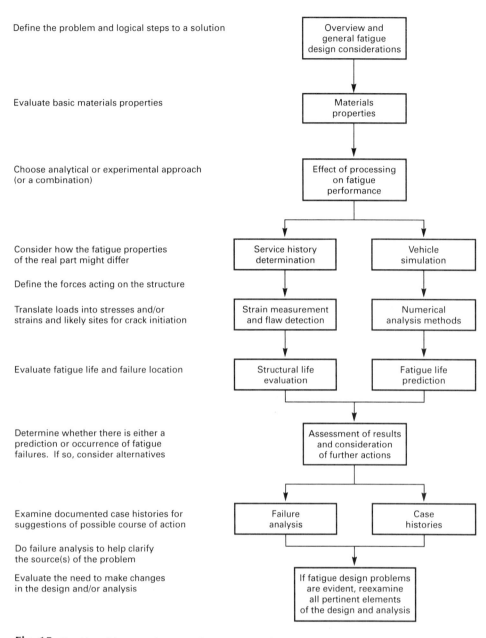

Fig. 15 Checklist of factors in fatigue analysis. Source: Ref 11

on AISI-SAE 4340 notched specimens that have been heat treated to a strength level of 1860 MPa (270 ksi) reduces the fatigue limit almost as much as a notch with $K_t = 3$.

When subjected to the same heat treatment as the core of the part, the decarburized surface layer is weaker and therefore less resistant to fatigue than the core. Hardening a part with a decarburized surface can also introduce residual tensile stresses, which reduce the fatigue limit of the material. Results of research studies have indicated that fatigue properties lost through decarburization can be at least partially regained by recarburization (carbon restoration in the surfaces).

Residual Stresses. The fatigue properties of a metal are significantly affected by the residual stresses in the metal. Compressive residual stresses at the surface of a part can improve its fatigue life; tensile residual stresses at the surface reduce fatigue life. Beneficial compressive residual stresses may be produced by surface alloying, surface hardening, mechanical (cold) working of the surface, or by a combination of these processes. In addition to introducing compressive residual stresses, each of these processes strengthens the surface layer of the material. Because most real components also receive significant bending and/or torsional loads, where the stress is highest at the surface, compressive surface stresses can provide significant benefit to fatigue.

Surface Alloying. Carburizing, carbonitriding, and nitriding are three processes for surface alloying. The techniques required to achieve these types of surface alloying are discussed in Volume 2 of the 8th Edition and Volume 4 of the 9th Edition of *Metals Handbook*. In these processes, carbon, nitrogen, or both elements are introduced into the surface layer of the steel part. The solute atoms strengthen the surface layer of the steel and increase its bulk relative to the metal below the surface. The case and core of a carburized steel part respond differently to the same heat treatment; because of its higher carbon content, the case is harder after quenching and harder after tempering. To achieve maximum effectiveness of surface alloying, the surface layer must be much thinner than the thickness of the part to maximize the effect of the residual stresses; however, the surface layer must be thick enough to prevent operating stresses from affecting the material just below the surface layer. Figure 19 shows the improvement in fatigue limit that can be achieved by nitriding. A particular advantage of surface alloying in the resistance to fatigue is that the alloyed layer closely follows the contours of the part.

Surface Hardening. Induction, flame, laser, and electron beam hardening selectively harden the surface of a steel part; the steel must contain sufficient carbon to permit hardening. In each operation, the surface of the part is rapidly heated, and the part is quenched either by externally applied quenchant or by internal mass effect. This treatment forms a surface layer of martensite that is bulkier than the steel beneath it. Further information on these processes may be found in Volume 2 of the 8th Edition and Volume 4 of the 9th Edition of *Metals Handbook*. Induction, flame, laser, and electron beam hardening can produce beneficial surface residual stresses that are compressive; by comparison, surface residual stresses resulting from through hardening are often tensile. Figure 20 compares the fatigue life of through-hardened, carburized, and induction-hardened transmission shafts.

Figure 21 shows the importance of the proper case depth on fatigue life; the hardened case must be deep enough to prevent operating stresses from affecting the steel beneath the case. However, it should be thin enough to maximize the effectiveness of the residual stresses. Three advantages of induction, flame, laser, or electron beam hardening in the resistance of fatigue are:

- The core may be heat treated to any appropriate condition
- The processes produce relatively little distortion
- The part may be machined before heat treatment

Mechanical working of the surface of a steel part effectively increases the resistance to fatigue. Shot peening and skin rolling are two methods for developing com-

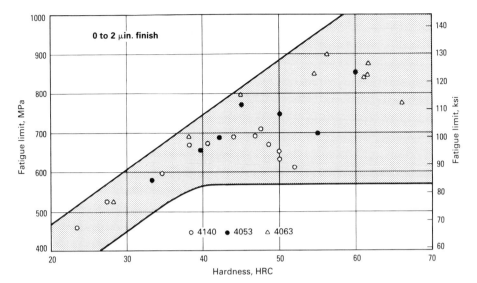

Fig. 16 Effect of carbon content and hardness on fatigue limit of through-hardened and tempered 4140, 4053, and 4063 steels. See the sections "Composition" and "Scatter of Data" in this article for additional discussions.

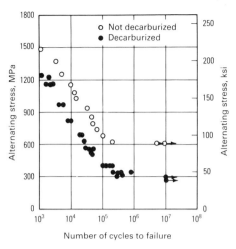

Fig. 18 Effect of decarburization on the fatigue behavior of a steel

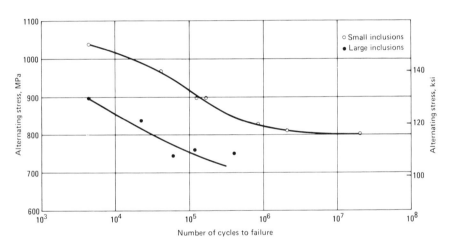

Fig. 17 Effect of nonmetallic inclusion size on fatigue. Steels were two lots of AISI-SAE 4340H; one lot (lower curve) contained abnormally large inclusions; the other lot (upper curve) contained small inclusions.

pressive residual stresses at the surface of the part. The improvement in fatigue life of a crankshaft that results from shot peening is shown in Fig. 19. Shot peening is useful in recovering the fatigue resistance lost through decarburization of the surface. Decarburized specimens similar to those described in Fig. 18 were shot peened, raising the fatigue limit from 275 MPa (40 ksi) after decarburizing to 655 MPa (95 ksi) after shot peening.

Tensile residual stresses at the surface of a steel part can severely reduce its fatigue limit. Such residual stresses can be produced by through hardening, cold drawing, welding, or abusive grinding. For applications involving cyclic loading, parts containing these residual stresses should be given a stress relief anneal if feasible.

Aggressive environments can substantially reduce the fatigue life of steels. In the absence of the medium causing corrosion, a previously corroded surface can substantially reduce the fatigue life of the steel, as shown in Fig. 8. Additional information on corrosion fatigue is contained in Volumes 8 and 13 of the 9th Edition of *Metals Handbook*.

Grain size of steel influences fatigue behavior indirectly through its effect on the strength and fracture toughness of the steel. Fine-grained steels have greater fatigue strength than do coarse-grained steels.

Composition. An increase in carbon content can increase the fatigue limit of steels, particularly when the steels are hardened to 45 HRC or higher (Fig. 16). Other alloying elements may be required to attain the desired hardenability, but they generally have little effect on fatigue behavior.

Microstructure. For specimens having comparable strength levels, resistance to fatigue depends somewhat on microstructure. A tempered martensite structure provides the highest fatigue limit. However, if the structure as-quenched is not fully martensitic, the fatigue limit will be lower (Fig. 22). Pearlitic structures, particularly those with coarse pearlite, have poor resistance to fatigue. *S-N* curves for pearlitic and spheroidized structures in a eutectoid steel are shown in Fig. 23.

Macrostructure differences typical of those seen when comparing ingot cast to continuously cast steels can have an effect on fatigue performance. While there is no inherent difference between these two types of steel after rolling to a similar reduction in area from the cast ingot, bloom, or billet, ingot cast steels will typically receive much larger reductions in area (with subsequent refinement of grain size and inclusions) than will continuously cast billets when rolled to a constant size. Therefore, the billet size of continuously cast steels becomes important to fatigue, at least as it relates to the size of the material from which the part was fabricated.

A significant amount of research has shown that for typical structural applications, strand cast reduction ratios should be above 3:1 or 5:1, although many designers of critical forgings still insist on reduction ratios greater than 10:1 or 15:1. These larger reduction ratio requirements will frequently preclude the use of continuously cast steels because the required caster size would be larger than existing equipment. While this may not be a major problem at this time, steel trends suggest that there will be very little domestic and almost no off-shore ingot cast material available at any cost within the next two decades. The problem will be reduced as larger and larger casters, approaching bloom and ingot sizes, are installed.

Creep-Fatigue Interaction. At temperatures sufficiently elevated to produce creep, creep-fatigue interaction can be a factor affecting fatigue resistance. Information on creep-fatigue interaction is contained in the article "Elevated-Temperature Properties of Ferritic Steels" in this Volume.

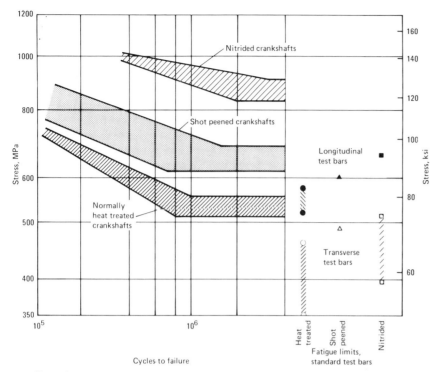

Fig. 19 Effect of nitriding and shot peening on fatigue behavior. Comparison between fatigue limits of crankshafts (S-N bands) and fatigue limits of separate test bars, which are indicated by plotted points at right. Steel was 4340.

Table 2 Improvement in the fatigue limits of SAE 4340 steel with the reduction of nonmetallic inclusions by vacuum melting compared to electric furnace melting

	Longitudinal fatigue limit(a)		Transverse fatigue limit(a)		Ratio of transverse to longitudinal	Hardness, HRC
	MPa	ksi	MPa	ksi		
Electric furnace melted	800	116	545	79	0.68	27
Vacuum melted	960	139	825	120	0.86	29

(a) Determined in repeated bending fatigue test ($R = 0$). Source: Ref 12

Application of Fatigue Data

The application of fatigue data in engineering design is complicated by the characteristic scatter of fatigue data; variations in surface conditions of actual parts; variations in manufacturing processes such as bending, forming, and welding; and the uncertainty of environmental and loading conditions in service. In spite of the scatter of fatigue data, it is possible to estimate service life under cyclic loading. It is essential to view such estimates for what they are, that is, estimates of the mean or average performance, and to recognize that there may be large discrepancies between the estimated and actual service lives.

Scatter of Data. Fatigue testing of test specimens and actual machine components produces a wide scatter of experimental results (see Fig. 25 and Ref 10 for examples). The data in Fig. 25 represent fatigue life simulated-service testing of 25 lots of 12 torsion bars each. In this program, the coefficient of variation, C_N, defined as the ratio of the standard deviation of the mean value, of fatigue life was 0.28. In Table 3, the range of values of the coefficient of variation for fatigue strength is compared with those for other mechanical properties.

For specimens tested near the fatigue limit, the probable range of fatigue life becomes so large that it is pointless to compute a coefficient of variation for fatigue life. Instead, values of C_N are calculated for the fatigue limit. Approximately 1000 fatigue specimens were made from a single heat of aircraft quality 4340 steel; all were taken parallel to the fiber axis of the steel. The specimens were heat treated to three different strength levels and polished to a surface roughness of 0 to 0.050 μm (0 to 2 μin.). Fatigue limits for these specimens are given in terms of the percent surviving 10 million cycles (Fig. 26). It should be noted that the scatter increases as the strength level is increased; a similar trend is shown in Fig. 16.

The orientation of cyclic stress relative to the fiber axis or rolling direction of a steel can affect the fatigue limit of the steel. Figure 24 shows the difference between the fatigue limit of specimens taken parallel to the rolling direction and those taken transverse to it. Any nonmetallic inclusions present will be elongated in the rolling direction and will reduce fatigue life in the transverse direction. The use of vacuum melting to reduce the number and size of nonmetallic inclusions therefore can have a beneficial effect on transverse fatigue resistance (Table 2).

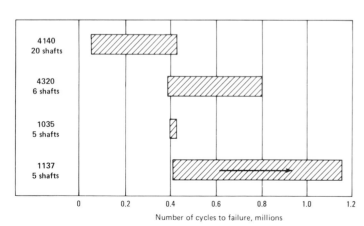

Steel	Surface hardness, HRC	Hardening process
4140	36–42	Through hardened
4320	40–46	Carburized to 1.0–1.3 m (0.040–0.050 in.)
1035	42–48	Induction hardened to 3 mm (0.120 in.) min effective depth (40 HRC)
1137	42–48	Induction hardened to 3 mm (0.120 in.) min effective depth (40 HRC)

Fig. 20 Effect of carburizing and surface hardening on fatigue life. Comparison of carburized, through-hardened, and induction-hardened transmission shafts tested in torsion. Arrow in lower bar on chart indicates that one shaft had not failed after the test was stopped at the number of cycles shown.

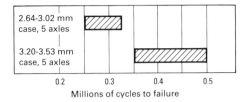

Fig. 21 Effect of case depth on fatigue life. Fatigue tests on induction-hardened 1038 steel automobile axle shafts 32 mm (1¼ in.) in diameter. Case depth ranges given on the chart are depths to 40 HRC. Shafts with lower fatigue life had a total case depth to 20 HRC of 4.5 to 5.2 mm (0.176 to 0.206 in.), and shafts with higher fatigue life, 6.4 to 7.0 mm (0.253 to 0.274 in.). Load in torsion fatigue was 2030 N · m (1500 ft · lbf), and surface hardness was 58 to 60 HRC after hardening.

Variation from heat to heat with the same steel is greater than variation within a single heat. Figure 27 shows the variations in fatigue limit among five heats of 8740 steel; all specimens were hardened and tempered to 39 HRC. Specimens taken from heat E were given a variety of heat treatments, all of which resulted in a hardness of 39 HRC. The variations in fatigue limit resulting from these heat treatments are also shown in Fig. 27.

Additional scatter of fatigue data is likely to result from variations in case depth, surface finish, dimensions of the part or specimen, or environmental or residual stresses. Axial load tests for fatigue properties are considered more conservative than rotating bending tests but have the advantage of obtaining information on fatigue properties at various mean stresses.

Estimating Fatigue Parameters. In the strain-based approach to fatigue, five parameters (σ_f', b, ϵ_f', c, and E) are used to describe fatigue behavior. These parameters can be determined experimentally; typical values (which should not be considered averages or minimums) obtained for several materials are given in Table 4. In the absence of experimentally determined values, these parameters have been estimated from uniaxial tension test results. The use of these parameters (either experimentally determined or estimated values) to predict fatigue behavior only approximates actual behavior and should never be substituted for full-scale testing of actual parts under service conditions.

Table 3 Coefficients of variation for mechanical properties

Mechanical property	Coefficient of variation, C_n(a)
Elastic modulus	0.03
Ultimate tensile strength	0.05
Brinell hardness	0.05
Tensile yield strength	0.07
Fracture toughness	0.07
Fatigue strength	0.08 to 1.0

(a) Coefficient of variation, C_n, is the standard deviation divided by the mean value. Source: Ref 12

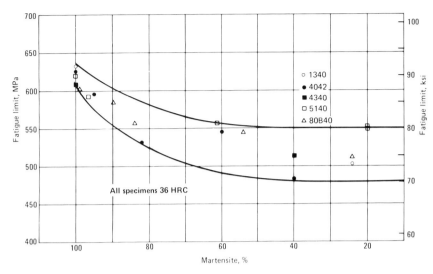

Fig. 22 Effect of martensite content on fatigue limit. Data are based on standard rotating-beam fatigue specimens of alloy steels 6.3 mm (0.250 in.) in diameter with polished surfaces.

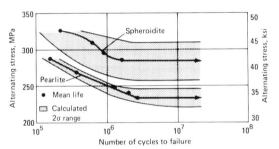

Fig. 23 Effect of microstructure on fatigue behavior of carbon steel (0.78% C, 0.27% Mn, 0.22% Si, 0.016% S, and 0.011% P)

Property	Spheroidite	Pearlite
Tensile strength, MPa (ksi)	641 (93)	676 (98)
Yield strength, MPa (ksi)	490 (71)(a)	248 (36)(b)
Elongation in 50 mm (2 in.), %	28.9	17.8
Reduction in area, %	57.7	25.8
Hardness, HB	92	89

(a) Lower yield point. (b) 0.1% offset yield strength

As described earlier, the fatigue strength coefficient, σ_f', is the intercept of the true stress amplitude-fatigue life plot at one reversal. The fatigue strength exponent, b, is the slope (always negative) of this line.

For steels with hardnesses below 500 HB, σ_f' may be approximated by:

$$\sigma_f' = S_u + 345 \quad \text{(Eq 10a)}$$

where σ_f' and S_u, the ultimate tensile strength, are given in MPa, or by:

$$\sigma_f' = S_u + 50 \quad \text{(Eq 10b)}$$

where σ_f' and S_u are given in ksi. If the tensile strength is not known, it may be approximated at 3.4 MPa (500 psi) times the Brinell hardness number.

The value of the fatigue strength exponent, b, is usually about -0.085. If the steel has been fully annealed, the value of b may be as high as -0.1. If the steel has been severely cold worked, the value of b may be as low as -0.05.

Steel	No. of tests(a)	Average tensile strength, MPa (ksi)	Hardness, HRC
Longitudinal tests			
4027	11	1179 (171)	37–39
4063	12	1682 (244)	47–48
4032	11	1627 (236)	46–48
Transverse tests			
4027	10	1130 (164)	34–39.5
4063	9	1682 (244)	47–48.5
4032	10	1254 (182)	47.5–48.5

(a) Number of fatigue specimens. For 4140 steel, 50 longitudinal and 50 transverse specimens were tested; for 4340 steel, 10 longitudinal and 10 transverse specimens were used.

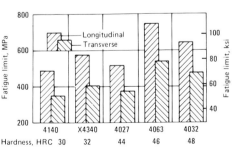

Fig. 24 Effect of specimen orientation on fatigue limit. Orientations are relative to the fiber axis resulting from hot working on the fatigue limit of low-alloy steels. Through-hardened and tempered specimens, 6.3 mm (0.250 in.) in diameter, were taken from production billets. Specimens for each grade were from the same heat of steel, but the tensile and fatigue specimens were heat treated separately, accounting for one discrepancy in hardness readings between the chart and the tabulation above. Fatigue limit is for 100×10^6 cycles.

684 / Service Characteristics of Carbon and Low-Alloy Steels

Table 4 Cyclic and monotonic properties of selected as-received and heat-treated steels
For a more complete, up-to-date listing of cyclic-fatigue properties, see Ref 13.

SAE steel Grade	Brinell hardness, HB	Condition(a)	Ultimate tensile strength MPa	ksi	Reduction in area, %	Modulus of elasticity GPa	10^6 psi	Yield strength MPa	ksi	Cyclic strain hardening exponent
1006	85	As-received	318	46.1	73	206	30	224	32.5	0.21
1018	106	As-received	354	51.3	...	200	29	236	34.2	0.27
1020	108	As-received	392	56.9	64	186	27	233	33.8	0.26
1030	128	As-received	454	65.8	59	206	30	248	36	0.29
1035	...	As-received	476	69.0	56	196	28.4	270	39	0.24
1045	...	As-received	671	97.3	44	216	31.3	353	51.2	0.22
1045	390	QT	1343	194.8	59	206	30	842	122	0.09
1045	450	QT	1584	229.7	55	206	30	1069	155	0.09
1045	500	QT	1825	265	51	206	30	1259	182.6	0.12
1045	595	QT	2240	325	41	206	30	1846	267.7	0.10
4142	380	QT	1412	205	48	206	30	966	140	0.14
4142	450	QT	1757	255	42	206	30	1160	168	0.11
4142	670	QT	2445	355	6	200	29	2238	324.6	0.07
4340	242	As-received	825	120	43	192	27.8	467	67.7	0.17
4340	409	QT	1467	213	38	200	29	876	127	0.13
SAE 950X	...	As-rolled	438	63.5	64	206	30	339	49.2	0.14
SAE 960X	...	As-rolled	480	70	...	206	30	417	60.5	0.14
SAE 980X	...	As-rolled	652	94.6	75	206	30	514	74.5	0.13

SAE steel Grade	Brinell hardness, HB	Cyclic strength coefficient MPa	ksi	Fatigue strength coefficient (σ_f') MPa	ksi	Fatigue strength exponent(b)	Fatigue ductility coefficient, ϵ_f'	Fatigue ductility exponent(c)
1006	85	813	118	756	109.6	−0.13	1.22	−0.67
1018	106	1259	182.6	782	113.4	−0.11	0.19	−0.41
1020	108	1206	175	850	123.2	−0.12	0.44	−0.51
1030	128	1545	224	902	130.8	−0.12	0.17	−0.42
1035	...	1185	172	906	131.4	−0.11	0.33	−0.47
1045	...	1402	203.3	1099	159.4	−0.11	0.52	−0.54
1045	390	1492	216.4	1408	204.2	−0.07	1.51	−0.85
1045	450	1874	271.8	1686	244.5	−0.06	0.97	−0.83
1045	500	2636	382.3	2165	314	−0.08	0.22	−0.66
1045	595	3498	507.3	3047	441.9	−0.10	0.13	−0.79
4142	380	2259	327.6	1820	264	−0.08	0.65	−0.76
4142	450	2359	342.1	2017	292.5	−0.08	0.85	−0.90
4142	670	3484	505.3	2727	395.5	−0.08	0.06	−1.47
4340	242	1384	200.7	1232	178.7	−0.10	0.53	−0.56
4340	409	1950	283	1898	275.3	−0.09	0.67	−0.64
SAE 950X	...	796	115.4	800	116	−0.10	1.23	−0.62
SAE 960X	...	969	140.5	895	130	−0.09	0.46	−0.65
SAE 980X	...	1135	164.6	1146	166.2	−0.09	1.10	−0.72

(a) QT, quenched and tempered. Source: Ref 10

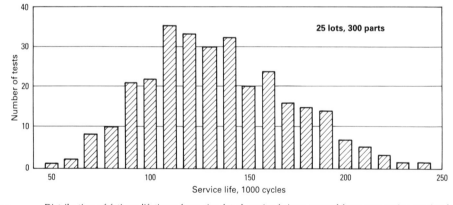

Fig. 25 Distribution of fatigue lifetimes from simulated service fatigue tests of front suspension torsion bar springs of 5160H steel. Size of hexagonal bar section was 32 mm (1.25 in.); mean service life, 134 000 cycles; standard deviation, 37 000 cycles; coefficient of variations, 0.28.

For a fatigue life of more than a million cycles, the use of these parameters in Eq 7 provides a slightly lower estimate of fatigue limit than the frequently used rule of thumb that the fatigue limit is half the ultimate tensile strength.

The fatigue ductility coefficient, ϵ_f', is approximated by the true fracture ductility, ϵ_f, which can be calculated from the reduction in area in a tension test by:

$$\epsilon_f' \approx \epsilon_f = \ln\left(\frac{100}{100 - \%RA}\right) \quad \text{(Eq 11)}$$

If the reduction in area (% RA) can be estimated from hardness levels, typical values of ϵ_f' can then be approximated by the use of Eq 11. For example:

- With hardness less than 200 HB, RA is approximately 65%, and $\epsilon_f' = 1.0$
- With hardness between 200 and 300 HB, RA is approximately 40%, and $\epsilon_f' = 0.5$
- With hardness greater than 400 HB, RA is approximately 10%, and $\epsilon_f' = 0.1$

The fatigue-ductility coefficient, ϵ_f', should be estimated from a measured percent of RA rather than obtained by using these approximate values, if possible.

The fatigue-ductility exponent, c, has approximately the same value (−0.6) for most ductile steels. Severe cold working may raise the value of c to −0.7; annealing or tempering at a high temperature may reduce c to about −0.5.

The elastic modulus (Young's modulus), E, is the slope of the elastic portion of the uniaxial stress-strain curve. For most steels, it has a value of about 200 GPa (29 × 10^6 psi). Further information on estimating these fatigue parameters may be found in Ref 10. As a check on estimating, the results should be compared with the data for a similar material in Table 4.

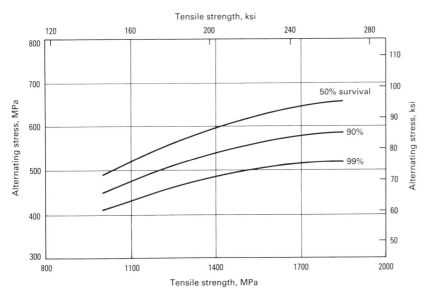

Fig. 26 Scatter of fatigue limit data. Based on the survival after 10 million cycles of approximately 1000 specimens, at one heat, of AISI-SAE 4340 steel with tensile strengths of 995, 1320, and 1840 MPa (144, 191, and 267 ksi). Rotating-beam fatigue specimens tested at 10 000 to 11 000 rev/min. Coefficients of variation, C_N, range from 0.17 to 0.20.

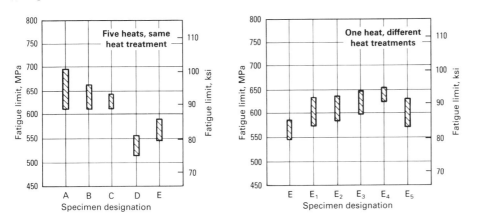

Specimen(a)	Hardness, HRC	Tensile strength MPa	ksi	Yield strength MPa	ksi	Elongation in 50 mm (2 in.), %	Reduction of area, %
Five heats, same heat treatment							
A	39.1	1250	181	1205	175	14.7	56.0
B	39.3	1225	178	1185	172	15.3	56.7
C	38.2	1235	179	1185	172	15.3	52.3
D	39.1	1235	179	1170	170	15.0	55.0
E	39.7	1270	184	1220	177	13.7	55.3
One heat, different heat treatments to produce the same hardness							
E	39.7	1270	184	1220	177	13.7	55.3
E_1(b)	40.3	1260	183	1250	181	13.0	55.7
E_2(c)	39.3	1270	184	1210	176	14.3	54.3
E_3(d)	38.7	1270	184	1220	177	15.7	54.3
E_4(e)	39.0	1275	185	1230	178	14.3	55.3
E_5(f)	37.8	1230	178	1170	170	14.7	58.3

(a) The letters A, B, C, D, and E indicate different heats of 8740 steel. Specimens were normalized at 900 °C (1650 °F) 1 h and air cooled; austenitized at 825 °C (1520 °F) 1 h and oil quenched; tempered 2 h. (b) Austenitized at 815 °C (1500 °F) ½ h and oil quenched; tempered 2 h. (c) Normalized at 900 °C (1650 °F) 1 h and air cooled; austenitized at 840 °C (1540 °F) 1¼ h and oil quenched; tempered 2 h. (d) Normalized at 900 °C (1650 °F) 1 h and air cooled; austenitized at 815 °C (1500 °F) ½ h and oil quenched; tempered 2 h. (e) Austenitized at 840 °C (1540 °F) 1¼ h and oil quenched; tempered 2 h. (f) Homogenized at 1150 °C (2100 °F) 24 h and air cooled; normalized at 900 °C (1650 °F) 1 hr and air cooled; austenitized at 825 °C (1520 °F) 1 h and oil quenched; tempered 2 h

Fig. 27 Variations in fatigue limit for different heats and heat treatments

Estimating Fatigue Life. Designers of machine components to be subjected to cyclic loading would like to be able to predict the fatigue life from basic materials parameters and anticipated loading patterns. However, the scatter of fatigue data is so great that the likelihood of accurate predictions is extremely low. The methods and approximations in this article and in Ref 10, 11, and 13 to 15 can provide some indication of fatigue life. Efforts to estimate fatigue life when service temperatures make creep-fatigue interaction a factor are discussed in the article "Elevated-Temperature Properties of Ferritic Steels" in this Volume.

In a specific situation, the assessment of the seriousness of fatigue is aided by a knowledge of the cyclic strains involved in fatigue at various lives. Certain generalizations are useful guidelines for ductile steels:

- If the peak localized strains are completely reversed and the total range of strain is less than S_u/E, fatigue failures are likely to occur in a large number of cycles or not at all
- If the total strain range is greater than 2% (amplitude ± 1%), fatigue failure will probably occur in less than 1000 cycles
- Part configurations that prevent the use of the ductility of the metal and metals that have limited ductility are highly susceptible to fatigue failures

With respect to long-life fatigue, the relative magnitude of the change in fatigue strength due to processing may be crudely estimated by the relative changes produced in ultimate tensile strength and in hardness. If the ductility change is also measured and if the qualitative effects of various processes on different types of metal are known, more refined estimates of the change in fatigue behavior can be made without resorting to extensive fatigue testing.

Fatigue life may be estimated by inserting a calculated strain amplitude and the appropriate materials parameters from Table 4 into Eq 9 and then solving for the number of cycles to failure, N_f. Where deformation is purely elastic, a calculated stress amplitude and Eq 7 may be used. The calculated fatigue life must be adjusted to compensate for stress concentrations, surface finish, and the presence of aggressive environments, as described in Fig. 8 and Ref 6 to 11. Alternatively, the calculated stress may be adjusted by using stress concentration factors such as those in Ref 16 and 17. Any of these calculations includes the assumption that the loading is fully reversed ($R = -1$).

Potter (Ref 18) has described a method for approximating a constant-lifetime fatigue diagram for unnotched specimens. Using this method, a series of points corresponding to different lifetimes are calculated and plotted along a diagonal line for $R = -1$. Each of these points is connected by a straight line to the point on another diagonal ($R = 1.0$) that corresponds to the ultimate tensile strength. The calculated lines correspond well with the experimental lines. Generally, the predicted lines represent lower stresses than the actual data. Estimating fatigue parameters from

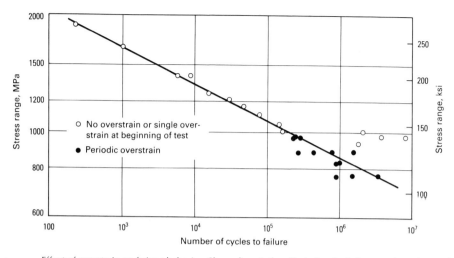

Fig. 28 Effect of overstrain on fatigue behavior. Shown here is the effect of periodic large strain cycles on the fatigue life of AISI-SAE 4340 steel hardened and tempered to a yield strength of 1100 MPa (160 ksi). Source: Ref 7

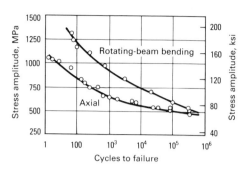

Fig. 29 Fatigue data under axial loading and rotating bending loading for 4340 steel. Source: Ref 21

the Brinell hardness number provides more conservative estimates. These results are only approximations, and the methods may not apply for every material.

While the likelihood of an accurate life prediction is relatively low, the use of these procedures is still valuable. There are very few "new" parts designed; most new parts are similar to a previously successful design, scaled up or down or operating at a slightly increased load. These procedures are very useful in estimating the change in life due to a change in design, load, processing, or material.

Cumulative Fatigue Damage. The data presented in this article, and most other published fatigue data, were obtained from constant-amplitude testing; all the load cycles in the test are identical. In actual service, however, the loading can vary widely during the lifetime of a part. There have been many approaches to evaluating the cumulative effects of variations in loading on the fatigue behavior of steels. References 7, 9, 10, 15, 18, and 19 describe methods of analyzing cumulative damage. A few overload cycles can reduce the fatigue life of steel, even though the mean load amplitude lies below the fatigue limit; this effect is shown in Fig. 28. The counting of each load cycle and the relative damage produced must be done with extreme accuracy and care. One method, rain flow counting (described in Ref 7, 9, 10, 15, 19, and 20), has been shown to be most effective. In this method, the cyclic stress-strain properties are applied such that the hysteresis behavior of the material (Fig. 9) is taken into account on each load excursion. Obviously, when there are millions of individual loads involved, the task becomes quite large. Frequently, a complex load or strain history will be simplified into a short block representing a fraction of the whole, and the damage in that block will be predicted. The block can then be repeated until failure, and component life can be predicted based on the fraction of the whole represented by the block. In any event, predicting fatigue behavior under these circumstances is difficult.

Notches. Fatigue failures in service nearly always start at the roots of notches. Because notches cannot always be avoided in design (though they should be avoided whenever possible), some allowance for notches must be made in calculating nominal stresses during the design process. A fatigue notch factor, K_f, should be introduced into the fatigue life calculations that use Eq 7:

$$S_a = \frac{\sigma_a}{K_f} = \frac{\sigma_f'}{K_f}(2N_f)^b \quad \text{(Eq 12)}$$

where σ_a is the amplitude of the true stress.

The appropriate value of K_f depends on the shape of the notch, fatigue strength, ductility of the metal, residual stress, and design life of the part. Its value varies between 1 (no notch effect) and the theoretical stress concentration factor, K_t. References 16 and 17 list many useful stress concentration factors.

First estimates of notched fatigue performance may be based on $K_f = K_t$, especially for moderately notched, heat-treated steel parts that are expected to withstand many cycles. A value of $K_f < K_t$ can be used if a more exact value of K_f is available.

For notched parts, the value of K_f may be estimated from the equation:

$$K_f = 1 + \frac{K_t - 1}{1 + a/r} \quad \text{(Eq 13)}$$

where r is the notch root radius; a is the material constant depending on strength and ductility.

For heat-treated steel, the following equation may be used to estimate a:

$$a = 0.025\left(\frac{2070}{S_u}\right)^{1.8} \quad \text{(Eq 14a)}$$

where a is given in millimeters and the ultimate tensile strength, S_u, is given in MPa. For steels, values of a range from 0.064 to 0.25 mm. Values of a may also be estimated by using the equation:

$$a = \left(\frac{300}{S_u}\right)^{1.8} \times 10^{-3} \quad \text{(Eq 14b)}$$

where a is given in inches and S_u in ksi.

When the required design life is relatively short, the effect of the notch will be even less than indicated by Eq 13 because of the large amount of inelastic strain at the root of the notch.

At low fatigue lives, a notch must be regarded as a strain concentration as well as a stress concentration. The product of the strain concentration factor, K_ϵ, and the stress concentration factor, K_σ, is equal to the square of the theoretical stress concentration factor:

$$K_\epsilon K_\sigma = (K_t)^2 \quad \text{(Eq 15)}$$

At long lives, the behavior is nominally elastic so that K_ϵ equals K_σ, and both K_ϵ and K_σ are equal to K_t. At short lives, where K_σ is nominally 1, the strain concentration factor is equal to $(K_t)^2$. Usually, however, K_t is replaced by an effective value of K_f, as in Eq 12. This relation may be rewritten as:

$$(K_f \times \Delta S)^2 = \sigma_a \epsilon_a E \quad \text{(Eq 16)}$$

where ΔS is the nominal calculated stress amplitude remote from the stress raiser, σ_a is the amplitude of the true stress (Eq 12), and ϵ_a is the amplitude of the true strain. The value of ϵ_a calculated can be used directly on a fatigue life diagram, such as Fig. 14, to estimate the fatigue life of an actual part.

Mean stresses may be introduced into the above equations by substituting the quantity $(\sigma_f - \sigma_0)$ for σ_f wherever it appears. Mean stresses affect fatigue behavior by increasing the amount of plastic strain whenever the algebraic sum of the mean and alternating stresses exceeds the yield strength.

Fatigue Resistance of Steels / 687

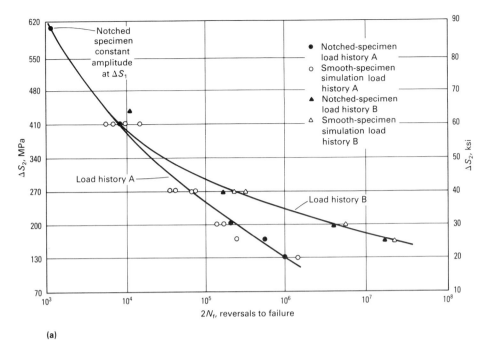

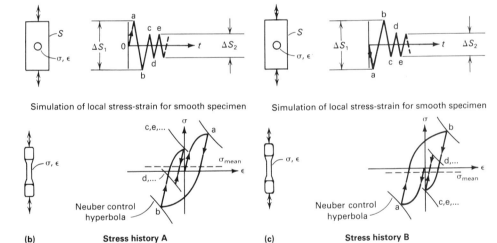

Fig. 30 Fatigue data (a) showing sequence effects for notched-specimen and smooth-specimen simulations (2024-T4 aluminum, $K_f = 2.0$). Load histories A and B have a similar cyclic load pattern (ΔS_2) but have slightly different initial transients (ΔS_1) with either (b) a tensile leading edge (first stress peak at $+\Delta S_1/2$) or (c) a compressive leading edge (first stress peak at $-\Delta S_1/2$). The sequence effect on fatigue life (a) becomes more pronounced as ΔS_2 becomes smaller. Source: Ref 21

Although residual stresses may be considered equivalent to mechanically imposed mean stresses when the cyclic stress is low, their effect on fatigue is less than that indicated by their initial value when the stress or strain is high, because the residual stress is "washed out" by repeated slip. The nominal maximum stress is defined as the algebraic sum of the alternating stress and the mean stress. When the nominal maximum stress is larger than the yield strength, S_y, there is little influence on the residual stress regardless of its original magnitude, and the behavior in fatigue will be similar to that of a stress-free member under fully reversed loading. Most change in residual stress occurs during the first few cycles.

Discontinuities. Many features of a material that are not reflected by the usual bulk mechanical properties may have a large influence on its fatigue resistance. Porosity and inclusions may have little effect on the fatigue behavior of a material, provided that they are less than a certain critical size and are not located in a highly stressed region. The critical size depends on the fracture toughness of the material, shape of the pore or inclusion, and stress intensity at the inclusion or pore. Pores and inclusions larger than this critical size can significantly reduce fatigue life, possibly causing failure during the first load application. Surface discontinuities such as folds, seams, score marks, cracks, and corrosion pits greatly influence fatigue behavior. The detrimental effect of these surface discontinuities on fatigue behavior can be somewhat reduced by surface treatments such as shot peening and surface rolling. Anisotropy of the microstructure can be detrimental to fatigue life, particularly if the tensile component of applied stress is nearly perpendicular to the long dimension of elongated grains or stringers. Because it is impractical to eliminate all discontinuities completely, the quantitative influence of discontinuities must be determined by fatigue tests involving the materials, manufacturing processes, and shapes of the parts in question.

Comparison of fatigue testing techniques can show large differences in the results of a life prediction. Socie (Ref 21) discusses four techniques:

- Load life
- Stress life
- Strain life
- Crack propagation

and the application of each. He makes the point that load life, while most accurate, is generally restricted to real parts and is difficult to apply to new designs. He points out that different load application methods, as in axial versus rotating bending, often make large differences in results (Fig. 29). Socie also cautions the user about sequence effects (Fig. 30). Load history A (Fig. 30b) and load history B (Fig. 30c) have similar-appearing strain histories with totally different stress-strain response and fatigue life (Fig. 30a).

Ultimately, the fatigue analyst will be required to include and correlate a number of material, shape, processing, and load factors in order to identify the critical locations within a part and to describe the local stress-strain response at those critical locations. The ability to anticipate pertinent factors will greatly affect the final accuracy of the life prediction.

Load data gathering is one remaining topic that must be included in any discussion of fatigue. Reference 21 discusses three load histories, suspension, transmission, and bracket vibration, that typify loads found in the ground vehicle industry. Additionally, there are vastly different histories unique to other industries, like the so-called ground-air-ground cycle in aeronautics. Without the ability to completely and accurately characterize anticipated and, occasionally, unanticipated customer use and resultant loads, the analyst will not be able to predict accurately the suitability of a new or revised design.

The last several years have seen a major change in the ability to gather customer or simulated customer load data. Testing methods have progressed from bulky, multichannel analogue tape recorders (where it took days or weeks before results were available) through portable frequency-modulated telemetry packages (where analysis could be performed immediately at a remote site) to hand-held packages capable of data acquisition and analysis on board the test vehicle in real time. Microelectronics is further reducing size and improving reliability to the point that data can be gathered from within small, complex, moving, hostile assemblies, such as engines.

REFERENCES

1. R.W. Hertzberg, *Deformation and Fracture Mechanics of Engineering Materials*, John Wiley & Sons, 1976
2. D.J. Wulpi, *Understanding How Components Fail*, American Society for Metals, 1985
3. Fatigue and Microstructure, in *Proceedings of the ASM Materials Science Seminar*, American Society for Metals, 1979
4. *Metallic Materials and Elements for Aerospace Vehicle Structures*, MIL-HDBK-5B, *Military Standardization Handbook*, U.S. Department of Defense, 1987
5. *Metallic Materials and Elements for Aerospace Vehicle Structures*, Vol 1, MIL-HDBK-5B, *Military Standardization Handbook*, U.S. Department of Defense, Sept 1971, p 2-29
6. R.C. Juvinall, *Engineering Considerations of Stress, Strain and Strength*, McGraw-Hill, 1967
7. N.E. Dowling, W.R. Brose, and W.K. Wilson, Notched Member Fatigue Life Predictions by the Local Strain Approach, in *Fatigue Under Complex Loading: Analyses and Experiments*, R.M. Wetzel, Ed., Society of Automotive Engineers, 1977
8. J.A. Graham, Ed., *Fatigue Design Handbook*, Society of Automotive Engineers, 1968
9. H.O. Fuchs and R.I. Stephens, *Metal Fatigue in Engineering*, John Wiley & Sons, 1980
10. Special Publication P-109, in *Proceedings of the SAE Fatigue Conference*, Society of Automotive Engineers, 1982
11. R.C. Rice, Ed., *Fatigue Design Handbook*, 2nd ed., Society of Automotive Engineers, 1988
12. J.T. Ransom, *Trans. ASM*, Vol 46, 1954, p 1254-1269
13. "Fatigue Properties," Technical Report, SAE J1099, Society of Automotive Engineers, 1977
14. P.H. Wirsching and J.E. Kempert, A Fresh Look at Fatigue, *Mach. Des.*, Vol 48 (No. 12), 1976, p 120-123
15. L.E. Tucker, S.D. Downing, and L. Camillo, Accuracy of Simplified Fatigue Prediction Methods, in *Fatigue Under Complex Loading: Analyses and Experiments*, R.M. Wetzel, Ed., Society of Automotive Engineers, 1977
16. R.E. Peterson, *Stress Concentration Design Factors*, John Wiley & Sons, 1974
17. R.J. Roark, *Formulas for Stress and Strain*, 4th ed., McGraw-Hill, 1965
18. J.M. Potter, Spectrum Fatigue Life Predictions for Typical Automotive Load Histories and Materials Using the Sequence Accountable Fatigue Analysis, in *Fatigue Under Complex Loading: Analyses and Experiments*, R.M. Wetzel, Ed., Society of Automotive Engineers, 1977
19. D.V. Nelson and H.O. Fuchs, Predictions of Cumulative Fatigue Damage Using Condensed Load Histories, in *Fatigue Under Complex Loading: Analyses and Experiments*, R.M. Wetzel, Ed., Society of Automotive Engineers, 1977
20. S.D. Downing and D.F. Socie, Simple Rainflow Counting Algorithms, *Int. J. Fatigue*, Jan 1981
21. D.F. Socie, "Fatigue Life Estimation Techniques," Technical Report 145, Electro General Corporation

Embrittlement of Steels

George F. Vander Voort, Carpenter Technology Corporation

IRON-BASE ALLOYS are susceptible to a number of embrittlement phenomena. Some of these affect a wide range of compositions, while others are specific to a rather narrow range of compositions. These problems promote brittle service failures that may be catastrophic in nature or may reduce the useful service life of a component. This article reviews these embrittlement problems and presents examples of their influence on mechanical properties. These problems arise from compositional, processing, or service conditions, or combinations of these three conditions.

If the embrittlement occurs during processing at the mill, it may be detected during routine testing depending on the degree of embrittlement and the nature of the testing program. The steelmaker is generally aware of the potential problems that particular grades are susceptible to and will check the various well-known parameters that can promote such problems. This starts with an examination of chemical composition, such as gas contents or impurity elements that are known to cause problems. For example, in the melting of ingots for pressure vessels or rotors where temper embrittlement is a potential problem, the selection of scrap for electric furnace melting is based on the scrap impurity level, and every effort is made to keep residual levels of phosphorus, antimony, tin, and arsenic as low as possible. Subsequent impact testing of coupons from the forgings is used to verify the initial toughness. Such information is compared to specification requirements and historical data base information to maintain quality. Special melting practices, such as vacuum carbon deoxidation, have also been adopted for critical composition control. Furthermore, additional knowledge is obtained by surveillance testing and by postmortems on retired forgings.

However, not all potential problems can be detected, or prevented, at the mill. Some arise from handling or fabrication problems. For example, low-carbon sheet steels that are not aluminum killed are roller leveled at the mill prior to shipment to suppress the yield point and prevent strain-age embrittlement. However, if this steel is not formed within a certain time period, the yield point will return, and discontinuous yielding may occur when the sheet is cold formed, resulting in cosmetically damaging Lüders bands on the product. For this and other fabrication problems, postfabrication inspection and testing programs must be properly planned and executed.

Service and environmental conditions can also cause a number of embrittlement problems. Engineers working with steel components that are susceptible to such operating-induced problems must be aware of the potential problems and must establish regular inspection programs to detect problems before they become critical. An excellent example of such programs is the on-site, *in situ* examination of steam piping in electric power generation plants; these examinations make extensive use of field metallography and replication to assess creep damage and predict remnant life.

Embrittlement of Iron

Before discussing the embrittlement of steels, this article will first examine the embrittlement of iron because the number of such studies are few compared to those for steels. Grain-boundary segregation of elements such as oxygen, sulfur, phosphorus, selenium, and tellurium is known to produce intergranular brittle fractures in iron at low temperatures. Studies of the effects of such impurities in pure iron have been greatly aided by the development of Auger electron spectroscopy (AES).

Intercrystalline fractures in iron at low temperatures occur when the carbon content is low. It has been assumed that the absence of carbon is more important than the presence of embrittling grain-boundary impurities. However, impurities must be present, and the role of carbon appears to be one of a competitor for grain-boundary sites when such impurities are present.

Oxygen. A study of the toughness of iron-oxygen alloys found that intergranular embrittlement and a rise in impact transition temperature in iron were produced by oxygen levels of 30 ppm or more (Ref 1). Figure 1 shows the Charpy V-notch impact transition curves from this study for a series of iron-oxygen alloys with increasing oxygen contents (from 10 to 2700 ppm). However, the sulfur levels of the heats were rather high, from 30 to 54 ppm for the alloys shown in Fig. 1, except for the 0.011% O heat,

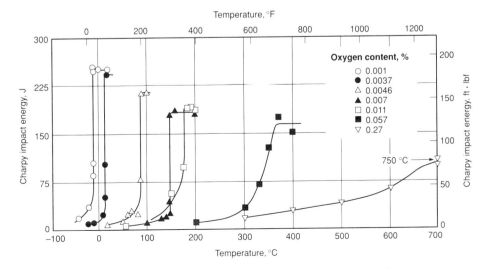

Fig. 1 Charpy V-notch impact energy curves for iron-oxygen alloys with varying oxygen content. Source: Ref 1

690 / Service Characteristics of Carbon and Low-Alloy Steels

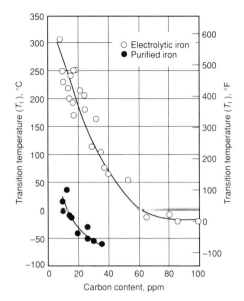

Fig. 2 Transition temperature, T_t, versus carbon content for two different high-purity irons, each containing 2000 ppm O. Source: Ref 4

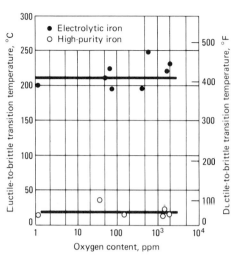

Fig. 3 Ductile-to-brittle transition temperatures (from tests using Charpy U-notch specimens) as a function of oxygen content for a decarburized electrolytic iron and a high-purity iron with 10 ppm C. Source: Ref 6

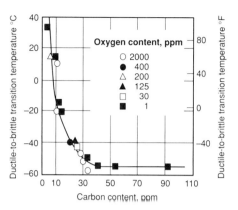

Fig. 4 Ductile-to-brittle transition temperatures of high-purity iron as a function of carbon content and oxygen content. Source: Ref 6

which had 16 ppm S. Because this work was conducted long before the introduction of AES, the true nature of the embrittling grain-boundary element(s) is open to question.

Work on high-purity iron found intergranular brittleness at low test temperatures when specimens were quenched from temperatures where carbon was in solution (Ref 2, 3). It was believed that this brittleness was due to oxygen. Lowering the quench temperature reduced the embrittlement. Titanium additions were not found to be helpful in reducing the intergranular embrittlement.

Reference 4 reviews the study from Ref 3 and reports on similar work performed in France that found that sulfur, not oxygen, was the cause of the embrittlement. Tests were done using electrolytic iron containing 35 ppm S and a purified iron containing less than 1 ppm S; each contained 2000 ppm O and varying carbon contents. The purified iron was found to be much less brittle than the electrolytic iron (Fig. 2). At 10 ppm C, the purified iron was free of intergranular fracture; the electrolytic iron fractured intergranularly below the ductile-to-brittle transition temperature (DBTT).

Reference 5, on the other hand, shows Auger analysis of intergranular fractures of the pure iron specimen from Ref 2 that contained 60 ppm S. Sulfur was not detected on the intergranular fracture, while carbon, nitrogen, and oxygen were. The authors did state that the oxygen peak could possibly be due to oxygen contamination after fracture in the high vacuum used for Auger work (fracture made inside the evacuated chamber).

It has been demonstrated that large variations in oxygen content have no influence on the brittleness of iron (Ref 6). The Charpy U-notch transition temperatures of electrolytic iron and high-purity iron with varying oxygen and carbon contents were determined, as in Ref 4. Figure 3 shows that the DBTTs for these two irons were constant over a wide range of oxygen contents (1 to 2000 ppm). The DBTT of the electrolytic iron (210 °C, or 410 °F) was consistently much higher than that of the high-purity iron (20 °C, or 70 °F). Also, fractures for the electrolytic iron specimens in the brittle range were fully intergranular, whereas those for the high-purity iron were by cleavage. Carbon and sulfur contents were 30 and 35 ppm, respectively, for the electrolytic iron and 10 and less than 3 ppm, respectively, for the high-purity iron. Manganese contents were 12 ppm for the electrolytic iron and less than 0.5 ppm for the high-purity iron, too little to tie up the sulfur completely in the electrolytic iron. Figure 4 shows the results when carbon and oxygen were both varied. When the carbon level was increased to 30 ppm, there was a large improvement in the DBTT, irrespective of the oxygen level. This study also demonstrated that the addition of elements that form sulfide precipitates improved the DBTT.

On the other hand, a study using irons with less than 2 ppm S and 0.5 ppm C and Auger analysis demonstrated oxygen grain-boundary enrichment of intergranular fractures broken by impact within the AES chamber (Ref 7). Segregation of elements such as sulfur, phosphorus, and nitrogen was not observed on the fracture, but oxygen was detected. For a specimen with a bulk oxygen content of 430 at. ppm, there were two monolayers of oxygen at the intergranular fracture surface. Carbon was not present at the fracture surface, consistent with the very low bulk carbon content. Another specimen with 235 at. ppm O exhibited about 15% cleavage fracture, along with areas that were predominantly intergranular. The study found that for the material with higher oxygen content, quenching specimens from temperatures of 1103 and 1123 °C (2017 and 2053 °F), near the solubility limit, gave predominantly intergranular low-temperature fractures. When the temperature was lowered from 973 to 773 °C (1783 to 1423 °F), below the solubility limit, the amount of intergranular fracture decreased. Also, if the specimen was slow cooled from temperatures near the solubility limit, low-temperature fractures were fully by cleavage. Therefore, the study concluded that intergranular brittle fracture of iron was due to the oxygen in solution.

Sulfur. The embrittlement of iron by sulfur has been demonstrated by several authors. Prior to Auger analysis, such results were assumed, but without direct proof. Auger work has, indeed, confirmed that sulfur is a potent embrittler of iron, even at bulk levels as low as 10 ppm. Furthermore, the displacement of sulfur on the grain boundaries when carbon is added has been proved by Auger analysis. Low-temperature impact tests performed on three heats of relatively pure iron obtained intergranular fractures, depending on the heat treatment used (Ref 8). Sulfur contents ranged from 14 to 100 ppm, carbon contents from less than 10 to 70 ppm, and oxygen contents from 8 to 420 ppm. Auger analysis revealed heavy segregation of sulfur to the grain boundaries. No clear evidence of oxygen on the intergranular fractures was obtained.

Other studies have shown the detrimental influence of small sulfur additions (Ref 9, 10). Figure 5, from these studies, shows the DBTT as a function of bulk sulfur content (≤60 ppm) and the beneficial influence of carbon additions (≤10 to 30 ppm) for iron containing 2000 ppm O. For less than or equal to 10 ppm C, the transition temperature increased from 0 to over 600 °C (30 to

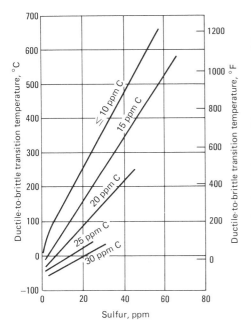

Fig. 5 Influence of sulfur on the transition temperature of purified iron containing 2000 ppm O, and the influence of carbon content on sulfur embrittlement. Increasing carbon content has the beneficial effect of decreasing sulfur embrittlement. Source: Ref 9, 10

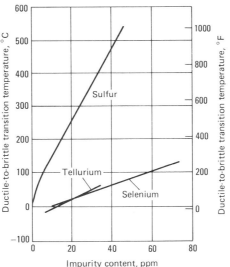

Fig. 6 Influence of sulfur, tellurium, and selenium on the transition temperature of purified iron containing up to 10 ppm carbon and approximately 2000 ppm O. Source: Ref 10

>1110 °F) with increasing sulfur content. However, with 25 or 30 ppm C, the influence of sulfur was small. Also demonstrated was the scavenging influence of a 0.5% Al addition, which suppressed embrittlement (constant DBTT) for the full range of sulfur (≤60 ppm) tested.

Selenium and tellurium are similar to sulfur and are potential embrittlers of pure iron. One study (Ref 10) showed that they do cause embrittlement, but to a lesser degree than sulfur (Fig. 6). In this work, the carbon content was less than 10 ppm, and 2000 ppm O was present. Although selenium and tellurium had less influence on the DBTT, these two elements reduced the absorbed energy values much more than did sulfur. Therefore, in terms of impact energy, the elements can be placed in order of increasing influence as follows: sulfur, selenium, and tellurium. Others have also reported the embrittlement of Fe-0.04% C by 0.02% Te (Ref 11).

Other Impurity Elements. Phosphorus has also been reported to embrittle pure iron (Ref 12, 13). However, both of these studies used materials with a significant sulfur content, and they were performed prior to the development of Auger analysis. However, radioactive tracer analysis demonstrated the segregation of phosphorus at the grain boundaries of an Fe-0.09% P alloy (Ref 12). Phosphorus was reported to be 50 times as prevalent at the grain boundaries as in the grain interiors. Phosphorus does substantially increase the strength of ferrite. Again, the addition of carbon was shown to reduce the influence of phosphorus on embrittlement.

Researchers have also studied phosphorus segregation in pure iron (Ref 14). Again, the specimens contained a significant amount of sulfur, but mechanical properties were not determined. However, a small amount of manganese was present that should precipitate the sulfur as a sulfide. The carbon content was reduced to below 10 ppm. Specimens were austenitized, water quenched, tempered at 850 °C (1560 °F) for 1 h, and then furnace cooled to the aging temperature. Specimens were fractured within the Auger chamber. Phosphorus was observed on the surface of intergranular fractures, but not on cleavage fractures. Auger analysis showed that the amount of phosphorus on the intergranular fractures increased with bulk phosphorus content. Also, as the aging temperature decreased, the grain-boundary phosphorus content increased, and the fracture became more intergranular. When carbon was added (≤80 ppm), the grain-boundary phosphorus concentration decreased. A deep-drawing steel containing 7 ppm C, 310 ppm P, and 0.36% Mn fractured intergranularly in the drawing direction, and phosphorus was detected on the grain boundaries (Ref 14). Similar steels with 14 ppm C, 80 ppm P, and 0.38% Mn did not fracture during deep drawing.

A study of the embrittlement of iron by phosphorus, phosphorus and sulfur, and antimony and sulfur demonstrated that the embrittlement was different from that of temper embrittlement in that it was not reversible (Ref 15). Segregation occurred at all temperatures in ferrite but was negligible or limited in austenite. Quenching from the austenite region produced specimens that fractured by cleavage. When quenched from the two-phase region, fractures did exhibit phosphorus at the grain boundaries. When an Fe-0.2P alloy was furnace cooled from the austenite region, the fracture surface exhibited a layer of nearly pure phosphorus at the grain boundary with a thickness of 1 to 1.5 nm (10 to 15 Å). The ternary alloys containing sulfur exhibited DBTTs of about 350 °C (660 °F). The study concluded that sulfur, even at much lower concentrations than phosphorus, is a more potent embrittler of iron.

Metalloids such as phosphorus, arsenic, antimony, and tin do not produce embrittlement of pure iron containing minor amounts of carbon in the same manner as sulfur, although they do in alloy steels (Ref 16). It has been demonstrated that such elements produce embrittlement of carbide/ferrite and surrounding ferrite/ferrite interfaces (Ref 17). This appears to be a nonequilibrium segregation problem, however.

The influence of tin on high-purity iron and low-carbon steel has been examined (Ref 18). Detailed chemical analyses of the pure iron specimens used were not given in the study, although it was stated that the base metal had a carbon content of 20 ppm and an oxygen content of 400 ppm. The addition of 0.5% Sn to the pure iron reduced the impact strength in the ductile region to such a degree that the absorbed energy was constant up to 70 °C (160 °F). Specimens water quenched from 650 °C (1200 °F) exhibited impact results similar to those of tin-free pure iron, while slowly cooled specimens were embrittled. The addition of 0.15% C to the Fe-0.5Sn alloy did reduce the embrittling influence of tin, and the alloy had better toughness than the pure iron specimen when water quenched from 650 °C (1200 °F). The addition of 0.15% P to the Fe-0.5Sn-0.15C alloy raised the DBTT about 20 °C (36 °F) and lowered the upper-shelf energy when water quenched from 650 °C (1200 °F). The examination of fractured specimens showed a change from transgranular cleavage to intergranular fracture as the tin content increased, particularly for the slowly cooled specimens.

This survey of the influence of impurities on the embrittlement of pure iron has demonstrated that the design of experiments and the interpretation of results are difficult. Many of the early studies did not recognize the significance of relatively minor amounts of sulfur in the high-purity irons used. It is clear that Auger analysis is required to determine the embrittling species. When sulfur is present in the absence of sulfide-forming elements, it has a dominating influence on properties and behavior. The addition of carbon above about 10 ppm will reduce or eliminate embrittlement effects.

Embrittlement in Carbon Steels and Alloy Steels

Several forms of embrittlement can occur during thermal treatment or elevated-temperature service of carbon and alloy steels.

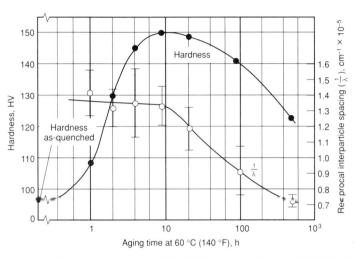
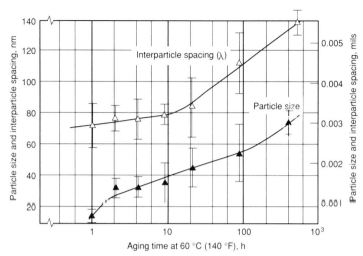

Fig. 7 Influence of aging time at 60 °C (140 °F) after quenching from 500 °C (930 °F) on the hardness, particle size, and interparticle spacing for an Fe-0.02N alloy. Source: Ref 25

These forms of embrittlement (and the types of steel that some forms specifically affect) are:

- Blue brittleness (carbon steels)
- Quench-age embrittlement (low-carbon steels)
- Strain-age embrittlement (low-carbon steels)
- Aluminum nitride embrittlement (carbon and alloy steels)
- Graphitization (carbon and alloy steels)

Blue Brittleness

Carbon steels generally exhibit an increase in strength and a reduction of ductility and toughness at temperatures around 300 °C (570 °F). Because such temperatures produce a bluish temper color on the surface of the specimen, this problem has been called blue brittleness (Ref 19-22). It is generally believed that blue brittleness is an accelerated form of strain-age embrittlement. Deformation in the blue-heat range followed by testing at room temperature produces an increase in strength that is greater than when the deformation is performed at ambient temperature. Blue brittleness can be eliminated if elements that tie up nitrogen are added to the steel, for example, aluminum or titanium.

Quench-Age Embrittlement

If a carbon steel is heated to a temperature slightly below its lower critical temperature and then quenched, it will become harder and stronger but less ductile (Ref 23-28). This problem has been called quench aging or quench-age embrittlement. The degree of embrittlement is a function of time at the aging temperature. Aging at room temperature requires several weeks to reach maximum embrittlement. Lowering the quenching temperature reduces the degree of embrittlement. Quenching from temperatures below about 560 °C (1040 °F) does not produce quench-age embrittlement. Carbon steels with a carbon content of 0.04 to 0.12% appear to be most susceptible to this problem.

Quench aging is caused by the precipitation of carbide and/or nitride from solid solution. One study has reviewed the quench aging of iron-nitrogen, iron-carbon, and iron-carbon-nitrogen alloys (Ref 25). Figure 7 shows the change in hardness, interparticle spacing, and particle size of precipitates in an Fe-0.02N alloy quenched from 500 °C (930 °F) and aged at 60 °C (140 °F) up to 500 h. The precipitates grew from about 30 nm (300 Å) in diameter after 2 h to about 80 nm (800 Å) in diameter after 500 h. With aging, the hardness increased rapidly to about 150 HV after 10-h aging at 60 °C (140 °F) and then decreased to about 120 HV with aging to 500 h.

Low-Carbon Steels. For low-carbon steels, quench aging is due mainly to carbon because the nitrogen level is usually too low to have a substantial effect. Results for a rimming steel containing 0.03% C (Fig. 8) show the increase in hardness with aging time at 60 °C (140 °F) for specimens quenched from 725 °C (1335 °F). The tensile strength increased rapidly to a maximum value after 10 h at 60 °C (140 °F) and then decreased slowly with further aging. The yield strength also increased rapidly and reached a maximum in about the same time but remained constant with continued aging. The elongation and reduction of area decreased as the strength increased; they reached a minimum after 10 h and then increased somewhat with continued aging.

Iron-Nitrogen and Iron-Carbon Alloys. For iron-nitrogen alloys, two types of nitrides can precipitate during quench aging:

- Face-centered cubic (fcc) Fe_4N platelets form at high temperatures and generally are found at grain boundaries
- Body-centered cubic (bcc) $Fe_{16}N_2$ nitrides with a circular disk shape precipitate at low temperature on dislocations

This latter type of nitride causes strengthening during quench aging in iron-nitrogen alloys (Ref 25). Iron-carbon alloys also have two types of carbides that can form during quench aging. Cementite forms at high temperatures, and a low-temperature carbide that is identical in morphology and habit to

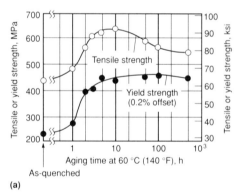

(a)

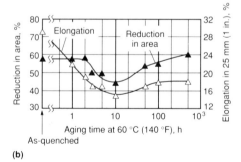

(b)

Fig. 8 Influence of aging time at 60 °C (140 °F) after quenching from 725 °C (1335 °F) on the tensile properties of an Fe-0.03C rimming alloy. (a) Tensile and yield strength. (b) Elongation and reduction in area. Source: Ref 25

Fig. 9 Example of stretcher-strain marks (Lüders bands) on a cold-formed steel part

$Fe_{16}N_2$ and may be isomorphous with it can also form. With aging, the low-temperature carbide will gradually be replaced by Fe_3C (Ref 25). The phase changes during aging of iron-nitrogen and iron-carbide quench aged steels are discussed in the literature (Ref 28).

Strain-Age Embrittlement

Strain aging occurs in low-carbon steels deformed certain amounts and then aged, which produces an increase in strength and hardness but a loss in ductility (Ref 25-27, 29-34). The degree of deformation, or cold work, is important. Generally, about a 15% reduction in thickness provides the maximum effect. The resulting brittleness varies with the aging temperature and time. Aging at room temperature is very slow, requiring several months to obtain maximum embrittlement. As the aging temperature is increased, the time for maximum embrittlement decreases. Certain coating treatments, such as hot dip galvanizing, can produce a high degree of embrittlement in areas that were cold worked the critical amount; this can lead to brittle fractures. To prevent this problem, the part can be annealed before galvanizing. Alternatively, the use of sheet steels containing elements that tie up nitrogen, for example, aluminum, titanium, zirconium, vanadium, or boron, will prevent strain-age embrittlement.

Strain aging may also lead to stretcher-strain marks (Lüders bands) on cold-formed low-carbon sheet steel components. These marks are cosmetic defects, rather than cracks, but their presence on formed parts is unacceptable (Fig. 9). During tensile loading, sheet steel that is susceptible to this defect will exhibit nonuniform yielding followed by uniform elongation. The elongation at maximum load and the total elongation are reduced, decreasing cold formability. In a nonaluminum-killed sheet steel, a small amount of deformation, typically 1% reduction, will suppress the yield point for several months. This process is referred to as roller levelling or temper rolling (Ref 31). This process is more effective in eliminating the sharp yield point and preventing strain aging than stretching the steel through the Lüders strain, which requires about 4 to 6% reduction. However, if the material is not formed within the safe period, discontinuous yielding will eventually return and impair formability.

Results of one study illustrate the influence of strain aging on mechanical properties (Ref 32). Three steels made by different processes were evaluated: Steel A, silicon and aluminum deoxidized steel; Steel B, capped open hearth steel; and Steel C, capped Bessemer steel. Steel C had the highest nitrogen content. Steels B and C had low aluminum contents, while steel A had sufficient aluminum to tie up the nitrogen. Strips of each were normalized and loaded in tension to a permanent strain of 10%. The strips were held at 25, 230, 480, and 650 °C (75, 450, 900, and 1200 °F) for various lengths of time (≤25 000 h at 25 and 230 °C, or 75 and 450 °F; ≤10 h at 480 °C, or 900 °F; and 2 h at 650 °C, or 1200 °F). Hardness, tensile properties, and impact properties (half-width Charpy V-notch specimens) were determined at different aging times.

Figure 10 shows the impact test results for steels A, B, and C strained 10% in tension and aged at room temperature up to 25 000 h. The impact curves are shifted with aging at room temperature for all three steels; steel A exhibits the best aged toughness, and steel C the poorest. Figure 11 shows the increase in hardness for steels A, B, and C aged for times up to 25 000 h at 25 °C (75 °F) and 230 °C (450 °F). Room-temperature aging produced a gradual increase in hardness with time. The maximum hardness was about the same and was reached quickest by steel C and slowest by steel A. The hardness increase with aging at 230 °C (450 °F) was constant for steel A and slowly decreased with aging for steels B and C.

In low-carbon steels, strain aging is caused chiefly by the presence of interstitial solutes (carbon and nitrogen), although hydrogen is known to produce a lesser effect at low temperatures. These interstitial solutes have high diffusion coefficients in iron and high interaction energies with dislocations. The change in mechanical properties of low-carbon rimming steels with different carbon and nitrogen contents that were prestrained 4% and aged various times at 60 °C (140 °F) has been demonstrated (Fig. 12) (Ref 34). This work clearly demonstrates the detrimental influence of higher carbon and nitrogen contents on strain aging. The solubilities of carbon and nitrogen in iron are quite different. Nitrogen solubility is high in the temperature range where rapid precipitation occurs; the solubility of carbon, in equilibrium with cementite, is very low. Therefore, strain aging that is due to carbon at temperatures below 100 °C (210 °F) is insignificant. However, above 100 °C (210 °F), ε carbide can redissolve and produce substantial strain aging (Ref 35). Strain aging attributable to nitrogen is caused by

694 / Service Characteristics of Carbon and Low-Alloy Steels

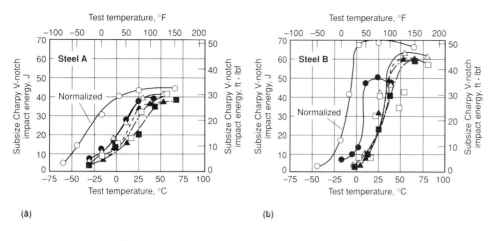

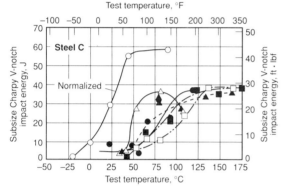

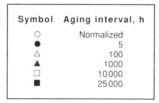

Fig. 10 Influence of straining in tension and aging at 24 °C (75 °F) on the Charpy V-notch (half width) impact strength for three steels. (a) Steel A, silicon and aluminum killed, 0.25% C with 0.013% Al and 0.011% N. (b) Steel B, capped open hearth steel, 0.07% C with 0.005% Al and 0.005% N. (c) Steel C, capped Bessemer steel, 0.08% C with 0.006% Al and 0.016% N. All three steels were strained 10% and aged. Source: Ref 32

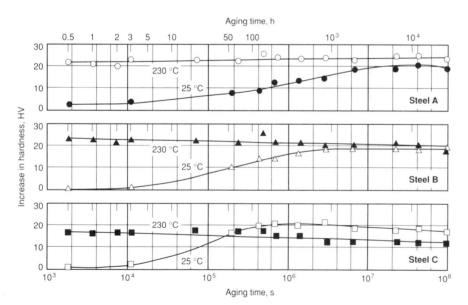

Fig. 11 Increase in hardness for steels A, B, and C from Fig. 10 after straining in tension (10%) and aging at 24 and 230 °C (75 and 450 °F) for up to 25 000 h. Source: Ref 32

nitrogen that is not tied up with strong nitride formers, for example, aluminum, titanium, zirconium, vanadium, or boron.

Dynamic Strain Aging. Strain aging can also occur dynamically, that is, aging occurs simultaneously with straining. In this case, the effective strain rate, that is, the dislocation velocity, controls the extent of aging of a particular steel. For normal tensile strain rates, dynamic strain aging occurs in the temperature range of 100 to 300 °C (210 to 570 °F) (which includes temperatures at which blue brittleness occurs). If the interstitial solute content is substantial, dynamic strain aging may be observed at room temperature. At very high strain rates, as in impact testing, dynamic strain aging is observed at temperatures above 400 °C (750 °F), up to about 670 °C (1240 °F). Reference 33 presents dynamic strain aging results for five carbon steels. Carbon and nitrogen are, again, the most important elements in dynamic strain aging. Nitrogen is more important than carbon because of the lower solubility of carbon.

Aluminum Nitride Embrittlement

It is well known that aluminum nitride precipitation in aluminum-killed steels can cause embrittlement and fracture. Several types of problems due to aluminum nitride precipitation have been observed: intergranular fractures in castings (Ref 36-43), panel cracking in ingots (Ref 44-47), and reduced hot ductility (Ref 48-53).

Intergranular fracture in castings, in both the as-cast and heat-treated conditions, have been sporadically observed for many years. The fractures occur at the primary austenite grain boundaries formed during solidification. In as-cast specimens, ferrite films are generally observed at these grain boundaries. The incidence of such cracking has been shown to increase with increases in aluminum and nitrogen contents and with slower cooling rates after casting.

It has been claimed that additions of titanium, zirconium, boron, sulfur, molybdenum, or copper decrease the tendency for cracking (Ref 36). The cooling rate between 1150 and 700 °C (2100 and 1290 °F) is important in controlling the amount of aluminum nitride precipitation. The minimum amount of aluminum nitride required to produce intergranular fracture is lower for alloy steels than for plain carbon steels (0.002 versus 0.004%) (Ref 37). Minimizing the nitrogen content, using the lowest possible amount of aluminum for deoxidation, and increasing the cooling rate after solidification are recommended, and it has been demonstrated that cracking can be prevented by deoxidizing with titanium or zirconium or by combined titanium and aluminum (Ref 37).

Some researchers have claimed that higher levels of phosphorus and sulfur reduce the susceptibility to aluminum nitride intergranular fractures (Ref 38). Nitrogen content has been shown to be more important than aluminum content because aluminum is always present in amounts greater than that required to tie up all of the nitrogen (Ref 41). Higher aluminum contents do, however, increase the solubility temperature of aluminum nitride and provide an additional driving force for precipitation.

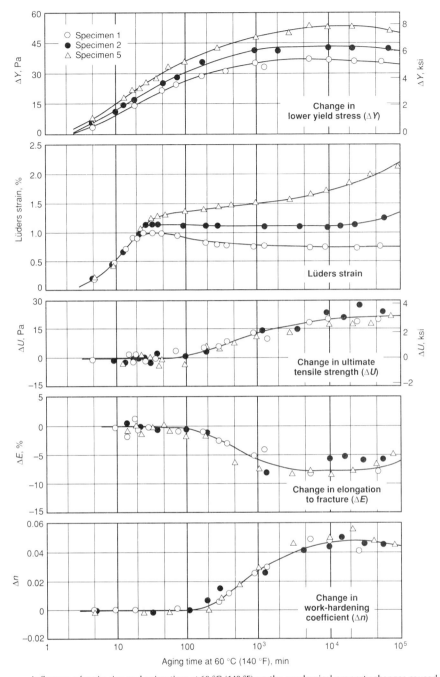

Fig. 12 Influence of grain size and aging time at 60 °C (140 °F) on the mechanical property changes caused by strain aging. A 0.038C-0.0042N-0.001Al steel was quenched from 200 °C (390 °F), prestrained 4%, and tested after different aging times. Grain sizes, in grains/mm² (ASTM number), were: specimen 1, 50 (2.7); specimen 2, 195 (4.7); specimen 5, 1850 (7.9). Source: Ref 34

Aluminum nitride is known to precipitate with one of two morphologies: plates or dendritic arrays (Ref 38). The dendritic form of aluminum nitride found on the intergranular fracture surfaces of aluminum nitride embrittled castings precipitates from the liquid near the conclusion of solidification (Ref 42). These aluminum nitride dendrites may be nucleation sites for platelike aluminum nitride that precipitates after solidification. The platelike aluminum nitride produces the small, shiny fracture surface facets that are generally observed (Ref 43).

Panel Cracking. Panel cracks are longitudinally oriented surface cracks on the side face of an ingot that generally form near the center of the face and can extend to the midradius of the ingot (Ref 44). Such cracks can occur on more than one ingot face and can run the entire length of the ingot. Panel cracks are observed in aluminum-killed steel ingots, particularly those with 0.4 to 0.7% C (plain carbon steels) or those with somewhat lower carbon contents for alloy steel ingots. These carbon contents lead to ferrite grain-boundary network films with predominantly pearlitic matrix structures. The susceptibility to panel cracking varies with melt practice: electric arc furnace steels are most prone; basic open hearth, basic oxygen furnace, and acid open hearth steels are less prone. Large ingots are more susceptible than small ingots. Stripping of the ingot at as hot a temperature as possible reduces susceptibility. Again, aluminum and nitrogen contents are very important. Panel cracking is not observed with less than 0.015% Al but occurs with increasing frequency with increasing aluminum content above this level. Also, for a given aluminum content, increasing the nitrogen above 0.005% increases panel cracking. Crack surfaces are oxidized but not decarburized. Because the intergranular cracks generally propagate along ferrite/pearlite interfaces, it has been suggested that cracking occurs at relatively low temperatures, probably below 850 °C (1560 °F) (Ref 44).

A statistical study of panel cracking found that the only significant variable was the level of the aluminum addition (Ref 45). It was suggested that cracking began internally and propagated to the surface because of cooling-induced stresses. In a study of panel cracking in two alloy steels, one containing 0.025% Al and 0.008% N and the other having 0.004% Al and 0.006% N, the former exhibited panel cracking, and the latter did not. Additions of aluminum and titanium to the crack-prone steel composition resulted in the elimination of grain-boundary ferrite networks and freedom from cracking (Ref 46).

A statistical study of panel cracking in low-carbon plate steels found that the steels did not exhibit grain-boundary ferrite networks, but rolled bloom surfaces were heavily cracked in some cases. Some of the observations in this study differ from those of other authors who have studied panel cracking. However, the statistical analysis showed that cracking increased with aluminum content and track time (time between ingot stripping and charging into the soaking pit). Cracking was minimized by holding the aluminum contents to 0.030% or less (Ref 47).

Reduced Hot Ductility. Numerous studies have demonstrated that increasing aluminum and nitrogen contents degrade hot ductility; this influence is most pronounced in the temperature range where aluminum nitride precipitation is greatest (Ref 48-53). One author found that 3.4% Ni raised the solution temperature of aluminum nitride in En 36 alloy steel by about 100 °C (210 °F) compared to plain carbon steels with the same aluminum and nitrogen levels (Ref 48). Another study evaluated the hot ductility of chromium-molybdenum-vanadium turbine rotor steels with 0.002 to 0.066% Al and 0.007 to 0.014% N. This study showed that nitrogen and residuals (copper, tin, antimony, and arsenic) reduced the hot ductility. The addition of titanium and/or boron improved the hot ductility in the test range (800 to 1000 °C, or 1470 to 1830 °F) where high nitrogen contents were detrimental (Ref 49).

In hot ductility tests on carbon-manganese steels containing 0.032 to 0.073% Al and 0.0073 to 0.0105% N, aluminum nitride was found to reduce deformability and increase the resistance to the deformation of austenite. These trends were enhanced as the aluminum nitride particle size decreased (Ref 50). Other hot ductility tests on low-carbon steels also showed a large decrease in hot ductility because of aluminum nitride precipitation, depending on the volume fraction and size of the precipitates (Ref 51). Five steels were tested containing different levels of soluble aluminum (aluminum not tied up with oxygen) and with nitrogen present as aluminum nitride. Figure 13 shows the reduction in area for hot tensile tests over a range of temperatures for five steels. Steels A and B showed a large decrease in percent reduction in area with increasing temperature; however, steel C did not exhibit such embrittlement. The difference between steels A, B, and C, which all had high levels of nitrogen in the form of aluminum nitride, was the particle size. The smaller aluminum nitride particles in steels A and B were detrimental to hot ductility, while the larger particles in steel C did not cause embrittlement; this agrees with the findings in Ref 50. Steel E contained 0.06% Ti and exhibited the best hot ductility.

An evaluation of the hot ductility of low-carbon killed steels found a substantial reduction in fracture strain between 700 and 800 °C (1290 and 1470 °F). This reduction was most pronounced between 750 and 775 °C (1380 and 1425 °F), where ferrite formed along the prior-austenite grain boundaries. The intergranular fracture surfaces exhibited aluminum nitride and manganese sulfide precipitates (Ref 53).

Hot ductility is also impaired by high residual impurity contents, chiefly those of copper and tin (Ref 54-60). Examinations of forging and rolling defects have frequently revealed concentrations of elemental copper, high levels of copper and tin, or copper in the scale. Tin residual levels are normally much lower than copper residual levels, but there appears to be a synergistic effect between copper and tin that enhances embrittlement (Ref 54). An examination of longitudinal cracks in medium- and high-carbon steels found that cracking occurred between 700 and 500 °C (1290 and 930 °F) and depended on high copper and tin contents. Aluminum was not added to these steels, but grain-boundary ferrite networks were present. The copper and tin were segregated to the ferrite networks (Ref 55).

Nickel, copper, tin, antimony, and arsenic often become enriched in the subscale layer at the surface of steels heated for forging and rolling in oxidizing atmospheres. It has been shown that tin, antimony, and arsenic residuals alter the solubility of copper in austenite during high-temperature soaking. Because the hot-working temperature is usually above the melting point of elemental copper, liquid copper is produced that will penetrate the austenite grain boundaries and cause cracking by liquid metal embrittlement. Nickel and molybdenum concentrate with copper and raise the melting point of copper; tin, antimony, and arsenic also concentrate at the scale/metal interface and lower the melting point of copper. If copper is not present, tin, antimony, and arsenic have little detrimental effect on hot workability (Ref 56).

It has been shown that tin reduces the solubility of copper in austenite, which is probably more important than its influence on the melting point of copper (Ref 56). Tin reduces the solubility of copper in austenite by a factor of three; therefore, when tin is present, molten copper can form at the surface at lower bulk copper contents. Nickel reduces copper-induced hot shortness, manganese slightly increases hot shortness, arsenic is slightly more detrimental than manganese, and tin and antimony are extremely detrimental to copper-induced hot shortness (Ref 58). Chromium decreases the solubility of copper in austenite and increases the susceptibility to copper-induced hot shortness, although its influence is small (Ref 59).

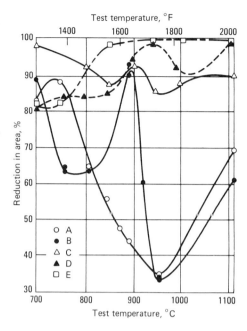

Fig. 13 Elevated-temperature tensile test results for five plain carbon steels containing various amounts of aluminum nitride. The nitrogen content (in ppm) of the steels in the form of aluminum nitride was: A, 80; B, 70; C, 72; D, 2; E, 1. Source: Ref 51

Low-carbon steel	Composition N, ppm(a)	Ti, %	Particle size nm	Å
Curve A	80	...	90	900
Curve B	70	...	80	800
Curve C	72	...	210	2100
Curve D	2	...	(b)	(b)
Curve E	1	0.06	(b)	(b)

(a) As AlN. (b) No data

Graphitization

In the early 1940s, several failures of welded joints in high-pressure steam lines occurred because of graphite formation in the region of the weld heat-affected zone that had been heated during welding to the critical temperature of the steel used (Ref 61-65). Extensive surveys of carbon and carbon-molybdenum steel specimens removed from various types of petroleum refining equipment revealed graphite in about one-third of the 554 specimens examined (Ref 61, 64). In most cases, graphite formation did not occur until about 40 000 h or more at temperatures from 455 to 595 °C (850 to 1100 °F). Aluminum-killed carbon steels were susceptible, but silicon-killed or low-aluminum killed steels were not. The C-0.5Mo steels were more resistant to graphitization than carbon steels, but they were similarly influenced by the nature of the deoxidation practice. Chromium additions and stress relieving at 650 °C (1200 °F) both retarded graphitization.

An examination of the graphitization susceptibility of a number of alloy steels showed that chromium-molybdenum steels used for steam piping, either annealed or normalized, were resistant to graphitization. Nickel and nickel-molybdenum steels did graphitize during high-temperature exposure. Chromium-bearing steels did not graphitize and appeared to be quite stable (Ref 65).

The deoxidation practice used in making the steels is the most important parameter influencing graphitization. High levels of aluminum deoxidation strongly promote graphitization for both carbon and carbon-molybdenum steels used for steam lines. While nitrogen appears to retard graphitization, high levels of aluminum remove nitrogen and thus promote graphitization (Ref 66). Molybdenum additions (0.5%) help stabilize cementite but do not fully offset the influence of high aluminum additions. Manganese and silicon both affect graphitization, but their influence is small at the levels used in these alloys. Chromium appears to be the best alloy addition for stabilizing carbides.

Tensile tests of affected steam piping indicate that the graphite present did not affect the tensile strength. Charpy V-notch impact strength, however, was reduced substantially. Localized graphitization near a welded joint appears to be much more damaging to pipe behavior than general, uniform graphitization. The localized graphitization apparently produces notches that concentrate stress and reduce load-bearing capability.

Overheating

The overheating of steels occurs when they are heated to excessively high temperatures prior to hot working (Ref 67-76). Heating at even higher temperatures will cause incipient grain-boundary melting, a problem known as burning. Thus, overheating occurs in the temperature range between the safe range normally used prior to hot working and the higher range where liquation begins. Hot working after burning generally results in the tearing or rupture of the steel due to the liquid in the grain boundaries. Hot working after overheating generally does not result in cracking; if sufficient hot reduction occurs, the influence of overheating may be minor or negligible. If the degree of hot reduction is small, mechanical properties, chiefly toughness and ductility, will be affected.

Fracture surfaces of overheated steels given limited hot reduction often exhibit a coarse-grain faceted appearance. Such features are most evident after quenching and tempering to develop optimum toughness. Other problems, such as aluminum nitride embrittlement, may also produce a faceted fracture surface. Additional tests are needed to distinguish among these problems (Ref 76).

Although the mechanical properties of burnt steels are always very poor, the mechanical properties of overheated steels show considerable scatter. For tensile tests, the elongation and reduction of area are most affected by overheating and decrease with increasing heating temperature. Fracture faceting and substantial decreases in tensile ductility normally are observed after severe overheating.

Impact properties are usually more sensitive to overheating than is tensile ductility. In examining impact test results, several interrelated features should be examined:

- Change in upper-shelf energy
- Impact strength transition temperature
- Presence of facets

Upper-shelf energy appears to be particularly sensitive to overheating. Figure 14 shows impact energy curves for En lll alloy steel specimens heated to a variety of temperatures from the normal soaking range to the burning range; no hot working was performed. The specimens were first heated to 950 °C (1740 °F) for 10 min, transferred to the high-temperature zone for 7 min, then transferred back to the furnace at 950 °C (1740 °F) and held for 50 min. They were oil quenched, tempered 1 h at 675 °C (1245 °F), and water quenched to minimize temper embrittlement (more information about temper embrittlement is given in the section "Temper Embrittlement in Alloy Steels" in

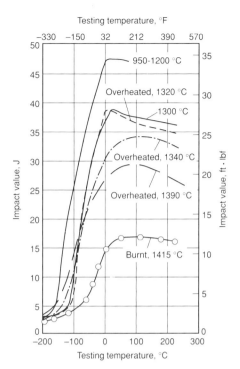

Fig. 14 Impact energy values versus test temperature for En 111 alloy steel specimens heated to the indicated temperature for 1 h, oil quenched, and tempered for 1 h at 675 °C (1245 °F). Source: Ref 68

this article). The use of temperatures up to 1200 °C (2190 °F) produced no change in impact energies, but temperatures above 1200 °C (2190 °F) produced a decrease in upper-shelf energy. The burnt specimens displayed a substantial loss in toughness. Because the pieces were not forged after the burning treatment, they did not fracture.

Presence of Facets. Many of the studies of overheating have been conducted without subsequent hot working after overheating and thus do not simulate actual commercial experience. These studies are of limited value. Numerous theories have been proposed to explain overheating. The examination of facets on fractures of overheated steels reveals that the facet surfaces are covered with fine ductile dimples, and small manganese sulfides can be found within the dimples (because two fracture surfaces are formed, a manganese sulfide particle will be found in either half of the mating dimples after fracture). The facet surfaces correspond to prior-austenite grain surfaces formed during overheating.

As the soaking temperature is raised, manganese sulfide in the steel dissolves (that is, the sulfur goes into solution in the austenite). Dissolved sulfur diffuses toward the austenite grain boundaries, where it reprecipitates. Therefore, overheating changes the size and distribution of sulfides in the steel. The cooling rate through the overheating range also affects the size and dispersion of the intergranular sulfides. In commercial practice, the size of the overheated piece, and any externally applied coolant during hot working, will control this cooling rate.

Steels containing less than 10 ppm S do not overheat, regardless of the heating temperature up to burning. However, this level of sulfur is difficult to obtain. Steels with relatively low sulfur contents, (for example, in the range of 0.001 to 0.005%) are being produced in greater quantities today because of the detrimental influence of sulfur on properties. However, it has been demonstrated that the minimum temperature at which overheating occurs in low-sulfur steels is lower than that for high-sulfur steels (>0.010% S). Overheating problems thus have been experienced in low-sulfur steels heated at temperatures that usually do not cause overheating (Ref 73-75). Additions of rare earth elements have been shown to prevent overheating by modifying the solubility of the sulfide formed. High-sulfur steels appear to require a greater degree of overheating to cause fracture faceting and impaired properties than do low-sulfur steels.

The problem of overheating is complex. Sulfide dissolution and reprecipitation at the prior-austenite grain boundaries causes fracture faceting and impairment of properties. However, when faceting is observed, mechanical properties may not be significantly affected. Overheating and its influence on properties depend on the sulfur content, the soaking temperature, grain size, the cooling rate through the overheating range, and the degree of hot reduction. Furthermore, the amount of faceting observed on the test fracture depends on the heat treatment (particularly the tempering temperature), the test temperature, the test specimen orientation, and the amount of deformation after sulfide reprecipitation.

Thermal Embrittlement of Maraging Steels

Maraging steels will fracture intergranularly at low impact energies if improperly processed after hot working. This problem, known as thermal embrittlement, occurs when maraging steels that have been heated above 1095 °C (2000 °F) are slowly cooled through, or held within, the temperature range of 980 to 815 °C (1800 to 1500 °F) (Ref 77-81). The embrittlement is caused by the precipitation of TiC and/or Ti(C,N) on the austenite grain boundaries during cooling through, or holding within, the critical temperature range. The degree of embrittlement increases with time within the critical range. Increased levels of carbon and nitrogen render maraging steels more susceptible to thermal embrittlement. Auger analysis has shown that embrittlement begins with the diffusion of titanium, carbon, and nitrogen to the grain boundaries, and observation of TiC or Ti(C,N) precipitates represents an advanced stage of embrittlement.

Results of a study on thermal embrittlement demonstrate its influence on an 18Ni(250) maraging steel (Ref 77). Plates 12.7 mm (0.5 in.) thick were rolled with finishing temperatures in the range of 1080 to 870 °C (1980 to 1600° F) and then cooled by three different methods (water, air, and vermiculite). Test results showed that the finishing temperature and cooling rate from the finishing temperature had minor effects on the tensile properties but a significant effect on fracture toughness. Figure 15 shows the plane-strain fracture toughness results as a function of finishing temperature (temperature at the end of rolling) and cooling rate for hot-rolled and aged material (Fig. 15a) and for annealed and aged material (Fig. 15b and 15c). In general, vermiculite cooling (slow) and a high finishing temperature produced the lowest toughness, except for specimens water quenched after rolling (Fig. 15b and 15c). Detailed information on maraging steels is available in the article "Maraging Steels" in this Volume.

Quench Cracking

The production of fully martensitic microstructures in steels requires a heat treatment cycle that involves quenching after austenitization. The composition of the steel, its size, and the desired depth of hardening dictate the required quench medium. Certain steels are known to be susceptible to cracking during or slightly after quenching. This is a relatively common problem for tool steels, particularly those that require liquid quenching (Ref 82).

Numerous factors can contribute to cracking susceptibility (Ref 82-87):

- Carbon content
- Hardenability
- M_s temperature (the temperature at which martensite starts to form)
- Part design
- Surface quality
- Furnace atmosphere
- Heat treatment practice

As the carbon content is raised, the M_s and M_f (temperature at which martensite formation ends) temperatures decrease, and the volumetric expansion and transformation stresses accompanying martensite formation increase. In general, steels with less than 0.35% C are free of quench cracking problems. Such low-carbon steels have higher M_s and M_f temperatures, which allow some stress relief to occur during the quench. Also, transformation stresses are lower, and the lower strength of the martensite formed (low-carbon lath martensite) can accommodate the strains more readily than can a higher-carbon steel.

Alloy steels with ideal critical diameters of 4 or greater are more susceptible to quench cracking than are lower-hardenability steels. Quench crack sensitivity also increases as the severity of the quench rate increases. Control of the austenitizing temperature is also important, particularly for high-carbon tool steels. Excessive retained austenite and coarse-grain structures promote quench cracking. Quench uniformity is important, particularly when liquid quenchants are employed. When high-carbon steels are quenched to form martensite, they are in a highly stressed condition. Therefore, tempering must be done immediately after quenching to relieve these stresses and minimize the risk of cracking. Surface quality is also very important because seams, laps, tool marks, stamp marks, and so on, act as stress concentrators to locate and enhance quench cracking susceptibility (Ref 82).

Quench cracking is a problem that often defies prediction and can be difficult to diagnose. Heat treaters have experienced short time periods in which cracking problems occur frequently and then stop for no apparent reason. Evidence also indicates that quench cracking can be more frequent for certain heats of steel, again for no obvious reason. Instances have also been documented (Ref 86) in which extensive quench cracking has occurred in material processed from the bottom portion of ingots.

Quench crack fractures are always intergranular. In quenched and tempered steels, proof of quench cracking is often obtained by opening a crack (if necessary) and visually looking for the temper color typical for the temperature used (Ref 88, 89). If the crack occurs during or after quenching but before tempering, and if the crack is open to the furnace atmosphere, a thin oxide layer will form on the surface. The color of the oxide layer depends on its thickness, which in turn depends on the tempering temperature and the steel composition. Quench cracks begin at the surface and propagate inward. They are usually oriented longitudinally or radially unless located by a change in section size, by surface imperfections, or by changes in surface microstructure (such as an interface between hardened and nonhardened areas).

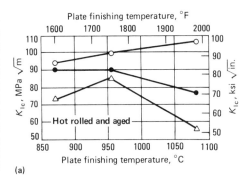
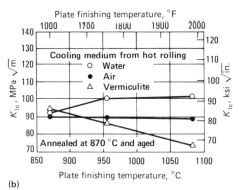
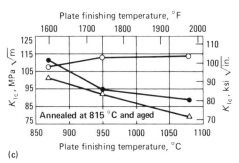

Fig. 15 Influence of mill finishing temperature and manner of cooling on the plane-strain fracture toughness (K_{Ic}) of 18Ni(250) maraging steel heat treated three ways. (a) Hot rolled and aged. (b) Annealed at 870 °C (1600 °F) and aged. (c) Annealed at 815 °C (1500 °F) and aged. Source: Ref 77

Temper Embrittlement in Alloy Steels

Temper embrittlement—also known as temper brittleness*, two-step temper embrittlement, or reversible temper embrittlement—is a problem associated with tempered alloy steels that are heated within, or slowly cooled through, a critical temperature range, generally 300 to 600 °C (570 to 1110 °F) for low-alloy steels. This treatment causes a decrease in toughness as determined with Charpy V-notch impact specimens (Ref 90-114). It is a particular problem for heavy-section components, such as pressure vessels and turbine rotors, that are slowly cooled through the embrittling range after tempering and also experience service at temperatures within the critical range.

Temper embrittled steels exhibit an increase in their DBTT and a change in fracture mode in the brittle test temperature range from cleavage to intergranular. The DBTT can be assessed in several ways, the most common being the temperature for 50% ductile and 50% brittle fracture (50% fracture appearance transition temperature, or FATT), or the lowest temperature at which the fracture is 100% ductile (100% fibrous criterion). Transition temperatures based on absorbed energy values are not normally employed. Temper embrittlement is reversible; that is, the toughness of embrittled steels can be restored by tempering them above the

---

*The term temper brittleness was first used to describe this problem by J.H.S. Dickenson in 1917.

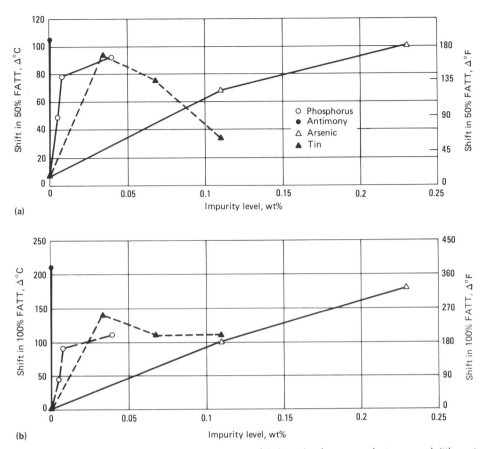

Fig. 16 Influence of phosphorus, antimony, arsenic, and tin impurity elements on the temper embrittlement susceptibility of nickel-chromium experimental steels based on the change in (a) 50% FATT and (b) 100% fibrous FATT after aging at 450 °C (840 °F) for 1000 h. Source: Ref 91

critical region followed by rapid cooling, for example, water quenching. This decreases the DBTT and changes the low-temperature (that is, below the 50% FATT) intergranular brittle appearance back to the cleavage mode.

Temper embrittlement does not occur in plain carbon steels, only in alloy steels. Also, the degree of embrittlement varies with alloy steel composition. Therefore, the alloying elements present, and their combinations and levels, are important. However, certain impurity elements must be present if temper embrittlement is to occur. The embrittling impurities are, in decreasing order of influence on a weight percent basis, antimony, phosphorus, tin, and arsenic. Of these elements, phosphorus is most commonly present in alloy steels and it has captured the most attention in research studies. Manganese and silicon also increase the susceptibility to embrittlement. Although alloy steels are ferritic in the tempered condition, fracture below the DBTT occurs along prior-austenite grain boundaries where both alloying elements and impurity elements are concentrated.

Effect of Composition on ΔFATT

The proof that antimony, phosphorus, tin and/or arsenic is an essential ingredient(s) for temper embrittlement was obtained in the late 1950s (Ref 90, 91). The change in 50% FATT and 100% fibrous FATT with isothermal aging at 450 °C (230 °F) for up to 1000 h for nickel-chromium and nickel-chromium-molybdenum laboratory heats with controlled compositions and impurity levels was determined. Figures 16 and 17 show some test results for the influence of antimony, phosphorus, tin, and arsenic and aging at 450 °C (840 °F) for 1000 h (Ref 90, 91). Embrittlement was greater for the nickel-chromium steels than for the nickel-chromium-molybdenum steels because of the beneficial influence of molybdenum. The nickel-chromium steels also showed substantially greater embrittlement from the manganese addition than did the nickel-chromium-molybdenum steels. The addition of about 0.7% Si had a smaller and similar embrittling influence for both grades.

The important role of alloying elements has been clearly demonstrated in tests that also used heats of controlled composition (Ref 92). The tests were performed using 0.4% C alloy steels containing nickel, chromium, or nickel and chromium, as well as a plain carbon steel. Controlled additions of antimony, phosphorus, tin, and arsenic were made to these compositions. The 50% FATT was evaluated for each composition after heat treatment (870 °C, or 1600 °F, for 1 h, oil quench; 625 °C, or 1155 °F, for 1 h, water quench) and after step-cool embrittlement. Figure 18 shows the results for additions of antimony to plain carbon and alloy steels of various analyses. The bars show the 50% FATT after tempering at 625 °C (1155 °F) (left end—not embrittled) and after step cooling (right end—embrittled); the value under the bar is the shift in 50% FATT. The addition of 600 to 800 ppm Sb to the 0.4C-3.5Ni-1.7Cr steel caused a shift in transition temperature of 695 °C (1285 °F). The same steel, but without carbon, exhibited a shift of 222 °C (432 °F), but its hardness was much lower (~80 HRB versus 27 HRC). The steels with only nickel and carbon or chromium and carbon and 600 to 800 ppm Sb exhibited much less embrittlement, while the plain carbon steel was not embrittled by antimony.

Figure 19 shows the results for additions of about 500 ppm each of phosphorus, tin, and antimony to the nickel-chromium-carbon, nickel-carbon, and chromium-carbon steels. The nonembrittled toughnesses of the nickel-chromium-carbon-phosphorus and chromium-carbon-phosphorus alloys were poorer than those of the other alloys shown, probably because of the segregation of phosphorus in austenite. Phosphorus also embrittled the chromium-carbon-phosphorus alloy much more than the nickel-carbon-phosphorus alloy. This is due to an interaction between chromium and phosphorus. Tin appears to embrittle the nickel-chromium-carbon alloy more than phosphorus in that the change in fracture appearance transition temperature (ΔFATT) was greater. However, the grain size of the nickel-chromium-carbon-phosphorus alloy was ASTM No. 8, while that of the nickel-chromium-carbon-tin alloy was ASTM No. 6. Also, the nonembrittled toughness of the alloy containing phosphorus was much poorer. The 50% FATT values for these two compositions are nearly identical and would be even closer if the grain sizes were the same. Otherwise, it appears that tin embrittled the nickel-carbon alloy more than phosphorus, while phosphorus embrittled the chromium-carbon alloy more than tin. Arsenic was a much weaker embrittler. The results in Fig. 18 and 19 clearly show that the combination of nickel and chromium resulted in much greater embrittlement, particularly for additions of antimony and tin. The results for phosphorus show that a strong interaction exists between chromium and phosphorus, while phosphorus causes little embrittlement in nickel steels not containing chromium.

The beneficial influence of molybdenum on phosphorus-induced temper embrittlement has been known for many years (Ref 93, 94). It has also been known that carbon steels are immune to temper embrittlement, but that substantial additions of manganese cause susceptibility to this problem. In ad-

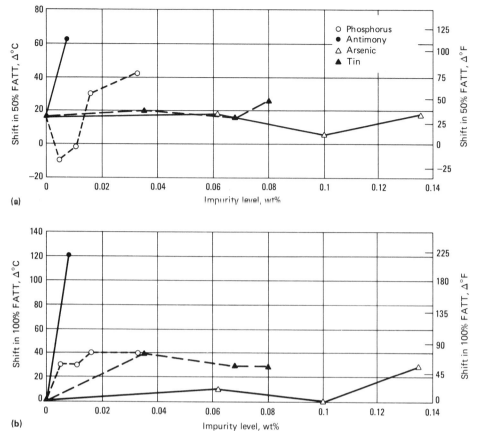

Fig. 17 Influence of phosphorus, antimony, arsenic, and tin impurity elements on the temper embrittlement susceptibility of nickel-chromium-molybdenum experimental steels based on the change in (a) 50% FATT and (b) 100% fibrous FATT after aging at 450 °C (840 °F) for 1000 h. Source: Ref 91

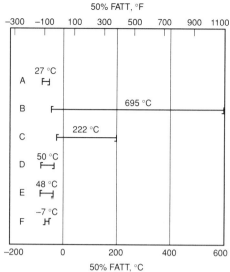

Fig. 18 Influence of alloying elements on the temper embrittlement of steels (compositions given in accompanying table) containing 600 to 800 ppm Sb. The left end of bar gives the nonembrittled DBTT; the right end of bar gives the DBTT after embrittlement (except for line F, which is reversed). Value between bar ends is the shift in 50% FATT. Source: Ref 92

Steel	Composition, wt%			
	Ni	Cr	C	Sb
Alloy				
Line A.........	3.5	1.7	0.4	...
Line B.........	3.5	1.7	0.4	0.06–0.08
Line C.........	3.5	1.7	...	>0.08
Line D.........	3.5	...	0.4	0.06–0.08
Line E.........	...	1.7	0.4	0.06–0.08
Plain carbon				
Line F.........	...	...	0.4	>0.08

dition, high levels of manganese in alloy steels have been known to render them more susceptible to temper embrittlement.

Reference 95 includes an evaluation of the addition of 0.6% Mo, along with controlled antimony, phosphorus, tin, arsenic, and other elements, to AISI 3340 (3.5Ni-1.7Cr-0.4C). The addition of molybdenum eliminated or greatly reduced embrittlement due to step cooling for additions of antimony, tin, and arsenic, but not for additions of phosphorus. The addition of 0.7% Mn to this steel produced substantial embrittlement, which was largely eliminated when 0.6% Mo was added. In this work, it was shown that phosphorus segregates to the austenite grain boundaries during austenitization; antimony does not do this. This work also clearly showed that manganese is an embrittling element, not merely an enhancer of embrittlement.

Later work was conducted to clarify these results (Ref 96). The earlier work had employed specimens with a very coarse grain size (Ref 95). Results with somewhat finer grain sizes, although still rather coarse, showed that 0.5 to 0.6% Mo additions would prevent temper embrittlement caused by phosphorus in 3.5Ni-1.7Cr-0.2C steels for aging times up to 1000 h at 475 and 500 °C (885 and 930 °F). The influence of molybdenum on the prevention of temper embrittlement appears to depend on how much of it is dissolved in the matrix as opposed to how much is tied up in carbides. As more molybdenum becomes tied up in the carbides, the beneficial influence of molybdenum decreases. Therefore, depending on the temperatures experienced and the presence of other strong carbide formers, molybdenum may or may not be able to suppress temper embrittlement.

A major advancement in understanding temper embrittlement was brought about by the development of Auger electron spectroscopy and its application to embrittlement studies in 1969 (Ref 97-99). This work permitted the direct chemical analysis of segregants on the intergranular fracture surfaces of embrittled specimens. Such work has shown that the embrittling impurity elements are segregated to within the first few monolayers of the embrittled grain boundaries. The degree of enrichment of these elements may be 100 to 10^3 times the bulk concentration. Alloy element segregation at these boundaries was also detected. However, the concentration of these alloying elements was found to be only 2 to 3 times that of the bulk concentration, and the concentration profile from the grain boundary into the grain interior was much shallower than for the impurity elements. Figure 20 shows an example of Auger analysis of antimony, sulfur, and phosphorus segregated to either fracture or free surfaces (Ref 99, 100). These results were obtained by alternate argon ion sputtering (depth profiling) and analysis.

The degree of embrittlement also depends upon the time at temperature within the critical region. Extensive isothermal embrittlement studies were performed on a heat of AISI/SAE 3140 alloy steel (0.39C-0.79Mn-0.30Si-0.028S-0.015P-1.26Ni-0.77Cr-0.02Mo, ASTM No. 8 grain size) (Ref 101). Specimens were austenitized at 900 °C (1650 °F) for 1 h, water quenched, and tempered at 675 °C (1245 °F) for 1 h, and water quenched. This resulted in a hardness of 23 HRC and a transition temperature (100% fibrous criterion) of −83 °C (−117 °F). Specimens were aged at temperatures from 325 to 650 °C (615 to 1200 °F) for times ranging from 4 min to 240 h. Figure 21 shows the time-temperature-embrittlement diagram developed from this work.

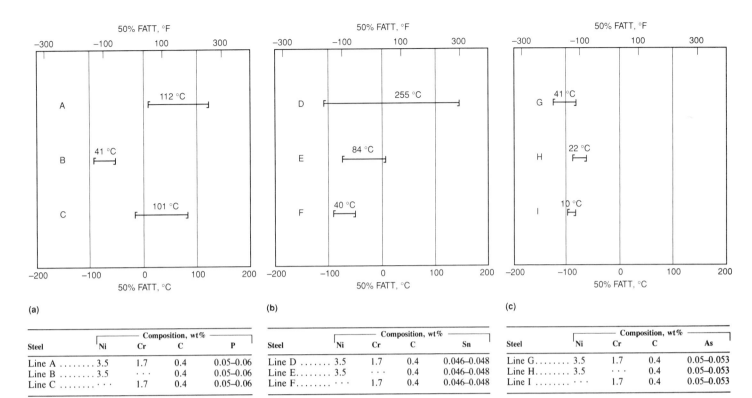

Fig. 19 Influence of alloying elements on the temper embrittlement of steels (compositions given in accompanying tables). (a) Steel containing 500 to 600 ppm P. (b) Steel containing 460 to 480 ppm Sn. (c) Steel containing 500 to 530 ppm As. The left end of bar gives the nonembrittled DBTT; the right end of bar gives the DBTT after embrittlement. Value between bar ends is the shift in 50% FATT. Source: Ref 92

Because of some minor differences in hardness and apparent inconsistencies in the results for aging at 525 °C (975 °F), additional work was performed (Ref 102). Minor corrections were made in the transition temperatures to account for hardness differences (Fig. 22). Only limited tests were done at temperatures above 575 °C (1065 °F), but the work performed indicated that temper embrittlement occurred at tempering temperatures up to 675 °C (1245 °F), which is close to the lower critical temperature. The diagrams exhibit the classic C-curve appearance. The nose of the curve is at 550 °C (1020 °F), but maximum embrittlement (ΔFATT ≈ 100 °C, or 210 °F) was obtained at 475 to 500 °C (885 to 930 °F) after 240 h aging (the longest time used). Tests done using step cooling showed that the degree of embrittlement that occurred was greater than would have been predicted from the isothermal data.

The above data was analyzed to predict the degree of grain-boundary phosphorus segregation in this steel (analysis for anti-

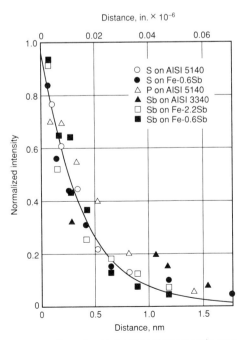

Fig. 20 Normalized intensities of Auger peaks (as a function of depth below the surface) from antimony, sulfur, and phosphorus segregated to grain boundaries or free surfaces (depth profiling by argon ion sputtering). Source: Ref 99

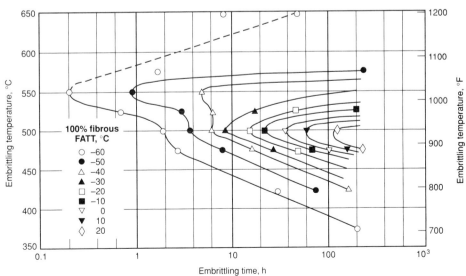

Fig. 21 Time-temperature diagram for isothermally temper-embrittled AISI/SAE 3140 alloy steel showing constant embrittlement levels (100% fibrous FATT) for quenched and tempered (675 °C, or 1245 °F, for 1 h) specimens. Source: Ref 101

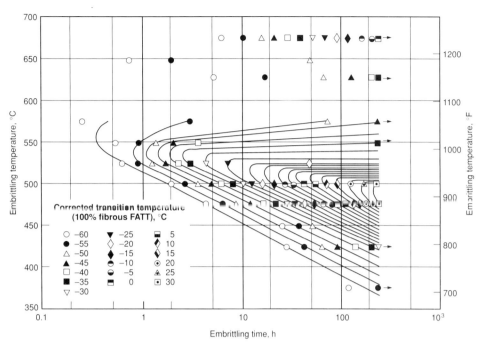

Fig. 22 Revised time-temperature diagram for temper-embrittled AISI/SAE 3140 alloy steel. Source: Ref 102

mony, tin, and arsenic was not performed, but the amount of these elements was assumed to be very low) (Ref 103). Auger analysis of similar steels was used to calculate the monolayers of phosphorus segregated to the prior-austenite grain boundaries (Fig. 23).

Microstructure and Grain Size. It is well known that microstructure influences the susceptibility to temper embrittlement and the resulting degree of embrittlement. Because the impurity and alloying elements segregate to the prior-austenite grain boundaries, grain size has an influence. As the grain size becomes larger, the grain-boundary surface area decreases. Therefore, for a fixed level of impurities and constant embrittling temperatures and time, there will be greater coverage of the grain boundaries in a coarse-grain steel than in a fine-grain steel. However, the distance over which the impurities must diffuse increases as the grain size becomes larger. Nevertheless, coarse-grain steels are recognized to be more severely embrittled than fine-grain steels. Figure 24 shows the results of aging a 0.33C - 0.59Mn - 0.03P - 0.031S - 0.27Si - 2.92Ni - 0.87Cr steel for various times at 500 °C (930 °F) followed by water quenching. Prior to aging, the specimens had been austenitized (at 850 °C, or 1560 °F, for fine-grain specimens; at 1200 °C, or 2190 °F, for coarse-grain specimens) and tempered at 650 °C (1200 °F) for 1 h; they were oil quenched after tempering. The coarse-grain specimens were embrittled to a greater extent than the fine-grain specimens. Similar results have been obtained by others (for example, Ref 105).

Matrix microstructures are also important because they control toughness, for both nonembrittled and embrittled steels. Most studies have evaluated temper embrittlement of martensitic specimens, but a few have compared results for a variety of microstructures as a function of hardness. In general, tempered martensite is more susceptible than tempered bainite to temper embrittlement, but tempered bainite is more susceptible than pearlitic-ferritic structures. This analysis is somewhat misleading, however, because nonembrittled tempered martensite is much tougher than nonembrittled bainite at the same hardness, and after embrittlement, tempered martensite is still tougher than tempered bainite. The same holds true when comparing bainitic and pearlitic-ferritic microstructures. This has been demonstrated for chromium-molybdenum-vanadium steels, as shown in Fig. 25 (Ref 106). The alloy composition was 0.3C-0.91Mn-0.27Si-0.15Ni-1.3Cr-1.2Mo-0.31V-0.025P-0.0045S-0.005As-0.0008Sb-0.027Sn. The toughness of embrittled tempered martensite was better than that of nonembrittled tempered bainite over the hardness range evaluated. Also, the shift in 50% FATT for tempered martensite increased with hardness and was greater than that for tempered bainite. Only one hardness level was obtained for the ferrite-pearlite condition, and the shift in 50% FATT because of embrittlement (step cooling) was only 5 °C (9 °F).

Embrittlement Predictive Equations. Much of the research on temper embrittlement has concentrated on the influence of composition on susceptibility to temper embrittlement for fixed embrittlement conditions. In general, these studies have concentrated on two basic steel grades, nickel-chromium-molybdenum-vanadium and 2¼Cr-1Mo, which are used for rotors and pressure vessels, respectively.

In a report on an ASTM study of vacuum carbon deoxidized nickel-chromium-molybdenum-vanadium rotor steels isothermally embrittled at 400 °C (750 °F) for 10 000 h (Ref 107), the shift in FATT (ΔFATT) in degrees centigrade was correlated to the impurity content and molybdenum concentration (all in weight percent) by:

$$\Delta\text{FATT} = 7524\text{P} + 7194\text{Sn} + 1166\text{As} - 52\text{Mo} - 450\,000(\text{P} \times \text{Sn}) \quad \text{(Eq 1)}$$

No significant influence was found for antimony. Equation 1 states that phosphorus, tin, and arsenic increased embrittlement, while molybdenum decreased it. Also, a phosphorus-tin interaction that decreased embrittlement was observed.

A correlation between the 50% FATT and impurity content (J factor) for both nickel-chromium-molybdenum-vanadium and 2¼Cr-1Mo steels has been demonstrated (Ref 108). The J factor equation is:

$$J = (\text{Mn} + \text{Si})(\text{P} + \text{Sn}) \times 10^4 \quad \text{(Eq 2)}$$

where all concentrations are in weight percent.

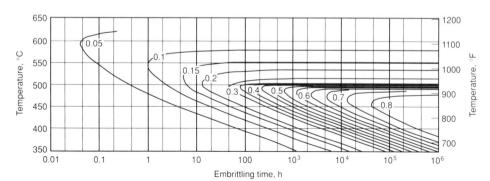

Fig. 23 Time-temperature diagram for the segregation of phosphorus in temper-embrittled AISI/SAE 3140 alloy steel. The numbers next to the curves describe the degree of phosphorus segregated during the embrittlement treatment (not including the 0.06 monolayers of phosphorus segregated prior to the isothermal aging treatments). Source: Ref 103

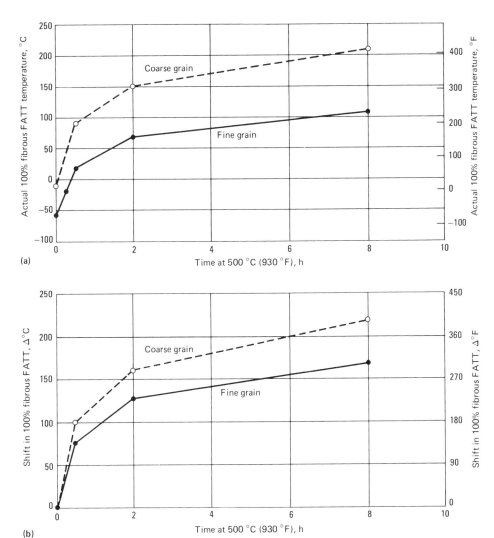

Fig. 24 Influence of prior-austenite grain size on the temper embrittlement of a nickel-chromium alloy steel that was heat treated to produce two levels of grain size. The alloy was tempered at 650 °C (1200 °F) and aged various times at 500 °C (930 °F). (a) Actual 100% fibrous FATT. (b) Change in 100% fibrous FATT. Source: Ref 104

A more detailed correlation has been provided for nickel-chromium steels doped with manganese, phosphorus, and tin (Ref 109). The equation combines the grain-boundary phosphorus and tin concentrations, the prior-austenite grain size, and the hardness level. Equation 3 was extended to a nickel-chromium-molybdenum-vanadium steel with hardnesses of 20 and 30 HRC, ASTM grain sizes of No. 3 and No. 7, and isothermal embrittlement at 480 °C (895 °F) for 6000 h (Ref 110). The resulting equation was:

$$\Delta \text{FATT} = 4.8P + 24.5\text{Sn} + 13.75(7 - \text{GS})$$
$$+ 2(\text{HRC} - 20) + 0.33(\text{HRC} - 20)$$
$$(P + \text{Sn}) + 0.036(7 - \text{GS})(\text{HRC} - 20)$$
$$(P + \text{Sn}) \quad \text{(Eq 3)}$$

where ΔFATT is in degrees centigrade, the concentrations of phosphorus and tin are the Auger peak-height ratios with respect to iron, HRC is the Rockwell hardness, and GS is the grain size number.

The above discussion shows that carbon steels are immune to temper embrittlement and that only alloy steels are susceptible to it. Both bulk composition and impurity levels are important, although without the latter, temper embrittlement will not occur. The most potent impurity elements are antimony, phosphorus, tin, and arsenic (in order of decreasing potency). Manganese and silicon also promote embrittlement. Alloys containing nickel and chromium in combination are more susceptible than those that contain nickel or chromium separately. Molybdenum additions are effective in retarding or eliminating temper embrittlement when impurities are present. However, to be effective, the molybdenum must be dissolved in the ferritic matrix, not tied up as carbides. Embrittlement occurs because of tempering or service time within a temperature range of 300 to 600 °C (570 to 1110 °F) or because of slow cooling through this range. Coarse-grain material is more susceptible than fine-grain material. The degree of embrittlement is greater for martensite than bainite and least for ferrite-pearlite. However, embrittled martensite is still tougher than nonembrittled bainite, while embrittled bainite is tougher than nonembrittled ferrite-pearlite. Additional information may be found in Ref 110 to 114.

Tempered Martensite Embrittlement

Tempered martensite embrittlement (also known as 350 °C, or 500 °F, embrittlement and as one-step temper embrittlement) of high-strength alloy steels occurs upon tempering in the range of 205 to 370 °C (400 to 700 °F) (Ref 115-132). It differs from temper embrittlement in the strength of the material and the temperature exposure range. In temper embrittlement, the steel is usually tempered at a relatively high temperature, producing lower strength and hardness, and embrittlement occurs upon slow cooling after tempering and during service at temperatures within the embrittlement range. In tempered martensite embrittlement, the steel is tempered within the embrittlement range, and service exposure is usually at room temperature. Therefore, temper embrittlement is often called two-step temper embrittlement, while tempered martensite embrittlement is often called one-step temper embrittlement.

Because alloy steels are heat treated to form martensite and then may be tempered within the embrittlement region, this problem has been widely referred to as tempered martensite embrittlement (TME). While some have claimed that structures other than martensite are not embrittled by tempering in this region, it is well established that lower bainite is embrittled when tempered in this region. Other structures, such as upper bainite and pearlite/ferrite, are not embrittled by tempering in this region.

While temper embrittlement is evaluated by the change in the ductile-to-brittle transition temperature, most studies of tempered martensite embrittlement have evaluated only the change in room-temperature impact energy. In general, when an as-quenched alloy steel is tempered, the toughness at room temperature increases with tempering temperature up to about 200 °C (390 °F). However, with further increases in tempering temperature, the toughness decreases. Then, with increasing tempering temperatures above about 400 °C (750 °F), the toughness increases again. This change in toughness with tempering temperature is not apparent when examining hardness or tensile strength data, which generally decrease with increasing tempering temperatures.

To illustrate the influence of tempering temperature on properties, Fig. 26 shows tension, hardness, and Charpy V-notch room-temperature tests for AISI 4340 alloy steel (Ref 115). Room-temperature impact

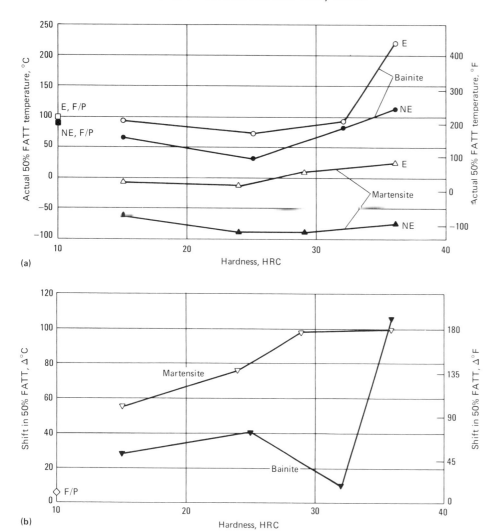

Fig. 25 Influence of microstructure on the temper embrittlement susceptibility of a chromium-molybdenum-vanadium alloy steel as a function of hardness. (a) Actual 50% FATT. (b) Change in 50% FATT. F/P, ferrite-30% pearlite structure; E, embrittled; NE, nonembrittled. Source: Ref 106

energy decreases with tempering in the critical region (205 to 370 °C, or 400 to 700 °F). The tensile and hardness results show no influence at room temperature. Figure 27 shows room-temperature Charpy V-notch impact energy for AISI 4340 at two different phosphorus levels, 0.003 and 0.03%, as a function of tempering temperature (Ref 128). The low-phosphorus specimens have consistently higher impact energies at all tempering temperatures, and both suffer a reduction in impact energy with tempering in the critical region. As a final example, Fig. 28 shows plane-strain fracture toughness and Charpy V-notch impact energy at room temperature for 300M alloy steel (a high-silicon modification of AISI 4340) as a function of tempering temperature (Ref 127). Both test procedures show somewhat similar results, although the tempering temperature ranges for embrittlement are slightly different.

Activating Mechanism. The mechanism causing TME is not as well defined as that which causes temper embrittlement. While many studies have shown that fractures are partly or substantially intergranular, particularly with tempering at about 350 °C (660 °F), other studies have observed only transgranular fractures. This difference may influence the respective interpretations of the TME mechanism. Also, studies have been made using controlled impurity and composition alloys or commercial alloys, and the factors that cause TME may differ for these two types of alloys.

The earlier studies of TME noted that embrittlement coincided with the initiation of cementite precipitation (Ref 115). Such low-temperature precipitated cementite is long and platelike and is present at the grain boundaries. Therefore, the shape and location of the cementite produced at low tempering temperatures was cited as the cause of TME (Ref 116).

The emphasis then shifted to the study of the influence of impurity elements, chiefly phosphorus (which segregates to the prior-austenite grain boundaries during austenitization). This shift was brought about by the observation that TME did not occur in high-purity steels but did occur in steels containing elements such as phosphorus (Ref 117, 118). Nitrogen was also observed to cause embrittlement, as were large amounts of manganese and silicon.

After these developments, several studies suggested that impurities segregated to prior-austenite grain boundaries and cementite precipitation along grain boundaries were both required for TME. The segregation of impurities was seen as a necessity; cementite precipitation was cited as an important factor, but not one that would cause TME by itself (Ref 120-122). At about the same time, transmission electron microscopy showed that martensite in as-quenched medium-carbon alloy steels has small amounts of retained austenite films present between the laths. Several studies attributed TME to the decomposition of interlath-retained austenite to interlath cementite. It was suggested that alloying elements that promote the decomposition of retained austenite to cementite at lower temperatures, such as manganese and chromium, enhance TME, while alloying elements that do not promote austenite decomposition at low temperatures, such as silicon, nickel, and aluminum, shift TME to higher temperatures. Most of these studies exhibited transgranular test fractures of controlled-composition alloys.

Other studies have observed cementite precipitation from interlath-retained austenite but have described a more complex process for TME. For example, a study of TME in AISI 4340 and 300M alloy steels stated that the magnitude of TME increased significantly when the volume fraction of interlath-retained austenite was high. A partial decomposition of the retained austenite to interlath cementite upon tempering was observed. Subsequent deformation during testing transformed the balance of the retained austenite that became destabilized because of the depletion of carbon during cementite precipitation. The study further concluded that the segregation of impurity elements at the prior-austenite grain boundaries contributes to the embrittlement process by providing an alternate weak path for fracture to follow. The study thus accounted for either transgranular or intergranular fracture modes.

The concept that a dual role played by interlath carbides is the primary problem and impurity segregation is a secondary influence has been verified by others (Ref 128-130). In an evaluation of TME in AISI 4340 containing 0.003 and 0.03% P, both steels exhibited TME (Ref 128). For the low-phosphorus steel, interlath cementite initiated cleavage across the martensite laths, while for the high-phosphorus steel, intergranular fracture resulted from the combined influence of carbide formation and phosphorus segregation. However, the

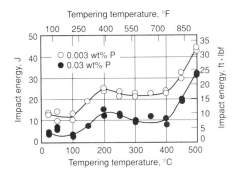

Fig. 27 Influence of phosphorus on the room-temperature Charpy V-notch impact energy of AISI 4340 alloy steel as a function of tempering temperature. Source: Ref 128

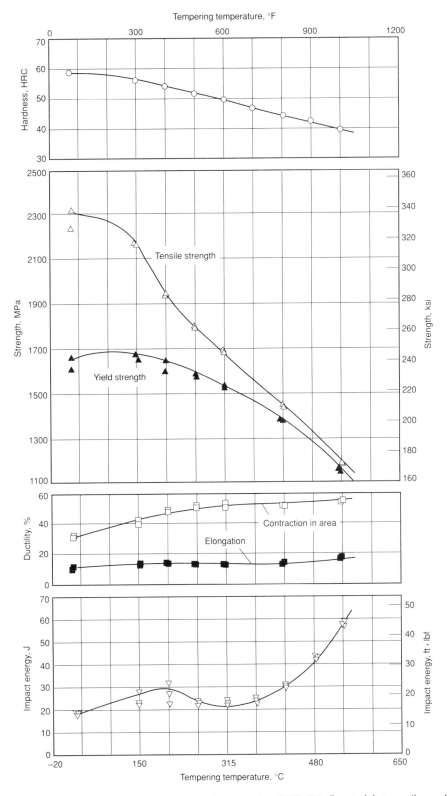

Fig. 26 Influence of tempering temperature on the properties of AISI 4340 alloy steel that was oil quenched and tempered 1 h. Source: Ref 115

plane-strain fracture toughness test results displayed TME only in the high-phosphorus steel; the Charpy V-notch results, on the other hand, showed TME in both steels. A study of AISI 4340 steel observed intergranular fractures in specimens with various levels of manganese, silicon, phosphorus, and sulfur. These fractures were attributed to carbide precipitation and impurity segregation at the prior-austenite grain boundaries (Ref 129). Another study of 4340 found substantial intergranular fracture only after a 350 °C (660 °F) temper (Ref 130). Cementite begins to precipitate at 250 °C (480 °F). With higher tempering temperatures, the number of precipitated carbides increases, and their thickness and length increase. Phosphorus segregates to the prior-austenite grain boundaries during austenitization but does not diffuse during these low-temperature tempering treatments. This study showed that the initial drop in impact energy with tempering at 250 to 300 °C (480 to 570 °F) was caused by the influence of cementite. However, with tempering at 350 °C (660 °F), the amount and size of the cementite at the grain boundaries was sufficient to cause intergranular fracture at the grain boundaries that were already weakened by the presence of impurities.

Instrumented Charpy V-notch testing of AISI 4130 alloy steel containing 0.002 and 0.02% P showed similar losses in impact energy for both steels when tempered in the critical region (Ref 131). As would be expected, the higher-phosphorus steel had lower impact strength at all tempering temperatures. However, the fractures were all transgranular, irrespective of phosphorus content or tempering temperature. The pre-maximum-load energy decrease with increasing tempering temperature in the range of 200 to 400 °C (390 to 750 °F) was attributed to a change in the work-hardening rate.

A recent study of TME in Fe-0.25C-10Cr-base martensitic steels concluded that TME was not due to the decomposition of interlath-retained austenite (Ref 132). This study instead suggested that TME was the result of the coarsening of interlath and intralath carbides.

As this review has shown, our understanding of tempered martensite embrittlement is not complete. Fractures of test specimens tempered within the critical region do vary with transgranular fractures observed with tempering between 200 and 300 °C (390 and 570 °F), and intergranular fractures usually, but not always, observed with tempering at about 350 °C (660 °F). These differences may be due to differences

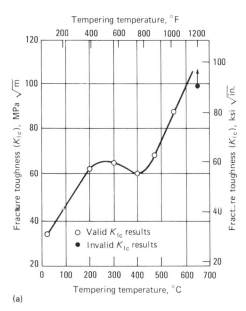

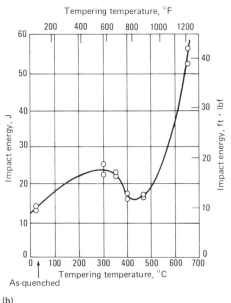

Fig. 28 Influence of tempering temperature on (a) the room-temperature plane-strain fracture toughness and (b) the Charpy V-notch impact energy of 300M alloy steel that was austenitized 1 h at 870 °C (1600 °F), oil quenched, and tempered for 1 h. Source: Ref 127

in carbon, alloy, and impurity content, as well as in strength level, test temperature, nature of the test, and grain size. Impurities appear to influence coarse-grain steels to a greater degree than fine-grain steels.

Sensitization

Austenitic stainless steels become susceptible to intergranular corrosion when subjected to temperatures in the range of 480 to 815 °C (900 to 1500 °F), generally from welding or service conditions. This susceptibility has been termed sensitization and has been attributed to the precipitation of $M_{23}C_6$ carbides on the austenite grain boundaries. The mechanism for sensitization was initially proposed as being due to the depletion of chromium at the grain boundaries when the carbides form (Ref 133, 134). The chromium content adjacent to the grain-boundary carbides was thought to drop below some critical limit, which rendered the alloy susceptible to severe localized attack by the corrosive environment.

This theory for the sensitization mechanism was widely accepted, and considerable indirect evidence for its validity was subsequently developed. However, when the electron microprobe was developed in the 1950s and researchers tried to detect the impoverished chromium zone, they were unable to do so (Ref 135, 136), or they could show chromium depletion only when the steel was carburized (Ref 137), or they detected only small regions of possible chromium depletion (Ref 138). Therefore, it was concluded that if such a zone exists, it must be less than 1 μm wide (40 μin.) and the concentration gradient must be rather steep. This prompted a number of other theories to be developed to explain sensitization, for example, electrochemical consideration of the nobility of the carbides (Ref 139, 140), strains at the carbide/austenite interface (Ref 139), and grain-boundary strain energy acting as the driving force for intergranular attack (Ref 141).

The picture was further complicated by a number of observations of intergranular corrosion in austenitic stainless steels under conditions unfavorable for sensitization, and in nonsensitized stainless steels with no detectable grain-boundary precipitation (Ref 142-149). The latter occurred in highly oxidizing solution, such as the nitric dichromate solution. Because of these difficulties, attention was focused on the influence of impurity elements, such as phosphorus, that are segregated to the austenite grain boundaries. It was subsequently proposed that intergranular corrosion of austenitic stainless steels was due to the presence of a continuous grain-boundary path of either carbides or solute segregated regions (Ref 142). Direct evidence for impurity segregation, chiefly of phosphorus, has been obtained from intergranular fracture surfaces of both sensitized and nonsensitized austenitic stainless steels (Ref 148-152).

The development of scanning transmission electron microscopes (STEM) with electron beam sizes of about 10 nm (100 Å) in diameter, coupled with energy-dispersive x-ray analysis, has provided direct proof of chromium depletion due to carbide precipitation at the grain boundaries (Ref 153-159). These studies have demonstrated that significant chromium depletion occurs after sensitization at grain boundaries adjacent to precipitated carbides. For equal times, the degree of chromium depletion for T316LN was greater with aging at 650 °C (1200 °F) than at 700 °C (1290 °F). Increasing the holding time caused the width of the depleted zone to increase. The width of chromium depletion along the grain boundaries around individual carbides was much greater than the width of the depletion zone into the grain interior (~3 μm, or 120 μin., versus ~0.15 μm, or 6 μin., for samples aged 100 h at 700 °C, or 1290 °F) (Ref 156). Evidence of molybdenum depletion has also been obtained. These direct measurements are in relative agreement with theoretical calculations of chromium levels in equilibrium with growing carbides during sensitization (Ref 160-163) and with empirical modifications of such theories (Ref 164).

Transmission electron microscopy studies of the precipitated carbides have demonstrated that the susceptibility to intergranular corrosion with sensitization temperature and time correlates well with the morphology of the grain-boundary carbide precipitates (Ref 165-169). These studies have demonstrated that the grain-boundary precipitate due to sensitization is always $(Cr,Fe)_{23}C_6$, that is, $M_{23}C_6$ carbide. The preferred sites for the nucleation of $M_{23}C_6$ are, in decreasing order of occurrence (Ref 166):

- Delta ferrite-austenite phase boundaries
- Austenite grain boundaries
- Incoherent twin boundaries
- Coherent twin boundaries

The carbides grow in the plane of the grain or twin boundaries. The morphology of the precipitated $M_{23}C_6$ depends on the type of boundary where precipitation occurs and the temperature (Ref 166). Precipitates that form at δ-ferrite-austenite phase boundaries or at austenite grain boundaries are dendritic or geometric in shape; the shape depends on the boundary orientation and misfit between the grains, the temperature, and the time at temperature. Those that form at incoherent twin boundaries look like ribbons of connected trapezoids while those that form at coherent twin boundaries have an equilateral thin triangular shape. Grain-boundary precipitates are classified into three categories:

- Dendritic shapes
- Separate geometric shapes
- Sheets of interconnected geometric particles

The sheets form at the lower temperatures; dendrites form at higher temperatures within the sensitization range, and the small, isolated geometric particles can form over the entire sensitization range and above, up to about 980 °C (1800 °F) (Ref 166). Sensitization is most severe for specimens sensitized at temperatures in the lower portion of the range where sheets of interconnected geometrically shaped carbides are formed. It is well known that healing occurs with very long times in the sensitization range; that is, long holding times result in reduced chromium depletion and reduced intergranular attack. When this occurs, the sheets of interconnected particles gradually separate into arrays of thick geometric particles (Ref 166).

Several approaches have been taken to minimize or prevent the sensitization of austenitic stainless steels. If sensitization results from welding heat and the component is small enough, solution annealing will dissolve the precipitates and restore immunity. However, in many cases this cannot be done because of distortion problems or the size of the component. In these cases, a low-carbon version of the grade or a stabilized composition should be used. Complete immunity requires a carbon content below about 0.015 to 0.02% (Ref 170). Additions of niobium or titanium to tie up the carbon are also effective in preventing sensitization as long as the ratio of these elements to the carbon content is high enough. Stabilizing heat treatments aimed at producing intergranular carbides are not very effective.

Few studies have been conducted on the influence of sensitization on mechanical properties. In general, carbide precipitation produces a slight increase in tensile strength and a minor reduction in tensile ductility (Ref 134). Heats of AISI 304 containing various levels of phosphorus and sulfur after sensitization were tested using half-size Charpy V-notch specimens fractured at liquid nitrogen temperature. Longitudinally oriented specimens were sensitized at 550 to 850 °C (1020 to 1560 °F) for 15 to 105 min. The decrease in impact energy, relative to the results before sensitization, were much greater in phosphorus-doped steels than in sulfur-doped steels. Fractures were intergranular for the phosphorus-doped steels and transgranular for the sulfur-doped steels. Figure 29 shows the reduction in half-size Charpy V-notch impact energy for phosphorus-doped steels aged at different temperatures for 5100 s (85 min); the specimens were tested after cooling in liquid nitrogen. For heats 1 and 6, embrittlement was greatest at 750 °C (1380 °F); somewhat higher or lower temperatures produced near-normal toughness. Embrittlement was greater for high-phosphorus heats, but these levels are greater than those encountered in commercial heats. The loss in toughness for the phosphorus-doped heats increased with holding times at temperatures between 650 and 825 °C (1200 and 1515 °F); somewhat higher or lower temperatures again produced near-normal toughness. The modified Strauss test produced higher corrosion rates for heats sensitized at 650 °C (1200 °F), and the corrosion rate increased linearly with increasing time and increasing impurity content.

Ferritic Stainless Steels. Sensitization and intergranular corrosion also occur in ferritic stainless steels (Ref 172-178). A wider range of corrosive environments can produce in-

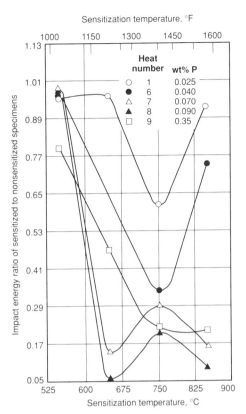

Fig. 29 Influence of phosphorus content and aging temperature on the relative loss in Charpy impact energy (tested using half-size specimens broken after cooling in liquid nitrogen) for sensitized AISI 304 stainless steel. Sensitization time, 5100 s. Source: Ref 171

tergranular attack in ferritic grades than is the case for austenitic grades. In the case of welds, the attacked region is usually larger for ferritic grades than for austenitic grades because temperatures above 925 °C (1700 °F) are involved in causing sensitization. Ferritic grades with less than 15% Cr are not susceptible, however. One study demonstrated that ferritic grades with 16 to 28% Cr were susceptible to intergranular corrosion when rapidly cooled from above 925 °C (1700 °F). This susceptibility was due to solution of carbides and nitrides followed by their reprecipitation in the grain boundaries. Subsequent annealing at 650 to 815 °C (1200 to 1500 °F) restored corrosion resistance (Ref 172). Therefore, the thermal processes causing intergranular corrosion in ferritic stainless steels are different from those for austenitic stainless steels. Reducing the carbon and nitrogen interstitial levels improves the intergranular corrosion resistance of ferritic stainless steels.

Sensitization can occur in titanium-stabilized ferritic stainless steels (Ref 177, 178). The thermal treatment that causes sensitization, however, is altered by the addition of titanium. First, high-temperature exposure requires a temperature in excess of 1050 °C (1920 °F) to dissolve the Ti(C,N) that reprecipitates upon cooling, even with water, forming grain-boundary precipitates of (Ti,Cr)(C,N) (Ref 177). The chromium-to-titanium (Cr/Ti) ratio in these precipitates is approximately 1/3. Aging at 480 to 550 °C (895 to 1020 °F) causes these precipitates to grow, and the Cr/Ti ratio increases to approximately 1/2. This depletes the grain-boundary zone around the precipitates of chromium, thereby increasing the susceptibility to intergranular corrosion. Again, long times at 480 to 550 °C (895 to 1020 °F) reduce the chromium gradient around the particles and restore corrosion resistance. Aging above 600 °C (1110 °F) also produces resistance to intergranular corrosion because the chromium in the (Ti,Cr)(C,N) precipitates is replaced by titanium, that is, the Cr/Ti ratio decreases. A titanium-stabilized 12% Cr ferritic grade was also found to sensitize under similar heat treatment conditions (Ref 178).

Duplex stainless steels are resistant to intergranular corrosion when aged in the region of 480 to 700 °C (895 to 1290 °F). It has been recognized for some time that duplex grades with 20 to 40 vol% ferrite exhibit excellent resistance to intergranular corrosion (Ref 179). A study of intergranular corrosion in AISI 308 stainless steel that was heat treated to produce 15% ferrite found that aging at 600 °C (1100 °F) caused the precipitation of $M_{23}C_6$ at austenite-ferrite boundaries (Ref 180). When this occurs, most of the chromium in the $M_{23}C_6$ comes from the ferrite grains; only a very small amount comes from the austenite grains. A chromium-depleted zone is formed at the austenite/carbide interface, which is very narrow, compared to those in fully austenitic sensitized stainless steels. Aging for 7 h at 600 °C (1110 °F) replenished the chromium-depleted zone and stopped the localized intergranular attack (ASTM A 262E test). Therefore, the healing of depleted zones is much more rapid in duplex grades.

The aging of duplex stainless steels produces a variety of phases in the ferrite (Ref 180). Aging at 480 °C (895 °F) does not produce $M_{23}C_6$, but it will produce an extremely small, finely distributed, chromium-rich α′ phase in the ferrite. Aging at 600 to 700 °C (1110 to 1290 °F) produces a complex series of transformations of the ferrite phase.

Another study of AISI 308 showed that for a given carbon content there is a critical amount and distribution of the ferrite/austenite interfacial boundary area (Ref 181). Above this critical level, the alloy is immune to intergranular corrosion with aging between 480 and 700 °C (895 and 1290 °F) for up to 1000 h. If the amount and distribution of these boundaries is below the critical level, two types of sensitization behavior can occur. The amount and distribution of the ferrite/austenite interfaces may be adequate to produce the rapid healing of chromium-depleted regions that form at the

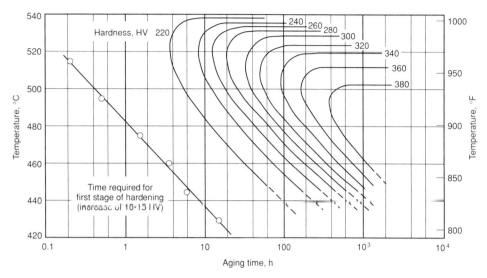

Fig. 30 Time-temperature-constant hardness curves for Fe-30Cr (Alloy 90) after aging done between approximately 430 and 540 °C (805 and 1005 °F), around the region of 475 °C embrittlement. Specimens rolled at 900 °C (1650 °F); starting hardness, 195 to 205 HV. Source: Ref 192

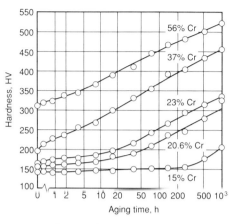

Fig. 31 Influence of aging time at 475 °C (885 °F) on the hardness of iron-chromium alloys with 15, 20.6, 23, 37, and 56% Cr. Source: Ref 192

475 °C Embrittlement

Iron-chromium alloys containing 13 to 90% Cr are susceptible to embrittlement when held within or cooled slowly through the temperature range of 550 to 400 °C (1020 to 750 °F). This phenomenon has been called 475 °C (885 °F) embrittlement and it produces an increase in tensile strength and hardness and a decrease in tensile ductility, impact strength, electrical resistivity, and corrosion resistance (Ref 182-203). Microstructure effects are minor: Grain boundaries etch more widely, and the grain interiors darken.

Numerous theories have been proposed to account for 475 °C embrittlement. The problem occurs with iron-chromium ferritic and duplex ferritic-austenitic stainless steels, and not with austenitic grades. The earliest theories suggested that embrittlement was due to precipitation of second phases such as phosphides (Ref 183), carbides or nitrides (Ref 184), or oxides (Ref 188). Others suggested that embrittlement was due to σ phase, which does form in iron-chromium alloys, or some transitional phase that precedes σ formation (Ref 185-187). However, σ forms at higher temperatures and has never been detected in 475 °C embrittled specimens. Japanese researchers (Ref 189, 190) suggested that 475 °C embrittlement was due to ordering, that is, from the formation of Fe_3Cr, $FeCr$, and $FeCr_3$ superlattices. However, subsequent experiments using neutron diffraction failed to detect evidence of ordering.

A 1953 study using transmission electron microscopy observed that 475 °C embrittlement caused the precipitation of a coherent chromium-rich bcc phase with a lattice parameter only slightly greater than the iron-rich bcc ferritic matrix phase. These precipitates were extremely small, for example, about 15 to 30 nm (150 to 300 Å) in diameter for an Fe-27% Cr alloy aged from 10 000 to 34 000 h at 480 °C (900 °F). The precipitates were nonmagnetic and contained about 80% Cr. The rate of growth of the precipitates was very slow, and they did not appear to over age (Ref 191). Other studies have confirmed these findings (Ref 192-199). The 1953 study was unable to explain its observations based on the existing iron-chromium phase diagram. Later work (Ref 192) concluded that 475 °C embrittlement was a precipitation-hardening phenomenon resulting from the presence of a miscibility gap in the iron-chromium system below 600 °C (1110 °F). The location of the miscibility gap was later refined (Ref 200).

Aging at 475 °C (885 °F) has been shown to cause a rapid rate of hardening with aging between about 20 and 120 h because of homogeneous precipitation. The rate of hardening is much slower with continued aging from 120 to 1000 h. During this aging period, the precipitates grow. Aging beyond 1000 h produces little increase in hardness because of the stability of the precipitates, which do not grow larger than about 30 nm (300 Å).

Precipitation of the chromium-rich α' phase in iron-chromium alloys can occur either by spinodal decomposition or by nucleation and growth, depending on the aging temperature and alloy composition. For example, an Fe-30% Cr alloy will decompose to chromium-rich precipitates in an iron-rich matrix (chromium-depleted) inside the spinodal at 475 °C (885 °F), forming spherical α', or outside the spinodal at 550 °C (1020 °F), forming disk-shaped α' (Ref 196). An Fe-20Cr alloy will decompose by nucleation and growth at 470 °C (880 °F), while Fe-30Cr, Fe-40Cr, and Fe-50Cr alloys will decompose spinodally at 470 °C (880 °F).

Even for a severely embrittled alloy, 475 °C embrittlement is reversible. Properties can be restored within minutes by reheating the alloy to 675 °C (1250 °F) or above (Ref 188, 194). The degree of embrittlement increases with chromium content; however, embrittlement is negligible below 13% Cr. Carbide-forming alloying additions, such as molybdenum, vanadium, titanium, and niobium, appear to increase embrittlement, particularly with higher chromium levels. Increased levels of carbon and nitrogen also enhance embrittlement and, of course, are detrimental to nonembrittled properties as well. Cold work prior to 475 °C (885 °F) exposure accelerates embrittlement, particularly for higher-chromium alloys.

Figure 30 demonstrates the C-curve nature of the increase in hardness due to aging for an Fe-30% Cr ferritic stainless steel (Ref 192). The nose of the curve decreases with time. Figure 31 shows the results for aging at 475 °C (885 °F) for up to 1000 h for iron-chromium alloys with 15, 20.6, 23, 37, and 56% Cr. As the chromium content increased, the time to the initial increase in hardness decreased. Over 200 h were required for the 15% Cr alloy, only about 4 h for the 20.6% Cr alloy, and less than 1 h for the 23% Cr alloy. The initial increase in hardness was nearly instantaneous for the 37 and 56% Cr alloys. Accompanying the increase in hardness with aging at 475 °C (885 °F) are an increase in tensile and yield strength and a decrease in tensile ductility and impact energy. Examples of such data can be found in Ref 185, 186, 199, and 201 to 203.

Sigma Phase Embrittlement

The existence of σ phase in iron-chromium alloys was first detected in 1907 by the observation of a thermal arrest in cooling curves (Ref 204). It was suggested that the thermal arrest could be due to the formation of an FeCr intermetallic compound. The first actual observation of σ in iron-chromium alloys was reported in 1927 (Ref 205). The phase was referred to as the brittle constituent (B constituent) and was reported to be corundum hard. In a discussion of this paper, it was suggested that the B constituent was the FeCr intermetallic phase detected in 1907. The σ phase was identified by x-ray diffraction in 1927 (Ref 206) and in 1931 (Ref 207). Numerous earlier investigators, however, had failed to detect the presence of σ; this failure caused considerable confusion.

After the existence of σ was firmly established, numerous studies were conducted to define the compositions and temperatures over which σ could be formed. This produced a series of refinements to the iron-chromium equilibrium diagram, as in the case of 475 °C embrittlement. Successive studies demonstrated that σ could form in alloys with lower and lower chromium contents under the proper conditions; one study demonstrated the formation of σ at 480 °C (900 °F) in an alloy with less than 12% Cr. In general, σ forms with long-time exposure in the range of 565 to 980 °C (1050 to 1800 °F), although this range varies somewhat with composition and processing. Sigma formation exhibits C-curve behavior with the shortest time for formation (nose) generally occurring between about 700 and 810 °C (1290 and 1490 °F); the temperature that produces the greatest amount of σ with time is usually somewhat lower.

The general characteristics of σ phase have been reviewed extensively (Ref 209-216). The name sigma stems from work done in 1936 (Ref 217). Sigma-type phases have since been identified in over fifty binary systems and in other commercial alloys (for example, nickel-base superalloys). Sigma phase has a tetragonal crystal structure with 30 atoms per unit cell and a c/a ratio of approximately 0.52 (Ref 216). Sigma in iron-chromium alloys has a hardness equivalent to approximately 68 HRC (940 HV). Because of its brittleness, σ often fractures during indentation. At room temperature, σ is nonmagnetic. Embrittlement effects due to σ are greatest at room temperature.

Austenitic and Ferritic Stainless Steels. In commercial alloys, silicon, even in small amounts, markedly accelerates the formation of σ. In general, all of the elements that stabilize ferrite promote σ formation. Molybdenum has an effect similar to that of silicon; aluminum has a lesser influence. Increasing the chromium content, of course, also favors σ formation. Small amounts of nickel and manganese also increase the rate of σ formation, although large amounts, which stabilize austenite,

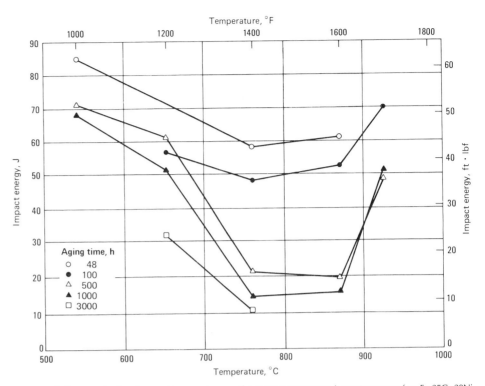

Fig. 32 Influence of aging time and temperature on the room-temperature impact energy of an Fe-25Cr-20Ni alloy. Annealed value, 89 J (66 ft · lbf). Source: Ref 225

retard σ formation. Carbon additions decrease σ formation by forming chromium carbides, thereby reducing the amount of chromium in solid solution. Additions of tungsten, vanadium, titanium, and niobium also promote σ formation. As might be expected, σ forms more readily in ferritic than in austenitic stainless steels. Coarse grain sizes from high solution-annealing temperatures retard σ formation, and prior cold working enhances it. The influence of cold work on σ formation depends on the amount of cold work and its effect on recrystallization. If the amount of cold work is sufficient to produce recrystallization at the service temperature, σ formation is enhanced. If recrystallization does not occur, the rate of σ formation may not be affected. Small amounts of cold work that do not promote recrystallization may actually retard σ formation (Ref 218).

The composition of σ in austenitic stainless steels is more complex than it is for simple iron-chromium ferritic grades. Several studies, particularly for AISI 316, have analyzed the composition of σ. These studies have used a traditional wet chemical analysis of bulk-extracted σ, wavelength-dispersive spectroscopy (WDS) with an electron microprobe, or energy-dispersive spectroscopy (EDS) with either a scanning transmission electron microscope (for thin foils) or a scanning electron microscope (SEM) (for bulk specimens) (Ref 219).

An analysis of σ in AISI 316 by the SEM-EDS approach obtained a composition of 11Mo-29Cr-55Fe-5Ni in a specimen aged 3000 h at 815 °C (1500 °F) (Ref 220). A WDS analysis using the electron microprobe of AISI 316 heated 60 h at 870 °C (1600 °F) obtained a σ composition of 26.4Cr-3.3Ni-53.7Fe-8.5Mo (Ref 221). An analysis of σ in a failed AISI 316 superheater tube using STEM-EDS of thin foils obtained a composition of 52.7Fe-37Cr-3.7Mo-4.8Ni-0.7Si-0.4Mn (average of several results).

Reference 223 presents an analysis of σ in three versions of AISI 310 (a standard version, a low-carbon version, and a high-silicon version) and titanium-stabilized AISI 316. The analysis was done using the electron microprobe. For the low-carbon AISI 310S, the chromium-to-iron (Cr/Fe) ratio in σ was constant and equal to 1 for all temperatures and times; the composition was 46Cr-46Fe-8Ni. For the AISI 310 and AISI 310Si, the composition of σ varied with temperature and time, and the Cr/Fe ratio of 1 was obtained after a certain time at any temperature used. The higher the temperature, the shorter the time required to obtain a Cr/Fe ratio of one. When the Cr/Fe ratio stabilized at 1 for 310 and 310Si, σ had the same composition as in the low-carbon version. The work with 316Ti showed that the molybdenum content in σ increased with temperature. As the molybdenum content in σ increased, the iron content increased, and the chromium and nickel contents decreased. The (Cr+Mo)/(Fe+Ni) ratio was constant, and the formu-

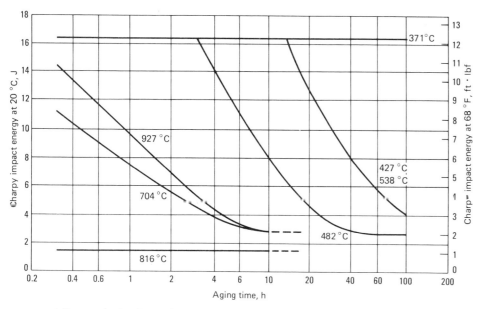

Fig. 33 Influence of aging time and temperature on the room-temperature Charpy impact energy of a low interstitial content 29Cr-4Mo ferritic stainless steel. Source: Ref 226

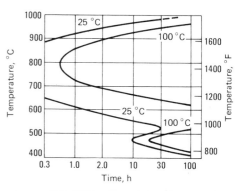

Fig. 34 Time-temperature relationships to produce 25 and 100 °C (75 and 212 °F) DBTTs for a 29Cr-4Mo ferritic stainless steel as a function of aging times that cover both the 475 °C (885 °F) embrittlement range and the σ phase embrittlement range. Source: Ref 226

la for σ was expressed as $(Cr,Mo)_{35}(Fe,Ni)_{65}$. Although the chemical composition of σ varied with temperature and time, the c/a ratio of the tetragonal unit cell was constant at 0.519.

The influence of molybdenum, which is known to promote σ formation, was examined using wrought 25Cr-20Ni alloys with additions of up to 8.2% Mo (Ref 224). Sigma was analyzed with the electron microprobe using specimens aged at 850 °C (1560 °F) for 525 h (the 8.2% Mo specimen was aged for 350 h). As the molybdenum content in the 25Cr-20Ni alloys increased to 8.2%, the chromium content in σ decreased from 42.6 to 31.0%, the iron content decreased from 43.3 to 38.8%, the nickel content remained constant at about 10%, and the molybdenum content increased to 14.3%. In addition, as the molybdenum content of the steels increased, the volume fraction of σ increased from 3 to 60% for these aging treatments.

Sigma formation in pure iron-chromium alloys is rather sluggish, which accounts for much of the confusion in early studies concerning its existence. Subsequent work showed that the formation of σ was dramatically accelerated by prior cold work and by silicon additions (Ref 217). In ferritic stainless steels, the addition of even minor amounts of other alloying elements expands the compositional range over which σ may form and increases the rate of formation. All of the ferritic stabilizing elements promote σ formation.

Sigma will also form in austenitic alloys. In fully austenitic alloys, σ forms from the austenite along grain boundaries. If δ-ferrite is present in the austenitic alloy, σ formation is more rapid and occurs in the δ-ferrite. Sigma will form more readily in austenitic alloys containing additions of ferrite-stabilizing elements such as molybdenum and titanium; the rate of formation can be quite rapid in these alloys.

The most sensitive room-temperature property for assessing the influence of σ is the impact strength. A study of the influence of σ on the toughness of AISI 310 shows the dramatic loss in toughness due to σ (Fig. 32). With increasing time at temperature, particularly in the range of 760 to 870 °C (1400 to 1600 °F), the toughness decreased by about 85%.

The influence of high-temperature exposure on the toughness of a low-interstitial 29Cr-4Mo ferritic stainless steel has been examined (Ref 226). Figure 33 shows the room-temperature impact strength trends for this alloy as a function of aging temperature and time. Aging at 371 °C (700 °F) produced no loss in toughness. However, aging at 427, 482, and 538 °C (800, 900, and 1000 °F), in the range of 475 °C embrittlement, produced a loss of toughness that was most pronounced at 482 °C (900 °F). Aging at higher temperatures, where σ was formed, produced more pronounced embrittlement, which was greatest at 816 °C (1500 °F). Figure 34, which is a summary of this data, shows the time at aging temperatures between 371 and 978 °C (700 and 1790 °F) required to produce a DBTT of 25 and 100 °C (77 and 212 °F). This produces a C-curve presentation of the time for embrittlement as a function of aging temperature. For σ formation, embrittlement was most rapid at about 775 °C (1425 °F); 475 °C embrittlement was slower with a maximum rate at about 480 °C (900° F). Chi phase was also observed, along with σ, after aging in the high-temperature range. Sigma formed over the range of 595 to 925 °C (1100 to 1700 °F). Embrittlement was most pronounced when intergranular σ films formed, producing intergranular tensile and impact fractures.

While some studies have demonstrated only a minor increase in hardness and strength because of σ formation, studies of some steels have demonstrated more substantial changes. Tensile ductility, like toughness, is generally substantially reduced. One study has demonstrated that high-chromium σ-hardenable alloys are useful in applications involving high-temperature erosion or wear, for example, exhaust valves (Ref 227). Such steels generally contain from 20 to 30% Cr, about 0.25 to 0.45% C, and additions of manganese and nickel to produce a duplex structure; they also generally contain additions of elements that promote σ formation, such as silicon and molybdenum. Such steels can be hardened to about 40 HRC by σ and retain their strength at high temperatures. Although the toughness of such steels is reduced about 35% by the presence of σ, they do perform well as long as the extent of σ-phase embrittlement is not severe. The toughness and ductility of σ-containing steels at high temperatures is considerably better than at room temperature; however, such steels are not useful at high temperatures if shock resistance is required.

An examination of the high-temperature properties of a 25Cr-20Ni-2Si (type 314) austenitic stainless steel aged between 650 and 980 °C (1200 and 1800 °F) showed that with the proper amount and distribution of σ, substantial increases in yield and tensile strengths result for test temperatures up to 760 °C (1400 °F) (Ref 228). For conditions involving slow strain rates, σ phase reduces creep resistance. Variations in austenite grain size, however, can exert an even greater effect on high-temperature tensile and creep properties. Fine grain sizes increase short-time high-temperature tensile strength but reduce long-time creep strength. The room-temperature ductility of these alloys is poor but is restored at temperatures above 540 to 650 °C (1000 to 1200 °F). A fine distribution of σ is detrimental to creep strength but minimizes loss of ductility at room temperature. Sigma has also been found to be detrimental in thermal-fatigue situations (Ref 229). High-temperature exposure can produce a variety of phases, and embrittlement is not always due solely to σ formation (Ref 230). Therefore, each situation must be carefully evaluated to determine the true cause of the degradation of properties.

Duplex Stainless Steels. Sigma phase is known to form quite rapidly in duplex stainless steels. One study, for example, observed σ after 15 min at 750 °C (1380 °F) and 2 min at 850 °C (1560 °F) in a 20Cr-10Ni-3Mo duplex alloy (Ref 231). Another study found that σ formed after 2 min at 900 °C (1650 °F) in a 21Cr-7Ni-2.4Mo-1.3Cu (UNS S32404) duplex stainless steel containing 33% ferrite (Ref 232). Sigma formed in a C-curve manner along with a number of other phases, but sigma was the worst embrittler and led to massive pit initiation in corrosion tests. Sigma formed in the ferrite, and pitting occurred in the chromium-molybdenum-depleted ferrite-σ regions.

An evaluation of the effects of alloying elements on σ formation in duplex stainless steels found that increasing chromium and molybdenum contents caused an increase in the rate of σ formation and in the maximum amount produced (Ref 233). Increasing the nickel content decreased the maximum amount of σ that could form but increased the rate of σ formation. Sigma formation occurred primarily by the decomposition of ferrite into σ and austenite.

The influence of σ on corrosion characteristics is rather complex. In many instances, little influence is observed in environments normally used with a particular alloy. Large σ particles appear to be rather harmless; a fine distribution of particles, particularly if present at the grain boundaries and in highly oxidizing solutions, is more harmful.

Hydrogen Damage

Of all the embrittlement problems discussed in this article, that due to hydrogen is the most widespread and influences the behavior and properties of nearly all ferrous alloys and many other metals. Indeed, the ubiquitous nature of hydrogen and the wide variety of forms of damage that it can produce make this subject difficult to address and summarize. Hydrogen can come from many sources. It can be retained in the steel:

- Upon solidification (supersaturated)
- From acid cleaning (pickling)
- From electroplating
- From contact with water or other hydrogen-containing liquids or gases
- From pure hydrogen gas

Table 1 Classification of hydrogen damage processes

Hydrogen embrittlement
   Hydrogen environmental embrittlement
   Hydrogen stress cracking
   Loss in tensile ductility
Hydrogen attack
Blistering
Shatter cracks, flakes, and fisheyes
Microperforation
Degradation in flow properties
Metal hydride formation

Source: Ref 235, 236

Hydrogen has no known beneficial effects on iron and steel, only detrimental effects. While the influence of hydrogen on high-strength steels is well known, even the softest irons are not immune from hydrogen damage.

Despite the thousands of research articles published on this subject, there is still a great deal that is not fully understood and considerable controversy over the mechanisms by which hydrogen exerts influence. The damage produced by hydrogen has been known since at least 1875 (Ref 234).

The general term hydrogen damage has been used to be consistent with a phenomenological classification scheme covering the various hydrogen degradation processes (Ref 235, 236). Hydrogen can cause a wide variety of problems, and it is helpful to categorize them (Table 1). Other reviews on hydrogen damage can be found in Ref 237-251.

Hydrogen environmental embrittlement, hydrogen stress cracking, and loss of tensile ductility are grouped under hydrogen embrittlement. Hydrogen environmental embrittlement refers to the degradation of mechanical properties in metals such as steels when they are deformed in a hydrogen-containing environment. Degradation is enhanced by slow strain rates and high hydrogen pressures.

Hydrogen stress cracking is characterized by the brittle fracture of an alloy that is normally ductile while being stressed below its yield strength in a hydrogen-containing environment. This problem is also called "static fatigue" and hydrogen-induced cracking. In general, there is a certain stress level (threshold) below which such cracking does not occur. Hydrogen stress cracking involves the absorption of hydrogen, followed by hydrogen diffusion to highly stressed regions, particularly those associated with notches. Thus, there is a delay time, or incubation time, before cracking occurs. H$_2$S stress cracking is a specific type of hydrogen stress cracking.

The loss of tensile ductility, one of the oldest forms of hydrogen damage, occurs in steels and other alloys exposed to hydrogen. It generally affects lower-strength alloys. Fracture does not occur, as it does with higher-strength alloys; these alloys experience only a decrease in tensile elongation and reduction in area. The phenomenon is sensitive to strain rate and hydrogen content. It is increased by decreases in strain rate and increases in hydrogen.

Hydrogen Environmental Embrittlement

Hydrogen attack occurs at high temperatures rather than at or near room temperature. Hydrogen reacts with carbon within the steel to form methane gas. This may simply decarburize the steel, lowering its strength, or may produce cracks or fissures. This problem does not occur below 200 °C (390 °F). Blistering is observed in low-strength steels and other metals. It is caused by atomic hydrogen diffusing to internal defects or inclusions where it precipitates as molecular (diatomic) hydrogen. This generates substantial pressure, which produces blisters. Shatter cracks, flakes, and fisheyes are terms that describe cracks or the surface appearance of a fracture in castings, forgings, wrought alloys, or weldments. These defects are caused by the presence of excessive hydrogen in the liquid melt prior to solidification. The solubility of hydrogen in the melt is much greater than after solidification; cracking can result if the hydrogen level is above a critical level and is not permitted to diffuse out of the material before cooling to ambient temperature. Microperforation occurs mainly when steels are exposed to very high-pressure hydrogen near room temperature. As the term implies, small fissures are formed in the metal. A degradation in flow properties occurs in steels at room temperature. Metal hydride formation does not occur in steels, but it is a problem in metals such as titanium, niobium, and zirconium.

The fact that hydrogen embrittlement increases with decreasing strain rate and increasing temperature above ambient, as well as the fact that these are mutually dependent variables, is directly opposite to the general trend for bcc metals that undergo a ductile-to-brittle transition behavior with increasing strain rate and decreasing temperature below ambient. This unique behavior arises because the phenomenon is controlled by the diffusion of hydrogen through the lattice. Hydrogen effects are further complicated by the chemical analytical limitations associated with hydrogen analyses. Although a bulk hydrogen analysis can be made, hydrogen in the lattice cannot be differentiated from hydrogen within voids or traps.

Figure 35 demonstrates the influence of hydrogen on the tensile reduction of area as a function of strength level (Ref 237). The AISI 4340 specimens, quenched and tempered to three different strength levels, were electrolytically charged with hydrogen using an aqueous 4% H$_2$SO$_4$ solution. Charging times were selected that lowered

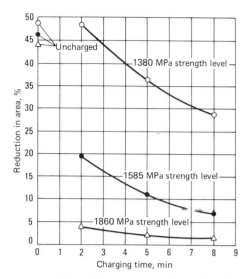

Fig. 35 Effect of charging time on the tensile reduction in area of AISI 4340 steel quenched and tempered to different strength levels. Source: Ref 237

the tensile ductility, but not to 0. Figure 35 shows that with increasing charging time, that is, with increased hydrogen pickup, the reduction of area decreased. As the tensile strength increased, the same degree of charging had a greater influence on decreasing the reduction in area.

Hydrogen embrittlement, in classic form, is reversible in that the damage should be eliminated by removing the hydrogen. However, hydrogen can also cause permanent damage. Figure 36 shows the recovery of tensile reduction in area for AISI 4340 specimens charged for 5 min and then aged for various times at room temperature before tensile testing (Ref 237). For all three strength levels (the same strength levels in Fig. 35), the tensile ductility returned to normal levels with time. About 100 h were required; however, there was no clear trend in recovery time with strength level.

Figure 37 shows the combined effects of reversible and permanent damage due to hydrogen (Ref 252). The AISI 1020 carbon steel tensile specimens were tested both uncharged and cathodically charged with hydrogen, and each group was further tested after aging for 20 h at 250 °C (480 °F) in air. Tensile tests were performed over a wide range of temperatures, from −196 °C (−321 °F) (liquid nitrogen temperature) to room temperature. The uncharged specimens, with or without aging, exhibited a gradual decrease in tensile reduction of area with decreasing test temperature, as would be expected. The aging treatment had no influence on the uncharged specimens. The charged specimens that were not aged exhibited much lower tensile ductility from room temperature to −160 °C (−255 °F), where they had about the same ductility as the uncharged specimens. However, at the liquid nitrogen temperature (−196 °C, or −321 °F), the reduction in area was much lower than that in the uncharged steel. The hydrogen-charged specimens that were aged at 250 °C (480 °F) exhibited the same high ductility as the uncharged specimens down to −160 °C (−255 °F), but at −196 °C (−321 °F) they exhibited the same very low percentage of reduction in area as the non-aged charged specimens. Aging apparently removed the hydrogen that caused the loss of ductility from 20 to −160 °C (70 to −255 °F), but it did not remove the low ductility at −196 °C (−321 °F). Further testing of aged, charged specimens where the surface layer was removed resulted in the same −196 °C (−321 °F) ductility as the uncharged specimens. It was shown that cathodic charging produced blisters and cracks at the surface (permanent damage) that made the charged specimens notch sensitive at −196 °C (−321 °F) whether hydrogen was present or not. The effects of combined reversible and permanent damage

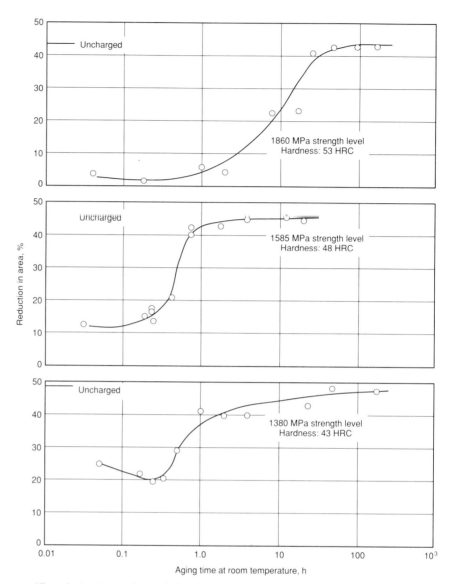

Fig. 36 Effect of aging time on the tensile ductility of heat-treated AISI 4340 at three different strength levels. Source: Ref 237

caused by hydrogen is difficult to separate, particularly in higher-strength materials.

Hydrogen Stress Cracking and Loss of Tensile Ductility

Another classic form of hydrogen embrittlement, hydrogen stress cracking (also called "static fatigue" or hydrogen-induced cracking) involves delayed fracture at stresses below the yield strength. Figure 38 illustrates the nature of these failures for high-strength hydrogen-charged steels (Ref 238). The notched tensile strength may be lower than for noncharged specimens. A wide range of applied stresses can cause delayed fractures, with the applied stress having only a minor influence on the time to failure. Below some level of applied stress, fracture does not occur. This critical level of stress is referred to as the "static endurance limit." Additional work has demonstrated that the time between the incubation

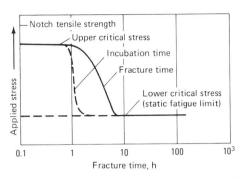

Fig. 38 Failure characteristics of a hydrogen-charged high-strength steel. Source: Ref 238

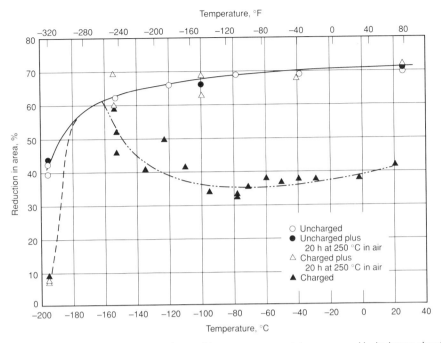

Fig. 37 Influence of test temperature and reversible versus permanent damage caused by hydrogen charging on the tensile ductility of AISI 1020 steel, with and without aging. Source: Ref 252

of cracking and failure decreases with decreasing test temperature until, at −46 °C (−50 °F), they are coincident (Ref 253). Resistivity measurements also showed that crack growth was discontinuous.

Figure 39 shows test results for hydrogen-charged notched specimens of AISI 4340 (cathodically charged 5 min in aqueous 4% H_2SO_4 and then cadmium plated, 25 min) that were aged at 150 °C (300 °F) for various times and then loaded in tension until fracture (Ref 238, 239). The notched tensile strength of the material was 2070 MPa (300 ksi), and the smooth bar tensile strength was 1585 MPa (230 ksi) before charging and plating. For any aging (baking) time, the time to fracture increased slightly with decreasing load until the critical applied stress at which fracture did not occur was reached. Also, as the aging time increased, the critical stress increased, and the times to fracture at loads above this level increased. With aging at 150 °C (300 °F), the critical stress level increased from 515 MPa (75 ksi) to about 1655 MPa (240 ksi), indicating a recovery brought about by the removal of hydrogen.

Work has been done that demonstrates the combined influence of strain rate and test temperature on the tensile ductility of AISI/SAE 1020 carbon steel (Ref 254, 255). Specimens were tested in tension at strain rates from 0.05 to 19 000 min^{-1} over the temperature range −196 to 100 °C (−320 to 212 °F). The test specimens were either spheroidize annealed before testing (for uncharged specimens) or cathodically charged. Figure 40 shows the results of these tests. For the uncharged specimens (Fig. 40a), the tensile ductility (natural log of the original cross-sectional area, a_0, to the final cross-sectional area, a_f) decreased to zero when the test temperature reached −195 °C (−320 °F). The variation in strain rate had little effect on the uncharged specimens. For the charged specimens (Fig. 40b), a marked reduction in tensile ductility is evident, with decreasing strain rate in the center of the test temperature range producing two intersecting surfaces (coded a and d in Fig. 40b). Surface a, in the lower-temperature test region, shows that ductility decreases with increasing temperature and decreasing strain rate. Surface d, in the higher-temperature test region, shows that ductility decreases with decreasing temperature and decreasing strain rate. As the strain rate increases, the temperature for lowest ductility (the intersection of surfaces a and d) increases from about −40 to 50 °C (−40 to 120 °F).

Figure 41 shows the test results for charged specimens as a function of charging time at the lowest strain rate, 0.05 min^{-1}. The graph shows the loss in fracture ductility with increased charging time and the

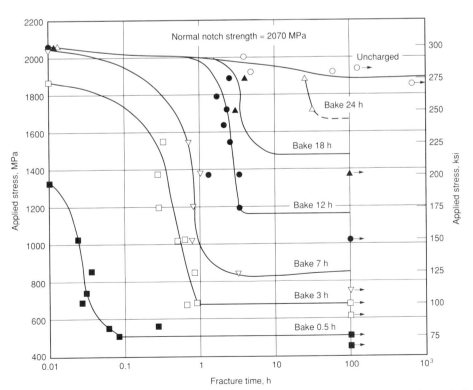

Fig. 39 Time to fracture as a function of applied stress and aging (baking) time at 150 °C (300 °F) for notched tensile specimens of AISI 4340 that were hydrogen charged and electroplated with cadmium. Source: Ref 238

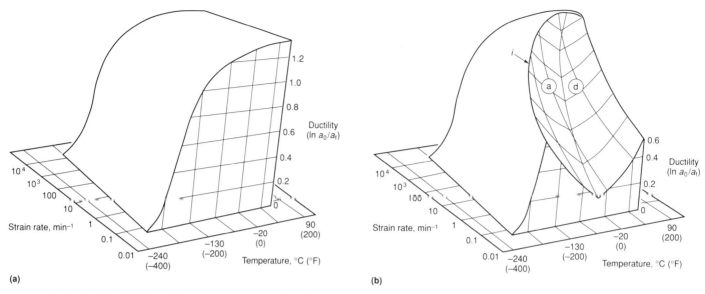

Fig. 40 Tensile ductility of AISI/SAE 1020 carbon steel as a function of strain rate and test temperature for (a) spheroidize-annealed specimens and (b) cathodically charged specimens. Curve i bounds the range of strain rates and temperatures where embrittlement was observed. Source: Ref 255

increase in hydrogen content. The loss in ductility becomes relatively constant after about 30 min charging time.

A correlation has also been demonstrated between total hydrogen content and tensile reduction in area (Ref 256). Test specimens were machined from the center, midpoint, and edge of a 127 × 127 mm (5 × 5 in.) square billet of AISI 1010 carbon steel known to have a rather high hydrogen content. Reduction in area results were 13.4, 23.0, and 33.1% with corresponding hydrogen contents of 5.7, 5.2, and 3 ppm. A midpoint specimen aged at 175 °C (350 °F) for 15 h had a 64% reduction in area and contained 1.4 ppm hydrogen. Fisheyes were observed on the nonaged specimen fractures but not on those that were aged. Eight months later, additional tests were made with better percentage reduction in area values and less of a ductility gradient across the billet. Figure 42 shows the results from these two test periods. The values on the graph indicated by the solid dots were from similar tests made on a 127 × 127 mm (5 × 5 in.) billet from another heat of the same grade.

Hydrogen-induced fractures are not always intergranular. In high-strength steels, cracking is often intergranular, but transgranular cleavage is also observed. For ductile low-strength steels, the fracture mode is ductile with changes in the dimple size. Prior temper embrittlement of high-strength alloy steels enhances hydrogen stress cracking (Ref 257-261). The threshold stress intensity for crack growth is lowered, and intergranular fractures occur. The influence of temper embrittlement on hydrogen stress cracking is particularly pronounced as the yield strength is decreased. Very low impurity content steels do show high threshold stress intensity values and freedom from intergranular fracture (Ref 258).

Hydrogen-assisted ductile fractures have been reported for a variety of steels, chiefly

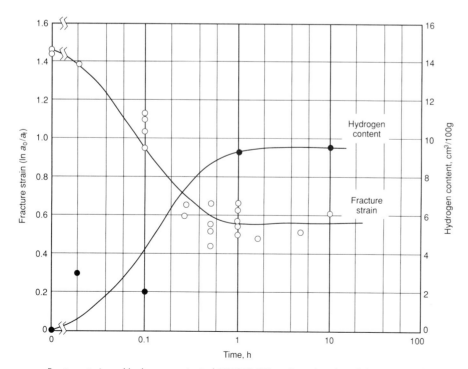

Fig. 41 Fracture strain and hydrogen content of AISI/SAE 1020 steel as a function of charging time for tensile tests conducted at room temperature, with a strain rate of 0.05 min^{-1}. Source: Ref 255

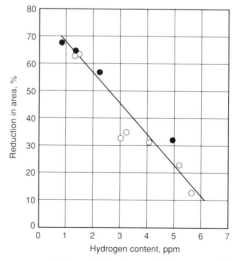

Fig. 42 Relationship between hydrogen content and tensile reduction in area for two heats of AISI 1010 steel. Source: Ref 256

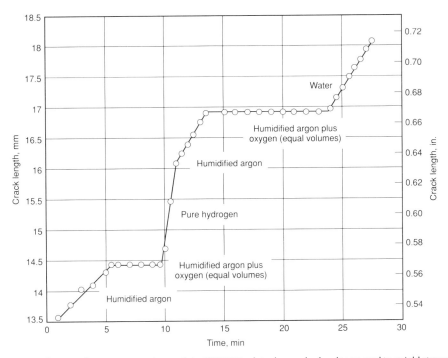

Fig. 43 Influence of oxygen on crack growth in AISI H11 tool steel quenched and tempered to a yield strength of 1585 MPa (230 ksi). Source: Ref 272

spheroidized carbon steels and austenitic stainless steels (Ref 262-266). The fractures exhibit ductile dimples (microvoid coalescence), and the dimple sizes in charged specimens have been reported to be either smaller or larger than those in uncharged specimens. The change in dimple size can be correlated with the loss in tensile reduction of area (Ref 262).

Austenitic Stainless Steels. For many years, it was believed that only bcc metals were susceptible to hydrogen embrittlement. However, it has since been shown that austenitic stainless steels, and several other fcc metals and alloys, can become embrittled by hydrogen, and their behavior is similar in nature to the bcc metals (Ref 242, 267). A number of studies have focused on hydrogen-embrittled AISI 304 or 304L. These studies have demonstrated that hydrogen charging partially transforms the austenite to α' (bcc) and ϵ (hexagonal close-packed) martensites (Ref 268). These phases can also be formed in 304 or 304L, and in other low-stability austenitic stainless steels, by cold working (uncharged) specimens at low temperatures. Tests of more stable austenitic grades, such as AISI 310, have produced conflicting results. While some studies showed no embrittlement (Ref 269), others detected embrittlement. The unstable 304 alloy exhibits transgranular cleavage fractures, and the stable 310 alloy exhibits ductile fractures, when tested in hydrogen gas environmental cells (Ref 270).

Tool Steels. Fracture mechanics approaches have also been used to study crack nucleation and growth in hydrogen-containing environments. Studies have demonstrated that high-strength steels are embrittled by exposure to hydrogen gas even at, or below, atmospheric pressure (Ref 271). For example, a study was conducted on the effect of hydrogen gas at atmospheric pressure on subcritical crack growth using AISI H11 tool steel (Ref 272). Although argon, helium, and nitrogen are inert environments, if they are humidified, subcritical crack growth occurs. The study also demonstrated that the addition of oxygen to the environment, even in amounts as low as 0.7% by volume, would terminate crack growth. The removal of the oxygen would restart crack growth. Oxygen dissolved in water, however, had no influence on the crack growth rate. Figure 43 shows results for H11, quenched and tempered to a yield strength of 1585 MPa (230 ksi), using a precracked specimen under load in various environments. For the first 5 min, the crack grew slowly in humidified argon. Then, an equal volume of oxygen was added to the environment, and cracking stopped for the next 5 min. Next, this environment was removed and replaced by pure hydrogen, and crack growth was very rapid. After about 1 min, the pure hydrogen was replaced by humidified argon, and the crack growth rate decreased. Again, an equal volume of oxygen was added to the humidified argon, and crack growth ceased. This atmosphere was removed and replaced by water, and the crack again began to grow at a constant rate.

Maraging Steels. The influence of hydrogen on slow crack growth in maraging steels has been examined by several researchers. One study determined the critical stress intensity for slow crack growth in hydrogen-precharged 18% Ni maraging steels and for 300M alloy steel (Ref 273). The maraging steel was tested at three yield strength levels: 1750, 1880, and 2020 MPa (254, 273, and 293 ksi); the 300M was tested at a yield strength of 1705 MPa (247 ksi). Specimens were charged with different levels of hydrogen, and the critical stress intensity for slow crack growth was determined (Fig. 44). Each of the four steels contained about 0.5 ppm H prior to charging, and the K_{Ic} values for the noncharged specimens are plotted at this value (dashed line). The data points for the charged specimens are the critical stress intensities where slow crack growth began. Note that the lowest-strength maraging specimens, with the highest K_{Ic}, had the highest critical stress intensities after charging. The critical stress intensities for the two higher-strength maraging steels were the same after charging to the same hydrogen levels despite their substantial differences in K_{Ic}. The 300M alloy, with the lowest yield strength (1705 MPa, or 245 ksi) and a K_{Ic} only slightly greater than the highest yield strength (2020 MPa, or 292 ksi) maraging grade, had much lower critical stress intensities and was more susceptible to hydrogen embrittlement.

Similar tests were performed using 18% Ni maraging steel at two different yield strengths: 1650 and 1915 MPa (239 and 278 ksi) in high-purity hydrogen at pressures from 11.5 to 133.3 kPa (86 to 1000 torr) (Ref 274). The crack growth rate as a function of the crack tip stress intensity, K, was determined for each steel as a function of temperature and hydrogen pressure. This information was used to determine the critical or threshold stress intensity, K_{th}, at which crack growth began. At K values above K_{th}, the crack growth increased (stage I). At intermediate K values, crack growth became relatively constant over a range of K values (stage II). At higher K values, crack growth became rapid (stage III) up to specimen failure. Figure 45 shows the data for the 1650 MPa (239 ksi) yield strength 18Ni(250) maraging steel tested in hydrogen at 133.3 kPa (1000 torr) for stage I and II crack growth. Over the temperature range of −60 to 23 °C (−75 to 73 °F), the apparent threshold stress intensity, K_{th}, varies only slightly from about 11 to 16.5 MPa $\sqrt{m}$ (10 to 15 ksi$\sqrt{in.}$). No detectable crack growth was observed at 50 or 80 °C (120 or 175 °F) for K levels up to 79 and 92 MPa$\sqrt{m}$ (72 and 84 ksi$\sqrt{in.}$), respectively. In the stage II regime, where the crack growth rate (da/dt) was constant, the highest crack growth rate occurred at 23 °C (73 °F). Figure 46 shows crack growth data for stage II crack growth as a function of test temperature and hydrogen pressure. The upper curve shows data for hydrogen at 133.3 kPa (1000 torr). At

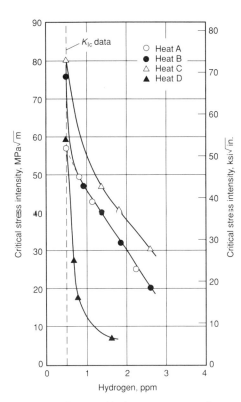

Fig. 44 Influence of hydrogen content on the critical stress at and below which cracks do not grow. Alloys A, B, and C are three different 18% Ni maraging steels at increasing levels of yield strength (1740, 1870, and 2020 MPa, or 252, 271, and 293 ksi); alloy D is 300M alloy steel at 1705 MPa (247 ksi). Source: Ref 273

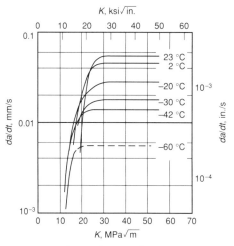

Fig. 45 Crack growth rates of 18Ni(250) maraging steel (1648 MPa, or 239 ksi, yield strength) in hydrogen at 133 kPa (1000 torr) as a function of test temperature and stress intensity range. Source: Ref 274

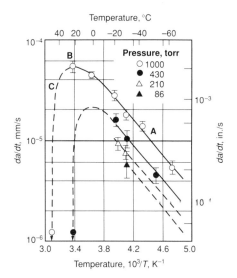

Fig. 46 Effect of test temperature on stage II crack growth rates for 18Ni(250) maraging steel in gaseous hydrogen at different pressures (133, 57, 28, and 11.5 kPa, or 1000, 430, 210, and 86 torr). Source: Ref 274

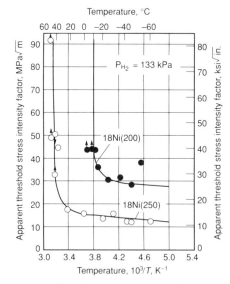

Fig. 47 Effect of test temperature on the apparent threshold stress intensity factor for two maraging steels tested in gaseous hydrogen at 133 kPa (1000 torr). Source: Ref 275

low temperatures, the crack growth rate increased with increasing temperature (region A) until a temperature range was reached where the crack growth did not vary much with temperature (region B), and the maximum crack growth rate was obtained. With further increases in temperature, the crack growth rates decrease rapidly (region C). Lowering the hydrogen pressure shifts the curves to lower crack growth rates, and the temperature for the maximum crack growth rate appears to decrease slightly.

Another study also tested 18% Ni maraging steels at two different yield strengths (18Ni(200) at 1270 MPa, or 184 ksi, and 18Ni(250) at 1650 MPa, or 239 ksi) in low-pressure hydrogen (Ref 275). Figure 47 shows a plot of the apparent threshold stress intensity level for the onset of slow crack growth, K_{th}, for these two grades in hydrogen at 133 kPa (1000 torr) and over the temperature range of −60 to 60 °C (−75 to 140 °F). The K_{th} is quite low at low temperatures (stage II) and increases slowly with increasing temperature until a temperature is obtained at which K_{th} increases rapidly and K_{Ic} is reached. The lower-strength steel has higher K_{th} values at low temperatures, and its critical temperature (where K_{th} increases rapidly) occurs at a much lower temperature than it does for the higher-strength grade.

Line Pipe Steels. Blisters form in steels when atomic hydrogen enters the steel from the surrounding environment. The hydrogen diffuses through the steel to an internal defect or inclusion, where it builds up as molecular hydrogen. This process produces high pressure, which causes a blister to form. The blister will grow as more hydrogen diffuses to it, deforming the surrounding metal until it ruptures. Blister formation is observed in environments containing hydrogen sulfide or where hydrogen is electrolytically charged. Blister formation is common in the more ductile steels, such as sour gas transmission line pipe steels (Ref 276-278).

Hydrogen-induced blistering of line pipe steels results from contact with wet environments containing H₂S. Hydrogen diffuses to inclusions and precipitates at the inclusion/matrix interface. A blister-crack array forms; cracking of the metal around the blisters is caused by the molecular hydrogen pressure and applied or residual stresses. These crack arrays are internal, rather than at the surface. Elongated manganese sulfides are the most common nucleation sites, but cracking can also occur at manganese silicates and at large complex carbonitrides.

It is well known that high sulfur contents are highly detrimental for sour gas applications. High-quality line pipe steels require very careful processing, which involves desulfurization to very low levels, sulfide shape control, controlled rolling, grain refinement, and segregation control.

Formation of Flakes in Steels. Flaking or hairline cracking in steels, particularly in forgings, heavy-section alloy plate steels, and railroad rails, has been an ongoing problem since the beginning of steelmaking (Ref 279-289). Flakes are caused by an excessive hydrogen content, that is, hydrogen above some safe level or threshold level. This safe level varies somewhat with steel composition, section thickness, cooling rate, inclusion content, and segregation and is generally in the range of 1 to 2 ppm. Some steel compositions, particularly low-carbon steels, appear to be rather free of flaking problems.

The solubility of hydrogen in liquid steel decreases substantially when it solidifies (to about 7 ppm in δ iron). With continued cooling, it decreases to about 5 ppm and then at the γ to α transformation it abruptly drops to about 2.5 ppm. With continued cooling, it decreases to less than 1 ppm at

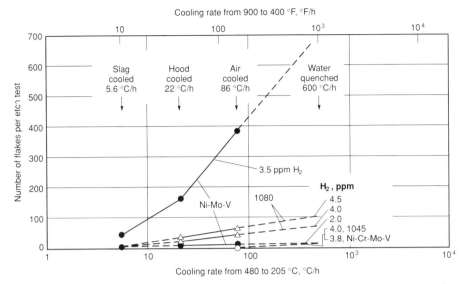

Fig. 48 Effect of cooling rate after hot working on the number of flakes in etch disks of AISI 1045 and 1080 carbon steels and nickel-molybdenum-vanadium and nickel-chromium-molybdenum-vanadium alloy steels. Source: Ref 287

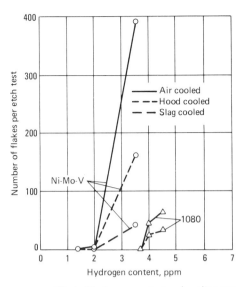

Fig. 49 Effect of hydrogen content and cooling rate after hot working on the number of flakes in etch disks of AISI 1080 carbon steel and nickel-molybdenum-vanadium alloy steel. Source: Ref 287

room temperature. Some of the hydrogen present in excess of the solubility limit at room temperature can be accommodated at microstructural traps, such as sulfide inclusions, and rendered harmless. However, if the hydrogen content is too high, cracking will occur after a certain delay time, usually less than 2 weeks, unless the steel is reheated to a temperature where it can safely diffuse out of the steel. In the years prior to the development of vacuum degassing (even today for some steels, such as rails), wrought or forged products susceptible to flaking were held at temperatures between about 500 and 650 °C (930 and 1200 °F) for a time that depended on the product thickness; this was done before the products were cooled to ambient temperature, and it allowed the hydrogen content to decrease to a safe level. With the drive to produce steels with lower sulfur contents, rail producers have discovered that these safe controlled-cooling practices have had to be extended for further hydrogen removal because of the decrease in hydrogen traps available. Similar flaking problems have occurred in low-sulfur plate steels that were free of such problems at the former higher-sulfur levels.

An interesting study was done of flake formation in plain carbon steels (AISI 1045 and 1080) and in nickel-molybdenum-vanadium and nickel-chromium-molybdenum-vanadium forging steels melted and teemed to produce a range of hydrogen contents (Ref 287). Five ingots of each grade were produced and rolled to 305 mm (12 in.) square blooms, and each was hot sheared to produce four 610 mm (24 in.) long sections. These sections were cooled at different rates and by different methods (water quench, air cool, hood cool, and slag cool). After a 2-week incubation period, to permit flaking to occur, transverse disks were cut from each block, macroetched, and examined. The total number of flakes were counted, except for disks that were quench cracked. Hydrogen analyses of the blocks were made immediately after cooling to a safe handling temperature.

The study showed that the number of flakes increased with the cooling rate (Ref 287). The flakes were generally located in a circumferential pattern between the surface and the center of the block and as the cooling rate decreased, they were closer to the center. In general, as the number of flakes decreased, their size increased.

Figure 48 shows data for the relationship between the number of flakes on the etch disks as a function of the cooling rate and hydrogen content for each grade. Quench cracking, which occurred in some of the water-quenched blocks, made the analysis more difficult by requiring the use of an estimated value of the number of flakes (dashed lines). In general, water quenching produced about twice as many flakes as did air cooling. The 1045 and nickel-chromium-molybdenum-vanadium steels had low susceptibilities to flaking at hydrogen levels of 4 and 3.8 ppm. The 1080 and nickel-molybdenum-vanadium steels were much more susceptible at hydrogen levels of 4.0 and 3.5 ppm. Based on the number of flakes, a relative susceptibility of 1 was assigned for 1045 and nickel-chromium-molybdenum-vanadium, and values of 8 and 75 were assigned for air-cooled blocks of 1080 and nickel-molybdenum-vanadium. The figure also shows the influence of hydrogen content on the nickel-molybdenum-vanadium steel blocks with hydrogen contents of 3.5 and 2.0 ppm. Reducing the hydrogen content of the steel from 3.5 to 2.0 reduced its susceptibility to a level similar to that of the 1045 and nickel-chromium-molybdenum-vanadium steels (indicated by the lower curve for nickel-molybdenum-vanadium). For the 1080 steel, the flaking index was 8 at 4.0 ppm; at 3.7 ppm it flaked only when water quenched, and at 2.2 ppm it did not flake even when water quenched. These results for nickel-molybdenum-vanadium and 1080 as a function of hydrogen content and cooling rate are better shown in Fig. 49, which clearly demonstrates the presence of a critical threshold level of hydrogen in these flake-sensitive steels. The threshold levels were approximately 3.5 ppm for 1045 and nickel-chromium-molybdenum-vanadium, 3.3 ppm for 1080, and 1.6 ppm for nickel-molybdenum-vanadium. The high susceptibility of the nickel-molybdenum-vanadium steel was attributed to the high transformation stresses associated with the formation of martensite from enriched austenite; the high susceptibility of 1080 was attributed to its low fracture toughness.

Metal-Induced Embrittlement

The term metal-induced embrittlement (MIE) applies to cases in which contact with a liquid metal embrittles a solid metal (liquid metal embrittlement, LME) or in which contact with a solid metal somewhat below its melting point (solid metal embrittlement, SME) embrittles a solid metal. Only recently has a distinction been made between these two forms of MIE. Historically, LME was the term used to describe all MIE problems.

Liquid Metal Embrittlement

Liquid metal embrittlement is a phenomenon in which the ductility or fracture stress of a solid metal is reduced by surface contact with a liquid metal. The first recorded recognition of this problem was made in 1914 when the embrittlement of β brass by mercury was observed (Ref 290). Since then, LME has been identified as the cause of numerous failures, and numerous reviews of LME have been published (Ref 291-301).

There are at least four distinct forms of liquid metal embrittlement (Ref 292):

- *Type 1*: Instantaneous fracture of a metal under an applied or residual tensile stress when in contact with certain liquid metals. This is the most common type of LME
- *Type 2*: Delayed failure of a metal when in contact with a specific liquid metal after a certain time interval at a static load below the ultimate tensile stress of the metal. This form involves grain-boundary penetration by the liquid metal and is less common than type 1
- *Type 3*: Grain-boundary penetration of a solid metal by a specific liquid metal, which causes the solid metal to eventually disintegrate. Stress does not appear to be a prerequisite for this type of LME in all observed cases
- *Type 4*: High-temperature corrosion of a solid metal by a liquid metal, causing embrittlement, an entirely different problem from types 1 to 3

In some respects, LME is similar to stress-corrosion cracking (SCC), which always requires a measurable incubation period before fracture occurs. Stress-corrosion cracking is more similar to type 2 LME than to the other types. This review will concentrate on the influence of MIE on the mechanical properties of ferrous alloys.

A number of observations have been made of interactions between liquid metal and solid metal. These observations can be classified as (Ref 291):

- No observed interaction
- Dissolution of the solid metal into the liquid metal. Diffusion of the liquid metal into the solid metal
- Formation of intermetallic compounds at the liquid/solid interface
- Intergranular penetration of the liquid metal into the solid metal without the presence of applied or residual stress
- Brittle premature failure of the solid metal caused by intergranular penetration of the liquid metal into the solid metal under the influence of applied or residual tensile stress. Failure occurs after a finite incubation period
- Brittle instantaneous failure of a stressed metal when wetted by a specific liquid metal. Grain boundary penetration is not necessarily observed

The last three observed interactions are the prime forms of liquid metal embrittlement. Most studies of LME have been concerned with the form that causes instantaneous failure of the solid metal, the most common form. This type of LME has been referred to as adsorption-induced embrittlement. Liquid metal embrittlement is often considered a special type of brittle fracture, and fracture mechanics approaches have been applied to its analysis.

Many different metals and alloys besides ferrous alloys can fail because of LME. Only specific liquid metals are known to cause LME of each specific metal or alloy. Indeed, this is one of the more perplexing aspects of LME, and some investigators refer to it as specificity (Ref 302). However, studies have generally used tests involving a limited number of conditions, for example, a single temperature, applied stress, grain size, and so on, and until a much wider range of variables has been tested, the validity of this concept cannot be proved. Also, many LME tests have been of the go/no go type rather than quantitative measurements of the degree of embrittlement.

Certain conditions are recognized as being required for, but not necessarily sufficient to cause, embrittlement. These conditions include (Ref 291):

- Adequate wetting of the solid-metal substrate by the liquid metal
- An applied or residual stress present in the solid metal
- A barrier to plastic flow existing at some point in the solid metal that is in contact with the liquid metal

Of course, for failure to occur, the liquid metal must be of the correct composition to cause LME of the solid metal.

If the solid metal is not notch sensitive, as in the case of fcc metals, the crack will propagate only when the liquid metal feeds the crack. However, for a notch-sensitive metal, such as bcc metals like iron, the nucleated crack may become unstable and propagate ahead of the liquid metal. Crack propagation can be extremely rapid; rates between 500 and 5000 mm/s (20 and 200 in./s) have been recorded (Ref 293). In most cases, crack paths are intergranular, but transgranular cracking has been observed, and a few cases of ductile fracture have even been reported. Factors that influence the fracture stress and ductility of metals subject to LME include:

- Composition of the solid and liquid metals
- Temperature
- Strain rate
- Grain size
- Thermal-mechanical history of the solid metal

Some of the earliest studies of LME did not recognize the important role of stress in LME (Ref 303). Some early studies, however, did demonstrate the significance of stress in producing LME (Ref 304-306). Tensile stresses, either applied or residual, are required, while the solid metal is wet by the liquid metal. Again, a certain critical stress (threshold) can be identified above which cracking occurs and below which it does not. Subsequent studies focused on the identification of specific liquid-solid metal couples that undergo LME.

Studies of specific LME couples have empirically identified certain trends that are usually, but not always, obeyed. For example, most LME couples have very little mutual solid solubility and exist as immiscible liquids in the liquid state. Solid metals that are highly soluble in the liquid metal and solid-liquid metal combinations that form intermetallic compounds are usually immune to LME. Such studies have demonstrated that good wetting of the substrate is required. It is interesting that the factors that promote the lowest interfacial energy and therefore the best wetting are high mutual solid solubility and intermetallic compound formation, both of which promote low susceptibility to LME. A number of studies have examined the effect of adding various solute elements to particular liquid metals, but no clear trends have been identified.

Several researchers have attempted to correlate the occurrence and severity of LME for particular couples with the electronegativities of the metals. For some couples, maximum embrittlement arises when the solid and liquid metals have similar electronegativities. As the difference in electronegativity increases, the degree of embrittlement decreases. While many examples support this relationship, exceptions also exist (Ref 294).

The presence of notches (stress concentrators) increases the severity of LME, and this has prompted the use of fracture mechanics concepts in the study of LME. As stress raisers are well known to be a detrimental influence in brittle fracture, their effect on LME is understandable.

Reference 301 reports the results of a very extensive literature survey concerning the nature and occurrence of LME. The following liquid metals have been reported to cause LME of ferrous alloys: aluminum (minor LME for pure iron, no LME for a 0.2% C steel), antimony, bismuth (both positive and negative results have been reported), cadmium, copper, gallium (both positive and negative results have been reported), indium, lead, lithium, mercury, sodium, tellurium, tin, and zinc. Solders, bearing metals, and brazing alloys have also been shown to produce LME in certain tested ferrous alloys. As might be expected, for each of the above-mentioned liquid metals only a limited range of solid ferrous alloys has been evaluated. Liquid metals that have not been found to cause LME of ferrous alloys include cesium, rubidium, selenium, silver (although silver-containing

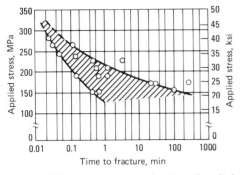

Fig. 50 Time to fracture as a function of applied load for AISI 4130 alloy steel specimens (at 44 HRC) wetted with molten lithium at 200 °C (390 °F). Source: Ref 291

brazing alloys do cause LME), and thallium. Reference 301 contains source data for these observations and should be consulted for further information.

The effect of test temperature, T, is complex and is not predictable. Generally, temperature increases above the melting point of the embrittling liquid decrease the degree of embrittlement. Maximum embrittlement often occurs near the melting point of the liquid, T_m. In many couples, embrittlement occurs below the melting point of the liquid. For example, the embrittlement of AISI 4140 alloy steel starts at temperatures below the melting point for cadmium ($T/T_m = 0.75$) and for lead and tin ($T/T_m = 0.85$) (Ref 307). This phenomenon is called solid metal embrittlement or solid metal induced embrittlement. In other cases, the liquid temperature must be above T_m, as for the embrittlement of austenitic stainless steel by zinc. It has been reported that the zinc embrittlement of austenitic stainless steels has not been observed below 750 °C (1380

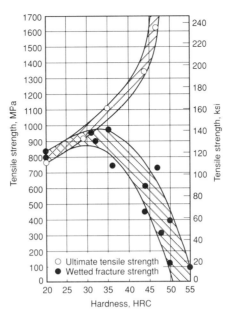

Fig. 51 Plot of hardness versus tensile strength for AISI 4130 alloy steel with and without wetting by molten lithium at 205 °C (400 °F). Source: Ref 291

°F), which is well above the melting point of zinc (419 °C, or 786 °F) (Ref 308).

Liquid metal embrittlement reduces tensile ductility and stress at fracture if the failure occurs at a stress below the usual yield strength of the metal. The time to fracture is usually directly related to the level of the applied stress. At some high level of applied stress, fracture can occur immediately after the specimen is wetted by the liquid metal. As the level of applied stress is reduced, the time to failure increases. This is shown in Fig. 50, which

shows test results for AISI 4130 steel at 44 HRC immersed in molten lithium at 200 °C (390 °F). As the applied stress decreased from 310 to 138 MPa (45 to 20 ksi), the time to failure increased from about 1 s to 1 200 min.

The strength of the material also influences LME results. Figure 51 demonstrates this by showing the normal tensile strength of AISI 4130 alloy steel quenched and tempered to hardnesses from 20 to 55 HRC and the corresponding fracture strength of specimens wet with liquid lithium at 205 °C (400 °F). At tensile strengths above 1035 MPa (150 ksi), catastrophic embrittlement occurs (Ref 291). Tensile ductility, however, was reduced at all of the strength levels.

While the majority of LME studies have been conducted using tensile specimens, some work has been done using fatigue loading conditions. Figure 52 shows axial load fatigue tests of AISI 4340 with a tensile strength of 1310 MPa (190 ksi) tested at a stress ratio of 0.75 with or without a coating with liquid mercury (Ref 291). As the maximum applied stress decreased, the coated specimens exhibited a proportionally lower fatigue life than did the uncoated specimens.

Fatigue crack growth rate measurements were performed using liquid lithium and AISI 304L stainless steel at temperatures ranging from 473 to 973 K (200 to 700 °C) (Ref 309). Figure 53 shows the crack growth data from these tests as a function of temperature for 304L specimens in argon and in liquid lithium using two different test frequencies and a stress intensity range of 14 MPa√m (12.7 ksi√in.). The crack growth rates in lithium were much greater than those in argon at all temperatures and load frequencies. In lithium, the crack growth rate decreased as the temperature increased from 473 to about 700 K (200 to 700 °C) and then increased strongly with further increases in temperature.

Crack propagation in argon was transgranular for all test conditions. At 573 K (300 °C) in lithium, a mixture of transgranular cleavage and intergranular fracture was observed that varied with load frequency and stress intensity range (Fig. 54). The percentage of intergranular fracture increased with decreasing stress intensity range and increasing frequency. The influence of load frequency at 573 and 873 K (300 and 600 °C) is shown in Fig. 55. The effect of frequency on the amount of intergranular fracture at 873 K (600 °C) is opposite to that at 573 K (300 °C). This difference was also accompanied by a difference in the appearance of the intergranular facets; at 873 K (600 °C), the facets were rounded, with numerous dimples; at 573 K (300 °C), the facets were sharp, and occasional tears and slip bands were observed.

The liquid metal embrittlement failures of steels caused by molten copper and copper-base alloys are common, and numerous

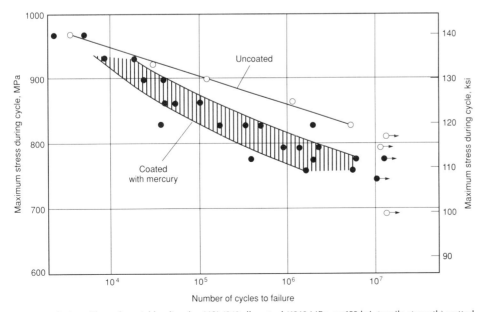

Fig. 52 Fatigue life under axial loading for AISI 4340 alloy steel (1310 MPa, or 190 ksi, tensile strength) wetted with mercury. Source: Ref 291

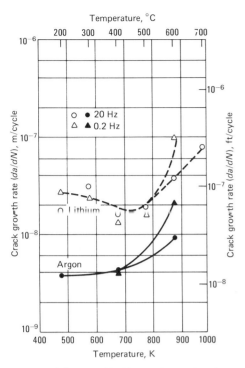

Fig. 53 Influence of test temperature and environment (argon versus molten lithium) on the crack growth rate using two different loading frequencies and a stress intensity factor range of 14 MPa√m (12.7 ksi√in.) for AISI 304L stainless steel. Source: Ref 309

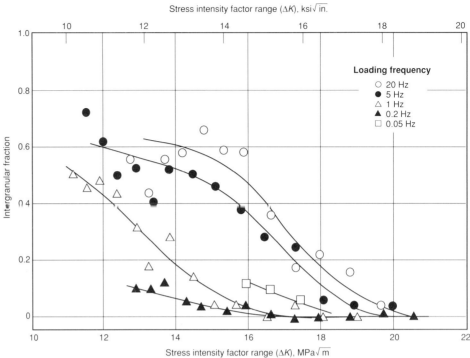

Fig. 54 Influence of loading frequency and stress intensity factor range on the percentage of intergranular fracture for AISI 304L in molten lithium at 573 K (300 °C). Source: Ref 309

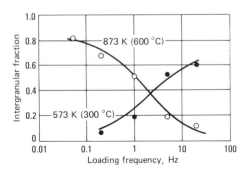

Fig. 55 Influence of loading frequency and temperature of lithium on the percentage of intergranular fracture for AISI 304L. Stress intensity factor range, ΔK, was 14 MPa√m (12.7 ksi√in.). Source: Ref 309

studies have been made. Such failures are characterized by grain-boundary penetration of liquid copper, which can be quite rapid. One of the earliest studies of LME of steel, in this case by molten brass, was made in 1927 (Ref 310). An evaluation of the influence of various liquid metals on plain carbon steel, silicon steel, and chromium steel found that these steels were embrittled at 1000 to 1200 °C (1830 to 2190 °F) by liquid tin, zinc, antimony, copper, 5% tin-bronze, and 10% zinc-brass; liquid bismuth, cadmium, lead, and silver caused little or no embrittlement (Ref 311).

A series of studies of LME of iron by liquid copper was done using tests performed under nonoxidizing conditions with notched tensile creep specimens at 1100 to 1130 °C (2010 to 2065 °F) and stresses of 8.3 and 11.0 MPa (1.2 and 1.6 ksi) (Ref 312-315). Liquid copper significantly altered the creep behavior of pure iron, causing premature failure. Embrittlement was of the delayed type controlled by grain-boundary copper penetration. The depth of surface cracking was controlled by the depth of copper penetration. Dihedral-angle measurements for liquid copper in the steel grain boundaries revealed that 34° was the most frequent angle for exposure at 1100 or 1130 °C (2010 or 2065 °F).

Liquid metal embrittlement of steels by copper has also been encountered during welding and joining processes. An examination of copper weldment deposition on ferritic and austenitic stainless steels in duplex ferrite-austenite grades found that stable ferrite reduced the penetration of copper; if ferrite was in excess of 30%, penetration did not occur. Also, copper deposited on fully ferritic grades did not cause cracking, but it did in austenitic grades. Wetting by molten copper was also examined. A contact (dihedral) angle of 92 to 100° was observed for ferritic grades (no wetting of the grain boundaries) and 22 to 28° for the austenitic grades (wetting) at 1100 °C (2010 °F) (Ref 316).

A series of studies has examined LME of steels by copper during welding (Ref 317-320). A study of HY-80 steel welded with or without copper-nickel filler metal found infiltration of the grain boundaries by the copper-nickel deposit under an applied strain field in the heat-affected zone while the steel was austenitic. Heat affected zone areas that remained ferritic were not penetrated (Ref 317). In another study involving copper deposition by a gas metal arc, alloy steels such as AISI 4340, 4140, and 304 were found to be extremely sensitive to copper penetration. In comparison, penetration depths for AISI 1340, 1050, Armco iron, and carburized Armco iron were only about one-third that of the alloy steels. A ferritic stainless steel, AISI 430, was almost completely immune to copper LME. The study concluded that, when stresses were absent, the ease of penetration by molten copper was a function of the alloy content of the steel. Grain-boundary copper penetration occurred under an applied stress, forming partially filled or open cracks in the steel. Steels that remained ferritic at the melting point of copper were not penetrated (Ref 318).

As reviewed earlier in the section "Aluminum Nitride Embrittlement," residual copper can detrimentally influence the hot workability of steel by internal LME. Reference 321 summarizes work on the influence of copper on the hot workability and scaling characteristics of steels. Elements such as tin, arsenic, and antimony dissolved in the steel increase the sensitivity of surface cracking that is due to copper.

Perhaps the most common LME failure is the fracture of railroad axles caused by copper penetration from overheated friction bearings. These failures are relatively common, even today, and have been ob-

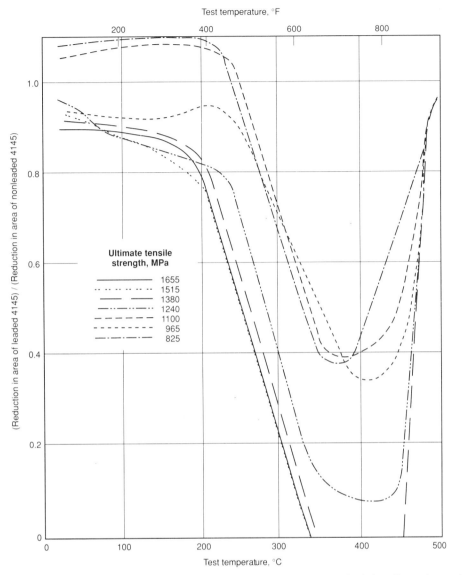

Fig. 56 Relative decrease in the percent reduction of area for leaded versus nonleaded tensile specimens of AISI 4145 alloy steel as a function of test temperature and ultimate tensile strength (825 to 1655 MPa, or 120 to 240 ksi). Source: Ref 329

served ever since the development of railroads. A friction bearing consists of a bronze shell lined with babbit metal bonded to the inner surface. The bearing is lubricated through an opening in the shell, which is packed with cotton waste. Oil from a reservoir is carried to the bearing through the cotton waste by a wicklike mechanism. In 1914, a study presented results for a Krupp axle that failed in service (Ref 322). The surface of the axle had been exposed to very high temperatures, as evidenced by the microstructural changes, and cracking was present. The heated region was the part of the axle within the support bearing. Bronze bearing metal was observed to be associated with the cracks and grain-boundary penetration. Such failures still occur and have recently been reviewed (Ref 323).

Solid Metal Embrittlement

Solid metal embrittlement occurs when the temperature is somewhat below the melting point of the embrittling metal (Ref 297, 300, 307, 324, 326). Its characteristics, except for this distinction, are quite similar to that of LME. Crack propagation rates for SME, however, are much slower than for LME, and multiple cracking is often observed in SME, but not in LME.

Iron-Base Alloys. Solid metal embrittlement of iron-base alloys can be caused by a number of low melting point metals. Solid metal embrittlement of steels by cadmium (Ref 307, 326, 327) was studied in depth as a result of the fracture of cadmium-plated nuts (Ref 327). The failures occurred in the temperature range of 200 to 300 °C (390 to 570 °F), which is well below the melting point of cadmium (321 °C, or 610 °F). Crack propagation rates increase with temperature (below the melting point, T_m), and a threshold stress level is required for cracking (Ref 328).

Leaded Alloy Steels. Solid metal embrittlement of leaded alloy steels, that is, internal rather than external embrittlement, has been studied extensively (Ref 329-333). Reference 329 contains the first discussion of the differences between liquid metal and solid metal embrittlement. Short-time high-temperature tensile specimens of AISI 4140 and 41L40 (leaded) were tested between room temperature and approximately 480 °C (900 °F). The melting point of lead is 327 °C (621 °F). Tensile test results for the leaded and nonleaded steels were compared. Figure 56 shows the ratio of the percentage reduction in area of the leaded steel to the unleaded steel for tensile strength levels from 825 to 1515 MPa (120 to 220 ksi) and temperatures from ambient to approximately 480 °C (900 °F). The ductility of the leaded steels was substantially reduced at temperatures well below the melting point of lead. The loss in ductility increased with tensile strength and was greatest at temperatures just above the melting point of lead. By 480 °C (900 °F), most of the ductility has been recovered. Similar tests of leaded and nonleaded Charpy V-notch impact specimens revealed that embrittlement effects were apparent beginning at about 28 °C (50 °F) below the melting point, suggesting a sensitivity to strain rate. Also, when comparing the ultimate tensile strength for leaded and nonleaded specimens as a function of hardness at the maximum embrittlement temperature (370 °C, or 700 °F), results were quite similar to those shown in Ref 291 for AISI 4130 in liquid lithium at 205 °C (400 °F) (Fig. 51). Catastrophic failure occurred for leaded steels at tensile strengths above 965 to 1140 MPa (140 to 165 ksi).

A fractographic examination was done of these test fractures (Ref 330). Five types of fractures were observed that related to the test temperature. Tensile specimens broken at temperatures up to 205 °C (400 °F) exhibited the classic cup-and-cone fracture and failed by microvoid coalescence (ductile). Specimens tested between 205 and 315 °C (400 and 600 °F) exhibited small shear lips and a mixture of intergranular fracture regions within a transgranular fracture. This was the range where embrittlement (SME only) began and became progressively worse. Tensiles broken between 315 and 345 °C (600 and 650 °F) showed severe embrittlement and numerous coarse facets (called "fisheyes" by the authors of the study). Fracture was a mixture of transgranular and intergranular modes. Tensile specimens broken between 345 and 480 °C (650 and 900 °F), covering the region of maximum embrittlement to recovery, exhibited

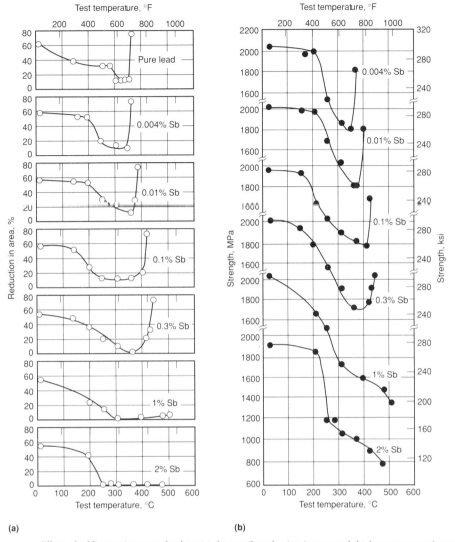

Fig. 57 Effect of adding antimony to lead on (a) the tensile reduction in area and (b) fracture strength as a function of test temperature for externally embrittled AISI 4145 steel heat treated to a yield strength of 1380 MPa (200 ksi). Source: Ref 333

large shear lips and a large faceted origin. The fractures were completely intergranular and were moving away from the origin toward the opposite shear lip. Inclusions were always observed to be associated with the intergranular facets. Above 480 °C (900 °F), the ductility was recovered, and there was a return to a cup-and-cone fracture.

Leaded Carbon and Alloy Steels. An evaluation of a variety of leaded carbon and alloy steels found that all were subject to SME and LME (Ref 331). Embrittlement began at about the same temperature and was most severe in the same temperature range, 315 to 370 °C (600 to 700 °F), but the recovery of ductility at higher temperatures varied somewhat, from about 400 to 480 °C (750 to 900 °F). It was thought that the composition of the lead could be affecting this brittle-to-ductile transition temperature, T_r.

Subsequent work compared externally applied liquid lead and lead alloys to the previous internal lead work (Ref 332, 333). Nonleaded AISI 4145 tensile specimens at a tensile strength of 1380 MPA (200 ksi) were externally wetted with high-purity lead and lead-antimony alloys. For the high-purity lead, embrittlement began at lower temperatures, and the recovery of embrittlement occurred at a lower temperature (370 °C, or 700 °F), than for the internally embrittled leaded steels. Additions of up to 2% Sb were made to the pure lead, and similar tests were run that showed that the onset of embrittlement, the temperature range for severe embrittlement, and the temperature for the recovery of ductility were all affected by the addition of antimony (Fig. 57) (Ref 332, 333). As the antimony content increased to 0.3%, T_r gradually increased. However, for the 1 and 2% Sb specimens, little or no evidence for recovery was obtained up to 480 °C (900 °F). Similar work was done for lead containing up to 9% Sn (Ref 333). Results were similar for tin except that on a weight percent basis, it enhanced lead embrittlement less than antimony additions did.

Combined SME and LME has been demonstrated for AISI 4140 alloy steel (1380 MPa, or 200 ksi, tensile strength) in contact with pure zinc, lead, cadmium, tin, and indium over the temperature range from ambient to the melting point of the pure, low melting point alloy (Ref 307). Figure 58 shows test results for 4140 in contact with these metals. All of the data have been normalized for ease of comparison. Temperature is expressed in homologous fashion (that is, by the ratio of the test temperature used to the melting point of the embrittler):

$$T_H = \frac{T}{T_m} \qquad (Eq\ 4)$$

where T_H is the homologous temperature. The true fracture strength (TFS) ratio is the ratio of the true fracture stress of the embrittlement couple to the true fracture stress of uncoated 4140 at the same temperature; the reduction in area (RA) ratio is defined in the same way, based on the percentage reduction in area values. As the melting point of the embrittler is approached, the TFS ratio and RA ratio decrease. For temperatures at the melting point of the embrittler, RA ratios vary from 0 to approximately 0.2, indicating severe embrittlement. All of the elements in this test were potent embrittlers when the test temperature was increased to about 75% of the melting temperature.

Times to failure are reported in Ref 326 for AISI 4140 (1380 MPa, or 200 ksi, tensile strength) in contact with solid zinc, lead, cadmium, tin, and indium at temperatures below the melting points of the embrittlers. The time to failure decreased as the temperature increased toward the melting point and decreased as the applied stress was increased toward the tensile strength. Crack propagation was intergranular along the prior-austenite grain boundaries.

As with other embrittlement mechanisms, the segregation of embrittling impurities (phosphorus, antimony, arsenic, and tin) to the prior-austenite grain boundaries can influence MIE susceptibility. In an examination of the influence of prior temper embrittlement of AISI 3340 alloy steel containing controlled additions of phosphorus, antimony, tin, and arsenic, the segregation of tin and antimony to the grain boundaries increased the susceptibility to MIE by lead or tin; however, the segregation of phosphorus and arsenic to the grain boundaries produced a minor decrease in MIE susceptibility in tests from room temperature to 425 °C (800 °F).

Neutron Irradiation Embrittlement

The neutron irradiation of nuclear reactor components can produce an increase in strength and in the ductile-to-brittle transi-

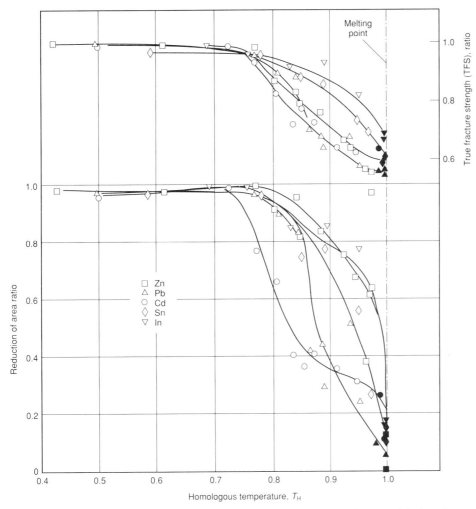

Fig. 58 Normalized true fracture strength ratio and reduction of area ratio as a function of the homologous temperature ($T_H = T/T_m$) for heat-treated AISI 4140 alloy steel in external contact with solid lead, cadmium, zinc, tin, and indium. Source: Ref 307

tion temperature. The degree of irradiation-induced embrittlement depends on the neutron dose, neutron spectrum, temperature, steel composition, and heat treatment. Tempered martensite is less susceptible to neutron irradiation embrittlement than tempered bainite or ferrite-pearlite microstructures. Impurity elements in the steels can influence embrittlement; for example, phosphorus levels above 0.015% and copper levels above 0.05% are detrimental.

Radiation produces swelling and void formation. Void density decreases as the irradiation temperature increases, but the average void size increases. The examination of fractures of irradiated ferritic materials tested below their DBTT reveals a change in fracture mode from transgranular cleavage to mixed cleavage and intergranular fracture. The examination of fracture made above the DBTT reveals a change in dimple size and depth. Additional information on this subject can be found in the article "Effect of Neutron Irradiation on Properties of Steels" in this Volume.

Stress-Corrosion Cracking

Stress-corrosion cracking is a generic term describing the initiation and propagation of cracks in a metal or alloy under the combined action of tensile stresses (applied and/or residual) and a corrosive environment. The history of SCC failures begins in the nineteenth century with the development of cold-drawn brass cartridges for ammunition. Similar failures of brass condenser tubes brought additional attention to this problem, which soon gained the name season cracking. These failures were due to residual tensile stresses, and the benefits of stress-relieving treatments were subsequently shown. Later, the detrimental influence of ammonia on the stress-corrosion cracking of brass was observed, which lead to the empirical rule of the specificity of the environment causing SCC for each alloy. This rule was cited for many years, but with subsequent research, particularly in the last 30 years, the list of environments that can cause SCC for specific alloys has grown dramatically, and the rule appears to be invalid. Also, in most cases the chemical species causing SCC need not be present in large amounts or high concentrations in the environment.

The caustic cracking of riveted boiler steels was also observed late in the nineteenth century. Caustic cracking involves a localized high concentration of free alkali at high temperatures, as would exist in a boiler. At the beginning of this century, SCC of aluminum alloys due to moisture was observed. Also, SCC of martensitic steels was encountered, but the nature of such cracks was not recognized until much later (additional information is available in the article "Wrought Stainless Steels" in this Volume). Cracking of plain carbon steels (mild steels) by nitrates was encountered in the chemical industry at the beginning of this century.

Stainless steels, first used extensively in the 1930s, were also observed to be susceptible to such failures, which were due to chlorides or caustic environments at high temperatures. Also in the 1930s, magnesium aircraft alloys were found to be susceptible to SCC caused by moisture. Cracking problems with titanium alloys were observed beginning in the 1950s.

Considerable research has been conducted on stress-corrosion cracking. Much of it has been centered on gaining an understanding of the conditions that cause it for commercial alloys and defining the influence of alloy composition, processing, and microstructure. Research has also focused on defining the mechanism(s) responsible for SCC and developing remedies to minimize or prevent SCC failures. This article concentrates on SCC of iron-base alloys, with emphasis on practical aspects rather than mechanisms. References 335 to 361 provide a wealth of additional information on stress-corrosion cracking, particularly with respect to iron-base alloys.

Properties and Conditions Producing Stress-Corrosion Cracking

A number of general characteristics of stress-corrosion cracking should be summarized before dealing with specific topics.

Alloy Susceptibility to SCC. First, the alloy must be susceptible to stress-corrosion cracking. Although it was formerly believed that pure metals are immune to SCC, recent experience has shown instances in which pure metals containing certain impurity elements were susceptible. It could be argued that such pure metals were not really pure. Also, tests of high-purity alloys have demonstrated greatly enhanced resistance to SCC even for alloys that are normally quite susceptible. Therefore, because pure is a relative term, and because the influence of impurity elements on SCC requires further study, it can be concluded that improving the purity of metals and alloys is very

Table 2 Partial list of aqueous environments known to cause stress-corrosion cracking in steels

Low-carbon steel and low-alloy ferritic steels	Austenitic stainless steels
Nitrates	Chlorides
Phosphates	Hydroxides
Sulfates	Fluorides
Carbonates	Bromides
Carbon monoxide/carbon dioxide	Water/oxygen
Hydrogen sulfide	Sulfates
Ferric chloride	Thiocyanates
Ammonia	Thiosulfates
Organic liquids	Tetrathionates
Water/oxygen	Polythionates
Cyanides	Sulfur dioxide
Hydroxides	Sulfurous acid

Source: Ref 338

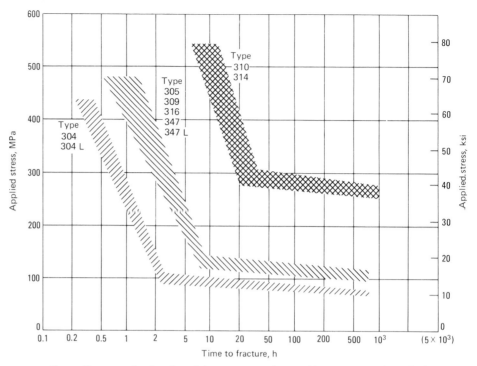

Fig. 59 Composite curves showing the relative stress-corrosion cracking resistance of standard austenitic stainless steels in boiling 42% MgCl$_2$. Source: Ref 362

beneficial, but it is not yet a commercially viable solution to the SCC problem.

Static Tensile Stresses. The metal or alloy must not only be susceptible to SCC, it must also be subjected to static tensile stresses for SCC to occur. These stresses may be applied service loads, or they may result from residual stresses, or both. Stresses can also arise from the wedging action of corrosion products growing in a crack. These stresses need not exceed the yield strength of the material and can be relatively low in comparison. Generally, but not always, some stress level will exist below which cracking does not occur but above which it does (that is, a threshold stress level discussed previously in the sections "Hydrogen Damage" and "Metal-Induced Embrittlement" in this article). Compressive residual stresses, such as from shot peening, are beneficial in reducing or preventing SCC.

SCC-Inducing Chemical Species. The alloy in question must not only be under the influence of an applied or residual tensile stress of sufficient magnitude, it must also be in contact with an environment that contains a chemical species that will cause SCC. An interesting aspect of SCC is that the alloy in question is normally nearly inert, from the standpoint of general corrosion, in the environment that causes SCC. This anomalous behavior was one of the biggest surprises to early researchers who studied SCC of stainless steels because these are alloys noted for their superior corrosion resistance. Indeed, SCC failures generally show very few corrosion effects. The damaging environment, however, may produce a film on the surface of the alloy. Such films can be tarnish, passivating layers, or dealloyed layers, depending on the system. These films retard general corrosion.

As mentioned, it was believed for many years that only a few such species could cause SCC for certain alloys or families of alloys. A partial list has been compiled of species in aqueous environments that cause SCC in iron-base alloys (Table 2). As Table 2 indicates, a great many damaging environments are known (and this is only a partial list); further research will probably add to the list. Also, as previously mentioned, the damaging species does not need to be present in large quantities or high concentrations. Indeed, even parts-per-million levels have been known to cause SCC failures. Only small quantities are necessary because the damaging species can become concentrated in the affected area. Also, an environment that causes SCC in one alloy may not produce SCC in another alloy (unless they are similar in nature and composition). Similarly, a particular environment may cause SCC of an alloy only under specific conditions, such as a particular temperature, degree of aeration, or pH. A particular heat treatment or strength of the alloy could also have an impact.

Parameters Affecting Stress-Corrosion Cracking

Variables such as temperature, pit geometry, intergranular corrosion, and slip dissolution processes can also promote stress-corrosion cracking, depending on their degree or severity.

Temperature is often a critical parameter in SCC. For some systems, failures occur only at temperatures above some specific level, and the severity or rapidity of failure increases with temperature above this limit. Chloride-induced SCC of austenitic stainless steels is known to follow this behavior.

Pit Geometry. For a smooth specimen or part under tension in an environment that can cause SCC, the process may begin with the formation of a corrosion pit. Such pits could be present on a part before it is placed in service if, for example, the part were previously excessively pickled (that is, acid cleaned). Pits may also form at inclusions that intersect the free surface of the part. Pit geometry is quite important; the pit depth must be much greater than the pit width by about a factor of ten, at a minimum. The environment within the pit must change if it is to act as a crack initiator. Not all pits initiate stress-corrosion cracks. It is possible that a preexisting pit may not develop the same electrochemical conditions as one that forms during exposure in the environment. The local chemistry in the pit appears to be important in SCC.

Intergranular Corrosion or Slip Dissolution. Stress-corrosion cracks can also initiate in the absence of pitting, for example, by slip dissolution (film rupture) processes or by intergranular corrosion. In intergranular corrosion-initiated SCC, the grain-boundary composition must differ from the bulk composition. This can occur because of the segregation of elements to the grain boundaries (or because of sensitization in austenitic stainless steels). The slip dissolution process is associated with localized corrosion on slip planes at the surface of low stacking fault energy austenitic alloys.

SCC Verification Procedures

Stress-corrosion cracking is a delayed failure process in which a certain incubation time is required to initiate cracking. The cracks propagate slowly, and propagation rates can vary from 0.1 to 10^{-9} mm/s (0.004 to 4.0×10^{-11} in./s) (Ref 358). Cracking continues until the stress exceeds the fracture strength of the remaining noncracked cross section. Crack initiation and propagation are usually divided into three stages:

- Crack initiation and stage I propagation (increasing growth rate with increasing stress intensity)
- Steady-state crack propagation (stage II: growth constant over intermediate stress intensities)
- Crack propagation to final fracture (stage III: rapid increase in crack growth rate at high stress intensities)

The threshold stress intensity necessary to produce SCC is called K_{ISCC}.

Historically, stress-corrosion cracking tests have been conducted on statically loaded smooth specimens at various stress levels, and the time to fracture is determined for specific environmental conditions. The resulting data are generally plotted on semilog paper, with the time to failure located on the log scale. Tests are conducted in relatively severe environments under the assumption that if cracking does not occur under the test conditions, the alloy is safe to use at the same stress level and temperature in a less severe environment.

Chloride cracking tests of stainless steels, for example, have been performed using boiling magnesium chloride ($MgCl_2$) solutions, which constitute a very severe environment. It is well known that the boiling point of aqueous magnesium chloride varies with the amount of magnesium chloride. This test was introduced in 1945 and is detailed in ASTM G 36. In this test, the composition is adjusted so that the solution boils at 155 ± 1 °C (311 ± 2 °F). The test must be carefully controlled because results do vary for different magnesium chloride contents and temperature.

U-Bend Testing. U-bend specimens have also been widely used to test for SCC. They exhibit large elastic and plastic strains and are a very severe test arrangement for a smooth specimen. ASTM G 30 describes the use of U-bend specimens for SCC testing.

Fracture mechanics methods are much more recent in development but are now widely used in research studies. Precracked specimens containing a sharp notch are used with either a constant applied load or a fixed crack opening displacement. Crack growth rates, da/dt, are measured as a function of the stress intensity factor, K. Tests are conducted to define the crack growth rate in stages I, II, and III and the critical threshold stress intensity level below which crack growth does not occur, K_{ISCC}. K_{ISCC} varies as a function of alloy chemistry, heat treatment, strength level, and environment.

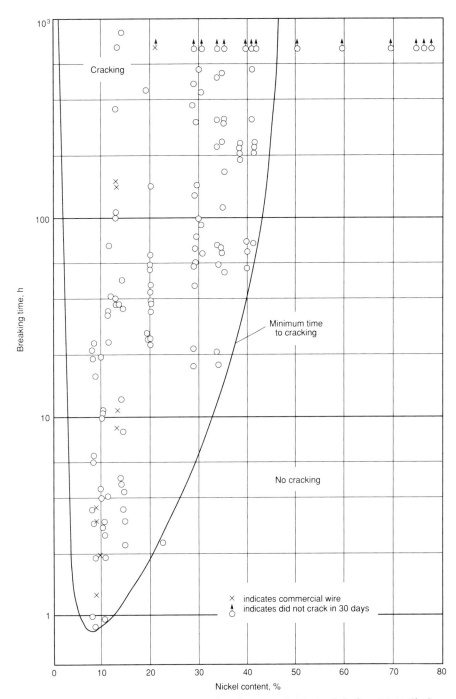

Fig. 60 Effect of nickel content on the time to cracking for Fe-18-20Cr-Ni wires in boiling 42% $MgCl_2$. Source: Ref 363

Slow strain rate tests are also used to evaluate the SCC tendency of alloys. Either precracked or smooth specimens are used in these tests, which are conducted by slowly increasing the load or strain on the alloy. Generally, results in the detrimental environment are compared to those in an inert environment, for example, by plotting the ratios of the strains to failure, the reduction in areas, or the ultimate tensile strengths. While these are useful tests for comparing different materials or environments, they are less useful than the fracture mechanics approach for predicting in-service behavior.

SCC Evaluation of Stainless Steels

Austenitic Stainless Steels. Figure 59 shows the time to failure by SCC in boiling magnesium chloride as a function of applied stress for a variety of austenitic stainless

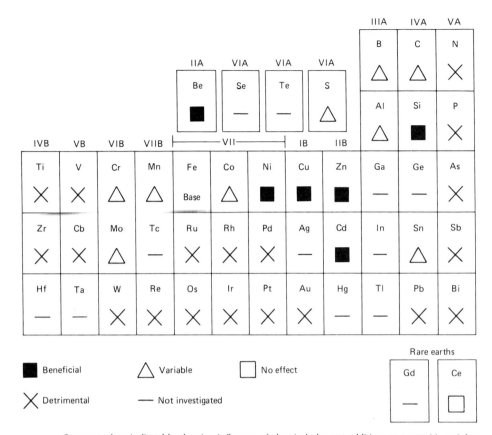

Fig. 61 Segment of periodic table showing influence of chemical element additions to austenitic stainless steels on their SCC behavior in chloride solutions. Source: Ref 350, 354

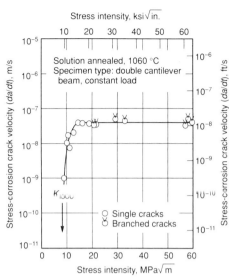

Fig. 62 Influence of stress intensity on crack growth rate and branching of stress-corrosion cracks for AISI 304L in 42% MgCl₂ at 130 °C (265 °F). Threshold stress intensity level, K_{ISCC}, was 8 MPa√m (7.3 ksi√in.). Source: Ref 265

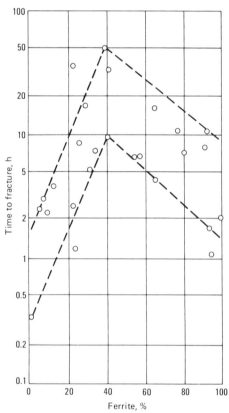

Fig. 63 Effect of ferrite content on the time to failure for duplex stainless steels in boiling 42% MgCl₂ under an applied stress of 240 MPa (34.8 ksi). Source: Ref 368

steels (Ref 362). The families of curves shift to greater SCC resistance with increased nickel content. The 304 and 304L alloys contained 9% Ni; the 305, 309, 316, 347, and 347L alloys had nickel contents ranging from 10.5 to 13.9%; and the 310 and 314 alloys had nickel contents of 24.5 and 19.7%, respectively.

The influence of nickel on stress-corrosion cracking of iron-chromium-nickel alloys has been demonstrated (Fig. 60) (Ref 363). Alloys with 8 to 77% Ni, 18 to 20% Cr, and a balance of iron were produced as wires (annealed, partially or fully hardened by drawing) and loaded in tension in boiling 42% MgCl₂ to determine the time to cracking with loads of 230 or 310 MPa (33 or 45 ksi). Cracking occurred most readily for alloys with about 8 to 12% Ni, and some of the 8% nickel wires broke within 1 h. The time to cracking increased with increasing nickel content; alloys with more than approximately 45% Ni were immune to cracking.

The effect of composition on SCC of austenitic stainless steels, particularly in magnesium chloride solutions, has been widely studied (Ref 344, 345, 349, 350, 352, 355, 363, 364). Multiple regression analysis was used to detect compositional influences for 19 austenitic heats tested in boiling 42% MgCl₂ (Ref 364). The study concluded that the most important alloying elements were carbon and molybdenum. As these elements increased in concentration, the time to fracture passed through a minimum at about 0.065 to 0.085% C and 1.5 to 2.1% Mo. Titanium additions had only a minor influence on the effect of carbon. For the rather small nickel range studied, nickel additions had a minor beneficial influence on the time to fracture.

References 350 and 354 summarize the influence of alloying-element additions to austenitic stainless steels on their SCC behavior in chloride solutions (Fig. 61). Many alloying elements are detrimental to chloride cracking, but some are beneficial; a few have variable effects. Nickel, cadmium, zinc, silicon, beryllium, and copper are listed as being beneficial. Nickel is beneficial for resisting chloride cracking in solutions other than magnesium chloride. However, for ferritic stainless steels, small additions of nickel are detrimental to SCC resistance in chlorides. Cadmium and zinc are listed as being beneficial, but they have not been studied systematically. The influences of silicon and beryllium have been studied in some depth. Silicon additions improve the resistance to SCC in magnesium chloride, but results in sodium chloride (NaCl) at higher temperatures showed little benefit. The beneficial influence of copper is minor.

Additions of boron, aluminum, and cobalt have variable effects. Small amounts appear to be detrimental, but larger amounts appear to be helpful. Tin and manganese additions have demonstrated either no influence or detrimental or beneficial effects, depending on the amounts present. Like carbon and molybdenum, chromium

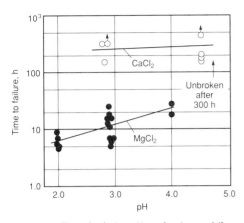

Fig. 64 Effect of solution pH on the time to failure for AISI 304 in MgCl$_2$ and CaCl$_2$ at 125 °C (255 °F) with an applied load of 345 MPa (50 ksi). Source: Ref 369

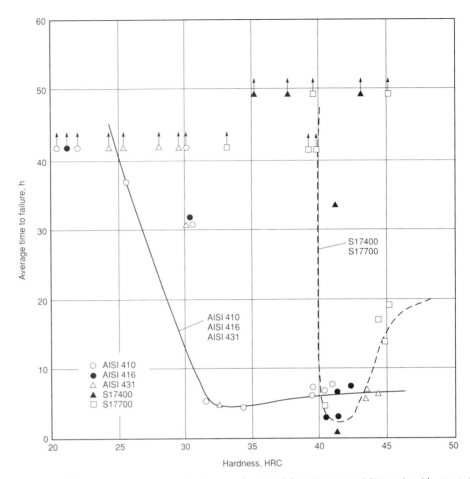

Fig. 65 Relationship between specimen hardness and time to failure in aqueous 0.5% acetic acid saturated with hydrogen sulfide for martensitic and precipitation-hardenable stainless steels. Source: Ref 370

additions exhibit a minimum in chloride cracking resistance, apparently at about 20%.

Crack growth rates for AISI 304L in magnesium chloride at 130 °C (265 °F) have been measured as a function of stress intensity, K (Ref 365). Figure 62 shows the results using precracked specimens. The threshold stress intensity level, K_{ISCC}, was about 8 MPa$\sqrt{m}$ (7.3 ksi$\sqrt{in.}$). Single cracks were observed in stage I, and branching cracks were observed in stage II.

Temperature has a very important influence on the chloride cracking of austenitic stainless steels. Experience has shown that it does not occur at room temperature. Intergranular attack was observed in severely sensitized alloys and in welded AISI 301 after a 5-year ambient temperature exposure in a marine atmosphere at Kure Beach, NC (Ref 366). Normally, chloride cracking is transgranular; however, these alloys cracked intergranularly, apparently because of the sensitization. Therefore, the failures may have been a form of intergranular stress-assisted attack, rather than true SCC (Ref 354). Chloride cracking does occur at temperatures above ambient, generally at temperatures above 60 °C (140 °F). As the temperature increases above this limit, the time to fracture decreases, other factors being constant. Additional information on the exposure of austenitic steels to marine atmospheres is available in the article "Wrought Stainless Steels" in this Volume.

Duplex stainless steels in the annealed condition are more resistant to SCC than are the common austenitic grades AISI 304 and 316. It is known that δ-ferrite in austenitic grades improves the resistance to chloride cracking. This has been attributed to the blocking of crack propagation by the δ-ferrite. Relatively high amounts (as in duplex grades) are required to be effective. Naturally, the morphology and distribution of the ferrite must be controlled in order to achieve such benefits.

An evaluation of the influence of small amounts of δ-ferrite (<10%) on SCC in 18% Cr-8% Ni grades in boiling 42% MgCl$_2$ found an increase in the time to fracture with increasing δ-ferrite content (Ref 367). The beneficial effect of ferrite in duplex stainless steels exposed to boiling magnesium chloride has been demonstrated (Ref 368). Figure 63 shows these results, which indicate a maximum in the time to fracture at about 40% ferrite.

Solution pH also has an influence on stress-corrosion cracking. The pH in the crack can be quite different from that of the bulk environment. Acidic conditions are required for chloride cracking, and an increase in acidity enhances SCC.

Austenitic Stainless Steels. Figure 64 shows the influence of pH on the time to fracture for AISI 304 in magnesium chloride and calcium chloride (CaCl$_2$) at 125 °C (255 °F). The specimens were loaded in tension at 345 MPa (50 ksi). Variations in the pH had a greater effect on magnesium chloride than on calcium chloride, and the former was a much more severe environment than the latter.

Heat-treated martensitic and precipitation-hardenable (PH) stainless steels can also fail by SCC. For these grades, the strength level has a strong influence on cracking tendency. Figure 65 shows this influence in a distilled water solution containing 0.5% acetic acid (CH$_3$COOH) saturated with hydrogen sulfide (H$_2$S), using U-bend specimens (Ref 370). For the standard martensitic grades, cracking occurred readily at hardnesses above 25 HRC; for the PH grades, cracking occurred readily above 40 HRC.

An evaluation of AISI 410 and three PH grades (UNS S17400, S13800, and S15500) was conducted using the National Association of Corrosion Engineers (NACE) sulfide stress cracking test (5% NaCl and 0.5% acetic acid solution purged with argon to remove oxygen and saturated with hydrogen sulfide) at loads of 172, 345, or 515 MPa (25, 50, or 75 ksi) in tension (Ref 371). Figure 66 shows the time to fracture as a function of yield strength for specimens loaded at 345 MPa (50 ksi). The tempering temperature (for AISI 410) and aging treatments (for the PH grades) are shown for each strength level and alloy. The numbers in parentheses are the amount of austenite

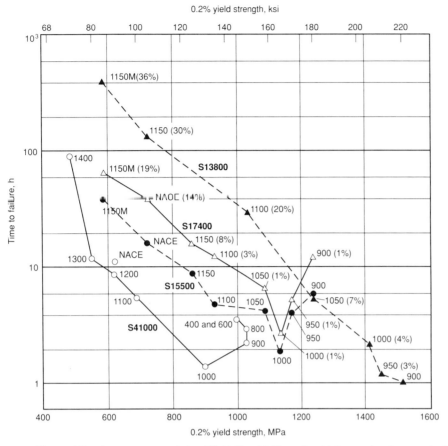

Fig. 66 Time to failure in an aqueous solution (5% NaCl and 0.5% acetic acid deaerated and saturated with hydrogen sulfide) as a function of yield strength for UNS S41000, S15500, S17400, and S13800 stainless steels loaded to 345 MPa (50 ksi) in tension. Tempering temperatures, in degrees Fahrenheit, or temper condition are shown by data points; values in parentheses are the approximate austenite contents. Source: Ref 371

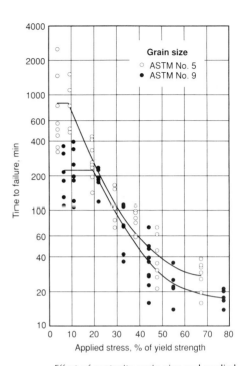

Fig. 67 Effect of austenite grain size and applied stress as a percentage of the yield strength (311.7 and 358.5 MPa, or 45.2 and 52 ksi) for specimens with grain sizes ASTM No. 5 and No. 9, respectively, on the time to failure for AISI 302 wires in boiling 42% MgCl₂. Source: Ref 372

present. The alloys exhibited increased cracking resistance in the order AISI 410, S15500, S17400, S13800.

Grain size differences can influence stress-corrosion cracking resistance. Studies have shown that increasing the grain size accelerates SCC, although the effect is not large, except at low applied stresses (Ref 372). Figure 67 shows the time to failure for AISI 302 wires in boiling magnesium chloride as a function of the applied stress, expressed as a percentage of the yield strength and for grain sizes of ASTM No. 5 and 9. The influence of grain size appears to be most important with regard to crack initiation, particularly for the slip dissolution mechanism.

REFERENCES

1. W.P. Rees and B.E. Hopkins, Intergranular Brittleness in Iron-Oxygen Alloys, *J. Iron Steel Inst.*, Vol 172, Dec 1952, p 403-409
2. C.J. McMahon, Jr., Intergranular Brittleness in Iron, *Acta Metall.*, Vol 14, July 1966, p 839-845
3. J.R. Rellick and C.J. McMahon, Jr., The Elimination of Oxygen-Induced Intergranular Brittleness in Iron by Addition of Scavengers, *Metall. Trans.*, Vol 1, April 1970, p 929-937
4. P. Jolly, Discussion of "The Elimination of Oxygen-Induced Intergranular Brittleness in Iron by Addition of Scavengers," *Metall. Trans.*, Vol 2, Jan 1971, p 341-342
5. J.R. Rellick et al., Further Information on Oxygen Induced Intergranular Brittleness in Iron, *Metall. Trans.*, Vol 2, Jan 1971, p 342-343
6. C. Pichard et al., The Influence of Oxygen and Sulfur on the Intergranular Brittleness of Iron, *Metall. Trans. A*, Vol 7A, Dec 1976, p 1811-1815
7. A. Kumar and V. Raman, Low Temperature Intergranular Brittleness of Iron, *Acta Metall.*, Vol 29, 1981, p 1131-1139
8. B.D. Powell et al., A Study of Intergranular Fracture in Iron Using Auger Spectroscopy, *Metall. Trans.*, Vol 4, Oct 1973, p 2357-2361
9. P. Jolly and C. Goux, Influence of Certain Impurities on Intercrystalline Embrittlement of Iron, *Mem. Sci. Rev. Met.*, Vol 66 (No. 9), 1969, p 605-617
10. C. Pichard et al., Influence of Sulfur Type Metalloid Impurities on the Intercrystalline Embrittlement of Pure Iron, *Mem. Sci. Rev. Met.*, Vol 70 (No. 1), 1973, p 13-22
11. J.R. Rellick et al., The Effect of Tellurium on Intergranular Cohesion in Iron, *Metall. Trans.*, Vol 2, May 1971, p 1492-1494
12. M.C. Inman and H.R. Tipler, Grain-Boundary Segregation of Phosphorus in an Iron-Phosphorus Alloy and the Effect Upon Mechanical Properties, *Acta Metall.*, Vol 6, Feb 1958, p 73-84
13. B.E. Hopkins and H.R. Tipler, The Effect of Phosphorus on the Tensile and Notch-Impact Properties of High-Purity Iron and Iron-Carbon Alloys, *J. Iron Steel Inst.*, Vol 188, March 1958, p 218-237
14. H. Erhart and H.J. Grabke, Equilibrium Segregation of Phosphorus at Grain Boundaries of Fe-P, Fe-C-P, Fe-Cr-P, and Fe-Cr-C-P Alloys, *Met. Sci.*, Vol 15, Sept 1981, p 401-408
15. P.V. Ramasubramanian and D.F. Stein, An Investigation of Grain-Boundary Embrittlement in Fe-P, Fe-P-S and Fe-Sb-S Alloys, *Metall. Trans.*, Vol 4, July 1973, p 1735-1742
16. C.J. McMahon, Jr., Strength of Grain Boundaries in Iron-Base Alloys, in *Grain Boundaries in Engineering Materials*, Claitor, 1975, p 525-552
17. J.R. Rellick and C.J. McMahon, Jr., Intergranular Embrittlement of Iron-Carbon Alloys by Impurities, *Metall. Trans.*, Vol 5, Nov 1974, p 2439-2450

18. C.J. Thwaites and S.K. Chatterjee, Effect of Tin on the Impact Behavior of Alloys Based on High-Purity Iron and Mild Steel, *J. Iron Steel Inst.*, Vol 210, Aug 1972, p 581-587
19. R.L. Kenyon and R.S. Burns, Testing Sheets for Blue Brittleness and Stability Against Changes Due to Aging, *Proc. ASTM*, Vol 34, 1934, p 48-58
20. E.O. Hall, The Deformation of Low-Carbon Steel in the Blue-Brittle Range, *J. Iron Steel Inst.*, Vol 170, April 1952, p 331-336
21. G. Mima and F. Inoko, A Study of the Blue-Brittle Behavior of a Mild Steel in Torsional Deformation, *Trans. JIM*, Vol 10, May 1969, p 227-231
22. T. Takeyama and H. Takahashi, Strength and Dislocation Structures of α-Irons Deformed in the Blue-Brittleness Temperature Range, *Trans. ISIJ*, Vol 13, 1973, p 293-302
23. A.L. Tsou et al., The Quench-Aging of Iron, *J. Iron Steel Inst.*, Vol 172, Oct 1952, p 163-171
24. T.C. Lindley and C.E. Richards, The Effect of Quench-Aging on the Cleavage Fracture of a Low-Carbon Steel, *Met. Sci. J.*, Vol 4, May 1970, p 81-84
25. A.S. Keh and W.C. Leslie, Recent Observations on Quench-Aging and Strain Aging of Iron and Steel, in *Materials Science Research*, Vol 1, Plenum Publishing, 1963, p 208-250
26. E.R. Morgan and J.F. Enrietto, Aging in Steels, in *AISI 1963 Regional Technical Meeting*, American Iron and Steel Institute, 1964, p 227-252
27. E. Stolfe and W. Heller, The State of Knowledge of the Aging of Steels: I, Fundamental Principles, *Stahl und Eisen*, Vol 90 (No. 16), 1970, p 861-868
28. G. Lagerberg and B.S. Lement, Morphological and Phase Changes During Quench-Aging of Ferrite Containing Carbon and Nitrogen, *Trans. ASM*, Vol 50, 1958, p 141-162
29. J.D. Baird, Strain Aging of Steel—A Critical Review, *Iron Steel*, Vol 36, 1963, p 186-192, 326-334, 368-374, 400-405, and 450-457
30. J.D. Baird, The Effects of Strain-Aging Due to Interstitial Solutes on the Mechanical Properties of Metals, *Met. Rev.*, Vol 16, Feb 1971, p 1-18
31. R.D. Butler and D.V. Wilson, The Mechanical Behavior of Temper Rolled Steel Sheets, *J. Iron Steel Inst.*, Vol 201, Jan 1963, p 16-33
32. F. Garofalo and G.V. Smith, The Effect of Time and Temperature on Various Mechanical Properties During Strain Aging of Normalized Low Carbon Steels, *Trans. ASM*, Vol 47, 1955, p 957-983
33. C.C. Li and W.C. Leslie, Effects of Dynamic Strain Aging on the Subsequent Mechanical Properties of Carbon Steels, *Metall. Trans. A*, Vol 9A, Dec 1978, p 1765-1775
34. D.V. Wilson and B. Russell, The Contribution of Precipitation to Strain Aging in Low Carbon Steels, *Acta Metall.*, Vol 8, July 1960, p 468-479
35. E.T. Stephenson and M. Cohen, The Effect of Prestraining and Retempering on AISI Type 4340, *Trans. ASM*, Vol 54, 1961, p 72-83
36. C.H. Lorig and A.R. Elsea, Occurrence of Intergranular Fracture in Cast Steels, *Trans. AFS*, Vol 55, 1947, p 160-174
37. B.C. Woodfine and A.G. Quarrell, Effects of Al and N on the Occurrence of Intergranular Fracture in Steel Castings, *J. Iron Steel Inst.*, Vol 195, Aug 1960, p 409-414
38. J.A. Wright and A.G. Quarrell, Effect of Chemical Composition on the Occurrence of Intergranular Fracture in Plain Carbon Steel Castings Containing Aluminum and Nitrogen, *J. Iron Steel Inst.*, Vol 200, April 1962, p 299-307
39. L. Barnard and R. Brook, Intergranular Fracture of Alloy Steels, *J. Iron Steel Inst.*, Vol 205, July 1967, p 756-762
40. N.E. Hannerz, Influence of Cooling Rate and Composition on Intergranular Fracture of Cast Steel, *Met. Sci. J.*, Vol 2, 1968, p 148-152
41. N.H. Croft, Use of Solubility Data to Predict the Effects of Aluminum and Nitrogen Contents on the Susceptibility of Steel Castings to Intergranular Embrittlement, *Met. Technol.*, Vol 10, Aug 1983, p 285-290
42. N.H. Croft et al., Origins of Dendritic AlN Precipitates in Aluminum-Killed-Steel Castings, *Met. Technol.*, Vol 10, April 1983, p 125-129
43. N.H. Croft et al., Intergranular Fracture of Steel Castings, in *Advances in the Physical Metallurgy and Applications of Steels*, Book 284, The Metals Society, 1982, p 286-295
44. S.C. Desai, Longitudinal Panel Cracking in Ingots, *J. Iron Steel Inst.*, Vol 191, March 1959, p 250-256
45. E. Colombo and B. Cesari, The Study of the Influence of Al and N on the Susceptibility to Crack Formation of Medium Carbon Steel Ingots, *Metall. Ital.*, Vol 59, 1967, p 71-75
46. L. Ericson, Cracking in Low Alloy Aluminum Grain Refined Steels, *Scand. J. Metall.*, Vol 6, 1977, p 116-124
47. R. Sussman et al., Occurrence and Control of Panel Cracking in Aluminum Containing Steel Heats, in *Mechanical Work. and Steel Processing*, Vol 17, American Institute of Mining, Metallurgical, and Petroleum Engineers, 1979, p 49-78
48. L.A. Erasmus, Effect of Aluminum Additions on Forgeability, Austenite Grain Coarsening Temperature, and Impact Properties, *J. Iron Steel Inst.*, Vol 202, Jan 1964, p 32-41
49. R. Harris and L. Barnard, Experiences of Hot Shortness in the Forging of Certain Low-Alloy Steels, in *Deformation Under Hot Working Conditions*, SR 108, Iron and Steel Institute, 1968, p 167-177
50. F. Vodopivec, Influence of Precipitation and Precipitates of Aluminum Nitride on Torsional Deformability of Low-Carbon Steel, *Met. Technol.*, Vol 5, April 1978, p 118-121
51. G.D. Funnell and R.J. Davies, Effect of Aluminum Nitride Particles on Hot Ductility of Steel, *Met. Technol.*, Vol 5, May 1978, p 150-153
52. G.D. Funnell, Observations on Effect of Aluminum Nitride on Hot Ductility of Steel, in *Hot Working and Forming Processes*, Book 264, The Metals Society, 1980, p 104-107
53. K. Yamanaka et al., Relation Between Hot Ductility and Grain-Boundary Embrittlement of Low-Carbon Killed Steels, *Trans. ISIJ*, Vol 20, 1980, p 810-816
54. S.L. Gertsman and H.P. Tardif, Tin and Copper in Steel: Both Are Bad, Together They're Worse, *Iron Age*, Vol 169 (No. 7), Feb 14, 1952, p 136-140
55. P. Bjornson and H. Nathorst, A Special Type of Ingot Cracks Caused by Certain Impurities, *Jernkontorets Ann.*, Vol 139 (No. 6), 1955, p 412-438
56. D.A. Melford, Surface Hot Shortness in Mild Steel, *J. Iron Steel Inst.*, Vol 200, April 1962, p 290-299
57. A. Nicholson and J.D. Murray, Surface Hot Shortness in Low-Carbon Steel, *J. Iron Steel Inst.*, Vol 203, Oct 1965, p 1007-1018
58. W.J.M. Salter, Effects of Alloying Elements on Solubility and Surface Energy of Copper in Mild Steel, *J. Iron Steel Inst.*, Vol 204, May 1966, p 478-488
59. W.J.M. Salter, Effect of Chromium on Solubility of Copper in Mild Steel, *J. Iron Steel Inst.*, Vol 205, Nov 1967, p 1156-1160
60. W.J.M. Salter, Effect of Mutual Additions of Tin and Nickel on the Solubility and Surface Energy of Copper in Mild Steel, *J. Iron Steel Inst.*, Vol 207, Dec 1969, p 1619-1623
61. H.J. Kerr and F. Eberle, Graphitization of Low-Carbon and Low-Carbon-Molybdenum Steels, *Trans. ASME*, Vol 67, 1945, p 1-46
62. S.L. Hoyt et al., Summary Report on the Joint E.E.I.-A.E.I.C. Investigation of Graphitization in Piping, *Trans. ASME*, Vol 68, Aug 1946, p 571-580

63. R.W. Emerson and M. Morrow, Further Observations of Graphitization in Aluminum-Killed Carbon-Molybdenum Steel Steam Piping, *Trans. AIME*, Vol 68, Aug 1946, p 597-607
64. J.G. Wilson, Graphitization of Steel in Petroleum Refining Equipment, *Weld. Res. Counc. Bull.*, No. 32, Jan 1957, p 1-10
65. A.B. Wilder *et al.*, Stability of AISI Alloy Steels, *Trans. AIME*, Vol 209, Oct 1957, p 1176-1181
66. E.J. Dulis and G.V. Smith, Roles of Aluminum and Nitrogen in Graphitization, *Trans. ASM*, Vol 46, 1954, p 1318-1330
67. A. Preece *et al.*, The Overheating and Burning of Steel, *J. Iron Steel Inst.*, Part I, Vol 153, 1946, p 237p-254p; and, Part III, Vol 164, 1950, p 37-45
68. I.S. Brammar, A New Examination of the Phenomena of Overheating and Burning of Steels, *J. Iron Steel Inst.*, Vol 201, Sept 1963, p 752-761
69. R.D.N. Lester, Overheating in Steels, *Steel Times*, Vol 193 (No. 513), 15 July 1966, p 96-102
70. G.D. Joy and J. Nutting, Influence of Intermetallic Phases and Non-Metallic Inclusions Upon the Ductility and Fracture Behavior of Some Alloy Steels, in *Effect of Second-Phase Particles on the Mechanical Properties of Steel*, Iron and Steel Institute, 1971, p 95-100
71. T.J. Baker and R. Johnson, Overheating and Fracture Toughness, *J. Iron Steel Inst.*, Vol 211, Nov 1973, p 783-791
72. D.R. Glue *et al.*, Effect of Composition and Thermal Treatment on the Overheating Characteristics of Low-Alloy Steels, *Met. Technol.*, Vol 2, Sept 1975, p 416-421
73. R.C. Andrew *et al.*, Overheating in Low-Sulphur Steels, *J. Australasian Inst. Met.*, Vol 21, June-Sept 1976, p 126-131
74. R.C. Andrew and G.M. Weston, The Effect of Overheating on the Toughness of Low Sulphur ESR Steels, *J. Aust. Inst. Met.*, Vol 22, Sept-Dec 1972, p 171-176
75. R.C. Andrew and G.M. Weston, The Effect of the Interaction Between Overheating and Tempering Temperature on the Notch Toughness of Two Low Sulphur Steels, *J. Aust. Inst. Met.*, Vol 22, Sept-Dec 1972, p 200-204
76. G.E. Hole and J. Nutting, Overheating of Low-Alloy Steels, *Int. Met. Rev.*, Vol 29, 1984, p 273-298
77. G.J. Spaeder *et al.*, The Effect of Hot Rolling Variables on the Fracture Toughness of 18Ni Maraging Steel, *Trans. ASM*, Vol 60, 1967, p 418-425
78. D. Kalish and H.J. Rack, Thermal Embrittlement of 18Ni(350) Maraging Steel, *Metall. Trans.*, Vol 2, Sept 1971, p 2665-2672
79. W.C. Johnson and D.F. Stein, A Study of Grain Boundary Segregants in Thermally Embrittled Maraging Steel, *Metall. Trans.*, Vol 5, March 1974, p 549-554
80. E. Nes and G. Thomas, Precipitation of TiC in Thermally Embrittled Maraging Steels, *Metall. Trans. A*, Vol 7A, July 1976, p 967-975
81. H.J. Rack and P.H. Holloway, Grain Boundary Precipitation in 18Ni-Maraging Steels, *Metall. Trans. A*, Vol 8A, Aug 1977, p 1313-1315
82. G.F. Vander Voort, Failures of Tools and Dies, in *Failure Analysis and Prevention*, Vol 11, 9th ed., *Metals Handbook*, American Society for Metals, 1986, p 563-585
83. L.D. Jaffee and J.R. Hollomon, Hardenability and Quench Cracking, *Trans. AIME*, Vol 167, 1946, p 617-626
84. M.C. Udy and M.K. Barnett, A Laboratory Study of Quench Cracking in Cast Alloy Steels, *Trans. ASM*, Vol 38, 1947, p 471-487
85. J.W. Spretnak and C. Wells, An Engineering Analysis of the Problem of Quench Cracking in Steel, *Trans. ASM*, Vol 42, 1950, p 233-269
86. C. Wells, Quench Cracks in Wrought Steel Tubes, *Met. Prog.*, Vol 65, May 1954, p 113-121
87. T. Kunitake and S. Sugisawa, The Quench-Cracking Susceptibility of Steel, *Sumitomo Search*, No. 5, May 1971, p 16-25
88. P. Gordon, The Temper Colors on Steel, *J. Heat Treat.*, Vol 1, June 1979, p 93
89. G.F. Vander Voort, Visual Examination and Light Microscopy, in *Fractography*, Vol 12, 9th ed., *Metals Handbook*, ASM INTERNATIONAL, 1987, p 91-165
90. K. Balajiva *et al.*, Effects of Trace Elements on Embrittlement of Steels, *Nature*, Vol 178, 1956, p 433
91. W. Steven and K. Balajiva, The Influence of Minor Elements on the Isothermal Embrittlement of Steels, *J. Iron Steel Inst.*, Vol 193, Oct 1959, p 141-147
92. J.R. Low *et al.*, Alloy and Impurity Effects on Temper Brittleness of Steel, *Trans. AIME*, Vol 242, Jan 1968, p 14-24
93. R.H. Greaves and J.A. Jones, Temper-Brittleness of Nickel-Chromium Steels, *J. Iron Steel Inst.*, Vol 102, 1920, p 171-222
94. R.H. Greaves and J.A. Jones, Temper-Brittleness of Steel: Susceptibility to Temper-Brittleness in Relation to Chemical Composition, *J. Iron Steel Inst.*, Vol 111, 1925, p 231-255
95. B.J. Schulz and C.J. McMahon, Jr., Alloy Effects in Temper Embrittlement, in *Temper Embrittlement of Alloy Steels*, STP 499, American Society for Testing and Materials, 1972, p 104-135
96. C.J. McMahon, Jr. *et al.*, The Influence of Mo on P-Induced Temper Embrittlement in Ni-Cr Steel, *Metall. Trans. A*, Vol 8A, July 1977, p 1055-1057
97. H.L. Marcus and P.W. Palmberg, Auger Fracture Surface Analysis of a Temper Embrittled 3340 Steel, *Trans. AIME*, Vol 245, July 1969, p 1664-1666
98. D.F. Stein *et al.*, Studies Utilizing Auger Electron Emission Spectroscopy on Temper Embrittlement in Low Alloy Steels, *Trans. ASM*, Vol 62, 1969, p 776-783
99. P.W. Palmberg and H.L. Marcus, An Auger Spectroscopic Analysis of the Extent of Grain Boundary Segregation, *Trans. ASM*, Vol 62, 1969, p 1016-1018
100. D.F. Stein, Reversible Temper Embrittlement, *Annu. Rev. Mater. Sci. 1977*, Vol 7, 1977, p 123-153
101. L.D. Jaffe and D.C. Buffum, Isothermal Temper Embrittlement, *Trans. ASM*, Vol 42, 1950, p 604-618
102. F.L. Carr *et al.*, Isothermal Temper Embrittlement of SAE 3140 Steel, *Trans. AIME*, Vol 197, Aug 1953, p 998
103. M.P. Seah, Grain Boundary Segregation and the T-t Dependence of Temper Brittleness, *Acta Metall.*, Vol 25, 1977, p 345-357
104. B.C. Woodfine, Some Aspects of Temper-Brittleness, *J. Iron Steel Inst.*, Vol 173, March 1953, p 240-255
105. J.M. Capus, Austenite Grain Size and Temper Brittleness, *J. Iron Steel Inst.*, Vol 200, Nov 1962, p 922-927
106. R. Viswanathan and A. Joshi, Effect of Microstructure on the Temper Embrittlement of Cr-Mo-V Steels, *Metall. Trans. A*, Vol 6A, Dec 1975, p 2289-2297
107. D.L. Newhouse *et al.*, Temper Embrittlement Study of Nickel-Chromium-Molybdenum-Vanadium Rotor Steels, I: Effects of Residual Elements, in *Temper Embrittlement of Alloy Steels*, STP 499, American Society for Testing and Materials, 1972, p 3-36
108. J. Watanabe and Y. Murakami, Prevention of Temper Embrittlement of Chromium-Molybdenum Steel Vessels by Use of Low-Silicon Forged Steels, *Proc. API Refin. Dept.*, Vol 60, 1981, p 216-224
109. S. Takayama *et al.*, The Calculation of Transition Temperature Changes in Steels Due to Temper Embrittlement, *Metall. Trans. A*, Vol 11A, Sept 1980,

110. R. Viswanathan and S.M. Bruemmer, In-Service Degradation of Toughness of Steam Turbine Rotors, *J. Eng. Mater. Technol. (Trans. ASME)*, Vol 107, Oct 1985, p 316-324
111. B.C. Woodfine, Temper-Brittleness: A Critical Review of the Literature, *J. Iron Steel Inst.*, Vol 173, March 1953, p 229-240
112. J.M. Capus, The Mechanism of Temper Brittleness, in *Temper Embrittlement in Steel*, STP 407, American Society for Testing and Materials, 1968, p 3-19
113. C.J. McMahon, Jr., Temper Brittleness—An Interpretive Review, in *Temper Embrittlement in Steel*, STP 407, American Society for Testing and Materials, 1968, p 127-167
114. B.L. Eyre et al., Physical Metallurgy of Reversible Temper Embrittlement, in *Advances in the Physical Metallurgy and Application of Steels*, Book 284, The Metals Society, 1982, p 246-258
115. L.J. Klingler et al., The Embrittlement of Alloy Steel at High Strength Levels, *Trans. ASM*, Vol 46, 1954, p 1557-1598
116. G. Delisle and A. Galibois, Tempered Martensite Brittleness in Extra-Low-Carbon Steels, *J. Iron Steel Inst.*, Vol 207, Dec 1969, p 1628-1634
117. M. Gensamer, "Study of the Effects of Vacuum Melting on 550°F Tempering Embrittlement," WADC Technical Report 57-85, ASTIA Document AD 130850, Wright Air Development Center, Sept 1957
118. J.M. Capus and G. Mayer, The Influence of Trace Elements on Embrittlement Phenomena in Low-Alloy Steels, *Metallurgia*, Vol 62, Oct 1960, p 133-138
119. E.B. Kula and A.A. Anctil, Tempered Martensite Embrittlement and Fracture Toughness in SAE 4340 Steel, *J. Mater.*, Vol 4, Dec 1969, p 817-841
120. S.K. Banerji et al., Intergranular Fracture in 4340-Type Steels: Effects of Impurities and Hydrogen, *Metall. Trans. A*, Vol 9A, Feb 1978, p 237-247
121. C.L. Briant and S.K. Banerji, Phosphorus Induced 350°C Embrittlement in an Ultra High Strength Steel, *Metall. Trans. A*, Vol 10A, Jan 1979, p 123-126
122. C.L. Briant and S.K. Banerji, Tempered Martensite Embrittlement in Phosphorus Doped Steels, *Metall. Trans. A*, Vol 10A, Nov 1979, p 1729-1737
123. G. Thomas, Retained Austenite and Tempered Martensite Embrittlement, *Metall. Trans. A*, Vol 9A, March 1978, p 439-450
124. H.K.D.H. Bhadeshia and D.V. Edmonds, Tempered Martensite Embrittlement: Role of Retained Austenite and Cementite, *Met. Sci.*, Vol 13, June 1979, p 325-334
125. M. Sarikaya et al., Retained Austenite and Tempered Martensite Embrittlement in Medium Carbon Steels, *Metall. Trans. A*, Vol 14A, June 1983, p 1121-1133
126. H. Kwon and C.H. Kim, Tempered Martensite Embrittlement in Fe-Mo-C and Fe-W-C Steel, *Metall. Trans. A*, Vol 14A, July 1983, p 1389-1394
127. R.M. Horn and R.O. Ritchie, Mechanisms of Tempered Martensite Embrittlement in Low Alloy Steels, *Metall. Trans. A*, Vol 9A, Aug 1978, p 1039-1053
128. J.P. Materkowski and G. Krauss, Tempered Martensite Embrittlement in SAE 4340 Steel, *Metall. Trans. A*, Vol 10A, Nov 1979, p 1643-1651
129. N. Bandyopadhyay and C.J. McMahon, Jr., The Micro-Mechanisms of Tempered Martensite Embrittlement in 4340-Type Steels, *Metall. Trans. A*, Vol 14A, July 1983, p 1313-1325
130. C.L. Briant, Role of Carbides in Tempered Martensite Embrittlement, *Mater. Sci. Technol.*, Vol 5, Feb 1989, p 138-147
131. F. Zia-Ebrahimi and G. Krauss, The Evaluation of Tempered Martensite Embrittlement in 4130 Steel by Instrumented Charpy V-Notch Testing, *Metall. Trans. A*, Vol 14A, June 1983, p 1109-1119
132. J.A. Peters et al., On the Mechanisms of Tempered Martensite Embrittlement, *Acta Metall.*, Vol 37, Feb 1989, p 675-686
133. B. Strauss et al., Carbide Precipitation in the Heat Treatment of Stainless Non-Magnetic Chromium-Nickel Steels, *Zh. Anorg. Allg. Chem.*, Vol 188, 1930, p 309-324
134. E.C. Bain et al., The Nature and Prevention of Intergranular Corrosion in Austenitic Stainless Steels, *Trans. ASST*, Vol 21, June 1933, p 481-509
135. K.G. Caroll et al., Chromium Distribution Around Grain Boundary Carbides Found in Austenitic Stainless Steel, *Nature*, Vol 184, 1959, p 1479-1480
136. C.W. Weaver, Grain-Boundary Precipitation in Nickel-Chromium-Base Alloys, *J. Inst. Met.*, Vol 90, 1961-1962, p 404
137. S. Alm and R. Kiessling, Chromium Depletion Around Grain-Boundary Precipitates in Austenitic Stainless Steel, *J. Inst. Met.*, Vol 91, 1962-1963, p 190
138. R.J. Hodges, Intergranular Corrosion in High Purity Ferritic Stainless Steels: Effect of Cooling Rate and Alloy Composition, *Corrosion*, Vol 27, March 1971, p 119-127
139. A.B. Kinzel, Chromium Carbide in Stainless Steel, *Trans. AIME*, Vol 194, May 1952, p 469-488
140. R. Stickler and A. Vinckier, Electron Microscope Investigation of the Intergranular Corrosion Fracture Surfaces in a Sensitized Austenitic Stainless Steel, *Corros. Sci.*, Vol 3, 1963, p 1-8
141. M.A. Streicher, General and Intergranular Corrosion of Austenitic Stainless Steels in Acids, *J. Electrochem. Soc.*, Vol 106, March 1959, p 161-180
142. K.T. Aust et al., Heat Treatment and Corrosion Resistance of Austenitic Type 304 Stainless Steel, *Trans. ASM*, Vol 59, 1966, p 544-556
143. K.T. Aust et al., Intergranular Corrosion and Electron Microscopic Studies of Austenitic Stainless Steels, *Trans. ASM*, Vol 60, 1967, p 360-372
144. K.T. Aust et al., Intergranular Corrosion and Mechanical Properties of Austenitic Stainless Steels, *Trans. ASM*, Vol 61, 1968, p 270-277
145. K.T. Aust, Intergranular Corrosion of Austenitic Stainless Steels, *Trans. AIME*, Vol 245, Oct 1969, p 2117-2126
146. J.S. Armijo, Impurity Adsorption and Intergranular Corrosion of Austenitic Stainless Steel in Boiling HNO_3-$K_2Cr_2O_7$ Solutions, *Corros. Sci.*, Vol 7, 1967, p 143-150
147. J.S. Armijo, Intergranular Corrosion of Nonsensitized Austenitic Stainless Steels, *Corrosion*, Vol 24, Jan 1968, p 24-30
148. T.M. Devine et al., Mechanism of Intergranular Corrosion of 316L Stainless Steel in Oxidizing Acids, *Scr. Metall.*, Vol 14, 1980, p 1175-1179
149. A. Joshi and D.F. Stein, Chemistry of Grain Boundaries and Its Relation to Intergranular Corrosion of Austenitic Stainless Steel, *Corrosion*, Vol 28, Sept 1972, p 321-330
150. C.L. Briant, The Effects of Sulfur and Phosphorus on the Intergranular Corrosion of 304 Stainless Steel, *Corrosion*, Vol 36, Sept 1980, p 497-509
151. C.L. Briant, The Effect of Alloying Elements on Impurity Induced Intergranular Corrosion, *Corrosion*, Vol 38, April 1982, p 230-232
152. C.L. Briant, Grain Boundary Segregation of Phosphorus and Sulfur in Types 304L and 316L Stainless Steel and Its Effect on Intergranular Corrosion in the Huey Test, *Metall. Trans. A*, Vol 18A, April 1987, p 691-699
153. C.S. Pande et al., Direct Evidence of Chromium Depletion Near the Grain Boundaries in Sensitized Stainless Steels, *Scr. Metall.*, Vol 11, 1977, p 681-684
154. P. Rao and E. Lifshin, Microchemical Analysis in Sensitized Austenitic

Steel, in *Proceedings of the 8th Annual Conference of the Microbeam Analysis Society*, 1977, p 118A-118F
155. R.A. Mulford et al., Sensitization of Austenitic Stainless Steels: II, Commercial Purity Alloys, *Corrosion*, Vol 39, April 1983, p 132-143
156. E.L. Hall and C.L. Briant, Chromium Depletion in the Vicinity of Carbides in Sensitized Austenitic Stainless Steels, *Metall. Trans. A*, Vol 15A, May 1984, p 793-811
157. C.L. Briant and E.L. Hall, A Comparison Between Grain Boundary Chromium Depletion in Austenitic Stainless Steel and Corrosion in the Modified Strauss Test, *Corrosion*, Vol 42, Sept 1986, p 522-531
158. S.M. Bruemmer and L.A. Charlot, Development of Grain Boundary Chromium Depletion in Type 304 and 316 Stainless Steels, *Scr. Metall.*, Vol 20, 1986, p 1019-1024
159. E.P. Butler and M.G. Burke, Chromium Depletion and Martensite Formation at Grain Boundaries in Sensitized Austenitic Stainless Steel, *Acta Metall.*, Vol 34, March 1986, p 557-570
160. C. Stawström and M. Hillert, An Improved Depleted-Zone Theory of Intergranular Corrosion of 18-8 Stainless Steel, *J. Iron Steel Inst.*, Vol 207, Jan 1967, p 77-85
161. C.S. Tedmon, Jr. et al., Intergranular Corrosion of Austenitic Stainless Steel, *J. Electrochem. Soc.*, Vol 118, Feb 1971, p 192-202
162. R.L. Fullman, A Thermodynamic Model of the Effects of Composition on the Susceptibility of Austenitic Stainless Steels to Intergranular Stress Corrosion Cracking, *Acta Metall.*, Vol 30, 1982, p 1407-1415
163. G.S. Was and R.M. Kruger, A Thermodynamic and Kinetic Basis for Understanding Chromium Depletion in Ni-Cr-Fe Alloys, *Acta Metall.*, Vol 33, May 1985, p 841-854
164. S.M. Bruemmer, Sensitization Development in Austenitic Stainless Steel. Measurement and Prediction of Thermomechanical History Effects, *Corrosion*, Vol 44, July 1988, p 427-434
165. E.M. Mahla and N.A. Nielsen, Carbide Precipitation in Type 304 Stainless Steel—An Electron Microscope Study, *Trans. ASM*, Vol 43, 1951, p 290-322
166. R. Stickler and A. Vinckier, Morphology of Grain-Boundary Carbides and Its Influence on Intergranular Corrosion of 304 Stainless Steel, *Trans. ASM*, Vol 54, 1961, p 362-380
167. F.R. Beckitt and B.R. Clark, The Shape and Mechanism of Formation of $M_{23}C_6$ Carbide in Austenite, *Acta Metall.*, Vol 15, Jan 1967, p 113-129
168. L.K. Singhal and J.W. Martin, The Growth of $M_{23}C_6$ Carbide on Grain Boundaries in an Austenitic Stainless Steel, *Trans. AIME*, Vol 242, May 1968, p 814-819
169. C. Da Casa et al., $M_{23}C_6$ Precipitation in Unstabilized Austenitic Stainless Steel, *J. Iron Steel Inst.*, Vol 207, Oct 1969, p 1325-1332
170. W.O. Binder et al., Resistance to Sensitization of Austenitic Chromium-Nickel Steels of 0.03% Max. Carbon Content, *Trans. ASM*, Vol 41, 1949, p 1301-1370
171. S. Danyluk et al., Intergranular Fracture, Corrosion Susceptibility, and Impurity Segregation in Sensitized Type 304 Stainless Steel, *J. Mater. Energy Syst.*, Vol 7, June 1985, p 6-15
172. R.A. Lula et al., Intergranular Corrosion of Ferritic Stainless Steels, *Trans. ASM*, Vol 46, 1954, p 197-230
173. A.P. Bond, Mechanisms of Intergranular Corrosion in Ferritic Stainless Steels, *Trans. AIME*, Vol 245, Oct 1969, p 2127-2134
174. R.J. Hodges, Intergranular Corrosion in High Purity Ferritic Stainless Steels: Isothermal Time-Temperature Sensitization Measurements, *Corrosion*, Vol 27, April 1971, p 164-167
175. J.J. Demo, Mechanism of High Temperature Embrittlement and Loss of Corrosion Resistance in AISI Type 446 Stainless Steel, *Corrosion*, Vol 27, Dec 1971, p 531-544
176. J.A. Davis et al., Intergranular Corrosion Resistance of a 26Cr-1Mo Ferritic Stainless Steel Containing Niobium, *Corrosion*, Vol 36, May 1980, p 215-220
177. T.M. Devine et al., Influence of Heat Treatment on the Sensitization of 18Cr-2Mo-Ti Stabilized Ferritic Stainless Steel, *Metall. Trans. A*, Vol 12A, Dec 1981, p 2063-2069
178. T.M. Devine and A.M. Ritter, Sensitization of 12 Wt Pct Chromium, Titanium-Stabilized Ferritic Stainless Steel, *Metall. Trans. A*, Vol 14A, Aug 1983, p 1721-1728
179. P. Payson, Prevention of Intergranular Corrosion in Corrosion-Resistant Chromium-Nickel Steels, *Trans. AIME*, Vol 100, 1932, p 306-333
180. T.M. Devine, Mechanism of Intergranular Corrosion and Pitting Corrosion of Austenitic and Duplex 308 Stainless Steel, *J. Electrochem. Soc.*, Vol 126, March 1979, p 374-385
181. T.M. Devine, Jr., Influence of Carbon Content and Ferrite Morphology on the Sensitization of Duplex Stainless Steel, *Metall. Trans. A*, Vol 11A, May 1980, p 791-800
182. F.M. Becket, On the Allotropy of Stainless Steels, *Trans. AIME*, Vol 131, 1938, p 15-36
183. G. Riedrich and F. Loib, Embrittlement of High Chromium Steels Within Temperature Range of 570-1100°F, *Arch. Eisenhüttenwes.*, Vol 15, Oct 1941, p 175-182
184. W. Dannöhl, Discussion of Ref 183 (and The Embrittlement of High-Alloy Chrome Steels in the Temperature Range About 500°, by G. Bandel and W. Tofaute, *Arch. Eisenhüttenwes.*, Vol 15, 1942, p 307-320), *Arch. Eisenhüttenwes.*, Vol 15, 1942, p 319
185. H.D. Newell, Properties and Characteristics of 27% Chromium Iron, *Met. Prog.*, Vol 49, May 1946, p 977-1028
186. J.J. Heger, 885°F Embrittlement of the Ferritic Chromium-Iron Alloys, *Met. Prog.*, Vol 60, Aug 1951, p 55-61
187. A.J. Lena and M.F. Hawkes, 475°C (885°F) Embrittlement in Stainless Steels, *Trans. AIME*, Vol 200, May 1954, p 607-615
188. C.A. Zapffe, Fractographic Pattern for 475°C Embrittlement in Stainless Steel, *Trans. AIME*, Vol 191, March 1951, p 247-248
189. H. Masumoto et al., The Anomaly of the Specific Heat at High Temperatures in α-Phase Alloys of Iron and Chromium, *Sci. Rep. Res. Inst., Tôhuko Univ. A*, Vol 5, 1953, p 203-207
190. S. Takeda and N. Nagai, Experimental Research on Superlattices in Iron-Chromium System, *Mem. Fac. Eng., Nagoya Univ.*, Vol 8, 1956, p 1-28
191. R.M. Fisher et al., Identification of the Precipitate Accompanying 885°F Embrittlement in Chromium Steels, *Trans. AIME*, Vol 197, May 1953, p 690-695
192. R.O. Williams and H.W. Paxton, The Nature of Aging of Binary Iron-Chromium Alloys Around 500°C, *J. Iron Steel Inst.*, Vol 185, March 1957, p 358-374
193. G.F. Tisinai and C.H. Samans, Some Observations of 885°F Embrittlement, *Trans. AIME*, Vol 209, Oct 1957, p 1221-1226
194. M.J. Blackburn and J. Nutting, Metallography of an Iron-21% Chromium Alloy Subjected to 475°C Embrittlement, *J. Iron Steel Inst.*, Vol 202, July 1964, p 610-613
195. M.J. Marcinkowski et al., Effect of 500°C Aging on the Deformation Behavior of an Iron-Chromium Alloy, *Trans. AIME*, Vol 230, June 1964, p 676-689
196. R. Lagneborg, Metallography of the 475°C Embrittlement in an Iron-30% Chromium Alloy, *Trans. ASM*, Vol 60, 1967, p 67-78
197. R. Lagneborg, Deformation in an Iron-30% Chromium Alloy Aged at 475°C, *Acta Metall.*, Vol 15, Nov 1967, p 1737-1745
198. T. DeNys and P.M. Gielen, Spinodal Decomposition in the Fe-Cr System,

Metall. Trans., Vol 2, May 1971, p 1423-1428
199. P.J. Grobner, The 885°F (475°C) Embrittlement of Ferritic Stainless Steels, Metall. Trans., Vol 4, Jan 1973, p 251-260
200. R.O. Williams, Further Studies of the Iron-Chromium System, Trans. AIME, Vol 212, Aug 1958, p 497-502
201. P.J. Grobner and R.F. Steigerwald, Effect of Cold Work on the 885°F (475°C) Embrittlement of 18Cr-2Mo Ferritic Stainless Steels, J. Met., Vol 29, July 1977, p 17-23
202. T.J. Nichol et al., Embrittlement of Ferritic Stainless Steels, Metall. Trans. A, Vol 11A, April 1980, p 573-585
203. W. Haoquan et al., Influence of Annealing and Aging Treatments on the Embrittlement of Type 446 Ferritic Stainless Steel, J. Mater. Eng., Vol 9, 1987, p 51-61
204. W. Trietschke and G. Tammnann, The Alloys of Iron and Chromium, Zh. Anorg. Chem., Vol 55, 1907, p 402-411
205. E.C. Bain and W.E. Griffiths, An Introduction to the Iron-Chromium-Nickel Alloys, Trans. AIME, Vol 75, 1927, p 166-213
206. P. Chevenard, Experimental Investigations of Iron, Nickel, and Chromium Alloys, Trav. Mem., Bur. Int. Poids et Mesures, Vol 17, 1927, p 90
207. F. Wever and W. Jellinghaus, The Two-Component System: Iron-Chromium, Mitt. Kaiser-Wilhelm Inst., Vol 13, 1931, p 143-147
208. D.C. Ludwigson and H.S. Link, Further Studies of the Formation of Sigma in 12 to 16 Per Cent Chromium Steels, in Advances in the Technology of Stainless Steels and Related Alloys, STP 369, American Society for Testing and Materials, 1965, p 299-311
209. J.H.G. Monypenny, The Brittle Phase in High-Chromium Steels, Metallurgia, Vol 21, 1939-1940, p 143-148
210. F.B. Foley, The Sigma Phase, Alloy Cast. Bull., July 1945, p 1-9
211. D.A. Oliver, The Sigma Phase in Stainless Steels, Met. Prog., Vol 55, May 1949, p 665-667
212. G.V. Smith, Sigma Phase in Stainless: What, When and Why, Iron Age, Vol 166, 30 Nov 1950, p 63-68; 7 Dec 1950, p 127-132
213. A.J. Lena, Sigma Phase—A Review, Met. Prog., Vol 66, July 1954, p 86-90; Aug 1954, p 94-99; Sept 1954, p 122-126, 128
214. F.B. Foley and V.N. Krivobok, Sigma Formation in Commercial Ni-Cr-Fe Alloys, Met. Prog., Vol 71, May 1957, p 81-86
215. F.B. Pickering, The Formation of Sigma in Austenitic-Stainless Steels, in Precipitation Processes in Steels, Special Report 64, The Iron and Steel Institute, 1959, p 118-124
216. E.O. Hall and S.H. Algie, The Sigma Phase, Metall. Rev., Vol 11, 1966, p 61-88
217. E.R. Jette and F. Foote, The Fe-Cr Alloy System, Met. Alloys, Vol 7, Aug 1936, p 207-210
218. A.J. Lena and W.E. Curry, The Effect of Cold Work and Recrystallization on the Formation of the Sigma Phase in Highly Stable Austenitic Stainless Steels, Trans. ASM, Vol 47, 1955, p 193-210
219. P. Duhaj et al., Sigma-Phase Precipitation in Austenitic Steels, J. Iron Steel Inst., Vol 206, Dec 1968, p 1245-1251
220. B. Weiss and R. Stickler, Phase Instabilities During High Temperature Exposure of 316 Austenitic Stainless Steel, Metall. Trans., Vol 3, April 1972, p 851-866
221. M.T. Shehata et al., A Quantitative Metallographic Study of the Ferrite to Sigma Transformation in Type 316 Stainless Steel, in Microstruct. Sci., Vol 11, Elsevier, 1983, p 89-99
222. J.K.L. Lai et al., Precipitate Phases in Type 316 Austenitic Stainless Steel Resulting From Long-Term High Temperature Service, Mater. Sci. Eng., Vol 49, 1981, p 19-29
223. J. Barcik and B. Brzycka, Chemical Composition of σ Phase Precipitated in Chromium-Nickel Austenitic Steels, Met. Sci., Vol 17, May 1983, p 256-260
224. T. Andersson and B. Lundberg, Effect of Mo on the Lattice Parameters and on the Chemical Composition of Sigma Phase and $M_{23}C_6$ Carbide in an Austenitic 25Cr-20Ni Steel, Metall. Trans. A, Vol 8A, May 1977, p 787-790
225. G.N. Emanuel, Sigma Phase and Other Effects of Prolonged Heating at Elevated Temperatures on 25 Per Cent Chromium-20 Per Cent Nickel Steel, in Symposium on the Nature, Occurrence, and Effects of Sigma Phase, STP 110, American Society for Testing and Materials, 1951, p 82-99
226. G. Aggen et al., Microstructures Versus Properties of 29-4 Ferritic Stainless Steel, in MiCon 78: Optimization of Processing, Properties, and Service Performance Through Microstructural Control, STP 672, American Society for Testing and Materials, 1979, p 334-366
227. J.J. Gilman, Hardening of High-Chromium Steels by Sigma Phase Formation, Trans. ASM, Vol 43, 1951, p 161-192
228. G.J. Guarnieri et al., The Effect of Sigma Phase on the Short-Time High Temperature Properties of 25 Chromium-20 Nickel Stainless Steel, Trans. ASM, Vol 42, 1950, p 981-1007
229. J.H. Jackson, The Occurrence of the Sigma Phase and Its Effect on Certain Properties of Cast Fe-Ni-Cr Alloys, in Symposium on the Nature, Occurrence, and Effects of Sigma Phase, STP 110, American Society for Testing and Materials, 1951, p 101-127
230. L.P. Stoter, Thermal Aging Effects in AISI Type 316 Stainless Steel, J. Mater. Sci., Vol 16, 1981, p 1039-1051
231. R.G. Ellis and G. Pollard, The Observation of Sigma Phase After Short Aging Times in a Duplex Steel, J. Iron Steel Inst., Vol 208, Aug 1970, p 783-784
232. H.D. Solomon and T.M. Devine, Influence of Microstructure on the Mechanical Properties and Localized Corrosion of a Duplex Stainless Steel, in MiCon 78: Optimization of Processing, Properties, and Service Performance Through Microstructural Control, STP 672, American Society for Testing and Materials, 1979, p 430-461
233. Y. Maehara et al., Effects of Alloying Elements on σ Phase Precipitation in δ-γ Duplex Phase Stainless Steels, Met. Sci., Vol 17, Nov 1983, p 541-547
234. W.H. Johnson, On Some Remarkable Changes Produced in Iron and Steel by the Action of Hydrogen and Acids, Proc. R. Soc., Vol 23 (No. 158), 1875, p 168-179; reprinted in Hydrogen Damage, American Society for Metals, 1977, p 1-12
235. B. Craig, Hydrogen Damage, in Corrosion, Vol 13, 9th ed., Metals Handbook, ASM INTERNATIONAL, 1987, p 163-189
236. J.P. Hirth and H.H. Johnson, Hydrogen Problems in Energy Related Technology, Corrosion, Vol 32, Jan 1976, p 3-26
237. R.P. Frohmberg et al., Delayed Failure and Hydrogen Embrittlement in Steel, Trans. ASM, Vol 47, 1955, p 892-925
238. A.R. Troiano, The Role of Hydrogen and Other Interstitials in the Mechanical Behavior of Metals, Trans. ASM, Vol 52, 1960, p 54-80
239. H.H. Johnson et al., Hydrogen, Crack Initiation, and Delayed Failure in Steel, Trans. AIME, Vol 212, Aug 1958, p 528-536
240. I.M. Bernstein, The Role of Hydrogen in the Embrittlement of Iron and Steel, Mater. Sci. Eng., Vol 6, 1970, p 1-19
241. M.R. Louthan, Jr. et al., Hydrogen Embrittlement of Metals, Mater. Sci. Eng., Vol 10, 1972, p 357-368
242. M.R. Louthan, Jr., Effects of Hydrogen on the Mechanical Properties of Low Carbon and Austenitic Steels, in Hydrogen in Metals, American Society for Metals, 1974, p 53-77
243. I.M. Bernstein et al., Effect of Dis-

solved Hydrogen on Mechanical Behavior of Metals, in *Effect of Hydrogen on Behavior of Materials*, The Metallurgical Society, 1976, p 37-58
244. I.M. Bernstein and A.W. Thompson, Effect of Metallurgical Variables on Environmental Fracture of Steels, *Int. Met. Rev.*, Vol 21, Dec 1976, p 269-287
245. A.W. Thompson, Effect of Metallurgical Variables on Environmental Fracture of Engineering Materials, in *Environment-Sensitive Fracture of Engineering Materials*, The Metallurgical Society, 1979, p 379-410
246. J.P. Hirth, Effects of Hydrogen on the Properties of Iron and Steel, *Metall. Trans. A*, Vol 11A, June 1980, p 861-890
247. A.W. Thompson and I.M. Bernstein, Microstructure and Hydrogen Embrittlement, in *Hydrogen Effects in Metals*, The Metallurgical Society, 1981, p 291-308
248. C.G. Interrante, Basic Aspects of the Problems of Hydrogen in Steels, in *Current Solutions to Hydrogen Problems in Steels*, American Society for Metals, 1982, p 3-17
249. G.M. Pressouyre, Current Solutions to Hydrogen Problems in Steel, in *Current Solutions to Hydrogen Problems in Steels*, American Society for Metals, 1982, p 18-34
250. H.G. Nelson, Hydrogen Embrittlement, in *Embrittlement of Engineering Alloys*, Vol 25, *Treatise on Materials Science and Technology*, Academic Press, 1983, p 275-359
251. H.H. Johnson, Keynote Lecture: Overview on Hydrogen Degradation Phenomena, in *Hydrogen Embrittlement and Stress Corrosion Cracking*, American Society for Metals, 1984, p 3-27
252. H.C. Rogers, Hydrogen Embrittlement in Engineering Materials, *Mater. Prot.*, Vol 1, April 1962, p 26, 28-30, 33
253. E.A. Steigerwald *et al.*, Discontinuous Crack Growth in Hydrogenated Steel, *Trans. AIME*, Vol 215, Dec 1959, p 1048-1052
254. J.T. Brown and W.M. Baldwin, Jr., Hydrogen Embrittlement of Steels, *Trans. AIME*, Vol 200, Feb 1954, p 298-303
255. T. Toh and W.M. Baldwin, Jr., Ductility of Steel With Varying Concentrations of Hydrogen, in *Stress Corrosion Cracking and Embrittlement*, John Wiley & Sons, 1956, p 176-186
256. S. Marshall *et al.*, Relationship Between Hydrogen Content and Ductility of Steels, in *Electric Furnace Steel Conference*, Vol 6, American Institute of Mining, Metallurgical, and Petroleum Engineers, 1949, p 59-73
257. C.L. Briant *et al.*, Embrittlement of a 5 Pct Nickel High Strength Steel by Impurities and Their Effects on Hydrogen-Induced Cracking, *Metall. Trans. A*, Vol 9A, May 1978, p 625-633
258. R. Viswanathan and R.I. Jaffie, "Clean Steels" to Control Hydrogen Embrittlement, in *Current Solutions to Hydrogen Problems in Steels*, American Society for Metals, 1982, p 275-278
259. N. Bandyopadhyay *et al.*, Hydrogen-Induced Cracking in 4340-Type Steel: Effects of Composition, Yield Strength, and H_2 Pressure, *Metall. Trans. A*, Vol 14A, May 1983, p 881-888
260. J. Kameda and C.J. McMahon, Jr., Solute Segregation and Hydrogen-Induced Intergranular Fracture in an Alloy Steel, *Metall. Trans. A*, Vol 14A, May 1983, p 903-911
261. H. Asahi *et al.*, Effects of Mn, P, and Mo on Sulfide Stress Cracking Resistance of High Strength Low Alloy Steels, *Metall. Trans. A*, Vol 19A, Sept 1988, p 2171-2177
262. A.W. Thompson, The Mechanism of Hydrogen Participation in Ductile Fracture, in *Effect of Hydrogen on Behavior of Materials*, American Institute of Mining, Metallurgical, and Petroleum Engineers, 1976, p 467-479
263. T.D. Lee *et al.*, Effect of Hydrogen on Fracture of U-Notched Bend Specimens of Spheroidized AISI 1095 Steel, *Metall. Trans. A*, Vol 10A, Feb 1979, p 199-208
264. H. Cialone and R.J. Asaro, The Role of Hydrogen in the Ductile Fracture of Plain Carbon Steels, *Metall. Trans. A*, Vol 10A, March 1979, p 367-375
265. R. Garber *et al.*, Hydrogen Assisted Ductile Fracture of Spheroidized Carbon Steels, *Metall. Trans. A*, Vol 12A, Feb 1981, p 225-234
266. H. Cialone and R.J. Asaro, Hydrogen Assisted Fracture of Spheroidized Plain Carbon Steels, *Metall. Trans. A*, Vol 12A, Aug 1981, p 1373-1387
267. M.B. Whiteman and A.R. Troiano, Hydrogen Embrittlement of Austenitic Stainless Steel, *Corrosion*, Vol 21, Feb 1965, p 53-56
268. M.L. Holzworth and M.R. Louthan, Jr., Hydrogen-Induced Phase Transformations in Type 304L Stainless Steels, *Corrosion*, Vol 24, April 1968, p 110-124
269. R.M. Vennett and G.S. Ansell, A Study of Gaseous Hydrogen Damage in Certain FCC Metals, *Trans. ASM*, Vol 62, 1969, p 1007-1013
270. N. Narita *et al.*, Hydrogen-Related Phase Transformations in Austenitic Stainless Steels, *Metall. Trans. A*, Vol 13A, Aug 1982, p 1355-1365
271. A.W. Thompson and J.A. Brooks, Hydrogen Performance of Precipitation-Strengthening Stainless Steels Based on A-286, *Metall. Trans. A*, Vol 6A, July 1975, p 1431-1442
272. G.G. Hancock and H.H. Johnson, Hydrogen, Oxygen, and Subcritical Crack Growth in a High Strength Steel, *Trans. AIME*, Vol 236, April 1966, p 513-516
273. D.P. Dautovich and S. Floreen, The Stress Intensities for Slow Crack Growth in Steels Containing Hydrogen, *Metall. Trans.*, Vol 4, Nov 1973, p 2627-2630
274. S.J. Hudak, Jr. and R.P. Wei, Hydrogen Enhanced Crack Growth in 18 Ni Maraging Steels, *Metall. Trans. A*, Vol 7A, Feb 1976, p 235-241
275. R.P. Gangloff and R.P. Wei, Gaseous Hydrogen Embrittlement of High Strength Steels, *Metall. Trans. A*, Vol 8A, July 1977, p 1043-1053
276. M. Iino, The Extension of Hydrogen Blister-Crack Array in Linepipe Steels, *Metall. Trans. A*, Vol 9A, Nov 1978, p 1581-1590
277. M. Iino, Influence of Sulfur Content on the Hydrogen-Induced Fracture in Linepipe Steels, *Metall. Trans. A*, Vol 10A, Nov 1979, p 1691-1698
278. B.E. Wilde *et al.*, Some Observations on the Role of Inclusions in the Hydrogen Induced Blister Cracking of Linepipe Steels in Sulfide Environments, *Corrosion*, Vol 36, Nov 1980, p 625-632
279. R.E. Cramer and E.C. Bast, The Prevention of Flakes by Holding Railroad Rails at Various Constant Temperatures, *Trans. ASM*, Vol 27, Dec 1939, p 923-934
280. C.A. Zapffe and C.E. Sims, Hydrogen, Flakes and Shatter Cracks, *Met. Alloys*, Vol 11, May 1940, p 145-151; June 1940, p 177-184; Vol 12, July 1940, p 44-51; Aug 1940, p 145-148
281. F.B. Foley, Flakes and Cooling Cracks in Forgings—A Problem in Ordnance, *Met. Alloys*, Vol 12, Oct 1940, p 442-445
282. C.A. Zapffe, Defects in Cast and Wrought Steel Caused by Hydrogen, *Met. Prog.*, Vol 42, Dec 1942, p 1051-1056
283. C.A. Zapffe, Sources of Hydrogen in Steel and Means for Its Elimination, *Met. Prog.*, Vol 43, March 1943, p 397-401
284. E.R. Johnson *et al.*, Flaking in Alloy Steels, in *Proceedings of the National Open Hearth Conference*, Vol 27, American Institute of Mining, Metallurgical, and Petroleum Engineers, 1944, p 358-377
285. A.W. Dana, Jr. *et al.*, Relation of Flake Formation in Steel to Hydrogen, Microstructure, and Stress, *Trans. AIME*, Vol 203, Aug 1955, p 895-905
286. W.L. Kerlie and J.H. Richards, Origin and Elimination of Hydrogen in Basic

286. Open-Hearth Steels, *Trans. AIME*, Vol 209, Dec 1957, p 1541-1548
287. J.M. Hodge et al., Effect of Hydrogen Content on Susceptibility to Flaking, *Trans. AIME*, Vol 230, Aug 1964, p 1182-1193
288. J.E. Ryall et al., The Effects of Hydrogen in Rolled Steel Products, *Met. Forum*, Vol 2 (No. 3), 1979, p 174-182
289. A.P. Lingras, Hydrogen Control in Steelmaking, in *Electric Furnace Steel Proceedings*, Vol 40, Iron and Steel Society, 1983, p 133-143
290. A.R. Huntington, discussion of paper by C.H. Desch, The Solidification of Metals From The Liquid State, *J. Inst. Met.*, Vol 11, 1914, p 108-109
291. W. Rostoker et al., *Embrittlement by Liquid Metals*, Reinhold, 1960
292. N.S. Stoloff, Liquid Metal Embrittlement, in *Surfaces and Interfaces, II*, Syracuse University Press, 1968, p 157-182
293. A.R.C. Westwood et al., Adsorption-Induced Brittle Fracture in Liquid-Metal Environments, in *Fracture*, Vol 3, Academic Press, 1971, p 589-644
294. M.H. Kamdar, Embrittlement by Liquid Metals, in *Progress in Materials Science*, Vol 15, Part 4, Pergamon Press, 1973, p 289-374
295. M.H. Kamdar, Mechanism of Embrittlement and Brittle Fracture in Liquid Metal Environment, in *Fracture 1977*, Vol 1, Pergamon Press, 1978, p 387-405
296. M.G. Nicholas and C.F. Old, Review: Liquid Metal Embrittlement, *J. Mater. Sci.*, Vol 14, Jan 1979, p 1-18
297. N.S. Stoloff, Recent Developments in Liquid-Metal Embrittlement, in *Environment-Sensitive Fracture of Engineering Materials*, American Institute of Mining, Metallurgical, and Petroleum Engineers, 1979, p 486-518
298. M.H. Kamdar, Liquid Metal Embrittlement, in *Embrittlement of Engineering Alloys*, Vol 25, *Treatise on Materials Science and Technology*, Academic Press, 1983, p 361-459
299. N.S. Stoloff, Liquid and Solid Metal Embrittlement, in *Atomistics of Fracture*, Plenum Press, 1983, p 921-953
300. N.S. Stoloff, Metal Induced Embrittlement—A Historical Perspective, in *Embrittlement by Liquid and Solid Metals*, American Institute of Mining, Metallurgical, and Petroleum Engineers, 1984, p 3-26
301. M.G. Nicholas, A Survey of Literature on Liquid Metal Embrittlement of Metals and Alloys, in *Embrittlement by Liquid and Solid Metals*, American Institute of Mining, Metallurgical, and Petroleum Engineers, 1984, p 27-50
302. F.A. Shunk and W.R. Warke, Specificity as an Aspect of Liquid Metal Embrittlement, *Scr. Metall.*, Vol 8, 1974, p 519-526
303. C.H. Desch, The Solidification of Metals From the Liquid State (2nd report), *J. Inst. Met.*, Vol 22, 1919, p 241-276
304. E. Heyn, Internal Strains in Cold-Wrought Metals, and Some Troubles Caused Thereby, *J. Inst. Met.*, Vol 12, 1914, p 3-37
305. H.S. Rawdon, The Use of Mercury Solutions for Predicting Season Cracking in Brass, *Proc. ASTM*, Vol 18 (No. 2), 1918, p 189-219
306. H. Moore and S. Beckinsale, The Removal of Internal Stresses in 70:30 Brass by Low Temperature Annealing, *J. Inst. Met.*, Vol 23, 1920, p 225-245
307. J.C. Lynn et al., Solid Metal Induced Embrittlement of Steel, *Mater. Sci. Eng.*, Vol 18, March 1975, p 51-62
308. J.M. Johnson et al., Zinc Embrittlement of Stainless Steel Welds, in *Embrittlement by Liquid and Solid Metals*, American Institute of Mining, Metallurgical, and Petroleum Engineers, 1984, p 415-434
309. D.L. Hammon et al., Embrittlement of Engineering Materials During High-Temperature Fatigue in a Liquid-Lithium Environment, in *Embrittlement by Liquid and Solid Metals*, American Institute of Mining, Metallurgical, and Petroleum Engineers, 1984, p 549-562
310. R. Genders, The Penetration of Mild Steel by Brazing Solder and Other Metals, *J. Inst. Met.*, Vol 37, 1927, p 215-221
311. H. Schottky et al., The Red-Shortening of Steels by Metals, *Arch. Eisenhüttenwes.*, Vol 4, 1931, p 541-547
312. R.R. Hough and R. Rolls, The High-Temperature Tensile Creep Behavior of Notched, Pure Iron Embrittled by Liquid Copper, *Scr. Metall.*, Vol 4, Jan 1970, p 17-24
313. R.R. Hough and R. Rolls, Creep Fracture Phenomena in Iron Embrittled by Liquid Copper, *J. Mater. Sci.*, Vol 6, 1971, p 1493-1498
314. R.R. Hough and R. Rolls, Copper Diffusion in Iron During High-Temperature Tensile Creep, *Metall. Trans.*, Vol 2, Sept 1971, p 2471-2475
315. R.R. Hough and R. Rolls, Some Factors Influencing the Effects of Liquid Copper on the Creep-Rupture Properties of Iron, *Scr. Metall.*, Vol 8, Jan 1974, p 39-44
316. E.A. Asnis and V.M. Prokhorenko, Mechanism of Cracking During the Welding or Deposition of Copper Onto Steel, *Weld. Prod. (USSR)*, Vol 12 (No. 11), 1965, p 15-17
317. S.J. Matthews and W.F. Savage, Heat-Affected Zone Infiltration by Dissimilar Liquid Weld Metal, *Weld. J.*, Vol 50, April 1971, p 174s-182s
318. W.F. Savage et al., Intergranular Attack of Steel by Molten Copper, *Weld. J.*, Vol 57, Jan 1978, p 9s-16s
319. W.F. Savage et al., Copper-Contamination Cracking in the Weld Heat-Affected Zone, *Weld. J.*, Vol 57, May 1978, p 145s-152s
320. W.F. Savage et al., Liquid-Metal Embrittlement of the Heat-Affected Zone by Copper Contamination, *Weld. J.*, Vol 57, Aug 1978, p 237s-245s
321. L. Habraken and J. Lecomte-Beckers, Hot Shortness and Scaling of Copper-Containing Steels, in *Copper in Iron and Steel*, John Wiley & Sons, 1982, p 45-81
322. B. Straub, Microscopic Steel Investigation, *Stahl Eisen*, Vol 34 (No. 50), Dec 1914, p 1814-1820
323. G.F. Vander Voort, Failures of Locomotive Axles, in *Failure Analysis and Prevention*, Vol 11, 9th ed., *Metals Handbook*, American Society for Metals, 1986, p 715-727
324. A.P. Druschitz and P. Gordon, Solid Metal-Induced Embrittlement of Metals, in *Embrittlement by Liquid and Solid Metals*, American Institute of Mining, Metallurgical, and Petroleum Engineers, 1984, p 285-316
325. J.C. Lynn and W.R. Warke, Delayed Failure of Steel Due to Solid Metal Induced Embrittlement, in *Embrittlement by Liquid and Solid Metals*, American Institute of Mining, Metallurgical, and Petroleum Engineers, 1984, p 343-365
326. Y. Asayama, Metal-Induced Embrittlement of Steels, in *Embrittlement by Liquid and Solid Metals*, American Institute of Mining, Metallurgical, and Petroleum Engineers, 1984, p 317-331
327. Y. Iwata et al., Delayed Failure of Cadmium Plated Steels at Elevated Temperatures, *Jpn. Inst. Met.*, Vol 31, 1967, p 77-83
328. D.N. Fager and W.F. Spurr, Solid Cadmium Embrittlement: Steel Alloys, *Corrosion*, Vol 27, Feb 1971, p 72-76
329. S. Mostovoy and N.N. Breyer, The Effects of Lead on the Mechanical Properties of 4145 Steel, *Trans. ASM*, Vol 61, 1968, p 219-232
330. R.D. Zipp et al., A Comparison of Elevated Temperature Tensile Fracture in Nonleaded and Leaded 4145 Steel, in *Electron Microfractography*, STP 453, American Society for Testing and Materials, 1969, p 111-133
331. W.R. Warke and N.N. Breyer, Effect of Steel Composition on Lead Embrittlement, *J. Iron Steel Inst.*, Vol 209, Oct 1971, p 779-784
332. N.N. Breyer, Some Effects of Certain Trace Elements on the Properties of High Strength Steels, in *Proceedings of the 31st Electric Furnace Confer-*

ence, Vol 31, American Institute of Mining, Metallurgical, and Petroleum Engineers, 1974, p 183-189
333. N.N. Breyer and K.L. Johnson, Liquid Metal Embrittlement of 4145 Steel by Lead-Tin and Lead-Antimony Alloys, *J. Test. Eval.*, Vol 2, Nov 1974, p 471-477
334. S. Dinda and W.R. Warke, The Effect of Grain Boundary Segregation on Liquid Metal Induced Embrittlement of Steel, *Mater. Sci. Eng.*, Vol 24, 1976, p 199-208
335. J.J. Heger, Stress-Corrosion of Stainless Steels, *Met. Prog.*, Vol 67, March 1955, p 109-116
336. H.H. Uhlig, New Perspectives in the Stress Corrosion Problem, in *Physical Metallurgy of Stress Corrosion Fractures*, Vol 4, AIME Metallurgical Society Conferences, Interscience, 1959, p 1-28
337. N.A. Nielsen, The Role of Corrosion Products in Crack Propagation in Austenitic Stainless Steel: Electron Microscopic Studies, in *Physical Metallurgy of Stress Corrosion Fractures*, Vol 4, AIME Metallurgical Society Conferences Interscience, 1959, p 121-154
338. J.G. Hines and R.W. Hugill, Metallographic and Crystallographic Examination of Stress Corrosion Cracks in Austenitic Cr-Ni Steels, in *Physical Metallurgy of Stress Corrosion Fractures*, Vol 4, AIME Metallurgical Society Conferences, Interscience, 1959, p 193-226
339. H.L. Logan, Stress Corrosion Cracking in Low Carbon Steel, in *Physical Metallurgy of Stress Corrosion Fractures*, Vol 4, AIME Metallurgical Society Conferences, Interscience, 1959, p 295-310
340. J.F. Bates and A.W. Loginow, Principles of Stress Corrosion Cracking as Related to Steels, *Corrosion*, Vol 20, June 1964, p 189t-197t
341. R.N. Parkins, Stress-Corrosion Cracking, *Metall. Rev.*, Vol 9 (No. 35), 1964, p 201-260
342. P.R. Swann and J.D. Embury, Microstructural Aspects of Stress-Corrosion Failure, in *High-Strength Materials*, John Wiley & Sons, 1965, p 327-362
343. H.L. Logan, *The Stress Corrosion of Metals*, John Wiley & Sons, 1966
344. R.M. Latanision and R.W. Staehle, Stress Corrosion Cracking of Iron-Nickel-Chromium Alloys, in *Fundamental Aspects of Stress Corrosion Cracking*, National Association of Corrosion Engineers, 1969, p 214-307
345. R.W. Staehle *et al.*, Effect of Alloy Composition on Stress Corrosion Cracking of Fe-Cr-Ni Base Alloys, *Corrosion*, Vol 26, Nov 1970, p 451-486
346. N.A. Nielsen, Observations and Thoughts on Stress Corrosion Mechanisms, *J. Mater.*, Vol 5, Dec 1970, p 794-829
347. R.N. Parkins, Stress Corrosion Cracking of Low-Strength Ferritic Steels, in *The Theory of Stress Corrosion Cracking in Alloys*, W.S. Maney & Sons, 1971, p 167-185
348. B.F. Brown, Stress Corrosion Cracking of High Strength Steels, in *The Theory of Stress Corrosion Cracking in Alloys*, W.S. Maney & Sons, 1971, p 186-204
349. R.W. Staehle, Stress Corrosion Cracking of the Fe-Cr-Ni Alloy System, in *The Theory of Stress Corrosion Cracking in Alloys*, W.S. Maney & Sons, 1971, p 223-288
350. A.J. Sedriks, Stress-Corrosion Cracking of Stainless Steels and Nickel Alloys, *J. Inst. Met.*, Vol 101, 1973, p 225-232
351. R.W. Staehle, Stress Corrosion Cracking (and Corrosion Fatigue), *Mater. Sci. Eng.*, Vol 25, 1976, p 207-215
352. G.J. Theus and R.W. Staehle, Review of Stress Corrosion Cracking and Hydrogen Embrittlement in the Austenitic Fe-Cr-Ni Alloys, in *Stress Corrosion Cracking and Hydrogen Embrittlement of Iron-Base Alloys*, National Association of Corrosion Engineers, 1977, p 845-892
353. G.E. Moller, Designing With Stainless Steels for Service in Stress Corrosion Environments, *Mater. Perform.*, Vol 16, May 1977, p 32-44
354. A.J. Sedriks, *Corrosion of Stainless Steels*, John Wiley & Sons, 1979, p 139-193
355. H.E. Hänninen, Influence of Metallurgical Variables on Environment-Sensitive Cracking of Austenitic Alloys, *Int. Met. Rev.*, Vol 24 (No. 3), 1979, p 85-135
356. B.M. Gordon, The Effect of Chloride and Oxygen on the Stress Corrosion Cracking of Stainless Steels: Review of Literature, *Mater. Perform.*, Vol 19, April 1980, p 29-38
357. J.E. Truman, Stress-Corrosion Cracking of Martensitic and Ferritic Stainless Steels, *Int. Met. Rev.*, Vol 26 (No. 6), 1981, p 301-349
358. F.P. Ford, Stress Corrosion Cracking of Iron-Base Alloys in Aqueous Environments, in *Embrittlement of Engineering Alloys*, Vol 25, *Treatise on Material Science and Technology*, Academic Press, 1983, p 235-274
359. R.D. Kane, Role of H_2S in Behavior of Engineering Alloys, *Int. Met. Rev.*, Vol 30 (No. 6), 1985, p 291-301
360. R.F. Hehemann, Stress Corrosion Cracking of Stainless Steels, *Metall. Trans. A*, Vol 16A, Nov 1985, p 1909-1923
361. R.H. Jones and R.E. Ricker, Stress-Corrosion Cracking, in *Corrosion*, Vol 13, 9th ed., *Metals Handbook*, ASM INTERNATIONAL, 1987, p 145-163, 187-188
362. E.E. Denhard, Jr., Effects of Composition and Heat Treatment on the Stress Corrosion Cracking of Austenitic Stainless Steels, *Corrosion*, Vol 16, July 1960, p 359t-369t
363. H.R. Copson, Effect of Composition on Stress Corrosion Cracking of Some Alloys Containing Nickel, in *Physical Metallurgy of Stress Corrosion Fracture*, Vol 4, AIME Metallurgical Society Conferences, Interscience, 1959, p 247-272
364. J.G. Hines and E.R.W. Jones, Some Effects of Alloy Composition on the Stress-Corrosion Behaviour of Austenitic Cr-Ni Steels, *Corros. Sci.*, Vol 1, 1961, p 88-107
365. M.O. Speidel, Stress Corrosion Crack Growth in Austenitic Stainless Steel, *Corrosion*, Vol 33, June 1977, p 199-203
366. K.L. Money and W.W. Kirk, Stress Corrosion Cracking Behavior of Wrought Fe-Cr-Ni Alloys in Marine Atmosphere, *Mater. Perform.*, Vol 17, July 1978, p 28-36
367. C. Edeleanu, Stress Corrosion Cracking in Austenitic Stainless Steels, in *Stress Corrosion Cracking and Embrittlement*, John Wiley & Sons, 1956, p 126-139
368. T. Suzuki *et al.*, Austenitic-Ferritic Stainless Steels With High Chromium Low-Nickel Content, *Nippon Kinzoku Gakkaishi*, Vol 32 (No. 11), 1968, p 1171-1177
369. K.C. Thomas *et al.*, Stress Corrosion of Type 304 Stainless Steel in Chloride Environments, *Corrosion*, Vol 20, March 1964, p 89t-92t
370. F.K. Bloom, Stress Corrosion Cracking of Hardenable Stainless Steels, *Corrosion*, Vol 11, Aug 1955, p 351t-361t
371. R.R. Gaugh, Sulfide Stress Cracking of Precipitation Hardening Stainless Steels, *Mater. Perform.*, Vol 16, Sept 1977, p 24-29
372. V.L. Barnwell *et al.*, Effect of Grain Size on Stress Corrosion of Type 302 Austenitic Stainless Steel, *Corrosion*, Vol 22, Sept 1966, p 261-264

Notch Toughness of Steels

G.J. Roe and B.L. Bramfitt, Bethlehem Steel Corporation

TOUGHNESS is an indication of the capacity of a steel to absorb energy and is dependent on strength as well as ductility. Notch toughness is an indication of the capacity of a steel to absorb energy when a stress concentrator or notch is present. All carbon and high-strength low-alloy (HSLA) steels undergo a ductile-to-brittle transition as the temperature is lowered. Depending on chemical composition, product processing, and service environment, this transition can occur at temperatures from several hundred degrees above to several hundred degrees below room temperature. A number of notch impact tests have been developed to screen and rate steel product toughness on a relative basis and to determine the ductile-to-brittle transition for a specific carbon or HSLA steel product. Examples of various notch toughness tests are:

Test	ASTM specification
Charpy V-notch	E 23
Drop-weight nil-ductility	E 208
Drop-weight tear	E 436
Dynamic tear	E 604

Of all the notched tests, the most widely applied is the Charpy V-notch test.

An illustration of a Charpy testing machine is shown in Fig. 1. The Charpy specimen shown in Fig. 2 is generally oriented so that the root of the notch lies perpendicular to the surface of the component being tested. The orientation (longitudinal or transverse) of the specimen is selected according to the appropriate product specification. The specimen is held for 10 min at the test temperature and then broken in the Charpy-type impact tester by a single blow of a freely swinging pendulum. Upon the breaking of the Charpy specimen, three criteria are commonly measured. The loss of energy in the pendulum swing provides the energy in terms of joules (foot-pounds of force) absorbed in breaking the specimen. The fracture appearance of the broken specimen can be rated in terms of ductile and brittle fracture modes. Also, the lateral expansion at the base of the fracture opposite the notch can be measured. Any of these three criteria can be plotted versus temperature to obtain a ductile-to-brittle transition curve. The most commonly used measurement is energy absorbed. For steel product specifications, a specific Charpy V-notch requirement can be negotiated between the customer and supplier (for example, a minimum energy absorbed at a specified test temperature).

Increased reliance on the Charpy test began during World War II when a number of Liberty ships experienced brittle fracture, primarily initiating at notches in the ship deck plate. The fractured plate removed from the ship met existing chemistry and strength requirements but experienced a high notch sensitivity at the ship operating temperatures. From this observation, a correlation was obtained that indicated that ship failures were not observed at temperatures where the Charpy V-notch had an absorbed energy of 20 J (15 ft · lbf) or greater. As a result of this work, the 20 J (15 ft · lbf) transition temperature requirement was established. Today Charpy requirements are used as a screening technique in many steel applications (for example, shipbuilding, linepipe, highway bridges, and pressure vessel applications) where brittle fracture is a concern.

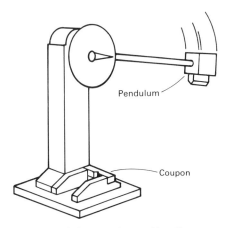

Fig. 1 Typical Charpy testing machine. The coupon is chilled to the desired temperature, then quickly placed into the anvil to be broken.

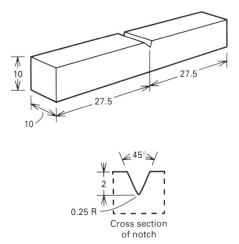

Fig. 2 Charpy V-notch specimen used for the evaluation of notch toughness (ASTM E 23). Dimensions given in millimeters

Ductile-to-Brittle Transition

In body-centered cubic metals such as plain carbon and low-alloy steels, a unique characteristic is found during impact testing over a temperature range of approximately 120 to −130 °C (250 to −200 °F). This characteristic is a ductile-to-brittle transition in toughness. At the higher end of the temperature range, the fracture behavior is ductile with an accompanying large degree of plastic deformation. The ductile mode of fracture is most commonly associated with microvoid coalescence. At the lower end of the range the fracture is brittle, and the mode of fracture is cleavage with little or no plastic deformation. Figure 3 is a Charpy curve showing a typical ductile-to-brittle transition in steel. The plot shows absorbed energy versus test temperature. The stable region at the higher temperatures is called the upper shelf, and the stable region at the lower temperatures is called the lower shelf. The region between the upper and lower shelves that displays a mixture of ductile and cleavage fracture is called the transition region.

Figure 4(a) shows an example of the fracture surface of a broken Charpy specimen from a low-alloy steel tested at the

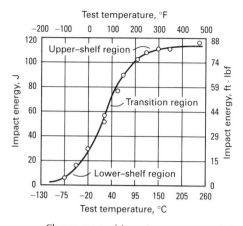

Fig. 3 Charpy curve of impact energy versus test temperature for a nickel-chromium-molybdenum steel

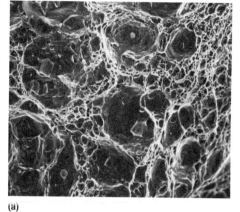
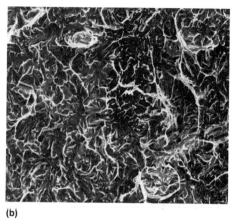

Fig. 4 SEM micrographs showing the fracture surfaces of broken Charpy V-notch specimens (nickel-chromium-molybdenum steel) tested in the (a) upper-shelf region and (b) lower-shelf region of the Charpy curve shown in Fig. 3. Both 670×

upper-shelf region. The absorbed energy was 117 J (86 ft · lbf). The fracture appears as a ductile, dimplelike surface with many shear lips. Inclusions, in this case titanium nitrides (square particles) and manganese sulfides (round particles), are usually located at the center of each large dimple. These inclusions initiated the formation of a void, which created the dimple. Figure 4(b) shows the typical cleavage fracture surface at the lower-shelf region for the same steel. The absorbed energy was 6.1 J (4.5 ft · lbf). In this scanning electron microscopy (SEM) photograph, no regions of ductile tearing or shear lips are present.

From the Charpy curve shown in Fig. 3 a transition temperature can be determined. There are a number of criteria for transition temperature, one being the median temperature of this transition range. In Fig. 3, this would be 25 °C (75 °F). The transition temperature could also be the temperature at a specified absorbed energy (20 J, or 15 ft · lbf, for example). Thus, in Fig. 3 the 20 J (15 ft · lbf) transition temperature would be −30 °C (−25 °F).

In addition to absorbed energy, other important measurements can be obtained from the Charpy test. For example, the lateral expansion can be measured on the broken Charpy specimen at the edge of the specimen opposite to the notch. Figure 5 shows the amount of lateral expansion one can expect at the upper- and lower-shelf regions of the Charpy curve. Lateral expansion is used as a measure of notch toughness similar to absorbed energy. Figure 6 shows a plot of lateral expansion versus test temperature for the same steel as in Fig. 3.

Another important measurement is fracture appearance transition temperature (FATT). To obtain this value, the percentage of shear (fibrous fracture) in the fracture surface is measured. The 50% FATT represents the temperature at 50% shear (50% fibrous). The 100% FATT would represent the temperature at 100% shear. Figure 7 shows a plot of percent shear versus test temperature for the same steel as in Fig. 3 and 6. From Fig. 7 the 50% FATT is 25 °C (75 °F). Note that this is the same temperature as determined from the data plotted in Fig. 3. In some steels the energy-established transition temperature and FATT may not be the same. In some steels it may be difficult to measure percent shear because of woody fracture surfaces caused by stringer-type inclusions, and so on. For these cases, it would be more appropriate to use the lateral expansion and absorbed-energy measurements to obtain a more accurate transition temperature.

Selection of the most appropriate method of measuring transition temperature for a given application is difficult and requires an understanding of both toughness testing and service behavior. Often, additional tests are needed to establish a correlation between the transition temperature determined using a certain method and the service behavior

Fig. 5 Fracture surfaces of the broken Charpy V-notch specimens in Fig. 4. These photographs show the amount of lateral expansion in nickel-chromium-molybdenum steel specimens tested in (a) the upper-shelf region (1.5 mm, or 0.059 mil) and (b) the lower-shelf region (0.05, or 0.002 mil) of the Charpy curve shown in Fig. 3.

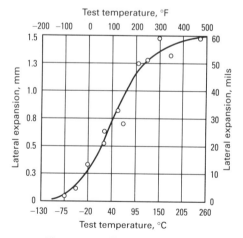

Fig. 6 Charpy curve of lateral expansion versus test temperature for the same nickel-chromium-molybdenum steel in Fig. 3

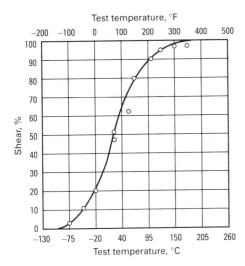

Fig. 7 Charpy curve of percent shear versus test temperature for the same nickel-chromium-molybdenum steel in Fig. 3 and 6

of a specific structure made of the same material. The method in which the impact energy at a given temperature must exceed a specific value is one method requiring such a correlation.

Despite the importance of notch toughness, other mechanical requirements must be considered, and often some compromise must be made when selecting a steel. For example, notch toughness usually decreases as carbon content, strength, and hardness are increased. Hard surfaces required for wear resistance also have an adverse effect on notch toughness.

The shape of the ductile-to-brittle transition can be represented by fitting a hyperbolic tangent curve to the data obtained from the Charpy test (Ref 1). This curve-fitting technique is now used by some design engineers to characterize and predict the toughness of a particular steel. Figure 8 shows an example of fitting a hyperbolic tangent to the set of Charpy impact data for the same steel as in Fig. 3. From this procedure, the median temperature in the transition range was calculated as 23 °C (74 °F), and the upper- and lower-shelf energies were 114 and 0.07 J (84 and 0.05 ft · lbf), respectively. Currently, the curve-fitting technique is only experimental; however, it is slowly becoming accepted by the metallurgical and materials community.

Effects of Composition

The composition of a steel, as well as its microstructure and processing history, significantly affects both the ductile-to-brittle transition temperature range and the energy absorbed during fracture at any particular temperature. The effect of the various alloying elements and those of microstructural and processing variables are intimately interrelated; in practice, it is difficult to change one variable without affecting another (Ref 2, 3). Each individual alloying element contributes to notch toughness to varying degrees.

Carbon. Increasing carbon content increases transition temperature and decreases upper-shelf fracture energy primarily as a result of increased strength and hardness. These effects, measured by Charpy V-notch

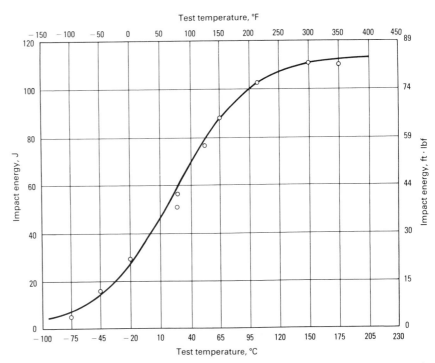

Fig. 8 Charpy curve plotted by fitting a hyperbolic tangent curve to the same test data shown for the nickel-chromium-molybdenum steel in Fig. 3

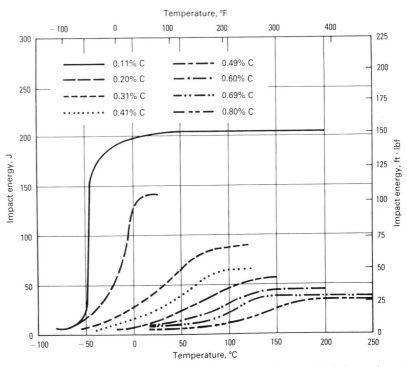

Fig. 9 Variation in Charpy V-notch impact energy with temperature for normalized plain carbon steels of varying carbon content. Source: Ref 4

impact tests, are shown in Fig. 9. Carbon is one of the most potent alloying elements in its effect on notch toughness and strength. Consequently, for maximum toughness, the carbon content should be kept as low as possible, consistent with strength requirements. Low-carbon steels tend to have very steep transition curves.

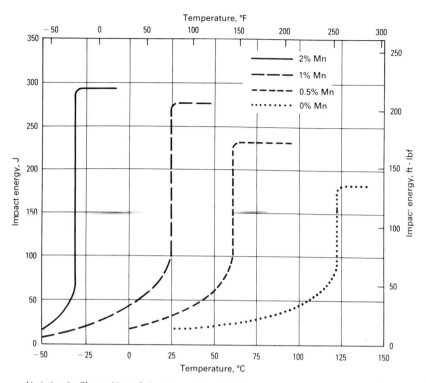

Fig. 10 Variation in Charpy V-notch impact energy with temperature for furnace-cooled Fe-Mn-0.05C alloys containing varying amounts of manganese. Source: Ref 5

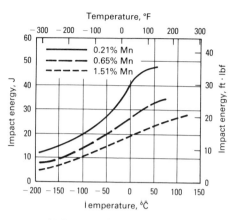

Fig. 12 Variation in Charpy V-notch impact energy with temperature for alloy steels containing 0.35% C, 0.35% Si, 0.80% Cr, 3.00% Ni, 0.30% Mo, 0.10% V, and the indicated amounts of manganese. The steels were hardened and tempered to a yield strength of approximately 1175 MPa (170 ksi). The microstructures of these steels contained tempered martensite. Source: Ref 7

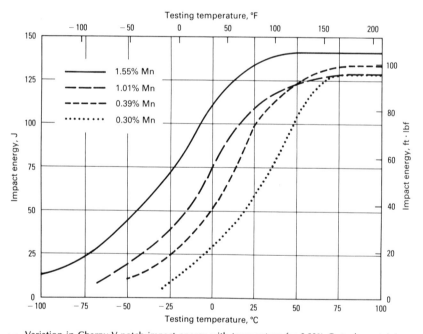

Fig. 11 Variation in Charpy V-notch impact energy with temperature for 0.30% C steels containing varying amounts of manganese. The specimens were austenitized at 900 °C (1650 °F) and cooled at approximately 14 °C/min (25 °F/min). The microstructures of these steels were pearlitic. Source: Ref 6

Manganese has a variety of effects on transition temperature. In low-carbon steels, it can substantially reduce the transition temperature, as shown in Fig. 10. In higher-carbon steels, manganese may be less beneficial. As illustrated in Fig. 11, increasing the manganese content of a normalized medium-carbon steel lowered the ductile-to-brittle fracture transition temperature, probably because the additional manganese reduced the pearlite interlamellar spacing (the spacing between the alternating plates of ferrite and cementite in pearlite). In a hardened and tempered steel, manganese can have the opposite effect, as illustrated in Fig. 12. Manganese can make the steel susceptible to temper embrittlement, and it may cause the formation of less tough upper bainite (rather than fine pearlite) during normalizing (see the article "Austenitic Manganese Steels" in this Volume).

Sulfur. The effect of sulfur on the notch toughness of steels is directly related to deoxidation practice. For rimmed, semikilled, and silicon-killed steels, sulfur in amounts up to about 0.04% has a negligible effect on notch toughness. Sulfur has a strong directional effect on Charpy results depending on the inclusion types present (for example, sulfides, oxides, and complex nonmetallics). Charpy tests taken perpendicular to the working or rolling direction have lower absorbed energies when manganese sulfide stringers are present. Steel ladle treatments, used to reduce sulfur and to provide inclusion shape control, minimize Charpy test directionality. For silicon-aluminum-killed steels, a reduction in sulfur content can substantially increase uppershelf energy, as shown in Fig. 13. This improvement in energy absorption results from a reduction in the number of sulfide stringers in the steel. Room-temperature Charpy results taken transverse to the rolling direction in plate steels show substantial improvement in energy absorbed only when sulfur levels are reduced below about 0.010% (Fig. 14).

Phosphorus has a strongly deleterious effect on the notch toughness of steel. It raises the 50% FATT about 7 °C (13 °F) for each 0.01% P and reduces upper-shelf energy. In addition, phosphorus increases the susceptibility of some steels to temper embrittlement.

Silicon, used in amounts of 0.15 to 0.30% to deoxidize steels, generally lowers the ductile-to-brittle fracture transition and raises upper-shelf energy. Compared to rimmed or semikilled steels, silicon-killed steels are cleaner and have more uniform ferrite grains. These effects are probably

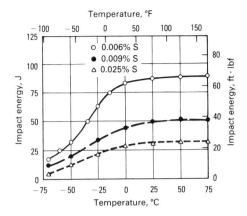

Fig. 13 Variation in transverse Charpy V-notch impact energy with temperature for HSLA steels containing varying amounts of sulfur. The steels were silicon-aluminum killed with a minimum yield strength of 450 MPa (65 ksi).

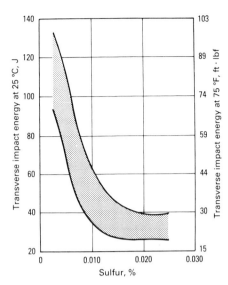

Fig. 14 Effect of sulfur content on transverse impact energy at room temperature in a silicon-aluminum-killed steel

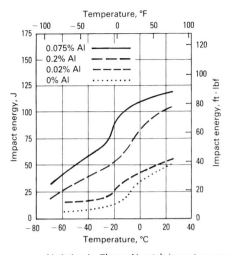

Fig. 15 Variation in Charpy V-notch impact energy with temperature for normalized and tempered medium-carbon steels containing varying amounts of aluminum

caused by variations in steelmaking practice characteristic of the deoxidation methods used, rather than by the silicon content.

Aluminum. The effect of aluminum on the notch toughness of a medium-carbon steel is illustrated in Fig. 15. Note that increasing the aluminum content above that needed for forming aluminum nitrides (≈0.075% Al) impairs notch toughness.

Nitrogen, by itself, lowers the upper-shelf energy and raises the transition temperature. However, most nitrogenized steels are deoxidized with silicon and aluminum, both of which combine with nitrogen. Aluminum nitrides formed during deoxidation serve to stabilize grain size and thus improve the notch toughness of these steels.

Nickel, like manganese, is useful for improving the notch toughness of steels at low temperatures. Nickel is less effective in improving the toughness of medium-carbon steels than low-carbon steels. Some high-nickel alloy steels, such as maraging steels and austenitic stainless steels (see the articles "Maraging Steels" and "Wrought Stainless Steels" in this Volume), do not exhibit the typical ductile-to-brittle transition (austenitic steels, being face-centered cubic, do not have a ductile-to-brittle transition). The high nickel content reduces upper-shelf fracture energy—but to a level that is still quite acceptable for most applications.

Chromium raises the transition temperature slightly. In steels having chromium contents in excess of 0.90%, it is very difficult to develop those microstructures and mechanical properties that are typical of plain carbon steels; therefore, impact test results are not comparable. Chromium is usually added to increase hardenability. The increase in hardenability is often sufficient to develop a martensitic microstructure, which provides high upper-shelf energy. Medium-carbon, straight chromium alloy steels, such as 5140, are susceptible to embrittlement when quenched to martensite and tempered between 370 and 575 °C (700 and 1070 °F).

Molybdenum in the typical quantities in alloy steels (up to about 0.40%) raises the 50% FATT. Molybdenum is frequently used to increase hardenability, and it influences notch toughness primarily through its effect on microstructure. About 0.5 to 1.0% Mo can be added to alloy steels to reduce their susceptibility to temper embrittlement, but it is effective only for relatively short heating times at embrittling temperatures. Molybdenum appears to delay rather than eliminate temper embrittlement, because steels containing small amounts of this element have become embrittled upon prolonged exposure within the embrittling temperature range.

Boron. For quenched and tempered steels, a practical way of improving toughness without reducing strength is to use a boron-containing grade of steel with a lower carbon content. As shown in Fig. 16, 10B21 steel has greater toughness than 1038 steel at all strength levels. However, the benefit of boron is applicable only to quenched and tempered steels; boron reduces the toughness of as-rolled, as-annealed, and as-normalized steels.

Copper in steels that have not been subjected to precipitation hardening appears to be moderately beneficial to low-temperature notch toughness. However, copper promotes precipitation hardening in steel and, as a result, may adversely affect notch toughness, particularly if the tempering temperature is between 400 and 565 °C (750 and 1050 °F).

Vanadium, niobium, and titanium are most often used in steels that receive controlled thermomechanical treatment. Consequently, the toughness of steels contain-

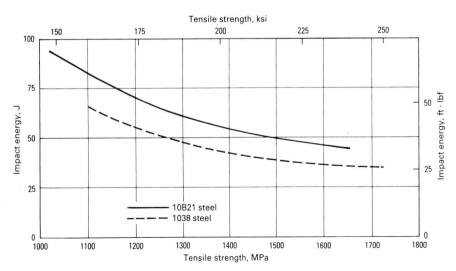

Fig. 16 Effect of boron content on notch toughness. Room-temperature Charpy V-notch impact energy varies with tensile strength for 10B21 and 1038 steels having tempered martensite structures.

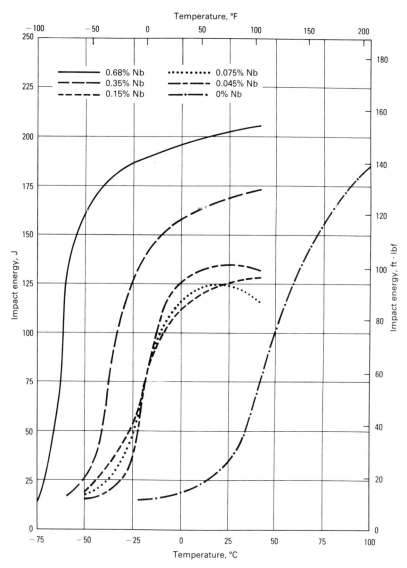

Fig. 17 Variation in Charpy V-notch impact energy with temperature for low-carbon steels containing varying amounts of niobium that were normalized from 955 °C (1750 °F). Source: Ref 8

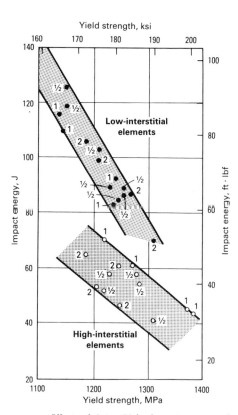

Fig. 18 Effect of interstitial elements on notch toughness. The notch toughness at −18 °C (0 °F) of 12% Ni maraging steel can be significantly raised by controlling the amount of interstitial alloying elements in the steel, regardless of the strength level. Numbers indicate plate thickness in inches.

ing these elements is largely a function of mill processing. When the steel is finished at temperatures below about 925 °C (1700 °F) (which is characteristic of certain HSLA steels), vanadium, niobium, and titanium improve toughness primarily by refining the ferrite grain size (see the articles "High-Strength Structural and High-Strength Low-Alloy Steels" and "High-Strength Low-Alloy Steel Forgings" in this Volume). At higher finishing temperatures, these elements may be detrimental to toughness. The effect of niobium on toughness in a low-carbon normalized steel is shown in Fig. 17.

Zirconium, titanium, calcium, and the rare earths can be used to control the shape of manganese sulfide inclusions, causing the inclusions to be spherical rather than elongated. Spherical inclusions raise upper-shelf energy and minimize the anisotropic nature of notch toughness; these effects are particularly useful in HSLA sheet and thin plate.

Interstitial elements, such as carbon, oxygen, nitrogen, and hydrogen, generally reduce the notch toughness of steels. This effect is especially evident in maraging steels, as shown in Fig. 18. Oxygen content in steels is usually determined by the deoxidation practice used in manufacturing the steel; rimmed steels have higher oxygen contents and higher transition temperatures than killed steels. Hydrogen reduces the notch toughness of steels; its chief deleterious effect occurs under conditions of slow or static loading. Specific effects are discussed in the sections "Carbon" and "Nitrogen" in this article.

Antimony, Arsenic, and Tin. In trace amounts, these elements reduce the notch toughness of steels and greatly increase the susceptibility of nickel- and chromium-alloy steels to temper embrittlement.

Interactive effects of alloying elements are very common, particularly between the interstitial elements carbon and nitrogen and the strong carbide or nitride formers, such as aluminum, vanadium, manganese, niobium, molybdenum, and titanium. Some interactive effects are described in the sections "Sulfur" and "Aluminum" in this article; other examples are shown in Fig. 19 and 20. Interactive effects of alloying elements and processing variables are discussed in the next section.

Effects of Manufacturing Practices

The notch toughness of a steel product is the result of a number of interactive effects, including composition, deoxidation and steelmaking practices, solidification, and rolling practices, as well as the resulting microstructure. This section focuses on the general influence of manufacturing practices and the interactive effects that simultaneously influence notch toughness.

Wrought Steels

Deoxidation Practice. As a general rule, notch toughness improves from rimmed through silicon-aluminum-killed deoxidation practices. Optimum toughness is obtained in either silicon-aluminum-killed or aluminum-killed steels.

The effect of deoxidation practice on the notch toughness of steels is directly traceable to the presence of those alloying ele-

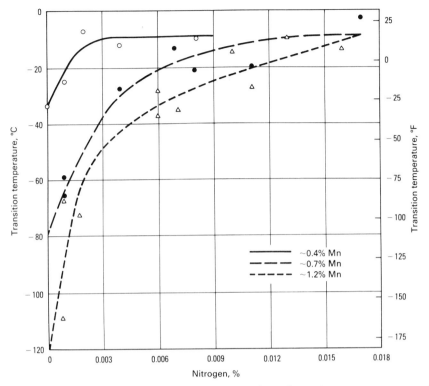

Fig. 19 Interactive effect of manganese and nitrogen on notch toughness. Fracture appearance transition temperature (50% shear FATT) in plain carbon steel (0.10% C) at three manganese levels (0.4, 0.7, and 1.2% Mn) varies with nitrogen content. The beneficial effect of manganese is particularly evident at low levels of nitrogen.

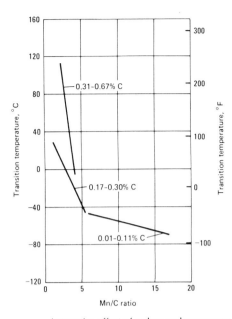

Fig. 20 Interactive effect of carbon and manganese on notch toughness. Manganese-to-carbon ratio affects the transition temperature of ferritic steels. Source: Ref 6

ments and impurities characteristic of the deoxidation practice. Rimmed steels typically contain appreciable quantities of oxygen and nitrogen; soundness and homogeneity of rimmed steel ingots are often poor. These characteristics of rimmed steels account for their poor notch toughness. Killed steels, particularly silicon-aluminum-killed steels, have lower transition temperatures and higher upper-shelf energy values than rimmed steels. Semikilled steels have toughness properties between those of rimmed and killed steels. A comparison of the notch toughness of these three types of steel is shown in Fig. 21. The combined effects of silicon-aluminum deoxidation and restriction of sulfur content are shown in Fig. 13. Upper-shelf energy is increased because silicon-aluminum killing reduces the incidence of detrimental silicate stringers in the steel; restriction of sulfur reduces the incidence of manganese sulfide inclusions. An aluminum killing practice reduces the inclusion levels and refines the ferrite grain size.

Hot Deformation Temperature. The effect of hot deformation on the notch toughness of steel can be directly related to the microstructure produced during hot rolling and forging. Steels deformed at temperatures

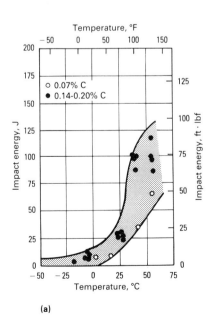
(a)

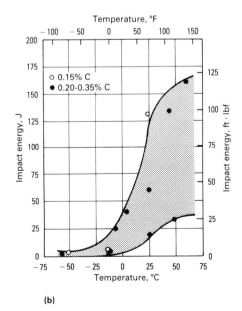
(b)

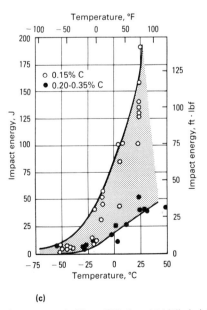
(c)

Fig. 21 Effects of deoxidation practice on notch toughness. Charpy V-notch impact energy varies with temperature for (a) rimmed, (b) semikilled, and (c) killed plain carbon steels.

744 / Service Characteristics of Carbon and Low-Alloy Steels

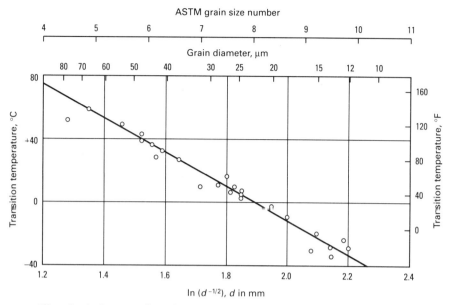

Fig. 22 Effect of grain size on notch toughness. Fracture appearance transition temperature varies with ferritic grain size for 0.11% C low-carbon steel. Transition temperature varies linearly with $\ln(d^{-1/2})$ and is lower for fine-grain steel. Source: Ref 9

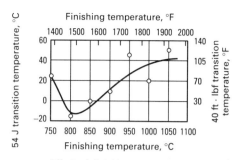

Fig. 23 Effect of finishing temperature on notch toughness. The 54 J (40 ft · lbf) Charpy V-notch transition temperature varies with hot-rolling finishing temperature for silicon-killed 0.24C-1.69Mn steel. Source: Ref 10

above about 980 °C (1800 °F) undergo considerable recrystallization and grain growth during rolling; the structure thus obtained is only slightly affected by the rolling process. The notch toughness of steels deformed at such high temperatures is largely determined by the size to which the austenite grains grow after recrystallization. The influence of grain size on transition temperature is shown in Fig. 22. When steels are deformed at lower temperatures, recrystallization and growth of austenite grains cannot proceed to the extent possible at higher deformation temperatures. Thus, the transition temperature can be significantly lowered by deforming in the lower portion of the austenite temperature range. Figure 23 shows the effect of finishing temperature during hot rolling on the 54 J (40 ft · lbf) transition temperature of a carbon-manganese steel.

The amount of deformation at specific temperatures is also important in determining the toughness of the steel. Figure 24 illustrates the importance of sufficient reduction during rolling; although extensive rolling reduces transition temperature, it also reduces upper-shelf fracture energy. The schedule of hot-rolling temperatures

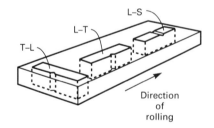

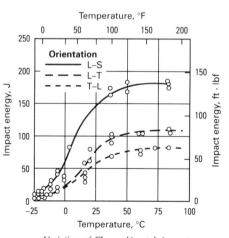

Fig. 24 Effect of hot deformation below 900 °C (1650 °F) on 50% FATT for an HSLA steel plate

Fig. 25 Variation of Charpy V-notch impact energy with notch orientation and temperature for steel plate containing 0.012% C. Source: Ref 11

and reductions is especially important for HSLA steels in order to ensure sufficient toughness and transition temperatures that are low enough. (The above discussion of hot-rolling temperature is applicable to other hot-working operations, such as extrusion or forging.)

Toughness Anisotropy. Steels can acquire strongly anisotropic microstructures as a result of working. Anisotropic microstructures are often indicative of the anisotropy of mechanical properties, particularly notch toughness. Anisotropy, therefore, is an important consideration in the design and fabrication of rolled, forged, drawn, or extruded steel products.

The effect of anisotropy on the notch toughness of as-rolled low-carbon steel plate is shown in Fig. 25. Specimens parallel to the rolling direction (orientations L-S and L-T) show higher impact energies throughout the ductile-to-brittle fracture transition temperature range than do specimens perpendicular to the rolling direction (orientation T-L). Orientation L-T is the standard longitudinal specimen, and orientation T-L is the standard transverse specimen referred to in ASTM E 23. Therefore, when a part is to be cut from plate, it is essential to specify the orientation of the part relative to the rolling direction.

The rolling schedule during the fabrication of plate affects anisotropy. For example, if the steel had been cross rolled so that it received about the same amount of hot reduction in both directions, the curves for orientation would nearly coincide at a position between the L-T and T-L curves for material reduced by conventional rolling.

Regardless of the amount of cross rolling, specimens that are notched parallel to the plate surface (orientation L-S) absorb greater amounts of energy than those notched at right angles to the plate surface (orientation L-T). In the experiment described in Fig. 25, the temperature range over which transition occurred (and also the shear-fracture transition temperature) was the same regardless of notch or specimen orientation. In other experiments, transition temperatures for both energy absorption and fracture appearance were higher for transverse

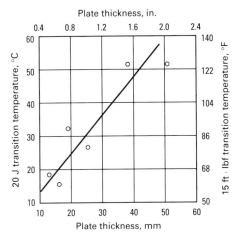

Fig. 26 Effect of plate thickness on notch toughness for aluminum semikilled steel (0.14C-1.25Mn-0.007S-0.020P-0.021Nb)

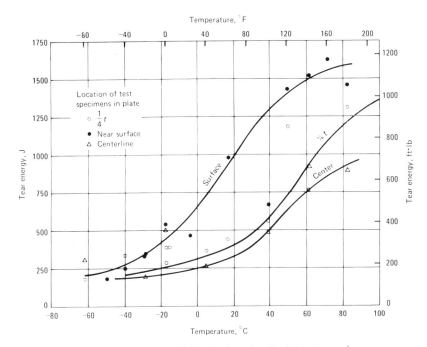

Fig. 27 Variation of tear test toughness with location in a plate. Variations in toughness were measured by drop-weight tear tests across the thickness of a 102 mm (4 in.) thick plate on ASTM A 387 (2.25Cr-1Mo) steel. The curve labeled ¼t was obtained from specimens taken approximately 25 mm (1 in.) from the top and bottom surfaces of the plate.

orientations than for longitudinal orientations.

Section and Part Size. Variations in fracture behavior can also result from differences in metallurgical structure between thin and thick stock of a given material. For example, the transition temperature of hot-rolled low-carbon steel varies with plate thickness, as shown in Fig. 26. In these tests, specimen size was constant, yet ductile-to-brittle energy transition still increased with increasing plate thickness. Reduction of toughness with increasing plate thickness is not limited to low-carbon steel but apparently applies to all steels. Because of the characteristics of normal commercial processing, the metallurgical structure of thick stock is different from that of thin stock, resulting in inherently lower toughness for the thicker stock. More important, the probability that a given part will contain a crack or a flaw of critical size (or greater) increases with increasing stock thickness or part size. The lower inherent toughness of thick stock and the higher probability that thick stock contains a large crack or flaw, combined with the plane-strain conditions inherent in thick members, account for the fact that large structures are more susceptible to brittle fracture than small structures. These factors are discussed in detail in the article "Fatigue Resistance of Steels" in this Volume.

The notch toughness of steels can vary through the thickness of a plate. Pellini (Ref 12) found that the transition temperature in a thick plate was far lower near the surface than near the center of the plate (Fig. 27). Aside from problems that might be caused by inhomogeneities in mechanical properties, such a variation in toughness would be especially serious if portions of the surface of the plate were cut away, exposing the low-toughness center section to impact loading.

In addition to the effects of the section size of the steel from which a part is made, the size of the part itself can influence fracture behavior. Large sections are more susceptible to brittle fracture than small sections because the large sections create conditions of plane strain under which cracks can propagate very readily.

Effects of Surface Condition on Notch Toughness. Nitriding and carburization adversely affect the notch toughness of carbon and alloy steels. Nitriding to even a relatively shallow depth (0.13 mm, or 0.005 in.) can affect notch toughness by making the surface harder and less ductile. Steel may be carburized accidentally during austenitizing under protective atmospheres; the resulting higher carbon content and hardness make the surface layer more susceptible to crack formation under impact loading.

The decrease in notch toughness that results from carburization was evaluated by heating test specimens of 1050 and 4130 steel in a carburizing atmosphere (dew point: 0 °C, or 32 °F) at 925 °C (1700 °F) for 30 min to 2 h. The specimens were quenched directly and tempered to a core hardness of 38 to 40 HRC. The decrease in notch toughness with increasing depth of carburization is shown in Fig. 28.

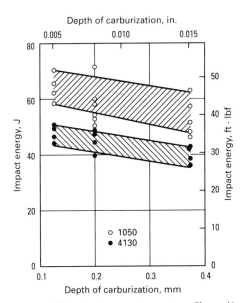

Fig. 28 Variation in room-temperature Charpy V-notch impact energy with depth of carburization. Specimens of 1050 and 4130 were carburized to the indicated case depth, then quenched and tempered to core hardness levels of 38 and 40 HRC, respectively.

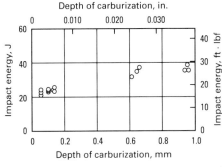

Fig. 29 Variation in room-temperature notch toughness with depth of decarburization. Specimens of 4340 steel were deliberately decarburized to the indicated depth, then hardened and tempered to 52 HRC.

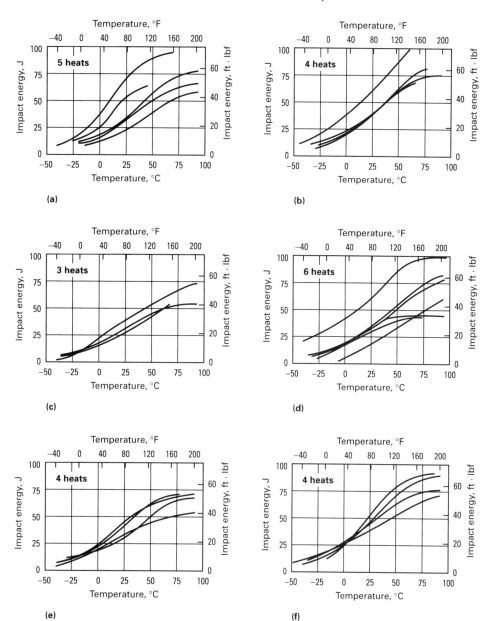

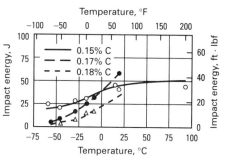

Fig. 32 Variation in Charpy V-notch impact energy with temperature for cast carbon steel containing different amounts of carbon, all normalized and tempered to 105 HB

Fig. 30 Effect of melting technique on notch toughness. Variation in Charpy V-notch impact energy with temperature for annealed (a, c, and e) and normalized (b, d, and f) cast carbon steels produced using three different melting techniques. (a) and (b) 0.27C-0.70Mn-0.43Si steel melted by acid electric technique. (c) and (d) 0.29C-0.72Mn-0.44Si steel melted by acid open hearth technique. (e) and (f) 0.33C-0.78Mn-0.38Si steel melted by basic open hearth technique. All steels were fully silicon-aluminum killed.

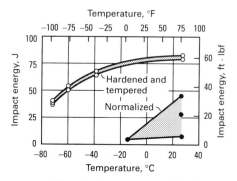

Fig. 31 Effect of microstructure and hardness on notch toughness of cast steels. Charpy V-notch impact energy varies with temperature for cast 4330 steel normalized to 228 HB or hardened and tempered to 269 HB.

Decarburization lowers surface hardness and permits plastic deformation at the root of a notch without crack initiation; thus, it will not lower, and may slightly increase, notch toughness. However, because decarburization adversely affects the fatigue resistance of steel, the decarburization of a notched member is rarely advisable. The effect of decarburization on notch toughness is shown in Fig. 29. Specimens of 4340 steel were deliberately decarburized during austenitizing at 820 °C (1510 °F), then quenched and tempered to a core hardness of 52 HRC. The observed values of impact energy increased slightly as the depth of decarburization increased.

Electroplating may impair notch toughness. Conventional hard plating, as with chromium or nickel, apparently has an adverse effect similar to that of carburized cases. Results from tests of specimens plated with these metals showed such wide scatter that no definite conclusions can be drawn. Hydrogen absorbed during plating may cause hydrogen embrittlement and thus contribute to the erratic results. The notch toughness of specimens plated with softer metals such as zinc and cadmium is not affected by the coating itself but may be impaired by hydrogen absorbed during plating.

Cast Steels

With the exception of working direction, most of the same chemical, microstructural, and manufacturing factors that influence the notch toughness of wrought steels also apply to cast steels. Direct comparisons are seldom made between wrought and cast steels because of differences in their chemical composition and microstructure (for example, cast steels usually contain more silicon). The porosity that sometimes occurs in cast steel may cause a greater scatter of test results than for wrought steel of similar composition. The notch toughness of cast steel is usually less than that of wrought steel measured parallel to the rolling direction but greater than that of wrought steel measured perpendicular to the rolling direction, provided the steels are otherwise similar.

To investigate the effect of melting method on the notch toughness of 0.30% C cast steels, Charpy V-notch specimens were machined from 355 × 355 × 38 mm (14 × 14 × 1½ in.) plates poured from 26 production heats—9 acid electric, 9 acid open hearth, and 8 basic open hearth. All steels were fully killed with silicon and aluminum. The results of these tests, shown in Fig. 30, indicate no significant differences in notch toughness among steels melted by the three different methods. Normalizing results in an increase in notch toughness over that of annealed materials.

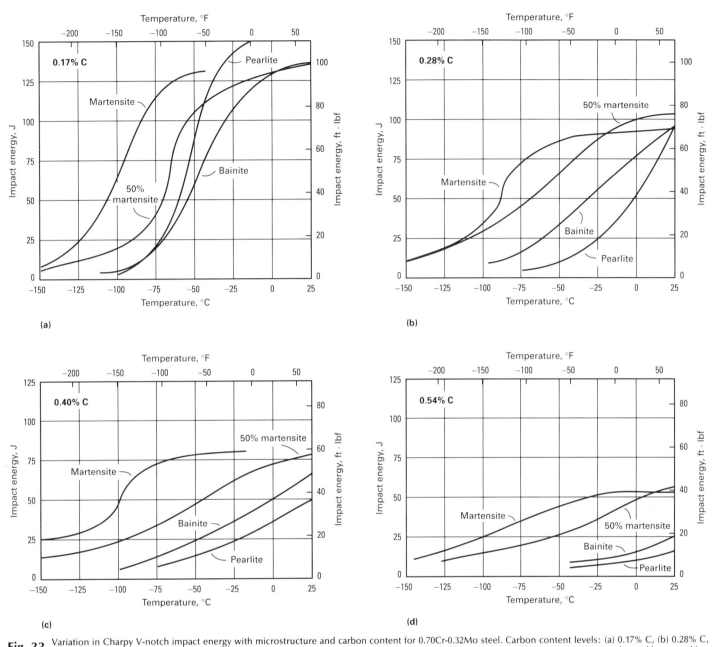

Fig. 33 Variation in Charpy V-notch impact energy with microstructure and carbon content for 0.70Cr-0.32Mo steel. Carbon content levels: (a) 0.17% C, (b) 0.28% C, (c) 0.40% C, and (d) 0.54% C. A pearlitic structure was formed by transformation at 650 °C (1200 °F). A structure with 50% martensite was formed by quenching in lead at 455 °C (850 °F) for (a) 10 s, (b) 19 s, (c) 35 s, and (d) 100 s. Fully martensitic structures were formed by quenching the 0.17 and 0.28% C grades in water and by oil quenching the grades containing 0.40 and 0.54% C. Bainite was formed by quenching in lead at 455 °C (850 °F) and holding 1 h, except that the 0.54% C grade was held 3 h. All specimens were tempered to the same hardness level.

Steel castings that have been quenched and tempered have higher notch toughness than similar castings in the as-cast, annealed, or normalized condition. The graph in Fig. 31 shows the difference in notch toughness between normalized 4330 steel and hardened and tempered 4330 steel. Despite the higher hardness of the quenched and tempered material, its upper-shelf energy is higher and its transition temperature is lower. Certain castings in the normalized condition (untempered) failed in service on excavating equipment when operated below −18 °C (0 °F).

Carbon steel castings are sensitive to minor variations in carbon content in the same way as wrought steels. The influence of small variations in carbon is shown in Fig. 32. Additional information on the notch toughness of cast steels is available in the article "Steel Castings" in this Volume.

Effects of Microstructure

Like most mechanical properties, the notch toughness of steel can usually be traced directly to microstructure. Because the microstructures of steel are readily observed and classified, it is convenient to attribute the various mechanical properties to the microstructure, even though the properties might also be attributed to the composition and manufacturing history of the steel.

Microstructural Constituents

In general, of the major microstructural constituents found in steels, pearlite and ferrite have the highest transition temperature, followed by upper bainite and, finally, tempered martensite or lower bainite. Values of notch toughness for similar steels having different carbon contents and microstructural constituents are shown in Fig. 33; to facilitate comparison of the different microstructures, these steels were tempered to uniform strength levels before testing. In practice, the cooling or quenching rate de-

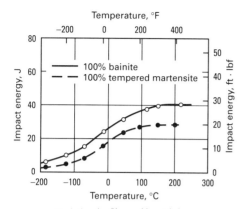

Fig. 34 Variation in Charpy V-notch impact energy with temperature for specimens of 4340 steel having 100% tempered martensite and 100% bainite microstructures. All specimens were austenitized for 30 min at 845 °C (1550 °F) in neutral salt. 100% bainite was produced by isothermal transformation for 1 h at 315 °C (600 °F) in agitated salt. 100% tempered martensite was produced by quenching in agitated oil at 50 to 55 °C (120 to 135 °F) and tempering at 315 °C (600 °F). All specimens had the same tensile strength. Source: Ref 13

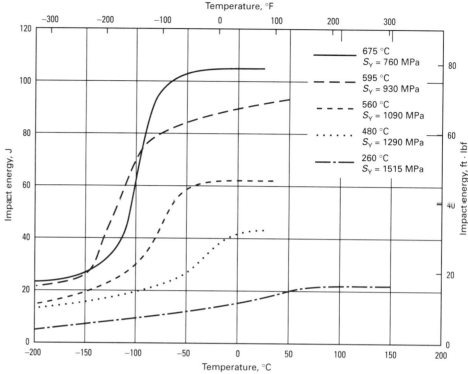

Fig. 35 Variation in Charpy V-notch impact energy with temperature for various tempering temperatures. Specimens of 4340 steel were tempered 1.5 h at the indicated temperatures. Yield strength, S_Y, obtained through each heat treatment is also indicated. Source: Ref 16

termines the resulting microstructure or mixture of microstructures in a particular steel. The transformation characteristics are controlled by the cooling rate, alloy content, austenitizing temperature, and austenite grain size.

Generally, treatments that produce microstructures with inferior room-temperature toughness also raise transition temperature. Precipitates and second-phase particles are detrimental to toughness, especially if located at grain boundaries. A spheroidization treatment of pearlitic steels can improve toughness by reducing strength and by eliminating ferrite lamellae (which have a platelike form and thus are paths of easy cleavage fracture). Spheroidization also improves toughness by changing the shape of the brittle cementite lamella of pearlite to innocuous spherical particles. In addition, steels can lose toughness because of various embrittlement phenomena such as temper embrittlement (see the section "Submicroscopic Structure" in this article).

Steels having a tempered martensitic or a lower bainitic structure offer an optimum balance of strength and toughness. When observed under a light microscope, these two microconstituents are indistinguishable. Only by transmission electron microscopy can one distinguish tempered martensite from lower bainite. Which of these two microconstituents provides better notch toughness has been a subject of controversy.

Some data indicate that lower bainite has better toughness than tempered martensite of the same strength (Fig. 34). Other data have shown the opposite effect (Ref 14). However, it is fairly well established that tempered martensite has better toughness than tempered lower bainite.

The presence of austenite inhibits the fast propagation of cleavage fracture in some ferritic and martensitic steels. For example, the presence of retained austenite in martensitic maraging steel significantly improves toughness by a process called transformation-induced plasticity. The means by which toughness is improved is not fully understood, but it is thought to occur as follows (Ref 15):

- When large quantities (of the order of 50% or more) of retained austenite are present, the austenite undergoes a strain-induced transformation to martensite in the plastic region ahead of the crack tip. This consumes energy, thus raising the total energy required for the fracture process
- When austenite is retained as a lamellar phase (in lesser quantities than for a strain-induced transformation), the retained-austenite lamellae block the growth of secondary cracks in the martensite matrix just ahead of the main crack front

Additional information on the effect of the austenite-to-martensite transformation on toughness is available in the article "Maraging Steels" in this Volume.

A continuum of microstructures ranging from martensite to spheroidized cementite can be achieved by tempering martensite at various combinations of time and temperature. The notch toughness of tempered steel varies widely, from excellent to very poor. As shown in Fig. 35, tempering at higher temperatures lowers the transition temperature and raises upper-shelf energy.

Grain Size

The effect of grain size on the notch toughness of steels is mentioned in the section "Hot Deformation Temperature" in this article. Hot deformation practices can significantly affect grain size and shape. The grain size of steels can also be affected by secondary heat treatments. As shown in Fig. 22, a decrease in grain size reduces the transition temperature of steel; therefore, hot deformation practices and secondary heat treatments are usually designed to produce fine grains. In addition to improving notch toughness, a fine grain size increases the strength of steel. Thus, refinement of grain size is the most important and most effective way to increase both strength and toughness.

When steel is heated, small austenite grains form at temperatures above the temperature at which austenite begins to form upon heating, Ac_1. Grain size continually increases with time at temperature, and higher temperatures result in faster grain growth. The strength, ductility, and impact toughness of coarse-grain metals are impaired not only by the large grain size but also by grain-boundary precipitation. Generally, when a metal has been overheated and incipient melting has not occurred, re-

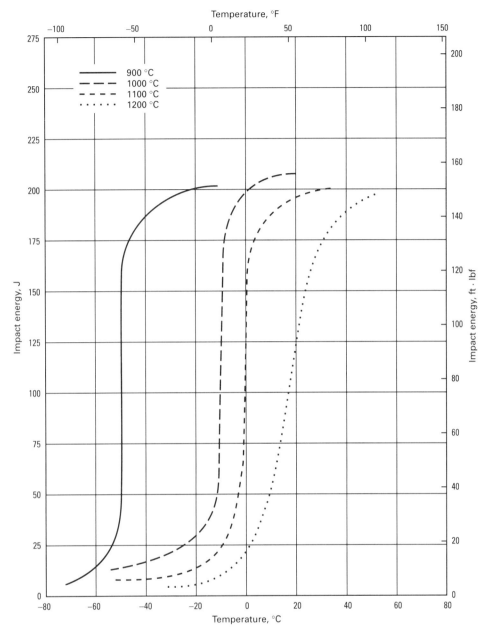

Fig. 36 Variation of Charpy V-notch impact energy with temperature for low-carbon steel normalized at various temperatures

heat treating at proper time and temperature will restore a desirable small-grain structure and satisfactory properties. The effect of normalizing at excessive temperatures is shown in Fig. 36.

Qualitatively, the embrittling effect of large grains in ferritic steel can be explained by stress concentration at the ends of slip bands and at grain boundaries. The larger the grains, the longer the slip bands and the greater the stress concentration. Severe stress concentration will induce nucleation of microcracks, which in turn may cause early and catastrophic fracture.

Submicroscopic Structure

Several structural features that are too small to resolve by light microscopy have significant effects on the notch toughness of steels. Among these are the precipitation that occurs during quench aging or strain aging, and the brittle particles and films sometimes formed during tempering. Low (Ref 17) has shown that the ductile-to-brittle transition temperature rises abruptly with time after quenching from 690 °C (1275 °F), as illustrated in Fig. 37. Although such a quenching treatment is not common practice, it might occur inadvertently; the critical aspect of quench aging is that it continues to raise the transition temperature for a period of several years, long after the steel might have passed all of the required impact tests.

Strain aging is common in sheet and plate steels that have been roller straightened or stretcher straightened; strains of only a few percent can produce noticeable strain aging. Both quench aging and strain aging can be controlled by alloying with those elements that form stable carbides and nitrides, that is, elements such as niobium, titanium, and aluminum.

Medium- or high-carbon steels are susceptible to two types of embrittlement associated with the tempering of martensitic structures: blue brittleness and temper embrittlement. Both are discussed below.

Blue Brittleness. Strain aging between 230 and 370 °C (450 and 700 °F) produces blue brittleness. Medium- or high-carbon steel tempered near 300 °C (575 °F) may absorb less energy during fracture than the same steel tempered at a lower temperature—or scarcely more energy than if not tempered at all. To avoid blue brittleness, these steels usually are not tempered between 230 and 370 °C (450 and 700 °F). Tetelman and McEvily (Ref 18) reported that the deformation or testing of low-carbon steels in the blue brittleness temperature range can cause a pronounced loss of toughness. This effect is an important consideration for steels to be used between 230 and 370 °C (450 and 700 °F).

Temper embrittlement can occur if a medium- or high-carbon alloy steel is tempered between 205 and 430 °C (400 and 800 °F) or cooled slowly through that range after tempering at a higher temperature. Steels containing nickel, chromium, manganese, phosphorus, tin, arsenic, or antimony are susceptible to temper embrittlement; Fig. 38 illustrates this effect in a straight chromium steel. Controlling the amount of residual elements present in the steel—particularly phosphorus, tin, arsenic, and antimony—is effective in preventing temper embrittlement.

Figure 39 shows the fracture surface of a broken Charpy bar from a temper-embrittled nickel-chromium-molybdenum steel. The facets shown in the fracture are due to the intergranular failure of prior-austenite grain boundaries. The residual elements mentioned above weaken the grain boundaries and cause embrittlement. An easy way to detect temper embrittlement is to examine the fracture surface of broken Charpy specimens using scanning electron microscopy.

Variability of Charpy Test Results

The Army Materials Testing Laboratory (formerly Watertown Arsenal Laboratory) conducted a closely controlled experiment that established the Charpy V-notch impact test as both reliable and reproducible.

Reproducibility of Test Results. A total of 1200 specimens from a single heat of aircraft quality 4340 steel were divided into three groups and heat treated to three different ranges of hardness: 43 to 46, 32.5 to 36.5,

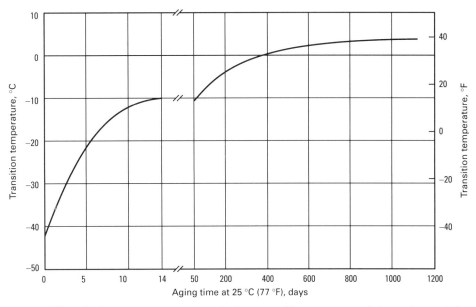

Fig. 37 Effect of aging at room temperature on the impact transition temperature of a low-carbon steel after quenching from 690 °C (1275 °F)

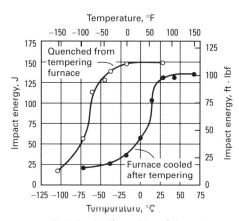

Fig. 38 Variation in Charpy V-notch impact energy with temperature for 5140 steel hardened and tempered at 620 °C (1150 °F). One series of specimens was quenched from tempering temperature; the other was furnace cooled. Slow cooling of susceptible steels causes temper embrittlement. Source: Ref 17

and 26 to 29 HRC. A total of 200 specimens at each hardness level were impact tested in each of two Charpy machines manufactured by two companies. The average impact energy values and distribution of results are shown in Fig. 40.

The test program described above clearly demonstrated the narrow spread of results that can be obtained under carefully controlled testing conditions. However, the experience of other laboratories indicates that even when the preparation and testing of impact specimens are closely controlled, a considerable spread of test results can still occur. When the effects of these variables are added to the inherent scatter that occurs among different heats of steel, the distribution of test results is broadened appreciably. Thus, judging notch toughness on the basis of one or two tests for a specific set of conditions is unwise without voluminous data on prior production heats of the material.

Some specifications designate that a specific number of specimens be tested at a particular temperature or over a particular range of temperatures. In other instances, the number of specimens with minimum or average values, or both, is negotiated. The accuracy and usefulness of results vary directly with the number of specimens tested.

Variability With Thickness and Grade. In addition to the variability in reproducibility of Charpy data during the test itself, there is variability in the steel being tested. The American Iron and Steel Institute (AISI)

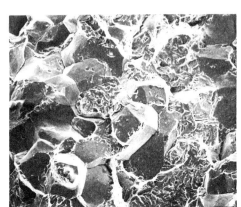

Fig. 39 SEM micrograph showing intergranular fracture in a nickel-chromium-molybdenum (HY80) steel. 315×

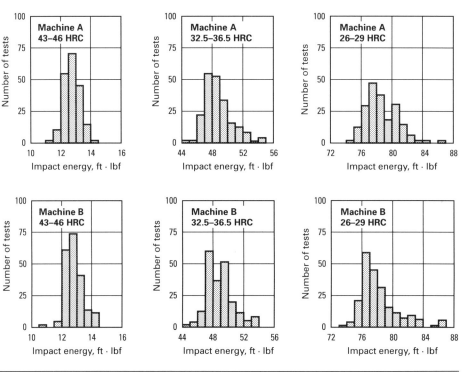

Machine	Charpy test average		
	43–46 HRC, ft · lbf	32.5–36.5 HRC, ft · lbf	26–29 HRC, ft · lbf
A	12.7	48.6	78.4
B	12.6	49.1	77.9

Fig. 40 Comparison of test results from two Charpy impact machines manufactured by two companies. All 1200 specimens were made from a single heat of aircraft quality 4340 steel. Specimens were hardened and tempered to three hardness levels: 43 to 46, 32.5 to 36.5, and 26 to 29 HRC. On each of the impact machines, 200 specimens at each of the three hardness levels were tested at 20 °C (70 °F). Source: Ref 19

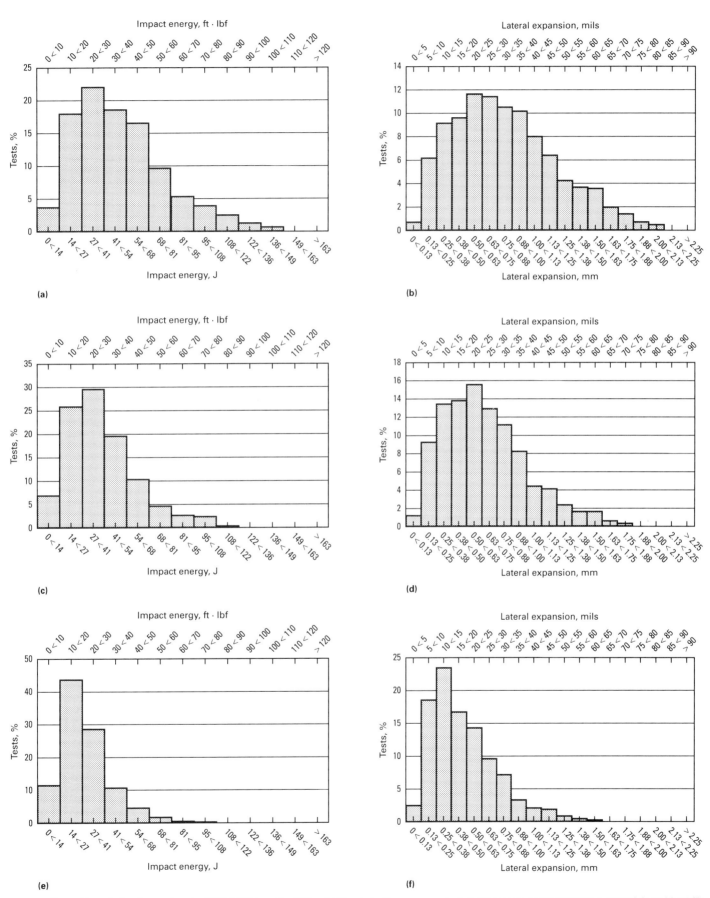

Fig. 41 Charpy V-notch data for steel grade ASTM A 572 from the AISI variability study. Coupons were tested at three selected temperatures. (a) and (b) 21 °C (70 °F). (c) and (d) 4 °C (40 °F). (e) and (f) −18 °C (0 °F)

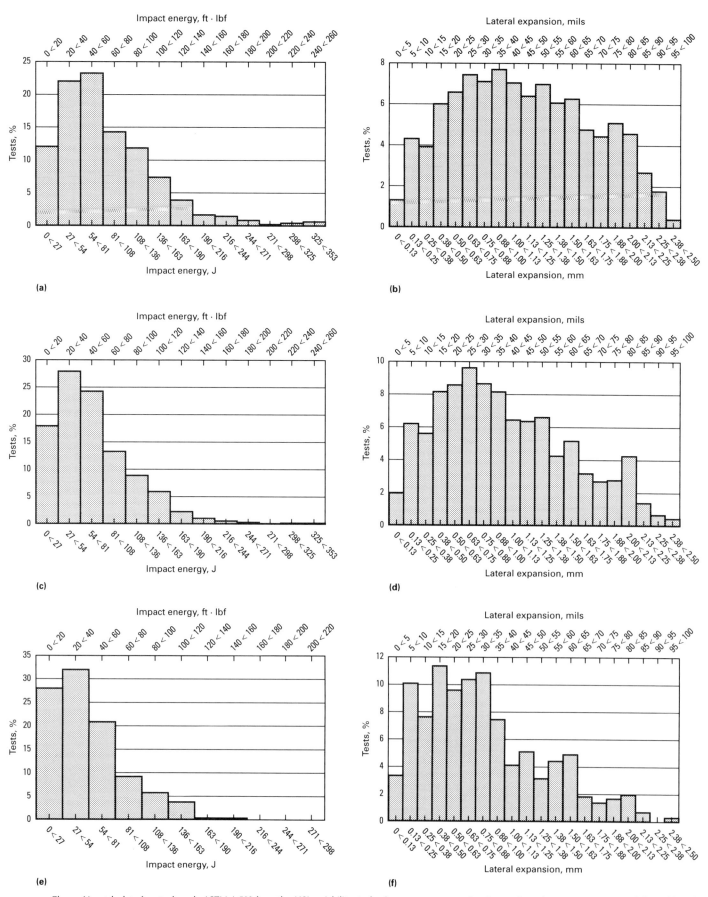

Fig. 42 Charpy V-notch data for steel grade ASTM A 588 from the AISI variability study. Coupons were tested at three selected temperatures. (a) and (b) 21 °C (70 °F). (c) and (d) 4 °C (40 °F). (e) and (f) −18 °C (0 °F)

Table 1 Statistical analysis of the process variability of ASTM A 572 and A 588 high-strength low-alloy steel plates using the Charpy impact test data in Fig. 41 and 42

Test temperature		Impact energy								Number of test locations
		Average value		Standard deviation		Maximum value		Minimum value		
°C	°F	J	ft · lbf	J	ft · lbf	J	ft · lbf	J	ft · lbf	
Grade A 572										
21	70	72.54	53.50	26.3	19.4	169.1	124.7	21.7	16.0	785
4	40	49.35	36.40	20.3	15.0	123.4	91.0	10.8	8.0	785
−18	0	28.70	21.17	15.2	11.2	104.4	77.0	6.4	4.7	785
Grade A 588										
21	70	115.54	85.22	61.3	45.2	347.1	256.0	15.3	11.3	417
4	40	85.31	62.92	53.6	39.5	393.2	290.0	7.2	5.3	417
−18	0	55.05	40.60	38.6	28.5	223.7	165.0	5.0	3.7	417

Test temperature		Lateral expansion								Number of test locations
		Average value		Standard deviation		Maximum value		Minimum value		
°C	°F	mm	mils	mm	mils	mm	mils	mm	mils	
Grade A 572										
21	70	1.16	45.79	0.37	14.7	2.4	92.7	0.3	13.0	785
4	40	0.82	32.33	0.33	12.9	1.8	71.3	0.2	9.0	785
−18	0	0.48	19.01	0.27	10.7	1.5	61.0	0.04	1.7	785
Grade A 588										
21	70	1.98	58.26	0.5	19.8	2.4	95.0	0.2	6.0	417
4	40	1.18	46.56	0.6	22.0	2.4	95.3	0.1	4.3	417
−18	0	0.82	32.28	0.5	20.5	2.4	95.0	0.01	0.5	417

has conducted two extensive programs to determine the variability of Charpy impact properties in steel plates (Ref 20, 21). The combined absorbed energy and lateral expansion data from both the 1979 and 1984 survey are shown in Fig. 41 for steel grade ASTM A 572 and in Fig. 42 for steel grade ASTM A 588. The thickness range includes plates from 9.53 to 101.6 mm (0.375 to 4.0 in.) thick. For grade A 572, 99 plates are included, and for grade A 588, 57 plates are included. At each of nine test locations, three Charpy specimens were tested at −18, 4, and 21 °C (0, 40, and 70 °F). The variability for all tests and thicknesses is summarized in Table 1.

Correlations of Notch Toughness With Other Mechanical Properties

The Charpy test is used worldwide to indicate the ductile-to-brittle transition of a steel. While Charpy results cannot be directly applied to structural design requirements, a number of correlations have been made between Charpy results and fracture toughness.

Charpy V-Notch Correlations to Fracture Mechanics. Fracture mechanics provides a calculation of tolerable crack size and shape for a specific material application. A designer can determine the allowable crack size a structure can tolerate at a specific design stress if the fracture toughness of a steel, the operating temperature, and loading rate are known. The design criteria for highway bridge and nuclear pressure vessel steels are partially based on Charpy correlations with fracture toughness. Examples of Charpy correlations with fracture toughness parameters are given in the article "Charpy Impact Testing" in Volume 8 of the 9th Edition of *Metals Handbook*.

For highway bridges, the American Association of State Highway and Transportation Officials (AASHTO) has adopted minimum Charpy energy requirements based on the minimum service temperatures expected for a bridge structure. For example, a 25 mm (1 in.) thick carbon steel ASTM A 36 plate would require 34 J (25 ft · lbf) at the following Charpy test temperatures:

- 21 °C (70 °F), zone 1; minimum bridge service temperature of −18 °C (0 °F) and above
- 4 °C (40 °F), zone 2; minimum bridge service temperature of −18 to −34 °C (−1 to −30 °F)
- −12 °C (10 °F), zone 3; minimum bridge service temperature of −35 to −51 °C (−31 to −60 °F)

A bridge constructed in Florida would be in zone 1, while a northern Minnesota bridge would require zone 3 testing. These AASHTO testing requirements include a temperature shift based on the difference in loading rate between the bridge structure and the Charpy test (Ref 22).

REFERENCES

1. W. Oldfield, Curve Fitting Impact Test Data, *ASTM Stand. News*, Vol 3 (No. 11), 1975, p 24-28
2. W.C. Leslie, *The Physical Metallurgy of Steels*, McGraw-Hill, 1981
3. F.B. Pickering, *Physical Metallurgy and the Design of Steels*, Applied Science, 1978
4. K.W. Burns and F.B. Pickering, Deformation and Fracture of Ferrite-Pearlite Structures, *J. Iron Steel Inst.*, Vol 202 (No. 11), Nov 1964, p 899-906
5. N.P. Allen et al., Tensile and Impact Properties of High-Purity Iron-Carbon and Iron-Carbon-Manganese Alloys of Low Carbon Content, *J. Iron Steel Inst.*, Vol 174, June 1953, p 108-120
6. J.A. Rineholt and W.J. Harris, Jr., Effect of Alloying Elements on Notch Toughness of Pearlitic Steels, *Trans. ASM*, Vol 43, 1951, p 1175-1214
7. C. Vishnevsky and E.A. Steigerwald, "Influence of Alloying Elements on the Toughness of Low-Alloy Martensitic High-Strength Steels," AAMRC CR-80-09(F), Army Materials and Mechanics Research Center, Nov 1968
8. R. Phillips, W.E. Duckworth, and F.E.L. Copley, Effect of Niobium and Tantalum on the Tensile and Impact Properties of Mild Steel, *J. Iron Steel Inst.*, Vol 202, July 1964, p 593-600
9. N.J. Petch, The Ductile-Cleavage Transition in Alpha-Iron, in *Fracture*, B.L. Averbach et al., Ed., Technology Press, 1959, p 54-67
10. R. Phillips and J.A. Chapman, Influence of Finish Rolling Temperature on the Mechanical Properties of Some Commercial Steels Rolled to 13/16 Diameter Bars, *J. Iron Steel Inst.*, Vol 204, 1966, p 615-622
11. P.P. Puzak, E.W. Eschbacher, and W.S. Pellini, Initiation and Propagation of Brittle Fracture in Structural Steels, *Weld. Res. Supp.*, Dec 1952, p 569s
12. W.S. Pellini, Evaluation of the Significance of Charpy Tests, in *Symposium on Effect of Temperature on the Brittle Behavior of Metals with Particular Reference to Low Temperatures*, STP 158, American Society for Testing and Materials, 1954, p 222; see also W.S. Pellini, "Evolution of Principles for Fracture-Safe Design of Steel Structures," NRL Report 6957, United States Naval Research Laboratory, Sept 1969, p 9
13. R.F. Hehemann, V.J. Luhan, and A.R. Troiano, The Influence of Bainite on Mechanical Properties, *Trans. ASM*, Vol 49, 1957, p 409-426
14. R.L. Bodnar, K.A. Taylor, K.S. Albano, and S.A. Heim, Improving the Toughness of 3½ NiCrMoV Steam Turbine Disk Forgings, *J. Eng. Mater. Technol. (Trans. ASME)*, Vol III, 1989, p 61
15. S.D. Antolovich, A. Saxens, and G.R. Chanani, Increased Fracture Toughness in a 300 Grade Maraging Steel as a Result of Thermal Cycling, *Metall. Trans.*, Vol 5, 1974, p 623
16. F.R. Larson and J. Nunes, Relation-

ships Between Energy, Fibrosity, and Temperature in Charpy Impact Tests on AISI 4340 Steel, *Proc. ASTM*, Vol 62, 1962, p 1192-1209
17. J.R. Low, Jr., The Effect of Quench-Aging on the Notch Sensitivity of Steel, *Weld. Res. Counc. Res. Rep.*, Vol 17, 1952, p 253s-256s
18. A.S. Tetelman and A.J. McEvily, Jr., *Fracture of Structural Materials*, John Wiley & Sons, 1967, p 512-514
19. D.E. Driscoll, Reproducibility of Charpy Impact Test, in *Symposium on Impact Testing*, STP 176, American Society for Testing and Materials, 1956, p 70-75
20. The Variations of Charpy V-Notch Impact Test Properties in Steel Plates, Publication SU/24, American Iron and Steel Institute, Jan 1979
21. The Variations in Charpy V-Notch Impact Properties in Steel Plates, Publication SU/27, American Iron and Steel Institute, Jan 1989
22. J.M. Barsom and S.T. Rolfe, *Fracture and Fatigue Control in Structures*, Prentice-Hall, 1987, p 526-537

Specialty Steels and Heat-Resistant Alloys

Wrought Tool Steels 757
P/M Tool Steels 780
Maraging Steels 793
Ferrous Powder Metallurgy Materials 801
Austenitic Manganese Steels 822
Wrought Stainless Steels 841
Cast Stainless Steels 908
Elevated-Temperature Properties of Stainless Steels 930
Wrought and P/M Superalloys...950
   Appendix: P/M Cobalt-Base Wear-Resistant Materials 977
Polycrystalline Cast Superalloys..981
Directionally Solidified and Single-Crystal Superalloys 995

Wrought Tool Steels

Revised by Alan M. Bayer, Teledyne Vasco, and Lee R. Walton, Latrobe Steel Company

A TOOL STEEL is any steel used to make tools for cutting, forming, or otherwise shaping a material into a part or component adapted to a definite use. The earliest tool steels were simple, plain carbon steels, but by 1868 and increasingly in the early 20th century, many complex, highly alloyed tool steels were developed. These complex alloy tool steels, which contain, among other elements, relatively large amounts of tungsten, molybdenum, vanadium, manganese, and chromium, make it possible to meet increasingly severe service demands and to provide greater dimensional control and freedom from cracking during heat treatment. Many alloy tool steels are also widely used for machinery components and structural applications in which particularly stringent requirements must be met, for example, high-temperature springs, ultrahigh-strength fasteners, special-purpose valves, and bearings of various types for elevated-temperature service.

In service, most tools are subjected to extremely high loads that are applied rapidly. The tools must withstand these loads a great number of times without breaking and without undergoing excessive wear or deformation. In many applications, tool steels must provide this capability under conditions that develop high temperatures in the tool. No single tool material combines maximum wear resistance, toughness, and resistance to softening at elevated temperatures. Consequently, the selection of the proper tool material for a given application often requires a trade-off to achieve the optimum combination of properties.

Most tool steels are wrought products, but precision castings can be used to advantage in some applications. The powder metallurgy (P/M) process is also used in making tool steels. It provides, first, a more uniform carbide size and distribution in large sections and, second, special compositions that are difficult or impossible to produce by melting and casting and then mechanically working the cast product.

For typical wrought tool steels, raw materials (including scrap) are carefully selected, not only for alloy content, but also for qualities that ensure cleanliness and homogeneity in the finished product. Tool steels are generally melted in relatively small-tonnage electric arc furnaces and refined in an argon oxygen decarburization (AOD) vessel to achieve composition tolerances at low cost, good cleanliness, and precise control of melting conditions. Special refining and secondary remelting processes have been introduced to satisfy particularly difficult demands regarding tool steel quality and performance. The medium-to-high alloy contents of many tool steels require careful control of forging and rolling, which often results in a large amount of process scrap. Semifinished and finished bars are given rigorous in-process and final inspection. This inspection can be so extensive that both ends of each bar may be inspected for macrostructure (etch quality), cleanliness, hardness, grain size, annealed structure, and hardening ability. Inspection may also require that the entire bar be subjected to magnetic and ultrasonic inspections for surface and internal discontinuities (see the articles "Magnetic Particle Inspection" and "Ultrasonic Inspection" in Volume 17 of the 9th Edition of *Metals Handbook*). It is important that finished tool steel bars have minimal decarburization within carefully controlled limits, which requires that annealing be done by special procedures under closely controlled conditions.

Such precise production practices and stringent quality controls contribute to the high cost of tool steels, as do the expensive alloying elements they contain. Insistence on quality in the manufacture of these specialty steels is justified, however, because tool steel bars generally are made into complicated cutting and forming tools worth many times the cost of the steel itself. Although some standard constructional alloy steels resemble tool steels in composition, they are seldom used for expensive tooling because, in general, they are not manufactured to the same rigorous quality standards as are tool steels.

The performance of a tool in service depends on the proper design of the tool, accuracy with which the tool is made, selection of the proper tool steel, and application of the proper heat treatment. A tool can perform successfully in service only when all four of these requirements have been fulfilled.

With few exceptions, all tool steels must be heat treated to develop specific combinations of wear resistance, resistance to deformation or breaking under high loads, and resistance to softening at elevated temperatures. Some tool steels are available as prehardened bar or other products. A few simple shapes may also be obtained directly from tool steel producers in correctly heat-treated condition. However, most tool steels are first formed or machined to produce the required shape and then heat treated by the tool manufacturer or ultimate user.

Classification and Characteristics

Table 1 gives composition limits for the tool steels most commonly used in 1989. Each group of tool steels of similar composition and properties is identified by a capital letter; within each group, individual tool steel types are assigned code numbers. Table 2 cross references U.S. tool steel designations with their foreign equivalents. Table 3 identifies tool steel types that have been dropped from active listings because they are no longer commonly used.

Tool steels are produced to various standards including several American Society for Testing and Materials (ASTM) specifications. Reference 3 contains much useful information that essentially represents the normal manufacturing practices of most of the tool steel producers. Frequently, more stringent chemical and/or metallurgical standards are invoked by the individual producers or consumers to achieve certain commercial goals. Where appropriate, standard specifications for tool steels—ASTM A 600, A 681, and A 686—may be used as a basis for procurement. ASTM A 600 sets forth standard requirements for both tungsten and molybdenum high-speed steels; A 681 is applicable to hot-work, cold-work, shock-resisting, special-purpose, and mold steels; A 686 covers water-hardening tool steels. In many instances, however, tool steels are purchased by trade name because

Table 1 Composition limits of principal types of tool steels

Designation AISI	UNS	C	Mn	Si	Cr	Ni	Mo	W	V	Co
Molybdenum high-speed steels										
M1	T11301	0.78–0.88	0.15–0.40	0.20–0.50	3.50–4.00	0.30 max	8.20–9.20	1.40–2.10	1.00–1.35	...
M2	T11302	0.78–0.88; 0.95–1.05	0.15–0.40	0.20–0.45	3.75–4.50	0.30 max	4.50–5.50	5.50–6.75	1.75–2.20	...
M3, class 1	T11313	1.00–1.10	0.15–0.40	0.20–0.45	3.75–4.50	0.30 max	4.75–6.50	5.00–6.75	2.25–2.75	...
M3, class 2	T11323	1.15–1.25	0.15–0.40	0.20–0.45	3.75–4.50	0.30 max	4.75–6.50	5.00–6.75	2.75–3.75	...
M4	T11304	1.25–1.40	0.15–0.40	0.20–0.45	3.75–4.75	0.30 max	4.25–5.50	5.25–6.50	3.75–4.50	...
M7	T11307	0.97–1.05	0.15–0.40	0.20–0.55	3.50–4.00	0.30 max	8.20–9.20	1.40–2.10	1.75–2.25	...
M10	T11310	0.84–0.94; 0.95–1.05	0.10–0.40	0.20–0.45	3.75–4.50	0.30 max	7.75–8.50	...	1.80–2.20	...
M30	T11330	0.75–0.85	0.15–0.40	0.20–0.45	3.50–4.25	0.30 max	7.75–9.00	1.30–2.30	1.00–1.40	4.50–5.50
M33	T11333	0.85–0.92	0.15–0.40	0.15–0.50	3.50–4.00	0.30 max	9.00–10.00	1.30–2.10	1.00–1.35	7.75–8.75
M34	T11334	0.85–0.92	0.15–0.40	0.20–0.45	3.50–4.00	0.30 max	7.75–9.20	1.40–2.10	1.90–2.30	7.75–8.75
M35	T11335	0.82–0.88	0.15–0.40	0.20–0.45	3.75–4.50	0.30 max	4.50–5.50	5.50–6.75	1.75–2.20	4.50–5.50
M36	T11336	0.80–0.90	0.15–0.40	0.20–0.45	3.75–4.50	0.30 max	4.50–5.50	5.50–6.50	1.75–2.25	7.75–8.75
M41	T11341	1.05–1.15	0.20–0.60	0.15–0.50	3.75–4.50	0.30 max	3.25–4.25	6.25–7.00	1.75–2.25	4.75–5.75
M42	T11342	1.05–1.15	0.15–0.40	0.15–0.65	3.50–4.25	0.30 max	9.00–10.00	1.15–1.85	0.95–1.35	7.75–8.75
M43	T11343	1.15–1.25	0.20–0.40	0.15–0.65	3.50–4.25	0.30 max	7.50–8.50	2.25–3.00	1.50–1.75	7.75–8.75
M44	T11344	1.10–1.20	0.20–0.40	0.30–0.55	4.00–4.75	0.30 max	6.00–7.00	5.00–5.75	1.85–2.20	11.00–12.25
M46	T11346	1.22–1.30	0.20–0.40	0.40–0.65	3.70–4.20	0.30 max	8.00–8.50	1.90–2.20	3.00–3.30	7.80–8.80
M47	T11347	1.05–1.15	0.15–0.40	0.20–0.45	3.50–4.00	0.30 max	9.25–10.00	1.30–1.80	1.15–1.35	4.75–5.25
M48	T11348	1.42–1.52	0.15–0.40	0.15–0.40	3.50–4.00	0.30 max	4.75–5.50	9.50–10.50	2.75–3.25	8.00–10.00
M62	T11362	1.25–1.35	0.15–0.40	0.15–0.40	3.50–4.00	0.30 max	10.00–11.00	5.75–6.50	1.80–2.10	...
Tungsten high-speed steels										
T1	T12001	0.65–0.80	0.10–0.40	0.20–0.40	3.75–4.50	0.30 max	...	17.25–18.75	0.90–1.30	...
T2	T12002	0.80–0.90	0.20–0.40	0.20–0.40	3.75–4.50	0.30 max	1.00 max	17.50–19.00	1.80–2.40	...
T4	T12004	0.70–0.80	0.10–0.40	0.20–0.40	3.75–4.50	0.30 max	0.40–1.00	17.50–19.00	0.80–1.20	4.25–5.75
T5	T12005	0.75–0.85	0.20–0.40	0.20–0.40	3.75–5.00	0.30 max	0.50–1.25	17.50–19.00	1.80–2.40	7.00–9.50
T6	T12006	0.75–0.85	0.20–0.40	0.20–0.40	4.00–4.75	0.30 max	0.40–1.00	18.50–21.00	1.50–2.10	11.00–13.00
T8	T12008	0.75–0.85	0.20–0.40	0.20–0.40	3.75–4.50	0.30 max	0.40–1.00	13.25–14.75	1.80–2.40	4.25–5.75
T15	T12015	1.50–1.60	0.15–0.40	0.15–0.40	3.75–5.00	0.30 max	1.00 max	11.75–13.00	4.50–5.25	4.75–5.25
Intermediate high-speed steels										
M50	T11350	0.78–0.88	0.15–0.45	0.20–0.60	3.75–4.50	0.30 max	3.90–4.75	...	0.80–1.25	...
M52	T11352	0.85–0.95	0.15–0.45	0.20–0.60	3.50–4.30	0.30 max	4.00–4.90	0.75–1.50	1.65–2.25	...
Chromium hot-work steels										
H10	T20810	0.35–0.45	0.25–0.70	0.80–1.20	3.00–3.75	0.30 max	2.00–3.00	...	0.25–0.75	...
H11	T20811	0.33–0.43	0.20–0.50	0.80–1.20	4.75–5.50	0.30 max	1.10–1.60	...	0.30–0.60	...
H12	T20812	0.30–0.40	0.20–0.50	0.80–1.20	4.75–5.50	0.30 max	1.25–1.75	1.00–1.70	0.50 max	...
H13	T20813	0.32–0.45	0.20–0.50	0.80–1.20	4.75–5.50	0.30 max	1.10–1.75	...	0.80–1.20	...
H14	T20814	0.35–0.45	0.20–0.50	0.80–1.20	4.75–5.50	0.30 max	...	4.00–5.25	...	...
H19	T20819	0.32–0.45	0.20–0.50	0.20–0.50	4.00–4.75	0.30 max	0.30–0.55	3.75–4.50	1.75–2.20	4.00–4.50
Tungsten hot-work steels										
H21	T20821	0.26–0.36	0.15–0.40	0.15–0.50	3.00–3.75	0.30 max	...	8.50–10.00	0.30–0.60	...
H22	T20822	0.30–0.40	0.15–0.40	0.15–0.40	1.75–3.75	0.30 max	...	10.00–11.75	0.25–0.50	...
H23	T20823	0.25–0.35	0.15–0.40	0.15–0.60	11.00–12.75	0.30 max	...	11.00–12.75	0.75–1.25	...
H24	T20824	0.42–0.53	0.15–0.40	0.15–0.40	2.50–3.50	0.30 max	...	14.00–16.00	0.40–0.60	...
H25	T20825	0.22–0.32	0.15–0.40	0.15–0.40	3.75–4.50	0.30 max	...	14.00–16.00	0.40–0.60	...
H26	T20826	0.45–0.55(b)	0.15–0.40	0.15–0.40	3.75–4.50	0.30 max	...	17.25–19.00	0.75–1.25	...
Molybdenum hot-work steels										
H42	T20842	0.55–0.70(b)	0.15–0.40	...	3.75–4.50	0.30 max	4.50–5.50	5.50–6.75	1.75–2.20	...
Air-hardening, medium-alloy, cold-work steels										
A2	T30102	0.95–1.05	1.00 max	0.50 max	4.75–5.50	0.30 max	0.90–1.40	...	0.15–0.50	...
A3	T30103	1.20–1.30	0.40–0.60	0.50 max	4.75–5.50	0.30 max	0.90–1.40	...	0.80–1.40	...
A4	T30104	0.95–1.05	1.80–2.20	0.50 max	0.90–2.20	0.30 max	0.90–1.40	...	...	...
A6	T30106	0.65–0.75	1.80–2.50	0.50 max	0.90–1.20	0.30 max	0.90–1.40	...	...	...
A7	T30107	2.00–2.85	0.80 max	0.50 max	5.00–5.75	0.30 max	0.90–1.40	0.50–1.50	3.90–5.15	...
A8	T30108	0.50–0.60	0.50 max	0.75–1.10	4.75–5.50	0.30 max	1.15–1.65	1.00–1.50	...	...
A9	T30109	0.45–0.55	0.50 max	0.95–1.15	4.75–5.50	1.25–1.75	1.30–1.80	...	0.80–1.40	...
A10	T30110	1.25–1.50(c)	1.60–2.10	1.00–1.50	...	1.55–2.05	1.25–1.75	...	...	...
High-carbon, high-chromium, cold-work steels										
D2	T30402	1.40–1.60	0.60 max	0.60 max	11.00–13.00	0.30 max	0.70–1.20	...	1.10 max	...
D3	T30403	2.00–2.35	0.60 max	0.60 max	11.00–13.50	0.30 max	...	1.00 max	1.00 max	...
D4	T30404	2.05–2.40	0.60 max	0.60 max	11.00–13.00	0.30 max	0.70–1.20	...	1.00 max	...
D5	T30405	1.40–1.60	0.60 max	0.60 max	11.00–13.00	0.30 max	0.70–1.20	...	1.00 max	2.50–3.50
D7	T30407	2.15–2.50	0.60 max	0.60 max	11.50–13.50	0.30 max	0.70–1.20	...	3.80–4.40	...
Oil-hardening cold-work steels										
O1	T31501	0.85–1.00	1.00–1.40	0.50 max	0.40–0.60	0.30 max	...	0.40–0.60	0.30 max	...
O2	T31502	0.85–0.95	1.40–1.80	0.50 max	0.50 max	0.30 max	0.30 max	...	0.30 max	...
O6	T31506	1.25–1.55(c)	0.30–1.10	0.55–1.50	0.30 max	0.30 max	0.20–0.30	...	...	...
O7	T31507	1.10–1.30	1.00 max	0.60 max	0.35–0.85	0.30 max	0.30 max	1.00–2.00	0.40 max	...

(continued)

(a) All steels except group W contain 0.25 max Cu, 0.03 max P, and 0.03 max S; group W steels contain 0.20 max Cu, 0.025 max P, and 0.025 max S. Where specified, sulfur may be increased to 0.06 to 0.15% to improve machinability of group A, D, H, M, and T steels. (b) Available in several carbon ranges. (c) Contains free graphite in the microstructure. (d) Optional. (e) Specified carbon ranges are designated by suffix numbers.

Table 1 (continued)

Designation		Composition(a), %								
AISI	UNS	C	Mn	Si	Cr	Ni	Mo	W	V	Co
Shock-resisting steels										
S1	T41901	0.40–0.55	0.10–0.40	0.15–1.20	1.00–1.80	0.30 max	0.50 max	1.50–3.00	0.15–0.30	...
S2	T41902	0.40–0.55	0.30–0.50	0.90–1.20	...	0.30 max	0.30–0.60	...	0.50 max	...
S5	T41905	0.50–0.65	0.60–1.00	1.75–2.25	...	0.50 max	0.20–1.35	...	0.35 max	...
S6	T41906	0.40–0.50	1.20–1.50	2.00–2.50	1.20–1.50	...	0.30–0.50	...	0.20–0.40	...
S7	T41907	0.45–0.55	0.20–0.90	0.20–1.00	3.00–3.50	...	1.30–1.80	...	0.20–0.30(d)	...
Low-alloy special-purpose tool steels										
L2	T61202	0.45–1.00(b)	0.10–0.90	0.50 max	0.70–1.20	...	0.25 max	...	0.10–0.30	...
L6	T61206	0.65–0.75	0.25–0.80	0.50 max	0.60–1.20	1.25–2.00	0.50 max	...	0.20–0.30(d)	...
Low-carbon mold steels										
P2	T51602	0.10 max	0.10–0.40	0.10–0.40	0.75–1.25	0.10–0.50	0.15–0.40	...	...	...
P3	T51603	0.10 max	0.20–0.60	0.40 max	0.40–0.75	1.00–1.50	...	...	...	...
P4	T51604	0.12 max	0.20–0.60	0.10–0.40	4.00–5.25	...	0.40–1.00	...	...	...
P5	T51605	0.10 max	0.20–0.60	0.40 max	2.00–2.50	0.35 max	...	...	...	...
P6	T51606	0.05–0.15	0.35–0.70	0.10–0.40	1.25–1.75	3.25–3.75	...	...	...	...
P20	T51620	0.28–0.40	0.60–1.00	0.20–0.80	1.40–2.00	...	0.30–0.55	...	...	...
P21	T51621	0.18–0.22	0.20–0.40	0.20–0.40	0.50 max	3.90–4.25	...	...	0.15–0.25	1.05–1.25Al
Water-hardening tool steels										
W1	T72301	0.70–1.50(e)	0.10–0.40	0.10–0.40	0.15 max	0.20 max	0.10 max	0.15 max	0.10 max	...
W2	T72302	0.85–1.50(e)	0.10–0.40	0.10–0.40	0.15 max	0.20 max	0.10 max	0.15 max	0.15–0.35	...
W5	T72305	1.05–1.15	0.10–0.40	0.10–0.40	0.40–0.60	0.20 max	0.10 max	0.15 max	0.10 max	...

(a) All steels except group W contain 0.25 max Cu, 0.03 max P, and 0.03 max S; group W steels contain 0.20 max Cu, 0.025 max P, and 0.025 max S. Where specified, sulfur may be increased to 0.06 to 0.15% to improve machinability of group A, D, H, M, and T steels. (b) Available in several carbon ranges. (c) Contains free graphite in the microstructure. (d) Optional. (e) Specified carbon ranges are designated by suffix numbers.

the user has found that a particular tool steel from a certain producer gives better performance in a specific application than does a tool steel of the same AISI type classification purchased from another source. Table 4 categorizes tool steels on the basis of specific machining applications.

High-Speed Steels

High-speed steels are tool materials developed largely for use in high-speed cutting tool applications. A chronology of some of the significant breakthroughs in high-speed tool steel technology is given in Table 5. There are two classifications of high-speed steel: molybdenum high-speed steels, or group M, and tungsten high-speed steels, or group T. Group M steels constitute greater than 95% of all high-speed steel produced in the United States. There is also a subgroup consisting of intermediate high-speed steels in the M group.

Group M and group T high-speed steels are equivalent in performance; the main advantage of the group M steels is lower initial cost (approximately 40% lower than that of similar group T steels). This difference in cost results from the lower atomic weight of molybdenum, about one-half that of tungsten. Based on weight percent, only about one-half as much molybdenum as tungsten is required to provide the same atom ratio.

Molybdenum high-speed steels and tungsten high-speed steels are similar in many other respects, including hardening ability. Typical applications for both categories include cutting tools of all sorts, such as drills, reamers, end mills, milling cutters, taps, and hobs (see the Section "Traditional Machining Processes" in Volume 16 of the 9th Edition of *Metals Handbook*). Some grades are satisfactory for cold-work applications, such as cold-header die inserts, thread-rolling dies, punches, and blanking dies. Steels of the M40 series are used to make cutting tools for machining modern, very tough, high-strength steels.

For die inserts and punches, high-speed steels frequently are under hardened, that is, quenched from austenitizing temperatures lower than those recommended for cutting tool applications, as a means of increasing toughness.

Molybdenum high-speed steels contain molybdenum, tungsten, chromium, vanadium, cobalt, and carbon as principal alloying elements. Group M steels have slightly greater toughness than group T steels at the same hardness. Otherwise, mechanical properties of the two groups are similar.

Increasing the carbon and vanadium contents of group M steels increases wear resistance; increasing the cobalt content improves red hardness (that is, the capability of certain steels to resist softening at temperatures high enough to cause the steel to emit radiation in the red part of the visible spectrum) but simultaneously lowers toughness. Type M2 and other grades in the M group have unusually high resistance to softening at elevated temperatures as a result of high alloy content (Fig. 1).

Because group M steels readily decarburize and can be damaged from overheating under adverse austenitizing environments, they are more sensitive than group T steels to hardening conditions, particularly austenitizing temperature and atmosphere. This is especially true of high-molybdenum, low-tungsten compositions.

Group M high-speed steels are deep hardening. They must be austenitized at temperatures lower than those for hardening group T steels to avoid incipient melting. Group M high-speed steels can develop full hardness when quenched from temperatures of 1175 to 1230 °C (2150 to 2250 °F).

The maximum hardness that can be obtained in group M high-speed steels varies with composition. For those with lower carbon contents, that is, types M1, M2, M10 (low-carbon composition), M30, M33, M34, and M36, maximum hardness is usually 65 HRC. For higher carbon contents, including types M3, M4, and M7, maximum hardness is about 66 HRC. Maximum hardness of the higher-carbon cobalt-containing steels, that is, types M41, M42, M43, M44, and M46, is 69 to 70 HRC. However, few industrial applications exist for steels of the M40 series at this maximum hardness. Usually, the heat treatment is adjusted to provide a hardness of 66 to 68 HRC.

Tungsten high-speed steels contain tungsten, chromium, vanadium, cobalt, and carbon as the principal alloying elements. Type T1 was developed partly as a result of the work of Taylor and White, who in the early 1900s, found that certain steels with more than 14% W, about 4% Cr, and about 0.3% V exhibited red hardness. In its earliest form, type T1 contained about 0.68% C, 18% W, 4% Cr, and 0.3% V. By 1920, the vanadium content had been increased to about 1.0%. Over a 30-year period, the

760 / Specialty Steels and Heat-Resistant Alloys

Table 2 Cross reference to tool steels. Similar specifications for tool steels established by the United States, West Germany, Japan, Great Britain, France, and Sweden are presented below. Exact chemical compositions for the non-U.S. tool steels can be found in Ref 1 and 2.

United States (AISI)	West Germany (DIN)(a)	Japan (JIS)(b)	Great Britain (B.S.)(c)	France (AFNOR)(d)	Sweden (SS$_{14}$)
Molybdenum high-speed steels (ASTM A 600)					
M1	1.3346	...	4659 BM1	A35-590 4441 Z85DCWV08-04-02-01	2715
M2, reg C	1.3341, 1.3343, 1.3345, 1.3553, 1.3554	G4403 SKH51 (SKH9)	4659 BM2	A35-590 4301 Z85WDCV06-05-04-02	2722
M2, high C	1.3340, 1.3342	...	...	A35-590 4302 Z90WDCV06-05-04-02	...
M3, class 1	...	G4403 SKH52	...	...	...
M3, class 2	1.3344	G4403 SKH53	...	A35-590 4360 Z120 WDCV06-05-04-03	(USA M3 class 2)
M4	...	G4403 SKH54	4659 BM4	A35-590 4361 Z130 WDCV06-05-04-04	...
M7	1.3348	G4403 SKH58	...	A35-590 4442 Z100DCWV09-04-02-02	2782
M10, reg C	...	...	...	...	...
M10, high C	...	...	...	...	...
M30	1.3249	...	4659 BM34	...	...
M33	1.3249	...	4659 BM34	...	...
M34	1.3249	...	4659 BM34	...	...
M35	1.3243	G4403 SKH55	...	A35-590 4371 Z85WDKCV06-05-05-04-02 A35-590 4372 Z90WDKCV06-05-05-04-02	...
M36	1.3243	G4403 SKH55 G4403 SKH56	...	A35-590 4371 Z85WDKCV06-05-05-04	2723
M41	1.3245, 1.3246	G4403 SKH55	...	A35-590 4374 Z110WKCDV07-05-04-04	2736
M42	1.3247	G4403 SKH59	4659 BM42	A35-590 4475 Z110DKCWV09-08-04-02	...
M43	...	...	...	A35-590 4475 Z110DKCWV09-08-04-02-01	...
M44	1.3207	G4403 SKH57	4659 (USA M44)	A35-590 4376 Z130KWDCV12-07-06-04-03	...
M46	1.3247	...	...	...	...
M47	1.3247	...	...	...	...
Intermediate high-speed steels					
M50	1.2369, 1.3551	...	...	A35-590 3551 Y80DCV42.16	(USA M50)
M52	...	...	...	...	...
Tungsten high-speed steels (ASTM A 600)					
T1	1.3355, 1.3558	G4403 SKH2	4659 BT1	A35-590 4201 Z80WCV18-04-01	...
T2	...	...	4659 BT2 4659 BT20	4203 18-0-2	...
T4	1.3255	G4403 SKH3	4659 BT4	A35-590 4271 Z80WKCV18-05-04-01	...
T5	1.3265	G4403 SKH4	4659 BT5	A35-590 4275 Z80WKCV18-10-04-02	(USA T5)
T6	1.3257	G4403 SKH4B	4659 BT6	...	...
T8	...	...	...	...	...
T15	1.3202	G4403 SKH10	4659 BT15	A35-590 4171 Z160WKVC12-05-05-04	(USA T15)
Chromium hot-work steels (ASTM A 681)					
H10	1.2365, 1.2367	G4404 SKD7	4659 BH10	A35-590 3451 32DCV28	...
H11	1.2343, 1.7783, 1.7784	G4404 SKD6	4659 BH11	A35-590 3431 FZ38CDV5	...
H12	1.2606	G4404 SKD62	4659 BH12	A35-590 3432 Z35CWDV5	...
H13	1.2344	G4404 SKD61	4659 BH13 4659 H13	A35-590 3433 Z40CDV5	2242
H14	1.2567	G4404 SKD4	...	3541 Z40WCV5	...
H19	1.2678	G4404 SKD8	4659 BH19	...	...
Tungsten hot-work steels (ASTM A 681)					
H21	1.2581	G4404 SKD5	4659 BH21 4659 H21A	A35-590 3543 Z30WCV9	2730
H22	1.2581	G4404 SKD5	...	...	...
H23	1.2625	...	...	...	...
H24	...	...	...	...	...
H25	...	...	...	...	...
H26	...	...	4659 BH26	...	...
Molybdenum hot-work steels (ASTM A 681)					
H42	...	...	...	3548 Z65WDCV6.05	...
Air-hardening, medium-alloy, cold-work steels (ASTM A 681)					
A2	1.2363	G4404 SKD12	4659 BA2	A35-590 2231 Z100CDV5	2260
A3	...	...	...	...	...
A4	...	...	...	...	...
A5	...	...	...	...	...
A6	...	...	4659 BA6	...	...
A7	...	...	...	...	...
A8	1.2606	G4404 SKD62	...	3432 Z38CDWV5	...
A9	...	...	...	...	...
A10	...	...	...	...	...
High-carbon, high-chromium, cold-work steels (ASTM A 681)					
D2	1.2201, 1.2379, 1.2601	G4404 SKD11	4659 (USA D2) 4659 BD2 4659 BD2A	A35-590 2235 Z160CDV12	2310
D3	1.2080, 1.2436, 1.2884	G4404 SKD1 G4404 SKD2	4659 BD3	A35-590 2233 Z200C12	...
D4	1.2436, 1.2884	G4404 SKD2	4659 (USA D4)	A35-590 2234 Z200CD12	2312

(continued)

(a) Deutsche Industrie Normen (German Industrial Standards). (b) Japanese Industrial Standard. (c) British Standard. (d) l'Association Française de Normalisation (French Standards Association). Source: Ref 1, 2

Table 2 (continued)

United States (AISI)	West Germany (DIN)(a)	Japan (JIS)(b)	Great Britain (B.S.)(c)	France (AFNOR)(d)	Sweden (SS₁₄)
High-carbon, high-chromium, cold-work steels (ASTM A 681) (continued)					
D5	1.2880	...	...	A35-590 2236 Z160CKDV12.03	...
D7	1.2378	...	...	2237 Z230CVA12.04	...
Oil-hardening cold-work steels (ASTM A 681)					
O1	1.2510	G4404 SKS21 G4404 SKS3 G4404 SKS93 G4404 SKS94 G4404 SKS95	4659 BO1	A35-590 2212 90 MWCV5	2140
O2	1.2842	...	4659 (USA O2) 4659 BO2	A35-590 2211 90MV8	...
O6	1.2206	...	...	A35-590 2132 130C3	...
O7	1.2414, 1.2419, 1.2442, 1.2516, 1.2519	G4404 SKS2	...	A35-590 2141 105WC13	...
Shock-resisting steels (ASTM A 681)					
S1	1.2542, 1.2550	G4404 SKS41	4659 BS1	A35-590 2341 55WC20	2710
S2	1.2103	...	4659 BS2	A35-590 2324 Y45SCD6	...
S5	1.2823	...	4659 BS5	...	...
S6	...	...	...	...	...
S7	...	...	...	...	...
Low-alloy special-purpose steels (ASTM A 681)					
L2	1.2235, 1.2241, 1.2242, 1.2243	G4404 SKT3 G4410 SKC11	...	A35-590 3335 55CNDV4	...
L6	1.2713, 1.2714	G4404 SKS51 G4404 SKT4	...	A35-590 3381 55NCDV7	...
Low-carbon mold steels (ASTM A 681)					
P2	...	...	...	...	...
P3	1.5713	...	...	2881 Y10NC6	...
P4	1.2341	...	...	...	(USA P4)
P5	...	...	...	...	...
P6	1.2735, 1.2745	G4410 SKC31	...	2882 10NC12	...
P20	1.2311, 1.2328, 1.2330	...	4659 (USA P20)	A35-590 2333 35CMD7	(USA P20)
P21	...	...	...	...	...
Water-hardening steels (ASTM A 686)					
W1	1.1525, 1.1545, 1.1625, 1.1654, 1.1663, 1.1673, 1.1744, 1.1750, 1.1820, 1.1830	G4401 SK1 G4401 SK2 G4401 SK3 G4401 SK4 G4401 SK5 G4401 SK6 G4401 SK7 G4410 SKC3	4659 (USA W1) 4659 BW1A 4659 BW1B 4659 BW1C	A35-590 1102 Y(1) 105 A35-590 1103 Y(1) 90 A35-590 1104 Y(1) 80 A35-590 1105 Y(1) 70 A35-590 1200 Y(2) 140 A35-590 1201 Y(2) 120 A35-596 Y75 A35-596 Y90	...
W2	1.1645, 1.2206, 1.2833	G4404 SKS43 G4404 SKS44	4659 BW2	A35-590 1161 Y120V A35-590 1162 Y105V A35-590 1163 Y90V A35-590 1164 Y75V A35-590 1230 Y(2) 140C A35-590 2130 Y100C2	(USA W2A) (USA W2B) (USA W2C)
W5	1.2002, 1.2004, 1.2056	...	...	A35-590 1232 Y105C	...

(a) Deutsche Industrie Normen (German Industrial Standards). (b) Japanese Industrial Standard. (c) British Standard. (d) l'Association Française de Normalisation (French Standards Association). Source: Ref 1, 2

carbon content was gradually increased to its present level of 0.75%.

Group T high-speed steels are characterized by high red hardness and wear resistance. They are so deep hardening that sections up to 76 mm (3 in.) in thickness or diameter can be hardened to 65 HRC or more by quenching in oil or molten salt. The high alloy and high carbon contents produce a large number of hard, wear-resistant carbides in the microstructure, particularly in those types containing more than 1.5% V and more than 1.0% C. Type T15 is the most wear-resistant steel of this group.

The combination of good wear resistance and high red hardness makes group T high-speed steels suitable for many high-performance cutting tool applications; their toughness allows them to outperform cemented carbides in delicate tools and interrupted-cut applications. Group T high-speed steels are primarily used for cutting tools such as bits, drills, reamers, taps, broaches, milling cutters, and hobs. These steels are also used for making dies, punches, and high-load high-temperature structural components such as aircraft bearings and pump parts.

Group T high-speed steels are all deep hardening when quenched from their recommended hardening temperatures of 1205 to 1300 °C (2200 to 2375 °F). They are seldom used to make hardened tools with section sizes greater than 76 mm (3 in.). Even very large cutting tools, such as drills 76 and 102 mm (3 and 4 in.) in diameter, have relatively small effective sections for hardening because metal has been removed to form the flutes. Some large-diameter solid tools are made from group T high-speed steels; these include broaches and cold extrusion punches as large as 102 to 127 mm (4 to 5 in.) in diameter.

As shown in Fig. 2, the difference between surface hardness and center hardness varies with bar size. The data in Fig. 2 are given to indicate the general trend of hardness variation rather than to provide specific values. The section size and total mass of a given tool often have an effect on its response to a given hardening treatment that is equal to or greater than the effect of the grade of tool steel selected. For tools of extremely large diameter or heavy section, it is relatively common practice to use an accelerated oil quench to provide full hardness. This practice may yield values of

762 / Specialty Steels and Heat-Resistant Alloys

Table 3 Compositions of tool steels no longer in common use

Type	_____ Composition, % _____					
	C	W	Mo	Cr	V	Others
High-speed steels						
M6	0.80	4.25	5.00	4.00	1.50	12.00 Co
M8	0.80	5.00	5.00	4.00	1.50	1.25 Nb
M15	1.50	6.50	3.50	4.00	5.00	5.00 Co
M45	1.25	8.00	5.00	4.25	1.60	5.50 Co
T3	1.05	18.00	...	4.00	3.00	...
T7	0.75	14.00	...	4.00	2.00	...
T9	1.20	18.00	...	4.00	4.00	...
Hot-work steels						
H15	0.40	...	5.00	5.00	...	...
H16	0.55	7.00	...	7.00	...	...
H20	0.35	9.00	...	2.00	...	...
H41	0.65	1.50	8.00	4.00	1.00	...
H43	0.55	...	8.00	4.00	2.00	...
Cold-work steels						
D1	1.00	...	1.00	12.00	...	...
D6(a)	...	...	...	...	...	...
A5	1.00	...	1.00	1.00	...	3.00 Mn
Shock-resisting steels						
S3	0.50	1.00	...	0.74	...	...
S4	0.55	...	...	...	...	2.00 Si, 0.80 Mn
Mold steel						
P1	0.10	...	...	...	...	...
Special-purpose steels						
L1	1.00	...	...	1.25	...	...
L3	1.00	...	...	1.50	0.20	...
L4	1.00	...	...	1.50	0.25	0.60 Mn
L5	1.00	...	0.25	1.00	...	1.00 Mn
L7	1.00	...	0.40	1.40	...	0.35 Mn
F1	1.00	1.25	...	...	...	...
F2	1.25	3.50	...	...	...	...
F3	1.25	3.50	...	0.75	...	...
Water-hardening tool steels						
W3	1.00	...	...	...	0.50	...
W4	0.60/1.40(b)	...	...	0.25	...	...
W6	1.00	...	...	0.25	0.25	...
W7	1.00	...	...	0.50	0.20	...

(a) Now included with D3 in Table 1. (b) Various carbon contents were available.

Rockwell C hardness only one or two points higher than those obtainable through hot-salt quenching or air cooling, which ordinarily produce full hardness in tools smaller than about 76 mm (3 in.) in diameter, but at such high hardnesses that a one- or two-point increase in Rockwell hardness may prove quite significant.

The maximum hardness of tungsten high-speed steels varies with carbon content and, to a lesser degree, with alloy content. A hardness of at least 64.5 HRC can be developed in any high-speed steel. Those types that have high carbon contents and hard carbides, such as T15, may be hardened to 67 HRC.

Hot-Work Steels

Many manufacturing operations involve punching, shearing, or forming of metals at high temperatures. Hot-work steels (group H) have been developed to withstand the combinations of heat, pressure, and abrasion associated with such operations. Table 6 gives data on resistance to softening after 100 h at temperatures from 480 to 760 °C (900 to 1400 °F) for four of these steels.

Group H tool steels usually have medium carbon contents (0.35 to 0.45%) and chromium, tungsten, molybdenum, and vanadium contents of 6 to 25%. H steels are divided into three subgroups: chromium hot-work steels (types H10 to H19), tungsten hot-work steels (types H21 to H26), and molybdenum hot-work steels (types H42 and H43).

Chromium hot-work steels (types H10 to H19) have good resistance to heat softening because of their medium chromium content and the addition of carbide-forming elements such as molybdenum, tungsten, and vanadium. The low carbon and low total alloy contents promote toughness at the normal working hardnesses of 40 to 55 HRC. Higher tungsten and molybdenum contents increase hot strength but slightly reduce toughness. Vanadium is added to increase resistance to washing (erosive wear) at high temperatures. An increase in silicon content improves oxidation resistance at temperatures up to 800 °C (1475 °F). The most widely used types in this group are H11, H12, H13, and, to a lesser extent, H19.

All of the chromium hot-work steels are deep hardening. The H11, H12, and H13 steels may be air hardened to full working hardness in section sizes up to 152 mm (6 in.); other group H steels may be air hardened in section sizes up to 305 mm (12 in.). The air-hardening qualities and balanced alloy contents of these steels result in low distortion during hardening. Chromium hot-work steels are especially well adapted to hot die work of all kinds, particularly dies for the extrusion of aluminum and magnesium, as well as die casting dies, forging dies, mandrels, and hot shears. Most of these steels have alloy and carbon contents low enough that tools made from them can be water cooled in service without cracking.

Tool steel H11 is used to make certain highly stressed structural parts, particularly in aerospace technology. Material for such demanding applications is produced by vacuum arc remelting of air-melted electrodes, which provides extremely low residual-gas content, excellent microcleanliness, and a high degree of structural homogeneity.

The chief advantage of H11 over conventional high-strength steels is its ability to resist softening during continued exposure to temperatures up to 540 °C (1000 °F) and at the same time provide moderate toughness and ductility at room-temperature tensile strengths of 1720 to 2070 MPa (250 to 300 ksi). In addition, because of its secondary hardening characteristic, H11 can be tempered at high temperatures, resulting in nearly complete relief of residual hardening stresses, which is necessary for maximum toughness at high strength levels. Other important advantages of H11, H12, and H13 steels for structural and hot-work applications include ease of forming and working, good weldability, relatively low coefficient of thermal expansion, acceptable thermal conductivity, and above-average resistance to oxidation and corrosion.

Tungsten Hot-Work Steels. The principal alloying elements of tungsten hot-work steels (types H21 to H26) are carbon, tungsten, chromium, and vanadium. The higher alloy contents of these steels make them more resistant to high-temperature softening and washing than H11 and H13 hot-work steels. However, high alloy content also makes them more prone to brittleness at normal working hardnesses (45 to 55 HRC) and makes it difficult for them to be safely water cooled in service.

Although tungsten hot-work steels can be air hardened, they are usually quenched in oil or hot salt to minimize scaling. When air hardened, they exhibit low distortion. Tungsten hot-work steels require higher hardening temperatures than do chromium

Table 4 Reference guide for tool steel selection

Application areas	High-speed tool steels, M and T	Hot-work tool steels, H	Cold-work tool steels, D, A, and O	Shock-resisting tool steels, S	Mold steels, P	Special-purpose tool steels, L	Water-hardening tool steels, W
Cutting tools Single-point types (lathe, planer, boring) Milling cutters Drills Reamers Taps Threading dies Form cutters	General-purpose production tools: M2, T1 For increased abrasion resistance: M3, M4, M10 Heavy-duty work calling for high hot hardness: T5, T15 Heavy-duty work calling for high abrasion resistance: M42, M44	...	Tools with sharp edges (knives, razors) Tools for operations in which no high speed is involved, yet stability in heat treatment and substantial abrasion resistance are needed	Pipe cutter wheels	...	...	Uses that do not require hot hardness or high abrasion resistance Examples with carbon content of applicable group: Taps (1.05–1.10% C) Reamers (1.10–1.15% C) Twist drills (1.20–1.25% C) Files (1.35–1.40% C)
Hot-forging tools and dies Dies and inserts Forging machine plungers and piercers	For combining hot hardness with high abrasion resistance: M2, T1	Dies for presses and hammers: H20, H21 For severe conditions over extended service periods: H22–H26	Hot trimming dies: D2	Hot-trimming dies Blacksmith tools Hot-swaging dies	...	...	Smith tools (0.65–0.70% C) Hot chisels (0.70–0.75% C) Drop forging dies (0.90–1.00% C) Applications limited to short-run production
Hot extrusion tools and dies Extrusion dies and mandrels Dummy blocks Valve extrusion tools	Brass extrusion dies: T1	Extrusion dies and dummy blocks: H21–H26 For tools that are exposed to less heat: H10–H14, H19	...	Compression molding: S1	...	...	...
Cold-forming dies Bending, forming, drawing, and deep-drawing dies and punches	Burnishing tools: M1, T1	Cold-heading die casings: H13	Drawing dies: O1 Coining tools: O1, D2 Forming and bending dies: A2 Thread rolling dies: D2	Hobbing and short-run applications: S1, S7 Rivet sets and rivet busters	...	Blanking, forming, and trimmer dies when toughness has precedence over abrasion resistance: L6	Cold-heading dies: W1 or W2 (C ~ 1.00%) Bending dies: W1 (C ~ 1.00%)
Shearing tools Dies for piercing, punching, and trimming Shear blades	Special dies for cold and hot work: T1 For work requiring high abrasion resistance: M2, M3	For shearing knives: H11, H12 For severe hot-shearing applications: H21, H25	Dies for medium runs: A2, A6, O1 Dies for long runs: D2, D3 Trimming dies (also for hot trimming): A2	Cold and hot shear blades Hot punching and piercing tools Boilermaker tools	...	Knives for work requiring high toughness: L6	Trimming dies (0.90–0.95% C) Cold-blanking and punching dies (1.00% C)
Die casting and molding dies	...	For aluminum and lead: H11, H13 For brass: H21	A2, A6, O1	...	Plastic molds: P2–P4, P20	...	...
Structural parts for severe service conditions	Roller bearings for high-temperature environment: T1 Lathe centers: M2, T1	For aircraft components (landing gears, arrester hooks, rocket cases): H11	Lathe centers: D2, D3 Arbors: O1 Bushings A4 Gages: D2	Pawls Clutch parts	...	Spindles and clutch parts (if high toughness is needed): L6	Spring steel (1.10–1.15% C)
Battering tools, hand and power	...	...	...	Pneumatic chisels for cold work: S5 For higher performance: S7	...	...	For intermittent use: W1 (0.80% C)

Source: Ref 4

hot-work steels, making the former more likely to scale when heated in an oxidizing atmosphere.

Although these steels have much greater toughness, in many characteristics they are similar to high-speed steels; in fact, type H26 is a low-carbon version of T1 high-speed steel. If tungsten hot-work steels are preheated to operating temperature before use, breakage can be minimized. These steels have been used to make mandrels and extrusion dies for high-temperature applications, such as the extrusion of brass, nickel alloys, and steel, and are also suitable for use in hot-forging dies of rugged design.

Molybdenum Hot-Work Steel. There are only two active molybdenum hot-work steels: type H42 and type H43. These alloys contain molybdenum, chromium, vanadium, carbon, and varying amounts of tungsten. They are similar to tungsten hot-work steels, having almost identical characteristics and uses. Although their compositions resemble those of various molybdenum high-speed steels, they have a low carbon content and greater toughness. The principal advantage of types H42 and H43 over tungsten hot-work steels is their lower initial cost. They are more resistant to heat checking than are tungsten hot-work steels but, in common with all high-molybdenum steels, require greater care in heat treatment, particularly with regard to decarburization and control of austenitizing temperature.

Cold-Work Steels

Cold-work tool steels, because they do not have the alloy content necessary to make them resistant to softening at elevated temperature, are restricted in application to those uses that do not involve prolonged or repeated heating above 205 to 260 °C (400 to 500 °F). There are three categories of cold-work steels: air-hardening steels, or group A; high-carbon, high-chromium steels, or group D; and oil-hardening steels, or group O.

Air-hardening, medium-alloy, cold-work steels (group A) contain enough alloying elements to enable them to achieve full hardness in sections up to about 102 mm (4

Table 5 Significant dates in the development of high-speed tool steels

Date	Development
1903	0.70% C, 14% W, 4% Cr prototype of modern high-speed tool steels
1904	0.30% V addition
1906	Introduction of electric furnace melting
1910	Introduction of first 18-4-1 composition (AISI T1)
1912	3–5% Co addition for improved hot hardness
1923	12% Co addition for increased cutting speeds
1939	Introduction of high-carbon, high-vanadium, super high speed tool steels (M4 and T15)
1940–1952	Increasing substitution of molybdenum for tungsten
1953	Introduction of sulfurized free-machining high-speed tool steel
1961	Introduction of high-carbon, high-cobalt, superhard high speed tool steels (M40 series)
1970	Introduction of powdered metal high-speed tool steels
1973	Addition of higher silicon/nitrogen content to M7 to increase hardness
1980	Development of cobalt-free super high speed tool steels
1982	Introduction of aluminum-modified high-speed tool steels for cutting tools

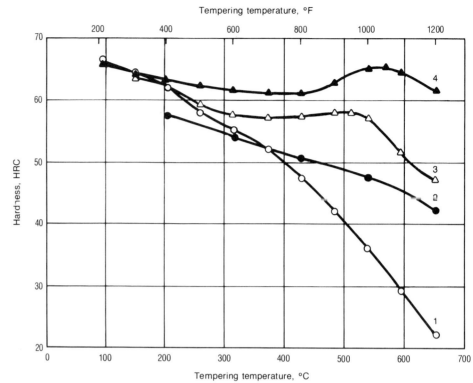

Fig. 1 Variation of hardness with tempering temperature for four typical tool steels. Curves are for 1 h at temperature. Curve 1 illustrates low resistance to softening as tempering temperatures increase, such as is exhibited by group W and group O tool steels. Curve 2 illustrates medium resistance to softening, such as is exhibited by type S1 tool steel. Curves 3 and 4 illustrate high and very high resistance to softening, respectively, such as are exhibited by the secondary hardening tool steels A2 and M2.

in.) in diameter upon air cooling from the austenitizing temperature. (Type A6 through hardens in sections as large as a cube 178 mm, or 7 in., on a side.) Because they are air hardening, group A tool steels exhibit minimum distortion and the highest safety (least tendency to crack) in hardening. Manganese, chromium, and molybdenum are the principal alloying elements used to provide this deep hardening. Types A2, A3, A7, A8, and A9 contain a high percentage of chromium (5%), which provides moderate resistance to softening at elevated temperatures (see curve 3 in Fig. 1 for a plot of hardness versus tempering temperature for type A2).

Types A4, A6, and A10 are lower in chromium content (1%) and higher in manganese content (2%). They can be hardened from temperatures about 110 °C (200 °F) lower than those required for the high-chromium types, further reducing distortion and undesirable surface reactions during heat treatment.

To improve toughness, silicon is added to type A8, and both silicon and nickel are added to types A9 and A10. Because of the high carbon and silicon contents of type A10, graphite is formed in the microstructure; as a result, A10 has much better machinability when in the annealed condition, and somewhat better resistance to galling and seizing when in the fully hardened condition, than other group A tool steels.

Typical applications for group A tool steels include shear knives, punches, blanking and trimming dies, forming dies, and coining dies. The inherent dimensional stability of these steels makes them suitable for

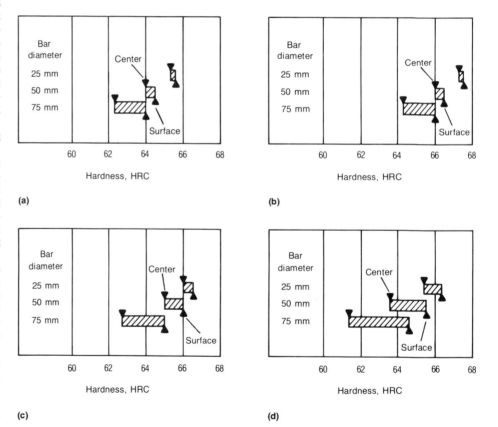

Fig. 2 Variation of surface and center hardness with bar diameter for four high-speed steels. (a) M2. (b) M3. (c) T1. (d) T2. Steels M2 and M3 were oil quenched from 1205 °C (2200 °F) and 1230 °C (2250 °F), respectively. Steels T1 and T2 were oil quenched from 1290 °C (2350 °F).

Table 6 Resistance of four hot-work steels to softening at elevated temperatures

Type	Original hardness, HRC	_____ Hardness(a), HRC, after 100 h at _____					
		480 °C (900 °F)	540 °C (1000 °F)	595 °C (1100 °F)	650 °C (1200 °F)	705 °C (1300 °F)	760 °C (1400 °F)
H13	50.2	48.7	45.3	29.0	22.7	20.1	13.9
	41.7	38.6	39.3	27.7	23.7	20.2	13.2
H21	49.2	48.7	47.6	37.2	27.4	19.8	15.2
	36.7	34.8	34.9	32.6	27.1	19.8	14.9
H23	40.8	40.0	40.6	40.8	38.6	33.2	25.8
	38.9	38.9	38.0	38.0	37.1	32.5	25.6
H26	51.0	50.6	50.3	47.1	38.4	26.9	21.3
	42.9	42.4	42.3	41.3	34.9	26.4	21.1

(a) At room temperature

gages and precision measuring tools. In addition, the extreme abrasion resistance of type A7 makes it suitable for brick molds, ceramic molds, and other highly abrasive applications.

The complex chromium or chromium-vanadium carbides in group A tool steels enhance the wear resistance provided by the martensitic matrix. Therefore, these steels perform well under abrasive conditions at less than full hardness. Although cooling in still air is adequate for producing full hardness in most tools, massive sections should be hardened by cooling in an air blast or by interrupted quenching in hot oil.

High-carbon, high-chromium, cold-work steels (group D) contain 1.50 to 2.35% C and 12% Cr; with the exception of type D3, they also contain 1% Mo. All group D tool steels except type D3 are air hardening and attain full hardness when cooled in still air. Type D3 is almost always quenched in oil (small parts can be austenitized in vacuum and then gas quenched); therefore, tools made of D3 are more susceptible to distortion and are more likely to crack during hardening.

Group D steels have high resistance to softening at elevated temperatures. These steels also exhibit excellent resistance to wear, especially type D7, which has the highest carbon and vanadium contents. All group D steels, particularly the higher-carbon types D3, D4, and D7, contain massive amounts of carbides, which make them susceptible to edge brittleness.

Typical applications of group D steels include long-run dies for blanking, forming, thread rolling, and deep drawing; dies for cutting laminations; brick molds; gages; burnishing tools; rolls; and shear and slitter knives.

Oil-hardening cold-work steels (group O) have high carbon contents, plus enough other alloying elements that small-to-moderate sections can attain full hardness when quenched in oil from the austenitizing temperature. Group O tool steels vary in type of alloy, as well as in alloy content, even though they are similar in general characteristics and are used for similar applications. Type O1 contains manganese, chromium, and tungsten. Type O2 is alloyed primarily with manganese. Type O6 contains silicon, manganese, and molybdenum; it has a high total carbon content that includes free carbon, as well as sufficient combined carbon to enable the steel to achieve maximum as-quenched hardness. Type O7 contains manganese and chromium and has a tungsten content higher than that of type O1.

The most important service-related property of group O steels is high resistance to wear at normal temperatures, a result of high carbon content. On the other hand, group O steels have a low resistance to softening at elevated temperatures.

The ability of group O steels to harden fully upon relatively slow quenching yields lower distortion and greater safety (less tendency to crack) in hardening than is characteristic of the water-hardening tool steels. Tools made from these steels can be successfully repaired or renovated by welding if proper procedures are followed. In addition, graphite in the microstructure of type O6 greatly improves the machinability of annealed stock and helps reduce galling and seizing of fully hardened steel.

Group O steels are used extensively in dies and punches for blanking, trimming, drawing, flanging, and forming. Surface hardnesses of 56 to 62 HRC, obtained through oil quenching followed by tempering at 175 to 315 °C (350 to 600 °F), provide a suitable combination of mechanical properties for most dies made from type O1, O2, or O6. Type O7, which has lower hardenability but better general wear resistance than any other group O tool steel, is more often used for tools requiring keen cutting edges. Oil-hardening tool steels are also used for machinery components (such as cams, bushings, and guides) and for gages (where good dimensional stability and wear resistance properties are needed).

The hardenability of group O steels can be measured effectively by the Jominy end-quench test. Hardenability bands for group O steels are shown in Fig. 3. Variation of hardness with diameter is shown in Fig. 4 for center, surface, and ¾ radius locations in oil-quenched bars of group O steels.

At normal hardening temperatures, group O steels retain greater amounts of undissolved carbides and thus do not harden as deeply as do steels that are lower in carbon

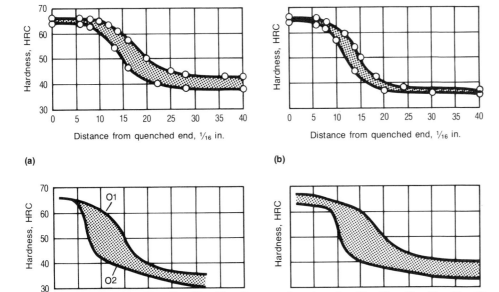

Fig. 3 End-quench hardenability bands for group O tool steels. (a) O1, source A. (b) O2, source A. (c) O1 and O2, source B. (d) O6. Hardenability bands from source B represent the data from five heats each for O1 and O2 tool steels. Data from source A were determined only on the basis of average hardness, not as hardenability bands. Data for O6 is for a spheroidized prior structure. Steels O1 and O6 were quenched from 815 °C (1500 °F); O2, from 790 °C (1450 °F).

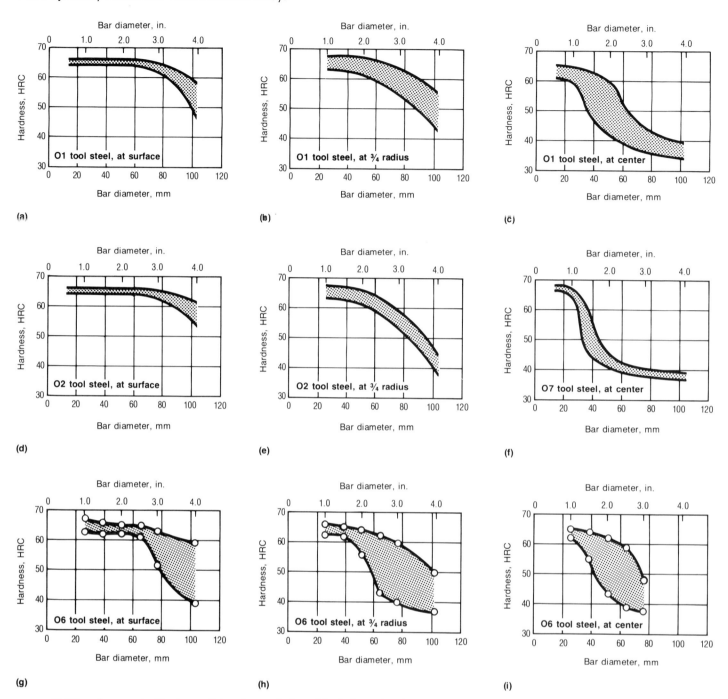

Fig. 4 Variation of as-quenched hardness with bar diameter for four oil-hardening tool steels. Data for (a), (b), (d), and (e) are from tests on 5 heats of each steel from source A. Data for (c) are from tests on 23 heats of O1 from source B. Data for (f) are from tests on 8 heats (source unknown). Information on the number of heats and source of data not available for (g), (h), and (i). Center hardness data not available for type O2; surface and ¾-radius data not available for type O7. Type O1 austenitized at 815 °C (1500 °F) in source A and at 775 °C (1425 °F) in source B. Type O2 austenitized at 790 °C (1450 °F). Austenitizing temperatures for types O6 and O7 not available

but similar in alloy content. On the other hand, group O steels attain higher surface hardness. Raising the hardening temperature increases grain size; increases solution of alloying elements; and dissolves more of the excess carbide, thereby increasing hardenability. However, raising the hardening temperature can have an adverse effect on certain mechanical properties, most notably ductility toughness, and also can increase the likelihood of cracking during hardening.

Shock-Resisting Steels

The principal alloying elements in shock-resisting, or group S, steels are manganese, silicon, chromium, tungsten, and molybdenum, in various combinations. Carbon content is about 0.50% for all group S steels, which produces a combination of high strength, high toughness, and low-to-medium wear resistance. Group S steels are used primarily for chisels, rivet sets, punches, driver bits, and other applications requiring high toughness and resistance to shock loading. Types S1 and S7 are also used for hot punching and shearing, which require some heat resistance.

Group S steels vary in hardenability from shallow hardening (S2) to deep hardening (S7). In these steels of intermediate alloy content, hardenability is controlled to a greater extent by composition than by the incidental effects of grain size and melting

Table 7 Nominal room-temperature mechanical properties of group L and group S tool steels

Type	Condition	Tensile strength MPa	ksi	0.2% yield strength MPa	ksi	Elongation(a), %	Reduction in area, %	Hardness, HRC	Impact energy J	ft · lbf
L2	Annealed	710	103	510	74	25	50	96 HRB	...	...
	Oil quenched from 855 °C (1575 °F) and single tempered at:									
	205 °C (400 °F)	2000	290	1790	260	5	15	54	28(b)	21(b)
	315 °C (600 °F)	1790	260	1655	240	10	30	52	19(b)	14(b)
	425 °C (800 °F)	1550	225	1380	200	12	35	47	26(b)	19(b)
	540 °C (1000 °F)	1275	185	1170	170	15	45	41	39(b)	29(b)
	650 °C (1200 °F)	930	135	760	110	25	55	30	125(b)	92(b)
L6	Annealed	655	95	380	55	25	55	93 HRB	...	...
	Oil quenched from 845 °C (1550 °F) and single tempered at:									
	315 °C (600 °F)	2000	290	1790	260	4	9	54	12(b)	9(b)
	425 °C (800 °F)	1585	230	1380	200	8	20	46	18(b)	13(b)
	540 °C (1000 °F)	1345	195	1100	160	12	30	42	23(b)	17(b)
	650 °C (1200 °F)	965	140	830	120	20	48	32	81(b)	60(b)
S1	Annealed	690	100	415	60	24	52	96 HRB	...	...
	Oil quenched from 925 °C (1700 °F) and single tempered at:									
	205 °C (400 °F)	2070	300	1895	275	...	...	57.5	249(c)	184(c)
	315 °C (600 °F)	2025	294	1860	270	4	12	54	233(c)	172(c)
	425 °C (800 °F)	1790	260	1690	245	5	17	50.5	203(c)	150(c)
	540 °C (1000 °F)	1680	244	1525	221	9	23	47.5	230(c)	170(c)
	650 °C (1200 °F)	1345	195	1240	180	12	37	42	...	...
S5	Annealed	725	105	440	64	25	50	96 HRB	...	...
	Oil quenched from 870 °C (1600 °F) and single tempered at:									
	205 °C (400 °F)	2345	340	1930	280	5	20	59	206(c)	152(c)
	315 °C (600 °F)	2240	325	1860	270	7	24	58	232(c)	171(c)
	425 °C (800 °F)	1895	275	1690	245	9	28	52	243(c)	179(c)
	540 °C (1000 °F)	1515	220	1380	200	10	30	48	188(c)	139(c)
	650 °C (1200 °F)	1035	150	1170	170	15	40	37	...	...
S7	Annealed	640	93	380	55	25	55	95 HRB	...	...
	Fan cooled from 940 °C (1725 °F) and single tempered at:									
	205 °C (400 °F)	2170	315	1450	210	7	20	58	244(c)	180(c)
	315 °C (600 °F)	1965	285	1585	230	9	25	55	309(c)	228(c)
	425 °C (800 °F)	1895	275	1410	205	10	29	53	243(c)	179(c)
	540 °C (1000 °F)	1820	264	1380	200	10	33	51	324(c)	239(c)
	650 °C (1200 °F)	1240	180	1035	150	14	45	39	358(c)	264(c)

(a) In 50 mm, or 2 in. (b) Charpy V-notch. (c) Charpy unnotched

practice, which are so important for group W steels. Group S steels require relatively high austenitizing temperatures to achieve optimum hardness; consequently, undissolved carbides are not a factor in the control of hardenability. Type S2 is normally water quenched; types S1, S5, and S6 are oil quenched; and type S7 is normally cooled in air, except for large sections, which are oil quenched.

Because group S steels exhibit excellent toughness at high strength levels, they are often considered for nontooling or structural applications. The nominal mechanical properties of S1, S5, and S7, in both annealed and hardened and tempered conditions, are presented in Table 7.

Low-Alloy Special-Purpose Steels

The low-alloy special-purpose, or group L, tool steels contain small amounts of chromium, vanadium, nickel, and molybdenum. At one time, seven steels were listed in this group, but because of falling demand, only types L2 and L6 remain. Type L2 is available in several carbon contents from 0.50 to 1.10%; its principal alloying elements are chromium and vanadium, which make it an oil-hardening steel of fine grain size. Type L6 contains small amounts of chromium and molybdenum, as well as 1.50% Ni for increased toughness.

Although both L2 and L6 are considered oil-hardening steels, large sections of L2 are often quenched in water. A type L2 steel containing 0.50% C is capable of attaining about 57 HRC as oil quenched, but it will not through harden in sections of more than about 12.7 mm (0.5 in.) thickness. Type L6, which contains 0.70% C, has an as-quenched hardness of about 64 HRC; it can maintain a hardness above 60 HRC through sections of 76 mm (3 in.) thickness.

Group L steels are generally used for machine parts such as arbors, cams, chucks, and collets, and for other special applications requiring good strength and toughness. Nominal mechanical properties of annealed and hardened-and-tempered L2 and L6 steels are given in Table 7.

Mold Steels

Mold steels, or group P, contain chromium and nickel as principal alloying elements. Types P2 and P6 are carburizing steels produced to tool steel quality standards. They have very low hardness and low resistance to work hardening in the annealed condition. These factors make it possible to produce a mold impression by cold hubbing. After the impression is formed, the mold is carburized, hardened, and tempered to a surface hardness of about 58 HRC. Types P4 and P6 are deep hardening; with type P4, full hardness in the carburized case can be achieved by cooling in air.

Types P20 and P21 normally are supplied heat treated to 30 to 36 HRC, a condition in which they can be machined readily into large, intricate dies and molds. Because these steels are prehardened, no subsequent high-temperature heat treatment is required, and distortion and size changes are avoided. However, when used for plastic molds, type P20 is sometimes carburized and hardened after the impression has been machined. Type P21 is an aluminum-containing precipitation-hardening steel that is supplied prehardened to 32 to 36 HRC. This steel is preferred for critical-finish molds because of its excellent polishability.

All group P steels have low resistance to softening at elevated temperatures, with the exception of P4 and P21, which have medium resistance. Group P steels are used almost exclusively in low-temperature die casting dies and in molds for the injection or compression molding of plastics. Plastic molds often require massive steel blocks up to 762 mm (30 in.) thick and weighing as much as 9 Mg (10 tons). Because these large die blocks must meet stringent requirements for soundness, cleanliness, and hardenability, electric furnace melting, vacuum degassing, and special deoxidation treatments have become standard practice in the production of group P tool steels. In addition, ingot casting and forging practices have been refined so that a high degree of homogeneity can be achieved.

Water-Hardening Steels

Water-hardening, or group W, tool steels contain carbon as the principal alloying element. Small amounts of chromium and vanadium are added to most of the group W steels—chromium to increase hardenability and wear resistance, and vanadium to maintain fine grain size and thus enhance toughness. Group W tool steels are made with various nominal carbon contents (~0.60 to 1.40%); the most popular grades contain approximately 1.00% C.

Group W tool steels are very shallow hardening and consequently develop a fully hardened zone that is relatively thin, even when quenched drastically. Sections more than about 13 mm (½ in.) thick generally have a hard case over a strong, tough, and resilient core.

Group W steels have low resistance to softening at elevated temperatures. They are suitable for cold heading, striking, coining, and embossing tools; woodworking tools; hard metal-cutting tools, such as taps and reamers; wear-resistant machine tool components; and cutlery.

This group of steels is made in as many as four different grades or quality levels for the same nominal composition. These quality levels, which have been given various names by different manufacturers, range from a clean carbon tool steel with precisely controlled hardenability, grain size, microstructure, and annealed hardness to a grade less carefully controlled but satisfactory for noncritical low-production applications.

The Society of Automotive Engineers (SAE) defines four grades of plain carbon tool steels as follows:

- *Special (grade 1)*: The highest-quality water-hardening tool steel. Hardenability is controlled, and composition is held to close limits. Bars are subjected to rigorous testing to ensure maximum uniformity in performance
- *Extra (grade 2)*: A high-quality water-hardening tool steel that is controlled for hardenability and is subjected to tests that ensure good performance in general applications
- *Standard (grade 3)*: A good-quality water-hardening tool steel that is not controlled for hardenability and that is recommended for applications in which some latitude in uniformity can be tolerated
- *Commercial (grade 4)*: A commercial-quality water-hardening tool steel that is neither controlled for hardenability nor subjected to special tests

Limits on manganese, silicon, and chromium generally are not required for special and extra grades. Instead, the Shepherd hardenability limits are prescribed:

Hardenability classification	Radial depth of hardening (P), 1/64 in.	Minimum fracture grain size, F
Carbon content, 0.70–0.95%		
Shallow	10 max	8
Regular	9–13	8
Deep	12 min	8
Carbon content, 0.95–1.30%		
Shallow	8 max	9
Regular	7–11	9
Deep	10–16	8

See the section "Testing of Tool Steels" in this article for more information on Shepherd hardenability.

The combined manganese, silicon, and chromium contents of SAE standard and commercial grades should not exceed 0.75%. Generally, both manganese and silicon are limited to 0.35% maximum in all standard and commercial grades; chromium is limited to 0.15% maximum in standard grades and to 0.20% maximum in commercial grades.

The ability of a group W tool steel to perform satisfactorily in many applications depends on the depth of the hardened zone. Depth of hardening in these steels is primarily controlled by the austenitic grain size, melting practice, alloy content, amount of excess carbide present at the quenching temperature and, to a lesser extent, initial structure of the steel prior to austenitizing for hardening.

Typical results in the Shepherd penetration-fracture (PF) test indicate an increase in P value of 0.8 mm (2/64 in.) for every increase in austenitic grain size of one ASTM number for the same grade. Increased amounts of undissolved carbides at the hardening temperature will reduce hardenability. This is doubly important in hyper-

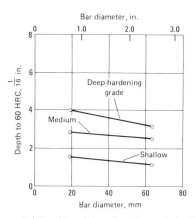

Fig. 5 Relationship of bar diameter and depth of hardened zone for shallow-, medium-, and deep-hardening grades of W1 tool steel containing 1% C

eutectoid grades, which are deliberately quenched to retain carbides undissolved at the austenitizing temperature in order to increase wear resistance. A fine lamellar microstructure prior to hardening, such as that obtained by normalizing, will result in fewer undissolved carbides at the normal austenitizing temperature than will a previously spheroidized microstructure. The presence of fewer carbides at the austenitizing temperature promotes deeper hardening because more carbon is dissolved in the austenite and there are fewer carbides to act as nucleation sites for nonmartensitic transformation products. Thus, normalized bars have deeper hardenability than do spheroidized bars of the same grade.

The addition of vanadium frequently decreases hardenability under normal hardening conditions because of the formation of many fine carbides that not only act as nucleation sites for nonmartensitic transformation products, but also refine the austenitic grain size. Austenitizing at higher-than-normal temperatures dissolves these excess carbides and thus increases the hardenability.

Group W steels with carbon contents lower than that of the eutectoid composition often have greater hardenability than do hypereutectoid grades. Grain coarsening resulting from the higher austenitizing temperatures used for hypoeutectoid grades is one cause of this, but the main cause is the absence of excess carbides at the austenitizing temperature.

Figure 5 shows a typical relationship between bar diameter and case depth (60 HRC or above) for three W1 tool steels that have the same carbon content (1% C) but different hardenabilities. Hardenability is varied by adjusting the manganese and silicon contents and altering the deoxidation procedure. This relationship illustrates the need for precise specification of hardenability in the selection of these grades: Group W tool steels purchased without hardenability requirements could vary widely enough in this

Table 8 Normalizing and annealing temperatures of tool steels

Type	Normalizing(a) °C	Normalizing(a) °F	Annealing(b) Temperature °C	Annealing(b) Temperature °F	Rate of cooling, maximum °C/h	Rate of cooling, maximum °F/h	Hardness, HB
Molybdenum high-speed steels							
M1, M10	Do not normalize		815–870	1500–1600	22	40	207–235
M2	Do not normalize		870–900	1600–1650	22	40	212–241
M3, M4	Do not normalize		870–900	1600–1650	22	40	223–255
M7	Do not normalize		815–870	1500–1600	22	40	217–255
M30, M33, M34, M35, M36, M41, M42, M46, M47	Do not normalize		870–900	1600–1650	22	40	235–269
M43	Do not normalize		870–900	1600–1650	22	40	248–269
M44	Do not normalize		870–900	1600–1650	22	40	248–293
M48	Do not normalize		870–900	1600–1650	22	40	285–311
M62	Do not normalize		870–900	1600–1650	22	40	262–285
Tungsten high-speed steels							
T1	Do not normalize		870–900	1600–1650	22	40	217–255
T2	Do not normalize		870–900	1600–1650	22	40	223–255
T4	Do not normalize		870–900	1600–1650	22	40	229–269
T5	Do not normalize		870–900	1600–1650	22	40	235–277
T6	Do not normalize		870–900	1600–1650	22	40	248–293
T8	Do not normalize		870–900	1600–1650	22	40	229–255
T15	Do not normalize		870–900	1600–1650	22	40	241–277
Intermediate high-speed steels							
M50	Do not normalize		830–845	1525–1550	22	40	197–235
M52	Do not normalize		830–845	1525–1550	22	40	197–235
Chromium hot-work steels							
H10, H11, H12, H13	Do not normalize		845–900	1550–1650	22	40	192–229
H14	Do not normalize		870–900	1600–1650	22	40	207–235
H19	Do not normalize		870–900	1600–1650	22	40	207–241
Tungsten hot-work steels							
H21, H22, H25	Do not normalize		870–900	1600–1650	22	40	207–235
H23	Do not normalize		870–900	1600–1650	22	40	212–255
H24, H26	Do not normalize		870–900	1600–1650	22	40	217–241
Molybdenum hot-work steels							
H41, H43	Do not normalize		815–870	1500–1600	22	40	207–235
H42	Do not normalize		845–900	1550–1650	22	40	207–235
High-carbon, high-chromium, cold-work steels							
D2, D3, D4	Do not normalize		870–900	1600–1650	22	40	217–255
D5	Do not normalize		870–900	1600–1650	22	40	223–255
D7	Do not normalize		870–900	1600–1650	22	40	235–262
Medium-alloy, air-hardening, cold-work steels							
A2	Do not normalize		845–870	1550–1600	22	40	201–229
A3	Do not normalize		845–870	1550–1600	22	40	207–229
A4	Do not normalize		740–760	1360–1400	14	25	200–241
A6	Do not normalize		730–745	1350–1375	14	25	217–248
A7	Do not normalize		870–900	1600–1650	14	25	235–262
A8	Do not normalize		845–870	1550–1600	22	40	192–223
A9	Do not normalize		845–870	1550–1600	14	25	212–248
A10	790	1450	765–795	1410–1460	8	15	235–269
Oil-hardening cold-work steels							
O1	870	1600	760–790	1400–1450	22	40	183–212
O2	845	1550	745–775	1375–1425	22	40	183–212
O6	870	1600	765–790	1410–1450	11	20	183–217
O7	900	1650	790–815	1450–1500	22	40	192–217
Shock-resisting steels							
S1	Do not normalize		790–815	1450–1500	22	40	183–229(c)
S2	Do not normalize		760–790	1400–1450	22	40	192–217
S5	Do not normalize		775–800	1425–1475	14	25	192–229
S7	Do not normalize		815–845	1500–1550	14	25	187–223
Mold steels							
P2	Not required		730–815	1350–1500	22	40	103–123
P3	Not required		730–815	1350–1500	22	40	109–137
P4	Do not normalize		870–900	1600–1650	14	25	116–128
P5	Not required		845–870	1550–1600	22	40	105–116
P6	Not required		845	1550	8	15	183–217
P20	900	1650	760–790	1400–1450	22	40	149–179
P21	900	1650	Do not anneal	Do not anneal			
Low-alloy special-purpose steels							
L2	871–900	1600–1650	760–790	1400–1450	22	40	163–197
L3	900	1650	790–815	1450–1500	22	40	174–201
L6	870	1600	760–790	1400–1450	22	40	183–212
Carbon-tungsten special-purpose steels							
F1	900	1650	760–800	1400–1475	22	40	183–207
F2	900	1650	790–815	1450–1500	22	40	207–235
Water-hardening steels							
W1, W2	790–925(d)	1450–1700(d)	740–790(e)	1360–1450(e)	22	40	156–201
W5	870–925	1600–1700	760–790	1400–1450	22	40	163–201

(a) Time held at temperature varies from 15 min for small sections to 1 h for large sizes. Cooling is done in still air. Normalizing should not be confused with low-temperature annealing. (b) The upper limit of ranges should be used for large sections, and the lower limit for smaller sections. Time held at temperature varies from 1 h for light sections to 4 h for heavy sections and large furnace charges of high-alloy steel. (c) For 0.25 Si type, 183 to 207 HB; for 1.00 Si type, 207 to 229 HB. (d) Temperature varies with carbon content: 0.60 to 0.75 C, 815 °C (1500 °F); 0.75 to 0.90 C, 790 °C (1450 °F); 0.90 to 1.10 C, 870 °C (1600 °F); 1.10 to 1.40 C, 870 to 925 °C (1600 to 1700 °F). (e) Temperature varies with carbon content: 0.60 to 0.90 C, 740 to 790 °C (1360 to 1450 °F); 0.90 to 1.40 C, 760 to 790 °C (1400 to 1450 °F).

Table 9 Hardening and tempering of tool steels

Type	Rate of heating	Preheat temperature °C	Preheat temperature °F	Hardening temperature °C	Hardening temperature °F	Time at temperature, min	Quenching medium(a)	Tempering temperature °C	Tempering temperature °F
Molybdenum high-speed steels									
M1, M7, M10	Rapidly from preheat	730–845	1350–1550	1175–1220	2150–2225(b)	2–5	O, A, or S	540–595(c)	1000–1100(c)
M2	Rapidly from preheat	730–845	1350–1550	1190–1230	2175–2250(b)	2–5	O, A, or S	540–595(c)	1000–1100(c)
M3, M4, M30, M33, M34, M35	Rapidly from preheat	730–845	1350–1550	1205–1230(b)	2200–2250(b)	2–5	O, A, or S	540–595(c)	1000–1100(c)
M36	Rapidly from preheat	730–845	1350–1550	1220–1245(b)	2225–2275(b)	2–5	O, A, or S	540–595(c)	1000–1100(c)
M41	Rapidly from preheat	730–845	1350–1550	1190–1215(b)	2175–2220(b)	2–5	O, A, or S	540–595(d)	1000–1100(d)
M42	Rapidly from preheat	730–845	1350–1550	1190–1210(b)	2175–2210(b)	2–5	O, A, or S	510–595(d)	950–1100(d)
M43	Rapidly from preheat	730–845	1350–1550	1190–1215(b)	2175–2220(b)	2–5	O, A, or S	510–595(d)	950–1100(d)
M44	Rapidly from preheat	730–845	1350–1550	1200–1225(b)	2190–2240(b)	2–5	O, A, or S	540–625(d)	1000–1160(d)
M46	Rapidly from preheat	730–845	1350–1550	1190–1220(b)	2175–2225(b)	2–5	O, A, or S	525–565(d)	975–1050(d)
M47	Rapidly from preheat	730–845	1350–1550	1175–1200(b)	2150–2200(b)	2–5	O, A, or S	525–595(d)	975–1100(d)
M48	Rapidly from preheat	730–845	1350–1550	1175–1200(b)	2150–2200(b)	2–5	O, A, or S	540–595(d)	1000–1100(d)
M62	Rapidly from preheat	730–845	1350–1550	1175–1200(b)	2150–2200(b)	2–5	O, A, or S	540–595(d)	1000–1100(d)
Tungsten high-speed steels									
T1, T2, T4, T8	Rapidly from preheat	815–870	1500–1600	1260–1300(b)	2300–2375(b)	2–5	O, A, or S	540–595(c)	1000–1100(c)
T5, T6	Rapidly from preheat	815–870	1500–1600	1275–1300(b)	2325–2375(b)	2–5	O, A, or S	540–595(c)	1000–1100(c)
T15	Rapidly from preheat	815–870	1500–1600	1205–1260(b)	2200–2300(b)	2–5	O, A, or S	540–650(d)	1000–1200(d)
Intermediate high-speed steels									
M50	Rapidly from preheat	730–845	1350–1550	1095–1120	2000–2050	2–5	O, A, or S	525–595	975–1100
M52	Rapidly from preheat	730–845	1350–1550	1120–1175	2050–2150	2–5	O, A, or S	525–595	975–1100
Chromium hot-work steels									
H10	Moderately from preheat	815	1500	1010–1040	1850–1900	15–40(e)	A	540–650	1000–1200
H11, H12	Moderately from preheat	815	1500	995–1025	1825–1875	15–40(e)	A	540–650	1000–1200
H13	Moderately from preheat	815	1500	995–1040	1825–1900	15–40(e)	A	540–650	1000–1200
H14	Moderately from preheat	815	1500	1010–1065	1850–1950	15–40(e)	A	540–650	1000–1200
H19	Moderately from preheat	815	1500	1095–1205	2000–2200	2–5	A or O	540–705	1000–1300
Tungsten hot-work steels									
H21, H22	Rapidly from preheat	815	1500	1095–1205	2000–2200	2–5	A or O	595–675	1100–1250
H23	Rapidly from preheat	845	1550	1205–1260	2200–2300	2–5	O	650–815	1200–1500
H24	Rapidly from preheat	815	1500	1095–1230	2000–2250	2–5	O	565–650	1050–1200
H25	Rapidly from preheat	815	1500	1150–1260	2100–2300	2–5	A or O	565–675	1050–1250
H26	Rapidly from preheat	870	1600	1175–1260	2150–2300	2–5	O, A, or S	565–675	1050–1250
Molybdenum hot-work steels									
H41, H43	Rapidly from preheat	730–845	1350–1550	1095–1190	2000–2175	2–5	O, A, or S	565–650	1050–1200
H42	Rapidly from preheat	730–845	1350–1550	1120–1220	2050–2225	2–5	O, A, or S	565–650	1050–1200
High-carbon, high-chromium, cold-work steels									
D1, D5	Very slowly	815	1500	980–1025	1800–1875	15–45	A	205–540	400–1000
D3	Very slowly	815	1500	925–980	1700–1800	15–45	O	205–540	400–1000
D4	Very slowly	815	1500	970–1010	1775–1850	15–45	A	205–540	400–1000
D7	Very slowly	815	1500	1010–1065	1850–1950	30–60	A	150–540	300–1000
Medium-alloy, air-hardening, cold-work steels									
A2	Slowly	790	1450	925–980	1700–1800	20–45	A	175–540	350–1000
A3	Slowly	790	1450	955–980	1750–1800	25–60	A	175–540	350–1000
A4	Slowly	675	1250	815–870	1500–1600	20–45	A	175–425	350–800
A6	Slowly	650	1200	830–870	1525–1600	20–45	A	150–500	300–800
A7	Very slowly	815	1500	955–980	1750–1800	30–60	A	150–540	300–1000
A8	Slowly	790	1450	980–1010	1800–1850	20–45	A	175–595	350–1100
A9	Slowly	790	1450	980–1025	1800–1875	20–45	A	510–620	950–1150
A10	Slowly	650	1200	790–815	1450–1500	30–60	A	175–425	350–800
Oil-hardening cold-work steels									
O1	Slowly	650	1200	790–815	1450–1500	10–30	O	175–260	350–500
O2	Slowly	650	1200	760–800	1400–1475	5–20	O	175–260	350–500
O6	Slowly	...	...	790–815	1450–1500	10–30	O	175–315	350–600
O7	Slowly	650	1200	790–830 845–885	W: 1450–1525 O: 1550–1625	10–30	O or W	175–290	350–550
Shock-resisting steels									
S1	Slowly	...	...	900–955	1650–1750	15–45	O	205–650	400–1200
S2	Slowly	650(f)	1200(f)	845–900	1550–1650	5–20	B or W	175–425	350–800
S5	Slowly	760	1400	870–925	1600–1700	5–20	O	175–425	350–800
S7	Slowly	650–705	1200–1300	925–955	1700–1750	15–45	A or O	205–620	400–1150
Mold steels									
P2		900–925(g)	1650–1700(g)	830–845(h)	1525–1550(h)	15	O	175–260	350–500
P3		900–925(g)	1650–1700(g)	800–830(h)	1475–1525(h)	15	O	175–260	350–500

(continued)

(a) O, oil quench; A, air cool; S, salt bath quench; W, water quench; B, brine quench. (b) When the high-temperature heating is carried out in a salt bath, the range of temperatures should be about 14 °C (25 °F) lower than given here. (c) Double tempering recommended for not less than 1 h at temperature each time. (d) Triple tempering recommended for not less than 1 h at temperature each time. (e) Times apply to open furnace heat treatment. For pack hardening, a common rule is to heat 1.2 min per mm (30 min per in.) of cross section of the pack. (f) Preferable for large tools to minimize decarburization. (g) Carburizing temperature. (h) After carburizing. (i) Carburized per case hardness. (j) P21 is a precipitation-hardening steel having a thermal treatment that involves solution treating and aging rather than hardening and tempering. (k) Recommended for large tools and tools with intricate sections

Table 9 (continued)

Type	Rate of heating	Hardening Preheat temperature °C	°F	Hardening temperature °C	°F	Time at temperature, min	Quenching medium(a)	Tempering temperature °C	°F
Mold steels (continued)									
P4	...	970–995(g)	1775–1825(g)	970–995(h)	1775–1825(h)	15	A	175–480	350–900
P5	...	900–925(g)	1650–1700(g)	845–870(h)	1550–1600(h)	15	O or W	175–260	350–500
P6	...	900–925(g)	1650–1700(g)	790–815(h)	1450–1500(h)	15	A or O	175–230	350–450
P20	...	870–900(h)	1600–1650(h)	815–870	1500–1600	15	O	480–595(i)	900–1100(i)
P21(j)	Slowly	Do not preheat		705–730	1300–1350	60–180	A or O	510–550	950–1025
Low-alloy special-purpose steels									
L2	Slowly	...	...	W: 790–845 O: 845–925	W: 1450–1550 O: 1550–1700	10–30	O or W	175–540	350–1000
L3	Slowly	...	...	W: 775–815 O: 815–870	W: 1425–1500 O: 1500–1600	10–30	O or W	175–315	350–600
L6	Slowly	...	...	790–845	1450–1550	10–30	O	175–540	350–1000
Carbon-tungsten special-purpose steels									
F1, F2	Slowly	650	1200	790–870	1450–1600	15	W or B	175–260	350–500
Water-hardening steels									
W1, W2, W3	Slowly	565–650(k)	1050–1200(k)	760–845	1400–1550	10–30	B or W	175–345	350–650

(a) O, oil quench; A, air cool; S, salt bath quench; W, water quench; B, brine quench. (b) When the high-temperature heating is carried out in a salt bath, the range of temperatures should be about 14 °C (25 °F) lower than given here. (c) Double tempering recommended for not less than 1 h at temperature each time. (d) Triple tempering recommended for not less than 1 h at temperature each time. (e) Times apply to open furnace heat treatment. For pack hardening, a common rule is to heat 1.2 min per mm (30 min per in.) of cross section of the pack. (f) Preferable for large tools to minimize decarburization. (g) Carburizing temperature. (h) After carburizing. (i) Carburized per case hardness. (j) P21 is a precipitation-hardening steel having a thermal treatment that involves solution treating and aging rather than hardening and tempering. (k) Recommended for large tools and tools with intricate sections

property to cause severe processing difficulties or actual tool failures.

With the very fast cooling rate required for the hardening of the W grades, there is a greater chance that a tool will crack during hardening. Consequently, most manufacturers prefer to use tool steels that can be satisfactorily hardened by quenching in oil or cooling in air to attempt to avoid the expense involved when a tool cracks during heat treatment.

Typical Heat Treatments and Properties

Condensed information on heat-treating specifications and on the processing and service characteristics of tool steels is presented in Tables 8 to 10. This information clarifies the problems involved in the selection, processing, and application of tool steels.

More detailed heat-treating information for each of these steels is available in Volume 4 of the 9th Edition "Heat Treating" of *Metals Handbook*. Additional detailed information on resistance to softening at elevated temperatures is summarized in Fig. 1, which presents curves of hardness versus tempering temperature. Similar curves for most of the tool steels covered in this article are presented in Ref 5.

Technical representatives of tool steel producers can supply more specific information on the properties developed by specific heat treatments in the steels produced by their companies. They should be consulted regarding the type of steel and heat treatment best suited to meet all service requirements at the least overall cost.

The physical properties, specifically, density, thermal expansion, and thermal conductivity, of selected tool steels are given in Tables 11 and 12.

Testing of Tool Steels

Because of the difficulty of obtaining reliable correlations between the properties of tool steels as measured by laboratory tests and the performance of these steels in service or in fabrication, these properties are usually presented as general comparisons rather than as specific data.

Performance in Service

The basic properties of tool steels that determine their performance in service are resistance to wear, deformation, and breakage; toughness; and, in many instances, resistance to softening at elevated temperatures. These properties are listed in Table 13 and compared in Fig. 6. Often, these characteristics can be measured by, or inferred from, the measurement of hardness. The hardness of tool steels is most commonly measured and reported on the Rockwell C scale (HRC) in the United States and on the Vickers scale (diamond pyramid hardness, or HV) in Great Britain and Europe. It is significant that the conversion from HRC to HV, or vice versa, is not linear (see Fig. 7). For example, an increase from 67 to 68 HRC corresponds to a 40-point increase on the HV scale, whereas increases from 57 to 58 HRC and from 49 to 50 HRC correspond, respectively, to 20-point and 10-point increases on the HV scale.

For a given tool steel at a given hardness, wear resistance may vary widely depending on the wear mechanism involved and the heat treatment used. It is important to note also that among tool steels with widely differing compositions but identical hardnesses, wear resistance may vary widely under identical wear conditions.

For all practical purposes, the resistance to elastic deformation (modulus of elasticity) of all tool steels in all conditions is about 210 GPa (30×10^6 psi) at room temperature. This decreases uniformly to about 185 GPa (27×10^6 psi) at 260 °C (500 °F) and about 150 GPa (22×10^6 psi) at 540 °C (1000 °F).

Except for special grades, the compositions and heat treatments of most tool steels are selected to provide very high resistance to plastic deformation. This course of action leaves the metal with very little ability to absorb deformation; in other words, it leaves the metal very brittle. Therefore, it is difficult to determine reliable values of strength at maximum hardness by tensile testing, even when specially designed clamping fixtures are used to provide accurate alignment.

Compression tests have been used to some extent to measure resistance to deformation. Bending tests using either three- or four-point supports can provide useful comparative information on tool steels with high hardness levels, but the results are often difficult to evaluate. Torsion tests have been used effectively to measure the toughness of tool steels, particularly those to be used in drills and other tools loaded in torsion during service.

The amount of energy absorbed when a notched bar of fully hardened tool steel (except for certain grades) is broken in impact (Charpy test) is so small that it is very difficult to measure the differences in toughness that may make it possible to predict service performance. Attempts have

772 / Specialty Steels and Heat-Resistant Alloys

Table 10 Processing and service characteristics of tool steels

AISI designation	Resistance to decarburization	Hardening and tempering			Approximate hardness(b), HRC	Fabrication and service			
		Hardening response	Amount of distortion(a)	Resistance to cracking		Machinability	Toughness	Resistance to softening	Resistance to wear
Molybdenum high-speed steels									
M1	Low	Deep	A or S, low; O, medium	Medium	60–65	Medium	Low	Very high	Very high
M2	Medium	Deep	A or S, low; O, medium	Medium	60–65	Medium	Low	Very high	Very high
M3 (class 1 and class 2)	Medium	Deep	A or S, low; O, medium	Medium	61–66	Medium	Low	Very high	Very high
M4	Medium	Deep	A or S, low; O, medium	Medium	61–66	Low to medium	Low	Very high	Highest
M7	Low	Deep	A or S, low; O, medium	Medium	61–66	Medium	Low	Very high	Very high
M10	Low	Deep	A or S, low; O, medium	Medium	60–65	Medium	Low	Very high	Very high
M30	Low	Deep	A or S, low; O, medium	Medium	60–65	Medium	Low	Highest	Very high
M33	Low	Deep	A or S, low; O, medium	Medium	60–65	Medium	Low	Highest	Very high
M34	Low	Deep	A or S, low; O, medium	Medium	60–65	Medium	Low	Highest	Very high
M35	Low	Deep	A or S, low; O, medium	Medium	60–65	Medium	Low	Highest	Very high
M36	Low	Deep	A or S, low; O, medium	Medium	60–65	Medium	Low	Highest	Very high
M41	Low	Deep	A or S, low; O, medium	Medium	65–70	Medium	Low	Highest	Very high
M42	Low	Deep	A or S, low; O, medium	Medium	65–70	Medium	Low	Highest	Very high
M43	Low	Deep	A or S, low; O, medium	Medium	65–70	Medium	Low	Highest	Very high
M44	Low	Deep	A or S, low; O, medium	Medium	62–70	Medium	Low	Highest	Very high
M46	Low	Deep	A or S, low; O, medium	Medium	67–69	Medium	Low	Highest	Very high
M47	Low	Deep	A or S, low; O, medium	Medium	65–70	Medium	Low	Highest	Very high
M48	Low	Deep	A or S, low; O, medium	Medium	65–70	Low	Low	Highest	Highest
M62	Low	Deep	A or S, low; O, medium	Medium	62–68	Medium	Low	Highest	Very high
Tungsten high-speed steels									
T1	High	Deep	A or S, low; O, medium	High	60–65	Medium	Low	Very high	Very high
T2	High	Deep	A or S, low; O, medium	High	61–66	Medium	Low	Very high	Very high
T4	Medium	Deep	A or S, low; O, medium	Medium	62–66	Medium	Low	Highest	Very high
T5	Low	Deep	A or S, low; O, medium	Medium	60–65	Medium	Low	Highest	Very high
T6	Low	Deep	A or S, low; O, medium	Medium	60–65	Low to medium	Low	Highest	Very high
T8	Medium	Deep	A or S, low; O, medium	Medium	60–65	Medium	Low	Highest	Very high
T15	Medium	Deep	A or S, low; O, medium	Medium	63–68	Low to medium	Low	Highest	Highest
Intermediate high-speed steels									
M50	Low	Deep	A or S, low; O, medium	Medium	58–63	Medium	Low	High	High
M52	Low	Deep	A or S, low; O, medium	Medium	58–64	Medium	Low	High	High
Chromium hot-work steels									
H10	Medium	Deep	Very low	Highest	39–56	Medium to high	High	High	Medium
H11	Medium	Deep	Very low	Highest	38–54	Medium to high	Very high	High	Medium
H12	Medium	Deep	Very low	Highest	38–55	Medium to high	Very high	High	Medium
H13	Medium	Deep	Very low	Highest	38–53	Medium to high	Very high	High	Medium
H14	Medium	Deep	Low	Highest	40–47	Medium	High	High	Medium
H19	Medium	Deep	A, low; O, medium	High	40–57	Medium	High	High	Medium to high
Tungsten hot-work steels									
H21	Medium	Deep	A, low; O, medium	High	36–54	Medium	High	High	Medium to high
H22	Medium	Deep	A, low; O, medium	High	39–52	Medium	High	High	Medium to high
H23	Medium	Deep	Medium	High	34–47	Medium	Medium	Very high	Medium to high
H24	Medium	Deep	A, low; O, medium	High	45–55	Medium	Medium	Very high	High
H25	Medium	Deep	A, low; O, medium	High	35–44	Medium	High	Very high	Medium
H26	Medium	Deep	A or S, low; O, medium	High	43–58	Medium	Medium	Very high	High
Molybdenum hot-work steels									
H42	Medium	Deep	A or S, low; O, medium	Medium	50–60	Medium	Medium	Very high	High
Air-hardening, medium-alloy, cold-work steels									
A2	Medium	Deep	Lowest	Highest	57–62	Medium	Medium	High	High
A3	Medium	Deep	Lowest	Highest	57–65	Medium	Medium	High	Very high
A4	Medium to high	Deep	Lowest	Highest	54–62	Low to medium	Medium	Medium	Medium to high
A6	Medium to high	Deep	Lowest	Highest	54–60	Low to medium	Medium	Medium	Medium to high
A7	Medium	Deep	Lowest	Highest	57–67	Low	Low	High	Highest
A8	Medium	Deep	Lowest	Highest	50–60	Medium	High	High	Medium to high
A9	Medium	Deep	Lowest	Highest	35–56	Medium	High	High	Medium to high
A10	Medium to high	Deep	Lowest	Highest	55–62	Medium to high	Medium	Medium	High
High-carbon, high-chromium, cold-work steels									
D2	Medium	Deep	Lowest	Highest	54–61	Low	Low	High	High to very high
D3	Medium	Deep	Very low	High	54–61	Low	Low	High	Very high
D4	Medium	Deep	Lowest	Highest	54–61	Low	Low	High	Very high
D5	Medium	Deep	Lowest	Highest	54–61	Low	Low	High	High to very high
D7	Medium	Deep	Lowest	Highest	58–65	Low	Low	High	Highest
Oil-hardening cold-work steels									
O1	High	Medium	Very low	Very high	57–62	High	Medium	Low	Medium
O2	High	Medium	Very low	Very high	57–62	High	Medium	Low	Medium
O6	High	Medium	Very low	Very high	58–63	Highest	Medium	Low	Medium

(continued)

(a) A, air cool; B, brine quench; O, oil quench; S, salt bath quench; W, water quench. (b) After tempering in temperature range normally recommended for this steel. (c) Carburized case hardness. (d) After aging at 510 to 550 °C (950 to 1025 °F). (e) Toughness decreases with increasing carbon content and depth of hardening. Source: Ref 3

Table 10 (continued)

AISI designation	Resistance to decarburization	Hardening and tempering			Approximate hardness(b), HRC	Fabrication and service			
		Hardening response	Amount of distortion(a)	Resistance to cracking		Machinability	Toughness	Resistance to softening	Resistance to wear
Oil-hardening cold-work steels (continued)									
O7	High	Medium	O, very low; W, high	W, low; O, very high	58–64	High	Medium	Low	Medium
Shock-resisting steels									
S1	Medium	Medium	Medium	High	40–58	Medium	Very high	Medium	Low to medium
S2	Low	Medium	High	Low	50–60	Medium to high	Highest	Low	Low to medium
S5	Low	Medium	Medium	High	50–60	Medium to high	Highest	Low	Low to medium
S6	Low	Medium	Medium	High	54–56	Medium	Very high	Low	Low to medium
S7	Medium	Deep	A, lowest; O, low	O, high; A, highest	45–57	Medium	Very high	High	Low to medium
Low-alloy special-purpose steels									
L2	High	Medium	O, medium; W, low	O, medium; W, high	45–63	High	Very high(c)	Low	Low to medium
L6	High	Medium	Low	High	45–62	Medium	Very high	Low	Medium
Low-carbon mold steels									
P2	High	Medium	Low	High	58–64(c)	Medium to high	High	Low	Medium
P3	High	Medium	Low	High	58–64(c)	Medium	High	Low	Medium
P4	High	High	Very low	High	58–64(c)	Low to medium	High	Medium	High
P5	High	…	O, low; W, high	High	58–64(c)	Medium	High	Low	Medium
P6	High	…	A, very low; O, low	High	58–61(c)	Medium	High	Low	Medium
P20	High	Medium	Low	High	28–37	Medium to high	High	Low	Low to medium
P21	High	Deep	Lowest	Highest	30–40(d)	Medium	Medium	Medium	Medium
Water-hardening steels									
W1	Highest	Shallow	High	Medium	50–64	Highest	High(e)	Low	Low to medium
W2	Highest	Shallow	High	Medium	50–64	Highest	High(e)	Low	Low to medium
W5	Highest	Shallow	High	Medium	50–64	Highest	High(e)	Low	Low to medium

(a) A, air cool; B, brine quench; O, oil quench; S, salt bath quench; W, water quench. (b) After tempering in temperature range normally recommended for this steel. (c) Carburized case hardness. (d) After aging at 510 to 550 °C (950 to 1025 °F). (e) Toughness decreases with increasing carbon content and depth of hardening. Source: Ref 3

been made to perform impact tests on unnotched tool steel bars, but excessive deformation of supporting fixtures makes it very difficult to obtain reproducible results. Torsion impact testing yields useful, reproducible data on the effects of variations in the composition and heat treatment of tool steels; however, it is difficult to correlate the results of such testing with service experience. Fatigue tests have been useful for research, but only in some instances have the results correlated well with field experience.

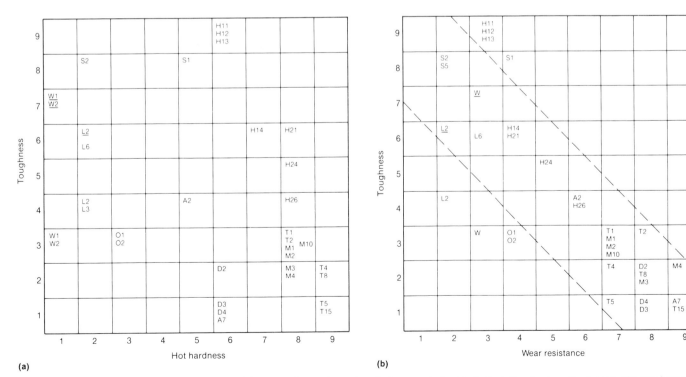

Fig. 6 Plots of toughness against (a) hot hardness and (b) wear resistance for tool steels. Types underlined indicate shallow-hardened tool steels. The area between the dashed lines in (b) represents average values.

774 / Specialty Steels and Heat-Resistant Alloys

Table 11 Density and thermal expansion of selected tool steels

Type	Density g/cm³	Density lb/in.³	Thermal expansion μm/m·K from 20 °C to 100 °C	200 °C	425 °C	540 °C	650 °C	Thermal expansion μin./in.·°F from 70 °F to 200 °F	400 °F	800 °F	1000 °F	1200 °F
W1	7.84	0.282	10.4	11.0	13.1	13.8(a)	14.2(b)	5.76	6.13	7.28	7.64(a)	7.90(b)
W2	7.85	0.283	...	...	14.4	14.8	14.9	...	...	8.0	8.2	8.3
S1	7.88	0.255	12.4	12.6	13.5	13.9	14.2	6.9	7.0	7.5	7.7	7.9
S2	7.79	0.281	10.9	11.9	13.5	14.0	14.2	6.0	6.6	7.5	7.8	7.9
S5	7.76	0.280	...	...	12.6	13.3	13.7	...	...	7.0	7.4	7.6
S6	7.75	0.279	...	...	12.6	13.3	...	...	...	7.0	7.4	...
S7	7.76	0.280	...	12.6	13.3	13.7(a)	13.3	...	7.0	7.4	7.6(a)	7.4
O1	7.85	0.283	...	10.6(c)	12.8	14.0(d)	14.4(d)	...	5.9(c)	7.1	7.8(d)	8.0(d)
O2	7.66	0.277	11.2	12.6	13.9	14.6	15.1	6.2	7.0	7.7	8.1	8.4
O6	7.70	0.277	...	11.2	12.6	12.9	13.7	...	6.2	7.0	7.2	7.6
O7	7.8	0.282	...	...	...	...	...	...	...	...	...	...
A2	7.86	0.284	10.7	10.6(c)	12.9	14.0	14.2	5.96	5.91(c)	7.2	7.8	7.9
A6	7.84	0.283	11.5	12.4	13.5	13.9	14.2	6.4	6.9	7.5	7.7	7.9
A7	7.66	0.277	...	...	12.4	12.9	13.5	...	...	6.9	7.2	7.5
A8	7.87	0.284	...	...	12.0	12.4	12.6	...	...	6.7	6.9	7.0
A9	7.78	0.281	...	...	12.0	12.4	12.6	...	...	6.7	6.9	7.0
A10	7.68	0.278	12.8	13.3	...	...	...	7.1	7.4	...	...	...
D2	7.70	0.278	10.4	10.3	11.9	12.2	12.2	5.8	5.7	6.6	6.8	6.8
D3	7.70	0.278	12.0	11.7	12.9	13.1	13.5	6.7	6.5	7.2	7.3	7.5
D4	7.70	0.278	...	...	12.4	...	...	...	...	6.9	...	...
D5	...	...	...	...	...	12.0	...	...	...	...	6.7	...
H10	7.81	0.281	...	...	12.2	13.3	13.7	...	...	6.8	7.4	7.6
H11	7.75	0.280	11.9	12.4	12.8	12.9	13.3	6.6	6.9	7.1	7.2	7.4
H13	7.76	0.280	10.4	11.5	12.2	12.4	13.1	5.8	6.4	6.8	6.9	7.3
H14	7.89	0.285	11.0	...	...	...	...	6.1	...	...	...	...
H19	7.98	0.288	11.0	11.0	12.0	12.4	12.9	6.1	6.1	6.7	6.9	7.2
H21	8.28	0.299	12.4	12.6	12.9	13.5	13.9	6.9	7.0	7.2	7.5	7.7
H22	8.36	0.302	11.0	...	11.5	12.0	12.4	6.1	...	6.4	6.7	6.9
H26	8.67	0.313	...	...	...	12.4	...	...	...	...	6.9	...
H42	8.15	0.295	...	...	...	11.9	...	...	...	...	6.6	...
T1	8.67	0.313	...	9.7	11.2	11.7	11.9	...	5.4	6.2	6.5	6.6
T2	8.67	0.313	...	...	...	...	...	...	...	...	...	...
T4	8.68	0.313	...	...	...	11.9	...	...	...	...	6.6	...
T5	8.75	0.316	11.2	...	...	11.5	...	6.2	...	...	6.4	...
T6	8.89	0.321	...	...	...	...	...	...	...	...	...	...
T8	8.43	0.305	...	...	...	...	...	...	...	...	...	...
T15	8.19	0.296	...	9.9	11.0	11.5	...	...	5.5(c)	6.1	6.4	...
M1	7.89	0.285	...	10.6(c)	11.3	12.0	12.4	...	5.9(c)	6.3	6.7	6.9
M2	8.16	0.295	10.1	9.4(c)	11.2	11.9	12.2	5.6	5.2(c)	6.2	6.6	6.8
M3, class 1	8.15	0.295	...	...	11.5	12.0	12.2	...	...	6.4	6.7	6.8
M3, class 2	8.16	0.295	...	...	11.5	12.0	12.8	...	...	6.4	6.7	7.1
M4	7.97	0.288	...	9.5(c)	11.2	12.0	12.2	...	5.3(c)	6.2	6.7	6.8
M7	7.95	0.287	...	9.5(c)	11.5	12.2	12.4	...	5.3(c)	6.4	6.8	6.9
M10	7.88	0.255	...	...	11.0	11.9	12.4	...	...	6.1	6.6	6.9
M30	8.01	0.289	...	...	11.2	11.7	12.2	...	...	6.2	6.5	6.8
M33	8.03	0.290	...	...	11.0	11.7	12.0	...	...	6.1	6.5	6.7
M36	8.18	0.296	...	...	...	...	...	...	...	...	...	...
M41	8.17	0.295	...	9.7	10.4	11.2	...	...	5.4	5.8	6.2	...
M42	7.98	0.288	...	...	...	...	...	...	...	...	...	...
M46	7.83	0.283	...	...	...	...	...	...	...	...	...	...
M47	7.96	0.288	10.6	11.0	11.9	...	12.6	5.9	6.1	6.6	...	7.0
L2	7.86	0.284	...	...	14.4	14.6	14.8	...	...	8.0	8.1	8.2
L6	7.86	0.284	11.3	12.6	12.6	13.5	13.7	6.3	7.0	7.0	7.5	7.6
P2	7.86	0.284	...	...	13.7	...	...	...	...	7.6	...	...
P5	7.80	0.282	...	...	...	...	...	...	...	...	...	...
P6	7.85	0.284	...	...	...	...	...	...	...	...	...	...
P20	7.85	0.284	...	...	12.8	13.7	14.2	...	...	7.1	7.6	7.9

(a) From 20 °C to 500 °C (70 °F to 930 °F). (b) From 20 °C to 600 °C (70 °F to 1110 °F). (c) From 20 °C to 260 °C (70 °F to 500 °F). (d) From 40 °C (100 °F)

In general, the ability of tool steels to withstand the rapid application of high loads without breaking increases with decreasing hardness. With hardness held constant, wide differences can be observed among tool steels of different compositions, or among steels of the same nominal composition made by different melting practices or heat treated according to different schedules.

The ability of a tool steel to resist softening at elevated temperatures is related to its ability to develop secondary hardening and to the amount of special phases, such as excess alloy carbides, in the microstructure. Useful information on the ability of tool steels to resist softening at elevated temperatures can be obtained from tempering curves such as those in Fig. 1. Hardness testing at elevated temperatures (see Fig. 8 and 9) also can provide useful information. Table 14 lists the hot hardness of selected high-speed and die steels.

Fabrication

The properties that influence the ease of fabrication of tool steels include machinability; grindability; weldability; hardenability; and extent of distortion, safety (freedom from cracking), and tendency to decarburize during heat treatment.

Machinability of tool steels can be measured by the usual methods applied to constructional steels. Results are reported as percentages of the machinability of water-hardening tool steels (see Table 15); 100% machinability in tool steels is equivalent to about 30% machinability in constructional steels, for which 100% machinability would be that of a free-machining constructional steel such as B1112. Improving the machinability of a tool steel by altering either the composition or preliminary heat treatment can be very important if a large amount of

Table 12 Thermal conductivity of selected tool steels

Temperature		Thermal conductivity	
°C	°F	W/m · K	Btu/ft · h · °F
Type W1			
95	200	48.3	27.9
260	500	41.5	24.0
400	750	38.1	22.0
540	1000	34.6	20.0
675	1250	29.4	17.0
815	1500	24.2	14.0
Type H11			
95	200	42.2	24.4
260	500	36.3	21.0
400	750	33.4	19.3
540	1000	31.5	18.2
675	1250	30.1	17.4
815	1500	28.6	16.5
Type H13			
215	420	28.6	16.5
350	660	28.4	16.4
475	890	28.4	16.4
605	1120	28.7	16.6
Type H21			
95	200	27.0	15.6
260	500	29.8	17.2
400	750	29.8	17.2
540	1000	29.4	17.0
675	1250	29.1	16.8
Type T1			
95	200	19.9	11.5
260	500	21.6	12.5
400	750	23.2	13.4
540	1000	24.7	14.3
Type T15			
95	200	20.9	12.1
200	500	24.1	13.9
400	750	25.4	14.7
540	1000	26.3	15.2
Type M2			
95	200	21.3	12.3
200	500	23.5	13.6
400	750	25.6	14.8
540	1000	27.0	15.6
675	1250	28.9	16.7

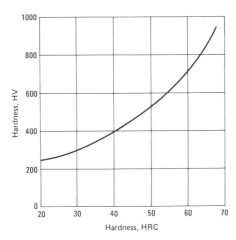

Fig. 7 Relationship of Vickers and Rockwell hardness scales

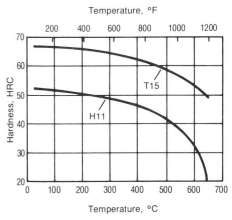

Fig. 8 Hot hardness of H11 and T15 tool steels. Type H11 has high resistance to softening at elevated temperatures; T15 has the highest resistance to softening. For these tests, H11 was air cooled from 1010 °C (1850 °F) and tempered 2 + 2 h at 565 °C (1050 °F); T15 was oil quenched from 1230 °C (2250 °F) and tempered 2 + 2 h at 550 °C (1025 °F). After hot-hardness testing at 650 °C (1200 °F), T15 had a room-temperature hardness of 63.4 HRC.

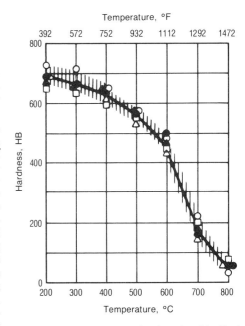

Fig. 9 Hot hardness (mutual indentation Brinell) of high-speed steel as a function of the temperature of testing. Average results of a series of tests on T1 tool steel. Ref 6

machining is required to form the tool and a large number of tools are to be made.

Grindability. One measure of grindability is the ease with which the excess stock on heat-treated tool steel can be removed using standard grinding wheels. The grinding ratio (grindability index) is the volume of metal removed per volume of wheel wear. The higher the grindability index, the easier the metal is to grind. The index is valid only for specific sets of grinding conditions. Table 16 gives grinding ratios for several high-speed steels. It should be noted that the grindability index does not indicate the susceptibility to cracking during or after grinding, the ability to produce the required surface (and subsurface) stress distribution, or the ease of obtaining the required surface smoothness.

Weldability. The ability to construct, alter, or repair tools by welding without causing the material to crack may be an important factor in the selection of a tool material, especially if the tool is large. It is only rarely of importance in selecting materials for small tools. Weldability is largely a function of composition, but welding method and procedure also influence weld soundness. Generally, tool steels that are deep hardening and that are classified as having relatively high safety in hardening are among the more readily welded tool steel compositions. These are generally the lower-alloy tool steel grades.

Hardenability includes both the maximum hardness obtainable when the quenched steel is fully martensitic and the depth of hardening obtained by quenching in a specific manner. In this context, depth of hardening must be defined, generally as a specific value of hardness or a specific microstructural appearance. As a very general rule, maximum hardness of a tool steel increases with increasing carbon content; increasing the austenitic grain size and the amount of alloying elements reduces the cooling rate required to produce maximum hardness (increases the depth of hardening). The Jominy end-quench test, which is applied extensively to measure hardenability of constructional steels (see the articles in the Section "Hardenability of Carbon and Low-Alloy Steels" in this Volume), has limited application to tool steels. This test gives useful information only for oil-hardening grades. Air-hardening grades are so deep hardening that the standard Jominy test is not sufficient to evaluate hardenability.

An air-hardenability test has been developed that is based on the principles involved in the Jominy test, but which uses only still-air cooling and a 152 mm (6 in.) diam end block to produce the very slow cooling rates of large sections. Such tests provide useful information for research but are of limited use for devising production heat treatments. By contrast, water-hardening grades of tool steel are so shallow hardening that the Jominy test is not sensitive enough. Special tests, such as the Shepherd PF test, are useful for research and for special applications of water-hardening tool steels.

In the Shepherd PF test, a bar 19 mm (¾ in.) in diameter, in the normalized condition, is brine quenched from 790 °C (1450 °F) and fractured; the case depth (penetration, P) is measured in 0.4 mm (¹⁄₆₄ in.) intervals, and the fracture grain size of the case (F) is determined by comparison with standard specimens. A PF value of 6 to 8 indicates a case depth of 2.4 mm (⁶⁄₆₄ in.) and a fracture grain size of 8. Fine-grain

776 / Specialty Steels and Heat-Resistant Alloys

Table 13 General properties of tool steels

AISI designation	Major factors(a)			Usual working hardness, HRC	Depth of hardening(d)	Minor factors		
	Wear resistance(b)	Toughness(c)	Hot hardness			Finest grain size at full hardness, Shepherd standard	As-quenched surface hardness, HRC	Core hardness (25 mm, or 1 in., diam round), HRC
Molybdenum high-speed steels								
M1	7	3	8	63–65	D	9½	64–66	64–66
M2	7	3	8	63–65	D	9½	64–66	64–66
M3, class 1	8	3	8	63–66	D	9½	64–66	64–66
M3, class 2	8	3	8	63–66	D	9½	64–66	64–66
M4	9	3	8	63–66	D	9½	65–67	65–67
M7	8	3	8	63–66	D	9½	64–66	64–66
M10	7	3	8	63–65	D	9½	64–66	64–66
M30	7	2	8	63–65	D	9½	64–66	64–66
M33	8	1	9	63–65	D	9½	64–66	64–66
M34	8	1	9	63–65	D	9½	64–66	64–66
M35	7	2	8	63–65	D	9½	64–66	64–66
M36	7	1	9	63–65	D	9½	64–66	64–66
M41	8	1	9	66–70	D	9½	63–65	63–65
M42	8	1	9	66–70	D	9½	63–65	63–65
M43	8	1	9	66–70	D	9½	63–65	63–65
M44	8	1	9	66–70	D	9½	63–65	63–65
M46	8	1	9	66–69	D	9½	63–65	63–65
M47	8	1	9	66–70	D	9½	63–65	63–65
Intermediate high-speed steels								
M50	6	3	6	61–63	D	8½	63–65	63–65
M52	6	3	6	62–64	D	8½	63–65	63–65
Tungsten high-speed steels								
T1	7	3	8	63–65	D	9½	64–66	64–66
T2	8	3	8	63–66	D	9½	65–67	65–67
T4	7	2	8	63–65	D	9½	63–66	63–66
T5	7	1	9	63–65	D	9½	64–66	64–66
T6	8	1	9	63–65	D	9½	64–66	64–66
T8	8	2	8	63–65	D	9½	64–66	64–66
T15	9	1	9	64–68	D	9½	65–68	65–68
Chromium hot-work steels								
H10	3	9	6	39–56	D	8	52–59	52–59
H11	3	9	6	38–55	D	8	53–55	53–55
H12	3	9	6	38–55	D	8	53–55	53–55
H13	3	9	6	40–53	D	8	51–54	51–54
H14	4	6	7	40–54	D	8	53–57	53–56
H19	5	6	7	40–55	D	8½	48–57	48–57
Tungsten hot-work steels								
H21	4	6	8	40–55	D	9	45–63	45–63
H22	5	5	8	36–54	D	9	48–56	48–56
H23	5	5	8	38–48	D	7	34–40	34–40
H24	5	5	8	40–55	D	9	52–56	52–56
H25	4	6	8	35–45	D	9	33–46	33–46
H26	6	4	8	50–58	D	9	51–59	51–59
Molybdenum hot-work steels								
H42	6	4	7	45–62	D	8½	54–62	54–62
Air-hardening, medium-alloy, cold-work steels								
A2	6	4	5	57–62	D	8½	63–65	63–65
A3	7	3	5	58–63	D	8½	63–65	63–65
A4	5	4	4	54–62	D	8½	61–63	61–63
A5	5	4	4	54–60	D	8½	60–62	60–62
A6	4	5	4	54–60	D	8½	60–62	60–62
A7	9	1	6	58–66	D	8½	64–66	64–66
A8	4	8	6	48–57	D	8	60–62	60–62
A9	4	8	6	40–56	D	8	55–57	55–57
A10	3	3	3	55–62	D	8	60–63	60–63
High-carbon, high-chromium, cold-work steels								
D2	8	2	6	58–64	D	7½	61–64	61–64
D3	8	1	6	58–64	D	7½	64–66	64–66
D4	8	1	6	58–64	D	7½	64–66	64–66
D5	8	2	7	58–63	D	7½	61–64	61–64
D7	9	1	6	58–66	D	7½	64–68	64–68
Oil-hardening cold-work steels								
O1	4	3	3	57–62	M	9	61–64	59–61
O2	4	3	3	57–62	M	9	61–64	59–61

(continued)

(a) Rating range from 1 (low) to 9 (high). (b) Wear resistance increases with increasing carbon content. (c) Toughness decreases with increasing carbon content and depth of hardening. (d) S, shallow; M, medium; and D, deep. (e) After carburizing. Source: Ref 6

Table 13 (continued)

AISI designation	Major factors(a) Wear resistance(b)	Toughness(c)	Hot hardness	Usual working hardness, HRC	Depth of hardening(d)	Minor factors Finest grain size at full hardness, Shepherd standard	As-quenched surface hardness, HRC	Core hardness (25 mm, or 1 in., diam round), HRC
Oil-hardening cold-work steels (continued)								
O6	3	3	2	58–63	M	9	65–67	50–55
O7	5	3	3	58–64	M	9	61–64	59–61
Shock-resisting steels								
S1	4	8	5	50–58	M	8	55–58	55–58
S2	2	8	2	50–60	M	8	61–63	56–60
S5	2	8	3	50–60	M	9	61–63	58–62
S6	2	8	3	50–56	M	8	56–58	56–58
S7	3	8	5	47–57	D	8	59–61	59–61
Low-alloy special-purpose steels								
L2	1	7	2	45–62	M	8½	56–62	54–58
L6	3	6	2	45–62	M	8	58–63	58–62
Low-carbon mold steels								
For hubbed and/or carburized cavities								
P2	1(e)	9	2(e)	58–64(e)	S	...	62–65(a)	15–21
P3	1(e)	9	2(e)	58–64(e)	S	...	62–64(a)	15–21
P4	1(e)	9	4(e)	58–64(e)	M	...	62–65(a)	33–35
P5	1(e)	9	2(e)	50–64(e)	S	...	62–65(a)	20–25
P6	1(e)	9	3(e)	58–61(e)	M	...	60–62(a)	35–37
For machined cavities								
P20	1(e)	8	2(e)	30–50	M	7½	52–54	45–50
P21	1	8	4	36–39	D	...	22–26	22–26
Water-hardening tool steels								
W1	2–4	3–7	1	58–65	S	9	65–67	38–43
W2	2–4	3–7	1	58–65	S	9	65–67	38–43
W5	3–4	3–7	1	58–65	S	9	65–67	38–43

(a) Rating range from 1 (low) to 9 (high). (b) Wear resistance increases with increasing carbon content. (c) Toughness decreases with increasing carbon content and depth of hardening. (d) S, shallow; M, medium; and D, deep. (e) After carburizing. Source: Ref 6

Table 14 Hot hardness of selected high-speed tool steels and die steels

AISI designation	Room temperature	Hardness, HRC — Hot hardness(a) 315 °C (600 °F)	425 °C (800 °F)	540 °C (1000 °F)	650 °C (1200 °F)
High-speed tool steels					
M1	65	61	58	54	32
M2	65	62	59	55	36
M3, class 1	65	63	60	56	36
M3, class 2	65	63	60	56	36
M4	66	63	60	56	37
M7	65	61	58	54	35
M10	65	60	57	52	33
M30	65	63	58	55	35
M33	65	64	60	57	40
M36	65	64	60	57	40
M42	68	66	65	62	44
M50	64	59	57	52	...
M52	64	60	57	53	...
T1	65	61	57	53	33
T4	65	61	59	55	38
T5	66	62	60	56	40
T15	68	64	61	57	42
Cold-work die steels					
A2	60	52	46	35	...
A8	58	55	52	45	...
D2	60	53	47	38	...
D4	62	52	46	37	...
Hot-work die steels					
A8	58	55	52	45	...
H11	54	49	47	42	22
H12	54	49	47	42	22
H13	55	49	47	42	22
H19	54	51	47	42	31
H21	54	52	49	45	29
H23	41	32	30	28	25
H26	58	54	50	46	31

(a) Small-diameter bars tested according to the recommended heat treatment. Source: Ref 7

water-hardening tool steels are those with fracture grain sizes (F values) of 8 or more. Deep-hardening steels of this type have P values of 12 or more; medium-hardening steels, 9 to 11; and shallow-hardening steels, 6 to 8.

Distortion and Safety in Hardening. Minimal distortion in heat treating is important for tools that must remain within close size limits. In general, the amount of distortion and the tendency to crack increase as the severity of quenching increases.

Table 15 Approximate machinability ratings for annealed tool steels

Type	Machinability rating
O6	125
W1, W2, W5	100(a)
A10	90
P2, P3, P4, P5, P6	75–90
P20, P21	65–80
L2, L6	65–75
S1, S2, S5, S6, S7	60–70
H10, H11, H13, H14, H19	60–70(b)
O1, O2, O7	45–60
A2, A3, A4, A6, A8, A9	45–60
H21, H22, H24, H25, H26, H42	45–55(b)
T1	40–50
M2	40–50
T4	35–40
M3, class 1	35–40
D2, D3, D4, D5, D7, A7	30–40
T15	25–30
M15	25–30

(a) Equivalent to approximately 30% of the machinability of B1112. (b) For hardness range 150 to 200 HB

778 / Specialty Steels and Heat-Resistant Alloys

Table 16 Typical grinding ratios for high-speed steels using three selected grinding wheels

Type	Hardness, HRC	Grinding ratio(a) 32A46-H8VBE	32A60-H8VBE	32A80-H8VBE
T15	65.7	0.49	0.62	0.51
M44	67.7	0.97	0.99	0.88
M41	68.7	1.2	1.6	1.4
M43	67.5	1.4	2.2	1.7
M42	68.8	4.8	6.5	3.8
M2	64.9	6.1	7.2	6.7
M1	64.9	7.8	8.0	11.9

(a) For the following conditions: work, 152 mm (6 in.) long by 38 mm (1.5 in.) wide; wheel size, 203 mm (8 in.) in diameter by 12.7 mm (0.5 in.) wide; wheel speed (idling), 30 m/s (6000 table speed, 0.3 m/s (60 sfm); unit crossfeed, 1.27 mm (0.050 in.) after each table traverse; unit downfeed, 0.025 mm (0.001 in.) after each complete cross feed; total down feed, 0.25 mm (0.010 in.) preceded by four unit down feeds to break wheel in after dressing with a diamond tool; grinding fluid, 1.25% water emulsion of general-purpose soluble oil

Table 17 Standard machining allowances for hot-rolled square and flat bars

Specified width mm	in.	Machining allowances(a) Top and bottom surfaces mm	in.	Edges mm	in.
Specified thickness, <12.7 mm (½ in.)					
0–12.7	0–½	0.64	0.025	0.64	0.025
>12.7–25.4	>½–1	0.64	0.025	0.89	0.035
>25.4–50.8	>1–2	0.76	0.030	1.02	0.040
>50.8–76.2	>2–3	0.89	0.035	1.27	0.050
>76.2–101.6	>3–4	1.02	0.040	1.65	0.065
>101.6–127.0	>4–5	1.14	0.045	2.03	0.080
>127.0–152.4	>5–6	1.27	0.050	2.41	0.095
>152.4–177.8	>6–7	1.40	0.055	2.67	0.105
>177.8–203.2	>7–8	1.52	0.060	3.05	0.120
>203.2–228.6	>8–9	1.52	0.060	3.30	0.130
>228.6–304.8	>9–12	1.52	0.060	3.56	0.140
Specified thickness, >12.7–25.4 mm (>½–1 in.)					
>12.7–25.4	>½–1	1.14	0.045	1.14	0.045
>25.4–50.8	>1–2	1.14	0.045	1.27	0.050
>50.8–76.2	>2–3	1.27	0.050	1.52	0.060
>76.2–101.6	>3–4	1.40	0.055	1.90	0.075
>101.6–127.0	>4–5	1.52	0.060	2.41	0.095
>127.0–152.4	>5–6	1.65	0.065	2.92	0.115
>152.4–177.8	>6–7	1.78	0.070	3.30	0.130
>177.8–203.2	>7–8	1.90	0.075	3.81	0.150
>203.2–228.6	>8–9	1.90	0.075	3.94	0.155
>228.6–304.8	>9–12	1.90	0.075	3.94	0.155
Specified thickness, >25.4–50.8 mm (>1–2 in.)					
>25.4–50.8	>1–2	1.65	0.065	1.65	0.065
>50.8–76.2	>2–3	1.65	0.065	1.78	0.070
>76.2–101.6	>3–4	1.78	0.070	2.16	0.085
>101.6–127.0	>4–5	1.78	0.070	2.67	0.105
>127.0–152.4	>5–6	1.90	0.075	3.18	0.125
>152.4–177.8	>6–7	2.03	0.080	3.68	0.145
>177.8–203.2	>7–8	2.03	0.080	4.19	0.165
>203.2–228.6	>8–9	2.41	0.095	4.32	0.170
>228.6–304.8	>9–12	2.54	0.100	4.32	0.170
Specified thickness, >50.8–76.2 mm (>2–3 in.)					
>50.8–76.2	>2–3	2.16	0.085	2.16	0.085
>76.2–101.6	>3–4	2.16	0.085	2.54	0.100
>101.6–127.0	>4–5	2.16	0.085	3.05	0.120
>127.0–152.4	>5–6	2.16	0.085	3.43	0.135
>152.4–177.8	>6–7	2.29	0.090	3.94	0.155
>177.8–203.2	>7–8	2.54	0.100	4.32	0.170
>203.2–228.6	>8–9	2.54	0.100	4.83	0.190
>228.6–304.8	>9–12	2.54	0.100	4.83	0.190
Specified thickness, >76.2–101.6 mm (>3–4 in.)					
>76.2–101.6	>3–4	2.92	0.115	2.92	0.115
>101.6–127.0	>4–5	2.92	0.115	3.18	0.125
>127.0–152.4	>5–6	2.92	0.115	3.56	0.140
>152.4–177.8	>6–7	2.92	0.115	4.32	0.170
>177.8–203.2	>7–8	3.18	0.125	4.83	0.190
>203.2–228.6	>8–9	3.18	0.125	4.83	0.190
>228.6–304.8	>9–12	3.18	0.125	4.83	0.190

(a) Minimum allowance per side for machining prior to heat treatment. Maximum decarburization limit, 80% of machining allowance

Resistance to decarburization is an important factor in determining whether a protective atmosphere is required during heat treating. In a decarburizing atmosphere, the rate of decarburization increases rapidly with increasing austenitizing temperature, and, for a given austenitizing temperature, the depth of decarburization increases in direct proportion to holding time. Some types of tool steel decarburize much more rapidly than others under the same conditions of atmosphere, austenitizing temperature, and time.

Machining Allowances

The standard machining allowance is the recommended total amount of stock that the user should remove from the as-supplied mill form to provide a surface free from imperfections that might adversely affect the response to heat treatment or the ability of tools to perform properly.

The decarburization resulting from oxidation at the exposed surfaces during the forging and rolling of the tool steel is a major factor in determining the amount of stock that should be removed. Although extra care is used in producing tool steels, scale, seams, and other surface imperfections that may be present must be removed.

Table 17 gives the standard machining allowances for various sizes of hot-rolled square and flat bars. Similar tables are available for other shapes and other methods of forming and finishing in ASTM specifications A 600, A 681, and A 686.

In addition to the standard machining allowance, sufficient stock must be provided to permit the cleanup of any decarburization or distortion that may occur during final heat treatment. The amount varies with the type of tool steel, the type of heat treating equipment, and the size and shape of the tool.

Group W and group O tool steels are considered highly resistant to decarburization. Group M steels, cobalt-containing group T steels, group D steels, and types H42, A2, and S5 are rated poor for resisting decarburization.

Decarburization during final heat treatment is undesirable because it alters the composition of the surface layer, thereby changing the response to heat treatment of this layer and usually adversely affecting the properties resulting from heat treatment. Decarburization can be controlled or avoided by heat treating in a salt bath or in a controlled atmosphere or vacuum furnace. When heat treating is accomplished in vacuum, a vacuum of 13 to 27 Pa (100 to 200 μm Hg) is satisfactory for most tools if the furnace is in good operating condition and has a very low leak rate. However, it is recommended that a vacuum of 7 to 13 Pa (50 to 100 μm Hg) be used wherever possible.

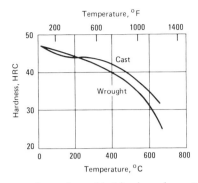

Fig. 10 Comparison of hot hardness for cast and wrought H13 tool steel. Source: Latrobe Steel Company

If special heat-treating equipment is not available, appreciable decarburization can be avoided by wrapping the tool in stainless steel foil. Type 321 stainless steel foil can be used at austenitizing temperatures up to about 1010 °C (1850 °F); either type 309 or type 310 foil is required at austenitizing temperatures from 1010 to 1205 °C (1850 to 2200 °F).

Precision Cast Hot-Work Tools

Precision casting of tools to nearly finished size offers important cost advantages through reductions in waste and machining. Casting is a particular advantage when pattern-making costs can be distributed over a large number of tools.

Experience with cast forging and extrusion dies has shown that cast tools are more resistant to heat checking. Minute cracks do occur, but they grow at much lower rates than in wrought material of the same grade and hardness. The slower propagation of thermal-fatigue cracks generally extends die life significantly. The mechanical testing of cast and wrought H13 indicates that yield and tensile strengths are virtually identical from room temperature to 595 °C (1100 °F), but that ductility is moderately lower in cast material. Hot hardness of cast H13 is higher than that of wrought H13 at temperatures above about 315 °C (600 °F); this hardness advantage increases with temperature, as illustrated in Fig. 10, and measures about eight points on the Rockwell C scale at 650 °C (1200 °F).

Because cast dies exhibit uniform properties in all directions, no problem of directionality (anisotropy) exists. The dimensional control of castings is very consistent after an initial die is made and any necessary corrections are incorporated in the pattern. Reasonable finishing allowances are 0.25 to 0.38 mm (0.010 to 0.015 in.) on the impression faces, 0.8 to 1.6 mm (1/32 to 1/16 in.) at the parting line of the mold, and 1.6 to 3.2 mm (1/16 to 1/8 in.) on the back and outside surfaces. The hot-work tool steels most commonly cast include H12, H13, H21, and H25.

Surface Treatments

In many applications, the service life of high-speed steel tools can be increased by surface treatments.

Oxide coatings, provided by treatment of the finish-ground tool in an alkali-nitrate bath or by steam oxidation, prevent or reduce adhesion of the tool to the workpiece. Oxide coatings have doubled tool life, particularly that of tools used to machine gummy materials such as soft copper and nonfree-cutting low-carbon steels.

Plating of finished high-speed steel tools with 0.0025 to 0.0125 mm (0.1 to 0.5 mil) of chromium also prolongs tool life by reducing adhesion of the tool to the workpiece. Chromium plating is relatively expensive, and precautions must be taken to prevent tool failure in service due to hydrogen embrittlement.

Carburizing is not recommended for high-speed steel cutting tools because the cases on such tools are extremely brittle. However, carburizing is useful for applications such as cold-work dies that require extreme wear resistance and that are not subjected to impact or highly concentrated loading. Carburizing is done at 1040 to 1065 °C (1900 to 1950 °F) for short periods of time (10 to 60 min) to produce a case 0.05 to 0.25 mm (0.002 to 0.010 in.) deep. The carburizing treatment also serves as an austenitizing treatment for the whole tool. A carburized case on high-speed steels has a hardness of 65 to 70 HRC, but does not have the high resistance to softening at elevated temperatures exhibited by normally hardened high-speed steel.

Nitriding successfully increases the life of all types of high-speed steel cutting tools. However, gas nitriding in dissociated ammonia produces a case that is too brittle for most applications. Liquid nitriding for about 1 h at 565 °C (1050 °F) provides a light case, increasing both surface hardness and resistance to adhesion. For nitrided high-speed steel taps, drills, and reamers used in machining annealed steel, fivefold increases in life have been reported, with average increases of 100 to 200%. Obviously, if this nitrided case is removed when the tool is reground, the tool must then be retreated, thereby reducing the cost advantage of the process.

Other special surface treatment processes, such as aerated nitriding baths, improve resistance to adhesive wear without producing excessive brittleness. Sulfur-containing nitriding baths provide a high-sulfur surface layer for additional resistance to seizing.

Titanium nitride coating is the most common of the newer types of wear-resistant coatings that are applied to tool steels. This shallow layer of titanium nitride, formed by physical vapor deposition process, has increased tool life in many instances by as much as 400%. This is primarily attributed to the increased lubricity of the coating due to a coefficient of friction that is one-third that of the bare metal surface of the tool. This increase in tool life justifies the application of the coating, despite the increase in cost. Additional information on the benefits of titanium nitride coatings used on tool steels is available in the article "High-Speed Tool Steels" in Volume 16 of the 9th Edition of *Metals Handbook*.

Sulfide Treatment. A low-temperature (190 °C, or 375 °F) electrolytic process using sodium and potassium thiocyanate provides a seizing-resistant iron sulfide layer. This process can be used as a final treatment for all types of hardened tool steels without great danger of overtempering.

REFERENCES

1. J.G. Gensure and D.L. Potts, *International Metallic Materials Cross-Reference*, 3rd Edition, Genium Publishing, 1988
2. C.W. Wegst, *Key to Steel*, Verlag Stahlschlüssel Wegst, 1989
3. "Tool Steels," Products Manual, American Iron and Steel Institute, March 1978
4. E. Orberg, F. Jones, and H. Horton, *Machinery's Handbook*, 23rd ed., H. Ryffel, Ed., Industrial Press, 1988
5. *Source Book on Industrial Alloys and Engineering Data*, American Society for Metals, 1978, p 251-292
6. G.A. Roberts and R.A. Gary, *Tool Steels*, 4th ed., American Society for Metals, 1980
7. "Tool Steel Guide," Product Literature, Teledyne Vasco, 1985

SELECTED REFERENCES

- P. Payson, *The Metallurgy of Tool Steels*, John Wiley & Sons, 1962
- R. Wilson, *Metallurgy and Heat Treatment of Tool Steels*, McGraw-Hill, 1975
- F.R. Palmer et al., *Tool Steel Simplified*, rev. ed., Chilton Book, 1978

P/M Tool Steels

K.E. Pinnow and W. Stasko, Crucible Materials Corporation, Crucible Research Center

POWDER METALLURGY (P/M) has become a major process for the manufacture of high-performance tool steels and tool steel products. The items now available include as-compacted or hot-worked billets and bars, semifinished parts, near-net shapes, and indexable cutting tool inserts. The P/M process has been used primarily for the production of advanced high-speed tool steels. However, it is now also being applied to the manufacture of improved cold-work and hot-work tool steels (Ref 1-5).

For most applications, the P/M tool steels offer distinct advantages over conventional tool steels. As a result of pronounced ingot segregation, conventional tool steels often contain a coarse, nonuniform microstructure accompanied by low transverse properties and problems with size control and hardness uniformity in heat treatment. Rapid solidification of the atomized powders used in P/M tool steels eliminates such segregation and produces a very fine microstructure with a uniform distribution of carbides and nonmetallic inclusions.

For high-speed tool steels, a number of important end-user properties have been improved by powder processing; notably, machinability, grindability, dimensional control during heat treatment, and cutting performance under difficult conditions where high edge toughness is essential (Ref 6). Several of these advantages also apply to P/M cold- and hot-work tool steels, which, compared to conventional tool steels, offer better toughness and ductility for cold-work tooling and better thermal fatigue life and greater toughness for hot-work tooling (Ref 2).

The alloying flexibility of the P/M process allows the production of new tool steels that cannot be made by conventional ingot processes, because of segregation-related hot-workability problems. Examples of these developments are the highly alloyed super-high-speed steels, such as CPM Rex 20, CPM Rex 76, and ASP 60, and the highly wear resistant cold-work tool steels, such as CPM 9V and CPM 10V.

As shown in Fig. 1, there are a large number of possibilities for powder production, shape making, and powder consolidation. The options shown in Fig. 1 are by no means complete because new processes are continuously being developed. However, because tool steels are highly stressed in service, the usual production routes applicable to P/M tooling are those that yield a fully dense, pore-free structure. Any remaining porosity in such products has been shown to act as a local stress raiser and to initiate premature failure (Ref 8, 9).

The basic production routes now in commercial use for P/M tool steels are summarized in Fig. 2. All these processes use gas- or water-atomized powders and either hot isostatic pressing (HIP), mechanical compaction (extrusion, forging, and so on), or vacuum sintering for densification (Ref 10). The basic difference among these processes is that the use of gas atomization will yield spherical particles, while water atomization will produce angular particles of significantly higher oxygen content. The angular particles can be cold pressed to provide a compact that has sufficient mechanical strength to be handled and processed directly, while the spherical gas-atomized powder must be encapsulated prior to densification.

The most widely used production practices for P/M tool steels utilize gas atomization and hot isostatic compaction and include the Anti-Segregation process (ASP), developed in Sweden by Stora Kopparberg and ASEA (Ref 11), and the Crucible Particle Metallurgy (CPM) process, developed in the United States by the Crucible Materials Corporation (Ref 12). Other processes using gas-atomized powders involve compaction at atmospheric pressure (CAP process), high-pressure compaction and hot working (STAMP process), and controlled spray deposition and hot working (CSD and Osprey) (Ref 13-16).

Most of the processes that use water-atomized powders utilize vacuum sintering for compaction and are used to produce complex tool shapes and indexable cutting tool inserts. They include the Powdrex pro-

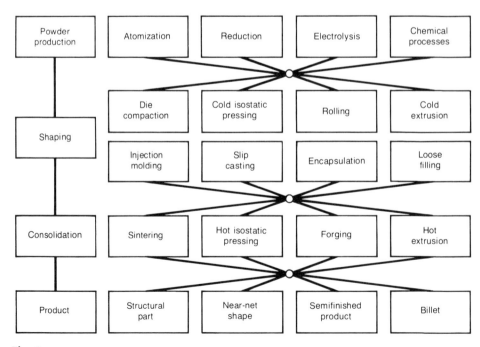

Fig. 1 P/M production methods. Source: Ref 7

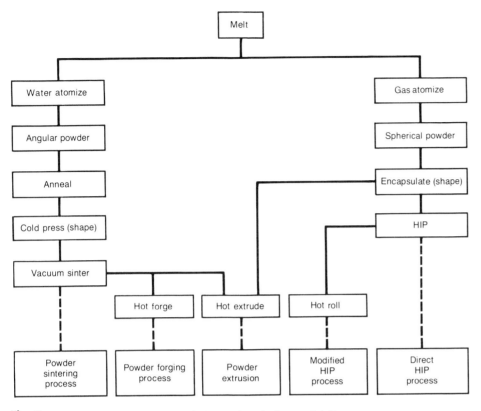

Fig. 2 Current manufacturing processes for P/M tool steels. Source: Ref 10

cess developed in the United Kingdom by Powdrex, Ltd. (Ref 17), and the full densification FULDENS process, developed in the United States by Consolidated Metallurgical Industries, Inc. (Ref 18) and currently licensed by Krupp Engineering, Inc. Additional information on the processes available for the production of P/M tool steels is given in the references at the end of this article and in Volumes 3, 7, and 16 of the 9th Edition of *Metals Handbook*.

P/M High-Speed Tool Steels

A large variety of P/M high-speed tool steels are available (Table 1). The P/M process has been applied primarily to standard American Iron and Steel Institute (AISI) high-speed tool steels (M2, M3, M4, M35, and so on) that are normally produced by conventional means but have significantly improved toughness and grindability when the P/M process is used. Highly resulfurized grades, such as M2HCHS and M3HCHS, also offer improved machinability when produced by this process. Other grades that are very difficult to produce by conventional means are readily produced by the P/M process, with an accompanying improvement in properties. Still others, such as CPM Rex 20, CPM Rex 76, and

Table 1 Commercial P/M tool steel compositions

Trade name	AISI designation	C	Cr	W	Mo	V	Co	S	Other	Hardness, HRC
High-speed tool steels(a)										
ASP 23	M3	1.28	4.20	6.40	5.00	3.10	...	...	...	65–67
ASP 30	...	1.28	4.20	6.40	5.00	3.10	8.5	...	...	66–68
ASP 60	...	2.30	4.00	6.50	7.00	6.50	10.50	...	...	67–69
CPM Rex M2HCHS	M2	1.00	4.15	6.40	5.00	2.00	...	0.27	...	64–66
CPM Rex M3HCHS	M3	1.30	4.00	6.25	5.00	3.00	...	0.27	...	65–67
CPM Rex M4	M4	1.35	4.25	5.75	4.50	4.00	...	0.06	...	64–66
CPM Rex M4HS	M4	1.35	4.25	5.75	4.50	4.00	...	0.22	...	64–66
CPM Rex M35HCHS	M35	1.00	4.15	6.00	5.00	2.00	5.0	0.27	...	65–67
CPM Rex M42	M42	1.10	3.75	1.50	9.50	1.15	8.0	...	...	66–68
CPM Rex 45	...	1.30	4.00	6.25	5.00	3.00	8.25	0.03	...	66–68
CPM Rex 45HS	...	1.30	4.00	6.25	5.00	3.00	8.25	0.22	...	66–68
CPM Rex 20	M62	1.30	3.75	6.25	10.50	2.00	...	...	...	66–68
CPM Rex 25	M61	1.80	4.00	12.50	6.50	5.00	...	...	...	67–69
CPM Rex T15	T15	1.55	4.00	12.25	...	5.00	5.0	0.06	...	65–67
CPM Rex T15HS	T15	1.55	4.00	12.25	...	5.00	5.0	0.22	...	65–67
CPM Rex 76	M48	1.50	3.75	10.0	5.25	3.10	9.00	0.06	...	67–69
CPM Rex 76HS	M48	1.50	3.75	10.0	5.25	3.10	9.00	0.22	...	67–69
HAP 10	...	1.35	5.0	3.0	6.0	3.8	...	...	...	64–66
HAP 40	...	1.30	4.0	6.0	5.0	3.0	8.0	...	...	...
HAP 50	...	1.50	4.0	8.0	6.0	4.0	8.0	...	...	...
HAP 60	...	2.00	4.0	10.0	4.0	7.0	12.0	...	...	...
HAP 70	...	2.00	4.0	12.0	10.0	4.5	12.0	...	...	...
KHA 33N	...	0.95	4.0	6.0	6.0	3.5	...	...	0.60N	65–66
Cold-work tool steels										
CPM 9V	...	1.78	5.25	...	1.30	9.00	...	0.03	...	53–55
CPM 10V	All	2.45	5.25	...	1.30	9.75	...	0.07	...	60–62
CPM 440V	...	2.15	17.50	...	0.50	5.75	...	...	...	57–59
Vanadis 4	...	1.50	8.00	...	1.50	4.00	...	...	...	59–63
Hot-work tool steels										
CPM H13	H13	0.40	5.00	...	1.30	1.05	...	...	...	42–48
CPM H19	H19	0.40	4.25	4.25	0.40	2.10	4.25	...	...	44–52
CPM H19V	...	0.80	4.25	4.25	0.40	4.00	4.25	...	...	44–56

(a) HCHS, high carbon, high sulfur; HS, high sulfur

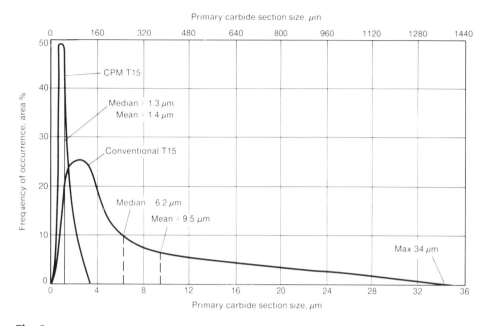

Fig. 3 Primary carbide size distributions in CPM and conventionally produced T15 high-speed tool steel

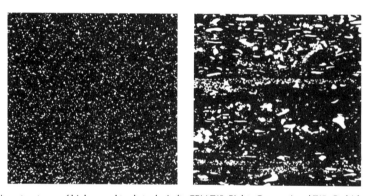

Fig. 4 Microstructures of high-speed tool steels. Left: CPM T15. Right: Conventional T15. Carbide segregation and its detrimental effects are eliminated with the CPM process, regardless of the size of the products. Courtesy of Crucible Materials Corporation

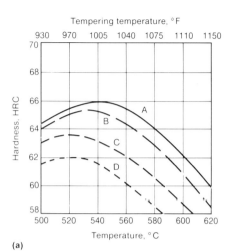

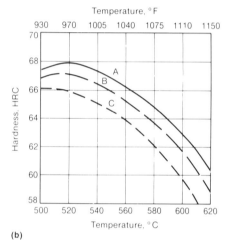

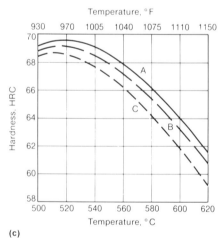

Fig. 5 Hardness of ASP steels after hardening and tempering a 25 mm (1 in.) diam specimen three times for 1 h. (a) ASP 23. (b) ASP 30. (c) ASP 60, cooled in salt bath. Hardening temperatures for the curves are: A, 1180 °C (2155 °F); B, 1150 °C (2100 °F); C, 1100 °C (2010 °F); D, 1050 °C (1920 °F)

ASP 60, represent superhigh-speed steels that are very difficult to produce by conventional means and are made only by P/M processing.

A distinguishing feature of P/M high-speed tool steels is the uniform distribution and small size of the primary carbides. The same is true of the sulfide inclusions in resulfurized grades. The size distributions of the primary carbides in P/M-produced and conventionally produced T15 have been measured, as shown in Fig. 3. Most of the carbides in CPM high-speed tool steels are less than about 3 μm (120 μin.), while those in the conventional product cover the entire size range up to about 34 μm (1360 μin.), with a median size of 6 μm (240 μin.). Microstructures of CPM and conventionally processed T15 are shown in Fig. 4.

Heat Treatment

Proper heat treatment of tool steels is essential for developing their properties. This is especially true of high-speed and high-alloy materials. Improper heat treatment of these steels can result in a tool with greatly reduced properties or even one that is unusable. Powder metallurgy tool steels utilize the same basic heat treatments as

Table 2 Austenitizing temperatures of ASP 23 steel

Hardness, HRC(a)	Temperature °C	Temperature °F	Salt bath(b) min/mm	Salt bath(b) min/in.	Other furnace, min(c)
58	1000	1830	0.59	15	30
60	1050	1925	0.47	12	25
62	1100	2010	0.39	10	20
64	1140	2085	0.31	8	15
66	1180	2155	0.24	6	10

(a) After triple temper at 560 °C (1040 °F); hardness values may vary by ±1%. (b) Total immersion time after preheating. (c) Holding time in minutes after tool has reached full temperature

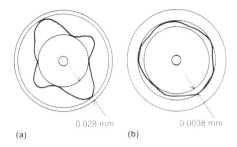

Fig. 6 Out-of-roundness measurements on test disks after hardening and tempering. Test disks machined from 102 mm (4 in.) diam bars. (a) AISI M2. (b) ASP 30

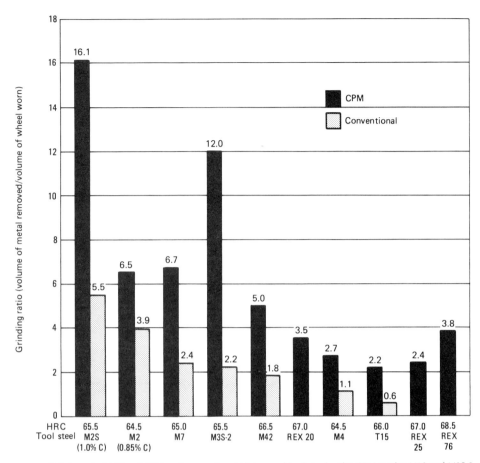

Fig. 7 Relative grindability of CPM and conventional high-speed tool steels. The CPM grades M2S and M3S-2 contain 0.25 to 0.30% S. Source: Crucible Materials Corporation

their conventional counterparts, but they tend to respond more rapidly and with better predictability to heat treatment because of their more uniform microstructure and finer carbide size. The basic heat treatments used include preheating, austenitizing, quenching, and tempering. However, optimum heat-treating temperatures may vary somewhat, even if chemical compositions are identical.

The following procedures are specific to ASP high-speed tool steels, but are generally applicable to all P/M high-speed tool steels. Deviations from these practices may be needed for some specific alloys and applications; in such cases, the recommendations by the manufacturer should be followed. The basic heat treatment steps are:

- *Annealing*: Heat to 850 to 900 °C (1560 to 1650 °F). Slow cool 10 °C/h (18 °F/h) to 700 °C (1290 °F). Typical annealed hardness values for ASP 23 are about 26 HRC maximum, 32 HRC for ASP 30, and 36 HRC for ASP 60
- *Stress relieving (before hardening)*: Hold for approximately 2 h at 600 to 700 °C (1110 to 1290 °F). Slow cool to 500 °C (930 °F) in furnace
- *Hardening*: Preheat in two steps, first at 450 to 500 °C (840 to 930 °F) and then at 850 to 900 °C (1560 to 1650 °F). Austenitize at 1050 to 1180 °C (1920 to 2155 °F) and quench, preferably in neutral salt baths. Cool to hand warmth. See Table 2 for recommended austenitizing temperatures
- *Tempering*: Raise temperature to 560 °C (1040 °F) or higher. Repeat two or three times for at least 1 h at full temperature. Cool to room temperature between tempers

The hardnesses of three P/M high-speed tool steels after hardening and tempering are shown in Fig. 5.

Three types of distortion are experienced metallurgically during heat treatment:

- Normal volume change due to phase transformations in the steel
- Variations in volume change in different parts of the tool due to the segregation in the steel
- Distortion due to residual stress caused by machining or nonuniform heating and cooling during heat treatment

Powder metallurgy grades, however, differ significantly from conventionally manufactured high-speed tool steels. Dimensional changes are more uniform in all directions. Because P/M high-speed tool steels are segregation free, variations in dimensional change are smaller. Therefore, the dimensional change that occurs during hardening can be predicted more accurately. Conventionally processed high-speed tool steels go out-of-round in a four-cornered pattern. The extent of distortion during heat treatment depends on the type and degree of segregation. In P/M high-speed tool steels, anisotropy is smaller, and out-of-roundness occurs in a close, circular pattern. Figure 6 shows typical results of measuring 102 mm (4 in.) diam disks after hardening and tempering. With P/M high-speed tool steels, cracking and variation of hardness are minimized because of their fine-grained, uniform structure.

Precautions must be taken to control distortion caused by residual stresses during heat treating. Mechanical stresses from rough machining can be eliminated by stress relieving prior to heat treating and finish machining.

Manufacturing Properties

During manufacturing, as much as 50% of a tool blank can be removed by sawing, turning, drilling, milling, grinding, and other machining operations. Consequently, the machinability and grindability of high-speed steels are important factors that affect alloy selection and the cost of a finished tool. In general, the machinability of the P/M high-speed tool steels in the annealed condition is comparable to and, in many cases, superior to that of conventional high-speed tool steels of similar composition. However, an important advantage of the P/M process is that the machinability of P/M tool steels can be improved by increasing their sulfur contents to much higher than conventional levels without sacrificing toughness or cutting performance. As indicated in Table 1, several P/M high-speed tool steels are available that contain as much as 0.30% S. Depending on specific conditions, the highly resulfurized P/M tool steels can offer as much as a 30% improvement in machinability over P/M tool steels with lower sulfur levels or conventional high-speed tool steels of similar composition.

The grindability of P/M high-speed steels is also superior to that of conventional high-speed steel of the same composition

784 / Specialty Steels and Heat-Resistant Alloys

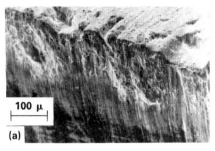 

Fig. 8 Comparison of cutting edge wear of a conventional high-speed tool steel and a P/M high-speed tool steel. (a) Cutting edge of tool made of conventional AISI M2 material showing severe microchipping. (b) Cutting edge of tool made of P/M-processed ASP 23 material showing no microchipping under the same service conditions. Courtesy of Speedsteel, Inc.

because of the small size and uniform distribution of carbides in P/M steels, regardless of bar size. The relative grindability of several conventional and P/M high-speed tool steels is illustrated in Fig. 7. The grinding ratio, or volume of metal removed to the volume of wheel worn, is clearly superior for the P/M tool steels, particularly those with high sulfur contents. As expected, the grinding ratios generally decrease for both the conventional and P/M tool steels as their alloy and carbon contents increase. The machining and grinding practices recommended for P/M high-speed steels are given in the article "Machining of P/M Tool Steels" in Volume 16 of the 9th Edition of *Metals Handbook*.

Cutting Tool Properties

The cutting performance of high-speed steels is primarily determined by their resistance to wear, by their resistance to tempering at operating temperatures (hot hardness), and by their toughness (Ref 6). Wear resistance is generally a function of hardness and of the type, volume, and shape of the primary carbides present in the tool. In this respect, the higher hardness attainable with P/M high-speed tool steels, along with the greater amount of alloy carbides that can be included in these steels, constitutes a significant advantage over conventional high-speed steels. Temper resistance, or hot hardness, is largely determined by the composition and growth of the secondary hardening carbides and is promoted by vanadium, molybdenum, and cobalt; these elements can be used in larger amounts in P/M high-speed steels than in conventional steels without degrading properties. Toughness of the high-speed steels is determined by the state of tempering of the matrix and the spatial and size distribution of the primary carbides. Here again, the uniform distribution and small size of the carbides in P/M high-speed steels represents an important toughness advantage.

The importance of toughness in high-speed tool steels is illustrated in Fig. 8. A cutting edge may suffer from repeated microchipping. As Fig. 8 shows, the cutting edge of ASP 23 steel displays minimal wear. The cutting edge of conventional M2, however, shows microchipping under the same service conditions. Microchipping blunts the cutting edge, increases stress, and accelerates other wear factors.

As shown in Fig. 9, toughness and hardness can be controlled by varying the hardening temperature. A low hardening temperature produces good toughness. A higher hardening temperature increases hardness, but lowers toughness.

Figure 10 summarizes the relative toughness, wear resistance, and red (hot) hardness characteristics of several P/M and conventional versions of various AISI high-speed tool steel grades. As Fig. 10 shows, the wear resistance and red hardness of a given grade of P/M high-speed tool steel are equal to those of the conventional versions of the same steel. However, wear resistance and red hardness generally increase with alloy content and are highest with the very highly alloyed grades, such as M4, T15, CPM Rex 45, M62 (CPM Rex 20), and M48 (CPM Rex 76). All of these grades are best produced or can be produced only by P/M methods. The toughness comparison shows that the P/M version of each grade is notably tougher than the conventional version.

Alloy Development

As indicated in Table 1, a number of P/M high-speed steels have been developed that cannot be made by conventional methods because of their high carbon, nitrogen, or alloy contents. Examples of these are CPM Rex 20, CPM Rex 25, CPM Rex 76, and ASP 60. CPM Rex 76 is a cobalt-rich high-speed steel with exceptional hot hardness and wear resistance, along with greatly increased tool life in difficult cutting operations. Its high alloy content (32.5%, compared to 27.8% for T15 and 25% for M42) renders this alloy unforgeable if produced by conventional methods.

CPM Rex 20 and CPM Rex 25 are examples of two high-performance P/M high-speed tool steels that were developed as cobalt-free equivalents of M42 or T15 (Ref 19). The latter two grades contain 8 and 5% Co, respectively, and are used in the machining of those difficult-to-machine superalloys and titanium alloys used by the aircraft industry. Cobalt increases the solidus

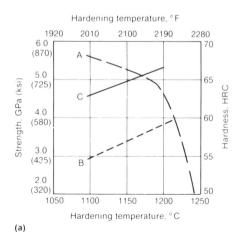 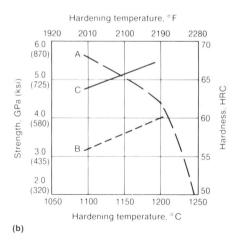 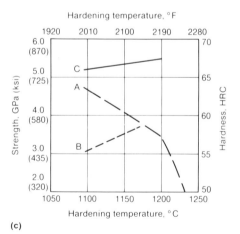

Fig. 9 Bend test results to determine the toughness of P/M-processed ASP high-speed tool steels. A, ultimate bend strength; B, bend yield strength; C, hardness (HRC). (a) Bend strength of a test bar of ASP 23 steel after hardening and tempering at 560 °C (1040 °F) (three times for 1 h). (b) Bend strength of a test bar of ASP 30 steel after hardening and tempering at 560 °C (1040 °F) (three times for 1 h). (c) Bend strength of a test bar of ASP 60 steel after hardening and tempering at 560 °C (1040 °F) (three times for 1 h). Ultimate bend strength may vary ±10%; bend yield strength may vary ±5%; hardness values may vary ±1%. Source: Speedsteel, Inc.

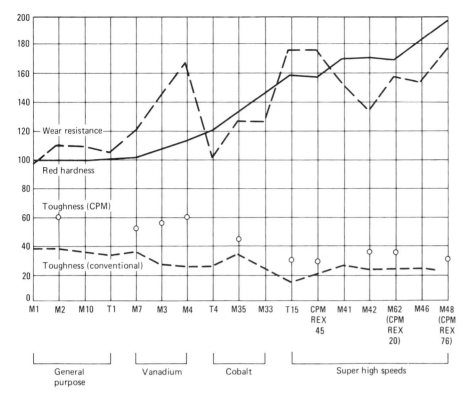

Fig. 10 Relative wear resistance, red (hot) hardness, and toughness of CPM and conventional high-speed tool steels. Source: Crucible Materials Corporation

temperature in high-speed tool steels, thus permitting the use of higher austenitizing temperatures to achieve greater solutioning of alloying elements. Cobalt also enhances the secondary hardening reaction, which results in a 1 to 2 HRC hardness advantage in the fully heat-treated condition. It also enhances hot hardness and temper resistance, thus allowing a tool to retain a sharp cutting edge at higher machining speeds that generate heat. Despite the advantages of adding cobalt, its high cost and occasional lack of availability have necessitated the development of cobalt-free alternatives.

Tables 3 and 4 show the results of temper resistance and hot hardness comparisons of specimens that were heat treated to full hardness. These results show that CPM Rex 20, CPM Rex M42, and conventional M42 are equivalent in both temper resistance and hot hardness. Table 5 compares the Charpy C-notch impact energy and bend fracture strength values obtained for CPM Rex 20, CPM Rex M42, and conventional M42. Both the Charpy C-notch impact energy and the bend fracture strength of CPM Rex 20 are equal to those of CPM Rex M42, but are notably higher than those of conventional M42. Table 6 gives the results of laboratory lathe tool tests in single-point turning on H13 tool steel and the P/M René 95 superalloy. The overall performance of CPM Rex 20 is comparable to that of CPM Rex M42.

Applications of P/M High-Speed Tool Steel

Milling. Milling cutters, such as those shown in Fig. 11, are a major application of P/M high-speed tool steels. Stock removal rates can usually be increased by raising the cutting speed and/or feed rate. In general, the feed per cutter tooth is increased in roughing operations, and the cutting speed is increased for finishing operations.

The performances of conventionally processed and P/M high-speed tool steel end mills in milling Ti-6Al-4V have been evaluated. In these tests, ASP 30 and ASP 60 were compared to M42. The cutting conditions used for this evaluation are given in Fig. 12, which shows tool life versus cutting speed. Both feed per tooth (0.203 mm, or 0.008 in.) and cutting speeds (>45.7 m/min, or 150 sfm) are higher than those used in production practice for machining aircraft parts. At a constant metal removal rate that corresponds to a cutting speed of 53.3 m/min (175 sfm), ASP 60 and ASP 30 lasted 8 times and 4.5 times longer, respectively, than the M42 end mill. Other materials machined by P/M high-speed tool steel milling cutters include tough, hardened steels, such as 4340; austenitic stainless steels, such as AISI type 316; and nickel-base superalloys, such as Nimonic 80.

Hole Machining. Reamers, taps, and drills (Fig. 13) are also made from P/M high-speed tool steels. In one application, the tool life of ¼-20 GH3 four-flute plug taps made from CPM Rex M4 and conventional M1, M7, and M42 were compared. The operation consisted of tapping a reamed 5.18 mm (0.204 in.) diam, 12.7 mm (0.500 in.) deep through hole in AISI 52100 steel at 32 to 34 HRC using a speed of 7.9 m/min (26 sfm) and chlorinated tapping oil. Eight to thirteen taps of each grade were tested. The CPM Rex M4 taps had an average tool life of 157 holes tapped before tool failure, compared to 35 holes for M1, 18 holes for M7, and 32 holes for M42. The tool life of the CPM Rex M4 in this application was about five times the life of conventional M42.

Broaching tools constitute another major application for P/M high-speed tool steels because tool life is often improved when P/M steels are used to broach difficult-to-cut materials, such as case-hardened steels and superalloys.

One application required broaching six ball tracks that are used in front-wheel-drive automobiles in constant velocity joint hubs made of a case-hardened steel. Figure 14 shows the joint hubs and the broaching tool used. In this broaching application,

Table 3 Temper resistance of CPM alloys

	Hardness, HRC				
	As-heat-treated at 1190 °C (2175 °F) + 550 °C (1025 °F) three times/2 h	As-heat-treated +		As-heat-treated +	
Alloy grade		595 °C (1100 °F)/2 h	595 °C (1100 °F)/2 + 2 h	650 °C (1200 °F)/ 2 h	650 °C (1200 °F)/ 2 + 2 h
CPM Rex 20. . . . . 67.5		66	65.5	60	57
CPM Rex M42. . . . . 67.5		65.5	65	59	55.5
Conventional M42. . . . . 67.5		65	65	59	55

Table 4 Hot hardness of CPM alloys

	Hot hardness, HRC				
Alloy grade	At room temperature before test	At 540 °C (1000 °F)	At 595 °C (1100 °F)	At 650 °C (1200 °F)	At room temperature after test
CPM Rex 20. . . . . 67.5		58.0	56.0	47.5	64.0
CPM Rex M42. . . . . 67.0		58.5	56.0	48.0	63.0
Conventional M42. . . . . 66.5		58.5	56.0	48.0	62.0

Table 5 Charpy C-notch impact and bend fracture strengths of two CPM alloys and one conventional alloy

Alloy grade	Austenitizing temperature(a) °C	°F	Hardness, HRC	Charpy C-notch impact energy J	ft · lbf	Bend fracture strength MPa	ksi
CPM Rex 20	1190	2175	67.5	16	12	4006	581
CPM Rex M42	1190	2175	67.5	16	12	4006	581
Conventional M42	1190	2175	67.5	7	5	2565	372

(a) 4-min soak in salt bath and oil quenched. Tempered at 550 °C (1025 °F) three times for 2 h

Table 6 Lathe tool test results on CPM alloys

Alloy grade	Austenitizing temperature °C	°F	Hardness, HRC	Tool life, minutes to 0.38 mm (0.015 in.) flank wear		
				Intermittent cut on H13 steel at 33 HRC	Continuous cut on H13 steel at 33 HRC	Continuous cut on P/M René 95 at 33 HRC
CPM Rex 20	1190	2175	67.5	8.5	14	31
CPM Rex M42	1190	2175	67	8	16	27
Test conditions						
Speed, m/s (sfm)				0.20 (40)	0.20 (40)	0.06 (12)
Feed, mm/rev (in./rev)				0.10 (0.004)	0.14 (0.0055)	0.18 (0.007)
Depth of cut, mm (in.)				1.57 (0.062)	1.57 (0.062)	1.57 (0.062)
Coolant				None	None	None

surface finish and form tolerance requirements are high because subsequent machining is not performed on the ball tracks. In broaching with low-carbon M35 tools with an 18.0 mm (0.709 in.) diameter and a 185.0 mm (7.283 in.) length, the total number of hubs machined per tool was 5600. The M35 tools experienced severe flank wear and developed a large built-up edge, which produced poor surface finishes. With an ASP 30 tool, 20 000 parts were produced.

Large broaching tools, such as those shown in Fig. 15, are also being made from P/M high-speed tool steels, such as CPM M3 and M4, to upgrade the broach material and its performance. In general, large rounds for broaches are not available in conventional high-speed tool steels in sizes above about 254 mm (10 in.), but the larger sizes are available in P/M high-speed tool steels. One application for these tools is the broaching of involute splines in bores of truck transmission gear blanks. Bores up to 305 mm (12 in.) in diameter and 1380 mm (54¼ in.) in length have been cut using such tools.

Gear Manufacturing. Gear hobs (Fig. 16) made from P/M high-speed tool steels can also provide substantial cost reductions by increasing machining rates. One application called for the hobbing of rear axle gears for heavy-duty trucks and tractor differentials. Hobs made of conventionally processed AISI M35 (65 HRC) and ASP 30 (67 HRC) were compared. Production results showed that the flank wear land on the hobs made of ASP 30 (0.44 mm, or 0.017 in.) was much less than on hobs made of M35 (0.71 mm, or 0.028 in.). Chipping of the edges was infrequent on the ASP 30 hobs, while the M35 hobs frequently displayed such damage.

Sintered Tooling

Figure 17 shows a number of P/M high-speed steel tools made by vacuum sintering cold isostatically pressed or mechanically pressed powders. Due to the higher temperatures used in processing, the primary carbides tend to be larger in vacuum-sintered, rather than hot isostatically pressed, high-speed steels. However, they are still smaller than in conventional high-speed steels of similar composition. Applications in which fully dense vacuum-sintered high-speed steels are currently in use include screw machine tooling, gear cutting tools, and indexable cutting tool inserts.

P/M Cold-Work Tool Steels

A number of improved, high-vanadium P/M tool steels designed for high-wear and cold-work applications are commercially available. The chemical compositions of four representative P/M cold-work grades are given in Table 1. As with P/M high-speed tool steels, the more uniform microstructure of the P/M cold-work steels yields

Fig. 11 Typical milling cutters made from P/M high-speed tool steels. Courtesy of Speedsteel Inc.

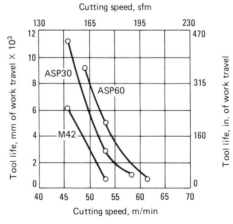

Cutter, mm (in.) .... 25 (1) diam end mills
Feed, mm/tooth (in./tooth) 0.203 (0.008)
Radial depth of cut, mm (in.) 6.35 (0.250)
Axial depth of cut, mm (in.) 25.4 (1.000)
Cutting fluid .... Soluble oil (1:20)
Tool life end point, mm (in.) ..... 0.5 (0.020) wear

Fig. 12 Results of end mill tests on Ti-6Al-4V. Hardness: 34 HRC

Fig. 13 Reamers, taps, and drills made from P/M high-speed tool steels. Courtesy of Crucible Materials Corporation

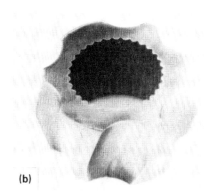

Fig. 14 Broaching application. (a) Tool made from P/M high-speed tool steel that was used to produce ball tracks on joint hub. (b) ASP 30 tools produced 20 000 parts compared to 5600 parts by tools made from conventional high-speed tool steel. Courtesy of Speedsteel Inc.

better toughness. This is very important in cold-work tooling because it allows higher hardnesses to be used with associated improvements in yield strength and wear resistance. Further, the use of higher vanadium contents in P/M cold-work tool steels than is practical in conventional cold-work tool steels has made for substantial improvements in wear resistance.

Both CPM 9V and CPM 10V are examples of two P/M cold-work tool steels with outstanding wear resistance and toughness (Ref 1). CPM 9V is capable of being heat treated to 58 to 60 HRC using austenitizing temperatures at or above 1149 °C (2100 °F). However, much better toughness is obtained in the hardness range of 46 to 55 HRC using austenitizing temperatures ranging from 1038 to 1121 °C (1900 to 2050 °F). The impact toughness and wear resistance values of CPM 9V, compared to those of a number of conventional and P/M hot- and cold-work tool steels, are shown in Fig. 18 and 19, respectively. In Fig. 19, the wear resistance of CPM 9V at 53 to 55 HRC, which is the maximum hardness level recommended for good toughness, is clearly superior to that of conventional D2 (62 HRC) and CPM M4 (64 HRC).

CPM 10V is favored over CPM 9V in cold-work applications where greater hardness, higher compressive strength, or better wear resistance is required. In practice, CPM 10V has proved to be more wear resistant than any commercially available high-alloy tool steel, including the most highly alloyed high-speed tool steels. The outstanding wear resistance properties of CPM 10V are illustrated in Fig. 20, which shows its laboratory wear test data compared to data for conventional D2, M2, and for CPM M4 at the maximum hardness levels typically used with these materials in cold-work tooling. Also included are wear data for D7, A7, and CPM T15. Although hardness has a major effect on wear resistance, its importance in this comparison is overshadowed by the very high vanadium carbide content of CPM 10V.

Figure 21 shows impact toughness data for CPM 10V compared to D2, M2, conventional M4, and CPM M4 at the same hardness levels used to generate the wear test data in Fig. 20. These results, confirmed by actual field experience, indicate that CPM 10V exhibits toughness comparable to that of conventional D2 and M4.

Typical applications for CPM 10V include punches and dies for cold-forming and stamping operations, powder metal compaction tooling, roll-forming rolls, woodworking tools, wear parts, and so on. CPM 10V should also be considered as a cost-effective replacement for tungsten carbide and the Ferro-Tic steel-bonded carbides, particularly where these materials are prone to chipping and breakage or where the cost of these materials is prohibitive. Some examples of tooling and wear part applications where CPM 10V has demonstrated its cost effectiveness follow.

Figure 22 shows a typical high-production-rate, progressive stamping die application in which CPM 10V punches were compared to

Fig. 15 Large broaching tool made from P/M high-speed tool steel that was used for broaching involute splines in bores of truck transmission gear blanks. Courtesy of Crucible Materials Corporation

Fig. 16 Gear hobs made from P/M high-speed tool steels. Courtesy of Speedsteel Inc.

788 / Specialty Steels and Heat-Resistant Alloys

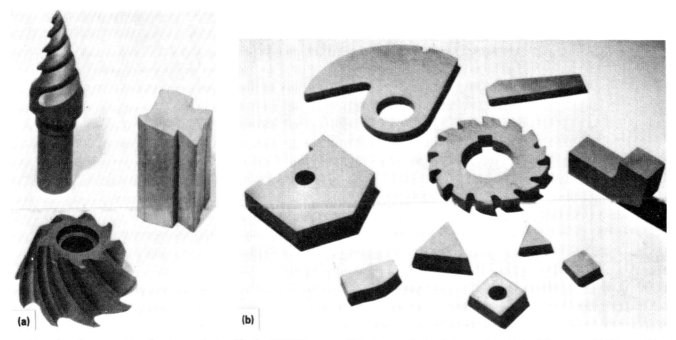

Fig. 17 Examples of vacuum-sintered parts manufactured by the FULDENS process. Note the complexity of shapes attainable by this process. (a) Using cold isostatic pressing. (b) Using mechanical pressing

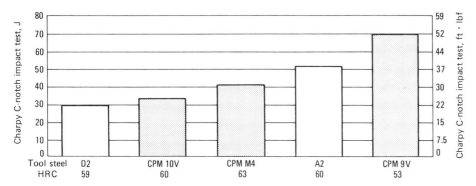

Fig. 18 Charpy C-notch impact properties of CPM 9V and other P/M and conventional tool steels at indicated hardnesses for cold-work applications

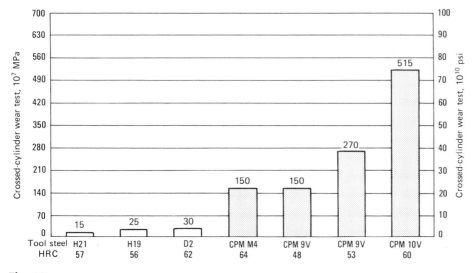

Fig. 19 Wear resistance of CPM 9V and other P/M and conventional tool steels at indicated hardnesses

D2 and CPM Rex M4. The material being stamped was a 0.381 mm (0.015 in.) thick copper-beryllium strip. In one operation, D2 piercing punches at 60 to 62 HRC averaged 75 000 parts before losing size; CPM Rex M4 at 64 HRC showed some signs of wear after 200 000 parts; and CPM 10V showed no wear after 400 000 parts and ultimately produced over 1 500 000 parts. In a coining operation in the same progressive die, CPM 10V was compared to D2 to produce a sharp radius bend indentation. D2 at 60 to 62 HRC required regrinding after 50 000 to 60 000 parts because of rounding of the punch nose radius, while the CPM 10V punch showed no signs of wear after 200 000 parts.

Figure 23 shows a unique wear part application in which CPM 10V was selected both for its high wear resistance and its ability to be manufactured economically into the component shown. This part is an automotive push-rod assembly for diesel engines. CPM 10V was selected to eliminate a localized wear problem on the ball ends. Manufacturing of the balls requires cold or warm upsetting of 6.35 mm (0.250 in.) diam slugs cut from coil stock to form approximately 9.53 mm (0.375 in.) diam balls that are subsequently rough ground, drilled, heat treated, finish ground, and resistance welded to the push-rod stems. This part could not be produced economically in carbide or other pressed and sintered materials.

CPM 440V is a high-vanadium, high-chromium tool steel for applications requiring both high wear resistance and good corrosion resistance. The composition of this material (Table 1) is essentially that of T440C martensitic stainless steel to which about 5.75% V

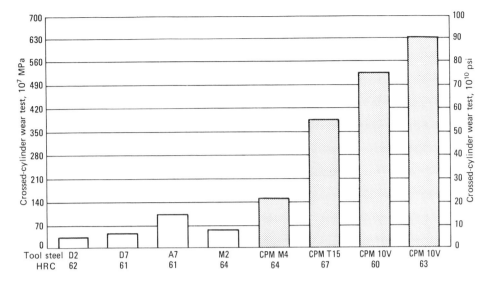

Fig. 20 Wear resistance of CPM 10V and other P/M and conventional tool steels at indicated hardnesses

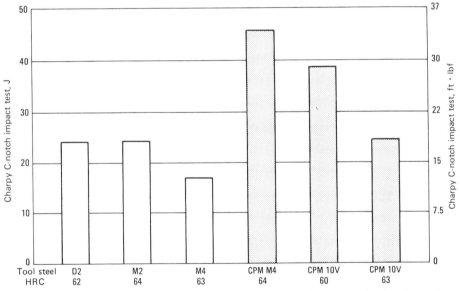

Fig. 21 Charpy C-notch impact properties of CPM 10V and other P/M and conventional tool steels at indicated hardnesses

and increased carbon have been added to improve wear resistance. The wear resistance and toughness properties of CPM 440V are compared with those of T440C, D2, and CPM 10V in Table 7. Although not as wear resistant as CPM 10V, CPM 440V exhibits outstanding wear resistance characteristics compared to conventional T440C and D2.

P/M Hot-Work Tool Steels

The absence of segregation in P/M tool steels makes them very attractive for hot-work tool and die applications because a frequent cause of premature failure of large die casting dies is thermal fatigue attributed to segregation and heterogeneous microstructures (Ref 20). As an example, H13 shot sleeves used to die cast four-cylinder engine blocks had an average service life of about 27 000 pieces when the starting material had little segregation, compared to about 13 000 pieces for material that showed clearly visible segregation in the starting material (Ref 21). During the past 10 years, efforts have been made to develop premium-quality, low-segregation tool and die steels by improved steelmaking processes such as vacuum degassing and vacuum arc or electroslag remelting.

Powder metallurgy processing offers an alternative method of producing segregation-free hot-work tool steels of both standard and improved compositions and offers near-net shape capability. The compositions of three P/M hot-work tool steels now commercially available are given in Table 1. Two are P/M versions of standard H13 and H19 hot-work tool steels, which, when made by P/M methods, have more uniform properties and equivalent or better toughness. The third is a high-vanadium modification of standard H19 with improved toughness and wear resistance.

Figure 24 compares the longitudinal microstructures of P/M H13, standard conventional H13, and premium-quality conventional H13. Both of the conventional H13 alloys show significant microbanding, or alloy segregation, while the P/M H13 material has a homogeneous, segregation-free structure.

Results of property determinations made on conventional and P/M H13 to compare their heat treatment response, size change, toughness, tensile strength, and thermal fatigue resistance characteristics are given in Tables 8 to 11. The hardness characteristics across large sections of conventional and P/M H13 after a standard heat treatment [that is, 1010 °C (1850 °F)/1 h, air cool, 593 °C (1100 °F)/2 + 2 h] are given in Table 8. These results show that the attainable hardnesses were the same but that the variation of hardness values across the P/M product was less than that of the conventional material, indicating greater uniformity. Table 9 gives the size change values for these products for the same heat-treated conditions. These results clearly show the isotropic nature of the P/M product.

Table 10 gives the standard Charpy V-notch impact strength and tensile strength values obtained at 21 and 538 °C (70 and 1000 °F). At both test temperatures, the uniform properties of the P/M product are shown by the equal values obtained for longitudinal and transverse specimens and for edge and center locations. Compared to the conventional H13, the P/M product has slightly lower longitudinal toughness but higher transverse toughness. The tension test results show that the P/M and conventional products have equal tensile properties.

Table 11 lists thermal fatigue test results obtained by alternately immersing specimens in molten lead at 621 °C (1150 °F) and water at 93 °C (200 °F) at a frequency of three cycles per minute. The average number of cycles to crack initiation was 50% greater for P/M H13.

Initial field trials of CPM H19V have been encouraging. In one application in which the material was used to punch holes in forged hammer heads at elevated temperatures, it yielded 1700 to 4500 pieces, while conventional H13 averaged only 900 pieces before failure.

Another major benefit of using the P/M process is the ability to produce near-net shape die cavities directly during HIP and thus minimize material input and subsequent machining. Figure 25 is an example of a die produced by ceramic core technology (Ref 22). In this process, a ceramic core material is cast in a reusable aluminum mold

Table 7 Comparative properties of CPM 440V, CPM 10V, conventional T440C, and D2 tool steel

Alloy grade	Hardness	Crossed-cylinder wear resistance 10^7 MPa	10^{10} psi	Charpy C-notch toughness J	ft · lbf
CPM 440V	59	276	40	16.3	12
CPM 440V	56	...	...	21.7	16
T440C	56.5	28	4	35.3	26
D2	59	28	4	31.2	23
CPM 10V	60	517	75	35.3	26

Table 8 Hardness of conventional and P/M H13

Product(a)	Section size mm	in.	Hardness taken at 1.6 mm (1/16 in.) intervals across section, HRC
P/M H13	152	6 round	47.3–48.1
H13	127	5 round	46.0–47.7
H13	152 × 406	6 × 16	46.0–48.6

(a) Preheated at 816 °C (1500 °F); austenitized at 1010 °C (1850 °F) for 1 h, air cooled; tempered at 593 °C (1100 °F) for 2 + 2 h

Table 9 Size change of conventional and P/M H13 after heat treatment

Product(a)	Section size mm	in.	Specimen location	Size change, 0.0001 in./in. Longitudinal direction	Transverse width	Transverse thickness
P/M H13	152	6 round	Edge center	+6	+6	+6
				+6	+6	+6
H13	127	5 round	Mid-radius	−5	+8	+14
H13	152 × 406	6 × 16	Edge center	−1	+4	+7
				−2	+6	+17

(a) Preheated at 816 °C (1500 °F); austenitized at 1010 °C (1850 °F) for 1 h, air cooled; tempered at 593 °C (1100 °F) for 2 + 2 h

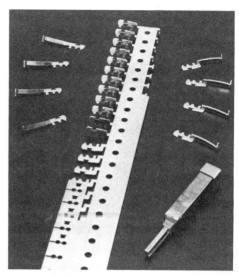

Fig. 22 CPM 10V punch and copper-beryllium blank used in a progressive stamping operation. Courtesy of Crucible Materials Corporation

ic core. Typical HIP parameters for hot-work die steels are 1149 °C (2100 °F) and 103 MPa (15 ksi). Following HIP, the ceramic cores can be removed by grit blasting and/or chemical leaching. The die is then ready for heat treatment and finish machining operations.

The same process can be used to place cooling water passages within the die. An example is shown in Fig. 27. With the P/M process, these passages can be designed and placed in a manner to make cooling most effective, rather than being restricted by machining limitations. With more effective cooling, die life would be expected to be enhanced.

Composite near-net shape dies can also be made by the P/M process. A highly alloyed material with high resistance to thermal fatigue or wear can be used in the die cavity while another lower cost or higher strength alloy is used in the remainder of the die. Figure 28 shows a P/M composite that uses Stellite 6 in the die cavity and H13 as the support material.

to the shape desired in the die cavity to be made. The ceramic core is then assembled with die steel powder in an outer steel container, as shown schematically in Fig. 26. The dies can be made in pairs in a single outer container by using a steel separating plate. After powder loading, the assembly is evacuated, sealed, and hot isostatically pressed to compact the powder to full density and form the powder around the ceram-

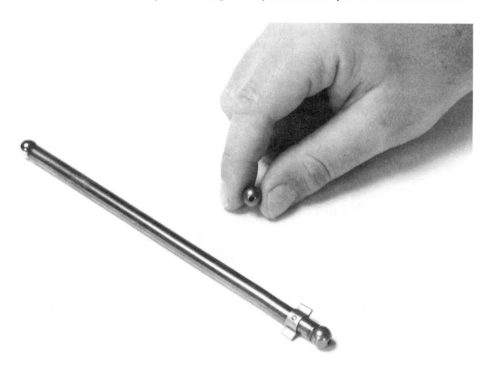

Fig. 23 Automotive push-rod assembly with wear-resistant spherical CPM 10V tips. Courtesy of Crucible Materials Corporation

REFERENCES

1. R. Dixon, "Advances in the Development of Wear-Resistant High-Vanadium Tool Steels for Both Tooling and Non-Tooling Applications," Paper 8201-085, presented at the ASM Metals Congress, St. Louis, American Society for Metals, 1982
2. E. Bayer, HIP Tool Materials, *Powder Metall.*, Vol 16 (No. 3), 1984, p 117-120
3. W. Stasko, V.K. Chandhok, and K.E. Pinnow, "Tool and Die Materials From Rapidly Solidified Powders," in *Rapidly Solidified Materials: Properties and Processing*, ASM INTERNATIONAL, 1988, p 49-57

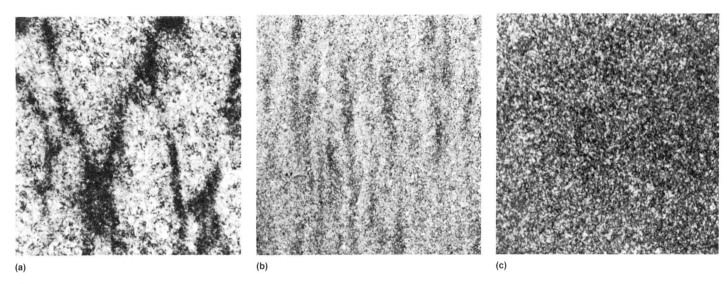

Fig. 24 Longitudinal microstructure of (a) standard conventional H13, (b) premium-quality conventional H13, and (c) P/M H13. Vilella's etch. 50×

4. H. Seilstorfer, PM Hot Work Tool Steels, *Metall*, Vol 42 (No. 2), Feb 1988, p 146-152 (in German)
5. V. Arnhold *et al.*, Cutting Tools From P/M High Speed Steels, *Powder Metall. Int.*, Vol 21 (No. 2), 1989, p 67-74 (in German)
6. R. Riedl *et al.*, Developments in High Speed Tool Steels, *Steel Res.*, Vol 58 (No. 8), 1987, p 339-352
7. P. Beiss, PM Methods for the Production of High Speed Steels, *Met. Powder Rep.*, Vol 38 (No. 4), April 1983, p 185-194
8. S. Karagoz and H. Fischmeister, Microstructure and Toughness in High Speed Tool Steels: The Influence of Hot Reduction and Austenitization Temperature, *Steel Res.*, Vol 58 (No. 8), 1987, p 360
9. H. Berns *et al.*, The Fatigue Behavior of Conventional and Powder Metallurgical High Speed Steels, *Powder Metall. Int.*, Vol 19 (No. 4), 1987, p 22-26
10. N. Kawai and H. Takigawa, Methods for Producing PM High-Speed Steels, *Met. Powder Rep.*, Vol 37 (No. 5), May 1982, p 237-240
11. P. Hellman *et al.*, The ASEA-Stora Process, in *Modern Developments in Powder Metallurgy*, Vol 4, Plenum Press, 1970, p 573-582
12. E.J. Dulis and T.A. Neumeyer, Particle Metallurgy of High Speed Tool Steel, in *Materials for Metal Cutting*, Publication 126, The Iron and Steel Institute, 1970, p 112-118
13. W.B. Kent, An Alternative Method of Processing High Speed Powder, in *Processing and Properties of High Speed Tool Steels*, Conference Proceedings, American Institute of Mining, Metallurgical, and Petroleum Engineers, 1980
14. M. Goransson *et al.*, Method of Special Steel Production Via the STAMP Process, *Met. Powder Rep.*, Vol 38 (No. 4), April 1983, p 205-209
15. B.A. Rickenson *et al.*, CSD: New Horizons for Special Steels, in *Towards Improved Performance of Tool Materials*, The Metals Society, 1982, p 73-77
16. R.G. Brooks *et al.*, The Osprey Process, *Powder Metall.*, Vol 20, 1977, p 100-102
17. J. Smart *et al.*, Pressing and Sintering Methods to Produce High Grade Tool Steels, *Met. Powder Rep.*, Vol 35 (No. 6), June 1980, p 241-244
18. M.T. Podob and R.P. Harvey, Advantages and Applications of CMI's FULDENS Process, in *Processing and Properties of High Speed Tool Steels*, The Metallurgical Society, 1980, p 181-195
19. F.R. Dax, W.T. Haswell, and W. Stasko, Cobalt-Free CPM High Speed Steels, in *Processing and Properties of High Speed Tool Steels*, The Metallurgical Society, 1980, p 148-158
20. K.E. Thelning, "How Far Can H-13 Die Casting Die Steel Be Improved," Paper 114, presented at the Sixth SDCE International Die Casting Congress, Cleveland, OH, Society of Die Casting Engineers, 1970

Table 10 Impact and tensile properties of conventional and P/M H13

Product(a)	Section size mm	in.	Specimen location(b)	Charpy V-notch impact strength J	ft · lbf	0.2% yield strength MPa	ksi	Tensile strength MPa	ksi	Elongation, %	Reduction of area, %
Tested at 21 °C (70 °F)											
P/M H13	152	6 round	A	13.6	10	1407	204	1682	244	11	42
			B	13.6	10	1407	204	1682	244	11	42
H13	152 × 406	6 × 16	C	13.6	10	1413	205	1696	246	12	43
			D	4	3	1400	203	1669	242	9	24
			E	12.2	9	...	...	...	...	...	...
Tested at 538 °C (1000 °F)											
P/M H13	152	6 round	A	24.4	18	945	137	1213	176	17	51
			B	24.4	18	945	137	1213	176	17	51
H13	152 × 406	6 × 16	C	24.4	18	910	132	1200	174	17	53
			D	16.3	12	924	134	1213	176	14	42
			E	40.7	30	...	...	...	...	...	...

(a) Preheated at 816 °C (1500 °F); austenitized at 1010 °C (1850 °F) for 1 h, air cooled; tempered at 593 °C (1100 °F) for 2 + 2 h. (b) A, longitudinal; B, transverse; C, longitudinal center; D, transverse edge; E, longitudinal edge

792 / Specialty Steels and Heat-Resistant Alloys

Table 11 Thermal fatigue resistance of conventional and P/M H13

Product(a)	Section size mm	in.	Average cycles to crack initiation
P/M H13	152	6 round	9000
H13	152 × 406	6 × 16	6000

(a) Preheated at 816 °C (1500 °F); austenitized at 1010 °C (1850 °F) for 1 h, air cooled; tempered at 593 °C (1100 °F) for 2 + 2 h

21. J.H. Stuhl and A.M. Schindler, "New Materials Study of 5% Chromium Type Steels for Use in Die Casting Dies," Paper G-T75-053, presented at the Eighth SDCE International Die Casting Congress, Detroit, MI, Society of Die Casting Engineers, 1975
22. V.K. Chandhok, J.H. Moll, and M.E. Ulitchny, Process for Producing Parts With Deep Pocketed Cavities Using P/M Shape Technology, in *Titanium Net Shape Technologies*, F.H. Froes and D. Eylon, Ed., The Metallurgical Society, 1980

SELECTED REFERENCES

- J.T. Berry, *High Performance High Hardness High Speed Steels*, Climax Molybdenum Company, 1970
- E.A. Carlson, J.E. Hansen, and J.C. Lynn, Characteristics of Full-Density P/M Tool Steel and Stainless Steel Parts, in *Modern Developments in Powder Metallurgy*, Vol 13, Metal Powder Industries Federation, 1980
- B.-A. Cehlin, "Improving Productivity With High Strength P/M High Speed Steel Cutting Tools," Paper MR82-948, presented at the Increasing Productivity With Advanced Machining Concepts Clinic, Los Angeles, CA, Society of Manufacturing Engineers, 1982
- P. Hellman, Wear Mechanism and Cutting Performance of Conventional and High-Strength P/M High-Speed Steels, *Powder Metall.*, Vol 25 (No. 2), 1982
- G. Hoyle, *High Speed Steels*, Butterworths, 1988
- A. Kasak and E.J. Dulis, Powder-Metallurgy Tool Steels, *Powder Metall.*, Vol 21 (No. 2), 1978, p 114-123
- A. Kasak, G. Steven, and T.A. Neumeyer, High-Speed Tool Steels by Particle Metallurgy, SAE Paper 720182, Society of Automotive Engineers, 1972, p 2-5
- O. Siegwarth, Higher Productivity With ASP Tooling Material, Technical Paper MF 81-137, Society of Manufacturing Engineers, p 1-22

Fig. 25 P/M H19 die produced using prealloyed powder and the ceramic core process. Courtesy of Crucible Materials Corporation

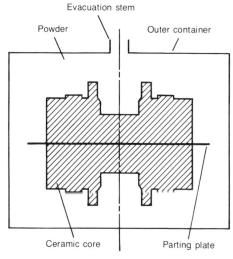

Fig. 26 Schematic diagram of an assembly for producing P/M dies by the ceramic core process

Fig. 27 P/M H13 die cut open to show cooling passages that were placed in the die using the ceramic core process. Courtesy of Crucible Materials Corporation

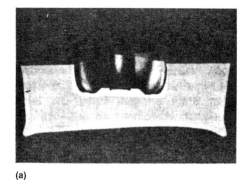
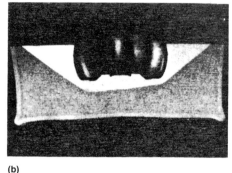

Fig. 28 Composite P/M die made of Stellite 6 and H13 tool steel. (a) About 6.5% Stellite. (b) About 38% Stellite. Source: Ref 4

Maraging Steels

Revised by Kurt Rohrbach and Michael Schmidt, Carpenter Technology Corporation

MARAGING STEELS comprise a special class of high-strength steels that differ from conventional steels in that they are hardened by a metallurgical reaction that does not involve carbon. Instead, these steels are strengthened by the precipitation of intermetallic compounds at temperatures of about 480 °C (900 °F). The term maraging is derived from martensite age hardening and denotes the age hardening of a low-carbon, iron-nickel lath martensite matrix.

Commercial maraging steels are designed to provide specific levels of yield strength from 1030 to 2420 MPa (150 to 350 ksi). Some experimental maraging steels have yield strengths as high as 3450 MPa (500 ksi). These steels typically have very high nickel, cobalt, and molybdenum contents and very low carbon contents. Carbon, in fact, is an impurity in these steels and is kept as low as commercially feasible in order to minimize the formation of titanium carbide (TiC), which can adversely affect strength, ductility, and toughness. Other varieties of maraging steel have been developed for special applications. Maraging steels are commercially produced by various steel companies in the United States and abroad.

The absence of carbon and the use of intermetallic precipitation to achieve hardening produce several unique characteristics that set maraging steels apart from conventional steels. Hardenability is of no concern. The low-carbon martensite formed after annealing is relatively soft—about 30 to 35 HRC. During age hardening, there are only very slight dimensional changes. Therefore, fairly intricate shapes can be machined in the soft condition and then hardened with a minimum of distortion. Weldability is excellent. Fracture toughness is considerably better than that of conventional high-strength steels. This characteristic in particular has led to the use of maraging steels in many demanding applications.

Physical Metallurgy

The unique characteristics of maraging steels have generated considerable interest, and extensive efforts have been devoted to gaining an understanding of the metallurgy of these steels. Much of this work has been summarized in review papers (Ref 1-3) and will not be recapitulated here. A brief description of the metallurgical characteristics of these steels is necessary, however, for an understanding of their behavior.

Maraging steels can be considered highly alloyed low-carbon, iron-nickel lath martensites. These alloys also contain small but significant amounts of titanium. Typical nominal compositions of maraging steels are given in Table 1. The phase transformations in these steels can be explained with the help of the two phase diagrams shown in Fig. 1, which depict the iron-rich end of the Fe-Ni binary system. Figure 1(a) is the metastable diagram plotting the austenite-to-martensite transformation upon cooling and the martensite-to-austenite reversion upon heating. Figure 1(b) is the equilibrium diagram showing that at higher nickel contents the equilibrium phases at low temperatures are austenite and ferrite.

The metastable diagram indicates the typical behavior of these steels during cooling from the austenitizing or solution annealing temperature. No phase transformations occur until the M_s temperature, the temperature at which martensite starts to transform from austenite, is reached. Even very slow cooling of heavy sections produces a fully martensitic structure, so there is no lack of hardenability in these alloys.

Alloying elements alter the M_s temperature significantly, but do not alter the characteristic that transformation is independent of cooling rate. In addition to nickel, the other alloy elements present in maraging steels generally lower the martensite transformation range, with the exception of cobalt, which raises it. One of the roles of cobalt in maraging steels is to raise the M_s temperature so that greater amounts of other alloying elements (that is, titanium and molybdenum, which lower the M_s temperature) can be added while still allowing complete transformation to martensite before the steel cools to room temperature.

Most grades of maraging steel have M_s temperatures of the order of 200 to 300 °C (390 to 570 °F) and are fully martensitic at room temperature. Therefore, retained austenite is generally not a problem in these alloys, and as a result, refrigeration treatments are not needed prior to aging. The martensite is normally a low-carbon, body-centered cubic (bcc) lath martensite containing a high dislocation density but no twinning. This martensite is relatively soft (~30 HRC), ductile, and machinable.

The age hardening of maraging steels is produced by heat treating for 3 to 9 h at temperatures of the order of 455 to 510 °C (850 to 950 °F). The metallurgical reactions that take place during such treatment can be

Table 1 Nominal compositions of commercial maraging steels

Grade	Composition, %(a)					
	Ni	Mo	Co	Ti	Al	Nb
Standard grades						
18Ni(200)	18	3.3	8.5	0.2	0.1	...
18Ni(250)	18	5.0	8.5	0.4	0.1	...
18Ni(300)	18	5.0	9.0	0.7	0.1	...
18Ni(350)	18	4.2(b)	12.5	1.6	0.1	...
18Ni(Cast)	17	4.6	10.0	0.3	0.1	...
12-5-3(180)(c)	12	3	...	0.2	0.3	...
Cobalt-free and low-cobalt bearing grades						
Cobalt-free 18Ni(200)	18.5	3.0	...	0.7	0.1	...
Cobalt-free 18Ni(250)	18.5	3.0	...	1.4	0.1	...
Low-cobalt 18Ni(250)	18.5	2.6	2.0	1.2	0.1	0.1
Cobalt-free 18Ni(300)	18.5	4.0	...	1.85	0.1	...

(a) All grades contain no more than 0.03% C. (b) Some producers use a combination of 4.8% Mo and 1.4% Ti, nominal. (c) Contains 5% Cr

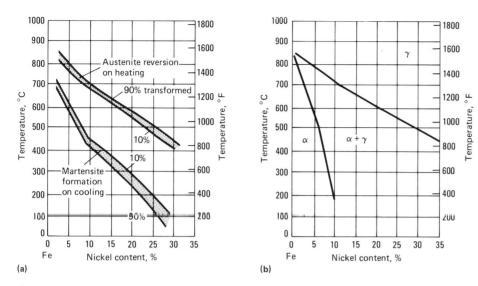

Fig. 1 Phase relationships in the iron-nickel system. (a) Metastable. (b) Equilibrium. Source: Ref 4

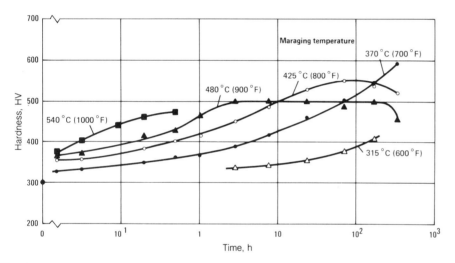

Fig. 2 Hardness of 18Ni(250) maraging steel versus aging time for various aging temperatures. Source: Ref 4

explained by using the equilibrium diagram (Fig. 1b).

With prolonged aging, the structure tends to revert to the equilibrium phases—primarily ferrite and austenite. Fortunately, the kinetics of the precipitation reactions that cause hardening are such that considerable age hardening—that is, approximately 20 HRC points (1035 MPa, or 150 ksi)—occurs before the onset of the reversion reactions that produce austenite and ferrite. With long aging times or high temperatures, however, hardness will reach a maximum and then will start to drop, as shown by the data in Fig. 2. Softening in these steels usually results not only from overaging in the usual sense of the term—that is, coarsening of the precipitate particles—but also from austenite reversion. The two processes are interlinked; dissolution of metastable nickel-rich precipitate particles in favor of equilibrium iron-rich precipitates locally enriches the matrix in nickel, which favors austenite formation. Very substantial amounts of austenite (of the order of 50%) can eventually be formed by overaging.

Maraging steels normally contain little or no austenite after standard maraging heat treatments. However, austenite is always present in the heat-affected zones around welds and is sometimes deliberately formed to enhance fabricability or service performance. For example, if maraging steel is to be used in an application where overaging in service is expected, such as in dies for aluminum die castings, it is deliberately overaged slightly before being put into service. This minimizes overaging in service, which often produces tensile stresses at the surface. Extreme overaging to form large amounts of austenite has also been employed as an intermediate treatment to enhance response to cold working or to minimize the effects of thermal gradients during hot working and subsequent storage of extraordinarily heavy sections.

Age hardening in maraging steels results primarily from the precipitation of intermetallic compounds. Precipitation takes place preferentially on dislocations and within the lath martensite to produce a fine uniform distribution of coherent particles. The major hardener is molybdenum, which upon aging initially forms Ni_3Mo, with an orthorhombic Cu_3Ti-type structure (Ref 5). The metastable Ni_3Mo phase forms initially because of its better lattice fit with the bcc martensitic matrix. Growth of the Ni_3Mo is restricted by coherency strains, and as such, further aging results in the *in situ* transformation of Ni_3Mo to the equilibrium Fe_2Mo phase, which has a hexagonal C14-type structure (Ref 6). Titanium, which is generally present in maraging steels, promotes additional age hardening through the precipitation of Ni_3Ti, which has a DO_{24} ordered hexagonal structure (Ref 6).

Cobalt does not directly participate in the age-hardening reaction, because this element does not form a precipitate with iron, nickel, molybdenum, or titanium in the 18Ni maraging alloy system. The main contribution of cobalt is to lower the solubility of molybdenum in the martensitic matrix and thus increase the amount of Ni_3Mo precipitate formed during age hardening. Some hardening also results from a short-range ordering reaction in the matrix that involves cobalt.

Molybdenum also plays the necessary supplemental role of minimizing localized grain-boundary precipitation by lowering the diffusion coefficients of a number of elements in solid solution. Precipitation of these grain-boundary phases severely impairs the toughness of most molybdenum-free ferrous alloys. Work by Schmidt (Ref 7) has shown that discrete particles of austenite are also present on the grain and subgrain boundaries in molybdenum-free 18Ni(300). It has been theorized that the precipitation of these discrete particles of austenite at the grain and subgrain boundaries results in a nickel-depleted zone, which adversely affects the toughness and ductility of the molybdenum-free 18Ni(300) alloy on a localized scale.

The precipitate particles are of a lattice size comparable to that of the martensite matrix and cause little distortion of the matrix. This characteristic, together with the absence of carbon, allows the steel to be age hardened to very high strength levels while minimizing changes in the shape of the part being hardened.

Another important precipitation reaction, which must be avoided if optimum toughness and ductility are to be achieved, can take place when the steel is in the austenitic condition. Maraging steels containing titanium are susceptible to the formation of TiC films at austenite grain boundaries. The solutioning of carbides at temperatures greater than 1150 °C (2100 °F), followed by holding at temperatures of the order of 900 to 1095 °C (1650 to 2000 °F), results in the reprecipitation of titanium carbide in the form of films on austenite grain boundaries. As discussed in the section "Hot Working" in this article, this problem first became evident in hot-worked billets and can be avoided by proper hot-working procedures. Prolonged annealing in this range should also be avoided.

Commercial Alloys

Table 1 lists the chemical compositions of the more common grades of maraging steel. The nomenclature that has become established for these steels is nominal yield strength (ksi units) in parentheses. Thus, for example, 18Ni(200) steel is normally age hardened to a yield strength of 1380 MPa (200 ksi). The first three steels in Table 1—18Ni(200), 18Ni(250), and 18Ni(300)—are the most widely used and most commonly available grades. The 18Ni(350) grade is an ultrahigh-strength variety made in limited quantities for special applications. Two 18Ni(350) compositions have been produced (see the footnote in Table 1). The 18Ni(Cast) grade was developed specifically as a cast composition.

Special varieties of maraging steels have been developed, including stainless grades, other cast grades, grades of other strength levels, a cobalt-free variety for nuclear service, a grade especially suited for heavy sections, and a grade with superior magnetic characteristics. Some of these steels have been made and used commercially, but only in limited amounts for specific applications.

A number of cobalt-free maraging steels and a low-cobalt bearing maraging steel have recently been developed. The driving force for the development of these particular alloys was the cobalt shortage and resultant price escalation of cobalt during the late 1970s and early 1980s. The nominal compositions for these alloys are also listed in Table 1.

The wrought steels in Table 1 are produced in various forms: forgings, plate, sheet, and bar stock are generally available. Most applications use bars or forgings. Thin strip, although it has been produced, is commercially uncommon. Special varieties can be produced in shapes suitable for specific applications.

Processing

Melting. Most grades of maraging steel are either air melted, then vacuum arc remelted, or vacuum induction melted, then vacuum arc remelted. Premium grades of maraging steels used in critical aircraft and aerospace applications, for which minimum residual element (carbon, manganese, sulfur, and phosphorus) and gas (O_2, N_2, and H_2) contents are required, are triple melted using the air, vacuum induction, and vacuum arc remelting processes.

The primary goals in melting are a clean microstructure, homogeneity, and low levels of deleterious residual elements, particularly sulfur and carbon. The sensitivity of maraging steels to the presence of inclusions is comparable to that of conventional steels of similar strength. Their sensitivity to residual elements, however, is different. Maraging steels show very little sensitivity to elements, such as arsenic, antimony, and phosphorus, that cause temper embrittlement in conventional quenched and tempered steels. As discussed in the section "Surface Treatment" in this article, maraging steels appear to be more resistant to hydrogen embrittlement than low-alloy steels. Carbon and sulfur are the most deleterious impurities in maraging steels because they tend to form brittle carbide, sulfide, carbonitride, and carbosulfide inclusions. These particles crack when the metal is strained, thus initiating fracture and lowering toughness and ductility.

Steels with high titanium and molybdenum contents are prone to microsegregation of these elements during solidification. If the ingot structure is not homogenized during hot working, a banded microstructure with highly anisotropic mechanical properties may persist. The only effective remedy for microsegregation is to adjust the ingot size and hot-working schedule to ensure elimination of the segregation. With the 18Ni(350) grade, ingot size may have to be smaller than normal, and the maximum hot-working temperature should be kept below 1230 °C (2250 °F).

Hot Working. Maraging steels can be hot worked by conventional steel mill techniques, even though allowances must be made for several unique characteristics. Steels with high titanium contents have greater hot strength than conventional steels and require higher hot-working loads or higher working temperatures. Working above about 1260 °C (2300 °F) should be avoided. To maximize their mechanical properties, maraging steels should be hot worked at the lowest temperatures that equipment power limitations permit.

The precipitation of TiC films at austenite grain boundaries must be avoided. This phenomenon first came to light in billets that had been worked at very high temperatures and then allowed to cool slowly through the temperature range of 750 to 1095 °C (1380 to 2000 °F) or to cool with inadvertent thermal arrests in this range. It is essential that long dwell times in this temperature range be avoided after working is completed so that the titanium and carbon remain in solution.

However, one should keep in mind that it is safe to heat into the 750 to 1095 °C (1380 to 2000 °F) range from room temperature because stable carbides will already have precipitated. This temperature range should be avoided only when cooling from temperatures above 1150 °C (2100 °F).

Cold Working. Maraging steels can be cold worked by any conventional technique when in the solution-annealed (unaged or as-transformed) condition. They have very low work-hardening rates and can be subjected to very heavy reductions (>50%) with only slight accompanying gains in hardness.

In maraging steels, both the strain-hardening exponent and tensile elongation are relatively low. Thus, cold-forming operations that depend on uniform tensile strain before necking (such as deep drawing) are generally limited to mild degrees of deformation. Uniform elongation can be significantly improved by overaging, which produces substantial amounts of austenite. This austenite is metastable and transforms to martensite during cold working, which greatly increases uniform elongation and deep-drawing capabilities.

Heavily cold worked structures can be softened by austenitizing or solution annealing at temperatures of about 815 °C (1500 °F). Cold-worked pieces can be directly age hardened, in which case the total strength includes the increase in strength produced by cold working plus that produced by precipitation hardening. Large amounts of prior cold work decrease toughness after direct aging, and considerable anisotropy in elastic modulus and in toughness may be present in unidirectionally worked structures.

Machining. Maraging steels can be machined by any conventional technique when in the solution-annealed or age-hardened condition. Machinability is generally as good as or better than that of conventional steels of the same hardness. Rigid equipment and firm tool supports are necessary. Tools should be flooded with cutting fluids free of lead, sulfur, and other low-melting components, because residues of these elements may cause surface embrittlement during subsequent heat treatment. The new generation of ceramic tooling provides superior cutting speeds and tool life when working with 18Ni maraging steels.

Heat Treating. The properties of conventionally heat treated maraging steels are given in Table 2 and the mechanical properites of 18Ni(250), cobalt-free 18Ni(250), and low-cobalt bearing 18Ni(250) are compared in Table 3. Maraging steels are usually solution annealed (austenitized) 1 h for each 25 mm (1 in.) of section size. It may be necessary to heat treat sheet products in dry hydrogen or dissociated ammonia to minimize surface damage. The cooling rate after annealing has a minimal effect on microstructure or properties. It is essential, however, to cool the steel completely to room temperature before aging. If this is not done, the steel may contain untransformed

796 / Specialty Steels and Heat-Resistant Alloys

Table 2 Heat treatments and typical mechanical properties of standard 18Ni maraging steels

Grade	Heat treatment(a)	Tensile strength MPa	ksi	Yield strength MPa	ksi	Elongation in 50 mm (2 in.), %	Reduction in area, %	Fracture toughness MPa√m	ksi√in.
18Ni(200)	A	1500	218	1400	203	10	60	155–240	140–220
18Ni(250)	A	1800	260	1700	247	8	55	120	110
18Ni(300)	A	2050	297	2000	290	7	40	80	73
18Ni(350)	B	2450	355	2400	348	6	25	35–50	32–45
18Ni(Cast)	C	1750	255	1650	240	8	35	105	95

(a) Treatment A: solution treat 1 h at 820 °C (1500 °F), then age 3 h at 480 °C (900 °F). Treatment B: solution treat 1 h at 820 °C (1500 °F), then age 12 h at 480 °C (900 °F). Treatment C: anneal 1 h at 1150 °C (2100 °F), age 1 h at 595 °C (1100 °F), solution treat 1 h at 820 °C (1500 °F) and age 3 h at 480 °C (900 °F)

Table 3 Comparison of the longitudinal, room-temperature mechanical properties of standard, cobalt-free, and low cobalt-bearing 18Ni(250) maraging steels

Heat treatment: solution heat 1 h at 815 °C (1500 °F), then age 5 h at 480 °C (900 °F). Testing was conducted on 63.5 × 88.9 mm (2.5 × 3.5 in.) billets produced from 200 mm (8 in.) diam vacuum induction melted/vacuum arc remelted ingots.

Grade	Ultimate tensile strength MPa	ksi	0.2% offset yield strength MPa	ksi	Elongation in 25 mm (1 in.), %	Reduction in area, %	Charpy V-notch impact toughness(a) J	ft · lbf	Plane-strain fracture toughness(a) MPa √m	ksi √in.
18Ni(250)	1870	271	1825	265	12	64.5	37	27	138	125
Cobalt-free 18Ni(250)	1895	275	1825	265	11.5	58.5	34	25	127	115
Low-cobalt 18Ni(250)	1835	266	1780	258	11	63.5	43	32	149	135

(a) Longitudinal-short transverse orientation tested (L-S orientation)

austenite and may be much softer than expected.

Aging is normally done at 480 °C (900 °F) for 3 to 6 h. For certain applications (such as die casting dies), aging at about 530 °C (985 °F) is employed. Figure 2 shows hardness versus aging time for 18Ni(250). Hardening is initially very rapid; several minutes at temperature will cause a substantial boost in strength in all grades of maraging steel. With very long times at temperature, hardness begins to drop due to coarsening of the precipitate particles and reversion of the martensite matrix to austenite. Austenite generally starts to form as rather small particles at both prior austenite boundaries and lath-martensite boundaries.

Thermal cycling of maraging steels between the M_f temperature (temperature at which martensite formation finishes during cooling) and a temperature considerably in excess of the solution-heating temperature can be used to refine the grain structure of coarse-grain maraging steels. For example, Saul, Roberson, and Adair (Ref 8) were able to refine an ASTM grain size of −1/1 to an ASTM grain size of 6/7 in 18Ni(300) following three thermal cycles between room temperature and 1025 °C (1880 °F). The shear strains produced by the diffusionless shear transformations of martensite to austenite and of austenite to martensite provide the driving force for recrystallization during these thermal cycles. One should recall that grain sizes finer than ASTM 6/7 cannot be achieved by this process and that the process becomes less effective as the starting grain size becomes finer.

The standard aging treatments listed in Table 2 produce contraction in length of 0.04% in 18Ni(200), 0.06% in 18Ni(250), and 0.08% in both 18Ni(300) and 18Ni(350). These very small dimensional changes during aging allow many maraging steel components to be finish machined in the annealed condition, then hardened. When precise dimensions must be held, an allowance for contraction can be made.

Much effort has been devoted to examining the properties of overaged maraging steels. The general belief is that a microstructure containing coarse precipitate particles and finely distributed austenite particles should have good resistance to both fracture and stress-corrosion cracking. In many instances, this has been found to be so, and very impressive improvements in plane-strain fracture toughness, K_{Ic}, or threshold stress intensity for stress-corrosion cracking, K_{Iscc}, have been achieved with only modest reductions in yield strength. Unfortunately, however, there appears to be more heat-to-heat variability when overaging heat treatments are used. This variability is the result of a greater sensitivity of austenite formation to minor changes in composition and processing. Even the most homogeneous-looking struc-

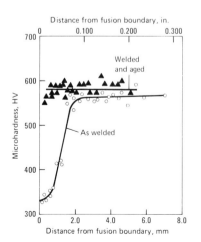

Fig. 3 Microhardness of a weld heat-affected zone in 18Ni(250) maraging steel. Source: Ref 4

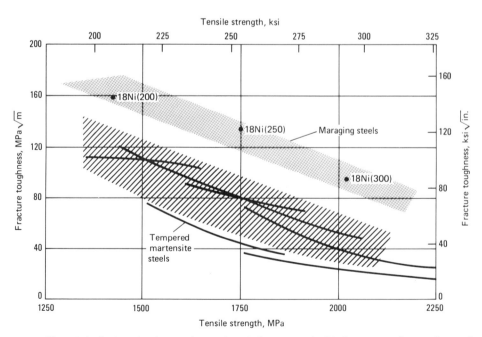

Fig. 4 Plane-strain fracture toughness of maraging steels compared with fracture toughness of several ultrahigh strength steels as a function of tensile strength. Source: Ref 2

Table 4 Typical compositions of filler wire for welding 18Ni maraging steels

Base metal grade	Composition of filler wire, %(a)				
	Ni	Mo	Co	Ti	Al
18Ni(200)	18	3.5	8	0.25	0.10
18Ni(250)	18	4.5	8	0.50	0.10
18Ni(300)	18	4.5	10	0.80	0.10
18Ni(350)	17	3.7	12.5	1.60	0.15

(a) All grades contain no more than 0.03% C.

tures contain minor alloy segregation that is manifested in the form of alloy-rich and alloy-lean bands. The alloy-rich bands tend to overage more rapidly than their alloy-lean counterparts. It is therefore difficult to recommend specific overaging heat treatments that will produce consistent mechanical properties. Generally, if a specific yield strength is required, it is better to use a maraging steel in which the required strength can be produced by conventional aging than to use an overaged steel of higher strength.

Surface Treatment. Grit blasting is the most efficient technique for removing oxide films formed by heat treatment. Maraging steels can be chemically cleaned by pickling in sulfuric acid or by duplex pickling in hydrochloric acid and then in nitric acid plus hydrofluoric acid. As with conventional steels, care must be taken to avoid overpickling. The sodium hydride cleaning of maraging steels should be avoided to minimize problems with crack formation. Grease and oils can be removed by cleaning in trichloroethane-type solutions.

Maraging steels can be nickel plated in chloride baths provided that proper surface-activation steps are followed. Heavy chromium deposits can be plated on top of nickel electrodeposits. Maraging steels are less susceptible to hydrogen embrittlement during plating than conventional quenched and tempered steels of comparable hardness. They are not immune to hydrogen, however, and baking after plating is recommended. Baking should be done at temperatures of about 150 to 205 °C (300 to 400 °F) for periods of 3 to 10 h, depending on size and baking temperature. Baking cannot be combined with age hardening, because considerable hydrogen remains in the steel after heat treating at the higher temperatures.

Considerable surface hardening can be achieved by nitriding maraging steels in dissociated ammonia. Hardness levels equivalent to 65 to 70 HRC can be achieved at depths of up to 0.15 mm (0.006 in.) after nitriding for 24 to 48 h at 455 °C (850 °F). Nitriding at this temperature allows age hardening to occur during nitriding; therefore, the two processes can be accomplished simultaneously. Salt bath nitriding for 90 min at 540 °C (1000 °F) has been done successfully. Such treatment must be very carefully controlled to avoid excessive

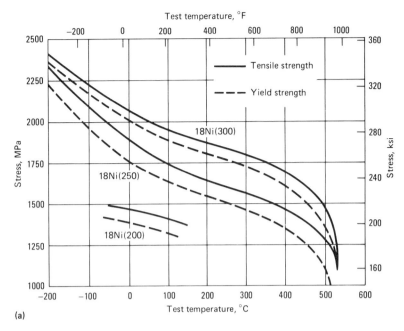
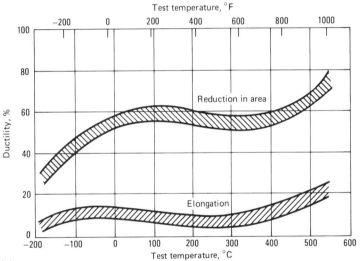
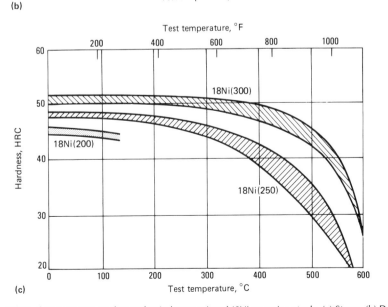

Fig. 5 Effect of temperature on the mechanical properties of 18Ni maraging steels. (a) Stress. (b) Ductility. (c) Hardness. Source: Ref 10

overaging. Both the fatigue strength and the wear resistance of maraging steels are improved by nitriding.

Welding. One of the main virtues of maraging steels is their excellent weldability. The extensive research and development work that has been done to optimize weldability is reviewed in Ref 4 and 9; only a brief summary of that work is given here.

Perhaps the characteristic that most enhances the weldability of maraging steels is that their low carbon content produces a soft, ductile martensite on cooling. A weld heat-affected zone in maraging steels can be divided into three regions. The region closest to the fusion line contains coarse martensite produced by solution annealing. Next is a narrow region containing reverted austenite produced by heating into the 595 to 805 °C (1100 to 1480 °F) range. Finally, there is a region where the maximum temperature reached during welding ranges from ambient temperature up to 595 °C (1100 °F); this region contains martensite that has been age hardened by various amounts. As shown in Fig. 3, the heat-affected zone in an as-welded structure is relatively soft. Because the metal in the area immediately surrounding the weld is soft and ductile, residual stresses are low, and the susceptibility to weld cracking is considerably less than in steels hardened by quenching. Subsequent local aging brings the hardness of the weld zone up to that of the base metal; this effect is also shown in Fig. 3. Resolution heat treating is advisable after welding if optimum properties are desired in the heat-affected weld zone.

All conventional welding processes have been used for welding maraging steel. Brazing can also be done. Most fabrication thus far, however, has employed inert-gas welding processes; the least work has been done with flux-shielded processes (Ref 4, 9).

Typical compositions of filler wire for welding maraging steels are listed in Table 4. These compositions are very similar to those of the steels with which they are used. After welding, an aging treatment is used to increase the strength of the weld joint. Weld metal strength does not depend significantly on the welding process used to make the joint; in most applications, joint efficiencies exceeding 90% can be achieved.

The toughness of the heat-affected zone after age hardening usually matches that of the unaffected base metal. The toughness of the weld metal, however, depends on the joining process. Gas tungsten arc welding produces the best toughness, gas metal arc welds have somewhat lower toughness, and the toughness of flux-shielded welds is poorer yet. In general, high-energy processes that produce coarser weld beads will produce weld metal of lower toughness. This is of special concern with submerged arc welding, where low toughness negates the advantage of the high welding speed associated with that process. The general

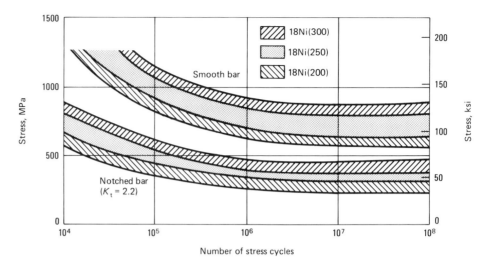

Fig. 6 Rotating-beam fatigue properties of three 18Ni maraging steels. Source: Ref 2

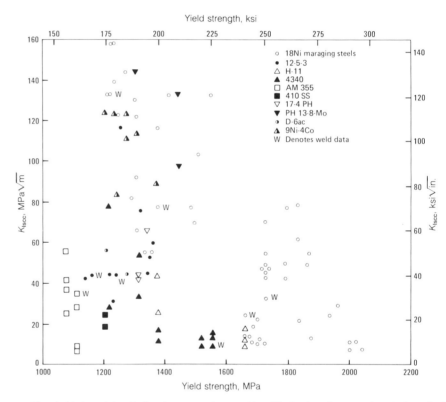

Fig. 7 Threshold stress intensity for stress-corrosion cracking (K_{Iscc}) values for maraging steels and other high-strength steels as a function of yield strength. Source: Ref 11

rules that follow are helpful for obtaining the best properties:

- Avoid prolonged dwell times at high temperatures
- Avoid preheat
- Keep interpass temperatures below 120 °C (250 °F)
- Minimize weld energy
- Avoid slow cooling rates
- Keep welds as clean as possible

Powder Metallurgy Products. Maraging steels having yield strengths of about 1585 MPa (230 ksi), combined with reasonable ductility, have been made by sintering elemental powders. Gas-atomized alloy powders that have been canned and extruded show excellent ductility at yield strengths up to 1900 MPa (275 ksi). However, during the atomization and collection of atomized maraging steel powder, care must be taken to minimize contamination of the surface of the powder particles. The presence of oxides or other nonmetallic inclusions on these surfaces can adversely affect the mechanical properties of parts consolidated from contaminated powder. Powder metallurgy parts made of maraging steel have been commercially produced only in limited quantities.

Properties of Maraging Steels

Mechanical Properties. Typical tensile properties and fracture toughness values of the conventional grades of maraging steel are listed in Table 2. One of the distinguishing features of maraging steels is superior toughness compared to conventional steels. Figure 4 compares K_{Ic} values for several maraging steels with those of quenched and tempered alloy steels as a function of tensile strength. The toughness of maraging steels is sensitive to purity level, and carbon and sulfur levels in particular should be kept low to obtain optimum fracture toughness.

The mechanical properties of 18Ni(250), cobalt-free 18Ni(250), and low-cobalt bearing 18Ni(250) are compared in Table 3. These data show that all three alloys exhibit comparable strength when aged at 480 °C (900 °F) for 5 h. However, these data also show that, in general, the cobalt-free 18Ni(250) material displays reduced localized ductility as measured by the percentage of reduction in area, and reduced plane-strain fracture toughness compared to the standard 18Ni(250) and low-cobalt bearing 18Ni(250) alloys. The reduced ductility and toughness of the cobalt-free 18Ni(250) alloy is related to the absence of cobalt in this material. The lack of cobalt eliminates the previously discussed cobalt/molybdenum interaction and thus necessitates a higher level of titanium, which is a more potent embrittling agent than either cobalt or molybdenum when present at levels significantly in excess of 1.3%.

The effects of temperature on the mechanical properties of maraging steels are illustrated in Fig. 5. Maraging steels can be used for prolonged service at temperatures up to approximately 400 °C (750 °F). Yield and tensile strengths at 400 °C (750 °F) are about 80% of the room-temperature values. Long-time stress-rupture failures can occur at 400 °C (750 °F), but the rupture stresses are fairly high. At temperatures above 400 °C (750 °F), reversion of the martensite matrix to austenite becomes dominant, and long-term load-carrying capacities decay fairly quickly.

At cryogenic temperatures, the strengths of maraging steels increase similarly to those of steels (Fig. 5); K_{Ic} values fall off roughly in a linear fashion with decreasing temperatures.

The fatigue properties of maraging steels are comparable to those of other high-strength steels. Smooth-bar and notched-bar fatigue properties for 18Ni(200), 18Ni(250), and 18Ni(300) grades are summarized in Fig. 6. Fatigue crack growth rates in maraging steels obey the $da/dN = (\Delta K)^m$ relationship commonly observed in steels, and the rates are similar to those of conventional steels. The bulk of the fatigue life in maraging steels is in the crack initiation stage. In general, cracks tend to initiate at noncoherent intermetallic inclusions, which once again suggests that low levels of residual elements and clean melt practices are important in this alloy system. Improved fatigue properties can be obtained by shot peening and by nitriding.

Resistance to Corrosion and Stress Corrosion. Maraging steels have slightly better corrosion resistance than tempered martensitic alloy steels. In industrial and marine atmospheres, the corrosion rates of maraging steels are about half those of conventional steels. In static and flowing seawater, maraging and conventional steels have essentially the same corrosion rates. In saline and acidic solutions, maraging steels show somewhat better corrosion resistance. Their hot oxidation resistance is noticeably better than that of tempered martensitic alloy steels because of the nickel and cobalt contents of these materials. In general, contact with more noble metals should be avoided to minimize the likelihood of galvanic corrosion.

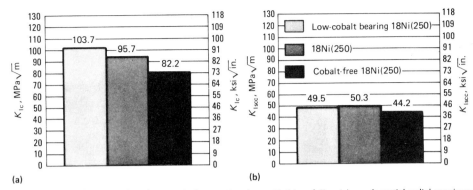

Fig. 8 Bar graphs comparing plane-strain fracture toughness K_{Ic} (a) and K_{Iscc} (circumferential-radial specimen orientation) (b) of low-cobalt bearing, standard, and cobalt-free 18Ni(250). K_{Iscc} testing was conducted in a marine atmosphere. Heat treatment: 815 °C (1500 °F), 1 h, air-cooled; 480 °C (900 °F), 5 h, air-cooled. Source: Ref 12

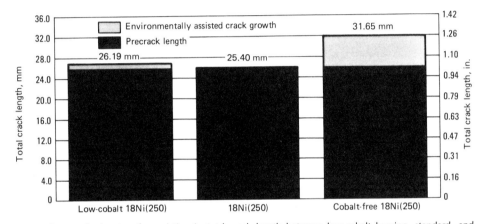

Fig. 9 Bar graph showing the variation in total crack length between low-cobalt bearing, standard, and cobalt-free 18Ni(250) maraging steels following K_{Iscc} testing in a marine atmosphere. Heat treatment: 815 °C (1500 °F), 1 h, air-cooled; 480 °C (900 °F), 5 h, air-cooled. Source: Ref 12

Maraging steels are susceptible to stress-corrosion cracking in most aqueous environments; their resistance to cracking increases significantly at lower yield strengths. Maraging steels have better resistance to stress-corrosion cracking than tempered martensitic steels of comparable strength. Figure 7 compares K_{Iscc} values of several high-strength steels. The resistance of maraging steels to stress-corrosion cracking generally parallels fracture toughness. Processing techniques that improve K_{Ic} (such as vacuum melting, proper hot working, and keeping residual impurity levels low) also improve resistance to stress-corrosion cracking. Surface treatments such as painting or shot peening have been successfully used to increase resistance. Cathodic protection can also be used, but must be employed very carefully to avoid hydrogen charging that could initiate cracking. Maraging steels can be used in aqueous environments; however, as with all high-strength steels, proper precautions to avoid stress-corrosion cracking must be taken.

A comparison of the environmentally assisted cracking (EAC) behavior of production scale lots of 18Ni(250), low-cobalt

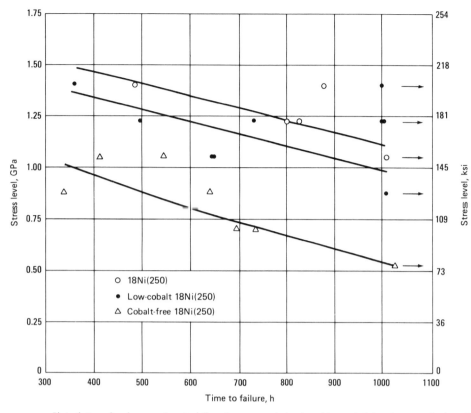

Fig. 10 Plot of stress level versus time to failure to compare behavior of low-cobalt bearing, standard, and cobalt-free 18Ni(250) maraging steels tested in stagnant 3.5% NaCl for 1000 h using proof-ring tensile specimens. Heat treatment: 815 °C (1500 °F), 1 h, air-cooled; 480 °C (900 °F), 5 h, air-cooled. Source: Ref 12

Table 5 Selected physical and mechanical properties of 18Ni(250) maraging steel

Density, g/cm³ (lb/in.³)	8.0 (0.289)
Coefficient of thermal expansion, μm/m · K (μin./in. · °F) at 24–284 °C (75–543 °F)	10.1 (5.6)
Thermal conductivity, W/m · K (Btu · in./ft³ · h · °F)	
20 °C (70 °F)	19.7 (137)
50 °C (120 °F)	20.1 (139)
100 °C (212 °F)	20.9 (145)
Electrical resistivity, μΩ · m	
Annealed	0.6–0.7
Aged	0.36–0.6
Melting temperature, °C (°F)	1430–1450 (2605–2640)
Length change during aging, %	−0.06
Poisson's ratio......................	0.30
Elastic modulus, GPa (10⁶ psi)	186 (27)
Shear modulus, GPa (10⁶ psi)	71 (10.3)
Magnetic coercive force, A · m⁻¹ (Oe)	
Annealed	1750–2700 (22–34)
Aged	1670–4300 (21–54)

18Ni(250), and cobalt-free 18Ni(250) is shown in Fig. 8 to 10. The 18Ni(250) and low-cobalt 18Ni(250) materials used for this comparison testing were produced by Carpenter Technology Corporation; the cobalt-free 18Ni(250) was produced by Teledyne Vasco Corporation. The K_{Ic} and K_{Iscc} data contained in Fig. 8 and 9 were produced by tests using compact tension specimens. The K_{Iscc} testing was conducted on samples mounted in a fiberglass rack that was 45° to the vertical and facing the prevailing southeast wind on a floating pier located in Galveston Bay. All samples were loaded to K_{init} of 66 MPa√m (60 ksi√in.), and the test duration was 1000 h. The proof-ring tensile data contained in Fig. 10 is from a test conducted in a stagnant 3.5% NaCl solution using Plexiglas vessels. Figures 8 and 9 show that both the 18Ni(250) and low-cobalt 18Ni(250) alloys display improved K_{Ic}, K_{Iscc}, and EAC growth compared to the cobalt-free 18Ni(250) alloy. Based on Fig. 10, the apparent threshold stress levels for 18Ni(250), low-cobalt 18Ni(250), and cobalt-free 18Ni(250) in stagnant 3.5% NaCl are 1035 MPa (150 ksi), 860 MPa (125 ksi), and 515 MPa (75 ksi), respectively.

Physical Properties. Table 5 summarizes the physical properties of 18Ni(250) maraging steel. Grade 18Ni(250) is the most widely used and most thoroughly evaluated maraging steel. Consequently, more extensive physical property data are available for this steel than for other maraging steels (Ref 13).

Applications

Maraging steels have been used in a wide variety of applications, including missile cases, aircraft forgings, structural parts, cannon recoil springs, Belleville springs, bearings, transmission shafts, fan shafts in commercial jet engines, couplings, hydraulic hoses, bolts, punches, and dies. Maraging steels have been extensively used in two general types of applications:

- Aircraft and aerospace applications, in which the superior mechanical properties and weldability of maraging steels are the most important characteristics
- Tooling applications, in which the excellent mechanical properties and superior fabricability (in particular, the lack of distortion during age hardening) are important

In many applications, even though maraging steels are more expensive than conventional steels in terms of alloy cost, finished parts made of maraging steel are less expensive because of significantly lower fabrication costs. Therefore, it is often economics rather than properties alone that determine the use of maraging steels.

REFERENCES

1. S. Floreen, *Metall. Rev.*, Vol 13 (No. 126), 1968
2. A. Magnee, J.M. Drapier, J. Dumont, D. Coutsouradis, and L. Habraken, "Cobalt-Containing High-Strength Steels," Cobalt Information Center, 1974
3. A.M. Hall and C.J. Slunder, "The Metallurgy, Behavior, and Application of the 18 Percent Nickel Maraging Steels," Report SP5051, National Aeronautics and Space Administration, 1968
4. F.H. Lang and N. Kenyon, Bulletin 159, Welding Research Council, 1971
5. W.B. Pearson, *Handbook of Lattice Spacings and Structure of Metals and Alloys*, Vol 2, Pergamon Press, 1967
6. W.B. Pearson, *Handbook of Lattice Spacings and Structure of Metals and Alloys*, Vol 1, Pergamon Press, 1958
7. M.L. Schmidt, *Maraging Steels: Recent Developments and Applications*, The Minerals, Metals & Materials Society, 1988, p 213-235
8. G. Saul, J.A. Roberson, and A.M. Adair, in *Source Book on Maraging Steels*, American Society for Metals, 1979
9. C.R. Weymueller, How to Weld Maraging Steels, *Weld. Des. Fabr.*, Feb 1989
10. "18% Ni Maraging Steel Data Bulletin," The International Nickel Company, Inc., 1964
11. D.P. Dautovich and S. Floreen, in *Proceedings of the International Conference on Stress Corrosion Cracking and Hydrogen Embrittlement*, National Association of Corrosion Engineers, 1976
12. M.L. Schmidt, Carpenter Technology Corporation, unpublished research
13. R.P. Wei, Ferrous Alloys, in *Aerospace Structural Metal Handbook*, Belfour Stulen, Inc., 1970

Ferrous Powder Metallurgy Materials

Revised by Leander F. Pease III, Powder-Tech Associates, Inc.

POWDER METALLURGY (P/M) in its simplest form consists of compressing metal powders in a shaped die to produce green compacts. These are then sintered, or diffusion bonded, at elevated temperatures in a furnace with a protective atmosphere. During sintering, the constituents usually do not melt, and the compacts become substantially strengthened by the development of bonds between individual particles.

For a specific metal powder and sintering condition, increased compact density results in improved mechanical properties. The density of sintered compacts may be increased by re-pressing. When re-pressing is performed primarily to increase dimensional accuracy rather than density, it is termed sizing. When re-pressing is intended to change the contour of the surface in contact with the punches, it is termed coining. For example, a sintered blank could be coined so that the surface is indented with small slots or letters and numbers. The re-pressing may be followed by resintering, which relieves the stresses due to cold work and may further strengthen the compact.

Alloy compacts can be formed from mixtures of metal powders that completely or partially diffuse during sintering. Alternatively, each individual particle may be completely prealloyed prior to compaction. The diffusion bonding process may be accelerated by sintering at a temperature at which one of the constituent metals is molten. For metals with relatively high melting points, such as iron or tungsten, a skeleton may be pressed, which is then infiltrated by a molten metal, such as copper or silver, having a melting point lower than that of the skeleton.

Process Capabilities. Theoretical density commonly refers to the density of a pore-free material of defined chemistry and thermomechanical history. For example, the theoretical density of iron or low-carbon steel is about 7.87 g/cm^3. Adding carbon or oxide inclusions to the solid steel lowers the theoretical density, even though no pores are formed.

By pressing and sintering only, parts are produced at 80 to 93% of theoretical density. By re-pressing, with or without resintering, the materials may be further densified to 85 to 96% of theoretical density. High-temperature sintering will also produce parts at these high densities. The density of pressed parts is limited by the size and shape of the compact. The most common P/M materials for structural parts are iron-copper-carbon, iron-nickel-carbon, and iron-carbon. Parts made from these materials respond to heat treatment with a defined hardenability band. Iron parts that are low in carbon and high in density can be carburized and quenched to form a definite, hard case.

Powder metallurgy parts are frequently competitive with forgings, castings, stampings, numerically controlled machined components, and fabricated assemblies. Within the limitations inherent to the P/M processes (discussed in detail in *Powder Metallurgy*, Volume 7 of the 9th Edition of *Metals Handbook*), parts can be fabricated to final or near-final shape, thereby eliminating or reducing scrap metal, secondary machining, and assembly operations. Although the unit cost of metal powder (about $0.73/kg, or $0.33/lb in 1989) is usually higher than that of steel bars, the savings achieved by eliminating fabricating operations and minimizing scrap losses often result in lower total costs for P/M parts. The pressing operation in an intricate set of tooling provides very complex geometry, at production rates of 8 to 60 pieces per minute.

Certain metal products can be produced only by powder metallurgy; among these products are materials whose porosity (number, distribution, and size of pores) is controlled. Two examples of controlled-porosity materials are filter elements and self-lubricating bearings. The porosity in a self-lubricating bearing is filled with oil, which lubricates the bearing/shaft interface as the temperature starts to rise.

Successful production by powder metallurgy depends on the proper selection and control of process variables:

- Powder characteristics
- Powder preparation
- Type of compacting press
- Design of compacting tools and dies
- Type of sintering furnace
- Composition of the sintering atmosphere
- Choice of production cycle, including sintering time and temperature
- Secondary operations and heat treatment

Metal Powder Characteristics and Control

The size and shape of powder particles have important effects on pressing and sintering characteristics, but the performance of each type of powder can be determined only empirically. The size and shape of iron powder particles are largely determined by the process used in producing the powder. Powder particles representative of the principal production processes are shown in Fig. 1.

To ensure uniformity in the handling and press performance of powders from batch to batch and in the strength of the compacts formed from them, laboratory tests are usually made on samples of powder from each batch. The results of quality control tests can be applied only to samples of the same basic composition made by the same method. This limitation is evident in Fig. 2, which shows typical compressibility values for three types of iron powders, atomized, prealloyed 4600V, and hydrogen-reduced sponge iron.

Test procedures are commonly specified for sampling and for the determination of particle size distribution, flow rate, apparent density, weight loss in hydrogen, compressibility, green strength, sintering characteristics, and chemical composition. A cross-index of standards for testing metal powders and ferrous P/M products is given in Table 1.

Sampling. Metal powders in containers are usually sampled by turning a specially designed auger-type sampler called a sampling thief vertically to the bottom of the container and removing the column of powder. (A straight, hollow tube does not provide a representative sample.) One or more of these samples are taken from each drum of powder, the number of samples depending on the drum size. Samples taken in this

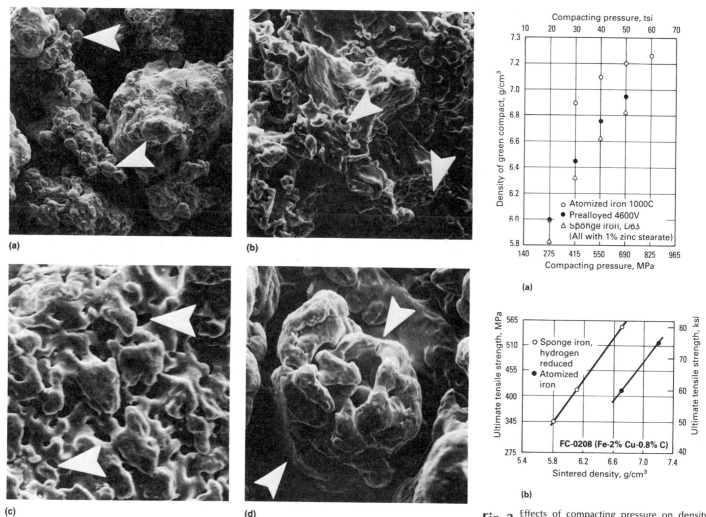

Fig. 1 Scanning electron micrographs of various iron powder particles. (a) Water-atomized and annealed iron powder (Ancorsteel 1000). Arrows indicate small fines that were agglomerated into larger particles. 190×. (b) Iron powder (Atomet 28). Arrows indicate porosity in the spongy regions. 750×. (c) Hydrogen-reduced sponge iron (Pyron 100). Arrows indicate pores opening into the spongy interior. 1000×. (d) Carbon-reduced iron ore (MH-100). Arrows indicate one particle with coarse internal porosity. 750×

Fig. 2 Effects of compacting pressure on density and mechanical properties. (a) Density versus compacting pressure for three types of iron powders. (b) Ultimate tensile strength versus density for FC-0208 (Fe-2%Cu-0.8%C).

manner are combined and reduced to the required size by means of a so-called sample splitter, or by rolling, coning, and quartering.

Particle Size Distribution. A complete screen analysis of the particle size distribution is determined by placing a 100 g (3.5 oz) sample of powder in the top screen of a "nest" of standard sieves having successively smaller openings, usually 80, 100, 140, 200, and 325 mesh. The assembly is shaken mechanically for 15 min. The powder that is retained on each mesh size and the powder that passes through the finest sieve are reported as weight percentages of the sample.

The powder passing 325-mesh can be further analyzed for its subsieve particle size distribution by use of an optical or electron microscope or by the scattering of laser light.

Flow Time. The standard (Hall) flowmeter for metal powders is a funnel-shape apparatus having a calibrated orifice. The time required for 50 g (1.75 oz) powder to flow unaided through the orifice is reported as the flow time of the powder.

Apparent Density. The Hall flowmeter is also used to determine the apparent density of free-flowing powders. For nonflowing powders, use is made of a baffled rectangular tower fed by a funnel having a large orifice (Scott volumeter). The Arnold meter measures apparent density and approximates the value that is found in actual press operation. All three instruments serve to supply a controlled flow of powder into a cup of known volume. The content of the full cup is weighed, and the apparent density is reported as g/cm³.

Oxide Content. A fairly accurate indication of the hydrogen-reducible oxide content of metal powders can be obtained by reducing a weighed sample of the powder in hydrogen under standard time-temperature conditions. The sample must be cooled to room temperature in an atmosphere of dry hydrogen. Loss in weight caused by reduction of the oxides present is expressed as a percentage of the initial sample weight. Direct determination of oxygen may also be used to estimate the oxide content of the powder.

Compressibility is the density to which a powder can be pressed at a given pressure. Alternatively, it can be expressed as the pressure required to attain a given density (see Fig. 2).

Green Strength. The strength of a green (unsintered) compact is usually determined by pressing a given weight of powder to a specified density in the shape of a rectangular bar of standard size. This test bar is supported at both ends as a simple beam and loaded transversely to fracture. The modulus of rupture is reported as the green strength of the powder.

Sintering Characteristics. A weighed quantity of a powder is compacted under controlled conditions in a die of known

Table 1 Test methods applicable to powder metallurgy

Methods applicable to metal powders	Designation ASTM	MPIF	ISO	Methods applicable to compacted and sintered materials	Designation ASTM	MPIF	ISO
Compressibility of metal powders	B 331	...	...	Density and interconnected porosity	B 328	42	2737, 2738
Compression testing of metallic materials at room temperature	E 9	...	...	Determination of apparent hardness of sintered metal powder products	...	43	4498/1
Density, apparent, of metal powders	B 212	4	3923/1	Determination of dimensional change of sintered metal powder specimens	B 610	44	4492
Density, apparent, of nonfree-flowing metal powders	B 417	28	3923/2	Determination of impact strength of sintered metal powder specimens	...	40	5754
Flow rate of metal powders	B 213	3	4490	Determination of transverse rupture or bending strength of sintered metal powder test specimens	B 528	41	3325
Hydrogen loss of metal powders	E 159	2	4491/3				
Insoluble matter in iron and copper powders	E 194	6	4496				
Iron content of iron powder	...	7	...				
Particle size average of metal powders by Fisher subsieve sizer	...	32	...	Determining the case hardness of powder metallurgy parts	...	37	4507
Sampling finished lots of metal powders	B 215	1	3954	Green strength of compacted metal powder specimens	B 312	15	3995
Sieve analysis of granular metal powders	B 214	5	4497	Determination of radial crushing strength of bearings	B 438	...	2739
Subsieve analysis of granular metal powders by air classification	B 293	12	...	Determination of Young's modulus	...	...	3312
Tension test specimens for pressed and sintered metal powders	E 8	10	2740	Determination of density of impermeable materials	B 311	...	3369
Recommended practice for analysis by microscopical methods for particle size distribution of particulate substances of subsieve sizes	E 20	...	...	Fatigue test pieces	...	...	3928
Terminology	B 438	...	2739	Determination of bubble test pore size in filters	...	39	4003
Tap density of metal powders	B 527	...	3953	Determination of fluid permeability	...	...	4022
Lubricant content of mixed powders	...	...	4495				
Compactability of metal powders	B 331	45	...				
Apparent density of Arnold meter	...	48	...				
Test of copper base powder for infiltrating	...	49	...				

dimensions. After sintering under controlled conditions, dimensional measurements are again taken. The difference in size is usually given as a percentage of the die dimension. Finally, the specimen is tested for mechanical properties. Sintering characteristics are generally reported in terms of dimensional changes, final density, and mechanical properties. The test is usually done by comparing the dimensional change or strength to a standard or reference bar processed at the same time.

Chemical Composition. In addition to the above tests, routine chemical analyses usually are run on all powders to determine the amount of alloying elements and impurities that might adversely affect the final product.

Powder Preparation

Powders are mixed or blended before use. This is done to obtain specific properties (for example, dimensional change or strength) in the finished product. Mixing or blending must be carefully controlled. Also, it is important to use equipment that mixes rapidly and produces a uniform distribution of the powders in the desired proportions in minimum time.

Excessive blending, especially in an overloaded blender, may work harden the powder, making it less compressible and thus more difficult to compact. The cascading of powder inside the blender will gradually round the edges of the particles, changing their shape and raising apparent density. As a result, the powder will exhibit different molding characteristics.

The double-cone blender is probably the most extensively used of the various types available. The final operation before compaction is to mix the powders with a lubricant, such as synthetic wax or stearic acid. The lubricant not only minimizes die friction and wear, but also reduces interparticle friction, allowing the particles to pack more closely and resulting in higher density for the lower range of molding pressure. Because of a hydraulic effect, densities above 7.0 g/cm^3 require less lubricant.

Small percentages of alloying additives such as graphite, copper, or nickel powders, if required, are also introduced during final mixing. Powder manufacturers are beginning to provide bonded mixes in which ⅛% of a polymer attaches the additives to the iron powder. This prevents segregation, increases flow rate, and decreases dimensional variation among parts.

Powder Compacting

Compacting powder serves several important functions: The powder is consolidated into the desired shape—a compact that must be strong enough to withstand subsequent processing; compacting controls the amount and type of porosity of the finished product; and compacting is largely responsible for the final dimensions of the part, subject to dimensional changes during sintering.

Of the numerous compacting methods, closed-die pressure compacting at room temperature is the method that is most commonly used in the production of ferrous powder metallurgy parts. A detailed discussion of the various compacting methods, their capabilities and limitations, and the equipment required by each method may be found in *Powder Metallurgy*, Volume 7 of the 9th Edition of *Metals Handbook* or in Ref 1 and 2.

Green strength is an important property of a green compact; the compact must be strong enough to avoid damage during ejection from the die and during normal handling between compacting and sintering. The green strength of a compact is affected by the type and composition of the powder, the distribution of powder particles with respect to both size and shape, the quantity and type of lubricant added to the powder, the configuration of the die, and the compacting pressure. If other compacting conditions are held constant, the green strength of powder metallurgy compacts will vary directly with compacting pressure. Also, as shown in Fig. 2, green density of compacts increases with compacting pressure. Green density is indicative of the density that can be expected after sintering.

Sintering

Sintering is the process by which a compact of metal powder is bonded by heating at a temperature below the melting point, or liquidus, of the major constituent. In iron-graphite mixes, no melting occurs during sintering. In the commonly used Fe-2Cu-0.8C mixes, the copper melts and diffuses into the iron. For an M-2 tool steel, sintering is done above the solidus temperature with 15 to 20% permanent liquid phase.

Densification does not occur in most commercially produced P/M parts. Sintering conditions and alloys are adjusted so that the final part is very close to the die size that molded it.

Sometimes a liquid phase forms and assists in sintering. This is true for the Fe-

Table 2 Compositions of ferrous P/M structural materials

Description	Designation(a) MPIF	ASTM	SAE	MPIF composition limits and ranges, %(b) C	Ni	Cu	Fe	Mo
P/M iron	F-0000	B 783	853, Cl 1	0.3 max	...	...	97.7–100	...
P/M steel	F-0005	B 783	853, Cl 2	0.3–0.6	...	...	97.4–99.7	...
P/M steel	F-0008	B 783	853, Cl 3	0.6–1.0	...	...	97.0–99.1	...
P/M copper iron	FC-0200	B 783	...	0.3 max	...	1.5–3.9	93.8–98.5	...
P/M copper steel	FC-0205	B 783	...	0.3–0.6	...	1.5–3.9	93.5–98.2	...
P/M copper steel	FC-0208	B 783	864, Gr 1, Cl 3	0.6–1.0	...	1.5–3.9	93.1–97.9	...
P/M copper steel	FC-0505	B 783	...	0.3–0.6	...	4.0–6.0	91.4–95.7	...
P/M copper steel	FC-0508	B 783	864, Gr 2, Cl 3	0.6–1.0	...	4.0–6.0	91.0–95.4	...
P/M copper steel	FC-0808	B 783	864, Gr 3, Cl 3	0.6–1.0	...	6.0–11.0	86.0–93.4	...
P/M copper steel	...	...	864, Gr 4, Cl 3	0.6–0.9	...	18.0–22.0	75.1 min	...
P/M iron-copper	FC-1000	B 783	862	0.3 max	...	9.5–10.5	87.2–90.5	...
P/M prealloyed steel	FL-4205	B 783	...	0.4–0.7	0.35–0.45	...	95.9–98.7	0.50–0.85
P/M prealloyed steel	FL-4605	B 783	...	0.4–0.7	1.70–2.00	...	94.5–97.5	0.40–0.80
P/M iron-nickel	FN-0200	B 783	...	0.3 max	1.0–3.0	2.5 max	92.2–99.0	...
P/M nickel steel	FN-0205	B 783	...	0.3–0.6	1.0–3.0	2.5 max	91.9–98.7	...
P/M nickel steel	FN-0208	B 783	...	0.6–0.9	1.0–3.0	2.5 max	91.6–98.4	...
P/M iron-nickel	FN-0400	B 783	...	0.3 max	3.0–5.5	2.0 max	90.2–97.0	...
P/M nickel steel	FN-0405	B 783	...	0.3–0.6	3.0–5.5	2.0 max	89.9–96.7	...
P/M nickel steel	FN-0408	B 783	...	0.6–0.9	3.0–5.5	2.0 max	89.6–96.4	...
P/M iron-nickel	FN-0700	...	...	0.3 max	6.0–8.0	2.0 max	87.7–94.0	...
P/M nickel steel	FN-0705	...	...	0.3–0.6	6.0–8.0	2.0 max	87.4–93.7	...
P/M nickel steel	FN-0708	...	...	0.6–0.9	6.0–8.0	2.0 max	87.1–93.4	...
P/M infiltrated steel	FX-1000	B 783	...	0–0.3	...	8.0–14.9	82.8–92.0	...
P/M infiltrated steel	FX-1005	B 783	...	0.3–0.6	...	8.0–14.9	80.5–91.7	...
P/M infiltrated steel	FX-1008	B 783	...	0.6–1.0	...	8.0–14.9	80.1–91.4	...
P/M infiltrated steel	FX-2000	B 783	870	0.3 max	...	15.0–25.0	70.7–85.0	...
P/M infiltrated steel	FX-2005	B 783	...	0.3–0.6	...	15.0–25.0	70.4–84.7	...
P/M infiltrated steel	FX-2008	B 783	872	0.6–1.0	...	15.0–25.0	70.0–84.4	...

(a) Designations listed are nearest comparable designations; ranges and limits may vary slightly between comparable designations. (b) MPIF standards require that the total amount of all other elements be less than 2.0%, except in infiltrated steels, for which the total amount of other elements must be less than 4.0%

2Cu-0.8C mixtures noted above. The Fe-0.45P and Fe-0.8P contain Fe$_3$P, which melts as a eutectic at 1048 °C (1918 °F) and forms a transient liquid phase. An example of a permanent liquid phase is the infiltration of iron with copper in which 10 to 20 wt% copper remains as a liquid inside the iron skeleton, prior to cooling. The iron-copper infiltrated parts do not shrink. The permanent liquid in the M-2 tool steel causes about 6% linear shrinkage and near-full densification.

The processes operating in sintering to form necks or bonds between the particles include surface, volume, and grain-boundary diffusion. In the initial stages of sintering, the atoms move over the surfaces of the particles toward the small radius of curvature regions, where the particles contact each other. During sintering, the compact initially shows increases in strength, thermal conductivity, and electrical conductivity. The particle contacts formed in pressing become larger. Later, the strength increases. Finally, toughness, ductility, and density increase with longer sintering time. The structure changes from an interlocked aggregate of powders to what may be thought of as a solid material containing small pores of various sizes and shapes. At longer sintering times, the densification results in the isolation of the pores. They are no longer connected to the surface. To obtain nearly fully density, it is necessary to sinter at temperatures approaching the melting point of the powder. A more detailed sintering theory is found in Volume 7 of the 9th Edition of *Metals Handbook*, p 309-321.

Sintering Techniques. The equipment and techniques used in sintering are diversified; there must be a means of heating the parts in a suitable protective atmosphere and a means of controlling the process variables. Heating and cooling rates, time and temperature of sintering, and composition of the atmosphere are the most critical variables. The dimensions of the finished parts can be affected by the sintering conditions, as well as by the compacting process and the properties of the powder itself. In the sintering of ferrous metals, the atmosphere serves a dual purpose: It both protects the metal from oxidation and helps to control the carbon content of the parts, especially the carbon content at the surface. Commercial ferrous powders, including prealloyed steel powders, generally contain very little carbon. The carbon content of the finished part is regulated by the amount of graphite mixed into the powder and the carbon potential of the sintering atmosphere. Particularly in alloy steels, the rate of cooling from the sintering temperature can determine which microconstituents are formed as a result of austenite decomposition, but the rate of cooling generally is slow enough to produce ferritic or pearlitic microstructures.

Sintering Equipment. Most iron parts are sintered in mesh belt conveyor furnaces at 1105 to 1120 °C (2020 to 2050 °F) in a mixture of nitrogen and 10% dissociated ammonia. The parts pass through a delubrication zone at 760 to 870 °C (1400 to 1600 °F) and then spend about 15 to 25 min at the high-heat temperature. These furnaces produce 90 to 360 kg/h (200 to 800 lb/h) of parts. They have been improved by the addition of preheat zones that heat rapidly using gas flames to provide an oxidizing environment, which prevents blistering and sooting of parts. Other furnaces have forced atmosphere gas recirculating coolers at the discharge end. These can give finer pearlite spacing or can even form some martensite or bainite, for increased strength.

For higher temperatures and improved toughness, parts are now being sintered in large-scale walking beam furnaces at 1230 to 1315 °C (2250 to 2400 °F). Yield can be up to 900 kg/h (2000 lb/h), with good dimensional control and decreased spread in mechanical properties, as well as increases in the mean properties. Mesh belts of SiC were introduced in 1989, allowing sintering at 1315 °C (2400 °F) and higher.

Stainless steel, silicon iron, and tool steels are usually sintered in vacuum furnaces. The lubricant is generally removed first in a separate furnace. Such delubrication furnaces have been linked to vacuum furnaces in such a way that trays of parts are automatically delubricated, transferred for sintering, and then transferred for gas pressure quenching. Rapid quenching in gas prevents the pickup of nitrogen in 316 stainless steel and promotes better corrosion resistance. Some furnaces are capable of 1315 to 1540 °C (2400 to 2800 °F). Furnaces with sweep gas systems can remove the

Table 3 Compositions of P/M stainless steels

MPIF designation	Fe	Cr	Ni	Mn	Si	S	C	P	Mo	N
SS-303N1, N2	rem	17.0–19.0	8.0–13.0	0–2.0	0–1.0	0.15–0.30	0–0.15	0–0.20	...	0.2–0.6
SS-303L	rem	17.0–19.0	8.0–13.0	0–2.0	0–1.0	0.15–0.30	0–0.03	0–0.20	...	...
SS-304N1, N2	rem	18.0–20.0	8.0–12.0	0–2.0	0–1.0	0–0.03	0–0.08	0–0.045	...	0.2–0.6
SS-304L	rem	18.0–20.0	8.0–12.0	0–2.0	0–1.0	0–0.03	0–0.03	0–0.045	...	...
SS-316N1, N2	rem	16.0–18.0	10.0–14.0	0–2.0	0–1.0	0–0.03	0–0.08	0–0.045	2.0–3.0	0.2–0.6
SS-316L	rem	16.0–18.0	10.0–14.0	0–2.0	0–1.0	0–0.03	0–0.03	0–0.045	2.0–3.0	...
SS-410	rem	11.5–13.0	...	0–1.0	0–1.0	0–0.03	0–0.25	0–0.04	...	0.2–0.6

binder in the vacuum furnace without the use of a separate atmosphere furnace.

Secondary Operations

In many applications, a P/M part that is made by compacting and sintering meets every performance requirement. In other instances, however, the functional requirements (mechanical properties, surface finish, and/or dimensional tolerances) for a part exceed the capabilities of compacted and sintered parts, and one or more secondary operations are required. Some of the common secondary operations are:

- *Sizing*: To tighten dimensional tolerances, usually in the radial direction, relative to the direction of compacting pressure
- *Coining*: To change axial dimensions and tolerances
- *Machining*: To obtain shapes that cannot be compacted, such as by tapping holes or cutting undercut grooves
- *Forming*: To change the shape of the part; can be done hot or cold
- *Re-pressing*: To reduce porosity and increase strength and ductility; may be accompanied by resintering
- *Infiltration*: To increase strength and decrease porosity
- *Heat treating*: To increase hardness and strength
- *Joining*: By sinter bonding, staking, brazing, infiltrating, or welding
- *Finishing*: Includes deburring, polishing, impregnating, and plating

When the application of a P/M part requires high levels of strength, toughness, or hardness, the mechanical properties can be improved or modified by infiltration, heat treatment, or a secondary mechanical forming operation such as cold (room-temperature) re-pressing or powder forging (hot forming). The effect of these secondary processes on P/M mechanical properties is discussed in later sections of this article.

Designation of P/M Materials

Powder metallurgy materials are customarily designated by the specifications or standards to which they are made, such as those listed in Tables 2 and 3. Comparable standards are published by ASTM, SAE, and MPIF (Metal Powder Industries Federation).

The MPIF designations for ferrous P/M materials, described in detail in Ref 3, include a prefix of one or more letters (the first of which is F to indicate an iron-base material), four numerals, and a suffix. The second letter in the prefix identifies the principal alloying element (if one is specified); the percentage of that element is indicated by the first two digits. The third and fourth digits indicate the amount of carbon in the compacted and sintered part; the code designation 00 indicates less than 0.3%, 05 indicates 0.3 to 0.6%, and 08 indicates 0.6 to 0.9%. The suffix is used to indicate the minimum 0.2% yield strength of as-sintered parts and the minimum ultimate tensile strength of heat-treated materials in units of 1000 psi (6.894 MPa). The letters HT designate heat treated.

Commercially produced iron-base powders often contain controlled amounts of alloying elements other than those specified by any of the designations listed in Table 2. Manganese and molybdenum may be added to improve strength and the response to heat treatment. Sulfur may be added to enhance machinability. Additions of 0.45 to 0.80% P can improve the toughness of the part and reduce magnetic hysteresis losses. These powders are usually identified by the trade name of the producer even though the amounts of alloy additions are small enough that the designations listed in Table 2 could be applied to the powders. Commercially produced iron-base powders usually contain very little carbon because carbon lowers compressibility and the amount of carbon in the finished part is readily controlled by the amount of admixed graphite and the composition of the sintering atmosphere.

Commercially produced stainless steels also contain controlled amounts of alloying elements, particularly chromium and nickel. Table 3 lists the designations and compositions of some typical stainless steel P/M materials.

Mechanical and Physical Properties of Sintered Ferrous P/M Materials

The mechanical and physical properties of compacted and sintered ferrous powder metallurgy materials depend on many fac-

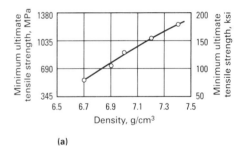

(a)

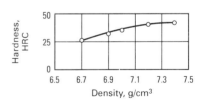

(b)

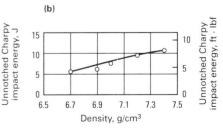

(c)

Fig. 3 Atomized iron powder with 0.3% graphite added to yield 0.1 to 0.2% combined carbon (6.7 g/cm³). Pressed at 410 to 480 MPa (30 to 35 tsi) and sintered 30 min at 1120 °C (2050 °F) in dissociated ammonia. White regions are ferrite. Arrows E surround a colony of eutectoid (pearlite). Arrow P points to a pore. 2% nital. 545×

Fig. 4 Effect of density on mechanical properties of heat-treated FN-0208 nickel steel. (a) Minimum ultimate tensile strength. (b) Hardness. (c) Impact toughness. Source: Ref 3

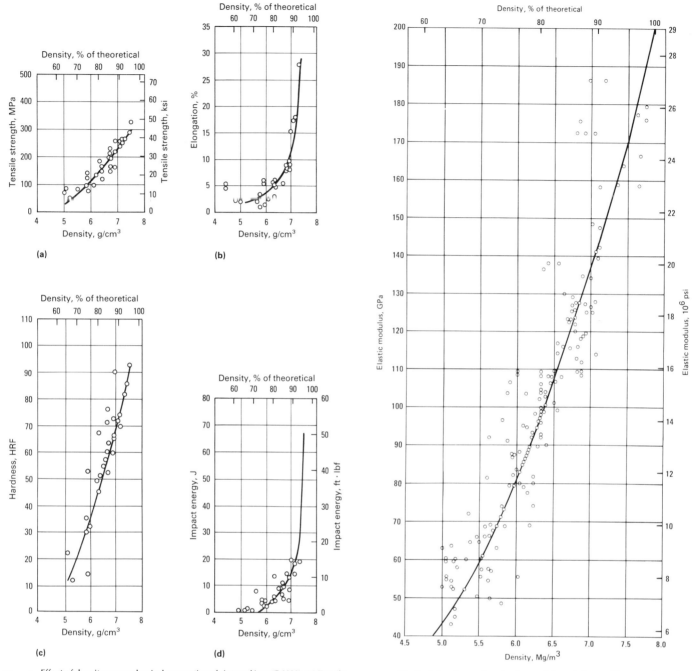

Fig. 5 Effect of density on mechanical properties of sintered iron (F-0000). (a) Tensile strength. (b) Elongation. (c) Hardness. (d) Unnotched Charpy impact energy

Fig. 6 Effect of density on the elastic modulus of sintered irons and steels after sintering to various densities

tors. Some factors, such as microstructure, chemical composition, and heat treatment, affect the properties of P/M materials just as they affect the properties of wrought or cast steels. However, these factors themselves are affected by conditions unique to the P/M process, such as type of iron powder, whether alloying additions are prealloyed or admixed, type and amount of lubricant added to the powder, variations in the compacting process, and sintering conditions. For example, prealloyed powders generally produce stronger and tougher parts than do powders mixed from elemental metal powders, assuming that the same compacted density is obtained with each powder. The variations in the P/M process, such as compacting pressure and sintering temperature, also affect the properties of P/M materials.

Porosity. The most obvious distinction between the microstructures of wrought and powder metallurgy materials is the porosity often found in powder metallurgy materials. This porosity originates as the spaces between powder particles and persists to some extent through sintering and subsequent secondary operations. The microstructure in Fig. 3 shows an example of a pore in a P/M material.

As was shown in Fig. 2, different base irons of the same density or porosity level have widely differing tensile properties. For that reason, a minimum density is no longer a requirement for a P/M material. P/M materials are specified in terms of the guaranteed minimum 0.2% tensile yield strength for as-sintered materials, and the minimum ultimate tensile strength for heat-treated materials. Nonetheless, for a specific base iron powder and degree of sintering, mechanical properties do increase with densi-

Table 4 Minimum and typical mechanical properties of ferrous P/M materials

Minimum strength values (in ksi) are specified by the suffix of the material designation code in the first column of the table. Typical values are given in the remaining columns.

Material designation code(a)	Ultimate strength MPa	ksi	0.2% offset yield strength MPa	ksi	Elongation in 25 mm (1 in.), %	Elastic modulus GPa	10^6 psi	Transverse rupture strength MPa	ksi	Impact energy(b) J	ft·lbf	Apparent hardness(c)	Fatigue strength(d) MPa	ksi
Iron and carbon steel														
F-0000-10(e)	125	18	90	13	1.5	96.5	14.0	248	36	4	3	40 HRF	48	7
F-0000-15(e)	172	25	125	18	2.5	117	17.0	345	50	8	6	60 HRF	70	10
F-0000-20(e)	262	38	172	25	7.0	141	20.5	655	95	47	35	80 HRF	96	14
F-0005-15(e)	165	24	125	18	<1.0	96.5	14.0	330	48	4	3	25 HRB	62	9
F-0005-20(e)	220	32	160	23	1.0	114	16.5	440	64	5.5	4	40 HRB	83	12
F-0005-25(e)	262	38	193	28	1.5	124	18.0	525	76	6.8	5	55 HRB	97	14
F-0005-50HT(f)	415	60	(g)	(g)	<0.5	114	16.5	725	105	4	3	20 HRC(h)	160	23
F-0005-60HT(f)	483	70	(g)	(g)	<0.5	120	17.5	827	120	4.7	3.5	22 HRC(h)	185	27
F-0005-70HT(f)	550	80	(g)	(g)	<0.5	130	19.0	965	140	5.5	4	25 HRC(h)	207	30
F-0008-20(e)	200	29	172	25	<0.5	83	12.0	350	51	3.5	2.5	35 HRB	75	11
F-0008-25(e)	240	35	207	30	<0.5	100	14.5	420	61	4	3	50 HRB	90	13
F-0008-30(e)	290	42	240	35	<1.0	114	16.5	510	74	5.5	4	60 HRB	110	16
F-0008-35(e)	393	57	275	40	1.0	130	19.0	690	100	6.8	5	70 HRB	150	22
F-0008-55HT(f)	448	65	(g)	(g)	<0.5	103	15.0	690	100	4	3	22 HRC(i)	172	25
F-0008-65HT(f)	517	75	(g)	(g)	<0.5	114	16.5	793	115	5.5	4	28 HRC(i)	200	29
F-0008-75HT(f)	585	85	(g)	(g)	<0.5	125	18.0	895	130	6	4.5	32 HRC(i)	220	32
F-0008-85HT(f)	655	95	(g)	(g)	<0.5	135	19.5	1000	145	6.8	5	35 HRC(i)	250	36
Iron-copper and copper steel														
FC-0200-15(e)	172	25	138	20	1.0	90	13.0	310	45	6	4.5	11 HRB	70	10
FC-0200-18(e)	193	28	160	23	1.5	103	15.0	350	51	6.8	5	18 HRB	75	11
FC-0200-21(e)	215	31	180	26	1.5	114	16.5	385	56	7.5	5.5	26 HRB	83	12
FC-0200-24(e)	235	34	200	29	2.0	124	18.0	435	63	8	6	36 HRB	90	13
FC-0205-30(e)	240	35	240	35	<1.0	90	13.0	415	60	<2.7	<2	37 HRB	90	13
FC-0205-35(e)	275	40	275	40	<1.0	103	15.0	517	75	4	3	48 HRB	103	15
FC-0205-40(e)	345	50	310	45	<1.0	117	17.0	655	95	6.8	5	60 HRB	130	19
FC-0205-45(e)	415	60	345	50	<1.0	134	19.5	793	115	11	8	72 HRB	160	23
FC-0205-60HT(f)	483	70	(g)	(g)	<0.5	100	14.5	655	95	3.5	2.5	19 HRC(h)	185	27
FC-0205-70HT(f)	550	80	(g)	(g)	<0.5	110	16.0	758	110	4.7	3.5	25 HRC(h)	207	30
FC-0205-80HT(f)	620	90	(g)	(g)	<0.5	121	17.5	827	120	6	4.5	31 HRC(h)	235	34
FC-0205-90HT(f)	690	100	(g)	(g)	<0.5	131	19.0	930	135	7.5	5.5	36 HRC(h)	262	38
FC-0208-30(e)	240	35	240	35	<1.0	83	12.0	415	60	<2.7	<2	50 HRB	90	13
FC-0208-40(e)	345	50	310	45	<1.0	103	15.0	620	90	2.7	2	61 HRB	130	19
FC-0208-50(e)	415	60	380	55	<1.0	117	17.0	860	125	6.8	5	73 HRB	160	23
FC-0208-60(e)	517	75	448	65	<1.0	138	20.0	1070	155	9.5	7	84 HRB	200	29
FC-0208-50HT(f)	450	65	(g)	(g)	<0.5	96.5	14.0	655	95	3.5	2.5	20 HRC(i)	172	25
FC-0208-65HT(f)	517	75	(g)	(g)	<0.5	107	15.5	760	110	4.7	3.5	27 HRC(i)	200	29
FC-0208-80HT(f)	620	90	(g)	(g)	<0.5	121	17.5	895	130	6	4.5	35 HRC(i)	235	34
FC-0208-95HT(f)	725	105	(g)	(g)	<0.5	134	19.5	1035	150	7.5	5.5	43 HRC(i)	275	40
FC-0505-30(e)	303	44	248	36	<0.5	83	12.0	530	77	4	3	51 HRB	117	17
FC-0505-40(e)	400	58	325	47	<0.5	103	15.0	703	102	6	4.5	62 HRB	152	22
FC-0505-50(e)	490	71	385	56	<1.0	117	17.0	855	124	6.8	5	72 HRB	185	27
FC-0508-40(e)	400	58	345	50	<0.5	86	12.5	690	100	4	3	60 HRB	152	22
FC-0508-50(e)	470	68	415	60	<0.5	103	15.0	827	120	4.7	3.5	68 HRB	180	26
FC-0508-60(e)	565	82	483	70	<1.0	121	17.5	1000	145	6	4.5	80 HRB	215	31
FC-0808-45(e)	380	55	345	50	<0.5	90	13.0	585	85	4	3	65 HRB	145	21
FC-1000-20(e)	207	30	180	26	<1.0	90	13.0	365	53	4.7	3.5	15 HRB	75	11
Iron-nickel and nickel steel														
FN-0200-15(e)	172	25	117	17	1.5	107	15.5	...	...	...	...	...	70	10
FN-0200-20(e)	240	35	172	25	4.0	134	19.5	550	80	26.5	19.5	75 HRF	90	13
FN-0200-25(e)	275	40	205	30	6.5	159	23.0	...	...	...	...	...	105	15
FN-0205-20(e)	275	40	172	25	1.5	107	15.5	450	65	8	6.0	44 HRB	105	15
FN-0205-25(e)	345	50	205	30	2.5	128	18.5	690	100	16.5	12.0	59 HRB	130	19
FN-0205-30(e)	415	60	240	35	4.0	152	22.0	860	125	28.5	21.0	69 HRB	160	23
FN-0205-35(e)	483	70	275	40	5.5	165	24.0	1035	150	46	34.0	78 HRB	185	27
FN-0205-80HT(f)	620	90	(g)	(g)	<0.5	107	15.5	827	120	4.5	3.5	23 HRC(j)	235	34
FN-0205-105HT(f)	827	120	(g)	(g)	<0.5	128	18.5	1100	160	6	4.5	29 HRC(j)	315	46
FN-0205-130HT(f)	1000	145	(g)	(g)	<0.5	145	21.0	1310	190	8	6.0	33 HRC(j)	380	55
FN-0205-155HT(f)	1100	160	(g)	(g)	<0.5	152	22.0	1480	215	9.5	7.0	36 HRC(j)	420	61
FN-0205-180HT(f)	1275	185	(g)	(g)	<0.5	165	24.0	1725	250	13	9.5	40 HRC(j)	480	70
FN-0208-30(e)	310	45	240	35	1.5	114	16.5	585	85	7.5	5.5	63 HRB	115	17
FN-0208-35(e)	380	55	275	40	1.5	128	18.5	725	105	11	8.0	71 HRB	145	21
FN-0208-40(e)	483	70	310	45	2.0	145	21.0	895	130	15	11.0	77 HRB	185	27
FN-0208-45(e)	550	80	345	50	2.5	159	23.0	1070	155	21.5	16.0	83 HRB	205	30
FN-0208-50(e)	620	90	380	55	3.0	165	24.0	1170	170	28.5	21.0	88 HRB	235	34

(continued)

(a) The suffix of the material designation codes represent either the minimum yield strength or the minimum ultimate strength in ksi. For example, the minimum yield strength of F-0000-10 is 10 ksi, while the minimum yield strength of F-0000-15 is 15 ksi. (b) Unnotched Charpy test. (c) Where applicable, the matrix (converted) hardness is also given in the footnotes. (d) Fatigue limit for 10^7 cycles from reverse-bending fatigue tests. (e) The suffix number represents the minimum yield strength (in ksi) for the material in the as-sintered condition. (f) The suffix number for heat-treated (HT) materials represents the minimum ultimate tensile strength in ksi. Tempering temperature for heat-treated materials is 175 °C (350 °F). (g) Yield strength and ultimate tensile strength are approximately the same for heat-treated materials. (h) Or a matrix (converted) hardness of 58 HRC. (i) Or a matrix (converted) hardness of 60 HRC. (j) Or a matrix (converted) hardness of 55 HRC. (k) Or a matrix (converted) hardness of 57 HRC. (l) All data based on single-pass infiltration. (m) Codes for the stainless steel designations: N1, nitrogen alloyed, with good strength and low elongation; N2, nitrogen alloyed, with high strength and medium elongation; L, low carbon, with lower strength and highest elongation; HT, martensitic grade, heat treated, and highest strength. Source: Ref 3

Table 4 (continued)

Material designation code(a)	Ultimate strength MPa	ksi	0.2% offset yield strength MPa	ksi	Elongation in 25 mm (1 in.), %	Elastic modulus GPa	10⁶ psi	Transverse rupture strength MPa	ksi	Impact energy(b) J	ft · lbf	Apparent hardness(c)	Fatigue strength(d) MPa	ksi
Iron-nickel and nickel steel (continued)														
FN-0208-80HT(f)	620	90	(g)	(g)	<0.5	114	16.5	827	120	5.5	4.0	26 HRC(k)	235	34
FN-0208-105HT(f)	827	120	(g)	(g)	<0.5	128	18.5	1035	150	6	4.5	31 HRC(k)	315	46
FN-0208-130HT(f)	1000	145	(g)	(g)	<0.5	134	19.5	1275	185	7.5	5.5	35 HRC(k)	380	55
FN-0208-155HT(f)	1170	170	(g)	(g)	<0.5	152	22.0	1515	220	9.5	7.0	39 HRC(k)	450	65
FN-0208-180HT(f)	1345	195	(g)	(g)	<0.5	165	24.0	1725	250	11	8.0	42 HRC(k)	510	74
FN-0405-25(e)	275	40	205	30	<1.0	96.5	14.0	450	65	6	4.5	49 HRB	105	15
FN-0405-35(e)	415	60	275	40	3.0	134	19.5	827	120	19.5	14.5	71 HRB	160	23
FN-0405-45(e)	620	90	345	50	4.5	165	24.0	1205	175	45.5	33.5	84 HRB	235	34
FN-0405-80HT(f)	585	85	(g)	(g)	<0.5	96.5	14.0	793	115	5.5	4.0	19 HRC(j)	220	32
FN-0405-105HT(f)	760	110	(g)	(g)	<0.5	121	17.5	1000	145	6.8	5.0	25 HRC(j)	290	42
FN-0405-130HT(f)	930	135	(g)	(g)	<0.5	134	19.5	1380	200	8.8	6.5	31 HRC(j)	350	51
FN-0405-155HT(f)	1100	160	(g)	(g)	<0.5	159	23.0	1690	245	13	9.5	37 HRC(j)	420	61
FN-0405-180HT(f)	1275	185	(g)	(g)	<0.5	165	24.0	1930	280	17.5	13.0	40 HRC(j)	480	70
FN-0408-35(e)	310	45	275	40	1.0	100	14.5	517	75	5.5	4.0	67 HRB	115	17
FN-0408-45(e)	450	65	345	50	1.0	128	18.5	793	115	10	7.5	78 HRB	170	25
FN-0408-55(e)	550	80	415	60	1.0	152	22.0	1035	150	15	11.0	87 HRB	205	30
Low-alloy steel														
FL-4205-80HT(f)	620	90	(g)	(g)	<0.5	117	17.0	930	135	4.7	3.5	28 HRC(i)	235	34
FL-4205-100HT(f)	760	110	(g)	(g)	<0.5	130	19.0	1100	160	5.5	4.0	32 HRC(i)	290	42
FL-4205-120HT(f)	895	130	(g)	(g)	<0.5	150	22.0	1275	185	5.5	4.0	36 HRC(i)	338	49
FL-4205-140HT(f)	1035	150	(g)	(g)	<0.5	172	25.0	1480	215	6	4.5	39 HRC(i)	393	57
FL-4605-80HT(f)	585	85	(g)	(g)	<0.5	115	16.5	895	130	4.7	3.5	24 HRC(i)	220	32
FL-4605-100HT(f)	760	110	(g)	(g)	<0.5	125	18.0	1135	165	6	4.5	29 HRC(i)	290	42
FL-4605-120HT(f)	895	130	(g)	(g)	<0.5	138	20.0	1345	195	8	6.0	34 HRC(i)	338	49
FL-4605-140HT(f)	1070	155	(g)	(g)	<0.5	148	21.5	1585	230	9.5	7.0	39 HRC(i)	405	59
Copper-infiltrated iron and steel(l)														
FX-1000-25(e)	350	51	220	32	7.0	110	16.0	910	132	34	25	65 HRB	130	19
FX-1005-40(e)	530	77	345	50	4.0	110	16.0	1090	158	17.5	13	82 HRB	200	29
FX-1005-110HT(f)	825	120	(g)	(g)	<0.5	110	16.0	1445	210	9.5	7	38 HRC(j)	315	46
FX-1008-50(e)	600	87	415	60	3.0	110	16.0	1145	166	13.5	10	89 HRB	225	33
FX-1008-110HT(f)	825	120	(g)	(g)	<0.5	110	16.0	1305	189	8.8	6.5	43 HRC(h)	315	46
FX-2000-25(e)	315	46	255	37	3.0	103	15.0	993	144	20	15	66 HRB	115	17
FX-2005-45(e)	515	75	415	60	1.5	103	15.0	1020	148	10.8	8	85 HRB	200	29
FX-2005-90HT(f)	690	100	(g)	(g)	<0.5	103	15.0	1180	171	9.5	7	36 HRC(j)	260	38
FX-2008-60(e)	550	80	483	70	1.0	103	15.0	1075	156	9.5	7	90 HRB	205	30
FX-2008-90HT(f)	690	100	(g)	(g)	<0.5	103	15.0	1095	159	6.8	5	36 HRC(h)	260	38
Stainless steel(m)														
SS-303N1-25(e)	270	39	220	32	0.5	...	...	593	86	4.7	3.5	62 HRB	...	...
SS-303N2-35(e)	380	55	290	42	5	...	...	675	98	26	19	63 HRB	...	...
SS-303L-12(e)	270	39	115	17	17.5	...	...	565	82	...	...	21 HRB	...	...
SS-304N1-30(e)	295	43	260	38	0.5	...	...	772	112	5.5	4	61 HRB	...	...
SS-304N2-33(e)	393	57	275	40	10	...	...	875	127	34	25	62 HRB	...	...
SS-304L-13(e)	295	43	125	18	23	...	...	...	...	...	...	...	...	...
SS-316N1-25(e)	283	41	235	34	0.5	...	...	745	108	6.8	5	59 HRB	...	...
SS-316N2-33(e)	415	60	270	39	10	...	...	860	125	38	28	62 HRB	...	...
SS-316L-15(e)	283	41	138	20	18.5	...	...	550	80	47	35	20 HRB	...	...
SS-410-90HT(f)	725	105	(g)	(g)	<0.5	...	...	780	113	3.5	2.5	23 HRC(j)	...	...

(a) The suffix of the material designation codes represent either the minimum yield strength or the minimum ultimate strength in ksi. For example, the minimum yield strength of F-0000-10 is 10 ksi, while the minimum yield strength of F-0000-15 is 15 ksi. (b) Unnotched Charpy test. (c) Where applicable, the matrix (converted) hardness is also given in the footnotes. (d) Fatigue limit for 10⁷ cycles from reverse-bending fatigue tests. (e) The suffix number represents the minimum yield strength (in ksi) for the material in the as-sintered condition. (f) The suffix number for heat-treated (HT) materials represents the minimum ultimate tensile strength in ksi. Tempering temperature for heat-treated materials is 175 °C (350 °F). (g) Yield strength and ultimate tensile strength are approximately the same for heat-treated materials. (h) Or a matrix (converted) hardness of 58 HRC. (i) Or a matrix (converted) hardness of 60 HRC. (j) Or a matrix (converted) hardness of 55 HRC. (k) Or a matrix (converted) hardness of 57 HRC. (l) All data based on single-pass infiltration. (m) Codes for the stainless steel designations: N1, nitrogen alloyed, with good strength and low elongation; N2, nitrogen alloyed, with high strength and medium elongation; L, low carbon, with lower strength and highest elongation; HT, martensitic grade, heat treated, and highest strength. Source: Ref 3

ty. Under the guaranteed minimum property system, the design engineer selects the tensile properties and chemistry he needs, and the parts fabricator chooses the density, base iron, sintering, and heat treatment to provide the properties. Table 4 shows all the guaranteed minimum tensile properties and related typical properties for the P/M materials that have been standardized by the MPIF.

The effects of density on the mechanical properties of heat-treated 2% Ni steel, FN-0208, are given in Fig. 4. The effects of density on the mechanical properties of sintered pure iron, F-0000, are shown in Fig. 5.

The apparent hardness of P/M materials is measured with the regular Rockwell hardness tester. The HRC, HRB, and HRF scales measure the resistance to indentation. They average the true hardness of the matrix material and the porosity and show lower readings than the corresponding material at full density. They are effective for quality control when applied to a specified region of the part, and an average of five readings is taken. Table 5 shows some typical Rockwell hardnesses and the precision that may be expected.

The abrasion or wear resistance of heat-treated materials is measured by the 100 g Knoop microhardness. Through the use of a microscope, these hardness measurements, made on the solid regions between the pores, show the intrinsic hardness of the steel matrix. The Knoop microhardness may be directly converted to HRC readings using the method described in Volume 7 of the 9th Edition of *Metals Handbook*, p 489. The precision or reproducibility between

Table 5 Precision of Rockwell apparent hardness readings on P/M parts

All laboratories tested identical coupons. If similar but different coupons were to be compared, variability would be increased, and larger differences between respective readings would be expected.

MPIF material designation	Density, g/cm³	Number of laboratories	Average apparent hardness	Repeatability(a) (95% confidence limits) One reading	Repeatability(a) (95% confidence limits) Average of six readings	Reproducibility(a) (95% confidence limits) One reading	Reproducibility(a) (95% confidence limits) Average of six readings
CZP-2002	7.92	9	82.5 HRC	1.7	0.7	2.2	0.9
F-0000	6.74	9	63.4 HRF	4.0	1.6	4.4	1.8
FC-0208	6.63	9	70.8 HRB	4.5	1.8	5.7	2.3
FX-2008	7.45	9	86.4 HRB	4.3	1.8	4.9	2.0
FL-4605-HT	6.90	8	107.2 HRB(b)	1.9	0.8	3.1	1.3
FL-4605-HT	6.90	8	34.6 HRC	2.2	0.9	3.1	1.3
FC-0208-HT	6.29	10	97.1 HRB(b)	3.1	1.3	4.4	1.8
FC-0208-HT	6.29	10	18.7 HRC	4.2	1.7	5.1	2.1
FN-0208-HT	6.89	10	105.3 HRB(b)	2.9	1.2	4.1	1.7
FN-0208-HT	6.89	10	30.5 HRC	3.8	1.5	4.6	1.9

(a) Repeatability and reproducibility defined according to ASTM E 691. (b) HRB scale with 1.6 mm (1/16 in.) diam carbide ball indentors

Table 6 Effect of density on elastic modulus and Poisson's ratio

Density range, g/cm³	Elastic modulus GPa	Elastic modulus 10⁶ psi	Poisson's ratio
5.6–6.0	72	10.5	0.18
6.0–6.4	90	13	0.20
6.4–6.8	110	16	0.21
6.8–7.2	130	19	0.23
7.2–7.6	160	23	0.26
7.86	205	30	0.28

laboratories of the measurement of microhardness on a 4650 heat-treated P/M steel is 3 HRC points when the laboratories take the average of six readings and use the conversion method noted above. Comparison of Knoop 100 g readings has a reproducibility between laboratories of 48 points, when the average of six readings are compared between laboratories.

In the range of densities normally encountered, the fatigue strength of a ferrous P/M material is approximately 25 to 40% of its tensile strength, regardless of composition. The effect of porosity on several other properties that are generally considered independent of variations in structure is given in Fig. 6 and in Table 6.

Composition. Alloying elements are added to ferrous P/M materials for the same reasons they are added to wrought or cast steels: primarily to improve strength and hardenability. The alloying elements most commonly used in ferrous P/M materials are carbon, copper, nickel, and molybdenum. The alloying elements are chosen for easy reduction in hydrogen. The amounts of alloying additions are important to the properties of the material, but so are the forms and sources of the alloying elements, particularly, whether they are added as elemental powders or as prealloyed powders. Elemental additions allow better compressibilities, but prealloying gives greater depth of hardening.

Carbon is the most important alloying element in P/M alloys because it strengthens iron and enables it to be heat treated. The effect of carbon content on the ultimate tensile strength of sintered iron is shown in Fig. 7. This strengthening effect can be traced to an increase in the amount of pearlite in the microstructure. As-sintered hypereutectoid P/M steels may have very low strength because proeutectoid cementite forms at austenite grain boundaries. The brittle cementite between the pores can fracture under relatively low levels of macroscopic stress. Combined carbon contents over 0.8% are not recommended.

Iron and prealloyed steel powders are usually made with very low carbon content to maximize their compressibilities; carbon is added to these powders as admixed graphite. The carbon content of the sintered compact is also affected by the composition of the sintering atmosphere. The amount of graphite that is retained in the sintered compact as combined carbon depends on the sintering atmosphere and temperature, the source and type of both the graphite and the metal powder, and the amount and types of oxides in the metal powder. The correct graphite addition is determined empirically for each lot of iron powder.

Copper increases the strength of ferrous P/M materials, as shown in Fig. 8, primarily through solid-solution strengthening. Copper is also thought to accelerate the sintering and homogenizing processes, particularly if elemental copper or a copper-rich master alloy exists in the liquid state during the early stages of sintering. Copper causes atomized iron-base compacts to expand during sintering; thus, additions of copper can be used to compensate for shrinkage in iron-carbon or iron-carbon-nickel alloys (Ref 4). Sponge iron blends with elemental copper do not expand.

Nickel strengthens ferrous P/M alloys by solid-solution strengthening and increases their hardenability. The effect of nickel con-

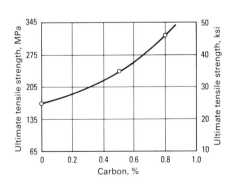

Fig. 7 Tensile properties of as-sintered iron-carbon alloys at 6.7 g/cm³ as a function of carbon content. Source: Ref 3

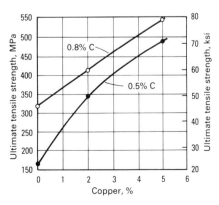

Fig. 8 Effect of copper and carbon content on the tensile strength of ferrous P/M materials. Source: Ref 3

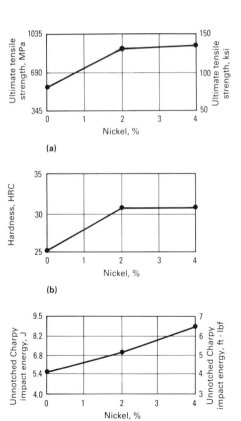

Fig. 9 Effect of nickel on the mechanical properties of heat treated P/M steels (7.0 g/cm³). (a) Ultimate tensile strength. (b) Hardness. (c) Impact toughness. Source: Ref 3

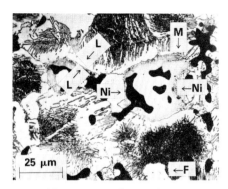

Fig. 10 Microstructure of sintered Fe-2Ni-0.5C alloy. Sintered for 30 min at 1120 °C (2050 °F). Arrows marked Ni outline nickel-rich particle. Arrow M, martensite or bainite at nickel-rich boundary. Arrows marked L, diffusion layer between nickel and pearlite. This is not unalloyed ferrite. Arrow F, ferrite. 4% nital etched

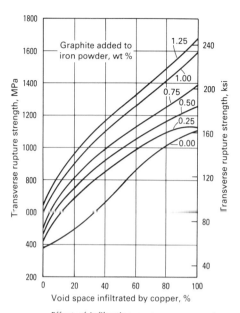

Fig. 11 Effect of infiltration on transverse rupture strength of iron-carbon alloys sintered to a density of 6.4 Mg/m³. Combined carbon in the alloys was about 80% of the amount of graphite added to the iron powder. The amount of copper infiltrant was adjusted to fill various fractions of void space. Source: Ref 7

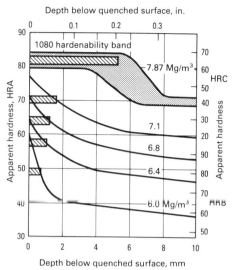

Fig. 12 Effect of porosity on hardenability. Compacts of F-0008 powder were pressed and sintered to various densities, then austenitized and end quenched. Apparent hardness traverses reflect both depth of hardening and density of compacts. Horizontal bars represent approximate distance over which the average amount of martensite in the microstructure exceeded 50%. Hardenability band for 1080 wrought steel, which has a higher manganese content, is included for comparison. Source: Ref 8

tent on the mechanical properties of P/M nickel steels is given in Fig. 9. The presence of nickel increases shrinkage during sintering; additions of copper help maintain dimensions during sintering. The heterogeneous microstructure with nickel-rich islands (see Fig. 10) decreases notch sensitivity and raises tensile strength and impact resistance of the heat-treated materials. The copper helps dissolve nickel in iron and increases hardenability.

Molybdenum and nickel are the elements most frequently added to iron powders to improve hardenability. The amounts of these elements are relatively small, as in the MPIF designations FL-4605 and FL-4205. These two commonly used prealloyed powders correspond in composition to 46xx and 42xx wrought steels except that they have a higher molybdenum content and a lower manganese content.

Phosphorus can reduce the temperature used for sintering. It forms a liquid phase, giving rapid pore rounding, strengthening, and toughening when used in amounts of 0.45 to 0.80% P. It also reduces the hysteresis losses in magnetic core applications. The hot shortness that phosphorus causes in wrought steels is generally not a problem in P/M materials. Some prealloyed steels with extra molybdenum and manganese, when blended with 2% Cu and eutectoid carbon, have enough hardenability to air harden during normal belt furnace sintering (Ref 5). After sintering, parts are tempered 1 h at 175 °C (350 °F).

Sulfur may be added to ferrous P/M materials to improve machinability. The amounts of sulfur used in P/M materials exceed those used in wrought steels; however, sulfur contents as high as 1.0% appear to have little effect on the mechanical properties of as-sintered compacts. Sulfur may be prealloyed to form MnS *in situ*, added as elemental sulfur to sponge iron, or added as manganese sulfide powder. The prealloyed form can decrease drill times by a factor of two (Ref 6). Small amounts of boron nitride can increase drill life by a factor of ten.

Alloying elements having oxides that are stable in typical sintering atmospheres are used sparingly, if at all, in low-alloy ferrous P/M materials. Manganese and chromium are sometimes used in very small amounts; vanadium and titanium are rarely used.

Infiltration

The infiltration of ferrous P/M materials is a common processing technique. Infiltrants, usually copper-base alloys, are placed in contact with the green compact, and the two are sintered together at a temperature above the melting point of the infiltrant. Capillary action draws the infiltrant into the ferrous compact, where it can promote the sintering of iron particles. In an alternative technique, the ferrous compact is sintered and then infiltrated in a second step. Traditionally, infiltration has allowed the manufacture of large parts on smaller presses because the green compacts can have a density as low as 6.0 to 6.2 g/cm³.

As indicated in Fig. 11 and Table 4, infiltration increases the strength, hardness, fatigue strength, and impact energy of ferrous P/M products compared to the iron-carbon matrix. Infiltration increases the density of the part and decreases variations in density. Porosity can be partly or completely eliminated by infiltration. As the infiltrant runs in under capillary action, it fills the higher-density fine-pore regions first. Recent proprietary techniques have increased unnotched impact toughness to more than 200 J (150 ft · lbf). Some of the disadvantages of infiltration are its high cost, the likelihood of growth of the part during infiltration, and the possibility of erosion of the surface of the compact by the infiltrant (Ref 7).

Heat Treatment of Ferrous P/M Materials

Ferrous P/M materials may be heat treated in the same manner as wrought or cast steels of comparable compositions. The porosity in P/M materials can cause complications, such as apparent loss of hardenability, greater depth of carburization or decarburization, and possible entrapment of quenchants. The effect of porosity on hardenability, which is exaggerated in measurements of apparent hardness, is shown in Fig. 12. The reduction in hardenability was caused by the reduced thermal conductivity of the porous sintered compact. The effect of porosity on depth of carbonitriding is shown in Fig. 13. The porosity of the sintered compact enabled the carbonitriding gas to penetrate well below the surface, hardening the compact to a greater depth. The use of vacuum carburizing for 2 to 10 min, followed by quenching, can form a definite case on low-carbon parts with a density as low as 6.8 g/cm³.

Mechanical properties of FN-0205 sintered steels hardened and tempered in are shown Fig. 9.

Steam treating is a heat treatment sometimes applied to ferrous P/M materials. A typical treatment might be 2 h in super-

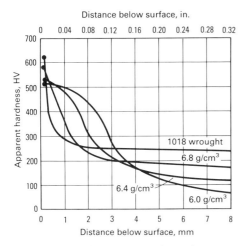

Fig. 13 Effect of porosity on carbonitriding. Compacts of F-0000 powder were pressed and sintered to various densities, then carbonitrided. Hardness traverses reflect both depth of carbonitrided case and density of compacts. Hardness traverse for a carbonitrided specimen of wrought 1018 steel is included for comparison. Source: Ref 8

Table 7 Effect of steam treating on the density and hardness of ferrous P/M materials

Material	Density, g/cm³ Sintered	Density, g/cm³ Steam treated	Apparent hardness Sintered	Apparent hardness Steam treated
F-0000-N	5.8	6.2	7 HRF	75 HRB
F-0000-P	6.2	6.4	32 HRF	61 HRB
F-0000-R	6.5	6.6	45 HRF	51 HRB
F-0008-M	5.8	6.1	44 HRB	100 HRB
F-0008-P	6.2	6.4	58 HRB	98 HRB
F-0008-R	6.5	6.6	60 HRB	97 HRB
FC-0700-N	5.7	6.0	14 HRB	73 HRB
FC-0700-P	6.35	6.5	49 HRB	78 HRB
FC-0700-R	6.6	6.6	58 HRB	77 HRB
FC-0708-N	5.7	6.0	52 HRB	97 HRB
FC-0708-P	6.3	6.4	72 HRB	94 HRB
FC-0708-R	6.6	6.6	79 HRB	93 HRB

heated steam at 540 to 600 °C (1000 to 1100 °F). In such a treatment, a layer of black iron oxide (Fe_3O_4) is formed on the surfaces of the sintered part, including the surfaces of pores that are connected to the surface. The process increases the density, hardness, compressive strength, and resistance to wear and corrosion of the part. The effect of steam treating on the density and hardness of several ferrous P/M materials is given in Table 7. Steam treating can reduce the tensile strength, elongation, and impact strength below the as-sintered values (Table 8).

Re-pressing

As a secondary mechanical forming operation performed at room temperature, re-pressing is done primarily to increase density, which increases mechanical and physical properties and hardness. Improvements in part dimensions can also be achieved by re-pressing. The amount of material deformation achieved with re-pressing is greater than in sizing because the forces used are greater than the sizing forces. The reduction in height of a ferrous part generally ranges from 3 to 5%. As with sizing, part tolerance after re-pressing depends on material type and part size.

Re-pressing generally refers to the application of high pressures on a sintered part at room temperature, while powder forging refers to processes in which a P/M preformed part is kept at an elevated temperature during the application of high pressure (see the section "Powder Forging"). At room temperature and at pressures as high as or higher than the compacting pressure, re-pressing increases the strength of a sintered P/M part by decreasing its porosity and by cold working the metal. The part is considerably strengthened, but at the expense of ductility. Resintering after re-pressing increases the ductility and toughness of the part without diminishing its strength. Those materials that are difficult to re-press after sintering usually can be re-pressed if the sintering is done at a low temperature at which alloying cannot take place; this low-temperature sintering is called presintering. For iron alloys, presintering is done at 845 °C (1550 °F).

The effect of re-pressing on the density of ferrous P/M materials is shown in Fig. 14. The density that is achieved by re-pressing depends on the density of the sintered or presintered compact, the re-pressing pressure and lubricant, and whether the powder used was prealloyed or mixed from elemental powders.

Table 8 Comparison of average mechanical properties of steam-blackened P/M materials with as-sintered properties and typical standard 35

Material	Blackening(a)	Ultimate tensile strength MPa	Ultimate tensile strength ksi	Average-density ultimate tensile strength MPa	Average-density ultimate tensile strength ksi	Elongation in 25 mm (1 in.), %	Transverse rupture strength MPa	Transverse rupture strength ksi	Impact energy(b) J	Impact energy(b) ft · lbf	0.2% offset yield strength MPa	0.2% offset yield strength ksi	Hardness, HRB
F-0000	None	138	20.0	42.7	6.20	6.4	330	47.9	5.4	4.0	...	...	37
	Light	154	22.3	45.0	6.52	1.8	367	53.2	2.7	2.0	...	...	88
	Heavy	152	22.1	44.6	6.47	1.2	434	63.0	2.7	2.0	...	...	85
Standard 35	None	124	18.0	...	...	1.5	248	36.0	4.1	3.0	690	10	40 HRF
F-0000	None	132	19.1	47.8	6.93	3.7	395	57.3	5.8	4.3	...	...	52
	Light	139	20.2	49.0	7.10	0.6	496	72.0	3.0	2.2	...	...	90
	Heavy	134	19.5	48.8	7.08	0.8	545	79.1	3.0	2.2	...	...	88
Standard 35	None	173	25.1	...	...	2.5	345	50.0	8.1	6.0	103	15	60 HRF
F-0008	None	211	30.6	43.2	6.26	1.5	453	65.7	3.6	2.7	...	...	69
	Light	134	19.4	45.6	6.61	0.4	434	63.0	2.3	1.7	...	...	108
	Heavy	116	16.8	45.4	6.59	0.4	450	65.2	2.4	1.8	...	...	106
Standard 35	None	241	35	...	...	<0.5	420	61	4.1	3	170	25	50
F-0008	None	272	39.5	46.7	6.78	1.6	665	96.4	4.3	3.2	...	...	87
	Light	192	27.9	48.0	6.96	0.3	656	95.1	3.0	2.2	...	...	110
	Heavy	193	28.0	48.1	6.97	0.8	752	109	3.1	2.3	...	...	108
Standard 35	None	345	50.0	...	...	<1.0	600	87.0	6.1	4.5	220	32	65
F-0208	None	331	48.0	43.2	6.27	1.2	684	99.2	4.7	3.5	...	...	87
	Light	208	30.2	45.1	6.54	0.4	606	87.9	2.7	2.0	...	...	112
	Heavy	205	29.8	44.9	6.51	0.7	738	107	2.7	2.0	...	...	110
Standard 35	None	345	50	...	...	<1.0	620	90	2.7	2.0	275	40	61
F-208	None	461	66.8	47.2	6.85	1.7	952	138	8.8	6.5	...	...	95
	Light	405	58.8	48.5	7.03	1.2	924	134	5.8	4.3	...	...	111
	Heavy	383	55.5	47.6	6.9	0.3	1040	151	5.7	4.2	...	...	108
Standard 35	None	434	63	...	...	<1.0	903	131	7.5	5.5	360	52	75

(a) None, as sintered; Light blackening, 2 h exposure in 538 °C (1000 °F) steam; heavy blackening, 4 h exposure in 538 °C (1000 °F) steam. Unnotched Charpy test at room temperature. Source: Ref 9

812 / Specialty Steels and Heat-Resistant Alloys

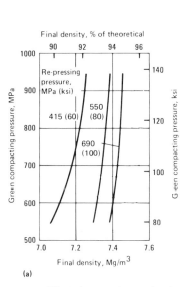
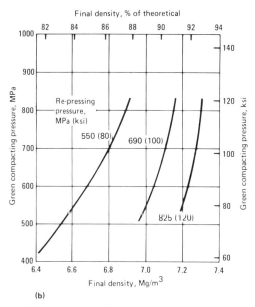

Fig. 14 Effect of re-pressing on density of powder metallurgy compacts. Alloy steel powders (4640 composition) were compacted at various pressures, then sintered, re-pressed, and resintered. For each specimen, the final density is indicated by the intersection between the curve that indicates the re-pressing pressure and the grid line that indicates the green compacting pressure. (a) Prealloyed powder. (b) Diffusional alloy made from elemental powders

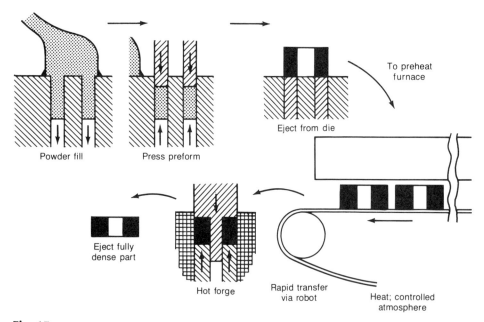

Fig. 15 The powder forging process

Powder Forging*

Powder forging is a process in which unsintered, presintered, or sintered powder metal preforms are hot formed in confined dies. The process is sometimes called P/M forging or P/M hot forming, or is simply referred to by the acronym P/F.

Powder forging is a natural extension of the conventional press and sinter (P/M)

---
*The text, photographs, and tables in this section on powder forging were adapted from the article "Powder Forging" in Volume 14 of the 9th Edition of *Metals Handbook*.

process, which has long been recognized as an effective technology for producing a great variety of parts to net or near-net shape. Figure 15 shows the powder forging process. In essence, a porous preform is densified by hot forging with a single blow. Forging is carried out in heated, totally enclosed dies, and virtually no flash is generated.

The shape, quantity, and distribution of porosity in P/M and P/F parts strongly influence their mechanical performance. Powder forging is a deformation processing technology aimed at increasing the density of P/M parts and thus their performance characteristics.

There are two basic forms of powder forging:

- Hot upsetting, in which the preform experiences a significant amount of lateral material flow
- Hot re-pressing, in which material flow during densification is mainly in the direction of pressing. This form of densification is sometimes referred to as hot restriking, or hot coining

While P/F parts are primarily used in automotive applications where they compete with cast and wrought products, parts have also been developed for military and off-road equipment.

The economics of powder forging have been reviewed by a number of authors (Ref 10-15). Some of the case histories included in the section "Applications of Powder Forged Parts" in this article compare the cost of powder forging with that of alternative forming technologies.

Material Considerations

The initial production steps of powder forging (preforming and sintering) are identical to those of the conventional press and sinter P/M process. Certain defined physical characteristics and properties are required in the powders used in these processes. In P/M parts, surface finish is related to the particle size distribution of the powder. In powder forging, however, the surface finish is directly related to the finish of the forging tools.

Typical pressing grades are −80 mesh with a median particle size of about 75 μm. The apparent density and flow are important for maintaining fast and accurate die filling. The chemistry affects the final alloy produced, as well as the compressibility.

Green strength and compressibility are more critical in P/M than they are in P/F applications. Although there is a need to maintain edge integrity in P/F preforms, there are rarely thin, delicate sections that require high green strength. Because P/F preforms do not require high densities (typically 6.2 to 6.8 g/cm³), the compressibility obtainable with prealloyed powders is sufficient. However, carbon is not prealloyed because it has an extremely detrimental effect on compressibility.

The two principal requirements for P/F materials are a capability to develop an appropriate hardenability that will guarantee strength and to control fatigue performance by microstructural features such as inclusions.

Hardenability. Nickel and molybdenum have the advantage that their oxides are reduced at conventional sintering temperatures. Alloy design is therefore a compromise, and the majority of atomized prealloyed powders in commercial use are nickel/molybdenum based, with manganese present in limited quantities. The compositions of three commercial P/M steels are:

Alloy	Composition, wt%(a)		
	Mn	Ni	Mo
P/F-4600	0.10–0.25	1.75–1.90	0.50–0.60
P/F-2000	0.25–0.35	0.40–0.50	0.55–0.65
P/F-1000	0.10–0.25	...	...

(a) All compositions contain balance of iron.

The higher cost of nickel and molybdenum, along with the higher cost of powder, compared with conventional wrought materials, is often offset by the higher material utilization inherent in the P/F process.

More recently, P/F parts have been produced from iron powders (0.10 to 0.25% Mn) with copper and/or graphite additions for parts that do not require the heat-treating response or high-strength properties achieved through the use of the low-alloy steels.

Inclusion Assessment. Because the properties of material powder forged to near-full density are strongly influenced by the composition, size distribution, and location of nonmetallic inclusions (Ref 16-18), a method has been developed for assessing the inclusion content of powders intended for P/F applications (Ref 19-22). Samples of powders intended for forging applications are re-press powder forged under closely controlled laboratory conditions. The resulting compacts are sectioned and prepared for metallographic examination. The inclusion assessment technique involves the use of automatic image analysis equipment. The compact used for inclusion assessment may also be used to measure the amount of unalloyed iron powder particles present.

Process Considerations

The development of a viable powder forging system requires the consideration of many process parameters. The mechanical, metallurgical, and economic outcomes depend to a large extent on operating conditions, such as temperature, pressure, flow/feed rates, atmospheres, and lubrication systems. Equally important consideration must be given to the types of processing equipment, such as presses, furnaces, dies, and robotics, and to secondary operations, in order to obtain the process conditions that are most efficient. This efficiency is maintained by optimizing the process line layout. Examples of effective equipment layouts for preforming, sintering, reheating, forging, and controlled cooling have been reviewed in the literature (Ref 10). Figure 16 shows a few of the many possible operational layouts. Each of these process stages is reviewed in the following sections.

Preforming. Preforms are manufactured from admixtures of metal powders, lubricants, and graphite. Compaction is predominantly accomplished in conventional P/M presses that use closed dies.

The control of weight distribution within preforms is essential to produce full density and thus maximize performance in the critical regions of the forged component. Excessive weight in any region of the preform may cause overload stresses that could lead to tool breakage at forging.

Successful preform designs have been developed by an iterative trial-and-error procedure, using prior experience to determine the initial shape. More recently, computer-aided design (CAD) has been used for preform design.

Preform design is intimately related to the design and dimensions of the forging tooling, the type of forging press, and the forging process parameters. Among the variables to be considered for the preforming tools are:

- Temperature, that is, preform temperature, die temperature, and, when applicable, core rod temperature
- Ejection temperature of the forged part
- Lubrication conditions, that is, influence on compaction/ejection forces and tooling temperatures
- Transfer time and handling of the preform from the preheat furnace to the forging die cavity

Correct preform design not only entails having the right amount of material in the various regions of the preform, but also is concerned with material flow between the regions and prevention of potential fractures and defects (Fig. 17).

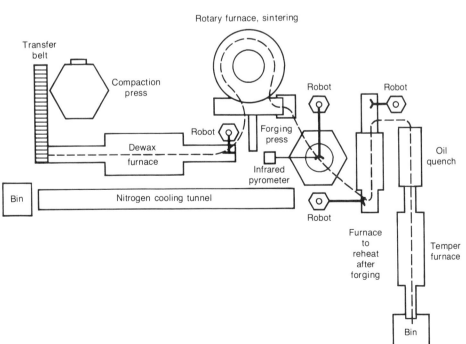

Fig. 16 A powder forging process line. Source: Ref 23

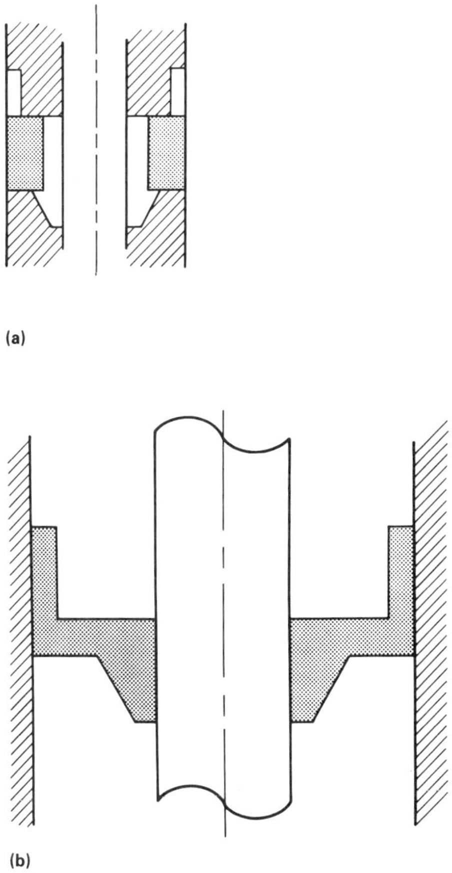

Fig. 17 Configuration for the ring preform (a) for forging the part shown in (b). (b) Cross section of the part under consideration for powder forging

Sintering and Reheating. Preforms may be forged directly from the sintering furnace; sintered, reheated, and forged; or sintered after the forging process. The basic requirements for sintering in a ferrous P/F system are:

- Lubricant removal
- Oxide reduction
- Carbon diffusion
- Development of particle contacts
- Heat for hot densification

Oxide reduction and carbon diffusion are the most important aspects of the sintering operations. For most ferrous powder forging alloys, sintering takes place at about 1120 °C (2050 °F) in a protective reducing atmosphere with a carbon potential to prevent decarburization. Typical P/M sintering has been performed at 1120 °C (2050 °F) for 20 to 30 min. Increases in temperature will reduce the time required for sintering by improving oxide reduction and increasing carbon diffusion. Chromium-manganese steels have been limited in their use because of the higher temperatures required to reduce their oxides and the greater care needed to prevent reoxidation.

Any of the furnaces used for sintering P/M parts, such as vacuum, pusher, belt, rotary hearth, walking beam, roller hearth, and batch/box, may be used for sintering or reheating P/F preforms. The sintered preforms may be forged directly from the sintering furnace; stabilized at lower temperatures and forged; or cooled to room temperature, reheated, and forged. All cooling, temperature stabilization, and reheating must occur under protective atmosphere to prevent oxidation.

Induction furnaces are often used to reheat axisymmetric preforms to the forging temperature because of the short time required to heat the material. Difficulties may be encountered in obtaining uniform heating throughout asymmetric shapes because of the variation in section thickness.

Powder forging involves removing heated preforms from a furnace, usually by robotic manipulators, and locating them in the die cavity for forging at high pressures (690 to 965 MPa, or 100 to 140 ksi). Preforms may be graphite coated to prevent oxidation during reheating and transfer to the forging die. Lubrication of the die and punches is usually accomplished by spraying a water-graphite suspension into the cavity.

The forging presses commonly used in conventional forging, including hammers, high energy rate forming machines, mechanical presses, hydraulic presses, and screw presses, have been evaluated for use in powder forging. The essential characteristics that differentiate presses are contact time, stroke velocity, available energy and load, stiffness, and guide accuracy.

Metal Flow in Powder Forging. Draft angles, which facilitate forging and ejection in conventional forging, are eliminated in P/F parts. This means that greater ejection forces—on the order of 15 to 20% of press capacity as a minimum—are required for the powder forging of simple shapes. However, the elimination of draft angles permits P/F parts to be forged closer to net shape.

Tool Design. In order to produce sound forged components, the forging tooling must be designed to take into account:

- Preform temperature
- Die temperature
- Forging pressure
- The elastic strain of the die
- The elastic/plastic strain of the forging
- The temperature of the part upon ejection
- The elastic strain of the forging upon ejection
- The contraction of the forging during cooling
- Tool wear

Secondary Operations. In general, the secondary operations applied to conventional components, such as plating and peening, may be applied to P/F components. The most commonly used secondary operations involve deburring, heat treating, and machining.

The heat treatment of P/M products is the same as that required for conventionally processed materials of similar composition. The most common heat-treating practices involve treatments such as carburizing, quench and temper cycles, and continuous-cooling transformation.

The amount of machining required for P/F components is less than the amount required for conventional forgings because of the improved dimensional tolerances. Standard machining operations may be used to achieve final dimensions and surface finish. One of the main economic benefits of powder forging is the reduced amount of machining required. Improved machinability can be accomplished by the addition of solid lubricants such as manganese sulfide.

Mechanical Properties

Wrought steel bar stock undergoes extensive deformation during cogging and rolling of the original ingot. This creates inclusion stringers and leads to planes of weakness, which affect the ductile failure of the material. The mechanical properties of wrought steels vary considerably with the direction test pieces are cut from the wrought billet. Powder forged materials, on the other hand, undergo relatively little material deformation, and their mechanical properties have been shown to be relatively isotropic.

The mechanical properties of P/F materials are usually intermediate to the transverse and longitudinal properties of wrought steels. The rotating-bending fatigue properties of P/F material have also been shown to fall between the longitudinal and transverse properties of wrought steel of the same tensile strength.

While the performance of machined laboratory test pieces follows the intermediate trend described above, in the case of actual components, P/F parts have been shown to have superior fatigue resistance. This has generally been attributed not only to the relative mechanical property isotropy of powder forgings, but also to their better surface finish and finer grain size.

This section reviews the mechanical properties of P/F materials. The data presented represent results obtained on machined standard laboratory test pieces. Data are reported for four primary materials. The first two material systems are based on prealloyed powders (P/F-4600 and P/F-4200; see the section "Hardenability" in this article). The third material, based on an iron-copper-carbon alloy, was used by Toyota in 1981 to make P/F connecting rods; Ford Motor Company introduced powder forged rods with a similar chemistry in 1986. Mechanical property data are therefore presented for copper and graphite powders mixed with an iron powder base to produce materials that generally contain 2% Cu. Some powder forged components are made from plain carbon steel. This is the fourth and final material for which mechanical property data are presented.

Forging Mode. It is well known that the forging mode has a major effect on the mechanical properties of components. With this in mind, the mechanical property data reported in this section were obtained on specimens that were either hot upset or hot re-press forged.

Heat Treatments. There were three heat treatments used in developing the properties of the prealloyed powder forged materials: case carburizing, blank carburizing, and through-hardening (quenching and tempering).

Hardenability. Jominy hardenability curves are presented in Fig. 18 for the P/F-4600, P/F-4200, and iron-copper-carbon alloys. Testing was carried out according to ASTM A 255. Specimens were machined from upset forged billets that had been sintered at 1120 °C (2050 °F) in dissociated ammonia.

Tensile, Impact, and Fatigue Properties. Tensile properties were determined on test pieces with a gage length of 25 mm (1 in.) and a gage diameter of 6.35 mm (0.25 in.). Testing was carried out according to ASTM E 8 using a crosshead speed of 0.5 mm/min (0.02 in./min). Room-temperature impact testing was carried out on standard Charpy V-notch specimens according to ASTM E 23. Rotating-bending fatigue (RBF) testing was performed using single-load, cantilever, rotating fatigue testers. The tensile, impact, and fatigue data for the various materials are summarized in Tables 9 to 11.

The iron-copper-carbon alloys were either still-air cooled or forced-air cooled

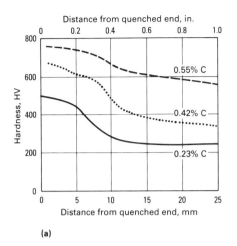
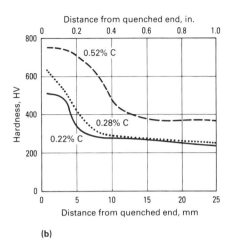
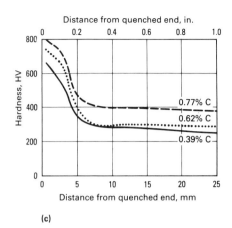

Fig. 18 Jominy hardenability curves for (a) P/F-4600, (b) P/F-4200, and (c) iron-copper-carbon materials at various forged-carbon levels. Vickers hardness was determined at a 30 kgf load.

from the austenitizing temperature of 845 °C (1550 °F). The austenitizing temperature influences core hardness. These iron-copper-carbon alloys are often used with manganese sulfide additions for enhanced machinability. The tensile, impact, and fatigue properties for a sample with a 0.35% manganese sulfide addition are compared with a material without sulfide additions in Table 12. Data from the samples with manganese sulfide and sulfurized powders are included for comparison. The manganese sulfide addition had little influence on tensile strength, whereas the sulfurization process degraded tensile properties.

Compressive Yield Strength. The 0.2% offset compressive yield strengths for P/F-4600 at various forged carbon levels after different heat treatments are summarized in Table 13.

Effect of Porosity on Mechanical Properties. The mechanical property data summarized in the previous sections are related to either hot re-press or hot upset forged pore-free material. The general effect of density on mechanical properties is presented in Table 14.

Quality Assurance for P/F Parts

Many of the quality assurance tests applied to wrought parts are similar to those used for powder forged parts. Among the parameters specified are part dimensions,

Table 9 Mechanical property and fatigue data for P/F-4600 materials
Sintered at 1120 °C (2050 °F) in dissociated ammonia unless otherwise noted

Forging mode	Carbon, %	Oxygen, ppm	Ultimate tensile strength MPa	ksi	0.2% offset yield strength MPa	ksi	Elongation in 25 mm (1 in.), %	Reduction of area, %	Room-temperature Charpy V-notch impact energy J	ft · lbf	Core hardness, HV30	Fatigue endurance limit MPa	ksi	Ratio of fatigue endurance to tensile strength
Blank carburized														
Upset	0.24	230	1565	227	1425	207	13.6	42.3	16.3	12.0	487	565	82	0.36
Re-press	0.24	210	1495	217	1325	192	11.0	34.3	12.9	9.5	479	550	80	0.37
Upset(a)	0.22	90	1455	211	1275	185	14.8	46.4	22.2	16.4	473	550	80	0.38
Re-press(a)	0.25	100	1455	211	1280	186	12.5	42.3	16.8	12.4	468	510	74	0.36
Upset(b)	0.28	600	1585	230	1380	200	7.8	23.9	10.8	8.0	513	590	86	0.37
Re-press(b)	0.24	620	1580	229	1305	189	6.8	16.9	6.8	5.0	464	455	66	0.29
Quenched and stress relieved														
Upset	0.38	270	1985	288	1505	218	11.5	33.5	11.5	8.5	554	...	...	...
Re-press	0.39	335	1960	284	1480	215	8.5	21.0	8.7	6.4	...	...	...	...
Upset	0.57	275	2275	330	...	...	3.3	5.8	7.5	5.5	655	...	...	...
Re-press	0.55	305	1945	282	...	...	0.9	2.9	8.1	6.0	...	...	...	...
Upset	0.79	290	940	136	...	...	...	...	1.4	1.0	712	...	...	...
Re-press	0.74	280	1055	153	...	...	...	...	2.4	1.8	...	...	...	...
Upset	1.01	330	800	116	...	...	...	...	1.3	1.0	672	...	...	...
Re-press	0.96	375	760	110	...	...	...	...	1.6	1.2	...	...	...	...
Quenched and tempered														
Upset(c)	0.38	230	1490	216	1340	194	10.0	40.0	28.4	21.0	473	...	...	...
Re-press(c)	...	...	1525	221	1340	194	8.5	32.3	...	...	...	...	...	...
Upset(d)	0.60	220	1455	211	1170	170	9.5	32.0	13.6	10.0	472	...	...	...
Re-press(d)	...	...	1550	225	1365	198	7.0	23.0	...	...	...	...	...	...
Upset(e)	0.82	235	1545	224	1380	200	8.0	16.0	8.8	6.5	496	...	...	...
Re-press(e)	...	...	1560	226	1340	194	6.0	12.0	...	...	...	...	...	...
Upset(f)	1.04	315	1560	226	1280	186	6.0	11.8	9.8	7.2	476	...	...	...
Re-press(f)	...	...	1480	215	1225	178	6.0	11.8	...	...	...	...	...	...
Upset(g)	0.39	260	825	120	745	108	21.0	57.0	62.4	46.0	269	...	...	...
Upset(g)	0.58	280	860	125	760	110	20.0	50.0	44.0	32.5	270	...	...	...
Upset(h)	0.80	360	850	123	600	87	19.5	46.0	24.4	18.0	253	...	...	...
Upset(i)	1.01	320	855	124	635	92	17.0	38.0	13.3	9.8	268	...	...	...

(a) Sintered at 1260 °C (2300 °F) in dissociated ammonia. (b) Sintered at 1120 °C (2050 °F) in endothermic gas atmosphere. (c) Tempered at 370 °C (700 °F). (d) Tempered at 440 °C (825 °F). (e) Tempered at 455 °C (850 °F). (f) Tempered at 480 °C (900 °F). (g) Tempered at 680 °C (1255 °F). (h) Tempered at 695 °C (1280 °F). (i) Tempered at 715 °C (1320 °F)

Table 10 Mechanical property data for P/F-4200 materials

Forging mode	Carbon, %	Oxygen, ppm	Ultimate tensile strength MPa	ksi	0.2% offset yield strength MPa	ksi	Elongation in 25 mm (1 in.), %	Reduction of area, %	Core hardness, HV(a)
Blank carburized									
Upset(b)	0.19	450	1205	175	...	...	10.0	37.4	390
Re-press(b)	0.23	720	1110	161	...	...	6.3	17.0	380
Upset(c)	0.25	130	1585	230	...	...	13.0	47.5	489
Re-press(c)	0.25	110	1460	212	...	...	11.3	36.1	466
Quenched and stress relieved									
Upset(b)	0.31	470	1790	260	...	...	9.0	27.3	532
Re-press(b)	0.32	700	1745	253	...	...	4.0	9.0	538
Upset(b)	0.54	380	2050	297	...	...	1.3	...	694
Re-press(b)	0.50	520	2160	313	...	...	2.0	...	653
Upset(c)	0.65	120	1605	233	...	...	...	...	710
Re-press(c)	0.67	130	1040	151	...	...	...	...	709
Upset(b)	0.73	270	1110	161	...	...	...	...	767
Re-press(b)	0.85	370	1345	195	...	...	...	...	727
Upset(b)	0.70	420	600	87	...	...	...	...	761
Re-press(b)	0.67	320	540	78	...	...	...	...	778
Upset(c)	0.91	120	910	132	...	...	...	...	820
Re-press(c)	0.86	120	840	122	...	...	...	...	825
Quenched and tempered									
Upset(d)	0.28	720	1050	153	895	130	10.6	42.8	336
Upset(c)	0.37	1200	1450	210	1385	201	10.2	33.0	447
Upset(e)	0.56	580	1680	244	7560	226	9.8	28.6	444
Upset(f)	0.70	760	1805	262	1565	227	5.0	11.8	531
Upset(g)	0.86	790	1425	207	1310	190	10.4	30.0	450
Upset(h)	0.26	920	835	121	705	102	22.6	57.6	269
Upset(i)	0.38	860	860	125	785	114	20.8	56.5	288
Upset(j)	0.55	840	917	133	820	119	17.8	49.5	305
Upset(k)	0.73	820	965	140	855	124	15.4	42.7	304
Upset(k)	0.87	920	995	144	850	123	15.6	33.9	318

(a) 30 kgf load. (b) Sintered in dissociated ammonia at 1120 °C (2050 °F). (c) Sintered in dissociated ammonia at 1260 °C (2300 °F). (d) Tempered at 175 °C (350 °F). (e) Tempered at 315 °C (600 °F). (f) Tempered at 345 °C (650 °F). (g) Tempered at 425 °C (800 °F). (h) Tempered at 620 °C (1150 °F). (i) Tempered at 650 °C (1200 °F). (j) Tempered at 660 °C (1225 °F). (k) Tempered at 675 °C (1250 °F)

Table 11 Mechanical property and fatigue data for iron-copper-carbon alloys

Sintered at 1120 °C (2050 °F) in dissociated ammonia, reheated to 980 °C (1800 °F) in dissociated ammonia, and forged

Forging mode	Carbon, %	Oxygen, ppm	Ultimate tensile strength MPa	ksi	0.2% offset yield strength MPa	ksi	Elongation in 25 mm (1 in.), %	Reduction of area, %	Room-temperature Charpy V-notch impact energy J	ft·lbf	Core hardness, HV30	Fatigue endurance limit MPa	ksi	Ratio of fatigue endurance to tensile strength
Upset(a)	0.39	250	670	97	475	69	15	37.8	4.1	3.0	228	...	...	...
Upset(b)	0.40	210	805	117	660	96	12.5	38.3	5.4	4.0	261	325	47	0.40
Re-press(a)	0.39	200	690	100	490	71	15	35.4	2.7	2.0	227	...	...	...
Re-press(b)	0.41	240	795	115	585	85	10	36.5	4.1	3.0	269	345	50	0.43
Upset(a)	0.67	170	840	122	750	109	10	22.9	2.7	2.0	267	...	...	...
Upset(b)	0.66	160	980	142	870	126	15	24.9	4.1	3.0	322	470	68	0.48
Re-press(a)	0.64	190	825	120	765	111	10	24.8	3.4	2.5	266	...	...	...
Re-press(b)	0.67	170	985	143	875	127	10	20.6	4.7	3.5	311	460	67	0.47
Upset(a)	0.81	240	1025	149	625	91	10	19.2	2.7	2.0	337	...	...	...
Upset(b)	0.85	280	1130	164	625	91	10	16.6	4.1	3.0	343	525	76	0.46
Re-press(a)	0.81	200	1040	151	640	93	10	16.2	2.7	2.0	335	...	...	...
Re-press(b)	0.82	220	1170	170	745	108	10	12.8	2.7	2.0	368	475	69	0.41

(a) Still-air cooled. (b) Forced-air cooled

Table 12 Mechanical property and fatigue data for iron-copper-carbon alloys with sulfur additions

Sintered at 1120 °F (2050 °F) in dissociated ammonia, reheated to 980 °C (1800 °F) in dissociated ammonia, and forged

Addition	Carbon, %	Oxygen, ppm	Sulfur, %	Ultimate tensile strength MPa	ksi	0.2% offset yield strength MPa	ksi	Elongation in 25 mm (1 in.), %	Reduction of area, %	Room-temperature Charpy V-notch impact energy J	ft·lbf	Core hardness, HV30	Fatigue endurance limit MPa	ksi	Ratio of fatigue endurance to tensile strength
Manganese sulfide	0.59	270	0.13	915	133	620	90	11	23.2	6.8	5.0	290	430	62	0.47
Sulfur	0.63	160	0.14	840	122	560	81	12	21.4	6.8	5.0	267	415	60	0.50
None	0.66	160	0.013	980	142	870	126	15	24.9	4.1	3.0	322	470	68	0.48

surface finish, magnetic particle inspection, composition, density, metallographic analysis, and nondestructive testing.

Part Dimensions and Surface Finish. Typical tolerances for P/F parts are summarized in Table 15. The as-forged surface finish of a P/F part is directly related to the surface finish of the forging tool. Surface finish is generally better than 0.8 μm (32 μin.) which is better than that obtained on wrought forged parts. This good surface finish is beneficial to the fatigue performance of P/F parts.

Magnetic particle inspection is used to detect surface blemishes such as cracks and laps.

Table 13 Compressive yield strengths of P/F-4600 materials
Sintered at 1120 °C (2050 °F) in dissociated ammonia

Forged carbon content, %	Forged oxygen content, ppm	Heat treatment	0.2% offset compressive yield strength	
			MPa	ksi
0.22	460	Stress relieved at 175 °C (350 °F)	1240	180
0.22	350	Tempered at 370 °C (700 °F)	1155	168
0.22	440	Tempered at 680 °C (1255 °F)	580	84
0.29	380	Stress relieved at 175 °C (350 °F)	1440	209
0.35	430	Stress relieved at 175 °C (350 °F)	1670	242
0.43	410	Stress relieved at 175 °C (350 °F)	1690	245
0.41	410	Tempered at 370 °C (700 °F)	1360	197
0.41	460	Tempered at 680 °C (1255 °F)	680	99
0.46	480	Stress relieved at 175 °C (350 °F)	1780	259
0.44	380	Tempered at 370 °C (700 °F)	1275	185
0.44	400	Tempered at 680 °C (1255 °F)	685	100
0.57	330	Stress relieved at 175 °C (350 °F)	1980	287
0.66	400	Tempered at 440 °C (825 °F)	1325	192
0.60	330	Tempered at 680 °C (1255 °F)	700	101
0.75	300	Stress relieved at 175 °C (350 °F)	2000	290
0.80	480	Tempered at 455 °C (850 °F)	1355	196
0.77	410	Tempered at 695 °C (1280 °F)	700	101

Density. Sectional density measurements are taken to ensure that sufficient densification has been achieved in critical areas. Displacement density checks are generally supplemented by microstructural examination to assess the residual porosity level. For a given level of porosity, the measured density will depend on the exact chemistry, thermomechanical condition, and microstructure of the sample. Parts may be specified to have a higher density in particular regions than is necessary in less critical sections of the same component.

Metallographic Analysis. Powder forged parts are subjected to extensive metallographic evaluation. The primary parameters of interest include those discussed below.

The extent of surface decarburization permitted in a forged part will generally be specified. The depth of decarburization may be estimated by metallographic examination, but it is best quantified using microhardness measurements as described in ASTM E 1077.

Surface finger oxides are defined as oxides that follow prior particle boundaries into the forged part from the surface and cannot be removed by physical means such as rotary tumbling. An example of surface finger oxides is shown in Fig. 19. Metallographic techniques are used to determine the maximum depth of surface finger oxide penetration.

Interparticle oxides follow prior particle boundaries. They may sometimes form a continuous three-dimensional network, but more often will, in a two-dimensional plane of polish, appear to be discontinuous. An example is presented in Fig. 19.

Unalloyed iron powder contamination in low-alloy powder forged parts can be quantified by means of the etching procedure described in the earlier section "Material Considerations."

The nonmetallic inclusion level in a P/F part may also be quantified using the image analysis technique described in the section "Material Considerations." However, if the section of a component selected for inclusion assessment is not pore-free, image analysis procedures are not applicable.

Nondestructive Testing. Although metallographic assessment of P/F parts is common, it is also useful to have a nondestructive method for evaluating the microstructural integrity of components. It has been demonstrated that this can be achieved with a magnetic bridge comparator.

Applications of Powder Forged Parts

Previous sections compared powder forging and drop forging and illustrated the range of mechanical property performance that can be achieved in powder forged material. The present section concentrates on two examples of P/F components and highlights some of the reasons for selecting P/F parts over those made by competing forming methods.

Example 1: Converter Clutch Cam. The automotive industry is the principal user of P/F parts, and components for automatic transmissions represent the major areas of application. One of the earliest powder forgings used in such an application is the converter clutch cam (Fig. 20). The primary reason that powder forging was chosen over competitive processes was that it reduced manufacturing costs by 58%, compared with the conventional process of machining a forged gear blank.

Powder forged cams are made from a water-atomized steel powder (P/F-4200) containing 0.6% Mo, 0.5% Ni, 0.3% Mn, and 0.3% graphite. Preforms weighing 0.33 kg (0.73 lb) are compacted to a density of 6.8 g/cm^3. The preforms are sintered at 1120 °C (2050 °F) in an endothermic gas atmosphere with a 2 °C (35 °F) dewpoint. The sintered preforms are graphite coated before being induction heated and forged to near-full density (less than 0.2% porosity) using both axial and lateral flow. After forging, the face of the converter clutch cam is ground, carburized to a depth of 1.75 mm (0.070 in.), and surface hardened by means of induction. The part requires a high density to withstand the high Hertzian stress experienced by the inner cam surface in service. Machining requires only one step on the P/F cam; seven machining operations were required for the conventionally processed cam.

Example 2: Powder Forged Connecting Rods. Connecting rods were among the components selected for a number of P/F development programs in the 1960s. However, it was not until 1976 that the first P/F connecting rod was produced commercially. This was the connecting rod for the Porsche 928 V-8 engine.

The connecting rod for the Porsche 928 engine was made from a water-atomized low-alloy steel powder (0.3 to 0.4% Mn, 0.1 to 0.25% Cr, 0.2 to 0.3% Ni, and 0.25 to

Table 14 Tensile and impact properties of P/F-4600 hot re-pressed at two temperatures

Re-pressing temperature		Re-pressing stress		Re-pressed density		0.2% offset yield strength		Ultimate tensile strength		Elongation, %	Reduction in area, %	Hardness, HV(a)	Charpy V-notch impact energy	
°C	°F	MPa	ksi	g/cm^3	lb/in.^{3}	MPa	ksi	MPa	ksi				J	ft · lbf
870	1600	406	59	7.65	0.276	1156	168	1634	237	2.6	2.8	519	2.9	2.13
870	1600	565	82	7.72	0.279	1243	180	1641	238	2.1	2.8	538	2.8	2.06
870	1600	741	107	7.78	0.281	1316	191	1702	247	2.4	2.4	564	3.1	2.29
870	1600	943	137	7.79	0.282	1349	196	1705	248	2.3	2.4	562	3.5	2.58
1120	2050	344	50	7.83	0.283	1364	198	1750	254	6.4	20.5	549	6.8	5.01
1120	2050	593	86	7.86	0.2840	1450	210	1777	258	6.7	17.3	566	6.2	4.57
1120	2050	856	124	7.87	0.2844	1592	231	1782	259	5.3	14.1	565	6.2	4.57
1120	2050	981	142	7.87	0.2844	1502	218	1788	260	5.5	12.3	572	6.0	4.42

(a) 30 kgf load

Table 15 Typical tolerances for P/F parts

Dimension or characteristic	Description	Typical tolerance mm/mm	Typical tolerance in./in.	Minimum tolerance mm	Minimum tolerance in.	Schematic of part
a	Linear dimension perpendicular to the press axis	0.0025	0.0025	0.08	0.003	
b	Linear dimensions parallel to the press axis	±0.25	±0.10	0.20	0.008	
c	Concentricity of holes to external dimensions	...	...	0.10	0.004	
d	Surface finish	...	...	Normally better than 0.8 μm (32 μin.)		

0.35% Mo) to which graphite was added to give a forged carbon content of 0.35 to 0.45%. The forgings were oil quenched and tempered to a core hardness of 28 HRC (ultimate tensile strength of 835 to 960 MPa, or 121 to 139 ksi), followed by shot peening to a surface finish of 11 to 13 on the Almen scale.

The preform was designed so that the P/F component had less than 0.2% porosity in the critical web region. The P/F connecting rod had considerably better fatigue properties than did conventional drop forged rods. Its weight control was good enough to allow a reduction in the size of the balance pads, resulting in a weight savings of about 10% (it weighed ~1 kg, or 2 lb). Powder forged connecting rods are currently used in both the Porsche 928 and 944 engines.

The first high-volume commercialization of P/F connecting rods was in the 1.9 L Toyota Camry engine. In this design, the balance pads were completely eliminated. Despite the publication of the results of development trials in 1972, it was not until the summer of 1981 that production rods were introduced.

Toyota selected a copper steel (Fe-0.55C-2Cu) based on a water-atomized iron powder to replace conventional forgings, which had been made from a quenched and tempered 10L55 free-machining steel. The preform, which has a preshaped partial I-beam web section, has an average green density of 6.5 g/cm³. The preform shape is such that forging is predominantly in the re-pressing mode. However, some lateral flow does take place where required in critical regions, such as the web.

Preforms are sintered for 20 min at 1150 °C (2100 °F) in an endothermic gas atmosphere in a specially designed rotary hearth furnace. During sintering, the preforms are supported on flat, ceramic plates. The preforms are allowed to stabilize at about 1010 °C (1850 °F) before closed-die forging.

Exposure of the preform to the atmosphere during transfer to the forging dies is limited to 4 to 5 s. An ion nitriding treatment is applied to the punches and dies in the regions at which forging deformation occurs. The connecting rods are forged at the rate of 10 per minute, and tool lives of over 100 000 pieces have been reported.

The forged rods are subjected to a thermal treatment after forging. This results in a ferrite/pearlite microstructure with a core hardness of 240 to 300 HV (30 kgf load). Subsequent operations include burr removal, shot peening, straightening, sizing, magnetic particle inspection, and finish machining.

Savings in material and energy are substantial for the P/F rods. The billet weight for a conventional forging is 1.2 kg (2.65 lb); the P/F preform weighs 0.7 kg (1.54 lb) and requires little machining. In addition to the benefits in process economics, the variability in fatigue performance for the P/F rods is reported to be half that of conventionally forged parts.

Ford Motor Company has introduced P/F connecting rods in the 1.9 L four-cylinder engine used in the Ford Escort model. Five million connecting rods were in service by June 1989. Figure 21 illustrates the Ford P/F connecting rod. Ford has also announced plans to use P/F rods in its modular engine, which is scheduled for production in 1992.

Metal Injection Molding (MIM) Technology

When 60 vol% of fine (10 μm, or 400 μin.) metal powder is blended with 40 vol% lubricant and binder, the resulting mixture can be injection molded, much like a conventional plastic. Any shape that can be molded in plastic can be molded in metal powder. The use of paraffin waxes and thermoplastics such as polyethylene or polypropylene provides the rheological basis for allowing the mixture to flow around corners and into undercuts in a way that is impossible in the uniaxial pressing of binder-free metal powders. The process leads to a whole new class of P/M geometries.

In the original 1971 Wiech process, the wax is removed with a liquid solvent. The remaining plastic binder is thermally removed during a very careful heat-up and sintering. The latter process takes 3 to 5 days. During the long debindering and sintering cycle at 1200 to 1300 °C (2200 to 2400

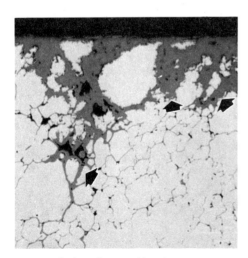

Fig. 19 Surface finger oxides (arrows at upper right) and interparticle oxide networks (arrow near lower left) in a powder forged material

Fig. 20 Powder forged converter clutch cam used in an automotive automatic transmission. Courtesy of Precision Forged Products Division, Federal Mogul Corporation

Fig. 21 Powder forged connecting rod for 1.9 L automobile engine. A similar rod will be used in the modular engine.

°F), near-theoretical density is reached. Following further development at Witec, the process (Wiech II) now consists of removal of the wax in air, at about 175 °C (350 °F), over a 1 to 3 day period. This oxidizes the part, resulting in a brownish color, while the oxide imparts extra strength for handling. Sintering takes place in a closed reaction vessel in a controlled atmosphere of argon and hydrogen. The 60% dense as-molded parts shrink 14 to 20% and achieve near-full density (>95% dense). The parts have very fine, noninterconnected porosity and much better elongation, toughness, and dynamic properties than conventionally pressed and sintered materials. Such parts are limited to section sizes of 9.5 to 13 mm (0.375 to 0.5 in.) and have tolerances of ±0.003 mm/mm (±0.003 in./in.). Because of the oxidation that occurs during wax removal, the process may be fundamentally limited to smaller cross sections. Also, the oxidation of sensitive elements such as chromium and manganese is only reversed at later sintering, with difficulty. Modification of the early Wiech process using solvent to debinder the wax, is now the most commonly used debindering technique. The residual thermoplastic can be handled in a vacuum furnace or atmosphere pusher furnace.

The Rivers process, licensed by Haynes International, uses methylcellulose as a binder, along with small amounts of water, glycerine, and boric acid. The methylcellulose dissolves in cold water to form a binder. Upon injection into a warm mold, it gels to form a fairly rigid part that can be removed from the tooling. The part is dried at 120 °C (250 °F) to remove water and then sintered by any convenient sintering process. This can be done in belt furnaces, pushers, or vacuum furnaces, at nearly any heat-up rate. Depending on the fineness of the starting powder and the sintering temperature, densities of 90 to 99% of full density are obtained. An injection molded and sintered 6.8 kg (15 lb) Stellite block has demonstrated the capability to make parts with a 75 mm (3 in.) thick section. This degree of binder removal is possible because the methyl cellulose breaks down easily upon heat-up and follows the channels created by the evaporated water.

Although there are some large-scale users of the Wiech-type batch sintering reactors in Japan (340 kg, or 750 lb, in a 60 h period), the trend in the United States is toward more continuous furnaces. The use of a 1315 °C (2400 °F) molybdenum wound pusher furnace has become widespread. A single-stage vacuum furnace to evaporate and thermally remove all binders and lubricants from MIM parts has been developed. It is an outgrowth of vacuum furnaces, which have long been used to dewax and liquid phase sinter carbide parts. The MIM parts are then heated to sintering temperatures of 1100 to 1315 °C (2000 to 2400 °F). Because the parts are not oxidized, the process works well with sensitive chromium- and manganese-containing materials, as well as low-alloy steel. A 45 kg (100 lb) load has a floor-to-floor time of 12 h or less and runs under microprocessor control.

Advantages of the MIM Process. The tolerance and shape factor advantages were noted above. The design engineer should think of using injection molded parts in mechanisms or as small enclosures. Some of the MIM fabricators produce runs of 2000 to 5000 pieces, particularly on more expensive parts. They can do a short run such as these because there is no danger to the tooling at setup, while in conventional P/M there is more risk. As the process becomes ever better understood and better controlled, small batches can be grouped together for sintering. Sintering to near-full density seems to reach an end point for shrinkage. This contributes to product uniformity, as long as the original molded part has uniform metal powder density.

Because the process uses fine powders (5 to 20 μm), alloys have been formed *in situ* by the diffusion of mixed elements or master alloys. Some stainless steel alloys have been made in this way, using an Fe-30Cr master alloy. It is convenient to make 4100 series alloys without the need for prealloyed

Fig. 22 Typical MIM microstructure with rounded, isolated areas of porosity. BASF grade OM carbonyl iron sintered for 1 h at 1315 °C (2400 °F) in vacuum. 94% of full density. Unetched. 180×

powders. The capability to blend an alloy allows a flexibility that does not exist in conventional powder metallurgy. There are only a limited number of prealloyed powders available.

Sintering to near-full density gives excellent toughness, elongation, and other dynamic properties. This is aided by the presence of fine spheroidized porosity versus the sharp, stress raiser porosity of conventional powder metallurgy (see Fig. 22 for a MIM microstructure). Without open pores, the parts can be plated, used in pressure-controlled environments, or used in food handling applications. The stainless steels thus produced have very low carbon contents and are more corrosion resistant than typical P/M materials.

The capital equipment (presses) for injection molding is more economical than that for large-scale P/M presses. Tool life is at least 300 000 pieces. These factors help offset the added short-term material cost.

Factors Impeding Growth of MIM Technology. In conventional press and sinter P/M, dimensional tolerances and quality control are very significant. Re-pressed or coined P/M parts are made to a tolerance of ±0.013 mm (±0.0005 in.) on a 25 mm (1 in.) diam part. As-sintered tolerances on well-behaved alloys are ±0.001 mm/mm (±0.001 in./in.). Most MIM producers are achieving ±0.003 mm/mm (±0.003 in./in.) with some

Table 16 Mechanical properties of metal injection molded materials

Alloy	Condition	Ultimate tensile strength MPa	ksi	0.2% offset yield strength MPa	ksi	Elongation in 25 mm (1 in.), %	Hardness	Processing
Fe-2Ni-0.43C	Sintered	470	68	255	37	21	74 HRB	Sinter
	Heat treated	1450	210	1095	159	4.0	38 HRC	Sinter, heat treat, then hold 1 h at 232 °C (450 °F)
	Heat treated	1125	163	980	142	5.5	31 HRC	Sinter, heat treat, then hold 1 h at 343 °C (650 °F)
	Heat treated	880	128	710	103	11.0	21 HRC	Sinter, heat treat, then hold 1 h at 427 °C (800 °F)
4600 + 0.45% C	Sintered	605	88	345	50	11.5	79 HRB	Sinter
	Heat treated	1620	235	1350	196	2.5	44 HRC	Sinter, heat treat, then hold 1 h at 232 °C (450 °F)
	Heat treated	1345	195	1115	162	4.5	78 HRC	Sinter, heat treat, then hold 1 h at 343 °C (650 °F)
	Heat treated	1090	158	938	136	6.0	31 HRC	Sinter, heat treat, then hold 1 h at 427 °C (800 °F)
316/304/L (18–20% Cr, 8–12% Ni)	Sintered	580	84	235	34	60	84 HRB	Sinter
316L duplex (20% Cr, 8% Ni, 2% Mo, 0.03% C max)	Sintered	580	84	262	38	40+	77 HRB	Sinter

Source: Brunswick Technetics Division of Brunswick Corporation

Fig. 23 Tensile bars specially developed for testing MIM. Courtesy of Omark Industries, Advanced Forming Technology, and Metal Injection Molding Association

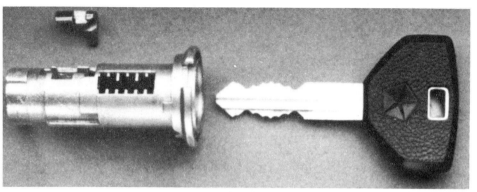

Fig. 24 MIM part (upper left) for an automobile ignition lock. The key forces the MIM part into contact with a security switch. Courtesy of SSI Technologies

companies offering 0.001 mm/mm (0.001 in./in.) on selected dimensions. The industry as a whole needs to reach the tolerance levels being offered by conventional powder metallurgy. Some MIM producers do re-press their parts to straighten them and achieve tolerances like those of conventional powder metallurgy.

The raw materials are expensive: in 1989, carbonyl Fe, reduced, is $8.60/kg ($3.90/lb). With binder added, ferrous material costs may by $7 to 10/kg ($3.20 to 4.50/lb). The total cost for low-alloy steel MIM parts is $30 to 40/kg ($14 to 18/lb). This explains why parts of less than 20 g (0.7 oz) have been the initial focus of sales activity. Work is underway on ways to liquid phase sinter mixtures of $0.70/kg ($0.32/lb) iron and a special graphite (Ref 24). At 1150 °C (2100 °F), some regions of the iron-graphite parts do sinter to full density, but the process is not yet successful. The development of sintering techniques for the coarser, less expensive powders will enormously impact the economics of the process. There are several Japanese and U.S. companies at work developing a way to make less expensive fine Fe powders or alloys.

MIM green parts can exhibit defects as they exit the press. Variations in metal loading or the segregation of metal and plastic will result in dimensional variations as full density is approached. Real-time x-ray equipment has disclosed voids and low density areas in injection molded ceramic parts with diameters of 0.1 mm (0.004 in.). It also may be possible to measure density variation and to look for cracks and pores in MIM parts with x-ray equipment.

Mechanical Properties. Typical mechanical properties for 95% dense MIM materials are given in Table 16. The Metal Injection Molding Association and ASTM Subcommittee B09.11 are currently developing the standardized minimum properties for the MIM materials that are in current use. These will be available in 1990. Figure 23 shows the new standard tensile test bar for metal injection molded materials. It is

(a)

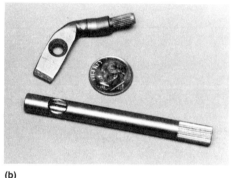
(b)

Fig. 25 Single-piece MIM part (a) that replaced a two-piece automobile turn signal lever assembly (b). The smaller MIM part in (a) was the first version, while the larger MIM part is the finished version that replaced the two-part assembly shown in (b). Courtesy of the Remington Arms Division of E.I. Du Pont de Nemours & Company, Inc.

pulled by inserting pins through the holes in each end and is self-aligning.

Applications. MIM parts have been used in a variety of production parts for automobiles. Other applications of MIM materials include the molding of threads on pressure manifolds and the fabrication of gun sight parts with a special nongalling cam locking mechanism that could not be machined on the parts.

The first two MIM production parts in automobiles were a part for an ignition lock (Fig. 24) and a single-part replacement (Fig. 25a) for a two-part turn signal lever assembly (Fig. 25b). Both have been in service since July 1988.

Figure 24 shows the entire ignition lock and the MIM subcomponent. As the key is inserted in the lock, the cam-shaped MIM part moves away and depresses an electrical switch, which is part of the security system. The initial design of the part was too small and complicated for the model shop to make and it was prototyped from the MIM tooling.

The turn signal indicator lever is an example of the replacement of a two-piece assembly (Fig. 25b) with a single MIM part. The lower portion of Fig. 25(a) shows the first version of the MIM part, and the upper view shows the final 19.0 g (0.670 oz) MIM part that replaced the assembly. The MIM material is iron with 2% Ni, sintered and then case hardened. It replaced AISI 4037 and SAE 1018 case hardened. The MIM part succeeded because of its superior strength compared to the two-piece assembly. The core properties of the materials are 415 MPa (60 ksi) tensile strength and 15% elongation at 60 HRB.

REFERENCES

1. R.M. German, *Powder Metallurgy Science*, Metal Powder Industries Federation, 1984
2. F.V. Lenel, *Powder Metallurgy Principles and Applications*, Metal Powder Industries Federation, 1980
3. P/M Materials Standards for P/M Structural Parts, MPIF Standard 35, Metal Powder Industries Federation, 1988
4. C. Durdaller, The Effect of Additions of Copper, Nickel and Graphite on the Sintered Properties of Iron-Base Sintered P/M Parts, *Progress in Powder Metallurgy*, Metal Powder Industries Federation, Vol 25, 1969, p 71-100
5. L.G. Roy and L.F. Pease III, Through Hardening of P/M Materials, in *Progress in Powder Metallurgy*, Vol 42, Metal Powder Industries Federation, 1986
6. A.F. deRege, G. l'Espérance, and L.F.

Pease III, Prealloyed MnS Powders for Improved Machinability in P/M Parts, in *Near Net Shaping Manufacturing*, P.W. Lee and B.L. Ferguson, Ed., ASM INTERNATIONAL, 1988, p 57-67
7. C. Durdaller, *Copper Infiltration of Iron-Base P/M Parts*, Hoeganaes Corporation, 1969
8. H. Ferguson, Heat Treatment of P/M Parts, *Met. Prog.*, Vol 107 (No. 6), June 1975, p 81-83; Vol 108 (No. 2), July 1975, p 66-69
9. L.F. Pease III, J.P. Collette, and D.A. Pease, Mechanical Properties of Steam Blackened P/M Materials, in *Modern Developments in Powder Metallurgy*, Vol 18-21, Metal Powder Industries Federation, 1988
10. G. Bockstiegel, Powder Forging—Development of the Technology and Its Acceptance in North America, Japan, and West Europe, in *Powder Metallurgy 1986—State of the Art*, Vol 2, Powder Metallurgy in Science and Practical Technology series, Verlag Schmid, 1986, p 239
11. P.K. Jones, The Technical and Economic Advantages of Powder Forged Products, *Powder Metall.*, Vol 13 (No. 26), 1970, p 114
12. Economic Aspects of P/M-Hot-Forming, *Mod. Dev. Powder Metall.*, Vol 7, 1974, p 91
13. J.W. Wisker and P.K. Jones, The Economics of Powder Forging Relative to Competing Processes—Present and Future, *Mod. Dev. Powder Metall.*, Vol 7, 1974, p 33
14. W.J. Huppmann and M. Hirschvogel, Powder Forging, Review 233, *Int. Met. Rev.*, No. 5, 1978, p 209
15. C. Tsumuti and I. Nagare, Application of Powder Forging to Automotive Parts, *Met. Powder Rep.*, Vol 39 (No. 11), 1984, p 629
16. R. Koos and G. Bockstiegel, The Influence of Heat Treatment, Inclusions and Porosity on the Machinability of Powder Forged Steel, *Prog. Powder Metall.*, Vol 37, 1981, p 145
17. B.L. Ferguson, H.A. Kuhn, and A. Lawley, Fatigue of Iron Base P/M Forgings, *Mod. Dev. Powder Metall.*, Vol 9, 1977, p 51
18. G.T. Brown and J.A. Steed, The Fatigue Performance of Some Connecting Rods Made by Powder Forging, *Powder Metall.*, Vol 16 (No. 32), 1973, p 405
19. W.B. James, The Use of Image Analysis for Assessing the Inclusion Content of Low Alloy Steel Powders for Forging Applications, in *Practical Applications of Quantitative Metallography*, STP 839, American Society for Testing and Materials, 1984, p 132
20. R. Causton, T.F. Murphy, C.-A. Blande, and H. Soderhjelm, Non-Metallic Inclusion Measurement of Powder Forged Steels Using an Automatic Image Analysis System, in *Horizons of Powder Metallurgy*, Part II, Verlag Schmid, 1986, p 727
21. W.B. James, "Quality Assurance Procedures for Powder Forged Materials," Technical Paper 830364, Society of Automotive Engineers, 1983
22. W.B. James, Automated Counting of Inclusions in Powder Forged Steels, *Mod. Dev. Powder Metall.*, Vol 14, 1981, p 541
23. W.B. James, New Shaping Methods for P/M Components, in *Powder Metallurgy 1986—State of the Art*, Vol 2, Powder Metallurgy in Science and Practical Technology series, Verlag Schmid GmbH, 1986, p 71
24. L.F. Pease, An Approach to Near Full Density P/M Automotive Components—Liquid Phase Sintering of Iron Graphite Materials, Technical Paper 870131, Society of Automotive Engineers, 1987

Austenitic Manganese Steels

Revised by D.K. Subramanyam,* Ergenics Inc.; A.E. Swansiger, ABC Rail Corporation; and H.S. Avery, Consultant

THE ORIGINAL AUSTENITIC manganese steel, containing about 1.2% C and 12% Mn, was invented by Sir Robert Hadfield in 1882. Hadfield's steel was unique in that it combined high toughness and ductility with high work-hardening capacity and, usually, good resistance to wear. Consequently, it rapidly gained acceptance as a very useful engineering material.

Hadfield's austenitic manganese steel is still used extensively, with minor modifications in composition and heat treatment, primarily in the fields of earthmoving, mining, quarrying, oil well drilling, steelmaking, railroading, dredging, lumbering, and in the manufacture of cement and clay products. Austenitic manganese steel is used in equipment for handling and processing earthen materials (such as rock crushers, grinding mills, dredge buckets, power shovel buckets and teeth, and pumps for handling gravel and rocks). Other applications include fragmentizer hammers and grates for automobile recycling and military applications such as tank track pads. Another important use is in railway trackwork at frogs, switches, and crossings, where wheel impacts at intersections are especially severe. Because austenitic manganese steel resists metal-to-metal wear, it is used in sprockets, pinions, gears, wheels, conveyor chains, wear plates, and shoes.

Austenitic manganese steel has certain properties that tend to restrict its use. It is difficult to machine and usually has a yield strength of only 345 to 415 MPa (50 to 60 ksi). Consequently, it is not well suited for parts that require close-tolerance machining or that must resist plastic deformation when highly stressed in service. However, hammering, pressing, cold rolling, or explosion shocking of the surface raises the yield strength to provide a hard surface on a tough core structure.

Composition

Many variations of the original austenitic manganese steel have been proposed, often in unexploited patents, but only a few have been adopted as significant improvements. These usually involve variations of carbon and manganese, with or without additional alloys such as chromium, nickel, molybdenum, vanadium, titanium, and bismuth. The most common of these compositions, as listed in ASTM A 128, are given in Table 1.

The available assortment of wrought grades is smaller and usually approximates ASTM composition B-3. Some wrought grades contain about 0.8% C and either 3% Ni or 1% Mo. Large heat orders are usually required for the production of wrought grades, while cast grades and their modifications are more easily obtained in small lots. A manganese steel foundry may have several dozen modified grades on its production list. Modified grades are usually produced to meet the requirements of application, section size, casting size, cost, and weldability considerations.

Carbon and Manganese. The ASTM A 128 compositions in Table 1 do not permit any austenite transformation when the alloys are water quenched from above the A_{cm} (that is, the temperature that corresponds to the boundary between the cementite-austenite and the austenite fields). However, this does not preclude lower ductility in heavy sections because of slower quenching rates. The effect is due to the formation of carbides along grain boundaries and other interdendritic areas and to some degree affects nearly all commercial castings except the very smallest. Figure 1 shows A_{cm} temperatures for 13% Mn steels containing between 0.6 and 1.4% C. Figure 2 shows the effects of carbon and manganese content on the M_s temperature, that is, the temperature at which martensite starts to form from austenite upon cooling, of a homogeneous austenite with all carbon and manganese in solid solution.

The mechanical properties of austenitic manganese steel vary with both carbon and manganese content. Figure 3 indicates that carbon increases strength up to the range of ASTM A 128, grade A. A pleateau is indicated at 1.05 to 1.35% C content. Any departure from this curve can be attributed to grain size unless good statistical evidence is found. The plateau at 827 MPa (120 ksi) is based on the 97-heat, 270-test scatter graph shown in Fig. 4. The data points in Fig. 4 were used to calculate the standard deviation, σ, data in Fig. 3. As carbon is increased it becomes increasingly difficult to retain all of the carbon in solid solution, which may account for reductions in tensile strength and ductility. Nevertheless, because abrasion resistance tends to increase with carbon, carbon content higher than the 1.20% midrange of grade A may be preferred even when ductility is lowered. Carbon content above 1.4% is seldom used because of the difficulty of obtaining an austenitic structure sufficiently free of grain boundary carbides, which are detrimental to strength and ductility. The effect can also be observed in 13% Mn steels containing less than 1.4% C because segregation may result in local variations of ±17% (±0.2%C) from the average carbon level determined by chemical analysis.

Table 1 Standard composition ranges for austenitic manganese steel castings

ASTM A 128 grade	C	Mn	Cr	Mo	Ni	Si (max)	P (max)
A	1.05–1.35	11.0 min	...	...	...	1.00	0.07
B-1	0.9–1.05	11.5–14.0	...	...	...	1.00	0.07
B-2	1.05–1.2	11.5–14.0	...	...	...	1.00	0.07
B-3	1.12–1.28	11.5–14.0	...	...	...	1.00	0.07
B-4	1.2–1.35	11.5–14.0	...	...	...	1.00	0.07
C	1.05–1.35	11.5–14.0	1.5–2.5	...	...	1.00	0.07
D	0.7–1.3	11.5–14.0	...	...	3.0–4.0	1.00	0.07
E-1	0.7–1.3	11.5–14.0	...	0.9–1.2	...	1.00	0.07
E-2	1.05–1.45	11.5–14.0	...	1.8–2.1	...	1.00	0.07
F	1.05–1.35	6.0–8.0	...	0.9–1.2	...	1.00	0.07

*Formerly with Abex Corporation

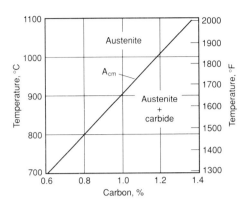

Fig. 1 Solubility of carbon in 13% Mn steels. Source: Ref 1

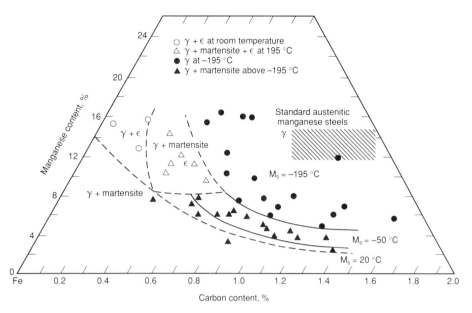

Fig. 2 Variation of M_s temperature with carbon and manganese contents. Source: Ref 2

The 0.7% C (min) of grades D and E-1 may be used to minimize carbide precipitation in heavy castings or in weldments, and similar low carbon contents are specified for welding filler metal. Carbides form in castings that are cooled slowly in the molds. In fact, carbides form in practically all as-cast grades containing more than 1.0% C, regardless of mold cooling rates. They form in heavy-section castings during heat treatment if quenching is ineffective in producing rapid cooling throughout the entire section thickness. Carbides can form during welding or during service at temperatures above about 275 °C (530 °F).

If carbon and manganese are lowered together, for instance to 0.53% C with 8.3% Mn or 0.62% C with 8.1% Mn, the work-hardening rate is increased because of the formation of strain-induced α (body-centered cubic, or bcc) martensite. However, this does not provide enhanced abrasion resistance (at least to high-stress grinding abrasion) as is often hoped (Ref 4).

Manganese contributes the vital austenite-stabilizing effect of delaying transformation (but not eliminating it). Thus, in a simple steel that contains 1.1% Mn, isothermal transformation at 370 °C (700 °F) begins about 15 s after the steel is quenched to that temperature, whereas in a 13% Mn steel, transformation at the same temperature does not begin until after 48 h (Ref 1). Below 260 °C (500 °F), phase changes and carbide precipitation are so sluggish that for all practical purposes they may be neglected, in the absence of deformation, if manganese content exceeds 10%.

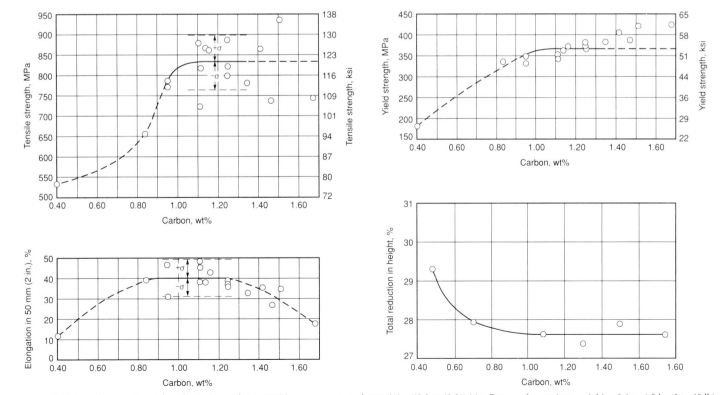

Fig. 3 Variation of properties with carbon content for austenitic manganese steel containing 12.2 to 13.8% Mn. Data are for castings weighing 3.6 to 4.5 kg (8 to 10 lb) and about 25 mm (1 in.) in section size that were water quenched from 1040 to 1095 °C (1900 to 2000 °F). Flow under impact is the total reduction in height sustained by a cylindrical specimen 25 mm (1 in.) in both diameter and length after absorbing 20 blows of 680 J (500 ft · lbf) each. Source: Abex Research Center

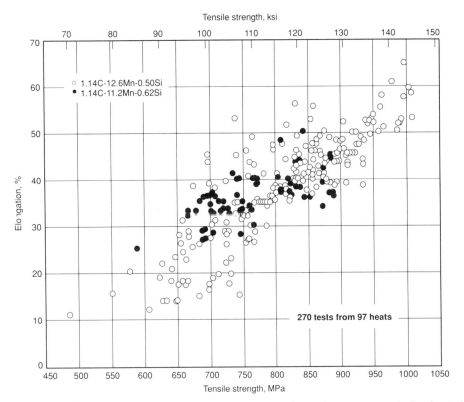

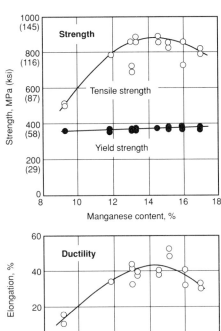

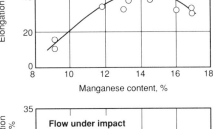

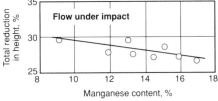

Fig. 4 Distribution of tensile strength and ductility values for 97 heats of manganese steel. The chemical compositions indicated are average for the specific data points plotted. Test specimens were 25 mm (1 in.) diam bars, austenitized and quench-annealed from 1010 °C (1850 °F) or above. Source: Ref 3

Fig. 5 Variation of properties with manganese content for austenitic manganese steel containing 1.15% C. Data are for castings weighing 3.6 to 4.5 kg (8 to 10 lb) and about 25 mm (1 in.) in section size that were water quenched from 1040 to 1095 °C (1900 to 2000 °F). Flow under impact is the total reduction in height sustained by a cylindrical specimen 25 mm (1 in.) in both diameter and length after absorbing 20 blows of 680 J (500 ft · lbf) each. Source: Abex Research Center

Figure 5 shows the influence of manganese content on the strength and ductility of cast austenitic steel that has been solution treated and water quenched. It confirms the observations of many investigators, including Sir Robert Hadfield (Ref 5), who studied the influence of manganese content up to about 22%. Manganese content has little effect on yield strength. In tensile testing, ultimate strength and ductility increase fairly rapidly with increasing manganese content up to about 12% and then tend to level off, although small improvements normally continue up to about 13% Mn.

Silicon and Phosphorus. As noted in Table 1, silicon and phosphorus are present in all ASTM A 128 grades of austenitic manganese steel. Silicon is seldom added except for steelmaking purposes. Silicon content exceeding 1% is uncommon, because foundries do not like to have the silicon pyramid in melts containing returned scrap. A silicon content of 1 to 2% might be used to increase yield strength to a moderate degree, but other elements are preferred for this effect. Loss of strength is abrupt above 2.2% Si, and Mn steel containing more than 2.3% Si may be worthless. On the other hand, silicon levels below 0.10% show decreased fluidity during casting.

The availability of low-phosphorus ferromanganese since about 1960 has enabled steelmakers to reduce phosphorus levels in manganese steel to a large extent. The preferred practice is to hold the phosphorus content below 0.04% even though 0.07% is permitted by ASTM A 128. Levels above 0.06%, which formerly were prevalent, contribute to hot shortness and low elongation at very high temperatures and frequently are the cause of hot tears in castings and underbead cracking in weldments. It is particularly advantageous to keep phosphorus at the lowest possible level in the grades that are welded, and in manganese steel welding electrodes, and in heavy section castings.

Common Alloy Modifications. The most common alloying elements are chromium, molybdenum, and nickel (see Table 1). Added to the usual carbon level of about 1.15%, both chromium and molybdenum increase yield strength (Fig. 6) and flow resistance under impact.

Chromium additions are less expensive for a given increase, and chromium grades (ASTM A 128, grade C, for instance) are probably the most common modifications. ASTM A 128, grade B, often contains some chromium also. The 2% chromium addition in grade C does not significantly lower toughness in light sections. However, in heavier sections, its effect is similar to that of raising the carbon level; the result is a decrease in ductility due to an increase in the volume fraction of carbides in the microstructure. Chromium additions have been used up to 6% for some applications, sometimes in combination with copper, but these grades no longer receive much attention. Chromium enhances resistance to both atmospheric corrosion and abrasion, although the latter effect is not always consistent and depends on the individual application. It is also used up to 18% in low-carbon electrodes for welding manganese steel. Because of the stabilizing effect of chromium on iron carbide, higher heat-treatment (solutionizing) temperatures are often necessary prior to water quenching.

Molybdenum additions, usually 0.5 to 2%, are made to improve the toughness and resistance to cracking of castings in the as-cast condition and to raise the yield strength (and possibly toughness) of heavy-section castings in the solution-treated and quenched condition. These effects occur because molybdenum in manganese steel is distributed partly in solution in the austenite and partly in primary carbides formed during solidification of the steel. The molybdenum in solution effectively suppresses the

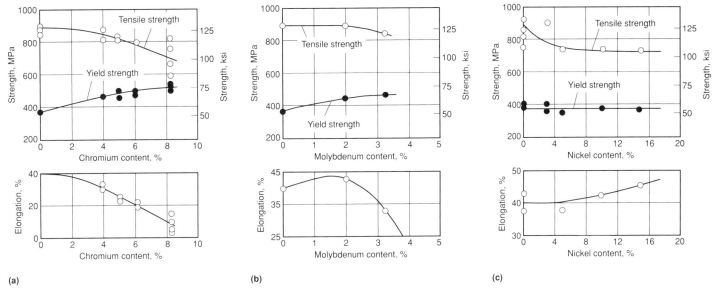

Fig. 6 Effects of (a) chromium, (b) molybdenum, and (c) nickel contents on the tensile properties of cast manganese steel. Steel was cast in 25 mm (1 in.) diam test bars, reheated to 1095 °C (2000 °F), and water quenched. Source: Ref 6

formation of both embrittling carbide precipitates and pearlite, even when the austenite is exposed to temperatures above 275 °C (530 °F) during welding or in service. The molybdenum in primary carbides tends to change the morphology from continuous envelopes around austenite dendrites to a less harmful nodular form, especially when the molybdenum content exceeds 1.5%.

The 1% molybdenum grades (ASTM A 128, grade E-1, and AWS A5.13, grade EFeMn-B) are resistant to the reheating effect that limits the usefulness of the standard B-2, B-3, and B-4 grades. Grade E-1 is adapted to heavy-section castings used in roll and impact crushers that are frequently reheated during weld buildup and overlays.

Grade E-2, which contains about 2% Mo, may be given a special heat treatment to develop a structure of finely dispersed carbides in austenite. This heat treatment entails a partial grain refinement (U.S. Patent 1,975,746) by pearlitizing near 595 °C (1105 °F) for 12 h and water quenching from 980 °C (1800 °F). This type of microstructure has been found to enhance abrasion resistance in crusher applications. The tensile properties of specimens removed from worn cone crusher parts ranged from 440 to 485 MPa (64 to 70 ksi) in yield strength, 695 to 850 MPa (100 to 125 ksi) in tensile strength, and 15 to 25% in elongation.

The addition of molybdenum in amounts greater than 1% can increase the susceptibility of the manganese steel to incipient fusion during heat treatment. Incipient melting refers to a liquation phenomenon that occurs because of the presence of low-melting constituents in interdendritic areas, both within individual grains and along grain boundaries. This tendency is aggravated by higher P levels (>0.05%), higher pouring temperatures (which promote segregation in the casting), and higher carbon levels (>1.3%) in the steel.

As a further use, molybdenum is added to the lean manganese steel grade F partly to suppress embrittlement in both as-cast and heat-treated conditions.

Nickel, in amounts up to 4%, stabilizes the austenite because it remains in solid solution. It is particularly effective for suppressing precipitates of carbide platelets, which can form between about 300 to 550 °C (570 to 1020 °F). Therefore, the presence of nickel helps retain nonmagnetic qualities in the steel, especially in the decarburized surface layers. Nickel additions increase ductility, decrease yield strength slightly, and lower the abrasion resistance of manganese steel. Nickel is used primarily in the lower-carbon or weldable grades of cast manganese steel and in wrought manganese steel products (including welding electrodes). In wrought products, nickel is sometimes used in conjunction with molybdenum. The stability of nickel-containing manganese steels when reheated is shown in Table 2. Table 3 contains compositional and tensile data for nickel manganese steels used in naval applications.

Manganese. ASTM A 128, grade F, has reduced manganese (6 to 8%) to make the austenite less stable, but this requires compensation with 1% Mo to gain acceptable properties. Work-hardening rates are reported to be higher than that of the standard 13% Mn grades, with some loss in toughness. This grade has been used in scoop lips, ball mill end liners, discharge grates, and grizzly screens for siliceous ore milling. One record indicates 45% longer life in ball mill discharge grates compared to pearlitic chromium-molybdenum steel used previously. Average properties reported were 415 MPa (60 ksi) yield strength, 585 MPa (85 ksi) tensile strength, and 12% elongation for carbon levels of 1.2 to 1.4%. ASTM A 128, grade F, is not adapted to heavy sections or to service involving temperatures above 315 °C (600 °F). It has poor weldability and should be avoided if a casting must be hardfaced or rebuilt.

Several other elements not listed in Table 1 (for example, vanadium, copper, titanium, and bismuth) are added to manganese steel for unique applications.

Vanadium is a strong carbide former, and its addition to manganese steels substantially increases yield strength, but with a corresponding decrease in ductility. Vanadium is used in precipitation-hardening manganese steels in amounts ranging from 0.5 to 2%. Because of the stability of vanadium carbonitrides, a higher solutionizing temperature 1120 to 1175 °C (2050 to 2150 °F) is recommended prior to aging (usually between 500 to 650 °C or 930 to 1200 °F). Yield strengths of over 700 MPa (100 ksi) are obtainable depending on the level of ductility that can be tolerated for a given application. Tests of an age-hardened manganese-nickel-molybdenum-vanadium austenitic alloy demonstrated that the abrasion resistance of this steel is not as good as that of the standard grades (Ref 7).

Copper. Like nickel, copper in amounts of 1 to 5% has been used in austenitic manganese steels to stabilize the austenite. The effects of copper on mechanical properties have not been clearly established. Scattered reports indicate that it may have an embrittling effect, which may be due to the limited solubility of copper in austenite.

Bismuth. Other elements such as bismuth and titanium are also added to standard

Table 2 Mechanical properties of three reheated austenitic manganese steels

Condition	Tensile strength(a) MPa	ksi	Yield strength(a) MPa	ksi	Elongation, %(a)	Reduction in area, %(a)	Hardness, HB (b)	Charpy V-notch impact strength(a) J	ft · lbf
12% Mn steel									
Solution treated	615	89	340	49	28	31	164	129	95.5
Reheated at 370 °C (700 °F)									
For 0.5 h	560	81	325	47	26	25	175	117	86.2
For 2 h	600	87	315	46	27	30	168	137	100.7
For 10 h	670	97	330	48	31	30	177	112	82.5
Reheated at 480 °C (900 °F)									
For 0.5 h	620	90	330	48	29	31	177	138	101.7
For 2 h	565	82	325	47	24	20	171	115	84.5
For 10 h	460	67	345	50	6	6	177	12	8.5
Reheated at 595 °C (1100 °F)									
For 0.5 h	395	57	325	47	4	7	173	12	8.8
For 2 h	475	69	350	51	1	1	182	5	3.8
For 10 h	540	78	340	49	1	1	180	3	2.3
12Mn-1Mo steel									
Solution treated	595	86	360	52	26	27	187	110	81.5
Reheated at 370 °C (700 °F)									
For 0.5 h	635	92	350	51	32	31	171	129	95.2
For 2 h	595	86	360	52	25	29	182	116	85.3
For 10 h	625	91	350	51	31	31	163	115	84.5
Reheated at 480 °C (900 °F)									
For 0.5 h	640	93	350	51	32	39	177	120	88.5
For 2 h	625	91	350	51	28	32	187	112	82.5
For 10 h	585	85	350	51	27	31	183	136	100.5
Reheated at 595 °C (1100 °F)									
For 0.5 h	540	78	345	50	21	26	182	103	76.0
For 2 h	405	59	350	51	5	6	182	19	13.8
For 10 h	525	76	380	55	1	2	227	5	4.0
12Mn-3.5Ni steel									
Solution treated	635	92	290	42	39	38	151	135	99.7
Reheated at 370 °C (700 °F)									
For 0.5 h	695	101	330	48	40	33	149	144	106
For 2 h	685	99	325	47	40	35	146	127	93.7
For 10 h	805	117	340	49	50	37	163	126	93
Reheated at 480 °C (900 °F)									
For 0.5 h	685	99	340	49	35	31	142	72	52.8
For 2 h	485	70	325	47	17	20	150	57	42.2
For 10 h	385	56	330	48	5	6	148	14	10.2
Reheated at 595 °C (1100 °F)									
For 0.5 h	460	67	325	47	14	11	146	30	21.8
For 2 h	435	63	330	48	7	6	142	15	10.8
For 10 h	395	57	315	46	4	3	145	6	4.2

(a) Average of two determinations. (b) 1000 kg load; average of six determinations

Table 3 Compositions and tensile properties of 14Mn-Ni steels meeting MIL-S-17758 (Ships)

Element or property	Alloy(a) Type 1	Type 2
Composition limits		
Carbon	0.70–0.90	0.70–0.90
Manganese	13–15	13–15
Silicon	0.50–1.00	0.50–1.00
Nickel	3.0–3.5	1.75–2.25
Chromium	0.50 max	0.50 max
Molybdenum	...	0.35–0.55
Phosphorus	0.07 max	0.07 max
Sulfur	0.05 max	0.05 max
Minimum tensile properties		
Tensile strength, MPa (ksi)	690 (100)	690 (100)
Yield strength, MPa (ksi)(b)	345 (50)	345 (50)
Elongation in 50 mm (2 in.), %	18	18
Elongation plus reduction in area, %	56	56

(a) These grades are intended primarily for nonmagnetic applications and have maximum surface permeability of 1.2, measured with a go/no go low permeability, μ, indicator, as described in MIL-I-17214 and MIL-N-17387. The comparable casting specification, MIL-S-17249(Ships), includes only the standard 1.00 to 2.35% C grade. (b) 0.2% offset

manganese steels. Bismuth was found to improve machinability, especially when coupled with higher manganese levels (>13%).

Titanium can reduce carbon in austenite by forming very stable carbides. The resulting properties may simulate those of a lower-carbon grade. Titanium may also somewhat neutralize the effect of excessive phosphorus; some European practice is apparently based on this idea. Microalloying additions (<0.1%) of titanium, vanadium, boron, zirconium, and nitrogen have been reported to promote grain refinement in manganese steels. The effect, however, is inconsistent. Higher levels of these elements can result in serious losses in ductility. Nitrogen in amounts greater than 0.20% can cause gas porosity in castings. An overall reduction in grain size lowers the susceptibility of the steel to hot tearing.

Sulfur. The sulfur content in manganese steels seldom influences its properties, because the scavenging effect of manganese operates to eliminate sulfur by fixing it in the form of innocuous, rounded, sulfide inclusions. The elongation of these inclusions in wrought steels may contribute to directional properties; in cast steels, such inclusions are harmless. However, it is best to keep sulfur as low as is practically possible to minimize the number of inclusions in the microstructure that would be potential sites for the nucleation of fatigue cracks in service.

Higher Manganese Content Steels

Austenitic steels with a higher manganese content (>15%) have recently been developed for applications requiring low magnetic permeability, μ, low temperature (cryogenic) strength, and low-temperature toughness. These applications stem from the development of superconducting technologies used in transportation systems and nuclear fusion research and to meet the

Table 4 Chemical compositions and mechanical properties of austenitic manganese steels for nonmagnetic and cryogenic applications

			Composition, %					Mechanical properties					
C	Si	Mn	P	S	Ni	Cr	Other	0.2% Yield strength, MPa (ksi)	Tensile strength, MPa (ksi)	Elongation, %	Reduction in area, %	Charpy V-notch impact energy, J (ft · lbf)	Remarks
0.50	0.42	18.0	<0.08	<0.025	<2.0	5.00	...	378 (54.8)	895 (130)	40	33	208 (154)	Hot-rolled bar
0.40	0.55	18.0	...	...	...	4.00	0.10N	358 (51.9)	908 (132)	71	50	258 (190)	Solution annealed
0.70	0.28	15.0	0.045	0.004	1.1	0.24	...	393 (57.0)	915 (133)	67	60	112 (82.6)	Forged, solution annealed
0.71	0.32	15.7	0.047	0.003	1.2	0.21	0.59V	498 (72.2)	1032 (150)	46	39	109 (80.4)	Age-hardened pipe
0.72	0.25	15.6	0.015	0.051	1.1	0.25	0.011Al	376 (54.5)	977 (142)	58	36	...	Free machining, weldable
0.60	0.21	24.6	0.008	0.048	...	6.00	...	340 (49.3)	810 (117)	74	67	238 (176)	Solution annealed, low coefficient of thermal expansion
0.60	0.30	14.0	...	...	2.0	2.00	...	363 (52.6)	794 (115)	68	...	181 (133)	200 mm (7.9 in.) plate, weldable
0.26	0.58	18.7	...	...	...	15.10	1.3V, 0.20N	...	...	...	...	...	Age hardenable, corrosion resistant
0.46	0.34	19.2	0.020	0.001	...	5.30	...	...	...	...	...	...	Weldable
0.90	0.43	14.2	0.050	0.005	...	...	...	...	...	...	...	...	...
0.22	0.46	22.5	0.006	0.064	...	5.2	0.14Pb, 0.001Ca	...	...	...	...	...	Free machining
0.50	0.43	18.45	0.014	0.006	1.22	2.48	...	420 (60.9)	940 (136)	60	57	216 (159)	Hot-rolled plate
0.16	0.21	24.50	0.025	0.006	1.07	5.00	...	397 (57.6)	739 (107)	56	...	141 (104)	...
0.27	0.44	24.50	0.026	0.008	...	1.65	...	398 (57.7)	940 (136)	55	60	182 (134)	Low coefficient of thermal expansion
0.74	0.40	15.90	0.024	0.005	...	1.82	...	...	...	...	...	...	...
0.60	1.00	17.00	...	...	...	6.00	2.00V	784 (114)	1054 (153)	29	...	...	Solutionized at 1125 °C (2060 °F) and aged at 525 °C (975 °F)
0.15	0.50	24.00	<0.02	<0.02	...	5.00	0.05N	264 (38.3)	692 (100)	59.5	...	174 (128)	Hot-rolled plate, stress corrosion cracking resistant
0.01	0.50	28.00	0.02	0.005	...	...	1.5Al, 0.50V	...	...	...	...	...	...
0.14	...	27.50	...	...	0.11	10.00	...	395 (57.3)	931 (135)	35	...	117–147 (86.3–108)(a)	Cast valve for liquified natural gas storage
0.008	...	28.60	0.004	0.001	5.30	15.10	...	...	...	...	...	...	...
0.019	...	10.07	0.003	0.011	14.95	15.16	5.18Mo, 0.16Ti	...	...	...	...	...	High strength
0.14	0.60	31.58	0.022	0.006	0.23	7.04	0.133N	...	...	...	...	...	Weldable
0.024	0.53	18.00	0.004	0.010	5.00	16.30	0.21N	...	...	...	...	...	...
0.05	...	25.00	...	...	1.00	15.00	0.21Cu, Nb, N	...	...	...	...	...	Corrosion resistant
0.20	0.50	24.00	...	...	...	7.00	2.0Al, 0.50V, 0.005Ca	...	...	...	...	...	High-temperature strength

(a) At −196 °C (−321 °F). Source: Ref 8

need for structural materials to store and transport liquefied gases.

Table 4 gives some typical chemical compositions of high-manganese austenitic steels. For low magnetic permeability, these alloys have a lower carbon content than the regular Hadfield steels. The corresponding loss in yield strength is compensated for by alloying with vanadium, nitrogen, chromium, molybdenum, and titanium. Chromium also imparts corrosion resistance, as required in some cryogenic applications.

The alloys are used in the heat-treated (solution-annealed and quenched) condition except for those that are age-hardenable. Wrought alloys are available in the hot-rolled condition. The microstructure is usually a mixture of γ (face-centered cubic, or fcc) austenite and ε (hexagonal close-packed, or hcp) martensite. Typical mechanical properties are included in Table 4. The alloys are characterized by good ductility and toughness, both especially desirable attributes in cryogenic applications. Further, the ductile-brittle transition is gradual, not abrupt. Because the stability of the austenite is composition dependent, a deformation-induced transformation can occur in service under certain conditions. This is usually undesirable because it is accompanied by a corresponding increase in magnetic permeability.

Additions of sulfur, calcium, and aluminum are made to enhance the machinability of these alloys where required. Because of their lower carbon content, most of these alloys are readily weldable by the shielded metal arc welding (SMAW), gas metal arc welding (GMAW), and electron beam welding (EBW) processes. The composition of the weld metal is similar to that of the base metal and tailored for low magnetic permeability. The phosphorus content is generally maintained below 0.02% to minimize the tendency for hot cracking.

Physical Properties. The coefficient of thermal expansion decreases with increasing manganese content and generally ranges from 8 to 17 × 10^{-6}/°C (4 to 9.5 × 10^{-6}/°F). The electrical resistivity of a 0.7C-15Mn-1Ni alloy is reported to be about 710 μΩ · mm. The thermal conductivity of the 32% Mn, 7% Cr alloy was determined to be 0.0314 cal$_{IT}$/cm · s · °C (~13 W/m · K) at room temperature.

Another class of austenitic steels with high manganese additions has been developed for cryogenic and for marine applications with good resistance to cavitation corrosion. Potential applications also include those that require high-temperature corrosion resistance. These alloys have been viewed as economical substitutes for conventional austenitic stainless steels because they contain aluminum and manganese instead of chromium and nickel. Some examples of these steels are given in Table 5.

828 / Specialty Steels and Heat-Resistant Alloys

Table 5 Composition and mechanical properties of high-manganese stainless steels for marine and cryogenic applications

Composition, %					Test temperature		Yield strength		Tensile strength		Elongation, %	Reduction in area, %	Miscellaneous
C	Mn	Al	Si	Other	°C	°F	MPa	ksi	MPa	ksi			
0.07	18.83	5.34	...	...	26	70	214	31.0	524	76.0	54	79	...
					593	1100	172	24.9	321	46.5	55	68	...
					732	1350	131	19.0	193	28.0	72	77	...
0.84	18.74	4.98	...	...	20	70	407	59.0	817	118	59	74	...
					593	1100	203	29.4	421	61.0	54	84	...
					732	1350	148	21.5	210	30.5	86	91	...
0.88	28.10	8.30	...	...	20	70	496	71.9	889	129	54	72	...
					593	1100	503	72.9	614	89.0	22	46	...
					732	1350	245	35.5	248	36.0	62	79	...
0.97	29.43	8.93	1.43	...	20	70	565	81.9	827	120	72	70	...
0.01	23.00	5.30	...	...	20	70	224	32.5	490	71.0	53	...	28 HRB
					−196	−320	456	66.1	915	133	82	...	190 J · cm
0.17	23.00	4.60	...	...	20	70	262	38.0	558	80.9	53	...	43 HRB
					196	320	500	04.1	1020	148	82	...	210 J · cm
1.01	30.50	10.40	1.31	0.007P, 0.006S, 0.056Re	20	70	...	...	605	87.7	10	...	40 HRC
0.50	22.30	7.20	0.30	0.005P, 0.003S, 0.038Re	20	70	...	...	545	79.0	14	...	80 HRB
0.23	29.20	7.40	1.25	0.008P, 0.003S, 0.061Re	20	70	...	...	550	79.8	11	...	84 HRB
0.13	20.70	5.60	0.28	0.007P, 0.005S, 0.063Re	20	70	...	...	510	74.0	15	...	86 HRB
0.76	34.50	10.00	...	...	20	70	717	104	917	133	56	61	...
					650	1200	407	59.0	510	74.0	49	79	...
1.00	30.00	8.00	1.50	...	20	70	496	71.9	862	125	62	57	...
					650	1200	400	58.0	552	80.0	12	19	...

Source: Ref 9–12

These alloys are generally of higher strength but lower ductility than conventional stainless steels such as type 304. Typical mechanical properties are also shown in Table 5. The microstructure in these alloys is a mixture of γ (fcc) austenite and ε (hcp) martensite, and in some cases (especially when the aluminum content exceeds about 5%) α (bcc) ferrite, too, and is obtained by a solution-annealing heat treatment. There is a tendency for an embrittling β-Mn phase to form in the high-manganese compositions during aging at elevated temperatures. The result is a significant decrease in ductility. The addition of aluminum to some extent suppresses the precipitation of this compound.

Fabrication. While these alloys require special precautions during melting due to their high aluminum contents, they are fairly easily hot worked and are usable in wrought form. Data on weldability is not readily available; it is suspected, however, that overall weldability is poor.

Oxidation. The oxidation resistance of these alloys is inferior to that of the chromium-containing grades of stainless steels because of the poor adhesion of aluminum oxide scale. This is especially evident under thermal cycling conditions.

Melt Practice

Austenitic manganese steels are most commonly produced in electric arc furnaces using a basic melting practice. Typical charge materials include carbon and manganese steel scrap, high-carbon ferromanganese, ferrosilicon, and silicon-manganese. Alloying elements such as chromium, molybdenum, and vanadium are usually added as ferroalloys, while elements such as nickel are used in a nearly pure metallic state. Deoxidation of the steel is accomplished with aluminum prior to pouring. In the case of most castings, the pouring temperature is regulated to less than 1470 °C (2680 °F) to prevent an excessively coarse grain size and minimize chemical segregation and other related casting defects. Sand castings are produced with olivine rather than silica sand to prevent mold-metal reaction. The grain size of wrought manganese steels is usually much lower than that of castings due to the recovery and recrystallization of the austenite grains during the hot rolling process. Further details of melting practice can be found in Ref 13.

As-Cast Properties

Although austenitic manganese steels in the as-cast condition are generally considered too brittle for normal use, Table 6 demonstrates that there are exceptions to this rule. Mechanical properties are listed for five grades of as-cast austenitic manganese steels of various thicknesses. These data indicate that lowering carbon content to less than 1.1% and/or adding about 1.0% Mo or about 3.5% Ni results in commercially acceptable as-cast ductilities in light and moderate section thicknesses. These data also apply to weld deposits, which are normally left in the as-deposited condition and therefore are essentially equivalent to material in the as-cast condition.

Adjustments in composition that limit carbide embrittlement and austenite transformation reduce or eliminate cracking of manganese steel castings during cooling in the molds or reheating for solution treatment. The steel is generally poured at temperatures just high enough to avoid misruns in the castings and excessive skulling of the metal in the ladle. This practice helps ensure rapid solidification of the metal in the molds, which in turn helps prevent an excessively coarse grain size. The final microstructure in most castings is not fully austenitic, but contains carbide precipitates and pearlite in an austenitic matrix.

Commercial use of castings in the as-cast condition results in cost and energy savings and eliminates the problems of decarburization of thin castings during solution treatment and warpage during water quenching. Full ductility or toughness is not required in certain applications of manganese steels. For example, as-cast manganese steels have been used successfully for pans on pan conveyors and for other light-section applications where solution-treated and water-quenched castings were prone to severe warpage.

Of the steels listed in Table 6, the 6Mn-1Mo grades are most susceptible to strain-induced martensite embrittlement because of their low manganese levels. However, even these grades can be used in applications for which 1% elongation (either determined in a tensile test or estimated from a transverse bend test) is considered adequate ductility. For example, the as-cast 6Mn-1Mo grade containing 0.8 to 1.0% C has been used successfully in grinding mill liners.

Table 6 Composition and mechanical properties of typical as-cast austenitic manganese steels

	Composition, %			Form	Section size mm	Section size in.	0.2% yield strength MPa	0.2% yield strength ksi	Tensile strength MPa	Tensile strength ksi	Elongation, %	Reduction in area, %	Charpy V-notch impact strength J	Charpy V-notch impact strength ft · lbf	Hardness, HB
C	Mn	Si	Other												
Plain manganese steels															
0.85	11.2	0.57	...	Round	25	1	...	...	440	64	14.5	...	...	...	...
0.95	13.0	0.51	...	Round	25	1	...	...	420	61	14	...	...	...	...
1.11	12.7	0.54	...	Round	25	1	360	52	450	65	4	...	...	...	...
1.27	11.7	0.56	...	Round	25	1	...	...	360	52	2	...	...	...	...
1.28	12.5	0.94	...	Keel block	102	4	...	...	330(a)	48(a)	1(a)	...	3.4	2.5	245
1.36	20.2	0.6	...	Y-block	51	2	...	...	425(a)	62(a)	1(a)	...	...	...	283
1% Mo manganese steels															
0.61	11.8	0.17	1.10 Mo	Round	25	1	315	46	710	103	27.5	23	...	...	163
0.75	13.9	0.58	0.90 Mo	Round	25	1	340	49	740	107	39.5	30	...	...	183
0.83	11.6	0.38	0.96 Mo	Round	25	1	345	50	695	101	30	29	...	...	163
0.89	14.1	0.54	1.00 Mo	Round	25	1	360	52	690	100	29.5	22	...	...	196
1.16	13.6	0.60	1.10 Mo	Round	25	1	400	58	560	81	13	15	...	...	185
0.93	13.6	0.67	0.96 Mo	Plate	25	1	365	53	510	74	11	16	72	53	188
0.99	12.6	0.6	0.87 Mo	Plate	25	1	...	...	460(a)	67(a)	6(a)	...	...	...	...
0.98	12.6	0.6	0.87 Mo	Plate	50	2	...	...	435(a)	63(a)	4(a)	...	...	...	...
0.95	12.6	0.6	0.87 Mo	Plate	102	4	345	50	385	56	4	4	...	...	...
1.30	13.1	0.78	0.99 Mo	Keel block	102	4	...	...	435(a)	63(a)	2(a)	...	8	6	230
1.33	19.8	0.6	0.99 Mo	Y-block	51	2	...	...	505(a)	73(a)	2.5(a)	...	...	...	231
2% Mo manganese steels															
0.52	14.3	1.47	2.4 Mo	Round	25	1	370	54	600	87	15.5	13	...	...	220
0.70	13.6	0.63	2.0 Mo	Round	25	1	360	52	785	114	41	29	...	...	180
0.75	14.1	0.99	2.0 Mo	Round	25	1	365	53	745	108	34.5	27	...	...	183
0.91	14.1	0.60	2.0 Mo	Round	25	1	395	57	705	102	27.5	21	...	...	196
1.24	14.1	0.64	3.0 Mo	Round	25	1	440	64	600	87	7.5	10	...	...	235
1.40	12.5	0.62	2.1 Mo	Round	25	1	420	61	550	80	3.5	5	...	...	228
1.34	12.0	0.43	2.2 Mo	Keel block	51	2	415	60	435	63	3.5	7	...	...	235
3.5% Ni manganese steels															
0.75	13.0	0.95	3.65 Ni	Round	25	1	295	43	655	95	36	26	...	...	150
0.80	13.5	0.53	3.61 Ni	Round	25	1	...	...	530	77	26	...	...	...	...
0.91	13.3	0.53	3.38 Ni	Round	25	1	...	...	510	74	24	...	...	...	...
6Mn-1Mo alloys															
0.90	5.8	0.37	1.46 Mo	Mill liner	102	4	325	47	340	49	2	...	9	7	181
1.00	6.0	0.43	1.03 Mo	Keel block	102	4	330	48	365	53	2	3	...	...	195
0.89	6.3	0.6	1.20 Mo	Plate	102	4	...	...	330(a)	48(a)	1(a)	...	...	...	...
1.27	6.1	0.42	1.07 Mo	Keel block	51	2	365	53	400	58	1	1	3	2	273

(a) Properties converted from transverse bend tests on 6.4 × 13 mm (¼ × ½ in.) bars cut from castings and broken by center loading across 25 mm (1 in.) span. Source: Ref 3

Heavy Sections. As section thickness increases, the rate at which castings cool in sand molds decreases. This increases the opportunity for embrittlement by carbide precipitation. Shapes that tend to develop high residual stresses, such as cylinders and cones, can be particularly affected. These stresses most probably result from volume changes accompanying the carbide precipitation and austenite transformation that occur during the normal cooling of castings.

Figure 7 shows the volume changes that occur during the isothermal decomposition of a 1.25C-12.8Mn steel at temperatures between 850 and 500 °C (1560 and 930 °F), the principal range within which embrittlement occurs when a casting is cooled in its mold or reheated for reaustenitization. Between 850 and about 705 °C (1560 and 1300 °F), only carbides are precipitated, primarily as envelopes around austenite grains and as lamellar-type patches within grains. The lamellar carbide patches have the appearance of coarse pearlite, but actually they are carbide plates in austenite. Below 705 °C (1300 °F), and particularly between 650 and 550 °C (1200 and 1020 °F), pearlite nodules, nucleated by previously precipitated carbides, grow at a relatively rapid rate.

Transgranular acicular carbides also tend to precipitate below about 600 °C (1110 °F), especially in austenite containing more than about 1.1% C. This precipitation can continue down to about 300 °C (570 °F) in a 1.2C-12Mn steel. It may be followed by the transformation of some of the carbon-depleted austenite to martensite as the ambient temperature is approached.

Heat Treatment

Heat treatment strengthens austenitic manganese steel so that it can be used safely and reliably in a wide variety of engineering applications. Solution annealing and quenching, the standard treatment that produces normal tensile properties and the desired toughness, involves austenitizing followed quickly by water quenching. Figure 8 shows the microstructures of a 76 mm (3 in.) section of austenitic manganese steel in the as-cast condition and after solution annealing and quenching. Variations of this treatment can be used to enhance specific desired properties such as yield strength and abrasion resistance. Usually, a fully austenitic structure, essentially free of carbides and reasonably homogeneous with respect to carbon and manganese, is desired in the as-quenched condition, although this is not always attainable in heavy sections or in steels containing carbide-forming elements such as chromium, molybdenum, va-

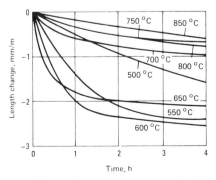

Fig. 7 Change in length of an austenitic 1.25C-12.8Mn steel during isothermal transformation. Source: Ref 14

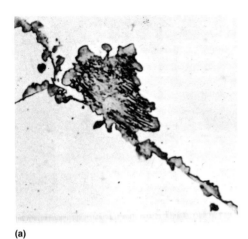

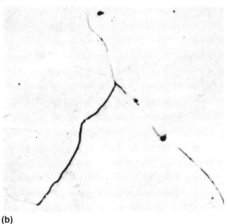

Fig. 8 Typical structures of as-cast and heat-treated ASTM A 128, grade B-3, manganese steel. Top: As-cast material 76 mm (3 in.) thick, with large carbides along grain boundaries. Bottom: 76 mm (3 in.) thick material heated to 1120 °C (2050 °F) and water quenched. Both samples etched in 2½% nital, rinsed in methanol, and reetched in 15% HCl. Both 500×.

nadium, and titanium. If carbides exist in the as-quenched structure, it is desirable for them to be present as relatively innocuous particles or nodules within the austenite grains rather than as continuous envelopes at grain boundaries.

Procedures. Full solution of carbides requires a solution-treating temperature that exceeds A_{cm} by about 30 to 50 °C (50 to 90 °F). Soaking for 1 to 2 h at temperature is usually adequate. Although it might appear that temperatures above 1095°C (2000 °F) would permit use of of 1.4 to 1.5% C, three factors discourage the use of very high temperatures:

- Incipient melting occurs in areas of carbon and phosphorus segregation
- Scaling and decarburization become excessive
- Commercial quenching rates are limited in their ability to retain high carbon concentrations in solution

The commercial heat treatment of manganese steel castings normally involves heating slowly to 1010 to 1090 °C (1850 to 2000 °F), soaking for 1 to 2 h per 25 mm (1 in.) of thickness at temperature, and then quenching in agitated water. There is some tendency for the austenite grains to grow during soaking, especially in wrought manganese steels, although final austenite grain size in castings is largely determined by pouring temperature and solidification rate.

For grade E-2 manganese steel (see Table 1), a modified heat treatment is often specified or recommended. This treatment consists of heating castings to about 595 °C (1100 °F) and soaking them 8 to 12 h at temperature, which causes substantial amounts of pearlite to form in the structure. The castings are then further heated to about 980 °C (1800 °F) to reaustenitize the structure. This step converts the pearlitic areas to fine-grain austenite containing a dispersion of small carbide particles, which remain undissolved as long as the austenitizing temperature does not exceed about 1010 °C (1850 °F). Quenching then results in a dispersion-hardened austenite, which is characterized by higher yield strength, higher hardness, and lower ductility than would be obtained if the same steel were given a full solution treatment at a higher austenitizing temperature. This dispersion-hardening heat treatment permits a relatively high carbon content which in turn can improve abrasion resistance.

Precautions. Speed of quenching is important, but it is difficult to increase beyond the rate of heat transfer from a hot surface to agitated water or beyond the rate fixed by the thermal conductivity of the metal. As a result, heavy-section castings have lower mechanical properties at the center than do thinner castings. Figure 9 shows the cooling rates that can be expected when metal plates of four different thicknesses are quenched in water. Table 7 lists average properties observed in castings of 1.11C-12.7Mn-0.5Si-0.043P steel water quenched from about 980 °C (1800 °F), which cooled the castings at the rates shown in Fig. 9.

Residual stresses from quenching, coupled with the lower properties of heavy sections, establish the usual maximum thickness of commercial castings at about 127 to 152 mm (5 to 6 in.), although castings with sections up to 406 mm (16 in.) thick have been produced.

There have been reports of two-step quenching cycles being used to decrease the overall thermal gradient and thereby to reduce the residual stresses associated with heat-treatment operations. However, this is not common practice in the industry and is probably restricted to specific products. Residual stresses in manganese steels are not a critical issue because of their inherent toughness.

The relatively high austenitizing temperature leads to marked surface decarburization by furnace gases and to some loss of manganese. Surface decarburization may

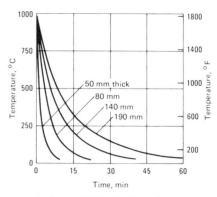

Fig. 9 Cooling curves for austenitic manganese steel of various thicknesses. Cooling curves are approximately equivalent to those for plate of the thicknesses indicated. Source: Ref 3

extend as much as 3.2 mm (⅛ in.) below the casting surface. Thus, the skin may be partly martensitic at times and usually exhibits properties less desirable than those of the underlying metal. This characteristic is not significant in parts subjected to abrasion, such as those used in crushing or grinding, because in these applications the skin is removed by normal wear. Tensile deformation in service sometimes produces numerous cracks in this inferior skin, which terminate where they reach the tough austenite of normal composition except along grain boundaries, which contain mostly continuous carbides (for example, due to slack quenching during heat treatment). Service performance is not seriously affected except under critical fatigue conditions or in very light sections; in such instances, premature failure may result. If considered necessary, a proprietary grade (equivalent to ASTM grade B-2 to which 6% Cr has been added) that is less prone to surface decarburization can be specified. Under certain conditions, sections such as wrought sheets may be protected with inert or reducing gas atmospheres, or with covers, metal envelopes, or organic or inorganic coatings, to minimize decarburization.

For applications that require nonmagnetic properties, alteration of the skin by furnace gases requires attention. If the affected skin, which is usually magnetic, is quite shallow, it may be possible to remove it by pickling. In heavy sections, where skin thickness may approach 3.2 mm (⅛ in.), the altered layer should be removed by grinding if a surface permeability below 1.3 is required. Metallographic procedures and microstructural interpretations are discussed in detail in Volume 9 of the 9th Edition of *Metals Handbook*.

Mechanical Properties After Heat Treatment

As the section size of manganese steel increases, tensile strength and ductility de-

Table 7 Average mechanical properties of 1.11C-12.7Mn-0.5Si-0.043P castings water quenched from 1040 °C (1900 °F)

Tension tests were performed on specimens 6.40 mm (0.252 in.) in diameter and 25 mm (1 in.) in gage length

Plate thickness		Type of grain	Yield strength		Tensile strength		Elongation, %(a)	Reduction in area, %	Izod V-notch impact strength	
mm	in.		MPa	ksi	MPa	ksi			J	ft · lbf
50	2	Coarse	338	49	635	92	37.0	35.7	137	101
		Fine	365	53	820	119	45.5	37.4	134	99
83	3¼	Coarse	345	50	620	90	25.0	34.5	133	98
		Fine	359	52	765	111	36.0	33.0	115	85
140	5½	Coarse	338	49	545	79	22.5	25.6	115	85
		Fine	352	51	705	102	32.0	28.3	100	74
190	7½	Coarse	324	47	455	66	18.0	25.1	77	57
		Fine	359	52	725	105	33.5	29.2	66	49

(a) In 25 mm or 1 in. Source: Ref 3

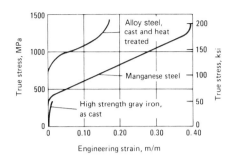

Alloy	Composition, %				
	C	Mn	Si	Cr	Other
Alloy steel, Q and T	0.29	1.30	0.52	0.37	0.36 Mo
Manganese steel	1.22	13.08	0.33	0.09	0.05 Al
Gray iron	2.79	0.75	1.32	0.10	...

Q and T, quenched and tempered

Fig. 10 True stress versus engineering strain for manganese steel, cast alloy steel (quenched and tempered) of similar tensile strength, and a high-strength gray iron. Source: Ref 3

crease substantially in specimens cut from heat-treated castings. This occurs because, except under specially controlled conditions, heavy sections do not solidify in the mold fast enough to prevent coarse grain size, a condition that is not altered by heat treatment. As shown in Table 7, fine-grain specimens may exhibit tensile strength and elongation as much as 30% greater than those of coarse-grain specimens. It is important to note that care should be taken during tensile tests to ensure that the gage section in cast specimens contains an adequate number of grains as required by the ASTM-E8 standard.

Grain size also is the chief reason for the differences between cast and wrought manganese steels (the latter are usually of fine grain size). For cast grade B-2, the standard deviations for tensile strength and elongation are about 69 MPa (10 ksi) and 9%, respectively. The midrange values of 825 MPa (120 ksi) and 40% apply to sound, medium-grain cast specimens that have been properly heat treated. The scatter bands for this grade extend from 620 to 1035 MPa (90 to 150 ksi) for tensile strength and from 13 to 67% for elongation.

Mechanical properties vary with section size. Tensile strength, tensile elongation, reduction in area, and impact strength are substantially lower in 102 mm (4 in.) thick sections than in 25 mm (1 in.) thick sections. Because section thicknesses of production castings are often from 102 to 152 mm (4 to 6 in.), this factor is an important consideration for proper grade specification.

Notched-bar impact test values can be exceptionally high. Charpy test specimens are sometimes bent and dragged through the machine rather than being fractured. Occasionally, observed values are biased because of the incorrect preparation of specimens. Notches should be cut by precision grinding to minimize work hardening at the apex of the notch.

Austenitic manganese steel remains tough at subzero temperatures above the M_s temperature. The steel is apparently immune to hydrogen embrittlement, although embrittlement has been produced in steels with low carbon content (<~0.02%) and high manganese content. There is a gradual decrease in impact strength with decreasing temperature. The transition temperature is not well defined because there is no sharp inflection in the impact strength-temperature curve down to temperatures as low as −85 °C (−120 °F). At a given temperature and section size, nickel and manganese additions are usually beneficial for enhancing impact strength, while higher carbon and chromium levels are not.

Resistance to crack propagation is high and is associated with very sluggish progressive failures. Because of this, any fatigue cracks that develop might be detected, and the affected part or parts removed from service before complete failure occurs, a capability that is a distinct advantage in railway track work. The fatigue limit of austenitic manganese steel has been reported as 270 MPa (39 ksi).

Yield strength and hardness vary only slightly with section size. The hardness of most grades is about 200 HB after solution annealing and quenching, but this value has little significance for estimating machinability or wear resistance. Hardness increases so rapidly because of work hardening during machining or while in service that austenitic manganese steels must be evaluated on some basis other than hardness.

The true tensile characteristics of manganese steel are better revealed by the stress-strain curves in Fig. 10, which compare manganese steel with gray iron and with a heat-treated, high-strength, low-alloy steel of about the same nominal tensile strength. The low yield strength is significant and may prevent the selection of this alloy where slight or moderate deformation is undesirable, unless the usefulness of the parts in question can be restored by grinding. However, if deformation is immaterial, the low yield values may be considered temporary, that is, deformation will produce a new, higher yield strength corresponding to the amount of strain that is absorbed locally.

Bend tests are often used as a qualitative indication of ductility in castings. A description of the bend test specimen and testing procedure is given in ASTM A 128 as a supplementary requirement. Typically, a separate test specimen is poured from the same heat and is heat treated with the batch of castings. This specimen is then tested by cold bending around a 25 mm (1 in.) diam mandrel without any further machining or grinding, except that required to remove surface decarburization. The ductility is judged to be acceptable in most cases if the specimen can be bent through an angle of 150° without breaking into two pieces.

Work Hardening

The approximate ranges of tensile properties produced in constructional alloy steels by heat treatment are developed in austenitic manganese steels by deformation-induced work hardening. In a tension test, yielding signifies the beginning of work hardening, and elongation is associated with its progress. Little or no necking occurs because work hardening is greatest at the point of greatest deformation. The increase in strength due to cold work stops further elongation, and deformation then occurs elsewhere in the gage section. This type of behavior is often compared to that exhibited by transformation-induced plasticity (TRIP) steels. Elongation thus occurs uniformly and without flow stress saturation, until failure.

A 1988 investigation (Ref 15) showed that fracture occurs because of a combination of microvoid coalescence (without shear localization) and surface cracking within regions of localized plastic flow. The nucleation of both voids and surface cracks was observed to be a function of carbides and nonmetallic inclusions in the steel, to some extent.

Plastic deformation is accompanied by the development of texture (Ref 16). X-ray diffraction studies have shown this texture to be [111] in tension and [110] in compression, similar to the behavior of 70-30 brass. Also, work-hardening rates tend to be higher in compression than in tension.

The mechanism of work hardening has been the subject of numerous investigations (Ref 17-20). Various mechanisms have been found to contribute to work hardening, depending on factors such as alloy composition (stacking fault energy, strain rate sensitivity), temperature, and strain rate. These mechanisms include twinning or pseudotwinning (Ref 16, 17), stacking fault formation (Ref 18), and dynamic strain aging (Ref 19, 20). However, it is well established that a deformation-induced transformation from austenite to α martensite (bcc) does not occur in ordinary Hadfield steels. The role of such a transformation in work hardening is more significant only at lower carbon and manganese levels.

On a macroscopic scale, the work-hardening rate has been observed to increase with increasing carbon content and decreasing grain size (Ref 21). Work-hardening rates of 1500 to 2500 MPa (218 to 363 ksi) have been measured in austenitic manganese steels at nominal strain rates of 10^{-4}/s at room temperature.

Determination of Work-Hardening Rate. The work-hardening rate is usually determined from ordinary tensile or compression tests as the slope of the true stress-true strain curve, which is mostly linear in the plastic region.

One other measure of the tendency of austenitic manganese steels to work harden is based on a determination of the so-called Meyer index or exponent. The technique employs a 10 mm diam Brinell ball indentor and a series of loads. The test loads are plotted against the diameters of the corresponding indentations on logarithmic scales. The result is expected to be a straight line that fit the equation:

$$P = A \cdot d^n \qquad (Eq\ 1)$$

where P represents the applied load; d, the diameter of the indentation; A, a constant; and n, a measure of the tendency of the metal to strain harden (also called the Meyer index or exponent). The Meyer index of a variety of austenitic manganese and stainless steels has been determined to be in the range of 2.17 to 2.60 (Ref 4).

Manganese steels are unequaled in their ability to work harden, exceeding even the metastable austenitic stainless steels in this feature. For example, a standard grade of manganese steel containing 1.0 to 1.4% C and 10 to 14% Mn can work harden from an initial level of 220 HV to a maximum of more than 900 HV. After extended service, the hardness at the wearing surfaces of railway frogs typically ranges from 495 to 535 HB. Maximum attainable hardness depends on many factors, including specified composition, service limitations, method of work hardening and preservice hardening procedures. It appears that rubbing under heavy pressure can produce higher values of maximum attainable hardness than can be produced by simple impact.

Service Limitations. In some cases, abrasion may remove surface metal before it can attain maximum hardness. In other instances, work hardening raises the elastic limit to the point that succeeding impact blows cannot cause more plastic flow and thus merely bounce off. However, if a rotating tool is applied with enough force and does not cut, accentuated work hardening can be expected. This is a common occurrence during drilling. A sharp twist drill of superior steel can drill 13% Mn steel provided a deep enough cut is taken, but cutting ceases if the drill becomes dull. When this happens, it is frequently futile to continue drilling, even with a sharp drill, because the bottom of the hole has become so hard that the cutting edge cannot penetrate it.

The low yield strength of manganese steel is sometimes a disadvantage in service. For example, plastic deformation due to wheel impact, as in railway frogs and crossings, increases yield strength to levels more resistant to flow, but the associated changes in dimensions are undesirable. With time, low spots develop at critically pounded locations, eventually requiring rebuilding with weld deposits. This problem is alleviated to some degree by hardening the running surface prior to in-service installation.

Low yield strength can also be a disadvantage when manganese steel is used in light armor and similar applications. Because much energy is absorbed during work hardening, manganese steel sheet is an effective light armor against slow-moving projectiles. However, manganese steel is relatively ineffective against high-velocity projectiles that shear through the armor with little accompanying deformation. One reason for this may be the negative strain rate sensitivity behavior exhibited by ordinary Hadfield steels. Heat-treated (quenched and tempered) alloy steels of higher yield strength are preferred for armor against high-velocity projectiles. However, most of the impact blows encountered in industrial service are of low velocity, and for this service, manganese steel is acceptable. It is the preferred choice for applications requiring high impact resistance, toughness, and absorption of energy.

Work-Hardening Methods. Work hardening is usually induced by impact, as from hammer blows. Light blows, even if they are of high velocity, cause shallow deformation with only superficial hardening even though the resulting surface hardness is ordinarily high. Heavy impact produces deeper hardening, usually with lower values of surface hardness. The course of flow under impact and the associated increase in hardness are shown in Fig. 11, which compares a standard 12% Mn

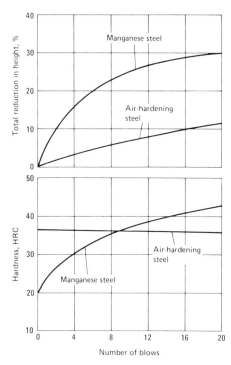

Fig. 11 Plastic flow (top) and work hardening (bottom) of a manganese steel and an air-hardening steel under repeated impact. Specimens 25 mm (1 in.) in both diameter and length were struck repeatedly on one end by blows with an impact energy of 680 J (500 ft · lbf). Composition and heat treatment of the manganese steel were 1.17C-12.8Mn-0.46Si, water quenched from 1010 °C (1850 °F). Composition and heat treatment of the air-hardening steel were 0.74C-0.88Mn-0.30Si-0.75Ni-1.40Cr-0.38Mo; air cooled from 900 °C (1650 °F), reheated to 705 °C (1300 °F), and air cooled. Source: Ref 3

steel with an air-hardening chromium-nickel-molybdenum alloy steel. Less well known is the fact that abrasion itself can produce work hardening.

Explosion hardening was developed as a substitute for hammer or press hardening to achieve hardening with less deformation. Pentaerythritol tetranitrate (3.1 mg/mm^2, or 0.0044 lb/in.^{2}) in the form of plastic explosive sheet 2.11 mm (0.083 in.) thick or mixtures of ammonium nitrate and trinitrotoluene (TNT) are cemented to the surface of the steel and detonated. Usually, three explosions are required to attain the desired hardness in railway track work. More explosions do not significantly change the pattern of hardening, but increase the possibility of cracking instead. The use of plastic explosive permits the hardening of areas such as track work flangeways and unsupported sections that cannot be hammered satisfactorily. Figure 12 compares the depth and intensity of hardening from three different explosion treatments and from one hammer peening operation applied to railway track work. Explosion hardening is now considered a satisfactory but expensive substitute for the hammer or press hardening of manganese steel trackwork castings.

Explosion treatment has also been applied to parts that are to be subjected to

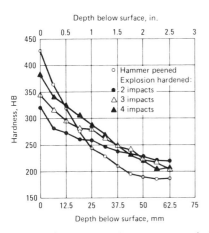

Fig. 12 Hardening patterns in manganese steel railway crossings. Measurements were made on sections taken 19 mm (¾ in.) from the flangeway on each sample crossing. Source: Ref 3

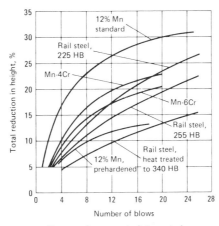

Fig. 13 Flow under repeated impact for several manganese steels and for rail steel at different hardnesses. Specimens 25 mm (1 in.) in both diameter and length were struck repeatedly by blows with an impact energy of 680 J (500 ft · lbf). Source: Ref 3

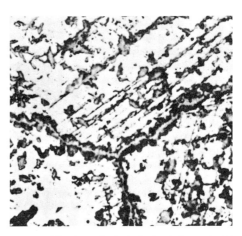

Fig. 14 Structure of a typical reheated manganese steel. ASTM A 128, grade B-3, steel annealed at 1120 °C (2050 °F), water quenched, and reheated above 315 °C (600 °F). Structure is austenite with extensive carbide precipitation along grain boundaries and within grains. Specimen was etched in 2.5% nital, rinsed in ethanol, and reetched in 15% HCl. 500×

abrasive wear. Initial reports were favorable, but have been reversed by subsequent experience; there is no solid evidence that explosion hardening is advantageous for service involving grinding or gouging abrasion.

Explosion hardening is accompanied by insignificant deformation in spite of the fact that work hardening is usually associated with plastic flow and deformation. Studies have centered on the premise that a different hardening mechanism, other than twinning, is involved in explosion hardening.

The addition of alloying elements such as vanadium, chromium, silicon, and molybdenum is also an effective means of raising yield strength, but vanadium, silicon, and chromium reduce ductility. The relative effects of alloying and of prehardening by deformation are compared in Fig. 13.

Reheating

Before manganese steel parts are reheated in the field, the effects of such reheating must be seriously considered. Unlike ordinary structural steels, which become softer and more ductile when reheated, manganese steels suffer reduced ductility when reheated enough to induce carbide precipitation or some transformation of the austenite. Figure 14 shows the microstructural effects of such heating. As a general rule, manganese steels should never be heated above 260 °C (500 °F), either intentionally or accidentally, unless such heating can be followed by standard solution annealing and quenching.

Time, temperature, and composition are variables in the embrittlement process. At lower temperatures, embrittlement takes longer to develop. The time-temperature relationship in 13Mn-1.2C-0.5Si steel is shown in Fig. 15, which presents data based on metallographic examination for structural changes that indicate the beginning of embrittlement. At 260 °C (500 °F), transformation requires more than 1000 h; reheating to as high as 425 °C (800 °F), even with close control of temperature, may be done for no longer than 1 h if transformation is to be avoided. Figure 16 shows the effect of composition on the magnitude of embrittlement.

Because large castings are sometimes mounted with backings of molten lead or zinc, the temperature of such castings should be carefully controlled during reheating. The time-temperature relationship should also be given due consideration for parts that must be welded.

When 12 to 14% Mn steels are to be heated above about 290 °C (550 °F) during service or welding, it is recommended that the carbon content be held below 1.0%, which will suppress embrittlement for at least 48 h at temperatures up to 370 °C (700 °F). The addition of 1.0% Mo will suppress embrittlement completely at temperatures up to 480 °C (900 °F) and will partly suppress it at temperatures of 480 to 595 °C (900 to 1100 °F). If the carbon content is held below about 0.9%, the addition of 3.5% Ni will completely suppress embrittlement up to 480 °C (900 °F) and will partly suppress it above this temperature. These rules can be expected to apply during heating periods of up to 100 h. For periods of 1000 h or more, limiting temperatures are substantially lower. It should be noted that localized carbon contents can still exceed 1.0% because of chemical segregation. Hence, these guidelines should be used with caution.

Table 2 lists properties of three 1.0% C manganese steels that were reheated for periods of up to 10 h at temperatures up to 595 °C (1100 °F). At 370 °C (700 °F), none of the three steels was embrittled; in fact, there were indications that the treatment had effected slight improvements in ductility. At 480 °C (900 °F), the plain manganese steel and the manganese-nickel steel were embrittled, but there was little change in the properties of the manganese-molybdenum steel. At 595 °C (1100 °F), all three steels were embrittled, but the plain manganese steel was more severely affected than the others.

Embrittlement, as revealed by microstructural investigation, was caused by the formation of acicular carbide and pearlite in the austenite grains of the plain manganese steel, by carbide nodules surrounded by pearlite within the austenite grains of the manganese-molybdenum steel, and by an envelope of proeutectoid cementite around each grain of the manganese-nickel steel. If the carbon content in the manganese-nickel steel had been lower, the steel might have been substantially less susceptible to embrittlement. In another investigation, a 0.9C-14.3Mn-1.75Si-3.4Ni steel did not become significantly embrittled when it was heated for 1½ h at 480 °C (900 °F), and its ductility was reduced by only 17 to 20% when it was heated for 1½ h at 595 to 760 °C (1100 to 1400 °F).

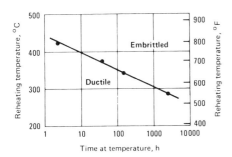

Fig. 15 Time-temperature relationship for embrittlement of 13Mn-1.2C-0.5Si steel. Prior to reheating, the alloy was annealed 2 h at 1095 °C (2000 °F) and water quenched. Source: Ref 3

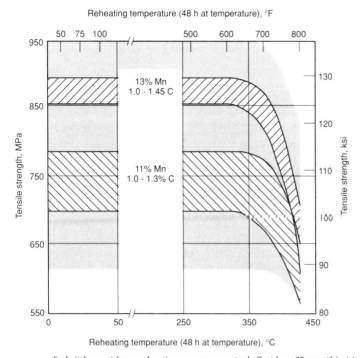
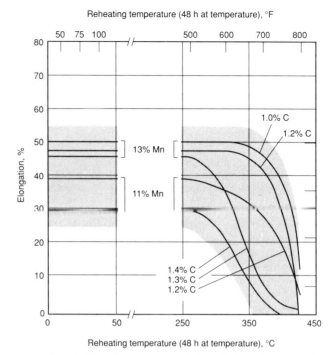

Fig. 16 Embrittlement from reheating manganese steel. Cast bars 25 mm (1 in.) in diameter were reheated 48 h at the temperatures indicated after solution annealing and quenching. Source: Ref 3

Wear Resistance

Compared to most other abrasion-resistant ferrous alloys, manganese steels are superior in toughness and moderate in cost, and it is primarily for these reasons that they are selected for a wide variety of abrasive applications. They are usually less resistant to abrasion than are martensitic white irons or martensitic high-carbon steels, but are often more resistant than pearlitic white irons or pearlitic steels.

The type of wear that is sustained has a major influence on the performance of manganese steels. They have excellent resistance to metal-to-metal wear, as in sheave wheels, crane wheels, and car wheels; good resistance to gouging abrasion, as in equipment for handling or crushing rock; intermediate resistance to high-stress (grinding) abrasion, as in ball mill and rod mill liners; and relatively low resistance to low-stress abrasion, as in equipment for handling loose sand or sand slurries.

Metal-to-Metal Contact. In applications involving metal-to-metal contact, the work hardening of manganese steel is a distinct advantage because it decreases the coefficient of friction and confers resistance to galling if temperatures are not excessive. Compressive loads, rather than impact, provide the deformation required, producing a smooth, hard surface that has good resistance to wear but that does not abrade the contacting part. Sheaves, wear plates, and castings for railway track work are common applications of this type. Manganese steel also has been used in some water-lubricated bearings.

Railway center plates, which are the bearing surfaces where trucks swivel under freight cars, provide a good example of the merit of manganese steel for metal-to-metal frictional wear. Initially lubricated, these plates soon are operating dry and are accessible to airborne grit. When plates are made of carbon steel, the mating surfaces become rough from galling (adhesive wear); the increased friction prevents easy motion, the trucks do not swivel properly on curves, and the lateral pressure is so accentuated that early wheel flange wear is induced. Poor swiveling may accentuate thrust loads at the ends of wheel bearings, generating excessive frictional heat, and center plates and mating bowls on trucks may wear severely. Many years ago, the substitution of 13% Mn steel center plates on heavily loaded ore cars demonstrated their superiority. More recently, various service tests have demonstrated that manganese steel not only wears less than carbon steel, but also develops a low-friction polished surface. The advantages of manganese steel center plates are most evident for cars that are very heavily loaded.

Manganese steel wear plates in a blooming mill housing were in good condition after 16 years, whereas carbon steel plates required replacement every 2 years. Manganese steel wear plates on ore bridge haulage drums were reported to be as good as new after 5½ years of service, whereas ordinary cast steel plates wore out in about 3 years. Manganese steel sheave wheels are expected to last about four times as long as cast carbon steel wheels and they do not groove and cause excessive rope wear, as do wheels made of alloys that do not work harden. In steel mill applications, welded manganese steel overlays on large mill-coupling boxes, pinions, spindles, and other items working under heavy impact loads perform satisfactorily.

Abrasion. The concept that manganese steel has poor wear resistance unless it has been work hardened is not a valid generalization. The misunderstanding has probably developed because, where significant impact and attendant work hardening are present, 12% Mn steel is so clearly superior to other metals that its performance is attributed to surface hardening. However, controlled abrasion tests have indicated that there are circumstances under which the abrasion resistance of austenitic manganese steel is modified little by preservice work hardening, and other circumstances under which this steel will outwear harder pearlitic white cast irons without work hardening.

In applications that involve heavy blows or high compressive and structural stresses, the very hard and abrasion-resistant martensitic cast irons may wear more slowly than manganese steel. However, these irons usually fail by early fracture with a considerable portion of the original cross section unworn, whereas manganese steel may become almost paper thin before fracturing.

Pearlitic white cast iron, which has a hardness of about 400 to 450 HB, is equally brittle but less resistant to wear. Comparative tests on log washer lugs indicated that manganese steel was about 25% worn out with no breakage, whereas in the same period white iron lugs wore to the point of

Table 8 Relative resistance of various materials to high-stress grinding abrasion by wet quartz sand

Material	Hardness, HB	Abrasion factor(a)
Cemented tungsten carbide	...	0.17
Martensitic Ni-Hard cast iron	550–750	0.25–0.60
Martensitic 4150 steel	715	0.60±
Bainitic 4150 steel	512	0.75±
Austenitic 12% Mn steel	200	0.75–0.85
Pearlitic 0.85% C steel	220–350	0.75–0.85
Alloyed white cast irons	400–600	0.70–1.00
Unalloyed white cast irons	400±	0.90–1.00
1020 steel (standard)	107	1.00
Gray cast irons	200±	1.00–1.50
Ferritic ingot iron	90	1.40

(a) Particle size of abrasive, 50 to 55 AFS grain fineness number, compressive stress on specimen about 370 KPa (54 psi), as determined at the Abex Research Center. Ratio of weight loss of sample to weight loss of the standard material, 1020 steel. High factors indicate poor wear resistance. Source: Ref 3

Table 9 Relative wear rates of various materials tested as 127 mm (5 in.) grinding balls in a ball mill

Material	Nominal composition, %					Hardness, HB(a)	Relative wear rate(b)	Order of toughness(c)
	C	Mn	Cr	Mo	Ni			
Martensitic Cr-Mo white iron	2.8	1.0	15.0	3.0	...	740	89	7
Martensitic high-chromium white iron	2.7	1.0	26.0	...	...	705	98	8
Martensitic high-carbon Cr-Mo steel (type 3)	1.0	0.8	6.0	1.0	...	615	100	6
Martensitic high-carbon Cr-Mo steel (type 2)	0.7	0.7	2.0	0.4	...	560	110	5
Martensitic Ni-Cr white iron	3.2	0.8	2.0	...	4.0	650	112	9
Austenitic 6Mn-1Mo steel	0.9	6.0	...	1.0	...	490	114	3
Martensitic medium-carbon Cr-Mo steel (type 1)	0.4	1.5	0.8	0.4	...	560	120	2
Pearlitic high-carbon Cr-Mo steel (type A)	0.8	0.8	2.5	0.4	...	380	127	4
Austenitic 12% Mn steel	1.2	12.0	...	...	...	410	138	1

(a) Hardnesses are converted from the average of HRC readings on the worn surfaces of the balls. Hardnesses below the cold-worked surface are normally lower than those given. (b) Relative wear rates are based on a nominally assigned factor of 100 for the high-carbon martensitic steel (type 3). Factors greater than 100 indicate higher rates of wear, and factors less than 100 indicate proportionately lower rates of wear. (c) Order of toughness listed here is a qualitative value, based partly on results of laboratory tests and partly on the basis of resistance to breakage and spalling that the respective materials have shown in service. Source: Ref 23

Table 10 Relative gouging abrasion resistance in the ASTM G 81 jaw crusher test

Alloy type(a)	Laboratory A		Laboratory B	
	Wear ratio(b)	Hardness, HB	Wear ratio(b)	Hardness, HB
T-1 type A steel (0.19% C), WQ and T (650 °C, or 1200 °F)	0.983	269	1.085±0.013	260
	1.024	269	1.085±0.013	260
	1.022	269	1.085±0.013	260
	1.027	269	1.085±0.013	260
Austenitic 12% Mn steel, WQ	0.279	...	0.279	199
Austenitic 12Mn-2Cr steel, WQ	0.247	...	0.249	232
4340 steel, OQ and T (650 °C, or 1200 °F)	0.788	321	0.716	340
4340 steel, OQ and T (205 °C, or 400 °F)	0.262	555	0.232	520
27% Cr white iron, AC and T (230 °C, or 445 °F)	0.166	653	0.144	662
15Cr-3Mo white iron, AC and T (230 °C, or 445 °F)	0.088	750+	0.076	816
20Cr-3Mo white iron, hardfacing alloy	0.201	(51.7 HRC)	0.170	...
0.30% C cast low-alloy steel, WQ and T (205 °C, or 400 °F)	0.286	514	0.288	499

(a) WQ, water quenched; T, tempered; OQ, oil quenched; AC, air-cooled. (b) Ratio of weight loss of sample to weight loss of the standard material, martensitic T-1 steel plate. High values indicate poor wear resistance

uselessness with 14% breakage. In clay crusher rolls, manganese steel lasted two to three times as long as white or chilled iron. In grinding barrel liners, cast irons lasted 2 to 3 years compared to 10 years for manganese steel. Part of the superiority of manganese steel over white cast iron is attributed to greater freedom from breakage and spalling, but some is probably due to better intrinsic wear resistance.

Manganese steel chain, with endless links cast in interlocking molds, also provides resistance to wear, lasting three to nine times as long as heat-treated steel chain in certain applications. Manganese steel is valuable in conveyors as well as in dragline chain subjected to abrasion and used for carrying heavy loads.

Manganese steel is not satisfactorily resistant to wear by a stream of airborne abrasive particles (impingement erosion), such as in sandblasting or gritblasting equipment, and consequently should not be selected for such service.

Abrasion Testing. A number of laboratory abrasion and wear testing procedures have been developed to simulate a variety of applications with differing degrees of severity. These include a jaw crusher test for gouging abrasion resistance (ASTM G 81), a dry-sand/rubber wheel low-stress abrasion test (ASTM G 65), testing for slurry abrasiveness (ASTM G 75), and tests for galling resistance (ASTM G 83 and G 98). Other procedures such as a wet-sand/rubber wheel low-stress abrasion test and a pin-on-disk high-stress abrasion test* are also nearing standardization at this writing.

Many variations of the pin test are used to measure wear resistance. One test method utilizes abrasive paper or cloth mounted on a revolving disk, the abraded flat tip of a small pin moving in a spiral path. The Climax lab used a milling machine platen with a revolving pin taking a zig-zag path. The Bureau of Mines and some workers in Australia have the pin making a spiral path against garnet paper or cloth fastened to a large revolving drum. The common feature of these test methods is a small pin specimen, but the pins are subjected to a variety of conditions: normal load, tangential force, abrasive type, abrasive grain size, abrasive angularity or roundness, temperature, abrasive concentration, and so on.

The rankings of a number of metals in various abrasion tests are given in Tables 8 to 10. A considerable amount of test data on abrasion resistance of a wide variety of austenitic manganese steels has been accumulated (see the article "Wear Resistance" in Volume 1 of the 9th edition of *Metals Handbook*).

There is little need for laboratory tests in the case of well authenticated applications, but such tests can be useful to judging suitability for new applications. The usefulness of these tests is dependent on their precision and reliability (versus statistical standards) and their validity as determined from comparisons with service tests. Crusher tests are slow and expensive, motivating attempts to correlate more convenient procedures, such as a pin on abrasive paper or cloth. Table 11 compares the abrasion resistance (abrasion factors) of a number of alloys as determined by four separate laboratory testing procedures.

The alloys in Table 11 are clustered towards the high end of the hardness range. Another jaw crusher test on 28 alloys (Ref 25) shows a more even hardness spread. The correlation coefficients in Table 11 highlight the fallacy of judging manganese steel by its initial (as-heat-treated) hardness, even though there is a direct relationship between this property and abrasion resistance for many carbon and low-alloy steels. One reason for the degradation in correlation coefficient due to the inclusion of manganese steel is that the actual hardness developed at the wear surface because of work hardening is significantly higher. Though 500 to 600 HB is usually considered the limit in service, laboratory studies have shown that the hardness can be as high as 900 HV (Ref 26).

The mediocre correlation coefficients and large standard errors emphasize that laboratory tests are often handicapped by large scatter. Often a certain amount of judgment is necessary to be able to apply the results of laboratory tests to new applications.

The abrasion resistance of austenitic 12 to 14% Mn steels with various carbon contents has been compared with the resistance of other steels and white irons in a jaw crusher abrasion test (see Fig. 17). The wear rate for

---

*The high stress term has been carelessly used and does not apply clearly to a definite condition. Originally used to characterize the AMSCO/Abex wet-sand grinding-abrasion test (Ref 4 and 22 and Table 8), it is not appropriate when applied to a pin being worn by an abrasive anchored on paper or cloth that can have a cushioning effect.

836 / Specialty Steels and Heat-Resistant Alloys

Table 11 Comparison of several abrasion tests involving manganese steel
Abrasion factors relative to A 514 steel. Data from U.S. Bureau of Mines Research Center

Material	Hardness, HB	Abrasion tests			
		Slurry	Dry sand/ rubber wheel	Pin-on-drum test	Jaw crusher
A 514 steel, 0.19% C	269	1.00	1.00	1.00	1.00
AISI 4340 steel, 0.42% C	595	0.33	0.37	0.58	0.16
AISI 52100 steel, 1.05% C	670	0.33	0.31	...	0.37
3Cr-0.5Mo steel, 0.91% C	550	0.60	0.51	0.44	0.31
Pearlitic cast iron, 3.2% C	440	...	0.37	0.61	0.19
6Mn-2Cr-1Mo steel, 1.29% C	512	0.44	0.50	0.35	0.24
Austenitic 13% Mn steel, 1.20% C	200	...	0.50	0.58	0.24
1.5Cr-4Ni white cast iron 3.13% C	601	0.19	0.12	0.31	0.26
9Cr-6Ni white cast iron, 2.81% C	700	0.21	0.21	0.34	0.13
18Cr-1Mo-1Ni cast iron 2.80% C	555	0.13	0.084	0.12	0.082

	Correlations of Brinell hardness with above tests			
	Steel without manganese		Steel with manganese	
Type of test	Correlation coefficient	Standard error on HB	Correlation coefficient	Standard error on HB
Slurry	0.849	0.140	...	...
Dry sand/rubber wheel	0.796	0.157	0.681	0.182
Pin on drum	0.772	0.158	0.665	0.177
Jaw crusher	0.732	0.177	0.492	0.215

Source: Ref 24

a quenched and tempered low-carbon low-alloy steel (ASTM 517, type B, at 269 HB), which was used as a comparative standard in each test, is also shown. When the relative wear rate (wear ratio) of each test material is plotted against increasing carbon content on a log-log scale, results for austenitic steels and irons tend to fall on a descending straight line, and results for martensitic steels and irons fall on a parallel line below the line for the austenitic alloys. A decrease in wear ratio represents a proportionate increase in abrasion resistance. Thus, Fig. 17 strongly supports the conclusions that the abrasion resistance of both austenitic and martensitic steels improves with increasing carbon content and that, for a given carbon content, martensitic steels have better abrasion resistance than do austenitic steels. However, martensitic steels and white irons have limited resistance to gouging abrasion because of their lack of toughness. The wear ratios of pearlitic steels, if plotted on Fig. 17, would lie above those of the austenitic steels. There is considerable scatter in the rates for pearlitic steels due to their wide variation in hardness for any given carbon content.

The same trends have been observed in ore milling tests and in sand slurry abrasion tests. Generally, austenitic 12 to 14% Mn steels have higher wear rates, and therefore poorer abrasion resistance than pearlitic or martensitic steels of equivalent carbon content in a low-stress slurry abrasion; a 6Mn-1Mo composition is usually intermediate between pearlitic and martensitic steels in wear resistance.

Corrosion

Manganese steel is not corrosion resistant; it rusts readily. Furthermore, where corrosion and abrasion are combined, as they frequently are in mining and manufacturing environments, the metal may deteriorate or be dissolved at a rate only slightly lower than that of carbon steel. If the toughness or nonmagnetic nature of manganese steel is essential for a marine application, protection by galvanizing is usually satisfactory.

Effects of Temperature

The excellent properties of 13% Mn steel between −45 and 205 °C (−50 and 400 °F) make it useful for all ambient-temperature applications, even in arctic climates. It is not recommended for hot-wear applications because of structural instability between 260 and 870 °C (500 and 1600 °F). At higher temperatures, it may lack the strength and ductility necessary to withstand severe welding stresses, and thus welding must be done under closely controlled conditions. It is not resistant to oxidation, and its creep rupture properties are inferior to those of iron-chromium-nickel austenites. Because the work-hardening rate decreases at higher temperatures, this steel is unsuitable for structural applications in the red-heat range.

The effects of temperature on mechanical properties (in both tension and compression) have been well documented (Ref 16, 17, 19). The general trend for changes in flow stress versus temperature is shown in Fig. 18 for both tension and compression. There is an increase in yield strength with decreasing temperature with a corresponding drop in ductility and ultimate tensile

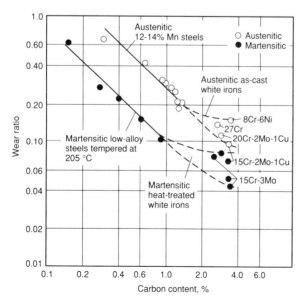

Fig. 17 Relative wear ratios of ferrous alloys in jaw crusher tests. Source: Ref 7

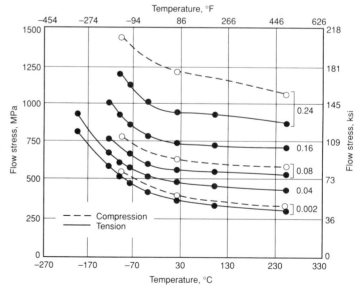

Fig. 18 Temperature dependence of flow stresses for equivalent true plastic strain $\bar{\epsilon}_p$, of 0.002, 0.04, 0.08, 0.16, and 0.24. Source: Ref 16

Table 12 Charpy V-notch impact strength

Composition, %				Impact strength(a)			
				At 24 °C (75 °F)		At −73 °C (−100 °F)(b)	
C	Mn	Si	Ni	J	ft · lbf	J	ft · lbf
1.03	12.9	0.52	...	128	94.5	71	52.5
1.18	13.0	0.50	...	144	106	79	58.5
1.19	14.6	0.50	...	141	104	79	58.5
0.84	12.5	0.48	3.46	136	100	108	80
1.17	12.7	0.53	3.56	142	104.5	119	88

(a) Average of duplicate tests. Temperature ± 3 °C (±5 °F). V-notches machine ground to avoid work hardening. Specimens water quenched from A_{cm} before machining. (b) Held in solid CO_2 and acetone until just before testing. Source: Ref 22

Table 13 Effects of temperature on the physical properties of 13% Mn steel

Temperature		Mean apparent specific heat,		Mean coefficient of thermal expansion from 0 °C (32 °F),		Electrical resistivity,	Thermal conductivity	
°C	°F	J/kg · K	Btu/lb · °F	10^{-6}/K	10^{-6}/°F	nΩ · m	W/m · K	Btu/ft · s · °F
0	32	494	0.118	...	...	6.65	13.2	0.00210
50	120	510	0.122	...	...	7.11	14.0	0.00225
100	212	527	0.126	18.01	10.01	7.57	14.9	0.00239
150	302	553	0.132	...	...	8.02	15.7	0.00252
200	390	573	0.137	19.37	10.77	8.47	16.5	0.00265
250	480	590	0.141	...	...	8.89	17.4	0.00279
300	570	603	0.144	20.71	11.51	9.31	18.0	0.00289
350	660	607	0.145	...	...	9.69	18.6	0.00298
350–650	660–1200	(a)	(a)	(a)	(a)	(a)	(a)	(a)
700	1290	...	...	20.49	11.39	11.53	21.8	0.00350
750	1380	...	...	...	...	11.80	21.8	0.00350
800	1470	...	...	21.86	12.15	12.11	22.2	0.00356
850	1560	...	...	...	...	12.40	22.4	0.00359

Composition: 1.22% C, 13.0% Mn, 0.22% Si, 0.038% P, 0.010% S, 0.07% Ni, 0.03% Cr, 0.07% Cu, 0.004% Al, 0.038% As. Condition: wrought steel, heated to 1050 °C (1925 °F) and air cooled. (a) Values depend on time at temperature and on amount of transformation. Source: Ref 27

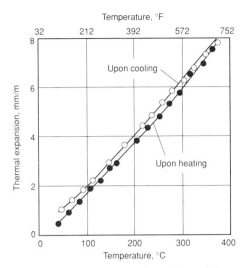

Fig. 19 Thermal expansion of a 13% Mn steel. Composition of steel was 1.18C-13Mn-0.5Si. Specimen was annealed 2 h at 1095 °C (2000 °F) and water quenched. Source: Ref 3

strength. The changes in strength and ductility, however, are not uniform, and the steel retains a major portion of its room-temperature ductility down to about −100 °C (−150 °F).

There has been some controversy with respect to the operative mechanism for work hardening in manganese steels at different temperatures. It currently appears that deformation twinning is predominant at the lower temperatures (below about 0 °C, or 30 °F). However, at temperatures above 0 °C (30 °F), strain hardening has been variously attributed to twinning, dynamic strain aging, and stacking fault formation. Some hardening due to the formation of Cottrell clusters and carbide precipitation has been reported at temperatures above 300 °C (570 °F).

At −75 °C (−105 °F), cast manganese steels retain from 50 to 85% of their room-temperature impact resistance (see Table 12). They are considerably more brittle at liquid-air temperature (−185 °C, or −300 °F), but at all atmospheric temperatures encountered by railway track work and mining and construction equipment, they have outstanding toughness that provides a valuable safety factor, compared to ferritic steels at subzero temperatures.

Associated with the embrittlement produced by reheating above 260 °C (500 °F) are changes in physical properties stemming from the same transformations that cause the loss in toughness. Because both composition and time at temperature influence these changes, erratic behavior and a considerable range in such properties as thermal and electrical conductivity may be expected above 315 °C (600 °F), as shown in Table 13.

Thermal-expansion characteristics of austenitic manganese steels are similar to those of other austenitic materials. The expected change in length upon heating is about 1½ times that of ferritic steels. A coefficient of linear thermal expansion of 18 × 10^{-6}/°C (10 × 10^{-6}/°F) is generally precise enough near room temperature (Fig. 19). Transformation to pearlite and the precipitation of carbide influence the values of the expansion coefficient in the range from 370 to 760 °C (700 to 1400 °F).

Magnetic Properties

The untransformed austenite of 13% Mn steel is virtually nonmagnetic, with a permeability of about 1.03 or less. This permits the use of the material where a strong, tough, nonmagnetic metal is required, as in magnet cover plates, collector shoes for traveling cranes, stator core parts for generators and motors, liner plates for storage bins holding materials that are handled by lifting magnets, magnetic-separator parts, instrument-testing devices, and furnace parts located in the magnetic fields of induction furnaces.

The changes that occur in the composition of the surface during heat treatment may produce a magnetic skin, one that is either a martensitic surface layer or a magnetic oxide. Permeability values of approximately 1.3 have been obtained on specimens that have this magnetic surface layer. Frequently, this layer does no harm, but if necessary, it may be removed by grinding or pickling or it may be prevented altogether by suitable (although often expensive) corrective measures during heat treatment.

Cast or wrought manganese steel is probably the most economical material for strong nonmagnetic parts if machining is not required. Fabrication costs and anticipated operating temperature are decisive factors in selection; operating temperature must not exceed 260 °C (500 °F).

A 20% Mn steel containing bismuth (U.S. Patent 3,010,823) has been developed for nonmagnetic parts that require machining. Laboratory tests indicate a machinability comparable to that of the more expensive type 304 stainless steel that this material was designed to replace. It can be lathe turned (horizontal force: 135 to 380 N at 1.35 m/s, or 30 to 85 lbf at 265 sfm), drilled, and tapped. This steel has not yet been exploited commercially.

Lack of magnetism is a disadvantage when austenitic manganese steel is used in components of material-handling systems that depend on magnetic separators to remove tramp iron from the process stream before it enters crushers, grinders, and other machinery. When there is the possibility that manganese steel parts may become detached from working equipment and fall into the process stream, it is advisable to cast mild steel inserts in the manganese steel parts. The inserts must be large enough to provide the level of ferromagnetism necessary for magnetic separators to detect and remove the lost parts so they cannot enter working machinery and cause damage.

Table 14 Typical electrodes for arc welding austenitic manganese steels

Type	American Welding Society class(a)	C	Mn	Ni	Cr	Mo	V	Si	P	Fe	Remarks
Mn-Ni	EFeMn-A	0.5–0.9	11–16	3–6	0.5 max	...	...	1.3 max	0.03 max	rem	...
Mn-Ni-Cr	(c)	0.6–0.85	14–17	2–4	2.5–4.0	...	...	0.2–0.7	0.02	rem	...
Mn-Mo	EFeMn-B	0.5–0.9	11–16	...	0.5 max	0.6–1.4	...	0.3–1.3	0.03 max	rem	...
Mn-Cr	(c)	0.3–0.6	14–15	1.0	14–15	0.3–1.7	0–0.6	0.2–0.5	...	rem	Corrosion resistant
Cr-Ni-Mn	(c)	0.5	4.5	10	20	1.4	...	0.6	...	rem	Corrosion resistant

(a) Covered electrode requirements are described in AWS A5.13–80, "Specification for Solid Surfacing Welding Rods and Electrodes." Bare and covered composite electrodes are described in AWS A5.21–80, "Specification for Composite Surfacing Welding Rods and Electrodes." (b) Composition given is for deposited weld metal with covered electrodes. (c) Proprietary electrodes. Source: Ref 28

Welding

Many of the common applications of austenitic manganese steel involve welding, either for fabrication or for repair. Consequently, it is important to understand that this material is unusually sensitive to the effects of reheating, often becoming embrittled to the point of losing its characteristic toughness. Oxyfuel gas welding is so likely to produce embrittlement that it is not accepted as a practical method of welding this alloy. When properly done, electric arc welding is the preferred method of joining or surfacing manganese steels. The phosphorus content is kept below 0.03% to minimize hot cracking. Welding is generally carried out after heat treatment to minimize cracking after deposition of the weld metal.

Arc Welding. Electrodes for arc welding austenitic manganese steel are commercially available in many compositions. They may be used for surfacing, for repair welds, and for joining manganese steel to itself or to carbon steels. Some of them are shown in Table 14. They have a lowered carbon content to minimize carbon precipitation as they cool from the welding temperatures. Though formulated to avoid embrittlement of the deposited filler metal, proper welding procedures still must be used to avoid damage in the heat-affected zone. It is generally recommended that the metal temperature adjacent to the weld not exceed 315 °C (600 °F) after an elapsed time of 1 min.

Electrodes of high manganese content, containing insignificant amounts of other alloying elements, are also available. Usually, these electrodes are recommended only for the buildup of worn areas, because they are inherently lower in toughness than more highly alloyed grades. High-alloy low-manganese electrodes have not been generally accepted as equivalent to high-manganese types. Austenitic stainless steel electrodes such as type 308 or 310 may be used for the repair of cracks or for the buildup of worn areas.

Factors that are frequently overlooked are the losses in carbon, manganese, and silicon that occur during welding. Although many electrode manufacturers compensate for these losses, improper welding techniques, such as use of excessive arc length and excessive puddling, may cause additional losses. The result is inferior properties in the weld deposit.

Frequently carbon steels are welded to high-manganese steel using austenitic stainless steel electrodes. Because the deposit tends to be a mixture or hybrid of the base and the filler metal, it can have quite different properties. Often it is air hardening, producing a martensitic zone as the weld cools. The ductility of the martensite is low, but the strength is high, and weldments are often satisfactory. The chief adverse factor may be cracks in the martensite. Cross-weld tensile properties of low-carbon 14Mn-1Mo steel plate welded to 1045 steel with EFeMn-A electrodes were 435 MPa (63 ksi) yield strength, 650 MPa (94 ksi) ultimate tensile strength, and 11% elongation with fracture in the 1045 steel. These properties are superior to those of many weldments of carbon steel.

Filler metal from 0.75% C, 15% Mn, 3.5% Ni, and 4.0% Cr seems to be superior to that of EFeMn-A and EFeMn-B.

Grades having chromium at levels above 14% are also useful for the joining of manganese steel and the buildup of worn parts such as trackwork, but because of their low carbon content, have sacrificed abrasion resistance. However, they are more machinable than the higher-carbon grades. If used to restore the dimensions of crusher parts, they should be overlaid with an effective hard-facing alloy.

Precautions. The primary consideration in welding austenitic manganese steel is minimum heating of the parent metal to avoid embrittling transformations or carbide precipitation. This precludes preheating. Under the most favorable circumstances, some precipitation is expected, and the resulting heat-affected zones seldom attain the toughness of normal parent metal. Because manganese steel work hardens in service, it may be assumed that any worn area requiring repair or rebuilding will have a work-hardened surface. This surface must be removed before welding in order to prevent cracking in the heat-affected zone. Also, any fatigue cracks that may have nucleated in the work-hardened region can cause delamination of the weld if not removed.

The low heat conductivity and high thermal expansion of manganese steel also cause difficulties, combining to produce steep thermal gradients and high residual stresses. Weld beads are subjected to tension as they cool. To minimize cracking, it is desirable to peen them while they are hot, producing plastic flow and changing the stress to compression. This hammering should be done promptly after 152 to 229 mm (6 to 9 in.) of weld bead has been deposited. Typical mechanical properties of austenitic manganese steel weld metal are shown in Tables 15 and 16.

Machining

Manganese steels are so tough and work harden at the point of a cutting tool to such an extent that frequently they are considered commercially unmachinable. However, these steels are regularly cut by adhering to generally accepted procedures. In addition, a new, more highly machinable grade of manganese steel has been developed that may be helpful in appropriate applications.

Table 15 Typical mechanical properties of austenitic manganese steel weld metal

Welding process(a)	Electrode type	Yield strength MPa	ksi	Tensile strength MPa	ksi	Elongation, %	Reduction of area, %	Hardness, HB
SMAW	Mn-Ni	442	64.1	836	121.3	47.0	37.6	207
SMAW	Mn-Ni-Cr	521	75.6	826	119.8	42.0	33.2	223
AW		542	78.7	843	122.3	37.0	31.6	235
SAW		547	79.4	831	120.6	38.0	34.0	207
SMAW	Mn-Mo	468	67.9	826	119.8	32.0	33.1	214
SMAW	Mn-Cr	827	120.0	1007	146.0	30.0	...	194

(a) SMAW, shielded metal arc welding; AW, semiautomatic without auxiliary shielding; SAW, submerged arc welding. Source: Ref 28

Table 16 Low-temperature Charpy V-notch impact strength of austenitic manganese-nickel steel weld metal

Test temperature °C	°F	Impact strength J	ft · lbf
25	75	160	118
−20	0	130	96
−60	−75	108	80
−100	−150	75	55

Source: Ref 28

Table 17 Feed forces required in lathe turning of austenitic manganese steels

Specimens were 31.7 mm (1.25 in) diam bars, toughened by water quenching. Roughing cuts 2.5 mm (0.098 in.) deep were taken using complex-carbide tools containing about 15% TiC + TaC (predominantly TiC) and about 7 to 10% Co. New cutting edges were used for each positive or negative rake. Cutting speed was 0.19 to 0.20 m/s (37 to 39 sfm).

	Negative 7° rake				Flat tool				Positive 6° rake			
	Horizontal		Vertical		Horizontal		Vertical		Horizontal		Vertical	
Type of steel	N	lbf	kN	lbf	N	lbf	kN	lbf	kN	lbf	kN	lbf
1.12C-13Mn	670	150	1.58	355	780	175	1.58	355	1.13	255	1.67	375
3Ni-13Mn	620	140	1.53	345	...	...	...	...	1.25	280	1.69	380
1Mo-13Mn	800	180	1.65	370	...	...	...	...	1.09	245	1.71	385
1.12C-13Mn leaded(b)	580	130	1.49	335	490	110	1.38	310	1.25	280	1.69	380
Wrought type 304 stainless(c)	690	155	1.16	260	890	200	1.25	280	Welded to tool		Welded to tool	
Cast CF-8 stainless(c)	670	150	1.13	255	...	...	...	...	2.25	505	2.56	575

(a) Recovery of 0.02% Pb from 0.35% added in ladle. The effect on machinability is inclusive. (b) Stainless steels suffer in comparison at this speed. They are more machinable at higher speeds and permit certain operations, such as drilling of 6.4 mm (¼-in.) diam holes, which are very difficult with austenitic manganese steel. The type 304 stainless steel was cold finished. Source: Ref 3

Table 18 Typical room-temperature properties of machinable manganese steel

Type	Treatment	Tensile strength		Yield strength		Elongation %	Reduction in area, %	Hardness, HB	Magnetic permeability, μ
		MPa	ksi	MPa	ksi				
Standard 13% Mn	Toughened	825	120	360	52	40	35	200	1.01
Machinable grade A, 20 Mn-0.6 C	Toughened	640–855	93–124	275–310	40–45	39–65	26–44	159–170	1.003
	As-cast	380–580	55–84	275–305	40–44	13–22	24	159–170	1.003

Source: Abex Research Center

Procedures. Although details of practice and tool design differ, there is general agreement on the procedures for machining manganese steels:

- Machine tools should be rigid and in good condition. Any factors that encourage chatter are undesirable
- Tools should be sharp. Dull tools cause excessive work hardening of the cut surface and accentuate the difficulty in machining
- Low speeds of about 9 to 12 m/min (30 to 40 sfm) should be used. High speeds are likely to create red-hot chips and to cause rapid tool breakdown
- Cobalt high-speed steel tools or tools with cemented carbide and ceramic inserts can be used. The latter are preferred
- The liberal use of a good grade of sulfur-bearing cutting oil is beneficial but not essential
- In castings, holes should be formed by cores in the foundry, rather than by machining, whenever possible
- Coolants are recommended for surface grinding operations

Various sources provide statements in favor of both positive-rake and negative-rake tools and both dry cutting and liquid coolants. Because high temperatures at the cutting edge are a large part of the problem, effective cooling seems desirable. Negative-rake tools are likely to require more force and thus to produce more heat. However, the thinner edge of a positive-rake tool is more vulnerable to heat. Comparative machining data are presented in Table 17.

Machinability is increased by the embrittlement that develops with reheating between about 540 and 650 °C (1000 and 1200 °F). Although not usually practicable, such a treatment may be useful if the part can subsequently be properly toughened. Milling usually is not considered practicable.

Machinable Grade. A 20Mn-0.6C steel was developed specifically for improved machinability. Table 18 gives the mechanical properties of this material. Even though the yield strength was deliberately reduced from 360 MPa (52 ksi) to a value between 240 and 310 MPa (35 and 45 ksi) to obtain improved machinability, the ultimate tensile strength exceeds 620 MPa (90 ksi), and elongation in small castings may reach 40%. The heat treatment of this steel involves water quenching from 1040 °C (1900 °F). As-cast properties are lower but are probably adequate for many applications.

This nonmagnetic modified grade can be lathe turned, drilled, tapped, and threaded; even holes 6.4 mm (¼ in.) in diameter can be drilled and tapped in this metal. In some machine shops, it is rated only slightly more difficult to drill than plain 1020 steel, and the quality of the tapped threads is considered very good. Typical machining data for this steel are presented in Table 19. Wear resistance has been sacrificed for machinability, and this grade has significantly less abrasion resistance than do the various types in ASTM A 128.

REFERENCES

1. E.C. Bain, E.S. Davenport, and W.S.N. Waring, The Equilibrium Diagram of Iron-Manganese-Carbon Alloys of Commercial Purity, *Trans. AIME*, Vol 100, 1932, p 228
2. C.H. Shih, B.L. Averbach, and M. Cohen, Work Hardening and Martensite Formation in Austenitic Manganese Alloys, Research Report, Massachusetts Institute of Technology, 1953
3. H.S. Avery, Austenitic Manganese Steel, *Metals Handbook*, Vol 1, 8th ed., American Society for Metals, 1961
4. H.S. Avery, Work Hardening in Relation to Abrasion Resistance, in *Proceedings of the Symposium on Materials for the Mining Industry*, published by Climax Molybdenum Company, 1974, p 43
5. *Manganese Steel*, Oliver and Boyd, for Hadfields Ltd., 1956
6. H.S. Avery and H.J. Chapin, Austenitic Manganese Steel Welding Electrodes, *Weld. J.*, Vol 33, 1954, p 459
7. F. Borik and W.G. Scholz, Gouging Abrasion Test for Materials Used in Ore and Rock Crushing, Part II, *J. Mater.*, Vol 6 (No. 3), Sept 1971, p 590
8. M. Fujikura, Recent Developments of

Table 19 Force requirements for single-point lathe turning of austenitic manganese steel

		Feed force(a)				Friction coefficient
		Horizontal		Vertical		
Type	Condition	N	lbf	N	lbf	
Standard 13% Mn	As-cast	535–670	120–150	1225	275	0.64
	Toughened	690	155	1310	295	0.76
Machinable grade A (20% Mn)	As-cast	155–290	35–65	890–980	200–220	0.31–0.48
	Toughened	180–380	40–85	955–1000	215–225	0.33–0.57

(a) Depth of cut, 3 mm (0.1 in.) on radius; feed, 0.16 mm/rev (0.0062 in./rev); turning speed, 1.35 m/s (265 ft/min); 6° positive-rake tool.
Source: Abex Research Center

Austenitic Manganese Steels for Non-Magnetic and Cryogenic Applications in Japan, The Manganese Center, Paris 1984
9. D.J. Schmatz, Structure and Properties of Austenitic Alloys Containing Aluminum and Silicon, *Trans. ASM*, Vol 52, 1960, p 898
10. J. Charles and A. Berghezan, Nickel-Free Austenitic Steels for Cryogenic Applications: The Fe-23% Mn-5% Al-0.2% C Alloys, *Cryogenics*, May 1981, p 278
11. R. Wang and F.H. Beck, New Stainless Steel Without Nickel or Chromium for Marine Applications, *Met. Prog.*, March 1983, p 72
12. J.C. Benz and H.W. Leavenworth, Jr., An Assessment of Fe-Mn-Al Alloys as Substitutes for Stainless Steels, *J. Met.*, March 1985, p 36
13. W.J. Jackson and M.W. Hubbard, Steelmaking for Steelfounders, Steel Castings Research and Trade Association, 1979, p 106
14. R. Castro and P. Garnier, Some Decomposition Structures of Austenitic Manganese Steels, *Rev. Métall., Cah. Inf. Tech.*, Vol 55, Jan 1958, p 17
15. D. Rittel and I. Roman, Tensile Fracture of Coarse-Grained Cast Austenitic Manganese Steels, *Metall. Trans. A*, Vol 19A, Sept 1988, p 2269-2277
16. P.H. Adler, G.B. Olson, and W.S. Owen, Strain Hardening of Hadfield Manganese Steel, *Metall. Trans. A*, Vol 17A, Oct 1986, p 1725
17. H.C. Doepken, Tensile Properties of Wrought Austenitic Manganese Steel in the Temperature Range from +100 °C to −196 °C, *J. Met., Trans. AIME*, Feb 1952, p 166
18. K.S. Raghavan, A.S. Sastri, and M.J. Marcinkowski, Nature of the Work Hardening Behaviour in Hadfield's Manganese Steel, *Trans. TMS-AIME*, Vol 245, July 1969, p 1569
19. Y.N. Dastur and W.C. Leslie, Mechanism of Work Hardening in Hadfield Manganese Steel, *Metall. Trans. A*, Vol 12A, May 1981, p 749
20. B.K. Zuidema, D.K. Subramanyam, and W.C. Leslie, The Effect of Aluminum on the Work Hardening and Wear Resistance of Hadfield Manganese Steel, *Metall. Trans. A*, Vol 18A, Sept 1987, p 1629
21. Abex Research Center, Abex Corporation, unpublished research, 1981-1983
22. H.S. Avery, Austenitic Manganese Steel, American Brakeshoe Company, 1949, condensed version in *Metals Handbook*, American Society for Metals, 1948, p 526-534
23. T.E. Norman, *Eng. Mining J.*, July, 1957, p 102
24. R. Blickensderfer, B.W. Madsen, and J.H. Tylczak, Comparison of Several Types of Abrasive Wear Tests, in *Wear of Materials 1985*, K.C. Ludema, Ed., American Society of Mechanical Engineers, p 313
25. D.E. Diesburg and F. Borik, Optimizing Abrasion Resistance and Toughness in Steels and Irons for the Mining Industry, in *Proceedings of the Symposium on Materials for the Mining Industry*, Climax Molybdenum Company, 1974, p 15
26. U. Bryggman, S. Hogmark, and O. Vingsbo, Abrasive Wear Studied in a Modified Impact Testing Machine, *Wear of Materials*, 1979, p 292
27. "The Physical Properties of a Series of Steels, Part II," Special Report 23, Alloy Steels Research Committee, British Iron and Steel Institute, Sept 1946
28. *Metals and Their Weldability*, Vol 4, 7th ed., *Welding Handbook*, American Welding Society, 1982, p 195

Wrought Stainless Steels

Revised by S.D. Washko and G. Aggen, Allegheny Ludlum Steel, Division of Allegheny Ludlum Corporation

STAINLESS STEELS are iron-base alloys containing at least 10.5% Cr. Few stainless steels contain more than 30% Cr or less than 50% Fe. They achieve their stainless characteristics through the formation of an invisible and adherent chromium-rich oxide surface film. This oxide forms and heals itself in the presence of oxygen. Other elements added to improve particular characteristics include nickel, molybdenum, copper, titanium, aluminum, silicon, niobium, nitrogen, sulfur, and selenium. Carbon is normally present in amounts ranging from less than 0.03% to over 1.0% in certain martensitic grades.

The selection of stainless steels may be based on corrosion resistance, fabrication characteristics, availability, mechanical properties in specific temperature ranges and product cost. However, corrosion resistance and mechanical properties are usually the most important factors in selecting a grade for a given application.

Original discoveries and developments in stainless steel technology began in England and Germany about 1910. The commercial production and use of stainless steels in the United States began in the 1920s, with Allegheny, Armco, Carpenter, Crucible, Firth-Sterling, Jessop, Ludlum, Republic, Rustless, and U.S. Steel being among the early producers.

Only modest tonnages of stainless steel were produced in the United States in the mid-1920s, but annual production has risen steadily since that time. Even so, tonnage has never exceeded about 1.5% of total production for the steel industry. Table 1 shows shipments of stainless steel over a recent 10-year period. Production tonnages are listed only for U.S. domestic production. France, Italy, Japan, Sweden, the United Kingdom, and West Germany produce substantial tonnages of steel, and data on production in these countries are also available. However, other free-world countries do not make their figures public, and production statistics are not available from the U.S.S.R. or other Communist nations, which makes it impossible to estimate accurately the total world production of stainless steel.

The development of precipitation-hardenable stainless steels was spearheaded by the successful production of Stainless W by U.S. Steel in 1945. Since then, Armco, Allegheny-Ludlum, and Carpenter Technology have developed a series of precipitation-hardenable alloys.

The problem of obtaining raw materials has been a real one, particularly in regard to nickel during the 1950s when civil wars raged in Africa and Asia, prime sources of nickel, and Cold War politics played a role because Eastern-bloc nations were also prime sources of the element. This led to the development of a series of alloys (AISI 200 type) in which manganese and nitrogen are partially substituted for nickel. These stainless steels are still produced today.

New refining techniques were adopted in the early 1970s that revolutionized stainless steel melting. Most important was the argon-oxygen-decarburization (AOD) process. The AOD and related processes, with different gas injections or partial pressure systems, permitted the ready removal of carbon without substantial loss of chromium to the slag. Furthermore, low carbon contents were readily achieved in 18% Cr alloys when using high-carbon ferrochromium in furnace charges in place of the much more expensive low-carbon ferrochromium. Major alloying elements could also be controlled more precisely, nitrogen became an easily controlled intentional alloying element, and sulfur could be reduced to exceptionally low levels when desired. Oxygen could also be reduced to low levels and, when coupled with low sulfur, resulted in marked improvements in steel cleanliness.

During the same period, continuous casting grew in popularity throughout the steel industry, particularly in the stainless steel segment. The incentive for continuous casting was primarily economic. Piping can be confined to the last segment to be cast such that yield improvements of approximately 10% are commonly achieved. Improvements in homogeneity are also attained.

Over the years, stainless steels have become firmly established as materials for cooking utensils, fasteners, cutlery, flatware, decorative architectural hardware, and equipment for use in chemical plants, dairy and food-processing plants, health and sanitation applications, petroleum and petrochemical plants, textile plants, and the pharmaceutical and transportation industries. Some of these applications involve exposure to either elevated or cryogenic temperatures; austenitic stainless steels are well suited to either type of service. Properties of stainless steels at elevated temperatures are discussed in the section "Elevated-Temperature Properties" of this article and more detailed information is available in the article "Elevated-Temperature Properties of Stainless Steels" in this Volume. Properties at cryogenic temperatures are discussed in the section "Subzero-Temperature Properties" of this article.

Modifications in composition are sometimes made to facilitate production. For instance, basic compositions are altered to make it easier to produce stainless steel tubing and castings. Similar modifications are made for the manufacture of stainless steel welding electrodes; here, combinations of electrode coating and wire composition are used to produce desired compositions in deposited weld metal.

Classification of Stainless Steels

Stainless steels are commonly divided into five groups: martensitic stainless steels, ferritic stainless steels, austenitic stainless steels, duplex (ferritic-austenitic) stainless steels, and precipitation-hardening stainless steels.

Martensitic stainless steels are essentially alloys of chromium and carbon that possess a distorted body-centered cubic (bcc) crystal structure (martensitic) in the hardened condition. They are ferromagnetic, hardenable by heat treatments, and are generally resistant to corrosion only to relatively mild environments. Chromium content is generally in the range of 10.5 to 18%, and carbon content may exceed 1.2%. The chromium and carbon contents are balanced to ensure a martensitic structure after hardening. Ex-

Table 1 Total U.S. shipments of stainless steel over the 10-year period from 1979 to 1988

Year	Shipments kt	1000 tons
1979	1234	1361
1980(a)	1022	1127
1981(a)	1055	1163
1982(a)	811	894
1983(a)	1032	1137
1984(a)	1132	1248
1985(a)	1135	1251
1986(a)	1077	1187
1987(a)	1287	1418
1988(b)	1439	1586

(a) Ref 1. (b) Ref 2

cess carbides may be present to increase wear resistance or to maintain cutting edges, as in the case of knife blades. Elements such as niobium, silicon, tungsten, and vanadium may be added to modify the tempering response after hardening. Small amounts of nickel may be added to improve corrosion resistance in some media and to improve toughness. Sulfur or selenium is added to some grades to improve machinability.

Ferritic stainless steels are essentially chromium containing alloys with bcc crystal structures. Chromium content is usually in the range of 10.5 to 30%. Some grades may contain molybdenum, silicon, aluminum, titanium, and niobium to confer particular characteristics. Sulfur or selenium may be added, as in the case of the austenitic grades, to improve machinability. The ferritic alloys are ferromagnetic. They can have good ductility and formability, but high-temperature strengths are relatively poor compared to the austenitic grades. Toughness may be somewhat limited at low temperatures and in heavy sections.

Austenitic stainless steels have a face-centered cubic (fcc) structure. This structure is attained through the liberal use of austenitizing elements such as nickel, manganese, and nitrogen. These steels are essentially nonmagnetic in the annealed condition and can be hardened only by cold working. They usually possess excellent cryogenic properties and good high-temperature strength. Chromium content generally varies from 16 to 26%; nickel, up to about 35%; and manganese, up to 15%. The $2xx$ series steels contain nitrogen, 4 to 15.5% Mn, and up to 7% Ni. The $3xx$ types contain larger amounts of nickel and up to 2% Mn. Molybdenum, copper, silicon, aluminum, titanium, and niobium may be added to confer certain characteristics such as halide pitting resistance or oxidation resistance. Sulfur or selenium may be added to certain grades to improve machinability.

Duplex stainless steels have a mixed structure of bcc ferrite and fcc austenite. The exact amount of each phase is a function of composition and heat treatment (see the article "Cast Stainless Steels" in this Volume). Most alloys are designed to contain about equal amounts of each phase in the annealed condition. The principal alloying elements are chromium and nickel, but nitrogen, molybdenum, copper, silicon, and tungsten may be added to control structural balance and to impart certain corrosion-resistance characteristics.

The corrosion resistance of duplex stainless steels is like that of austenitic stainless steels with similar alloying contents. However, duplex stainless steels possess higher tensile and yield strengths and improved resistance to stress-corrosion cracking than their austenitic counterparts. The toughness of duplex stainless steels is between that of austenitic and ferritic stainless steels.

Precipitation-hardening stainless steels are chromium-nickel alloys containing precipitation-hardening elements such as copper, aluminum, or titanium. Precipitation-hardening stainless steels may be either austenitic or martensitic in the annealed condition. Those that are austenitic in the annealed condition are frequently transformable to martensite through conditioning heat treatments, sometimes with a subzero treatment. In most cases, these stainless steels attain high strength by precipitation hardening of the martensitic structure.

Standard Types. A list of standard types of stainless steels, similar to those originally published by the American Iron and Steel Institute (AISI), appears in Table 2. The criteria used to decide which types of stainless steel are standard types have been rather loosely defined but include tonnage produced during a specific period, availability (number of producers), and compositional limits. Specification-writing organizations such as ASTM and SAE include these standard types in their specifications. In referring to specific compositions, the term type is preferred over the term grade. Some specifications establish a series of grades within a given type, which makes it possible to specify properties more precisely for a given nominal composition.

In each of the three original groups of stainless steels—austenitic, ferritic, and martensitic—there is one composition that represents the basic, general-purpose alloy. All other compositions derive from this basic alloy, with specific variations in composition being made to impart very specific properties. The so-called family relationships for these three groups are summarized in Fig. 1 to 3. Type 329 is a duplex stainless steel (about 80% ferrite, 20% austenite as annealed) and is listed separately in Table 2.

Nonstandard Types. In addition to the standard types, many proprietary stainless steels are used for specific applications. Compositions of the more popular, nonstandard stainless steels are given in Table 3; some of the nonstandard grades are identified by AISI type numbers.

A cooperative study of ASTM and SAE resulted in the Unified Numbering System (UNS) for designation and identification of metals and alloys in commercial use in the United States. In UNS listings, stainless steels are identified by the letter S, followed by five digits. A few stainless alloys are classified as nickel alloys in the UNS system (identification letter N) because of their high nickel and low iron (less than 50%) contents.

Use of UNS numbers and AISI standard-type numbers ensures that a consumer can obtain suitable material time after time even from different producers or suppliers. Nevertheless, some variation in fabrication and service characteristics can be expected, even with material obtained from a single producer.

Factors in Selection

The first and most important step toward successful use of a stainless steel is selection of a type that is appropriate for the application. There are a large number of standard types that differ from one another in composition, corrosion resistance, physical properties, and mechanical properties; selection of the optimum type for a specific application is the key to satisfactory performance at minimum total cost.

The characteristics and properties of individual types discussed in this article and elsewhere in this Volume provide some of the information useful in steel selection. For a more detailed discussion, the reader is referred to *Design Guidelines for the Selection and Use of Stainless Steel*, published by the Committee of Stainless Steel Producers and available through AISI.

A checklist of characteristics to be considered in selecting the proper type of stainless steel for a specific application includes:

- Corrosion resistance
- Resistance to oxidation and sulfidation
- Strength and ductility at ambient and service temperatures
- Suitability for intended fabrication techniques
- Suitability for intended cleaning procedures
- Stability of properties in service
- Toughness
- Resistance to abrasion and erosion
- Resistance to galling and seizing
- Surface finish and/or reflectivity
- Magnetic properties
- Thermal conductivity
- Electrical resistivity
- Sharpness (retention of cutting edge)
- Rigidity

Corrosion resistance is frequently the most important characteristic of a stainless steel, but often is also the most difficult to assess for a specific application. General corrosion resistance to pure chemical solu-

Table 2 Compositions of standard stainless steels

Type	UNS designation	Composition, %(a)							
		C	Mn	Si	Cr	Ni	P	S	Other
Austenitic types									
201	S20100	0.15	5.5–7.5	1.00	16.0–18.0	3.5–5.5	0.06	0.03	0.25 N
202	S20200	0.15	7.5–10.0	1.00	17.0–19.0	4.0–6.0	0.06	0.03	0.25 N
205	S20500	0.12–0.25	14.0–15.5	1.00	16.5–18.0	1.0–1.75	0.06	0.03	0.32–0.40 N
301	S30100	0.15	2.00	1.00	16.0–18.0	6.0–8.0	0.045	0.03	...
302	S30200	0.15	2.00	1.00	17.0–19.0	8.0–10.0	0.045	0.03	...
302B	S30215	0.15	2.00	2.0–3.0	17.0–19.0	8.0–10.0	0.045	0.03	...
303	S30300	0.15	2.00	1.00	17.0–19.0	8.0–10.0	0.20	0.15 min	0.6 Mo(b)
303Se	S30323	0.15	2.00	1.00	17.0–19.0	8.0–10.0	0.20	0.06	0.15 min Se
304	S30400	0.08	2.00	1.00	18.0–20.0	8.0–10.5	0.045	0.03	...
304H	S30409	0.04–0.10	2.00	1.00	18.0–20.0	8.0–10.5	0.045	0.03	...
304L	S30403	0.03	2.00	1.00	18.0–20.0	8.0–12.0	0.045	0.03	...
304LN	S30453	0.03	2.00	1.00	18.0–20.0	8.0–12.0	0.045	0.03	0.10–0.16 N
302Cu	S30430	0.08	2.00	1.00	17.0–19.0	8.0–10.0	0.045	0.03	3.0–4.0 Cu
304N	S30451	0.08	2.00	1.00	18.0–20.0	8.0–10.5	0.045	0.03	0.10–0.16 N
305	S30500	0.12	2.00	1.00	17.0–19.0	10.5–13.0	0.045	0.03	...
308	S30800	0.08	2.00	1.00	19.0–21.0	10.0–12.0	0.045	0.03	...
309	S30900	0.20	2.00	1.00	22.0–24.0	12.0–15.0	0.045	0.03	...
309S	S30908	0.08	2.00	1.00	22.0–24.0	12.0–15.0	0.045	0.03	...
310	S31000	0.25	2.00	1.50	24.0–26.0	19.0–22.0	0.045	0.03	...
310S	S31008	0.08	2.00	1.50	24.0–26.0	19.0–22.0	0.045	0.03	...
314	S31400	0.25	2.00	1.5–3.0	23.0–26.0	19.0–22.0	0.045	0.03	...
316	S31600	0.08	2.00	1.00	16.0–18.0	10.0–14.0	0.045	0.03	2.0–3.0 Mo
316F	S31620	0.08	2.00	1.00	16.0–18.0	10.0–14.0	0.20	0.10 min	1.75–2.5 Mo
316H	S31609	0.04–0.10	2.00	1.00	16.0–18.0	10.0–14.0	0.045	0.03	2.0–3.0 Mo
316L	S31603	0.03	2.00	1.00	16.0–18.0	10.0–14.0	0.045	0.03	2.0–3.0 Mo
316LN	S31653	0.03	2.00	1.00	16.0–18.0	10.0–14.0	0.045	0.03	2.0–3.0 Mo; 0.10–0.16 N
316N	S31651	0.08	2.00	1.00	16.0–18.0	10.0–14.0	0.045	0.03	2.0–3.0 Mo; 0.10–0.16 N
317	S31700	0.08	2.00	1.00	18.0–20.0	11.0–15.0	0.045	0.03	3.0–4.0 Mo
317L	S31703	0.03	2.00	1.00	18.0–20.0	11.0–15.0	0.045	0.03	3.0–4.0 Mo
321	S32100	0.08	2.00	1.00	17.0–19.0	9.0–12.0	0.045	0.03	5 × %C min Ti
321H	S32109	0.04–0.10	2.00	1.00	17.0–19.0	9.0–12.0	0.045	0.03	5 × %C min Ti
330	N08330	0.08	2.00	0.75–1.5	17.0–20.0	34.0–37.0	0.04	0.03	...
347	S34700	0.08	2.00	1.00	17.0–19.0	9.0–13.0	0.045	0.03	10 × %C min Nb
347H	S34709	0.04–0.10	2.00	1.00	17.0–19.0	9.0–13.0	0.045	0.03	8 × %C min – 1.0 max Nb
348	S34800	0.08	2.00	1.00	17.0–19.0	9.0–13.0	0.045	0.03	0.2 Co; 10 × %C min Nb; 0.10 Ta
348H	S34809	0.04–0.10	2.00	1.00	17.0–19.0	9.0–13.0	0.045	0.03	0.2 Co; 8 × %C min – 1.0 max Nb; 0.10 Ta
384	S38400	0.08	2.00	1.00	15.0–17.0	17.0–19.0	0.045	0.03	...
Ferritic types									
405	S40500	0.08	1.00	1.00	11.5–14.5	...	0.04	0.03	0.10–0.30 Al
409	S40900	0.08	1.00	1.00	10.5–11.75	0.50	0.045	0.045	6 × %C min – 0.75 max Ti
429	S42900	0.12	1.00	1.00	14.0–16.0	...	0.04	0.03	...
430	S43000	0.12	1.00	1.00	16.0–18.0	...	0.04	0.03	...
430F	S43020	0.12	1.25	1.00	16.0–18.0	...	0.06	0.15 min	0.6 Mo(b)
430FSe	S43023	0.12	1.25	1.00	16.0–18.0	...	0.06	0.06	0.15 min Se
434	S43400	0.12	1.00	1.00	16.0–18.0	...	0.04	0.03	0.75–1.25 Mo
436	S43600	0.12	1.00	1.00	16.0–18.0	...	0.04	0.03	0.75–1.25 Mo; 5 × %C min – 0.70 max Nb
439	S43035	0.07	1.00	1.00	17.0–19.0	0.50	0.04	0.03	0.15 Al; 12 × %C min – 1.10 Ti
442	S44200	0.20	1.00	1.00	18.0–23.0	...	0.04	0.03	...
444	S44400	0.025	1.00	1.00	17.5–19.5	1.00	0.04	0.03	1.75–2.50 Mo; 0.025 N; 0.2 + 4 (%C + %N) min – 0.8 max (Ti + Nb)
446	S44600	0.20	1.50	1.00	23.0–27.0	...	0.04	0.03	0.25 N
Duplex (ferritic-austenitic) type									
329	S32900	0.20	1.00	0.75	23.0–28.0	2.50–5.00	0.040	0.030	1.00–2.00 Mo
Martensitic types									
403	S40300	0.15	1.00	0.50	11.5–13.0	...	0.04	0.03	...
410	S41000	0.15	1.00	1.00	11.5–13.5	...	0.04	0.03	...
414	S41400	0.15	1.00	1.00	11.5–13.5	1.25–2.50	0.04	0.03	...
416	S41600	0.15	1.25	1.00	12.0–14.0	...	0.06	0.15 min	0.6 Mo(b)
416Se	S41623	0.15	1.25	1.00	12.0–14.0	...	0.06	0.06	0.15 min Se
420	S42000	0.15 min	1.00	1.00	12.0–14.0	...	0.04	0.03	...
420F	S42020	0.15 min	1.25	1.00	12.0–14.0	...	0.06	0.15 min	0.6 Mo(b)
422	S42200	0.20–0.25	1.00	0.75	11.5–13.5	0.5–1.0	0.04	0.03	0.75–1.25 Mo; 0.75–1.25 W; 0.15–0.3 V
431	S43100	0.20	1.00	1.00	15.0–17.0	1.25–2.50	0.04	0.03	...
440A	S44002	0.60–0.75	1.00	1.00	16.0–18.0	...	0.04	0.03	0.75 Mo
440B	S44003	0.75–0.95	1.00	1.00	16.0–18.0	...	0.04	0.03	0.75 Mo
440C	S44004	0.95–1.20	1.00	1.00	16.0–18.0	...	0.04	0.03	0.75 Mo
Precipitation-hardening types									
PH 13-8 Mo	S13800	0.05	0.20	0.10	12.25–13.25	7.5–8.5	0.01	0.008	2.0–2.5 Mo; 0.90–1.35 Al; 0.01 N
15-5 PH	S15500	0.07	1.00	1.00	14.0–15.5	3.5–5.5	0.04	0.03	2.5–4.5 Cu; 0.15–0.45 Nb
17-4 PH	S17400	0.07	1.00	1.00	15.5–17.5	3.0–5.0	0.04	0.03	3.0–5.0 Cu; 0.15–0.45 Nb
17-7 PH	S17700	0.09	1.00	1.00	16.0–18.0	6.5–7.75	0.04	0.04	0.75–1.5 Al

(a) Single values are maximum values unless otherwise indicated. (b) Optional

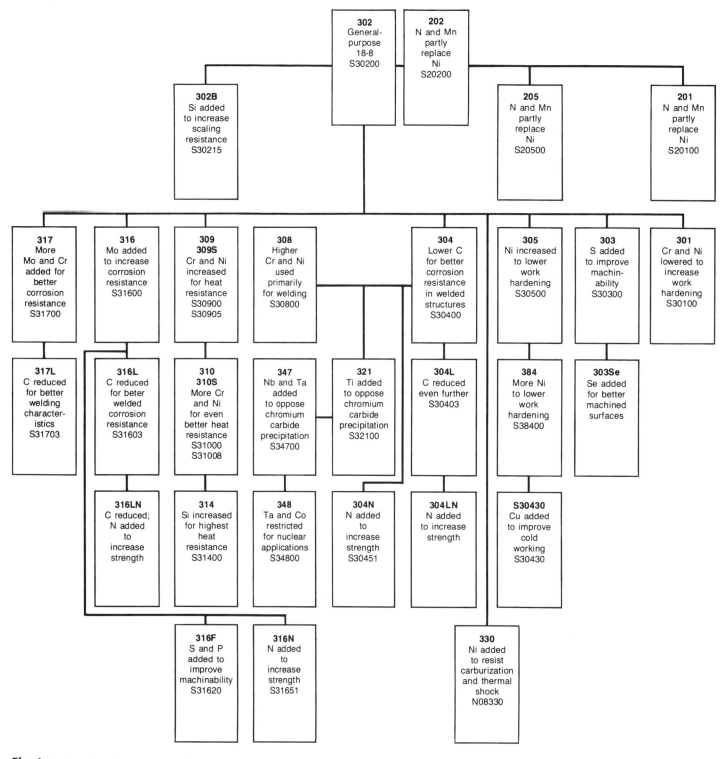

Fig. 1 Family relationships for standard austenitic stainless steels

tions is comparatively easy to determine, but actual environments are usually much more complex. Tables 4 and 5 show resistance of standard types to various common media.

General corrosion is often much less serious than localized forms such as stress-corrosion cracking, crevice corrosion in tight spaces or under deposits, pitting attack, and intergranular attack in sensitized material such as weld heat-affected zones (HAZ). Such localized corrosion can cause unexpected and sometimes catastrophic failure while most of the structure remains unaffected, and therefore must be considered carefully in the design and selection of the proper grade of stainless steel. Corrosive attack can also be increased dramatically by seemingly minor impurities in the medium that may be difficult to anticipate but that can have major effects, even when present in only parts-per-million concentrations; by heat transfer through the steel to or from the corrosive medium; by contact

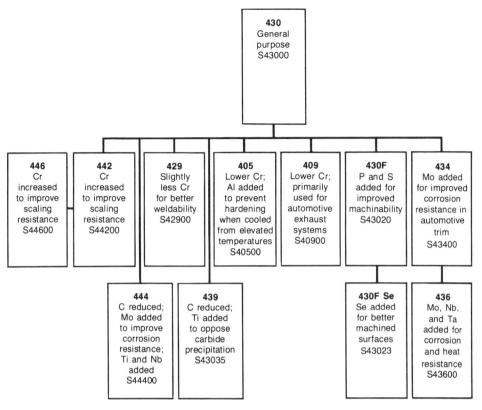

Fig. 2 Family relationships for standard ferritic stainless steels

with dissimilar metallic materials; by stray electrical currents; and by many other subtle factors. At elevated temperatures, attack can be accelerated significantly by seemingly minor changes in atmosphere that affect scaling, sulfidation, or carburization.

Despite these complications, a suitable steel can be selected for most applications on the basis of experience, perhaps with assistance from the steel producer. Laboratory corrosion data can be misleading in predicting service performance. Even actual service data have limitations, because similar corrosive media may differ substantially because of slight variations in some of the corrosion factors listed above. For difficult applications, the extensive study of comparative data may be necessary, sometimes followed by pilot plant or in-service testing.

More detailed information is available in the section "Corrosion Properties" in this article.

Mechanical properties at service temperature are obviously important, but satisfactory performance at other temperatures must be considered also. Thus, a product for arctic service must have suitable properties at subzero temperatures even though steady-state operating temperature may be much higher; room-temperature properties after extended service at elevated temperature can be important for applications such as boilers and jet engines, which are intermittently shut down.

Fabrication and Cleaning. Frequently a particular stainless steel is chosen for a fabrication characteristic such as formability or weldability. Even a required or preferred cleaning procedure may dictate the selection of a specific type. For instance, a weldment that is to be cleaned in a medium such as nitric-hydrofluoric acid, which attacks sensitized stainless steel, should be produced from stabilized or low-carbon stainless steel even though sensitization may not affect performance under service conditions.

Experience in the use of stainless steels indicates that many factors can affect their corrosion resistance. Some of the more prominent factors are:

- Chemical composition of the corrosive medium, including impurities
- Physical state of the medium—liquid, gaseous, solid, or combinations thereof
- Temperature
- Temperature variations
- Aeration of the medium
- Oxygen content of the medium
- Bacteria content of the medium
- Ionization of the medium
- Repeated formation and collapse of bubbles in the medium
- Relative motion of the medium with respect to the steel
- Chemical composition of the metal
- Nature and distribution of microstructural constituents
- Continuity of exposure of the metal to the medium
- Surface condition of the metal
- Stresses in the metal during exposure to the medium
- Contact of the metal with one or more dissimilar metallic materials
- Stray electric currents
- Differences in electric potential
- Marine growths such as barnacles
- Sludge deposits on the metal
- Carbon deposits from heated organic compounds
- Dust on exposed surfaces
- Effects of welding, brazing, and soldering

Surface Finish. Other characteristics in the stainless steel selection checklist are vital for some specialized applications but of little concern for many applications. Among these characteristics, surface finish is important more often than any other except corrosion resistance. Stainless steels are sometimes selected because they are available in a variety of attractive finishes. Surface finish selection may be made on the basis of appearance, frictional characteristics, or sanitation. The effect of finish on sanitation sometimes is thought to be simpler than it actually is, and tests of several candidate finishes may be advisable. The selection of finish may in turn influence the selection of the alloy because of differences in availability or durability of the various finishes for different types. For example, a more corrosion-resistant stainless steel will maintain a bright finish in a corrosive environment that would dull a lower-alloy type. Selection among finishes is described in more detail in this article in the section "Surface Finishing of Stainless Steel."

Product Forms

Stainless steels are available in the form of plate, sheet, strip, foil, bar, wire, semifinished products, pipes, tubes, and tubing.

Plate

Plate is a flat-rolled or forged product more than 250 mm (10 in.) in width and at least 4.76 mm (0.1875 in.) in thickness. Stainless steel plate is produced in most of the types shown in Table 2. Exceptions include highly alloyed ferritic stainless steels, some of the martensitic stainless steels, and a few of the free-machining grades. Plate is usually produced by hot rolling from slabs that have been directly cast or rolled from ingots and that usually have been conditioned to improve plate surface. Some plate may be produced by direct rolling from ingot. This plate is referred to as sheared plate or sheared mill plate when rolled between horizontal rolls and trimmed on all edges, and as universal plate or universal mill plate when rolled between horizontal and vertical rolls and

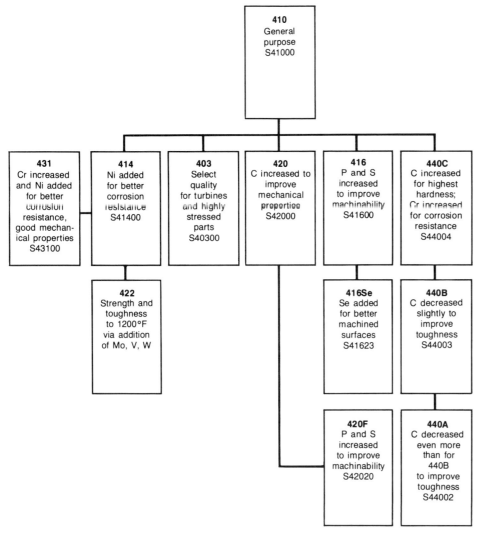

Fig. 3 Family relationships for standard martensitic stainless steels

trimmed only on the ends. Universal plate is sometimes rolled between grooved rolls.

Stainless steel plate is generally produced in the annealed condition and is either blast cleaned or pickled. Blast cleaning is generally followed by further cleaning in appropriate acids to remove surface contaminants such as particles of steel picked up from the mill rolls. Plate can be produced with mill edge and uncropped ends.

Sheet

Sheet is a flat-rolled product in coils or cut lengths at least 610 mm (24 in.) wide and less than 4.76 mm (0.1875 in.) thick. Stainless steel sheet is produced in nearly all types shown in Table 2 except the free-machining and certain martensitic grades. Sheet from the conventional grades is almost exclusively produced on continuous mills. Hand mill production is usually confined to alloys that cannot be produced economically on continuous mills, such as certain high-temperature alloys.

The steel is cast in ingots, and the ingots are rolled on a slabbing mill or a blooming mill into slabs or sheet bars. The slabs or sheet bars are then conditioned prior to being hot rolled on a finishing mill. Alternatively, the steel may be continuous cast directly into slabs that are ready for hot rolling on a finishing mill. The current trend worldwide is toward greater production from continuous cast slabs.

Sheet produced from slabs on continuous rolling mills is coiled directly off the mill. After they are descaled, these hot bands are cold rolled to the required thickness, and coils off the cold mill are either annealed and descaled or bright annealed. Belt grinding to remove surface defects is frequently required at hot bands or at an intermediate stage of processing. Full coils or lengths cut from coils may then be lightly cold rolled on either dull or bright rolls to produce the required finish. Sheet may be shipped in coils, or cut sheets may be produced by shearing lengths from a coil and flattening them by roller leveling or stretcher leveling.

Sheet produced on hand mills from sheet bars is rolled in lengths and then annealed and descaled. It may be subjected to additional operations, including cold reduction, annealing, descaling, light cold rolling for finish, or flattening.

A specified minimum tensile strength, minimum yield strength, or hardness level higher than that normally obtained on sheet in the annealed condition, or a combination thereof, can be attained by controlled cold rolling.

Sheet made of chromium-nickel stainless steel (often type 301) or of chromium-nickel-manganese stainless steel (often type 201) is produced in the following cold-rolled tempers:

Temper	Minimum tensile strength		Minimum yield strength	
	MPa	ksi	MPa	ksi
¼ hard	860	125	515	75
½ hard	1035	150	760	110
¾ hard	1205	175	930	135
Full hard	1275	185	965	140

Strip

Strip is a flat-rolled product, in coils or cut lengths, less than 610 mm (24 in.) wide and 0.13 to 4.76 mm (0.005 to 0.1875 in.) thick. Cold finished material 0.13 mm (0.005 in.) thick and less than 610 mm (24 in.) wide fits the definitions of both strip and foil and may be referred to by either term.

Cold-rolled stainless steel strip is manufactured from hot-rolled, annealed, and pickled strip (or from slit sheet) by rolling between polished rolls. Depending on the desired thickness, various numbers of cold-rolling passes through the mill are required for effecting the necessary reduction and securing the desired surface characteristics and mechanical properties.

Hot-rolled stainless steel strip is a semi-finished product obtained by hot-rolling slabs or billets and is produced for conversion to finished strip by cold rolling.

Heat Treatment. Strip of all types of stainless steel is usually either annealed or annealed and skin passed, depending on requirements. When severe forming, bending, and drawing operations are involved, it is recommended that such requirements be indicated so that the producer will have all the information necessary to ensure that he supplies the proper type and condition. When stretcher strains are objectionable in ferritic stainless steels such as type 430, they can be minimized by specifying a No. 2 finish. Cold-rolled strip in types 410, 414, 416, 420, 431, 440A, 440B, and 440C can be produced in the hardened and tempered condition.

Strip made of chromium-nickel stainless steel (often type 301) or of chromium-nickel-manganese stainless steel (often type 201) is produced in the same cold-rolled tempers in which sheet is produced.

Table 3 Compositions of nonstandard stainless steels

Designation(a)	UNS designation	C	Mn	Si	Cr	Ni	P	S	Other
Austenitic stainless steels									
Gall-Tough	S20161	0.15	4.00–6.00	3.00–4.00	15.00–18.00	4.00–6.00	0.040	0.040	0.08–0.20 N
203 EZ (XM-1)	S20300	0.08	5.0–6.5	1.00	16.0–18.0	5.0–6.5	0.040	0.18–0.35	0.5 Mo; 1.75–2.25 Cu
Nitronic 50 (XM-19)	S20910	0.06	4.0–6.0	1.00	20.5–23.5	11.5–13.5	0.040	0.030	1.5–3.0 Mo; 0.2–0.4 N; 0.1–0.3 Nb; 0.1–0.3 V
Tenelon (XM-31)	S21400	0.12	14.5–16.0	0.3–1.0	17.0–18.5	0.75	0.045	0.030	0.35 N
Cryogenic Tenelon (XM-14)	S21460	0.12	14.0–16.0	1.00	17.0–19.0	5.0–6.0	0.060	0.030	0.35–0.50 N
Esshete 1250	S21500	0.15	5.5–7.0	1.20	14.0–16.0	9.0–11.0	0.040	0.030	0.003–0.009 B; 0.75–1.25 Nb; 0.15–0.40 V
Type 216 (XM-17)	S21600	0.08	7.5–9.0	1.00	17.5–22.0	5.0–7.0	0.045	0.030	2.0–3.0 Mo; 0.25–0.50 N
Type 216 L (XM-18)	S21603	0.03	7.5–9.0	1.00	17.5–22.0	7.5–9.0	0.045	0.030	2.0–3.0 Mo; 0.25–0.50 N
Nitronic 60	S21800	0.10	7.0–9.0	3.5–4.5	16.0–18.0	8.0–9.0	0.040	0.030	0.08–0.18 N
Nitronic 40 (XM-10)	S21900	0.08	8.0–10.0	1.00	19.0–21.5	5.5–7.5	0.060	0.030	0.15–0.40 N
21-6-9 LC	S21904	0.04	8.00–10.00	1.00	19.00–21.50	5.50–7.50	0.060	0.030	0.15–0.40 N
Nitronic 33 (18-3-Mn)	S24000	0.08	11.50–14.50	1.00	17.00–19.00	2.50–3.75	0.060	0.030	0.20–0.40 N
Nitronic 32 (18-2-Mn)	S24100	0.15	11.00–14.00	1.00	16.50–19.50	0.50–2.50	0.060	0.030	0.20–0.45 N
18-18 Plus	S28200	0.15	17.0–19.0	1.00	17.5–19.5	...	0.045	0.030	0.5–1.5 Mo; 0.5–1.5 Cu; 0.4–0.6 N
303 Plus X (XM-5)	S30310	0.15	2.5–4.5	1.00	17.0–19.0	7.0–10.0	0.020	0.25 min	0.6 Mo
MVMA(c)	S30415	0.05	0.60	1.30	18.5	9.50	...	...	0.15 N; 0.04 Ce
304BI(d)	S30424	0.08	2.00	0.75	18.00–20.00	12.00–15.00	0.045	0.030	0.10 N; 1.00–1.25 B
304 HN (XM-21)	S30452	0.04–0.10	2.00	1.00	18.0–20.0	8.0–10.5	0.045	0.030	0.16–0.30 N
Cronifer 1815 LCSi	S30600	0.018	2.00	3.7–4.3	17.0–18.5	14.0–15.5	0.020	0.020	0.2 Mo
RA 85 H(c)	S30615	0.20	0.80	3.50	18.5	14.50	...	...	1.0 Al
253 MA	S30815	0.05–0.10	0.80	1.4–2.0	20.0–22.0	10.0–12.0	0.040	0.030	0.14–0.20 N; 0.03–0.08 Ce; 1.0 Al
Type 309 S Cb	S30940	0.08	2.00	1.00	22.0–24.0	12.0–15.0	0.045	0.030	10 × %C min to 1.10 max Nb
Type 310 Cb	S31040	0.08	2.00	1.50	24.0–26.0	19.0–22.0	0.045	0.030	10 × %C min to 1.10 max Nb + Ta
254 SMO	S31254	0.020	1.00	0.80	19.50–20.50	17.50–18.50	0.030	0.010	6.00–6.50 Mo; 0.50–1.00 Cu; 0.180–0.220 N
Type 316 Ti	S31635	0.08	2.00	1.00	16.0–18.0	10.0–14.0	0.045	0.030	5 × %(C + N) min to 0.70 max Ti; 2.0–3.0 Mo; 0.10 N
Type 316 Cb	S31640	0.08	2.00	1.00	16.0–18.0	10.0–14.0	0.045	0.030	10 × %C min to 1.10 max Nb + Ta; 2.0–3.0 Mo; 0.10 N
Type 316 HQ	...	0.030	2.00	1.00	16.00–18.25	10.00–14.00	0.030	0.015	3.00–4.00 Cu; 2.00–3.00 Mo
Type 317 LM	S31725	0.03	2.00	1.00	18.0–20.0	13.5–17.5	0.045	0.030	4.0–5.0 Mo; 0.10 N
17-14-4 LN	S31726	0.03	2.00	0.75	17.0–20.0	13.5–17.5	0.045	0.030	4.0–5.0 Mo; 0.10–0.20 N
Type 317 LN	S31753	0.03	2.00	1.00	18.0–20.0	11.0–15.0	0.030	0.030	0.10–0.22 N
Type 370	S37000	0.03–0.05	1.65–2.35	0.5–1.0	12.5–14.5	14.5–16.5	0.040	0.010	1.5–2.5 Mo; 0.1–0.4 Ti; 0.005 N; 0.05 Co
18-18-2 (XM-15)	S38100	0.08	2.00	1.5–2.5	17.0–19.0	17.5–18.5	0.030	0.030	...
19-9 DL	S63198	0.28–0.35	0.75–1.50	0.03–0.8	18.0–21.0	8.0–11.0	0.040	0.030	1.0–1.75 Mo; 0.1–0.35 Ti; 1.0–1.75 W; 0.25–0.60 Nb
20Cb-3	N08020	0.07	2.00	1.00	19.0–21.0	32.0–38.0	0.045	0.035	2.0–3.0 Mo; 3.0–4.0 Cu; 8 × %C min to 1.00 max Nb
20Mo-4	N08024	0.03	1.00	0.50	22.5–25.0	35.0–40.0	0.035	0.035	3.50–5.00 Mo; 0.50–1.50 Cu; 0.15–0.35 Nb
20Mo-6	N08026	0.03	1.00	0.50	22.00–26.00	33.00–37.20	0.03	0.03	5.00–6.70 Mo; 2.00–4.00 Cu
Sanicro 28	N08028	0.02	2.00	1.00	26.0–28.0	29.5–32.5	0.020	0.015	3.0–4.0 Mo; 0.6–1.4 Cu
AL-6X	N08366	0.035	2.00	1.00	20.0–22.0	23.5–25.5	0.030	0.030	6.0–7.0 Mo
AL-6XN	N08367	0.030	2.00	1.00	20.0–22.0	23.50–25.50	0.040	0.030	6.00–7.00 Mo; 0.18–0.25 N
JS-700	N08700	0.04	2.00	1.00	19.0–23.0	24.0–26.0	0.040	0.030	4.3–5.0 Mo; 8 × %C min to 0.5 max Nb; 0.5 Cu; 0.005 Pb; 0.035 S
Type 332	N08800	0.01	1.50	1.00	19.0–23.0	30.0–35.0	0.045	0.015	0.15–0.60 Ti; 0.15–0.60 Al
904L	N08904	0.02	2.00	1.00	19.0–23.0	23.0–28.0	0.045	0.035	4.0–5.0 Mo; 1.0–2.0 Cu
Cronifer 1925 hMo	N08925	0.02	1.00	0.50	24.0–26.0	19.0–21.0	0.045	0.030	6.0–7.0 Mo; 0.8–1.5 Cu; 0.10–0.20 N
Cronifer 2328	...	0.04	0.75	0.75	22.0–24.0	26.0–28.0	0.030	0.015	2.5–3.5 Cu; 0.4–0.7 Ti; 2.5–3.0 Mo
Ferritic stainless steels									
18-2 FM (XM-34)	S18200	0.08	1.25–2.50	1.00	17.5–19.5	...	0.040	0.15 min	1.5–2.5 Mo
Type 430 Ti	S43036	0.10	1.00	1.00	16.0–19.5	0.75	0.040	0.030	5 × %C min to 0.75 max Ti
Type 441	S44100	0.03	1.00	1.00	17.5–19.5	1.00	0.040	0.040	0.3 + 9 × (%C) min to 0.90 max Nb; 0.1–0.5 Ti; 0.03 N
E-Brite 26-1	S44627	0.01	0.40	0.40	25.0–27.0	0.50	0.020	0.020	0.75–1.5 Mo; 0.05–0.2 Nb; 0.015 N; 0.2 Cu
MONIT (25-4-4)	S44635	0.025	1.00	0.75	24.5–26.0	3.5–4.5	0.040	0.030	3.5–4.5 Mo; 0.2 + 4 (%C + %N) min to 0.8 max (Ti + Nb); 0.035 N
Sea-Cure (SC-1)	S44660	0.025	1.00	1.00	25.0–27.0	1.5–3.5	0.040	0.030	2.5–3.5 Mo; 0.2 + 4 (%C + %N) min to 0.8 max (Ti + Nb); 0.035 N
AL 29-4C	S44735	0.030	1.00	1.00	28.0–30.0	1.00	0.040	0.030	3.60–4.20 Mo; 0.20–1.00 Ti + Nb and 6 (%C + %N) min Ti + Nb; 0.045 N
AL 29-4-2	S44800	0.01	0.30	0.20	28.0–30.0	2.0–2.5	0.025	0.020	3.5–4.2 Mo; 0.15 Cu; 0.02 N; 0.025 max (%C + %N)
18 SR(c)	...	0.04	0.30	1.00	18.0	...	...	...	2.0 Al; 0.4 Ti
12 SR(c)	...	0.02	...	0.50	12.0	...	...	...	1.2 Al; 0.3 Ti
406	...	0.06	1.00	0.50	12.0–14.0	0.50	0.040	0.030	2.75–4.25 Al; 0.6 Ti
408 Cb	...	0.03	0.2–0.5	0.2–0.5	11.75–12.25	0.45	0.030	0.020	0.75–1.25 Al; 0.65–0.75 Nb; 0.3–0.5 Ti; 0.03 N

(continued)

(a) XM designations in this column are ASTM designations for the listed alloy. (b) Single values are maximum values unless otherwise indicated. (c) Nominal compositions. (d) UNS designation has not been specified. This designation appears in ASTM A 887 and merely indicates the form to be used.

Table 3 (continued)

Designation(a)	UNS designation	Composition, %(b)							
		C	Mn	Si	Cr	Ni	P	S	Other
Ferritic stainless steels (continued)									
ALFA IV	...	0.03	0.50	0.60	19.0–21.0	0.45	0.035	0.005	4.75–5.25 Al; 0.005–0.035 Ce; 0.03 N
Sealmet 1	...	0.08	0.5–0.8	0.3–0.6	28.0–29.0	0.40	0.030	0.015	0.04 N
Duplex stainless steels									
44LN	S31200	0.030	2.00	1.00	24.0–26.0	5.50–6.50	0.045	0.030	1.20–2.00 Mo; 0.14–0.20 N
DP-3	S31260	0.030	1.00	0.75	24.0–26.0	5.50–7.50	0.030	0.030	2.50–3.50 Mo; 0.20–0.80 Cu; 0.10–0.30 N; 0.10–0.50 W
3RE60	S31500	0.030	1.20–2.00	1.40–2.00	18.00–19.00	4.25–5.25	0.030	0.030	2.50–3.00 Mo
2205	S31803	0.030	2.00	1.00	21.0–23.0	4.50–6.50	0.030	0.020	2.50–3.50 Mo; 0.08–0.20 N
2304	S32304	0.030	2.50	1.0	21.5–24.5	3.0–5.5	0.040	0.040	0.05–0.60 Mo; 0.05–0.60 Cu; 0.05–0.20 N
Uranus 50	S32404	0.04	2.00	1.0	20.5–22.5	5.5–8.5	0.030	0.010	2.0–3.0 Mo; 1.0–2.0 Cu; 0.20 N
Ferralium 255	S32550	0.04	1.50	1.00	24.0–27.0	4.50–6.50	0.04	0.03	2.00–4.00 Mo; 1.50–2.50 Cu; 0.10–0.25 N
7-Mo PLUS	S32950	0.03	2.00	0.60	26.0–29.0	3.50–5.20	0.035	0.010	1.00–2.50 Mo; 0.15–0.35 N
Martensitic stainless steels									
Type 410S	S41008	0.08	1.00	1.00	11.5–13.5	0.60	0.040	0.030	...
Type 410 Cb (XM-30)	S41040	0.15	1.00	1.00	11.5–13.5	...	0.040	0.030	0.05–0.20 Nb
E4	S41050	0.04	1.00	1.00	10.5–12.5	0.60–1.1	0.045	0.030	0.10 N
CA6NM	S41500	0.05	0.5–1.0	0.60	11.5–14.0	3.5–5.5	0.030	0.030	0.5–1.0 Mo
416 Plus X (XM-6)	S41610	0.15	1.5–2.5	1.00	12.0–14.0	...	0.060	0.15 min	0.6 Mo
Type 418 (Greek Ascolloy)	S41800	0.15–0.20	0.50	0.50	12.0–14.0	1.8–2.2	0.040	0.030	2.5–3.5 W
TrimRite	S42010	0.15–0.30	1.00	1.00	13.5–15.0	0.25–1.00	0.040	0.030	0.40–1.00 Mo
Type 420 F Se	S42023	0.3–0.4	1.25	1.00	12.0–14.0	...	0.060	0.060	0.15 min Se; 0.6 Zr; 0.6 Cu
Lapelloy	S42300	0.27–0.32	0.95–1.35	0.50	11.0–12.0	0.50	0.025	0.025	2.5–3.0 Mo; 0.2–0.3 V
Type 440 F	S44020	0.95–1.20	1.25	1.00	16.0–18.0	0.75	0.040	0.10–0.35	0.08 N
Type 440 F Se	S44023	0.95–1.20	1.25	1.00	16.0–18.0	0.75	0.040	0.030	0.15 min Se; 0.60 Mo
Precipitation-hardening stainless steels									
PH 14-4 Mo	S14800	0.05	1.00	1.00	13.75–15.0	7.75–8.75	0.015	0.010	2.0–3.0 Mo; 0.75–1.50 Al
PH 15-7 Mo (Type 632)	S15700	0.09	1.00	1.00	14.0–16.0	6.5–7.75	0.040	0.030	2.0–3.0 Mo; 0.75–1.5 Al
AM-350 (Type 633)	S35000	0.07–0.11	0.5–1.25	0.50	16.0–17.0	4.0–5.0	0.040	0.030	2.5–3.25 Mo; 0.07–0.13 N
AM-355 (Type 634)	S35500	0.10–0.15	0.5–1.25	0.50	15.0–16.0	4.0–5.0	0.040	0.030	2.5–3.25 Mo; 0.07–0.13 N
Custom 450 (XM-25)	S45000	0.05	1.00	1.00	14.0–16.0	5.0–7.0	0.030	0.030	1.25–1.75 Cu; 0.5–1.0 Mo; 8 × %C min Nb
Custom 455 (XM-16)	S45500	0.05	0.50	0.50	11.0–12.5	7.5–9.5	0.040	0.030	1.5–2.5 Cu; 0.8–1.4 Ti; 0.1–0.5 Nb; 0.5 Mo

(a) XM designations in this column are ASTM designations for the listed alloy. (b) Single values are maximum values unless otherwise indicated. (c) Nominal compositions. (d) UNS designation has not been specified. This designation appears in ASTM A 887 and merely indicates the form to be used.

For strip, edge condition is often important—more important than it usually is for sheet. Strip can be furnished with various edge specifications:

- Mill edge (as produced, condition unspecified)
- No. 1 edge (edge rolled, rounded, or square)
- No. 3 edge (as slit)
- No. 5 edge (square edge produced by rolling or filing after slitting)

Mill edge is the least expensive edge condition, and is adequate for many purposes. No. 1 edge provides improved width tolerance over mill edge plus a cold-rolled edge condition; rounded edges are preferred for applications requiring the lowest degree of stress concentration at corners. No. 3 and No. 5 edges give progressively better width tolerance and squareness over No. 1 edge.

Foil

Foil is a flat-rolled product, in coil form, up to 0.13 mm (0.005 in.) thick and less than 610 mm (24 in.) wide. Foil is produced in slit widths with edge conditions corresponding to No. 3 and No. 5 edge conditions for strip.

Foil is made from types 201, 202, 301, 302, 304, 304L, 305, 316, 316L, 321, 347, 430, and 442, as well as from certain proprietary alloys.

The finishes, tolerances, and mechanical properties of foil differ from those of strip because of limitations associated with the way in which foil is manufactured. Nomenclature for finishes, and for width and thickness tolerances, vary among producers.

Finishes for foil are described by the finishing operations employed in their manufacture. However, each finish in itself is a category of finishes, with variations in appearance and smoothness that depend on composition, thickness, and method of manufacture. Chromium-nickel and chromium-nickel-manganese stainless steels have a characteristic appearance different from that of straight chromium types for corresponding finish designations.

Mechanical Properties. In general, mechanical properties of foil vary with thickness. Tensile strength is increased somewhat, and ductility is lowered, by a decrease in thickness.

Bar

Bar is a product supplied in straight lengths; it is either hot or cold finished and is available in various shapes, sizes, and surface finishes. This category includes small shapes whose dimensions do not exceed 75 mm (3 in.) and, second, hot-rolled flat stock at least 3.2 mm (0.125 in.) thick and up to 250 mm (10 in.) wide.

Hot-finished bar is commonly produced by hot rolling, forging, or pressing ingots to blooms or billets of intermediate size, which are subsequently hot rolled, forged, or extruded to final dimensions. Whether rolling, forging, or extrusion is selected as the finishing method depends on several factors, including composition and final size.

Following hot rolling or forging, hot-finished bar may be subjected to various operations, including:

- Annealing or other heat treatment
- Descaling by pickling, blast cleaning, or other methods
- Surface conditioning by grinding or rough turning
- Machine straightening

Cold-finished bar is produced from hot-finished bar or rod by additional operations such as cold rolling or cold drawing, which result in the close control of dimensions, a smooth surface finish, and higher tensile and yield strengths. Sizes and shapes of cold reduced stock classified as bar are essentially the same as for hot-finished bar, except that all cold reduced flat stock less

Table 4 Resistance of standard types of stainless steel to various classes of environments

Type	Mild atmospheric and fresh water	Atmospheric Industrial	Atmospheric Marine	Salt water	Chemical Mild	Chemical Oxidizing	Chemical Reducing
Austenitic stainless steels							
201	x	x	x	...	x	x	...
202	x	x	x	...	x	x	...
205	x	x	x	...	x	x	...
301	x	x	x	...	x	x	...
302	x	x	x	...	x	x	...
302B	x	x	x	...	x	x	...
303	x	x	...	...	x	...	...
303Se	x	x	...	...	x	...	...
304	x	x	x	...	x	x	...
304H	x	x	x	...	x	x	...
304L	x	x	x	...	x	x	...
304N	x	x	x	...	x	x	...
S30430	x	x	x	...	x	x	...
305	x	x	x	...	x	x	...
308	x	x	x	...	x	x	...
309	x	x	x	...	x	x	...
309S	x	x	x	...	x	x	...
310	x	x	x	...	x	x	...
310S	x	x	x	...	x	x	...
314	x	x	x	...	x	x	...
316	x	x	x	x	x	x	x
316F	x	x	x	x	x	x	x
316H	x	x	x	x	x	x	x
316L	x	x	x	x	x	x	x
316N	x	x	x	x	x	x	x
317	x	x	x	x	x	x	x
317L	x	x	x	x	x	x	x
321	x	x	x	...	x	x	...
321H	x	x	x	...	x	x	...
329	x	x	x	x	x	x	x
330	x	x	x	x	x	x	x
347	x	x	x	...	x	x	...
347H	x	x	x	...	x	x	...
348	x	x	x	...	x	x	...
348H	x	x	x	...	x	x	...
384	x	x	x	...	x	x	...
Ferritic stainless steels							
405	x	...	...	...	x	...	...
409	x	...	...	...	x	...	...
429	x	x	...	...	x	x	...
430	x	x	...	...	x	x	...
430F	x	x	...	...	x	...	...
430FSe	x	x	...	...	x	...	...
434	x	x	x	...	x	x	...
436	x	x	x	...	x	x	...
442	x	x	...	...	x	x	...
446	x	x	x	...	x	x	...
Martensitic stainless steels							
403	x	...	...	...	x	...	...
410	x	...	...	...	x	...	...
414	x	...	...	...	x	...	...
416	x	...	...	...	...	...	...
416Se	x	...	...	...	...	...	...
420	x	...	...	...	...	...	...
420F	x	...	...	...	...	...	...
422	x	...	...	...	...	...	...
431	x	x	x	...	x	...	...
440A	x	...	...	...	x	...	...
440B	x	...	...	...	...	...	...
440C	x	...	...	...	...	...	...
501	...	...	...	...	...	...	...
502	...	...	...	...	...	...	...
503	...	...	...	...	...	...	...
504	...	...	...	...	...	...	...
Precipitation-hardening stainless steels							
PH 13-8 Mo	x	x	...	...	x	x	...
15-5 PH	x	x	x	...	x	x	...
17-4 PH	x	x	x	...	x	x	...
17-7 PH	x	x	x	...	x	x	...

An x notation indicates that the specific type may be considered for application in the corrosive environment.

than 4.76 mm (0.1875 in.) thick and over 9.5 mm (0.375 in.) wide is classified as strip.

Cold-finished round bar is commonly machine straightened; afterward, it can be centerless ground or centerless ground and polished. Centerless grinding and polishing do not alter the mechanical properties of cold-finished bar and are used only to improve surface finish or provide closer tolerances. Some increase in hardness, more marked at the surface and particularly in 2xx and 3xx stainless steels, results from machine straigthening. The amount of increase varies chiefly with composition, size, and amount of cold work necessary to straighten the bar.

Cold-finished bars that are square, flat, hexagonal, octagonal, or of certain special shapes are produced from hot-finished bars by cold drawing or cold rolling.

When cold-finished bar is required to have high strength and hardness, it is cold drawn or heat treated, depending on composition, section size, and required properties. Round sections can be subsequently centerless ground or centerless ground and polished.

Free-machining wire is a bar commodity used for making parts in automatic screw machines or other types of machining equipment. The principal types used are 303, 303Se, 416, 416Se, 420F, 430F, and 430FSe. Free-machining wire is commonly produced with a cold drawn or centerless ground finish and with selected hardnesses, depending on the machining operation involved.

Structural Shapes. Hot-rolled, bar-size structural shapes are produced in angles, channels, tees, and zees. They can be purchased in various conditions:

- Hot rolled
- Hot rolled and annealed
- Hot rolled, annealed, and blast cleaned
- Hot rolled, annealed, and chemically cleaned
- Hot rolled, annealed, blast cleaned, and chemically cleaned

Wire

Wire is a coiled product derived by cold finishing hot-rolled and annealed rod. Cold finishing imparts excellent dimensional accuracy, good surface smoothness, a fine finish, and specific mechanical properties. Wire is produced in several tempers and finishes.

Wire is customarily referred to as round wire when the contour is completely cylindrical and as shape wire when the contour is other than cylindrical. For example, wires that are half round, half oval, oval, square, rectangular, hexagonal, octagonal, or triangular in cross section are all referred to as shape wire. Shape wire is cold finished either by drawing or by a combination of drawing and rolling.

Table 5 Relative corrosion resistance of AISI stainless steels for different grade applications

Environment	Grades(a)
Acids	
Hydrochloric acid	Stainless is not generally recommended except when solutions are very dilute and at room temperature (pitting may occur).
Mixed acids	There is usually no appreciable attack on type 304 or 316 as long as sufficient nitric acid is present.
Nitric acid	Type 304L and 430 and some higher-alloy stainless grades have been used.
Phosphoric acid	Type 304 is satisfactory for storing cold phosphoric acid up to 85% and for handling concentrations up to 5% in some unit processes of manufacture. Type 316 is more resistant and is generally used for storing and manufacture if the fluorine content is not too high. Type 317 is somewhat more resistant than type 316. At concentrations ≤85%, the metal temperature should not exceed 100 °C (212 °F) with type 316 and slightly higher with type 317. Oxidizing ions inhibit attack.
Sulfuric acid	Type 304 can be used at room temperature for concentrations >80 to 90%. Type 316 can be used in contact with sulfuric acid ≤10% at temperatures ≤50 °C (120 °F) if the solutions are aerated; the attack is greater in air-free solutions. Type 317 may be used at temperatures as high as 65 °C (150 °F) with ≤5% concentration. The presence of other materials may markedly change the corrosion rate. As little as 500 to 2000 ppm of cupric ions make it possible to use type 304 in hot solutions of moderate concentration. Other additives may have the opposite effect.
Sulfurous acid	Type 304 may be subject to pitting, particularly if some sulfuric acid is present. Type 316 is usable at moderate concentrations and temperatures.
Bases	
Ammonium hydroxide, sodium hydroxide, caustic solutions	Steels in the 300 series generally have good corrosion resistance at virtually all concentrations and temperatures in weak bases, such as ammonium hydroxide. In stronger bases, such as sodium hydroxide, there may be some attack, cracking, or etching in more concentrated solutions and/or at higher temperatures. Commercial-purity caustic solutions may contain chlorides, which will accentuate any attack and may cause pitting of type 316, as well as type 304.
Organics	
Acetic acid	Acetic acid is seldom pure in chemical plants but generally includes numerous and varied minor constituents. Type 304 is used for a wide variety of equipment including stills, base heaters, holding tanks, heat exchangers, pipelines, valves, and pumps for concentrations ≤99% at temperatures ≤ ~50 °C (120 °F). Type 304 is also satisfactory—if small amounts of turbidity or color pickup can be tolerated—for room temperature storage of glacial acetic acid. Types 316 and 317 have the broadest range of usefulness, especially if formic acid is also present or if solutions are unaerated. Type 316 is used for fractionating equipment, for 30–99% concentrations where type 304 cannot be used, for storage vessels, pumps, and process equipment handling glacial acetic acid, which would be discolored by type 304. Type 316 is likewise applicable for parts having temperatures >50 °C (120 °F), for dilute vapors, and for high pressures. Type 317 has somewhat greater corrosion resistance than type 316 under severely corrosive conditions. None of the stainless steels has adequate corrosion resistance to glacial acetic acid at the boiling temperature or at superheated vapor temperatures.
Aldehydes	Type 304 is generally satisfactory.
Amines	Type 316 is usually preferred to type 304.
Cellulose acetate	Type 304 is satisfactory for low temperatures, but type 316 or type 317 is needed for high temperatures.
Formic acids	Type 304 is generally acceptable at moderate temperatures, but type 316 is resistant to all concentrations at temperatures up to boiling.
Esters	With regard to corrosion, esters are comparable to organic acids.
Fatty acids	Type 304 is resistant to fats and fatty acids ≤ ~150 °C (300 °F), but type 316 is needed at 150–260 °C (300–500 °F), and type 317, at higher temperatures.
Paint vehicles	Type 316 may be needed if exact color and lack of contamination are important.
Phthalic anhydride	Type 316 is usually used for reactors, fractionating columns, traps, baffles, caps, and piping.
Soaps	Type 304 is used for parts such as spray towers, but type 316 may be preferred for spray nozzles and flake-drying belts to minimize off-color product.
Synthetic detergents	Type 316 is used for preheat, piping, pumps, and reactors in catalytic hydrogenation of fatty acids to give salts of sulfonated high-molecular alcohols.
Tall oil (pulp and paper industry)	Type 304 has only limited use in tall-oil distillation service. High rosin acid streams can be handled by type 316L with a minimum molybdenum content of 2.75%. Type 316 can also be used in the more corrosive high fatty acid streams at temperatures ≤245 °C (475 °F), but type 317 will probably be required at higher temperatures.
Tar	Tar distillation equipment is almost all type 316 because coal tar has a high chloride content; type 304 does not have adequate resistance to pitting.
Urea	Type 316L is generally required.
Pharmaceuticals	Type 316 is usually selected for all parts in contact with the product because of its inherent corrosion resistance and greater assurance of product purity.

(a) The stainless steels mentioned may be considered for use in the indicated environments. Additional information or corrosion expertise may be necessary prior to use in some environments; for example, some impurities may cause localized corrosion (such as chlorides causing pitting or stress-corrosion cracking of some grades). Source: Ref 3

In the production of wire, rod (which is a coiled hot-rolled product approximately round in cross section) is drawn through the tapered hole of a die or a series of dies. The smallest size of hot-rolled rod commonly made is 5.5 mm (0.218 in.). Rod smaller than this is produced by cold work, the number of dies employed depending on the finished diameter required.

Round stainless steel wire is commonly produced within the approximate size range 0.08 to 15.9 mm (0.003 to 0.625 in.). Shape wire, except cold-finished flat wire, is commonly produced within the approximate size range of 1.12 to 12.7 mm (0.044 to 0.500 in.), although the particular shape governs the specific sizes that can be produced.

Tempers of Wire. There are four classifications of wire temper: annealed-temper, soft-temper, intermediate-temper, and spring-temper.

Annealed temper describes soft wire that has undergone no further cold drawing after the last annealing treatment. Wire in this temper is made by annealing in open-fired furnaces or molten salt, and annealing ordinarily is followed by pickling that produces a clean, gray, matte finish. It is also made with a bright finish by annealing in a protective atmosphere and sometimes is described as bright annealed wire.

Soft-temper wire is given a single light draft following the final annealing operation and generally is produced to a defined upper limit of tensile strength or hardness. Wire in this temper is produced with various dry-drawn finishes, including lime soap, lead, copper, and oxide. It may also be given a bright finish produced by oil or grease drawing.

Intermediate-temper wire is drawn one or more drafts after annealing as required to produce a specific minimum strength or hardness. The properties of this wire can vary between the properties of soft-temper wire and properties approaching those of spring-temper wire. Intermediate-temper wire is usually produced with one of the dry-drawn finishes.

Spring-temper wire is drawn several drafts as required to produce high tensile strengths.

Special Wire Commodities. There are many classes of stainless steel wire that have been developed for specific components or for particular applications. The unique properties of each of these individual wire commodities are developed by employing a particular combination of composition, steel quality, process heat treatment, and cold drawing practice. The details of manufacture may vary slightly from one wire manufacturer to another, but the finished wire will fulfill the specified requirements.

Cold-heading wire is produced in any of the various types of stainless steel. In all instances, cold-heading wire is subjected to special testing and inspection to ensure satisfactory performance in cold-heading and cold-forging operations.

Of the chromium-nickel group, types 305 and 302Cu are used for cold-heading wire and generally are necessary for severe upsetting. Other grades commonly cold formed include 304, 316, 321, 347, and 384.

Of the 4xx series, types 410, 420, 430, and 431 are used for a variety of cold-headed products. Types 430 and 410 are commonly used for severe upsetting and for recessed-head screws and bolts. Types 416, 416Se, 430F, and 430FSe are intended primarily for free cutting and are not recommended for cold heading.

Cold-heading wire is manufactured using a closely controlled annealing treatment that produces optimum softness and still permits a very light finishing draft after pickling. The purposes of the finishing draft are to provide a lubricating coating that will aid the cold-heading operation and to produce a kink-free wire coil having more uniform dimensions.

Cold-heading wire is produced with a variety of finishes, all of which have the function of providing proper lubrication in the header dies. The finish or coating should be suitably adherent to prevent galling and excessively rapid die wear. A copper coating, which is applied after the annealing treatment and just prior to the finishing draft, is available; the copper-coated wire is then lime coated and drawn, using soap as the drawing lubricant. Coatings of lime and soap or of oxide and soap are also employed.

Spring wire is drawn from annealed rod and is subjected to mill tests and inspection that ensures the quality required for extension and compression springs. The types of stainless steel of which spring wire is commonly produced include 302, 304, and, for additional corrosion resistance, 316, and UNS N08020.

Spring wire in large sizes can be furnished in a variety of finishes, such as dry-drawn lead, copper, lime and soap, and oxide and soap. Fine sizes are usually wet drawn, although they can be dry drawn.

Tensile strength ranges or minimums for types 302, 304, 305, and 316 spring wire in various sizes are given in Table 6.

The torsional modulus for stainless steel spring wire may range from 59 to 76 GPa (8.5 to 11 × 10⁶ psi), depending on alloy and wire size. Magnetic permeability is extremely low compared to that of carbon steel wire. Springs made from stainless steel wire retain their physical and mechanical properties at temperatures up to about 315 °C (600 °F).

Rope wire is used to make rope, cable, and cord for a variety of uses, such as aircraft control cable, marine rope, elevator cable, slings, and anchor cable. Because of special requirements for fatigue strength, rope wire is produced from specially selected and processed material.

Rope wire is made of type 302 or type 304 unless a higher level of corrosion resistance is required, in which case type 316 is generally selected. Special nonmagnetic characteristics may be required, which necessitate the selection of grades that have little or no ferrite or martensite in the microstructure and the use of special drawing practices to limit or avoid deformation-induced transformation to martensite.

Tensile properties of regular rope wire are slightly lower than those of stainless steel spring wire. Finishes for rope wire vary from a gray matte finish to a bright finish and include a series of bright to dark soap finishes. Soap finishes afford some lubrication that facilitates laying up of rope and also to some extent aids in-service use.

Weaving wire is used in the weaving of screens for many different applications in coal mines, sand-and-gravel pits, paper mills, chemical plants, dairy plants, oil refineries, and food-processing plants. Annealing and final drawing must be carefully controlled to maintain uniform temper and finish throughout each coil or spool. Because weaving wire must be ductile, it is usually furnished in the annealed temper with a bright annealed finish, or in the soft temper with either a lime-soap finish or an oil- or grease-drawn finish.

Most types of stainless steel are available in weaving wire; the most widely used types are 302, 304, 309, 310, 316, 410, and 430. Annealed wire in the 3xx series commonly has a tensile strength of 655 to 860 MPa (95 to 125 ksi) and an elongation (in 50 mm, or 2 in.) of 35 to 60%. Soft-temper wire, which is commonly specified for sizes over 0.75 mm (0.030 in.), averages 860 to 1035 MPa (125 to 150 ksi) in tensile strength and exhibits 15 to 40% elongation. For annealed wire in types 410 and 430, tensile strength averages 495 to 585 MPa (72 to 85 ksi), and elongation averages 17 to 23%.

Armature binding wire is produced in types 302 or 304 stainless steel of a composition that is balanced to produce high tensile and yield strengths and low magnetic permeability. Minimum tensile strength of 1515 MPa (220 ksi), minimum yield strength (0.2% offset) of 1170 MPa (170 ksi), and maximum permeability of 4.0 at 16 kA · m⁻¹ (200 oersteds) are usually specified. The

Table 6 Room-temperature tensile strength of stainless steel spring wire

Diameter		Tensile strength	
mm	in.	MPa	ksi
Types 302 and 304			
≤0.23	≤0.009	2241–2448	325–355
>0.23–0.25	>0.009–0.010	2206–2413	320–350
>0.25–0.28	>0.010–0.011	2192–2399	318–348
>0.28–0.30	>0.011–0.012	2179–2385	316–346
>0.30–0.33	>0.012–0.013	2165–2372	314–344
>0.33–0.36	>0.013–0.014	2151–2358	312–342
>0.36–0.38	>0.014–0.015	2137–2344	310–340
>0.38–0.41	>0.015–0.016	2124–2330	308–338
>0.41–0.43	>0.016–0.017	2110–2317	306–336
>0.43–0.46	>0.017–0.018	2096–2303	304–334
>0.46–0.51	>0.018–0.020	2068–2275	300–330
>0.51–0.56	>0.020–0.022	2041–2248	296–326
>0.56–0.61	>0.022–0.024	2013–2220	292–322
>0.61–0.66	>0.024–0.026	2006–2206	291–320
>0.66–0.71	>0.026–0.028	1993–2192	289–318
>0.71–0.79	>0.028–0.031	1965–2172	285–315
>0.79–0.86	>0.031–0.034	1944–2137	282–310
>0.86–0.94	>0.034–0.037	1930–2124	280–308
>0.94–1.04	>0.037–0.041	1896–2096	275–304
>1.04–1.14	>0.041–0.045	1875–2068	272–300
>1.14–1.27	>0.045–0.050	1841–2034	267–295
>1.27–1.37	>0.050–0.054	1827–2020	265–293
>1.37–1.47	>0.054–0.058	1800–1993	261–289
>1.47–1.60	>0.058–0.063	1779–1965	258–285
>1.60–1.78	>0.063–0.070	1737–1937	252–281
>1.78–1.90	>0.070–0.075	1724–1917	250–278
>1.90–2.03	>0.075–0.080	1696–1896	246–275
>2.03–2.21	>0.080–0.087	1668–1868	242–271
>2.21–2.41	>0.087–0.095	1641–1848	238–268
>2.41–2.67	>0.095–0.105	1600–1806	232–262
>2.67–2.92	>0.105–0.115	1565–1772	227–257
>2.92–3.18	>0.115–0.125	1531–1744	222–253
>3.18–3.43	>0.125–0.135	1496–1710	217–248
>3.43–3.76	>0.135–0.148	1448–1662	210–241
>3.76–4.12	>0.148–0.162	1413–1620	205–235
>4.12–4.50	>0.162–0.177	1365–1572	198–228
>4.50–4.88	>0.177–0.192	1338–1551	194–225
>4.88–5.26	>0.192–0.207	1296–1517	188–220
>5.26–5.72	>0.207–0.225	1255–1475	182–214
>5.72–6.35	>0.225–0.250	1207–1413	175–205
>6.35–7.06	>0.250–0.278	1158–1365	168–198
>7.06–7.77	>0.278–0.306	1110–1324	161–192
>7.77–8.41	>0.306–0.331	1069–1282	155–186
>8.41–9.20	>0.331–0.362	1020–1241	148–180
>9.20–10.01	>0.362–0.394	979–1193	142–173
>10.01–11.12	>0.394–0.438	931–1138	135–165
>11.12–12.70	>0.438–0.500	862–1069	125–155
Types 305 and 316			
≤0.25	≤0.010	1689–1896	245–275
>0.25–0.38	>0.010–0.015	1655–1862	240–270
>0.38–1.04	>0.015–0.041	1620–1827	235–265
>1.04–1.19	>0.041–0.047	1586–1723	230–260
>1.19–1.37	>0.047–0.054	1551–1758	225–255
>1.37–1.58	>0.054–0.062	1517–1724	220–250
>1.58–1.85	>0.062–0.072	1482–1689	215–245
>1.85–2.03	>0.072–0.080	1448–1655	210–240
>2.03–2.34	>0.080–0.092	1413–1620	205–235
>2.34–2.67	>0.092–0.105	1379–1586	200–230
>2.67–3.05	>0.105–0.120	1344–1551	195–225
>3.05–3.76	>0.120–0.148	1276–1482	185–215
>3.76–4.22	>0.148–0.166	1241–1448	180–210
>4.22–4.50	>0.166–0.177	1172–1379	170–200
>4.50–5.26	>0.177–0.207	1103–1310	160–190
>5.26–5.72	>0.207–0.225	1069–1276	155–185
>5.72–6.35	>0.225–0.250	1034–1241	150–180
>6.35–7.92	>0.250–0.312	931–1138	135–165
>7.92–12.68	>0.312–0.499	793–1000	115–145
>12.68	>0.499	Consult producer	

wire must be strong enough to withstand the centrifugal forces encountered in use, yet ductile enough to withstand being bent sharply back on itself without cracking when a hook is formed to hold the armature wire during the binding operation. Armature binding wire is furnished on spools and has a smooth, tightly adherent tinned coating that facilitates soldering.

Slide forming wire is produced in all standard types, particularly in types 302, 304, 316, 410, and 430. It can be produced in any temper suitable for forming any of the numerous shapes made on slide-type wire forming machines.

Wool wire is designed for the production of wool by shredding. It is commonly furnished in an intermediate temper and produced to rigid standards so that it will perform satisfactorily in the wool-cutting operation. Wool wire usually is made from type 430 and has a lime-soap finish.

Reed wire is high-quality wire produced for the manufacture of dents for reeds that, once assembled, are used in weaving textiles and other products. Dents are made by rolling the round reed wire into a flat section, and then machining and polishing the edges to a very smooth and accurate contour before cutting the wire into individual dents. Accuracy in size and shape are necessary because of the various processes that the wire must undergo.

Reed wire is usually made from type 430 in an intermediate temper that must be uniform in properties throughout each coil and each shipment. The finish also must be uniform and bright.

Lashing wire is designed for lashing electric power transmission lines to support cables. Lashing wire is usually made from type 430. It is furnished in the annealed temper with a bright finish and has a maximum tensile strength of 655 MPa (95 ksi) and minimum elongation of 17% in 255 mm (10 in.). It is normally furnished on coreless spools.

Cotter pin wire is approximately half-round wire designed for fabricating cotter pins. It is generally produced by rolling round wire between power-driven rolls, by drawing it between power-driven rolls, or by drawing it through a die or Turk's-head roll. To facilitate the spreading of the cotter pin ends, it is desirable that the flat side of the wire have a small radius rather than sharp corners at the edges.

Cotter pin wire is commonly furnished in vibrated or hank-wound coils with the flat side of the wire facing inward. Ordinarily it is produced in the soft temper to prevent undesirable springback in the legs of formed cotter pins. Usually it is furnished with a bright finish, but it is also available with a metallic coating.

Stainless welding wire is available for many grades to provide good weldability with optimized mechanical properties and corrosion resistance of the weldment. For example, the weldability of austenitic stainless steels is enhanced by controlling unwanted residual elements or balancing the wire composition to provide a small amount of ferrite in the as-deposited weld metal. Also, the composition of duplex stainless weld wire is generally controlled to produce levels of austenite and ferrite in the weld metal that will optimize mechanical properties and corrosion resistance.

Stainless steel weld wire is produced in layer-level wound spools, straight lengths (both included in the American Welding Society AWS A5.9) and coated electrodes (AWS A5.4).

Semifinished Products

Blooms, billets, and slabs are hot rolled, hot forged, or hot pressed to approximate cross-sectional dimensions and generally have rounded corners. Round billets are also produced, typically for extrusion or closed-die forging. These semifinished products, as well as tube rounds, are produced in random lengths or are cut to specified lengths or to specified weights. There are no invariable criteria for distinguishing between the terms bloom and billet, and often they are used interchangeably.

Dimensions. The nominal cross-sectional dimensions of blooms, billets, and slabs are designated in inches and fractions of an inch. The size ranges commonly listed as hot-rolled stainless steel blooms, billets, and slabs include square sections 100 × 100 mm (4 × 4 in.) and larger, and rectangular sections at least 10 300 mm^2 (16 in.^{2}) in cross-sectional area.

Stainless Types. Blooms, billets, and slabs made of 4xx stainless steels that are highly hardenable (types 414, 420, 420F, 422, 431, 440A, 440B, and 440C) are annealed before shipment to prevent cracking. Other hardenable types, such as 403, 410, 416, and 416Se, also may be furnished in the annealed condition, depending on composition and size.

Processing. In general practice, blooms, billets, and slabs are cut to length by hot shearing. Hot sawing and flame cutting are also used. When the end distortion or burrs normally encountered in regular mill cutting are not acceptable, ends can be prepared for subsequent operations by any method that does not leave distortion or burrs. Usually, this is grinding. Blooms, billets, tube rounds, and slabs are surface conditioned by grinding or turning prior to being processed by hot rolling, hot forging, hot extruding, or hot piercing. Material can be tested by ultrasonic and macroetching techniques in the as-worked condition; however, a more critical evaluation is possible after the material has been conditioned. At the time an order is placed, producer and customer should come to an agreement regarding the manner in which testing or inspection is to be conducted and results interpreted.

Pipe, Tubes, and Tubing

Pipe, tubes, and tubing are hollow products made either by piercing rounds or by rolling and welding strip. They are used for conveying gases, liquids, and solids, and for various mechanical and structural purposes. (Cylindrical forms intended for use as containers for storage and shipping purposes and products cast to tubular shape are not included in this category.) The number of terms used in describing sizes and other characteristics of stainless steel tubular products has grown with the industry, and in some cases terms may be difficult to define or to distinguish from one another. For example, the terms pipe, tubes, and tubing are distinguished from one another only by general use, not by clear-cut rules. Pipe is distinguished from tubes chiefly by the fact that it is commonly produced in relatively few standard sizes. Tubing is generally made to more exacting specifications than either pipe or tubes, regarding dimensions, finish, chemical composition, and mechanical properties.

Stainless steel tubular products are classified according to intended service, as described in the following paragraphs and tabular matter.

Stainless Steel Tubing for General Corrosion-Resisting Service. Straight chromium (ferritic or martensitic) types are produced in the annealed or heat-treated condition, and chromium-nickel (austenitic) types are produced in the annealed or cold-worked condition. Austenitic types are inherently tougher and more ductile than ferritic types for similar material conditions or tempers.

ASTM specifications A 268 and A 269 apply to stainless steel tubing for general service: A 268 applies to ferritic grades, and A 269, to austenitic grades. Most ferritic grades are also covered by ASME SA268, which sets forth the same material requirements as does ASTM A 268.

Stainless steel pressure pipe is made from straight chromium and chromium-nickel types and is governed by the specifications:

Specifications		
ASTM	ASME	Description
A 312	SA312	Seamless and welded pipe
A 358	SA358	Electric fusion welded pipe for high-temperature service
A 376	SA376	Seamless pipe for high-temperature central-station service
A 409	··· ········	Large-diameter welded pipe for corrosion or high-temperature service
A 790	SA790	Seamless and welded ferritic/austenitic stainless steel pipe

Stainless steel pressure tubes include boiler, superheater, condenser, and heat-exchanger tubes, which commonly are manufactured from chromium-nickel types;

requirements are set forth in the specifications:

| Specifications | | Description |
ASTM	ASME	
A 213	SA213	Ferritic and austenitic alloy seamless tubes for boilers, superheaters, and heat exchangers
A 249	SA249	Austenitic alloy welded tubes for boilers, superheaters, heat exchangers, and condensers
A 271	SA271	Austenitic alloy seamless still tubes for refinery service
A 498	...	Ferritic and austenitic alloy seamless and welded tubes with integral fins
A 688	SA688	Welded austenitic stainless steel feedwater heater tubes
A 789	SA789	Seamless and welded ferritic/austenitic stainless steel tubing

Stainless steel sanitary tubing is used extensively in the dairy and food industries, where cleanliness and exceptional corrosion resistance are important surface characteristics. In many instances, even the slight amounts of corrosion that result in tarnishing or in release of a few ppm of metallic ions into the process stream are objectionable. Sanitary tubing may be polished on the outside or the inside, or both, to provide smooth, easily cleanable surfaces. Special finishes and close dimensional tolerances for special fittings are sometimes required. ASTM A 270 is in common use for this tubing.

Stainless steel mechanical tubing is produced in round, square, rectangular, and special-shape cross sections. It is used for many different applications, most of which do not require the tubing to be pressurized. Mechanical tubing is used for bushings; small cylinders; bearing parts; fittings; various types of hollow, cylindrical or ringlike formed parts; and structural members such as furniture frames, machinery frames, and architectural members. ASTM A 511 and A 554 apply to seamless and welded mechanical tubing, respectively.

Stainless steel aircraft tubing, produced from various chromium-nickel types, has many structural and hydraulic applications in aircraft construction because of its high resistance to both heat and corrosion. Work-hardened tubing can be used in high-strength applications, but it is not recommended for parts that may be exposed to certain corrosive substances or to certain combinations of corrosive static or fluctuating stress. Low-carbon types or compositions stabilized by titanium or by niobium with or without tantalum are commonly used when welding is to be done without subsequent heat treatment.

Aircraft tubing is made to close tolerances and with special surface finishes, special mechanical properties, and stringent requirements for testing and inspection. It is used for structural components of aircraft fuselages, engine mounts, engine oil lines, landing gear components, and engine parts and is finding increasing application in parts for hydraulic, fuel-injection, exhaust, and heating systems.

Aircraft structural tubing is both seamless and welded stainless steel tubing in sizes larger than those referred to as aircraft tubing. It is commonly used in exhaust systems (including stacks), cross headers, collector rings, engine parts, heaters, and pressurizers. Sometimes, stainless steel aircraft structural tubing is produced especially for parts that are to be machined. Stabilized types are used for welded and brazed structures.

Seamless and welded stainless steel aircraft structural tubing is made in sizes ranging from 1.6 to 125 mm ($\frac{1}{16}$ to 5 in.) in outside diameter and from 0.25 to 6.35 mm (0.010 to 0.250 in.) in wall thickness. It is ordinarily produced to the federal and Aerospace Material Specification (AMS) specifications listed below. However, because the U.S. government has embarked on a program of replacing military (MIL) specifications with AMS and ASTM specifications, the MIL specifications listed may no longer apply.

Specification	UNS number, composition, and condition
Seamless tubing	
AMS 5560	S30400; 19Cr-9Ni; annealed
AMS 5561	S21900; 21Cr-6Ni-9Mn; annealed
AMS 5570	S32100; 18Cr-11Ni (Ti stabilized); annealed
AMS 5571	S34700; 18Cr-11Ni (Nb + Ta stabilized); annealed
AMS 5572	S31008; 25Cr-20Ni; annealed
AMS 5573	S31600; 17Cr-12.5Ni-2.5Mo; annealed
AMS 5574	S30908; 23Cr-13.5Ni; annealed
AMS 5578	S45500; 12.5Cr-8.5Ni-0.03 (Nb + Ta)-1.1Ti-2.0Cu; annealed
Welded tubing	
MIL-T-6737	18-8 (stabilized); annealed
AMS 5565	S30400; 19Cr-9Ni; annealed
AMS 5575	S34700; 18Cr-11Ni (Nb + Ta stabilized); annealed
AMS 5576	S32100; 18Cr-10Ni (Ti stabilized); annealed
AMS 5577	S31008; 25Cr-20Ni; annealed
Seamless and welded tubing	
MIL-T-5695	18-8; hardened (cold worked)
MIL-T-8506	S30400; 18-8; annealed
MIL-T-8686	18-8 (stabilized); annealed

Aircraft Hydraulic-Line Tubing. Stainless steel tubing is used widely in aircraft and aerospace vehicles for fuel-injection lines and hydraulic systems. Most of the tubing used for such applications is relatively small; types 304, 304L, 321, 347, and 21-6-9 are most often specified. Aircraft hydraulic-line tubing must have high strength, high ductility, high fatigue resistance, high corrosion resistance, and good cold-working qualities. The ability to be flared for use with standard flare fittings, the ability to be bent without excessive distortion or fracture, and cleanliness of the inside surface are important requirements.

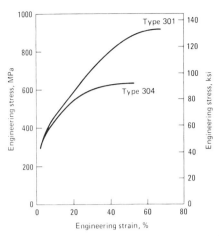

Fig. 4 Typical stress-strain curves for types 301 and 304 stainless steel

Stainless steel aircraft hydraulic-line tubing is produced in either the annealed or the cold-worked ($\frac{1}{8}$ hard) condition. The $\frac{1}{8}$ hard temper is used wherever possible to save weight. Specifications for stainless steel aircraft hydraulic-line tubing, either seamless or welded, are:

Specification	UNS number, or type, and condition
MIL-T-6845	S30400; $\frac{1}{8}$ hard
MIL-T-8504	S30400; annealed
MIL-T-8808	321 or 347; annealed
AMS 5556	S34700; annealed
AMS 5557	S32100; annealed
AMS 5560	S30400; annealed
AMS 5566	S30400; $\frac{1}{8}$ hard

Tensile Properties

Mechanical properties of most stainless steels, especially ductility and toughness, are higher than the same properties of carbon steels. Strength and hardness can be raised by cold work for ferritic and austenitic types, and by heat treatment for precipitation-hardening and martensitic types. Certain ferritic stainless steels can also be hardened slightly by heat treatment.

Austenitic Types. Basic room-temperature properties of standard austenitic stainless steels and of several nonstandard austenitic stainless steels are given in Tables 7 and 8. Additional specifications for austenitic stainless steels include:

Product form	ASTM specification
Wire	A 313, A 492, A 493, A 555, B 471, B 475
Bar and wire	B 649
Wire, bar, shapes	A 479
Billet and bar	A 314, B 472
Flanges, fittings, and/or values, and so on	A 182, A 403, B 462, A 403
Bolting	A 193
Nuts	A 194

Table 7 Minimum room-temperature mechanical properties of austenitic stainless steels

Product form(a)	Condition	Tensile strength MPa	ksi	0.2% yield strength MPa	ksi	Elongation, %	Reduction in area, %	Hardness, HRB	ASTM specification
Type 301 (UNS S30100)									
B	Annealed	620	90	205	30	40	...	95 max	A 666
B, P, Sh, St	Annealed	515	75	205	30	40	...	92 max	A 167
B, P, Sh, St	¼ hard	860	125	515	75	25	...	...	A 666
B, P, Sh, St	½ hard	1030	150	760	110	18	...	...	A 666
B, P, Sh, St	¾ hard	1210	175	930	135	12	...	...	A 666
B, P, Sh, St	Full hard	1280	185	965	140	9	...	...	A 666
Type 302 (UNS S30200)									
B, F	Hot finished and annealed	515	75	205	30	40	50	...	A 276, A 473
B	Cold finished(b) and annealed	620	90	310	45	30	40	...	A 276
B	Cold finished(c) and annealed	515	75	205	30	30	40	...	A 276
W	Annealed	515	75	205	30	35(d)	50(d)	...	A 580
W	Cold finished	620	90	310	45	30(d)	40	...	A 580
P, Sh, St	Annealed	515	75	205	30	40	...	92 max	A 167, A 240, A 666
B, P, Sh, St	High tensile, ¼ hard	860	125	515	75	10	...	...	A 666
B, P, Sh, St	High tensile, ½ hard	1030	150	760	110	10	...	...	A 666
B, P, Sh, St	High tensile, ¾ hard	1205	175	930	135	6	...	...	A 666
B, P, Sh, St	Full hard	1275	185	965	140	4	...	...	A 666
Type 302B (UNS S30215)									
B, F	Hot finished and annealed	515	75	205	30	40	50	...	A 276, A 473
B	Cold finished(b) and annealed	620	90	310	45	30	40	...	A 276
B	Cold finished(c) and annealed	515	75	205	30	30	40	...	A 276
W	Annealed	515	75	205	30	35(d)	50(d)	...	A 580
W	Cold finished	620	90	310	45	30(d)	40	...	A 580
P, Sh, St	Annealed	515	75	205	30	40	...	95 max	A 167
Type 302Cu (UNS S30430)									
W(e)	Annealed	550	80	...	...	...	...	...	A 493
W(e)	Lightly drafted	585	85	...	...	...	...	...	A 493
Types 303 (UNS S30300) and 303Se (UNS S30323)									
F	Annealed	515	75	205	30	40	50	...	A 473
W	Annealed	585–860	85–125	...	...	...	...	...	A 581
W	Cold worked	790–1000	115–145	...	...	...	...	...	A 581
Type 304 (UNS S30400)									
B, F(f)	Hot finished and annealed	515	75	205	30	40	50	...	A 276, A 473
B	Cold finished(b) and annealed	620	90	310	45	30	40	...	A 276
B	Cold finished(c) and annealed	515	75	205	30	30	40	...	A 276
W	Annealed	515	75	205	30	35(d)	50(d)	...	A 580
W	Cold finished	620	90	310	45	30(d)	40	...	A 580
P, Sh, St	Annealed	515	75	205	30	40	...	92 max	A 167
B, P, Sh, St	⅛ hard	690	100	380	55	35	...	...	A 666
B, P, Sh, St	¼ hard	860	125	515	75	10	...	...	A 666
B, P, Sh, St	½ hard	1035	150	760	110	7	...	...	A 666
Type 304L (UNS S30403)									
F	Annealed	450	65	170	25	40	50	...	A 473
B	Hot finished and annealed	480	70	170	25	40	50	...	A 276
B	Cold finished(b) and annealed	620	90	310	45	30	40	...	A 276
B	Cold finished(c) and annealed	480	70	170	25	30	40	...	A 276
W	Annealed	480	70	170	25	35(d)	50(d)	...	A 580
W	Cold finished	620	90	310	45	30(d)	40	...	A 580
P, Sh, St	Annealed	480	70	170	25	40	...	88 max	A 167, A 240
Type 304B4 (UNS S30424)									
P, Sh, St grade A	Annealed	515	75	205	30	27	...	95 max	A 887
P, Sh, St grade B	Annealed	515	75	205	30	16	...	95 max	A 887
Type 305 (UNS S30500)									
B, F	Hot finished and annealed	515	75	205	30	40	50	...	A 276, A 473
B	Cold finished(b) and annealed	260	90	310	45	30	40	...	A 276
B	Cold finished(c) and annealed	515	75	205	30	30	40	...	A 276
W	Annealed	515	75	205	30	35(d)	50(d)	...	A 580
W	Cold finished	620	90	310	45	30(d)	40	...	A 580
P, Sh, St	Annealed	480	70	170	25	40	...	88 max	A 167
B, W	High tensile(d)	1690	245	...	...	...	...	...	...
Cronifer 18-15 LCSi (UNS S30600)									
P, Sh, St	Annealed	540	78	240	35	40	...	...	A 167, A 240

(continued)

(a) B, bar; F, forgings; P, plate; Pi, pipe; Sh, sheet; St, strip; T, tube; W, wire. (b) Up to 13 mm (0.5 in.) thick. (c) Over 13 mm (0.5 in.) thick. (d) For wire 3.96 mm (5/32 in.) and under, elongation and reduction in area shall be 25 and 40%, respectively. (e) 4 mm (0.156 in.) in diameter and over. (f) For forged sections 127 mm (5 in.) and over, the tensile strength shall be 485 MPa (70 ksi). (g) For information only, not a basis for acceptance or rejection

Table 7 (continued)

Product form(a)	Condition	Tensile strength MPa	ksi	0.2% yield strength MPa	ksi	Elongation, %	Reduction in area, %	Hardness, HRB	ASTM specification
Type 308 (UNS S30800)									
B, F	Hot finished and annealed	515	75	205	30	40	50	...	A 276, A 473
B	Cold finished(b) and annealed	620	90	310	45	30	40	...	A 276
B	Cold finished(c) and annealed	515	75	205	30	30	40	...	A 276
W	Annealed	515	75	205	30	35(d)	50(d)	...	A 580
W	Cold finished	620	90	310	45	30(d)	40	...	A 580
P, Sh, St	Annealed	515	75	205	30	40	...	88 max	A 167
Types 309 (UNS S30900), 309S (UNS S30908), 310 (UNS S31000) and 310S (UNS S31008)									
B, F	Hot finished and annealed	515	75	205	30	40	50	...	A 276, A 473
B	Cold finished(b) and annealed	620	90	310	45	30	40	...	A 276
B	Cold finished(c) and annealed	515	75	205	30	30	40	...	A 276
W	Annealed	515	75	205	30	35(d)	50(d)	...	A 580
W	Cold finished	620	90	310	45	30(d)	40	...	A 580
P, Sh, St	Annealed	515	75	205	30	40	...	95 max	A 167
310Cb (UNS S31040)									
P, Sh, St	Annealed	515	75	205	30	40	...	95	A 167, A 240
B, Shapes	Hot finished and annealed	515	75	205	30	40	50	...	A 276
B, Shapes	Cold finished(b) and annealed	620	90	310	45	30	40	...	A 276
B, Shapes	Cold finished(c) and annealed	515	75	205	30	30	40	...	A 276
W	Annealed	515	75	205	30	35(d)	50(d)	...	A 580
W	Cold finished	620	90	310	45	30(d)	40	...	A 580
Type 314 (UNS S31400)									
B, F	Hot finished and annealed	515	75	205	30	40	50	...	A 276, A 473
B	Cold finished(b) and annealed	620	90	310	45	30	40	...	A 276
B	Cold finished(c) and annealed	515	75	205	30	30	40	...	A 276
W	Annealed	515	75	205	30	35(d)	50(d)	...	A 580
W	Cold finished	620	90	310	45	30(d)	40	...	A 580
Type 316 (UNS S31600)									
B, F(f)	Hot finished and annealed	515	75	205	30	40	50	...	A 276, A 473
B	Cold finished(b) and annealed	620	90	310	45	30	40	...	A 276
B	Cold finished(c) and annealed	515	75	205	30	30	40	...	A 276
W	Annealed	515	75	205	30	35(d)	50(d)	...	...
W	Cold finished	620	90	310	45	40(d)	40	...	A 580
P, Sh, St	Annealed	515	75	205	30	40	...	95 max	A 167, A 240
Type 316L (UNS S31603)									
F	Annealed	450	65	170	25	40	50	...	A 473
B	Hot finished and annealed	480	70	170	25	40	50	...	A 276
B	Cold finished(b) and annealed	620	90	310	45	30	40	...	A 276
B	Cold finished(c) and annealed	480	70	170	25	30	40	...	A 276
W	Annealed	480	70	170	25	35(d)	50(d)	...	A 580
W	Cold finished	620	90	310	45	30(d)	40	...	A 580
P, Sh, St	Annealed	485	70	170	25	40	...	95 max	A 167, A 240
Type 316Cb (UNS S31640)									
P, Sh, St	Annealed	515	75	205	30	30	...	95	A 167, A 240
B, Shapes	Hot finished and annealed	515	75	205	30	40	50	...	A 276
B, Shapes	Cold finished(b) and annealed	620	90	310	45	30	40	...	A 276
B, Shapes	Cold finished(c) and annealed	515	75	205	30	30	40	...	A 276
W	Annealed	515	75	205	30	35(d)	50(d)	...	A 580
W	Cold finished	620	90	310	45	30(d)	40	...	A 580
Type 317 (UNS S31700)									
B, F	Hot finished and annealed	515	75	205	30	40	50	...	A 276, A 473
B	Cold finished(b) and annealed	620	90	310	45	30	40	...	A 276
B	Cold finished(c) and annealed	515	75	205	30	30	40	...	A 276
W	Annealed	515	75	205	30	35(d)	50(d)	...	A 580
W	Cold finished	620	90	310	45	30(d)	40	...	A 580
P, Sh, St	Annealed	515	75	205	30	35	...	95 max	A 167, A 240
Type 317L (UNS S31703)									
B	Annealed	585(g)	85(g)	240(g)	35(g)	55(g)	65(g)	85 max(g)	...
P, Sh, St	Annealed	515	75	205	30	40	...	95 max	A 167
Type 317LM (UNS S31725)									
B, P	Annealed	515	75	205	30	40	...	...	A 276
P, Sh, St	Annealed	515	75	205	30	40	...	96 max	A 167
Types 321 (UNS S32100) and 321H (UNS 32109)									
B, F	Hot finished and annealed	515	75	205	30	40	50	...	A 276, A 473
B	Cold finished(b) and annealed	620	90	310	45	30	40	...	A 276
B	Cold finished(c) and annealed	515	75	205	30	30	40	...	A 276

(continued)

(a) B, bar; F, forgings; P, plate; Pi, pipe; Sh, sheet; St, strip; T, tube; W, wire. (b) Up to 13 mm (0.5 in.) thick. (c) Over 13 mm (0.5 in.) thick. (d) For wire 3.96 mm (5/32 in.) and under, elongation and reduction in area shall be 25 and 40%, respectively. (e) 4 mm (0.156 in.) diameter and over. (f) For forged sections 127 mm (5 in.) and over, the tensile strength shall be 485 MPa (70 ksi). (g) For information only, not a basis for acceptance or rejection

Table 7 (continued)

Product form(a)	Condition	Tensile strength MPa	ksi	0.2% yield strength MPa	ksi	Elongation, %	Reduction in area, %	Hardness, HRB	ASTM specification
Types 321 (UNS S32100) and 321H (UNS 32109) (continued)									
W	Annealed	515	75	205	30	35(d)	50(d)	...	A 580
W	Cold finished	620	90	310	45	30(d)	40	...	A 580
P, Sh, St	Annealed	515	75	205	30	40	...	95 max	A 167, A 240
Types 347 (UNS S34700) and 348 (UNS S34800)									
B, F	Hot finished and annealed	515	75	205	30	40	50	...	A 276, A 473
B	Cold finished(b) and annealed	620	90	310	45	30	40	...	A 276
B	Cold finished(c) and annealed	515	75	205	30	30	40	...	A 276
W	Annealed	515	75	205	30	35(d)	50(d)	...	A 580
W	Cold finished	620	90	310	45	30(d)	40	...	A 580
P, Sh, St	Annealed	515	75	205	30	40	...	92 max	A 167, A 240
18 18 2 (UNS S38100)									
P, Sh, St	Annealed	515	75	205	30	40	...	95 max	A 167, A 240
Type 384 (UNS S38400)									
W(e)	Annealed	550	80	...	...	...	...	...	A 493
W(e)	Lightly drafted	585	85	...	...	...	...	...	A 493
20Cb-3 (UNS N08020), 20Mo-4 (UNS N08024), and 20Mo-6 (UNS N08026)									
B, W	Annealed	550	80	240	35	30	50	...	B 473
Shapes	Annealed	550	80	240	35	15	50	...	B 473
B, W	Annealed and strain hardened	620	90	415	60	15	40	...	B 473
W	Annealed and cold finished	620–830	90–120	...	...	...	...	...	B 473
P, Sh, St	Annealed	550	80	240	35	30	...	95 max	B 463
Pi, T	Annealed	550	80	240	35	30	...	...	B 464, B 468, B 474, B 729
Sanicro 28 (UNS N08028)									
P, Sh, St	Annealed	500	73	215	31	40	...	70–90(g)	B 709
Seamless Tube	Annealed	500	73	215	31	40	...	...	B 668
Type 330 (UNS N08330)									
B	Annealed	485	70	210	30	30	...	...	B 511
P, Sh, St	Annealed	485	70	210	30	30	...	70–90(g)	B 536
Pi	Annealed	485	70	210	30	30	...	70–90(g)	B 535, B 546
AL-6X (UNS N08366)									
B, W	Annealed	515	75	210	30	30	...	...	B 691
P, Sh, St	Annealed	515	75	240	35	30	...	95 max	B 688
Pi, T	Annealed	515	75	210	30	30	...	...	B 675, B 676, B 690
Welded T	Cold worked	515	75	210	30	10	...	...	B 676
JS-700 (UNS N08700)									
B, W	Annealed	550	80	240	35	30	50	...	B 672
P, Sh, St	Annealed	550	80	240	35	30	...	75–90(g)	B 599
Type 332 (UNS N08800)									
Pi, T	Annealed	515	75	210	30	30	...	...	B 163, B 407, B 514, B 515
Seamless Pi, T	Hot finished	450	65	170	25	30	...	...	B 407
B	Hot worked	550	80	240	35	25	...	...	B 408
B	Annealed	515	75	210	30	30	...	...	B 408
P	Hot rolled	550	80	240	35	25	...	...	B 409
P, Sh, St	Annealed	515	75	210	30	30	...	...	B 409
Type 904L (UNS N08904)									
B	Annealed	490	71	220	31	35	...	...	B 649
W	Cold finished	620–830	90–120	...	...	...	...	...	B 649
Pi, T	Annealed	490	71	220	31	35	...	...	B 673, B 674, B 677
P, Sh, St	Annealed	490	71	220	31	35	...	70–90(g)	B 625

(a) B, bar; F, forgings; P, plate; Pi, pipe; Sh, sheet; St, strip; T, tube; W, wire. (b) Up to 13 mm (0.5 in.) thick. (c) Over 13 mm (0.5 in.) thick. (d) For wire 3.96 mm (5/32 in.) and under, elongation and reduction in area shall be 25 and 40%, respectively. (e) 4 mm (0.156 in.) in diameter and over. (f) For forged sections 127 mm (5 in.) and over, the tensile strength shall be 485 MPa (70 ksi). (g) For information only, not a basis for acceptance or rejection

Certain austenitic stainless steels—the so-called metastable types—can develop higher strengths and hardnesses than other stable types for a given amount of cold work. In metastable austenitic stainless steels, deformation triggers the transformation of austenite to martensite. The effect of this transformation on strength is shown in Fig. 4, which compares the stress-strain curve for stable type 304 with that for metastable type 301. The parabolic shape of the curve for type 304 indicates that strain hardening occurs throughout the duration of the application of stress, but that the amount of strain hardening for a given increment of stress decreases as stress increases.

On the other hand, type 301 continues to strain harden well into the plastic range. The extended strain hardening is the result of the deformation-induced transformation of austenite to martensite.

Ferritic types of stainless steel are defined as those that contain at least 10.5% Cr and that have microstructures of ferrite plus carbides. These steels are lower in toughness than the austenitic types. Basic room-temperature mechanical properties of ferritic stainless steels are given in Table 9. Strength is enhanced only moderately by cold working. Additional specifications for ferritic stainless steels include:

Product form	ASTM specification
Bar and wire	A 493
Billet and bar	A 314
Flanges, fittings and valves, and so on	A 182

Duplex (Austenite/Ferrite) Types. Most wrought duplex stainless steels contain

Table 8 Minimum mechanical properties of high-nitrogen austenitic stainless steels

Product form(a)	Condition	Tensile strength MPa	ksi	0.2% yield strength MPa	ksi	Elongation, %	Reduction in area, %	Hardness, HRB	ASTM specification
Type 201 (UNS S20100)									
B	Annealed	515	75	275	40	40	45	...	A 276
P, Sh, St	Annealed	655	95	310	45	40	...	100 max	A 276, A 666
Sh, St	¼ hard	860	125	515	75	25	...	...	A 666
Sh, St	½ hard	1030	150	760	110	18	...	...	A 666
Sh, St	¾ hard	1210	175	930	135	12	...	...	A 666
Sh, St	Full hard	1280	185	965	140	9	...	...	A 666
Type 202 (UNS S20200)									
B	Annealed	515	75	275	40	40	45	...	A 276
P, Sh, St	Annealed	620	90	260	38	40	...	...	A 666
Sh, St	¼ hard	860	125	515	75	12	...	...	A 660
Type 205 (UNS S20500)									
B, P, Sh, St	Annealed	790	115	450	65	40	...	100 max	A 666
Nitronic 50 (UNS S20910)									
B	Annealed	690	100	380	55	35	55	...	A 276
W	Annealed	690	100	380	55	35	55	...	A 580
Sh, St	Annealed	725	105	415	60	30	...	100 max	A 240
P	Annealed	690	100	380	55	35	...	100 max	A 240
Cryogenic Tenelon (UNS S21460)									
B, P, Sh, St	Annealed	725	105	380	55	40	...	...	A 666
Types 216 (UNS S21600) and 216L (UNS S21603)									
Sh, St	Annealed	690	100	415	60	40	...	100 max	A 240
P	Annealed	620	90	345	50	40	...	100 max	A 240
Nitronic 40 (UNS S21900)									
B, W	Annealed	620	90	345	50	45	60	...	A 276, A 580
21-6-9 LC (XM-11) (UNS S21904)									
B, W, shapes	Annealed	620	90	345	50	45	60	...	A 276, A 580
Sh, St	Annealed	690	100	415	60	40	...	...	A 666
P	Annealed	620	90	345	50	45	...	...	A 666
Nitronic 33 (UNS S24000)									
B, W	Annealed	690	100	380	55	30	50	...	A 276, A 580
Sh, St	Annealed	690	100	415	60	40	...	100 max	A 240
P	Annealed	690	100	380	55	40	...	100 max	A 240
Nitronic 32 (UNS S24100)									
B, W	Annealed	690	100	380	55	30	50	...	A 276, A 580
Type 304N (UNS S30451)									
B	Annealed	550	80	240	35	30	...	...	A 276
P, Sh, St	Annealed	550	80	240	35	30	...	92 max	A 240
Type 340HN (UNS S30452)									
B	Annealed	620	90	345	50	30	50	...	A 276
Sh, St	Annealed	620	90	345	50	30	...	100 max	A 240
P	Annealed	585	85	275	40	30	...	100 max	A 240
Type 304LN (UNS 30453)									
B	Annealed	515	75	205	30	...	...	...	A 276
P, Sh, St	Annealed	515	75	205	30	40	...	92 max	A 167, A 240
253 MA (UNS S30815)									
P, Sh, St	Annealed	600	87	310	45	40	...	95 max	A 167, A 240
B, shapes	Annealed	600	87	310	45	40	50	...	A 276
254 SMO (UNS S31254)									
P, Sh, St	Annealed	650	94	300	44	35	...	96	A 167, A 240
B, shapes	Annealed	650	95	300	44	35	50	...	A 276
Type 316N (UNS S31651)									
B	Annealed	550	80	240	35	30	...	...	A 276
P, Sh, St	Annealed	550	80	240	35	35	...	95 max	A 240
17-14-4 LN (UNS S31726)									
P, Sh, St	Annealed	550	80	240	35	40	...	96	A 167, A 240
B, shapes	Annealed	550	80	240	35	40	...	...	A 276
317LN (UNS S31753)									
P, Sh, St	Annealed	550	80	240	35	40	...	95	A 167, A 240

(continued)

(a) B, bar; P, plate; Pi, pipe; Sh, sheet; St, strip; T, tube; W, wire

Table 8 (continued)

Product form(a)	Condition	Tensile strength MPa	ksi	0.2% yield strength MPa	ksi	Elongation, %	Reduction in area, %	Hardness, HRB	ASTM specification
AL 6XN (UNS N08367)									
B, W	Annealed	715	104	315	46	30	...	...	B691
P, Sh, St	Annealed	715	104	315	46	30	...	100	B688
Flanges, fittings, valves, and so on	Annealed	715	104	315	46	30	50	...	B462
Seamless Pi, T	Annealed	715	104	315	46	30	...	...	B690
Welded Pi	Annealed	715	104	315	46	30	...	...	B676
Welded T	Solution treated and annealed	715	104	315	46	30	...	...	B676
Welded T	Cold worked	...	...	...	...	10	...	...	B676
Cronifer 1925 hMO (UNS N08925)									
B, W	Annealed	600	87	300	43	40	...	...	B649
Seamless Pi, T	Annealed	600	87	300	43	40	...	...	B677
Welded Pi	Annealed	600	87	300	43	40	...	...	B673
Welded T	Annealed	600	87	300	43	40	...	...	B674

(a) B, bar; P, plate; Pi, pipe; Sh, sheet; St, strip; T, tube; W, wire

about 50% austenite-50% ferrite because of the balancing of elements that stabilize austenite (carbon, nitrogen, nickel, copper, and manganese) and ferrite (chromium, molybdenum, and silicon). Low carbon is maintained in most grades to minimize intergranular carbide precipitation. The austenite-ferrite balance provides wrought material with the optimum levels of mechanical properties and corrosion resistance. Because typically less austenite is present as-cast, welding consumables with enriched nickel are generally used to maintain austenite in the weld metal at levels generally similar to those in the base material.

Yield strengths approximately twice that of type 316 can be obtained with annealed duplex stainless steels. Basic room-temperature mechanical properties of duplex stainless steels are given in Table 10. Strength levels can be enhanced by cold working. Lower transverse ductility and impact strength can be expected because of the directional nature of the wrought microstructure (typically elongated austenite islands in a ferrite matrix).

Martensitic types are iron-chromium steels with or without small additions of other alloying elements. They are ferritic in the annealed condition, but are martensitic after rapid cooling in air or a liquid medium from above the critical temperature. Steels in this group usually contain no more than 14% Cr—except types 440A, 440B, and 440C, which contain 16 to 18% Cr—and an amount of carbon sufficent to permit hardening. If other elements are present, the total concentration is usually no more than 2 to 3%. Martensitic stainless steels may be hardened and tempered in the same manner as alloy steels. They have excellent strength and are magnetic. Basic room-temperature properties of the martensitic types are given in Table 11 and Fig. 5. Additional specifications for martensitic stainless steels are:

Product form	ASTM specification
Bar	A 582
Bolting	A 193
Nuts	A 194

Martensitic stainless steels harden when cooled off the mill after hot processing; therefore, they are often given a process anneal at 650 to 760 °C (1200 to 1400 °F) for about 4 h. Process annealing differs from full annealing, which is done by heating at 815 to 870 °C (1500 to 1600 °F), cooling in the furnace at a rate of 40 to 55 °C/h (75 to 100 °F/h) to about 540 °C (1000 °F), and then cooling in air to room temperature. Occasionally, martensitic types are purchased in the tempered condition; this condition is achieved by cooling directly off the mill to harden the steel and then reheating to a tempering temperature of 540 to 650 °C (1000 to 1200 °F), or by reheating the steel to a hardening temperature of 1010 to 1065 °C (1850 to 1950 °F), cooling it, and then tempering it. The influence of tempering temperature on the properties of hardened martensitic stainless steels is shown in Fig. 6. In heat treating martensitic stainless steels, temperatures up to about 480 °C (900

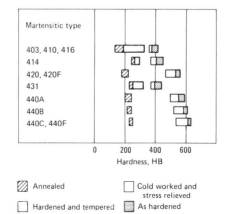

Fig. 5 Typical hardnesses of selected martensitic stainless steels

°F) are referred to as stress-relieved temperatures because little change in tensile properties occurs upon heating hardened material to these temperatures. Temperatures of 540 to 650 °C (1000 to 1200 °F) are referred to as tempering temperatures, and temperatures of 650 to 760 °C (1200 to 1400 °F) are called annealing temperatures.

Precipitation-hardening types are generally heat treated to final properties by the fabricator. Table 12 summarizes the minimum properties that can be expected in both material as received from the mill and material that has been properly heat treated.

The precipitation-hardening stainless steels are of two general classes: single-treatment alloys and double-treatment alloys. Single-treatment alloys, such as Custom 450, 17-4 PH, and 15-5 PH, are solution annealed at about 1040 °C (1900 °F) to dissolve the hardening agent. Upon cooling to room temperature, the structure transforms to martensite that is supersaturated with respect to the hardening agent. A single tempering treatment at about 480 to 620 °C (900 to 1150 °F) is all that is required to precipitate a secondary phase to strengthen the alloy. As listed in Table 12, different tempering temperatures within this range produce different properties.

Double-treatment alloys such as 17-7 PH are solution treated at about 1040 °C (1900 °F) and then water quenched to retain the hardening agent in solution in an austenitic structure. The austenite is conditioned by heating to 760 °C (1400 °F) to precipitate carbides and thereby unbalance the austenite so that it transforms to martensite upon cooling to a temperature below 15 °C (60 °F); this treatment produces condition T. Alternatively, the austenite may be conditioned at a higher temperature, 925 °C (1700 °F), at which fewer carbides precipitate, and then may be transformed to martensite by cooling to room temperature, followed by refrigerating to −75 °C (−100 °F); this treatment produces condition R. Transfor-

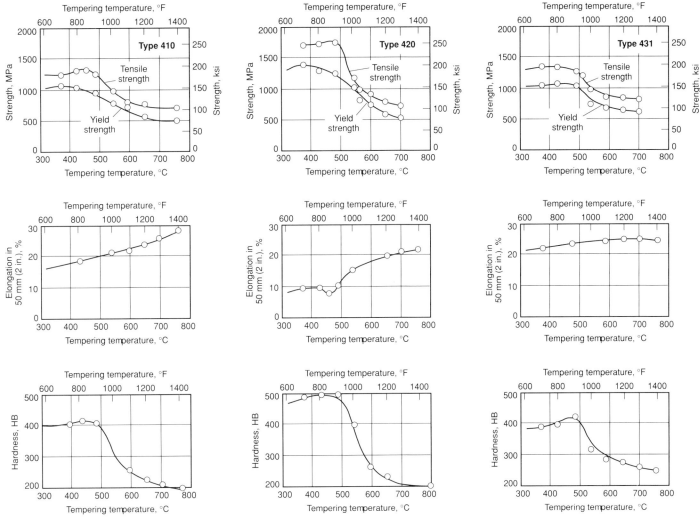

Fig. 6 Variation of tensile properties and hardness with tempering temperature for three martensitic stainless steels

mation can also be effected by severe cold work (about 60 to 70% reduction); such treatment produces condition C. Once the structure has been transformed to martensite by one of these three processes, tempering at 480 to 620 °C (900 to 1150 °F) induces precipitation of a secondary metallic phase, which strengthens the alloy. Properties developed by typical TH, RH, and CH treatments are given in Table 12 for 17-7 PH.

Notch Toughness and Transition Temperature

Notched-bar impact testing of stainless steels is likely to show a wide scatter in test results, regardless of type or test conditions. Because of this wide scatter, only general behavior of the different classes can be described.

Austenitic types have good notched-bar impact resistance. Charpy impact energies of 135 J (100 ft · lb) or greater are typical of all types at room temperature. Cryogenic temperatures have little or no effect on notch toughness; ordinarily, austenitic stainless steels maintain values exceeding 135 J even at very low temperatures. On the other hand, cold work lowers the resistance to impact at all temperatures.

Martensitic and ferritic stainless steels exhibit a decreasing resistance to impact with decreasing temperature, and the fracture appearance changes from a ductile mode at mildly elevated temperatures to a brittle mode at low temperatures. This fracture transition is characteristic of martensitic and ferritic materials. Both the upper-shelf energy and the lower-shelf energy are not greatly influenced by heat treatment in these stainless steels. However, the temperature range over which transition occurs is affected by heat treatment, minor variations in composition, and cold work. Heat treatments that result in high hardness move the transition range to higher temperatures, and those that result in low hardness move the transition range to lower temperatures. As indicated in Fig. 7, transition generally occurs in the range of −75 to 95 °C (−100 to 200 °F), which is the temperature range in which martensitic stainless steels are ordinarily used. Consequently,

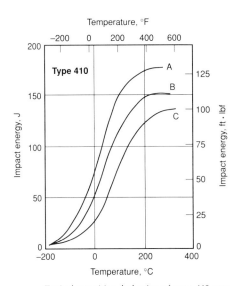

Fig. 7 Typical transition behavior of type 410 martensitic stainless steel. All data from Charpy V-notch tests: A represents material tempered at 790 °C (1450 °F); final hardness, 95 HRB. B represents material tempered at 665 °C (1225 °F); final hardness, 24 HRC. C represents material tempered at 595 °C (1100 °F); final hardness, 30 HRC.

Table 9 Minimum mechanical properties of ferritic stainless steels

Product form(a)	Condition	Tensile strength MPa	ksi	0.2% yield strength MPa	ksi	Elongation, %	Reduction in area, %	Hardness, HRB	ASTM specification
Type 405 (UNS S40500)									
B	Annealed	415	60	170	25	20	45	...	A 479
F	Annealed	415	60	205	30	20	45	...	A 473
W	Annealed	480	70	280	40	20	45	...	A 580
P, Sh, St	Annealed	415	60	170	25	20	...	88 max	A 176, A 240
Type 409 (UNS S40900)									
P, Sh, St	Annealed	380	55	205	30	20	...	80 max	A 240
P, Sh, St	Annealed	380	55	205	30	22(c)	...	80 max	A 176
Type 429 (UNS S42900)									
B	Annealed	480	70	275	40	20	45	...	A 276
P, Sh, St	Annealed	450	65	205	30	22(c)	...	88 max	A 176, A 240
Type 430 (UNS S43000)									
B	Annealed	415	60	205	30	20	45	...	A 276
W	Annealed	480	70	275	40	20	45	...	A 580
P, Sh, St	Annealed	450	65	205	30	22(c)	...	88 max	A 176, A 240
Type 430F (UNS S43020)									
F	Annealed	485	70	275	40	20	45	...	A 473
W	Annealed	585–860	85–125	...	...	...	...	...	A 581
Type 439 (UNS S43035)									
B	Annealed	485	70	275	40	20	45	...	A 479
P, Sh, St	Annealed	450	65	205	30	22	...	88 max	A 240
Type 430Ti (UNS S43036)									
B	Annealed	515(b)	75(b)	310(b)	45(b)	30(b)	65(b)	...	...
Type 434 (UNS S43400)									
W	Annealed	545(b)	79(b)	415(b)	60(b)	33(b)	78(b)	90 max(b)	...
Sh	Annealed	530(b)	77(b)	365(b)	53(b)	23(b)	...	83 max(b)	...
Type 436 (UNS S43600)									
Sh, St	Annealed	530(b)	77(b)	365(b)	53(b)	23(b)	...	83 max(b)	...
Type 442 (UNS S44200)									
B	Annealed	550(b)	80(b)	310(b)	45(b)	20(b)	40(b)	90 max(b)	...
P, Sh, St	Annealed	515	75	275	40	20	...	95 max	A 176
Type 444 (UNS S44400)									
P, Sh, St	Annealed	415	60	275	40	20	...	95 max	A 176
Type 446 (UNS S44600)									
B	Annealed, hot finished	480	70	275	40	20	45	...	A 276
B	Annealed, cold finished	480	70	275	40	16	45	...	A 276
W	Annealed	480	70	275	40	20	45	...	A 580
W	Annealed, cold finished	480	70	275	40	16	45	...	A 580
P, Sh, St	Annealed	515	75	275	40	20	...	95 max	A 176
18 SR									
Sh, St	Annealed	620(b)	90(b)	450(b)	65(b)	25(b)	...	90 min(b)	...
E-Brite 26-1 (UNS S44627)									
B	Annealed, hot finished	450	65	275	40	20	45	...	A 276
B	Annealed, cold finished	450	65	275	40	16	45	...	A 276
P, Sh, St	Annealed	450	65	275	40	22(c)	...	90 max	A 176, A 240
MONIT (UNS S44635)									
P, Sh, St	Annealed	620	90	515	75	20	...	...	A 176, A 240
Sea-Cure/SC-1 (UNS S44660)									
P, Sh, St	Annealed	585	85	450	65	18	...	100 max	A 176, A 240
29-4C (UNS S44735)									
P, Sh, St	Annealed	550	80	415	60	18	...	...	A 276, A 240
29-4-2 (UNS S44800)									
P, Sh, St	Annealed	550	80	415	60	20	...	98 max	A 176, A 240
B	Hot finished	480	70	380	55	20	40	...	A 276
B	Cold finished	520	75	415	60	15	30	...	A 276
B		480	70	380	55	20	40	...	A 479

(a) B, bar; F, forgings; W, wire; P, plate; Sh, sheet; St, strip. (b) Typical values. (c) 20% reduction for 1.3 mm (0.050 in.) and under in thickness

it may be necessary to investigate fracture behavior thoroughly before specifying a martensitic or ferritic stainless steel for a particular application.

Fracture toughness data are not available for many of the standard types of stainless steel. Most testing has been concentrated on the high-strength precipitation-hardening stainless steels because these materials have been used in critical applications where fracture toughness testing has been found most useful for evaluating materials. Table 13 lists

Table 10 Minimum mechanical properties of duplex stainless steels
Minimum values unless otherwise indicated

Product form(a)	Condition	Tensile strength MPa	Tensile strength ksi	0.2% yield strength MPa	0.2% yield strength ksi	Elongation, %	Reduction in area, %	Maximum hardness, HRC	ASTM specification
44LN (UNS S31200)									
F	Annealed	690–900	100–130	450	65	25	50	...	A 182
P, Sh, St	Annealed	690	100	450	65	25	...	220 HB	A 240
T	Annealed	690	100	450	65	25	...	280 HB	A 789
Pi	Annealed	690	100	450	65	25	...	280 HB	A 790
DP-3 (UNS S31260)									
P, Sh, St	Annealed	690	100	485	70	20	...	290 HB	A 240
T	Annealed	690	100	450	65	25	...	30.5	A 789
Pi	Annealed	690	100	450	65	25	...	...	A 790
3RE60 (UNS S31500)									
T	Annealed	630	92	440	64	30	...	30.5	A 789
Pi	Annealed	630	92	440	64	30	...	30.5	A 790
2205 (UNS S31803)									
F	Annealed	620	90	450	65	25	45	...	A 182
P, Sh, St	Annealed	620	90	450	65	25	...	32	A 240
B, Shapes	Annealed	620	90	448	65	25	...	290 HB	A 276
T	Annealed	620	90	450	65	25	...	30.5	A 789
Pi	Annealed	620	90	450	65	25	...	30.5	A 790
2304 (UNS S32304)									
T	Annealed	600	87	400	58	25	...	30.5	A 789
Pi	Annealed	600	87	400	58	25	...	30.5	A 790
Ferralium 255 (UNS S32550)									
P, Sh, St	Annealed	760	110	550	80	15	...	32	A 240
B, Shapes	Annealed	760	110	550	80	15	...	297 HB	A 479
T	Annealed	760	110	550	80	15	...	31.5	A 789
Pi	Annealed	760	110	550	80	15	...	31.5	A 790
Type 329 (UNS S32900)									
P, Sh, St	Annealed	620	90	485	70	15	...	28	A 240
T	Annealed	620	90	485	70	20	...	28	A 789
Pi	Annealed	620	90	485	70	20	...	28	A 790
7-Mo PLUS (UNS S32950)									
P, Sh, St	Annealed	690	100	480	70	15	...	31	A 240
B, Shapes	Annealed	690	100	480	70	15	...	297 HB	A 479
T	Annealed	690	100	480	70	20	...	30.5	A 789
Pi	Annealed	690	100	480	70	20	...	30.5	A 790

(a) B, bar; W, wire; P, plate; Sh, sheet; St, strip; T, tubing; Fl, flanges, fittings, valves, and parts for high-temperature service; Pi, pipe

typical fracture toughness for several of the high-strength stainless steels for which this property has been determined.

Fatigue Strength

Three types of fatigue tests are used to develop data on the fatigue behavior of stainless steels:

- Rotating-beam test, the most commonly used, which most closely approximates the kind of loading to which shafts and axles are subjected
- Flexural fatigue test (used to evaluate the behavior of sheet), which most closely simulates the action of leaf springs, which are expected to flex without deforming or breaking
- Axial-load fatigue test, which subjects a fatigue specimen to unidirectional loading that can range from full reversal (tension-compression) to tension-tension loading and can have virtually any conceivable ratio of maximum stress to minimum stress

Fatigue data can be given in the form of stress-number of cycles (S-N) curves (Fig. 8) or constant-life diagrams (Fig. 9). Data from any of the three types of tests can be presented as S-N curves, but only data from flexural fatigue and axial fatigue tests can be presented in the form of a constant-life diagram. In analyzing fatigue data, and particularly in selecting materials on the basis of fatigue life, it is important to understand the influence of the stress ratio on fatigue life. In general, fatigue conditions involving tension-compression loading (stress ratio, R, between 0 and -1) lead to shorter fatigue lives than conditions involving tension-tension loading (stress ratio, R, between 0 and $+1$) at the same value of maximum stress.

Elevated-Temperature Properties

Many stainless steels—particularly the austenitic types 304, 309, 310, 316, 321, and 347 and certain precipitation-hardening types such as PH 15-7 Mo, 15-5 PH, 17-4 PH, 17-7 PH, AM-350, and AM-355—are used extensively for elevated-temperature applications such as chemical processing equipment, high-temperature heat exchangers, and superheater tubes for power boilers. For more details on the elevated-temperature properties of selected types, see the article "Elevated-Temperature Properties of Stainless Steels" in this Volume.

Extended service at elevated temperature can result in the embrittlement of stainless steels or in sensitization, which degrades the ability of the material to withstand corrosion, particularly in acid

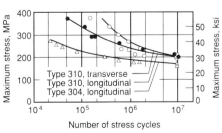

Fig. 8 Typical rotating-beam fatigue behavior of types 304 and 310 stainless steel

Table 11 Minimum mechanical properties of martensitic stainless steels

Product form(a)	Condition	Tensile strength MPa	ksi	0.2% yield strength MPa	ksi	Elongation, %	Reduction in area, %	Rockwell hardness	ASTM specification
Type 403 (UNS S40300)									
B, F	Annealed, hot finished	485	70	275	40	20	45	...	A 276, A 473, A 479
B	Annealed, cold finished	485	70	275	40	16	45	...	A 276
B	Intermediate temper, hot finished	690	100	550	80	15	45	...	A 276
B	Intermediate temper, cold finished	690	100	550	80	12	40	...	A 276
B	Hard temper, hot or cold finished	825	120	620	90	12	40	...	A 276
W	Annealed	485	70	275	40	20	45	...	A 580
W	Annealed, cold finished	485	70	275	40	16	45	...	A 580
W	Intermediate temper, cold finished	690	100	550	80	12	40	...	A 580
W	Hard temper, cold finished	825	120	620	90	12	40	...	A 580
P, Sh, St	Annealed	485	70	205	30	25(b)	...	88 HRB max	A 176
Type 410 (UNS S41000)									
B, F	Annealed, hot finished	485	70	275	40	20	45	...	A 276, A 473, A 479
B	Annealed, cold finished	485	70	275	40	16	45	...	A 276
B	Intermediate temper, hot finished	690	100	550	80	15	45	...	A 276
B	Intermediate temper, cold finished	690	100	550	80	12	40	...	A 276
B	Hard temper, hot or cold finished	825	120	620	90	12	40	...	A 276
W	Annealed	485	70	275	40	20	45	...	A 580
W	Annealed, cold finished	485	70	275	40	16	45	...	A 580
W	Intermediate temper, cold finished	690	100	550	80	12	40	...	A 580
W	Hard temper, cold finished	825	120	620	90	12	40	...	A 580
P, Sh, St	Annealed	450	65	205	30	22(b)	...	95 HRB max	A 176
P, Sh, St	Annealed	450	65	205	30	20	...	95 HRB max	A 240
Type 410S (UNS S41008)									
F	Annealed	450	65	240	35	22	45	...	A 473
P, Sh, St	Annealed	415	60	205	30	22(b)	...	88 HRB max	A 176, A 240
Type 410Cb (UNS S41040)									
B	Annealed, hot finished	485	70	275	40	13	45	...	A 276, A 479
B	Annealed, cold finished	485	70	275	40	12	35	...	A 276, A 479
B	Intermediate temper, hot finished	860	125	690	100	13	45	...	A 276, A 479
B	Intermediate temper, cold finished	860	125	690	100	12	35	...	A 276, A 479
E-4 (UNS S41050)									
P, Sh, St	Annealed	415	60	205	30	22	...	88 HRB max	A 276, A 240
Type 414 (UNS S41400)									
B	Intermediate temper, cold or hot finished	795	115	620	90	15	45	...	A 276, A 479
W	Annealed, cold finished	1030 max	150 max	...	...	...	...	...	A 580
CA6NM (UNS S41500)									
P, Sh, St	Tempered	795	115	620	90	15	...	32 HRC max	A 176, A 240
B, F	Tempered	795	115	620	90	15	45	...	A 276, A 473, A 479
Types 416 (UNS S41600) and 416Se (UNS S41623)									
F	Annealed	485	70	275	40	20	45	...	A 473
W	Annealed	585–860	85–125	...	...	...	...	...	A 581
W	Intermediate temper	795–1000	115–145	...	...	...	...	...	A 581
W	Hard temper	965–1210	140–175	...	...	...	...	...	A 581
Type 416 plus X (UNS S41610)									
W	Annealed	585–860	85–125	...	...	...	...	...	A 581
W	Intermediate temper	795–1000	115–145	...	...	...	...	...	A 581
W	Hard temper	965–1210	140–175	...	...	...	...	...	A 581
Type 418 (UNS S41800)									
B, F	Tempered at 620 °C (1150 °F)	965	140	760	110	15	45	...	A 565
Type 420 (UNS S42000)									
B	Tempered at 204 °C (400 °F)	1720	250	1480(c)	215(c)	8(c)	25(c)	52 HRC(c)	...
W	Annealed, cold finished	860 max	125 max	...	...	...	...	...	A 580
P, Sh, St	Annealed	690	100	...	...	15	...	96 HRB max	A 176
TrimRite (UNS S42010)									
W	Annealed	690 max	100 max	...	...	...	...	...	A 493
W	Lightly drafted	725 max	105 max	...	...	...	...	...	A 493
Type 422 (UNS S42200)									
B, F	Tempered at 675 °C (1250 °F)	825	120	585	85	17(d)	35	...	A 565
B, F	Tempered at 620 °C (1150 °F)	965	140	760	110	13	30	...	A 565
Lapelloy (UNS S42300)									
B, F	Tempered at 620 °C (1150 °F)	965	140	760	110	8	20	...	A 565

(continued)

(a) B, bar; F, forgings; P, plate; Sh, sheet; St, strip; W, wire. (b) 20% elongation for 1.3 mm (0.050 in.) and under in thickness. (c) Typical values. (d) Minimum elongation of 15% for forgings

Table 11 (continued)

Product form(a)	Condition	Tensile strength MPa	ksi	0.2% yield strength MPa	ksi	Elongation, %	Reduction in area, %	Rockwell hardness	ASTM specification
Type 431 (UNS S43100)									
F	Intermediate temper	795	115	620	90	15	...	...	A 473
F	Hard temper	1210	175	930	135	13	...	...	A 473
W	Annealed, cold finished	965 max	140 max	...	...	...	...	...	A 580
W	Annealed	760	110	...	...	...	...	...	A 493
W	Lightly drafted	795	115	...	...	...	...	...	A 493
Type 440A (UNS S44002)									
B	Annealed	725(c)	105(c)	415(c)	60(c)	20(c)	...	95 HRB(c)	...
B	Tempered at 315 °C (600 °F)	1790(c)	260(c)	1650(c)	240(c)	5(c)	20(c)	51 HRC(c)	...
W	Annealed, cold finished	965 max	140 max	...	...	...	...	...	A 580
Type 440B (UNS S44003)									
B	Annealed	740(c)	107(c)	425(c)	62(c)	18(c)	...	96 HRB(c)	...
B	Tempered at 315 °C (600 °F)	1930(c)	280(c)	1860(c)	270(c)	3(c)	15(c)	55 HRC(c)	...
W	Annealed, cold finished	965 max	140 max	...	...	...	...	...	A 580
Type 440C (UNS S44004)									
B	Annealed	760(c)	110(c)	450(c)	65(c)	14(c)	...	97 HRB(c)	...
B	Tempered at 315 °C (600 °F)	1970(c)	285(c)	1900(c)	275(c)	2(c)	10(c)	57 HRC(c)	...
W	Annealed, cold finished	965 max	140 max	...	...	...	...	...	A 580

(a) B, bar; F, forgings; P, plate; Sh, sheet; St, strip; W, wire. (b) 20% elongation for 1.3 mm (0.050 in.) and under in thickness. (c) Typical values. (d) Minimum elongation of 15% for forgings

media. Most often, such degradation is caused by the precipitation of secondary phases such as carbides, α′ phase, or σ phase. Precipitation depends on both time and temperature; longer times at temperature and higher temperatures within the precipitation temperature range promote more extensive precipitation. The problems arising from embrittlement and sensitization and the remedies that can help combat them are discussed in detail in the section "Fabrication Characteristics" in this article.

Subzero-Temperature Properties

Austenitic stainless steels have been used extensively for subzero applications to −270 °C (−450 °F). These steels contain sufficient amounts of nickel and manganese to depress the temperature at which martensite starts to form from austenite upon cooling, M_s, into the subzero range. Thus they retain face-centered cubic (fcc) crystal structures upon cooling from hot working or annealing temperatures. Yield and tensile strengths of austenitic stainless steels increase substantially as testing temperature is decreased, and these steels retain good ductility and toughness at −270 °C (−450 °F). Most austenitic stainless steels may be readily fabricated by welding, but sometimes the welding heat causes sensitization that reduces corrosion resistance in the weld area. The strength of austenitic steels can be increased by cold rolling or cold drawing. Cold working at −195 °C (−320 °F) is more effective in increasing strength than cold working at room temperature. For metallurgically unstable stainless steels such as 301, 304, and 304L, plastic deformation at subzero temperatures causes partial transformation to martensite, which increases strength. For some cryogenic applications, it is desirable to use a stable stainless steel such as type 310.

Compositions of austenitic stainless steels of interest are presented in Table 3. Small amounts of nitrogen increase the strengths of these steels. Manganese additions are used in some steels to replace some of the nickel. Type 416 is a martensitic chromium stainless steel that is usually used in the quenched and tempered condition. It is included in this series because there are applications in rotating pumps and other machinery in which a magnetic material is needed to activate counters.

Types 301 and 310 have been used in the form of extrahard cold-rolled sheet to provide high strength in such applications as the liquid oxygen and liquid hydrogen tanks for Atlas and Centaur rockets. Joining was done by butt fusion welding, and reinforcing strips were spot welded to the tank along the weld joint. In another method for producing high-strength cylindrical tanks, welded preform tanks are fabricated from annealed type 301 stainless steel, submerged in liquid nitrogen while in a cylindrical die, and expanded (cryoformed) by pressurizing until the preform fits the die. The amount of strengthening depends on the amount of plastic deformation incurred in expanding the preform to the size of the die. Strengthening results from the dual effects of the cold working of the austenite and the partial transformation of the austenite to martensite.

Type 304 stainless steel is usually used in the annealed condition for tubing, pipes, and valves employed in the transfer of cryogens; for Dewar flasks and storage tanks; and for structural components that do not require high strength.

Types 310 and 310S are considered metallurgically stable for all conditions of cryogenic exposure. Therefore, these steels are

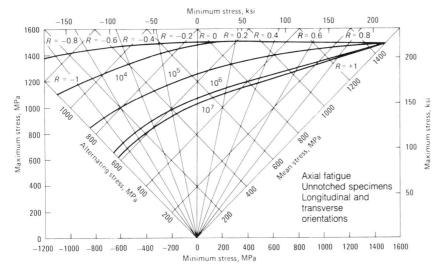

Fig. 9 Constant-life fatigue diagram for PH 13-8 Mo stainless steel, condition H1000

864 / Specialty Steels and Heat-Resistant Alloys

Table 12 Minimum mechanical properties of precipitation-hardening stainless steels

Product form(a)	Condition	Tensile strength MPa	Tensile strength ksi	Yield strength MPa	Yield strength ksi	Elongation, %	Reduction in area, %	Hardness, HRC min	Hardness, HRC max	ASTM specification
PH 13-8 Mo (UNS S13800)										
B, F	H950	1520	220	1410	205	10	45;35(b)	45	...	A 564, A 705
B, F	H1000	1410	205	1310	190	10	50;40(b)	43	...	A 564, A 705
B, F	H1025	1275	185	1210	175	11	50;45(b)	41	...	A 564, A 705
B, F	H1050	1210	175	1140	165	12	50;45(b)	40	...	A 564, A 705
B, F	H1100	1030	150	930	135	14	50	34	...	A 564, A 705
B, F	H1150	930	135	620	90	14	50	30	...	A 564, A 705
B, F	H1150M	860	125	585	85	16	55	26	...	A 564, A 705
P, Sh, St	H950	1520	220	1410	205	6–10(c)	...	45	...	A 693
P, Sh, St	H1000	1380	200	1310	190	6–10(c)	...	43	...	A 693
15-5 PH (UNS S15500)										
B, F	H900	1310	190	1170	170	10;6(b)	35;15(b)	40	...	A 564, A 705
B, F	H925	1170	170	1070	155	10;7(b)	38;20(b)	38	...	A 564, A 705
B, F	H1025	1070	155	1000	145	12;8(b)	45;27(b)	35	...	A 564, A 705
B, F	H1075	1000	145	860	125	13;9(b)	45;28(b)	32	...	A 564, A 705
B, F	H1100	965	140	795	115	14;10(b)	45;29(b)	31	...	A 564, A 705
B, F	H1150	930	135	725	105	16;11(b)	50;30(b)	28	...	A 564, A 705
B, F	H1150M	795	115	515	75	18;14(b)	55;35(b)	24	...	A 564, A 705
P, Sh, St	H900	1310	190	1170	170	5–10(c)	...	40	48	A 693
P, Sh, St	H1100	965	140	790	115	5–14(c)	...	29	40	A 693
PH 15-7 Mo (UNS S15700)										
B, F	RH950	1380	200	1210	175	7	25	...	...	A 564, A 705
B, F	TH1050	1240	180	1100	160	8	25	...	...	A 564, A 705
P, Sh, St	Annealed	1035 max	150 max	450 max	65 max	25 min	...	...	...	A 693
P, Sh, St	RH950(d)	1550	225	1380	200	1–4(c)	...	45–46	...	A 693
P, Sh, St	TH1050(d)	1310	190	1170	170	2–5(c)	...	40	...	A 693
P, Sh, St	Cold rolled condition C	1380	200	1210	175	1	...	41	...	A 693
P, Sh, St	CH900	1650	240	1590	230	1	...	46	...	A 693
17-4 PH (UNS S17400)										
B, F	H900(d)	1310	190	1170	170	10	40;35(e)	40	...	A 564, A 705
B, F	H925(d)	1170	170	1070	155	10	44;38(e)	38	...	A 564, A 705
B, F	H1025(d)	1070	155	1000	145	12	45	35	...	A 564, A 705
B, F	H1075(d)	1000	145	860	125	13	45	32	...	A 564, A 705
B, F	H1100(d)	965	140	795	115	14	45	31	...	A 564, A 705
B, F	H1150(d)	930	135	725	105	16	50	28	...	A 564, A 705
B, F	H1150M(d)	795	115	515	75	18	55	24	...	A 564, A 705
P, Sh, St	H900	1310	190	1170	170	5–10(c)	...	40	48	A 693
P, Sh, St	H1100	965	140	790	115	5–14(c)	...	29	40	A 693
17-7 PH (UNS S17700)										
B, F	RH950(d)	1275	185	1030	150	6	10	41	...	A 564, A 705
B, F	TH1050(d)	1170	170	965	140	6	25	38	...	A 564, A 705
P, Sh, St	RH950	1450(c)	210(c)	1310(c)	190(c)	1–6(c)	...	43(c)	44(c)	A 693
P, Sh, St	TH1050	1240(c)	180(c)	1030(c)	150(c)	3–7(c)	...	38	...	A 693
P, Sh, St	Cold rolled condition C	1380	200	1210	175	1	...	41	...	A 693
P, Sh, St	CH900	1650	240	1590	230	1	...	46	...	A 693
W	Cold drawn condition C	1400–2035(c)	203–295(c)	...	...	...	...	...	...	A 313
W	CH900	1585–2515(c)	230–365(c)	...	...	...	...	...	...	A 313
AM-350 (UNS S35000)										
P, Sh, St	Annealed	1380 max	200 max	585–620 max(c)	85–90 max(c)	8–12(c)	...	...	30	A 693
P, Sh, St	H850	1275	185	1030	150	2–8(c)	...	42	...	A 693
P, Sh, St	H1000	1140	165	1000	145	2–8(c)	...	36	...	A 693
AM-355 (UNS S35500)										
F	H1000	1170	170	1070	155	12	25	37	...	A 705
P, Sh, St	H850	1310	190	1140	165	10	...	...	...	A 693
P, Sh, St	H1000	1170	170	1030	150	12	...	37	...	A 693
Custom 450 (UNS S45000)										
B, shapes	Annealed	895(f)	130(f)	655	95	10	40	...	32	A 564(f)
F, shapes	Annealed	860(f)	125(f)	655	95	10	40	...	33	A 705(f)
B, F, shapes	H900	1240(g)	180(g)	1170	170	6;10(b)	20;40(b)	39	...	A 564(g), A 705(g)
B, F, shapes	H950	1170(g)	170(g)	1100	160	7;10(b)	22;40(b)	37	...	A 564(g), A 705(g)
B, F, shapes	H1000	1100(g)	160(g)	1030	150	8;12(b)	27;45(b)	36	...	A 564(g), A 705(g)
B, F, shapes	H1025	1030(g)	150(g)	965	140	12	45	34	...	A 564(g), A 705
B, F, shapes	H1050	1000(g)	145(g)	930	135	9;12(b)	30;45(b)	34	...	A 564(g), A 705
B, F, shapes	H1100	895(g)	130(g)	725	105	11;16(b)	30;50(b)	30	...	A 564(g), A 705
B, F, shapes	H1150	860(g)	125(g)	515	75	12–18(h)	35–55(h)	26	...	A 564(g), A 705

(continued)

(a) B, bar; F, forgings; P, plate; Sh, sheet; St, strip; W, wire. (b) Higher value is longitudinal; lower value is transverse. (c) Values vary with thickness or diameter. (d) Longitudinal properties only. (e) Higher values are for sizes up to and including 75 mm (3 in.); lower values are for sizes over 75 mm (3 in.) up to and including 200 mm (8 in.). (f) Tensile strengths of 860 to 140 MPa (125 to 165 ksi) for sizes up to 13 mm (½ in.). (g) Tensile strength only applicable up to sizes of 13 mm (½ in.). (h) Varies with section size and test direction. (i) Up to and including 150 mm (6 in.)

Table 12 Minimum mechanical properties of precipitation-hardening stainless steels

Product form(a)	Condition	Tensile strength MPa	ksi	Yield strength MPa	ksi	Elongation, %	Reduction in area, %	Hardness, HRC min	max	ASTM specification
Custom 450 (UNS S45000) (continued)										
P, Sh, St	Annealed	895–1205	130–165	620–1035	90–150	4 min	...	25	33	A 693
P, Sh, St	H900	1240	180	1170	170	3–5(c)	...	40	...	A 693
P, Sh, St	H1000	1105	160	1035	150	5–7(c)	...	36	...	A 693
P, Sh, St	H1150	860	125	515	75	8–10(c)	...	26	...	A 693
Custom 455 (UNS S45500)										
B, F, shapes	H900(i)	1620	235	1520	220	8	30	47	...	A 564(g), A 705(g)
B, F, shapes	H950(i)	1520	220	1410	205	10	40	44	...	A 564(g), A 705(g)
B, F, shapes	H1000(i)	1410	205	1280	185	10	40	40	...	A 564(g), A 705(g)
P, Sh, St	H950	1530	222	1410	205	≤4	...	...	...	A 693

(a) B, bar; F, forgings; P, plate; Sh, sheet; St, strip; W, wire. (b) Higher value is longitudinal; lower value is transverse. (c) Values vary with thickness or diameter. (d) Longitudinal properties only. (e) Higher values are for sizes up to and including 75 mm (3 in.); lower values are for sizes over 75 mm (3 in.) up to and including 200 mm (8 in.). (f) Tensile strengths of 860 to 140 MPa (125 to 165 ksi) for sizes up to 13 mm (½ in.). (g) Tensile strength only applicable up to sizes of 13 mm (½ in.). (h) Varies with section size and test direction. (i) Up to and including 150 mm (6 in.)

used for structural components in which maximum stability and a high degree of toughness are required at cryogenic temperatures.

Type 316 stainless steel is less stable than type 310, but tensile specimens of type 316 pulled to 0.2% offset (at the yield load) at −270 °C (−450 °F) showed no indication of martensite formation in the deformed regions (Ref 4). However, when tensile specimens of type 316 were pulled to fracture at −270 °C (−450 °F), the metallographic structures in the areas of the fractures transformed to approximately 50% martensite (Ref 5). Type 316 stainless steel is an important candidate material for structural components of superconducting and magnetic fusion machinery.

For higher-strength components of cryogenic structures, there are several stainless steels that contain significant amounts of manganese in place of some of the nickel, along with small additions of nitrogen and other elements that increase strength. Among these stainless steels are 21-6-9, Pyromet 538, Nitronic 40, and Nitronic 60.

Tensile Properties. Typical tensile properties of annealed 300 series austenitic stainless steels at room temperature and at subzero temperatures are presented in Table 14, and tensile properties of cold worked 300 series stainless steels are given in Table 15. Cold working substantially increases yield and tensile strengths and reduces ductility, but ductility and notch toughness of the cold-worked alloy are often sufficient for cryogenic applications. Tensile properties of other stainless steels are presented in Table 16. For the annealed alloys, the greatest effect of the nitrogen addition is to produce an increase in yield strength at cryogenic temperatures. The data for cold-worked AISI 202, a nitrogen-strengthened stainless steel, indicates how this alloy can be strengthened by cold working that results in reduced ductility. Because of its low ductility, alloy 416 is not recommended for use below −196 °C (−320 °F) except in nonstressed applications.

Results of tensile tests on stainless steel weldments at subzero temperatures, given in Table 17, may be significant in selecting stainless steels for cryogenic applications. Results of ultrasonic determinations of Young's modulus and Poisson's ratio for three stainless steels, shown in Fig. 10 and 11, serve to supplement the tensile data.

Fracture Toughness. Fracture toughness data for stainless steels are limited because steels of this type that are suitable for use at cryogenic temperatures have very high toughness. The fracture toughness data that are available were obtained by the J-integral method and converted to $K_{Ic}(J)$ values. Such data for base metal and weldments are shown in Table 18. Fracture toughness of base metals are relatively high even at −269 °C (−452 °F); fracture toughness of fusion zones (FZ) of welds may be lower or higher than that of the base metal.

Fracture Crack Growth Rates. Available data for determining fatigue crack growth rates at room temperature and at subzero temperatures for austenitic stainless steels and weldments are presented in Table 19. The fatigue crack growth rates of the base metals are generally higher at room temperature than at subzero temperatures, or about equal at room temperature and at subzero temperatures, except for 21-6-9 stainless steel. For 21-6-9, fatigue crack growth rates are higher at −269 °C (−452 °F) than at room temperature. A log-log plot of the da/dN data for type 304 stainless steel is shown in Fig. 12. For this steel, fatigue crack growth rates are nearly the same, at the same values of ΔK, for room-temperature and cryogenic-temperature tests. Fatigue crack growth rates in the fusion zones of welds tend to be higher than in the base metal.

Fatigue Strength. The results of flexural and axial fatigue tests at 10^6 cycles on austenitic stainless steels at room temperature and at subzero temperatures are presented in Table 20. Fatigue strength increases as exposure temperature is decreased. Notched specimens have substantially lower fatigue strengths than corresponding unnotched specimens at all testing temperatures. Reducing the surface roughness of unnotched specimens improves fatigue strength.

Influence of Product Form on Properties

The mechanical properties of cast or wrought stainless steels vary widely from

Table 13 Longitudinal fracture toughness of precipitation-hardening (PH) stainless steels

Designation	Condition(a)	Hardness, HRC	Fracture toughness MPa√m	ksi√in.
17-4 PH	H900	44	53	48
17-7 PH	RH950	44	76	69
Custom 450	Aged at 480 °C (900 °F)	43	81	74
Custom 455	Aged at 480 °C (900 °F)	50	47	43
	Aged at 510 °C (950 °F)	48	80	73
	Aged at 540 °C (1000 °F)	44	110	100
PH 13-8 Mo	H950	47	99	90
	H1000	46	121	110
PH 15-7 Mo	TH1080	42	55	50

(a) H, hardened; RH, refrigeration hardened; TH, transformation hardened

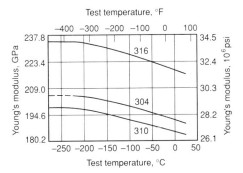

Fig. 10 Young's modulus for three austenitic stainless steels as determined ultrasonically. Source: Ref 23

Table 14 Typical tensile properties of annealed type 300 austenitic stainless steels

Temperature °C	°F	Tensile strength MPa	ksi	Yield strength MPa	ksi	Elongation, %	Reduction, in area, %	Notch tensile strength(a) MPa	ksi	Young's modulus GPa	10^6 psi
303 bar, longitudinal orientation											
24	75	730	106	425	61.4	67	70	...	...	...	...
−78	−108	1190	172	435	63.3	43	60	...	...	...	...
−196	−320	1660	240	465	67.3	36	54	...	...	...	...
−253	−423	2060	298	570	82.6	33	...	...	...	...	...
−269	−452	1830	266	...	...	30	37	...	...	...	...
304 sheet, longitudinal orientation											
24	75	660	95.5	295	42.5	75	...	715	104	...	...
−196	−320	1625	236	380	55.0	42	...	1450	210	...	...
−253	−423	1800	261	425	62.0	31	...	1160	168	...	...
−269	−452	1700	247	570	82.5	30	...	1230	178	...	...
304 plate, longitudinal orientation											
24	75	590	85.9	330	47.6	64	...	...	...	...	...
−253	−423	1720	250	410	59.4	...	...	...	...	...	...
304 bar, longitudinal orientation											
24	75	640	92.8	235	33.9	76	82	710	103	...	...
−78	−108	1150	167	300	43.2	50	76	...	...	...	...
−196	−320	1520	221	280	40.9	45	66	1060	153	...	...
−253	−423	1860	270	420	60.6	27	54	1120	162	...	...
−269	−452	1720	250	400	58.2	30	55	...	...	...	...
304L sheet, longitudinal orientation											
24	75	660	95.9	295	42.8	56	...	730	106	...	...
−78	−108	980	142	250	36.0	43	...	1030	150	...	...
−196	−320	1460	212	275	39.6	37	...	1420	206	...	...
−253	−423	1750	254	305	44.5	33	...	1290	187	...	...
−269	−452	1590	230	405	58.5	29	...	1460	212	...	...
304L sheet, transverse orientation											
−269	−452	1540	223	410	59.5	35	...	...	...	...	...
304L bar, longitudinal orientation											
24	75	660	95.5	405	58.9	78	81	...	...	190	27.6
−78	−108	1060	153	435	62.8	70	74	...	...	...	...
−196	−320	1510	219	460	66.6	43	66	...	...	205	29.7
−253	−423	1880	273	525	75.8	42	41	...	...	...	...
−269	−452	1660	241	545	79.4	34	56	...	...	200	29.2
310 sheet, longitudinal orientation											
24	75	570	83.0	240	35.0	50	...	645	93.9	...	...
−196	−320	1080	156	545	79.1	68	...	1070	155	...	...
−253	−423	1300	188	715	104	56	...	1250	182	...	...
−269	−452	1230	178	770	112	58	...	...	...	...	...
310 sheet, transverse orientation											
24	75	600	86.8	240	34.8	46	...	630	91.6	...	...
−269	−452	1280	186	800	116	58	...	...	...	...	...
310 bar, longitudinal orientation											
24	75	585	84.8	340	49.1	50	76	770	112	...	...
−78	−108	740	107	305	43.9	72	68	...	...	...	...
−196	−320	1090	158	520	75.5	68	50	...	...	205	29.9
−253	−423	1390	202	855	124	44	48	1305	189	...	...
−269	−452	1300	189	715	104	50	41	...	...	205	29.9
310S forging, transverse orientation											
24	75	585	84.8	260	37.9	54	71	800	116	...	...
−196	−320	1100	159	605	87.6	72	52	1350	196	...	...
−269	−452	1300	189	815	118	64	45	1600	232	...	...
316 sheet, longitudinal orientation											
24	75	595	86.4	275	39.8	60	...	...	...	...	...
−253	−423	1580	229	664	96.6	55	...	...	...	...	...
321 sheet, longitudinal orientation											
24	75	620	89.6	225	32.4	55	...	625	90.4	180	26.0
−196	−320	1380	200	315	45.6	46	...	1520	220	205	29.5
−253	−423	1650	239	375	54.5	36	...	1460	212	210	30.7
321 bar, longitudinal orientation											
24	75	675	97.6	430	62.2	55	79	...	...	...	...
−78	−108	1060	153	385	55.9	46	73	...	...	...	...
−196	−320	1540	223	450	65.4	38	60	...	...	...	...
−253	−423	1860	270	405	58.5	35	44	...	...	...	...
347 sheet, longitudinal orientation											
24	75	650	94	255	37	52	...	...	...	...	...
−196	−320	1365	198	420	61	47	...	...	...	...	...
−253	−423	1610	234	435	63	35	...	...	...	...	...
347 bar											
24	75	670	97.4	340	49.3	57	76	...	...	...	...
−78	−108	995	144	475	68.8	51	71	...	...	...	...
−196	−320	1470	214	430	62.2	43	60	...	...	...	...
−253	−423	1850	268	525	76.4	38	45	...	...	...	...

(a) Stress concentration factor, K_t, is 5.2 for 304 and 304L sheet, 14 for 304 bar, 6.3 for 310 sheet, 6.4 for 310 bar; K_t is 10 for 310S forging; K_t is 3.5 for 321 sheet. Source: Ref 6–12

Table 15 Typical tensile properties of cold-worked type 300 austenitic stainless steel sheet

Temperature °C	°F	Tensile strength MPa	ksi	Yield strength MPa	ksi	Elongation, %	Notch tensile strength(a) MPa	ksi	Young's modulus GPa	10⁶ psi
301, hard, cold rolled (42–60% reduction), longitudinal orientation										
24	75	1310	190	1200	174	18	1390	201	...	...
−78	−108	1560	226	1130	164	23	1460	212	...	...
−196	−320	2020	293	1380	200	19	1660	241	...	...
−253	−423	2110	306	1610	233	14	1830	265	...	...
301, hard, cold rolled (42–60% reduction), transverse orientation										
24	75	1310	190	1060	153	10	1430	207	...	...
−78	−108	1560	226	1070	155	28	1430	208	...	...
−196	−320	2060	299	1310	190	28	1670	243	...	...
−253	−423	1900	275	1570	227	8	1360	197	...	...
301, extra hard, cold rolled (>60% reduction), longitudinal orientation										
24	75	1500	217	1370	198	9	1600	232	175	25.6
−78	−108	1710	248	1400	203	22	1680	244	180	26.3
−196	−320	2220	322	1610	234	22	1940	282	180	26.2
−253	−423	2220	322	1810	262	13	1890	274	190	27.6
−269	−452	2140	310	1930	280	2	...	...	...	...
301, extra hard, cold rolled (>60% reduction), transverse orientation										
24	75	1590	230	1280	186	8	1520	220	...	...
−78	−108	1770	257	1250	181	18	1590	230	...	...
−196	−320	2190	318	1560	226	18	1680	244	...	...
−253	−423	2180	316	1830	266	5	1340	194	...	...
304, hard, cold rolled, longitudinal orientation										
24	75	1320	191	1190	173	3	1460	212	180	25.9
−78	−108	1470	213	1300	188	10	1590	231	185	26.9
−196	−320	1900	276	1430	208	29	1910	277	200	29.1
−253	−423	2010	292	1560	226	2	2160	313	210	30.5
304, hard, cold rolled, transverse orientation										
24	75	1440	209	1180	171	5	1200	174	195	28.0
−78	−108	1600	232	1330	193	7	1400	203	200	28.9
−196	−320	1870	271	1480	214	23	1690	245	205	30.0
−253	−423	2160	313	1560	226	1	1900	276	215	31.1
304L, 70% cold reduced, longitudinal orientation										
24	75	1320	192	1080	156	3	...	...	...	...
−196	−320	1770	256	1530	222	14	...	...	...	...
−253	−423	1990	288	1770	256	2	...	...	...	...
304L, 70% cold reduced, transverse orientation										
24	75	1440	209	1220	177	4	...	...	...	...
−196	−320	1890	274	1630	236	12	...	...	...	...
−253	−423	2230	324	1940	282	1	...	...	...	...
310, 75% cold reduced, longitudinal orientation										
24	75	1180	171	1100	160	3	1360	197	175	25.4
−78	−108	1410	204	1290	187	4	1530	222	175	25.5
−196	−320	1720	249	1540	223	10	1900	276	180	26.4
−253	−423	2000	290	1790	259	10	2230	324	195	28.3
310, 75% cold reduced, transverse orientation										
24	75	1370	199	1110	161	4	1370	199	195	28.1
−78	−108	1540	224	1290	187	8	1640	238	190	27.6
−196	−320	1880	272	1520	221	10	2050	297	195	28.2
−253	−423	2140	311	1790	260	9	2190	318	200	29.1

(a) K_t = 6.3. Source: Ref 6, 10, 13

group to group, vary less widely from type to type within groups, and may vary with product form for a given type. Because of the wide variation from group to group, one must first decide whether a martensitic, ferritic, austenitic, duplex, or precipitation-hardening stainless steel is most suitable for a given application. Once the appropriate group is selected, the method of fabrication or service conditions may then dictate which specific type is required.

Before typical properties of the various product forms are discussed, it is important that two key points about stainless steels be recognized. First, many stainless steels are manufactured and/or used in a heat-treated condition, that is, in some thermally treated condition other than process annealed or, typically, mill processed. When this is the case, a tabulation of typical properties may not give all the required information. Second, in many products strain hardening during fabrication is a very important consideration. All stainless steels strain harden to some degree depending on structure, alloy content, and amount of cold working. Consequently, for applications in which the service performance of the finished product depends on the enhancement of properties during fabrication, it is essential that the manufacturer determine this effect independently for each individual product. Here, techniques such as statistical-reliability testing are invaluable.

Cast Structures. Whether produced as ingot, slab, or billet in a mill or as shape castings in a foundry, cast structures can exhibit wide variations in properties. Because of the possible existence of large dendritic grains, inter-

Table 16 Typical tensile properties of stainless steels other than type 300 series steels

Temperature °C	°F	Tensile strength MPa	ksi	Yield strength MPa	ksi	Elongation, %	Reduction, in area, %	Notch tensile strength(a) MPa	ksi	Young's modulus GPa	10⁶ psi
202 sheet, annealed, longitudinal orientation											
24	75	705	102	325	47.1	57	...	...	...	...	...
−73	−100	1080	156	485	70.2	41	...	...	...	...	...
−196	−320	1590	231	610	88.3	52	...	...	...	...	...
−268	−450	1420	206	765	111	25	...	...	...	...	...
202 sheet, cold reduced 50%, longitudinal orientation											
24	75	1080	156	965	140	21	...	...	...	...	...
−196	−320	1970	286	1070	155	28	...	...	...	...	...
−268	−450	1950	283	1240	180	20	...	...	...	...	...
21-6-9 plate, longitudinal orientation(b)											
24	75	705	102	385	55.9	54	80	...	...	...	...
−78	−108	895	130	590	85.4	60	75	...	...	...	...
−196	−320	1510	219	970	141	41	33	...	...	...	...
−253	−423	1660	241	1220	177	16	26	...	...	...	...
−269	−452	1700	247	1350	196	22	30	...	...	...	...
Pyromet 538 plate, longitudinal orientation(c)											
24	75	675	97.9	340	49.0	75	81	...	...	...	...
−196	−320	1370	199	800	116	76	73	...	...	...	...
−269	−452	1490	216	1010	147	52	59	...	...	...	...
Nitronic 40 plate, electroslag remelted; as-rolled											
24	75	1010	146	840	122	35	72	...	...	...	...
−73	−100	1170	169	945	137	36	71	...	...	...	...
−196	−320	1830	266	1540	223	31	64	...	...	...	...
Nitronic 60 bar, annealed, longitudinal orientation											
24	75	750	109	400	58.1	66	79	1080	157	165	24.0
−73	−100	1020	148	535	77.9	70	81	1480	215	165	24.2
−196	−320	1500	218	695	101	60	66	1900	275	170	24.8
−253	−423	1410	204	860	125	24	27	1870	271	170	24.8
416 bar, longitudinal orientation(d)											
24	75	1400	203	1200	174	15	53	...	...	...	...
−78	−108	1500	218	1260	183	15	52	...	...	...	...
−196	−320	1800	261	1600	232	9	24	...	...	...	...
−253	−423	2020	293	2020	293	0.4	2	...	...	...	...

(a) $K_t = 7$ for Nitronic 60 bar. (b) Annealed 1 h at 1065 °C (1950 °F), water quenched. (c) Annealed 1 h at 1095 °C (2000 °F), water quenched. (d) Heat treatment: 1 h at 980 °C (1800 °F), oil quenched, tempered 4 h at 370 °C (700 °F), air cooled. Source: Ref 6, 10, 11, 14–20

granular phases, and alloy segregation, typical mechanical properties cannot be stated precisely and generally are inferior to those of any wrought structure. Detailed information on the composition and properties of cast stainless steels is given in the article "Cast Stainless Steels" in this Volume.

Hot Processing. The initial purpose of hot rolling or forging an ingot, slab, or billet is to refine the cast structure and improve mechanical properties. Hot reduced products and hot reduced and annealed products exhibit coarser grain structures and lower strengths than cold processed products. Grain size and shape depend chiefly on start and finish temperatures and on the method of hot reduction. For instance, cross-rolled hand mill plate will exhibit a more equiaxed grain structure than continuous hot rolled strip.

Hot reduction may be a final sizing operation, as in the case of hot-rolled bar, billet, plate, or bar flats, or it may be an intermediate processing step for products such as cold-finished bar, rod, and wire, and cold-rolled sheet and strip.

Typical properties of hot processed products and of hot processed and annealed products are different from those of either cast or cold reduced products. Hot processed products tend to have coarser grain sizes than cold reduced products.

Cold Reduced Products. When strained at ambient temperatures, all stainless steels tend to work harden, as shown in Fig. 13. Because recrystallization does not occur during cold working, the final properties of thermally treated products depend on:

- Amount of cold reduction (which helps determine the number of potential recrystallization sites)
- Type of mill thermal treatment (subcritical annealing, normalizing, or solution treatment)
- Time at any given temperature

Wrought products that have been cold reduced and annealed generally have finer grain sizes, which produce higher strengths than hot processed products. Cold reduced products sometimes exhibit greater differences between transverse and longitudinal properties than hot processed products.

Cold finishing is generally done to improve dimensional tolerances or surface finish or to raise mechanical strength. Cold-finished products—whether they have been previously hot worked and annealed or have been hot worked, cold worked, and annealed—have higher mechanical strength and slightly lower ductility than their process-annealed counterparts.

Physical Properties

There are relatively few applications for stainless steels in which physical properties

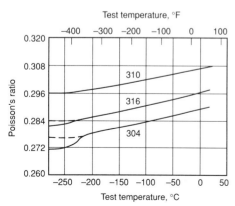

Fig. 11 Poisson's ratios for three austenitic stainless steels as determined ultrasonically. Source: Ref 23

Table 17 Typical tensile properties of stainless steel weldments

Alloy condition	Welding process	Filler	Form	Base metal orientation(a)	Test temperature °C	Test temperature °F	Yield strength MPa	Yield strength ksi	Tensile strength MPa	Tensile strength ksi	Elongation, %	Reduction in area, %	Notch tensile strength(b) MPa	Notch tensile strength(b) ksi
Type 301, cold rolled 60%; tested as welded	GTA	None	Sheet	L	24	75	...	...	1034	150	7	...	...	...
					−78	−108	...	...	1489	216	13	...	...	...
					−196	−320	...	...	2006	291	16	...	...	...
					−253	−423	...	...	1675	243	6	...	...	...
Type 310, ¾ hard; tested as welded	GTA	310	Sheet	L	24	75	380	55.1	530	76.8	4	...	...	...
					−78	−108	523	75.9	723	105	4	...	...	...
					−196	−320	752	109	1026	149	4	...	...	...
AISI, 310S, annealed	SMA	310S	Plate	...	24	75	334	48.5	582	84.4	40	76	841	122
					−196	−320	660	96.6	1066	155	46	67	1428	207
					−269	−452	829	120	1102	160	26	24	1672	242
21-6-9, annealed	SMA	Inconel 625	Plate	Weld(c)	−269	−452	878	127	1276	185	31	27	...	...
				HAZ(c)	−269	−452	1728	251	1873	272	21	33	...	...
	GTA	Inconel 625	Plate	Weld(c)	−269	−452	951	138	1222	177	18	20	...	...
				HAZ(c)	−269	−452	1740	252	1921	279	17	37	...	...
	GMA	Inconel 625	Plate	Weld(c)	−269	−452	833	121	1087	158	19	27	...	...
				HAZ(c)	−269	−452	1689	245	1866	271	15	27	...	...
	GTA	Pyromet 538	Plate	...	24	75	414	60.0	725	105	51	74	1238	180
					−196	−320	1009	146	1456	211	48	61	2119	307
					−269	−452	1240	180	1646	239	31	24	1841	267
	GMA	IN-182	Plate	...	24	75	413	59.9	729	106	53	75	1018	148
					−196	−320	800	116	1045	152	6	37	1416	205
					−269	−452	805	117	1086	158	6	40	1419	206

(a) L, longitudinal. (b) $K_t = 10$. (c) Weld parallel with specimen axis; weld specimens were all weld metal; HAZ specimens contained HAZ plus some weld metal and some base metal. Source: Ref 6, 15, 17, 18, 21, 22

are the determining factors in selection. However, there are many applications in which physical properties are important in product design. For instance, stainless steels are used for many elevated-temperature applications, often in conjunction with steels of lesser alloy content. Because austenitic stainless steels have higher coefficients of thermal expansion and lower thermal conductivities than carbon and alloy steels, these characteristics must be taken into account in the design of stainless steel-to-carbon steel or stainless steel-to-alloy steel products such as heat exchangers. In such products, differential thermal expansion imposes stresses on the unit that would not be present were the unit made entirely of carbon or alloy steel; also, if the heat-transfer surface is made of stainless steel, it must be larger than if it were made of carbon or alloy steel.

Typical physical properties of selected grades of annealed wrought stainless steels are given in Table 21. Physical properties may vary slightly with product form and size, but such variations are usually not of critical importance to the application.

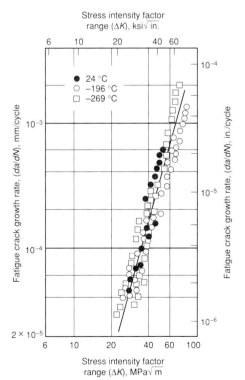

Fig. 12 Fatigue crack growth rate data for type 304 austenitic stainless steel (annealed) at room temperature and at subzero temperatures. Source: Ref 27

Corrosion Properties

Stainless steels are susceptible to several forms of localized corrosive attack. The avoidance of such localized corrosion is the focus of much of the effort involved in selecting stainless steel. Furthermore, the corrosion performance of stainless steels can be strongly affected by practices of design, fabrication, surface conditioning, and maintenance.

The selection of a grade of stainless steel for a particular application involves the consideration of many factors, but always begins with corrosion resistance. It is first necessary to characterize the probable service environment. It is not enough to consider only the design conditions. It is also necessary to consider the reasonably anticipated excursions or upsets in service conditions. The suitability of various grades can be estimated from laboratory tests or from documentation of field experience in comparable environments. Once the grades with adequate corrosion resistance have been identified, it is then appropriate to consider mechanical properties, ease of fabrication, the types and degree of risk present in the application, the availability of the necessary product forms, and cost.

Mechanism of Corrosion Resistance

The mechanism of corrosion protection for stainless steels differs from that for carbon steels, alloy steels, and most other metals. In these other cases, the formation of a barrier of true oxide separates the metal from the surrounding atmosphere. The degree of protection afforded by such an oxide is a function of the thickness of the oxide layer, its continuity, its coherence and adhesion to the metal, and the diffusivities of oxygen and metal in the oxide. In high-temperature oxidation, stainless steels use a generally similar model for corrosion protection. However, at low temperatures, stainless steels do not form a layer of true oxide. Instead, a passive film is formed. One mechanism that has been suggested is the formation of a film of hydrated oxide, but there is not total agreement on the nature of the oxide complex on the metal surface. However, the oxide film should be continuous, nonporous, insoluble, and self-healing if broken in the presence of oxygen.

Passivity exists under certain conditions for particular environments. The range of conditions over which passivity can be maintained depends on the precise environment and on the family and composition of

Table 18 Fracture toughness of austenitic stainless steels and weldments for compact tension specimens

Alloy and condition(a)	Form	Room-temperature yield strength		Orientation	Fracture toughness, (K_{Ic}), J, at					
					24 °C (75 °F)		−196 °C (−320 °F)		−269 °C (−452 °F)	
		MPa	ksi		MPa$\sqrt{m}$	ksi$\sqrt{in.}$	MPa$\sqrt{m}$	ksi$\sqrt{in.}$	MPa$\sqrt{m}$	ksi$\sqrt{in.}$
Type 310S, annealed	Plate	261	37.9	T-L	...	...	...	...	262	236
	Weldment	...	...		...	...	...	...	118	106
Pyromet 538, STQ	Plate	338	49	T-L	...	...	275	250	182	165
	Weldment	...	...		...	...	...	...	82.4	74.4
	Weldment	...	...		...	...	...	...	176	159

(a) STQ, solution treated and quenched. Filler wires for 310S: E 310-16; For Pyromet 538: 21-6-9. Source: Ref 17, 24–26

Table 19 Fatigue crack growth rate (da/dN) data for compact tension specimens of austenitic stainless steels

Alloy and condition	Orientation(a)	Frequency, Hz	Stress ratio, R	Test temperature or temperature range		C(b)		n(b)	Estimated range for ΔK	
				°C	°F	da/dN:mm/cycle ΔK:MPa$\sqrt{m}$	da/dN:in./cycle ΔK:ksi$\sqrt{in.}$		MPa$\sqrt{m}$	ksi$\sqrt{in.}$
Type 304 annealed plate	T-L	20–28	0.1	24 to −269	75 to −452	2.7×10^{-9}	1.4×10^{-10}	3.0	22–80	20–73
Type 304L annealed plate	T-L	20–28	0.1	24	75	2.0×10^{-10}	1.2×10^{-11}	4.0	22–54	20–49
				−196, −269	−320, −452	3.4×10^{-11}	2.0×10^{-12}	4.0	26–80	24–73
Type 310S annealed plate	T-L	20–28	0.1	24	75	3.5×10^{-11}	2.1×10^{-12}	4.4	24–35	22–32
				24	75	4.7×10^{-9}	2.4×10^{-10}	3.0	35–60	32–55
				−196, −269	−320, −452	1.1×10^{-10}	6.1×10^{-12}	3.7	25–80	23–73
	...	10	0.1	−196, −269	−320, −452	1.4×10^{-10}	7.9×10^{-12}	3.75	24–71	22–65
Type 310S, SMA weld with E310-16 filler	...	10	0.1	−196, −269	−320, −452	7.8×10^{-13}	5.0×10^{-14}	5.15	27–66	25–60
Type 316 annealed plate	T-L	20–28	0.1	24 to −269	75 to −452	2.1×10^{-10}	1.2×10^{-11}	3.8	19–16	17–14
21-6-9 annealed plate	T-L	20–28	0.1	24, −196	75, −320	1.9×10^{-10}	1.1×10^{-11}	3.7	25–80	23–73
				−269	−452	3.6×10^{-11}	2.2×10^{-12}	4.4	25–70	23–64
Pyromet 538, GTA weld in annealed plate using 21-6-9 filler	T-L	10	0.1	24	75	1.8×10^{-10}	9.9×10^{-12}	3.7	26–55	24–50
				−196, −269	−320, −452	7.6×10^{-14}	5.47×10^{-15}	6.36	24–44	22–40
Pyromet 538, SMA weld in annealed plate using Inconel 182 filler	T-L	10	0.1	24 to −269	75 to −452	2.5×10^{-12}	1.6×10^{-13}	5.13	25–55	23–50

(a) T, transverse; L, longitudinal. (b) C and n are constants from $da/dN = C (\Delta K)^n$; ΔK, stress intensity factor range. Source: Ref 17, 26

Table 20 Results of fatigue life tests on austenitic stainless steels

Alloy and condition	Stressing mode	Stress ratio, R	Cyclic frequency, Hz	K_t	Fatigue strengths at 10^6 cycles					
					24 °C (75 °F)		−196 °C (−320 °F)		−253 °C (−423 °F)	
					MPa	ksi	MPa	ksi	MPa	ksi
Type 301 sheet, extra full hard	Flex	−1.0	29, 86	1	496	72	793	115	669	97
				3.1	172	25	303	44	...	...
Type 304L bar, annealed	Axial	−1.0	...	1	269	39	483	70	552(a)	80(a)
				3.1	193	28	207	30	228(a)	33(a)
Type 310 sheet, annealed	Flex(b)	−1.0	...	1	186	27	455	66	597	84
	Flex(c)	−1.0	...	1	213	31	490	71	662	96
Type 310 bar, annealed	Axial	−1.0	...	1	255	37	469	68	607(a)	88(a)
				3.1	186	27	234	34	352(a)	51(a)
Type 321 sheet, annealed	Axial	−1.0	...	1	221	32	303	44	372	54
				3.5	124	18	154	22.3	181	26.3
	Flex(b)	−1.0	30–40	1	172	25	303	44	358	52
Type 347 sheet, annealed	Flex(b)	−1.0	30–40	1	221	32	421	61	386	56
	Flex(c)	−1.0	30–40	1	241	35	469	68	510	74

(a) Tested at −269 °C (−452 °F). (b) Surface finish 64 rms. (c) Surface finish 11 rms. Source: Ref 6, 28–31

the stainless steel. When conditions are favorable for maintaining passivity, stainless steels exhibit extremely low corrosion rates. If passivity is destroyed under conditions that do not permit the restoration of the passive film, stainless steel will corrode much like a carbon or low-alloy steel.

The presence of oxygen is essential to the corrosion resistance of a stainless steel. The corrosion resistance of stainless steel is at its maximum when the steel is boldly exposed and the surface is maintained free of deposits by a flowing bulk environment. Covering a portion of the surface, for example, by biofouling, painting, or installing a gasket, produces an oxygen-depleted region under the covered region. The oxygen-de-

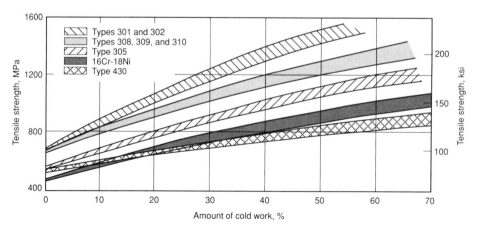

Fig. 13 Typical effect of cold rolling on the tensile strength of selected stainless steels

Table 21 Typical physical properties of wrought stainless steels, annealed condition

Type	UNS number	Density g/cm³ (lb/in.³)	Elastic modulus GPa (10⁶ psi)	Mean CTE from 0°C (32°F) to: 100°C (212°F) μm/m·°C (μin./in.·°F)	315°C (600°F) μm/m·°C (μin./in.·°F)	538°C (1000°F) μm/m·°C (μin./in.·°F)	Thermal conductivity at 100°C (212°F) W/m·K (Btu/ft·h·°F)	at 500°C (932°F) W/m·K (Btu/ft·h·°F)	Specific heat(a) J/kg·K (Btu/lb·°F)	Electrical resistivity, nΩ·m	Magnetic permeability(b)	Melting range °C (°F)
201	S20100	7.8 (0.28)	197 (28.6)	15.7 (8.7)	17.5 (9.7)	18.4 (10.2)	16.2 (9.4)	21.5 (12.4)	500 (0.12)	690	1.02	1400–1450 (2550–2650)
202	S20200	7.8 (0.28)	...	...	17.5 (9.7)	18.4 (10.2)	19.2 (10.7)	16.2 (9.4)	21.6 (12.5)	500 (0.12)	690	1.02	1400–1450 (2550–2650)
205	S20500	7.8 (0.28)	197 (28.6)	...	17.9 (9.9)	19.1 (10.6)	...	...	500 (0.12)	...	...	...
301	S30100	8.0 (0.29)	193 (28.0)	17.0 (9.4)	17.2 (9.6)	18.2 (10.1)	16.2 (9.4)	21.5 (12.4)	500 (0.12)	720	1.02	1400–1420 (2550–2590)
302	S30200	8.0 (0.29)	193 (28.0)	17.2 (9.6)	17.8 (9.9)	18.4 (10.2)	16.2 (9.4)	21.5 (12.4)	500 (0.12)	720	1.02	1400–1420 (2550–2590)
302B	S30215	8.0 (0.29)	193 (28.0)	16.2 (9.0)	18.0 (10.0)	19.4 (10.8)	15.9 (9.2)	21.6 (12.5)	500 (0.12)	720	1.02	1375–1400 (2500–2550)
303	S30300	8.0 (0.29)	193 (28.0)	17.2 (9.6)	17.8 (9.9)	18.4 (10.2)	16.2 (9.4)	21.5 (12.4)	500 (0.12)	720	1.02	1400–1420 (2550–2590)
304	S30400	8.0 (0.29)	193 (28.0)	17.2 (9.6)	17.8 (9.9)	18.4 (10.2)	16.2 (9.4)	21.5 (12.4)	500 (0.12)	720	1.02	1400–1450 (2550–2650)
304L	S30403	8.0 (0.29)	...	...	...	...	...	...	...	...	1.02	1400–1450 (2550–2650)
302Cu	S30430	...	193 (28.0)	17.2 (9.6)	17.8 (9.9)	...	11.2 (6.5)	21.5 (12.4)	500 (0.12)	720	1.02	1400–1450 (2550–2650)
304N	S30451	8.0 (0.29)	196 (28.5)	...	...	...	...	...	500 (0.12)	720	1.02	1400–1450 (2550–2650)
305	S30500	8.0 (0.29)	193 (28.0)	17.2 (9.6)	17.8 (9.9)	18.4 (10.2)	16.2 (9.4)	21.5 (12.4)	500 (0.12)	720	1.02	1400–1450 (2550–2650)
308	S30800	8.0 (0.29)	193 (28.0)	17.2 (9.6)	17.8 (9.9)	18.4 (10.2)	15.2 (8.8)	21.6 (12.5)	500 (0.12)	720	...	1400–1420 (2550–2590)
309	S30900	8.0 (0.29)	200 (29.0)	15.0 (8.3)	16.6 (9.2)	17.2 (9.6)	15.6 (9.0)	18.7 (10.8)	500 (0.12)	780	1.02	1400–1450 (2550–2650)
310	S31000	8.0 (0.29)	200 (29.0)	15.9 (8.8)	16.2 (9.0)	17.0 (9.4)	14.2 (8.2)	18.7 (10.8)	500 (0.12)	780	1.02	1400–1450 (2550–2650)
314	S31400	7.8 (0.28)	200 (29.0)	...	15.1 (8.4)	...	17.5 (10.1)	20.9 (12.1)	500 (0.12)	770	1.02	...
316	S31600	8.0 (0.29)	193 (28.0)	15.9 (8.8)	16.2 (9.0)	17.5 (9.7)	16.2 (9.4)	21.5 (12.4)	500 (0.12)	740	1.02	1375–1400 (2500–2550)
316L	S31603	8.0 (0.29)	...	...	...	...	...	...	...	...	1.02	1375–1400 (2500–2550)
316N	S31651	8.0 (0.29)	196 (28.5)	...	...	...	...	...	500 (0.12)	740	1.02	1375–1400 (2500–2550)
317	S31700	8.0 (0.29)	193 (28.0)	15.9 (8.8)	16.2 (9.0)	17.5 (9.7)	16.2 (9.4)	21.5 (12.4)	500 (0.12)	740	1.02	1375–1400 (2500–2550)
317L	S31703	8.0 (0.29)	200 (29.0)	16.5 (9.2)	...	18.1 (10.1)	14.4 (8.3)	...	500 (0.12)	790	...	1375–1400 (2500–2550)
321	S32100	8.0 (0.29)	193 (28.0)	16.6 (9.2)	17.2 (9.6)	18.6 (10.3)	16.1 (9.3)	22.2 (12.8)	500 (0.12)	720	1.02	1400–1425 (2550–2600)
329	S32900	7.8 (0.28)	...	...	...	...	...	...	460 (0.11)	750	...	...
330	N08330	8.0 (0.29)	196 (28.5)	14.4 (8.0)	16.0 (8.9)	16.7 (9.3)	16.1 (9.3)	22.2 (12.8)	460 (0.11)	1020	1.02	1400–1425 (2550–2600)
347	S34700	8.0 (0.29)	193 (28.0)	16.6 (9.2)	17.2 (9.6)	18.6 (10.3)	16.2 (9.4)	21.5 (12.4)	500 (0.12)	730	1.02	1400–1425 (2550–2600)
384	S38400	8.0 (0.29)	193 (28.0)	17.2 (9.6)	17.8 (9.9)	18.4 (10.2)	16.2 (9.4)	21.5 (12.4)	500 (0.12)	790	1.02	1400–1450 (2550–2650)
405	S40500	7.8 (0.28)	200 (29.0)	10.8 (6.0)	11.6 (6.4)	12.1 (6.7)	27.0 (15.6)	...	460 (0.11)	600	...	1480–1530 (2700–2790)
409	S40900	7.8 (0.28)	...	11.7 (6.5)	...	...	...	...	...	...	...	1480–1530 (2700–2790)
410	S41000	7.8 (0.28)	200 (29.0)	9.9 (5.5)	11.4 (6.3)	11.6 (6.4)	24.9 (14.4)	28.7 (16.6)	460 (0.11)	570	700–1000	1480–1530 (2700–2790)
414	S41400	7.8 (0.28)	200 (29.0)	10.4 (5.8)	11.0 (6.1)	12.1 (6.7)	24.9 (14.4)	28.7 (16.6)	460 (0.11)	700	...	1425–1480 (2600–2700)
416	S41600	7.8 (0.28)	200 (29.0)	9.9 (5.5)	11.0 (6.1)	11.6 (6.4)	24.9 (14.4)	28.7 (16.6)	460 (0.11)	570	700–1000	1480–1530 (2700–2790)
420	S42000	7.8 (0.28)	200 (29.0)	10.3 (5.7)	10.8 (6.0)	11.7 (6.5)	24.9 (14.4)	...	460 (0.11)	550	...	1450–1510 (2650–2750)
422	S42200	7.8 (0.28)	...	...	11.2 (6.2)	11.4 (6.3)	11.9 (6.6)	23.9 (13.8)	27.3 (15.8)	460 (0.11)	...	...	1470–1480 (2675–2700)
429	S42900	7.8 (0.28)	200 (29.0)	10.3 (5.7)	...	...	25.6 (14.8)	...	460 (0.11)	590	...	1450–1510 (2650–2750)
430	S43000	7.8 (0.28)	200 (29.0)	10.4 (5.8)	11.0 (6.1)	11.4 (6.3)	26.1 (15.1)	26.3 (15.2)	460 (0.11)	600	600–1100	1425–1510 (2600–2750)
430F	S43020	7.8 (0.28)	200 (29.0)	10.4 (5.8)	11.0 (6.1)	11.4 (6.3)	26.1 (15.1)	26.3 (15.2)	460 (0.11)	600	...	1425–1510 (2600–2750)
431	S43100	7.8 (0.28)	200 (29.0)	10.2 (5.7)	12.1 (6.7)	...	20.2 (11.7)	...	460 (0.11)	720	...	...
434	S43400	7.8 (0.28)	200 (29.0)	10.4 (5.8)	11.0 (6.1)	11.4 (6.3)	...	26.3 (15.2)	460 (0.11)	600	600–1100	1425–1510 (2600–2750)
436	S43600	7.8 (0.28)	200 (29.0)	9.3 (5.2)	...	...	23.9 (13.8)	26.0 (15.0)	460 (0.11)	600	600–1100	1425–1510 (2600–2750)
439	S43035	7.7 (0.28)	200 (29.0)	10.4 (5.8)	11.0 (6.1)	11.4 (6.3)	24.2 (14.0)	...	460 (0.11)	630	...	...
440A	S44002	7.8 (0.28)	200 (29.0)	10.2 (5.7)	...	...	24.2 (14.0)	...	460 (0.11)	600	...	1370–1480 (2500–2700)
440C	S44004	7.8 (0.28)	200 (29.0)	10.2 (5.7)	...	...	24.2 (14.0)	...	460 (0.11)	600	...	1370–1480 (2500–2700)
444	S44400	7.8 (0.28)	200 (29.0)	10.0 (5.6)	10.6 (5.9)	11.4 (6.3)	26.8 (15.5)	...	420 (0.10)	620	...	...
446	S44600	7.5 (0.27)	200 (29.0)	10.4 (5.8)	10.8 (6.0)	11.2 (6.2)	20.9 (12.1)	24.4 (14.1)	500 (0.12)	670	400–700	1425–1510 (2600–2750)
PH 13-8 Mo	S13800	7.8 (0.28)	203 (29.4)	10.6 (5.9)	11.2 (6.2)	11.9 (6.6)	14.0 (8.1)	22.0 (12.7)	460 (0.11)	1020	...	1400–1440 (2560–2625)
15-5 PH	S15500	7.8 (0.28)	196 (28.5)	10.8 (6.0)	11.4 (6.3)	...	17.8 (10.3)	23.0 (13.1)	420 (0.10)	770	95	1400–1440 (2560–2625)
17-4 PH	S17400	7.8 (0.28)	196 (28.5)	10.8 (6.0)	11.6 (6.4)	...	18.3 (10.6)	23.0 (13.1)	460 (0.11)	800	95	1400–1440 (2560–2625)
17-7 PH	S17700	7.8 (0.28)	204 (29.5)	11.0 (6.1)	11.6 (6.4)	...	16.4 (9.5)	21.8 (12.6)	460 (0.11)	830	...	1400–1440 (2560–2625)

CTE, coefficient of thermal expansion. (a) At 0 to 100 °C (32 to 212 °F). (b) Approximate values

pleted region is anodic relative to the well-aerated boldly exposed surface, and a higher level of alloy content in the stainless steel is required to prevent corrosion.

With appropriate grade selection, stainless steel will perform for very long times with minimal corrosion, but an inadequate grade can corrode and perforate more rapidly than a plain carbon steel will fail by uniform corrosion. The selection of the appropriate grade of stainless steel, then, is a balancing of the desire to minimize cost and the risk of corrosion damage by excursions of environmental conditions during operation or downtime.

Confusion exists regarding the meaning of the term passivation. It is not necessary to chemically treat a stainless steel to obtain the passive film; the film forms spontaneously in the presence of oxygen. Most frequently, the function of passivation is to remove free iron, oxides, and other surface contamination. For example, in the steel mill, the stainless steel may be pickled in an acid solution, often a mixture of nitric and hydrofluoric acids (HNO_3-HF), to remove oxides formed in heat treatment. Once the surface is cleaned and the bulk composition of the stainless steel is exposed to air, the passive film forms immediately.

Effects of Composition

Chromium is the one element essential in forming the passive film. Other elements can influence the effectiveness of chromium in forming or maintaining the film, but no other element can, by itself, create the properties of stainless steel.

Chromium. The film is first observed at about 10.5% Cr, but it is rather weak at this composition and affords only mild atmospheric protection. Increasing the chromium content to 17 to 20%, typical of the austenitic stainless steels, or to 26 to 29%, as possible in the newer ferritic stainless steels, greatly increases the stability of the passive film. However, higher chromium may adversely affect mechanical properties, fabricability, weldability, or suitability for applications involving certain thermal exposures. Therefore, it is often more efficient to improve corrosion resistance by altering other elements, with or without some increase in chromium.

Nickel, in sufficient quantities, will stabilize the austenitic structure; this greatly enhances mechanical properties and fabrication characteristics. Nickel is effective in promoting repassivation, especially in reducing environments. Also, it is particularly

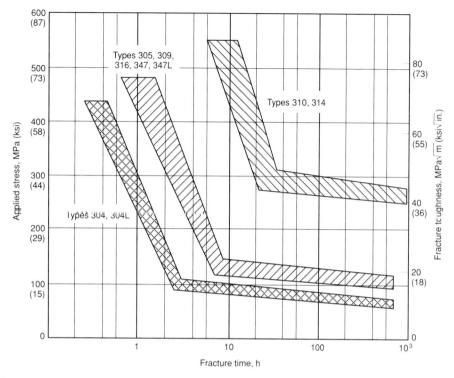

Fig. 14 Relative SCC behavior of austenitic stainless steels in boiling magnesium chloride. Source: Ref 35

useful in resisting corrosion in mineral acids. Increasing nickel content to about 8 to 10% decreases resistance to stress-corrosion cracking (SCC), but further increases begin to restore SCC resistance. Resistance to SCC in most service environments is achieved at about 30% Ni. In the newer ferritic grades, in which the nickel addition is less than that required to destabilize the ferritic phase, there are still substantial effects. In this range, nickel increases yield strength, toughness, and resistance to reducing acids, but makes the ferritic grades susceptible to SCC in concentrated magnesium chloride ($MgCl_2$) solutions.

Most chloride cracking testing has been carried out in accelerated test media such as boiling $MgCl_2$ solution (boiling point: 154 °C, or 309 °F) (Ref 32-34). All austenitic stainless steels are susceptible to chloride cracking (Fig. 14). It is noteworthy, however, that the higher-nickel types 310 and 314 were appreciably more resistant than the others. Although this solution causes rapid cracking, it does not necessarily simulate the cracking observed in field applications.

Manganese in moderate quantities and in association with nickel additions will perform many of the functions attributed to nickel. However, total replacement of nickel by manganese is not practical. Very high-manganese steels have some unusual and useful mechanical properties, such as resistance to galling. Manganese interacts with sulfur in stainless steels to form manganese sulfides. The morphology and composition of these sulfides can have substantial effects on corrosion resistance, especially pitting resistance.

Molybdenum in combination with chromium is very effective in terms of stabilizing the passive film in the presence of chlorides. Molybdenum is especially effective in increasing resistance to the initiation of pitting and crevice corrosion.

Carbon is useful to the extent that it permits hardenability by heat treatment, which is the basis of the martensitic grades, and provides strength in the high-temperature applications of stainless steels. In all other applications, carbon is detrimental to corrosion resistance through its reaction with chromium. In the ferritic grades, carbon is also extremely detrimental to toughness.

Nitrogen is beneficial to austenitic stainless steels in that it enhances pitting resistance, retards the formation of the chromium-molybdenum σ phase, and strengthens the steel. Nitrogen is essential in the newer duplex grades for increasing the austenite content, diminishing chromium and molybdenum segregation, and raising the corrosion resistance of the austenitic phase. Nitrogen is highly detrimental to the mechanical properties of the ferritic grades and must be treated as comparable to carbon when a stabilizing element is added to the steel.

Forms of Corrosion of Stainless Steels

The various forms of corrosive attack will be briefly discussed in this section. Detailed information on each of these forms of corrosion is available in the Section "Forms of Corrosion" in Volume 13 of the 9th Edition of *Metals Handbook*.

General (uniform) corrosion of a stainless steel suggests an environment capable of stripping the passive film from the surface and preventing repassivation. Such an occurrence could indicate an error in grade selection. An example is the exposure of a lower-chromium ferritic stainless steel to moderate concentration of hot sulfuric acid (H_2SO_4).

Galvanic corrosion results when two dissimilar metals are in electrical contact in a corrosive medium. As a highly corrosion-resistant metal, stainless steel can act as a cathode when in contact with a less noble metal, such as steel. The corrosion of steel parts, for example, steel bolts in a stainless steel construction, can be a significant problem. However, the effect can be used in a beneficial way for protecting critical stainless steel components within a larger steel construction. In the case of stainless steel connected to a more noble metal, the active-passive condition of the stainless steel must be considered. If the stainless steel is passive in the environment, galvanic interaction with a more noble metal is unlikely to produce significant corrosion. If the stainless steel is active or only marginally passive, galvanic interaction with a more noble metal will probably produce sustained rapid corrosion of the stainless steel without repassivation. The most important aspect of galvanic interaction for stainless steels is the need to select fasteners and weldments of adequate corrosion resistance relative to the bulk material, which is likely to have a much larger exposed area.

Pitting is a localized attack that can produce the penetration of a stainless steel with almost negligible weight loss to the total structure. Pitting is associated with a local discontinuity of the passive film. It can be a mechanical imperfection, such as an inclusion or surface damage, or it can be a local chemical breakdown of the film. Chloride is the most common agent for the initiation of pitting. Once a pit is formed, it in effect becomes a crevice; the local chemical environment is substantially more aggressive than the bulk environment. Therefore, very high flow rates over a stainless steel surface tend to reduce pitting corrosion; a high flow rate prevents the concentration of corrosive species in the pit. The stability of the passive film with respect to resistance to pitting initiation is controlled primarily by chromium and molybdenum. Minor alloying elements can also have an important effect by influencing the amount and type of inclusions (for example, sulfides) in the steel that can act as pitting sites.

Pitting initiation can also be influenced by surface condition, including the presence of deposits, and by temperature. For a particular environment, a grade of stainless steel may be characterized by a single temperature, or a very narrow range of temperatures, above which pitting will initiate and below which pitting will not initiate. It is therefore possible to select a grade that will not be subject to pitting attack if the chemical environment and temperature do not exceed the critical levels. If the range of operating conditions can be accurately characterized, a meaningful laboratory evaluation is possible. The formation of deposits in service can reduce the pitting temperature.

Although chloride is known to be the primary agent of pitting attack, it is not possible to establish a single critical chloride limit for each grade. The corrosivity of a particular concentration of chloride solution can be profoundly affected by the presence or absence of various other chemical species that may accelerate or inhibit corrosion. Chloride concentration may increase where evaporation or deposits occur. Because of the nature of pitting attack—rapid penetration with little total weight loss—it is rare for any significant amount of pitting to be acceptable in practical application.

Crevice corrosion can be considered a severe form of pitting. Any crevice, whether the result of a metal-to-metal joint, a gasket, fouling, or deposits, tends to restrict oxygen access, resulting in attack. In practice, it is extremely difficult to prevent all crevices, but every effort should be made to do so. Higher-chromium, and especially higher-molybdenum, grades are more resistant to crevice attack. Just as there is a critical pitting temperature for a particular environment, there is a critical crevice temperature. This temperature is specific to the geometry and nature of the crevice and to the precise corrosion environment for each grade. The critical crevice temperature can be useful in selecting an adequately resistant grade for a particular application.

Intergranular corrosion is a preferential attack at the grain boundaries of a stainless steel. It is generally the result of sensitization. This condition occurs when a thermal cycle leads to grain-boundary precipitation of a carbide, nitride, or intermetallic phase without providing sufficient time for chromium diffusion to fill the locally depleted region. A grain-boundary precipitate is not the point of attack; instead, the low-chromium region adjacent to the precipitate is susceptible.

Sensitization is not necessarily detrimental unless the grade is to be used in an environment capable of attacking the region. For example, elevated-temperature applications for stainless steel can operate with sensitized steel, but concern for intergranular attack must be given to possible corrosion during downtime when condensation might provide a corrosive medium. Because chromium provides corrosion resistance, sensitization also increases the susceptibility of chromium-depleted regions to other forms of corrosion, such as pitting, crevice corrosion, and stress-corrosion cracking (SCC). The thermal exposures required to sensitize a steel can be relatively brief, as in high-temperature service.

Stress-corrosion cracking is a corrosion mechanism in which the combination of a susceptible alloy, sustained tensile stress, and a particular environment leads to cracking of the metal. Stainless steels are particularly susceptible to SCC in chloride environments; temperature and the presence of oxygen tend to aggravate chloride SCC of stainless steels. Most ferritic and duplex stainless steels are either immune or highly resistant to SCC. All austenitic grades, especially AISI types 304 and 316, are susceptible to some degree. The highly alloyed austenitic grades are resistant to sodium chloride (NaCl) solutions, but crack readily in $MgCl_2$ solutions. Although some localized pitting or crevice corrosion probably precedes SCC, the amount of pitting or crevice attack may be so small that it is undetectable. Stress corrosion is difficult to detect while in progress, even when pervasive, and can lead to rapid catastrophic failures of pressurized equipment.

It is difficult to alleviate the environmental conditions that lead to SCC. The level of chlorides required to produce stress-corrosion cracking is very low. In operation, there can be evaporative concentration or a concentration in the surface film on a heat-rejecting surface. Temperature is often a process parameter, as in the case of a heat exchanger. Tensile stress is one parameter that might be controlled. However, the residual stresses associated with fabrication, welding, or thermal cycling, rather than design stresses, are often responsible for SCC, and even stress-relieving heat treatments do not completely eliminate these residual stresses.

Erosion-Corrosion. Corrosion of a metal or alloy can be accelerated when there is an abrasive removal of the protective oxide layer. This form of attack is especially significant when the thickness of the oxide layer is an important factor in determining corrosion resistance. In the case of a stainless steel, erosion of the passive film can lead to some acceleration of attack.

Oxidation. Because of their high chromium contents, stainless steels tend to be very resistant to oxidation. Important factors to be considered in the selection of stainless steel grades for high-temperature service are the stability of the composition and microstructure upon thermal exposure and the adherence of the oxide scale upon thermal cycling. Because many of the stainless steels used for high temperatures are austenitic grades with relatively high nickel contents, it is also necessary to be alert to the possibility of sulfidation attack.

Corrosion in Specific Environments

The selection of a suitable stainless steel for a specific environment requires consideration of several criteria. The first is corrosion resistance. Alloys are available that provide resistance to mild atmospheres (for example, type 430) or to many food-processing environments (for example, type 304 stainless). Chemicals and more severe corrodents require type 316 or a more highly alloyed material, such as 20Cb-3 (UNS N08020). Factors that affect the corrosivity of an environment include the concentration of chemical species, pH, aeration, flow rate (velocity), impurities (such as chlorides), and temperature, including effects from heat transfer.

The second criterion is mechanical properties, or strength. High-strength materials often sacrifice resistance to some form of corrosion, particularly SCC.

Third, fabrication must be considered, including such factors as the ability of the steel to be machined, welded, or formed. Resistance of the fabricated article to the environment must be considered, for example, the ability of the material to resist attack in crevices that cannot be avoided in the design.

Fourth, total cost must be estimated, including initial alloy price, installed cost, and the effective life expectancy of the finished product. Finally, consideration must be given to product availability.

Many applications for stainless steels, particularly those involving heat exchangers, can be analyzed in terms of a process side and a water side. The process side is usually a specific chemical combination that has its own requirements for a stainless steel grade. The water side is common in many applications. This section will discuss the corrosivity of various environments for stainless steels.

Atmospheric Corrosion. The atmospheric contaminants most often responsible for the rusting of structural stainless steels are chlorides and metallic iron dust. Chloride contamination may originate from the calcium chloride ($CaCl_2$) used to make concrete or from exposure in marine or industrial locations. Iron contamination may occur during fabrication or erection of the structure. Contamination should be minimized, if possible.

The corrosivity of different atmospheric exposures can vary greatly and can dictate application of different grades of stainless steel. Rural atmospheres, uncontaminated by industrial fumes or coastal salt, are extremely mild in terms of corrosivity for stainless steel, even in areas of high humidity. Industrial or marine environments can be considerably more severe.

Table 22 demonstrates that resistance to staining can depend on the specific exposure. For example, several 300-series stainless steels showed no rust during long-term exposures in New York City. On the other hand, staining was observed after much shorter exposures at Niagara Falls in a

Table 22 Atmospheric corrosion of austenitic stainless steels at two industrial sites

Type(a)	New York City (industrial) Exposure time, years	Specimen surface evaluation	Niagara Falls (industrial-chemical) Exposure time, years	Specimen surface evaluation
302	5	Free from rust stains	<$\frac{2}{3}$	Rust stains
302	26	Free from rust stains	...	...
304	26	Free from rust stains	<1	Rust stains
304	...	...	6	Covered with rust spots; pitted
347	26	Free from rust stains	...	...
316	23	Free from rust stains	<$\frac{2}{3}$	Slight stains
316	...	...	6	Slight rust spots, slightly pitted
317	...	...	<$\frac{2}{3}$	Slight stains
317	...	...	6	Slight stains
310	...	...	<1	Rust stains
310	...	...	6	Rust spots; pitted

(a) Solution-annealed sheet, 1.6 mm ($\frac{1}{16}$ in.) thick

Table 23 Corrosion of AISI 300-series stainless steels in a marine atmosphere
Based on 15-year exposures 250 m (800 ft) from the ocean at Kure Beach, NC

AISI type	Average corrosion rate mm/year	mils/year	Average depth of pits mm	mils	Appearance(a)
301	<2.5 × 10^{-5}	<0.001	0.04	1.6	Light rust and rust stain on 20% of surface
302	<2.5 × 10^{-5}	<0.001	0.03	1.2	Spotted with rust stain on 10% of surface
304	<2.5 × 10^{-5}	<0.001	0.028	1.1	Spotted with slight rust stain on 15% of surface
321	<2.5 × 10^{-5}	<0.001	0.067	2.6	Spotted with slight rust stain on 15% of surface
347	2.5 × 10^{-5}	0.001	0.086	3.4	Spotted with moderate rust stain on 20% of surface
316	<2.5 × 10^{-5}	<0.001	0.025	1.0	Extremely slight rust stain on 15% of surface
317	<2.5 × 10^{-5}	<0.001	0.028	1.1	Extremely slight rust stain on 20% of surface
308	<2.5 × 10^{-5}	<0.001	0.04	1.6	Spotted with rust stain on 25% of surface
309	<2.5 × 10^{-5}	<0.001	0.028	1.1	Spotted with slight rust stain on 25% of surface
310	<2.5 × 10^{-5}	<0.001	0.01	0.4	Spotted with slight rust stain on 20% of surface

(a) All stains easily removed to reveal bright surface. Source: Ref 36

Table 24 SCC of U-bend test specimens 25 m (80 ft) from the ocean at Kure Beach, NC

Alloy	Final heat treatment	Hardness, HRC	Specimen orientation	Time to failure of each specimen, days(a)
Custom 450	Aged at 480 °C (900 °F)	42	Transverse	NF, NF, NF, NF, NF
Type 410	Tempered at 260 °C (500 °F)	45	Longitudinal	379, 379, 471
	Tempered at 550 °C (1025 °F)	35	Longitudinal	4, 4
Alloy 355	Tempered at 540 °C (1000 °F)	38	Longitudinal	NF, NF, NF
15Cr-7Ni-Mo	Aged at 510 °C (950 °F)	49	Longitudinal	1, 1, 1
17Cr-4Ni	Aged at 480 °C (900 °F)	42	Longitudinal	93, 129, NF
	Aged at 620 °C (1150 °F)	32	Longitudinal	93, 129, NF
14Cr-6Ni	Aged at 480 °C (900 °F)	39	Longitudinal	93, 872, NF

(a) NF, no failure in over 4400 days for Custom 450 and in 1290 days for the other materials. Source: Ref 38

severe industrial-chemical environment near plants producing chlorine and HCl.

Although marine environments can be severe, stainless steels often provide good resistance. Table 23 compares several AISI 300-series stainless steels after a 15-year exposure to a marine atmosphere 250 m (800 ft) from the ocean at Kure Beach, NC. Materials containing molybdenum exhibited only extremely slight rust stain, and all grades were easily cleaned to reveal a bright surface. Type 304 stainless steel may provide satisfactory resistance in many marine applications, but more highly alloyed grades are often selected when the stainless is sheltered from washing by the weather and is not cleaned regularly.

Types 302 and 304 stainless steels have had many successful architectural applications. Type 430 stainless steel has been used in many locations, but there have been problems. For example, this stainless steel rusted in sheltered areas after only a few months' exposure in an industrial environment. It was replaced by type 302, which provided satisfactory service. In more aggressive environments, such as marine or severely contaminated atmospheres, type 316 stainless steel is especially useful.

Stress-corrosion cracking is generally not a concern when austenitic or ferritic stainless steels are used in atmospheric exposures. Several austenitic stainless steels were exposed to a marine atmosphere at Kure Beach, NC. Annealed and quarter-hard wrought AISI types 201, 301, 302, 304, and 316 stainless steels were not susceptible to SCC. In the as-welded condition, only type 301 stainless steel experienced failure. Following sensitization at 650 °C (1200 °F) for 1.5 h and furnace cooling, failures were obtained only for materials with carbon contents of 0.043% or more (Ref 37).

Stress-corrosion cracking must be considered when quench-hardened martensitic stainless steels or precipitation-hardening grades are used in marine environments or in industrial locations in which chlorides are present. Several hardenable stainless grades were exposed as U-bends 25 m (80 ft) from the ocean at Kure Beach, NC. Most samples were cut longitudinally, and two alloys received different heat treatments to produce different hardness or strength levels. The results of the study (Table 24) indicated that Custom 450 stainless and stainless alloy 355 resisted cracking. Stainless alloy 355 failed in this type of test when fully hardened; resistance was imparted by the 540 °C (1000 °F) temper. Precipitation-hardenable grades are expected to exhibit improved corrosion resistance when higher aging temperatures (lower strengths) are used.

Resistance to SCC is of particular interest in the selection of high-strength stainless steels for fastener applications. Cracking of high-strength fasteners is possible and often results from hydrogen generation due to corrosion or contact with a less noble material, such as aluminum. Resistance to SCC can be improved by optimizing the heat treatment, as noted above.

Fasteners for atmospheric exposure have been fabricated from a wide variety of alloys. Type 430 and unhardened type 410 stainless steels have been used when moderate corrosion resistance is required in a lower-strength material. Better-than-average corrosion resistance has been obtained by using type 305 and Custom Flo 302HQ stainless steels when lower strength is acceptable.

Corrosion in Water. Water may vary from extremely pure to chemically treated water to highly concentrated chloride solutions, such as brackish water or seawater, which can be further concentrated by recycling. This chloride content poses the danger of pitting or crevice attack of stainless steels. When the application involves moderately increased temperatures, even as low as 45 °C (110 °F), and particularly where there is heat transfer into the chloride-containing medium, there is the possibility of SCC. It is useful to consider water with two general levels of chloride content: fresh water, which can have chloride levels up to approximately 600 ppm, and seawater, which encompasses brackish and severely contaminated waters. The corrosivity of a particular level of chloride can be strongly affected by the other chemical constituents present, making the water either more or less corrosive.

Permanganate ion (MnO_4^-), which is associated with the dumping of chemicals, has been related to the pitting of type 304 stainless steel. The presence of sulfur compounds and oxygen or other oxidizing agents can affect the corrosion of copper and copper alloys, but does not have very

Table 25 Crevice corrosion indexes of several alloys in tests in filtered seawater

Mill-finished panels exposed for 30, 60, and 90 days to seawater at 30 °C (85 °F) flowing at less than 0.1 m/s (0.33 ft/s); crevice washers tightened to 2.8 or 8.5 N · m (25 or 75 lbf · in.)

Alloy	UNS designation	Number of sides (S) attacked(a)	Maximum pit depth (D) mm	Maximum pit depth (D) mils	Crevice corrosion index (S × D)
AL-29-4C	S44735	0	nil	nil	0
MONIT	S44635	3	0.01	0.4	0.03
Ferralium 255	S32550	1	0.09	3.5	0.09
Alloy 904L	N08904	3	0.37	14.6	1.1(b)
254SMO	S31254	6	0.19	7.5	1.1
Sea-Cure	S44660	14	0.11	4.3	1.5
AL-6X	N08366	8	0.34	13.4	2.7
JS777	...	6	2.3	90.6	14(b)
JS700	N08700	14	1.8	70.9	24
AISI type 329	...	17	1.6	63	28(c)
Nitronic 50	S20910	17	1.2	47.2	20

(a) Total number of sides was 18. (b) Also showed tunneling attack perpendicular to the upper edge, or attack at edges. (c) Perforated by attack from both sides. Source: Ref 40

significant effects on stainless steels at ambient or slightly elevated temperatures (up to approximately 260 °C, or 500 °F).

In fresh water, type 304 stainless steel has provided excellent service for such items as valve parts, weirs, fasteners, and pump shafts in water and wastewater treatment plants. Custom 450 stainless steel has been used as shafts for large butterfly valves in potable water. The higher strength of a precipitation-hardenable stainless steel permits reduced shaft diameter and increased flow. Type 201 stainless steel has seen service in revetment mats to reduce shoreline erosion in fresh water. Type 316 stainless steel has been used as wire for microstrainers in tertiary sewage treatment and is suggested for waters containing minor amounts of chloride.

Seawater is a very corrosive environment for many materials. Stainless steels are more likely to be attacked in low-velocity seawater or at crevices resulting from equipment design or at attachments of barnacles. Types 304 and 316 stainless steels suffer deep pitting if the seawater flow rate decreases to below about 1.5 m/s (5 ft/s) because of the crevices produced by fouling organisms. However, in one study, type 316 stainless steel provided satisfactory service as tubing in the heat recovery section of a desalination test plant with relatively high flow rates (Ref 39).

The choice of stainless steel for seawater service can depend on whether stagnant conditions can be minimized or eliminated. For example, boat shafting of 17Cr-4Ni stainless steel has been used for trawlers where stagnant exposure and the associated pitting would not be expected to be a problem. When seagoing vessels are expected to lie idle for extended periods of time, more resistant boat shaft materials, such as 22Cr-13Ni-5Mn stainless steel, are considered. Boat shafts with intermediate corrosion resistance are provided by 18Cr-2Ni-12Mn and high-nitrogen type 304 (type 304HN) stainless steels.

The most severe exposure conditions are often used in seawater test programs. In one example of such data, flat-rolled specimens of 11 commercially available alloys with several mill finishes were exposed to seawater (Table 25). Triplicate samples were prepared with plastic multiple-crevice washers, each containing 20 plateaus or crevices. These washers were affixed to both sides of each panel by using a torque of either 2.8 or 8.5 N · m (25 or 75 lbf · in.). The panels were exposed for up to 90 days in filtered seawater flowing at a velocity of less than 0.1 m/s (0.33 ft/s).

Table 25 gives the number of sides that experienced crevice attack and the maximum attack depth at any crevice for that alloy. A crevice corrosion index (CCI) was calculated by multiplying the maximum attack depth by the number of sides attacked. This provided a ranking system that accounts for both initiation and growth of attack. Lower values of the CCI imply improved resistance.

Attack in the above test does not mean that materials with high CCIs cannot be used in seawater. For example, 22Cr-13Ni-5Mn stainless steel with a CCI of 20 has proved to be a highly resistant boat shaft alloy. Some of the more resistant materials in the above tests have been used for utility condenser tubing. These alloys include MONIT, AL-29-4C, 254SMO, Sea-Cure, and AL-6XN.

The possibility of galvanic corrosion must be considered if stainless steel is to be used in contact with other metals in seawater. Figure 15 provides corrosion potentials in flowing seawater for several materials. Preferably, only those materials that exhibit closely related electrode potentials should be coupled to avoid attack of the less noble material. Galvanic differences have been used to advantage in the cathodic protection of stainless steel in seawater. Crevice corrosion and pitting of austenitic types 302 and 316 stainless steels have been prevented by cathodic protection, but types 410 and 430 stainless steels develop hydrogen blisters at current densities below those required for complete protection.

Other factors that should be noted for applications of stainless steels in seawater include the effects of high velocity, aeration, and temperature. Stainless steels generally show excellent resistance to high velocities, impingement attack, and cavitation in seawater. Also, stainless steels provide optimum service in aerated seawater because a lack of aeration at a specific site often leads to crevice attack. Very little oxygen is required to maintain the passive film on a clean stainless surface. Increasing the temperature from ambient to about 50 °C (120 °F) often reduces the attack of stainless steels, possibly because of differences in the amount of dissolved oxygen, changes in the surface film, or changes in the resistance of the boldly exposed sample area (Ref 42). Further temperature increases can result in increased corrosion, such as SCC.

Corrosion in Chemical Environments. The selection of stainless steels for service in chemical environments requires consideration of all forms of corrosion, as well as impurity levels and degree of aeration. When an alloy with sufficient general corrosion resistance has been selected, care must be taken to ensure that the material will not fail by pitting or SCC due to chloride contamination. Aeration may be an important factor in corrosion, particularly in cases of borderline passivity. If dissimilar-metal contact or stray currents occur, the possibility of galvanic attack or hydrogen embrittlement must be considered.

Alloy selection also depends on fabrication and operation details. If a material is to be used in the as-welded or stress-relieved condition, it must resist intergranular attack in service after these thermal treatments. In chloride environments, the possibility of crevice corrosion must be considered when crevices are present because of equipment design or the formation of adherent deposits. Higher flow rates may prevent the formation of deposits, but in extreme cases may also cause accelerated attack because of erosion or cavitation. Increased operating temperatures generally increase corrosion. In heat-transfer applications, higher metal wall temperatures result in higher rates than expected from the lower temperature of the bulk solution. These and other items may require consideration in the selection of stainless steels, yet suitable materials continue to be chosen for a wide variety of chemical plant applications (see the article "Corrosion in the Chemical Processing Industry" in Volume 13 of the 9th Edition of *Metals Handbook*).

Some generalizations can be made regarding the performance of various categories of stainless steels in certain types of chemical environments. These observations

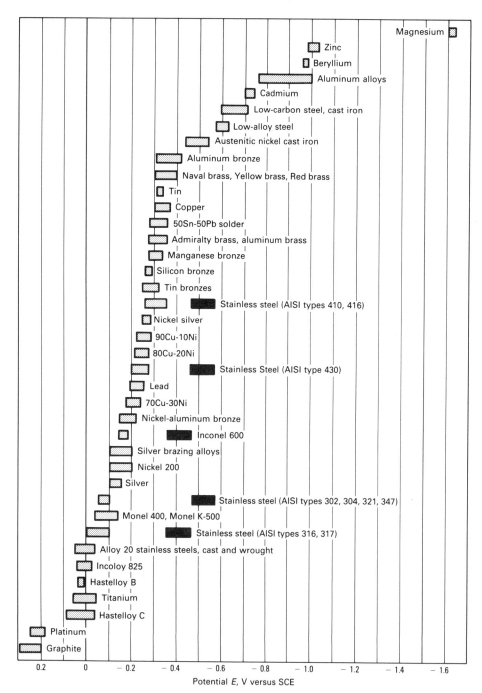

Fig. 15 Corrosion potentials of various metals and alloys in flowing seawater at 10 to 25 °C (50 to 80 °F). Flow rate was 2.5 to 4 m/s (8 to 13 ft/s); alloys are listed in order of the potential versus saturated calomel electrode (SCE) that they exhibited. Those metals and alloys indicated by black bar may become active and exhibit a potential near −0.5 V versus SCE in low velocity or poorly aerated water and in shielded areas. Source: Ref 41

pH of the acid because HNO_3 is highly oxidizing and forms a passive film because of the chromium content of the alloy. On the other hand, stainless steels are rapidly attacked by strong HCl because a passive film is not easily attained. Even in strong HNO_3, stainless steels can be rapidly attacked if they contain sufficient amounts of carbon and are sensitized. Oxidizing species, such as ferric salts, result in reduced general corrosion in some acids, but can cause accelerated pitting attack if chloride ions (Cl^-) are present.

In nitric acid (HNO_3), as noted above, stainless steels have broad applicability, primarily because of their chromium content. Most AISI 300-series stainless steels exhibit good or excellent resistance in the annealed condition in concentrations from 0 to 65% up to the boiling point. Figure 16 shows the good resistance of type 304 stainless steel, particularly when compared with the lower-chromium type 410 stainless steel. More severe environments at elevated temperatures require alloys with higher chromium. In HNO_3 cooler-condensers, such stainless alloys as 7-Mo PLUS (UNS S32950) and 2RE10 (UNS S31008) are candidates for service.

In sulfuric acid, stainless steels can approach the borderline between activity and passivity. Conventional ferritic grades, such as type 430, have limited use in H_2SO_4, but the newer ferritic grades containing higher chromium and molybdenum (for example, 28% Cr and 4% Mo) with additions of at least 0.25% Ni have shown good resistance in boiling 10% H_2SO_4 (Ref 45), but corrode rapidly when acid concentration is increased.

The conventional austenitic grades exhibit good resistance in very dilute or highly concentrated H_2SO_4 at slightly elevated temperatures. Acid of intermediate concentration is more aggressive, and conventional grades have very limited use. The resistance of several stainless steels in up to about 50% H_2SO_4 is shown in Fig. 17. Aeration or the addition of oxidizing species can significantly reduce the attack of stainless steels in H_2SO_4. This occurs because the more oxidizing environment is better able to maintain the chromium-rich passive oxide film.

Improved resistance to H_2SO_4 has been obtained by using austenitic grades containing high levels of nickel and copper, such as 20Cb-3 stainless steel. In addition to reducing general corrosion, the increased nickel provides resistance to SCC. Because of its resistance to these forms of corrosion, 20Cb-3 stainless steel has been used for valve springs in H_2SO_4 service.

In phosphoric acid (H_3PO_4), conventional straight-chromium stainless steels have very limited general corrosion resistance and exhibit lower rates only in very dilute or more highly concentrated solutions. Con-

relate to the compositions of the grades. For example, the presence of nickel and copper in some austenitic grades greatly enhances the resistance to H_2SO_4 compared to the resistance of the ferritic grades. However, combinations of chemicals that are encountered in practice can be either more or less corrosive than might be expected from the corrosivity of the individual components. Testing in actual or simulated environments is always recommended as the best procedure for selecting a stainless steel grade. Additional information describing service experience is available from alloy suppliers.

Mineral Acids. The resistance of stainless steel to acids depends on the hydrogen ion (H^+) concentration and the oxidizing capacity of the acid, along with such material variables as chromium content, nickel content, carbon content, and heat treatment (Ref 43). For example, annealed stainless steel resists strong HNO_3 in spite of the low

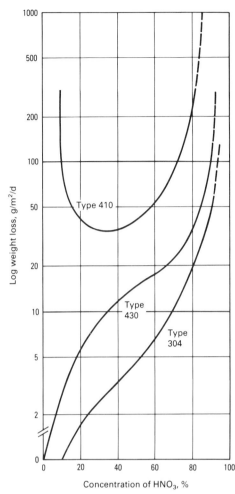

Fig. 16 Corrosion rates of various stainless steels in boiling HNO₃. Source: Ref 44

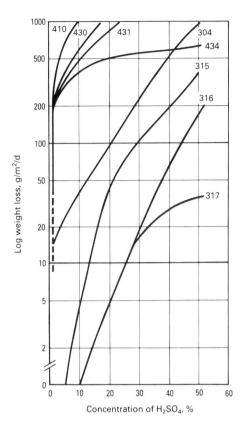

Fig. 17 Corrosion rates of various stainless steels in underaerated H₂SO₄ at 20 °C (70 °F). Source: Ref 44

Table 26 Corrosion of austenitic stainless steels in boiling glacial acetic acid
Data are from averaged results of 11-, 12-, and 21-day field tests.

AISI type	Corrosion rate mm/year	mils/year
304	0.46	18
321	1.19	47
347	1.04	41
308	1.35	53
310	0.99	39
316	0.015	0.6

Source: Ref 46

ventional austenitic stainless steels provide useful general corrosion resistance over the full range of concentrations up to about 65 °C (150 °F); use at temperatures up to the boiling point is possible for acid concentrations up to about 40%.

In commercial applications, however, wet-process H_3PO_4 environments include impurities derived from the phosphate rock, such as chlorides, fluorides, and H_2SO_4. These three impurities accelerate corrosion, particularly pitting or crevice corrosion in the presence of the halogens. Higher-alloyed materials than the conventional austenitic stainless steels are required to resist wet-process H_3PO_4. Candidate materials include alloy 904L, alloy 28, 20Cb-3, 20Mo-4, and 20Mo-6 stainless steels.

Hydrochloric acid service is generally not an application for stainless steels, except perhaps for very dilute solutions at room temperature. Stainless materials can be susceptible to accelerated general corrosion, SCC, and pitting in HCl environments.

Sulfurous acid (H_2SO_3) is a reducing agent; several stainless steels have provided satisfactory service in H_2SO_3 environments. Conventional austenitic stainless steels have been used in sulfite digesters, and type 316, type 317, 20Cb-3, and cast Alloy Casting Institute (ACI) alloys CF-8M and CN-7M stainless steels have seen service in wet sulfur dioxide (SO_2) and H_2SO_3 environments. Cast stainless steels are discussed in the article "Cast Stainless Steels" in this Volume. Service life is improved by eliminating crevices, including those formed from the settling of suspended solids, or by using molybdenum-containing grades. In some environments, SCC is also a possibility.

Organic acids and compounds are generally less aggressive than mineral acids because they do not ionize as completely, but they can be corrosive to stainless steels, especially when impurities are present. The presence of oxidizing agents in the absence of chlorides can reduce corrosion rates.

For pure acetic acid, corrosion resistance has been obtained by using types 316 and 316L stainless steels over all concentrations up to the boiling point. Type 304 stainless steel may be considered in all concentrations below about 90% at temperatures up to the boiling point. Corrosion rates for several stainless steels in acetic acid are listed in Table 26. Impurities present in the manufacture of acetic acid, such as acetaldehyde, formic acid, chlorides, and propionic acid, are expected to increase the attack of stainless steels. Chlorides may cause pitting or SCC.

Formic acid is one of the more aggressive organic acids, and corrosion rates can be higher in the condensing vapor than in the liquid. Type 304 stainless steel has been used at moderate temperatures. However, type 316 stainless steel or higher alloys, such as 20Cb-3, are often preferred, and high-alloy ferritic stainless steels containing 26% Cr and 1% Mo or 29% Cr and 4% Mo also show some promise.

The corrosivity of propionic and acrylic acids at a given temperature is generally similar to that of acetic acid. Impurities are important and may strongly affect the corrosion rate. In citric and tartaric acids, type 304 stainless steel has been used for moderate temperatures, and type 316 has been suggested for all concentrations up to the boiling point.

Most dry organic halides do not attack stainless steels, but the presence of water allows halide acids to form and can cause pitting or SCC. Therefore, care should be exercised when using stainless steels in organic halides to ensure that water is excluded.

Type 304 stainless steel has generally been satisfactory in aldehydes, in cellulose acetate at lower temperatures, and in fatty acids up to about 150 °C (300 °F); at higher temperatures, these chemicals require types 316 or 317. Type 316 stainless steel is also used in amines, phthalic anhydride, tar, and urea service.

Stainless steels have been used in the plastics and synthetic fiber industries. Types 420 and 440C stainless steels have been used as plastic mold steels. More resistant materials, such as Custom 450, have been used for extruding polyvinyl chloride (PVC) pipe. Spinnerettes, pack parts, and metering pumps of Custom 450 and Custom 455 stainless steels have been used in the synthetic fiber industry to produce nylon, rayon, and polyesters.

Alkalies. All stainless steels resist general corrosion by all concentrations of sodium hydroxide (NaOH) up to about 65 °C (150 °F). Types 304 and 316 stainless steels exhibit low rates of general corrosion in boil-

Table 27 Generally accepted maximum service temperatures in air for AISI stainless steels

AISI type	Maximum service temperature			
	Intermittent service		Continuous service	
	°C	°F	°C	°F
Austenitic grades				
201	815	1500	845	1550
202	815	1500	845	1550
301	840	1545	900	1650
302	870	1600	925	1700
304	870	1600	925	1700
308	925	1700	980	1795
309	980	1795	1095	2000
310	1035	1895	1150	2100
316	870	1600	925	1700
317	870	1600	925	1700
321	870	1600	925	1700
330	1035	1895	1150	2100
347	870	1600	925	1700
Ferritic grades				
405	815	1500	705	1300
406	815	1500	1035	1895
430	870	1600	815	1500
442	1035	1895	980	1795
446	1175	2145	1095	2000
Martensitic grades				
410	815	1500	705	1300
416	760	1400	675	1250
420	735	1355	620	1150
440	815	1500	760	1400

Source: Ref 48

occurs. Table 27 lists generally accepted maximum safe service temperatures for wrought stainless steels. Maximum temperatures for intermittent service are lower for the austenitic stainless steels, but are higher for most of the martensitic and ferritic stainless steels listed.

Contamination of the air with water and CO_2 often increases corrosion at elevated temperatures. Increased attack can also occur because of sulfidation as a result of SO_2, H_2S, or sulfur vapor.

Carburization of stainless steels can occur in carbon monoxide (CO), methane (CH_4), and other hydrocarbons. Carburization can also occur when stainless steels contaminated with oil or grease are annealed without sufficient oxygen to burn off the carbon. This can occur during vacuum or inert-gas annealing, as well as in open-air annealing of oily parts with shapes that restrict air access. Chromium, silicon, and nickel are useful in combating carburization.

Nitriding can occur in dissociated NH_3 at high temperatures. Resistance to nitriding depends on alloy composition as well as NH_3 concentration, temperature, and pressure. Stainless steels are readily attacked in pure NH_3 at about 540 °C (1000 °F).

Liquid Metals. The 18-8 stainless steels are highly resistant to liquid sodium or sodium-potassium alloys. Mass transfer is not expected up to 540 °C (1000 °F) and remains at moderately low levels up to 870 °C (1600 °F). The accelerated attack of stainless steels in liquid sodium occurs with oxygen contamination, with a noticeable effect occurring at about 0.02% oxygen by weight (Ref 43).

Exposure to molten lead under dynamic conditions often results in mass transfer in common stainless alloy systems. Particularly severe corrosion can occur in strongly oxidizing conditions. Stainless steels are generally attacked by molten aluminum, zinc, antimony, bismuth, cadmium, and tin.

Corrosion in Various Applications

Every industry features a variety of applications encompassing a range of corrosion environments. This section characterizes the experience of each industry according to the corrosion problems most frequently encountered and suggests appropriate grade selections.

Food and Beverage Industry. Stainless steels have been relied upon in these applications because of the lack of corrosion products that could contaminate the process environment and because of the superior cleanability of the stainless steels. The corrosion environment often involves moderately to highly concentrated chlorides on the process side, often mixed with significant concentrations of organic acids. The water side can range from steam heating to brine cooling. Purity and sanitation standards require excellent resistance to pitting and crevice corrosion.

Foods such as vegetables represent milder environments and can generally be handled by using type 304 stainless steel. Sauces and pickle liquors, however, are more aggressive and can pit even type 316 stainless steel. For improved pitting resistance, such alloys as 22Cr-13Ni-5Mn, 904L, 20Mo-4, 254SMO, AL-6XN, and MONIT stainless steels should be considered.

At elevated temperatures, materials must be selected for resistance to pitting and SCC in the presence of chlorides. Stress corrosion must be avoided in heat transfer applications, such as steam jacketing for cooking or processing vessels or in heat exchangers. Cracking may occur from the process or water side or may initiate outside the unit under chloride-containing insulation. Brewery applications of austenitic stainless steels have been generally successful except for a number of cases of the SCC of high-temperature water lines. The use of ferritic or duplex stainless steels is an appropriate remedy for the SCC. More information on this subject is available in the article "Corrosion in the Brewery Industry" in Volume 13 of the 9th Edition of *Metals Handbook*.

Stainless steel equipment should be cleaned frequently to prolong its service life. The equipment should be flushed with fresh water, scrubbed with a nylon brush and detergent, and rinsed. On the other hand, consideration should be given to the effect of very aggressive cleaning procedures on the stainless steels, as in the chemical sterilization of commercial dishwashers. In some cases, it may be necessary to select a more highly alloyed stainless steel grade to deal with these brief exposures to highly aggressive environments.

Conventional AISI grades provide satisfactory service in many food and beverage applications. Type 304 stainless steel is widely used in the dairy industry, and type 316 finds application as piping and tubing in breweries. These grades, along with type 444 and Custom 450 stainless steels, have been used for chains to transfer food through processing equipment. Machined parts for beverage-dispensing equipment have been fabricated from types 304, 304L, 316, 316L, and 303Al MODIFIED, 302HQ-FM, and 303BV stainless steels. When the free-machining grades are used, it is important to passivate and rinse properly before service in order to optimize corrosion resistance.

Food-handling equipment should be designed without crevices in which food can become lodged. In more corrosive food products, extralow-carbon stainless steels should be used when possible. Improved results have been obtained when equipment is finished with a 2B (general-purpose cold-rolled) finish rather than No. 4 (general-purpose polished) finish. Alternatively, an electropolished surface may be considered.

ing NaOH up to nearly 20% concentration. Stress-corrosion cracking of these grades can occur at about 100 °C (212 °F). Good resistance to general corrosion and SCC in 50% NaOH at 135 °C (275 °F) is provided by E-Brite and 7-Mo stainless steels (Ref 47). In ammonia (NH_3) and ammonium hydroxide (NH_4OH), stainless steels have shown good resistance at all concentrations up to the boiling point.

Salts. Stainless steels are highly resistant to most neutral or alkaline nonhalide salts. In some cases, type 316 is preferred for its resistance to pitting, but even the higher-molybdenum type 317 stainless steel is readily attacked by sodium sulfide (Na_2S) solutions.

Halogen salts are more corrosive to stainless steels because of the ability of the halide ions to penetrate the passive film and cause pitting. Pitting is promoted in aerated or mildly acidic oxidizing solutions. Chlorides are generally more aggressive than the other halides in their ability to cause pitting.

Gases. At lower temperatures, most austenitic stainless steels resist chlorine or fluorine gas if the gas is completely dry. The presence of even small amounts of moisture results in accelerated attack, especially pitting, and possibly SCC.

At elevated temperatures, stainless steels resist oxidation primarily because of their chromium content. Increased nickel minimizes spalling when temperature cycling

Pharmaceutical Industry. The production and handling of drugs and other medical applications require exceedingly high standards for preserving the sterility and purity of process streams. Process environments can include complex organic compounds, strong acids, chloride solutions comparable to seawater, and elevated processing temperatures. Higher-alloy grades, such as type 316 or higher, may be necessary instead of type 304 in order to prevent even superficial corrosion. Electropolishing may be desirable in order to reduce or prevent adherent deposits and the possibility of underdeposit corrosion. Superior cleanability and ease of inspection make stainless steel the preferred material.

The 18-8 stainless grades have been used for a wide variety of applications from pill punches to operating tables. However, care is required in selecting stainless steels for pharmaceutical applications because small amounts of contamination can be objectionable. For example, stainless steel has been used to process vitamin C, but copper must be eliminated because copper in aqueous solutions accelerates the decomposition of vitamin C. Also, stainless steel is not used to handle vitamin B_6 hydrochloride, even though corrosion rates may be low, because trace amounts of iron are objectionable (see the article "Stainless Steels in Corrosion Service" in Volume 3 of the 9th Edition of *Metals Handbook*).

The effects of temperature and chloride concentration must be considered. At ambient temperature, chloride pitting of 18Cr-8Ni stainless steel may occur, but SCC is unlikely. At about 65 °C (150 °F) or above, the SCC of austenitic grades must be considered. Duplex alloys, such as 7-Mo PLUS, alloy 2205, and Ferralium 255, possess improved resistance to SCC in elevated-temperature chloride environments. Ferritic grades with lower nickel content, such as 18Cr-2Mo stainless steel, provide another means of avoiding chloride SCC. Additional information on the use of stainless steels in the pharmaceutical industry is available in the article "Corrosion in the Pharmaceutical Industry" in Volume 13 of the 9th Edition of *Metals Handbook*.

Stainless steels have also found application as orthopedic implants. Material is required that is capable of moderately high strength and resistance to wear and fretting corrosion and to pitting and crevice attack. Vacuum-melted type 316 stainless steel has been used for temporary internal fixing devices, such as bone plates, screws, pins, and suture wire. Higher purity improves electropolishing and increased chromium (17 to 19%) improves corrosion resistance.

In permanent implants, such as artificial joints, very high strength and resistance to wear, fatigue, and corrosion are essential. Cobalt- or titanium-base alloys are used for these applications. More information on this subject is available in the article "Corrosion of Metallic Implants and Prosthetic Devices" in Volume 13 of the 9th Edition of *Metals Handbook*.

Oil and Gas Industry. Stainless steels were not frequently used in oil and gas production until the tapping of sour reservoirs (those containing hydrogen sulfide, H_2S) and the use of enhanced recovery systems in the mid-1970s. Sour environments can result in sulfide stress cracking (SSC) of susceptible materials. This phenomenon generally occurs at ambient or slightly elevated temperatures; it is difficult to establish an accurate temperature maximum for all alloys. Factors affecting SSC resistance include material variables, pH, H_2S concentration, total pressure, maximum tensile stress, temperature, and time. A description of some of these factors, along with information on materials that have demonstrated resistance to SSC, is available in Ref 49.

The resistance of stainless steels to SSC improves with reduced hardness. Conventional materials, such as types 410, 430, and 304 stainless steels exhibit acceptable resistance at hardnesses below 22 HRC. Specialized grades, such as 22Cr-13Ni-5Mn, Custom 450, 20Mo-4, and some duplex stainless steels, have demonstrated resistance at higher hardnesses. Duplex alloy 2205 has been used for its strength and corrosion resistance as gathering lines for CO_2 gas before gas cleaning. Custom 450 and 22Cr-13Ni-5Mn stainless steels have seen service as valve parts. Other grades used in these environments include 254SMO and alloy 28, particularly for chloride and sulfide resistance, respectively.

In addition to the lower-temperature SSC, resistance to cracking in high-temperature environments is required in many oil field applications. Most stainless steels, including austenitic and duplex grades, are known to be susceptible to elevated-temperature cracking, probably by a mechanism similar to chloride SCC. Failure appears to be accelerated by H_2S and other sulfur compounds. Increased susceptibility is noted in material of higher yield strength, for example, because of the high residual tensile stresses imparted by some cold-working operations.

The above discussion is pertinent to the production phase of a well. However, drilling takes place in an environment of drilling mud, which usually consists of water, clay, weighting materials, and an inhibitor (frequently an oxygen scavenger). Chlorides are also present when drilling through salt formations. Austenitic stainless steels containing nitrogen have found use in this environment as nonmagnetic drill collars, as weight for the drill bit, and as housings for measurement-while-drilling (MWD) instruments. Nonmagnetic materials are required for the operation of these instruments, which are used to locate the drill bit in directional-drilling operations. Stainless steels used as drill collars or measurement-while-drilling components include the standard 316LN type stainless steel (Table 2) and several nonstandard stainless steels. More information on corrosion during petroleum production is available in the article "Corrosion in Petroleum Production Operations" in Volume 13 of the 9th Edition of *Metals Handbook*.

In refinery applications, the raw crude contains such impurities as sulfur, water, salts, organic acids, and organic nitrogen compounds. These and other corrosives and their products must be considered in providing stainless steels for the various refinery steps.

Raw crude is separated into materials from petroleum gas to various oils by fractional distillation. These materials are then treated to remove impurities, such as CO_2, NH_3, and H_2S, and to optimize product quality. Refinery applications of stainless steels often involve heat exchangers. Duplex and ferritic grades have been used in this application for their improved SCC resistance. Type 430 and type 444 stainless steel exchanger tubing has been used for resisting hydrogen, chlorides, and sulfur and nitrogen compounds in oil refinery streams. The article "Corrosion in Petroleum Refining and Petrochemical Operations" in Volume 13 of the 9th Edition of *Metals Handbook* contains detailed information on the corrosion of materials in these applications.

Power Industry. Stainless steels are used in the power industry for generator components, feedwater heaters, boiler applications, heat exchangers, condenser tubing, flue gas desulfurization (FGD) systems, and nuclear power applications.

Generator blades and vanes have been fabricated of modified 12% Cr stainless steel, such as ASTM types 615 (UNS S41800) and 616 (UNS S42200). In some equipment, Custom 450 has replaced AISI type 410 and ASTM type 616 stainless steels.

Heat Exchangers. Stainless steels have been widely used in tubing for surface condensers and feedwater heaters. Both of these are shell and tube heat exchangers that condense steam from the turbine on the shell side. In these heat exchangers, the severity of the corrosion increases with higher temperatures and pressures. Stainless steels resist failure by erosion and do not suffer SCC in NH_3 (from the decomposition of boiler feedwater additives), as do some nonferrous materials.

Stainless steels must be chosen to resist chloride pitting. The amount of chloride that can be tolerated is expected to be higher with a higher pH and cleaner stainless steel surfaces, that is, the absence of

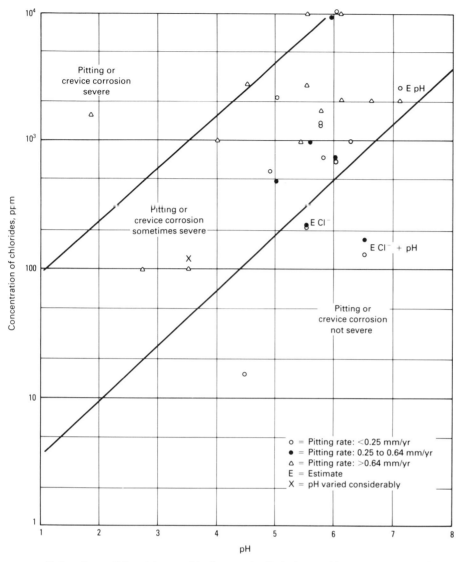

Fig. 18 Pitting of type 316L stainless steel in flue gas desulfurization scrubber environments. Solid lines indicate zones of differing severity of corrosion; because the zones are not clearly defined, the lines cannot be precisely drawn. Source: Ref 52

deposits. For example, type 304 stainless steel may resist pitting in chloride levels of 1000 ppm or higher in the absence of fouling, crevices, or stagnant conditions. The presence of one or more of these conditions can allow chlorides to concentrate at the metal surface and initiate pits. Several high-performance stainless steels have been used to resist chloride pitting in brackish water or seawater. High-performance austenitic grades have been useful in feedwater heaters, although duplex stainless steels may also be considered because of their high strength. Ferritic stainless steels have proved to be economically competitive in exchangers and condensers. High-performance austenitic and ferritic grades have been satisfactory for seawater-cooled units. These grades include MONIT, AL-29-4C, Usinor 290 Mo, Sea-Cure, AL-6X, AL-6XN, and 254SMO stainless steels.

Compatibility of materials and good installation practice are required. Tubes of such materials as those listed above have been installed in tubesheets fabricated of alloy 904L, 20Mo-4, and 254SMO stainless steels. Crevice corrosion can occur when some tube materials are rolled into type 316 stainless steel tubesheets (Ref 50). Appropriate levels of cathodic protection have been identified (Ref 51).

Flue Gas Desulfurization. A wide variety of alloys have been used in scrubbers, which are located between the boiler and smokestack of fossil fuel power plants to treat effluent gases and to remove SO_2 and other pollutants. Typically, fly ash is removed, and the gas travels through an inlet gas duct and then the quencher section. Next, SO_2 is removed in the absorber section, most often using either a lime or limestone system. A mist eliminator is employed to remove suspended droplets, and the gas proceeds to the treated-gas duct, reheater section, and the stack.

Two important items for consideration in selecting stainless steels for resistance to pitting in scrubber environments are pH and chloride level. Stainless steels are more resistant to higher pH and lower chloride levels, as shown in Fig. 18 for type 316L stainless steel. Environments that cause the pitting or crevice attack of type 316 stainless steel can be handled by using higher-alloy materials, for example, those with increased molybdenum and chromium.

Some of the materials being considered and specified for varying chloride levels are given in Ref 53. Other materials can also provide good resistance, as evidenced by the results given in Table 28 for samples exposed to several scrubber environments. The maximum depth of localized corrosion and pit density are given for the stainless steels tested. Exposure at the quencher spray header (above slurry) was more severe than expected, probably because of wet-dry concentration effects. Severe attack also occurred in the outlet duct. Samples in this area were exposed to high chlorides, high temperatures, and low pH during the 39 days on bypass operation. More information on corrosion in fossil fuel power generation is available in the article "Corrosion in Fossil Fuel Power Plants" in Volume 13 of the 9th Edition of *Metals Handbook*; corrosion in FGD systems is also discussed in the article "Corrosion of Emission-Control Equipment" in Volume 13 of the 9th Edition of *Metals Handbook*.

Nuclear Power Applications. Type 304 stainless steel piping has been used in boiling-water nuclear reactor plants. The operating temperatures of these reactors are about 290 °C (550 °F), and a wide range of conditions can be present during startup, operation, and shutdown. Because these pipes are joined by welding, there is a possibility of sensitization. This can result in intergranular SCC in chloride-free high-temperature water that contains small amounts of oxygen, for example, 0.2 to 8 ppm. Nondestructive electrochemical tests have been used to evaluate weldments for this service (Ref 55).

Type 304 stainless steel with additions of boron (about 1%) has been used to construct spent-fuel storage units, dry storage casks, and transportation casks. The high boron level provides neutron-absorbing properties. More information on nuclear applications is available in the article "Corrosion in the Nuclear Power Industry" in Volume 13 of the 9th Edition of *Metals Handbook*.

Pulp and Paper Industry. In the kraft process, paper is produced by digesting wood chips with a mixture of Na_2S and NaOH (white liquor). The product is transferred to the brown stock washers to remove the

Table 28 Pitting of stainless steel spool test specimens in an FGD system
The slurry contained 7000 ppm dissolved Cl⁻; test duration was 6 months, with 39 days on bypass.

Spool location(a)	pH	Maximum temperature °C	Maximum temperature °F	Maximum chloride concentration, ppm	Maximum pit depth, mm (mils), and pit density							
					Type 304	Type 316L	Type 317L	Type 317LM	JS700	JS777	904L	20Mo-6
Wet/dry line at inlet duct	1–2(b)	60–170	140–335	7 000(b)	>1.24 (>49) Profuse	>0.91 (>36) Profuse	0.53 (21) Sparse	0.53 (21) Sparse	0.33 (13) Sparse	0.33 (13) Profuse	0.43 (17) Sparse	(c)
Quencher sump (submerged; 1.8 m, or 6 ft, level)	4.4	60	140	7 000	>1.19 (>47) Sparse	>0.91 (>36) Sparse	0.28 (11)	0.1 (4) Single	nil	nil	nil	nil
Quencher sump (submerged; 3.4 m, or 11 ft level)	4.4	60	140	7 000	>1.2 (>48) Profuse	>0.9 (>36) Sparse	<0.03 (<1)	0.05 (2)	nil	nil	nil	nil
Quencher spray header, above slurry	4.4	60	140	100	>1.19 (>47) Profuse	0.58 (23) Profuse	0.61 (24) Profuse	0.46 (18) Profuse	0.33 (13) Sparse	0.61 (24) Profuse	0.25 (10) Sparse	0.15 (6) Sparse
Absorber, spray area	6.2	60	140	100	0.58 (23) Sparse	0.10 (4)	nil	nil	nil	nil	nil	nil
Outlet duct	2–4(d)	55	130(d)	100(d)	>1.19 (>47) Profuse	>0.91 (>36) Profuse	0.58 (23) Profuse	0.58 (23) Profuse	0.18 (7) Single	0.51 (20) Profuse	0.53 (21) Profuse	0.36 (14) IG etch
	1.5(e)	170	335(e)	82 000(e)								

(a) Slurry contained 7000 ppm dissolved Cl⁻. Deposits in the quencher, inlet duct, absorber, and outlet ducting contained 3000 to 4000 ppm Cl⁻ and 800 to 1900 ppm F⁻. (b) Present as halide gases. (c) Not tested. (d) During operation. (e) During bypass. Bypass condition gas stream contained SO_2, SO_3, HCl, HF, and condensate. Source: Ref 54

liquor (black liquor) from brown pulp. After screening, the pulp may go directly to the paper mill to produce unbleached paper or may be directed first to the bleach plant to produce white paper.

The digester vapors are condensed, and the condensate is pumped to the brown stock washers. The black liquor from these washers is concentrated and burned with sodium sulfate (Na_2SO_4) to recover sodium carbonate (Na_2CO_3) and Na_2S. After dissolution in water, this green liquor is treated with calcium hydroxide ($Ca(OH)_2$) to produce NaOH to replenish the white liquor. Pulp bleaching involves treating with various chemicals, including chlorine, chlorine dioxide (ClO_2), sodium hypochlorite (NaClO), calcium hypochlorite ($Ca(ClO)_2$), peroxide, caustic soda, quicklime, and oxygen.

The sulfite process uses a liquor in the digester that is different from that used in the kraft process. This liquor contains free SO_2 dissolved in water, along with SO_2 as a bisulfite. The compositions of the specific liquors differ, and the pH can range from 1 for an acid process to 10 for alkaline cooking. Sulfur dioxide for the cooking liquor is produced by burning elemental sulfur, cooling rapidly, absorbing the SO_2 in a weak alkaline solution, and fortifying the raw acid.

Various alloys are selected for the wide range of corrosion conditions encountered in pulp and paper mills. Paper mill headboxes are typically fabricated from type 316L stainless steel plate with superior surface finish and are sometimes electropolished to prevent scaling, which may affect pulp flow. The blades used to remove paper from the drums are fabricated from types 410 and 420 stainless steels and from cold reduced 22Cr-13Ni-5Mn stainless steel.

Evaporators and reheaters must deal with corrosive liquors and must minimize scaling to provide optimum heat transfer. Type 304 stainless steel ferrite-free welded tubing has been used in kraft black-liquor evaporators. Cleaning is often performed with HCl, which attacks ferrite. In the sulfite process, type 316 (>2.75% Mo) and type 317 stainless steels have been used in black-liquor evaporators. Digester-liquor heaters in the kraft and sulfite processes have used 7-Mo stainless for resistance to caustic or chloride SCC.

Bleach plants have used types 316 and 317 stainless steels and are upgrading to austenitic grades containing 4.5 and 6% Mo in problem locations. The tightening of environmental regulations has generally increased temperature, chloride level, and acidity in the plant, and this requires grades of stainless steel that are more highly alloyed than those used in the past. Tall oil units have shifted from types 316 and 317 stainless steels to such candidate alloys as 904L and 20Mo-4 stainless steels and most recently to 254SMO and 20Mo-6 stainless steels.

Tests including higher-alloyed materials have been coordinated by the Metals Subcommittee of the Corrosion and Materials Engineering Committee of the Technical Association of the Pulp and Paper Industry (TAPPI). Racks of test samples, which included crevices at polytetrafluoroethylene (PTFE) spacers, were submerged in the vat below the washer in the C (chlorination), D (chlorine dioxide), and H (hypochlorite) stages of several paper mills. The sum of the maximum attack depth on all samples for each alloy—at crevices and remote from crevices—is shown in Fig. 19. It should be noted that the vertical axes are different in Fig. 19(a), (b), and (c). Additional information on corrosion in this industry is available in the article "Corrosion in the Pulp and Paper Industry" in Volume 13 of the 9th Edition of *Metals Handbook*.

Transportation Industry. A wide range of functional and decorative components for transportation vehicles are fabricated from stainless steels.

Automobile. Bright automobile parts, such as trim, fasteners, wheel covers, mirror mounts, and windshield wiper arms, have generally been fabricated from 17Cr or 18Cr-8Ni stainless steel or similar grades. Example alloys include types 430, 434, 304, and 305 stainless steels. Type 302HQ-FM remains a candidate for such applications as wheel nuts, and Custom 455 stainless has been used as wheel lock nuts. The use of type 301 stainless steel for wheel covers has diminished with the weight reduction programs of the automotive industry.

Stainless steels also serve many nondecorative functions in automotive design. Small-diameter shafts of type 416 and, occasionally, type 303 stainless steels have been used in connection with power equipment, such as windows, door locks, and antennas. Solenoid grades, such as type 430FR stainless steels, have also found application. Type 409 stainless steel has been used for mufflers and catalytic converters for many years, but it is now being employed through the exhaust system. The article "Corrosion in the Automotive Industry" in Volume 13 of the 9th Edition of *Metals Handbook* contains detailed information on corrosion in the automotive environment.

Railroad Cars and Large Vehicles. In railroad cars, external and structural stainless steels provide durability, low-cost

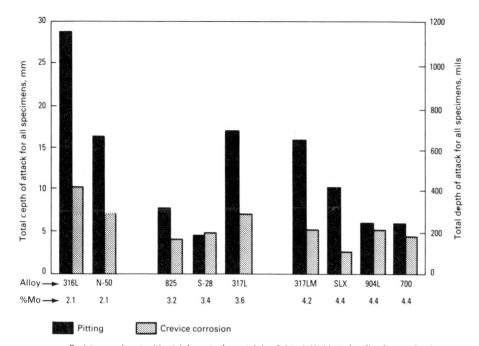

Fig. 19(a) Resistance of austenitic stainless steels containing 2.1 to 4.4% Mo to localized corrosion in a paper mill bleach plant environment. Total depth of attack has been divided by 4 because there were four crevice sites per specimen. See also Fig. 19(b) and 19(c). Source: Ref 56

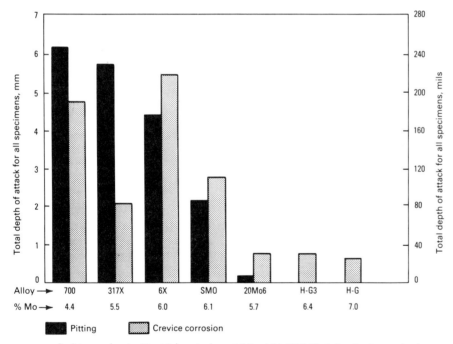

Fig. 19(b) Resistance of austenitic stainless steels containing 4.4 to 7.0% Mo to localized corrosion in a paper mill bleach plant environment. Total depth of attack has been divided by 4 because there were four crevice sites per specimen. See also Fig. 19(a) and 19(c). Source: Ref 56

maintenance, and superior safety through crashworthiness. The fire resistance of stainless steel is a significant safety advantage. Modified type 409 stainless steel is used as a structural component in buses. Types 430 and 304 are used for exposed functional parts on buses. Type 304 stainless steel has provided economical performance in truck trailers. For tank trucks, type 304 has been the most frequently used stainless steel, but type 316 and higher-alloyed grades have been used where appropriate to carry more corrosive chemicals safely over the highways.

Ships. Stainless steels are used for seagoing chemical tankers with types 304, 316, 317, and alloy 2205 being selected according to the corrosivity of the cargoes being carried. Conscientious adherence to cleaning procedures between cargo changeovers has allowed these grades to give many years of service with a great variety of corrosive cargoes.

Aerospace. In aerospace, quench-hardenable and precipitation-hardenable stainless steels have been used in varying applications. Heat treatments are chosen to optimize fracture toughness and resistance to SCC. Stainless steel grades 15-5PH and PH13-8Mo have been used in structural parts, and PH3-8Mo stainless steel has served as fasteners. Parts in cooler sections of the engine have been fabricated from type 410 stainless steel. Custom 455, 17-4PH, 17-7PH, and 15-5PH stainless steels have been used in the space shuttle program (see the articles "Corrosion in the Aircraft Industry" and "Corrosion in the Aerospace Industry" in Volume 13 of the 9th Edition of *Metals Handbook*).

Architectural Applications. Typically, types 430 or 304 have been used in architectural applications. In bold exposure, these grades are generally satisfactory; however, in marine and industrially contaminated atmospheres, type 316 is often suggested and has performed well (see the article "Corrosion in Structures" in Volume 13 of the 9th Edition of *Metals Handbook*).

In all applications, but particularly in the cases listed above in which appearance is important, it is essential that any chemical cleaning solutions be thoroughly rinsed from the metal.

Corrosion Testing

The physical and financial risks involved in selecting stainless steels for particular applications can be reduced through consideration of corrosion tests. However, care must be taken when selecting a corrosion test. The test must relate to the type of corrosion possible in the application. The steel should be tested in the condition in which it will be applied. The test conditions should be representative of the operating conditions and all reasonably anticipated excursions of operating conditions.

Corrosion tests vary in their degree of simulation of operation in terms of the design of the specimen and the selection of medium and test conditions. Standard tests use specimens of defined nature and geometry exposed in precisely defined media and conditions. Standard tests can confirm that a particular lot of steel conforms to the level of performance of standard and proprietary grades. The relevance of test results to performance in particular applications increases as the specimen is made to resemble more closely the final fabricated structure, for example, bent, welded, stressed, or creviced. Relevance also increases as the test medium and conditions more closely approach the most severe operating conditions. However, many types of failures occur only after extended exposures to operating cycles. Therefore, there is often an

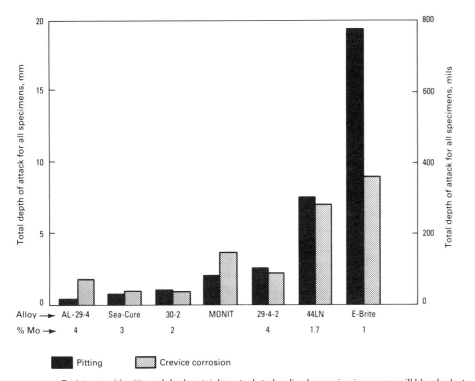

Fig. 20 Assembled crevice corrosion test specimen. Source: Ref 59

Fig. 19(c) Resistance of ferritic and duplex stainless steels to localized corrosion in a paper mill bleach plant environment. Total depth of attack has been divided by 4 because there were four crevice sites per specimen. See also Fig. 19(a) and 19(b). Source: Ref 56

Figure 21 shows one of several frequently used specimens with a multiple-crevice assembly. The presence of many separate crevices helps to deal with the statistical nature of corrosion initiation. The severity of the crevices can be regulated by means of a standard crevice design and the use of a selected torque in its application.

Laboratory media do not necessarily have the same response of corrosivity as a function of temperature as do engineering environments. For example, the ASTM G 48 solution is thought to be roughly comparable to seawater at ambient temperatures. However, the corrosivity of $FeCl_3$ increases steadily with temperature. The response of seawater to increasing temperature is quite complex, relating to such factors as concentration of oxygen and biological activity. Also, the various families of stainless steels will be internally consistent, but will differ from one another in response to a particular medium.

effort to accelerate testing by increasing the severity of one or more environmental factors, such as temperature, concentration, aeration, and pH. Care must be taken that the altered conditions do not give spurious results. For example, an excessive temperature may either introduce a new failure mode or prevent a failure mode relevant to the actual application. The effects of minor constituents or impurities on corrosion are of special concern in simulated testing.

Pitting and crevice corrosion are readily tested in the laboratory by using small coupons and controlled-temperature conditions. A procedure for such tests using 6% $FeCl_3$ (10% $FeCl_3 \cdot 6H_2O$) is described in ASTM G 48 (Ref 57). This procedure is performed in 3 days. The coupon may be evaluated in terms of weight loss, pit depth, pit density, and appearance. Several suggestions for methods of pitting evaluation are given in ASTM G 46 (Ref 58). Reference 57 also describes the construction of a crevice corrosion coupon (Fig. 20). It is possible to determine a temperature below which crevice corrosion is not initiated for a particular material and test environment. This critical crevice temperature (CCT) provides a useful ranking of stainless steels. For the CCT to be directly applicable to design, it is necessary to determine that the test medium and conditions relate to the most severe conditions to be encountered in service.

Pitting and crevice corrosion may also be evaluated by electrochemical techniques. When immersed in a particular medium, a metal coupon will assume a potential that can be measured relative to a standard reference electrode. It is then possible to impress a potential on the coupon and observe the corrosion as measured by the resulting current. Various techniques of

Fig. 21 Multiple-crevice cylinders for use in crevice corrosion testing. Source: Ref 59

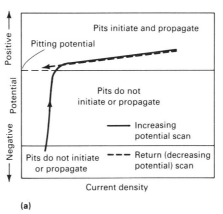

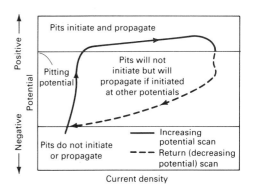

Fig. 22 Schematics showing how electrochemical tests can indicate the susceptibility to pitting of a material in a given environment. (a) Specimen has good resistance to pitting. (b) Specimen has poor resistance to pitting. In both cases, attack occurs at the highest potentials. Source: Ref 59

Table 29 ASTM standard tests for susceptibility to intergranular corrosion in stainless steels

ASTM test method	Test medium and duration	Alloys	Phases detected
A 262, practice A	Oxalic acid etch; etch test	AISI types 304, 304L, 316, 316L, 317L, 321, 347 casting alloys	Chromium carbide
A 262, practice B	$Fe_2(SO_4)_3$–H_2SO_4; 120 h	Same as above	Chromium carbide, σ phase(a)
A 262, practice C	HNO_3 (Huey test); 240 h	Same as above	Chromium carbide, σ phase(b)
A 262, practice D	HNO_3–HF; 4 h	AISI types 316, 316L, 317, 317L	Chromium carbide
A 262, practice E	$CuSO_4$–16% H_2SO_4, with copper contact; 24 h	Austenitic stainless steels	Chromium carbide
A 708 (formerly A 393)	$CuSO_4$–16% H_2SO_4, without copper contact; 72 h	Austenitic stainless steels	Chromium carbide
G 28	$Fe_2(SO_4)_3$–H_2SO_4; 24–120 h	20Cb-3	Carbides and/or intermetallic phases(c)
A 763, practice X	$Fe_2(SO_4)_3$–H_2SO_4; 24–120 h	AISI types 403 and 446; E-Brite, 29-4, 29-4-2	Chromium carbide and nitride intermetallic phases(d)
A 763, practice Y	$CuSO_4$–50% H_2SO_4; 96–120 h	AISI types 446, XM27(e), XM33(f), 29-4, 29-4-2	Chromium carbide and nitride
A 763, practice Z	$CuSO_4$–16% H_2SO_4; 24 h	AISI types 430, 434, 436, 439, 444	Chromium carbide and nitride

(a) There is some effect of σ phase in type 321 stainless steel. (b) Detects σ phase in AISI types 316, 316L, 317, 317L, and 321. (c) Carbides and perhaps other phases detected. (d) Detects γ and σ phases, which do not cause intergranular attack in unstabilized iron-chromium-molybdenum alloys. (e) UNS S44625. (f) UNS S44626. Source: Ref 59

scanning the potential range provide extremely useful data on corrosion resistance. Figure 22 demonstrates a simplified view of how these tests may indicate the corrosion resistance for various materials and media.

The nature of intergranular sensitization has been discussed earlier in this article in the section "Forms of Corrosion in Stainless Steels." There are many corrosion tests for detecting susceptibility to preferential attack at the grain boundaries. The appropriate media and test conditions vary widely for the different families of stainless steels. Table 29 summarizes the ASTM tests for intergranular sensitization. Figure 23 shows that electrochemical techniques may also be used, as in the electrochemical potentiostatic reactivation (EPR) test.

Stress-corrosion cracking covers all types of corrosion involving the combined action of tensile stress and corrodent. Important variables include the level of stress, the presence of oxygen, the concentration of the corrodent, temperature, and the conditions of heat transfer. It is important to recognize the type of corrodent likely to produce cracking in a particular family of steel. For example, austenitic stainless steels are susceptible to chloride SCC (Table 30). Martensitic and ferritic grades are susceptible to cracking related to hydrogen embrittlement.

It is important to realize that corrosion tests are designed to single out one particular corrosion mechanism. Therefore, determining the suitability of a stainless steel for a particular application usually requires consideration of more than one type of test. No single chemical or electrochemical test has been shown to be an all-purpose measure of corrosion resistance. More information on corrosion testing is available in the Section "Corrosion Testing and Evaluation" in Volume 13 of the 9th Edition of *Metals Handbook*.

Surface Finishing of Stainless Steel

Stainless steels have the capability of maintaining bright finished surfaces in many ambient environments. Lacquers and other protective coatings are seldom recommended, because unprotected stainless steel surfaces are more serviceable and easier to clean than coated surfaces. (Stainless steels may be painted provided that their surfaces are first chemically cleaned or etched, or both, for good paint adherence.) Any of several surface textures and degrees of finish can be specified, depending on the particular requirements of the application.

After fabrication, stainless steels may be bright finished by any of the following methods: mechanical grinding and polishing, Tampico brushing, buffing, electrolytic polishing, barrel finishing, or wet or dry blasting (see Volume 5 of the 9th Edition of *Metals Handbook* for detailed information on finishing and cleaning of stainless steels). Sometimes, two or more of these methods are used in combination. In addition, any of several processes that involve combinations of heating and chemical treatment can be used to produce dull black surfaces having good abrasion resistance; this is done most often on small parts. The most widely used finishing process, however, is grinding or grinding and polishing.

When fabrication does not involve extensive machining, severe deformation, or general surface marring, sheet and round bar are often polished before fabrication. Welds in prepolished parts may be finished by grinding the welds flush and then polishing contrasting stripes over the weld seams. Alternatively, the entire surface may be given a final polish in the direction of previous polishing to render weld areas virtually invisible.

Grinding, Polishing, and Buffing. The term grinding applies to rough, often dry, finishing operations in which significant amounts of metal are removed using loose abrasives or abrasive belts coarser than 100 grit. Polishing refers to finishing operations using abrasives finer than 100 grit suspended in lubricants. Buffing is a special type of polishing that is done with high-speed cloth wheels charged with extremely fine abrasives.

Table 30 Stress-corrosion cracking resistance of stainless steels

	Stress-corrosion cracking test(a)		
Grade	Boiling 42% $MgCl_2$	Wick test	Boiling 25% NaCl
AISI type 304	F(b)	F	F
AISI type 316	F	F	F
AISI type 317	F	[P(c) or F](d)	(P or F)
Type 317LM	F	(P or F)	(P or F)
Alloy 904L	F	(P or F)	(P or F)
AL-6XN	F	P	P
254SMO	F	P	P
20Mo-6	F	P	P
AISI type 409	P	P	P
Type 439	P	P	P
AISI type 444	P	P	P
E-Brite	P	P	P
Sea-Cure	F	P	P
MONIT	F	P	P
AL 29-4	P	P	P
AL 29-4-2	F	P	P
AL 29-4C	F	P	P
3RE60	F	NT	NT
2205	F	NT	(P or F)(e)
Ferralium	F	NT	(P or F)(e)

(a) U-bend tests, stressed beyond yielding. (b) Fails, cracking observed. (c) Passes, no cracking observed. (d) Susceptibility of grade to SCC determined by variation of composition within specified range. (e) Susceptibility of grade to SCC determined by variation of thermal history. Source: Ref 60

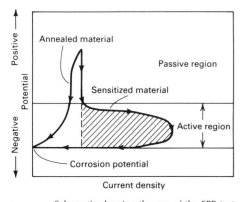

Fig. 23 Schematic showing the use of the EPR test to evaluate sensitization. The specimen is first polarized up to a passive potential at which the metal resists corrosion. Potential is then swept back through the active region, where corrosion may occur. Source: Ref 59

Although equipment used for grinding and polishing stainless steel is the same as standard equipment used for grinding and polishing other metals, the characteristics of stainless steel necessitate some modifications in technique. The suggestions given below for austenitic chromium-nickel types apply to straight-chromium types as well, even though the latter have certain characteristics similar to those of carbon steels. Variations in procedure may be necessary because of differing characteristics of the various types. The chromium-nickel types have about half the thermal conductivity and nearly twice the coefficient of expansion of carbon steels. This combination of characteristics means that austenitic stainless steels tend to heat up locally when ground or polished. When this occurs, they expand excessively in the heated area. It is very important, therefore, that polishing heads not be allowed to dwell on one spot, and excessive polishing pressure should be avoided. Attempts to hasten polishing will aggravate overheating and may cause the part to warp or buckle.

Straight-chromium stainless steels tend to load the abrasive, and this can result in scoring or galling of the polished surface. Frequently, chromium stainless steel appears slightly rougher than chromium-nickel stainless steel when the same polishing grits have been used. A slight color difference exists between polished finishes on chromium and those on chromium-nickel types, although this difference is not usually noticeable, unless pieces are adjacent and in the same general viewing plane.

Polishing in different directions in different areas of the same surface produces contrasts in appearance. A pleasing contrast can be obtained by polishing one area in a direction perpendicular to the polishing direction for another area. However, there sometimes is little or no contrast when different areas are polished in directions opposite (180°) to each other. If uniformity is desired, grit lines should be parallel over the entire surface; long, even strokes are required when hand tools are used.

Buffing, when used to produce a fine, scratch-free, mirror finish, usually consists of two operations: the cutting operation, in which a high wheel speed and a cutting compound are used; and the coloring operation, in which a slightly lower wheel speed and a color compound are employed. A satin finish, which is really the result of fine polishing rather than a true buffed finish, is produced by final buffing with satin finish compounds. Usually, grinding and polishing are necessary to obtain a surface suitable for buffing.

Grinding of welds using solid grinding wheels of suitable grit will remove excess weld metal and prepare weld surfaces for subsequent polishing operations. Welds should be ground in the direction of the weld bead without allowing the wheel to dwell in one spot. Grinding across the weld is objectionable, because cross grinding tends to cut into parent metal surfaces. Small-diameter wheels are controlled more easily than large-diameter wheels. It is advisable to stop rough grinding before the weld reinforcement is entirely removed, thereby leaving some stock for finishing. When only one surface of a weld must be finished, the side opposite that from which the weld was made usually has less surplus metal and requires less grinding. Light-gage sheet or strip should be rigidly backed up to provide a firm base for grinding.

Subsequent polishing and buffing can then be done in the usual manner. Where feasible, welds in mill-polished sheet should run in the direction of the polishing lines. This makes it easier to polish the weld to match the existing finish.

Electropolishing provides, at low cost, the combination of an attractive high-luster finish plus deburring. In electropolishing, which is the reverse of electroplating, a small amount of surface metal is removed. Wire and bar products, stampings, and small forgings have been electropolished with good results. Recesses in intricately shaped pieces may be readily brightened, making electropolishing particularly suitable for springs, wire racks, and the inside surfaces of deep-drawn bowls and pans.

Because it removes only a small amount of metal, electropolishing does not eliminate heavy scratches, deep die scoring, or embedded nonmetallic particles. The process is not recommended for descaling, although it removes discoloration such as that resulting from spot welding. An electropolishing finish does not simulate a mechanically polished finish.

Mill Finishes

Sheet, strip, plate, bar, and wire made of stainless steel all have different designations of mill finish, each representing a standardized appearance that is characteristic of the process used to impart final mechanical properties. Although the various mill finishes are standardized, there is sufficient variability in mill processing that exact matching of color and reflectivity cannot be expected from lot to lot. Even wider differences can be expected between mill products from different producers.

Sheet finishes are designated by a system of numbers: No. 1, 2D, 2B, and 2BA for unpolished finishes; and No. 3, 4, 6, 7, and 8 for polished finishes.

Each of the unpolished finishes is in itself a category of finishes, with variations in appearance and smoothness, depending on composition, sheet thickness, and method of manufacture. Generally, the thinner the sheet, the smoother the surface. Chromium-nickel and chromium-nickel-manganese stainless steel are characteristically different in appearance from straight-chromium types having the same finish. Furthermore, sheet produced continuously (in coil form) generally differs in appearance from sheet produced as individual pieces on hand mills.

The appearance or "color" of polished finishes may differ slightly among 2xx-, 3xx-, and 4xx-series stainless steels. Sheet can be produced with one or both sides polished; when only one side is polished, the other side is often rough ground to obtain better flatness.

No. 1 finish is produced by hot rolling followed by annealing and descaling. It is generally used in industrial applications where smoothness of finish is not particularly important, such as equipment for elevated-temperature or corrosion service.

No. 2D finish is a dull cold-rolled finish produced by cold rolling, annealing, and descaling. The dull finish may result from descaling or pickling or it may be developed by a final light cold rolling pass using dull rolls. No. 2D finish is favorable for retention of lubricants in deep drawing and is generally preferred for deep-drawn articles that will be polished after fabrication.

No. 2B finish is a bright cold-rolled finish commonly produced in the same manner as No. 2D, except that the final light cold rolling pass is done using polished rolls. No. 2B is a general-purpose cold-rolled finish commonly used for all but exceptionally difficult deep-drawing applications. It is more readily polished to a high luster than is a No. 1 or No. 2D finish.

No. 2BA finish is a mirrorlike appearance produced by cold rolling, followed by bright annealing or double bright annealing. The final appearance is developed by a single light skin pass through a cold mill over highly polished rolls, but is dependent on additional millwork, such as grinding of the surface at an intermediate gage. A No. 2BA finish is often specified for architectural applications and for other uses where a highly reflective surface is desired on the as-fabricated part.

No. 3 finish is a polished finish obtained with abrasives approximately 100 mesh in particle size. It is used for articles that may or may not receive additional polishing during fabrication.

No. 4 finish is a general-purpose polished finish widely used for restaurant equipment, kitchen equipment, storefronts, and dairy equipment. Following initial grinding with coarser abrasives, final finishing is generally done with abrasives having a particle size of approximately 120 to 150 mesh.

No. 6 finish is a dull satin finish having lower reflectivity than No. 4 finish. It is produced by Tampico brushing No. 4 finish sheet in a medium of abrasive and oil. It is used in architectural and ornamental applications where high luster is undesirable or for contrast with brighter finishes.

No. 7 finish has a high degree of reflectivity. It is produced by buffing finely ground surfaces, but not to the extent that existing grit lines are removed. It is used chiefly for architectural and ornamental parts.

No. 8 finish, the most reflective finish that is commonly produced on sheet, is obtained by polishing with successively finer abrasives and buffing extensively with very fine buffing rouges. The surface is essentially free of grit lines from preliminary grinding operations. No. 8 finish is most widely used for press plates, small mirrors, and reflectors.

Strip Finishes. Only three unpolished finishes (No. 1, No. 2, and bright annealed) and one polished finish (mill buffed) are commonly supplied on stainless steel strip. As with finishes on stainless steel sheet, each unpolished strip finish comprises a category of finishes that vary in appearance and smoothness depending on composition, thickness, and method of manufacture. Generally, the thinner the strip, the smoother the surface. Chromium-nickel and chromium-nickel-manganese stainless steels are characteristically different in appearance from straight-chromium types having the same finish.

No. 1 finish is produced by cold rolling, annealing, and pickling. Appearance varies from dull gray matte to fairly reflective, depending largely on stainless steel type. This finish is used for severely drawn or formed parts, as well as for applications where the brighter No. 2 finish is not required, such as parts to be used at high temperatures. No. 1 finish for strip approximates No. 2D finish for sheet in corresponding chromium-nickel or chromium-nickel-manganese types.

No. 2 finish is produced by the same treatment used for No. 1 finish, followed by a final light cold rolling pass, which is generally done using highly polished rolls. This final pass produces a smoother and more reflective surface, the appearance of which varies with stainless steel type. No. 2 finish for strip is a general-purpose finish widely used for household appliances, automotive trim, tableware, and utensils. No. 2 finish for strip approximates No. 2B finish for sheet in corresponding chromium-nickel or chromium-nickel-manganese stainless steels.

Bright annealed finish is a bright, cold-rolled, highly reflective finish retained by final annealing in a controlled-atmosphere furnace. The purpose of atmospheric control is to prevent scaling or oxidation during annealing. The atmosphere usually consists of either dry hydrogen or dissociated ammonia. Bright annealed strip is used most extensively for automotive trim.

Mill-buffed finish is a highly reflective finish obtained by subjecting either No. 2 or bright annealed coiled strip to a continuous buffing pass. The purpose of mill buffing is to provide a finish uniform in color and reflectivity. It also can provide a surface receptive to chromium plating. This type of finish is used chiefly for automotive trim, household trim, tableware, utensils, fire extinguishers, and plumbing fixtures.

Plate Finishes. Stainless steel plate can be produced in a variety of conditions and surface finishes:

Condition and finish	Description and remarks
Hot rolled	Scale not removed; not heat treated; plate not recommended for final use in this condition(a)
Hot or cold rolled, annealed or heat treated	Scale not removed; use of plate in this condition generally confined to heat-resisting applications; scale impairs corrosion resistance(a)
Hot or cold rolled, annealed or heat treated, blast cleaned or pickled	Condition and finish commonly preferred for corrosion-resisting and most heat-resisting applications
Hot or cold rolled, annealed, descaled and temper passed	Smoother finish for specialized applications
Hot rolled, annealed, descaled, cold rolled, annealed, descaled, optionally temper passed	Smooth finish with greater freedom from surface imperfections than any of the above
Hot or cold rolled, annealed or heat treated, surface cleaned and polished	Polished finishes similar to the polished finishes on sheet

(a) Surface inspection is not practicable for plate that has not been pickled or otherwise descaled.

Plate is often conditioned by localized grinding to remove surface imperfections on either one or both surfaces; ground areas are well flared, and the thickness is not reduced below the allowable tolerance in any of these areas.

Bar Finishes. Stainless steel bar is produced in a number of conditions and surface finishes (it is important that both condition and finish be specified, because each finish is applicable only to certain conditions):

Condition	Surface finish
Hot worked only	Scale not removed (except for spot conditioning)
	Rough turned(a)(b)
	Blast cleaned
Annealed or otherwise heat treated	Scale not removed (except for spot conditioning)
	Rough turned(a)
	Pickled or blast cleaned and pickled
	Cold drawn or cold rolled
	Centerless ground(a)
	Polished(a)
Annealed and cold worked to high tensile strength(c)	Cold drawn or cold rolled
	Centerless ground(a)
	Polished(a)

(a) Applicable to round bar only. (b) Bar of 4xx series stainless steels that are highly hardenable, such as types 414, 420, 420F, 431, 440A, 440B and 440C, are annealed before rough turning. Other hardenable types, such as types 403, 410, 416, and 416Se, also may require annealing depending on composition and size. (c) Produced only in mill orders; made predominantly in types 301, 302, 303Se, 304, 304N, 316, and 316N

Wire Finishes. Finishes used on wires can be classified as:

- Oil drawn or grease drawn
- Diamond drawn
- Copper coated
- Tinned
- Lead coated

Oil- or grease-drawn finish is a special bright finish for wire intended for uses, such as racks and handles, where the finish supplied is to be the final finish of the end product. In producing this finish, lower drawing speeds are necessary, and additional care in processing is needed to provide a surface with few scratches and with only a very light residue of lubricant.

Diamond-drawn finish is a very bright finish generally limited to wet-drawn stainless steel wire in fine sizes. Drafting speeds are necessarily reduced to obtain the desired brightness.

Copper-coated wire is supplied when a special finish is required for lubrication in an operation such as spring coiling or cold heading. Generally, copper-coated wire is drawn after coating, the amount depending on the desired cold-worked temper of the wire.

Tinned wire is coated by passing single strands through a bath of molten tin. Tinned wire is used in soldering applications. The temper of the finished wire is controlled by processing prior to tinning.

Lead-coated wire is coated by passing single strands through, or immersing bundles of wire in, a bath of molten lead. The wire is then drawn to final size, with the lead forming a thin coating over the entire surface. This coating is useful on wire for coil springs, where it serves as a lubricant during coiling operations.

Interim Surface Protection

Finishes on stainless steel often require protection during shipment of mill products to fabrication plants. Otherwise, it is inevitable that scratches, dings, and other evidence of material-handling operations will mar the appearance of some end products. Furthermore, if a mill finish is intended to give the end product its appearance, the finish must be protected from incidental damage during fabrication of parts. Although finishes on completed parts and assemblies can be protected by proper packaging, such packaging can be made much simpler in design if the need to protect the finish from certain kinds of incidental damage during shipment can be reduced.

All of these considerations have led to the development of various masking materials for the protection of finishes on stainless steel surfaces prior to end use. Masking materials used in mills are most often obtained in the form of rolls of adhesive-backed protective film that can be applied

Table 31 Typical methods of cleansing stainless steel surfaces

Cleansing problem	Cleansing agent(a)	Method of application(b)
Routine cleansing	Soap, ammonia, detergent, water	Sponge with cloth, then rinse with clear water and wipe dry. Satisfactory for use on all finishes
Smears and fingerprints	Arcal 20, Lac-O-Nu, Lumin Wash, O'Cedar Cream Polish, Stainless Shine, Wind-O-Shine	Rub with cloth as directed on the package. Satisfactory for use on all finishes. Provides barrier film to minimize prints
Stubborn spots, stains, and other light discolorations	Allchem Concentrated Cleaner, Samae, Cameo Copper Cleaner, Cooper's Stainless Steel Cleaner, Revere Stainless Steel Cleaner, Paste NuSteel, DuBois Temp, Aerogroom Household cleansers such as Old Dutch, Bab-O, Sapolio, Bon Ami, Ajax, Comet; grade F Italian pumice, Steel Bright, Lumin Cleaner, Zud, Restoro, Sta-Clean, Highlite, Penny-Brite, Copper-Brite, DuBois Stainless Steel Polish	Apply with sponge or cloth. Satisfactory for use on all finishes. Use in direction of polish lines on No. 4 finish. Use light pressure on No. 2, 7, and 8 finishes. May scratch No. 2, 7, and 8 finishes
Burnt-on foods and grease, fatty acids, milk-stone (where swabbing or rubbing is not practical)	Easy-Off, De-Grease-It, 4–6% hot solution of such agents as trisodium phosphate or sodium tripolyphosphate, 5–15% caustic soda solution	Apply generous coating, allow to stand for 10–15 min, rinse. Repeated application may be necessary. Satisfactory for use on all finishes
Tenacious deposits, rusty discolorations, industrial atmospheric stains	Oakite No. 33, Dilac, Flash-Flenz, Caddy Cleaner, Turco Scale 4368, Permag 57	Swab and soak with clean cloth, let stand 15 minutes or more according to directions on package, then rinse and dry. Satisfactory for use on all finishes
Hard-water spots and scale	Vinegar, 5% oxalic acid, 5% sulfamic acid, 5–10% phosphoric acid, Dilac, Oakite No. 33, Texo 12	Swab or soak with cloth, let stand 10–15 min. Always follow with neutralizer rinse and dry. Satisfactory for use on all finishes
Grease and oil	Organic solvents, detergents, caustic cleaners	Rub with cloth. (Organic solvents may be flammable and/or toxic.) Satisfactory for use on all finishes

(a) Use of proprietary names is intended to indicate type of cleanser and does not constitute an endorsement. Omission of any proprietary cleanser does not imply inadequacy. All products should be used in strict accordance with instructions on package. (b) In all applications, stainless steel wool, a sponge, or a fibrous brush or pad is recommended for scouring stainless steel. Use of ordinary steel wool or steel brushes will leave a residue and result in corrosion and/or rust staining.

by hand or machine to stainless steel mill products. In addition, water-base and solvent-base strippable coatings are available; these are applied by dip-dry methods. Sometimes, special adhesive-backed protective coverings of the customer's choosing can be applied at the mill prior to shipment.

Most often, adhesives are based on latex or proprietary organic compounds. Masking materials include paper, rubber, and plastics. Plastics such as polyvinyl chloride, polyolefin, polyethylene, and polypropylene are the most popular. Coverings range in thickness from about 0.06 to 0.13 mm (2.5 to 5 mils) and are available in several colors (generally white, clear, black, and light blue). Sometimes, paper coverings are printed with a company logo or a proprietary message such as instructions for in-service care of the finish.

Adhesion to stainless steel is generally measured as the force per unit width required to remove the covering from sheet having a No. 4 finish. Other data available from the masking material manufacturer include tensile strength, elongation, and unwinding force.

Plastic and rubber coverings have sufficient elongation and adhesion to enable them to survive all but the most severe fabrication operations, including drawing, bending, and roll forming.

Of interest to the end-user, of course, is the ease with which the masking material can be removed. Often, a compromise must be reached between the degree of protection afforded the finish and the ease of removal. Usually, adhesive-backed coverings are easiest to remove when the coated steel is first received. They often become more difficult to remove if they are exposed to temperatures significantly above or below normal room ambient, to ultraviolet light, or to outdoor (weathering) environments. Excessively prolonged indoor storage can also make it quite difficult to remove adhesive-backed coverings.

In-Service Care

Despite the fact that stainless steel surfaces are generally considered nontarnishing, they still need a certain amount of care to maintain a given surface appearance under normal conditions of service. Table 31 summarizes methods of cleansing stainless steel for a wide variety of applications.

Architectural Applications. For exposed exterior surfaces in inland, light industrial areas, minimum maintenance is needed. Ordinarily, normal rainfall is adequate to maintain the desired appearance, and only sheltered areas such as entryways need occasional washing with a scrub brush or a pressurized stream of water. In marine atmospheres and heavy industrial areas, periodic cleaning with detergents and water to remove salt and dirt deposits is advisable. Heavy or stubborn deposits may have to be removed with strong industrial cleaners.

For interior surfaces, only occasional cleaning with detergent and water is required for maintenance of finish. Where fingerprints are a problem, a commercial glass cleaner or wax is suggested. Often, a No. 4 sheet finish is specified to minimize the effect of fingerprints on appearance.

Food-Handling Applications. Stainless steel is widely specified for food-handling equipment because of its excellent bacterial cleanability. In many instances, strong sanitizing sterilizing solutions are used for cleaning the equipment to prevent bacterial contamination of the food products being processed. Where this is done, it should be standard practice to monitor exposure time and thoroughly flush the cleaned surfaces with water. Burnt-on foods and grease spots can be removed by soaking in hot water and detergent. Stubborn spots can be removed by scrubbing with a nonabrasive cleanser and a fiber brush, a sponge, or a pad of stainless steel wool or nickel-silver wool.

Chemical, textile, and drug applications often require high purity in the product being processed. Stainless steel is used in equipment for these industries not only because it is chemically inert to the products, which effectively eliminates corrosion as a possible source of low-level contamination, but also because the surfaces of stainless steel equipment are easy to clean and sterilize, which effectively eliminates bacteria and residues as sources of contamination. Equipment is usually cleaned with strong chemical cleaners and then repeatedly rinsed with water. To facilitate cleaning, equipment usually has rounded corners and fillets, welded construction is used instead of mechanical seams, and all welds and other protrusions are ground flush and polished.

Fabrication Characteristics

Wrought stainless steels come from the melting furnaces in the form of either ingot or continuously cast slabs. Ingots require a roughing or primary hot working, which the other form commonly bypasses. All then go through fabricating and finishing operations such as hot and cold forming, welding, rolling, machining, spinning, and polishing.

No stainless steel is excluded from any of the common industrial processes because of its special properties; yet all stainless steels require certain modifications of technique.

Hot working is influenced by the heat-resisting characteristic of many of the stainless steels. These alloys are stronger at elevated temperatures than ordinary steel. Therefore they require greater roll and forge pressure, and perhaps lesser reductions per pass or per blow. The austenitic steels are particularly heat resistant.

Machining and forming processes are adaptable to all grades, with these major precautions: First, the stainless steels are generally stronger and tougher than carbon steel, so that more power and rigidity are needed in tooling. Second, the powerful work-hardening effect gives the austenitic grades the property of being instantaneously strengthened upon the first touch of the tool or pass of the roll. Machine tools must therefore bite surely and securely, with care taken to ensure that they do not ride the piece. Difficult forming operations warrant careful attention to variations in grade and in heat treatment to accomplish end purposes without unnecessary working problems. Spinning, for example, has a type 305 modification of 18-8 that greatly favors the operation: at the opposite extreme is type 301 or 17-7, when strengthening due to cold working is an advantage.

Welding is influenced by another aspect of the high-temperature resistance of these metals: the resistance to scaling. Oxidation during service at high temperatures does not become catastrophic with most stainless steel because the steel immediately forms a hard and protective scale. However, this means that welding must be conducted under conditions that protect the metal from such reactions with the environment. This can be done with specially prepared coatings on electrodes, under cover of fluxes, using shielding gas, or in a vacuum.

Sheet Formability

Stainless steels are blanked, pierced, formed, and drawn in basically the same press tools and machines as those used for other metals. However, because stainless steels have higher strength and are more prone to galling than low-carbon steels and because they have a surface finish that often must be preserved, the techniques used in the fabrication of sheet metal parts from stainless steels are more exacting than those used for low-carbon steels. In contrast to carbon steels, stainless steels usually have:

- Greater strength
- Greater susceptibility to work hardening
- Higher propensity to weld or gall to tooling
- Lower heat conductivity

General ratings of the relative suitability of the commonly used austenitic, martensitic, and ferritic types of stainless steels to various methods of forming are given in Table 32. These ratings are based on formability and on the power required for forming.

As Table 32 shows, the austenitic and ferritic steels are, almost without exception, well suited to all of the forming methods listed. Of the martensitic steels, however, only types 403, 410, and 414 are generally recommended for cold-forming applications. Because the higher carbon content of the remaining martensitic types severely limits their cold formability, these steels are sometimes formed warm. Duplex types have properties intermediate between those of typical austenitic and ferritic grades. Warm forming can also be used to advantage with other stainless steels in difficult applications.

The characteristics of stainless steel that affect its formability include yield strength, tensile strength, ductility (and the effect of work hardening on these properties), and the plastic-strain ratio, r, value. The composition of stainless steel is also an important factor in formability. Figure 24 compares the effect of cold work on the tensile strength and yield strength of type 301 (an austenitic alloy), types 409 and 430 (both ferritic alloys), and 1008 low-carbon steel sheet.

Formability of Austenitic Types. Type 301 stainless steel has the lowest nickel and chromium contents of the standard austenitic types; is also has the highest tensile strength in the annealed condition. The extremely high rate of work hardening of type 301 results in appreciable increases in tensile strength and yield strength with each increase in the amount of cold working, as measured by cold reduction (Fig. 24). This response to work hardening is particularly important for structural parts, including angles and channel sections, which, after fabrication, are expected to have additional strength and stiffness. On the other hand, for deep-drawing applications, a lower rate of work hardening is usually preferable and can be obtained in the austenitic alloys that have higher nickel contents, notably, types 304, 304L, and 305.

Table 32 Relative suitability of stainless steels for various methods of forming

Suitability ratings are based on the comparison of steels within any one class; therefore, it should not be inferred that a ferritic steel with an A rating is more formable than an austenitic steel with a C rating for a particular method. Ratings are A, excellent; B, good; C, fair; D, not generally recommended

Steel	0.2% yield strength, 6.89 MPa (1 ksi)	Blanking	Piercing	Press-brake forming	Deep drawing	Spinning	Roll forming	Coining	Embossing
Austenitic steels									
201	55	B	C	B	A–B	C–D	B	B–C	B–C
202	55	B	B	A	A	B–C	A	B	B
301	40	B	C	B	A–B	C–D	B	B–C	B–C
302	37	B	B	A	A	B–C	A	B	B
302B	40	B	B	B	B–C	C	...	C	B–C
303, 303(Se)	35	D	D	D(a)	D	D	D	C–D	C
304	35	B	B	A	A	B	A	B	B
304L	30	B	B	A	A	B	A	B	B
305	37	B	B	A	B	A	A	A–B	A–B
308	35	B	...	B(a)	D	D	...	D	D
309, 309S	40	B	B	A(a)	B	C	B	B	B
310, 310S	40	B	B	A(a)	B	B	A	B	B
314	50	B	B	A(a)	B–C	C	B	B	B–C
316	35	B	B	A(a)	B	B	A	B	B
316L	30	B	B	A(a)	B	B	A	B	B
317	40	B	B	A(a)	B	B–C	B	B	B
321, 347, 348	35	B	B	A	B	B–C	B	B	B
Martensitic steels									
403, 410	40	A	A–B	A	A	A	A	A	A
414	95	A	B	A(a)	B	C	C	B	C
416, 416(Se)	40	B	A–B	C(a)	D	D	D	D	C
420	50	B	B–C	C(a)	C–D	D	C–D	C–D	C
431	95	C–D	C–D	C(a)	C–D	D	C–D	C–D	C–D
440A	60	B–C	...	C(a)	C–D	D	C–D	D	C
440B	62	...	...	...	...	D	...	D	D
440C	65	...	...	...	...	D	...	D	D
Ferritic steels									
405	40	A	A–B	A(a)	A	A	A	A	A
409	38	A	A–B	A(a)	A	A	A	A	A
430	45	A	A–B	A(a)	A–B	A	A	A	A
430F, 430F(Se)	55	B	A–B	B–C(a)	D	D	D	C–D	C
442	...	A	A–B	A(a)	B	B–C	A	B	B
446	50	A	B	A(a)	B–C	C	B	B	B

(a) Severe sharp bends should be avoided.

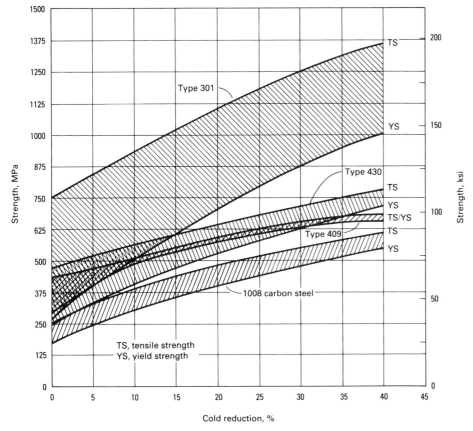

Fig. 24 Comparison of work-hardening qualities of type 301 austenitic stainless steel, types 409 and 430 ferritic stainless steels, and 1008 low-carbon steel

In general, the austenitic alloys are more difficult to form as the nickel content or both the nickel and the chromium contents are lowered, as in type 301. Such alloys show increased work-hardening rates and are less suitable for deep drawing or multiple forming operations. The presence of the stabilizing elements niobium, titanium, and tantalum, as well as higher carbon contents, also exerts an adverse effect on the forming characteristics of the austenitic stainless steels. Therefore, the forming properties of types 321 and 347 stainless steel are less favorable than those of types 302, 304, and 305.

Formability of Ferritic Types. The range between yield strength and tensile strength of types 409 and 430 narrows markedly as cold work increases, as shown in Fig. 24. This response, typical of the ferritic alloys, limits their formability (ductility) in comparison with the austenitic alloys. Nevertheless, types 409 and 430, although lacking the formability of type 302, are widely used in applications that require forming by blanking, bending, drawing, or spinning. One of the most important applications for type 430 stainless steel is in automotive trim or molding. Type 409 stainless steel has found wide acceptance as the material of choice in automotive exhaust systems.

Formability of Duplex Types. Duplex stainless steels have about twice the yield strength of most austenitic stainless steels. Their elongation, toughness, and work-hardening rates are generally intermediate to those of the usual austenitic and ferritic grades.

Duplex stainless steels can be cold formed and expanded. Their higher strength relative to their austenitic counterparts necessitates greater loads in cold-forming operations. Because elongation is lower, they should be formed to more generous radii than fully austenitic materials. Heavily cold-formed (>~25%) sections should be fully annealed and quenched whenever applications to the alloy present the possibility of stress-corrosion cracking in the service environment. Full annealing is conducted in the temperature range of 1010 to 1100 °C (1850 to 2010 °F), followed by a rapid cooling.

Hot forming operations are usually performed in the temperature range of 980 to 1260 °C (1800 to 2300 °F); the preferred temperature range is dictated by the specific alloy composition. The temperature range of about 370 to 925 °C (700 to 1700 °F) should be avoided to preclude the precipitation of such deleterious phases as σ and α′. These phases can adversely affect mechanical properties and corrosion resistance.

Duplex stainless steels, especially in light gages, can be made to exhibit superplastic behavior at elevated temperatures. This property may be used to advantage in certain operations involving the continuous line annealing of strip and sheet.

Comparison With Carbon Steel. The curves for 1008 low-carbon steel are included in Fig. 24 as a reference for the evaluation of stainless steels. The decrease in formability of 1008 steel with cold work appears to fall between that of types 409/430 and that of the more formable type 301. Figure 24 also shows that cold work does not increase the strength of 1008 as rapidly as it does that of type 301 and the ferritic alloys.

Stress-Strain Relationships. Figure 25 shows load-elongation curves for six types of stainless steel: four austenitic (202, 301, 302, and 304), one martensitic (410), and one ferritic (430). The figure also shows that the type of failure in the cup drawing of the austenitic types was different from that of types 410 and 430, as shown in Fig. 25. The austenitic types broke in a fairly clean line near the punch nose radius, almost as though the bottom of the drawn cup were blanked out; types 410 and 430 broke in the sidewall in sharp jagged lines, showing extreme brittleness as a result of the severe cold work.

As suggested by the data in Fig. 25, the power required to form type 301 exceeds that required by the other austenitic alloys. In addition, type 301 will develop maximum elongation before failing. Types 410 and 430 require considerably less power to form, but fail at comparatively low elongation levels.

Power requirements for forming stainless steel, because of the high yield strength, are greater than those for low-carbon steel; generally, twice as much power is used in forming stainless steel. Because the austenitic steels work harden rapidly in cold-forming operations, the need for added power after the start of initial deformation is greater than that for the ferritic steels. The ferritic steels behave much like plain carbon steels once deformation begins, although higher power is also needed to start plastic deformation.

Forgeability

Stainless steels, based on forging pressure and load requirements, are considerably more difficult to forge than carbon or low-alloy steels, primarily because of the greater strength of stainless steels at elevated temperatures and the limitations on the maximum temperatures at which stainless steels can be forged without incurring microstructural damage. Forging load requirements and forgeability vary widely among stainless steels of different types and compositions; the most difficult alloys to forge are those with the greatest strength at elevated temperatures.

890 / Specialty Steels and Heat-Resistant Alloys

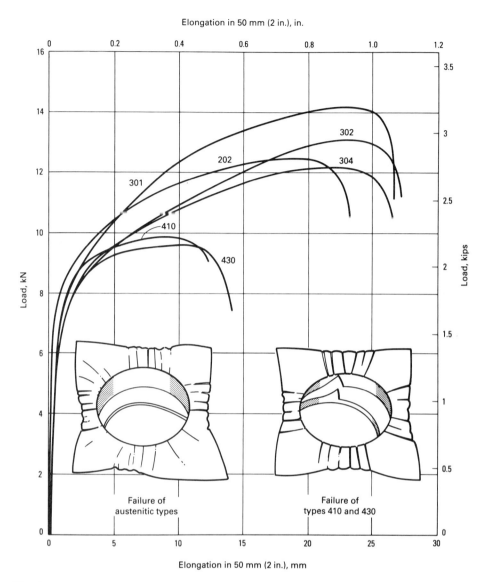

Fig. 25 Comparison of ductility of six stainless steels and of the types of failure resulting from deep drawing

Ingot Breakdown. In discussing the forgeability of the stainless steels, it is critical to understand the types of primary mill practices available to the user of semifinished billet or bloom product.

Primary Forging and Ingot Breakdown. Most stainless steel ingots destined for the forge shop are melted by the electric furnace argon-oxygen decarburization process. They usually weigh between 900 and 13 500 kg (2000 to 30 000 lb), depending on the shop and the size of the finished piece. Common ingot shapes are round, octagonal, and fluted; less common ingot shapes include squares. Until recently, all of these ingots would have been top poured. Increasing numbers of producers are switching to the bottom-poured ingot process. This process is slightly more expensive to implement in the melt shop, but it more than pays for itself in extended mold life and greatly improved ingot surface.

Some stainless steel grades used in the aircraft and aerospace industries are double melted. The first melt is done with the electric furnace and argon-oxygen decarburization, and these electrodes are then remelted by a vacuum arc remelting (VAR) or electroslag remelting (ESR) process. This remelting under a vacuum (VAR) or a slag (ESR) tends to give a much cleaner product with better hot workability. For severe forging applications, the use of remelt steels can sometimes be a critical factor in producing acceptable parts. These double-melted ingots are round in shape and vary in diameter from 460 to 915 mm (18 to 36 in.), and in some cases, they weigh in excess of 11 000 kg (25 000 lb). The breakdown of ingots is usually done on large hydraulic presses (13 500 kN, or 1500 tonf). A few shops, however, still use large hammers, and the four-hammer radial forging machine is being used increasingly for ingot breakdown.

Heating is the single most critical step in the initial forging of ingots. The size of the ingot and the grade of the end-product stainless steel dictate the practice necessary to reduce thermal shock and to avoid unacceptable segregation levels. It is essential to have accurate and programmable control of the furnaces used to heat stainless steel ingots and large blooms.

Primary forging or breakdown of an ingot is usually achieved using flat dies. However, some forgers work the ingot down as a round, using V or swage dies. Because of the high hot hardness of stainless steel and the narrow range of working temperatures for these alloys, light reductions, or saddening is the preferred initial step in the forging of the entire surface of the ingot. Saddening is an operation in which an ingot is given a succession of light reductions in a press or rolling mill or under a hammer in order to break down the skin and overcome the initial fragility due to a coarse crystalline structure preparatory to reheating prior to heavier reductions.

After the initial saddening of the ingot surface is complete, normal reductions of 50 to 100 mm (2 to 4 in.) can be taken. If the chemistry of the heat is in accordance with specifications and if heating practices have been followed and minimum forging temperatures observed, no problems should be encountered in making the bloom and other semifinished product.

If surface tears occur, the forging should be stopped, and the workpiece conditioned. Some forgers use hot powder scarfing, but this presents environmental problems. The most common method is to grind out the defect. The ferritic, austenitic, and nitrogen-strengthened austenitic stainless steels can be air cooled, ground, and reheated for reforging. The martensitic and precipitation-hardening grades must be slow cooled and overaged before grinding and reheating. The ingot surface is important, and many producers find it advantageous to grind the ingots before forging to ensure good starting surfaces.

Billet and Bloom Product. Forgers buy bars, billets, or blooms of stainless steel for subsequent forging on hammers and presses. Forged stainless steel billet and bloom products tend to have better internal integrity than rolled product, especially with larger-diameter sections (>180 mm, or 7 in.). Correctly conditioned billet and bloom product should yield acceptable finished forgings if good heating practices are followed and if attention is paid to the minimum temperature requirements. Special consideration must be given to sharp corners and thin sections, because these tend to cool off very rapidly. Precautions should be taken when forging precipitation-hardening or nitrogen-strengthened austenitic grades.

Closed-Die Forgeability. The relative forging characteristics of stainless steels can be

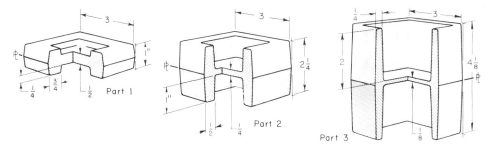

Fig. 26 Three degrees of forging severity. Dimensions given in inches

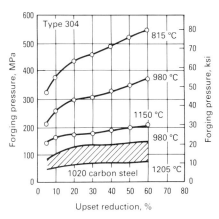

Fig. 28 Effect of upset reduction on forging pressure for various temperatures. Source: Ref 61

most easily depicted through examples of closed-die forgings. The forgeability trends these examples establish can be interpreted in light of the grade, type of part, and forging method to be used.

Stainless steels of the 300 and 400 series can be forged into any of the hypothetical parts illustrated in Fig. 26. However, the forging of stainless steel into shapes equivalent in severity to part 3 may be prohibited by shortened die life (20 to 35% of that obtained in forging such a shape from carbon or low-alloy steel) and by the resulting high cost. For a given shape, die life is shorter in forging stainless steel than in forging carbon or low-alloy steel.

Forgings of mild severity, such as part 1 in Fig. 26, can be produced economically from any stainless steel with a single heating and about five blows. Forgings approximating the severity of part 2 can be produced from any stainless steel with a single heating and about ten blows. For any type of stainless steel, die life in the forging of part 1 will be about twice that in the forging of part 2.

Part 3 represents the maximum severity for forging all stainless steels and especially those with high strength at elevated temperature, namely, types 309, 310, 314, 316, 317, 321, and 347. Straight-chromium types 403, 405, 410, 416, 420, 430, 431, and 440 are the easiest to forge into a severe shape such as part 3 (although type 440, because of its high carbon content, would be the least practical). Types 201, 301, 302, 303, and 304 are intermediate between the two previous groups.

One forge shop has reported that part 3 would be practical and economical to produce in the higher-strength alloys if the center web were increased from 3.2 to 6.4 mm (⅛ to ¼ in.) and if all fillets and radii were increased in size. It could then be forged with 15 to 20 blows and one reheating, dividing the number of blows about equally between the first heat and the reheat.

Hot Upsetting. Stainless steel forgings of the severity represented by hypothetical parts 4, 5, and 6 in Fig. 27 can be hot upset in one blow in a steel die. However, the conditions are similar to those encountered in hot-die forging. First, with a stainless steel, die wear in the upsetting of part 6 will be several times as great as in the upsetting of part 4. Second, die wear for the forming of any shape will increase as the elevated-temperature strength of the alloy increases. Therefore, type 410, with about the lowest strength at high temperature, would be the most economical stainless steel to be formed of any of the parts, particularly part 6. Conversely, type 310 would be the least economical.

Upset Reduction Versus Forging Pressure. The effect of percentage of upset reduction (upset height versus original height) on forging pressure for low-carbon steel and for type 304 stainless steel at various temperatures is shown in Fig. 28. Temperature has a marked effect on the pressure required for any given percentage of upset, and at any given forging temperature and percentage of upset, type 304 stainless requires at least twice the pressure required for 1020 steel.

The effects of temperature on forging pressure are further emphasized in Fig. 29(a). These data, based on an upset reduction of 10%, show that at 760 °C (1400 °F) type 304 stainless steel requires only half as much pressure as A-286 (an iron-base heat-resistant alloy), although the curves for forging pressure for the two metals converge at 1100 °C (2000 °F). However, at a forging temperature of 1100 °C (2000 °F), the pressure required for a 10% upset reduction on type 304 is more than twice that required for a carbon steel (1020) and about 60% more than that required for 4340 alloy steel. Differences in forgeability, based on percentage of upset reduction and forging pressure for type 304 stainless steel, 1020, and 4340 at the same temperature (980 °C, or 1800 °F), are plotted in Fig. 29(b).

Austenitic stainless steels are more difficult to forge than the straight-chromium types, but are less susceptible to surface defects. Most of the austenitic stainless steels can be forged over a wide range of temperatures above 925 °C (1700 °F) and because they do not undergo major phase transformation at elevated temperature, they can be forged at higher temperatures than the martensitic types (Table 33). Exceptions to the above statements occur when the composition of the austenitic

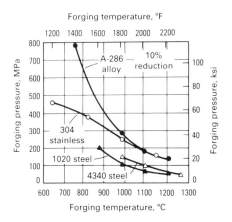

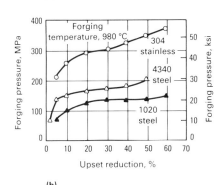

Fig. 29 Forging pressure required for upsetting versus (a) forging temperature and (b) percentage of upset reduction. Source: Ref 62

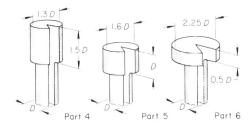

Fig. 27 Three degrees of upsetting severity

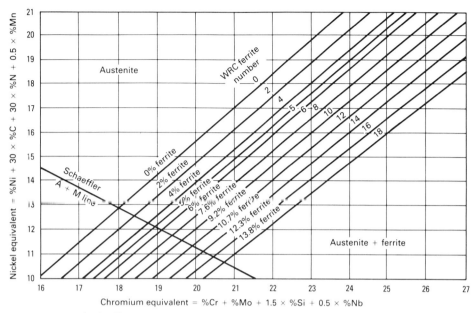

Fig. 30 Revised Schaeffler (constitution) diagram used to predict the amount of δ-ferrite that will be obtained during elevated-temperature forging or welding of austenite/ferritic stainless steels. A, austenite; M, martensite. WRC, Welding Research Council. Source: Ref 64

Table 33 Typical forging temperature ranges of stainless steels

Stainless steel	Temperature °C	°F
More difficult to hot work		
UNS N08020	980–1230	1800–2245
Pyromet 355	925–1150	1700–2100
Type 440C	925–1150	1700–2100
19-9DL/19DX	870–1150	1600–2100
Types 347 and 348	925–1230	1700–2245
Type 321	925–1260	1700–2300
Type 440B	925–1175	1700–2145
Type 440A	925–1200	1700–2200
Type 310	980–1175	1800–2145
Type 310S	980–1175	1800–2145
17-4 PH	1095–1175	2000–2145
15-5 PH	1095–1175	2000–2145
13-8 Mo	1095–1175	2000–2145
Type 317	925–1260	1700–2300
Type 316L	925–1260	1700–2300
Type 316	925–1260	1700–2300
Type 309S	980–1175	1800–2145
Type 309	980–1175	1800–2145
Type 303	925–1260	1700–2300
Type 303Se	925–1260	1700–2300
Type 305	925–1260	1700–2300
Type 329	925–1095	1700–2000
Easier to hot work		
Types 302 and 304	925–1260	1700–2300
Nitronic 60	1095–1175	2000–2145
Carpenter No. 10 (Type 384)	925–1230	1700–2245
Lapelloy	1040–1150	1900–2100
AMS 5616 (Greek Ascoloy)	955–1175	1750–2145
Type 431	900–1200	1650–2200
Type 414	900–1200	1650–2200
Type 420F	900–1200	1650–2200
Type 420	900–1200	1650–2200
Type 416	925–1230	1700–2245
Type 410	900–1200	1650–2200
Type 404	900–1150	1650–2100
Type 446	900–1120	1650–2050
Type 443	900–1120	1650–2050
Type 430F	815–1150	1560–2100
Type 430	815–1120	1500–2050

Source: Ref 63

stainless steel promotes the formation of δ-ferrite, as in the case of the 309S, 310S, or 314 grades. At temperatures above 1100 °C (2000 °F), these steels, depending on their composition, may form appreciable amounts of δ-ferrite. Figure 30 depicts these compositional effects in terms of nickel equivalent (austenitic-forming elements) and chromium equivalent. Delta-ferrite formation adversely affects forgeability, and compensation for the amount of ferrite present can be accomplished with forging temperature restrictions.

Equally important restrictions in forging the austenitic stainless steels apply to the finishing temperatures. All but the stabilized types (321, 347, 348) and the extralow-carbon types should be finished at temperatures above the sensitizing range (~815 to 480 °C, or 1500 to 900 °F) and cooled rapidly from 870 °C (1600 °F) to a black heat. The highly alloyed grades, such as 309, 310, and 314, are also limited with regard to finishing temperature, because of their susceptibility at lower temperatures to hot tearing and σ formation. A final annealing by cooling rapidly from about 1065 °C (1950 °F) is generally advised for nonstabilized austenitic stainless steel forgings in order to retain the chromium carbides in solid solution.

Finishing temperatures for austenitic stainless steels become more critical when section sizes increase and ultrasonic testing requirements are specified. During ultrasonic examination, coarse-grain austenitic stainless steels frequently display sweep noise that can be excessive due to a coarse-grain microstructure. The degree of sound attenuation normally increases with section size and may become too great to permit detection of discontinuities. Careful control of forging conditions, including final forge reductions of at least 5%, can assist in the improvement of ultrasonic penetrability.

The stabilized or extralow-carbon austenitic stainless steels, which are resistant to sensitization, are sometimes strain hardened by small reductions at temperatures well below the forging temperature. Strain hardening is usually accomplished at 535 to 650 °C (1000 to 1200 °F) (referred to as warm working or hot-cold working). When minimum hardness is required, the forgings are solution annealed.

Sulfur or selenium can be added to austenitic stainless steel to improve machinability. Selenium is preferred because harmful stringers are less likely to exist. Type 321, stabilized with titanium, may also contain stringers of segregate that will open as the surface ruptures when the steel is forged. Type 347, stabilized with niobium, is less susceptible to stringer segregation and is the stabilized grade that is usually specified for forgings.

When heating the austenitic stainless steels, it is especially desirable that a slightly oxidizing furnace atmosphere be maintained. A carburizing atmosphere or an excessively oxidizing atmosphere will impair corrosion resistance, either by harmful carbon pickup or by chromium depletion. In types 309 and 310, chromium depletion can be especially severe.

Nitrogen-strengthened austenitic stainless steels are iron-base alloys containing chromium and manganese. Varying amounts of nickel, molybdenum, niobium, vanadium, and/or silicon are also added to achieve specific properties. Nitrogen-strengthened austenitic stainless steels provide high strength, excellent cryogenic properties and corrosion resistance, low magnetic permeability (even after cold work or subzero temperature), and higher elevated-temperature strengths, compared to the 300-series stainless steels. A partial list of these alloys includes:

- UNS S24100 (Nitronic 32) ASTM XM-28. High work hardening while remaining nonmagnetic plus twice the yield strength of type 304 with equivalent corrosion resistance
- UNS S24000 (Nitronic 33) ASTM XM-29. Twice the yield strength of type 304, low magnetic permeability after severe cold work, high resistance to wear and galling compared to standard austenitic stainless steels, and good cryogenic properties
- UNS S21904 (Nitronic 40) ASTM XM-11. Twice the yield strength of type 304 with good corrosion resistance, low magnetic permeability after severe cold working, and good cryogenic properties
- UNS S20910 (Nitronic 50) ASTM XM-19. Corrosion resistance greater than that of

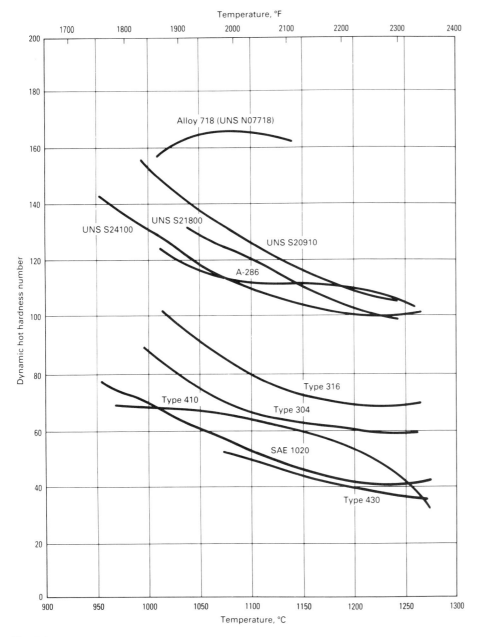

Fig. 31 Comparative dynamic hot hardness versus temperature (forgeability) for various ferrous alloys

type 316L with twice the yield strength, good elevated and cryogenic properties, and low magnetic permeability after severe cold work
- UNS S21800 (Nitronic 60). Galling resistance with corrosion resistance equal to that of type 304 and twice the yield strength and good oxidation resistance

A forgeability comparison, as defined by dynamic hot hardness, is provided in Fig. 31.

Martensitic stainless steels have high hardenability to the extent that they are generally air hardened. Therefore, precautions must be taken when cooling forgings of martensitic steels, especially those with high carbon content, in order to prevent cracking. The martensitic alloys are generally cooled slowly to about 595 °C (1100 °F), either by burying in an insulating medium or by temperature equalizing in a furnace. Direct water sprays, such as might be employed to cool dies, should be avoided, because they would cause cracking of the forging.

Forgings of the martensitic steels are often tempered in order to soften them for machining. They are later quench hardened and tempered.

Maximum forging temperatures for these steels are low enough to avoid the formation of δ-ferrite. If δ-ferrite stringers are present at forging temperatures, cracking is likely to occur. Delta-ferrite usually forms at temperatures from 1095 to 1260 °C (2000 to 2300 °F). Care must be exercised to keep the temperature below this level during forging and to avoid rapid metal movement that might result in local overheating. Surface decarburization, which promotes ferrite formation, must be minimized.

The δ-ferrite formation temperature decreases with increasing chromium content, and small amounts of δ-ferrite reduce forgeability significantly. As the δ-ferrite increases above about 15% (Fig. 30), forgeability improves gradually until the structure becomes entirely ferritic. Finishing temperatures are limited by the allotropic transformation, which begins near 815 °C (1500 °F). However, forging of these steels is usually stopped at about 925 °C (1700 °F) because the metal is difficult to deform at lower temperatures.

Sulfur or selenium can be added to type 410 to improve machinability. However, these elements can cause forging problems, particularly when they form surface stringers that open and form cracks. This can sometimes be overcome by adjusting the forging temperature or the procedure. With sulfur additions, it may be impossible to eliminate all cracking of this type. Therefore, selenium additions are preferred.

Ferritic Stainless Steels. The ferritic straight-chromium stainless steels exhibit virtually no increase in hardness upon quenching. They will work harden during forging; the degree of work hardening depends on the temperature and the amount of metal flow. Cooling from the forging temperature is not critical.

The ferritic stainless steels have a broad range of forgeability, which is restricted somewhat at higher temperature because of grain growth and structural weakness and is closely restricted in finishing temperature only for type 405. Type 405 requires special consideration because of the grain-boundary weakness resulting from the development of a small amount of austenite. The other ferritic stainless steels are commonly finished at any temperature down to 705 °C (1300 °F). For type 446, the final 10% reduction should be made below 870 °C (1600 °F) to achieve grain refinement and room-temperature ductility. Annealing after forging is recommended for ferritic steels.

Precipitation-Hardening Stainless Steels. The semiaustenitic and martensitic precipitation-hardening stainless steels can be heat treated to high hardness through a combination of martensite transformation and precipitation. They are the most difficult to forge and will crack if temperature schedules are not accurately maintained. The forging range is narrow, and the steel must be reheated if the temperature falls below 980 °C (1800 °F). They have the least plasticity (greatest stiffness) at forging temperature of any of the classes and are subject to grain growth and δ-ferrite formation. Heavier equipment and a greater number of blows are required to achieve metal flow equivalent to that of the other types.

During trimming, the forgings must be kept hot enough to prevent the formation of flash-like cracks. To avoid these cracks, it is often necessary to reheat the forgings slightly between the finish forging and the trimming operations. Cooling, especially the cooling of the martensitic grades, must be controlled to avoid cracking.

Duplex Stainless Steels. Because of their higher strength, duplex stainless steels are generally stiff when hot worked, relative to many other stainless grades. Rolling and forging equipment must have sufficient power to reduce the material while it is in the temperature range for optimum hot workability. The surface of finished forgings can be optimized by controlling the initial surface, material composition, phase balance, and hot-working temperature. Wrought structures generally produce a better surface after reforging than does cast material after the initial forging operation.

Machinability

Because of the wide variety of stainless steels available, a simple characterization of their machinability can be somewhat misleading. As shown in later sections of this article, the machinability of stainless steels varies from low to very high, depending on the final choice of alloy. In general, however, stainless steels are considered more difficult to machine than other metals, such as aluminum or low-carbon steels. Stainless steels have been characterized as gummy during cutting, showing a tendency to produce long, stringy chips, which seize or form a built-up edge on the tool. This may result in reduced tool life and degraded surface finish. These general characteristics are due to the following properties possessed by stainless steels to varying degrees (Ref 65, 66) (see also the article "Machining of Stainless Steel" in Volume 16 of the 9th Edition of *Metals Handbook*):

- High tensile strength (Fig. 24)
- Large spread between yield strength (YS) and ultimate tensile strength (UTS) (Fig. 24)
- High ductility and toughness
- High work-hardening rate (Fig. 24)
- Low thermal conductivity (Fig. 32)

Despite these properties, stainless steels can be machined under the appropriate conditions. In general, more power is required to machine stainless steels than carbon steels, cutting speeds must often be lower, a positive feed must be maintained, tooling and fixtures must be rigid, chip breakers or curlers may be needed on the tools, and care must be taken to ensure good lubrication and cooling during cutting (Ref 68).

These and other practices are discussed in more detail in the sections on individual conventional machining techniques in the article "Machining of Stainless Steels" in Volume 16 of the 9th Edition of *Metals Handbook*.

Significant differences in machinability exist between different alloy systems and alloy families, including the various free-machining alloys. This section will discuss machinability both within and among the five basic families of stainless steels.

Machinability of Ferritic and Martensitic Alloys. Free-machining ferritic alloys (such as S43020) and annealed, low-carbon, free-machining martensitic alloys (such as S41600) are the easiest to machine of the stainless steels (Ref 66, 69-72). In fact, their machinability ratings approach and in some cases are comparable to those of certain free-machining carbon steels (Ref 69-71). The nonfree-machining lower-chromium ferritic alloys (S40500, S43000) and annealed, low-carbon, straight-chromium martensitic alloys (S40300, S41000) are also generally easier to machine than most other nonfree-machining alloys (Ref 69-72). The higher-chromium ferritic alloys, such as S44600, are considered by some to be somewhat more difficult to machine than the lower-chromium alloys because of gumminess and stringy chips (Ref 73, 74).

Other than the presence or lack of a free-machining additive, the machining characteristics of martensitic stainless steels are influenced by certain variables:

- Hardness level
- Carbon content
- Nickel content
- Phase balance, that is, the percentage of free or δ-ferrite in the martensitic matrix

Increasing the hardness level for a particular alloy results in a decrease in machinability as measured by various criteria (tool life, drillability, and so on) (Ref 66, 72, 75, 76). Within certain limits, however, surface finish can be improved by machining harder material (Ref 72, 75).

In the martensitic grades, machinability decreases as the carbon content increases from S41000 to S42000 to S44004 or from S41600/S41623 to S42020/S42023 to S44020/S44023. With higher carbon levels, there also tends to be a smaller difference in machinability between the corresponding free-machining and nonfree-machining versions. These effects are primarily due to the increasing quantities of abrasive chromium carbides present as carbon level increases in this series of alloys. As a further detriment to machinability, annealed hardness level increases with increasing carbon level (Ref 72-75).

Nickel content also influences machinability by increasing annealed hardness levels. Consequently, alloys such as S41400 and S43100 will be more difficult to machine

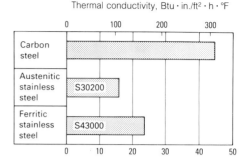

Fig. 32 Comparison of thermal conductivities for carbon steel, S30200 austenitic stainless steel, and S43000 ferritic stainless steel. Source: Ref 67

than S41000 in the annealed condition (Ref 70, 71).

Changing phase balance has been used to improve the machining characteristics of S41600. It has generally been found that increasing free or δ-ferrite content results in improved machinability, including tool life and surface finish (Ref 72, 75, 76-79). The introduction of a higher ferrite content also results in a decreasing hardness capability.

Machinability of Austenitic Alloys. The difficulties in machining attributed to stainless steels in general are more specifically attributable to the austenitic stainless steels (Ref 69, 71-74, 79). Compared to ferritic and martensitic alloys, typical austenitic alloys have a higher work-hardening rate, a wider spread between yield and ultimate tensile strengths, and higher toughness and ductility. When machining austenitic stainless steels, particularly the nonfree-machining alloys, several factors become more pronounced:

- Tools will run hotter, with more tendency to form a large built-up edge
- Chips will be stringier, with a tendency to tangle, making their removal difficult
- Chatter will be more likely if tool rigidity is inadequate or marginal
- Cut surfaces will be work hardened and more difficult to machine if cutting is interrupted or if the feed rate is too low

Because of these factors, the general precautions for machining stainless steels are particularly important for austenitic alloys.

Although there have been differing opinions (Ref 80), a moderate amount of cold work has been regarded as beneficial to the overall machining characteristics of austenitic stainless steels (Ref 72, 74). The cold working reduces the ductility of the material, which results in cutting with a cleaner chip and less tendency for a built-up edge. This produces a better-machined surface finish but with some loss of tool life due to the higher hardness level (Ref 72).

Automatic screw machine testing has shown that the effects of cold working and hardness are variable and may or may not

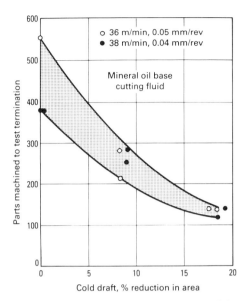

Fig. 33 Effect of percent cold draft on machinability in a screw machine test for an enhanced-machining version of S30400. Termination is defined as a 0.075 mm (0.003 in.) increase in the diameter of the part being cut.

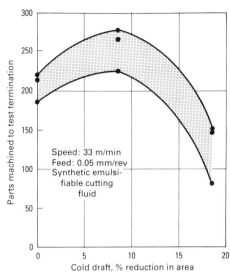

Fig. 34 Effect of percent cold draft on machinability in a screw machine test for an enhanced-machining version of S30400. Termination is defined as a 0.075 mm (0.003 in.) increase in the diameter of the part being cut.

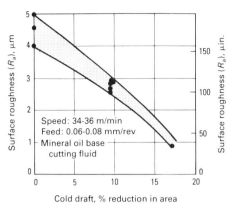

Fig. 35 Effect of percent cold draft on machined surface finish in a screw machine test for an enhanced-machining version of S31600

be seen, depending on the type of alloy and the machining conditions. In such testing, tool life has been lowered by an increasing level of cold work for both free-machining (S30300) and nonfree-machining (S30400, S31600) austenitic stainless steels. This effect is shown in Fig. 33 for S30400. On the other hand, there have also been indications under different cutting conditions of an optimum level of tool life at an intermediate level of cold work (Fig. 34).

Machined surface finish can be improved by an increasing level of cold work for nonfree-machining alloys (S30400, S31600). Figure 35 shows this effect for S31600. A decreasing tendency for tool chatter with increasing cold work has also been seen for these alloys. On the other hand, the use of cold-drawn bar does not consistently benefit the machined surface finish of a free-machining alloy (S30300).

Additions of manganese or copper can increase the machinability (Fig. 36) and decrease the high work-hardening rate of the lower-alloy austenitic stainless steels (Ref 79, 81-84). Austenitic free-machining alloys that have additions of manganese and/or copper include S20300, S30310, S30330, and S30431. Although higher alloy content generally reduces the work-hardening rate, it may not necessarily benefit machinability. Highly alloyed austenitic stainless steels, such as S30900, S31000, and N08020, tend to be more difficult to machine (Ref 70, 71).

Carbon and nitrogen can affect work-hardening rate and will increase the strength and hardness of austenitic stainless steels. Higher levels of either or both elements will decrease machinability (Fig. 37). Conse-

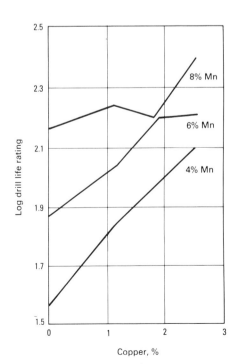

Fig. 36 Effect of copper and manganese contents on machinability in a drill test for a free-machining chromium-manganese-nickel austenitic stainless steel. Source: Ref 81

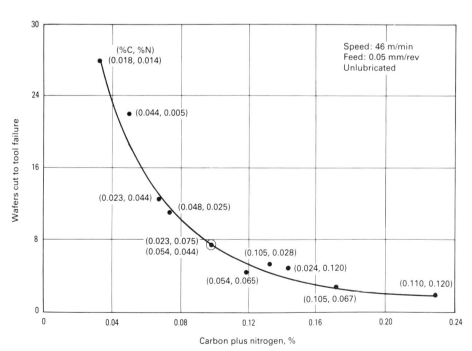

Fig. 37 Effect of carbon and nitrogen contents on machinability in a tool life test for a free-machining 18Cr-9Ni-3Mn austenitic stainless steel. Source: Ref 85

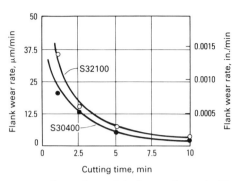

Fig. 38 Comparison of tool wear for austenitic stainless steels with (S32100) and without (S30400) titanium carbide inclusions. Source: Ref 86

quently, the high-nitrogen austenitic alloys, such as S20910 and S28200, are more difficult to machine than the standard lower-nitrogen austenitic alloys (Ref 70).

Strong carbide/nitride-forming elements, including titanium and niobium, are used in stainless steels such as S32100 and S34700 to prevent grain-boundary carbide, which can reduce intergranular corrosion resistance. However, the carbide/nitride inclusions are abrasive and will increase tool wear (Fig. 38).

Machinability of Duplex Alloys. The machinability of duplex stainless steels is limited by their high annealed strength level. Figures 39 and 40 compare the machinability of a duplex alloy, S32950, with that of a high-nitrogen austenitic alloy, S20910, and a conventional austenitic alloy, S31600, in standard (0.004% S) and enhanced-machining (0.027% S) versions. The duplex alloy (S32950) has a hardness level comparable to that of the high-nitrogen austenitic alloy (S20910), but provides better machinability. However, it does not machine as well as either the standard or the enhanced-machining S31600 alloy.

Other nitrogen-bearing duplex alloys are expected to machine similarly to S32950. No enhanced-machining versions of duplex alloys are available.

Machinability of Precipitation-Hardenable Alloys. The machinability of precipitation-hardenable stainless steels depends on the type of alloy and its hardness level. Martensitic precipitation-hardened stainless steels are often machined in the solution-treated condition; therefore, only a single aging treatment is required afterward to reach the desired strength level. In this condition, the relatively high hardness limits machinability. Most of these alloys machine comparably to, or somewhat worse than, a standard austenitic alloy such as S30400. Alloy S17400 is available in enhanced-machining versions that allow machining at higher speeds with a significantly reduced tendency toward chatter.

Martensitic precipitation-hardenable stainless steels can also be machined in an aged condition so that the heat treating can be avoided and closer tolerances maintained. The ease of cutting generally varies with the hardness or heat-treated condition (Table 34).

In the annealed, austenitic condition, semiaustenitic alloys can be expected to machine with difficulty, somewhat worse than an alloy such as S30200, which has a high work-hardening rate. Alloys S35000 and S35500 can be supplied in an equalized and overtempered condition, which provides the best machinability. As with the martensitic precipitation-hardenable alloys, machining difficulties increase with aged hardness level.

Austenitic precipitation-hardenable alloys, such as S66286, machine quite poorly,

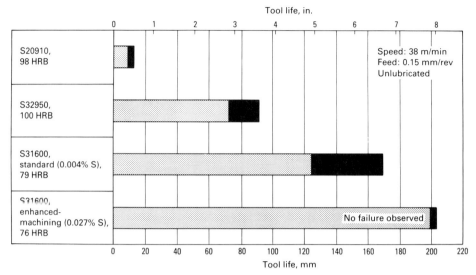

Fig. 39 Comparison of tool life for a duplex stainless steel (S32950), a high-nitrogen austenitic stainless steel (S20910), and a lower-nitrogen austenitic stainless steel (S31600). Tool life is measured as the distance traveled along a 25 mm (1.0 in.) diam bar until tool failure. Shaded areas represent distance to failure.

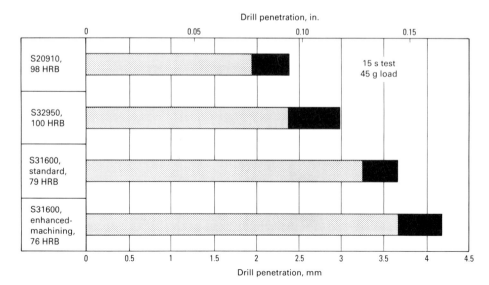

Fig. 40 Comparison of machinability in a drill penetration test for a duplex stainless steel (S32950), a high-nitrogen austenitic stainless steel (S20910), and a lower-nitrogen austenitic stainless steel (S31600)

Table 34 Relative machinability of 17-4 PH (S17400) in various heat-treated conditions

Condition	Typical hardness, HRC
Improved machinability (higher cutting speed)	
H1150M	27
H1150	33
H1075	36
A (solution treated)	34
H1025	38
H900	44
Improved surface finish	

Source: Ref 87

Wrought Stainless Steels / 897

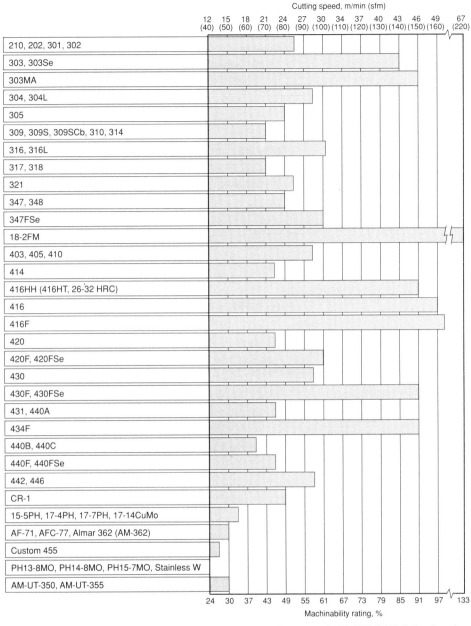

Fig. 41 General comparison of machinability of stainless steels compared with AISI B1112. Rating based on 100% for AISI B1112 using high-speed steel tools. Source: Ref 88

requiring slower cutting rates than even the highly alloyed austenitic stainless steels (Ref 70). Machining in an aged condition requires even slower speeds.

General Guidelines. The characteristics of stainless steels that have a large influence on machinability include:

- Relatively high tensile strength
- High work-hardening rate, particularly for the austenitic alloys
- High ductility

These factors explain the tendency of the material to form a built-up edge on the tool during traditional machining operations. The chips removed in machining exert high pressures on the nose of the tool; these pressures, when combined with the high temperature at the chip/tool interface, cause pressure welding of portions of the chip to the tool. In addition, the low thermal conductivity of stainless steels contributes to a continuing heat buildup.

Figure 41 compares the machinability ratings of selected stainless steels using AISI B1112 as the reference. The difficulties involved in the traditional machining of stainless steels can be minimized by observing the following points:

- Because more power is generally required to machine stainless steels, equipment should be used only up to about 75% of the rating for carbon steels
- To avoid chatter, tooling and fixtures must be as rigid as possible. Overhang or protrusion of either the workpiece or the tool must be minimized. This applies to turning tools, drills, reamers, and so on
- To avoid glazed, work-hardened surfaces, particularly with austenitic alloys, a positive feed must be maintained. In some cases, increasing the feed and reducing the speed may be necessary. Dwelling, interrupted cuts, or a succession of thin cuts should be avoided
- Lower cutting speeds may be necessary, particularly for nonfree-machining austenitic alloys, precipitation-hardenable stainless steels, or higher-hardness martensitic alloys. Excessive cutting speeds result in tool wear or tool failure and shutdown for tool regrinding or replacement. Slower speeds with longer tool life are often the answer to higher output and lower costs
- Tools, both high-speed steel and carbides, must be kept sharp, with a fine finish to minimize friction with the chip. A sharp cutting edge produces the best surface finish and provides the longest tool life. To produce the best cutting edge on high-speed steel tools, 60-grit roughing should be followed by 120- and 150-grit finishing. Honing produces an even finer finish
- Cutting fluids must be selected or modified to provide proper lubrication and heat removal. Fluids must be carefully directed to the cutting area at a sufficient flow rate to prevent overheating

Weldability

The metallurgical features of each group generally determines the weldability characteristics of the steels in that group. The weldability of martensitic stainless steels is greatly affected by hardenability that can result in cold cracking. Welded joints in ferritic stainless steel have low ductility as a result of grain coarsening that is related to the absence of allotropic (phase) transformation. The weldability of austenitic stainless steels is governed by their susceptibility to hot cracking, as is the case with other single-phase alloys with a face-centered cubic (fcc) crystal structure. With the precipitation-hardening stainless steels, weldability is related to the mechanisms associated with the transformation (hardening) reactions (Ref 89).

Stainless steels can be joined by most welding processes, but with some restrictions. In general, those steels that contain aluminum or titanium, or both, can be arc welded only with the gas-shielded processes. The weld joint efficiency depends upon the ability of the welding process and procedures to produce nearly uniform mechanical properties in the weld metal, heat-affected zone, and base metal in the as-welded or postweld heat-treated condition.

These properties can vary considerably with ferritic, martensitic, and precipitation-hardening steels (Ref 89).

Weldability and various suitability-for-service conditions, including temperature, pressure, creep, impact, and corrosion environments (see the article "Corrosion of Weldments" in Volume 13 of the 9th Edition of *Metals Handbook*), require careful evaluation because of the complex metallurgical aspects of stainless steels.

Shielded metal arc (SMAW), submerged arc (SAW), gas metal arc (GMAW), gas tungsten arc (GTAW), and plasma arc welding (PAW) are used extensively for joining stainless steels. Flux cored arc welding (FCAW) is also used, but to a lesser extent. The remainder of the article discusses the weldability of the various grades and the suitability of arc welding processes for specific conditions and requirements.

Austenitic Stainless Steels. Differences in composition among the standard austenitic stainless steels affect weldability and performance in service. For example, types 302, 304, and 304L differ primarily in carbon content, and consequently there is a difference in the amount of carbide precipitation that can occur in the heat-affected zone (HAZ) after the heating and cooling cycle encountered in welding. Types 303 and 303Se contain 0.20% P (max) plus 0.15% Se or S for free machining. These elements are detrimental to weldability and can cause severe hot cracking in the weld metal. Types 316 and 317 contain molybdenum for increased corrosion resistance and higher creep strength at elevated temperatures. However, unless controlled by extralow carbon content, as in type 316L, carbide precipitation occurs in the HAZ during welding. Types 318, 321, 347, and 348 are stabilized with titanium, or niobium-plus-tantalum, to prevent the intergranular precipitation of chromium carbides when the steels are heated to a temperature in the sensitizing range, as during welding.

Welding Characteristics. The austenitic stainless steels, except for the free-machining grades, are the easiest to weld and produce welded joints that are characterized by a high degree of toughness, even in the as-welded condition. Serviceable joints can be readily produced if the composition and the physical and mechanical properties are well suited to the welding process and condition. The heat of welding, contamination, carbide precipitation, cracking, and porosity must be considered before, during, and after welding stainless steels.

Heat of Welding. Excessive heat input may result in weld cracking, loss of corrosion resistance, warping, and undesirable changes in mechanical properties. Welds in stainless steels generally require 20 to 30% less heat input than welds in carbon grades because stainless steels have lower thermal conductivity and higher electrical resistance. Because of low thermal conductivity, heat remains near the weld, so that more heat is available to melt the material, which may produce detrimental results. Excessive heat produces large thermal gradients across the joint, which can cause distortion. Because heat dissipates slowly in stainless steel, it may lower corrosion resistance and change strength. These effects can be minimized with chill bars and less heat input. Weld metal cracking, another problem resulting from excessive welding heat, is discussed later in this article.

The high electrical resistivity of stainless steel makes it suitable for welding with low heat inputs. With reduced heat, good penetration and fusion result because low thermal conductivity retains heat in the weld area. Comparative electrical resistivities are 25 to 50 $\mu\Omega \cdot$ in. for carbon steel and 175 to 200 $\mu\Omega \cdot$ in. for austenitic grades. Heat input can be reduced by using low amperages, low voltages (short arc lengths), high travel speeds, and stringer beads. With GMAW and GTAW processes, heat input can be affected by the type of shielding gas. Argon produces a cool, stable arc, while helium produces a hot arc that is somewhat unstable. For manual processes, pure argon is generally best. When working with automatic welding equipment that offers good control of amperage, voltage, and travel speed, mixed gases can be used without risking damage from high heat. Finally, pulsed arc welding techniques can be used to lower heat input.

Weld Contamination. Contaminants not only hinder successful welding, but may prevent an apparently sound weld from functioning satisfactorily. A contaminated weld has inferior corrosion resistance and strength, and the weldment may fail prematurely. The stainless steel itself may also contain the contaminant. Free-machining stainless steels frequently contain sulfur or selenium. Both elements can make the steel unweldable. Similarly, high concentrations of carbon in high-strength stainless steel can inhibit weld serviceability.

External sources of contamination include carbon, nitrogen, oxygen, iron, and water. Carbon is often picked up from shop dirt, grease, forming lubricants, paint, marking materials, and tools; consequently, steel parts should be cleaned before welding and during welding. Otherwise, carbon contamination can cause welds to crack, change the mechanical properties, and lower the corrosion resistance in weld areas. Although iron contamination generally does not affect weldability, it can lower serviceability. Flakes of iron on surfaces rust, thereby speeding localized corrosion. The welder may unknowingly cause the contamination by grinding stainless steel with a wheel previously used on carbon steel. Clean Al_2O_3 grinding wheels, preferably those not used for grinding other alloys, should be used.

One of the most troublesome types of contamination is stainless steel surface contamination by copper, bronze, lead, or zinc from hammers, hold-down fingers on seamers, or tools used in fabrication. Small amounts of these materials on the surface of the stainless steel can lead to cracking in the high-temperature HAZ of the weld. This type of cracking generally occurs in the HAZ, where the contaminant attacks the grain boundaries.

Effect of Carbide Precipitation on Corrosion Resistance of Welded Joints. The precipitation of intergranular chromium carbides is accelerated by an increase in temperature within the sensitizing range and by an increase in time at temperature. When intergranular chromium carbides are precipitated at welded joints, resistance to intergranular corrosion and stress corrosion markedly decreases. The decrease in corrosion resistance is attributed to the presence of the chromium-rich carbides at the grain boundaries and the depletion of chromium in the adjacent matrix material. Although intergranular carbide precipitation generally occurs between 425 and 870 °C (800 and 1600 °F), sensitization is restricted to a narrow range by the fairly rapid heating and cooling that usually occur in welding. The narrower range varies with time at temperature and steel composition, but is approximately 650 to 870 °C (1200 to 1600 °F).

The base metal immediately adjacent to the weld is annealed or solution treated by the heat of welding and, because it generally is cooled rapidly enough to hold the dissolved carbides in solution, this zone usually exhibits normal resistance to corrosive attack. A short distance from the weld, about 3.2 mm (⅛ in.) (the distance depending on the thermal cycle and material thickness), there is a narrow zone in which lower heating and cooling rates prevail. In this HAZ, intergranular precipitation of chromium carbides is most likely to take place. Harmful carbide precipitation can be overcome or prevented by the use of:

- Postweld solution annealing
- Extralow-carbon, that is, 0.03% C (max), alloy
- Stabilized alloy containing preferential carbide-forming elements, such as niobium-plus-tantalum or titanium

Solution annealing puts carbides back into solution and restores normal corrosion resistance, but is generally inconvenient. The solution-annealing temperature range is very high, 1040 °C (1900 °F) minimum, and unless stainless steel is protected from air at these temperatures, it oxidizes rapidly, forming adherent oxide scale. Thin sections, unless adequately supported, may sag or be severely distorted at these temperatures or during rapid cooling from them. Rapid cooling in solution annealing may present other problems. Water quenching,

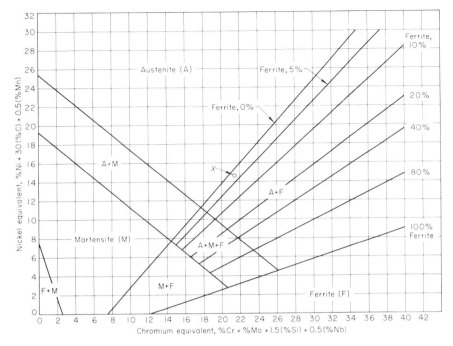

Fig. 42 Schaeffler diagram. See Fig. 30 for revised Schaeffler diagram.

although effective, is seldom feasible except for small workpieces of simple shape. Unless adequate safeguards are available, water quenching of large workpieces from the solution-annealing temperature is hazardous. Often solution annealing is impractical because the workpiece is too large for available furnace and cooling facilities.

Extralow-carbon stainless steels (types 304L and 316L) are resistant to carbide precipitation in the 425 to 870 °C (800 to 1600 °F) range and can thus undergo normal welding without reduction in corrosion resistance. Carbides precipitate in significant quantities when extralow-carbon steels are heated and held in the sensitizing temperature range for an extended period, as in service. These steels are generally recommended for use below 425 °C (800 °F).

Stabilized Steels. Compared with the extralow-carbon steels, the stabilized steels exhibit higher strength at elevated temperature. For service in a corrosive environment in the sensitizing temperature range of 425 to 870 °C (800 to 1600 °F), an austenitic steel stabilized with niobium-plus-tantalum or titanium is needed. The filler metal used for welding should also be of a stabilized composition. Because an inert shielding gas is used, GTAW and GMAW are suitable for titanium-stabilized steel without oxidizing the titanium. Under certain conditions, stabilized stainless steel weldments are susceptible to sensitization, which occurs in narrow zones of the base metal immediately adjacent to the line of weld fusion. During welding, stabilized carbides are dissolved and, as a result of rapid cooling, are retained in solution. Subsequent reheating to about 650 °C (1200 °F) results in preferential precipitation of chromium carbides in a narrow zone that exhibits less than normal corrosion resistance.

Microfissuring in Welded Joints. Interdendritic cracking in the weld area that occurs before the weld cools to room temperature is known as hot cracking or microfissuring. The occurrence of microfissuring is related to the:

- Microstructure of the weld metal as solidified
- Composition of the weld metal, especially the content of certain residual or trace elements
- Amounts of stress developed in the weld as it cools
- Ductility of the weld metal at high temperatures
- Presence of notches

Susceptibility to microfissuring is highly dependent on the microstructure of the weld metal. Weld metal with a wholly austenitic microstructure is considerably more susceptible to microfissuring than weld metal with a duplex structure of δ-ferrite in austenite. The content of alloying elements and residual elements strongly influences the susceptibility of fully austenitic stainless steel weld metal to microfissuring. Susceptibility can be reduced by a small increase in carbon or nitrogen content or by a substantial increase in manganese content. Residual or trace elements that contribute to microfissuring are boron, phosphorus, sulfur, selenium, silicon, niobium, and tantalum.

The amount of stress imposed on austenitic stainless weld metal as it cools from the solidus down to about 980 °C (1800 °F) should be minimized. In this temperature range, the weld metal is most susceptible to microfissuring, and if the level of stress is high, the fissures propagate to form visible cracks. Peening is not an effective method of preventing this type of cracking because it can seldom be applied early enough to reduce stress buildup.

Prevention of Microfissuring. To obtain duplex-structured weld metal that has a controlled ferrite content of at least 3 to 5 ferrite number (FN), a filler metal of suitable composition is selected. The ferrite number is a magnetically determined scale of ferrite measurement. The Welding Research Council Advisory Subcommittee on Welding Stainless Steels determined that the ferrite number of a weld metal, at least from 0 to 6 FN, approximates the average value of percent ferrite assigned by laboratories applying metallographic measurements of ferrite to a given weld metal.

Microfissuring can be prevented or minimized by the proper control of ferrite in the weld metal. Wide use has been made of the Schaeffler diagram (Fig. 42) to determine the approximate amount of ferrite that will be obtained in the austenitic weld metal of a given composition. Point X indicates the equivalent composition of a type 318 (316Cb) weld deposit containing 0.07C-1.55Mn-0.57Si-18.02Cr-11.87Ni-2.16Mo-0.80Nb. To determine the chromium and nickel equivalents, each percentage was multiplied by the potency factor indicated for the respective element along the axes of the diagram. When these were plotted as point X, the constitution of the weld was indicated as austenite plus 0 to 5% ferrite. Magnetic analysis of an actual sample revealed an average 2 FN. For austenite-plus-ferrite structures, the diagram predicts ferrite within 4% for stainless steel types 308, 309, 309Cb, 310, 312, 316, 317, and 318. Actual measurements of ferrite content can be made conveniently with the aid of a magnetic analysis device. American Welding Society (AWS) A4.2-74, "Standard Procedures for Calibrating Magnetic Instruments to Measure the Delta Ferrite Content of Austenitic Stainless Steel Weld Metal," and A5.4-81, "Specification for Covered Corrosion-Resisting Chromium and Chromium-Nickel Steel Welding Electrodes," discuss this measurement.

Because many heats of austenitic stainless steel contain appreciable amounts of nitrogen (a very strong austenitizer), a revised constitution diagram for austenitic stainless steel weld metal has been developed to include nitrogen in the nickel equivalent (Fig. 30). Compared to Fig. 42, Fig. 30 is modified in shape and slope to improve the accuracy of ferrite estimation for types 309, 309Cb, 316, 316L, 317, 317L, and 318. In addition, ferrite calculation for types 308 and 347 weld metal is improved on samples with either high or low nitrogen content. For use in the diagram, actual nitrogen

Table 35 Suggested filler metals for stainless steels

AISI No.	Recommended filler metal(a) First choice	Second choice	Remarks
Chromium-nickel-manganese austenitic nonhardenable			
201	308	308L	Substitute for 301
202	308	308L	Substitute for 302
301	308	308L	
302	308	308L	
302B	308	309	High silicon
303	...	...	Free-machining stainless steel: welding not recommended—312
303Se	...	...	Free-machining stainless steel: welding not recommended—312
Chromium-nickel austenitic nonhardenable			
304	308	308L	
304L	308L	347	Extralow carbon
305	308	...	
308	308	...	
309	309	...	
309S	309	...	Low carbon
310	310	...	
310S	310	...	Low carbon
314	310	...	
316	316	309Cb	
316L	316L	309Cb	Extralow carbon
317	317	309Cb	
321	347	308L	
347	347	308L	Difficult to weld in heavy sections
348	347	...	
Chromium martensitic hardenable			
403	410	...	
410	410	430	
414	410	...	
416	410	...	410-15 should be used
416Se	...	...	Free-machining: welding not recommended
420	410	...	High carbon
431	430	...	
440A	...	...	High carbon: welding not recommended
440B	...	...	High carbon: welding not recommended
440C	...	...	High carbon: welding not recommended
Chromium ferritic nonhardenable			
405	410	405Cb	
430	430	309	
430F	...	...	Free machining: welding not recommended
430FSe	...	...	Free machining: welding not recommended
446	309	310	

(a) Use E, electrode, or R, welding rod, prefix. Source: Ref 90

content is preferred. If it is not available, the following nitrogen value shall be used: 0.12% for self-shielding flux cored electrode GMAW welds, 0.08% for other GMAW welds, and 0.06% for welds of other processes.

When the weld metal must be wholly austenitic (that is, when the metal must be nonmagnetic or when specific corrosive environments that selectively attack δ-ferrite will be encountered), the content of crack-promoting residual elements must be stringently controlled, and the composition of the weld metal must be adjusted to increase crack resistance. Crack resistance can be increased by modifying the carbon, manganese, sulfur, phosphorus, silicon, and nitrogen contents of the weld metal. However, even with optimum compositions and the most favorable welding procedures, wholly austenitic weld deposits are more crack sensitive than those of a duplex structure.

Underbead cracking can occur in the heat-affected zones adjacent to welds in austenitic stainless steel, particularly when the weld zone is heavily restrained or the section thickness is greater than 19 mm (¾ in.). Such cracking is most common in type 347 because of the strain-induced precipitation of niobium carbides, but it has been reported in other types as well. Weld restraint is the most important factor in the control of underbead cracking. Preheating is of little value in preventing cracking of austenitic stainless steel and can cause other problems, such as increased carbide precipitation.

Selection of Filler Metals. Electrodes and welding rods suitable for use as filler metal in the welding of austenitic stainless steels are shown in Table 35. These filler metals, with AWS standard composition specifications, are for GMAW, SAW, and SMAW. The selection of filler metals for welding austenitic stainless steels requires consideration of the microstructural constituents of the as-deposited weld metal. Ultimately, these microstructural constituents determine the mechanical properties, crack sensitivity, and corrosion resistance of the weld. The constituents of principal concern are austenite, δ-ferrite, and precipitated carbides.

Some filler metals, such as types 309Cb and 310, invariably deposit a fully austenitic weld metal. In these alloys, the ratio of ferrite formers to austenite formers cannot, within permissible limits, be raised high enough to produce any δ-ferrite in the austenite. Consequently, when these filler metals are applied to restrained joints or to base metals containing additions of elements such as phosphorus, sulfur, selenium, and silicon, only those procedures proved suitable by experience should be used.

The compositions of most filler metals are adjusted by the manufacturers to produce weld deposits that have ferrite-containing microstructures. Thus, ferrite-forming elements, such as chromium and molybdenum, are maintained at the high side of their allowable ranges, and austenite-forming elements, such as nickel, are kept low. The amount of ferrite in the structure of the weld metal depends on the ratio or balance of these elements. At least 3 or 4 FN δ-ferrite is needed in the as-deposited weld metal for effective suppression of hot cracking. With the proper techniques, however, types 316 and 316L can be welded with as little as 0.5 FN. Ferrite-containing weld metal may have certain disadvantages in a welded austenitic stainless steel. Ferrite is ferromagnetic, and the increased magnetic permeability of the weld metal may be objectionable in applications that require nonmagnetic properties. When exposed to service at elevated temperature, the ferrite in some weld metals may transform to σ phase and adversely affect mechanical properties and corrosion resistance, a problem that has been encountered in power plant applications.

When a joint is arc welded without the addition of filler metal, the structure of the weld metal is determined by the composition of the base metal. Sometimes this leads to unfavorable results, because wrought base metals may not have the compositional limits required for good weld metal.

Preheating. In general, no benefit is derived from preheating austenitic stainless steels. In some applications, preheating can increase carbide precipitation, cause shape distortion of the workpiece, or increase hot-cracking tendencies.

Postweld Stress Relieving. Although the effects of residual stress from welding on the properties of austenitic stainless steels are limited in comparison to the effects of cold working, residual stress may significantly affect the mechanical properties. Because the effective yield strength varies from point to point, the application of further stresses at later stages of fabrication can cause excessive distortion and even premature failure. Nonuniform heating, which relieves some local residual stress, may also contribute to distortion. For these reasons, stress relieving may be required to ensure dimensional stability.

Stress relieving can be performed over a wide range of temperatures, depending on the amount of relaxation required. Time at temperature ranges from about 1 h per inch of section thickness at temperatures above 650 °C (1200 °F) to 4 h per inch of section thickness at temperatures below 650 °C (1200 °F). Because of the high coefficient of expansion and the low thermal conductivity of austenitic stainless steels, cooling from the stress-relieving temperature must be slow. The stress-relieving temperature selected must be compatible with the extent of carbide precipitation acceptable and with the corrosion resistance desired. Nonstabilized stainless steels cannot be stress relieved in the sensitizing temperature range without sacrifice of corrosion resistance. Extralow-carbon stainless steels are affected much less, because carbide precipitation in these steels is sluggish. Stabilized stainless steels exhibit minimal chromium carbide precipitation tendencies.

For austenitic stainless steels, the estimated percentages of residual stress relieved at various temperatures, for the times previously noted, are:

Temperature		Stress relief, %
°C	°F	
845–900	1550–1650	85
540–650	1000–1200	35

Ferritic Stainless Steels. The ferritic stainless steels are generally less weldable than the austenitic stainless steels and produce welded joints having lower toughness because of grain coarsening that occurs at the high welding temperatures. The standard ferritic stainless steels are:

- Type 446 (25% Cr)
- Types 430, 430F, and 430F-Se (17% Cr)
- Types 405 and 409 (13% Cr)

Type 409 is ferritic because it has a low carbon content (0.08% max) and a minimum titanium content equal to six times the carbon content. Type 405, which also contains only 0.08% C (max), contains an average of 0.20% Al, which promotes ferrite formation.

There are a number of ferritic stainless steels that contain very low amounts of carbon and nitrogen. These low-interstitial ferritic stainless steels (including the 26Cr-1Mo and 29Cr-4Mo alloys) can be welded so that they do not lose any of the base metal ductility in the weld area, while retaining grain growth. The key to the successful welding of these steels is to prevent any carbon, nitrogen, or oxygen contamination during welding. Thus, the part and filler material must be clean before welding, and both the molten weld metal and the hot weld-area metal must be fully shielded from the atmosphere. All moisture must be excluded from the weld area before and during welding.

Effect of Welding Heat on Ductility and Grain Size. Although most ferritic stainless steels have compositions that ensure a ductile ferritic structure at room temperature, variations in composition within the standard compositional limits can result in the formation of small amounts of austenite during heating to elevated temperature. Upon cooling, the austenite transforms to martensite, resulting in a duplex structure of ferrite and a small amount of martensite. The martensite reduces both the ductility and toughness of the steel. Annealing transforms the martensite and restores normal ferritic properties, but annealing increases costs and can result in an excessive amount of distortion, particularly in parts that were previously formed by a cold-working process.

All ferritic stainless steel mill products are normally annealed at the mill to transform any martensite that may be present to a softer structure of ferrite and carbides. In this condition, the steel can be readily cold formed. Only when the steel is heated near or above the transformation temperature (approximately 870 °C, or 1600 °F), as during welding, does the risk of austenite formation and subsequent transformation to martensite arise. In addition, heating to temperatures above 955 °C (1750 °F) results in enlargement of the ferrite grain size, which also reduces the ductility and toughness of the steel. Although martensite can be eliminated by annealing, coarsened ferrite grains remain unaffected. Because martensite responds to annealing and inhibits ferrite grain growth, some applications may benefit from martensite formation, provided that the workpiece can be annealed after welding. In a 17% Cr steel, martensite formation is promoted by lowering the chromium content to 15 to 16%. When this practice is adopted, it is usually necessary to preheat before welding or to select a steel of lower carbon content to guard against cracking in the HAZ.

When postweld annealing is not feasible, the ductility of the welded joint can be controlled by the selection of a stainless steel base metal containing a substantial amount of strong ferrite-former, such as aluminum, niobium, or titanium. One such steel, which has the commercial designation of type 430Ti, is a 17% Cr steel containing 0.12% C (max) and a minimum titanium content equal to six times the carbon content. The metallurgical functions of the titanium in this steel are to form stable titanium carbides and to promote the formation of ferrite. When a completely ferritic steel is welded, no martensite is formed in the HAZ, although some grain coarsening may occur. Grain coarsening can be controlled, to some extent, by minimizing heat input during welding and by avoiding slow cooling from the weld temperature.

Effect of Temperature on Notch Toughness. For the 17% Cr steels, the temperature range is just above room temperature for transition from a tough shear-type fracture at the higher temperature to a brittle, cleavage-type fracture at the lower temperature, upon impact at a notch. Under impact loading at room temperature and below, these steels are notch sensitive, and impact test values of less than 20 J (15 ft · lbf) are usual. At 95 to 120 °C (200 to 250 °F), the impact test values increase to approximately 40 to 70 J (30 to 50 ft · lbf).

This relationship between notch toughness and temperature is important in the selection of joint design and welding conditions for ferritic stainless steel. In service, a weldment designed primarily to withstand static load may be subjected to accidental or unforeseen impact loading. Furthermore, a weldment with low notch toughness may not withstand an appreciable number of cyclical stresses even under a low rate of loading. Because multiaxial residual stresses are often developed during welding (especially when welding heavy sections), notches and points of stress concentration that might cause failure in service must be avoided whenever possible. Preheating before welding is often useful in preventing cracking during welding.

Preheating. The recommended preheating temperature range for ferritic stainless steels is 150 to 230 °C (300 to 450 °F). The need for preheating is determined largely by the composition, mechanical properties, and thickness of the steel being welded. Steels less than 6.4 mm (¼ in.) thick are much less likely to crack during welding than those of greater thickness. The type of joint, joint location, restraints imposed by clamping and jigging, welding process, and rate of cooling from the welding temperature can also affect weld cracking.

Postweld Annealing. The temperature range for postheating or postweld annealing of ferritic stainless steels is 790 to 845 °C (1450 to 1550 °F), which is safely below the temperatures for austenitic formation and grain coarsening. Annealing transforms a mixed structure to a wholly ferritic structure and restores the mechanical properties and corrosion resistance that may have been adversely encountered in welding. Thus, except for its inability to refine coarsened ferrite grains, annealing is generally beneficial. Annealing has two major disadvantages: the time and cost of the treatment, and the need to prevent the formation of the oxide scale or to remove it if it is already present. Annealing may also require the use of elaborate fixturing to prevent sagging or distortion of the weldment.

Cooling ferritic stainless steel from the annealing temperature may be done by air or water quenching. To minimize distortion from handling, weldments are often allowed to cool to about 595 °C (1100 °F) before they are removed from the furnace. Slow cooling through the temperature range of 565 to 400 °C (1050 to 750 °F) must be avoided because it produces brittleness in the steel. Susceptibility to this type of embrittlement, known as 475 °C (885 °F) embrittlement, normally increases as chromium content increases. Heavy sections may require forced cooling or a spray quench to bring them safely through this embrittlement range.

Selection of Filler Metal. As shown in Table 35, both ferritic and austenitic stainless steel filler metals are used in the arc welding of ferritic stainless steel. Ferritic stainless steel filler metals offer the advantages of having the same color and appearance, the same coefficient of thermal expansion, and essentially the same corrosion resistance as the base metal. However, austenitic stainless steel filler metals are often used to obtain more ductile weld metal in the as-welded condition.

Although austenitic stainless steel weld metal does not prevent grain growth or martensite formation in the HAZ, the ductility of austenitic weld metal improves the ductility of the welded joint. The selection of austenitic stainless steel filler metal, however, should be carefully related to the specific application to determine whether differences in color or in the physical corrosion and mechanical properties of the weld metal and the base metal cause difficulty.

For weldments that are to be annealed after welding, the use of austenitic filler metal can introduce several problems. The normal range of annealing temperature for ferritic stainless steels falls within the sensitizing temperature range for austenitic steels. Consequently, unless the austenitic weld metal is of extralow-carbon content or is stabilized with niobium or titanium, its corrosion resistance may be seriously impaired. If the annealing treatment is intended to relieve residual stress in the weldment, it cannot be fully effective because of the difference in the coefficients of thermal expansion of the weld metal and the base metal.

Corrosion Resistance. Ferritic stainless steels usually exhibit less corrosion resistance than austenitic stainless steels. Any condition of the welded joint that might therefore impair corrosion resistance must be avoided. The presence of martensite or the precipitation of σ phase at the grain boundaries can cause severe intergranular corrosion in the HAZ. Completely ferritic steels, such as types 430Ti and 446, display little or no susceptibility to intergranular attack at the weld joint. Annealing of any welded ferritic steel eliminates the unfavorable structural conditions that promote corrosive attack.

Martensitic Stainless Steels. The standard martensitic stainless steels are types 403, 410, 414, 416, 416Se, 420, 431, 440A, 440B, and 440C. These steels derive their corrosion resistance from chromium, which they contain in proportions ranging from 11.5 to 18%. Martensitic stainless steels are the most difficult stainless steels to weld because they are chemically balanced to become harder, stronger, and less ductile through thermal treatment. These same metallurgical changes occur from the heat of welding. As a result of welding, these changes are restricted to the weld area and are not uniform over the entire section. The nonuniform metallurgical condition of the part makes it susceptible to cracking when subjected to the high stresses from welding. Increasing carbon content in martensitic stainless steels generally results in increased hardness and reduced ductility. Thus, the three type 440 stainless steels are seldom considered for applications that require welding, and filler metals of the type 440 compositions are not readily available.

Modifications of the standard martensitic steels contain additions of elements such as nickel, molybdenum, vanadium, and tungsten, primarily to raise the allowable service temperature above the 595 °C (1100 °F) limit for the standard steels. When these elements are added, carbon content is increased, and the problem of avoiding cracking in the hardened HAZ of weldments becomes more serious.

Martensitic stainless steels can be welded in the annealed, hardened, and hardened and tempered conditions. Regardless of the prior condition of the steel, welding produces a hardened martensitic zone adjacent to the weld. The hardness of the HAZ depends primarily on the carbon content of the base metal. As hardness increases, toughness decreases, and the zone becomes more susceptible to cracking. Preheating and control of the interpass temperature are the most effective means of avoiding cracking. Postweld heat treatment is required to obtain optimum properties.

Preheating and Postweld Heat Treating. The usual preheating temperature range of martensitic steels is 205 to 315 °C (400 to 600 °F). The carbon content of the steel is the most important factor in determining whether preheating is necessary. On the basis of carbon content alone, a steel containing not more than 0.10% C seldom requires preheating, and one with more than 0.10% C requires preheating to prevent cracking. Other factors that determine the need for preheating are the mass of the joint, degree of restraint, presence of a notch effect, and composition of the filler metal. Correlations of preheating and postweld heat-treating practice with carbon contents and welding characteristics of martensitic stainless steels can be used:

- Carbon below 0.10%: Neither preheating nor postweld annealing is generally required; steels with carbon contents this low are not standard
- Carbon 0.10 to 0.20%: Preheat to 260 °C (500 °F); weld at this temperature; cool slowly
- Carbon 0.20 to 0.50%: Preheat to 260 °C (500 °F); weld at this temperature; anneal
- Carbon over 0.50%: Preheat to 260 °C (500 °F); weld with high heat input; anneal

If the weldment is to be hardened and tempered immediately after welding, annealing may be omitted. Otherwise, the weldment should be annealed immediately after welding, without cooling to room temperature.

The functions of a postweld heat treatment are:

- To temper or anneal the weld metal and HAZ to optimize hardness, toughness, and strength for the intended application
- To decrease residual stresses associated with welding

Postweld heat treatments normally used for martensitic stainless steels are subcritical annealing and full annealing (Ref 89).

The necessity for a postweld heat treatment depends on the composition of the steel, the filler metal, and the service requirements. Full annealing transforms a multiple-phase weld zone to a wholly ferritic structure. This annealing procedure requires proper control of the complete thermal cycle. It should not be used unless maximum softness is required because of the formation of coarse carbides in the microstructure that take longer to dissolve at the austenitizing temperature. Typical postweld annealing temperatures are given in Table 36.

Precipitation-Hardening Stainless Steels. Steels are divided into three groups on the dual basis of characteristic alloying additions, particularly the elements added to promote precipitation hardening, and the matrix structures of the steels in the solution-annealed and aged condition. Because differences among the steels have a direct bearing on the behavior of the steels in heat treatment and welding, the metallurgical characteristics of each are considered separately.

Martensitic Precipitation-Hardening Steels. These steels have a predominantly austenitic structure at the solution-annealing temperature of approximately 1040 to 1065 °C (1900 to 1950 °F), but they undergo an austenite-to-martensite transformation when cooled to room temperature. The

Table 36 Annealing treatments for martensitic stainless steels

Type	Subcritical annealing temperature range, °C (°F)(a)	Full annealing temperature range, °C (°F)(b)
403, 410, 416	650–760 (1200–1400)	830–885 (1525–1625)
414	650–730 (1200–1350)	Not recommended
420	675–760 (1250–1400)	830–885 (1525–1625)
431	620–705 (1150–1300)	Not recommended
440A, 440B, 440C	675–760 (1250–1400)	845–900 (1550–1650)
CA-6NM	595–620 (1100–1150)	790–815 (1450–1500)
CA-15, CA-40	620–650 (1150–1200)	845–900 (1550–1650)

(a) Air cool from temperature; lowest hardness is obtained by heating near the top of the range. (b) Furnace cool to 595 °C (1100 °F); weldment can then be air-cooled. Source: Ref 89

temperature at which martensite starts to form from austenite upon cooling, M_s, is usually in the range of 95 to 150 °C (200 to 300 °F). When martensite is reheated to 480 to 595 °C (900 to 1100 °F), precipitation hardening and strengthening occur, promoted by the presence of one or more alloying additions. Molybdenum, copper, titanium, niobium, and aluminum (and their compounds) are dissolved during annealing and retained in solid solution by rapid cooling, producing precipitate (usually submicroscopic particles) that increases both the strength and the hardness of the martensitic matrix.

The compositional balance in the martensitic precipitation-hardening (PH) steels is critical, because relatively slight variations can lead to the formation of excessive amounts of δ-ferrite during solution annealing. If the austenite is too stable, large amounts of austenite can also be retained at room temperature after solution annealing. Either of these two conditions prevents full hardening during aging. Carbon and nitrogen contents can significantly affect this balance. Increased carbon or nitrogen may result in contamination. Typical sources of these contaminants are shop dirt and the atmosphere.

These steels can be readily welded. The welding procedures resemble those ordinarily used for the 300-series stainless steels, despite differences in composition and structure between the two classes. The formation of martensite, which occurs during cooling from elevated temperatures, as in welding, does not result in full hardening. These steels are not sensitive to cracking and do not require preheating.

The selection of filler metal depends on the properties required for the welded joint (Table 37). If strength comparable to that of the base metal is not required at the welded joint, a tough 300-series stainless steel filler metal may be adequate. When a weld having mechanical properties comparable to those of the hardened base metal is desired, the filler metal must be of comparable composition, although slight modifications are permissible to obtain better weldability.

When welds are deposited in a single pass, the weld metal and the HAZ usually respond uniformly to a postweld precipitation-hardening heat treatment. There is seldom any significant variation in hardness across the joint. Multiple-pass welds, however, exhibit less uniform response to the same heat treatment because successive applications of heat during welding result in marked variations in the structure of weld metal, HAZ, and base metal. Annealing eliminates these variations and provides a more uniform microstructure capable of responding uniformly to precipitation hardening.

Semiaustenitic Precipitation-Hardening Steels. Unlike martensitic PH steels, semiaustenitic PH steels are soft enough in the annealed condition to permit cold working. When cooled rapidly from the annealing temperature to room temperature, they retain their austenitic structure, which displays good toughness and ductility in cold-forming operations. The M_s temperatures for these steels are well below room temperature, but they vary depending on composition and annealing temperature.

To obtain hardening and strengthening, the austenitic structure must be transformed to an essentially martensitic one. This can be accomplished by treating the steel before subjecting it to the precipitation-hardening heat treatment by:

- Heating the steel in the range of 650 to 870 °C (1200 to 1600 °F) to precipitate carbides and other compounds, thereby depleting the matrix of enough austenite-stabilizing elements to allow transformation of austenite to martensite when the steel is cooled to room temperature
- Refrigerating the steel to a temperature well below the M_s point (−75 °C, or −100 °F, for example)
- Cold working the steel enough so that the austenite transforms to martensite

After transformation to martensite, the semiaustenitic PH steels, like the martensitic PH steels, respond to precipitation hardening in the temperature range of 455 to 595 °C (850 to 1100 °F). Whether a precipitate forms or a tempering reaction takes place depends on the steel composition.

The M_s temperature of the semiaustenitic PH steels is controlled by the solution-annealing temperature, as well as by composition. For example, when AM-350 steel is solution annealed at temperatures below 925 °C (1700 °F), incomplete carbide solution raises the M_s temperature above room temperature. On the other hand, when the solution-annealing temperature is raised above 925 °C (1700 °F), the M_s temperature drops precipitously. In practice, the solution-annealing temperature is not permitted to exceed about 1050 °C (1925 °F), because higher temperatures promote the formation of δ-ferrite.

The semiaustenitic PH steels are normally welded in the annealed condition. The tough austenitic structure imparts welding characteristics similar to those of 300-series stainless steels. The semiaustenitic PH steels are not susceptible to cracking when welded, even when welded after transformation to martensite, because the low-carbon martensite developed is not of high hardness or low ductility. Also, cold cracking does not occur in the base metal adjacent to the weld because the HAZ is austenitized during welding and remains substantially austenitic as the joint cools to room temperature.

The choice of filler metal depends largely on the weld properties desired. The filler metal can be an alloy of precipitation-hardening composition capable of developing mechanical properties comparable to those of the base metal (Table 37). If high strength is not a requisite, the filler metal can be a 300-series austenitic stainless steel. When these steels are welded in the annealed condition, certain microstructural relations are generally obtained as a result of relatively rapid heating and cooling at the joint:

- Weld metal contains small amounts of ferrite in an essentially austenitic matrix; hardness is approximately 90 HRB
- Base metal immediately adjacent to the weld displays high-temperature annealed (austenitic) structure; hardness is approximately 90 HRB
- Base metal in the narrow zone just beyond the annealed zone next to the weld is hardened slightly; hardness is approximately 90 to 98 HRB

If the welded assembly is given the customary double-aging heat treatment, the three areas identified above, as well as the unaffected base metal, transform and precipitation harden uniformly to a hardness range commensurate with the precipitation-hardening temperature. The weld metal may be somewhat less tough than the wrought base metal, as measured by the results of tensile-elongation and bend tests, depending on the type of joint, the welding process, and the hardening temperature.

Higher precipitation-hardening temperatures ensure good weld toughness with little sacrifice of strength. Maximum toughness requires the annealing of the weldment prior to the transformation and hardening treatments. Although other variations in the welding and heat-treating sequence are possible and may be desirable at times, the choice of the sequence should ensure that, after welding, the weld metal and the HAZ are in the annealed (austenitic) condition. To harden these areas, both the transforma-

Table 37 Recommended filler metals for welding precipitation-hardening stainless steels

Type	UNS designation	Covered electrodes	Bare welding wire	Dissimilar PH stainless steels
Martensitic				
17-4 PH	S 17400	AMS 5827B (17-4 PH) or E308	AMS 5826 (17-4 PH) or ER308	E309 or ER309, E309Cb or ER309Cb
15-5 PH	S 15500	AMS 5827B (17-4 PH) or E308	AMS 5826 (17-4 PH) or ER308	E309 or ER309, E309Cb or ER309Cb
Semiaustenitic				
17-7 PH	S 17700	AMS 5827B (17-4 PH), E308, or E309	AMS 5824A (17-7 PH)	E310 or ER 310, ENiCrFe-2, or ERNiCr-3
PH 15-7 Mo	S 15700	E308 or E309	AMS 5812C (PH 15-7 Mo)	E309 or ER309, E310 or ER310
AM350	S 35000	AMS 5775A (AM350)	AMS 5774B (AM 350)	E308 or ER308, E309 or ER309
AM355	S 35500	AMS 5781A (AM355)	AMS 5780A (AM 355)	E308 or ER308, E309 or ER309

(a) See AWS A5.11-85. (b) See AWS A5.14-89. Source: Ref 89

tion and the precipitation-hardening heat treatments must be applied. If the components are given the transformation treatment before welding, the precipitation-hardening treatment alone, after welding, produces no significant hardening in either the weld metal or the HAZ.

Austenitic Precipitation-Hardening Steels. The alloy content of these steels is high enough to maintain an austenitic structure after annealing and after any aging or hardening treatment. The precipitation-hardening phase is soluble at the annealing temperature of 1095 to 1120 °C (2000 to 2050 °F), and it remains in solution during rapid cooling from the annealing temperature. When these steels are reheated to about 650 to 760 °C (1200 to 1400 °F), precipitation occurs, and the hardness and strength of the austenitic structure increase. The hardness attained is lower than that of the martensitic or semiaustenitic PH steels, but the nonmagnetic properties are retained. Although the austenitic PH steels remain austenitic during all phases of forming, welding, and heat treatment, some contain alloying elements (for precipitation-hardening purposes) that greatly affect behavior in welding.

The austenitic precipitation-hardening stainless steels can be welded using the arc welding techniques described earlier in the section "Weldability" for the austenitic stainless steels. The major difference is that these steels are usually heat treated after welding to achieve the required mechanical properties, which is usually unnecessary with austenitic PH stainless steels. Austenitic precipitation-hardening stainless steels may be welded with matched or dissimilar filler metals or without filler metals, as is the case with most stainless steels. There is a wide variety of hardenable filler metals available for these PH grades through the manufacturer of consumables. The most commonly used grade is the 630 alloy, the only one currently included in the AWS specifications. Its composition is shown in Table 38.

Duplex Stainless Steels (Ref 90-92). The physical and mechanical properties of duplex stainless steels affect the welding process. Because of their better stress-corrosion cracking resistance and appreciably higher yield and tensile strengths, these steels are currently used as direct substitutes for austenitic stainless steels when service above about 260 to 315 °C (500 to 600 °F) is not required.

Duplex stainless steels are magnetic and have yield strengths typically about double that of type 316L, with tensile strengths considerably higher than those of standard austenitic grades. Duplex stainless steels also generally have a higher thermal conductivity and a lower coefficient of thermal expansion than austenitic stainless steels. Because of these factors, duplex stainless steels generally exhibit less distortion during welding than do austenitic stainless steels.

The general welding characteristics of the duplex alloy steels are very similar to those of austenitic stainless steels. They can be welded by any of the conventional arc welding processes. The normal arc welding processes, shielded metal arc, gas tungsten arc, gas metal arc, plasma arc, and submerged arc welding can all be used. In addition, electron beam and laser welding are used, as well as resistance welding. Heat input should be low enough to minimize intergranular carbide precipitation. Surface cleanliness is a must when welding duplex stainless steels. It is necessary to eliminate any source of hydrogen in the welding operation. For the gas-shielded processes, particularly on pipe, argon-helium purge gas should be used.

Two differences between the characteristics of duplex stainless steels and those of austenitic stainless steels can be discerned:

- Because duplex materials are appreciably stiffer than austenitic grades, greater forces are required for tube rolling into tube sheets, and so on
- Because duplex alloys are highly sensitive to 475 °C (885 °F) and σ phase embrittlement, control over interpass temperature is imperative

It is good practice to weld austenitic grades at a maximum interpass temperature of approximately 200 °C (390 °F), but, in cases such as multipass orbital GTAW of pipe, interpass temperatures well above this level can often be tolerated with no significant adverse effect. These situations should be regarded with caution for duplex alloys, in view of embrittlement following prolonged exposure to temperatures much above 300 °C (570 °F), and a maximum interpass temperature of 200 °C (390 °F) is suggested.

Wrought duplex stainless steels typically contain elongated islands of austenite in a ferrite matrix, as seen in Fig. 43. Generally, weld metal contains less austenite than parent metal of the same composition. As-welded properties may be reduced because of this change in structure. Welding parameters and/or consumables are chosen to optimize the mechanical properties and corrosion resistance of the weld. Useful properties have been obtained using duplex stainless consumables with increased nickel to maintain sufficient austenite in the weld (Fig. 44). Slightly slower cooling rates after welding also permit more time for austenite formation and can improve as-welded mechanical properties and corrosion resistance. Weldments may also be annealed to optimize properties, but this is not always practical. It is essential that thorough cleaning be done after welding.

Table 38 Chemical compositions of welding consumables for precipitation-hardening stainless steels
All are maximum percentages.

AWS classification	C	Cr	Ni	Mo	Nb + Ta	Mn	Si	P	S	Cu
E630(a), ER630(b)	0.05	16.0–16.75	4.5–5.0	0.75	0.15–0.30	0.25–0.75	0.75	0.04	0.03	3.25–4.00

(a) Undiluted weld metal composition. (b) Consumable composition. Source: AWS A5.4-81, A5.9-81

Fig. 43 Photomicrograph of a duplex stainless steel showing elongated austenite islands in the ferrite matrix. The mill-annealed 19.1 mm (0.752 in.) thick plate sample is a longitudinal section etched using 15 ml HCl in 100 ml ethyl alcohol. 200×

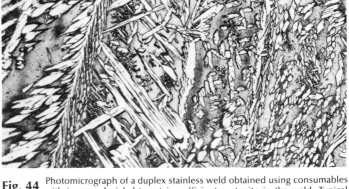

Fig. 44 Photomicrograph of a duplex stainless weld obtained using consumables with increased nickel to retain sufficient austenite in the weld. Typical weld metal microstructure for gas metal arc weld in 19.1 mm (0.752 in.) thick plate. The light-etching phase is austenite. The micrograph shows a cross section etched in Groesbeck's reagent. 200×

Test weldments may be evaluated in the laboratory before vessel fabrication and service exposure. The welding techniques, filler metal, and plate thickness studied in these evaluations are typically the same as those employed in field fabrication. Corrosion evaluation of weldments should employ environments and stress levels based on service expectations.

Filler metal selection is critical. A filler metal with matching composition may result in a higher ferrite content than that of the base metal. Gooch (Ref 91) and others have shown that filler metals of duplex stainless steel with higher nickel and/or nitrogen are preferred. Weld metal cracking in duplex steels has rarely been identified. Because extensive austenite-to-ferrite and ferrite-to-austenite transformations occur in the HAZ, along with grain growth, the welding procedure must be carefully controlled. The extent of the above transformations depends to a great extent on the composition and the precise weld thermal cycle experienced by the HAZ during welding. Because of their high chromium level, these steels are prone to σ phase and 475 °C (885 °F) embrittlement. Though the weld thermal cycle is too short for σ-phase formation of 475 °C (885 °F) embrittlement to occur, care must be exercised in welding heavy-section steels of this type.

REFERENCES

1. Metal Statistics: 1988, *American Metal Market*, Fairchild Publications, 1988
2. 1988 Annual Statistical Report, American Iron and Steel Institute, 1989
3. D.J. De Renzo, Ed., *Corrosion Resistant Materials Handbook*, Noyes Data Corporation, 1985
4. R.L. Tobler, R.P. Reed, and D.S. Burkhalter, "Temperature Dependence of Yielding in Austenitic Stainless Steels," National Bureau of Standards, U.S. Department of Commerce
5. D.C. Larbalestier and H.W. King, Austenitic Stainless Steels at Cryogenic Temperatures, 1-Structural Stability and Magnetic Properties, *Cryogenics*, Vol 13 (No. 3), March 1973, p 160-168
6. K.R. Hanby et al., "Handbook on Materials for Superconducting Machinery," MCIC-HB-04, Metals and Ceramics Information Center, Battelle Columbus Laboratories, Jan 1977
7. A.J. Nachtigall, Strain Cycling Fatigue Behavior of Ten Structural Metals Tested in Liquid Helium, Liquid Nitrogen, and Ambient Air, in *Properties of Materials for Liquified Natural Gas Tankage*, STP 579, American Society for Testing and Materials, 1975, p 378-396
8. W. Weleff, H.S. McQueen, and W.F. Emmons, Cryogenic Tensile Properties of Selected Aerospace Materials, in *Advances in Cryogenic Engineering*, Vol 10, K.D. Timmerhaus, Ed., Plenum Press, 1965, p 14-15
9. L.P. Rice, J.E. Campbell, and W.F. Simmons, Tensile Behavior of Parent-Metal and Welded 5000-Series Aluminum Alloy Plate at Room and Cryogenic Temperatures, in *Advances in Cryogenic Engineering*, Vol 7, K.D. Timmerhaus, Ed., Plenum Press, 1962, p 478-489
10. K.A. Warren and R.P. Reed, *Tensile and Impact Properties of Selected Materials from 20° to 300 °K*, Monograph 63, National Bureau of Standards, U.S. Department of Commerce, June 1963
11. C.J. Guntner and R.P. Reed, Mechanical Properties of Four Austenitic Steels at Temperatures Between 300° and 20 °K, in *Advances in Cryogenic Engineering*, Vol 6, K.D. Timmerhaus, Ed., Plenum Press, 1961, p 565-576
12. C.J. Guntner and R.P. Reed, The Effect of Experimental Variables Including the Martensitic Transformation on the Low-Temperature Mechanical Properties of Austenitic Stainless Steels, *Trans. ASM*, Vol 55, Sept 1962, p 399-419
13. J.F. Watson and J.L. Christian, Low Temperature Properties of Cold-Rolled AISI Types 301, 302, 304ELC, and 310 Stainless Steel Sheet, in *Low-Temperature Properties of High-Strength Aircraft and Missile Materials*, STP 287, American Society for Testing and Materials, 1961, p 170-193
14. J.H. Bolton, L.L. Godby, and B.L. Taft, Materials for Use at Liquid Hydrogen Temperature, in *Low-Temperature Properties of High-Strength Aircraft and Missile Materials*, STP 287, American Society for Testing and Materials, 1961, p 108-120
15. H.L. Martin et al., "Effects of Low Temperature on the Mechanical Properties of Structural Metals," NASA SP-5012 (01), Office of Technology Utilization, National Aeronautics and Space Administration, 1968
16. D.T. Read and H.M. Ledbetter, Temperature Dependencies of the Elastic Constants of Precipitation-Hardened Aluminum Alloys 2014 and 2219, in *J. Eng. Mater. Technol. (Trans. ASME)*, Series H, Vol 99 (No. 2), April 1977, p 181-184
17. J.M. Wells, W.A. Logsdon, and R. Kossowsky, Evaluations of Weldments in Austenitic Stainless Steels for Cryogenic Applications, in *Advances in Cryogenic Engineering*, Vol 24, K.D. Timmerhaus et al., Ed., Plenum Press, 1978, p 150-159
18. R.R. Vandervoort, Mechanical Properties of Inconel 625 Welds in 21-6-9 Stainless Steel, *Cryogenics*, Vol 18 (No. 8), Aug 1979, p 448-452
19. J.W. Montano, "The Stress Corrosion

20. J.E. Campbell and L.P. Rice, Properties of Some Precipitation-Hardening Stainless Steels at Very Low Temperatures, in *Low-Temperature Properties of High-Strength Aircraft and Missile Materials*, STP 287, American Society for Testing and Materials, 1961, p 158-167
21. R.W. Finger, "Proof Test Criteria for Thin-Walled 2219 Aluminum Pressure Vessels," Vol I, NASA CR-135036, Vol II, NASA CR-135037, The Boeing Aerospace Company, Aug 1976
22. J.F. Watson and J.L. Christian, Mechanical Properties of High-Strength 301 Stainless Steel Sheet at 70, -320, and -423 F in Base Metal and Welded Joint Configuration, in *Low-Temperature Properties of High-Strength Aircraft and Missile Materials*, STP 287, American Society for Testing and Materials, 1961
23. H.M. Ledbetter, W.F. Weston, and E.R. Naimon, Low-Temperature Elastic Properties of Four Austenitic Stainless Steels, *J. Appl. Phys.*, Vol 6 (No. 9), Sept 1975, p 3855-3860
24. R.L. Tobler et al., Low Temperature Fracture Behavior of Iron Nickel Alloy Steels, in *Properties of Materials for Liquified Natural Gas Tankage*, STP 579, American Society for Testing and Materials, Sept 1975, p 261-287
25. W.A. Logsdon, J.M. Wells, and R. Kossowsky, Fracture Mechanics Properties of Austenitic Stainless Steels for Advanced Applications, in *Proceedings of the Second International Conference on Mechanical Behavior of Materials*, American Society for Metals, 1976, p 1283-1289
26. R.P. Reed, R.L. Tobler, and R.P. Mikesell, The Fracture Toughness and Fatigue Crack Growth Rate of an Fe-Ni-Cr Superalloy at 298, 76, and 4K, in *Advances in Cryogenic Engineering*, Vol 22, K.D. Timmerhaus et al., Ed., Plenum Press, 1977, p 68-79
27. R.L. Tobler and R.P. Reed, Fatigue Crack Growth Resistance of Structural Alloys at Cryogenic Temperatures, in *Advances in Cryogenic Engineering*, Vol 24, K.D. Timmerhaus et al., Ed., Plenum Press, 1978, p 82-90
28. T.F. Kiefer, R.D. Keys, and F.R. Schwartzberg, "Determination of Low-Temperature Fatigue Properties of Structural Metal Alloys," Final Report, The Martin Company, Oct 1965
29. E.H. Schmidt, "Fatigue Properties of Sheet, Bar, and Cast Metallic Materials for Cryogenic Applications," Report R-7564, Rocketdyne Division, North American Rockwell Corporation, Aug 1968
30. D.N. Gideon et al., The Fatigue Behavior of Certain Alloys in the Temperature Range from Room Temperature to -423 F, in *Advances in Cryogenic Engineering*, Vol 7, K.D. Timmerhaus, Ed., Plenum Press, 1962, p 503-508
31. D.N. Gideon et al., "Investigation of Notch Fatigue Behavior of Certain Alloys in the Temperature Range of Room Temperature to -423 F," ASD-TR-62-351, Battelle Memorial Institute, Aug 1962
32. B.F. Brown, *Stress Corrosion Cracking Control Measures*, Monograph 156, National Bureau of Standards, U.S. Department of Commerce, June 1977
33. R.M. Latanision and R.W. Staehle, Stress Corrosion Cracking of Iron-Nickel-Chromium Alloys, in *Proceedings of Conference on Fundamental Aspects of Stress Corrosion Cracking*, National Association of Corrosion Engineers, 1969, p 214-307
34. S.W. Dean, Review of Recent Studies on the Mechanism of Stress Corrosion Cracking in Austenitic Stainless Steels, in *Stress Corrosion—New Approaches*, STP 610, H.L. Craig, Jr., Ed., American Society for Testing and Materials, 1976, p 308-337
35. G. Fontana, *Corrosion Engineering*, 3rd ed., McGraw-Hill, 1986
36. *Corrosion Resistance of the Austenitic Chromium-Nickel Stainless Steels in Atmospheric Environments*, The International Nickel Company, Inc., 1963
37. K.L. Money and W.W. Kirk, Stress Corrosion Cracking Behavior of Wrought Fe-Cr-Ni Alloys in Marine Atmosphere, *Mater. Perform.*, Vol 17, July 1978, p 28-36
38. M. Henthorne, T.A. DeBold, and R.J. Yinger, "Custom 450—A New High Strength Stainless Steel," Paper 53, presented at Corrosion/72, National Association of Corrosion Engineers, 1972
39. *The Role of Stainless Steels in Desalination*, American Iron and Steel Institute, 1974
40. M.A. Streicher, Analysis of Crevice Corrosion Data from Two Sea Water Exposure Tests on Stainless Alloys, *Mater. Perform.*, Vol 22, May 1983, p 37-50
41. A.H. Tuthill and C.M. Schillmoller, *Guidelines for Selection of Marine Materials*, The International Nickel Company, Inc., 1971
42. R.M. Kain, "Crevice Corrosion Resistance of Austenitic Stainless Steels in Ambient and Elevated Temperature Seawater," Paper 230, presented at Corrosion/79, National Association of Corrosion Engineers, 1979
43. F.L. LaQue and H.R. Copson, Ed., *Corrosion Resistance of Metals and Alloys*, Reinhold, 1963, p 375-445
44. J.E. Truman, in *Corrosion: Metal/Environment Reactions*, Vol 1, L.L. Shreir, Ed., Newness-Butterworths, 1976, p 352
45. M.A. Streicher, Development of Pitting Resistant Fe-Cr-Mo Alloys, *Corrosion*, Vol 30, 1974, p 77-91
46. H.O. Teeple, Corrosion by Some Organic Acids and Related Compounds, *Corrosion*, Vol 8, Jan 1952, p 14-28
47. T.A. DeBold, J.W. Martin, and J.C. Tverberg, Duplex Stainless Offers Strength and Corrosion Resistance, in *Duplex Stainless Steels*, R.A. Lula, Ed., American Society for Metals, 1983, p 169-189
48. L.A. Morris, in *Handbook of Stainless Steels*, D. Peckner and I.M. Bernstein, Ed., McGraw-Hill, 1977, p 17-1
49. "Material Requirements: Sulfide Stress Cracking Resistant Metallic Materials for Oil Field Equipment," MR-01-84, National Association of Corrosion Engineers
50. J.R. Kearns, M.J. Johnson, and J.F. Grubb, "Accelerated Corrosion in Dissimilar Metal Crevices," Paper 228, presented at Corrosion/86, National Association of Corrosion Engineers, 1986
51. L.S. Redmerski, J.J. Eckenrod, and K.E. Pinnow, "Cathodic Protection of Seawater-Cooled Power Plant Condensers Operating with High Performance Ferritic Stainless Steel Tubing," Paper 208, presented at Corrosion/85, National Association of Corrosion Engineers, 1985
52. E.C. Hoxie and G.W. Tuffnell, A Summary of INCO Corrosion Tests in Power Plant Flue Gas Scrubbing Processes, in *Resolving Corrosion Problems in Air Pollution Control Equipment*, National Association of Corrosion Engineers, 1976
53. *Effective Use of Stainless Steel in FGD Scrubber Systems*, American Iron and Steel Institute, 1978
54. G.T. Paul and R.W. Ross, Jr., "Corrosion Performance in FGD Systems at Laramie River and Dallman Stations," Paper 194, presented at Corrosion/83, National Association of Corrosion Engineers, 1983
55. A.P. Majidi and M.A. Streicher, "Four Non-Destructive Electrochemical Tests for Detecting Sensitization in Type 304 and 304L Stainless Steels," Paper 62, presented at Corrosion/85, National Association of Corrosion Engineers, 1985
56. A.H. Tuthill, Resistance of Highly Alloyed Materials and Titanium to Localized Corrosion in Bleach Plant Environments, *Mater. Perform.*, Vol 24, Sept 1985, p 43-49
57. "Standard Test Methods for Pitting and Crevice Corrosion Resistance of Stainless Steels and Related Alloys by the Use of Ferric Chloride Solution," G 48,

Annual Book of ASTM Standards, American Society for Testing and Materials
58. "Standard Recommended Practice for Examination and Evaluation of Pitting Corrosion," G 46, *Annual Book of ASTM Standards*, American Society for Testing and Materials
59. T.A. DeBold, Which Corrosion Test for Stainless Steels, *Mater. Eng.*, Vol 2 (No. 1), July 1980
60. R.M. Davison et al., *A Review of Worldwide Developments in Stainless Steels in Specialty Steels and Hard Materials*, Pergamon Press, 1983, p 67-85
61. A.M. Sabroff, F.W. Boulger, and H.J. Henning, *Forging Materials and Practices*, Reinhold, 1968
62. H.J. Henning, A.M. Sabroff, and F.W. Boulger, *A Study of Forging Variables*, Report ML-TDR-64-95, U.S. Air Force, 1964
63. *Open Die Forging Manual*, 3rd ed., Forging Industry Association, 1982, p 106-107
64. *ASME Boiler and Pressure Vessel Code*, Section III, Division I, Figure NB-2433.1-1, American Society of Mechanical Engineers, 1986
65. Machining and Abrasive Wheel Grinding of Carpenter Stainless Steels, in *Carpenter Stainless Steels, Selection, Alloy Data, Fabrication*, Carpenter Technology Corporation, 1987, p 240-241
66. V.A Tipnis, Machining of Stainless Steels, *Wire*, Aug 1971, p 153-161
67. Metallurgy of Welding Stainless Steels, in *Stainless Steels*, American Society for Metals, 1978, p 11-1 to 11-22
68. R.A. Lula, Fabrication of Stainless Steels—Machining, in *Stainless Steels*, American Society for Metals, 1986, p 112-114
69. "Free-Machining Stainless Steels," American Iron and Steel Institute, 1975
70. *Guide to Machining Stainless Steels and Other Specialty Metals*, Carpenter Technology Corporation, 1985
71. D.M. Blott, Machining Wrought and Cast Stainless Steels, in *Handbook of Stainless Steels*, McGraw-Hill, 1977, p 24-2 to 24-30
72. C.A. Divine, Jr., What to Consider in Choosing an Alloy, *Met. Prog.*, Feb 1968, p 19-23
73. L. Colombier and J. Hochmann, Manufacturing, Forming and Finishing Techniques—Machining, in *Stainless and Heat Resisting Steels*, St. Martin's Press, 1968, p 508-514
74. Machining Operations, in *Stainless Steel Fabrication*, Allegheny Ludlum Steel Corporation, 1959, p 223-259
75. W.C. Clarke, Which Free-Machining Chromium Stainless?, *Metalwork. Prod.*, Sept 1964, p 68-71
76. A. Moskowitz et al., Free-Machining Stainless Steels, U.S. Patent 3,401,035, 1968
77. F.M. Richmond, A Decade of Progress in Machinability, Finishing and Forming, *Met. Prog.*, Aug 1967, p 85-86
78. C.W. Kovach and A. Moskowitz, How to Upgrade Free-Machining Properties, *Met. Prog.*, Aug 1967, p 173-180
79. J.R. Blank et al., Improved and More Consistent Steels for Machining, in *Influence of Metallurgy on Machinability of Steel*, Proceedings of an International Symposium, American Society for Metals, 1977, p 397-419
80. W.C. Clarke, Which Free-Machining Stainless?, *Metalwork. Prod.*, 27 May 1964, p 43-45
81. J.A. Ferree, Jr., Free Machining Austenitic Stainless Steel, U.S. Patent 3,888,659, 1975
82. W.C. Clarke, Jr., Free-Machining Stainless Steel and Method, U.S. Patent 2,697,035, 1954
83. J.J. Eckenrod and C.W. Kovach, Effects of Manganese on Austenitic Stainless Steels, *Met. Eng. Q.*, Feb 1972, p 5-10
84. R.P. Ney, Sr., Free Machining, Cold Formable Austenitic Stainless Steel, U.S. Patent 4,444,588, 1984
85. J.J. Eckenrod et al., Low Carbon Plus Nitrogen, Free-Machining Austenitic Stainless Steel, U.S. Patent 4,613,367, 1986
86. P.K. Wright and A. Bagchi, Wear Mechanisms That Dominate Tool-Life in Machining, *J. Appl. Metalwork.*, Vol 1 (No. 4), 1981, p 15-23
87. "Carpenter Custom 630 (17Cr-4Ni)," Carpenter Technology Corporation, 1971
88. D. Pechner and I.M. Bernstein, *Handbook of Stainless Steels*, McGraw-Hill, 1977
89. *Welding Handbook*, Vol 4, *Metals and Their Weldability*, 9th ed., American Welding Society, 1982
90. H.B. Cary, *Modern Welding Technology*, 2nd ed., Prentice-Hall, 1989
91. T.G. Gooch, Weldability of Duplex Ferritic-Austenitic Stainless Steels, in *Duplex Stainless Steels*, R.A. Lula, Ed., Proceedings of Conference on Duplex Austenitic-Ferritic Stainless Steel, American Society for Metals, Oct 1982
92. S.A. David, Welding of Stainless Steels, in *Encyclopedia of Materials Science and Engineering*, Vol 7, M.B. Bever, Ed., The MIT Press, p 5316-5320

SELECTED REFERENCES

- R.Q. Barr, Ed., *Stainless Steel '77*, Climax Molybdenum Company, 1978
- *Book on Metals and Alloys in the Unified Numbering System*, Society of Automotive Engineers, Inc., and American Society for Testing and Materials
- A.B. Kinzel et al., *The Alloys of Iron and Chromium*, 2 vol., McGraw-Hill, 1937
- R.A. Lula, Ed., New Developments in Stainless Steel Technology, in *Proceedings of International Conference on New Developments in Stainless Steel Technology*, American Society for Metals, 1985
- H.E. McGannon, Ed., *The Making, Shaping and Treating of Steel*, 9th ed., United States Steel Corporation, 1971
- J.H.G. Monypenny, *Stainless Iron and Steel*, 2 vol., Chapman and Hall, 1951
- J.G. Parr and A. Hanson, *An Introduction to Stainless Steel*, American Society for Metals, 1965
- D. Peckner and I.M. Bernstein, *Handbook of Stainless Steels*, McGraw-Hill, 1977
- F.B. Pickering, Ed., *The Metallurgical Evolution of Stainless Steels*, American Society for Metals, 1979
- A.O. Schaefer, Ed., *Elevated Temperature Properties in Austenitic Stainless Steels*, American Society of Mechanical Engineers, 1974
- A.J. Sedriks, *Corrosion of Stainless Steels*, Wiley-Interscience, 1979
- *Source Book on Stainless Steels*, American Society for Metals, 1976
- *Stainless Steel for Architectural Use*, STP 454, American Society for Testing and Materials, 1969
- *Stainless Steels '87*, The Institute of Metals, 1988
- E.E. Thum, Ed., *The Book of Stainless Steels*, American Society for Metals, 1935
- C.A. Zapffe, *Stainless Steels*, American Society for Metals, 1949

Cast Stainless Steels

Revised by Malcolm Blair, Steel Founders' Society of America

STAINLESS STEELS are a class of chromium-containing steels widely used for their corrosion resistance in aqueous environments and for service at elevated temperatures. Stainless steels are distinguished from other steels by the enhanced corrosion and oxidation resistance created by chromium additions. Chromium imparts passivity to ferrous alloys when present in amounts of more than about 11%, particularly if conditions are strongly oxidizing. Consequently, steels with more than 10 or 12% Cr are sometimes defined as stainless steels.

This article reviews the properties of cast steels that are specified either for liquid corrosion service at temperatures below 650 °C (1200 °F) or for service at temperatures above 650 °C (1200 °F). The cast steels suitable for these applications are often high-alloy compositions because carbon and low-alloy steels do not provide sufficient corrosion resistance and/or strength at elevated temperatures. These high-alloy cast steels generally have more than 10% Cr and primarily consist of stainless steel compositions.

Stainless steel castings are usually classified as either corrosion-resistant castings (which are used in aqueous environments below 650 °C, or 1200 °F) or heat-resistant castings (which are suitable for service temperatures above 650 °C, or 1200 °F). However, this line of demarcation in terms of application is not always distinct, particularly for steel castings used in the range from 480 to 650 °C (900 to 1200 °F). The usual distinction between heat-resistant and corrosion-resistant cast steels is based on carbon content.

In general, the cast and wrought stainless steels possess equivalent resistance to corrosive media and they are frequently used in conjunction with each other. Important differences do exist, however, between some cast stainless steels and their wrought counterparts. One significant difference is in the microstructure of cast austenitic stainless steels. There is usually a small amount of ferrite present in austenitic stainless steel castings, in contrast to the single-phase austenitic structure of the wrought alloys. The presence of ferrite in the castings is desirable for facilitating weld repair, but ferrite also increases resistance to stress-corrosion cracking. There have been only a few stress-corrosion cracking failures with cast stainless steels in comparison to the approximately equivalent wrought compositions. The principal reasons for this resistance are apparently:

- Silicon added for fluidity gives added benefit from the standpoint of stress-corrosion cracking
- Sand castings are usually tumbled or sandblasted to remove molding sand and scale; this probably tends to put the surface in compression

Wrought and cast stainless steels may also differ in mechanical properties, magnetic properties, and chemical content. Because of the possible existence of large dendritic grains, intergranular phases, and alloy segregation, typical mechanical properties of cast stainless steels may vary more and generally are inferior to those of any wrought structure.

Grade Designations and Compositions

Cast stainless steels are most often specified on the basis of composition using the designation system of the High Alloy Product Group of the Steel Founders' Society of America. (The High Alloy Product Group has replaced the Alloy Casting Institute, or ACI, which formerly administered these designations.) The first letter of the designation indicates whether the alloy is intended primarily for liquid corrosion service (C) or high-temperature service (H). The second letter denotes the nominal chromium-nickel type of the alloy (Fig. 1). As nickel content increases, the second letter of the designation is changed from A to Z. The numeral or numerals following the first two letters indicate maximum carbon content (percentage × 100) of the alloy. Finally, if further alloying elements are present, these are indicated by the addition of one or more letters as a suffix. Thus, the designation of CF-8M refers to an alloy for corrosion-resistant service (C) of the 19Cr-9Ni type

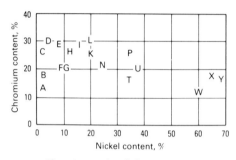

Fig. 1 Chromium and nickel contents in ACI standard grades of heat- and corrosion-resistant steel castings. See text for details.

(Fig. 1), with a maximum carbon content of 0.08% and containing molybdenum (M).

Some of the high-alloy cast steels exhibit many of the same properties of cast carbon and low-alloy steels (see the article "Steel Castings" in this Volume). Some of the mechanical properties of these grades (for example, hardness and tensile strength) can be altered by a suitable heat treatment. The cast high-alloy grades that contain more than 20 to 30% Cr plus Ni, however, do not show the phase changes observed in plain carbon and low-alloy steels during heating or cooling between room temperature and the melting point. These materials are therefore nonhardenable, and their properties depend on composition rather than heat treatment. Therefore, special consideration must be given to each grade of high-alloy cast steel with regard to casting design, foundry practice, and subsequent thermal processing (if any).

Compositions of C-Type (Corrosion-Resistant) Steel Castings. The C-type steel castings for liquid corrosion service are often classified on the basis of composition, although it should be recognized that classification by composition often involves microstructural distinctions (see the section "Composition and Microstructure" in this article). Table 1 lists the compositions of the commercial cast corrosion-resistant alloys. Alloys are grouped as:

- Chromium steels
- Chromium-nickel steels, in which chromium is the predominant alloying element

Table 1 Compositions and typical microstructures of ACI corrosion-resistant cast steels

ACI type	Wrought alloy type(a)	ASTM specifications	Most common end-use microstructure	C	Mn	Si	Cr	Ni	Others(c)
Chromium steels									
CA-15	410	A 743, A 217, A 487	Martensite	0.15	1.00	1.50	11.5–14.0	1.0	0.50Mo(d)
CA-15M	...	A 743	Martensite	0.15	1.00	0.65	11.5–14.0	1.0	0.15–1.00Mo
CA-40	420	A 743	Martensite	0.40	1.00	1.50	11.5–14.0	1.0	0.5Mo(d)
CA-40F	...	A 743	Martensite	0.2–0.4	1.00	1.50	11.5–14.0	1.0	...
CB-30	431, 442	A 743	Ferrite and carbides	0.30	1.00	1.50	18.0–22.0	2.0	...
CC-50	446	A 743	Ferrite and carbides	0.30	1.00	1.50	26.0–30.0	4.0	...
Chromium-nickel steels									
CA-6N	...	A 743	Martensite	0.06	0.50	1.00	10.5–12.5	6.0–8.0	...
CA-6NM	...	A 743, A 487	Martensite	0.06	1.00	1.00	11.5–14.0	3.5–4.5	0.4–1.0Mo
CA-28MWV	...	A 743	Martensite	0.20–0.28	0.50–1.00	1.00	11.0–12.5	0.50–1.00	0.9–1.25Mo; 0.9–1.25W; 0.2–0.3V
CB-7Cu-1	...	A 747	Martensite, age hardenable	0.07	0.70	1.00	15.5–17.7	3.6–4.6	2.5–3.2Cu; 0.20–0.35Nb; 0.05N max
CB-7Cu-2	...	A 747	Martensite, age hardenable	0.07	0.70	1.00	14.0–15.5	4.5–5.5	2.5–3.2Cu; 0.20–0.35 Nb; 0.05N max
CD-4MCu	...	A 351, A 743, A 744, A 890	Austenite in ferrite, age hardenable	0.04	1.00	1.00	25.0–26.5	4.75–6.0	1.75–2.25Mo; 2.75–3.25Cu
CE-30	312	A 743	Ferrite in austenite	0.30	1.50	2.00	26.0–30.0	8.0–11.0	...
CF-3(e)	304L	A 351, A 743, A 744	Ferrite in austenite	0.03	1.50	2.00	17.0–21.0	8.0–12.0	...
CF-3M(e)	316L	A 351, A 743, A 744	Ferrite in austenite	0.03	1.50	1.50	17.0–21.0	9.0–13.0	2.0–3.0Mo
CF-3MN	...	A 743	Ferrite in austenite	0.03	1.50	1.50	17.0–21.0	9.0–13.0	2.0–3.0Mo; 0.10–0.20N
CF-8(e)	304	A 351, A 743, A 744	Ferrite in austenite	0.08	1.50	2.00	18.0–21.0	8.0–11.0	...
CF-8C	347	A 351, A 743, A 744	Ferrite in austenite	0.08	1.50	2.00	18.0–21.0	9.0–12.0	Nb(f)
CF-8M	316	A 351, A 743, A 744	Ferrite in austenite	0.08	1.50	2.00	18.0–21.0	9.0–12.0	2.0–3.0Mo
CF-10	...	A 351	Ferrite in austenite	0.04–0.10	1.50	2.00	18.0–21.0	8.0–11.0	...
CF-10M	...	A 351	Ferrite in austenite	0.04–0.10	1.50	1.50	18.0–21.0	9.0–12.0	2.0–3.0Mo
CF-10MC	...	A 351	Ferrite in austenite	0.10	1.50	1.50	15.0–18.0	13.0–16.0	1.75–2.25Mo
CF-10SMnN	...	A 351, A 743	Ferrite in austenite	0.10	7.00–9.00	3.50–4.50	16.0–18.0	8.0–9.0	0.08–0.18N
CF-12M	316	...	Ferrite in austenite or austenite	0.12	1.50	2.00	18.0–21.0	9.0–12.0	2.0–3.0Mo
CF-16F	303	A 743	Austenite	0.16	1.50	2.00	18.0–21.0	9.0–12.0	1.50Mo max; 0.20–0.35Se
CF-20	302	A 743	Austenite	0.20	1.50	2.00	18.0–21.0	8.0–11.0	...
CG-6MMN	...	A 351, A 743	Ferrite in austenite	0.06	4.00–6.00	1.00	20.5–23.5	11.5–13.5	1.50–3.00Mo; 0.10–0.30Nb; 0.10–30V; 0.20–40N
CG-8M	317	A 351, A 743, A 744	Ferrite in austenite	0.08	1.50	1.50	18.0–21.0	9.0–13.0	3.0–4.0Mo
CG-12	...	A 743	Ferrite in austenite	0.12	1.50	2.00	20.0–23.0	10.0–13.0	...
CH-8	...	A 351	Ferrite in austenite	0.08	1.50	1.50	22.0–26.0	12.0–15.0	...
CH-10	...	A 351	Ferrite in austenite	0.04–0.10	1.50	2.00	22.0–26.0	12.0–15.0	...
CH-20	309	A 351, A 743	Austenite	0.20	1.50	2.00	22.0–26.0	12.0–15.0	...
CK-3MCuN	...	A 351, A 743, A 744	Ferrite in austenite	0.025	1.20	1.00	19.5–20.5	17.5–19.5	6.0–7.0V; 0.18–0.24N; 0.50–1.00Cu
CK-20	310	A 743	Austenite	0.20	2.00	2.00	23.0–27.0	19.0–22.0	...
Nickel-chromium steel									
CN-3M	...	A 743	Austenite	0.03	2.00	1.00	20.0–22.0	23.0–27.0	4.5–5.5Mo
CN-7M	...	A 351, A 743, A 744	Austenite	0.07	1.50	1.50	19.0–22.0	27.5–30.5	2.0–3.0Mo; 3.0–4.0Cu
CN-7MS	...	A 743, A 744	Austenite	0.07	1.50	3.50(g)	18.0–20.0	22.0–25.0	2.5–3.0Mo; 1.5–2.0Cu
CT-15C	...	A 351	Austenite	0.05–0.15	0.15–1.50	0.50–1.50	19.0–21.0	31.0–34.0	0.5–1.5V

(a) Type numbers of wrought alloys are listed only for nominal identification of corresponding wrought and cast grades. Composition ranges of cast alloys are not the same as for corresponding wrought alloys; cast alloy designations should be used for castings only. (b) Maximum unless a range is given. The balance of all compositions is iron. (c) Sulfur content is 0.04% in all grades except: CG-6MMN, 0.030% S (max); CF-10SMnN, 0.03% S (max); CT-15C, 0.03% S (max); CK-3MCuN, 0.010% S (max); CN-3M, 0.030% S (max), CA-6N, 0.020% S (max); CA-28MWV, 0.030% S (max); CA-40F, 0.20–0.40% S; CB-7Cu-1 and -2, 0.03% S (max). Phosphorus content is 0.04% (max) in all grades except: CF-16F, 0.17% P (max); CF-10SMnN, 0.060% P (max); CT-15C, 0.030% P (max); CK-3MCuN, 0.045% P (max); CN-3M, 0.030% P (max); CA-6N, 0.020% P (max); CA-28MWV, 0.030% P (max); CB-7Cu-1 and -2, 0.035% P (max). (d) Molybdenum not intentionally added. (e) CF-3A, CF-3MA, and CF-8A have the same composition ranges as CF-3, CF-3M, and CF-8, respectively, but have balanced compositions so that ferrite contents are at levels that permit higher mechanical property specifications than those for related grades. They are covered by ASTM A 351. (f) Nb, 8 × %C min (1.0% max); or Nb + Ta × %C (1.1% max). (g) For CN-7MS, silicon ranges from 2.50 to 3.50%.

- Nickel-chromium steels, in which nickel is the predominant alloying element

The serviceability of cast corrosion-resistant steels depends greatly on the absence of carbon, and especially precipitated carbides, in the alloy microstructure. Therefore, cast corrosion-resistant alloys are generally low in carbon (usually lower than 0.20% and sometimes lower than 0.03%).

All cast corrosion-resistant steels contain more than 11% chromium, and most contain from 1 to 30% nickel (a few have less than 1% Ni). About two-thirds of the corrosion-resistant steel castings produced in the United States are of grades that contain 18 to 22% Cr and 8 to 12% Ni.

In general, the addition of nickel to iron-chromium alloys improves ductility and impact strength. An increase in nickel content increases resistance to corrosion by neutral chloride solutions and weakly oxidizing acids.

The addition of molybdenum increases resistance to pitting attack by chloride solutions. It also extends the range of passivity in solutions of low oxidizing characteristics.

The addition of copper to duplex (ferrite in austenite) nickel-chromium alloys produces alloys that can be precipitation hardened to higher strength and hardness. The addition of copper to single-phase austenitic alloys greatly improves their resistance to corrosion by sulfuric acid. In all iron-chromium-nickel stainless alloys, resistance to corrosion by environments that cause intergranular attack can be improved by lowering the carbon content. Information on the corrosion characteristics and mechanical properties of the C-type steel castings is provided in the sec-

Table 2 Compositions of ACI heat-resistant casting alloys

ACI designation	UNS number	ASTM specifications(a)	C	Cr	Ni	Si (max)
HA	...	A 217	0.20 max	8–10	...	1.00
HC	J92605	A 297, A 608	0.50 max	26–30	4 max	2.00
HD	J93005	A 297, A 608	0.50 max	26–30	4–7	2.00
HE	J93403	A 297, A 608	0.20–0.50	26–30	8–11	2.00
HF	J92603	A 297, A 608	0.20–0.40	19–23	9–12	2.00
HH	J93503	A 297, A 608, A 447	0.20–0.50	24–28	11–14	2.00
HI	J94003	A 297, A 567, A 608	0.20–0.50	26–30	14–18	2.00
HK	J94224	A 297, A 351, A 567, A 608	0.20–0.60	24–28	18–22	2.00
HK30	...	A 351	0.25–0.35	23.0–27.0	19.0–22.0	1.75
HK40	...	A 351	0.35–0.45	23.0–27.0	19.0–22.0	1.75
HL	J94604	A 297, A 608	0.20–0.60	28–32	18–22	2.00
HN	J94213	A 297, A 608	0.20–0.50	19–23	23–27	2.00
HP	...	A 297	0.35–0.75	24–28	33–37	2.00
HP-50WZ(c)	...	...	0.45–0.55	24–28	33–37	2.50
HT	J94605	A 297, A 351, A 567, A 608	0.35–0.75	13–17	33–37	2.50
HT30	...	A 351	0.25–0.35	13.0–17.0	33.0–37.0	2.50
HU	...	A 297, A 608	0.35–0.75	17–21	37–41	2.50
HW	...	A 297, A 608	0.35–0.75	10–14	58–62	2.50
HX	...	A 297, A 608	0.35–0.75	15–19	64–68	2.50

(a) ASTM designations are the same as ACI designations. (b) Rem Fe in all compositions. Manganese content: 0.35 to 0.65% for HA, 1% for HC, 1.5% for HD, and 2% for the other alloys. Phosphorus and sulfur contents: 0.04% (max) for all but HP-50WZ. Molybdenum is intentionally added only to HA, which has 0.90 to 1.20% Mo; maximum for other alloys is set at 0.5% Mo. HH also contains 0.2% N (max). (c) Also contains 4 to 6% W, 0.1 to 1.0% Zr, and 0.035% S (max) and P (max)

tion "Corrosion-Resistant Steel Castings" in this article.

Compositions of H-Type (Heat-Resistant) Steel Castings. Castings are classified as heat resistant if they are capable of sustained operation while exposed, either continuously or intermittently, to operating temperatures that result in metal temperatures in excess of 650 °C (1200 °F). Heat-resistant steel castings resemble high-alloy corrosion-resistant steels except for their higher carbon content, which imparts greater strength at elevated temperature. The higher carbon content and, to a lesser extent, alloy composition ranges distinguish cast heat-resistant steel grades from their wrought counterparts. Table 2 summarizes the compositions of standard cast heat-resistant grades and three grade variations (HK30, HK40, HT30) specified in ASTM A 351 for elevated-temperature and corrosive service of pressure-containing parts.

The three principal categories of H-type cast steels, based on composition, are:

- Iron-chromium alloys
- Iron-chromium-nickel alloys
- Iron-nickel-chromium alloys

Information on the properties of H-type grades of steel castings is contained in the section "Heat-Resistant Cast Steels" in this article.

Composition and Microstructure

As shown in Table 1, cast stainless steels can also be classified on the basis of microstructure. Structures may be austenitic, ferritic, martensitic, or ferritic-austenitic (duplex).

The structure of a particular grade is primarily determined by composition. Chromium, molybdenum, and silicon promote the formation of ferrite (magnetic), while carbon, nickel, nitrogen, and manganese favor the formation of austenite (nonmagnetic). For example, a cast extra-low-carbon grade such as 0.03% C (max) cannot be completely nonmagnetic unless it contains 12 to 15% Ni. The wrought grades of these alloys normally contain about 13% Ni. They are made fully austenitic to improve rolling and forging characteristics.

Chromium (a ferrite and martensite promoter), nickel, and carbon (austenite promoters) are particularly important in determining microstructure (see the section "Ferrite Control" in this article). In general, straight chromium grades of high-alloy cast steel are either martensitic or ferritic, the chromium-nickel grades are either duplex or austenitic, and the nickel-chromium steels are fully austenitic.

Ferrite in Cast Austenitic Stainless Steels. Cast austenitic alloys usually have from 5 to 20% ferrite distributed in discontinuous pools throughout the matrix, the percent of ferrite depending on the nickel, chromium, and carbon contents (see the section "Ferrite Control"). The presence of ferrite in austenite may be beneficial or detrimental, depending on the application.

Ferrite is beneficial and intentionally present in various corrosion-resistant cast steels (see some of the CF grades in Table 1, for example) to improve weldability and to maximize corrosion resistance in specific environments. Ferrite is also used for strengthening duplex alloys. The section "Austenitic-Ferritic (Duplex) Alloys" in this article gives further information.

Ferrite can be beneficial in terms of weldability because fully austenitic stainless steels are susceptible to a weldability problem known as hot cracking, or microfissuring. The intergranular cracking occurs in the weld deposit and/or in the weld heat-affected zone and can be avoided if the composition of the filler metal is controlled to produce about 4% ferrite in the austenitic weld deposit. Duplex CF grade alloy castings are immune to this problem.

The presence of ferrite in duplex CF alloys improves the resistance to stress-corrosion cracking (SCC) and generally to intergranular attack. In the case of SCC, the presence of ferrite pools in the austenite matrix is thought to block or make more difficult the propagation of cracks. In the case of intergranular corrosion, ferrite is helpful in sensitized castings because it promotes the preferential precipitation of carbides in the ferrite phase rather than at the austenite grain boundaries, where they would increase susceptibility to intergranular attack. The presence of ferrite also places additional grain boundaries in the austenite matrix, and there is evidence that intergranular attack is arrested at austenite-ferrite boundaries. It is important to note, however, that not all studies have shown ferrite to be unconditionally beneficial to the general corrosion resistance of cast stainless steels. Some solutions attack the austenite phase in heat-treated alloys, whereas others attack the ferrite. For instance, calcium chloride solutions attack the austenite. On the other hand, a 10° Baumé cornstarch solution, acidified to a pH of 1.8 with sulfuric acid and heated to a temperature of 135 °C (275 °F), attacks the ferrite. Whether corrosion resistance is improved by ferrite and to what degree depends on the specific alloy composition, the heat treatment, and the service conditions (environment and stress state).

Ferrite can be detrimental in some applications. One concern may be the reduced toughness from ferrite, although this is not a major concern, given the extremely high toughness of the austenite matrix. A much greater concern is for applications that require exposure to elevated temperatures, usually 315 °C (600 °F) and higher, where the metallurgical changes associated with the ferrite can be severe and detrimental. In applications requiring that these steels be heated in the range from 425 to 650 °C (800 to 1200 °F), carbide precipitation occurs at the edges of the ferrite pools in preference to the austenite grain boundaries. When the steel is heated above 540 °C (1000 °F), the ferrite pools transform to a χ or σ phase. If these pools are distributed in such a way that a continuous network is formed, embrittlement or a network of corrosion penetration may result. Also, if the amount of ferrite is too great, the ferrite may form continuous stringers where corrosion can take place, producing a condition similar to grain boundary attack.

In the lower end of this temperature range, the reductions in toughness observed have been attributed to carbide precipitation or reactions associated with 475 °C (885 °F) embrittlement. The 475 °C embrittlement is caused by the precipitation of an intermetallic phase with a composition of approximately 80Cr-20Fe. The name derives from the fact that this embrittlement is most severe and rapid when it occurs at approximately 475 °C (885 °F).

At 540 °C (1000 °F) and above, the ferrite phase may transform to a complex iron-chromium-nickel-molybdenum intermetallic compound known as σ phase, which reduces toughness, corrosion resistance, and creep ductility. The extent of the reduction increases with time and temperature to about 815 °C (1500 °F) and may persist to 925 °C (1700 °F). In extreme cases, Charpy V-notch energy at room temperature may be reduced 95% from its initial value (Ref 1, 2).

At temperatures above 540 °C (1000 °F), austenite also has better creep resistance than ferrite. The weaker ferrite phase may lend better plasticity to the alloy, but after long exposure at temperatures in the 540 to 760 °C (1000 to 1400 °F) range, it may transform to σ or χ phase, which reduces resistance to impact. In some instances, the alloy is deliberately aged to form the σ or χ phase and thus increase strength. Austenite can transform directly to σ or χ without going through the ferrite phase.

In weld deposits, the presence of σ or χ phase is extremely detrimental to ductility. When welding for service at room temperature or up to 540 °C (1000 °F), 4 to 10% ferrite may be present and will greatly reduce the tendency toward weld cracking. However, for service at temperatures between 540 and 815 °C (1000 and 1500 °F), the amount of ferrite in the weld must be reduced to less than 5% to avoid embrittlement from excessive σ or χ phase.

Ferrite Control. From the preceding discussion, it is apparent that ferrite in predominantly austenitic cast stainless steels can offer property advantages in some steels (notably the CF alloys) and disadvantages in other cases (primarily at elevated temperatures). The underlying causes for the dependence of ferrite content on composition are found in the phase equilibria for the iron-chromium-nickel system. These phase equilibria have been exhaustively documented and related to commercial stainless steels.

The major elemental components of cast stainless steels are in competition to promote austenite or ferrite phases in the alloy microstructure. Chromium, silicon, molybdenum, and niobium promote the presence of ferrite in the alloy microstructure; nickel, carbon, nitrogen, and manganese promote the presence of austenite. By balancing the contents of ferrite- and austenite-forming elements within the specified ranges for the elements in a given alloy, it is possible to control the amount of ferrite present in the austenite matrix. The alloy can usually be made fully austenitic or with ferrite contents up to 30% or more in the austenite matrix.

The relationship between composition and microstructure in cast stainless steels permits the foundryman to predict and control the ferrite content of an alloy, as well as its resultant properties, by adjusting the composition of the alloy. This is accomplished with the Schoefer constitution diagram for cast chromium-nickel alloys (Fig. 2). This diagram was derived from an earlier diagram developed by Schaeffler for stainless steel weld metal (Ref 1). The use of Fig. 2 requires that all ferrite-stabilizing elements in the composition be converted into chromium equivalents and that all austenite-stabilizing elements be converted into nickel equivalents by means of empirically derived coefficients representing the ferritizing or austenitizing power of each element. A composition ratio is then obtained from the total chromium equivalent, Cr_e, and nickel equivalent, Ni_e, calculated for the alloy composition by:

$$Cr_e = \%Cr + 1.5(\%Si) + 1.4(\%Mo) + \%Nb - 4.99 \quad \text{(Eq 1)}$$

$$Ni_e = \%Ni + 30(\%C) + 0.5(\%Mn) + 26(\%N - 0.02) + 2.77 \quad \text{(Eq 2)}$$

where the elemental concentrations are given in weight percent. Although similar expressions have been derived that take into account additional alloying elements and different compositional ranges in the iron-chromium-nickel alloy system, use of the Schoefer diagram has become standard for estimating and controlling ferrite content in stainless steel castings.

The Schoefer diagram possesses obvious utility for casting users and the foundryman. It is helpful for estimating or predicting ferrite content if the alloy composition is known and for setting nominal values for individual elements when calculating the furnace charge for an alloy in which a specified ferrite range is desired.

Limits of Ferrite Control. Although ferrite content can be estimated and controlled on the basis of alloy composition only, there are limits to the accuracy with which this can be done. The reasons for this are many. First, there is an unavoidable degree of uncertainty in the chemical analysis of an alloy (note the scatter band in Fig. 2). Second, in addition to composition, the ferrite content depends on thermal history, although to a lesser extent. Third, ferrite contents at different locations in individual castings can vary considerably, depending on section size, ferrite orientation, presence of alloying-element segregation, and other factors.

Both the foundryman and the user of stainless steel castings should recognize that the factors mentioned above place significant limits on the degree to which ferrite content (either as ferrite number or ferrite

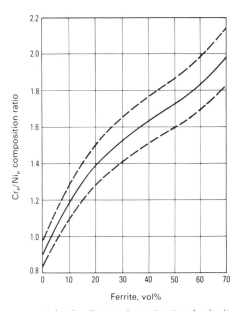

Fig. 2 Schoefer diagram for estimating the ferrite content of steel castings in the composition range of 16 to 26% Cr, 6 to 14% Ni, 4% Mo (max), 1% Nb (max), 0.2% C (max), 0.19% N (max), 2% Mn (max), and 2% Si (max). Dashed lines denote scatter bands caused by the uncertainty of the chemical analysis of individual elements. See text for equations used to calculate Cr_e and Ni_e. Source: Ref 1

percentage) can be specified and controlled. In general, the accuracy of ferrite measurement and the precision of ferrite control diminish as the ferrite number increases. As a working rule, it is suggested that the ±6 about the mean or desired ferrite number be viewed as a limit of ferrite control under ordinary circumstances, with ±3 possible under ideal circumstances.

Heat Treatment

The heat treatment of stainless steel castings is very similar in purpose and procedure to the thermal processing of comparable wrought materials (see the article "Heat Treating of Stainless Steels" in Volume 4 of the 9th Edition of *Metals Handbook*). However, some differences warrant separate consideration here.

Homogenization. Alloy segregation and dendritic structures may occur in castings and may be particularly pronounced in heavy sections. Because castings are not subjected to the high-temperature mechanical reduction and soaking treatments involved in the mill processing of wrought alloys, it is frequently necessary to homogenize some alloys at temperatures above 1095 °C (2000 °F) to promote uniformity of chemical composition and microstructure. The full annealing of martensitic castings results in recrystallization and maximum softness, but it is less effective than homogenization in eliminating segregation. Homogenization is a common procedure in the heat treatment of precipitation-hardening castings.

Sensitization and Solution Annealing of Austenitic and Duplex Alloys. When austenitic or duplex (ferrite in austenite matrix) stainless steels are heated in or cooled slowly through a temperature range of about 425 to 870 °C (800 to 1600 °F), chromium-rich carbides form at grain boundaries in austenitic alloys and at ferrite-austenite interfaces in duplex alloys. These carbides deplete the surrounding matrix of chromium, thus diminishing the corrosion resistance of the alloy. In small amounts, these carbides may lead to localized pitting in the alloy, but if the chromium-depleted zones are extensive throughout the alloy or heat-affected zone (HAZ) of a weld, the alloy may disintegrate intergranularly in some environments.

An alloy in this condition of reduced corrosion resistance due to the formation of chromium carbides is said to be sensitized, a situation that is most pronounced for austenitic alloys. In austenitic structures, the complex chromium carbides precipitate preferentially along the grain boundaries. This microstructure is susceptible to intergranular corrosion, especially in oxidizing solutions. In partially ferritic alloys, carbides tend to precipitate in the discontinuous carbide pools; thus, these alloys are less susceptible to intergranular attack.

Solution annealing of austenitic and duplex stainless steels makes these alloys less susceptible to intergranular attack by ensuring the complete solution of the carbides in the matrix. Depending on the specific alloy in question, temperatures between 1040 and 1205 °C (1900 and 2200 °F) will ensure the complete solution of all carbides and phases, such as σ and χ, that sometimes form in highly alloyed stainless steels. Alloys containing relatively high total alloy content, particularly high molybdenum content, often require the higher solution treatment temperature. Water quenching from the temperature range of 1040 to 1205 °C (1900 to 2200 °F) normally completes the solution treatment. Solution-annealing procedures for all austenitic alloys require holding for a sufficient amount of time to accomplish the complete solution of carbides and quenching at a rate fast enough to prevent reprecipitation of the carbides, particularly while cooling through the range of 870 to 540 °C (1600 to 1000 °F).

A two-step heat-treating procedure can be applied to the niobium-containing CF-8C alloy. The first treatment consists of solution annealing. This is followed by a stabilizing treatment at 870 to 925 °C (1600 to 1700 °F), which precipitates niobium carbides, prevents the formation of damaging chromium carbides, and provides maximum resistance to intergranular attack.

Because of their low carbon content, CF-3 and CF-3M as-cast do not contain enough chromium carbide to cause selective intergranular attack; therefore, these alloys can be used in some environments in this condition. However, for maximum corrosion resistance, these grades require solution annealing.

If the usual quenching treatment is difficult or impossible, holding for 24 to 48 h at 870 to 980 °C (1600 to 1800 °F) and air cooling is helpful for improving the resistance of castings to intergranular corrosion. However, except for alloys of very low carbon content and castings with thin sections, this treatment fails to produce material with as good a resistance to intergranular corrosion as properly quench-annealed material.

Corrosion-Resistant Steel Castings

As previously mentioned, various high-alloy steel castings are classified as corrosion resistant (Table 1). These corrosion-resistant cast steels are widely used in chemical processing and power-generating equipment that requires corrosion resistance in aqueous or liquid-vapor environments at temperatures normally below 315 °C (600 °F). These alloys are also used in special applications with temperatures up to 650 °C (1200 °F).

Compositions

The chemical compositions of various corrosion-resistant cast steels are given in Table 1. These cast steels are specified in the ASTM standards listed in Table 1.

Straight chromium stainless steels contain 10 to 30% Cr and little or no nickel. Although about two-thirds of the corrosion-resistant steel castings produced in the United States are of grades that contain 18 to 22% Cr and 8 to 12% Ni, the straight chromium compositions are also produced in considerable quantity, particularly the steel with 11.5 to 14.0% Cr. Corrosion resistance improves as chromium content is increased. In general, intergranular corrosion is less of a concern in the straight chromium alloys (which are typically ferritic), especially those containing 25% Cr or more. This is attributed to the high bulk chromium contents and the rapid diffusion rates of chromium in ferrite.

Iron-chromium-nickel alloys have found wide acceptance and constitute about 60% of total production of high-alloy castings. They generally are austenitic with some ferrite. The most popular alloys of this type are CF-8 and CF-8M. These alloys are nominally 18-8 stainless steels and are the cast counterparts of wrought types 304 and 316, respectively. The carbon content of each is maintained at 0.08% (max).

Effects of Molybdenum on Corrosion Resistance. Alloys CF-3M and CF-8M are modifications of CF-3 and CF-8 containing 2 to 3% Mo to enhance general corrosion resistance. Their passivity under weakly oxidizing conditions is more stable than that of CF-3 and CF-8. The addition of 2 to 3% Mo increases resistance to corrosion by seawater and improves resistance to many chloride-bearing environments. The presence of 2 to 3% Mo also improves crevice corrosion and pitting resistance compared to the CF-8 and CF-3 alloys. The CF-8M and CF-3M alloys have good resistance to such corrosive media as sulfurous and acetic acids and are more resistant to pitting by mild chlorides. These alloys are suitable for use in flowing seawater, but will pit under stagnant conditions.

Alloy CG-8M is slightly more highly alloyed than the CF-8M alloys, with the primary addition being increased molybdenum (3 to 4%). The increased amount of molybdenum provides superior corrosion resistance to halide-bearing media and reducing acids, particularly H_2SO_3 and H_2SO_4 solutions. The high molybdenum content, however, renders CG-8M generally unsuitable in highly oxidizing environments.

Molybdenum-bearing alloys are generally not as resistant to highly oxidizing environments (this is particularly true for boiling HNO_3), but for weakly oxidizing environments and reducing environments, Mo-bearing alloys are generally superior. Molybdenum may also produce detrimental catalytic reactions. For example, the residual molybdenum in CF-8 alloy must be held below 0.5% in the presence of hydrazine.

Effects of Chromium, Carbon, and Silicon on Corrosion Resistance. In alloys of the CF type, the effects of composition on rates of general corrosion attack have been studied, and certain definite relationships have been established. Through the use of the Huey test (five 48 h periods of exposure to boiling 65% nitric acid, as described in practice C of ASTM A 262), it has been shown that, in this standardized environment, carbide-free quench-annealed alloys of various nickel, chromium, silicon, carbon, and manganese contents have corrosion rates directly related to these contents.

Figure 3 shows the influences on corrosion rate exerted by various elements in a 19Cr-9Ni casting alloy. Variations in nickel, manganese, and nitrogen contents for the ranges shown have relatively slight influences, but variations in chromium, carbon, and silicon have marked effects. The relationship between composition and corrosion rate for properly heat-treated CF alloys in boiling 65% nitric acid is summarized in the nomograph presented in Fig. 4.

Iron-Nickel-Chromium Alloys. For some types of service, extensive use is made of iron-nickel-chromium alloys that contain more nickel than chromium. Most important among this group is alloy CN-7M, which has a nominal composition of 28% Ni, 20% Cr, 3.5% Cu, 2.5% Mo, and 0.07% C (max). In effect, this alloy is made by adding 20% Ni and 3.5% Cu to alloy CF-

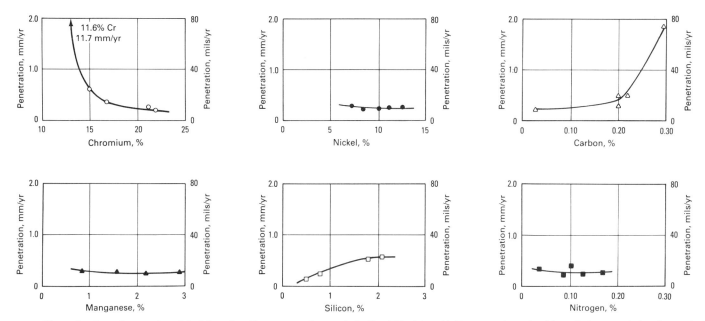

Fig. 3 Effects of various elements in a 19Cr-9Ni casting alloy on corrosion rate in boiling 65% nitric acid. Data were determined for solution-annealed and quenched specimens. Composition of base alloy was 19Cr, 9Ni, 0.09C, 0.8Mn, 1.0Si, 0.04P (max), 0.03S (max), 0.06N.

8M, which greatly improves resistance to hot, concentrated, weakly oxidizing solutions such as sulfuric acid and also improves resistance to severely oxidizing media. Alloys of this type can withstand all concentrations of sulfuric acid at temperatures up to 65 °C (150 °F) and many concentrations up to 80 °C (175 °F). They are widely used in nitric-hydrofluoric pickling solutions; phosphoric acid; cold dilute hydrochloric acid; hot acetic acid; strong, hot caustic solutions; brines; and many complex plating solutions and rayon spin baths.

Results of in-plant corrosion testing of CF-8, CF-8M, and CN-7M alloys are shown in Table 3. These tests give the specific effect of molybdenum on 19Cr-9Ni alloys in reducing selective attack and pitting, and the overall corrosion rate computed from loss in weight. The higher nickel plus copper and molybdenum in the CN-7M alloy reduces the rate of corrosion to a rate lower than that of the CF-8M alloy.

Corrosion From Chlorine. The influence of contaminants is one of the most important considerations in selecting an alloy for a particular process application. Ferric chloride in relatively small amounts, for example, will cause concentration cell corrosion and pitting. The buildup of corrosion products in a chloride solution may increase the iron concentration to a level high enough to be destructive. Thus, chlorine salts, wet chlorine gas, and unstable chlorinated organic compounds cannot be handled by any of the iron-base alloys, creating a need for nickel-base alloys.

Microstructures

Although corrosion-resistant cast steels are usually classified on the basis of composition, it should be recognized that classification by composition also often involves microstructural distinctions. Table 1 shows the typical microstructures of various corrosion-resistant cast steels. As noted previously, straight chromium grades of high-alloy cast steel are either martensitic or ferritic, the chromium-nickel grades are either duplex or austenitic, and the nickel-chromium steels are fully austenitic.

Martensitic grades include Alloys CA-15, CA-40, CA-15M, and CA-6NM. The CA-15 alloy contains the minimum amount of chromium necessary to make it essentially rustproof. It has good resistance to atmospheric corrosion, as well as to many organic media in relatively mild service. A higher-carbon modification of CA-15, CA-40 can be heat treated to higher strength and hardness levels. Alloy CA-15M is a molybdenum-containing modification of CA-15 that provides improved elevated-temperature strength. Alloy CA-6NM is an iron-chromium-nickel-molybdenum alloy of low carbon content.

Austenitic grades include CH-20, CK-20, and CN-7M. The CH-20 and CK-20 alloys are high-chromium, high-carbon, wholly austenitic compositions in which the chromium content exceeds the nickel content. The more highly alloyed CN-7M, as described earlier in the section "Iron-Nickel-Chromium Alloys," has excellent corrosion resistance in many environments and is often used in sulfuric acid environments. The CN-7MS alloy has a corrosion resistance similar to that of CN-7M. The CN-7MS alloy has outstanding resistance to corrosion from high-strength (>90%) nitric acid.

Ferritic grades include CB-30 and CC-50. Alloy CB-30 is practically nonhardenable by heat treatment. As this alloy is normally made, the balance among the elements in the composition results in a wholly ferritic structure similar to wrought AISI type 442 stainless steel. Alloy CC-50 has substantially more chromium than CB-30 and has relatively high resistance to localized corrosion in many environments.

Austenitic-ferritic (duplex) alloys include CE-30, CF-3, CF-3A, CF-8, CF-8A, CF-20, CF-3M, CF-3MA, CF-8M, CF-8C, CF-16F, and CG-8M. The microstructures of these alloys usually contain 5 to 40% ferrite, depending on the particular grade and the balance among the ferrite-promoting and austenite-promoting elements in the chemical composition (see the section "Ferrite

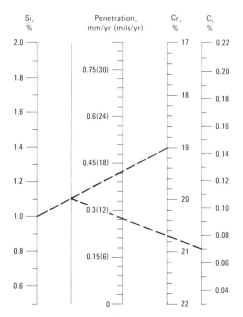

Fig. 4 Nomograph for determining corrosion rate in boiling 65% nitric acid for solution-annealed and quenched type CF casting alloys

Table 3 Results of in-plant corrosion testing of CF-8, CF-8M, and CN-7M alloys

Type and composition of corroding solution	Temperature of solution, °C	°F	Alloy	Metal loss on surface μm/yr	mils/yr	Surface condition by visual examination	Remarks
Neutralizer after formation of ammonium sulfate: ammonium sulfate plus small excess of sulfuric acid, ammonia vapor, and steam	100	212	CF-8 CF-8M CN-7M	665 28 18	26.2 1.1 0.7	Very heavy etch(a) Light tarnish(b) Bright	CF-8M was installed for low corrosion tolerance equipment in this service and performed satisfactorily
Settling tank after neutralizer: ammonium sulfate plus excess of sulfuric acid	50	122	CF-8 CF-8M CN-7M	385 10 2.5	15.2 0.4 0.1	Very heavy etch(a) Slight tarnish Bright(b)	CF-8 in service showed excessive corrosion rate plus heavy concentration cell attack
Ammonium sulfate processing solution: ammonium sulfate at pH of 8.0	50	122	CF-8 CF-8M CN-7M	685 175 50	27.0 6.8 2.0	Heavy etch Moderate etch Light etch	CF-8M had too high a corrosion rate in service for good valve life, although suitable for equipment of greater corrosion tolerance. CN-7M was installed in this service
99 to 100% fuming nitric acid	20	60	CF-8 CN-7M CF-8M	245 79 345	9.6 3.1 13.5	Moderate etch Light etch Moderate etch	CF-8 was satisfactory except for low-tolerance equipment such as valves. CN-7M valves performed satisfactorily in service
Saturated solution of sodium chloride plus 15% sodium sulfate; pH of 4.5	60	140	CF-8M CF-8	2.5 240	0.1 9.5	Bright Concentration cell corrosion at various small areas of specimen	CF-8M was installed for valves in service

(a) Concentration cell attack under insulating washer. (b) Slight concentration cell attack under insulating washer

Control" in this article). Duplex alloys offer superior strength, corrosion resistance, and weldability.

The use of duplex cast steels has focused primarily on the CF grades, particularly by the power generation industry. Strengthening in the cast CF grade alloys is limited essentially to that which can be gained by incorporating ferrite into the austenite matrix phase. These alloys cannot be strengthened by thermal treatment, as can the cast martensitic alloys, nor by hot or cold working, as can the wrought austenitic alloys. Strengthening by carbide precipitation is also out of the question because of the detrimental effect of carbides on corrosion resistance in most aqueous environments. Thus, the alloys are effectively strengthened by balancing the alloy composition to produce a duplex microstructure consisting of ferrite (up to 40% by volume) distributed in an austenite matrix. It has been shown that the incorporation of ferrite into 19Cr-9Ni cast steels improves yield and tensile strengths without substantial loss of ductility or impact toughness at temperatures below 425 °C (800 °F). The magnitude of this strengthening effect for CF-8 and CF-8M alloys at room temperature is shown in Fig. 5. Table 4 shows the effect of ferrite content on the tensile properties of 19Cr-9Ni alloys at room temperature and at 355 °C (670 °F). Table 5 shows the effect of ferrite content on impact toughness.

Other duplex alloys of interest include CD-4MCu and Ferralium. Alloy CD-4MCu is the most highly alloyed duplex alloy. Ferralium was developed by Langley Alloys and is essentially CD-4MCu with about 0.15% N added. With high levels of ferrite (about 40 to 50%) and low nickel, the duplex alloys have better resistance to stress-corrosion cracking (SCC) than CF-3M. Alloy CD-4MCu, which contains no nitrogen and has a relatively low molybdenum content, has only slightly better resistance to local-

Table 4 Effect of ferrite content on tensile properties of 19Cr-9Ni alloys

Ferrite content, %	Tensile strength MPa	ksi	Yield strength at 0.2% offset MPa	ksi	Elongation in 50 mm (2 in.), %	Reduction in area, %
Tested at room temperature						
3	465	67.4	216	31.3	60.5	64.2
10	498	72.2	234	34.0	61.0	73.0
20	584	84.7	296	43.0	53.5	58.5
41	634	91.9	331	48.0	45.5	47.9
Tested at 355 °C (670 °F)						
3	339	49.1	104	15.1	45.5	63.2
10	350	50.8	109	15.8	43.0	69.7
20	457	66.3	183	26.5	36.5	47.5
41	488	70.8	188	27.3	33.8	49.4

Table 5 Charpy V-notch impact energy, ferrite content, and Cr_e/Ni_e ratio of duplex cast steels

Alloy	Charpy V-notch energy J	ft·lbf	Ferrite content, % Calculated	MG(a)	FS(b)	Cr_e/Ni_e ratio(c)
CF 3M	197	145	28.5	20	20	1.5
CF 3C	183	135	20.7	12.5	14	1.4
CG 8M	216	159	18	9	10	1.34
CF 3C	>358	>264	15	13	15	1.29
CF 3M	>358	>264	7.7	6	7	1.12

(a) MG, magna gage. (b) FS, ferrite scope. (c) See Eq 1 and 2 for formulas to compute Cr_e and Ni_e.

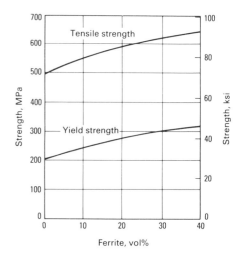

Fig. 5 Yield strength and tensile strength versus percentage of ferrite for CF-8 and CF-8M alloys. Curves are mean values for 277 heats of CF-8 and 62 heats of CF-8M. Source: Ref 3

ized corrosion than CF-3M. Ferralium, which has nitrogen and slightly higher molybdenum than CD-4MCu, exhibits better localized corrosion resistance than either CF-3M or CD-4MCu.

Improvements in stainless steel production practices (for example, electron beam refining, vacuum and argon-oxygen decarburization, and vacuum induction melting) have also created a second generation of duplex stainless steels. These steels offer excellent resistance to pitting and crevice corrosion, significantly better resistance to chloride SCC than the austenitic stainless steels, good toughness, and yield strengths two to three times higher than those of type 304 or 316 stainless steels.

First-generation duplex stainless steels, for example, AISI type 329 and CD-4MCu, have been in use for many years. The need for improvement in the weldability and corrosion resistance of these alloys resulted in the second-generation alloys, which are characterized by the addition of nitrogen as an alloying element.

Second-generation duplex stainless steels are usually about a fifty-fifty blend of ferrite and austenite. The new duplex alloys combine the near immunity to chloride SCC of the ferritic grades with the toughness and ease of fabrication of the austenitics. Among the second-generation duplexes, Alloy 2205 seems to have become the general-purpose stainless. Table 6 lists the nominal compositions of first- and second-generation duplex alloys.

Precipitation-Hardening Alloys. Corrosion-resistant alloys capable of being hardened by low-temperature treatment to obtain improved mechanical properties are usually duplex-structure alloys with much more chromium than nickel. The addition of copper enables these alloys to be strengthened by precipitation hardening. These alloys are significantly higher in strength than the other corrosion-resistant alloys even without hardening.

The alloys CB-7Cu-1 and CB-7Cu-2 have corrosion resistances between those of CA-15 and CF-8. They are widely used for structural components requiring moderate corrosion resistance, as well as for components requiring resistance to erosion and wear.

The alloy CD-4MCu is widely used in many applications where its good corrosion resistance (which often equals or even exceeds that of CF-8M) and excellent resistance to erosion make it the most desirable alloy. The steel CD-4MCu has outstanding resistance to nitric acid and mixtures of nitric acid and organic acids, as well as excellent resistance to a wide range of corrosive chemical process conditions. This alloy is normally used in the solution-annealed condition, but it can be precipitation hardened for carefully selected applications when lower corrosion resistance can be tolerated and when there is no potential for stress-corrosion cracking.

Corrosion Characteristics

Table 7 compares the general corrosion resistance of the C-type (corrosion-resistant in liquid service) cast steels. Additional information on the corrosion resistance of cast steels is contained below and in Volume 13 of the 9th Edition of *Metals Handbook*.

General Corrosion of Martensitic Alloys. The martensitic grades include CA-15, CA-15M, CA-6NM, CA-6NM-B, CA-40, CB-7Cu-1, and CB-7Cu-2. These alloys are generally used in applications requiring high strength and some corrosion resistance.

Table 6 Nominal compositions of first- and second-generation duplex stainless steels

UNS designation	Common name	Composition, %(a)					
		Cr	Ni	Mo	Cu	N	Others
First generation steels							
S31500	3RE60	18.5	4.7	2.7	...	...	1.7Si
S32404	Uranus 50	21	7.0	2.5	1.5	...	...
S32900	Type 329	26	4.5	1.5	...	...	...
J93370	CD-4MCu	25	5	2	3	...	...
Second generation steels							
S31200	44LN	25	6	1.7	...	0.15	...
S31260	DP-3	25	7	3	0.5	0.15	0.3W
S31803	Alloy 2205	22	5	3	...	0.15	...
S32550	Ferralium 255	25	6	3	2	0.20	...
S32950	7-Mo PLUS	26.5	4.8	1.5	...	0.20	...
J93404	Atlas 958, COR 25	25	7	4.5	...	0.25	...

(a) All compositions contain balance of iron.

Table 7 Summary of applications for various corrosion-resistant cast steels

Alloy	Characteristics
CA-15	Widely used in mildly corrosive environments; hardenable; good erosion resistance
CA-40	Similar to CA-15 at higher strength level
CA-6NM	Improved properties over CA-15, especially improved resistance to cavitation
CA-6N	Outstanding combinations of strength, toughness, and weldability with moderately good corrosion resistance
CB-30	Improved performance in oxidizing environments compared to CA-15; excellent resistance to corrosion by nitric acid, alkaline solutions, and many organic chemicals
CB-7Cu-1	Hardenable with good corrosion resistance
CB-7Cu-2	Superior combination of strength, toughness, and weldability with moderately good corrosion resistance
CC-50	Used in highly oxidizing media (hot HNO_3, acid mine waters)
CD-4MCu	Similar to CF-8 in corrosion resistance, but higher strength, hardness, and stress-corrosion cracking resistance; excellent resistance to environments involving abrasion or erosion-corrosion; usefully employed in handling both oxidizing and reducing corrodents
CE-30	Similar to CC-50, but Ni imparts higher strength and toughness levels. A grade available with controlled ferrite
CF-3, CF-8, CF-20, CF-3M, CF-8M, CF-8C, CF-16F	CF types: most widely used corrosion-resistant alloys at ambient and cryogenic temperatures M variations: enhanced resistance to halogen ion and reducing acids C and F variations: used where application does not permit postweld heat treat A grades available with controlled ferrite
CG-8M	Greater resistance to pitting and corrosion in reducing media than CF-8M; not suitable for nitric acids or other strongly oxidizing environments
CH-20	Superior to CF-8 in specialized chemical and paper applications in resistance to hot H_2SO_3, organic acids, and dilute H_2SO_4; the high nickel and chromium contents also make this alloy less susceptible to intergranular corrosion after exposure to carbide-precipitating temperatures
CK-20	Improved corrosion resistance compared to CH-20
CN-7M	Highly resistant to H_2SO_4, H_3PO_4, H_2SO_3, salts, and seawater. Good resistance to hot chloride salt solutions, nitric acid, and many reducing chemicals

Alloy CA-15 typically exhibits a microstructure of martensite and ferrite. This alloy contains the minimum amount of chromium to be considered a stainless steel (11 to 14% Cr) and as such may not be used in aggressive environments. It does, however, exhibit good atmospheric-corrosion resistance and it resists staining by many organic environments. Alloy CA-15M may contain slightly more molybdenum than CA-15 (up to 1% Mo) and therefore may have improved general corrosion resistance in relatively mild environments. Alloy CA-6NM is similar to CA-15M except that it contains more nickel and molybdenum, thereby improving its general corrosion resistance. Alloy CA-6NM-B is a lower-carbon version of this alloy. The lower strength level promotes resistance to sulfide stress cracking. Alloy CA-40 is a higher-strength version of CA-15 and it, too, exhibits excellent atmospheric-corrosion resistance after a normalize and temper heat treatment. Microstructurally, the CB-7Cu alloys usually consist of mixed martensite and ferrite and, because of the increased chromium and nickel levels compared to the other martensitic alloys, they offer improved corrosion resistance to seawater and some mild acids. These alloys also have good atmospheric-corrosion resistance. The CB-7Cu alloys are hardenable and offer the possibility of increased strength and improved corrosion resistance among the martensitic alloys.

General Corrosion of Ferritic Alloys. Alloys CB-30 and CC-50 are higher-carbon and higher-chromium alloys than are the CA alloys mentioned above. Each alloy is predominantly ferritic, although a small amount of martensite may be found in CB-30. Alloy CB-30 contains 18 to 21% Cr and is used in chemical processing and oil refining applications. The chromium content is sufficient to have good corrosion resistance to many acids, including nitric acid (HNO_3).

General Corrosion of Austenitic and Duplex Alloys. Alloy CF-8 may be fully austenitic, but it more commonly contains some residual ferrite (3 to 30%) in an austenite matrix. In the solution-annealed condition, this alloy has excellent resistance to a wide variety of acids. It is particularly resistant to highly oxidizing acids, such as boiling HNO_3. The duplex nature of the microstructure of this alloy imparts additional resistance to SCC compared to its wholly austenitic counterparts. Alloy CF-3 is a reduced-carbon version of CF-8 with essentially identical corrosion resistance except that CF-3 is much less susceptible to sensitization. For applications in which the corrosion resistance of the weld HAZ may be critical, CF-3 is a common material selection.

Alloys CF-8A and CF-3A contain more ferrite than their CF-8 and CF-3 counterparts. Because the higher ferrite content is achieved by increasing the chromium/nickel equivalent ratio, the CF-8A and CF-3A alloys may have slightly higher chromium or slightly lower nickel contents than the low-ferrite equivalents. In general, the corrosion resistance is very similar, but the strength increases with ferrite content. Because of the high ferrite content, service should be restricted to temperatures below 400 °C (750 °F) because of the possibility of severe embrittlement. Alloy CF-8C is the niobium-stabilized grade of the CF-8 alloy class. This alloy contains small amounts of niobium, which tend to form carbides preferentially over chromium carbides and improve intergranular corrosion resistance in applications involving relatively high service temperatures.

Alloy CF-16F is a selenium-bearing free-machining grade of cast stainless steel. Because CF-16F nominally contains 19% Cr and 10% Ni, it has adequate corrosion resistance to a wide range of corrosive materials but the large number of selenide inclusions makes surface deterioration and pitting definite possibilities.

Alloy CE-30 is a nominally 27Cr-9Ni alloy that normally contains 10 to 20% ferrite in an austenite matrix. The high carbon, high ferrite content provides relatively high strength. The high chromium content and duplex structure act to minimize corrosion because of the formation of chromium carbides in the microstructure. This particular alloy is known for good resistance to sulfurous acid and sulfuric acid and is used extensively in the pulp and paper industry (see the article "Corrosion in the Pulp and Paper Industry" in Volume 13 of the 9th Edition of *Metals Handbook*).

Alloy CD-4MCu is the most highly alloyed material in this group of alloys, and a microstructure containing approximately equal amounts of ferrite and austenite is common. The low carbon content and high chromium content render the alloy relatively immune to intergranular corrosion. High chromium and molybdenum provide a high degree of localized corrosion resistance (crevices and pitting), and the duplex microstructure provides SCC resistance in many environments. This alloy can be precipitation hardened to provide strength and is also relatively resistant to abrasion and erosion-corrosion.

Fully Austenitic Alloys. Alloys CH-10 and CH-20 are fully austenitic and contain 22 to 26% Cr and 12 to 15% Ni. The high chromium content minimizes the tendency toward the formation of chromium-depleted zones due to sensitization. These alloys are used for handling paper pulp solutions and are known for good resistance to dilute H_2SO_4 and HNO_3.

Alloy CK-20 contains 23 to 27% Cr and 19 to 22% Ni and is less susceptible than CH-20 to intergranular corrosion attack in many acids after brief exposures to the chromium carbide formation temperature range. Maximum corrosion resistance is achieved by solution treatment. Alloy CK-20 possesses good corrosion resistance to many acids and, because of its fully austenitic structure, can be used at relatively high temperatures.

Alloy CN-7M exhibits excellent corrosion resistance in a wide variety of environments and is often used for H_2SO_4 service. Relatively high resistance to intergranular corrosion and SCC make this alloy attractive for many applications. Although CN-7M is relatively highly alloyed, its fully austenitic structure may lead to SCC susceptibility for some environments and stress states.

Alloy CF-20 is a fully austenitic, relatively high-strength corrosion-resistant alloy. The 19% Cr content provides resistance to many types of oxidizing acids, but the high carbon content makes it imperative that this alloy be used in the solution-treated condition for environments known to cause intergranular corrosion.

Intergranular Corrosion. Ferritic alloys may also be sensitized by the formation of extensive chromium carbide networks, but because of the high bulk chromium content and rapid diffusion rates of chromium in ferrite, the formation of carbides can be tolerated if the alloy has been slowly cooled from a solutionizing temperature of 780 to 900 °C (1435 to 1650 °F). The slow cooling allows replenishment of the chromium adjacent to the carbides. Martensitic alloys normally do not contain sufficient bulk chromium to be used in applications in which intergranular corrosion is likely to be a concern.

Austenitic and duplex stainless steels use solution annealing for the prevention or reduction of intergranular corrosion (see "Sensitization and Solution Annealing of Austenitic and Duplex Alloys" in this article). Failure to solution treat a particular alloy or an improper solution treatment may seriously compromise the observed corrosion resistance in service.

If solution treatment of the alloy after casting and/or welding is impractical or impossible, the metallurgist has several tools from which to choose to minimize potential intergranular corrosion problems. The low-carbon grades CF-3 and CF-3M are commonly used when heat treatment is impractical or as a solution to the sensitization incurred during welding. The low carbon content, that is, 0.03% C (max), of these alloys precludes the formation of an extensive number of chromium carbides. In addition, these alloys normally contain 3 to 30% ferrite in an austenitic matrix. By virtue of rapid carbide precipitation kinetics at ferrite/austenite interfaces compared to austenite/austenite interfaces, carbide precipitation is confined to ferrite-austenite boundaries in alloys containing a minimum of about 3 to 5% ferrite (Ref 4, 5). If the ferrite network is discontinuous in the austenite matrix (depending on the amount, size, and distribution of ferrite pools), extensive intergranular corrosion will not be a problem in most of the environments to which these alloys will be subjected.

The niobium-modified grade of 18-8, known as CF-89C, is produced for similar applications in which heat treatment is impractical. Niobium-containing alloys that have been heated to sensitizing temperatures around 650 °C (1200 °F) are not susceptible to intergranular corrosion. However, they are more susceptible to overall corrosion when tested in nitric acid, compared to the niobium-free, quench-annealed alloys of the same nickel, chromium, and carbon contents. Addition of niobium to molybdenum-containing type CF alloys has also been found unsatisfactory for castings. When both niobium and molybdenum are present, the ferrite phase tends to form as an interconnected network and is especially likely to transform into the brittle σ phase. As a result, castings in the as-cast condition become embrittled and have a tendency to crack.

When the niobium-bearing grade CF-8C is in the as-cast condition, most of its carbon is in the form of niobium carbide, precluding chromium carbide precipitation in the critical temperature range from 425 to 870 °C (800 to 1600 °F) and particularly from 565 to 650 °C (1050 to 1200 °F). The alloy CF-8C is solution treated at 1120 °C (2050 °F), quenched to room temperature, and then reheated to 870 to 925 °C (1600 to 1700 °F), at which temperature precipitation of niobium carbide occurs. An alternative method is solution treating at 1120 °C (2050 °F), cooling to the 870 to 925 °C (1600 to 1700 °F) range, and then holding at this temperature before cooling to room temperature. For maximum corrosion resistance, it is recommended that this alloy be solution treated before being stabilized.

Weld crack sensitivity of CF alloys containing niobium (CF-8C) is more pronounced in the fully austenitic grade. Cracking may be alleviated through the introduction into the weld deposit of a small amount of ferrite, usually between 4 and 10%. However, appreciable amounts of ferrite in niobium-bearing corrosion-resistant steels will transform, at least partly, to the σ or χ phase upon heating to between 540 and 925 °C (1000 and 1700 °F).

Stress-Corrosion Cracking. The SCC of cast stainless steels has been investigated for only a limited number of environments, heat treatments, and test conditions. From the limited information available, the following generalizations apply.

First, SCC resistance seems to improve as the composition is adjusted to provide increasingly greater amounts of ferrite in an austenitic matrix. This trend continues to a certain level, apparently near 50% ferrite (Fig. 6). Second, a lower nickel content tends to improve SCC resistance in cast duplex alloys, possibly because of its effect on ferrite content (Ref 6). Third, ferrite appears to be involved in a keying action in discouraging SCC. At low and medium stress levels, the ferrite tends to block the propagation of stress-corrosion cracks. This may be due to a change in composition and/or crystal structure across the austenite/ferrite boundary. As the stress level increases, crack propagation may change from austenite/ferrite boundaries to transgranular propagation (Ref 6, 7). Finally, reducing the carbon content of cast stainless alloys, thereby reducing the susceptibility to sensitization, improves SCC resistance. This is also true for wrought alloys.

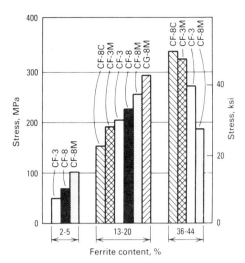

Fig. 6 Stress required to produce stress-corrosion cracking in several corrosion-resistant cast steels with varying amounts of ferrite

Mechanical Properties of Corrosion-Resistant Cast Steels

The importance of mechanical properties in the selection of corrosion-resistant cast steels is established by the casting application. The paramount basis for alloy selection is normally the resistance of the alloy to the specific corrosive media or environment of interest. The mechanical properties of the alloy are usually, but not always, secondary considerations in these applications.

Room-Temperature Mechanical Properties. Representative room-temperature tensile properties, hardness, and Charpy impact values for corrosion-resistant cast steels are given in Fig. 7. These properties are representative of the alloys rather than the specification requirements. Minimum specified mechanical properties for these alloys are given in ASTM standards A 351, A 743, A 744, and A 747. A wide range of mechanical properties are attainable depending on the selection of alloy composition and heat treatment. Tensile strengths ranging from 475 to 1310 MPa (69 to 190 ksi) and hardnesses from 130 to 400 HB are available among the cast corrosion-resistant alloys. Similarly, wide ranges exist in yield strength, elongation, and impact toughness.

The straight chromium steels (CA-15, CA-40, CB-30, and CC-50) possess either martensitic or ferritic microstructures in the end-use condition (Table 1). The CA-15 and CA-40 alloys, which contain nominally 12% Cr, are hardenable through heat treatment by means of the martensite transformation and are often selected as much or more for their high strength as for their comparatively modest corrosion resistance.

The higher-chromium CB-30 and CC-50 alloys, on the other hand, are fully ferritic alloys that are not hardenable by heat treatment. These alloys are generally used in the annealed condition and exhibit moderate tensile properties and hardness. Like most ferritic alloys, CB-30 and CC-50 possess limited impact toughness, especially at low temperatures.

Three chromium-nickel alloys, CA-6NM, CB-7Cu, and CD-4MCu, are exceptional in their response to heat treatment and in the resultant mechanical properties. Alloy CA-6NM is balanced compositionally for martensitic hardening response. This alloy was developed as an alternative to CA-15 and has improved impact toughness and weldability. The CB-7Cu and CD-4MCu alloys both contain copper and can be strengthened by age hardening. These alloys are initially solution heat treated and then cooled rapidly (usually by quenching in oil or water); thus, the phases that would normally precipitate at slow cooling rates cannot form. The casting is then heated to an intermediate aging temperature at which the precipitation reaction can occur under controlled conditions until the desired combination of strength and other properties is achieved. The CB-7Cu alloy possesses a martensitic matrix, while the CD-4MCu alloy possesses a duplex microstructure, consisting of approximately 40% austenite in a ferritic matrix. Alloy CB-7Cu is applied in the aged condition to obtain the benefit of its excellent combination of strength and corrosion resistance, but alloy CD-4MCu is seldom applied in the aged condition because of its relatively low resistance to SCC in this condition compared to its superior corrosion resistance in the solution-annealed condition.

The CE, CF, CG, CH, CN, and CK alloys are essentially not hardenable by heat treatment. To ensure maximum corrosion resistance, however, it is necessary that castings of these grades receive a high-temperature solution anneal (see "Sensitization and Solution Annealing of Austenitic and Duplex Alloys" in this article). By virtue of their microstructures, which are fully austenitic or duplex without significant carbide precipitation, the alloys exhibit generally excellent impact toughness at low temperatures. The tensile strength range represented by these alloys typically extends from 475 to 670 MPa (69 to 97 ksi). As indicated earlier in the section "Austenitic-Ferritic (Duplex) Alloys" in this article, the alloys with duplex structures can be strengthened by balancing the composition for higher ferrite levels (Fig. 5). The tensile and yield strengths of CF alloys with a ferrite number of 35 are typically 150 MPa (22 ksi) higher than those of fully austenitic alloys. Tensile ductility (Table 4) and impact toughness (Table 5) are lowered with increasing ferrite content.

Effects From High Temperatures. Cast corrosion-resistant high-alloy steels are used extensively at moderately elevated temperatures (up to 650 °C, or 1200 °F). Elevated-temperature properties are important selec-

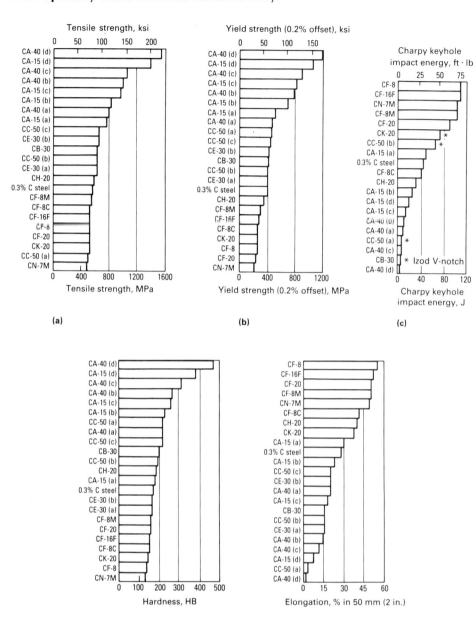

Fig. 7 Mechanical properties of various cast corrosion-resistant steels at room temperature. (a) Tensile strength. (b) 0.2% offset yield strength. (c) Charpy keyhole impact energy. (d) Brinell hardness. (e) Elongation. Also given are the heat treatments used for test materials: AC, air cool; FC, furnace cool; WQ, water quench; A, anneal; T, temper.

tion criteria for these applications. Table 8 gives the tensile properties of a corrosion-resistant cast steel at various test temperatures. In addition, mechanical properties after long-term exposure at elevated temperatures are increasingly considered because of the aging effect that these exposures may have. For example, cast alloys CF-8C, CF-8M, CE-30A, and CA-15 are currently used in high-pressure service at temperatures up to 540 °C (1000 °F) in sulfurous acid environments in the petrochemical industry. Other uses are in the power-generating industry at temperatures up to 565 °C (1050 °F).

Room-temperature properties after exposure to elevated service temperatures may differ from those in the as-heat-treated condition because of the microstructural changes that may take place at the service temperature. Microstructural changes in iron-nickel-chromium-(molybdenum) alloys may involve the formation of carbides and such phases as σ, χ, and η (Laves). The extent to which these phases form depends on the composition, as well as the time at elevated temperature.

The martensitic alloys CA-15 and CA-6NM are subject to minor changes in mechanical properties and SCC resistance in NaCl and polythionic acid environments upon exposure for 3000 h at up to 565 °C (1050 °F). In CF-type chromium-nickel-(molybdenum) steels, only negligible changes in ferrite content occur during 10 000 h exposure at 400 °C (750 °F) and during 3000 h exposure at 425 °C (800 °F). Carbide precipitation, however, does occur at these temperatures, and noticeable Charpy V-notch energy losses have been reported.

Above 425 °C (800 °F), microstructural changes in chromium-nickel-(molybdenum) alloys take place at an increased rate. Carbides and σ phase form rapidly at 650 °C (1200 °F) at the expense of ferrite. Tensile ductility and Charpy V-notch impact energy (Fig. 8) are prone to significant losses under these conditions. Density changes, resulting in contraction, have been reported as a result of these high-temperature exposures.

Fatigue Properties and Corrosion Fatigue. The resistance of cast stainless steels to fatigue depends on a sizable number of material, design, and environmental factors. For example, design factors of importance include the stress distribution within the casting (residual and applied stresses), the location and severity of stress concentrators (surface integrity), and the environment and service temperatures. Material factors of importance include strength and microstructure. It is generally found that fatigue strength increases with the tensile strength of a material. Both fatigue strength and tensile strength usually increase with decreasing temperature. Under equivalent conditions of stress, stress concentration,

Table 8 Short-time tensile properties of peripheral-welded cylinders of CF-8 alloy

Cylinders were 38 mm (1½ in.) thick; specimens were machined with longitudinal axes perpendicular to welded seam and with seam at middle of gage length.

Testing temperature		Tensile strength		Yield strength at 0.2% offset		Proportional limit(a)		Reduction in area, %	Elongation in 50 mm (2 in.), %	Modulus of elasticity(a)		Location of final rupture
°C	°F	MPa	ksi	MPa	ksi	MPa	ksi			GPa	10⁶ psi	
Base metal												
Keel block(b)		500	72.5	238	34.5	...	...	59.0	49	...	...	...
	Room	500	72.5	261	37.8	179	26	62.1	58	186	27	...
315	600	330	47.8	169	24.5	90	13	54.9	33.5	152	22	...
425	800	339	49.2	167	24.2	59	8.5	58.6	37.5	134	19.5	...
540	1000	291	42.2	140	20.3	55	8	60.8	32.5	117	17	...
595	1100	279	40.4	130	18.8	45	6.5	59.1	38	110	16	...
Welded joint												
	Room	490	71.0	247	35.8	148	21.5	70.8	42	186	27	Base metal
315	600	341	49.5	199	28.8	72	10.5	58.3	15.5	152	22	Base metal
425	800	355	51.5	171	24.8	69	10	46.3	24.5	131	19	Base metal
540	1000	326	47.3	188	27.3	62	9	62.8	23.5	114	16.5	Base metal
595	1100	272	39.4	134	19.5	55	8	70.4	31	107	15.5	Base metal

(a) Values of proportional limit and modulus of elasticity at elevated temperatures are apparent values because creep occurs. (b) Separately cast from same heat as cylinders

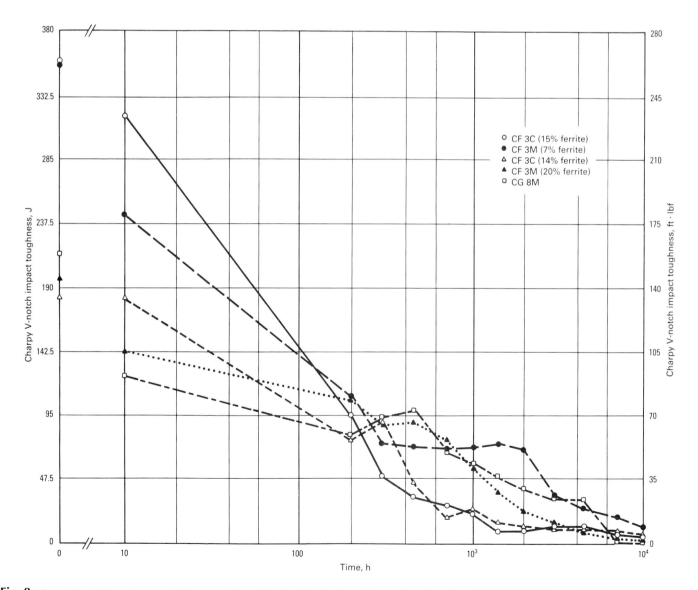

Fig. 8 Charpy V-notch impact energy of three corrosion-resistant cast steels at room temperature after aging at 594 °C (1100 °F). Source: Ref 8

920 / Specialty Steels and Heat-Resistant Alloys

Table 9 General corrosion characteristics of heat-resistant cast steels and typical limiting creep stress values at indicated temperatures

Alloy	Corrosion characteristics	Creep test temperature °C	°F	Limiting creep stress (0.0001%/h) MPa	ksi
HA	Good oxidation resistance to 650 °C (1200 °F); widely used in oil refining industry	650	1200	21.5	3.1
HC	Good sulfur and oxidation resistance up to 1095 °C (2000 °F); minimal mechanical properties; used in applications where strength is not a consideration or for moderate load bearing up to 650 °C (1200 °F)	870	1600	5.15	0.75
HD	Excellent oxidation and sulfur resistance plus weldability	980	1800	6.2	0.9
HE	Higher temperature and sulfur resistant capabilities than HD	980	1800	9.5	1.4
HF	Excellent general corrosion resistance to 815 °C (1500 °F) with moderate mechanical properties	870	1600	27	3.9
HH(a)	High strength; oxidation resistant to 1090 °C (2000 °F); most widely used	980	1800	7.5 (type I) 14.5 (type II)	1.1 (type I) 2.1 (type II)
HI	Improved oxidation resistance compared to HH	980	1800	13	1.9
HK	Because of its high temperature strength, widely used for stressed parts in structural applications up to 1150 °C (2100 °F); offers good resistance to corrosion by hot gases, including sulfur-bearing gases, in both oxidizing and reducing conditions (although HC, HE, and HI are more resistant in oxidizing gases); used in air, ammonia, hydrogen, and molten neutral salts; widely used for tubes and furnace parts	1040	1900	9.5	1.4
HL	Improved sulfur resistance compared to HK; especially useful where excessive scaling must be avoided	980	1800	15	2.2
HN	Very high strength at high temperatures; resistant to oxidizing and reducing flue gases	1040	1900	11	1.6
HP	Resistant to both oxidizing and carburizing atmospheres at high temperatures	980	1800	19	2.8
HP-50WZ	Improved creep rupture strength at 1090 °C (2000 °F) and above compared to HP	1090	2000	4.8	0.7
HT	Widely used in thermal shock applications; corrosion resistant in air, oxidizing and reducing flue gases, carburizing gases, salts, and molten metals; performs satisfactorily up to 1150 °C (2100 °F) in oxidizing atmospheres and up to 1095 °C (2000 °F) in reducing atmospheres, provided that limiting creep stress values are not exceeded	980	1800	14	2.0
HU	Higher hot strength than HT and often selected for its superior corrosion resistance	980	1800	15	2.2
HW	High hot strength and electrical resistivity; performs satisfactorily to 1120 °C (2050 °F) in strongly oxidizing atmospheres and up to 1040 °C (1900 °F) in oxidizing or reducing products of combustion that do not contain sulfur; resistant to some salts and molten metals	980	1800	9.5	1.4
HX	Resistant to hot-gas corrosion under cycling conditions without cracking or warping; corrosion resistant in air, carburizing gases, combustion gases, flue gases, hydrogen, molten cyanide, molten lead, and molten neutral salts at temperatures up to 1150 °C (2100 °F)	980	1800	11	1.6

(a) Two grades: type I (ferrite in austenite) and type II (wholly austenitic), per ASTM A 447

and strength, evidence suggests that austenitic materials are less notch sensitive than martensitic or ferritic materials.

Corrosion fatigue is highly specific to the environment and alloy. The martensitic materials are degraded the most in both absolute and relative terms. If left to corrode freely in seawater, they have very little resistance to corrosion fatigue. This is remarkable in view of their very high strength and fatigue resistance in air.

Properties can be protected if suitable cathodic protection is applied. However, because these materials are susceptible to hydrogen embrittlement, cathodic protection must be carefully applied. Too large a protective potential will lead to catastrophic hydrogen stress cracking.

Austenitic materials are also severely degraded in corrosion fatigue strength under conditions conducive to pitting, such as in seawater. However, they are easily cathodically protected without fear of hydrogen embrittlement and perform well in fresh waters. The corrosion fatigue behavior of duplex alloys has not been widely studied.

Heat-Resistant Cast Steels

As previously mentioned, castings are classified as heat resistant if they are capable of sustained operation while exposed, either continuously or intermittently, to operating temperatures that result in metal temperatures in excess of 650 °C (1200 °F). Cast steels for this type of service include iron-chromium (straight chromium), iron-chromium-nickel, and iron-nickel-chromium alloys. In applications of heat-resistant alloys, considerations include:

- Resistance to corrosion at elevated temperatures
- Stability (resistance to warping, cracking, or thermal fatigue)
- Creep strength (resistance to plastic flow)

Table 9 briefly compares the various H-type grades of heat-resistant steel castings in terms of general corrosion resistance and creep values.

Commercial applications of heat-resistant castings include metal treatment furnaces, gas turbines, aircraft engines, military equipment, oil refinery furnaces, cement mill equipment, petrochemical furnaces, chemical process equipment, power plant equipment, steel mill equipment, turbochargers, and equipment used in manufacturing glass and synthetic rubber. Alloys of the iron-chromium and iron-chromium-nickel groups are of the greatest commercial importance.

General Properties

General corrosion and creep properties of heat-resistant steel castings are compared in Table 8. The compositions of these heat-resistant cast steels are given in Table 2. These heat-resistant cast steels resemble corrosion-resistant cast steels (Table 1) except for their higher carbon content, which imparts greater strength at elevated temperatures. Typical tensile properties of heat-resistant cast steels at room temperature are given in Table 10 and (at elevated temperatures) in Table 11.

Iron-chromium alloys contain 10 to 30% Cr and little or no nickel. These alloys are useful chiefly for resistance to oxidation; they have low strength at elevated temperatures. Use of these alloys is restricted to conditions, either oxidizing or reducing, that involve low static loads and uniform heating. Chromium content depends on anticipated service temperature.

Iron-chromium-nickel alloys contain more than 13% Cr and more than 7% Ni (always more chromium than nickel). These austenitic alloys are ordinarily used under oxidizing or reducing conditions similar to those withstood by the ferritic iron-chromium alloys, but in service they have greater strength and ductility than the straight chromium alloys. They are used, therefore, to withstand greater loads and moderate changes of temperature. These alloys also are used in the presence of oxidizing and reducing gases that are high in sulfur content.

Table 10 Typical room-temperature properties of ACI heat-resistant casting alloys

Alloy	Condition	Tensile strength MPa	ksi	Yield strength MPa	ksi	Elongation, %	Hardness, HB
HC	As-cast	760	110	515	75	19	223
	Aged(a)	790	115	550	80	18	...
HD	As-cast	585	85	330	48	16	90
HE	As-cast	655	95	310	45	20	200
	Aged(a)	620	90	380	55	10	270
HF	As-cast	635	92	310	45	38	165
	Aged(a)	690	100	345	50	25	190
HH, type 1	As-cast	585	85	345	50	25	185
	Aged(a)	595	86	380	55	11	200
HH, type 2	As-cast	550	80	275	40	15	180
	Aged(a)	635	92	310	45	8	200
HI	As-cast	550	80	310	45	12	180
	Aged(a)	620	90	450	65	6	200
HK	As-cast	515	75	345	50	17	170
	Aged(b)	585	85	345	50	10	190
HL	As-cast	565	82	360	52	19	192
HN	As-cast	470	68	260	38	13	160
HP	As-cast	490	71	275	40	11	170
HT	As-cast	485	70	275	40	10	180
	Aged(b)	515	75	310	45	5	200
HU	As-cast	485	70	275	40	9	170
	Aged(c)	505	73	295	43	5	190
HW	As-cast	470	68	250	36	4	185
	Aged(d)	580	84	360	52	4	205
HX	As-cast	450	65	250	36	9	176
	Aged(c)	505	73	305	44	9	185

(a) Aging treatment: 24 h at 760 °C (1400 °F), furnace cool. (b) Aging treatment: 24 h at 760 °C (1400 °F), air cool. (c) Aging treatment: 48 h at 980 °C (1800 °F), air cool. (d) Aging treatment: 48 h at 980 °C (1800 °F), furnace cool

Iron-nickel-chromium alloys contain more than 25% Ni and more than 10% Cr (always more nickel than chromium). These austenitic alloys are used for withstanding reducing as well as oxidizing atmospheres, except where sulfur content is appreciable. (In atmospheres containing 0.05% or more hydrogen sulfide, for example, iron-chromium-nickel alloys are recommended.) In contrast with iron-chromium-nickel alloys, iron-nickel-chromium alloys do not carburize rapidly or become brittle and do not take up nitrogen in nitriding atmospheres. These characteristics become enhanced as nickel content is increased, and in carburizing and nitriding atmospheres casting life increases with nickel content. Austenitic iron-nickel-chromium alloys are used extensively under conditions of severe temperature fluctuations such as those encountered by fixtures used in quenching and by parts that are not heated uniformly or that are heated and cooled intermittently. In addition, these alloys have characteristics that make them suitable for electrical resistance heating elements.

Metallurgical Structures

The structures of chromium-nickel and nickel-chromium cast steels must be wholly austenitic, or mostly austenitic with some ferrite, if these alloys are to be used for heat-resistant service. Depending on the chromium and nickel content (see the section "Composition and Microstructure" in this article), the structures of these iron-base alloys can be austenitic (stable), ferritic (stable, but also soft, weak, and ductile) or martensitic (unstable); therefore, chromium and nickel levels should be selected to achieve good strength at elevated temperatures combined with resistance to carburization and hot-gas corrosion.

A fine dispersion of carbides or intermetallic compounds in an austenitic matrix increases high-temperature strength considerably. For this reason, heat-resistant cast steels are higher in carbon content than are corrosion-resistant alloys of comparable chromium and nickel content. By holding at temperatures where carbon diffusion is rapid (such as above 1200 °C) and then rapidly cooling, a high and uniform carbon content is established, and up to about 0.20% C is retained in the austenite. Some chromium carbides are present in the structures of alloys with carbon contents greater than 0.20%, regardless of solution treatment, as described in the section "Sensitization and Solution Annealing of Austenitic and Duplex Alloys" in this article.

Castings develop considerable segregation as they freeze. In standard grades, either in the as-cast condition or after rapid cooling from a temperature near the melting point, much of the carbon is in supersaturated solid solution. Subsequent reheating precipitates excess carbides. The lower the reheating temperature, the slower the reaction and the finer the precipitated carbides. Fine carbides increase creep strength and decrease ductility. Intermetallic compounds such as Ni_3Al, if present, have a similar effect.

Reheating material containing precipitated carbides in the range between 980 and 1200 °C (1800 and 2200 °F) will agglomerate and spheroidize the carbides, which reduces creep strength and increases ductility. Above 1100 °C (2000 °F), so many of the fine carbides are dissolved or spheroidized that this strengthening mechanism loses its importance. For service above 1100 °C (2000 °F), certain proprietary alloys of the iron-nickel-chromium type have been de-

Table 11 Representative short-term tensile properties of cast heat-resistant alloys at elevated temperatures

	Property at indicated temperature													
	760 °C (1400 °F)					870 °C (1600 °F)					980 °C (1800 °F)				
	Ultimate tensile strength		Yield strength at 0.2% offset		Elongation, %	Ultimate tensile strength		Yield strength at 0.2% offset		Elongation, %	Ultimate tensile strength		Yield strength at 0.2% offset		Elongation, %
Alloy	MPa	ksi	MPa	ksi		MPa	ksi	MPa	ksi		MPa	ksi	MPa	ksi	
HA	462(a)	67(a)	220(b)	32(b)	...	...	...	...	...	...	...	...	...	...	...
HD	248	36	...	...	14	159	23	...	...	18	103	15	...	...	40
HF	262	38	172	25	16	145	21	107	15.5	16	...	...	...	...	...
HH (type I)(c)	228	33	117	17	18	127	18.5	93	13.5	30	62	9	43	6.3	45
HH (type II)(c)	258	37.4	136	19.8	16	148	21.5	110	16	18	75	10.9	50	7.3	31
HI	262	38	...	...	6	179	26	...	...	12	...	...	...	...	...
HK	258	37.5	168	24.4	12	161	23	101	15	16	85.5	12.4	60	8.7	42
HL	345	50	...	...	...	210	30.5	...	...	...	129	18.7	...	...	...
HN	...	...	...	...	...	140	20	100	14.5	37	83	12	66	9.6	51
HP	296	43	200	29	15	179	26	121	17.5	27	100	14.5	76	11	46
HT	240	35	180	26	10	130	19	103	15	24	76	11	55	8	28
HU	275	40	...	...	...	135	19.5	...	...	20	69	10	43	6.2	28
HW	220	32	158	23	...	131	19	103	15	...	69	10	55	8	40
HX	310(d)	45(d)	138(d)	20(d)	8(d)	141	20.5	121	17.5	48	74	10.7	47	6.9	40

(a) In this instance, test temperature was 540 °C (1000 °F). (b) Test temperature was 590 °C (1100 °F). (c) Type I and II per ASTM A 447. (d) Test temperature was 650 °C (1200 °F).

veloped. Alloys for this service contain tungsten to form tungsten carbides, which are more stable than chromium carbides at these temperatures.

Aging at a low temperature, such as 760 °C (1400 °F), where a fine, uniformly dispersed carbide precipitate will form, confers a high level of strength that is retained at temperatures up to those at which agglomeration changes the character of the carbide dispersion (overaging temperatures). Solution heat treatment or quench annealing, followed by aging, is the treatment generally employed to attain maximum creep strength.

Ductility is usually reduced when strengthening occurs; but in some alloys the strengthening treatment corrects an unfavorable grain-boundary network of brittle carbides, and both properties benefit. However, such treatment is costly and may warp castings excessively. Hence, this treatment is applied to heat-resistant castings only for the small percentage of applications for which the need for premium performance justifies the high cost.

Carbide networks at grain boundaries are generally undesirable in iron-base heat-resistant alloys. Grain-boundary networks usually occur in very-high-carbon alloys or in alloys that have cooled slowly through the high-temperature ranges in which excess carbon in the austenite is rejected as grain-boundary networks rather than as dispersed particles. These networks confer brittleness in proportion to their continuity.

Carbide networks also provide paths for selective attack in some atmospheres and in certain molten salts. Therefore, it is advisable in some salt bath applications to sacrifice the high-temperature strength imparted by high carbon content and gain resistance to intergranular corrosion by specifying that carbon content be no greater than 0.08%.

Straight Chromium Heat-Resistant Castings

Iron-chromium alloys, also known as straight chromium alloys, contain either 9 or 28% Cr. HC and HD alloys are included among the straight chromium alloys, although they contain low levels of nickel.

HA alloy (9Cr-1Mo), a heat treatable material, contains enough chromium to provide good resistance to oxidation at temperatures up to about 650 °C (1200 °F). The 1% molybdenum is present to provide increased strength. HA alloy castings are widely used in oil refinery service. A higher-chromium modification of this alloy (12 to 14% Cr) is widely used in the glass industry.

HA alloy has a structure that is essentially ferritic; carbides are present in pearlitic areas or as agglomerated particles, depending on prior heat treatment. Hardening of the alloy occurs upon cooling in air from temperatures above 815 °C (1500 °F). In the normalized and tempered condition, the alloy exhibits satisfactory toughness throughout its useful temperature range.

HC alloy (28% Cr) resists oxidation and the effects of high-sulfur flue gases at temperatures up to 1100 °C (2000 °F). It is used for applications in which strength is not a consideration, or in which only moderate loads are involved, at temperatures of about 650 °C (1200 °F). It is also used where appreciable nickel cannot be tolerated, as in very-high-sulfur atmospheres, or where nickel may act as an undesirable catalyst and destroy hydrocarbons by causing them to crack.

HC alloy is ferritic at all temperatures. Its ductility and impact strength are very low at room temperature and its creep strength is very low at elevated temperatures unless some nickel is present. In a variation of HC alloy that contains more than 2% Ni, substantial improvement in all three of these properties is obtained by increasing the nitrogen content to 0.15% or more.

HC alloy becomes embrittled when heated for prolonged periods at temperatures between 400 and 550 °C (750 and 1025 °F), and it shows low resistance to impact. The alloy is magnetic and has a low coefficient of thermal expansion, comparable to that of carbon steel. It has about eight times the electrical resistivity and about half the thermal conductivity of carbon steel. Its thermal conductivity, however, is roughly double the value for austenitic iron-chromium-nickel alloys.

HD alloy (28Cr-5Ni) is very similar in general properties to HC, except that its nickel content gives it somewhat greater strength at high temperatures. The high chromium content of this alloy makes it suitable for use in high-sulfur atmospheres.

HD alloy has a two-phase, ferrite-plus-austenite structure that is not hardenable by conventional heat treatment. Long exposure at 700 to 900 °C (1300 to 1650 °F), however, may result in considerable hardening and severe loss of room-temperature ductility through the formation of σ phase. Ductility may be restored by heating uniformly to 980 °C (1800 °F) or higher and then cooling rapidly to below 650 °C (1200 °F).

Iron-Chromium-Nickel Heat-Resistant Castings

Heat-resistant ferrous alloys in which the chromium content exceeds the nickel content are made in compositions ranging from 20Cr-10Ni to 30Cr-20Ni.

HE alloy (28Cr-10Ni) has excellent resistance to corrosion at elevated temperatures. Because of its higher chromium content, it can be used at higher temperatures than HF alloy and is suitable for applications up to 1100 °C (2000 °F). This alloy is stronger and more ductile at room temperature than the straight chromium alloys.

In the as-cast condition, HE alloy has a two-phase, austenite-plus-ferrite structure containing carbides. HE castings cannot be hardened by heat treatment; however, as with HD castings, long exposure to temperatures near 815 °C (1500 °F) will promote formation of σ phase and consequent embrittlement of the alloy at room temperature. The ductility of this alloy can be improved somewhat by quenching from about 1100 °C (2000 °F).

Castings of HE alloy have good machining and welding properties. Thermal expansion is about 50% greater than that of either carbon steel or the Fe-Cr alloy HC. Thermal conductivity is much lower than for HD or HC, but electrical resistivity is about the same. HE alloy is weakly magnetic.

HF alloy (20Cr-10Ni) is the cast version of 18-8 stainless steel, which is widely used for its outstanding resistance to corrosion. HF alloy is suitable for use at temperatures up to 870 °C (1600 °F). When this alloy is used for resistance to oxidation at elevated temperatures, it is not necessary to keep the carbon content at the low level specified for corrosion-resistant castings. Molybdenum, tungsten, niobium, and titanium are sometimes added to the basic HF composition to improve elevated-temperature strength.

In the as-cast condition, HF alloy has an austenitic matrix that contains interdendritic eutectic carbides and, occasionally, a lamellar constituent presumed to consist of alternating platelets of austenite and carbide or carbonitride. Exposure at service temperatures usually promotes precipitation of finely dispersed carbides, which increases room-temperature strength and causes some loss of ductility. If improperly balanced, as-cast HF may be partly ferritic. HF is susceptible to embrittlement due to σ-phase formation after long exposure at 760 to 815 °C (1400 to 1500 °F).

HH Alloy (26Cr-12Ni). Alloys of this nominal composition comprise about one-third of the total production of iron-base heat-resistant castings. Alloy HH is basically austenitic and holds considerable carbon in solid solution, but carbides, ferrite (soft, ductile, and magnetic) and σ (hard, brittle, and nonmagnetic) may also be present in the microstructure. The amounts of the various structural constituents present depend on composition and thermal history. In fact, two distinct grades of material can be obtained within the stated chemical compositional range of the cast alloy HH. These grades are defined as type I (partially ferritic) and type II (wholly austenitic) in ASTM A 447.

The partially ferritic (type I) alloy HH is adapted to operating conditions that are subject to changes in temperature level and applied stress. A plastic extension in the weaker, ductile ferrite under changing load tends to occur more readily than in the stronger austenitic phase, thereby reducing

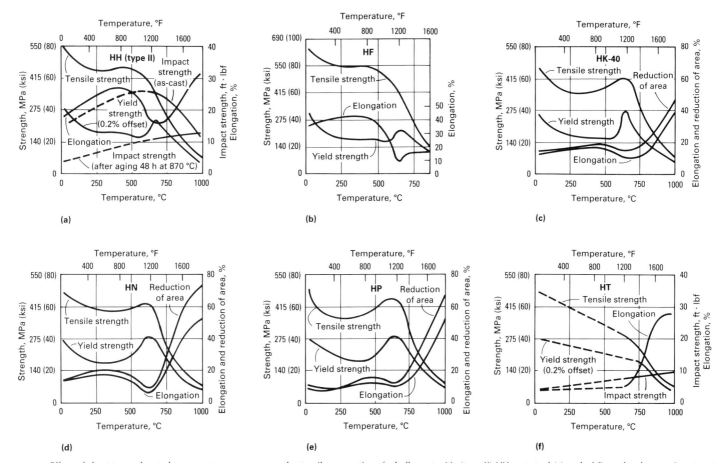

Fig. 9 Effect of short-term elevated-temperature exposure on the tensile properties of wholly austenitic (type II) HH cast steel (a) and of five other heat-resistant cast steels: (b) HF cast steel, (c) HK-40 cast steel, (d) HN cast steel, (e) HP cast steel, and (f) HT cast steel. Long-term elevated-temperature exposure reduces the strengthening effects between 500 to 750 °C (900 to 1400 °F) in (c), (d), and (e). Tensile properties of alloy HT in (f) include extrapolated data (dotted lines) below 750 °C but should be similar to alloy HN in terms of yield and tensile strengths. Source: Ref 9

unit stresses and stress concentrations and permitting rapid adjustment to suddenly applied overloads without cracking. Near 870 °C (1600 °F), the partially ferritic alloys tend to embrittle from the development of σ phase, while close to 760 °C (1400 °F), carbide precipitation may cause comparable loss of ductility. Such possible embrittlement suggests that 930 to 1090 °C (1700 to 2000 °F) is the best service temperature range, but this is not critical for steady temperature conditions in the absence of unusual thermal or mechanical stresses.

To achieve maximum strength at elevated temperatures, the HH alloy must be wholly austenitic. Where load and temperature conditions are comparatively constant, the wholly austenitic (type II) alloy HH provides the highest creep strength and permits the use of maximum design stress. The stable austenitic alloy is also favored for cyclic temperature service that might induce σ-phase formation in the partially ferritic type. When HH alloy is heated to between 650 and 870 °C (1200 and 1600 °F), a loss in ductility may be produced by either of two changes within the alloy: precipitation of carbides or transformation of ferrite to σ. When the composition is balanced so that the structure is wholly austenitic, only carbide precipitation normally occurs. In partly ferritic alloys, both carbides and σ phase may form.

The wholly austenitic (type II) HH alloy is used extensively in high-temperature applications because of its combination of relatively high strength and oxidation resistance at temperatures up to 1100 °C (2000 °F). Typical tensile properties and impact toughness of the type II HH alloy at elevated temperatures are shown in Fig. 9(a). The HH alloy (type I or II) is seldom used for carburizing applications because of embrittlement from carbon absorption. High silicon content (over 1.5%) will fortify the alloy against carburization under mild conditions, but will promote ferrite formation and possible σ embrittlement.

For the wholly austenitic (type II) HH alloy, composition balance is critical in achieving the desired austenitic microstructure (see "Composition and Microstructure" in this article). An imbalance of higher levels of ferrite-promoting elements compared to levels of austenite-promoting elements may result in substantial amounts of ferrite which improves ductility, but decreases strength at high temperatures. If a balance is maintained between ferrite-promoting elements (such as chromium and silicon) and austenite-promoting elements (such as nickel, carbon, and nitrogen), the desired austenitic structure can be obtained. In commercial HH alloy castings, with the usual carbon, nitrogen, manganese, and silicon contents, the ratio of chromium to nickel necessary for a stable austenitic structure is expressed by the inequality:

$$\frac{\%Cr - 16(\%C)}{\%Ni} < 1.7 \qquad (Eq\ 3)$$

Silicon and molybdenum have definite effects on the formation of σ phase. A silicon content in excess of 1% is equivalent to a chromium content three times as great, and any molybdenum content is equivalent to a chromium content four times as great.

Before HH alloy is selected as a material for heat-resistant castings, it is advisable to consider the relationship between chemical composition and operating-temperature range. For castings that are to be exposed continuously at temperatures appreciably above 870 °C (1600 °F), there is little danger of severe embrittlement from either the

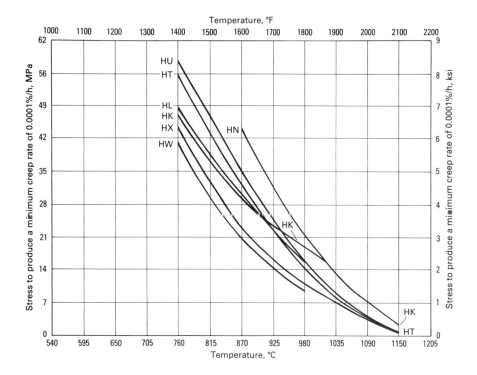

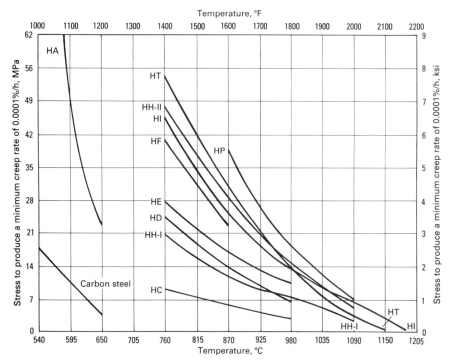

Fig. 10 Creep strength of heat-resistant alloy castings (HT curve is included in both graphs for ease of comparison). Source: Ref 10

Short-time tensile testing of fully austenitic HH alloys shows that tensile strength and elongation depend on carbon and nitrogen contents. For maximum creep strength, HH alloy should be fully austenitic in structure (Fig. 10). In design of load-carrying castings, data concerning creep stresses should be used with an understanding of the limitations of such data. An extrapolated limiting creep stress for 1% elongation in 10 000 h cannot necessarily be sustained for that length of time without structural damage. Stress-rupture testing is a valuable adjunct to creep testing and a useful aid in selecting section sizes to obtain appropriate levels of design stress.

Because HH alloys of wholly austenitic structure have greater strength at high temperatures than partly ferritic alloys of similar composition, measurement of ferrite content is recommended. Although a ratio calculated from Eq 3 that is less than 1.7 indicates wholly austenitic material, ratios greater than 1.7 do not constitute quantitative indications of ferrite content. It is possible, however, to measure ferrite content by magnetic analysis after quenching from about 1100 °C (2000 °F). The magnetic permeability of HH alloys increases with ferrite content. This measurement of magnetic permeability, preferably after holding 24 h at 1100 °C (2000 °F) and then quenching in water, can be related to creep strength, which also depends on structure.

HH alloys are often evaluated by measuring percentage elongation in room-temperature tension testing of specimens that have been held 24 h at 760 °C (1400 °F). Such a test may be misleading because there is a natural tendency for engineers to favor compositions that exhibit the greatest elongation after this particular heat treatment. High ductility values are often measured for alloys that have low creep resistance, but, conversely, low ductility values do not necessarily connote high creep resistance.

HI alloy (28Cr-15Ni) is similar to HH but contains more nickel and chromium. The higher chromium content makes HI more resistant to oxidation than HH, and the additional nickel serves to maintain good strength at high temperatures. Exhibiting adequate strength, ductility, and corrosion resistance, this alloy has been used extensively for retorts operating with an internal vacuum at a continuous temperature of 1175 °C (2150 °F). It has an essentially austenitic structure that contains carbides and that, depending on the exact composition balance, may or may not contain small amounts of ferrite. Service at 760 to 870 °C (1400 to 1600 °F) results in precipitation of finely dispersed carbides, which increases strength and decreases ductility at room temperature. At service temperatures above 1100 °C (2000 °F), however, carbides remain in solution, and room-temperature ductility is not impaired.

precipitation of carbide or the formation of σ phase, and composition should be 0.50% C (max) (0.35 to 0.40% preferred), 10 to 12% Ni, and 24 to 27% Cr. On the other hand, castings to be used at temperatures from 650 to 870 °C (1200 to 1600 °F) should have compositions of 0.40% C (max), 11 to 14% Ni, and 23 to 27% Cr. For applications involving either of these temperature ranges, that is, 650 to 870 °C (1200 to 1600 °F), or appreciably above 870 °C (1600 °F), composition should be balanced to provide an austenitic structure. For service from 650 to 870 °C (1200 to 1600 °F), for example, a combination of 11% Ni and 27% Cr is likely to produce σ phase and its associated embrittlement, which occurs most rapidly around 870 °C (1600 °F). It is preferable, therefore, to avoid using the maximum chromium content with the minimum nickel content.

Table 12 Approximate rates of corrosion for ACI heat-resistant casting alloys in air and in flue gas

Alloy	Oxidation rate in air, mm/yr			Corrosion rate, mm/yr, at 980 °C (1800 °F) in flue gas with sulfur content of:			
	870 °C (1600 °F)	980 °C (1800 °F)	1090 °C (2000 °F)	0.12 g/m³		2.3 g/m³	
				Oxidizing	Reducing	Oxidizing	Reducing
HB	0.63–	6.25–	12.5–	2.5+	12.5	6.25–	12.5
HC	0.25	1.25	1.25	0.63–	0.63+	0.63	0.63–
HD	0.25–	1.25–	1.25–	0.63–	0.63–	0.63–	0.63–
HE	0.13–	0.63–	0.88–	0.63–	0.63–	0.63–	0.63–
HF	0.13–	1.25+	2.5	1.25+	2.5+	1.25+	6.25
HH	0.13–	0.63–	1.25	0.63–	0.63	0.63	0.63–
HI	0.13–	0.25+	0.88–	0.63–	0.63–	0.63–	0.63–
HK	0.25–	0.25–	0.88–	0.63–	0.63–	0.63–	0.63–
HL	0.25+	0.63–	0.88	0.63–	0.63–	0.63	0.63–
HN	0.13	0.25+	1.25–	0.63–	0.63–	0.63	0.63
HP	0.63–	0.63	1.25	0.63–	0.63–	0.63–	0.63–
HT	0.13–	0.25+	1.25	0.63	0.63–	0.63	2.5
HU	0.13–	0.25–	0.88–	0.63–	0.63–	0.63–	0.63
HW	0.13–	0.25–	0.88	0.63	0.63–	1.25–	6.25
HX	0.13–	0.25–	0.88–	0.63–	0.63–	0.63–	0.63–

Source: Ref 11

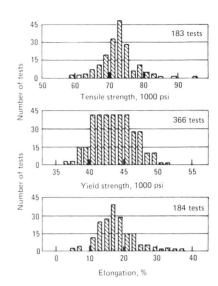

Fig. 11 Statistical spread in mechanical properties of HK alloy. Data are for 183 heats of HK alloy produced in a single foundry. Tests were performed at room temperature on as-cast material.

HK alloy (26Cr-20Ni) is somewhat similar to wholly austenitic HH alloy in general characteristics and mechanical properties. Although less resistant to oxidizing gases than HC, HE, or HI (Table 12), HK alloy contains enough chromium to ensure good resistance to corrosion by hot gases, including sulfur-bearing gases, under both oxidizing and reducing conditions. The high nickel content of this alloy helps make it one of the strongest heat-resistant casting alloys at temperatures above 1040 °C (1900 °F). Accordingly, HK alloy castings are widely used for stressed parts in structural applications at temperatures up to 1150 °C (2100 °F). As normally produced, HK is a stable austenitic alloy over its entire range of service temperatures. The as-cast microstructure consists of an austenitic matrix containing relatively large carbides in the form of either scattered islands or networks. After the alloy has been exposed to service temperature, fine, granular carbides precipitate within the grains of austenite and, if the temperature is high enough, undergo subsequent agglomeration. These fine, dispersed carbides contribute to creep strength. A lamellar constituent that resembles pearlite, but that is presumed to be carbide or carbonitride platelets in austenite, is also frequently observed in HK alloy.

Unbalanced compositions are possible within the standard composition range for HK alloy, and hence some ferrite may be present in the austenitic matrix. Ferrite will transform to brittle σ phase if the alloy is held for more than a short time at about 815 °C (1500 °F), with consequent embrittlement upon cooling to room temperature. Direct transformation of austenite to σ phase can occur in HK alloy in the range of 760 to 870 °C (1400 to 1600 °F), particularly at lower carbon levels (0.20 to 0.30%). The presence of σ phase can cause considerable scatter in property values at intermediate temperatures.

The minimum creep rate and average rupture life of HK are strongly influenced by variations in carbon content. Under the same conditions of temperature and load, alloys with higher carbon content have lower creep rates and longer lives than lower-carbon compositions. Room-temperature properties after aging at elevated temperatures are affected also: The higher the carbon, the lower the residual ductility. For these reasons, three grades of HK alloys with carbon ranges narrower than the standard HK alloy in Table 2 are recognized: HK-30, HK-40, and HK-50. In these designations, the number indicates the midpoint of a 0.10% C range. HK-40 (Table 2) is widely used for high-temperature processing equipment in the petroleum and petrochemical industries.

Figure 9(c) shows the effect of short-term temperature exposure on an HK-40 alloy. Figure 11 indicates the statistical spread in room-temperature mechanical properties obtained for an HK alloy. These data were obtained in a single foundry and are based on 183 heats of the same alloy.

HL alloy (30Cr-20Ni) is similar to HK; its higher chromium content gives it greater resistance to corrosion by hot gases, particularly those containing appreciable amounts of sulfur. Because essentially equivalent high-temperature strength can be obtained with either HK or HL, the superior corrosion resistance of HL makes it especially useful for service in which excessive scaling must be avoided. The as-cast and aged microstructures of HL alloy, as well as its physical properties and fabricating characteristics, are similar to those of HK.

Iron-Nickel-Chromium Heat-Resistant Castings

Iron-nickel-chromium alloys generally have more stable structures than those of iron-base alloys in which chromium is the predominant alloying element. There is no evidence of an embrittling phase change in iron-nickel-chromium alloys that would impair their ability to withstand prolonged service at elevated temperature. Experimental data indicate that composition limits are not critical; therefore, the production of castings from these alloys does not require the close composition control necessary for making castings from iron-chromium-nickel alloys.

The following general observations should be considered in the selection of iron-nickel-chromium alloys:

- As nickel content is increased, the ability of the alloy to absorb carbon from a carburizing atmosphere decreases
- As nickel content is increased, tensile strength at elevated temperatures decreases somewhat, but resistance to thermal shock and thermal fatigue increases
- As chromium content is increased, resistance to oxidation and to corrosion in chemical environments increases
- As carbon content is increased, tensile strength at elevated temperatures increases
- As silicon content is increased, tensile strength at elevated temperatures decreases, but resistance to carburization increases somewhat

HN alloy (25Ni-20Cr) contains enough chromium for good high-temperature corrosion resistance. HN has mechanical properties somewhat similar to those of the much more widely used HT alloy, but has better ductility (see Fig. 9d and 9f for a comparison of HN and HT tensile properties above 750 °C, or 1400 °F). It is used for highly stressed components in the temperature range of 980 to 1100 °C (1800 to 2000 °F). In several specialized applications (notably, brazing fixtures), it has given satisfactory

926 / Specialty Steels and Heat-Resistant Alloys

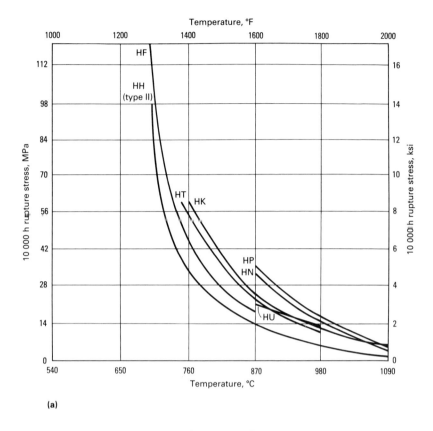

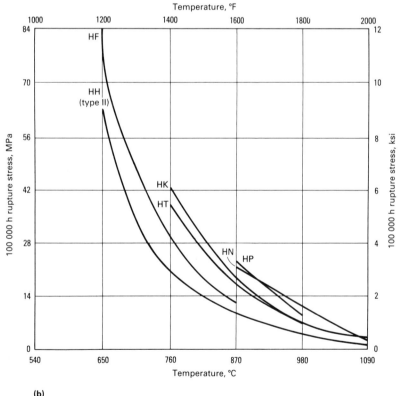

Fig. 12 Stress-rupture properties of several heat-resistant alloy castings. (a) 10 000 h rupture stress. (b) 100 000 h rupture stress. Source: Ref 10

service at temperatures from 1100 to 1150 °C (2000 to 2100 °F). HN alloy is austenitic at all temperatures: Its composition limits lie well within the stable austenite field. In the as-cast condition it contains carbide areas, and additional fine carbides precipitate with aging. HN is not susceptible to σ phase formation, and increases in its carbon content are not especially detrimental to ductility.

HP, HT, HU, HW, and HX alloys make up about one-third of the total production of heat-resistant alloy castings. When used for fixtures and trays for heat treating furnaces, which are subjected to rapid heating and cooling, these five high-nickel alloys have exhibited excellent service life. Because these compositions are not as readily carburized as iron-chromium-nickel alloys, they are used extensively for parts of carburizing furnaces. Because they form an adherent scale that does not flake off, castings of these alloys are also used in enameling applications in which loose scale would be detrimental.

Four of these high-nickel alloys (HT, HU, HW, and HX) also exhibit good corrosion resistance with molten salts and metal. They have excellent corrosion resistance to tempering and to cyaniding salts and fair resistance to neutral salts, with proper control. With molten metal, these alloys exhibit excellent resistance to molten lead, good resistance to molten tin to 345 °C (650 °F), and good resistance to molten cadmium to 410 °C (775 °F). The alloys have poor resistance to antimony, babbitt, soft solder, and similar metal. In many respects, there are no sharp lines of demarcation among the HP, HT, HU, HW, and HX alloys with respect to service applications.

HP alloy (35Ni-26Cr) is related to HN and HT alloys, but is higher in alloy content. It contains the same amount of chromium but more nickel than HK, and the same amount of nickel but more chromium than HT. This combination of elements makes HP resistant to both oxidizing and carburizing atmospheres at high temperatures. It has creep-rupture properties that are comparable to, or better than, those of HK-40 and HN alloys (Fig. 12).

HP alloy is austenitic at all temperatures, and is not susceptible to σ-phase formation. Its microstructure consists of massive primary carbides in an austenitic matrix; in addition, fine secondary carbides are precipitated within the austenite grains upon exposure to elevated temperatures. This precipitation of carbides is responsible for the strengthening between 500 and 750 °C (900 and 1400 °F) in Fig. 9(e). This strengthening, which is reduced after long-term exposure at high temperatures, also occurs for the cast stainless steels shown in Fig. 9(c) and (d).

HT alloy (35Ni-17Cr) contains nearly equal amounts of iron and alloying elements. Its high nickel content enables it to resist the thermal shock of rapid heating and cooling. In addition, HT is resistant to high-temperature oxidation and carburization and has good strength at the temperatures ordinarily used for heat treating steel. Except in high-sulfur gases, and provided that limiting creep-stress values are not exceeded, it performs satisfactorily in oxidizing atmospheres at temperatures up to 1150 °C (2100 °F) and in reducing atmospheres at temperatures up to 1100 °C (2000 °F).

HT alloy is widely used for highly stressed parts in general heat-resistant applications. It has an austenitic structure containing carbides in amounts that vary with carbon content and thermal history. In the as-cast condition, it has large carbide areas at interdendritic boundaries; but fine carbides precipitate within the grains after exposure to service temperature, causing a decrease in room-temperature ductility. Increases in carbon content may decrease the high-temperature ductility of the alloy. A silicon content above about 1.6% provides additional protection against carburization, but at some sacrifice in elevated-temperature strength. HT can be made still more resistant to thermal shock by the addition of up to 2% niobium.

HU alloy (39Ni-18Cr) is similar to HT, but its higher chromium and nickel contents give it greater resistance to corrosion by either oxidizing or reducing hot gases, including those that contain sulfur in amounts up to 2.3 g/m^3 (see Table 12). Its high-temperature strength and resistance to carburization are essentially the same as those of HT and thus its superior corrosion resistance makes it especially well suited for severe service involving high stress and/or rapid thermal cycling, in combination with an aggressive environment.

HW alloy (60Ni-12Cr) is especially well suited for applications in which wide and/or rapid fluctuations in temperature are encountered. In addition, HW exhibits excellent resistance to carburization and high-temperature oxidation. HW alloy has good strength at steel-treating temperatures, although it is not as strong as HT. HW performs satisfactorily at temperatures up to about 1120 °C (2050 °F) in strongly oxidizing atmospheres and up to 1040 °C (1900 °F) in oxidizing or reducing products of combustion, provided that sulfur is not present in the gas. The generally adherent nature of its oxide scale makes HW suitable for enameling furnace service, where even small flakes of dislodged scale could ruin the work in process.

HW alloy is widely used for intricate heat-treating fixtures that are quenched with the load and for many other applications (such as furnace retorts and muffles) that involve thermal shock, steep temperature gradients, and high stresses. Its structure is austenitic and contains carbides in amounts that vary with carbon content and thermal history. In the as-cast condition, the microstructure consists of a continuous interdendritic network of elongated eutectic carbides. Upon prolonged exposure at service temperatures, the austenitic matrix becomes uniformly peppered with small carbide particles except in the immediate vicinity of eutectic carbides. This change in structure is accompanied by an increase in room-temperature strength, but there is no change in ductility.

HX alloy (66Ni-17Cr) is similar to HW, but contains more nickel and chromium. Its higher chromium content gives it substantially better resistance to corrosion by hot gases (even sulfur-bearing gases), which permits it to be used in severe service applications at temperatures up to 1150 °C (2100 °F). However, it has been reported that HX alloy decarburizes rapidly at temperatures from 1100 to 1150 °C (2000 to 2100 °F). High-temperature strength (Table 11), resistance to thermal fatigue, and resistance to carburization are essentially the same as for HW; hence HX is suitable for the same general applications in which corrosion must be minimized. The as-cast and aged microstructures of HX, as well as its mechanical properties and fabricating characteristics, are similar to those of HW.

Properties of Heat-Resistant Alloys

Elevated-Temperature Tensile Properties. The short-term elevated-temperature test, in which a standard tension test bar is heated to a designated uniform temperature and then strained to fracture at a standardized rate, identifies the stress due to a short-term overload that will cause fracture in uniaxial loading. The manner in which the values of tensile strength and ductility change with increasing temperature is shown in Fig. 9 for selected alloys. Representative tensile properties at temperatures between 760 and 980 °C (1400 and 1800 °F) are given in Table 11 for several heat-resistant cast steel grades.

Creep and Stress-Rupture Properties. Creep is defined as the time-dependent strain that occurs under load at elevated temperature and is operative in most applications of heat-resistant high-alloy castings at the normal service temperatures. In time, creep may lead to excessive deformation and even fracture at stresses considerably below those determined in room-temperature and elevated-temperature short-term tension tests.

When the rate or degree of deformation is the limiting factor, the design stress is based on the minimum creep rate and design life after allowing for initial transient creep. The stress that produces a specified minimum creep rate of an alloy or a specified amount of creep deformation in a given time (for example, 1% total creep in 100 000 h) is referred to as the limiting creep strength, or limiting stress. Table 9 lists the creep strength of various H-type castings at specific temperatures. Figure 10 shows creep rates as a function of temperature.

Stress-rupture testing is a valuable adjunct to creep testing and is used to select the section sizes necessary to prevent creep rupture of a component. Figure 12 compares the creep-rupture strength of various H-type steel castings at 10 000 and 100 000 h. It should be recognized that long-term creep and stress-rupture values (for example, 100 000 h) are often extrapolated from shorter-term tests. Whether these property values are extrapolated or determined directly often has little bearing on the operating life of high-temperature parts. The actual material behavior is often difficult to predict accurately because of the complexity of the service stresses relative to the idealized, uniaxial loading conditions in the standardized tests and because of the attenuating factors such as cyclic loading, temperature fluctuations, and metal loss from corrosion. The designer should anticipate the synergistic effects of these variables.

Thermal fatigue failure involves cracking caused by heating and cooling cycles. Very little experimental thermal fatigue information is available on which to base a comparison of the various alloys, and no standard test as yet has been adopted. Field experience indicates that resistance to thermal

Table 13 Thermal conductivity and mean coefficient of linear thermal expansion of ACI heat-resistant cast steels at various temperatures

Alloy	Mean coefficient of linear thermal expansion for a temperature change				Thermal conductivity, W/m · K, at:		
	From 21 to 540 °C (700 to 1000 °F)		From 21 to 1090 °C (70 to 2000 °F)		100 °C (212 °F)	540 °C (1000 °F)	1090 °C (2000 °F)
	mm/mm/°C × 10^{-6}	in./in./°F × 10^{-6}	mm/mm/°C × 10^{-6}	in./in./°F × 10^{-6}			
HA	12.8	7.1	...	...	26.0	27.2	...
HC	11.3	6.3	13.9	7.7	21.8	31.0	41.9
HD	13.9	7.7	16.6	9.2	21.8	31.0	41.9
HE	17.3	9.6	20.0	11.1	14.7	21.5	31.5
HF	17.8	9.9	19.3	10.7	14.4	21.3	...
HH (type I)(a)	17.1	9.5	19.3	10.7	14.2	20.8	30.3
HH (type II)(a)	17.1	9.5	19.3	10.7	14.2	20.8	30.3
HI	17.8	9.9	19.4	10.8	14.2	20.8	30.3
HK	16.9	9.4	18.7	10.4	13.7	20.4	32.2
HL	16.6	9.2	18.2	10.1	14.2	21.1	33.4
HN	16.7	9.3	18.4	10.2	13.0	19.0	29.4
HP	16.6	9.2	19.1	10.6	13.0	19.0	29.4
HT	15.8	8.8	18.0	10.0	12.1	18.7	28.2
HU	15.8	8.8	17.5	9.7	12.1	18.7	28.2
HW	14.2	7.9	16.7	9.3	12.5	19.2	29.4
HX	14.0	7.8	17.1	9.5	12.5	19.2	29.4

(a) Type I and II specified per ASTM A 447. Source: Ref 10

fatigue is usually improved with an increase in nickel content. Niobium-modified alloys have been employed successfully when a high degree of thermal fatigue resistance is desired such as in reformer outlet headers.

Thermal Shock Resistance. Thermal shock failure may occur as a result of a single, rapid temperature change or as a result of rapid cyclic temperature changes, which induce stresses that are high enough to cause failure. Thermal shock resistance is influenced by the coefficient of thermal expansion and the thermal conductivity of materials. Increases in the thermal expansion coefficient or decreases in thermal conductivity reduce the resistance against thermal shock. Table 13 lists the thermal conductivities and expansion coefficients for heat-resistant castings at various temperatures. The HA, HC, and HD alloys, because of their predominately ferritic microstructure, have the lowest thermal expansion coefficients and the highest thermal conductivities.

Resistance to Hot-Gas Corrosion. The atmospheres most commonly encountered by heat-resistant cast steel are air, flue gases, and process gases; such gases may be either oxidizing or reducing and may be sulfidizing or carburizing if sulfur or carbon is present. The corrosion of heat-resistant alloys by the environment at elevated temperatures varies significantly with alloy type, temperature, velocity, and the nature of the specific environment to which the part is exposed. Table 14 presents a general ranking of the standard cast, heat-resistant grades in various environments at 980 °C (1800 °F). Corrosion rates at other temperatures are given in Table 12.

Manufacturing Characteristics

Foundry practices for cast high-alloy steels for corrosion resistance or heat resistance are essentially the same as those used for cast plain carbon steels. Details on melting practice, metal treatment, and foundry practices, including gating, risering, and cleaning of castings, are available in Volume 15 of the 9th Edition of *Metals Handbook*.

Iron-base alloys can be cast from heats melted in electric arc furnaces that have either acid or basic linings. When melting is done in acid-lined furnaces, however, chromium losses are high and silicon content is difficult to control, and thus acid-lined furnaces are seldom used. Alloys that contain appreciable amounts of aluminum, titanium, or other reactive metals are melted by induction or electron beam processes under vacuum or a protective atmosphere prior to casting.

Welding. As the alloy content of steel castings is increased to produce a fully austenitic structure, welding without cracking becomes more difficult. The fully austenitic low-carbon grades tend to form microfissures adjacent to the weld. This tendency toward microfissuring increases as nickel and silicon contents increase and carbon content decreases. Microfissuring is most evident in coarse-grain alloys with a carbon content of approximately 0.10 to 0.20% and a nickel content exceeding 13%. The microfissuring is reduced by an extremely low sulfur content. In welding these grades, low interpass temperatures, low heat inputs, and peening of the weld to relieve mechanical stresses are all effective. If strength is not a great factor, an initial weld deposit or "buttering of the weld" is also occasionally used.

Welding of corrosion-resistant steel castings can be done by shielded metal arc welding, gas tungsten arc welding, gas metal arc welding, and electroslag (submerged arc) welding. Austenitic castings are normally welded without preheat, and are solution annealed after welding. Martensitic castings require preheating to avoid cracking during welding and are given an appropriate postweld heat treatment. Specific conditions for welding specific alloys are listed in Table 15. When welds are properly made, tensile and yield strengths of the welded joint are similar to those of the unwelded castings (Table 8). Elongation is generally lower for specimens taken perpendicular to the weld bead.

Most of the corrosion-resistant cast steels, such as the CF-8 or CF-8M grade, are readily weldable, especially if their microstructures contain small percentages of δ-ferrite. Because stainless steels can become sensitized and lose their corrosion resistance if subjected to temperatures above 425 °C (800 °F), great care must be taken in welding to make certain that the casting or fabricated component is not heated excessively. For this reason, many stainless steels are almost never preheated. In many cases, the weld is cooled with a water spray between passes to reduce the interpass temperature to 150 °C (300 °F) or below.

Any welding performed on the corrosion-resistant grades will affect the corrosion resistance of the casting, but for many services the castings will perform satisfactorily in the as-welded condition. Where extremely corrosive conditions exist or where SCC may be a problem, complete reheat treatment may be required after welding. Heating the casting above 1065 °C (1950 °F) and then cooling it rapidly redissolves the carbides precipitated during the welding operation and restores corrosion resistance.

When maximum corrosion resistance is desired and postweld heat treatment (solution annealing) cannot be performed, alloying elements can be added to form stable carbides. Although niobium and titanium both form stable carbides, titanium is readily oxidized during the casting operation and therefore is seldom used. The niobium-stabilized grade CF-8C is the most commonly used cast grade. The stability of the niobium carbides prevents the formation of chromium carbides and the consequent chromium depletion of the base metal. This grade may therefore be welded without postweld heat treatment. Another approach to take when postweld heat treatment is undesirable or impossible is to keep the carbon content below 0.03%, as in the CF-3 and CF-3M grades. At this low carbon level, the depletion of the chromium due to carbide precipitation is so slight that the corrosion resistance of the grade is unaffected by the welding operation.

Galling

Stainless steel castings are susceptible to galling and seizing when dry surfaces slide or chafe against each other. However, the

Table 14 Corrosion resistance of heat-resistant cast steels at 980 °C (1800 °F) in 100 h tests in various atmospheres

Alloy	Air	Oxidizing flue gas(b)	Reducing flue gas(b)	Reducing flue gas(c)	Reducing flue gas (constant temperature)(d)	Reducing flue gas cooled to 150 °C (300 °F) every 12 h(d)
HA	U	U	U	U	U	U
HC	G	G	G	S	G	G
HD	G	G	G	S	G	G
HE	G	G	G	...	G	...
HF	S	G	S	U	S	S
HH	G	G	G	S	G	G
HI	G	G	G	S	G	G
HK	G	G	G	U	G	G
HL	G	G	G	S	G	G
HN	G	G	G	U	S	S
HP	G	G	G	G	G	G
HT	G	G	G	U	S	U
HU	G	G	G	U	S	U
HW	G	G	G	U	U	U
HX	G	G	G	S	G	U

(a) G, good (corrosion rate $r < 1.27$ mm/yr, or 50 mils/yr); S, satisfactory ($r < 2.54$ mm/yr, or 100 mils/yr); U, unsatisfactory ($r > 2.54$ mm/yr, or 100 mils/yr). (b) Contained 2 g of sulfur/m^3 (5 grains S/100 ft^3). (c) Contained 120 g S/m^3 (300 grains S/100 ft^3). (d) Contained 40 g S/m^3 (100 grains S/100 ft^3)

Table 15 Welding conditions for corrosion-resistant steel castings

ACI designation	Type of electrodes used(a)	Preheat °C	Preheat °F	Postweld heat treatment
CA-6NM	Same composition	100–150	212–300	590–620 °C (1100–1150 °F)
CA-15	410	200–315	400–600	610–760 °C (1125–1400 °F), air cool
CA-40	410 or 420	200–315	400–600	610–760 °C (1125–1400 °F), air cool
CB-7Cu	Same composition or 308	Not required		480–590 °C (900–1100 °F), air cool
CB-30	442	315–425	600–800	790 °C (1450 °F) min, air cool
CC-50	446	200–700	400–1300	900 °C (1650 °F), air cool
CD-4MCu	Same composition	Not required		Heat to 1120 °C (2050 °F), cool to 1040 °C (1900 °F), quench
CE-30	312	Not required		Quench from 1090–1120 °C (2000–2050 °F)
CF-3	308L	Not required		Usually unnecessary
CF-8	308	Not required		Quench from 1040–1120 °C (1900–2050 °F)
CF-8C	347	Not required		Usually unnecessary
CF-3M	316L	Not required		Usually unnecessary
CF-8M	316	Not required		Quench from 1070–1150 °C (1950–2100 °F)
CF-12M	316	Not required		Quench from 1070–1150 °C (1950–2100 °F)
CF-16F	308 or 308L	Not required		Quench from 1090–1150 °C (2000–2100 °F)
CF-20	308	Not required		Quench from 1090–1150 °C (2000–2100 °F)
CG-8M	317	Not required		Quench from 1040–1120 °C (1900–2050 °F)
CH-20	309	Not required		Quench from 1090–1150 °C (2000–2100 °F)
CK-20	310	Not required		Quench from 1090–1180 °C (2000–2150 °F)
CN-7M	320	200	400	Quench from 1120 °C (2050 °F)

Note: Metal arc, inert-gas arc, and electroslag welding methods can be used. Suggested electrical settings and electrode sizes for various section thicknesses are:

Section thickness, mm (in.)	Electrode diameter, mm (in.)	Current, A	Maximum arc voltage, V
3.2–6.4 (1/8–1/4)	2.4 (3/32)	45–70	24
3.2–6.4 (1/8–1/4)	3.2 (1/8)	70–105	25
3.2–6.4 (1/8–1/4)	4.0 (5/32)	100–140	25
6.4–13 (1/4–1/2)	4.8 (3/16)	130–180	26
≥13 (1/2)	6.4 (1/4)	210–290	27

(a) Lime-coated electrodes are recommended.

surfaces of the castings can be nitrided so that they are hard and wear resistant. Tensile properties are not impaired. Nitriding reduces resistance to corrosion by concentrated nitric or mixed acids.

Parts such as gate disks for gate valves are usually furnished in the solution-treated condition, but may be nitrided to reduce susceptibility to seizing in service. Similar results are obtained by hardfacing with cobalt-chromium-tungsten alloys.

Magnetic Properties

The magnetic properties of high-alloy castings depend on microstructure. The straight chromium types are ferritic and ferromagnetic. All other grades are mainly austenitic, with or without minor amounts of ferrite, and are either weakly magnetic or wholly nonmagnetic.

Cast nonmagnetic parts for applications in radar and in minesweepers require close control of ferrite content. Thicker sections have higher permeability than thinner sections. Therefore, to ensure low magnetic permeability in all areas of a casting, magnetic permeability checks should be made on the thicker sections.

REFERENCES

1. M. Prager, Cast High Alloy Metallurgy, in *Steel Casting Metallurgy*, J. Svoboda, Ed., Steel Founders' Society of America, 1984, p 221-245
2. C.E. Bates and L.T. Tillery, *Atlas of Cast Corrosion-Resistant Alloy Microstructures*, Steel Founders' Society of America, 1985
3. F. Beck, E.A. Schoefer, E. Flowers, and M. Fontana, New Cast High Strength Alloy Grades by Structure Control, in *Advances in the Technology of Stainless Steels and Related Alloys*, STP 369, American Society for Testing and Materials, 1965, p 159-174
4. T.M. Devine, Mechanism of Intergranular Corrosion and Pitting Corrosion of Austenitic and Duplex 308 Stainless Steel, *J. Electrochem. Soc.*, Vol 126 (No. 3), 1979, p 374
5. E.E. Stansbury, C.D. Lundin, and S.J. Pawel, Sensitization Behavior of Cast Stainless Steels Subjected to Simulated Weld Repair, in *Proceedings of the 38th SFSA Technical and Operating Conference*, Steel Founders' Society of America, 1983, p 223
6. S. Shimodaira et al., Mechanisms of Transgranular Stress Corrosion Cracking of Duplex and Ferrite Stainless Steels, in *Stress Corrosion Cracking and Hydrogen Embrittlement in Iron Base Alloys*, NACE Reference Book 5, National Association of Corrosion Engineers, 1977
7. P.L. Andersen and D.J. Duquette, The Effect of Cl- Concentration and Applied Potential on the SCC Behavior of Type 304 Stainless Steel in Deaerated High Temperature Water, *Corrosion*, Vol 36 (No. 2), 1980, p 85-93
8. S.B. Shendye, "Effect of Long Term Elevated Temperature Exposure on the Mechanical Properties and Weldability of Cast Duplex Steels," Master's thesis, Oregon Graduate Center, 1985
9. High Alloy Data Sheet, Heat Series, in *Steel Castings Handbook Supplement 9*, Steel Founders' Society of America
10. "Heat and Corrosion-Resistant Castings," The International Nickel Company, 1978
11. A. Brasunas, J.T. Glow, and O.E. Harder, Resistance of Fe-Ni-Cr Alloys to Corrosion in Air at 1600 to 2200 °F, in *Proceedings of the ASTM Symposium for Gas Turbines*, American Society for Testing and Materials, 1946, p 129-152

Elevated-Temperature Properties of Stainless Steels

STAINLESS STEELS are widely used at elevated temperatures when carbon and low-alloy steels do not provide adequate corrosion resistance and/or sufficient strength at these temperatures. Carbon and low-alloy steels are generally more economical than stainless steels and are often used in applications with temperatures below about 370 °C (700 °F). Several low-alloy steels with moderate chromium contents (between 1 and 10%) and improved high-temperature strength are also widely used at elevated temperatures above 370 °C (700 °F). These steels include the creep-resistant chromium-molybdenum ferritic steels discussed in the article "Elevated-Temperature Properties of Ferritic Steels" in this Volume. Carbon steels may even be suitable for temperatures above 370 °C (700 °F), if high strength and oxidation are not concerns.

This article deals with the wrought stainless steels used for high-temperature applications (see the article "Cast Stainless Steels" in this Volume for the elevated-temperature properties of cast stainless steels). Corrosion resistance is often the first criterion used to select stainless steel for a particular application. However, strength is also a significant factor in a majority of elevated-temperature applications and may even be the key factor governing the choice of a stainless steel. The stainless steels used in applications in which high-temperature strength is important are sometimes referred to as heat-resistant steels.

Table 1 gives some typical compositions of wrought heat-resistant stainless steels, which are grouped into ferritic, martensitic, austenitic, and precipitation-hardening grades. Of these steels, the austenitic grades offer the highest strength at high temperatures (Fig. 1). The precipitation-hardening steels have the highest strength at lower temperatures (Fig. 1), but they weaken considerably at temperatures above about 425 °C (800 °F).

Production of Steel

In recent years, the melting and refining of wrought heat-resistant alloys have become more and more sophisticated. Traditionally, special melting techniques have been applied to high-strength, high temperature resistant alloys, which often contain large amounts of reactive elements. However, even for low-alloy steels and other lower-strength materials, innovations in melting have steadily increased as melting and casting techniques have been optimized.

Low-alloy steels are generally melted in electric arc furnaces or in the basic oxygen furnace. Primary melting may be followed by a further refining procedure such as ladle treatments, vacuum arc remelting (VAR), or electroslag remelting (ESR). An ingot or billet product from the electric furnace is used as an electrode in these remelting operations. Vacuum degassing may also be used to remove gases, particularly hydrogen. These alloys may be cast directly into ingots or, in some cases, may be continuously cast. The VAR and ESR steels have fewer segregations and a finer grain size.

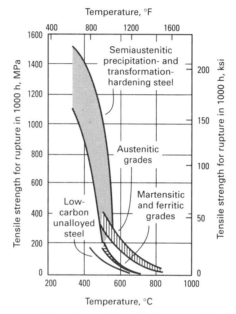

Fig. 1 General comparison of the hot-strength characteristics of austenitic, martensitic, and ferritic stainless steels with those of low-carbon unalloyed steel and semiaustenitic precipitation and transformation-hardening steels

Ferritic stainless steels are electric furnace melted; melting is followed by argon-oxygen deoxidation (AOD) for controlled oxidation of impurities such as sulfur and carbon (when used for reducing carbon, the process is often termed argon-oxygen decarburization). Because nitrogen is less expensive, it is often used to replace most of the argon in this process; nitrogen levels in stainless steels are higher today than 20 years ago because of its use in the AOD process. Basic open hearth melting is also used occasionally, with successful results. Vacuum induction melting and electron beam hearth refining have been used for melting some grades in order to control interstitial elements.

The austenitic 300 and 200 series stainless steels are usually produced by the AOD process. The old process for making stainless steel required the use of expensive low-carbon ferrochromium to produce these low-carbon alloys. This is unnecessary with the AOD process, and lower-cost high-carbon ferrochromium can be used. The ultimate product is not only less expensive, but is also of better quality. In the AOD process, the carbon monoxide partial pressure over the bath is lowered with argon and/or nitrogen, thus enhancing the removal of carbon by the carbon-oxygen reaction. One disadvantage is the high consumption of expensive argon, but nitrogen can be used to replace most of the argon.

The yield of forgeable stock per ingot is generally higher with vacuum melting because of increased purity. Yield may be increased 10% or more, depending on the alloy. Alloys used for forged gas turbine rotors are manufactured on a large scale by the consumable-electrode vacuum arc process and the electroslag process. In both methods, arc melting is accomplished directly in a water-cooled crucible. Mechanical properties, especially the transverse ductility of forgings made from these ingots, are higher than in forgings made from conventional ingots poured in iron molds.

Product Forms

Wrought stainless steel alloys are manufactured in all the forms common to the

Table 1 Nominal compositions of wrought iron-base heat-resistant alloys

Designation	UNS number	C	Cr	Ni	Mo	N	Nb	Ti	Other
Ferritic stainless steels									
405	S40500	0.15 max	13.0	...	...	...	...	...	0.2 Al
406	...	0.15 max	13.0	...	...	...	...	...	4.0 Al
409	S40900	0.08 max	11.0	0.5 max	...	...	...	6 × C min	...
429	S42900	0.12 max	15	...	...	...	...	...	...
430	S43000	0.12 max	16.0	...	...	...	...	...	...
434	S43400	0.12 max	17.0	...	1.0	...	...	...	...
439	S43035	0.07 max	18.25	...	...	...	...	12 × C min	1.10 Ti max
18 SR	...	0.05	18.0	...	...	...	...	0.40 max	2.0 Al max
18Cr-2Mo	S44400	...	18.5	...	2.0	...	(a)	(a)	0.8 (Ti + Nb) max
446	S44600	0.20 max	25.0	...	...	0.25	...	...	...
E-Brite 26-1	S44627	0.01 max	26.0	...	1.0	0.015 max	0.1	...	...
26-1Ti	S44626	0.04	26.0	...	1.0	...	...	10 × C min	...
29Cr-4Mo	S44700	0.01 max	29.0	...	4.0	0.02 max	...	...	...
Quenched and tempered martensitic stainless steels									
403	S40300	0.15 max	12.0	...	...	...	...	...	...
410	S41000	0.15 max	12.5	...	...	...	...	...	...
410Cb	S41040	0.15 max	12.5	...	...	...	0.12	...	...
416	S41600	0.15 max	13.0	...	0.6(b)	...	...	...	0.15 min S
422	S42200	0.20	12.5	0.75	1.0	...	...	...	1.0 W, 0.22 V
H-46	...	0.12	10.75	0.50	0.85	0.07	0.30	...	0.20 V
Moly Ascoloy	...	0.14	12.0	2.4	1.80	0.05	...	...	0.35 V
Greek Ascoloy	S41800	0.15	13.0	2.0	...	...	...	...	3.0 W
Jethete M-152	...	0.12	12.0	2.5	1.7	...	...	...	0.30 V
Almar 363	...	0.05	11.5	4.5	...	...	...	10 × C min	...
431	S43100	0.20 max	16.0	2.0	...	...	...	...	...
Lapelloy	S42300	0.30	11.5	...	2.75	...	...	...	0.25 V
Precipitation-hardening martensitic stainless steels									
Custom 450	...	0.05 max	15.5	6.0	0.75	...	8 × C min	...	1.5 Cu
Custom 455	...	0.03	11.75	8.5	...	...	0.30	1.2	2.25 Cu
15-5 PH	S15500	0.07	15.0	4.5	...	...	0.30	...	3.5 Cu
17-4PH	S17400	0.04	16.5	4.25	...	...	0.25	...	3.6 Cu
PH 13-8 Mo	S13800	0.05	12.5	8.0	2.25	...	...	...	1.1 Al
Precipitation-hardening semiaustenitic stainless steels									
AM-350	S35000	0.10	16.5	4.25	2.75	0.10	...	...	...
AM-355	S35500	0.13	15.5	4.25	2.75	0.10	...	...	...
17-7 PH	S17700	0.07	17.0	7.0	...	...	...	...	1.15 Al
PH 15-7 Mo	S15700	0.07	15.0	7.0	2.25	...	...	...	1.15 Al
Austenitic stainless steels									
304	S30400	0.08 max	19.0	10.0	...	...	...	...	...
304H	S30409	0.04–0.10	19.0	10.0	...	...	...	...	...
304L	S30403	0.03 max	19.0	10.0	...	...	...	...	...
304N	S30451	0.08 max	19.0	9.25	...	0.13	...	...	...
309	S30900	0.20 max	23.0	13.0	...	...	...	...	...
309H	S30909	0.04–0.10	23.0	13.0	...	...	...	...	...
310	S31000	0.25 max	25.0	20.0	...	...	...	...	...
310H	S31009	0.04–0.10	25.0	20.0	...	...	...	...	...
316	S31600	0.08 max	17.0	12.0	2.5	...	...	...	...
316L	S31603	0.03 max	17.0	12.0	2.5	...	...	...	...
316N	S31651	0.08 max	17.0	12.0	2.5	0.13	...	...	...
316H	S31609	0.04–0.10	17.0	12.0	2.5	...	...	...	...
316LN	S31653	0.035 max	17.0	12.0	2.5	0.13	...	...	...
317	S31700	0.08 max	19.0	13.0	3.5	...	...	...	...
317L	S31703	0.035 max	19.0	13.0	3.5	...	...	...	...
321	S32100	0.08 max	18.0	10.0	...	...	...	5 × C min, 0.70 max	...
321H	S32109	0.04–0.10	18.0	10.0	...	...	...	4 × C min, 0.60 max	...
347	S34700	0.08 max	18.0	11.0	...	...	10 × C min(c)	...	1.0 (Nb + Ta) max
347H	S34709	0.04–0.10	18.0	11.0	...	...	8 × C min(c)	...	1.0 (Nb + Ta) max
348	S34800	0.08 max	18.0	11.0	...	...	10 × C min(c)	...	0.10 Ta max, 1.0 (Nb + Ta) max
348H	S34809	0.04–0.10	18.0	11.0	...	...	8 × C min(c)	...	0.10 Ta max, 1.0 (Nb + Ta) max
19-9 DL	K63198	0.30	19.0	9.0	1.25	...	0.4	0.3	1.25 W
19-9 DX	K63199	0.30	19.2	9.0	1.5	...	...	0.55	1.2 W
17-14-CuMo	...	0.12	16.0	14.0	2.5	...	0.4	0.3	3.0 Cu
201	S20100	0.15 max	17	4.2	...	0.25 max	...	...	...
202	S20200	0.09	18.0	5.0	...	0.10	...	...	8.0 Mn
205	S20500	0.18	17.2	1.4	...	0.36	...	...	...
216	S21600	0.05	20.0	6.0	2.5	0.35	...	...	8.5 Mn
21-6-9	S21900	0.04 max	20.25	6.5	...	0.30	...	...	9.0 Mn
Nitronic 32	S24100	0.10	18.0	1.6	...	0.34	...	...	12.0 Mn
Nitronic 33	S24000	0.08 max	18.0	3.0	...	0.30	...	...	13.0 Mn
Nitronic 50	...	0.06 max	21.0	12.0	2.0	0.30	0.20	...	5.0 Mn
Nitronic 60	S21800	0.10 max	17.0	8.5	2.0	...	...	...	8.0 Mn, 0.20 V, 4.0 Si
Carpenter 18-18 Plus	S28200	0.10	18.0	<0.50	1.0	0.50	...	...	16.0 Mn, 0.40 Si, 1.0 Cu

(a) Ti + Nb = (0.20 + 4C + 4N) min. (b) Optional. (c) Minimum for Nb + Ta

metal industry. A partial list of American Society for Testing and Materials (ASTM) specifications for stainless steel products used at elevated temperatures includes:

- Heat-resisting stainless steel plate, sheet, and strip in ASTM A 240
- Heat-resisting stainless steel bars and shapes in ASTM A 479
- Heat-resisting stainless steel forgings in ASTM A 473
- Stainless steel tube in ASTM A 213, A 249, A 268, A 269, A 511, A 632, A 688, A 771, and A 791
- Stainless steel pipe in ASTM A 312, A 376, A 409, A 430, A 731, A 813, and A 814

These specifications include numerous types and variations of stainless steel compositions besides those listed in Table 1. Table 1 does not list silicon contents, which may vary according to product form. For example, 316 stainless steel forgings (ASTM A 473) and bar (ASTM A 479) have silicon contents of 1.0% max, while 316 stainless steel pipe (ASTM A 312) and pressure vessel plate (ASTM A 240) have silicon contents of 0.75% max.

The elevated-temperature properties of any of these materials are influenced to some extent by the form of the product; these properties depend largely on the specific alloy characteristics, such as oxidation resistance, type of oxide scale, thermal conductivity, and thermal expansion. Time-temperature exposure and the duration and type of loading are also significant factors, as are differences in properties among different product forms.

For alloys that form thin, tenacious scales at elevated temperature, the stress-rupture properties of bar and sheet of the same alloy will be about the same. On the other hand, for alloys that are less resistant to oxidation, rupture values are likely to be significantly lower for sheet than for the same alloy in bar form because of the greater ratio of surface area to volume, which causes greater interaction between the environment and the substrate metal. In the case of oxidation, a fixed depth of oxidation (such as 75 to 125 μm, or 3 to 5 mils) will more drastically affect properties in 1.3 mm (50 mil) sheet than in 6.5 mm (250 mil) bar stock.

The high-temperature strength of the heat-resistant alloys can be increased by such cold-working processes as rolling, swaging, or hammering. The increased strength is retained, however, only up to the recrystallization temperature. Figure 2, for example, shows the effect of temperature on the tensile properties of cold-worked 301 stainless steel. In particular, cold-worked products have poor resistance to creep, which generally occurs at temperatures slightly above the recrystallization temperature of the metal. During long-term high-temperature exposure, the benefit of cold

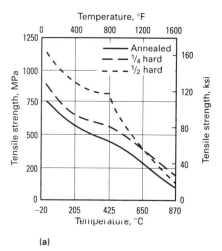

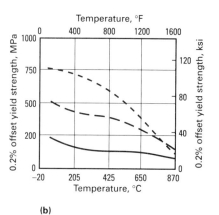

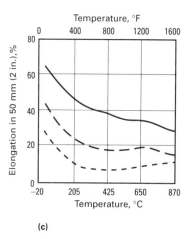

Fig. 2 Effect of short-term elevated temperature on tensile properties of cold-worked 301 stainless steel. (a) Tensile strength. (b) Yield strength. (c) Elongation

working is lost, and stress-rupture strength may even fall below annealed strength.

Mechanical Property Considerations

For service at elevated temperatures, the first property considered is the tensile strength during short-term exposure at ele-

Table 2 Data sources for the mechanical properties of various steels at elevated temperatures

Steel	Data source	Year
Carbon steel	ASTM DS 11S1	1970
	ASTM STP 503	1972
Carbon-molybdenum steel	ASTM DS 47	1971
½Cr-½Mo, 1Cr-½Mo, 1¼Cr-½Mo	ASTM DS 50	1973
2¼Cr-1Mo steel	ASTM DS 6S2	1971
3 to 9Cr-Mo steel	ASTM DS 58	1975
12 to 27 Cr steel	ASTM DS 59	1980
Types 304, 316, 321, and 347 stainless steels	ASTM DS 5S2	1969
	ORNL 5285	Oct 1977
9Cr-1Mo-V	Code Case 1943 of the ASME Boiler Code	...

vated temperatures (Fig. 1). For applications involving short-term exposure to temperatures below about 480 °C (900 °F), the short-time tensile properties are usually sufficient in the mechanical design of steel components. For temperatures above 480 °C (900 °F), the design process must include other properties such as creep rate, creep-rupture strength, creep-rupture ductility, and creep-fatigue interaction. Mechanical data of various steels at elevated temperatures are available in the ASTM data series (DS) listed in Table 2.

Various methods, depending on the application, are used to establish the design criteria for using materials at elevated temperatures. One method, for example, develops allowable stresses by multiplying tensile strengths, yield strengths, creep strength, and/or rupture strength with safety factors (Fig. 3). This method is illustrated in Fig. 3, where various safety factors are used to establish allowable stresses for 18-8 austenitic stainless steel at various temperatures. This method does not take into account environmental interactions, aging effects from long-term temperature exposure, or the potential of creep-fatigue interaction. Another design basis is to define maximum allowable temperatures. Table 3, for example, lists maximum allowable service temperatures of various materials for tube and plate products used in refinery or boiler applications.

Creep and Stress Rupture. Creep is defined as the time-dependent strain that occurs under load at elevated temperature. Creep is operative in most applications when metal temperatures exceed 480 °C (900 °F). In time, creep may lead to excessive deformation and even fracture at stresses considerably below those determined in room-temperature and elevated-temperature short-term tension tests. The designer must usually determine whether the serviceability of the component in question is limited by the rate or the degree of deformation.

Elevated-Temperature Properties of Stainless Steels / 933

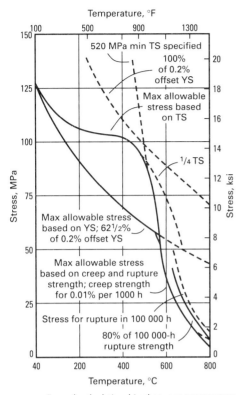

Fig. 3 Example of relationships between temperature and high-temperature strengths previously used in the ASME Boiler Code (1974) to establish maximum allowable stresses in tension for type 18-8 austenitic stainless steel. The current code uses two-thirds of the yield strength instead of 62.5% of the yield strength.

Table 3 Suggested maximum temperatures for continuous service based on creep or rupture data

Material	Maximum temperature based on creep rate °C	°F	Maximum temperature based on rupture °C	°F
Carbon steel	450	850	540	1000
C-0.5Mo steel	510	950	595	1100
2.25Cr-1Mo steel	540	1000	650	1200
Type 304 stainless steel	595	1100	815	1500
Alloy C-276 nickel-base alloy	650	1200	1040	1900

When the rate or degree of deformation is the limiting factor, the design stress is based on the minimum creep rate and design life after allowing for initial transient creep. The stress that produces a specified minimum creep rate of an alloy or a specified amount of creep deformation in a given time (for example, 1% total creep in 100 000 h) is referred to as the limiting creep strength or limiting stress. Of the various types of stainless steels, the austenitic types provide the highest limiting creep strength. Figure 4 plots typical creep rates of various austenitic stainless steels. The original data for Fig. 4 were generated on steels with carbon contents greater than 0.04%; the steels were solution annealed at sufficiently high temperatures to meet H grade requirements (see the section "H Grades" in this article). Today, the 300-series stainless steels are usually low carbon, unless an H grade is specified.

When fracture is the limiting factor, stress-to-rupture values can be used in design. Typical stress-to-rupture values of various austenitic stainless steels are shown in Fig. 5. The values were generated from steels meeting H grade requirements.

It should be recognized that long-term creep and stress-rupture values (for example, 100 000 h) are often extrapolated from shorter-term tests, which are conducted at high stresses and in which creep is dislocation controlled. Whether these property values are extrapolated or determined directly often has little bearing on the operating life of high-temperature parts, where operating stresses are lower and where the mechanisms of creep are diffusion controlled. The actual material behavior can also be difficult to predict accurately because of the complexity of the service stresses relative to the idealized, uniaxial loading conditions in the standardized tests and because of attenuating factors such as cyclic loading, temperature fluctuations, and metal loss from corrosion.

Rupture Ductility. While creep strength and rupture strength are given considerable attention as design and failure parameters, rupture ductility is an important mechanical property when stress concentrations and localized defects such as notches are a factor in design. Rupture ductility, which varies inversely with creep and rupture strength, influences the growth of cracks or defects and thus affects notch sensitivity. This general effect of rupture ductility on rupture strength is shown conceptually in Fig. 6. When smooth parts are tested, the rupture strength is higher for the steel with lower ductility (steel A). However, when a notch is introduced, the rupture strength of steel A plummets; the rupture strength of

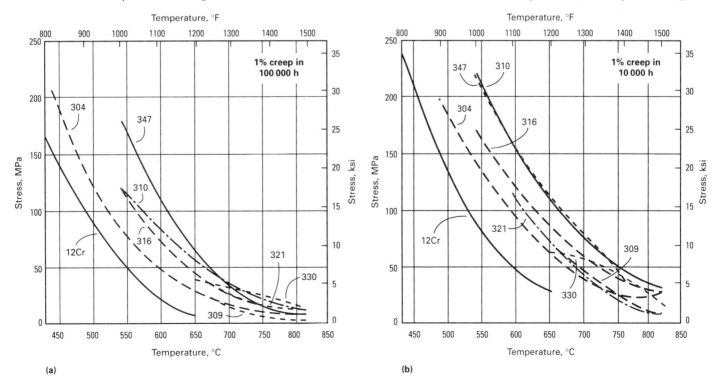

Fig. 4 Creep rate curves for several annealed H-grade austenitic stainless steels. (a) 1% creep in 100 000 h. (b) 1% creep in 10 000 h. Source: Ref 1

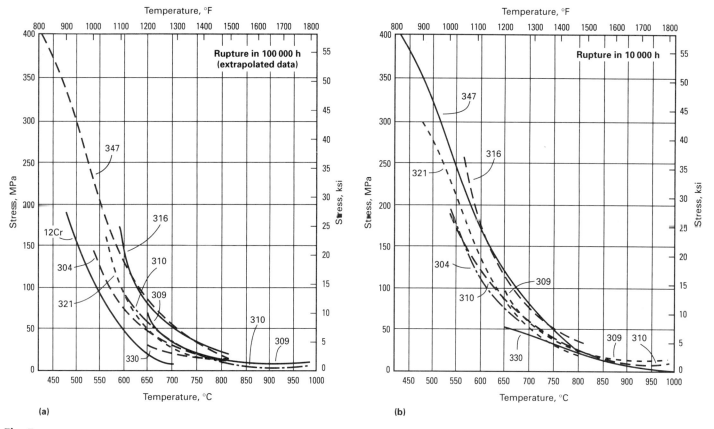

Fig. 5 Stress-rupture curves for several annealed H-grade austenitic stainless steels. (a) Extrapolated data for rupture in 100 000 h. (b) Rupture in 10 000 h. Source: Ref

steel B, on the other hand, is less notch sensitive because of its higher ductility. It is clear from Fig. 6 that for low-stress, long-term applications, steel B would be preferred to steel A. This is true even though steel B is weaker than steel A, as shown by results of smooth-bar rupture tests.

In many service conditions, the amount of deformation is not critical, and relatively high rupture ductility can be used in design. Under such conditions, with the combined uncertainties of actual stress, temperature, and strength, it may be important that failure not occur without warning and that the metal retain high elongation and reduction in area throughout its service life. In the oil and chemical industries, for instance, many applications of tubing under high pressure require high long-time ductility, and impending rupture will be evident from the bulging of the tubes.

Values of elongation and reduction in area obtained in rupture tests are used in judging the ability of metal to adjust to stress concentration. The requirements are not well defined and are controversial. Most engineers are reluctant to use alloys with elongations of less than 5%, and this limit is sometimes considerably higher. Low ductility in a rupture test almost always indicates high resistance to the relaxation of stress by creep, and possible sensitivity to stress concentrations. There is also ample evidence that rupture ductility has a major influence on creep-fatigue interaction (see the section below). Large changes in elongation with increasing fracture time usually indicate extensive changes in metallurgical structure or surface corrosion.

Creep-fatigue interaction is a phenomenon that can have a detrimental effect on the performance of metal parts or components operating at elevated temperature. When temperatures are high enough to produce creep strains and when cyclic (that is, fatigue) strains are present, the two can interact with one another. For example, it has been found that creep strains can seriously reduce fatigue life and that fatigue strains can seriously reduce creep life. This effect occurs in both stainless steels and low-alloy or carbon steels when temperatures are in the creep range (see the article "Elevated-Temperature Properties of Ferritic Steels" in this Volume).

Creep-fatigue interaction causes a reduction in fatigue life when either the frequency of the cycling stress is reduced or the cycling waveform has a tensile (and sometimes compressive) hold time (Fig. 7). Early studies (Ref 3-8) on AISI type 316 stainless steel showed that tensile hold periods in the temperature region from 550 to 625 °C (1020 to 1160 °F) were very damaging, as shown in Fig. 7. Because the strain ranges were fairly high and the hold periods were short, failures were dominated by fatigue. More recent results at lower strain ranges and longer hold periods have revealed that creep-dominated failures also occur in stainless steels (Ref 9-11). Creep-dominated failures have been observed for tensile hold times up to 16 h at 600 °C (1110 °F) (Ref 9) and in tests at 625 °C (1160 °F) with tensile hold times up to 48 h (Ref 10).

Some investigators have observed a saturation in the detrimental effects of tensile hold periods, that is, a recovery of the endurance occurring at longer hold periods. This has been attributed to microstructural changes leading to increases in ductility. Aging and the concomitant precipitation and growth of large carbides prior to testing

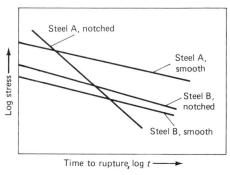

Fig. 6 Notched-bar rupture behavior for a creep-brittle steel (A) and a creep-ductile steel (B)

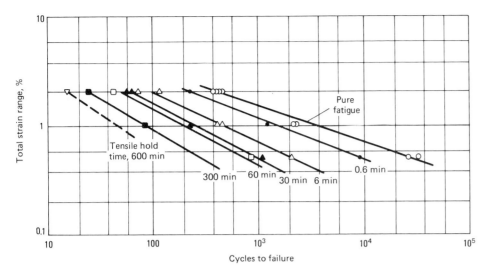

Fig. 7 Effect of tensile hold time on fatigue endurance of type 316 stainless steel. Source: Ref 2

have been shown to eliminate creep-fatigue effects altogether in type 316 stainless steel at 650 °C (1200 °F) (Ref 12). Compressive holds have been found to have an effect similar to that in chromium-molybdenum-vanadium steels; that is, they nullified the detrimental effects caused by tensile holds. The effect of slow-fast cycles, in which the strain increased slowly during the tension cycle but increased rapidly during the compression-going cycle, on the endurance of type 304, type 316, and other stainless steels has been investigated (Ref 13-16). In other tests, lower strain rates in the tension cycle were found to reduce endurance. Data on creep-fatigue interaction for various stainless steels are given in the article "Creep-Fatigue Interaction" in Volume 8 of the 9th Edition of *Metals Handbook*.

Life reduction due to tensile hold has been observed to be related to stress-rupture ductility, leading to heat-to-heat variations (Ref 11, 17). The influence of rupture ductility on the creep fracture component of creep-fatigue interaction is negligible in continuous-cycle and high-frequency or short hold time fatigue tests (where fracture is fatigue dominated). However, as frequency is decreased or as hold time is increased, the effect of rupture ductility becomes more pronounced, as illustrated in Fig. 8(b). The hold time effects on the fatigue lives of two rotors are compared in Fig. 8(b) in terms of their rupture-ductility behavior. The fatigue life of the low-ductility rotor steel is much more adversely affected by hold time than that of the high-ductility rotor steel. The rupture ductilities diverge with increasing time to rupture (Fig. 8a), which is correspondingly reflected in the long hold time tests. Endurance data for several ferritic steels and austenitic stainless steels in relation to the range of rupture ductility exhibited by them are presented in Ref 16. The lower the ductility, the lower the creep-fatigue endurance. In addition, long hold periods, small strain ranges, and low ductility favor creep-dominated failures, whereas short hold periods, intermediate strain ranges, and high creep ductility favor creep-fatigue interaction failures.

Corrosion Considerations

The corrosion and oxidation resistance of wrought stainless steels is similar to that of cast stainless steels with comparable compositions. However, austenitic cast stainless steels may have a more pronounced duplex (ferrite-in-austenite) structure than wrought austenitic stainless steels. This ferrite-in-austenitestainless steel is beneficial in preventing intergranular stress-corrosion cracking (see the article "Cast Stainless Steels" in this Volume for the general corrosion resistance of stainless steels at elevated temperatures).

Cast and wrought steels with similar composition, heat treatment, and microstructure (as noted above) exhibit about the same corrosion resistance in a given environment. The temperature limits in air for various stainless steels are given in the article "Wrought Stainless Steels" in this Volume. Because oxidation resistance is affected by many factors, including temperature, time, type of service (cyclic or continuous), and atmosphere, selection of material for a specific application should be based on tests that duplicate anticipated conditions as closely as possible. Figure 9 compares the oxidation resistance of type 430, type 446, and several martensitic and austenitic grades in 1000-h continuous exposure to water-saturated air at temperatures from 815 to 1095 °C (1500 to 2000 °F).

Sulfur attack is second only to air oxidation in frequency of occurrence in many

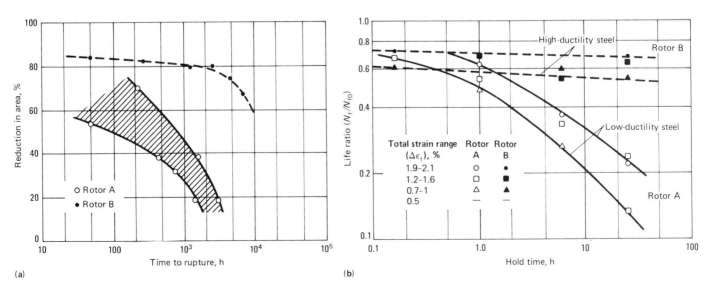

Fig. 8 Effects of creep-rupture ductility (a) on hold time effects (b) during low-cycle fatigue testing of a 1Cr-molybdenum-vanadium steel at 500 °C (930 °F). N_{f0} = fatigue life with zero hold time. Source: Ref 18

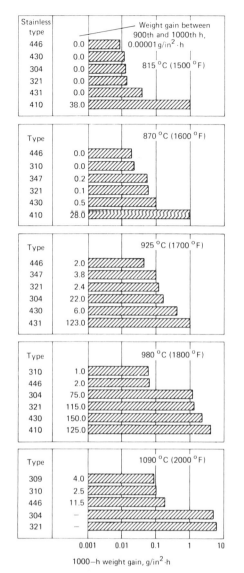

Fig. 9 The 1000-h oxidation resistance of selected stainless steels

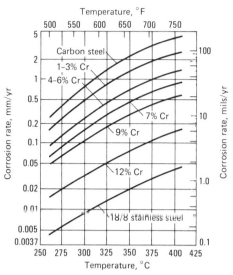

Fig. 10 Average high-temperature sulfur corrosion rates in a hydrogen-free environment compiled from an American Petroleum Institute survey. Source: Ref 19

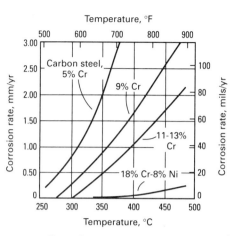

Fig. 11 Elevated-temperature corrosion rates of steels in hydrogen and hydrogen sulfide environments (H_2S concentrations above 1 mole %). Source: Ref 20

industries and is likely to be more severe. Figure 10 shows average corrosion rates at high temperatures from sulfur corrosion in a hydrogen-free environment. Figure 11 shows corrosion from hydrogen sulfide. Additional information on corrosion is given in the sections below and in the article "Wrought Stainless Steels" in this Volume.

Ferritic Stainless Steels

The main advantage of ferritic stainless steels for high-temperature use is their good oxidation resistance, which is comparable to that of austenitic grades. Ferritic stainless steels can also be more resistant to liquid metal attack than austenitic stainless steels. They are therefore used for some applications in the lead and copper metal industries.

In view of their lower alloy content and lower cost, ferritic steels should be used in preference to austenitic steels, stress conditions permitting. Ferritic stainless steels can also be more desirable than austenitic steels in applications with thermal cycling because ferritic steels have higher thermal conductivities and lower thermal expansion coefficients than austenitic steels. Ferritic steels may therefore allow reductions in thermal stresses and improved thermal-fatigue resistance.

For the purposes of this article, ferritic stainless steels are classified into two categories:

- Ferritic stainless steels without vanadium (Table 1)
- High-chromium ferritic steels with vanadium (Table 4)

These two categories are described in the next two sections. An important structural characteristic of all ferritic stainless steels is precipitation of alpha prime (α'), a chromium-rich ferrite, when the steel is exposed to temperatures in the range from 370 to 540 °C (700 to 1000 °F). This precipitation results in an increase in hardness and a drastic reduction in room-temperature toughness, which is known as 475 °C (885 °F) embrittlement (see the article "Embrittlement of Steels" in this Volume). This embrittlement occurs in all ferritic grades that have chromium contents above approximately 13%, and its severity increases at higher chromium levels. This characteristic has to be taken into consideration for applications involving exposure to temperatures in the range from 370 to 540 °C (700 to 1000 °F) because subsequent room-temperature ductility will be severely impaired. In the higher-chromium alloys such as 18Cr-2Mo, type 446, 26-1, and 29-4 (Table 1), σ phase is encountered at temperatures above 565 °C (1050 °F). The χ phase will also form in 26-1Ti and 29Cr-4Mo. The high-molybdenum steels such as 29Cr-4Mo will also form a χ phase, which has an embrittling effect similar to that of the σ phase. The χ phase is only formed in high-molybdenum steels. Titanium has little, if any, effect on the formation of χ phase.

Ferritic Stainless Steels Without Vanadium

Many stainless steels of the 400 series (Table 1) have essentially ferritic structures at all temperatures. Types 405, 430, 434, and 446 form a certain amount of austenite when heated to high temperatures. Type 409 may also form some austenite, particularly if the titanium content is relatively low, but the other steels listed are completely ferritic at all temperatures.

The amount of chromium added for corrosion and oxidation resistance varies from 11% in type 409 to 29% in 29Cr-4Mo. Titanium is used to tie up carbon and nitrogen for structure control and resistance to intergranular corrosion. Molybdenum is used to improve corrosion resistance, whereas aluminum and silicon are added for resistance to oxidation.

Tensile and yield strengths of ferritic stainless steels in Table 1 in the annealed condition are shown in Fig. 12. At room temperature, these properties are nearly equivalent to those of austenitic stainless steels. At higher temperatures, however, ferritic steels are much lower in strength. The rupture strength and creep strength of types 430 and 446 are compared with the same properties of various austenitic stainless steels and two nickel-base alloys (Inconel and Incoloy) in Fig. 13. Long-time and short-time high-temperature strengths of ferritic steels are relatively low compared to those of austenitic steels. Data on the cyclic oxidation resistance of ferritic stainless steels in air containing 10% water vapor are presented in Table 5. At 705 and 815 °C (1300 and 1500 °F), all the alloys listed are

Table 4 High-chromium heat-resistant ferritic steels

| Grade | \---- Nominal chemical composition, wt% ---- | | | | | | | | | Heat treatment, °C(a) | 100 000-h creep rupture strength at: | | | | | |
| | C | Mn | Si | Ni | Cr | Mo | W | V | Nb | Other | | 550 °C (1020 °F) | | 600 °C (1110 °F) | | 650 °C (1200 °F) | |
												MPa	ksi	MPa	ksi	MPa	ksi
High-chromium heat-resistant steels																	
H-46	0.16	0.6	0.4	...	11.5	0.65	...	0.3	0.3	0.05 N	1150 OQ/650 AC	308	44.7	118	17	61	8.8
FV448	0.13	1.0	0.5	...	10.5	0.75	...	0.15	0.45	0.05 N	1150 OQ/650 AC	293	42.5	139	20.2	60	8.7
AISI 422	0.23	0.6	0.4	0.7	12.5	1.0	1.0	0.25	...	...	1060 OQ/650 AC	257	37.3	130	19	64	9.3
A.L. type 419	0.25	1.0	0.3	0.5	11.5	0.5	2.5	0.4	...	0.10 N	1100 OQ/650 AC	...	...	...	...	...	...
Lapelloy	0.30	1.0	0.25	0.3	12.0	2.75	...	0.25	...	...	1100 OQ/650 AC	258	37.4	127	18.4	...	...
HT-9	0.20	0.6	0.4	0.5	11.5	1.0	0.5	0.3	...	...	1050 OQ/760 AC	200	29	100	14.5	...	...
EM-12	0.10	1.0	0.4	...	9.5	2.0	...	0.3	0.4	...	1080 OQ/785 AC	210	30.5	120	17.4	60	8.7
TAF	0.18	0.5	0.3	...	10.5	1.5	...	0.2	0.15	0.3 B	1150 OQ/700 AC	373	54	216	31.3	137	20.0
Advanced 10–12% Cr steels for steam turbine rotors																	
G.E.	0.19	0.65	0.3	0.6	10.5	1.0	...	0.20	0.085	0.06 N	1050 OQ/570 AC/620 AC	...	...	...	...	...	...
TR1100	0.14	0.50	0.05	0.6	10.2	1.5	...	0.17	0.055	0.04 N, 0.002 Al	1050 OQ/570 AC/680 AC	...	...	118	17.1	64	9.3
TR1150	0.13	0.50	0.05	0.7	10.2	0.4	1.8	0.17	0.055	0.045 N, 0.005 Al	1050 OQ/570 AC/680 AC	...	...	157	22.8	83	12.0
TR1200	0.12	0.50	0.05	0.8	11.2	0.3	1.8	0.20	0.055	0.055 N, 0.005 Al	1020 OQ/570 AC/710 AC	...	...	...	...	...	...
Boiler tube materials																	
9Cr-1Mo	0.10	0.4	0.05	...	9.0	1.0	...	...	...	0.02 N	...	...	...	39	5.7	20	2.9
Mod 9Cr-1Mo	0.10	0.45	0.35	<0.2	8.75	0.95	...	0.21	0.08	0.05 N	...	...	...	98	14.2	49	7.1
Mod NSCR9	0.08	0.5	0.05	0.1	9.0	1.6	...	0.16	0.05	0.03 N, 0.003 B	...	...	...	128	18.6	69	10
T-1	0.05	0.5	0.3	...	10.0	2.0	...	0.1	0.05	0.02 N	1050 AC/700 AC	...	...	...	...	...	...
T-2	0.10	0.5	0.3	...	10.0	2.0	...	0.1	0.05	0.02 N	1050 AC/700 AC	...	...	...	...	...	...
TB9	0.08	0.5	0.05	0.1	9.0	0.5	1.8	0.2	0.05	0.05 N	...	...	...	196	28.4	98	14.2
TB12	0.08	0.5	0.05	0.1	12.0	0.5	1.8	0.2	0.05	0.05 N, 0.003 B	...	...	...	206	30	108	15.7
AISI 304	0.08	1.5	0.6	10	18.5	...	...	...	...	0.05 N	...	...	...	118	17.1	69	10
AISI 347	0.06	1.7	0.5	12	17.5	...	...	...	...	0.02 N	...	...	...	128	18.6	78	11.3

(a) OQ, oil quench; AC, air cool. Source: Ref 21

resistant to oxidation. At 980 °C (1800 °F), the lower-alloy types 409, 430, and 304 exhibit high oxidation. At 1090 °C (2000 °F), alloys 18 SR and Inconel 601 (shown in Table 5) and alloys such as 446 and 310 have adequate oxidation resistance. The cycling oxidation resistances of E-Brite 26-1, type 310, and Incoloy 800 are compared in Fig. 14.

Type 409, the lowest-alloy stainless steel with a nominal chromium content of 11.0%, is used extensively because of its good fabricating characteristics, including weldability and formability, and its availability. Its best-known high-temperature applications are in automotive exhaust systems; metal temperatures in catalytic converters exceed 540 °C (1000 °F). Type 409 is also used for exhaust ducting and silencers in gas turbines. Type 405 is used in stationary vanes and spacers in steam turbines and in various furnace components. Types 430 and 439 are used for heat exchangers, hot-water tanks, condensers, and furnace parts. Type 18 SR, like type 446, is used in industrial ovens, blowers, exhaust systems, furnace equipment, annealing boxes, kiln liners, and pyrometer tubes.

Heat-Resistant High-Chromium Ferritic Steels

Table 4 lists various precipitation-strengthened high-chromium ferritic steels. These steels contain vanadium and perhaps other carbide formers such as niobium or tungsten to strengthen the steel by precipitation hardening during tempering or elevated-temperature exposure.

Although several steels in Table 4 are not strictly considered stainless steels, the chromium contents are high enough to exhibit corrosion resistance comparable to that of typical stainless steels. Some of these steels, such as the modified 9Cr-1Mo steel, are also substitutes for austenitic steels in boiler tube applications, as described below. The modified 9Cr-1Mo steel, which is designated ASTM grade 91, is specified in ASTM standards for boiler tubes (ASTM A 213), forgings (ASTM A 336), seamless pipe (ASTM A 335), and forged and bored pipe (ASTM A 369). High-chromium ferritic steels are also used for steam turbine rotors and bolting materials.

Historical Background. High-chromium heat-resistant ferritic steels were first developed for use in gas and steam turbine applications. In the early 1940s, a need for improved high-strength, corrosion-resistant materials for gas turbine disks and steam turbine blades, operating at temperatures near 540 °C (1000 °F), led metallurgists in England to increase development work on 12% Cr heat-resistant steels. This research produced two steels (H-46 and FV448 in Table 4) and other alloy steels in the 1950s with improved creep-rupture strengths. During the same time period, other 12% Cr alloys such as AISI 422 (UNS S42200) and Lapelloy (listed in Table 4) were introduced

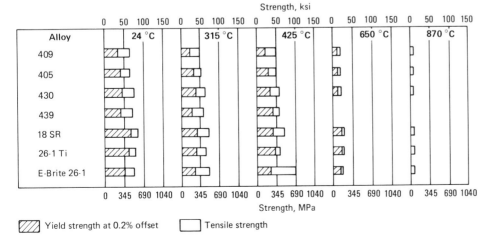

Fig. 12 Room-temperature and high-temperature tensile properties of selected ferritic stainless steels from Table 1. All alloys in the annealed condition: fast cooled from 815 to 925 °C (1500 to 1700 °F)

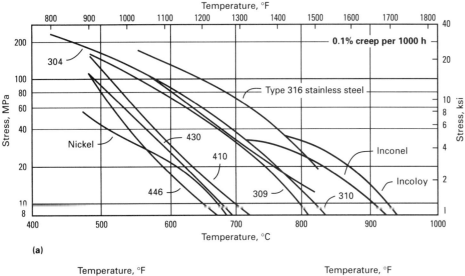

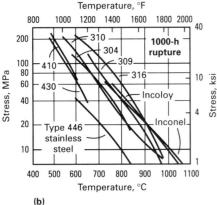 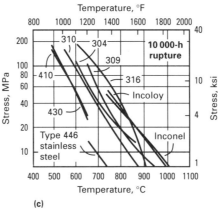

Fig. 13 Creep and rupture behavior of selected heat-resistant alloys as a function of temperature. (a) Stresses for a creep rate of 0.1% in 1000 h. (b) Stresses for rupture in 1000 h. (c) Stresses for rupture in 10 000 h

in the United States; they have been used for steam turbine blades and bolting materials for turbine cylinders. Work in Japan led to the development in 1956 of TAF steel (Table 4), which is twice as strong as H-46 at 650 °C (1200 °F).

Turbine Rotors. The elevated-temperature strength of turbine rotors is a major factor that influences the allowable temperature conditions of turbines.

The heat-resistant high-chromium ferritic steels developed in the 1950s, such as H-46 and AISI 422, have good creep-rupture strengths and are widely used in steam and gas turbines. Also, General Electric's rotor steel (Table 4), with improved creep-rupture strength at 570 °C (1060 °F), was developed in the early 1960s for application in supercritical steam power plants.

In the early 1980s, joint research by Kobe Steel and Mitsubishi Heavy Industry produced modified versions of ferritic TAF steel for operation at 620 °C (1150 °F). The new rotor steels, TR1100, TR1150, and TR1200 (Table 4) improve creep-rupture strength substantially over the General Electric rotor steel (which exhibits a marked decrease in creep-rupture strength over extended times). These improved rotor steels are based on chemistry modifications. The carbon and nitrogen contents of TR1100 are lowered to 0.13% and 0.04%, respectively, and the steel contains 1.5% Mo for solid-solution strengthening and stabilization of $M_{23}C_6$ and M_6C carbides. In addition, tests show that tungsten has a stronger high-temperature strengthening effect than molybdenum at elevated temperatures. Based on these data, TR1150 and TR1200 steels are being developed with increased tungsten content within a molybdenum equivalence (Mo% + ½ W%) range of 1.2 to 1.5% Mo.

The rotor steels, such as TR1100 and TR1150, are expected to be used in applications with steam conditions of 595 °C (1100 °F) and 620 °C (1150 °F), respectively, with a rotor design stress of 120 to 130 MPa (17.5 to 18.8 ksi). The TR1200 rotor steel in development is intended for steam conditions of 34 MPa (4.9 ksi) and 650 °C (1200 °F).

Boiler tubing requires a combination of high-temperature strength, good formability, and weldability. Various types of high-chromium heat-resistant ferritic steels for boiler tubes are listed in Table 4, along with two austenitic stainless steels (AISI 304 and 347) for comparison. In addition, the HT-9 and EM-12 alloys in Table 4 are widely used in Europe for boiler tubing.

The 10Cr-2Mo boiler tube steels with niobium and vanadium additions include the T-1 and T-2 steels in Table 4. These steels contain 0.05% and 0.10% C, respectively; this level of carbon provides high-temperature strength as well as the good formability

Table 5 Cyclic oxidation resistance of ferritic stainless steels
Specimens were exposed for 100 h in air containing 10% water vapor, cooled to room temperature every 2 h, then reheated to test temperature.

Alloy	Weight change (scale not removed), g/m² at:			
	705 °C (1300 °F)	815 °C (1500 °F)	980 °C (1800 °F)	1090 °C (2000 °F)
409	+0.1	+0.8	+1430(a)	−10 000(b)
430	+0.4	+1.3	−1660(c)	−10 000(b)
18 SR	+0.1	+0.3	+2.5	+7.4
304	+0.2	+1.7	−3400	−10 000(d)
309	+0.2	+2.7	−120	−910
201	+0.8	+3.1	+10	−150
Incoloy 800	+0.3	+3.2	+8.6	−560
Inconel 601	+0.1	+1.2	+10	−2.1

(a) Removed after 36 h. (b) Removed after 12 h. (c) Removed after 30 h. (d) Removed after 24 h

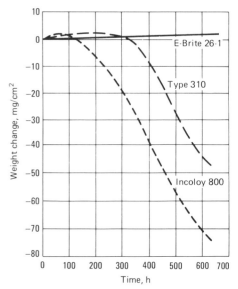

Fig. 14 Cyclic oxidation behavior of three iron-base heat-resistant alloys at 980 °C (1800 °F)

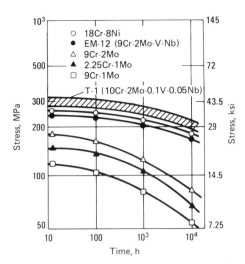

Fig. 15 Creep-rupture strengths of various boiler tube steels at 600 °C (1110 °F). Source: Ref 21

and weldability required for boiler tubing. A chromium content of 10% provides very good high-temperature corrosion resistance at about 600 °C (1100 °F), and other alloying elements produce a 20 to 30% δ-ferrite content, which improves weldability.

Increasing the molybdenum content enhances solution hardening and strengthening by $M_{23}C_6$ and M_6C carbide precipitation and intermetallic compound (Fe_2Mo) formation. However, molybdenum is fixed at 2% for the best corrosion resistance, formability, toughness, and δ-ferrite content.

Small amounts of vanadium and niobium affect the high-temperature strength of this steel significantly. Although the optimum content of vanadium is 0.25% for a carbon content of 0.2%, the addition of this amount to a 0.05% carbon steel causes only V_4C_3 precipitation. Therefore, a maximum vanadium content of 0.10 to 0.15% is necessary to raise high-temperature strength by the gradual precipitation of various carbides such as $M_{23}C_6$, M_6C and NbC. The vanadium content is fixed at 0.10% for good weldability.

The niobium content must be relatively low (0.02 to 0.05%) in these steels so that NbC can dissolve in the matrix when normalized at 1050 °C (1900 °F). However, at the low end of the range, the complete dissolution of carbides results in grain growth at the normalizing temperature with a reduction in notch toughness; thus, niobium content is fixed at 0.05%.

The 0.05% C in T-1 steel is shared to form carbides in the following manner: 0.02% for V_4C_3, 0.005% for NbC, and 0.025% for $M_{23}C_6$ and M_6C. In addition, the equation $(V/51) + (Nb/93) < (C/12)$ must be satisfied so that $M_{23}C_6$ and M_6C can precipitate along with V_4C_3 and NbC.

The heat-resistant steel designed in this manner is strengthened by the precipitation of very fine V_4C_3 and NbC during temper-

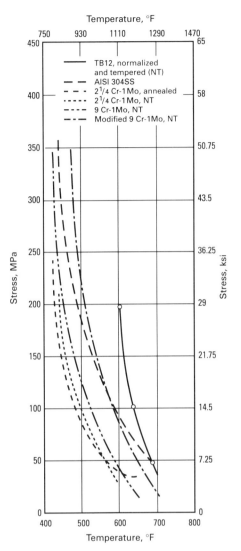

Fig. 16 100 000-h creep-rupture strength of various steels used in boiler tubes. TB12 steel has as much as five times the 100 000-h creep-rupture strength of conventional ferritic steels at 600 °C (1110 °F). This allows an increase in boiler tube operating temperature of 120 to 130 °C (215 to 235 °F). Source: Ref 21

ing and in the early stages of creep, followed by $M_{23}C_6$ and M_6C precipitation. Figure 15 compares the creep-rupture strength of T-1 at 600 °C (1110 °F) with that of other boiler tube materials.

Modified 9 and 12Cr-Mo Boiler Tube Steels. Further investigation of the effects of molybdenum, vanadium plus niobium, chromium, nickel, and tungsten additions led to the development of modified chromium-molybdenum steels. These steels include the TB9 and TB12 alloys (Table 4) developed in Japan and the vanadium-niobium modified 9Cr-1Mo steel developed by Oak Ridge National Laboratory. The modified 9Cr-1Mo steel, previously covered in Code Case 1943 of the ASME Boiler Code, is now covered in the regular code. For boiler tubes, these steels offer a promising alternative to 2.25Cr-1Mo steel and types 304, 316, 321, and 347 austenitic stainless steels. Figure 16 compares the 100 000-h creep-rupture strengths of these steels as a function of temperature.

Quenched and Tempered Martensitic Stainless Steels

Quenched and tempered martensitic stainless steels are essentially martensitic, and harden when air cooled from the austenitizing temperature. These alloys offer good combinations of mechanical properties, with usable short-time strength up to 590 °C (1100 °F), and relatively good corrosion resistance. The strength levels at temperatures up to 590 °C (1100 °F) that can be attained in these alloys through heat treatment are considerably higher than those attainable in ferritic stainless steels, but the martensitic alloys have inferior corrosion resistance. Also, the martensitic stainless steels are not very tough.

These alloys are normally purchased in the annealed or fully treated (hardened and tempered) condition. They are used in the hardened and tempered condition. For best long-time thermal stability, these alloys should be tempered at a temperature that is 110 to 165 °C (200 to 300 °F) above the expected service temperature.

Properties. Quenched and tempered martensitic stainless steels can be grouped according to increasing strength and heat resistance:

- *Group 1* (lowest strength and heat resistance): types 403, 410, and 416
- *Group 2*: Greek Ascoloy and type 431
- *Group 3*: Moly Ascoloy (Jethete M-152)
- *Group 4* (highest strength and heat resistance): H-46 and type 422

A general comparison of mechanical property data is presented in Fig. 17 for some of these alloys. Data for type 410 are typical of group 1 alloys. Data for Greek Ascoloy are typical of data for type 431. Data for Moly Ascoloy are typical of data for group 3 alloys (the composition of Jethete M-152 is very similar to that of Moly Ascoloy). Although H-46 and type 422 are similar in strength, their compositions are somewhat different; therefore, data are shown for both alloys.

The short-time tensile and rupture data shown in Fig. 17 were generated in tests of material that had been given austenitizing treatments typical for the specific alloys tested. Because these alloys are normally used at service temperatures near 540 °C (1000 °F), although they may be used up to 590 °C (1100 °F), data are shown for a relatively high tempering temperature of 650 °C (1200 °F), which results in good thermal stability in these alloys at 540 °C (1000 °F).

It should be noted that the group 1 alloys, of which type 410 is typical, show the

940 / Specialty Steels and Heat-Resistant Alloys

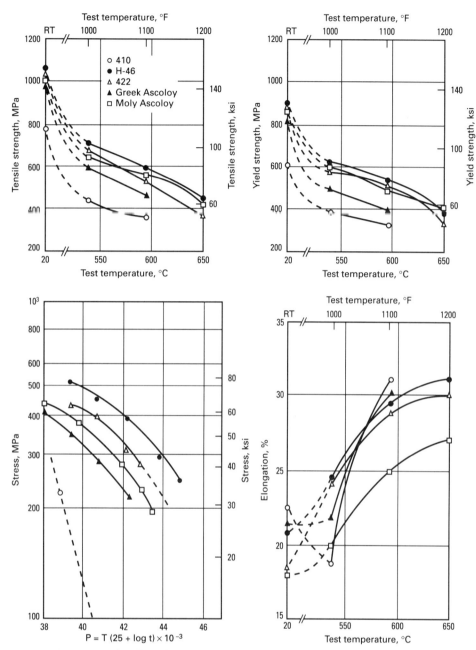

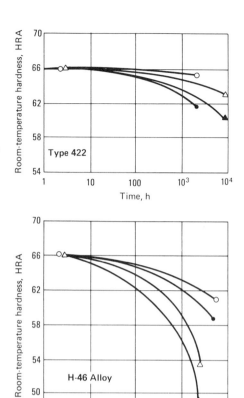

Fig. 17 Comparison of mechanical properties of martensitic stainless steels. Heat treating schedules were as follows. Type 410: 1 h at 980 °C (1800 °F), oil quench; 2 h at 650 °C (1200 °F), air cool. H-46: 1 h at 1150 °C (2100 °F), air cool; 2 h at 650 °C (1200 °F), air cool. Type 422: 1 h at 1040 °C (1900 °F), oil quench; 2 h at 650 °C (1200 °F), air cool. Greek Ascoloy: 1 h at 955 °C (1750 °F), oil quench; then 2 h at 650 °C (1200 °F), air cool. Moly Ascoloy: 30 min at 1050 °C (1925 °F), oil quench; then 2 h at 650 °C (1200 °F), air cool

Fig. 18 Approximate effects of time and stress on tempering of types 422 and H-46. Circles indicate specimens heated to 1150 °C (2100 °F) and rapidly cooled, tempered 2 h at 705 °C (1300 °F), and tested (to fracture) at a temperature of 540 °C (1000 °F) and a stress of 380 MPa (55 ksi). Triangles indicate specimens heated to 980 °C (1800 °F) and rapidly cooled, tempered 2 h at 705 °C, and tested (to fracture) at 540 °C (1000 °F) and 275 MPa (40 ksi). Open symbols represent data taken at the unstressed specimen shoulder; solid symbols represent data taken within the stressed gage length.

lowest values of strength capability as a function of test temperature. Greek Ascoloy is considerably stronger than type 410, with a yield strength (0.2% offset) of 480 MPa (70 ksi) and a tensile strength of 585 MPa (85 ksi) at 540 °C (1000 °F). The tensile strength capabilities of H-46, Moly Ascoloy, and type 422 are fairly similar and are the highest in this group of alloys. Tensile elongation data for all these alloys are similar—from about 20% elongation at 21 °C (70 °F) to about 30% at 650 °C (1200 °F).

Stress-rupture data for alloys typical of each subgroup are compared in Fig. 17 by means of a master stress-rupture plot. It should be noted that the niobium-containing H-46 alloy has the highest stress-rupture capability, with type 422, Moly Ascoloy, and Greek Ascoloy, in that order, having increasingly lower rupture capabilities. Type 410 has a very low stress-rupture capability and is the weakest of all the martensitic stainless alloys being considered. The niobium-containing alloys such as H-46 usually show an advantage in stress-rupture capability (creep resistance) for short testing times (100 to 1000 h) but lose their strength advantage when tested for periods of about 10 000 h or more. The favorable effects of niobium additions on short-time stress-rupture properties are attributed to a finely dispersed precipitation of NbC. The favorable effects tend to diminish as tempering temperature is increased, and a coarsely dispersed precipitate is formed. The effect of tempering in service is shown by the hardness data in Fig. 18 for type 422 and H-46. The H-46 alloy shows a larger hardness drop for extended thermal exposure at a testing temperature of 540 °C (1000 °F) than does type 422.

Applications. Quenched and tempered martensitic stainless steels find their greatest application in steam and gas turbines, where they are used in blading at temperatures up to 540 °C (1000 °F). Other uses include steam valves, bolts, and miscellaneous parts requiring corrosion resistance and good strength up to 540 °C (1000 °F).

Type 410 is the basic, general-purpose steel, used for steam valves, pump shafts, bolts, and miscellaneous parts requiring

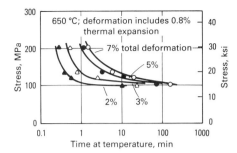
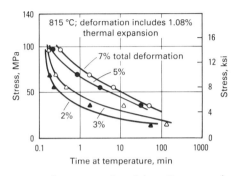

Fig. 19 Stress-versus-time deformation curves for type 410 stainless steel sheet, showing effect of time at temperature on total deformation at specific stress levels. Design curves in the chart at left represent a heating rate of 90 °C/s (160 °F/s) to 650 °C (1200 °F). Those at right represent a heating rate of 105 °C/s (190 °F/s) to 815 °C (1500 °F). Room-temperature properties of the sheet used in these tests were: tensile strength, 650 to 695 MPa (94.5 to 101 ksi); yield strength at 0.2% offset, 555 to 565 MPa (80.7 to 82.3 ksi); and 9.6 to 16% elongation in 50 mm (2 in.) after air cooling from the normalizing temperature of 955 °C (1750 °F) (AFTR 6731, Part 4)

corrosion resistance and moderate strength up to 540 °C (1000 °F). Type 403 is similar to 410, but the chemical composition is adjusted to prevent formation of δ-ferrite in heavy sections. It is used extensively for steam turbine rotor blades and gas turbine compressor blades operating at temperatures up to 480 °C (900 °F). For this type of application the steel is tempered at 590 °C (1100 °F) or above, after which embrittlement is negligible in the service temperature range of 370 to 480 °C (700 to 900 °F).

A satisfactory heat treatment for these steels is to austenitize at 950 to 980 °C (1750 to 1800 °F), cool rapidly in air or oil, and temper. Cooling from the hot-rolling temperature and tempering without intermediate austenitizing is sometimes practical but may result in a structure that contains free ferrite, which is detrimental to transverse properties. Warm or cold work after tempering sets up residual stresses that can be relieved by heating to approximately 620 °C (1150 °F).

Stress-versus-time deformation curves for type 410 sheet are given in Fig. 19. These values are useful for special applications where heating rates are high.

Greek Ascoloy, type 431, and type 422 are variants of 410, modified by the addition of such elements as nickel, tungsten, aluminum, molybdenum, and vanadium. Nickel serves a useful purpose by causing the steel to be entirely austenitic at conventional heating temperatures when the carbon and chromium contents are such that a two-phase structure would exist if nickel were absent. The tempering temperature for Greek Ascoloy may be 55 °C (100 °F) or more, higher than that for type 410 for equivalent strength and hardness. Type 422 develops the highest mechanical properties and at 650 °C (1200 °F) has a tensile strength equivalent to that of type 403 at 590 °C (1100 °F). The rupture strength of type 422 at 540 °C (1000 °F) is considerably higher than those of the other steels in this series (Fig. 20).

Types 430 and 446, which have nominal chromium contents of 16 and 25%, respectively, are not hardenable by conventional quenching and tempering treatments. Nitrogen to 0.25% (max) is added to type 446 for grain refinement. These grades are generally used in the annealed condition, but they can be cold worked to increase strength and decrease ductility. They have greater oxidation resistance than the 12% Cr steels and may be used without excessive scaling to temperatures of 840 °C (1550 °F) for type 430 and 1090 °C (2000 °F) for type 446 (Fig. 9). The rate of oxidation and the character of the scale are affected greatly by variations in the chemical and physical nature of the air and gases in the environment, and tests under actual service conditions are required for meaningful evaluation.

These alloys age harden, with a loss of ductility, when held for prolonged periods at 370 to 540 °C (700 to 1000 °F); this age hardening is due to 885 °F embrittlement. A typical hardness-temperature curve for a 27% Cr alloy was obtained on a temperature gradient bar (Fig. 21). The specimen was heated in a specially designed furnace so that one end only was subjected to a maximum and uniform temperature. The opposite end reached about 95 °C (200 °F), and the area between showed a uniform increase in temperature from the colder end to the hotter end. Thermocouples were inserted through the colder end of the bar to different depths so that temperatures could be measured at specified distances from the hot end.

The increase in hardness with time at 475 °C (885 °F) and the effect of the increase in chromium content are shown in Fig. 22. Notch sensitivity is greatly increased by the increase in hardness. The embrittled condition can be alleviated, and the original hardness restored, by heating at 590 °C (1100 °F). This embrittlement is significant only for applications in the range from 425 to 540 °C (800 to 1000 °F).

Type 430 is used for heat exchange equipment, condensers, and piping for nitric acid, as well as for furnace parts and retorts operating at temperatures up to 840 °C (1550 °F). It can be stamped, spun, and formed more easily than can type 446.

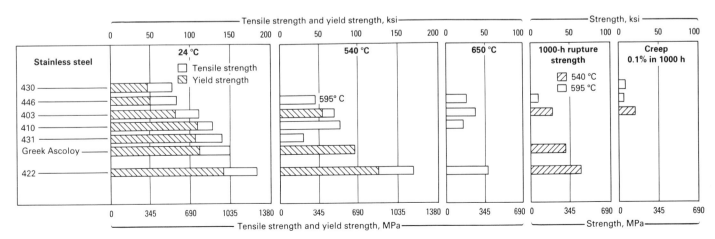

Fig. 20 Tensile, yield, rupture, and creep strengths for seven ferritic and martensitic stainless steels. Types 430 and 446 were annealed. Type 403 was quenched from 870 °C (1600 °F) and tempered at 620 °C (1150 °F). Type 410 was quenched from 955 °C (1750 °F) and tempered at 590 °C (1100 °F). Type 431 was quenched from 1025 °C (1875 °F) and tempered at 590 °C (1100 °F). Greek Ascoloy was quenched from 955 °C (1750 °F) and tempered at 590 °C (1100 °F). Type 422 was quenched from 1040 °C (1900 °F) and tempered at 590 °C (1100 °F).

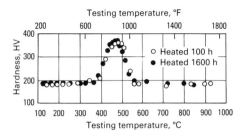

Fig. 21 Hardness values along a temperature gradient bar of type 446 stainless steel (0.19C-0.73Mn-0.54Si-27.31Cr-0.16Ni-0.15N) after exposure for times indicated

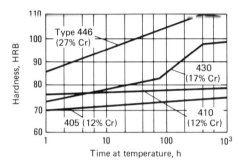

Fig. 22 Effect of time at 475 °C (885 °F) on age-hardening characteristics of chromium steels containing 12, 17, and 27% Cr

Type 446 is used for furnace parts, oil burners, heat exchangers, kiln liners, glass molds, and stationary soot blowers in steam boilers. Its elevated-temperature strength is low, which is indicated by the creep and rupture-strength values given in Fig. 13. These plots compare elevated-temperature properties of types 410, 430, and 446 with those of several other materials, including the standard austenitic stainless steels, Incoloy, and Inconel, a nickel-base alloy.

Welding. The primary factor in welding martensitic steels is their hardenability. The heat-affected zones surrounding the weld harden upon cooling, setting up stresses that can give rise to cracking. If the filler metal is similar to the parent metal, the weld itself will also harden upon cooling and become very brittle. These difficulties can be avoided by two precautions:

- Using austenitic filler metal, which will remain ductile and will absorb the stresses set up by hardening in the heat-affected zone
- Preheating the work gradually before welding and postheating after welding to avoid quenching stresses and cracks. Preheating should be carried out at 205 to 315 °C (400 to 600 °F); postheating at 590 to 760 °C (1100 to 1400 °F)

Precipitation-Hardening Martensitic Stainless Steels

The precipitation-hardening (PH) martensitic stainless (maraging) steels fill an important position between the chromium-free 18% Ni maraging steels and the 12% Cr, low-nickel quenched and tempered martensitic stainless alloys. These PH alloys contain 12 to 16% Cr for corrosion resistance and scaling resistance at elevated temperatures and have the highest tensile strengths at temperatures below about 450 °C (850 °F), as shown in Fig. 1.

These alloys are normally purchased in the solution-annealed condition. Depending on the application, they may be used in the annealed condition or in the annealed plus age-hardened condition. In some cases, material will be supplied in an overaged condition to facilitate the forming of parts. The formed parts are then solution annealed following fabrication.

Properties. The maraging alloys listed in Table 1 include Custom 450, 17-4 PH, 15-5 PH, Custom 455, and PH 13-8 Mo. Property data for these alloys are shown in Fig. 23. Data for 15-5 PH are not shown separately because the properties of this alloy are very similar to those of 17-4 PH.

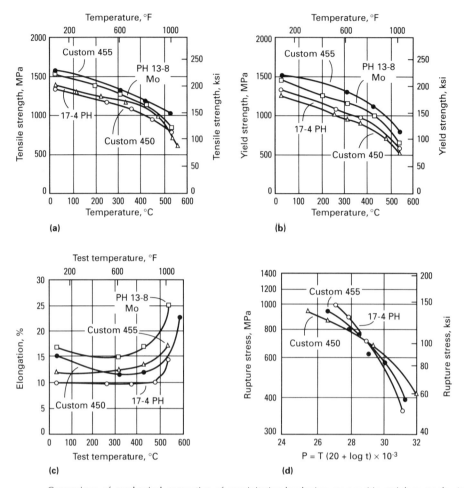

Fig. 23 Comparison of mechanical properties of precipitation-hardening martensitic stainless steels. (a) Tensile strength. (b) Yield strength. (c) Elongation. (d) Rupture strength. Heat treating schedules were as follows. Custom 450: 1 h at 1040 °C (1900 °F), water quench; then 4 h at 480 °C (900 °F), air cool. 17-4 PH: 30 min at 1040 °C (1900 °F), oil quench; then 4 h at 480 °C (900 °F), air cool. Custom 455: 30 min at 815 °C (1500 °F), water quench; then 4 h at 510 °C (950 °F), air cool. PH 13-8 Mo: oil quenched from 925 °C (1700 °F); then 4 h at 540 °C (1000 °F), air cool

Short-time tensile data indicate that Custom 455 and PH 13-8 Mo have higher strengths than Custom 450, 17-4 PH, or 15-5 PH. For all of these alloys, tensile and yield strengths drop rapidly at temperatures above 425 °C (800 °F), and tensile elongation is greater than 10% over the temperature range from ambient to 540 °C (1000 °F).

Stress-rupture data are compared in Fig. 23 by means of a master parameter plot. Data were developed at testing temperatures of 425 and 480 °C (800 and 900 °F) during time periods of 100 and 1000 h. It should be noted that 17-4 PH appears to have better stress-rupture strength at 425 °C (800 °F), whereas Custom 450 is superior in this respect at 480 °C (900 °F). The only stress-rupture data available for Custom 455 appears to indicate that the alloy is intermediate in rupture strength between 17-4 PH and Custom 450.

It is possible to produce a wide variety of useful properties in a given alloy by varying the aging temperature. An example of this can be seen in Fig. 24, where tensile

al and reduces the possibility of crack formation in weld metal and in the heat-affected zone.

Precipitation-Hardening Semiaustenitic Stainless Steels

The precipitation-hardening semiaustenitic heat-resistant stainless steels are modifications of standard 18-8 austenitic stainless steels. Nickel contents are lower, and such elements as aluminum, copper, molybdenum, and niobium are added. These steels are used at temperatures up to 480 °C (900 °F).

Heat Treatment. Typical schedules for heat treating precipitation-hardening alloys are given in Table 6. These alloys are solution annealed above 1040 °C (1900 °F) and in this condition can be formed, stamped, stretched, and otherwise cold worked to about the same extent as can 18-8 alloys, although they are less ductile and may require intermediate annealing.

All the semiaustenitic stainless steels can also be used in the cold-worked condition, in either sheet or wire form. Cold working causes partial transformation of the rather unstable austenite to martensite because of plastic deformation. Aging or tempering is performed after cold working.

Mechanical Properties. Typical short-time tensile, rupture, and creep properties of several precipitation-hardening semiaustenitic alloys are compared in Fig. 25. Different hardening heat treatments may produce a wide variety of useful properties for the same alloy. For example, the PH 15-7 Mo alloy treated to condition RH950 has a higher rupture strength than the same alloy treated to TH1050 (Fig. 26). Strengths also degrade after long-term exposure at elevated temperature because of overaging (coarsening) of precipitates.

Compressive and tensile yield strengths are approximately equal for all precipitation-hardened steels. For sheet, yield strengths of specimens taken transverse and parallel to the direction of rolling may vary appreciably. The magnitude of this effect varies from grade to grade and with the heat treatment for a given grade.

Welding. The precipitation-hardening stainless steels in either the annealed or hardened condition can be welded by any of the processes used for welding 18-8 (300-series) stainless steel. These steels are less ductile and more notch sensitive than conventional 18-8 grades and require more care to prevent stress concentrations at corners and notches.

The gas-shielded tungsten arc process has been satisfactory for welding large assemblies of sheet and forgings that are subsequently aged. Postweld annealing is desirable.

A complete cycle of heat treatment, including high-temperature solution anneal-

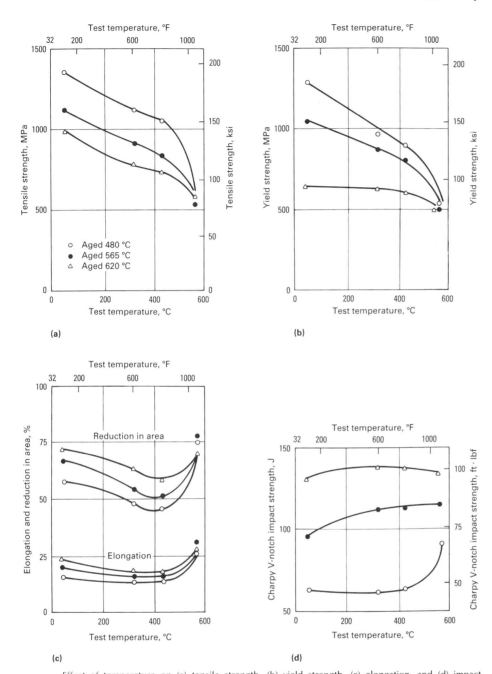

Fig. 24 Effect of temperature on (a) tensile strength, (b) yield strength, (c) elongation, and (d) impact toughness of Custom 450. Material used for testing was round bar stock 25 mm (1 in.) in diameter that had been solution treated by heating 1 h at 1040 °C (1900 °F) and water quenching

strength, yield strength (0.2% offset), ductility, and impact data are shown for Custom 450 at three different aging temperatures. Aging at 480 °C (900 °F) can produce significant strengthening at testing temperatures as high as 450 °C (850 °F), but it also results in lower toughness values (as measured by Charpy V-notch testing) than aging at either of the two higher temperatures.

Precipitation-hardening martensitic stainless steels are used for short-time elevated-temperature exposures in industrial and military applications for which resistance to corrosion and high mechanical properties at temperatures up to 425 °C (800 °F) are necessary. Typical uses include valve parts, ball bearings, forgings, turbine blades, mandrels, conveyor chain, miscellaneous hardware, and mechanical and structural components for aircraft.

Welding. All of these alloys are martensitic in the annealed condition and, because of their low carbon levels, are readily weldable with minimal danger of cracking. Any of the standard welding procedures, such as gas tungsten arc, gas metal arc, covered electrode, and resistance welding, may be used. No preheating is required because the very low carbon content of these alloys restricts the hardness of rapidly cooled met-

944 / Specialty Steels and Heat-Resistant Alloys

Table 6 Heat-treating schedules for precipitation-hardening semiaustenitic stainless steels

Alloy	Mill heat treatment (solution anneal)	Fabrication	Conditioning and hardening treatments	Aging or tempering treatment
17-7 PH	1065 °C (1950 °F), air cool	Forming, welding	10 min at 955 °C (1750 °F), air cool, 8 h at −75 °C (−100 °F)	1 h at 510, 565, or 620 °C (950, 1050, or 1150 °F)
			1½ h at 760 °C (1400 °F), air cool to 15 °C (60 °F), hold ½ h	1 h at 510, 565, or 620 °C (950, 1050, or 1150 °F)
15-7 Mo	1065 °C (1950 °F), air cool	Forming, welding	10 min at 955 °C (1750 °F), air cool, 8 h at −75 °C (−100 °F)	1 h at 510, 565, or 620 °C (950, 1050, or 1150 °F)
			1½ h at 790 °C (1450 °F), air cool to 15 °C (60 °F), hold ½ h	1 h at 510, 565, or 620 °C (950, 1050, or 1150 °F)
AM-350	1040–1080 °C (1900–1975 °F), air cool	Forming, welding	930 °C (1710 °F), air cool, 3 h at −75 °C (−100 °F)	3 h at 455 or 540 °C (850 or 1000 °F)
			3 h at 745 °C (1375 °F), air cool to 27 °C (80 °F) max	3 h at 455 °C (850 °F)
AM-355	3 h at 775 °C (1425 °F), oil or water quench to 27 °C (80 °F) max, 3 h at 580 °C (1075 °F), air cool	Machining and other	1040 °C (1900 °F), water quench, 3 h at −75 °C (−100 °F), reheat to 955 °C (1750 °F), air cool, 3 h at −75 °C (−100 °F)	3 h at 455 or 540 °C (850 or 1000 °F)

ing, is desirable after welding. Strength and ductility in fully heat-treated fusion welds are superior to these properties in material hardened before welding and not heat treated after welding. Joint efficiencies near 90% have been reported for welded joints fully heat treated after welding.

The precipitation-hardening steels can be spot welded by the same technique used for 18-8 steels. This may be done before or after precipitation hardening. The minimum shear requirements for stainless steel at 1035 MPa (150 ksi) can be obtained.

Applications. Precipitation-hardening steels are used for industrial and military applications that require resistance to corrosion as well as high mechanical properties at temperatures up to 425 °C (800 °F). Typical uses of these steels include landing-gear hooks, poppet valves, fuel tanks, hydraulic lines, hydraulic fittings, compressor casings, miscellaneous hardware, and structural components for aircraft. The higher-carbon grade (0.13%) is used for compressor blades, spacers, frames and casings for gas turbines, oil well drill rods, and rocket casings.

Precipitation-hardening steels can be cold rolled and tempered. Work-hardening rates are higher than for type 301 stainless steel and can be varied by regulating the annealing temperature. Compared with cold-rolled 301, cold-rolled precipitation-hardened steels have higher ductility at a given strength level, higher modulus of elasticity in compression in the rolling direction, and less reduction in strength with increasing temperature.

Austenitic Stainless Steels

Austenitic stainless steels comprise a group of iron-base alloys that contain 16 to 25% Cr and residual to 20% Ni. Some alloys may contain as much as 18% Mn. These stainless steels are not hardenable by heat treatment but can be hardened by cold work. However, the effect of cold work on strength is lost after elevated-temperature exposure (Fig. 2) because of recrystallization.

The austenitic stainless steels listed in Table 1 can be grouped into three categories, based primarily on composition:

- *300-series alloys*, which are essentially chromium-nickel and chromium-nickel-molybdenum austenitic stainless steels to which small amounts of other elements have been added
- *19-9 DL, 19-9 DX, and 17-14-CuMo*, all of which contain 1.25 to 2.5% Mo and 0.3 to 0.55% Ti. Other elements used include 1.25% W and 3% Cu in 17-14-CuMo
- *Chromium-nickel-molybdenum alloys*, which include types 201 and 202; 21-6-9; Nitronics 32, 33, 50, and 60; and Carpenter 18-18 Plus. These alloys contain 5 to 18% Mn and 0.10 to 0.50% N

Austenitic stainless steels are noted for high strength and for exceptional toughness, ductility, and formability. As a class, they exhibit considerably better corrosion resistance than martensitic or ferritic steels and also have excellent strength and oxidation resistance at elevated temperatures.

Solution heat treatment of these alloys is done by heating to about 1095 °C (2000 °F), followed by rapid cooling. Carbides that are dissolved at these temperatures may precipitate at grain boundaries upon exposure to temperatures from 425 to 870 °C (800 to 1600 °F), causing chromium depletion in grain-boundary regions. In this condition,

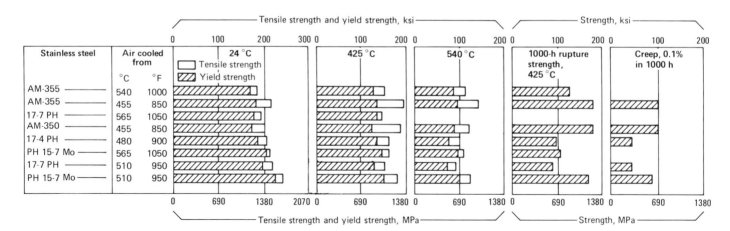

Fig. 25 Short-time tensile, rupture, and creep properties of precipitation-hardening stainless steels. AM-355 was finish hot worked from a maximum temperature of 980 °C (1800 °F), reheated to 930 to 955 °C (1710 to 1750 °F), water quenched, treated at −75 °C (−100 °F), and aged at 540 and 455 °C (1000 and 850 °F). 17-7 PH and PH 15-7 Mo were solution treated at 1040 to 1065 °C (1900 to 1950 °F). 17-7 PH (TH1050) and PH 15-7 Mo (TH1050) were reheated to 760 °C (1400 °F), air cooled to 15 °C (60 °F) within 1 h, and aged 90 min at 565 °C (1050 °F). 17-7 PH (RH950) and PH 15-7 Mo (RH950) were reheated to 955 °C (1750 °F) after solution annealing, cold treated at −75 °C (−100 °F), and aged at 510 °C (950 °F). 17-4 PH was aged at 480 °C (900 °F) after solution annealing. AM-350 was solution annealed at 1040 to 1065 °C (1900 to 1950 °F), reheated to 930 °C (1710 °F), air cooled, treated at −75 °C (−100 °F), and then aged at 455 °C (850 °F).

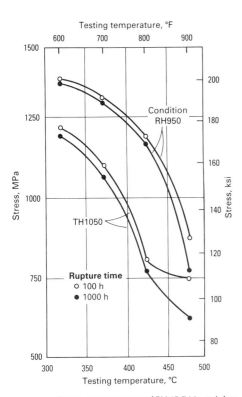

Fig. 26 Stress-rupture curves of PH 15-7 Mo stainless steel in the TH1050 and RH950 condition

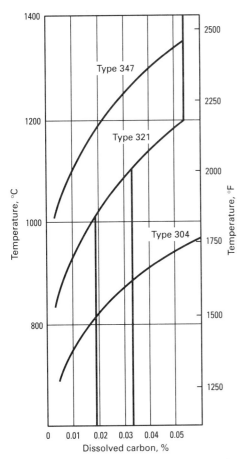

Fig. 27 Dissolved carbon under equilibrium conditions calculated from the solubility products for a type 347 (0.054% C, 0.76% Nb), a type 321 (0.054% C, 0.42% Ti), and an unstabilized 18Cr-8Ni stainless steel

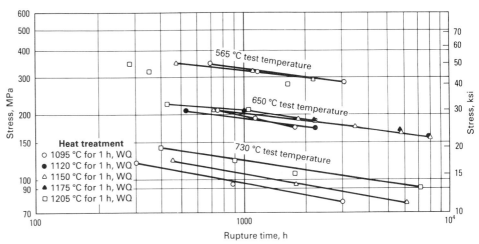

Fig. 28 Stress-rupture strength of type 347H stainless steel treated at different solution-annealing temperatures

the metal is sensitive to intergranular corrosion in oxidizing acids. The precipitation of chromium carbides can be controlled by reducing carbon content, as in types 304L and 316L, or by adding the stronger carbide formers titanium and niobium, as in types 321 and 347. These alloys are normally purchased and used in the annealed condition. The reduced carbon in solution in the low-carbon (304L, 316L) and stabilized grades (321, 347) results in reduced creep strength and creep-rupture strength.

H Grades. For the best creep strength and creep-rupture strength, the H grades of austenitic stainless steels are specified. These steels have carbon contents of 0.04 to 0.10% (Table 1) and are solution annealed at temperatures high enough to produce improved creep properties. The minimum annealing temperatures specified in ASTM A 312 and A 240 for H grades in general corrosion service are:

- 1040 °C (1900 °F) for hot-finished or cold-worked 304H and 316H steels
- 1095 °C (2000 °F) for cold-worked 321H, 347H, and 348H steels
- 1050 °C (1095 °F) for hot-finished 321H, 347H, and 348H steels

The stabilized grades (such as 321H, 347H, and 348H) have additions of strong carbide-forming elements, which lower the amount of dissolved carbon at a given annealing temperature (Fig. 27). The carbide formers in the stabilized grades, such as niobium in 347H, increase the resistance to intergranular corrosion by making less dissolved carbon available for chromium carbide formation, thereby preventing the depletion of chromium in grain-boundary regions.

When intergranular corrosion is of concern, annealing temperatures must be low enough to keep dissolved carbon at low levels. Type 347H tube, for example, at the minimum annealing temperature of 1095 °C (2000 °F) (specified in ASME SA213) would have only about 0.01% soluble carbon (Fig. 27), and the alloy should be stabilized against chromium depletion in the grain boundaries. However, annealing temperatures above 1065 °C (1950 °F) may still impair the intergranular corrosion resistance of stabilized grades such as 321, 321H, 347, 347H, 348, and 348H. When type 321H and 347H are used in applications where intergranular corrosion may be a problem, it is possible to apply an additional stabilizing treatment at a temperature near 900 °C (1650 °F) to reduce free carbon content by carbide precipitation.

Lower annealing temperatures improve resistance to intergranular corrosion but also reduce creep strength and creep-rupture strength. Figure 28 shows the creep-rupture strength of 347H tube treated at different annealing temperatures. In applications where intergranular corrosion is not a concern, better creep properties can be obtained with higher annealing temperatures. Types 321H and 347H, for example, are solution annealed at temperatures above 1120 °C (2050 °F) and 1150 °C (2100 °F), respectively, in order to put carbides in solution and to coarsen the grain structure, thereby ensuring the best creep strength and creep-rupture strength.

Tensile Properties. Typical mechanical property data for austenitic stainless steels are given in Fig. 29. Room-temperature tensile properties of annealed 300-series alloys are similar. At higher testing temperatures (425 and 650 °C, or 800 and 1200 °F), types 321, 347, and 309 appear to have yield strengths somewhat higher than those of types 304, 310, and 316. Types 309, 310, and 316 have the highest tensile strengths at 650 °C (1200 °F).

Tensile and yield strengths of 19-9 DL and 19-9 DX are higher than those of any 300-series alloy. However, 19-9 DL and 19-9 DX are heat treated at 705 °C (1300 °F),

946 / Specialty Steels and Heat-Resistant Alloys

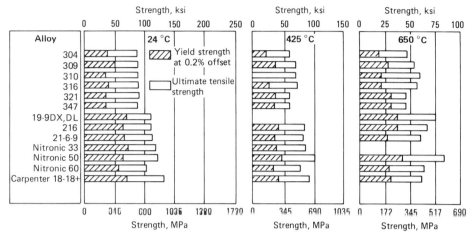

Fig. 29 Effect of testing temperature on tensile properties of austenitic stainless steels. Heat treating schedules were as follows. Type 304: 1065 °C (1950 °F), water quench. Type 309: 1090 °C (2000 °F), water quench. Type 310: 1120 °C (2050 °F), water quench. Type 316: 1090 °C (2000 °F), water quench. Type 321: 1010 °C (1850 °F), water quench. Type 347: 1065 °C (1950 °F), water quench. 19-9 DX and 19-9 DL: 705 °C (1300 °F), air cool. Type 216: 1065 °C (1950 °F), water quench. 21-6-9: 1065 °C (1950 °F), water quench. Nitronic 33: 1065 °C (1950 °F), water quench. Nitronic 50: 1090 °C (2000 °F), water quench. Nitronic 60: 1065 °C (1950 °F), water quench. Carpenter 18-18 Plus: 1065 °C (1950 °F), water quench

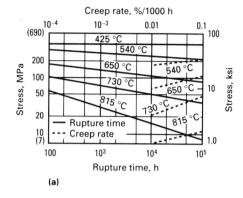

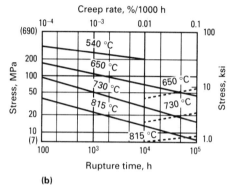

Fig. 30 Stress-rupture times and creep rates for annealed (a) type 347 stainless steel and (b) type 316 stainless steel. Source: Ref 22

compared with an average of 1065 °C (1950 °F) for 300-series alloys. Also, 19-9 DL and 19-9 DX are normally strengthened by controlled amounts of hot and cold work.

Tensile and yield strengths of chromium-nickel-molybdenum alloys are higher than those of 300-series alloys at both room and elevated temperatures. Carpenter 18-18 Plus exhibits the highest room-temperature tensile strength, whereas Nitronic 50 has the highest tensile strength at 650 °C (1200 °F).

Stress-rupture properties of various stainless steels in the 300 series are shown in Fig. 5. The H grades of type 347 and 316 appear to be the two strongest alloys over a range of temperatures. Stress-rupture times and creep rates of these two austenitic stainless steels are shown in Fig. 30.

The 19-9 DL and 19-9 DX alloys have rupture strengths superior to those of all 300-series alloys over the limited temperature and time range for which rupture data are available (1000-h rupture strength at 540 to 815 °C, or 1000 to 1500 °F) (Fig. 31). At longer times or higher temperatures, the 300 series may be superior. For the time-temperature range in Fig. 31, the chromium-nickel-molybdenum alloys have higher stress-rupture capabilities than do 300-series alloys, with the following exceptions: type 316 is superior to 21-6-9, and both 316 and 347 are stronger than Nitronic 60. The spread in rupture strength capability among these alloys is greater at the lower testing temperatures (540 to 700 °C, or 1000 to 1300 °F) and becomes progressively smaller as temperature is increased to approximately 980 °C (1800 °F), where all the alloys exhibit 1000-h rupture stresses of about 7 to 10 MPa (1 to 1.5 ksi). Types 304 and 310 have the lowest stress-rupture strengths.

Aging and the degradation of mechanical properties occur in austenitic steel because of two principal factors:

- Precipitation reactions that occur during prolonged exposure at elevated temperatures
- Environmental effects, such as corrosion or nuclear irradiation

The precipitation reactions in austenitic stainless steels that occur during prolonged exposure at elevated temperatures are complex, but some general guidance to the precipitates formed can be gained by reference to constitutive diagrams in Ref 23. The precipitates formed at temperatures in the range of about 500 to 600 °C (930 to 1110 °F) are predominantly carbides, while at higher

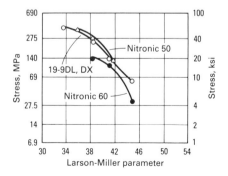
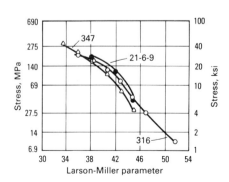
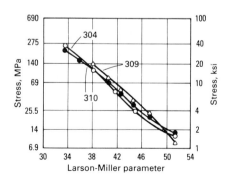

Fig. 31 Stress-rupture plots for various austenitic stainless steels. Heat treating schedules were as follows. Type 304: 1065 °C (1950 °F), water quench. Type 309: 1090 °C (2000 °F), water quench. Type 310: 1120 °C (2050 °F), water quench. Type 316: 1090 °C (2000 °F), water quench. Type 347: 1065 °C (1950 °F), water quench. 21-6-9: 1065 °C (1950 °F), water quench. 19-9 DX and 19-9 DL: for tests above 705 °C (1300 °F), 1065 °C (1950 °F) and water quench, then 705 °C (1300 °F) and air cool; for tests below 705 °C (1300 °F), 705 °C (1300 °F) and air cool. Nitronic 50: 1090 °C (2000 °F), water quench. Nitronic 60: 1065 °C (1950 °F), water quench. Larson-Miller parameter = $T/1000$ $(20 + \log t)$ where T is temperature in °R and t is time in h. All data taken from 1000-h tests

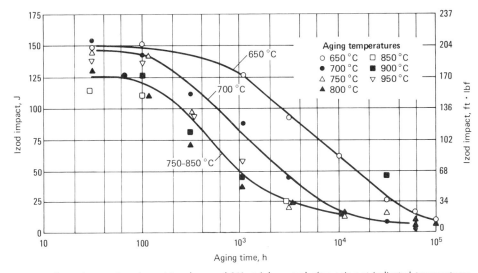

Fig. 32 Room-temperature impact toughness of 316 stainless steel after aging at indicated temperatures. Source: Ref 23

temperatures they are in the form of intermetallic phases. Although about 30 phases have been identified in stainless steels, the predominant precipitates found in plant-serviced alloys and weld metals exposed to temperatures of about 550 °C (1020 °F) are $M_{23}C_6$ and MC carbides in unstabilized and stabilized steels, respectively. At temperatures above 600 °C (1110 °F), σ phase and Fe_2Mo are also formed. The aging process also tends to occur more rapidly in weld metals that contain δ-ferrite (Ref 23).

Environmental effects on aging can be classified as either surface effects or bulk effects. Surface effects include oxidation and carburization, while bulk effects include changes in properties from nuclear irradiation (see the article "Effect of Neutron Irradiation on Properties of Steels" in this Volume).

Impact Toughness. Solution-annealed austenitic stainless steels are very tough, and their impact energies are very high. However, following elevated-temperature thermal aging, the impact energy decreases with increasing time at temperature. The impact energy of AISI type 316 stainless steel (Fig. 32) shows a continuous fall with increasing exposure time and temperature in the range of 650 to 850 °C (1200 to 1560 °F). Service-exposed type 316 stainless steel exhibited impact energies of 80 and 300 J (60 and 220 ft · lbf) in the serviced and resolution heat-treated conditions, respectively, when tested at room temperature (Ref 23).

Duplex stainless steels, with δ-ferrite in an austenite matrix, have improved resistance to stress-corrosion cracking and increased yield and tensile strengths. A further advantage of the presence of δ-ferrite is that it causes grain refinement, which produces additional strengthening. Further refining of the grain size can be achieved by a controlled-rolling treatment, using hot working in the range of 900 to 950 °C (1650 to 1740 °F) or at even lower temperatures. This causes a very fine dispersion of ferrite and austenite, in approximately equal proportions, which can give yield strengths in excess of 450 MPa (65 ksi). Duplex steels, however, require a careful balance of the ferrite- and austenite-forming elements (see the article "Cast Stainless Steels" in this Volume). The duplex steels are susceptible to 885 °F embrittlement at 370 to 480 °C (700 to 900 °F) and to σ formation at 650 to 815 °C (1200 to 1500 °F). For this reason, they are not very suitable for high-temperature applications.

Fatigue properties at elevated temperatures are dependent on several variables including strain range, temperature, cyclic frequency, hold times, and the environment. The fatigue design curves in Fig. 33(a) show the simple case of pure fatigue (that is, continuous cycles without hold times) for 304 and 316 stainless steel. These design curves (from Code Case N-47 in the ASME Boiler Code) have a built-in factor of safety and are established by applying a safety factor of 2 with respect to strain range or a factor of 20 with respect to the number of cycles, whichever gives the lower value. The creep-life fraction is determined by the time-life fraction per cycle using assumed stresses 1.1 times the applied stress and the minimum stress-rupture curves incorporated in the code. The total damage must not exceed the envelope defined by the bilinear damage curve shown in Fig. 34.

The design curves in Fig. 33(a) are based on a strain rate of 1×10^{-3}/s. If the strain rate decreases, fatigue life also decreases. In Fig. 35, for example, the fatigue lives of several stainless steels are shown for continuous cycling at two different strain rates. Fatigue life is reduced with a lower strain rate, while grain size has little effect on fatigue life when life is determined from pure fatigue (or continuous cycling).

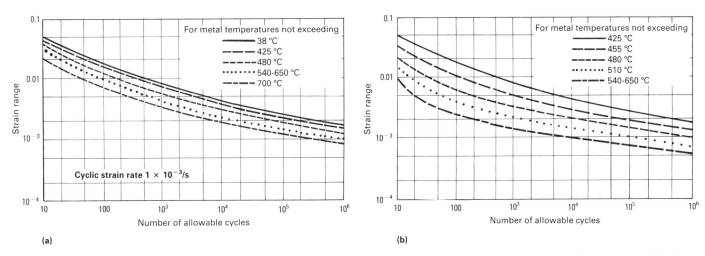

Fig. 33 Design fatigue-strain range curves for 304 and 316 stainless steel. (a) Design curves with continuous cycling (pure fatigue). (b) Design curves with hold times (creep-fatigue interaction)

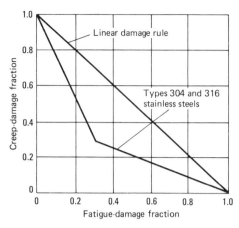

Fig. 34 Comparison of linear damage rule of creep-fatigue interaction with design envelopes in ASME Code Case N-47 for 304 and 316 stainless steel. Creep-damage fraction = time/time-to-rupture (multiplied by a safety factor). Fatigue-damage fraction = number of cycles/cycles to failures (multiplied by a safety factor).

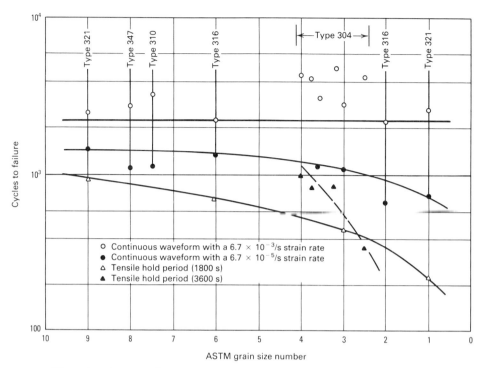

Fig. 35 Effect of strain rate and grain size on the fatigue life of various stainless steels at elevated temperatures. Grain size has the greatest influence on fatigue life when hold times are increased. Test conditions: total strain range = 1.0%; test temperature, 593 to 600 °C (1100 to 1110 °F). Source: Ref 24

When hold times are introduced, a different set of design curves is used (Fig. 33b) to determine the allowable fatigue-life fraction (creep-life fraction is determined the same way as for continuous cycling). These allowable fatigue-life curves are a more conservative set of curves than those of Fig. 33(a). They incorporate the effect of creep damage by applying a fatigue life reduction factor, which includes hold time effects in addition to the factor of safety (2 in strength and 20 in cycles, whichever gives the lower value). Figure 36 compares the 540 to 650 °C (1000 to 1200 °F) design curve in Fig. 33(b) with actual fatigue life results from testing 316 stainless at 593 °C (1100 °F) and various hold times. When hold times are introduced, the influence of grain size may also be more pronounced (Fig. 35).

Fatigue Crack Growth. Although S-N curves have been used in the past as the basic design tool against fatigue, their limitations have become increasingly obvious. One of the more serious limitations is that they do not distinguish between crack initiation and crack propagation. Particularly in the low-stress regions, a large fraction of the life of a component may be spent in crack propagation, allowing crack tolerance over a large portion of the life. Engineering structures often contain flaws or cracklike imperfections that may altogether eliminate the crack initiation step. A methodology that quantitatively describes crack growth as a function of the loading variables is, therefore, of great value in design and in assessing the remaining lives of components.

Because fatigue crack growth rates are obtained at various ΔK and temperature ranges, it is difficult to compare the various types of materials directly. At a constant ΔK (arbitrarily chosen as 30 MPa$\sqrt{m}$, or 27 ksi$\sqrt{in.}$), a clear trend of crack growth rate increase with increasing temperature can be seen (Fig. 37). In this figure it can be seen that at temperatures up to about 50% of the melting point (550 to 600 °C, or 1020 to 1110 °F), the growth rates are relatively insensitive to temperature, but sensitivity increases rapidly at higher temperatures. The crack growth rates for all the materials at temperatures up to 600 °C (1110 °F) relative to the room-temperature rates can be estimated by a maximum correlation factor of 5 (2 for ferritic steels).

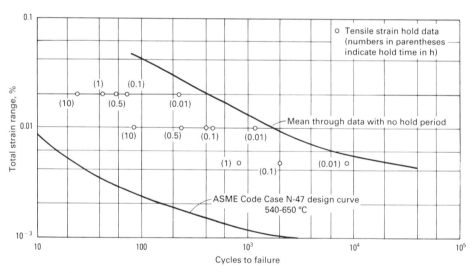

Fig. 36 Influence of tensile hold times at peak strain on failure life of a single heat of type 316 stainless steel tested at 593 °C (1100 °F). Source: Ref 24

Applications that use the heat-resisting capabilities of austenitic stainless steels to advantage include furnace parts, heat exchanger tubing, steam lines, exhaust systems in reciprocating engines and gas turbines, afterburner parts, and similar parts that require strength and oxidation resistance.

Type 304 has good resistance to atmospheric corrosion and oxidation. Types 309 and 310 rank higher in these properties because of their higher nickel and chromium contents. Type 310 is useful where intermittent heating and cooling are encoun-

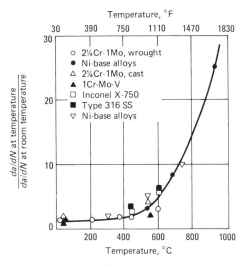

Fig. 37 Variation of fatigue crack growth rates as a function of temperature at $\Delta K = 30$ MPa$\sqrt{m}$ (27 ksi$\sqrt{in.}$). Source: Ref 2

tered, because it forms a more adherent scale than does type 309. Types 309 and 310 are used for parts such as firebox sheets, furnace linings, boiler baffles, thermocouple wells, aircraft cabin heaters, and jet engine burner liners.

The austenitic stainless steels may become susceptible to intergranular corrosion in some environments. This occurs when these steels are exposed to temperatures between 500 and 900 °C (950 and 1650 °F) and carbon diffuses to the grain boundaries to form chromium carbides. The formation of these carbides can significantly reduce the amount of free chromium in the alloy, thus impairing the corrosion resistance of the grain boundaries. The intergranular corrosion that occurs in the heat-affected zones of welds is often called weld knife edge attack. Types 321 and 347 (when correctly heat treated to form titanium and niobium carbides, respectively) and the naturally low-carbon grades 304L and 316L are resistant to intergranular corrosion. Intergranular attack is usually limited to aqueous environments and is covered in more detail in Volume 13 of the 9th Edition of *Metals Handbook*. When types 321H and 347H are used in applications where intergranular corrosion may be a problem, it is possible to apply an additional stabilizing heat treatment at a temperature near 900 °C (1650 °F) to produce stable titanium and niobium carbides and thus reduce the free carbon content.

Types 321 and 347 can be used where solution treatment after welding is not feasible, such as in steam lines and superheater tubes and exhaust systems in reciprocating engines and gas turbines that operate at temperatures from 425 to 870 °C (800 to 1600 °F). The low-carbon types 304L and 316L are used for similar applications but are more susceptible to intergranular attack during long exposure to high temperature.

Type 316 has better mechanical properties than 304 or 321 and is more resistant to corrosion in some media, such as fatty acids at elevated temperature and mild sulfuric acid solutions. The tensile and yield strengths of 304, 304L, 316, and 316L can be increased by alloying these grades with nitrogen. These modifications are designated as 304N (UNS S30451), 304LN (UNS S30454), 316N (UNS S31651), and 316LN (UNS S31653).

ACKNOWLEDGMENT

ASM INTERNATIONAL would like to thank R. Lula and W.C. Mack of Babcock & Wilcox, and V.K. Sikka of Oak Ridge National Laboratory for their helpful comments and suggestions in the preparation of this article.

REFERENCES

1. W.F. Simmons and J.A. Van Echo, "Report on the Elevated-Temperature Properties of Stainless Steels," ASTM Data Series, Publication DS-5-S1 (formerly STP 124), American Society for Testing and Materials
2. R. Viswanathan, *Damage Mechanisms and Life Assessment of High-Temperature Components*, ASM INTERNATIONAL, 1989
3. J. Wareing, *Met. Trans. A*, Vol 8A, 1977, p 711-721
4. C.R. Brinkman, G.E. Korth, and R.R. Hobbins, *Nucl. Tech.*, Vol 16, 1972, p 299-307
5. Y. Asada and S. Mitsuhaski, in *Fourth International Conference on Pressure Vessel Technology*, Vol 1, 1980, p 321
6. C.R. Brinkman and G.E. Korth, *Met. Trans.*, Vol 5, 1974, p 792
7. J. Wareing, *Met. Trans. A*, Vol 6A, 1975, p 1367
8. J. Wareing, H.G. Vaughan, and B. Tomkins, Report NDR-447S, United Kingdom Atomic Energy Agency, 1980
9. I.W. Goodall, R. Hales, and D.J. Walters, in *Proceedings of IUTAM 103*, International Union of Theoretical and Applied Mechanics, 1980
10. D.S. Wood, J. Wynn, A.B. Baldwin, and P. O'Riordan, *Fatigue Eng. Mater. Struct.*, Vol 3, 1980, p 89
11. J. Wareing, *Fatigue Eng. Mater. Struct.*, Vol 4, 1981, p 131
12. C.E. Jaske, M. Mindlin, and J.S. Perrin, Development of Elevated Temperature Fatigue Design Information for Type 316 Stainless Steel, International Conference on Creep and Fatigue, Conference Publication 13, Institute of Mechanical Engineers, 1973, p 163.1-163.7
13. S. Majumdar and P.S. Maiya, *J. Eng. Mater. Technol. (Trans. ASME)*, Vol 102 (No. 1), 1980, p 159
14. V.B. Livesey and J. Wareing, *Met. Sci.*, Vol 17, 1983, p 297
15. D. Gladwin and D.A. Miller, *Fatigue Eng. Mater. Struct.*, Vol 5, 1982, p 275-286
16. D.A. Miller, R.H. Priest, and E.G. Ellison, A Review of Material Response and Life Prediction Techniques Under Fatigue-Creep Loading Conditions, *High Temp. Mater. Process*, Vol 6 (No. 3 and 4), 1984, p 115-194
17. J.K. Lai and C.P. Horton, Report RD/L/R/200S, Central Electricity Generating Board, 1979
18. Y. Kadoya et al., Creep Fatigue Life Prediction of Turbine Rotors, in *Life Assessment and Improvement of Turbogenerator Rotors for Fossil Plants*, R. Viswanathan, Ed., Pergamon Press, 1985, p 3.101-3.114
19. H.E. McCoy, *Corrosion*, Vol 21, 1965, p 84
20. J.D. McCoy, "Corrosion Rates for H_2S at Elevated Temperatures in Refinery Hydrosulfurization Processes," Paper 128, National Association of Corrosion Engineers, 1974
21. T. Fujita, Advanced High-Chromium Ferritic Steels for High Temperatures, *Met. Prog.*, Vol 130, Aug 1986, p 33
22. W.F. Simmons and H.C. Cross, "Report on the Elevated Temperature Properties of Chromium Steels, 12 to 27 Percent," STP 228, American Society for Testing and Materials, 1952
23. P. Marshall, *Austenitic Stainless Steels Microstructure and Mechanical Properties*, Elsevier, 1984
24. C.R. Brinkman, High-temperature time-dependent fatigue behavior of several engineering structural alloys, *International Metals Review*, Vol 30 (No. 5), 1985, p 235-258

SELECTED REFERENCES

- *Austenitic Chromium-Nickel Stainless Steels: Engineering Properties at Elevated Temperatures*, INCO Europe Limited, 1963
- *High Temperature Characteristics of Stainless Steels*, American Iron and Steel Institute, 1979
- R.A. Lula, *Source Book on the Ferritic Stainless Steels*, American Society for Metals, 1982
- P. Marshall, *Austenitic Stainless Steels Microstructure and Mechanical Properties*, Elsevier, 1984
- G.V. Smith, *Evaluations of the Elevated Temperature Tensile and Creep Rupture Properties of 12 to 27 Percent Chromium Steels*, DS 59, American Society for Testing and Materials, 1980
- R. Viswanathan, *Damage Mechanisms and Life Assessment of High-Temperature Components*, ASM INTERNATIONAL, 1989

Wrought and P/M Superalloys

N.S. Stoloff, Rensselaer Polytechnic Institute

SUPERALLOYS are heat-resisting alloys based on nickel, nickel-iron, or cobalt that exhibit a combination of mechanical strength and resistance to surface degradation. Superalloys are primarily used in gas turbines, coal conversion plants, and chemical process industries, and for other specialized applications requiring heat and/or corrosion resistance. The modern high-performance aircraft (jet) engine could not operate without the major advances made in superalloy development over the past 50 years. A noteworthy feature of nickel-base alloys is their use in load-bearing applications at temperatures in excess of 80% of their incipient melting temperatures, a fraction that is higher than for any other class of engineering alloys.

This article focuses on the properties of conventional wrought superalloys based on nickel, iron, and cobalt, as well as on the properties of alloys produced from powder. The powder metallurgy (P/M) category includes alloys that were originally developed as casting alloys; new alloy compositions developed specifically to benefit from powder processing; and oxide dispersion strengthened (ODS) alloys (particularly those produced by mechanical alloying). The ODS alloys based on nickel and iron have been commercialized, whereas those based on cobalt have not. Other types of superalloys are described in the articles "Polycrystalline Cast Superalloys" and "Directionally Solidified and Single-Crystal Superalloys" in this Volume.

Applications of superalloys are categorized below; the bulk of tonnage is used in gas turbines:

- *Aircraft gas turbines*: disks, combustion chambers, bolts, casings, shafts, exhaust systems, cases, blades, vanes, burner cans, afterburners, thrust reversers
- *Steam turbine power plants*: bolts, blades, stack gas reheaters
- *Reciprocating engines*: turbochargers, exhaust valves, hot plugs, valve seat inserts
- *Metal processing*: hot-work tools and dies, casting dies
- *Medical applications*: dentistry uses, prosthetic devices
- *Space vehicles*: aerodynamically heated skins, rocket engine parts
- *Heat-treating equipment*: trays, fixtures, conveyor belts, baskets, fans, furnace mufflers
- *Nuclear power systems*: control rod drive mechanisms, valve stems, springs, ducting
- *Chemical and petrochemical industries*: bolts, fans, valves, reaction vessels, piping, pumps
- *Pollution control equipment*: scrubbers
- *Metals processing mills*: ovens, afterburners, exhaust fans
- *Coal gasification and liquefaction systems*: heat exchangers, reheaters, piping

Many technical considerations, such as formability, strength, creep resistance, fatigue strength, and surface stability, must be evaluated when selecting a superalloy for any of the applications identified above. Unfortunately, those compositional and microstructural variables that benefit one property may result in undesirable performance in another area. For example, fine grain size is desirable for low-temperature tensile strength, fatigue crack initiation resistance, and high-temperature formability, but creep resistance is usually adversely affected. Similarly, high chromium contents in nickel alloys improve the resistance to oxidation and hot corrosion, but result in lower tensile and creep strengths and promote the formation of σ phase. Further, the more temperature resistant the alloy, the more likely it is to be segregation prone and, perhaps, brittle, and thus formable only by casting to shape or by using powder processing. For these and other compelling reasons, the interplay between composition, microstructure, consolidation method, mechanical properties, and surface stability is emphasized in this article.

Wrought Nickel Alloys

Nickel alloys in commercial service and under development range from single-phase alloys to precipitation-hardened superalloys and oxide dispersion strengthened alloys and composites, the latter of which is described in the section "P/M Alloys" in this article. Nickel-base superalloys are the most complex, the most widely used for the hottest parts, and, to many metallurgists, the most interesting of all superalloys. They currently constitute over 50% of the weight of advanced aircraft engines. Their use in cast form extends to the highest homologous temperature of any common alloy system (see the article "Polycrystalline Cast Superalloys" in this Volume).

The principal characteristics of nickel as an alloy base are the high phase stability of the face-centered cubic (fcc) nickel matrix and the capability to be strengthened by a variety of direct and indirect means. Further, the surface stability of nickel is readily improved by alloying with chromium and/or aluminum. In order to adequately describe mechanical behavior, however, it is first necessary to consider the composition and microstructure of the various classes of nickel alloys.

Chemical Composition

The compositions of many representative nickel-base wrought alloys are listed in Table 1. They can be categorized as nickel-iron-base alloys, in which nickel is the major solute element, or nickel-base, in which at least 50% Ni is present. The nickel-iron alloys are discussed in detail in a later section. The nickel-base superalloys discussed below are considered to be complex because they incorporate as many as a dozen elements. In addition, deleterious elements such as silicon, phosphorus, sulfur, oxygen, and nitrogen must be controlled through appropriate melting practices. Other trace elements, such as selenium, bismuth, and lead, must be held to very small (ppm) levels in critical parts.

Many wrought nickel-base superalloys contain 10 to 20% Cr, up to about 8% Al and Ti combined, 5 to 15% Co, and small amounts of boron, zirconium, magnesium, and carbon. Other common additions are molybdenum, niobium, and tungsten, all of which play dual roles as strengthening solutes and carbide formers. Chromium and aluminum are also necessary to improve surface stability through the formation of

Table 1 Nominal compositions of wrought nickel-base alloys

Alloy	Ni	Cr	Co	Mo	W	Nb	Al	Ti	Fe	Mn	Si	C	B	Zr	Other
Astroloy	55.0	15.0	17.0	5.3	...	...	4.0	3.5	...	...	...	0.06	0.030	...	...
Cabot 214	75.0	16.0	...	...	...	...	4.5	...	2.5	...	...	...	...	...	0.01 Y
D-979	45.0	15.0	...	4.0	...	...	1.0	3.0	27.0	0.3	0.2	0.05	0.010	...	...
Hastelloy C-22	51.6	21.5	2.5	13.5	4.0	...	...	...	5.5	1.0	0.1	0.01	...	...	0.3 V
Hastelloy C-276	...	15.5	2.5	16.0	3.7	...	...	...	5.5	1.0	0.1	0.01	...	...	0.3 V
Hastelloy G-30	42.7	29.5	2.0	5.5	2.5	0.8	...	...	15.0	1.0	1.0	0.03	...	...	2.0 Cu
Hastelloy S	67.0	15.5	...	14.5	...	...	0.3	...	1.0	0.5	0.4	...	0.009	...	0.05 La
Hastelloy X	47.0	22.0	1.5	9.0	0.6	...	...	...	18.5	0.5	0.5	0.10	...	...	...
Haynes 230	57.0	22.0	...	2.0	14.0	...	0.3	...	...	0.5	0.4	0.10	...	...	0.02 La
Inconel 587(a)	bal	28.5	20.0	...	...	0.7	1.2	2.3	...	...	...	0.05	0.003	0.05	...
Inconel 597(a)	bal	24.5	20.0	1.5	...	1.0	1.5	3.0	...	...	...	0.05	0.012	0.05	0.02 Mg
Inconel 600	76.0	15.5	...	...	...	...	...	...	8.0	0.5	0.2	0.08	...	...	...
Inconel 601	60.5	23.0	...	...	...	...	1.4	...	14.1	0.5	0.2	0.05	...	...	...
Inconel 617	54.0	22.0	12.5	9.0	...	...	1.0	0.3	...	...	...	0.07	...	...	...
Inconel 625	61.0	21.5	...	9.0	...	3.6	0.2	0.2	2.5	0.2	0.2	0.05	...	...	...
Inconel 706	41.5	16.0	...	...	...	2.9	0.2	1.8	40.0	0.2	0.2	0.03	...	...	...
Inconel 718	52.5	19.0	...	3.0	...	5.1	0.5	0.9	18.5	0.2	0.2	0.04	...	...	...
Inconel X750	73.0	15.5	...	...	...	1.0	0.7	2.5	7.0	0.5	0.2	0.04	...	...	...
M-252	55.0	20.0	10.0	10.0	...	...	1.0	2.6	...	0.5	0.5	0.15	0.005	...	...
Nimonic 75	76.0	19.5	...	...	...	...	...	0.4	3.0	0.3	0.3	0.10	...	...	...
Nimonic 80A	76.0	19.5	...	...	...	...	1.4	2.4	...	0.3	0.3	0.06	0.003	0.06	...
Nimonic 90	59.0	19.5	16.5	...	...	...	1.5	2.5	...	0.3	0.3	0.07	0.003	0.06	...
Nimonic 105	53.0	15.0	20.0	5.0	...	...	4.7	1.2	...	0.3	0.3	0.13	0.005	0.10	...
Nimonic 115	60.0	14.3	13.2	...	...	...	4.9	3.7	...	...	...	0.15	0.160	0.04	...
Nimonic 263	51.0	20.0	20.0	5.9	...	...	0.5	2.1	...	0.4	0.3	0.06	0.001	0.02	...
Nimonic 942(a)	bal	12.5	...	6.0	...	...	0.6	3.7	37	0.2	0.30	0.03	0.010	...	...
Nimonic PE.11(a)	bal	18.0	...	5.2	...	...	0.8	2.3	35	0.20	0.30	0.05	0.03	0.2	...
Nimonic PE.16	43.0	16.5	1.0	1.1	...	...	1.2	1.2	33.0	0.1	0.1	0.05	0.020	...	...
Nimonic PK.33	56.0	18.5	14.0	7.0	...	...	2.0	2.0	0.3	0.1	0.1	0.05	0.030	...	...
Pyromet 860	43	12.6	4.0	6.0	...	...	1.25	3.0	30.0	0.05	0.05	0.05	0.010	...	...
René 41	55.0	19.0	11.0	1.0	...	...	1.5	3.1	...	...	...	0.09	0.005	...	...
René 95	61.0	14.0	8.0	3.5	3.5	3.5	3.5	2.5	...	...	...	0.15	0.010	0.05	...
Udimet 400(a)	bal	17.5	14.0	4.0	...	0.5	1.5	2.5	...	...	...	0.06	0.008	0.06	...
Udimet 500	54.0	18.0	18.5	4.0	...	...	2.9	2.9	...	...	...	0.08	0.006	0.05	...
Udimet 520	57.0	19.0	12.0	6.0	1.0	...	2.0	3.0	...	...	...	0.05	0.005	...	...
Udimet 630(a)	bal	18.0	...	3.0	3.0	6.5	0.5	1.0	18.0	...	...	0.03	...	...	...
Udimet 700	55.0	15.0	17.0	5.0	...	...	4.0	3.5	...	...	...	0.06	0.030	...	...
Udimet 710	55.0	18.0	15.0	3.0	1.5	...	2.5	5.0	...	...	...	0.07	0.020	...	...
Udimet 720	55.0	17.9	14.7	3.0	1.3	...	2.5	5.0	...	...	...	0.03	0.033	0.03	...
Unitemp AF2-1DA6	60.0	12.0	10.0	2.7	6.5	...	4.0	2.8	...	...	...	0.04	0.015	0.10	1.5 Ta
Waspaloy	58.0	19.5	13.5	4.3	...	...	1.3	3.0	...	...	...	0.08	0.006	...	...

(a) Ref 2 (1984 data). Bal, balance. Source: Ref 1

Table 2 Role of elements in superalloys

Effect	Iron base	Cobalt base	Nickel base
Solid-solution strengtheners	Cr, Mo	Nb, Cr, Mo, Ni, W, Ta	Co, Cr, Fe, Mo, W, Ta
Fcc matrix stabilizers	C, W, Ni	Ni	...
Carbide form			
MC type	Ti	Ti, Ta, Nb	W, Ta, Ti, Mo, Nb
M_7C_3 type	...	Cr	Cr
$M_{23}C_6$ type	Cr	Cr	Cr, Mo, W
M_6C type	Mo	Mo, W	Mo, W
Carbonitrides			
M(CN) type	C, N	C, N	C, N
Forms γ' Ni_3 (Al, Ti)	Al, Ni, Ti	...	Al, Ti
Retards formation of hexagonal η (Ni_3Ti)	Al, Zr	...	...
Raises solvus temperature of γ'	...	...	Co
Hardening precipitates and/or intermetallics	Al, Ti, Nb	Al, Mo, Ti(a), W, Ta	Al, Ti, Nb
Forms γ'' (Ni_3Nb)	...	...	Nb
Oxidation resistance	Cr	Al, Cr	Al, Cr
Improves hot corrosion resistance	La, Y	La, Y, Th	La, Th
Sulfidation resistance	Cr	Cr	Cr
Increases rupture ductility	B	B, Zr	B(b), Zr
Causes grain-boundary segregation	...	...	B, C, Zr
Facilitates working	...	Ni_3Ti	...

(a) Hardening by precipitation of Ni_3Ti also occurs if sufficient Ni is present. (b) If present in large amounts, borides are formed. Source: Ref 3

Cr_2O_3 and Al_2O_3, respectively. The functions of the various elements in nickel alloys are summarized in Table 2, where in addition, they are compared to iron- and cobalt-base alloys. Other alloys that have been developed primarily for low-temperature service, often in corrosive environments (refer to the Hastelloy series and Inconel 600 shown in Table 1), are likely to contain chromium, molybdenum, iron, or tungsten in solution, with little or no second phase present.

Microstructure

The major phases that may be present in nickel-base alloys are:

- *Gamma matrix*, γ, in which the continuous matrix is an fcc nickel-base nonmagnetic phase that usually contains a high percentage of solid-solution elements such as cobalt, iron, chromium, molybdenum, and tungsten. All nickel-base alloys contain this phase as the matrix
- *Gamma prime*, γ', in which aluminum and titanium are added in amounts required to precipitate fcc γ' (Ni_3Al,Ti), which precipitates coherently with the austenitic gamma matrix. Other elements, notably niobium, tantalum, and chromium, also enter γ'. This phase is required for high-temperature strength and creep resistance.
- *Gamma double prime*, γ'', in which nickel and niobium combine in the presence of iron to form body-centered tetragonal (bct) Ni_3Nb, which is coherent with the gamma matrix, while inducing large mismatch strains of the order of 2.9%. This phase provides very high strength at low

Table 3 Solution treatments for selected wrought nickel-base superalloys
As used in blades or in creep-limiting applications

Alloy	Solution temperature(a) °C	°F	Time, h
Inconel X750	1150	2100	4
Nimonic 90	1080	1975	8
Nimonic 105	1125–1150	2060–2100	4
Udimet 500	1175	2150	2
Udimet 700	1175	2150	4
Waspaloy	1080	1975	4

(a) All materials air cooled after solution treatment

to intermediate temperatures, but is unstable at temperatures above about 650 °C (1200 °F). This precipitate is found in nickel-iron alloys.
- *Grain boundary* γ', a film of γ' along the grain boundaries in the stronger alloys, produced by heat treatments and service exposure. This film is believed to improve rupture properties
- *Carbides*, in which carbon that is added in amounts of about 0.02 to 0.2 wt% combines with reactive elements, such as titanium, tantalum, hafnium, and niobium, to form metal carbides (MC). During heat treatment and service, these MC carbides tend to decompose and generate other carbides, such as $M_{23}C_6$ and/or M_6C, which tend to form at grain boundaries. Carbides in nominally solid-solution alloys may form after extended service exposures
- *Borides*, a relatively low density of boride particles formed when boron segregates to grain boundaries
- *Topologically close-packed (TCP)-type phases*, which are platelike or needle-like phases such as σ, μ, and Laves that may form for some compositions and under certain conditions. These cause lowered rupture strength and ductility. The likelihood of their presence increases as the solute segregation of the ingot increases

In solid-solution alloys such as Inconel 600 and Hastelloy C (Table 1), only the gamma matrix is present. In nickel-base superalloys, however, the above phases, except γ'', are generally present. Solution treatments are designed to dissolve most precipitated phases (see Table 3). On the other hand, several nickel-iron superalloys, such as Inconel 706 and Inconel 718, contain γ'' Ni_3Nb as the principal precipitate, as well as γ'. Further, oxide dispersion strengthened alloys contain a few volume percent of a dispersed phase such as Y_2O_3 in a γ-γ' matrix, while composites (mechanically incorporated) contain tungsten or tungsten alloy fibers.

The Gamma Matrix (γ). Pure nickel does not display an unusually high elastic modulus or low diffusivity (two factors that promote creep rupture resistance), but the gamma matrix is readily strengthened for the most severe temperature and time conditions. Some superalloys can be used at 0.85 T_m (melting point) and, for times up to 100 000 h, at somewhat lower temperatures. These conditions can be tolerated because of three factors (Ref 4):

- The high tolerance of nickel for solutes without phase instability, because of its nearly filled $d-$ shell
- The tendency, with chromium additions, to form Cr_2O_3 having few cation vacancies, thereby restricting the diffusion rate of metallic elements outward and the rate of oxygen, nitrogen, and sulfur inward (Ref 5)
- The additional tendency, at high temperatures, to form Al_2O_3 barriers, which display exceptional resistance to further oxidation

Gamma prime is an intermetallic compound of nominal composition Ni_3Al, which is stable over a relatively narrow range of compositions (Fig. 1). It precipitated as spheroidal particles in early nickel-base alloys, which tended to have a low volume fraction of particles (see Fig. 2a). Later, cuboidal precipitates were noted in alloys

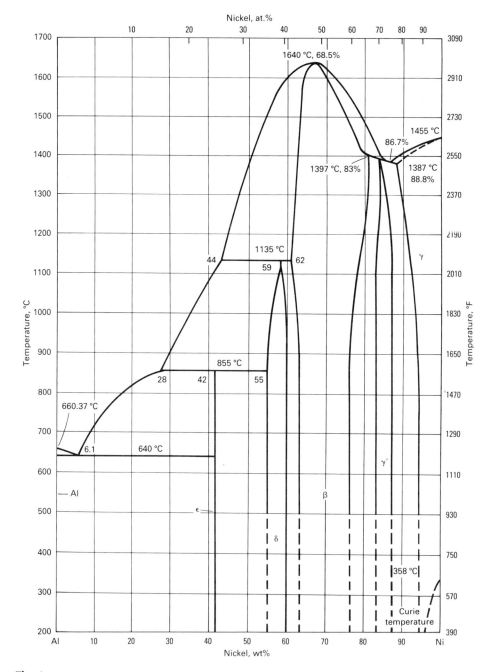

Fig. 1 Nickel-aluminum phase diagram

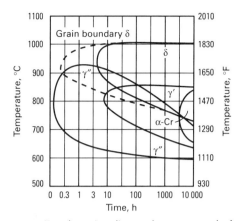

Fig. 4 Transformation diagram for vacuum-melted and hot-forged Inconel 718 bar. Source: Ref 8

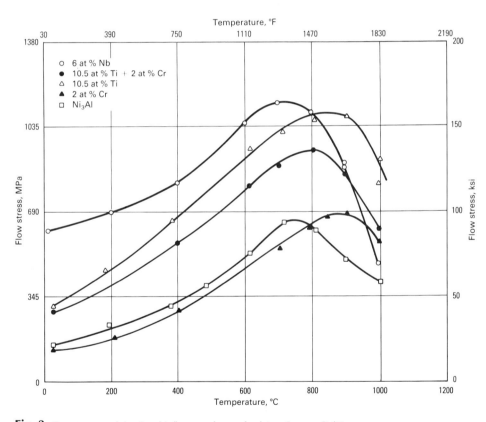

Fig. 2 Microstructure of (a) fully heat-treated Waspaloy showing MC and $M_{23}C_6$ carbides. 3400×. (b) Fully heat-treated Udimet 700 showing cubical γ'. 6800×. Source: Ref 6

Fig. 3 Flow stress peak in γ' and influence of several solutes. Source: Ref 7

with higher aluminum and titanium contents (Fig. 2b). The change in morphology is related to a matrix-precipitate mismatch. It is observed that γ' occurs as spheres for 0 to 0.2% mismatches, becomes cuboidal for mismatches of 0.5 to 1%, and is platelike at mismatches above about 1.25%.

To understand fully the vital role of γ' in the nickel-base superalloys, it is necessary to consider the structure and properties of this phase in some detail. Gamma prime is a superlattice that possesses the Cu_3Au ($L1_2$)-type structure, which exhibits long-range order to its melting point of 1385 °C (2525 °F). It exists over a fairly restricted range of composition, but alloying elements may substitute to a considerable degree for either of its constituents. In particular, most nickel-base alloys are strengthened by a precipitate in which up to 60% of the aluminum can be substituted by titanium and/or niobium. Also, nickel sites in the superlattice may be occupied by iron or cobalt atoms.

Both single crystals and polycrystals of unalloyed γ' exhibit a startling, reversible increase in flow stress between −196 and about 800 °C (−320 and 1470 °F), which is highly dependent on solute content (Ref 7), as shown in Fig. 3. Several other superlattices, such as Ni_3Si, Co_3Ti, Ni_3Ge, and Ni_3Ga, all of $L1_2$ structure, display increasing strength over a temperature range comparable to that of Ni_3Al.

The magnitude and temperature position of the peak in flow stress of γ' may be shifted by alloying elements such as titanium, chromium, and niobium (Fig. 3). There is no simple relation between the magnitude of flow stress increase and the change in the temperature of the peak. Tantalum, niobium, and titanium are effective solid-solution hardeners of γ' at room temperature. Tungsten and molybdenum are strengtheners at both room and elevated temperatures, while cobalt does not solid-solution strengthen γ'.

Gamma double prime (γ''), a bct coherent precipitate of composition Ni_3Nb, precipitates in nickel-iron-base alloys such as Inconel 706 and Inconel 718. In the absence of iron, or at temperatures and times shown in the transformation diagram of an iron-containing alloy (Fig. 4), an orthorhombic precipitate of the same Ni_3Nb composition (delta phase) forms instead. The latter is invariably incoherent and does not confer strength when present in large quantities. However, small amounts of delta phase can be used to control and refine grain size, resulting in improved tensile properties, fatigue resistance, and creep rupture ductility. Careful heat treatment is required to ensure precipitation of γ'' instead of δ.

Gamma double prime often precipitates together with γ' in Inconel 718, but γ'' is the principal strengthening phase under such circumstances. Unlike γ', which causes strengthening through the necessity to disorder the particles as they are sheared, γ'' strengthens by virtue of high coherency

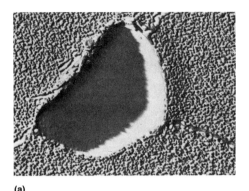
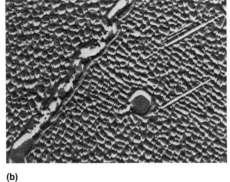

Fig. 5 Udimet 700 nickel-base heat-resistant alloy. (a) Udimet 700 solution annealed at 1177 °C (2150 °F) for 4 to 6 h and then aged 5000 h at 760 °C (1400 °F). Replica electron micrograph shows large particle of MC at grain-boundary intersection and γ' in grains of γ matrix. 4500×. (b) Udimet 700 solution annealed as for (a) and aged for 5000 h at 815 °C (1500 °F). Replica electron micrograph shows acicular sigma, carbide ($M_{23}C_6$) at grain boundary, and γ' within grains of the γ' matrix. 4700×. Source: Ref 9

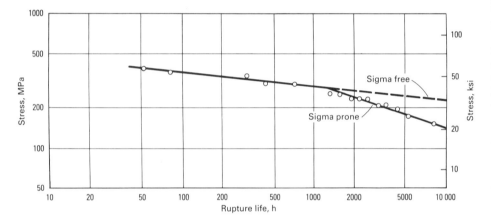

Fig. 6 Log stress versus log rupture life at 815 °C (1500 °F) for Udimet 700. Source: Ref 12

strains in the lattice. More detailed descriptions of the physical metallurgy of γ/γ" alloys appears in the section "Iron-Base Superalloys" in this article.

Grain-Boundary Chemistry. The improvement of creep properties by very small additions of boron and zirconium is a notable feature of nickel-base superalloys. Improved forgeability and better properties have also resulted from magnesium additions of 0.01 to 0.05% (Ref 5). It is believed that this is due primarily to the tying up of sulfur, a grain-boundary embrittler, by the magnesium.

Mechanisms for these property effects are unclear. However, it is believed that boron and zirconium segregate to grain boundaries because of their large size misfit with nickel. Because, at higher temperatures, cracks in superalloys usually propagate along grain boundaries, the importance of grain-boundary chemistry is apparent. Although early work on coarse-grain materials suggested that boron and zirconium influence rupture properties because of their effects on carbide and γ' distribution, recent work on P/M superalloys has revealed no such effects with ultrafine grain size.

Boron and zirconium also improve the rupture life of γ'-free alloys, cobalt alloys, and stainless steels, so that microstructural alterations cannot, in any case, apply to all systems. Boron may also reduce carbide precipitation at grain boundaries by releasing carbon into the grains. Magnesium may have a similar effect in a nickel-chromium-titanium-aluminum alloy in which intergranular MC has been noted (Ref 5). Finally, the segregation of misfitting atoms to grain boundaries may reduce grain-boundary diffusion rates. Direct evidence of a lowering of grain-boundary diffusivity by 0.11% Zr in Ni-20% Cr alloys over the range of 800 to 1200 K has recently been reported (Ref 10). This effect was accompanied by the precipitation of several precipitates containing zirconium at grain boundaries. The creep strength increased significantly, and the fracture mode changed from intergranular to transgranular ductile rupture, with a corresponding substantial increase in ductility.

Carbides in superalloys serve a number of functions. They often precipitate at grain boundaries in nickel alloys, whereas in cobalt and iron superalloys, intergranular sites are common. Early work suggested that some grain-boundary carbides were detrimental to ductility, but most investigators now believe that discrete carbides (as well as magnesium) exert a beneficial effect on rupture strength at high temperatures.

Carbide Types and Typical Morphologies. The common nickel-base alloy carbides are MC, $M_{23}C_6$, and M_6C. MC usually exhibits a coarse, random, cubic (Fig. 5a), or script morphology. The carbide $M_{23}C_6$ is found primarily at grain boundaries (Fig. 5b) and usually occurs as irregular, discontinuous, blocky particles, although plates and regular geometric forms have been observed. The M_6C carbides also can precipitate in blocky form in grain boundaries and less often in a Widmanstätten intragranular morphology. Although data are sparse, it appears that continuous grain-boundary $M_{23}C_6$ and Widmanstätten M_6C, caused by an improper choice of processing or heat-treatment temperatures, are to be avoided for best ductility and rupture life.

MC carbides, fcc in structure, usually form in superalloys during freezing. They are distributed heterogeneously through the alloy, both in intergranular and transgranular positions, often interdendritically. Little or no orientation relation with the alloy matrix has been noted. MC carbides are a major source of carbon for subsequent phase reactions during heat treatment and service (Ref 5). In some alloys, such as Incoloy 901 and A286, MC films may form along grain boundaries and reduce ductility.

These carbides, for example, TiC and HfC, are among the most stable compounds in nature. The preferred order of formation (in order of decreasing stability) in superalloys for these carbides is HfC, TaC, NbC, and TiC. This order is not the same as that of thermodynamic stability, which is HfC, TiC, TaC, and NbC. In these carbides, M atoms can readily substitute for each other, as in (Ti, Nb)C. However, the less reactive elements, principally molybdenum and tungsten, can also substitute in these carbides. For example, (Ti,Mo)C is found in Udimet 500, M-252, and René 77. It appears that the change in stability order cited above is due to the molybdenum or tungsten substitution, which weakens the binding forces in MC carbides to such an extent that degeneration reactions, discussed later, can occur. This typically leads to the formation of the more stable compounds $M_{23}C_6$- and M_6C-type carbides in the alloys during processing or after heat treatment and/or service. Additions of niobium and tantalum tend to counteract this effect. Recent alloys with high niobium and tantalum contents contain MC carbides that do not break down easily during processing or solution treatment in the range of 1200 to 1260 °C (2190 to 2300 °F).

The carbides $M_{23}C_6$ readily form in alloys with moderate to high chromium content. They form during lower-temperature heat treatment and service (that is, 760 to 980 °C,

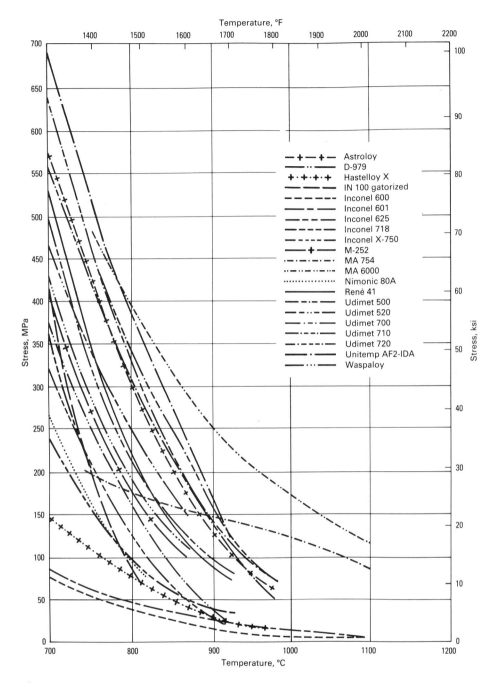

Fig. 7(a)   1000-h rupture strengths of selected wrought nickel-base superalloys

The M_6C carbides have a complex cubic structure. They form when the molybdenum and/or tungsten content is more than 6 to 8 at.%, typically in the range of 815 to 980 °C (1500 to 1800 °F). The M_6C forms with $M_{23}C_6$ in René 80, René 41, and AF 1753. Typical formulas for M_6C are $(Ni,Co)_3Mo_3C$ and $(Ni,Co)_2W_4C$, although a wider range of compositions has been reported for Hastelloy X. Therefore, M_6C carbides are formed when molybdenum or tungsten acts to replace chromium in other carbides; unlike the more rigid $M_{23}C_6$, the compositions can vary widely. Because M_6C carbides are stable at higher levels than are $M_{23}C_6$ carbides, M_6C is more commercially important as a grain-boundary precipitate for controlling grain size during the processing of wrought alloys.

Carbide Reactions. MC carbides are a major source of carbon in most nickel-base superalloys below 980 °C (1800 °F). However, MC decomposes slowly during heat treatment and service, releasing carbon for several important reactions.

The principal carbide reaction in many alloys is believed to be the formation of $M_{23}C_6$ (Ref 5):

$MC + \gamma \rightarrow M_{23}C_6 + \gamma'$ or

$(Ti,Mo)C + (Ni,Cr,Al,Ti) \rightarrow Cr_{21}Mo_2C_6$
$ + Ni_3(Al,Ti)$ (Eq 1)

The carbide M_6C can form in a similar manner.

Also, M_6C and $M_{23}C_6$ interact, forming one from the other (Ref 4):

$M_6C + M' \rightarrow M_{23}C_6 + M''$ (Eq 2)

or

$Mo_3(Ni,Co)_3C + Cr \leftrightarrows Cr_{21}Mo_2C_6$
$ + (Ni,Co,Mo)$ (Eq 3)

depending on the alloy. For example, René 41 and M-252 can be heat treated to generate MC and M_6C initially, with long-time exposure causing the conversion of M_6C to $M_{23}C_6$ (Ref 4).

These reactions lead to carbide precipitation in various locations, but typically at grain boundaries. Perhaps the most beneficial reaction (for high creep resistance applications), one that is controlled in many heat treatments, is that shown in Eq 1. Both the blocky carbides and the γ' produced are important in that they may inhibit grain-boundary sliding. In many cases, the γ' generated by this reaction coats the carbides, and the grain boundary becomes a relatively ductile, creep-resistant region.

Borides. Small additions of boron are essential to improved creep-rupture resistance of superalloys. Borides are hard particles, blocky to half moon in appearance, that are observed at grain boundaries. The boride found in superalloys is of the form M_3B_2, with a tetragonal unit cell. At least two types of borides have been observed in

or 1400 to 1800 °F) both from the degeneration of MC carbide and from soluble residual carbon in the alloy matrix. Although usually seen at grain boundaries, they occasionally occur along twin bands, stacking faults, and at twin ends. The carbides $M_{23}C_6$ have a complex cubic structure, which, if the carbon atoms were removed, would closely approximate the structure of the TCP σ phase. In fact, σ plates often nucleate on $M_{23}C_6$ particles.

When tungsten or molybdenum is present, the approximate composition of $M_{23}C_6$ is $Cr_{21}(Mo,W)_2C_6$, although it also has been shown that appreciable nickel can substitute in the carbide. It is also possible for small amounts of cobalt or iron to substitute for chromium.

The $M_{23}C_6$ particles strongly influence the properties of nickel alloys. Rupture strength is improved by the presence of discrete particles, apparently through the inhibition of grain-boundary sliding. Eventually, however, failure can initiate either by fracture of particles or by decohesion of the carbide/matrix interface. In some alloys, cellular structures of $M_{23}C_6$ have been noted. These can cause premature failures, but can be avoided by proper processing and/or heat treatment.

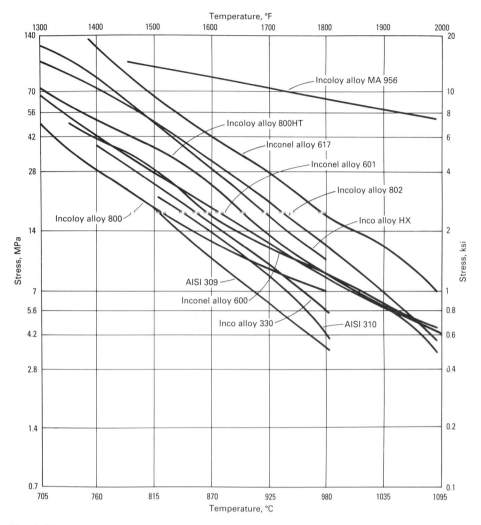

Fig. 7(b) 10 000-h rupture strengths of nickel-base alloys and stainless steels

Table 4 Incipient melting temperatures of selected wrought superalloys

Alloy	Incipient melting temperature	
	°C	°F
Hastelloy X (Ni)	1250	2280
L-605 (Co)	1329	2425
Haynes 188 (Co)	1302	2375
Incoloy 800 (Ni)	1357	2475
Incoloy 825 (Fe)	1370	2500
Incoloy 617 (Ni)	1333	2430
Inconel 625 (Ni)	1288	2350
Inconel X750 (Ni)	1393	2540
Nimonic 80A (Ni)	1360	2480
Nimonic 90 (Ni)	1310	2390
Nimonic 105 (Ni)	1290	2354
René 41 (Ni)	1232	2250
Udimet 500 (Ni)	1260	2300
Udimet 700 (Ni)	1216	2220
Waspaloy (Ni)	1329	2425

Source: Ref 3

Udimet 700; the type observed depends upon the thermal history of the alloy. These boride types are $(Mo_{0.48}Ti_{0.07}Cr_{0.39}Ni_{0.3}Co_{0.3})_3 B_2$ and $(Mo_{0.31}Ti_{0.07}Cr_{0.49}Ni_{0.08}Co_{0.07})_3 B_2$. However, when Phacomp procedures are used to estimate long-term alloy stability, the composition $(Mo_{0.5}Ti_{0.15}Cr_{0.25}Ni_{0.10})_3 B_2$ is usually assumed. Because the level of boron added rarely exceeds 0.03 wt%, and because it is often substantially less than the solubility limit of 0.01%, the volume fraction of borides tends to be quite small. In fact, direct observation has been made of boron segregation to grain boundaries in Udimet 700 containing 0.03 wt% B, but grain boundaries decorated with fine borides were difficult to find in these observations (Ref 11).

TCP Phases. In some alloys, if composition has not been carefully controlled, undesirable phases can form either during heat treatment or, more commonly, during service. These precipitates, known as TCP phases, are composed of close-packed layers of atoms parallel to {111} planes of the γ matrix. Usually harmful, they may appear as long plates or needles, often nucleating on grain-boundary carbides. Nickel alloys are especially prone to the formation of σ and μ. The formula for σ is $(Fe,Mo)_x(Ni,Co)_y$, where x and y can vary from 1 to 7. Alloys containing a high level of body-centered cubic (bcc) transition metals (tantalum, niobium, chromium, tungsten, and molybdenum) are most susceptible to TCP formation.

The σ hardness and its platelike morphology cause premature cracking, leading to low-temperature brittle failure, although yield strength is unaffected. However, the major effect is on elevated-temperature rupture strength, as shown for Udimet 700 in Fig. 6. Sigma formation must deplete refractory metals in the γ matrix, causing loss of strength of the matrix. Also, high-temperature fracture can occur along σ plates rather than along the normal intergranular path, resulting in sharply reduced rupture life. Platelike μ can form also, but little is known about its detrimental effects.

Heat Treatment

Wrought alloys are, first, solution treated to dissolve nearly all γ' and carbides other than the very stable MC carbides. Typical solution treatments (for creep-limited applications) are in the range of 1050 to 1200 °C (1920 to 2190 °F) (Table 3) and may be followed by a second solution treatment at lower temperature. Some γ' can form upon air cooling from the solution treatment temperature. Aging is then carried out in several steps to coarsen the γ' that is formed upon cooling, as well as to precipitate additional γ'.

A two-step aging treatment is commonly used, with the first treatment in the range of 850 to 1100 °C (1560 to 2010 °F) over a period of up to 24 h. Aging at one or more lower temperatures, for example at 760 °C (1400 °F) for 16 h, completes the precipitation of γ'. The finer γ' produced in the second aging treatment is advantageous for tensile strength as well as for rupture life. Both solution and aging anneals are followed by air cooling.

Carbide distribution also is controlled by the heat treatment schedule. Modifications to the γ' heat treatment procedure often are required to avoid problems with carbide films at grain boundaries. For example, a solution treatment of René 41 at 1175 °C (2150 °F) leads to subsequent precipitation of a grain-boundary film of $M_{23}C_6$, with deleterious effects on mechanical properties. Therefore, a lower solution treatment temperature (about 1075 °C, or 1970 °F) is used to preserve the fine-grain as-worked structure with well-dispersed M_6C (Ref 4).

Additional heat treatments may be carried out in connection with the application of diffusion or overlay coatings, although in some cases the coating treatment coincides with an aging treatment.

Alloying for Surface Stability

Low-Temperature Corrosion. High chromium contents are required in nickel-base alloys for good resistance to corrosive media such as aqueous solutions and acids at

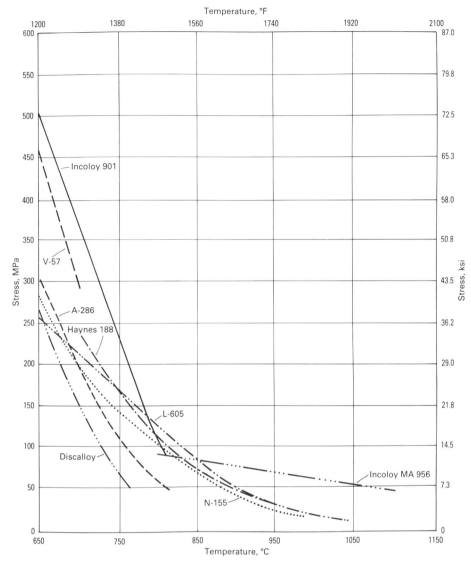

Fig. 8 1000-h stress-rupture curves of wrought cobalt-base (Haynes 188 and L-605) and wrought iron-base superalloys

low temperatures. A series of Hastelloy and Inconel alloys has been developed for such applications. Hastelloy C and Inconel 600 are typical: the former contains 16.5% Cr and 17% Mo as principal components for corrosion resistance, while Inconel 600 contains 15.5% Cr and 8% Fe. Other Hastelloy alloys contain up to 28% Mo, sometimes with small additions of tungsten.

Oxidation and Hot Corrosion. At elevated temperatures, oxidation resistance is provided by Al_2O_3 or Cr_2O_3 protective films. Accordingly, nickel-base alloys must contain one or both of these elements even where strength is not a principal factor. For example, Hastelloy X, one of the most oxidation and (hot) corrosion resistant of all nickel-base alloys, contains 22% Cr, 9% Mo, and 18.5% Fe as principal solutes (Table 1). Because Hastelloy X is essentially a solid-solution alloy when placed into service (carbides precipitate after long-term exposure), the alloy is much weaker than superalloys containing γ' or γ'' as strengthening precipitates.

Because chromium is known to degrade the high-temperature strength of γ' (see Fig. 3), there has been a strong incentive to lower chromium content in modern superalloys. Thus, the level of chromium decreased from 20% in earlier wrought alloys to as little as 9% in modern cast alloys. Unfortunately, this compositional change degraded hot corrosion resistance to the point that superalloys used in gas turbines had to be coated.

Further, as turbine blade temperatures exceed 1000 °C (1830 °F), Cr_2O_3 tends to decompose to CrO_3, which is more volatile and therefore less protective. To some extent, the loss of oxidation resistance has been compensated for by raising aluminum contents, although aluminum resides primarily in γ'. (Aluminum in small quantities promotes the formation of Cr_2O_3.) However, Al_2O_3 is less protective than Cr_2O_3 under sulfidizing conditions, making coatings indispensible in aircraft turbines and, more recently, in industrial turbines.

Other elements that contribute to oxidation and hot corrosion resistance are tantalum, yttrium, and lanthanum. The rare earths appear to improve oxidation resistance by preventing spalling of the oxide, while the mechanism for improvement with tantalum is not known. Yttrium is now widely used in overlay coatings of the NiCrAlY type.

Finally, it must be pointed out that molybdenum and tungsten are considered to be the most deleterious solutes in terms of hot corrosion resistance. Nevertheless, one or both of these elements are required for strength (for example, most of the γ'-strengthened alloys in Tables 1 and 9), so that alloying for improved surface stability is often in conflict with alloying for strength. The two most prominent solutes that provide both strength and surface stability are aluminum and tantalum.

Protection Against Oxidation and Corrosion. Superalloys are often diffusion or overlay coated in order to improve their corrosion resistance. The overlay coatings used are based on iron-chromium-aluminum-yttrium, cobalt-chromium-aluminum-yttrium, and nickel-chromium-aluminum-yttrium alloys. The earlier, less protective diffusion coatings are based on the reaction of aluminum with the substrate to form one or more aluminum-rich intermetallics (NiAl and/or Ni_2Al_3).

An important consideration for coated superalloys is the reduction in incipient melting temperature of the system (coating/base metal) that may result from the change in composition caused by the diffusion of coating components inward from the surface. Incipient melting (Table 4) reduces grain-boundary strength and ductility and thus reduces stress-rupture capabilities. Once an alloy has been heated above its incipient melting point, the alloy properties cannot be restored by heat treatment. Generally, a cast alloy should not be used for a structural application at any temperature higher than the point about 125 °C (225 °F) below its incipient melting temperature. Oxidation behavior and strength will determine how close actual metal temperatures may approach this suggested upper limit. Wrought alloys are used for applications, such as turbine disks and rotating seals, that typically operate at much lower temperatures. Therefore, coatings are much less likely to be used for such applications.

An important difference between nickel- and cobalt-base superalloys is related to the superior hot corrosion resistance claimed for cobalt-base alloys in atmospheres containing sulfur, sodium salts, halides, vanadium oxides, and lead oxide, all of which can be found in fuel-burning systems. In part, this apparent superiority may arise from the higher chromium content characteristic of cobalt-base alloys. Nickel forms low melting point eutectics with nickel sul-

Table 5 Ultimate tensile strengths of wrought nickel-, iron-, and cobalt-base superalloys

Alloy	Form	21 °C (70 °F) MPa	ksi	540 °C (1000 °F) MPa	ksi	650 °C (1200 °F) MPa	ksi	760 °C (1400 °F) MPa	ksi	870 °C (1600 °F) MPa	ksi	Condition of test material(a)
Nickel base												
Astroloy	Bar	1415	205	1240	180	1310	190	1160	168	775	112	1095 °C (2000 °F)/4 h/OQ + 870 °C (1600 °F)/8 h/AC + 980 °C (1800 °F)/4 h/AC + 650 °C (1200 °F) 24 h/AC + 760 °C (1400 °F)/8 h/AC
Cabot 214	...	915	133	715	104	675	98	560	84	440	64	1120 °C (2050 °F)
D-979	Bar	1410	204	1295	188	1105	160	720	104	345	50	1040 °C (1900 °F)/1 h/OQ + 845 °C (1550 °F)/6 h/AC + 705 °C (1300 °F)/16 h/AC
Hastelloy C-22	Sheet	800	116	625	91	585	85	525	76	...	...	1120 °C (2050 °F)/RQ
Hastelloy G-30	Sheet	690	100	490	71	...	...	...	...	...	...	1175 °C (2150 °F)/RAC-WQ
Hastelloy S	Bar	845	130	775	112	720	105	575	84	340	50	1065 °C (1950 °F)/AC
Hastelloy X	Sheet	785	114	650	94	570	83	435	63	255	37	1175 °C (2150 °F)/1 h/RAC
Haynes 230	...	870	126	720	105	675	98	575	84	385	56	1230 °C (2250 °F)/AC
Inconel 587(b)	Bar	1180	171	1035	150	1005	146	830	120	525	76	...
Inconel 597(b)	Bar	1220	177	1140	165	1060	154	930	135	...	...	...
Inconel 600	Bar	660	96	560	81	450	65	260	38	140	20	1120 °C (2050 °F)/2 h/AC
Inconel 601	Sheet	740	107	725	105	525	76	290	42	160	23	1150 °C (2100 °F)/2 h/AC
Inconel 617	Bar	740	107	580	84	565	82	440	64	275	40	1175 °C (2150 °F)/AC
Inconel 617	Sheet	770	112	590	86	590	86	470	68	310	45	1175 °C (2150 °F)/0.2 h/AC
Inconel 625	Bar	965	140	910	132	835	121	550	80	275	40	1150 °C (2100 °F)/1 h/WQ
Inconel 706	Bar	1310	190	1145	166	1035	150	725	105	...	...	980 °C (1800 °F)/1 h/AC + 845 °C (1550 °F)/3 h/AC + 720 °C (1325 °F)/8 h/FC + 620 °C (1150 °F)/8 h/AC
Inconel 718	Bar	1435	208	1275	185	1228	178	950	138	340	49	980 °C (1800 °F)/1 h/AC + 720 °C (1325 °F)/8 h/FC + 620 °C (1150 °F)/18 h/AC
Inconel 718 Direct Age	Bar	1530	222	1350	196	1235	179	...	...	...	...	735 °C (1325 °F)/8 h/SC + 620 °C (1150 °F)/8 h/AC
Inconel 718 Super	Bar	1350	196	1200	174	1130	164	...	...	...	...	925 °C (1700 °F)/1 h/AC + 735 °C (1325 °F)/8 h/SC + 620 °C (1150 °F)/8 h/AC
Inconel X750	Bar	1200	174	1050	152	940	136	...	...	...	...	1150 °C (2100 °F)/2 h/AC + 845 °C (1550 °F)/24 h/AC + 705 °C (1300 °F)/20 h/AC
M-252	Bar	1240	180	1230	178	1160	168	945	137	510	74	1040 °C (1900 °F)/4 h/AC + 760 °C (1400 °F)/16 h/AC
Nimonic 75	Bar	745	108	675	98	540	78	310	45	150	22	1050 °F (1925 °F)/1 h/AC
Nimonic 80A	Bar	1000	145	875	127	795	115	600	87	310	45	1080 °C (1975 °F)/8 h/AC + 705 °C (1300 °F)/16 h/AC
Nimonic 90	Bar	1235	179	1075	156	940	136	655	95	330	48	1080 °C (1975 °F)/8 h/AC + 705 °C (1300 °F)/16 h/AC
Nimonic 105	Bar	1180	171	1130	164	1095	159	930	135	660	96	1150 °C (2100 °F)/4 h/AC + 1060 °C (1940 °F)/16 h/AC + 850 °C (1560 °F)/16 h/AC
Nimonic 115	Bar	1240	180	1090	158	1125	163	1085	157	830	120	1190 °C (2175 °F)/1.5 h/AC + 1100 °C (2010 °F)/6 h/AC
Nimonic 263	Sheet	970	141	800	116	770	112	650	94	280	40	1150 °C (2100 °F)/0.2 h/WQ + 800 °C (1470 °F)/8 h/AC
Nimonic 942(b)	Bar	1405	204	1300	189	1240	180	900	131	...	...	...
Nimonic PE.11(b)	Bar	1080	157	1000	145	940	136	760	110	...	...	...
Nimonic PE.16	Bar	885	128	740	107	660	96	510	74	215	31	1040 °C (1900 °F)/4 h/AC + 800 °C (1470 °F)/2 h/AC + 700 °C (1290 °F)/16 h/AC
Nimonic PK.33	Sheet	1180	171	1000	145	1000	145	885	128	510	74	1100–1115 °C (2010–2040 °F)/0.25 h/AC + 850 °C (1500 °F)/4 h/AC
Pyromet 860(b)	Bar	1295	188	1255	182	1110	161	910	132	...	...	1095 °C (2000 °F)/2 h/WQ + 830 °C (1525 °F)/2 h/AC + 760 °C (1400 °F)/24 h/AC
René 41	Bar	1420	206	1400	203	1340	194	1105	160	620	90	1065 °C (1950 °F)/4 h/AC + 760 °C (1400 °F)/16 h/AC
René 95	Bar	1620	235	1550	224	1460	212	1170	170	...	...	900 °C (1650 °F)/24 h + 1105 °C (2025 °F)/1 h/OQ + 730 °C (1350 °F)/64 h/AC
Udimet 400(b)	Bar	1310	190	1185	172	...	...	...	...	...	...	...
Udimet 500	Bar	1310	190	1240	180	1215	176	1040	151	640	93	1080 °C (1975 °F)/4 h/AC + 845 °C (1550 °F)/24 h/AC + 760 °C (1400 °F)/16 h/AC
Udimet 520	Bar	1310	190	1240	180	1175	170	725	105	515	75	1105 °C (2025 °F)/4 h/AC + 845 °C (1550 °F)/24 h/AC + 760 °C (1400 °F)/16 h/AC
Udimet 630(b)	Bar	1520	220	1380	200	1275	185	965	140	...	...	...
Udimet 700	Bar	1410	204	1275	185	1240	180	1035	150	690	100	1175 °C (2150 °F)/4 h/AC + 1080 °C (1975 °F)/4 h/AC + 845 °C (1550 °F)/24 h/AC + 760 °C (1400 °F)/16 h/AC
Udimet 710	Bar	1185	172	1150	167	1290	187	1020	148	705	102	1175 °C (2150 °F)/4 h/AC + 1080 °C (1975 °F)/4 h/AC + 845 °C (1550 °F)/24 h/AC + 760 °C (1400 °F)/16 h/AC
Udimet 720	Bar	1570	228	...	...	1455	211	1455	211	1150	167	1115 °C (2035 °F)/2 h/AC + 1080 °C (1975 °F)/4 h/OQ + 650 °C (1200 °F)/24 h/AC + 760 °C (1400 °F)/8 h/AC
Unitemp AF2–1DA6	Bar	1560	226	1480	215	1400	203	1290	187	...	...	1150 °C (2100 °F)/4 h/AC + 760 °C (1400 °F)/16 h/AC
Waspaloy	Bar	1275	185	1170	170	1115	162	650	94	275	40	1080 °C (1975 °F)/4 h/AC + 845 °C (1550 °F)/24 h/AC + 760 °C (1400 °F)/16 h/AC
Iron base												
A-286	Bar	1005	146	905	131	720	104	440	64	...	...	980 °C (1800 °F)/1 h/OQ + 720 °C (1325 °F)/16 h/AC
Alloy 901	Bar	1205	175	1030	149	960	139	725	105	...	...	1095 °C (2000 °F)/2 h/WQ + 790 °C (1450 °F)/2 h/AC + 720 °C (1325 °F)/24 h/AC
Discaloy	Bar	1000	145	865	125	720	104	485	70	...	...	1010 °C (1850 °F)/2 h/OQ + 730 °C (1350 °F)/20 h/AC + 650 °C (1200 °F)/20 h/AC
Haynes 556	Sheet	815	118	645	93	590	85	470	69	330	48	1175 °C (2150 °F)/AC
Incoloy 800(b)	Bar	595	86	510	74	405	59	235	34	...	...	...
Incoloy 801(b)	Bar	785	114	660	96	540	78	325	47	...	...	...

(continued)

(a) OQ, oil quench; AC, air cool; RQ, rapid quench; RAC-WQ, rapid air cool-water quench; FC, furnace cool; SC, slow cool; CW, cold worked. (b) Ref 13. (c) Ref 14. (d) Annealed. (e) Precipitation hardened. (f) Ref 15. (g) Ref 3. (h) Work strengthened and aged. (i) At 700 °C (1290 °F). (j) At 900 °C (1650 °F). Source: Ref 1, except as noted

Table 5 (continued)

Alloy	Form	Ultimate tensile strength at										Condition of test material(a)
		21 °C (70 °F)		540 °C (1000 °F)		650 °C (1200 °F)		700 °C (1400 °F)		870 °C (1600 °F)		
		MPa	ksi	MPa	ksi	MPa	ksi	MPa	ksi	MPa	ksi	
Iron base (continued)												
Incoloy 802(b)	Bar	690	100	600	87	525	76	400	58	195	28	...
Incoloy 807(b)	Bar	655	95	470	68	440	64	350	51	220	32	...
Incoloy 825(c)(d)	...	690	100	~590	~86	~470	~68	~275	~40	~140	~20	...
Incoloy 903	Bar	1310	190	...	...	1000	145	...	...	...	...	845 °C (1550 °F)/1 h/WQ + 720 °C (1325 °F)/8 h/FC + 620 °C (1150 °F)/8 h/AC
Incoloy 907(c)(e)	...	~1365	~198	~1205	~175	~1035	~150	~655	~95	...	...	...
Incoloy 909	Bar	1310	190	1160	168	1025	149	615	89	...	...	980 °C (1800 °F)/1 h/WQ + 720 °C (1325 °F)/8 h/FC + 620 °C (1150 °F)/8 h/AC
N-155	Bar	815	118	650	94	545	79	428	62	260	38	1175 °C (2150 °F)/1 h/WQ + 815 °C (1500 °F)/4 h/AC
V-57	Bar	1170	170	1000	145	895	130	620	90	...	...	980 °C (1800 °F)/2–4 h/OQ + 730 °C (1350 °F)/16 h/AC
19-9 DL(f)	...	815	118	615	89	517	75	...	...	...	...	...
16-25-6(f)	...	980	142	...	...	620	90	415	60	...	...	...
Cobalt base												
AirResist 213(g)	...	1120	162	...	...	960	139	485	70	315	46	...
Elgiloy(g)	...	690(d)–2480(h)	100(d)–360(h)	...	...	...	...	...	...	...	...	...
Haynes 188	Sheet	960	139	740	107	710	103	635	92	420	61	1175 °C (2150 °F)/1 h/RAC
L-605	Sheet	1005	146	800	116	710	103	455	66	325	47	1230 °C (2250 °F)/1 h/RAC
MAR-M 918	Sheet	895	130	...	...	...	...	...	...	...	...	1190 °C (2175 °F)/4 h/AC
MP35N	Bar	2025	294	...	...	...	...	...	...	...	...	53% CW + 565 °C (1050 °F)/4 h/AC
MP159	Bar	1895	275	1565	227	1540	223	...	...	...	...	48% CW + 665 °C (1225 °F)/4 h/AC
Stellite 6B(g)	Sheet	1010	146	...	...	...	...	...	...	385	56	2 mm (0.063 in.) sheet heat treated at 1232 °C (2250 °F) and RAC
Haynes 150(g)	...	925	134	...	...	325(i)	47	...	...	155(j)	22.8	...

(a) OQ, oil quench; AC, air cool; RQ, rapid quench; RAC-WQ, rapid air cool-water quench; FC, furnace cool; SC, slow cool; CW, cold worked. (b) Ref 13. (c) Ref 14. (d) Annealed. (e) Precipitation hardened. (f) Ref 15. (g) Ref 3. (h) Work strengthened and aged. (i) At 700 °C (1290 °F). (j) At 900 °C (1650 °F). Source: Ref 1, except as noted

fide; in sulfur-bearing gases, the attack on nickel alloys may be devastating. Generalizations about the comparative oxidation and hot corrosion resistance of nickel- and cobalt-base alloys must be treated with some caution because resistance to corrosion varies widely within each alloy group. Small variations in service conditions (temperature, fuel impurity content, gas flow rates) can sharply alter results.

Property Data

Mechanical and physical properties for nickel-, iron-, and cobalt-base wrought superalloys are defined in the tables and graphs briefly described here. Although much data are presented comprehensively in this section, descriptions of iron- and cobalt-base superalloys are provided independently in respective sections, along with representative data. Ultimate tensile strength values at a variety of temperatures are given in Table 5, while tensile yield strengths and elongations are provided in Table 6. Rupture strengths measured at 1000 h are given in Table 7, while stress-rupture curves for selected superalloys are shown in Fig. 7 and 8. Physical properties of wrought superalloys are given in Table 8.

Iron-Base Superalloys

Microstructure. Iron-base superalloys evolved from austenitic stainless steels and are based on the principle of combining a close-packed fcc matrix with (in most cases) both solid-solution hardening and precipitate-forming elements. The austenitic matrix is based on nickel and iron, with at least 25% Ni needed to stabilize the fcc phase. Other alloying elements, such as

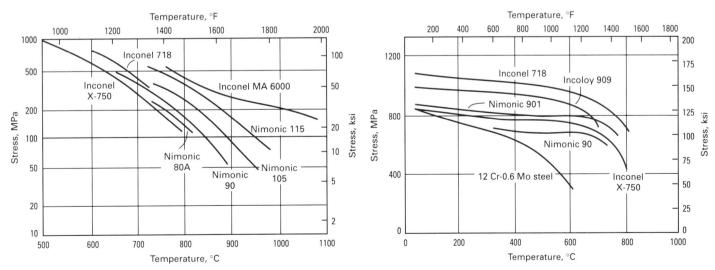

Fig. 9 1000-h creep rupture strength of turbine rotor and compressor blade alloys. Source: Ref 14

Fig. 10 Yield strength at 0.2% offset of five candidate compressor blade alloys compared with that of 12Cr-0.6Mo steel. Source: Ref 14

960 / Specialty Steels and Heat-Resistant Alloys

Table 6 Additional tensile properties of wrought nickel-, iron-, and cobalt-base superalloys

Alloy	Form	Yield strength at 0.2% offset										Tensile elongation, %				
		At 21 °C (70 °F)		At 540 °C (1000 °F)		At 650 °C (1200 °F)		At 760 °C (1400 °F)		At 870 °C (1600 °F)		At 21 °C (70 °F)	At 540 °C (1000 °F)	At 650 °C (1200 °F)	At 760 °C (1400 °F)	At 870 °C (1600 °F)
		MPa	ksi	MPa	ksi	MPa	ksi	MPa	ksi	MPa	ksi					
Nickel base																
Astroloy	Bar	1050	152	965	140	965	140	910	132	690	100	16	16	18	21	25
Cabot 214	...	560	81	510	74	505	73	495	72	310	45	38	19	14	9	11
D-979	Bar	1005	146	925	134	980	129	655	95	305	44	15	15	21	17	18
Hastelloy C-22	Sheet	405	59	275	40	250	36	240	35	...	...	57	61	65	63	...
Hastelloy G-30	Sheet	315	46	170	25	...	...	...	...	...	...	64	75	...	...	...
Hastelloy S	Bar	455	65	340	49	320	47	310	45	220	32	49	50	56	70	47
Hastelloy X	Sheet	360	52	290	42	275	40	260	38	180	26	43	45	37	37	50
Haynes 230	(b)(c)	390	57	275	40	270	39	285	41	225	32	48	56	55	46	59
Inconel 587(a)	Bar	705	102	620	90	615	89	605	88	400	58	28	22	21	20	16
Inconel 597(a)	Bar	760	110	720	104	675	98	665	96	...	...	15	15	15	16	...
Inconel 600	Bar	285	41	220	32	205	30	180	26	40	6	45	41	49	70	80
Inconel 601	Sheet	455	66	350	51	310	45	220	32	55	8	40	34	33	78	128
Inconel 617	Bar	295	43	200	29	170	25	180	26	195	28	70	68	75	84	118
Inconel 617	Sheet	345	50	230	33	220	32	230	33	205	30	55	62	61	59	73
Inconel 625	Bar	490	71	415	60	420	61	415	60	275	40	50	50	34	45	125
Inconel 706	Bar	1005	146	910	132	860	125	660	96	...	...	20	19	24	32	...
Inconel 718	Bar	1185	172	1065	154	1020	148	740	107	330	48	21	18	19	25	88
Inconel 718 Direct Age	Bar	1365	198	1180	171	1090	158	...	...	...	...	16	15	23	...	...
Inconel 718 Super	Bar	1105	160	1020	148	960	139	...	...	...	...	16	18	14	...	...
Inconel X750	Bar	815	118	725	105	710	103	...	...	...	...	27	26	10	...	...
M-252	Bar	840	122	765	111	745	108	720	104	485	70	16	15	11	10	18
Nimonic 75	Bar	285	41	200	29	200	29	160	23	90	13	40	40	46	67	68
Nimonic 80A	Bar	620	90	530	77	550	80	505	73	260	38	39	37	21	17	30
Nimonic 90	Bar	810	117	725	105	685	99	540	78	260	38	33	28	14	12	23
Nimonic 105	Bar	830	120	775	112	765	111	740	107	490	71	16	22	24	25	27
Nimonic 115	Bar	865	125	795	115	815	118	800	116	550	80	27	18	23	24	16
Nimonic 263	Sheet	580	84	485	70	485	70	460	67	180	26	39	42	27	21	25
Nimonic 942(a)	Bar	1060	154	970	141	1000	145	860	125	...	...					
Nimonic PE.11(a)	Bar	720	105	690	100	670	97	560	81	...	...					
Nimonic PE.16(a)	Bar	530	77	485	70	485	70	370	54	140	20	37	26	30	42	80
Nimonic PK.33(a)	Sheet	780	113	725	105	725	105	670	97	420	61	30	30	26	18	24
Pyromet 860(a)	Bar	835	121	840	122	850	123	835	121	...	...	22	15	17	18	...
René 41	Bar	1060	154	1020	147	1000	145	940	136	550	80	14	14	14	11	19
René 95	Bar	1310	190	1255	182	1220	177	1100	160	...	...	15	12	14	15	...
Udimet 400(a)	Bar	930	135	830	120	...	...	...	...	...	...	30	26	...	...	...
Udimet 500	Bar	840	122	795	115	760	110	730	106	495	72	32	28	28	39	20
Udimet 520	Bar	860	125	825	120	795	115	725	105	520	75	21	20	17	15	20
Udimet 630(a)	Bar	1310	190	1170	170	1105	160	860	125	...	...	15	15	7	5	...
Udimet 700	Bar	965	140	895	130	855	124	825	120	635	92	17	16	16	20	27
Udimet 710	Bar	910	132	850	123	860	125	815	118	635	92	7	10	15	25	29
Udimet 720	Bar	1195	173	...	...	1130	164	1050	152	...	...	13	...	17	9	...
Unitemp AF2-1DA6	Bar	1015	147	1040	151	1020	148	995	144	...	...	20	19	18	16	...
Waspaloy	Bar	795	115	725	105	690	100	675	98	520	75	25	23	34	28	35
Iron base																
A-286	Bar	725	105	605	88	605	88	430	62	...	...	25	19	13	19	...
Alloy 901(a)	Bar	895	130	780	113	760	110	635	92	...	...	14	14	13	19	...
Discaloy	Bar	730	106	650	94	630	91	430	62	...	...	19	16	19	...	...
Haynes 556	Sheet	410	60	240	35	225	33	220	32	195	29	48	54	52	49	53
Incoloy 800(a)	Bar	250	36	180	26	180	26	150	22	...	...	44	38	51	83	...
Incoloy 801(a)	Bar	385	56	310	45	305	44	290	42	...	...	30	28	26	55	...
Incoloy 802(a)	Bar	290	42	195	28	200	29	200	29	150	22	44	39	25	15	38
Incoloy 807(a)	Bar	380	55	255	37	240	35	225	32.5	185	26.5	48	40	35	34	71
Incoloy 825(d)(e)	...	310	45	~234	~34	~220	~32	180	~26	~105	~15	45	~44	~35	~86	~100
Incoloy 903	Bar	1105	160	...	...	895	130	...	...	...	...	14	...	18	...	...
Incoloy 907(d)(f)	...	~1110	~161	~960	~139	~895	~130	~565	~82	...	...	~12	~11	~10	~20	...
Incoloy 909	Bar	1020	148	945	137	870	126	540	78	...	...	16	14	24	34	...
N-155	Bar	400	58	340	49	295	43	250	36	175	25	40	33	32	32	33
V-57	Bar	830	120	760	110	745	108	485	70	...	...	26	19	22	34	...
19-9 DL(g)	...	570	83	395	57	360	52	...	...	...	...	43	30	30	...	...
16-25-6(g)	...	770	112	...	...	517	75	345	50	255	37	23	...	12	11	9
Cobalt base																
AirResist 213(h)	...	625	91	...	...	425	66	385	56	220	32	14	...	28	47	55
Elgiloy(i)	Sheet	480(e)–2000(i)	70–290	...	...	...	...	...	...	...	...	34	...	...	...	...
Haynes 188	Sheet	485	70	305	44	305	44	290	42	260	38	56	70	61	43	73
L-605	Sheet	460	67	250	36	240	35	260	38	240	35	64	59	35	12	35
MAR-M 918	Sheet	895	130	...	...	...	...	...	...	...	...	48	...	...	...	...
MP35N	Bar	1620	235	...	...	...	...	...	...	...	...	10	...	...	...	...
MP159	Bar	1825	265	1495	217	1415	205	...	...	...	...	8	8	7	...	...
Stellite 6B(h)	Sheet	635	92	...	...	...	...	...	...	270	39	11	...	...	...	18
Haynes 150(g)	...	317	46	...	...	160	23	...	...	...	...	8	...	...	...	...

(continued)

(a) Ref 13. (b) Cold-rolled and solution-annealed sheet, 1.2 to 1.6 mm (0.048 to 0.063 in.) thick. (c) Ref 16. (d) Ref 14. (e) Annealed. (f) Precipitation hardened. (g) Ref 15. (h) Ref 3. (i) Work strengthened and aged. (j) Data for bar, rather than sheet. Source: Ref 1, except as noted

Table 6 (continued)

Alloy	Form	At 21 °C (70 °F) GPa	10⁶ psi	At 540 °C (1000 °F) GPa	10⁶ psi	At 650 °C (1200 °F) GPa	10⁶ psi	At 760 °C (1400 °F) GPa	10⁶ psi	At 870 °C (1600 °F) GPa	10⁶ psi
Nickel base											
D-979	Bar	207	30.0	178	25.8	167	24.2	156	22.6	146	21.2
Hastelloy S	Bar	212	30.8	182	26.4	174	25.2	166	24.1	...	...
Hastelloy X	Sheet	197	28.6	161	23.4	154	22.3	146	21.1	137	19.9
Haynes 230	(b)(c)	211	30.6	184	26.4	177	25.3	171	24.1	164	23.1
Inconel 587	Bar	222	32.1	...	...	...	...	...	...	...	...
Inconel 596	Bar	186	27.0	...	...	...	...	...	...	...	...
Inconel 600	Bar	214	31.1	184	26.7	176	25.5	168	24.3	157	22.8
Inconel 601(j)	Sheet	207	30.0	175	25.4	166	24.1	155	22.5	141	20.5
Inconel 617	Bar	210	30.4	176	25.6	168	24.4	160	23.2	150	21.8
Inconel 625	Bar	208	30.1	179	25.9	170	24.7	161	23.3	148	21.4
Inconel 706	Bar	210	30.4	179	25.9	170	24.7	...	...	...	...
Inconel 718	Bar	200	29.0	171	24.8	163	23.7	154	22.3	139	20.2
Inconel X750	Bar	214	31.0	184	26.7	176	25.5	166	24.0	153	22.1
M-252	Bar	206	29.8	177	25.7	168	24.4	156	22.6	145	21.0
Nimonic 75	Bar	221	32.0	186	27.0	176	25.5	170	24.6	156	22.6
Nimonic 80A	Bar	219	31.8	188	27.2	179	26.0	170	24.6	157	22.7
Nimonic 90	Bar	226	32.7	190	27.6	181	26.3	170	24.7	158	22.9
Nimonic 105	Bar	223	32.3	186	27.0	178	25.8	168	24.4	155	22.5
Nimonic 115	Bar	224	32.4	188	27.2	181	26.3	173	25.1	164	23.8
Nimonic 263	Sheet	222	32.1	190	27.5	181	26.2	171	24.8	158	22.9
Nimonic 942	Bar	196	28.4	166	24.1	158	22.9	150	21.8	138	20.0
Nimonic PE.11	Bar	198	28.7	166	24.0	157	22.8	...	...	...	...
Nimonic PE.16	Bar	199	28.8	165	23.9	157	22.7	147	21.3	137	19.9
Nimonic PK.33	Sheet	222	32.1	191	27.6	183	26.5	173	25.1	162	23.5
Pyromet 860	Bar	200	29.0	...	...	...	...	...	...	...	...
René 95	Bar	209	30.3	183	26.5	176	25.5	168	24.3	...	...
Udimet 500	Bar	222	32.1	191	27.7	183	26.5	173	25.1	161	23.4
Udimet 700	Bar	224	32.4	194	28.1	186	27.0	177	25.7	167	24.2
Udimet 710	Bar	222	32.1	...	...	...	...	...	...	...	...
Waspaloy	Bar	213	30.9	184	26.7	177	25.6	168	24.3	158	22.9
Iron base											
A-286	Bar	201	29.1	162	23.5	153	22.2	142	20.6	130	18.9
Alloy 901(a)	Bar	206	29.9	167	24.2	153	22.1	...	...	...	...
Discaloy	Bar	196	28.4	154	22.3	145	21.0	...	...	...	...
Haynes 556	Sheet	203	29.5	165	23.9	156	22.6	146	21.1	137	19.9
Incoloy 800	Bar	196	28.4	161	23.4	154	22.3	146	21.1	138	20.0
Incoloy 801	Bar	208	30.1	170	24.7	162	23.5	154	22.3	144	20.9
Incoloy 802	Bar	205	29.7	169	24.5	161	23.4	156	22.6	152	22.0
Incoloy 807	Bar	184	26.6	155	22.4	146	21.2	137	19.9	128	18.5
Incoloy 903	Bar	147(e)	21.3	152(f)	22.1	...	...	...	...	...	...
Incoloy 907(d)(f)	...	165(e)	23.9	165(f)	23.9	159(e)	23	...	...	...	...
N-155	Bar	202	29.3	167	24.2	159	23.0	149	21.6	138	20.0
V-57	Bar	199	28.8	163	23.6	153	22.2	144	20.8	130	18.9
19-9 DL(g)	...	203	29.5	...	...	152	22.1	...	...	...	...
16-25-6(g)	...	195	28.2	...	...	123	17.9	...	...	...	...
Cobalt base											
Haynes 188	Sheet	207	30	...	...	...	...	...	...	...	...
L-605	Sheet	216	31.4	...	...	185	26.8	...	...	166	24.0
MAR-M 918	Sheet	225	32.6	186	27.0	176	25.5	168	24.3	159	23.0
MP35N	Bar	231(e)	33.6	...	...	...	...	...	...	...	...
Haynes 150(g)	...	217(e)	31.5	...	...	...	...	...	...	...	...

(a) Ref 13. (b) Cold-rolled and solution-annealed sheet, 1.2 to 1.6 mm (0.048 to 0.063 in.) thick. (c) Ref 16. (d) Ref 14. (e) Annealed. (f) Precipitation hardened. (g) Ref 15. (h) Ref 3. (i) Work strengthened and aged. (j) Data for bar, rather than sheet. Source: Ref 1, except as noted

chromium, partition primarily to the austenite for solid-solution hardening. The strengthening precipitates are primarily ordered intermetallics, such as γ' Ni$_3$Al, η Ni$_3$Ti, and γ'' Ni$_3$Nb, although carbides and carbonitrides may also be present. Elements that partition to grain boundaries, such as boron and zirconium, perform a function similar to that which occurs in nickel-base alloys; that is, grain-boundary fracture is suppressed under creep rupture conditions, resulting in significant increases in rupture life. Compositions of wrought iron-nickel alloys are listed in Table 9. A summary of the function of various alloying elements is found in Table 2.

Several groupings of iron-nickel alloys based on composition and strengthening mechanisms have been established. Alloys that are strengthened by ordered fcc γ', such as V-57 and A-286, and contain 25 to 35 wt% Ni, represent one subgroup. The γ' phase is titanium-rich in these alloys, and care must be taken to avoid an excessively high titanium-to-aluminum ratio, resulting in the replacement of fcc γ' by hexagonal close-packed (hcp) η(Ni$_3$Ti), a less effective strengthener.

A second iron-rich subgroup, of which Inconel X750 and Incoloy 901 are examples, contains at least 40% Ni, as well as higher levels of solid-solution strengthening and precipitate-forming elements.

Another iron-rich group, based on the iron-nickel-cobalt system strengthened by fcc γ', combines low thermal expansion coefficients and relatively high strength to a temperature of 650 °C (1200 °F). These alloys are typified by Incoloy 903, 907, 909, Pyromet CTX-1, and Pyromet CTX-3, which do not have any chromium and oxidize and spall readily. The unusually low thermal expansion coefficients result from the elimination of ferrite-stabilizing elements. In each of these alloys, a small quantity of titanium is present to allow the formation of metastable γ'(Ni$_3$Ti,Al) in the austenite matrix during aging. Another phase that may occur after extended aging

Table 7 1000-h rupture strengths of wrought nickel-, cobalt-, and iron-base superalloys

Alloy	Form	At 650 °C (1200 °F) MPa	ksi	At 760 °C (1400 °F) MPa	ksi	At 870 °C (1600 °F) MPa	ksi	At 980 °C (1800 °F) MPa	ksi
Nickel base									
Astroloy	Bar	770	112	425	62	170	25	55	8
Cabot 214	...	...	...	...	...	30	4	15	2
D-979	Bar	515	75	250	36	70	10	...	...
Hastelloy S	Bar	...	...	90	13	25	4	...	...
Hastelloy X	Sheet	215	31	105	15	40	6	15	2
Haynes 230	...	...	...	125	18	55	8	15	2
Inconel 587(a)	Bar	...	...	285	41	...	...	...	...
Inconel 597(a)	Bar	...	...	340	49	...	...	...	...
Inconel 600	Bar	...	...	...	...	30	4	15	2
Inconel 601	Sheet	195	28	60	9	30	4	15	2
Inconel 617	Bar	360	52	165	24	60	9	30	4
Inconel 617	Sheet	...	...	160	23	60	9	30	4
Inconel 625	Bar	370	54	160	23	50	7	20	3
Inconel 706	Bar	580	84	...	...	...	...	...	...
Inconel 718	Bar	595	86	195	28	...	...	...	...
Inconel 718 Direct Age	Bar	405	59	...	...	...	...	...	...
Inconel 718 Super	Bar	600	87	...	...	...	...	...	...
Inconel X750	Bar	470	68	...	...	50	7	...	...
M-252	Bar	565	82	270	39	95	14	...	...
Nimonic 75	Bar	170	25	50	7	5	1	...	...
Nimonic 80A	Bar	420	61	160	23	...	...	...	...
Nimonic 90	Bar	455	66	205	30	60	9	...	...
Nimonic 105	Bar	...	...	330	48	130	19	30	4
Nimonic 115	Bar	...	...	420	61	185	27	70	10
Nimonic 942(a)	Bar	520	75	270	39	...	...	...	...
Nimonic PE.11(a)	Bar	335	49	145	21	...	...	...	...
Nimonic PE.16	Bar	345	50	150	22	...	...	...	...
Nimonic PK.33	Sheet	655	95	310	45	90	13	...	...
Pyromet 860(a)	Bar	545	79	250	36	...	...	...	...
René 41	Bar	705	102	345	50	115	17	...	...
René 95	Bar	860	125	...	...	...	...	...	...
Udimet 400(a)	Bar	600	87	305	44	110	16	...	...
Udimet 500	Bar	760	110	325	47	125	18	...	...
Udimet 520	Bar	585	85	345	50	150	22	...	...
Udimet 700	Bar	705	102	425	62	200	29	55	8
Udimet 710	Bar	870	126	460	67	200	29	70	10
Udimet 720	Bar	670	97	...	...	...	...	...	...
Unitemp AF2-1DA6	Bar	885	128	360	52	...	...	...	...
Waspaloy	Bar	615	89	290	42	110	16	...	...
Iron base									
A-286	Bar	315	46	105	15	...	...	...	...
Alloy 901	Sheet	525	76	205	30	...	...	...	...
Discaloy	Bar	275	40	60	9	...	...	...	...
Haynes 556	Sheet	275	40	125	18	55	8	20	3
Incoloy 800(a)	Bar	165	24	66	9.5	30	4.4	13	1.9
Incoloy 801(a)	Bar	...	...	...	...	...	...	...	...
Incoloy 802(a)	Bar	170	25	110	16	69	10	24	3.5
Incoloy 807(a)	Bar	...	...	105	15	43	6.2	19	2.7
Incoloy 903	Bar	510	74	...	...	...	...	...	...
Incoloy 909	Bar	345	50	...	...	...	...	...	...
N-155	Bar	295	43	140	20	70	10	20	3
V-57	Bar	485	70	...	...	...	...	...	...
Cobalt base									
Haynes 188	Sheet	...	...	165	24	70	10	30	4
L-605	Sheet	270	39	165	24	75	11	30	4
MAR-M 918	Sheet	...	...	60	9	20	3	5	1
Haynes 150(b)	...	...	...	40(c)	5.8	...	...	...	...

(a) Ref 13. (b) Ref 15. (c) At 815 °C (1500 °F). Source: Ref 1, except as noted

of the Pyromet alloys is η(Ni₃Ti), a plate-like, stable hcp phase that contributes less to strengthening than does γ'. The alloys are used for shafts, rings, and casings to permit the reduction of clearances between rotating and static components.

Alloys With High Nickel Content. Two important classes of iron-nickel alloys actually contain more nickel than iron. The first group, typified by Incoloy 706 and Inconel 718 (Table 1), are strengthened by a coherent bct phase known as γ" (Ni_3Nb). These alloys contain 3 and 5 wt% Nb, respectively. The iron acts principally as a catalyst for the formation of γ", which is metastable. These alloys also contain smaller quantities of aluminum and titanium, thereby leading to the formation of γ' Ni_3Al,Ti. Improper heat treatment can lead to the formation of a stable orthorhombic δ phase with the composition Ni_3Nb. Because proper heat treatment procedures are essential for these alloys, time-temperature-transformation diagrams such as that shown in Fig. 4 are used to establish appropriate heat treatment schedules.

The maximum-use temperature of Inconel 718 is about 650 °C (1200 °F) because of instability of the γ" precipitate. Other problems that have arisen with this alloy are notch sensitivity from about 525 to 750 °C (975 to 1380 °F) when tested in air (generally caused by improper processing) and a significantly higher rate of crack growth in air than in vacuum. Recent work has shown that lowering the carbon content can result in equal or better properties than those of standard Inconel 718 (Ref 17). Further, increasing the aluminum and niobium content and the aluminum-to-titanium ratio in Inconel 718 enhances its mechanical properties by producing more of the stable γ' phase and less of the undesirable δ phase (Ref 18).

A second set of nickel-rich alloys displays little or no precipitation strengthening. However, second phases may be seen after long-term exposure. Alloys in this group include Hastelloy X and N-155.

Boron in quantities of 0.003 to 0.03 wt% and, less frequently, small additions of zirconium are added to improve stress-rupture properties and hot workability. Zirconium also forms the MC carbide ZrC. Another MC carbide, NbC, is found in alloys that contain niobium, such as Inconel 706 and Inconel 718. Vanadium also is added in small quantities to iron-nickel alloys to improve both notch ductility at service temperatures and hot workability. Manganese and rare earth elements may be present as deoxidizers; rare earths also have been added to Hastelloy X to improve oxidation resistance.

Generally, the iron-nickel alloys described above have useful strengths to about 650 °C (1200 °F). Inconel 718 is one of the strongest (at low temperatures) and most widely used of all superalloys, but it rapidly loses strength in the range of 650 to 815 °C (1200 to 1500 °F). Figures 9 and 10 show creep strength and yield strength, respectively. This is probably due to the high lattice misfit associated with the precipitation of γ" in the austenitic matrix.

The carbide and carbonitride strengthened alloys have useful strength to at least 815 °C (1500 °F). Hastelloy X, a very weak alloy at all temperatures, is widely used as a combustion chamber liner, operating at temperatures in excess of 1050 °C (1920 °F) because of its oxidation resistance (Fig. 11). Other physical and mechanical properties are given in Tables 5 to 8 and Fig. 7 and 8.

Cobalt-Base Superalloys

Wrought cobalt-base alloys, unlike other superalloys, are not strengthened by a co-

Table 8 Physical properties of selected wrought superalloys

Designation	Form	Density, g/cm³	Melting range °C	Melting range °F	Specific heat capacity At 21 °C (70 °F) J/kg·K	At 21 °C (70 °F) Btu/lb·°F	At 538 °C (1000 °F) J/kg·K	At 538 °C (1000 °F) Btu/lb·°F	At 871 °C (1600 °F) J/kg·K	At 871 °C (1600 °F) Btu/lb·°F
Nickel base										
Astroloy	Bar	7.91	...	...	...	...	...	...	...	...
D-979	Bar	8.19	1200–1390	2225–2530	...	...	...	...	...	...
Hastelloy X	Sheet	8.21	1260–1355	2300–2470	485	0.116	...	...	700	0.167
Hastelloy S	Bar	8.76	1335–1380	2435–2515	405	0.097	495	0.118	595	0.142
Inconel 597	Bar	8.04	...	...	...	...	...	...	...	...
Inconel 600	Bar	8.41	1355–1415	2470–2575	445	0.106	555	0.132	625	0.149
Inconel 601	Bar	8.05	1300–1370	2375–2495	450	0.107	590	0.140	680	0.162
Inconel 617	Bar	8.36	1330–1375	2430–2510	420	0.100	550	0.131	630	0.150
Inconel 625	Bar	8.44	1290–1350	2350–2460	410	0.098	535	0.128	620	0.148
Inconel 690	Bar	8.14	1345–1375	2450–2510	...	...	...	...	...	...
Inconel 706	Bar	8.08	1335–1370	2435–2500	445	0.106	580	0.138	670	0.159
Inconel 718	Bar	8.22	1260–1335	2300–2435	430	0.102	560	0.133	645	0.153
Inconel X750	Bar	8.25	1395–1425	2540–2600	430	0.103	545	0.130	715	0.171
Haynes 230	...	8.8	1300–1370	2375–2500	397	0.095	473	0.112	595	0.145
M-252	Bar	8.25	1315–1370	2400–2500	...	...	...	...	...	...
Nimonic 75	Bar	8.37	...	...	460	0.11	...	...	...	...
Nimonic 80A	Bar	8.16	1360–1390	2480–2535	460	0.11	...	...	...	...
Nimonic 81	Bar	8.06	...	...	460	0.11	585	0.14	670	0.16
Nimonic 90	Bar	8.19	1335–1360	2435–2480	460	0.11	585	0.14	670	0.16
Nimonic 105	Bar	8.00	...	...	420	0.10	545	0.13	670	0.16
Nimonic 115	Bar	7.85	...	...	460	0.11	...	...	...	...
Nimonic 263	Sheet	8.36	...	...	460	0.11	...	...	...	...
Nimonic 942	Bar	8.19	1240–1300	2265–2370	420	0.10	...	...	...	...
Nimonic PE.11	Bar	8.02	1280–1350	2335–2460	420	0.10	585	0.14	...	...
Nimonic PE.16	Bar	8.02	...	...	545	0.13	...	...	...	...
Nimonic PK.33	Sheet	8.21	...	...	420	0.10	545	0.13	670	0.16
Pyromet 860	Bar	8.21	...	...	...	...	...	...	...	...
René 41	Bar	8.25	1315–1370	2400–2500	...	...	545	0.13	725	0.173
René 95	Bar	...	...	...	...	...	...	...	...	...
Udimet 500	Bar	8.02	1300–1395	2375–2540	...	...	...	...	...	...
Udimet 520	Bar	8.21	1260–1405	2300–2560	...	...	...	...	...	...
Udimet 700	Bar	7.91	1205–1400	2200–2550	...	...	575	0.137	590	0.141
Unitemp AF2-1DA	Bar	8.26	...	...	420	0.100	...	...	...	...
Waspaloy	Bar	8.19	1330–1355	2425–2475	...	...	...	...	...	...
Iron base										
Alloy 901	Bar	8.21	1230–1400	2250–2550	...	...	...	...	...	...
A-286	Bar	7.91	1370–1400	2500–2550	460	0.11	...	...	...	...
Discaloy	Bar	7.97	1380–1465	2515–2665	475	0.113	...	...	...	...
Haynes 556	Sheet	8.23	...	...	450	0.107	...	...	...	...
Incoloy 800	Bar	7.95	1355–1385	2475–2525	455	0.108	...	...	...	...
Incoloy 801	Bar	7.95	1355–1385	2475–2525	455	0.108	...	...	...	...
Incoloy 802	Bar	7.83	1345–1370	2450–2500	445	0.106	...	...	...	...
Incoloy 807	Bar	8.32	1275–1355	2325–2475	...	...	...	...	...	...
Incoloy 825(a)	...	8.14	1370–1400	2500–2550	440	0.105	...	...	...	...
Incoloy 903	Bar	8.14	1320–1395	2405–2540	435	0.104	...	...	...	...
Incoloy 904	Bar	8.12	...	...	460	0.11	...	...	...	...
Incoloy 907(a)	...	8.33	1335–1400	2440–2550	431	0.103	...	...	...	...
Incoloy 909(a)	...	8.30	1395–1430	2540–2610	427	0.102	...	...	...	...
N-155	Bar	8.19	1275–1355	2325–2475	430	0.103	...	...	...	...
19-9 DL(b)	...	7.9	1425–1430	2600–2610	...	...	...	...	...	...
16-25-6(b)	...	8.0	...	...	...	...	...	...	...	...
Cobalt base										
Haynes 188	Sheet	9.13	1300–1330	2375–2425	405	0.097	510	0.122	565	0.135
L-605	Sheet	9.13	1330–1410	2425–2570	385	0.092	...	...	...	...
Stellite 6B(a)	...	8.38	1265–1354	2310–2470	421	0.101	...	...	...	...
Haynes 150(b)	...	8.05	1395	2540	...	...	...	...	...	...
MP35N(b)	...	8.41	1315–1425	2400–2600	...	...	...	...	...	...
Elgiloy(b)	...	8.3	1495	2720	...	...	...	...	...	...

Designation	Form	Thermal conductivity At 21 °C (70 °F) W/m·K	At 21 °C (70 °F) Btu/ft²·in.·h·°F	At 538 °C (1000 °F) W/m·K	At 538 °C (1000 °F) Btu/ft²·in.·h·°F	At 871 °C (1600 °F) W/m·K	At 871 °C (1600 °F) Btu/ft²·in.·h·°F	Mean coefficient of thermal expansion, 10⁻⁶/K At 538 °C (1000 °F)	At 871 °C (1600 °F)	Electrical resistivity, nΩ·m
Nickel base										
Astroloy	Bar	...	...	...	...	...	...	13.9	16.2	...
D-979	Bar	12.6	87	18.5	128	...	...	14.9	17.7	...
Hastelloy X	Sheet	9.1	63	19.6	136	26.0	180	15.1	16.2	1180(c)
Hastelloy S	Bar	...	...	20.0	139	26.1	181	13.3	14.9	...
Inconel 597	Bar	...	...	18.2	126	...	...	...	...	...
Inconel 600	Bar	14.8	103	22.8	158	28.8	200	15.1	16.4	1030(c)
Inconel 601	Bar	11.3	78	20.0	139	25.7	178	15.3	17.1	1190(d)

(continued)

(a) Ref 14. (b) Ref 12. (c) Ref 33. (d) At 21 to 93 °C (70 to 200 °F). (e) At 25 to 427 °C (77 to 800 °F). (f) At 705 °C (1300 °F). (g) At 980 °C (1800 °F). Source: Ref 13, unless otherwise noted

Table 8 (continued)

Designation	Form	Thermal conductivity At 21 °C (70 °F) W/m·K	Btu/ft²·in.·h·°F	At 538 °C (1000 °F) W/m·K	Btu/ft²·in.·h·°F	At 871 °C (1600 °F) W/m·K	Btu/ft²·in.·h·°F	Mean coefficient of thermal expansion, 10^{-6}/K At 538 °C (1000 °F)	At 871 °C (1600 °F)	Electrical resistivity, $n\Omega \cdot m$
Nickel base (continued)										
Inconel 617	Bar	13.6	94	21.5	149	26.7	185	13.9	15.7	1220(d)
Inconel 625	Bar	9.8	68	17.5	121	22.8	158	14.0	15.8	1290(c)
Inconel 690	Bar	13.3	95	22.8	158	27.8	193	...	...	148(c)
Inconel 706	Bar	12.6	87	21.2	147	...	...	15.7	...	...
Inconel 718	Bar	11.4	79	19.6	136	24.9	173	14.4	...	1250(d)
Inconel X750	Bar	12.0	83	18.9	131	23.6	164	14.6	16.8	1220(d)
Haynes 230	...	8.9	62	18.4	133	24.4	179	14.0	15.2	1250
M-252	Bar	11.8	82	...	...	...	...	13.0	15.3	...
Nimonic 75	Bar	...	...	...	...	...	...	14.7	17.0	1090(d)
Nimonic 80A	Bar	8.7	60	15.9	110	22.5	156	13.9	15.5	1240(d)
Nimonic 81	Bar	10.8	75	19.2	133	25.1	174	14.2	17.5	1270(d)
Nimonic 90	Bar	9.8	68	17.0	118	...	...	13.9	16.2	1180(d)
Nimonic 105	Bar	10.8	75	18.6	129	24.0	166	13.9	16.0	1310(d)
Nimonic 115	Bar	10.7	74	17.6	124	22.6	154	13.3	16.4	1390(d)
Nimonic 263	Sheet	11.7	81	20.4	141	26.2	182	13.7	16.2	1150(d)
Nimonic 942	Bar	...	...	...	...	...	...	14.7	16.5	...
Nimonic PE.11	Bar	...	...	...	...	...	...	15.2	...	...
Nimonic PE.16	Bar	11.7	81	20.2	140	26.4	183	15.3	18.5	1100(d)
Nimonic PK.33	Sheet	10.7	74	19.2	133	24.7	171	13.1	16.2	1260(d)
Pyromet 860	Bar	...	...	...	...	...	...	15.4	16.4	...
René 41	Bar	9.0	62	18.0	125	23.1	160	13.5	15.6	1308(c)
René 95	Bar	8.7	60	17.4	120	...	...	...	...	...
Udimet 500	Bar	11.1	77	18.3	127	24.5	170	14.0	16.1	1203(c)
Udimet 700	Bar	19.6	136	20.6	143	27.7	192	13.9	16.1	...
Unitemp AF2-1DA	Bar	10.8	75	16.5	114	19.5	135	12.4	14.1	...
Waspaloy	Bar	10.7	74	18.1	125	24.1	167	14.0	16.0	1240
Iron base										
Alloy 901	Bar	13.3	92	...	...	...	...	15.3	...	...
A-286	Bar	12.7	88	22.5	156	...	...	17.6	...	...
Discaloy	Bar	13.3	92	21.1	146	...	...	17.1	...	...
Haynes 556	Sheet	11.6	80	17.5	121	...	...	16.2	17.5	...
Incoloy 800	Bar	11.6	80	20.1	139	...	...	16.4	18.4	989
Incoloy 801	Bar	12.4	86	20.7	143	25.6	177	17.3	18.7	...
Incoloy 802	Bar	11.9	82	19.8	137	24.2	168	16.7	18.2	...
Incoloy 807	Bar	...	...	...	...	...	...	15.2	17.6	...
Incoloy 825(a)	...	11.1	76.8	...	...	...	...	14.0(d)	...	1130
Incoloy 903	Bar	16.8	116	20.9	145	...	...	8.6	...	610
Incoloy 904	Bar	16.8	116	22.4	155	...	...	...	...	...
Incoloy 907(a)	...	14.8	103	...	...	...	...	7.7(e)	...	697
Incoloy 909(a)	...	14.8	103	...	...	...	...	7.7(e)	...	728
N-155	Bar	12.3	85	19.2	133	...	...	16.4	17.8	...
19-9 DL(b)	...	...	...	...	...	...	...	17.8	...	...
16-25-6(b)	...	...	...	15	104	...	...	16.9	...	...
Cobalt base										
Haynes 188	Sheet	...	...	19.9	138	25.1	174	14.8	17.0	922
L-605	Sheet	9.4	65	19.5	135	26.1	181	14.4	16.3	890
Stellite 6B(c)	...	14.7	101	...	...	...	...	15.0	16.9	910
Haynes 150(b)	...	...	...	0.75(f)	5.2	...	...	...	16.8(g)	810
MP35N(b)	...	...	...	...	...	...	...	...	15.7(g)	1010
Elgiloy(b)	...	1.0	7.2	1.4	10	...	...	...	15.8(g)	995

(a) Ref 14. (b) Ref 12. (c) Ref 33. (d) At 21 to 93 °C (70 to 200 °F). (e) At 25 to 427 °C (77 to 800 °F). (f) At 705 °C (1300 °F). (g) At 980 °C (1800 °F). Source: Ref 13, unless otherwise noted

herent, ordered precipitate. Rather, they are characterized by a solid solution strengthened austenitic (fcc) matrix in which a small quantity of carbides is distributed. (Cast cobalt alloys rely upon carbide strengthening to a much greater extent.) Cobalt crystallizes in the hcp structure below 417 °C (780 °F). At higher temperatures, it transforms to fcc. To avoid this transformation during service, virtually all cobalt-base alloys are alloyed with nickel in order to stabilize the fcc structure between room temperature and the melting point.

Melting point is not a reliable guide to the temperature capability of iron-, nickel-, and cobalt-base superalloys. Although nickel possesses the lowest melting point of the three alloy bases, it has by far the highest temperature capability under moderate-to-high stresses. Further, the incipient melting temperatures of nickel- and cobalt-base alloys are very similar (see Table 4). Cobalt-base alloys, with a flatter rupture stress-temperature relationship, may actually display better creep rupture properties above about 1000 °C (1830 °F) than the other superalloys. This may best be seen by means of comparative Larson-Miller plots for superalloys (Fig. 12). Also, cobalt-base alloys display superior hot corrosion resistance at high temperatures, probably a consequence of the considerably higher chromium contents that are characteristic of these alloys. A list of wrought cobalt-base alloys and their compositions appears in Table 10, and the function of various solutes is summarized in Table 2.

Cobalt-base alloys generally exhibit better weldability and thermal-fatigue resistance than do nickel-base alloys. Another advantage of cobalt-base alloys is the capability to be melted in air or argon, in contrast to the vacuum melting required for nickel-base and iron-nickel-base alloys containing the reactive metals aluminum and titanium. However, unlike nickel-base alloys, which have a high tolerance for alloy-

Table 9 Nominal compositions of wrought iron-base alloys

Alloy	Ni	Cr	Co	Mo	W	Nb	Al	Ti	Fe	Mn	Si	C	B	Other
A-286(a)	26.0	15.0	...	1.3	...	...	0.2	2.0	54.0	1.3	0.5	0.05	0.015	...
Discaloy(a)	26.0	13.5	...	2.7	...	...	0.1	1.7	54.0	0.9	0.8	0.04	0.005	...
Alloy 901(a)	42.5	12.5	...	5.7	...	...	0.2	2.8	36.0	0.1	0.1	0.05	0.015	...
Haynes 556(a)	20.0	22.0	20.0	3.0	2.5	0.1	0.3	...	29.0	1.5	0.4	0.10	...	0.2 N, 0.02 La, 0.9 Ta
Incoloy 800(b)	32.5	21.0	...	...	...	...	0.4	0.4	46	0.8	0.5	0.05	...	...
Incoloy 801(b)	32.0	20.5	...	...	...	...	...	1.1	44.5	0.8	0.5	0.05	...	...
Incoloy 802(b)	32.5	21.5	...	...	...	...	...	...	46	0.8	0.4	0.4	...	...
Incoloy 807(b)	40.0	20.5	8.0	0.1	5.0	...	0.2	0.3	25	0.50	0.40	0.05	...	...
Incoloy 825(c)	38–46	19.5–23.5	...	2.5–3.5	...	...	0.2	0.6–1.2	22	1.0	0.5	0.05	...	1.5–3 Cu, 0.03 S
Incoloy 903(a)	38.0	...	15.0	...	...	3.0	0.7	1.4	41.0	...	...	...	...	...
Incoloy 907(c)	38	...	13	...	...	4.7	0.03	1.5	42	...	0.15	...	...	...
Incoloy 909(a)	38.0	...	13.0	...	...	4.7	...	1.5	42.0	...	0.4	0.01	0.001	...
N-155(a)	20.0	21.0	20.0	3.0	2.5	1.0	...	...	30.0	1.5	0.5	0.15	...	0.15 N
V-57(a)	27.0	14.8	...	1.3	...	...	0.3	3.0	52.0	0.3	0.7	0.08	0.010	...
19-9 DL(d)	9.0	19.0	0.4	...	1.3	...	...	0.3	bal	1.0	0.50	0.3	...	...
16-25-6(d)	25.5	16.25	...	6.0	...	...	...	...	bal	2.0	1.0	0.10	...	...
Pyromet CTX-1(e)	37.7	0.1	16.0	0.1	...	3.0	1.0	1.7	39.0	...	...	0.03	...	...
Pyromet CTX-3	38.3	0.2	13.6	...	...	4.9	0.1	1.6	bal	...	0.15	0.05	0.007	...
17-14CuMo(e)	14.0	16.0	...	2.5	...	0.4	...	0.3	62.4	0.75	0.50	0.12	...	3.0 Cu
20-Cb3(e)	34.0	20.0	...	2.5	...	1.0	...	...	42.4	...	...	0.07	...	3.5 Cu

(a) Ref 1. (b) Ref 13. (c) Ref 14. (d) Ref 15. (e) Ref 3

Table 10 Nominal compositions of wrought cobalt-base alloys

Alloy	Ni	Cr	Co	Mo	W	Ta	Nb	Al	Fe	Mn	Si	C	Zr	Other
AirResist 213(a)	...	19	66	...	4.7	6.5	...	3.5	...	...	...	0.18	0.15	0.1 Y
Elgiloy(b)	15	20	40	7	...	...	...	...	bal	2	...	0.1	...	0.04 Be
Haynes 188(c)	22.0	22.0	39.2	...	14.0	...	...	...	3.0	...	...	0.10	...	...
L-605(c)	10.0	20.0	52.9	...	15.0	...	...	...	...	...	...	0.05	...	...
MAR-M 918(c)	20.0	20.0	52.5	...	...	7.5	...	...	...	...	...	0.05	0.10	...
MP35N(c)	35.0	20.0	35.0	10.0	...	...	...	...	...	...	...	...	...	...
MP159(c)	25.5	19.0	35.7	7.0	...	...	0.6	0.2	9.0	...	...	...	...	3.0 Ti
Stellite 6B(a)	3.0	30	bal	1.5	4.5	...	...	...	3.0	2.0	2.0	1.1	...	...
Haynes 150(b)	...	28	50.5	...	...	...	...	...	bal	...	0.75	...	...	0.02 P, 0.002 S
S-816(b)	20.0	20.0	bal	4.0	4.0	...	4.0	...	3.0	1.20	...	0.40	...	...
V-36(b)	20.0	25.0	bal	4.0	...	...	2.3	...	2.4	1.0	...	0.32	...	...

(a) Ref 3. (b) Ref 15. (c) Ref 1

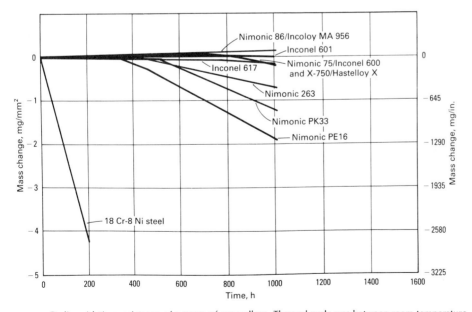

Fig. 11 Cyclic oxidation resistance of a range of superalloys. Thermal cycle was between room temperature and 1000 °C (1830 °F) (except for Inconel 601 and 617); 15 min heating, 5 min cooling. For Inconel alloys 601 and 617, cycle was between room temperature and 1095 °C (2000 °F). Source: Ref 14

ing elements in solid solution, cobalt-base alloys are more likely to precipitate undesirable platelike σ, Laves, and similar TCP phases.

Microstructure. Virtually all cobalt alloys are based on an fcc matrix obtained by alloying with 10% or more nickel. Iron, manganese, and carbon additions also stabilize the fcc phase, while nickel and iron additions improve workability. Exerting the opposite hcp stabilizing tendency are other common alloying elements, such as tungsten, added primarily for solid-solution strengthening, as in L-605, and chromium, added primarily for oxidation and hot corrosion resistance. Tungsten is favored over molybdenum as a strengthener, even though the latter is more effective per atomic percent because tungsten, alone among potential solutes in cobalt, raises the melting temperature of cobalt. Tantalum has been used as a replacement for tungsten in sheet alloys MAR-M 918 and S-57, while molybdenum contents of up to 10 wt% are found in the work-hardened Multiphase (MP) alloys.

Improved oxidation and corrosion resistance with 5 wt% Al have been noted in a few cobalt-base alloys (Ref 19). Titanium additions also have been made in order to precipitate coherent, ordered Co_3Ti as a strengthening phase. Unfortunately, this phase is stable only to about 700 °C (1290 °F), which is much lower than for γ′ Ni_3Al,Ti in Ni-base alloys.

Increased susceptibility to Laves-phase precipitation has been noted in the presence of silicon in tungsten-containing alloys such

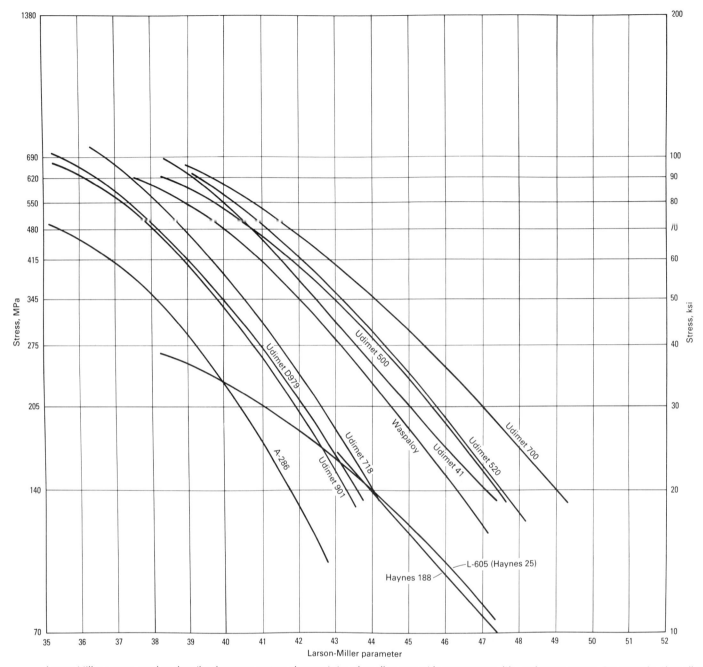

Fig. 12 Larson-Miller curves, used to describe the stress-rupture characteristics of an alloy over wide temperature, life, and stress ranges. For example, if an alloy selected for an application requires a minimum of 100 h of exposure at 925 °C (1700 °F) under a stress of 170 MPa (25 ksi), then the Larson-Miller formula of $P + T(20 + \log t) \times 10^3$ gives the parameter 47.5, where P = LM parameter, T = temperature, and t = time in hours. Comparing 47.5 on the abscissa with a stress of 170 MPa (25 ksi) on the ordinate, it is evident that Udimet 700 is the only alloy that falls above the required point.

as L-605. Haynes 188 was developed with reduced tungsten, increased nickel, and controlled silicon contents in order to avoid this problem.

As in the case of nickel-base alloys, a variety of carbides have been found in cobalt alloys. These include $M_{23}C_6$, M_6C, and MC carbides. In both L-605 and Haynes 188, M_6C transforms into $M_{23}C_6$ during exposure to temperatures in the range of 816 to 927 °C (1500 to 1700 °F) for 3000 h. MC carbides are found only in alloys containing tantalum, niobium, zirconium, titanium, or hafnium. Among wrought cobalt-base alloys, MAR-M 918 contains tantalum, while CM-7, a Belgian modification of L-605, contains a small amount of aluminum and titanium. Therefore, these two alloys will display MC carbides.

In addition to carbides, small quantities of intermetallic phases such as Co_3W, Co_2W, and Co_7W_6 have been found in L-605. Other alloys display the compounds CoAl, Co_3Ti, and $Co_2(Ta,Nb,Ti)$. However, it is unlikely that these phases contribute to the strengthening of the γ matrix. On the contrary, Co_7W_6 and $Co_2(Ta,Nb,Ti)$ are TCP phases that are likely to cause the deterioration of mechanical properties. A uniform dispersion of $Co_3(Ti,Al)$ has been achieved in CM-7 by solution treatment at 1200 °C (2190 °F) and aging at 800 °C (1470 °F). However, this microstructure is unstable at temperatures of 815 °C (1500 °F) and above for times over 1000 h. Figure 13 shows the distribution of phases after various heat treatments for two age-hardened cobalt-base alloys, Haynes 25 and Haynes 188.

An additional source of strengthening, in the Multiphase alloys, arises from the strain-induced transformation of the γ ma-

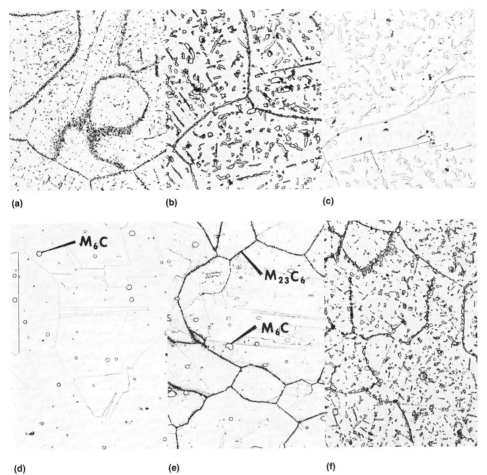

Fig. 13 Microstructures of Haynes 25 and 188 after various heat treatments. (a) Haynes 25, solution annealed at 1204 °C (2200 °F) and aged for 3400 h at 816 °C (1500 °F). Structure is made up of precipitates of M_6C and Co_2W intermetallic compound in an fcc matrix. (b) Haynes 25, solution annealed at 1204 °C (2200 °F) and aged for 3400 h at 871 °C (1600 °F). Structure is same as (a). (c) Haynes 25, solution annealed at 1204 °C (2200 °F) and aged for 3400 h at 927 °C (1700 °F). Structure is same as (a), but M_6C is considered primary, while Co_2W is considered secondary. (d) Haynes 188, cold rolled, 20% solution annealed at 1177 °C (2150 °F) for 10 min before water quenching. Fully annealed structure is M_6C particles in an fcc matrix. (e) Haynes 188, solution annealed at 1177 °C (2150 °F) and aged at 649 °C (1200 °F) for 3400 h. Microstructure is particles of M_6C and $M_{23}C_6$ in an fcc matrix. (f) Haynes 188, solution annealed at 1177 °C (2150 °F) and aged at 871 °C (1600 °F) for 6244 h. Structure is $M_{23}C_6$, Laves phase, and probably M_6C in an fcc matrix. All 500×

Table 11 Oxidation role of alloying additions to cobalt-base materials

Alloying element	Probable effect of addition on the oxidation behavior of a Co(20–30)Cr base
Titanium	Innocuous at low levels
Zirconium	Innocuous at low levels
Carbon	Slightly deleterious; ties up chromium
Vanadium	Harmful, even at 0.5%
Niobium	Harmful, even at 0.5%
Tantalum	Beneficial to moderate levels (<5%)
Molybdenum	Harmful; forms volatile oxides
Tungsten	Innocuous below ~1000 °C (1800 °F); harmful >1000 °C (1800 °F); forms volatile oxides
Yttrium	Beneficial; improves scale adherence
Nickel	May be slightly deleterious
Manganese	Beneficial; induces the formation of spinels
Iron	Tends to induce spinel formation

Source: Ref 19

trix to an hcp structure. This transformation is closely linked to the occurrence of stacking faults in cobalt-base alloys. The lower the stacking fault energy of the γ matrix (as in low nickel, high refractory element content alloys), the more readily will faulting occur.

Applications and Properties. Various wrought cobalt-base alloys can be grouped according to use:

- Alloys for use primarily at high temperatures from 650 to 1150 °C (1200 to 2100 °F), including S-816, Haynes 25, Haynes 188, Haynes 556, and UMCo-50
- Fastener alloys MP35N and MP159, for use to about 650 °C (1200 °F)
- Wear-resistant Stellite 6B

All alloys in the heat-treated and softened condition have fcc crystal structures. However, alloys MP35N and MP159 develop controlled amounts of close-packed hexagonal (cph) structure during the thermomechanical processing recommended before service applications, which are typically used as fasteners for service to 650 °C (1200 °F). Stellite 6B, when heat treated between 650 and 1060 °C (1200 and 1900 °F), and Haynes 25, when exposed for 1000 h or more at temperatures near 650 °C (1200 °F), may partly transform to a cph structure.

None of the cobalt-base superalloys are complete solid-solution alloys because all contain secondary phases in the form of carbides (M_6C, $M_{23}C_6$, M_7C_3, or MC) or intermetallic compounds. Aging causes additional second-phase precipitation, which generally results in some loss of room-temperature ductility (refer to Fig. 11). Of the high-temperature group, alloy S-816 originally was used extensively in turbochargers, as well as gas turbine wheels, blades, and vanes, but has been largely replaced by higher-strength, lower-density nickel-base alloys with improved resistance to adverse environments.

Haynes 25 is perhaps the best-known wrought cobalt-base alloy and has been widely used for hot sections of gas turbines; components of nuclear reactors; devices for surgical implants; and, in the cold-worked condition, for fasteners and wear pads.

Haynes 188 is an alloy that was specially designed for sheet-metal components, such as combustors and transition ducts, in gas turbines. The basic composition, provided that lanthanum, silicon, aluminum, and manganese contents are judiciously controlled, provides excellent qualities, such as oxidation resistance at temperatures up to 1100 °C (2000 °F), hot corrosion resistance, creep resistance, room-temperature formability, and ductility after long-term aging at service temperatures.

Figure 14 compares the static oxidation resistance of Haynes 188 with that of several other alloys. It should be noted that Haynes 188 has excellent oxidation resistance in dry air (Fig. 14a), but displays inferior properties in moist air (Fig. 14b). Haynes 188, like Haynes 25, MP35N, and MP159, can be work hardened to relatively high hardness and tensile strength; after 50% cold reduction, Haynes 188 has a tensile strength of 1690 MPa (245 ksi) at room temperature and 1585 MPa (230 ksi) at 540 °C (1000 °F).

UMCo-50, which contains about 21% iron, is not as strong as Haynes 25 or 188. It is not used extensively in the United States, and especially not in gas turbine applications. In Europe, on the other hand, it is used extensively for furnace parts and fixtures.

The last type of wrought cobalt alloy consists of a single alloy, Stellite 6B, which is characterized by high hot hardness and relatively good resistance to oxidation. The latter property is derived chiefly from its high chromium content (about 30%), whereas its hot hardness property is obtained

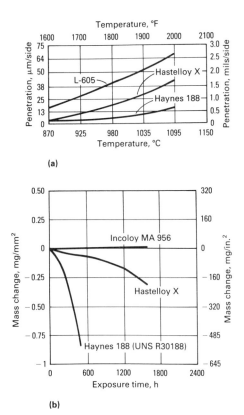

Fig. 14 Oxidation resistance. (a) In dry air for Haynes 188 versus Hastelloy X and L-605 alloys showing continuous penetration from original thickness. (b) Static values at 1100 °C (2010 °F) in air with 5% water vapor

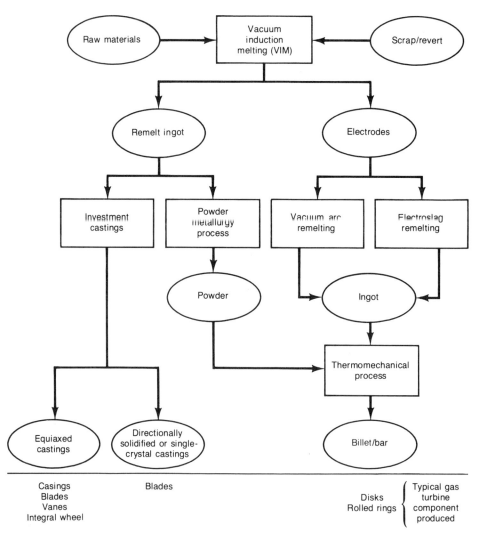

Fig. 15 Flow diagram of processes widely used to produce superalloy components. Source: Ref 22

through the formation of complex carbides of the Cr_7Co_3 and $M_{23}C_6$ types. Stellite 6B is widely used for erosion shields in steam turbines, for wear pads in gas turbines, and for bends in tube systems carrying particulate matter at high temperatures and high velocities.

Several cobalt casting alloys have been successfully prepared as powder product. Both X-40 and MAR-M 509 have exhibited superior mechanical properties under some circumstances, due either to effects of grain size or to carbide size and distribution. Property data and corresponding information on a family of cobalt-chromium-tungsten-carbon alloys that use P/M processing are provided in an appendix to this article. Powder processing also has been used to prepare oxide dispersion strengthened cobalt alloys with improved creep rupture strengths, but no commercial alloys have yet been developed.

Oxidation and Hot Corrosion. The influence of solute elements on the oxidation of cobalt-base alloys has been summarized by Sims (Ref 19) and Beltran (Ref 20) (see Table 11). Among refractory metals, only tantalum is considered beneficial, but tantalum is not added to wrought alloys. Tungsten, molybdenum, vanadium, and niobium are decidedly harmful. Caution must be exercised because carbide distribution may also play a role in oxidation resistance. Beltran (Ref 20) has suggested that the same refractory metals that degrade oxidation resistance play a similar role in hot corrosion. Niobium is cited as being harmful to both the oxidation and corrosion resistance of Wi-52, a cast alloy, but less so for wrought S-816, even though the latter contains twice the niobium content.

Pettit and Giggins (Ref 21) note that cobalt and cobalt-aluminum alloys with chromium and aluminum contents below the level needed to form external scales of Cr_2O_3 or Al_2O_3 are susceptible to a form of catastrophic hot corrosion known as basic fluxing. This type of attack is inhibited by higher levels of chromium or aluminum. In general, cobalt-chromium alloys are less susceptible than nickel-aluminum and nickel-aluminum-X alloys, where X = Mo, W, or V, to two other forms of hot corrosion: alloy-induced acidic degradation and sulfur-induced degradation.

Melting and Consolidation of Wrought Alloys

Superalloy ingots must be melted and cast with due regard for the volatility and reactivity of the elements present. Vacuum melting processes are a necessity for many nickel- and iron-nickel-base alloys because of the presence of aluminum and titanium as solutes. Cobalt-base alloys, on the other hand, do not usually contain these elements and, therefore, may be melted in air.

Melt Processes (Ref 22)

Superalloy melt processes must effectively consolidate raw materials into a product that meets chemistry specifications, mechanical property requirements, microstructure and macrostructure standards, and ultrasonic inspection requirements. The traditional processes used to produce superalloy components are identified in Fig. 15. Of these processes, vacuum induction melting (VIM), vacuum arc remelting (VAR), and electroslag remelting (ESR) are

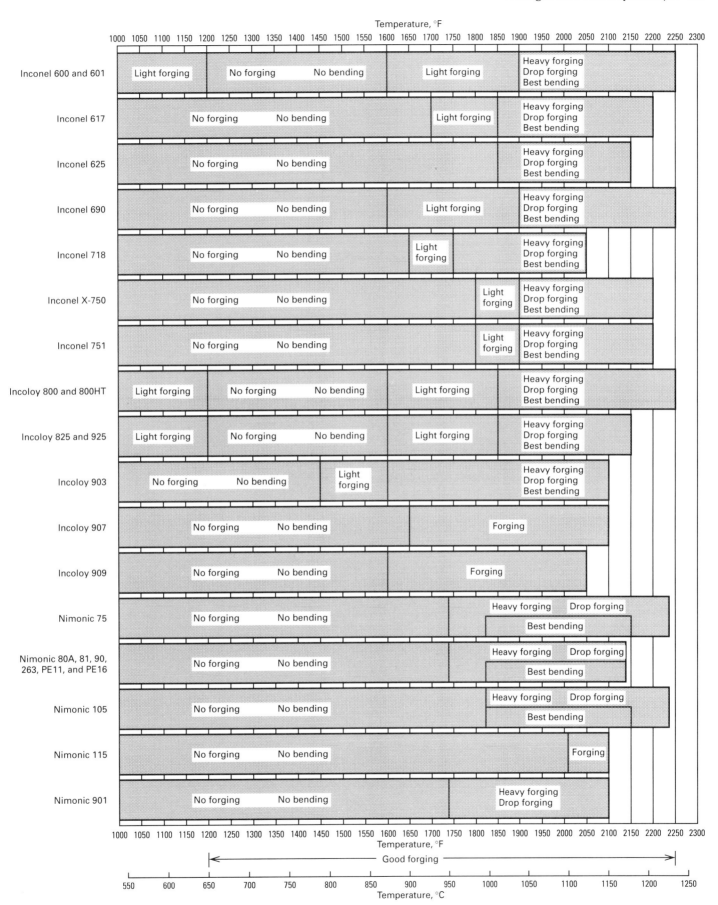

Fig. 16 Temperature ranges for the hot forming of various superalloys. Source: INCO Alloys International, Inc.

Table 12 Chemical compositions of some nickel-base superalloys produced by powder metallurgy

Alloy	C	Ni	Cr	Co	Mo	W	Ta	Nb	Hf	Al	Ti	V	B	Zr
IN-100	0.07	bal	12.4	18.5	3.2	...	...	...	...	5.0	4.3	0.8	0.02	0.06
LC Astroloy	0.023	bal	15.1	17.0	5.2	...	...	...	...	4.0	3.5	...	0.024	<0.01
Waspaloy	0.04	bal	19.3	13.6	4.2	...	...	...	...	1.3	3.6	...	0.005	0.048
NASA II B-7	0.12	bal	8.9	9.1	2.0	7.6	10.1	...	1.0	3.4	0.7	0.5	0.023	0.080
René 80	0.20	bal	14.5	10.0	3.8	3.8	...	...	...	3.1	5.1	...	0.014	0.05
Unitemp AF2-1DA	0.35	bal	12.2	10.0	3.0	6.2	1.7	...	...	4.6	3.0	...	0.014	0.12
MAR-M 200	0.15	bal	9.0	10.0	...	12.0	...	1.0	...	5.0	2.0	...	0.015	0.05
IN-713 LC	0.05	bal	12.0	0.08	4.7	...	...	(2.0)	...	6.2	0.8	...	0.005	0.1
IN-738	0.17	bal	16.0	8.5	1.7	2.6	1.7	0.9	...	3.4	3.4	...	0.01	0.1
IN-792 (PA 101)	0.12	bal	12.4	9.0	1.9	3.8	3.9	...	...	3.1	4.5	...	0.02	0.10
AF-115	0.045	bal	10.9	15.0	2.8	5.7	...	1.7	0.7	3.8	3.7	...	0.016	0.05
MERL 76	0.025	bal	12.2	18.2	3.2	...	...	1.3	0.3	5.0	4.3	...	0.02	0.06
René 95	0.08	bal	12.8	8.1	3.6	3.6	...	3.6	...	3.6	2.6	...	0.01	0.053
Modified MAR-M 432	0.14	bal	15.4	19.6	...	2.9	0.7	1.9	0.7	3.1	3.5	...	0.02	0.05
New alloys														
RSR 103	...	bal	...	...	15.0	...	...	...	...	8.4	...	...	...	...
RSR 104	...	bal	...	...	18.0	...	...	...	...	8.0	...	...	...	...
RSR 143	...	bal	...	...	14.0	...	6.0	...	...	6.0	...	...	...	...
RSR 185	0.04	bal	...	...	14.4	6.1	...	...	...	6.8	...	...	...	...

Source: Ref 2

briefly described below; additional details are provided in Ref 22 and in *Castings*, Volume 15 of the 9th Edition of *Metals Handbook*. Both the thermomechanical and powder metallurgy processes shown in Fig. 15 are described in more detail within this article in the sections "Thermomechanical Processing" and "P/M Alloys." Investment castings, as mentioned earlier, are discussed in the articles "Polycrystalline Cast Superalloys" and "Directionally Solidified and Single-Crystal Superalloys" in this Volume.

The VIM process produces liquid metal under vacuum in an induction-heated crucible. It is used as a primary melting step in the route to producing wrought and cast products, as well as near-net shape. Before being melted, the raw material can be refined and purified and its composition can be controlled. Vacuum induction melting has been widely used in the manufacture of superalloys, which must be melted under vacuum or in an inert gas atmosphere because of their reactivity with atmospheric oxygen and nitrogen. A recent development, the installation of programmable control, in combination with a process computer for automation, allows better melt reproducibility and the capability to meet cleanliness and homogeneity demands.

The VAR process, a secondary melting technique, converts VIM-processed electrodes into ingots whose chemical and physical homogeneity have been significantly improved. In this process, a stub is welded to one end of an electrode, which is then suspended over a water-cooled copper crucible. Next, an arc is struck between the end of the electrode and the crucible bottom. Maintaining the arc generates the heat required to melt the electrode, which drips into the crucible and can subsequently be poured into molds. Many inclusions can be removed by flotation or chemical and physical processes before the molten material solidifies.

The ESR process, another secondary melting technique, may appear to be similar to the VAR process, but there are a number of differences. Remelting does not occur by striking an arc under vacuum. Instead, an ingot is built up in a water-cooled mold by melting a consumable electrode that is immersed in a slag, which is superheated by means of resistance heating. Rather than operating in a vacuum, the process is conducted in air under the molten slag. During melting, metal droplets fall through the molten slag, and chemical reactions reduce sulfur and nonmetallic inclusions. Both ESR and VAR processes allow directional solidification of an ingot from bottom to top, yielding high density and homogeneity in its macrostructure, as well as an absence of segregation and shrinkage cavities.

Future Options. Modifications of existing processes and the development of alternatives in an effort to provide cleaner and more uniform microstructures are ongoing. Future options in processing sequences may involve argon/oxygen degassing prior to vacuum induction melting in the primary melting step. Secondary melting techniques may include electron beam cold hearth refining and plasma cold hearth refining, in addition to VAR and ESR techniques. The electron beam technique appears promising because it is refractoryless, has demonstrated an ability to remove oxides, and offers a greater degree of process control. Triple melting alternatives may be represented by the VAR and vacuum arc double electrode remelt (VADER) processes. The latter is unique because the semicontinuous process can accept ultraclean electrodes and process the material in a refractoryless environment. Excellent descriptions of these options are provided in Ref 22.

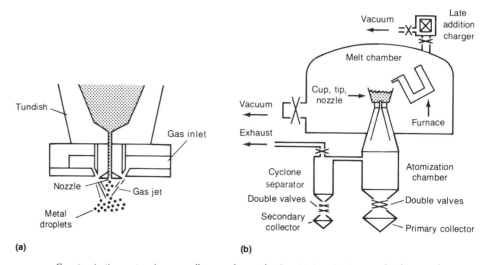

Fig. 17 Gas atomization system for superalloy powder production. (a) Atomization nozzle. (b) Typical system. Source: Ref 28

Table 13 Powder production methods

Step	Inert gas atomization(a)	Soluble gas process	Rotating electrode process(b)	Plasma rotating electrode process	Centrifugal atomization with forced convective cooling (RSR)(c)
Melting 1	VIM; ceramic crucible	VIM; ceramic crucible	VIM, VAR, ESR	VIM, VAR, ESR	VIM; ceramic crucible
Melting 2	...	...	Argon arc	Plasma	...
Melt disintegration system/environment	Nozzle; argon stream	Expansion of dissolved hydrogen against vacuum and Ar + H_2 mixture	Rotating consumable electrode; argon or helium	Rotating consumable electrode; argon	Rotating disk; forced helium convective cooling

(a) VIM, vacuum induction melting. (b) VAR, vacuum arc remelting; ESR, electroslag remelting. (c) RSR, rapid solidification rate. Source: Ref 2

Alloy Types Versus Melt Process. The austenitic high-nickel alloys and nickel-base solid-solution alloys are generally electric furnace melted. Inconels and Incoloys are electric furnace melted, followed by argon-oxygen decarburization (AOD processing). Hastelloys are generally electric furnace melted and then electroslag remelted, although vacuum induction melting followed by electroslag remelting may also be employed. Some other alloys may be simply electric arc furnace melted.

Precipitation-hardening iron-base alloys are electric furnace or vacuum induction melted and then vacuum arc or electroslag remelted. Double vacuum induction melting may be employed when critical applications are involved. In general, vacuum melting improves cleanliness and trace element-sensitive properties such as fatigue strength, ductility, and impact strength. Its effect on tensile strength is negligible.

The precipitation-hardening nickel-base alloys are generally double vacuum induction melted and then vacuum arc or electroslag remelted. Vacuum induction melting lowers the gas content (hydrogen, nitrogen, and oxygen) and evaporates trace elements (such as lead, bismuth, cadmium, tellurium, arsenic, antimony, titanium, and selenium), the presence of which can adversely affect the mechanical properties and workability of ingots. Several of these elements, even in trace amounts, can adversely affect service life. Secondary remelting using vacuum arc and electroslag processes further refines the metal by eliminating gases, nonmetallic and metallic impurities, and inclusions. Secondary remelting also produces larger ingots of uniform composition and dense homogeneous structure. Electroslag remelting is done by arc melting under a cover of slag, and the removal of sulfur is quite easy compared to vacuum melting processes. High sulfur levels are deleterious to the workability and properties of wrought heat-resisting alloys, particularly high-nickel alloys.

Forming Processes

Wrought heat-resistant alloys are manufactured in all mill forms common to the metal industry. Iron-base, cobalt-base, and nickel-base superalloys are produced conventionally as bar, billet, extrusions, plate, sheet, strip, wire, and forgings by primary mills. Inconels and Hastelloys also are available as rod, bar, plate, sheet, strip, tube, pipe, shapes, wire, forging stock, and specialty items from secondary converters.

Cast structure is refined by a working operation known as cogging (Ref 23). In this process, hydraulic presses are used with open dies. Uniform billet structure and improved surface finish are additional objectives of this process. Initial ingot breakdown is followed by methods such as rolling, forging, or extrusion.

Sheet, bar, and ring rolling are commonly employed for secondary hot working of superalloys. Working temperature ranges for many nickel-base alloys are listed in Fig. 16. Edge cracking is a problem that is minimized by rapid handling. Forgings are used in both the turbine and compressor sections of gas tubes. Rates of die closure vary from 0.3 mm/s (0.012 in./s) for hydraulic presses to 7.5 m/s (295 in./s) for hammers. While most forging is carried out with steel tooling heated in the range of 200 to 430 °C (390 to 805 °F), isothermal forging is now widely used for near-net shape processing. Superalloy or molybdenum alloy dies are heated to the same temperature as the forging, in the range of 650 to 980 °C (1200 to 1800 °F). Isothermal forging produces a uniform microstructure while requiring less material, thereby lowering machining costs. For temperatures at which superplastic behavior of superalloys occurs, less powerful presses or hammers and slower die closure rates can be used. The benefits outweigh the increased costs of hot die tooling for expensive input materials.

Extrusion is used for the conversion of ingot to billet, especially for the stronger, crack-prone alloys, as well as for powder-processed alloys. Canning in mild steel or stainless steel is required to avoid chilling and surface cracking. Glass is the most common lubricant (Ref 24). Seamless tubing is also produced by this method.

Sheet and other semifinished products are produced by rolling, often with many reheats, frequent conditioning, and possibly encasing of the alloy in a can. Open-die forging is carried out in flat or swaging dies. Closed-die forging is used to produce shapes that match the impressions of dies attached to the ram and anvil. Both hammers and presses are used; for the hammers, stresses required for deformation are higher because of the higher strain rates employed.

A few superalloys can be cold formed on high-capacity equipment by drawing, extrusion, pressing and deep drawing, spinning, or rolling. For all processes except rolling, a lubricant is usually required, and speeds are relatively low. The uniform, fine-grain microstructures resulting from cold working and reannealing lead to improved mechanical properties.

Thermomechanical Processing

Grain structure may be controlled by thermomechanical processing in several iron-nickel-base alloys that have two precipitates present, the primary strengthening precipitate (γ'' Ni_3Nb in Inconel 718 and γ' Ni_3Ti in Inconel 901) and a secondary precipitate (δ in Inconel 718 and η Ni_3Ti in Inconel 901) (Ref 24, 25). The secondary precipitate is produced first, by an appropriate heat treatment (8 h at 900 °C, or 1650 °F, for 901), followed by working at about 950 °C (1740 °F), below the η solvus. Final working is carried out below the recrystallization temperature, and the alloy is subsequently recrystallized below the η solvus. Finally, the alloy is aged by standard procedures. The result is a fine-grain alloy with higher tensile strength and improved fatigue resistance.

The critical warm-working temperature range for Inconel 718 is 955 to 995 °C (1750 to 1820 °F). The upper limit avoids grain coarsening at higher temperatures due to re-solution of δ, while the lower limit is established to avoid an excessively high flow stress during working. Delta and γ'' precipitates compete for the available niobium. Therefore, any factor suppressing δ tends to favor γ'' formation, and vice versa. Delta does not strengthen Inconel 718, but it reduces the room-temperature ductility. However, when some δ phase is precipitated prior to or during working, the grain size can be reduced substantially, leading to increased tensile and fatigue strength.

Other thermomechanical working schedules are used to produce a double necklace

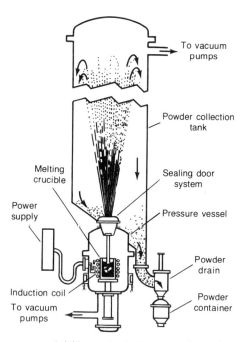

Fig. 18 Soluble gas atomization system for producing superalloy powder. Source: Ref 28

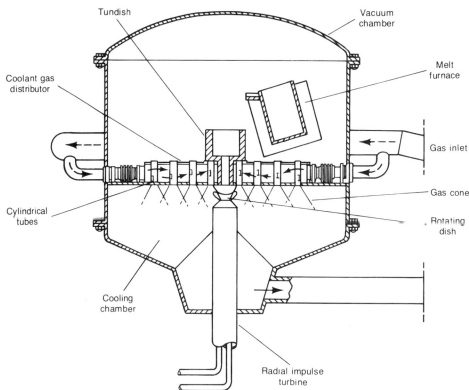

Fig. 19 Schematic of Pratt and Whitney powder rig for producing rapidly solidified powders. Source: Ref 2

structure of fine grains surrounding the large grains formed during high-temperature recrystallization. Reductions of 25 to 50% are needed in the final working operations at 1080 to 1110 °C (1975 to 2030 °F) to produce the small recrystallized grains in cast/wrought René 95.

P/M Alloys

Aircraft gas turbine disks, designed to operate at temperatures of about 650 °C (1200 °F) in current high-performance engines, require forgeable alloys with high yield strength (to tolerate overspeed without burst), high creep resistance, and good damage tolerance. The crack growth rate must be kept low even under conditions of environmental attack and hold times under stress.

Several P/M superalloys have replaced forged alloys as turbine disks, including MERL 76, LC Astroloy, IN-100, and René 95. Table 12 gives compositions. In general, the strength of these alloys is a direct function of the γ' content. Powder processing permits the attainment of a fine grain size, which lends the alloys superplastic forming capability, as in the Pratt and Whitney Gatorizing process. The alloys are characterized by a high homogeneous concentration of both solid-solution strengthening elements and the γ'-forming elements aluminum and titanium. These factors would limit forgeability of conventionally cast and wrought alloys.

Other P/M alloys, strengthened by a small volume percent of fine oxide particles, have been developed for turbine blade and vane applications. These alloys will be discussed in the section "Oxide Dispersion Strengthened Alloys."

The advantages of powder-processed disk alloys are:

- More uniform composition and phase distribution
- Finer grain size
- Reduced carbide segregation
- Higher material yields
- Increased flexibility in alloy design

However, several problems arise directly from powder techniques:

- Increased residual gas content
- Carbon contamination
- Ceramic inclusions
- Formation of prior particle boundary oxide and/or carbide films

These problems can lead directly to inferior mechanical properties. For example, oxide and other inclusions more readily cause crack nucleation and greater variability in fatigue life than is the case with wrought alloys. For these reasons major efforts have been made to produce a cleaner master melt and to reduce contamination during the preparation of powders. The semicontinuous carbide films on prior particle boundaries cannot be altered by simple heat treatment, but can be affected by proper hot-working schedules. These boundaries may also be minimized by the proper selection and balance of solute elements.

Powder Production. Various means of producing superalloy powders are summarized in Table 13; detailed descriptions have been published (Ref 2, 27, 28). Inert gas atomization is the most common technique (Fig. 17). An ingot is first cast, typically by vacuum induction melting, in order to minimize oxygen and nitrogen contents. In some cases remelting may be carried out by electron beam heating, arc melting under argon, or plasma heating. Atomization is carried out by pouring master melt through a refractory orifice; a high-pressure inert gas stream (typically argon) breaks up the alloy into liquid droplets, which are solidified at a rate of about 10^2 K/s.

The spherical powder is collected at the outlet of the atomization chamber. The maximum particle diameter resulting from this process depends on the surface tension, γ, viscosity, η, and density, ρ, of the melt, as well as the velocity, v, of the atomizing gas. The principal factor is gas velocity. Oxygen contents are of the order of 100 ppm, depending on particle size. Finer particle sizes are obtained by screening. Generally, spherical fine particles are desired for further processing (Ref 2).

Another important powder production method, the soluble gas process, is based on the rapid expansion of gas-saturated molten metal, resulting in a fine spray of molten droplets that form as the dissolved gas, usually hydrogen, is suddenly released (Ref 28) (Fig. 18). The droplets solidify at a rate

Table 14 Oxide dispersion strengthened superalloys

Alloy	Density, g/cm³	Ni	Cr	Y₂O₃	Ti	Al	C	Co	Fe	Other
Inconel MA 754	8.3	bal	20	0.6	0.5	0.3	0.05	...	...	...
Inconel MA 758	...	bal	30	0.6	0.5	0.3	0.05	...	...	...
Incoloy MA 956	7.2	...	20	0.5	0.5	4.5	...	...	bal	...
Incoloy MA 957	...	...	14	0.25	1.0	...	...	0.3	bal	...
Inconel MA 6000	8.1	bal	15	1.1	2.5	4.5	0.05	2.0	...	2 Mo, 4 W, 2 Ta, 0.15 Zr, 0.1 B
Inconel MA 760	...	bal	20	0.95	...	6	0.05	...	...	2 Mo, 3.5 W, 0.15 Zr, 0.01 B

Source: Ref 3

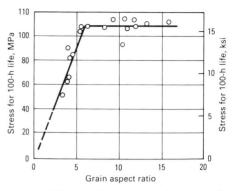

Fig. 20 Influence of grain aspect ratio on stress for 100-h life for MA 753 at 1040 °C (1900 °F). Source: Ref 13

of about 10³ K/s, and the cooled powder is collected under vacuum in another chamber, which is sealed and backfilled with an inert gas. This method is capable of atomizing up to 1000 kg (2200 lb) of superalloy in one heat and produces spherical powder that can be made very fine. This method has been successfully employed for LC Astroloy, MERL 76, and IN-100.

The third method of powder preparation is based on centrifugal atomization. The melt is accelerated and disintegrated by rotating under vacuum or in a protective atmosphere. One example of this method is the rotating electrode process (REP) used in the early production of IN-100 and René 95 powder. In this process, a bar of the desired composition, 15 to 75 mm (0.6 to 3 in.) in diameter, serves as a consumable electrode. The face of this positive electrode, which is rotated at high speed, is melted by a direct current electric arc between the consumable electrode and a stationary tungsten negative electrode. The process is carried out in helium. Centrifugal force causes spherical molten droplets to fly off the rotating electrode. These droplets freeze and are collected at the bottom of the tank, which is filled with helium or argon. A major advantage of this process is the elimination of ceramic inclusions and the lack of any increase in the gas content of the powder relative to the alloy electrode.

A variant on the REP process is the plasma rotating electrode process (PREP). Instead of a tungsten electrode, a plasma arc is used to melt the superalloy electrode surface. Cooling rates are higher, up to 10⁵ K/s for IN-100 powder. On average, particle sizes are nearly twice as large in these processes as in gas atomization. Neither REP nor PREP processes are currently in active production for superalloys (Ref 28).

When solidification rates of powder exceed 10^5 K/s, the process is referred to as rapid solidification rate (RSR). For superalloys, the objective of the high rates is to obtain a microcrystalline alloy rather than an amorphous material. Apart from extremely fine grain size, such powders display nonequilibrium solubilities and very uniform compositions because of the very fine dendritic arm spacings resulting from rapid solidification. Both conventional superalloys and new alloys based on Ni-Al-Mo-X alloys, where X = Ta or W, have been prepared by RSR.

Rapid solidification processing can be done by centrifugal atomization with forced convective cooling, as in the method shown in Fig. 19. In this method, the alloy is vacuum induction melted in the upper part of a chamber. The chamber is then backfilled with helium, and the alloy is poured in a preheated tundish. The liquid is poured through the tundish nozzle onto a rotor that is spun at 24 000 rev/min. The melt is accelerated to rim speed and then ejected longitudinally as droplets. Further atomization and cooling of the droplets is accomplished by the injection of helium gas through annular nozzles. Spherical powder in the 10 to 100 μm (400 to 4000 μin.) diam range is produced; the cooling rate typically varies between 10^5 and 10^7 K/s, the higher rates being achieved with the smaller particles.

Rapidly solidified powders may also be prepared from melt spun ribbon that is pulverized after solidification. The ribbon is produced by pouring the melt through an orifice under pressure and impinging it on a rotating wheel of, for example, copper, that acts as a heat sink. Typical cooling rates are approximately 10^6 K/s, and ribbon thicknesses are less than 25 μm (1000 μin.). The ribbon must be mechanically pulverized, and this method is generally limited to small quantities of experimental alloys.

Powder Consolidation. Although virtually every P/M consolidation technique has been applied to superalloys, production is presently limited to hot isostatic pressing (HIP) and related processes and to hot compaction followed by extrusion (Ref 28), in which extrusion is carried out to achieve full density.

A key feature of any consolidation process is the necessity to minimize contamination, especially from absorbed surface gases and organic material mixed in with the powder. Therefore, powders are packed into sheet metal containers under dynamic vacuum (either warm or cold). The evacuated container is then sealed, heated to the desired temperature, and then compacted, either isostatically under gas pressure or in a closed die.

The advantage of hot isostatic pressing is that a fully dense product can be obtained without retained prior particle boundaries (PPBs). The densification mode is either plastic deformation or creep, depending on the HIP cycle time, temperature, and pressure. Grain size control during hot isostatic pressing is achieved by choosing a temperature either above or below the γ′ solvus.

The advantage of extrusion is greater certainty in breaking up reaction areas that are due to organic contamination. Extrusion has been used for a significant portion of consolidation in production. The hot compaction that precedes extrusion is usually carried out below the solvus.

Recently, both glassy and ceramic containers have been used for hot isostatic pressing (Ref 2). Glassy containers are weak and cannot support the weight of a large part. However, a ceramic container inserted into a large steel container allows the production of intricate shapes. During the HIP cycle, the outer metal jacket can be subjected to the autoclave gas pressure. The pressure is transmitted to the ceramic mold through a bed of fine alumina powder.

Other powder consolidation techniques related to hot isostatic pressing are the consolidation by atmospheric pressure (CAP) process and the fluid die process (Ref 2). Neither technique requires an expensive HIP unit. The CAP process resembles a vacuum sintering process, assisted by low pressure exerted on the surface of a shaped glass container. The fluid die process incorporates a cavity surrounded by a dense, incompressible mass of material. The higher mass of these containers, compared to sheet containers, allows greater vibration during filling and sealing, ensuring more complete filling and uniform tap density. The outer material softens appreciably at the compaction temperature, allowing pressure to be transmitted to the powder. Convectional die forging equipment is capable of much higher ram pressures than those possible with HIP autoclaves; full consolidation occurs in less than 1 s.

Table 15 Heat treatments, grain size, and tensile properties of René 95 forms

Heat treatment/property	Extruded and forged(a)	Hot isostatic pressing(b)	Cast and wrought(c)
Heat treatment	1120 °C (2050 °F)/1 h AC + 760 °C (1400 °F)/8 h AC	1120 °C (2050 °F)/1 h AC + 760 °C (1400 °F)/8 h AC	1220 °C (2230 °F)/1 h AC + 1120 °C (2050 °F)/1 h AC + 760 °C (1400 °F)/8 h AC
Grain size, μm (mils)	5 (0.2) (ASTM No. 11)	8 (0.3) (ASTM No. 8)	150 (6) (ASTM No. 3-6)
40 °C (100 °F) tensile properties			
0.2% yield strength MPa (ksi)	1140 (165.4)	1120 (162.4)	940 (136.4)
Ultimate tensile strength, MPa (ksi)	1560 (226.3)	1560 (226.3)	1210 (175.5)
Elongation, %	8.6	16.6	8.6
Reduction in area, %	19.6	19.1	14.3
650 °C (1200 °F) tensile properties			
0.2% yield strength MPa (ksi)	1140 (165.4)	1100 (159.5)	930 (134.7)
Ultimate tensile strength, MPa (ksi)	1500 (217.6)	1500 (217.6)	1250 (181.3)
Elongation, %	12.4	13.8	9.0
Reduction in area, %	16.2	13.4	13.0

(a) AC, air cooled. Processing: −150 mesh powder, extruded at 1070 °C (1900 °F) to a reduction of 7 to 1 in area, isothermally forged at 1100 °C (2012 °F) to 80% height reduction. (b) Processing: −150 mesh powder, HIP processed at 1120 °C (2050 °F) at 100 MPa (15 ksi) for 3 h. (c) Processing: cross rolled plate, heat treated at 1218 °C (2225 °F) for 1 h. Source: Ref 28, 31

Thermomechanical Working (Ref 2, 24, 25). Processing by thermomechanical means achieves hot deformation that can neutralize the detrimental effects of the contamination of superalloy powder by ceramic inclusions, rubber, cloth, mill scale, and similar debris. In addition, the superplastic behavior of some P/M superalloys, when grain size is in the range of 1 to 10 μm (40 to 400 μin.), leads to improved formability so that near-net shape parts can be more readily produced. Thermomechanical processing also can modify grain shape, size, and orientation; dislocation substructure; and grain-boundary morphology, in addition to producing necklace structures, which may result in improved fatigue crack growth behavior because of their microstructure.

In the Pratt and Whitney Gatorizing process, initially fine-grain P/M preforms are isothermally forged to improve structural uniformity relative to other consolidation methods. A high volume fraction of γ' is required to stabilize the grain size during forging. The fine grain structure is achieved by extrusion of previously compacted preforms or by direct powder extrusion. These alloys exhibit high strain rate sensitivity, making strain rate control during forging critical.

Oxide Dispersion Strengthened Alloys. An important application of P/M techniques is the production of oxide dispersion strengthened superalloys. Commercial alloy compositions are listed in Table 14 and are based on either nickel or iron-chromium-aluminum matrices.

While a variety of methods has been used to incorporate particles in a ductile matrix, the only commercial technique currently being employed is mechanical alloying. In this proprietary process, a controlled mixture of alloy powder and about 1 vol% oxides (typically Y_2O_3) are charged into a high-energy ball mill. The metal particle size initially is in the 2 to 200 μm (80 to 8000 μin.) range, while the oxide particles are less than 10 μm (<400 μin.) in size. The milling operation, carried out dry, causes the superalloy particles to weld repeatedly to the oxide particles and then break apart. The resultant acicular or platelike powders are composites with extremely fine, homogeneous microstructures. Consolidation is carried out by placing the powders in steel cylinders, aligning the ends closed, and either extruding to bar or rolling to plate or sheet.

Hard particles produced by mechanical alloying are strengthened by impeding dislocation motion. An additional factor affecting the strength of these alloys is the elongated grain structure resulting from extrusion or rolling. The Y_2O_3 particles in the mechanically alloyed materials listed in Table 14 are fine and quite uniformly dispersed. The hardening due to these factors must be added to the strengthening effects of grain boundaries, subboundaries, and solid-solution additions. An additional factor is the grain aspect ratio (GAR), or the ratio of grain length, D, to width, d. At high temperatures, tensile strength varies approximately linearly with GAR (Ref 29).

Creep and stess rupture behaviors also correlate well with GAR, as shown in Fig. 20 for MA 753 at 1040 °C (1900 °F) (Ref 30). Wilcox and Clauer (Ref 29) concluded that when grains are elongated, the GAR effect swamps any contribution of grain size. As shown in Fig. 7(b), the strength of MA 6000 is greater than that of most cast or cast and wrought alloys at temperatures above 1000 °C (1830 °F).

A further advantage of mechanical alloying is a much higher ratio of fatigue resistance to tensile strength relative to that of precipitation-hardened alloys.

The thermomechanical processing of ODS alloys is carried out by standard su-

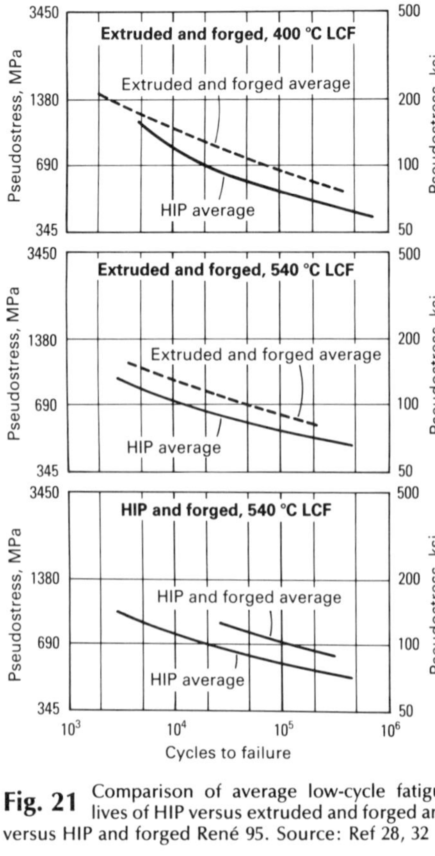

Fig. 21 Comparison of average low-cycle fatigue lives of HIP versus extruded and forged and versus HIP and forged René 95. Source: Ref 28, 32

Table 16 Powder metallurgy René 95 tensile properties at 650 °C (1200 °F)

Alloy	ASTM grain size	Ultimate tensile strength MPa	Ultimate tensile strength ksi	0.2% offset yield strength MPa	0.2% offset yield strength ksi	Elongation, %	Reduction in area, %	Stress-rupture life at 965 MPa (140 ksi), h
HIP René 95(a)	8	1505	218	1140	165	13.8	15	87 150
HIP René 95 with supersolvus heat treatment(b)	6	1415	205	1000	145	16	19	244 298

(a) Processing: −150 mesh powder, HIP processed at 1120 °C (2050 °F), 103 MPa (15 ksi)/3 h + 1120 °C (2050 °F)/1 h + 870 °C (1600 °F)/1 h + 650 °C (1200 °F)/24 h. (b) Processing: −150 mesh powder, HIP processed at 1120 °C (2050 °F), 103 MPa (15 ksi)/3 h + 1200 °C (2200 °F)/1 h + 1120 °C (2050 °F)/1 h + 870 °C (1600 °F)/1 h + 650 °C (1200 °F)/24 h. Source: Ref 28

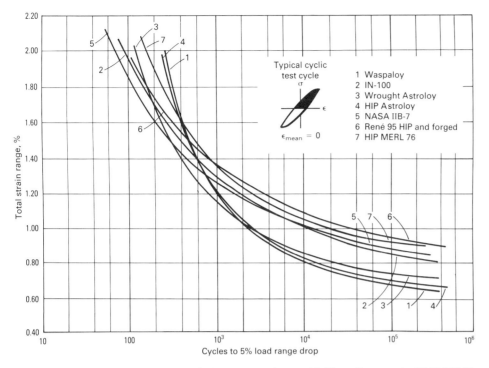

Fig. 22 Comparison of cyclic strain control LCF properties of seven nickel-base alloys tested at 650 °C (1200 °F). Source: Ref 33

peralloy procedures because grain size is fine and alloys such as MA 6000 can be deformed superplastically (Ref 27). A critical level of warm deformation (815 to 1100 °C, or 1500 to 2010 °F) is required for subsequent recrystallization at higher temperature (<1200 °C, or 2190 °F) to produce coarse, elongated grains with the desired cube-on-edge texture. Finally, heat treatment is carried out to control γ' size and carbide distribution.

The ODS alloys have applications in hot sections of aircraft gas turbines, in powder generation equipment, in furnaces, and in the chemical process industries. For example, Inconel MA 6000, which is also strengthened by γ', is suitable for turbine blades, while the solid-solution alloy Inconel MA 754 is a turbine vane alloy and has applications in which resistance to impure helium (in gas-cooled nuclear reactors) or to corrosive liquids such as molten glass is required. The iron-base sheet alloy MA 956 has been developed for service at temperatures above 1100 °C (2010 °F) in corrosive atmospheres.

Mechanical Properties (Ref 28). Monotonic P/M superalloy properties are directly related to the composition and structure of the alloy. In turn, the structure is related to powder particle size, consolidation process, and heat treatment. For example, the effects of grain size and heat treatment on the tensile properties of René 95 are obvious in Table 15, which shows that tensile strength and ductility increase as grain size becomes finer (Ref 28, 31). However, when René 95 is heat treated above the γ' solvus temperature, both grain size and stress-rupture life increase, while tensile yield decreases, as shown in Table 16.

Controlling the cyclic mechanical properties of superalloys, particularly low-cycle fatigue (LCF), is important to gas turbine engine performance and reliability improvements. Low-cycle fatigue is primarily controlled by defects in the material that result in the initiation of fatigue cracks by any mechanism that is noncrystallographic in nature. The effects of defects on low-cycle fatigue depend on the size and location of the defects in the LCF specimen. (The defects themselves, in terms of size and distribution, are dependent on powder particle size and consolidation process.) A defect of a given size is less detrimental when located internally rather than externally. This is primarily due to the absence, when internal, of adverse environmental interaction (Ref 28).

The effects of thermomechanical working on low-cycle fatigue also are important. As shown in Fig. 21, extruded and forged material showed longer life than the average HIP or HIP and forged materials (Ref 28, 32). The effects of forging are that defects are dispersed during processing, their size may be reduced, and the grain size is further refined. Controlling the thermomechanical process can reduce or even eliminate the effects of PPB defects. The LCF life of a thermomechanically processed P/M superalloy does appear to be limited by small ceramic defects (Ref 28).

Test conditions and alloy chemistry also are factors in fatigue resistance. This may

Table 17 Selected P/M superalloy properties

Property	MERL 76(a)	RSR 185(b)	IN-100(c)	LC Astroloy(d)	Astroloy(c)	René 95(e)	IN-718(f)
Condition	Gatorized	As-extruded	...	HIPed(g)	...	HIPed at 1121 °C (2050 °F)(g)	RS(h)
Grain size, μm	20	16–20	5	...	5	...	...
Properties at 25 °C (77 °F)							
0.2% yield strength, MPa (ksi)	1035 (150)	1380 (200)	940 (136)	932 (135)	936 (135)	1215 (176)	1240 (180)
Ultimate tensile strength, MPa (ksi)	1505 (218)	1860 (270)	1130 (164)	1380 (200)	1393 (202)	1636 (237)	1450 (210)
Elongation, %	38	8	8	26	...	...	18
Reduction in area, %	30	...	...	...	...	15	33.5
Properties at 649 °C (1200 °F)							
0.2% yield strength, MPa (ksi)	1050 (152)	...	1080 (157)	863 (125)	1025 (149)	1120 (162)	1035 (150)
Ultimate tensile strength, MPa (ksi)	1276 (185)	...	1290 (187)	1290 (187)	1300 (189)	1514 (220)	1173 (170)
Elongation, %	20	...	16	25	25	17	20
Reduction in area, %	20	...	...	...	...	...	32
Properties at 704 °C (1300 °F)							
0.2% yield strength, MPa (ksi)	1050 (152)	1104 (160)	1065 (154)	...	1030 (149)	...	...
Ultimate tensile strength, MPa (ksi)	1320 (191)	1310 (190)	1270 (184)	...	1160 (168)	...	...
Elongation, %	16	...	20	...	24	...	...
Reduction in area, %	23	...	...	...	...	...	...

(a) Source: Ref 34. (b) Source: Ref 35. (c) Source: Ref 36. (d) Source: Ref 37. (e) Source: Ref 38. (f) Source: Ref 39. (g) HIP, hot isostatic pressing. (h) RS, rapidly solidified

Table 18 Material properties of oxide dispersion strengthened superalloys

Property	Inconel MA 754(a)	Incoloy MA 956(b)	Inconel MA 6000(c)
Mechanical			
Ultimate tensile strength, MPa (ksi)			
At 21 °C (70 °F), longitudinal	965 (140)	645 (94)	1295 (188)
At 540 °C (1000 °F), longitudinal	760 (110)	370 (54)	1155 (168)
At 1095 °C (2000 °F)			
Longitudinal(d)	148 (21.5)	91 (13.2)	222 (32.2)
Transverse(d)	131 (19)	90 (13.0)	177 (25.7)
Yield strength, 0.2% offset			
At 21 °C (70 °F), longitudinal	585 (85)	555 (80)	1285 (186)
At 540 °C (1000 °F), longitudinal	515 (75)	285 (41)	1010 (147)
At 1095 °C (2000 °F)			
Longitudinal	134 (19.5)	84.8 (12.3)	192 (27.8)
Transverse	121 (17.5)	82.7 (12.0)	170 (24.7)
Elongation, %			
At 21 °C (70 °F), longitudinal	21	10	4
At 540 °C (1000 °F), longitudinal	19	20	6
At 1095 °C (2000 °F)			
Longitudinal	12.5	3.5	9.0
Transverse	3.5	4.0	2.0
Reduction in area at 1095 °C (2000 °F), %			
Longitudinal	24	...	31.0
Transverse	1.5	...	1.0
1000-h rupture strength, MPa (ksi)			
At 650 °C (1200 °F)	255 (37)	110(e) (16)	...
At 980 °C (1800 °F)	130 (19)	65 (10)	185 (27)
Physical			
Melting range, °C (°F)	...	1480(f)(g) (2700)	...
Specific heat capacity at 21 °C (70 °F), J/kg · K (Btu/lb · °F)	...	469(g) (0.112)	...
Thermal conductivity at 21 °C (70 °F), W/m · K (Btu/ft^2 · in. · h · °F)	...	10.9(g) (76)	...
Mean coefficient of thermal expansion at 538 °C (1000 °F), 10^{-6}/K	...	11.3(g)	...
Electrical resistivity, nΩ · m	...	1310(g)	...

(a) Data for bar form. Condition of test material was 1315 °C (2400 °F)/1 h/AC. (b) Data for sheet. Condition of test material was 1300 °C (2375 °F)/1 h/AC. (c) Data for bar. Condition of test material was 1230 °C (2250 °F)/0.5 h/AC + 955 °C (1750 °F)/2 h/AC + 845 °C (1550 °F)/24 h/AC. (d) Ref 28. (e) At 760 °C (1400 °F). (f) Approximate solidus temperature. (g) Ref 14

best be seen in Fig. 22, which compares the strain-controlled fatigue lives of seven superalloys. At low strain amplitude, the strongest alloys (for example, René 95 and MERL 76) are the most fatigue resistant, while at high strain amplitudes, the weakest alloys (Waspaloy and HIP Astroloy) display the longest lives. The crossover in ranking of the various alloys occurs because strength is the most important variable at low strain amplitudes, while ductility dominates at high strain amplitudes.

For alloys that are highly oxidation resistant, the stress-rupture properties of bar and sheet of the same alloy differ little. However, less oxidation-resistant alloys are likely to exhibit lower properties for sheet than for bar of the same composition. Similarly, elevated-temperature fatigue properties are very susceptible to changes in environment. Oxygen and sulfur are particularly harmful with respect to crack growth at elevated temperatures. The application of a coating may actually decrease fatigue resistance at high stress or strain levels because of alterations in microstructure induced by the coating treatment. Long-time exposures of coated materials, however, should yield superior fatigue properties.

Selected P/M superalloy mechanical properties are provided in Table 17, while those of ODS materials are given in Table 18. It should be noted that mechanical properties can change significantly with changes in processing conditions and chemical compositions. In particular, changes in grain size and precipitate distribution that result from processing can have a substantial influence on strength.

REFERENCES

1. Appendix B, compiled by T.P. Gabb and R.L. Dreshfield, in *Superalloys II*, C.T. Sims, N.S. Stoloff, and W.C. Hagel, Ed., John Wiley & Sons, 1987, p 575-596
2. G.H. Gessinger, *Powder Metallurgy of Superalloys*, Butterworths, 1984
3. F.R. Morral, Ed., Wrought Superalloys, in *Properties and Selection: Stainless Steels, Tool Materials and Special-Purpose Metals*, Volume 3, 9th ed., Metals Handbook, American Society for Metals, 1980
4. E.W. Ross and C.T. Sims, in *Superalloys II*, C.T. Sims, N.S. Stoloff, and W.C. Hagel, Ed., John Wiley & Sons, 1987, p 97
5. R.F. Decker, "Strengthening Mechanisms in Nickel-Base Superalloys," Paper presented at the Climax Molybdenum Company Symposium, Zurich, May 1969
6. M.J. Donachie, *Superalloy Source Book*, American Society for Metals, 1984, p 105
7. P.H. Thornton et al., *Metall. Trans.*, Vol 1, 1970, p 207
8. J.W. Brook and P.J. Bridges, in *Superalloys 1988*, The Metallurgical Society, 1988, p 33-42
9. Volume 7, 8th Edition, *Metals Handbook*, American Society for Metals, 1972, p 170
10. J.H. Schneibel, C.L. White, and M.H. Yoo, *Metall. Trans. A*, Vol 16A, 1985, p 651
11. J.M. Walsh and D.H. Kear, *Metall. Trans. A*, Vol 6A, 1975, p 226-229
12. D. Moon and F. Wall, in *Proceedings of the Symposium on Structural Stability in Superalloys*, American Institute of Mining, Metallurgical, and Petroleum Engineers, 1968, p 115
13. "High-Temperature High-Strength Nickel Base Alloys," Inco Alloys International Ltd., distributed by Nickel Development Institute
14. "Product Handbook," Publication 1A1-38, Inco Alloys International, Inc., 1988
15. *Materials Selector 1988*, Penton, 1987
16. Alloy 230 Product Literature, Haynes International
17. J.M. Moyer, in *Proceedings of Superalloys 1984 Conference*, The Metallurgical Society, 1984, p 445
18. J.P. Collier, A.O. Selius, and J.K. Tien, in *Proceedings of Superalloys 1988 Conference*, The Metallurgical Society, 1988, p 43
19. C.T. Sims, *J. Met.*, Vol 21 (No. 12), 1969, p 27
20. A.M. Beltran, *Cobalt*, No. 46, 1970, p 3
21. F.S. Pettit and C.S. Giggins, in *Superalloys II*, C.T. Sims, N.S. Stoloff, and W.C. Hagel, Ed., John Wiley & Sons, 1987, p 327
22. G.E. Maurer, Primary and Secondary Melt Processing—Superalloys, in *Superalloys, Supercomposites, and Superceramics*, Academic Press, 1989, p 64-96
23. W.H. Couts, Jr., and T.E. Howson, in *Superalloys II*, C.T. Sims, N.S. Stoloff, and W.C. Hagel, Ed., John Wiley & Sons, 1987, p 441
24. L.A. Jackman, in *Proceedings of the Symposium on Properties of High Temperature Alloys*, Electrochemical Society, 1976, p 42
25. N.A. Wilkinson, *Met. Technol.*, July 1977, p 346
26. E.E. Brown and D.R. Muzyka, in *Superalloys II*, C.T. Sims, N.S. Stoloff, and W.C. Hagel, Ed., John Wiley & Sons, 1987, p 165
27. G.H. Gessinger, *Powder Metall. Int.*, Vol 13, 1981, p 93

28. S. Reichman and D.S. Chang, in *Superalloys II*, C.T. Sims, N.S. Stoloff, and W.C. Hagel, Ed., John Wiley & Sons, 1987, p 459
29. B.A. Wilcox and A.H. Clauer, in *Oxide Dispersion Strengthening*, Gordon & Breach, 1968, p 323
30. J.S. Benjamin and M.J. Bonford, *Metall. Trans.*, Vol 5, 1974, p 416
31. R.V. Miner and J. Gayda, *Int. J. Fatigue*, Vol 6 (No. 3), 1984, p 189
32. G.I. Friedman and G.S. Ansell, *The Superalloys*, C.T. Sims and W.C. Hagel, Ed., John Wiley & Sons, 1972, p 427
33. B.A. Cowles, D.L. Sims, and J.R. Warren, NASA-CR-159409, National Aeronautics and Space Administration, 1978
34. R.H. Caless and D.F. Paulonis, in *Superalloys 1988*, The Metallurgical Society, 1988, p 101-110
35. D.B. Miracle, K.A. Williams, and H.A. Lipsitt, in *Rapid Solidification Processing, Principles and Technologies III*, 1982, p 234-239
36. C. Ducrocq, A. Lasalmonie, and Y. Honnorat, in *Superalloys 1988*, The Metallurgical Society, 1988, p 63-72
37. R.M. Pelloux and J.S. Huang, in *Creep-Fatigue Environment Interactions*, The Metallurgical Society of AIME, 1980, p 151-164
38. S.J. Choe, S.V. Golwalker, D.J. Duquette, and N.S. Stoloff, in *Superalloys 1984*, The Metallurgical Society of AIME, 1984, p 309-318
39. J.F. Radovich and D.J. Myers, in *Superalloys 1984*, The Metallurgical Society of AIME, 1984, p 347-356

Appendix: P/M Cobalt-Base Wear-Resistant Materials

R.J. Dawson and E.M. Foley, Deloro Stellite, Inc.

Alloys based on cobalt chromium-tungsten-carbon, such as the Stellite family of alloys, are extremely versatile, wear-resistant materials (Ref 1-6). They are particularly effective when high temperature or corrosive conditions are combined with abrasive, erosive, or metal-on-metal wear conditions. These materials retain their hardness at elevated temperatures and excel in applications in which lubrication cannot be used.

Full density P/M components can now be manufactured by hot isostatic pressing or extruding operations (Ref 7, 8). Powder metallurgy processing offers the advantages of economical automation, near-net shape processing, and resulting refined microstructures.

Material Descriptions and Applications. Table 1 gives the nominal chemical analysis for the most popular Stellite P/M materials. The microstructure of Stellite alloys contains complex combinations of M_7C_3, M_6C, and $M_{23}C_6$ carbides embedded in a cobalt-chromium-tungsten superalloy matrix. Figure 1 shows a range of structures that can be tailored to give a desired combination of wear, corrosion, and mechanical properties.

For highly abrasive conditions, alloys such as Stellite 3, 98M2, and Star J are preferable because they contain a large volume fraction of carbides. The lower carbon grades are more ductile and are suitable for applications involving shock loading or impact.

Figure 2 shows a range of Stellite P/M components. Typical applications include precision balls, bearing race blanks, valve seat inserts, saw tips, cutters, spacer bushings, and wear pads. Figure 3 shows stainless steel needle holders with Stellite 3 serrated-jaw inserts, which are widely used in the medical surgical field. Because Stellite 3 is more ductile than is tungsten carbide, the serrations are less prone to chip off. The corrosion resistance of Stellite 3 makes it more compatible with body fluids and easier to clean. Brazing of the jaw inserts is also much easier with Stellite 3 because the thermal expansion characteristics of Stellite 3 are closer to those of the stainless steel forging.

Material Properties. Tables 2 and 3 show the physical and mechanical properties of Stellite materials. The physical properties were measured on castings, but are thought to be representative of P/M parts. It should be noted that Stellite materials are only weakly magnetic under most conditions.

The excellent hot hardness of the materials is particularly important to prevent the dulling of cutting tools where frictional heating is typically extreme at the sharp edge. It may be noted that some grades, such as Stellite 31, are relatively soft and ductile. This grade is used when high-temperature strength, shock loading, fatigue, and fretting are concerns.

Service Performance Tests. Stellite 12 P/M has proved to be exceptional for cutting timber. Table 4 compares the dulling of

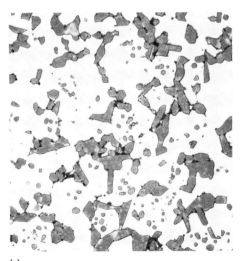
(a)

(b)
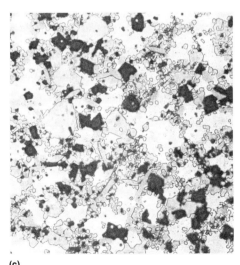
(c)

Fig. 1 Microstructure of Stellite P/M materials. Chromium and tungsten alloy carbides are supported by a cobalt-base superalloy matrix. Carbide level varies from 5 to 60 vol%. (a) Stellite 12. (b) Stellite 31. (c) Stellite 98M2. 500×

Table 1   Nominal chemical analysis of most widely used Stellite P/M materials

Alloy	Composition, %									
	Cr	W	Ni	Mo(a)	Si(a)	Mn(a)	Fe(a)	C	B(a)	Other(a)
Stellite 3	31	12.5	3(a)	...	1	1	3	2.4	1	1
Stellite 6	29	4.5	3(a)	1.5	1.5	1	3	1.2	1	2
Stellite 12	30	8.5	3(a)	...	1	1	3	1.5	1	3
Stellite 19	31	10.5	3(a)	...	1	1	3	1.9	1	2
Stellite 31	25.5	7.5	10.5	...	1	1	2	0.5	...	2
Stellite 190	26	14	3(a)	1	1	1	5	3.1	1	2
Stellite 98M2	30	18.5	3.5	0.8	1	1	5	2	1	2; 4.2 V
Stellite Star J	32.5	17.5	3(a)	...	1	1	3	2.5	1	2

(a) Maximum; all materials cobalt base

Fig. 2 Typical Stellite P/M components

Table 2   Physical properties of Stellite materials

Property	Stellite trade designation						
	3	6	12	19	31	98M2	Star J
Density, g/cm³(a)	8.40	8.20	8.28	8.35	8.45	8.45	8.58
Melting range, °C (°F)	1213–1285 (2215–2345)	1260–1357 (2300–2475)	1255–1341 (2290–2445)	1239–1299 (2260–2370)	1340–1396 (2445–2545)	1224–1275 (2235–2325)	1215–1299 (2220–2370)
Coefficient of thermal expansion, 10^{-6}/K							
At 0–100 °C (32–212 °F)	12.3	13.9	11.9	12.8	14.0	10.8	12.0
At 0–400 °C (32–750 °F)	12.9	14.7	13.4	13.8	14.5	11.7	12.3
At 0–700 °C (32–1290 °F)	13.4	15.8	14.4	14.8	16.0	13.0	13.2
At 0–1000 °C (32–1830 °F)	16	17.4	15.8	16.9	17.0	14.1	15.1
Magnetic permeability at 200 Oersteds and 22 °C (72 °F)	<1.2	<1.2	<1.2	<1.2	...	<1.2	<1.2

(a) P/M materials are typically 97 to 100% dense

tungsten carbide and Stellite 12 P/M for four types of wood. The dulling of the teeth was measured at various stages of wood cutting. Corrosion by wood acids at high temperature is the main cause of dulling. The primary reason that Stellite alloys outperform tungsten carbide is that the wear of Stellite 12 P/M teeth is insensitive to the wood species being cut.

Bearing performance was simulated in a Rolltact test, as shown in Fig. 4. Test results are shown in Table 5. It is worth noting that the P/M balls outperformed their cast counterparts. This was attributed to the fine P/M structure. The wear of the cups and separators was also reduced. The crush strength of the P/M balls was measured and found to be nearly twice that of the cast balls.

Heat Treatment. Stellite P/M materials are not generally considered heat treatable, nor is heat treatment required. Stress relief treatment involves a slow heat-up and soaking at 900 °C (1650 °F) for at least 2 h, followed by furnace cooling. Stellite 6 can be solution annealed by soaking at 1200 °C (2190 °F) for 2 h followed by rapid cooling. This maximizes the corrosion resistance and ductility of Stellite 6. Stellite 31 can be age hardened using a 2-day aging treatment at 800 °C (1470 °F).

Joining. Stellite materials can be joined to other materials by brazing or welding. Flame brazing is typically done with silver solder and an oxyacetylene flame. Vacuum furnace brazing uses a nickel-base brazing alloy such as AMS 4777.

Gas tungsten arc and gas metal arc resistance welding are commonly used for Stellite P/M materials. For the fusion processes, a preheat of 810 °C (1490 °F) is required for the harder grades, such as Stellite 3, 19, 98M2, and Star J. The softer grades, such as Stellite 6 and 31, should be preheated to 540 °C (1005 °F). Furnace cooling is required for the harder grades, whereas cooling in still air is satisfactory for the softer grades. Stellite 25 or Nistelle W filler metals are recommended for joining Stellite material to mild steel or stainless steel.

Machining. With the proper machine setup, Stellite alloys can be rough machined. Alloys having hardnesses of less than 55 HRC can usually be machined, whereas alloys having hardnesses greater than 60 HRC are normally ground. For better surface finishes, any of the Stellite wear-resistant alloys should be ground.

These alloys are usually machined with tungsten carbide tools. Tools for turning should have a 5° end relief, 10° side relief, and a side cutting edge angle of 45°. Tools for facing and boring are essentially the same, except for greater clearance where needed. For best results in drilling, the drill web should be kept as thin as possible, using carbide-tipped drills. In reaming, a 45° side cutting edge angle should be used. The tapping of holes is not recommended for any of the harder alloys, but threads can be produced by electrical discharge machining techniques.

Table 6 is a guide to machining these alloys; it has been written to cover the machining of all the alloys and does not necessarily represent the optimum conditions for each. In general, the harder alloys should be machined, starting at the lower end of the speed and feed ranges shown in Table 6. Softer alloys or more rigid setups will allow higher speeds and feeds.

Table 3 Mechanical properties of Stellite materials

Property	\multicolumn{8}{c}{Stellite P/M trade designation}							
	3	6	12	19	31	190	98M2	Star J
Transverse rupture strength, MPa (ksi)								
At 22 °C (72 °F)	966 (140)	1725 (250)	...	1898 (275)	...	932 (135)	1035 (150)	863 (125)
Ultimate tensile strength, MPa (ksi)								
At 22 °C (72 °F)	863 (125)	897 (130)	...	1035 (150)	828 (120)	621 (90)	794 (115)	523 (75)
At 1000 °C (1830 °F)	725 (105)	828 (120)	...	...	676 (98)	518 (75)	725 (105)	539 (78)
At 1200 °C (2190 °F)	690 (100)	766 (110)	...	...	614 (89)	518 (75)	690 (100)	569 (82.5)
At 1400 °C (2550 °F)	621 (90)	518 (75)	...	...	545 (79)	518 (75)	656 (95)	573 (83)
Elongation, %								
At 22 °C (72 °F)	<1	<1	...	<1	4	<1	<1	<1
At 1000 °C (1830 °F)	<1	1	...	...	14	<1	<1	<1
At 1200 °C (2190 °F)	<1	1	...	...	13	<1	<1	<1
At 1400 °C (2550 °F)	1	10	...	...	16	<1	<1	<1
Hardness, HRC								
At 22 °C (72 °F)	54	40	44	49	32	58	58	56
At 1000 °C (1830 °F)	46	37	...	...	...	54	...	52
At 1200 °C (2190 °F)	39	30	...	...	...	46	...	43
At 1400 °C (2550 °F)	28	15	...	...	...	34	...	31

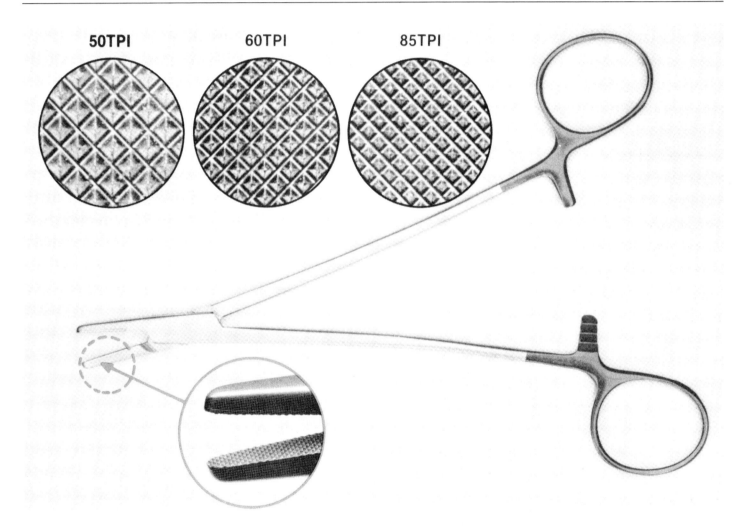

Fig. 3 Stellite P/M needle holder inserts manufactured from one of the harder wear-resistant cobalt-chromium-tungsten-carbon alloys

Table 4 Tool wear rate of saw teeth when cutting various green woods

Wood	Stellite 12 P/M μm/km	Stellite 12 P/M μin./mile	K6 tungsten carbide μm/km	K6 tungsten carbide μin./mile
Douglas fir	0.28	18	0.47	30
Western red cedar	0.29	19	0.87	56
White spruce	0.28	18	0.50	32
Southern pine	0.31	20	0.40	26

Grinding. Whenever close tolerances are required, grinding is recommended for finishing Stellite wear-resistant alloys. Recommended wheels and coolants are listed in Table 7. Grinding speeds should be kept between 14 and 31 m/s (45 and 100 ft/s). Alloys being ground dry should not be quenched because this may cause surface checking. Care should be taken in grinding Stellite 98M2, Star J, and Stellite 3 P/M because they are more sensitive to thermal shock than are the other alloys, and flood cooling is recommended.

REFERENCES

1. G.A. Fritzlen and J.K. Elbaum, Cobalt-Chromium-Tungsten-Molybdenum

Table 5 Rolltact test component wear

Material Parameter		Component wear(a)											
		Drive ball		Alternate balls		Separators (17-4PH)		Test cup (Haynes 25)		Balls only		All components	
Drive ball	Alternate balls	mg	gr	mg	gr	mg	gr	mg	gr	mg	gr	mg	gr
Cast 3	Cast 3	1.8	115	6.3	410	1.6	105	0.4	25	8.1	525	10.1	655
P/M 3	P/M 3	0.1	6.5	0.5	32	0.3	19	0.2	13	0.6	39	1.1	71
19	Cast 3	1.0	65	4.6	300	6.2	400	1.5	97	5.6	365	13.3	860
19	P/M 3	0.3	19	3.9	250	1.6	105	0.3	19	4.2	270	6.1	395

(a) In terms of average weight loss from three tests. Gr, grain

Table 6 Guide for machining Stellite alloys

Operation	Speed		Roughing				Finishing			
			Feed		Depth of cut		Feed		Depth of cut	
	mm/min	ft/min	mm/rev	mils/rev	mm	mils	mm/rev	mils/rev	mm	mils
Turning(a)	9–15	30–50	0.203–0.305	8–12	0.127–5.08	5–205	0.152–0.254	6–10	≤0.635	≤25
Facing(a)	9–15	30–50	0.203–0.305	8–12	0.127–2.54	5–100	0.152–0.254	6–10	≤0.635	≤25
Boring(a)	6–15	20–50	0.127–0.203(b)	5–8	0.635–2.54(b)	25–100	0.127–0.203	5–8	≤0.635	≤25
Drilling(c)	3–8	10–26	...	...	...	...	0.025–0.127(d)	1–5	...	...
Reaming(e)	3–9	10–30	0.025–0.127(f)	1–5	0.762–1.52	30–60	0.025–0.076(f)	1–3	0.254–0.635	10–25
Threading			Use single-point C-3 tungsten carbide tools at 3–8 m/min (10–26 ft/min)							

(a) C-3 tungsten carbide tools should be used. Coolant is water-base fluid diluted 15 parts water to one part fluid. Tools for facing and boring are basically the same as turning tools, except for greater clearances where needed. (b) Depends on bar and size of tool. (c) C-2 tungsten carbide twist drills should be used, with drill web kept as thin as possible. Coolant is same as in (a). (d) Depends on size of drills. (e) C-3 tungsten carbide tools should be used. Reamers should have a 45° side cutting edge angle. Coolant is same as in (a). (f) Per tooth

Wear-Resistant Alloys, in Vol 1, 8th ed., *Metals Handbook,* American Society for Metals, 1961, p 669-671
2. K.C. Anthony, Wear Resistant Cobalt-Base Alloys, *J. Met.,* Feb 1983, p 52-60
3. F.H. Stott, C.W. Stevenson, and G.C. Wood, Friction and Wear of Stellite 31 at Temperatures From 293 to 1073 °K, *Met. Technol.,* Feb 1977, p 66-74
4. W.L. Silence, Effect of Structure on Wear Resistance of Co-, Fe- and Ni-Base Alloys, in *Wear of Materials 1977,* American Society of Mechanical Engineers, 1977, p 77-85
5. H.J. Wagner and A.M. Hall "The Physical Metallurgy of Cobalt-Base Superalloys" Report 171, Defense Materials Information Centre, Battelle Memorial Institute, July 1962
6. G.G. Gould, Cavitation Erosion of Stellite and Other Metallic Materials in *Proceedings of the Third International Conference on Rain Erosion,* Royal Aircraft Establishment, United Kingdom, 1970, p 881-906
7. K.M. Kulkarni, P.J. Walsh, and E.M. Foley, Full Density P/M Parts of Stellite Alloys, *Met. Powder Rep.,* Dec 1983, p 681-684
8. E.M. Foley, Powder Metallurgy Processing Produces Wear and Corrosion Resistant Cobalt-Base Alloys With Improved Properties, in *Proceedings of the P/M in Defense Technology Seminar,* Metal Powder Industries Federation and American Powder Metallurgy Institute, Dec 1986

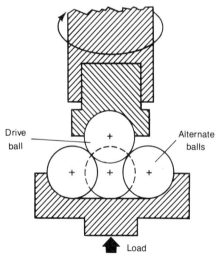

Fig. 4 Schematic of Rolltact test, in which component wear indicates expected wear of a rolling contact bearing operated under similar conditions. The drive ball and cup, alternate balls, and test separator are representative of a full-scale bearing. Test conditions are: spindle speed, 2000 rev/min; contact angle, 30°; contact load, 1380 MPa (200 ksi); test duration, 20 h; environment, deionized water at 22 °C (72 °F) in air

Table 7 General recommendations for grinding wheels

Purpose	Recommended wheels	Coolant
Peripheral wheels		
Stellite alloy deposits	A60JV	Emulsifying oil
Form grinding (Finer grits recommended for sharper forms)	A100GV A120KV A150GV	Emulsifying oil
Ring wheels		
Mixed Stellite alloy and steel	A46DZB	3.8 L (1 gal) good grade of water-soluble oil plus 190 L (50 gal) water
Stellite alloy only	A50BB	
Segmental wheels		
Mixed Stellite alloy and steel	A46DZB	0.45 kg (1 lb) Sal Soda to 190 L (50 gal) water
Stellite alloy only	A50BB	
Tool and cutter grinding		
Cup wheels	A46JV	3.8 L (1 gal) good grade of water-soluble oil plus 190 L (50 gal) water
Peripheral wheels (see above)		
Internal grinding		
Bores >~100 mm (4 in.) diameter	A54JV, A54MV	3.8 L (1 gal) good grade of water-soluble oil plus 190 L (50 gal) water
Centerless grinding <~115 mm (4.6 in.) diameter	A60L-V	3.8 L (1 gal) good grade of water-soluble oil plus 190 L (50 gal) water
Off-hand grinding	A46MV	Dry
Cutting off Stellite alloy	A46QB	1 part water-soluble oil to 60 parts water
Surface grinding		
General Stellite alloy grinding		
For wheels ≤305 mm (12 in.)	A60JV	1 part water-soluble oil to 60 parts water
For wheels >305 mm (12 in.)	A46JV	

Note: Wheels carrying the same marking from different manufacturers do not necessarily have the same cutting action. Wheel descriptions based on ANSI B74.13-1970 and BS 4481

Polycrystalline Cast Superalloys

Gary L. Erickson, Cannon-Muskegon Corporation

SUPERALLOYS are a group of nickel-, iron-nickel-, and cobalt-base materials that exhibit outstanding strength and surface stability at temperatures up to 85% of their melting points ($0.85\ T_M$). They are generally used at temperatures above 540 °C (1000 °F). Superalloys were initially developed for use in aircraft piston engine turbosuperchargers, and their development over the last 50 years has been paced by the demands of advancing gas turbine engine technology.

The initial cast superalloy developments in the United States centered on cobalt-base materials. Cast Vitallium (Co-27Cr-5.5Mo-2.5Ni-0.25C) turbosupercharger blades were produced by the Austenal Company in 1942 in response to an overloaded forging industry and forgeability problems experienced with the early nickel-chromium-iron solid-solution wrought superalloys. This work ultimately led to the successful manufacture of investment cast components for the first U.S. production gas turbine engine in 1945 (Ref 1).

Nickel-base and nickel-iron-base superalloys owe their high-temperature strength potential to their gamma prime (γ') (Ni_3Al,Ti) content. The first reference to aluminum or titanium additions to the 80-20 Ni-Cr system occurred in a patent filed by Heraeus Vacuumschmelze A.G. in 1926, in which as much as 6% Al was added to a nickel-chromium-iron alloy for increased tensile yield strength. Not until later in the decade, however, did a French patent application recognize the occurrence of precipitation hardening in nickel-chromium alloys. In 1929, Pilling and Merica concurrently filed a number of patent applications in the United States for precipitation-hardening nickel-base alloys containing aluminum and titanium, and in 1931 the first British patent applications covering aluminum plus titanium additions to nickel-base alloys were filed.

Although age hardening was recognized, it was not until the development of the Whittle engine in Great Britain, along with gas turbine engine developments in Germany, that material creep strength was a recognized consideration. The requirement for creep resistance led to an understanding of the relationship between age hardening and creep, eventually pacing the rate of development of γ'-strengthened nickel-base alloys.

The first commercial nickel-base alloy developmental work, undertaken by the British in the early 1940s, led to the wrought Nimonic 75 and 80 alloys. Increased operating-temperature requirements for U.S. aircraft engines resulted in the use of aluminum plus titanium strengthened wrought materials during the same period of time. Component forgeability problems, however, led to the use of cast Vitallium until the shortages of cobalt supply experienced during the Korean War caused further research on nickel-base alloys.

Cast nickel-base alloy developments outpaced cobalt-base developmental work by the late 1950s because of their superior strengthening potential, that is, stable, coherent intermetallic compound γ' phase introduction. The introduction of commercial vacuum induction melting (VIM) and vacuum investment casting in the early 1950s provided further potential for γ' exploitation. Many nickel-base alloy developments resulted, continuing through the 1960s (Fig. 1). The compositions of cast nickel- and cobalt-base superalloys are listed in Tables 1 and 2, respectively; Tables 3 to 6 list physical and mechanical properties of cast nickel- and cobalt-base superalloys.

The development of new polycrystalline alloys continued through the 1970s, howev-

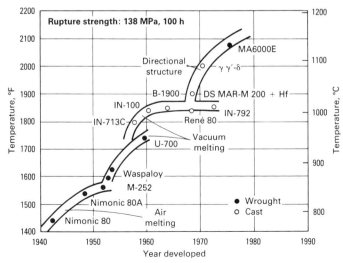

Fig. 1 Progress in the high-temperature capabilities of superalloys since the 1940s. Source: Ref 2

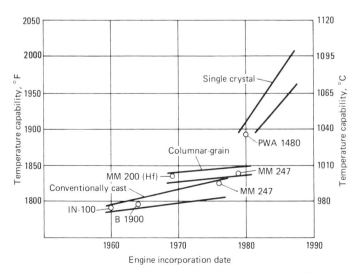

Fig. 2 Advances in turbine blade materials and processes since 1960. Source: Ref 4

982 / Specialty Steels and Heat-Resistant Alloys

Table 1 Nominal compositions and densities of selected cast nickel-base superalloys

Alloy	C	Cr	Co	Mo	W	Ta	Nb	Al	Ti	Hf	Zr	B	Ni	Other	Density, g/cm³
IN-718	0.04	18.5	...	3.0	...	...	5.1	0.5	0.9	...	...	...	bal	18.5 Fe	8.22
René 200	0.03	19.0	12.0	3.2	...	3.1	5.1	0.5	1.0	...	...	...	bal	...	...
IN-625	0.06	21.5	...	8.5	...	...	4.0	0.2	0.2	...	...	...	bal	2.5 Fe	...
IN-713C	0.12	12.5	...	4.2	...	...	2.0	6.1	0.8	...	0.10	0.012	bal	...	8.25
IN-713LC	0.05	12.0	...	4.5	...	...	2.0	5.9	0.6	...	0.10	0.01	bal	...	8.00
IN-713 Hf (MM 004)	0.05	12.0	...	4.5	...	...	2.0	5.9	0.6	1.3	0.10	0.01	bal	...	...
IN-100	0.18	10.0	15.0	3.0	...	...	...	5.5	4.7	...	0.06	0.014	bal	1.0 V	7.75
IN-738C	0.17	16.0	8.5	1.75	2.6	1.75	0.9	3.4	3.4	...	0.10	0.01	bal	...	8.11
IN-738LC	0.11	16.0	8.5	1.75	2.6	1.75	0.9	3.4	3.4	...	0.04	0.01	bal	...	...
IN-792	0.21	12.7	9.0	2.0	3.9	3.9	...	3.2	4.2	...	0.10	0.02	bal	...	8.25
IN-939	0.15	22.4	19.0	...	2.0	1.4	1.0	1.9	3.7	...	0.10	0.009	bal	...	8.2
B-1900	0.10	8.0	10.0	6.0	...	4.3	...	6.0	1.0	...	0.08	0.015	bal	...	8.2
B-1900 Hf (MM 007)	0.10	8.0	10.0	6.0	...	4.3	...	6.0	1.0	1.5	0.08	0.015	bal	...	8.25
B-1910	0.10	10.0	10.0	3.0	7.0	...	...	6.0	1.0	...	0.10	0.015	bal	...	...
MM 002	0.15	9.0	10.0	...	...	2.5	...	5.5	1.5	1.5	0.05	0.015	bal	...	...
MAR-M 200	0.15	9.0	10.0	...	12.5	...	1.8	5.0	2.0	...	0.05	0.015	bal	...	8.53
MAR-M 200 Hf (MM 009)	0.14	9.0	10.0	...	12.5	...	1.0	5.0	2.0	2.0	...	0.015	bal	...	...
MAR-M 246	0.15	9.0	10.0	2.5	10.0	1.5	...	5.5	1.5	...	0.05	0.015	bal	...	8.44
MAR-M 246 Hf (MM 006)	0.15	9.0	10.0	2.5	10.0	1.5	...	5.5	1.5	1.4	0.05	0.015	bal	...	...
MAR-M 247 (MM 0011)	0.16	8.5	10.0	0.65	10.0	3.0	...	5.6	1.0	1.4	0.04	0.015	bal	...	8.53
CM 247LC	0.07	8.1	9.3	0.5	9.5	3.0	...	5.6	0.7	1.4	0.01	0.015	bal	...	...
René 41	0.08	19.0	10.5	9.5	...	...	...	1.7	3.2	...	...	0.005	bal	...	...
René 77	0.08	15.0	18.5	5.2	...	...	...	4.25	3.5	...	...	0.015	bal	...	7.91
René 80	0.17	14.0	9.5	4.0	4.0	...	...	3.0	5.0	...	0.03	0.015	bal	...	8.16
René 80 Hf	0.15	14.0	9.5	4.0	4.0	...	...	3.0	4.7	0.8	0.01	0.015	bal	...	...
René 100	0.15	9.5	15.0	3.0	...	...	...	5.5	4.2	...	0.06	0.015	bal	1.0 V	7.75
René 125 Hf (MM 005)	0.10	9.0	10.0	2.0	7.0	3.8	...	4.8	2.6	1.6	0.05	0.015	bal	...	...
Nimocast 75	0.12	20.0	...	...	...	...	...	...	0.5	...	...	...	bal	...	8.44
Nimocast 80	0.05	19.5	...	...	...	...	...	1.4	2.3	...	...	...	bal	1.5 Fe	8.17
Nimocast 90	0.06	19.5	18.0	...	...	...	...	1.4	2.4	...	...	...	bal	1.5 Fe	8.18
Nimocast 95	0.07	19.5	18.0	...	...	...	...	2.0	2.9	...	0.02	0.015	bal	...	...
Nimocast 100	0.20	11.0	20.0	5.0	...	...	...	5.0	1.5	...	0.03	0.015	bal	...	...
Udimet 500	0.08	18.5	16.5	3.5	...	...	...	3.0	3.0	...	...	0.006	bal	...	8.02
Udimet 700	0.08	14.3	14.5	4.3	...	...	...	4.25	3.5	...	0.02	0.015	bal	...	...
Udimet 710	0.13	18.0	15.0	3.0	1.5	...	...	2.5	5.0	...	0.08	...	bal	...	8.08
C 130	0.04	21.5	...	10.0	...	...	...	0.8	2.6	...	...	...	bal	...	...
C 242	0.30	20.0	10.0	10.3	...	...	...	0.1	0.2	...	...	...	bal	...	...
C 263	0.06	20.0	20.0	5.9	...	...	...	0.45	2.15	...	0.02	0.001	bal	...	...
C 1023	0.15	15.5	10.0	8.0	...	...	...	4.2	3.6	...	...	0.006	bal	...	...
Hastelloy X	0.08	21.8	1.5	9.0	0.6	...	...	...	...	...	...	...	bal	18.5 Fe, 0.5 Mn, 0.3 Si	...
Hastelloy S	0.01	16.0	...	15.0	...	...	...	0.40	...	...	...	0.009	bal	3.0 Fe, 0.02 La, 0.65 Si, 0.55 Mn	...
Waspaloy	0.06	19.0	12.3	3.8	...	...	...	1.2	3.0	...	0.01	0.005	bal	0.45 Mn	...
NX 188	0.04	...	...	18.0	...	...	...	8.0	...	...	...	...	bal	...	...
SEL	0.08	15.0	26.0	4.5	...	...	...	4.4	2.4	...	...	0.015	bal	...	...
CMSX-2(a)	...	8.0	4.6	0.6	8.0	6.0	...	5.6	1.0	...	...	...	bal	...	8.6
GMR-235	0.15	15.0	...	4.8	...	...	...	3.8	2.0	...	...	0.05	bal	0.3 Mn, 0.4 Si, 11.0 Fe	8.0
CMSX-3(a)	...	8.0	4.6	0.6	8.0	6.0	...	5.6	1.0	0.10	...	...	bal	...	8.6
CMSX-4(a)	...	6.4	9.6	0.6	6.4	6.5	...	5.6	1.0	0.10	...	...	bal	3.0 Re	8.7
CMSX-6(a)	...	9.9	5.0	3.0	...	2.0	...	4.8	4.7	0.05	...	...	bal	...	7.98
GMR-235	0.15	15.0	...	4.8	...	...	...	3.5	2.5	...	...	0.05	bal	4.5 Fe	8.04
SEL-15	0.07	11.0	14.5	6.5	1.5	...	0.5	5.4	2.5	...	...	0.015	bal	...	8.7
UDM 56	0.02	16.0	5.0	1.5	6.0	...	...	4.5	2.0	...	0.03	0.070	bal	0.5 V	8.2
M-22	0.13	5.7	...	2.0	11.0	3.0	...	6.3	...	...	0.60	...	bal	...	8.63
IN-731	0.18	9.5	10.0	2.5	...	...	...	5.5	4.6	...	0.06	0.015	bal	1.0 V	7.75
MAR-M 421	0.14	15.8	9.5	2.0	3.8	...	...	4.3	1.8	...	0.05	0.015	bal	...	8.08
MAR-M 432	0.15	15.5	20.0	...	3.0	2.0	2.0	2.8	4.3	...	0.05	0.015	bal	...	8.16
MC-102	0.04	20.0	...	6.0	2.5	0.6	6.0	...	...	...	...	...	bal	0.25 Si, 0.30 Mn	...
Nimocast 242	0.34	20.5	10.0	10.5	...	...	...	0.2	0.3	...	...	...	bal	1.0 Fe, 0.3 Mn, 0.3 Si	8.40
Nimocast 263	0.06	20.0	20.0	5.8	...	...	...	0.5	2.2	...	0.04	0.008	bal	0.5 Fe, 0.5 Mn	8.36

(a) Single crystal

er, at a more moderate rate. Attention was concentrated instead on process development, with specific interest directed toward grain orientation and directional-solidification (DS) turbine blade and vane casting technology (Fig. 2).

Applied to turbine blades and vanes, the DS casting process results in the alignment of all component grain boundaries such that they are parallel to the blade/vane stacking fault axis, essentially eliminating transverse grain boundaries (Fig. 3). Because turbine blades/vanes encounter major operating stress in the direction which is near normal to the stacking fault axis, transverse grain boundaries provide relatively easy fracture paths. The elimination of these paths provides increased strain elasticity by virtue of the lower ⟨001⟩ elastic modulus, thereby creating opportunities for further exploitation of the nickel-base alloy potential.

The logical progression to grain-boundary reduction is the total elimination thereof. Thus, single-crystal turbine blade/vane casting technology soon developed, providing further opportunity for nickel-base alloy design innovation.

The late 1970s and the 1980s have, therefore, been a productive development period for nickel-base alloys designed specifically for directionally solidified columnar-grain and single-crystal cast components. These new process technologies, which are more fully discussed in the article "Directionally Solidified and Single-Crystal Superalloys" in this Volume, have contributed to dramat-

Table 2 Nominal compositions of selected cast cobalt-base superalloys

Alloy	C	Cr	Ni	W	Ta	Nb	Mo	Ti	B	Zr	Fe	Co	Other	Density, g/cm³
HS-21 (MOD Vitallium)	0.25	27.0	3.0	...	...	...	5.0	...	...	...	1.0	bal	...	...
HS-31 (X-40)	0.50	25.0	10.0	7.5	...	...	...	...	...	0.17	1.5	bal	0.4 Si	...
HS-25 (L-605)	0.10	20.0	10.0	15.0	...	...	...	...	...	...	...	bal	...	...
ML-1700	0.2	25.0	...	15.0	...	...	...	...	0.4	...	...	bal	...	...
WI-52	0.42	21.0	1.0 max	11.0	...	2.0	...	...	...	...	2.0	bal	...	8.88
MAR-M 302	0.85	21.5	...	10.0	9.0	...	...	0.2	0.005	...	1.5 max	bal	...	9.21
MAR-M 322	1.0	21.5	...	9.0	4.5	...	...	0.75	...	2.25	0.75	bal	...	8.91
MAR-M 509	0.60	24.0	10.0	7.0	7.5	...	...	0.2	...	...	1.0	bal	...	8.85
AiResist 13	0.45	21.0	...	11.0	...	2.0	...	...	...	...	2.5 max	bal	3.4 Al, 0.1 Y	8.43
AiResist 215	0.35	19.0	0.5	4.5	7.5	...	...	...	...	0.13	...	bal	4.3 Al, 0.1 Y	8.47
F 75	0.25	28.0	1.0 max	...	...	...	5.5	...	...	...	...	bal	...	...
FSX-414	0.25	29.5	10.5	7.0	...	...	...	...	0.012	...	2.0 max	bal	...	8.3
X-45	0.25	25.5	10.5	7.0	...	...	...	...	0.010	...	2.0 max	bal	...	...

ic improvements in gas turbine engine operating efficiency.

Superalloy Design

Nickel-base superalloys have microstructures consisting of an austenitic face-centered cubic (fcc) matrix (γ) dispersed intermetallic fcc γ' Ni$_3$(Al,Ti) precipitates coherent with the matrix (0 to 0.5% lattice mismatch), and carbides, borides, and other phases distributed throughout the matrix and along the grain boundaries. These complex alloys generally contain more than ten different alloying constituents. Various combinations of carbon, boron, zirconium, hafnium, cobalt, chromium, aluminum, titanium, vanadium, molybdenum, tungsten, niobium, tantalum, and rhenium result in the commercial alloys used in today's gas turbine engines.

Some alloying elements have single-function importance, whereas others provide multiple functions. For example, chromium is primarily added to nickel-base alloys for sulfidation resistance (Cr$_2$O$_3$ protective-scale formation), whereas aluminum not only is a strong γ' former but also helps provide oxidation resistance when present in sufficient quantity by forming a protective Al$_2$O$_3$ scale.

Many of the other alloying elements also have multiple roles. Titanium, while primarily partitioning to the γ', also participates in

Table 3 Physical properties of cast nickel-base and cobalt-base alloys

Alloy	Density, g/cm³	Melting range °C	Melting range °F	Specific heat At 21 °C (70 °F) J/kg·K	Btu/lb·°F	At 538 °C (1000 °F) J/kg·K	Btu/lb·°F	At 1093 °C (2000 °F) J/kg·K	Btu/lb·°F	Thermal conductivity At 93 °C (200 °F) W/m·K	Btu·in./h·ft²·°F	At 538 °C (1000 °F) W/m·K	Btu·in./h·ft²·°F	At 1093 °C (2000 °F) W/m·K	Btu·in./h·ft²·°F	Mean coefficient of thermal expansion, 10⁻⁶/K(a) At 93 °C (200 °F)	At 538 °F (1000 °C)	At 1093 °C (2000 °F)
Nickel base																		
IN-713 C	7.91	1260–1290	2300–2350	420	0.10	565	0.135	710	0.17	10.9	76	17.0	118	26.4	183	10.6	13.5	17.1
IN-713 LC	8.00	1290–1320	2350–2410	440	0.105	565	0.135	710	0.17	10.7	74	16.7	116	25.3	176	10.1	15.8	18.9
B-1900	8.22	1275–1300	2325–2375	...	...	...	...	...	...	(10.2)	(71)	16.3	113	...	...	11.7	13.3	16.2
Cast alloy 625	8.44	...	...	...	...	...	...	...	...	...	...	...	...	...	...	...	...	...
Cast alloy 718	8.22	1205–1345	2200–2450	...	...	...	...	...	...	...	...	...	...	...	...	...	...	...
IN-100	7.75	1265–1335	2305–2435	...	...	480	0.115	605	0.145	...	...	17.3	120	...	...	13.0	13.9	18.1
IN-162	8.08	1275–1305	2330–2380	...	...	...	...	...	...	...	...	...	...	...	...	12.2	14.1	...
IN-731	7.75	...	...	...	...	...	...	...	...	...	...	...	...	...	...	...	...	...
IN-738	8.11	1230–1315	2250–2400	420	0.10	565	0.135	710	0.17	...	...	17.7	123	27.2	189	11.6	14.0	...
IN-792	8.25	...	...	...	...	...	...	...	...	...	...	...	...	...	...	...	...	...
M-22	8.63	...	...	...	...	...	...	...	...	...	...	...	...	...	...	12.4	13.3	...
MAR-M 200	8.53	1315–1370	2400–2500	400	0.095	420	0.10	565	0.135	13.0	90	15.2	110	29.7	206	...	13.1	17.0
MAR-M 246	8.44	1315–1345	2400–2450	...	...	...	...	...	...	...	...	18.9	131	30.0	208	11.3	14.8	18.6
MAR-M 247	8.53	...	...	...	...	...	...	...	...	...	...	...	...	...	...	...	...	...
MAR-M 421	8.08	...	...	...	...	...	...	...	...	...	...	19.1	137	32.0	229	...	14.9	19.8
MAR-M 432	8.16	...	...	...	...	...	...	...	...	...	...	...	...	...	...	...	14.9	19.3
MC-102	8.84	...	...	...	...	...	...	...	...	...	...	...	...	...	...	12.8	14.9	...
Nimocast 75	8.44	1410(b)	2570(b)	...	...	...	...	...	...	...	...	...	...	...	...	12.8	14.9	...
Nimocast 80	8.17	1310–1380	2390–2515	...	...	...	...	...	...	...	...	...	...	...	...	12.8	14.9	...
Nimocast 90	8.18	1310–1380	2390–2515	...	...	...	...	...	...	...	...	...	...	...	...	12.3	14.8	...
Nimocast 242	8.40	1225–1340	2235–2445	...	...	...	...	...	...	...	...	...	...	...	...	12.5	14.4	...
Nimocast 263	8.36	1300–1355	2370–2470	...	...	...	...	...	...	...	...	...	...	...	...	11.0	13.6	...
René 77	7.91	...	...	...	...	...	...	...	...	...	...	...	...	...	...	...	...	...
René 80	8.16	...	...	...	...	...	...	...	...	...	...	...	...	...	...	...	...	...
Udimet 500	8.02	1300–1395	2375–2540	...	...	...	...	...	...	...	...	...	...	...	...	13.3	...	...
Udimet 710	8.08	...	...	...	...	...	...	...	...	12.1	84	18.1	126	...	...	...	...	...
Cobalt base																		
FSX-414	8.3	...	...	...	...	...	...	...	...	...	...	...	...	...	...	...	...	...
Haynes 1002	8.75	1305–1420	2380–2590	420	0.10	530	0.126	645	0.154	11.0	76	21.8	151	32.1	222	12.2	14.4	...
MAR-M 302	9.21	1315–1370	2400–2500	...	...	...	...	...	...	18.7	130	22.2	154	...	...	...	13.7	16.6
MAR-M 322	8.91	1315–1360	2400–2475	...	...	...	...	...	...	...	...	...	...	...	...	...	...	...
MAR-M 509	8.85	...	...	...	...	...	...	...	...	...	...	27.9	194	44.6	310	9.8	15.9	18.2
WI-52	8.88	1300–1355	2425–2475	420	0.10	...	...	...	...	24.8	172	27.4	190	40.3	280	...	14.4	17.5
X-40	8.60	...	...	...	...	...	...	...	...	11.8	82	21.6	150	...	...	...	15.1	...

(a) From room temperature to indicated temperature. (b) Liquidus temperature. Source: Nickel Development Institute.

Table 4 Mechanical properties of cast nickel-base and cobalt-base alloys

	Ultimate tensile strength						0.2% yield strength						Tensile elongation, %			Dynamic modulus of elasticity					
	At 21 °C (70 °F)		At 538 °C (1000 °F)		At 1093 °C (2000 °F)		At 21 °C (70 °F)		At 538 °C (1000 °F)		At 1093 °C (2000 °F)		At 21 °C (70 °F)	At 538 °C (1000 °F)	At 1093 °C (2000 °F)	At 21 °C (70 °F)		At 538 °C (1000 °F)		At 1093 °C (2000 °F)	
Alloy	MPa	ksi	MPa	ksi	MPa	ksi	MPa	ksi	MPa	ksi	MPa	ksi				GPa	10⁶ psi	GPa	10⁶ psi	GPa	10⁶ psi
Nickel base																					
IN-713 C	850	123	860	125	...	...	740	107	705	102	...	...	8	10	...	206	29.9	179	26.2	...	...
IN-713 LC	895	130	895	130	...	...	750	109	760	110	...	...	15	11	...	197	28.6	172	25.0	...	...
B-1900	970	141	1005	146	270	38	825	120	870	126	195	28	8	7	11	214	31.0	183	27.0	...	...
IN-625	710	103	510	74	...	...	350	51	235	34	...	...	48	50	...	...	...	...	...	...	...
IN-718	1090	158	...	...	...	...	915	133	...	...	...	...	11	...	...	...	...	...	...	...	...
IN-100	1018	147	1090	150	(380)	(55)	850	123	885	128	(240)	(35)	9	9	...	215	31.2	187	27.1	...	...
IN-162	1005	146	1020	148	...	...	815	118	795	115	...	...	7	6.5	...	197	28.5	172	24.9	...	...
IN-731	835	121	...	...	275	40	725	105	...	...	170	25	6.5	...	...	...	...	...	...	...	...
IN-738	1095	159	...	...	...	...	950	138	...	...	...	...	...	...	...	201	29.2	175	25.4	...	...
IN-792	1170	170	...	...	...	...	1060	154	...	...	...	...	4	...	...	...	...	...	...	...	...
M-22	730	106	780	113	...	...	685	99	730	106	...	...	5.5	4.5	...	...	...	...	...	...	...
MAR-M 200	930	135	945	137	325	47	840	122	880	128	...	...	7	5	...	218	31.6	184	26.7	...	...
MAR-M 246	965	140	1000	145	345	50	860	125	860	125	...	...	5	5	...	205	29.8	178	25.8	145	21.1
MAR-M 247	965	140	1035	150	...	...	815	118	825	120	...	...	7	...	...	...	...	...	...	...	...
MAR-M 421	1085	157	995	147	...	...	930	135	815	118	...	...	4.5	3	...	203	29.4	...	...	141	20.4
MAR-M 432	1240	180	1105	160	...	...	1070	155	910	132	...	...	6	...	...	...	...	...	...	...	...
MC-102	675	98	655	95	...	...	605	88	540	78	...	...	5	9	...	...	...	...	...	...	...
Nimocast 75	500	72	...	...	...	...	179	26	...	...	...	...	39	...	...	...	...	...	...	...	...
Nimocast 80	730	106	...	...	...	...	520	75	...	...	...	...	15	...	...	...	...	...	...	...	...
Nimocast 90	700	102	595	86	...	...	520	75	420	61	...	...	14	15	...	...	...	...	...	...	...
Nimocast 242	460	67	...	...	...	...	300	44	...	...	...	...	8	...	...	...	...	...	...	...	...
Nimocast 263	730	106	...	...	...	...	510	74	...	...	...	...	18	...	...	...	...	...	...	...	...
René 77	...	...	...	...	...	...	...	...	...	...	...	...	...	...	...	...	...	...	...	...	...
René 80	...	...	...	...	...	...	...	...	...	...	...	...	...	...	...	208	30.2	...	...	...	...
Udimet 500	930	135	895	130	...	...	815	118	725	105	...	...	13	13	...	...	...	...	...	...	...
Udimet 710	1075	156	...	...	240	35	895	130	...	...	170	25	8	...	...	...	...	...	...	...	...
CMSX-2(a)(b)	1185	172	1295(c)	188(c)	...	...	1135	165	1245(c)	181(c)	...	...	10	17(c)	...	...	...	...	...	...	...
GMR-235(b)	710	103	...	...	...	...	640	93	...	...	...	...	3	...	18(d)	...	...	...	...	...	...
IN-939(b)	1050	152	915(c)	133(c)	325(d)	47(d)	800	116	635(c)	92(c)	205(d)	30(d)	5	7(c)	25(d)	...	...	...	...	...	...
MM 002(b)(e)	1035	150	1035(c)	150(c)	550(d)	80(d)	825	120	860(c)	125(c)	345(d)	50(d)	7	5(c)	12(d)	...	...	...	...	...	...
IN-713 Hf(b)(f)	1000	145	895(c)	130(c)	380(d)	55(d)	760	110	620(c)	90(c)	240(d)	35(d)	11	6(c)	20(d)	...	...	...	...	...	...
René 125 Hf(b)(g)	1070	155	1070(c)	155(c)	550(d)	80(d)	825	120	860(c)	125(c)	345(d)	50(d)	5	5(c)	12(d)	...	...	...	...	...	...
MAR-M 246 Hf(b)(h)	1105	160	1070(c)	155(c)	565(d)	82(d)	860	125	860(c)	125(c)	345(d)	50(d)	6	7(c)	14(d)	...	...	...	...	...	...
MAR-M 200 Hf(b)(i)	1035	150	1035(c)	150(c)	540(d)	78(d)	825	120	860(c)	125(c)	345(d)	50(d)	5	5(c)	10(d)	...	...	...	...	...	...
PWA-1480(a)(b)	...	...	1130(c)	164(c)	685(d)	99(d)	895	130	905(c)	131(c)	495(d)	72(d)	4	8(c)	20(d)	...	...	...	...	...	...
SEL(b)	1020	148	875(c)	127(c)	...	...	905	131	795(c)	115(c)	...	...	6	7(c)	...	...	...	...	...	...	...
UDM 56(b)	945	137	945(c)	137(c)	...	...	850	123	725(c)	105(c)	...	...	3	5(c)	...	...	...	...	...	...	...
SEL-15(b)	1060	154	1090(c)	158(c)	...	...	895	130	815(c)	118(c)	...	...	9	5(c)	...	...	...	...	...	...	...
Cobalt base																					
AiResist 13(j)	600	87	420(c)	61(c)	...	...	530	77	330(c)	48(c)	...	...	1.5	4.5(c)	...	...	...	...	...	...	...
AiResist 215(j)	690	100	570(k)	83(k)	...	...	485	70	315(k)	46(k)	...	...	4	12(k)	...	...	...	...	...	...	...
FSX-414	...	...	...	...	...	...	...	...	...	...	...	...	...	...	...	...	...	...	...	...	...
Haynes 1002	770	112	560	81	115	17	470	68	345	50	95	14	6	8	28	210	30.4	173	25.1	...	...
MAR-M 302	930	135	795	115	150	22	690	100	505	73	150	22	2	...	21	...	...	...	...	...	...
MAR-M 322(j)	830	120	595(c)	86(c)	...	...	630	91	345(c)	50(c)	...	...	4	6.5(c)	...	...	...	...	...	...	...
MAR-M 509	785	114	570	83	...	...	570	83	400	58	...	...	4	6	...	225	32.7	...	...	...	...
WI-52	750	109	745	108	160	23	585	85	440	64	105	15	5	7	35	...	...	...	...	...	...
X-40	745	108	550	80	...	...	525	76	275	40	...	...	9	17	...	...	...	...	...	...	...

(a) Single crystal [001]. (b) Data from Ref 3. (c) At 760 °C (1400 °F). (d) At 980 °C (1800 °F). (e) RR-7080. (f) MM 004. (g) M 005. (h) MM 006. (i) MM 009. (j) Data from Volume 3, 9th Edition, *Metals Handbook*, 1980. (k) At 650 °C (1200 °F). Source: Nickel Development Institute, except as noted

---

the formation of primary (MC) carbides, the hexagonal close-packed (hcp) eta (η) phase, and undesirable nitride and carbosulfide formation. Molybdenum, tungsten, tantalum, rhenium, cobalt, and chromium additions promote solid-solution strengthening, but it is known that tantalum, tungsten, and rhenium may also partition to the γ' to varying degrees and that tantalum and rhenium may also be beneficial to environmental resistance properties.

Vanadium is a γ' partitioner, but it also promotes the formation of M_3B_2-type borides. Niobium forms the intermetallic phases delta (δ) (orthorhombic Ni_3Nb) and γ" (body-centered tetragonal Ni_3Nb), but it is also involved in the formation of Laves (Fe,Ni_2Nb) phase, carbides, borides, and/or nitrides. Hafnium is a strong carbide former that is added to polycrystalline alloys to improve grain-boundary ductility. However, at the same time, it increases the volume fraction of γ/γ' eutectic and increases oxidation resistance. Carbon, boron, and zirconium are used at varying levels for grain-boundary strengthening.

All of these constituents interact in various ways to provide high tensile, creep, and fatigue strengths, plus oxidation and sulfidation resistance. Proper control of the cast microstructure and subsequent solutioning and aging treatments generally result in satisfactory component performance.

Under the extreme temperature/stress conditions in which superalloy components operate, however, microstructural features change, often with attendant property changes. The microstructural instabilities that may occur include:

- Intermetallic phase precipitation (σ, μ, Laves)
- Phasial decomposition (carbides, borides, nitrides)
- Phase coalescence and coarsening (γ')
- Phasial solutioning and reprecipitation (γ')
- Order-disorder transition
- Material oxidation
- Stress-corrosion cracking

The formation of topologically close-packed phases (σ, μ, and so on) generally decreases rupture properties. Their occurrence is controlled through chemistry adjustment and is fairly predictable through

Table 5 Stress-rupture strengths for selected polycrystalline nickel-base superalloys

	Rupture stress					
	At 815 °C (1500 °F)		At 870 °C (1600 °F)		At 980 °C (1800 °F)	
	100 h	1000 h	100 h	1000 h	100 h	1000 h
Alloy	MPa (ksi)	MPa (ksi)	MPa (ksi)	MPa (ksi)	MPa (ksi)	MPa (ksi)
IN-713 LC	425 (62)	325 (47)	295 (43)	240 (35)	140 (20)	105 (15)
IN-713 C	370 (54)	305 (44)	305 (44)	215 (31)	130 (19)	70 (10)
IN-738 C	470 (68)	345 (50)	330 (48)	235 (34)	130 (19)	90 (13)
IN-738 LC	430 (62)(a)	315 (46)	295 (43)(a)	215 (31)	140 (20)(a)	90 (13)
IN-100	455 (66)	365 (53)	360 (52)	260 (38)	160 (23)	90 (13)
MAR-M 247 (MM 0011)	585 (85)	415 (60)	455 (66)	290 (42)	185 (27)	125 (18)
MAR-M 246(a)	525 (76)	435 (62)	440 (63)	290 (42)	195 (28)	125 (18)
MAR-M 246 Hf (MM 006)	530 (77)	425 (62)	425 (62)	285 (41)	205 (30)	130 (19)
MAR-M 200	495 (72)(a)	415 (60)(a)	385 (56)(a)	295 (43)(a)	170 (25)	125 (18)
MAR-M 200 Hf (MM 009)(b)	...	...	...	305 (44)	...	125 (18)
B-1900	510 (74)	380 (55)	385 (56)	250 (36)	180 (26)	110 (16)
René 77(a)	...	...	310 (45)	215 (31.5)	130 (19)	62 (9.0)
René 80	...	...	350 (51)	240 (35)	160 (23)	105 (15)
IN-625(a)	130 (19)	110 (16)	97 (14)	76 (11)	34 (5)	28 (4)
IN-162(a)	505 (73)	370 (54)	340 (49)	255 (37)	165 (24)	110 (16)
IN-731(a)	505 (73)	365 (53)	...	...	165 (24)	105 (15)
IN-792(a)	515 (75)	380 (55)	365 (53)	260 (38)	165 (24)	105 (15)
M-22(a)	515 (75)	385 (56)	395 (57)	285 (41)	200 (29)	130 (19)
MAR-M 421(a)	450 (65)	305 (44)	310 (46)	215 (31)	125 (18)	83 (12)
MAR-M 432(a)	435 (63)	330 (48)	295 (40)	215 (31)	140 (20)	97 (14)
MC-102(a)	195 (28)	145 (21)	145 (21)	105 (15)	...	...
Nimocast 90(a)	160 (23)	110 (17)	125 (18)	83 (12)	...	...
Nimocast 242(a)	110 (16)	83 (12)	90 (13)	59 (8.6)	45 (6.5)	...
Udimet 500(a)	330 (48)	240 (35)	230 (33)	165 (24)	90 (13)	...
Udimet 710(a)	420 (61)	325 (47)	305 (44)	215 (31)	150 (22)	76 (11)
CMSX-2(b)	...	...	...	345 (50)	...	170 (25)
GMR-235(b)	...	...	...	180 (26)	...	75 (11)
IN-939(b)	...	...	...	195 (28)	...	60 (9)
MM 002(b)	...	...	...	305 (44)	...	125 (18)
IN-713 Hf (MM 004)(b)	...	...	...	205 (30)	...	90 (13)
René 125 Hf (MM 005)(b)	...	...	...	305 (44)	...	115 (17)
SEL-15(b)	...	...	...	295 (43)	...	75 (11)
UDM 56(b)	...	...	...	270 (39)	...	125 (18)

(a) Ref 3. (b) Ref 1

Table 6 Stress-rupture strengths for selected polycrystalline cobalt-base superalloys

	Rupture stress							
	At 815 °C (1500 °F)		At 870 °C (1600 °F)		At 980 °C (1800 °F)		At 1095 °C (2000 °F)	
	100 h	1000 h	100 h	1000 h	100 h	1000 h	100 h	1000 h
Alloy	MPa (ksi)	MPa (ksi)	MPa (ksi)	MPa (ksi)	MPa (ksi)	MPa (ksi)	MPa (ksi)	MPa (ksi)
HS-21	150 (22)	95 (14)	115 (17)	90 (13)	60 (9)	50 (7)	...	...
X-40 (HS-31)	180 (26)	140 (20)	130 (19)	105 (15)	75 (11)	55 (8)	...	...
MAR-M 509	270 (39)	225 (33)	200 (29)	140 (20)	115 (17)	90 (13)	55 (8)	41 (6)
FSX-414	150 (22)	115 (17)	110 (16)	85 (12)	55 (8)	35 (5)	21 (3)	...
WI-52	...	195 (28)	175 (25)	150 (22)	90 (13)	70 (10)	...	...

use of the commonly accepted methods of calculating the so-called electron vacancy number, N_v, of the given alloys. Different calculation methods exist; however, all provide a useful key to the prediction of σ formation when proper reference points are known.

Although it occurs during both solidification and heat treatment, carbide precipitation is generally promoted during component heat treatment to effect an optimum grain-boundary carbide morphology and population. Discrete, blocky $M_{23}C_6$ particles distributed in a discontinuous fashion are preferred. High-temperature, stressed exposure tends to cause carbide degeneration, often resulting in grain-boundary overload and compromised rupture strength.

MC-type carbides generally occur during alloy solidification. They are titanium-rich (MC-1) or tantalum-rich (MC-2) and may partially degenerate with high-temperature exposure to form hafnium-rich (MC-3) carbides and/or $M_{23}C_6$, M_7C_3, and M_6C carbides (secondary carbides); the specific type depends upon alloy chemistry and exposure temperature. The chromium-rich $M_{23}C_6$ generally forms at the grain boundaries in polycrystalline materials; when present as discrete, discontinuous particles, it provides the grain-boundary strength and resistance to fracture needed to prolong service life.

On the other hand, carbide degeneration also releases titanium and tantalum to the solid-solution matrix, resulting in further matrix saturation. Oversaturation can result in the formation of undesirable secondary phases such as μ (tungsten- and/or molybdenum-rich), α-W, α-Cr, and or M_6C carbides, making chemistry balancing and controlled thermal treatment necessary for ultimate success.

Superalloys are, indeed, complex. However, careful alloy design and processing will provide the desired results. Simply stated, superalloy property attainment is principally a function of the amount and morphology of the γ', grain size and shape, and carbide distribution. Early superalloys contained less than 25 vol% γ'. However, commercial vacuum induction refining and casting provided the opportunity for greater γ' volume fraction, to the extent that today's commercial superalloys generally contain approximately 60 vol% γ'.

This increased level of γ' results in greater alloy creep strength (Fig. 4), but it can be fully exploited only in single-crystal components, where full γ' solutioning is generally possible. For polycrystalline superalloy components, high-temperature strength is affected by the condition of the grain boundaries and, in particular, the grain-boundary carbide morphology and distribution. Optimized properties can be achieved if solutioning and aging treatments are developed to attain discrete, globular carbide formation along the grain boundaries in conjunction with the optimized γ' volume fraction/morphology and component grain structure. Representative stress-rupture curves for selected nickel-base superalloys are shown in Fig. 5. Table 5 also provides stress-rupture data.

Cobalt-base alloys (see Table 2) are designed around a cobalt-chromium matrix with chromium contents ranging from 18 to 35 wt%. The high chromium content contributes to oxidation and sulfidation resistance, but also participates in carbide formation (Cr_7C_3 and $M_{23}C_6$) and solid-solution strengthening. Carbon content generally ranges from 0.25 to 1.0%, with nitrogen occasionally substituting for carbon.

Cobalt-base alloys are often designed with significant levels of both nickel and tungsten. The addition of nickel helps to stabilize the desired fcc matrix, while tungsten provides solid-solution strengthening and promotes carbide formation. Molybdenum also contributes to solid-solution strengthening but is less effective and potentially more deleterious than a tungsten addition. Other alloying elements contributing to the solid solution and/or carbide formation are tantalum, niobium, zirconium, vanadium, and titanium.

These additions provide strength by means of solid-solution and second-phase strengthening. No intermetallic precipitated phase has been discovered to equal the benefit imparted by γ' in nickel-base superalloys. Solid-solution strengthening results principally from the chromium, tantalum, niobium, and tungsten additions, while second-phase strengthening is obtained primarily from the carbides and carbonitrides formed with chromium. The multiple-composition complex carbides may be present as MC, M_6C, M_7C_3, $M_{23}C_6$, and M_2C_3.

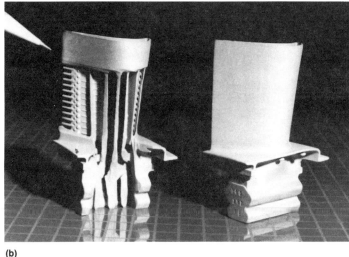

Fig. 3 The evolution of the processing of nickel-base superalloy turbine blades. (a) From left, equiaxed, directionally solidified, and single-crystal blades. (b) An exposed view of the internal cooling passages of an aircraft turbine blade. Source: Ref 5

As with nickel-base superalloys, carbides must be precipitated at grain boundaries to control gross grain-boundary sliding and migration. Optimum mechanical properties are obtained through the careful balancing of the carbides at grain boundaries and within the matrix. When carbides are present in sufficient quantity, the skeletal carbide network that results can contribute to component strength much like the strengthening that is achieved in a composite (Ref 9).

Most cobalt-base alloy aerospace castings are not heat treated apart from a relatively low-temperature stress-relief anneal. Carbide distribution is, therefore, determined during solidification, thereby highlighting the need for stringent control of the alloy pouring temperature and cooling rate during and after solidification. Exceptions to this may be found in medical applications, where cobalt-base alloy castings for orthopedic implants are sometimes solution treated. Representative stress-rupture curves for selected cobalt-base alloys are shown in Fig. 5(c) and 5(e). Table 6 also provides stress-rupture data.

Vacuum Induction Melting of Superalloys

Commercial vacuum induction melting was developed in the early 1950s, having been stimulated by the need to produce superalloys containing reactive elements within an evacuated atmosphere. The process is relatively flexible, featuring the independent control of time, temperature, pressure, and mass transport through melt stirring. As such, VIM offers more control over alloy composition and homogeneity than all other vacuum melting processes.

The primary purification reaction occurring in the process is the removal of melt-contained oxygen by means of a reaction with carbon to form carbon monoxide (CO). The reaction occurs most readily at or near the melt surface with the reaction kinetics being affected by crucible geometry and melt stirring. The removal of oxygen from the melt as CO is favored by decreased melt chamber pressure, elevated bath temperature, and increased carbon activity (Ref 10).

The melting crucible material is not inert and is actually another source of oxygen and other impurities, depending on refractory type and condition. Therefore, both melt refining temperature and refining duration are carefully scrutinized. Proper melt stirring is integral to the deoxidation process and must be optimized through proper furnace power frequency and application procedure to prevent refractory lining erosion, a potential problem particularly during the controlled but more vigorous CO boiling portion of the process.

Vacuum induction melting deoxidation, that is, the generation of CO gas, proceeds as CO bubbles are nucleated heterogeneously along the walls and, sometimes, bottom of the melt/lining-refractory interface. This occurs preferentially at small crevices existing in the lining, with the bubbles growing during movement toward the molten metal/vacuum interface (Ref 11-13). Actual bubble formation is dependent on the number of gas molecules present; the pressure in the liquid at the level of the bubble; the temperature of the gas; and, for very small bubbles, the interfacial tension between the gas and the liquid metal. Figure 6 shows bubble formation during the VIM process.

Following formation, bubble growth and mass transport within the liquid toward the liquid/vacuum interface is dependent on:

- The quantity of the dissolved gas
- The decreased pressure exerted on the bubble as it rises in the melt
- The bath temperature
- The time it takes for the bubble to rise through the melt to the surface, which, in turn, is a function of melt stirring
- The pressure above the melt
- The interfacial tension between the bubble and the liquid metal

The relatively vigorous, but controlled, portion of the boiling process results in the greatest CO removal. Concurrently, a slight nitrogen loss is realized because of scavenging associated with the CO bubbles, and a slight sulfur reduction may occur during the CO supersaturation stage via sulfur dioxide (SO_2) evolution. Minor tramp elements such as lead, silver, bismuth, selenium, and tellurium, which are deleterious to alloy elevated-temperature rupture strength and ductility (Ref 14, 15), are partially evaporat-

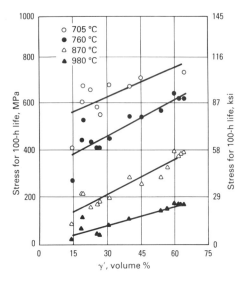

Fig. 4 The relationship between γ' volume percent and stress-rupture strength for nickel-base superalloys. Source: Ref 6

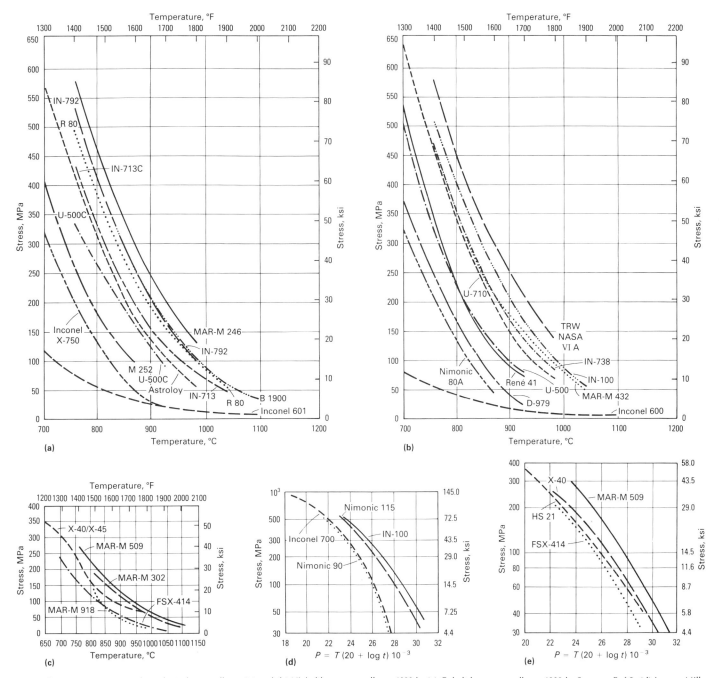

Fig. 5 Stress-rupture curves for selected superalloys. (a) and (b) Nickel-base superalloys. 1000 h. (c) Cobalt-base superalloys. 1000 h. Source: Ref 3. (d) Larson-Miller stress-rupture curves for selected nickel-base superalloys. Source: Ref 7. (e) Larson-Miller stress-rupture curves for selected cobalt-base superalloys. Source: Ref 8

ed during this period as well as throughout the entire refining process (Ref 16, 17). Some undesirable elements, however, such as arsenic and tin, must be controlled through raw material selection because they are not removed by vacuum refining. Figure 7 shows the effects of VIM time and temperature on tramp element concentration. Once the boiling subsides, surface desorption of additional CO occurs, and it is during this nonboiling period that nitrogen removal (desorption) is most effective (Ref 18).

Refractory Materials. Superalloy melting is generally undertaken in a relatively unreactive, high-bond strength, high-purity $MgO-Al_2O_3$ spinel refractory lining. Refractories may be monolithic or brick and mortar, with the former providing the greater potential for alloy quality. By minimizing extremes in thermal cycling; optimizing alloy sequencing; and refining temperature, time, and pressure, alloys practically void of any lining-related nonmetallics can be produced.

The types of raw materials and the melt procedure vary depending on the quality of alloy being produced. Alloys destined for critical application, the components of which may be difficult to cast, are produced using the highest-quality raw materials commercially available, in conjunction with sophisticated melt processing. Lower-quality raw materials, such as GMR-235 or IN-713 C, are used for commercial application, such as nonrotating, noncritical, or nonaerospace use, for example, turbocharger wheels, because a higher-quality alloy product is not necessary. The more sophisticated alloy systems are generally produced to premium-quality or integral wheel quality levels, with special attention given not only to cleanliness but also to specific micro-

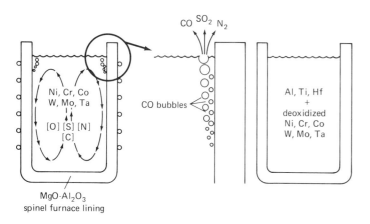

Fig. 6 Vacuum induction refining process

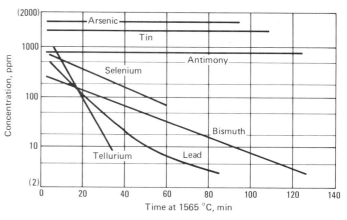

Fig. 7 Evaporation of elements from an 80Ni-20Cr alloy during VIM

structural characteristics and/or chemistry requirements to assist castability, weldability, and component mechanical properties.

The Melt Process. Base charge materials are layered in the relatively warm furnace, in a manner which recognizes and accommodates the elemental melting point of the material and bridging tendency. Only those materials with oxides that are relatively easily reduced for the encountered melt conditions are placed in the initial furnace charge along with a small, controlled carbon addition. Also, those elements that have a particularly strong affinity for nitrogen may be withheld from the base charge because they lower the activity of the dissolved nitrogen.

Following furnace evacuation and particular heat-up cycles that ensure proper closure of any refractory lining cracks prior to metal liquation, optimum temperature and vacuum pressure, consistent with promoting a somewhat vigorous CO boil, is attained. Bath refining is undertaken at a temperature and duration long enough to reach the so-called system equilibrium conditions, the assurance of which is provided by the attainment of consistent furnace leak-up rates. At this point, those elements that were held from the base charge because of their relative reactivity toward oxygen, for example, aluminum, titanium, zirconium, and hafnium, are added with an associated solutioning and homogenization procedure.

Dip sample alloy chemistry is checked in a relatively short time period with any necessary chemistry adjustment undertaken. A similar analytical check is undertaken prior to pouring.

Pouring proceeds once the correct chemistry is ensured, the bath is properly solutioned/homogenized, and the proper pour temperature is attained. It is generally undertaken under high vacuum condition and proceeds from the furnace crucible into a relatively sophisticated multicompartment tundish, thereby ensuring that extensive time for flotation is achieved and also that laminar flow conditions prevail in the final separation and pour compartments. The flow-rate controlling, high-alumina tundish system results in a relatively slow pouring rate, effectively maximizing alloy cleanliness.

Filters. One of the most critical stages with respect to cleanliness is the pouring of the melt. Ceramic foam filters are used in some master metal operations to remove relatively large melt inclusions by means of entrapment. Foam filters are most effective where extremely high pour rate conditions and gross cleanliness problems prevail. Filter performance often varies because of the occasional use of filters with poor mechanical strength and/or thermal shock resistance. Foam cell particle breakage often results from handling during shipment or tundish installation and, if undetected, results in filter particulate in the alloy bar stock and subsequently cast components. Optimized VIM technology and practice without filters provide a clean alloy, without the inherent risks associated with filter use when applied to master metal production.

Alternative Melt Techniques. Electron beam, cold hearth refining systems have been shown to be effective for nonmetallic-inclusion removal from superalloys. Initial problems occurring during the process that are associated with analytical control, for example, chromium evaporation, have almost been resolved. However, the process is not currently cost effective for virgin cast superalloy production, although use may be found for powder metallurgy (P/M) superalloy production and/or revert scrap reclamation in the future. Vacuum arc skull melting and casting technology may also be used for superalloy castings.

A plasma melting/refining facility for the production of superalloy powder metal is currently being commissioned. Plasma melt investment casting trials have been undertaken, but no data are currently available. Skull casting development would also help pace the interest in this secondary refining technology. Each of the above-mentioned techniques is described in detail in *Casting*, Volume 15 of the 9th Edition of *Metals Handbook*.

Quality Considerations

Cast cobalt-base superalloys have large application in the aerospace, medical, and chemical industries. Nickel-base superalloys are used in a wide range of applications also, although most are related to the aerospace industry. For aerospace applications, the quality level required for static components differs tremendously from that required for critical, highly stressed rotating components. Thus, it is useful to classify superalloy products according to quality level.

Superalloy quality can be categorized as commercial grade, aerospace quality, or premium grade. Commercial-grade alloys may be produced with select materials, foundry revert, and/or lower-quality elemental raw materials. Aerospace quality materials may also be produced with select, revert, and/or virgin materials. However, the resulting quality, as measured by cleanliness and gas and tramp element content, is greater. Premium grades are produced for blade/vane airfoil, wheel, and structural-part applications using top-quality raw materials and sophisticated VIM procedures. Typical gas and tramp element level data for the three grades are presented in Table 7.

Stringent control of alloy gas and tramp element content is paramount to achieving satisfactory foundry performance and component mechanical property response. Low gas content is desired because oxygen in the alloy is present as nonmetallic stable oxide inclusions, which affect weldability and mechanical properties. Nitrogen also is controlled because it induces microporosity in castings and, when sufficiently high, causes agglomerated microporosity. Nitrogen promotes alloy/crucible wetting during precision investment casting operations and may also be involved in deleterious titanium carbonitride particle formation.

Table 7 Quality level (allowable tramp element concentrations) for selected nickel-base superalloys

Element	Commercial grade Tooling applications	Other	IN-718/MAR-M 247, typical alloy concentration		Aerospace quality		Premium quality	
N, ppm	20+	60–100	60		5–15		10–25	1
O, ppm	5+	5–10	<5		<5		2	1
Si, wt%	0.05+	0.10–0.30 wt%	0.05–0.10		0.02–0.04		<0.02	0.008
Mn, wt%	0.01+	0.05+	<0.02		<0.002		<0.002	<0.002
S, ppm	15+	10–40	10–30		5–15		10	<5
Zr, wt%	...	<0.01	0.001		...		<10 ppm	...
Fe, wt%	0.10+	...	...		0.05–0.10		...	0.03
Cu, wt%	0.002+	0.08	0.01–0.05		0.002–0.005		<0.001	<0.001
P, wt%	0.002	0.005	0.005		<0.005		0.001–0.002	<0.001
Pb, ppm	<1	1–5	<1		<1		<1	<0.5
Ag, ppm	<0.5	<1	<1		<0.5		<0.5	<0.5
Bi, ppm	<0.3	<0.5	<0.5		<0.3		<0.2	<0.2
Se, ppm	<0.5	<1	<1		<0.5		<0.5	<0.5
Te, ppm	<0.3	<0.5	<0.5		<0.2		<0.2	<0.2
Tl, ppm	<0.3	<0.5	<0.5		<0.2		<0.2	<0.2
Sn, ppm	<5	15–40	<10		<5		<10	<5
Sb, ppm	<2	2+	<2		<1		<2	<1
As, ppm	<2	5	<2		<2		<2	<1
Zn, ppm	<2	2+	<2		<1		<2	<1

(a) + = or higher

Superalloy silicon content is controlled to low levels because it adversely affects mechanical properties and may cause hot-short cracking during welding operations. In the case of IN-718, silicon is avoided because it also impedes the rate of Laves phase transformation. Low sulfur content is also desired because it tends to migrate to grain boundaries, thereby decreasing hot ductility and promoting cracking. In addition, it can be deleterious to fatigue strength and contribute to increased alloy/crucible wetting during precision casting.

Zirconium is added to superalloys to control grain-boundary strength. However, it is minimized in alloys that undergo extensive welding because it increases the tendency for cracking in the weld and heat-affected zone of the base material.

Investment Casting of Superalloys (Ref 19)

A number of casting processes can provide near-net shape superalloy cast parts, but essentially all components are produced by investment casting (Fig. 8). The characteristic physical and mechanical properties and complex, hollow shape-making capabilities of investment casting have made it ideal for amplifying the unusual high-temperature properties of superalloys.

Cast superalloys are made in a wider range of compositions than are wrought alloys. Creep and rupture properties of a given superalloy composition are maximized by the casting and heat-treatment processes. Ductility and fatigue properties of polycrystalline materials are generally lower in castings than in their wrought counterparts of similar composition. The gap, however, is being reduced by new technological developments to eliminate casting defects and refine grain size.

Investment Casting Process

Patterns, Cores, and Molds. The first step in the investment casting process is to produce an exact replica or pattern of the part in wax, plastic, or a combination thereof. Pattern dimensions must compensate for wax, mold, and metal shrinkage during processing. If the product contains internal passages, a preformed ceramic core is inserted in the die cavity, around which the pattern material is injected. Except for large or complex castings, a number of patterns may be assembled in a cluster and held in position in order to channel the molten metal into the various mold cavities. Design and positioning of the runners and gating is critical to achieving sound, metallurgically acceptable castings. Today the molds are produced by first immersing the pattern assembly in an aqueous ceramic slurry (Fig. 9). A dry, granular ceramic stucco is applied immediately after dipping to strengthen the shell. These steps are repeated several times to develop a rigid shell. After slow, thorough drying, the wax is melted out of the shell, and the mold is fired to increase substantially its strength for handling and storage. An insulating blanket is tailored to the mold configuration to minimize heat loss during the casting operation and to control solidification. More information on the production of patterns, cores, and shells for investment casting is available in the article "Investment Casting" in Volume 15 of the 9th Edition of *Metals Handbook*.

To make equiaxed-grain castings, the mold is preheated to enhance mold filling, control solidification, and develop the proper microstructure. For vacuum casting, the alloy charge is melted in an isolated chamber before the preheated mold is inserted, and the pressure is maintained at about 1 μm for pouring. After casting, exothermic material is applied as a hot top for feeding purposes, and the mold is allowed to cool. A different procedure is followed in the production of directionally solidified (DS) and single-crystal (SC) superalloy castings; this is described in the article "Directionally Solidified and Single-Crystal Superalloys" in this Volume and in the article "New and Emerging Processes" in Volume 15 of the 9th Edition of *Metals Handbook*.

The Casting Process. Most superalloys are cast in vacuum to avoid the oxidation of reactive elements in their compositions. Some cobalt-base superalloys are cast in air using induction or indirect arc rollover furnaces. The vacuum casting of equiaxed-grain products is usually done in a furnace divided into two major chambers, each held

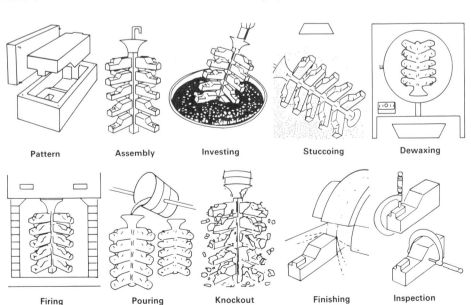

Fig. 8 Basic steps in the investment casting process. See Fig. 9(a) for a close-up of an automated slurry coating process. Source: Ref 4

 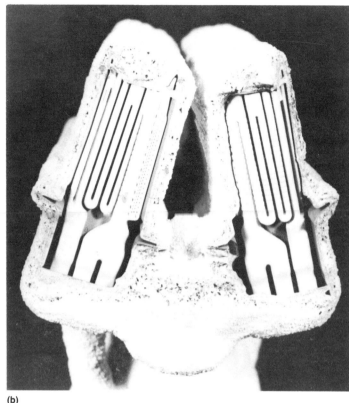

Fig. 9 (a) Automated slurry coating of an investment casting mold. (b) Cutaway view of a shell mold for an air-cooled turbine blade casting. Source: Ref 5

under vacuum and separated by a large door or valve. The upper chamber contains an induction-heated reusable ceramic crucible in which the alloy is melted. Zirconia crucibles are commonly employed; single-use silica liners may be specified when alloy cleanliness is especially critical.

The preweighed charge is introduced through a lock device and is melted rapidly to a predetermined temperature, usually 85 to 165 °C (150 to 300 °F) above the liquidus temperature. Precise optical measurement of this temperature is crucial. Metal temperature during casting is much more critical than mold temperature in controlling grain size and orientation; it also strongly affects the presence and location of microshrinkage. When the superheat condition is satisfied, the preheated mold is rapidly transferred from the preheat furnace to the lower chamber, which is then evacuated. The mold is raised to the casting position, and the molten superalloy is quickly poured into the cavity; speed and reproducibility are essential in order to achieve good fill without cold shuts and other related imperfections. Precise mold positioning and pour rates also are imperative. For maximum consistency, melting and casting are automated with programmed closed-loop furnace control. The filled mold is lowered and removed from the furnace.

Shrinkage during solidification is minimized in part by maintaining a head of molten metal to feed the casting; this is achieved by adding an exothermic material immediately after mold removal from the furnace.

Because of thermal expansion differences, the shell mold usually fractures upon cooling, facilitating its removal by mechanical or hydraulic means. Before grit- and sand-blasting operations, the individual castings are separated from the cluster by abrasive cutoff. After shell removal, the cluster is checked by one of several commercially available emission or x-ray fluorescence instruments to verify the alloy identification.

A major portion of the casting cost is in the finishing operations, which remain labor intensive. Superficial surface defects are blended out abrasively within specified limits, and the castings may require mechanical straightening operations before and after heat treatment to satisfy dimensional requirements.

Control of Casting Microstructure (Ref 20)

The solidification of investment cast superalloy components is precisely controlled so that the microstructure, which ultimately determines mechanical properties, remains consistent. For example, once the process for a particular component has been defined, the production of these components does not deviate from the agreed-upon steps for the entire production run, which may last many years without proper approval. If steps are changed, it must be shown that the new steps do not cause a degradation in the properties of the component.

To control the solidification of equiaxed-grain castings, the investment caster has several tools at his disposal: facecoats that encourage grain nucleation, pour temperature of the metal, preheat temperature of the shell, shell thickness, part orientation, part spacing, gating locations, insulation to wrap the shells, pouring speed, and shell agitation. However, the investment caster must first fill the shell cavity, prevent hot tears or other cracks, and minimize porosity. If the first two objectives can be met, the investment caster has some freedom to produce the desired structure. If the desired structure still cannot be made, other more complex techniques may be employed, including changing the thermal conductivity of the shell.

Dendrites are probably the most visible microstructural feature in superalloy castings. Primary and secondary dendrite arm spacings are controlled by the cooling rate. As the dendrite arm spacing is reduced, segregation in the dendrite core and interdendritic regions is also reduced, thereby benefitting mechanical properties.

Carbides. Conventional equiaxed-grain nickel-base superalloys typically have 0.05

to 0.20 wt% C, while cobalt-base alloys contain up to about 1.0% C. Both alloy systems may use carbon to increase grain-boundary strength. Cobalt-base alloys require more because internal carbides are one of the primary strengthening mechanisms.

Carbide morphology is controlled by solidification or composition. For example, by increasing the cooling rate, more discrete, blocky-type MC carbides are formed in IN-713 C, and this often results in an improvement of at least two times to low-cycle fatigue (LCF) properties (Ref 21, 22). If it is not possible to influence the cooling rate of a casting significantly, adding small amounts of magnesium, calcium, cerium, or other rare earth metals acting as nucleating agents will assist in carbide shape control.

Eutectic Segregation. By the very nature of solidification, segregation is introduced into the component. Important segregants of interest in cast superalloys are eutectics, which often are found in interdendritic or intergranular regions. In nickel-base alloys, eutectic pools are the last constituents to solidify and have a cellular appearance. The composition of the eutectic pools varies, but they typically contain excess γ', carbides, borides, and low melting point phases. Control of the eutectic pool is done primarily through composition. However, it has been shown that while the volume fraction of eutectic remained constant near 0.10 vol% in IN-713, the size of the eutectic pool increased from 11 to 19 µm as the cooling rate decreased from 0.56 to 0.036 °C/s (1 to 0.065 °F/s) (Ref 20).

In cobalt-base alloys, the eutectics typically form lamellar γ and $M_{23}C_6$ pools or colonies. Heat treatment between 1150 and 1230 °C (2100 and 2250 °F) for 4 h resolutions these eutectic colonies, redistributing much of the carbon.

Porosity. It is important to minimize the porosity in castings because the pores serve as initiation sites for fracture, especially fatigue cracks. There are three primary sources of porosity in superalloy investment castings: undissolved gas, microshrinkage caused by poor feeding between dendrites, and macroshrinkage caused by inadequate gating. Undissolved gas is gas that has come out of solution but, with today's vacuum technology, is seldom experienced. Usually made up of oxygen, nitrogen, or hydrogen, this gas can form spherical voids up to two or more times the diameter of the dendrite arm spacing. Gas porosity can be essentially eliminated by maintaining a vacuum during remelting and casting.

Microshrinkage (microporosity) is inherent to castings that experience dendritic solidification. The pores are spherical, but they typically have a diameter less than the dendrite spacing. Microshrinkage forms just ahead of the advancing solidus interface because liquid metal feeding is impeded by the tortuous path through and around the secondary dendrite arms (a fluid flow problem).

The 2 to 6% shrinkage experienced upon solidification by superalloy castings makes macroshrinkage (solidification shrinkage) a problem. This type of porosity tends to be confined within the thickest section of the casting, where the last solidification takes place.

The investment caster can control solidification shrinkage to a great extent by gating or feeding those areas that are the last to solidify. With complex geometries and the desire to produce net-shape castings, the size and placement of gating is based on experience, which necessitates experimentation before the process can be defined. Modeling with computers, however, is changing this practice. Recently, it has become possible to model the solidification process of simple castings by taking into account the thermal properties of the metal and the shell. Thus, the areas of solidification shrinkage can be predicted. Once the

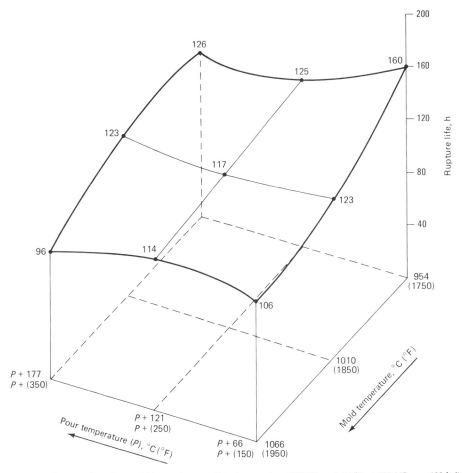

Fig. 10 Influence of casting variables on intermediate-temperature (760 °C, or 1400 °F, at 690 MPa, or 100 ksi) stress-rupture properties of a cast nickel-base superalloy. Pour temperature and mold temperature affect solidification and thus grain size of the component. Source: Ref 19

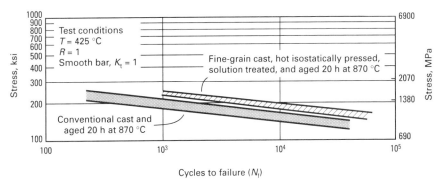

Fig. 11 Effect of grain size control and HIP processing on the strain-controlled (axial) low-cycle fatigue of CM 247 LC

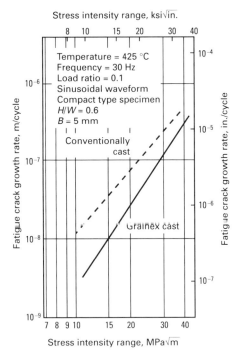

Fig. 12 Influence of grain refinement on fatigue crack growth rate. Source: Ref 19

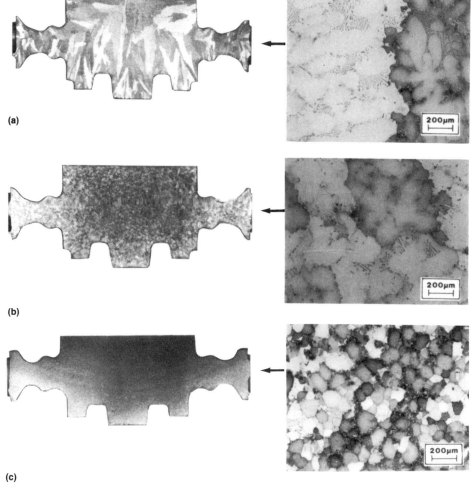

Fig. 13 Structure of conventionally cast turbine wheel (a) compared to wheels cast using the Grainex (b) and Microcast-X (c) processes. Courtesy of Howmet Turbine Corporation

shrinkage areas are located, various gating schemes can be evaluated until the shrink within the part is pulled into the gate. At this point, the model is verified by an experiment, significantly reducing the overall time it takes to design gating configurations. More information on the use of modeling to predict solidification shrinkage and other casting variables is available in the Section "Computer Applications in Metal Casting" in Volume 15 of the 9th Edition of *Metals Handbook*.

Grain Size. The control of grain size is an important means for developing and maintaining both physical and mechanical properties. Generally, a number of randomly oriented equiaxed grains in a given cross section is preferred to provide consistent properties, but often this is difficult to achieve in thin sections. To meet this objective, mold facecoat nucleants, mold and metal temperature, and other parameters are chosen to accelerate grain nucleation and solidification.

Finer grain size generally improves tensile, fatigue, and creep properties at low-to-intermediate temperatures (Fig. 10 to 12). The finer grain size produced by relatively rapid solidification is accompanied by a finer distribution of γ' particles and a tendency to form blocky carbide particles. The latter morphology is preferred to the script-type carbides produced by slow solidification rates, particularly for a fatigue-sensitive environment. Under these conditions, the carbide particles do not contribute to superalloy properties. As the service temperature increases, they impart important grain-boundary strengthening, provided that continuous films or necklaces are avoided.

For high-temperature rupture performance, slower solidification and cooling rates are preferred to coarsen both the grain size during solidification and the γ' precipitated during cooling. While this benefits high-temperature strength through a reduction in grain-boundary content, more property scatter can be expected due to (random) crystallographic orientation effects. For turbine blades, the desired microstructure is difficult to achieve because the thin airfoils operating at the highest temperatures should have coarse grains, and the heavier-section root attachment area, being less rupture dependent, should have a fine-grain microstructure. Where conventional practice fails, a gate, or gutter, along the airfoil edges may be employed, through which metal is caused to flow, thereby creating deliberate hot spots to retard the local solidification rate.

A significant foundry advancement has been the development of processes to produce fine-grain superalloy castings. In the late 1960s, experiments were conducted on a grain refinement technique for integral turbine wheels. The technique used the mechanical motion of a mold to shear dendrites from the solidifying metal. These dendrites then acted as nucleation sites for additional grains. However, the process was not commercially introduced because it produced castings with unacceptable levels of porosity.

In the mid-1970s, developmental work on this process resumed when it was realized that hot isostatic pressing could be used to eliminate residual casting porosity. The Howmet Corporation process that developed from this work, known as Grainex, results in ASTM grain sizes as fine as No. 2. A further Howmet Corporation development, the Microcast-X process, has led to a greater refinement in grain size (ASTM No. 3 to 5). Figure 13 compares the microstructures of grain-refined rotors with those of a conventionally cast part. References 21 to 25 provide additional information on grain size control/property relationships and on fine-grain casting process development work performed by others in the precision investment casting industry.

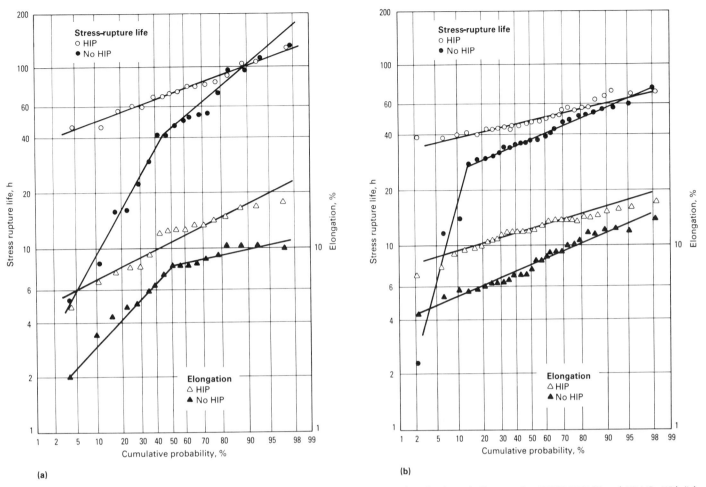

Fig. 14 Effect of hot isostatic pressing on stress-rupture properties of cast IN-738. Test material was hot isostatically pressed at 1205 °C (2200 °F) and 103 MPa (15 ksi) for 4 h. (a) Test conditions: 760 °C (1400 °F) and 586 MPa (85 ksi). (b) Test conditions: 980 °C (1800 °F) and 152 MPa (22 ksi). Source: Howmet Corporation

Postcasting Processing

Heat Treatment. The heat treatment of cast superalloys in the traditional sense was not employed until the mid-1960s. Before the use of shell molds, the heavy-walled investment mold dictated a slow cooling rate with its associated aging effect on the casting. As faster cooling rates with shell molds developed, the aging response varied with section size and the many possible casting variables. These factors, coupled with significant γ′ alloying additions, provided the opportunity to minimize property scatter by heat treatment. The combination of hot isostatic pressing plus heat treatment has also greatly enhanced properties.

Generally, heat treating cast superalloys involves homogenization and solution heat treatments or aging heat treatments. A stress-relief heat treatment may also be performed in order to reduce residual casting, welding, or machining stresses. Detailed information can be found in Ref 20.

Cast cobalt-base superalloys are not usually solutioned. However, they may be given stress relief and/or aging treatments. When required, aging is generally done at 760 °C (1400 °F) to promote the formation of discrete $Cr_{23}C_6$ particles. Higher-temperature aging can result in acicular and/or lamellar precipitate formation.

Cobalt-base alloy heat treatment may be done in an air atmosphere unless unusually high-temperature treatments are required, in which case vacuum or inert gas environments are used. Conversely, nickel-base alloys are always heat treated in a vacuum or in an inert gas medium.

Polycrystalline cast nickel-base superalloys may or may not be given solution treatment. Because alloys respond differently to γ′ solutioning, some are given only aging treatment. For those that do respond to partial solutioning, the treatment is performed at a temperature safely below the alloy's incipient melting point, for times ranging from 2 to 6 h at temperature. Rapid cooling from high temperature is necessary to ensure that the γ′ precipitate is fine sized, thereby maximizing strength potential.

Solution heat-treating procedures must be optimized to stabilize the carbide morphology. High-temperature exposure may cause extensive carbide degeneration, resulting in grain-boundary carbide overload and compromised mechanical properties. Many polycrystalline materials are used in the as-cast plus aged condition.

A typical aging cycle involves heating to 980 °C (1800 °F), holding for 5 h, and air cooling, followed by heating to 870 °C (1600 °F), holding for 20 h, and air cooling. An alternative is heating to 1080 °C (1975 °F), holding for 4 h, and air cooling, followed by heating to 870 °C (1600 °F), holding for 20 h, and air cooling. The 980 °C (1800 °F) and 1080 °C (1975 °F) exposures are carried out in conjunction with protective coating diffusion treatments. It is important that the cooling rate from the 980 °C (1800 °F) and 1080 °C (1975 °F) treatments be rapid to maintain the optimum γ′ size. However, a furnace cool from 870 °C (1600 °F) treatment is acceptable.

Hot Isostatic Pressing. Hot isostatic pressing (HIP) subjects a cast component to both elevated-temperature and isostatic gas pressure in an autoclave. The most widely used pressurizing gas is argon. For the processing of castings, argon is applied at pressures between 103 and 206 MPa (15 and 30 ksi), with 103 MPa (15 ksi) being the most common. Process temperatures of 1200 to 1220

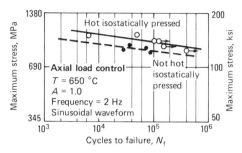

Fig. 15 Improvement of fatigue properties by the elimination of microporosity through HIP processing. Source: Ref 19

°C (2200 to 2225 °F) are common for polycrystalline superalloy castings.

When castings are hot isostatically pressed, the simultaneous application of heat and pressure virtually eliminates internal voids and microporosity through a combination of plastic deformation, creep, and diffusion. The elimination of internal defects leads to improved nondestructive testing ratings, increased mechanical properties, and reduced data scatter (Fig. 14 and 15).

In the past 15 years, hot isostatic pressing has become an integral part of the manufacturing process for high-integrity aerospace castings. The growth of hot isostatic pressing has paralleled the introduction of advanced nickel-base superalloys and increasingly complex casting designs, both of which tend to increase levels of microporosity. In addition, to optimize mechanical properties, turbine engine manufacturers have become more stringent in allowances for microporosity. The requirement for reduced porosity levels and increased mechanical properties has been achieved in many cases through the use of hot isostatic pressing.

In selecting HIP process parameters for a particular alloy, the primary objective is to use a combination of time, temperature, and pressure that is sufficient to achieve closure of internal voids and microporosity in the casting. There are also material considerations for avoiding such deleterious effects as incipient melting, grain growth, and the degradation of constituent phases such as carbides.

If encountered, incipient melting can be avoided by pre-HIP homogenization heat treatments or by lowering HIP temperatures. If the temperature is lowered, an increase in processing pressure may be required to obtain closure in certain alloys. For example, hafnium-bearing nickel-base superalloys such as C 101 and MAR-M 247, when cast with heavy sections (for example, integral wheels), have been found to undergo incipient melting when hot isostatically pressed at 1205 °C (2200 °F) and 103 MPa (15 ksi) for 4 h. To prevent incipient melting and still obtain closure, the HIP parameters were changed to 1185 °C (2165 °F) and 172 MPa (25 ksi) for 4 h. This trade-off between temperature and pressure can also be used to prevent grain growth and to prevent or limit carbide degradation while obtaining closure of microporosity.

Time at temperature and pressure will obviously affect processing cost. For most alloys, 2 to 4 h is sufficient. Exceptions are massive section sizes, which require additional thermal soaking time.

REFERENCES

1. R.W. Fawley, Superalloy Progress, in *The Superalloys*, C.T. Sims and W.C. Hagel, Ed., John Wiley & Sons, 1972, p 12
2. R.F. Decker, Superalloys—Does Life Renew at 50?, in *Proceedings of the Fourth International Symposium on Superalloys*, American Society for Metals, 1980, p 2
3. Appendix B: Superalloy Data, in *Superalloys II*, C.T. Sims, N.S. Stoloff, and W.C. Hagel, Ed., John Wiley & Sons, 1987, p 575-597
4. M. Gell and D.N. Duhl, The Development of Single-Crystal Superalloy Turbine Blades, in *Advanced High-Temperature Alloys: Processing and Properties*, American Society for Metals, 1986, p 41-49
5. L.E. Dardi, R.P. Dalal, and C. Yaker, Metallurgical Advancements in Investment Casting Technology, in *Advanced High-Temperature Alloys: Processing and Properties*, American Society for Metals, 1986, p 25-39
6. R.F. Decker, "Strengthening Mechanisms in Nickel-Base Superalloys," Paper presented at the Steel Strengthening Mechanisms Symposium, Zurich, 1969
7. W. Betteridge, *Nickel and Its Alloys*, Ellis Horwood, 1984
8. W. Betteridge, *Cobalt and Its Alloys*, Ellis Horwood, 1982
9. M.J. Donachie, Introduction to Superalloys, in *Superalloys—Source Book*, American Society for Metals, 1984, p 9
10. D.R. Gaskell, *Introduction to Metallurgical Thermodynamics*, Scripta Publishing, 1973, p 268-273
11. D. Winkler, Thermodynamics and Kinetics in Vacuum Metallurgy, in *Vacuum Metallurgy*, O. Winkler and R. Bakish, Ed., Elsevier, 1971, p 42-54
12. J.S. Foster, "Liquid Metal-Gas Systems and Kinetics of Metal Degassing," Metallurgical Kinetics, Michigan Technological University course material, 1974
13. J.F. Elliot, "Metal Refractory Reactions in Vacuum Processing of Steel and Superalloys," Paper presented at the AIME Electric Furnace Conference, American Institute of Mining, Metallurgical, and Petroleum Engineers, 1971
14. D.R. Wood and R.M. Cook, Effects of Trace Elements on the Creep Rupture Properties of Nickel Base Alloys, *Metallurgia*, Vol 7, 1963, p 109
15. W.B. Kent, Trace Element Effects in Vacuum Melted Alloys, *J. Vac. Sci. Technol.*, Vol 11 (No. 6), Nov/Dec 1974, p 1038-1046
16. Evaporation of Elements From 80/20 Nickel-Chromium During Vacuum Induction Melting, in *Transactions of the Vacuum Metallurgy Conference*, American Vacuum Society, 1963
17. R.E. Schwer, M.J. Gray, and S.F. Morykwas, "Trace Element Refining of Ni-Base Superalloys by Vacuum Induction Melting," Paper presented at the Vacuum Metallurgy Conference, American Vacuum Society, Battelle Memorial Institute, Columbus, OH, June 1975
18. V.M. Antipov, Refining of High-Temperature Nickel Alloy in Vacuum Induction Furnaces, *Stal'*, Feb 1968, p 117-120
19. W.R. Freeman, Jr., Chapter 15 in *Superalloys II*, C.T. Sims, N.S. Stoloff, and W.C. Hagel, Ed., John Wiley & Sons, 1987, p 411-439
20. G.K. Bouse and J.R. Mihalisin, Metallurgy of Investment Cast Superalloy Components, in *Superalloys, Supercomposites and Superceramics*, Academic Press, 1989, p 99-148
21. G.L. Erickson, K. Harris, and R.E. Schwer, "Optimized Superalloy Manufacturing Process for Critical Investment Cast Components," Cannon-Muskegon Corporation, 1982, p 6-9
22. M. Lamberigts, S. Ballarati, and J.M. Drapier, Optimization of the High Temperature, Low Cycle Fatigue Strength of Precision Cast Turbine Wheels, in *Proceedings of the Fifth International Symposium on Superalloys*, American Institute of Mining, Metallurgical, and Petroleum Engineers, 1984, p 13-22
23. B.A. Ewing and K.A. Green, Polycrystalline Grain Controlled Castings for Rotating Compressor and Turbine Components, in *Proceedings of the Fifth International Symposium on Superalloys*, American Institute of Mining, Metallurgical, and Petroleum Engineers, 1984, p 33-42
24. M.J. Woulds and H. Benson, Development of a Conventional Fine Grain Casting Process, in *Proceedings of the Fifth International Symposium on Superalloys*, American Institute of Mining, Metallurgical, and Petroleum Engineers, 1984, p 3-12
25. S.J. Veeck, L.E. Dardi, and J.A. Butzer, "High Fatigue Strength, Investment Cast Integral Rotors for Gas Turbine Applications," Paper presented at the TMS-AIME Annual Meeting, The Metallurgical Society, Dallas, Feb 1982

Directionally Solidified and Single-Crystal Superalloys

K. Harris, G.L. Erickson, and R.E. Schwer, Cannon-Muskegon Corporation

THE PRIMARY GOALS in the continuing development of the aircraft gas turbine are increased operating temperatures and improved efficiencies. A more efficient turbine is required to achieve lower fuel consumption. Higher turbine inlet temperature and increased stage loading result in fewer parts, shorter engine lengths, and reduced weight. Engine operating costs can be reduced if higher temperatures are possible without increasing part life-cycle costs.

Critical turbine components include high-pressure turbine blades, vanes, and disks. During the last 15 years, turbine inlet temperatures have increased by 278 °C (500 °F). About half of this increase is due to a more efficient design for the air cooling of turbine blades and vanes, while the other half is due to improved superalloys and casting processes (Ref 1). The cooling that is now possible with serpentine cores and multiple shaped-hole film cooling (Fig. 1) enables high-pressure turbine blades and vanes to operate with turbine inlet temperatures of typically 1343 °C (2450 °F), which is above the melting point of the superalloy materials. Turbine inlet temperatures as high as 1571 °C (2860 °F) are current parameters for several advanced fighter engines (Ref 2). It is forecast that by the mid to late 1990s, fabricated single-crystal airfoils with ultra-efficient transpiration cooling schemes will be capable of operating in gas temperatures greater than 1649 °C (3000 °F) with sufficient durability and reliability for man-rated flight turbine engines. These advanced superalloy single-crystal diffusion-bonded airfoils will present a major challenge to the emerging ceramic composite component technology.

For the past 28 years, high-pressure turbine blades and vanes have been made from cast nickel-base superalloys. The higher-strength alloys are hardened by a combination of approximately 60 vol% γ' [$Ni_3(Al,Ti)$] precipitated in a γ matrix, with solid-solution strengthening provided by the powerful strengtheners tantalum, tungsten, and molybdenum. The γ' phase, which has an ordered face-centered cubic structure, is coherent with the γ matrix, their lattice parameters being almost identical (<1% mismatch). This allows homogeneous nucleation of the precipitate with low surface energy and long-time stability at temperature, ensuring the potential usefulness of the alloys at elevated temperatures up to 0.85 T_m (melting point) for extended periods of time. Tantalum, tungsten, and hafnium substitute for some of the aluminum and titanium in the γ', thus stiffening this phase because of their relatively large atomic size. Initially, the blades were made as isotropic polycrystal or equiaxed castings. Under aerospace turbine engine operating conditions, failure of these equiaxed-grain components usually occurred at the grain boundaries from a combination of creep, thermal fatigue, and oxidation.

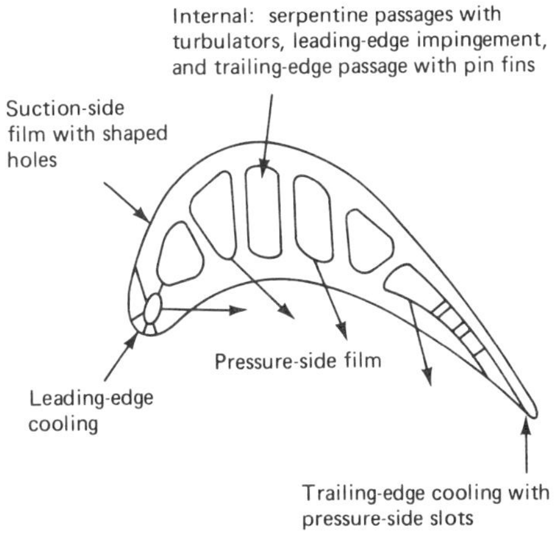

Development of the directional solidification (DS) casting process to produce blades and vanes with low-modulus (100)-oriented columnar grains aligned parallel to the longitudinal, or principal-stress, axis (Fig. 2) resulted in significant improvements in creep strength and ductility as well as in thermal fatigue resistance (5× improvement). Pratt and Whitney Aircraft (PWA)

Fig. 1 Shaped holes, turbulators, pin fins, and other techniques used in turbine rotor blade cooling

Fig. 2 Directionally solidified turbine blade CM 247 LC

pioneered this process (Ref 3, 4), as well as its turbine engine application, and has accumulated 18 years of production experience with over 25 million flight hours with DS blades and vanes (Ref 5).

There has been recent interest in DS blades not only for small- to medium-size airfoils for industrial turbines that burn natural gas but also for large base-load electricity-generating machines. Improved fuel efficiency requirements, along with the desire for high-temperature exhaust gases from the gas turbine (to produce steam suitable for co-generation electricity production), have resulted in the development and application engineering of DS blades with component lengths in the range of 305 to 635 mm (12 to 25 in.).

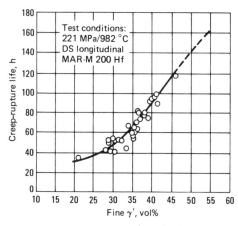

Fig. 3 Rupture life versus volume fraction of fine γ' at a fixed total amount of fine and coarse γ' for DS MAR-M 200 Hf alloy

Single-crystal (SX) casting technology was pioneered in the mid-1960s by PWA (Ref 5, 6). However, there was limited interest in the development of single-crystal blades because the conventional heat treatments being applied to MAR-M 200-type single-crystal components did not produce improvements in creep strength, thermal fatigue strength, and oxidation resistance that significantly exceeded the results achieved with the DS columnar-grain MAR-M 200 Hf. Only ductility and transverse creep resistance were improved. Around 1975, the beneficial role of γ' solutioning heat treatment applied to DS MAR-M 200 Hf was shown by PWA (Ref 7). It was found that creep strength was a direct function of the volume fraction of solutioned and reprecipitated fine γ' (Fig. 3). Experimental work by PWA showed that the elimination of grain-boundary strengthening elements (boron, hafnium, zirconium, and carbon) resulted in a substantial increase in the incipient melting temperature of the alloy (Ref 6). Consequently, the complete solutioning of the γ' phase, with appreciable solutioning of the γ-γ' eutectic phase, became possible without provoking incipient melting of the alloy.

Single-crystal alloy PWA 1480 (Table 1) offered a 25 to 50 °C (45 to 90 °F) temperature capability improvement in terms of time-to-1% creep, compared to the extensively used DS MAR-M 200 Hf alloy (Ref 5). The creep property improvement, which increases with temperature, depended on optimized single-crystal microstructures with full solutioning of the as-cast coarse γ'. The PWA 1480 alloy was developed to utilize the relatively low thermal gradient, single-crystal casting facilities already available as DS production units, without the freckling problems of alloy 444 (single-crystal MAR-M 200 with no carbon, boron, hafnium, zirconium, or cobalt) (Ref 5). Alloy PWA 1480, with its high tantalum (12%) and low tungsten (4%) contents, proved to be unique with this castability feature. Multistep homogenization/solutioning treatments with tight temperature control were developed to completely solution the γ' PWA 1480 without inducing incipient melting. Since 1982, PWA has had more than 5 million flight hours of successful experience using turbine blade and vane parts of single-crystal alloy PWA 1480 in commercial and military engines (Ref 8).

Directionally solidified and single-crystal superalloys and process technology are contributing to significant advances in turbine engine efficiency and durability. Further appreciable gains are forecast, particularly from single-crystal technology over the next 10 years. These gains are expected to arise from the development of higher creep strength and improved oxidation-resistant SX alloy compositions as well as from the development of SX casting and fabrication technology to utilize advanced transpiration-cooling schemes.

Directionally Solidified Superalloys

Chemistry and DS Castability. Early work with directionally solidified columnar-grain turbine blades in the 1960s involved the superalloys used for conventionally cast, equiaxed blades containing approximately 60 vol% γ', such as IN 100 and MAR-M 200. The problems encountered ranged from little longitudinal stress-rupture improvement with IN 100 to the lack of transverse ductility and DS grain-boundary cracking with MAR-M 200.

Pioneering work by Martin Metals resulted in the addition of hafnium to conventionally cast equiaxed superalloys to improve 760 °C (1400 °F) stress-rupture ductility and castability. For directional solidification, PWA added hafnium to MAR-M 200, which reduced DS grain-boundary cracking and increased transverse ductility. Although hafnium levels of up to 2% and greater in MAR-M 200 Hf combatted DS grain-boundary cracking, increasing levels of hafnium also increased the DS airfoil component rejection rate and the number of quality assurance problems. This was due to the occurrence of HfO inclusions that usually resulted from hafnium-ceramic reactions (core, shell-mold). Other first-generation DS alloys that were successfully and extensively adopted by turbine engine companies included René 80H (René 80 + Hf) by GE, MAR-M 002 by Rolls-Royce, and MAR-M 247 by Garrett. Both MAR-M 002 and MAR-M 247 were originally developed by Martin Metals to contain hafnium for optimized equiaxed turbine blade mechanical properties and castability. The nominal compositions of these first-generation DS superalloys are listed in Table 2. Directionally solidified superalloy turbine blades employed in large commercial turbofan engines for long-distance flights have been used for up to 15 000 h with high reliability.

Continuing improvements in airfoil cooling techniques have usually led to significant gains in gas turbine operating efficiencies. However, these cooling techniques often result in very complex cored, thin-wall (0.5 to 1 mm, or 0.02 to 0.04 in.) airfoil

Table 1 First-generation single-crystal superalloys

Alloy	Cr	Co	Mo	W	Ta	V	Nb	Al	Ti	Hf	Ni	Density, g/cm³
PWA 1480	10	5	...	4	12	...	...	5.0	1.5	...	bal	8.70
René N-4	9	8	2	6	4	...	0.5	3.7	4.2	...	bal	8.56
SRR 99	8	5	...	10	3	...	...	5.5	2.2	...	bal	8.56
RR 2000	10	15	3	...	...	1	...	5.5	4.0	...	bal	7.87
AM1	7	8	2	5	8	...	1	5.0	1.8	...	bal	8.59
CMSX-2	8	5	0.6	8	6	...	...	5.6	1.0	...	bal	8.56
CMSX-3	8	5	0.6	8	6	...	...	5.6	1.0	0.1	bal	8.56
CMSX-6	10	5	3	...	2	...	...	4.8	4.7	0.1	bal	7.98

Table 2 First-generation DS superalloys with extensive turbine engine airfoil applications

Alloy	C	Cr	Co	Mo	W	Nb	Ta	Al	Ti	B	Zr	Hf	Ni
MAR-M 200 Hf	0.13	8	9	...	12	1	...	5.0	1.9	0.015	0.03	2	bal
René 80H	0.16	14	9	4	4	...	...	3.0	4.7	0.015	0.01	0.8	bal
MAR-M 002	0.15	8	10	...	10	...	2.6	5.5	1.5	0.015	0.03	1.5	bal
MAR-M 247	0.15	8	10	0.6	10	...	3.0	5.5	1.0	0.015	0.03	1.5	bal

Directionally Solidified and Single-Crystal Superalloys / 997

Table 3 Second-generation DS and SX superalloys

Alloy	Nominal composition, wt%													Density, g/cm³
	C	Cr	Co	Mo	W	Ta	Re	Al	Ti	B	Zr	Hf	Ni	
DS alloy														
CM 247 LC	0.07	8	9	0.5	10	3.2	...	5.6	0.7	0.015	0.010	1.4	bal	8.54
SX alloys														
PWA 1484 (Ref 8)	...	5	10	2	6	9	3	5.6	...	...	...	0.1	bal	8.95
CMSX-4 (Ref 10)	...	6	9	0.6	6	7	3	5.6	1.0	...	...	0.1	bal	8.70

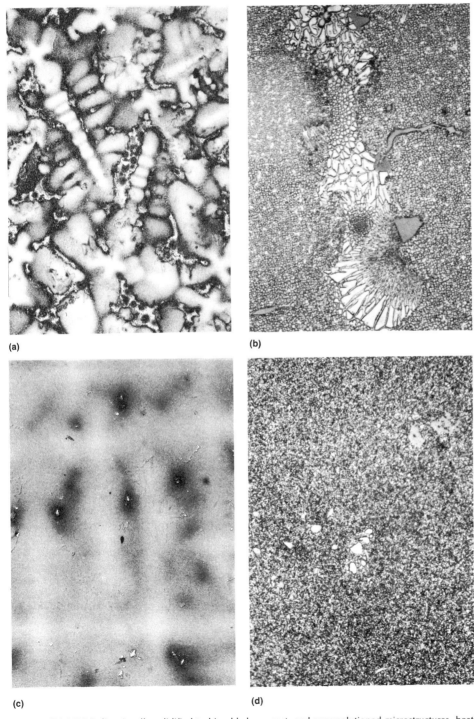

Fig. 4 CM 247 LC directionally solidified turbine blade, as-cast, and supersolutioned microstructures, heat V6692. Micrographs taken from airfoil, transverse orientation. (a) As-cast. 90×. (b) As-cast. 905×. (c) Supersolutioned. 90×. (d) Supersolutioned. 905×

designs, which can be susceptible to grain-boundary cracking during the DS casting of high-creep-strength alloys, particularly with modern high-thermal-gradient casting processes. Thus, the need for improved DS castability resulted in the development by the Cannon-Muskegon Corporation of CM 247 LC, a second-generation alloy from the MAR-M 247 composition (Ref 9). The nominal composition of this superalloy, which is also known as René 108, is given in Table 3. The CM 247 LC alloy has particularly excellent resistance to DS grain-boundary cracking and is capable of essentially 100% γ' solutioning to maximize creep strength without incipient melting or deleterious M_6C platelet formation upon subsequent high-temperature stress exposure, but with adequate transverse ductility retention.

With respect to DS grain-boundary cracking, zirconium and silicon are generally known to be bad actors. Small amounts of a brittle, hafnium-rich eutectic phase containing high concentrations of zirconium and silicon have been found in DS crack-prone tests (Ref 9). It was observed that very small reductions in zirconium and titanium contents, combined with a very tight control of silicon and sulfur, dramatically reduced the DS grain-boundary cracking tendency of a high-creep-strength superalloy such as MAR-M 247 (Ref 11). The major microstructural effect of the lower titanium content in CM 247 LC, compared to MAR-M 247, is to significantly reduce the size of the γ/γ' eutectic nodules as well as to lower the volume fraction of the eutectic from approximately 4 vol% in MAR-M 247 to 3 vol% in CM 247 LC DS components. This factor is also believed to be significant in reducing the DS grain-boundary cracking tendency of CM 247 LC.

Heat Treatment and Mechanical Properties. Multistep solutioning techniques based on a slow temperature increase between steps and temperatures up to 1254 °C (2290 °F) are used to supersolution heat treat CM 247 LC DS airfoil components to attain microstructures such as those shown in Fig. 4. Resultant stress-rupture property improvements are illustrated in Fig. 5.

The advent of single-crystal technology is not likely to preempt the need for DS airfoil components in the intermediate term. Directionally solidified airfoils will continue to be used for vane segments and low-pressure

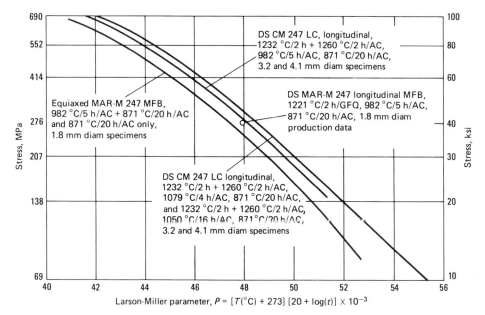

Fig. 5 Larson-Miller stress-rupture strength of DS CM 247 LC versus DS and equiaxed MAR-M 247. MFB, machined from blade; GFQ, gas fan quenched; AC, air cooled

blades in advanced turbine engines because of the producibility of the components, which makes them cost effective.

Several third-generation DS superalloys containing rhenium have been developed that have stress-rupture strength values close to those of the first-generation single-crystal alloys (Ref 12). These new alloys are particularly useful for DS vanes where load-bearing capability, such as to support a bearing, is an important design consideration.

Single-Crystal Superalloys

The greatest advance in the metal temperature capability of turbine blades in the last 25 years has been the single-crystal superalloy and process technology pioneered by PWA (Fig. 6 and 7). The dramatic improvement in the durability of the f-100 fighter engine turbine, as evidenced by the service performance of the f-220 version, is largely due to the PWA 1480 superalloy single-crystal first- and second-stage blades and vanes (Fig. 8).

Other pioneering single-crystal alloy development work resulted in the derivation of several single-crystal compositions from MAR-M 247 during the Garrett/NASA Materials for Advanced Technology Engines (MATE) program, which began in 1977 (Ref 14, 15). The two alloys studied extensively were NASAIR 100 and NASAIR Alloy 3; the latter contained a minor hafnium addition.

The compositions of the first-generation single-crystal superalloys, many of which are being used in turbine engine applications, are shown in Table 1: René N-4 developed by General Electric (Ref 16); SRR 99 and RR 2000 by Rolls-Royce plc (Ref 17); AM1 by the Office National d'Etudes et de Recherches Aerospatiales (ONERA) and Snecma (Ref 18); and CMSX-2, CMSX-3 (Ref 19), and CMSX-6 (Ref 20) by Cannon-Muskegon Corporation. These alloys are characterized by approximately the same creep-rupture strength (density corrected) but have differing SX castabilities, grain qualities, solution heat treatment windows, propensities for recrystallization upon solution treatment, environmental oxidation and hot corrosion properties, and densities. Typical stress-rupture properties are shown in Fig. 9 and 10.

Chemistry and SX Castability

Alloy CMSX-2 was developed in 1979 from the MAR-M 247 composition using some of the experience of the Garrett/NASA MATE program (Ref 14). A multidimensional development approach was used to achieve a high level of balanced properties (Fig. 11). The chemistry modifications applied to MAR-M 247 to develop CMSX-2 (Table 1) are summarized below with respect to function and objectives:

- Grain-boundary strengthening elements (boron, hafnium, zirconium, and carbon) were removed to achieve a very high incipient melting temperature (1335 °C, or 2435 °F)
- Partial substitution of tantalum for tungsten (CMSX-2 has 6% Ta) for good single-crystal castability, high γ' volume fraction (68%), improved γ' precipitate strength, microstructural stability (freedom from α-tungsten and tungsten, molybdenum-rich μ phases), good oxidation resistance, and coating stability
- Cobalt maintained to increase solid solubility and microstructural stability
- Chemistry balance designed to ensure a wide and practical solution heat treatment temperature range, or window (difference between the γ' solvus and the incipient melting temperature), of at least 22 °C (40 °F)
- Phacomp control of the chemistry of the alloy to avoid the occurrence of deleterious topologically close-packed phases

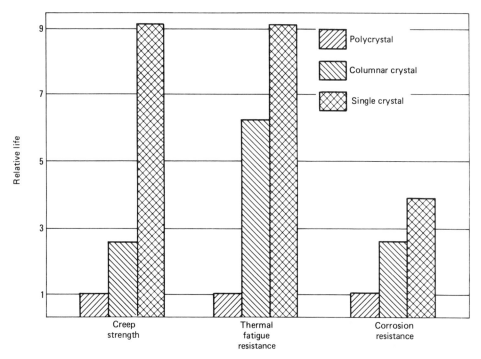

Fig. 6 Comparative properties of polycrystal, DS columnar-crystal, and single-crystal superalloys. Source: Ref 13

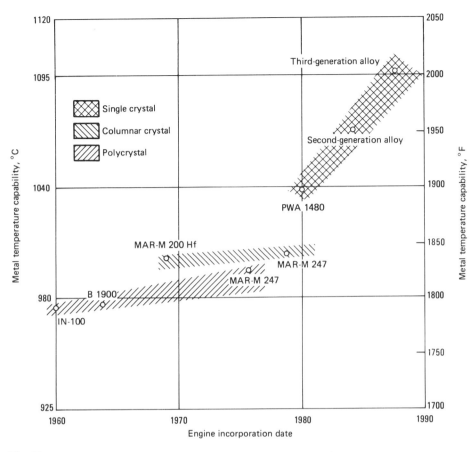

Fig. 7 Progress in turbine airfoil metal temperature capability. Source: Ref 13

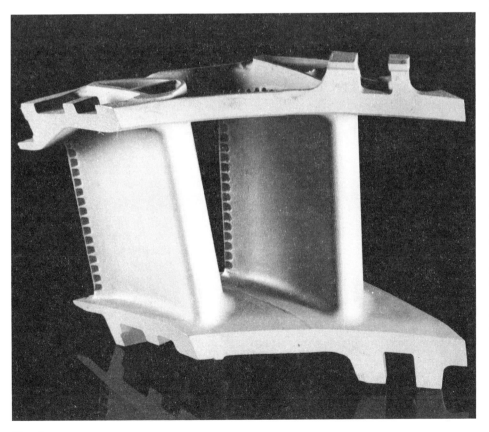

Fig. 8 f-100 paired single-crystal vane cast in PWA 1480

Figure 12 shows the relative potency of tantalum, tungsten, and molybdenum as solid-solution strengtheners in binary nickel alloys, where tantalum is the most powerful strengthener on an atomic percent basis. An increase in the lattice parameter of the γ phase due to alloy additions increases the solid-solution strengthening. Tantalum also partitions strongly to the γ' phase, increasing the volume fraction and stiffening the γ' due to its relatively large atomic size. The strength of the γ' phase is important in superalloys with a high volume fraction of γ' (>50%) because γ' shearing is the primary strengthening mechanism. With the mean free edge-to-edge distance in the γ matrix between the precipitates being smaller than the average precipitate size itself, dislocation shearing of the γ' particle is favored over Orowan dislocation looping around the γ' particles.

Detailed transmission electron microscopy studies of dislocation movement in cast high-strength superalloys, such as MAR-M 002 (Table 2) and its single-crystal derivative SRR 99 (Table 1), have shown the importance of ensuring that the antiphase boundary (APB) energy is high, so that the stacking fault mode of creep deformation occurs at temperatures up to 850 °C (1562 °F), thus ensuring high creep strength (Ref 17). Tantalum additions raise the APB energy relative to the stacking fault energy (Ref 17), leading to the increased tendency for stacking faults to be formed at lower temperatures.

The CMSX-2 alloy is designed to provide good SX foundry performance because castability is a crucial alloy performance criterion for any complex, thin-wall turbine blade or vane component, a characteristic sometimes given limited attention in alloy design. It affects not only the yield and cost of components but also the defect level and therefore component performance. Single-crystal casting defects of concern are:

- *Freckling*: A spiral of equiaxed grains caused by elemental segregation in the liquid state
- *Slivers*: Moderate-angle grain defects
- *Microporosity*: A uniform distribution of interdendritic micropores
- *Spurious grains*: High-angle grain boundaries
- *Stable oxide inclusions*: Al_2O_3
- *Carbides*: TiC

The partial substitution of tantalum for tungsten in the CMSX-2 alloy, compared to the MAR-M 247 chemistry, helps overcome the freckling problems inherent in the low-tantalum, high-tungsten single-crystal alloys. The strong γ'-forming elements, aluminum and titanium, which are also low density, tend to segregate to the last liquid to solidify in the interdendritic spaces during the SX solidification process. This can create density changes and consequential

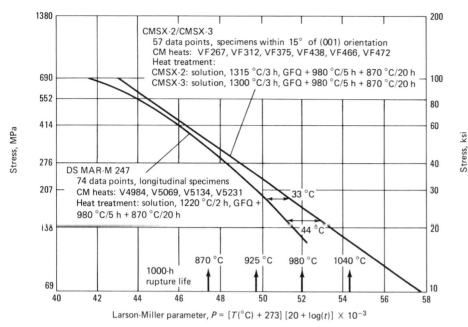

Fig. 9 Larson-Miller stress-rupture strength of CMSX-2/CMSX-3 versus DS MAR-M 247, using 1.8 mm (0.070 in.) specimens machined from blades

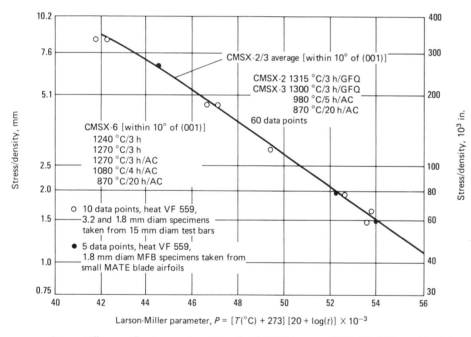

Fig. 10 Larson-Miller specific stress-rupture strength of CMSX-6 versus CMSX-2/3. MFB, machined from blade; GFQ, gas fan quenched; AC, air cooled

flow in the liquid metal close to the solidification front, which can nucleate freckle trails of equiaxed grains. This can occur particularly under conditions of low or changing thermal gradients. Tantalum, which is a strong γ′-forming element of high density, also tends to segregate to the last liquid to solidify in the interdendritic spaces and thus evens out these density changes in the liquid, or mushy, zone and reduces freckling tendencies.

Several studies undertaken in the United States, Europe, and Japan confirm that high [N] and [O] levels in single-crystal superalloy ingot adversely affect SX casting grain yield, supporting the importance for low [N] and [O] levels in the master alloy. Carbon, sulfur, and [O] master alloy impurities are shown to transfer nonmetallic inclusions, such as Al_2O_3, (Ti,Ta) C/N, and $(Ti,Ta)_x$ S, to SX parts (Ref 22). Grain defects can nucleate on these inclusions.

Several second-generation, rhenium-containing, single-crystal superalloys have been developed for turbine engine applications. Two typical compositions are given in Table 3. Rhenium partitions mainly to the γ matrix; this retards coarsening of the γ′-strengthening phase and increases γ/γ′ misfit (Ref 23). Atom-probe microanalysis of rhenium-containing modifications of the PWA 1480 and CMSX-2 alloys reveals the occurrence of short-range order in the matrix with small rhenium clusters (~1.0 nm, or 10 Å, in size) detected in the γ in the alloys (Ref 24). The rhenium clusters can act as more efficient obstacles against dislocation movement compared to isolated solute atoms in the γ solid solution; therefore, they play a significant role in improving the creep strength. The Larson-Miller stress-rupture comparison of CMSX-4 and CMSX-2/3 is shown in Fig. 13. The stress-rupture temperature capability advantage of CMSX-4 over CMSX-2/3 is 27 °C (48 °F) (density corrected) in the 248 MPa/982 °C (36 ksi/1800 °F) region. In the 103 MPa/1121 °C (15 ksi/2050 °F) region, the stress-rupture temperature capability advantage is 30 °C (54 °F) (density corrected). The data also indicate that CMSX-4 has a potential peak-use temperature under stress of at least 1149 °C (2100 °F).

Single-Crystal Casting Techniques. A variety of single-crystal airfoil component-casting techniques have been developed to production status around the world in the last 10 years. Most involve a withdrawal-type vacuum induction casting furnace with mold susceptor heating. Cooling plate sizes range in diameter from 140 to 610 mm (5½ to 24 in.). Some of the developed SX casting techniques are presented in Ref 12, 13, 25, and 26.

The modern helicopter engine turbine vane shown in Fig. 14 represents a difficult cored configuration. The large shrouds and core make this vane susceptible to shrinkage, grain nucleation, and recrystallization during solution heat treatment. Single-crystal casting processes developed by the Allison Gas Turbine Division of General Motors Corporation result in high yields for this vane in CMSX-3. Similar yields have been demonstrated with CMSX-4 using the same Allison casting process.

Single-Crystal Heat Treatment and Microstructures. With regard to solutioning, the latest multistep ramped cycles developed for single-crystal components are designed to completely solution the γ′ and most of the γ/γ′ eutectic without incipient melting. An additional benefit of the high-temperature cycles is the element homogenization effect, as shown in Fig. 15. Alloy CMSX-4, which is solutioned at a maximum temperature of 1321 °C (2410 °F) in commercial vacuum heat treatment furnaces, readily attains the 99%+ (<1% remnant γ/γ′ eutectic) solutioned microstructure, as illustrated in Fig. 16.

With regard to aging, the weight fraction of γ′ in CMSX-2 is approximately 68% with chemistry, as shown in Table 4, both being

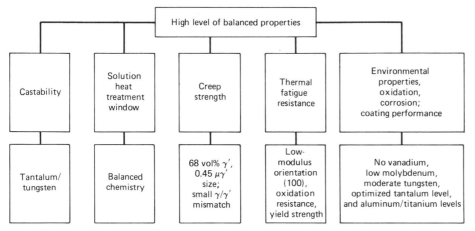

Fig. 11 CMSX-2 alloy development goal

Table 4 Chemical composition of the γ' phase in CMSX-2

Element	Composition, wt%
Nickel	69.25
Cobalt	3.15
Chromium	2.05
Molybdenum	0.30
Tungsten	7.25
Aluminum	7.55
Titanium	1.30
Tantalum	9.15

Source: Ref 27

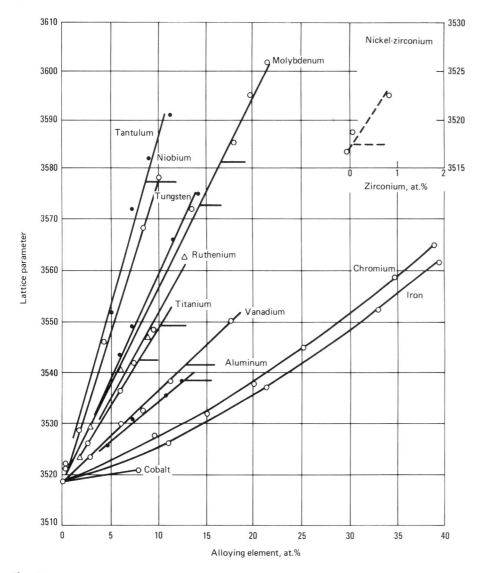

Fig. 12 Influence of alloying elements on the lattice parameter of binary nickel alloys. Source: Ref 21

independent of the aging treatments used. The measured lattice parameter of the γ' is 0.35865 nm (3.5865 Å), with a γ/γ' misfit at room temperature of 0.14% (Ref 27) and −0.33% at 1050 °C (1922 °F) (Ref 28).

It has been reported by ONERA that a high-temperature aging heat treatment (T_2) [16 h/1050 °C (1922 °F) air cooled (AC)] following solution treatment, with subsequent intermediate temperature aging [20 h/871 °C (1600 °F) or 48 h/850 °C (1562 °F)], gives CMSX-2 cuboidal γ' with a mean size of 0.45 μm (18 μin.), which optimizes creep response (Ref 27, 29). Similar γ' morphology and size are obtained with a 4 h/1079 °C (1975 °F) AC postsolution treatment. The morphology of the γ' in CMSX-2 with this ONERA-type aging treatment is shown in Fig. 17(a), which should be compared to the conventional irregularly shaped γ' particles with a mean size of 0.3 μm (12 μin.) shown in Fig. 17(b). The particles shown in Fig. 17(b) result from a 5 h/982 °C (1800 °F) AC + 48 h/850 °C (1562 °F) aging (T_1). Specimens of the T_2 type at 760 °C (1400 °F) deform in a homogeneous manner in the early stage of creep (Fig. 18). The homogeneous nature of the deformation leads to a rapid strain hardening of the material, causing a decrease in the creep rate. The T_1 heat treatment, which produces smaller, irregularly shaped particles, favors inhomogeneous deformation within the specimen due to the precipitate shearing during the early stages of creep (Fig. 19). In this case, the amplitude of primary creep is high, and the strain hardening of the material is achieved at a much later stage, compared with that of the T_2-type heat-treated specimens.

During creep at high temperature, the γ' precipitates coarsen in the form of rafts perpendicular to the stress axis. The kinetics of raft formation depend on the testing temperature, among other factors. At 1050 °C (1922 °F) under a stress of 120 MPa (17.4 ksi), the rafts form within a few hours (Fig. 20). The rafts have a high aspect ratio in the T_2-type heat-treated specimens in which the cuboidal γ' precipitates are already aligned. The lateral extension of the γ' phase in the form of rafts causes the specimen to creep at a much lower rate, compared with the creep rate of the material in which the γ' phase coalesces irregularly. The CMSX-2 and CMSX-3 alloys that show this type of rafted γ' morphology possess very long rupture lives at high temperatures. In these alloys, the misfit between the γ and γ' phases is found to be negative at high temperatures (Ref 30).

Work by ONERA and Ishikawajima-Harima Heavy Industries Company, Ltd. (IHI) shows some interesting effects of the crystal orientation and heat treatments on the creep behavior and strength of several single-

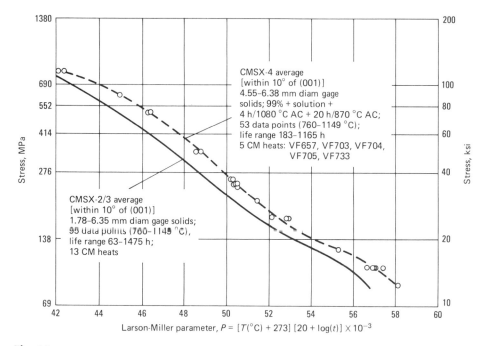

Fig. 13 Larson-Miller stress-rupture strength of CMSX-4 versus CMSX-2/3

crystal superalloys (Ref 31). The salient features can be summarized as follows:

- At intermediate temperatures (760 to 849 °C, or 1400 to 1560 °F), the creep behavior of nickel-base single-crystal superalloys is extremely sensitive to crystal orientation and γ' precipitate size. For a γ' size in the range of 0.35 to 0.5 μm (14 to 20 μin.), the highest creep strength is obtained near [001], while orientations near the [111]-[011] boundary of the standard stereographic triangle exhibit very short creep lives. When the γ' size decreases to 0.2 μm (8 μin.), the longest creep lives are exhibited, in decreasing order, by the crystals oriented near [111], [001], and [110]. The anisotropy in creep between the [001] and [111] orientations can therefore be reduced by appropriate precipitation heat treatments. The creep strengths, however, remain poor near the [011] orientation

- At high temperatures (982 to 1049 °C, or 1800 to 1920 °F), the creep behavior of the single-crystal superalloys is much less sensitive to crystal orientation and γ' size than it is at intermediate temperatures. The [001]-oriented single crystals develop a rafted γ' structure normal to the tensile stress axis, while the γ' precipitates coarsen irregularly in the [111] specimens

Fatigue. An important property that must be considered when selecting single-crystal superalloys for turbine blade applications is fatigue strength. Single crystals of CMSX-2 have been cast both under low- and high-gradient conditions and then subjected to high-cycle fatigue tests in the repeated tension mode at 870 °C (1598 °F) (Ref 29); the results are reported in Fig. 21. The fatigue resistance of specimens cast under a very high temperature gradient (laboratory conditions) is much superior to that of material cast under industrial conditions, primarily because of the very small pore size (<10 μm, or 40 μin.) inherent in the high-gradient specimens. The single crystals cast under industrial conditions have a more heterogeneous structure where the interdendritic spacing and the level of porosity vary along the length of the bar. Specimens corresponding to the beginning of solidification exhibit better fatigue resistance than those corresponding to the end of solidification. Some fatigue tests were also performed on specimens in which a rafted γ' morphology was developed prior to testing. It is interesting to note that the fatigue behavior is not significantly affected by the rafted γ' morphology (Ref 29).

Strain-controlled, fully reversed low-cycle fatigue tests performed at 760 °C (1400 °F) confirm the much better fatigue behavior of single crystals cast under a high gradient (Fig. 22). In this type of test, the higher the deviation from the [001] orientation, the shorter the fatigue life. It can be seen in Fig. 22 that for a total strain range of 1.2%, the fatigue life is decreased by an order of magnitude when the crystal orientation, relative to the [001], moves away from 6 to 22°. The decrease in fatigue life is a consequence of the increase in stress level through the increase of elastic modulus. Because the plastic strain component at 760 °C (1400 °F) is small, the results can be plotted as total stress versus the number of cycles to failure (Fig. 22). In Fig. 22, the effect of crystalline orientation on the fatigue life of the industrially processed single crystals is not apparent, and all the results of low-gradient single crystals can be represented by a single curve.

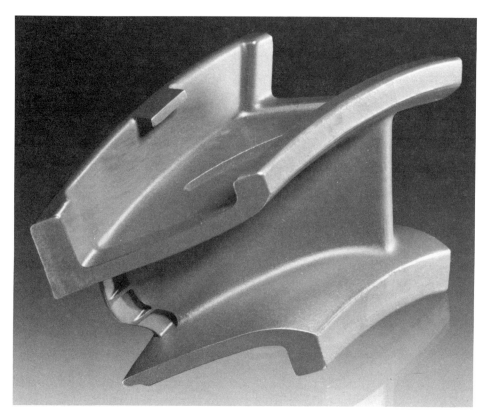

Fig. 14 SX turbine vane cast in CMSX-4

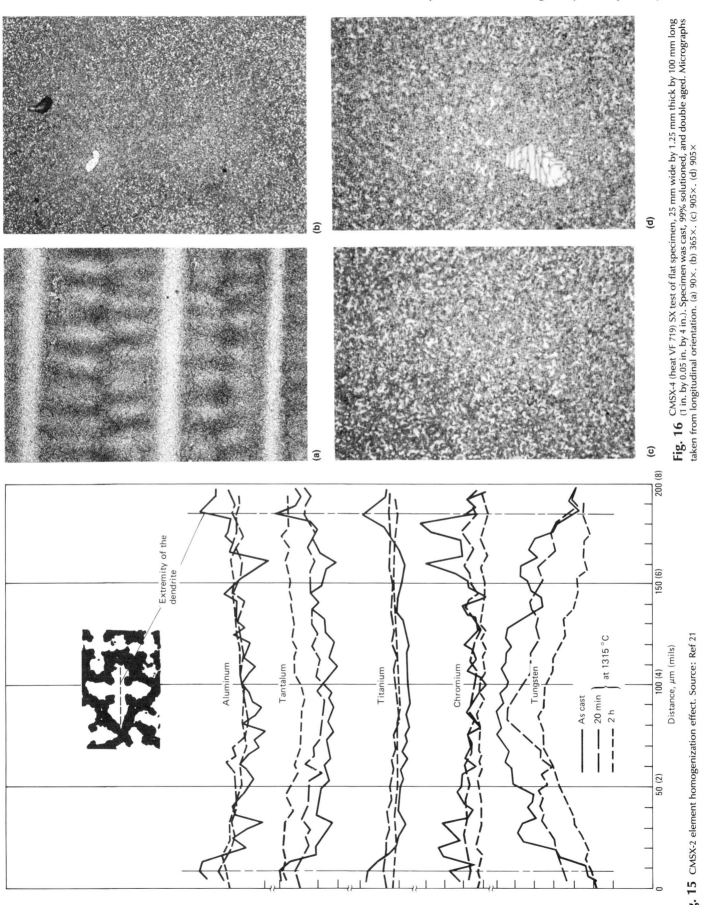

Fig. 16 CMSX-4 (heat VF 719) SX test of flat specimen, 25 mm wide by 1.25 mm thick by 100 mm long (1 in. by 0.05 in. by 4 in.). Specimen was cast, 99% solutioned, and double aged. Micrographs taken from longitudinal orientation. (a) 90×. (b) 365×. (c) 905×. (d) 905×

Fig. 15 CMSX-2 element homogenization effect. Source: Ref 21

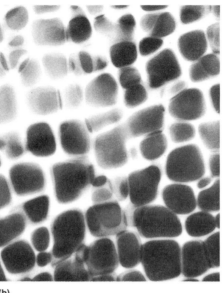

Fig. 17 Morphology of γ' precipitates in CMSX-2 alloy. (a) After T_2 heat treatment. (b) After T_1 heat treatment. Source: Ref 29

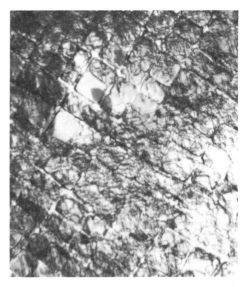

Fig. 18 Homogeneous deformation in CMSX-2 (T_2 heat treatment) after 0.16% creep strain at 760 °C (1400 °F). Source: Ref 29

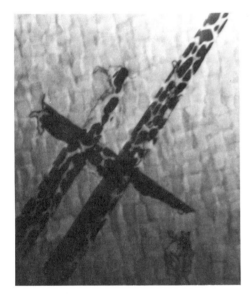

Fig. 19 Inhomogeneous deformation in CMSX-2 (T_1 heat treatment) during primary creep at 760 °C (1400 °F). Source: Ref 29

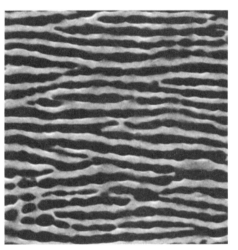

Fig. 20 Oriented coalescence of the γ' phase in CMSX-2 after 20 h of creep at 1050 °C (1920 °F) under 120 MPa (17.4 ksi). Tensile stress axis is [001]. Source: Ref 29

An examination of fracture surfaces shows that the cracks are initiated at microporosity, which indicates that these defects (microporosity) are of primary importance in determining the fatigue life of CMSX-2. The size of microporosity in industrial single crystals can be as large as 50 to 80 μm (2000 to 3200 μin.), but is rarely more than 10 μm (400 μin.), in single crystals cast under very high temperature gradients (laboratory conditions). The adverse effect of microporosity is also confirmed by the results obtained after hot isostatic pressing [1000 bars/1315 °C (2400 °F)] CMSX-2 single-crystal bars solidified under the low-temperature gradient. The fatigue strength of single crystals cast under low gradients after hot isostatic pressing can be improved to that of the high-gradient-processed specimens (Fig. 22). The techniques developed to hot isostatic press CMSX-2 ensure that no recrystallization occurs internally within the test bars during the hot isostatic press cycle.

Oxidation and Hot Corrosion. Coatings perform well with single-crystal alloys, particularly when they are optimized to the base alloy system, as shown by the work with PWA 1480 alloy in Fig. 23. The absence of grain boundaries, rosette clusters of carbides, and elemental segregation in heat-treated single-crystal alloys contribute to improved coating performance.

Cyclic oxidation testing at 900 and 1050 °C (1652 and 1922 °F) shows uncoated CMSX-4 and CMSX-2 to have excellent and similar performance (Ref 32). This work also shows uncoated CMSX-4 to have superior molten-salt hot corrosion characteristics compared to CMSX-2, most probably because of its lower tungsten content. Ultrahigh-temperature cyclic burner rig oxidation studies undertaken by Allison at 1177 °C (2150 °F) show that uncoated, modified CMSX-3 (designated CMSX-3 Mod A) has excellent performance similar to uncoated MAR-M 247 (which contains 1.4% Hf), with little or no attack after 140 h of testing (Fig. 24). Alloy CMSX-4 Mod A has also shown similar results.

REFERENCES

1. F.E. Pickering, Advances in Turbomachinery, Cliff Garrett Award Lecture, *Aerosp. Eng.*, Jan 1986, p 30-35
2. Snecma Advances M88 Demonstrator, *Flight Int.*, 22 March 1986, p 26
3. B.J. Piearcey and F.L. VerSnyder, *J. Aircr.*, Vol 3 (No. 5), 1966, p 390
4. F.L. VerSnyder and M.E. Shank, *Mater. Sci. Eng.*, Vol 6 (No. 4), 1970, p 321
5. M. Gell, The Science & Technology of Single Crystal Superalloys, in *Proceedings of Japan-U.S. Seminar on Superalloys*, International Iron and Steel Institute, 1984
6. M. Gell, D.N. Duhl, and A.F. Giamei, The Development of Single Crystal Superalloy Turbine Blades, in *Proceedings of the Fourth International Symposium on Superalloys* (Seven Springs, PA), American Society for Metals, 1980, p 205-214

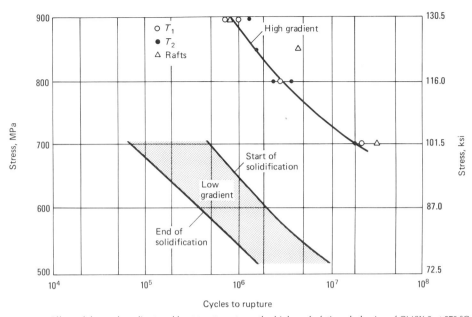

Fig. 21 Effect of thermal gradient and heat treatments on the high-cycle fatigue behavior of CMSX-2 at 870 °C (1598 °F) with frequency of 50 Hz. Source: Ref 29

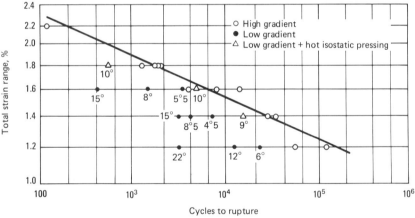

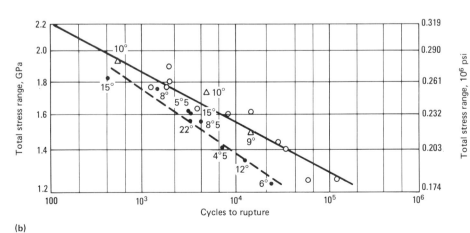

Fig. 22 Effect of thermal gradient, orientation, and hot isostatic pressing on the strain-controlled low-cycle fatigue behavior of CMSX-2 (fully reversed, with frequency of 0.33 Hz) at 760 °C (1400 °F). Numbers represent the deviation, in degrees, from the [001] orientation. (a) Strain versus cycles to failure. (b) Stress versus cycles to failure. Source: Ref 29

7. J.J. Jackson, M.J. Donachie, R.J. Henricks, and M. Gell, The Effects of Volume % of Fine γ' on Creep in DS MAR M 200 Hf, *Metall. Trans. A*, Vol 8A (No. 10), 1977, p 1615

8. A.D. Cetel and D.N. Duhl, Second-Generation Nickel-Base Single Crystal Superalloy, in *Sixth International Symposium on Superalloys* (Seven Springs, PA), The Metallurgical Society, 1988, p 235-244

9. J.J. Burke, H.L. Wheaton, and J.R. Feller, Paper presented at the Annual TMS-AIME Meeting, Denver, CO, The Metallurgical Society, 1978

10. K. Harris, G.L. Erickson, and R.E. Schwer, Development of CMSX-4 for Small Gas Turbines, Paper presented at the TMS-AIME Fall Meeting, Philadelphia, The Metallurgical Society, Oct 1983

11. K. Harris, G.L. Erickson, and R.E. Schwer, MAR M 247 Derivations: CM 247 LC DS Alloy, CMSX Single Crystal Alloys, Properties and Performance, in *Proceedings of the Fifth International Symposium on Superalloys* (Seven Springs, PA), The Metallurgical Society, 1984, 221-230

12. K. Harris, G.L. Erickson, and R.E. Schwer, CMSX Single Crystal, CM DS and Integral Wheel Alloys: Properties and Performances, in *Cost 50/501 Conference on High Temperature Alloys for Gas Turbines and Other Applications* (Liege), Reidel, 1986

13. M. Gell, D.N. Duhl, D.K. Gupta, and K.D. Sheffler, Advanced Superalloy Airfoils, *J. Met.*, July 1987, p 11-15

14. T.E. Strangman, G.S. Hoppin III *et al.*, Development of Exothermically Cast Single Crystal Mar M 247 and Derivative Alloys, in *Proceedings of the Fourth International Symposium on Superalloys* (Seven Springs, PA), American Society for Metals, 1980, p 215-224

15. G.S. Hoppin III and W.P. Danesi, Manufacturing Processes for Long Life Gas Turbines, *J. Met.*, July 1986

16. C.S. Wukusick, Final Report, Contract N62269-78-C-0315, Naval Air Systems Command, 25 Aug 1980

17. D.A. Ford and R.P. Arthey, Development of Single Crystal Alloys for Specific Engine Applications, in *Proceedings of the Fifth International Symposium on Superalloys* (Seven Springs, PA), The Metallurgical Society, 1984, 115-124

18. E. Bachelet and G. Lamanthe, AM1I, High Temperature Superalloy for Turbine Blades, Paper presented at the National Symposium on Single Crystal Superalloys, Villard-de-Lans, France, Feb 1986

19. K. Harris, G.L. Erickson and R.E. Schwer, "Development of the Single Crystal Alloys CMSX-2 & CMSX-3 for

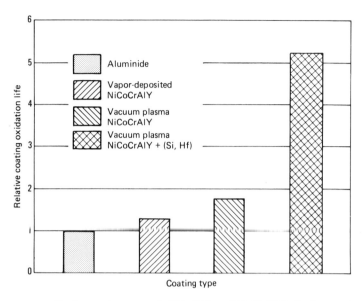

Fig. 23 Coating performance of PWA 1480 at 1149 °C (2100 °F), burner rig oxidation. Source: Ref 13

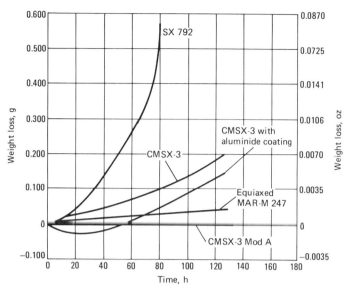

Fig. 24 Dynamic oxidation, 1177 °C (2150 °F), cyclic. Source: Allison Gas Turbine Division, General Motors Corporation

Advanced Technology Turbines," Paper 83-GT-244, American Society of Mechanical Engineers
20. J. Wortmann, R. Wege, K. Harris, G.L. Erickson, and R.E. Schwer, Low Density Single Crystal Superalloy CMSX-6, in *Proceedings of the Seventh World Conference on Investment Casting* (Munich, West Germany), European Investment Casters' Federation, 1988
21. I.I. Kornilov and A.Y. Snetkov, Lattice Parameter Limitations of Some Solid Solution Elements in Nickel, *Izv. Akad. Nauk.*, 1960, p 106-111
22. S. Isobe et al., "The Effects of Impurities on Defects in Single Crystals of NASAIR 100," Paper presented at the International Gas Turbine Congress, Tokyo, International Iron and Steel Institute, 1983
23. A.F. Giamei and D.L. Anton, *Metall. Trans. A*, Vol 16A, 1985, p 1997
24. D. Blavette, P. Caron, and T. Khan, *Scr. Metall.*, Vol 20 (No. 10), Oct 1986
25. M.J. Goulette, P.D. Spilling, and R.P. Arltey, Cost Effective Single Crystals, in *Proceedings of the Fifth International Symposium on Superalloys* (Seven Springs, PA), The Metallurgical Society, 1984, p 167-176
26. E. Staub, B. Walser, and J. Wortmann, An Alternative Single Crystal Manufacturing Process of Significant Economic and Quality Improvement Potential, in *Proceedings of the Symposium on Advanced Materials and Processing Techniques for Structural Applications* (Paris), ASM INTERNATIONAL, 1987
27. T. Khan and P. Caron, The Effect of Processing Conditions and Heat Treatments on the Mechanical Properties of a Single Crystal Superalloy, The Institute of Metals, 1985
28. A. Fredholm and J.L. Strudel, On the Creep Resistance of Some Nickel Base Single Crystals, in *Proceedings of the Fifth International Symposium on Superalloys* (Seven Springs, PA), The Metallurgical Society, 1984, p 220-221
29. T. Khan, P. Caron, D. Fournier, and K. Harris, "Single Crystal Superalloys for Turbine Blades: Characterization and Optimization of CMSX-2 Alloy," Paper presented at the 11th Symposium on Steels & Special Alloys for Aerospace, Paris Air Show—LeBourget, L'Association Aéronautique et Astronautique de France, June 1985
30. T. Khan and P. Caron, Effect of Heat Treatment on the Creep Behavior of a Ni-Base Single Crystal Superalloy, in *Fourth RISO International Symposium on Metallurgy and Material Sciences*, Office National d'Etudes et de Recherches Aerospatiales, 1983, p 173
31. T. Khan, P. Caron, Y.G. Nakagawa, and Y. Ohta, Creep Deformation Anisotropy in Single Crystal Superalloy, in *Proceedings of the Sixth International Symposium on Superalloys* (Seven Springs, PA), The Metallurgical Society, 1988, p 215-224
32. M. Matsubara, A. Nitta, and K. Kuwabara, Paper presented at the International Gas Turbine Congress, Tokyo, International Iron and Steel Institute, Oct 1987

Special Engineering Topics

Strategic Materials Availability and Supply 1009
   Appendix: Manganese Availability 1021
Recycling of Iron, Steel, and Superalloys 1023

Strategic Materials Availability and Supply

Joseph R. Stephens, National Aeronautics and Space Administration, Lewis Research Center

THE SUPERALLOYS are critical to the economic survival of the United States aerospace industry. Thus, it is imperative that the raw material resources required for superalloy production be readily available to U.S. producers. During the initial years of superalloy development, resources for iron, nickel, and to some extent chromium were available within the United States or from neighboring countries such as Canada and Cuba. However, in more recent years, superalloy compositions have become more complex, requiring ten or more different elements in a single alloy; environmental restrictions and labor costs have pushed sources offshore; and political and military changes have made resources supplies from countries that were once dependable suppliers unreliable or unavailable.

Because of these changes in the world market, superalloy producers have sought to ensure a reliable supply of imported materials. At the same time, they are exploring the suitability of alternative alloys in the event of the future unavailability of any imported alloying elements currently used in superalloys. One approach to alternative alloys is to substitute readily available elements for imported alloying elements. Obviously, this is not a simple solution because the compositions of superalloys are not only complex, their microstructures (which are dependent upon a critical balance of alloying constituents) must be maintained in order to achieve desired properties.

Because of shortages or limited availability of alloying elements over the years, for example, cobalt in the 1950s, chromium in the 1970s, and cobalt and other elements in the late 1970s and early 1980s, the Lewis Research Center of the National Aeronautics and Space Administration (NASA) undertook a program to address this continuing problem. This article reviews some of the trends in superalloy development; defines the term strategic materials; summarizes the current status of U.S. resources and reserves; discusses the supply sources and availability of strategic materials; and, finally, concentrates on the results achieved by the NASA Conservation of Strategic Aerospace Materials (COSAM) program (Ref 1).

Reserves and Resources

As already stated, a thriving superalloy industry within the United States requires a readily available supply of alloying ingredients. The ideal situation would be for U.S. mining companies to act as primary suppliers to the alloy producers. Unfortunately, the United States does not have ample reserves and resources for some required alloying elements and therefore has become more and more dependent on foreign sources for a number of strategic materials.

Planning for future aerospace materials needs requires an assessment of U.S reserves and resources. The principal distinction between reserves and resources is that reserves are based on current economical availability. They are known, identified deposits of mineral-bearing rock from which mineral(s) can be extracted profitably with existing technologies and under current economic conditions. On the other hand, resources include not only these reserves, but other mineral deposits that may eventually become available. The latter can be either known deposits that are not economically or technologically recoverable at present or unknown deposits that have not yet been discovered, but whose existence may have been inferred (Ref 2, 3).

Table 1 summarizes the estimated reserve and resource status of the United States for 18 metals. The United States ranks first for molybdenum reserves and second for copper. In comparison, the U.S.S.R. ranks either first or second for 8 of the 18 metals, and South Africa ranks either first or second for 5 of the 18. The U.S. domestic position presented in Table 1 and supported by further information indicates negligible reserves of chromium, cobalt, niobium (columbium), tantalum, and manganese, all vital to the aerospace and steel industries.

Although the United States does not have reserves of a number of elements, it is nonetheless a leading producer because it imports the necessary minerals and converts them into metals and alloys. Table 2 shows that the United States ranks first or second in 6 metals: aluminum, copper, magnesium, molybdenum, titanium, and rhenium. The U.S.S.R. holds this favorable position in 12 metals: rhenium, iron, manganese, nickel, aluminum, magnesium, titanium, tungsten, vanadium, gold, chromium, and platinum.

One other important consideration is the domestic consumption and production of the metals needed for the U.S. economy. Table 3 gives information on this factor for all 18 elements. It should be noted that in 1988 foreign purchases were required for the four elements (chromium, cobalt, niobium, and tantalum) designated as strategic metals in the COSAM program, as well as for manganese and nickel. In contrast, there is a fairly good matchup of production and consumption for iron, aluminum, copper, magnesium, titanium, and rhenium, and molybdenum production greatly exceeds consumption.

Strategic Materials

The United States has adequate supplies of metals such as copper, iron, and molybdenum. Stable/friendly foreign countries are sources for others, for example, nickel (Canada), titanium (Australia for rutile), aluminum (Jamaica, for bauxite), and tungsten (Canada) (Ref 5). However, by examining our import dependence on other metals, as shown in Fig. 1, it is apparent that we are a have-not nation for many important metals. Of particular concern is the aerospace industry because it is highly dependent on imports for several key metals that are considered strategic materials. For the purposes of this article, strategic materials are defined as predominately or wholly imported metals that are contained in the alloys used in aerospace components that

Table 1 World and U.S. reserves of 18 metals and U.S. resources of them, 1988

Metal	Reserves				Primary location of reserves		U.S. resources	
	World		United States					
	10^6 kg	10^6 ton	10^6 kg	10^6 ton	No. 1	No. 2	10^6 kg	10^6 ton
Re	2.7	0.003	0.36	0.0004	Chile	U.S.S.R.	4.5	0.005
Au	54	0.059	5.2	0.0057	South Africa	U.S.S.R.	5.7	0.0063
Pt	68	0.075	0.30	0.00033	South Africa	U.S.S.R.	0.907	0.001
Ta	22	0.024	...	...	Thailand	Australia	...	...
Ag	340	0.375	38	0.042	U.S.S.R.	Canada	50	0.0542
W	2 650	2.922	150	0.165	China	Canada	210	0.231
Co	3 310	3.650	...	...	Zaire	Cuba	860	0.950
Nb	3 445	3.800	...	...	Brazil	Canada	...	...
Mo	5 530	6.100	2 720	3	United States	Chile	5 350	5.9
V	4 265	4.705	136	0.150	U.S.S.R.	South Africa	2 175	2.4
Ni	48 980	54	1 270	1.4	Cuba	Canada	2 540	2.8
Ti(a)	85	0.094	0.30	0.0003	Brazil	Australia	1 450	1.6
Cu	349 200	385	56 870	62.7	Chile	United States	87 790	99
Mg	Unlimited	Unlimited	Unlimited	Unlimited	...	...	Unlimited	Unlimited
Cr	102 750	1 133	...	...	South Africa	U.S.S.R.	...	...
Mn	907 000	1 000	...	...	South Africa	U.S.S.R.	...	...
Al(b)	21 750 000	23 980	37 915	41.8	Guinea	Australia	39 900	44
Fe	65 300 000	72 000	3 355 900	3 700	U.S.S.R.	Canada	5 350 000	5 900

(a) Rutile. (b) Bauxite. Source: Ref 4

Table 2 World production in 1988

Metal	Production		Primary producers	
	10^6 kg	10^6 ton	No. 1	No. 2
Re	0.245	270×10^{-6}	United States	U.S.S.R.
Pt	0.335	371×10^{-6}	South Africa	U.S.S.R.
Ta	0.285	315×10^{-6}	Brazil	Australia
Au	2.08	2291×10^{-6}	South Africa	U.S.S.R.
Nb	13.5	0.0149	Brazil	Canada
Ag	16.3	0.01792	Mexico	Peru
Co	46.0	0.05068	Zaire	Zambia
V	30.4	0.0335	South Africa	U.S.S.R.
W	41.0	0.04524	China	U.S.S.R.
Ti	85.3	0.094	U.S.S.R.	United States
Mo	85.7	0.0945	United States	Chile
Mg	330	0.364	United States	U.S.S.R.
Ni	795	0.876	Canada	U.S.S.R.
Cu	8 480	9.350	United States	Chile
Cr	11 260	12.419	South Africa	U.S.S.R.
Al	16 860	18.590	United States	U.S.S.R.
Mn	23 220	25.600	U.S.S.R.	South Africa
Fe	832 630	918.000	U.S.S.R.	Brazil

Source: Ref 4

Table 3 United States production and consumption in 1988

Metal	Production		Consumption		Top two foreign suppliers	
	10^6 kg	10^6 ton	10^6 kg	10^6 ton	No. 1	No. 2
Re	0.0100	11×10^{-6}	0.0073	8×10^{-6}	Chile	Germany
Au	0.250	275×10^{-6}	0.120	133×10^{-6}	Canada	Switzerland
Pt	W	W	0.085	92×10^{-6}	South Africa	United Kingdom
Ta	...	...	0.042	46×10^{-6}	Thailand	Brazil
Nb	...	...	3.40	0.00375	Brazil	Canada
Ag	1.700	1875×10^{-6}	4.75	0.00525	Mexico	Canada
V	W	W	4.40	0.00484	South Africa	Chile
Co	...	...	7.3	0.008	Zaire	Zambia
W	0.225	0.00025	10	0.011	China	Bolivia
Ti	21.8	0.024	20	0.023	Japan	...
Mo	34.0	0.0375	16	0.018	Chile	Canada
Mg	140	0.155	137	0.151	Norway	Canada
Ni	...	...	180	0.200	Canada	Norway
Cr	...	...	525	0.579	South Africa	Turkey
Mn	...	...	750	0.825	South Africa	France
Cu	1 435	1.584	2 275	2.508	Canada	Chile
Al	3 890	4.290	5 390	5.940	Canada	Japan
Fe	52 970	58.400	68 115	75.100	Canada	Brazil

(a) W, Withheld. Source: Ref 4

are essential to the future economic health of the U.S. aerospace industry. In the study of strategic materials, two approaches have been used to identify the strategic materials most critical to the aerospace industry, an index method and a survey method.

The index method was used to define a vulnerability index for metals used in superalloys. Metals of greater vulnerability were identified as strategic metals. This study was undertaken by Stalker et al. (Ref 6). The index described 18 metals, listed in Table 1. Each metal was ranked 21 different ways, for example, its importance in relation to U.S. needs (in a peace economy and in a war economy), in relation to U.S. reserves, and in relation to cost in dollars per pound.

An analysis of the data from this study shows that the 18 metals generally fall in three groupings. The most strategic metals have an index of about 8 or greater on a scale from 1 to 10, with 10 being the most strategic. The midgroup metals have numbers between about 5 and 8; and the least strategic metals have indexes lower than 5. The breakdown of the 18 metals is shown below:

Most strategic	Midgroup	Least strategic
Ta	V	Al
Cr	Re	Cu
Pt	Ti	Mo
Nb	W	Mg
Mn	Ag	Fe
Co	Ni	
	Au	

It should be noted that although this ranking is in order of decreasing index for each of the three groups, the absolute rank within a group can be altered by using other data, such as a more complete breakdown of resources and reserves (to be discussed later). In addition, price volatility can affect known economic reserves as well as the weighting factors used. Regardless of these variables, it is doubtful that the most strategic list will change in composition.

The output of the study, with data from all 21 categories, is summarized in Table 4. The rankings are presented in terms of peacetime and wartime scenarios, based on such factors as reserves, consumption, pro-

Strategic Materials Availability and Supply / 1011

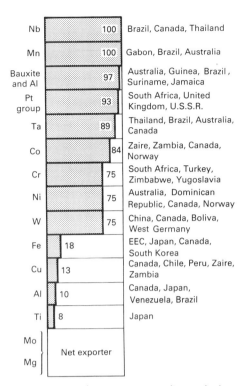

Fig. 1 Estimated 1988 net import reliance of selected nonfuel mineral materials as a percent of apparent consumption, where net import reliance equals imports minus exports plus adjustments for government and industry stock changes. EEC, European Economic Community

duction costs, and recyclability, in addition to subjective judgments regarding the likelihood of a mineral cartel. Consideration of the needs of a wartime economy yielded slightly different normalized scores, but with major concern for the same 6 elements.

In a further refinement, subjective weighting factors were applied to obtain a still more realistic appraisal. Weighting yielded significant increases in importance for manganese, copper, and aluminum and a decrease in importance for gold. However, the overall picture remained fairly constant. The availability and strategic nature of manganese are described in an appendix to this article.

Each of the most strategic metals has special capabilities such that the U.S. economy would not function well without them. Because the United States has a very limited reserve of each, attention must be given to the dependence on foreign supplies. Each element is unique and therefore requires a careful review of its role in superalloys in order to develop short-range and long-range plans for continuing supplies.

Survey Method. The second approach to identification of the metals most strategic to the United States involved meetings with the American Society of Mechanical Engineers (ASME) Gas Turbine Panel in 1979 and a survey of aerospace companies in 1980, which led to the recognition of the need to focus primarily on the aircraft engine industry. Based on the findings of these meetings and further discussions with several aircraft engine manufacturers, four elements that were of particular concern were identified (Ref 7). The alloys used to build the critical high-temperature components for aircraft propulsion systems require the use of three refractory metals, chromium, tantalum, and niobium, plus a fourth strategic metal, cobalt. These metals are contained in superalloys used in engine compressors, turbines, and combustors and are among the six elements having the highest strategic indexes of the 18 metals evaluated in Ref 6. Although the remaining two elements were considered, along with less strategic metals, it was decided to focus on the four aforementioned metals in the NASA COSAM program.

The Superalloys

Superalloys are the major materials of construction for today's high-temperature gas turbine engines used for both commercial and military aircraft. Nickel-base superalloys, along with iron-base and cobalt-base superalloys, are used throughout the engines in wrought, cast, powder metallurgy (P/M), and cast single-crystal forms to meet the requirements imposed by the aircraft industry.

Nickel-base superalloys were created at approximately the turn of the century with the addition of 20 wt% Cr in an 80 wt% Ni alloy for electrical heating elements (Ref 8). In the late 1920s, small amounts of aluminum and titanium were added to the 80Ni-20Cr alloy, with a significant gain in creep strength at elevated temperatures. It soon became apparent that iron and cobalt alloys could be more effectively strengthened by solid-solution additions, while nickel alloys could be strengthened by a coherent phase, γ'. Concurrent with these additions, the carbon present in the alloys was identified as having a strengthening effect when combined with other alloying elements to form M_6C and $M_{23}C_6$ carbides. Other grain-boundary formers, such as boron and zirconium, were added to polycrystalline materials to hold the material together.

Table 4 Strategic metals index analysis for peacetime and wartime economies

Ranking in relation to:	Economy	Al	Cr	Co	Nb	Cu	Au	Fe	Mg	Mn	Mo	Ni	Pt	Re	Ag	Ta	Ti	W	V	Weighting factor
U.S. needs	Peace	17	15	13	7	12	2	18	6	11	10	16	4	1	3	5	14	9	8	1
	War	17	14	12	11	7	5	18	2	8	13	16	4	1	3	6	15	10	9	2
World reserves	...	3	5	12	11	6	17	2	1	4	10	8	16	18	14	15	7	13	9	1
North American reserves	...	7	18	11	9	2	15	1	3	5	8	6	17	16	13	14	4	10	12	2
U.S. reserves	...	4	15	8	17	3	12	2	1	18	5	6	14	13	11	16	7	9	10	5
World production	Peace	3	4	12	14	5	15	1	7	2	8	6	17	18	13	16	9	10	11	1
	War	3	4	12	10	5	15	1	13	2	7	6	17	18	16	14	8	9	11	1
North American production	Peace	2	18	13	9	3	15	1	6	4	7	5	16	17	11	14	8	10	12	1
	War	2	13	11	9	3	18	1	6	5	7	4	16	17	15	14	8	10	12	1
U.S. production	Peace	2	16	14	17	3	11	1	4	15	5	7	13	12	10	18	6	9	8	2
	War	2	17	14	16	3	11	1	4	15	5	7	13	12	10	18	6	9	8	5
Availability in U.S. consumption	Peace	2	5	11	14	3	17	1	7	11	4	6	16	18	13	15	9	10	12	1
	War	2	14	10	13	3	17	1	4	11	5	6	12	18	16	15	8	7	9	1
Reliability of supply source	Peace	4	14	15	18	6	9	5	3	16	1	12	13	11	2	17	8	10	7	1
	War	4	18	16	14	6	8	5	3	15	1	9	13	7	2	17	12	10	11	2
Stockpile versus goal	War	17	10	9	12	16	3	2	1	8	4	18	11	5	6	13	14	7	15	5
Price	Peace	3	5	12	13	4	17	1	6	2	10	7	18	16	14	15	8	11	9	1
Recyclability	Peace	6	15	9	14	1	2	5	10	16	11	7	4	18	3	17	8	12	13	1
	War	6	12	11	15	1	2	5	10	16	9	7	4	18	3	17	8	13	14	2
Probability of mineral cartel	Peace	6	15	14	12	10	3	1	2	17	4	5	18	9	8	13	11	7	16	3
	War	6	17	15	12	10	3	1	2	16	4	5	18	9	8	13	11	7	14	5
Ranking point totals, normalized(a)	...	2.7	8.5	8.1	8.6	2.5	6.7	1.0	2.1	6.4	3.5	4.8	8.9	8.8	5.7	10	5.5	6.1	7.2	...
Weighted ranking point totals, normalized(a)	...	3.7	9.1	7.8	8.9	3.5	5.2	1.0	1.5	8.2	3.2	5.2	9.0	7.3	5.4	10	5.9	5.7	7.4	...

(a) Normalized on a 1 to 10 sliding scale; the lower the number, the better the ranking

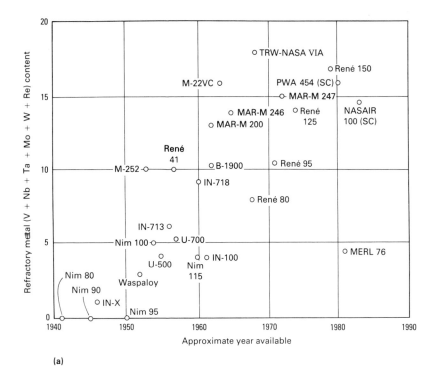

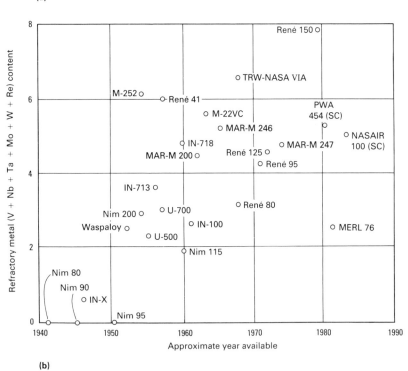

Fig. 2 Increased use of refractory metals in nickel-base superalloys. (a) Weight percent. (b) Atomic percent

came available in about 1960. In the early 1960s, tungsten and tantalum were widely accepted for alloying in nickel-base alloys. Finally, the demonstration of the effectiveness of rhenium additions in nickel-base alloys occurred in the late 1960s.

The original 20 wt% Cr level in superalloys was increased to 25 wt% or higher in some alloys to gain oxidation resistance, but because of its perceived deleterious effect on strength, chromium was reduced to as low as 10 wt% in favor of aluminum for oxidation protection (Ref 9). However, reducing chromium led to the onset of hot corrosion-enhanced oxidation resulting from sodium and sulfur in the fuel and exhaust gas stream. Ingestion of seawater spray into helicopter engines used in the Vietnam war wreaked havoc in low-chromium turbine blades, leading to a reevaluation of the use of chromium in superalloys.

The trend toward the increased use of refractory metals is shown in Fig. 2 and Table 5. It is apparent that, based on weight (Fig. 2a), the refractory metal content of nickel-base alloys tended to increase steadily from the mid-1940s to about 1980. Figure 2(b) shows that, based on atoms, the use of refractory metals increased from 1 to about 6 at.% in less than a decade. With this increasing use of refractory metals in superalloys, the concern for the availability of those determined to be strategic, that is, chromium, niobium, and tantalum, becomes apparent.

Cobalt is used in a variety of both cobalt-base and nickel-base superalloys. Its largest usage, in terms of pounds consumed, is in nickel-base alloys. Several nickel-base and cobalt-base superalloys are listed in Table 6, which shows the range of cobalt content in these alloys. It was the sharp rise in the cost of cobalt, more than any other factor, that brought on the need for the COSAM program. The cost of cobalt which sold for approximately $12.15/kg ($5.50/lb) in 1977, increased to over $66/kg ($30/lb) in 1979, with spot prices as high as $121/kg ($55/lb). As a rule-of-thumb, the price of cobalt has historically been higher than that of nickel by a factor of two to three. In 1980, that factor was in excess of seven. Primarily because of the spiraling cost of cobalt, the United States experienced a decline in cobalt use. In 1978, 9×10^6 kg (20×10^6 lb) of cobalt were consumed. In 1980, use was down to 7.3×10^6 kg (16×10^6 lb) (Ref 10). During this same time period, the use of cobalt to produce superalloys, primarily for aircraft engines, increased from 1.8×10^6 kg (4×10^6 lb) in 1978 to 3.3×10^6 kg (7.2×10^6 lb) in 1980. This proportionately high increase in cobalt use in superalloys can be attributed to the increased orders of aircraft during this time period. In 1988, of 6.8×10^6 kg (15×10^6 lb) consumed, 2.5×10^6 kg (5.6×10^6 lb) was used in superalloys. Because the well-being of the aircraft industry was

In the early development time period (about 1926), Heraeus Vacuumschmelze A.G. received a patent for a nickel-chromium alloy that contained up to 15 wt% W and 12 wt% Mo, thereby introducing the refractory metals into superalloy compositions. The purpose of adding refractory metals was to increase the high-temperature strength of the nickel-base alloy. By the 1930s there were two iron-base "heat-resisting alloys" containing either tungsten or molybdenum additions, and the use of these two metals was widely accepted in cobalt-base alloys. In the early 1950s, alloys containing about 5 wt% Mo were introduced in the United States. The commercial exploitation of Mo additions took place in 1955. A cast alloy containing 2 wt% Nb was available in the late 1950s. Only one commercially significant alloy uses vanadium; it be-

Table 5 Refractory metal content of selected nickel-base superalloys and year of availability

Alloy	Cr	Mo	W	Nb	Ta	Re	V	Year(a)
Nimonic 80A	20	0	0	0	0	0	0	1942
Waspaloy	19	4.4	0	0	0	0	0	1951
Nimonic 100	11	5	0	0	0	0	0	1953
M-252	20	4.0	0	0	0	0	0	1953
Inconel 713C	12	4.5	0	2.0	0	0	0	1956
Inconel 718	19	3	0	5.0	0	0	0	1960
TRW-NASA VIA	6.1	2	5.8	0.5	9	0	0.5	1968
René 150	5	1	5	0	6	2.2	3	1978
PWA 454	10	4	0	0	12	0	0	1980

(a) Approximate year of availability

Table 6 Cobalt content of typical superalloys

Alloy designation	Cobalt, nominal %
Cast	
HS-31	55
MAR-M 509	55
IN-100	15
B-1900	10
IN-738	8
Wrought	
L-605	55
S-816	45
HA-188	39
Udimet 700	17
Waspaloy	13
MAR-M 247	10
Powder metallurgy	
1056	19
Udimet 700	19
R-95	8

considered very important, the COSAM program, as well as industry activities, was undertaken.

Table 7 presents a list of several superalloys that have been used in gas turbine engines or that are emerging as replacements because of the promise of increased operating temperatures and higher efficiencies for the aircraft of the future. These alloys are used in a variety of forms, such as turbine blades, vanes, and disks; compressor components; and ducting components, and they serve a multitude of needs. The next section of this article focuses on the NASA COSAM program and some of its results.

COSAM Program Approach

The COSAM program had three general objectives: First, to contribute basic scientific information to the turbine engine "technology bank" in order to maintain U.S. national security in the event of constriction or interruption of strategic material supply lines; second, to help reduce the dependence of U.S. military and civilian gas turbine engines on disruptive, worldwide supply/price fluctuations in regard to strategic materials; and finally, by virtue of these research contributions, to help minimize the acquisition costs and optimize the performance of these engines in order to contribute to U.S. preeminence in world gas turbine markets.

To achieve these objectives, a three-pronged approach was undertaken, consisting of research on strategic element substitution, advanced processing concepts, and alternate materials (Ref 11). The intent was to achieve conservation, and reduced dependence on strategic metals by systematically examining the effects of replacing cobalt, niobium, and tantalum with less strategic elements in current, high-use engine alloys. This would help guide future material specifications if one or more of these metals were ever in short supply and would create a powerful base of understanding to benefit future advanced alloy development.

Conservation through advanced processing concepts research can be achieved by creating the means for using dual-alloy and multiple-alloy tailored structures that can minimize strategic material input requirements, by using them only where mandatory, and thus lowering total use.

In the long term, the development (higher risk) of readily available alternate materials to replace most strategic metals could lead to a dramatic reduction in U.S. dependence on foreign sources. The last two technologic areas could help conserve all four strategic metals: cobalt, tantalum, niobium, and chromium.

The various efforts of the COSAM program were conducted under the management of NASA Lewis Research Center, where some work was conducted in-house. In addition, there were cooperative programs between NASA Lewis, industry, and various universities in triparty projects to utilize the expertise of each and to produce synergistic results.

Typically, the research conducted at NASA involved oxidation/corrosion, low-cycle fatigue, thermal fatigue, and coatings; the research done by industry involved fabricability; and that done by universities involved mechanical properties, microstructures, and microchemistry. Varying levels of support flowed from one group to another, depending on the specific project. Highlights of the results obtained are described in the remainder of the article.

Table 7 Compositions of selected Fe-, Ni-, and Co-base superalloys

Alloy	Fe	Ni	Co	Cr	V	Nb	Ta	Mo	W	Re	Zr	Al	Ti	B	C	Hf
Fe-base alloys																
A-286	53	26	...	15	0.2	...	...	1.25	...	...	...	0.2	2.15	...	0.05	...
N-155	30	20	20	21	...	1	...	3	2.5	...	...	...	...	...	0.15	...
CG-27	38	38	...	13	...	0.6	...	5.5	...	...	...	1.5	2.5	0.01	0.05	...
Ni-base alloys																
Inconel 718	19	53	...	19	...	5.2	...	3	...	...	...	0.6	0.8	0.006	0.05	...
MAR-M 247	...	62	10	8.2	...	...	3	0.6	10	...	0.09	5.5	1.4	0.001	0.006	...
Udimet 700	...	53	19	15	...	...	...	5.2	...	...	...	4.3	3.5	0.03	0.08	...
CMSX-2	...	66	4.6	8	...	...	5.8	0.6	7.9	...	...	5.6	0.9	...	0.005	...
Inconel 713C	...	74	...	12	...	2	...	4.2	...	...	0.1	6.1	0.8	0.012	0.12	...
PWA 1480	...	63	5	10	...	...	12	...	4	...	...	5	1.5	...	...	...
Waspaloy	...	58	13	19	...	...	...	4.3	...	...	0.06	1.3	3	0.006	0.08	...
N-4	...	63	7.5	9.2	...	0.5	4	1.6	6	...	...	3.77	4.25	...	0.005	...
René 150	...	58	12	5	3	...	6	1	5	2.2	0.03	5.5	...	0.015	0.06	1.5
Co-base alloys																
HS-188	3	22	39	22	...	...	...	14	...	...	...	...	...	...	0.1	...
X-40	...	10	54	25	...	...	...	...	7.5	...	...	...	...	...	0.5	...

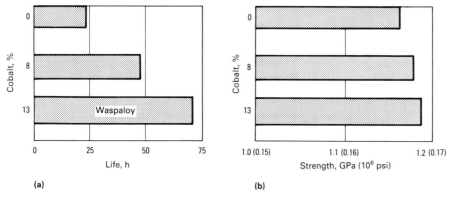

Fig. 3 Effect of cobalt content in Waspaloy on (a) rupture life at 730 °C (1345 °F) and 550 MPa (80 ksi) and (b) tensile strength at 535 °C (995 °F)

Table 8 Effects of removing cobalt from Waspaloy

Property or characteristic	Result
Hot workability	
Heating	No change
Cooling	Decrease
Tensile strength	Slight reduction
Tensile ductility	No change
Stress-rupture life	Major decrease
Creep rate	Sixfold increase
γ' solvus temperature	No change
γ' volume fraction	Slight decrease (18–16%)
γ' chemistry	Decrease of Cr and Ti and increase of Al
Carbides	
Chemistry	More metal carbide as-rolled; more $M_{23}C_6$ 843 °C (1550 °F) aging
Morphology	Coarser

Substitution

The reduction and/or replacement of cobalt, cobalt and tantalum, tantalum, and niobium in certain alloys is described below.

Cobalt in Waspaloy and Udimet 700. Waspaloy, which contains 13% Co, is used in turbine disks. Because of the size and weight of a turbine disk, a major portion of the cobalt used in gas turbine engine components is used in this application. Udimet 700, containing 17% Co, can be used for both disks and blades, depending on the processing history and the heat treatment used.

The effects of reducing cobalt in Waspaloy are reported in Ref 12. Highlights of that study are shown in Fig. 3. Tensile strength decreased only slightly as the amount of cobalt in the alloy decreased. However, rupture life decreased substantially with decreasing amounts of cobalt. Table 8 summarizes the major findings of this study. In addition to the slight decrease in the amount of γ' (Ni_3Al, the major strengthening phase in nickel-base superalloys) in Waspaloy, the major effects that removing cobalt had on mechanical properties were attributed to a possible higher stacking fault energy of the matrix and to changes in carbide partitioning in grain boundaries.

Barrett (Ref 13) examined the effect of cobalt on the oxidation resistance of Waspaloy. Results, shown in Fig. 4, indicate that, based on specific weight change data up to 1100 °C (2010 °F), cyclic oxidation resistance is essentially independent of cobalt content.

A further study of the reduced cobalt composition Waspaloy alloys was conducted at Purdue University by Durako (Ref 14). This investigation focused on the microstructure of the alloys and on metallographic studies of extracted γ' and carbide precipitates. The effect of the removal of cobalt in Waspaloy on mechanical properties was attributed by Durako to be due in

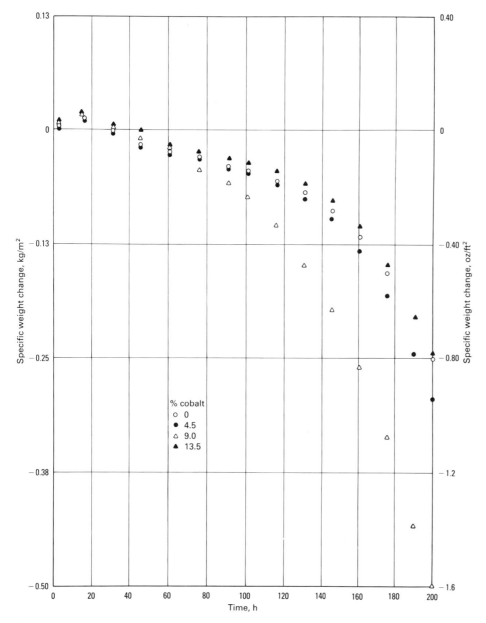

Fig. 4 Effect of cobalt on cyclic oxidation resistance of Waspaloy at 1100 °C (2010 °F) and 1 h/cycle

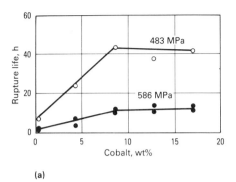
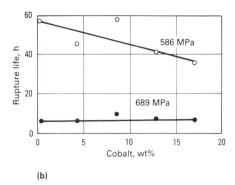

Fig. 5 Stress-rupture life of Udimet 700 at 760 °C (1400 °F) versus cobalt content. (a) Disk heat treatment. (b) Blade heat treatment

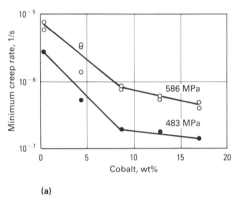
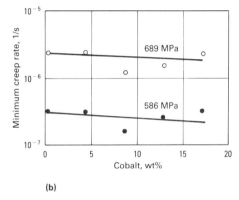

Fig. 6 Creep rates of Udimet 700 at 760 °C (1400 °F) as a function of cobalt content. (a) Disk heat treatment. (b) Blade heat treatment

part to the decrease in volume percent γ′, in agreement with Maurer (Ref 12); to the reduction in γ/γ′ mismatch, with increasing dislocation mobility; and to an indirect increase in the matrix stacking fault energy resulting from matrix chromium depletion caused by the formation of massive $M_{23}C_6$ chromium-rich carbides. Both Durako and Maurer suggested that alloy modifications might allow the reduction or removal of cobalt from Waspaloy.

The effects of the removal of cobalt from wrought Udimet 700 alloy were studied extensively as part of a cooperative program involving Special Metals Corporation, Columbia and Purdue Universities, and NASA Lewis Research Center. Fabricability was investigated by Jackman and Maurer (Ref 12) and Sczerenie and Maurer (Ref 15) of Special Metals Corporation, while mechanical and metallurgical properties were studied by Jarrett and Tien (Ref 16) of Columbia University. Fabricability, based on Gleeble and high strain rate tensile tests corresponding to rolling temperatures in the 1000 to 1100 °C (1830 to 2010 °F) range show no cobalt effect on high-temperature ductilities. Of particular interest is the work described in Ref 16 on the effect of the disk (partial γ′ solutioning) and blade (complete γ′ solutioning) heat treatments on stress-rupture and creep properties. Rupture life as a function of cobalt content is shown in Fig. 5 for the two heat-treated Udimet 700 conditions. The disk heat treatment resulted in a reduction in rupture life when the cobalt content was less than 9%. In the blade heat-treated condition, specimens exhibited an increase in rupture life with decreasing cobalt content at the lower stress level and were insensitive to cobalt content at a higher stress level. Creep rates, as expected, showed similar trends with cobalt content and heat treatment (Fig. 6). The results of the work of Jarrett and Tien (Ref 16) are summarized below.

Room-temperature tensile yield strength and tensile strength were only slightly decreased in the disk alloys and were basically unaffected in the blade alloys as cobalt was removed. Creep and stress-rupture resistance at 760 °C (1400 °F) were found to be unaffected by the cobalt level in the blade alloys and decreased sharply only when the cobalt level was reduced below about 9% in the disk alloys.

The microstructure was found to be very heat-treatment sensitive. After the fine grain, disk heat treatment, the fine strengthening γ′ precipitates fraction decreased as cobalt was removed because of a corresponding increase in undissolved γ′ fraction. No such change occurred after the higher-temperature, coarse grain heat-treatment, during which all γ′ particles were initially dissolved.

Cobalt was observed through scanning transmission electron microscopy and energy dispersive spectroscopy to partition mostly to the γ matrix phase. Cobalt also changed the relative stability of the various carbides and destabilized rather than stabilized the alloy with respect to σ-phase formation after long-term aging. It did not significantly alter γ′ coarsening kinetics.

The correlation of the detailed microstructural and microchemistry information with yield strength and creep rate formulas especially developed for particle-strengthened systems showed that the slight decrease in yield strength was due to γ′ fraction and antiphase boundary energy considerations. The significant drop in creep and stress-rupture resistance in the low-cobalt and cobalt-free disk alloys is due to a change in the fine γ′ volume fraction and is relatively unaffected by matrix composition or stacking fault energy factors.

Harf (Ref 17) of NASA Lewis conducted a parallel program on hot isostatic pressed (HIP) powder metallurgy (P/M) Udimet 700. Initial results confirmed the previous results of Ref 16 on the cast-plus-wrought (CW) material. Harf then focused on modifying the disk heat treatment to improve the creep-rupture properties of the 0% Co alloy (Ref 18).

In the original concept of comparing the properties of Udimet 700 alloys with decreased cobalt levels, the comparison was made with minimal change in the heat treatment of the various compositions. A major compromise in the disk-type heat treatments had been to adjust the partial solutioning temperature to maintain a nearly constant temperature difference from the γ′ solvus, particularly in the HIP-P/M alloys. However, because the γ′ solvus increased with decreasing cobalt content, this meant that the thermal gap between the partial solutioning temperature and the subsequent aging temperatures (which were the same for all cobalt contents) increased with decreasing cobalt content.

Harf (Ref 18) modified the aging temperatures to keep the thermal gap for the 0% Co alloy similar to that used for the 17% Co Udimet 700 alloy. His results showed that this technique was successful in improving the rupture life and creep resistance of the 0% Co alloy at 650 °C (1200 °F). He attributed this improvement to the microstructure, which contained an increased quantity of ultra fine, 20 nm (200 Å) γ′ particles, as shown in Fig. 7. The standard heat treatment (Fig. 7a) has fewer particles than the modified heat treatment (Fig. 7b). In the CW alloy, too, substantial improvements in creep-rupture properties were observed at 760 °C (1400 °F) as a result of a similar modification in heat treatment for the 0% Co alloy.

Barrett (Ref 13) has also investigated the cyclic oxidation resistance of the low- and/or no-cobalt Udimet 700 alloys. Results of this study are shown in Fig. 8. At 1100 °C

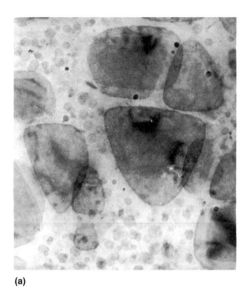
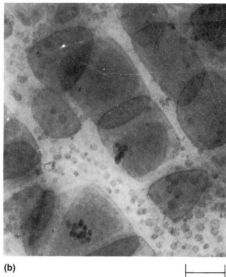
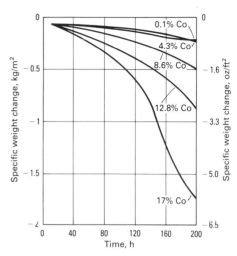

Fig. 7 Transmission electron micrographs comparing ultrafine particles in Udimet 700-type alloys with 0% cobalt content. (a) Standard heat treatment. (b) Modified heat treatment

Fig. 8 Effect of cobalt on cyclic oxidation resistance of Udimet 700 at 1100 °C (2010 °F) and 1 h/cycle

(2010 °F), the removal of cobalt from Udimet 700 improved the cyclic oxidation resistance based on specific weight change data. Hot corrosion resistance of the low- and/or no-cobalt Udimet 700 alloys was also investigated. Results by Deadmore (Ref 19) from tests using NaCl-doped flames in a Mach 0.3 burner rig indicate that corrosion resistance increases with decreasing cobalt content. Photographs of exposed specimens are shown in Fig. 9, where the improved corrosion resistance for the lower cobalt concentrations is evident.

In contrast, Zaplatynsky (Ref 20) found that the alloys with an aluminide coating exhibited improved oxidation resistance with increasing cobalt content based on a weight loss criteria during testing in the Mach 0.3 burner rig. Leis et al. (Ref 21), of the Battelle-Columbus Laboratories, investigated the creep fatigue behavior of low-cobalt P/M and CW Udimet 700 alloys and saw no correlation between fatigue resistance and cobalt content. It is concluded that an alloy based on Udimet 700 in which nickel has been substituted for all the cobalt is a viable superalloy for use in turbine applications. This statement applies to both the cast plus wrought and the hot isostatically pressed prealloyed powder processed alloys. Jarrett et al. (Ref 16) had previously reported that the alloy, when given a different heat treatment, might also qualify for use in turbine blades. It is suggested that this alloy be considered for future use in aerospace and land-based turbine applications.

Cobalt and Tantalum in MAR-M 247. MAR-M 247, an advanced nickel-base superalloy used in polycrystalline, directionally solidified, and single-crystal form, contains nominally 10 wt% Co and 3 wt% Ta. The effects of removing cobalt from MAR-M 247 have been investigated as part of a cooperative program involving TRW Inc., Teledyne CAE, Case Western Reserve University, and NASA Lewis. The potential industrial application was related to an integral, cast rotor. Therefore, the casting mold and pouring temperatures were selected by Teledyne CAE to simulate blade and hub conditions. Major findings by McLaughlin (Ref 22) of Teledyne and Kortovich (Ref 23) of TRW are summarized in Table 9. A parallel in-depth study on the effects of cobalt on the mechanical properties of MAR-M 247 was undertaken by Nathal (Ref 24). This study explored the mechanisms associated with the effects of cobalt content on the mechanical properties of polycrystalline materials. Nathal postulated that a reduction in γ' weight fraction and carbide formation as a grain-boundary film were responsible for the deleterious effects on creep-rupture properties. It was proposed that reducing the carbon level in the 5% Co alloy may result in an alloy with properties comparable to MAR-M 247, while conserving 50% of the cobalt normally used in this alloy. Nathal also showed that, based on weight change data, the removal of cobalt from MAR-M 247 improves the cyclic oxidation resistance of this alloy at 1100 °C (2010 °F). As in Udimet 700 testing, hot corrosion testing of alloys based on MAR-M 247 chemistry revealed that reducing cobalt also improved corrosion resistance (Ref 19).

Nathal et al. (Ref 25) further showed that in single-crystal form, removing cobalt from MAR-M 247 appears to increase rupture life and decrease creep rate, trends opposite to those observed for the polycrystalline material. The single-crystal findings by Nathal supported previous results reported by Strangman et al. (Ref 26) in which 0% Co levels in single-crystal alloys had longer rupture lives than the 10% cobalt MAR-M 247 single crystals. However, a 5% Co level was required for alloy stability with respect to the formation of the μ phase, a topologically close-packed, hard compound.

Nathal and Ebert (Ref 27) studied the influence of composition on the tensile and

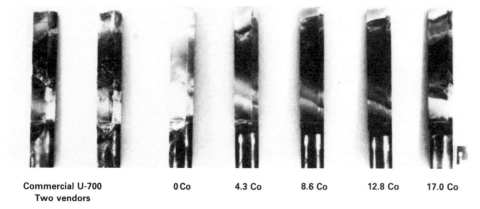

Fig. 9 Effect of cobalt on hot corrosion; modified Udimet 700, 170 1-h cycles at 900 °C (1650 °F), with 0.5 ppm by weight Na as NaCl, Mach 0.5

Table 9 Effects of removing cobalt from MAR-M 247 blade and hub

Property or characteristic item	Result	
	Blade	Hub
Yield strength	Slight decrease	Slight decrease
Ultimate tensile strength	Decrease	Decrease
Tensile ductility	Decrease	Slight decrease
Stress-rupture life	Decrease	Decrease
Oxidation resistance	No change	No change
Thermal shock	No change	No change
Fracture mode		
Tensile	From transcolony	To intercolony
Stress rupture	From transcolony	To intercolony

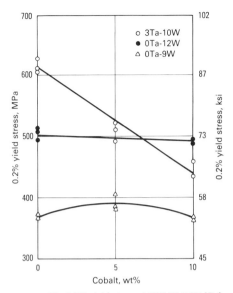

Fig. 10 The 0.2% yield stress at 1000 °C (1830 °F) for single-crystal MAR-M 247 alloys

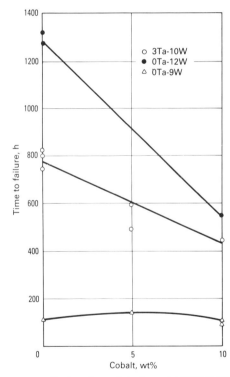

Fig. 11 Rupture lives of single-crystal MAR-M 247 alloys at 1000 °C (1830 °F) and 148 MPa (22 ksi)

creep strength of [001]-oriented nickel-base superalloy single crystals at temperatures near 1000 °C (1830 °F). Cobalt, tantalum, and tungsten concentrations were varied according to a matrix of compositions based on the single-crystal version of MAR-M 247. For alloys with the baseline refractory metal level of 3 wt% Ta and 10 wt% W, decreasing the cobalt level from 10 to 0 wt% resulted in increased tensile and creep strength. Substitution of 2 wt% W for 3 wt% Ta resulted in decreased creep life at high stresses, but improved creep life at low stresses. Substitution of nickel for tantalum caused large reductions in tensile strength and creep resistance and corresponding increases in ductility. For these alloys with low tantalum plus tungsten totals, strength was independent of the cobalt level. Figures 10 and 11 show the yield stress and creep-rupture properties of the reduced-cobalt and reduced-tantalum alloys. The results of the extensive studies of Nathal and Ebert on microstructure and mechanical properties are summarized in the following paragraphs.

The removal of tantalum and tungsten from the baseline 3Ta-10W alloys to form the 0Ta-9W alloys caused large reductions in γ' solvus temperature and γ' volume fraction. The substitution of tungsten for tantalum to form the 0Ta-12W alloys resulted in intermediate reductions in solvus temperature and volume fraction. The amount of γ' was independent of the cobalt level, although the γ' solvus temperature increased significantly as the cobalt content was reduced from 10 to 0%. The partitioning of elements between the γ and γ' phases did not vary appreciably as the alloy composition varied. Tantalum and titanium partitioned almost totally to γ'; aluminum and tungsten partitioned preferentially to γ'; and cobalt, chromium, and molybdenum partitioned preferentially to γ.

The γ' lattice parameter was independent of the cobalt content, but increased as the total refractory metal level increased. At the 0% Co level, the 3Ta-10W alloy exhibited a room-temperature lattice mismatch, where the difference was −0.0037 and the 0Ta-12W alloy exhibited a difference of −0.002. The 0Ta-9W alloys and all alloys with 5 and 10% Co possessed mismatch values below the detection limit.

For the alloys with γ' that remained coherent during aging, the unstressed γ' coarsening rate increased as the cobalt level was reduced from 10 to 0%. The alloys with a high lattice mismatch possessed a γ' that became semicoherent during aging and exhibited anomalously low coarsening rates.

Oriented γ' coarsening, which resulted in lamellae perpendicular to the applied stress, was very prominent during creep. Alloys with higher magnitudes of lattice mismatch exhibited faster directional coarsening rates and a finer spacing of misfit dislocations at the γ/γ' interfaces.

The substitution of nickel for cobalt caused large increases in creep resistance for alloys with high tantalum plus tungsten totals. This was consistent with an increase in γ/γ' lattice mismatch. High values of lattice mismatch resulted in a finer dislocation network at the γ/γ' interface, providing a more effective barrier for dislocation motion. The substitution of nickel for tantalum and tungsten to form the 0Ta-9W alloys caused large reductions in creep resistance, which were related to the decreases in γ' volume fraction, γ/γ' mismatch, and solid-solution hardening. The substitution of tungsten for tantalum to form the 0Ta-12W alloys resulted in a decrease in creep resistance at high stresses and an increase in creep strength at low stresses. This crossover in creep resistance between the 3Ta-10W and 0Ta-12W alloys was not easily explained. The decreased creep life of the 0Ta-12W alloys at high stresses was attributed to the slight decreases in γ' volume fraction and γ/γ' mismatch, although the reason that tungsten appears to be a more effective solid-solution strengthener at low stresses remains unclear.

Decreases in the cobalt level from 10 to 0% caused significant increases in the 1000 °C (1830 °F) yield and ultimate tensile strengths of the 3Ta-10W alloys, but cobalt had much less effect on alloys with other refractory metals contents. The influence of cobalt on the strength of the 3Ta-10W alloys was attributed to the coherency strain hardening associated with the increased lattice mismatch as the cobalt level decreased. The reduction of tantalum and tungsten content to form the 0Ta-9W alloys caused large reductions in tensile strength, and the substitution of tungsten for tantalum caused intermediate decreases in strength. These changes in tensile strength with refractory metal levels were related to the increases in γ' volume fraction and solid-solution hardening, which resulted from high tantalum plus tungsten totals.

Based on these various studies, it appears that reducing the cobalt content by 50% in MAR-M 247 may be feasible in the event of a cobalt shortage and tungsten may be used

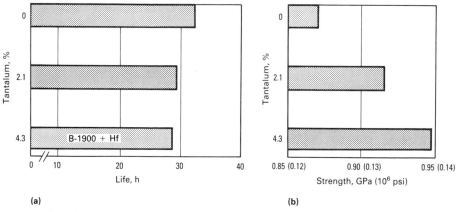

Fig. 12 Effect of tantalum on mechanical properties of B-1900+Hf at 760 °C (1400 °F). (a) Rupture life at 648 MPa (94 ksi). (b) Tensile strength

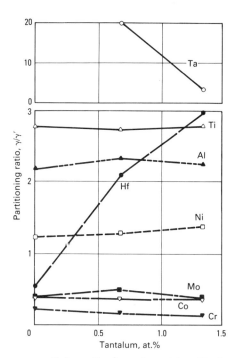

Fig. 13 Various γ'/γ phase elemental partitioning ratios for B-1900+Hf alloys as a function of tantalum content

as a substitute for a significant amount of tantalum in this alloy.

Tantalum in B-1900+Hf. The alloy B-1900+Hf containing 10 wt% Co and 4 wt% Ta is a high-strength, nickel-base superalloy used in turbine blade applications. Kortovich (Ref 28) conducted some independent studies of the role of tantalum on mechanical properties and microstructure of B-1900+Hf. His results indicated that tensile strength decreased with decreasing tantalum content upon testing at room temperature and at 760 °C (1400 °F). Stress-rupture testing at 760 °C/650 MPa (1400 °F/95 ksi) indicated that the rupture life increased with decreasing tantalum content, while at 980 °C/200 MPa (1795 °F/30 ksi), rupture exhibited a maximum at a 50% reduction in the normal tantalum content. TRW stress-rupture results and tensile results at 760 °C (1400 °F) are shown in Fig. 12.

Janowski (Ref 29) studied the microstructure of B-1900+Hf alloys tested by Kortovich. Figure 13 shows the γ'/γ partitioning ratios for the various tantalum levels. The partitioning ratios of cobalt, chromium, and nickel are constant with tantalum variations, consistent with MAR-M 247 results. In addition, the partitioning ratios for aluminum, molybdenum, and titanium are also independent of tantalum content. However, the hafnium and tantalum partitioning ratios are very sensitive to tantalum content; the partitioning ratios of hafnium and tantalum decrease and increase, respectively, as tantalum is removed from the alloy. The change in the hafnium distribution was postulated to be a consequence of the increase in available hafnium in the matrix phases as a result of the replacement of part of the hafnium contained in the metal carbides (MCs) by tantalum. Further work will be required to determine whether an alloy with reduced-tantalum content will be a viable substitute for B-1900+Hf.

Niobium in Inconel 718. Inconel 718 is used for disks in gas turbine engines and, by weight, is the most widely used superalloy. Ziegler and Wallace (Ref 30) explored a series of Inconel 718 alloys with reduced niobium contents. The alloys had niobium contents of 5.1 (Inconel 718 composition), 3.0, and 1.1 wt%. Substitutions of 3.0% W, 3.0% W plus 0.95% V, and 3 to 5.8% Mo were investigated. Two additional alloys, one containing 3.49% Nb plus 1.10% Ti and a second containing 3.89% Nb and 1.27% Ti, were also studied. Additions of solid-solution elements to a reduced niobium alloy had no significant effect on the properties of the alloys under either process condition. The solution and aged alloys with substitutions of 1.27% Ti at 3.89% Nb had tensile properties similar to those of the original alloy and stress-rupture properties superior to the original alloy. The improved stress-rupture properties were the result of significant precipitation in the alloy, of Ni_3Ti-γ', which is more stable than γ'' at the elevated temperatures. At lower temperatures this modified alloy benefits from the γ'' strengthening. Much more information is needed to characterize the modified alloy fully, but these initial results suggest that with more precise control and proper processing, the reduced niobium direct-age alloy could substitute for Inconel 718 in high-strength applications.

Advanced Processing

Advanced processing, that is, tailored fabrication and dual-alloy concepts, were to be developed under the COSAM program initially, but very little work in these areas actually occurred because of funding limitations and the focus on substitution. However, industry has independently investigated dual-alloy concepts to improve properties.

A turbine disk is a good candidate for the dual-alloy concept because of its weight and operating characteristics. The rim of a disk operates at higher temperatures than the hub of the disk and requires a nickel-base superalloy for creep resistance. The hub of a disk operates at lower temperatures where fatigue resistance is important, and thus an iron-base superalloy with lower strategic metal content may suffice. The key to this dual-alloy concept is the interface between two alloys of widely different compositions.

Harf (Ref 31) has studied the interface properties of alloy 901 combined with three nickel-base superalloys. His preliminary results showed that the alloys could be joined by the HIP-P/M process and could form a strong, integral bond. Stress-rupture properties of the alloy combinations, with heat treatments for each of the alloys, are shown in Fig. 14. Failure always occurred in alloy 901, indicating the superior strength of the bond. In addition, the rupture lives of the combinations were always equal to or greater than that for alloy 901 alone. These results imply that a dual-alloy concept is viable for enhancing the local properties of gas turbine components and for conserving costly strategic materials. Dual-alloy disks are being investigated for current engines. However, the goal is to achieve optimum properties, not to conserve strategic materials.

Alternate Materials

The third major COSAM program thrust, alternate materials, has the potential of making a major contribution to the conservation of the four strategic elements studied, including chromium, which is critical to currently used superalloys because of the oxidation/corrosion resistance it provides. The materials being investigated in this part of the program are described below.

Intermetallic Compounds. This effort emphasizes a basic research approach toward

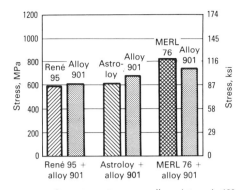

Fig. 14 Stress-to-rupture superalloy mixtures in 100 h at 650 °C (1200 °F); heat treatment sequence

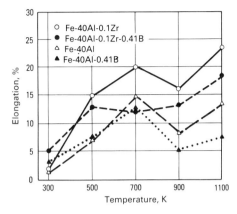

Fig. 15 Ductility of Fe-40Al alloys

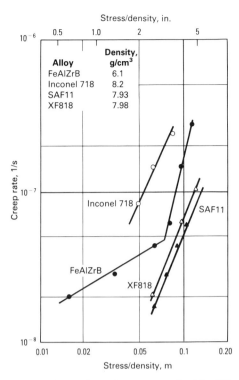

Fig. 16 Comparison of creep strength of Fe-40Al-0.1Zr-0.41B alloy and commercial iron-nickel-base superalloys

understanding the deformation mechanisms that control high-temperature creep, as well as those that control the lack of room-temperature ductility. By necessity, this is a long-term, high-risk effort, but it offers the potential of a high payoff if materials are evolved that permit the conservation of all four currently identified strategic metals.

A research program that focused on the equiatomic iron and nickel aluminides (FeAl and NiAl) as potential alternatives to nickel-base superalloys was conducted in-house at NASA Lewis Research Center and at Dartmouth College, Stanford University, Texas A & M University, and Case Western Reserve University (Ref 32). These binary aluminides have a number of advantages: They exist over a wide range of compositions and have a large solubility for substitutional third-element additions; they have a cubic crystal structure; they have very high melting points (except for FeAl, which has a somewhat lower melting point); they contain inexpensive, readily available elements; and they possess the potential for self-protection in oxidizing environments. Their chief disadvantage is the lack of room-temperature ductility. Cobalt aluminide has been used for comparative purposes because it has unusual mechanical properties and phase equilibria.

Some of the highlights of FeAl are shown in Fig. 15 and 16 (Ref 33, 34). Figure 15 shows the room-temperature tensile elongation of Fe-40Al to be about 1 or 2%. However, 5 to 6% elongation was achieved with the addition of zirconium and boron. The fracture for alloys without boron was intergranular, while alloys containing boron exhibited transgranular fracture. The creep strength of the Fe-40Al-0.1Zr-0.41B alloy is compared with several commercial iron-nickel-base superalloys in Fig. 16. Results indicate that this alloy has promising properties at elevated temperatures.

Schulson has described the effects of grain size on the tensile ductility of NiAl (Ref 35). His work was based on the models of Cottrell (Ref 36) and Petch (Ref 37), which stated that the stress required to nucleate microcracks in coarse-grained materials is more than enough to propagate them. In contrast, the stress required to nucleate cracks in fine-grain materials is less than that required to propagate them. In the first case, coarse-grain materials will fail in a brittle manner; in the second, fine-grained material must undergo permanent deformation and work hardening prior to failure. The conclusion is, therefore, that a critical grain size should exist for plastic flow. This does indeed appear to be the case for Ni-49Al at 295 °C (565 °F), as illustrated by the results shown in Fig. 17, where the critical grain size is ~20 μm (~785 μin.). At lower temperatures, the critical grain size becomes even smaller.

Vedula *et al.* (Ref 38) have shown that tantalum-rich precipitates in a Ni-48Al-2Ta alloy can produce an increase in creep strength of nearly two orders of magnitude over the simple, equiatomic, binary NiAl. Dislocations interacting with the precipitates during creep testing are shown in Fig. 18. This alloy is comparable to some nickel-base superalloys at 1025 °C (1875 °F). Inter-

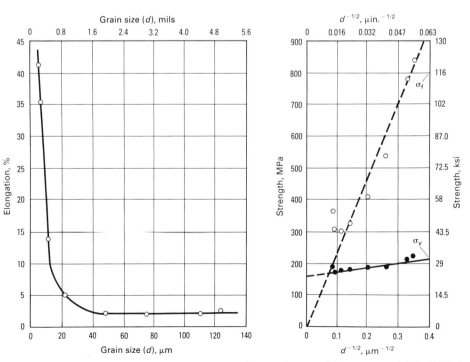

Fig. 17 Effect of grain size (d) on tensile elongation, yield strength (σ_y), and fracture strength (σ_f) of NiAl at 400 °C (750 °F)

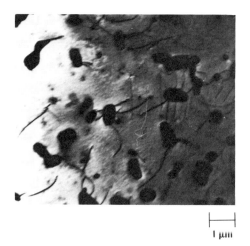

Fig. 18 Transmission electron micrograph of NiAl-2 at.% Ta specimen deformed in hot compression at 1300 K to about 7% strain

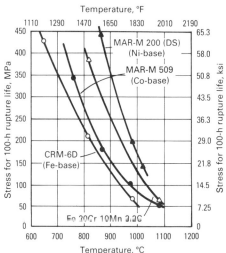

Fig. 19 Stress-rupture potential of Fe-20Cr-10Mn-3.2C alloy

metallic compounds are currently under extensive investigation by NASA, the U.S. Department of Defense, the U.S. Department of Energy, industry, and universities, both as monolithic materials and as matrices for composites.

Iron-Base Alloys. With the successful development of high-strength nickel-base superalloys (and to some extent cobalt-base superalloys) over the last 30 years, there has been little recent interest in developing iron-base alloys for the higher-temperature gas turbine engine components. However, with the threat of strategic material supply disruptions or interruptions, iron-base alloys with low strategic metal contents are now attractive as alternative materials for U.S. industrial consideration.

A program was initiated to investigate iron-base superalloys with aligned carbides for further strengthening as potential alternatives to current high strategic element content nickel- and cobalt-base superalloys. This was a joint program involving the University of Connecticut, United Technologies Research Center, and NASA Lewis.

The potential of these iron-chromium-manganese-aluminum alloys is shown in Fig. 19, where rupture lives determined by Lemkey and Bailey (Ref 39) compare favorably with other iron-, nickel-, and cobalt-base alloys. Some of these alloys are leading candidates for Stirling engine applications.

Composites. A third area of alternate-materials technology conducted in-house at NASA Lewis sought to determine the potential of silicon carbide reinforced, low strategic element content, iron-base matrix composites. The program focused on understanding matrix/fiber interface compatibility in the service range from 760 to 900 °C (1400 to 1650 °F) for turbine engine components (Ref 40). This concept offers the potential not only of conserving strategic materials, but also of either reducing component weight because of the potential strength of the fibers and their high volume fraction or of maintaining weight and extending service life. The results of this program can be summarized as follows: First, a low-temperature fabrication process, that is, hollow-cathode sputtering, can be used successfully to produce single-filament composites from B_4C-B and SiC filaments and iron-base matrix alloys while retaining high fractions of the filament strength. No evidence of filament/matrix reaction due to processing was observed. Second, single-fiber composites of B_4C-B and SiC filament-reinforced iron-base alloys have stress-rupture strengths at 870 °C (1600 °F) that are superior to those of the strongest superalloys. The 1000-h rupture strength projected for a 50 vol% B_4C-B filament reinforced iron-base alloy composite at 870 °C (1600 °F) is 455 MPa (66 ksi), which represents a 30% increase in strength over a single-crystal CMSX-2. Much more impressive, however, is the 1000-h rupture strength-to-density ratio at 870 °C (1600 °F) projected for the 50 vol% fiber content B_4C-B iron-base alloy composite. This is twice that of CMSX-2, as shown in Fig. 20. Finally, the potential for B_4C-B and SiC filament reinforced iron-base alloy composites for use at intermediate temperatures appears promising if similar processing technologies can be developed for multi-ply composites. The successful result for the development of this technology for turbine components would be the reduction of turbine blade weight. A 40% weight reduction could be obtained in a composite blade using this material, as well as increased blade life. In addition, with a 50 vol% fiber content, a significant savings in strategic materials would be achieved, compared to nickel-base superalloys.

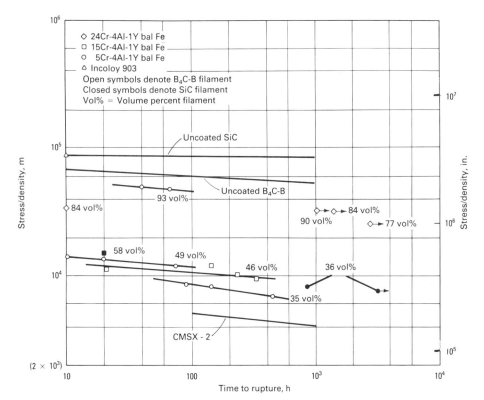

Fig. 20 Ratio of stress rupture to density for uncoated and coated filaments compared to that of CMSX-2 at 870 °C (1600 °F)

COSAM Program Results

The COSAM program began in 1980 and continued through 1983. Because of the drop in the cost of cobalt, niobium, and tantalum, and the lessening of any immedi-

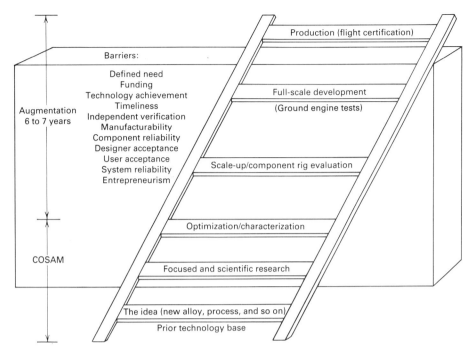

Fig. 21 Technology transfer ladder for strategic material substitution

ate threat of a strategic materials shortage, funding for the program was terminated in 1984. However, research initiated under the COSAM program continues at this writing (funded under other NASA programs) because of the technical promise shown in certain areas, especially that of alternate materials.

Comparing the results that were achieved in the short time that the COSAM program existed with the initial objectives stated earlier in the section "COSAM Program Approach," it is evident that a significant technology bank has been established in the area of a substitution for strategic materials in nickel-base superalloys. This is especially true for cobalt, where knowledge of the physical metallurgy of low- and no-cobalt alloys was characterized in terms of composition, microstructure, heat treatment, and thermomechanical processing. Mechanical and environmental properties were described in detail. A similar conclusion for tantalum substitution can also be reached, while niobium substitution was much less clearly defined.

The transfer of this technology to flight hardware must overcome significant barriers between laboratory feasibility and deployment, as discussed by Stephens and Tien (Ref 41). A commitment of both time and money is required to overcome these barriers in the course of the acquisition process, as shown in Fig. 21. Experience in the aircraft engine industry has shown that a 10-year period of development for a new material or concept to reach application, with a minimal expenditure of $1 million per year, is not uncommon. Because a strategic materials shortage is currently not a threat to aircraft engine producers, such a commitment to move one or more of the lower strategic metal content alloys into flight operation will not in all probability be forthcoming.

Because the COSAM program was not carried to fruition, it cannot be claimed that a reduced dependence on potentially unreliable foreign sources was achieved. However, the on-the-shelf knowledge that was gained may contribute to a reduced impact of future supply or cost-derived disruptions.

The objective of reducing costs and optimizing the performance of aircraft engines may well be realized in future advanced aircraft for which lightweight, high-temperature materials are of utmost importance. The continuing alternate materials research on the aluminides and metal matrix and, more recently, intermetallic matrix composites may well lead to improved engines for such vehicles as hypersonic, supersonic, and subsonic transport aircraft. Because of the technical requirements of such aircraft, materials with low or no strategic metal content that have the potential for meeting these needs may overcome the transfer barriers from research to flight hardware if their advantages can be realized and their disadvantages overcome. In addition, potential terrestrial applications for the heater head of experimental automotive Stirling engines (Ref 42) represent a spin-off of research on the iron-base alloys.

Appendix: Manganese Availability

As discussed previously under the section "Strategic Materials" in this article, manganese falls under the first category, that of most strategic, of the three categories of the 18 metals examined by Stalker et al. (Ref 6). Manganese is essential for the production of virtually all steels, and steelmaking constitutes its principal use. It is primarily used to counteract the adverse effects of sulfur and prevent hot shortness or brittleness, which otherwise would develop in steels that contain sulfur. In addition, manganese is used as a deoxidizer to inhibit the formation of grain-boundary carbides, and to impart increased strength, toughness, hardness, and hardenability to steels. Manganese is also used for similar reasons in the production of iron castings. It is used as an alloying element in aluminum, magnesium, and copper. About 90% of the U.S. consumption is for the production of iron and steel. In nonmetallurgical applications, manganese is used in the production of dry cell batteries and as an oxidant or pigmentation agent in various chemicals, paints, and pharmaceutical products.

Strategic Nature. The reliance of the United States on the importation of manganese has been 100% for a number of years. Imports in the form of manganese ore over the 1983 to 1986 time period have been primarily from Gabon (43%) and Brazil (24%). Ferromanganese also constitutes a major source of manganese to the United States and has been imported from South Africa (33%) and France (29%). During the 1983 to 1986 period, the ratio of manganese (imported as ferromanganese) to ore was approximately 1/6. Because the cost of manganese has remained relatively low during the 1980s, a suitable substitute for steelmaking at a similar cost has not been identified. Because most steels have a low manganese content, recycling specifically for manganese recovery is insignificant. Based on these factors, stockpiling seems to be the best hedge against supply disruption for the United States.

REFERENCES

1. J.R. Stephens, in *Proceedings, U.S. Department of Commerce Public Workshop on Critical Materials Needs in the Aerospace Industry*, NBSIR81-2305, U.S. Department of Commerce, 1981, p T20-11 to T20-33

2. D.A. Brobst and P.P. Walden, Professional Paper 820, United States Mineral Resources, U.S. Geological Survey, 1973
3. Circular 831, U.S. Geological Survey, 1980
4. "Mineral Commodity Summaries 1989," Bureau of Mines, U.S. Department of Interior, 1989
5. "Mineral Commodity Summaries 1986," Bureau of Mines, U.S. Department of Interior, 1986
6. K.W. Stalker, C.C. Clark, J.A. Ford, F.M. Richmond, and J.R. Stephens, *Met. Prog.*, Vol 126 (No. 10), 1984, p 55-65
7. J.R. Stephens, NASA TM-82662, National Aeronautics and Space Administration, 1981
8. J.R. Stephens, R.L. Dreshfield, and M.V. Nathal, in *Refractory Alloying Elements in Superalloys*, J.K. Tien and S. Reichman, Ed., American Society for Metals, 1984, p 31-42
9. C.T. Sims, in *The Superalloys*, C. Sims and W. Hagel, Ed., Wiley Interscience, 1972, p 145-174
10. "Mineral Industry Surveys, Cobalt in 1978, 1979, 1980 and 1988," Bureau of Mines, U.S. Department of Interior
11. J.R. Stephens, NASA TM-82852, National Aeronautics and Space Administration, 1982
12. G.E. Maurer, L.A. Jackman, and J.A. Domingoe, in *Superalloys 1980*, J. Tien, Ed., American Society for Metals, 1980, p 43-52
13. C.A. Barrett, in "COSAM (Conservation of Strategic Aerospace Materials) Program Overview," NASA TM-83006, National Aeronautics and Space Administration, 1982, p 89-94
14. W.J. Durako, Jr., in *Senior Project Reviews*, Purdue University, 1981
15. F.E. Sczerenie and G.E. Maurer, in "COSAM (Conservation of Strategic Aerospace Materials) Program Overview," NASA TM-83006, National Aeronautics and Space Administration, 1982, p 21-35
16. R.N. Jarrett and J.K. Tien, *Metall. Trans. A*, Vol 13, p 1021-1032
17. F.H. Harf, *Metall. Trans. A*, Vol 16, 1985, p 993-1003
18. F.H. Harf, *J. Mater. Sci.*, Vol 21, 1986, p 2497-2508
19. D.L. Deadmore, NASA TP-2338, National Aeronautics and Space Administration, 1984
20. I. Zaplatynsky, NASA TM-87173, National Aeronautics and Space Administration, 1985
21. B.N. Leis, R. Rungta, and T.A. Hopper, NASA CR-168260, National Aeronautics and Space Administration, 1983
22. J. McLaughlin, Paper presented at the Cobalt Substitution Workshop, NASA Lewis Research Center, National Aeronautics and Space Administration, 1981
23. C.S. Kortovich, Paper presented at the Cobalt Substitution Workshop, NASA Lewis Research Center, National Aeronautics and Space Administration, 1981
24. M.V. Nathal, NASA CR-165384, National Aeronautics and Space Administration, 1981
25. M.V. Nathal, R.D. Maier, and L.J. Ebert, *Metall. Trans. A*, Vol 13, 1982, p 1767-1783
26. J.E. Strangman, G.S. Hoppin III, C.M. Phipps, K. Harris, and R.E. Schwer, in *Superalloys 1980*, J.K. Tien, S.T. Wlodek, H. Morrow III, M. Gell, and G.E. Maurer, Ed., American Society for Metals, 1980, p 215-224
27. M.V. Nathal and L.J. Ebert, *Metall. Trans. A*, Vol 16, 1985, p 1849-1870
28. C.S. Kortovich, in "COSAM (Conservation of Strategic Aerospace Materials) Program Review," NASA TM-83006, National Aeronautics and Space Administration, 1982, p 125-131
29. G.M. Janowski, NASA CR-174847, National Aeronautics and Space Administration, 1985
30. K.R. Ziegler and J.F. Wallace, NASA CR-174841, National Aeronautics and Space Administration, 1985
31. F.H. Harf, NASA TM-86987, National Aeronautics and Space Administration, 1985
32. J.R. Stephens, in *High-Temperature Ordered Intermetallic Alloys*, C.C. Koch, C.T. Liu, and N.S. Stoloff, Ed., Elsevier, 1984, p 381-396
33. D.J. Gaydosh and M.V. Nathal, NASA TM-87290, National Aeronautics and Space Administration, 1986
34. N. Mantravadi, K. Vedula, O.J. Gaydosh, and R.H. Titran, NASA TM-87293, National Aeronautics and Space Administration, 1986
35. E.M. Schulson, in *High-Temperature Ordered Intermetallic Alloys*, C.C. Koch, C.T. Liu, and N.S. Stoloff, Ed., Elsevier, 1984, p 193-204
36. A.H. Cottrell, *Trans. AIME*, Vol 212, 1958, p 192-203
37. N.J. Petch, *Phil. Mag.*, Vol 3, 1958, p 1089-1097
38. K. Vedula, V. Pathare, I. Aslanidis, and R.H. Titran, in *High-Temperature Ordered Intermetallic Alloys*, C.C. Koch, C.T. Liu, and N.S. Stoloff, Ed., Elsevier, 1984, p 411-422
39. F.D. Lemkey and M.L. Bailey, in "COSAM (Conservation of Strategic Aerospace Materials) Program Overview," NASA TM-83006, National Aeronautics and Space Administration, 1982, p 209-214
40. D.W. Petrasek, NASA TM-87223, National Aeronautics and Space Administration, 1986
41. J.R. Stephens and J.K. Tien, NASA TM-83395, National Aeronautics and Space Administration, 1983
42. J.R. Stephens and R.H. Titran, in *Materials for Future Energy Systems*, American Society for Metals, 1985, p 317-328

Recycling of Iron, Steel, and Superalloys

Thomas A. Phillips, United States Department of the Interior, Bureau of Mines

RECYCLING can be simply defined as the use of a material over and over again. The many benefits of recycling include the conservation of natural resources and reductions in energy consumption and in the amount of disposable waste. Recycling within the metals industry involves collecting and reusing scrap metal and metallic wastes to produce new metal or goods such as chemicals. Recycled scrap is a major raw material for the metals industry, supplying up to 100% of the feedstock for some products. Consequently, the scrap metal recycling industry is very important to the economy, providing tens of thousands of jobs across the United States. In addition, the significant quantities of scrap that are exported by the United States improve its balance of trade.

Scrap metal can be divided into two categories, home scrap and purchased scrap, based on its origin. Home scrap is generated within the facility of a producer and recycled directly back into the production process. Purchased scrap, on the other hand, is bought by a dealer or broker and resold to potential users. Purchased scrap is further categorized as either prompt scrap or obsolete scrap. Prompt scrap is scrap generated during fabrication and manufacturing processes. Examples of prompt scrap include turnings, borings, stampings, and even rejected parts. As indicated by the word prompt, this type of scrap can usually be recycled quickly and, in many cases, with minimal processing. Obsolete scrap refers to finished goods that are worn out, broken, or otherwise no longer useful. Automobiles, appliances (white goods), ships, aircraft, machinery, and even cans collected from municipal garbage are examples of obsolete scrap.

Home scrap is usually recycled in the form in which it is generated. If the scrap cannot be used as generated, it is usually sold to a scrap dealer and becomes purchased scrap. Companies that deal in scrap can be categorized as either processors or brokers. Processors purchase iron and steel scrap and process, or upgrade, the scrap to meet customer needs. Brokers act as intermediaries, purchasing the scrap from a source and selling it unaltered to a consumer. There are more scrap processors than brokers. However, many processors act as brokers when the scrap is suitable and thus represent both categories.

Although scrap metal is an important raw material for most metal-producing operations, scrap dealers generally operate independently from the metal-producing companies they buy from and sell to. However, a few steel scrap dealers now own steelmaking facilities that utilize the scrap they collect as their primary feedstock material. Scrap metal dealers also tend to specialize in one type of metal, or they have specialized divisions that service the various metals markets. Iron and steel scrap is usually handled separately from stainless steel or superalloy scrap. Therefore, iron and steel recycling is discussed separately from stainless steel and superalloy recycling in this article. Stainless steels and superalloys are discussed together because their technologies and operations are the same; however, they are generally handled at separate facilities, even when the facilities are owned by one company.

This article focuses primarily on the methods and technology used by scrap dealers to recycle purchased scrap. Discussion about the supply and use of this scrap is limited to how processing influences the availability and handling of the scrap. The sections "Recycling Iron and Steel Scrap" and "Recycling Stainless Steel and Superalloy Scrap" in this article describe the size of the scrap recycling industry, scrap use by industry, scrap demand, and the scrap supply.

Recycling Iron and Steel Scrap

Scrap iron and steel are the major raw materials in the production of new iron and steel. In 1987, 62×10^6 Mg (68.3×10^6 tons) of iron and steel, or 58% of the total amount produced, was supplied by recycled scrap (Ref 1). Of this amount, about 38% was home scrap, with the remaining 62% consisting of purchased scrap. The delivered value of the purchased scrap was about $3.6 billion. An additional 9.4×10^6 Mg (10.4×10^6 tons) of scrap, with a value of about $967 million, was exported. Thus, the companies that handle this scrap constitute an important industry, one that has a significant impact on the economies of the United States and the world.

Raw steel production consumes about 75% of the purchased iron and steel scrap recycled. An additional 24% is used by foundries that produce cast iron and steel. The remaining 1% is consumed in small quantities for use in ferroalloys, for iron chemical production, and for the precipitation of copper. Of the 24% consumed by foundries, iron foundries take 89%; steel foundries use the remaining 11%.

Scrap Use by Industry

Most recycled iron and steel scrap is used to produce new steel. Three basic steelmaking processes are used today, each utilizing scrap. Most steel, about 60%, is currently produced in a basic oxygen furnace (BOF), which requires about 25 to 30% scrap. In the steelmaking process, scrap acts as a coolant in addition to providing part of the required iron. The metallic charge to a BOF is comprised of hot metal, iron from a blast furnace, and scrap. Oxygen is then blown into the furnace; the oxygen reacts with carbon, impurities, and a small amount of the iron to produce heat, which melts the scrap. Without the scrap, the furnace would have to be cooled by other means. Thus, the use of scrap simplifies the operation of the furnace and improves the overall energy balance by using heat that would otherwise be lost. Forms of cold iron, such as direct-reduced iron (DRI), could be used, but the current cost of DRI favors the use of scrap.

Open-hearth furnaces can be used to produce steel in the same manner as the BOF, but they have been almost totally replaced in the steelmaking process by the BOF process.

Fig. 1 Electric arc furnace being charged with baled scrap. Courtesy of the American Iron and Steel Institute

This is due to the higher thermal efficiency and shorter cycle time offered by the BOF. In 1987, scrap consumed in open-hearth furnaces amounted to only 1×10^6 Mg (1.1×10^6 tons), or about 2% of the total.

Steelmaking capacity using the electric arc furnace (EAF) has increased significantly in the recent past (Fig. 1). In 1950, EAFs accounted for only 6.2% of domestic steel production (Ref 2). By 1987, this had increased to 38%. The EAF can be charged with up to 100% scrap as feedstock; the average charge in 1987 was 99.2% scrap (Ref 1).

Electric furnaces have several advantages that have led to their increased use. The first is that an EAF steelmaking facility requires much less capital to build than an integrated steel production facility. Another is that coke is not required, eliminating the cost and environmental problems associated with coke production. In addition, because only scrap melting is taking place, the energy requirement per ton of steel produced is lower for an EAF facility than it is for an integrated steel producer.

Iron and steel foundries that produce cast products are the other major users of iron and steel scrap. Although the primary alternative to scrap is pig iron, current economics favor the use of scrap. Iron and steel foundries used scrap for over 92% of their feedstock in 1987 (Ref 1).

A small market for iron scrap exists for use in the cementation of copper from heap leach solutions. The switch in the copper industry from cementation to solvent extraction and electrowinning technology has substantially reduced this market. Other relatively small markets for iron scrap also exist. One provides iron for the production of chemicals such as pigments; another provides the iron content of some ferroalloys. In total, these minor uses account for about 1% of the iron and steel scrap consumed.

Factors Influencing Scrap Demand

Demand for iron and steel scrap is a function of both the quantity of iron and steel produced and the technology used to produce it. The biggest short-term influence on scrap demand is the general health of the iron and steel industry. Because this industry consumes 99% of the scrap, any change in its production rate affects the demand for scrap.

Changes in steelmaking technology also affect scrap demand. One example of a technology change is the switch from open-hearth furnaces to BOFs at the facilities of integrated steel producers. This switch has reduced the demand for scrap because the BOF uses a lower percentage of scrap in its feedstock. This change in technology is almost complete; very few open-hearth furnaces are still in operation. Additional impact on demand will therefore be negligible.

A second factor influencing the demand for scrap is the steady increase in steel production by mini-mills, which use EAFs to produce steel for specialized markets. These mills also use 98 to 99% scrap as feedstock, which greatly increases the total demand for scrap. Because mini-mills are projected to continue to increase their market share for steel products, the demand for scrap is expected to increase.

Another factor influencing scrap demand is the continuing effort by the iron and steel industries to lower their production costs by investing in newer, more efficient equipment. The adaptation of continuous casting technology is an example of this trend. Current estimates are that 60% of domestic steel production is based on continuous casting. The net result is a more efficient conversion of raw steel to finished product and the generation of less home scrap. Evidence of the trend toward more efficient production technology is the rising percentage of raw steel being converted into finished product. During the past 10 years this percentage has increased from 72 to 87% (Ref 3). To offset the corresponding decrease in home scrap production, steel producers are purchasing additional scrap.

Quality issues also affect demand, especially for high-quality scrap. Manufacturers such as automobile makers are tightening quality standards for their steel purchases. This, in turn, requires tighter control of the quality of the scrap used to produce the steel. High-quality scrap is therefore in short supply, while there can be surpluses of lower-quality materials such as ferrous scrap from municipal garbage.

Purchased Scrap Supply

As the generation of home scrap has diminished, the demand for purchased scrap per ton of steel produced has increased. The supply of prompt scrap is generally dependent on the production of iron and steel products and is relatively fixed. To satisfy the rising demand, scrap processors are increasing the quantity of obsolete scrap being recycled. Fortunately, a large amount of obsolete scrap has accu-

Table 1 Ferrous scrap specifications

ISRI code number	Item	Specification
Basic oxygen, electric furnace, and blast furnace grades		
200	No. 1 heavy melting steel	Wrought iron and/or steel scrap 6.4 mm (¼ in.) and more in thickness. Individual pieces not more than 1.5 × 0.60 m (60 × 24 in.) (charging box size), prepared in a manner to ensure compact charging
203	No. 2 heavy melting steel	Wrought iron and steel scrap, black and galvanized, 3.2 mm (⅛ in.) and more in thickness. Charging box size to include material not suitable as No. 1 heavy melting steel. Prepared to ensure compact charging
207	No. 1 busheling	Clean steel scrap, not exceeding 305 mm (12 in.) in any dimensions, including new factory busheling (for example, sheet clippings, stampings, and so on). May not include old auto body and fender stock. Free of metal-coated, limed, vitreous-enameled, and electrical sheet containing over 0.5% Si
208	No. 1 bundles	New black steel sheet scrap, clippings, or skeleton scrap, compressed or hand bundled to charging box size, with a density not less than 1.2 g/cm³ (75 lb/ft³). (Hand bundles are tightly secured for handling with a magnet.) May include Stanley balls or mandrel-wound bundles or skeleton reels, tightly secured. May include chemically detinned material. May not include old auto body or fender stock. Free of metal-coated, limed, vitreous-enameled, and electrical sheet containing over 0.5% Si
209	No. 2 bundles	Old black and galvanized steel sheet scrap, hydraulically compressed to charging box size, with a density not less than 1.2 g/cm³ (75 lb/ft³). May not include tin- and lead-coated material or vitreous-enameled material
210	Shredded scrap	Homogeneous iron and steel scrap, magnetically separated, originating from automobiles, unprepared No. 1 and No. 2 steel, and miscellaneous baling and sheet scrap. Average density, 1.12 g/cm³ (70 lb/ft³)
213	Shredded tin can for remelting	Tin-coated or tin-free shredded steel cans. May include aluminum tops but must be free of aluminum cans, nonferrous metals except those used in can construction, and nonmetallics of any kind
217	Bundled No. 1 steel	Wrought iron and/or steel scrap 3.2 mm (⅛ in.) or more in thickness, compressed to charging box size, with a density not less than 1.2 g/cm³ (75 lb/ft³), free of all metal-coated material
221	Shoveling turnings	Clean short steel or wrought iron turnings, drillings, or screw cuttings. May include any such material whether resulting from crushing, raking, or other processes. Free of springy, bushy, tangled, or matted material; also free of lumps, iron borings, nonferrous metals in a free state, grindings, or excessive oil
222	Shoveling turnings and iron borings	Same as shoveling turnings, but including iron borings
223	Iron borings	Clean cast iron or malleable iron borings and drillings. Free of steel turnings, scale, lumps, and excessive oil
Electric furnace casting and foundry grades		
231	Plate and structural steel, 1.5 m (5 ft) and under	Cut structural and plate scrap, 1.5 m (5 ft) and under. Clean, open-hearth steel plates, structural shapes, crop ends, shearings, or broken steel tires. Dimensions not less than 6.4 mm (¼ in.) thick; not more than 1.5 m (5 ft) long and 455 mm (18 in.) wide. Phosphorus or sulfur not over 0.05%
Electric furnace casting and foundry grades (continued)		
235	Electric furnace bundles	New black steel sheet scrap hydraulically compressed into bundles of size and weight as specified by consumer
236	Cut structural and plate scrap, 0.9 m (3 ft) and under	Clean, open-hearth steel plates, structural shapes, crop ends, shearings, or broken steel tires. Dimensions not less than 6.4 mm (¼ in.) thick; not more than 0.9 m (3 ft) long and 455 mm (18 in.) wide. Phosphorus or sulfur not over 0.05%
242	Foundry steel, 0.6 m (2 ft) and under	Steel scrap 3.2 mm (⅛ in.) and more in thickness, not more than 0.6 m (2 ft) long or 455 mm (18 in.) wide. Individual pieces free from attachments. May not include nonferrous metals, cast or malleable iron, cable, or vitreous-enameled or metal-coated material
Specially processed grades to meet consumer requirements		
254	Heavy breakable cast	Cast iron scrap over charging box size or weighing more than 225 kg (500 lb). May include cylinders and driving wheel centers. May include steel that does not exceed 10% of the casting by weight
259	Clean auto cast	Clean auto blocks. Free of all steel parts except camshafts, valves, valve springs, and studs. Free of nonferrous and nonmetallic parts
260	Unstripped motor blocks	Automobile or truck motors from which steel and nonferrous fittings may or may not have been removed. Free from driveshafts and all parts of frames
261	Drop broken machinery cast	Clean, heavy, cast iron machinery scrap that has been broken under a drop. All pieces must be of cupola size, not more than 0.60 × 0.75 m (24 × 30 in.) with no piece more than 70 kg (150 lb) in weight
Railroad ferrous scrap		
24	Melting steel, railroad No. 1	Clean wrought iron or steel scrap, 6.4 mm (¼ in.) and more in thickness, not more than 455 mm (18 in.) wide, and not more than 1.5 m (5 ft) long. May include pipe ends and material 3.2–6.4 mm (⅛–¼ in.) thick, not more than 380 × 380 mm (15 × 15 in.). Individual pieces cut so as to lie reasonably flat in charging box
27	Rail, steel No. 1	Standard section tee rails, original weight of 25 kg/m (50 lb/yd) or more; 3 m (10 ft) long and more. Suitable for rerolling into bars and shapes. Free from bent and twisted rails; frog, switch, and guard rails; or rails with split heads and broken flanges. Continuous-welded rail may be included, provided no weld is more than 230 mm (9 in.) from the end of the piece of rail
28A	Rail, steel No. 2, cropped rail ends	Standard section, original weight of 25 kg/m (50 lb/yd) and more; 455 mm (18 in.) long and less
28B	Rail, steel No. 2, cropped rail ends	Standard section, original weight of 25 kg/m (50 lb/yd) and more; 0.6 m (2 ft) long and less

Source: Ref 4

mulated over the years, and the nation has a substantial scrap inventory.

A problem that might affect the use of some forms of obsolete scrap is the presence of environmentally hazardous components in the scrap products, many of which eventually enter the waste products of the scrap processor. These substances are now regulated as hazardous wastes, which greatly increases disposal costs. Examples of the hazardous materials encountered are cadmium-plated bolts in automobiles, polychlorinated biphenyls in pre-1980 appliances, chromium pigments in paints, and sodium azide in automobile airbags. Under current law, the recycler inherits responsibility for these wastes by processing the scrap. If the cost or liability incurred in the disposal of these wastes becomes excessive, the processor may stop recycling these products; this not only reduces the supply of obsolete scrap but also increases the negative impact of these hazardous materials.

The Scrap Processor

The scrap processor collects scrap and converts it into a usable form, as defined by the potential customer. A major concern is maximizing the quality of all the scrap collected to increase its total value. Scrap quality depends on many factors, including size, shape, density, chemical composition, and cleanliness.

The scrap processor identifies and stores the scrap by grade. If necessary, the scrap can also be upgraded. Because iron and steel scrap is a relatively low value and high volume commodity, economically viable processing methods are limited. The scrap may have to be separated to remove contaminating items. In addition, large scrap may have to be reduced in size, while loose scrap may have to be compacted.

Collection. Iron and steel scrap is collected—that is, purchased—from a variety of sources, such as salvage yards, automobile wreckers, steel producers, and manufacturers. The value of most scrap is determined by the grade under which it is sold. The Institute of Scrap Recycling Industries (ISRI) has developed specifications for the recognized scrap grades; specifications for some common grades are listed in Table 1. Scrap generators who keep the scrap segregated by grade receive the highest value for their materials because they have eliminated the need for additional processing.

The collected scrap must be transported to the processing facilities. Because of the relatively low unit value of iron and steel scrap, transportation costs can become significant if the material has to be transported over a great distance. Most scrap processors and brokers can usually bid competitively for scrap within a few hundred miles radius from their facilities, or essentially a 1-day trip by truck. A major exception to this is the export market; the lack of adequate scrap supplies overseas justifies high transportation costs.

Separation and Sorting. The separation of iron and steel scrap from other metals serves two purposes. First, it controls impurities that will alter the properties of the iron or steel produced from the scrap. This is particularly important to foundries because they typically do not use ladle refining technology. The other purpose served by separation is the recovery of additional products that are valuable and can be recycled. Two separation methods commonly used for iron and steel scrap are hand sorting and magnetic separation.

Hand sorting simply involves the removal of components from the scrap by hand; it is therefore a labor-intensive process. Hand sorting is most advantageous when used to remove miscellaneous items from the scrap, or when handling the scrap is unavoidable, such as when loading or off-loading small scrap shipments. Items being sorted are usually identified visually.

Magnetic separation is used when large quantities of iron and steel scrap must be separated from other materials. There are two basic types of magnets, permanent magnets and electromagnets. Permanent magnets are metal alloys that, once magnetized, retain their magnetic properties. An electromagnet uses an electric current passed through copper windings to generate a magnetic field. Permanent magnets are generally less expensive, but electromagnets can be operated at higher magnetic field strengths and can be turned on and off to pick up and drop items.

Electromagnets can be used to separate scrap in stationary piles by picking up the magnetic portion of the material, but they are generally used to move scrap from one point to another. Permanent magnets are used in conjunction with magnetic drum and belt separators, the most common type of separation equipment. In a drum separator, a permanent magnet is located inside a rotating stainless steel shell. Magnetic material on a conveyor passing under the drum is attracted to the magnet and picked up off the conveyor. As the shell turns, it carries the magnetic material out of the magnetic field, allowing it to drop free of the drum to be collected. A magnetic belt separator works on the same principle, except that the magnet is located between two pulleys, around which a continuous belt travels.

Magnetically separated scrap can still contain impurities for several reasons. Nonmagnetic material attached in some way to the magnetic material will be collected. Frequently, a piece of nonmagnetic material caught between the magnet and the magnetic scrap will be carried along with the magnetic material. In addition, other undesired magnetic metals, such as nickel and certain stainless steels, will be collected as well. The most common method of removing these materials is hand sorting because they are usually identifiable by sight. The additional cost of hand sorting is often offset by the value of the nonferrous metal or stainless steel collected.

Size Reduction and Compaction. Large items, such as obsolete ships, railroad cars, truck bodies, structural steel, and large castings, must first be cut to facilitate handling and to enable them to be charged into a furnace. Shears, hand-held torches, and crushers are required for this labor-intensive task. Shears function by forcing a cutter head past a stationary anvil to cut the scrap. Two types of shears are used, alligator and guillotine shears. The alligator shears operate like large scissors with the lower edge fixed. Guillotine shears use hydraulic pressure to force the cutter head down past the anvil. Although cutting an object with a torch is slow and laborious, it is essential for very large items (Fig. 2).

Other forms of large scrap can be reduced in size by shredding. Scrap shredders are becoming increasingly important to scrap processors for the recycling of obsolete scrap such as automobiles and white goods. Inside a shredder, steel hammers attached to a drum are rotated past a fixed anvil. As an item is fed into the shredder, it is struck by the hammers and broken, either by direct impact or by being driven against the anvil. Brittle items are shattered, which liberates many of the components previously joined together. An automobile fed into a shredder (Fig. 3) will be reduced to fist-size chunks in less than 1 min. The shredded scrap can then be magnetically separated into magnetic iron and steel fractions and nonmagnetic fractions. Shredded scrap has many desirable qualities, including good density and size uniformity. Other products, such as nonferrous metals and plastic, can be recovered from the nonmagnetic fractions. Automobiles, white goods, and similar items can also be pressed into logs and then sheared to produce a regularly sized product.

Loose scrap that has a high surface area and low density is normally compacted by baling or briquetting. Examples of this type of scrap are lathe turnings, punchings, and surplus sheet metal from stampings. Baling involves pressing the scrap into cubic bundles. A baler is a very heavy piece of equipment that uses up to three hydraulic rams to compress the scrap against a fixed wall. In a briquetter, small scrap, such as turnings, is compacted into pockets as it passes between two counterrotating drums. Baled and briquetted scrap has a higher value due to the advantages it gives the furnace operator. In particular, compacted scrap is easier to handle and oxidizes less during melting, which results in better metal recoveries. In addition, for electric furnace operators, the increased density means fewer scrap charges are required per heat.

Detinning. Recycling tin plate scrap generated during the production of steel food and beverage cans requires specialized processors, commonly called detinners. In the detinning process, the scrap is leached with a hot alkaline solution that usually contains sodium hydroxide and sodium nitrate. Metallic tin dissolves quickly, while the tin that has alloyed with the iron (referred to as hardhead) takes additional time. Complete dissolution of the tin is not practical because of the time required; however, the tin content of the steel is reduced to less than 0.03% by this process, which is acceptable

Fig. 2 Large tank being cut down to size with a torch. Courtesy of the Institute of Scrap Recycling Industries, Inc.

Fig. 3 Junked automobile being fed into a shredder. Courtesy of the Institute of Scrap Recycling Industries, Inc.

for most uses (Ref 5). The steel plate is relatively unaffected by the leach and is washed and baled for recycling. Tin is precipitated from the alkaline solution and recovered by an electrochemical process. Tin cans collected from municipal refuse can also be detinned, but this has not been common practice because the cost of collection and transportation is prohibitive. The presence of aluminum in municipal scrap, especially from bimetallic cans, is often a problem. Aluminum dissolves and consumes reagent, and interferes in subsequent tin recovery steps.

Blending. When levels of impurities are considered to be low, or when the total quantity of contaminated scrap is small, a processor can upgrade the scrap by blending it with scrap of a higher purity. Scrap collected from municipal refuse is frequently blended with higher-grade scrap to dilute the tin and other impurities in the municipal scrap to tolerable levels. Care must be taken when blending scrap to ensure that its final composition is acceptable to the metal producer. If the final impurity levels are too high, the value of all the scrap will be reduced.

Incineration is used by some scrap processors to remove combustible materials. These materials can include an oil or grease coating on the scrap as well as wood or paper mixed in with the scrap. Incineration can also be used to remove volatile metals such as lead and zinc. Rotary furnaces are commonly used by scrap processors.

Recycling Stainless Steel and Superalloy Scrap

Stainless steels are iron-base alloys that contain at least 10% Cr. Other metals are frequently added to stainless steel, including nickel, molybdenum, manganese, cobalt, and vanadium. The American Iron and Steel Institute has three general categories of stainless steel, grouped in numbered series.

The 400 series represents straight chromium steel, which contains 10 to 30% chromium, with the balance consisting of iron. A major characteristic of these alloys is that most are magnetic. Stainless steel of the 300 series, on the other hand, is generally nonmagnetic. This series contains nickel and is frequently called 18-8, which signifies the percentages of chromium and nickel in type 304, the most common grade in the series. Additional elements added to stainless steels of this series enable them to attain a wide variety of properties. The 200 series was developed as a low-nickel replacement for the 300 series. In general, the 200 series contains about 4 to 6% Ni, 16 to 20% Cr, and 5 to 16% Mn. The manganese is added to replace nickel. Like the 300 series, the 200 series is nonmagnetic.

Superalloys are those alloys developed for high-temperature service where relatively high stresses (tensile, thermal, vibratory, and shock) are encountered, and where resistance to oxidation is frequently required. These alloys are generally either nickel-iron or nickel- or cobalt-base with varying amounts of additional elements, which can include chromium, molybdenum, niobium, tungsten, titanium, manganese, tantalum, and silicon. Since 1976, the most predominant superalloy has been Inconel 718, with a nominal composition of 52.5% Ni, 19% Cr, 18.5% Fe, 3% Mo, and 5.1% Nb, plus tantalum. Applications of superalloys range from chemically resistant tanks and heat exchanger tubes to components for jet engines.

Scrap Use By Industry

Stainless steel scrap is almost always used to produce more stainless steel. In 1980, 350×10^3 Mg (386 000 tons) of stainless steel was recycled in the United States, supplying about 34% of the stainless steel market (Ref 2). Because the process for making stainless steel is similar to the EAF steelmaking process, the role of scrap is also similar. A wide variety of alloys, including stainless steels and superalloys, are produced by specialty foundries that can use scrap if it matches the composition of the alloy being produced.

In 1986, approximately 25×10^3 Mg (55×10^6 lb) of superalloy scrap was processed, of which 70% was recycled into the same alloy and about 20% was downgraded. The remaining 10% of the recycled scrap was sold to nickel refiners; this is also considered to be a form of downgrading because many valuable elements are refined out of the nickel product and not recovered. This breakdown of the 25×10^3 Mg (55×10^6 lb) includes the 2×10^3 Mg (4.3×10^6 lb) that was exported.

Downgrading refers to the practice of including an alloy in the feedstock of a less critical alloy. For example, a nickel-base superalloy can be blended with other scrap to produce a less critical superalloy or stainless steel. A disadvantage of this practice is that some components of the superalloy can be lost; that is, low levels of the element can be tolerated in the alloy being produced but will have no real function. This represents an effective loss of these valuable metals to the economy.

Superalloys can be roughly divided into two broad categories, air-melted and vacuum-melted alloys. Air melting offers cost advantages for certain alloys. Recycled scrap is acceptable for most air-melted superalloys. Vacuum melting was developed to prevent the oxidation of alloying elements such as aluminum and titanium. It also allows additional purification of the alloy during production. Product specifications often prohibit the use of recycled feed in vacuum-melted alloys to reduce the chance that detrimental impurities will be included in the final product. This is especially important for critical components such as jet engine parts.

Stainless steels and superalloys are recycled in much the same manner and for the same purpose as iron and steel. The primary difference is that the volume of material recycled is less, while the value per pound is greater. As with iron and steel scrap,

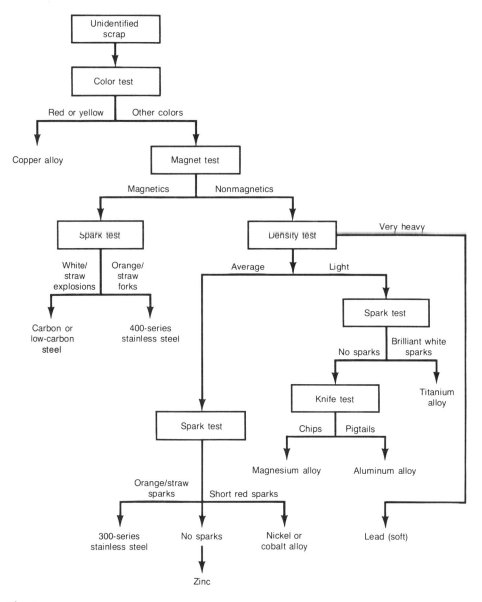

Fig. 4 Simple flowsheet for presorting scrap. Source: Ref 6

stainless steel and superalloy scrap is recycled as either home scrap or purchased scrap. Purchased scrap is handled by either processors or brokers. Processors alter the form of the scrap to that required by the metal producer, while brokers buy and sell scrap without changing its form. Although the companies that recycle iron and steel are similar to those that recycle stainless steel and superalloys, they are usually independent operations.

Scrap Demand

The demand for stainless steel scrap or superalloy scrap is largely a function of the demand for the stainless steel or superalloy itself. A general increase over time is therefore expected as manufacturers increase their use of stainless steels and superalloys to improve product performance. However, the general industrial production rate, which is driven by the overall economy, is the primary factor that influences demand for both the metals and scrap. High demand for stainless steel in 1987 through 1989, following a period of low demand, resulted in a sharply increased demand for stainless steel scrap. Demand is heightened by tight world markets for both chromium and nickel, the use of which can only be replaced by an increased use of scrap. Fortunately, increased demand also raises the value of scrap, which enables more to be recycled. As a major producer and consumer of stainless steel, the United States usually has excess stainless steel scrap, which is exported.

Processing Stainless Steel and Superalloy Scrap

Stainless steels and superalloys are recycled to produce the same metal whenever possible. For example, type 316 stainless

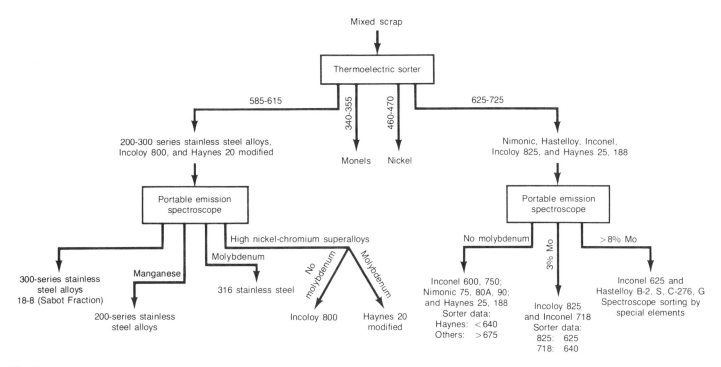

Fig. 5 Instrumental identification technique for sorting stainless steels, nickel alloys, and superalloys. Source: Ref 7

steel is recycled to produce more type 316. It is the responsibility of the scrap processor to ensure that the metal producer receives the desired scrap, as needed, and in the desired form.

A scrap processor collects and receives scrap from a number of sources. It is then analyzed and sorted with scrap of the same composition. Mixed scrap must be separated by alloy type. Most of the sorting is done by hand.

Accurate identification of the alloys being recycled is very important because the processor is responsible for the quality of the scrap sold. Scrap that does not meet specifications will be returned by the customer. Superalloy scrap identification utilizes the same procedures as stainless steel scrap identification, except that each piece of superalloy not only has to be identified but also its composition must be certified before it is sold. Scrap that is mixed and impractical to separate will be downgraded.

Collection. Stainless steel and superalloy scrap has a much higher unit value than iron and steel scrap, which reduces the relative impact of transportation costs on the overall recycling cost. Scrap dealers therefore compete for stainless steel and, especially, superalloy scrap on a national and international basis. In addition, scrap may be handled at several locations before it is finally sold for use. Generators of scrap will receive the highest value for their scrap if they keep it separated by alloy type. However, most scrap is mixed and must be separated to be useful.

Separation and Sorting. The separation of the scrap by individual alloy is one of the most important tasks for the scrap processor. Most will handle a wide variety of materials, which makes the ability to identify the scrap a prime concern. Although stainless steel may represent the largest volume of material processed, metals such as superalloys, titanium, nickel alloys, copper, and brasses provide significant income to the processing facility. Identification and sorting utilize a mix of simple and relatively complex procedures, based on any of the physical or chemical properties of the scrap.

Identifying scrap of completely unknown origin usually requires several steps, the choice of which depends on the results of the preceding step (Fig. 4). Initially, physical properties such as color, density, and relative hardness can be used to quickly separate certain classes of material. For example, copper and brass can be identified by their reddish color, while lead can be recognized by both its density and relative softness.

Differentiating between alloys of similar composition, such as between grades of stainless steel, is more difficult. For these separations, magnetic testing, spark testing, chemical spot testing, optical and x-ray spectroscopic analysis, thermoelectric analysis, or quantitative chemical analysis is used (Fig. 5).

Magnetic testing is a simple test used to determine whether or not a material is fer-

Table 2 Spark characteristics of various metals and alloys

Material	Spark characteristic
Normal carbon steel	Heavy, dense sparks 455–610 mm (18–24 in.) long that travel completely around the grinding wheel. Sparks are white to straw colored with main burst throughout.
400-series chromium stainless steel	Sparks are not as heavy or dense as those of normal carbon steel. Sparks are 355–455 mm (14–18 in.) long, travel completely around the grinding wheel, and are orange to straw colored, ending with a forked tongue. Preliminary bursts and few main bursts.
300-series stainless steel	Sparks are not as heavy or as dense as those of normal carbon steel. Sparks are 305–455 mm (12–18 in.) long, travel completely around the grinding wheel, and are orange to straw colored, ending in a straight line with few if any bursts.
310-series 25-20 stainless steel	The spark stream is thin and 100–150 mm (4–6 in.) long. Sparks are orange to red in color and do not travel around the grinding wheel; there are no bursts.
Nickel and cobalt high-temperature alloys	The spark stream is thin and about 50 mm (2 in.) long. The sparks are dark red in color and do not travel around the grinding wheel; there are no bursts.

Source: Ref 6

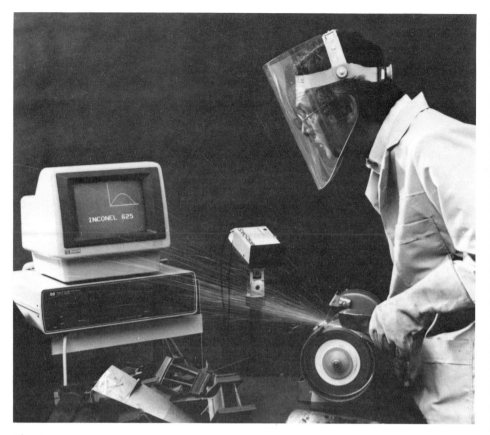

Fig. 6 Identification of scrap by experimental spark analysis equipment

romagnetic. Iron, nickel, and cobalt are ferromagnetic metals. Low-alloy (mild) and ferritic stainless steels, which include much of the 400 series, are also ferromagnetic. Magnetic testing is therefore a potential means of determining to which series a stainless steel belongs, but it cannot be used to determine the actual alloy within the series.

Spark testing is a scrap identification technique that is indispensable for most processors. A piece of scrap is ground on a high-speed abrasive grinding wheel to produce sparks. A stationary wheel can be used to grind small items, while portable wheels can be used for larger items. The sparks are generated when the abrasive wheel tears off small pieces of the alloy being tested, which are then ignited by the heat generated during this process. The burning particle is seen as the spark. The color and length of the spark are determined by the composition of the piece of metal and the speed at which it burns. Carbides in the alloy will cause secondary bursts, or splitting of the particles. The number and shape of these bursts also help identify the alloy being tested. Metal fragments will tend to stick to the grinding wheel in varying degrees, depending on the alloy being tested. This tendency is also helpful in identifying the alloy. The characteristics of the sparks emitted by several alloys are presented in Table 2.

The reliability of spark testing depends on the ability of the test operator to visually differentiate the various spark patterns produced. Skilled personnel are therefore required for dependable results. Test conditions, including lighting conditions, background (preferably dark), and the pressure used to hold the specimen against the wheel, should be uniform. In addition, the wheel must be frequently dressed to remove the metal particles that adhere to it during tests. The use of standards, or samples of known composition, is desirable if positive identification of the alloy is required.

Equipment is available to automatically measure the spectrum given off by the sparks (Fig. 6). The data collected are then compared to known standards by a computer to identify the alloy being tested. The advantages of this system are that a skilled tester is not required and that spectral differences that cannot be detected by eye can be determined by the equipment. However, truly portable equipment has yet to be developed, and the equipment that does exist is not sufficiently proven. As a result, this system is not in general use by the recycling industry.

Chemical spot tests are used as part of a general separation scheme or as the final step in identifying a piece of scrap. The test can be as simple as placing a drop of acid on a clean portion of the testpiece and observing the resulting reaction or lack thereof. A more complicated test involves dissolving a portion of the metal to be identified in a reagent. The resulting solution is then reacted with other reagents and the results observed.

The simplest method of conducting the latter test is to place a few drops of reagent on a clean portion of the surface of the unidentified metal. After an appropriate time has elapsed, the resulting solution is placed on a supporting surface, such as glass, or on a porous surface, such as a piece of filter paper. Other reagents are then mixed with the solution to complete the test. A second approach, an electrographic method, requires sandwiching an electrolyte-impregnated filter cloth between the unidentified metal and an aluminum or platinum cathode. With the piece of scrap acting as the anode, an electric current dissolves a portion of the scrap into the electrolyte. The filter paper can then be removed and tested.

Because chemical spot testing is based on the chemical reactions between the reagents used and the unknown metals being analyzed, test accuracy requires considerable expertise in both the procedures used and the interpretation of the results. The analyzer must be able to differentiate between shades of colors in the solutions or precipitates formed and in the speed at which the reaction occurs. It is very important to use standardized reagents to produce uniform results. An example of an identification technique based on chemical testing is shown in Fig. 7.

Spectroscopic Testing. As the number and complexity of stainless steel and other metallic alloys have increased, the need of the scrap metal processor to more accurately quantify and qualify the metals present has also increased. Spectroscopic analysis has been developed as an analytical tool and is in general use by scrap processors. The basic principle of this analytical technique is that an element, upon being energized, will reemit energy in characteristic wavelengths. Measuring these emissions from the sample and comparing them to a standard permits a quick and accurate analysis of most metals.

In optical spectroscopes, an electrical discharge is used to vaporize a small quantity of metal from the surface of a specimen. The heat provided is sufficient to cause the vaporized metal to luminesce as well. The light emitted is broken down into individual wavelengths, and the individual elements are then identified by the presence of their characteristic wavelengths. Although this type of analysis is very useful in discovering the presence of an element, it is not always accurate in determining the precise quantity that is present.

X-ray spectroscopy (Fig. 8) is another technique for identifying the elements present in an alloy; it is also useful in quantify-

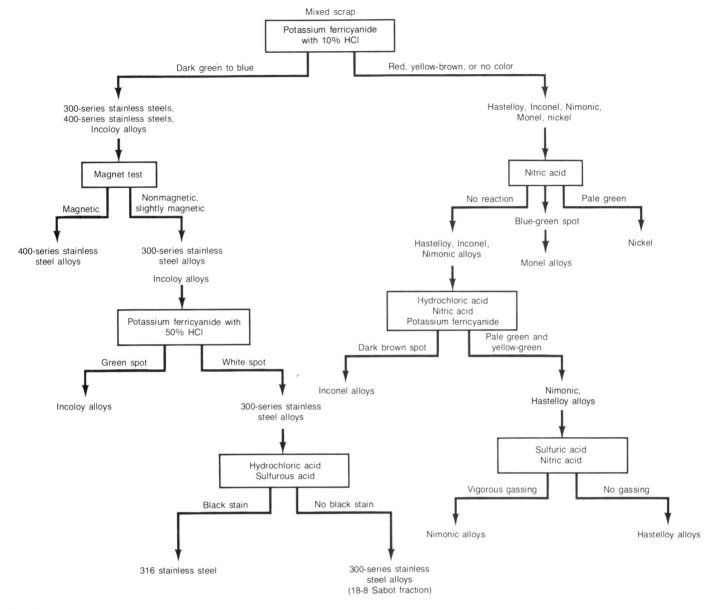

Fig. 7 Chemical test identification technique for nickel-base and stainless steel alloys

ing the elements present. In principle, a sample is bombarded with x-rays and then radiates back characteristic x-rays that can be detected and used to identify the elements present by comparing the emissions with emissions from known standards. A potential problem occurs when elements with close atomic numbers are being analyzed because resolution at low concentrations is often difficult, although not always impossible.

Available spectroscopic equipment ranges from portable, simple-to-use instruments to highly sophisticated and precise equipment. This equipment is most frequently used by scrap processors to identify and sort scrap with a high alloy content; the unit value of the scrap must be high enough to warrant the use of such equipment. Considerable experience and knowledge of the equipment are required. The results, however, can be obtained relatively quickly, which helps make this type of analytical equipment practical.

Thermoelectric Testing. When two different metals at two different temperatures are in contact, they generate an electric potential known as the Seebeck effect. Thermoelectric devices utilize this effect to identify and sort scrap. These devices contain two probes made of the same metal, one heated and one at ambient temperature. When they contact another piece of metal, the scrap in this case, a potential difference between the two probes is generated and is measured by a voltmeter. The potential difference is characteristic of the metal being tested. Calibration against standards enables various alloys to be identified.

Thermoelectric testing is not possible in all cases because some metals give readings that are too similar to differentiate. Another problem is that the potential difference measured is also affected by the physical condition of the alloy; that is, different voltage readings will be obtained depending on the structure and surface condition of the testpiece. Heat-treated and work-hardened alloys will also produce different results. This kind of testing is generally used in combination with other testing procedures for accurate identification of an alloy.

Quantitative Chemical Analysis. Complete and positive identification of a metal may be required, especially to obtain accurate standards upon which to identify unknown items using less intensive tests. Either standard wet chemical analysis or spectral analysis (or both) can be used for

Fig. 8 Portable x-ray spectroscope. Source: Ref 7

such an analysis. Wet chemical analysis involves dissolving the metal, usually into an acid, and then applying classical chemical analysis techniques to determine the quantity of each element of interest. Optical or x-ray spectroscopy can also be used, either on the sample itself or on solutions derived from it. This type of analysis is often provided by contract laboratories because the equipment and personnel required are costly and specialized.

Size Reduction and Compaction. Large items must be cut to manageable sizes by a processor. The equipment used is the same as that used to size iron and steel scrap. Massive items are cut with torches, while lighter items can be cut with shears. Although shredders are rarely used on stainless steel and superalloy scrap, balers are used to compress suitable scrap in the same manner as for steel scrap.

Degreasing. Some of the metals to be recycled may be coated with oil or grease, such as cutting lubricants or surface-protecting greases, which may have to be removed to facilitate metal identification or to meet product specifications. Solvents such as trichloroethylene are commonly used to remove these contaminants from the scrap. Usually, the scrap to be degreased is loaded into a basket or drum and tumbled or otherwise agitated in the solvent. The oils or grease dissolve into the solvent, and the loaded solvent can be drained away from the metals. To remove any adhering solvent, the clean metal will be washed. The solvent is usually recovered for reuse by selective distillation in an automated solvent regeneration system.

Blending. Many forms of mixed scrap cannot be separated or are too small or otherwise impractical to sort by the techniques described. Much of this, but primarily stainless steel, can still be recycled using a technique known as blending. First, various lots of mixed scrap are extensively sampled and analyzed. The analysis of the entire lot can then be determined using computerized statistics programs. By combining lots of known analysis, a mixture that meets the specifications of the stainless steel producer can be obtained. Careful sampling and the use of a computer to perform the many requisite calculations are necessary. The scrap processor using this technique must guarantee that the blended scrap will meet the requirements of the customer.

Secondary nickel refining is another recycling method used for scrap that is deemed undesirable in its current form. Examples of the kind of scrap that can be treated by this operation are metals coated with undesirable materials, bimetallic scrap, spills, slags, very oily turnings or grindings, unsortable mixed scrap, or scrap that is physically bound to dissimilar materials. In the alloys processed through such an operation, nickel is the metal of primary economic importance. Other metals such as cobalt, chromium, molybdenum, iron, and copper can also be recovered, but as nickel-base alloys.

The specific technology used by secondary nickel refiners is proprietary. Generally, scrap of known analysis is fed into an electric furnace and melted, and impurities such as oil, grease, and plastic coatings burn off. Volatile metals such as lead or zinc also vaporize and can be collected as baghouse dust. Refractory elements such as chromium or tungsten can be oxidized and collected in the slag by controlling the furnace conditions. The final product will be a nickel (or nickel alloy) pig or shot that is uniform and of known analysis. By carefully analyzing and adjusting the scrap feedstock in advance, specific alloys that become prime products can be produced.

Metallurgical Wastes. Stainless steel and superalloy producers and product fabricators generate significant quantities of waste materials such as furnace flue dust, mill scale, and grinding swarf. All these materials contain valuable metal constituents, but their analyses will vary depending on the alloy being produced and the technology utilized. These materials are difficult to recycle because they are generally finely divided and often oxidized. They are also often contaminated with a variety of other materials, including slag, grinding abrasives, and other foreign materials.

The technology used to recycle these materials requires that they be pelletized. To facilitate this, the large portions of the wastes must be separated on screens and then ground to a suitable size. Equipment such as a rod mill, a rotating cylinder in which steel rods tumble to grind the material, is commonly used for this step. The fine material is then mixed with a reducing agent (such as coal) and a binder (such as cement). Water is then added, and the mix is fed to a pelletizing disk or drum. A pelletizing disk is a rotating disk that is tilted at an angle. If the water content of the feedstock is correct, the mix will roll on the disk, forming small balls, or pellets. When the pellets reach the desired size, they are automatically discharged from the disk. A pelletizing drum operates in the same way, except that it consists of a tilted rotating cylinder in which the feedstock tumbles until discharged from one end. The formed pellets are then dried.

A few stainless steel producers recycle some of their own waste by feeding the dry pellets directly into the EAF as part of the furnace charge. Once in the furnace the added reductant reacts with the oxidized metals in the pellets to form metals that become part of the stainless steel. Other components of the pellets are incorporated into the furnace slag.

At other facilities, however, the pellets are used to produce a master alloy that can be recycled. Pellets that have simply been dried cannot withstand much handling. Therefore, the dried pellets are indurated, a process that involves heating the pellets in a furnace to a temperature where the individual components fuse together. This hardens the pellets, which are then fed to an EAF where they are melted. Once again, the reductant converts the oxidized metals back to metal. If the feedstock materials used to produce the pellets are blended correctly, a usable alloy is produced, which is then cast into ingots for recycling.

REFERENCES

1. R.E. Brown, Iron and Steel Scrap, in *BuMines Minerals Yearbook*, Bureau of Mines, 1987, p 519-535
2. Iron and Steel, in *Recycled Metals in the 1980's*, National Association of Recycling Industries, 1982, p 89-97

3. C.J. Labee, *Iron Steel Eng.*, Vol 66 (No. 2), Feb 1989, p D11-D32
4. "ISRI Guidelines for Ferrous Scrap: FS-88," Scrap Specifications Circular, Institute of Scrap Recycling Industries, 1988
5. W. Germain, H.P. Wilson, and V.E. Archer, Tin and Tin Alloys (Detinning), in *Encyclopedia of Chemical Technology*, Vol 21, 3rd ed., John Wiley & Sons, p 37
6. W.D. Riley, R.D. Brown, Jr., and J.M. Larrain, Large-Scale Metals Identification and Sorting Using Instrumental Techniques, *J. Test. Eval.*, Vol 15 (No. 4), July 1987, p 239-247
7. R.D. Brown, Jr., W.D. Riley, and C.A. Zieba, "Rapid Identification of Stainless Steel and Superalloy Scrap," RI 8858, Bureau of Mines, 1984, p 16-17
8. R. Newell, R.E. Brown, D.M. Soboroff, and H.V. Makar, "A Review of Methods for Identifying Scrap Metals," IC 8902, Bureau of Mines, 1982, p 5

Metric Conversion Guide

This Section is intended as a guide for expressing weights and measures in the Système International d'Unités (SI). The purpose of SI units, developed and maintained by the General Conference of Weights and Measures, is to provide a basis for worldwide standardization of units and measure. For more information on metric conversions, the reader should consult the following references:

- "Standard for Metric Practice," E 380, *Annual Book of ASTM Standards*, American Society for Testing and Materials, 1916 Race Street, Philadelphia, PA 19103
- "Metric Practice," ANSI/IEEE 268–1982, American National Standards Institute, 1430 Broadway, New York, NY 10018
- *The International System of Units*, SP 330, 1986, National Institute of Standards and Technology. Order from Superintendent of Documents, U.S. Government Printing Office, Washington, DC 20402-9325
- *Metric Editorial Guide*, 4th ed. (revised), 1985, American National Metric Council, 1010 Vermont Avenue NW, Suite 1000, Washington, DC 20005-4960
- *ASME Orientation and Guide for Use of SI (Metric) Units*, ASME Guide SI 1, 9th ed., 1982, The American Society of Mechanical Engineers, 345 East 47th Street, New York, NY 10017

Base, supplementary, and derived SI units

Measure	Unit	Symbol
Base units		
Amount of substance	mole	mol
Electric current	ampere	A
Length	meter	m
Luminous intensity	candela	cd
Mass	kilogram	kg
Thermodynamic temperature	kelvin	K
Time	second	s
Supplementary units		
Plane angle	radian	rad
Solid angle	steradian	sr
Derived units		
Absorbed dose	gray	Gy
Acceleration	meter per second squared	m/s^2
Activity (of radionuclides)	becquerel	Bq
Angular acceleration	radian per second squared	rad/s^2
Angular velocity	radian per second	rad/s
Area	square meter	m^2
Capacitance	farad	F
Concentration (of amount of substance)	mole per cubic meter	mol/m^3
Conductance	siemens	S
Current density	ampere per square meter	A/m^2
Density, mass	kilogram per cubic meter	kg/m^3
Electric charge density	coulomb per cubic meter	C/m^3
Electric field strength	volt per meter	V/m
Electric flux density	coulomb per square meter	C/m^2
Electric potential, potential difference, electromotive force	volt	V
Electric resistance	ohm	Ω
Energy, work, quantity of heat	joule	J
Energy density	joule per cubic meter	J/m^3
Entropy	joule per kelvin	J/K
Force	newton	N
Frequency	hertz	Hz
Heat capacity	joule per kelvin	J/K
Heat flux density	watt per square meter	W/m^2
Illuminance	lux	lx
Inductance	henry	H
Irradiance	watt per square meter	W/m^2
Luminance	candela per square meter	cd/m^2
Luminous flux	lumen	lm
Magnetic field strength	ampere per meter	A/m
Magnetic flux	weber	Wb
Magnetic flux density	tesla	T
Molar energy	joule per mole	J/mol
Molar entropy	joule per mole kelvin	J/mol · K
Molar heat capacity	joule per mole kelvin	J/mol · K
Moment of force	newton meter	N · m
Permeability	henry per meter	H/m
Permittivity	farad per meter	F/m
Power, radiant flux	watt	W
Pressure, stress	pascal	Pa
Quantity of electricity, electric charge	coulomb	C
Radiance	watt per square meter steradian	W/m^2 · sr
Radiant intensity	watt per steradian	W/sr
Specific heat capacity	joule per kilogram kelvin	J/kg · K
Specific energy	joule per kilogram	J/kg
Specific entropy	joule per kilogram kelvin	J/kg · K
Specific volume	cubic meter per kilogram	m^3/kg
Surface tension	newton per meter	N/m
Thermal conductivity	watt per meter kelvin	W/m · K
Velocity	meter per second	m/s
Viscosity, dynamic	pascal second	Pa · s
Viscosity, kinematic	square meter per second	m^2/s
Volume	cubic meter	m^3
Wavenumber	1 per meter	1/m

Conversion factors

To convert from	to	multiply by
Angle		
degree	rad	1.745 329 E − 02
Area		
in.^{2}	mm^2	6.451 600 E + 02
in.^{2}	cm^2	6.451 600 E + 00
in.^{2}	m^2	6.451 600 E − 04
ft^2	m^2	9.290 304 E − 02
Bending moment or torque		
lbf · in.	N · m	1.129 848 E − 01
lbf · ft	N · m	1.355 818 E + 00
kgf · m	N · m	9.806 650 E + 00
ozf · in.	N · m	7.061 552 E − 03
Bending moment or torque per unit length		
lbf · in./in.	N · m/m	4.448 222 E + 00
lbf · ft/in.	N · m/m	5.337 866 E + 01
Current density		
A/in.^{2}	A/cm^2	1.550 003 E − 01
A/in.^{2}	A/mm^2	1.550 003 E − 03
A/ft^2	A/m^2	1.076 400 E + 01
Electricity and magnetism		
gauss	T	1.000 000 E − 04
maxwell	μWb	1.000 000 E − 02
mho	S	1.000 000 E + 00
Oersted	A/m	7.957 700 E + 01
Ω · cm	Ω · m	1.000 000 E − 02
Ω circular-mil/ft	μΩ · m	1.662 426 E − 03
Energy (impact, other)		
ft · lbf	J	1.355 818 E + 00
Btu (thermochemical)	J	1.054 350 E + 03
cal (thermochemical)	J	4.184 000 E + 00
kW · h	J	3.600 000 E + 06
W · h	J	3.600 000 E + 03
Flow rate		
ft^3/h	L/min	4.719 475 E − 01
ft^3/min	L/min	2.831 000 E + 01
gal./h	L/min	6.309 020 E − 02
gal./min	L/min	3.785 412 E + 00
Force		
lbf	N	4.448 222 E + 00
kip (1000 lbf)	N	4.448 222 E + 03
tonf	kN	8.896 443 E + 00
kgf	N	9.806 650 E + 00
Force per unit length		
lbf/ft	N/m	1.459 390 E + 01
lbf/in.	N/m	1.751 268 E + 02
Fracture toughness		
ksi$\sqrt{in.}$	MPa $\sqrt{m}$	1.098 800 E + 00
Heat content		
Btu/lb	kJ/kg	2.326 000 E + 00
cal/g	kJ/kg	4.186 800 E + 00

To convert from	to	multiply by
Heat input		
J/in.	J/m	3.937 008 E + 01
kJ/in.	kJ/m	3.937 008 E + 01
Length		
Å	nm	1.000 000 E − 01
μin.	μm	2.540 000 E − 02
mil	μm	2.540 000 E + 01
in.	mm	2.540 000 E + 01
in.	cm	2.540 000 E + 00
ft	m	3.048 000 E − 01
yd	m	9.144 000 E − 01
mile	km	1.609 300 E + 00
Mass		
oz	kg	2.834 952 E − 02
lb	kg	4.535 924 E − 01
ton (short, 2000 lb)	kg	9.071 847 E + 02
ton (short, 2000 lb)	kg × 10^3(a)	9.071 847 E − 01
ton (long, 2240 lb)	kg	1.016 047 E + 03
Mass per unit area		
oz/in.^{2}	kg/m^2	4.395 000 E + 01
oz/ft^2	kg/m^2	3.051 517 E − 01
oz/yd^2	kg/m^2	3.390 575 E − 02
lb/ft^2	kg/m^2	4.882 428 E + 00
Mass per unit length		
lb/ft	kg/m	1.488 164 E + 00
lb/in.	kg/m	1.785 797 E + 01
Mass per unit time		
lb/h	kg/s	1.259 979 E − 04
lb/min	kg/s	7.559 873 E − 03
lb/s	kg/s	4.535 924 E − 01
Mass per unit volume (includes density)		
g/cm^3	kg/m^3	1.000 000 E + 03
lb/ft^3	g/cm^3	1.601 846 E − 02
lb/ft^3	kg/m^3	1.601 846 E + 01
lb/in.^{3}	g/cm^3	2.767 990 E + 01
lb/in.^{3}	kg/m^3	2.767 990 E + 04
Power		
Btu/s	kW	1.055 056 E + 00
Btu/min	kW	1.758 426 E − 02
Btu/h	W	2.928 751 E − 01
erg/s	W	1.000 000 E − 07
ft · lbf/s	W	1.355 818 E + 00
ft · lbf/min	W	2.259 697 E − 02
ft · lbf/h	W	3.766 161 E − 04
hp (550 ft · lbf/s)	kW	7.456 999 E − 01
hp (electric)	kW	7.460 000 E − 01
Power density		
W/in.^{2}	W/m^2	1.550 003 E + 03
Press capacity		
See **Force**		
Pressure (fluid)		
atm (standard)	Pa	1.013 250 E + 05
bar	Pa	1.000 000 E + 05
in. Hg (32 °F)	Pa	3.386 380 E + 03

To convert from	to	multiply by
in. Hg (60 °F)	Pa	3.376 850 E + 03
lbf/in.^{2} (psi)	Pa	6.894 757 E + 03
torr (mm Hg, 0 °C)	Pa	1.333 220 E + 02
Specific heat		
Btu/lb · °F	J/kg · K	4.186 800 E + 03
cal/g · °C	J/kg · K	4.186 800 E + 03
Stress (force per unit area)		
tonf/in.^{2} (tsi)	MPa	1.378 951 E + 01
kgf/mm^2	MPa	9.806 650 E + 00
ksi	MPa	6.894 757 E + 00
lbf/in.^{2} (psi)	MPa	6.894 757 E − 03
MN/m^2	MPa	1.000 000 E + 00
Temperature		
°F	°C	5/9 · (°F − 32)
°R	°K	5/9
Temperature interval		
°F	°C	5/9
Thermal conductivity		
Btu · in./s · ft^2 · °F	W/m · K	5.192 204 E + 02
Btu/ft · h · °F	W/m · K	1.730 735 E + 00
Btu · in./h · ft^2 · °F	W/m · K	1.442 279 E − 01
cal/cm · s · °C	W/m · K	4.184 000 E + 02
Thermal expansion		
in./in. · °C	m/m · K	1.000 000 E + 00
in./in. · °F	m/m · K	1.800 000 E + 00
Velocity		
ft/h	m/s	8.466 667 E − 05
ft/min	m/s	5.080 000 E − 03
ft/s	m/s	3.048 000 E − 01
in./s	m/s	2.540 000 E − 02
km/h	m/s	2.777 778 E − 01
mph	km/h	1.609 344 E + 00
Velocity of rotation		
rev/min (rpm)	rad/s	1.047 164 E − 01
rev/s	rad/s	6.283 185 E + 00
Viscosity		
poise	Pa · s	1.000 000 E − 01
stokes	m^2/s	1.000 000 E − 04
ft^2/s	m^2/s	9.290 304 E − 02
in.^{2}/s	mm^2/s	6.451 600 E + 02
Volume		
in.^{3}	m^3	1.638 706 E − 05
ft^3	m^3	2.831 685 E − 02
fluid oz	m^3	2.957 353 E − 05
gal. (U.S. liquid)	m^3	3.785 412 E − 03
Volume per unit time		
ft^3/min	m^3/s	4.719 474 E − 04
ft^3/s	m^3/s	2.831 685 E − 02
in.^{3}/min	m^3/s	2.731 177 E − 07
Wavelength		
Å	nm	1.000 000 E − 01

(a) kg × 10^3 = 1 metric ton or 1 megagram (Mg)

SI prefixes—names and symbols

Exponential expression	Multiplication factor	Prefix	Symbol
10^{18}	1 000 000 000 000 000 000	exa	E
10^{15}	1 000 000 000 000 000	peta	P
10^{12}	1 000 000 000 000	tera	T
10^{9}	1 000 000 000	giga	G
10^{6}	1 000 000	mega	M
10^{3}	1 000	kilo	k
10^{2}	100	hecto(a)	h
10^{1}	10	deka(a)	da
10^{0}	1	BASE UNIT	
10^{-1}	0.1	deci(a)	d
10^{-2}	0.01	centi(a)	c
10^{-3}	0.001	milli	m
10^{-6}	0.000 001	micro	μ
10^{-9}	0.000 000 001	nano	n
10^{-12}	0.000 000 000 001	pico	p
10^{-15}	0.000 000 000 000 001	femto	f
10^{-18}	0.000 000 000 000 000 001	atto	a

(a) Nonpreferred. Prefixes should be selected in steps of 10^3 so that the resultant number before the prefix is between 0.1 and 1000. These prefixes should not be used for units of linear measurement, but may be used for higher order units. For example, the linear measurement, decimeter, is nonpreferred, but square decimeter is acceptable.

Abbreviations, Symbols, and Tradenames

Abbreviations and Symbols

a crack length; edge length in crystal structure
a_f final cross-sectional area
a_0 original cross-sectional area
A ampere
A area; ratio of alternating stress to mean stress
Å angstrom
AAR Association of American Railroads
AASHTO American Association of State Highway and Transportation Officials
ABS American Bureau of Shipping
ac alternating current
AC air cooled
Ac_{cm} in hypereutectoid steel, temperature at which cementite completes solution in austenite
Ac_1 temperature at which austenite begins to form on heating
Ac_3 temperature at which transformation of ferrite to austenite is completed on heating
Ae_{cm}, Ae_1, Ae_3 equilibrium transformation temperatures in steel
ACD annealed cold drawn
ACI Alloy Casting Institute
ADI austempered ductile iron
AES auger electron spectroscopy
AFNOR Assocation Francaise de Normalisation
AFS American Foundrymen's Society
AISI American Iron and Steel Institute
AKDQ aluminum-killed drawing quality
AMS Aerospace Material Specification
ANSI American National Standards Institute
AOD argon oxygen decarburization
AP armor piercing
APB antiphase boundary
API American Petroleum Institute
AQ as quenched

Ar_{cm} temperature at which cementite begins to precipitate from austenite on cooling
Ar_1 temperature at which transformation to ferrite or to ferrite plus cementite is completed on cooling
Ar_3 temperature at which transformation of austenite to ferrite begins on cooling
AREA American Railway Engineering Association
ASCE American Society of Civil Engineers
ASME American Society of Mechanical Engineers
ASP antisegregation process
ASTM American Society for Testing and Materials
at.% atomic percent
atm atmospheres (pressure)
at. ppm atomic parts per million
AWG American wire gage
AWS American Welding Society
bal balance
bcc body-centered cubic
BCIRA British Cast Iron Research Association
bct body-centered tetragonal
BID Brinell indentation diameter
BOF basic oxygen furnace
BOP basic oxygen process
BS British Standard
BWG Birmingham wire gage
c edge length in crystal structure
CAP consolidation by atmospheric pressure
CC combined carbon
CCI crevice corrosion index
CCR conventional controlled rolling
CCT continuous cooling transformation
CE carbon equivalent
CG compacted graphite
cm centimeter
CP cathodic protection
CQ commercial quality

CSA Canadian Standards Association
CSD controlled spray deposition
CSP compact strip production
CT compact tension; continuous transformation
CTOD crack tip opening displacement
CVD chemical vapor deposition
CVN Charpy V-notch (impact test or specimen)
d diameter
D diameter
da/dN fatigue crack growth rate
dB decibel
DBTT ductile-to-brittle transition temperature
dc direct current
DCB double cantilever beam
DIN Deutsche Industrie-Normen
DIS Ductile Iron Society
DP dual phase
DQ drawing quality
DQSK drawing quality special killed
DRCR dynamic recrystallization controlled rolling
DRI direct reduced iron
DS directional solidification
DT dynamic tear
e natural log base, 2.71828
E Young's modulus
EAC environmental assisted cracking
EAF electric arc furnace
EBW electron beam welding
EDS energy-dispersive spectroscopy
EDXA energy-dispersive x-ray analysis
ELI extralow interstitial
Eq equation
ESR electroslag remelting
et al. and others
eV electron volt
FAD failure assessment diagram
FATT fracture-appearance transition temperature
FC furnace cool
FCAW flux-cored arc welding

fcc face-centered cubic
FG flake graphite
FGD flue gas desulfurization
FGHAZ fine grain heat-affected zone
Fig. figure
FLD forming limit diagram
FN ferrite number
ft foot
FZ fusion zone
g gram
G gauss
G modulus of rigidity
gal gallon
GAR grain aspect ratio
GC grain-coarsened
GCHAZ grain-coarsened heat-affected zone
GMAW gas metal arc welding
GPa gigapascal
gr grain
Gr graphite
GS grain size
h hour
HAZ heat-affected zone
HB Brinell hardness
HCF high-cycle fatigue
hcp hexagonal close-packed
HIC hydrogen-induced cracking
HIP hot isostatic pressing
HK Knoop hardness
hp horsepower
HR Rockwell hardness (requires scale designation, such as HRC for Rockwell C hardness)
HSLA high-strength low-alloy (steel)
HSS high-speed steel
HTLA heat-treatable low-alloy (steel)
HV Vickers hardness
Hz hertz
ICHAZ intercritical heat-affected zone
ID inside diameter
IF interstitial-free (steel)
IFI Industrial Fasteners Institute
IGA inert-gas atomization
IIW International Institute of Welding
in. inch
IR injection refining
IRGCHAZ intercritically reheated grain-coarsened heat-affected zone
ISCC intergranular stress-corrosion cracking
ISO International Organization for Standardization
ISRI Institute of Scrap Recycling Industries
IT isothermal transformation
J joule

J_{ec} Jominy equivalent cooling
J_{eh} Jominy equivalent hardness
JIS Japanese Industrial Standard
K Kelvin
K stress intensity factor
K_{Ic} plane-strain fracture toughness
K_{ISCC} threshold stress intensity to produce stress corrosion cracking
K_f fatigue notch factor
K_t theoretical stress-concentration factor
K_{th} critical or threshold stress intensity
K-BOP Kawasaki basic oxygen process
kg kilogram
km kilometer
KMS Kloeckner metallurgy scrap (process)
kN kilonewton
kPa kilopascal
ksi kips (1000 lbf) per square inch
kV kilovolt
kW kilowatt
ℓ length
L longitudinal; liter
L length
LAST lowest anticipated service temperature
lb pound
LBE Lance bubbling equilibrium
lbf pound force
LBZ local brittle zone
LCF low-cycle fatigue
LDH limiting dome height
LDR limiting draw ratio
LF ladle furnace
LiMCA liquid metal cleanness analyzer
LME liquid-metal embrittlement
ln natural logarithm (base e)
LNG liquefied natural gas
log common logarithm (base 10)
LVDT linear variable differential transformer
m meter
m strain rate sensitivity factor
M_f temperature at which martensite formation finishes during cooling
M_s temperature at which martensite starts to form from austenite on cooling
mA milliampere
MC metal carbide
mg milligram
Mg megagram (metric tonne)
MIE metal-induced embrittlement
MIL military
MIM metal injection molding
min minute; minimum
mL milliliter
mm millimeter

MPa megapascal
mph miles per hour
MPIF Metal Powder Industries Federation
ms millisecond
mV millivolt
MWG music wire gage
n strain-hardening exponent
N newton
N_f number of cycles to failure
NASA National Aeronautics and Space Administration
NDT nil ductility temperature
NDTT nil ductility transition temperature
nm nanometer
NMTP nonmartensitic transformation product
No. number
ns nanoseconds
NTHM net tonne of hot metal
OBM oxygen blown method
OD outside diameter
ODS oxide dispersion strengthened
Oe oersted
ONERA Office Nationale d'Etudes et de Recherches Aerospatiales
OQ oil quenched
OQ & T oil quenched and tempered
OSTE one-step temper embrittlement
oz ounce
p page
P applied load
Pa pascal
PAW plasma arc welding
pH negative logarithm of hydrogen-ion activity
PH precipitation hardenable
P/M powder metallurgy
ppb parts per billion
PPB prior particle boundary
ppm parts per million
PQ physical quality
PREP plasma rotating electrode process
psi pounds per square inch
PTFE polytetrafluoroethylene
PVA polyvinyl alcohol
PVP polyvinylpyrrolidone
PWHT post-weld heat treatment
q fatigue notch sensitivity
Q-BOP quick-quiet basic oxygen process
r plastic strain ratio; radius
R stress (load) ratio; radius; gas constant
RA reduction in area
RCR recrystallization controlled rolling
RD rolling direction
RE rare earth
Ref reference

rem remainder
REP rotating electrode process
RH Rurstahl Hereaus (degasser)
rms root mean square
RSR rapid solidification rate
RTE reversible temper embrittlement
s second
S Siemens
S_a alternating stress amplitude
S_m mean stress
S_r stress range
SACD spheroidized annealed cold drawn
SAE Society of Automotive Engineers
SAW submerged arc welding
SC single crystal
SCC stress-corrosion cracking
SCE saturated calomel electrode
SCF stress concentration factor
SCHAZ subcritical heat-affected zone
SEM scanning electron microscopy
sfm surface feet per minute
SG spheroidal graphite
SGA soluble-gas atomization
SI Système International d'Unités
SMAW shielded metal arc welding
SME solid-metal embrittlement
SMIE solid-metal-induced embrittlement
SNECMA Societe Nationale d'Etude et de Construction de Moteurs
SPC statistical process control
SQ structural quality
SRGHAZ subcritically reheated grain-coarsened heat-affected zone
SS Swedish Standard
SSC sulfide stress cracking
ST short transverse
STB Sumitomo top and bottom blowing (process)
STEM scanning transmission electron microscope
STQ solution treated and quenched
SWG steel wire gage
t thickness; time
T temperature
T_H homologous temperature
T_m melting temperature
T_{nr} no recrystallization temperature
TAPPI Technical Association of the Pulp and Paper Industry
TC total carbon
TCP topologically close packed
TEM transmission electron microscopy
TFS true fracture stress
TH transformation hardened
TIG tungsten inert gas (welding)
TIR total indicator reading
TMCP thermomechanical controlled processing
TME tempered martensite embrittlement
TMP thermomechanical processing
TRIP transformation induced plasticity
tsi tons per square inch
TTT time-temperature transformation
UNI Ente Nazionale Italiano di Unificazione
UNS Unified Numbering System
UTS ultimate tensile strength
USSWG United States steel wire gage
V volt
VAD vacuum arc degassing
VADER vacuum arc double-electrode remelting
VAR vacuum arc remelting
V-D vacuum degassing
VIM vacuum induction melting
vol volume
vol% volume percent
W watt
W width; weight
WDS wavelength dispersive spectroscopy
wt% weight percent
WQ water quenched
WRC Welding Research Council
yr year
° angular measure; degree
°C degree Celsius (centigrade)
°F degree Fahrenheit
⇆ direction of reaction
÷ divided by
= equals
≅ approximately equals
≠ not equal to
≡ identical with
> greater than
≫ much greater than
≥ greater than or equal to
∞ infinity
∝ is proportional to; varies as
∫ integral of
< less than
≪ much less than
≤ less than or equal to
± maximum deviation
− minus; negative ion charge
× diameters (magnification); multiplied by
· multiplied by
/ per
% percent
+ plus; positive ion charge
√ square root of
~ approximately; similar to
∂ partial derivative
Δ change in quantity; an increment; a range
ε strain
ε̇ strain rate
μ friction coefficient; magnetic permeability
μin. microinch
μm micron (micrometer)
μs microsecond
ν Poisson's ratio
π pi (3.141592)
ρ density
σ stress
Σ summation of
τ shear stress
Ω ohm

Greek Alphabet

A, α alpha
B, β beta
Γ, γ gamma
Δ, δ delta
E, ε epsilon
Z, ζ zeta
H, η eta
Θ, θ theta
I, ι iota
K, κ kappa
Λ, λ lambda
M, μ mu
N, ν nu
Ξ, ξ xi
O, ο omicron
Π, π pi
P, ρ rho
Σ, σ sigma
T, τ tau
Υ, υ upsilon
X, χ chi
Ψ, ψ psi
Ω, ω omega

Tradenames

AF-56 is a registered tradename of Allison Gas Turbine, Division of General Motors Corporation.
AL-6X, AL-6XN, AL 29-4C, AL 29-4-2, AL 904L, AL 2205, ALFA IV, E-Brite 26-1, Sealmet, and **203 EZ** are registered tradenames of Allegheny Ludlum Steel, Division of Allegheny Ludlum Corporation.
AM1 is a registered tradename of SNECMA/ONERA.
CM 247 LC and **CMSX** are registered tradenames of Cannon-Muskegon Corporation.
Cronifer is a registered tradename of Vereingte Deutsche Metallwerks.
Cryogenic Tenelon and **Tenelon** are registered tradenames of USS, Division of USX Corporation.

Custom 450, Custom 455, Gall-Tough, Pyromet, TrimRite, 7-Mo PLUS, 18-18 PLUS, 20Cb-3, 20Mo-4, and **20Mo-6** are registered tradenames of Carpenter Technology Corporation.

Discaloy is a registered tradename of Westinghouse Electric Corporation.

DP3 is a registered tradename of Sumitomo Metal America, Inc.

Esshete is a registered tradename of British Steel Corporation.

Ferralium is a registered tradename of Bonar Langley Alloy Ltd.

Hastelloy and **Haynes** are registered tradenames of Haynes International, Inc.

Incoloy, Inconel, Nimocast, and **Nimonic** are registered tradenames of INCO Alloys International, Inc.

JS700 is a registered tradename of Jessop Steel Company.

MAR-M is a registered tradename of Martin Marietta Corporation.

Monit is a registered tradename of Uddeholms Aktiebolag.

MP (Multiphase) is a registered tradename of Standard Pressed Steel Company.

Nitronic and **PH 13-8 Mo** are registered tradenames of Baltimore Specialty Steels Corporation.

PH 15-7 MO, 12SR, 15-5 PH, 17-4 PH, 18 SR, and **21-6-9** are registered tradenames of Armco Advanced Materials Corporation.

PWA 1484 is a registered tradename of Pratt & Whitney Aircraft.

RA85H is a registered tradename of Rolled Alloys, Inc.

René is a registered tradename of General Electric Company.

René 41 is a registered tradename of Allvac Metals Company, a Teledyne Company.

RR 2000 and **SRR 99** are registered tradenames of Rolls Royce, Inc.

Sanicro and **3RE60** are registered tradenames of Sandvik, Inc.

Sea-Cure is a registered tradename of Crucible, Inc.

Stellite is a registered tradename of Deloro Stellite, Inc.

Udimet is a registered tradename of Special Metals Corporation.

Unitemp is a registered tradename of Universal Cyclops Steel Corporation.

Uranus is a registered tradename of Compagnie des Ateliers et Forges de la Loire.

Vitallium is a registered tradename of Pfizer Hospital Products Group, Inc.

Waspaloy is a registered tradename of United Technologies, Inc.

253MA and **254SMO** are registered tradenames of Avesta Stainless, Inc.

Index

A

Abbreviations, symbols, and tradenames 1038–1041
Abrasion-resistant cast irons. *See* Alloy cast irons.
Acicular ferrite steels 148, 400, 404–405
ADI. *See* Austempered ductile iron.
AF1410 steel 431, 446–447
   heat treatment for 447
   properties of 445, 446, 447
AFNOR (French) standards for steels 158
   compositions of 186–189
   cross-referenced to SAE-AISI steels ...166–174
Aggressive environments, and fatigue resistance 677, 681
Aging 641–642, 946–947
   effect on toughness of austenitic stainless steel 947
Agricultural machinery components, hardened steel for 456
Aircraft cord wire 286
Aircraft gas turbine, development of 995
Aircraft quality 254
   of low-alloy steel 209
Aircraft quality plates 237
Aircraft structural quality, of low-alloy steel 209
Air hardenability test 465–466
Air hardening of low-alloy steels 644–645, 646
Air-hardening steels, medium alloy, cold work tool steels 763–765
AISI designations. *See* SAE-AISI designations.
Alloy cast irons 11, 85–104
   abrasion-resistant cast irons 11, 90–98
      abrasion resistance 96–98
      annealing, effect of......................92
      compositions 85, 91
      hardness, conversions....................95
      hardness, of microconstituents 97
      heat treatment 91–92
      mechanical properties 94
      microstructure 92–94, 95
      physical properties 96
      relative toughness 96
      tensile strength 95–96
      transverse strength 96
      wear rates 97–98
   alloying elements, effects of..........11, 86–90
      carbon................................86, 88
      chromium 86, 88–90, 100
      copper 89
      on depth of chill 87
      manganese 87
      molybdenum 89, 90
      nickel 89
      phosphorus 87–88
      on rate of growth 101, 103
      on scaling 101–102, 103
      silicon 86–87, 88, 100
      sulfur 87
      vanadium 89–90
   for automotive service 103–104
   classification of 5
      corrosion-resistant irons 86
      heat-resistant irons 86
      white cast irons 85–86
   corrosion-resistant cast irons.......11, 98–100
      alloying elements, effects of 98
      compositions 85
      high-chromium irons...............99–100
      high-nickel irons 100
      high-silicon irons 98–99
      mechanical properties 99
      physical properties 96
   heat-resistant cast irons 11, 100–104
      alloy ductile irons 103
      alloying elements, effects of 100
      composition 85
      creep 102
      growth 100–101
      high-aluminum irons 103
      high-chromium irons 103
      high-nickel irons 100, 103
      high-silicon irons 102–103
      high-temperature strength 102
      mechanical properties 99
      physical properties 96
      scaling.....................100, 101–102
   inoculants, effects of 90
   oxidation of 101
Alloy ductile irons 103
Alloying
   effect of, on hardenability and tempering of steel 392–394, 395, 396, 468–469
   to modify as-cast properties in gray iron....................................28–29
   for surface stability of superalloys 953, 955, 956–957, 958–959
Alloy steel. *See also* Low-alloy steel.
   alloying elements in 144–147, 456–457
   bulk formability of 581–590
   composition 152–153
   definition of 149
   distortion in heat treatment.........369–370
   embrittlement
      aluminum nitride 694–696
      blue brittleness 692
      graphitization 696–697
      quench-age 692–693
      strain-age 693–694, 695
   fabrication of parts and assemblies 463
   hardenable alloy steels....................453
   induction and flame hardening 463
   machinability of through-hardening ...600–601
   mechanical properties 457–458
   temper embrittlement in 698–703
   tempering 458–459, 462
   weldability of 609
Alloy steel bars 245–246
   aircraft quality and magnaflux quality 246
   axle shaft quality 246
   ball and roller bearing quality and bearing quality 246
   cold-shearing quality 246
   cold-working quality 246
   quality descriptors 253–254
   regular quality 246
   structural quality 246
Alloy steel rod 275
   qualities and commodities for............275
   special requirements for 275–276
Alloy steel spring wire........................287
Alloy wire 286–287
Aluminized wire 281
Aluminum
   and austenitic grain growth 227
   in cast iron 5, 6, 8
   in compacted graphite iron 56
   effect of, on notch toughness 741
   effect of, on steel 146, 577
   in ferrite.............................404, 408
   in malleable iron 10
   resistance of, to liquid-metal corrosion 635–636
Aluminum-coated fence wire 285
Aluminum-coated strand wire 282
Aluminum coating
   base metal and formability 219
   corrosion resistance 219
   handling and storage 220
   heat reflection 219–220
   heat resistance 219
   mechanical properties 218, 219
   painting 220
   for threaded steel fasteners 295
   weldability 220
Aluminum conductor steel reinforced wire 283
Aluminum-killed steels 6, 578
Aluminum nitride embrittlement 694–696
   intergranular fracture in castings 694–695
   panel cracking............................695
   reduced hot ductility.................695–696
Aluminum-zinc alloy coatings 220–221
AMS designations of carbon and alloy steels 153–154, 159, 160–162
Anisotropy
   effect of, on notch toughness 744–745
   in high-strength steel 343, 344, 345
Annealed low-carbon manufacturers' wire 282
Annealed spring wire, characteristics of ...307–308
Annealed temper 850
Annealing 122–123, 132–133, 272
   batch 122
   black 280
   bright 280
   of cold-rolled steel products 132–133
   continuous 122–123
   of dual-phase steels 424–425
   of ductile iron 41
   of gray iron 23–24
   of hot-rolled steel bars 241
   lime bright 280
   of low-alloy steel sheet/strip..............209
   of malleable iron 72–73
   of powder metallurgy high-speed tool steels 783
   salt 280
   solution annealing, austenitic stainless steels 898–899, 912, 945
   spheroidize 209, 280
   of steel wire 280
   strand 280
Antimony
   in cast iron 5, 8
   effects of, on notch toughness 742
   resistance of, to liquid-metal corrosion 635
Anti-segregation process (ASP) 780
API specifications, for steel tubular products 328, 329, 330, 332
Applications
   for alloy cast irons 103–104
   for austenitic manganese steels...........822
   for austenitic stainless steels 948–949
   for carbon and low-alloy steel sheet and strip 200

1044 / Index

Applications (continued)
  for cobalt-base superalloys......965, 967–968
  for compacted graphite iron..............70
  for ductile iron.......................35–38
  for gray iron............................12
  for high-temperature bearing steels...384–386
  for HSLA steels................399, 415–423
  for malleable irons............73, 74, 83, 84
  for precipitation-hardening semiaustenitic
    stainless steels.......................944
  for quenched and tempered martensitic
    stainless steels...................940–942
  for steel plate........................226
  for superalloys........................950
Argon
  crack propagation in...............719, 720
  in stainless steels.....................930
**Argon-oxygen decarburization (AOD)
  process**........................841, 970
Argon-oxygen deoxidation (AOD)..........930
Armature binding wire................851–852
Arsenic
  in cast iron...........................5, 8
  in compacted graphite iron..............59
  effects of, on notch toughness..........742
As-quenched hardness.............471, 476
As-rolled pearlitic steels..............399, 404
  for steel tubular products..328, 329, 330, 331,
    332, 333, 334
ASTM specifications........150, 154, 156, 162,
  163–164, 334
  for alloy steel pressure pipe and
    pressure tubes...................331, 334
  for aluminum coatings.............218–219
  for aluminum-zinc alloy coatings........220
  for austenitic grain size................274
  of carbon and alloy steel pipe..........332
  for carbon and alloy steel pressure tubes...333
  for carbon and alloy steel structural and
    mechanical tubing....................335
  for carbon and low-alloy steels for
    elevated-temperature services.........618
  for carbon steel rod...................274
  for carburizing steels..................483
  for chromate passivation..............215
  for cold-finished steel bars.............251
  for concrete reinforcement rod.........274
  for ductile iron................34, 36, 40, 41
  for flake graphite......................13
  for fracture toughness.................341
  for gray iron...............15, 16, 17, 22
  on hardenability......................464
  for high-carbon steels.................483
  for high-strength carbon and low-alloy
    steels................................390
  for hot-rolled steel bars and shapes...240, 241,
    242, 243, 244, 245, 246, 247
  for low-alloy steel.................208, 209
  for machinability testing for screw
    machines............................593
  for nonmetallic inclusion testing........274
  for notch toughness...................753
  for stainless steel products.............932
  for steel castings..364, 365, 366, 368, 370, 377,
    378, 379
  for steel tubular products..............335
  for structural quality steel plate.....230, 236
  for structural steel....................664
  for terne coatings.....................221
  for threaded fasteners...289, 290, 294, 295, 296
  for tin coatings.......................221
  for tool steels....................757, 759
  for weathering steels..................399
  for wrought tool steels................757
  for zinc coatings..................213–215
Auger analysis, of thermal embrittlement of
  maraging steels...................697–698
Auger electron spectroscopy (AES).........689
Austempered ductile iron (ADI)....34, 35, 37–38
  advantages.........................37–38
  effect of austempering temperature on
    strength and ductility.................40
  heat treatment for..................40, 42
  mechanical property requirements......34

  properties............................42
  specifications........................36
Austempering........................455, 457
Austenite
  in abrasion-resistant cast irons........93–94
  formation of, during intercritical
    annealing...........................425
  in maraging steels....................794
  transformation of, after intercritical
    annealing...........................425
Austenite grains
  influence of size of, on hardenability...392–393
  size of, and steel plate production....227–228
  transformation of, to ferrite...........586
Austenitic-ferritic (duplex) alloys. See Duplex
  stainless steels.
Austenitic grades, of corrosion-resistant steel
  castings.............................913
Austenitic manganese steels.........822–840
  applications.........................822
  as-cast properties................828, 829
    commercial use of castings......828, 829
    heavy sections.....................829
  composition of......................822
    bismuth.......................825–826
    carbon........................822–824
    common alloy modifications.....824–825
    copper............................825
    manganese.............822–824, 825
    phosphorus.................822, 825
    silicon.......................822, 824
    sulfur.............................826
    titanium..........................826
    vanadium.........................825
  corrosion of.........................836
    effect of temperature on........836–837
    arc welding........................838
    magnetic properties...............837
    precautions.......................838
    welding......................837–838
  heat treatment for
    precautions...................830, 831
    procedures...................822, 830
  higher manganese content steels...826–827
    fabrication........................828
    oxidation.........................828
    physical properties.............44–45
  machining......................838–839
    machinable grade..................839
    procedures........................839
  mechanical properties after heat
    treatment.......................830–831
  melt practice........................828
  reheating............................833
  wear resistance......................834
    abrasion testing................835–836
    metal-to-metal contact.........834–835
  work hardening.................831–832
    determination of rate..............832
    methods of....................832–833
    service limitations.................832
Austenitic stainless steels. See also Cast stainless
  steels; Wrought stainless steel.
  aging............................946–947
  applications.....................948–949
  categories of..................931, 944–945
  compositions of..............843, 847–848
  elevated-temperature properties...944–949
  fatigue crack growth.............948, 949
  fatigue properties...............947–948
  forgeability of...................891–893
  formability of...................888–889
  H grades.......................931, 945
  hydrogen embrittlement in............715
  impact toughness....................947
  machinability of.................894–896
  resistance of, to stress-corrosion
    cracking........................726–728
  sensitization of, to intergranular
    corrosion.......................706–707
  sigma phase embrittlement in......709–711
  tensile properties...........934, 945–946
  void swelling in..................655–656
  weldability of...................897–905

Automotive applications
  of alloy cast irons................103–104
  of gray cast irons......................19
  of high-strength low-alloy steels........417
  of stainless steels....................881
Axial-load fatigue test...................861
Axle shaft quality steel..................253

B

Bainite.............................128, 129
Baling wire............................282
**Ball and roller bearing quality and bearing
  quality**........................253–254
Ball and roller bearings, alloy steel wire
  for..............................286–287
Bar, steel. See Alloy steel bars; Carbon steel
  bars; Cold finished steel bars; Hot-rolled
  steel bars and shapes.
Basic oxygen process (BOP)......110, 111, 112
  Kawasaki basic oxygen process.........111
  quick-quiet basic oxygen process.......112
Bearing quality, of low-alloy steel.........209
Bearing steels...................149, 380–388
  carburizing...................381–382, 383
  composition
    carburizing steels..................382
    corrosion-resistant steels...........388
    high-carbon steels.................381
    high-temperature steels............387
  heat treatment, effect of......383, 384, 385
  high-carbon......................381, 382
  high- or low-carbon steels for...24–25, 380–381
  induction-hardened..............380, 381
  mechanical properties
    hardness......................381, 387
    impact strength....................381
    nonmetallic inclusion rating........385
    tensile strength....................381
  microstructure characteristics...381–382, 383
    carburizing...................381–382, 383
    high-carbon....................381, 382
  quality of...................382–384, 385
  special-purpose.................384–388
Bend test.....................582, 583, 610
Birmingham wire gage (BWG) system......277
Bismuth
  in alloy cast irons......................90
  in austenitic manganese steel......825–826
  in cast iron...........................5, 8
  in malleable iron......................10
  resistance of, to liquid-metal corrosion...636
Black annealing........................280
Blackheart malleable iron.................74
Blast furnace......................107–108
  current technology for...........108–109
Blue brittleness........................692
Boiler tube steels..................616, 937
  maximum-use temperature of..........617
Bolts. See also Threaded steel fasteners.
  for elevated-temperature service..296, 620, 631
  relaxation tests.......................624
  selection of steel for..........292, 293, 295
Borides.......................952, 955–956
Boron
  effect of, on hardenability of
    steels........................395, 469–470
  effect of, on notch toughness of steels....741
  at elevated-temperature service.........641
  in high-strength low-alloy steels.........408
  in malleable iron......................10
  in steel..............................145
  in superalloys...................954, 984
Boron steels............................208
Brass-plated wire.......................281
Brazing, of malleable iron............76, 83–84
Bright annealing.......................280
Brinell hardness, of gray iron..........16–17
Brinell hardness number (HB)............40
Brinell indentation diameter (BID).......40
Brinell test
  for ductile iron........................40
  for gray iron....................18–19, 30
British standards (BS) for steels..........158

Index / 1045

compositions of 182–186
cross-referenced to SAE-AISI steels ...166–174
Broaching, applications of P/M
 high-speed tool steel for 785–786
Bronze-coated wire 281
Bulk formability of steels 581–590
 characteristics of
  bulk versus sheet formability 581
  of carbon and alloy steels 581
  tests for 581
 flow localization 584–585
 flow stress and forging pressure 585
 formability characteristics 581
 formability tests 581–584
  bend test........................582, 583
  compression test 582
  ductility testing 582
  hot twist testing 583, 584
  nonisothermal upset test 584
  notched-bar upset test 584
  partial-width indentation test 583
  plane-strain compression test 582–583
  ring compression test 583
  secondary-tension test 583
  sidepressing test 583–584
  tension test 581–582
  torsion test 582
  truncated-cone indentation test 584
  wedge-forging test 583, 584
 microalloyed steels 585–586
  comparison of microalloyed plate
   and bar products 588, 589
  processing of microalloyed bars 587–588
  processing of microalloyed forging
   steels 588–589, 590
  processing of microalloyed plate
   steels 586–587
 stainless steels 889–894

C

Cadmium
 in cast iron 8
 resistance of, to liquid-metal corrosion ...635
Cadmium coatings, for threaded steel
  fasteners 295
Calcium
 in cast iron 5
 in compacted graphite iron 56
 effect of, on machinability of carbon
  steels 599
 effects of, on notch toughness 742
Capped steels 141, 143
Carbide degeneration 985
Carbides 952, 954–955
 in austenitic stainless steels 946–947
 in cobalt-base alloys 986
 formation of in 2¼Cr-1Mo
  steel 632–633, 638, 641–642
 hardness of 394
 solubility in austenite 407
 spheroidization 642
Carbon
 in alloy cast irons 86, 88
 in austenitic manganese steel 822–824
 in cast iron 5
 content, microstructures, and properties
  of steels 127–128, 144, 576
 in ductile iron 40, 43
 effect of, on cast stainless steel corrosion
  resistance 912, 913
 effect of, on hardenability ..392, 465, 467–468
 effect of, on notch toughness 739
 at elevated-temperature service 640
 in ferrite 406
 in nickel-base superalloys 984
 in P/M alloys 809
 solubility in ferrite vs temperature 132
 in wrought stainless steels 872
Carbon and low-alloy steels. *See also* Alloy
  steel; Carbon steel; Low-alloy steel.
 alloy designations and specifications
  for elevated-temperature service ...617–618
 alloying elements, effects of 144–147

aluminum 146
boron 145
carbon 144
chromium 145–146
copper 145
lead 145
manganese 144
molybdenum 146
nickel 146
niobium 146
phosphorus 144
silicon 145
sulfur 144–145
titanium 146
zirconium 147
 chemical analysis 141–142
  heat and product analysis 141
  residual elements.....................141
  silicon content 141
 classification of 140–141
 deoxidation practice 142–143
  capped steel 143
  killed steel 142
  rimmed steel 143
  semikilled steel 142–143
 quality descriptors 143–144
 spheroidization and graphitization in 644
Carbon and low-alloy steel plate. *See* Steel
 plate.
Carbonitrides, precipitation of, in steel ..115–116
Carbon-manganese cast steels 373
Carbon-manganese steels
 composition of 151
 hot ductility tests on 696
 sheet and strip 206, 208
Carbon-manganese structural steels,
  hot-rolled 390, 391
Carbon restoration 261, 263–264
Carbon-silicon steels 208
Carbon steel. *See also* Hardenable steels.
 alloying elements in 144–147, 456–457
 AMS designations 153–154, 159, 160–162
 ASTM specifications 150, 154, 156, 162,
   163–164
 bulk formability of 581
 carbon contents of 147–148, 454–456
 castings and 363
 classification of 147
 composition of 147, 149–151, 363
 definition of 147
 deoxidation practice 148
 distortion in heat treatment 369–370
 elevated-temperature properties 618, 619
 embrittlement
  aluminum nitride 694–696
  blue brittleness 692
  graphitization 696–697
  quench-age 692–693
  strain-age 693–694
 fabrication of parts and assemblies 463
 hardenability 451–454
 induction and flame hardening 463
 international designations and
  specifications for 156–159, 166–194
  British (BS) steel
   compositions......158, 166–174, 182–184
  French (AFNOR) steel
   compositions 158, 166–174, 186–188
  German (DIN) steel
   compositions 157, 166–174, 175–178
  Italian (UNI) steel
   compositions 159, 166–174, 190–191
  Japanese (JIS) steel
   compositions 157–158, 166–174, 180
  Swedish (SS) steel
   compositions 159, 166–174, 193
 machinability of 595–597
  resulfurized 597–599
  with calcium 599
  with lead 599–600
  with nitrogen 599
  with phosphorus 599
  with selenium 599
  with tellurium 599

mechanical properties ...202, 205, 206, 457–458
in pressure vessel fabrication 618
physical properties 195–199
quality descriptors 201, 203
SAE-AISI designations 149–151
sheet formability of................573–579
tempering 458–459, 462
UNS designations..................151, 153
weldability of 608–609
Carbon steel bars, for specific applications. *See
 also* Cold-finished steel bars; Hot-rolled
 steel bars and shapes.
 axle shaft quality 245
 cold-shearing quality 245
 cold-working quality 245
 structural quality 245
Carbon steel plate. *See also* Steel
 plate. 226, 227, 232–233, 235
Carbon steel rod 272
 mechanical properties of 275–276
 qualities and commodities of 272–274
 special requirements for 274
Carbon steel sheet and strip............200–208
 application of 200
 control of flatness 205–208
 direct casting methods 211
 mechanical properties of carbon
  steels 202, 205, 206
 mill heat treatment of cold-rolled steel
  products 202–204
 modified low-carbon steel sheet and
  strip 206, 207
 production of 200–201, 203, 204
 quality descriptors for carbon
  steels 201–202, 203
  commercial quality 201
  drawing quality 201–202
  structural quality 202
 surface characteristics 204–205
 strain aging 204–205
 stretcher strains 204
Carbon steel spring wire, characteristics of ..307
Carburization, at elevated-temperature
  service 643
Carburized hardenability test 464–465, 466
Carburized steels
 characteristics of 380–381
 machinability of 600
Carburizing, as surface treatment in wrought
  tool steels 779
Carburizing bearing steels 381–382, 383
Castability
 of compacted graphite iron 57
 of gray iron 12–13
Cast austenitic stainless steels,
  ferrite in 909, 910–911
Cast cobalt-base superalloys. *See also*
  Polycrystalline cast superalloys.
 compositions of 983
 design of 985–986
 physical properties 983
 stress-rupture properties 985–987
 tensile properties 984
Casting methods, direct 211
Castings, intergranular fractures in 694–695
Cast iron. *See also* Alloy cast irons; Compacted
  graphite iron; Ductile iron; Gray iron;
  Malleable iron. 3–104
 alloying elements, graphitization
  potential of 6
 automotive applications of gray 19
 basic metallurgy of 3–11
 carbon content of 3, 5
 classification of 3–11
 common 3
 compacted graphite irons 3, 8–9
 composition ranges 5, 6
 definition of 3, 12
 ductile iron 3, 7–8
 graphite shape 3, 6
 gray iron 3, 4–7
 heat treatment 7
 liquid treatment of 7
 malleable irons 3, 9–11

Cast iron (continued)
  matrix 3
  microstructures and processing for
    obtaining common commercial...........4
  mottled iron 3
  principles of metallurgy of 3–4
  special 3, 11
  spheroidal graphite iron 7–8
  tensile strength 6, 7
  white iron 3
Cast nickel-base superalloys. *See also*
  Polycrystalline cast superalloys.
  compositions of 982
  design of 983–985
  directionally solidified alloys 996–998
    castability 996–997
    compositions of 996
    heat treatment 997–998
    stress-rupture properties 998
  heat treatment 993
  hot isostatic pressing 993–994
    effect on fatigue properties ...991, 992, 994
  investment casting 989–990
  melting practice 986–988
  microstructures 990–992
    carbides 990–991
    dendrites 990
    eutectic segregation 991
    grain size 992
    porosity 991–992
  single-crystal alloys 998–1006
    castability 998
    compositions 996, 997
    fatigue properties 1002, 1004, 1005
    heat treatment 1000–1002
    microstructure 1000–1002
    oxidation of 1004, 1006
    stress-rupture properties ..1000, 1002, 1004
  stress-rupture
    properties 985, 986, 987, 991, 993
  tensile properties 984
Cast stainless steels 908–929
  composition and microstructure of ...909, 910
  ferrite control 911
  ferrite in cast austenitic stainless
    steels 909, 910–911
  corrosion-resistant steel castings.....909, 912
    compositions 909, 912–913
    corrosion characteristics 915–917
    mechanical properties 909, 914, 917–920
    microstructures 909, 913–915
  grade designations and compositions 908
  C-type (corrosion-resistant) steel
    castings 908–910
  H-type (heat-resistant) steel castings ...910
  heat-resistant cast steels 920
    galling 928–929
    general properties 909, 919, 920–921
    iron-chromium-nickel 922–925
    iron-nickel-chromium 921, 923, 925–927
    magnetic properties 929
    manufacturing characteristics...919, 928, 929
    metallurgical structures 921–922
    properties of heat-resistant
      alloys......920, 921, 924, 925, 926, 927–928
    straight chromium heat-resistant
      castings 922
  heat treatment of.......................911
  homogenization 911
  sensitization and solution annealing of
    austenitic alloys 912
Cast steels. *See also* Steel castings. 363–379
  microalloyed steel castings, applications
    compositions 420
  notch toughness of 746–747
Cavities, formation of, during irradiation 654
Cementite, proeutectoid 127, 129–130
Centrifugal atomization 972–973
Cerium
  in cast iron 5
  in compacted graphite iron 56
  effect of, on steel composition and
    formability 577
  in ferrite 408

  in malleable iron 10
Chain link fence wire 285
Chain quality rod 273
Charpy V-notch test 610–611, 737, 753
  correlation of, to fracture mechanics 753
  for steel castings 367
  variability of results 749–753
Chemical analysis, for classifying steels 141
Chemical composition, effect of, on
  weldability 606, 609
Chemical spot tests 1030, 1031
Chemistry, and single-crystal
  castability 998–1003
Chilling tendency, of compacted graphite
  iron 57
Chill tests, in ductile iron 39
Chloride cracking tests, of stainless steels ...725
Chlorine, corrosion of corrosion-resistant
  steel castings from 913
Chromate passivation 214–215
Chromium
  in abrasion-resistant cast irons...........115
  in alloy cast irons 86, 88–89, 100
  in austenitic manganese steel........824, 825
  in cast iron 6, 28
  at elevated-temperature service 640
  effect of, on corrosion resistance ...912, 913
  effect of, on hardenability 395, 468
  effect of, on notch toughness 741
  in ferrite 408
  in gray iron 22
  in nickel-base superalloys 984
  in P/M alloys 810
  in steel 145–146, 577
  in wrought stainless steels............871
Chromium heat-resistant castings 922
Chromium hot-work steels 762
Chromium-molybdenum cast steels.........374
**Chromium-molybdenum heat-resistant
  steels** 619–630
  ASTM specifications 157
  compositions of 158
  definition of 149
Chromium-molybdenum steels ...149–150, 618–620
  0.5Mo steel 619–620
  1.0Cr-0.5Mo steel 620
  2.25Cr-1Mo steel 620, 645–647
  9Cr-1Mo steel 620, 622, 623, 625, 937
  allowable stresses 625
  compositions 618
  creep-rupture strength 622, 937
  creep strengths 620
  modified chromium-molybdenum
    steels 621, 939
  room-temperature tensile properties 618
  tensile properties at elevated
    temperatures 624
Chromium-molybdenum-vanadium steels,
  for elevated-temperature
  service 619, 620–621, 624, 937, 939
Chromium-silicon steel spring wire and strip,
  characteristics of 306–307
Chromium-silicon steel VSQ wire,
  characteristics of 307
Chromium stainless steel, composition of 912
**Chromium-vanadium steel spring wire
  and strip,** characteristics of 306–307
Chromium-vanadium steel wire,
  characteristics of 307
Circle grid analysis, for steel sheet 575–576
Clamping forces, of threaded steel
  fasteners 300–301
Classification of cast iron 3
Classification of steel 140–141
Clean bright wire finish, for steel wire 279
Cleanliness, and fatigue
  resistance 678–679, 681, 682
Closed-die forgings 337–357
  allowance for machining 352
  decarburization 352
  design for tooling economy 352–353
  design of hot extrusion forgings...354, 355, 356
  machining allowance 357
  mechanical properties 356, 357

  mismatch tolerances 356–357
  design of hot upset forgings 353
  design of specific parts 353–354, 355
  machining stock allowances 353, 354
  tolerances 353
  design stress calculations ...342, 343–346, 347
  fundamentals of hammer and press
    forgings 346
    draft 346, 347, 348
    fillets and radii 347–348, 349
    holes and cavities 348
    lightening holes in webs 348–349
    minimum web thickness 348, 349
    parting line 346, 347, 348
    ribs and bosses 346–347
    scale control 349
  material control 338–339
    combined specifications 339
    critical forging 339
    ductility and amount of forging
      reduction 340, 341
    end-grain exposure 342
    fatigue strength 341, 342
    fracture toughness 341–342
    grain flow 341
    grain size and microconstituents 341
    identification 339
    material specification 339
    quality assurance and quality control ...339
    residual stress 342
    routine production 339
    test plans 339, 340
    tests and test coupons 339–340
    wrought structure and ductility 340
  mechanical properties 342
    anisotropy in high-strength
      steel 343, 344, 345
    grain flow and anisotropy 342–343
  selection of steel for 337
    cost 338
    design requirements 338
    forgeability 338
    microalloyed high-strength low-alloy
      (HSLA) steels 337
    precipitation-hardenable stainless
      steels 337–338
  tolerances 349, 350
    broaching allowance 352
    die wear 351
    draft 351
    flash 351
    hot shearing 351
    length 349–350, 351
    piercing 351–352
    shift or mismatch tolerance 349, 350
    trimming 351
  types of 337
Coatings. *See also* specific coatings.
  for elevated temperatures 296
  tests and designations 212–214, 215
  weight 218–219
Cobalt
  in age hardening in maraging steels.......794
  in MAR-M 247...................1016–1018
  in nickel-base superalloys 984
  in Udimet 700 1014–1016
  in Waspaloy 1014–1016
Cobalt-base alloy
  design for 983, 985–986
  heat treatment for 993
Cobalt-base superalloys. *See* specific types.
  cast cobalt-base superalloys......983, 985–987
  powder metallurgy (P/M) cobalt-base
    alloys 977–980
  wrought cobalt-base superalloys ..950, 962–968
Coefficient of thermal expansion, of gray iron...31
Cogging 971
Cokemaking 107
Cold-drawn steel, machinability of..........601
Cold-finished steel bars 248–271
  bar sizes 248
  classifications 248
  commercial grades 248–249
  heat treatment 260

carbon restoration 261, 263–264
machinability 264–265
mechanical properties 254, 257
  hardness 258
  impact properties 259
  tensile and yield strengths 257–258
product quality descriptors 252
  alloy steel quality descriptors 253–254
  carbon steel quality descriptors 252–253
product types 248
  cold-drawn bars 250–251
  machined bars 249–250
  turning versus cold drawing 251
residual stresses 259
  straightening........................259
  stress relieving 259–261
special die drawing 268–269
  drawing at elevated temperatures ..269–271
  heavy drafts 269
  strength considerations 265–266, 268
Cold finishing, for steel tubular
  products 328–329
Cold finishing quality rod...................273
Cold-forming strip, high-strength low-alloy
  steels for 418–419
Cold heading quality alloy steel rod 275
Cold-heading wire 851
**Cold-rolled high-strength low-alloy
  steels**............................420–421
Cold-rolled steel
  effects of steelmaking practices on
    formability of 577–578
  tensile properties and formability factors
    of 398, 420
Cold-rolled steel products, mill heat
  treatment of 202–204
Cold rolling, in processing of solid
  steel 121–122, 123
Cold working, of maraging steels 795
Cold-work tool steels
  powder metallurgy 786–789
  wrought 763–766
    air-hardening, medium-alloy 763–765
    high-carbon, high-chromium 765
    oil-hardening 765
Commercial quality, of carbon steels 201
Compacted graphite iron 3, 8–9, 56–70
  advantages 70
  applications 70
  castability
    chilling tendency 57
    fluidity 57
    shrinkage characteristics 57
  composition of 5, 8–9, 56–57
  cooling rate 9
  corrosion resistance 68, 69
  damping capacity 69–70
  ferritization tendency 57
  graphite morphology 56
  heat treatment for 9
  liquid treatment of 9
  machinability 68–69
  mechanical properties at elevated
    temperature 63–64
    growth and scaling 63, 66
    tensile properties 63, 65
    thermal fatigue 63–64, 66, 67
  mechanical properties at room
    temperature 57–63
    compressive
      properties 58, 60–61, 62, 63, 64
    fatigue strength 58, 62–63, 65
    impact properties 61–62, 63, 64
    modulus of elasticity 58, 61, 62, 63
    shear properties 61
    tensile properties and
      hardness 57, 58, 59–60, 62
    thermal conductivity 58
  microstructure 56, 61
  physical properties 66–70
    damping capacity 69–70
    sonic and ultrasonic properties ...67–68, 69
    thermal conductivity 58, 64, 66–68, 69
    thermal expansion 67, 68

Compression, flow stress in 585
Compression springs 302, 319–320, 322
Compressive properties
  of compacted
    graphite iron 58, 60–61, 62, 63, 64
  of ductile iron 42, 45
Compressive strength
  of gray iron 17–18, 19
  of malleable irons 82
Concrete-reinforcing bars 246–247
Concrete-reinforcing wire 282–283
Conduit pipe 331
**Conservation of Strategic Aerospace Materials
  (COSAM) program** 1009
**Consolidation at atmospheric pressure
  (CAP process)** 780, 973
Constant-amplitude tests of smooth bars ...369–370
Constant-force springs 302
Constant-lifetime diagram, and fatigue
  resistance 675
**Constant load amplitude fatigue crack
  growth** 370, 371, 372
**Continuous-welded cold-finished mechanical
  tubing** 335
Controlled cooling 272
Controlled rolling 115, 117–118, 131,
                         408–409, 587–588
  conventional controlled rolling 117, 409
  dynamic recrystallization controlled
    rolling 117–118, 409
  mechanical properties of control-rolled
    steel 409
  of microalloyed bar................587–588
  of microalloyed plate 586–587
  recrystallization controlled rolling ...117, 409
  temperature-time schedules 130, 131
**Controlled spray deposition and hot working
  (CSD and Osprey)** 780
Controlled thermal severity test 612–613
Control-rolled steels......................148
  mechanical properties of 409
**Conventional controlled rolling
  (CCR)** 117, 408–409
Conversion guide, metric 1035–1037
Cooling and coiling system
  in processing of solid steel 118–119, 121
  precipitation 119–120, 123
Cooling rate
  of compacted graphite irons 9
  of ductile iron 8
  effect on precipitation strengthening......402
  of gray iron 6–7
  of malleable iron 10
Copper
  in abrasion-resistant cast irons...........115
  in alloy cast irons 89
  in austenitic manganese steel 825
  in cast iron 6, 28
  in compacted graphite iron 57, 59
  in ductile iron 44
  effect of, on hardenability 393, 395
  effect of, on notch toughness 741
  in ferrite 406–407
  in P/M alloys 809
  precipitation strengthening with 411
  in steel 145, 577
Copper-bearing cast steels 374
Coppered finish and liquor finishes, for steel
  wire 279
Copper steels 208
Core hardness........................481–482
Corrosion
  and alloying for surface
    stability 953, 956–957, 958–959
  of austenitic manganese steel 836
  of cobalt-base alloys 968
  resistance of maraging steels to 799–800
  of stainless steels 869–884, 935–936
Corrosion fatigue, of corrosion-resistant steel
  castings 918, 920
Corrosion protection, for threaded
  steel fasteners 291, 296–297
Corrosion resistance 581, 584
  of aluminum coatings 219

of cast steels 376
of compacted graphite iron...........68, 69
effects of carbon on 912, 913
effects of chromium on 912, 913
effects of molybdenum on 912
effects of silicon on 912, 913
of ferritic malleable iron 76
of high-phosphorus weathering steel 400
Corrosion-resistant cast irons. See Alloy cast
  irons.
Corrosion-resistant cast steels, mechanical
  properties of 909, 914, 917–920
Corrosion-resistant steel castings ...908–910, 912
Corten A, corrosion resistance of 400
COSAM program approach 1013
  advanced processing 1018
  alternate materials.................1018–1020
  results 1020–1021
  substitution....................1014–1018
Cotter pin wire 852
Couper-Gorman curves, and
  elevated-temperature service 632
Crack propagation, in argon and liquid
  lithium 720
**Crack tip opening displacement (CTOD)
  tests** 611, 662–663
Crane and vehicle applications, high-strength
  low-alloy steels for 419
Creep. See also Elevated-temperature
  properties; Elevated-temperature service.
  of cast stainless steels 920, 924, 927
  classical behavior 629
  definition of................622, 927, 932
  of gray iron 27, 102
  in heat-resistant alloys 920, 924, 927
  relationship with hardness 640
  of wrought stainless steels 932
Creep embrittlement, and elevated-temperature
  service 626
Creep-fatigue interaction
  and chromium-molybdenum
    steels 625, 632, 633
  effect of ductility on 633, 935
  and wrought stainless steels 934–935
Creep-rupture strength. See also Stress rupture.
  of 1Cr-1Mo-0.25V steel 629
  of 2¼Cr-1Mo steel 622, 623, 636, 647
  of 9Cr-1Mo steel 622, 623
  austenitic stainless steels 934
    304 stainless steel 622
    effect of solution annealing temperature
      on 945
  carbon steel 629
  chromium-molybdenum steels 622
  chromium-molybdenum-vanadium steels ..619
  effect of heat treatment on
    2¼Cr-1Mo 641, 642
    austenitic stainless steel..............945
  effect of microstructure on 638
  effect of spheroidization on 644
  ferritic steels 622, 939, 941
  maximum-use temperatures based on 617
Creep strength
  of 2¼Cr-1Mo steel.............635, 636, 647
  of carbon steel 629
  of chromium-molybdenum
    steels 620, 623, 629
  chromium and molybdenum, effect of 643
  compared with hardness 640
  of ductile iron 48, 49, 50
  effects of chromium on 643
  maximum-use temperatures based on 617
  of wrought stainless steels 932–933
Critical temperatures 115–116, 126–127, 130
Crucible Particle Metallurgy (CPM) process..780
CSA specifications
  of carbon and alloy steel pipe 332
  of steel tubular products ...328, 329, 330, 332
Cupping and drawing, for steel tubular
  products 328
Cutting speed 591–592
Cycle annealing 261

D

D-6a/6ac steel 436
   heat treatment for 436
   properties of 436–437, 438
Damping capacity
   of compacted graphite iron 69–70
   of ductile iron 45–46
   of gray iron 31–32
   of malleable iron 82, 84
Decarburization
   at elevated-temperature service 642–643
   and fatigue resistance 679–680, 681
   in hot-rolled steel bars 241
   and notch toughness in wrought steels 746
   and steel plate imperfections 230
   of steel springs 308–309
Decarburization limits
   for alloy steel rod 274
   for carbon steel rod 274
Density
   of cast steel 374
   of ductile iron 50
   of gray iron 31
Deoxidation practice
   capped steel 143
   influence of, on graphitization 696
   killed steel 142
   rimmed steel 143
   semikilled steel 142–143
   steel plate production 226–227
Design stress 316–317, 319
Desulfurization 109, 110, 228
Die springs 302
Dimensional stability, of gray iron 26–28
DIN (German) standards for steels 157
   compositions of 175–179
   cross-referenced to SAE-AISI steels ...166–174
Direct casting methods, of carbon and low-alloy steel sheet and strip 211
Direct-cooled forging microstructures,
   processing of 137, 138
Directional properties, of steel plate 238
Directionally solidified superalloys. See also Cast nickel-base superalloys.
   chemistry and DS castability 996–997
   heat treatment and mechanical
     properties 997–998
Directional solidification (DS) casting process ..995
Distortion and safety in hardening of wrought tool steels 777
Double-wall brazed tubing 333–334
Downgrading 1028
Drawing quality
   of carbon steels 201–202
   of low-alloy steel 208–209
Drop-weight test 610
   for steel castings 367–368
Dry-drawn finish 279
Dual-phase steels 148, 400, 405, 424–429
   cold-forming strip 417
   heat treatment of
     annealing techniques and steel
       compositions 424–425
     austenite transformation after intercritical
       annealing 425
     ferrite phase changes during intercritical
       annealing 425–426
     formation of austenite during intercritical
       annealing 425
   forming properties 398, 418, 420
   mechanical properties
     ductility 427
     tempering and strain aging 427–428
     work hardening and yield
       behavior 424, 426
     yield and tensile strength 417, 426–427
   new advances in 428, 429
Ductile iron 3, 7–8, 33–55
   advantages of 33–34, 37
   alloys 34
   annealing 41
   applications for 34–38
   austempered 34, 35, 36, 37–38, 42
   compared to gray or malleable iron 33–34
   composition of 33, 34, 35
   formation of graphite during solidification ..33
   graphite shape and distribution 38–39
   hardenability 48–50
   heat treatment 38, 40, 41–42
   machinability of 52–53, 54
   mechanical properties 36, 40, 42–43
     compressive properties 42, 45
     composition, effect of 40, 43–44
     creep strength 48, 49, 50
     damping capacity 45–46
     at elevated temperatures 48, 49
     fatigue strength 39, 46–47, 48
     fracture toughness 46, 47
     graphite shape, effect of 44–45
     hot-tensile properties 48, 52
     impact properties 40, 43, 44, 45, 46
     notch sensitivity 48
     section size, effect of 45
     strain rate sensitivity 47
     stress-rupture properties 48, 51, 52
     tensile properties 37, 42, 45
     torsional properties 42, 45
   need for risers 33–35
   normalizing 41–42
   physical properties
     density 50
     electrical and thermal relationship ..50–51, 52
     electrical resistivity 51, 53
     magnetic properties 51–52
     thermal properties 49, 50
   quenching 38, 41–42
   shrinkage allowance 34
   specifications 34–35, 36
   strength and toughness of 33
   stress relieving 41
   surface hardening 42
   tempering 41–42
   testing and inspection 37, 38, 39–41
   welding of 53–55
Ductile-to-brittle transition 737–739
Ductility
   in closed-die forgings 340–341
   of dual-phase steels 427
   and elevated-temperature
     service 626–627, 633, 634
   and fatigue resistance 678
   of steel castings 365
Ductility testing 582
Duplex stainless steels. See also Cast stainless steels; Wrought stainless steels.
   compositions of 843, 847–848
   elevated-temperature properties of 947
   forgeability of 894
   formability of 889
   machinability of 896
   resistance of, to stress-corrosion cracking ..727
   sensitization in 707–708
   sigma phase embrittlement in 711
   stress-corrosion cracking resistance to boiling magnesium chloride 727
   tensile properties of 856, 858
   weldability of 904–905
Dynamic recrystallization controlled rolling (DRCR) 117–118
Dynamic tear properties
   of compacted graphite iron 61, 64
   of ductile iron 40, 43, 45, 46

E

8640 steel 438
   heat treatments for 439
   properties of 439
   processing of 438
Elastic constants, of steel castings 374
Electrical conductivity, and thermal conductivity of ductile iron,
   relationship between 50–51, 52
Electrical or conductor applications, wire for ..283
Electrical properties
   of gray iron 31
   of steel castings 374
Electrical resistivity, of ductile iron 51, 53
Electric furnace steelmaking 111
Electric resistance welding, for steel tubular products 327–328
Electrogalvanizing 212, 217
Electroplating, and notch toughness in wrought steels 746
Electroslag remelting (ESR) ...930, 968–969, 970
Elevated-temperature properties. See also Elevated-temperature service.
   of austenitic stainless steels 944–949
   of carbon steels 619, 629, 640
     creep-rupture strength 629
     creep strength 629
     hardness 640
     maximum-use temperatures 617–618
     stress rupture 619
   of cast steels 376–377
   of chromium-molybdenum steels 617–652
   of compacted graphite iron 63–66
   creep-fatigue
     interaction 625, 632, 633, 934–935
   effect of composition on 639–641
   effect of heat treatment on 638–639
   effect of microstructure on 638
   of ferritic stainless steels 936–937
   of gray iron 22, 26
   of heat-resistant cast alloys 921, 923, 927
   of low-alloy ferritic steels 620–636
     creep behavior of 2¼Cr-1Mo
       steel 635, 636, 647
     creep strengths 620, 623
     creep rupture of 2¼Cr-1Mo and 9Cr-1Mo
       steel 622
     effect of ductility on fatigue endurance ...633
     elastic and shear modulus of 2¼Cr-1Mo
       steel 628
     fatigue properties 624–626, 633, 649
     relaxation strengths 631
     stress rupture 619, 623, 629, 630,
                           636, 641, 642, 644
     tensile strengths......................624
   of martensitic stainless steels 939–942
   of precipitation-hardening steels 942–944
   of steel plate 238
Elevated-temperature service
   for 2¼Cr-1Mo steel 618, 645–647
   bolt steels 296, 620
   carbon and low-alloy steels for 617–618
     alloy designations and specifications ...618
     maximum-use temperatures 617
   carbon steels 618, 619, 629, 640
   chromium-molybdenum steels ...619–620, 623
   chromium-molybdenum-vanadium
     steels 620–621, 624
   corrosion 629–636
     hydrogen damage 632–634, 639
     oxidation 617, 629–630, 636
     resistance to liquid-metal..........634–636
     sulfidation 630–632, 636, 637
   creep-resistant low-alloy steels 619–621
   data presentation and
     analysis 627–629, 634, 635
   ductility and toughness 626–627
   fatigue 624–626, 632, 633
     creep-fatigue interaction 625, 632, 633
     fatigue-crack growth 625, 633
     load frequency, effect of 624
     thermal 626
   long-term exposure 623, 627
   long-term tests 622–624, 629
     creep strength 620, 623
     relaxation tests 624, 631
     stress rupture ..620, 622, 623–624, 629, 630
   mechanical properties 621–622, 636–645
   modified chromium-molybdenum
     steels 621, 622, 625, 626
   short-term tests 622, 628
   thermal expansion and
     conductivity 647, 651, 652
Embrittlement of iron 689–691
   other impurities, effect of 691
   oxygen, effect of 689–690
   selenium and tellurium, effect of 691

sulfur, effect of 690–691
Embrittlement of steels. See also Temper embrittlement in alloy steels. 689–736
  475 °C embrittlement 708
  in alloy steels 691–697
    aluminum nitride embrittlement 694–696
    blue brittleness 692
    graphitization 696–697
    quench-age embrittlement 692–693
    strain-age embrittlement 693–694
  aqueous environments causing stress-corrosion cracking...............724
  in austenitic stainless steels
    hydrogen-stress cracking and loss of tensile ductility 715
    sensitization 706–707
    sigma phase embrittlement 709–711
    solution pH.........................727
    stress-corrosion cracking resistance to boiling magnesium chloride......725–727
  in carbon steels 691–697
    aluminum nitride embrittlement 694–696
    blue brittleness 692
    graphitization 696–697
    quench-age embrittlement 692–693
    strain-age embrittlement 693–694
  creep embrittlement 626
  in duplex stainless steels
    sensitization 707–708
    sigma phase embrittlement 711
    stress-corrosion cracking to boiling magnesium chloride 727
  in ferritic stainless steels
    sensitization 707
    sigma phase embrittlement 709–711
  formation of flakes in steels.........716–717
  in heat-treated martensitic stainless steels, solution pH......................727–728
  hydrogen damage processes 711
  hydrogen environmental embrittlement......................711–712
  in iron-base alloys, solid-metal embrittlement.........................721
  in iron-nitrogen and iron-carbon alloys, quench-age embrittlement 692–693
  in leaded alloy steels, solid-metal embrittlement.........................721–722
  in line pipe steels, hydrogen-stress cracking and loss of tensile ductility 716
  liquid-metal embrittlement 717–721
  in low-carbon steels
    quench-age embrittlement 692
    strain-age embrittlement 693–694
  in maraging steels
    hydrogen-stress cracking and loss of tensile ductility.....................715–716
    thermal embrittlement 697–698
  neutron irradiation embrittlement 722–723
  overheating............................697
  presence of facets 697
  upper shelf energy 697
  parameters affecting stress-corrosion cracking............................724
    intergranular corrosion or slip dissolution.........................724
    pit geometry 724
    temperature 724
  in precipitation-hardenable stainless steels, solution pH......................727–728
  properties and conditions producing stress-corrosion cracking..........723–724
    alloy susceptibility to stress-corrosion cracking.........................723–724
    static tensile stresses 724
    stress-corrosion-cracking-inducing chemical species..............................724
  quench cracking 698
  stress-corrosion cracking verification procedures........................724–725
    chloride cracking tests 725
    fracture mechanics methods 725
    slow strain rate tests 725
    U-bend testing 725
  in tool steels, hydrogen-stress cracking and loss of tensile ductility 715
End-grain exposure, in closed-die forgings...342
End hooks, stresses in 321
End-quench hardenability test...452, 454, 470, 471
Engineering properties, of steel castings ..376–378
Eutectics, types of 3
Extension springs 302, 320–321
Extra smooth clean bright wire finish, for steel wire 279

F

475 °C embrittlement 708
4130 steel
  heat treatments for 431
  properties of........................431–432
4140 steel 432
  heat treatments for 432
  properties of...........................432
4340 steel.............................432–433
  heat treatments for 433
  properties of 432, 433–434
Fabrication
  of steel plate.........................238–239
  of threaded steel fasteners 300
  of wrought tool steels................774–778
    distortion and safety in hardening 777
    grindability 775
    hardenability 775–777
    machinability 774–775
    resistance to decarburization 778
    weldability 775
Fabrication characteristics of stainless steels........................887–905
  forgeability 889–894
  machinability 894–897
  sheet formability 888–889
  weldability.........................897–905
Fabrication characteristics of steels
  bulk formability of carbon and low-alloy steels 581–590
  machinability of carbon and low-alloy steels 591–602
  sheet formability of carbon and low-alloy steels 573–580
  weldability of carbon and low-alloy steels 603–613
Fabrication weldability tests...........611–613
Failure analyses, for matching steel properties to requirements 338
Fasteners, wire for 284
Fastener tests 296
Fatigue
  causes of 673
  definition of 673
  and elevated-temperature service 624–626, 632, 633
  environmental effects on 624–625, 643, 645, 677, 681
  prevention of 673–674
Fatigue characteristics, of HSLA steels 413
Fatigue-crack growth
  and elevated-temperature service 625, 633
  in structural steel 663–664, 665, 666
Fatigue crack initiation, causes and prevention of, in threaded steel fasteners 298, 300
Fatigue failures, for threaded steel fasteners 297–299, 300, 301
Fatigue limit
  and fatigue resistance....................675
  of ferritic malleable iron 75
  of gray iron.........................19–22
Fatigue notch factor, and fatigue resistance ..675
Fatigue notch sensitivity
  and fatigue resistance................675–676
  in gray iron........................21–22
  of steel plate 369–370
Fatigue properties
  of corrosion-resistant steel castings ..918, 920
  of steel castings 369–370
Fatigue resistance of steels...........673–688
  application of fatigue data 673–688
  comparison of fatigue testing techniques 687
  cumulative fatigue damage 677, 686
  discontinuities 687
  estimating fatigue life 677, 684, 685–686
  estimating fatigue parameters.......683–684
  load data gathering 687–688
  mean stresses 686–687
  notches 686
  scatter of data 682–683, 684
  metallurgical variables of fatigue behavior 678
    aggressive environments 677, 681
    cleanliness................678–679, 681, 682
    composition 681
    creep-fatigue interaction 681
    ductility 678
    grain size 681
    macrostructure differences 681
    microstructure 681, 683
    orientation of cyclic stress 682, 683
    residual stresses 591, 680–681, 682
    strength level 676, 678, 679, 681
    surface conditions 677, 679
    tensile residual stresses 681
  strain-based approach to fatigue ...677–678, 679
  stress-based approach to fatigue ..675, 676, 677
    correction factors for test data 676–677
  symbols and definitions 674
    applied stresses 674
    constant-lifetime diagram 675
    fatigue limit 675
    fatigue notch factor 675
    fatigue notch sensitivity 675–676
    fatigue strength 675
    nominal axial stresses 674
    S-N curves 674–675
    stress concentration factor 675
    stress ratio 674
Fatigue strength
  in closed-die forgings 341, 342
  and fatigue resistance 675
  of ductile iron 39, 46–47, 48
  of steel plate 238
  of threaded fasteners 298, 300
Ferrite
  in cast austenitic stainless steels ...909, 910–911
  proeutectoid 129–130
  strengthening mechanisms in high-strength low-alloy steels..............401, 402, 403, 404, 405, 406–408
    precipitation strengthening 403
    solid solution strengthening 400
  transformation of austenite grains to 586
Ferrite-pearlite microstructures, processing of..............................127, 130–131
Ferrite phase, changes in, during intercritical annealing........................425–426, 427
Ferritic grades, of corrosion-resistant steel castings 913
Ferritic malleable iron 72, 75–76
  alloying elements 73, 74, 75
  brazing 76
  corrosion resistance 76
  elevated-temperature properties of 76
  fatigue limit 75
  fracture toughness 75–76, 80
  graphite content 75
  heat treatment 75
  mechanical properties...............75–76
  microstructure 72
  modulus of elasticity 75
  stress-rupture plot 75
  tensile properties...................73, 75
  welding 76
Ferritic microstructures, processing of 127, 131–133
Ferritic stainless steels 842, 936
  categories of 845
  compositions of 843, 847–848
  elevated-temperature properties 936–937
  forgeability of 893
  formability of 889
  machinability of 894

Ferritic stainless steels (continued)
  notch toughness of 859
  production of 930
  sensitization 707
  sigma phase embrittlement in 709–711
  tensile properties of 856, 860
  weldability of 900, 901–902, 903

Ferritic steels
  elevated-temperature properties of ...617–652
  formability of 889
  heat-resistant high-chromium 937
    boiler tubing 937, 938–939
    historical background 937–938
    turbine rotors 937, 938
    without vanadium 931, 936–938
  thermal expansion and conductivity
    of 647, 651, 652
  void swelling in 656–657

Ferroalloy/deoxidizer additions, in
  steelmaking......................111–112

Ferrosilicon, in alloy cast irons 90

Ferrosilicon-based inoculant 39, 90

Ferrous powder metallurgy materials 801–821
  applications of powder forged parts ..817–818
  designation of P/M materials 804, 805
  heat treatment of 809, 810–811
  infiltration 807–808, 810
  material considerations 812
    hardenability 812–813
    inclusion assessment 813
  mechanical and physical properties of
    sintered ferrous P/M materials 805–806
    composition 809–810
    porosity 802, 805, 806, 807–809
  mechanical properties 814
    compressive yield strength 815, 817
    effect of porosity on mechanical
      properties 815, 817
    forging mode 814
    hardenability 814
    heat treatments 814
    tensile, impact, and fatigue
      properties 814–815, 816
  metal injection molding (MIM)
    technology 818–819
    advantages of 819
    applications 820
    factors impeding growth 819–820
    mechanical properties 820
  metal powder characteristics and
    control 801, 802
    apparent density 802
    chemical composition 803
    compressibility....................802
    flow time 802
    green strength 802, 803
    oxide content 802
    particle size distribution 802
    sampling 801–802
    sintering characteristics 802–803
  powder compacting 802, 803
  powder forging......................812
  powder preparation 803
  process capabilities 801
  process considerations 813
    metal flow in powder forging 814
    powder forging....................814
    preforming 813
    secondary operations 814
    sintering and reheating..............814
    tool design 814
  quality assurance for P/M parts 815–816
    density 817, 819
    magnetic particle inspection 816
    metallographic analysis 817
    nondestructive testing 817
    part dimensions and surface finish...816, 818
  re-pressing 811, 812
  secondary operations 805
  sintering 803–804
    equipment 804–805
    techniques 804

Fine steel wire 286
Fine wire quality rod 273

Finishes 279–280
Flake graphite 13
Flakes, formation of, in steels 716–717
Flatness, control of, in carbon steel sheet
  and strip 205–206
Flat springs 302, 305, 306, 307
Flat wire 277
Flexural fatigue test 861
Flow localization 584–585
Flow stress, and forging pressure 585
Fluidity
  of compacted graphite iron 57
  of gray iron........................12–13
Foil
  finishes for 848
  mechanical properties 848
  stainless steel 848
Forgeability of steels. *See* Bulk formability of
  steels.
Forged parts 455
Forgings. *See also* Closed-die forgings.
  alloy steel wire for 287
  definition of 337
  direct-cooled forging 137, 419
  microalloyed forgings ..137, 358–362, 419, 588
    applications and compositions of......420
    carbon contents of 585
    control of properties 588
    forging temperatures, effects of 588
    generations of 359–360
    properties of 360–361, 588, 599
Forging pressure, and flow stress 585
Forging quality plates 236, 237
Forging temperature, effects of 588–589
Formability. *See also* Sheet formability of steel.
  of aluminum coatings 219
  bulk. *See* Bulk formability of steels.
  of cold-rolled steels 420
  definition of 573
  of dual-phase steels 398, 420
  of interstitial-free steel 398
  of preprimed sheet 222
  of stainless steel 888–889
  sheet. *See* Sheet formability of steel.
  of steel plate........................238
Forming properties of sheet steels ..398, 888–889
Foundry practices, for corrosion-resistant steel
  castings 928
**Fracture appearance transition temperature
  (FATT)** 738
  effect of composition on....699–700, 701, 702
Fracture mechanics, correlation of Charpy
  V-notch to 753
Fracture mechanics tests
  for steel castings 368–369
  for stress-corrosion cracking 725
Fracture resistance, assessment of 662–663
Fracture testing, for carbon steel rod 274
Fracture toughness
  of closed-die forgings 341–342
  of ductile iron 46, 47
  of ferritic malleable iron 75–76
  of pearlitic and martensitic malleable
    iron 74, 80, 82
  of steel castings 377–378
  of structural
    steel........397–398, 663, 664, 666–667, 669
  of welded structures ...667–669, 670, 671, 672
Fretting, and fatigue resistance 679
FULDENS process 781
Furnaces 108–109, 110
  basic oxygen furnace 110
  blast furnace........................109

G

Galfan 218
Galling, of stainless steel castings 928–929
Gallium, resistance of, to liquid-metal
  corrosion 636
Galvanized bridge wire 282
Galvanized rope wire 284
Galvanized wire 281

Galvanizing
  definition of 212
  electrogalvanizing................212, 217
  hot dip 212, 216–217
Gamma double prime 951–952, 953–954
Gamma matrix 951, 952
Gamma prime 951, 952–953
Gas pipelines, HSLA steels for 416–417
Gatorizing 974
Gear manufacturing, P/M high-speed tool steel
  for 786
General spring quality wire,
  characteristics of 303–304, 305–307
Gleeble unit 581
Gouging wear ratio, of abrasion-resistant cast
  iron 119
Gouging wear test, of abrasion-resistant cast
  iron 120
Grain boundary 952
Grain-boundary chemistry 954
Grain flow, in closed-die forgings ..341, 342–343
Grain refinement, in ferrite 404, 406
Grain size
  in closed-die forgings 341
  effect of, on hardenability of steel ...393, 470
  effect of, on notch
    toughness 115, 744, 748–749
  effect of, on precipitation
    strengthening 402, 403
  effect of, on strength of ferrite 115
  effect of, on strength of martensite.......393
  effect of, on stress-corrosion cracking 728
  effect of, on temper embrittlement in
    alloy steels 700–702
Graphite
  in compacted graphite iron 56, 57, 61
  in ductile iron 38–39, 44–45
  in gray iron.......................24–26
Graphitization 696–697
  at elevated-temperature service 642
Gray iron 3, 4–7, 12–32, 85, 100–104
  alloying elements for 6, 100–104
  alloying to modify as-cast properties ...28–29
  applications of 12
  automotive applications of 19, 104
  castability of.......................12–13
  classes of 12
  composition of 4–6, 19
  cooling rate for 6–7
  creep in 102
  dimensional stability 26–28
    creep 27, 102
    growth 26–27, 101
    machining practice 27–28
    residual stresses 27, 28
    scaling 27, 101–102
    temperature, effect of..............26–27
  elevated-temperature properties...........26
  fatigue limit in reversed bending 19–22
  fatigue notch sensitivity 21–22
  fluidity............................12–13
  heat treatment 7, 21, 29–31
    hardenability 29, 30
    localized hardening 31
    mechanical properties 29–31
  impact resistance.....................23
  liquid treatment of 7
  machinability of....................23–24
    adhering sand......................23
    annealing 23–24
    chill 23
    machinability rating 23
    shifted castings 23
    shrinks............................23
    swells.............................23
  mechanical properties of..............19, 20
  microstructure of....................13–14
  physical properties 31–32
    coefficient of thermal expansion 31
    damping capacity 31–32
    density 31
    electrical and magnetic properties 31
    thermal conductivity 31
  pressure tightness 22–23

prevailing sections 16
room-temperature structure 14
scuffing resistance 24-25
   chemical composition, effect of 25
   graphite structure, effect of 24-25
   matrix structure, effect of 25
   surface finish effects 25
section sensitivity 14-16
   section size, effects of...............14, 15
   volume/area ratios 15-16
shakeout practice, effect of 28
solidification 13-14
tensile strength
   influence of CE on 5
   influence of composition and cooling rate
     on 6
test bar properties 16-19
   compressive strength 17-18, 19
   elongation 18
   hardness 18-19, 21
   modulus of elasticity 18, 20, 21
   tensile strength 18
   testing precautions 17
   torsional shear strength 18, 20
   transverse strength and deflection........18
   typical specifications 17, 18, 19
   usual tests 16-17, 18
wear.....................................24
   abrasive wear 24
   adhesive wear 24
   corrosive wear 24
   cutting wear 24
wear resistance 25-26
   graphite structure, effect of 25-26
   matrix microstructure, effect of 26
Grindability, of wrought tool steels 775
Grit blasting, of maraging steels 797-798
Ground-air-ground cycle in aeronautics 687
Growth and scaling, for compacted graphite
   iron 63, 66

H

H11 die steels, at elevated temperatures 621
H11 modified steel 439
   heat treatment for 440-441
   processing of 439-440
   properties of 440, 441
H13 steel 441-442
   heat treatments for 442-443
   processing of 442
   properties of 431, 442, 443-444
HA alloy 922
Hafnium, in nickel-base superalloys 984
Hall-Petch relations 115
**Hammer and press forgings, fundamentals
   of** 346
   draft 346, 347, 348
   fillets and radii 347-348, 349
   holes and cavities 348
   lightening holes in webs 348-349
   minimum web thickness 348, 349
   parting line 346, 347, 348
   ribs and bosses 346-347
   scale control 349
Hammers, hardened steel for 456
Hand cutting tools, hardened steel for 456
Hard-drawn spring wire,
   characteristics of..........303-304, 305-306
Hardenable steels. See also Hardenability
   of carbon and low-alloy steels. 451-463
   compositions of 452, 453
     alloy steels 453
     carbon and carbon-boron steels 452
   distortion during heat treating 462
   effect of alloying on quenching 456-457
   effect of carbon content on
     hardenability 454-456
     low-carbon content 454
     medium-carbon content 454-455
     high-carbon content 455-456
   flame and induction hardening 463
   selection of alloy H-steels 460-461
   tempering of 458-459

Hardenability
   of ductile iron 48-50
   effect of, on weldability 603-604, 606
   of gray iron 29, 30
   of wrought tool steels................775-777
Hardenability curves 485-570
**Hardenability of carbon and low-alloy
   steels**..................................464-484
   alloying elements 394, 395, 468-469
   boron 469-470
   calculation of hardenability 467
   carbon content, effect of ...465, 467-468, 469
   determining hardenability requirements
     as-quenched hardness 471, 476
     depth of hardening 471
     hardenability versus size and
       shape........465, 467-468, 469, 473, 477
     quenching media 471-473
   general hardenability selection
     charts 473, 474-476, 478, 479, 480
     estimating hardenability 479
     rectangular or hexagonal bars
       and plate 478-479, 480, 481
     scaled rounds 478, 480
     tubular parts 479
   grain size, effect of 393, 470
   H-steels 474-476, 480-481
   low-hardenability steels 466
     hot-brine test 466, 467
     SAC test 466-467, 468
   steel castings 483
   steels for case hardening 481
     applications 482
     core hardness 481-482
   testing of 464
     air hardenability test 465-466, 467
     carburized hardenability test ..464-465, 466
     continuous-cooling-transformation
       diagrams 466
     Jominy end-quench test 464, 466
   use of charts 479-480, 482, 483
   use of hardenability limits 481
   variations within heats 459, 470, 471, 472
   hot working 470-471, 473
Hardening by quenching, for hot-rolled steel
   bars 241
Hardening wrenches, hardened steel for 456
Hardness
   of carbon steel
     as a function of carbon content 127
     as a function of temperature 640
   of gray iron 18-19, 21
Hardness test
   for gray iron 16-17, 18-19, 21
   for hot-rolled steel bars..............242, 243
Hastelloy alloys 971
Haynes 25 967
Haynes 188 967
HC alloy 922
HD alloy 922
Heading, alloy steel wire for 287
**Heat-affected zone microstructure,
   effect of,** on weldability 605-606, 608
Heat analysis, for classifying steels 141
Heat-resistant cast irons. See Alloy cast irons.
Heat-resistant cast steels
   galling 928-929
   general properties 909, 919, 920-921
   iron-chromium-nickel 922-925
   iron-nickel-chromium 921, 923, 925-927
   magnetic properties 929
   manufacturing characteristics ...919, 928, 929
   metallurgical structures 921-922
   properties of heat-resistant alloys ...920, 921,
       923, 924, 925, 926, 927-928
   straight chromium heat-resistant castings ...922
Heat-treatable low-alloy steels, weldability
   of 609
Heat treatment
   of abrasion-resistant cast iron 91-92
   for austenitic manganese steel 829-831
   of compacted graphite irons 9
   distortion in 462
   of ductile iron 8, 9, 38, 40, 41-42

effect of, on bearing steels......383, 384, 385
effect of, at elevated-temperature
   service 638-639, 642
of ferritic malleable iron 75
of ferrous powder metallurgy
   materials.................809, 810-811
of gray iron 7, 21, 29-31
of hot-rolled steel bars 241
of malleable iron 10-11, 75, 76-80
of maraging steels 795-797
of pearlitic and martensitic malleable
   iron...................................76-80
powder metallurgy high-speed tool
   steels 782-783
   annealing 783
   hardening 783
   stress-relieving (before tempering) 783
   tempering 783
of stainless steel castings............911-912
of steel castings 367, 370-371
of steel plate.....................230-232
Heavy machinery parts, hardened steel
   for...............................455-456
HF alloy 922
HH alloy 922-924
HI alloy 924
High-aluminum irons 103
High-carbon, high chromium, cold-work
   wrought tool steels 765
High-carbon bearing steels 381, 382
High-carbon cast steels 372
High-carbon steels. See also Carbon steel.
   for bearings 24-25, 380-381
   definition of 148
   weldability of 609
High-carbon wire for mechanical tensioning ..283
High-chromium irons 103
High-nickel irons 103
**High-nickel steels for low-temperature
   service**..................392, 396-397, 398
**High-pressure compaction and hot working
   (STAMP process)** 780
High-silicon irons 102-103
High-speed tool steels
   powder metallurgy 781-786
     alloy development 784-785
     applications 785-786
     cutting tool properties 784
     heat treatment 782-783
     manufacturing properties 783-784
     sintered tooling 786
   wrought............................759-762
     molybdenum........................759
     tungsten 759-762
High-strength carbon and low-alloy steels 389
   quenched and tempered low-alloy
     steel 391-392
   effects of alloying elements
     on 392-394, 395, 396
   high-nickel steels for low-temperature
     service 392, 396-397, 398
   mechanical properties ...389, 392, 396, 397
   microalloyed quenched and tempered
     grades 394-396
   structural carbon steels 389
   high-strength structural carbon
     steels 390-391
   hot-rolled carbon-manganese structural
     steels 390, 391
   mild steels 390
High-strength cast steels 374
High-strength low-alloy (HSLA) steels. See also
   Microalloyed steel. 148, 151, 154,
        358-362, 397-398, 399
   applications of 399, 415-423
     automotive 416-417
     castings 419, 420
     cold-forming strip 418
     forgings 420
     offshore structures 417-418
     oil and gas pipelines 416
     railway tank cars 419
     shipbuilding 418
     structural 399, 401, 418, 420, 421

1052 / Index

High-strength low-alloy (HSLA) steels (continued)
  categories and specifications 398–405
    ASTM specifications 399, 406, 411
    SAE categories 401
  classification of 148
    acicular-ferrite 148, 399, 404–405
    as-rolled pearlitic 404
    dual-phase steels 148, 398, 405, 424–429
    hydrogen-induced cracking resistant
      steels 400, 416
    inclusion-shape controlled
      steels 400, 405, 412–413
    microalloyed ferrite-pearlite
      steels 400–404, 585–588
    weathering steels 400
  cold-rolled 419–420
  compositions 401, 406, 410
    in ASTM specifications 406
    of normalized European HSLA steels . . 410
    in SAE specifications 401
  control of properties 405–411, 588
    with alloying elements 406–408
    by controlled rolling 408–409, 586–588
    cooling, effect of 402
  definition of 148
  fatigue characteristics of 413
  forgings 137, 358–362, 419, 420, 588
  forming of 402, 408, 413–414
  forming properties of 398, 418
    bending radii 413, 414
  mechanical properties 410–413
    of acicular ferrite steels 404–405
    of cold-rolled sheet 420
    of cold-forming strip 417
    compared with carbon steel 389
    of control-rolled steels 409
    directionality of properties 412–413
    of dual-phase steels 398
    effect of manganese on 402
    of HSLA forgings 360–361, 588, 599
    of microalloyed ferrite-pearlite
      steels 411–412
    of normalized HSLA steels 409–410
    of weathering steels 400
  metallurgical effects 359–361
    first-generation microalloy steels 359
    second-generation microalloy
      steels 359–360
    third-generation microalloy steels . . 360–361
  microalloying elements . . 358–359, 400–404, 419
    molybdenum 358, 403, 407
    niobium 358, 402–403, 407, 419
    titanium 231–232, 359, 403–404, 408
    vanadium 358, 401–402, 403, 408, 419
  processing methods 130–131, 398,
                      408–410, 586, 587, 588
  selection guidelines 415–416
  steelmaking 405–406
  welding of 414–415, 609
High-strength low-alloy steel bars 246, 588
High-strength low-alloy steel
  forgings 137, 358–362, 419, 420, 588
High-strength low-alloy steel plate . . 235–236, 587
High-strength structural carbon steels . . . 390–391
High-tensile hard-drawn wire,
  characteristics of 307
HK alloy 925
HL alloy 925
HN alloy 923, 925–926
Hole machining, applications of P/M
  high-speed tool steel for 785
Home scrap, recycling of 1023
Homogenization, of stainless steel castings . . 911
Hooke's Law 318
Hose-reinforcing wire 286
Hot-brine test 466, 467
Hot cracking 608
Hot deformation, effect of, on notch
  toughness 741, 742–744
Hot dip galvanizing 212, 216–217, 218
Hot ductility tests, on carbon-manganese
  steels 696
Hot extrusion, for steel tubular products 328
Hot extrusion forgings, design of 355, 356

  machining allowance 357
  mechanical properties 356, 357
  mismatch tolerances 356–357
Hot-gas corrosion, resistance to, in
  heat-resistant alloys 925, 928
Hot isostatic pressing (HIP) 973, 993–994
Hot metal desulfurization 108, 109
Hot mill finishing temperature,
  effects of 588, 590
Hot-rolled carbon-manganese structural
  steels 390, 391
Hot-rolled steel, effects of steelmaking
  practices on formability of 577
Hot-rolled steel bars and shapes 240–247
  allowance for surface imperfections in
    machining applications 241
  alloy steel bars 245–246
    aircraft quality and magnaflux quality . . 246
    axle shaft quality 246
    ball and roller bearing quality and
      bearing quality 246
    cold-shearing quality 246
    cold-working quality 246
    regular quality 246
    structural quality 246
  carbon steel bars for specific applications . . 245
    axle shaft quality 245
    cold-shearing quality 245
    cold-working quality 245
    structural quality 245
  decarburization 241
  dimensions and tolerances 240
  heat treatment 241
    annealing for specified microstructures . . 241
    hardening by quenching 241
    normalizing 241
    ordinary annealing 241
    stress relieving 241
    tempering 241
  merchant quality bars 243
    grades 243
    sizes 243
  product categories 242–243
  product requirements 241–242
  special quality bars 244–245
  special shapes 247
  structural shapes 247
  surface imperfections 240
    laps 240
    seams 240
    slivers 240–241
  surface treatment 241
Hot rolling 115–120
  controlled rolling 115, 117–118,
                        130–131, 408–409
  critical temperatures . . . 115–116, 126–127, 130
Hot-tensile properties, of ductile iron 48, 52
Hot twist testing 583, 584
Hot working, of maraging steels 795
Hot-work tool steels
  powder metallurgy 789–790
  wrought 762–763
    chromium 762
    molybdenum 763
    tungsten 762–763
Hot-wound springs 315
HP-9-4-30 steel 431, 444–445
  heat treatments for 445–446
  properties of 444, 446
HP alloy 926
H-steels 474, 480–481
HT alloy 926–927
HU alloy 926, 927
HW alloy 926, 927
HX alloy 921, 926, 927
Hydrogen attack, and elevated-temperature
  service 633–634, 639
Hydrogen blistering, and elevated-temperature
  service 634
Hydrogen damage
  effect of M_3C carbide in 632–633
  and elevated-temperature
    service 632–634, 639

  hydrogen environmental
    embrittlement 711–712, 713
  hydrogen-stress cracking and loss of
    tensile ductility 712–717
Hydrogen embrittlement
  in austenitic stainless steels 715
  and elevated-temperature service 634
  and formation of flakes in steels 716–717
  in line pipe steels 716
  in maraging steels 715–716
  in tool steels 715
Hydrogen-induced cracking 606–607
Hydrogen-induced cracking resistant
  steels 400, 416
Hydrogen relief treatment, of steel springs . . 312
Hydrogen-stress cracking 342
  characteristics of 711
  and loss of tensile ductility 712–717
Hypoeutectic iron, solidification in 13–14

I

Impact properties 61–62, 63, 64
  of ductile iron 40, 43, 44, 45, 46
Impact resistance
  of gray iron 23
  of steel castings 365, 367–369
Inclusion shape controlled
  steels 400, 405, 412–413
Indium, resistance of, to liquid-metal
  corrosion 636
Industrial Fasteners Institute (IFI) 289
Industrial or standard-quality wire 282
Industrial quality rod 273
Inoculants, effect of, on alloy cast irons 90
In roller leveling, to control flatness in carbon
  steel sheet 206
Institute of Scrap Recycling Industries (ISRI),
  scrap specifications of 1026
In temper rolling, to control flatness in carbon
  steel sheet 205–206
Interactive effects of alloying elements,
  on notch toughness 742, 743
Intergranular corrosion
  of austenitic stainless steels 912, 945
  effect on stress-corrosion cracking . . . 724, 725
  of ferritic alloys 916–917
Intergranular fractures, effect of, on
  stress-corrosion cracking 724
Intermediate-temper wire 850
International designations and
  specifications 156–159, 166–174, 174–194
  British (BS) standards 158
    compositions of BS carbon steels . . 182–184
    compositions of BS alloy steels 185–186
    cross-referenced to SAE-AISI
      steels 166–174
  French (AFNOR) standards 158–159
    composition of AFNOR carbon
      steels 186–188
    composition of AFNOR alloy
      steels 189–190
    cross-referenced to SAE-AISI
      steels 166–174
  German (DIN) standards 157
    compositions of DIN carbon steels . . 175–178
    compositions of DIN alloy steels . . . 178–179
    cross-referenced to SAE-AISI
      steels 166–174
  Italian (UNI) standards 159
    compositions of UNI carbon steels . . 190–191
    compositions of UNI alloy steels . . 192–193
    cross-referenced to SAE-AISI
      steels 166–174
  Japanese (JIS) standards 157–158
    compositions of JIS carbon steels 180
    compositions of JIS alloy steels 181
    cross-referenced to SAE-AISI
      steels 166–174
  Swedish (SS_{14}) standards 159
    compositions of SS_{14} carbon steels . . . 193
    compositions of SS_{14} alloy steels 194
    cross-referenced to SAE-AISI
      steels 166–174

International Organization for Standardization (ISO), specifications for threaded fasteners 289
International System of grade designation, specifications for ductile iron 35, 36
Interstitial elements, effects of, on notch toughness 742
Interstitial-free steel ..112–113, 131–132, 405, 578
    cold-rolled strip 417
    composition of 417
    deep-drawing properties of 398
    effects of steelmaking on formability of ...578
    production of 112–113, 131–132
    tensile and yield strengths of 417
Investment cast superalloys. *See* Polycrystalline cast superalloys.
Iron, embrittlement of 689–691
    by oxygen 689–690
    by selenium 691
    by sulfur 690–691
    by tellurium 691
Iron-base alloys
    solid-metal embrittlement of 721
    susceptibility to embrittlement 689
Iron-base superalloys
    alloying elements, effect of 951
    compositions of 965
    dispersion-strengthened alloys
        compositions 973
        properties 976
    microstructure 959, 961–962
    with high nickel content 959
    physical properties 963–964
    stress-rupture properties 962
    tensile properties 958–959, 960–961
Iron-carbon alloys, quench-aging in 692–693
Iron-carbon phase diagram 126–127
Iron-chromium alloys, properties of 920
Iron-chromium-nickel alloys
    composition of 912
    properties of 920
Iron-chromium-nickel heat-resistant castings 922–925
Ironmaking 107, 109
Iron-nickel-chromium alloys
    composition of 912–913
    properties of 920–921
Iron-nitrogen alloys, quench-aging in ...692–693
Iron phosphate coatings 222
Iron scrap, recycling of 1023
    factors influencing scrap demand 1024
    purchased scrap supply 1024–1026
    scrap use by industry 1023–1024
Isothermal forging 971

J

Jaw crusher test, of abrasion-resistant cast iron 97
JIS (Japanese) standards for steels 157–158
    compositions of 180–181
    cross-referenced to SAE-AISI steels ...166–174
Jominy end-quench test 452, 464, 466

K

Kawasaki basic oxygen process (K-BOP) operation 111
Killed steel 142, 226, 227
Kloeckner metallurgy scrap (KMS) process ...111

L

Ladle refining 480
Ladle steelmaking 112–113
Ladle treatments 930
Lamellar cracking 607–608, 609
Lanthanum
    in compacted graphite iron 56
    in ferrite 408
Laps, in hot-rolled steel bars 240
Larson-Miller Parameter, and elevated-temperature service 627–628

Lashing wire 852
Lead
    in cast iron 5, 8
    corrosion resistance of 221
    effect of, on machinability of carbon steels 599–600
    effect of, on steel 145
    resistance of, to liquid-metal corrosion ...636
Leaded alloy steels, solid-metal embrittlement of 719, 721–722
Leaf springs 321–322
    types of 323–324
Lehigh restraint test 612
Lime bright annealing 280
Limiting dome height (LDH) test 576
Line pipe steels, hydrogen embrittlement in ..716
Linze-Donovitz (LD) method 111
Liquid-metal corrosion, resistance to ...634–636
Liquid-metal embrittlement 635, 717–721
Liquid processing of steel 107, 108
    blast furnace stove use 107–108
    cokemaking 107
    future technology for 114
Liquid treatment
    of compacted graphite irons 9
    of ductile iron 8, 9
    of gray iron 7
    of malleable iron 10
Lithium, resistance of, to liquid-metal corrosion 635
Long-term exposure, and elevated-temperature service 623, 627
Low-alloy cast steels 372–374, 375
Low-alloy special-purpose tool steels 767
Low-alloy steel. *See also* Alloy steel. 201, 207, 208–211
    air-hardening of 644–645, 646
    alloying elements in 144–147
    castings and 363–364
    classification of 149
    composition of 152–153
    creep-resistance of 619–621
    definition of 149
    direct castings methods 211
    for elevated-temperature service 618
    forgings. *See* High-strength, low-alloy steel forgings.
    hardenable low-alloy steels 453
    International designations and specifications for 156–159, 166–194
        British (BS) steel compositions 158, 166–174, 185
        French (AFNOR) steel compositions 158, 166–174, 188
        German (DIN) steel compositions 157, 166–174, 178–179
        Italian (UNI) steel compositions 159, 166–174, 192
        Japanese (JIS) steel compositions 157, 166–174, 181
        Swedish (SS) steel compositions 159, 166–174, 194
    mechanical properties 209–211, 396
    mill heat treatment 209
    physical properties of 195–200
    plate. *See* Steel plate.
    production of sheet and strip 208
    production of 930
    quality descriptors 208–209
    quenched and tempered 391–397
    SAE-AISI designations 152–153
    sheet and strip 207–208
Low-carbon bainite 404–405
Low-carbon cast steels 364, 371–372
Low-carbon quenched and tempered steels ...149
Low-carbon steels. *See also* Carbon steel.
    for bearings 24–25, 480–481
    definition of 147
    quench-age embrittlement 692
    weldability of 608
Low-carbon steel wire, for general use 282
Low-cycle fatigue tests 626
Low-hardenability steels 466

Low-temperature impact energy, of steel plate 238
Low-temperature properties of structural steel 662–672
    advances in steel technology 665–666, 668, 669
    assessment of fracture resistance 662–663
    design and failure criteria 662
    fatigue crack growth in structural steel 663–664, 665, 666
    fracture toughness characteristics 664, 666–667, 669
    fracture toughness of welded structures 667–669, 670, 671, 672
    fracture toughness requirements for ..663, 664
    structural steel specifications 664–665, 667, 668
Low-temperature service, high-nickel steels for 392, 396–397, 398
Low-temperature toughness, of steel castings 376, 377–378
Lüders lines 574

M

Machinability 68–69, 240, 591–602
    of austenitic manganese steel 838–839
    of carbon steels 595–597
    of carbon steels with other additives 599
        calcium 599
        nitrogen 599
        phosphorus 599
        selenium and tellurium 599
    of carburizing steels 600
    of cold-drawn steel 601
    of compacted graphite iron 68–69
    of ductile iron 52–53, 54
    of gray iron 23–24
    of leaded carbon and resulfurized steels 599–600
    of maraging steels 795
    measures of 591–592
        cutting speed 591–592
        machinability testing for screw machines 593
        power consumption 591, 592
        quality of surface finish 592–593
        tool life 591–592
    microstructure and 595, 596
    resulfurized carbon steels 597–599
        control and effect of sulfide morphology 598
        economic 598–599
        manganese content 597
    scatter in machinability ratings 593
    of stainless steels 894–897
    of steel castings 378
    of steel plate 238
    of through-hardening alloy steels 600–601
    of wrought tool steels 774–775
Machinability ratings
    of gray iron 23
    of steels 593–595
Machinability testing, for screw machines ...593
Machined bars 249–250
Machining allowances for wrought tool steels 778–779
Macroetch testing 275
    for carbon steel rod 274
Macrostructure differences, and fatigue resistance 681
Magnaflux quality 254
Magnesium
    in cast iron 5
    in compacted graphite iron 56
    in malleable iron 10
    in production of spheroidal graphite 7
Magnesium chloride
    stress-corrosion cracking resistance to boiling 725–727
    stress-corrosion cracking verification procedures 725
Magnesium-containing alloys 39

1054 / Index

Magnetic particle and eddy current testing of steel springs 309
Magnetic-particle inspection 274–275
Magnetic properties
  of austenitic manganese steel 837
  of ductile iron 51–52
  of gray iron 31
  of high-alloy castings 929
  of steel castings 374
Magnetic separation 1026
Magnetic testing 1029–1030
Malleable iron 9–11, 71–84
  alloying elements 10, 71–72
  annealing of 71, 72–73
  applications 73, 74, 83, 84
  blackheart malleable iron 74
  compared to ductile iron 71
  composition of 5, 9–10, 71–72
  control of mottle 71
  control of nodule count 73–74
  cooling rate of 10
  damping capacity 82, 84
  ferritic malleable iron 75–76
    alloying elements 75
    corrosion resistance 76
    fatigue limit 75
    fracture toughness 75–76, 80
    graphite content 75
    heat treatment 75
    microstructure 72
    modulus of elasticity 75
    stress-rupture plot 75
    tensile properties 73, 75
    welding and brazing of 76
  grades of 73, 74
  heat treatment of 10–11, 75, 76–80
  liquid treatment of 10
  melting practices 72
  metallurgical factors 71
  microstructure 72, 76, 77
  pearlitic-martensitic malleable irons 76–84
    Charpy V-notch impact energy 81
    compressive strength 82
    fatigue properties 81
    fracture toughness 74, 80, 82
    hardness 80–82
    heat treatment 76–80
    mechanical properties at elevated temperatures 80, 82
    microstructure 76, 77
    modulus of elasticity 82
    shear strength 82
    stress-rupture plot 81
    tempering times 80
    tensile properties 73, 78, 79, 82
    torsional strength 80, 82
    unnotched fatigue limits 81, 83
    wear resistance 83, 84
    welding and brazing of 83–84
  properties of 73, 74–83
  solidification of 72
  types of 74–75
  whiteheart malleable iron 74
Manganese
  in alloy cast irons 87
  in austenitic manganese steel 822–824, 825
  availability of 1021
  in cast iron 5, 28
  in compacted graphite iron 59
  in ductile iron 43
  in ferrite 402, 406
  effect of, on hardenability 393, 394, 395
  effect of, on machinability of carbon steels 598
  effect of, on notch toughness 740
  at elevated-temperature service 640
  in malleable iron 10
  in P/M alloys 810
  in steel 144, 576–577
  in structural steels 407
  in wrought stainless steels 872
Manganese-molybdenum cast steels 373
Manganese-nickel-chromium-molybdenum cast steels 373

Manufacturing practices, effects of, on notch toughness 742
  cast steels 746–747
  wrought steels 741, 742–746
Maraging steels 793–800
  age hardening 793–794
  applications 800
  commercial alloys 795
  hydrogen-stress cracking and loss of tensile ductility 715–716
  mechanical properties 799
  physical metallurgy 793–795
  physical properties 800
  processing
    cold working 795
    heat treating 795–797
    hot working 795
    machining 795
    melting 795
    powder metallurgy products 798–799
    surface treatment 797–798
    welding 798
  resistance to corrosion and stress corrosion 799
  thermal embrittlement of 697–698
Martempering 455, 457
Martensite 127, 133–134
  grain size effect on strength of 393
  hardness of 394
  tempering of 134–136, 137
Martensitic grades, of corrosion-resistant steel castings 913
Martensitic malleable irons. See Pearlitic-martensitic malleable iron.
Martensitic stainless steel. See also Cast stainless steels; Wrought stainless steels 841–842
  compositions of 843, 847–848
  elevated-temperature properties 939–942
  forgeability 892, 893
  machinability of 894
  notch toughness of 859
  tensile properties of 858, 862–863
  weldability of 902
Matrix structure, in gray iron 25
McConomy curves, and elevated-temperature service 630, 631
McQuaid-Ehn test, to determine austenitic grain size 241
Mechanically capped steel 143
Mechanical plating, of steel springs 312
Mechanical prestressing, of leaf springs 325–326
Mechanical properties
  of abrasion-resistant cast irons 94
  of carbon steels 202, 205, 206
  of compacted graphite iron 57–66
  of corrosion-resistant cast irons 99
  of corrosion-resistant steel castings 909, 914, 917–920
  of ductile iron 40, 42–43, 48, 49
  factors affecting, in high-temperature service 636–644
  of ferritic malleable iron 75–76
  and formability 573–575
  of gray iron 29–31
  of heat-resistant cast irons 99
  of leaf springs 325
  long-term elevated-temperature tests 620, 622–624, 629, 630
  of low-alloy cast steels 374
  of low-alloy steel sheet/strip 209–210
  of pearlitic and martensitic malleable iron 74, 81
  of powder metallurgy (P/M) superalloys 974, 975–976
  of quenched and tempered alloy steels 389, 392, 396, 397
  short-term elevated-temperature tests 622, 627, 628
  of steel castings 367, 375, 376
  of steel plate 237, 238
  of wrought stainless steels 930, 931, 932–935
Mechanical spring wire
  for general use 284–285

  for special applications 285
Mechanical steel tubing 334–336
Medium-carbon cast steels 364, 372, 373
Medium-carbon low-alloy steels 430–431
Medium-carbon steels
  definition of 148
  weldability of 608–609
Medium-carbon ultrahigh-strength steels. See also Ultrahigh-strength steels.
  classification of 149
  compositions of 157
Melting
  of malleable iron 72
  of maraging steels 795
  in steel plate production 228
  of superalloys 968, 970–971, 986–988
Merchant quality hot-rolled carbon steel bars 243
  grades of 243
  sizes of 243
Merchant quality steels, compositions of 150
Merchant wire 282
Mercury, resistance of, to liquid-metal corrosion 635
Metal-induced embrittlement of steels 717–722
  of liquid metal 717–721
  of solid metal 719, 721–722
Metal injection molding (MIM) technology 818–819
  advantages of 819
  applications 820
  factors impeding growth of 819–820
  mechanical properties of 820
Metallic coatings, effect of, on formability 579–580
Metallurgical control, in production of ductile iron 38
Metallurgical factors affecting weldability of steels 603–606, 607
  chemical composition effect 606
  hardenability and weldability 603–604
  heat-affected zone microstructure 605–606
  preweld and postweld heat treatments 606
  weld metal microstructure 604–605
Metallurgical structures, of heat-resistant cast steels 921–922
Metallurgical variables of fatigue behavior 678
Metallurgy, of cast iron 3–4
Metric conversion guide 1035–1037
Metric fasteners, property class designations of 289–290, 291
Meyer index of austenitic manganese and stainless steel 832
Microalloyed steel. See also High-strength low-alloy (HSLA) steels.
  ASTM specifications of 399, 406, 411
  compositions 406
  mechanical properties 411
  brittle fracture of 412
  comparison of plate and bar products 588
  control of properties 405–410
  controlled rolling of 115, 117–118, 131, 408–409, 586–588
  directionality of properties 412–413
  elements 358–359
  metallurgical effects 359–362
  microalloyed bars 246, 419–420
    alternative strengthening mechanisms in 587–588
    applications and compositions 420
    high-strength low-alloy bar products 588
    processing of 587–588
  microalloyed castings, applications and compositions 420
  microalloyed forgings 137, 358–362, 419, 588
    applications 420
    carbon contents of 585
    control of properties 588
    forging temperature, effects of 588
    generations of 359–360
    properties of 360–361, 588, 599
  microalloyed plates
    compositions 406, 410
    mechanical properties 409, 410, 411

Index / 1055

processing of 586–587
microalloyed quenched and tempered
 grades 394–396
microalloying elements ..358–359, 400–404, 419
 effects on properties 358
 molybdenum 358, 403, 407
 niobium 358, 402–403, 407, 419
 titanium 231–232, 359, 403–404, 408
 vanadium 358, 401–402, 403, 408, 419
notch toughness
 of acicular ferrite................404–405
 compared with carbon steel 389
 effect of reheating on 414
 of microalloyed ferrite-pearlite........412
 of microalloyed forgings 360, 361
processing of ..130–131, 408–410, 586, 587, 588
 control of properties 408, 588
 coiling temperature and nitrogen content,
  effect of..........................412
 forging temperature, effect of.....588–589
 hot mill finishing temperature,
  effect of 412, 588, 590
strengthening mechanisms of 405, 586–587
 grain refinement 405
 precipitation strengthening ...403, 586–587
Microalloyed steel bars............246, 419–420
 applications 420
 processing of 587–588
 alternative strengthening mechanisms
  in 587–588
 high-strength low-alloy bar products....588
Microcast-X process 992
Microconstituents, in low-carbon steel 579
Microstructure
 of abrasion-resistant cast iron 92–94, 95
 of bainite 128, 129
 of carburizing bearing steels 381–382, 383
 of cast stainless steels 909, 910
 of corrosion-resistant steel
  castings 909, 913–915
 direct-cooled forging 137, 138
 effect of, at elevated-temperature
  service 638, 641
 effect of, on notch
  toughness 744, 747–749, 750
 effect of, on weldability 604–606, 607, 608
 of ferrite-pearlite 127, 130–131
 ferritic 127, 131–133
 and formability, correlation between ...578–579
 of gray iron 13–14
 of high-carbon bearing steels 381, 382, 383
 influence of, on temper embrittlement
  in alloy steels 702
 and machinability of steels 595, 596
 of malleable iron 72, 76, 77
 of martensite 127, 133–134
  tempering of 134–136, 137
 of pearlite 127, 128–129
 proeutectoid ferrite and
  cementite 127, 129–130
 properties associated with..............126
 quenched and tempered 136–137, 138
 of wrought cobalt-base superalloys ...965–967
 of wrought iron-base
  superalloys 951, 959, 961–962
 of wrought nickel-base superalloys ...951–956
Mild steels 390
Mill heat treatment
 of cold-rolled steel products 202–204
 of low-alloy steel 209
Milling, applications of P/M high-speed tool
  steel for 785
**Modified low-carbon steel sheet
 and strip** 206, 208
Modulus of elasticity 58, 61, 62, 63
 of ferritic malleable iron 75
 of gray iron 18, 20, 21
 for threaded steel fasteners 295–296
 of pearlitic malleable iron 82
Mold metallurgy........................114
Mold steels 767–768
Molybdenum
 in age hardening in maraging steels.......794
 in alloy cast irons 89, 90

in austenitic manganese steel 824–825
in cast iron 6, 28–29
in cobalt-base alloys 985
in compacted graphite iron 57, 59
in ductile iron........................45
effect of, on corrosion resistance 912
effect of, on hardenability 395, 413, 468
effect of, on notch toughness 741
at elevated-temperature service 640
in ferrite 408
in gray iron 22
influence of, on phosphorus-induced temper
 embrittlement 699–700
influence of, on sigma phase
 embrittlement 710
in microalloy steel....................358
in nickel-base superalloys 984
in P/M alloys 810
in steel 146, 577
in wrought stainless steels 872
Molybdenum high-speed tool steels 759, 764
Molybdenum hot-work tool steels 763
Motor springs 302
**MPIF (Metal Powder Industries Federation)
 designations for ferrous P/M materials** ..805
Multipass weldments..................603, 605
Music spring steel wire 285
Music wire
 characteristics of 306
 size of 277
Music wire gage (MWG) 277

N

Neutron irradiation 653–661
 damage to steels 653
 irradiation damage processes 653
  displacement damage 653–654
  transmutation helium 654–655
 mechanical properties
  elevated-temperature tensile
   behavior 657–658
  fatigue 660
  irradiation-assisted SCC 660–661
  irradiation creep 660
  irradiation embrittlement ...658–660, 722–723
  low-temperature tensile behavior.......657
  thermal creep 660
  void swelling 655–657
**Neutron irradiation
 embrittlement**............658–660, 722–723
Nickel
 in alloy cast irons 89
 in austenitic manganese steel 825
 in cast iron 6, 28
 in cobalt-base alloys 985
 in compacted graphite iron 59
 in ductile iron.....................43–44
 effect of, on hardenability 393, 395, 468
 effect of, on notch toughness 741
 in ferrite 408
 in gray iron 22
 in P/M alloys 809–810
 in steels 146, 577
 in wrought stainless steels...........871–872
Nickel-base alloys
 mechanical properties of 984
 physical properties of 983
 stress-rupture strengths for 985
Nickel-base nodulizers 39
Nickel-base superalloys. *See* specific types.
 cast nickel-base
  superalloys...........981–985, 986–994
 directionally solidified
  superalloys............995, 996–998
 powder metallurgy (P/M) superalloys ..972–976
 single-crystal superalloys ..995–996, 998–1006
 wrought nickel-base
  superalloys 950–959, 968–972
Nickel cast steels 373–374
Nickel-chromium-molybdenum cast steels 374
Nickel-manganese cast steels 374
Nickel-vanadium cast steels 374
Nicrosilal 103

Ni-Hard cast irons 89
**Nil-ductility transition temperature
 (NDTT)** 367–368, 412
Niobium
 in cobalt-base alloys.....................985
 effect of, on hardenability 413, 419
 effect of, on notch toughness 741–742
 at elevated-temperature service 640–641
 in ferrite 403, 408
 in high-strength low-alloy steel ..358, 402–403
 in Inconel 718 1018
 in steel 146, 577
Niobium-molybdenum microalloyed steels 403
Nipples, pipe for 331
Ni-Resist 103
Nitriding, as surface treatment in wrought
 tool steels 779
Nitrogen
 effect of, on machinability of carbon
  steels 599
 effect of, on notch toughness 741
 effect of, on steel composition and
  formability 577
 in ferrite 406, 408
 in stainless steels 930
 in wrought stainless steels.............872
Nitrogenized steels 208
**Nitrogen-strengthened austenitic stainless
 steels** 892–893
Nodule count, control of, in malleable
  iron 73–74
Nodular iron. *See* Ductile iron.
Nominal axial stresses 674
 and fatigue resistance 674
Nondestructive inspection, of steel
  castings 378–379
Nonisothermal upset test 584
Nonmetallic-inclusion testing, for carbon steel
  rod 274
Normalizing
 of cold-rolled steel products 204
 for ductile iron 41
 effect on toughness....................390
 of high-strength structural carbon
  steels 390, 391
 of low-alloy steel sheet/strip 209
 of steel plate 231
Notched-bar upset test 584
Notch sensitivity 48
Notch toughness of steels 737–754
 of acicular ferrite 404–405
 comparison of a low-carbon steel
  and an HSLA steel 389, 404–405
 composition, effect of 668, 739–742
 correlations of, with other mechanical
  properties 753
 ductile-to-brittle transition 737–739
 manganese, effect of 390
 manufacturing practices, effect of 742
 normalizing 390
 wrought steels 741, 742–746
 microstructure, effects of 747
  grain size 744, 748–749
  microstructural constituents 747–748
  reheating on 414
  submicroscopic structure 749, 750
 of microalloyed forgings 360, 361
 of microalloyed forgings 412
 variability of Charpy test results 749–753
Nut steels 291, 292–294, 295

O

Obsolete scrap 1023
Offshore applications, high-strength low-alloy
  steels for 417–418
Oil country tubular goods 329, 330, 332–333
Oil-hardening cold-work tool steels 765
Oil pipelines, high-strength low-alloy steels
  for 416–417
Oil-tempered wire 280–281
 characteristics of 306
Olsen-Erichsen cup test 576
Optical spectroscopy 1030

1056 / Index

Organic composite coatings 223
Ostwald ripening 637
Overheating, of steels.....................697
    presence of facets 697
    upper-shelf energy 697
Oxidation
    of cast irons 101
    of chromium-molybdenum
      steels 617, 629–630, 636
    of single-crystal superalloys 1004, 1006
    of wrought cobalt-base
      superalloys...................965–966, 968
    of wrought nickel-base
      superalloys...................957, 959, 965
      alloying for surface stability 956–957
      protection against oxidation 957, 959
Oxygen
    basic oxygen process 110, 111, 112
    effect of, on steel composition and
      formability 577
    embrittlement of iron by 689–690

P

Packaging and container applications, wire
    for.....................................282
Panel cracking, and aluminum nitride
    embrittlement 695
Paper clip wire 286
Paris equation.............................663
Partial-width indentation test..............583
Pearlite 127, 128–129
Pearlite/ferrite ratio, in compacted graphite
    iron 57, 59
Pearlite-reduced steels.....................148
Pearlitic-martensitic malleable iron 76–84
    brazing...............................83–84
    Charpy V-notch impact energy 81
    compressive strength 82
    fracture toughness 74, 80, 82
    heat treatment 76–80
    mechanical properties 74, 78, 79, 80–83
      shear strength 82
      tensile properties......................82
      torsional strength 80, 82
      unnotched fatigue limits 81, 83
    microstructure 76, 77
    modulus of elasticity 82
    rehardened-and-tempered malleable iron ...79
    selective surface hardening 84
    stress-rupture plot 81
    tempering times 80
    wear resistance 83–84
    welding 83–84
Phosphate coatings.........................222
Phosphorus
    in alloy cast irons 87–88
    in austenitic manganese steel........822, 824
    in cast iron 5
    effect of, on machinability of carbon
      steels 599
    effect of, on notch toughness 740
    at elevated-temperature service 640
    in embrittlement of iron 691
    in ferrite 401, 407–408
    in gray iron 22
    in P/M alloys 810
    in steel 144, 577
Physical properties
    of abrasion-resistant cast iron 96
    of carbon and alloy steels 195–199
    of cast stainless steels 927
    of cast steels 374, 376
    of cast superalloys 983
    of compacted graphite iron 66–69
    of corrosion-resistant cast iron 96
    of ductile iron 49, 50
    of gray iron.........................31–32
    of heat-resistant cast iron 96
    of steel castings 374, 376
    of wrought stainless steels 871
    of wrought superalloys 963–964
    of wrought tool steels 774, 775
Piling pipe 331

Pin test, of abrasion-resistant cast iron 97
Pipe sizes and specifications, for steel
    tubular products 328, 329–331, 333
Pit geometry, effect of, on stress-corrosion
    cracking..............................724
Planar anisotropy 575
Plane-strain compression test 582–583
Plane-strain deformation 121
Plasma rotating electrode process (PREP)....973
Plastic-strain ratio 575
Plating
    of springs 311–312
    as surface treatment in tool steels 779
Polycrystalline cast superalloys. *See also* Cast
    cobalt-base superalloys; Cast nickel-base
    superalloys. 981–994
    age hardening..........................981
    application of 981
    composition and density 982
      cobalt-base 983
      nickel-base 982
    control of casting microstructure 990–992
      carbides..........................990–991
      dendrites............................990
      eutectic segregation 991
      grain size 992
      porosity.........................991–992
    design of 983–986
      cobalt-base 985–986
      nickel-base 983–985
    heat treatment 993
    hot isostatic pressing 993–994
      effect on fatigue strength 992, 994
    investment casting of 989–990
    mechanical properties of 984
      fatigue properties 991, 992, 994
      stress-rupture properties 985, 986,
                             987, 991, 993
      tensile properties.....................984
    physical properties of 983
    quality considerations 988–989
      tramp element content 989
    stress-rupture strengths for selected 985
    vacuum induction melting of 986–988
Polymer quenchants 455
Powder forging, and ferrous powder
    metallurgy materials..................812
Powder metallurgy (P/M) 801
    in making tool steels 757
**Powder metallurgy (P/M) cobalt-base
    alloys** 977–980
    compositions 978
    grinding of 979–980
    machining of 978, 980
    mechanical properties 977, 978
    physical properties 977, 978
Powder metallurgy products 798–799
Powder metallurgy (P/M) superalloys. *See also*
    Powder metallurgy (P/M) cobalt-base
    alloys.
    mechanical properties of 974–976
      fatigue properties 974, 975
      stress-rupture properties 974
      tensile properties................974, 975
    oxide dispersion strengthened
      alloys 972, 973, 974–975, 976
      compositions of 973
      physical properties 976
      stress-rupture properties 976
      tensile properties.....................976
    powder consolidation 973
    powder production 970, 972–973
    thermomechanical working 973–974
Powder metallurgy tool steels 780–792
    advantages over conventional tool steels ..780
    applications of high-speed tool steels ...785–786
      broaching 785–786
      gear manufacturing 786
      hole machining 785
      milling 785
    classification......................781–792
      cold-work steels 786–789
      high-speed steels 781–786
      hot-work steels 789–790

      cold-work tool steels 786–789
    composition 781
    heat treatment of high-speed tool
      steels.............................782–783
      annealing............................783
      austenitizing temperature of ASP 23 782
      hardening 783
      stress relieving (before hardening) 783
      tempering 783
    heat treatment of H13, effect on size
      change of 790
    high-speed tool steels 781–786
      alloy development 784–786
      applications 785–786
      cutting tool properties 784
      heat treatment 782–783
      manufacturing properties 783–784
      sintered tooling 786
    hot-work tool steels 789–790
    machinability, tool life of CPM alloys 786
    mechanical properties of CPM alloys
      bend fracture strength 786
      Charpy C-notch toughness 786, 790
      hot hardness 785
      temper resistance 785
      wear resistance 790
    mechanical properties of H13
      Charpy V-notch impact strength 791
      hardness 790
      tensile strength......................791
      thermal fatigue resistance 792
Powdrex process 780–781
Power consumption 592
Praseodymium
    in compacted graphite iron 56
    in ferrite 408
Pratt and Whitney Gatorizing process 974
Preaging...................................259
Precipitation-hardening stainless steels. *See also*
    Wrought stainless steels.
    compositions of 843, 847–848
    elevated-temperature properties 942–944
    forgeability of 893–894
    fracture toughness properties 865
    machinability of 896–897
    tensile properties of 864, 865
    weldability of 902–904
Precipitation strengthening
    in copper 411
    dependence on precipitate size 402, 403
    effect of cooling on 119, 402
    in elevated-temperature service 637
    in ferrite 402, 404, 406
Precision cast hot-work wrought tool steels ...779
Precoated steels, weldability of 609–610
Precoated steel sheet 212–225
    aluminum coatings 218–220
      base metal and formability 219
      coating weight 218–219
      corrosion resistance 219
      handling and storage 220
      heat reflection 219–220
      heat resistance 219
      mechanical properties 218, 219
      painting 220
      weldability 220
    aluminum-zinc alloy coatings 220–221
    organic composite coatings 223
    phosphate coatings 222
    prepainted sheet 223–224
      design considerations 224
      packaging and handling 224
      selection of paint system 224
      shop practices 224
    preprimed sheet 222–223
      formability 222
      zinc chromate primers 222
      zinc-rich primers 222–223
    terne coatings 221–222
    tin coatings 221
    zinc coatings 212–218
      chromate passivation 214–215
      coating tests and designations ...212–214, 215
      corrosion resistance 212, 581, 584

electrogalvanizing 217
hot dip galvanizing 216–217, 218
packaging and storage 215–216
painting 215
zinc alloy 217–218
Zincrometal 217
zinc spraying 218
Prepainted sheet 223–224
design considerations 224
packaging and handling 224
selection of paint system 224
shop practices 224
Preprimed sheet 222–223
formability of 222
zinc chromate primers 222
zinc-rich primers 222–223
Pressure steel tubes 327, 333–334
Pressure tightness, in gray iron 22–23
Pressure vessel plate 234–235, 236, 237
Prestressed concrete, wire for 283
Preweld and postweld heat treatments, effect of, on weldability 606
Primary purification reaction 986–987
Processing of solid steel 115–124
annealing 122–123
cold rolling 121–122, 123
conventional controlled rolling (CCR) 117
cooling and coiling system 118–119, 121
dynamic recrystallization controlled rolling (DRCR) 117–118
hot rolling 115
precipitation during cooling and coiling 119–120, 123
precipitation of carbonitrides and sulfides 115–116
recrystallization controlled rolling (RCR) 117, 120
Stelco coil box 118, 120
warm rolling 120, 124
Product analysis, for classifying steels 141
Proeutectoid ferrite 129
Prompt scrap 1023
Proof stress
of bolt or stud 296–297
of nut 297
Purchased scrap 1023

Q

Quality assurance and quality control, in closed-die forgings 339
Quality descriptors
for carbon steels 201, 203
for steel 143–144, 146
Quality of surface finish 592–593
Quantitative chemical analysis 1031–1032
Quench-age embrittlement 692–693
for iron-nitrogen and iron-carbon alloys 692–693
for low-carbon steels 692
Quench cracking 698
Quenched and tempered steels
H-steels 474–476, 480–481
low-alloy steels 391–392, 396
martensitic stainless steels 939–942
microstructures and processing of....136–137
structural carbon steels 390–391
weldability of 609
Quenching
alloying elements in 395, 456–457
for ductile iron 38, 41–42
of high-strength structural carbon steels 389, 391
of low-alloy steel sheet/strip 209
of steel plate 231
Quenching media 471–473
Quick-quiet basic oxygen process (Q-BOP) ...111

R

Radiation damage. *See* Neutron irradiation.
Radiation-induced segregation 654
Railway tank car applications, high-strength low-alloy steels for 419–420

Rapid solidification rate (RSR) 973
Rare-earth elements
effect of, on notch toughness 742
in ferrite 408
Recrystallization controlled rolling (RCR) 117, 120
Recycling 1023–1033
definition of 1023
of home scrap 1023
of iron and steel scrap 1023
factors influencing scrap demand ...1024
purchased scrap supply 1024–1026
scrap use by industry 1023–1024
of stainless steel and superalloy scrap 1027–1028
blending..........................1032
collection 1029
degreasing 1032
metallurgical wastes 1032
processing stainless steel and superalloy scrap 1028–1032
scrap demand 1028
scrap use by industry 1028
secondary nickel refining 1032
separation.................1028, 1029–1032
size reduction and compaction 1032
scrap processor 1026
blending 1027
collection 1025, 1026
detinning 1026–1027
incineration......................1027
separation and sorting 1026
size reduction and compaction ..1026, 1027
Reed wire 852
Regular quality of low-alloy steel 208
Rehardened-and-tempered malleable iron 79
Reheating, for austenitic manganese steel ...833
Relaxation, in threaded steel fasteners 295
effect of thread design on 296
strengths 631
Relaxation tests, and elevated-temperature service 624, 631
Rephosphorized steels 208
Residual elements, in steels 141
Residual stress
in closed-die forgings 342
and fatigue resistance 680–681, 682, 683
for steel springs 309, 311
Resistance to decarburization in wrought tool steels 777
Resulfurized carbon steels, machinability of 597–599
control and effect of sulfide morphology ..598
economic 598–599
manganese content 597
Reversible temper embrittlement 698
Rhenium, in nickel-base superalloys 984
Rimmed steels 6, 141, 143, 145, 578
Ring compression test 583
Rivers process.........................819
Roller leveling 693
of carbon steel sheet 206
Rolling-element bearings.................380
Roll threading, alloy steel wire for 287
Room-temperature structure, of gray iron 14
Rope wire 283–284, 851
Rotary piercing, for steel tubular products ...328
Rotating-beam test......................861
Rotating electrode process (REP) 973
Rotors, for elevated-temperature service 620–621
Rubber-wheel test, of abrasion-resistant cast iron...............................97
Rupture ductility 933–934

S

6150 steel 437–438
heat treatments for 438
properties of 438, 439
SAC (surface area center) test 150–151, 152–153, 454

SAE-AISI designations
cross-referenced to international steel specifications
British (BS) steel specifications 166–174
French (AFNOR) steel specifications 166–174
German (DIN) steel specifications ...166–174
Italian (UNI) steel specifications 166–174
Japanese (JIN) steel specifications ...166–174
Swedish (SS) steel specifications ...166–174
for ductile iron 35, 36
for former standard steels 155
for free-cutting steels 151
for free-cutting (resulfurized) steels 150
for high-strength low-alloy steels 154
for low-alloy (alloy) steels 152, 153
for merchant quality steels 150
for potential standard steels 153
for steel castings 364, 365, 366
for threaded steel fasteners ..289, 290, 292, 296
for threaded fasteners 289, 290, 292, 296
steels, formerly listed 151, 155–156
Safety pin wire 286
Scaling, in gray iron 27, 100, 101–102
Scatter, in machinability ratings 593
Scrapless nut quality rod 273
Screw machine tests 591, 593
Scuff resistance, in gray iron 24–25
Seamless mechanical tubing 335
Seamless processes, for steel tubular products 328
Seams
in hot-rolled steel bars 240
in steel plate 230
Secondary hardening
in elevated-temperature service 637–638, 640, 641
during tempering 396, 641
Secondary-tension test 583
Section sensitivity, of gray iron 14–16
Section size, effect of
in compacted graphite iron 60, 61, 62
in gray iron 15
Segregation, and steel plate imperfections ...230
Selective surface hardening, of pearlitic malleable iron 84
Selenium
effect of, on machinability of carbon steels 599
embrittlement of iron by 691
Semiductile cast iron. *See* Compacted graphite iron.
Semikilled steel 142–143, 226, 227
Sensitization
in austenitic stainless steels 706–707, 912
in ferritic stainless steels 707–708
SG iron 56
Shakeout practice, effect on gray iron........28
Shear and torsional strength, of malleable irons 82
Shear properties........................61
Sheet finishes 885–886
Sheet formability of steel. *See also* Formability. 573–580, 888–889
circle grid analysis 575–576
correlation between microstructure and formability 578–579
grain shape 579
grain size.........................578
microconstituents 579
effect of metallic coatings on formability 579–580
effects of steel composition on formability 576–577
aluminum 577
carbon 576
cerium 577
chromium, nickel, molybdenum, and vanadium 577
copper 577
manganese 576
niobium 577
nitrogen 577
oxygen 577

1058 / Index

Sheet formability of steel (continued)
　phosphorus and sulfur 577
　silicon 577
　titanium 577
　effects of steelmaking practices on
　　formability 577–578
　　aluminum-killed steels 578
　　cold-rolled steel 577–578
　　hot-rolled steel 577
　　interstitial-free steel 578
　　rimmed steels 578
　　surface finish 578
　mechanical properties and
　　formability 573–575
　　planar anisotropy 575
　　plastic strain ratio 575
　　strain-hardening exponent 575
　　total elongation 573–574
　　uniform elongation 574
　　yield point elongation 574–575
　　yield strength 573
　selection of steel sheet 580
　simulative forming tests 576
　of stainless steel 888–889
Sheet steels. See also Carbon steel sheet and
　strip.
　cold-rolled HSLA steel 420
　forming properties of various types ..398, 418
　HSLA sheet steels, tensile properties of ..411
　interstitial-free steels ...112–113, 131–132, 405
　　composition 417
　　deep-drawing properties of 398
　　production of 112–113, 131–132, 578
　mechanical properties and formability. See
　　also Sheet formability of steel.
　　planar anisotropy 575
　　plastic-strain ratio 575
　　strain-hardening exponent 575
　　total elongation 573–574
　　uniform elongation 574
　　yield point elongation 574–575
　　yield strength 573
　precoated. See Precoated steel sheet.
　selection of 580
　simulative forming tests 576
　stainless steel 888–889
Shipbuilding, high-strength low-alloy steels
　for 419
Shock-resisting tool steels 766–767
Shot peening, of springs 313–314
Shrinkage, in compacted graphite iron 57
Sidepressing test 583–585
Sigma phase embrittlement 708–709
　in austenitic and ferritic stainless
　　steels 709–711
　in duplex stainless steels 711
Silal 103
Silicon
　in austenitic manganese steel 822, 824
　in cast iron 5, 6, 86–87, 88, 100
　in ductile iron 43
　effect of, on corrosion resistance 912, 913
　effect of, on hardenability 393, 395
　effect of, on notch toughness 740–741
　at elevated-temperature service 640
　in ferrite 401, 406
　influence of, on microstructure of white
　　iron 94
　in steel 141–142, 145, 577
Simulative forming tests, for steel sheet 576
Single-crystal superalloys 995–996, 998–1006
　casting of 998–1000
　chemistry control 998–999
　compositions of 996, 997
　fatigue properties 1002, 1004, 1005, 1006
　　heat treatment, effect of 1005
　　hot isostatic pressing, effect of 1005
　heat treatment 1000–1002
　hot corrosion 1004
　microstructure 1000–1002
　　heat treatment, effect of 1000, 1003
　　precipitate formation 1001
　oxidation of 1004, 1006
　protective coatings 1006

Single-pass weldments 603, 604
Sintered high-speed tool steels, powder
　metallurgy 786
Sintering
　equipment 804–805
　and ferrous powder metallurgy
　　materials 803–804
　techniques 804
Slide forming wire 852
Slivers, in hot-rolled steel bars 240–241
Slow strain rate test, for stress-corrosion
　cracking 725–726
S-N curves, and fatigue resistance 674–675
Society of Automotive Engineers (SAE). See
　SAE-AISI designations.
Sodium alloys, resistance of, to liquid-metal
　corrosion 635
Sodium-potassium alloys, resistance of, to
　liquid-metal corrosion 635
Soft-temper wire 850
Soil corrosion, of cast steels 376
Solidification
　of gray iron 13–14
　of malleable iron 72
Solid-metal embrittlement 721
　in iron-base alloys 721
　in leaded alloy steels 719, 721–722
　in leaded carbon and alloy steels 722
Solid-solution strengthening
　effect of alloying elements on 400
　in elevated-temperature service 637, 639
Solid steel
　processing of 114, 118
　solubility of carbonitrides in austenite 407
Soluble-gas process 972
Solution annealing, of austenitic stainless
　steels 898–899, 912, 945
　effect on creep-rupture strength 945
　effect on intergranular corrosion 912, 945
Sonic and ultrasonic properties 67–68, 69
Spark testing 1030
Special cast iron 11
Special quality hot-rolled carbon steel
　bars 244–245
Spectroscopic testing 1030–1031, 1032
Spheroidal graphite iron. See Ductile iron.
Spheroidization, at elevated-temperature
　service 642, 644
Spheroidize annealing 272, 280
　of low-alloy steel sheet/strip 209
Spray forming, of carbon steel sheet and
　strip 210, 211
Spring-temper wire 850
Spring wire 851
Stabilizing 259
Stainless steel. See Cast stainless steels;
　Wrought stainless steels.
Stainless steel spring wire, characteristics of ...308
Stainless welding wire 852
Standard pipe 331
Static tensile stresses, role of in
　stress-corrosion cracking 724
Steel bars. See Alloy steel bars; Carbon steel
　bars; Cold-finished steel bars; Hot-rolled
　steel bars and shapes.
Steel castings 363–379, 483
　classifications and
　　specifications 363–364, 365, 366
　engineering properties 376–378
　　corrosion resistance 376
　　elevated-temperature properties 376–377
　　low-temperature toughness 377–378
　　machinability 378
　　soil corrosion 376
　　wear resistance 376
　　weldability 378
　high-carbon cast steels 372
　low-alloy cast steels 367, 372–374, 375
　low-carbon cast steels 364, 371–372
　mechanical properties 365, 367–371
　　ductility 365
　　fatigue properties 369–370
　　heat treatment 367, 370–371
　　section size and mass effects 370, 373

　　specimens 370
　　tensile and yield strengths 365
　　toughness and impact
　　　resistance 365, 367–369
　medium-carbon cast steels 364, 372, 373
　nondestructive inspection 378–379
　physical properties 374, 376
　　density 374
　　elastic constants 374
　　electrical properties 374
　　magnetic properties 374
　　volumetric changes 374, 376
Steelmaking. See also Steel processing
　technology.
　effect of practices on formability 577–578
　electric furnace steelmaking 111
　ferroalloy/deoxidizer additions 111–112
　first-stage refining 110, 111, 112
　of HSLA steel 405–406
　ladle steelmaking 112–113
　mold metallurgy 114
　second-stage refining and technology
　　advances 111
　steel plate 226–228
　stainless steels 930
　third-stage refining 113
　tundish metallurgy and continuous
　　casting 113–114
Steel plate 226–239
　applications of 226
　fabrication considerations 238–239
　　formability 238
　　machinability 238
　　weldability 238–239
　heat treatment of 230–232
　　normalizing 231
　　quenching 231
　　stress relieving 231–232
　　tempering 231
　high-strength low-alloy steel plate
　　compositions 401, 406, 410
　　controlled rolling of 408–409, 586–587
　　mechanical properties
　　　of 401, 409, 411, 586–587
　　normalized HSLA steel plate 409–410
　　specifications 399, 401
　imperfections 230
　　decarburization 230
　　seams 230
　　segregation 230
　mechanical properties 237–238
　　directional properties 238
　　elevated-temperature properties 238
　　fatigue strength 238
　　low-temperature impact energy 238
　　static tensile properties 227, 228, 229,
　　　230, 232–233, 235–236, 237–238
　platemaking practices 228, 230
　quality of 226, 234–235, 236–237, 741
　quenched and tempered carbon steel 391
　low-alloy steel compositions 392
　tensile properties 391, 396
　steelmaking practices 226–228
　　austenitic grain size 227–228
　　deoxidation practice 226–227
　　melting practices 228
　types of 226, 227–228, 232–237
　　aircraft quality 237
　　carbon 232–233
　　forging quality 237
　　high-strength low-alloy 235–236
　　low-alloy 233
　　pressure vessel 237
　　regular quality 236
　　structural quality 236
Steel processing technology 107–125
　basic oxygen process (BOP) 110, 111, 112
　Kawasaki basic oxygen process 111
　quick-quiet basic oxygen process 112
　controlled rolling of microalloyed steels ..115,
　　117–118, 130–131, 408–409, 587–588
　conventional controlled rolling 117, 409
　dynamic recrystallization controlled
　　rolling 117–118, 409

recrystallization controlled rolling ...117, 409
furnaces 108–109, 110
   basic oxygen furnace 110
   blast furnace 109
hot metal desulfurization 109
   current technology 108, 109
ironmaking 107–109
   blast furnace stove use 107–108
   cokemaking 107
   current blast furnace technology ...108–109
liquid processing 107–114
   desulfurization 109–110
   future technology for 114
   ironmaking 107–109
   steelmaking.....................110–114
of solid steel 114–123
   annealing 122–123, 132–133
   cold rolling 121–122, 132–133
   hot rolling 115–120
   warm rolling 120
steelmaking 110–114, 226–228, 930
   effects on formability 577–588
   electric furnace steelmaking 111
   ferroalloy/deoxidizer additions ...111–112
   first-stage refining 110, 111, 112
   ladle steelmaking 112–113
   mold metallurgy.....................114
   second-stage refining and technology
      advances 111
   temperature-time schedules for various
      steel processing technologies 130, 131
   third-stage refining 113
   tundish metallurgy and continuous
      casting 113–114
Steel production
   raw steel production by type of furnace,
      steel grade, and casting technique 147
   raw steel production by various
      countries......................154, 165
Steel scrap, recycling of 1023
   factors influencing scrap demand 1024
   purchased scrap supply 1024–1026
   scrap use by industry 1023–1024
Steel sheet. See Sheet steel.
Steel springs 302–326
   characteristics of spring steel grade
      annealed spring wire..............307–308
      stainless steel spring wire 308
      valve-spring quality (VSQ) wire 307
   compression springs............319–320, 322
      active coils 320
      modulus change, effect of 320
      solid heights...................320, 322
   costs 317–318, 320
   design 318
   life 319, 321
   stress range 308, 309, 310, 319
   Wahl corrections 318–319, 320, 321
   extension springs...................320–321
      end hooks 321
   fatigue 307, 312
      shot peening 313–314
      stress range 308, 309, 310, 311, 312–313
   hot-wound springs....................315
      hardenability requirements 315–316, 317
      surface quality.................316, 318
   leaf springs 321–324
      mechanical prestressing 325–326
      mechanical properties 325
      steel grades 322, 325
      surface finishes and protective coatings ..326
      for vehicle suspension 322, 325
   mechanical properties 302–305
   flat springs 305, 306, 307
   plating of springs 311–312
      hydrogen relief treatment 312
      mechanical plating 312
   residual stresses for 309, 311
      stress relieving 307, 311
   temperature, effect
      of 303–304, 312, 313, 314–315
   types of 302
   wire quality 303–304, 308
      decarburization 308–309

magnetic particle and eddy current
    testing 309
  seams 308
Steel technology, advances in ...665–666, 668, 669
Steel tubular products 327–336
  cold finishing for 328–329
  common types of pipe 331–333
    conduit pipe 331
    oil country tubular goods ..329, 330, 332–333
    piling pipe 331
    pipe nipples 331
    standard pipe 331
    transmission or line pipe 331
    water main pipe 331–332
    water well pipe 329, 332, 333
  maximum-use temperatures of boiler tube
    steels 617
  mechanical tubing 334–336
    continuous-welded cold-finished
      mechanical tubing 335
    seamless mechanical tubing 335
    square, rectangular, and special-shape
      sections 335–336
    welded mechanical tubing 334–335
  pipe sizes and specifications
    for 328, 329–331, 333
  pressure tubes 327, 333–334
    double-wall brazed tubing 333–334
    structural tubing 334
  product classification 327
  seamless processes for 328
    cupping and drawing 328
    hot extrusion 328
    rotary piercing 328
  welding processes
    continuous welding 328
    double submerged arc welding 328
    electric resistance welding 327–328
    fusion welding 328
Steel wire 277–288
  configurations and sizes 277
  mechanical properties of round wire versus
    flat wire 287–288
  metallic coated wire 281–282
  for packaging and container applications ..282
  quality descriptions and commodities 282
    alloy wire 286–287
    aluminum conductor steel reinforced
      wire 283
    for electrical or conductor applications .283
    fine steel wire 286
    low-carbon steel wire for general usage ..282
    mechanical spring wire for general
      use 284–285
    mechanical spring wire for special
      applications 285
    for packaging and container
      applications 282
    for prestressed concrete 283
    structural applications (not prestressed
      concrete) 282
    upholstery spring construction wire 285
    wire for fasteners 284
    wire for other specific applications ..285–296
  specification wire 281
  wiremaking practices
    cleaning and coating 280
    lubricants 279
    thermal treatments 280–281
    welds 279
Steel wire gage (SWG) system 277
Steel wire rod 272–276
  alloy steel rod 275
    qualities and commodities for..........275
    special requirements for 275–276
  carbon steel rod......................272
    mechanical properties 275–276
    special requirements for 274
  cleaning and coating 272
  configurations and sizes of 272
  heat treatment of 272
  wiremaking practices 277
  wiredrawing 277–279
Stelco coil box 118, 120

Strain-age embrittlement
  in carbon and alloy steels 693–694, 695
  dynamic strain aging...................694
  low-carbon steels 693–694
Strain aging, in carbon steel sheet and
  strip 204–205
Strain annealing 259, 280
Strain-based approach to fatigue ...677–678, 679
Strain-hardening exponent 575
Strain rate sensitivity, of ductile iron 47
Strain relieving 259
Strand annealing 280
Strapping wire 282
Strategic materials availability and
  supply 1009–1022
  COSAM program approach 1013
    advanced processing 1018
    alternate materials............1018–1020
    results 1020–1021
    substitution 1014–1018
  reserves and resources 1009, 1010
  strategic materials 1009–1011
    index method 1010
    survey method 1011
  superalloys 1011–1013
Strength level, and fatigue
  resistance 674, 678, 679, 681
Stress-based approach to fatigue ...675, 676, 677
Stress concentration factor, and fatigue
  resistance 675
Stress corrosion, resistance of maraging
  steels to 799–800
Stress-corrosion cracking
  (SCC) 299–300, 723–728
  of cast stainless steels 917
  evaluation in stainless
    steels 724, 725–728, 873
  parameters affecting 724
  properties and conditions producing ..723–724
  verification procedures 724–725
Stress-corrosion cracking test 611
Stress ratio, and fatigue resistance 674
Stress-relief cracking 607
Stress-relieved uncoated high-carbon wire 283
Stress relieving
  for ductile iron 41
  for hot-rolled steel bars 241
  for powder metallurgy high-speed tool
    steels 783
  of steel plate 231–232
Stress rupture. See also Creep-rupture strength.
  of carbon and low-alloy
    steels 620, 622, 623–624, 629
  of stainless steels 932–933, 934, 937,
        938, 939, 942, 945, 946
Stress-rupture properties. See specific material
  type.
Stretcher leveling, of carbon steel sheet 206
Stretcher strains 574
  in carbon steel sheet and strip 204
Strip. See also Carbon steel sheet and strip.
  cold-rolled...........................846
  heat treatment 846, 848
  high-strength low-alloy steel strip
    cold-forming strip 417
    compositions 406, 417
    forming properties 418, 420
    specifications.................399, 401
    yield and tensile strengths............417
  hot-rolled...........................846
  stainless steel 846, 848
Strip finishes 886
Structural carbon steels 389
  high-strength 390–391
  hot-rolled carbon-manganese structural
    steels 390, 391
  mild steels..........................390
Structural HSLA steels
  applications 399, 401, 420
  compared with carbon steel 389
  compositions 401, 406, 410, 420
  tensile properties
    cold-rolled.........................420
    hot-rolled 411

Structural HSLA steels (continued)
  normalized 410
Structural quality carbon steels 202
Structural steel
  fatigue crack growth in 663–664, 665, 666
  fracture toughness characteristics
    of 664, 666–667, 669
  high-strength low-alloy steels for 419
  low-temperature properties of 662–672
  notch toughness of 412
  specifications 664–665, 667, 668
Structural steel tubing 334, 335
Stud steels 292
Sulfidation, and elevated-temperature
  service 630–632, 636, 637
Sulfides, precipitation of, in processing of
  solid steel 115–116
Sulfide treatment, as surface treatment in
  wrought tool steels 779
Sulfur
  in alloy cast irons 109
  in austenitic manganese steel 826
  effect of, on machinability of carbon
    steels 597–599
  effect of, on notch toughness........740, 741
  effect of, on steel composition and
    formability 577
  at elevated-temperature service 640
  embrittlement of iron by 690–691
  in P/M alloys 810
  in steel..............................144–145
**Sumitomo top and bottom blowing (STB)
  process** 111
Superalloys. *See also* specific types.
  cast cobalt-base superalloys.....983, 985–987
  cast nickel-base superalloys ..981–985, 986–994
  directionally solidified
    superalloys.................995, 996–998
  iron-base superalloys 950, 958–962, 965
  powder metallurgy (P/M) cobalt-base
    alloys............................977–980
  powder metallurgy (P/M) superalloys ..972–976
  single-crystal superalloys ..995–996, 998–1006
  wrought cobalt-base superalloys ..950, 962–968
  wrought nickel-base
    superalloys 950–959, 968–972
Superalloy scrap, recycling of 1027–1028
  blending..............................1032
  collection 1029
  degreasing 1032
  demand 1028
  by industry 1028
  of metallurgical wastes..................1032
  processing 1028–1032
  secondary nickel refining................1032
  separation and sorting 1028, 1029–1032
  size reduction and compaction 1032
Support wire 283
Surface alloying, and fatigue resistance 680
Surface finishes and protective coatings,
  for leaf springs 326
Surface finishing, of stainless steel 884–885
Surface hardening
  of ductile iron 42
  and fatigue resistance.................680–681
  for hot-rolled steel bars 241
  of maraging steels 797–798
Surface treatments for wrought tool steels ...779
  carburizing 779
  nitriding 779
  oxide coatings 779
  plating 779
  sulfide treatment 779
  titanium nitride 779
Swedish standards for steels 159
  compositions of.....................193, 194
  cross-referenced to SAE-AISI steels ...166–174
Swift cup test 576
**Symbols, abbreviations, and
  tradenames** 1038–1041

T

300M steel 435–436

heat treatments for 435
  properties of......................435–436
Tantalum
  in cobalt-base alloys 985
  in MAR-M 247..................1016–1018
  in nickel-base superalloys 984
Taylor constant 591
TCMP steels 666
  strength in...........................666
TCP phases 956
Telephone and telegraph wire 283
Tellurium
  in alloy cast irons 112
  in cast iron 8
  effect of, on machinability of carbon
    steels 599
  embrittlement of iron by 691
  in malleable iron 10
Temper 281
Temperature
  effect of, on austenitic manganese
    steel 836–837
  effect of, on steel
    springs........303–304, 312, 313, 314–315
  effect of, on stress-corrosion
    cracking.......................724–725
  effect of high, on corrosion-resistant
    cast steel 917–918
Temper brittleness 698
Tempered martensite embrittlement 703–706
  activating mechanism 704–706
Temper embrittlement in alloy steels. *See also*
  Tempered martensite
    embrittlement....................698–703
  composition, effect of 699–700
  grain size, effect of 700–702
  microstructure, effect of 702
Tempering
  of dual-phase steels 427–428
  for ductile iron.........................41–42
  effects of alloys on 393–394, 395, 396
  effects of chromium on 641
  for hardened steels 458–459, 462
  of high-strength structural carbon
    steels 389, 391
  of hot-rolled steel bars 241
  of low-alloy steel sheet/strip 209
  of martensite 134–136, 137
  of powder metallurgy high-speed tool
    steels 783
  of steel plate..........................231
Tempering times 80
Temper rolling 693
  of carbon steel sheet 205–206
Tensile and hardness tests 242
Tensile ductility, loss of, and hydrogen-stress
  cracking.........................712–717
Tensile properties. *See* specific material type.
Tensile residual stresses, and fatigue
  resistance 681
Tensile test, for hot-rolled steel bars 242, 243
Tension leveling, of carbon steel sheet 206
Tension test 581–582
  for gray iron 16
  for hot-rolled steel bars 242, 243
Terne coatings 221–222
Terneplate 221
Test bar properties, in gray iron.........16–19
Testing of wrought tool steels 771–778
  fabrication........................774–778
    distortion and safety in hardening 777
    grindability 775
    hardenability 775–777
    machinability 774–775
    resistance to decarburization 778
    weldability 775
  performance in service 771–774
Thermal conductivity 58, 64, 66–68, 69
  of compacted graphite iron.........64, 66–67
  of gray iron............................31
  relationship between electrical conductivity
    and, of ductile iron 50–51, 52
Thermal embrittlement, of maraging
  steels 697–698

Thermal expansion 67, 68
  of ferritic steels 647, 651, 652
Thermal exposure and aging, at
  elevated-temperature service 638, 641
Thermal fatigue
  of compacted graphite iron 63–64, 66, 67
  and elevated-temperature service 626
  in heat-resistant alloys 927–928
Thermal properties, of ductile iron 49, 50
Thermal shock resistance, in heat-resistant
  alloys 928
Thermal treatments, for steel wire 280–281
Thermoelectric testing 1031
Thermomechanical fatigue (TMF) tests 626
Thermomechanical processing. *See also*
  Controlled rolling.
  accelerated cooling of high-strength
    low-alloy steels 398
  interpass cooling 409
  of microalloyed steels 585–589
  temperature time schedules 130, 131
Thin slab casting........................211
Thin strip casting 210, 211
Threaded steel fasteners 289–301
  clamping forces of 300–301
  corrosion protection for 291
    aluminum coatings 295
    cadmium coatings 295
    zinc coating 295
  fabrication 300
    platings and coatings 300
  fastener performance at elevated
    temperatures 295
  bolt steels for elevated
    temperatures 296, 620
  coatings for elevated temperatures 296
  effect of thread design on relaxation 296
  relaxation strengths 631
  time- and temperature-related
    factors 295–296
  fastener tests 296
    proof stress of a bolt or stud 296–297
    proof stress of nut 297
    wedge tensile test of bolts 296
  mechanical properties........291, 297–300, 301
    fatigue failures 297–299, 300, 301
    hardness versus tensile strength 297
    strengths with static loads 297
    stress-corrosion cracking (SCC) 299–300
  specifications and selection 289, 290
  steels for 290–291, 292, 293
    bolt steels 290–291, 294
    nut steels 291, 292–294, 295
    selection of steel for bolts and studs 292
    stud steels 292
  strength grades and property classes ...289–290
Through-hardening alloy steels, machinability
  of 600–601
Time-dependent fatigue behavior, methods for
  predicting 629
Tin
  in cast iron 6
  in compacted graphite iron 57, 59
  effects of, on notch toughness 742
  resistance of, to liquid-metal corrosion 636
Tin coatings.............................221
Tinned wire 282
Tire bead wire 286
Titanium
  in austenitic manganese steel 826
  in cast iron 8
  in cobalt-base alloys 985
  in compacted graphite iron 56
  effect of, on hardenability 395, 413, 470
  effect of, on notch toughness 741–742
  in ferrite 404, 408
  in maraging steels 794–795
  in malleable iron 10
  in microalloy steel 359
  in nickel-base superalloys 983–984
  in steel 146, 577
Titanium-microalloyed steels 403–404
Titanium-niobium microalloyed steels 404

Index / 1061

Titanium nitride, as surface treatment in
   wrought tool steels 779
Tool life 591–592
Tool life tests 591
Tool steels
  hydrogen-stress cracking and loss of tensile
    ductility 715
  powder metallurgy
    advantages over conventional tool
     steels 780
    applications of high-speed tool
     steels 785–786
    classification 781–792
    cold-work tool steels 786–789
    heat treatment of high-speed tool
     steels 782–783
    high-speed tool steels 781–786
    hot-work tool steels 789–790
  wrought 757–779
    classification and characteristics ...757–771
    cold-work steels 763–766
    fabrication of 774–778
    high-speed steels 759–762
    hot-work steels 762–763
    low-alloy special-purpose steels 767
    machining allowances 778–779
    mold steels 767–768
    precision cast hot-work 779
    shock-resisting steels 766–767
    surface treatments 779
    testing of 771–778
    typical heat treatments and properties ..771
    water-hardening steels 768–771
Topologically close-packed (TCP)-type
  phases 952
Torsional properties, of ductile iron 42, 45
Torsional shear strength, of gray iron.....18, 20
Torsion springs 302
Torsion test 582
Toughness and impact resistance
  of austenitic stainless steel after aging 947
  comparison of mild steel, HSLA steel, and
   heat-treated low-alloy steel 389
  of nickel-chromium-molybdenum
   steel 396–397
  of steel castings 365, 367–369
Tradenames, abbreviations, and
  symbols 1038–1041
Transgranular acicular carbides, precipitation
  of 829
Transmission or line pipe 331
Tranverse strength
  of abrasion-resistant cast iron 96
  of gray iron..........................18
Transverse test, for gray iron 16
Truncated-cone indentation test 584
Tube reducing and swaging, for steel tubular
  products 328
Tungsten
  in cast iron 6
  in cobalt-base alloys 985
  effect of, on hardenability 395, 413
  at elevated-temperature service 641
  in nickel-base superalloys 984
Tungsten tool steels
  high-speed steels 759–762
  hot-work steels 762–763
Turbine casing, for elevated-temperature
  service 621
Turbine rotor steels 619, 620–621, 937, 938
Two-step temper embrittlement 698
Tying wire 282

U

U-bend testing, to verify stress-corrosion
  cracking 725
Ultrahigh-carbon steels. See also Carbon steel.
  definition of 148
Ultrahigh-strength steels. See also specific
  types. 430–448
  high fracture toughness steels 444
    AF1410 steel 431, 445, 446–447
    HP-9-4-30 steel 431, 444–446

medium-alloy air-hardening steels....431, 439
   H11 modified 439–441
medium-carbon low-alloy steels 430–431
   300M steel 434–436
   4130 steel 431–432
   4140 steel 432
   4340 steel 432–434
   6150 steel 437–438, 439
   8640 steel 438–439
   D-6a and D-6ac steel 436–437, 438
   H13 steel 431, 441–444
Underbead cracking, in austenitic stainless
  steels 898
Unified Numbering System (UNS) designations.
  See also SAE/AISI
  designation. 151, 153, 842
UNI (Italian) standards for steels 159
  compositions of 190–193
  cross-referenced to SAE-AISI steels ...166–174
United States steel wire gage (USSWG) 277
Unnotched fatigue limits, of tempered pearlitic
  malleable irons 83
Upgraded cast iron. See Compacted graphite
  iron.
Upholstery spring construction wire 285

V

Vacuum arc remelting (VAR) 930, 968, 970
Vacuum degassing....................228, 930
Vacuum induction melting (VIM).......970, 981
  of superalloys 986–988
   alternative melt techniques 988
   filters 988
   melt process 988
   primary purification reaction 986–987
   refractory materials 987–988
Valve-spring quality (VSQ) wire, characteristics
  of 307
Valve-spring wire, testing of 309
Vanadium
  in alloy cast irons 89–90
  in austenitic manganese steel 825
  in cast iron 6, 28
  in cobalt-base alloys 985
  in compacted graphite iron 59
  effect of, on hardenability 395, 413, 419
  effect of, on notch toughness 741–742
  effect of, on steel composition and
   formability 577
  at elevated-temperature service 640–641
  in ferrite 408
  ferritic stainless steels without 936–937
  in microalloy steel 358
  in nickel-base superalloys 984
Vanadium-niobium microalloyed steels 403
Vanadium-nitrogen microalloyed steels 403
Varestraint test 612
Variable load amplitude fatigue tests 370, 372
Vehicle suspension, leaf springs for 322, 325
Vermicular graphite cast iron. See Compacted
  graphite iron.
Volume/area ratios, of gray iron 15–16
Volumetric changes, in steel castings ...374, 376
Volute springs..........................302

W

Wahl corrections, for steel
  springs 318–319, 320, 321
Warm rolling, in processing solid steel ...120, 124
Waspaloy, cobalt in..................1014–1016
Water-hardening steels 768–771
Water main pipe 331–332
Wear, in gray iron 24, 25–26
Wear resistance
  of austenitic manganese steel 834
  of cast iron 376
  of gray iron.....................25–26
  of pearlitic and martensitic malleable
   iron 83–84
Weathering steels 148, 399, 400
  corrosion resistance of 400
  specifications of 399

Weaving wire 851
Wedge-forging test 583, 584
Wedge tensile test of bolts 296
Weldability 603–613
  of aluminum coatings 220
  of heat-treatable low-alloy steels 609
  of high-carbon steels 609
  of high-strength low-alloy steel 609
  of low-carbon steels 608
  of medium-carbon steels 608–609
  metallurgical factors affecting ...603–606, 607
   chemical composition effect......606, 609
   hardenability and weldability ..603–604, 606
   heat-affected zone
    microstructure 605–606, 608
   preweld and postweld heat treatments ..606
   weld metal microstructure 604–605, 607
  of precoated steels 609–610
  of quenched and tempered steels.........609
  of stainless steels 897–905
  of steel castings 378
  of steel plate....................238–239
  of steels 603
  weld cracking 606–608
   hot cracking 608
   hydrogen-induced cracking 606–607
   inclusions 608
   lamellar cracking 607–608, 609
   stress-relief cracking 607–608
  of wrought tool steels 775
Weldability tests 610–613
  bend test 610
  Charpy V-notch test 610–611
  crack tip opening displacement test 611
  drop-weight test 610
  fabrication 611–613
   controlled thermal severity test 612–613
   Lehigh restraint test 612
   Varestraint test 612
  stress-corrosion cracking test 611
  weld tension test 610
Weld cracking 606–608
  hot 608
  hydrogen-induced 606–607
  inclusion 608
  lamellar 607–608
  stress-relief 607
Welded chains, alloy steel wire for 287
Welded mechanical tubing 334–335
Welded structures, fracture toughness
  of 667–669, 670, 671, 672
Welding
  of austenitic manganese steel 837–838
  of corrosion-resistant steel
   castings 919, 928, 929
  of ductile iron 53–55
  of ferritic malleable iron 76
  of HSLA steels 414–415
  of maraging steels 798
  of pearlitic-martensitic malleable irons ...83, 84
  of PH martensitic stainless steels 946
  of PH semiaustenitic stainless steels ..943–944
  of quenched and tempered martensitic
   stainless steels 942
  for steel tubular products 327–328
Welding-quality rod 273–274, 275
Weld metal microstructure, effect of, on
  weldability 604–605, 607
Welds, characteristic features of ...603, 604, 605
  multipass weldments 603, 605
  single-pass weldments 603, 604
Weld tension test 610
Wet drawing, for steel wire 279
White cast irons 107–108
Whiteheart malleable iron 74
White iron 3
Wire
  special commodities 851–852
  stainless steel 849–852
  tempers of 850
Wiredrawing 277–279
Wire forms 302
Wire rod, steel. See Steel wire rod.
Wood screw quality rod 273

Wool wire 852
Workability. *See* Bulk formability of steels.
Work hardening
  in austenitic manganese steel 831–832
  in dual-phase steels 424, 426
Wrought cobalt-base superalloys
  alloying elements, effect of 951
  applications for 950, 967–968
  compositions of 965
  mechanical properties
    stress-rupture properties 957, 962, 967
    tensile properties 958–959, 960–961
  melting (incipient) temperatures 956
  microstructure 965–967
  oxidation and hot corrosion 968
  physical properties 963–964
Wrought nickel-base superalloys
  alloying for corrosion resistance 956–957
  alloying elements, effect of 951
  applications for 950
  coatings for oxidation and corrosion
    resistance 957, 959
  compositions of 950–951
  forming of 971
  hot forming temperatures 969
  heat treatment 956
  low-temperature corrosion 956
  mechanical properties
    stress-rupture properties 954, 955, 956,
      957, 959, 962, 966
    tensile properties 958–959, 960–961
  melting processes
    alloy type versus melting process 971
    electroslag remelting 970
    vacuum arc remelting 970
    vacuum double electrode remelting 970
    vacuum induction melting 970
  melting (incipient) temperatures 856
  microstructure 951–956
    borides 955–956
    carbides 954–955
    gamma double prime 953–954
    gamma matrix 952–953
    gamma prime 952–953
    grain-boundary chemistry 954
    topologically close-packed phases 956
  physical properties 963–964
  thermomechanical processing 963–964
Wrought stainless steels 841–907
  classification of 841–842
    nonstandard types 842, 847–848
    standard types 842, 843, 844, 845, 846
  compositions of 843, 847–848
  corrosion in specific environments 873
    atmospheric corrosion 873–874
    corrosion in chemical
      environments 875–878
    corrosion in water 874–875, 876
    corrosion in various applications 878–882
      architectural 882
      food and beverage industries 878
      oil and gas industry 879–880
      pharmaceutical industry 879
      power industry 879–880
      pulp and paper industry 880–881
      transportation industry 881–882
  corrosion properties 869–884
  corrosion testing 882–883
    pitting and crevice corrosion 883–884
  creep rupture of 304 stainless steel 622
  effect of composition on corrosion ... 871–872
    carbon content 872
    chromium content 871
    manganese content 872
    molybdenum content 872
    nickel content 871–872
    nitrogen content 872
  elevated-temperature properties 861–863,
    930–949
    of austenitic stainless steels 944–949
    corrosion 935–936
    creep and stress rupture 932–933
    creep-fatigue interaction 934–935
    of ferritic stainless steels 936–937
    of martensitic stainless steels 939–942
    of precipitation-hardening alloys ... 942–944
    rupture ductility 933–934
  embrittlement (475 °C) 708
  fabrication characteristics 887–895
  factors in selection 842
    corrosion resistance ... 842, 844–845, 849, 850
    fabrication and cleaning 845
    mechanical properties 845
    surface finish 845
  fatigue strength 861, 863
  forgeability of 889–894
    austenitic stainless steels 891–893
    closed-die forgeability 890–891
    ingot breakdown 890
    duplex stainless steels 894
    ferritic stainless steels 893
    martensitic stainless steels 892, 893
    precipitation-hardening stainless
      steels 893–894
  formability of 888–889
    austenitic stainless steels 888–889
    duplex stainless steels 889
    ferritic stainless steels 889
  forms of corrosion 869–873
    crevice 873
    erosion-corrosion 873
    galvanic 872
    general 872
    intergranular 873
    oxidation 873
    pitting 872–873
    stress-corrosion cracking 873
  hydrogen damage 715
  influence of product form on
    properties 865, 867
    cast structures 867–868
    cold reduced products 868, 869
    hot processing 868
  in-service care 887
  interim surface protection 886–887
  machinability of 894–897
    austenitic stainless steels 894–896
    duplex stainless steels 896
    ferritic and martensitic stainless steels ..894
    precipitation-hardening alloys 896–897
  mechanism of corrosion resistance ... 870–871
  mill finishes 885–886
    plate finishes 886
    sheet finishes 885–886
    strip finishes 886
    wire finishes 886
  notch toughness and transition
    temperature 859–861, 865
  physical properties 868–869, 871
  product forms 845–853, 930, 932
    bar 848–849
    foil 848
    pipe, tubes, and tubing 852–853
    plate 843, 845–846
    semifinished products 852
    sheet 843, 846
    strip 846, 848
    wire 849–852
  production tonnage 841
  recycling 1027–1028
    blending 1032
    collection 1029
    degreasing 1032
    demand 1028
    by industry 1028
    processing 1028–1032
    secondary nickel refining 1032
    separation and sorting 1028, 1029–1032
    size reduction and compaction 1032
  recycling of metallurgical wastes 1032
  sensitization 706–708
    of austenitic stainless steels 706–707
    of duplex stainless steels 707–708
  sheet formability 888–889
    power requirements 889
    stress-strain relationships 889
  shipments of 841, 842
  sigma-phase embrittlement 708–711
    of austenitic stainless steels 708–711
    of duplex stainless steels 711
    of ferritic stainless steels 709–711
  stress-corrosion cracking 725–728, 873
    of austenitic stainless steels 725–726
    of duplex stainless steels 727
  subzero-temperature
    properties 847–848, 863–865
    fatigue strength 865
    fracture crack growth rates ... 865, 869, 870
    fracture toughness 865, 867
    tensile properties .. 865, 866, 867, 868, 869
  surface finishing of 884–885
    electropolishing 885
    grinding, polishing, and buffing 884–885
  tensile properties 853
    austenitic types 853–858
    duplex (austenite/ferrite) types 856, 858
    ferritic types 856, 860
    martensitic types 858, 862–863
    precipitation-hardening
      types 858–859, 864–865
  weldability of 897–905
    austenitic stainless steels 898–901
    duplex stainless steels 904–905
    ferritic stainless steels 901–902
    martensitic stainless steels 902
    precipitation-hardening steels 902–904
Wrought steels
  decarburization 745, 746
  deoxidation practice 741, 742–743
  effects of surface condition on notch
    toughness in 745
  electroplating 746
  hot deformation temperature 743–744
  section and part size 745
  toughness anisotropy 744–745
**Wrought structure and ductility, in closed-die
  forgings** 340
Wrought superalloys. *See* Wrought cobalt-base
  superalloys; Wrought nickel-base
  superalloys.
Wrought tool steels. *See also* specific
  types. 757–779
  applications, selection guide 763
  classification and characteristics 757–771
    cold-work steels 763–766
    cross-reference of United States and
      foreign designations 760
    high-speed steels 759–762
    hot-work steels 762–763
    low-alloy special-purpose steels 767
    mold steels 767–768
    shock-resisting steels 766–767
    water-hardening steels 768–771
  cold-work steels 763–766
    air-hardening, medium alloy 763–765
    high-carbon, high-chromium 765
    oil-hardening 765–766
  composition
    of nonstandard steels 762
    of standard steels 758–759
  fabrication of 774–778
    distortion and safety in hardening 777
    grindability 775
    hardenability 775–777
    machinability 774–775
    resistance to decarburization 778
    weldability 775
  general properties
    factors in tool selection 776–777
    processing and service
      characteristics 772–773
  heat treating
    hardening and tempering 770
    normalizing and annealing
      temperatures 769
  high-speed steels 759–762
    molybdenum 759
    tungsten 759–762
  hot-work steels 762–763
    chromium 762
    molybdenum 763
    tungsten 762–763

low-alloy special-purpose steels 767
machinability
   of annealed tool steels 777
   grindability index 778
machining allowances 778–779
   hot-rolled square and flat bars 778
mechanical properties
   hot hardness of die steels 777
   hot hardness of high-speed steels 787
   of low-alloy special-purpose steels at
     room temperature 767
   of shock-resisting steels at room
     temperature 767
mold steels 767–768
precision cast hot-work 779
shock-resisting steels 766–767
surface treatments 779
   carburizing 779
   nitriding 779
   oxide coatings 779
   plating 779
   sulfide treatment 779
   titanium nitride 779
testing of 771–778
   performance in service 771–774
thermal properties
   resistance to softening of hot-work steels
     at elevated temperatures 765
   thermal conductivity 775
   thermal expansion 774
typical heat treatments and properties 771
water-hardening steels 768–771

X

X-ray spectroscopy 1030–1031, 1032

Y

Yield point elongation 574–575
Yield strength, of dual-phase steels 426–427
   of steel castings 365

Z

Zinc, resistance of, to liquid-metal corrosion ... 635
Zinc alloy coated steels 217–218
Zinc chromate primers 222
Zinc-coated fence wire 285
Zinc-coated strand wire 282

Zinc coating
   chromate passivation 214–215
   coating tests and designations ... 212–214, 215
   corrosion resistance 581, 584
   electrogalvanizing 217
   hot dip galvanizing 216–217, 218
   packaging and storage 215–216
   painting 215
   for threaded steel fasteners 295
   Zincrometal 217
   zinc spraying 218
Zinc phosphate coatings 222
Zinc-rich primers 222–223
Zincrometal 217, 223
Zinc spraying 212, 218
Zirconium
   in cast iron 8
   in cobalt-base alloys 985
   effect of, on hardenability of steel 413, 470
   effects of, on notch toughness 742
   in ferrite 408
   in malleable iron 10
   in nickel-base superalloys 984
   in steel 147
   in superalloys 954, 989